VIRTUAL BIOLOGY LABORATORY 3.0

FROM LAB MATERIALS TO PRINT AND AUDIO TOOLS . . . RESOURCES THAT HELP YOU SUCCEED

This online tool includes 14 modules with 130 activities that give you virtual experience gathering data and performing experiments through engaging simulations. Change parameters to see what happens in each simulation, generate your own data, and write up results. Each experiment includes a general introduction followed by a series of interactive laboratory activities, each with its own set of questions. With an easy-to-use design and unparalleled flexibility, **Virtual Biology Laboratory 3.0** will make you feel like you're in a real lab!

The following experimental modules are available for purchase at www.thomsonedu.com/biology:

Choose from the following lab modules:

- Biochemistry
- Cell Chemistry
- Cell Division
- Cell Membranes
- Cell Respiration
- Cell Structure
- Ecology
- Evolution
- Genetics
- Microscopy
- Molecular Biology
- Pedigree Analysis
- Photosynthesis
- Population Biology

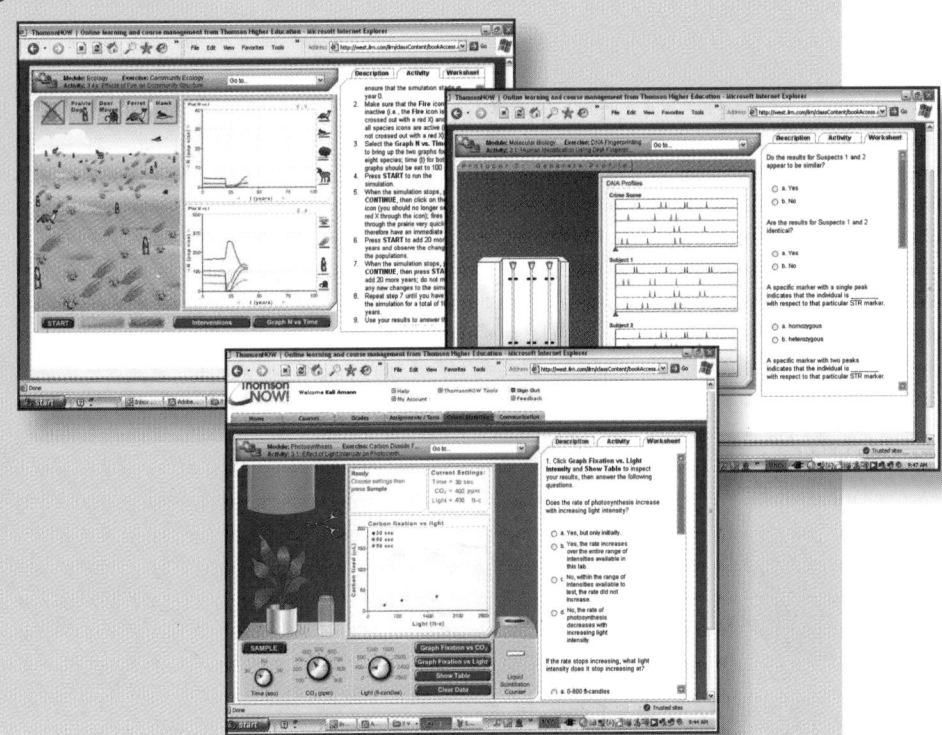

AVAILABLE

Audio Study Tools

Download these study aids containing concept reviews, key terms, questions, clarifications of common misconceptions, and study tips for each chapter.

Study Guide

This interactive workbook pairs text-specific concepts with questions, illustrations, and exercises that promote active learning, as well as topic maps and study strategies to help you study more efficiently.

BIOLOGY

the dynamic science

Peter J. Russell

Stephen L. Wolfe

Paul E. Hertz

Cecie Starr

Beverly McMillan

THOMSON

BROOKS/COLE

™ Australia · Brazil · Canada · Mexico · Singapore · Spain · United Kingdom · United States

THOMSON

BROOKS/COLE

Biology: The Dynamic Science, First Edition

Peter J. Russell, Stephen L. Wolfe, Paul E. Hertz, Cecie Starr, Beverly McMillan

Vice President, Editor in Chief: Michelle Julet

Publisher: Yolanda Cossio

Managing Editor: Peggy Williams

Senior Development Editors: Mary Arbogast,
Shelley Parlante

Development Editor: Christopher Delgado

Assistant Editor: Jessica Kuhn

Editorial Assistant: Rose Barlow

Technology Project Managers: Keli Amann, Kristina Razmara,
Melinda Newfarmer

Marketing Manager: Kara Kindstrom

Development Project Manager: Terri Mynatt

Production Manager: Shelley Ryan

Creative Director: Rob Hugel

Art Directors: John Walker, Lee Friedman

Art Developers: Steve McEntee, Dragonfly Media Group

Print Buyer: Karen Hunt

Permissions Editor: Sarah D'Stair

Production Service: Graphic World Inc.

Production Service Manager: Suzanne Kastner

Text Designer: Jeanne Calabrese

Photo Researchers: Linda Sykes, Robin Samper

Copy Editor: Christy Goldfinch

Illustrators: Dragonfly Media Group, Steve McEntee,
Precision Graphics, Dartmouth Publishing, Inc.

Cover Designer: Jeremy Mendes

Cover Image: © Leonardo Papini/Getty Images®

Cover Printer: Transcontinental Printing/Interglobe

Compositor: Graphic World Inc.

Printer: Transcontinental Printing/Interglobe

Printed in Canada
1 2 3 4 5 6 7 12 11 10 09 08 07

Library of Congress Control Number: 2007931665

ISBN-13: 978-0-534-24966-3
ISBN-10: 0-534-24966-3

For more information about our products, contact us at:
Thomson Learning Academic Resource Center
1-800-423-0563

For permission to use material from this text or product, submit a request online at http://www.thomsonrights.com.
Any additional questions about permissions can be submitted by e-mail to thomsonrights@thomson.com.

Thomson Higher Education
10 Davis Drive
Belmont, CA 94002-3098
USA

About the Authors

PETER J. RUSSELL received a B.Sc. in Biology from the University of Sussex, England, in 1968 and a Ph.D. in Genetics from Cornell University in 1972. He has been a member of the Biology faculty of Reed College since 1972; he is currently a Professor of Biology. He teaches a section of the introductory biology course, a genetics course, an advanced molecular genetics course, and a research literature course on molecular virology. In 1987 he received the Burlington Northern Faculty Achievement Award from Reed College in recognition of his excellence in teaching. Since 1986, he has been the author of a successful genetics textbook; current editions are *iGenetics: A Mendelian Approach, iGenetics: A Molecular Approach,* and *Essential iGenetics.* He wrote nine of the BioCoach Activities for The Biology Place. Peter Russell's research is in the area of molecular genetics, with a specific interest in characterizing the role of host genes in pathogenic RNA plant virus gene expression; yeast is used as the model host. His research has been funded by agencies including the National Institutes of Health, the National Science Foundation, and the American Cancer Society. He has published his research results in a variety of journals, including *Genetics, Journal of Bacteriology, Molecular and General Genetics, Nucleic Acids Research, Plasmid,* and *Molecular and Cellular Biology.* He has a long history of encouraging faculty research involving undergraduates, including cofounding the biology division of the Council on Undergraduate Research (CUR) in 1985. He was Principal Investigator/Program Director of an NSF Award for the Integration of Research and Education (AIRE) to Reed College, 1998–2002.

STEPHEN L. WOLFE received his Ph.D. from Johns Hopkins University and taught general biology and cell biology for many years at the University of California, Davis. He has a remarkable list of successful textbooks, including multiple editions of *Biology of the Cell, Biology: The Foundations, Cell Ultrastructure, Molecular and Cellular Biology,* and *Introduction to Cell and Molecular Biology.*

PAUL E. HERTZ was born and raised in New York City. He received a bachelor's degree in Biology at Stanford University in 1972, a master's degree in Biology at Harvard University in 1973, and a doctorate in Biology at Harvard University in 1977. While completing field research for the doctorate, he served on the Biology faculty of the University of Puerto Rico at Rio Piedras. After spending 2 years as an Isaac Walton Killam Postdoctoral Fellow at Dalhousie University, Hertz accepted a teaching position at Barnard College, where he has taught since 1979. He was named Ann Whitney Olin Professor of Biology in 2000, and he received The Barnard Award for Excellence in Teaching in 2007. In addition to his service on numerous college committees, Professor Hertz was Chair of Barnard's Biology Department for 8 years. He has also been the Program Director of the Hughes Science Pipeline Project at Barnard, an undergraduate curriculum and research program funded by the Howard Hughes Medical Institute, since its inception in 1992. The Pipeline Project includes the Intercollegiate Partnership, a program for local community college students that facilitates their transfer to 4-year colleges and universities. He teaches one semester of the introductory sequence for Biology majors and preprofessional students as well as lecture and laboratory courses in vertebrate zoology and ecology. Professor Hertz is an animal physiological ecologist with a specific research interest in the thermal biology of lizards. He has conducted fieldwork in the West Indies since the mid-1970s, most recently focusing on the lizards of Cuba. His work has been funded by the National Science Foundation, and he has published his research in such prestigious journals as *The American Naturalist, Ecology, Nature,* and *Oecologia.*

CECIE STARR is the author of best-selling biology textbooks. Her books include multiple editions of *Unity and Diversity of Life, Biology: Concepts and Applications,* and *Biology Today and Tomorrow.* Her original dream was to be an architect. She may not be building houses, but with the same care and attention to detail, she builds incredible books: *"I invite students into a chapter through an intriguing story. Once inside, they get the great windows that biologists construct on the world of life. Biology is not just another house. It is a conceptual mansion. I hope to do it justice."*

BEVERLY McMILLAN has been a science writer for more than 20 years and is coauthor of a college text in human biology, now in its seventh edition. She has worked extensively in educational and commercial publishing, including 8 years in editorial management positions in the college divisions of Random House and McGraw-Hill. In a multifaceted freelance career, Bev also has written or coauthored six trade books and numerous magazine and newspaper articles, as well as story panels for exhibitions at the Science Museum of Virginia and the San Francisco Exploratorium. She has worked as a radio producer and speechwriter for the University of California system and as a media relations advisor for the College of William and Mary. She holds undergraduate and graduate degrees from the University of California, Berkeley.

Preface

Welcome to *Biology: The Dynamic Science*. The title of our book reflects an explosive growth in the knowledge of living systems over the past few decades. Although this rapid pace of discovery makes biology the most exciting of all the natural sciences, it also makes it the most difficult to teach. How can college instructors—and, more important, college students—absorb the ever-growing body of ideas and information? The task is daunting, especially in introductory courses that provide a broad overview of the discipline.

Our primary goal in this text is to convey fundamental concepts while maintaining student interest in biology
In this entirely new textbook, we have applied our collective experience as college teachers, science writers, and researchers to create a readable and understandable introduction to our field. We provide students with straightforward explanations of fundamental concepts presented from the evolutionary viewpoint that binds all of the biological sciences together. Having watched our students struggle to navigate the many arcane details of college-level introductory biology, we have constantly reminded ourselves and each other to "include fewer facts, provide better explanations, and maintain the narrative flow," thereby enabling students to see the big picture. Clarity of presentation, a high level of organization, a seamless flow of topics within chapters, and spectacularly useful illustrations are central to our approach.

One of the main goals in this book is to sustain students' fascination with the living world instead of burying it under a mountain of disconnected facts. As teachers of biology, we encourage students to appreciate the dynamic nature of science by conveying our passion for biological research. We want to amaze students with *what* biologists know about the living world and *how* we know it. We also hope to excite them about the opportunities they will have to expand that knowledge. Inspired by our collective effort as teachers and authors, some of our students will take up the challenge and become biologists themselves, asking important new questions and answering them through their own innovative research. For students who pursue other career paths, we hope that they will leave their introductory—and perhaps only—biology courses armed with the knowledge and intellectual skills that allow them to evaluate future discoveries with a critical eye.

We emphasize that, through research, our understanding of biological systems is alive and constantly changing
In this book, we introduce students to a biologist's "ways of knowing." Scientists constantly integrate new observations, hypotheses, experiments, and insights with existing knowledge and ideas. To do this well, biology instructors must not simply introduce students to the current state of our knowledge. We must also foster an appreciation of the historical context within which that knowledge developed and identify the future directions that biological research is likely to take.

To achieve these goals, we explicitly base our presentation and explanations on the research that established the basic facts and principles of biology. Thus, a substantial proportion of each chapter focuses on studies that define the state of biological knowledge today. We describe recent research in straightforward terms, first identifying the question that inspired the work and relating it to the overall topic under discussion. Our research-oriented theme teaches students, through example, how to ask scientific questions and pose hypotheses, two key elements of the "scientific process."

Because advances in science occur against a background of past research, we also give students a feeling for how biologists of the past uncovered and formulated basic knowledge in the field. By fostering an appreciation of such discoveries, given the information and theories that were available to scientists in their own time, we can help students to better understand the successes and limitations of what we consider cutting edge today. This historical perspective also encourages students to view biology as a dynamic intellectual endeavor, and not just a list of facts and generalities to be memorized.

One of our greatest efforts has been to make the science of biology come alive by describing how biologists formulate hypotheses and evaluate them using hard-won data, how data sometimes tell only part of a story, and how studies often end up posing more questions than they answer. Although students often prefer to read about the "right" answer to a question, they must be encouraged to embrace "the unknown," those gaps in our knowledge that create opportunities for further research. An appreciation of what we *don't* know will draw more students into the field. And by defining *why* we don't understand interesting phenomena, we encourage students to follow paths dictated by their own curiosity. We hope that this approach will encourage students to make biology a part of their daily lives—to have informal discussions about new scientific discoveries, just as they do about politics, sports, or entertainment.

Special features establish a story line in every chapter and describe the process of science
In preparing this book, we developed several special features to help students broaden their understanding of the material presented and of the research process itself.

- The chapter openers, entitled *Why It Matters,* tell the story of how a researcher arrived at a key insight or how biological research solved a major societal problem or shed light on a fundamental process or phenomenon. These engaging, short vignettes are designed to capture students' imagination and whet their appetite for the topic that the chapter addresses.

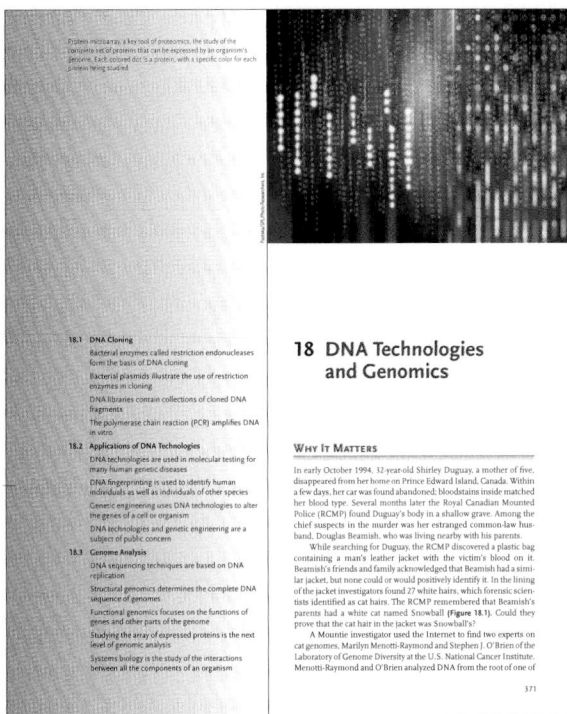

- To complement this historical or practical perspective, each chapter closes with a brief essay, entitled *Unanswered Questions,* often prepared by an expert in the field. These essays identify important unresolved issues relating to the chapter topic and describe cutting-edge research that will advance our knowledge in the future.

- Each chapter also includes a short boxed essay, entitled *Insights from the Molecular Revolution,* which describes how molecular technologies allow scientists to answer questions that they could not have even posed 20 or 30 years ago. Each *Insight*

focuses on a single study and includes sufficient detail for its content to stand alone.

- Almost every chapter is further supplemented with one or more short boxed essays that *Focus on Research.* Some of these essays describe seminal studies that provided a new perspective on an important question. Others describe how basic research has solved everyday problems relating to health or the environment. Another set introduces model research organisms—such as *E. coli, Drosophila, Arabidopsis, Caenorhabditis,* and *Anolis*—and explains why they have been selected as subjects for in-depth analysis.

Spectacular illustrations enable students to visualize biological processes, relationships, and structures

Today's students are accustomed to receiving ideas and information visually, making the illustrations and photographs in a textbook more important than ever before. Our illustration program provides an exceptionally clear supplement to the narrative in a style that is consistent throughout the book. Graphs and anatomical drawings are annotated with interpretative explanations that lead students through the major points they convey.

Three types of specially designed Research Figures provide more detailed information about how biologists formulate and test specific hypotheses by gathering and interpreting data.

- *Research Method* figures provide examples of important techniques, such as gel electrophoresis, the use of radioisotopes, and cladistic analysis. Each *Research Method* figure leads a student through the technique's purpose and protocol and describes how scientists interpret the data it generates.

- *Observational Research* figures describe specific studies in which biologists have tested hypotheses by comparing systems under varying natural circumstances.

- *Experimental Research* figures describe specific studies in which researchers used both experimental and control treatments—either in the laboratory or in the field—to test hypotheses by manipulating the system they study.

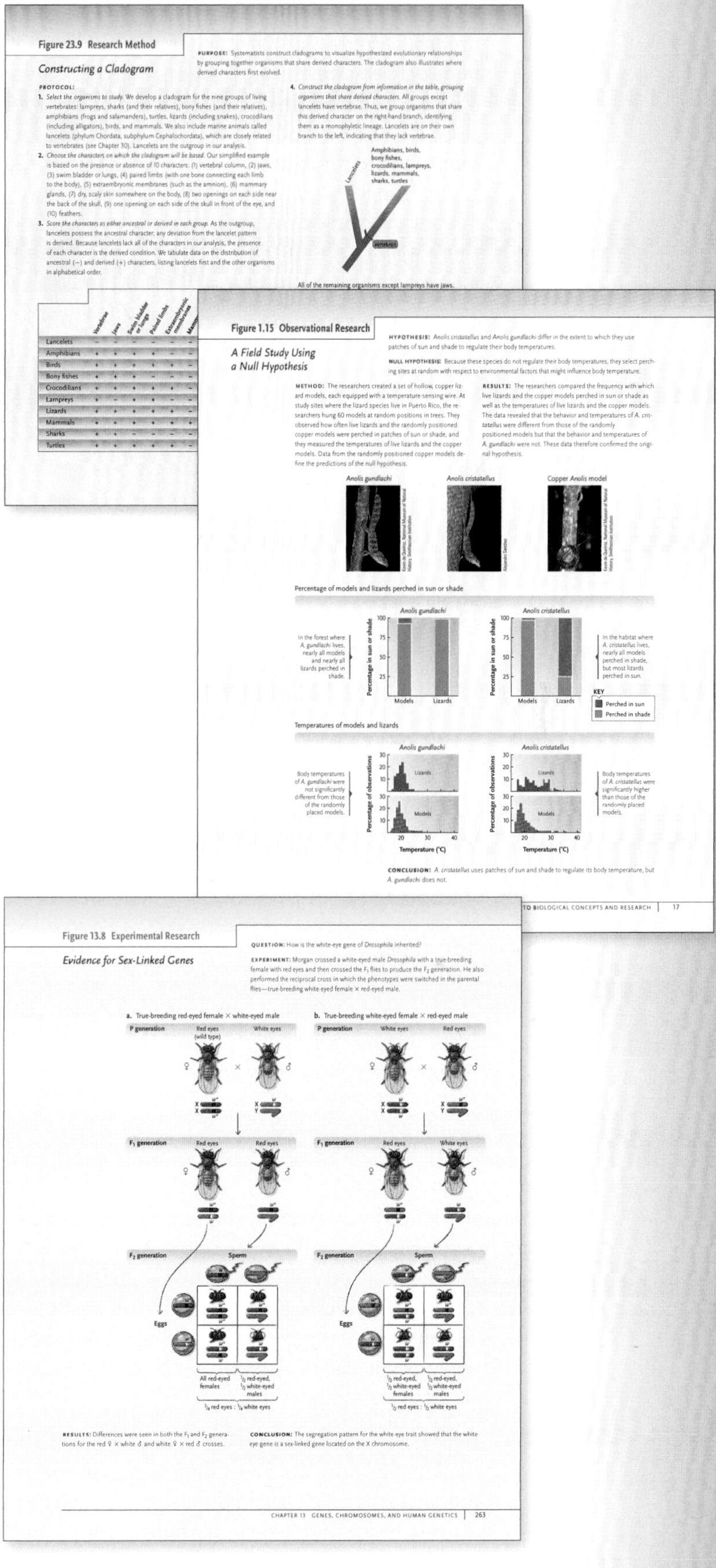

Chapters are structured to emphasize the big picture and the most important concepts

As authors and college teachers, we know how easily students can get lost within a chapter that spans 15 pages or more. When students request advice about how to approach such a large task, we usually suggest that, after reading each section, they pause and quiz themselves on the material they have just encountered. After completing all of the sections in a chapter, they should quiz themselves again, even more rigorously, on the individual sections and, most important, on how the concepts developed in different sections fit together. To assist these efforts, we have adopted a structure for each chapter that will help students review concepts as they learn them.

- The organization within chapters presents material in digestible chunks, building on students' knowledge and understanding as they acquire it. Each major section covers one broad topic. Each subsection, titled with a declarative sentence that summarizes the main idea of its content, explores a narrower range of material.

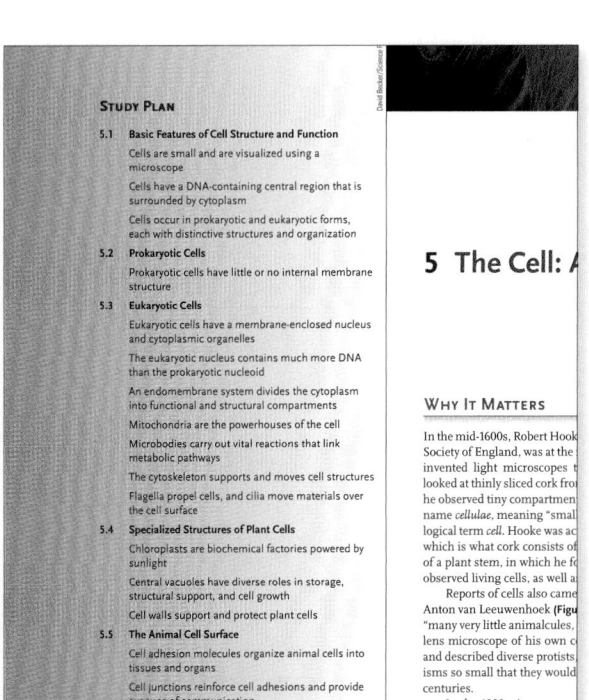

- Whenever possible, we include the derivation of unfamiliar terms so that students will see connections between words that share etymological roots. Mastery of the technical language of biology will allow students to discuss ideas and processes precisely. At the same time, we have minimized the use of unnecessary jargon as much as possible.

- Sets of embedded *Study Break* questions follow every major section. These questions encourage students to pause at the end of a section and review what they have learned before going on to the next topic within the chapter. Short answers to these questions appear in an appendix.

- Several open-ended *Questions for Discussion* emphasize concepts, the interpretation of data, and practical applications of the material.

- A question on *Experimental Analysis* asks students to consider how they would develop and test hypotheses about a situation that relates to the chapter's main topic.

- An *Evolution Link* question relates the subject of the chapter to evolutionary biology.

- The *How Would You Vote?* exercise allows students to weigh both sides of an issue by reading pro/con articles, and then making their opinion known through an online voting process.

We hope that, after reading parts of this textbook, you agree that we have developed a clear, fresh, and well-integrated introduction to biology as it is understood by researchers today. Just as important, we hope that our efforts will excite students about the research process and the new discoveries it generates.

Acknowledgments

We are grateful to the many people who have generously fostered the creation of this text

The creation of a new textbook is a colossal undertaking, and we could never have completed the task without the kind assistance of many people.

Jack Carey first conceived of this project and put together the author team. Michelle Julet and Yolanda Cossio have provided the support and encouragement necessary to move it forward to completion. Peggy Williams has served as the extraordinarily able coordinator of the authors, editors, reviewers, contributors, artists, and production team—we like to think of Peggy as the "cat herder."

Developmental Editors play nearly as large a role as the authors, interpreting and deconstructing reviewer comments and constantly making suggestions about how we could tighten the narrative and stay on course. Mary Arbogast has done banner service as a Developmental Editor, patiently working on the project since its inception. Shelley Parlante has provided very helpful guidance as the manuscript matured. Jody Larson and Catherine Murphy have offered useful comments on many of the chapters.

We are grateful to Christopher Delgado and Jessica Kuhn for coordinating the print supplements, and our Editorial Assistant Rose Barlow for managing all our reviewer information.

Many thanks to Keli Amann, Kristina Razmara, and Christopher Delgado, who were responsible for partnering with our technology authors and media advisory board in creating tools to support students in learning and instructors in teaching.

We appreciate the help of the production staff led by Shelley Ryan and Suzanne Kastner at Graphic

End-of-chapter material encourages students to review concepts, test their knowledge, and think analytically

Supplementary materials at the end of each chapter help students review the material they have learned, assess their understanding, and think analytically as they apply the principles developed in the chapter to novel situations. Many of the end-of-chapter questions also serve as good starting points for class discussions or out-of-class assignments.

- A brief *Review* that references figures and tables in the chapter provides an outline summary of important ideas developed in the chapter. The *Reviews* are much too short to serve as a substitute for reading the chapter. Instead, students may use them as an outline of the material, filling in the details on their own.

- Each chapter also closes with a set of 10 multiple choice *Self-Test* questions that focus on factual material.

World. We thank our Creative Director Rob Hugel, Art Director John Walker.

The outstanding art program is the result of the collaborative talent, hard work, and dedication of a select group of people. The meticulous styling and planning of the program is credited to Steve McEntee and Dragonfly Media Group, led by Craig Durant and Mike Demaray. The DMG group created hundreds of complex, vibrant art pieces. Steve's role was crucial in overseeing the development and consistency of the art program; he was also the illustrator for the unique Research features.

We appreciate Kara Kindstrom, our Marketing Manager, and Terri Mynatt, our Development Project Manager, whose expertise ensured that you would know all about this new book.

Peter Russell thanks Stephen Arch of Reed College for valuable discussions and advice during the writing of the Unit Six chapters on Animal Structure and Function. Paul E. Hertz thanks Hilary Callahan, John Glendinning, and Brian Morton of Barnard College for their generous advice on many phases of this project, and John Alcock of Arizona State University and James Danoff-Burg of Columbia University for their contributions to the discussions of Animal Behavior and Conservation Biology, respectively. Paul would also like to thank Jamie Rauchman, for extraordinary patience and endless support as this book was written, and his thousands of past students, who have taught him at least as much as he has taught them.

We would also like to thank our advisors and contributors:

Media Advisory Board

Scott Bowling, Auburn University
Jennifer Jeffery, Wharton County Junior College
Shannon Lee, California State University, Northridge
Roderick M. Morgan, Grand Valley State University
Debra Pires, University of California, Los Angeles

Art Advisory Board

Lissa Leege, Georgia Southern University
Michael Meighan, University of California, Berkeley
Melissa Michael, University of Illinois at Urbana–Champaign
Craig Peebles, University of Pittsburgh
Laurel Roberts, University of Pittsburgh

Accuracy Checkers

Brent Ewers, University of Wyoming
Richard Falk, University of California, Davis
Michael Meighan, University of California, Berkeley
Michael Palladino, Monmouth University

End-of-Chapter Questions

Patricia Colberg, University of Wyoming
Elizabeth Godrick, Boston University

Student Study Guide

Carolyn Bunde, Idaho State University
William Kroll, Loyola University Chicago
Mark Sheridan, North Dakota State University
Jyoti Wagle, Houston Community College

Instructor's Resource Manual

Benjie Blair, Jacksonville State University
Nancy Boury, Idaho State University
Mark Meade, Jacksonville State University
Debra Pires, University of California, Los Angeles
James Rayburn, Jacksonville State University

Test Bank

Scott Bowling, Auburn University
Laurie Bradley, Hudson Valley Community College
Jose Egremy, Northwest Vista College
Darrel L. Murray, University of Illinois, Chicago
Jacalyn Newman, University of Pittsburgh
Mark Sugalski, Southern Polytechnic State University

Technology Authors

Catherine Black, Idaho State University
David Byres, Florida Community College, Jacksonville
Kevin Dixon, University of Illinois
Albia Dugger, Miami Dade College
Mary Durant, North Harris College
Brent Ewers, University of Wyoming
Debbie Folkerts, Auburn University
Stephen Kilpatrick, University of Pittsburgh
Laurel Roberts, University of Pittsburgh
Thomas Sasek, University of Louisiana, Monroe
Bruce Stallsmith, University of Alabama–Huntsville

Workshop and Focus Group Participants

Karl Aufderheide, *Texas A&M University*

Bob Bailey, *Central Michigan University*

John Bell, *Brigham Young University*

Catherine Black, *Idaho State University*

Hessel Bouma III, *Calvin College*

Scott Bowling, *Auburn University*

Bob Brick, *Blinn College, Bryan*

Randy Brooks, *Florida Atlantic University*

Nancy Burley, *University of California, Irvine*

Genevieve Chung, *Broward Community College*

Allison Cleveland, *University of South Florida*

Patricia Colberg, *University of Wyoming*

Jay Comeaux, *Louisiana State University*

Sehoya Cotner, *University of Minnesota*

Joe Cowles, *Virginia Tech*

Anita Davelos-Baines, *University of Texas, Pan American*

Donald Deters, *Bowling Green State University*

Kevin Dixon, *University of Illinois at Urbana–Champaign*

Jose Egremy, *Northwest Vista College*

Diana Elrod, *University of North Texas*

Zen Faulkes, *University of Texas–Pan American*

Elizabeth Godrick, *Boston University*

Barbara Haas, *Loyola University Chicago*

Julie Harless, *Montgomery College*

Jean Helgeson, *Collin County Community College*

Mark Hunter, *University of Michigan*

Andrew Jarosz, *Michigan State University, Montgomery College*

Jennifer Jeffery, *Wharton County Junior College*

John Jenkin, *Blinn College, Bryan*

Wendy Keenleyside, *University of Guelph*

Steve Kilpatrick, *University of Pittsburgh at Johnstown*

Gary Kuleck, *Loyola Marymount University*

Allen Kurta, *Eastern Michigan University*

Mark Lyford, *University of Wyoming*

Andrew McCubbin, *Washington State University*

Michael Meighan, *University of California, Berkeley*

John Merrill, *Michigan State University*

Richard Merritt, *Houston Community College, Northwest*

Melissa Michael, *University of Illinois at Urbana–Champaign*

James Mickle, *North Carolina State University*

Betsy Morgan, *Kingwood College*

Kenneth Mossman, *Arizona State University*

Darrel Murray, *University of Illinois, Chicago*

Jacalyn Newman, *University of Pittsburgh*

Dennis Nyberg, *University of Illinois–Chicago*

Bruce Ostrow, *Grand Valley State University–Allendale*

Craig Peebles, *University of Pittsburgh*

Nancy Pencoe, *University of West Georgia*

Mitch Price, *Pennsylvania State University*

Kelli Prior, *Finger Lakes Community College*

Laurel Roberts, *University of Pittsburgh*

Ann Rushing, *Baylor University*

Bruce Stallsmith, *University of Alabama–Huntsville*

David Tam, *University of North Texas*

Franklyn Te, *Miami Dade College*

Nanette Van Loon, *Borough of Manhattan Community College*

Alexander Wait, *Missouri State University*

Lisa Webb, *Christopher Newport University*

Larry Williams, *University of Houston*

Michelle Withers, *Louisiana State University*

Denise Woodward, *Pennsylvania State University*

Class Test Participants

Tamarah Adair, *Baylor University*

Idelissa Ayala, *Broward Community College–Central*

Tim Beagley, *Salt Lake Community College*

Catherine Black, *Idaho State University*

Laurie Bradley, *Hudson Valley Community College*

Mirjana Brockett, *Georgia Tech*

Carolyn Bunde, *Idaho State University*

John Cogan, *Ohio State University*

Anne M. Cusic, *University of Alabama–Birmingham*

Ingeborg Eley, *Hudson Valley Community College*

Brent Ewers, *University of Wyoming*

Miriam Ferzli, *North Carolina State University*

Debbie Folkerts, *Auburn University*

Mark Hens, *University of North Carolina, Greensboro*

Anna Hill, *University of Louisiana, Monroe*

Anne Hitt, *Oakland University*

Jennifer Jeffery, *Wharton County Junior College*

David Jones, *Dixie State College*

Wendy Keenleyside, *University of Guelph*

Brian Kinkle, *University of Cincinnati*

Brian Larkins, *University of Arizona*

Shannon Lee, *California State University, Northridge*

Harvey Liftin, *Broward Community College*

Jim Marinaccio, *Raritan Valley Community College*

Monica Marquez-Nelson, *Joliet Junior College*

Kelly Meckling, *University of Guelph*

Richard Merritt, *Houston Community College–Town and Country*

Russ Minton, *University of Louisiana, Monroe*

Necia Nichols, *Calhoun State Community College*

Nancy Rice, *Western Kentucky University*

Laurel Roberts, *University of Pittsburgh*

John Russell, *Calhoun State Community College*

Pramila Sen, *Houston Community College*

Jacquelyn Smith, *Pima County Community College*

Bruce Stallsmith, *University of Alabama–Huntsville*

Joe Steffen, *University of Louisville*

Gail Stewart, *Camden County College*

Mark Sugalski, *Southern Polytechnic State University*

Marsha Turrell, *Houston Community College*

Fil Ventura-Smolenski, *Santa Fe Community College*

Beth Vlad, *College of DuPage*

Alexander Wait, *Missouri State University*

Matthew Wallenfang, *Barnard College*

David Wolfe, *American River College*

Reviewers

Heather Addy, *University of Calgary*

Adrienne Alaie-Petrillo, *Hunter College–CUNY*

Richard Allison, *Michigan State University*

Terry Allison, *University of Texas–Pan American*

Deborah Anderson, *Saint Norbert College*

Robert C. Anderson, *Idaho State University*

Andrew Andres, *University of Nevada–Las Vegas*

Steven M. Aquilani, *Delaware County Community College*

Jonathan W. Armbruster, *Auburn University*

Peter Armstrong, *University of California, Davis*

John N. Aronson, *University of Arizona*

Joe Arruda, *Pittsburg State University*

Karl Aufderheide, *Texas A&M University*

Charles Baer, *University of Florida*

Gary I. Baird, *Brigham Young University*

Aimee Bakken, *University of Washington*

Marica Bakovic, *University of Guelph*

Michael Baranski, *Catawba College*

Michael Barbour, *University of California, Davis*

Edward M. Barrows, *Georgetown University*

Anton Baudoin, *Virginia Tech*

Penelope H. Bauer, *Colorado State University*

Kevin Beach, *University of Tampa*

Mike Beach, *Southern Polytechnic State University*

Ruth Beattie, *University of Kentucky*

Robert Beckmann, *North Carolina State University*

Jane Beiswenger, *University of Wyoming*

Andrew Bendall, *University of Guelph*

Catherine Black, *Idaho State University*

Andrew Blaustein, *Oregon State University*

Anthony H. Bledsoe, *University of Pittsburgh*

Harriette Howard-Lee Block, *Prairie View A&M University*

Dennis Bogyo, *Valdosta State University*

David Bohr, *University of Michigan*

Emily Boone, *University of Richmond*

Hessel Bouma III, *Calvin College*

Nancy Boury, *Iowa State University*

Scott Bowling, *Auburn University*

Laurie Bradley, *Hudson Valley Community College*

William Bradshaw, *Brigham Young University*

J. D. Brammer, *North Dakota State University*

G. L. Brengelmann, *University of Washington*

Randy Brewton, *University of Tennessee–Knoxville*

Bob Brick, *Blinn College, Bryan*

Mirjana Brockett, *Georgia Tech*

William Bromer, *University of Saint Francis*

William Randy Brooks, *Florida Atlantic University–Boca Raton*

Mark Browning, *Purdue University*

Gary Brusca, *Humboldt State University*

Alan H. Brush, *University of Connecticut*

Arthur L. Buikema, Jr., *Virginia Tech*

Carolyn Bunde, *Idaho State University*

E. Robert Burns, *University of Arkansas for Medical Sciences*

Ruth Buskirk, *University of Texas–Austin*

David Byres, *Florida Community College, Jacksonville*

Christopher S. Campbell, *University of Maine*

Angelo Capparella, *Illinois State University*

Marcella D. Carabelli, *Broward Community College–North*

Jeffrey Carmichael, *University of North Dakota*

Bruce Carroll, *North Harris Montgomery Community College*

Robert Carroll, *East Carolina University*

Patrick Carter, *Washington State University*

Christine Case, *Skyline College*

Domenic Castignetti, *Loyola University Chicago–Lakeshore*

Jung H. Choi, *Georgia Tech*

Kent Christensen, *University Michigan School of Medicine*

John Cogan, *Ohio State University*

Linda T. Collins, *University of Tennessee–Chattanooga*

Lewis Coons, *University of Memphis*

Joe Cowles, *Virginia Tech*

George W. Cox, *San Diego State University*

David Crews, *University of Texas*

Paul V. Cupp, Jr., *Eastern Kentucky University*

Karen Curto, *University of Pittsburgh*

Anne M. Cusic, *University of Alabama–Birmingham*

David Dalton, *Reed College*

Frank Damiani, *Monmouth University*

Peter J. Davies, *Cornell University*

Fred Delcomyn, *University of Illinois at Urbana–Champaign*

Jerome Dempsey, *University of Wisconsin–Madison*

Philias Denette, *Delgado Community College–City Park*

Nancy G. Dengler, *University of Toronto*

Jonathan J. Dennis, *University of Alberta*

Daniel DerVartanian, *University of Georgia*

Donald Deters, *Bowling Green State University*

Kathryn Dickson, *CSU Fullerton*

Kevin Dixon, *University of Illinois at Urbana–Champaign*

Gordon Patrick Duffie, *Loyola University Chicago–Lakeshore*

Charles Duggins, *University of South Carolina*

Carolyn S. Dunn, *University North Carolina–Wilmington*

Roland R. Dute, *Auburn University*

Melinda Dwinell, *Medical College of Wisconsin*

Gerald Eck, *University of Washington*

Gordon Edlin, *University of Hawaii*

William Eickmeier, *Vanderbilt University*

Ingeborg Eley, *Hudson Valley Community College*

Paul R. Elliott, *Florida State University*

John A. Endler, *University of Exeter*

Brent Ewers, *University of Wyoming*

Daniel J. Fairbanks, *Brigham Young University*

Piotr G. Fajer, *Florida State University*

Richard H. Falk, *University of California, Davis*

Ibrahim Farah, *Jackson State University*

Jacqueline Fern, *Lane Community College*

Daniel P. Fitzsimons, *University of Wisconsin–Madison*

Daniel Flisser, *Camden County College*

R. G. Foster, *University of Virginia*

Dan Friderici, *Michigan State University*

J. W. Froehlich, *University of New Mexico*

Paul Garcia, *Houston Community College–SW*

Umadevi Garimella, *University of Central Arkansas*

Robert P. George, *University of Wyoming*

Stephen George, *Amherst College*

John Giannini, *St. Olaf College*

Joseph Glass, *Camden County College*

John Glendinning, *Barnard College*

Elizabeth Godrick, *Boston University*

Judith Goodenough, *University of Massachusetts Amherst*

H. Maurice Goodman, *University of Massachusetts Medical School*

Bruce Grant, *College of William and Mary*

Becky Green-Marroquin, *Los Angeles Valley College*

Christopher Gregg, *Louisiana State University*

Katharine B. Gregg, *West Virginia Wesleyan College*

John Griffin, *College of William and Mary*

Samuel Hammer, *Boston University*

Aslam Hassan, *University of Illinois at Urbana–Champaign, Veterinary Medicine*

Albert Herrera, *University of Southern California*

Wilford M. Hess, *Brigham Young University*

Martinez J. Hewlett, *University of Arizona*

Christopher Higgins, *Tarleton State University*

Phyllis C. Hirsch, *East Los Angeles College*

Carl Hoagstrom, *Ohio Northern University*

Stanton F. Hoegerman, *College of William and Mary*

Ronald W. Hoham, *Colgate University*

Margaret Hollyday, *Bryn Mawr College*

John E. Hoover, *Millersville University*

Howard Hosick, *Washington State University*

William Irby, *Georgia Southern*

John Ivy, *Texas A&M University*

Alice Jacklet, *SUNY Albany*

John D. Jackson, *North Hennepin Community College*

Jennifer Jeffery, *Wharton County Junior College*

John Jenkin, *Blinn College, Bryan*

Leonard R. Johnson, *University Tennessee College of Medicine*

Walter Judd, *University of Florida*

Prem S. Kahlon, *Tennessee State University*

Thomas C. Kane, *University of Cincinnati*

Peter Kareiva, *University of Washington*

Gordon I. Kaye, *Albany Medical College*

Greg Keller, *Eastern New Mexico University*

Stephen Kelso, *University of Illinois–Chicago*

Bryce Kendrick, *University of Waterloo*

Bretton Kent, *University of Maryland*

Jack L. Keyes, *Linfield College Portland Campus*

John Kimball, *Tufts University*

Hillar Klandorf, *West Virginia University*

Michael Klymkowsky, *University of Colorado–Boulder*

Loren Knapp, *University of South Carolina*

Ana Koshy, *Houston Community College–NW*

Kari Beth Krieger, *University of Wisconsin–Green Bay*

David T. Krohne, *Wabash College*

William Kroll, *Loyola University Chicago–Lakeshore*

Josepha Kurdziel, *University of Michigan*

Allen Kurta, *Eastern Michigan University*

Howard Kutchai, *University of Virginia*

Paul K. Lago, *University of Mississippi*

John Lammert, *Gustavus Adolphus College*

William L'Amoreaux, *College of Staten Island*

Brian Larkins, *University of Arizona*

William E. Lassiter, *University of North Carolina–Chapel Hill*

Shannon Lee, *California State University, Northridge*

Lissa Leege, *Georgia Southern University*

Matthew Levy, *Case Western Reserve University*

Harvey Liftin, *Broward Community College–Central*

Tom Lonergan, *University of New Orleans*

Lynn Mahaffy, *University of Delaware*

Alan Mann, *University of Pennsylvania*

Kathleen Marrs, *Indiana University Purdue University Indianapolis*

Robert Martinez, *Quinnipiac University*

Joyce B. Maxwell, *California State University, Northridge*

Jeffrey D. May, *Marshall University*

Geri Mayer, *Florida Atlantic University*

Jerry W. McClure, *Miami University*

Andrew G. McCubbin, *Washington State University*

Mark McGinley, *Texas Tech University*

F. M. Anne McNabb, *Virginia Tech*

Mark Meade, *Jacksonvile State University*

Bradley Mehrtens, *University of Illinois at Urbana–Champaign*

Michael Meighan, *University of California, Berkeley*

Catherine Merovich, *West Virginia University*

Richard Merritt, *Houston Community College–Town and Country*

Ralph Meyer, *University of Cincinnati*

James E. "Jim" Mickle, *North Carolina State University*

Hector C. Miranda, Jr., *Texas Southern University*

Jasleen Mishra, *Houston Community College–SW*

David Mohrman, *University of Minnesota Medical School*

John M. Moore, *Taylor University*

David Morton, *Frostburg State University*

Alexander Motten, *Duke University*

Alan Muchlinski, *California State University, Los Angeles*

Michael Muller, *University of Illinois–Chicago*

Richard Murphy, *University of Virginia*

Darrel L. Murray, *University of Illinois–Chicago*

Allan Nelson, *Tarleton State University*

David H. Nelson, *University of South Alabama*

Jacalyn Newman, *University of Pittsburgh*

David O. Norris, *University of Colorado*

Bette Nybakken, *Hartnell College, California*

Tom Oeltmann, *Vanderbilt University*

Diana Oliveras, *University of Colorado–Boulder*

Alexander E. Olvido, *Virginia State University*

Karen Otto, *University of Tampa*

William W. Parson, *University of Washington School of Medicine*

James F. Payne, *University of Memphis*

Craig Peebles, *University of Pittsburgh*

Joe Pelliccia, *Bates College*

Susan Petro, *Rampao College of New Jersey*

Debra Pires, *University of California, Los Angeles*

Thomas Pitzer, *Florida International University*

Roberta Pollock, *Occidental College*

Jerry Purcell, *San Antonio College*

Kim Raun, *Wharton County Junior College*

Tara Reed, *University of Wisconsin–Green Bay*

Lynn Robbins, *Missouri State University*

Carolyn Roberson, *Roane State Community College*

Laurel Roberts, *University of Pittsburgh*

Kenneth Robinson, *Purdue University*

Frank A. Romano, *Jacksonville State University*

Michael R. Rose, *University of California, Irvine*

Michael S. Rosenzweig, *Virginia Tech*

Linda S. Ross, *Ohio University*

Ann Rushing, *Baylor University*

Linda Sabatino, *Suffolk Community College*

Tyson Sacco, *Cornell University*

Peter Sakaris, *Southern Polytechnic State University*

Frank B. Salisbury, *Utah State University*

Mark F. Sanders, *University of California, Davis*

Andrew Scala, *Dutchess Community College*

John Schiefelbein, *University of Michigan*

Deemah Schirf, *University of Texas–San Antonio*

Kathryn J. Schneider, *Hudson Valley Community College*

Jurgen Schnermann, *University Michigan School of Medicine*

Thomas W. Schoener, *University California, Davis*

Brian Shea, *Northwestern University*

Mark Sheridan, *North Dakota State University–Fargo*

Dennis Shevlin, *College of New Jersey*

Richard Showman, *University of South Carolina*

Bill Simcik, *Tomball College*

Robert Simons, *University of California, Los Angeles*

Roger Sloboda, *Dartmouth College*

Jerry W. Smith, *St. Petersburg College*

Nancy Solomon, *Miami University*

Bruce Stallsmith, *University of Alabama–Huntsville*

Karl Sternberg, *Western New England College*

Pat Steubing, *University of Nevada–Las Vegas*

Karen Steudel, *University of Wisconsin–Madison*

Richard D. Storey, *Colorado College*

Michael A. Sulzinski, *University of Scranton*

Marshall Sundberg, *Emporia State University*

David Tam, *University of North Texas*

David Tauck, *Santa Clara University*

Jeffrey Taylor, *Slippery Rock University*

Franklyn Te, *Miami Dade College*

Roger E. Thibault, *Bowling Green State University*

Megan Thomas, *University of Nevada–Las Vegas*

Patrick Thorpe, *Grand Valley State University–Allendale*

Ian Tizard, *Texas A&M University*

Robert Turner, *Western Oregon University*

Joe Vanable, *Purdue University*

Linda H. Vick, *North Park University*

J. Robert Waaland, *University of Washington*

Douglas Walker, *Wharton County Junior College*

James Bruce Walsh, *University of Arizona*

Fred Wasserman, *Boston University*

Edward Weiss, *Christopher Newport University*

Mark Weiss, *Wayne State University*

Adrian M. Wenner, *University of California, Santa Barbara*

Adrienne Williams, *University of California, Irvine*

Mary Wise, *Northern Virginia Community College*

Charles R. Wyttenbach, *University of Kansas*

Robert Yost, *Indiana University Purdue University Indianapolis*

Xinsheng Zhu, *University of Wisconsin–Madison*

Adrienne Zihlman, *University of California–Santa Cruz*

Unanswered Questions Contributors

CHAPTERS 1–18
Peter J. Russell
Reed College

CHAPTER 19
Douglas J. Futuyma
Stony Brook University

CHAPTER 20
Mohammed Noor
Duke University

CHAPTER 21
Jerry Coyne
University of Chicago

CHAPTER 22
Elena M. Kramer
Harvard University

CHAPTER 23
Rich Glor
University of Rochester

CHAPTER 24
Andrew Pohorille
NASA

CHAPTER 25
Peter J. Russell
Reed College

CHAPTER 26
Geoff McFadden
University of Melbourne

CHAPTER 27
Amy Litt
New York Botanical Garden

CHAPTER 28
Peter J. Russell
Reed College

CHAPTER 29
William S. Irby
Georgia Southern University

CHAPTER 30
Marvalee H. Wake
University of California, Berkeley

CHAPTER 31
Marianne Hopkins
University of Waterloo

Susan Lolle
University of Waterloo

CHAPTER 32
Beverly McMillan

CHAPTER 33
Beverly McMillan

CHAPTER 34
Ravi Palanivelu
University of Arizona

CHAPTER 35
Susan Lolle
University of Waterloo

CHAPTER 36
R. Daniel Rudic
Medical College of Georgia

CHAPTER 37
Paul Katz
Georgia State University

CHAPTER 38
Peter J. Russell
Reed College

CHAPTER 39
Rona Delay
University of Vermont

CHAPTER 40
Peter J. Russell
Reed College

CHAPTER 41
Buel (Dan) Rodgers
Washington State University

CHAPTER 42
Russell Doolittle
University of California, San Diego

CHAPTER 43
Peter J. Russell
Reed College

CHAPTER 44
Ralph Fregosi
University of Arizona

CHAPTER 45
Mark Sheridan
North Dakota State University

CHAPTER 46
Paul H. Yancey
Whitman College

CHAPTER 47
Paul H. Yancey
Whitman College

CHAPTER 48
Laura Carruth
Georgia State University

CHAPTER 49
David Reznick
University of California, Riverside

CHAPTER 50
Anurag Agrawal
Cornell University

CHAPTER 51
Kevin Griffin
*Lamont-Doherty Earth Observatory of
Columbia University*

CHAPTER 52
Camille Parmesan
University of Texas–Austin

CHAPTER 53
Diego Vázquez
*Instituto Argentino de Investigaciones de las
Zonas Áridas*

CHAPTER 54
Gene E. Robinson
University of Illinois at Urbana–Champaign

CHAPTER 55
Michael J. Ryan
University of Texas–Austin

Brief Contents

Contents

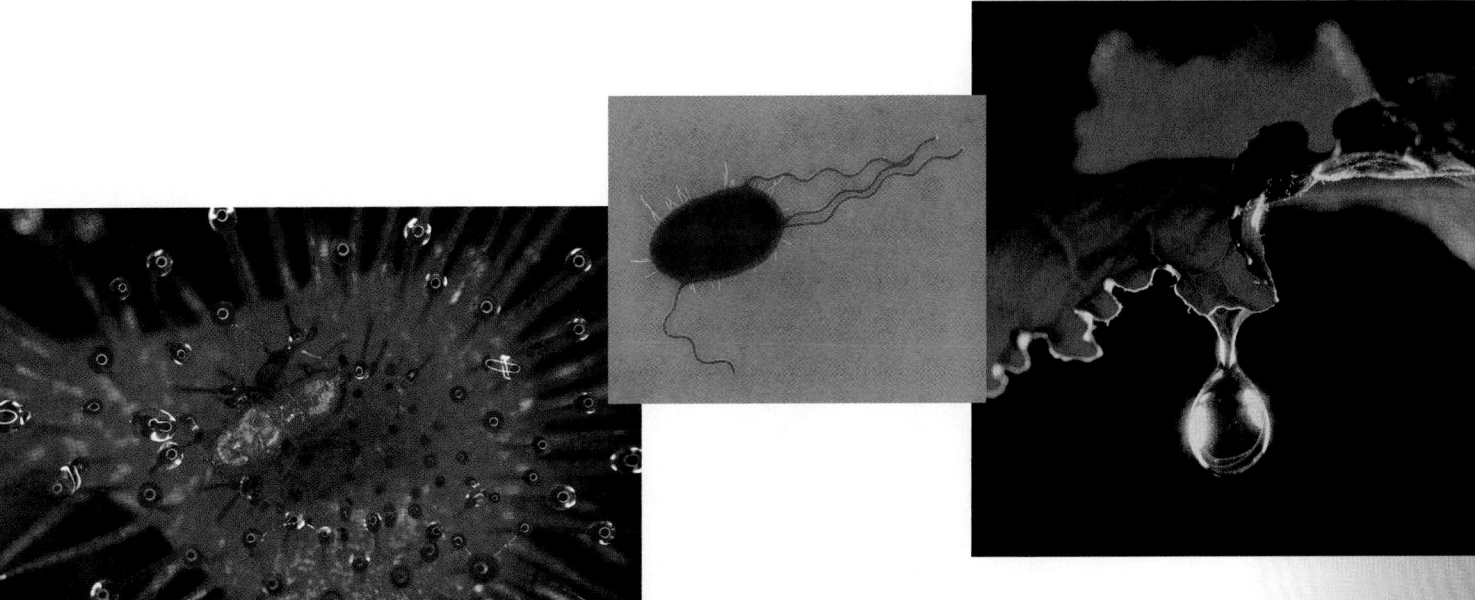

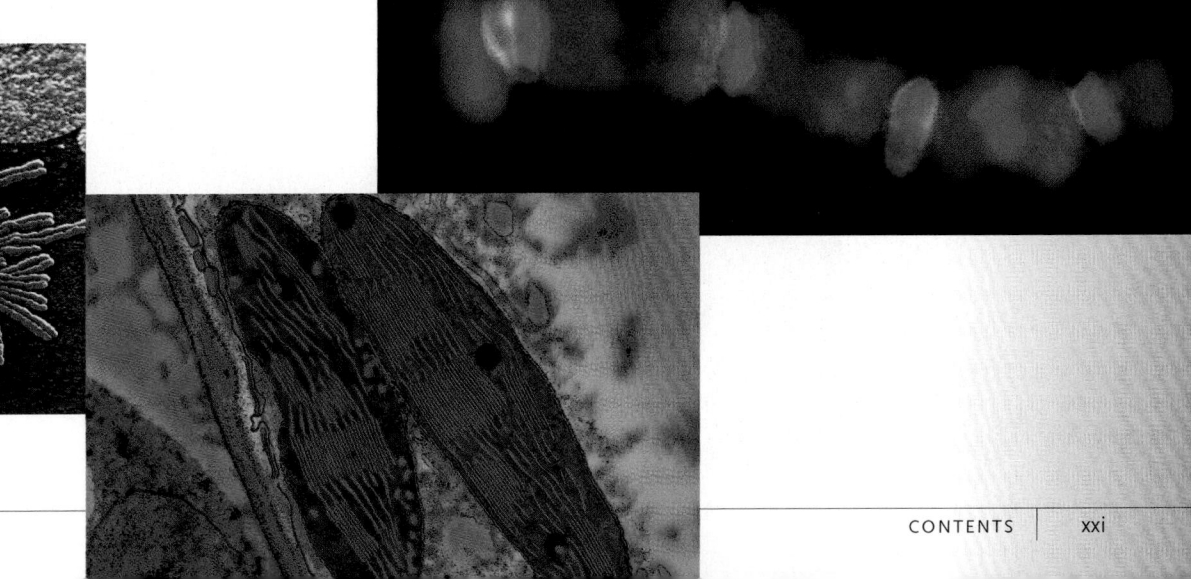

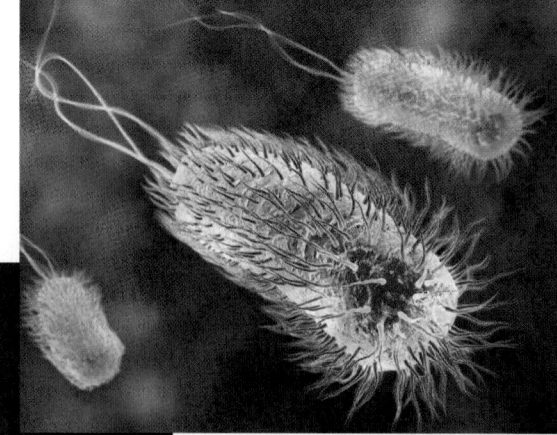

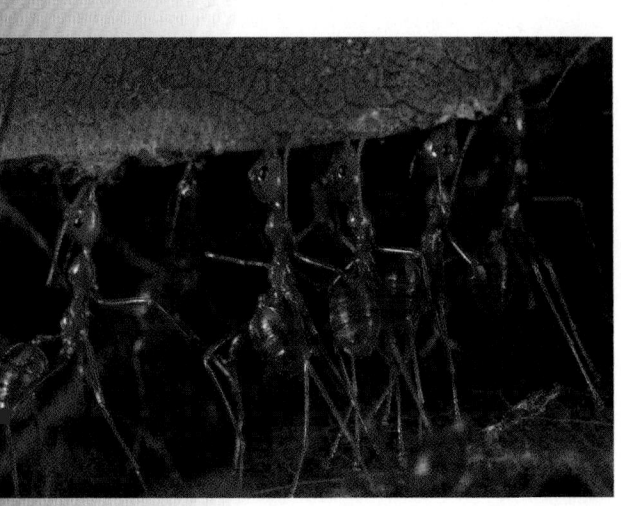

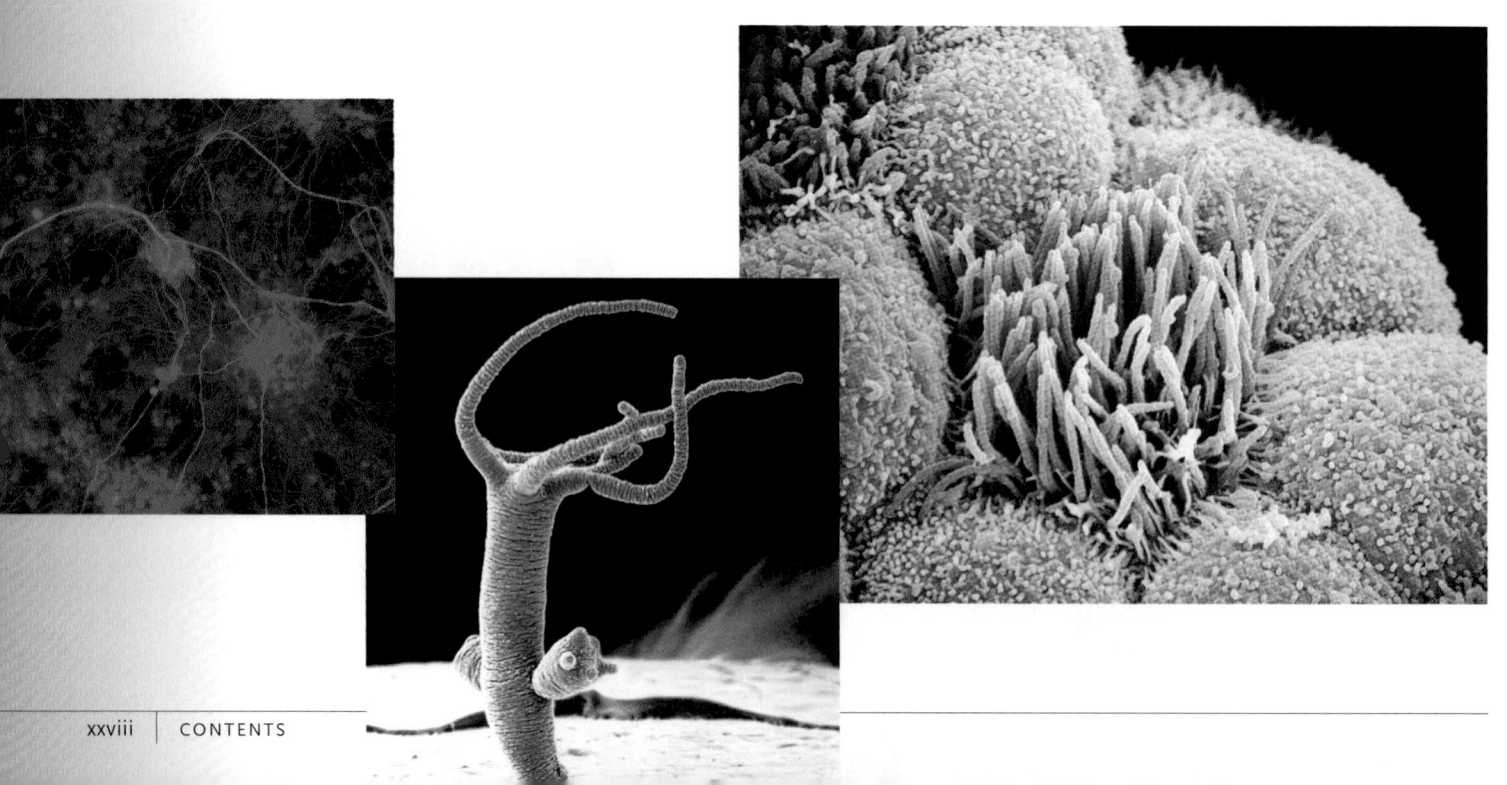

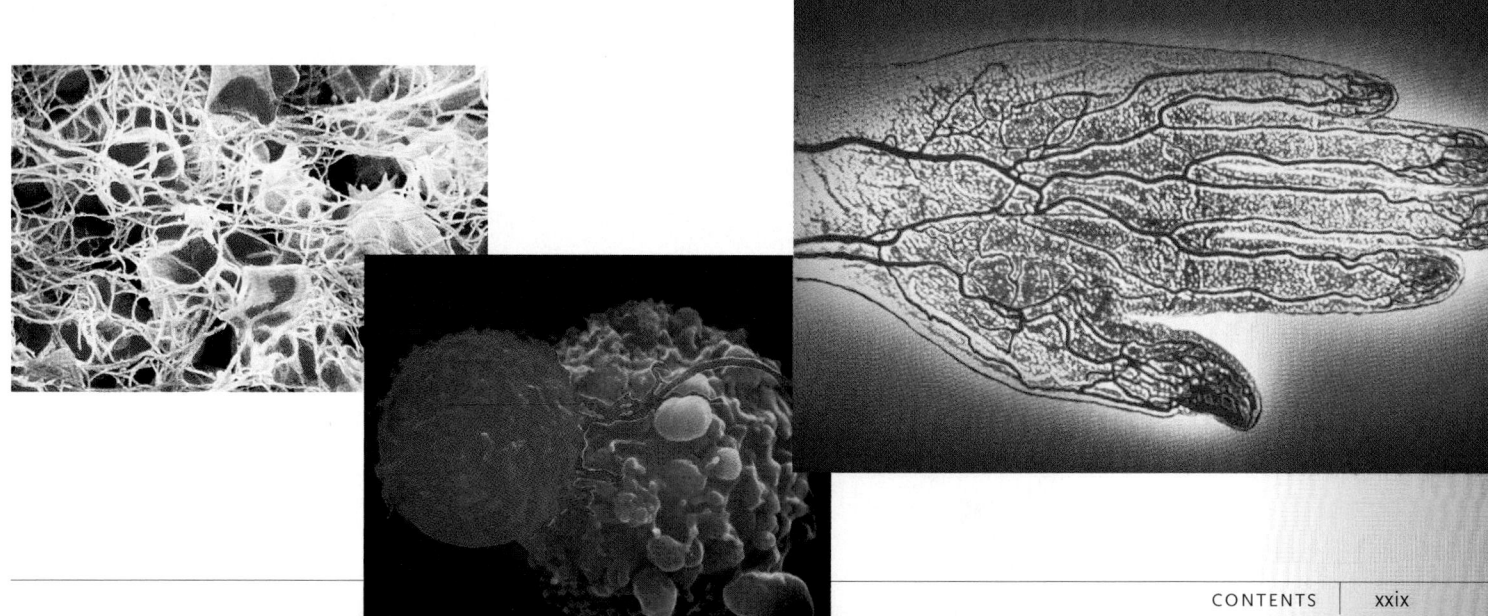

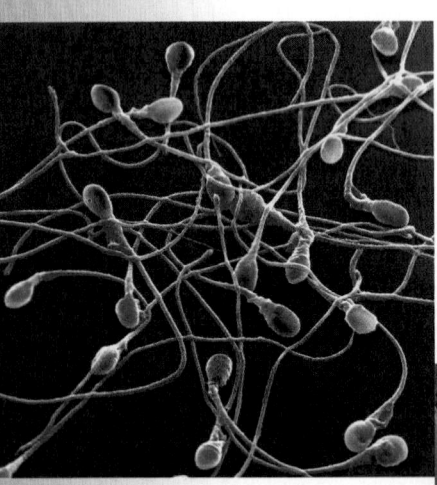

Earth, a planet teeming with life, is seen here in a satellite photograph.

Bryan Allen/Corbis

1 Introduction to Biological Concepts and Research

WHY IT MATTERS

Life abounds in almost every nook and cranny on Earth. A lion creeps through the brush of an African plain, ready to spring at a zebra. The leaves of a sunflower in Kansas turn slowly through the day, keeping their surfaces fully exposed to the sun's light. Fungi and bacteria in the soil of a Canadian forest obtain nutrients from decomposing organisms. A child plays in a park in Madrid, laughing happily as his dog chases a tennis ball. In one room of a nearby hospital, a mother hears the first cry of her newborn baby; in another room, an elderly man sighs away his last breath. All over the world, countless organisms are born, live, and die every second of every day. How did life originate, how does it persist, and how is it changing? Biology, the science of life, provides scientific answers to these questions.

What *is* life? Offhandedly, you might say that although you cannot define it, you know it when you see it. The question has no simple answer, because the story of life has been unfolding for billions of years, ever since ancient events assembled nonliving materials into the first organized, living cells. Clearly, any list of criteria for the living state only hints at the meaning of "life." Deeper insight requires a

1

wide-ranging examination of the characteristics of life, which is what this book is all about.

Over the next semester or two, you will encounter examples of how organisms are constructed, how they function, where they live, and what they do. The examples provide evidence in support of concepts that will greatly enhance your appreciation and understanding of the living world, including its fundamental unity and striking diversity. This chapter provides a brief overview of these basic concepts. It also describes some of the ways in which biologists conduct research: the process in which they observe nature, formulate explanations of their observations, and test their ideas. It is through research that we further our knowledge of living systems.

1.1 What Is Life? Characteristics of Living Systems

Picture a lizard on a rock, slowly shifting its head to follow the movements of another lizard nearby **(Figure 1.1).** You know that the lizard is alive and that the rock is not. If you examine both at the atomic and molecular levels, however, you will find that the differences between them blur. Lizards, rocks, and all other things are composed of atoms and molecules, which behave according to the same physical laws. Nevertheless, living systems share a set of characteristics that collectively set them apart from nonliving matter.

The differences between a lizard and a rock depend not only on the kinds of atoms and molecules present but also on their organization and their interactions. Individual organisms are at the middle of a hierarchy that ranges from the atoms and molecules within their bodies to the assemblages of organisms that occupy Earth's environments. Within every individual, certain biological molecules contain instructions for building other molecules, which, in turn, are assembled into complex structures. Living organisms must gather energy and materials from their surroundings to build new

Kevin Schafer

Figure 1.1
Living organisms and inanimate objects. All living organisms, such as this lizard *(Iguana iguana),* have characteristics that are fundamentally different from those of inanimate objects, such as the rock on which it is sitting.

biological molecules, grow in size, maintain and repair their parts, and produce offspring. They must also respond to environmental changes by altering their chemistry and activity in ways that allow them to survive. Finally, the structure and function of living systems often change from one generation to the next.

Living Systems Are Organized in a Hierarchy, with Each Level of Organization Having Its Own Emergent Properties

The organization of life extends through several levels of a hierarchy **(Figure 1.2).** Complex biological molecules exist at the lowest level of organization, but by themselves, these molecules are not alive. The properties of life do not appear until they are organized into cells. A **cell** consists of an organized chemical system, including many specialized molecules, surrounded by a membrane. A cell is the lowest level of biological organization that can survive and reproduce—as long as it has access to a usable energy source, the necessary raw materials, and appropriate environmental conditions. However, a cell is alive only as long as it is organized as a cell; if broken into its component parts, a cell is no longer alive even if the parts themselves are unchanged. Characteristics that depend on the level of organization of matter, but do not exist at lower levels of organization, are called **emergent properties.** Life is thus an emergent property of the organization of matter into cells.

Many single cells, such as bacteria and protozoans, exist as **unicellular organisms.** By contrast, plants and animals are **multicellular organisms.** Their cells live in tightly coordinated groups and are so interdependent that they cannot survive on their own. For example, human cells cannot live by themselves in nature because they must be bathed in body fluids and supported by the activities of other cells. Like individual cells, multicellular organisms have emergent properties that their individual components lack; for example, humans can learn biology.

The next, more inclusive level of organization is the **population,** a group of unicellular or multicellular organisms of the same kind that live together in the same place. The humans who occupy the island of Tahiti, a colony of penguins in Antarctica, or a group of bacteria in a laboratory flask are examples of populations. Like multicellular organisms, populations have emergent properties that do not exist at lower levels of organization. For example, a population has characteristics such as its birth or death rate—that is, the number of individuals who are born or die over a period of time—that do not exist for single cells or individual organisms.

Working our way up the biological hierarchy, all the populations of different organisms that live in the same place form a **community.** The bacteria, penguins, fishes, seals, whales, and other organisms that

Biosphere

All regions of Earth's crust, waters, and atmosphere that sustain life

Bryan Allen/Corbis

Ecosystem

Group of communities interacting with their shared physical environment

Jamie and Judy Wild/Danita Delimont.com

Community

Populations of all species that occupy the same area

Ron Sefton/Bruce Coleman USA

Population

Group of individuals of the same kind (that is, the same species) that occupy the same area

Jamie and Judy Wild/Danita Delimont.com

Multicellular organism

Individual consisting of interdependent cells

Edward Snow/Bruce Coleman USA

Cell

Smallest unit with the capacity to live and reproduce, independently or as part of a multicellular organism

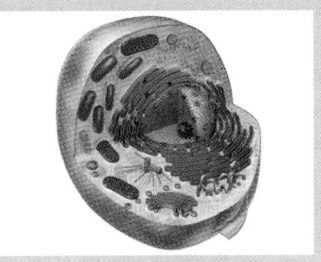

Figure 1.2
The hierarchy of life. Each level in the hierarchy of life exhibits emergent properties that do not exist at lower levels. The middle four photos depict a rocky intertidal zone on the coast of Washington State.

live along the coast of Antarctica, taken together, make up a community. The next highest level, the **ecosystem**, includes the community *and* the nonliving environmental factors with which it interacts. For example, a forest ecosystem comprises a community of living organisms, as well as soil, air, water, minerals, and the input of the energy in sunlight. The highest level, the **biosphere**, encompasses all the ecosystems of Earth's waters, crust, and atmosphere. Communities, ecosystems, and the biosphere also have emergent properties. For example, communities can be described in terms of their *diversity*—the number and types of different populations they contain—and their *stability*—the degree to which the populations within the community remain the same through time.

Living Systems Contain Chemical Instructions That Govern Their Structure and Function

The most fundamental and important molecule that distinguishes living systems from nonliving matter is **deoxyribonucleic acid (DNA; Figure 1.3).** DNA is a large, double-stranded, helical molecule that contains instructions for assembling a living organism from simpler molecules. The organisms that we recognize as bacteria, trees, fishes, or humans include some striking differences in their DNA. Thus, molecular building blocks are arranged differently in each of these organisms, producing differences in their appearance and function. (Some nonliving systems, notably certain viruses [see Chapter 25 for a discussion of viruses], also contain DNA, but biologists do not consider viruses to be alive because they cannot reproduce independently of the organisms they infect.)

DNA functions similarly in all living organisms. The information in DNA is copied into molecules of a related substance, **ribonucleic acid (RNA),** which then directs the production of different protein molecules **(Figure 1.4). Proteins** carry out most of the activities of life, including the synthesis of all other biologi-

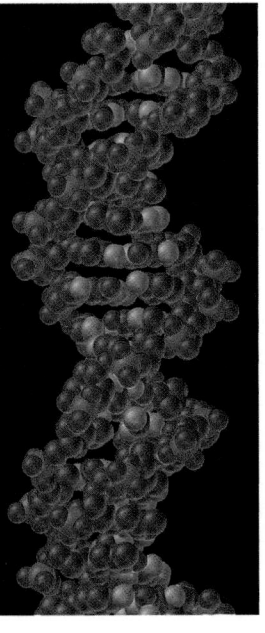

Figure 1.3
Deoxyribonucleic acid (DNA).
A computer-generated model of DNA illustrates that it is made up of two strands twisted into a double helix.

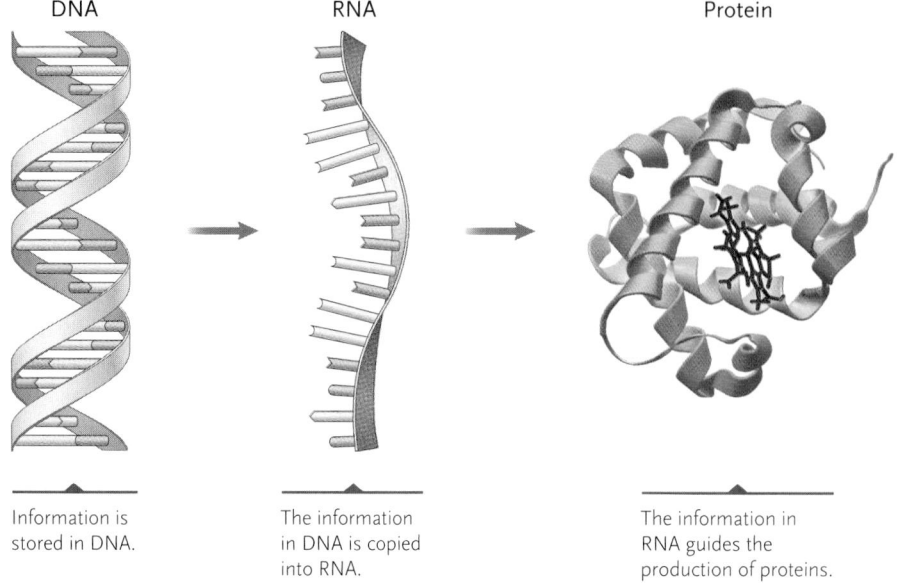

DNA   RNA   Protein

Information is stored in DNA.

The information in DNA is copied into RNA.

The information in RNA guides the production of proteins.

Figure 1.4

The pathway of information flow in living organisms. Information stored in DNA is copied into RNA, which then directs the construction of protein molecules. The protein shown here is one of four subunits of hemoglobin, an oxygen-carrying protein found inside red blood cells.
(PDB ID: 1BBB; Silva, M. M., Rogers, P. H., Amone, A. A third quaternary structure of hemoglobin A at 1.7-Å resolution, *J Biol Chem,* 267, p. 17248, 1992.)

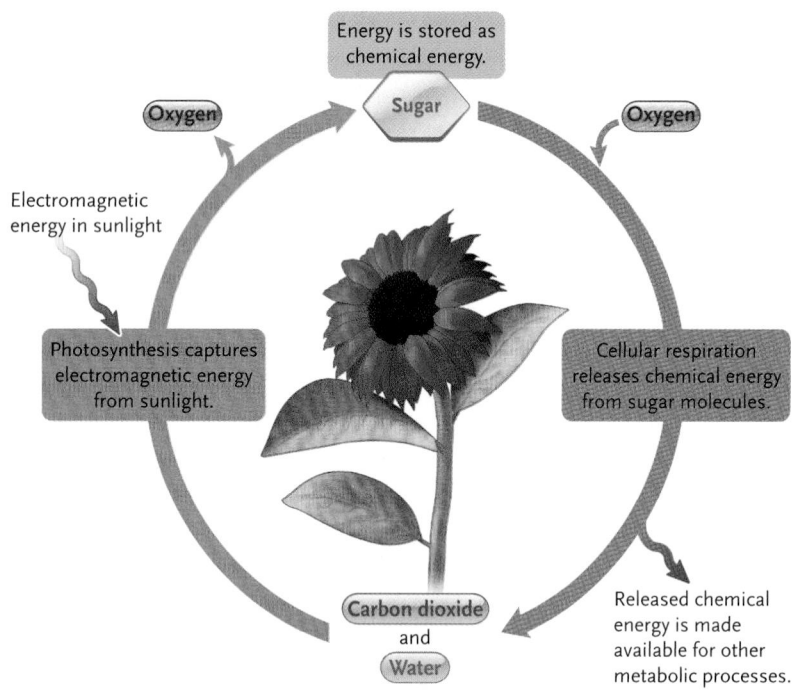

Figure 1.5

Metabolic activities. Photosynthesis uses the energy in sunlight to build sugar molecules from carbon dioxide and water, releasing oxygen as a by-product of the reaction. The process converts the electromagnetic energy in sunlight into chemical energy, which is stored in sugars and starches. Cellular respiration uses oxygen to break down sugar molecules, releasing their chemical energy and making it available for other metabolic processes.

cal molecules. This information pathway is preserved from generation to generation by the ability of DNA to direct its own replication so that offspring receive the same basic molecular instructions as their parents.

Living Organisms Engage in Metabolic Activities

Metabolism is a key property of living cells and organisms. **Metabolism** describes the ability of a cell or organism to extract energy from its surroundings and use that energy to maintain itself, grow, and reproduce. As a part of metabolism, cells carry out chemical reactions that assemble, alter, and disassemble molecules **(Figure 1.5).** For example, a growing sunflower plant carries out **photosynthesis**, in which the electromagnetic energy in sunlight is absorbed and converted into chemical energy. The cells of the plant store some chemical energy in sugars and starches, and they use the rest to manufacture other biological molecules from simple raw materials obtained from the environment.

Sunflowers concentrate some of their energy reserves in seeds from which more sunflower plants may grow. The chemical energy stored in the seeds also supports other organisms, such as insects, birds, and humans, that eat them. Most organisms, including sunflower plants, tap stored chemical energy through another metabolic process, **cellular respiration**, in which complex biological molecules are broken down with oxygen, releasing some of their energy content for cellular activities.

Figure 1.6

Energy flow and nutrient recycling. Within most ecosystems, energy flows from the sun to producers to consumers to decomposers. In this illustration, the sun provides energy to grasses (producers); the zebras (consumers) then feed on the grasses before being eaten by a lion (another consumer); and fungi (decomposers) absorb nutrients and energy from the digestive wastes of animals and from the remains of dead animals and plants. All of the energy that enters an ecosystem is ultimately lost from the system as heat. Nutrients move through the same pathways, but they are conserved and recycled.

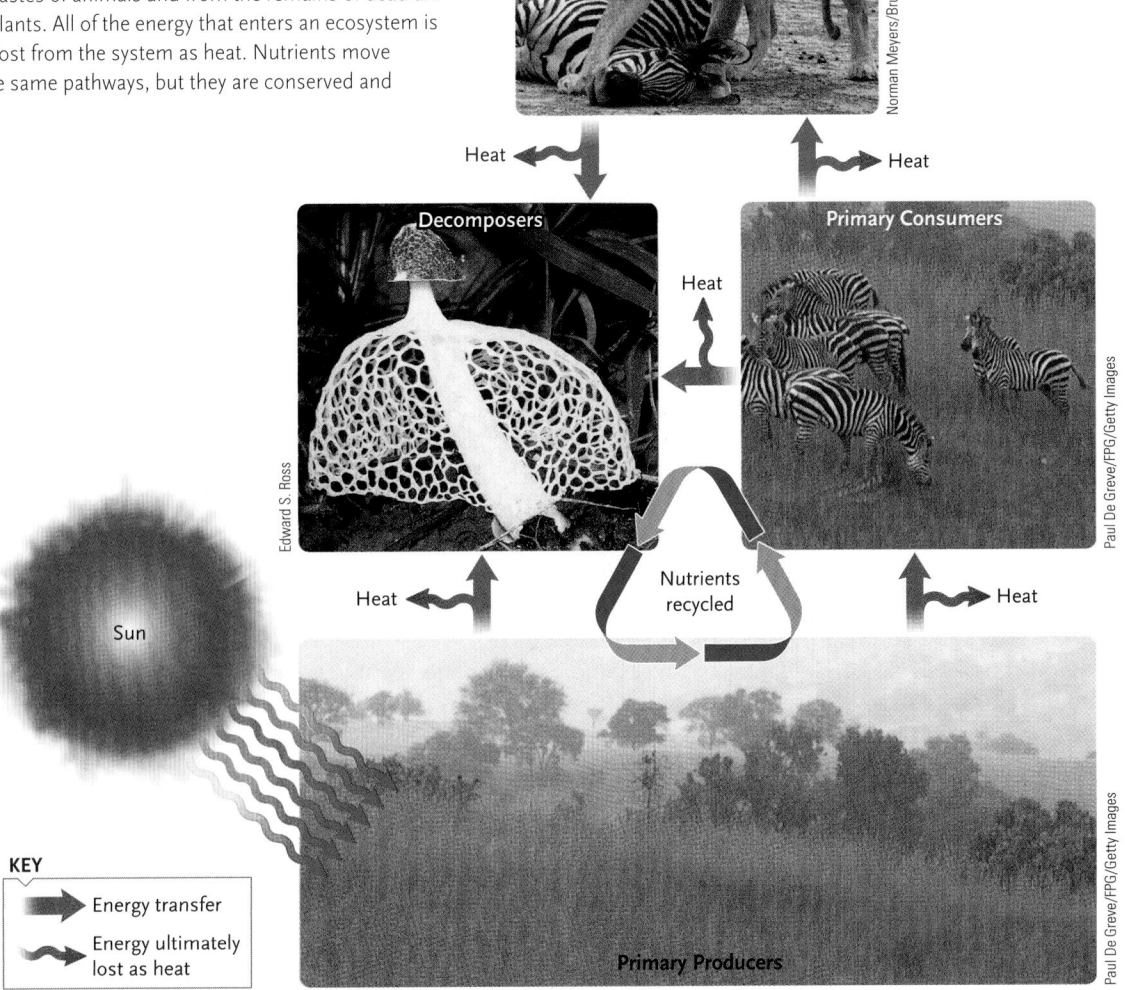

Energy Flows and Matter Cycles through Living Systems

With few exceptions, energy from sunlight supports life on Earth. Plants and other photosynthetic organisms absorb energy from sunlight and convert it into chemical energy, which they use to assemble complex molecules, such as sugars, from simple raw materials, such as water and carbon dioxide. As such, photosynthetic organisms are the **primary producers** of the food on which all other organisms rely. By contrast, animals are **consumers**: directly or indirectly, they feed on the complex molecules manufactured by plants **(Figure 1.6)**. For example, zebras tap directly into the molecules of plants when they eat grass, and lions tap into it indirectly when they eat zebras. Certain bacteria and fungi are **decomposers**: they feed on the remains of dead organisms, breaking down complex biological molecules into simpler raw materials, which may then be recycled by the producers.

Some of the energy that photosynthetic organisms trap from sunlight *flows* within and between populations, communities, and ecosystems. But because the biological processes that transfer energy from one organism to another are not 100% efficient, some of the energy is lost as heat. Although heat energy can be used by some animals to maintain body temperature, it cannot sustain other life processes. By contrast, matter—nutrients such as carbon and nitrogen—*cycles* between living organisms and the nonliving components of the biosphere, to be used again and again (see Figure 1.6).

Living Organisms Compensate for Changes in the External Environment

All objects, whether living or nonliving, respond to changes in the environment; for example, a rock warms up on a sunny day and cools at night. But only living systems have the capacity to detect environmental changes and *compensate* for them through controlled responses. They do so by means of diverse and varied *receptors*—molecules or larger structures, located on individual cells and body surfaces, that are able to detect changes in external and internal conditions. When stimulated, the receptors trigger reactions that produce a compensating response.

For example, your internal body temperature remains reasonably constant, even though the environment in which you live is usually either cooler or warmer than you are. Your body compensates for these environmental variations and maintains its internal temperature at about 37° Celsius (C). When environmental temperatures decrease significantly, receptors in your skin detect the temperature change and transmit that information to your brain. Your brain may send a signal to your muscles, causing you to shiver; the muscular activity of shivering releases heat that keeps your body temperature from dropping below its optimal level. When environmental temperatures increase significantly, glands in your skin secrete sweat, which evaporates, cooling the skin and its underlying blood supply. The cooled blood circulates internally and keeps your body temperature from rising above 37°C. People also compensate behaviorally by dressing warmly on a cold winter day or jumping into a swimming pool in the heat of summer. Maintaining your body's internal temperature within a narrow tolerable range is one example of **homeostasis**—a steady internal condition maintained by responses that compensate for changes in the external environment. All organisms have mechanisms that help maintain homeostasis in relation to temperature, blood chemistry, or other important factors.

Living Organisms Reproduce and Undergo Development

Humans and all other organisms are part of an unbroken chain of life that began billions of years ago. This chain continues today through **reproduction,** the process in which parents produce offspring. Offspring generally resemble their parents because the parents pass copies of their DNA—with all the accompanying instructions for virtually every life process—to their offspring. The transmission of DNA (that is, genetic information) from one generation to the next is called **inheritance.** For example, the eggs produced by storks hatch into little storks, not into pelicans, because they inherited stork DNA, which is different from pelican DNA.

Multicellular organisms also undergo a process of **development**, a series of programmed changes encoded in DNA, through which a fertilized egg divides into many cells that ultimately are transformed into an adult, which is itself capable of reproduction. As an example, consider the development of a moth **(Figure 1.7).** This insect begins its life as a tiny egg that contains all the instructions necessary for its development into an adult moth. Following these instructions, the egg first hatches into a caterpillar, a larval form adapted for feeding and rapid growth. The caterpillar increases in size until internal chemical signals indicate that it is time to spin a cocoon and become a pupa. Inside its cocoon, the pupa undergoes profound developmental changes that remodel its body completely. Some cells die, and others multiply and become organized in different patterns. When these transformations are complete, the adult moth emerges from the cocoon. It is equipped with structures and behaviors, quite different from those of the caterpillar, that enable it to reproduce.

The sequential stages through which individuals develop, grow, maintain themselves, and reproduce are known collectively as the **life cycle** of an organism. The moth's life cycle includes egg, larva, pupa, and adult stages. Adult moths, through reproduction, continue the cycle by producing the sperm and eggs that unite to form the fertilized egg, which starts the next generation.

Figure 1.7
Life cycle of a giant silkworm moth (family Saturniidae).

a. Egg   **b.** Larva   **c.** Pupa   **d.** Recently emerged adult   **e.** Adult

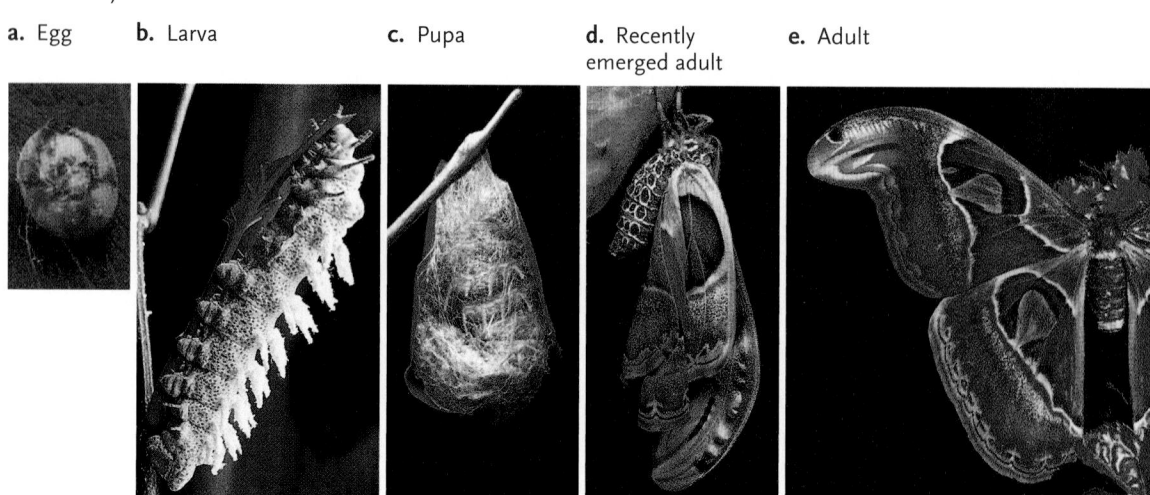

Photographs by Jack de Coningh

Populations of Living Organisms Change from One Generation to the Next

Although offspring generally resemble their parents, individuals with unusual characteristics sometimes suddenly appear in a population. Moreover, the features that distinguish these oddballs often are inherited by their offspring. Our awareness of the inheritance of unusual characteristics has had an enormous impact on human history because it allows plant and animal breeders to produce crops and domesticated animals with especially desirable characteristics.

Biologists have observed that similar changes also take place under natural conditions. In other words, populations of all organisms change from one generation to the next, because some individuals experience changes in their DNA and they pass those modified instructions along to their offspring. We consider this fundamental process, **biological evolution**, in the next section.

STUDY BREAK

1. List the major levels in the hierarchy of life and identify one emergent property of each level.
2. What do living organisms do with the energy they collect from the external environment?
3. What is a life cycle?

1.2 Biological Evolution

All research in biology—ranging from analyses of the precise structure of biological molecules to energy flow through the biosphere—is undertaken with the knowledge that biological evolution has shaped life on Earth. Our understanding of the evolutionary process reveals several truths about the living world: (1) all populations change through time, (2) all organisms are related through a shared ancestry, and (3) evolution has produced the spectacular diversity of life that we see around us. Evolution is the unifying theme that links all the subfields of the biological sciences.

Darwin and Wallace Explained How Populations of Organisms Change through Time

How do evolutionary changes take place? One important mechanism was first explained in the mid-nineteenth century by two British naturalists, Charles Darwin and Alfred Russel Wallace. On a 5-year voyage around the world, Darwin observed many strange and wondrous organisms. He also found fossils of species that are now extinct (that is, all members of the species are dead). The extinct forms often resembled living species in some traits but differed in others. Originally a believer in special creation—the idea that living organisms were placed on Earth in their present numbers and kinds and have not changed since their creation—Darwin became convinced that organisms do not remain constant with the passage of time, but change from one form into another. Wallace came to the same conclusion through his observations of the great variety of plants and animals in the Amazon basin and the region now called Malaysia.

Darwin also studied the process of evolution through observations and experiments on domesticated animals. Pigeons were among his favorite experimental subjects. Domesticated pigeons exist in a variety of sizes, colors, and shapes **(Figure 1.8).** Darwin noted that pigeon breeders who wished to promote a certain characteristic, such as elaborately curled tail feathers, selected individuals with the most curl in their feathers as parents for the next generation. By permitting only these birds to mate, the breeders fostered the desired characteristic and gradually eliminated or reduced other traits. The same practice is still used today to increase the frequency of desired traits in tomatoes, dogs, and other domesticated plants and animals. Darwin called this practice **artificial selection.** He termed the equivalent process that occurs in nature **natural selection.**

In 1858, Darwin and Wallace formally summarized their observations and conclusions explaining biological evolution. (1) Most organisms can produce

Wild rock dove

Figure 1.8
Artificial selection. Using artificial selection, pigeon breeders have produced more than 300 varieties of domesticated pigeons from ancestral wild rock doves (*Columba livia*).

Photographs courtesy Derrell Fowler, Tecumseh, Oklahoma

numerous offspring, but environmental factors limit the number that actually survive and reproduce. (2) Heritable variations allow some individuals to compete more successfully for space, food, and mates. (3) These successful individuals somehow pass the favorable characteristics to their offspring. (4) As a result, the favorable traits become more common in the next generation, and less successful traits become less common. This process of natural selection results in evolutionary change. Today, evolutionary biologists recognize that natural selection is just one of several potent evolutionary processes.

Over many generations, the evolutionary changes in a population may become extensive enough to produce a new kind of organism. These new types are distinct from their ancestors and cannot interbreed with them. Nevertheless, parental and descendant species often share many characteristics, allowing researchers to understand their relationships and reconstruct their shared evolutionary history. Starting with the first organized cells, this aspect of evolutionary change has contributed to the diversity of life that exists today.

Darwin and Wallace described evolutionary change largely in terms of how natural selection changes the commonness or rarity of particular variations over time. Their intellectual achievement was remarkable for its time. Although Darwin and Wallace understood the central importance of variability among organisms to the process of evolution, they could not explain how new variations arose or how they were passed to the next generation.

Mutations in DNA Are the Raw Materials That Allow Evolutionary Change

Today, we know that both the origin and the inheritance of new variations arise from the structure and variability of DNA, which is organized into functional units called **genes**. Each gene contains the code for (that is, the instructions for building) a protein molecule or one of its parts. Proteins determine all the structural and functional characteristics of an organism.

Variability among individuals—the raw material molded by evolutionary processes—arises ultimately through **mutations**, random changes in the structure, number, or arrangement of DNA molecules. When mutations occur in the DNA of reproductive cells, they may change the instructions for the development of offspring that the reproductive cells produce. Many mutations are of neutral value to individuals bearing them, and some turn out to be harmful. On rare occasions, however, a mutation is beneficial under the prevailing environmental conditions. Beneficial mutations increase the likelihood that individuals carrying the mutation will survive and reproduce. Thus, through the persistence and spread of beneficial mutations among individuals and their descendants, the genetic makeup of a population will change from one generation to the next.

Adaptations Enable Organisms to Survive and Reproduce in the Environments Where They Live

Favorable mutations may produce **adaptations**, characteristics that help an organism survive longer or reproduce more under a particular set of environmental conditions. To convey the sense of how organisms benefit from adaptations, consider an example from the recent literature on *cryptic coloration* (camouflage) in animals. Many animals have skin, scales, feathers, or fur that matches the color and appearance of the background in their environment, enabling them to blend into their surroundings. Camouflage makes it harder for predators to identify and then catch them—an obvious advantage to survival. Animals that are not camouflaged are often just sitting ducks.

The rock pocket mouse *(Chaetodipus intermedius)*, which lives in the deserts of the southwestern United States, is mostly nocturnal (that is, active at night). At most desert localities, the rocks are pale brown, and rock pocket mice have sandy-colored fur on their backs. However, at several sites, the rocks—remnants of lava flows from now-extinct volcanoes—are black. At these localities, rock pocket mice have black fur on their backs. Thus, like the sandy-colored mice in other areas, they, too, are camouflaged in their habitats **(Figure 1.9)**. Camouflage appears to be important to these mice because owls, which locate prey using their exceptionally keen eyesight, frequently eat nocturnal desert mice.

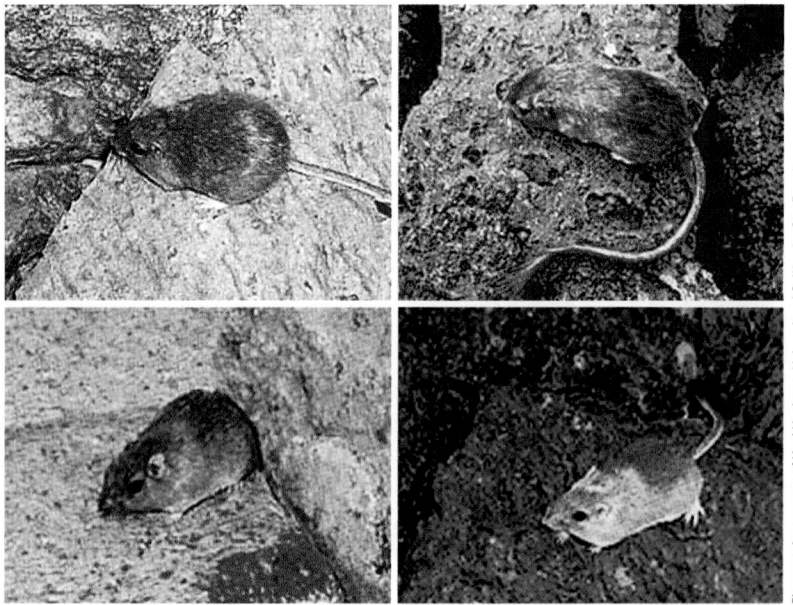

Photographs courtesy of Hopi Hoekstra, University of California, San Francisco

Figure 1.9
Camouflage in rock pocket mice *(Chaetodipus intermedius)*. Sandy-colored mice are well camouflaged on pale rocks, and black mice are well camouflaged on dark rocks (top); but mice with fur that does not match their backgrounds (bottom) are easy to see.

Examples of cryptic coloration are well documented in the scientific literature, and biologists generally interpret them as adaptations that reduce the likelihood of being captured by a predator. Nevertheless, few researchers have been able to identify precisely the genetic mutations that produced these adaptations. Michael W. Nachman, Hopi E. Hoekstra, and their colleagues at the University of Arizona tackled this problem in a study of the genetic and evolutionary basis for the color difference between rock pocket mice that live on light and dark backgrounds. In an article published in 2003, they reported the results of an analysis of mice sampled at six sites in southern Arizona and New Mexico. In two regions (Pinacate, AZ, and Armendaris, NM), both light and dark rocks were present, allowing the researchers to compare mice that lived on differently colored backgrounds. Two other sites had only light rocks and sandy-colored mice.

Nachman and his colleagues found that nearly all of the mice they captured on dark rocks had dark fur and that nearly all of the mice they captured on light rocks had light fur **(Figure 1.10)**. The researchers then studied the structure of *Mc1r*, a gene known to influence fur color in laboratory mice. The 17 black mice from Pinacate all shared certain mutations in their *Mc1r* gene, which established four specific changes in the structure of the Mc1r protein. However, none of the 12 sandy-colored mice from Pinacate carried these mutations. The exact match between the presence of the mutations and the color of the mouse strongly suggests that these mutations in the *Mc1r* gene are responsible for the dark fur in the mice from Pinacate. Thus, data on the distributions of light and dark mice coupled with analyses of their DNA suggest that the color difference is the product of specific mutations that were favored by natural selection.

Nachman's team then analyzed the *Mc1r* gene in the dark and light mice from Armendaris and in the light mice at two intermediate sites. Because the mice in these regions also closely matched the color of their environments, the researchers expected to find the *Mc1r* mutations in the dark mice but not in the light mice. However, none of the mice from Armendaris shared any of the mutations that contributed to the dark color of mice from Pinacate. Apparently, mutations in some other gene or genes, which the researchers have not yet identified, are responsible for the camouflaging black coloration of mice that live on black rocks in Armendaris.

The example of an adaptation provided by the rock pocket mice illustrates the observation that genetic differences often develop between populations. Sometimes these differences become so great that the organisms develop different appearances and adopt different ways of life, and biologists regard them as distinct types. Over immense spans of time, evolutionary processes have produced many types of organisms, which

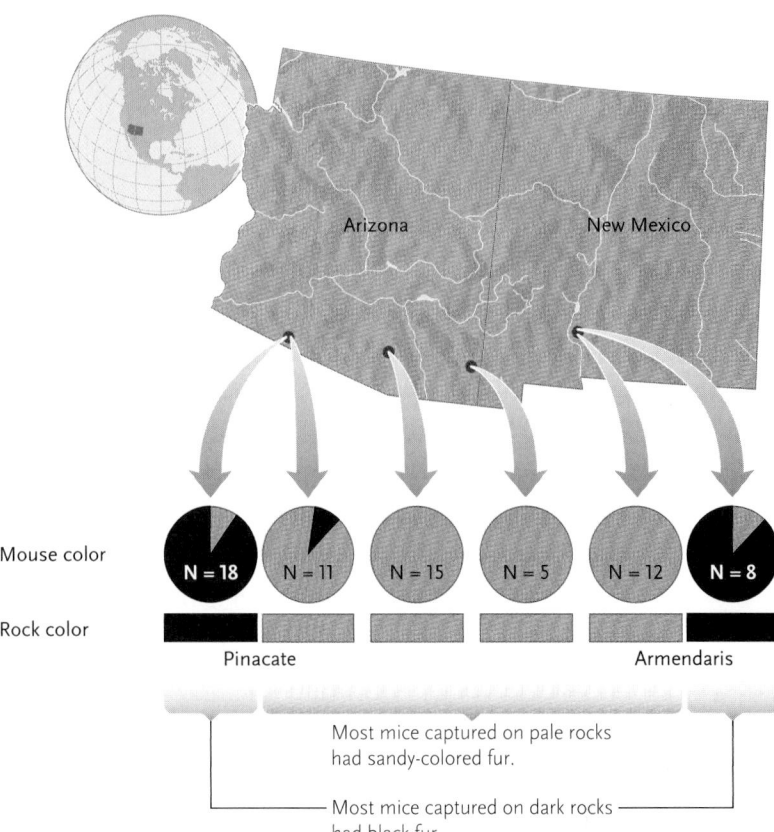

Figure 1.10

Distributions of rock pocket mice with light and dark fur. At six sites in Arizona and New Mexico, mouse fur color closely matched the colors of the backgrounds where they lived. The pie charts below the map show the proportion of mice with sandy-colored or black fur. N indicates the number of mice sampled at each site. The bars beneath the pie charts indicate the rock color.

constitute the diversity of life on Earth. In the next section, we survey this diversity and consider how it is classified for study by biologists.

STUDY BREAK

1. What is the difference between artificial selection and natural selection?
2. How do random changes in the structure of DNA affect the characteristics of organisms?
3. What is the usefulness of being camouflaged in natural environments?

1.3 Biodiversity

The great diversity of life, the product of evolution, represents the many different ways in which the common elements of life's organization have combined to provide new and successful ways to survive and reproduce.

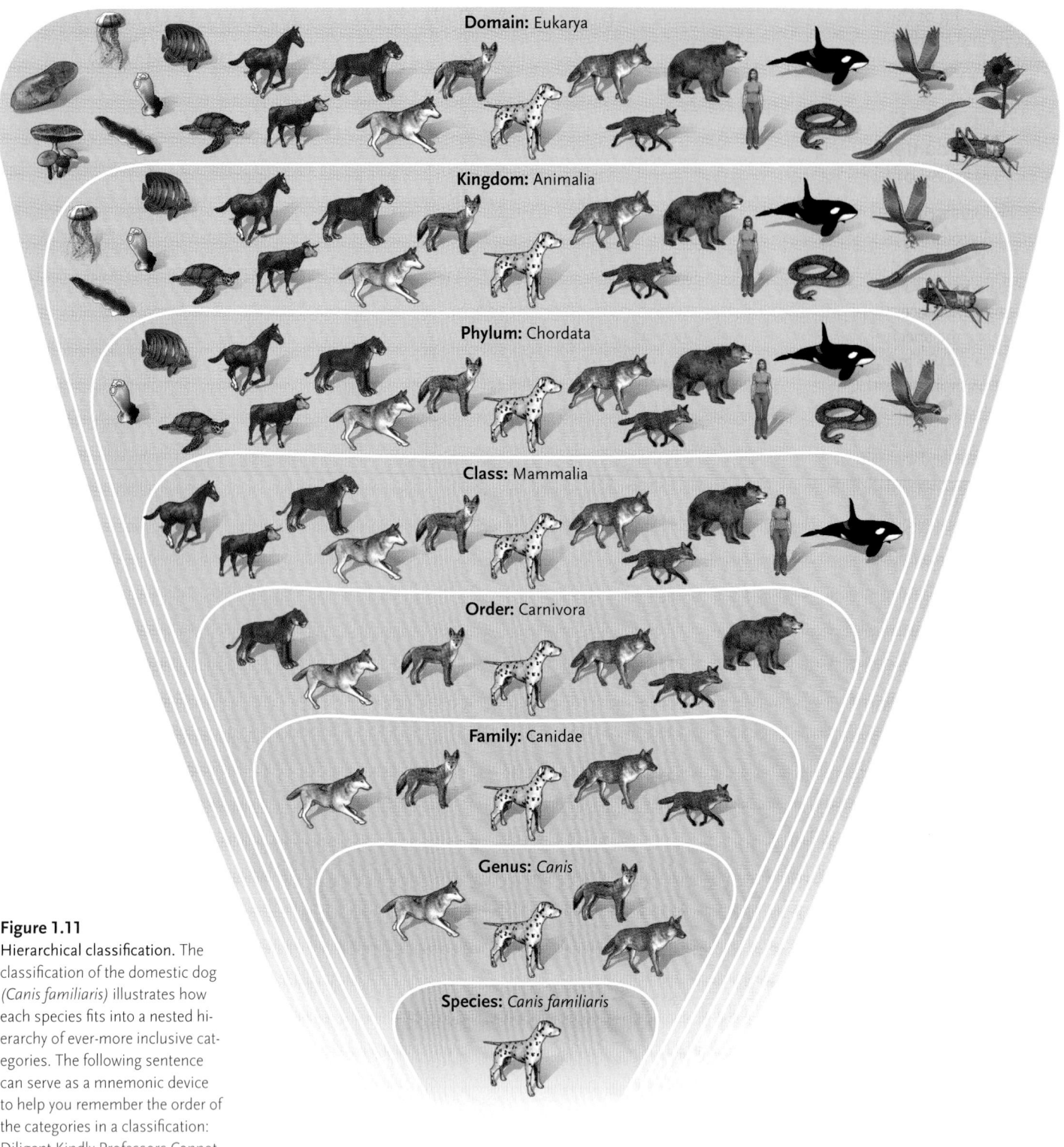

Figure 1.11
Hierarchical classification. The classification of the domestic dog *(Canis familiaris)* illustrates how each species fits into a nested hierarchy of ever-more inclusive categories. The following sentence can serve as a mnemonic device to help you remember the order of the categories in a classification: Diligent Kindly Professors Cannot Often Fail Good Students.

Millions of different kinds of organisms live on Earth. Many millions more existed in the past and became extinct. To make sense of the past and present diversity of life on Earth, scientists have developed classification systems that attempt to arrange organisms, living and dead, into groups that reflect their relationships and evolutionary origins. Although scientists traditionally relied on similarities and differences in external appearance to understand these evolutionary relationships, they now use analyses of proteins and DNA in this effort. The task is so daunting that there is no consensus on the numbers and kinds of divisions and categories to use; the classification system also changes as investigators learn more about extinct and living organisms. The attempt is worth the effort, however, because the classification of life leads to greater understanding of the relationships between living organisms and sheds light on the pathways of evolutionary change.

Biologists Consider the Species to Be a Fundamental Unit in a Hierarchy of Categories

Most biologists consider the species to be the most fundamental grouping in the diversity of life. A **species** is a group of populations in which the individuals are so closely related in structure, biochemistry, and behavior that they can successfully interbreed. At the level directly above the species, biologists recognize the **genus** (plural, *genera*), a group of similar species that share recent common ancestry. Species in the same genus usually also share many characteristics. For example, a group of closely related four-legged mammals that have elongated faces, large piercing teeth at the front of the mouth, slicing teeth behind them, and crushing teeth at the back of the mouth are classified together in the genus *Canis,* commonly known as dogs.

Each species is assigned a two-part **scientific name:** the first part identifies the genus to which it belongs, and the second part designates a particular species within that genus. In the genus *Canis,* for example, *Canis familiaris* is the scientific name of the domesticated dog; *Canis lupus,* the gray wolf, and *Canis latrans,* the coyote, are two other species in the same genus. Scientific names are always written in italics, and only the genus name is capitalized. After its first mention in a discussion, the genus name is frequently abbreviated to its first letter, as in *C. familiaris* and *C. lupus.*

At successively more inclusive levels **(Figure 1.11)**, related genera are placed in the same **family**, related families in the same **order**, and related orders in the same **class**. Related classes are grouped into a **phylum** (plural, *phyla*), and related phyla are assigned to a **kingdom**. In recent years, biologists have added the **domain** as the most inclusive group.

Biologists Classify Organisms into Three Domains and Several Kingdoms

Biologists distinguish three domains—Bacteria, Archaea, and Eukarya. Each domain represents a group of cellular organisms with characteristics that set it apart as a major branch of the evolutionary tree. Two of the three domains, Bacteria and Archaea, are described as **prokaryotes** (*pro* = before; *karyon* = nucleus) because they exhibit a relatively simple organization of their DNA and cell structures **(Figure 1.12a)**. In these organisms, the DNA is suspended in the cell interior without separation from other cellular components. By contrast, the Eukarya are described as **eukaryotes** (*eu* = typical) because their DNA is enclosed in a nucleus, a separate structure within cells **(Figure 1.12b)**. The nucleus and other specialized internal compartments of eukaryotic cells are called **organelles** ("little organs").

The Domain Bacteria. The Domain Bacteria **(Figure 1.13a)** comprises unicellular organisms that are generally visible only under the microscope. These prokaryotes live as producers or decomposers almost everywhere on Earth, utilizing metabolic processes that are the most varied of any organisms. They share a relatively simple cellular organization of DNA and internal structures with the archaeans, but bacteria have some structural molecules and mechanisms of photosynthesis that are unique and found only in this domain.

The Domain Archaea. Like bacteria, species in the Domain Archaea **(Figure 1.13b)** are unicellular, microscopic organisms that live as producers or decomposers. However, many archaeans inhabit extreme environments—hot springs, extremely salty ponds, or habitats with little or no oxygen—that other organisms cannot tolerate. They are distinguished by some structural molecules and by a primitive form of photosynthesis that are unique to their domain. Although archaeans are pro-

a. *Escherichia coli*, a prokaryote

DNA

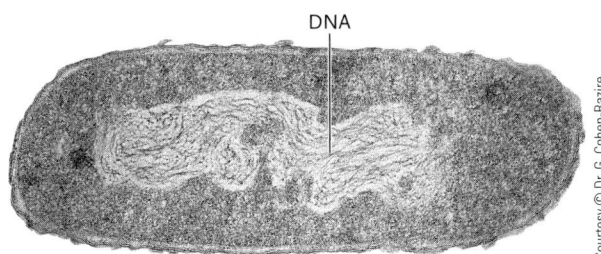

Courtesy © Dr. G. Cohen-Bazire

b. *Paramecium aurelia*, a eukaryote

Nucleus with DNA

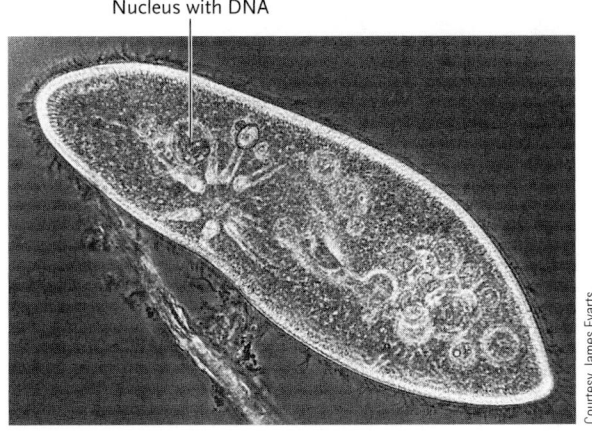

Courtesy James Evarts

Figure 1.12
Prokaryotic and eukaryotic cells. **(a)** *Escherichia coli*, a prokaryote, lacks the complex internal structures apparent in **(b)** *Paramecium aurelia*, a eukaryote. Most eukaryotic cells are 25 to 50 times larger than prokaryotic cells.

karyotic, they have some molecular and biochemical characteristics that are typical of eukaryotes, including features of DNA and RNA organization and processes of protein synthesis.

The Domain Eukarya. All the remaining organisms on Earth, including the familiar plants and animals, are members of the Domain Eukarya **(Figure 1.13c).** The organisms of this domain, all eukaryotic in cell structure, are divided into four kingdoms: Protoctista, Plantae, Fungi, and Animalia.

1. **The Kingdom Protoctista.** The Kingdom Protoctista forms a large and diverse group of single-celled and multicellular eukaryotic species. Most researchers divide the Protoctista into several kingdoms, but they do not yet agree on a classification. Protozoans, which are primarily unicellular, and algae, which range from single-celled, microscopic species to large, multicellular seaweeds, are the most familiar protoctistans. Protozoans live as consumers and decomposers, but almost all algae are producers because they carry out photosynthesis.

2. **The Kingdom Plantae.** Members of the Kingdom Plantae are multicellular organisms that, with few exceptions, carry out photosynthesis; they therefore function as producers in ecosystems. Except for the reproductive cells (pollen and seeds) of some species, plants do not move from place to place. The kingdom includes the familiar flowering plants, conifers, and mosses.

3. **The Kingdom Fungi.** The Kingdom Fungi includes a highly varied group of unicellular and multicellular species, among them the yeasts and molds.

a. Domain Bacteria

c. Domain Eukarya

Kingdom Protoctista

Kingdom Fungi

b. Domain Archaea

Kingdom Plantae

Kingdom Animalia

Figure 1.13

Three domains of life. **(a)** This member of the Domain Bacteria *(Helicobacter pylori)* causes ulcers in the digestive systems of humans. **(b)** This example from the Domain Archaea *(Methanosarcina* species) lives in the oxygen-free muck of swamps and bogs. **(c)** The Domain Eukarya is divided into four kingdoms in this book. The Kingdom Protoctista is represented by a trichomonad *(Trichonympha* species) that lives in the gut of a termite. Coast redwoods *(Sequoia sempervirens)* are among the largest members of the Kingdom Plantae; the picture shows a young tree with the trunk of an older tree behind it. The Kingdom Fungi includes the big laughing mushroom *(Gymnophilus* species), which lives on the forest floor. Members of the Kingdom Animalia are consumers, as illustrated by the fishing spider *(Dolomedes* species), which is feasting on a minnow it has captured.

Most fungi live as decomposers by breaking down and then absorbing biological molecules from dead organisms. No fungi carry out photosynthesis.

4. **The Kingdom Animalia.** Members of the Kingdom Animalia are multicellular organisms that live as consumers by ingesting organisms in all three domains. One of the distinguishing features of animals is their motility, the ability to move actively from one place to another during some stage of their life cycles. The kingdom encompasses a great range of organisms, including groups as varied as sponges, worms, insects, fishes, amphibians, reptiles, birds, and mammals.

Now that we have introduced the characteristics of living systems, basic concepts of evolution, and biological diversity, we turn our attention to the ways in which biologists examine the living world to make new discoveries and gain new insights about life on Earth.

STUDY BREAK

1. What is a major difference between prokaryotic and eukaryotic organisms?
2. In which domain and in which kingdom are humans classified?

1.4 Biological Research

The entire content of this book—every observation, experimental result, and generality—is the product of **biological research,** the collective effort of countless individuals who have worked to understand how living systems function. This section describes how researchers working today define and answer questions about biology.

People have been adding to our knowledge of living systems ever since our distant ancestors first thought about gathering food or hunting game. However, beginning about 500 years ago in Europe, inquisitive people began to understand that direct observation is the most reliable and productive way to study natural phenomena. By the nineteenth century, researchers were using the **scientific method,** an investigative approach to acquiring knowledge in which scientists make observations about the natural world, develop working explanations about what they observe, and then test those explanations by collecting more information.

Grade school teachers often describe the scientific method as a strict, stepwise procedure for observing and explaining the world around us, but it is really more of an attitude—an attitude of inquiry and skepticism. Successful scientists question the current state of our knowledge and challenge old concepts with new ideas and new observations. Scientists like to be shown

why an idea is correct, rather than simply being told that it is. They refuse to accept explanations of natural phenomena unless they are backed up by objective evidence rooted in observation and measurement. Most important, scientists develop ideas that can be confirmed or refuted by different researchers testing them in different settings.

Biologists Confront the Unknown by Conducting Basic and Applied Research

Although nonscientists may be intimidated by natural processes they do not understand, scientists embrace the "unknown." To a scientist, unexplained phenomena provide opportunities to apply creative thinking to important problems. As you read this book, at first you may be uncomfortable discovering how many fundamental questions have not been answered. How and where did life begin? How exactly do genes govern the growth and development of an organism? What triggers the signs of aging? To help you develop an appreciation of how exciting it is to enter unknown territory, most chapters close with a discussion of Unanswered Questions. Although the concepts and facts that you will learn about biological systems are profoundly interesting, you will discover that the unanswered questions are even more exciting. In many cases, we do not even know exactly how you and other scientists of your generation will answer these questions.

Research science is often broken down into two complementary activities—basic research and applied research—that constantly inform one another. Biologists who conduct **basic research** search for explanations about natural phenomena to satisfy their own curiosity and to advance our collective knowledge of living systems. Sometimes, they may not have a specific practical goal in mind. For example, some biologists study how lizards control their body temperatures in different environments. At other times, basic research is inspired by specific practical concerns. For example, understanding how certain bacteria attack the cells of larger organisms might someday prove useful for the development of a new antibiotic (that is, a bacteria-killing agent). Many chapters in this book include a *Focus on Research,* which describes particularly elegant or insightful basic research that advanced our knowledge.

Other scientists conduct **applied research,** with the goal of solving specific practical problems. For example, biomedical scientists conduct applied research to develop new drugs and to learn how illnesses spread from animals to humans or through human populations. Similarly, agricultural scientists try to develop varieties of important crop plants that are more productive and more pest-resistant than the varieties currently in use. Examples of applied research are presented throughout this book, some of them described in detail as a *Focus on Research.*

Biologists Conduct Research by Collecting Observational and Experimental Data

Biologists generally use one of two complementary approaches—or a combination of the two—to advance our knowledge. In many cases, they collect **observational data,** basic information on biological structures or the details of biological processes. This approach, which is sometimes called *descriptive science,* provides information about systems that have not yet been well studied. For example, biologists are now collecting observational data about the precise chemical structure of the DNA in different species of organisms. When conducting descriptive research, a scientist must make detailed observations and describe the methods of observation as carefully and accurately as possible so that other researchers can repeat and verify those observations at a later time.

In other cases, researchers collect **experimental data,** information that describes the result of a careful manipulation of the system under study. Experimental science often answers questions about why or how systems work as they do. For example, a biologist who wonders whether a particular snail species influences the distribution of algae on a rocky shoreline might remove the snail from some enclosed patches of shoreline and examine whether the distribution of algae changes as a result. Similarly, a geneticist who wants to understand the role of a particular gene in the functioning of an organism might make mutations in the gene and examine the consequences.

Researchers Often Test Hypotheses with Controlled Experiments

Research on a previously unexplored system usually starts with basic observations. Once a solid base of carefully observed and described facts is established, scientists may develop a **hypothesis,** a "working explanation" of the observed facts. And whenever scientists create a hypothesis, they simultaneously define—either explicitly or implicitly—a **null hypothesis,** a statement of what they would see if the hypothesis being tested is not correct.

The development of a scientific hypothesis is a creative activity that is constrained by one crucial requirement: it must be *falsifiable* by experimentation or further observation. In other words, scientists must describe an idea in such a way that, if it is wrong, they will be able to demonstrate that it is wrong. The principle of falsifiability helps scientists define testable, focused hypotheses. Hypotheses that are testable and falsifiable fall within the realm of science, whereas those that cannot be falsified—although possibly valid and true—do not fall within the realm of science.

Hypotheses generally explain the relationship between **variables,** environmental factors that may differ among places or organismal characteristics that may differ among individuals. Thus, hypotheses yield testable **predictions,** statements about what the researcher expects to happen to one variable if another variable changes. And if data from just one experiment refute a scientific hypothesis (that is, demonstrate that its predictions are incorrect), the scientist must modify the hypothesis and test it again or abandon it altogether. However, no amount of data can *prove* beyond a doubt that a hypothesis is correct; there may always be a contradictory example somewhere on Earth, and it is impossible to test every imaginable example. That is why scientists say that positive results *are consistent with, support,* or *confirm* a hypothesis.

To make these ideas more concrete, consider a simple example of hypothesis creation and testing. Say that a friend gives you a plant that she grew on her windowsill. Under her loving care, the plant flowered profusely. You place the plant on your windowsill and water it regularly, but the plant never blooms. You know that your friend always gave fertilizer to the plant—your observation—and you wonder whether fertilizing the plant would make it flower. In other words, you create a hypothesis with a specific prediction: "This type of plant will produce flowers if it receives fertilizer." This is a good scientific hypothesis because it is falsifiable. To test the hypothesis, you would simply give the plant fertilizer. If it blooms, the data—the fact that it flowers—confirm your hypothesis. If it does not bloom, the data force you to reject or revise your hypothesis.

One problem with this experiment is that the hypothesis does not address other possible reasons that the plant did not flower. Maybe it received too little water. Maybe it did not get enough sunlight. Maybe your windowsill was too cold. All of these explanations could be the basis of **alternative hypotheses,** which a conscientious scientist always considers when designing experiments. You could easily test any of these hypotheses by providing more water, more hours of sunlight, or warmer temperatures to the plant.

But even if you provide each of these necessities in turn, your efforts will not definitively confirm or refute your hypothesis unless you introduce a control treatment. The **control,** as it is often called, tells what we would see in the absence of the experimental manipulation. For example, your experiment would need to compare plants that received fertilizer (the experimental treatment) with plants grown without fertilizer (the control treatment). The presence or absence of fertilizer is the **experimental variable,** and in a controlled experiment, everything except the experimental variable—the flower pots, the soil, the amount of water, and exposure to sunlight—is exactly the same, or as close to exactly the same as possible. Thus, if your experiment is well controlled **(Figure 1.14),** any difference in flowering pattern observed between plants that receive the experimental treatment (fertilizer) and those that receive the control treatment (no fertilizer) can be attributed to the experimental variable. If the plants that receive fertilizer did not flower more than the control plants, you would reject your initial hypothesis.

Figure 1.14 Experimental Research

Hypothetical Experiment Illustrating the Use of Control Treatment and Replicates

OBSERVATION: Your friend fertilizes a plant that she grows on her windowsill, and it flowers profusely. After she gives you the plant, you put it on your windowsill, but you do not give it any fertilizer and it does not flower.

Friend added fertilizer

You did not add fertilizer

HYPOTHESIS: The plant requires fertilizer to produce flowers.

METHOD: Establish six replicates of an experimental treatment (identical plants grown with fertilizer) and six replicates of a control treatment (identical plants grown without fertilizer).

Experimental Treatment

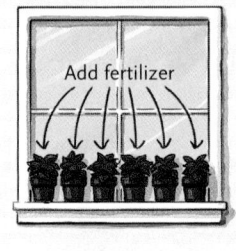
Add fertilizer

Control Treatment

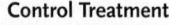
No fertilizer

POSSIBLE RESULT 1: Neither experimental nor control plants flower.

Experimentals	Controls

CONCLUSION: Fertilizer alone does not cause the plant to flower. Consider alternative hypotheses and conduct additional experiments, each testing a different experimental treatment, such as the amount of water or sunlight the plant receives or the temperature to which it is exposed.

POSSIBLE RESULT 2: Plants in the experimental group flower, but plants in the control group do not.

Experimentals	Controls

CONCLUSION: The application of fertilizer induces flowering in this type of plant, confirming your original hypothesis. Pat yourself on the back and apply to graduate school in plant biology.

The elements of a typical experimental approach, as well as our hypothetical experiment, are summarized in Figure 1.14. Figures that present observational and experimental research using this basic format are provided throughout this book.

Notice that in the preceding discussion we discussed plants (plural) that received fertilizer and plants that did not. Nearly all experiments in biology include **replicates**, multiple subjects that receive either the same experimental treatment or the same control treatment. Scientists use replicates in experiments because individuals typically vary in genetic makeup, size, health, or other characteristics—and because accidents sometimes disrupt a couple of replicates. By exposing multiple subjects to both treatments, we can compare the average result of the experimental treatment with the average result of the control treatment, giving us more confidence in the overall result. Thus, in the fertilizer experiment we described, we might expose six or more individual plants to each treatment and compare the results obtained for the experimental group with those obtained for the control group. We would also try to ensure that the individuals included in the experiment were as similar as possible. For example, we might specify that they all must be the same species and the same age or size.

When Controlled Experiments Are Unfeasible, Researchers Use Null Hypotheses to Evaluate Observational Data

In some fields of biology, especially ecology and evolution, the systems under study may be too large or complex for experimental manipulation. In such cases, biologists can use a null hypothesis to evaluate observational data. For example, Paul E. Hertz of Barnard College studies temperature regulation in lizards. As in many other animals, a lizard's body temperature can vary substantially as environmental temperatures change. Research on many lizard species has demonstrated that they often compensate for fluctuations in environmental temperature—that is, maintain thermal homeostasis—by perching in the sun to warm up or in the shade when they feel hot. Previous observations of two closely related lizard species in Puerto Rico, *Anolis cristatellus* and *Anolis gundlachi,* suggested that the two species respond differently to variations in environmental temperature. Based on this prior work, Hertz's working hypothesis was that *A. gundlachi* almost never tries to regulate its body temperature, whereas *A. cristatellus* often does, particularly when environmental temperatures are low.

To discover whether these two lizard species differed in their thermoregulatory behaviors, Hertz needed to determine what he would see if lizards were *not* trying to control their body temperatures. In other words, he needed to know the predictions of a null hypothesis that states: "Lizards do not regulate their body temperature, and they select perching sites at random with respect to factors that influence body temperature" **(Figure 1.15)**. Of course, it would be impossible to force a natural population of lizards to perch in places that define the null hypothesis. Instead, he and his students created a population of artificial lizards, copper models that served as lizard-sized, lizard-shaped, and lizard-colored thermometers. Each hollow copper model was equipped with a built-in temperature-sensing wire that can be connected to an electronic thermometer. After constructing the copper models, Hertz and his students verified that the models reached the same internal temperatures as live lizards under various laboratory conditions. They then traveled to Puerto Rico and hung 60 models at randomly selected positions in the habitats where the two lizard species lived.

How did the copper models allow Hertz and his students to interpret their data? Because the researchers placed these inanimate objects at random positions in the lizards' habitats, the percentages of mod-

els observed in sun and in shade provided a measure of how sunny or shady a particular habitat was. In other words, the copper models established the null hypothesis about the percentage of lizards that would perch in sunlit spots just by chance. Similarly, the temperatures of the models provided a null hypothesis about what the temperatures of lizards would be if they perched at random in their habitats. Hertz and his students gathered data on the use of sunny perching places and temperatures from both the copper models and live lizards. By comparing the behavior and temperatures of live lizards with the random "behavior" and random temperatures of the copper models, they demonstrated that *A. cristatellus* did, in fact, regulate its body temperature but that *A. gundlachi* did not (see Figure 1.15).

Biologists Often Use Model Organisms to Study Fundamental Biological Processes

Certain species or groups of organisms have become favorite subjects for laboratory and field studies because their characteristics make them relatively easy subjects of research. In most cases, biologists began working with these **model organisms** because they have rapid development, short life cycles, and small adult size. Thus, researchers can rear and house large numbers of them in the laboratory. Also, as fuller portraits of their genetics and other aspects of their biology emerge, their appeal as research subjects grows because biologists have a better understanding of the biological context within which specific processes occur.

Because many forms of life share similar molecules, structures, and processes, research on these small and often simple organisms provides insights into biological processes that operate in larger and more complex organisms. For example, early analyses of inheritance in a fruit fly *(Drosophila melanogaster)* established our basic understanding of genetics in all eukaryotic organisms. Research in the mid-twentieth century with the bacterium *Escherichia coli* demonstrated the mechanisms that control whether the information in any particular gene is used to manufacture a protein molecule, fueling additional work on this important subject in both prokaryotes and eukaryotes. In fact, the body of research with *E. coli* formed the foundation that now allows scientists to make and clone (that is, produce multiples copies of) DNA molecules. Similarly, research on a tiny mustard plant *(Arabidopsis thaliana)* is providing information about the genetic and molecular control of development in all plants. Other model organisms facilitate research in ecology and evolution. For example, the *Anolis* lizards described earlier are just 2 of more than 400 *Anolis* species. The geographic distribution of these species allows researchers to study general processes and interactions that affect the ecology and evolution of all forms of life. You will read about eight of the organisms most

Figure 1.15 Observational Research

A Field Study Using a Null Hypothesis

HYPOTHESIS: *Anolis cristatellus* and *Anolis gundlachi* differ in the extent to which they use patches of sun and shade to regulate their body temperatures.

NULL HYPOTHESIS: Because these species do not regulate their body temperatures, they select perching sites at random with respect to environmental factors that might influence body temperature.

METHOD: The researchers created a set of hollow, copper lizard models, each equipped with a temperature-sensing wire. At study sites where the lizard species live in Puerto Rico, the researchers hung 60 models at random positions in trees. They observed how often live lizards and the randomly positioned copper models were perched in patches of sun or shade, and they measured the temperatures of live lizards and the copper models. Data from the randomly positioned copper models define the predictions of the null hypothesis.

RESULTS: The researchers compared the frequency with which live lizards and the copper models perched in sun or shade as well as the temperatures of live lizards and the copper models. The data revealed that the behavior and temperatures of *A. cristatellus* were different from those of the randomly positioned models but that the behavior and temperatures of *A. gundlachi* were not. These data therefore confirmed the original hypothesis.

Anolis gundlachi

Kevin de Queiroz, National Museum of Natural History, Smithsonian Institution

Anolis cristatellus

Alejandro Sanchez

Copper *Anolis* model

Kevin de Queiroz, National Museum of Natural History, Smithsonian Institution

Percentage of models and lizards perched in sun or shade

In the forest where *A. gundlachi* lives, nearly all models and nearly all lizards perched in shade.

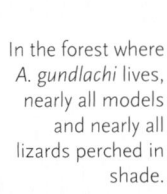
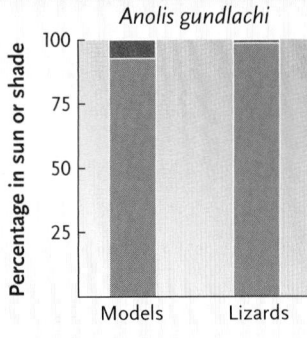

Anolis gundlachi

Percentage in sun or shade

Models   Lizards

Anolis cristatellus

Percentage in sun or shade

Models   Lizards

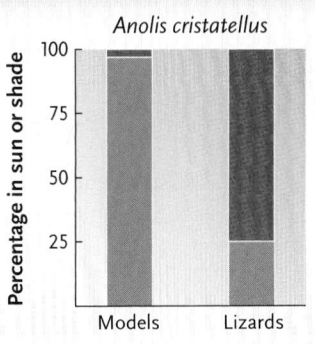

In the habitat where *A. cristatellus* lives, nearly all models perched in shade, but most lizards perched in sun.

KEY

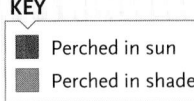

Perched in sun
Perched in shade

Temperatures of models and lizards

Body temperatures of *A. gundlachi* were not significantly different from those of the randomly placed models.

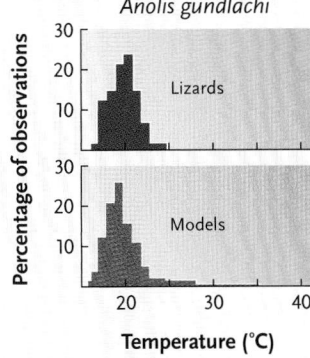

Anolis gundlachi

Percentage of observations

Lizards

Models

Temperature (°C)

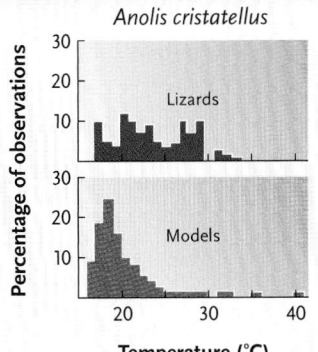

Anolis cristatellus

Percentage of observations

Lizards

Models

Temperature (°C)

Body temperatures of *A. cristatellus* were significantly higher than those of the randomly placed models.

CONCLUSION: *A. cristatellus* uses patches of sun and shade to regulate its body temperature, but *A. gundlachi* does not.

frequently used in research in some of the *Focus on Research* boxes distributed throughout this book.

Molecular Techniques Have Revolutionized Biological Research

In 1941, George Beadle and Edward Tatum used a simple bread mold *(Neurospora crassa)* as a model organism to demonstrate that genes provide the instructions for constructing certain proteins. Their work represents the beginning of the molecular revolution. In 1953, James Watson and Francis Crick determined the structure of DNA, giving us a molecular vision of what a gene is. In the years since those pivotal discoveries, our understanding of the molecular aspects of life has increased exponentially, because many new techniques allow us to study life processes at the molecular level. For example, we can isolate individual genes and study them in detail—even manipulate them—in the test tube. We can modify organisms by replacing or adding genes. We can explore the interactions that each individual protein in the cell has with other proteins. We can identify and characterize each of the genes in an organism and learn its exact structure. The list of experimental possibilities is nearly endless.

This molecular revolution has made it possible to answer questions about biological systems that we could not even ask just a few years ago. For example: What specific DNA changes are responsible for genetic diseases? How is development controlled at the molecular level? What genes do humans and chimpanzees share? In particular, the continued unraveling of the structure of DNA in many organisms is fueling a new intensity of scientific enquiry focused on the role of whole genomes (all of the DNA of an organism) in directing biological processes. To give you a sense of the exciting impact of molecular research on all areas of biology, most chapters in this book include a box dedicated to *Insights from the Molecular Revolution*.

Advances in molecular biology have also revolutionized applied research. DNA "fingerprinting" allows forensic scientists to identify individuals who left molecular traces at crime scenes. **Biotechnology**, the manipulation of living organisms to produce useful products, has also revolutionized the pharmaceutical industry. For example, insulin—a protein used to treat the metabolic disorder diabetes—is now routinely produced by bacteria into which the gene coding for this protein has been inserted. Current research on gene therapy and the cloning of stem cells also promises great medical advances in the future.

Scientific Theories Are Grand Ideas That Have Withstood the Test of Time

When a hypothesis stands up to repeated experimental tests, it is gradually accepted as an accurate explanation of natural events. This acceptance may take many years, and it usually involves repeated experimental confirmations. When every conceivable test has confirmed a hypothesis that addresses many broad questions, it may become regarded as a **scientific theory**. In chemistry and physics, long-established theories of fundamental importance are called *laws* of science. However, living systems are so variable that we do not really recognize many overarching biological laws.

Most scientific theories are supported by exhaustive experimentation; thus, scientists usually regard them as established truths that are not likely to be contradicted by future research. Note that this use of the word *theory* is quite different from its informal meaning in everyday life. In common usage, the word *theory* most often labels ideas that are either speculative or downright suspect, as in the expression "It's only a theory." But when scientists talk about theories, they do so with reverence for ideas that have withstood the test of many experiments.

Because of the difference between the scientific and common usage of the word *theory,* many people fail to appreciate the extensive evidence that supports most scientific theories. For example, nearly every scientist accepts the theory of evolution as fully supported scientific truth: all species change with time, new species are formed, and older species eventually die off. Although evolutionary biologists debate the details of how evolutionary processes bring about these changes, very few scientists doubt that the theory of evolution is essentially correct. Moreover, *no scientist who has tried to cast doubt on the theory of evolution has ever devised or conducted a study that disproves any part of it.* Unfortunately, the confusion between the scientific and common usage of the word *theory* has led to endless public debate about supposed faults and inadequacies in the theory of evolution.

Curiosity and the Joy of Discovery Motivate Scientific Research

What drives scientists in their quest for knowledge? The motivations of scientists are as complex as those driving people toward any goal. Intense curiosity about ourselves, our fellow creatures, and the chemical and physical objects of the world and their interactions is a basic ingredient of scientific research. The discovery of information that no one knew before is as exciting to a scientist as finding buried treasure. There is also an element of play in science, a joy in the manipulation of scientific ideas and apparatus, and the chase toward a scientific goal. Biological research also has practical motivations—for example, to cure disease or improve agricultural productivity. In all this research, one strict requirement of science is honesty—without honesty in the gathering and reporting of results, the work of science is meaningless. Dishonesty is actually rare in science, not least because repetition of experiments by others soon exposes any funny business.

Whatever the level of investigation or the motivation, the work of every scientist adds to the fund of knowledge about us and our world. For better or worse, the scientific method—that inquiring and skeptical attitude—has provided knowledge and technology that have revolutionized the world and improved the quality of human life immeasurably. This book presents the fruits of the biologists' labors in the most important and fundamental areas of biological science—cell and molecular biology, genetics, evolution, systematics, physiology, developmental biology, ecology, and behavior.

STUDY BREAK

1. In your own words, explain the most important requirement of a scientific hypothesis.
2. What information did the copper lizard models provide in the study of temperature regulation described earlier?
3. Why do biologists often use model organisms in their research?
4. How would you respond to a nonscientist who told you that Darwin's ideas about evolution were "just a theory"?

Review

Go to **ThomsonNOW**™ at www.thomsonedu.com/login to access quizzing, animations, exercises, articles, and personalized homework help.

1.1 What Is Life? Characteristics of Living Systems

- Living systems are organized in a hierarchy, each level having its own emergent properties (Figure 1.2). Cells, which represent the lowest level of organization that is alive, are organized into unicellular or multicellular organisms. At the next level of organization, populations are groups of organisms of the same kind that live together in the same area. An ecological community comprises all the populations living in an area, and ecosystems include communities that interact through their shared physical environment. At the highest level, the biosphere includes all of Earth's ecosystems.

- Living organisms have complex structures established by instructions coded in their DNA (Figure 1.3). The information in DNA is copied into RNA, which guides the production of protein molecules (Figure 1.4). Proteins carry out most of the activities of life.

- Living cells and organisms engage in metabolism, the activity of obtaining energy and using it to maintain themselves, grow, and reproduce. The two primary metabolic processes are photosynthesis and cellular respiration (Figure 1.5).

- Energy that flows through the hierarchy of life is eventually released as heat, which cannot be used by living systems. By contrast, matter is recycled within the biosphere (Figure 1.6).

- Cells and organisms use receptors to detect changes in their environment. Detection of an environmental change triggers a compensating reaction that allows the organism to survive.

- Organisms reproduce, and their offspring develop into mature, reproductive adults (Figure 1.7).

- Populations of living organisms undergo evolutionary change as generations replace one another over time.

Animation: Life's levels of organization

Animation: One-way energy flow and materials cycling

Animation: Insect development

1.2 Biological Evolution

- The structure, function, and types of organisms change with time. According to the theory of evolution by natural selection, certain characteristics allow some organisms to survive better and reproduce more than others in their population. If the instructions that produce those characteristics are coded in DNA, successful characteristics will become more common in later generations. As a result, the average characteristics of the offspring generation differ from those of the parent generation (Figure 1.8).

- The instructions for many characteristics are coded by segments of DNA called genes, which are passed through reproduction from parents to offspring.

- Mutations—changes in the structure, number, or arrangement of DNA molecules—create variability among individuals. Variability is the raw material of natural selection and other processes that cause biological evolution.

- Over many generations, the accumulation of favorable characteristics may produce adaptations, which enable individuals to survive longer or reproduce more (Figures 1.9 and 1.10).

- Over long spans of time, the accumulation of different adaptations and other genetic differences between populations has produced the diversity of life on Earth.

1.3 Biodiversity

- Scientists classify organisms in a hierarchy of categories. The species is the most basic category, followed by genus, family, order, class, phylum, and kingdom as increasingly inclusive categories (Figure 1.11).

- Most biologists organize the kingdoms into three domains—Bacteria, Archaea, and Eukarya—based on fundamental characteristics of cell structure. The Bacteria and Archaea each include one kingdom; the Eukarya is divided into four kingdoms: Protoctista, Plantae, Fungi, and Animalia (Figures 1.12 and 1.13).

Animation: Life's diversity

1.4 Biological Research

- Biologists conduct basic research to advance our knowledge of living systems and applied research to solve practical problems.

- Scientists may collect observational data, which describe biological organisms or the details of biological processes, or experimental data, which describe the results of an experimental manipulation.

- Scientists develop hypotheses—working explanations about the relationships between variables. Scientific hypotheses must be falsifiable.

- A well-designed experiment considers alternative hypotheses and includes control treatments and replicates (Figure 1.14). When experiments are unfeasible, biologists often use null hypotheses, explanations of what they would see if their hypothesis was wrong, to evaluate observational or experimental data (Figure 1.15).

- Model organisms, which are easy to maintain in the laboratory, have been the subject of much research.
- Molecular techniques allow detailed analysis of the DNA of many species and the manipulation of specific genes in the laboratory.

- A scientific theory is a set of broadly applicable hypotheses that have been completely supported by repeated tests under many conditions and in many different situations. The theory of evolution by natural selection is of central importance to biology because it explains how life evolved through natural processes.

Animation: Sample size and accuracy

Animation: How do scientists use random samples to test hypotheses?

Questions

Self-Test Questions

1. What is the lowest level of biological organization that biologists consider to be alive?
   a. a protein
   b. DNA
   c. a cell
   d. a multicellular organism
   e. a population of organisms

2. Which category falls immediately below "class" in the systematic hierarchy?
   a. species
   b. order
   c. family
   d. genus
   e. phylum

3. Which of the following represents the application of the "scientific method"?
   a. comparing one experimental subject to one control subject
   b. believing an explanation that is too complex to be tested
   c. using controlled experiments to test falsifiable hypotheses
   d. developing one testable hypothesis to explain a natural phenomenon
   e. observing a once-in-a-lifetime event under natural conditions

4. Houseflies develop through a series of programmed stages from egg, to pupa, to larva, to flying adult. This series of stages is called:
   a. artificial selection.
   b. respiration.
   c. homeostasis.
   d. a life cycle.
   e. metabolism.

5. Which structure allows living organisms to detect changes in the environment?
   a. a protein
   b. a receptor
   c. a gene
   d. RNA
   e. a nucleus

6. Which of the following is *not* a component of Darwin's theory as he understood it?
   a. Some individuals in a population survive longer than others.
   b. Some individuals in a population reproduce more than others.
   c. Heritable variations allow some individuals to compete more successfully for resources.
   d. Mutations in genes produce new variations in a population.
   e. Some new variations are passed to the next generation.

7. What role did the copper lizard models play in the field of study on temperature regulation?
   a. They attracted live lizards to the study site.
   b. They measured the temperatures of live lizards.
   c. They established null hypotheses about basking behavior and temperatures.
   d. They scared predators away from the study site.
   e. They allowed researchers to practice taking lizard temperatures.

8. Which of the following questions best exemplifies basic research?
   a. How did life begin?
   b. How does alcohol intake affect aging?
   c. How fast does avian flu spread among humans?
   d. How can we reduce hereditary problems in pure bred dogs?
   e. How does the consumption of soft drinks promote obesity?

9. When researchers say that a scientific hypothesis must be falsifiable, they mean that:
   a. the hypothesis must be proved correct before it is accepted as truth.
   b. the hypothesis has already withstood many experimental tests.
   c. they have an idea about what will happen to one variable if another variable changes.
   d. appropriate data can prove without question that the hypothesis is correct.
   e. if the hypothesis is wrong, scientists must be able to demonstrate that it is wrong.

10. Which of the following characteristics would *not* qualify an animal as a model research organism?
   a. It has rapid development.
   b. It has small adult size.
   c. It has a rapid life cycle.
   d. It has unique genes and unusual cells.
   e. It is easy to raise in the laboratory.

Questions for Discussion

1. Viruses are infectious agents that contain either DNA or RNA surrounded by a protein coat. They cannot reproduce on their own, but they can take over the cells of the organisms they infect and force those cells to produce more virus particles. Based on the characteristics of living organisms described in this chapter, should viruses be considered living organisms?

2. While walking through the woods, you discover a large rock covered with a gelatinous, sticky substance. What tests could you perform to determine whether the substance is inanimate, alive, or the product of a living organism?

3. Explain why control treatments are a necessary component of well-designed experiments.

Experimental Analysis

Design an experiment to test the hypothesis that the color of farmed salmon is produced by pigments in their food.

Evolution Link

When a biologist first tested a new pesticide on a population of insects, she found that only 1% of the insects survived their exposure to the poison. She allowed the survivors to reproduce and discovered that 10% of the offspring survived exposure to the same concentration of pesticide. One generation later, 50% of the insects survived this experimental treatment. What is a likely explanation for the increasing survival rate of these insects over time?

Life as we know it would be impossible without water, a small inorganic compound with unique properties.

Nigel Cattlin/Holt Studios International/Science Photo Library/Photo Researchers, Inc.

STUDY PLAN

2 Life, Chemistry, and Water

WHY IT MATTERS

We—like all plants, animals, and other organisms—are collections of atoms and molecules linked together by chemical bonds. Our chemical nature makes it impossible to understand biology without knowledge of basic chemistry and chemical interactions.

For example, the element selenium is a natural ingredient of rocks and soil. In minute amounts it is necessary for the normal growth and survival of humans and many other animals, but high concentrations of selenium are toxic. In 1983, thousands of dead or deformed waterfowl were discovered at the Kesterson Wildlife Refuge in the San Joaquin Valley of California. The deaths and deformities were traced to high concentrations of selenium in the environment, alerting the public and scientists alike to the dangers of this element to all animals. Decades of irrigation had washed selenium-containing chemicals from the soil into the water of the refuge. With the problem identified, engineers have diverted agricultural drainage water from the area, and the Kesterson refuge is now being restored.

The study of selenium and its biological effects has suggested a possible way to prevent it from accumulating in the environment. In 1996, Norman Terry and his coworkers at the University of California at Berkeley started a large-scale experiment designed to test natural methods for removing excess selenium from contaminated soils. Terry found that some wetland plants could remove up to 90% of the selenium in wastewater from a gasoline refinery. The plants convert much of the selenium into a relatively nontoxic gas, methyl selenide, which can pass into the atmosphere without harming plants and animals.

To test further the ability of plants to remove selenium, Terry and his coworkers grew wetland plants in 10 experimental plots watered by runoff from agricultural irrigation **(Figure 2.1).** The researchers measured how much selenium remained in the soil of the plots, how much was incorporated into plant tissues, and how much escaped into the air as a gas. Terry's results indicate that before the runoff trickles through to local ponds, the plants in his plots reduce selenium to nontoxic levels, less than 2 parts per billion. Such applications of chemical and biological knowledge to decontaminate polluted environments are known as **bioremediation.** They could help safeguard our food supplies, our health, and the environment.

The selenium example shows the importance of understanding and applying chemistry in biology. However, reactions involving selenium are only a few of the many thousands of chemical reactions that take place inside living organisms. Decades of research have taught us much about these reactions and have confirmed that the same laws of chemistry and physics govern both living and nonliving things. We can therefore apply with confidence information from chemical experiments in the laboratory to the processes inside living organisms. An understanding of the relationship between the structure of chemical substances and their behavior is the first step toward learning biology,

Figure 2.1
Researcher Norman Terry in an experimental wetlands plot in Corcoran, California.
Terry is testing the ability of cattails, bulrushes, and marsh grasses to reduce selenium contamination in water draining from irrigated fields.

and this knowledge will help you appreciate the benefits and risks of applying chemistry to human affairs.

2.1 The Organization of Matter: Elements and Atoms

Selenium is an example of an **element**—a pure substance that cannot be broken down into simpler substances by ordinary chemical or physical techniques. All **matter** of the universe—anything that occupies space and has mass—is composed of elements and combinations of elements. Ninety-two different elements occur naturally on Earth, and more than fifteen artificial elements have been synthesized in the laboratory.

Living Organisms Are Composed of about 25 Key Elements

Four elements—carbon, hydrogen, oxygen, and nitrogen—make up more than 96% of the weight of living organisms. Seven other elements—calcium, phosphorus, potassium, sulfur, sodium, chlorine, and magnesium—contribute most of the remaining 4%. Several other elements occur in organisms in quantities so small (<0.01%) that they are known as **trace elements. Figure 2.2** compares the relative proportions of different elements in a human, a plant, Earth's crust, and seawater, and lists the most important trace elements in a human. The proportions of elements in living organisms, as represented by the human and the plant, differ markedly from those of Earth's crust and seawater; these differences reflect the highly ordered chemical structure of living organisms.

Trace elements are vital to normal biological functions. For example, iodine makes up only about 0.0004% of a human's weight. However, a lack of iodine in the human diet severely impairs the function of the thyroid gland, which produces hormones that regulate metabolism and growth. Symptoms of iodine deficiency include lethargy, apathy, and sensitivity to cold temperatures. Prolonged iodine deficiency causes a *goiter,* a condition in which the thyroid gland enlarges so much that the front of the neck swells grotesquely. Once a common condition, goiter has almost been eliminated by adding iodine to table salt, especially in regions where soils are iodine-deficient.

Elements Are Composed of Atoms, Which Combine to Form Molecules

Elements are composed of individual **atoms**—the smallest units that retain the chemical and physical properties of an element. Any given element has only one type of atom. Several million atoms arranged side by side would be needed to equal the width of the period at the end of this sentence.

Atoms are identified by a standard one- or two-letter symbol. The element carbon is identified by the single letter *C*, which stands for both the carbon atom and the element; iron is identified by the two-letter symbol *Fe* (*ferrum* = iron). **Table 2.1** lists the chemical symbols of these and other atoms common in living organisms.

Atoms combined chemically in fixed numbers and ratios form the **molecules** of living and nonliving matter. For example, the oxygen we breathe is a molecule formed from the chemical combination of two oxygen atoms; a molecule of the carbon dioxide that we exhale contains one carbon atom and two oxygen atoms. The name of a molecule is written in chemical shorthand as a **formula**, using the standard symbols for the elements and using subscripts to indicate the number of atoms of each element in the molecule. The subscript is omitted for atoms that occur only once in a molecule. For example, the formula for an oxygen molecule is written as O_2; for a carbon dioxide molecule the formula is CO_2.

Molecules whose component atoms are different (such as carbon dioxide) are called **compounds.** The chemical and physical properties of compounds are typically distinct from those of their atoms or elements. For example, we all know that water is a liquid at room temperature. We also know that water does not burn. However, the properties of the individual elements of water—hydrogen and oxygen—are quite different. Hydrogen and oxygen are gases at room temperature, and both are highly reactive.

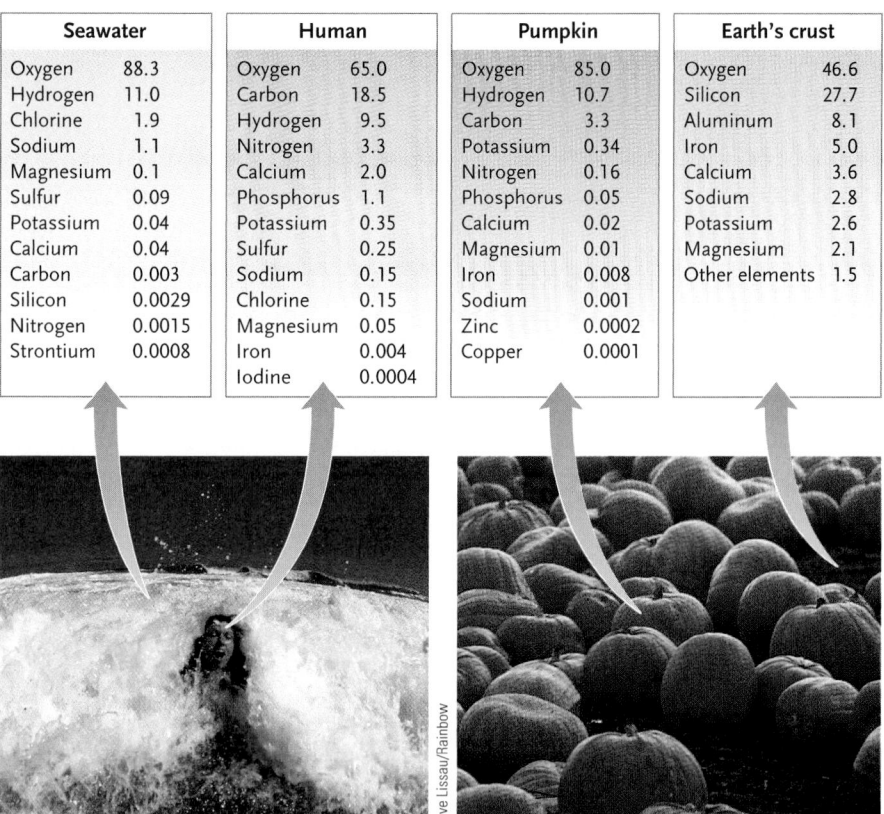

Seawater	
Oxygen	88.3
Hydrogen	11.0
Chlorine	1.9
Sodium	1.1
Magnesium	0.1
Sulfur	0.09
Potassium	0.04
Calcium	0.04
Carbon	0.003
Silicon	0.0029
Nitrogen	0.0015
Strontium	0.0008

Human	
Oxygen	65.0
Carbon	18.5
Hydrogen	9.5
Nitrogen	3.3
Calcium	2.0
Phosphorus	1.1
Potassium	0.35
Sulfur	0.25
Sodium	0.15
Chlorine	0.15
Magnesium	0.05
Iron	0.004
Iodine	0.0004

Pumpkin	
Oxygen	85.0
Hydrogen	10.7
Carbon	3.3
Potassium	0.34
Nitrogen	0.16
Phosphorus	0.05
Calcium	0.02
Magnesium	0.01
Iron	0.008
Sodium	0.001
Zinc	0.0002
Copper	0.0001

Earth's crust	
Oxygen	46.6
Silicon	27.7
Aluminum	8.1
Iron	5.0
Calcium	3.6
Sodium	2.8
Potassium	2.6
Magnesium	2.1
Other elements	1.5

Steve Lissau/Rainbow

Jack Carey

Figure 2.2
The proportions by mass of different elements in seawater, the human body, a fruit, and Earth's crust. Trace elements in humans include boron, chromium, cobalt, copper, fluorine, iodine, iron, manganese, molybdenum, selenium, tin, vanadium, and zinc, as well as variable traces of other elements.

Table 2.1	Atomic Number and Mass Number of the Most Common Elements in Living Organisms		
Element	Symbol	Atomic Number	Mass Number of the Most Common Form
Hydrogen	H	1	1
Carbon	C	6	12
Nitrogen	N	7	14
Oxygen	O	8	16
Sodium	Na	11	23
Magnesium	Mg	12	24
Phosphorus	P	15	31
Sulfur	S	16	32
Chlorine	Cl	17	35
Potassium	K	19	39
Calcium	Ca	20	40
Iron	Fe	26	56
Iodine	I	53	127

STUDY BREAK

Distinguish between an element and an atom and between a molecule and a compound.

2.2 Atomic Structure

Each element consists of one type of atom. However, all atoms share the same basic structure **(Figure 2.3)**. Each atom consists of an **atomic nucleus,** surrounded by one or more smaller, fast-moving particles called **electrons.** Although the electrons occupy more than 99.99% of the space of an atom, the nucleus makes up more than 99.99% of its total mass.

a. Hydrogen   **b.** Carbon

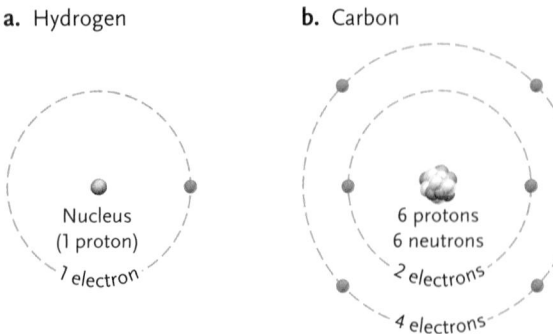

Figure 2.3

Atomic structure. The nucleus of an atom contains one or more protons and, except for the most common form of hydrogen, a similar number of neutrons. Fast-moving electrons, in numbers equal to the protons, surround the nucleus. **(a)** The most common form of hydrogen, the simplest atom, has a single proton in its nucleus and a single electron. **(b)** Carbon, a more complex atom, has a nucleus surrounded by electrons at two levels. The electrons in the outer level follow more complex pathways than shown here.

The Atomic Nucleus Contains Protons and Neutrons

All atomic nuclei contain one or more positively charged particles called **protons.** The number of protons in the nucleus of each kind of atom is referred to as the **atomic number.** This number does not vary and thus specifically identifies the atom. The smallest atom, hydrogen, has a single proton in its nucleus, so its atomic number is 1. The heaviest naturally occurring atom, uranium, has 92 protons in its nucleus and therefore has an atomic number of 92. Similarly, carbon with six protons, nitrogen with seven protons, and oxygen with eight protons have atomic numbers of 6, 7, and 8, respectively (see Table 2.1).

With one exception, the nuclei of all atoms also contain uncharged particles called **neutrons,** which occur in variable numbers approximately equal to the number of protons. The single exception is the most common form of hydrogen, which has a nucleus that contains only a single proton. There are two less common forms of hydrogen as well. One form, named deuterium, has a neutron in its nucleus in addition to a single proton. The other form, named tritium, has two neutrons and a single proton.

Other atoms also have common and less common forms with different numbers of neutrons. For example, the most common form of the carbon atom has six protons and six neutrons in its nucleus, but about 1% of carbon atoms have six protons and seven neutrons in their nuclei and an even smaller percentage has six protons and eight neutrons.

These distinct forms of the atoms of an element, all with the same number of protons but different numbers of neutrons, are called **isotopes (Figure 2.4).** The various isotopes of an atom differ in mass and other physical characteristics, but all have essentially the same chemical properties. Therefore, organisms can use any hydrogen or carbon isotope, for example, without a change in their chemical reactions.

Isotopes of hydrogen

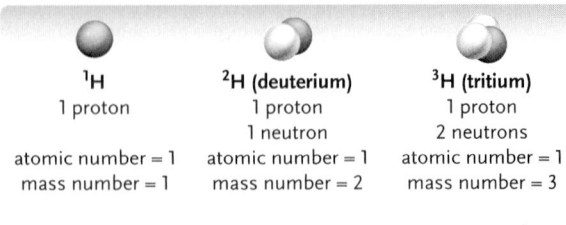

Isotopes of carbon

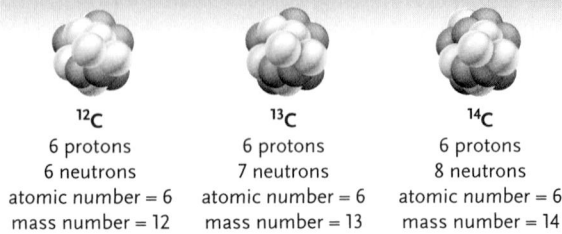

Figure 2.4

The atomic nuclei of hydrogen and carbon isotopes. Note that isotopes of an atom have the same atomic number but different mass numbers.

A neutron and a proton have almost the same mass, about 1.66×10^{-24} grams (g). This mass is defined as a standard unit, the **dalton,** named after John Dalton, a nineteenth-century English scientist who contributed to the development of atomic theory. Atoms are assigned a **mass number** based on the total number of protons and neutrons in the atomic nucleus (see Table 2.1). Electrons are ignored in determinations of atomic mass because the mass of an electron, at only 1/1800 of the mass of a proton or neutron, does not contribute significantly to the mass of an atom. Thus, the mass number of the hydrogen isotope with one proton in its nucleus is 1, and its mass is 1 dalton. The mass number of the hydrogen isotope deuterium is 2, and the mass number of tritium is 3. The carbon isotope with six protons and six neutrons in its nucleus has a mass number of 12; the isotope with six protons and seven neutrons has a mass number of 13, and the isotope with six protons and eight neutrons has a mass number of 14 (see Figure 2.4). These carbon mass numbers are written as ^{12}C, ^{13}C, and ^{14}C, or carbon-12, carbon-13, and carbon-14, respectively. However, all the carbon isotopes have the same atomic number of 6, because this number reflects only the number of protons in the nucleus.

You might wonder about the meaning of mass as compared to weight. Mass is the amount of matter in an object, whereas weight measures the pull of gravity on an object. Mass is constant, but the weight of an object may vary because of differences in gravity. For example, the mass of a piece of lead is the same on Earth and in outer space, but the same piece of lead that weighs 1 kilogram (kg) on Earth is weightless in an orbiting spacecraft, even though its mass remains the same. However, as long as an object is on Earth's surface, its mass and weight are equivalent. Thus, we can weigh an object in the laboratory and be assured that its weight accurately reflects its mass.

The Nuclei of Some Atoms Are Unstable and Tend to Break Down to Form Simpler Atoms

The nuclei of some isotopes are unstable and break down, or *decay*, giving off particles of matter and energy that can be detected as **radioactivity.** The decay transforms the unstable, radioactive isotope—called a **radioisotope**—into an atom of another element. The decay continues at a steady, clocklike rate, with a constant proportion of the radioisotope breaking down at any instant. The rate of decay is not affected by chemical reactions or environmental conditions such as temperature or pressure. For example, the carbon isotope ^{14}C is unstable and undergoes radioactive decay in which one of its neutrons splits into a proton and an electron. The electron is ejected from the nucleus, but the proton is retained, giving a new total of seven protons and seven neutrons, which is characteristic of the most common form of nitrogen. Thus, the decay transforms the carbon atom into an atom of nitrogen.

Because unstable isotopes decay at a clocklike rate, they can be used to estimate the age of organic material, rocks, or fossils that contain them. These techniques have been vital in dating animal remains and tracing evolutionary lineages, as described in Chapter 22. Isotopes are also used in biological research as **tracers** to label molecules so that they can be tracked as they pass through biochemical reactions. Radioactive isotopes of carbon (^{14}C), phosphorus (^{32}P), and sulfur (^{35}S) can be traced easily by their radioactivity. A number of stable, nonradioactive isotopes, such as ^{15}N (called heavy nitrogen), can be detected by their mass differences and have also proved valuable as tracers in biological experiments. *Focus on Research* describes some applications of radioisotopes in research and medicine.

The Electrons of an Atom Occupy Orbitals around the Nucleus

In an atom, the number of electrons surrounding the nucleus is equal to the number of protons in the nucleus. An electron carries a negative charge that is exactly equal and opposite to the positive charge of a proton, balancing the positive and negative charges and making the total structure of an atom electrically neutral.

An atom is often drawn in a simple way with electrons orbiting the nucleus similar to planets orbiting a sun. The reality is different. Electrons are in constant motion around the nucleus, moving at speeds that approach the speed of light. At any instant, an electron may be in any location with respect to its nucleus, from the immediate vicinity of the nucleus to practically infinite space. An electron moves so fast that it almost occupies all the locations at the same time; however, it passes through some locations much more frequently than others. The locations where an electron occurs most frequently around the atomic nucleus define a path called an orbital. An **orbital** is essentially the region of space where the electron "lives" most of the time. Although either one or two electrons may occupy an orbital, the most stable and balanced condition occurs when an orbital contains a pair of electrons.

Electrons are maintained in their orbitals by a combination of attraction to the positively charged nucleus and mutual repulsion because of their negative charge. The orbitals take different shapes depending on their distance from the nucleus and their degree of repulsion by electrons in other orbitals.

Under certain conditions, electrons may pass from one orbital to another within an atom, enter orbitals shared by two or more atoms, or pass completely from orbitals in one atom to orbitals in another. As discussed later in this chapter, the ability of electrons to move from one orbital to another underlies the chemical reactions that combine atoms into molecules.

Orbitals Occur in Discrete Layers around an Atomic Nucleus

Within an atom, electrons are found in regions of space called **energy levels**, or more simply, **shells**. Within each energy level, electrons are grouped into orbitals. The lowest energy level of an atom, the one nearest the nucleus, may be occupied by a maximum of two electrons in a single orbital **(Figure 2.5a).** This orbital, which has a spherical shape, is called the 1s orbital. (The "1" signifies that the orbital is in the energy level closest to the nucleus, and the "s" signifies the shape of the orbital, in this case, spherically symmetric around the nucleus.) Hydrogen has one electron in this orbital, and helium has two.

Atoms with atomic numbers between 3 (lithium) and 10 (neon) have two energy levels, with two electrons in the 1s orbital and one to eight electrons in orbitals at the next highest energy level. The electrons at this second energy level occupy one spherical orbital, called the 2s orbital **(Figure 2.5b)**, and as many as three orbitals that are pushed into a dumbbell shape by repulsions between electrons, called 2p orbitals **(Figure 2.5c)**. **Figure 2.5d** shows the orbitals for neon.

Larger atoms have more energy levels. The third energy level, which may contain as many as 18 electrons in 9 orbitals, includes the atoms from sodium (11 electrons) to argon (18 electrons). (**Figure 2.6** shows the 18 elements that have electrons in the lowest three energy levels only.) The fourth energy level may contain as many as 32 electrons in 16 orbitals. In all cases, the total number of electrons in the orbitals is matched by the number of protons in the nucleus. However, no matter what the size of an atom, the outermost energy level typically contains one to eight electrons occupying a maximum of four orbitals.

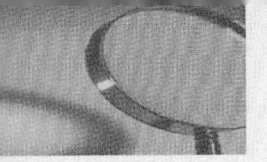
Basic Research: Using Radioisotopes to Trace Reactions and Save Lives

In 1896, the French physicist Henri Becquerel wrapped a rock containing uranium in paper and tucked it into a desk drawer on top of a case containing an unexposed photographic plate. When he opened the case containing the plate a few days later, he was surprised to find an image of the rock on the plate—apparently caused by energy emitted from the rock. One of his coworkers, Marie Curie, named the phenomenon "radioactivity." Although radioactivity can be dangerous to life (more than one researcher, including Marie Curie, has died from its effects), it has been harnessed and put to highly productive use for scientific and medical purposes.

The radiation released by unstable isotopes can be detected by placing a photographic film over samples containing the isotopes (as Becquerel discovered) or by using an instrument known as a *scintillation counter*. These techniques allow researchers to use isotopes as tracers in chemical reactions. Typically, organisms are exposed to a reactant chemical that has been "labeled" with a radioactive isotope such as ^{14}C or ^{3}H. After being exposed to the tracer, the chemical products in which the isotope appears, and their sequence of appearance, can be detected by their radioactivity and identified.

For example, algae and plants use carbon dioxide (CO_2) as a raw material in photosynthesis. To trace the reactions of photosynthesis, Melvin Calvin and his coworkers grew algal cells in a medium with CO_2 that contained the radioisotope ^{14}C. Then they extracted various substances from the cells at intervals, separated them

on a piece of paper based on their different solubilities in particular solvents, and placed the paper on a photographic film. The particular substances that exposed spots on the film because they were radioactive, as well as their order of appearance in the cells, allowed the researchers to piece together the sequence of reactions in photosynthesis, as described and illustrated in the *Focus on Research* in Chapter 9.

Radioisotopes are widely used in medicine to diagnose and cure disease, to produce images of diseased body organs, and, as in biological research, to trace the locations and routes followed by individual substances marked for identification by radioactivity. One example of their use in diagnosis is in the evaluation of thyroid gland disease. The thyroid is the only structure in the body that absorbs iodine in quantity. The size and shape of the thyroid, which reflect its health, are measured by injecting a small amount of a radioactive iodine isotope into the patient's bloodstream. After the isotope is concentrated in the thyroid, the gland is then scanned by an apparatus that uses the radioactivity to produce an image of the gland on a photographic film. Examples of what the scans may show are presented in the figure. Another application uses the fact that radioactive thallium is not taken up by regions of the heart muscle with poor circulation to detect coronary artery disease. Other isotopes are used to detect bone injuries and defects, including injured, arthritic, or abnormally growing segments of bone.

Treatment of disease with radioisotopes takes advantage of the fact that radioactivity in large doses can kill cells (radiation generates highly reactive chemical groups that break and disrupt biological molecules). Dangerously overactive thyroid glands are treated by giving patients a dose of radioactive iodine calculated to destroy just enough thyroid cells to reduce activity of the gland to normal levels. In radiation therapy, cancer cells are killed by bombarding them with radiation emitted by radium-226 or cobalt-60. As much as is possible, the radiation is focused on the tumor to avoid destroying nearby healthy tissues. In some forms of chemotherapy for cancer, patients are given radioactive substances at levels that kill cancer cells without also killing the patient.

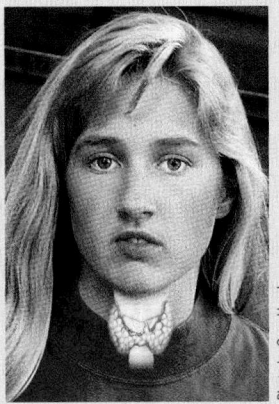
Photo by Gary Head

normal   enlarged   cancerous

Scans of human thyroid glands after iodine-123 was injected into the bloodstream. The radioactive iodine becomes concentrated in the thyroid gland.

The Number of Electrons in the Outermost Energy Level of an Atom Determines Its Chemical Activity

The electrons in an atom's outermost energy level are known as **valence electrons** (*valentia* = power or capacity). Atoms in which the outermost energy level is not completely filled with electrons tend to be chemi-

cally reactive; those with a completely filled outermost energy level are nonreactive, or inert. For example, hydrogen has a single, unpaired electron in its outermost and only energy level, and it is highly reactive; helium has two valence electrons filling its single orbital, and it is inert. For atoms with two or more energy levels, only those with unfilled outer energy levels are reactive. Those with eight electrons completely fill-

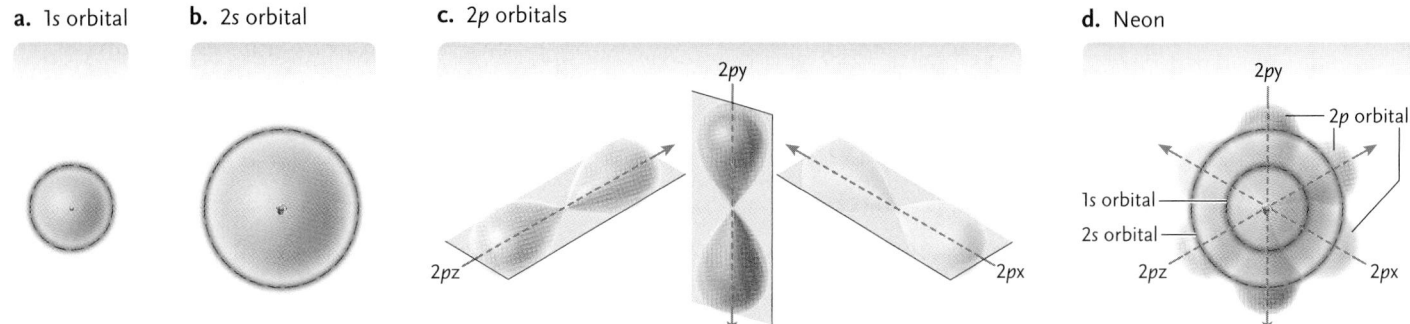

a. 1s orbital **b.** 2s orbital **c.** 2p orbitals **d.** Neon

Figure 2.5
Electron orbitals. **(a)** The single 1s orbital of hydrogen and helium approximates a sphere centered on the nucleus. **(b)** The 2s orbital. **(c)** The 2p orbitals lie in the three planes x, y, and z, each at right angles to the others. **(d)** In atoms with two energy levels, such as neon, the lowest energy level is occupied by a single 1s orbital as in hydrogen and helium. The second, higher energy level is occupied by a maximum of four orbitals—a spherical 2s orbital and three dumbbell-shaped 2p orbitals.

ing the four orbitals of the outer energy level, such as neon and argon, are stable and chemically unreactive (see Figure 2.6).

Atoms with outer energy levels that contain electrons near the stable numbers tend to gain or lose electrons to reach the stable configuration. For example, sodium has two electrons in its first energy level, eight in the second, and one in the third and outermost level (see Figure 2.6). The outermost electron is readily lost to another atom, giving the sodium atom a stable second energy level (now the outermost level) with eight electrons. Chlorine, with seven electrons in its outermost energy level, tends to take up an electron from another atom to attain the stable number of eight electrons.

Atoms that differ from the stable configuration by more than one or two electrons tend to attain stability by *sharing* electrons in joint orbitals with other atoms rather than by gaining or losing electrons completely. Among the atoms that form biological molecules, electron sharing is most characteristic of carbon, which has four electrons in its outer energy level and thus falls at the midpoint between the tendency to gain or lose electrons. Oxygen, with six electrons in its outer level, and nitrogen, with five electrons in its outer level, also share electrons readily. Hydrogen may either share or lose its single electron. The relative tendency to gain, share, or lose valence electrons underlies the chemical bonds and forces that hold the atoms of molecules together.

Figure 2.6
The atoms with electrons distributed in one, two, or three energy levels. The atomic number of each element (shown in boldface in each panel) is equivalent to the number of protons in its nucleus.

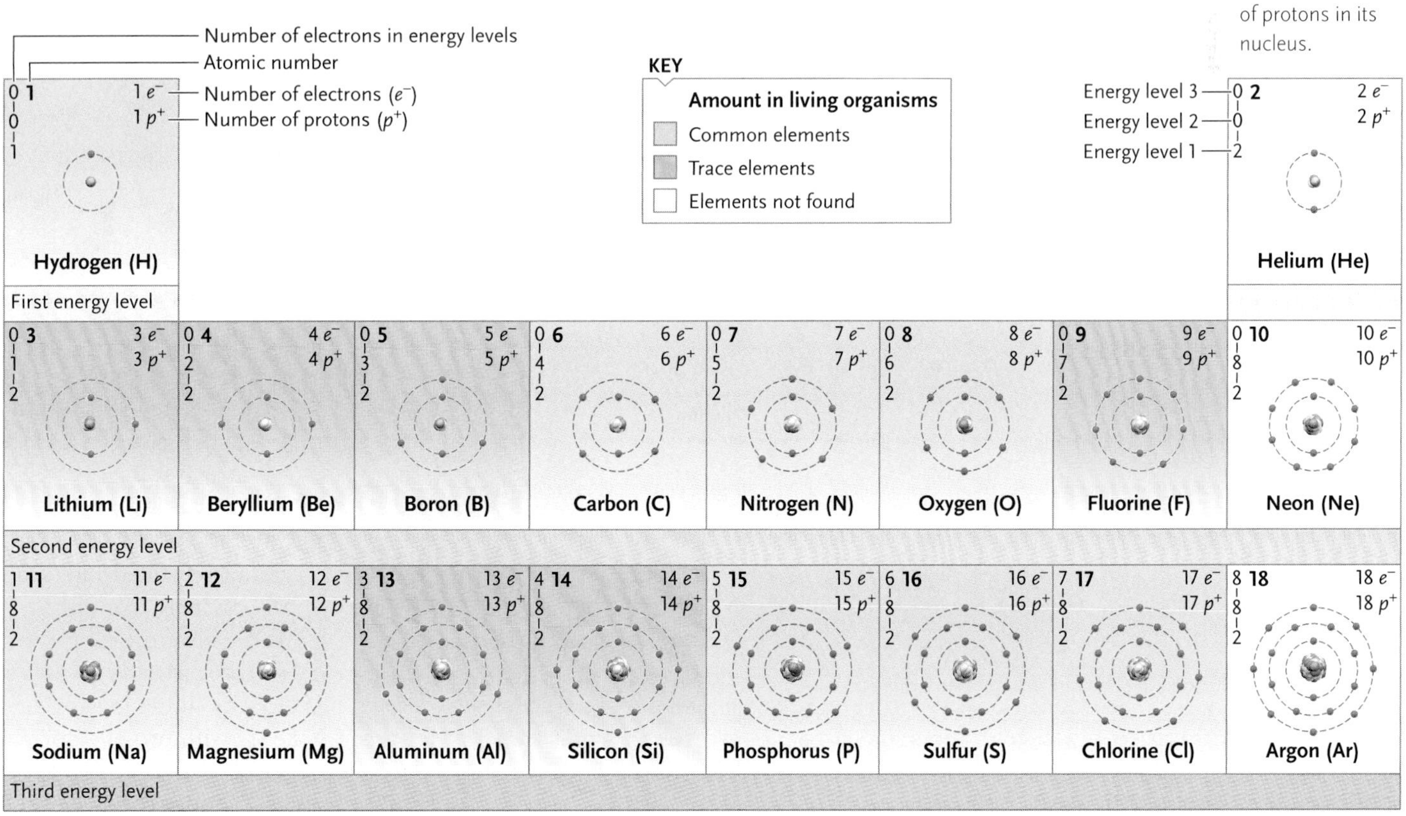

Number of electrons in energy levels
Atomic number
Number of electrons (e^-)
Number of protons (p^+)

0 **1**	1 e^-
0	1 p^+
1	

Hydrogen (H)

First energy level

KEY

Amount in living organisms
- Common elements
- Trace elements
- Elements not found

Energy level 3 — 0 **2**   2 e^-
Energy level 2 — 0   2 p^+
Energy level 1 — 2

Helium (He)

0 **3** 3 e^-	0 **4** 4 e^-	0 **5** 5 e^-	0 **6** 6 e^-	0 **7** 7 e^-	0 **8** 8 e^-	0 **9** 9 e^-	0 **10** 10 e^-
1 3 p^+	2 4 p^+	3 5 p^+	4 6 p^+	5 7 p^+	6 8 p^+	7 9 p^+	8 10 p^+
2	2	2	2	2	2	2	2
Lithium (Li)	Beryllium (Be)	Boron (B)	Carbon (C)	Nitrogen (N)	Oxygen (O)	Fluorine (F)	Neon (Ne)

Second energy level

1 **11** 11 e^-	2 **12** 12 e^-	3 **13** 13 e^-	4 **14** 14 e^-	5 **15** 15 e^-	6 **16** 16 e^-	7 **17** 17 e^-	8 **18** 18 e^-
8 11 p^+	8 12 p^+	8 13 p^+	8 14 p^+	8 15 p^+	8 16 p^+	8 17 p^+	8 18 p^+
2	2	2	2	2	2	2	2
Sodium (Na)	Magnesium (Mg)	Aluminum (Al)	Silicon (Si)	Phosphorus (P)	Sulfur (S)	Chlorine (Cl)	Argon (Ar)

Third energy level

1. Where are protons, electrons, and neutrons found in an atom?
2. The isotopes carbon-11 and oxygen-15 do not occur in nature, but they can be made in the laboratory. Both are used in a medical imaging procedure called positron emission tomography. Give the number of protons and neutrons in carbon-11 and in oxygen-15.
3. What determines the chemical reactivity of an atom?

2.3 Chemical Bonds

Atoms of inert elements, such as helium, neon, and argon, occur naturally in uncombined forms, but atoms of reactive elements tend to combine into molecules by forming **chemical bonds.** The four chemical linkages that are important in biological molecules are **ionic bonds**, resulting from electrical attractions between atoms that have lost or gained electrons; **covalent bonds**, formed by electron sharing between atoms; **hydrogen bonds**, noncovalent bonds formed by unequal electron sharing between hydrogen atoms and oxygen, nitrogen, or sulfur atoms; and **van der Waals forces**, weak molecular attractions over short distances.

Ionic Bonds Are Multidirectional and Vary in Strength

Ionic bonds form between atoms that gain or lose valence electrons completely. A sodium atom (Na) readily loses a single electron to achieve a stable outer energy level, and chlorine (Cl) readily gains an electron:

$$Na\cdot \; + \; .\overset{..}{\underset{..}{Cl}}: \; \rightarrow \; Na^+ \; :\overset{..}{\underset{..}{Cl}}:^-$$

(The dots in the preceding formula represent the electrons in the outermost energy level.) After the transfer, the sodium atom, now with 11 protons and 10 electrons, carries a single positive charge. The chlorine atom, now with 17 protons and 18 electrons, carries a single negative charge. In this charged condition, the sodium and chlorine atoms are called **ions** instead of atoms and are written as Na^+ and Cl^- **(Figure 2.7).** A positively charged ion such as Na^+ is

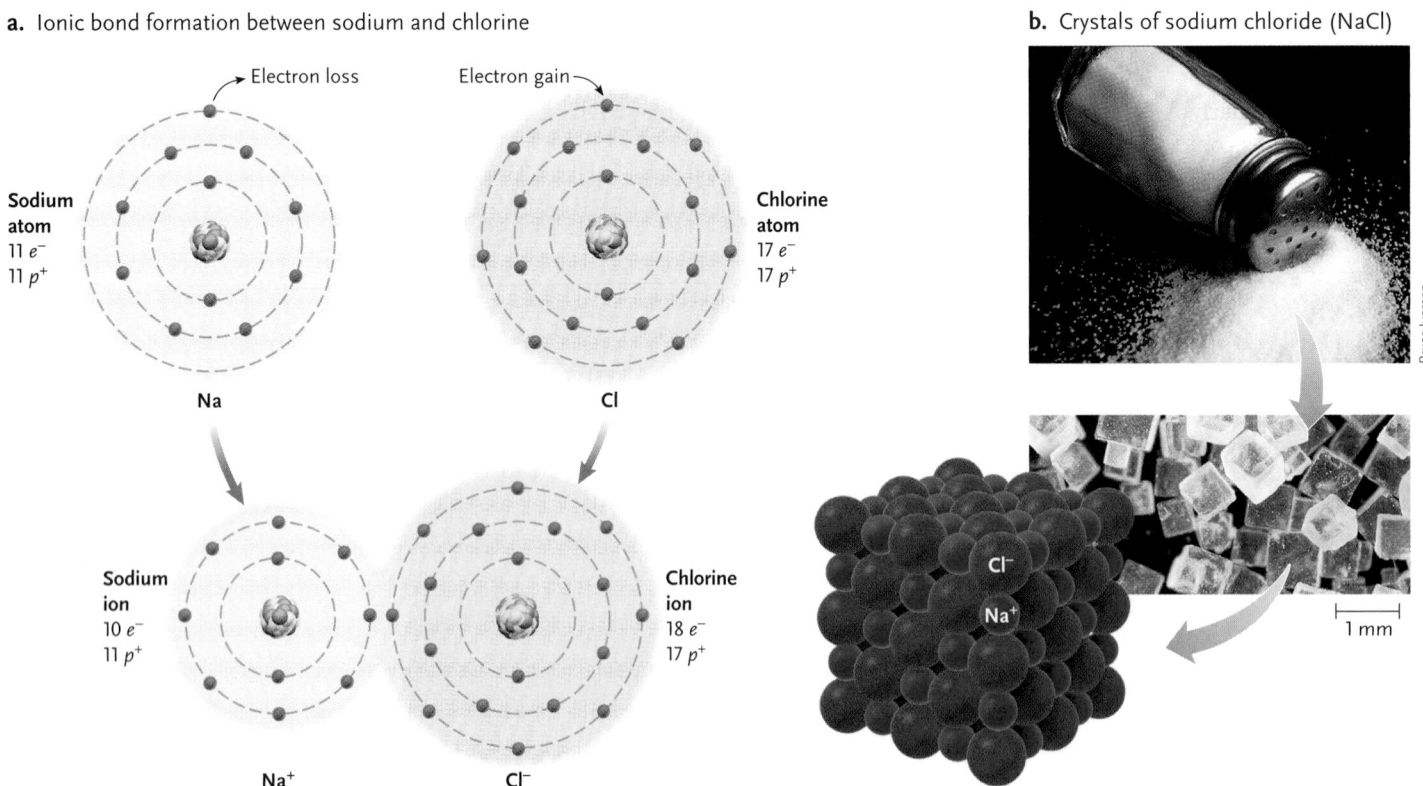

a. Ionic bond formation between sodium and chlorine

Electron loss

Electron gain

Sodium atom
11 e^-
11 p^+

Na

Chlorine atom
17 e^-
17 p^+

Cl

Sodium ion
10 e^-
11 p^+

Na^+

Chlorine ion
18 e^-
17 p^+

Cl^-

b. Crystals of sodium chloride (NaCl)

Bruce Iverson

Cl^-

Na^+

1 mm

Figure 2.7

Formation of an ionic bond. **(a)** Sodium, with one electron in its outermost energy level, readily loses that electron to attain a stable state in which its second energy level, with eight electrons, becomes the outer level. Chlorine, with seven electrons in its outer energy level, readily gains an electron to attain the stable number of eight. The transfer creates the ions Na^+ and Cl^-. **(b)** The combination forms sodium chloride (NaCl), common table salt.

called a **cation**, and a negatively charged ion such as Cl^- is called an **anion**. The difference in charge between cations and anions creates an attraction—the ionic bond—that holds the ions together in solid NaCl (sodium chloride).

Many other atoms that differ from stable outer energy levels by one electron, including hydrogen, can gain or lose electrons completely to form ions and ionic bonds. When a hydrogen atom loses its single electron to form a hydrogen ion (H^+), it consists of only a proton and is often simply called a proton to reflect this fact. A number of atoms with outer energy levels that differ from the stable number by two or three electrons, particularly metallic atoms such as calcium (Ca^{2+}), magnesium (Mg^{2+}), and iron (Fe^{2+} or Fe^{3+}), also lose their electrons readily to form cations and to join in ionic bonds with anions.

Ionic bonds are common among the forces that hold ions, atoms, and molecules together in living organisms because these bonds have three key features: (1) they exert an attractive force over greater distances than any other chemical bond, (2) their attractive force extends in all directions, and (3) they vary in strength depending on the presence of other charged substances. That is, in some systems, ionic bonds form in locations that exclude other charged substances, setting up strong and stable attractions that are not easily disturbed. For example, iron ions are stabilized by ionic bonds in the interior of the large biological molecule hemoglobin, where the ions are key to the distinctive chemical properties of that molecule. In other systems, particularly at molecular surfaces exposed to water molecules, ionic bonds are relatively weak, allowing ionic attractions to be established or broken quickly. For example, as part of their activity in speeding biological reactions, many enzymatic proteins bind and release molecules by forming and breaking relatively weak ionic bonds.

Covalent Bonds Are Formed by Electrons in Shared Orbitals

Covalent bonds form when atoms share a pair of valence electrons rather than gaining or losing them. The formation of molecular hydrogen, H_2, by two hydrogen atoms is the simplest example of the sharing mechanism. If two hydrogen atoms collide, the single electron of each atom may join in a new, combined two-electron orbital that surrounds both nuclei. The two electrons fill the orbital; thus, the hydrogen atoms tend to remain linked stably together. The linkage formed by the shared orbital is a covalent bond.

In molecular diagrams, a covalent bond is designated by a pair of dots or a single line that represents a pair of shared electrons. For example, in H_2, the covalent bond that holds the molecule together is represented as H:H or H—H.

Unlike ionic bonds, which extend their attractive force in all directions, the shared orbitals that form covalent bonds extend between atoms at discrete angles and directions, giving covalently bound molecules distinct, three-dimensional forms. For biological molecules such as proteins, which are held together primarily by covalent bonds, the three-dimensional form imparted by these bonds is critical to their functions.

Carbon, with four unpaired outer electrons, typically forms four covalent bonds to complete its outermost energy level. An example is methane, CH_4 **(Figure 2.8a, b),** the main component of natural gas. The four covalent bonds formed by the carbon atom are fixed at an angle of 109.5° from each other, forming a tetrahedron. The tetrahedral arrangement of the bonds allows carbon "building blocks" **(Figure 2.8c)** to link to each other in both branched and unbranched chains and rings **(Figure 2.8d).** Such structures form the backbones of an almost unlimited variety of molecules. Carbon can also form double bonds, in which atoms share two pairs of electrons, and triple bonds, in which atoms share three pairs of electrons.

Oxygen, hydrogen, nitrogen, and sulfur also share electrons readily to form covalent linkages, and they commonly combine with carbon in biological molecules. In these linkages, oxygen typically forms two covalent bonds; hydrogen, one; nitrogen, three; and sulfur, two.

Unequal Electron Sharing Results in Polarity

Electronegativity is the measure of an atom's attraction for the electrons it shares in a chemical bond with another atom. The more electronegative an atom is, the more strongly it attracts shared electrons. Among atoms, electronegativity increases as the number of protons in the nucleus increases and as the distance of electrons from the nucleus increases.

Although all covalent bonds involve the sharing of valence electrons, they differ widely in the degree of sharing. Depending on the difference in electronegativity between the bonded atoms, the covalent bonds are classified as **nonpolar covalent bonds** or **polar covalent bonds.** In a nonpolar covalent bond, electrons are shared equally, whereas in a polar covalent bond, they are shared unequally. When electron sharing is unequal, as in polar covalent bonds, the atom that attracts the electrons more strongly carries a partial negative charge and the atom deprived of electrons carries a partial positive charge. The atoms carrying partial charges may give the whole molecule partially positive and negative ends; in other words, the molecule is *polar,* hence the name given to the bond.

Nonpolar covalent bonds are characteristic of molecules that contain atoms of one kind, such as hydrogen (H_2) and oxygen (O_2), although there are some exceptions. Polar covalent bonds are characteristic of molecules that contain atoms of different types.

a. Shared orbitals of methane (CH₄)

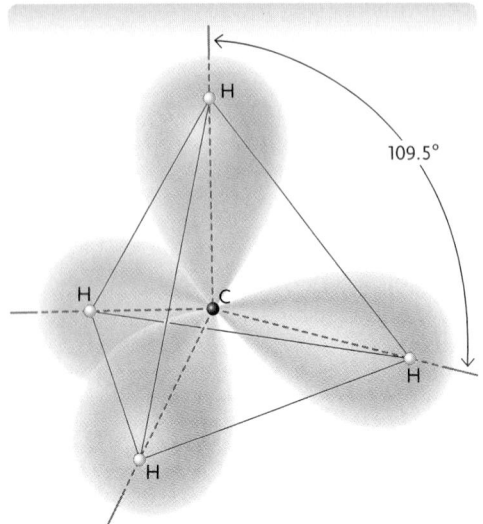

b. Space-filling model of methane

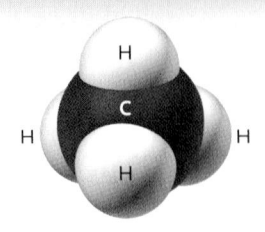

c. A carbon "building block" used to make molecular models

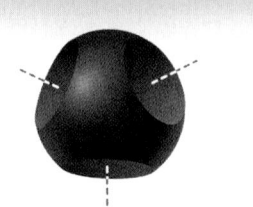

d. Cholesterol

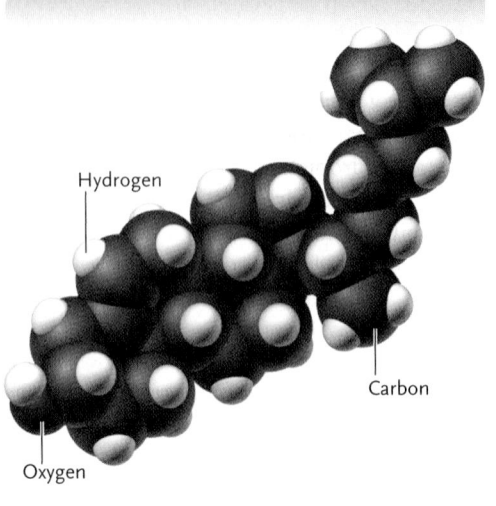

Hydrogen

Carbon

Oxygen

Figure 2.8

Covalent bonds shared by carbon. **(a)** The four covalent bonds of carbon in methane (CH₄) are shown as shared orbitals. The bonds extend outward from the carbon nucleus at angles of 109.5° from each other (dashed lines). The red lines connecting the hydrogen nuclei form a regular tetrahedron with four faces. **(b)** Space-filling model of methane, in which the diameter of the sphere representing an atom shows the approximate limit of its electron orbitals. **(c)** A tetrahedral carbon "building block." One of the four faces of the block is not visible. **(d)** Carbon atoms assembled into rings and chains forming a complex molecule.

For example, in water, an oxygen atom forms polar covalent bonds with two hydrogen atoms. Because the oxygen nucleus with its eight protons attracts electrons much more strongly than the hydrogen nuclei do, the bonds are strongly polar **(Figure 2.9)**. In addition, the water molecule is asymmetric, with the oxygen atom located on one side and the hydrogen atoms on the other. This arrangement gives the entire molecule an unequal charge distribution, with the hydrogen end partially positive and the oxygen end partially negative, and makes water molecules strongly polar. In fact, water is the primary biological example of a polar molecule. The polar nature of water underlies its ability to adhere to ions and weaken their attractions.

Oxygen, nitrogen, and sulfur, which all share electrons unequally with hydrogen, are located asymmetrically in many biological molecules. Therefore, the presence of —OH, —NH, or —SH groups tends to make regions in biological molecules containing them polar.

Although carbon and hydrogen share electrons somewhat unequally, these atoms tend to be arranged symmetrically in biological molecules. Thus, regions that contain only carbon–hydrogen chains are typically nonpolar. For example, the C—H bonds in methane are located symmetrically around the carbon atom (see Figure 2.8), so their partial charges cancel each other and the molecule as a whole is nonpolar.

Polar Molecules Tend to Associate with Each Other and Exclude Nonpolar Molecules

Polar molecules attract and align themselves with other polar molecules and with charged ions and molecules. These **polar associations** create environments that tend to exclude nonpolar molecules. When present in quantity, the excluded nonpolar molecules tend to clump together in arrangements called **nonpolar associations**; these nonpolar associations reduce the surface area exposed to the surrounding polar environment. Polar molecules that associate readily with water are identified as **hydrophilic** (*hydro* = water; *philic* = preferring). Nonpolar substances that are excluded by water and other polar molecules are identified as **hydrophobic** (*phobic* = avoiding).

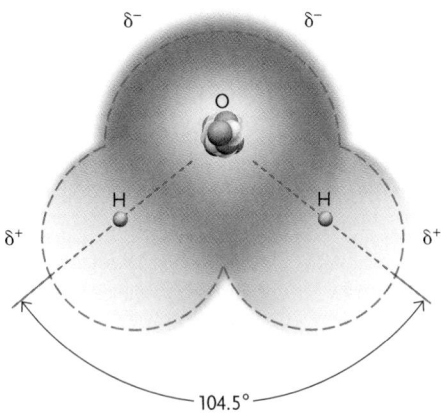

Figure 2.9

Polarity in the water molecule, created by unequal electron sharing between the two hydrogen atoms and the oxygen atom and the asymmetric shape of the molecule. The unequal electron sharing gives the hydrogen end of the molecule a partial positive charge, δ⁺ ("delta plus"), and the oxygen end of the molecule a partial negative charge, δ⁻ ("delta minus"). Regions of deepest color indicate the most frequent locations of the shared electrons. The orbitals occupied by the electrons are more complex than the spherical forms shown here.

Polar and nonpolar associations can be demonstrated with an apparatus no more complex than a bottle containing water and vegetable oil. If the bottle has been placed at rest for some time, the nonpolar oil and polar water form separate layers, with the oil on top. If you shake the bottle, the oil becomes suspended as spherical droplets in the water; the harder you shake, the smaller the oil droplets become (the spherical form of the oil droplets exposes the least surface area per unit volume to the watery polar surroundings). If you place the bottle at rest, the oil and water quickly separate again into distinct polar and nonpolar layers.

Hydrogen Bonds Also Involve Unequal Electron Sharing

When hydrogen atoms are made partially positive by sharing electrons unequally with oxygen, nitrogen, or sulfur, they may be attracted to nearby oxygen, nitrogen, or sulfur atoms made partially negative by unequal electron sharing in a different covalent bond **(Figure 2.10a).** This attractive force is the hydrogen bond, illustrated by a dotted line in structural diagrams of molecules. Hydrogen bonds may be intramolecular (between atoms in the same molecule) or intermolecular (between atoms in different molecules).

Individual hydrogen bonds are weak compared with ionic and covalent bonds. However, large biological molecules may offer many opportunities for hydrogen bonding, both within and between molecules. When numerous, hydrogen bonds are collectively strong and lend stability to the three-dimensional structure of molecules such as proteins **(Figure 2.10b).** Hydrogen bonds between water molecules are responsible for many of the properties that make water uniquely important to life (see Section 2.4 for a more detailed discussion).

The weak attractive force of hydrogen bonds makes them much easier to break than covalent and ionic bonds, particularly when elevated temperature increases the movements of molecules. Hydrogen bonds begin to break extensively as temperatures rise above 45°C and become practically nonexistent at 100°C. The disruption of hydrogen bonds by heat—for instance, the bonds in proteins—is one of the primary reasons most organisms cannot survive temperatures much greater than 45°C. Thermophilic (temperature-loving) organisms, which live at temperatures higher than 45°C, some at 120°C or more, have different molecules from those of organisms that live at lower temperatures. For example, proteins in thermophiles are stabilized at high temperatures by van der Waals forces and other noncovalent interactions.

Van der Waals Forces Are Weak Attractions over Very Short Distances

Van der Waals forces are even weaker than hydrogen bonds. These forces develop between nonpolar molecules or regions of molecules when, through their con-

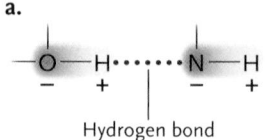

a.

Hydrogen bond

b.

Hydrogen bond

Hydrogen bonds stabilize the protein molecule into a helical shape.

Figure 2.10

Hydrogen bonds. **(a)** A hydrogen bond (dotted line) between the hydrogen of an —OH group and a nearby nitrogen atom, which also shares electrons unequally with another hydrogen. Regions of deepest blue indicate the most likely locations of electrons. **(b)** Multiple hydrogen bonds stabilize the backbone chain of a protein molecule into a spiral called the alpha helix. The spheres labeled *R* represent chemical groups of different kinds.

stant motion, electrons accumulate by chance in one part of a molecule or another. This process leads to zones of positive and negative charge, making the molecule polar. If they are oriented in the right way, the polar parts of the molecules are attracted electrically to one another and cause the molecules to stick together briefly. Although an individual bond formed with van der Waals forces is weak and transient, the formation of many bonds of this type can stabilize the shape of a large molecule, such as a protein.

A striking example of the collective power of van der Waals forces concerns the ability of geckos, a group of tropical lizard species, to cling to and walk up vertical smooth surfaces **(Figure 2.11).** The toes of the gecko are covered with millions of hairs, called *setae* (pronounced "see-tea"), that are about 100 micrometers (μm; 0.004 inch) long. At the tip of each hair are hundreds of thousands of pads, each about 200 nanometers (nm;

Figure 2.11
An example of van der Waals forces in biology. **(a)** A Tokay gecko *(Gekko gecko)* climbing while inverted on a glass plate. **(b)** Gecko toe. **(c)** Setae on a toe. **(d)** Pads (spatulae) on an individual seta.

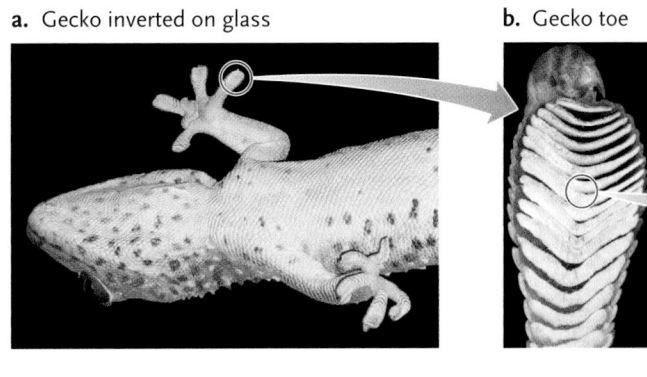

a. Gecko inverted on glass   b. Gecko toe   c. Setae on toe   d. Pads on a seta

Dr. Kellar Autumn, Autumn Lab, Lewis & Clark College

0.000008 inch) wide—smaller than the wavelength of visible light. Each pad forms a weak interaction—using van der Waals forces—with molecules on the surface. Magnified by the huge number of pads involved, the attractive forces are 1000 times greater than necessary for the gecko to hang on a vertical wall. To climb a wall, the animal rolls the hairs onto the surface and then peels them off like a piece of tape. Understanding the gecko's remarkable ability to climb has led to the development of gecko tape, a superadhesive prototype tape capable of holding a 3-kg weight with a 1-cm^2 (centimeter squared) piece.

Bonds Form and Break in Chemical Reactions

Chemical reactions occur when atoms or molecules interact to form new chemical bonds or break old ones. As a result of bond formation or breakage, atoms are added to or removed from molecules, or the linkages of atoms in molecules are rearranged. When any of these alterations occur, molecules change from one type to another, usually with different chemical and physical properties. In biological systems, chemical reactions are accelerated by molecules called *enzymes* (which are discussed in more detail in Chapter 4).

The atoms or molecules entering a chemical reaction are called the **reactants,** and those leaving a reaction are the **products.** A chemical reaction is written with an arrow showing the direction of the reaction; reactants are placed to the left of the arrow, and products are placed to the right. Both reactants and products are usually written in chemical shorthand as formulas.

For example, the overall reaction of photosynthesis, in which carbon dioxide and water are combined to produce sugars and oxygen (see Chapter 9), is written as follows:

$$6\ CO_2 + 6\ H_2O + light \rightarrow C_6H_{12}O_6 + 6\ O_2$$

carbon dioxide   water   a sugar   molecular oxygen

The number in front of each formula indicates the number of molecules of that type among the reactants and products (the number 1 is not written). Notice that there are as many atoms of each element to the left of the arrow as there are to the right, even though the products

are different from the reactants. This balance reflects the fact that in such reactions, atoms may be rearranged but not created or destroyed. Chemical reactions written in balanced form are known as **chemical equations.**

With the information about chemical bonds and reactions provided thus far, you are ready to examine the effects of chemical structure and bonding, particularly hydrogen bonding, in the production of the unusual properties of water, the most important substance to life on Earth.

STUDY BREAK

1. Explain how an ionic bond forms.
2. Explain how a covalent bond forms.
3. What is electronegativity, and how does it relate to nonpolar covalent bonds and polar covalent bonds?
4. What is a chemical reaction?

2.4 Hydrogen Bonds and the Properties of Water

All living organisms contain water, and many kinds of organisms live directly in water. Even those that live in dry environments contain water in all their structures—different organisms range from 50% to more than 95% water by weight. The water inside organisms is crucial for life: it is required for many important biochemical reactions and plays major roles in maintaining the shape and organization of cells and tissues. The properties of water molecules that make them so important to life depend to a great extent on their polar structure and their ability to link to each other by hydrogen bonds.

A Lattice of Hydrogen Bonds Gives Water Unusual Properties

Hydrogen bonds form readily between water molecules in both liquid water and ice. In liquid water, each water molecule establishes an average of 3.4 hydrogen bonds with its neighbors, forming an arrangement known as the **water lattice (Figure 2.12a).** In liquid wa-

a. Hydrogen-bond lattice of liquid water

b. Hydrogen-bond lattice of ice

KEY

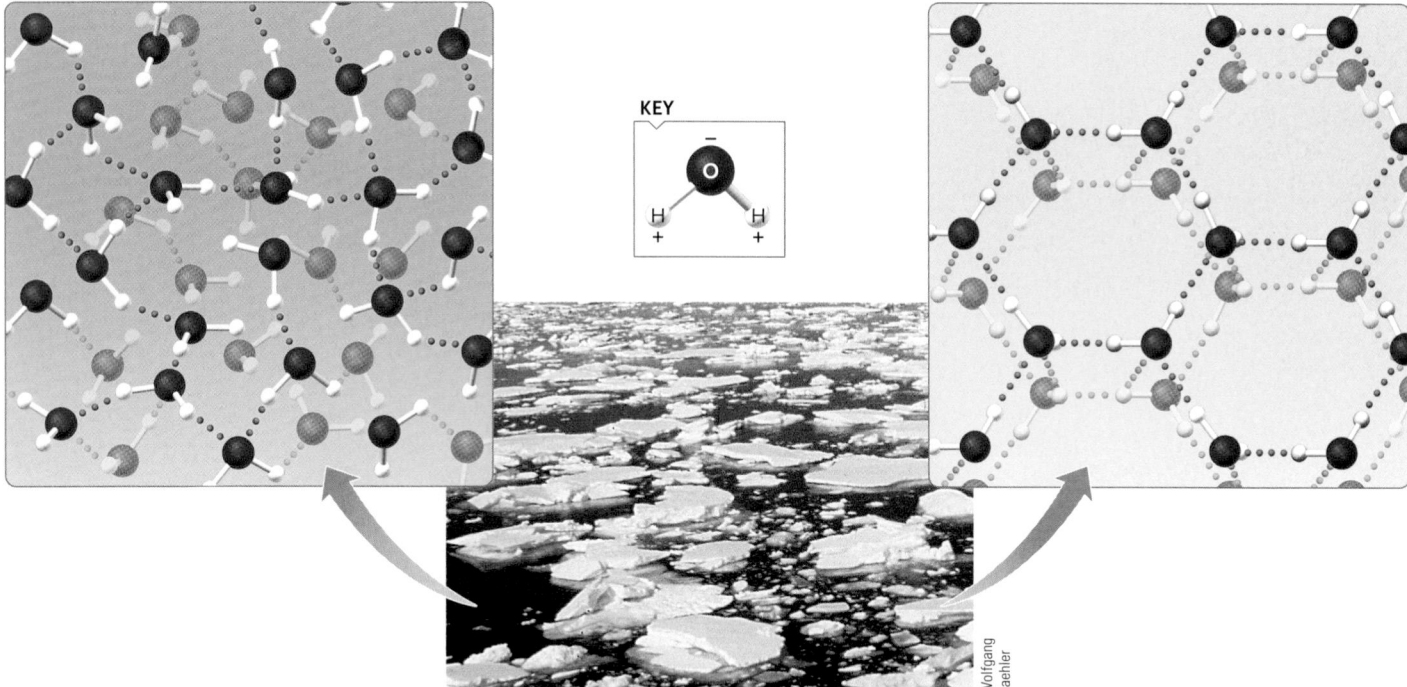

Wolfgang
Kaehler

ter, the hydrogen bonds that hold the lattice together constantly break and reform, allowing the water molecules to break loose from the lattice, slip past one another, and reform the lattice in new positions.

In ice, the water lattice is a rigid, crystalline structure in which each water molecule forms four hydrogen bonds with neighboring molecules **(Figure 2.12b).** The rigid **ice lattice** spaces the water molecules farther apart than the water lattice. Because of this greater spacing, water has the unusual property of being about 10% less dense when solid than when liquid. (Almost all other substances are denser in solid form than in liquid form.) Hence, ice cubes are a little larger than the water volume poured into the ice tray, and water filling a closed glass vessel will break the vessel when the water freezes. At atmospheric pressure, water reaches its greatest density at a temperature of 4°C, while it is still a liquid.

Because it is less dense than liquid water, ice forms at the surface of a body of water and remains floating at the surface. The ice creates an insulating layer that helps keep the water below from freezing. If ice were denser than liquid water, it would sink to the bottom as it freezes, continually exposing liquid water at the surface to freezing. Under those conditions, most bodies of water would freeze entirely solid, making life difficult or impossible for aquatic plants and animals.

The Hydrogen-Bond Lattice of Water Contributes to Polar and Nonpolar Environments in and around Cells

The hydrogen-bond lattice and the polarity of water molecules give water other properties that make it unique and ideal as a life-sustaining medium. In liquid

water, the lattice resists invasion by other molecules unless the invading molecule also contains polar or charged regions that can form competing attractions with water molecules. If present, the competing attractions open the water lattice, creating a cavity into which the polar or charged molecule can move. By contrast, nonpolar molecules are unable to disturb the water lattice. The lattice thus excludes nonpolar substances, forcing them to form the nonpolar associations that expose the least surface area to the surrounding water—such as the spherical droplets of oil that form when oil and water are shaken.

The distinct polar and nonpolar environments created by water are critical to the organization of cells. For example, biological membranes, which form boundaries around and inside cells, consist of lipid molecules with dual polarity: one end of each molecule is polar, and the other end is nonpolar. (Lipids are described in more detail in Chapter 3.) The membranes are surrounded on both sides by strongly polar water molecules. Exclusion by the water molecules forces the lipid molecules to associate into a double layer, a **bilayer**, in which only the polar ends of the surface molecules are exposed to the water **(Figure 2.13).** The nonpolar ends of the molecules associate in the interior of the bilayer, where they are not exposed to the water. Exclusion of their nonpolar regions by water is all that holds membranes together.

The membrane at the surface of cells prevents the watery solution inside the cell from mixing directly with the watery solution outside the cell. By doing so, the surface membrane, kept intact by nonpolar exclusion by water, maintains the internal environment and organization necessary for cellular life.

Figure 2.12
Hydrogen bonds and water. **(a)** In liquid water, hydrogen bonds between molecules (dotted lines) form and break rapidly, allowing the molecules to slip past each other easily. **(b)** In ice, water molecules are fixed into a rigid lattice.

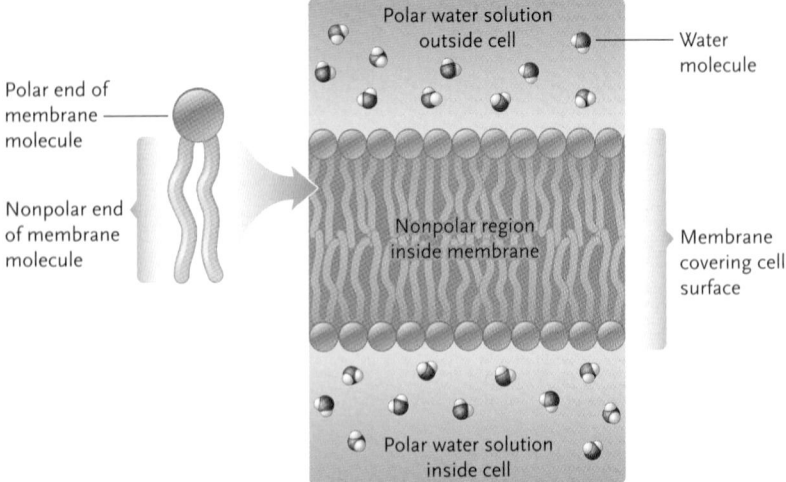

Figure 2.13
Double layer of lipid molecules that forms a membrane covering the cell surface.
Exclusion by polar water molecules forces the nonpolar ends of the surface molecules
to associate into the thin, double layer—the bilayer—that forms the membrane.

The Small Size and Polarity of Its Molecules Makes Water a Good Solvent

Because water molecules are small and strongly polar, they readily penetrate or coat the surfaces of other polar and charged molecules and ions. The surface coat, called a **hydration layer**, reduces the attraction between the molecules or ions and promotes their separation and entry into a **solution**, where they are suspended individually, surrounded by water molecules. Once in solution, the hydration layer prevents the polar molecules or ions from reassociating. In such a solution, water is called the **solvent**, and the molecules of a substance dissolved in water are called the **solute**.

For example, when a teaspoon of table salt is added to water, water molecules quickly form hydration layers around the Na^+ and Cl^- ions in the salt crystals, reducing the attraction between the ions so much that they separate from the crystal and enter the surrounding water lattice as individual ions **(Figure 2.14)**. If the water evaporates, the hydration layer is eliminated, exposing the strong positive and negative charges of the ions. The opposite charges attract and reestablish the ionic bonds that hold the ions in salt crystals. As the last of the water evaporates, all of the Na^+ and Cl^- ions relink into the solid, crystalline form.

In the cell, chemical reactions depend on solutes dissolved in aqueous solutions. To understand these reactions, you need to know the number of atoms and molecules involved. **Concentration** is the number of molecules or ions of a substance in a unit volume of space, such as a milliliter (mL) or liter (L). The number of molecules or ions in a unit volume cannot be counted directly but can be calculated indirectly by using the mass number of atoms as the starting point. The same method is used to prepare a solution with a known number of molecules per unit volume.

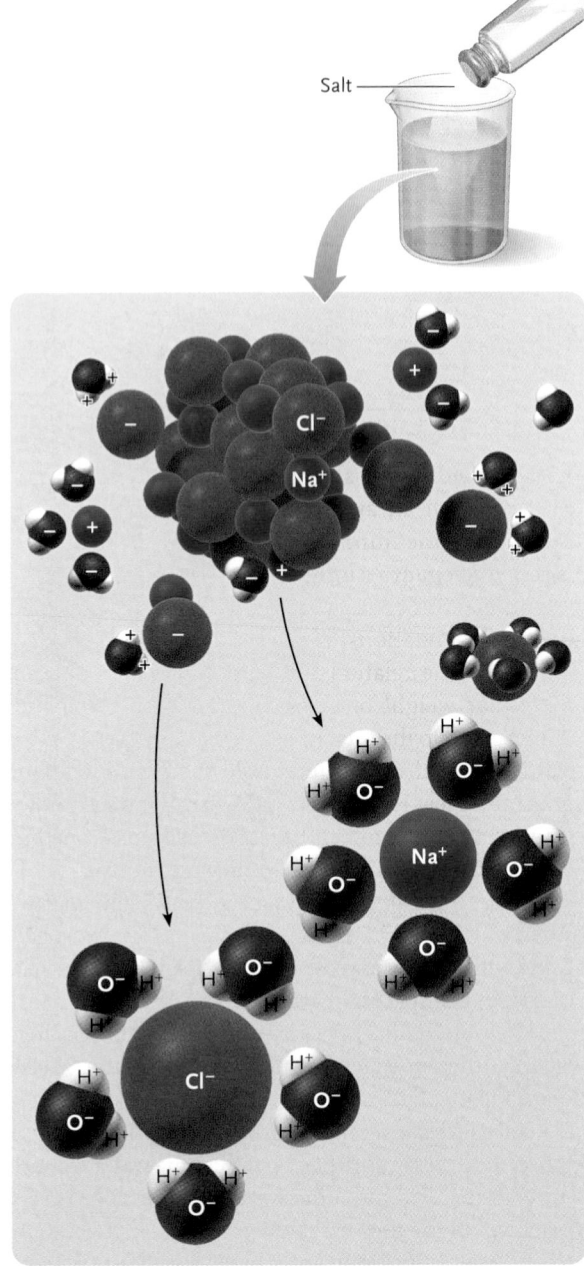

Figure 2.14
Water molecules forming a hydration layer around Na^+ and Cl^- ions, which promotes their separation and entry into solution.

The mass number of an atom is equivalent to the number of protons and neutrons in its nucleus. From the mass number, and the fact that neutrons and protons are approximately the same weight (that is, 1.66×10^{-24} g), you can calculate the weight of an atom of any substance. For an atom of the most common form of carbon, with 6 protons and 6 neutrons in its nucleus, the total weight is:

$$12 \times (1.66 \times 10^{-24}\,g) = 1.992 \times 10^{-23}\,g$$

For an oxygen atom, with 8 protons and 8 neutrons in its nucleus, the total weight is:

$$16 \times (1.66 \times 10^{-24}\,g) = 2.656 \times 10^{-23}\,g$$

Dividing the total weight of a sample of an element by the weight of a single atom gives the number of atoms in the sample. Suppose you have a carbon sample that weighs 12 g—a weight in grams equal to the atom's mass number. (A weight in grams equal to the mass number is known as an **atomic weight** of an element.) Dividing 12 g by the weight of one carbon atom gives:

$$\frac{12}{(1.992 \times 10^{-23}\,g)} = 6.022 \times 10^{23}\text{ atoms}$$

If you divide the atomic weight of oxygen (16 g) by the weight of one oxygen atom, you get the same result:

$$\frac{16}{(2.656 \times 10^{-23}\,g)} = 6.022 \times 10^{23}\text{ atoms}$$

In fact, dividing the atomic weight of any element by the weight of an atom of that element always produces the same number: 6.022×10^{23}. This number is called **Avogadro's number** after Amedeo Avogadro, the nineteenth-century Italian chemist who first discovered the relationship.

The same relationship holds for molecules. The **molecular weight** of any molecule is the weight in grams equal to the total mass number of its atoms. For NaCl, the total mass number is $23 + 35 = 58$ (a sodium atom has 11 protons and 12 neutrons, and a chlorine atom has 17 protons and 18 neutrons). The weight of an NaCl molecule is therefore:

$$58 \times (1.66 \times 10^{-24}\,g) = 9.628 \times 10^{-23}\,g$$

Dividing a molecular weight of NaCl (58 g) by the weight of a single NaCl molecule gives:

$$\frac{58}{(9.628 \times 10^{-23}\,g)} = 6.022 \times 10^{23}\text{ molecules}$$

When concentrations are described, the atomic weight of an element or the molecular weight of a compound—the amount that contains 6.022×10^{23} atoms or molecules—is known as a **mole** (abbreviated **mol**). The number of moles of a substance dissolved in 1 L of solution is known as the **molarity** (abbreviated **M**) of the solution. This relationship is highly useful in chemistry and biology because we know that two solutions having the same volume and molarity but composed of different substances will contain the same number of molecules of the substances.

The Hydrogen-Bond Lattice Gives Water Other Life-Sustaining Properties as Well

The hydrogen-bond lattice gives water other unique properties that make it a medium suitable for the molecules and reactions of life. Compared with substances that have a similar molecular structure, such as H_2S (hydrogen sulfide):

- Water has an unusual ability to resist changes in temperature by absorbing or releasing heat, plus an unusually high boiling point.

- Water has an unusually high internal cohesion and surface tension.

The Boiling Point and Temperature-Stabilizing Effects of Water. The hydrogen-bond lattice of liquid water retards the escape of individual water molecules as the water is heated. As a result, relatively high temperatures and the addition of considerable heat are required to break enough hydrogen bonds to make water boil. The high boiling point maintains water as a liquid over the wide temperature range of 0° to 100°C. Similar molecules that do not form an extended hydrogen-bond lattice, such as H_2S, have much lower boiling points and are gases rather than liquids at room temperature. The properties of these related substances indicate that without its hydrogen-bond lattice, water would boil at $-81°C$. If this were the case, most of the water on Earth would be in gaseous form and life as described in this book could not exist.

As a result of water's stabilizing hydrogen-bond lattice, it also has a relatively high **specific heat**—that is, the amount of heat required to increase the temperature of a given quantity of water. As heat flows into water, much of it is absorbed in the breakage of hydrogen bonds. As a result, the temperature of water, reflected in the average motion of its molecules, increases relatively slowly as heat is added. For example, a given amount of heat increases the temperature of water by only half as much as that of an equal quantity of ethyl alcohol. High specific heat allows water to absorb or release relatively large quantities of heat without undergoing extreme changes in temperature; this gives it a moderating and stabilizing effect on both living organisms and their environments.

The specific heat of water is measured in **calories.** This unit, used both in the sciences and in dieting, is the amount of heat required to raise 1 g of water by 1°C (technically, from 14.5 to 15.5°C at one atmosphere of pressure). This amount of heat is known as a "small" calorie and is written with a small c. The unit most familiar to dieters, equal to 1000 small calories, is written with a capital C as a **Calorie;** the same 1000-calorie unit is known scientifically as a **kilocalorie (kcal).** A 300-Calorie candy bar therefore really contains 300,000 calories.

A large amount of heat, 586 calories per gram, must be added to give water molecules enough energy of motion to break loose from liquid water and form a gas. This required heat, known as the **heat of vaporization,** allows humans and many other organisms to cool off when hot. In humans, water is released onto the surface of the skin by more than 2.5 million sweat glands; the heat absorbed by this water as it evaporates cools the skin and the underlying blood vessels. The heat loss helps keep body temperature from increasing when environmental temperatures are high. Plants use a similar cooling mechanism as water evaporates from their leaves.

Cohesion and Surface Tension. The high resistance of water molecules to separation, provided by the hydrogen-bond lattice, is known as internal **cohesion.** For example, in land plants, cohesion holds water molecules in unbroken columns in microscopic conducting tubes that extend from the roots to the highest leaves. As water evaporates from the leaves, water molecules in the columns, held together by cohesion, move upward through the tubes to replace the lost water. This movement raises water from roots to the tops of the tallest trees (see discussion in Chapter 32). Maintenance of the long columns of water in the tubes is aided by **adhesion,** in which molecules "stick" to the walls of the tubes by forming hydrogen bonds with charged and polar groups in molecules that form the walls of the tubes.

Water molecules at surfaces facing air can form hydrogen bonds with water molecules beside and below them but not on the sides that face the air. This unbalanced bonding produces a force that places the surface water molecules under tension, making them more resistant to separation than the underlying water molecules **(Figure 2.15a).** The force, called **surface tension,** is strong enough to allow small insects such as water striders to walk on water **(Figure 2.15b).** Similarly, the surface tension of water will support a sewing needle placed carefully on the surface, even though the needle is about 10 times denser than the water. Surface tension also causes water to form water droplets; the surface tension pulls the water in around itself to produce the smallest possible area, which is a spherical bead or droplet.

Water has still other properties that contribute to its ability to sustain life, the most important being that its molecules separate into ions. These ions help maintain an environment inside living organisms that promotes the chemical reactions of life.

STUDY BREAK

1. How do hydrogen bonds between water molecules contribute to the properties of water?
2. Distinguish between a solute, a solvent, and a solution.

2.5 Water Ionization and Acids, Bases, and Buffers

The most critical property of water that is unrelated to its hydrogen-bond lattice is its ability to separate, or **dissociate,** to produce positively charged *hydrogen ions* (H^+, or protons) and *hydroxide ions* (OH^-):

$$H_2O \rightleftharpoons H^+ + OH^-$$

(The double arrow means that the reaction is **reversible**— that is, depending on conditions, it may go from left to right or from right to left.) The proportion of water molecules that dissociates to release protons and hydroxide ions is small. However, because of the dissociation, water always contains some H^+ and OH^- ions.

Substances Act as Acids or Bases by Altering the Concentrations of H^+ and OH^- Ions in Water

In pure water, the concentrations of H^+ and OH^- ions are equal. However, adding other substances may alter the relative concentrations of H^+ and OH^-, making them unequal. Some substances, called **acids,** are proton donors that release H^+ (and anions) when they are dissolved in water, effectively increasing the H^+ concentration. For example, hydrochloric acid (HCl) dissociates into H^+ and Cl^- when dissolved in water:

$$HCl \rightleftharpoons H^+ + Cl^-$$

Other substances, called **bases,** are proton acceptors that reduce the H^+ concentration of a solution. Most bases dissociate in water into a hydroxide ion (OH^-) and a cation. The hydroxide ion can act as a base by accepting a proton (H^+) to produce water. For example,

a.

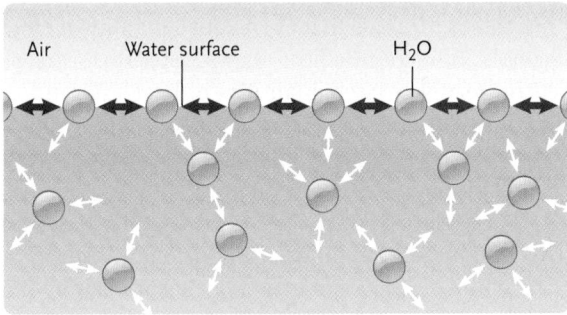

Air   Water surface H_2O

b.

H. Eisenbeiss/Frank Lane Picture Agency

Figure 2.15

Surface tension in water. **(a)** The unbalanced hydrogen bonding that places water molecules under lateral tension where a water surface faces the air. **(b)** A water strider (*Gerris* species) supported by the surface tension of water.

sodium hydroxide (NaOH) separates into Na^+ and OH^- ions when dissolved in water:

$$NaOH \rightarrow Na^+ + OH^-$$

The excess OH^- combines with H^+ to produce water:

$$OH^- + H^+ \rightarrow H_2O$$

thereby reducing the H^+ concentration.

Other bases do not dissociate to produce hydroxide ions directly. For example, ammonia (NH_3), a poisonous gas, acts as a base when dissolved in water, directly accepting a proton from water to produce an ammonium ion and releasing a hydroxide ion:

$$NH_3 + H_2O \rightarrow NH_4^+ + OH^-$$

The concentration of H^+ ions in a water solution, as compared with the concentration of OH^- ions, determines the **acidity** of the solution. Scientists measure acidity using a numerical scale from 0 to 14, called the **pH scale.** Because the number of H^+ ions in solution increases exponentially as the acidity increases, the scale is based on logarithms of this number to make the values manageable:

$$pH = -\log_{10}[H^+]$$

In this formula, the brackets indicate concentration in moles per liter. The negative of the logarithm is used to give a positive number for the pH value. For example, in a water solution that is *neutral*—neither acidic nor basic—the concentration of *both* H^+ and OH^- ions is $1 \times 10^{-7}\,M$ (0.0000001 M). The $\log_{10}$ of 1×10^{-7} is -7. The negative of the logarithm -7 is 7. Thus, a neutral water solution with an H^+ concentration of $1 \times 10^{-7}\,M$ has a pH of 7. *Acidic* solutions have pH values less than 7, with pH 0 being the value for the highly acidic 1 M hydrochloric acid (HCl); *basic* solutions have pH values greater than 7, with pH 14 being the value for the highly basic 1 M sodium hydroxide (NaOH) (basic solutions are also called *alkaline* solutions). Each whole number on the pH scale represents a value 10 times greater or less than the next number. Thus, a solution with a pH of 4 is 10 times more acidic than one with a pH of 5, and a solution with a pH of 6 is 100 times more acidic than a solution with a pH of 8. (The pH of many familiar solutions is shown in **Figure 2.16.**)

Acidity is important to cells because even small changes, on the order of 0.1 or even 0.01 pH unit, can drastically affect biological reactions. In large part, this effect reflects changes in the structure of proteins that occur when the water solution surrounding the proteins has too few or too many hydrogen ions. Consequently, all living organisms have elaborate systems that control their internal acidity by regulating H^+ concentration near the neutral value of pH 7.

Acidity is also important to the environment in which we live. Where the air is unpolluted, rainwater is only slightly acidic. However, in regions where certain pollutants are released into the air in large quantities by industry and automobile exhaust, the polluting chemicals combine with atmospheric water to produce "acid rain" with a pH as low as 3, about the same pH as that of vinegar. Acid rain can sicken and kill wildlife such as fishes and birds, as well as plants and trees (**Figure 2.17;** see also discussion in Chapter 53). Humans are also affected; acid rain and acidified water vapor in the air can contribute to human respiratory diseases such as bronchitis and asthma.

Buffers Help Keep pH under Control

Living organisms control the internal pH of their cells with **buffers,** substances that compensate for pH changes by absorbing or releasing H^+. When H^+ ions are released in excess by biological reactions, buffers combine

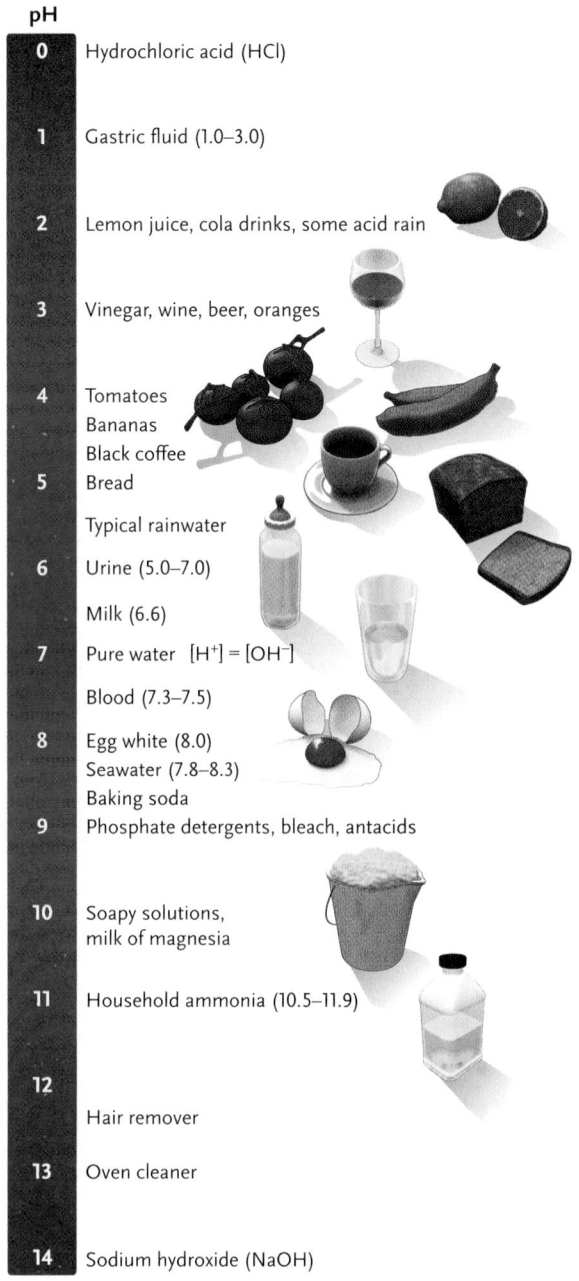

pH

0	Hydrochloric acid (HCl)
1	Gastric fluid (1.0–3.0)
2	Lemon juice, cola drinks, some acid rain
3	Vinegar, wine, beer, oranges
4	Tomatoes / Bananas / Black coffee
5	Bread / Typical rainwater
6	Urine (5.0–7.0) / Milk (6.6)
7	Pure water $[H^+] = [OH^-]$ / Blood (7.3–7.5)
8	Egg white (8.0) / Seawater (7.8–8.3) / Baking soda
9	Phosphate detergents, bleach, antacids
10	Soapy solutions, milk of magnesia
11	Household ammonia (10.5–11.9)
12	Hair remover
13	Oven cleaner
14	Sodium hydroxide (NaOH)

Figure 2.16
The pH scale, showing the pH of substances commonly encountered in the environment.

Figure 2.17

Forest affected by acid rain and other forms of air pollution in the Great Smoky Mountains National Park. The trees are susceptible to drought, disease, and insect pests.

Frederica Georgia/Photo Researchers, Inc.

with them and remove them from the solution; if the concentration of H^+ decreases greatly, buffers release additional H^+ to restore the balance. Most buffers are weak acids or bases, or combinations of these substances, that dissociate reversibly in water solutions to release or absorb H^+ or OH^-. (Weak acids, such as acetic acid, or weak bases, such as ammonia, are substances that release relatively few H^+ or OH^- ions in a water solution. Strong acids or bases are substances that dissociate extensively in a water solution. HCl is a strong acid; NaOH is a strong base.)

The buffering mechanism that maintains blood pH near neutral values is a primary example. In humans and many other animals, blood pH is buffered by a chemical system based on carbonic acid (H_2CO_3), a weak acid. In water solutions, carbonic acid dissociates readily into bicarbonate ions (HCO_3^-) and H^+:

$$H_2CO_3 \rightleftharpoons HCO_3^- + H^+$$

The reaction is reversible. If H^+ is present in excess, the reaction is pushed to the left—the excess H^+ ions combine with bicarbonate ions to form H_2CO_3. If the H^+ concentration declines below normal levels, the reaction is pushed to the right—H_2CO_3 dissociates into HCO_3^- and H^+, restoring the H^+ concentration. The back-and-forth adjustments of the buffer system help keep human blood close to its normal pH of 7.4.

The effects of hyperventilation highlight the importance of the system that buffers blood pH. Hyperventilation, caused by breathing too fast, drastically reduces the CO_2 concentration in blood **(Figure 2.18)**. Carbon dioxide is the primary source of carbonic acid in the bloodstream ($CO_2 + H_2O \rightarrow H_2CO_3$); removing too much CO_2 causes the amount of carbonic acid in the blood to decrease. If the amount of blood CO_2 drops so low that the carbonic acid buffer is no longer able to maintain pH at normal levels, a series of internal reac-

Unanswered Questions

Can arsenic be removed from the soil by bioremediation?

In the Why It Matters section, we learned that bioremediation of selenium in wastewater is possible using plants. Research is showing that bioremediation can also be used to remove other toxic chemicals in the environment, including perchlorate and arsenic. For example, arsenic-contaminated soils and sediments are the major sources of arsenic contamination in surface water and groundwater, which leads to contamination of foods. In some parts of the world, the drinking water is contaminated. Arsenic poses serious health risks to humans and other animals; for example, some cancers have been correlated with high levels of arsenic. Arsenic contamination is a worldwide concern, with arsenic levels in the environment in some parts of the world being tens of thousands of times higher than the maximum contaminant level set in the United States.

One research group at LaTrobe University, Melbourne, Australia, led by Joanne Santini, is exploring whether bacteria can be used for arsenic bioremediation in contaminated wastewater on mining sites and from groundwater in Bangladesh and West Bengal, India. Their approach has been to study 13 rare bacteria isolated from gold mines, a typical place to find arsenic. Arsenic is present in water in two toxic forms; one of these forms is easy and safe to get rid of, but the other is not. Santini's group has identified a bacterium that can "eat" the difficult-to-get-rid-of form of arsenic and convert it to the easy-to-get-rid-of form. Potentially, this bacterium could be developed for use in bioremediation of arsenic in contaminated locations.

Is food irradiation effective for killing microorganisms?

Radioisotopes are widely used to answer questions in biological research and as tools in medicine. Radioisotopes are also used to irradiate foods with the goal of killing microorganisms capable of causing disease. In most instances, the irradiation of food is done using the ra-

dioactive element cobalt-60 as a source of high-energy gamma rays. The energy of the gamma rays is sufficient to dislodge electrons from some food molecules, converting them to ions. But, there is insufficient energy to affect the neutrons in the nuclei of those molecules, so the food is not rendered radioactive by the treatment.

The effectiveness of food irradiation is tested in the laboratory. Researchers perform experiments to determine the dosage needed to kill a population of various pathogens in food. They have shown that irradiation kills many bacteria and parasites and destroys some viruses in food; moreover, they have not seen the development of radiation resistance in the microbial strains and species tested. However, some viruses and spore-forming bacteria are not destroyed by irradiation.

While many organizations such as the World Health Organization (WHO), U.S. Food and Drug Administration (FDA), and Institute of Food Science and Technology have concluded that irradiation of food is safe and can be effective in killing microbial contaminants, questions remain in some quarters, including with some consumers. For example, Does irradiation destroy vitamins? and Are toxic products produced by irradiation?

Researchers have shown that although vitamins in solution can be degraded by irradiation, they are less sensitive to irradiation when present in the complex chemical organization of food. There is some evidence, though, that irradiation sometimes causes chemical changes in food similar to those produced during cooking. Evidence from studies with laboratory animals indicated no adverse health effects when irradiated foods containing these compounds were consumed. However, some concerns remain about the generation of potentially harmful chemical compounds if the food is irradiated in its final packaging. More research needs to be done to determine how serious this concern is to human health.

Peter J. Russell

| Rapid breathing | → | Blood CO_2 concentration decreases | → | Blood carbonic acid level decreases | → | Blood pH changes from normal levels | → | Adverse physiological effects, such as dizziness, visual impairment, fainting, seizures, or death |

Figure 2.18
Effects of hyperventilation.

tions occurs that can produce dizziness, visual impairment, fainting, seizures, or even death.

This chapter examined the basic structure of atoms and molecules and discussed the unusual properties of water that make it ideal for supporting life. The next chapter looks more closely at the structure and properties of carbon and at the great multitude of molecules based on this element.

STUDY BREAK

1. Distinguish between acids and bases. What are their properties?
2. Why are buffers important for living organisms?

Review

Go to ThomsonNOW™ at www.thomsonedu.com/login to access quizzing, animations, exercises, articles, and personalized homework help.

2.1 The Organization of Matter: Elements and Atoms

- Matter is anything that occupies space and has mass. Matter is composed of elements, each consisting of atoms of the same kind.
- Atoms combine chemically in fixed numbers and ratios to form the molecules of living and nonliving matter. Compounds are molecules in which the component atoms are different.

2.2 Atomic Structure

- Atoms consist of an atomic nucleus that contains protons and neutrons surrounded by one or more electrons traveling in orbitals. Each orbital can hold a maximum of two electrons (Figure 2.3).
- All atoms of an element have the same number of protons, but the number of neutrons is variable. The number of protons in an atom is designated by its atomic number; the number of protons plus neutrons is designated by the mass number (Figure 2.4 and Table 2.1).
- Isotopes are atoms of an element with differing numbers of neutrons. The isotopes of an atom differ in physical but not chemical properties (Figure 2.4).
- Electrons surround an atomic nucleus in orbitals occupying energy levels that increase in discrete steps (Figures 2.5 and 2.6).
- The chemical activities of atoms are determined largely by the number of electrons in the outermost energy level. Atoms that have the outermost level filled with electrons are nonreactive, whereas atoms in which that level is not completely filled with electrons are reactive. Atoms tend to lose, gain, or share electrons to fill the outermost energy level.

  Video: Isotopes of hydrogen

  Animation: Electron arrangements in atoms

  Animation: The shell model of electron distribution

  Practice: Predicting the number of bonds of elements

2.3 Chemical Bonds

- An ionic bond forms between atoms that gain or lose electrons in the outermost energy level completely, that is, between a positively charged cation and a negatively charged anion (Figure 2.7).
- A covalent bond is established by a pair of electrons shared between two atoms. If the electrons are shared equally, the covalent bond is nonpolar (Figure 2.8).
- If electrons are shared unequally in a covalent bond, the atoms carry partial positive and negative charges and the bond is polar (Figure 2.9).
- Polar molecules tend to associate with other polar molecules and to exclude nonpolar molecules. Polar molecules that associate readily with water are hydrophilic; nonpolar molecules excluded by water are hydrophobic.
- A hydrogen bond is a weak attraction between a hydrogen atom made partially positive by unequal electron sharing and another atom—usually oxygen, nitrogen, or sulfur—made partially negative by unequal electron sharing (Figure 2.10).
- Van der Waals forces, bonds even weaker than hydrogen bonds, can form when natural changes in the electron density of molecules produce regions of positive and negative charge, which cause the molecules to stick together briefly.
- Chemical reactions occur when molecules form or break chemical bonds. The atoms or molecules entering into a chemical reaction are the reactants, and those leaving a reaction are the products.

  Animation: How atoms bond

2.4 Hydrogen Bonds and the Properties of Water

- The hydrogen-bond lattice formed by polar water molecules makes it difficult for nonpolar substances to penetrate the lattice. The distinct polar and nonpolar environments created by water are critical to the organization of cells (Figures 2.12 and 2.13).
- The polar properties of water allow it to form a hydration layer over the surfaces of polar and charged biological molecules, particularly proteins. Many chemical reactions depend on the special molecular conditions created by the hydration layer (Figure 2.14).
- The polarity of water allows ions and polar molecules to dissolve readily in water, making it a good solvent.

- The hydrogen-bond lattice gives water unusual properties that are vital to living organisms, including high specific heat, boiling point, cohesion, and surface tension (Figure 2.15).

  Animation: **Structure of water**

  Animation: **Spheres of hydration**

2.5 Water Ionization and Acids, Bases, and Buffers

- Acids are substances that increase the H^+ concentration by releasing additional H^+ as they dissolve in water; bases are substances that decrease the H^+ concentration by gathering H^+ or releasing OH^- as they dissolve.

- The relative concentrations of H^+ and OH^- in a water solution determine the acidity of the solution, which is expressed quantitatively as pH on a number scale ranging from 0 to 14. Neutral solutions, in which the concentrations of H^+ and OH^- are equal, have a pH of 7. Solutions with pH less than 7 have H^+ in excess and are acidic; solutions with pH greater than 7 have OH^- in excess and are basic or alkaline (Figure 2.16).

- The pH of living cells is regulated by buffers, which absorb or release H^+ to compensate for changes in H^+ concentration.

  Animation: **The pH scale**

Questions

Self-Test Questions

1. Which of the following statements about the mass number of an atom is *incorrect*?
   a. It has a unit defined as a dalton.
   b. On Earth, it equals the atomic weight.
   c. Unlike the atomic weight of an atom, it does not change when gravitational forces change.
   d. It equals the number of electrons in an atom.
   e. It is the sum of the protons and neutrons in the atomic nucleus.

2. To make 5 L of a 0.2 *M* aqueous solution of glucose, how many grams of glucose ($C_6H_{12}O_6$) do you need? Atomic masses are carbon 12, hydrogen 1, oxygen 16.
   a. 18.1   b. 180   c. 181   d. 905   e. 9.05

3. The chemical activity of an atom:
   a. depends on the electrons in the outermost energy level.
   b. is increased when the outermost energy level is filled with electrons.
   c. depends on its 1*s* but not its 2*s* or 2*p* orbitals.
   d. is increased when valence electrons completely fill the outer orbitals.
   e. of oxygen prevents it from sharing its electrons with other atoms.

4. When electrons are shared equally, this represents a (an):
   a. polar covalent bond.   d. hydrogen bond.
   b. nonpolar covalent bond.   e. van der Waals force.
   c. ionic bond.

5. Which of the following is *not* a property of water?
   a. It has a low boiling point compared with other molecules.
   b. It has a high heat of vaporization.
   c. Its molecules resist separation, a property called cohesion.
   d. It has the property of adhesion, the ability to stick to charged and polar groups in molecules.
   e. It can hydrogen bond to molecules below but not above its surface.

6. Which of the following would *not* represent a hydrophilic body fluid?
   a. blood   d. oil
   b. sweat   e. saliva
   c. tears

7. The water lattice:
   a. is formed from hydrophobic bonds.
   b. causes ice to be denser than water.
   c. reduces water's ability as a solvent.
   d. excludes polar substances.
   e. contributes to polar and nonpolar spaces around cells.

8. A hydrogen bond is:
   a. a strong attraction between hydrogen and another atom.
   b. a bond between a hydrogen atom already covalently bound to one atom and made partially negative by unequal electron sharing with another atom.
   c. a bond between a hydrogen atom already covalently bound to one atom and made partially positive by unequal electron sharing with another atom.
   d. weaker than van der Waals forces.
   e. exemplified by the two hydrogens covalently bound to oxygen in the water molecule.

9. If the water in a pond has a pH of 5, the hydroxide concentration would be
   a. 10^{-5} *M*.   d. 10^9 *M*.
   b. 10^{-10} *M*.   e. 10^{-9} *M*.
   c. 10^5 *M*.

10. Because of a sudden hormonal imbalance, a patient's blood was tested and shown to have a pH of 7.46. What does this pH value mean?
    a. This is more acidic than normal blood.
    b. It represents a weak alkaline fluid.
    c. This is caused by a release of large amounts of hydrogen ions into the system.
    d. The reaction $H_2CO_2 \rightarrow HCO_3^- + H^+$ is pushed to the left.
    e. This is probably caused by excess CO_2 in the blood.

Questions for Discussion

1. Detergents allow particles of oil to mix with water. From the information presented in this chapter, how do you think detergents work?

2. What would living conditions be like on Earth if ice were denser than liquid water?

3. You place a metal pan full of water on the stove and turn on the heat. After a few minutes, the handle is too hot to touch but the water is only warm. How do you explain this observation?

4. You are studying a chemical reaction accelerated by an enzyme. H^+ forms during the reaction, but the enzyme's activity is lost at low pH. What could you include in the reaction mix to keep the enzyme's activity at high levels? Explain how your suggestion might solve the problem.

Experimental Analysis

You know that adding NaOH to HCl results in the formation of common table salt, NaCl. You have a 0.5 *M* HCl solution. What weight of NaOH would you need to add to convert all of the HCl to NaCl? (**Note:** Chemical reactions have the potential to be dangerous. Please do not attempt to perform this reaction.)

Evolution Link

What properties of water made the evolution of life possible?

The lipoproteins HDL and LDL, cholesterol-transporting molecules composed of protein and lipid units, which are found in the bloodstream (computer illustration).

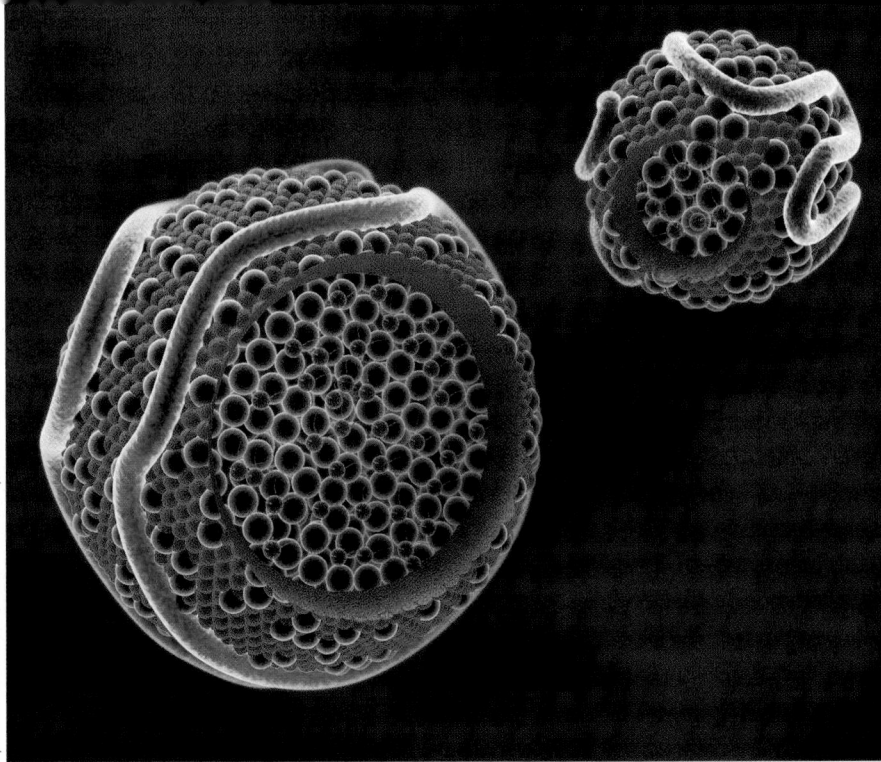

3 Biological Molecules: The Carbon Compounds of Life

WHY IT MATTERS

High in the mountains of the Pacific Northwest, vast forests of coniferous trees have survived another cold winter **(Figure 3.1).** With the arrival of spring, rising temperatures and water from melting snow stimulate renewed growth. Carbon dioxide (CO_2) from the air enters the needle-like leaves of the trees through microscopic pores. Using energy from sunlight, the trees combine the water and carbon dioxide into sugars and other carbon-based compounds through the process known as photosynthesis. The lives of plants, and almost all other organisms, depend directly or indirectly on the products of photosynthesis.

The amount of CO_2 in the atmosphere is critical to photosynthesis. Researchers have been studying the atmospheric concentration of CO_2 since the early 1950s. Among other things, they found that the concentration shifts with the seasons. It declines during spring and summer, when plants and other photosynthetic organisms withdraw large amounts of the gas from the air and convert it into sugars and other complex carbon compounds. It increases during autumn and winter, when global photosynthesis decreases and decomposers that release the gas as a metabolic by-product increase. Great quantities of CO_2 are also added to the atmosphere by forest fires and by the

Figure 3.1

Conifers in winter on Silver Star Mountain in Washington State. As is true of all other organisms, the structure, activities, and survival of these trees start with the carbon atom and its diverse molecular partners in organic compounds.

burning of coal, oil, gasoline, and other fossil fuels in automobiles, aircraft, trains, power plants, and other industries. The resulting increase in atmospheric CO_2 contributes to global warming, which may have profound effects on life in years to come.

The importance of atmospheric CO_2 to food production and world climate are just two examples of how carbon and its compounds are fundamental to the entire living world, from the structures and activities of single cells to physical effects on a global scale. Carbon compounds form the structures of living organisms and take part in all biological reactions. They also serve as sources of energy for living organisms and as an energy resource for much of the world's industry—for example, coal and oil are the fossil remains of long-dead organisms. This chapter outlines the structures and functions of biological carbon compounds.

3.1 Carbon Bonding

Carbon Chains and Rings Form the Backbones of All Biological Molecules

Carbon's central role in life arises from its bonding properties: it can assemble into an astounding variety of chain and ring structures that form the backbones of all biological molecules. Collectively, molecules based on carbon are known as **organic molecules**. All other substances, that is, those without carbon atoms in their structures, are **inorganic molecules**. A few of the smallest carbon-containing molecules that occur in the environment as minerals or atmospheric gases, such as CO_2, are also considered inorganic molecules.

In organic molecules, carbon atoms bond covalently to each other and to other atoms, chiefly hydrogen, oxygen, nitrogen, and sulfur, in molecular structures that range in size from a few to thousands or even millions of atoms. Molecules consisting of carbon linked only to hydrogen atoms are called **hydrocarbons** (*hydro-* refers to hydrogen, not to water).

As discussed in Section 2.3, carbon has four unpaired outer electrons that it readily shares to complete its outermost energy level, forming four covalent bonds. The simplest hydrocarbon, CH_4 (methane), consists of a single carbon atom bonded to four hydrogen atoms (see Figure 2.8a). Removing one hydrogen from methane leaves a methyl group, which occurs in many biological molecules:

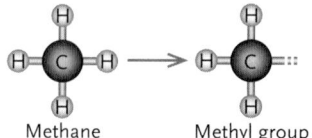

Now imagine bonding two methyl groups together. Removing a hydrogen atom from the resulting structure, ethane, produces an ethyl group:

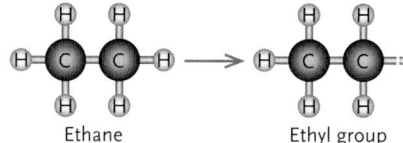

Repeating the process builds a linear hydrocarbon chain:

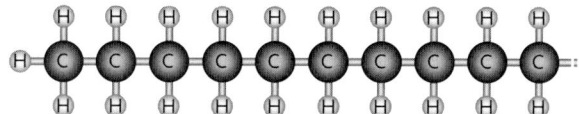

Branches can be added to produce a branched hydrocarbon chain:

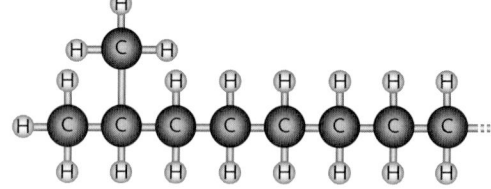

A chain can loop back on itself to form a ring. For example, cyclohexane is C_6H_{12}, with single covalent bonds between each pair of carbon atoms and two hydrogen atoms attached to each carbon atom:

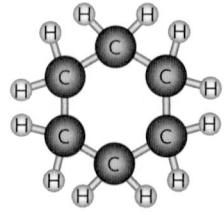

C_6H_{12}, cyclohexane

Hydrocarbons gain added complexity when neighboring carbon atoms form double or triple bonds. Because each carbon atom can form a maximum of four bonds, the number of hydrogen atoms in a molecule decreases as the number of bonds between any two carbon atoms increases:

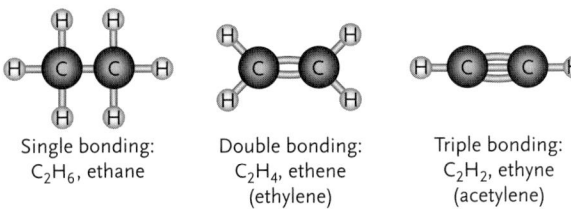

Single bonding:
C_2H_6, ethane

Double bonding:
C_2H_4, ethene
(ethylene)

Triple bonding:
C_2H_2, ethyne
(acetylene)

Double bonds between carbon atoms are also found in carbon rings:

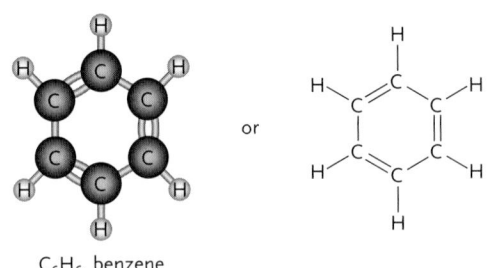

C_6H_6, benzene

We will also use this depiction of a carbon ring in figures:

Many carbon rings can join together to produce larger molecules, as in the string of sugar molecules that makes up a polysaccharide chain:

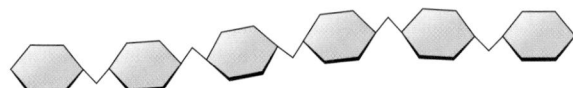

There is almost no limit to the number of different hydrocarbon structures that carbon and hydrogen can form. As you will learn in the next section, the molecules of living systems typically contain other elements in addition to carbon and hydrogen. These other elements confer functional properties on organic molecules. Subsequent sections detail the four major classes of organic molecules—*carbohydrates, lipids, proteins,* and *nucleic acids*—that form almost the entire substance of living organisms.

STUDY BREAK

1. Distinguish between hydrocarbons and other organic molecules.
2. What is the maximum number of bonds that a carbon atom can form?

3.2 Functional Groups in Biological Molecules

Carbohydrates, lipids, proteins, and nucleic acids are synthesized and degraded in living organisms through interactions between small, reactive groups of atoms attached to the organic molecules. The atoms in these reactive groups, called **functional groups,** occur in positions in which their covalent bonds are more readily broken or rearranged than the bonds in other parts of the molecules.

The functional groups that enter most frequently into biological reactions are the *hydroxyl, carbonyl, carboxyl, amino, phosphate,* and *sulfhydryl* groups **(Table 3.1).** The unconnected covalent bonds written to the left of each structure link these functional groups to other atoms in biological molecules, usually carbon atoms. A double bond, such as that in the carbonyl group, indicates that two pairs of electrons are shared between the carbon and oxygen atoms.

In many of the reactions that involve functional groups, the components of a water molecule, —H and —OH, are removed from or added to the groups as they interact. When the components of a water molecule are *removed* during a reaction, usually as part of the assembly of a larger molecule from smaller subunits, the reaction is called a **dehydration synthesis reaction** or **condensation reaction (Figure 3.2a).** For example, this type of reaction occurs when individual sugar molecules combine to form a starch molecule. In **hydrolysis,** the reverse reaction, the components of a water molecule are *added* to functional groups as molecules are broken into smaller subunits **(Figure 3.2b).** For example, the breakdown of a protein molecule into individual amino acids occurs by hydrolysis in the digestive processes of animals.

The Hydroxyl Group Is a Key Component of Alcohols

A **hydroxyl group** (—OH) consists of an oxygen atom linked to a hydrogen atom on one side and to a carbon chain on the other side. Hydroxyl groups readily enter dehydration synthesis reactions, and they are formed as part of hydrolysis reactions. Hydroxyl groups are polar, and they give a polar nature to parts of the molecules that contain them (see Section 2.3 for a discussion of polarity).

The hydroxyl group is a key component of **alcohols.** Alcohols take the form R—OH, in which *R* indicates a chain of one or more carbon atoms. In the R chain of an alcohol, the carbon atoms are all linked to hydrogen atoms, as in ethyl alcohol (see Table 3.1). Ethyl alcohol (ethanol) is the alcohol found in beer, wine, and spirits, and it is used to precipitate DNA from solutions in molecular biology experiments. The hydroxyl group enables an alcohol to form linkages to other organic molecules through dehydration synthesis reactions (see Figure 3.2a).

Table 3.1 — Common Functional Groups of Organic Molecules

Functional Group	Major Classes of Molecules	Example
Hydroxyl —C—OH	Alcohols	$H-\underset{\underset{H}{\mid}}{\overset{\overset{H}{\mid}}{C}}-\underset{\underset{H}{\mid}}{\overset{\overset{H}{\mid}}{C}}-OH$ Ethyl alcohol (in alcoholic beverages)
Carbonyl —C—C=O with H	Aldehydes	Acetaldehyde
—C—C=O with C	Ketones	Acetone (a solvent)
Carboxyl —C—COOH or —C—C(=O)OH	Organic acids	Acetic acid (in vinegar)
Amino —C—NH₂ or —C—N(H)(H)	Amino acids	Alanine (an amino acid)
Phosphate —C—O—PO₃²⁻ or —C—O—P(=O)(O⁻)(O⁻)	Nucleotides, nucleic acids, many other cellular molecules	Glyceraldehyde-3-phosphate (product of photosynthesis)
Sulfhydryl —C—SH	Many cellular molecules	HO—C—C—SH Mercaptoethanol

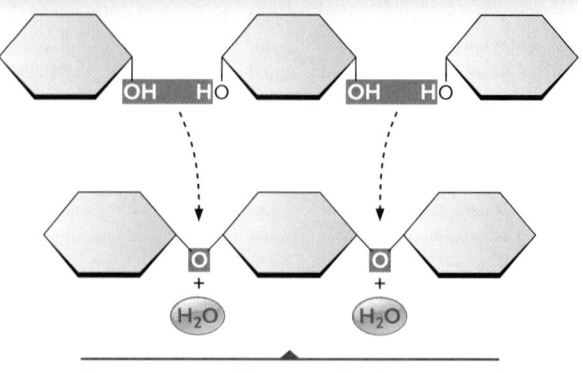

The components of a water molecule are removed as subunits join into a larger molecule.

b. Hydrolysis

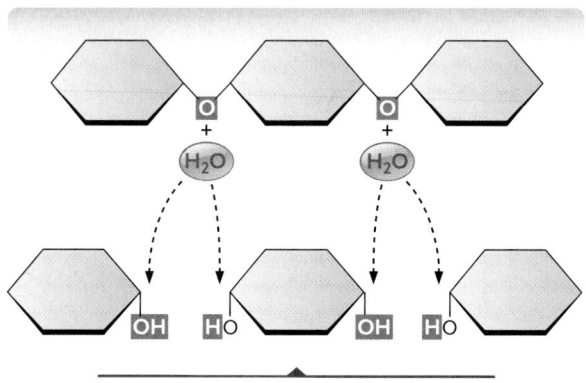

The components of a water molecule are added as molecules are split into smaller subunits.

Figure 3.2
Dehydration synthesis and hydrolysis reactions.

Carbonyl groups are the reactive parts of aldehydes and ketones, molecules that are important building blocks of carbohydrates and that also take part in the reactions supplying energy for cellular activities. In an **aldehyde,** the carbonyl group is linked to a carbon atom at the end of a carbon chain, along with a hydrogen atom, as in acetaldehyde (see Table 3.1). In a **ketone,** the carbonyl group is linked to a carbon atom in the interior of a carbon chain, as in acetone (see Table 3.1).

The Carboxyl Group Forms Organic Acids

Carbonyl and hydroxyl groups combine to form a **carboxyl group** (—COOH), the characteristic functional group of **organic acids** (also called *carboxylic acids*); an example is acetic acid (see Table 3.1). The carboxyl group gives organic molecules acidic properties because the —OH group readily releases its hydrogen as a proton (H^+) in water solutions (see Section 2.5):

$$R-\overset{\overset{\displaystyle O}{\parallel}}{C}-OH \;\rightleftharpoons\; R-\overset{\overset{\displaystyle O}{\parallel}}{C}-O^- \;+\; H^+$$

The Carbonyl Group Is the Reactive Part of Aldehydes and Ketones

A **carbonyl group** C=O consists of an oxygen atom linked to a carbon atom by a double bond. The oxygen atom of a carbonyl group is highly reactive, especially with substances that act as bases (see Section 2.5 for a discussion of acids and bases).

The carboxyl group readily enters into dehydration synthesis reactions, giving up its hydroxyl group as organic molecules combine into larger assemblies (see Figure 3.17). Many organic acids, such as citric acid and acetic acid, are central components of energy-generating reactions in living organisms.

The Amino Group Acts as an Organic Base

The **amino group** ($-NH_2$) consists of a nitrogen atom bonded on one side to two hydrogen atoms and on the other side to a carbon chain, as in the amino acid alanine (see Table 3.1) and all other amino acids. It readily acts as a base by accepting H^+ (a proton) in water solutions:

$$R-N\begin{smallmatrix}H\\ \\H\end{smallmatrix} + H^+ \rightleftharpoons R-N\begin{smallmatrix}H\\|\\H\end{smallmatrix}-H^+$$

The amino group also readily enters dehydration synthesis reactions, releasing a hydrogen ion as it links subunits into larger molecules (see Figure 3.17).

The Phosphate Group Is a Reactive Jack-of-All-Trades

The **phosphate group** ($-OPO_3^{2-}$) consists of a central phosphorus atom held in four linkages. Two of the linkages bind $-OH$ groups to the central phosphorus atom; a third linkage, formed by a double bond, binds an oxygen atom to the central phosphorus atom. The remaining bond links the phosphate group to an oxygen atom, which, in turn, binds to a carbon chain. An example is glyceraldehyde-3-phosphate, a product of photosynthesis (see Table 3.1).

Phosphate groups give molecules that contain them the ability to react as weak acids because one or both $-OH$ groups readily release their hydrogens as H^+:

$$-O-\underset{\underset{O}{\|}}{\overset{\overset{OH}{|}}{P}}-OH \rightleftharpoons -O-\underset{\underset{O}{\|}}{\overset{\overset{O^-}{|}}{P}}-O^- + 2H^+$$

A phosphate group can also form a chemical bridge that links two organic building blocks into a larger structure:

$$\text{Organic subunit}-O-\underset{\underset{O}{\|}}{\overset{\overset{O^-}{|}}{P}}-O-\text{Organic subunit}$$

Among the large biological molecules linked together by phosphate groups is the nucleic acid DNA, the genetic material of all living organisms. When acting as a linking bridge, a phosphate group still has one $-OH$ group that can dissociate to release a hydrogen ion (shown in dissociated form above as O^-).

Phosphate groups are also added to or removed from biological molecules as part of reactions that con-serve or release energy. In addition, they control biological activity—the activity of many proteins is turned on or off by the addition or removal of phosphate groups.

The Sulfhydryl Group Works as a Molecular Fastener

In the **sulfhydryl group** ($-SH$), a sulfur atom is linked on one side to a hydrogen atom and on the other side to a carbon chain, as in mercaptoethanol (see Table 3.1). The sulfhydryl group is easily converted into a covalent linkage, in which it loses its hydrogen atom as it binds. In many of these linking reactions, two sulfhydryl groups interact to form a **disulfide linkage** ($-S-S-$):

$$R-SH + HS-R \rightarrow R-\underset{\substack{\text{disulfide}\\\text{linkage}}}{S-S}-R + 2H^+ + 2\,\text{electrons}$$

In many proteins, the disulfide linkage forms a sort of molecular fastener that holds proteins in their folded form or links protein subunits into larger structures (see Figure 3.16).

The hydroxyl, carbonyl, carboxyl, amino, phosphate, and sulfhydryl functional groups provide most of the reactive sites on biological molecules. We now turn to the arrangement of these groups and carbon chains in the four classes of organic molecules—carbohydrates, lipids, proteins, and nucleic acids.

STUDY BREAK

1. Distinguish between a dehydration synthesis reaction (condensation reaction) and hydrolysis.
2. Explain whether carboxyl groups, amino groups, and phosphate groups act as acids or bases.

3.3 Carbohydrates

Carbohydrates, the most abundant organic molecules in the world, serve many functions. Together with fats, they act as the major fuel substances providing chemical energy for cellular activities. Table sugar is an example of a carbohydrate consumed in large quantities as an energy source in the human diet. For example, athletic activity is partly fueled by carbohydrates. Energy-providing carbohydrates are stored in plant cells as **starch** and in animal cells as **glycogen**, both consisting of long chains of repeating carbohydrate subunits linked end to end. Chains of carbohydrate subunits also form many structural molecules, such as **cellulose**, one of the primary constituents of plant cell walls.

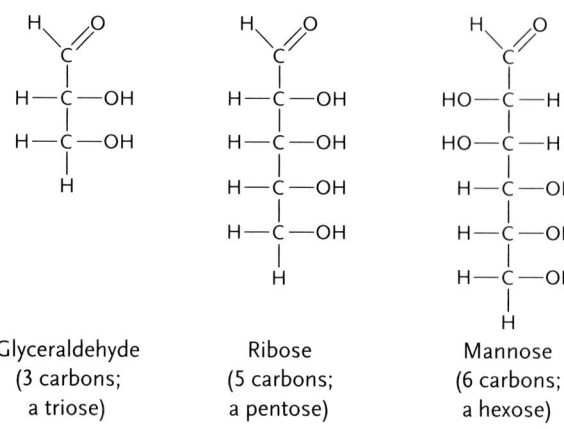

Glyceraldehyde
(3 carbons;
a triose)

Ribose
(5 carbons;
a pentose)

Mannose
(6 carbons;
a hexose)

Figure 3.3

Some representative monosaccharides. The triose, glyceraldehyde, takes part in energy-yielding reactions and photosynthesis. The pentose, ribose, is a component of RNA and of molecules that carry energy. The hexose, mannose, is a fuel substance and a component of glycolipids and glycoproteins.

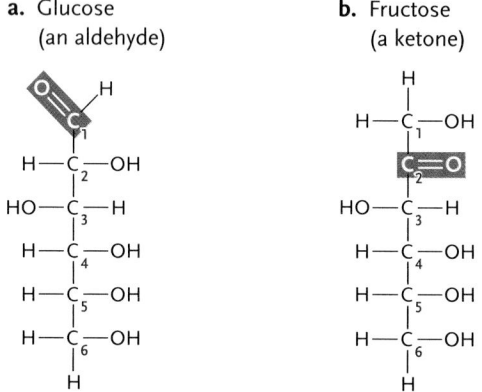

a. Glucose
(an aldehyde)

b. Fructose
(a ketone)

Figure 3.4

The aldehyde and ketone positions for the carbonyl group (shaded regions) in monosaccharides. **(a)** In the aldehyde position, the carbonyl group is located at the end of the carbon chain. **(b)** In the ketone position, the carbonyl group is located inside the carbon chain. For convenience, the carbons of monosaccharides are numbered, with 1 being the carbon at the end nearest the carbonyl group.

Carbohydrates contain only carbon, hydrogen, and oxygen atoms, in an approximate ratio of 1 carbon : 2 hydrogens : 1 oxygen (CH_2O). The names of many carbohydrates end in -*ose*. The smallest carbohydrates, the **monosaccharides** (*mono* = one; *saccharum* = sugar), contain three to seven carbon atoms. For example, the monosaccharide glucose consists of a chain of six carbons and has the molecular formula $C_6H_{12}O_6$. Two monosaccharides combine to form a *disaccharide* such as sucrose, common table sugar. Chains with more than 10 linked monosaccharide subunits are called **polysaccharides** (*poly* = many). Starch, glycogen, and cellulose are common polysaccharides.

Monosaccharides Are the Structural Units of Carbohydrates

Carbohydrates occur either as monosaccharides or as chains of monosaccharide units linked together. Mono-

saccharides are soluble in water, and most have a distinctly sweet taste. Of the monosaccharides, those that contain three carbons (*trioses*), five carbons (*pentoses*), and six carbons (*hexoses*) are most common in living organisms **(Figure 3.3)**.

Linear and Ring Forms of the Monosaccharides. All monosaccharides can occur in the linear form shown in Figure 3.3. In this form, each carbon atom in the chain except one has both an —H and an —OH group attached to it. The remaining carbon is part of a carbonyl group, which may be located at the end of the carbon chain in the aldehyde position (as in glucose in **Figure 3.4a**) or inside the chain in the ketone position (as in fructose in **Figure 3.4b**).

Monosaccharides with five or more carbons can fold back on themselves to assume a ring form. Folding into a ring occurs through a reaction between two functional groups in the same monosaccharide, as occurs in glucose **(Figure 3.5)**. The ring form of most five- and six-carbon sugars is much more common in cells than the linear form.

Isomers of the Monosaccharides. Typically, one or more of the carbon atoms in a monosaccharide links to four different atoms or chemical groups. Carbons linked in this way are called *asymmetric* carbons; they have important effects on the structure of a monosaccharide because they can take either of two fixed positions with respect to other carbons in a carbon chain. For example, the middle carbon of the three-carbon sugar glyceraldehyde is asymmetric because it shares electrons in covalent bonds with four different atoms or groups: —H, —OH, —CHO, and —CH$_2$OH. The —H and —OH groups can take either of two positions, with the —OH extending to either the left or right of the carbon chain relative to the —CHO and —CH$_2$OH groups:

CHO   CHO

H—C—OH HO—C—H

CH$_2$OH   CH$_2$OH

D-Glyceraldehyde L-Glyceraldehyde

Note that the two forms of glyceraldehyde have the same chemical formula, $C_3H_6O_3$. The difference between the two forms is similar to the difference between your two hands. Although both hands have four fingers and a thumb, they are not identical; rather, they are mirror images of each other. That is, when you hold your right hand in front of a mirror, the reflection looks like your left hand and vice versa.

Two or more molecules with the same chemical formula but different molecular structures are called **isomers**. Isomers that are mirror images of each other, like the two forms of glyceraldehyde, are called **enantiomers**, or **optical isomers**. One of the enantiomers—the one in which the hydroxyl group extends

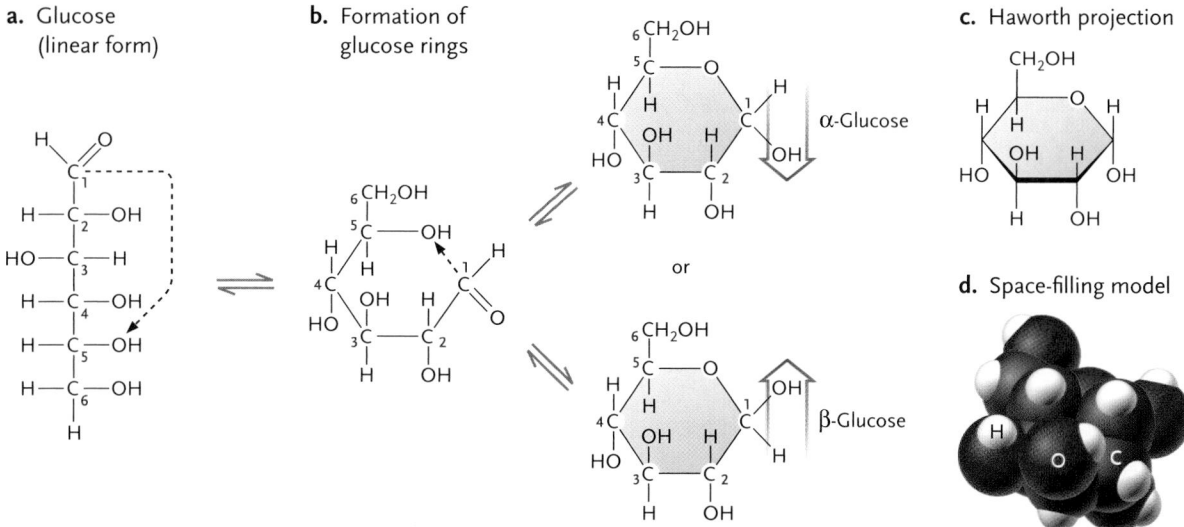

a. Glucose (linear form)

b. Formation of glucose rings

α-Glucose

or

β-Glucose

c. Haworth projection

CH₂OH

d. Space-filling model

Figure 3.5

Ring formation by glucose. **(a)** Glucose in linear form. **(b)** The ring form of glucose is produced by a reaction between the aldehyde group at the 1 carbon and the hydroxyl group at the 5 carbon. The reaction produces two alternate glucose enantiomers, α- and β-glucose. If the ring is considered to lie in the plane of the page, the —OH group points below the page in α-glucose and upward from the page in β-glucose. For simplicity, the group at the 6 carbon is shown as CH₂OH in this and later diagrams. **(c)** A commonly used, simplified representation of the glucose ring, in which the C's designating carbons of the ring are omitted. The thicker lines along one side indicate that the ring lies in a flat plane with the thickest edge closest to the viewer. **(d)** A space-filling model of glucose, showing the volumes occupied by the atoms. Carbon atoms are shown in black, oxygen in red, and hydrogen in white.

to the left in the view just shown—is called the L-form (*laevus* = left). The other enantiomer, in which the —OH extends to the right, is called the D-form (*dexter* = right). The difference between L- and D-enantiomers is critical to biological function. Typically, one of the two forms enters much more readily into cellular reactions; just as your left hand does not fit readily into a right-hand glove, enzymes (proteins that accelerate chemical reactions in living organisms) fit best to one of the two forms of an enantiomer. For example, most of the enzymes that catalyze the biochemical reactions of monosaccharides react more rapidly with the D-form, making this form much more common among cellular carbohydrates than L-forms. Many other kinds of biological molecules besides carbohydrates form enantiomers; an example is the amino acids.

In the ring form of many five- or six-carbon monosaccharides, including glucose, the carbon at the 1 position of the ring is asymmetric because its four bonds link to different groups of atoms. This asymmetry allows monosaccharides such as glucose to exist as two different enantiomers. The glucose enantiomer with an —OH group pointing below the plane of the ring is known as *alpha-glucose,* or *α-glucose;* the enantiomer with an —OH group pointing above the plane of the ring is known as *beta-glucose,* or *β-glucose* (see Figure 3.5b). Other five- and six-carbon monosaccharide rings have similar α- and β-configurations.

The α- and β-rings of monosaccharides can give the polysaccharides assembled from them vastly dif-ferent chemical properties. For example, starches, which are assembled from α-glucose units, are biologically reactive polysaccharides easily digested by animals; cellulose, which is assembled from β-glucose units, is relatively unreactive and, for most animals, completely indigestible.

Another form of isomerism is found in monosaccharides, as well as in other molecules. Two molecules with the same chemical formula but atoms that are arranged in different ways are called **structural isomers.** The sugars glucose and fructose are examples of structural isomers (see Figure 3.4).

Two Monosaccharides Link to Form a Disaccharide

Disaccharides typically are assembled from two monosaccharides linked together by a dehydration synthesis reaction. For example, the disaccharide maltose is formed by the linkage of two α-glucose molecules **(Figure 3.6a)** with oxygen as a bridge between the number 1 carbon of the first glucose unit and the 4 carbon of the second glucose unit. Bonds of this type, which commonly link monosaccharides into chains, are known as **glycosidic bonds.** A glycosidic bond between a 1 carbon and a 4 carbon is written in chemical shorthand as a 1→4 linkage; 1→2, 1→3, and 1→6 linkages are also common in carbohydrate chains. The linkages are designated as α or β depending on the orientation of the —OH group at the 1 carbon that forms the bond.

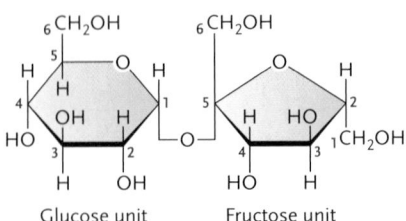

a. Formation of maltose

Glucose + Glucose → Maltose + H_2O

b. Sucrose

Glucose unit   Fructose unit

c. Lactose

Galactose unit   Glucose unit

Figure 3.6

Disaccharides. **(a)** Combination of two glucose molecules by a dehydration synthesis reaction to form the disaccharide maltose. The components of a water molecule (in blue) are removed from the monosaccharides as they join. **(b)** Sucrose, assembled from glucose and fructose. **(c)** Lactose, assembled from galactose and glucose.

In maltose, the —OH group is in the α position. Therefore, the link between the two glucose subunits of maltose is written as an α(1→4) linkage.

Maltose, sucrose, and lactose are common disaccharides. Maltose is present in germinating seeds and is a major sugar used in the brewing industry. Sucrose, which contains a glucose and a fructose unit **(Figure 3.6b)**, is transported to and from different parts of leafy plants. It is probably the most plentiful sugar in nature. Table sugar is made by extracting and crystallizing sucrose from plants, such as sugar cane and sugar beets. Lactose, assembled from a glucose and a galactose unit **(Figure 3.6c)**, is the primary sugar of milk.

Monosaccharides Link in Longer Chains to Form Polysaccharides

Polysaccharides are longer chains formed by end-to-end linkage of monosaccharides through dehydration synthesis reactions. A polysaccharide is a type of **macromolecule**, meaning a very large molecule assembled by the covalent linkage of smaller subunit molecules. The subunit for a polysaccharide is the monosaccharide.

The dehydration synthesis reactions that assemble polysaccharides from monosaccharides are examples of **polymerization**, in which identical or nearly identical subunits, called the **monomers** of the reaction, join like links in a chain to form a larger molecule called a **polymer.** Linkage of a relatively small number of non-

identical subunits can create highly diverse and varied biological molecules. Many kinds of polymers are found in cells, not just polysaccharides. DNA is a primary example of a highly diverse polymer assembled from various combinations of only four different types of monomers.

The most common polysaccharides—the plant starches, glycogen, and cellulose—are all assembled from hundreds or thousands of glucose units. Other polysaccharides are built up from a variety of different sugar units. Polysaccharides may be linear, unbranched molecules, or they may contain one or more branches in which side chains of sugar units attach to a main chain.

Figure 3.7 shows four common polysaccharides. Plant starches include both linear, unbranched forms such as amylose (Figure 3.7a) and branched forms such as amylopectin. Glycogen (Figure 3.7b), a more highly branched polysaccharide than amylopectin, can be assembled or disassembled readily to take up or release glucose; it is stored in large quantities in the liver and muscle tissues of many animals.

Cellulose (Figure 3.7c), probably the most abundant carbohydrate on Earth, is an unbranched polysaccharide assembled from glucose units bound together by β-linkages. It is the primary structural fiber of plant cell walls; in this role, cellulose has been likened to the steel rods in reinforced concrete. Its tough fibers enable the cell walls of plants to withstand enormous weight and stress. Fabrics such as cotton and linen are made from cellulose fibers extracted from plant cell walls. Animals such as mollusks, crustaceans, and insects synthesize an enzyme that digests the cellulose they eat. In ruminant mammals, such as cows, microorganisms in the digestive tract break down cellulose. Cellulose passes unchanged through the human digestive tract as indigestible fibrous matter. Many nutritionists maintain that the bulk provided by cellulose fibers helps maintain healthy digestive function.

Chitin (Figure 3.7d), another tough and resilient polysaccharide, is assembled from glucose units modified by the addition of nitrogen-containing groups. Similar to the subunits of cellulose, the modified glucose units of chitin are held together by β-linkages. Chitin is the main structural fiber in the external skeletons and other hard body parts of arthropods such as insects, crabs, and spiders. It is also a structural

a.

Amylose, formed from α-glucose units joined end to end in α(1→4) linkages. The coiled structures are induced by the bond angles in the α-linkages.

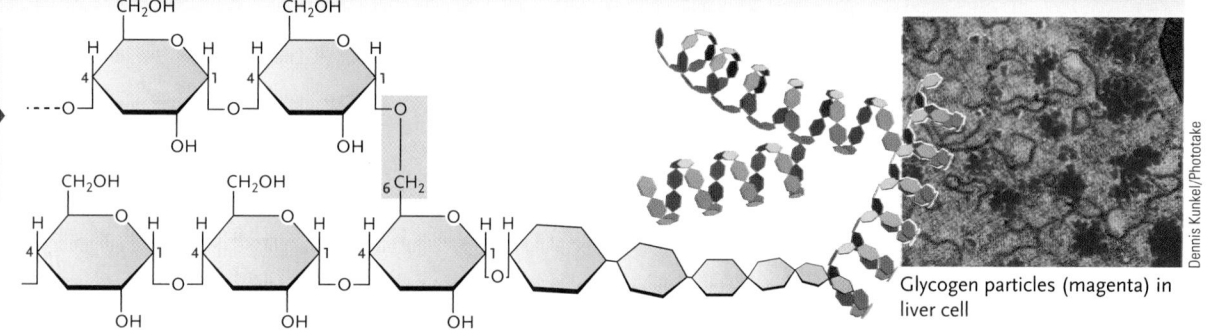

Amylose grains (purple) in plant root tissue

Ed Reschke/Peter Arnold, Inc.

b.

Glycogen, formed from glucose units joined in chains by α(1→4) linkages; side branches are linked to the chains by α(1→6) linkages (boxed in blue).

Glycogen particles (magenta) in liver cell

Dennis Kunkel/Phototake

c.

Cellulose, formed from glucose units joined end to end by β(1→4) linkages. Hundreds to thousands of cellulose chains line up side by side, in an arrangement reinforced by hydrogen bonds between the chains, to form cellulose microfibrils in plant cells.

Glucose subunit

Cellulose molecule

Cellulose microfibril

Cellulose microfibrils in plant cell wall

© Biophoto Associates/Photo Researchers, Inc.

d.

Chitin, formed from β-linkages joining glucose units modified by the addition of nitrogen-containing groups. The external body armor of the tick is reinforced by chitin fibers.

David Scharf/Peter Arnold, Inc.

Figure 3.7

Four common polysaccharides: **(a)** amylose, a plant starch; **(b)** glycogen; **(c)** cellulose, the primary fiber in plant cell walls; and **(d)** chitin, a reinforcing fiber in the external skeleton of arthropods and the cell walls of some fungi.

material in the cell walls of fungi such as mushrooms and yeasts. Unlike cellulose, chitin is digested by enzymes that are widespread among microorganisms, plants, and many animals. In plants and animals, including humans and other mammals, chitin-digesting enzymes occur primarily as part of defenses against fungal infections. However, humans cannot digest chitin as a food source.

Polysaccharides also occur on the surfaces of cells, particularly in animals. These surface polysaccharides are attached to both the protein and lipid molecules in membranes. They help hold the cells of animals together and serve as recognition sites between cells.

STUDY BREAK

Distinguish among a monosaccharide, a disaccharide, and a polysaccharide. Give examples of each.

3.4 Lipids

Lipids are a diverse group of water-insoluble, primarily nonpolar biological molecules composed mostly of hydrocarbons. Some are large molecules, but they are not large enough to be considered macromolecules. As a result of their nonpolar character, lipids typically dissolve much more readily in nonpolar solvents, such as acetone and chloroform, than in water, the polar solvent of living organisms. Their insolubility in water underlies their ability to form cell membranes, the thin molecular films that create boundaries between and within cells.

In addition to forming membranes, some lipids are stored and used in cells as an energy source. Other lipids serve as hormones that regulate cellular activities. Three types of lipid molecules—*neutral lipids,* *phospholipids,* and *steroids*—occur most commonly in living organisms.

Neutral Lipids Are Familiar as Fats and Oils

Neutral lipids, commonly found in cells as energy-storage molecules, are called "neutral" because at cellular pH they have no charged groups; they are therefore nonpolar. There are two types of neutral lipids: **oils** and **fats.** Oils are liquid at biological temperatures, and fats are semisolid. Generally, neutral lipids are insoluble in water. Almost all neutral lipids consist of a three-carbon backbone chain formed from glycerol, an alcohol, with each carbon of the glycerol backbone linked to a side chain consisting of a *fatty acid*.

Fatty Acids. A **fatty acid** contains a single hydrocarbon chain with a carboxyl group (—COOH) linked at one end **(Figure 3.8).** The carboxyl group gives the fatty acid its acidic properties. The fatty acids in living organisms contain 4 or more carbons in their hydrocarbon chain, with the most common forms having even-numbered chains of 14 to 22 carbons. Only the shortest fatty acid chains are water-soluble. As chain length increases, fatty acids become progressively less water-soluble and become oily.

If the hydrocarbon chain of a fatty acid binds the maximum possible number of hydrogen atoms, so that only single bonds link the carbon atoms, the fatty acid is said to be **saturated** with hydrogen atoms (as in stearic acid in Figure 3.8a). If one or more double bonds link the carbons (see Figure 3.8b, arrow), reducing the number of hydrogen atoms bound, the fatty acid is **unsaturated.** Fatty acids with one double bond are **monounsaturated;** those with more than one double bond are **polyunsaturated.**

Unsaturated fatty acid chains tend to bend or "kink" at a double bond (see Figures 3.8b and 3.12c). The kink makes the chains more disordered and thus more fluid at biological temperatures. Consequently, unsaturated fatty acids—and lipids that contain them—melt at lower temperatures than saturated fatty acids of the same length, and they generally have oily rather than fatty characteristics.

In foods, saturated fatty acids are usually found in solid animal fat, such as butter, whereas unsaturated fatty acids are usually found in vegetable oils, such as liquid canola oil. Nonetheless, both solid animal fat and liquid vegetable oils contain some saturated and some unsaturated fatty acids.

Glycerol and Triglyceride Formation. The glycerol unit that forms the backbone of neutral lipids has three —OH groups at which fatty acids may link **(Figure 3.9a).** In its free state, glycerol is a polar, water-soluble, sweet-tasting substance with the properties of an alcohol. If a fatty acid binds by a dehydration synthesis reaction at each of glycerol's three —OH-bearing sites, the polar groups are eliminated, producing a nonpolar compound known as a **triglyceride** (see Figure 3.9). Most

Figure 3.8
Fatty acids, one of two components of a neutral lipid. **(a)** Stearic acid, a saturated fatty acid. **(b)** Oleic acid, an unsaturated fatty acid. An arrow marks the "kink" introduced by the double bond.

a. Stearic acid, $CH_3(CH_2)_{16}COOH$

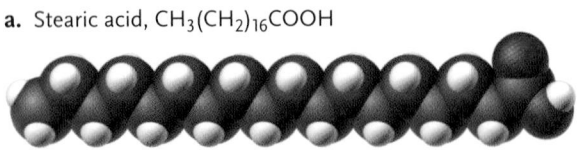

b. Oleic acid, $CH_3(CH_2)_7CH=CH(CH_2)_7COOH$

a. Formation of a triglyceride

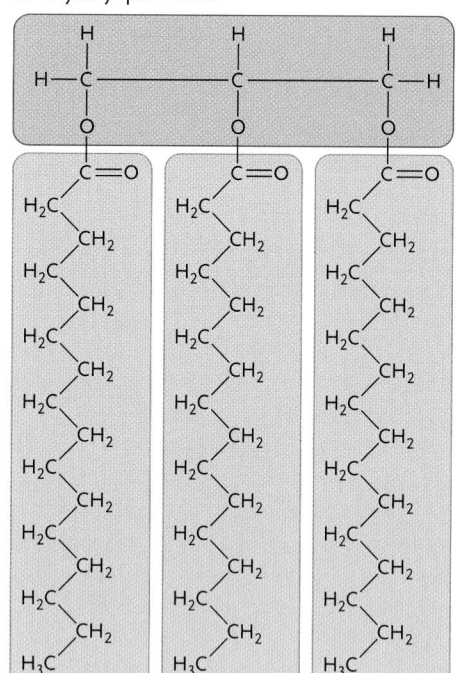

Glycerol

Fatty acids

Triglyceride

3 H_2O

b. Glyceryl palmitate

c. Triglyceride model

Figure 3.9

Triglycerides. **(a)** Formation of a triglyceride by dehydration synthesis of glycerol with three fatty acids. The R groups represent the hydrocarbon chains of the fatty acids. The components of a water molecule (in blue) are removed from the glycerol and fatty acids in each of the three bonds formed. **(b)** Chemical structure and **(c)** space-filling model of glyceryl palmitate, a triglyceride.

lipids stored as an energy reserve in living systems are triglycerides.

The fatty acids linked to glycerol may be different or the same. Different organisms usually have distinctive combinations of fatty acids in their triglycerides. As with individual fatty acids, triglycerides generally become less fluid as the length of their fatty acid chains increases; those with shorter chains remain liquid as oils at biological temperatures, and those with longer chains solidify as fats. The degree of saturation of the fatty acid chains also affects the fluidity of triglycerides—the more saturated, the less fluid the triglyceride. Plant oils are converted commercially to fats by *hydrogenation*—that is, adding hydrogen atoms to increase the degree of saturation, as in the conversion of vegetable oils to margarines and shortening.

Triglycerides are used widely as stored energy in animals. Gram for gram, they yield more than twice as much energy as carbohydrates by weight. Therefore, fats are an excellent source of energy in the diet. Storing the equivalent amount of energy as carbohydrates rather than fats would add more than 100 pounds to the weight of an average man or woman. A layer of fatty tissue just under the skin also serves as an insulating blanket in humans, other mammals, and birds. Triglycerides secreted from special glands in waterfowl and other birds help make feathers water repellent (as in the penguins shown in **Figure 3.10**).

Unsaturated fats are considered healthier than saturated fats in the human diet. Saturated fats have been implicated in the development of atherosclerosis (see the *Focus on Research*), a disease in which arteries, particularly those serving the heart, become clogged with fatty deposits.

Clem Haagner/Ardea, London

Figure 3.10

Penguins of the Antarctic, one of several animals that have a thick, insulating layer of fatty tissue that contains triglycerides under the skin. Penguins also use their face and bill to spread oil, secreted by a gland near their tail, over their feathers. The oily coating keeps their feathers watertight and dry.

Applied Research: Fats, Cholesterol, and Coronary Artery Disease

Butter! Bacon and eggs! Ice cream! Cheesecake! Possibly you think of such foods as irresistible, off limits, or both. After all, who doesn't know about animal fats, cholesterol, and hardening of the arteries? Hardening of the arteries, or *atherosclerosis*, is a condition in which deposits of lipid and fibrous material called plaque build up in the walls of arteries, the vessels that supply oxygenated blood to body tissues. Plaque reduces the internal diameter of the arteries, restricting or even completely blocking the flow of blood. One of the most serious consequences occurs when atherosclerosis narrows or blocks the coronary arteries that supply oxygenated blood to the heart muscle (see figure). This condition can severely impair heart function, as in coronary heart disease, and, in extreme cases, can lead to destruction of heart muscle tissue, as occurs in a heart attack.

Your body requires a certain amount of cholesterol, but the liver normally makes enough to meet this demand. Additional cholesterol is made from fats taken in as food.

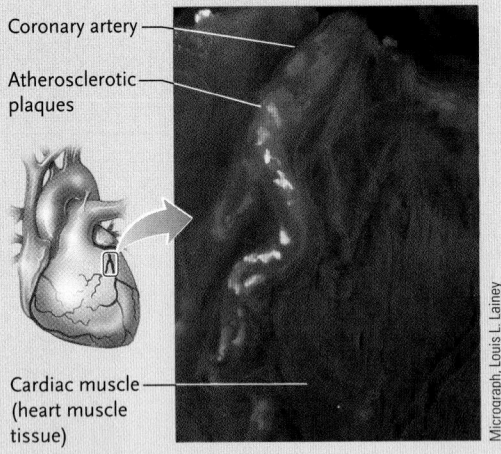

Coronary artery

Atherosclerotic plaques

Cardiac muscle (heart muscle tissue)

Micrograph, Louis L. Lainey

Atherosclerotic plaques (bright areas) in the coronary arteries of a patient with heart disease.

Cholesterol is found in the blood bound to low-density lipoprotein (LDL) and high-density lipoprotein (HDL). LDL cholesterol is considered "bad" because clinical studies have shown a positive correlation between its level in the blood and the risk for coronary heart disease. LDL cholesterol contributes to plaque formation as atherosclerosis proceeds. In contrast, HDL cholesterol is "good" because clinical studies have shown that high levels of this form appear to provide some protection against coronary heart disease. Simplifying, HDL cholesterol removes excess cholesterol from plaques in arteries, thereby reducing plaque buildup. The cholesterol that has been removed is transported by the HDL cholesterol to the liver where it is broken down.

Fats in food affect cholesterol levels in the blood. Diets high in saturated fats raise LDL cholesterol levels, but levels of HDL cholesterol appear not to be affected by such a diet. Foods of animal origin typically contain saturated fats, and foods of plant origin typically contain unsaturated fats.

In the food industry, unsaturated vegetable oils are often processed to solidify the fats. The process, partial hydrogenation, adds hydrogen atoms to unsaturated sites, eliminating many double bonds and generating substances known as trans fatty acids (or trans fats). Usually the hydrogen atoms at a double bond are positioned on the same side of the carbon chain, producing a cis (Latin, "on the same side") fatty acid:

$$
\begin{array}{cc}
H & H \\
| & | \\
-C & = C- \\
\end{array}
$$

but in a trans (Latin, "across") fatty acid, the hydrogen atoms are on

different sides of the chain at some double bonds:

$$
\begin{array}{cc}
H & \\
| & \\
-C & = C- \\
& | \\
& H \\
\end{array}
$$

Trans fatty acids are found in many vegetable shortenings, some margarines, cookies, cakes, doughnuts, and other foods made with or fried in partially hydrogenated fats.

Research from human feeding studies has shown that trans fatty acids raise LDL cholesterol levels nearly as much as saturated fatty acids do. More seriously, intake of trans fatty acids at levels found in a typical U.S. diet also appears to reduce HDL cholesterol levels. In addition, clinical studies have demonstrated a positive correlation between the intake of trans fatty acids and the occurrence of coronary heart disease. A regulation to add the trans fatty acid content to nutritional labels went into effect in the United States in January 2006. A number of federal and state agencies are considering legislation to ban trans fatty acids in food.

Many questions about dietary cholesterol still remain. For example, people in some cultures consume large quantities of fatty foods of the "wrong" kind yet rarely develop atherosclerosis. For example, atherosclerosis was once virtually nonexistent in Inuits, whose diet in their native culture contained more than 90% animal fat; however, atherosclerosis developed in that same population when they adopted a "civilized" diet and lifestyle. In France, the incidence of atherosclerosis is relatively low even though cheese and other dairy products are diet staples. Of course, the French say that wine keeps them healthy!

Waxes. Fatty acids may also combine with long-chain alcohols or hydrocarbon structures to form **waxes**, which are harder and less greasy than fats. Insoluble in water, waxy coatings help keep skin, hair, or feathers of animals protected, lubricated, and pliable. In humans, earwax lubricates the outer ear canal and protects the eardrum. Honeybees use a wax secreted by glands in their abdomen to construct the comb in which larvae are raised and honey is stored **(Figure 3.11a)**.

Many plants secrete waxes that form a protective exterior layer, which greatly reduces water loss from plants and resists invasion by infective agents such as bacteria and viruses. This waxy covering gives cherries, apples, and many other fruits their shiny appearance **(Figure 3.11b)**.

Phospholipids Provide the Framework of Biological Membranes

Phosphate-containing lipids called **phospholipids** are the primary lipids of cell membranes. In the most common phospholipids, glycerol forms the backbone of the molecule as in triglycerides, but only two of its binding sites are linked to fatty acids **(Figure 3.12)**. The third site is linked to a polar phosphate group, which binds to yet another polar unit. The end of the molecule containing the fatty acids is nonpolar and hydrophobic, and the end with the phosphate group is polar and hydrophilic.

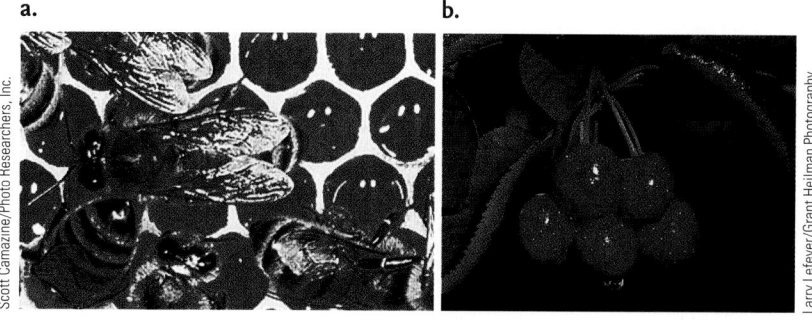

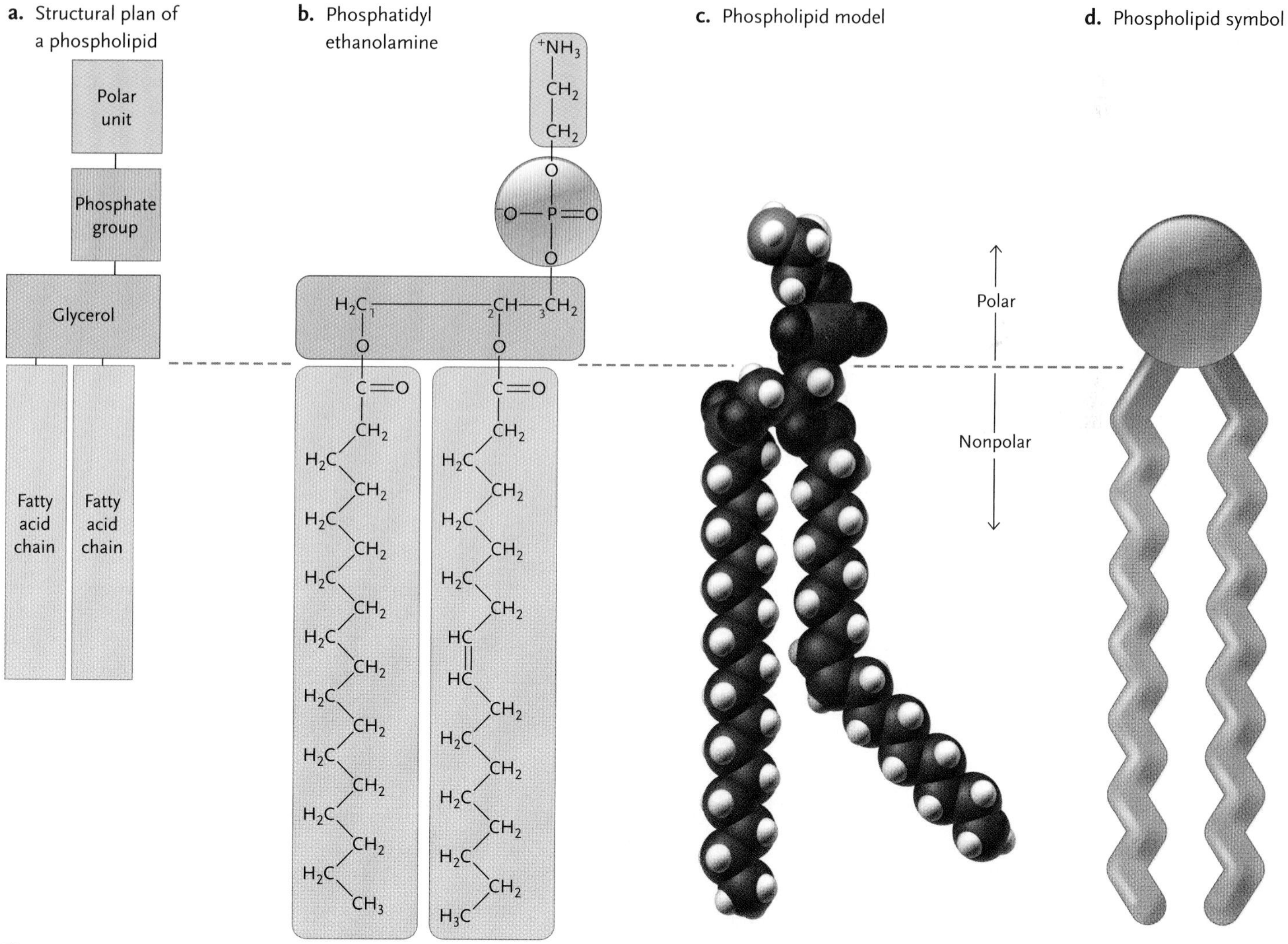

Figure 3.11

Waxy structures in nature. **(a)** The comb constructed by honeybees is made from a wax secreted by abdominal glands. **(b)** Beads of water on the waxy cuticle of cherries.

Scott Camazine/Photo Researchers, Inc.

Larry Lefever/Grant Heilman Photography

a. Structural plan of a phospholipid

Polar unit

Phosphate group

Glycerol

Fatty acid chain | Fatty acid chain

b. Phosphatidyl ethanolamine

$^+NH_3$
|
CH_2
|
CH_2
|
O
|
$O-P=O$
|
O

H_2C — CH — CH_2
| |
O   O
| |
C=O   C=O
| |
CH_2 CH_2
H_2C H_2C
CH_2 CH_2
H_2C H_2C
CH_2 CH_2
H_2C H_2C
CH_2 HC
H_2C ‖
CH_2 HC
H_2C CH_2
CH_2 H_2C
H_2C CH_2
CH_2 H_2C
H_2C CH_2
CH_3 H_2C
                     CH_2
                     H_3C

c. Phospholipid model

Polar

Nonpolar

d. Phospholipid symbol

Figure 3.12

Phospholipid structure. **(a)** The arrangement of components in phospholipids. **(b)** Phosphatidyl ethanolamine, a common membrane phospholipid. **(c)** Space-filling model of phosphatidyl ethanolamine. The kink in the fatty acid chain on the right reflects a double bond at this position. **(d)** Diagram widely used to depict a phospholipid molecule in cell membrane diagrams. The sphere represents the polar end of the molecule, and the zigzag lines represent the nonpolar fatty acid chains.

a. Arrangement of carbon rings in a steroid

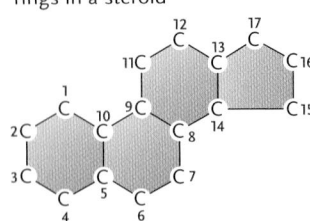

b. Cholesterol, a sterol

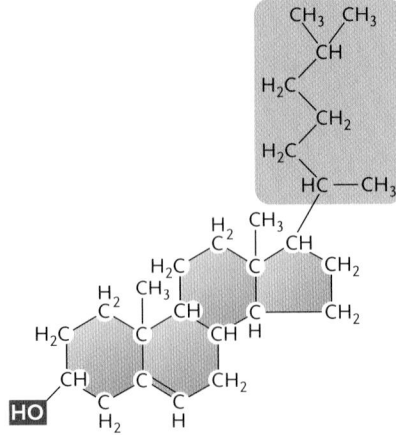

c. Cholesterol model

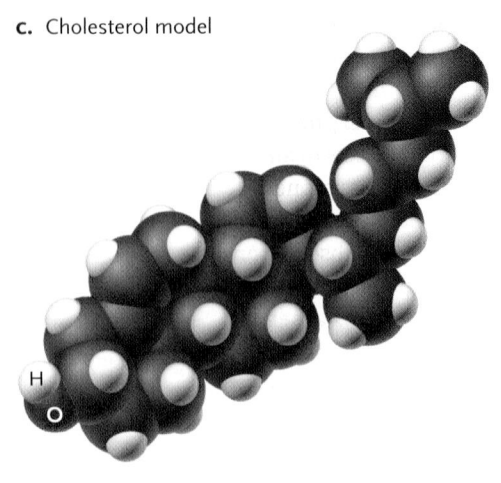

Figure 3.13

Steroids. **(a)** Typical arrangement of four carbon rings in a steroid molecule. **(b)** A sterol, cholesterol. Sterols have a hydrocarbon side chain linked to the ring structure at one end and a single —OH group at the other end (boxed in red). The —OH group makes its end of a sterol slightly polar. The rest of the molecule is nonpolar. **(c)** A space-filling model of cholesterol.

In polar environments, such as a water solution, phospholipids assume arrangements in which only their polar ends are exposed to the water; their nonpolar ends collect together in a region that excludes water. One of these arrangements, the *bilayer,* is the structural basis of membranes, the organizing boundaries of all living cells (see Figure 2.13). In a bilayer, formed by a film of phospholipids just two molecules thick, the phospholipid molecules are aligned so that the polar groups face the surrounding water molecules at the surfaces of the bilayer. The hydrocarbon chains of the phospholipids are packed together in the interior of the bilayer, where they form a nonpolar, hydrophobic region that excludes water. The bilayer remains stable because, if disturbed, the hydrophobic, nonpolar hydrocarbon chains of the phospholipids become exposed to the surrounding watery solution, and the molecule returns to its normal arrangement.

Steroids Contribute to Membrane Structure and Work as Hormones

Steroids are a group of lipids with structures based on a framework of four carbon rings **(Figure 3.13a)**. Small differences in the side groups attached to the rings distinguish one steroid from another. The most abundant steroids, the **sterols**, have a single polar —OH group linked to one end of the ring framework and a complex, nonpolar hydrocarbon chain at the other end **(Figure 3.13b)**. Although sterols are almost completely hydrophobic, the single hydroxyl group gives one end of the molecules a slightly polar, hydrophilic character. As a result, sterols also have dual solubility properties and, like phospholipids, tend to assume positions that satisfy these properties. In biological membranes, they line up beside the phospholipid molecules with their polar —OH group facing the membrane surface and their nonpolar ends buried in the nonpolar membrane interior.

Cholesterol (see **Figure 3.13b, c**) is an important component of the boundary membrane surrounding animal cells; similar sterols, called **phytosterols**, occur in plant cell membranes. Deposits derived from cholesterol also collect inside arteries in atherosclerosis (see the *Focus on Research*).

Other steroids, the *steroid hormones,* are important regulatory molecules in animals; they control development, behavior, and many internal biochemical pro-

a. Estradiol, an estrogen

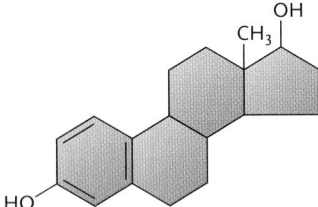

b. Testosterone

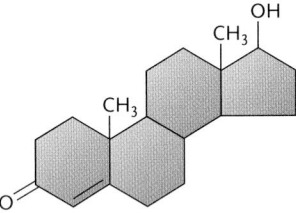

c.

Female wood duck   Male wood duck

Tim Davis, Photo Researchers, Inc.

Figure 3.14

Steroid sex hormones and their effects. The female sex hormone, estradiol **(a)**, and the male sex hormone, testosterone **(b)**, differ only in substitution of an —OH group for an oxygen and the absence of one methyl group (—CH₃) in the estrogen. Although small, these differences greatly alter sexual structures and behavior in animals, such as humans, and the wood ducks *(Aix sponsa)* shown in **(c)**.

cesses. The sex hormones that control differentiation of the sexes and sexual behavior are primary examples of steroid hormones **(Figure 3.14)**. Small differences in the functional groups of steroid hormones have vastly different effects in animals. For instance, the male and female sex hormones are almost identical, except that the female sex hormone contains a single hydrogen atom that is absent from the male sex hormone, and the male sex hormone contains a single methyl group (—CH$_3$) that is absent from the female sex hormone.

Bodybuilders and other athletes sometimes use hormonelike steroids (anabolic-androgenic steroids) to increase their muscle mass (see the *Focus on Research* in Chapter 40). Unfortunately, these substances also produce numerous side effects, including elevated cholesterol, elevated blood pressure, and acne. Other steroids occur as poisons in the venoms of toads and other animals.

Several other lipid types have structures unrelated to triglycerides, phospholipids, or steroids. Among these are *chlorophylls* and *carotenoids,* pigments that absorb light and participate in its conversion to chemical energy in plants (see Chapter 9). Lipid groups also combine with carbohydrates to form *glycolipids* and with proteins to form *lipoproteins*. Both glycolipids and lipoproteins form parts of cell membranes, where they perform vital structural and functional roles.

STUDY BREAK

What are the three most common lipids in living organisms? Distinguish between their structures.

3.5 Proteins

Proteins perform many vital functions in living organisms **(Table 3.2)**: as structural molecules, they provide much of the supporting framework of cells; as **enzymes**, perhaps the most important type of protein, they accelerate the rate of cellular reactions; and as motile molecules, they impart movement to cells and cellular structures. Proteins also transport substances across biological membranes, serve as recognition and recep-

Table 3.2	**Major Protein Functions**	
Protein Type	Function	Examples
Structural	Support	Microtubule and microfilament proteins, which form supporting fibers inside cells; collagen and other proteins that surround and support animal cells; cell wall proteins of plant cells
Enzymatic	Speed biological reactions	Among thousands of examples, DNA polymerase, the enzyme that speeds the duplication of DNA molecules; RuBP (ribulose 1,5-bisphosphate) carboxylase, which speeds the first synthetic reactions of photosynthesis; digestive enzymes such as lipases and proteases, which speed the breakdown of fats and proteins, respectively
Membrane transport	Speed movement of substances across biological membranes	Ion transporters, which move ions such as Na$^+$, K$^+$, and Ca^{2+} across membranes; glucose transporters, which move glucose into cells; aquaporins, which allow water molecules to move across membranes
Motile	Produce cellular movements	Myosin, which acts on microfilaments (called thin filaments in muscle) to produce muscle movements; dynein, which acts on microtubules to produce the whipping movements of sperm tails, flagella, and cilia (the last two are whiplike appendages on the surfaces of many eukaryotic cells); kinesin, which acts on microtubules of the cytoskeleton, a three-dimensional structure in the cytoplasm of eukaryotic cells responsible for cellular movement, cell division, and the organization of organelles
Regulatory	Promote or inhibit the activity of other cellular molecules	Nuclear regulatory proteins, which turn genes on or off to control the activity of DNA; protein kinases, which add phosphate groups to other proteins to modify their activity
Receptor	Bind molecules at cell surface or within cell; some trigger internal cellular responses	Hormone receptors, which bind hormones at the cell surface or within cells and trigger cellular responses; cellular adhesion molecules, which help hold cells together by binding molecules on other cells; LDL receptors, which bind cholesterol-containing particles to cell surfaces
Hormones	Carry regulatory signals between cells	Insulin, which regulates sugar levels in the bloodstream; growth hormone, which regulates cellular growth and division
Antibodies	Defend against invading molecules and organisms	Recognize, bind, and help eliminate essentially any protein of infecting bacteria and viruses, and many other types of molecules, both natural and artificial
Storage	Hold amino acids and other substances in stored form	Ovalbumin, a storage protein of eggs; apolipoproteins, which hold cholesterol in stored form for transport through the bloodstream
Venoms and toxins	Interfere with competing organisms	Ricin, a castor-bean protein that stops protein synthesis; bungarotoxin, a snake venom that causes muscle paralysis

Figure 3.15

The 20 amino acids used by cells to make proteins. The side group of each amino acid is boxed in brown. The amino acids are shown in the ionic forms in which they are found at the pH within the cell; the amino group becomes —NH₃⁺, and the carboxyl group becomes —COO⁻. Three-letter and one-letter abbreviations commonly used for the amino acids appear below each diagram. All amino acids assembled into proteins are in the L-form, one of two possible enantiomers.

Nonpolar amino acids

Alanine
Ala
A

Valine
Val
V

Leucine
Leu
L

Isoleucine
Ile
I

Glycine
Gly
G

Cysteine
Cys
C

Phenylalanine
Phe
F

Tryptophan
Trp
W

Methionine
Met
M

Proline
Pro
P

Uncharged polar amino acids

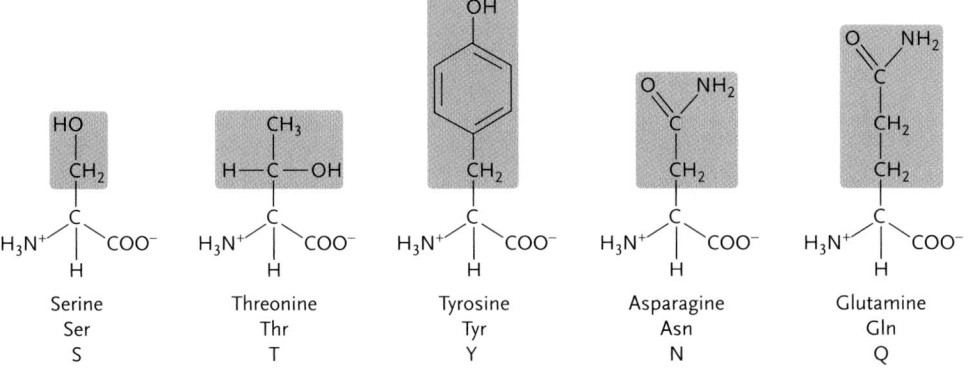

Serine
Ser
S

Threonine
Thr
T

Tyrosine
Tyr
Y

Asparagine
Asn
N

Glutamine
Gln
Q

Negatively charged (acidic) polar amino acids

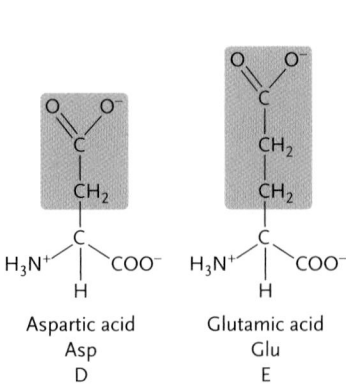

Aspartic acid
Asp
D

Glutamic acid
Glu
E

Positively charged (basic) polar amino acids

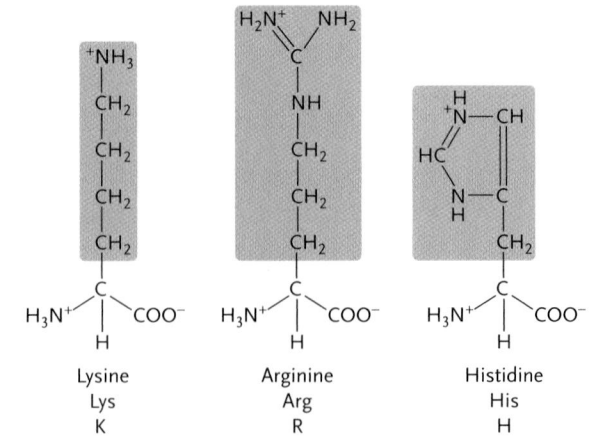

Lysine
Lys
K

Arginine
Arg
R

Histidine
His
H

tor molecules at cell surfaces, and regulate the activity of other proteins and DNA.

Proteins are also released to the cell exterior. Some form parts of extracellular structures, such as cell walls in plants and tendons, bone, cartilage, hair, hooves, and claws in animals. Other proteins released by animals work as hormones, digestive enzymes, or antibodies. (Antibodies are protein molecules that recognize and inactivate foreign material, such as infectious microorganisms.) Many toxins and venoms are based on proteins. For example, botulinum toxin, which is produced by the bacterium *Clostridium botulinum,* is one of the most toxic substances known, with a lethal dose to humans of about 200 to 300 pg/kg (picogram [pg] = 10^{-12} gram).

All of the protein molecules that carry out these and other functions are fundamentally similar in structure. All are macromolecules—polymers consisting of one or more unbranched chains of monomers called amino acids. An **amino acid** is a chemical that contains both an amino and a carboxyl group. Although the most common proteins contain 50 to 1000 amino acids, some proteins found in nature have as few as 3 or as many as 50,000 amino acid units. Proteins range in shape from globular or spherical forms to elongated fibers, and they vary from soluble to completely insoluble in water solutions. Some proteins have single functions, whereas others have multiple functions.

Cells Assemble 20 Kinds of Amino Acids into Proteins by Forming Peptide Bonds

The cells of all organisms use 20 different amino acids as the initial building blocks of proteins. Of these 20 amino acids, 19 have the same structural plan **(Figure 3.15).** In this plan, a central carbon atom is attached to an amino group (—NH$_2$), a carboxyl group (—COOH), and a hydrogen atom:

$$\overset{\displaystyle R}{\underset{\displaystyle H}{H_2N-\overset{|}{\underset{|}{C}}-COOH}}$$

The remaining bond of the central carbon is linked to 1 of 19 different side groups represented by the R (see shaded regions in Figure 3.15), ranging from a single hydrogen atom to complex carbon chains or rings. The remaining amino acid, proline, differs slightly in that it has a ring structure that includes the central carbon atom; the central carbon bonds to a —COOH group on one side and to an *imino* (=NH) group that forms part of the ring at the other side (see Figure 3.15). Although they are called acids, all 20 of the amino acids can act as either acids or bases—depending on cellular conditions, the amino (or imino) group can produce a basic reaction by accepting H$^+$, or the carboxyl group can produce an acidic reaction by releasing H$^+$.

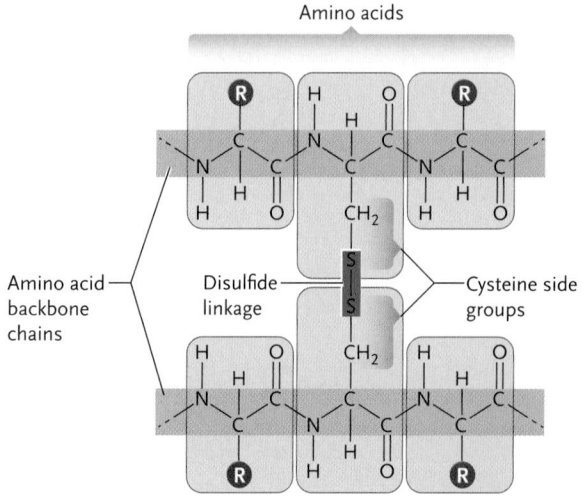

Figure 3.16
A disulfide linkage between two amino acid chains or two regions of the same chain. The linkage is formed by a reaction between the sulfhydryl groups (—SH) of cysteines. The circled R's indicate the side groups of other amino acids in the chains. Figure 3.19 shows disulfide linkages in a real protein.

Differences in the side groups give the amino acids their individual properties. Some side groups are polar, and some are nonpolar; among the polar side groups, some carry a positive or negative charge and some act as acids or bases (see Figure 3.15). Many of the side groups contain reactive functional groups, such as —NH$_2$, —OH, —COOH, or —SH, which may interact with atoms located elsewhere in the same protein or with molecules and ions outside the protein.

The sulfhydryl group (—SH) in the amino acid cysteine is particularly important in protein structure. The sulfhydryl groups of cysteines located in different regions of the same protein, or in different proteins, can interact to produce disulfide linkages (—S—S—). The linkages fasten amino acid chains together **(Figure 3.16)** and help hold proteins in their three-dimensional shape.

Overall, the varied properties and functions of proteins depend on the types and locations of the different amino acid side groups in their structures. The variations in the number and types of amino acids mean that the total number of possible proteins is extremely large.

Covalent bonds link amino acids into the chains of subunits that make proteins. The link, a **peptide bond,** is formed by a dehydration synthesis reaction between the —NH$_2$ group of one amino acid and the —COOH group of a second **(Figure 3.17).** An amino acid chain always has an —NH$_2$ group at one end, called the **N-terminal end,** and a —COOH group at the other end, called the **C-terminal end.** In cells, amino acids are added only to the —COOH end of another amino acid.

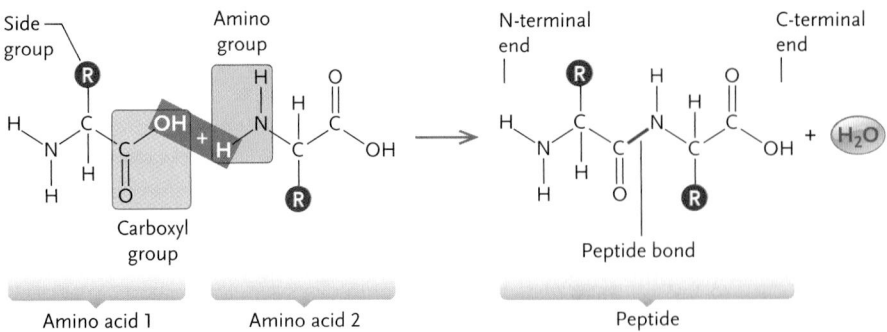

Figure 3.17

A peptide bond formed by reaction of the carboxyl group of one amino acid with the amino group of a second amino acid. The reaction is a typical dehydration synthesis reaction.

a. Primary structure: the sequence of amino acids in a protein

b. Secondary structure: regions of alpha helix, beta strand, or random coil in a polypeptide chain

c. Tertiary structure: overall three-dimensional folding of a polypeptide chain

Heme group

β-Globin polypeptide

β-Globin polypeptide

d. Quaternary structure: the arrangement of polypeptide chains in a protein that contains more than one chain

α-Globin polypeptide

α-Globin polypeptide

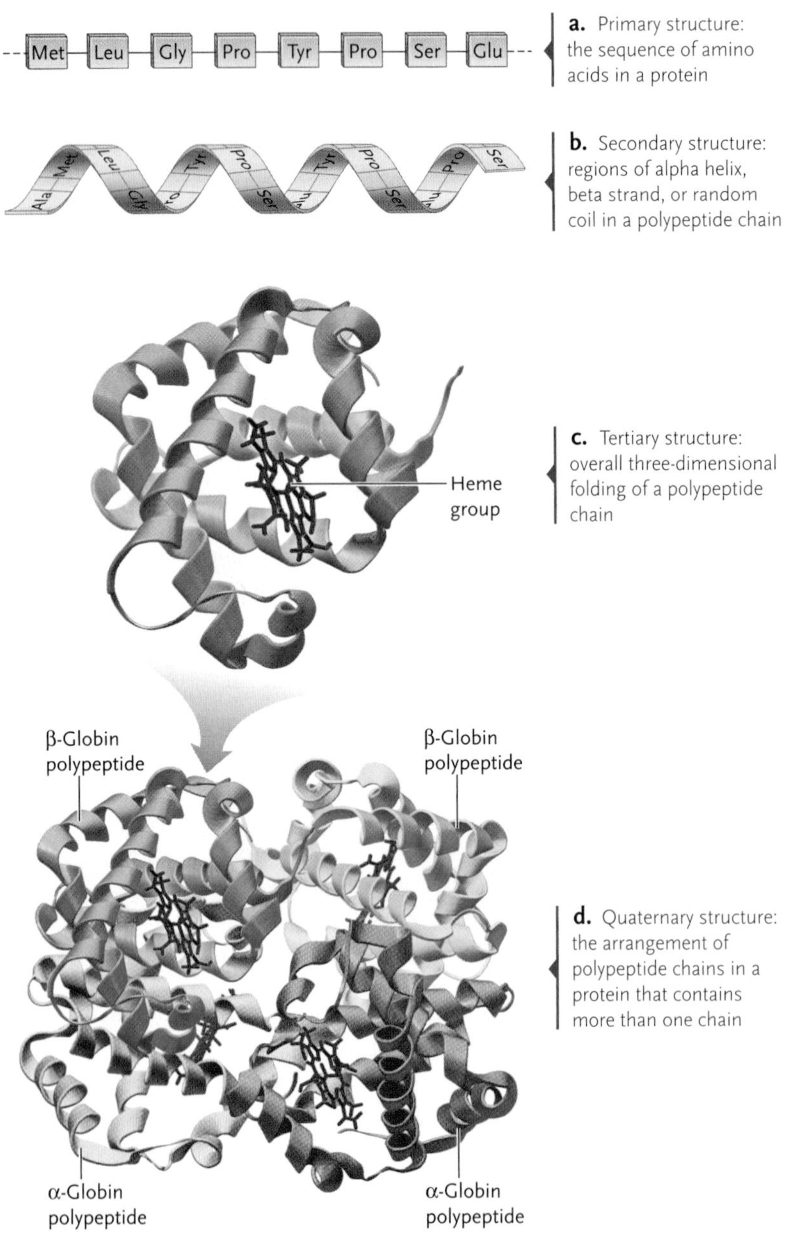

Figure 3.18

The four levels of protein structure. The protein shown in **(c)** is one of the subunits of a hemoglobin molecule; the heme group (in red) is an iron-containing group that binds oxygen. **(d)** A complete hemoglobin molecule.

The chain of amino acids formed by sequential peptide bonds, that is, a **polypeptide**, is only part of the complex structure of proteins. Once assembled, an amino acid chain may fold in various patterns, and more than one chain may combine to form a finished protein, adding to the structural and functional variability of proteins.

Proteins Have as Many as Four Levels of Structure

Proteins potentially have four levels of structure, with each level imparting different characteristics and degrees of structural complexity to the molecule **(Figure 3.18)**. **Primary structure** is the particular and unique sequence of amino acids forming a polypeptide; **secondary structure** is produced by the twists and turns of the amino acid chain. **Tertiary structure** is the folding of the amino acid chain, with its secondary structures, into the overall three-dimensional shape of a protein. All proteins have primary, secondary, and tertiary structures. **Quaternary structure**, when present, refers to the arrangement of amino acid chains in a protein that is formed from more than one chain.

Primary Structure Is the Fundamental Determinant of Protein Form and Function

The primary structure of a protein—that is, the sequence in which amino acids are linked—underlies the other, higher levels of structure. Changing even a single amino acid of the primary structure alters the secondary, tertiary, and quaternary structures to at least some degree and, by so doing, can alter or even destroy the biological functions of a protein. For example, substitution of a single amino acid in the blood protein hemoglobin produces an altered form responsible for sickle-cell disease (see Chapter 12); many other blood disorders are caused by single amino acid substitutions in other parts of the protein.

Because primary structure is so fundamentally important, many years of intensive research have been devoted to determining the amino acid sequence of proteins. Initial success came in 1953, when the English biochemist Frederick Sanger deduced the amino acid sequence of insulin, a protein-based hormone, from samples obtained from cows **(Figure 3.19)**. Now, the amino acid sequences of literally thousands of proteins have been determined, and more are constantly being added to the list. Knowledge of the primary structure of proteins often allows their three-dimensional structure and functions to be predicted and reveals relationships among proteins.

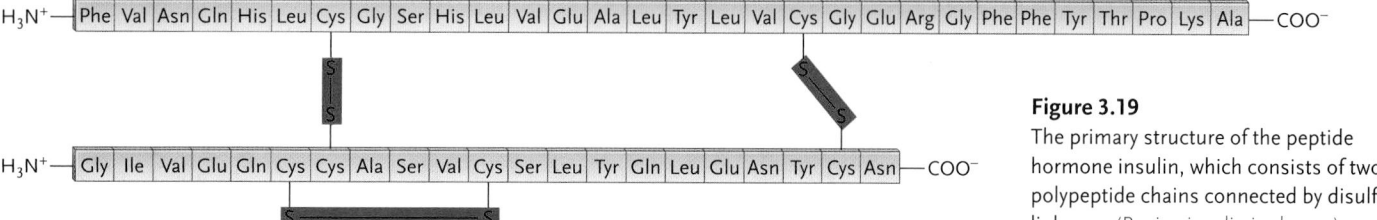

Figure 3.19
The primary structure of the peptide hormone insulin, which consists of two polypeptide chains connected by disulfide linkages. (Bovine insulin is shown.)

Twists and Other Arrangements of the Amino Acid Chain Form the Secondary Structure of a Protein

The amino acid chain of a protein, rather than being stretched out in linear form, is folded into arrangements that form the protein's secondary structure. Two highly regular secondary structures, the *alpha helix* and the *beta strand,* are particularly stable and make an amino acid chain resistant to bending. A third, less regular arrangement, the *random coil* or *loop,* provides flexible regions that allow sections of amino acid chains containing them to bend. Most proteins have segments of all three arrangements.

The Alpha Helix. In the alpha (α) helix, first identified by Linus Pauling and Robert Corey at the California Institute of Technology in 1951, the backbone of the amino acid chain is twisted into a regular, right-hand spiral **(Figure 3.20).** The amino acid side groups extend outward from the twisted backbone. The structure is stabilized by regularly spaced hydrogen bonds (see dotted lines in Figure 3.20) between atoms in the backbone.

Most proteins contain segments of α helix, which are rigid and rodlike, in at least some regions. Globular proteins usually contain several short α-helical segments that run in different directions, connected by segments of random coil. Fibrous proteins, such as the collagens, a major component of tendons, bone, and other extracellular structures in animals, typically contain one or more α-helical segments that run the length of the molecule, with few or no bendable regions of random coil.

The Beta Strand. Pauling and Corey were also the first to identify the beta strand as a major secondary protein structure **(Figure 3.21a).** In a beta (β) strand, the amino acid chain zigzags in a flat plane rather than twisting into a coil.

In many proteins, β strands are aligned side by side in the same or opposite directions to form a structure known as a **beta (β) sheet (Figure 3.21b).** Hydrogen bonds between adjacent β strands stabilize the sheet, making it a highly rigid structure. Beta sheets may lie in a flat plane or may twist into propeller- or barrel-like structures.

Beta strands and sheets occur in many proteins, usually in combination with α-helical segments. One notable exception is in the silk protein secreted by silk-

worms, which contains only β sheets. This exceptionally stable structure, reinforced by an extensive network of hydrogen bonds, underlies the unusually high tensile strength of silk fibers.

The Random Coil. In a **random coil** the amino acid chain has an irregularly folded arrangement. The amino acid proline is often present in random-coil structures. Its ring form does not fit into an α helix or β sheet, and it has no sites available for formation of stabilizing hydrogen bonds.

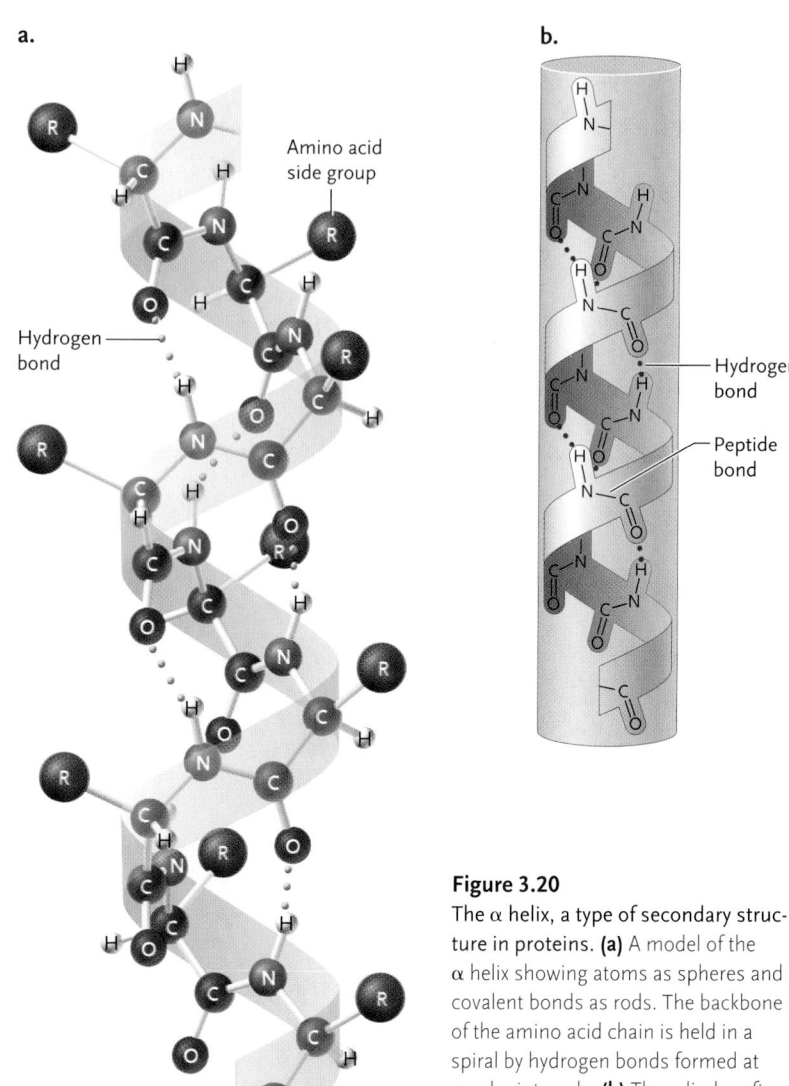

a.

Amino acid side group

Hydrogen bond

b.

Hydrogen bond

Peptide bond

Figure 3.20
The α helix, a type of secondary structure in proteins. **(a)** A model of the α helix showing atoms as spheres and covalent bonds as rods. The backbone of the amino acid chain is held in a spiral by hydrogen bonds formed at regular intervals. **(b)** The cylinder often is used to depict an α helix in protein diagrams, with peptide and hydrogen bonds also shown.

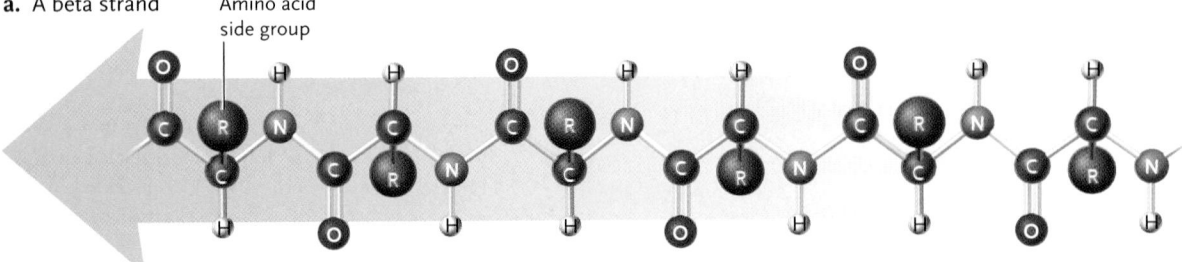

a. A beta strand

Amino acid side group

b. Two beta strands forming a beta sheet

The beta strands run in opposite directions in this sheet; strands may also run in the same direction in a beta sheet.

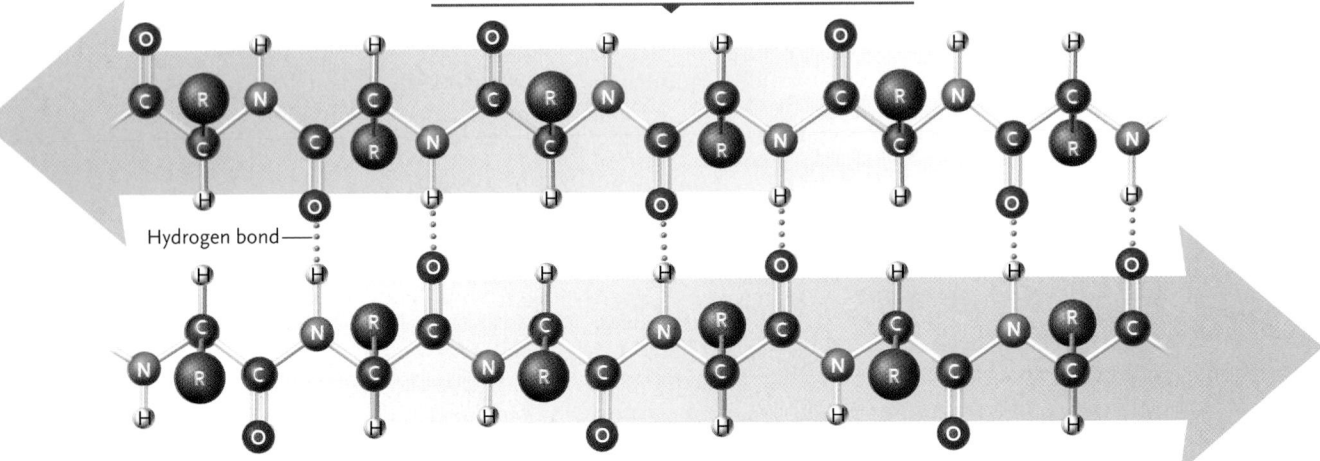

Hydrogen bond

Figure 3.21

The β strand, a type of secondary structure in proteins. **(a)** A single β strand; the arrow points in the direction of the C-terminal end of the amino acid chain. Arrows alone often are used to represent β strands in protein diagrams. **(b)** A β sheet formed by side-by-side alignment of two β strands, held together stably by hydrogen bonds.

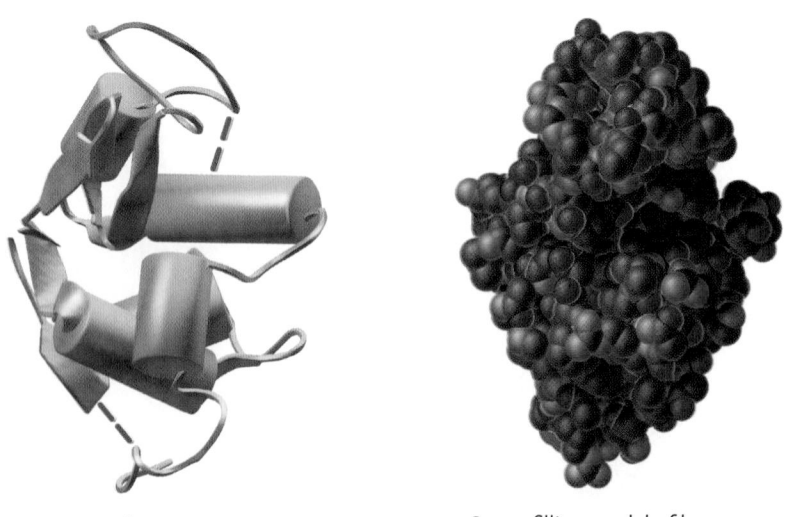

Lysozyme   Space-filling model of lysozyme

Figure 3.22

Tertiary structure of the protein lysozyme, with α helices shown as cylinders, β strands as arrows, and random coils as lines. Lysozyme is an enzyme found in nasal mucus, tears, and other body secretions; it destroys the cell walls of bacteria by breaking down molecules in the wall. Disulfide bonds are shown in yellow. A space-filling model of lysozyme is shown for comparison.

Segments of random coil provide flexible sites that allow α-helical or β-strand segments to bend or fold back on themselves. The fold-back loops of random coils often occur at the surfaces of proteins, at points where they link segments of α helix or β strand located deeper in the protein. Segments of random coil also commonly act as "hinges" that allow major parts of proteins to move with respect to one another.

The Tertiary Structure of a Protein Is Its Overall Three-Dimensional Conformation

The content of α-helical, β-strand, and random-coil segments, together with the number and position of disulfide linkages and hydrogen bonds, folds each protein into its tertiary structure—that is, its overall three-dimensional shape, or **conformation (Figure 3.22).** Attractions between positively and negatively charged side groups and polar or nonpolar associations also contribute to the tertiary structure.

A protein's tertiary structure buries some amino acid side groups in its interior and exposes others at the surface. The distribution and three-dimensional arrangement of the side groups, in combination with their chemical properties, determine the overall chemical activity of a protein. For example, the tertiary structure of the antibacterial enzyme lysozyme (see Figure

Getting Good Vibrations from Proteins

Many functions of proteins—their ability to speed biochemical reactions, obtain energy from sugars, transport molecules in and out of cells, move the limbs of animals, and even produce your thoughts as you read this page—depend on their ability to undergo changes in conformation (shape).

While conformational changes can produce major effects, the molecular movements that underlie these actions are often so minute that they are extremely difficult for scientists to detect. The information is well worth the quest, however, because detecting the exact instant that a protein's shape changes can lead to answers about the part of the protein that produces an effect. Through these answers, researchers can unearth the molecular processes responsible for activities such as muscle contraction or cellular transport.

A number of methods exist for detecting conformational changes in proteins. Atomic force sensing, developed by investigators at the Jerusalem Hebrew University and the Weizmann Institute in Israel, is a technique that can detect protein motion directly, by watching the "wiggles" of a microscopic glass fiber touching the surface of a protein. As a test object, the researchers chose bacteriorhodopsin, a protein found in some members of the prokaryotic domain Archaea (introduced in Chapter 1). They knew that, when bacteriorhodopsin absorbs light at a certain wavelength, it undergoes a conformational change that pumps H^+ from one side of a membrane to the other, initiating a primitive form of photosynthesis. A related animal protein, rhodopsin, undergoes similar changes when it absorbs light as part of the visual process.

The investigators isolated bacteriorhodopsin from the archaean *Halobacterium*, together with lipid molecules from the surface membrane of the organism. They created a film only a few molecules thick on the surface of a glass slide and positioned a curved, microscopic glass probe so that its tip just touched the surface of the film. By shining a microscopic laser beam at the glass fiber, the investigators could detect minute changes in its position by noting changes in the direction of the light reflected from the fiber.

When the researchers directed a brief pulse of light toward the protein film at the wavelength absorbed by bacteriorhodopsin (532 nm), the glass fiber wiggled directionally for a few thousandths of a second, indicating that the protein's conformation changed. Then, they directed a second pulse at a different wavelength known to reverse the pumping action of bacteriorhodopsin (410 nm), and the fiber wiggles also reversed their direction. These results provided further evidence that light causes a conformational change in the protein.

This experiment also produced some novel results. The wiggles showed some motions of bacteriorhodopsin that had never before been detected by measurement of changes in light absorption. These newly detected motions, as the investigators pointed out, may lead to new hypotheses and novel findings about how bacteriorhodopsin, as well as its rhodopsin cousin in animals, functions in living cells.

3.22) has a cleft that binds a polysaccharide found in bacterial cell walls; hydrolysis of the polysaccharide is accelerated by the enzyme.

Tertiary structure also determines the solubility of a protein. Water-soluble proteins have mostly polar or charged amino acid side groups exposed at their surfaces, whereas nonpolar side groups are clustered in the interior. Proteins embedded in nonpolar membranes are arranged in patterns similar to phospholipids, with their polar segments facing the surrounding watery solution and their nonpolar surfaces embedded in the nonpolar membrane interior. These dual-solubility proteins perform many important functions in membranes, such as transporting ions and molecules into and out of cells.

The tertiary structure of most proteins is flexible, allowing them to undergo limited alterations in three-dimensional shape known as **conformational changes**. These changes contribute to the function of many proteins, particularly those working as enzymes, in cellular movements or in the transport of substances across cell membranes. *Insights from the Molecular Revolution* describes a method that allows researchers to detect directly movements produced by the conformational changes of proteins.

Extreme conditions can unfold a protein from its conformation, causing **denaturation**, a loss of both the structure and function of the protein **(Figure 3.23).** For example, excessive heat can break the hydrogen bonds holding a protein in its natural conformation, causing it to unfold and lose its biological activity. Denaturation is one of the major reasons few living organisms can tolerate temperatures greater than 45°C. Extreme changes in pH, which alter the charge of amino acid side groups and weaken or destroy ionic bonds, can also cause protein denaturation.

For some proteins, denaturation is permanent. A familiar example of a permanently denatured protein is a cooked egg white. In its natural form, the egg white protein albumin dissolves in water to form a clear solution. The heat of cooking denatures it permanently into an insoluble, whitish mass.

For other proteins (for example, the enzyme in Figure 3.23), denaturation is reversible: the proteins can return to their natural, functional form if the temperature or pH returns to normal values. Disulfide linkages in an enzyme help limit protein denaturation by preventing amino acid chains from unfolding completely.

One of the active research areas of biology concerns the process by which proteins fold into their tertiary structure as they are made inside cells. Results indicate that proteins fold gradually as they are made—as successive amino acids are linked into the primary structure, the chain folds into increasingly complex structures. As the final amino acids are added to the sequence, the protein completes its folding to the final three-dimensional form. One nagging question about this process is how proteins assume their correct tertiary structure among the different possibilities that may exist for a given amino acid sequence. For many proteins, "guide" proteins called **chaperone proteins** or **chaperonins** solve this problem; they bind temporarily with newly synthesized proteins, directing their conformation toward the correct tertiary structure and inhibiting incorrect arrangements as the new proteins fold **(Figure 3.24).**

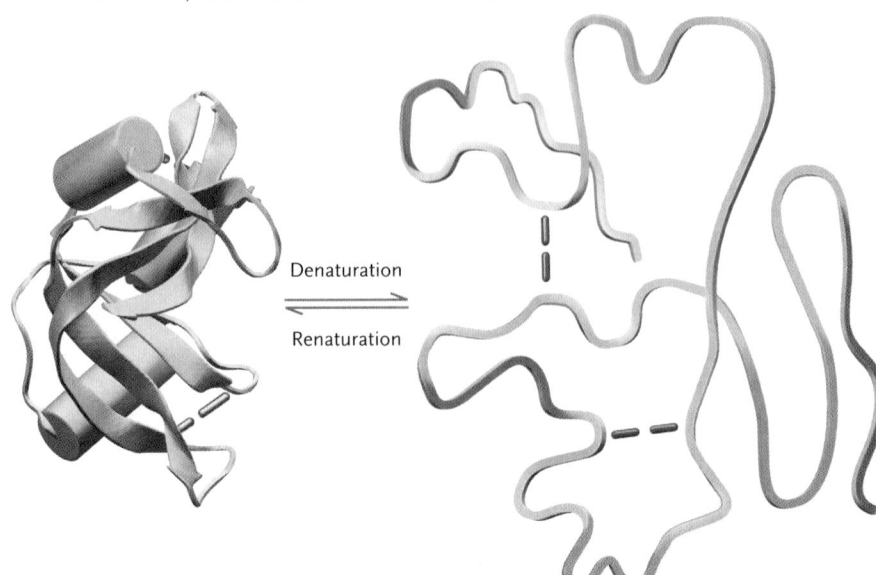

a. Ribonuclease A, natural form   **b.** Denatured form

Denaturation

Renaturation

Figure 3.23

Denaturation and renaturation of ribonuclease A, an enzyme that is released into the digestive tract. Note that all segments of the α helix and β strand are lost when the protein is denatured. Disulfide bonds (in yellow) help the protein return to its natural form during renaturation. (Not all of the disulfide bonds are shown.)

Multiple Amino Acid Chains Form Quaternary Structure

Some complex proteins, such as hemoglobin and antibody molecules, have quaternary structure—that is, the presence and arrangement of two or more amino acid chains (see Figure 3.18d). The same bonds and forces that fold single amino acid chains into tertiary structures, including hydrogen bonds, polar and nonpolar attractions, and disulfide linkages, also hold the multiple polypeptide chains together. During the assembly of multichain proteins, chaperonins also promote correct association of the individual amino acid chains and inhibit incorrect formations.

Combinations of Secondary, Tertiary, and Quaternary Structure Form Functional Domains in Many Proteins

In many proteins, folding of the amino acid chain produces distinct, large structural subdivisions called **domains (Figure 3.25a, b).** Often, one domain of a protein is connected to another by a segment of random coil. The hinge formed by the flexible random coil allows domains to move with respect to one another. Hinged domains of this type are typical of proteins that produce motion and also occur in many enzymes.

Many proteins have multiple functions. For instance, the sperm surface protein SPAM1 (sperm adhesion molecule 1) plays

Figure 3.24
Role of a chaperonin in folding a polypeptide. The three parts of the chaperonin are the top and bottom, which form a cylinder, and the cap.

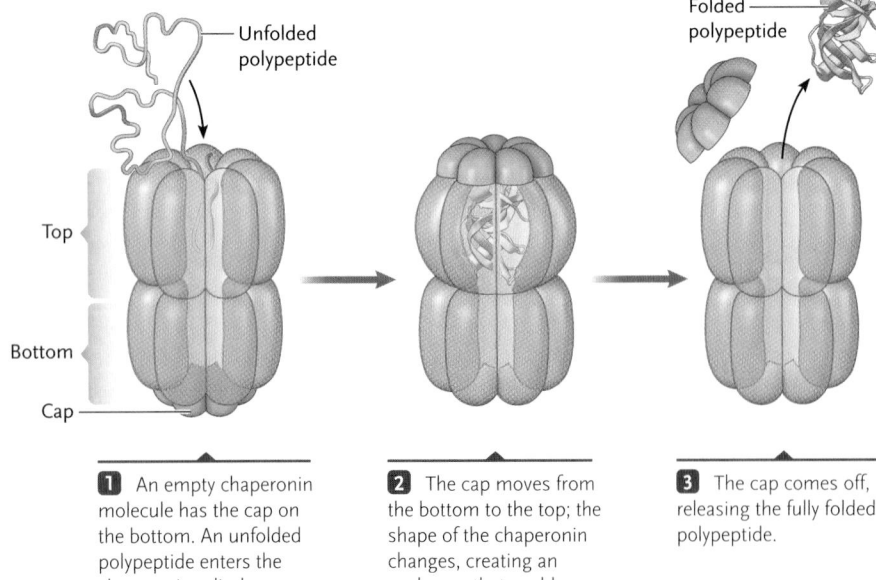

Unfolded polypeptide

Folded polypeptide

Top

Bottom

Cap

1 An empty chaperonin molecule has the cap on the bottom. An unfolded polypeptide enters the chaperonin cylinder at the top.

2 The cap moves from the bottom to the top; the shape of the chaperonin changes, creating an enclosure that enables the polypeptide to fold.

3 The cap comes off, releasing the fully folded polypeptide.

multiple roles in mammalian fertilization. In proteins with multiple functions, individual functions are often located in different domains (see Figure 3.25), meaning domains are functional as well as structural subdivisions. Different proteins often share one or more domains with particular functions. For example, a type of domain that releases energy to power biological reactions appears in similar form in many enzymes and motile proteins. The appearance of similar domains in different proteins suggests that the proteins may have evolved through a mechanism that mixes existing domains into new combinations.

The three-dimensional arrangement of amino acid chains within and between domains also produces highly specialized regions called **motifs**. Several types of motifs, each with a specialized function, occur in proteins. For example, a structural motif called the *leucine zipper* **(Figure 3.25c, d)** holds together proteins that become functional when they join into pairs. The amino acid sequence of the α helix forming each half of the zipper has leucine at every seventh position. The rows of leucine side groups, which project from the α helices, are the "teeth" of the zipper. When two zipper halves come together, as on proteins that join into a pair, they line up by hydrophobic associations (see Section 2.3) into a stable, closed zipper that links the proteins. Many other types of motifs occur in proteins, including some that fit perfectly to a segment of a DNA molecule; for example, the *helix-turn-helix motif* **(Figure 3.25e)** is found in many proteins that regulate DNA activity.

Proteins Combine with Units Derived from Other Classes of Biological Molecules

We have already mentioned the linkage of proteins to lipids to form lipoproteins. Proteins also link with carbohydrates to form *glycoproteins,* which function as

a. Two domains in an enzyme that assembles DNA molecules

Domain a   Domain b

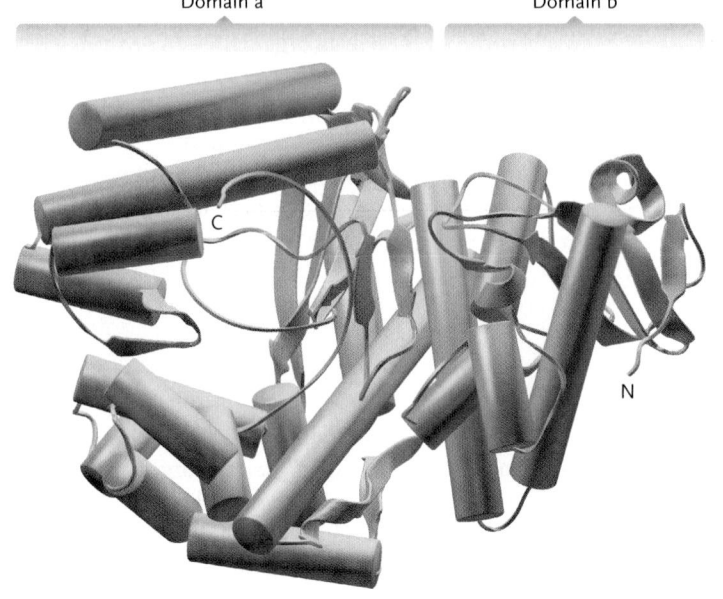

b. The same protein, showing the domain surfaces

Domain a   Domain b

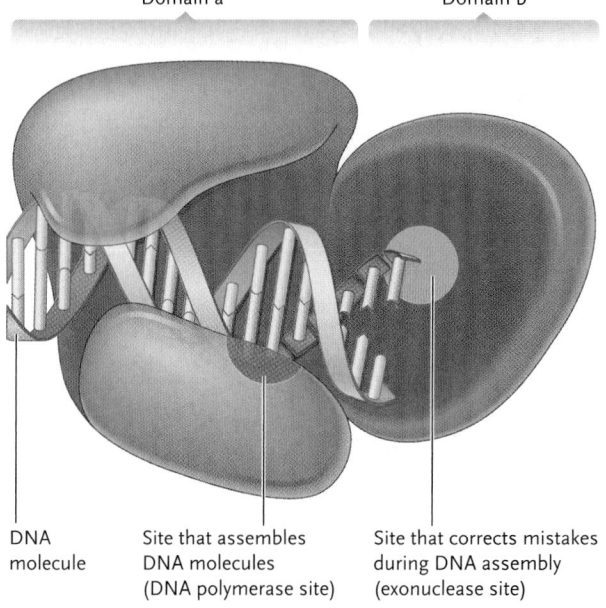

DNA molecule | Site that assembles DNA molecules (DNA polymerase site) | Site that corrects mistakes during DNA assembly (exonuclease site)

c. Leucine zipper, unzipped   **d.** Leucine zipper, zipped

e. Helix-turn-helix motif

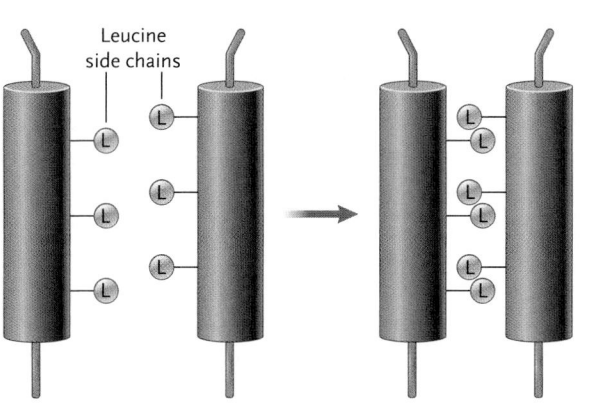

Leucine side chains

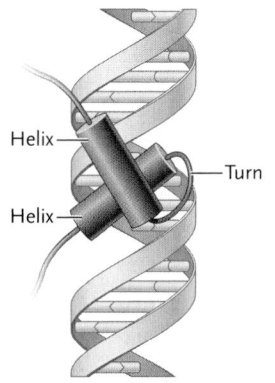

Helix
Turn
Helix

Figure 3.25
Domains and motifs in proteins. **(a)** Two domains in part of an enzyme that assembles DNA molecules in the bacterium *Escherichia coli*, showing α helices as cylinders and β strands as arrows. **(b)** The same view of the protein as in **(a)**, showing only the domain surfaces. **(c, d)** The leucine zipper motif, which holds proteins together in active pairs. **(e)** The helix-turn-helix motif, found in regulatory proteins, which fits precisely into the side of a DNA molecule.

enzymes, antibodies, recognition and receptor molecules at the cell surface, and parts of extracellular supports such as collagen. In fact, most of the known proteins located at the cell surface or in the spaces between cells are glycoproteins. Linkage of proteins to nucleic acids produces *nucleoproteins,* which form such vital structures as *chromosomes,* the structures that organize DNA inside cells.

This section has demonstrated the importance of the amino acid sequence to the structure and function of proteins and highlighted the great variability in proteins produced by differences in their amino acid sequence. The next section considers the nucleic acids, which store

and transmit the information required to arrange amino acids into particular sequences in proteins.

STUDY BREAK

1. What gives amino acids their individual properties?
2. What is a peptide bond, and what type of reaction forms it?
3. What are functional domains of proteins, and how are they formed?

3.6 Nucleotides and Nucleic Acids

Nucleic acids are another class of macromolecules, in this case, long polymers assembled from repeating monomers called *nucleotides.* Two types of nucleic acids exist: DNA and RNA. **DNA (deoxyribonucleic acid)** stores the hereditary information responsible for inherited traits in all eukaryotes and prokaryotes and in many viruses. **RNA (ribonucleic acid)** is the hereditary molecule of some viruses; in all organisms, one type of RNA carries the instructions for assembling proteins from DNA to the sites where the proteins are made inside cells. Another type of RNA forms part of ribosomes, the structural units that assemble proteins, and a third type of RNA functions to bring amino acids to the ribosome for their assembly into proteins (see Chapter 15).

Nucleotides Consist of a Nitrogenous Base, a Five-Carbon Sugar, and One or More Phosphate Groups

A **nucleotide**, the monomer of nucleic acids, consists of three parts linked together by covalent bonds: (1) a **nitrogenous base** (a nitrogen-containing molecule with the property of a base), formed from rings of carbon and nitrogen atoms; (2) a five-carbon, ring-shaped sugar; and (3) one to three phosphate groups **(Figure 3.26).** The two types of nitrogenous bases are **pyrimidines**, with one carbon-nitrogen ring, and **purines**, with two rings **(Figure 3.27).** Three pyrimidine bases—uracil (U), thymine (T), and cytosine (C)—and two purine bases—adenine (A) and guanine (G)—form parts of nucleic acids in cells.

In nucleotides, the nitrogenous bases link covalently to a five-carbon sugar, either **deoxyribose** or **ribose.** The carbons of the two sugars are numbered with a prime symbol—1′, 2′, 3′, 4′, and 5′ (see Figure 3.26). The prime symbols are added to distinguish the carbons in the sugars from those in the nitrogenous bases, which are written without primes. The two sugars differ only in the chemical group bound to the 2′ carbon (boxed in red in Figure 3.26b): deoxyribose has

Figure 3.26
Nucleotide structure.

a. Overall structural plan of a nucleotide

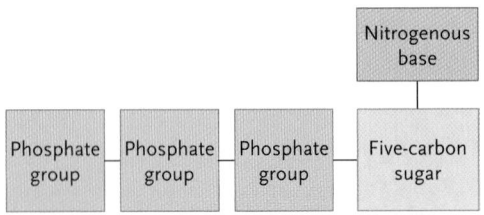

b. Chemical structures of nucleotides

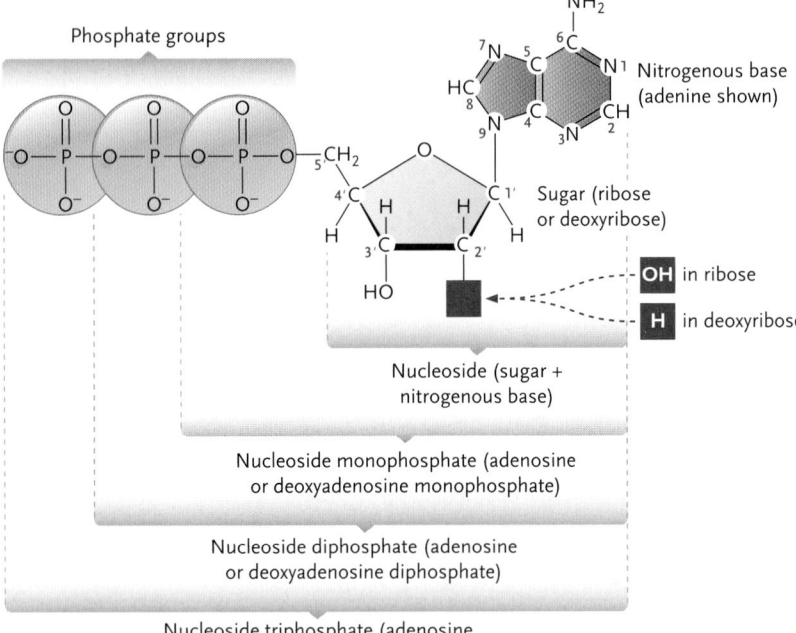

Nucleoside (sugar + nitrogenous base)

Nucleoside monophosphate (adenosine or deoxyadenosine monophosphate)

Nucleoside diphosphate (adenosine or deoxyadenosine diphosphate)

Nucleoside triphosphate (adenosine or deoxyadenosine triphosphate)

Other nucleotides:

Containing guanine: Guanosine or deoxyguanosine monophosphate, diphosphate, or triphosphate

Containing cytosine: Cytidine or deoxycytidine monophosphate, diphosphate, or triphosphate

Containing thymine: Thymidine monophosphate, diphosphate, or triphosphate

Containing uracil: Uridine monophosphate, diphosphate, or triphosphate

Pyrimidines

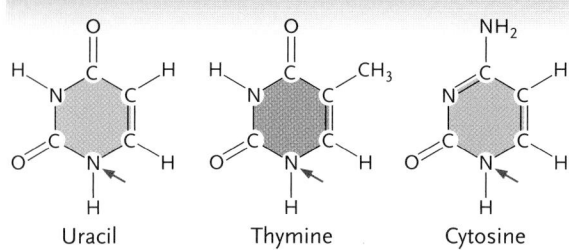

Uracil   Thymine   Cytosine

Purines

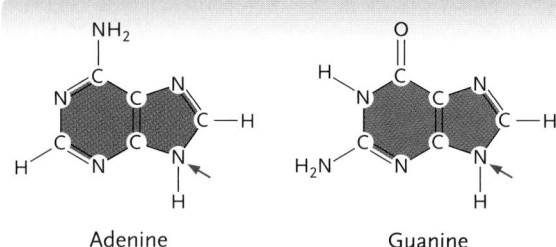

Adenine   Guanine

Figure 3.27

Pyrimidine and purine bases of nucleotides and nucleic acids. Red arrows indicate where the bases link to ribose or deoxyribose sugars in the formation of nucleotides.

an —H at this position and ribose has an —OH group. The prefix *deoxy-* in deoxyribose reflects the oxygen that is absent at this position in the DNA sugar. In nucleotides in the free, unlinked form, a chain of one, two, or three phosphate groups links to the ribose or deoxyribose sugar at the 5′ carbon; nucleotides are called monophosphates, diphosphates, or triphosphates according to the length of this phosphate chain.

A structure that contains only a nitrogenous base and a five-carbon sugar is called a *nucleoside* (see Figure 3.26b). Thus, nucleotides are *nucleoside phosphates.* For example, the nucleoside containing adenine and ribose is called *adenosine.* Adding one phosphate group to this structure produces *adenosine monophosphate (AMP),* adding two phosphate groups produces *adenosine diphosphate (ADP),* and adding three produces *adenosine triphosphate (ATP).* The corresponding adenine–deoxyribose complexes are named *deoxyadenosine monophosphate (dAMP), deoxyadenosine diphosphate (dADP),* and *deoxyadenosine triphosphate (dATP).* The lowercase *d* in the abbreviations indicates that the nucleoside contains the deoxyribose form of the sugar. Equivalent names and abbreviations are used for the other nucleotides (see Figure 3.26b). Whether a nucleotide is a monophosphate, diphosphate, or triphosphate has fundamentally important effects on its activities.

Nucleotides perform many functions in cells in addition to serving as the building blocks of nucleic acids. Two nucleotides in particular, ATP and guanosine triphosphate (GTP), are the primary molecules that transport chemical energy from one reaction system to another; the same nucleotides function to regu-

late and adjust cellular activity. Molecules derived from nucleotides play important roles in biochemical reactions by delivering reactants or electrons from one system to another.

Nucleic Acids DNA and RNA Are the Informational Molecules of All Organisms

DNA and RNA consist of chains of nucleotides, *polynucleotide chains,* with one nucleotide linked to the next by a bridging phosphate group between the 5′ carbon of one sugar and the 3′ carbon of the next sugar in line; this linkage is called a **phosphodiester bond (Figure 3.28).** This arrangement of alternating sugar and phosphate groups forms the backbone of a nucleic acid chain. The nitrogenous bases of the nucleotides project from this backbone.

Each nucleotide of a DNA chain contains deoxyribose and one of the four bases A, T, G, or C. Each nucleotide of an RNA chain contains ribose and one of the four bases A, U, G, or C. Thymine and uracil differ only in a single methyl (—CH₃) group linked to the ring in T but replaced by a hydrogen in U (see Figure 3.27). The differences in sugar and pyrimidine bases

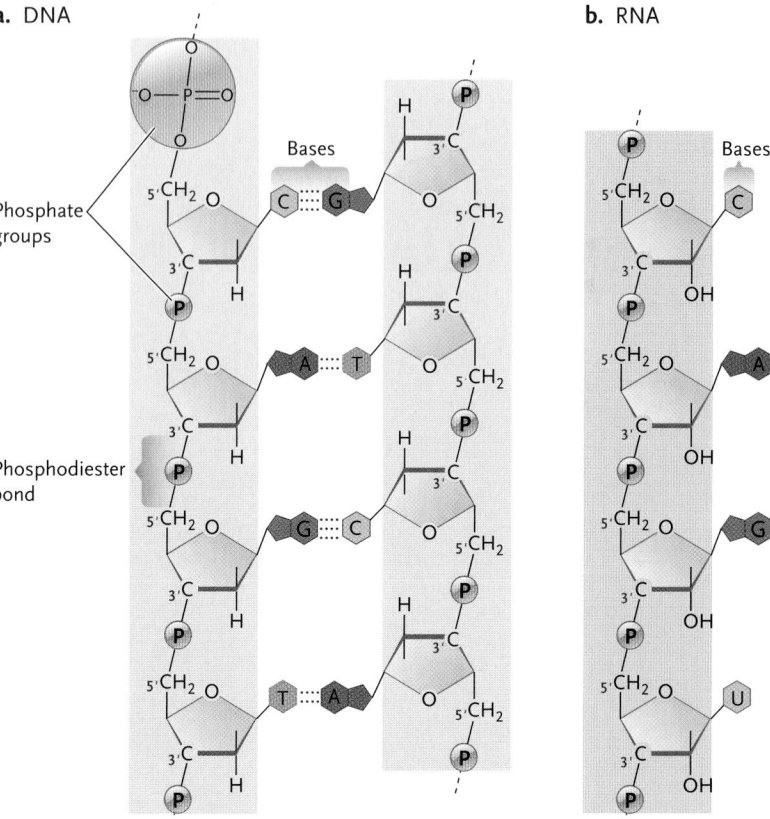

a. DNA   **b.** RNA

Figure 3.28

Linkage of nucleotides to form the nucleic acids DNA and RNA. P is a phosphate group (see Figure 3.26). **(a)** In DNA, the bases adenine (A), thymine (T), cytosine (C), or guanine (G) are bound at the positions marked "base." **(b)** In RNA, A, G, C, or uracil (U) may occur at these sites.

between DNA and RNA account for important differences in the structure and functions of these nucleic acids inside cells.

DNA Molecules Consist of Two Nucleotide Chains Wound Together

In cells, DNA takes the form of a **double helix**, first discovered by James D. Watson and Francis H. C. Crick in 1953, in collaboration with Maurice Wilkins and Rosalind Franklin (see Chapter 14 for details of their discovery). The double helix they described consists of two nucleotide chains wrapped around each other in a spiral that resembles a twisted ladder **(Figure 3.29).** The sides of the ladder are the sugar–phosphate backbones of the two chains, which twist around each other in a right-handed direction to form the double spiral. The rungs of the ladder are the nitrogenous bases, which extend inward from the sugars toward the center of the helix. Each rung consists of a pair of nitrogenous bases held in a flat plane roughly perpendicular to the long axis of the helix. The two nucleotide chains of a DNA double helix are held together primarily by hydrogen bonds between the base pairs. Slightly more than 10 base pairs are packed into each turn of the double helix. A DNA double-helix molecule is also referred to as double-stranded DNA.

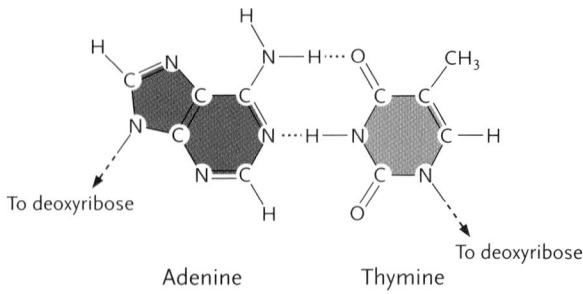

Adenine   Thymine

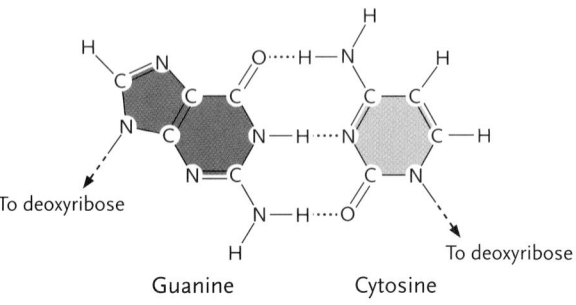

Guanine   Cytosine

Figure 3.30
The DNA base pairs A–T (adenine–thymine) and G–C (guanine–cytosine), as seen from one end of a DNA molecule. Dotted lines designate hydrogen bonds.

The space separating the sugar–phosphate backbones of a DNA double helix is just wide enough to accommodate a base pair that consists of one purine and one pyrimidine. Purine–purine base pairs are too wide and pyrimidine–pyrimidine pairs are too narrow to fit this space exactly. More specifically, of the possible purine–pyrimidine pairs, only two combinations, adenine with thymine, and guanine with cytosine, can form stable hydrogen bonds so that the base pair fits precisely within the double helix **(Figure 3.30).** An adenine–thymine (A–T) pair forms two stabilizing hydrogen bonds; a guanine–cytosine (G–C) pair forms three.

As Watson and Crick pointed out in the initial report of their discovery, the formation of A–T and G–C pairs allows the sequence of one nucleotide chain to determine the sequence of its partner in the double helix. Thus, wherever a T occurs on one chain of a DNA double helix, an A occurs opposite it on the other chain; wherever a C occurs on one chain, a G occurs on the other side (see Figure 3.28). That is, the nucleotide sequence of one chain is said to be *complementary* to the nucleotide sequence of the other chain. The complementary nature of the two chains underlies the processes when DNA molecules are copied—replicated—to pass hereditary information from parents to offspring and when RNA copies are made of DNA molecules to transmit information within cells. In DNA replication, one nucleotide chain is used as a **template** for the assembly of a complementary chain according to the A–T and G–C base-pairing rules **(Figure 3.31).**

a. DNA double helix, showing arrangement of sugars, phosphate groups, and bases

b. Space-filling model of DNA double helix

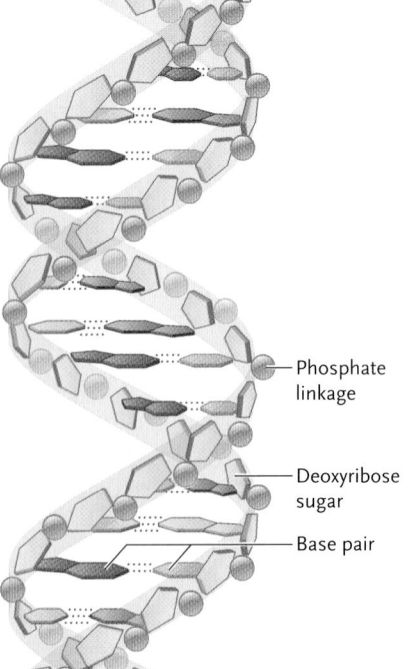

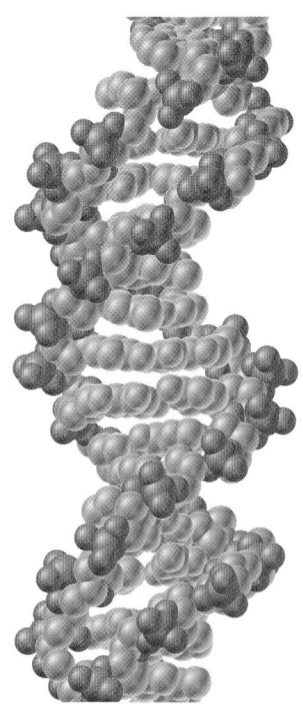

Phosphate linkage

Deoxyribose sugar

Base pair

Figure 3.29
The DNA double helix. **(a)** Arrangement of sugars, phosphate groups, and bases in the DNA double helix. **(b)** Space-filling model of the DNA double helix. The paired bases, which lie in flat planes, are seen on the edge in this view.

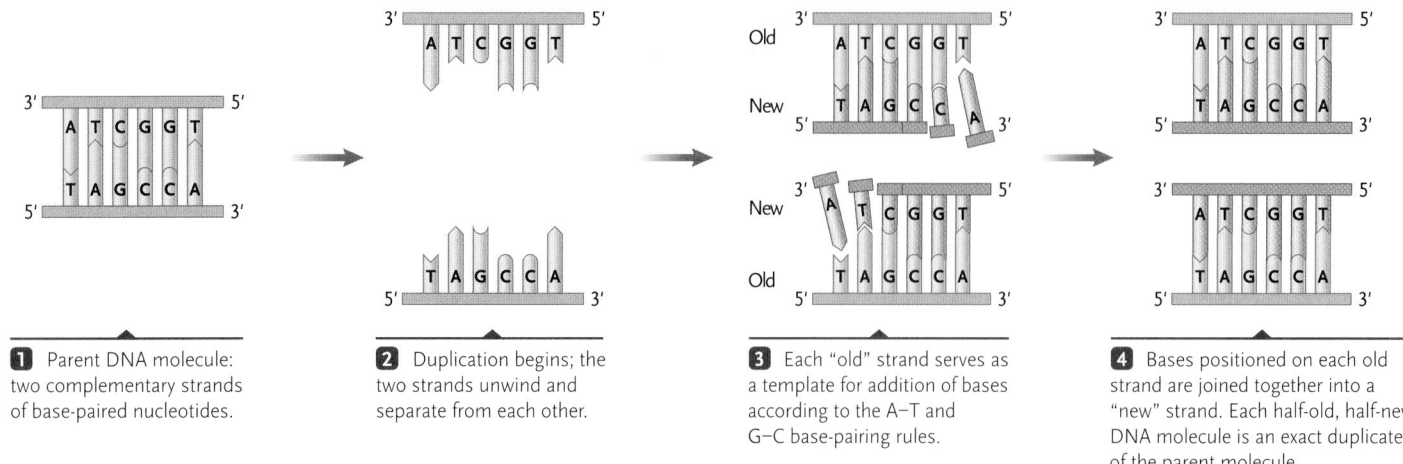

1 Parent DNA molecule: two complementary strands of base-paired nucleotides.

2 Duplication begins; the two strands unwind and separate from each other.

3 Each "old" strand serves as a template for addition of bases according to the A–T and G–C base-pairing rules.

4 Bases positioned on each old strand are joined together into a "new" strand. Each half-old, half-new DNA molecule is an exact duplicate of the parent molecule.

Figure 3.31
How complementary base pairing allows DNA molecules to be replicated precisely.

RNA Molecules Are Usually Single Nucleotide Chains

In contrast to DNA, RNA molecules exist largely as single, rather than double, nucleotide chains in living cells. That is, RNA is typically single-stranded. However, RNA molecules can fold and twist back on themselves to form double-helical regions. The patterns of these fold-back double helices are as vital to RNA function as the folding of amino acid chains is to protein function. "Hybrid" double helices, which consist of an RNA chain paired with a DNA chain, are formed temporarily when RNA copies are made of DNA chains. In the RNA–RNA or hybrid RNA–DNA helices, U in RNA takes over the pairing functions of T, forming A–U rather than A–T base pairs.

Unanswered Questions

Much of biological investigation has become molecular in nature. Here are a few highlights of the extensive research to answer questions about biological molecules.

How is the synthesis of cholesterol and fatty acids regulated?

The regulation of cholesterol and fatty acid synthesis is important because of the link between LDL cholesterol levels and the formation of plaques in arteries. Nobel Prize winners Michael Brown and Joseph Goldstein (University of Texas, Southwestern) recently identified sterol regulatory element-binding proteins (SREBPs), regulatory proteins that control the expression of genes involved in cholesterol and fatty acid synthesis. If lipid is at a low concentration in the cell, the cell needs to make more lipids. To do so, SREBPs enter the nucleus of the cell and activate genes required for the synthesis of cholesterol and fatty acids. In the initial steps of the pathway, SREBPs are escorted by SCAP, another protein. If cholesterol levels in cells are high, the genes must be turned off to prevent cholesterol overproduction. In this case, the accumulation of cholesterol in the cell causes SCAP to change its conformation. As a result, SREBPs cannot follow the pathway for activating the cholesterol and fatty acid genes, and those genes remain inactive.

Brown and Goldstein's lab has been characterizing the SREBPs and the genes that encode them. Current research focuses on understanding how the steps in the pathway for activating the cholesterol and fatty acid genes are regulated, particularly how physiological changes affect the pathway. Their research approaches include molecular analysis of the genes involved, biochemical analysis of the steps in the pathway, protein crystallography to characterize the structures and functions of the proteins involved, cell biology studies to examine the process in living cells, and animal physiology studies to investigate the system at the whole organism level.

What is the role of chaperonins in protein folding?

Chaperonins are crucial in the folding of proteins into their final and functional forms, and properly folded proteins are key to the life of a cell.

Many research groups are studying how chaperonins do their job. One group, headed by Martin Carden at the University of Kent (UK), is studying the structure and function of the human chaperonin CCT. CCT is a barrel-shaped, multiprotein ring that folds actin and tubulin proteins into their final shapes. Actins and tubulins help give eukaryotic cells their shape and provide mechanical strength, among other key properties.

The mechanism, roles, and cellular interactions of CCT are poorly understood. For example, what functions does each of the eight different proteins in the CCT molecule have? Carden's group is studying to what extent CCT separates into individual proteins in the living cell, whether the individual proteins have specialized roles in the cell, and whether they interact in other ways from those already characterized for the folding of actin and tubulin. Using structural information about the individual proteins in the chaperonin, they are building possible models of CCT that can be tested to determine the form of the chaperonin in the living cell. Understanding CCT's structure and function more completely may contribute to research investigating a variety of human diseases caused by protein misfolding. Examples of such diseases are Alzheimer disease, Parkinson disease, and non–insulin-dependent (type 2) diabetes.

Peter J. Russell

The description of nucleic acid molecules in this section, with the discussions of carbohydrates, lipids, and proteins in earlier sections, completes our survey of the major classes of organic molecules found in living organisms. The next chapter discusses the functions of molecules in one of these classes, the enzymatic proteins, and the relationships of energy changes to the biological reactions speeded by enzymes.

STUDY BREAK

1. What is the monomer of a nucleic acid macromolecule?
2. What are the chemical differences between DNA and RNA?

Review

Go to **ThomsonNOW** at www.thomsonedu.com/login to access quizzing, animations, exercises, articles, and personalized homework help.

3.1 Carbon Bonding

- Carbon atoms readily share electrons, allowing each carbon atom to form four covalent bonds with other carbon atoms or atoms of other elements. The resulting extensive chain and ring structures form the backbones of diverse organic compounds.

3.2 Functional Groups in Biological Molecules

- The structure and behavior of organic molecules, as well as their linkage into larger units, depend on the chemical properties of functional groups (Table 3.1).
- Particular combinations of functional groups determine whether an organic molecule is an alcohol, aldehyde, ketone, or acid (Table 3.1).
- In a dehydration synthesis reaction, the components of a water molecule are removed as subunits assemble. In hydrolysis, the components of a water molecule are added as subunits are broken apart (Figure 3.2).

Animation: Functional groups

Animation: Dehydration synthesis and hydrolysis

3.3 Carbohydrates

- Carbohydrates are molecules in which carbon, hydrogen, and oxygen occur in the approximate ratio $1 : 2 : 1$.
- Monosaccharides are carbohydrate subunits that contain three to seven carbons (Figures 3.3–3.5).
- Monosaccharides have D and L enantiomers. Typically, one of the two forms is used in cellular reactions because it has a molecular shape that can be recognized by the enzyme accelerating the reaction, whereas the other form does not.
- Two monosaccharides join to form a disaccharide; greater numbers form polysaccharides (Figures 3.6 and 3.7).
- In polymerization reactions, monomers link to form the polymer. Polysaccharides, proteins, and nucleic acids are assembled by polymerization reactions.

Animation: Structure of starch and cellulose

3.4 Lipids

- Lipids are hydrocarbon-based, water-insoluble, nonpolar molecules. Biological lipids include neutral lipids, phospholipids, and steroids.
- Neutral lipids, which are primarily energy-storing molecules, have a glycerol backbone and three fatty acid chains (Figures 3.8 and 3.9).

- Phospholipids are similar to neutral lipids except that a phosphate group and a polar organic unit substitute for one of the fatty acids (Figure 3.12).
- In polar environments (such as a water solution), phospholipids orient with their polar end facing the water and their nonpolar ends clustered in a region that excludes water. This orientation underlies the formation of bilayers, the structural framework of biological membranes.
- Steroids, which consist of four carbon rings carrying primarily nonpolar groups, function chiefly as components of membranes and as hormones in animals (Figures 3.13 and 3.14).
- Lipids link with carbohydrates to form glycolipids and with proteins to form lipoproteins, both of which play important roles in cell membranes.

Animation: Structure of a phospholipid

3.5 Proteins

- Proteins are assembled from 20 different amino acids. Amino acids have a central carbon to which is attached an amino group, a carboxyl group, a hydrogen atom, and a side group that differs in each amino acid (Figure 3.15).
- Peptide bonds between the amino group of one amino acid and the carboxyl group of another amino acid link amino acids into chains (Figure 3.17).
- A protein may have four levels of structure. Its primary structure is the linear sequence of amino acids in a polypeptide chain; secondary structure is the arrangement of the amino acid chain into α helices, β strands and sheets, or random coils; tertiary structure is the protein's overall conformation. Quaternary structure is the number and arrangement of polypeptide chains in a protein (Figures 3.18–3.22).
- In many proteins, combinations of secondary, tertiary, and quaternary structure form functional domains.
- Proteins combine with lipids to produce lipoproteins, with carbohydrates to produce glycoproteins, and with nucleic acids to form nucleoproteins.

Animation: Structure of an amino acid

Animation: Peptide bond formation

Animation: The primary and secondary structure of proteins

Animation: Secondary and tertiary structure

Animation: Globin and hemoglobin structure

3.6 Nucleotides and Nucleic Acids

- A nucleotide consists of a nitrogenous base, a five-carbon sugar, and one to three phosphate groups (Figures 3.26 and 3.27).
- Nucleotides are linked into nucleic acid chains by covalent bonds between their sugar and phosphate groups. The alternat-

ing sugar and phosphate groups form the backbone of a nucleic acid chain (Figure 3.28).

- There are two nucleic acids: DNA and RNA. DNA contains nucleotides with the nitrogenous bases adenine (A), thymine (T), guanine (G), or cytosine (C) linked to the sugar deoxyribose; RNA contains nucleotides with the nitrogenous bases adenine, uracil (U), guanine, or cytosine linked to the sugar ribose (Figures 3.26–3.28).

- In a DNA double helix, two nucleotide chains wind around each other like a twisted ladder, with the sugar–phosphate backbones of the two chains forming the sides of the ladder and the nitrogenous bases forming the rungs of the ladder (Figure 3.29).

- A-T and G-C base pairs mean that the sequences of the two nucleotide chains of a DNA double helix are complements of each other. The complementary pairing underlies the processes that replicate DNA and copy RNA from DNA (Figures 3.30 and 3.31).

Questions

Self-Test Questions

1. Which functional group has a double bond?
   a. carboxyl
   b. amino
   c. hydroxyl
   d. methyl
   e. sulfhydryl

2. Which of the following characteristics is *not* common to carbohydrates, lipids, and proteins?
   a. They are composed of a carbon backbone with functional groups attached.
   b. Monomers of these molecules undergo dehydration synthesis to form polymers.
   c. Their polymers are broken apart by hydrolysis.
   d. The backbones of the polymers are primarily polar molecules.
   e. The molecules are held together by covalent bonding.

3. Cellulose is to carbohydrate as:
   a. amino acid is to protein.
   b. lipid is to fat.
   c. keratin is to protein.
   d. nucleic acid is to DNA.
   e. nucleic acid is to RNA.

4. Maltose, sucrose, and lactose differ from one another:
   a. because not all contain glucose.
   b. because not all of them exist in ring form.
   c. in the number of carbons in the sugar.
   d. in the number of hexose monomers involved.
   e. by the linkage of the monomers.

5. Lipids that are liquid at room temperature:
   a. are fats.
   b. contain more hydrogen atoms than lipids that are solids at room temperature.
   c. if polyunsaturated, contain several double bonds in their fatty acid chains.
   d. lack glycerol.
   e. are not stored in cells as triglycerides.

6. Which of the following statements about steroids is *false*?
   a. They are classified as lipids because, like lipids, they are nonpolar.
   b. They can act as regulatory molecules in animals.
   c. They are composed of four interlocking rings.
   d. They are highly soluble in water.
   e. Their most abundant form is as sterols.

7. The term *secondary structure* refers to a protein's:
   a. sequence of amino acids.
   b. structure that results from local interactions between different amino acids in the chain.
   c. interactions with a second protein chain.
   d. interaction with a chaperonin.
   e. interactions with carbohydrates.

8. The first and major effect in denaturation of proteins is that:
   a. peptide bonds break.
   b. α helices unwind.
   c. β sheet structures unfold.
   d. tertiary structure is changed.
   e. quaternary structures disassemble.

9. In living systems:
   a. proteins rarely combine with other macromolecules.
   b. enzymes are always proteins.
   c. proteins are composed of 24 amino acids.
   d. chaperonins inhibit protein movement.
   e. a protein domain refers to the place in the cell where proteins are synthesized and function.

10. RNA differs from DNA because:
   a. RNA may contain the pyrimidine uracil, and DNA does not.
   b. RNA is always single-stranded when functioning, and DNA is always double-stranded.
   c. the pentose sugar in RNA has one less O atom than the pentose sugar in DNA.
   d. RNA is more stable and is broken down by enzymes less easily than DNA.
   e. RNA is much a much larger molecule than DNA.

Questions for Discussion

1. Identify the following structures as a carbohydrate, fatty acid, amino acid, or polypeptide:

   a. $H_3N^+-\underset{\underset{H}{|}}{\overset{\overset{R}{|}}{C}}-COO^-$ (The *R* indicates an organic group.)

   b. $C_6H_{12}O_6$

   c. $(glycine)_{20}$

   d. $CH_3(CH_2)_{16}COOH$

2. Cholesterol from food or synthesized in the liver is too hydrophobic to circulate in the blood; complexes of proteins and lipids ferry it around. Low-density lipoprotein (LDL) transports cholesterol out of the liver and into cells. High-density lipoprotein (HDL) ferries the cholesterol that is released from dead cells back into the liver.

   High LDL levels are implicated in atherosclerosis, heart problems, and strokes. The main protein in LDL is called ApoA1 (apolipoprotein A1). A mutant form of ApoA1 has the wrong amino acid (cysteine instead of arginine) at one place in its primary sequence. Carriers of this LDL mutation have very low levels of HDL, which is typically predictive of heart disease. Yet, the carriers have no heart problems. When medical investigators gave some heart patients injections of the mutant LDL, it acted like a drain cleaner, quickly reducing the size of cholesterol deposits in the patients' arteries.

   Soon, such a treatment may reverse years of damage. However, many researchers caution that a low-fat, low-cholesterol diet is still the best assurance of long-term health. Would you choose artery-cleansing treatments over a healthy diet? Explain your choice.

3. The shapes of a protein's domains often give clues to its functions. For example, protein HLA (human leukocyte antigen) is a type of recognition protein on the outer surface of all vertebrate body cells. Certain cells of the immune system use HLAs to distinguish self (the body's own cells) from nonself (invading cells). Each HLA protein has a jawlike region that can bind to molecular parts of an invader. It thus alerts the immune system that the body has been invaded. Speculate on what might happen if a mutation makes the jawlike region misfold.

4. Explain how polar and nonpolar groups are important in the structure and functions of lipids, proteins, and nucleic acids.

Experimental Analysis

A clerk in a health food store tells you that natural vitamin C extracted from rose hips is better for you than synthetic vitamin C. Given your understanding of the structure of organic molecules, how would you respond? Design an experiment to test whether the rose hips and synthetic vitamin C preparations differ in their effects.

Evolution Link

How do you think the primary structure (amino acid) sequence of proteins could inform us about the evolutionary relationships of proteins?

How Would You Vote?

Scientists have discovered vast reservoirs of methane (natural gas, an important fossil fuel) under sediments covering the seafloor. It occurs in a highly unstable form that can cause immense explosions if the temperature rises or the pressure falls slightly. Should we work toward developing these vast undersea methane deposits as an energy source, given that the environmental costs and risks to life are unknown? Go to www.thomsonedu.com/login to investigate both sides of the issue and then vote.

Leaf of a roundleaf sundew leaf *(Drosera rotundifolia)*. Enzymes secreted by the hairs on the leaf digest trapped insects, providing nutrients to the plant.

Carolina Biological Supply/Phototake, Inc.

4 Energy, Enzymes, and Biological Reactions

WHY IT MATTERS

The rotting trunk of a fallen tree is a reminder that death comes to all organisms. Yet, the rotten hulk is crowded with living organisms. Various insects and fungi live on the organic matter of the decaying tree, and with a microscope, you could see that it is also teeming with bacteria and other microorganisms.

If the tree fell in a forest in eastern North America, one of the fungi you might find would be the "old man of the woods" mushroom, known scientifically as *Strobilomyces floccopus* **(Figure 4.1).** The cap and stalk are the most visible parts of the mushroom, but the fungus also includes slender filaments that thread into the rotting tree. Collectively, the filaments represent the mycelium, the part of the fungus devoted to securing nutrients. Similar to that of other fungi, the mycelium of the old man of the woods secretes enzymes for the extracellular digestion of complex compounds—in this case, those of the tree—and absorbs the simple molecules that are produced, converting them into simpler molecules that can be absorbed for use as raw materials and as an energy source. If you look on the underside of the mushroom's mottled brown cap, you will see hundreds of minute tubes holding the mushroom's reproductive spores. They will be re-

Figure 4.1
Old man of the woods mushroom (*Strobilomyces floccopus*) growing on a rotting tree trunk. Enzymes produced by the fungus help convert the wood to sugars that can be used as an energy source.

David Work

leased, producing other mushrooms of the same type if they fall into a favorable environment.

Thus, in death, the fallen tree becomes a basis for new life. The energy and raw materials derived from its organic molecules allow other organisms to grow, maintain their highly organized state, and reproduce.

You would have arrived at the same fundamental understanding of the connection between energy and life if you had focused your attention on a living tree in the forest, a robin, a squirrel, an earthworm, a fly, or any other living organism. **Metabolism**—the biochemical modification and use of organic molecules and energy to support the activities of life—happens only in living organisms. Metabolism comprises thousands of biochemical reactions that accomplish the special activities we associate with life, such as growth, reproduction, movement, and the ability to respond to stimuli. Metabolism depends on enzymes, that is, proteins that speed the rate of cellular chemical reactions. Without enzymes, the pace of life would be very slow indeed.

Understanding how biological reactions occur and how enzymes work requires knowledge of the basic laws of chemistry and physics. All reactions, whether they occur inside living organisms or in the outside, inanimate world, obey the same chemical and physical laws that operate everywhere in the universe. These fundamental laws are the subject of this chapter, which is our starting point for exploring the nature of energy and how cells use it to conduct their activities.

4.1 Energy, Life, and the Laws of Thermodynamics

Life, like all chemical and physical activities, is an energy-driven process. Energy cannot be measured or weighed directly. We can detect it only through its effects on matter, including its ability to move objects against opposing forces, such as friction, gravity, or pressure, or to push chemical reactions toward completion. Therefore, **energy** is most conveniently defined as *the capacity to do work*. Even when you are asleep, cells of your muscles, brain, and other parts of your body are at work and using energy.

Energy Exists in Different Forms and States

Energy takes several different forms, including heat, chemical, electrical, mechanical, and radiant energy. Visible light, infrared and ultraviolet light, gamma rays, and X-rays are all types of radiant energy. Although the forms of energy are different, energy can be converted readily from one form to another. For example, chemical energy is transformed into electrical energy in a flashlight battery, and electrical energy is transformed into light and heat energy in the flashlight bulb. In green plants, the radiant energy of sunlight is converted into chemical energy in the form of complex sugars and other organic molecules.

Kinetic and Potential Energy. All forms of energy can exist in one of two states: kinetic and potential. **Kinetic energy** (*kinetikos* = putting in motion) is the energy of motion, for example, of waves, electrons, atoms, molecules, substances, and objects. Electrical energy, radiant energy, thermal energy, sound, and motion energy are forms of kinetic energy. For instance, a moving object can transfer some of its energy to other objects, as when a baseball is hit with a bat. **Potential energy** is stored energy; it is energy present in a nonmoving location or in the specific arrangement of atoms. Chemical energy, nuclear energy, gravitational energy, and stored mechanical energy are forms of potential energy. Heavy snow located high on a mountainside represents an example of potential energy because it is readily converted into the kinetic energy of an avalanche if it begins to slide downward. A compressed spring provides another example of potential energy; it converts its potential energy to kinetic form when it is released. The reverse conversion, from kinetic to potential energy, also takes place readily. For example, a cyclist converts kinetic energy to potential energy when pushing a bicycle uphill.

Energy Conversions in Living Organisms: Catabolic and Anabolic Reactions. Conversions between potential and kinetic energy also occur in living organisms. For example, sugar has potential energy in the form of the complex arrangement of atoms and chemical bonds in the sugar molecules. All living organisms break down sugar molecules to convert their potential energy into kinetic energy; they then use the kinetic energy to do the metabolic work of life.

Cellular reactions that break down complex molecules such as sugar to make their energy available for cellular work are called **catabolic reactions** (*cata* = downward, as in the sense of a rock releasing energy as it rolls down a hill). Metabolic reactions of the opposite type, which require energy to assemble simple substances into more complex molecules, are termed **anabolic reactions** (*ana* = upward, as in the sense of using energy to push a rock up a hill). An example of an anabolic reaction is the assembly of proteins from amino acids. Typically, living organisms use energy released in catabolic reactions to drive their anabolic reactions.

The Laws of Thermodynamics Describe the Energy Flow in Natural Systems

The study of the energy flow during chemical and physical reactions, including the catabolic and anabolic reactions of living organisms, is called **thermodynamics.** The results of this study are summarized in two fundamental laws of thermodynamics, which allow us to predict whether reactions of any kind, including biological reactions, can occur. That is, if particular groups of molecules are placed together, are they likely to react chemically and change into a different group of molecules? The laws also give us the information necessary to trace energy flows in biological reactions: they allow us to estimate the amount of energy released or required as a reaction proceeds.

The group of reacting molecules studied in thermodynamics is called a *system.* A system is whatever we define it to be; it can be as small as a single molecule or as large as the universe. Everything outside a system is its *surroundings.* In undergoing any type of change, such as a chemical reaction, a system goes from an *initial* state before the reaction begins to a *final* state when the reaction ends. There are two main types of systems: *closed* and *open.* Closed systems **(Figure 4.2a)** can exchange energy but not matter with their surroundings, whereas open systems **(Figure 4.2b)** can exchange both energy and matter with their surroundings. Living organisms are open systems because they constantly exchange matter with their surroundings. However, within a living organism, many individual biochemical reactions operate as closed systems.

The First Law of Thermodynamics Addresses the Energy Content of Systems and Their Surroundings

The **first law of thermodynamics,** also called the principle of the conservation of energy, states that *energy can be transferred and transformed but it cannot be created or destroyed.* That is, in any process that involves an energy change, the *total amount of energy in a system and its surroundings remains constant.*

If energy can be neither created nor destroyed, what is the ultimate source of the energy we and other living organisms use? For almost all organisms, the ultimate source is the sun **(Figure 4.3).** Plants capture the kinetic energy of the light radiating from the sun by absorbing it and converting it to the potential chemical energy of complex organic molecules—primarily sugars, starches, and lipids. These substances are used as fuels by the plants themselves, by animals that feed

on plants, and by organisms (such as fungi and bacteria) that break down the bodies of dead organisms. The potential energy stored in sugars and other organic molecules is used for growth, reproduction, and other work of living organisms.

Eventually, most of the solar energy absorbed by green plants is converted into heat energy as the activities of life take place. Heat energy (a form of kinetic energy) is largely unusable by living organisms; as a result, most of the heat released by the reactions of living organisms radiates to their surroundings, and then from Earth into space.

a. Closed system

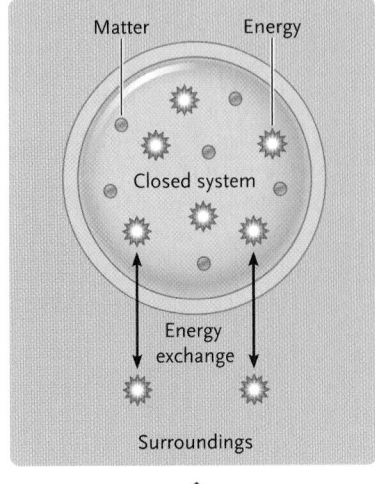

A closed system exchanges energy with its surroundings.

b. Open system

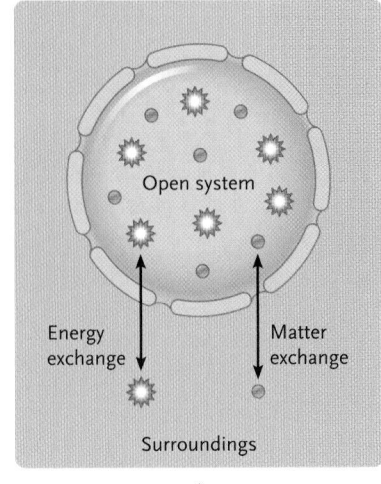

An open system exchanges both energy and matter with its surroundings.

Figure 4.2
Closed and open systems in thermodynamics.

NASA

Radiant energy lost from the sun

Light energy absorbed by photosynthetic organisms and converted into potential chemical energy

Manfred Kage/Peter Arnold, Inc.

Heat energy lost from photosynthetic organisms

Figure 4.3
Energy flow from the sun to photosynthetic organisms (colonies of the green alga *Volvox*), which capture the kinetic radiant energy of sunlight and convert it to potential chemical energy in the form of complex organic molecules.

How does the principle of conservation of energy apply to biochemical reactions? Molecules have both kinetic and potential energy. Kinetic energy for molecules above absolute zero ($-273°C$) is reflected in the constant motion of the molecules, whereas potential energy for molecules is the energy contained in the arrangement of atoms and chemical bonds. The energy content of reacting systems provides part of the information required to predict the likelihood and direction of chemical reactions. Usually, the energy content of the reactants in a chemical reaction (see Section 2.3) is larger than the energy content of the products. Thus, reactions usually progress to a state in which the products have *minimum energy content*. When this is the case, the difference in energy content between reactants and products in the reacting system is released to the surroundings.

The Second Law of Thermodynamics Considers Changes in the Degree of Order in Reacting Systems

The second law of thermodynamics explains why, as any energy change occurs, the objects (matter) involved in the change typically become more disordered. (Your room and the kitchen at home are probably the best examples of this phenomenon.) You know from experience that it takes energy to straighten out (decrease) the disorder (as when you have to clean up your room).

The **second law of thermodynamics** states this tendency toward disorder formally, in terms of a system and its surroundings: in any process in which a system changes from an initial to a final state, *the total disorder of the system and its surroundings always increases*. In thermodynamics, disorder is called **entropy**. If the system and its surroundings are defined as the entire universe, the second law means that as changes occur anywhere in the universe, the total disorder or entropy of the universe constantly increases. As the first law of thermodynamics asserts, however, the total energy in the universe does not change.

At first glance, living organisms appear to violate the second law of thermodynamics. As a fertilized egg develops into an adult animal, it becomes more highly ordered (decreases its entropy) as it synthesizes organic molecules from less complex substances. However, the entropy of the whole system—the surroundings, as well as the organism—must be considered as growth proceeds. For a fertilized egg—the initial state—its surroundings include all the carbohydrates, fats, and other complex organic molecules that the developing animal uses to develop into an adult. When development is complete—the final state—the surroundings include the animal's waste products (water, carbon dioxide, and many relatively simple organic molecules), which are collectively much less complex than the organic molecules used as fuels. When the total reactants, including all the nutrients, and the total products, including all the waste materials, are included, the total change satisfies both laws of thermodynamics—the total energy content remains constant, and the entropy of the system and its surroundings increases.

Applying the first and second laws of thermodynamics together allows us to predict whether any particular chemical or physical reaction will occur without outside help. Such reactions are called **spontaneous reactions** in thermodynamics. In this usage, the word *spontaneous* means only that a reaction will occur. It does not describe the rate of a reaction; indeed, spontaneous reactions may proceed very slowly, such as the formation of rust on a nail, or very quickly, such as a match bursting into flame. This concept is important in biology, because enzymes cannot make a reaction take place if it is not already spontaneous—*enzymes can only make spontaneous reactions go faster.*

Change in Free Energy Indicates Whether a Reaction Is Spontaneous

Free energy is the energy in a system that is available to do work. In living organisms, free energy accomplishes the chemical and physical work involved in activities such as the synthesis of molecules, movement, and reproduction.

A free energy equation combines the energy and entropy changes in a system going from initial to final states (such as reactants to products):

$$\Delta G = \Delta H - T\Delta S$$

in which Δ (pronounced delta) means "change in." ΔG is the change in free energy in the system (where the G recognizes physicist Josiah Willard Gibbs, the creator of the concept), ΔH is the change in energy content (considered as heat, H), T is the absolute temperature in degrees Kelvin (K, where K = °C + 273.16), and ΔS is the change in entropy. The equation states that *the free energy change as a system goes from initial to final states is the sum of the changes in energy content and entropy.* For a reaction to be spontaneous, ΔG must be negative. This negative value indicates that the free energy released by the reaction is lost from the system and is gained by the surroundings as the reaction goes from the initial to the final state. Overall, entropy has increased.

Reactions that have a negative ΔG because they release free energy are termed **exergonic reactions** (*ergon* = work) **(Figure 4.4a).** For biological systems, the free energy released by exergonic reactions accomplishes growth, movement, and all the other activities of life. If ΔG is positive, a reaction proceeds only if free energy is added to the system. Reactions that can proceed only if free energy is supplied are termed **endergonic reactions (Figure 4.4b).** Typically, the free energy for such reactions is supplied from other, exergonic, reactions.

In practical applications of the free energy equation, free energy changes (ΔG) are determined under standard conditions with the results given in kilocalories per mole (kcal/mol) of reactants converted to products. The value obtained allows the energy released or required by one reacting system to be compared directly with another.

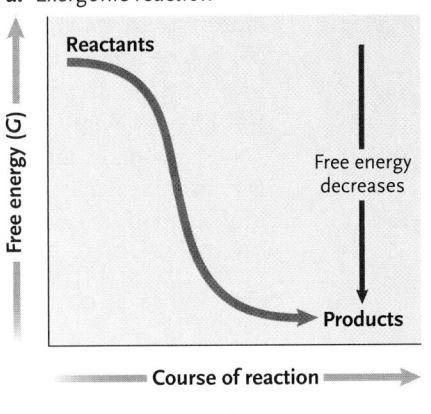

a. Exergonic reaction

Free energy (G) · Course of reaction

Reactants · Free energy decreases · Products

In an exergonic reaction, free energy is released. The products have less free energy than was present in the reactants, and the reaction proceeds spontaneously.

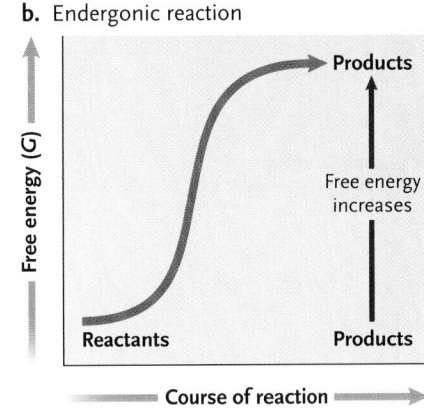

b. Endergonic reaction

Free energy (G) · Course of reaction

Products · Free energy increases · Reactants · Products

In an endergonic reaction, free energy is gained. The products have more free energy than was present in the reactants. An endergonic reaction is not spontaneous: it proceeds only if energy is supplied by an exergonic reaction.

Figure 4.4
Exergonic **(a)** and endergonic **(b)** reactions.

The calculated free energy can determine the likelihood of a reaction occurring. For example, if sucrose is placed in a test tube with water, will it break down (hydrolyze) into glucose and fructose? Or, if glucose and fructose are placed together with water in a test tube, will they combine to form sucrose? The hydrolysis reaction has a ΔG of -5.5 kcal/mol, which means that for each mole of sucrose hydrolyzed, 5.5 kcal of energy is *released*. By contrast, the synthesis reaction has a ΔG of $+5.5$ kcal/mol; 5.5 kcal of energy must be *added* to convert 1 mole of reactants into 1 mole of products. Therefore, the hydrolysis of sucrose to glucose and fructose can proceed spontaneously because ΔG is negative for this reaction, but the synthesis of sucrose from glucose and fructose cannot proceed spontaneously because it has a positive ΔG. However, plants in particular perform this synthesis reaction on a regular basis. How do they do it? The next section describes how plants, and in fact all living organisms, carry out synthetic reactions without violating the laws of thermodynamics.

STUDY BREAK

1. Distinguish between kinetic and potential energy.
2. Distinguish between catabolic and anabolic reactions.
3. In thermodynamics, what is meant by an open system and a closed system?

4.2 How Living Organisms Couple Reactions to Make Synthesis Spontaneous

Many individual reactions of living organisms, particularly those that involve the assembly of complex molecules from less complex building blocks, have a positive ΔG and therefore are not spontaneous. For example, cells commonly carry out reactions in which ammonia (NH_3) is added to glutamic acid, an amino acid with one amino group, to produce glutamine, an amino acid with two amino groups:

$$\text{glutamic acid} + NH_3 \rightarrow \text{glutamine} + H_2O$$
$$\Delta G = +3.4 \text{ kcal/mol}$$

The glutamine is used in the assembly of proteins and is a donor of nitrogen for other reactions in the cell. The positive value for ΔG shows that the reaction cannot proceed spontaneously.

How, then, do cells carry out this reaction? They join it to another reaction with a large negative ΔG. The combined reaction, called a **coupled reaction**, has a negative ΔG, which indicates that it is spontaneous and will release free energy. In effect, the coupling system works by joining an exergonic reaction to the endergonic reaction, producing an overall reaction that is exergonic. All the endergonic reactions of living organisms, including those of growth, reproduction, movement, and response to stimuli, are made possible by coupling reactions in this way.

ATP Is the Primary Coupling Agent in All Living Organisms

All cells, from bacteria to those of plants and animals, use the nucleotide **ATP** as the primary agent that couples exergonic and endergonic reactions. ATP provides an injection of free energy that does biological work.

ATP consists of a five-carbon sugar, ribose, linked to the nitrogenous base adenine and a chain of three phosphate groups **(Figure 4.5a)**. Much of the potential energy of ATP is associated with the arrangement of the three phosphate groups, which carry a strongly negative charge in the cellular environment. Because of the close alignment of the three

a. Chemical structure of ATP

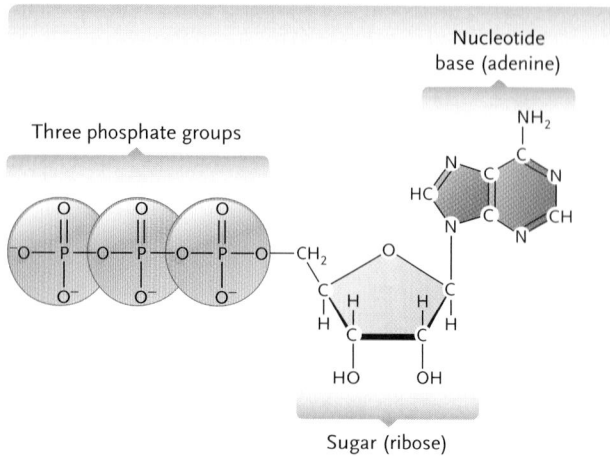

Three phosphate groups

Nucleotide base (adenine)

Sugar (ribose)

b. Adenine nucleotides

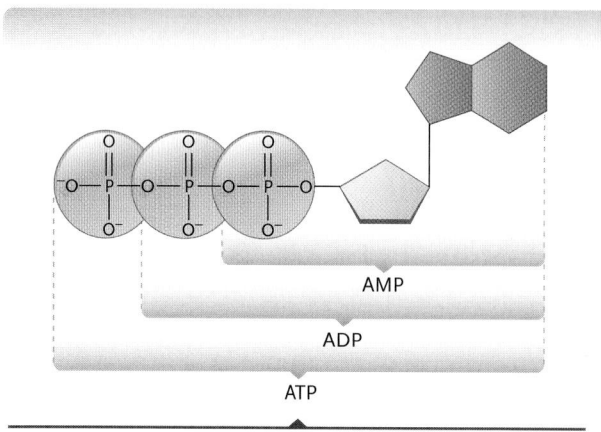

AMP

ADP

ATP

With one phosphate group, the molecule is known as AMP; with two phosphates, the molecule is called ADP. Each added phosphate packs additional potential chemical energy into the molecular structure.

c. Hydrolysis reaction removing a phosphate group from ATP

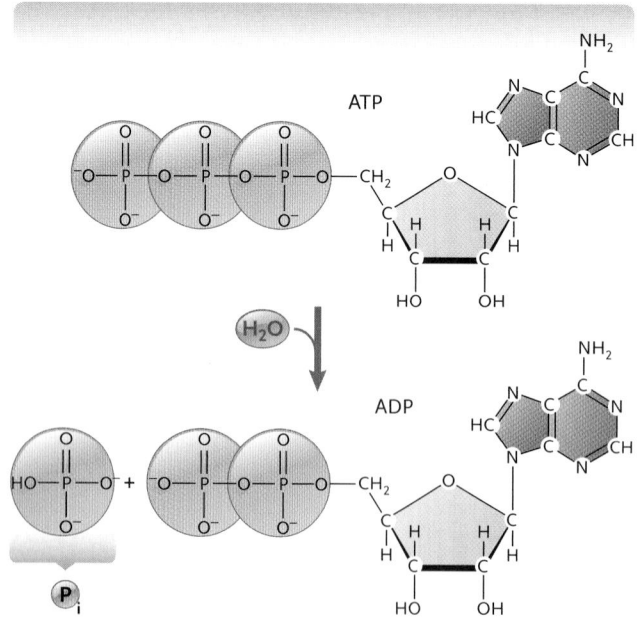

ATP

H_2O

ADP

P_i

Figure 4.5

ATP, the primary molecule that couples energy-requiring reactions to energy-releasing reactions in living organisms. (P_i is the symbol used in this book for inorganic phosphate.)

phosphate groups, the negative charges repel each other strongly. Removal of one or two of the three phosphate groups is a spontaneous reaction that relieves the repulsion and releases large amounts of free energy (**Figure 4.5b**).

For example, removal of just one phosphate group (a hydrolysis reaction; see **Figure 4.5c**) releases a large increment of free energy:

$$ATP + H_2O \rightarrow ADP + P_i$$
$$\Delta G = -7.3 \text{ kcal/mol}$$

The products of this reaction are ADP and inorganic phosphate (P_i). Removal of two of the phosphate groups produces adenosine monophosphate (AMP) and almost doubles the amount of free energy released.

Cells Couple Reactions Directly by Linking Phosphate Groups from ATP to Other Molecules

Although ATP hydrolysis releases a large burst of free energy, cells do not use ATP hydrolysis directly as a mechanism to couple reactions and release energy. Instead, the ADP or phosphate group produced by ATP breakdown is temporarily linked to one of the reacting molecules or to an enzyme that accelerates a coupled reaction. In effect, the linkage transfers potential chemical energy to the molecule binding the ADP or phosphate group and, in this way, conserves much of the free energy released by ATP hydrolysis.

The reactions that couple ATP breakdown to the synthesis of glutamine from glutamic acid illustrate the process. As a first step, the phosphate group removed from ATP is transferred to glutamic acid, forming glutamyl phosphate:

glutamic acid + ATP → glutamyl phosphate + ADP

The addition of a phosphate group to a molecule is called **phosphorylation**. The ΔG for this reaction is negative, making the reaction spontaneous, but much less free energy is released than in the hydrolysis of ATP to ADP + P_i. In the second step, glutamyl phosphate reacts with NH_3:

glutamyl phosphate + NH_3 → glutamine + P_i

This second reaction also has a negative value for ΔG and is spontaneous.

Even though the reaction proceeds in two steps, it is usually written for convenience as one reaction, with a combined negative value for ΔG:

$$\text{glutamic acid} + NH_3 + ATP \rightarrow$$
$$\text{glutamine} + ADP + P_i$$
$$\Delta G = -3.9 \text{ kcal/mol}$$

Because ΔG is negative, the coupled reaction is spontaneous and releases energy. The difference between -3.9 kcal/mol and the -7.3 kcal/mol released by hydrolyzing ATP to ADP + P_i represents potential chemical energy transferred to the glutamine molecules produced by the reaction.

Cells also Couple Reactions to Replenish Their ATP Supplies

How do cells replace the ATP used in coupling reactions? The reaction has a positive ΔG and is therefore endergonic:

$$ADP + P_i \rightarrow ATP + H_2O$$
$$\Delta G = +7.3 \text{ kcal/mol}$$

Cells accomplish this feat by coupling reactions that link ATP synthesis to catabolic reactions such as the breakdown of energy-rich sugar molecules. For example, if glucose is simply burned by igniting it in air, large quantities of free energy are released:

$$C_6H_{12}O_6 + 6 O_2 \rightarrow 6 CO_2 + 6 H_2O$$
$$\Delta G = -686 \text{ kcal/mol}$$

Rather than burning glucose directly, cells couple the reaction breaking down glucose to the synthesis of ATP from ADP and P_i:

$$C_6H_{12}O_6 + 30 ADP + 30 P_i + 6 O_2 \rightarrow 6 CO_2$$
$$+ 6 H_2O + 30 ATP$$
$$\Delta G = -476 \text{ kcal/mol}$$

The coupled reaction, shown here in greatly simplified form, is spontaneous and releases free energy, but much less than when glucose is burned in air; the difference represents energy conserved in ATP.

ATP thus cycles between reactions that release free energy and reactions that require free energy **(Figure 4.6a)**. Adding or removing phosphate groups in the ATP/ADP/AMP system is similar to compressing or releasing a spring. Adding phosphate groups, up to a limit of three, compresses the spring and stores potential energy in the molecule. Removing one or two phosphate groups releases the spring and makes free energy available for cellular work. Examples of cellular events driven by ATP hydrolysis are shown in **Figure 4.6b**; additional examples appear in many other chapters of this book.

The discussion up to this point has assumed that spontaneous reactions go to completion—that is, that reactants are converted completely into products. However, as the next section shows, other factors can oppose completion, stopping the progress of a reaction at a point where both reactants and products are present and remain in the system.

STUDY BREAK

1. How are coupled reactions important to cell function?
2. Explain the composition of the ATP molecule. What happens to ATP when it is used in a phosphorylation reaction?

a. The ATP/ADP cycle, which couples reactions that release free energy to reactions that require free energy

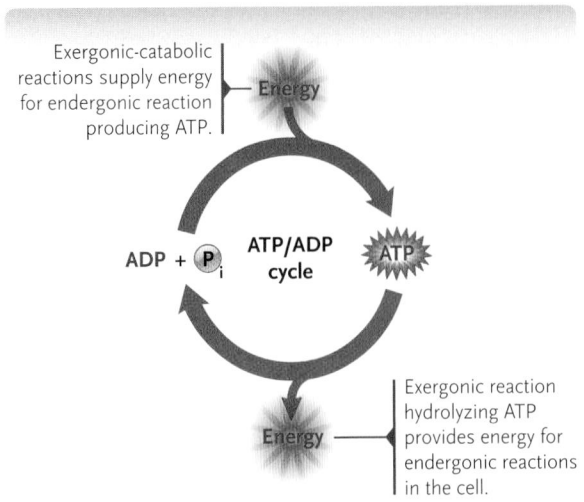

Figure 4.6

Formation and hydrolysis of ATP, which is centrally important for biological reactions. **(a)** ATP/ADP cycle. This cycle couples reactions that release free energy to reactions that require free energy. **(b)** Examples of cellular events driven by ATP hydrolysis.

b. Examples of cellular events driven by ATP hydrolysis

Anabolic reactions:

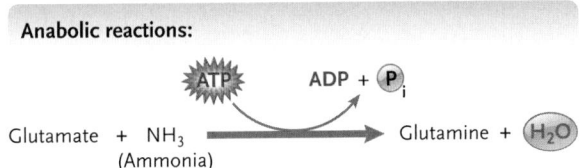

Regulation of protein activity:

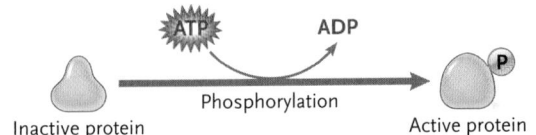

Transport of solutes:

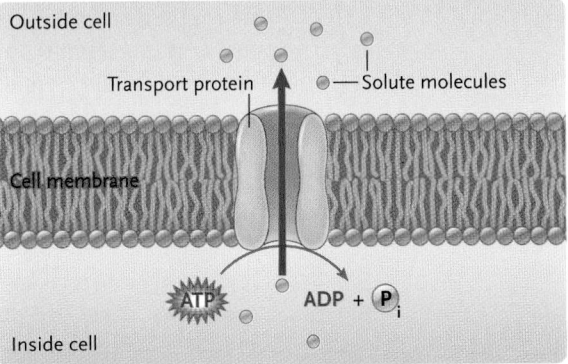

4.3 Thermodynamics and Reversible Reactions

Several conditions can oppose the completion of spontaneous biological reactions, even though the reactions have a negative ΔG. Instead, the reactions run in the direction of completion (toward reactants or toward products) until they reach the **equilibrium point**, a state of balance between the opposing factors pushing the reaction in either direction. At the equilibrium point, both reactants and products are present and the reactions typically are reversible.

The Concentration of Reactants and Products Often Establishes an Equilibrium Point

The relative concentrations (concentration = number of molecules per unit volume) of reactant and product molecules in the solution containing a reaction can oppose completion of the reaction. A solution containing reactants and products of a reaction is at a state of maximum entropy (disorder) when all the molecules are evenly distributed in equal concentrations. In terms of your room, this situation is equivalent to having all your books and clothing in a complete state of disorder on the floor. As a reaction runs past the point when the concentrations of reactants and products are equal, it begins to reduce entropy as it decreases the number of reactant molecules and adds additional product molecules, which is equivalent to beginning to hang clothes on hooks and place books on shelves.

This entropy reduction requires energy, and it begins to use some of the free energy released by the reaction. As the reaction continues, eventually the free energy released by the reaction is no longer sufficient to reduce entropy further—the equilibrium point has been reached **(Figure 4.7)**. At this balance point, reactant molecules constantly change into products, and products change into reactants, at equal rates. In other words, the rates of the forward and backward reactions are equal at the equilibrium point. (For chemical reactions of all types, *rate* means the number of reactant molecules converted to products, or products to reactants, per unit time.)

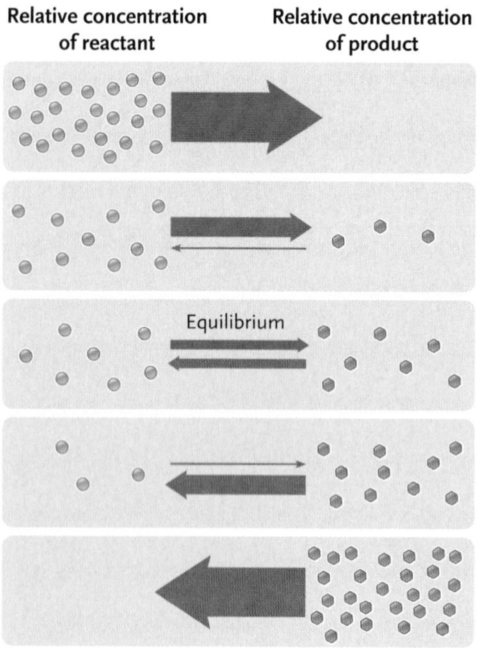

The concentrations of reactants and products at the equilibrium point are not necessarily equal. Generally, the greater the negative value of ΔG, the further a reaction will proceed toward completion, with proportionately greater numbers of product molecules than reactant molecules at the equilibrium point.

At the equilibrium point, small changes in conditions can push a reaction in either direction, toward reactants or toward products. Thus, reactions that reach an equilibrium point are **reversible** (see Figure 4.7). Reversible reactions are written with a double arrow to indicate this feature:

$$A + B \rightleftharpoons C + D$$
reactants   products

Many Biological Reactions Keep Running Because They Never Reach Equilibrium

Most biological reactions have an equilibrium point and are reversible. However, many individual reactions in living organisms never reach an equilibrium point because they are parts of a *metabolic pathway*—a series of sequential reactions in which the products of one reaction are used immediately as the reactants for the next reaction in the series. (An example of a pathway is shown in Figure 4.18.) This immediate use of reactants keeps the individual reactions, and the entire metabolic pathway, running as long as the final products do not accumulate in excess. Metabolic pathways may be anabolic, synthesizing complex molecules from simpler substances, or catabolic, degrading complex molecules to simpler forms.

Like all biological reactions, each reversible reaction of a metabolic pathway is speeded by an enzyme. The role of enzymes in biological reactions is described in the next section.

STUDY BREAK

What is the relation between ΔG and the concentrations of reactants and products at the equilibrium point of a reaction?

4.4 Role of Enzymes in Biological Reactions

Many reactions, although spontaneous, proceed so slowly that their rate is essentially zero at the temperatures characteristic of living organisms. Enzymes increase the rate of biological reactions to levels that sustain the activities of life. For reversible reactions, enzymes speed progress toward the equilibrium point. For most enzymes, the increase ranges from a minimum of about a million to as much as a trillion times faster than the same reaction would proceed without its enzyme.

Figure 4.7

The equilibrium point of a reaction. Conditions opposing completion of spontaneous reactions, such as relative concentrations of reactants and products, stop the progress of reactions at an equilibrium point. At this point, the number of reactant molecules being converted to products equals the number of product molecules being converted back to reactants. The reaction at the equilibrium point is reversible; it may be made to run to the right (forward) by adding more reactants or to the left (backward) by adding more products.

Relative concentration of reactant

Relative concentration of product

Equilibrium

The majority of enzymes have names ending in -*ase*. The rest of the name typically relates to the substrate of the enzyme or to the type of reaction with which the enzyme is associated. For example, enzymes that break down proteins are called *proteinases* or *proteases*.

Enzymes are not the only biological molecules capable of accelerating reaction rates. Some RNA molecules (see Section 4.6) also have this capacity. To distinguish between the two types of molecules, most biologists reserve the term *enzyme* for protein molecules that can accelerate reaction rates and call the RNA molecules with this capacity *ribozymes*. We follow this usage in this book.

Many inorganic substances, particularly metallic ions, function as catalysts. One common example is platinum, which is used in the catalytic converter of automobiles to speed the breakdown of smog-forming substances in the exhaust. Substances with the ability to accelerate spontaneous reactions without being changed by the reactions, including enzymes, ribozymes, and their inorganic counterparts, are called **catalysts**—that is, they *catalyze* reactions. The acceleration of a reaction by a catalyst is called *catalysis*.

Enzymes Accelerate Reactions by Reducing Activation Energy

Enzymes and other catalysts accelerate reactions by reducing the **activation energy** of a reaction—that is, the initial input of energy required to start a reaction. Even though a reaction is spontaneous, with a negative ΔG, it may not actually start unless a relatively small boost of energy is added **(Figure 4.8a)**. After it starts, the reaction becomes self-sustaining as it releases more than enough free energy to compensate for the original boost.

A rock resting in a depression at the top of a hill provides a physical example of activating energy **(Figure 4.8b)**. The rock will not roll downhill spontaneously, even though its position represents considerable potential energy and the total "reaction"—the downward movement of the rock—is spontaneous and releases free energy. (Trying to stop the rock halfway down the hill would give an idea of the free energy being released.) In this physical example, the activation energy is the effort required to raise the rock over the rim of the depression and start its downhill roll. For chemical reactions, the activation energy is the energy required to disturb the existing bonds in the reactants enough to begin the conversion of reactants to products.

What provides the activation energy in chemical reactions? The molecules that participate in chemical reactions are in constant motion at temperatures above absolute zero. Although the average amount of kinetic energy may be less than the amount required for activation, collisions between the moving molecules may raise some of them to the energy level required for the reaction to proceed.

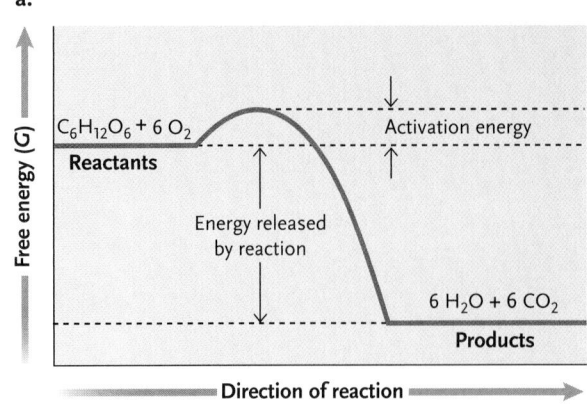

a.

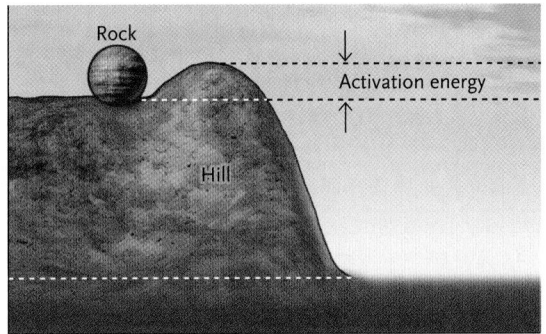

b.

Figure 4.8
Activation energy. **(a)** The activation energy for the oxidation of glucose is an energy barrier over which glucose molecules must be raised before they can react to form H_2O and CO_2. **(b)** In an analogous physical situation, a rock poised in a depression at the top of a hill will not roll downhill unless activating energy is added to raise it over the rim of the depression.

For nonbiological reactions, heat is often added to reacting systems to supply activation energy. The addition of heat increases both the speed of individual molecules and the rate of their collisions, making it more likely that molecules will gain enough energy to react. Heating biological reactions enough to make them self-sustaining would be an unsatisfactory condition for living organisms. Instead, enzymes decrease the activation energy **(Figure 4.9)**, greatly increasing the probability that molecules will gain enough energy to react at the rela-

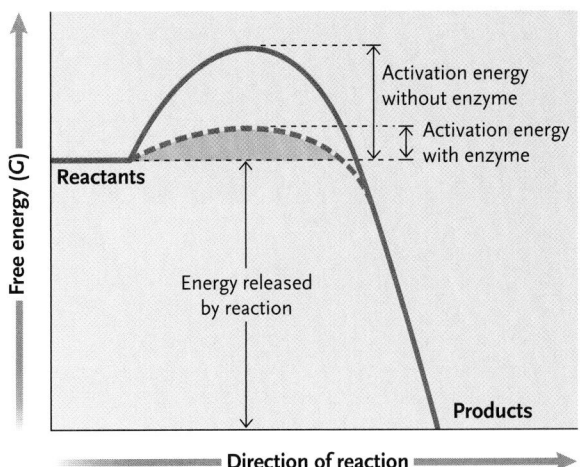

Figure 4.9
Effect of enzymes in reducing the activation energy. The reduction allows biological reactions to proceed rapidly at the relatively low temperatures that can be tolerated by living organisms.

tively low temperatures tolerated by living organisms. For the rock resting on the side of a hill, the role of an enzyme would be equivalent to reducing the depth of the depression so that the rock is finely balanced and only the slightest push is needed to start its journey downhill.

Enzymes Combine with Reactants and Are Released Unchanged

In catalysis, an enzyme combines briefly with reacting molecules and is released unchanged when the reaction is complete. For example, the enzyme in **Figure 4.10,** hexokinase, catalyzes the following reaction:

$$\text{glucose} + \text{ATP} \xrightarrow{\text{hexokinase}} \text{glucose-6-phosphate} + \text{ADP}$$

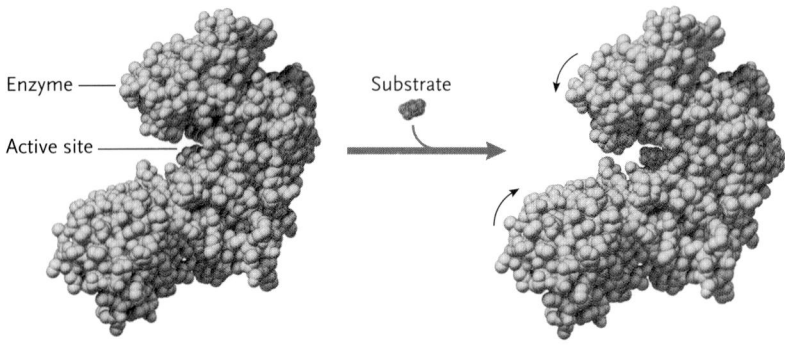

Figure 4.10
Space-filling models showing the combination of an enzyme, hexokinase (in blue), with its substrate, glucose (in yellow). Hexokinase catalyzes the phosphorylation of glucose to form glucose-6-phosphate. The phosphate group that enters the reaction is not shown. Note how the enzyme undergoes a conformational change, closing the active site more tightly as it binds the substrate.

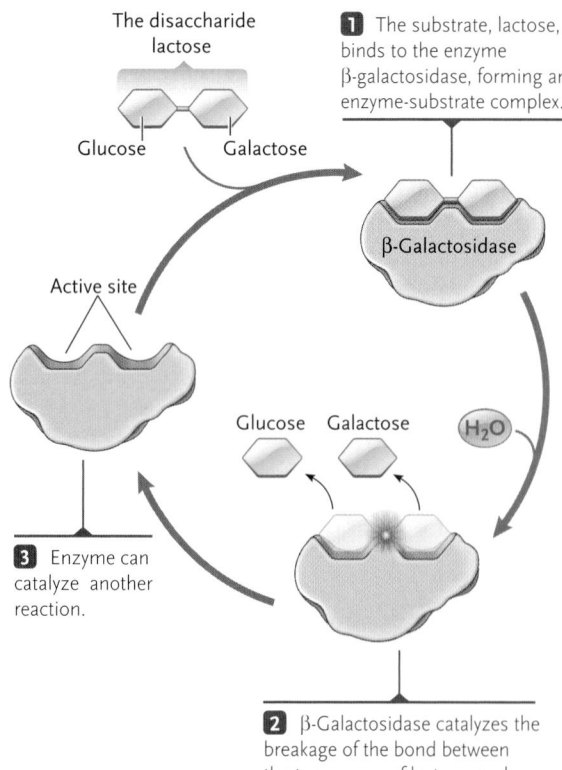

Figure 4.11
The catalytic cycle of enzymes. Shown is the enzyme β-galactosidase, which cleaves the sugar lactose to produce glucose and galactose.

The disaccharide lactose

Glucose   Galactose

1 The substrate, lactose, binds to the enzyme β-galactosidase, forming an enzyme-substrate complex.

β-Galactosidase

Active site

Glucose   Galactose H_2O

3 Enzyme can catalyze another reaction.

2 β-Galactosidase catalyzes the breakage of the bond between the two sugars of lactose, and the products are released.

Writing the enzyme name (for example, hexokinase) above the reaction arrow indicates that it is required but not involved as a reactant or a product.

Because enzymes are released unchanged after a reaction, enzyme molecules cycle repeatedly through reactions, combining with reactants and releasing products **(Figure 4.11).** Depending on the enzyme, the rate at which reactants are bound and catalyzed and at which products are released varies from 100 times to 10 million times per second. These astoundingly high rates of catalysis mean that a small number of enzyme molecules can catalyze large numbers of reactions.

Each type of enzyme catalyzes the reaction of only a single type of molecule or group of closely related molecules. This characteristic is known as **enzyme specificity.** The particular reacting molecule or molecular group that an enzyme catalyzes is known as the **substrate.** The region of an enzyme that recognizes and combines with a substrate molecule is the **active site.** In most enzymes, the active site is located in a cavity or pocket on the enzyme surface (as in the active site in Figure 4.10).

Cells have thousands of different enzymes. They vary from relatively small molecules, with single polypeptide chains containing as few as 100 amino acids, to large complexes that include many polypeptide chains totaling thousands of amino acids. Different enzymes are found in all areas of the cell, from the aqueous cell solution to the cell membranes. Other enzymes are released to catalyze reactions outside the cell. For example, enzymes that catalyze reactions breaking down food molecules are released from cells into the digestive cavity in all animals.

Many enzymes include a **cofactor,** an inorganic or organic nonprotein group that is necessary for catalysis to take place. Cofactors function in a variety of ways. Inorganic cofactors, which are all metallic ions, include iron, copper, magnesium, zinc, potassium, and manganese. Organic cofactors, also called **coenzymes,** are complex chemical groups of various kinds; in higher animals, many coenzymes are derived from vitamins.

Enzymes Reduce Activation Energy by Inducing the Transition State

The central question of enzyme activity is: How do enzymes reduce the activation energy to speed biological reactions? Evidence from many years of experiments indicates that enzymes reduce activation energy by altering the reacting molecules to a form known as the transition state of a reaction. The **transition state** is an intermediate arrangement of atoms and bonds that both the reactants and the products of a reaction can assume. It is an activated state that is highly unstable and can move forward toward products or backward toward reactants with relatively little change in energy. The *Focus on Research* outlines an experiment showing that a transition state is involved in enzymatic catalysis.

Basic Research: Testing the Transition State

One of the most inventive researchers of the twentieth century, American biochemist Linus Pauling, first proposed the idea that pushing molecules toward the transition state might be the mechanism that underlies enzymatic catalysis. Another biochemist, W. P. Jencks of Brandeis University, proposed a way of using antibodies to test Pauling's hypothesis. Animals produce proteins called antibodies when they are exposed to a foreign substance called an *antigen;* as part of their structure, antibodies contain a binding site that exactly fits the antigen. Combining an antibody with its antigen leads to the destruction or removal of the antigen from the body (as described in Chapter 43). Jencks reasoned that if antibodies

could be made with a binding site that, like an enzyme, can fit the transition state of a reaction, they might act as enzymes and speed the rate of the reaction. If this occurred, the experiment would provide strong support for the idea that the transition state is part of the mechanism of enzyme action. In 1986, two groups working independently, one led by R. A. Lerner and the other by P. G. Schultz, successfully performed Jencks' proposed experimental test.

Lerner's group, at Scripps Research Institute, studied a common biological interaction called an acyl transfer reaction, in which an acyl group (shown in red in the figure) is transferred from one organic side group to another:

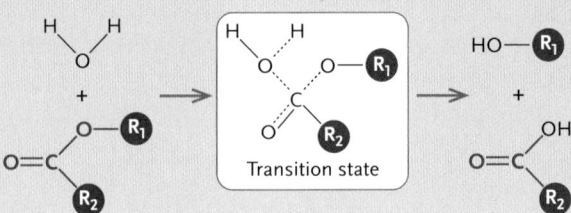

Transition state

(R_1 and R_2 designate the organic side groups.) The transition state for this reaction is mimicked by a group of stable, unrelated molecules called phosphonate esters:

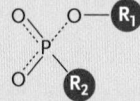

By injecting phosphonate esters into test animals, Lerner's group induced formation of antibodies tailored to fit the transition state for the acyl transfer reaction. These antibodies, as predicted by the Pauling–Jencks hypotheses, acted as enzymes speeding the rate of acyl transfer reactions.

The results directly support the proposal that achievement of the transition state is an important part of enzymatic catalysis. The technique also opened an entire new field of chemistry, the manufacture of "designer enzymes"—artificial enzymes made by developing antibodies that bind the transition state for a reaction desired in research, medicine, or industry.

Because enzymes bind the transition state, they can bind either the reactants or products of a reaction and can catalyze reversible reactions in either direction **(Figure 4.12).** However, the binding does not alter the equilibrium point of a reversible reaction; the enzyme simply increases the rate at which reversible reactions reach equilibrium.

Research has shown that at least three mechanisms contribute to the formation of the transition state. One of the key mechanisms enzymes use to induce the transition state is *bringing the reacting molecules into close proximity.* Reacting molecules can assume the transition state only when they collide; binding to an enzyme's active site brings the reactants so close together that a collision is almost certain to occur.

A second mechanism enzymes use is *orienting the reactants in positions that favor the transition state.* Binding at the active site positions the substrate molecules so that they are much more likely to collide at exactly the correct sites and angles required for achievement of the transition state.

A third mechanism is *exposing the reactant molecules to altered environments that promote their interaction.* For example, in some reactions, the active site of

the enzyme may contain ionic groups with positive or negative charges that help distort reactants toward the transition state.

Many conditions and factors alter the rates at which enzymes catalyze their reactions, and enzymes rarely work at their maximum possible rates inside cells. Instead, their rates are regulated and adjusted to match the requirements of a cell for the products of the reactions they catalyze. The next section describes conditions and factors that alter enzyme activity and outlines some of the most important regulatory mechanisms that key enzymatic catalysis to cellular requirements.

Figure 4.12
Fit of the active site to reactants, products, and the transition state. The strongest binding is to the transition state.

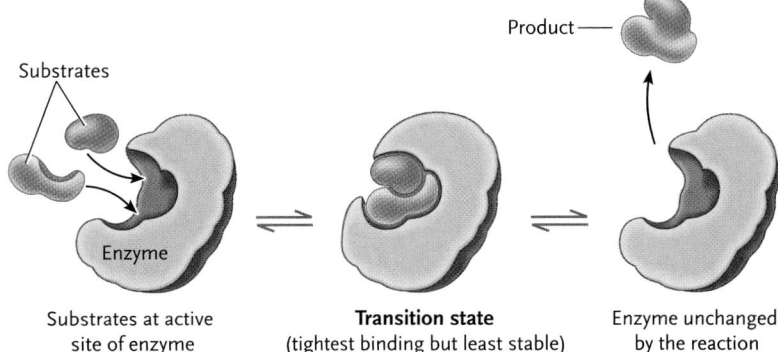

Substrates at active site of enzyme | **Transition state** (tightest binding but least stable) | Enzyme unchanged by the reaction

Explain how enzymes accelerate reactions.

4.5 Conditions and Factors That Affect Enzyme Activity

Several conditions can alter enzyme activity, including changes in temperature and pH and changes in the concentration of substrate and other molecules that can bind to enzymes. The activity of enzymes is also regulated by control mechanisms that modify enzyme activity, thereby adjusting reaction rates to meet a cell's requirements for chemical products.

Most Enzymes Reach Maximum Activity within a Narrow Range of Temperature and pH

The activity of most enzymes is strongly altered by changes in pH and temperature. Characteristically, enzymes reach maximal activity within a narrow range of temperature or pH; at levels outside this range, enzyme activity drops off. These effects produce a typically peaked curve when enzyme activity is plotted, with the peak where temperature or pH produces maximal activity.

Effects of Temperature Changes. The effects of temperature changes on enzyme activity reflect two distinct processes. First, temperature has a general effect on chemical reactions of all kinds. As the temperature rises, the rate of chemical reactions typically increases. This effect reflects increases in the kinetic motion of all molecules, with more frequent and stronger collisions as the tem-

perature rises. Second, temperature has an effect on all proteins, including enzymes. As the temperature rises, the kinetic motions of the amino acid chains of an enzyme increase, along with the strength and frequency of collisions between enzymes and surrounding molecules. At some point, these disturbances become strong enough to denature the enzyme: the hydrogen bonds and other forces that maintain its three-dimensional structure break, making the enzyme unfold and lose its function.

The two effects of temperature act in opposition to each other to produce characteristic changes in the rate of enzymatic catalysis **(Figure 4.13).** In the range of 0° to about 40°C, the reaction rate doubles for every 10°C increase in temperature. Above 40°C, the increasing kinetic motion begins to unfold the enzyme, reducing the rate of increase in enzyme activity. At some point, as temperature continues to rise, the unfolding causes the reaction rate to level off at a peak. Further increases cause such extensive unfolding that the reaction rate decreases rapidly to zero. For most enzymes, the peak in activity lies between 40° and 50°C; the drop-off becomes steep at 55°C and falls to zero at about 60°C. Thus, the rate of an enzyme-catalyzed reaction peaks at a temperature at which kinetic motion is greatest but no significant unfolding of the enzyme has occurred.

Although most enzymes have a temperature optimum between 40° and 50°C, some have activity peaks below or above this range. For example, the enzymes of maize (corn) pollen function best near 30°C and undergo steep reductions in activity above 32°C. As a result, environmental temperatures above 32°C can seriously inhibit the growth of corn crops. Many animals living in frigid regions have enzymes with much lower temperature optima than average. For example, the enzymes of arctic snow fleas are most active at −10°C. At the other extreme are the enzymes of archaeans that live in hot springs, which are so resistant to denaturation that they remain active at temperatures of 85°C or more.

Effects of pH Changes. Typically, each enzyme has an optimal pH where it operates at peak efficiency in speeding the rate of its biochemical reaction **(Figure 4.14).** On either side of this pH optimum, the rate of the catalyzed reaction decreases because of the resulting alterations in charged groups. The effects on the structure and function of the active site become more extreme at pH values farther from the optimum, until the rate drops to zero.

Most enzymes have a pH optimum near the pH of the cellular contents, about pH 7. Enzymes that are secreted from cells may have pH optima farther from neutrality. An example is pepsin, a protein-digesting enzyme secreted into the stomach. This enzyme's pH optimum is 1.5, close to the acidity of stomach contents. Similarly, trypsin, also a protein-digesting enzyme, has a pH optimum at about pH 8, allowing it to function well in the somewhat alkaline contents of the intestine where it is secreted.

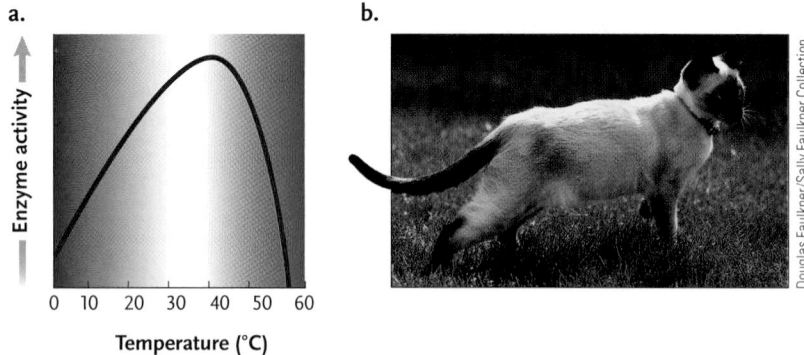

a.

Enzyme activity

Temperature (°C)
0 10 20 30 40 50 60

b.

Douglas Faulkner/Sally Faulkner Collection

Figure 4.13

Effect of temperature on enzyme activity. **(a)** As the temperature rises, the rate of the catalyzed reaction increases proportionally until the temperature reaches the point at which the enzyme begins to denature. The rate drops off steeply as denaturation progresses and becomes complete. **(b)** Visible effects of environmental temperature on enzyme activity in Siamese cats. The fur on the extremities—ears, nose, paws, and tail—contains more dark brown pigment (melanin) than the rest of the body. A heat-sensitive enzyme controlling melanin production is denatured in warmer body regions, so dark pigment is not produced and fur color is lighter.

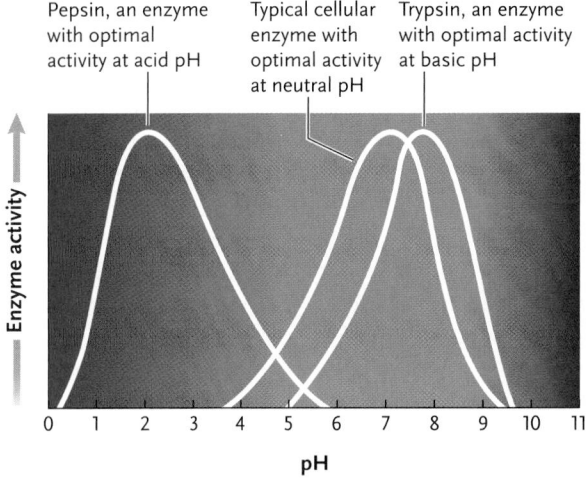

Pepsin, an enzyme with optimal activity at acid pH

Typical cellular enzyme with optimal activity at neutral pH

Trypsin, an enzyme with optimal activity at basic pH

Figure 4.14
Effects of pH on enzyme activity. An enzyme typically has an optimal pH at which it is most active; at pH values above or below the optimum, the rate of enzyme activity drops off. At extreme pH values, the rate drops to zero.

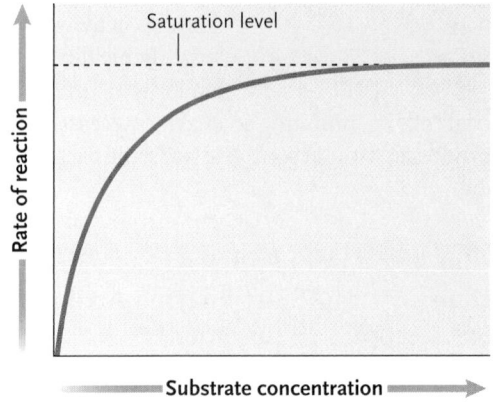

Figure 4.15
Effect of increasing substrate concentration on the rate of an enzyme-catalyzed reaction. At saturation (horizontal dashed line), further increases in substrate concentration do not increase the rate of the reaction.

Enzyme-Catalyzed Reactions Reach a Saturation Level beyond Which Increasing Substrate Concentration Does Not Increase the Reaction Rate

When substrate concentration is altered experimentally from low to high and the temperature and concentration of enzyme molecules are held constant, the rate of enzyme catalysis eventually levels off **(Figure 4.15).** At very low concentrations, substrate molecules collide so infrequently with enzyme molecules that the reaction proceeds slowly. As the substrate concentration increases, the reaction rate initially increases as enzyme and substrate molecules collide more frequently. But, as the enzyme molecules approach the maximum rate at which they can combine with reactants and release products, increasing substrate concentration has a smaller and smaller effect and the rate of reaction eventually levels off. When the enzymes are cycling as rapidly as possible, further increases in substrate concentration have no effect on the reaction rate. At this point, the enzymes are said to be **saturated,** and the reaction rate remains constant at the saturation level (see the horizontal dashed line in Figure 4.15).

By contrast, uncatalyzed reactions do not reach a saturation level. Therefore, if researchers perform the type of experiment presented in Figure 4.15 and observe that saturation occurs, they will conclude that an enzyme catalyzes the reaction.

Enzyme Inhibitors Have Characteristic Effects on Enzyme Activity

The rate at which enzymes catalyze reactions is reduced by *enzyme inhibitors,* substances that reduce enzyme activity by combining with enzyme molecules. Some inhibitors work by combining with the active site of an enzyme; others combine with critical sites located elsewhere in the structure of an enzyme **(Figure 4.16).**

Figure 4.16
Actions of competitive and noncompetitive inhibitors of enzyme activity.

a. Normal substrate binding to enzyme active site

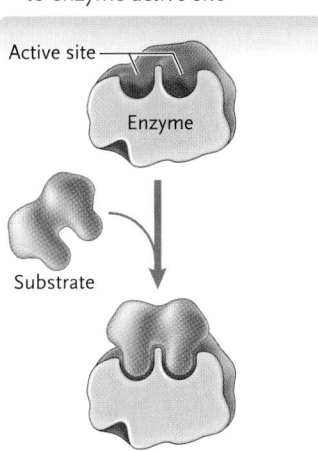

b. Competitive inhibition

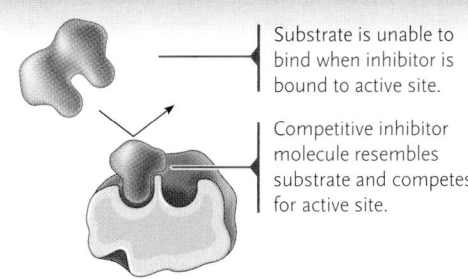

Substrate is unable to bind when inhibitor is bound to active site.

Competitive inhibitor molecule resembles substrate and competes for active site.

c. Noncompetitive inhibition

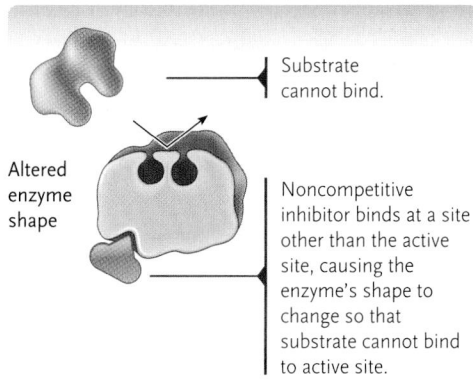

Substrate cannot bind.

Noncompetitive inhibitor binds at a site other than the active site, causing the enzyme's shape to change so that substrate cannot bind to active site.

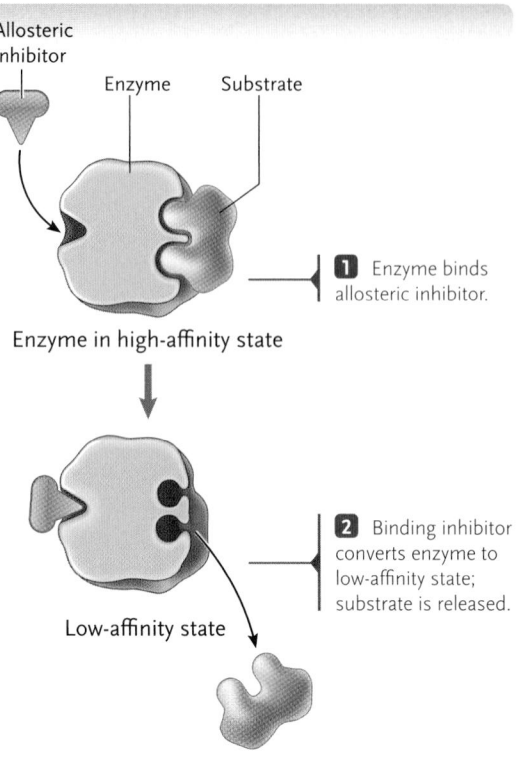

Allosteric activation

Allosteric activator

Allosteric site

Active site

Substrate

Enzyme in low-affinity state

1 Enzyme binds allosteric activator.

High-affinity state

2 Binding activator converts enzyme to high-affinity state.

High-affinity state

3 In high-affinity state, enzyme binds substrate.

Allosteric inhibition

Allosteric inhibitor

Enzyme

Substrate

Enzyme in high-affinity state

1 Enzyme binds allosteric inhibitor.

Low-affinity state

2 Binding inhibitor converts enzyme to low-affinity state; substrate is released.

Figure 4.17
Allosteric regulation.

Inhibitors that combine with the active site have molecular structures that resemble the normal substrate closely enough to fit into and occupy the site, thereby blocking access for the normal substrate and slowing the reaction rate. If the concentration of the inhibitor is high enough, the reaction may stop completely. Inhibition of this type is called **competitive inhibition** because the inhibitor *competes* with the normal substrate for binding to the active site. Competitive inhibitors are useful in enzyme research because their structure helps identify the region of a normal substrate that binds to an enzyme.

Inhibitors that combine with enzymes at locations other than the active site often alter the conformation of the enzyme. The alterations reduce the ability of the active site to bind the normal substrate and thus induce the transition state in the substrate. Because such inhibitors do not compete directly with the substrate for binding to the active site, their pattern of inhibition is called **noncompetitive inhibition.**

Both competitive inhibitors and noncompetitive inhibitors bind to enzymes with varying strength, depending on type. Some bind so strongly that their linkage is essentially permanent and the enzyme becomes completely disabled. Others bind more loosely, so their attachment is reversible.

Some foreign molecules, such as poisons and toxins, act as inhibitors of enzyme activity. Such molecules often bind so strongly to enzymes that their inhibitory effects are essentially irreversible. For example, cyanide is a potent poison because it binds strongly to and inhibits cytochrome oxidase, the enzyme that catalyzes the use of oxygen in cellular metabolism. Humans and other animals die quickly if exposed to cyanide because of the almost instant and complete inhibition of cytochrome oxidase by the poison. Many antibiotics are toxins that inhibit enzyme activity in bacteria. Penicillin, a toxin made by a fungus, kills bacterial cells by inhibiting an enzyme necessary for cell wall synthesis; without complete walls, the bacteria burst and die.

Cellular Regulatory Pathways Use Several Mechanisms to Adjust Enzyme Activity to Meet Metabolic Requirements

Cells adjust the activity of many enzymes upward or downward to meet their needs for reaction products. Several mechanisms are used in this regulation, including competitive and noncompetitive inhibition, a form of noncompetitive control called *allosteric regulation,* and covalent modification of enzyme structure by the addition or removal of chemical groups.

Regulation by Inhibitors. Many cellular enzymes are regulated by natural inhibitors, including inhibitors that work either competitively or noncompetitively. Typically, the combination between these inhibitors and the enzyme is fully reversible. If the concentration of the inhibitor increases, it combines with the enzymes in greater numbers, thereby interfering with enzyme activity and decreasing the rate of the reaction. If the concentration of the inhibitor decreases, its combination with enzymes decreases proportionately and the rate of the reaction increases. Control by the inhibitors changes enzyme activity precisely to meet the needs of the cell for the products of the reaction catalyzed by the enzyme.

For example, the specialized control mechanism called **allosteric regulation** (*allo* = different; *stereo* =

shape) occurs by the reversible combination of a regulatory molecule with the **allosteric site**, a location on the enzyme outside the active site. The mechanism, first discovered in 1965 by French biologist Jacques Monod and his colleagues J. P. Changeux and J. Wyman at the Pasteur Institut, Paris, France, may either slow or accelerate enzyme activity.

Enzymes controlled by allosteric regulation typically have two alternate conformations controlled from the allosteric site. In one conformation, called the *high-affinity state* (the active form), the enzyme binds strongly to its substrate; in the other conformation, the *low-affinity state* (the inactive form), the enzyme binds the substrate weakly or not at all. Binding with regulatory substances may induce either state: binding an **allosteric inhibitor** converts an allosteric enzyme from the high- to low-affinity state, and binding an **allosteric activator** converts it from the low- to high-affinity state **(Figure 4.17)**. Because allosteric inhibitors work by binding to sites separate from the active site, their action is noncompetitive.

Frequently, allosteric inhibitors are a product of the metabolic pathway that they regulate. If the product accumulates in excess, its effect as an inhibitor automatically slows or stops the enzymatic reaction producing it, typically by inhibiting the enzyme that catalyzes the first reaction of the pathway. If the product becomes too scarce, the inhibition is reduced and its production increases. Regulation of this type, in which the product of a reaction acts as a regulator of the reaction, is termed **feedback inhibition** (also called **end-product inhibition**). Feedback inhibition prevents cellular resources from being wasted in the synthesis of molecules made at intermediate steps of the pathway.

For instance, a biochemical pathway that makes the amino acid isoleucine from threonine proceeds in five steps, each catalyzed by an enzyme **(Figure 4.18)**. The end product of the pathway, isoleucine, is an allosteric inhibitor of the first enzyme of the pathway, threonine deaminase. If the cell makes more isoleucine than it needs, isoleucine combines reversibly with threonine deaminase at the allosteric site, converting the enzyme to the low-affinity state and inhibiting its ability to combine with threonine, the substrate for the first reaction in the pathway. If isoleucine levels drop too low, the allosteric site of threonine deaminase is vacated, the enzyme is converted to the high-affinity state, and isoleucine production increases.

Regulation by Chemical Modification. Many key enzymes are regulated by chemical linkage to other substances, typically ions, functional groups such as phosphate or methyl groups, or units derived from nucleotides. The regulatory substances induce folding changes in the enzyme that adjust its activity upward or downward.

For example, chemical modification by the addition or removal of phosphate groups is a highly significant mechanism of cellular regulation that is used by

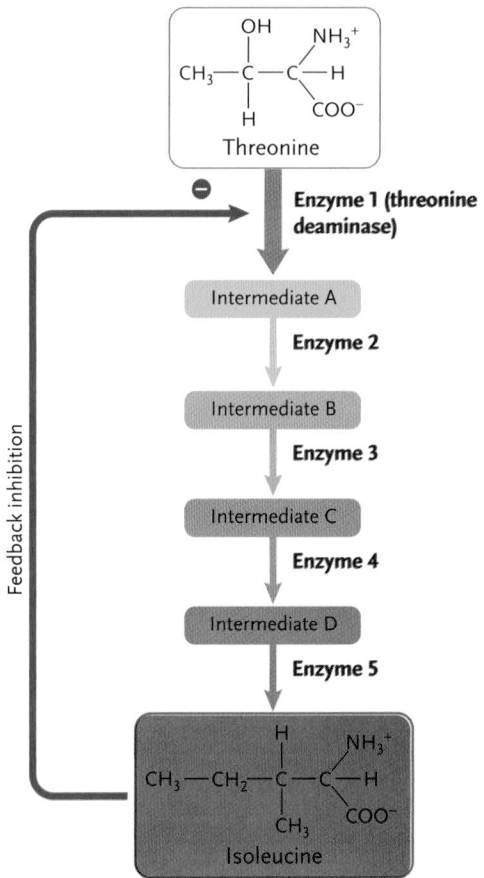

Figure 4.18

Feedback inhibition in the pathway that produces isoleucine from threonine. If the product of the pathway, isoleucine, accumulates in excess, it slows or stops the pathway by acting as an allosteric inhibitor of the enzyme that catalyzes the first step in the pathway.

all organisms from bacteria to humans. Typically, regulatory phosphate groups derived from ATP or other nucleotides are added to the regulated enzymes by other enzymes known as *protein kinases*. The addition of a phosphate group (phosphorylation) either increases or decreases enzyme activity or activates or deactivates the enzyme, depending on the particular enzyme and where the phosphate group is added to the enzyme.

Regulatory phosphate groups are removed (a process called dephosphorylation), reversing the effects of the protein kinases, by a different group of enzymes called *protein phosphatases*. The balance between phosphorylation and dephosphorylation of the enzymes modified by the kinases and protein phosphatases closely regulates cellular activity, often as a part of the response to external signal molecules (see Chapter 7).

Enzymes have been actively investigated since their discovery in the late 1800s. The word *enzyme* means "in yeast," in reference to the discovery of these protein-based catalysts in extracts of yeast cells. A hundred years of intensive research after the discovery of enzymatic proteins gave no hint that other molecules could act as biological catalysts. Thus, it came as a big surprise when RNA-based catalysts were discovered.

Ribozymes Take the First Step in Protein Synthesis

Harry Noller's experiment showed that if proteins were removed from ribosomes, the remaining RNA molecules could still catalyze the central reaction of protein synthesis, linkage of amino acids into chains via peptide bonds. However, his work did not eliminate the possibility that undetectable small amounts of ribosomal proteins in the preparations might be catalyzing peptide bond formation.

Billiang Zhang and Thomas R. Cech performed a definitive experiment that eliminated the possibility of protein contamination. They synthesized RNA molecules artificially, in solutions that had never been exposed to ribosomal proteins, and then tested the ability of the artificial RNA to catalyze formation of peptide bonds.

As a first step in their experiments, Zhang and Cech synthesized a large pool of artificial RNA molecules. Part of the nucleotide sequence was the same in every molecule and part differed randomly from molecule to molecule, but all were the same length. The investigators then linked an amino acid, phenylalanine, to one end of each RNA molecule by a disulfide (—S—S—) bond.

To that pool, they next added the amino acid methionine linked to the nucleotide AMP. In the cell, single amino acids linked to AMP are used in the pathway that makes proteins. The methionine–AMP combination was "tagged" by combining it with *biotin*, a small organic molecule, so that it could be identified in the reaction solution.

All the ingredients were mixed together and allowed to react. If any of the RNA molecules could act as ribozymes, catalyzing formation of a peptide bond, some of the tagged methionine should become linked to the phenylalanine at the end of the ribozyme. To find out if this had happened, the investigators poured the reaction mixture through a column packed with plastic beads that could bind to the biotin tag. Binding between the beads and the biotin tag trapped any RNA molecules that were able to catalyze linkage of the two amino acids, whereas unreactive RNA molecules flowed out of the bottom of the column. The RNA molecules with the biotin tag were then washed from the column and separated from the linked amino acids by adding a reagent that breaks disulfide bonds. Chemical tests showed that peptide bonds formed between the amino acids, the same type of linkage that joins amino acids in natural proteins.

The RNA molecules that had functioned successfully as ribozymes were then separated from their dipeptide product for further study and refinement. Eventually, the researchers obtained ribozymes that catalyzed peptide bond formation at rates 100,000 times faster than the same reaction without a catalyst.

Zhang and Cech's experiments confirmed a feature of ribozyme activity that is critical to the role proposed for these RNA-based catalysts in the primitive RNA world—their ability to catalyze formation of the fundamental linkage tying amino acids together in proteins. Thus, during the evolution of life, proteins could have been made first in quantity by RNA, with no requirement for either DNA or enzymatic proteins.

STUDY BREAK

1. Explain why the activity of an enzyme will eventually decrease to zero as the temperature rises.
2. Why do enzyme-catalyzed reactions reach a saturation level when substrate concentration is increased?
3. Distinguish between competitive and noncompetitive inhibition.

4.6 RNA-Based Biological Catalysts: Ribozymes

Ribozymes Catalyze Certain Biological Reactions

In 1981, biochemist Thomas R. Cech of the University of Colorado, Boulder, discovered a group of RNA molecules that appeared to be capable of accelerating the rate of certain biological reactions without being changed by the reactions. Further work demonstrated that these RNA-based catalysts, now called **ribozymes**, are part of the biochemical machinery of all cells. Cech and another scientist, Yale University biochemist Sidney Altman, received the Nobel Prize in 1989 for their research establishing that ribozymes are essential cellular catalysts.

Most of the known ribozymes speed the cutting and splicing reactions that remove surplus segments from RNA molecules as part of their conversion into finished form. Some have other functions, however. For example, Harry F. Noller and his coworkers at the University of California at Santa Cruz found that ribosomes, the cell structures that assemble amino acids into proteins, can still link amino acids together even if their proteins are removed. Only RNA molecules are left in the ribosomes after the proteins are extracted, suggesting that ribozymes might catalyze this central reaction of protein synthesis. After Noller's discovery, Cech and his colleague, Billiang Zhang, confirmed that ribozymes can actually catalyze this reaction (see the *Insights from the Molecular Revolution* for an outline of Cech and Zhang's experiment).

Many biological processes rely on enzymes to catalyze key reactions. A complete understanding of those processes requires knowledge about the structure and function of the enzymes involved. Much research continues to be done to elucidate enzyme structure and function.

How does protein structure relate to enzyme function?

Many researchers are studying protein structure and its relation to protein function. For example, Janet Smith at the University of Michigan uses X-ray crystallography to determine the structures of proteins. The patterns of diffraction of X-rays shone at a protein crystal give information about how the protein's atoms are organized. The crystal structure is "solved" once a model for the protein's structure is achieved in this way.

Smith's group uses information about the structure of solved proteins to predict the functions of other proteins. Even though it is possible to solve protein structures rapidly, it is not practical to solve the structures of all proteins involved in important biological processes. Instead, Smith, as well as other researchers, draws on the current understanding of the evolution of proteins. In particular, genes for useful proteins often have been duplicated during evolution and the duplicate copy adapted to a new function. Therefore, proteins can be related in an evolutionary sense. An understanding of the molecular mechanisms of particular enzymes may then be transferable to other proteins, which is an underlying theme of Smith's research.

How does ribozyme structure relate to function, and how might ribozymes be used as therapeutic agents?

Ribozymes are catalytic RNA molecules. Various types of ribozymes exist, each type differing in its three-dimensional structure and mechanism of catalysis.

Researcher John Burke at the University of Vermont and his group are studying hairpin ribozymes and hammerhead ribozymes, which are catalytically active once they fold into those two shapes (the hammerhead shape is similar to that of the head of a hammerhead shark). Their research has four directions: determining the molecular structure of ribozymes, characterizing RNA conformational changes during catalysis, elucidating the mechanisms of catalysis, and exploring ways to use ribozymes as therapeutic agents.

For example, Burke's group has shown that the hairpin ribozyme undergoes a dramatic conformational change when the substrate binds to the active site. Furthermore, they have engineered hairpin ribozymes that can inhibit viral replication in mammalian cells. The particular viruses targeted have RNA genomes and include HIV-1 (the causative agent of AIDS) and hepatitis B virus. To achieve their goal, they had to identify appropriate target sites within the viral RNA molecules and to express the engineered ribozymes efficiently within the cell. Current research focuses on optimizing the inhibition of viral replication by the ribozymes, determining the mechanism of antiviral activity, and extending this technology to develop therapeutic approaches for significant infectious diseases such as AIDS and hepatitis B.

Peter J. Russell

Ribozymes provide a possible solution to a longstanding "chicken-or-egg" paradox about the evolution of life: Did proteins or nucleic acids come first in evolution? It is difficult to understand how DNA could exist without the enzymatic proteins required for its duplication. At the same time, it is difficult to understand how enzymes could exist without nucleic acids, which contain the information required to make them. Ribozymes offer a way around this dilemma because they could have acted as *both* enzymes and informational molecules when cellular life first appeared. The earliest forms of life therefore might have inhabited an "RNA world" in which neither DNA nor proteins played critical roles (see discussion in Chapter 24). If so, ribozymes—the most recently discov-ered biological catalysts—may have existed for the longest time!

This chapter concludes our survey of the chemical underpinnings of biology. In the next chapter, we survey the structure of cells, the fundamental units into which biological molecules are organized and where molecules interact to produce the characteristics of life.

STUDY BREAK

What is a ribozyme, and how does it fit the definition of an enzyme?

Review

Go to **ThomsonNOW** at www.thomsonedu.com/login to access quizzing, animations, exercises, articles, and personalized homework help.

4.1 Energy, Life, and the Laws of Thermodynamics

- Energy, the capacity to do work, exists in kinetic and potential states. Kinetic energy is the energy of motion; potential energy is energy represented in the nonmoving location of matter or the specific arrangement of atoms. Energy may be readily converted between potential and kinetic states.

- Metabolism is the biochemical modification and use of energy in the synthesis and breakdown of organic molecules. Catabolic reactions release the potential energy of complex molecules to do cellular work. Anabolic reactions convert simple substances into more complex forms.

- Thermodynamics is the study of energy flow between a system and its surroundings during chemical and physical reactions. A system that exchanges energy but not matter with its surroundings is a closed system. A system that exchanges both energy and matter with its surroundings is an open system (Figure 4.2).

- The first law of thermodynamics states that the total amount of energy in a system and its surroundings remains constant. The second law states that in any process involving a spontaneous (possible) change from an initial to a final state, the total entropy (disorder) of the system and its surroundings always increases.

- Energy released by reactions that move spontaneously to the final state is free energy, that is, energy available to do work. The free energy equation, $\Delta G = \Delta H - T\Delta S$, states that the free energy change, ΔG, is the sum of the changes in energy content and entropy of the system as a reaction goes to completion.

- Reactions with a negative ΔG are spontaneous; they release free energy and are known as exergonic reactions. Reactions with a positive ΔG require free energy and are known as endergonic reactions (Figure 4.4).

4.2 How Living Organisms Couple Reactions to Make Synthesis Spontaneous

- Cells carry out endergonic reactions by using ATP to couple them to exergonic reactions, producing an overall reaction that proceeds spontaneously. In the coupled reactions, ATP is hydrolyzed to ADP and P_i, and one of these molecules is temporarily linked to reactants or the enzyme (Figure 4.5).

- The ATP used in coupling reactions is replenished by reactions that link ATP synthesis to catabolic reactions. ATP thus cycles between reactions that release free energy and reactions that require free energy (Figure 4.6).

Animation: Structure of ATP

Animation: Active transport

4.3 Thermodynamics and Reversible Reactions

- Factors that oppose the completion of spontaneous reactions, such as the relative concentrations of reactants and products, produce an equilibrium point at which reactants are converted to products, and products are converted back to reactants, at equal rates. Small changes in reaction conditions can easily reverse the overall progress of the reaction (Figure 4.7).

Animation: Chemical equilibrium

4.4 Role of Enzymes in Biological Reactions

- Enzymes are catalysts; they greatly speed the rate at which spontaneous reactions occur, and for reversible reactions, they increase the rate at which a reaction reaches equilibrium.

- Enzymes usually are specific: they catalyze reactions of only a single type of molecule or a group of closely related molecules.

- The active site of an enzyme combines briefly with the reactants (the substrates); the enzyme is released unchanged when the reaction is complete.

- Many enzymes include a cofactor, which is an inorganic ion or an organic nonprotein group called a coenzyme that is necessary for catalysis to occur.

- Enzymes work by decreasing the activation energy required for a chemical reaction to proceed. They reduce the activation energy by inducing the transition state of the reaction, from which the reaction can move easily in the direction of either products or reactants (Figures 4.8–4.12).

- Several mechanisms contribute to enzymatic catalysis by helping to induce the transition state. They include bringing the reactant molecules into close proximity, orienting the reactants in positions that favor the transition state, and exposing the reactants to altered environments that promote their interaction.

Animation: Activation energy

Animation: How catalase works

Animation: Enzymes and their role in lowering activation energy

4.5 Conditions and Factors That Affect Enzyme Activity

- Typically, enzymes have optimal activity at a certain temperature and a certain pH; at temperature and pH values above and below the optimum, reaction rates fall off (Figures 4.13 and 4.14).

- At high substrate concentrations, enzymes become saturated with reactants, and further increases in substrate concentration do not increase the rate of the reaction (Figure 4.15).

- Enzymes may be inhibited by nonsubstrate molecules. Competitive inhibitors interfere with reaction rates by combining with the active site of an enzyme; noncompetitive inhibitors combine with sites elsewhere on the enzyme (Figure 4.16).

- Many cellular enzymes are regulated by inhibitors. A special type of regulation, allosteric regulation, resembles noncompetitive inhibition, except that regulatory molecules may either increase or decrease enzyme activity. Allosteric regulation often carries out feedback inhibition, in which a product of an enzyme-catalyzed pathway acts as an allosteric inhibitor of the first enzyme in the pathway (Figures 4.17 and 4.18).

- Enzymes also are regulated by chemical modification, in many cases by reversible addition or removal of phosphate groups.

Animation: Allosteric activation

Animation: Allosteric inhibition

Interaction: Feedback inhibition

Interaction: Enzymes and temperature

4.6 RNA-Based Biological Catalysts: Ribozymes

- RNA-based catalysts called ribozymes speed some types of biological reactions; these include cutting and splicing reactions in which surplus segments are removed from RNA molecules and linking reactions that combine amino acids into polypeptide chains.

Questions

Self-Test Questions

1. The capacity to do work best defines:
   a. a metabolic pathway.
   b. entropy.
   c. kinetic or potential energy.
   d. a catabolic reaction.
   e. thermodynamics.

2. When two glucose molecules combine:
   a. the reaction represents a negative ΔG.
   b. free energy had to be available to allow the reaction to proceed.
   c. the reaction is exothermic.
   d. it supports the second law of thermodynamics, which states there is tendency of the universe toward disorder.
   e. the resulting product has less potential energy than the reactants.

3. When glucose is converted to glucose-6-phosphate:
   a. the synthesis of glucose-6-phosphate is exergonic.
   b. ADP is at a higher energy level than ATP.
   c. glucose-6-phosphate is at a higher energy level than glucose.
   d. because ATP donates a phosphate to glucose, this is not a coupled reaction.
   e. this is a spontaneous reaction.

4. In the following graph:

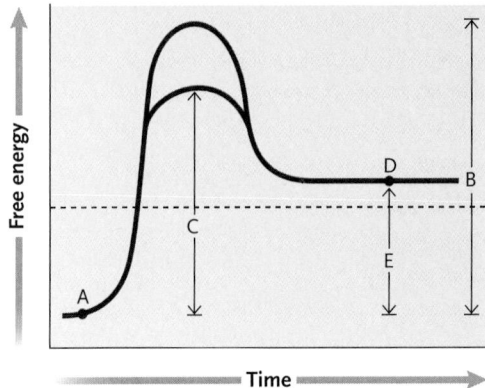

   a. A represents the product.
   b. B represents the energy of activation when enzymes are present.
   c. C is the free energy difference between A and D.
   d. C is the energy of activation without enzymes.
   e. E is the difference in free energy between the reactant and the products.

5. Subtilisin, a component in many laundry detergents, removes chocolate (which contains protein) from clothes in hot water. If used regularly on your silk shirt, the silk shirt might emulsify. From this information, which of the following is *not* a reasonable deduction?
   a. Subtilisin must be heat-stable.
   b. Subtilisin is an enzyme that must have a broad range of activity.
   c. Chocolate is composed of an α helix, and silk has a β-sheet structure; thus, subtilisin probably attacks a certain amino acid group linkage rather than a specifically shaped molecule.
   d. Subtilisin is not a protein.
   e. Be careful eating chocolate if wearing a silk shirt.

6. Which of the following methods is *not* used by enzymes to increase the rate of reactions?
   a. covalent bonding with the substrate at their active site
   b. bringing reacting molecules into close proximity
   c. orienting reactants into positions to favor transition states
   d. changing charges on reactants to hasten their reactivity
   e. increasing fit of enzyme and substrate that reduces the energy of activation

7. In an enzymatic reaction:
   a. the enzyme leaves the reaction chemically unchanged.
   b. if the enzyme molecules approach maximal rate, and the substrate is continually increased, the rate of the reaction does not reach saturation.
   c. in the stomach, enzymes would have an optimal activity at a neutral pH.
   d. increasing temperature above the optimal value slows the reaction rate.
   e. the least important level of organization for an enzyme is its tertiary structure.

8. Which of the following statements about the allosteric site is true?
   a. The allosteric site is a second active site on a substrate in a metabolic pathway.
   b. The allosteric site on an enzyme can allow the product of a metabolic pathway to inhibit that enzyme and stop the pathway.
   c. When the allosteric site of an enzyme is occupied, the reaction is irreversible and the enzyme cannot react again.
   d. An allosteric activator prevents binding at the active site.
   e. An enzyme that possesses allosteric sites does not possess an active site.

9. Which of the following statements about inhibition is true?
   a. Allosteric inhibitors and allosteric activators are competitive for a given enzyme.
   b. If an inhibitor binds the active site, it is considered noncompetitive.
   c. If an inhibitor binds to a site other than the active site, this is competitive inhibition.
   d. A noncompetitive inhibitor is believed to change the shape of the enzyme, making its active site inoperable.
   e. Competitive inhibition is usually not reversible.

10. Which of the following statements is *incorrect*?
    a. Ribozymes can link amino acids to form protein.
    b. Ribozymes can act as enzymes.
    c. Ribozymes can act as informational molecules.
    d. Ribozymes are suggested as the first molecules of life.
    e. Ribozymes are proteins.

Questions for Discussion

1. Trees become more complex as they develop spontaneously from seeds to adults. Does this process violate the second law of thermodynamics? Why or why not?

2. Trace the flow of energy through your body. What products increase the entropy of you and your surroundings?

3. You have found a molecular substance that accelerates the rate of a particular reaction. What kind of information would you need to demonstrate that this molecular substance is an enzyme?

4. The addition or removal of phosphate groups from ATP is a fully reversible reaction. In what way does this reversibility facilitate the use of ATP as a coupling agent for cellular reactions?

5. Researchers once hypothesized that an enzyme and its substrate fit together like a lock and key but that the products do not fit the enzyme. Examine this idea with respect to reversible reactions.

Experimental Analysis

Succinate dehydrogenase is part of the cellular biochemical machinery for breaking down sugars, fatty acids, and amino acids into carbon dioxide and water, with the capture of their chemical energy as ATP. Suppose you are measuring the activity of this enzyme extracted from cells in test-tube reactions. You find that the rate of the reaction converting succinate to fumarate catalyzed by succinate dehydrogenase is inhibited by the addition of malonate to the reaction mixture. Design an experiment that will tell you whether malonate is acting as a competitive or a noncompetitive inhibitor.

Evolution Link

If RNA appeared first in evolution, establishing an RNA world, which do you think would evolve next: DNA or proteins? Why?

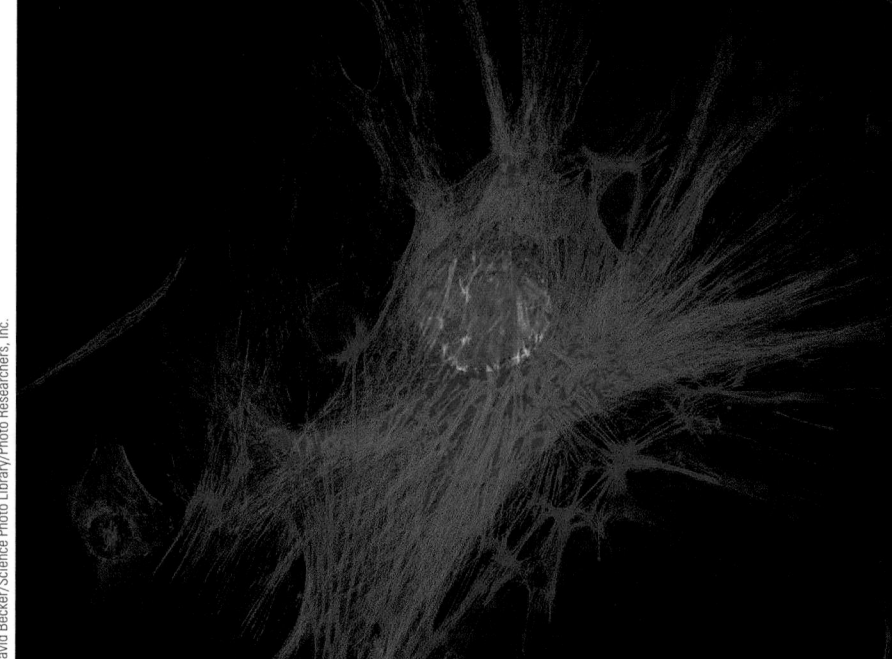

Cells fluorescently labeled to visualize their internal structure (confocal light micrograph). Cell nuclei are shown in blue and parts of the cytoskeleton in red and green.

5 The Cell: An Overview

WHY IT MATTERS

In the mid-1600s, Robert Hooke, Curator of Instruments for the Royal Society of England, was at the forefront of studies applying the newly invented light microscopes to biological materials. When Hooke looked at thinly sliced cork from a mature tree through a microscope, he observed tiny compartments **(Figure 5.1a).** He gave them the Latin name *cellulae,* meaning "small rooms"—hence, the origin of the biological term *cell.* Hooke was actually looking at the walls of dead cells, which is what cork consists of. Hooke also looked at the central pith of a plant stem, in which he found cells "fill'd with juices." Thus, he observed living cells, as well as dead and empty ones.

Reports of cells also came from other sources. By the late 1600s, Anton van Leeuwenhoek **(Figure 5.1b),** a Dutch shopkeeper, observed "many very little animalcules, very prettily a-moving," using a single-lens microscope of his own construction. Leeuwenhoek discovered and described diverse protists, sperm cells, and even bacteria, organisms so small that they would not be seen by others for another two centuries.

In the 1820s, improvements in microscopes brought cells into sharper focus. Robert Brown, an English botanist, noticed a discrete,

a. Hooke's microscope

b. Leeuwenhoek and microscope

National Library of Medicine

Armed Forces Institute of Pathology

Figure 5.1

Investigations leading to the first descriptions of cells. **(a)** The cork cells drawn by Robert Hooke and the compound microscope he used to examine them. **(b)** Anton van Leeuwenhoek holding his microscope, which consisted of a single, small sphere of glass fixed in a holder. He viewed objects by holding them close to one side of the glass sphere and looking at them through the other side.

Thus, by the middle of the nineteenth century, microscopic observations had yielded three profound generalizations, which together constitute what is now known as the **cell theory:**

1. All organisms are composed of one or more cells.
2. The cell is the smallest unit that has the properties of life.
3. Cells arise only from the growth and division of preexisting cells.

These tenets were fundamental to the development of biological science.

This chapter provides an overview of our current understanding of the structure and functions of cells, emphasizing both the similarities among all cells and some of the most basic differences among cells of various organisms. The variations in cells that help make particular groups of organisms distinctive are discussed in later chapters. This chapter also introduces some of the modern microscopes that transport us more deeply into the spectacular worlds of cells "fill'd with juices" and enable us to learn more about cell structure.

5.1 Basic Features of Cell Structure and Function

As the basic structural and functional units of all living organisms, cells carry out the essential processes of life. They contain highly organized systems of molecules, including the nucleic acids DNA and RNA, which carry hereditary information and direct the manufacture of cellular molecules. Cells use organic fuel molecules as energy sources for their activities. They use that energy to generate movements, and can alter their internal reactions in response to changes in their external environment. Cells can also duplicate and pass on their hereditary information as part of cellular reproduction. All these activities occur in cells that, in most cases, are invisible to the naked eye.

Some types of organisms, including bacteria, archaeans, and some protists such as the protozoa, are

spherical body inside some cells; he called it a *nucleus.* In 1838, a German botanist, Matthias Schleiden, speculated that the nucleus had something to do with the development of a cell. The following year, the zoologist Theodor Schwann of Germany expanded Schleiden's idea to propose that all animals and plants consist of cells that contain a nucleus. He also proposed that even when a cell forms part of a larger organism, it has an individual life of its own. However, an important question remained: Where do cells come from? A decade later, the German physiologist Rudolf Virchow answered this question. From his studies of cell growth and reproduction, Virchow proposed that cells arise only from preexisting cells by a process of division.

Figure 5.2

Examples of the varied kinds of cells. **(a)** A bacterial cell with flagella. **(b)** A trichomonad, a protist living in a termite's gut. **(c)** Two cells of *Micrasterias,* an alga. **(d)** Cells of a surface layer in the human kidney. **(e)** Cells in the leaf of a kidney bean plant (*Phaseolus*).

a. Bacterium

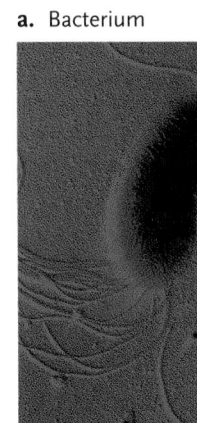

Tony Brain/SPL/Photo Researchers, Inc.

b. Protozoan

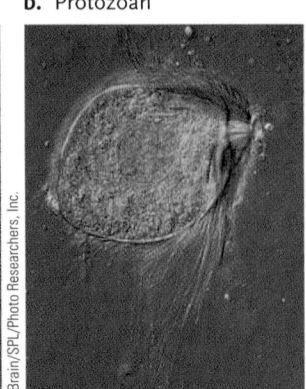

M. Abbey/Visuals Unlimited

c. Algae

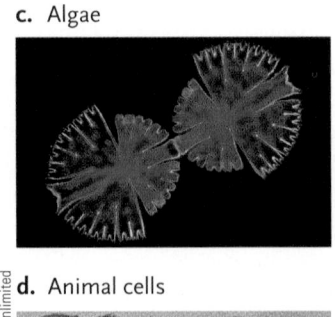

Wim van Egmond

d. Animal cells

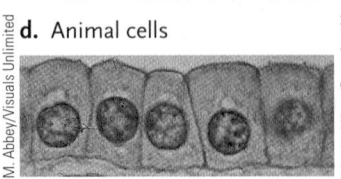

Manfred Kage/Peter Arnold

e. Plant cells

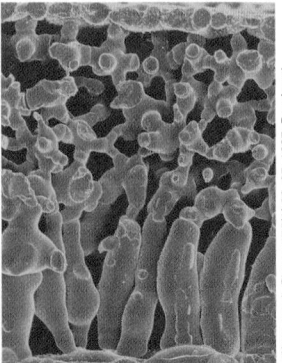

C. E. Jeffree, et al. *Planta.* 172(1):20–37. 1987. Reprinted.

unicellular. Each cell is a functionally independent organism capable of carrying out all life activities. In more complex multicellular organisms, including plants and animals, the activities of life are divided among varying numbers of specialized cells. However, individual cells of multicellular organisms are potentially capable of surviving by themselves if placed in a chemical medium that can sustain them.

If cells are broken open, the property of life is lost: they are unable to grow, reproduce, or respond to outside stimuli in a coordinated, potentially independent fashion. This fact confirms the second tenet of the cell theory: life as we know it does not exist in units more simple than individual cells. *Viruses,* which consist only of a nucleic acid molecule surrounded by a protein coat, cannot carry out all of the activities of life. Their only capacity is to infect living cells and direct them to make more virus particles of the same kind. (Viruses are discussed in Chapters 17 and 25.)

Cells Are Small and Are Visualized Using a Microscope

Cells assume a wide variety of forms in different microorganisms, plants, and animals **(Figure 5.2).** Individual cells range in size from tiny bacteria to an egg yolk, a single cell that can be several centimeters in diameter. Yet, all cells are organized according to the same basic plan, and all have structures that perform similar activities.

Most cells are too small to be seen by the unaided eye: humans cannot see objects smaller than about 0.1 mm in diameter. The smallest bacteria have diameters of about 0.5 μm (a micrometer is 1000 times smaller than a millimeter). The cells of multicellular animals range from about 5 to 30 μm in diameter. Your red blood cells are 7 to 8 μm across—a string of 2500 of these cells is needed to span the width of your thumbnail. Plant cells range from about 10 μm to a few hundred micrometers in diameter. **(Figure 5.3** explains the units of measurement used in biology to study molecules and cells.)

To see cells and the structures within them we use **microscopy,** a technique for producing visible images of objects, biological or otherwise, that are too small to be seen by the human eye **(Figure 5.4).** The instrument of microscopy is the **microscope.** The two common types of microscopes are **light microscopes,** which use light to illuminate the specimen (the object being viewed), and **electron microscopes,** which use electrons to illuminate the specimen. Different types of microscopes give different magnification and resolution of the specimen. Just as for a camera or a pair of binoculars, **magnification** is the ratio of the object as viewed to its real size, usually given as something like 1200×. **Resolution** is the minimum distance two points in the specimen can be separated and still be seen as two points. Resolution depends primarily on the wave-

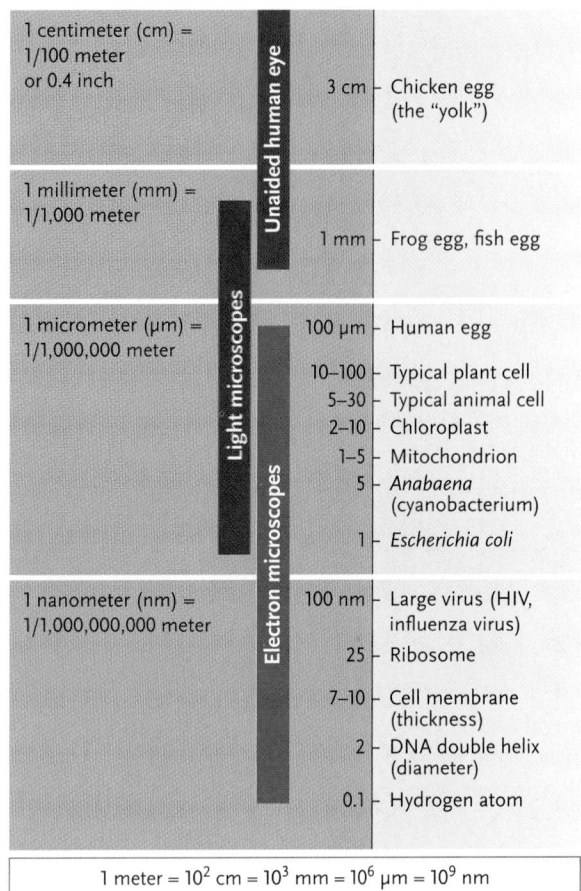

length of light or electrons used to illuminate the specimen; the shorter the wavelength, the better the resolution. Hence, electron microscopes have higher resolution than light microscopes. Biologists choose the type of microscopy technique based on what they need to see in the specimen; selected examples are shown in Figure 5.4.

Why are most cells so small? The answer depends partly on the change in the surface area-to-volume ratio of an object as its size increases **(Figure 5.5).** For example, doubling the diameter of a cell increases its volume by eight times but increases its surface area by only four times. The significance of this relationship is that the volume of a cell determines the amount of chemical activity that can take place within it, whereas the surface area determines the amount of substances that can be exchanged between the inside of the cell and the outside environment. Nutrients must constantly enter cells, and wastes must constantly leave; however, past a certain point, increasing the diameter of a cell gives a surface area that is insufficient to maintain an adequate nutrient–waste exchange for its entire volume.

Some cells increase their ability to exchange materials with their surroundings by flattening or by developing surface folds or extensions that increase their surface area. For example, human intestinal cells have closely packed, fingerlike extensions that increase their

Figure 5.3
Units of measure and the ranges in which they are used in the study of molecules and cells. The vertical scale in each box is logarithmic.

Figure 5.4 Research Method

Light and Electron Microscopy

PURPOSE: Microscopy is a widely used technique in biology to view organisms, cells, and structures within cells in their natural state or after being treated (stained) so that specific structures can be seen more clearly. All of the photographs of cells and cell structures in this book were made using microscopy.

PROTOCOL: A light microscope uses a beam of light to illuminate the specimen and forms a magnified image of the specimen with glass lenses. An electron microscope uses a beam of electrons to illuminate the specimen and forms a magnified image with magnetic fields. Electron microscopy provides higher resolution and higher magnification than light microscopy.

Light microscopy
Micrographs are of the protist *Paramecium*.

Electron microscopy
Micrographs are of the green alga *Scenedesmus*.

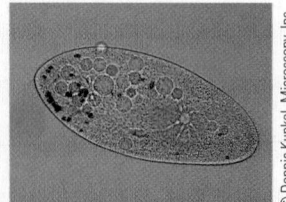

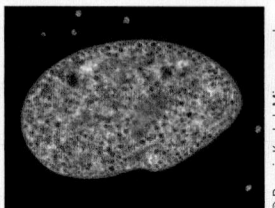

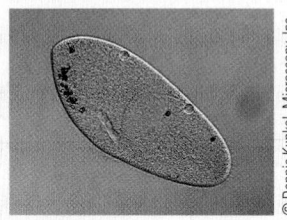

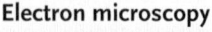

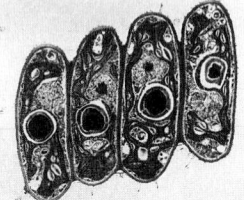

© Dennis Kunkel, Microscopy, Inc.

Jeremy Pickett-Heaps, University of Colorado

Bright field microscopy: Light passes directly through the specimen. Many cell structures have insufficient contrast to be discerned. Staining with a dye is used to enhance contrast in a specimen as shown here, but this treatment usually fixes and kills the cells.

Dark field microscopy: Light illuminates the specimen at an angle, and only light scattered by the specimen reaches the viewing lens of the microscope. This gives a bright image of the cell against a black background.

Phase-contrast microscopy: Differences in refraction (the way light is bent) caused by variations in the density of the specimen are visualized as differences in contrast. Otherwise invisible structures are revealed with this technique, and living cells in action can be photographed or filmed.

Transmission electron microscopy (TEM): A beam of electrons is focused on a thin section of a specimen in a vacuum. Electrons that pass through form the image; structures that scatter electrons appear dark. TEM is used primarily to examine structures within cells. Various staining and fixing methods are used to highlight structures of interest.

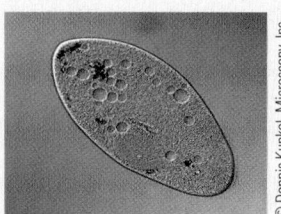

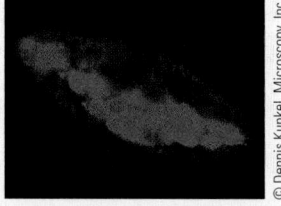

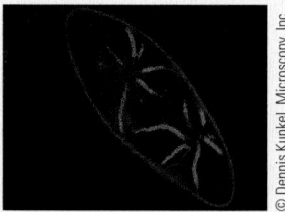

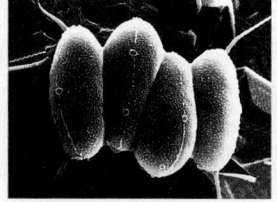

© Dennis Kunkel, Microscopy, Inc.

Jeremy Pickett-Heaps, University of Colorado

Nomarski (differential interference contrast): Similar to phase-contrast microscopy, special lenses enhance differences in density, giving a cell a 3D appearance.

Fluorescence microscopy: Different structures or molecules in cells are stained with specific fluorescent dyes. The stained structures or molecules fluoresce when the microscope illuminates them with ultraviolet light, and their locations are seen by viewing the emitted visible light.

Confocal laser scanning microscopy: Lasers scan across a fluorescently stained specimen, and a computer focuses the light to show a single plane through the cell. This provides a sharper 3D image than other light microscopy techniques.

Scanning electron microscopy (SEM): A beam of electrons is scanned across a whole cell or organism, and the electrons excited on the specimen surface are converted to a 3D-appearing image.

INTERPRETING THE RESULTS: Different techniques of light and electron microscopy produce images that reveal different structures or functions of the specimen. A micrograph is a photograph of an image formed by a microscope.

surface area, which greatly enhances their ability to absorb digested food molecules.

Cells Have a DNA-Containing Central Region That Is Surrounded by Cytoplasm

All cells have a central region that contains DNA molecules, which store hereditary information. The hereditary information is organized in the form of *genes*—segments of DNA that code for individual proteins. The central region also contains proteins that help maintain the DNA structure and enzymes that duplicate DNA and copy its information into RNA.

All the parts of the cell that surround the central region comprise the **cytoplasm.** The cytoplasm consists of the **cytosol,** which is an aqueous (water) solution containing ions and various organic molecules, and **organelles** ("little organs"), which are small, organized structures important for cell function. The outer limit of the cytoplasm is the **plasma membrane,** a bilayer made of lipids with embedded protein molecules **(Figure 5.6).**

The lipid bilayer of the plasma membrane is a hydrophobic barrier to the passage of water-soluble substances, but selected water-soluble substances can penetrate cell membranes through transport protein channels. The selective movement of ions and water-soluble molecules through the transport proteins maintains the specialized internal ionic and molecular environments required for cellular life. (Membrane structure and functions are discussed further in Chapter 6.)

Many of the cell's vital activities occur in the cytoplasm, including the synthesis and assembly of most of the molecules required for growth and reproduction (except those made in the central region) and the conversion of chemical and light energy into forms that can be used by cells. The cytoplasm also conducts stimulatory signals from the outside into the cell interior and carries out chemical reactions that respond to these signals.

Cells Occur in Prokaryotic and Eukaryotic Forms, Each with Distinctive Structures and Organization

Organisms fall into two fundamental groups, prokaryotes and eukaryotes, based on the organization of their cells. **Prokaryotes** (*pro* = before; *karyon* = nucleus) make up two domains of organisms, the Bacteria and the Archaea. The central region of prokaryotic cells, the **nucleoid,** has no boundary membrane separating it from the cytoplasm. Prokaryotic membranes are limited to the plasma membrane and, in some cases, simple saclike membranes in the cytoplasm.

The **eukaryotes** (*eu* = true) make up the domain Eukarya, which includes all the remaining organisms. The central region of eukaryotic cells, a true **nucleus,**

is separated by membranes from the surrounding cytoplasm. The cytoplasm of eukaryotic cells contains membrane systems that form organelles with their own distinct environments and specialized functions. As in prokaryotes, a plasma membrane surrounds eukaryotic cells as the outer limit of the cytoplasm.

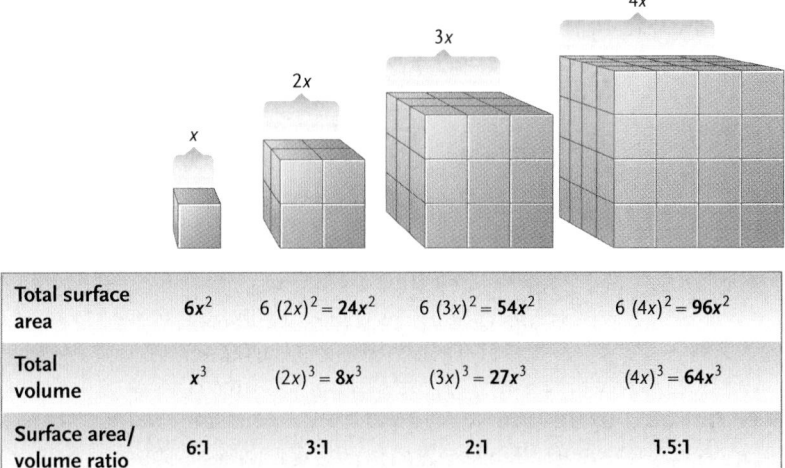

Total surface area	$6x^2$	$6\,(2x)^2 = \mathbf{24x^2}$	$6\,(3x)^2 = \mathbf{54x^2}$	$6\,(4x)^2 = \mathbf{96x^2}$
Total volume	x^3	$(2x)^3 = \mathbf{8x^3}$	$(3x)^3 = \mathbf{27x^3}$	$(4x)^3 = \mathbf{64x^3}$
Surface area/ volume ratio	6:1	3:1	2:1	1.5:1

Figure 5.5
Relationship between surface area and volume. The surface area of an object increases as a square of the linear dimension, whereas the volume increases as a cube of that dimension.

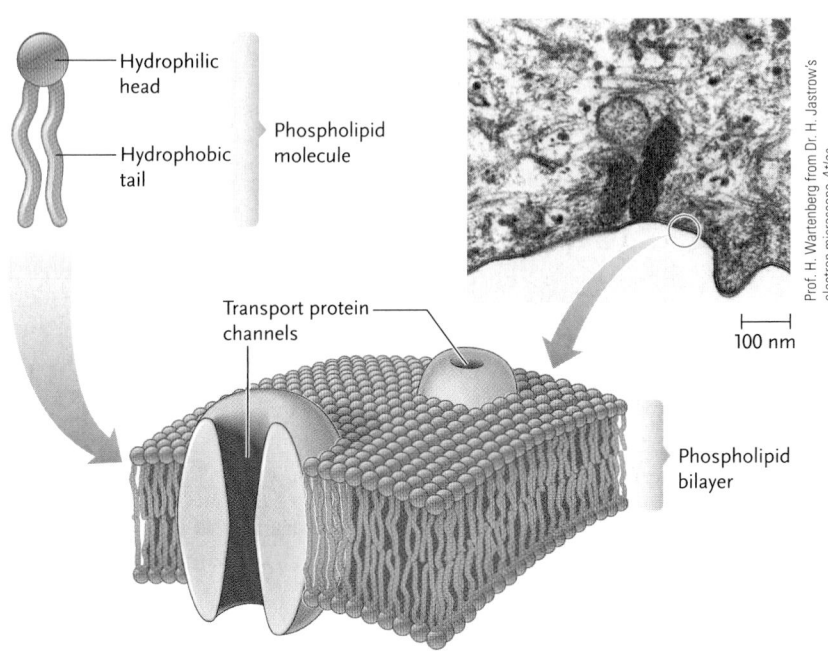

Figure 5.6
The plasma membrane, which forms the outer limit of a cell's cytoplasm. The plasma membrane consists of a phospholipid bilayer, an arrangement of phospholipids two molecules thick, which provides the framework of all biological membranes. Water-soluble substances cannot pass through the phospholipid part of the membrane. Instead, they pass through protein channels in the membrane; two proteins that transport substances across the membrane are shown. Other types of proteins are also associated with the plasma membrane. *(Inset)* Electron micrograph of part of an animal cell, showing the plasma membrane (circled).

Table 5.1 — Components of Prokaryotic and Eukaryotic Cells

Cell Component	Major Functions	Prokaryotes		Eukaryotes			
		Bacteria	Archaea	Protists	Fungi	Plants	Animals
Nucleoid	DNA replication and RNA transcription	■	■				
Nucleus	DNA replication and RNA transcription			■	■	■	■
Nuclear envelope	Separation of nucleus from cytoplasm			■	■	■	■
Nucleolus	Ribosomal RNA synthesis and assembly of ribosomal subunits			■	■	■	■
Plasma membrane	Regulation of substances moving into and out of cells	■	■	■	■	■	■
Cell wall	Cell protection and support	■	■	Some	■	■	
Ribosomes	Protein synthesis	■	■	■	■	■	■
Endoplasmic reticulum	Synthesis, transport, and initial modification of membrane proteins, lipids, and secreted proteins			■	■	■	■
Golgi complex	Final modification, sorting, and distribution of membrane lipids, proteins, and secreted proteins			■	■	■	■
Lysosome	Digestion of biological molecules and structures			■	Some	■	■
Mitochondrion	Conversion of energy associated with glucose into ATP			■	■	■	■
Microbody	Housing of reactions that link major pathways			?	?	■	■
Chloroplast	Conversion of light energy to chemical energy of organic molecules			Some		■	
Central vacuole	Storage, cell growth, and support					■	
Microfilament	Reinforcement of cell shape, motility			■	■	■	■
Microtubule	Reinforcement of cell shape, motility			■	■	■	■
Intermediate filament	Reinforcement of cell shape			■			■
Flagellum or cilium with 9 + 2 system of microtubules	Cell motility			Some	Some	Some	■

■ Bullets denote presence of cell component in designated group.

The remainder of this chapter surveys the components of prokaryotic and eukaryotic cells in more detail. **Table 5.1** summarizes these cellular components and notes the organisms in which they appear.

STUDY BREAK

What is the plasma membrane, and what are its main functions?

5.2 Prokaryotic Cells

Prokaryotic Cells Have Little or No Internal Membrane Structure

Prokaryotic cells **(Figure 5.7)** are relatively small, usually not much more than a few micrometers in length and a micrometer or less in diameter; they have little or no internal membrane structure. In almost all prokaryotes, the plasma membrane is surrounded by a rigid external layer of material, the **cell wall**, which ranges in thickness from 15 to 100 nm or more (a nanometer is one-billionth of a meter). In many prokaryotic cells, the wall is coated with an external layer of sticky or slimy polysaccharides called a **capsule**. The cell wall provides rigidity to prokaryotic cells and, with the capsule, protects the cell from physical damage.

The plasma membrane performs several vital functions in prokaryotes. Besides transporting materials into and out of the cells, it contains most of the molecular systems that metabolize food molecules into the chemical energy of ATP. In photosynthetic prokaryotes, the molecules that absorb light energy and convert it to the chemical energy of ATP are also associated with the plasma membrane or with internal, saclike membranes derived from the plasma membrane.

In an electron microscope, the nucleoid of a prokaryotic cell is seen to contain a folded mass of DNA (see Figure 5.7). For most species, the DNA is a single,

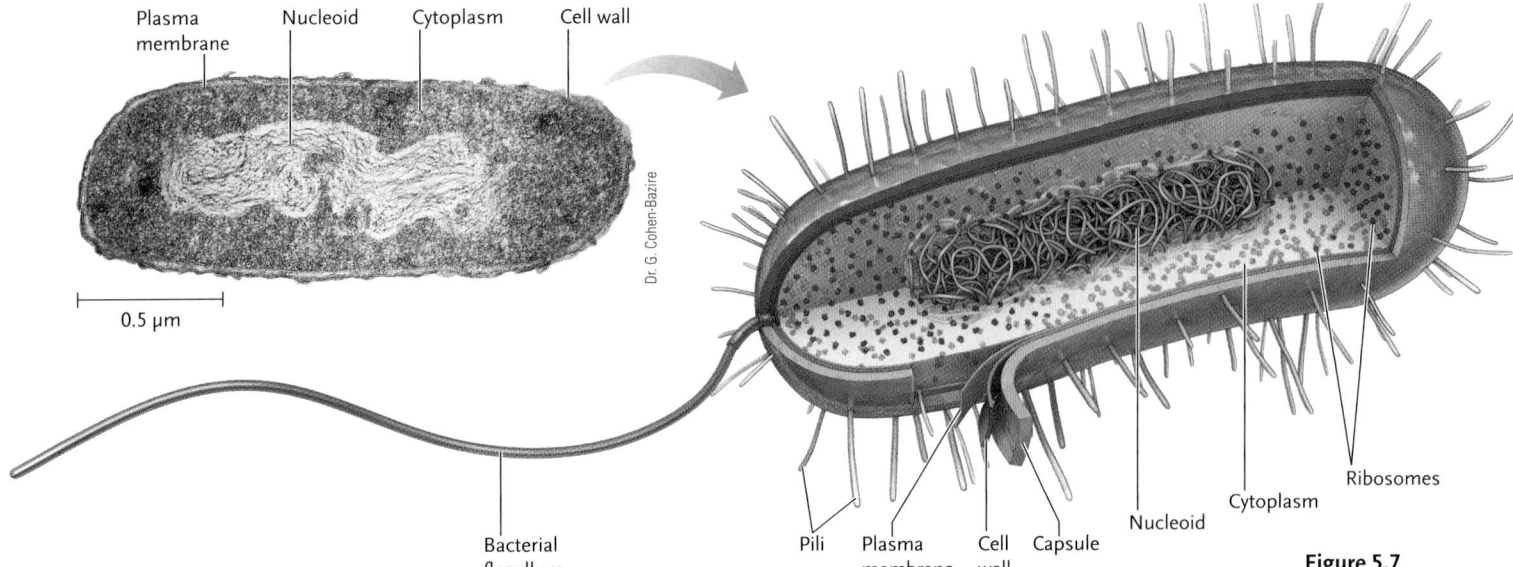

Plasma membrane   Nucleoid   Cytoplasm   Cell wall

Dr. G. Cohen-Bazire

0.5 μm

Bacterial flagellum

Pili   Plasma membrane   Cell wall   Capsule   Nucleoid   Cytoplasm   Ribosomes

Figure 5.7
Prokaryotic cell structure. An electron micrograph (left) and a diagram (right) of the bacterium *Escherichia coli*. The pili extending from the cell wall attach bacterial cells to other cells of the same species or to eukaryotic cells as a part of infection. A typical *E. coli* has four flagella.

circular molecule that unfolds when released from the cell. This DNA molecule is the **prokaryotic chromosome.** (Chapter 17 discusses the genetics of prokaryotes.)

Individual genes in the DNA molecule encode the information required to make proteins. This information is copied into a type of RNA molecule called *messenger RNA*. Small, roughly spherical particles in the cytoplasm, the **ribosomes,** are organelles that use the information in the messenger RNA to assemble amino acids into proteins. A prokaryotic ribosome consists of a large and a small subunit, each formed from a combination of *ribosomal RNA* and protein molecules. In all, each prokaryotic ribosome contains three types of ribosomal RNA molecules, which are also copied from the DNA, and more than 50 proteins.

Many prokaryotes swim by means of long, thread-like protein fibers called **flagella** (singular, *flagellum*), which extend from the cell surface (see Figure 5.2a). The **prokaryotic flagellum,** which is helically shaped, rotates in a socket in the plasma membrane and cell wall to push the cell through a liquid medium (see Section 25.1). Prokaryotic flagella are fundamentally different from the much larger and more complex flagella of eukaryotic cells, which are described in Section 5.3.

Although prokaryotic cells appear relatively simple, their simplicity is deceptive. Most can use a variety of substances as energy and carbon sources, and they are able to synthesize almost all of their required organic molecules from simple inorganic raw materials. In many respects, prokaryotes are more versatile biochemically than eukaryotes. Their small size and metabolic versatility are reflected in their abundance; prokaryotes vastly outnumber all other types of organisms and live successfully in almost all regions of Earth's surface. (Chapter 25 outlines the diversity of prokaryotes and extends the discussion of prokaryotic structure.)

The two domains of the prokaryotes, the Bacteria and the Archaea, share many biochemical and molecular features. However, the archaeans also share some features with eukaryotes and have other characteristics that are unique to their group. *Insights from the Molecular Revolution* describes the discovery of features that support the classification of the Archaea as a separate domain.

STUDY BREAK

Where in a prokaryotic cell is DNA found? How is that DNA organized?

5.3 Eukaryotic Cells

Eukaryotic Cells Have a Membrane-Enclosed Nucleus and Cytoplasmic Organelles

The domain of the eukaryotes, Eukarya, is divided into four major groups: the protists, fungi, animals, and plants. The cells of all eukaryotes have a true nucleus enclosed by membranes. The cytoplasm surrounding the nucleus contains a remarkable system of membranous organelles, each specialized to carry out one or more major functions of energy metabolism and molecular synthesis, storage, and transport. The cytosol, the cytoplasmic solution surrounding the organelles, participates in energy metabolism and molecular synthesis and performs specialized functions in support and motility.

The eukaryotic plasma membrane carries out various functions through embedded proteins. Proteins that form channels through the membrane transport

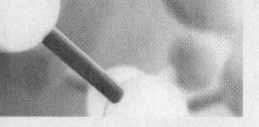

An Old Kingdom in a New Domain

In 1996, Carol J. Bult, Carl R. Woese, J. Craig Venter, and 37 other scientists at the Institute for Genomic Research published the complete DNA sequence of *Methanococcus jannaschii*, a member of the prokaryotic domain Archaea. Information obtained from the DNA sequence clearly supports the conclusion that archaeans are as different from the Bacteria, the other prokaryotic domain, as they are from the eukaryotes.

Many archaeans live in extreme environments that can be tolerated by no other organisms, suggesting that they might belong in a distinct domain. For example, *Methanococcus* was first found in an oceanic hot water vent at a depth of more than 2600 m (8500 feet). It can live at temperatures as high as 94°C, which is almost the temperature of boiling water, and can tolerate pressures as high as 200 times the pressure of air at sea level!

The complete DNA sequence of *Methanococcus* was obtained using techniques outlined in Chapter 18. Using computer algorithms, the scientists compared the final sequence with the already known sequences of several bacteria and of brewer's yeast (*Saccharomyces cerevisiae*), the first eukaryote to be sequenced completely. They found genes coding for 1738 proteins in the *Methanococcus* DNA. Of these, only 38% were related to genes coding for known proteins in either bacteria or eukaryotes. The remaining 62%, representing sequences with no known relatives in organisms of the other domains, demonstrated the unique character of the archaeans.

Some features of *Methanococcus* DNA are typically prokaryotic. Its single, circular chromosome is in a nucleoid, which is not bounded by a membrane. Its genes are organized into functional groups called *operons*, each having several genes copied as a unit into a single messenger RNA molecule (see discussion in Section 16.1). By contrast, each gene in eukaryotes is copied into a separate messenger RNA molecule. Some of the proteins encoded in *Methanococcus* DNA, including enzymes active in energy metabolism, membrane transport, and cell division, are similar to those of bacteria. Other proteins encoded in the *Methanococcus* DNA are similar to those of eukaryotes, including enzymes and other proteins that carry out DNA replication and the copying of genes into messenger RNA.

Thus, *Methanococcus* has a majority of genes that are unique, some that are typically bacterial, and some that are typically eukaryotic. This finding supports the proposal, first advanced by Woese, that *Methanococcus* and its archaean relatives are a separate domain of life, with the Bacteria and the Eukarya as the other domains. Woese's three-domain system is used in this book.

---

substances into and out of the cell. Other proteins in the plasma membrane act as receptors; they recognize and bind specific signal molecules in the cellular environment and trigger internal responses. In some eukaryotes, particularly animals, other plasma membrane proteins recognize and adhere to molecules on the surfaces of other cells. Other plasma membrane proteins are important markers in the immune system, labeling cells as "self," that is, belonging to the organism. Therefore, the immune system can identify cells without those markers as being foreign, most likely *pathogens* (disease-causing organisms or viruses).

A supportive cell wall surrounds the plasma membrane of fungal, plant, and many cells of protists. Because the cell wall lies outside the plasma membrane, it is an *extracellular* structure (*extra* = outside). Although animal cells do not have cell walls, they also form extracellular material with supportive and other functions.

Figure 5.8 show where the nucleus, cytoplasmic organelles, and other structures are located in representative animal cells. **Figure 5.9** show their locations in plant cells. The following sections discuss the structure and function of eukaryotic cell parts in more detail, beginning with the nucleus.

The Eukaryotic Nucleus Contains Much More DNA Than the Prokaryotic Nucleoid

The nucleus (see Figures 5.8 and 5.9) is separated from the cytoplasm by the **nuclear envelope**, which consists of two membranes, one layered just inside the other and separated by a narrow space **(Figure 5.10)**. **Nuclear pores** form openings through both membranes. The pores are made of protein structures that control the movement of large molecules, such as RNA and proteins, between the nucleus and cytoplasm. A network of protein filaments called *lamins* lines and reinforces the inner surface of the nuclear envelope in animal cells. Other, unrelated reinforcing proteins line the inner surface of the nuclear envelope in many protists, fungi, and plants.

The liquid or semiliquid substance within the nucleus is called the **nucleoplasm.** Most of the space inside the nucleus is filled with **chromatin**, a combination of DNA and proteins. By contrast with most prokaryotes, most of the hereditary information of a eukaryote is distributed among several to many linear DNA molecules in the nucleus. Each individual DNA molecule with its associated proteins is a **eukaryotic chromosome.** The terms *chromatin* and *chromosome* are similar but have distinct meanings. *Chromatin*

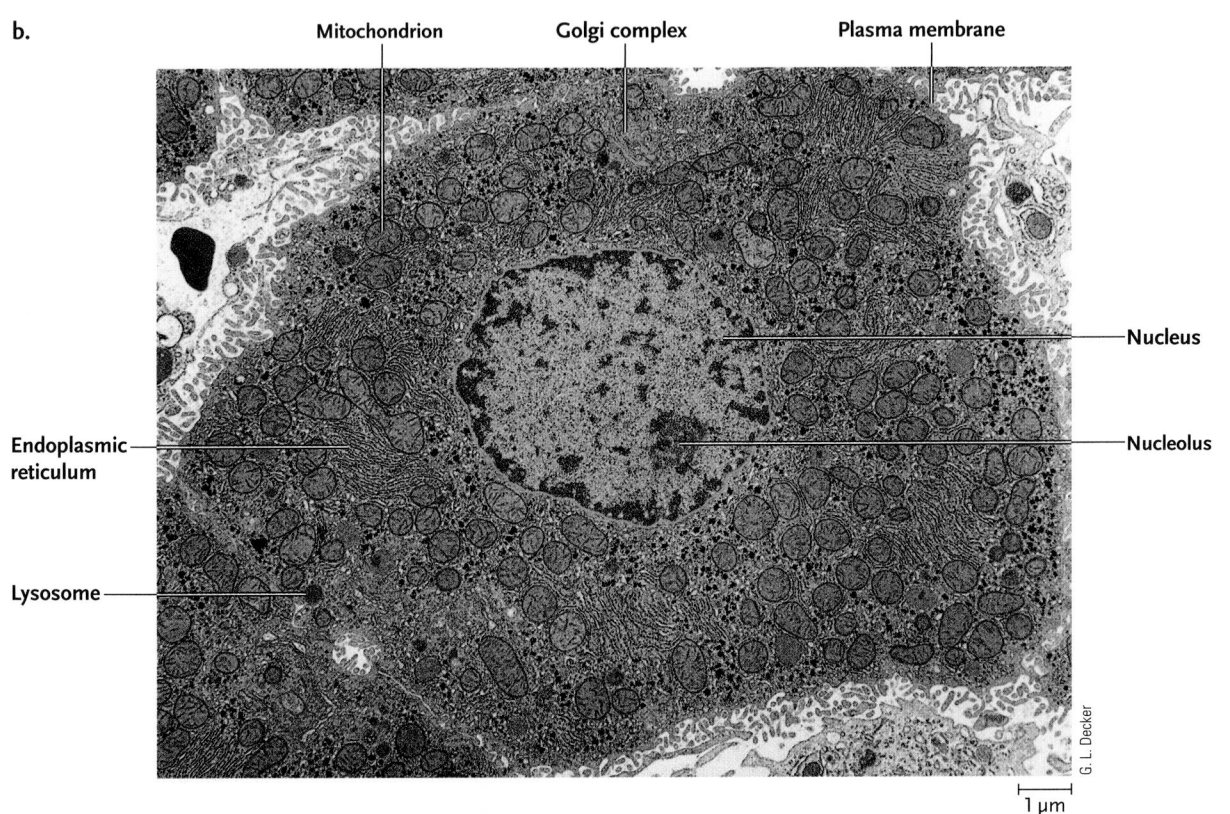

a.

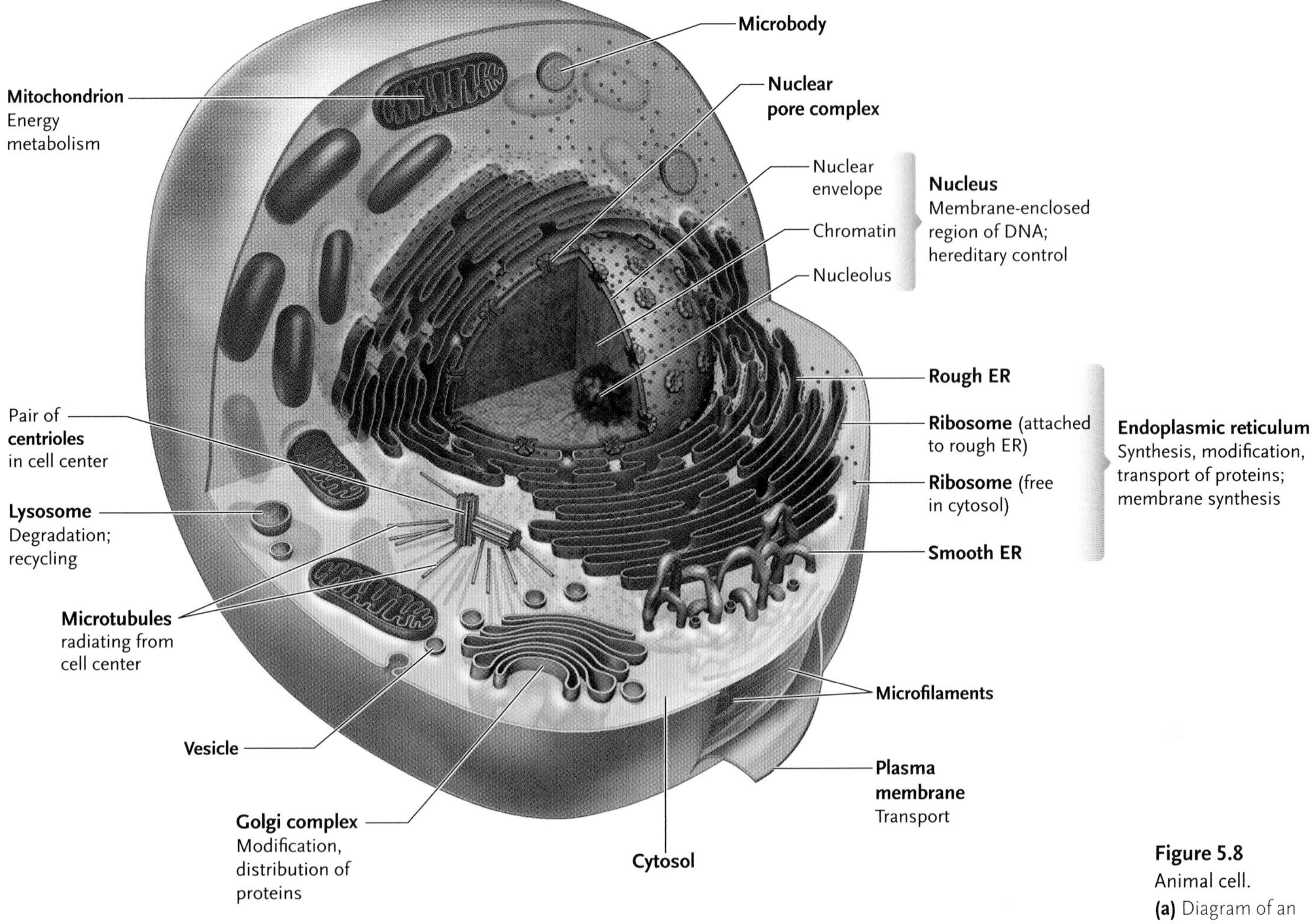

Microbody

Mitochondrion
Energy
metabolism

Nuclear
pore complex

Nuclear
envelope

Nucleus
Membrane-enclosed
region of DNA;
hereditary control

Chromatin

Nucleolus

Rough ER

Ribosome (attached
to rough ER)

Endoplasmic reticulum
Synthesis, modification,
transport of proteins;
membrane synthesis

Ribosome (free
in cytosol)

Smooth ER

Pair of
centrioles
in cell center

Lysosome
Degradation;
recycling

Microtubules
radiating from
cell center

Microfilaments

**Plasma
membrane**
Transport

Vesicle

Golgi complex
Modification,
distribution of
proteins

Cytosol

Figure 5.8
Animal cell.
(a) Diagram of an
animal cell high-
lighting the major
organelles and
their primary
locations.
(b) Electron
micrograph of a
rat liver cell.

b.

Mitochondrion

Golgi complex

Plasma membrane

Nucleus

Nucleolus

Endoplasmic
reticulum

Lysosome

G. L. Decker

1 μm

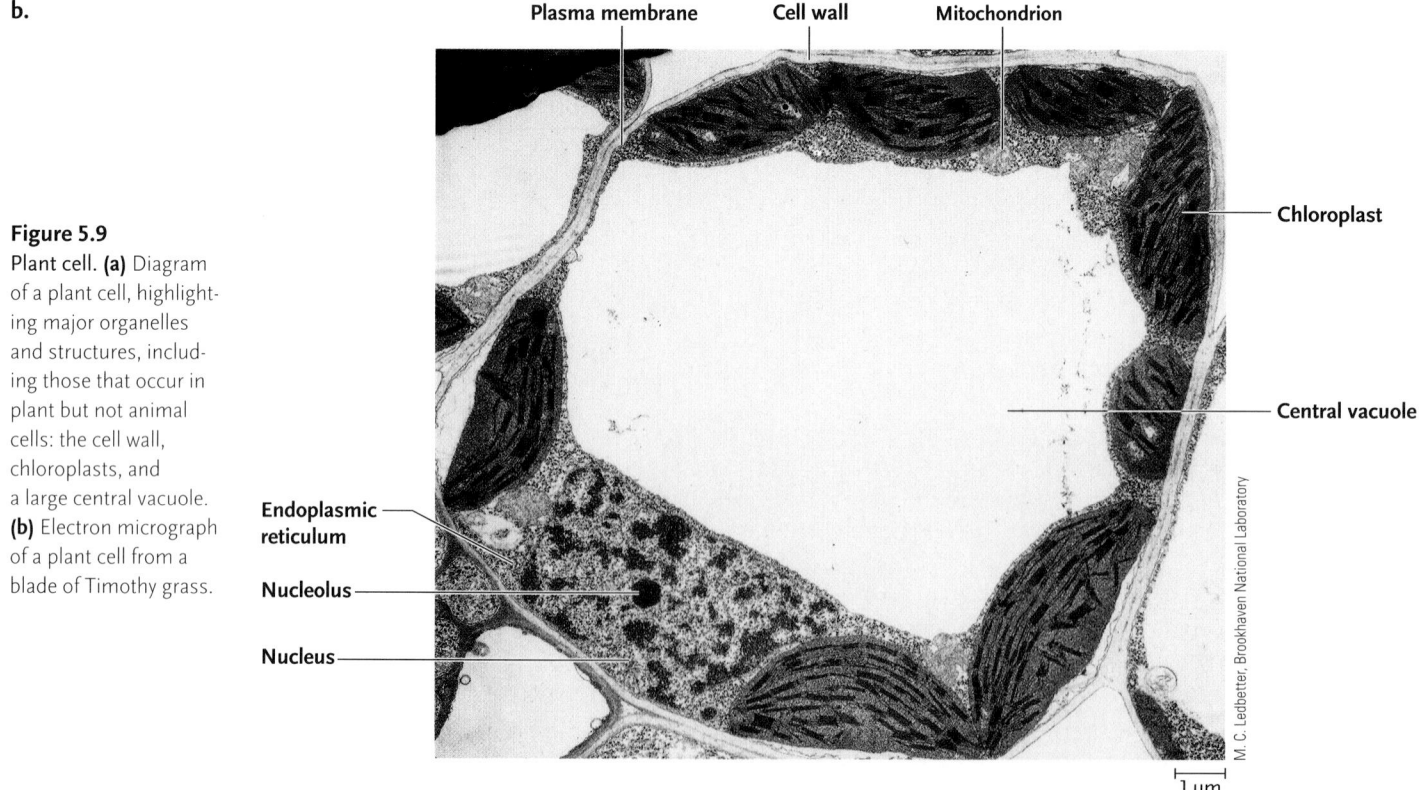

a.

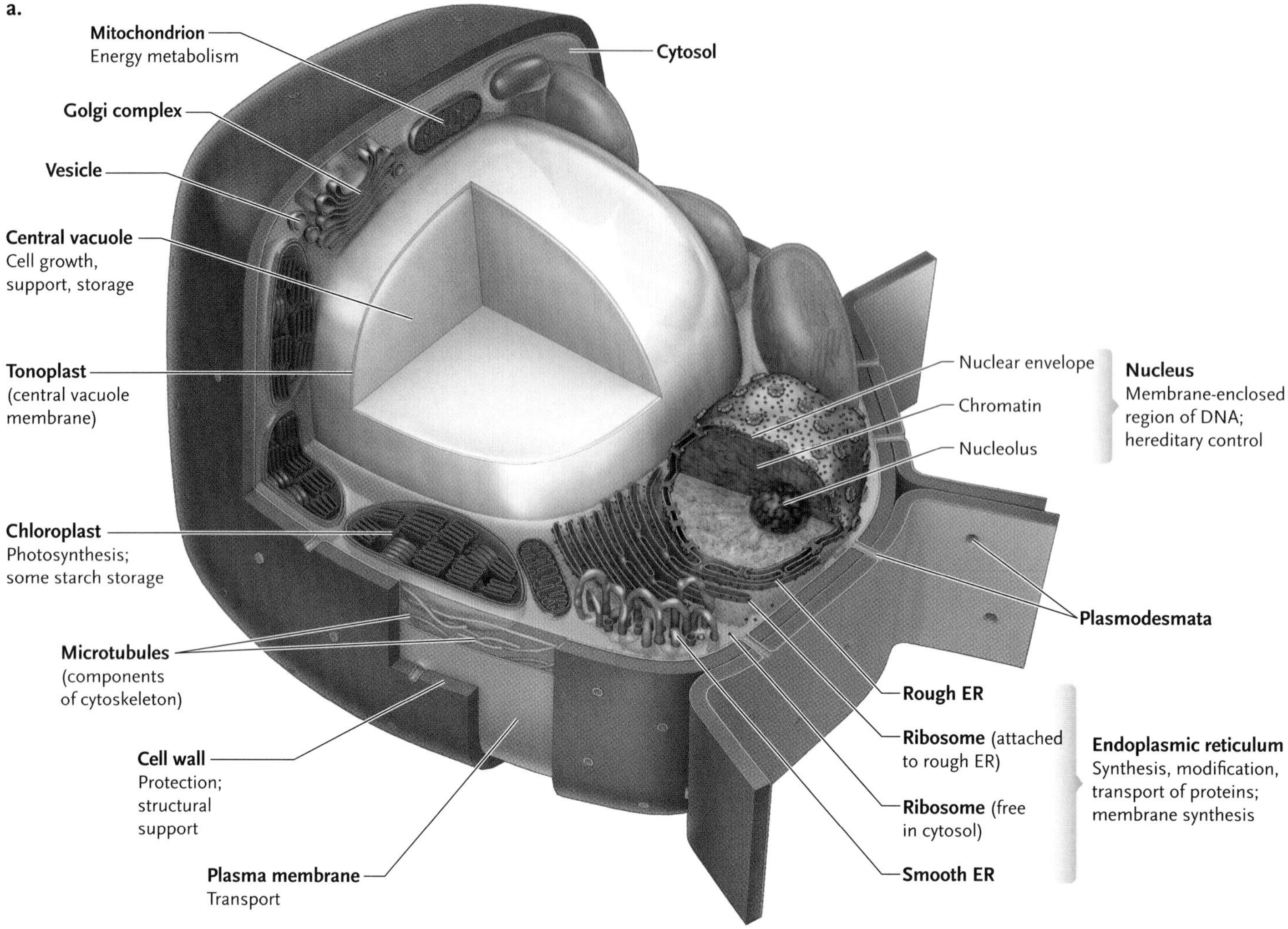

Mitochondrion
Energy metabolism

Golgi complex

Vesicle

Central vacuole
Cell growth,
support, storage

Tonoplast
(central vacuole
membrane)

Chloroplast
Photosynthesis;
some starch storage

Microtubules
(components
of cytoskeleton)

Cell wall
Protection;
structural
support

Plasma membrane
Transport

Cytosol

Nuclear envelope

Chromatin

Nucleolus

Nucleus
Membrane-enclosed
region of DNA;
hereditary control

Plasmodesmata

Rough ER

Ribosome (attached
to rough ER)

Ribosome (free
in cytosol)

Smooth ER

Endoplasmic reticulum
Synthesis, modification,
transport of proteins;
membrane synthesis

b.

Plasma membrane   Cell wall   Mitochondrion

Chloroplast

Central vacuole

Endoplasmic
reticulum

Nucleolus

Nucleus

Figure 5.9
Plant cell. **(a)** Diagram
of a plant cell, highlight-
ing major organelles
and structures, includ-
ing those that occur in
plant but not animal
cells: the cell wall,
chloroplasts, and
a large central vacuole.
(b) Electron micrograph
of a plant cell from a
blade of Timothy grass.

M. C. Ledbetter, Brookhaven National Laboratory

1 μm

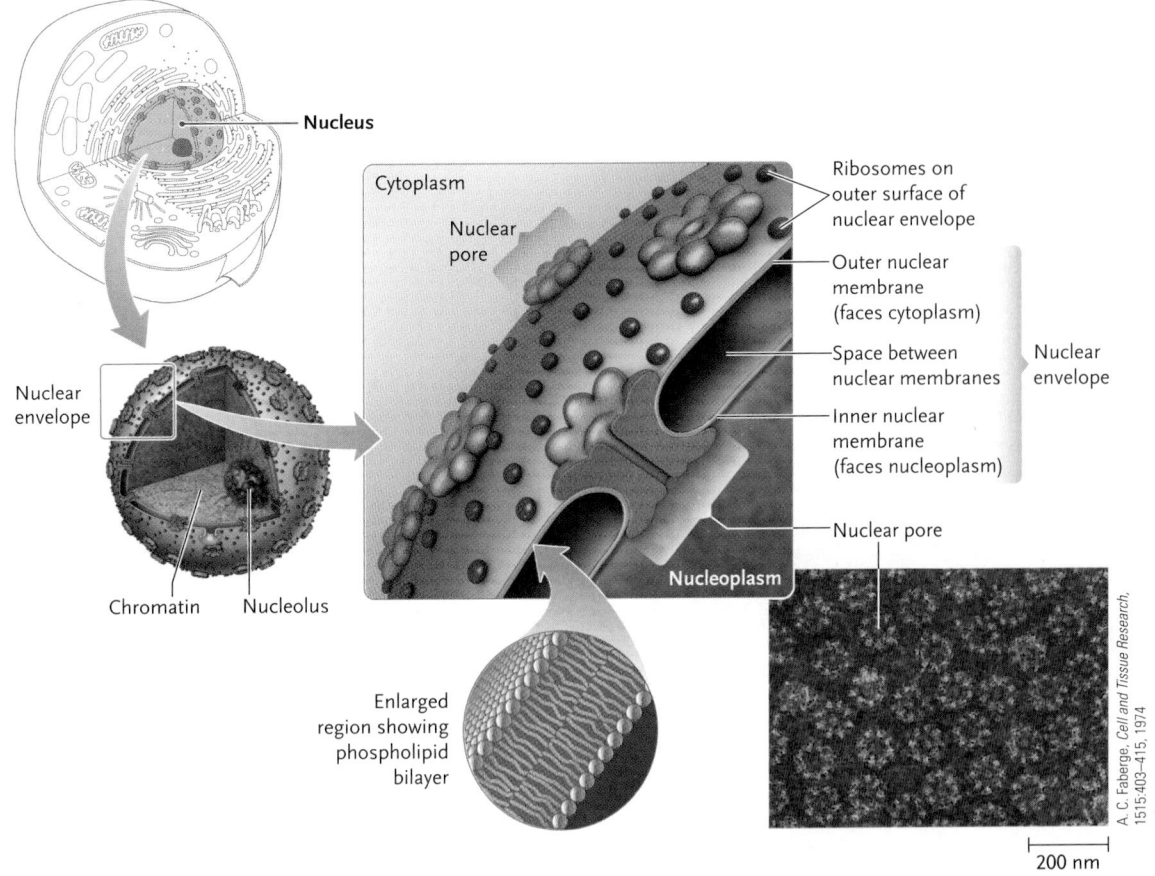

Figure 5.10
The nuclear envelope, which consists of a system of two concentric membranes perforated by nuclear pores. The electron micrograph shows nuclear pores; each pore is an organized cluster of membrane proteins that spans the membrane and facilitates transport of molecules between the nucleus and cytoplasm.

refers to any collection of eukaryotic DNA molecules with their associated proteins. *Chromosome* refers to one complete DNA molecule with its associated proteins.

Eukaryotic nuclei contain much more DNA than do prokaryotic nucleoids. For example, the entire complement of 46 chromosomes in the nucleus of a human cell has a total DNA length of about 2 meters (m), compared with about 1500 μm in prokaryotic cells with the most DNA. Some eukaryotic cells contain even more DNA; for example, a single frog or salamander nucleus, although of microscopic diameter, is packed with about 10 m of DNA!

A eukaryotic nucleus also contains one or more **nucleoli** (singular, *nucleolus*), which look like irregular masses of small fibers and granules in the electron microscope (see Figures 5.8b and 5.9b). These structures form around the genes coding for the ribosomal RNA molecules of ribosomes. Within the nucleolus, the information in ribosomal RNA genes is copied into the ribosomal RNA molecules, which combine with proteins to form ribosomal subunits. The ribosomal subunits then leave the nucleoli and exit the nucleus through the nuclear pores to enter the cytoplasm, where they join to form complete ribosomes.

The genes for most of the proteins that the organism can make are found within the chromatin, as are the genes for specialized RNA molecules such as ribosomal RNA molecules. Expression of these genes is carefully controlled as required for the function of each cell. (The other proteins in the cell are specified by DNA in the mitochondria and chloroplasts.)

An Endomembrane System Divides the Cytoplasm into Functional and Structural Compartments

Eukaryotic cells are characterized by an **endomembrane system** (*endo* = within), a collection of interrelated internal membranous sacs that divide the cell into functional and structural compartments. The endomembrane system has a number of functions, including the synthesis and modification of proteins and their transport into membranes and organelles or to the outside of the cell, the synthesis of lipids, and the detoxification of some toxins. The membranes of the system are connected either directly in the physical sense or indirectly by **vesicles**, which are small membrane-bound compartments that transfer substances between parts of the system.

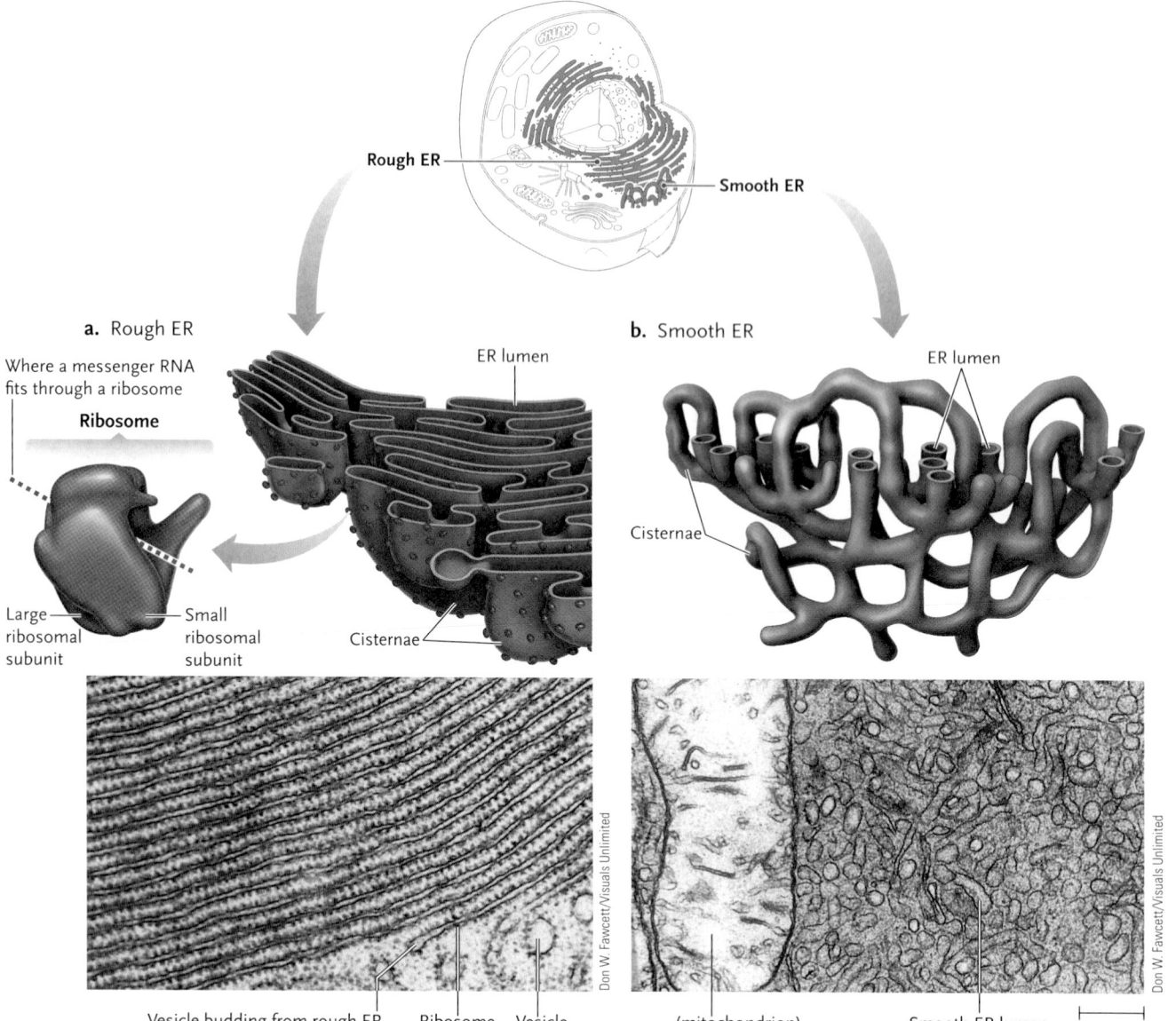

Rough ER

Smooth ER

a. Rough ER

Where a messenger RNA fits through a ribosome

Ribosome

ER lumen

Large ribosomal subunit

Small ribosomal subunit

Cisternae

b. Smooth ER

ER lumen

Cisternae

Don W. Fawcett/Visuals Unlimited

Vesicle budding from rough ER   Ribosome   Vesicle

(mitochondrion)   Smooth ER lumen   0.5 μm

Don W. Fawcett/Visuals Unlimited

Figure 5.11
The endoplasmic reticulum.
(a) Rough ER, showing the ribosomes that stud the membrane surfaces facing the cytoplasm. The structure of a single ribosome is shown on the top left. **(b)** Smooth ER membranes.

The components of the endomembrane system include the endoplasmic reticulum, Golgi complex, nuclear envelope, lysosomes, vesicles, and plasma membrane. The plasma membrane and the nuclear envelope are discussed earlier in this chapter. The functions of the other organelles are described in the following sections.

Endoplasmic Reticulum. The **endoplasmic reticulum (ER)** is an extensive interconnected network (*reticulum* = little net) of membranous channels and vesicles called **cisternae** (singular, *cisterna*). Each cisterna is formed by a single membrane that surrounds an enclosed space called the **ER lumen (Figure 5.11).** The ER occurs in two forms: rough ER and smooth ER, each with specialized structure and function.

The **rough ER** gets its name from the many ribosomes that stud its outer surface. Like a prokaryotic ribosome, a eukaryotic ribosome consists of a large and a small subunit (see Figure 5.11a). Eukaryotic ri-

bosomes are larger than prokaryotic ribosomes and contain four types of ribosomal RNA molecules and more than 80 proteins. Their function is identical to that of prokaryotic ribosomes: they use the information in messenger RNA to assemble amino acids into proteins.

The proteins made on ribosomes attached to the ER enter the ER lumen, where they fold into their final form. Chemical modifications of these proteins, such as addition of carbohydrate groups to produce glycoproteins, occur in the lumen. The proteins are then delivered to other regions of the cell within small vesicles that pinch off from the ER, travel through the cytosol, and join with the organelle that performs the next steps in their modification and distribution. For most of the proteins made on the rough ER, the next destination is the Golgi complex, which packages and sorts them for delivery to their final destinations.

The outer membrane of the nuclear envelope is closely related in structure and function to the rough

ER, to which it is often connected. This membrane is also a "rough" membrane, covered with ribosomes attached to the surface facing the cytoplasm. The proteins made on these ribosomes enter the space between the two nuclear envelope membranes. From there, the proteins can move into the ER and on to other cellular locations.

Proteins made on ribosomes that are freely suspended in the cytosol may remain in the cytosol, pass through the nuclear pores to enter the nucleus, or become parts of mitochondria, chloroplasts, the cytoskeleton, or other cytoplasmic structures. Proteins that enter the nucleus become part of chromatin or remain in solution in the nucleoplasm.

The **smooth ER** is so called because its membranes have no ribosomes attached to their surfaces (see Figure 5.11b). The smooth ER has various functions in the cytoplasm, including synthesis of lipids that become part of cell membranes. In some cells, such as those of the liver, smooth ER membranes contain enzymes that convert drugs, poisons, and toxic by-products of cellular metabolism into substances that can be tolerated or more easily removed from the body.

The rough and smooth ER membranes are often connected, making the entire ER system a continuous network of interconnected channels in the cytoplasm. The relative proportions of rough and smooth ER reflect cellular activities in protein and lipid synthesis. Cells that are highly active in making proteins to be released outside the cell, such as pancreatic cells that make digestive enzymes, are packed with rough ER but have relatively little smooth ER. By contrast, cells that primarily synthesize lipids or break down toxic substances are packed with smooth ER but contain little rough ER.

Golgi Complex. Camillo Golgi, a late-nineteenth-century Italian neuroscientist and Nobel laureate, discovered the Golgi complex. The **Golgi complex** consists of a stack of flattened, membranous sacs without attached ribosomes. In most cells, the complex looks like a stack of cupped pancakes **(Figure 5.12).** The Golgi complex is usually located near concentrations of rough ER membranes, between the ER and the plasma membrane.

The Golgi complex receives proteins that were made in the ER and transported to the complex in vesicles. Within the Golgi complex, further chemical modifications of the proteins occur, for example, removal of segments of the amino acid chain, addition of small functional groups, or addition of lipid or carbohydrate units. The modified proteins then are sorted into vesicles that pinch off from the margins of Golgi sacs on the side of the complex that faces the plasma membrane.

The Golgi complex regulates the movement of several types of proteins. Some are secreted from the cell, others become embedded in the plasma membrane as integral membrane proteins, and yet others are placed in lysosomes. For instance, proteins secreted from the cell are transported to the plasma membrane by

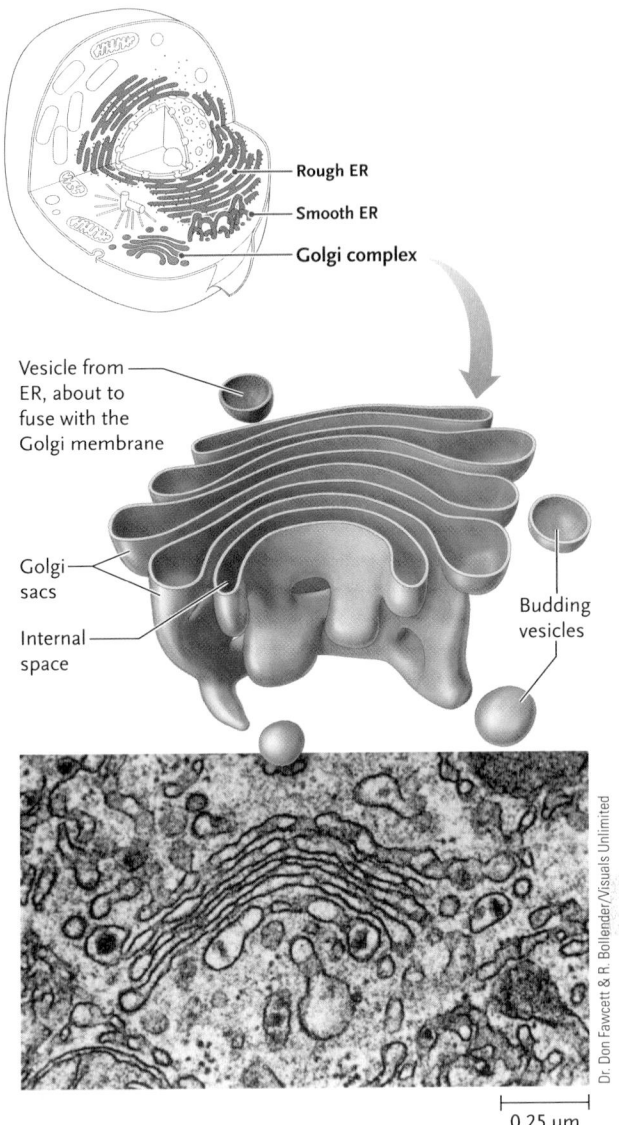

Rough ER

Smooth ER

Golgi complex

Vesicle from ER, about to fuse with the Golgi membrane

Golgi sacs

Internal space

Budding vesicles

0.25 μm

Dr. Don Fawcett & R. Bollender/Visuals Unlimited

Figure 5.12
The Golgi complex.

secretory vesicles, which release their contents to the exterior by **exocytosis (Figure 5.13a).** In this process, a secretory vesicle fuses with the plasma membrane and spills the vesicle contents to the outside.

Vesicles also may form by the reverse process, called **endocytosis,** which brings molecules into the cell from the exterior **(Figure 5.13b).** In this process, the plasma membrane forms a pocket, which bulges inward and pinches off into the cytoplasm as an **endocytic vesicle.** Once in the cytoplasm, endocytic vesicles, which contain segments of the plasma membrane as well as proteins and other molecules, are carried to the Golgi complex or to other destinations such as lysosomes in animal cells. The substances carried to the Golgi complex are sorted and placed into vesicles for routing to other locations, which may include lysosomes. Those routed to lysosomes are digested into molecular subunits that may be recycled as the building blocks for the biological molecules of the cell. Exo-

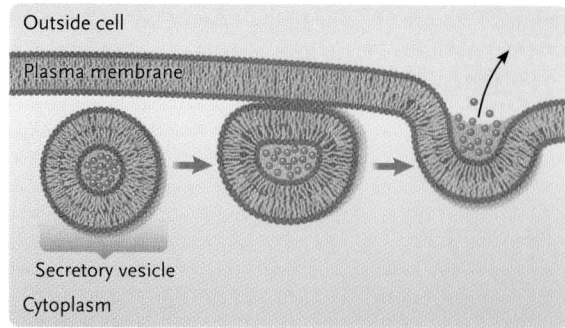

a. Exocytosis

Outside cell

Plasma membrane

Secretory vesicle

Cytoplasm

In exocytosis, a secretory vesicle fuses with the plasma membrane, releasing the vesicle contents to the cell exterior. The vesicle membrane becomes part of the plasma membrane.

b. Endocytosis

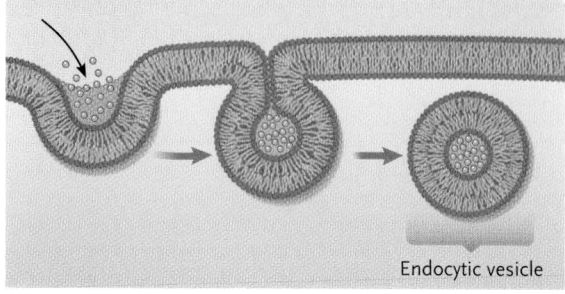

In endocytosis, materials from the cell exterior are enclosed in a segment of the plasma membrane that pockets inward and pinches off as an endocytic vesicle.

Endocytic vesicle

Figure 5.13
Exocytosis and endocytosis.

cytosis and endocytosis are discussed in more detail in Chapter 6.

Lysosomes. Lysosomes are membrane-bound vesicles that contain more than 30 hydrolytic enzymes for the digestion of many complex molecules, including proteins, lipids, nucleic acids, and polysaccharides **(Figure 5.14).** The cell recycles the subunits of these molecules.

Lysosomes are formed by budding from the Golgi complex. Their hydrolytic enzymes are synthesized in the rough ER, modified in the lumen of the ER to identify them as being bound for a lysosome, transported to the Golgi complex in a vesicle, and then packaged in the budding lysosome.

The pH within lysosomes is acidic (pH ~5) and is significantly lower than the pH of the cytosol (pH ~7.2). The hydrolytic enzymes in the lysosomes function optimally at the acidic pH within the organelle, but they do not function well at the pH of the cytosol; this difference reduces the risk to the viability of the cell should the en-

Figure 5.14
A lysosome.

Lysosome

Lysosome containing ingested material

Don Fawcett/Photo Researchers, Inc.

zymes be released from the vesicle.

Lysosomal enzymes can digest several types of materials. They digest food molecules entering the cell by endocytosis when an endocytic vesicle fuses with a lysosome. In a process called *autophagy,* they digest organelles that are not functioning correctly. A membrane surrounds the defective organelle, forming a large vesicle that fuses with one or more lysosomes; the organelle then is degraded by the hydrolytic enzymes. They also play a role in **phagocytosis**, a process in which some types of cells engulf bacteria or other cellular debris to break them down. These cells include the white blood cells known as *phagocytes,* which play an important role in the immune system (see Chapter 43). Phagocytosis produces a large vesicle that contains the engulfed materials until lysosomes fuse with the vesicle and release the hydrolytic enzymes necessary for degrading them.

In certain human genetic diseases known as *lysosomal storage diseases,* one of the hydrolytic enzymes normally found in the lysosome is absent. As a result, the substrate of that enzyme accumulates in the lysosomes, and this accumulation eventually interferes with normal cellular activities. An example is Tay–Sachs disease, which is a fatal disease of the central nervous system caused by the failure to synthesize the enzyme needed for hydrolysis of fatty acid derivatives found in brain and nerve cells.

Summary. In summary, the endomembrane system is a major traffic network for proteins and other substances within the cell. The Golgi complex in particular is a key distribution station for membranes and proteins **(Figure 5.15).** From the Golgi complex, lipids and proteins may move to storage or secretory vesicles, and from the secretory vesicles, they may move to the cell exterior by exocytosis. Membranes and proteins may also move between the nuclear envelope and the endomembrane system. Proteins and other materials that enter cells by endocytosis also enter the endomembrane system to travel to the Golgi complex for sorting and distribution to other locations. Details of how proteins are routed within cells to their final destinations are presented in Chapter 7.

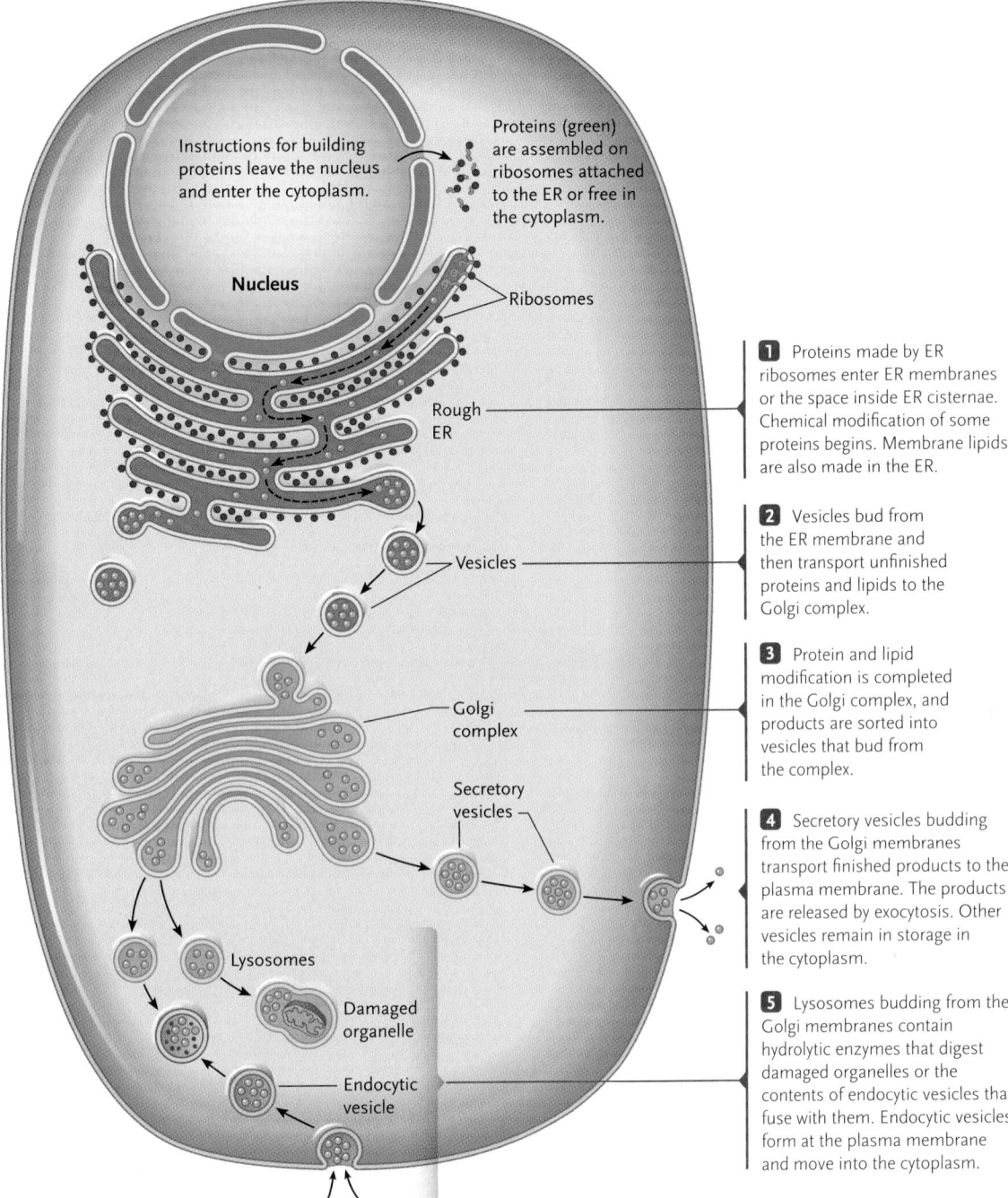

Instructions for building proteins leave the nucleus and enter the cytoplasm.

Proteins (green) are assembled on ribosomes attached to the ER or free in the cytoplasm.

Nucleus

Ribosomes

Rough ER

1 Proteins made by ER ribosomes enter ER membranes or the space inside ER cisternae. Chemical modification of some proteins begins. Membrane lipids are also made in the ER.

Vesicles

2 Vesicles bud from the ER membrane and then transport unfinished proteins and lipids to the Golgi complex.

Golgi complex

3 Protein and lipid modification is completed in the Golgi complex, and products are sorted into vesicles that bud from the complex.

Secretory vesicles

4 Secretory vesicles budding from the Golgi membranes transport finished products to the plasma membrane. The products are released by exocytosis. Other vesicles remain in storage in the cytoplasm.

Lysosomes

Damaged organelle

Endocytic vesicle

5 Lysosomes budding from the Golgi membranes contain hydrolytic enzymes that digest damaged organelles or the contents of endocytic vesicles that fuse with them. Endocytic vesicles form at the plasma membrane and move into the cytoplasm.

Figure 5.15
Vesicle traffic in the cytoplasm. The ER and Golgi complex are part of the endomembrane system, which releases proteins and other substances to the cell exterior and gathers materials from outside the cell.

Mitochondria Are the Powerhouses of the Cell

Mitochondria (singular, *mitochondrion*) are the membrane-bound organelles in which cellular respiration occurs. *Cellular respiration* is the process by which energy-rich molecules such as sugars, fats, and other fuels are broken down to water and carbon dioxide by mitochondrial reactions, with the release of energy. Much of the energy released by the breakdown is captured in ATP. Mitochondria require oxygen for this process—when you breathe, you are taking in oxygen primarily for your mitochondrial reactions (see Chapter 8). Mitochondria are frequently called the powerhouses of the cell because of their ATP-generating activities.

Mitochondria are enclosed by two membranes **(Figure 5.16)**. The **outer mitochondrial membrane** is smooth and covers the outside of the organelle. The surface area of the **inner mitochondrial membrane** is expanded by folds called **cristae** (singular, *crista*). Both membranes surround the innermost compartment of the mitochondrion, called the **mitochondrial matrix**. The ATP-generating reactions of mitochondria occur in the cristae and matrix.

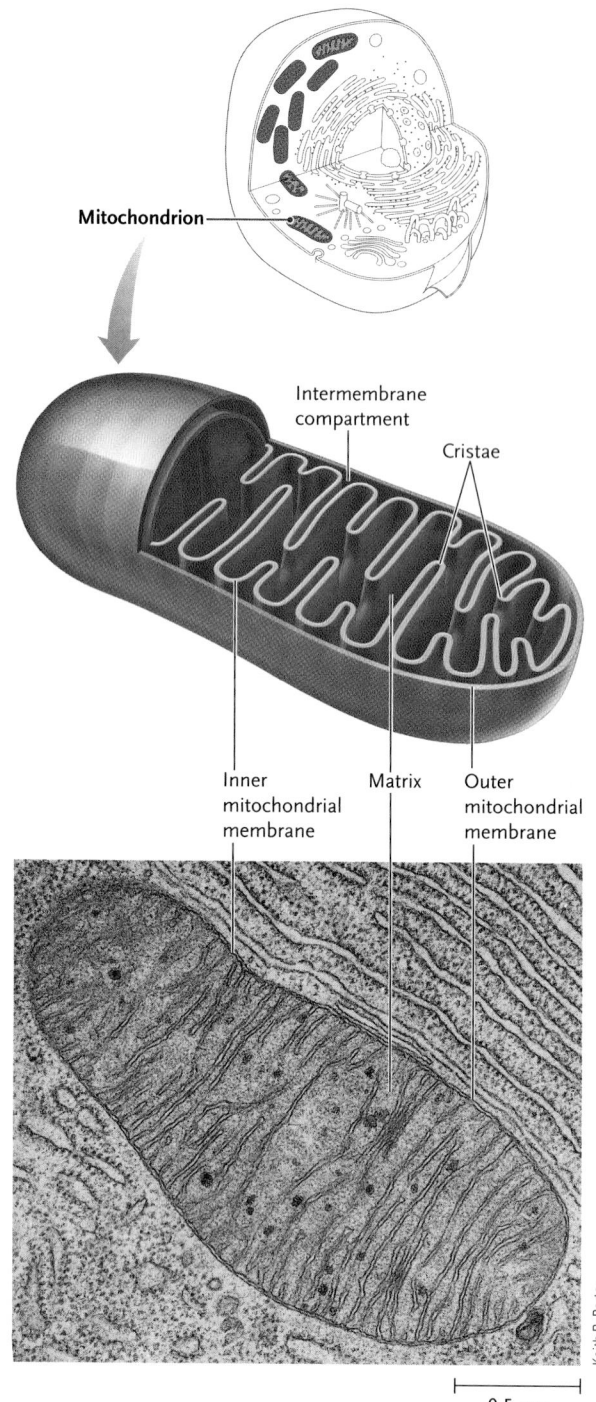

Figure 5.16

Mitochondria. The electron micrograph shows a mitochondrion from bat pancreas, surrounded by cytoplasm that contains rough ER. Cristae extend into the interior of the mitochondrion as folds from the inner mitochondrial membrane. The darkly stained granules inside the mitochondrion are probably lipid deposits.

Mitochondrion

Intermembrane compartment

Cristae

Inner mitochondrial membrane

Matrix

Outer mitochondrial membrane

0.5 μm

Keith R. Porter

The mitochondrial matrix also contains DNA and ribosomes that resemble the equivalent structures in bacteria. These and other similarities suggest that mitochondria originated from ancient bacteria that became permanent residents of the cytoplasm during the evolution of eukaryotic cells (see Chapter 24 for further discussion).

Microbodies Carry Out Vital Reactions That Link Metabolic Pathways

Microbodies are small, relatively simple membrane-bound organelles found in various forms in essentially all eukaryotic cells. They consist of a single boundary membrane that encloses a collection of enzymes and other proteins **(Figure 5.17)**. Recent research has shown that the ER is involved in microbody production. Proteins and phospholipids are continuously imported into microbodies. The phospholipids are used for new membrane synthesis, leading to growth of the microbody. Division of a microbody then produces new microbodies.

Microbodies have various functions that are often specific to an organism or cell type. Commonly, microbodies contain enzymes that conduct preparatory or intermediate reactions linking major biochemical pathways. For example, the series of reactions that allows cells to use fats as an energy source begins in microbodies and continues in mitochondria. Beginning or intermediate steps in the breakdown of some amino acids and alcohols also take place in microbodies, including about half of the ethyl alcohol that humans consume. Many types of microbodies produce as a by-product the toxic substance hydrogen peroxide (H_2O_2), which is broken down into water and oxygen by the enzyme *catalase*. Microbodies with this reaction are often termed **peroxisomes.**

Microbodies in plants convert oils or fats to sugars that can be used directly for energy-releasing reactions in mitochondria or for reactions that require sugars as chemical building blocks. These microbody reactions are particularly important in plant embryos that develop from oily seeds, such as those of the peanut or soybean. Depending on the particular reaction pathways they carry out, plant microbodies are called peroxisomes, *glyoxysomes*, or *glycosomes*.

The Cytoskeleton Supports and Moves Cell Structures

The characteristic shape and internal organization of each type of cell is maintained in part by its **cytoskeleton**, the interconnected system of protein fibers and tubes that extends throughout the cytoplasm. The cytoskeleton also reinforces the plasma membrane and functions in movement, both of structures within the cell and of the cell as a whole. It is most highly developed in animal cells, in which it fills and supports the cytoplasm from the plasma membrane to the nuclear envelope **(Figure 5.18)**. Although cytoskeletal structures are also present in plant cells, the fibers and tubes of the system are less prominent; much of cellular support in plants is provided by the cell wall and a large central vacuole (described in Section 5.4).

The cytoskeleton of animal cells contains structural elements of three major types: microtubules, intermediate filaments, and microfilaments. Plant cytoskeletons contain only microtubules and microfilaments. **Microtubules (Figure 5.19a)** are microscopic tubes about 25 nm in diameter; they function much like the tubes used by human engineers to construct supportive structures. **Intermediate filaments (Figure 5.19b)** are fibers with diameters of about 8 to 12 nm. These fibers occur singly, in parallel bundles, and in interlinked networks, either alone or in combination with microtubules, microfilaments, or both. **Microfilaments (Figure 5.19c)** are thin fibers 5 to 7 nm in diameter that consist of two rows of protein subunits wound around each other in a long spiral.

Each cytoskeletal element is assembled from proteins—microtubules from *tubulins,* intermediate filaments from a large and varied group of *intermediate filament proteins,* and microfilaments from *actins* (see Figure 5.19). The keratins of animal hair, nails, and claws contain a common form of intermediate filament proteins known as the *cytokeratins.* For example, human hair consists of thick bundles of cytokeratin fibers extruded from hair follicle cells. The lamins that line the inner surface of the nuclear envelope in animal cells are also assembled from intermediate filament proteins.

Many of the cytoskeletal microtubules in animal cells are formed and radiate outward from a site near the nucleus termed the **cell center** or **centrosome** (see Figure 5.8a). At its midpoint are two short, barrel-shaped structures also formed from microtubules called the **centrioles** (see Figure 5.23). Often, intermediate filaments extend from the cell center as well, apparently held in the same radiating pattern by linkage to microtubules. Microtubules that radiate from the cell center anchor the ER, Golgi complex, lysosomes, secretory vesicles, and at least some mitochondria in

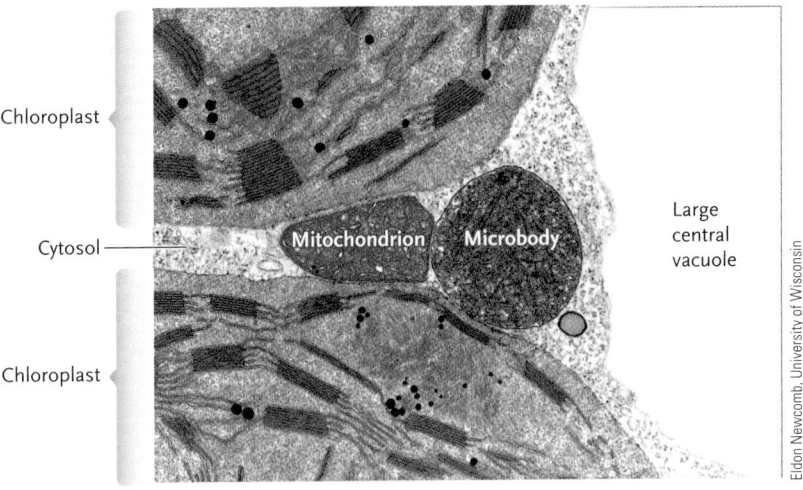

Figure 5.17

A microbody in the cytoplasm of a tobacco leaf cell. The EM has been colorized to make the structures easier to identify.

position. The microtubules also provide tracks along which vesicles move from the cell interior to the plasma membrane and in the reverse direction. The intermediate filaments probably add support to the microtubule arrays.

Eukaryotic cell movements are generated by "motor" proteins that push or pull against microtubules or microfilaments, much as our muscles produce body movements by acting on bones of the skeleton. One end of a motor protein is firmly fixed to a cell structure such as a vesicle or to a microtubule or microfilament. The other end has reactive groups that "walk" along another microtubule or microfilament by making an attachment, forcefully swiveling a short distance, and then releasing **(Figure 5.20)**. ATP supplies the energy

a. Microtubules   **b.** Intermediate filaments   **c.** Microfilaments

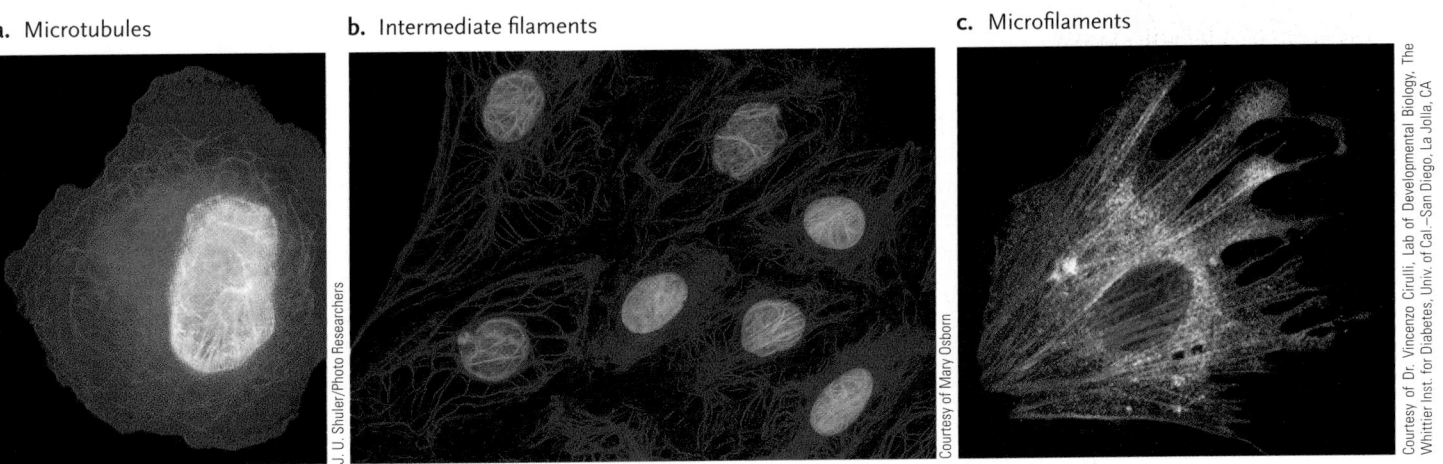

Figure 5.18

Cytoskeletons of eukaryotic cells, as seen in cells stained for light microscopy. **(a)** Microtubules (yellow) and microfilaments (red) in a pancreatic cell. **(b)** Intermediate filaments assembled from keratin proteins in cells of the kangaroo rat. The nucleus is stained blue in these cells. **(c)** Microfilaments (red) in a migrating mammalian cell.

a. Microtubule **b.** Intermediate filament **c.** Microfilament

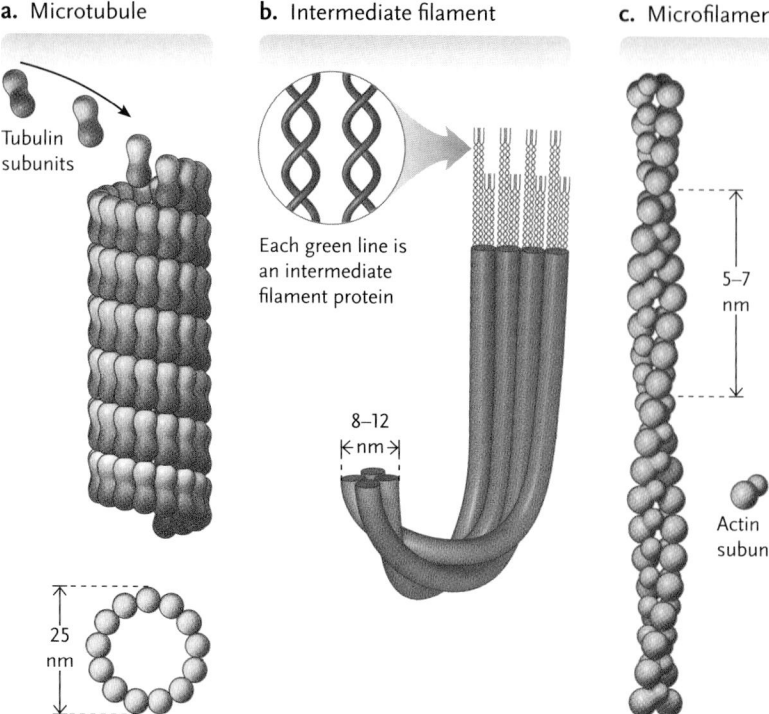

Tubulin subunits

Each green line is an intermediate filament protein

8–12 nm

25 nm

5–7 nm

Actin subunit

Figure 5.19

The major components of the cytoskeleton. **(a)** A microtubule, assembled from tubulin proteins. **(b)** An intermediate filament. Eight protein chains wind together to form each subunit shown as a green cylinder. **(c)** A microfilament, assembled from two rows of actin proteins, wound around each other into a double helix.

for the walking movements. The motor proteins that walk along microfilaments are called *myosins,* and the ones that walk along microtubules are called *dyneins* and *kinesins.*

Some cell movements, such as the whipping motions of sperm tails, depend entirely on microtubules and their motor proteins. Microfilaments are solely responsible for other types of movements, including *amoeboid motion,* the actively flowing motion of cytoplasm called *cytoplasmic streaming,* and the contraction of muscle cells (the roles of myosin and microfilaments in muscle contraction are discussed further in Chapter 41). When animal cells divide, both microtubules and microfilaments are active—the chromosomes are separated and moved by microtubules, and the cytoplasm is divided by microfilaments (see Chapter 10 for further discussion).

a. "Walking" end of a kinesin molecule

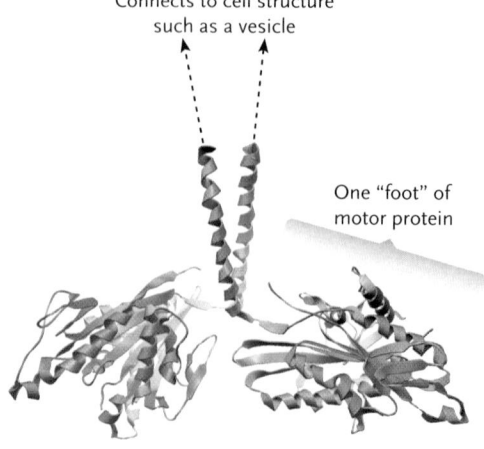

Connects to cell structure such as a vesicle

One "foot" of motor protein

b. How a kinesin molecule "walks"

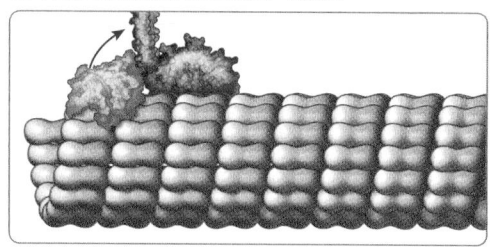

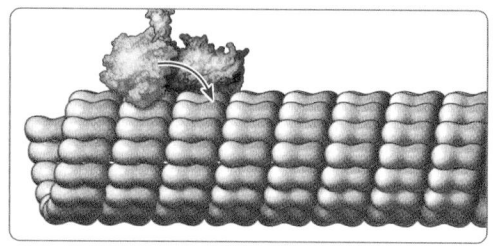

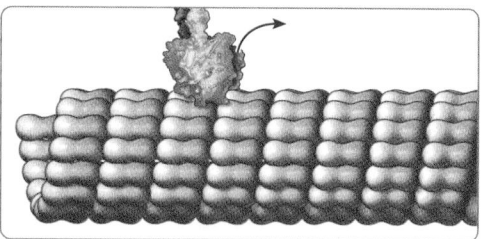

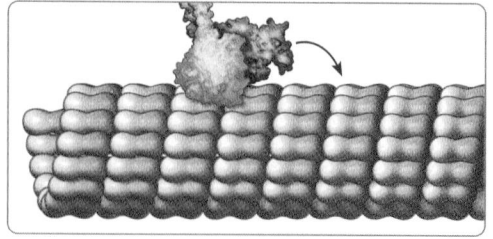

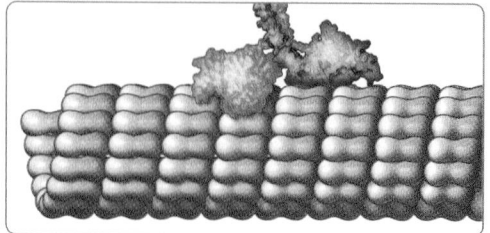

Figure 5.20

The microtubule motor protein kinesin. **(a)** Structure of the end of a kinesin molecule that "walks" along a microtubule, with α-helical segments shown as spirals and β strands as flat ribbons. **(b)** How a kinesin molecule walks along the surface of a molecule by alternately attaching and releasing its "feet."

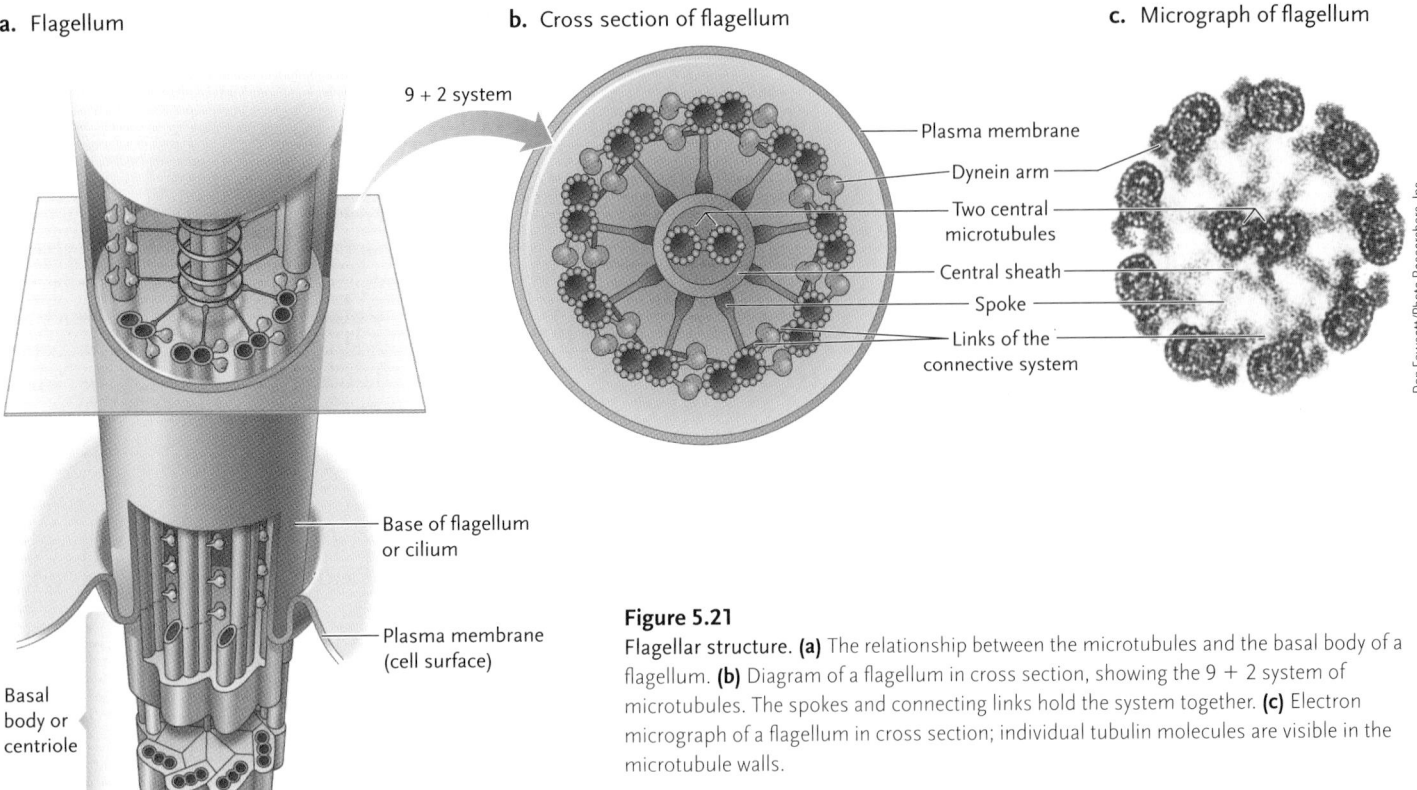

a. Flagellum   **b.** Cross section of flagellum   **c.** Micrograph of flagellum

9 + 2 system

Plasma membrane

Dynein arm

Two central microtubules

Central sheath

Spoke

Links of the connective system

Base of flagellum or cilium

Plasma membrane (cell surface)

Basal body or centriole

Don Fawcett/Photo Researchers, Inc.

Figure 5.21
Flagellar structure. (a) The relationship between the microtubules and the basal body of a flagellum. **(b)** Diagram of a flagellum in cross section, showing the 9 + 2 system of microtubules. The spokes and connecting links hold the system together. **(c)** Electron micrograph of a flagellum in cross section; individual tubulin molecules are visible in the microtubule walls.

Flagella Propel Cells, and Cilia Move Materials over the Cell Surface

Flagella and **cilia** (singular, *cilium*) are elongated, slender, motile structures that extend from the cell surface. They are identical in structure except that cilia are usually shorter than flagella and occur on cells in greater numbers. Whiplike or oarlike movements of a flagellum propel a cell through a watery medium, and cilia move fluids over the cell surface.

A bundle of microtubules extends from the base to the tip of a flagellum or cilium **(Figure 5.21)**. In the bundle, a circle of nine double microtubules surrounds a central pair of single microtubules, forming what is known as the 9 + 2 complex. Dynein motor proteins slide the microtubules of the 9 + 2 complex over each other to produce the movements of a flagellum or cilium **(Figure 5.22)**.

Flagella and cilia arise from the centrioles. These barrel-shaped structures contain a bundle of microtubules similar to the 9 + 2 complex, except that the central pair of microtubules is missing and the outer circle is formed from a ring of nine triple rather than double microtubules (compare Figure 5.21 and **Figure 5.23**). During the formation of a flagellum or cilium, a centriole moves to a position just under the plasma membrane. Then two of the three microtubules of each triplet grow outward from one end of the centriole to form the ring of nine double microtubules. The two central microtubules of the 9 + 2 complex also grow

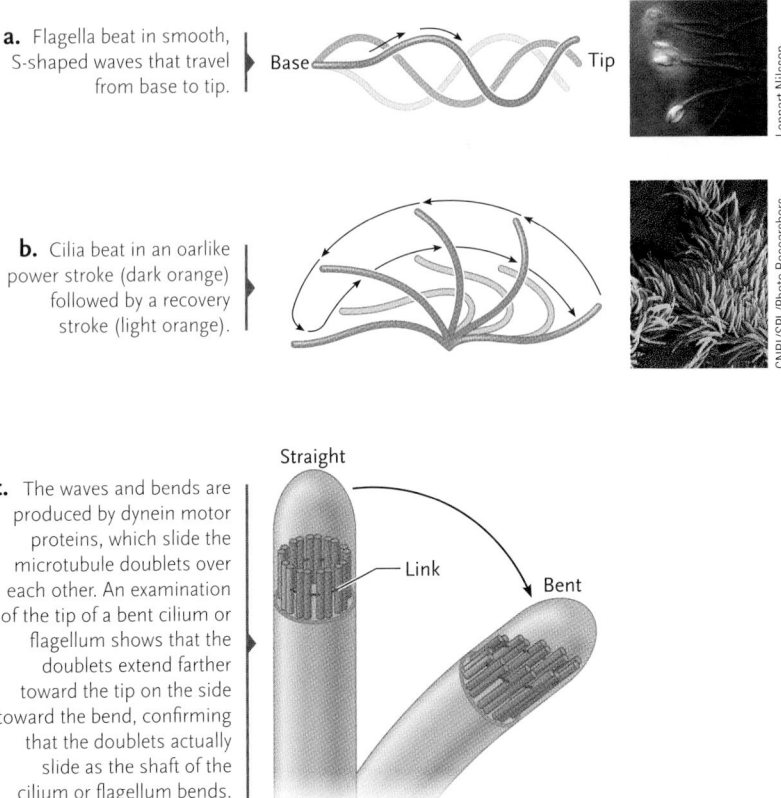

a. Flagella beat in smooth, S-shaped waves that travel from base to tip.

Base   Tip

Lennart Nilsson

b. Cilia beat in an oarlike power stroke (dark orange) followed by a recovery stroke (light orange).

CNRI/SPL/Photo Researchers

Straight

c. The waves and bends are produced by dynein motor proteins, which slide the microtubule doublets over each other. An examination of the tip of a bent cilium or flagellum shows that the doublets extend farther toward the tip on the side toward the bend, confirming that the doublets actually slide as the shaft of the cilium or flagellum bends.

Link

Bent

Figure 5.22
Flagellar and ciliary beating patterns. The micrographs show a few human sperm, each with a flagellum (top), and cilia from the lining of an airway in the lungs (bottom).

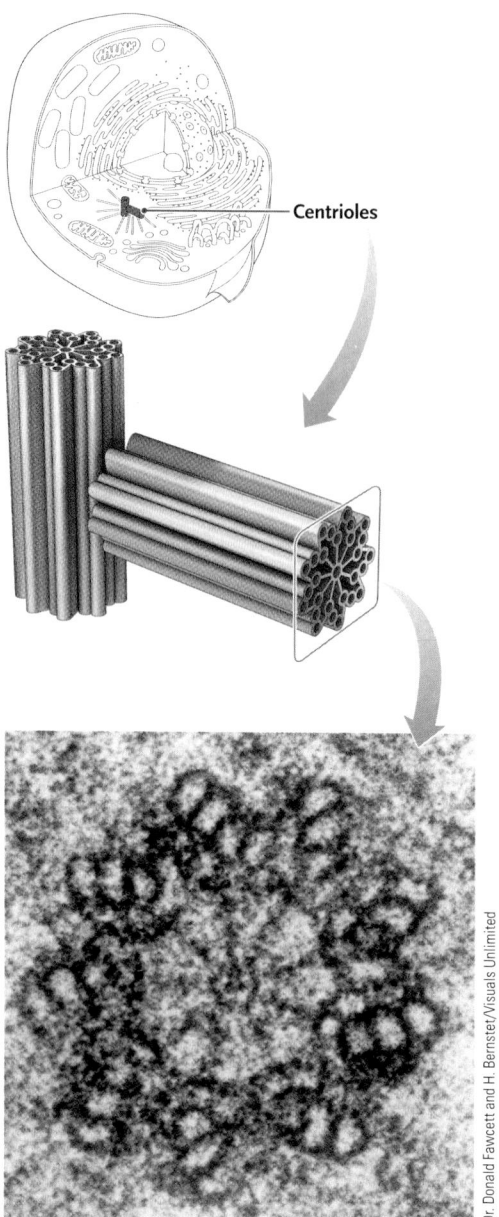

Centrioles

Figure 5.23
Centrioles. The two centrioles of the pair at the cell center usually lie at right angles to each other as shown. The electron micrograph shows a centriole from a mouse cell in cross section. A centriole gives rise to the 9 + 2 system of a flagellum and persists as the basal body at the inner end of the flagellum.

Dr. Donald Fawcett and H. Bernster/Visuals Unlimited

Although the purpose of the eukaryotic flagellum is the same as that of prokaryotic flagella, the genes that encode the components of the flagellar apparatus of cells of Bacteria, Archaea, and Eukarya are different in each case. Thus, the three types of flagella are analogous, not homologous, structures, and they must have evolved independently.

With a few exceptions, the cell structures described so far in this chapter occur in all eukaryotic cells. The major exception is intermediate filaments, which appear to be restricted to animal cells. The next section describes three additional structures that are characteristic of plant cells.

STUDY BREAK

1. Where in a eukaryotic cell is DNA found? How is that DNA organized?
2. What is the nucleolus, and what is its function?
3. Explain the structure and function of the endomembrane system.
4. What is the structure and function of a mitochondrion?
5. What is the structure and function of the cytoskeleton?

5.4 Specialized Structures of Plant Cells

Chloroplasts, a large and highly specialized central vacuole, and cell walls give plant cells their distinctive characteristics, but these structures also occur in some other eukaryotes—chloroplasts in algal protists and cell walls in algal protists and fungi. They are shown in Figure 5.9 and described in the following sections.

Chloroplasts Are Biochemical Factories Powered by Sunlight

Chloroplasts (Figure 5.24), like mitochondria, are usually lens- or disc-shaped and are surrounded by a smooth **outer boundary membrane,** and an **inner boundary membrane,** which lies just inside the outer membrane. These two boundary membranes completely enclose an inner compartment, the **stroma.** Within the stroma is a third membrane system that consists of flattened, closed sacs called **thylakoids.** In higher plants, the thylakoids are stacked, one on top of another, forming structures called **grana** (singular, *granum*).

Chloroplasts are the sites of photosynthesis in plant cells. The thylakoid membranes contain molecules that absorb light energy and convert it to chemical energy. The primary molecule absorbing light is

from the end of the centriole, but without direct connection to any centriole microtubules. The centriole remains at the innermost end of a flagellum or cilium when its development is complete as the **basal body** of the structure (see Figure 5.21).

Cilia and flagella are found in protozoa and algae, and many types of animal cells have flagella—the tail of a sperm cell is a flagellum—as do the reproductive cells of some plants. In humans, cilia cover the surfaces of cells lining cavities or tubes in some parts of the body. For example, cilia on cells lining the ventricles (cavities) of the brain circulate fluid through the brain, and cilia in the oviducts conduct eggs from the ovaries to the uterus. Cilia covering cells that line the air passages of the lungs sweep out mucus containing bacteria, dust particles, and other contaminants.

chlorophyll, a green pigment that is present in all chloroplasts. The chemical energy is used by enzymes in the stroma to make carbohydrates and other complex organic molecules from water, carbon dioxide, and other simple inorganic precursors. The organic molecules produced in chloroplasts, or from biochemical building blocks made in chloroplasts, are the ultimate food source for most organisms. (The physical and biochemical reactions of chloroplasts are described in Chapter 9.)

Chloroplasts are members of a family of plant organelles known collectively as **plastids**. Other members of the family include amyloplasts and chromoplasts. **Amyloplasts** (*amylo* = starch) are colorless plastids that store starch, a product of photosynthesis. They occur in great numbers in the roots or tubers of some plants, such as the potato. **Chromoplasts** (*chromo* = color) contain red and yellow pigments and are responsible for the colors of ripening fruits or autumn leaves.

The chloroplast stroma contains DNA and ribosomes that resemble those of certain photosynthetic bacteria. Because of these similarities, chloroplasts, like mitochondria, are believed to have originated from ancient prokaryotes that became permanent residents of the eukaryotic cells ancestral to the plant lineage (see Chapter 24 for further discussion).

Central Vacuoles Have Diverse Roles in Storage, Structural Support, and Cell Growth

Central vacuoles (see Figure 5.9) are large vesicles that are identified as distinct organelles of plant cells because they perform specialized functions unique to plants. In a mature plant cell, 90% or more of the cell's volume may be occupied by one or more large central vacuoles. The remainder of the cytoplasm and the nucleus of these cells are restricted to a narrow zone between the central vacuole and the plasma membrane. The pressure within the central vacuole supports the cells.

The membrane that surrounds the central vacuole, the **tonoplast**, contains transport proteins that move substances into and out of the central vacuole. As plant cells mature, they grow primarily by increases in the pressure and volume of the central vacuole.

Central vacuoles conduct other vital functions. They store salts, organic acids, sugars, storage proteins, pigments, and, in some cells, waste products. Pigments concentrated in the vacuoles produce the colors of many flowers. Enzymes capable of breaking down biological molecules are present in some central vacuoles, giving them some of the properties of lysosomes. Molecules that provide chemical defenses against pathogenic organisms also occur in the central vacuoles of some plants.

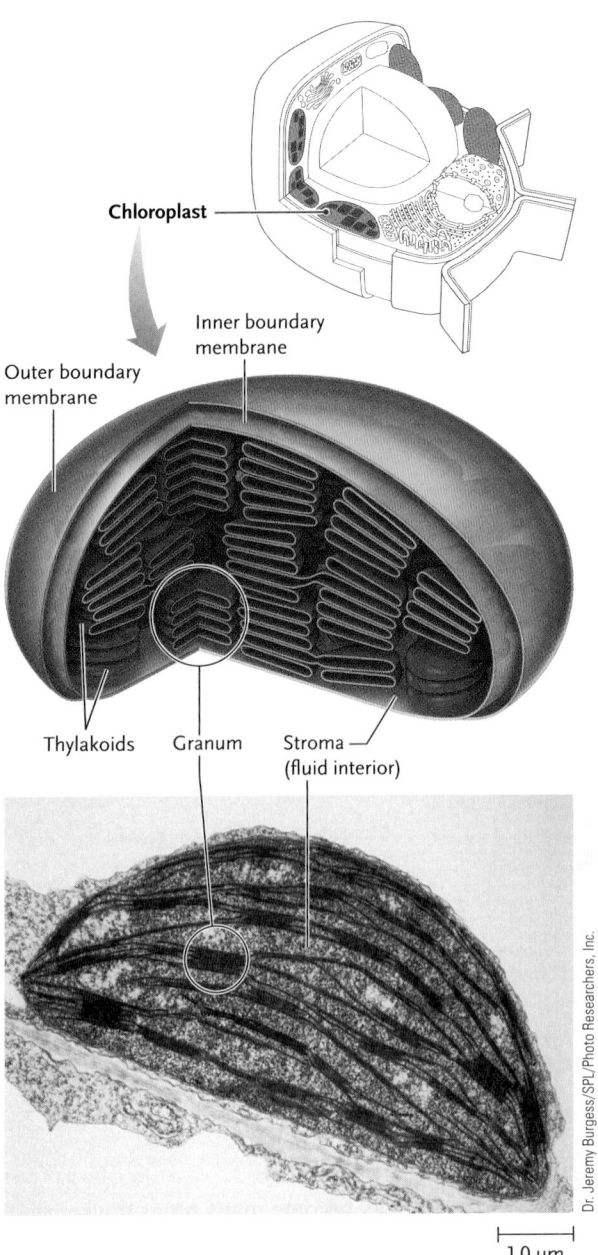

Outer boundary membrane · Inner boundary membrane · Chloroplast · Thylakoids · Granum · Stroma (fluid interior)

1.0 μm

Dr. Jeremy Burgess/SPL/Photo Researchers, Inc.

Figure 5.24
Chloroplast structure. The electron micrograph shows a maize (corn) chloroplast.

Cell Walls Support and Protect Plant Cells

The cell walls of plants are extracellular structures because they are located outside the plasma membrane **(Figure 5.25)**. Cell walls provide support to individual cells, contain the pressure produced in the central vacuole, and protect cells against invading bacteria and fungi.

Cell walls consist of cellulose fibers (see Figure 3.7c), which give tensile strength to the walls, embedded in a network of highly branched carbohydrates. The initial cell wall laid down by a plant cell, the **primary cell wall**, is relatively soft and flexible. As the cell grows and matures, the primary wall expands and additional layers of cellulose fibers and branched carbohydrates are laid down between the primary wall and the plasma membrane. The added wall layer, which is

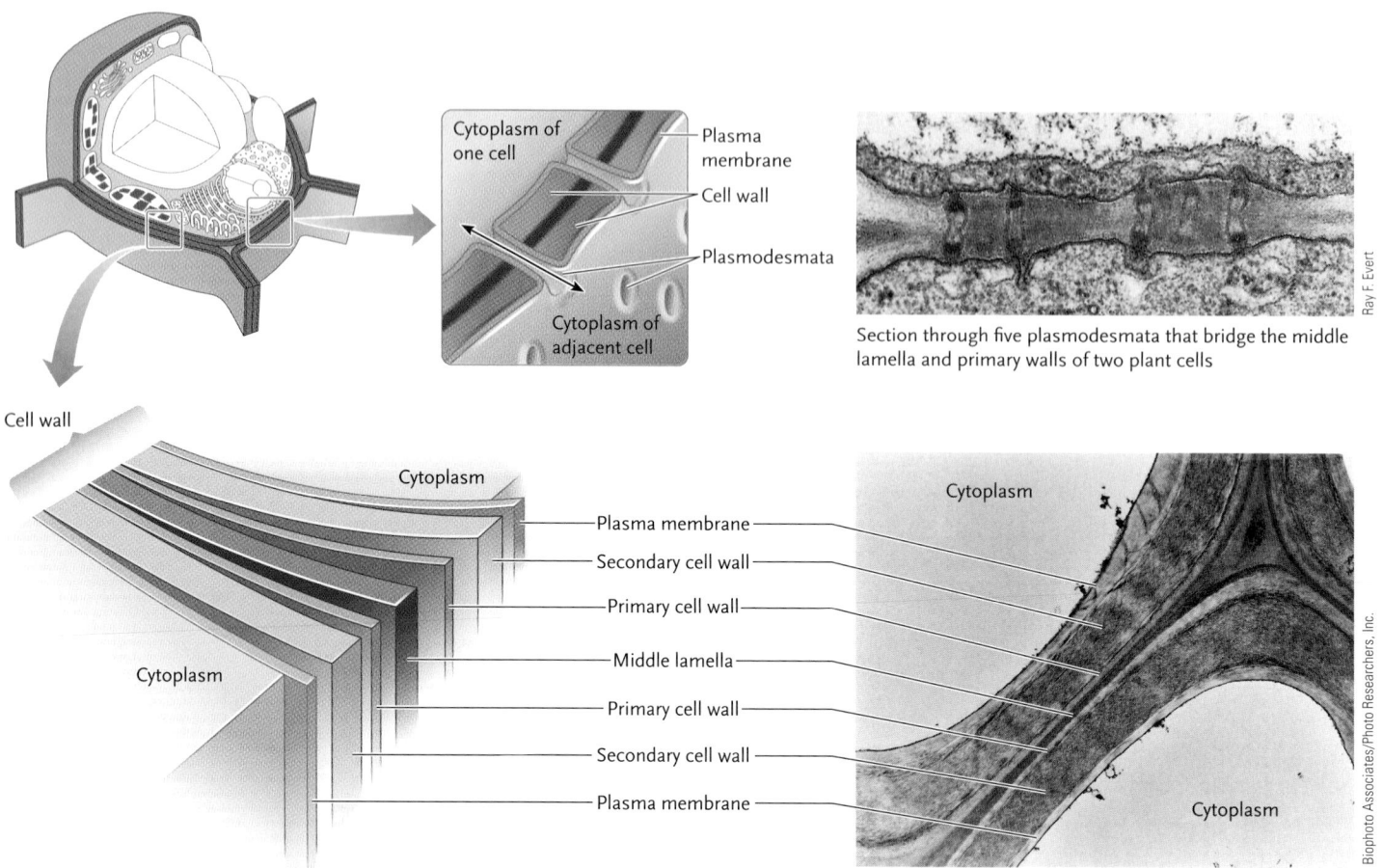

Labels in upper right diagram:
Cytoplasm of one cell — Plasma membrane — Cell wall — Plasmodesmata — Cytoplasm of adjacent cell

Ray F. Evert

Section through five plasmodesmata that bridge the middle lamella and primary walls of two plant cells

Labels in lower diagram:
Cell wall
Cytoplasm
Plasma membrane
Secondary cell wall
Primary cell wall
Middle lamella
Primary cell wall
Secondary cell wall
Plasma membrane
Cytoplasm
Cytoplasm

Biophoto Associates/Photo Researchers, Inc.

Figure 5.25

Cell wall structure. The upper right diagram and electron micrograph show plasmodesmata, which form openings in the cell wall that directly connect the cytoplasm of adjacent cells. The lower diagram and electron micrograph show the successive layers in the cell wall between two plant cells that have laid down secondary wall material.

more rigid and may become many times thicker than the primary wall, is the **secondary cell wall.** In woody plants and trees, secondary cell walls are reinforced by *lignin,* a hard, highly resistant substance assembled from complex alcohols, surrounding the cellulose fibers. Lignin-impregnated cell walls are actually stronger than reinforced concrete by weight; hence, trees can grow to substantial size, and the wood of trees is used extensively in human cultures to make many structures and objects, including houses, tables, and chairs.

The walls of adjacent cells are held together by a layer of gel-like polysaccharides called the **middle lamella,** which acts as an intercellular glue (see Figure 5.25). The polysaccharide material of the middle lamella, called *pectin,* is extracted from some plants and used to thicken jams and jellies.

Both primary and secondary cell walls are perforated by minute channels, the **plasmodesmata** (singular, *plasmodesma;* see Figure 5.25). These chan-

nels, lined by plasma membranes, contain extensions of the cytoplasm that directly connect adjacent plant cells. Plasmodesmata allow ions and small molecules to move directly from one cell to another, without having to penetrate the plasma membranes or cell walls.

Cell walls also surround the cells of fungi and algal protists. Carbohydrate molecules form the major framework of cell walls in most of these organisms, as they do in plants. In some, the wall fibers contain chitin (see Figure 3.7d) instead of cellulose. Details of cell wall structure in the algal protists and fungi, as well as in different subgroups of the plants, are presented in later chapters devoted to these organisms.

As noted earlier, animal cells do not form rigid, external, layered structures equivalent to the walls of plant cells. However, most animal cells secrete extracellular material and have other structures at the cell surface that play vital roles in the support and organization of animal body structures. The next section describes these and other surface structures of animal cells.

STUDY BREAK

1. What is the structure and function of a chloroplast?
2. What is the function of the central vacuole in plants?

5.5 The Animal Cell Surface

Animal cells have specialized structures that help hold cells together, produce avenues of communication between cells, and organize body structures. Molecular systems that perform these functions are organized at three levels: individual **cell adhesion molecules** bind cells together, more complex **cell junctions** seal the spaces between cells and provide direct communication between cells, and the **extracellular matrix (ECM)** supports and protects cells and provides mechanical linkages, such as those between muscles and bone.

Cell Adhesion Molecules Organize Animal Cells into Tissues and Organs

Cell adhesion molecules are glycoproteins embedded in the plasma membrane. They help maintain body form and structure in animals ranging from sponges to the most complex invertebrates and vertebrates. Rather than acting as a generalized intercellular glue, cell adhesion molecules bind to specific molecules on other cells. Most cells in solid body tissues are held together by many different cell adhesion molecules.

Cell adhesion molecules make initial connections between cells early in embryonic development, but then attachments are broken and remade as individual cells or tissues change position in the developing embryo. As an embryo develops into an adult, the connections become permanent and are reinforced by cell junctions. Cancer cells typically lose these adhesions, allowing them to break loose from their original locations, migrate to new locations, and form additional tumors.

Some bacteria and viruses—such as the virus that causes the common cold—target cell adhesion molecules as attachment sites during infection. Cell adhesion molecules are also partially responsible for the ability of cells to recognize one another as being part of the same individual or foreign. For example, rejection of organ transplants in mammals results from an immune response triggered by the foreign cell-surface molecules.

Cell Junctions Reinforce Cell Adhesions and Provide Avenues of Communication

Three types of cell junctions are common in animal tissues **(Figure 5.26)**. **Anchoring junctions** form buttonlike spots, or belts, that run entirely around cells, "welding" adjacent cells together. For some anchoring junctions known as **desmosomes**, intermediate filaments anchor the junction in the underlying cytoplasm; in other anchoring junctions known as **adherens junctions**, microfilaments are the anchoring cytoskeletal component. Anchoring junctions are most common in tissues that are subject to stretching, shear, or other mechanical forces—for example, heart muscle, skin, and the cell layers that cover organs or line body cavities and ducts.

Tight junctions, as the name indicates, are regions of tight connections between membranes of adjacent cells. The connection is so tight that it can keep particles as small as ions from moving between the cells in the layers.

Tight junctions seal the spaces between cells in the cell layers that cover internal organs and the outer surface of the body, or the layers that line internal cavities and ducts. For example, tight junctions between cells that line the stomach, intestine, and bladder keep the contents of these body cavities from leaking into surrounding tissues.

A tight junction is formed by direct fusion of proteins on the outer surfaces of the two plasma membranes of adjacent cells. Strands of the tight junction proteins form a complex network that gives the appearance of stitch work holding the cells together. Within a tight junction, the plasma membrane is not joined continuously; instead, there are regions of intercellular space. Nonetheless, the network of junction proteins is sufficient to make the tight cell connections characteristic of these junctions.

Gap junctions open direct channels that allow ions and small molecules to pass directly from one cell to another (see Figure 5.26). Hollow protein cylinders embedded in the plasma membranes of adjacent cells line up and form a sort of pipeline that connects the cytoplasm of one cell with the cytoplasm of the next. The flow of ions and small molecules through the channels provides almost instantaneous communication between animal cells, similar to the communication that plasmodesmata provide between plant cells.

In vertebrates, gap junctions occur between cells within almost all body tissues, but not between cells of different tissues. These junctions are particularly important in heart muscle tissues and in the smooth muscle tissues that form the uterus, where their pathways of communication allow cells of the organ to operate as a coordinated unit. Although most nerve tissues do not have gap junctions, nerve cells in dental pulp are connected by gap junctions; they are responsible for the

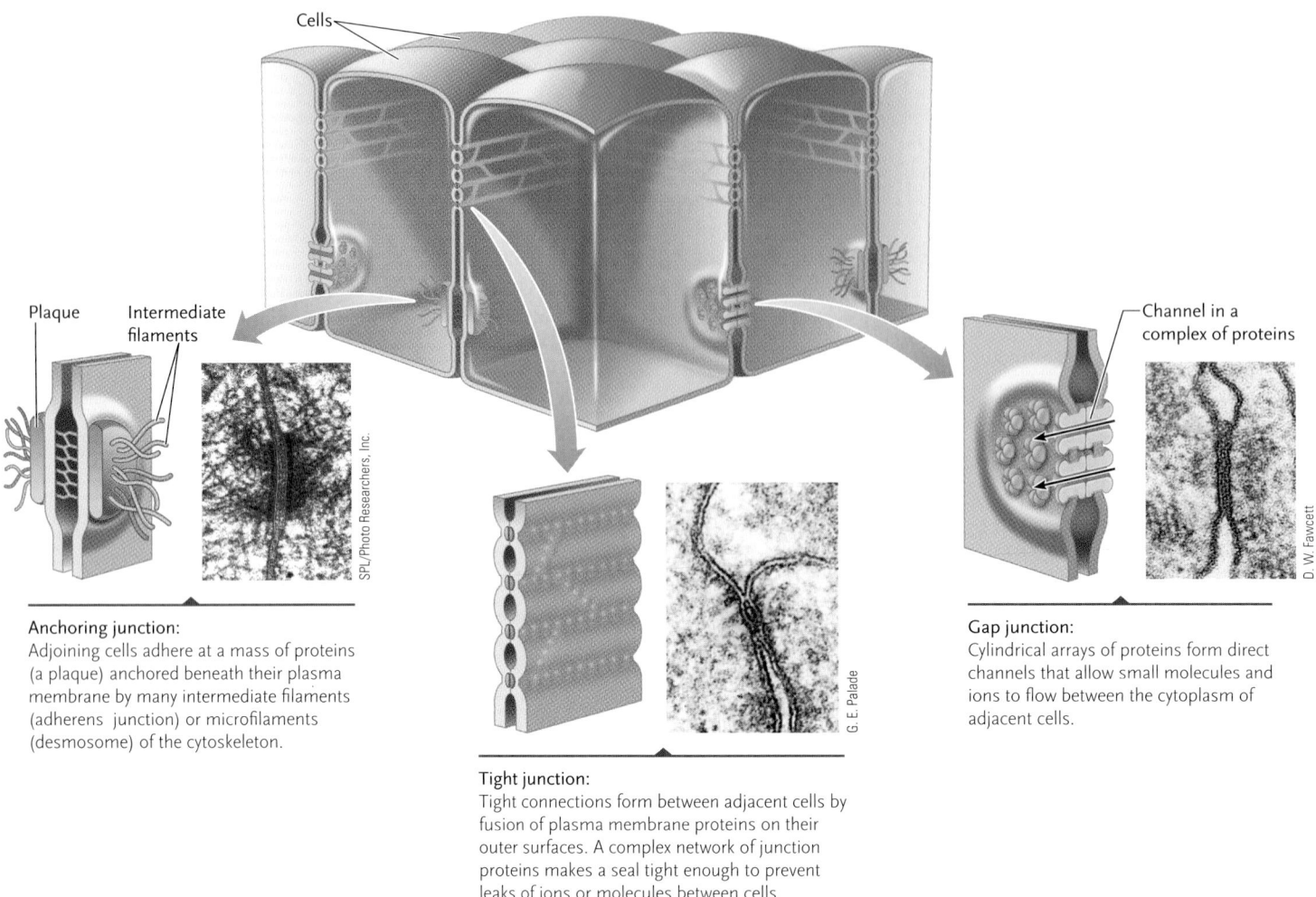

Cells

Plaque **Intermediate filaments**

SPL/Photo Researchers, Inc.

Anchoring junction:
Adjoining cells adhere at a mass of proteins (a plaque) anchored beneath their plasma membrane by many intermediate filaments (adherens junction) or microfilaments (desmosome) of the cytoskeleton.

G. E. Palade

Tight junction:
Tight connections form between adjacent cells by fusion of plasma membrane proteins on their outer surfaces. A complex network of junction proteins makes a seal tight enough to prevent leaks of ions or molecules between cells.

Channel in a complex of proteins

D. W. Fawcett

Gap junction:
Cylindrical arrays of proteins form direct channels that allow small molecules and ions to flow between the cytoplasm of adjacent cells.

Figure 5.26
Anchoring junctions, tight junctions, and gap junctions, which connect cells in animal tissues. Anchoring junctions reinforce the cell-to-cell connections made by cell adhesion molecules, tight junctions seal the spaces between cells, and gap junctions create direct channels of communication between animal cells.

discomfort you feel if your teeth are disturbed or damaged, or when a dentist pokes a probe into a cavity.

The Extracellular Matrix Organizes the Cell Exterior

Many types of animal cells are embedded in an ECM that consists of proteins and polysaccharides secreted by the cells themselves **(Figure 5.27)**. The primary function of the ECM is protection and support. The ECM forms the mass of skin, bones, and tendons; it also forms many highly specialized extracellular structures such as the cornea of the eye and filtering networks in the kidney. The ECM also affects cell division, adhesion, motility, and embryonic development, and it takes part in reactions to wounds and disease.

Glycoproteins are the main component of the ECM. In most animals, the most abundant ECM glycoprotein is *collagen,* which forms fibers with great tensile strength and elasticity. In vertebrates, the collagens of tendons, cartilage, and bone are the most abundant proteins of the body, making up about half of the total body protein by weight. (Collagens and their roles in body structures are described in further detail in Chapter 36.)

The consistency of the matrix, which may range from soft and jellylike to hard and elastic, depends on a network of proteoglycans that surrounds the collagen fibers. *Proteoglycans* are glycoproteins that consist of small proteins noncovalently attached to long polysaccharide molecules. Matrix consistency depends on the number of interlinks in this network, which determines how much water can be trapped in it. For example, cartilage, which contains a high proportion of interlinked glycoproteins, is relatively soft. Tendons, which are almost pure collagen, are tough and elastic. In bone, the glycoprotein network that surrounds collagen fibers is impregnated with mineral crystals, producing a dense and hard—but still elastic—structure that is about as strong as fiberglass or reinforced concrete.

Yet another class of glycoproteins is *fibronectins,* which aid in organizing the ECM and help cells attach to it. Fibronectins bind to receptor proteins called *integrins* that span the plasma membrane. On the cytoplasmic side of the plasma membrane, the integrins bind to microfilaments of the cytoskeleton. Integrins integrate changes outside and inside the cell by communicating changes in the ECM to the cytoskeleton.

Having laid the groundwork for cell structure and function in this chapter, we next take up further details of individual cell structures, beginning with the roles of cell membranes in transport in the next chapter.

STUDY BREAK

1. Distinguish between anchoring junctions, tight junctions, and gap junctions.
2. What is the structure and function of the extracellular matrix?

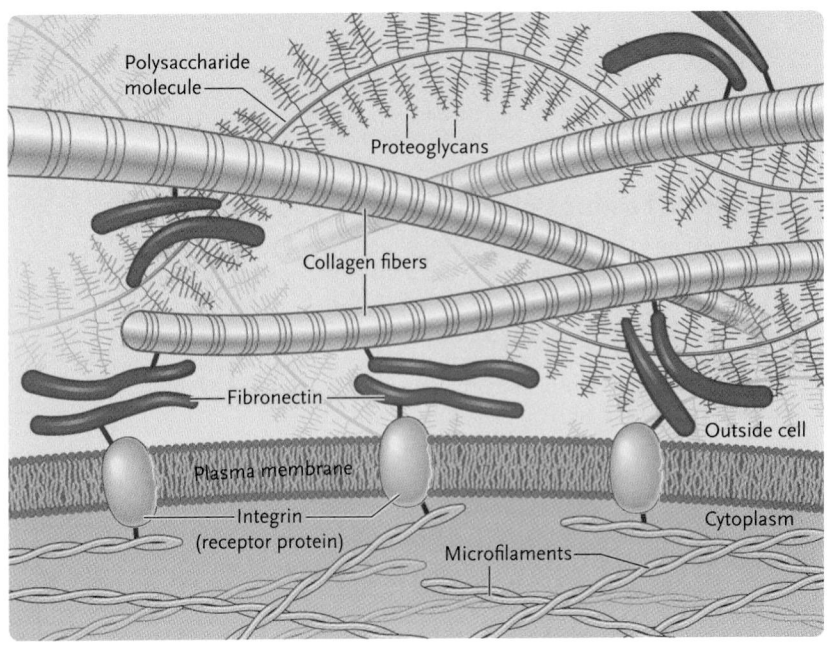

Figure 5.27
Components of the extracellular matrix.

UNANSWERED QUESTIONS

The study of cell structure and function is the focus of the field of cell biology. Research in cell biology includes an analysis of the physiological properties of cells, such as their structure at the whole-cell and subcellular levels, the reactions they conduct, their division, and their interactions with their environments. Understanding the structure and functions of cells is of core importance to all aspects of biology. Research on this topic is often closely allied with genetics, molecular biology, developmental biology, and biochemistry.

Many research questions are being addressed for both prokaryotes and eukaryotes. In prokaryotes, investigators are studying the nature of proteins that hold DNA in its structural conformations in the nucleoid, and the types of molecules found in many prokaryotes for which no function is currently known, including certain vesicles and molecular deposits. In eukaryotes, researchers are asking questions about every major eukaryotic structure described in this chapter.

What are the molecular mechanisms for protein insertion into lipid bilayers?

Stephen High's research group at the University of Manchester, England, is studying how proteins are inserted into lipid bilayers to form biologically functional membranes. They work with the ER, where many integral membrane proteins are synthesized. Using molecular approaches, High's team has identified several protein components of the ER that help regulate the integration of membrane proteins into the lipid bilayer of the ER. The Sec61 protein complex in particular plays a central role in this process, and High's group is studying the molecular mechanisms by which this protein works. They are also studying how the ER performs "quality control"—that is, how it deals with misfolded integral membrane proteins that have been inserted into the ER. Such proteins do not function properly, and it is important to remove them. Research on this problem is focused on identifying the ER components that recognize the aberrant proteins and on elucidating the molecular pathways by which those proteins are removed from the lipid bilayer and degraded.

How does the actin cytoskeleton regulate cell shape?

The ability to change shape and to move is critical for the function of a variety of cells, including single-celled amoebas and cells of the human immune system. The cellular structure responsible for shape changes and movement is the actin cytoskeleton; hence, many research groups are studying this cell component. For example, Matt Welch's research team at the University of California, Berkeley, is studying how the assembly of actin filaments in the cytoskeleton is initiated and regulated, how the actin cytoskeleton interacts with other cellular forces to drive shape change and movement, and how the actin cytoskeleton is targeted specifically by bacterial and viral pathogens to enhance their spread during infections in the organism.

Welch's work with pathogens draws from the fact that various bacterial and viral pathogens target the actin cytoskeleton of eukaryotic cells during infection. Determination of the mechanisms involved in these attacks is necessary to understand how pathogens infect cells and how to combat those infections. For instance, the spotted fever group of Rickettsia (Gram-negative, intracellular bacteria), which cause diseases such as Rocky Mountain spotted fever, enter the host cells and cause them to assemble actin filaments at their surfaces. The energy produced from this actin assembly powers the movement of the pathogen within the cell, and then its cell-to-cell spread. Currently, Welch's research team is trying to determine the mechanism by which these proteins initiate actin assembly and organize the actin into distinctive networks within infected cells.

Peter J. Russell

Review

Go to ThomsonNOW™ at www.thomsonedu.com/login to access quizzing, animations, exercises, articles, and personalized homework help.

5.1 Basic Features of Cell Structure and Function

- According to the cell theory: (1) all living organisms are composed of cells, (2) cells are the functional units of life, and (3) cells arise only from preexisting cells by a process of division.

- Cells of all kinds are divided internally into a central region containing the genetic material, and the cytoplasm, which consists of the cytosol and organelles and is bounded by the plasma membrane.

- The plasma membrane is a lipid bilayer in which transport proteins are embedded (Figure 5.6).

- In the cytoplasm, proteins are made, most of the other molecules required for growth and reproduction are assembled, and energy absorbed from the surroundings is converted into energy usable by the cell.

  Animation: Overview of cells

  Animation: Surface-to-volume ratio

  Animation: Cell membranes

5.2 Prokaryotic Cells

- Prokaryotic cells are surrounded by a plasma membrane and, in most groups, are enclosed by a cell wall. The genetic material, typically a single, circular DNA molecule, is located in the nucleoid. The cytoplasm contains masses of ribosomes (Figure 5.7).

  Animation: Typical prokaryotic cell

5.3 Eukaryotic Cells

- Eukaryotic cells have a true nucleus, which is separated from the cytoplasm by the nuclear envelope perforated by nuclear pores. A plasma membrane forms the outer boundary of the cell. Other membrane systems enclose specialized compartments as organelles in the cytoplasm (Figures 5.8 and 5.9).

- The eukaryotic nucleus contains chromatin, a combination of DNA and proteins. A specialized segment of the chromatin forms the nucleolus, where ribosomal RNA molecules are made and combined with ribosomal proteins to make ribosomes. The nuclear envelope is perforated by pores that open channels between the nucleus and the cytoplasm (Figure 5.10).

- Eukaryotic cytoplasm contains ribosomes, an endomembrane system, mitochondria, microbodies, the cytoskeleton, and some organelles specific to certain organisms. The endomembrane system includes the ER, Golgi complex, nuclear envelope, lysosomes, vesicles, and plasma membrane.

- The endoplasmic reticulum (ER) occurs in two forms, as rough and smooth ER. The ribosome-studded rough ER makes proteins that become part of cell membranes or are released from the cell. Smooth ER synthesizes lipids and breaks down toxic substances (Figure 5.11).

- The Golgi complex chemically modifies proteins made in the rough ER and sorts finished proteins to be secreted from the cell, embedded in the plasma membrane, or included in lysosomes (Figures 5.12, 5.13, and 5.15).

- Lysosomes, specialized vesicles that contain hydrolytic enzymes, digest complex molecules such as food molecules that enter the cell by endocytosis, cellular organelles that are no longer functioning correctly, and engulfed bacteria and cell debris (Figure 5.14).

- Mitochondria carry out cellular respiration, the conversion of fuel molecules into the energy of ATP (Figure 5.16).

- Microbodies conduct the initial steps in fat breakdown and other reactions that link major biochemical pathways in the cytoplasm (Figure 5.17).

- The cytoskeleton is a supportive structure built from microtubules, intermediate filaments, and microfilaments in animal cells, but from only microtubules and microfilaments in plants. Motor proteins walking along microtubules and microfilaments produce most cell movements (Figures 5.18–5.20).

- Motor protein-controlled sliding of microtubules generates the movements of flagella and cilia. Flagella and cilia arise from centrioles (Figures 5.21–5.23).

  Animation: Common eukaryotic organelles

  Animation: Nuclear envelope

  Animation: The endomembrane system

  Practice: Structure of a mitochondrion

  Animation: Cytoskeletal components

  Animation: Motor proteins

  Animation: Flagella structure

5.4 Specialized Structures of Plant Cells

- Plant cells contain all the eukaryotic structures found in animal cells except for intermediate filaments. They also contain three structures not found in animal cells: chloroplasts, a large central vacuole, and a cell wall (Figure 5.9).

- Chloroplasts contain pigments and molecular systems that absorb light energy and convert it to chemical energy. The chemical energy is used inside the chloroplasts to assemble carbohydrates and other organic molecules from simple inorganic raw materials (Figure 5.24).

- The large central vacuole, which consists of a tonoplast enclosing an inner space, develops pressure that supports plant cells, accounts for much of cellular growth by enlarging as cells mature, and serves as a storage site for substances including waste materials (Figure 5.9).

- A cellulose cell wall surrounds plant cells, providing support and protection. Plant cell walls are perforated by plasmodesmata, channels that provide direct pathways of communication between the cytoplasm of adjacent cells (Figure 5.25).

  Practice: Structure of a chloroplast

  Animation: Plant cell walls

5.5 The Animal Cell Surface

- Animal cells have specialized surface molecules and structures that function in cell adhesion, communication, and support.

- Cell adhesion molecules bind to specific molecules on other cells. The adhesions organize and hold together cells of the same type in body tissues.

- Cell adhesions are reinforced by various junctions. Anchoring junctions hold cells together. Tight junctions seal together the plasma membranes of adjacent cells, preventing ions and molecules from moving between the cells. Gap junctions open direct channels between the cytoplasm of adjacent cells (Figure 5.26).

- The extracellular matrix, formed from collagen proteins embedded in a matrix of branched glycoproteins, functions primarily in cell and body protection and support but also affects cell division, motility, embryonic development, and wound healing (Figure 5.27).

Animation: Animal cell junctions

Questions

Self-Test Questions

1. A cell found on the surface of your textbook contains ribosomes, DNA, a plasma membrane, a cell wall, and mitochondria. What type of cell is it?
   a. lung cell
   b. bacterium
   c. sperm cell
   d. plant cell
   e. fingernail

2. A prokaryote converts food energy to ATP on/in its:
   a. chromosome.
   b. flagella.
   c. ribosomes.
   d. cell wall.
   e. plasma membrane.

3. Which of the following structures does *not* require an immediate source of energy to function?
   a. central vacuoles
   b. cilia
   c. microtubules
   d. microfilaments
   e. microbodies

4. Which of the following structures is *not* used in eukaryotic protein manufacture and secretion?
   a. ribosome
   b. lysosome
   c. rough ER
   d. smooth ER
   e. Golgi complex

5. When a person has an infection, white blood cells are summoned to roll, stick, and squeeze through the inner surface of blood vessels. The major components for this action are:
   a. plasmodesmata.
   b. desmosomes.
   c. cell adhesion molecules.
   d. flagella.
   e. cilia.

6. An electron micrograph shows that a cell has extensive amounts of rough ER throughout. One can deduce from this that the cell is:
   a. synthesizing and metabolizing carbohydrates.
   b. synthesizing and secreting proteins.
   c. synthesizing ATP.
   d. contracting.
   e. resting metabolically.

7. Which of the following contributes to the sealed lining of the digestive tract to keep food inside it?
   a. A central vacuole stores proteins.
   b. Tight junctions form a hollow tube for transport of molecules.
   c. Gap junctions communicate between cells of the stomach lining and its muscular wall.
   d. Desmosomes form buttonlike spots or a belt to keep cells joined together.
   e. Plasmodesmata help cells communicate their activities.

8. Which of the following structures are found in the same organelle?
   a. stroma and vacuole
   b. basal body and flagellum
   c. matrix and cristae
   d. DNA and ribosomes
   e. cytosol and plasma membrane

9. Which of the following statements about proteins is correct?
   a. Proteins are transported to the rough ER for use within the cell.
   b. Lipids and carbohydrates are added to proteins by the Golgi complex.
   c. Proteins are transported directly into the cytosol for secretion from the cell.
   d. Proteins that are to be stored by the cell are moved to the rough ER.
   e. Proteins are synthesized in vesicles.

10. Which of the following is *not* a component of the cytoskeleton?
   a. microtubules
   b. microfilaments
   c. cytokeratins
   d. actins
   e. cilia

Questions for Discussion

1. Many compound microscopes have a filter that eliminates all wavelengths except that of blue light, thereby allowing only blue light to pass through the microscope. Use the spectrum of visible light (see Figure 9.4) to explain why the filter improves the resolution of light microscopes.

2. Explain why aliens invading Earth are not likely to be giant cells the size of humans.

3. An electron micrograph of a cell shows the cytoplasm packed with rough ER membranes, a Golgi complex, and mitochondria. What activities might this cell concentrate on? Why would large numbers of mitochondria be required for these activities?

4. Assuming that mitochondria evolved from bacteria that entered cells by endocytosis, what are the likely origins of the outer and inner mitochondrial membranes?

5. Researchers have noticed that some men who were sterile because their sperm cells were unable to move also had chronic infections of the respiratory tract. What might be the connection between these two symptoms?

Experimental Analysis

The unicellular alga *Chlamydomonas reinhardtii* has two flagella assembled from tubulin proteins. If a researcher changes the pH from approximately neutral (their normal growing condition) to pH 4.5, *Chlamydomonas* cells spontaneously lose their flagella. After the cells are returned to neutral pH, they regrow the flagella—a process called reflagellation. Assuming that you have deflagellated *Chlamydomonas* cells, devise experiments to answer the following questions:

1. Do new tubulin proteins need to be made for reflagellation to occur, or is there a reservoir of proteins in the cell?

2. Is the production of new messenger RNA for the tubulin proteins necessary for reflagellation?

3. What is the optimal pH for reflagellation?

Evolution Link

What aspects of cell structure suggest that prokaryotes and eukaryotes share a common ancestor in their evolutionary history?

How Would You Vote?

Researchers are modifying prokaryotes to identify what it takes to "be alive." They are creating "new" organisms by removing genes from living cells, one at a time. What are the potential advantages or bioethical pitfalls of this kind of research? Go to www.thomsonedu.com/login to investigate both sides of the issue and then vote.

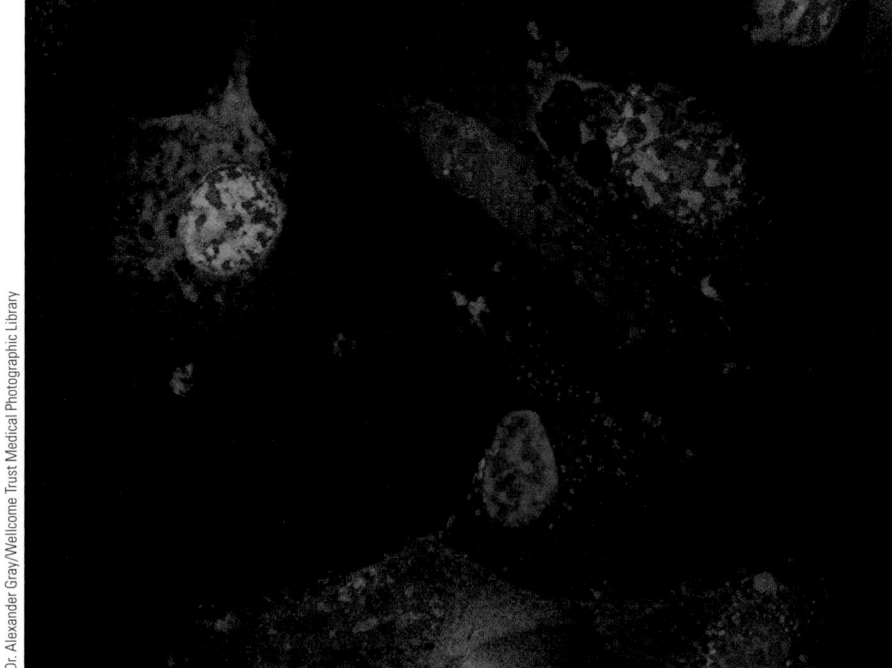

Endocytosis in cancer cells (confocal micrograph). The red spots are fluorescent spheres used to follow the process of endocytosis; some of the spheres have been taken up by cells.

6 Membranes and Transport

WHY IT MATTERS

All organisms encounter environmental factors that could disrupt their water content and internal concentrations of ions and molecules, but some species face dramatic challenges. Consider the striped bass *(Morone saxatilis)*, which migrates between the ocean and freshwater streams in North America **(Figure 6.1)**. Seawater is more salty than the fluids inside the fish. In this situation, water tends to leave the body of the fish and enter the seawater, and salt ions from the water tend to enter the fish. When the bass migrates into freshwater streams, now the inside of the fish is more salty than the surrounding freshwater, and its cells must keep its ions in and excess water out. If the cellular systems that regulate the balance fail in either situation, death is likely.

The challenge is not unique to organisms migrating between freshwater and the oceans—all living things must constantly bring in some molecules and ions and keep out others to maintain their internal environment. The **plasma membrane**—the exceedingly thin layer of lipids and proteins that covers the surface of all cells—makes this possible.

The plasma membrane is the primary zone of contact between a cell and its environment. It forms a barrier that keeps the cell con-

Figure 6.1
A striped bass, an organism that tolerates both saltwater and freshwater environments.

Andrew Martinez/Photo Researchers, Inc.

tents from mixing freely with molecules outside the cell. Only selected ions can move across the barrier to enter or leave the cytoplasm. Within eukaryotic cells, membranes surrounding internal organelles play similar roles, creating environments that differ from the surrounding cytoplasm.

The structure and function of biological membranes are the focus of this chapter. We first consider the structure of membranes and then examine how membranes selectively transport substances in and out of cells and organelles. Other roles of membranes, including recognition of molecules on other cells, adherence to other cells or extracellular materials, and reception of molecular signals such as hormones, are the subjects of Chapters 7, 40, and 43 in this book.

6.1 Membrane Structure

A watery fluid medium—or aqueous solution—bathes both surfaces of all biological membranes. The membranes are also fluid, but they are kept separate from

their surroundings by the properties of the lipid and protein molecules from which they are formed.

Biological Membranes Contain Both Lipid and Protein Molecules

Biological membranes consist of lipids and proteins assembled into a thin film. The proportions of lipid and protein molecules in membranes vary, depending on the functions of the membranes in the cells.

Membrane Lipids. *Phospholipids* and *sterols* are the two major types of lipids in membranes (see Section 3.4). Phospholipids have nonpolar fatty acid chains at one end; at the other end, phospholipids have a phosphate group linked to one of several alcohols or amino acids, making this end polar **(Figure 6.2a)**. The polar end is hydrophilic—it "prefers" being in an aqueous environment—and the nonpolar end is hydrophobic—it "prefers" being in an environment from which water is excluded. In other words, phospholipids have dual solubility properties.

In an aqueous medium, phospholipid molecules satisfy their dual solubility characteristics by assembling into a **bilayer**—a layer two molecules thick **(Figure 6.2b)**. In a bilayer, the polar ends of the phospholipid molecules are located at the surfaces, where

a. Phospholipid molecule

Polar end (hydrophilic)

Nonpolar end (hydrophobic)

CH₃
H₃C—N⁺—CH₃
H₂C
CH₂ — Polar alcohol

O
-O-P=O — Phosphate group
O

H₂C—CH—CH₂ — Glycerol
C=O   C=O

H₂C CH₂
 CH₂ H₂C
H₂C CH₂
 CH₂ H₂C
H₂C CH₂
 CH₂ H₂C — Hydrophobic tail
H₂C CH₂
 CH₂ H₂C
H₂C CH₂
 CH₂ H₂C
CH₃ H₃C

b. Fluid bilayer

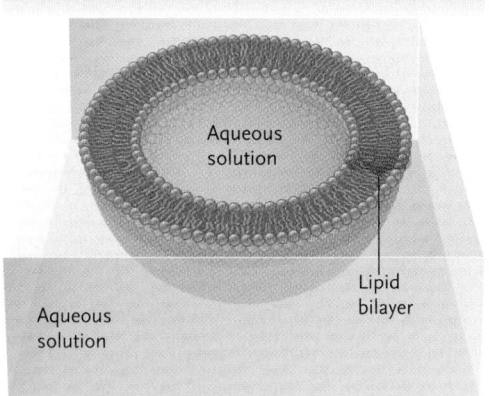

Aqueous solution

Aqueous solution

c. "Frozen" bilayer

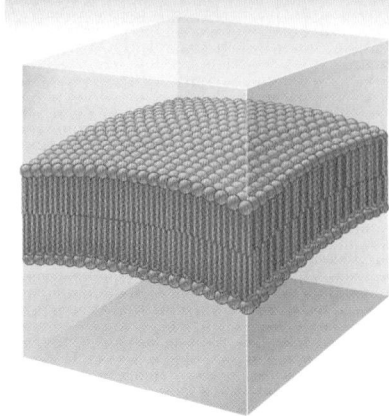

d. Bilayer vesicle

Aqueous solution

Aqueous solution

Lipid bilayer

Figure 6.2
Phospholipid bilayers. **(a)** Phospholipid molecule (phosphatidyl choline). Within the circle at the top representing the polar end of the molecule, the polar alcohol (choline) is shown in blue, the phosphate group in orange, and the glycerol unit in pink.
(b) Phospholipid bilayer in the fluid state, in which individual molecules are free to flex, rotate, and exchange places. **(c)** A bilayer frozen in a semisolid, gel-like state; note the close alignment of the hydrophobic tails compared with part **(b)**. **(d)** Phospholipid bilayer forming a vesicle.

they face the surrounding aqueous medium. The nonpolar fatty acid chains arrange themselves end to end in the membrane interior, in a nonpolar region that excludes water. At low temperatures, the phospholipid bilayer becomes frozen to produce a semisolid, gel-like state **(Figure 6.2c).** When a phospholipid bilayer sheet is shaken in water, it breaks and spontaneously forms small *vesicles* **(Figure 6.2d).** Vesicles consist of a spherical shell of phospholipid bilayer enclosing a small droplet of water.

Membrane sterols also have dual solubility characteristics. As explained in Section 3.4, these molecules have nonpolar carbon rings with a nonpolar side chain at one end and a single polar group (an —OH group) at the other end. In biological membranes, sterols pack into membranes alongside the phospholipid hydrocarbon chains, with only the polar end extending into the polar membrane surface **(Figure 6.3).** The predominant sterol of animal cell membranes is **cholesterol,** which is important for maintaining membrane fluidity. A variety of sterols, called *phytosterols,* is found in plants.

Membrane Proteins. Membrane proteins also have hydrophilic and hydrophobic regions that give them dual solubility properties. The hydrophobic regions of membrane proteins are formed by segments of the amino acid chain with hydrophobic side groups. These hydrophobic segments are often wound into alpha helices, which span the membrane bilayer **(Figure 6.4).** The hydrophobic segments are connected by loops of hydrophilic amino acids that extend into the polar regions at the membrane surfaces (for example, see Figure 6.4).

Each type of membrane has a characteristic group of proteins that is responsible for its specialized functions. **Transport proteins** form channels that allow selected polar molecules and ions to pass across a membrane. **Recognition proteins** in the plasma membrane identify a cell as part of the same individual or as foreign. **Receptor proteins** recognize and bind molecules from other cells that act as chemical signals, such as the peptide hormone insulin in animals. **Cell adhesion proteins** bind cells together by recognizing and binding receptors or chemical groups on other cells or on the extracellular

Cholesterol

Figure 6.3
The position taken by cholesterol in bilayers. The hydrophilic —OH group at one end of the molecule extends into the polar regions of the bilayer; the ring structure extends into the nonpolar membrane interior.

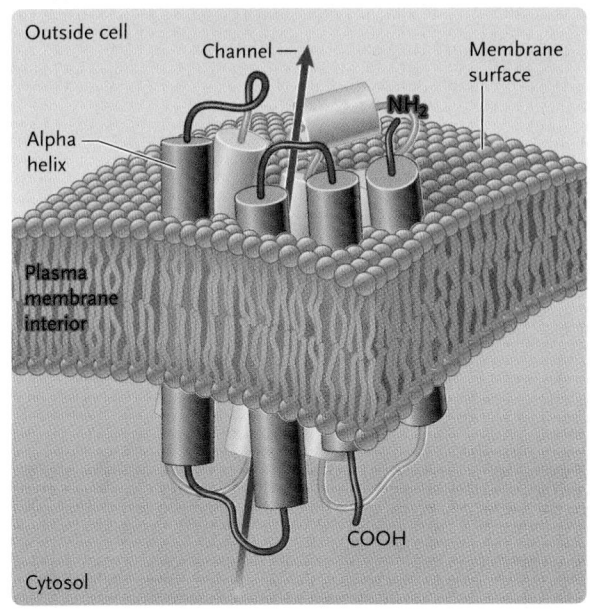

a. Typical membrane protein

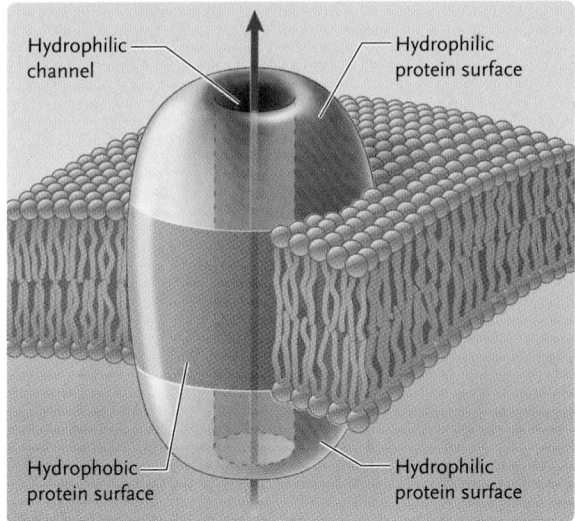

b. Hydrophilic and hydrophobic surfaces

Figure 6.4
Structure of membrane proteins. **(a)** Typical membrane protein, bacteriorhodopsin, showing the membrane-spanning alpha-helical segments (blue cylinders), connected by flexible loops of the amino acid chain at the membrane surfaces. **(b)** The same protein as in **(a)** in a diagram that shows hydrophilic (blue) and hydrophobic (orange) surfaces and the membrane-spanning channel created by this protein. Bacteriorhodopsin absorbs light energy in plasma membranes of photosynthetic archaeans.

matrix. Still other proteins are enzymes that speed chemical reactions carried out by membranes, such as the reactions in mitochondria that create ATP.

Membrane Glycolipids and Glycoproteins. Many of the phospholipids and proteins in membranes have carbohydrate groups linked to them, forming glycolipids and glycoproteins **(Figure 6.5)**. In the plasma membrane, the carbohydrate groups, which are polar, are attached covalently to parts of membrane lipid and protein molecules that face the exterior membrane surface. They are so abundant on the exterior surface that they give cells a "sugar coating" or **glycocalyx** (*glykys* = sweet; *calyx* = cup or vessel).

The Fluid Mosaic Model Explains Membrane Structure

The current view of membrane structure is based on the fluid mosaic model (see Figure 6.5). S. Jonathan Singer and Garth L. Nicolson at the University of California, San Diego, advanced this model in 1972. The **fluid mosaic model** proposes that the membrane consists of a fluid phospholipid bilayer in which proteins are embedded and float freely.

The "fluid" part of the fluid mosaic model refers to the phospholipid molecules, which vibrate, flex back and forth, spin around their long axis, move sideways, and exchange places within the same bilayer half. Only rarely does a phospholipid flip-flop between the two layers. Phospholipids exchange places within a layer millions of times a second, making the phospholipid molecules in the membrane highly dynamic. Membrane fluidity is critical to the functions of membrane proteins and allows membranes to accommodate, for example, cell growth, motility, and surface stresses.

Membranes remain fluid at a relatively wide range of temperatures. Low temperatures can be detrimental to membrane structure, and therefore membrane function, because at a sufficiently low temperature the phospholipid molecules become closely packed and the membrane becomes a nonfluid gel. A common modification that helps keep membranes fluid at low temperatures is increasing the proportion of unsaturated fatty acid chains in membrane phospholipids. The double bonds in the unsaturated fatty acid chains produce physical kinks in the chain that interfere with the packing of the phospholipids at low temperatures, thereby reducing the temperature at which the bilayer becomes a nonfluid gel. In a paradoxical way, cholesterol also helps protect against the adverse effects of low temperatures. Cholesterol in the membrane decreases membrane fluidity at moderate to high temperatures because of

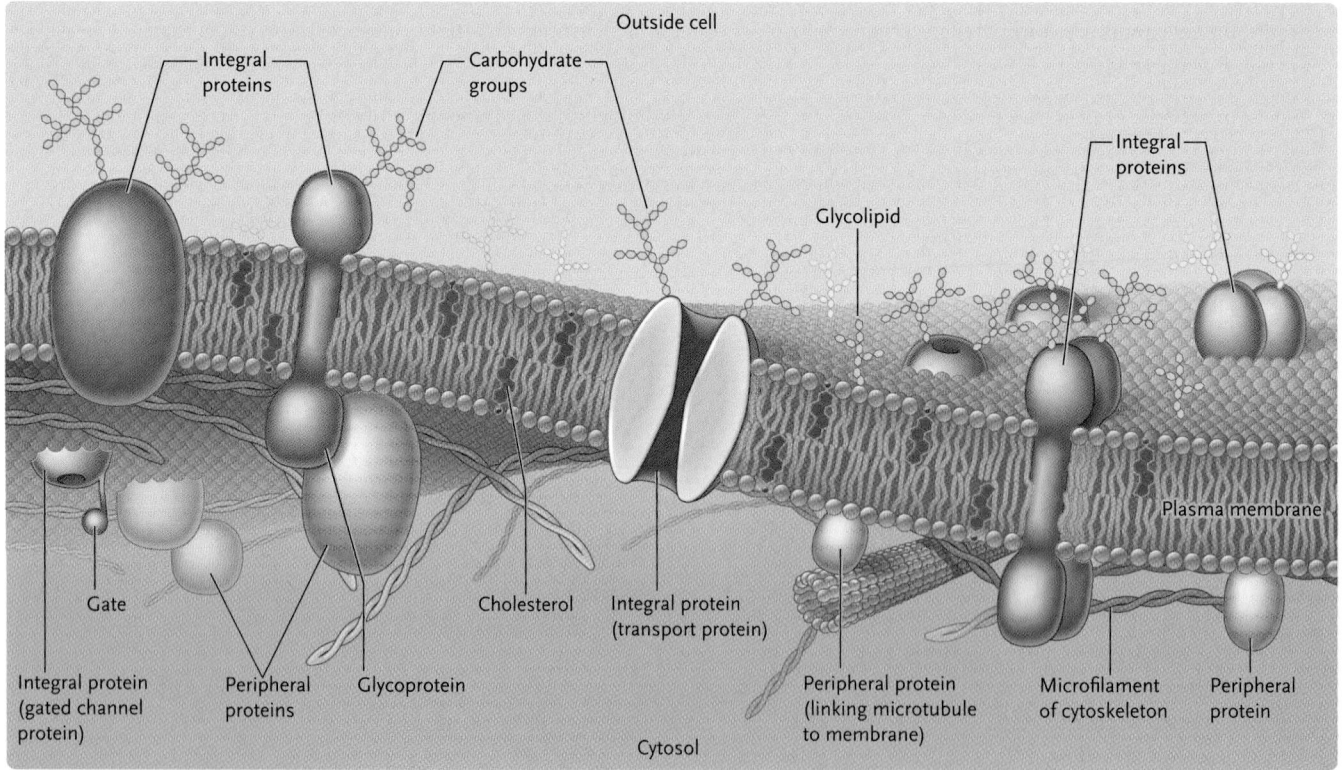

Figure 6.5

Membrane structure according to the fluid mosaic model, in which integral membrane proteins are suspended individually in a fluid bilayer. Peripheral proteins are attached to integral proteins or membrane lipids mostly on the cytoplasmic side of the membrane (shown only on the inner surface in the figure). In the plasma membrane, carbohydrate groups of membrane glycoproteins and glycolipids face the cell exterior.

Basic Research: Keeping Membranes Fluid at Cold Temperatures

The fluid state of biological membranes is critical to membrane function and absolutely vital to cellular life. When membranes freeze, researchers have shown that the phospholipids form a semisolid gel in which they are unable to move (see Figure 6.2c), and proteins become locked in place. Freezing can kill cells by impeding vital membrane functions such as transport.

Many eukaryotic organisms, including algae, higher plants, protozoa, and animals, adapt to colder temperatures by changing membrane lipids. Experiments comparing membrane composition have shown that, in animals with body temperatures that fluctuate with environmental temperature, such as fish, amphibians, and reptiles, both the proportion of double bonds in membrane phospholipids and the cholesterol content are increased at lower temperatures. How do these changes affect membrane fluidity? Double bonds in unsaturated fatty acids introduce "kinks" in their hydrocarbon chain (see Figure 3.12); the kinks help bilayers stay fluid at lower temperatures by interfering with packing of the hydrocarbons. Cholesterol depresses the freezing point by interfering with close packing of membrane phospholipids.

All of these membrane changes also occur in mammals that enter hibernation in cold climates. When mammals enter hibernation, their body temperature may fall to as low as 5°C without freezing their membranes. The resistance to freezing allows the nerve cells of a hibernating mammal to remain active so that the animal can maintain basic body functions and respond, although sluggishly, to external stimuli. In active, nonhibernating mammals, membranes freeze into the gel state at about 15°C.

---

interactions between the rigid cholesterol rings and the membrane phospholipids (see Figure 6.3). However, at the high concentrations found in eukaryotic membranes, the disruption of the ordered packing of phospholipids by cholesterol helps slow the transition of the membrane to the nonfluid gel state when temperatures drop. (See *Focus on Research* for a description of other strategies that organisms use to keep their membranes from freezing at low temperatures.)

At high temperatures, membranes can become too fluid and will become leaky, allowing ions to cross in an uncontrolled manner. This leaking disrupts the function of the cell; thus, it is likely to die. As described previously, cholesterol reduces membrane fluidity at high temperatures, thereby providing some protection.

The "mosaic" part of the fluid mosaic model refers to the membrane proteins, most of which float individually in the fluid lipid bilayer, like icebergs in the sea. Membrane proteins are larger than membrane lipids, and those that move do so much more slowly than do lipids. A number of membrane proteins are attached to the cytoskeleton. These proteins either are immobile or move in a directed fashion, perhaps along cytoskeletal filaments.

Membrane proteins are oriented across the membrane so that particular functional groups and active sites face either the inside or the outside membrane surface. The inside and outside halves of the bilayer also contain different mixtures of phospholipids. These differences make biological membranes *asymmetric* and give their inside and outside surfaces different functions.

Proteins that are embedded in the phospholipid bilayer are termed **integral proteins** (see Figure 6.5). Essentially all transport, receptor, recognition, and cell adhesion proteins that give membranes their specific functions are integral membrane proteins.

Other proteins, called **peripheral proteins** (see Figure 6.5), are held to membrane surfaces by noncovalent bonds—hydrogen bonds and ionic bonds—formed with the polar parts of integral membrane proteins or membrane lipids. Most peripheral proteins are on the cytoplasmic side of the membrane. Some peripheral proteins are parts of the cytoskeleton, such as microtubules, microfilaments, or intermediate filaments, or proteins that link the cytoskeleton together. These structures hold some integral membrane proteins in place. For example, this anchoring constrains many types of receptors to the sides of cells that face body surfaces, cavities, or tubes.

The Fluid Mosaic Model Is Fully Supported by Experimental Evidence

The novel ideas of a fluid membrane and a flexible mosaic arrangement of proteins and lipids challenged an accepted model in which a relatively rigid, stable membrane was coated on both sides with proteins arranged like jam on bread. Researchers tested the new model with an intensive burst of research. The experimental evidence from that research completely supports every major hypothesis of the model: that membrane lipids are arranged in a bilayer, that the bilayer is fluid, that proteins are suspended individually in the bilayer, and that the arrangement of both membrane lipids and proteins is asymmetric.

Evidence That Membranes Are Fluid. In a now-classic study carried out in 1970, L. David Frye and Michael A. Edidin grew human cells and mouse cells separately in tissue culture. Then they added antibodies that bound to either human or mouse membrane proteins **(Figure 6.6)**. The anti-human antibodies were

Figure 6.6 Experimental Research

The Frye-Edidin Experiment Demonstrating That the Phospholipid Bilayer Is Fluid

QUESTION: Do membrane proteins move in the phospholipid bilayer?

EXPERIMENT: Frye and Edidin labeled membrane proteins on cultured human and mouse cells with fluorescent dyes, red for human proteins and green for mouse proteins. Human and mouse cells were then fused and the pattern of fluorescence was followed under a microscope.

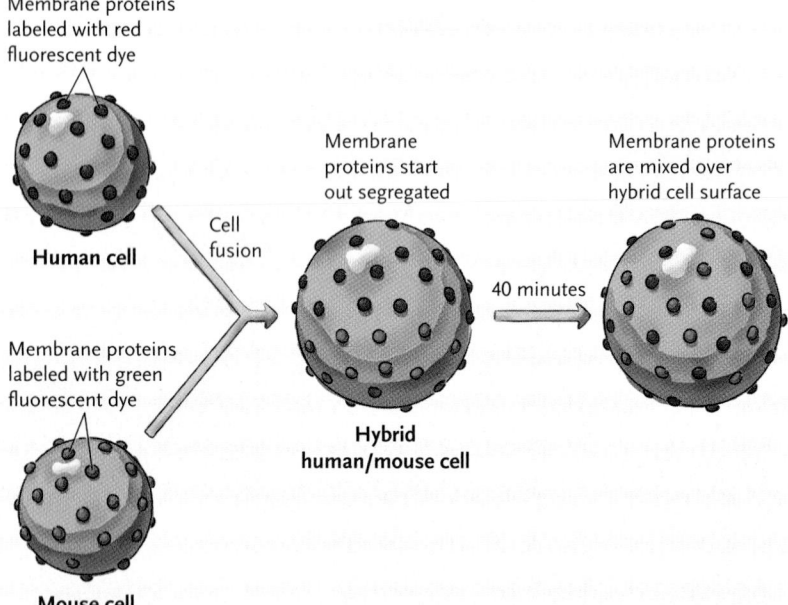

CONCLUSION: The rapid mixing of membrane proteins in the hybrid human/mouse cells showed that membrane proteins move in the phospholipid bilayer, indicating that the membrane is fluid.

attached to dye molecules that fluoresce with a red color under ultraviolet light; the anti-mouse antibodies were linked to dye molecules that fluoresce green. Next, they fused the human and mouse cells. Immediately after fusion, the cells were half red and half green, with a clear dividing line between the colors. Within a few minutes, the colors began to mix along the dividing line. In 40 minutes, the colors were completely intermixed on fused cells, indicating the mouse and human proteins had moved around in the fused membranes. In other words, the experiment showed that membrane proteins move in membranes; this movement occurs because the membranes are fluid.

Based on the measured rates at which molecules mix in biological membranes, the membrane bilayer appears to be about as fluid as light machine oil, such as the lubricants you might use around the house to oil a door hinge, the wheels of a skateboard, or a bicycle.

Evidence for Membrane Asymmetry and Individual Suspension of Proteins. An experiment that used membranes prepared for electron microscopy by the **freeze-fracture technique** confirmed that the membrane is a bilayer with proteins suspended in it individually and that the arrangement of membrane lipids and proteins is asymmetric **(Figure 6.7)**. In this technique, experimenters freeze a block of cells rapidly by dipping it in, for example, liquid nitrogen. Then they fracture the block by giving it a blow from a microscopically sharp knife edge. Often, the fracture splits bilayers into inner and outer halves, exposing the hydrophobic membrane interior. In the electron microscope, the split membranes appear as smooth layers in which individual particles the size of proteins are embedded (see Figure 6.7c).

STUDY BREAK

1. Describe the fluid mosaic model for membrane structure.
2. Give two examples each of integral proteins and peripheral proteins.

6.2 Functions of Membranes in Transport: Passive Transport

The primary function of cellular membranes is **transport,** the controlled movement of ions and molecules from one side of a membrane to the other. The membrane proteins are the molecules responsible for transport. The movement is typically *directional;* that is, some ions and molecules consistently move into cells, whereas others move out of cells. Transport is also *specific;* that is, only certain ions and molecules move directionally across membranes. Transport is critical to the ionic and molecular organization of cells, and with it, the maintenance of cellular life.

Transport occurs by two mechanisms. The first mechanism, **passive transport,** depends on concentration differences on the two sides of a membrane (concentration = number of molecules or ions per unit volume). In this mechanism, ions and molecules move across the membrane from the side with the higher concentration to the side with the lower concentration (that is, *with* the gradient). The difference in concentration provides the energy for this form of transport.

The second mechanism, **active transport,** moves ions or molecules *against* the concentration gradient; that is, from the side with the lower concentration to the side with the higher concentration. Active transport

Figure 6.7 Research Method

Freeze Fracture

PURPOSE: Quick-frozen cells are fractured to split apart lipid bilayers for analysis of the membrane interior.

PROTOCOL:

1. The specimen is frozen quickly in liquid nitrogen and then fractured by a sharp blow by a knife edge.

2. The fracture may travel over membrane surfaces as it passes through the specimen, or it may split membrane bilayers into inner and outer halves as shown here.

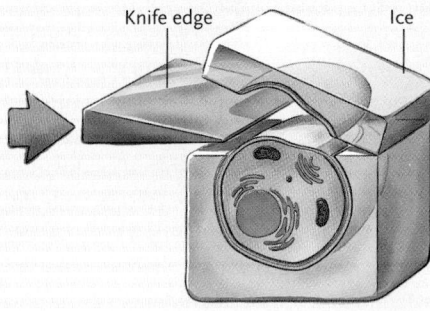

Knife edge   Ice

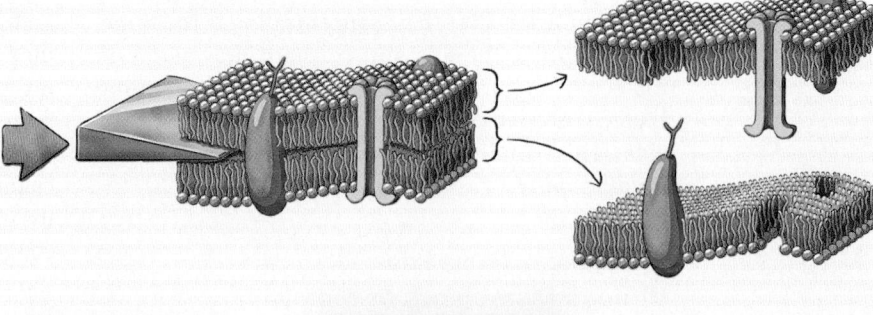

INTERPRETING THE RESULTS: The image of a freeze-fractured plasma membrane is visualized using the electron microscope. The particles visible in the exposed membrane interior are integral membrane proteins.

Don W. Fawcett/Photo Researchers, Inc.

Outer membrane surface

Exposed membrane interior

---

uses energy directly or indirectly obtained by breaking down ATP. The properties of passive and active transport are compared in **Table 6.1**.

Passive Transport Is Based on Diffusion

Passive transport is a form of **diffusion**, the net movement of ions or molecules from a region of higher concentration to a region of lower concentration. If you add a drop of food dye to a container of clear water, the dye molecules, and therefore the color, will spread or *diffuse* from their initial center of high concentration until they are distributed evenly. At this point, the water has an even color. Diffusion depends on the constant motion of ions or molecules at temperatures above absolute zero ($-273°C$). The constant motion gradually mixes the dye molecules and water molecules until they are distributed uniformly.

The concentration difference that drives diffusion, a **concentration gradient**, is a form of potential energy. At the initial state, when molecules are more concentrated in one region of a solution, as when a dye is dropped into one side of a container of water, the molecules are highly organized and at a state of minimum entropy. In the final state, when they are distributed evenly throughout the solution, they are less organized and at a state of maximum entropy. As the distribution proceeds to the state of maximum dis-

Table 6.1	Characteristics of Transport Mechanisms		
Characteristic	Passive Transport		Active Transport
	Simple Diffusion	Facilitated Diffusion	
Membrane component responsible for transport	Lipids	Proteins	Proteins
Binding of transported substance	No	Yes	Yes
Energy source	Concentration gradients	Concentration gradients	ATP hydrolysis or concentration gradients
Direction of transport	With gradient of transported substance	With gradient of transported substance	Against gradient of transported substance
Specificity for molecules or molecular classes	Nonspecific	Specific	Specific
Saturation at high concentrations of transported molecules	No	Yes	Yes

order, the molecules release free energy that can accomplish work (see Section 4.1 for a discussion of entropy and free energy).

Diffusion involves a *net* movement of molecules or ions. Molecules and ions actually move in all directions at all times in a solution. But when molecules or ions exist in a concentration gradient, more of them move from the area of higher concentration to areas of lower concentration than in the opposite direction. Even after their concentration is the same in all regions, there is still constant movement of molecules or ions from one space to another, but there is no net change in concentration on either side. This condition is an example of a *dynamic equilibrium*.

Substances Move Passively through Membranes by Simple or Facilitated Diffusion

Hydrophobic (nonpolar) molecules are able to dissolve in the lipid bilayer of a membrane and move through it freely. By contrast, hydrophilic molecules such as ions and polar molecules are impeded in their movement through the membrane by the hydrophobic core; thus, their passage is slow. Charged atoms and molecules are mostly blocked from moving through the membrane because of the hydrophobic core. Membranes that affect diffusion in this way are said to be **selectively permeable.**

Transport by Simple Diffusion. A few small substances diffuse through the lipid part of a biological membrane. With one major exception—water—these substances are nonpolar inorganic gases such as O_2, N_2, and CO_2 and nonpolar organic molecules such as steroid hormones. This type of transport, which depends solely on molecular size and lipid solubility, is **simple diffusion** (see Table 6.1).

Water is a strongly polar molecule. Nevertheless, water molecules are small enough to slip through momentary spaces created between the hydrocarbon tails of phospholipid molecules as they flex and move in a fluid bilayer. However, this type of water movement across the membrane is relatively slow.

Transport by Facilitated Diffusion. Many polar and charged molecules such as water, amino acids, sugars, and ions diffuse across membranes with the help of transport proteins, a mechanism termed **facilitated diffusion.** In essence, the transport proteins enable polar and charged molecules to avoid interaction with the hydrophobic lipid bilayer (see Table 6.1).

Facilitated diffusion is specific: the membrane proteins involved transport certain polar and charged molecules, but not others. Facilitated diffusion is also dependent on concentration gradients: proteins aid the transport of polar and charged molecules through membranes, but a favorable concentration gradient provides the energy for transport. Transport stops if the gradient falls to zero.

Two Groups of Transport Proteins Carry Out Facilitated Diffusion

The proteins that carry out facilitated diffusion are integral membrane proteins that extend entirely through the membrane. There are two types of transport proteins involved in facilitated diffusion. One type, **channel proteins,** forms hydrophilic channels in the membrane through which water and ions can pass **(Figure 6.8a).** The channel "facilitates" the diffusion of molecules through the membrane by providing an avenue. For example, facilitated diffusion of water through membranes occurs through specialized water channels called **aquaporins** (see Figure 6.8a). A billion molecules of water per second can move through an aquaporin channel. How the molecules move is fascinating. Each water molecule is severed from its hydrogen-bonded neighbors as it is handed off to a succession of hydrogen-bonding sites on the aquaporin protein in the channel.

Other channel proteins facilitate the transport of ions such as sodium (Na^+), potassium (K^+), calcium (Ca^{2+}), and chlorine (Cl^-). Most of these ion transporters, which occur in all eukaryotes, are **gated channels;** that is, they switch between open, closed, or intermediate states. The gates may be opened or closed by changes in voltage across the membrane, for instance, or by binding signal molecules. In animals, voltage-gated ion channels are used in nerve conduction and the control of muscle contraction. *Insights from the Molecular Revolution* describes molecular experiments showing the conformational changes that open channel gates.

Gated ion channels perform functions that are vital to survival, as illustrated by the effects of hereditary defects in the channels. For example, the hereditary disease *cystic fibrosis* results from a fault in a gated Cl^- channel. The faulty channel allows unusually high levels of Cl^-, as well as Na^+, to pass into extracellular fluids and from sweat glands. Abnormally sticky mucus accumulates in the respiratory tract leading to chronic lung infections that, with other effects of the Cl^- transport deficiency, are typically fatal by age 30 for persons born with the disease.

Carrier proteins are the second type of transport proteins; they also form passageways through the lipid bilayer **(Figure 6.8b).** Carrier proteins each bind a specific single solute, such as glucose or an amino acid, and transport it across the lipid bilayer (glucose is also transported by active transport, as described in the next section). Because a single solute is transferred in this carrier-mediated fashion, the transfer is called *uniport transport*. In performing the transport step, the carrier

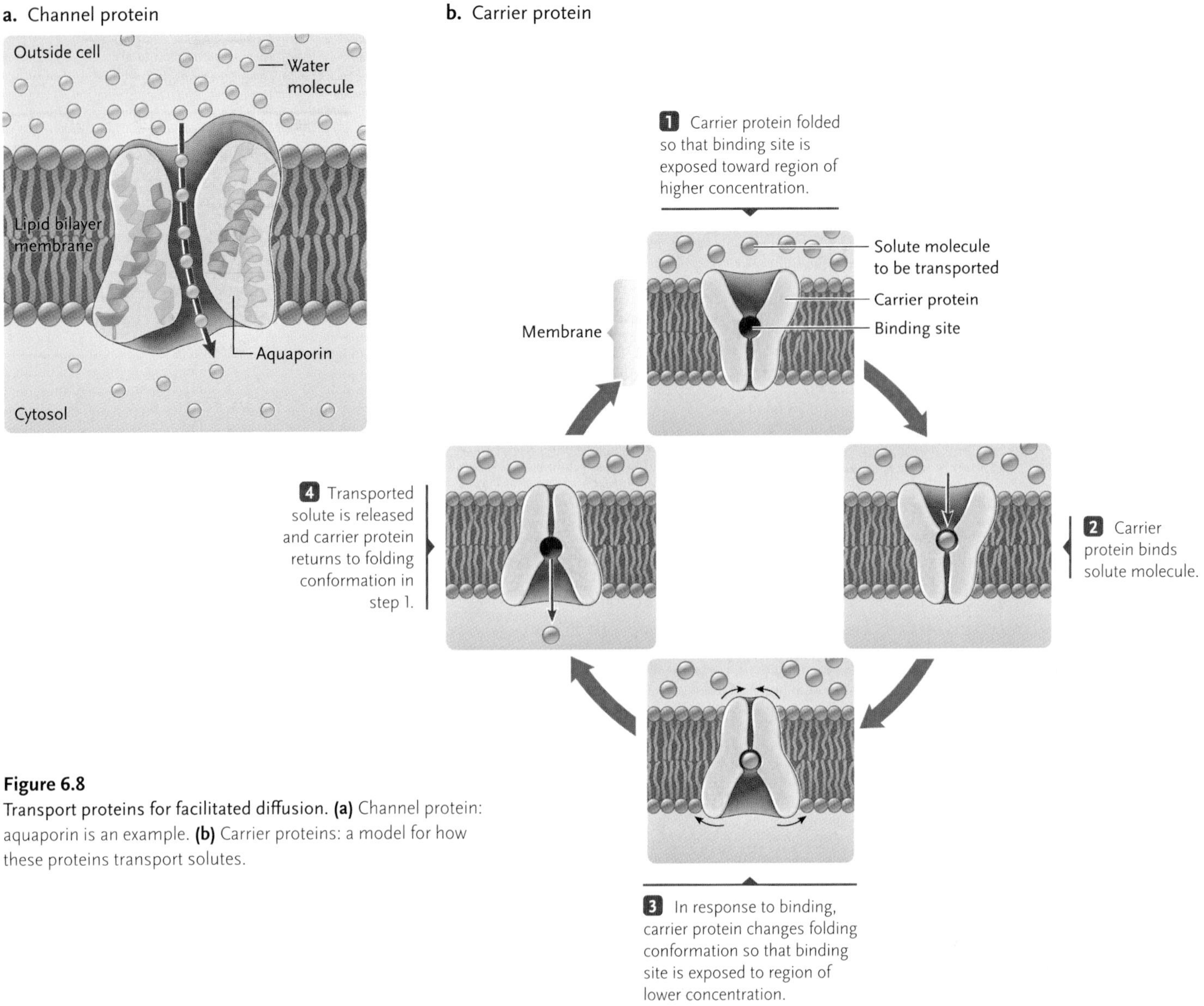

a. Channel protein

b. Carrier protein

1 Carrier protein folded so that binding site is exposed toward region of higher concentration.

Solute molecule to be transported
Carrier protein
Binding site

Membrane

2 Carrier protein binds solute molecule.

4 Transported solute is released and carrier protein returns to folding conformation in step 1.

3 In response to binding, carrier protein changes folding conformation so that binding site is exposed to region of lower concentration.

Figure 6.8
Transport proteins for facilitated diffusion. **(a)** Channel protein: aquaporin is an example. **(b)** Carrier proteins: a model for how these proteins transport solutes.

protein undergoes conformational changes that progressively move the solute-binding site from one side of the membrane to the other, thereby transporting the solute. This property distinguishes carrier protein function from channel protein function.

Facilitated diffusion by carrier proteins can become *saturated* when there are not enough transport proteins to handle all the solute molecules. For example, if glucose is added at higher and higher concentrations to the solution that surrounds an animal cell, the rate at which it passes through the membrane at first increases proportionately with the increase in concentration. However, at some point, as the glucose concentration is increased still further, the increase in the rate of transport slows. Eventually, further increases in concentration cause no additional rise in the rate of transport—the transport mechanism is saturated. By contrast, saturation does not occur for simple diffusion.

Because the proteins that perform facilitated diffusion are specific, cells can control the kinds of molecules and ions that pass through their membranes by regulating the types of transport proteins in their membranes. As a result, each type of cellular membrane, and each type of cell, has its own group of transport proteins and passes a characteristic group of substances by facilitated diffusion. The kinds of transport proteins present in a cell ultimately depend on the activity of genes in the cell nucleus.

STUDY BREAK

1. What is the difference between passive and active transport?
2. What is the difference between simple and facilitated diffusion?

Tracking Gating Movements in a Channel Protein

Passive transport of ions often occurs through gated channels, and in animals, such channels operate to generate nerve signals and to stimulate muscle contraction. Knowledge of the structures of the gates and how each structural component plays a role in its function is required to understand this type of transport mechanism. For instance, researchers have studied voltage-gated ion transport proteins in nerve cells to determine such structure–function relations. As part of nerve transmission, a gated sodium channel opens to allow sodium ions to flow inward and a gated potassium channel opens to allow potassium ions to flow outward, both with their concentration gradients. The proteins of both channels have six alpha-helical segments, designated S1 to S6, that zigzag back and forth across the plasma membrane and form the passage through which ions move when the channel opens. Do these segments also participate in opening and closing the channels? To answer this question, scientists investigated whether any of the six helices responds to the voltage changes that stimulate opening and closing of the channels, and whether the helix moves as part of the gating.

Several experiments had implicated S4 as the critical helix. For example, in one experiment, the researchers substituted one amino acid for another at different sites in the channel proteins; of these substitutions, those in S4 had the greatest effect on the ability of the channels to respond to voltage changes in the membrane.

Lidia M. Mannuzzu, Mario M. Morrone, and Ehud Y. Isacoff of the University of California at Berkeley performed an experiment that confirmed that movement of S4 is the first step in voltage gating. The investigators tagged S4 with a dye molecule so they could trace movements of the helix in voltage-gated potassium channels in the plasma membranes of egg cells of the African clawed frog, *Xenopus laevis*.

Mannuzzu and her colleagues used molecular techniques to make five different versions of the potassium channel, each with the amino acid cysteine substituted at a different position in helix S4. They combined the cysteine with a particular dye that fluoresces—emits light—with a different wavelength (color) when it is in a nonpolar or polar environment.

Before the voltage change was made, all the different versions of the channel protein emitted light at a wavelength characteristic of a nonpolar environment. This result indicated that, before the channel opens, S4 is buried in the nonpolar interior of the plasma membrane. Immediately after the voltage change, the emitted light wavelength changed to that of a polar environment, indicating that S4 moves from the channel interior to the polar membrane surface as the channel gate opens.

The investigators were even able to find out whether S4 moves to the outside or the inside of the plasma membrane by applying a substance to the outside of the membrane that would "quench" emission of the fluorescent dye. Since quenching did occur, they concluded that helix S4 moves to the outside surface of the plasma membrane when the channel gate opens.

Taken together, the results of the experiments show that, in the first response of the gated channel to a change in membrane voltage, S4 moves from the interior to the external surface of the channel protein, and the channel gates open. This result provides the first direct confirmation that S4 actually moves as a part of the response. The techniques used in the experiments also provide a new way to track further gating changes that, until now, had been invisible to scientists investigating membrane channels.

6.3 Passive Water Transport and Osmosis

As discussed earlier, water can also follow concentration gradients and diffuse passively across membranes in response. It diffuses both directly through the membrane and through aquaporins. The passive transport of water, called **osmosis**, occurs constantly in living cells. Inward or outward movement of water by osmosis develops forces that can cause cells to swell and burst or shrink and shrivel up. Much of the energy budget of many cell types, particularly in animals, is spent counteracting the inward or outward movement of water by osmosis.

Osmosis Can Be Demonstrated in a Purely Physical System

The apparatus shown in **Figure 6.9a** is a favorite laboratory demonstration of osmosis. It consists of an inverted thistle tube (so named because its shape resembles a thistle flower) tightly sealed at its lower end by a sheet of cellophane. The tube is filled with a solution of glucose molecules in water and is suspended in a beaker of distilled water. The cellophane film acts as a selectively permeable membrane because its pores are large enough to admit water molecules but not glucose. At the start of the experiment, the position of the tube is set so the level of the liquid in the tube is at same level as the distilled water in the beaker. Almost

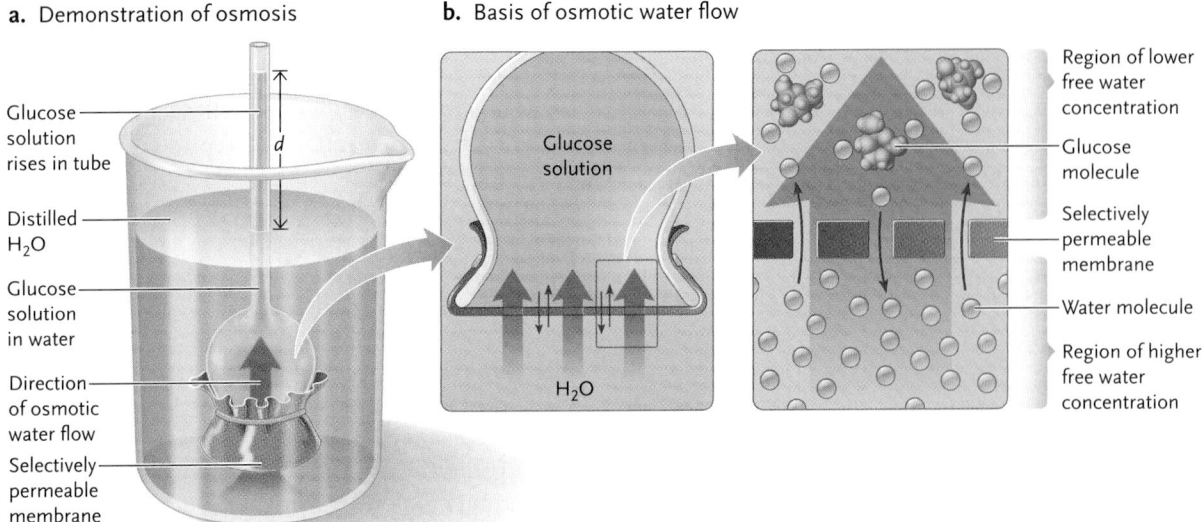

a. Demonstration of osmosis

b. Basis of osmotic water flow

Glucose solution rises in tube

Distilled H₂O

Glucose solution in water

Direction of osmotic water flow

Selectively permeable membrane

Glucose solution

H₂O

Region of lower free water concentration

Glucose molecule

Selectively permeable membrane

Water molecule

Region of higher free water concentration

Figure 6.9

Osmosis. **(a)** An apparatus demonstrating osmosis. The fluid in the tube rises due to the osmotic flow of water through the cellophane membrane, which is permeable to water but not to glucose molecules. Osmotic flow continues until the weight of the water in column *d* develops enough pressure to counterbalance the movement of water molecules into the tube. **(b)** The basis of osmotic water flow. The pure water solution on the left is separated from the glucose solution on the right by a membrane permeable to water but not to glucose. The free water concentration on the glucose side is lower than on the water-only side because water molecules are associated with the glucose molecules. That is, water molecules are in greater concentration on the bottom than on the top. Although water molecules move in both directions across the membrane (small red arrows), there is a net upward movement of water (blue arrows), with the water's concentration gradient.

immediately, the level of the solution in the tube begins to rise, eventually reaching a maximum height above the liquid in the beaker.

The liquid rises in the tube because water moves by osmosis from the beaker into the thistle tube. The movement occurs passively, in response to a concentration gradient in which the water molecules are higher in concentration in the beaker than inside the thistle tube. The basis for the gradient is shown in **Figure 6.9b.** The glucose molecules are more concentrated on one side of the selectively permeable membrane. On this side, association of water molecules with those solute molecules reduces the amount of water available to cross the membrane. Thus, although initially there is an equal apparent water concentration on each side of the membrane, there is a difference in the *free water* concentration—that is, the water available to move across the membrane. Specifically, the concentration of free water molecules is lower on the glucose side than on the pure water side. In response, a net movement of water occurs from the pure water side to the glucose solution side. Osmosis is the net diffusion of water molecules through a selectively permeable membrane in response to a gradient of this type.

The solution stops rising in the tube when the pressure created by the weight of the raised solution exactly balances the tendency of water molecules to move from the beaker into the tube in response to the concentration gradient. This pressure is the **osmotic pressure** of the solution in the tube. At this point, the system is in a state of dynamic equilibrium and no further net movement of water molecules occurs.

A formal definition for osmosis is *the net movement of water molecules across a selectively permeable membrane by passive diffusion, from a solution of lesser solute concentration to a solution of greater solute concentration* (the *solute* is the substance dissolved in water). For osmosis to occur, the selectively permeable membrane must allow water molecules, but not molecules of the solute, to pass. Pure water does not need to be on one side of the membrane; osmotic water movement also occurs if a solute is at different concentrations on the two sides. Because osmosis occurs in response to a concentration gradient, it releases free energy and can accomplish work.

The Free Energy Released by Osmosis May Work for or against Cellular Life

Osmosis occurs in cells because they contain a solution of proteins and other molecules that are retained in the cytoplasm by a membrane impermeable to them but freely permeable to water. The resulting osmotic movement of water is used as an energy source for some of the activities of life. However, it can also create a disturbance that cells must counteract by ex-

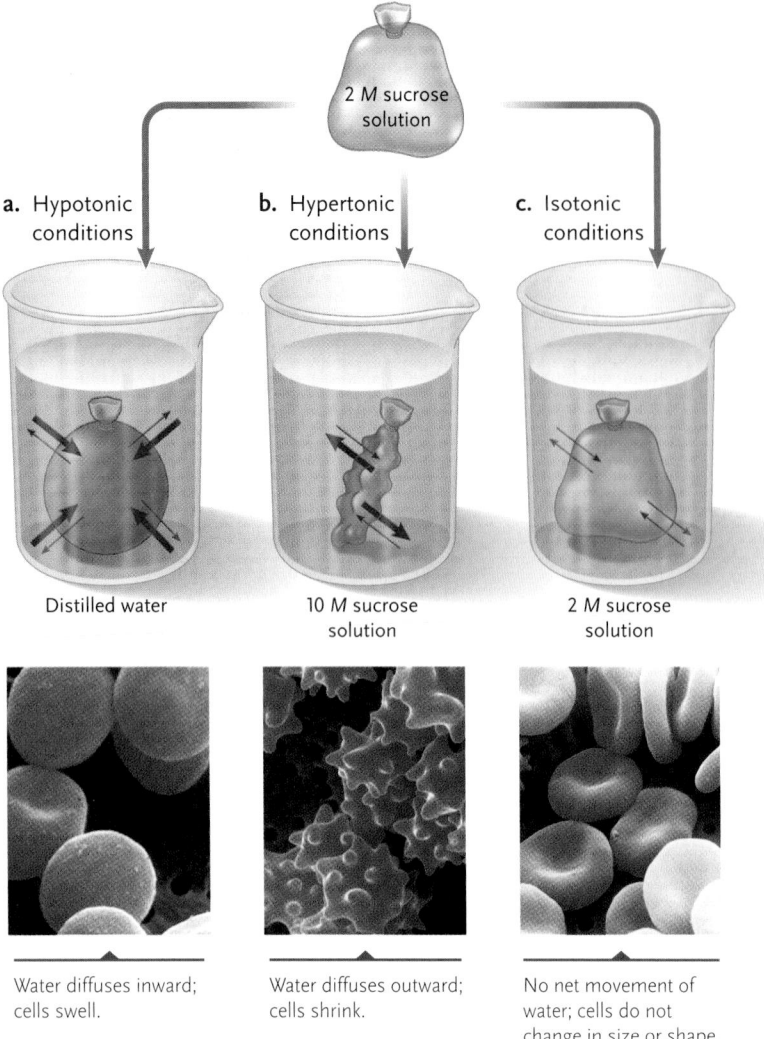

a. Hypotonic
conditions

b. Hypertonic
conditions

c. Isotonic
conditions

2 M sucrose
solution

Distilled water

10 M sucrose
solution

2 M sucrose
solution

Water diffuses inward;
cells swell.

Water diffuses outward;
cells shrink.

No net movement of
water; cells do not
change in size or shape.

Figure 6.10

Tonicity and osmotic water movement. The diagrams show what
happens when a cellophane bag filled with a 2 M sucrose solution is placed in a **(a)** hypotonic, **(b)** hypertonic, or **(c)** isotonic
solution. The cellophane is permeable to water but not to sucrose
molecules. The width of the arrows shows the amount of water
movement. In the first beaker, the distilled water is hypotonic to
the solution in the bag; net movement of water is into the bag. In
the second beaker, the 10 M solution is hypertonic to the solution
in the bag; net movement of water is out of the bag. In the third
beaker, the solutions inside and outside the bag are isotonic;
there is no net movement of water in or out of the bag. The
animal cell micrographs show the corresponding effects on red
blood cells placed in **(a)** hypotonic, **(b)** hypertonic, or **(c)** isotonic
solutions.

(Micrographs, M. Sheetz, R. Painter, and S. Singer. *Journal of Cell Biology,*
70:493, 1976. By permission of Rockefeller University Press.)

pending energy. If the solution that surrounds a cell
contains dissolved substances at lower concentrations
than the cell, the solution is said to be **hypotonic** to
the cell (*hypo* = under or below; *tonos* = tension or
tone). When a cell is in a hypotonic solution, water
enters by osmosis and the cell tends to swell **(Figure
6.10a).** Animal cells in a hypotonic solution may actually swell to the point of bursting. However, in most

plant cells, strong walls prevent the cells from bursting in a hypotonic solution. In most land plants, the
cells at the surfaces of roots are surrounded by almost
pure water, which is hypotonic to the cells and tissues
of the root. As a result, water flows from the surrounding soil into the root cells by osmosis. The osmotic pressure developed by the inward flow contributes part of the force required to raise water from the
roots to the leaves of the plant. Water also normally
moves into cells of the stems and leaves of plants by
osmosis. The resulting osmotic pressure, called **turgor pressure**, pushes the cells tightly against their
walls and supports the softer tissues against the force
of gravity **(Figure 6.11a).**

Organisms living in surroundings that contain
salts or other molecules at higher concentrations than
their own bodies must constantly expend energy to
replace water lost by osmosis. In this situation, the outside solution is said to be **hypertonic** to the cells
(*hyper* = over or above; see **Figure 6.10b**). If the outward
osmotic movement exceeds the capacity of cells to replace the lost water, both animal and plant cells will
shrink. In plants, the shrinkage and loss of internal
osmotic pressure under these conditions causes stems
and leaves to wilt. In extreme cases, plant cells shrink
so much that they retract from their walls, a condition
known as **plasmolysis (Figure 6.11b).**

In animals, ions, proteins, and other molecules
are concentrated in extracellular fluids, as well as inside cells, so that the concentration of water inside and
outside cells is usually equal or **isotonic** (*iso* = the
same; see **Figure 6.10c**). To keep fluids on either side
of the plasma membrane isotonic, animal cells must
constantly use energy to pump Na^+ from inside to
outside by active transport (see Section 6.4); otherwise,
water would move inward by osmosis and cause the
cells to burst. For animal cells, an isotonic solution is
usually optimal, whereas for plant cells, an isotonic
solution results in some loss of turgor **(Figure 6.11c).**
The mechanisms by which animals balance their water content by regulating osmosis are discussed in
Chapter 46.

Passive transport, driven by concentration gradients, accounts for much of the movement of water,
ions, and many types of molecules into or out of cells.
In addition, all cells transport some ions and molecules
against their concentration gradients by active transport (see the next section).

STUDY BREAK

1. What conditions are required for osmosis to
   occur?
2. Explain the effect of a hypertonic solution that
   surrounds animal cells.

6.4 Active Transport

Many substances are pushed across membranes against their concentration gradients by active transport "pumps." Active transport concentrates molecules such as sugars and amino acids inside cells and pushes ions in or out of cells. Ion transport may contribute to a voltage difference across a membrane, called a *membrane potential*. It may also control osmotic pressures.

Active Transport Requires a Direct or Indirect Input of Energy Derived from ATP Hydrolysis

There are two kinds of active transport: primary and secondary. In **primary active transport**, the same protein that transports a substance also hydrolyzes ATP to power the transport directly. In **secondary active transport**, the transport is indirectly driven by ATP hydrolysis. That is, the transport proteins do not break down ATP; instead, the transporters use a favorable concentration gradient of ions, built up by primary active transport, as their energy source for active transport of a different ion or molecule.

Other features of active transport resemble facilitated diffusion (listed in Table 6.1). Both processes depend on membrane transport proteins, both are specific, and both can be saturated. The transport proteins are carrier proteins that change their conformation as they function.

Primary Active Transport Moves Positively Charged Ions across Membranes

The primary active transport pumps all move positively charged ions—H^+, Ca^{2+}, Na^+, and K^+—across membranes **(Figure 6.12)**. The gradients of positive ions established by primary active transport pumps underlie functions that are absolutely essential for cellular life. For example, H^+ **pumps** (also called **proton pumps**) move hydrogen ions across membranes and push hydrogen ions across the plasma membrane from the cytoplasm to the cell exterior. The pumps of the plasma membrane (see Figure 6.12) temporarily bind a phosphate group removed from ATP during the pumping cycle. Proton pumps are not common in animals.

The Ca^{2+} **pump** (or **calcium pump**) is distributed widely among eukaryotes. It pushes Ca^{2+} from the cytoplasm to the cell exterior, and also from the cytosol into the vesicles of the endoplasmic reticulum (ER). As a result, Ca^{2+} concentration is typically high outside cells and inside ER vesicles and low in the cytoplasmic solution. This Ca^{2+} gradient is used universally among eukaryotes as a regulatory control of cellular activities as diverse as secretion, microtubule assembly, and muscle contraction; the latter is discussed further in Chapters 7 and 41.

a. Hypotonic conditions: normal turgor pressure

b. Hypertonic conditions: plasmolysis

c. Isotonic conditions: weakened turgor pressure

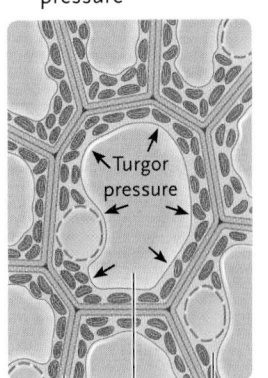
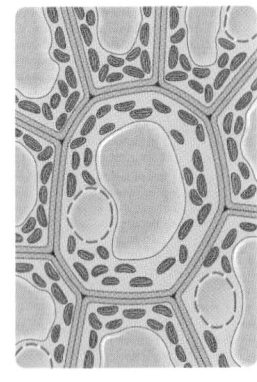

Cell wall　Vacuole　Cytoplasm

Figure 6.11

Effects of turgor pressure and plasmolysis in plants. **(a)** Plant cells developing normal turgor pressure, which keeps the cytoplasmic contents pressed against the cell walls. The pressure is developed by osmotic water flow into the large central vacuole. **(b)** Plant cells in plasmolysis, in which the cells have lost so much water due to outward osmotic flow that they have shrunk away from their walls. **(c)** Plant cells in an isotonic solution, which results in decreased water flow into the cell, shrinkage of the central vacuole, and some loss of turgor.

The Na^+/K^+ **pump** (or **sodium-potassium pump**), located in the plasma membrane, pushes 3 Na^+ out of the cell and 2 K^+ into the cell in the same pumping cycle. As a result, positive charges accumulate in excess outside the membrane, and the inside of the cell becomes negatively charged with respect to the outside. Voltage—an electrical potential difference—across the plasma membrane results in part from this difference in charge. It also results from the unequal distribution of ions across the membrane created by passive transport. The voltage across a membrane is called a **membrane potential**; it measures from about -50 to -200 millivolts (mV; 1 millivolt = 1/1000th of a volt), with the minus sign indicating that the charge inside the cell is negative versus the outside. In summary, there is both a concentration difference (of the ions) and an electrical charge difference on the two sides of the membrane, constituting what is called an **electrochemical gradient**. Electrochemical gradients store energy that is used for other transport mechanisms. For instance, the electrochemical gradient across the membrane is involved with the movement of ions associated with nerve impulse transmission (see Chapter 37).

Secondary Active Transport Moves Both Ions and Organic Molecules across Membranes

As noted earlier, secondary active transport pumps use the concentration gradient of an ion established by a primary pump as their energy source. For example, the

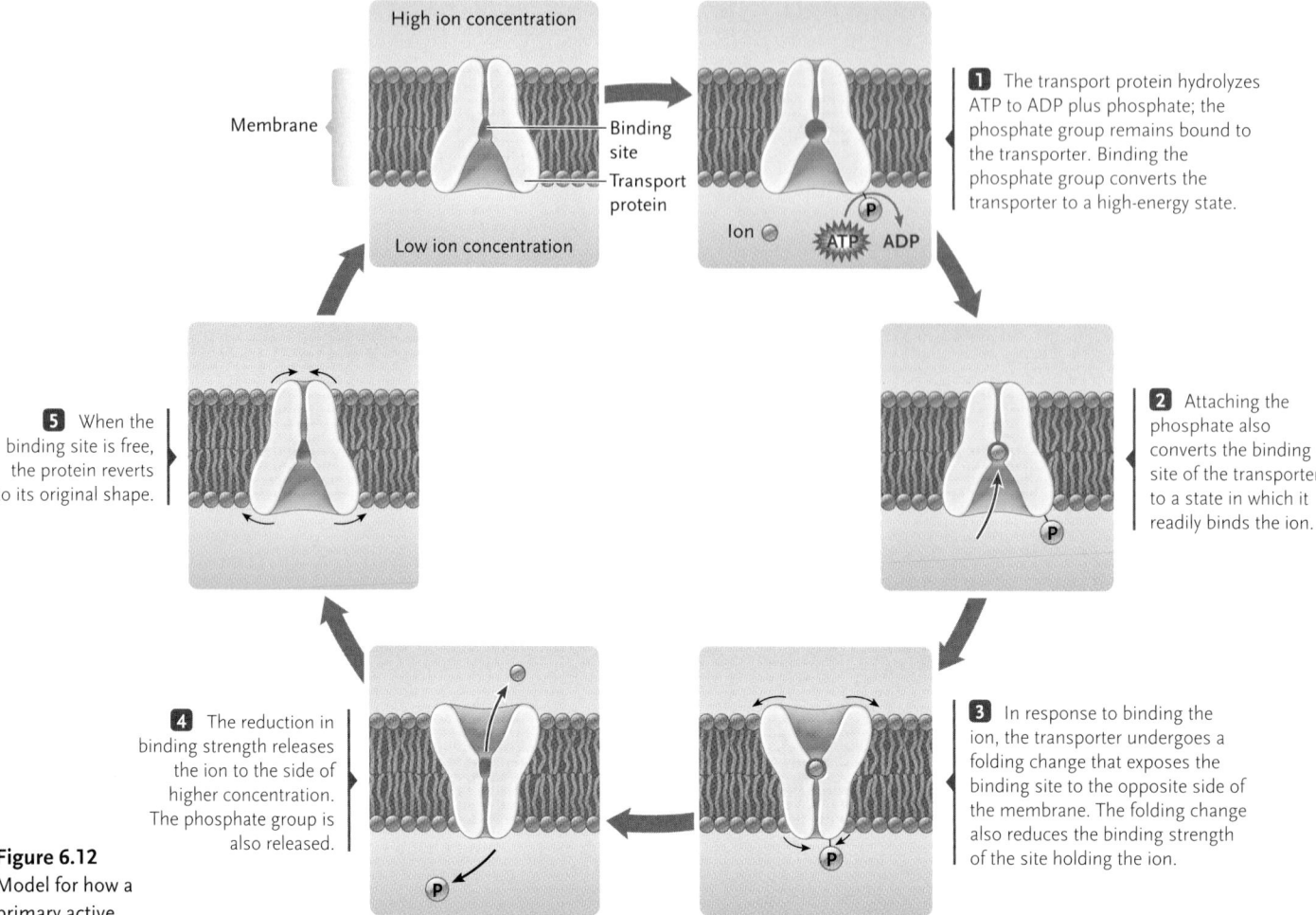

High ion concentration

Membrane

Binding
site

Transport
protein

Low ion concentration

1 The transport protein hydrolyzes ATP to ADP plus phosphate; the phosphate group remains bound to the transporter. Binding the phosphate group converts the transporter to a high-energy state.

Ion

ATP ADP

2 Attaching the phosphate also converts the binding site of the transporter to a state in which it readily binds the ion.

3 In response to binding the ion, the transporter undergoes a folding change that exposes the binding site to the opposite side of the membrane. The folding change also reduces the binding strength of the site holding the ion.

4 The reduction in binding strength releases the ion to the side of higher concentration. The phosphate group is also released.

5 When the binding site is free, the protein reverts to its original shape.

Figure 6.12
Model for how a primary active transport pump operates.

driving force for most secondary active transport in animal cells is the high outside/low inside Na$^+$ gradient created by the Na$^+$/K$^+$ pump. Also, in secondary active transport, the transfer of the solute across the membrane always occurs coupled with transfer of the ion that supplies the driving force.

Secondary active transport occurs by two mechanisms known as *symport* and *antiport* **(Figure 6.13).** In **symport** (also called **cotransport**), the solute moves through the membrane channel in the same direction as the driving ion. Sugars, such as glucose, and amino acids are examples of molecules actively transported into cells by symport. In **antiport** (also known as **exchange diffusion**), the driving ion moves through the membrane channel in one direction, providing the energy for the active transport of another molecule through the membrane in the opposite direction. In many cases, ions are exchanged by antiport. For example, in red blood cells, antiport is the mechanism used for the coupled movement of chloride and bicarbonate ions through a membrane channel; depending on the conditions, either chloride ions enter and bicarbonate ions leave the cells, or bicarbonate ions enter and chloride ions leave.

Active and passive transport move ions and smaller hydrophilic molecules across cellular membranes. Cells can also move much larger molecules or aggregates of molecules from inside to outside, or in the reverse direction, by including them in the inward or outward vesicle traffic of the cell. The mechanisms that carry out this movement—exocytosis and endocytosis—are discussed in the next section.

STUDY BREAK

1. What is active transport? What is the difference between primary and secondary active transport?
2. How is a membrane potential generated?

6.5 Exocytosis and Endocytosis

The largest molecules transported through cellular membranes by passive and active transport are in the size range of amino acids or monosaccharides such as glucose. Eukaryotic cells import and export larger molecules by exocytosis and endocytosis (introduced in Section 5.3). The export of materials by exocytosis primarily carries secretory proteins and some waste materials from the cytoplasm to the cell exterior. Import by endocytosis may carry proteins, larger aggregates of molecules, or even whole cells from the outside into the cytoplasm. Exocytosis and endocytosis also contribute to the back-and-forth flow of membranes between the endomembrane system and the plasma membrane. Both exocytosis and endocytosis require energy; thus, both processes stop if the ability of a cell to make ATP is inhibited.

Exocytosis Releases Molecules to the Outside by Means of Secretory Vesicles

In exocytosis, secretory vesicles move through the cytoplasm and contact the plasma membrane **(Figure 6.14a)**. The vesicle membrane fuses with the plasma membrane, releasing the contents of the vesicle to the cell exterior.

All eukaryotic cells secrete materials to the outside through exocytosis. For example, in animals, glandular cells secrete peptide hormones or milk proteins, and cells that line the digestive tract secrete mucus and digestive enzymes. Plant cells secrete carbohydrates by exocytosis to build a strong cell wall.

Endocytosis Brings Materials into Cells in Endocytic Vesicles

In endocytosis, proteins and other substances are trapped in pitlike depressions that bulge inward from the plasma membrane. The depression then pinches off as an endocytic vesicle. Endocytosis occurs in most eukaryotic cells by one of two distinct but related pathways. In the simplest of these mechanisms, **bulk-phase endocytosis** (sometimes called **pinocytosis**, meaning "cell drinking"), extracellular water is taken in together with any molecules that happen to be in solution in the water **(Figure 6.14b)**. No binding by surface receptors occurs.

In the second endocytic pathway, **receptor-mediated endocytosis**, the target molecules to be taken in are bound to the outer cell surface by receptor proteins **(Figure 6.14c, d)**. The receptors, which are integral proteins of the plasma membrane, recognize and bind only certain molecules—primarily proteins or other molecules carried by proteins—from the solution that surrounds the cell. After binding their target molecules, the receptors collect into a depression in the plasma membrane; this depression is called a **coated pit** because of the network of proteins (called **clathrin**) that coat and reinforce the cytoplasmic side. With the target molecules attached, the pits

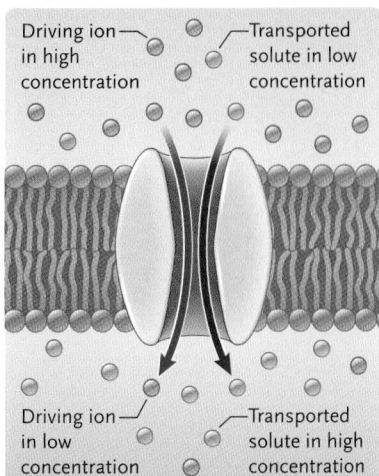
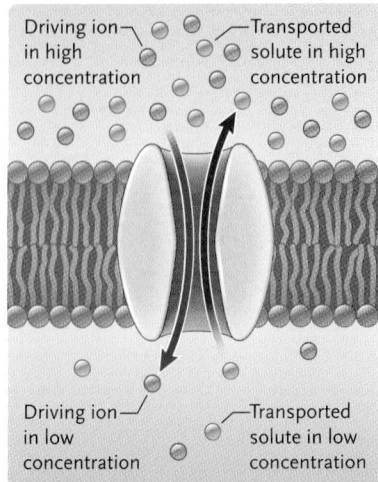

Figure 6.13

Secondary active transport, in which a concentration gradient of an ion is used as the energy source for active transport of a solute. **(a)** In symport, the transported solute moves in the same direction as the gradient of the driving ion. **(b)** In antiport, the transported solute moves in the direction opposite from the gradient of the driving ion.

deepen and pinch free of the plasma membrane to form endocytic vesicles. Once in the cytoplasm, an endocytic vesicle rapidly loses its clathrin coat and may fuse with a lysosome. The enzymes within the lysosome then digest the contents of the vesicle, breaking them down into smaller molecules useful to the cell. These molecular products—for example, amino acids and monosaccharides—enter the cytoplasm by crossing the vesicle membrane via transport proteins. The membrane proteins are recycled to the plasma membrane.

Mammalian cells take in many substances by receptor-mediated endocytosis, including peptide hormones, antibodies, and blood proteins. The receptors that bind these substances to the plasma membrane are present in thousands to hundreds of thousands of copies. For example, a mammalian cell plasma membrane has about 20,000 receptors for *low-density lipoprotein (LDL)*. LDL, a complex of lipids and proteins, is the way cholesterol moves through the bloodstream. When LDL binds to its receptor on the membrane, it is taken into the cell by receptor-mediated endocytosis; then, by the steps described earlier, the LDL is broken down within the cell and cholesterol is released into the cytoplasm.

Some cells, such as certain white blood cells *(phagocytes)* in the bloodstream, or protists such as *Amoeba proteus,* can take in large aggregates of molecules, cell parts, or even whole cells by a process related to receptor-mediated endocytosis. The process, called **phagocytosis** (meaning "cell eating"), begins when surface receptors bind molecules on the substances to be

Figure 6.14

Exocytosis and endocytosis. **(a)** Exocytosis. **(b)** Bulk-phase endocytosis. **(c)** Diagram and **(d)** electron micrographs of receptor-mediated endocytosis.

a. Exocytosis

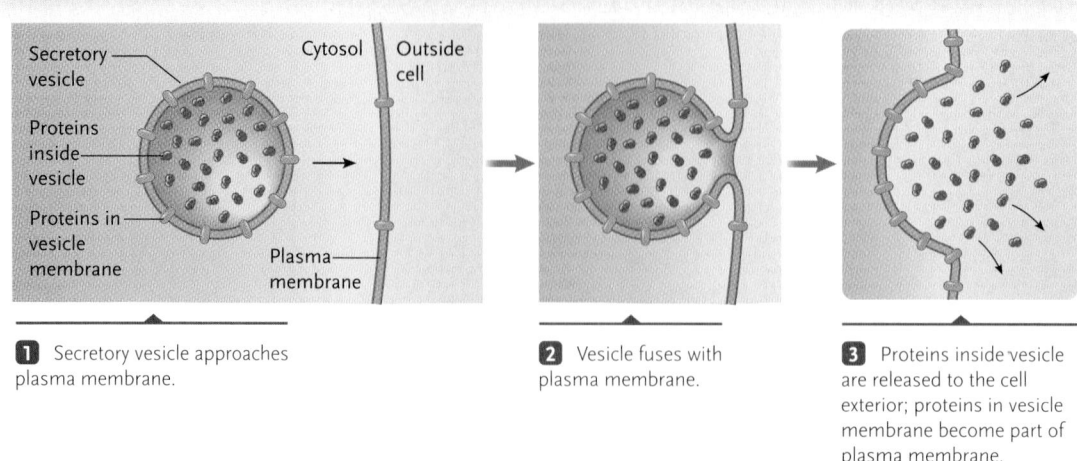

1 Secretory vesicle approaches plasma membrane.

2 Vesicle fuses with plasma membrane.

3 Proteins inside vesicle are released to the cell exterior; proteins in vesicle membrane become part of plasma membrane.

b. Bulk-phase endocytosis (pinocytosis)

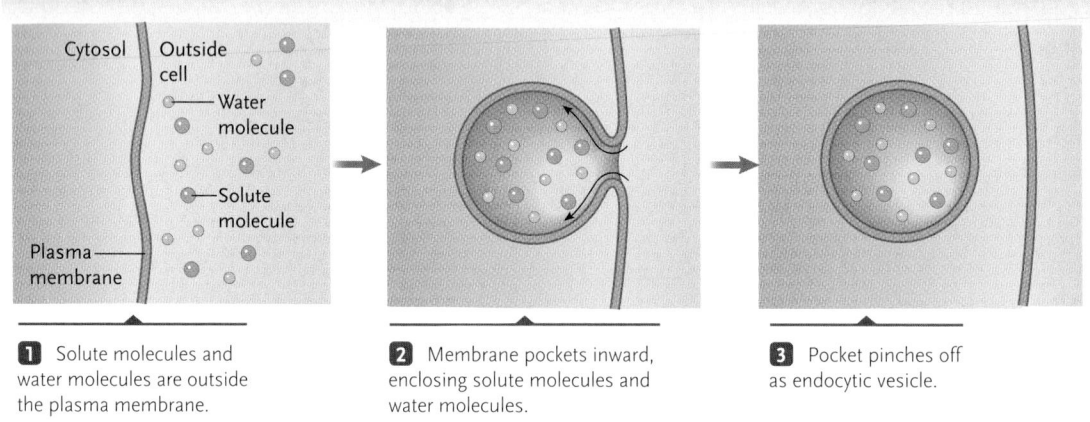

1 Solute molecules and water molecules are outside the plasma membrane.

2 Membrane pockets inward, enclosing solute molecules and water molecules.

3 Pocket pinches off as endocytic vesicle.

c. Receptor-mediated endocytosis

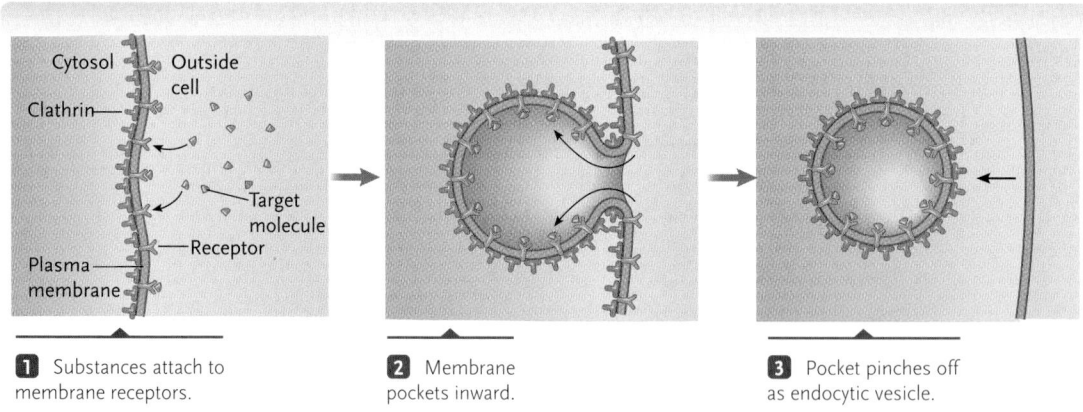

1 Substances attach to membrane receptors.

2 Membrane pockets inward.

3 Pocket pinches off as endocytic vesicle.

d. Micrographs of stages of receptor-mediated endocytosis shown in c

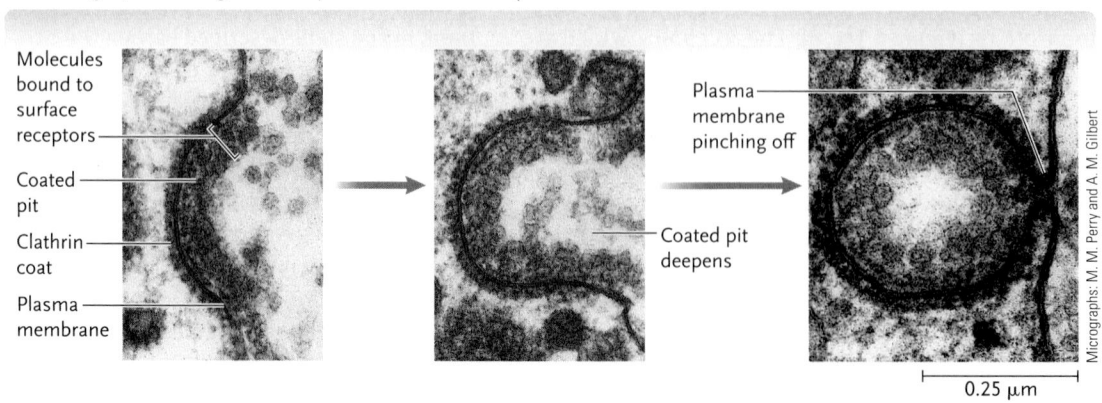

0.25 μm

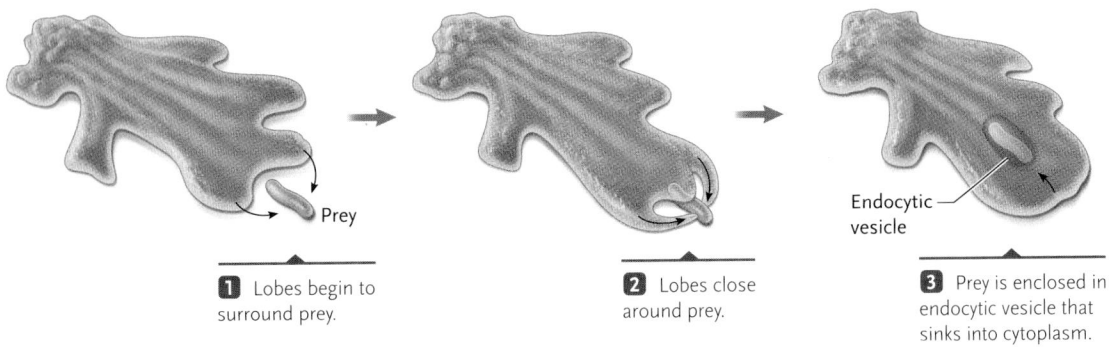

1 Lobes begin to surround prey.

2 Lobes close around prey.

3 Prey is enclosed in endocytic vesicle that sinks into cytoplasm.

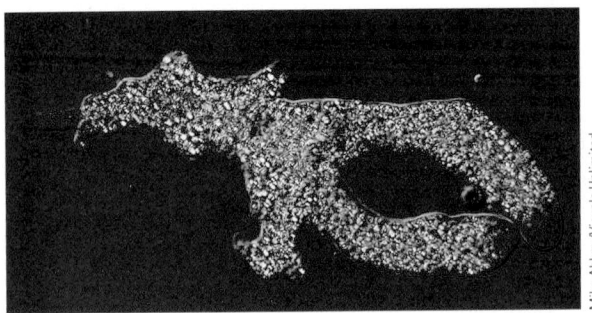

Mike Abbey/Visuals Unlimited

Figure 6.15

Phagocytosis, in which lobes of the cytoplasm extend outward and surround a cell targeted as prey. The micrograph shows the protist *Chaos carolinense* engulfing a single-celled alga *(Pandorina)* by phagocytosis (corresponding to step 2 in the diagram); white blood cells called phagocytes carry out a similar process in mammals.

taken in **(Figure 6.15)**. Cytoplasmic lobes then extend, surround, and engulf the materials, forming a pit that pinches off and sinks into the cytoplasm as a large endocytic vesicle. The materials then are digested within the cell as in receptor-mediated endocytosis, and any remaining residues are sequestered permanently into storage vesicles or are expelled from cells by exocytosis as wastes.

The combined workings of exocytosis and endocytosis constantly cycle membrane segments between the internal cytoplasm and the cell surface. The balance of the two mechanisms maintains the surface area of the plasma membrane at controlled levels.

Thus, through the combined mechanisms of passive transport, active transport, exocytosis, and endocytosis, cells maintain their internal concentrations of ions and molecules and exchange larger molecules such as proteins with their surroundings. The next chapter explores cell communication through intercellular chemical messengers. Many of these messengers act through binding to specific proteins embedded in the plasma membrane.

STUDY BREAK

1. What is the mechanism of exocytosis?
2. What is the difference between bulk-phase endocytosis and receptor-mediated endocytosis?

Research continues into the structure and assembly of membranes and the transport of substances across membranes by various mechanisms.

How do aquaporin channels function?

Peter Agre at Johns Hopkins University School of Medicine in Baltimore received a Nobel Prize in 2003 for his discovery of aquaporins. Aquaporins are specific channels for water transport across cell membranes. Problems with aquaporin function are associated with various human diseases, such as congenital cataracts, a form of diabetes, congestive heart failure, and brain edema (swelling caused by excess fluid). Therefore, a better understanding of aquaporin function could help facilitate the development of drugs to treat those diseases.

The ability to absorb or release water varies considerably among cells and tissues of an organism and between organisms. Since Agre's discovery, more than 200 different aquaporins have been identified in tissues from mammals, nonmammalian vertebrates, invertebrates, plants, and various microorganisms. Variation in aquaporin structure among these forms is likely responsible for their differences in function. Agre's research group is pursuing this issue by characterizing the structures of various aquaporins from humans, yeast, and bacteria to produce high-resolution models. Such models will be informative for designing experiments to further our understanding of the function of these channel molecules. Agre's group is also studying the regulation of the aquaporin genes to characterize tissue-specific production of aquaporins. The results of this line of investigation will provide a valuable piece of the puzzle concerning variation in water uptake.

Can endocytosis of nanotubes deliver therapeutic agents into cells?

The goal of many research groups has been the use of endocytosis to deliver therapeutic agents to diseased cells, such as cancer cells. There are many possible ways to deliver therapeutic agents to cells. Hongjie Dai, a physical chemist at Stanford University in California, has been working with carbon nanotubes, which are cylindrical carbon molecules with a diameter of just a few nanometers (about 50,000 times smaller than the width of a human hair) and up to several centimeters in length.

Dai's research team has shown that carbon nanotubes can carry proteins and DNA into cells. How are the carbon nanotubes taken into the cells? Knowing the route is important for determining what kinds of chemical bonds will be needed to attach therapeutic agents to the carbon nanotubes. For example, endocytosis produces vesicles that can fuse with lysosomes. Therefore, if carbon nanotubes are taken up by endocytosis, then the drug or DNA being delivered could be attached to the nanotubes by disulfide bonds because those bonds would readily be broken by the acidic environment of the lysosome, thereby releasing the agent.

Dai's group has evidence that carbon nanotubes are taken into cells by endocytosis. Endocytosis requires energy in the form of either ATP or heat, and when they cooled the cell cultures or treated them with an inhibitor that stopped ATP production, the cells could no longer take in carbon nanotubes.

Future research will focus on use of carbon nanotubes to deliver anticancer agents specifically to cancer cells in tissue culture. Undoubtedly, much work will be needed to produce an efficient method for that delivery, as well as an effective way to release and activate the anticancer agent within the cell. If success is forthcoming with tissue culture systems, the protocols will be moved to model organisms for cancer and eventually to humans for clinical trials.

Peter J. Russell

Review

Go to **ThomsonNOW** at www.thomsonedu.com/login to access quizzing, animations, exercises, articles, and personalized homework help.

6.1 Membrane Structure

- Both membrane phospholipids and membrane proteins have hydrophobic and hydrophilic regions, giving them dual solubility properties.

- Membranes are based on a fluid phospholipid bilayer, in which the polar regions of the phospholipids lie at the surfaces of the bilayer and their nonpolar tails associate together in the interior (Figures 6.2–6.5).

- Membrane proteins are suspended individually in the bilayer, with their hydrophilic regions at the membrane surfaces and their hydrophobic regions in the interior (Figures 6.4 and 6.5).

- The lipid bilayer forms the structural framework of membranes and serves as a barrier that prevents the passage of most water-soluble molecules.

- Proteins embedded in the phospholipid bilayer perform most membrane functions, including transport of selected hydrophilic substances, recognition, signal reception, cell adhesion, and metabolism.

- Integral membrane proteins are embedded deeply in the bilayer and cannot be removed without dispersing the bilayer. Peripheral membrane proteins associate with membrane surfaces (Figure 6.5).

- Membranes are asymmetric—that is, different proportions of phospholipid types occur in the two bilayer halves.

Animation: Lipid bilayer organization

Animation: Cell membranes

6.2 Functions of Membranes in Transport: Passive Transport

- Passive transport depends on diffusion, the net movement of molecules with a concentration gradient, from a region of higher concentration to a region of lower concentration. Passive transport does not require cells to expend energy (Table 6.1).

- Simple diffusion is the passive transport of substances across the lipid portion of cellular membranes with their concentration gradients. It proceeds most rapidly for small molecules that are soluble in lipids (Table 6.1).

- Facilitated diffusion is the passive transport of substances at rates higher than predicted from their lipid solubility. It depends on membrane proteins, follows concentration gradients, is specific for certain substances, and becomes saturated at high concentrations of the transported substance (Figure 6.8 and Table 6.1).
- Most proteins that carry out facilitated diffusion of ions are controlled by "gates" that open or close their transport channels (Figure 6.8).

Interaction: Selective permeability

Animation: Passive transport

6.3 Passive Water Transport and Osmosis

- Osmosis is the net diffusion of water molecules across a selectively permeable membrane in response to differences in the concentration of solute molecules (Figure 6.9). Water moves from hypotonic (lower concentrations of solute molecules) to hypertonic solutions (higher concentrations of solute molecules). When the solutions on each side are isotonic, there is no net osmotic movement of water in either direction (Figure 6.10).

Animation: Solute concentration and osmosis

Interaction: Tonicity and water movement

6.4 Active Transport

- Active transport moves substances against their concentration gradients and requires cells to expend energy. It depends on membrane proteins, is specific for certain substances, and becomes saturated at high concentrations of the transported substance (Table 6.1).

- Active transport proteins are either primary transport pumps, which directly use ATP as their energy source, or secondary transport pumps, which use favorable concentration gradients of positively charged ions, created by primary transport pumps, as their energy source for transport (Figure 6.12).
- Secondary active transport may occur by symport, in which the transported substance moves in the same direction as the concentration gradient used as the energy source, or by antiport, in which the transported substance moves in the direction opposite to the concentration gradient used as the energy source (Figure 6.13).

Animation: Active transport

6.5 Exocytosis and Endocytosis

- Large molecules and particles are moved out of and into cells by exocytosis and endocytosis. The mechanisms allow substances to leave and enter cells without directly passing through the plasma membrane (Figure 6.14).
- In exocytosis, a vesicle carrying secreted materials contacts and fuses with the plasma membrane on its cytoplasmic side. The fusion introduces the vesicle membrane into the plasma membrane and releases the vesicle contents to the cell exterior (Figure 6.14a).
- In endocytosis, materials on the cell exterior are enclosed in a segment of the plasma membrane that pockets inward and pinches off on the cytoplasmic side as an endocytic vesicle. Endocytosis occurs in two overall forms, bulk-phase (pinocytosis) and receptor-mediated endocytosis. Most of the materials that enter cells are digested into molecular subunits small enough to be transported across the vesicle membranes (Figures 6.14b–d).

Animation: Phagocytosis

Questions

Self-Test Questions

1. In the fluid mosaic model:
   a. plasma membrane proteins orient their hydrophilic sides toward the internal bilayer.
   b. phospholipids often flip-flop between the inner and outer layers.
   c. the mosaic refers to proteins attached to the underlying cytoskeleton.
   d. the fluid refers to the phospholipid bilayer.
   e. the mosaic refers to the symmetry of the internal membrane proteins and sterols.

2. Which of the following statements is *false*? Proteins in the plasma membrane can:
   a. transport proteins.
   b. synthesize polypeptides.
   c. recognize self versus foreign molecules.
   d. allow adhesion between the same tissue cells or cells of different tissues.
   e. combine with lipids or sugars to form complex macromolecules.

3. The freeze-fracture technique demonstrated:
   a. that the plasma membrane is a bilayer with individual proteins suspended in it.
   b. that the plasma membrane is fluid.
   c. the different functions of membrane proteins.
   d. that proteins are bound to the cytoplasmic side but not embedded in the lipid bilayer.
   e. the direction of movement of solutes through the membrane.

4. In the following diagram, assume that the setup was left unattended. Which of the following statements is correct?

Selectively permeable membrane	
Inside a cell	**Outside fluids**
Solvent 95%	Solvent 98%
Solute 5%	Solute 2%

   a. The relation of the cell to its environment is isotonic.
   b. The cell is in a hypertonic environment.
   c. The net flow of solvent is into the cell.
   d. The cell will soon shrink.
   e. Diffusion can occur here but not osmosis.

5. Which of the following statements is true for the diagram in question 4?
   a. The net movement of solutes is into the cell.
   b. There is no concentration gradient.
   c. There is a potential for plasmolysis.
   d. The solvent will move against its concentration gradient.
   e. If this were a plant cell, turgor pressure would be maintained.

6. Using the principle of diffusion, a dialysis machine removes waste solutes from a patient's blood. Imagine blood runs through a cylinder wherein diffusion can occur across an artificial selectively permeable membrane to a saline solution

on the other side. Which of the following statements is correct?
a. Solutes move from lower to higher concentration.
b. The concentration gradient is lower in the patient's blood than in the saline solution wash.
c. The solutes are transported through a symport in the blood cell membrane.
d. The saline solution has a lower concentration gradient of solute than the blood.
e. The waste solutes are actively transported from the blood.

7. A characteristic of carrier molecules in a primary active transport pump is that:
a. they cannot transport a substance and also hydrolyze ATP.
b. they retain their same shape as they perform different roles.
c. their primary role is to move negatively charged ions across membranes.
d. they move Na^+ into a neural cell and K^+ out of the same cell.
e. They act to establish an electrochemical gradient.

8. A driving ion moving through a membrane channel in one direction gives energy to actively transport another molecule in the opposite direction. What is this process called?
a. facilitated diffusion
b. exchange diffusion
c. symport transport
d. primary active transport pump
e. cotransport

9. Phagocytosis illustrates which phenomenon?
a. receptor-mediated endocytosis
b. bulk-phase endocytosis
c. exocytosis
d. pinocytosis
e. cotransport

10. Place in order the following events of receptor-mediated endocytosis.
(1) Clathrin coat disappears.
(2) Receptors collect in a coated pit covered with clathrin on the cytoplasmic side.
(3) Receptors recognize and bind specific molecules.
(4) Endocytic vesicle may fuse with lysosome while receptors are recycled to the cell surface.
(5) Pits deepen and pinch free of plasma membrane to form endocytic vesicles.
a. 4, 1, 2, 5, 3   d. 4, 1, 5, 2, 3
b. 2, 1, 3, 5, 4   e. 3, 1, 2, 4, 5
c. 3, 2, 5, 1, 4

Questions for Discussion

1. The bacterium *Vibrio cholerae* causes cholera, a disease characterized by severe diarrhea that may cause infected people to lose up to 20 *L* of fluid in a day. The bacterium enters the body when someone drinks contaminated water. It adheres to the intestinal lining, where it causes cells of the lining to release sodium and chloride ions. Explain how this release is related to the massive fluid loss.

2. Irrigation is widely used in dryer areas of the United States to support agriculture. In those regions, the water evaporates and leaves behind deposits of salt. What problems might these salt deposits cause for plants?

3. In hospitals, solutions of glucose with a concentration of 0.3 *M* can be introduced directly into the bloodstream of patients without tissue damage by osmotic water movement. The same is true of NaCl solutions, but these must be adjusted to 0.15 *M* to be introduced without damage. Explain why one solution is introduced at 0.3 *M* and the other at 0.15 *M*.

Experimental Analysis

Design an experiment to determine the concentration of NaCl (table salt) in water that is isotonic to potato cells. Use only the following materials: a knife, small cookie cutters, and a balance.

Evolution Link

What evidence would convince you that membranes and active transport mechanisms evolved from an ancestor common to both prokaryotes and eukaryotes?

How Would You Vote?

The ability to detect mutant genes that cause severe disorders raises bioethical questions. Should we encourage the mass screening of prospective parents for mutant genes that cause cystic fibrosis? Should society encourage women to give birth only if their child will not develop severe medical problems? Go to www.thomsonedu.com/login to investigate both sides of the issue and then vote.

A B cell and a T cell communicating by direct contact in the human immune system (computer image). Cell communication coordinates the cellular defense against disease.

© Russell Kightley Media

7 Cell Communication

WHY IT MATTERS

Hundreds of aircraft, ranging from small private planes to huge passenger jets, approach and leave airports in Southern California. In addition to the large terminals in Los Angeles and San Diego, dozens of smaller airports are located in the vicinity. The aircraft that approach these airports are traveling at various speeds, entering from all points of the compass, and flying at different altitudes. Airplanes are also leaving the same airports with routes distributed over the same directions, speeds, and altitudes. A wrong turn, ascent, or descent by any one of the hundreds of planes could lead to disaster. Yet, disasters are extremely rare. How are all these aircraft kept separate, and routed to and from their airports safely and efficiently? The answer lies in a highly organized system of controllers, signals, and receivers.

As the aircraft thread their way along the various approach and departure routes, they follow directions issued by air traffic controllers. Each aircraft has a radio receiver tuned to a frequency that has been assigned by the controllers. By speaking on a transmitter tuned to the frequency assigned to Piper 4879Z, a slow-moving two-seater headed for Montgomery Field near San Diego, and using its identifier ("seven niner Zulu"), controllers can keep this plane's path separate from that of "five-two heavy," a passenger jet, leaving the main San Diego air terminal. The flow of directing signals, followed individually by each aircraft in the vicinity, keeps the traffic unscrambled and moving safely.

The principle of the air control system is nothing new. An equivalent system of signals and tuned receivers evolved hundreds of millions of years ago, as one of the developments that made multicellular life possible. Within a multicellular organism, the activities of individual cells are directed by molecular signals, such as hormones, that are released by controlling cells. Although the controlling cells release many signals, each receiving cell has receptors that are "tuned" to recognize only one or a few of the many signal molecules that circulate in its vicinity; other signals pass by without effect because the cell has no receptors for them.

When a cell binds a signal molecule, it modifies its internal activities in accordance with the signal, coordinating its functions with the activities of other cells of the organism. The responses of the receiving cell may include changes in gene activity, protein synthesis, transport of molecules across the plasma membrane, metabolic reactions, secretion, movement, and division. In some cases, the response to a signal may be "suicide"—that is, the programmed death of the receiving cell **(Figure 7.1)**. As part of its response, a cell may itself become a controller and thus contribute to the organizational network by releasing signal molecules that modify the activity of other cell types. The total network of signals and responses allows multicellular organisms to grow, develop, reproduce, and compensate for environmental changes in an internally coordinated fashion.

This chapter describes the major pathways that form parts of the cell communication system based on both surface and internal receptors, including the links that tie the different response pathways into fully integrated networks. (Communication pathways based on neurons—nerve cells—in animals are discussed in Chapter 37.) This chapter concentrates primarily on the systems working in animals, particularly in mammals, from which most of our knowledge of cell communication has been developed. Nonetheless, the prin-

ciples of cell communication illustrated by these pathways apply to most multicellular eukaryotic organisms, including plants, protists, and fungi, and to single-celled eukaryotic organisms such as yeast. (The plant communication and control systems are described in more detail in Chapter 35.) This discussion begins with a few fundamental principles that underlie the often complex networks of cell communication.

7.1 Cell Communication: An Overview

Communication is critical for the function and survival of cells that compose a multicellular animal. For example, the ability of cells to communicate with one another in a regulated way is responsible for the controlled growth and development of an animal, as well as the integrated activities of its tissues and organs.

Cells communicate with one another in three ways. Adjacent cells use direct channels of communication. In this rapid means of communication, small molecules and ions exchange directly between the two cytoplasms. In animal cells, the direct channels of communication are *gap junctions,* the specialized connections between the cytoplasms of adjacent cells (see Section 5.5 for a detailed discussion of gap junctions). The main role of gap junctions is to synchronize metabolic activities or electronic signals between cells in a tissue. For example, gap junctions play a key role in the spread of electrical signals from one cell to the next in cardiac muscle. In plant cells, the direct channels of communication are plasmodesmata (see discussion in Section 5.4). Small molecules moving between adjacent cells in plants include plant hormones that regulate growth. In this way, responses triggered by plant hormones are spread to other cells.

Cells also communicate through *specific contact between cells.* Certain cells have molecules on their surfaces that allow them to interact directly with other cells. For example, some cells of a mammal's immune system use their surface molecules to recognize particular molecules on the surfaces of invading pathogens that signal them as foreign. The host cell then engulfs the invader. Cells also have on their surfaces *cell adhesion molecules,* integral membrane proteins that allow the cells to bind to other cells or to the extracellular matrix. There are many important functions of cell adhesion molecules, including roles in coordinating tissue and organ formation as an embryo develops.

Finally, cells communicate through *intercellular ("between cell") chemical messengers.* This method is the most common means of cell communication. Here, one cell, the *controlling cell,* synthesizes a specific molecule that acts as a *signaling molecule* to affect the activity of another cell, the *target cell.* The target cell is not in contact with the cell that synthesizes the signaling molecule; rather, it is either nearby or at a distance away in the organism.

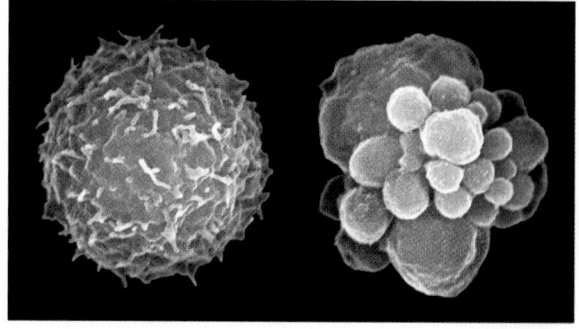

Figure 7.1
A normal cell (left) and a cell undergoing apoptosis (programmed cell death) (right).

Surviving Something Bad by Taking a Risk

Programmed cell death, called *apoptosis,* is a natural part of many developmental pathways. For example, human embryos initially have webbed fingers and toes; as the embryo develops, cells in the webbing die and the fingers and toes become fully separated. The instructions for apoptosis are transmitted to the condemned cells by extracellular signal molecules and executed by internal response pathways. The pathways activate enzymes that break down DNA and proteins, fragmenting the chromosomes and destroying vital cellular processes. The cell quickly dies and disintegrates.

Programmed cell death also occurs as the brain develops. Nerve cells that fail to make normal connections are marked for death and cleared from the brain. Apoptosis in neurons occurs via a surface-receptor–regulated signal transduction pathway that activates a death protein called BAD, which sets off the killing reactions. Not all of the marked neurons die; some survive if a signal molecule called *BDNF (brain-derived neurotrophic factor)* is bound

by its receptor on the cell surface. Molecular research by Michael E. Greenberg and his coworkers at Harvard Medical School, Cambridge, Massachusetts, has shown how BDNF counteracts the BAD effects and rescues cells marked for death.

Greenberg and his colleagues set out to discover the signal transduction pathway that starts with the BDNF signal and ends with inactivation of BAD. In their experiments, they activated this pathway in nerve cells, then broke open the cells and identified the activated signal transduction pathway proteins. To identify the proteins, they used specific marker proteins that bound to BDNF pathway proteins only if they were in their activated, phosphorylated state.

Using this method, Greenberg and his colleagues discovered that BDNF-triggered cell rescue occurred by activating a Ras G protein, which, in turn, activated a phosphorylation cascade that involved MAP kinases (see Figure 7.13). The phosphorylation cascade activates another protein kinase called Rsk, which then phos-

phorylates (inactivates) BAD and saves the cell.

To test this hypothesis, Greenberg and coworkers introduced genes that encode MAP kinase, Rsk, and BAD into cultures of a cell type unrelated to neurons. After activation of MAP kinase, the cells were broken open and tested with the antibody that reacts with inactivated BAD. The antibody reacted positively, showing that the death protein was inactivated. However, if a gene encoding a mutant form of Rsk, unable to phosphorylate BAD, was introduced into the cells instead of the normal form, BAD remained unphosphorylated. These results supported their hypothesis. In summary, Greenberg's laboratory had evidence for the following model for the cellular response pathway that saves neurons from apoptosis:

BDNF → receptor → unknown steps → Ras → MAP kinase series → Rsk → inactive BAD

Therefore, preventing BAD things does indeed involve some Rsk.

For example, in response to stress, cells of a mammal's adrenal glands (located on top of the kidneys)—the controlling cells—secrete the hormone epinephrine into the bloodstream. Epinephrine acts on target cells to increase the amount of glucose in the blood.

Cell communication through intercellular chemical messengers is the focus of this chapter, and the epinephrine example is used to illustrate the principles involved. In the 1950s, Earl Sutherland and his research team at Case Western Reserve University, Cleveland, Ohio, began investigating this cell communication system. Sutherland discovered that the hormone epinephrine acts by activating an enzyme, glycogen phosphorylase, which catalyzes the production of glucose from glycogen. That is, the result of the secretion of epinephrine into the blood by adrenal gland cells is an increase in the amount of glucose in the blood. Sutherland's experiments showed that enzyme activation did not involve epinephrine directly but did require an unknown (at the time) cellular substance. Sutherland called the hormone the *first messenger* in the system and the unknown cellular substance the *second messenger.* He proposed that the following chain of reactions was involved: epinephrine (the first messenger) leads to the forma-

tion of the second messenger, which activates the enzyme for conversion of glycogen to glucose.

Sutherland's work was the foundation for research that developed our current understanding of this type of cell communication. In brief, a controlling cell releases a signal molecule that causes a response (affects the function) of target cells. Target cells process the signal in the following three sequential steps **(Figure 7.2):**

1. **Reception.** Reception is the binding of a signal molecule with a specific receptor of target cells. Target cells have receptors that are specific for the signal molecule, which distinguishes them from cells that do not respond to the signal molecule. The signals themselves may be polar (charged, hydrophilic) molecules or nonpolar (hydrophobic) molecules, and their receptors are shaped to recognize and bind them specifically **(Figure 7.3).** Receptors for polar signal molecules are embedded in the plasma membrane with a binding site for the signal molecule on the cell surface (see Figures 7.2 and 7.3a). Epinephrine, the first messenger in Sutherland's research, is a peptide hormone, a polar molecule that is recognized by a surface recep-

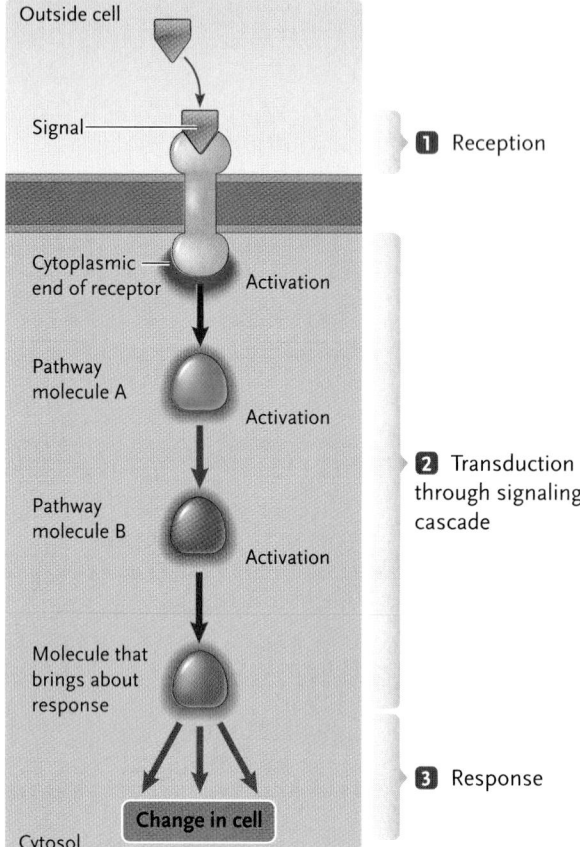

Figure 7.2
The three stages of signal transduction: reception, transduction, and response (shown for a system using a surface receptor).

a. Reception by a cell-surface receptor

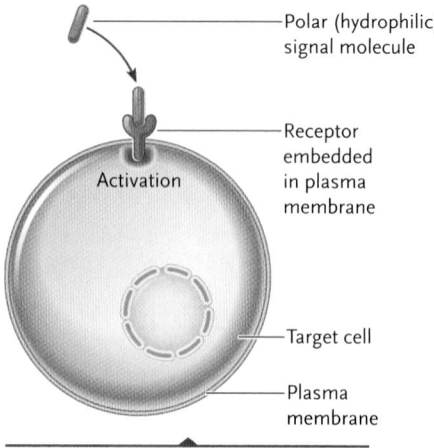

Polar (hydrophilic) signal molecule

Receptor embedded in plasma membrane

Activation

Target cell

Plasma membrane

Polar signal molecules cannot pass through the plasma membrane. In this case they bind to a receptor on the surface.

b. Reception by a receptor within cell

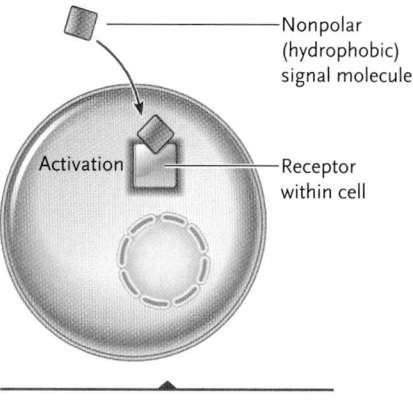

Nonpolar (hydrophobic) signal molecule

Activation

Receptor within cell

Nonpolar signal molecules pass through the plasma membrane and bind to their receptors in the cell.

Figure 7.3
Reception **(a)** of a polar (hydrophilic) signal molecule by a receptor on the cell surface and **(b)** of a nonpolar (hydrophobic) signal molecule by a receptor in the cell.

tor on target cells. Receptors for nonpolar molecules are located within the cell (see Figure 7.3b). In this case, the nonpolar signal molecule passes freely through the plasma membrane and interacts with its receptor within the cell. Steroid hormones such as testosterone and estrogen are examples of nonpolar signal molecules.

2. **Transduction.** Transduction is the process of changing the signal into the form necessary to cause the cellular response (see Figure 7.2). In other words, the binding of a signal molecule to its receptor is not directly responsible for the response. Transduction may occur in a single step, although more often it involves a cascade of reactions that include several different molecules, often referred to as a *signaling cascade.* For example, in Sutherland's work, after epinephrine bound to its surface receptor, the signal was transmitted through the plasma membrane into the cell, where transduction by a signaling cascade activated a molecule that triggered a cellular response. This molecule was Sutherland's *second messenger.*

3. **Response.** In the third and last stage, the transduced signal causes a specific cellular response. That response depends on the signal and the receptors on the target cell. In Sutherland's work, the response was the activation of the enzyme glycogen phosphorylase; the active enzyme catalyzed the conversion of stored glycogen to glucose.

The whole series of events from reception to response is called **signal transduction.** As explained in subsequent sections, signal transduction occurs by different mechanisms, depending on the receptor type. Earl Sutherland was awarded a Nobel Prize in 1971 for his research on the mechanisms of action of hormones.

STUDY BREAK

What accounts for the specificity of a cellular response in signal transduction?

7.2 Characteristics of Cell Communication Systems with Surface Receptors

Cell communication systems based on surface receptors have three components: (1) the extracellular signal molecules released by controlling cells, (2) the surface receptors on target cells that receive the signals, and (3) the internal response pathways triggered when receptors bind a signal.

Peptide Hormones and Neurotransmitters Are Extracellular Signal Molecules Recognized by Surface Receptors in Animals

Surface receptors in mammals and other vertebrates recognize and bind two major types of extracellular signal molecules: *peptide hormones* and *neurotransmitters*. These signal molecules are polar, water-soluble molecules that are released by control cells and enter the fluids that surround cells, including the blood circulation in animals with a circulatory system.

Peptide hormones are small proteins with a few to more than 200 amino acids. As a group, they affect all body systems. For example, they regulate sugar levels in blood, pigmentation, and ovulation. A special class of peptide hormones, the *growth factors,* affects cell growth, division, and differentiation.

Cells that release peptide hormones are called gland cells. They may form part of distinct, individual organs such as the thyroid or pituitary gland, or they may be distributed among the cells of organs with other functions, such as the stomach and intestines, heart, brain, liver, and kidneys in humans and other mammals. For example, gland cells scattered through the lining of the human stomach and small intestine secrete peptide hormones that regulate digestive functions. (Peptide hormones and growth factors are discussed in further detail in Chapter 40.)

Neurotransmitters are molecules released by neurons that trigger activity in other neurons or other cells in the body; they include small peptides, individual amino acids or their derivatives, and other chemical substances. Some neurotransmitters affect only one or a few cells in the immediate vicinity of the neuron that releases the signal molecule, whereas others are released into the body circulation and act essentially as hormones, affecting many types of tissues. (Neurotransmitters are discussed in further detail in Chapter 37.)

Once signal molecules are released into the body's circulation, they remain for only a certain time. They are either broken down at a steady rate by enzymes in their target cells or in organs such as the liver, or they are excreted by the kidneys. The removal process ensures that the signal molecules are active only as long as controlling cells are secreting them.

Surface Receptors Are Integral Membrane Glycoproteins

The surface receptors that recognize and bind signal molecules are all glycoproteins—proteins with attached carbohydrate chains (see Section 3.5). They are integral membrane proteins that extend entirely through the plasma membrane **(Figure 7.4)**. The signal-binding site of the receptor, which extends from the outer membrane surface, is folded in a way that closely fits the signal molecule. The fit, similar to the fit of an enzyme to its substrate, is specific, so a particular receptor binds only one type of signal molecule or a closely related group of signal molecules.

A signal molecule brings about specific changes in cells to which it binds. When a signal molecule binds to a surface receptor, the molecular structure of that receptor is changed so that it transmits the signal through the plasma membrane, activating the cytoplasmic end of the receptor. The activated receptor then initiates the first step in a cascade of molecular events—the signal transduction pathway—that triggers the cellular response (see Figure 7.2).

Animal cells typically have hundreds to thousands of surface receptors that represent many receptor types. Receptors for a specific peptide hormone may number from 500 to as many as 100,000 or more per cell. Different cell types contain distinct combinations of receptors, allowing them to react individually to the

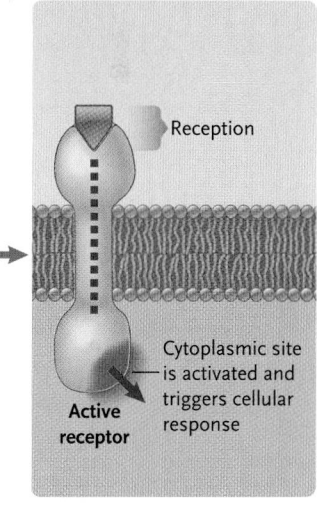

a. Surface receptor

b. Activation of receptor by binding of a specific signal molecule

A surface receptor has an extracellular segment with a site that recognizes and binds a particular signal molecule.

When the signal molecule is bound, a conformational change is transmitted through the transmembrane segment that activates a site on the cytoplasmic segment of receptor. The activation triggers a reaction pathway that results in the cellular response.

Figure 7.4
The mechanism by which a surface receptor responds when it binds a signal molecule.

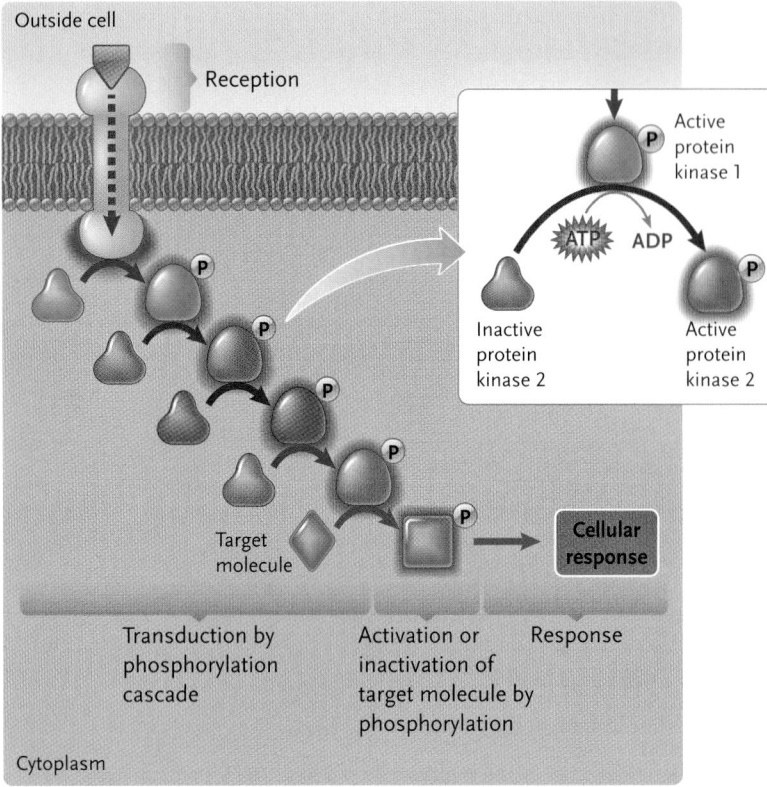

Figure 7.5

Phosphorylation, a key reaction in many signaling pathways.

teins **(Figure 7.5;** see Section 4.5). These phosphorylated proteins are known as *target proteins* because they are the proteins modified by signaling pathways. The added phosphate groups either stimulate or inhibit the activity of the target proteins; the change in the target proteins' activity leads directly or indirectly to the cellular response. Often, protein kinases act in a chain, called a *protein kinase cascade,* to pass along a signal. The first kinase catalyzes phosphorylation of the second, which then becomes active and phosphorylates the third kinase, and so on. The proteins that bring about the cellular response may be parts of the reaction pathways, enzymes of other cellular reactions, end targets of the signal transduction pathways (such as transport proteins), or at the most fundamental level, proteins that regulate gene transcription.

The effects of protein kinases in the signal transduction pathways are balanced or reversed by another group of enzymes called **protein phosphatases,** which remove phosphate groups from target proteins. Unlike the protein kinases, which are active only when a surface receptor binds a signal molecule, most of the protein phosphatases are continuously active in cells. By continually removing phosphate groups from target proteins, the protein phosphatases quickly shut off a signal transduction pathway if its signal molecule is no longer bound at the cell surface.

Two scientists, Edwin Krebs and Edmond Fischer at the University of Washington, Seattle, first discovered that protein kinases add phosphate groups to control the activities of key proteins in cells and provided evidence showing that protein phosphatases reverse these phosphorylations. Krebs and Fischer, who began their experiments in the 1950s, received a Nobel Prize in 1992 for their discoveries concerning reversible protein phosphorylation.

A third characteristic of signal transduction pathways involving surface receptors is **amplification**—an increase in the magnitude of each step as a signal transduction pathway proceeds **(Figure 7.6).** Amplification occurs because many of the proteins that carry out individual steps in the pathways, including the protein kinases, are enzymes. Once activated, each enzyme can activate hundreds of proteins, including other enzymes, that enter the next step in the pathway. Generally, the more enzyme-catalyzed steps in a response pathway, the greater the amplification. As a result, just a few extracellular signal molecules binding to their receptors can produce a full internal response. For similar reasons, amplification also occurs for signal transduction pathways that involve internal receptors.

As signal transduction runs its course, the receptors and their bound signal molecules are removed from the cell surface by endocytosis. Both the receptor and its bound signal molecule may be degraded in lysosomes after entering the cell. Alternatively, the

hormones and growth factors circulating in the extracellular fluids. The combination of surface receptors on particular cell types is not fixed but rather changes as cells develop. Changes also occur as normal cells are transformed into cancer cells.

The Signaling Molecule Bound by a Surface Receptor Triggers Response Pathways within the Cell

Signal transduction pathways triggered by surface receptors are common to all animal cells. At least parts of the pathways are also found in protists, fungi, and plants. In all cases, binding of a signal molecule to a surface receptor is sufficient to trigger the cellular response—the signal molecule does not have to enter the cell. For example, experiments have shown that (1) a signal molecule produces no response if it is injected directly into the cytoplasm, and (2) unrelated molecules that mimic the structure of the normal extracellular signal molecule can trigger a full cellular response as long as they can bind to the recognition site of the receptor.

Another typical characteristic of signal transduction is that the signal is relayed inside the cell by **protein kinases,** enzymes that transfer a phosphate group from ATP to one or more sites on particular pro-receptors may be separated from the signal molecules and recycled to the cell surface, whereas only the signal molecules are degraded. Thus, surface receptors participate in an extremely lively cellular "conversation" with moment-to-moment shifts in the information.

The next two sections discuss two large families of surface receptors: the receptor tyrosine kinases and the G-protein–coupled receptors.

STUDY BREAK

1. What are protein kinases, and how are they involved in signal transduction pathways?
2. How is amplification accomplished in a signal transduction pathway?

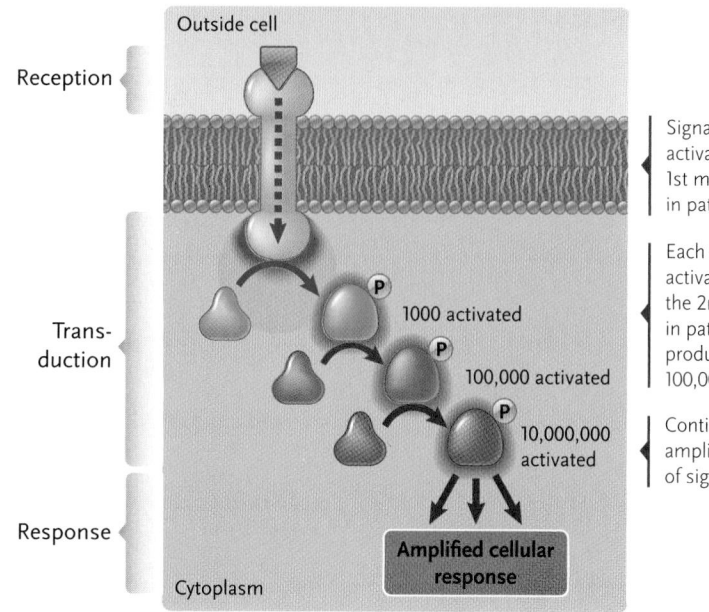

Figure 7.6
Amplification in signal transduction.

7.3 Surface Receptors with Built-in Protein Kinase Activity: Receptor Tyrosine Kinases

In the simplest form of signal transduction, the receptor itself has a protein kinase site at its cytoplasmic end. For this type of receptor, initiation of transduction occurs when two receptor molecules each bind a signal molecule in the reception step, move together in the membrane, and assemble into a pair called a *dimer* **(Figure 7.7).** Dimer assembly activates the receptor's protein kinase, which adds phosphate groups to sites on the receptor itself, a process known as *autophosphorylation.* Target proteins recognize and bind to the phosphorylated sites on the receptor and are then activated by being phosphorylated themselves. The total effect of the phosphorylations is to initiate the signal transduction pathway controlled by the receptor.

In autophosphorylation, the phosphate groups are added to tyrosine amino acids on the receptor. The protein kinase activity of the activated receptors also adds phosphate groups to tyrosines in the amino acid chains of target proteins. Because of this specificity of phosphorylation, the receptors in this group are called **receptor tyrosine kinases**. More than 50 receptor tyrosine kinases are known. In mammals, receptor tyrosine kinases fall into 14 different families, all related to one another in structure and amino acid sequence. Relatives of the mammalian receptors have been discovered in yeasts, *Drosophila,* and higher plants, indicating that the origin of the receptor tyrosine kinases is a single ancestral type that must have appeared before the evolutionary splits that led to the fungi, plants, and animals.

The cellular responses triggered by receptor tyrosine kinases are among the most important processes of animal cells. For example, the receptor tyrosine kinases binding the peptide hormone *insulin,* a regulator of carbohydrate metabolism, triggers diverse cellular responses, including effects on glucose uptake, the rates of many metabolic reactions, and cell growth and division. (The insulin receptor is exceptional because it is permanently in the dimer form.) Other receptor tyrosine kinases bind growth factors, including *epidermal growth factor, platelet-derived growth factor,* and *nerve growth factor,* which are all important peptide hormones that regulate cell growth and division in higher animals.

Hereditary defects in the insulin receptor are responsible for some forms of *diabetes,* a disease in which glucose accumulates in the blood because it cannot be absorbed in sufficient quantity by body cells. The inherited defects may impair the ability of the receptor to bind insulin or block its ability to trigger a cellular response. In either case, the cell is unresponsive to insulin and does not add sufficient glucose transporters to take up glucose. (The role of insulin in glucose metabolism and diabetes is discussed further in Chapter 40.)

STUDY BREAK

1. How does a receptor tyrosine kinase become activated?
2. How is the insulin receptor different from general receptor tyrosine kinases?

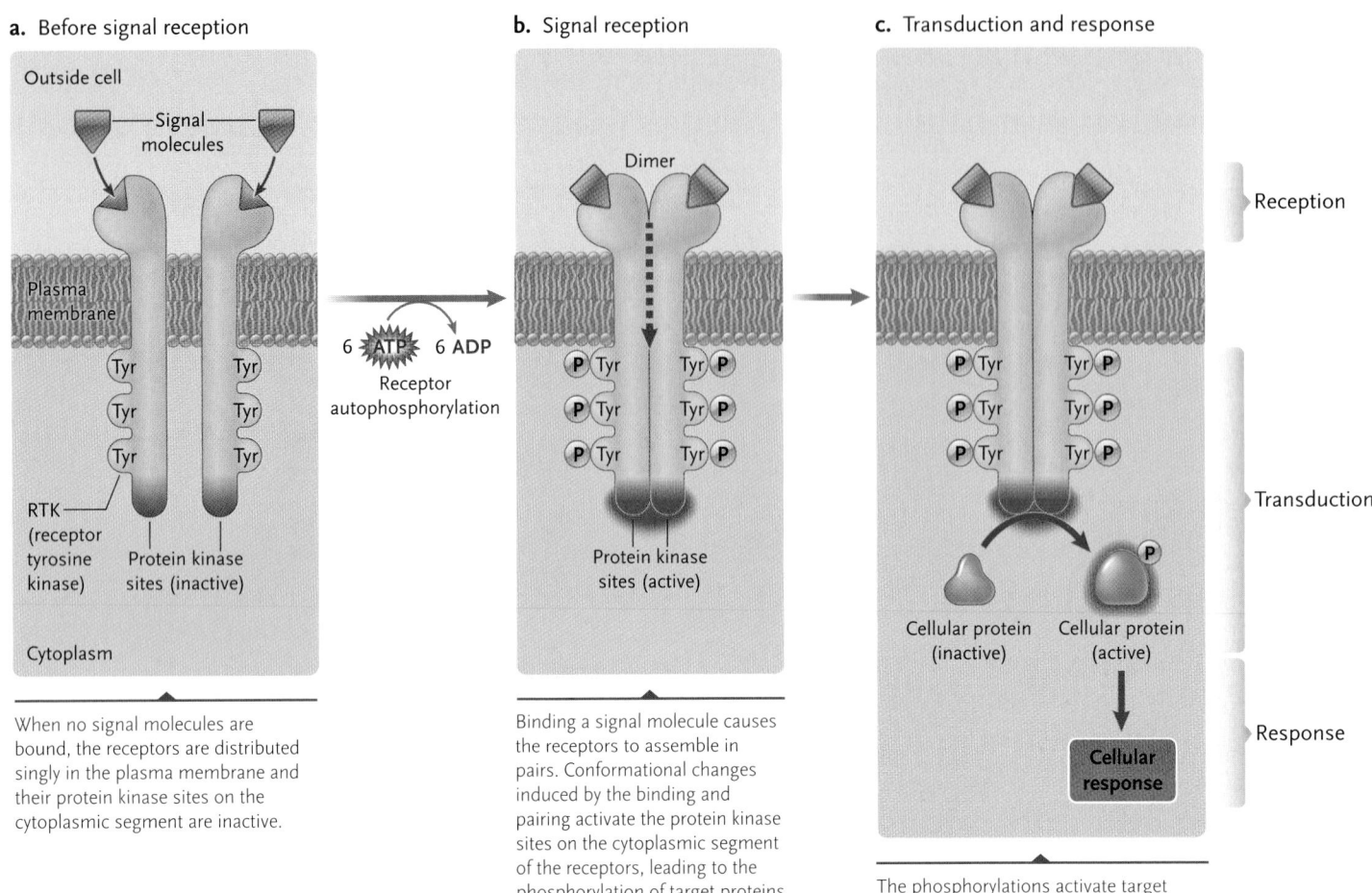

a. Before signal reception

Outside cell

Signal molecules

Plasma membrane

Tyr Tyr

Tyr Tyr

Tyr Tyr

RTK (receptor tyrosine kinase)

Protein kinase sites (inactive)

Cytoplasm

When no signal molecules are bound, the receptors are distributed singly in the plasma membrane and their protein kinase sites on the cytoplasmic segment are inactive.

b. Signal reception

Dimer

6 ATP → 6 ADP

Receptor autophosphorylation

P Tyr   Tyr P
P Tyr   Tyr P
P Tyr   Tyr P

Protein kinase sites (active)

Binding a signal molecule causes the receptors to assemble in pairs. Conformational changes induced by the binding and pairing activate the protein kinase sites on the cytoplasmic segment of the receptors, leading to the phosphorylation of target proteins and of receptors themselves.

c. Transduction and response

Reception

P Tyr   Tyr P
P Tyr   Tyr P
P Tyr   Tyr P

Transduction

Cellular protein (inactive)   Cellular protein (active) P

Cellular response

Response

The phosphorylations activate target proteins and initiate the cellular response.

Figure 7.7

The action of receptors with built-in protein kinase activity leading to the phosphorylation of the receptors themselves and the subsequent phosphorylation of target proteins. These receptors are called receptor tyrosine kinases because they add phosphate groups to tyrosines in target proteins. These receptors combine into pairs (dimers) when they bind signal molecules; the assembly into a dimer transmits the signal that activates the cytoplasmic end of the receptors.

7.4 G-Protein–Coupled Receptors

A second large family of surface receptors, known as the **G-protein–coupled receptors**, respond to a signal by activating an inner membrane protein called a G protein, which is closely associated with the cytoplasmic end of the receptor. About 1000 different G-protein–coupled receptors have been identified in mammals; several hundred types are involved in recognizing and binding odor molecules as part of the mammalian sense of smell. Almost all of the receptors of this group are large glycoproteins built up from a single polypeptide chain anchored in the plasma membrane by seven segments of the amino acid chain that zigzag back and forth across the membrane seven times **(Figure 7.8)**.

Unlike receptor tyrosine kinases, these receptors lack built-in protein kinase activity.

G Proteins Are Key Molecular Switches in Second-Messenger Pathways

The extracellular signal molecule in signal transduction pathways controlled by G-protein–coupled receptors is termed the **first messenger.** Binding the first messenger by the receptor activates a site on the cytoplasmic end of the receptor (**Figure 7.9,** step 1). The active site of the receptor then activates the G protein next to it by inducing the G protein to bind GTP, replacing the GDP that was bound to it (step 2). The G protein is an example of a *molecular switch* protein because it changes between inactive and active states. If GDP is bound to the G protein, the G protein is inactive, whereas if GTP is bound, it is active. In fact, G proteins are named because they use GDP and GTP to control their activities. The role of a switched-on G protein is to activate a plasma membrane–associated enzyme called the **effector** (step 3). In turn, the effector generates one or

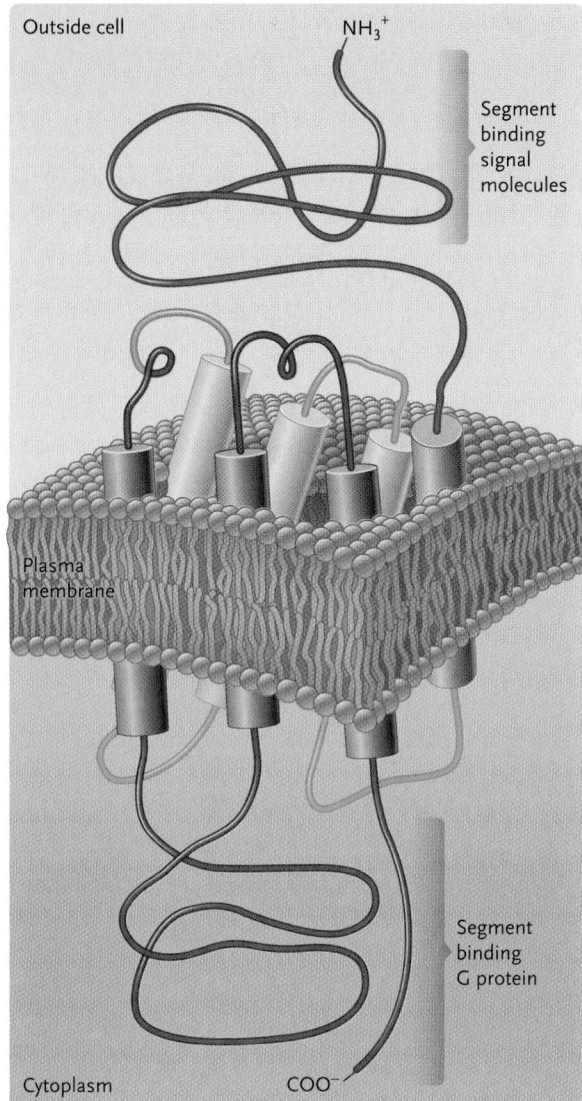

Figure 7.8

Structure of the G-protein–coupled receptors, which activate separate protein kinases. These receptors have seven transmembrane α-helical segments (shown as cylinders) that zigzag across the plasma membrane. Binding of a signal molecule at the cell surface, by inducing changes in the positions of some of the helices, activates the cytoplasmic end of the receptor.

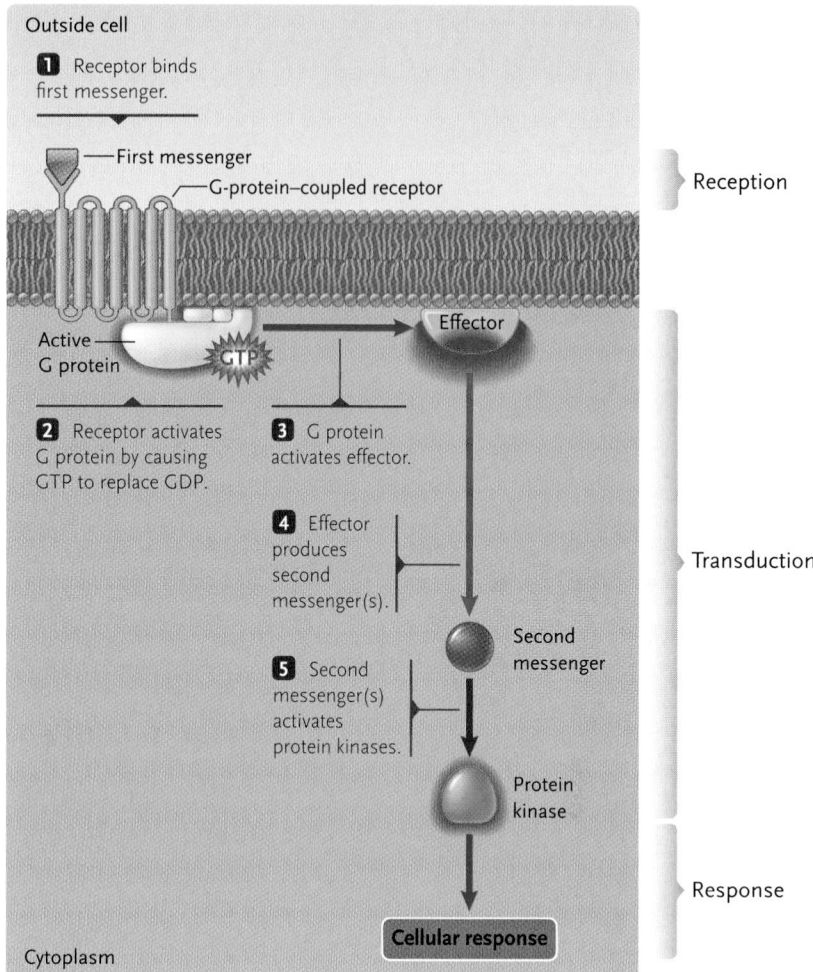

Figure 7.9

Response pathways activated by G-protein–coupled receptors, in which protein kinase activity is separate from the receptor. The signal molecule is called the first messenger; the effector is an enzyme that generates one or more internal signal molecules called second messengers. The second messengers directly or indirectly activate the protein kinases of the pathway, leading to the cellular response.

more internal, nonprotein signal molecules called **second messengers** (step 4). The second messengers directly or indirectly activate protein kinases, which elicit the cellular response by adding phosphate groups to specific target proteins (step 5). Thus, the entire control pathway operates through the following sequence:

first messenger → receptor → G proteins → effector → protein kinases → target proteins

The separate protein kinases of these pathways all add phosphate groups to serine or threonine amino acids in their target proteins, which are typically:

- Enzymes catalyzing steps in metabolic pathways
- Ion channels in the plasma and other membranes
- Regulatory proteins that control gene activity and cell division

The pathway from first messengers to target proteins is common to all G-protein–coupled receptors.

As long as a G-protein–coupled receptor is bound to a first messenger, the receptor keeps the G protein active. The activated G protein, in turn, keeps the effector active in generating second messengers. If the first messenger is released from the receptor, or if the receptor is taken into the cell by endocytosis, GTP is hydrolyzed to GDP, which inactivates the G protein. As a result, the effector becomes inactive, turning "off" the response pathway.

Cells can make a variety of G proteins, with each type activating a different cellular response. Alfred G. Gilman at the University of Virginia, Charlottesville, and Martin Rodbell at the National Institutes of Health,

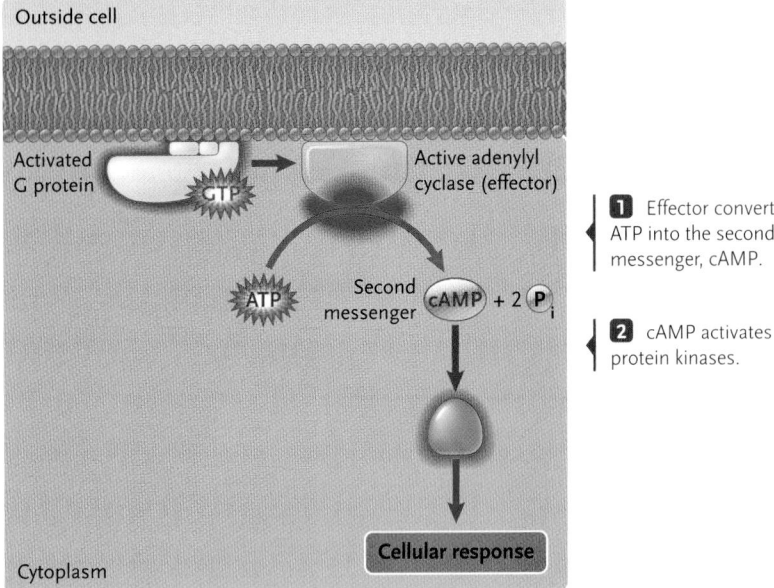

Figure 7.10

The operation of cAMP receptor–response pathways. The second messenger of the pathway, cAMP, activates one or more cAMP-dependent protein kinases, which add phosphate groups to target proteins to initiate the cellular response.

1 Effector converts ATP into the second messenger, cAMP.

2 cAMP activates protein kinases.

Two Major G-Protein–Coupled Receptor–Response Pathways Involve Different Second Messengers

Activated G proteins bring about a cellular response through two major receptor–response pathways in which different effectors generate different second messengers. One pathway involves the second messenger **cyclic AMP (cAMP)**, a relatively small, water-soluble molecule derived from ATP **(Figure 7.10)**. The effector that produces cAMP is the enzyme *adenylyl cyclase,* which converts ATP to cAMP **(Figure 7.11)**. cAMP diffuses through the cytoplasm and activates protein kinases that add phosphate groups to target proteins. The other pathway involves two second messengers: **inositol triphosphate (IP$_3$)** and **diacylglycerol (DAG)**. The effector of this pathway, an enzyme called *phospholipase C,* produces both of these second messengers by breaking down a membrane phospholipid **(Figure 7.12)**. IP$_3$ is a small, water-soluble molecule that diffuses rapidly through the cytoplasm. DAG is hydrophobic; it remains and functions in the plasma membrane.

The primary effect of IP$_3$ in animal cells is to activate transport proteins in the endoplasmic reticulum (ER), which release Ca^{2+} stored in the ER into the cytoplasm. The released Ca^{2+}, either alone or in combination with DAG, activates a protein kinase cascade that brings about the cellular effect. Techniques designed to detect Ca^{2+} release inside cells are among the most important tools of researchers studying cell signaling (see the *Focus on Research* for a description of two of these techniques.)

Both major G-protein–coupled receptor–response pathways are balanced by reactions that constantly eliminate their second messengers. For example, cAMP is quickly degraded by *phosphodiesterase,* an enzyme that is continuously active in the cytoplasm (see Figure 7.11). The rapid elimination of the second messengers provides another highly effective off switch for the pathways, ensuring that protein kinases are inactivated quickly if the receptor becomes inactive. Still another off switch is provided by protein phosphatases that remove the phosphate groups added to proteins by the protein kinases.

Bethesda, Maryland, received a Nobel Prize in 1994 for their discovery of G proteins and their role in signal transduction in cells.

The importance of G proteins to cellular metabolism is underscored by the fact that they are targets of toxins released by some infecting bacteria. The cholera toxin produced by *Vibrio cholerae,* the pertussis toxin that causes whooping cough produced by *Bordetella pertussis,* and a toxin produced by a disease-causing form of *Escherichia coli* are all enzymes that modify the G proteins, making them continuously active and keeping their response pathways turned "on" at high levels. For example, the cholera toxin prevents a G protein from hydrolyzing GTP, keeping the G protein switched on and the pathway in a permanently active state. Among other effects, the pathway opens ion channels in intestinal cells, causing severe diarrhea through a massive release of salt and water from the body into the intestinal tract. Unless the resulting dehydration of the body is relieved, death can result quickly. The *E. coli* toxin, which has similar but milder effects, is the cause of many cases of traveler's diarrhea.

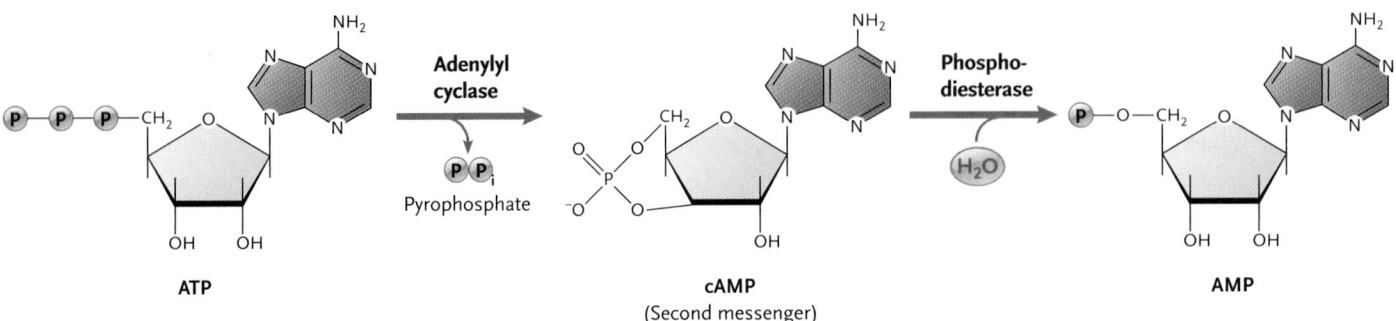

Figure 7.11

cAMP. The second messenger, cAMP, is made from ATP by adenylyl cyclase and is broken down to AMP by phosphodiesterase.

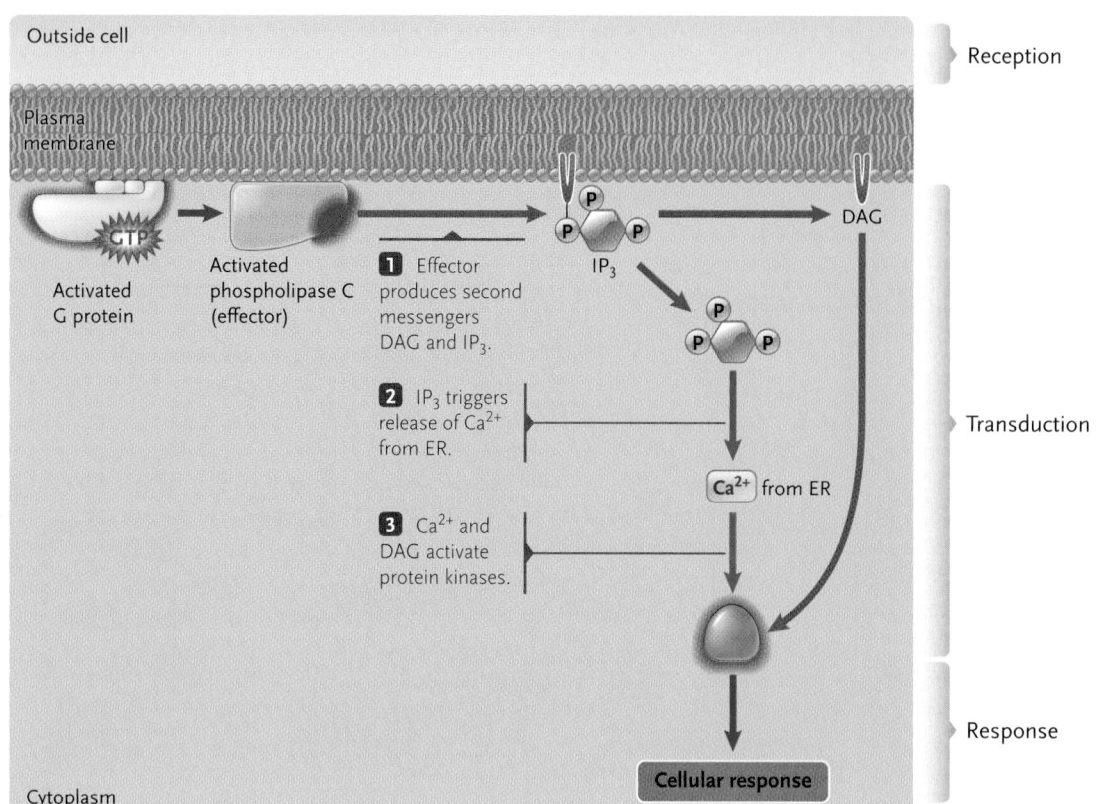

Figure 7.12
The operation of IP_3/DAG receptor–response pathways. Two second messengers, IP_3 and DAG, are produced by the pathway. IP_3 opens Ca^{2+} channels in ER membranes, releasing the ion into the cytoplasm. The Ca^{2+}, with DAG in some cases, directly or indirectly activates the protein kinases of the pathway, which add phosphate groups to target proteins to initiate the cellular response.

Labels in figure:
Outside cell
Plasma membrane
Reception
Activated G protein
Activated phospholipase C (effector)
1 Effector produces second messengers DAG and IP_3.
IP_3
DAG
2 IP_3 triggers release of Ca^{2+} from ER.
Transduction
Ca^{2+} from ER
3 Ca^{2+} and DAG activate protein kinases.
Response
Cellular response
Cytoplasm

As in the receptor tyrosine kinase pathways, the activities of the pathways controlled by cAMP and IP_3/DAG second messengers are also stopped by endocytosis of receptors and their bound extracellular signals. As with all cell signaling pathways, cells vary in their response to cAMP or IP_3/DAG pathways depending on the type of G-protein–coupled receptors on the cell surface and the kinds of protein kinases present in the cytoplasm.

The cAMP pathway is limited to animals and some fungi. The IP_3/DAG pathway is universally distributed among eukaryotic organisms, including both vertebrate and invertebrate animals, fungi, and plants. In plants, IP_3 releases Ca^{2+} primarily from the large central vacuole rather than from the ER.

Specific Examples of Cyclic AMP Pathways. Many peptide hormones act as first messengers for cAMP pathways in mammals and other vertebrates. The receptors that bind these hormones control such varied cellular responses as the uptake and oxidation of glucose, glycogen breakdown or synthesis, ion transport, the transport of amino acids into cells, and cell division.

For example, a cAMP pathway is involved in the regulation of the level of glucose, the fundamental fuel of cells. When the level of blood glucose falls too low in mammals, cells in the pancreas release the peptide hormone glucagon. Binding of the hormone by a G-protein–coupled glucagon receptor on the surface of liver cells triggers the cAMP receptor–response pathway (see Figure 7.10). The cAMP produced activates a pro-

tein kinase cascade that amplifies the effects of the pathway at each step. Two enzymes are end targets of the protein kinase cascades. One enzyme is *glycogen phosphorylase,* which catalyzes the breakdown of glycogen into glucose units that pass from the liver cells into the bloodstream and increase the glucose level in the blood; it is activated by the cascades. The other enzyme is *glycogen synthase,* which adds glucose units to glycogen; it is inactivated by the cascades, ensuring that glucose is not converted back into glycogen in the liver cells.

Specific Examples of IP_3/DAG Pathways. The IP_3/DAG-response–pathways are also activated by a large number of peptide hormones (including growth factors) and neurotransmitters, leading to responses as varied as sugar and ion transport, glucose oxidation, cell growth and division, and movements such as smooth muscle contraction.

Among the mammalian hormones that activate the pathways are vasopressin, angiotensin, and norepinephrine. Vasopressin, also known as antidiuretic hormone, helps the body conserve water by reducing the output of urine. Angiotensin helps maintain blood volume and pressure. Norepinephrine, together with epinephrine, brings about the fight-or-flight response in threatening or stressful situations.

Many growth factors operate through IP_3/DAG pathways. Defects in the receptors or other parts of the pathways that lead to higher-than-normal levels of DAG in response to growth factors are often associated with the progression of some forms of cancer. This is

Basic Research: Detecting Calcium Release in Cells

Because calcium ions are used as a control element in all eukaryotic cells, it was important to develop techniques for detecting Ca^{2+} when it is released into the cytosol. One of the most interesting techniques uses substances that release a burst of light when they bind the ion. One of these substances is *aequorin*, a protein produced by jellyfish, ctenophores, and many other luminescent organisms. Aequorin is injected into the cytoplasm of cells using microscopic needles, and it releases light when IP_3 opens Ca^{2+} channels in the ER, causing an increase in cytosolic Ca^{2+} concentration.

Artificially made, water-soluble molecules called *fura-2* and *quin-2* are also used as indicators of Ca^{2+} release. These molecules fluoresce (emit light) when exposed to ultraviolet (UV) light. The wavelength of UV light that causes the fluorescence is different depending on whether the molecules are bound to or free of Ca^{2+}. Therefore, the amount of Ca^{2+} released into the cytosol can be quantified by measuring how much fluorescence occurs at each of the two wavelengths. Rather than injecting fura-2 or quin-2 into cells, investigators combine them with a hydrophobic organic molecule that allows them to pass directly through the plasma membrane. After fura-2 or quin-2 are inside the cell, enzymes normally found in the cytoplasm remove the added organic group, releasing the Ca^{2+} indicators into the cytosol. This approach sidesteps injection by microneedles, which is a technically demanding technique that can damage the cells being studied.

In a typical experiment designed to follow steps in the IP_3/DAG pathway, an investigator might want to know whether a given hormone triggers the pathway in a group of cells. The investigator first adds aequorin or quin-2 to the cells, and then the hormone. If the cells emit a bright flash of light after they are exposed to the hormone, it is a good indication that the hormone triggers the IP_3/DAG pathway.

Other techniques allow investigators to study the effects of Ca^{2+} in cells independently, without complications introduced by the addition of hormones or other first messengers. One commonly used technique involves adding substances called *calcium ionophores* to the plasma membrane. The ionophores are open Ca^{2+} channels, produced naturally as antibiotics by some microorganisms, that can bury themselves in the plasma membrane of targeted cells. In the membrane, they allow Ca^{2+} to flow into the cytoplasm from the outside. After adding an ionophore to cells in a calcium-free medium, an investigator can detect any cellular activities controlled by Ca^{2+} simply by adding the ion to the medium.

Experiments using these methods have revealed the many cellular processes controlled by Ca^{2+} concentration inside cells, including cellular response pathways, cell movements, assembly and disassembly of the cytoskeleton, secretion, and endocytosis.

---

because DAG, in turn, causes an overactivity of the protein kinases responsible for stimulating cell growth and division. Also, plant substances in a group called *phorbol esters* resemble DAG so closely that they can promote cancer in animals by activating the same protein kinases.

IP_3/DAG pathways have also been linked to mental disease, particularly *bipolar disorder* (previously called *manic depression*), in which patients experience periodic changes in mood. Lithium has been used for many years as a therapeutic agent for bipolar disorder. Recent research has shown that lithium reduces the activity of IP_3/DAG pathways that release neurotransmitters, among them some that take part in brain function. Lithium also relieves cluster headaches and premenstrual tension, suggesting that they may be related to IP_3/DAG pathways as well.

In plants, IP_3/DAG pathways control responses to conditions such as water loss and changes in light intensity or salinity. Plant hormones—relatively small, nonprotein molecules such as *auxin* (a derivative of the amino acid tryptophan) and the *cytokinins* (derivatives of the nucleotide base adenine)—act as first messengers activating some of the IP_3/DAG pathways of these organisms.

Example of a Signaling Pathway That Combines a Receptor Tyrosine Kinase with a G Protein. Some pathways important in gene regulation link certain receptor tyrosine kinases to a specific type of G protein called Ras. When the receptor tyrosine kinase receives a signal (**Figure 7.13,** step 1), it activates by autophosphorylation (step 2). Adapter proteins then bind to the phosphorylated receptor and bridge to Ras, stimulating the activation of Ras (step 3). Like other G proteins, Ras is activated by binding GTP. The activated Ras sets in motion a phosphorylation cascade that involves a series of three enzymes known as *mitogen-activated protein kinases* (MAP kinases; step 4). The last MAP kinase in the cascade, when activated, enters the nucleus (step 5) and phosphorylates other proteins, which then change the expression of certain genes, particularly activating those involved in cell division (step 6). (A *mitogen* is a substance that controls cell division, hence the name of the kinases.) Changes in gene expression can have far-reaching effects on the cell, such as determining whether a cell divides or how frequently it divides. The Ras proteins are of major interest to investigators because of their role in linking receptor tyrosine kinases to gene regulation, as well as their major roles in the development of many types of cancer when their function is altered.

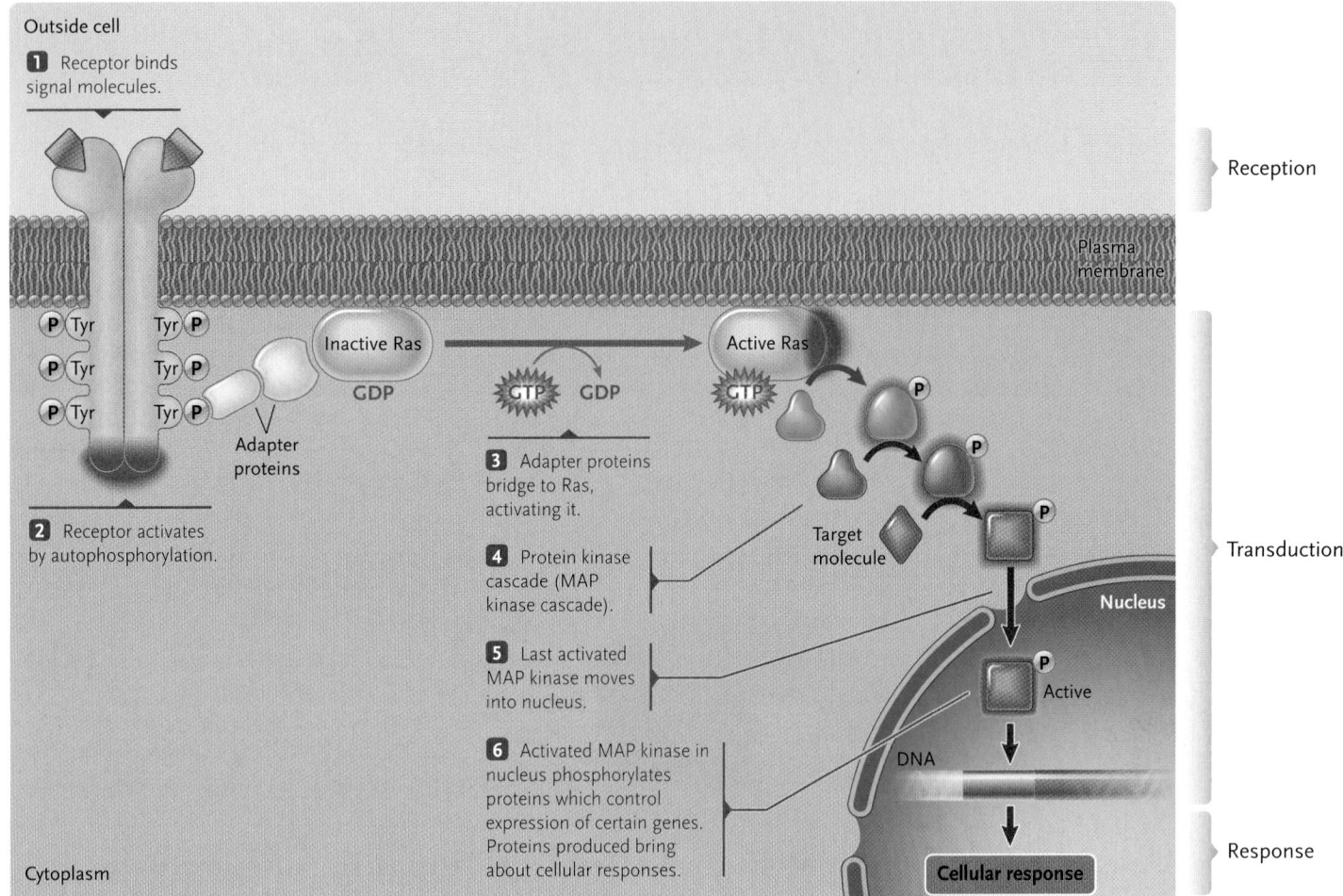

Outside cell

1 Receptor binds signal molecules.

2 Receptor activates by autophosphorylation.

P Tyr   Tyr P
P Tyr   Tyr P
P Tyr   Tyr P

Adapter proteins

Inactive Ras
GDP

GTP → GDP

Active Ras
GTP

3 Adapter proteins bridge to Ras, activating it.

4 Protein kinase cascade (MAP kinase cascade).

5 Last activated MAP kinase moves into nucleus.

6 Activated MAP kinase in nucleus phosphorylates proteins which control expression of certain genes. Proteins produced bring about cellular responses.

Target molecule

Plasma membrane

Nucleus

Active

DNA

Cellular response

Cytoplasm

Reception

Transduction

Response

Figure 7.13
The pathway from receptor tyrosine kinases to gene regulation, including the G protein, Ras, and MAP kinase.

In this section, we have surveyed major response pathways linked to surface receptors that bind peptide hormones, growth factors, and neurotransmitters. We now turn to the other major type of signal receptor: the internal receptors binding signal molecules—primarily steroid hormones—that penetrate through the plasma membrane.

STUDY BREAK

1. What is the role of the first messenger in a G-protein–coupled receptor-controlled pathway?
2. What is the role of the effector?
3. For a cAMP second-messenger pathway, how is the pathway turned off if no more signal molecules are present in the extracellular fluids?

7.5 Pathways Triggered by Internal Receptors: Steroid Hormone Receptors

Cells of many types have internal receptors that respond to signals arriving from the cell exterior. Unlike the signal molecules that bind to surface receptors,

these signals, primarily steroid hormones, penetrate through the plasma membrane to trigger response pathways inside the cells. The internal receptors, called **steroid hormone receptors**, are typically control proteins that turn on specific genes when they are activated by binding a signal molecule.

Steroid Hormones Have Widely Different Effects That Depend on Relatively Small Chemical Differences

Steroid hormones are relatively small, nonpolar molecules derived from cholesterol, with a chemical structure based on four carbon rings (see Figure 3.14). Steroid hormones combine with hydrophilic carrier proteins that mask their hydrophobic groups and hold them in solution in the blood and extracellular fluids. When a steroid hormone–carrier protein complex collides with the surface of a cell, the hormone is released and penetrates directly through the nonpolar part of the plasma membrane. On the cytoplasmic side, the hormone binds to its internal receptor.

The various steroid hormones differ only in the side groups attached to their carbon rings. Although the differences are small, they are responsible for highly distinctive effects. For example, the male and

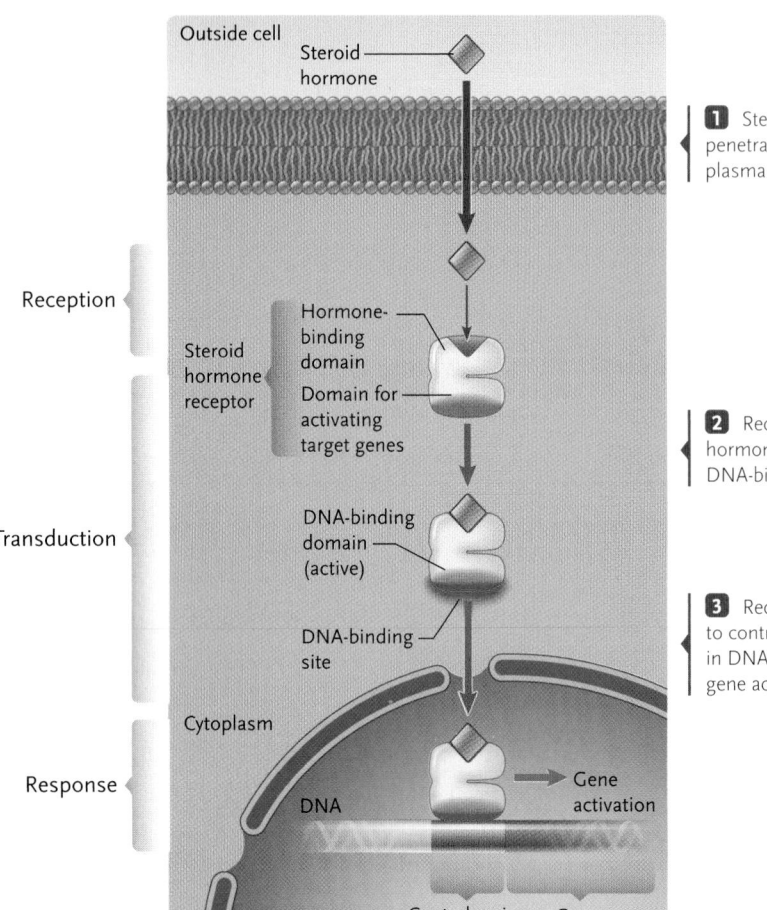

Figure 7.14

Pathway of gene activation by steroid hormone

Outside cell

Steroid hormone

Reception

1 Steroid hormone penetrates through plasma membrane.

Steroid hormone receptor

Hormone-binding domain

Domain for activating target genes

2 Receptor binds hormone, activating DNA-binding site.

Transduction

DNA-binding domain (active)

DNA-binding site

3 Receptor binds to control sequence in DNA, leading to gene activation.

Cytoplasm

Response

Gene activation

DNA

Control region of gene   Gene

Nucleus

carried by the organism's circulation to other cells. Whether a cell responds to a steroid hormone depends on whether it has an internal receptor for the hormone within the cell. The type of response depends on the genes that are recognized and turned on by an activated receptor. Depending on the receptor type and the particular genes it recognizes, even the same steroid hormone can have highly varied effects on different cells. (The effects of steroid hormones are described in more detail in Chapter 40.)

Taken together, the various types of receptor tyrosine kinases, G-protein–coupled receptors, and steroid hormone receptors prime cells to respond to a stream of specific signals that continuously fine-tune their function. How are the signals integrated within the cell and organism to produce harmony rather than chaos? The next section shows how the various signal pathways are integrated into a coordinated response.

female sex hormones of mammals, testosterone and estrogen, respectively, which are responsible for many of the structural and behavioral differences between male and female mammals, differ only in minor substitutions in side groups at two positions (see Figure 3.14). The differences cause the hormones to be recognized by different receptors, which activate specific group of genes leading to development of individuals as males or females.

The Response of a Cell to Steroid Hormones Depends on Its Internal Receptors and the Genes They Activate

Steroid hormone receptors are proteins with two major domains **(Figure 7.14).** One domain recognizes and binds a specific steroid hormone. The other domain interacts with the regions of target genes that control their expression. When a steroid hormone combines with the hormone-binding domain, the gene activation domain changes shape, thus enabling the complex to bind to the control regions of the target genes that the hormone affects. For most steroid hormone receptors, binding of the activated receptor to a gene control region activates that gene.

Steroid hormones, like peptide hormones, are released by cells in one part of an organism and are

STUDY BREAK

1. What distinguishes a steroid receptor from an receptor tyrosine kinase receptor or a G-protein–coupled receptor?
2. By what means does a specific steroid hormone result in a specific cellular response?

7.6 Integration of Cell Communication Pathways

Cells are under the continual influence of many simultaneous signal molecules. The cell signaling pathways may communicate with one another to integrate their responses to cellular signals. The interpathway interaction is called **cross-talk;** a conceptual example that involves two second-messenger pathways is shown in **Figure 7.15.** For example, a protein kinase in one pathway might phosphorylate a site on a target protein in another signal transduction pathway, activating or inhibiting that protein, depending on the site of the phosphorylation. The cross-talk can be extensive, resulting in a complex network of interactions between cell communication pathways.

Cross-talk often leads to modifications of the cellular responses controlled by the pathways. Such modifications fine-tune the effects of combinations of signal molecules binding to the receptors of a cell. For example, cross-talk between second-messenger pathways is involved in particular types of olfactory (smell) signal transduction in rats, and probably in many animals. The two pathways involved are activated upon stimulation with distinct odors. One pathway involves cAMP as the second messenger, and the other involves IP$_3$. However, the two olfactory second-messenger pathways do not work independently; rather, they operate in an antagonistic way. That is, experimentally blocking key enzymes of one signal transduction cascade inhibits that pathway, while simultaneously augmenting the activity of the other pathway. The cross-talk may be a way to refine the animal's olfactory sensory perception by helping discriminate different odor molecules more effectively.

Cross-talk networks may also involve inputs from other cellular response systems, such as those triggered by cell adhesion molecules as a result of specific contact between cells. Cell adhesion molecules are receptor-like glycoproteins in plasma membranes; they link cells together or bind them to molecules of the extracellular matrix. Many of these surface molecules also trigger cellular responses. For example, when the surface molecule *integrin* binds to another cell or to a molecule of the extracellular matrix, such as collagen, it triggers a cellular response, often including cross-talk steps that link the reactions to the cAMP and IP$_3$/DAG pathways. The responses triggered by the cell adhesion molecules include changes in the rate of cell division and gene activity and alterations in cell motility, development, and differentiation.

Direct channels of communication may also be involved in a cross-talk network. For example, gap junctions between the cytoplasms of adjacent cells admit ions and small molecules, including the Ca^{2+}, cAMP, and IP$_3$ second messengers released by the receptor–response pathways. (Gap junctions are discussed in further detail in Section 5.5.) Thus, one cell that receives a signal through its surface receptors can transmit the signal to other cells in the same tissue via the connecting gap junctions, thereby coordinating the functions of those cells. For instance, cardiac muscle cells are connected by gap junctions, and the Ca^{2+} flow regulates coordinated muscle fiber contractions.

The entire system integrating cellular response mechanisms, tied together by many avenues of cross-talk between individual pathways, creates a sensitively balanced control mechanism that regulates and coordinates the activities of individual cells into the working unit of the organism.

In this chapter you have seen that proteins in the plasma membrane play an important role in many aspects of cell communication. The next two chapters take up another vital role of membranes in cells of all kinds—their participation in fundamentally important reactions of energy metabolism.

STUDY BREAK

What cell communication pathways might be integrated in a cross-talk network?

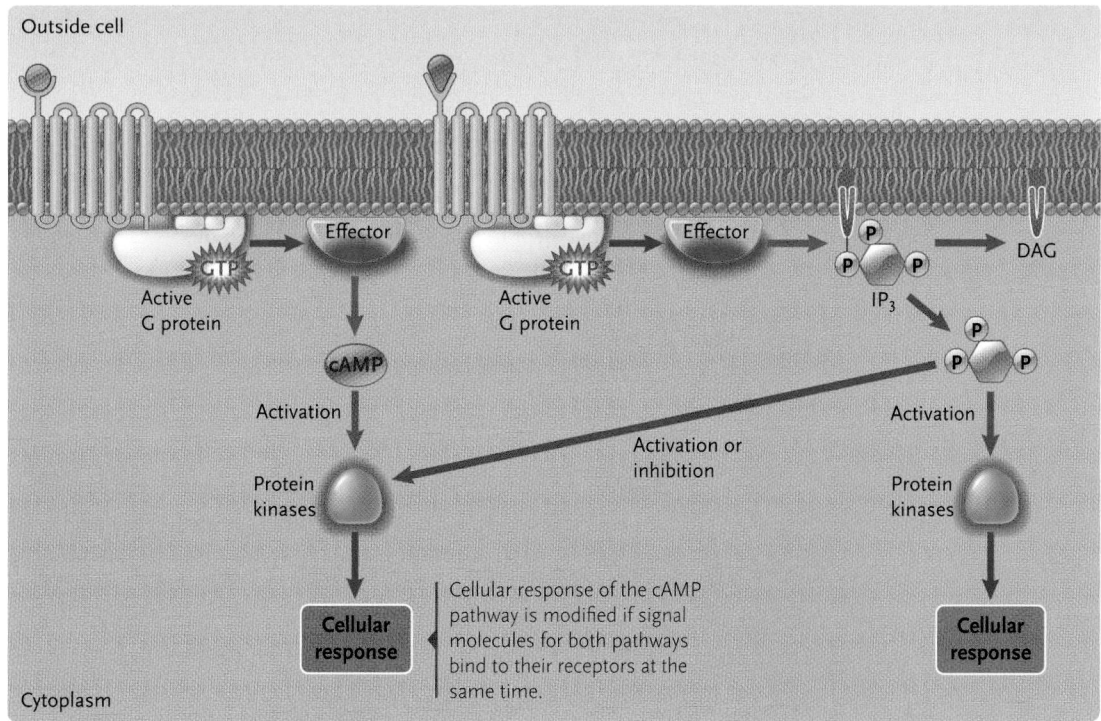

Figure 7.15
Cross-talk, the interaction between cell communication pathways to integrate the responses to signal molecules.

Intercellular signal molecules control many cellular activities; therefore, it is not surprising that many laboratories are researching extensively the mechanisms involved. Experimental goals include determining the molecular details of the receptor structures and how they interact with and change when a signal molecular binds, identifying and characterizing all of the components of the transduction steps, detailing how the final activated component of the transduction steps triggers the cellular responses, and understanding the regulation of signal transduction pathways.

What are the prospects for treating human diseases caused by signal transduction pathway malfunctions?

Receptor tyrosine kinase–mediated signaling is critical for cell growth, division, differentiation, and development. Some human diseases and developmental abnormalities result from mutations in the genes for receptor tyrosine kinases and from overexpression of those genes. Examples are dwarfism, heritable cancer susceptibility, vein malformations, and piebaldism. Researchers are determining the exact nature of the receptor gene mutations in order explore how the mutations cause the malfunctions of the signal transduction pathways. They have found that some mutations affect the ability of the receptor to form a dimer when the signal molecule binds, and others affect the kinase activity of the cytoplasmic side of the receptor. In fact, there are a surprisingly large number of different mutations that affect receptor tyrosine kinases, meaning that there are many ways that their functions can be affected. In terms of treating human diseases resulting from receptor tyrosine kinase mutations, research is at a relatively early stage. Prospects for therapeutic approaches to treat these diseases include developing anti-tyrosine kinase drugs. Clearly an increased understanding of receptor tyrosine kinases' signaling and function is crucial for prog-

ress to be made in diagnosis and treatment of human diseases resulting from mutations that cause abnormal regulation of receptor tyrosine kinase function.

How do steroid hormones act in the brain to modify brain function and behavior?

Research is being done on the cellular processes by which steroid hormones involved in mammalian reproductive behavior act in neurons. During the estrous cycle of female rats, estradiol and progesterone (ovarian hormones) regulate the expression of reproductive behaviors by binding to steroid hormone receptors in neurons.

The model presented in this chapter is that a steroid receptor becomes activated when the steroid hormone binds to it. However, several groups have now shown that neurotransmitters can activate steroid hormone receptors in the absence of hormone. In addition, experiments have demonstrated that when a male rat mates with a female rat, the mating stimulates the female's neural steroid hormone receptors. This activation causes neuronal changes in the brain, which result in changes in behavior and physiology. These changes are similar to those induced by steroid hormones. That is, how a male behaves toward a female alters neurotransmitters in her brain and creates events, many of which are the same as those caused by hormone secretion from the female's ovaries. Jeffrey Blaustein's research group at the University of Massachusetts (Amherst) is studying a number of questions in this area: How does the hormone-independent steroid hormone receptor activation occur? In which neurons do these events occur? What regulates the process? The results will give valuable insights into the mechanisms of steroid hormone action in the brain.

Peter J. Russell

Review

7.1 Cell Communication: An Overview

- Cells communicate with one another through direct channels of communication, specific contact between cells, and intercellular chemical messengers.

- In communication that involves an intercellular chemical messenger, a controlling cell releases a signal molecule that causes a response of target cells. The target cell processes the signal in three steps: reception, transduction, and response. The series of events from reception to response is called signal transduction (Figures 7.2 and 7.3).

7.2 Characteristics of Cell Communication Systems with Surface Receptors

- Cell communication systems based on surface receptors have three components: (1) extracellular signal molecules, (2) surface receptors that receive the signals, and (3) internal response pathways triggered when receptors bind a signal.

- The systems based on surface receptors respond to peptide hormones and neurotransmitters.

- Peptide hormones are small proteins. A special class of peptide hormones is the growth factors, which affect cell growth, division, and differentiation. Neurotransmitters include small peptides, individual amino acids or their derivatives, and other chemical substances.

- Surface receptors are integral membrane proteins that extend entirely through the plasma membrane. Binding a signal molecule induces a molecular change in the receptor that activates its cytoplasmic end (Figure 7.4).

- Cellular response pathways operate by activating protein kinases. Phosphate groups added by the protein kinases stimulate or inhibit the activities of the target proteins, thereby accomplishing the cellular response. The response is reversed by protein phosphatases that remove phosphate groups from target proteins. In addition, receptors are removed by endocytosis when signal transduction has run its course (Figure 7.5).

- Each step of a response pathway catalyzed by an enzyme is amplified, because each enzyme can activate hundreds or thousands of proteins that enter the next step in the pathway. Amplification allows a full cellular response when a few signal molecules bind to their receptors (Figure 7.6).

Animation: Signal transduction

7.3 Surface Receptors with Built-In Protein Kinase Activity: Receptor Tyrosine Kinases

- When receptor tyrosine kinases bind a signal molecule, the protein kinase site becomes active and adds phosphate groups to tyrosines in the receptor itself and to target proteins. The phosphate groups added to the cytoplasmic end of the receptor are recognition sites for proteins that are activated by binding to the receptor (Figure 7.7).

7.4 G-Protein–Coupled Receptors

- In the pathways activated by G-protein-coupled receptors, binding of the extracellular signal molecule (the first messenger) activates a site on the cytoplasmic end of the receptor (Figure 7.8).

- An activated receptor turns on a G protein, which acts as a molecular switch. The G protein is active when it is bound to GTP and inactive when it is bound to GDP (Figure 7.9).

- When a G protein is active, it switches on the effector of the pathway, an enzyme that generates small internal signal molecules called second messengers. The second messengers activate the protein kinases of the pathway (Figure 7.9).

- In one of the two major pathways triggered by G-protein–coupled receptors, the effector, adenylyl cyclase, generates cAMP as second messenger. cAMP activates specific protein kinases (Figures 7.10 and 7.11).

- In the other major pathway, the activated effector, phospholipase C, generates two second messengers, IP_3 and DAG. IP_3 activates transport proteins in the ER, which release stored Ca^{2+} into the cytoplasm. The released Ca^{2+}, alone or in combination with DAG, activates specific protein kinases that add phosphate groups to their target proteins (Figure 7.12).

- Both the cAMP and IP_3/DAG pathways are balanced by reactions that constantly eliminate their second messengers. Both pathways are also stopped by protein phosphatases that continually remove phosphate groups from target proteins and by endocytosis of receptors and their bound extracellular signals.

- Mutated systems can turn on the pathways permanently, contributing to the progression of some forms of cancer.

- Some pathways important in gene regulation link certain receptor tyrosine kinases to a specific G protein called Ras.

When the receptor binds a signal molecule, it phosphorylates itself and adapter proteins then bind, bridging to Ras, activating it. Activated Ras turns on the MAP kinase cascade. The last MAP kinase in the cascade, when activated, phosphorylates target proteins in the nucleus, activating them to turn on specific genes. Many of those genes control cell division (Figure 7.13).

Practice: Response pathways activated by G-protein–coupled receptors

7.5 Pathways Triggered by Internal Receptors: Steroid Hormone Receptors

- Steroid hormones penetrate through the plasma membrane to bind to receptors within the cell. The internal receptors are regulatory proteins that turn on specific genes when they are activated by binding a signal molecule, thereby producing the cellular response (Figure 7.14).

- Steroid hormone receptors have a domain that recognizes and binds a specific steroid hormone and a domain that interacts with the controlling regions of target genes (Figure 7.14).

- Whether a cell responds to a steroid hormone depends on whether it has an internal receptor for the hormone; within the cell, the type of response depends on the genes that are recognized and turned on by an activated receptor.

Animation: Pathway of gene activation by steroid hormone receptors

7.6 Integration of Cell Communication Pathways

- In cross-talk, cell signaling pathways communicate with one another to integrate responses to cellular signals. Cross-talk may result a complex network of interactions between cell communication pathways (Figure 7.15).

- Cross-talk often results in modifications of the cellular responses controlled by the pathways, fine-tuning the effects of combinations of signal molecules binding to the receptors of a cell.

- In animals, inputs from other cellular response systems, including cell adhesion molecules, as well as molecules arriving through gap junctions, also can become involved in the cross-talk network.

Animation: Animal cell junctions

Questions

Self-Test Questions

1. In signal transduction, which of the following is *not* a target protein?
   a. proteins that regulate gene activity
   b. hormones that activate the receptor
   c. enzymes of pathways
   d. transport proteins
   e. enzymes of cell reactions

2. Which of the following could *not* elicit a signal transduction response?
   a. a signal molecule injected directly into the cytoplasm
   b. a virus mimicking a normal signal molecule
   c. a peptide hormone
   d. a steroid hormone
   e. a neurotransmitter

3. A cell that responds to a signal molecule is distinguished from a cell that does not respond by the fact that it has:
   a. a cell adhesion molecule.
   b. cAMP.
   c. a first messenger molecule.

   d. a receptor.
   e. a protein kinase.

4. The mechanism to activate an immune cell to make an antibody involves signal transduction using tyrosine kinases. Place in order the following series of steps to activate this function.
   (1) The activated receptor phosphorylates cytoplasmic proteins.
   (2) Conformational change occurs in the receptor tyrosine kinase.
   (3) Cytoplasmic protein crosses the nuclear membrane to activate genes.
   (4) An immune hormone signals the immune cell.
   (5) Activation of protein kinase site(s) adds phosphates to the receptor to activate it.
   a. 2, 1, 4, 3, 5
   b. 5, 3, 4, 2, 1
   c. 4, 1, 5, 2, 3
   d. 4, 2, 5, 1, 3
   e. 2, 5, 3, 4, 1

5. Which of the following is the ability of enzymes, requiring few receptors, to activate thousands of molecules in a stepwise pathway?
   a. autophosphorylation
   b. second-messenger enhancement
   c. amplification
   d. ion channel regulation
   e. G protein turn-on

6. Which of the following is *incorrect* about pathways activated by G-protein–coupled receptors?
   a. The extracellular signal is the first messenger.
   b. When activated, plasma membrane–bound G protein can switch on an effector.
   c. Second messengers enter the nucleus.
   d. ATP converts to cAMP to activate protein kinases.
   e. Protein kinases phosphorylate molecules to change cellular activity.

7. Which of the following would *not* inhibit signal transduction?
   a. Phosphate groups are removed from proteins.
   b. Endocytosis acts on receptors and their bound signals.
   c. Receptors and signals separate.
   d. Receptors and bound signals enter lysosomes.
   e. Autophosphorylation targets the cytoplasmic portion of the receptor.

8. An internal receptor binds both a signal molecule and controlling region of a gene. What type of receptor is it?
   a. protein   d. receptor tyrosine kinase
   b. steroid   e. switch protein
   c. IP_3/DAG

9. Place in order the following steps for the normal activity of a Ras protein.
   (1) Ras turns on the MAP kinase cascade.
   (2) Adaptor proteins connect phosphorylated tyrosine on a receptor to Ras.
   (3) GTP activates Ras by binding to it, displacing GDP.
   (4) The last MAP kinase in the cascade phosphorylates proteins in the nucleus that activate genes.
   (5) Receptor tyrosine kinase binds a signal molecule and is activated.
   a. 1, 2, 3, 4, 5   d. 2, 3, 1, 5, 4
   b. 2, 3, 5, 1, 4   e. 4, 1, 5, 3, 2
   c. 5, 2, 3, 1, 4

10. Cross-talk is best exemplified as:
    a. second messenger activates protein kinases.
    b. effector protein produces second messengers DAG and IP_3.
    c. an MAP kinase in a cascade that activates DNA.
    d. a protein kinase in a pathway that activates or inactivates a protein in another pathway.
    e. steroid hormones can move across the plasma membrane.

Questions for Discussion

1. Describe the possible ways in which a G-protein–coupled receptor pathway could become defective and not trigger any cellular responses.

2. Is providing extra insulin an effective cure for an individual who has diabetes caused by a hereditary defect in the insulin receptor? Why or why not?

3. There are molecules called GTP analogs that resemble GTP so closely that they can be bound by G proteins. However, they cannot be hydrolyzed by cellular GTPases. What differences in effect would you expect if you inject GTP or a nonhydrolyzable GTP analog into a liver cell that responds to glucagon?

4. Why do you suppose cells evolved internal response mechanisms using switching molecules that bind GTP instead of ATP?

Experimental Analysis

How would you set up an experiment to determine whether a hormone receptor is located on the cell surface or inside the cell?

Evolution Link

Based on their distributions among different groups of organisms, which signaling pathway is the oldest?

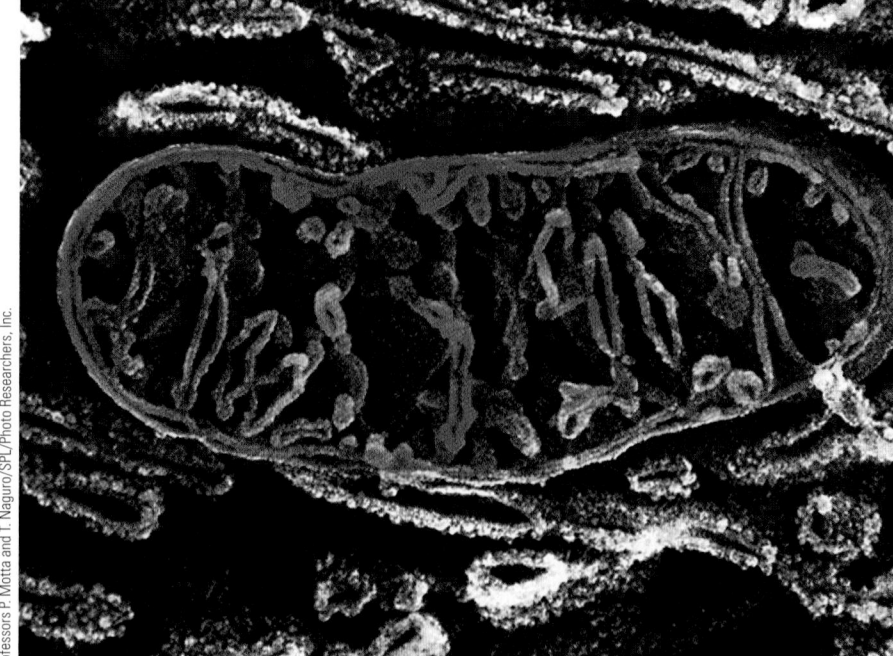

Mitochondrion (colorized SEM). Mitochondria are the sites of cellular respiration.

Professors P. Motta and T. Naguro/SPL/Photo Researchers, Inc.

8 Harvesting Chemical Energy: Cellular Respiration

WHY IT MATTERS

In the early 1960s, Swedish physician Rolf Luft mulled over some odd symptoms of a patient. The young woman felt weak and too hot all the time. Even on the coldest winter days, she never stopped perspiring and her skin was always flushed. She was also thin, despite a huge appetite.

Luft inferred that his patient's symptoms pointed to a metabolic disorder. Her cells seemed to be active, but much of their activity was being dissipated as metabolic heat. He decided to order tests to measure her metabolic rates. The patient's oxygen consumption was the highest ever recorded!

Luft also examined a tissue sample from the patient's skeletal muscles. Using a microscope, he found that her muscle cells contained many more mitochondria—the ATP-producing organelles of the cell—than are normal; also, her mitochondria were abnormally shaped. Other studies showed that the mitochondria were engaged in cellular respiration—their prime function—but little ATP was being generated.

The disorder, now called *Luft syndrome*, was the first disorder to be linked directly to a defective cellular organelle. By analogy, someone

157

with this mitochondrial disorder functions like a city with half of its power plants shut down. Skeletal and heart muscles, the brain, and other hardworking body parts with the highest energy demands are hurt the most. More than 100 mitochondrial disorders are now known.

Defective mitochondria also contribute to many age-related problems, including type 1 diabetes, atherosclerosis, amyotrophic lateral sclerosis (ALS, also called Lou Gehrig disease), as well as Parkinson, Alzheimer, and Huntington diseases.

Clearly, human health depends on mitochondria that are sound structurally and functioning properly. More broadly, every animal, plant, and fungus and most protists depend on mitochondria that are functioning correctly to grow and survive.

In mitochondria, ATP forms as part of the reactions of cellular respiration. The **cellular respiration** pathway breaks down food molecules to produce energy in the form of ATP, releasing water and carbon dioxide in the process. ATP fuels nearly all of the reactions that keep cells, and organisms, metabolically active. Respiration powers metabolism in most eukaryotes and many prokaryotes. This chapter discusses the reactions of cellular respiration.

Photosynthesis, the ultimate source of the chemical energy used by most organisms, is described in Chapter 9. Photosynthesis captures energy from light by splitting water molecules, and hydrogen from the water is combined with carbon dioxide to synthesize carbohydrates. A major by-product of photosynthesis is oxygen, a molecule needed for cellular respiration. Photosynthesis occurs in most plants, many protists, and some prokaryotes.

Respiration and photosynthesis are the major biological steps of the carbon cycle, the global movement of carbon atoms. The physiological connection between respiration and photosynthesis is a consequence of evolution.

8.1 Overview of Cellular Energy Metabolism

Electron-rich food molecules synthesized by plants are used by the plants themselves, and by animals and other eukaryotes. The electrons are removed from fuel substances, such as sugars, and donated to other molecules, such as oxygen, that act as electron acceptors. In the process, some of the energy of the electrons is released and used to drive the synthesis of ATP. ATP provides energy for most of the energy-consuming activities in the cell. Thus, life and its systems are driven by a cycle of electron flow powered by light in photosynthesis and oxidation in cellular respiration.

Coupled Oxidation and Reduction Reactions Produce the Flow of Electrons for Energy Metabolism

The removal of electrons (e^-) from a substance is termed an **oxidation**, and the substance from which the electrons are removed is said to be **oxidized.** The addition of electrons to a substance is termed a **reduction**, and the substance that receives the electrons is said to be **reduced.** A simple mnemonic to remember the direction of electron transfer is OIL RIG—Oxidation Is Loss (of electrons), Reduction Is Gain (of electrons). The term *oxidation* was originally used to describe the reaction that occurs when fuel substances are burned in air, in which oxygen directly accepts electrons removed from the fuels. However, although oxidation suggests that oxygen is involved in electron removal, most cellular oxidations occur without the direct participation of oxygen. The term *reduction* refers to the decrease in positive electrical charge that occurs when electrons, which are negatively charged, are added to a substance. Although reduction suggests that the energy level of molecules is decreased when they accept electrons, molecules typically gain energy from added electrons.

Oxidation and reduction *invariably* are coupled reactions that remove electrons from a donor molecule and simultaneously add them to an acceptor molecule. In such coupled oxidation–reduction reactions, also called **redox reactions**, electrons release some of their energy as they pass from a donor to an acceptor molecule. This free energy is available for cellular work, such as ATP synthesis.

Frequently, protons (hydrogen atoms stripped of electrons, symbolized as H^+) are also removed from a molecule during oxidation. (Recall from Chapter 2 that a hydrogen atom, H, consists of a proton and an electron: $H = H^+ + e^-$.) The molecules that accept electrons may also combine with protons, as oxygen does when it is reduced to form water.

The gain or loss of an electron in a redox reaction is not always complete. That is, depending on the redox reaction, electrons are transferred completely from one atom to another, or alternatively, the degree of electron sharing in covalent bonds changes. The latter condition is said to involve a relative loss or gain of electrons; most redox reactions in the electron transfer system discussed later in the chapter are of this type. The redox reaction between methane and oxygen (the burning of natural gas in air) that produces carbon dioxide and water illustrates a change in the degree of electron sharing. The dots in **Figure 8.1** indicate the positions of the electrons involved in the covalent bonds of the reactants and products.

Compare the reactant methane with the product carbon dioxide. In methane, the covalent electrons are shared essentially equally between bonded C and H atoms because C and H are almost equally electronegative. In carbon dioxide, electrons are closer to the O atoms than to the C atom in the C=O bonds because O atoms are highly electronegative. Overall, this means that the C atom has partially "lost" its shared electrons in the reaction. In short, methane has been oxidized. Now compare the oxygen reactant with the product water. In the oxygen molecule, the two O atoms share their electrons equally. The oxygen reacts with the hydrogen from methane, producing water, in which the electrons are closer to the O atom than to the H atoms. This means that each O atom has partially "gained" electrons; in short, oxygen has been reduced.

The movement of electrons away from an atom requires energy. The more electronegative an atom is, the greater the force that holds the electrons to that atom and therefore the greater the energy required to remove an electron. The changes in electron positions in a redox reaction consequently change the amount of chemical energy in the reactants and products. In our example of methane burning in oxygen, electrons are held more tightly in the product molecules (by being closer to the highly electronegative O atoms) than in the reactant molecules. Therefore, in this redox reaction, the potential energy of the reactants has dropped and chemical energy that can be used for cellular work is released.

Electrons Flow from Fuel Substances to Final Electron Acceptors

The energy of the electrons removed during cellular oxidations originates in the reactions of photosynthesis **(Figure 8.2a)**. During photosynthesis, electrons derived from water are pushed to very high energy levels using energy from the absorption of light. The high-energy electrons, together with H^+ from water, are combined with carbon dioxide to form sugar molecules and then are removed by the oxidative reactions that release energy for cellular activities **(Figure 8.2b)**. As electrons pass to acceptor molecules, they lose much

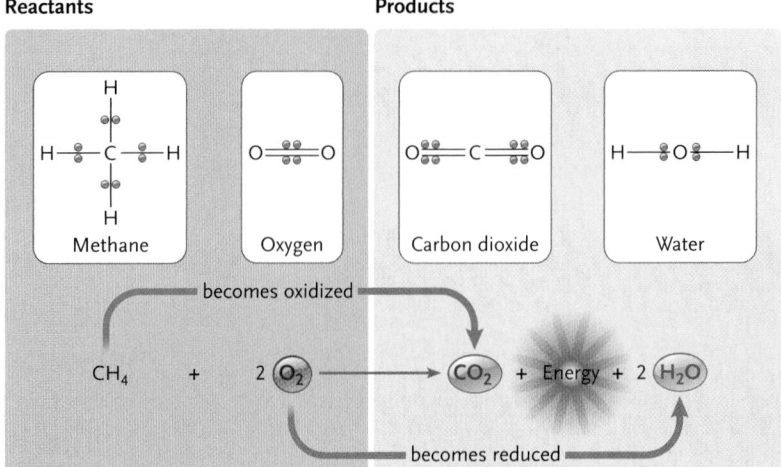

Reactants **Products**

Figure 8.1

Relative loss and gain of electrons in a redox reaction, the burning of methane (natural gas) in oxygen. Compare the positions of the electrons in the covalent bonds of reactants and products. In this redox reaction, methane is oxidized and oxygen is reduced.

of their energy; some of this energy drives the synthesis of ATP from ADP and P_i (a phosphate group from an inorganic source) (see Section 4.2).

The total amount of energy obtained from electrons flowing through cellular oxidative pathways depends on the difference between their high energy level in fuel substances and the lower energy level in the molecule that acts as the *final acceptor* for electrons, that is, the last molecule reduced in cellular pathways. The lower the energy level in the final acceptor, the greater the yield of energy for cellular activities. Oxygen is the final acceptor in the most efficient and highly developed form of cellular oxidation: cellular respiration (see Figure 8.2b). The very low energy level of the electrons added to oxygen allows a maximum output of energy for ATP synthesis. As part of the final reduction, oxygen combines with protons and electrons to form water.

Figure 8.2
Flow of energy from sunlight to ATP. **(a)** Photosynthesis occurs in plants, many protists, and some prokaryotes; **(b)** cellular respiration occurs in all eukaryotes, including plants, and in some prokaryotes.

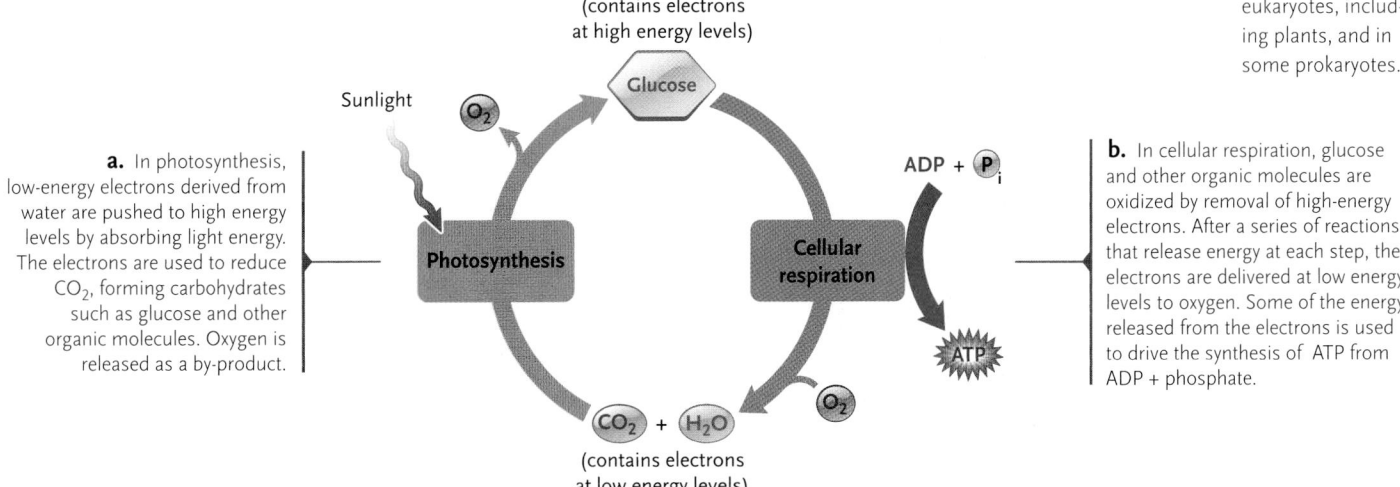

a. In photosynthesis, low-energy electrons derived from water are pushed to high energy levels by absorbing light energy. The electrons are used to reduce CO_2, forming carbohydrates such as glucose and other organic molecules. Oxygen is released as a by-product.

b. In cellular respiration, glucose and other organic molecules are oxidized by removal of high-energy electrons. After a series of reactions that release energy at each step, the electrons are delivered at low energy levels to oxygen. Some of the energy released from the electrons is used to drive the synthesis of ATP from ADP + phosphate.

In Cellular Respiration, Cells Make ATP by Oxidative Phosphorylation

Cellular respiration includes both the reactions that transfer electrons from organic molecules to oxygen and the reactions that make ATP. These reactions are often written in a summary form that uses glucose ($C_6H_{12}O_6$) as the initial reactant:

$$C_6H_{12}O_6 + 6\ O_2 + 32\ ADP + 32\ P_i \rightarrow$$
$$6\ H_2O + 6\ CO_2 + 32\ ATP$$

In this overall reaction, electrons and protons are transferred from glucose to oxygen, forming water, and the carbons left after this transfer are released as carbon dioxide. How we derive the 32 ATP molecules is explained later in this chapter.

ATP synthesis is the key part of this reaction. As discussed in Section 4.2, phosphorylation is a reaction that adds a phosphate group to a substance such as ADP. The process by which ATP is synthesized using the energy released by electrons as they are transferred to oxygen is called oxidative phosphorylation.

The entire process of cellular respiration can be divided into three stages **(Figure 8.3):**

1. In **glycolysis**, enzymes break down a molecule of glucose (containing six carbon atoms) into two molecules of pyruvate (an organic compound with a backbone of three C atoms). Some ATP is synthesized during glycolysis.

2. In **pyruvate oxidation**, enzymes convert the three-carbon molecule pyruvate into a two-carbon acetyl group, which enters the **citric acid cycle**, where it is completely oxidized to carbon dioxide. Some ATP is synthesized during the citric acid cycle.

3. In the **electron transfer system**, high-energy electrons produced from glycolysis, pyruvate oxidation, and the citric acid cycle are delivered to oxygen by a sequence of electron carriers. Free energy released by the electron flow generates an H^+ gradient. In **oxidative phosphorylation**, the enzyme **ATP synthase** uses the H^+ gradient built by the electron transfer system as the energy source to make ATP.

In eukaryotes, most of the reactions of cellular respiration occur in various regions of the mitochondrion **(Figure 8.4);** only glycolysis is located in the cytosol. Pyruvate oxidation and the citric acid cycle take place in the mitochondrial matrix. The inner mitochondrial membrane houses the electron transfer system and the ATP synthase enzymes. Transport proteins, concentrated primarily in the inner membrane, control the substances that enter and leave mitochondria.

The locations of the reactions in mitochondria were determined by studies of mitochondria that had been isolated from cells by **cell fractionation**—a technique that divides cells into fractions containing a single type of organelle, such as mitochondria or chloroplasts, or other structures, such as ribosomes **(Figure 8.5).** The

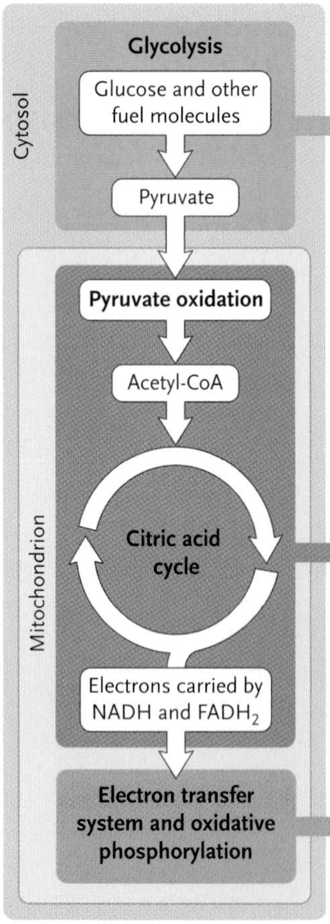

Figure 8.3

The three stages of cellular respiration: (1) glycolysis, (2) pyruvate oxidation and the citric acid cycle, and (3) the electron transfer system and oxidative phosphorylation.

Glycolysis

Glucose and other fuel molecules

ATP — Substrate-level phosphorylation

Pyruvate

Pyruvate oxidation

Acetyl-CoA

Citric acid cycle

ATP — Substrate-level phosphorylation

Electrons carried by NADH and FADH$_2$

Electron transfer system and oxidative phosphorylation

ATP — Oxidative phosphorylation

Cytosol

Mitochondrion

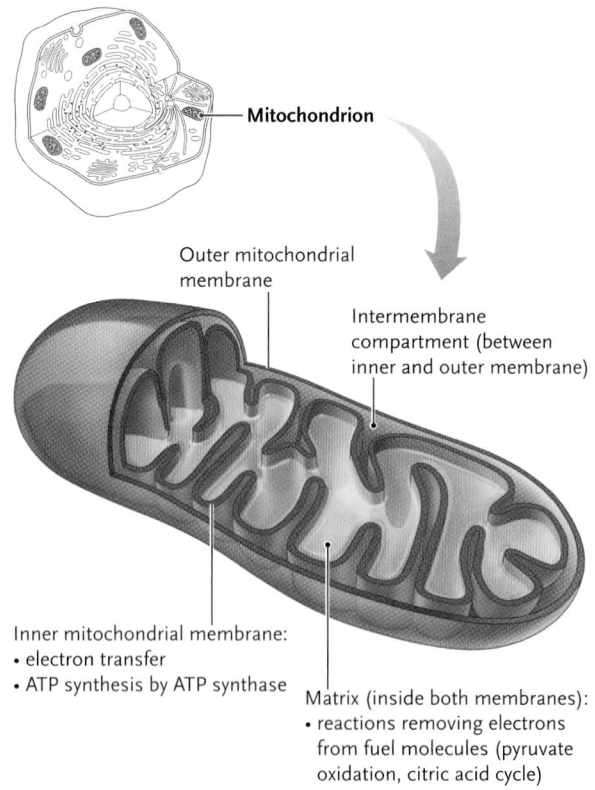

Mitochondrion

Outer mitochondrial membrane

Intermembrane compartment (between inner and outer membrane)

Inner mitochondrial membrane:
• electron transfer
• ATP synthesis by ATP synthase

Matrix (inside both membranes):
• reactions removing electrons from fuel molecules (pyruvate oxidation, citric acid cycle)

Figure 8.4

Membranes and compartments of mitochondria. Label lines that end in a dot indicate a compartment enclosed by the membranes.

Figure 8.5 Research Method

Cell Fractionation

PROTOCOL:

1. Break open intact cells by sonication (high-frequency sound waves), grinding in fine glass beads, or exposure to detergents that disrupt plasma membranes.

PURPOSE: Cell fractionation breaks cells into fractions containing a single cell component, such as mitochondria or ribosomes. Once isolated, the cell component can be disassembled by the same general techniques to analyze its structure and function. This example shows the isolation and subfractionation of mitochondria.

2. Use sequential centrifugations at increasing speeds to separate and purify cell structures. The spinning centrifuge drives cellular structures to bottom of tube at a rate that depends on their shape and density. With each centrifugation, the largest and densest components are isolated and concentrated into a pellet; the remaining solution, the supernatant, is drawn off and can be centrifuged again at higher speed.

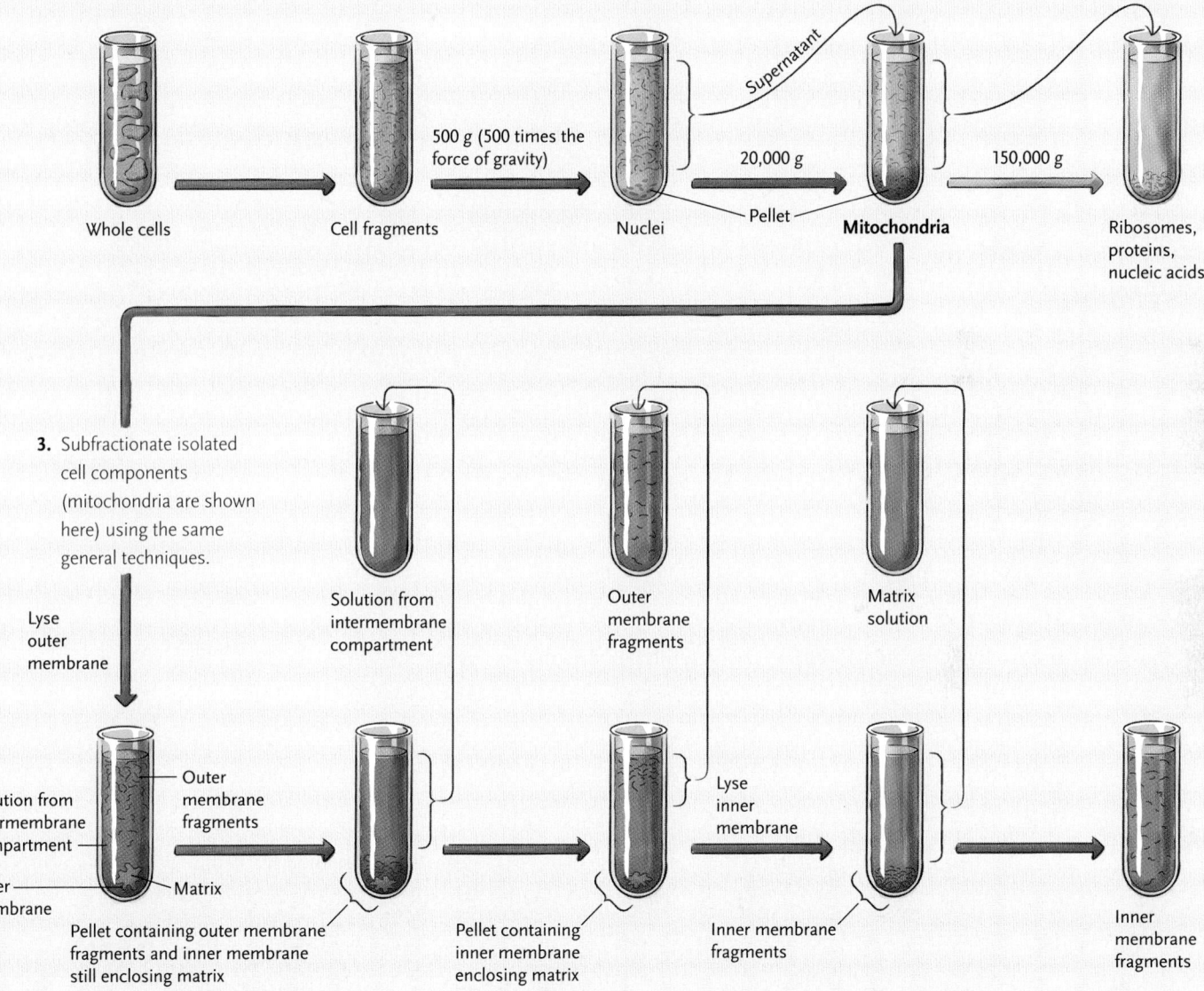

3. Subfractionate isolated cell components (mitochondria are shown here) using the same general techniques.

4. Centrifuge to concentrate outer membrane fragments and inner membrane enclosing matrix into a pellet.

5. Resuspend pellet and centrifuge it to separate outer membrane fragments and inner membrane enclosing matrix.

6. Sonicate pellet to break inner membrane and release matrix contents. Centrifuge to separate inner membrane fragments and matrix solution.

INTERPRETING THE RESULTS: Many of the cell or organelle subfractions generated by cell fractionation retain their biological activity, making them useful in studies of various cellular processes. For example, mitochondrial subfractions were used to work out the structure and function of the electron transfer system. Cell fractionation is still used to determine the cellular location of a protein or biological reaction, such as whether it is free in the cytosol or associated with a membrane.

collected mitochondria were, in turn, fractionated into different subfractions using experimental treatments. For example, the outer and inner mitochondrial membranes react differently to particular detergents, permitting each membrane, as well as the solutions in the matrix and intermembrane compartment, to be purified individually and then studied in detail. Each subfraction was then analyzed to identify the locations of the individual reactions of cellular respiration.

In prokaryotes, glycolysis, pyruvate oxidation, and the citric acid cycle are all located in the cytosol. The other reactions of cellular respiration occur in the plasma membrane.

The following three sections examine the three stages of cellular respiration in turn.

STUDY BREAK

1. Distinguish between oxidation and reduction.
2. Distinguish between cellular respiration and oxidative phosphorylation.

8.2 Glycolysis

Glycolysis, the first series of oxidative reactions that remove electrons from cellular fuel molecules, takes place in the cytosol of all organisms. In glycolysis (*glykys* = sweet; *lysis* = breakdown), sugars such as glucose are partially oxidized and broken down into smaller molecules, and a relatively small amount of ATP is produced. Glycolysis is also known as the Embden–Meyerhof pathway in honor of Gustav Embden and Otto Meyerhof, two German physiological chemists who (separately) made the most important contributions to determining the sequence of reactions in the pathway. Meyerhof received a Nobel Prize in 1922 for his work.

Glycolysis starts with the six-carbon sugar glucose and produces two molecules of the three-carbon organic substance *pyruvate* or *pyruvic acid* in 10 sequential enzyme-catalyzed reactions. (The *-ate* suffix indicates the ionized form of organic acids such as pyruvate, in which the carboxyl group —COOH dissociates to —COO$^-$ + H$^+$, as is usual under cellular conditions.) Pyruvate still contains many electrons that can be removed by oxidation, and it is the primary fuel substance for the second stage of cellular respiration.

The Reactions of Glycolysis Include Energy-Requiring and Energy-Releasing Steps

The initial steps of glycolysis (red in **Figure 8.6**) are energy-requiring reactions—2 ATP are hydrolyzed; they convert glucose into an unstable phosphorylated derivative. In the subsequent energy-releasing part of

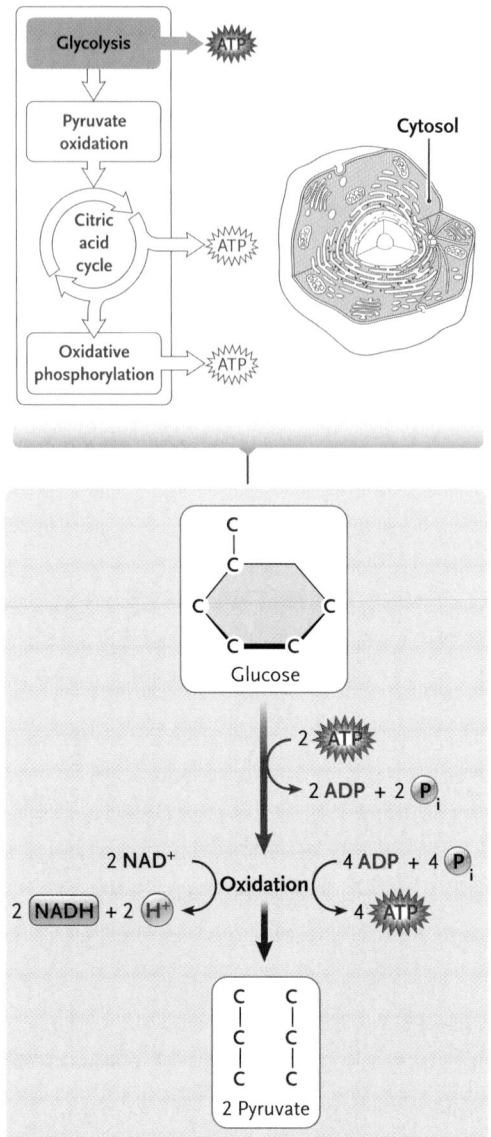

Figure 8.6

Overall reactions of glycolysis. Glycosis splits glucose (six carbons) into pyruvate (three carbons) and yields ATP and NADH.

glycolysis (blue in Figure 8.6), electrons are removed from the phosphorylated derivatives of glucose and 4 ATP are produced, giving a net gain of 2 ATP. Two molecules of pyruvate are generated in the final reaction of the pathway.

The electrons removed from fuel molecules in glycolysis are accepted by the electron carrier molecule *nicotinamide adenine dinucleotide* (**Figure 8.7**). The oxidized form of this electron carrier is NAD$^+$; the reduced form, NADH, carries a pair of electrons and a proton removed from fuel molecules. Nicotinamide adenine dinucleotide is one of many nucleotide-based carriers that shuttle electrons, protons, or metabolic products between major reaction systems (nucleotides are discussed in Section 3.6).

The reactions of glycolysis are shown in **Figure 8.8**. The major oxidation of glycolysis, which occurs in reac-

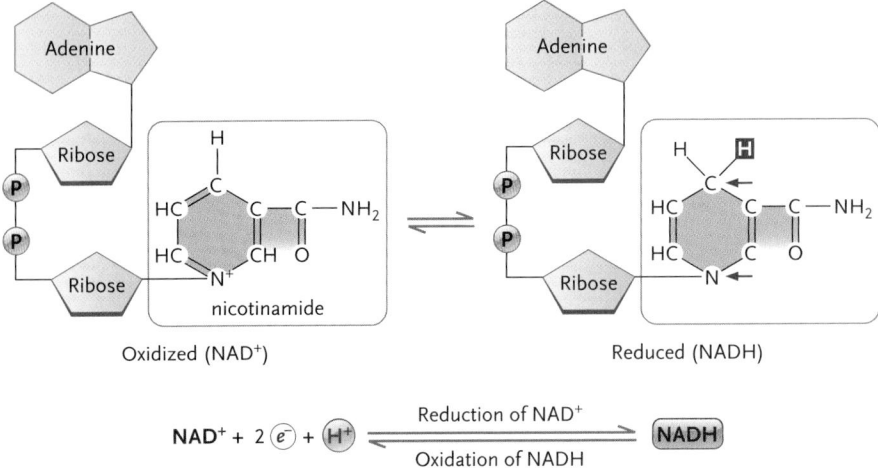

$$NAD^+ + 2 \, \textcircled{e^-} + \textcircled{H^+} \xrightleftharpoons[\text{Oxidation of NADH}]{\text{Reduction of NAD}^+} \boxed{\text{NADH}}$$

Figure 8.7

Electron carrier NAD^+. As the carrier is reduced to NADH, an electron is added at each of the two positions marked by a red arrow; a proton is also added at the position boxed in red. The nitrogenous base (blue) that adds and releases electrons and protons is nicotinamide, which is derived from the vitamin niacin (nicotinic acid).

tion 6, removes two electrons and two protons from the three-carbon substance *glyceraldehyde-3-phosphate* (G3P). Both electrons and one proton are picked up by NAD^+ to form NADH (see Figure 8.7). The other proton is released into the cytosol.

For each molecule of glucose that enters the pathway (see Figure 8.8), reactions 1 to 5 generate 2 molecules of G3P using 2 ATP, and reactions 6 to 10 convert the 2 molecules of G3P to 2 molecules of pyruvate, producing 4 ATP and 2 NADH. The net reactants and products of glycolysis are:

glucose + 2 ADP + 2 P_i + 2 $NAD^+ \rightarrow$
    2 pyruvate + 2 NADH + 2 H^+ + 2 ATP

The total of six carbon atoms in the two molecules of pyruvate is the same as in glucose; no carbons are released as CO_2 by glycolysis.

Each ATP molecule produced in the energy-releasing steps of glycolysis—steps 8 and 10 (see Figure 8.8)—results from **substrate-level phosphorylation**, an enzyme-catalyzed reaction that transfers a phosphate group from a substrate to ADP **(Figure 8.9)**.

Glycolysis Is Regulated at Key Points

The rate of sugar oxidation by glycolysis is closely regulated by several mechanisms to match the cell's need for ATP. For example, if excess ATP is present in the cytosol, it binds to *phosphofructokinase,* the enzyme that catalyzes reaction 3 in Figure 8.8, inhibiting its action. This is an example of feedback inhibition (introduced in Section 4.5). The resulting decrease in the concentration of the product of reaction 3, fructose-1,6-bisphosphate, slows or stops the subsequent reactions of glycolysis. Thus, glycolysis does not oxidize

fuel substances needlessly when ATP is in adequate supply.

If energy-requiring activities then take place in the cell, ATP concentration decreases and ADP concentration increases in the cytosol. As a result, ATP is released from phosphofructokinase, relieving inhibition of the enzyme. In addition, ADP activates the enzyme. Therefore, the rates of glycolysis and ATP production increase proportionately as cellular activities convert ATP to ADP.

NADH also inhibits phosphofructokinase. This inhibition slows glycolysis if excess NADH is present, such as when oxidative phosphorylation has been slowed by limited oxygen supplies. The systems that regulate phosphofructokinase and other enzymes of glycolysis closely balance the rate of the pathway to produce adequate supplies of ATP and NADH without oxidizing excess quantities of glucose and other sugars.

Our discussion of the oxidative reactions that supply electrons now moves from the cytosol to mitochondria, the locale of pyruvate oxidation and the citric acid cycle. These reactions complete the breakdown of fuel substances into carbon dioxide and provide most of the electrons that drive electron transfer and ATP synthesis.

STUDY BREAK

1. What are the energy-requiring and energy-releasing steps of glycolysis?
2. Why is phosphofructokinase a target for inhibition by ATP?

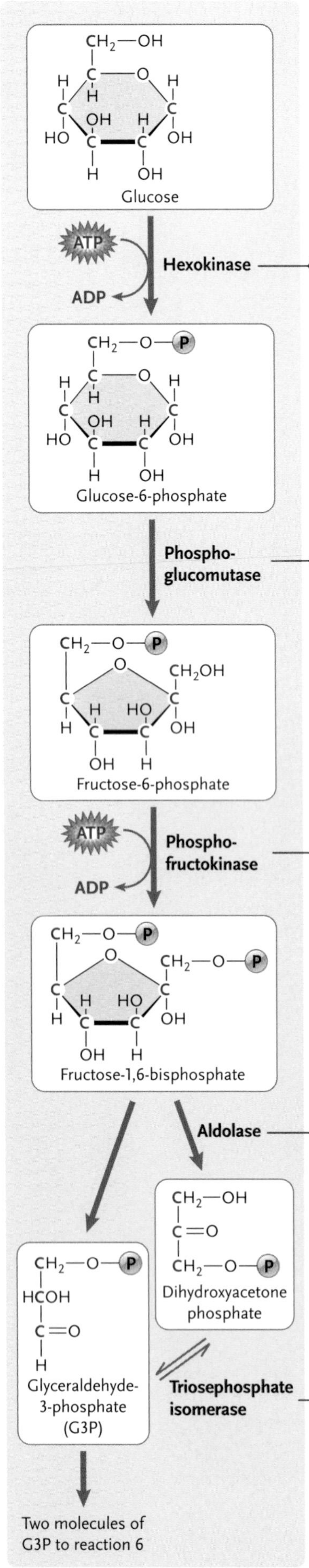

Glucose

ATP **Hexokinase**
ADP

1 Glucose receives a phosphate group from ATP, producing glucose-6-phosphate. (phosphorylation reaction)

Glucose-6-phosphate

Phospho-glucomutase

2 Glucose-6-phosphate is rearranged into its isomer, fructose-6-phosphate. (isomerization reaction)

Fructose-6-phosphate

ATP **Phospho-fructokinase**
ADP

3 Another phosphate group derived from ATP is attached to fructose-6-phosphate, producing fructose-1,6-bisphosphate. (phosphorylation reaction)

Fructose-1,6-bisphosphate

Aldolase

4 Fructose-1,6-bisphosphate is split into glyceraldehyde-3-phosphate (G3P) and dihydroxyacetone phosphate (DAP). (hydrolysis reaction)

Dihydroxyacetone phosphate

Glyceraldehyde-3-phosphate (G3P)

Triosephosphate isomerase

5 The DAP produced in reaction 4 is converted into G3P, giving a total of two of these molecules per molecule of glucose. (isomerization reaction)

Two molecules of G3P to reaction 6

Continued from reaction 5

G3P
(2 molecules)

NAD⁺ **Triosephosphate dehydrogenase**
NADH **P_i**
H⁺

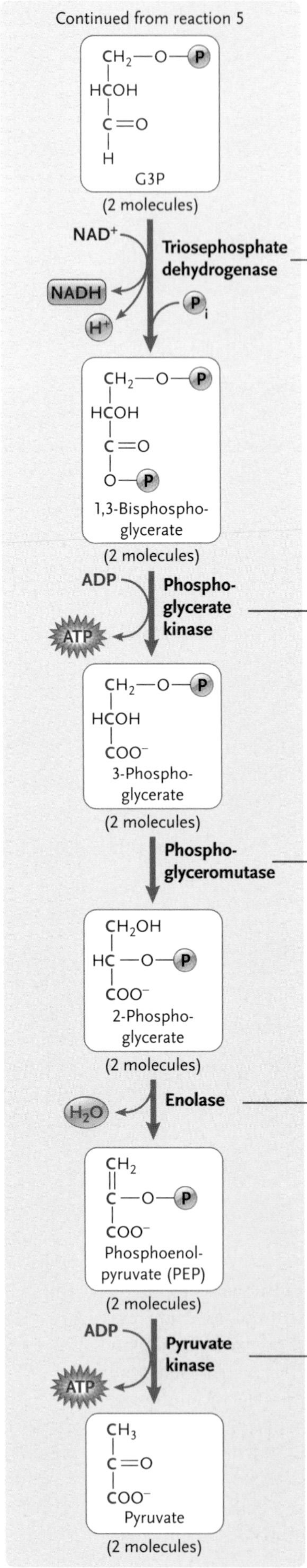

6 Two electrons and two protons are removed from G3P. Some of the energy released in this reaction is trapped by the addition of an inorganic phosphate group from the cytosol (not derived from ATP). The electrons are accepted by NAD⁺, along with one of the protons. The other proton is released to the cytosol. (redox reaction)

1,3-Bisphospho-glycerate
(2 molecules)

ADP **Phospho-glycerate kinase**
ATP

7 One of the two phosphate groups of 1,3-bisphosphoglycerate is transferred to ADP to produce ATP. (substrate-level phosphorylation reaction)

3-Phospho-glycerate
(2 molecules)

Phospho-glyceromutase

8 3-Phosphoglycerate is rearranged, shifting the phosphate group from the 3 carbon to the 2 carbon to produce 2-phosphoglycerate. (mutase reaction—shifting of a chemical group to another within same molecule)

2-Phospho-glycerate
(2 molecules)

H₂O **Enolase**

9 Electrons are removed from one part of 2-phosphoglycerate and delivered to another part of the molecule. Most of the energy lost by the electrons is retained in the product, phosphoenolpyruvate. (redox reaction)

Phosphoenol-pyruvate (PEP)
(2 molecules)

ADP **Pyruvate kinase**
ATP

10 The remaining phosphate group is removed from phosphoenolpyruvate and transferred to ADP. The reaction forms ATP and the final product of glycolysis, pyruvate. (substrate-level phosphorylation reaction)

Pyruvate
(2 molecules)

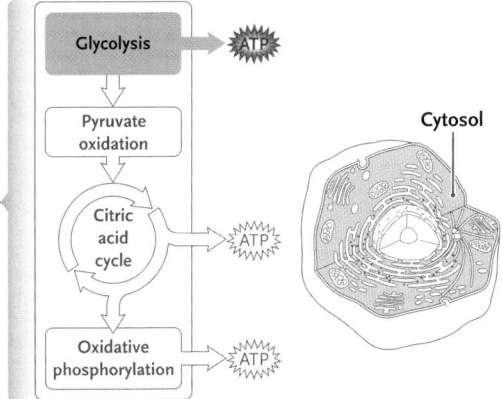

Figure 8.8
Reactions of glycolysis, which occur in the cytosol. Because two molecules of G3P are produced in reaction 5, all the reactions from 6 to 10 are doubled (not shown). The names of the enzymes that catalyze each reaction are in rust.

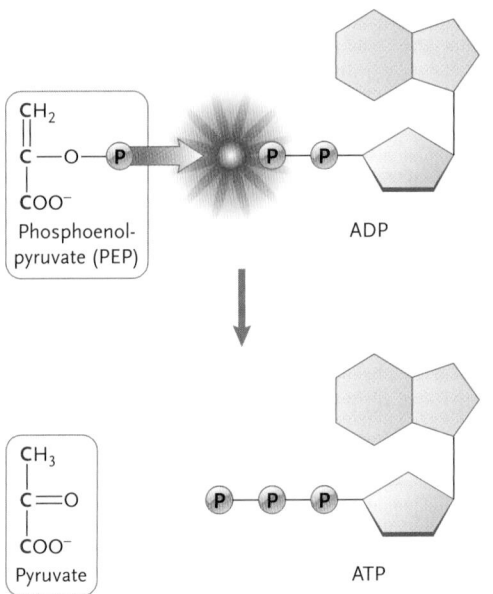

Figure 8.9
Mechanism that synthesizes ATP by substrate-level phosphorylation. A phosphate group is transferred from a high-energy donor directly to ADP, forming ATP.

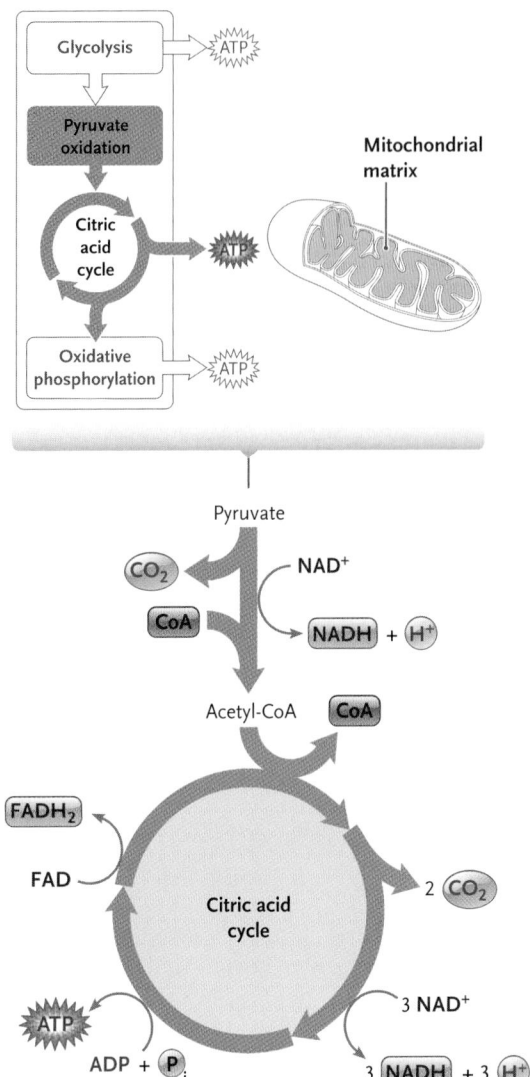

Figure 8.10
Overall reactions of pyruvate oxidation and the citric acid cycle. Each turn of the cycle oxidizes an acetyl group of acetyl-CoA to 2 CO_2. Acetyl-CoA, NAD^+, FAD, and ADP enter the cycle; CoA, NADH, $FADH_2$, ATP, and CO_2 are released as products.

8.3 Pyruvate Oxidation and the Citric Acid Cycle

Glycolysis produces pyruvate molecules in the cytosol, and an active transport mechanism moves them into the mitochondrial matrix, where pyruvate oxidation and the citric acid cycle proceed. An overview of these two processes is presented in **Figure 8.10.** Oxidation of pyruvate generates CO_2, *acetyl-coenzyme A* (acetyl-CoA), and NADH. The acetyl group of acetyl-CoA enters the citric acid cycle. As the citric acid cycle turns, every available electron carried into the cycle from pyruvate oxidation is transferred to NAD^+ or to another nucleotide-based molecule, *flavin adenine dinucleotide*

(FAD; the reduced form is $FADH_2$). With each turn of the cycle, substrate-level phosphorylation produces 1 ATP. The combined action of pyruvate oxidation and the citric acid cycle oxidizes the three-carbon products of glycolysis completely to carbon dioxide. The NADH and $FADH_2$ produced during this stage carry high-energy electrons to the electron transfer system in the mitochondrion.

Pyruvate Oxidation Produces the Two-Carbon Fuel of the Citric Acid Cycle

In pyruvate oxidation (also called **pyruvic acid oxidation**), a multienzyme complex removes the $—COO^-$ from pyruvate as CO_2 and then oxidizes the remaining two-carbon fragment of pyruvate to an acetyl group ($CH_3CO—$) **(Figure 8.11).** Two electrons and two protons are released by these reactions; the electrons and

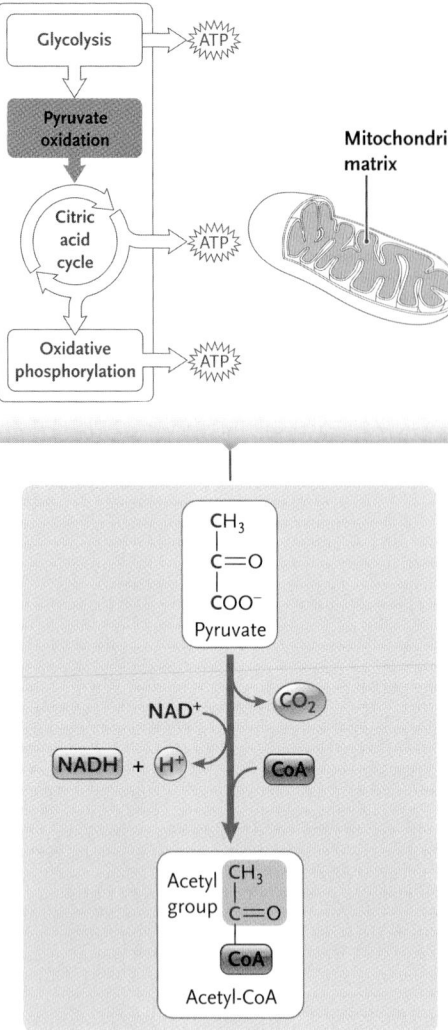

Figure 8.11

Reactions of pyruvate oxidation. Pyruvate (three carbons) is oxidized to an acetyl group (two carbons), which is carried from the cycle by CoA. The third carbon is released as CO_2. NAD^+ accepts two electrons and one proton removed in the oxidation. The acetyl group carried from the reaction by CoA is the fuel for the citric acid cycle.

from citrate, the product of the first reaction of the cycle. It is also called the **tricarboxylic acid cycle** or **Krebs cycle,** the latter after Hans Krebs, a German-born scientist who worked out the majority of the reactions in the cycle in research he conducted in England beginning in 1932. Using slices of fresh liver and kidney tissue, he tested various compounds thought to be important in cellular energy metabolism and discovered that a number of organic acids, including citrate, succinate, fumarate, and acetate, are oxidized rapidly. Several other scientists pieced together segments of the reaction series, but Krebs found the key reaction that linked the series into a cycle (see reaction 1 in Figure 8.12). Krebs was awarded a Nobel Prize in 1953 for his elucidation of the citric acid cycle.

The citric acid cycle has eight reactions, each catalyzed by a specific enzyme. All of the enzymes are located in the mitochondrial matrix except the enzyme for reaction 6, which is bound to the inner mitochondrial membrane on the matrix side. In a complete turn of the cycle, one two-carbon acetyl unit is consumed and two molecules of CO_2 are released (at reactions 3 and 4), thereby completing the conversion of all the C atoms originally in glucose to CO_2. The CoA molecule that carried the acetyl group to the cycle is released and participates again in pyruvate oxidation to pick up another acetyl group. Electron pairs are removed at each of four oxidations in the cycle (reactions 3, 4, 6, and 8). Three of the oxidations use NAD^+ as the electron acceptor, producing 3 NADH, and one uses FAD, producing 1 $FADH_2$. Substrate-level phosphorylation generates 1 ATP as part of reaction 5. Therefore, the net reactants and products of one turn of the citric acid cycle are:

$$1 \text{ acetyl-CoA} + 3 \text{ NAD}^+ + 1 \text{ FAD}$$
$$+ 1 \text{ ADP} + 1 \text{ P}_i + 2 \text{ H}_2\text{O} \rightarrow 2 \text{ CO}_2 + 3 \text{ NADH}$$
$$+ 1 \text{ FADH}_2 + 1 \text{ ATP} + 3 \text{ H}^+ + 1 \text{ CoA}$$

Because one molecule of glucose is converted to two molecules of pyruvate by glycolysis and each molecule of pyruvate is converted to one acetyl group, all the reactants and products in this equation are doubled when the citric acid cycle is considered as a continuation of glycolysis and pyruvate oxidation.

Most of the energy released by the four oxidations of the cycle is associated with the high-energy electrons carried by the 3 NADH and 1 $FADH_2$. These high-energy electrons enter the electron transfer system, where their energy is used to make most of the ATP produced in cellular respiration.

Like glycolysis, the citric acid cycle is regulated at several steps to match its rate to the cell's requirements for ATP. For example, the enzyme that catalyzes the first reaction of the citric acid cycle, *citrate synthase,* is inhibited by elevated ATP concentrations. The inhibitions automatically slow or stop the cycle when ATP production exceeds the demands of the cell and, by doing so, conserve cellular fuels.

one proton are accepted by NAD^+, reducing it to NADH, and the other proton is released as free H^+. The acetyl group is transferred to the nucleotide-based carrier *coenzyme A* (CoA). As acetyl-CoA, it carries acetyl groups to the citric acid cycle.

In summary, the pyruvate oxidation reaction is:

$$\text{pyruvate} + \text{CoA} + \text{NAD}^+ \rightarrow$$
$$\text{acetyl-CoA} + \text{NADH} + \text{H}^+ + \text{CO}_2$$

Because each glucose molecule that enters glycolysis produces two molecules of pyruvate, all the reactants and products in this equation are doubled when pyruvate oxidation is considered as a continuation of glycolysis.

The Citric Acid Cycle Oxidizes Acetyl Groups Completely to CO_2

The reactions of the citric acid cycle **(Figure 8.12)** oxidize acetyl groups completely to CO_2 and synthesize some ATP molecules. The citric acid cycle gets its name

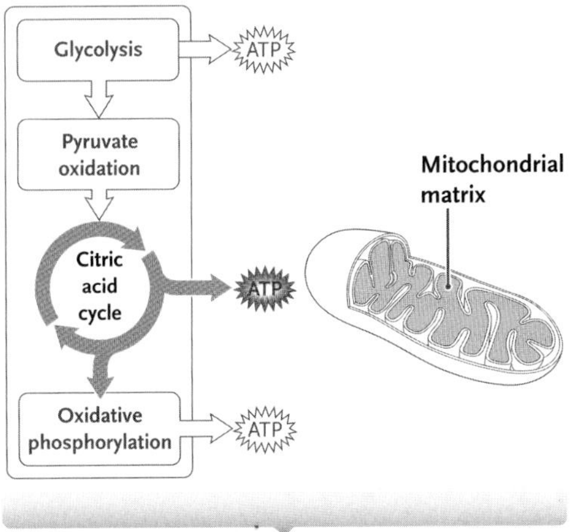
Glycolysis
ATP
Pyruvate oxidation
Citric acid cycle
ATP
Oxidative phosphorylation
ATP
Mitochondrial matrix

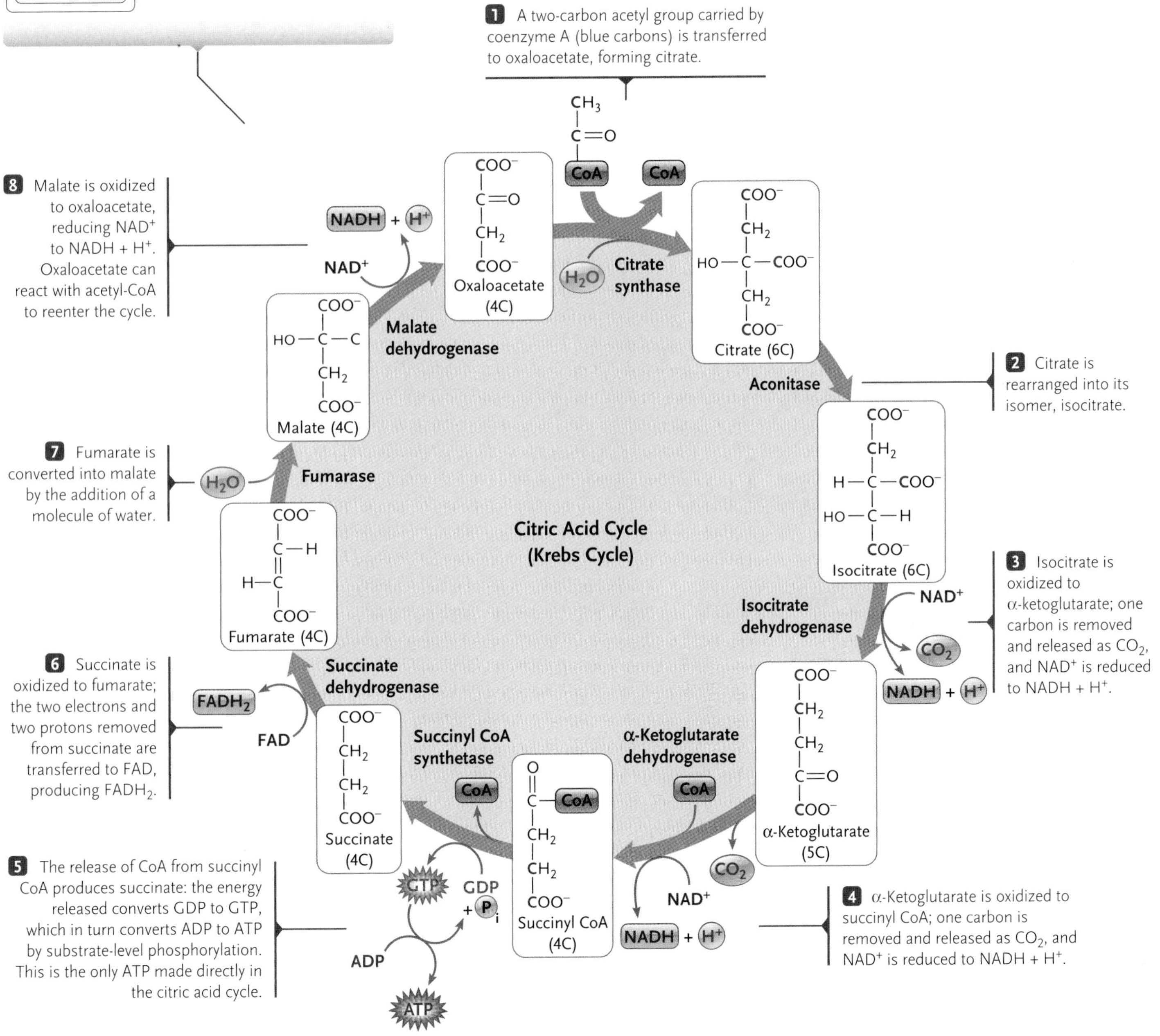
1 A two-carbon acetyl group carried by coenzyme A (blue carbons) is transferred to oxaloacetate, forming citrate.
2 Citrate is rearranged into its isomer, isocitrate.
3 Isocitrate is oxidized to α-ketoglutarate; one carbon is removed and released as CO_2, and NAD^+ is reduced to NADH + H^+.
4 α-Ketoglutarate is oxidized to succinyl CoA; one carbon is removed and released as CO_2, and NAD^+ is reduced to NADH + H^+.
5 The release of CoA from succinyl CoA produces succinate: the energy released converts GDP to GTP, which in turn converts ADP to ATP by substrate-level phosphorylation. This is the only ATP made directly in the citric acid cycle.
6 Succinate is oxidized to fumarate; the two electrons and two protons removed from succinate are transferred to FAD, producing $FADH_2$.
7 Fumarate is converted into malate by the addition of a molecule of water.
8 Malate is oxidized to oxaloacetate, reducing NAD^+ to NADH + H^+. Oxaloacetate can react with acetyl-CoA to reenter the cycle.
Citric Acid Cycle (Krebs Cycle)
CoA
CoA
H_2O
Citrate synthase
Oxaloacetate (4C)
Citrate (6C)
Aconitase
Isocitrate (6C)
Isocitrate dehydrogenase
NAD^+
CO_2
NADH + H^+
α-Ketoglutarate (5C)
α-Ketoglutarate dehydrogenase
CoA
CO_2
NAD^+
NADH + H^+
Succinyl CoA (4C)
Succinyl CoA synthetase
CoA
GTP
GDP + P_i
ADP
ATP
Succinate (4C)
Succinate dehydrogenase
FAD
$FADH_2$
Fumarate (4C)
Fumarase
H_2O
Malate (4C)
Malate dehydrogenase
NAD^+
NADH + H^+

Figure 8.12

Reactions of the citric acid cycle. Acetyl-CoA, NAD^+, FAD, and ADP enter the cycle; CoA, NADH, $FADH_2$, ATP, and CO_2 are released as products. The CoA released in reaction 1 can cycle back for another turn of pyruvate oxidation. Enzyme names are in rust.

Ralph Pleasant/FPG/Getty Images

Carbohydrates, Fats, and Proteins Can Function as Electron Sources for Oxidative Pathways

In addition to glucose and other six-carbon sugars, reactions leading from glycolysis through pyruvate oxidation also oxidize a wide range of carbohydrates, lipids, and proteins, which enter the reaction pathways at various points. **Figure 8.13** summarizes the cellular pathways involved; it shows the central role of CoA in funneling acetyl groups from different pathways into the citric acid cycle and of the mitochondrion as the site where most of these groups are oxidized.

Carbohydrates such as sucrose and other disaccharides are easily broken into monosaccharides such as glucose and fructose, which enter glycolysis at early steps. Starch (see Figure 3.7a) is hydrolyzed by digestive enzymes into individual glucose molecules, which enter the first reaction of glycolysis. Glycogen, a more complex carbohydrate that consists of glucose subunits (see Figure 3.7b), is broken down and converted by enzymes into glucose-6-phosphate, which enters glycolysis at reaction 2 of Figure 8.8.

Among the fats, triglycerides (see Figure 3.9) are major sources of electrons for ATP synthesis. Before entering the oxidative reactions, they are hydrolyzed into glycerol and individual fatty acids. The glycerol is converted to G3P and enters glycolysis at reaction 6 of Figure 8.8, in the ATP-producing portion of the pathway. The fatty acids—and many other types of lipids—are split into two-carbon fragments, which enter the citric acid cycle as acetyl-CoA. The energy released by the oxidation of fats, by weight, is comparatively high—about twice the energy yield of carbohydrates. This fact explains why fats are an excellent source of energy in the diet.

Proteins are hydrolyzed to amino acids before oxidation. The amino group ($-NH_2$) is removed, and the remainder of the molecule enters the pathway of carbohydrate oxidation as either pyruvate, acetyl units carried by CoA, or intermediates of the citric acid cycle. For example, the amino acid alanine is converted into pyruvate; leucine, into acetyl units; and phenylalanine, into fumarate, which enters the citric acid cycle at reaction 7 of Figure 8.12.

STUDY BREAK

Summarize the fate of pyruvate molecules produced by glycolysis.

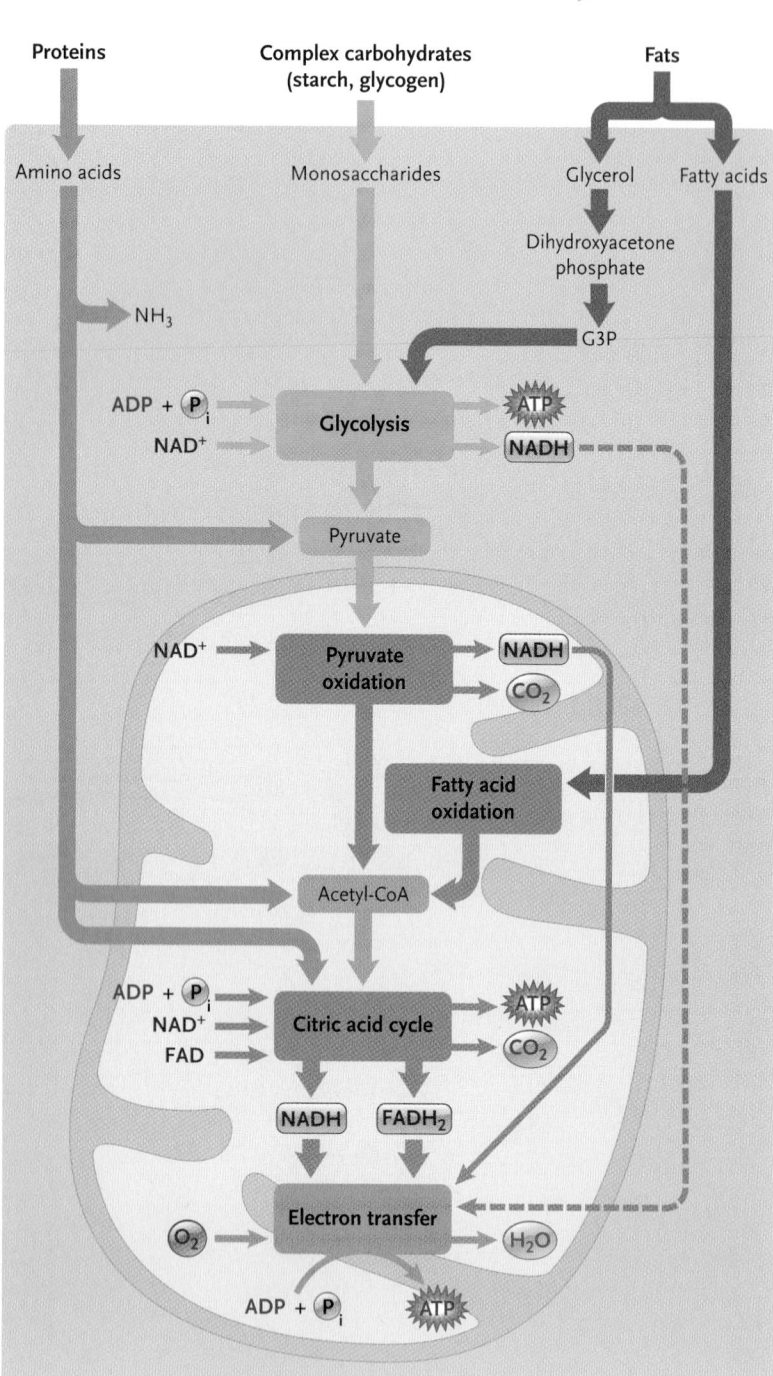

Figure 8.13

Major pathways that oxidize carbohydrates, fats, and proteins. Reactions that occur in the cytosol are shown against a tan background; reactions that occur in mitochondria are shown inside the organelle. CoA funnels the products of many oxidative pathways into the citric acid cycle.

8.4 The Electron Transfer System and Oxidative Phosphorylation

From the standpoint of ATP synthesis, the most significant products of glycolysis, pyruvate oxidation, and the citric acid cycle are the many high-energy electrons removed from fuel molecules and picked up by the carrier molecules NAD^+ or FAD. These electrons are released by the carriers into the electron transfer system of mitochondria.

The **mitochondrial electron transfer system** consists of a series of electron carriers that alternately pick up and release electrons, ultimately transferring them to their final acceptor, oxygen. As the electrons flow through the system, they release free energy, which is used to build a gradient of H^+ across the inner mitochondrial membrane. The gradient goes from a high H^+ concentration in the intermembrane compartment to a low concentration in the matrix. The H^+ gradient supplies the energy that drives ATP synthesis by mitochondrial ATP synthase.

In the Electron Transfer System, Electrons Flow through Protein Complexes in the Inner Mitochondrial Membrane

The mitochondrial electron transfer system includes three major protein complexes, numbered I, III, and IV, which serve as electron carriers **(Figure 8.14)**. These protein complexes are integral membrane proteins located in the inner mitochondrial membrane. In addition, a smaller complex, complex II, is bound to the inner mitochondrial membrane on the matrix side. Associated with the system are two small, highly mobile electron carriers, *cytochrome c* and *ubiquinone* (also known as coenzyme Q, or CoQ), which shuttle electrons between the major complexes. (**Cytochromes** are proteins with a heme prosthetic group that contains an iron atom. The iron atom accepts and donates electrons.)

Electrons flow through the major complexes as shown in Figure 8.14. Complex I picks up high-energy electrons from NADH on its side facing the mitochondrial matrix and conducts them via two electron carriers within the mitochondrial membrane, FMN (flavin mononucleotide) and an Fe/S (iron–sulfur) protein, to ubiquinone molecules. Complex II also contributes high-energy electrons to ubiquinone. Complex II is a succinate dehydrogenase complex that catalyzes two reactions. One is reaction 6 of the citric acid cycle, the conversion of succinate to fumarate (see Figure 8.12). In that reaction, FAD accepts two protons and two electrons and is reduced to $FADH_2$. The other reaction is the transfer to ubiquinone of two electrons obtained from the oxidation of $FADH_2$ to FAD. Two protons are also released in this reaction, and they are released back into the mitochondrial matrix. Electrons that pass to ubiquinone by the complex II reaction bypass complex I of the electron transfer system. Complex III accepts electrons from ubiquinone and transfers them through the electron carriers in the complex—cytochrome *b*, an Fe/S protein, and cytochrome c_1—to cytochrome *c*, which diffuses freely in the intermembrane space. Complex IV accepts electrons from cytochrome *c* and delivers them

via electron carriers cytochromes *a* and a_3 to oxygen. Four protons are added to a molecule of O_2 as it accepts four electrons, forming 2 H_2O.

The gas carbon monoxide inhibits complex IV activity, leading to abnormalities in mitochondrial function. In this way, the carbon monoxide in tobacco smoke contributes to the development of diseases associated with smoking.

Ubiquinone and the Three Major Electron Transfer Complexes Pump H^+ across the Inner Mitochondrial Membrane

Ubiquinone and the proteins of complexes I, III, and IV pump (actively transport) H^+ (protons) using energy from electron flow. Complex II, which does not pump H^+, works primarily as an entry point for the electrons removed from succinate.

The electron transfer system pumps protons from the matrix to the intermembrane compartment, resulting in an H^+ gradient with a high concentration in the intermembrane compartment and a low concentration in the matrix. Because protons carry a positive charge, the asymmetric distribution of protons generates an electrical and chemical gradient across the inner mitochondrial membrane, with the intermembrane compartment more positively charged than the matrix. The combination of a proton gradient and voltage gradient across the membrane produces stored energy known as the **proton-motive force.** This force contributes energy for ATP synthesis, as well as for cotransport of substances to and from mitochondria (see Section 6.4).

In our explanation of cellular respiration, electrons have been transferred to oxygen and the H^+ gradient has been generated across the inner mitochondrial membrane. We now focus on the use of this gradient to power the synthesis of ATP.

Chemiosmosis Powers ATP Synthesis by a Proton Gradient

Within the mitochondrion, ATP is synthesized by ATP synthase, an enzyme embedded in the inner mitochondrial membrane. In 1961, British scientist Peter Mitchell of Glynn Research Laboratories proposed that mitochondrial electron transfer produces an H^+ gradient and that the gradient powers ATP synthesis by ATP synthase. He called this pioneering model the **chemiosmotic hypothesis;** the process is commonly called *chemiosmosis* (see Figure 8.14). At the time, this hypothesis was a radical proposal because most researchers thought that the energy of electron transfer was stored as a high-energy chemical intermediate. No such intermediate was ever found, and eventually, Mitchell's hypothesis was supported by the results of many experiments. Mitchell received a Nobel Prize in 1978 for his model and supporting research.

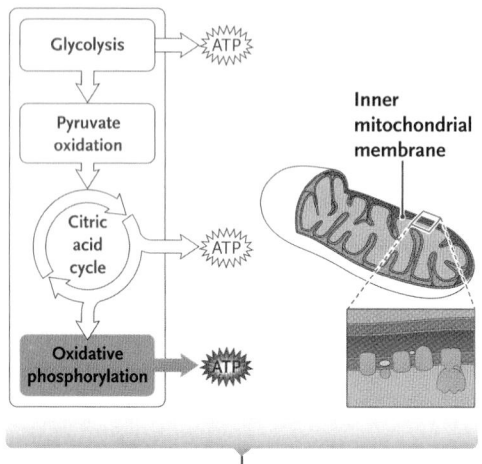

Figure 8.14

Mitochondrial electron transfer system and oxidative phosphorylation. The electron transfer system includes three major complexes, I, III, and IV. Two smaller electron carriers, ubiquinone and cytochrome *c*, act as shuttles between the major complexes, and succinate dehydrogenase (complex II) passes electrons to ubiquinone, bypassing complex I. Blue arrows indicate electron flow; red arrows indicate H⁺ movement. H⁺ is pumped from the matrix to the intermembrane compartment as electrons pass through complexes I, III, and IV. Oxidative phosphorylation involves the ATP synthase–catalyzed synthesis of ATP using the energy of the H⁺ gradient across the inner mitochondrial membrane—that is, by chemiosmosis. H⁺ moves through the membrane between the ATP synthase's basal unit and the membrane-embedded part of the stator. Sites in the headpiece convert ADP to ATP.

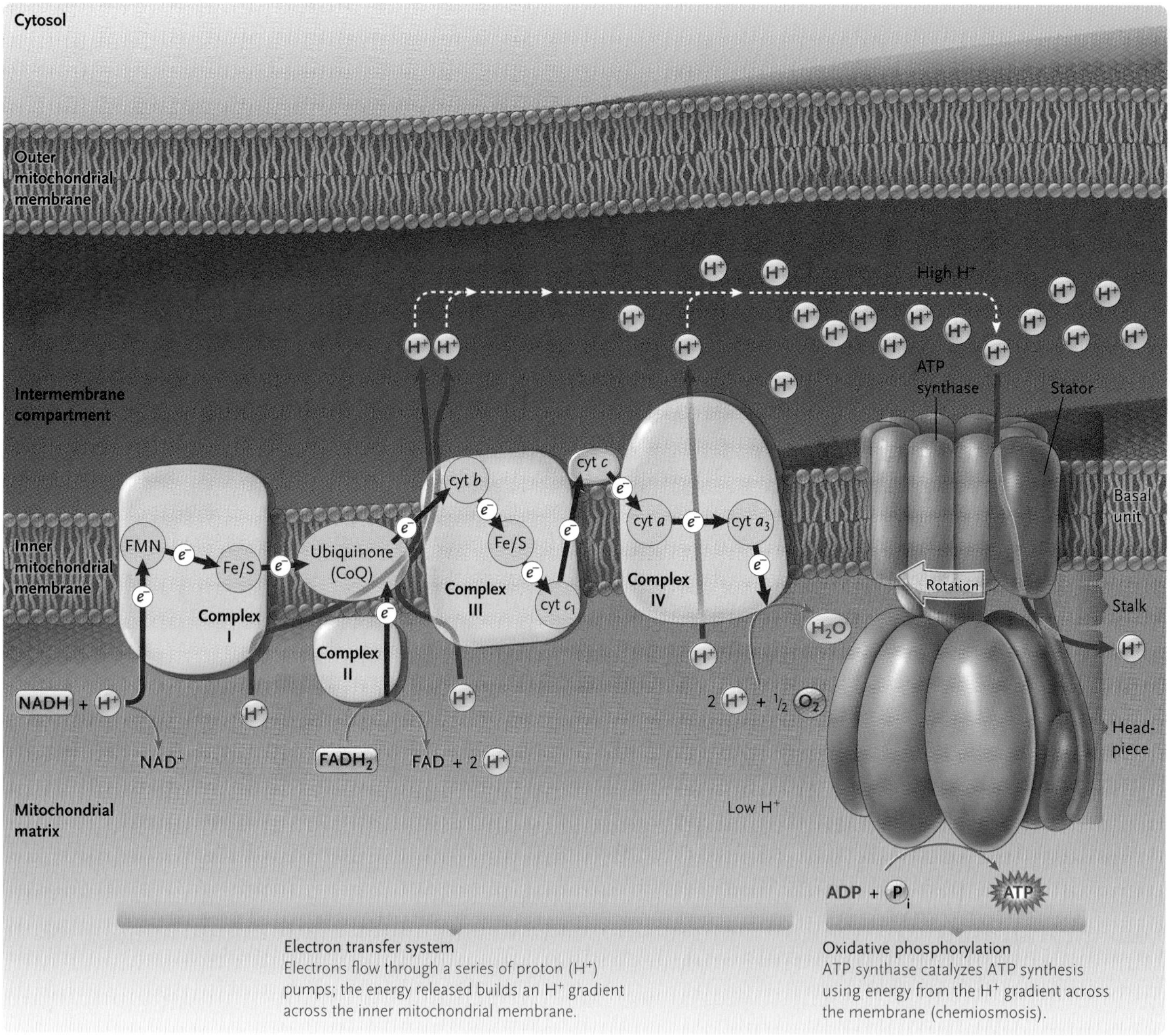

Cytosol

Outer mitochondrial membrane

High H⁺

Intermembrane compartment

ATP synthase

Stator

Inner mitochondrial membrane

FMN — e^- — Fe/S — e^- — Ubiquinone (CoQ)

cyt *b*

cyt *c*

Basal unit

Complex I

Complex III

Fe/S

cyt c_1

cyt *a*

cyt a_3

Complex IV

Rotation

Stalk

Complex II

H_2O

$NADH$ + H^+

NAD^+

$FADH_2$

$FAD + 2 H^+$

$2 H^+ + \frac{1}{2} O_2$

Low H⁺

Headpiece

Mitochondrial matrix

$ADP + P_i$

ATP

Electron transfer system
Electrons flow through a series of proton (H⁺) pumps; the energy released builds an H⁺ gradient across the inner mitochondrial membrane.

Oxidative phosphorylation
ATP synthase catalyzes ATP synthesis using energy from the H⁺ gradient across the membrane (chemiosmosis).

How does ATP synthase use the H^+ gradient to power ATP synthesis in chemiosmosis? ATP synthase consists of a *basal unit,* which is embedded in the inner mitochondrial membrane, connected to a *headpiece* by a *stalk,* and with a peripheral stalk called a *stator* bridging the basal unit and headpiece (see Figure 8.14). The headpiece extends into the mitochondrial matrix. Protons move between the basal unit and the membrane-embedded part of the stator. ATP synthase functions like an active transport ion pump. In Chapter 6, we described active transport pumps that use the energy created by hydrolysis of ATP to ADP and P_i to transport ions across membranes against their concentration gradients (see Figure 6.11). However, if the concentration of an ion is very high on the side toward which it is normally transported, the pump runs in reverse—that is, the ion is transported backward through the pump, and the pump adds phosphate to ADP to generate ATP. That is how ATP synthase operates in mitochondrial membranes. Proton-motive force moves protons in the intermembrane space through the channel in the enzyme's basal unit down their concentration gradient into the matrix. The flow of protons powers ATP synthesis by the headpiece; this phosphorylation reaction is oxidative phosphorylation. ATP synthase occurs in similar form and works in the same way in mitochondria, chloroplasts, and prokaryotes capable of oxidative phosphorylation.

Many details of the chemiosmotic mechanism are still being investigated. Paul D. Boyer of UCLA, one of the major contributors to this research, proposed the novel idea that passage of protons through the channel of the basal unit makes the stalk and headpiece spin like a top, just as the flow of water makes a waterwheel turn. The turning motion cycles each of three catalytic sites on the headpiece through sequential conformational changes that pick up ADP and phosphate, combine them, and release the ATP product. Another researcher, John Walker of the Laboratory of Molecular Biology (Cambridge, United Kingdom) used X-ray diffraction to create a three-dimensional picture of ATP synthase that clearly verified Boyer's model by showing the head in different rotational positions as ATP synthesis proceeds. Boyer and Walker jointly received a Nobel Prize in 1997 for their research into the mechanisms by which ATP synthase makes ATP.

Thirty-Two ATP Molecules Are Produced for Each Molecule of Glucose Completely Oxidized to CO_2 and H_2O

How many ATP molecules are produced as electrons flow through the mitochondrial electron transfer system? The most recent research indicates that approximately 2.5 ATP are synthesized as a pair of electrons released by NADH travels through the entire electron transfer pathway to oxygen. The shorter pathway, followed by an electron pair released from $FADH_2$ by

complex II to oxygen, synthesizes about 1.5 ATP. (Some accounts of ATP production round these numbers to 3 and 2 molecules of ATP, respectively.)

These numbers allow us to estimate the total amount of ATP that would be produced by the complete oxidation of glucose to CO_2 and H_2O if the entire H^+ gradient produced by electron transfer is used for ATP synthesis (**Figure 8.15**). During glycolysis, substrate-level phosphorylation produces 2 ATP. Glycolysis also produces 2 NADH, which leads to 5 ATP (see earlier discussion). In pyruvate oxidation, 2 NADH are produced from the two molecules of pyruvate, again leading to 5 ATP. In summary, glycolysis and pyruvate oxidation together yield 2 ATP, 4 NADH, and 2 CO_2 and, in the end, are responsible for 12 of the ATP produced by oxidation of glucose.

The subsequent citric acid cycle turns twice for each molecule of glucose that enters glycolysis, yielding a total of 2 ATP produced by substrate-level phosphorylation, as well as 6 NADH, 2 $FADH_2$, and 4 CO_2. The 6 NADH lead to 15 ATP, and the 2 $FADH_2$ lead to 3 ATP, for a total of 20 ATP from the citric acid cycle. With the ATP from glycolysis and pyruvate oxidation,

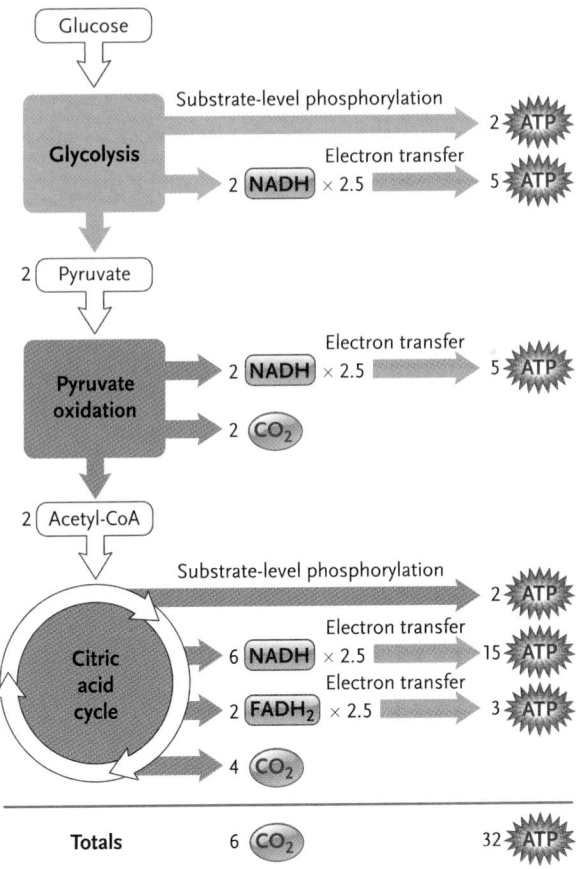

Figure 8.15

Summary of ATP production from the complete oxidation of a molecule of glucose. The total of 32 ATP assumes that electrons carried from glycolysis by NADH are transferred to NAD^+ inside mitochondria. If the electrons from glycolysis are instead transferred to FAD inside mitochondria, total production will be 30 ATP.

the total yield is 32 ATP from each molecule of glucose oxidized to carbon dioxide and water.

The combination of glycolysis, pyruvate oxidation, and the citric acid cycle has the following summary reaction:

$$\text{glucose} + 4\,\text{ADP} + 4\,P_i + 10\,\text{NAD}^+ + 2\,\text{FAD} \rightarrow$$
$$4\,\text{ATP} + 10\,\text{NADH} + 10\,\text{H}^+ + 2\,\text{FADH}_2 + 6\,CO_2$$

The total of 32 ATP assumes that the two pairs of electrons carried by the 2 NADH reduced in glycolysis each drive the synthesis of 2.5 ATP when traversing the mitochondrial electron transfer system. However, because NADH cannot penetrate the mitochondrial membranes, its electrons are transferred inside by one of two shuttle systems. The more efficient shuttle mechanism transfers the electrons to NAD^+ as the acceptor inside mitochondria. These electron pairs, when passed through the electron transfer system, result in the synthesis of 2.5 ATP each, producing the grand total of 32 ATP. The less efficient shuttle transfers the electrons to FAD as the acceptor inside mitochondria. These electron pairs, when passed through the electron transfer system, result in the synthesis of only 1.5 ATP each and produce a grand total of 30 ATP instead of 32.

Which shuttle predominates depends on the particular species and cell types involved. For example, heart, liver, and kidney cells in mammals use the more efficient shuttle; skeletal muscle and brain cells use the less efficient shuttle. Regardless, the numbers are ideal, because mitochondria also use the H^+ gradient to drive cotransport; any of the energy in the gradient used for this activity would reduce ATP production proportionally.

Cellular Respiration Conserves More Than 30% of the Chemical Energy of Glucose in ATP

Cellular respiration is not 100% efficient in converting the chemical energy of glucose to ATP. Using the estimate of 32 ATP produced for each molecule of glucose oxidized under ideal conditions, we can estimate the overall efficiency of cellular glucose oxidation—that is, the percentage of the chemical energy of glucose conserved as ATP energy.

Under standard conditions, including neutral pH (pH = 7) and a temperature of 25°C, the hydrolysis of ATP to ADP yields about 7.0 kilocalories per mole (kcal/mol). Assuming that complete glucose oxidation produces 32 ATP, the total energy conserved in ATP production would be about 224 kcal/mol. By contrast, if glucose is simply burned in air, it releases 686 kcal/mol. On this basis, the efficiency of cellular glucose oxidation would be about 32% (224/686 × 100 = about 32%). This value is considerably better than that of most devices designed by human engineers—for example, an automobile extracts only about 25% of the energy in the fuel it burns.

The chemical energy released by cellular oxidations that is not captured in ATP synthesis is released as heat. In mammals and birds, this source of heat maintains body temperature at a constant level. In certain mammalian tissues, including *brown fat* (see Chapter 46), the inner mitochondrial membranes contain *uncoupling proteins* (UCPs) that make the inner mitochondrial membrane "leaky" to H^+. As a result, electron transfer runs without building an H^+ gradient or synthesizing ATP and releases all the energy extracted from the electrons as heat. Brown fat with UCPs occurs in significant quantities in hibernating mammals and in very young offspring, including human infants. (*Insights from the Molecular Revolution* describes research showing that some plants also use UCPs in mitochondrial membranes to heat tissues.)

STUDY BREAK

1. What distinguishes the four complexes of the mitochondrial electron transfer system?
2. Explain how the proton pumps of complexes I, III, and IV relate to ATP synthesis.

8.5 Fermentation

Fermentation Keeps ATP Production Going When Oxygen Is Unavailable

When oxygen is plentiful, electrons carried by the 2 NADH produced by glycolysis are passed to the electron transfer system inside mitochondria, and the released energy drives the synthesis of ATP. If, instead, oxygen is absent or in short supply, the electrons may be used in fermentation. In **fermentation**, electrons carried by NADH are transferred to an organic acceptor molecule rather than to the electron transfer system. This transfer converts the NADH to NAD^+, which is required to accept electrons in reaction 6 of glycolysis (see Figure 8.8). As a result, glycolysis continues to supply ATP by substrate-level phosphorylation.

Two types of fermentation reactions exist: lactate fermentation and alcoholic fermentation **(Figure 8.16)**. **Lactate fermentation** converts pyruvate into lactate (Figure 8.16a). This reaction occurs in the cytosol of muscle cells in animals whenever vigorous or strenuous activity calls for more oxygen than breathing and circulation can supply. For example, significant quantities of lactate accumulate in the leg muscles of a sprinter during a 100-meter race. The lactate temporarily stores electrons, and when the oxygen content of the muscle cells returns to normal levels, the reverse of the reaction in Figure 8.16a regenerates pyruvate and NADH. The pyruvate can be used in the second stage

Keeping the Potatoes Hot

Mammals use several biochemical and molecular processes to maintain body heat. One process is shivering; the muscular activity of shivering releases heat that helps keep body temperature at normal levels. Another mechanism operates through uncoupling proteins (UCPs), which eliminate the mitochondrial H^+ gradient by making the inner mitochondrial membrane leaky to protons. Electron transfer and the oxidative reactions then run at high rates in mitochondria without trapping energy in ATP. The energy is released as heat that helps maintain body temperature.

Until recently, production of body heat by UCPs was thought to be confined to animals. But research by Maryse Laloi and her colleagues at the Max Planck Institute for Molecular Plant Physiology in Germany shows that some tissues in plants may use the same process to generate heat. The research team used molecular techniques to show that po-

tato plants (Solanum tuberosum) have a gene with a DNA sequence similar to that of a mammalian UCP gene. The potato gene encodes a protein of the same size as the two known UCPs of mammalian mitochondria. Enough sequence similarities exist to indicate that the potato and mammalian proteins are related and have the same overall three-dimensional structure.

The investigators then used the DNA of the potato UCP gene to probe for the presence of messenger RNA (mRNA), the molecules that serve as instructions for making proteins in the cytoplasm. This test determined whether the UCP genes were actually active in the potatoes. Potato plants grown at 20°C showed a low level of UCP mRNA in leaves and tubers, a moderate level in stems and fruits, and a very high level in roots and flowers. These results indicate that the gene encoding the plant UCP is active at different levels in various plant tissues, sug-

gesting that certain tissues naturally need warming for optimal function.

Laloi and her coworkers then used the same method to test whether exposing potato plants to cold temperatures could induce greater synthesis of the UCP mRNA. After potato plants were kept for 1 to 3 days at 4°C, the UCP mRNA in leaves rose to a level comparable with the high level found in the flowers of plants kept at 20°C.

The research indicates that although potato plants cannot shiver to keep warm, they probably use the mitochondrial uncoupling process to warm tissues when they are stressed by low temperatures. Thus, mechanisms for warming body tissues, once thought to be the province only of animals, appear to be much more widespread. In particular, UCPs, which were believed to have evolved in relatively recent evolutionary times with the appearance of birds and mammals, may be a much more ancient development.

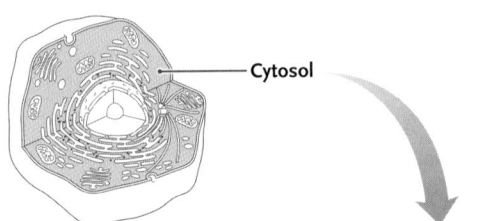

a. Lactate fermentation

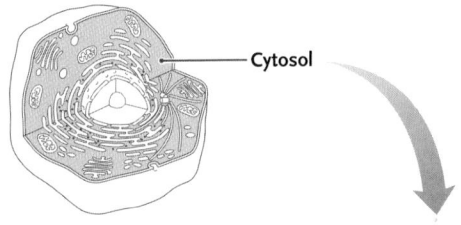

b. Alcoholic fermentation

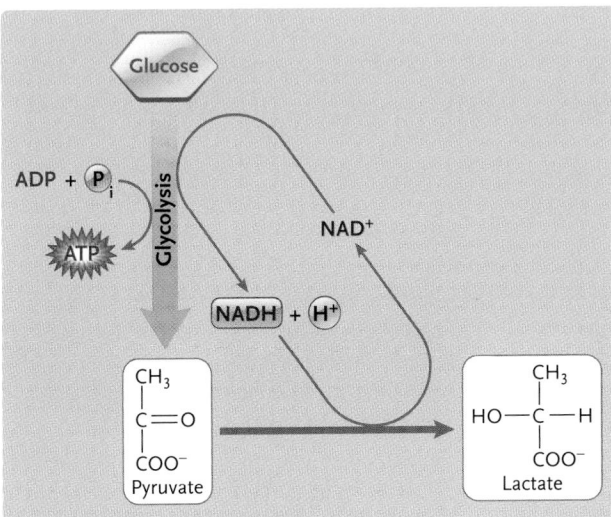

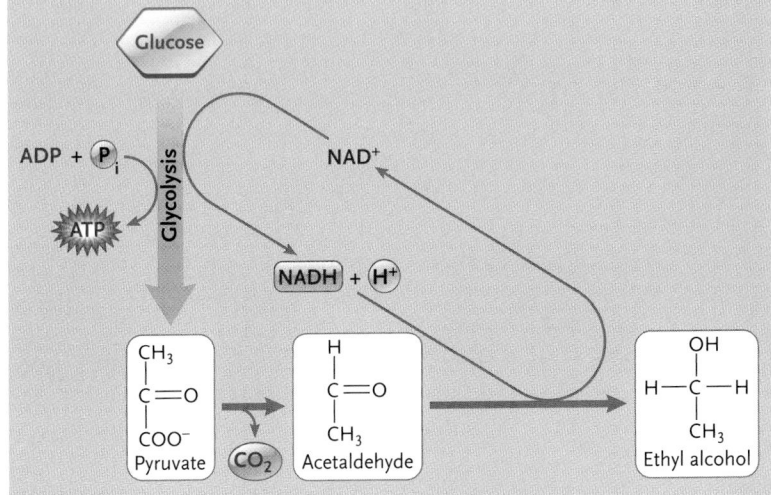

Figure 8.16

Fermentation reactions that produce (a) lactate and (b) ethyl alcohol. The fermentations, which occur in the cytosol, convert NADH to NAD^+, allowing the electron carrier to cycle back to glycolysis. This process keeps glycolysis running, with continued production of ATP.

of cellular respiration, and the NADH contributes its electron pair to the electron transfer system. Some bacteria also produce lactate as their fermentation product; the sour taste of buttermilk, yogurt, and dill pickles is a sign of their activity.

Alcoholic fermentation (Figure 8.16b) occurs in microorganisms such as yeasts, which are single-celled fungi. In this reaction, pyruvate is converted into ethyl alcohol (which has two carbons) and CO_2 in a two-step series that also converts NADH into NAD^+. Alcoholic fermentation by yeasts has widespread commercial applications. Bakers use the yeast *Saccharomyces cerevisiae* to make bread dough rise. They mix the yeast with a small amount of sugar and blend the mixture into the dough where oxygen levels are low. As the yeast cells convert the sugar into ethyl alcohol and carbon dioxide, the gaseous CO_2 expands and creates bubbles that cause the dough to rise. Oven heat evaporates the alcohol and causes further expansion of the bubbles, producing a light-textured product. Alcoholic fermentation is also the mainstay of beer and wine brewing. Fruits are a natural home to wild yeasts **(Figure 8.17)**; for example, winemakers rely on a mixture of wild and cultivated yeasts to produce wine. Alcoholic fermenta-

Figure 8.17
Alcoholic fermentation in nature: wild yeast cells, visible as a dust-like coating on grapes.

David M. Phillips/Visuals Unlimited

tion also occurs naturally in the environment; for example, overripe or rotting fruit frequently will start to ferment, and birds that eat the fruit may become too drunk to fly.

Fermentation is a lifestyle for some organisms. In bacteria and fungi that lack the enzymes and factors to carry out oxidative phosphorylation, fermentation is the only source of ATP. These organisms are called **strict anaerobes** (*an* = without; *aero* = air; *bios* = life). In general, these organisms require an oxygen-free environment; they cannot utilize oxygen as a final electron acceptor. Among these organisms are the bacteria that cause botulism, tetanus, and some other serious diseases. For example, the bacterium that causes botulism thrives in the oxygen-free environment of canned foods that prevents the growth of most other microorganisms.

Other organisms, called **facultative anaerobes**, can switch between fermentation and full oxidative pathways, depending on the oxygen supply. Facultative anaerobes include *Escherichia coli*, the bacterium that inhabits the digestive tract of humans; the *Lactobacillus* bacteria used to produce buttermilk and yogurt; and *S. cerevisiae*, the yeast used in brewing

UNANSWERED QUESTIONS

Glycolysis and energy metabolism are crucial for the normal functioning of an animal. Research of many kinds is being conducted in this area, such as characterizing the molecular components in detail and determining how the reactions are regulated. The goal is to generate comprehensive models of cellular respiration and its regulation. Following are two specific examples of ongoing research related to human disease caused by defects in cellular respiration.

How do mitochondrial proteins change in patients with Alzheimer disease?

Alzheimer disease (AD) is an age-dependent, irreversible, neurodegenerative disorder in humans. Symptoms include a progressive deterioration of cognitive functions and, in particular, a significant loss of memory. Reduced brain metabolism occurs early in the onset of AD. One of the mechanisms for this physiological change appears to be damage to or reduction of key mitochondrial components, including enzymes of the citric acid cycle and the oxidative phosphorylation system. However, the complete scope of mitochondrial protein changes has not been established, nor have detailed comparisons been made in mitochondrial protein changes among AD patients. Currently, Gail Breen at the University of Texas, Dallas, is performing research to detail qualitatively and quantitatively all mitochondrial proteins and their levels in healthy and AD brains. A mouse model of AD is being used for this research. Breen's group hopes that the information they obtain will provide a better understanding of how mitochondrial dysfunction contributes to AD. With such information in hand, it may be possible to develop interventions that slow or halt the progression of AD in humans.

How are the oxidative phosphorylation complexes in the mitochondrion assembled?

Defects in oxidative phosphorylation may cause disorders in which several systems of the human body are adversely affected. Often, these disorders involve the nervous system and the skeletal and cardiac muscles. The enzyme complexes of the oxidative phosphorylation system consist of about 80 different protein subunits, some of which are encoded by nuclear genes and some by mitochondrial genes. The protein subunits are assembled into complexes in the mitochondria. This assembly process requires a large number of accessory proteins, and many important mitochondrial diseases are caused by defects in the assembly protein genes.

Eric Shoubridge of McGill University in Canada is studying the molecular genetics of assembly of oxidative phosphorylation complexes. His focus is identifying and characterizing the assembly genes with long-term goals of understanding how the complexes are assembled and how defects in complex assembly lead to disease. Shoubridge's group has identified mutations in four different assembly genes in infants with a fatal disease caused by cytochrome *c* deficiency (a defect in the assembly of complex IV). They have also identified complex I assembly proteins, and they were the first to show an association between a defect in one of the proteins and a human disease. Unexpectedly, the biochemical deficiencies caused by the mutant assembly proteins tend to be tissue-specific, even though the assembly protein genes are expressed in all tissues. As a result, clinical symptoms caused by defective assembly proteins vary based on the extent of the enzyme deficiencies in different tissues. Understanding how the tissue-specific differences occur and how they are regulated will be important in developing therapies for patients with the diseases.

Peter J. Russell

and baking. Many cell types in higher organisms, including vertebrate muscle cells, are also facultative anaerobes.

Some prokaryotic and eukaryotic cells are **strict aerobes**—that is, they have an absolute requirement for oxygen to survive and are unable to live solely by fermentations. Vertebrate brain cells are key examples of strict aerobes.

This chapter traced the flow of high-energy electrons from fuel molecules to ATP. As part of the process, the fuels are broken into molecules of carbon di-oxide. The next chapter shows how photosynthetic organisms use these inorganic raw materials to produce organic molecules through a process that pushes the electrons back to high energy levels by absorbing the energy of sunlight.

STUDY BREAK

What is fermentation, and when does it occur?
What are the two types of fermentation?

Review

Go to **ThomsonNOW**™ at www.thomsonedu.com/login to access quizzing, animations, exercises, articles, and personalized homework help.

8.1 Overview of Cellular Energy Metabolism

- Oxidation–reduction reactions, called redox reactions, partially or completely transfer electrons from donor to acceptor atoms; the donor is oxidized as it releases electrons, and the acceptor is reduced (Figure 8.1).

- Plants and almost all other organisms obtain energy for cellular activities through cellular respiration, the process of transferring electrons from donor organic molecules to a final acceptor molecule such as oxygen; the energy that is released drives ATP synthesis (Figure 8.2).

- Cellular respiration occurs in three stages: (1) In glycolysis, glucose is converted to two molecules of pyruvate through a series of enzyme-catalyzed reactions; (2) in pyruvate oxidation and the citric acid cycle, pyruvate is converted to an acetyl compound that is oxidized completely to carbon dioxide; and (3) in the electron transfer system and oxidative phosphorylation, high-energy electrons produced from the first two stages pass through the transfer system, with much of their energy being used to establish an H^+ gradient across the membrane that drives the synthesis of ATP from ADP and P_i (Figure 8.3).

- In eukaryotes, most of the reactions of cellular respiration occur in mitochondria (Figure 8.4).

Animation: The functional zones in mitochondria

8.2 Glycolysis

- In glycolysis, which occurs in the cytosol, glucose (six carbons) is oxidized into two molecules of pyruvate (three carbons each). Electrons removed in the oxidations are delivered to NAD^+, producing NADH. The reaction sequence produces a net gain of 2 ATP, 2 NADH, and 2 pyruvate molecules for each molecule of glucose oxidized (Figures 8.6 and 8.8).

- ATP molecules produced in the energy-releasing steps of glycolysis result from substrate-level phosphorylation, an enzyme-catalyzed reaction that transfers a phosphate group from a substrate to ADP (Figure 8.9).

Animation: The overall reactions of glycolysis

Practice: Recreating the reactions of glycolysis

8.3 Pyruvate Oxidation and the Citric Acid Cycle

- In pyruvate oxidation, which occurs inside mitochondria, 1 pyruvate (three carbons) is oxidized to 1 acetyl group (two carbons) and 1 CO_2. Electrons removed in the oxidation are ac-cepted by 1 NAD^+ to produce 1 NADH. The acetyl group is transferred to coenzyme A, which carries it to the citric acid cycle (Figure 8.11).

- In the citric acid cycle, acetyl groups are oxidized completely to CO_2. Electrons removed in the oxidations are accepted by NAD^+ or FAD, and substrate-level phosphorylation produces ATP. For each acetyl group oxidized by the cycle, 2 CO_2, 1 ATP, 3 NADH, and 1 $FADH_2$ are produced (Figure 8.12).

Animation: Pyruvate oxidation and the citric acid cycle

Animation: Major pathways oxidizing carbohydrates, fats, and proteins

8.4 The Electron Transfer System and Oxidative Phosphorylation

- Electrons are passed from NADH and $FADH_2$ to the electron transfer system, which consists of four protein complexes and two smaller shuttle carriers. As the electrons flow from one carrier to the next through the system, some of their energy is used by the complexes to pump protons across the inner mitochondrial membrane (Figure 8.14).

- Ubiquinone and the three major protein complexes (I, III, and IV) pump H^+ from the matrix to the intermembrane compartment, generating an H^+ gradient with a high concentration in the intermembrane compartment and a low concentration in the matrix (Figure 8.14).

- The H^+ gradient produced by the electron transfer system is used by ATP synthase as an energy source for synthesis of ATP from ADP and P_i. The ATP synthase is embedded in the inner mitochondrial membrane together with the electron transfer system (Figure 8.14).

- An estimated 2.5 ATP are synthesized as each electron pair travels from NADH to oxygen through the mitochondrial electron transfer system; about 1.5 ATP are synthesized as each electron pair travels through the system from $FADH_2$ to oxygen. Using these totals gives an efficiency of more than 30% for the utilization of energy released by glucose oxidation if the H^+ gradient is used only for ATP production (Figure 8.15).

Animation: The mitochondrial electron transfer system and oxidative phosphorylation

8.5 Fermentation

- Fermentations are reaction pathways that deliver electrons carried from glycolysis by NADH to organic acceptor molecules, thereby converting NADH back to NAD^+. The NAD^+ can accept electrons generated by glycolysis, allowing glycolysis to supply ATP by substrate-level phosphorylation (Figure 8.16).

Animation: The fermentation reactions

Questions

Self-Test Questions

1. In glycolysis:
   a. free oxygen is required for the reactions to occur.
   b. ATP is used when glucose and fructose-6-phosphate are catabolized, and ATP is synthesized when 3-phosphoglycerate and pyruvate are formed.
   c. the enzymes that move phosphate groups on and off the molecules are uncoupling proteins.
   d. the product with the highest potential energy in the pathway is pyruvate.
   e. the end product of glycolysis moves to the electron transfer system.

2. Which of the following statements about phosphofructokinase is *false*?
   a. It is located and has its main activity on the inner mitochondrial membrane.
   b. It catalyzes a reaction to form a product with the highest potential energy in the pathway.
   c. It can be inactivated by ATP at an inhibitory site on its surface.
   d. It can be activated by ADP at an excitatory site on its surface.
   e. It can cause ADP to form.

3. Which of the following statements is *false*? Imagine that you ingested three chocolate bars just before sitting down to study this chapter. Most likely:
   a. your brain cells are using ATP.
   b. there is no deficit of the initial substrate to begin glycolysis.
   c. the respiratory processes in your brain cells are moving atoms from glycolysis through the citric acid cycle to the electron transfer system.
   d. after a couple of hours, you change position and stretch to rest certain muscle cells, which removes lactate from these muscles.
   e. after 2 hours, your brain cells are oxygen-deficient.

4. If ADP produced throughout the respiratory reactions is in excess, this excess ADP will:
   a. bind glucose to turn off glycolysis.
   b. bind glucose-6-phosphate to turn off glycolysis.
   c. bind phosphofructokinase to turn on or keep glycolysis turned on.
   d. cause lactate to form.
   e. increase oxaloacetate binding to increase NAD^+ production.

5. Which of the following statements is *false*? In cellular respiration:
   a. one molecule of glucose can produce about 32 ATP.
   b. oxygen unites directly with glucose to form carbon dioxide.
   c. a series of energy-requiring reactions is coupled to a series of energy-releasing reactions.
   d. NADH and $FADH_2$ allow H^+ to be pumped across the inner mitochondrial membrane.
   e. the electron transfer system occurs on the inner mitochondrial membrane.

6. You are reading this text while breathing in oxygen and breathing out carbon dioxide. The carbon dioxide arises from:
   a. glucose in glycolysis.
   b. NAD^+ redox reactions in the mitochondrial matrix.
   c. NADH redox reactions on the inner mitochondrial membrane.
   d. $FADH_2$ in the electron transfer system.
   e. the oxidation of pyruvate, isocitrate, and α-ketoglutarate in the citric acid cycle.

7. In the citric acid cycle:
   a. NADH and H^+ are produced when α-ketoglutarate is both produced and metabolized.
   b. ATP is produced by oxidative phosphorylation.
   c. to progress from a four-carbon molecule to a six-carbon molecule, CO_2 enters the cycle.
   d. $FADH_2$ is formed when succinate is converted to oxaloacetate.
   e. for each molecule of glucose metabolized, the cycle "turns" once.

8. For each NADH produced from the citric acid cycle, about how many ATP are formed?
   a. 38   b. 36   c. 32   d. 2.5   e. 2.0

9. In the 1950s, a diet pill that had the effect of "poisoning" ATP synthase was tried. The person taking it could not use glucose and "lost weight"—and ultimately his or her life. Today, we know that the immediate effect of poisoning ATP synthase is:
   a. ATP would not be made at the electron transfer system.
   b. there would be an increase in H^+ movement across the inner mitochondrial membrane.
   c. more than 32 ATP could be produced from a molecule of glucose.
   d. ADP would be united with phosphate more readily in the mitochondria.
   e. ATP would react with oxygen.

10. Amino acids and fats enter the respiration pathway:
    a. by joining to NADH.
    b. by joining to glucose.
    c. at the citric acid cycle.
    d. on the inner mitochondrial membrane.
    e. on the electron transfer system.

Questions for Discussion

1. Why do you think nucleic acids are not oxidized extensively as a cellular energy source?

2. A hospital patient was regularly found to be intoxicated. He denied that he was drinking alcoholic beverages. The doctors and nurses made a special point to eliminate the possibility that the patient or his friends were smuggling alcohol into his room, but he was still regularly intoxicated. Then, one of the doctors had an idea that turned out to be correct and cured the patient of his intoxication. The idea involved the patient's digestive system and one of the oxidative reactions covered in this chapter. What was the doctor's idea?

Experimental Analysis

There are several ways to measure cellular respiration experimentally. For example, CO_2 and O_2 gas sensors measure changes over time in the concentration of carbon dioxide or oxygen, respectively. Design two experiments to test the effects of changing two different variables or conditions (one per experiment) on the respiration of a research organism of your choice.

Evolution Link

Which of the two phosphorylation mechanisms, oxidative phosphorylation or substrate-level phosphorylation, is likely to have appeared first in evolution? Why?

How Would You Vote?

Developing new drugs is costly. There is little incentive for pharmaceutical companies to target ailments that affect relatively few individuals, such as Luft syndrome. Should the federal government allocate some funds to private companies that search for cures for diseases affecting a relatively small number of people? Go to www.thomsonedu.com/login to investigate both sides of the issue and then vote.

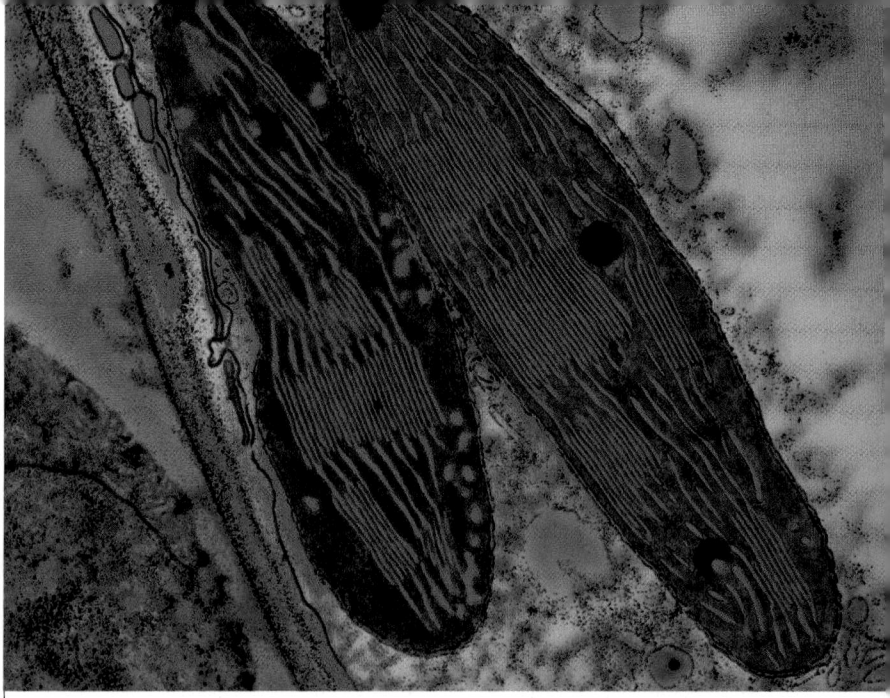

Dr. Kari Lounatmaa/Science Photo Library/Photo Researchers, Inc.

Chloroplasts in the leaf of the pea plant *Pisum sativum* (colorized TEM). The light-dependent reactions of photosynthesis take place within the thylakoids of the chloroplasts (thylakoid membranes are shown in yellow).

9 Photosynthesis

WHY IT MATTERS

By the late 1880s, scientists realized that green algae and plants use light as a source of energy to make organic molecules. This conversion of light energy to chemical energy in the form of sugar and other organic molecules is called **photosynthesis.** The scientists also knew that these organisms release oxygen as part of their photosynthetic reactions. Among these scientists was a German botanist, Theodor Engelmann, who was curious about the particular colors of light used in photosynthesis. Was green light the most effective in promoting photosynthesis, as you might expect from looking at a plant, or were other colors used more?

Engelmann used only a light microscope and a glass prism to find the answer to this question. Yet his experiment stands today as a classic, both for the fundamental importance of his answer and for the simple but elegant methods he used to obtain it. Engelmann placed a strand of a green alga, *Spirogyra,* on a glass microscope slide, along with water containing bacteria that require oxygen to survive. He adjusted the prism so that it split a beam of light into its separate colors, which spread like a rainbow across the strand **(Figure 9.1).** After a short time, he noticed that the bacteria had begun to cluster around the algal

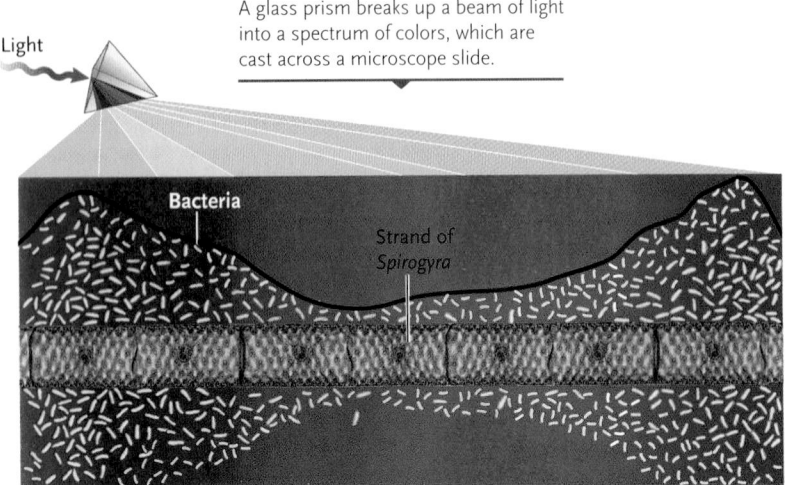

A glass prism breaks up a beam of light into a spectrum of colors, which are cast across a microscope slide.

Light

Bacteria

Strand of *Spirogyra*

Figure 9.1

Engelmann's 1882 experiment revealing the action spectrum of light used in photosynthesis by *Spirogyra*, a green alga. The aerobic bacteria clustered along the algal strand in the regions where oxygen was released in greatest quantity—the regions in which photosynthesis proceeded at the greatest rate. Those regions corresponded to the colors (wavelengths) of light being absorbed most effectively by the alga—in this case, violet and red.

strand in different locations. The largest clusters were under the blue and violet light at one end of the strand and the red light at the other end. Very few bacteria were found in the green light. Evidently, violet, blue, and red light caused the most oxygen to be released, and Engelmann concluded that these colors of light—rather than green—were used most effectively in photosynthesis.

Engelmann used the distribution of bacteria to construct a curve called an *action spectrum* for the wavelengths of light falling on the *Spirogyra;* it shows the relative effect of each color of light on photosynthesis (black curve in Figure 9.1). Engelmann's results were so accurate that an action spectrum obtained from *Spirogyra* with modern equipment fits closely with his bacterial distribution. However, his results were so advanced for his time that they remained controversial for some 60 years, until instruments that directly measure the effects of different wavelengths of light became available.

Scientists now know that photosynthetic organisms, which include plants, some protists (the algae), and some archaeans and bacteria, absorb the radiant energy of sunlight and convert it into chemical energy. The organisms use the chemical energy to convert simple inorganic raw materials—water, carbon dioxide (CO_2) from the air, and inorganic minerals from the soil—into complex organic molecules. Photosynthesis is still not completely understood, so it remains a subject of active research today.

This chapter begins with an overview of the photosynthetic reactions. We then examine light and light absorption and the reactions that use absorbed energy to make organic molecules from inorganic substances. This chapter focuses primarily on photosynthesis in plants and green algae; other eukaryotic photosynthesizers have individual variations on the process (see Chapter 26). Prokaryotic photosynthesis is described in Chapter 25.

9.1 Photosynthesis: An Overview

Plants and other photosynthetic organisms are the *primary producers* of Earth; they convert the energy of sunlight into chemical energy and use it to assemble simple inorganic raw materials into complex organic molecules. Primary producers use some of the organic molecules they make as an energy source for their own activities. But they also serve—directly or indirectly—as a food source for *consumers*, the animals that live by eating plants or other animals. Eventually, the bodies of both primary producers and consumers provide chemical energy for bacteria, fungi, and other *decomposers*.

Photosynthesizers and other organisms that use energy to make all of their required organic molecules from CO_2 and other inorganic sources such as water are called **autotrophs** (*autos* = self; *trophos* = feeding). Autotrophs that use light as the energy source to make organic molecules by photosynthesis are called **photoautotrophs**. Consumers and decomposers, which need a source of organic molecules to survive, are called **heterotrophs** (*hetero* = different).

As the pathway of energy flow proceeds from primary producers to decomposers, the organic molecules made by photosynthesis are broken down into inorganic molecules again, and the chemical energy captured in photosynthesis is released as heat. Thus the energy required for life flows from the sun through plants, animals, and decomposers, and finally is released as heat. Because the reactions capturing light energy are the first step in this pathway, photosynthesis is the vital link between the energy of sunlight and the vast majority of living organisms.

Electrons Play a Primary Role in Photosynthesis

Photosynthesis proceeds in two stages, each involving multiple reactions. In the first stage, the **light-dependent reactions**, the energy of sunlight is absorbed and converted into chemical energy in the form of two substances: ATP and NADPH. ATP is the main energy source for plant cells, and NADPH (nicotinamide adenine dinucleotide phosphate) carries electrons pushed to high energy levels by absorbed light. In the second stage of photosynthesis, the **light-independent reactions** (also called the *Calvin cycle*), these electrons are used as a source of energy to convert inorganic CO_2 to an organic form, a process called **CO_2 fixation.** The conversion is a reduction, in which electrons are added to CO_2;

as part of the reduction, protons are also added to CO_2 (reduction and oxidation are discussed in Section 8.1). With the added electrons and protons (H^+), CO_2 is converted to a carbohydrate, with carbon, hydrogen, and oxygen atoms in the ratio $1\ C : 2\ H : 1\ O$. Carbohydrate units are often symbolized as $(CH_2O)_n$, with the "n" indicating that different carbohydrates are formed from different multiples of the carbohydrate unit.

In plants, algae, and one group of photosynthetic bacteria (the cyanobacteria), the source of electrons and protons for CO_2 fixation is the most abundant substance on Earth: water. Oxygen generated from the splitting of water is released to the environment as a by-product:

$$2\ H_2O \rightarrow 4\ H^+ + 4\ e^- + O_2$$

Thus plants, algae, and cyanobacteria use three resources that are readily available—sunlight, water, and CO_2—to produce almost all the organic matter on Earth and to supply the oxygen of our atmosphere.

In the organisms able to split water, the two reactions shown above are combined and multiplied by 6 to produce a six-carbon carbohydrate such as glucose:

$$6\ CO_2 + 12\ H_2O \rightarrow C_6H_{12}O_6 + 6\ O_2 + 6\ H_2O$$

Note that water appears on both sides of the equation; it is both consumed as a reactant and generated as a product in photosynthesis.

The water-splitting reaction probably developed even before oxygen-consuming organisms appeared, evolving first in photosynthetic bacteria that resembled present-day cyanobacteria. The oxygen released by the reaction profoundly changed the atmosphere, allowing aerobic respiration, in which oxygen serves as the final acceptor for electrons removed in cellular oxidations. The existence of all animals depends on the oxygen provided by the water-splitting reaction of photosynthesis.

Glucose is often shown as the only product of photosynthesis. Glucose is the major product of photosynthesis; other monosaccharides, disaccharides, polysaccharides, lipids, and amino acids are also produced. In fact, all the organic molecules of plants are assembled as direct or indirect products of photosynthesis.

Originally, investigators thought that the O_2 released by photosynthesis came from the CO_2 entering the process. The fact that it comes from water was first established in the 1940s, when researchers used a heavy isotope of oxygen, ^{18}O, to trace the pathways of the atoms through photosynthesis. A substance containing heavy ^{18}O can be distinguished readily from the same substance containing the normal isotope, ^{16}O. When a photosynthetic organism was supplied with water containing ^{18}O, the heavy isotope showed up in the O_2 given off in photosynthesis. However, if the organisms were supplied with carbon dioxide con-

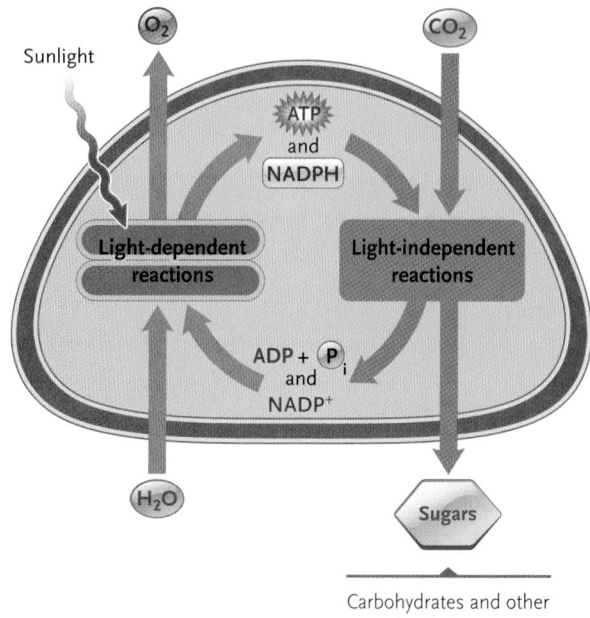

Figure 9.2

The light-dependent and light-independent reactions of photosynthesis, and their interlinking reactants and products. Both series of reactions occur in the chloroplasts of plants and algae.

taining ^{18}O, the heavy isotope showed up in the carbohydrate and water molecules assembled during the reactions—but not in the oxygen gas. This experiment, and other similar experiments using different isotopes, revealed where each atom of the reactants end up in products:

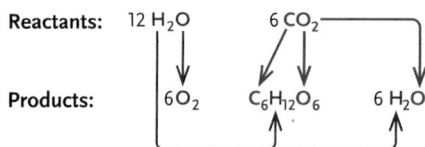

Figure 9.2 summarizes the relationships of the light-dependent and light-independent reactions. Notice that the ATP and NADPH produced by the light-dependent reactions, along with CO_2, are the reactants of the light-independent reactions. The ADP, inorganic phosphate (P_i), and $NADP^+$ produced by the light-independent reactions, along with H_2O, are the reactants for the light-dependent reactions. The light-dependent and light-independent reactions thus form a cycle in which the net inputs are H_2O and CO_2, and the net outputs are organic molecules and O_2.

In Eukaryotes, Photosynthesis Takes Place in Chloroplasts

In eukaryotes, the photosynthetic reactions take place in the chloroplasts of plants and algae; in cyanobacteria, the reactions are distributed between the plasma membrane and the cytosol.

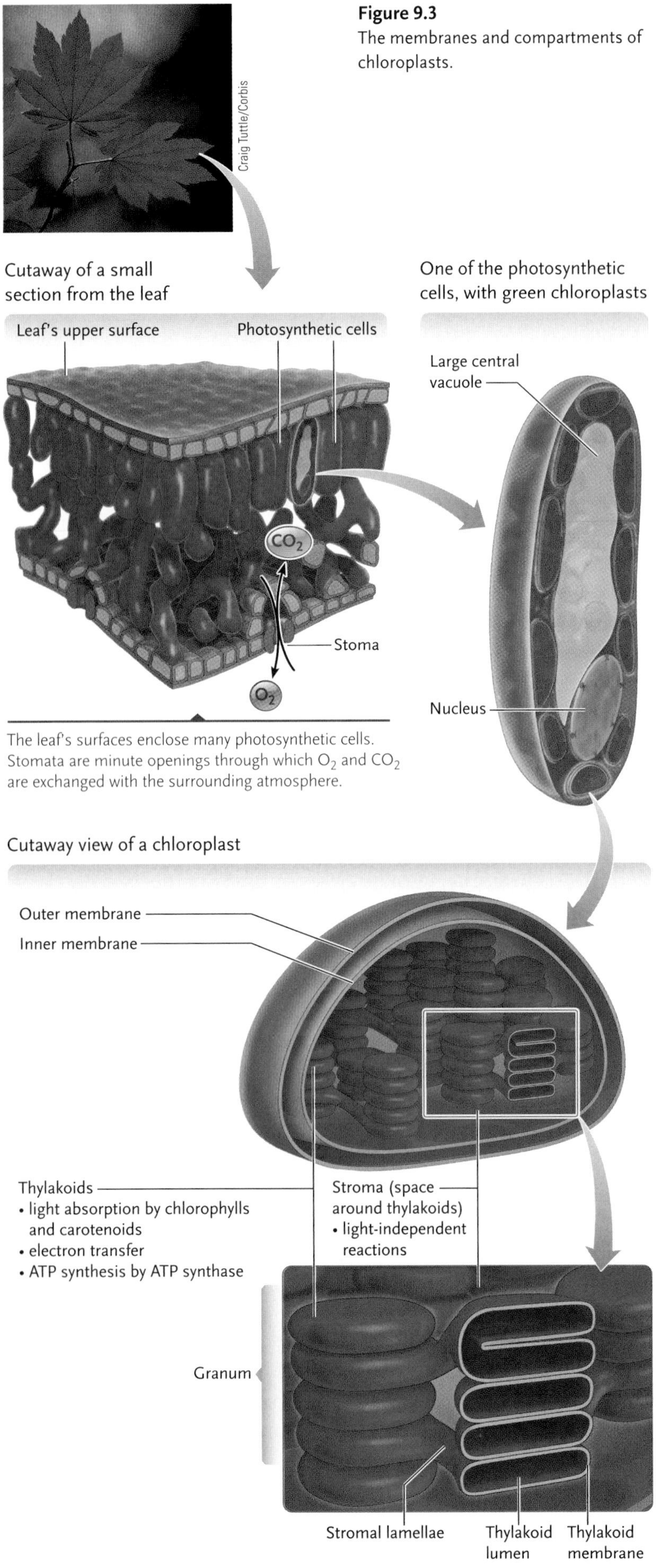

Figure 9.3
The membranes and compartments of chloroplasts.

Craig Tuttle/Corbis

Chloroplasts from individual algal and plant groups differ in structural details. The chloroplasts of plants and green algae are formed from three membranes that enclose three compartments inside the organelles (**Figure 9.3;** chloroplast structure is also described in Section 5.4). An *outer membrane* covers the entire surface of the organelle. An *inner membrane* lies just inside the outer membrane. Between the outer and inner membranes is an *intermembrane compartment*. The fluid within the inner membrane is the *stroma.* Within the stroma is the third membrane system, the *thylakoid membranes,* which form flattened, closed sacs called *thylakoids.* The space enclosed by a thylakoid is called the *thylakoid lumen.*

In green algae and higher plants, thylakoids are arranged into stacks called *grana* (singular, *granum;* shown in Figure 9.3). The grana are interconnected by flattened, tubular membranes called *stromal lamellae.* The stromal lamellae probably link the thylakoid lumens into a single continuous space within the stroma.

The thylakoid membranes and stromal lamellae house the molecules that carry out the light-dependent reactions of photosynthesis, including the pigments, electron transfer carriers, and ATP synthase enzymes for ATP production. The light-independent reactions are concentrated in the stroma.

In higher plants, the CO_2 required for photosynthesis diffuses to cells containing chloroplasts after entering the plant through *stomata,* minute "air valves" in leaves and stems. The O_2 produced in photosynthesis diffuses from the cells and exits through the stomata, as does water. Water and minerals required for photosynthesis are absorbed by the roots and transported to cells containing chloroplasts through tubular conductive vessels; the organic products of photosynthesis are distributed to all parts of the plant by other vessels (see Chapter 32).

Cutaway of a small section from the leaf

Leaf's upper surface   Photosynthetic cells

CO_2

Stoma

O_2

The leaf's surfaces enclose many photosynthetic cells. Stomata are minute openings through which O_2 and CO_2 are exchanged with the surrounding atmosphere.

One of the photosynthetic cells, with green chloroplasts

Large central vacuole

Nucleus

Cutaway view of a chloroplast

Outer membrane
Inner membrane

Thylakoids
• light absorption by chlorophylls and carotenoids
• electron transfer
• ATP synthesis by ATP synthase

Stroma (space around thylakoids)
• light-independent reactions

Granum

Stromal lamellae   Thylakoid lumen   Thylakoid membrane

STUDY BREAK

1. What are the two stages of photosynthesis?
2. In which organelle does photosynthesis take place in plants? Where in that organelle are the two stages of photosynthesis carried out?

9.2 The Light-Dependent Reactions of Photosynthesis

In this section we discuss the light-dependent reactions (also referred to more simply as the light reactions), in which light energy is converted to chemical energy. The light-dependent reactions involve two main processes: (1) light absorption and (2) synthesis of NADPH and ATP. We will describe these processes in turn. Through the discussion, it may be useful for you to refer to the summary Figure 9.2 periodically to keep the bigger picture in perspective.

Electrons in Pigment Molecules Absorb Light Energy in Photosynthesis

The first process in photosynthesis is light absorption. What is light? Visible light is a form of radiant energy. It makes up a small part of the **electromagnetic spectrum (Figure 9.4),** which ranges from radio waves to gamma rays. The various forms of electromagnetic radiation differ in *wavelength*—the horizontal distance between the crests of successive waves. Radio waves have wavelengths in the range of 10 meters to hundreds of kilometers, and gamma rays have wavelengths in the range of one hundredth to one millionth of a nanometer. The average wavelength for an FM radio station, for example, is 3 m. Generally, the shorter the wavelength, the greater the energy of the radiation.

The radiation we detect as visible light has wavelengths between about 700 nm, seen as red light, and 400 nm, seen as blue light. We see the entire spectrum of wavelengths from 700 to 400 nm, combined together, as white light. Although radiated in apparently continuous beams that follow a wave path through space, the energy of light interacts with matter in discrete units called *photons*. Each photon contains a fixed amount of energy that is inversely proportional to its wavelength: the shorter the wavelength, the greater the energy of a photon.

In photosynthesis, light is absorbed by molecules of green pigments called **chlorophylls** (*chloros* = yellow-green; *phyllon* = leaf) and yellow-orange pigments called **carotenoids** (*carota* = carrot). These pigment molecules are embedded in the thylakoid membranes of chloroplasts.

Pigment molecules such as chlorophyll appear colored to an observer because they absorb the energy of visible light at certain wavelengths and transmit or reflect other wavelengths. The color of a pigment is produced by the transmitted or reflected light. Plants look green because chlorophyll absorbs blue and red light most strongly and transmits or reflects most of the wavelengths in between; we see the reflected light as green. This green light, as demonstrated by Engel-

a. Visible spectrum

400-nm wavelength

700-nm wavelength

Barker Blankenship/FPG/Getty Images

Figure 9.4

The electromagnetic spectrum and the visible wavelengths used as the energy source for photosynthesis. **(a)** Examples of wavelengths, showing the difference between the longest and shortest wavelengths of visible light. **(b)** The entire electromagnetic spectrum, ranging from gamma rays to radio waves; visible light and the wavelengths used for photosynthesis occupy only a narrow band of the spectrum.

b. Range of the electromagnetic spectrum

The shortest, most energetic wavelengths

Most of the radiation that reaches Earth's surface is in this range.

Heat that escapes into space from Earth's surface is in this range.

The longest, lowest-energy wavelengths

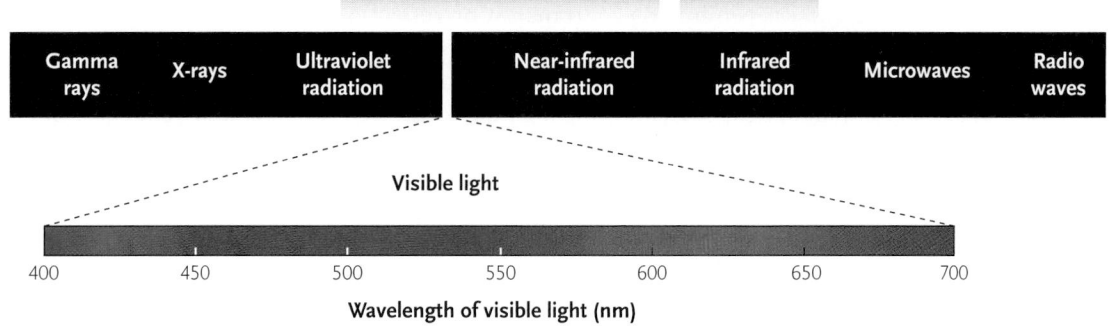

Visible light

Wavelength of visible light (nm)

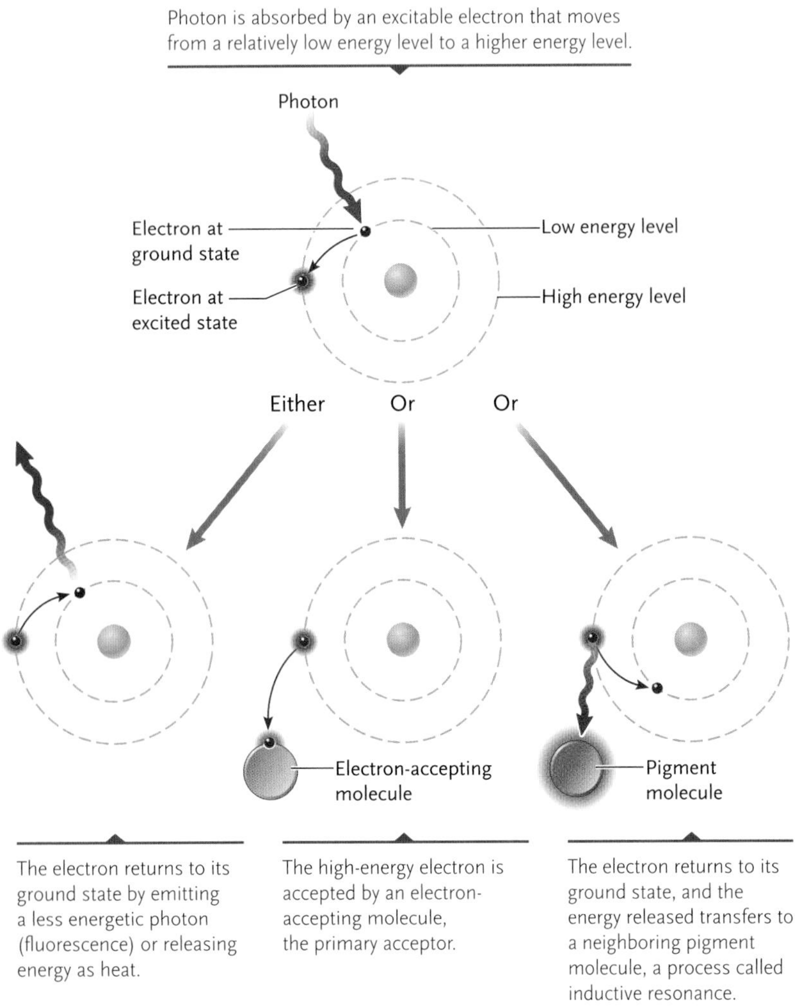

Photon is absorbed by an excitable electron that moves from a relatively low energy level to a higher energy level.

Photon

Electron at ground state

Low energy level

Electron at excited state

High energy level

Either   Or   Or

Electron-accepting molecule

Pigment molecule

The electron returns to its ground state by emitting a less energetic photon (fluorescence) or releasing energy as heat.

The high-energy electron is accepted by an electron-accepting molecule, the primary acceptor.

The electron returns to its ground state, and the energy released transfers to a neighboring pigment molecule, a process called inductive resonance.

Figure 9.5
Alternative effects of light absorbed by a pigment molecule.

mann's experiment described in the introduction to this chapter, is the combination of wavelengths *not used* by the plants in photosynthesis.

Light is absorbed in a pigment molecule by excitable electrons occupying certain energy levels (shells) in the atoms (see Section 2.2). When not absorbing light, these electrons are at a relatively low energy level known as the *ground state*. If an electron in the pigment absorbs the energy of a photon, it jumps to a higher energy level farther from the atomic nucleus called the *excited state* **(Figure 9.5)**. The difference in energy level between the ground state and the excited state is equivalent to the energy of the photon of light that was absorbed.

One of three events then occurs, depending on the atom and other molecules in the vicinity. The electron may return to its ground state, releasing its energy either as heat or as an emission of light of a longer wavelength than the absorbed light, a process called *fluorescence*. Alternatively, the high-energy electron is transferred from the pigment molecule to a nearby electron-accepting molecule called a *primary acceptor*. In green algae and plants, chlorophyll is the

pigment molecule from which excited electrons transfer to stable orbitals in acceptor molecules. In the transfer, chlorophyll is oxidized because it loses an electron, and the primary acceptor is reduced because it gains an electron. In the third way, the energy of the excited electron, but not the electron itself, is transferred to a neighboring pigment molecule, a process called *inductive resonance*. This transfer excites the second molecule, while the first molecule returns to its ground state. Very little energy is lost in this energy transfer.

Chlorophylls and Carotenoids Cooperate in Light Absorption

Chlorophylls are the major photosynthetic pigments in plants, green algae, and cyanobacteria. They absorb photons and transfer excited electrons to stable orbitals in primary acceptors. Closely related molecules, the *bacteriochlorophylls,* carry out the same functions in other photosynthetic bacteria. Carotenoids absorb light energy and pass it on to the chlorophylls by inductive resonance in both eukaryotes and bacteria. Chlorophylls and carotenoids are bound to proteins in photosynthetic membranes.

Molecules of the chlorophyll family **(Figure 9.6a)** have a carbon ring structure, to which is attached a long, hydrophobic side chain. A magnesium atom is bound at the center of the ring structure. The most important kind of chlorophyll is chlorophyll *a,* which is found in plants, green algae, and cyanobacteria. A second kind, chlorophyll *b,* is found only in plants and green algae. Chlorophyll *a* and chlorophyll *b* differ only in one side group attached to a carbon of the ring structure (shown in Figure 9.6a).

A chlorophyll molecule contains a network of electrons capable of absorbing light (shaded in orange in Figure 9.6a). The amount of light absorbed at each wavelength is represented by a curve called an **absorption spectrum**, in which the height of the curve at any wavelength indicates the amount of light absorbed. **Figure 9.7a** shows the absorption spectra for chlorophylls *a* and *b.*

The carotenoids are built on a long backbone that typically contains 40 carbon atoms **(Figure 9.6b)**. Carotenoids expand the range of wavelengths used for photosynthesis because they absorb different wavelengths than chlorophyll does. Carotenoids transmit or reflect other wavelengths in combinations that appear yellow, orange, red, or brown, depending on the type of carotenoid. The carotenoids contribute to the red, orange, and yellow colors of vegetables and fruits and to the brilliant colors of autumn leaves, in which the green color is lost when the chlorophylls break down.

The light absorbed by the carotenoids and chlorophylls, acting in combination, drives the reactions

a. Chlorophyll structure

b. Carotenoid structure

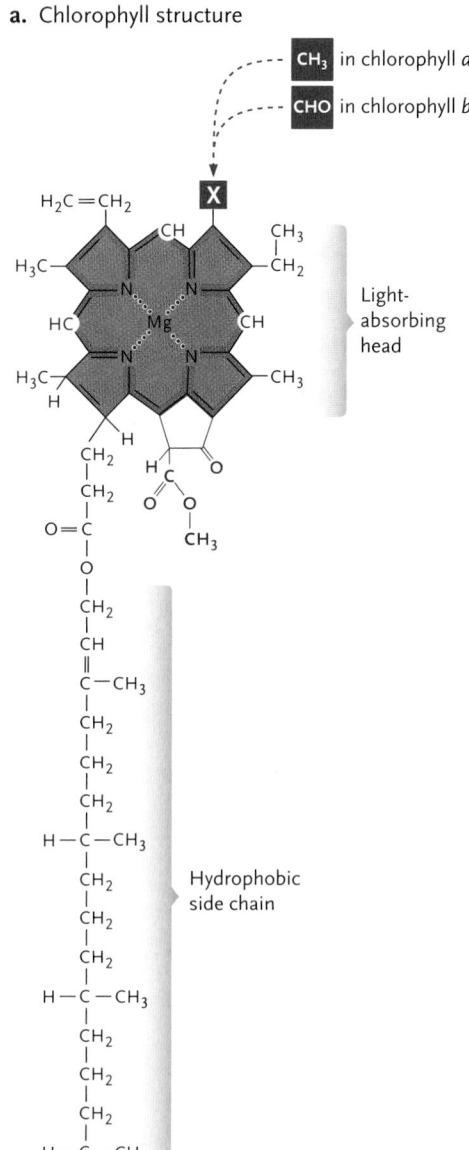

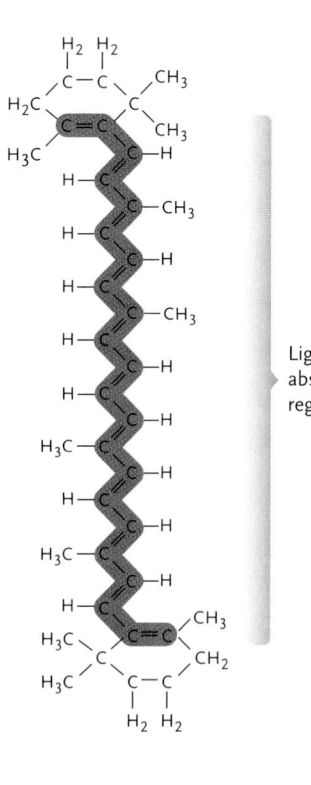

Figure 9.6
Pigment molecules used in photosynthesis. **(a)** Chlorophylls *a* and *b*, which differ only in the side group attached at the X. Light-absorbing electrons are distributed among the bonds shaded in orange. The chlorophylls are similar in structure to the cytochromes, which occur in both the chloroplast and mitochondrial electron transfer systems. **(b)** Carotenoids. The electrons absorbing light are distributed in a series of alternating double and single bonds in the backbone of these pigments.

nance to the specialized chlorophyll *a* molecules that are directly involved in transforming light into chemical energy.

The Photosynthetic Pigments Are Organized into Photosystems in Chloroplasts

The light-absorbing pigments are organized with proteins and other molecules into large complexes called **photosystems (Figure 9.8),** which are embedded in thylakoid membranes and stromal lamellae. The photosystems are the sites at which light is absorbed and converted into chemical energy.

Plants, green algae, and cyanobacteria have two types of these complexes, called **photosystems I** and **II,** which carry out different parts of the light-dependent reactions. Each consists of two closely associated components: an **antenna complex** (also called a *light-harvesting complex*), and a **reaction center.**

The antenna complex contains an aggregate of many chlorophyll pigments and a number of carotenoid pigments. The chlorophyll molecules are anchored in the complex by being bound to specific membrane proteins. In this form, they are efficiently arranged to optimize the capture of light energy.

The reaction center contains a pair of specialized chlorophyll *a* molecules complexed with proteins. The specialized chlorophyll *a* at the reaction center of pho-

of photosynthesis. Plotting the effectiveness of light at each wavelength in driving photosynthesis produces a graph called the **action spectrum** of photosynthesis (**Figure 9.7b** shows the action spectrum of higher plants). The action spectrum is usually determined by measuring the amount of O_2 released by photosynthesis at different wavelengths of visible light, as Engelmann did indirectly in the experiment described in the introduction to this chapter (compare Figures 9.1 and 9.7b).

In all eukaryotes, a specialized chlorophyll *a* molecule passes excited electrons to stable orbitals in the primary acceptor. Other chlorophyll molecules, along with carotenoids, act as *accessory pigments* that pass their energy to chlorophyll *a*. Light energy absorbed by the entire collection of chlorophyll and carotenoid molecules in chloroplasts is passed by inductive reso-

a. The absorption spectra of chlorophylls *a* and *b* and carotenoids

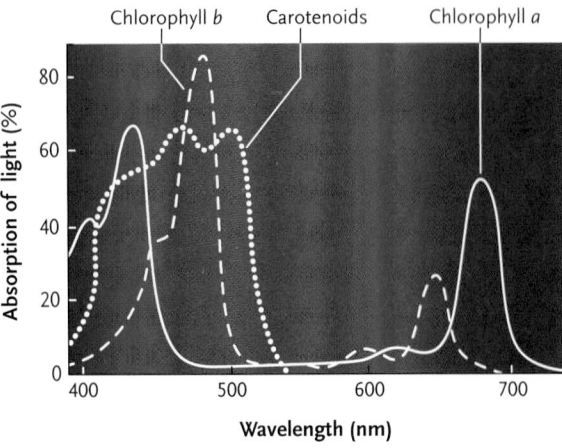

b. The action spectrum in higher plants, representing the combined effects of chlorophylls and carotenoids

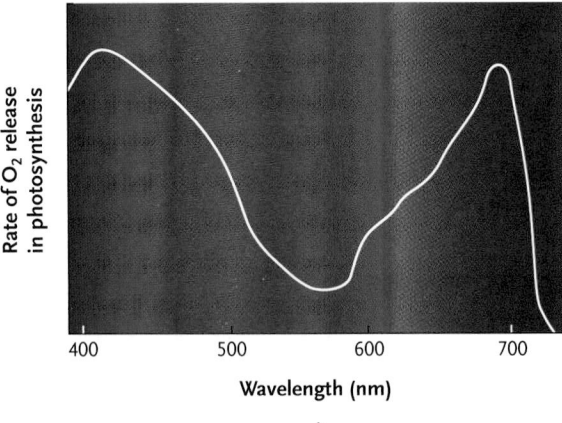

The peaks in the action spectrum are typically broader than those for the individual pigments, reflecting both their combined effects and changes in the absorption spectra of individual pigments by their combination with proteins in chloroplasts.

Figure 9.7
The absorption spectra of the photosynthetic pigments **(a)** and the action spectrum of photosynthesis **(b)** in higher plants. The absorption spectra in **(a)** were made from pigments that were extracted from cells and purified.

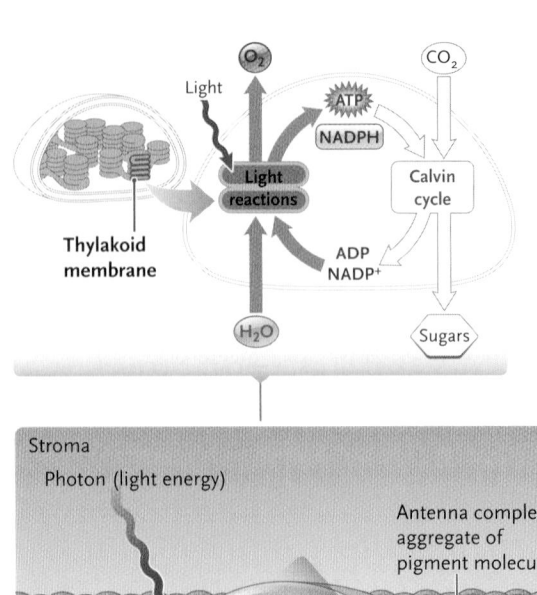

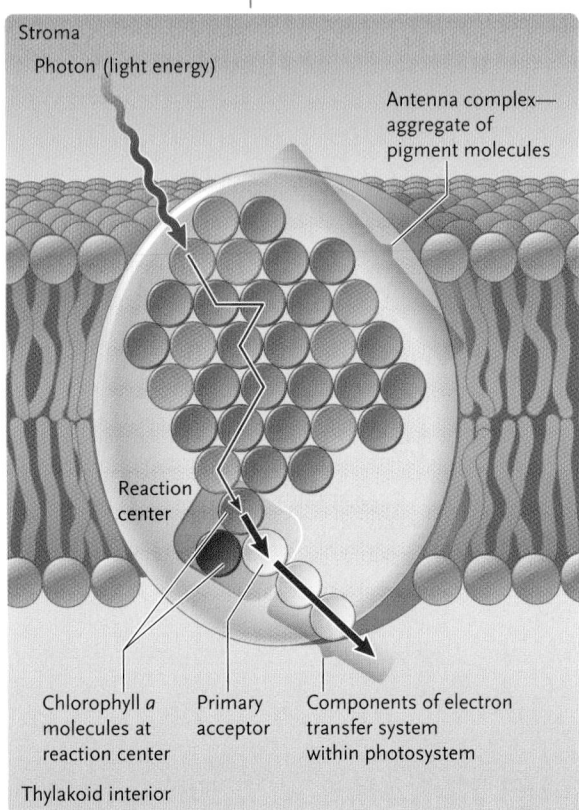

Figure 9.8
Major components of a photosystem: a group of pigments forming an antenna complex (light-harvesting complex) and a reaction center. Some components of the electron transfer system are located within the photosystems. Light energy absorbed anywhere in the antenna complex is conducted by inductive resonance to specialized chlorophyll *a* molecules in the reaction center. The absorbed light is converted to chemical energy when an excited electron is transferred to a stable orbital in a primary acceptor, also in the reaction center. High-energy electrons are conducted out of the photosystem by the components of the electron transfer system. The blue arrows show the path of energy flow.

tosystem I is called *P700* (P = pigment) because it absorbs light optimally at a wavelength of 700 nm. The reaction center of photosystem II contains a different specialized chlorophyll *a, P680,* which absorbs light optimally at a wavelength of 680 nm. P700 and P680 are structurally identical to other chlorophyll *a* molecules; their specific light absorption patterns result from interactions with particular proteins in the photosystems.

Light energy in the form of photons is absorbed by the pigment molecules of the antenna complex. This absorbed light energy reaches P700 and P680 in the reaction center by inductive resonance. On arrival, the energy is captured quickly in the form of an excited electron passed to a stable orbital in a primary acceptor molecule. That electron is passed to

the electron transfer system, which has some components within the photosystems and other components separate.

The electron transfer system within a photosystem carries electrons away from the primary acceptor. Photosystem II, in addition, is closely linked to a group of enzymes that carries out the initial reaction splitting water into electrons, protons, and oxygen.

Electrons Flow from Water to Photosystem II to Photosystem I to NADP$^+$ Leading to the Synthesis of NADPH and ATP

In the second main process of the light-dependent reactions, the electrons obtained from the splitting of water (two electrons per molecule of water; see Section 9.1) are used for the synthesis of NADPH and ATP. These electrons, which were pushed to higher levels by the energy from light, pass through an electron transfer system consisting of a series of electron carriers that are alternately reduced and oxidized as they pick up and release electrons in sequence. The electron carriers are embedded in a thylakoid membrane in eukaryotes and in the plasma membrane in prokaryotes.

As in all electron transfer systems, the electron carriers of the photosynthetic system consist of nonprotein organic groups that pick up and release the electrons traveling through the system. The carriers include the same types that act in mitochondrial electron transfer—cytochromes, quinones, and iron-sulfur centers (discussed in Section 8.4). Most of the carriers are organized with proteins into larger complexes, which are distributed among the thylakoid membranes and stromal lamellae of chloroplasts.

The electron carriers of photosynthesis are arranged in a chain first deduced by Robert Hill and Fay Bendall of Cambridge University **(Figure 9.9)**. Electrons from water first flow through photosystem II, becoming excited to a higher energy level in P680 through energy absorbed from light. The electrons then flow "downhill" in energy level through an electron transfer system connecting photosystems II and I. (Note: Photosystem I is so named because it was discovered first; the systems were given their numbers before their order of use in the pathway was worked out.) The electron transfer system consists of a pool of molecules of the electron carrier plastoquinone, a cytochrome complex, and the protein plastocyanin. The electrons release free energy at each transfer from a donor to an acceptor molecule as they pass through the system; some of this energy is used to create a gradient of H$^+$ across the membrane. The gradient provides the energy source for ATP synthesis, just as it does in mitochondria.

The electrons then pass to photosystem I, where they are excited a second time in P700 through energy absorbed from light. The high-energy electrons enter a short electron transfer system leading to the final acceptor of the chloroplast system, NADP$^+$. The enzyme NADP$^+$ reductase reduces NADP$^+$ to NADPH, using two electrons and two protons from the surrounding water solution and releasing one proton.

This pathway is frequently called **noncyclic electron flow** because electrons travel in a one-way direction from H$_2$O to NADP$^+$; it is sometimes called the *Z scheme* because of the sawtooth changes in electron energy level.

NADPH has the same primary role in all eukaryotes—to deliver high-energy electrons to synthetic reactions that require a reduction. In photosynthesis, the reaction requiring a reduction, the fixation of CO$_2$, takes place in the second stage of photosynthesis, the light-independent reactions.

Figure 9.10 shows how the electron transfer and ATP synthesis systems for the light-dependent reactions are organized in the thylakoid membrane. Let us follow the noncyclic electron pathway using this figure.

1. **Excitation in P680.** Electrons entering the pathway from the water-splitting reaction system associated with photosystem II are accepted one at a time by a P680 chlorophyll *a* in the reaction center of photosystem II. As P680 accepts the electrons, they are raised to the excited state, using energy passed to the reaction center from the light-absorbing pigment molecules in the antenna complex. The excited electrons are immediately transferred to the primary acceptor of photosystem II, which is a modified form of chlorophyll *a* without magnesium.

2. **Movement to the Plastoquinone Pool.** From the primary acceptor the electrons flow through a short chain of carriers within the photosystem and then transfer to a *plastoquinone,* which forms the first carrier of the electron transfer system linking photosystem II to photosystem I. The plastoquinones, analogous in structure and function to the ubiquinones of the mitochondrial electron system (shown in Figure 8.14), form a "pool" of molecules within the thylakoid membranes.

3. **H$^+$ Pumping by Plastoquinones and the Cytochrome Complex.** Electrons then pass from the plastoquinones to the next carrier, the *cytochrome complex,* in a structure that is closely related to complex III of the mitochondrial electron transfer system. As it accepts and releases electrons, the cytochrome complex pumps H$^+$ from the stroma into the thylakoid lumen. Those protons drive ATP synthesis (see step 7).

4. **Shuttling by Plastocyanin.** From the cytochrome complex, electrons pass to the mobile carrier *plastocyanin,* which shuttles electrons between the cytochrome complex and photosystem I.

5. **The Second Excitation in P700.** Electrons pass from plastocyanin to a P700 chlorophyll *a* in the reaction center of photosystem I, where they are excited to high energy levels again by absorbing more light energy. The excited electrons are transferred from P700 to the primary acceptor of this

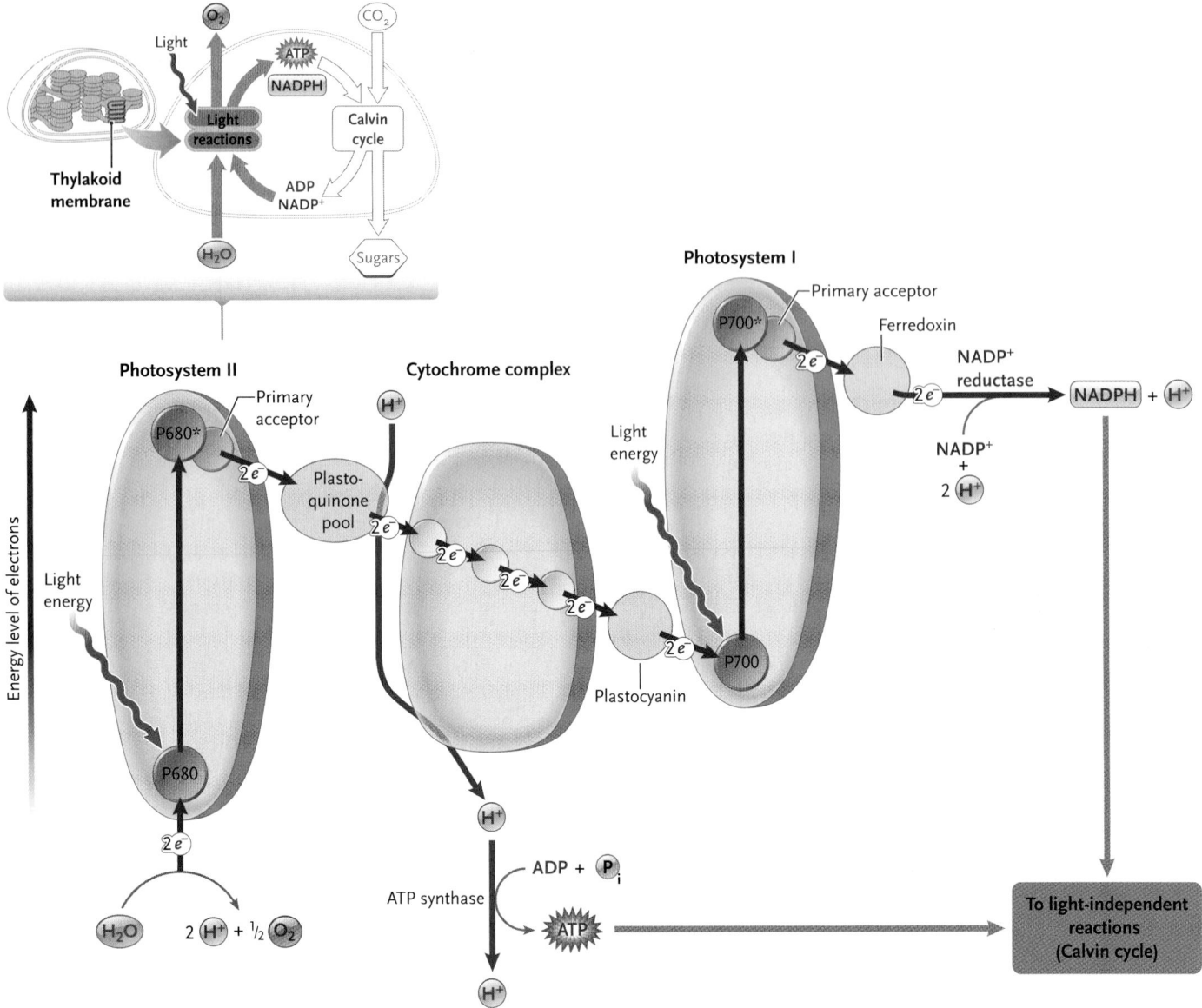

Figure 9.9

The pathway of the light-dependent reactions, noncyclic electron flow. Electrons (e^- in the figure) derived from water absorb light energy in photosystem II and, after transfer to a primary acceptor, travel through the electron transfer system to reach photosystem I. As they travel, some of their energy is tapped off to drive ATP synthesis. In photosystem I, the electrons absorb a second boost of energy and then, after transfer to a primary acceptor, are delivered to the final electron acceptor, $NADP^+$. As $NADP^+$ accepts the electrons, it combines with two protons to form NADPH and a proton. The asterisks indicate the excited forms of P680 and P700.

photosystem, formed by another specialized chlorophyll *a* molecule.

6. **Transfer to $NADP^+$ by Ferredoxin.** After passage through a short sequence of carriers within photosystem I, the electrons are transferred to *ferredoxin,* an iron-sulfur protein that acts as another mobile electron carrier of the pathway. The ferredoxin transfers the electrons, still at very high energy levels, to $NADP^+$, the final acceptor of the noncyclic pathway. $NADP^+$ is reduced to NADPH by $NADP^+$ reductase.

7. **ATP Synthesis.** Proton pumping by the plastoquinones and the cytochrome complex, as described in step 3, creates a concentration gradient of H^+ with the high concentration within the thylakoid lumen and the low concentration in the stroma. The gradient is enhanced by the addition of two protons to the lumen for each water molecule split, and by the removal of one proton from the stroma for each NADPH molecule synthesized. Because protons carry a positive charge, an electrical gradient forms across the thylakoid mem-

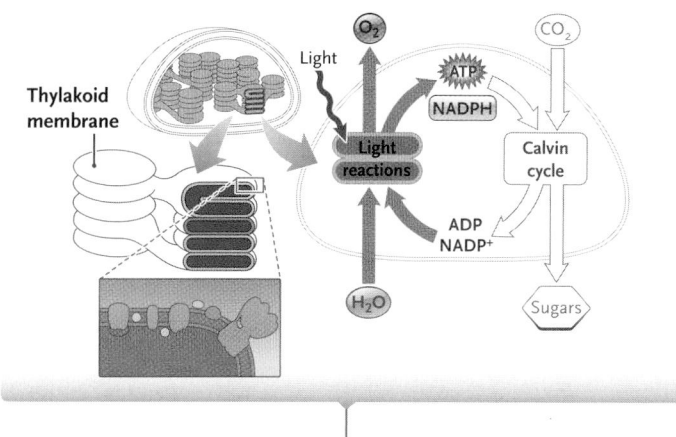

Figure 9.10

The components of the electron transfer and ATP synthesis systems in the thylakoid membrane, illustrating the synthesis of NADPH and ATP by the noncyclic electron flow pathway. The electron transfer system is organized into four complexes and two individual electron carriers. Photosystems II and I, both of which are embedded in the membrane, form two of the complexes. One of the remaining complexes is the membrane-embedded cytochrome complex. The other is ferredoxin, which is on the stromal surface of the membrane alongside the membrane-embedded $NADP^+$ reductase, which catalyzes the reduction of $NADP^+$ to NADPH. Plastoquinone is dissolved as a pool of molecules in the thylakoid membrane interior; plastocyanin is located on the membrane surface facing the thylakoid lumen. The enzyme for ATP synthesis by chemiosmosis, ATP synthase, is embedded in the same membrane.

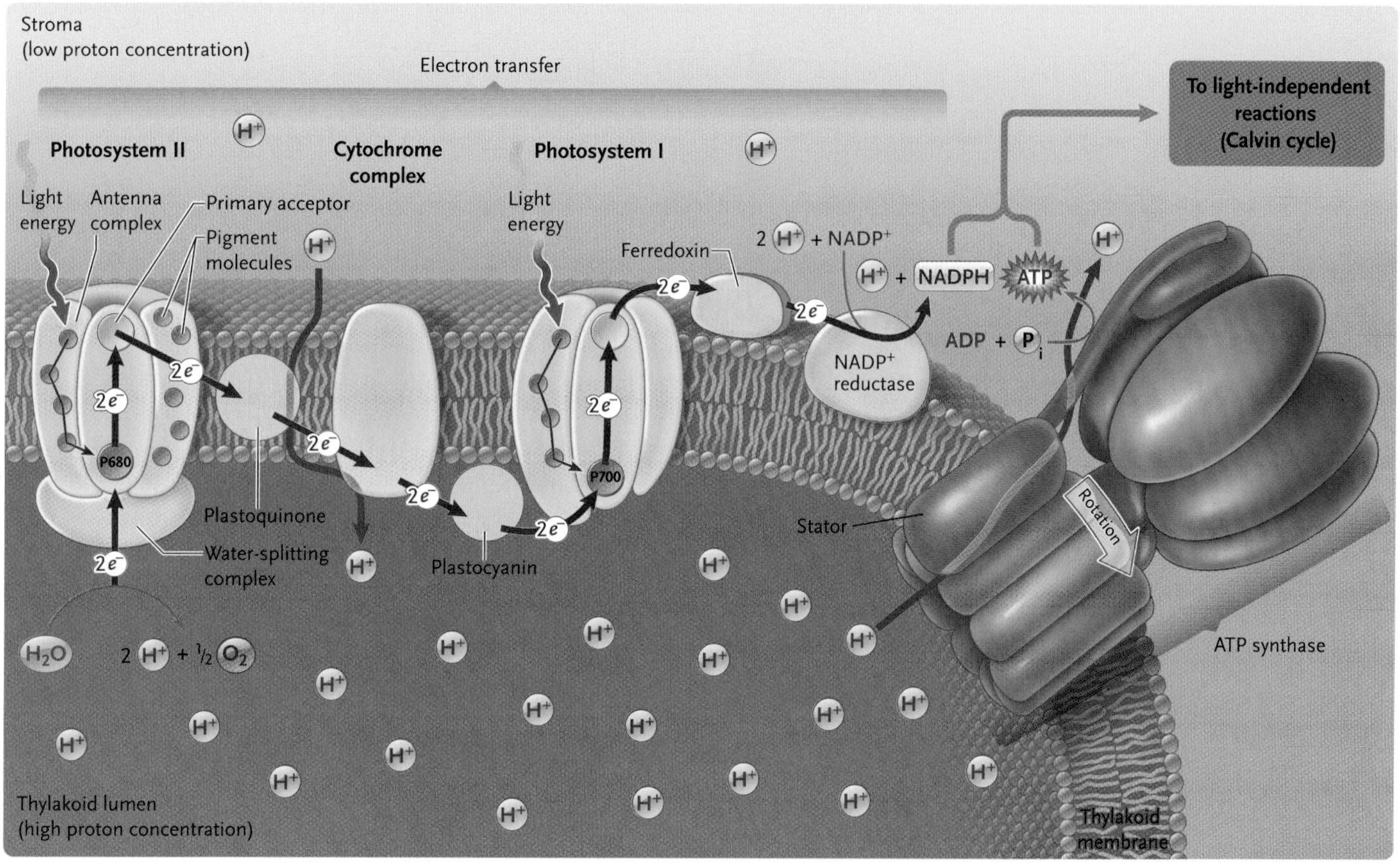

brane, with the lumen more positively charged than the stroma. The combination of a proton gradient and a voltage gradient across the membrane produces stored energy known as the *proton-motive force* (also discussed in Section 8.4), which contributes energy for ATP synthesis by ATP synthase. Just as for the mitochondrial ATP synthase, the chloroplast enzyme is embedded in the same membranes as the electron transfer system. Protons flow through a membrane channel from the thylakoid lumen to the stroma along their concentration gradient (see Figure 9.10). Free energy is released as H^+ moves through the channel, and it powers synthesis of ATP from ADP and P_i by the ATP synthase. This process of using an H^+ gradient to power ATP synthesis, called *chemiosmosis,* is the same as that used for ATP synthesis in mitochondria (see Section 8.4).

The overall yield of the noncyclic electron flow pathway is one molecule of NADPH and one molecule of ATP for each pair of electrons produced from the splitting of water. The synthesis of ATP coupled to the transfer of electrons energized by photons of light is called **photophosphorylation.** This process is analogous to oxidative phosphorylation in mitochondria (see Section 8.4), except that in chloroplasts light provides the energy for establishing the proton gradient.

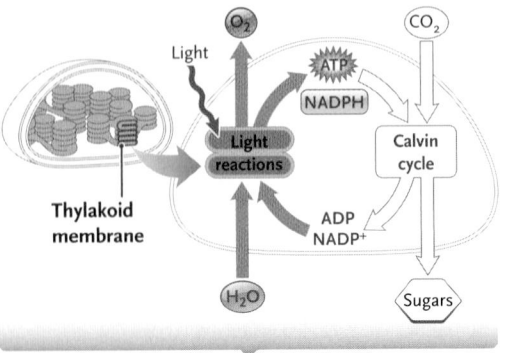

Figure 9.11

Cyclic electron flow around photosystem I. Electrons move in a circular pathway from ferredoxin back to the cytochrome complex, then to plastocyanin, through photosystem I, and back to ferredoxin again. The cycle pumps additional H⁺ each time electrons flow through the cytochrome complex. The H⁺ drive ATP synthesis as described for the noncyclic flow pathway.

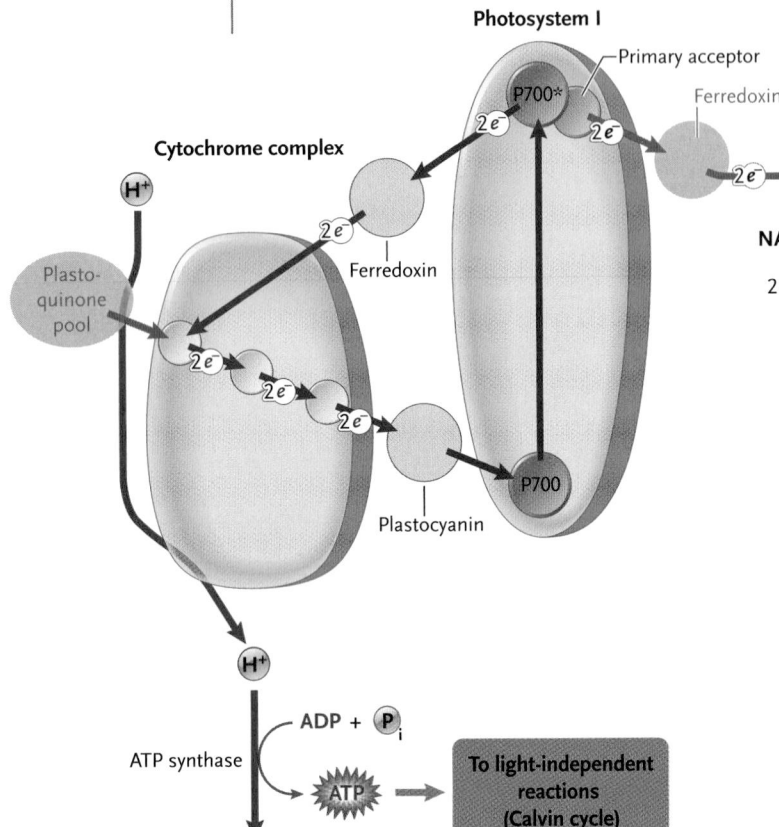

transfer system are distributed among both thylakoids and stromal lamellae.

Electrons Can Also Drive ATP Synthesis by Flowing Cyclically around Photosystem I

Photosystem I can work independently of photosystem II in a circular process called **cyclic electron flow (Figure 9.11).** In this process, electrons pass through the cytochrome complex and plastocyanin to the P700 chlorophyll *a* in the reaction center of photosystem I where they are excited by light energy. The electrons then flow from photosystem I to ferredoxin, but rather than being used for NADP⁺ reduction by NADP⁺ reductase, they flow back to the cytochrome complex. The electrons again pass to plastocyanin and on to photosystem I where they receive another energy boost, and so the cycle continues. Each time electrons flow around the cycle, more H⁺ is pumped across the thylakoid membranes, driving ATP synthesis in the way already described. The net result of cyclic electron flow is that the energy absorbed from light is converted into the chemical energy of ATP *without* reduction of NADP⁺ to NADPH.

The cyclic electron flow pathway is an important part of photosynthesis. The light-independent reactions require more ATP molecules than NADPH molecules, and the additional ATP molecules are provided by cyclic electron flow. Other energy-requiring reactions in the chloroplast also depend on ATP produced by cyclic electron flow.

Comparing the noncyclic pathway with the mitochondrial electron transfer system (shown in Figure 8.14) reveals that the pathway from the plastoquinones through plastocyanin in chloroplasts is essentially the same as the pathway from the ubiquinones through cytochrome *c* in mitochondria. The similarities between the two pathways indicate that the electron transfer system is a very ancient evolutionary development that became adapted to both photosynthesis and oxidative phosphorylation.

The elements of the noncyclic pathway are not located in fixed, organized assemblies as Figure 9.10 might suggest. Instead, photosystem II is located almost exclusively in thylakoid membranes, in regions where one thylakoid membrane is fused to the next in the stacks of grana; photosystem I is located primarily in stromal lamellae. Other components of the electron

Experiments with Chloroplasts Helped Confirm the Synthesis of ATP by Chemiosmosis

Our present understanding of the connection between electron transfer and ATP synthesis was first proposed for mitochondria in Mitchell's chemiosmotic hypothesis (discussed in Section 8.4). Several experiments have shown that the same mechanism operates in chloroplasts.

One of these experiments was carried out in 1966 by Andre T. Jagendorf and Ernest Uribe at Johns Hopkins University **(Figure 9.12)**. The two scientists placed a solution containing intact chloroplasts (isolated from cells by cell fractionation: see Figure 8.5) in darkness, thereby eliminating light absorption and electron transfer as a source of energy for photosynthesis. They next created a surplus of H^+ inside the chloroplasts by adding an organic acid to the solution, which lowered the pH of the solution inside the stroma and thylakoids to pH 4. The chloroplasts, still in darkness, were then transferred to a second solution at a basic pH (pH 8). This process created an H^+ gradient, high inside the thylakoid lumen and low in the stroma. As H^+ moved from the thylakoid lumen to the stroma in response to the gradient, ATP was synthesized in the chloroplasts. Because the darkness eliminated electron transfer as an energy source, the observed ATP synthesis could have been powered only by the H^+ gradient.

Our description of photosynthesis to this point shows how the light-dependent reactions generate NADPH and ATP, which provide the reducing power and chemical energy required to produce organic molecules from CO_2. The next section follows NADPH and ATP through the light-independent reactions and shows how the organic molecules are produced.

STUDY BREAK

1. What is the difference between the chlorophyll *a* molecules in the antenna complexes and the chlorophyll *a* molecules in the reaction centers of the photosystems?
2. How is NADPH made in the noncyclic electron flow pathway?
3. What is the difference between the noncyclic electron flow pathway and the cyclic electron flow pathway?

9.3 The Light-Independent Reactions of Photosynthesis

The electrons carried from the light-dependent reactions by NADPH retain much of the energy absorbed from sunlight. These electrons provide the reducing

Figure 9.12 Experimental Research

Demonstration That an H⁺ Gradient Drives ATP Synthesis in Chloroplasts

QUESTION: Does chemiosmosis power ATP synthesis by a proton gradient in chloroplasts?

EXPERIMENT: Jagendorf and Uribe placed chloroplasts in darkness in an acidic medium, which allowed H^+ to penetrate inside, including into the thylakoid lumen.

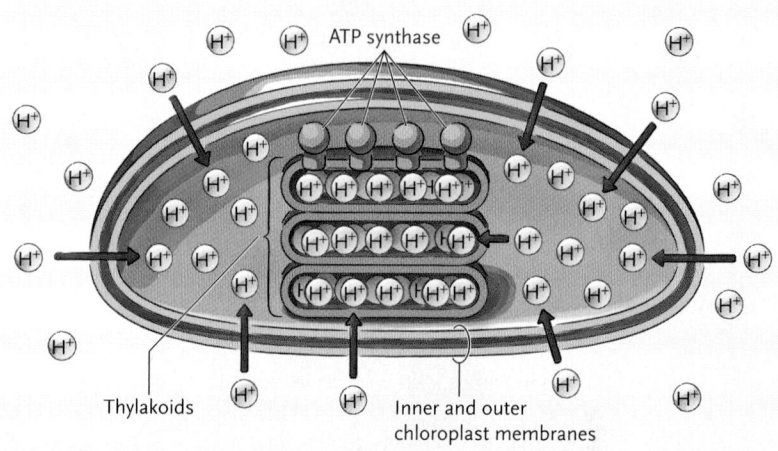

The chloroplasts were then placed in a medium at a basic pH, causing H^+ to move out of the thylakoid lumen in response to the gradient.

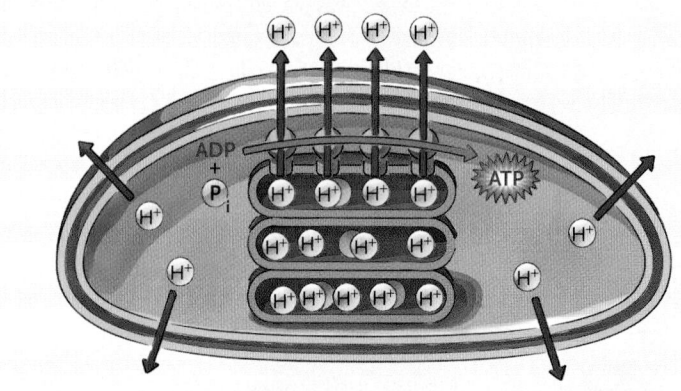

RESULT: ATP was synthesized by the chloroplast.

CONCLUSION: Because chloroplasts in darkness cannot use electron transfer as an energy source, chemiosmosis must power ATP synthesis by an H^+ gradient.

power required to fix CO_2 into carbohydrates and other organic molecules in the light-independent reactions. Additional energy for the light-independent reactions is supplied by the ATP generated in the light-dependent reactions. The reactions using NADPH and ATP to fix CO_2 occur in a circuit known as the **Calvin cycle,** named for its discoverer, Melvin

Basic Research: Two-Dimensional Paper Chromatography and the Calvin Cycle

The first significant progress in unraveling the light-independent reactions was made in the 1940s, when newly developed radioactive compounds became available to biochemists. One substance, CO_2 labeled with the radioactive carbon isotope ^{14}C (discussed in the *Focus on Research* in Chapter 2), was critical to this research.

Beginning in 1945, Melvin Calvin, Andrew A. Benson, and their colleagues at the University of California, Berkeley, combined ^{14}C-labeled CO_2 with a widely used technique called *two-dimensional paper chromatography* to trace the pathways of the light-independent reactions in a green alga, *Chlorella*. The researchers exposed actively photosynthesizing *Chlorella* cells to the labeled carbon dioxide. Then, at various times, cells were removed and placed in hot alcohol, which instantly stopped all the photosynthetic reactions of the algae. Radioactive carbohydrates were then extracted from the cells and, to identify them chemically, a drop of the extract was placed at one corner of a piece of paper and dried. The paper was placed with its edge touching a solvent (step 1 in the figure); Calvin used a water solution of butyl alcohol and propionic acid for this step. The compounds in the dried spot dissolved and were carried upward by the solvent through the paper (step 2 in the figure), at rates that varied according to their molecular size and solubility. This line of spots was the first dimension of the two-dimensional technique.

The paper was then dried, turned 90°, and touched to a second solvent (Calvin used a water solution of phenol for this part of the experiment). As this solvent moved through the paper, the compounds again migrated upward from the spots produced by the first dimension, but at rates that were different from their mobility in the first solvent (step 3 in the figure). This step, the second dimension of the two-dimensional technique, separated molecules that, although different, had produced a single spot in the first solvent because they had migrated at the same rate. After all the molecules had migrated through the second dimension, the individual spots were identified by comparing their locations with the positions of spots made by known molecules when the "knowns" were run through the same procedure.

In a final step, the dried paper was covered with a sheet of photographic film. The radioactive compounds exposed spots on the film (step 4 in the figure), which was developed and compared with the spots on the paper to identify compounds that were radioactive. By comparing the labeled compounds revealed by the two-dimensional chromatography technique in extracts prepared from *Chlorella* cells under different conditions, Calvin and his colleagues were able to reconstruct the reactions of the Calvin cycle.

In carbohydrate extracts made within a few seconds after the cells were exposed to the labeled CO_2, most of the radioactivity was found in 3PGA, indicating that it is one of the earliest products of photosynthesis. In extracts made after longer periods of exposure to the label, radioactivity showed up in G3P and in more complex substances including a variety of six-carbon sugars, sucrose, and starch.

In other experiments, Calvin reduced the amount of CO_2 available to the *Chlorella* cells so that photosynthesis worked slowly even in bright light. Under these conditions, RuBP accumulated in the cells, suggesting that it is the first substance to react with CO_2 in the light-independent reactions, and that it accumulates if CO_2 is in short supply. By similar methods, most of the intermediate compounds between CO_2 and six-carbon sugars were identified.

Using this information, Calvin and his colleagues pieced together the light-independent reactions of photosynthesis and showed that they formed a continuous cycle. Melvin Calvin was awarded a Nobel Prize in 1961 for his work on the assimilation of carbon dioxide in plants.

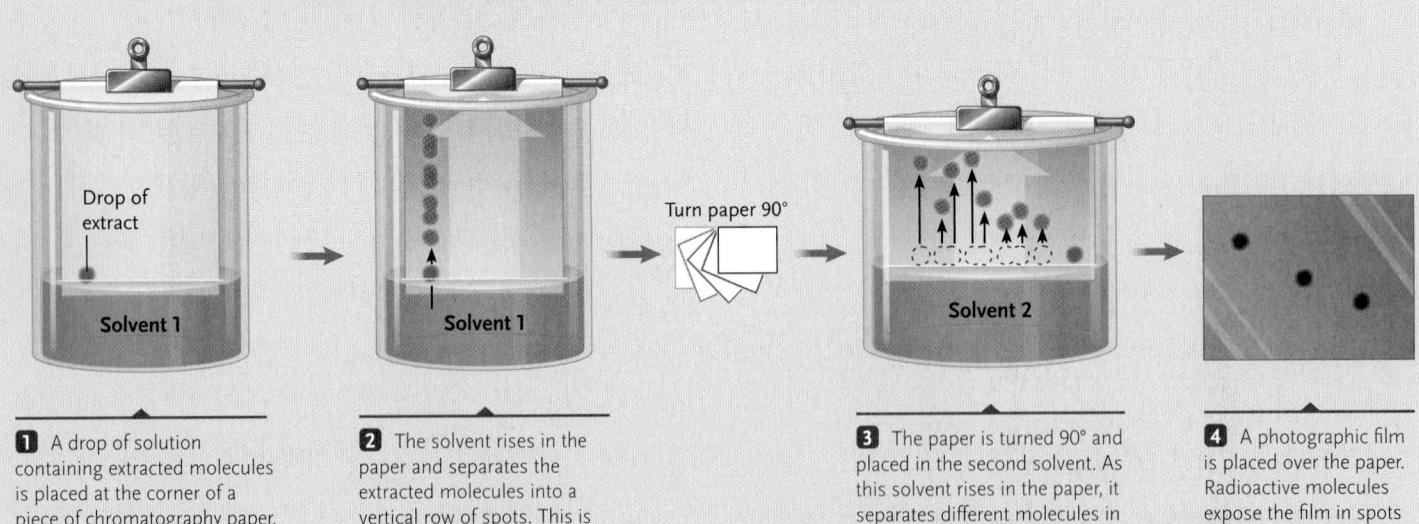

1 A drop of solution containing extracted molecules is placed at the corner of a piece of chromatography paper. The edge of the paper is placed in a solvent.

2 The solvent rises in the paper and separates the extracted molecules into a vertical row of spots. This is the first dimension of the technique.

Turn paper 90°

3 The paper is turned 90° and placed in the second solvent. As this solvent rises in the paper, it separates different molecules in the first row into vertical rows of spots. This is the second dimension of the technique.

4 A photographic film is placed over the paper. Radioactive molecules expose the film in spots over their locations in the paper. Developing the film reveals the locations of the radioactive spots.

Calvin. *Focus on Research* describes the experiments Calvin and his colleagues used to elucidate the light-independent reactions.

The Calvin Cycle Uses NADPH, ATP, and CO_2 to Generate Carbohydrates

The light-independent reactions of the Calvin cycle use CO_2, ATP, and NADPH as inputs. As products, the cycle releases ADP; $NADP^+$; the three-carbon carbohydrate molecule glyceraldehyde-3-phosphate (G3P), already familiar as part of glycolysis; and inorganic phosphate (outlined in **Figure 9.13a**). The Calvin cycle takes place entirely in the chloroplast stroma.

Figure 9.13a focuses primarily on tracking the carbon atoms through the cycle. In phase 1 of the cycle, *carbon fixation*, a carbon atom from CO_2 is added to ribulose 1,5-bisphosphate (RuBP), a five-carbon sugar, to produce two three-carbon molecules of 3-phosphoglycerate (3PGA). In phase 2, *reduction*, reactions using NADPH and ATP from the light-dependent reactions convert 3PGA into G3P, another three-carbon molecule. After several rounds of the Calvin cycle, two molecules of G3P leave the cycle and are used to form the products of the cycle, the six-carbon sugar glucose and other organic compounds. In phase 3, *regeneration*, some G3P molecules are used to produce the five-carbon RuBP with the help of energy from ATP. The cycle then begins again.

Now let us consider the reactions in more detail. **Figure 9.13b** shows the chemical structures, reactions, and enzymes of the cycle. The key reaction of the cycle is the first, carbon fixation, in which CO_2 combines with RuBP, forming a transient six-carbon molecule that is cleaved to form 3PGA. This reaction, which fixes CO_2 into organic form, is catalyzed by the key enzyme of the Calvin cycle, **RuBP carboxylase/ oxygenase** (abbreviated as **rubisco**). In the next two reactions (reactions 2 and 3 in Figure 9.13b, shown in two parallel paths because two molecules of 3PGA are being processed), the three-carbon molecules are raised in energy level by the addition of phosphate groups transferred from ATP and electrons from NADPH (the ATP and NADPH are products of the light-dependent reactions). The G3P generated by reaction 3 is the carbohydrate product of the Calvin cycle.

Most of the G3P produced by the reactions is used to regenerate the RuBP entering in the first reaction of the cycle. However, some G3P is released as a net product; it serves as the primary building block for reactions producing glucose and many other organic molecules in chloroplasts.

The G3P used to regenerate RuBP enters a complex series of reactions (reaction series 4 in Figure 9.13b) that yields the five-carbon sugar ribulose 5-phosphate. In the final reaction of the cycle (reaction 5), a phosphate group is transferred from ATP to regenerate the RuBP used in the first reaction, and the cycle is ready to turn again.

Three Turns of the Calvin Cycle Are Needed to Make One Net G3P Molecule

If the Calvin cycle is run through one turn, the cycle cannot turn again if a molecule of G3P is taken away. The remaining G3P, with three carbons in its structure, cannot supply the five carbons needed to regenerate the RuBP molecule required for another turn. In fact, the cycle must run through *three* turns before enough G3P molecules are made so that one can be released.

Here's how it works. Three turns of the Calvin cycle produce six molecules of G3P (totaling 18 carbons) and use three molecules of RuBP (totaling 15 carbons) and three molecules of CO_2 (totaling three carbons). Of the six G3P molecules, five (totaling 15 carbons) go back into the cycle to regenerate the three RuBP molecules (15 carbons) used in the three turns. Thus, the cycle can generate one surplus molecule of G3P (three carbons) after three turns. The leftover G3P is free to enter reaction pathways that yield glucose, sucrose, starch, and other complex organic substances.

Another way to look at it is to consider that one turn of the Calvin cycle takes up one molecule of CO_2 and generates one (CH_2O) unit of carbohydrate. On this basis, you can understand that three turns are required to make enough (CH_2O) units to assemble one surplus molecule of G3P. Providing enough (CH_2O) units to make a six-carbon carbohydrate such as glucose requires six turns of the cycle.

This approach allows us to total all the inputs and outputs of the Calvin cycle. For each turn of the cycle, 2 ATP and 2 NADPH are used in reactions 2 and 3, and one additional ATP is used in reaction 5, for a total of 3 ATP and 2 NADPH for each turn. Although one of the phosphates derived from ATP is attached to G3P, this phosphate is eventually released when G3P is converted into other substances. As net reactants and products, one complete turn of the cycle therefore includes:

$$CO_2 + 2\,NADPH + 3\,ATP \rightarrow$$
$$(CH_2O) + 2\,NADP^+ + 3\,ADP + 3\,P_i$$

Rubisco Is the Key Enzyme of the World's Food Economy

Rubisco, the enzyme that catalyzes the first reaction of the Calvin cycle, is unique to photosynthetic organisms. By catalyzing CO_2 fixation, it provides the source of organic molecules for most of the world's organisms—the enzyme converts about 100 billion tons of CO_2 into carbohydrates annually. There are so many rubisco mole-

a. Overall phases of the Calvin Cycle

CO_2
P—C—C—C—C—C—P
RuBP
P—C—C—C
C—C—C—P
2 molecules of 3PGA
Phase 1
Fixation
of CO_2
Phase 3
Regeneration
of RuBP
Calvin
Cycle
Phase 2
Reduction
and sugar
production
ADP
ATP
ATP
ADP
2 NADPH
2 $NADP^+$
Organic
compounds
(glucose)
P—C—C—C
C—C—C—P
2 molecules of G3P
2 P_i

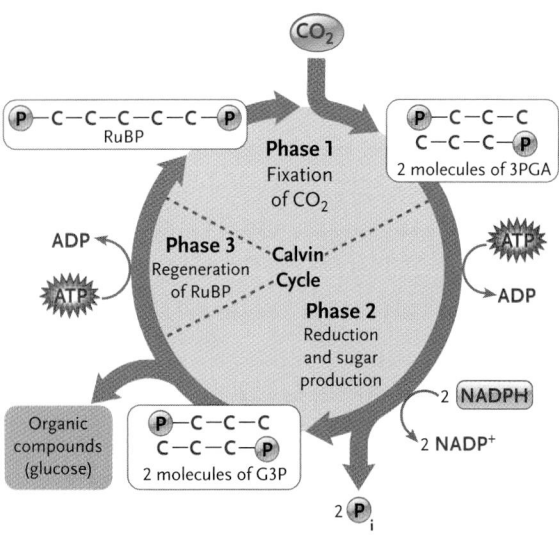

Figure 9.13

The Calvin cycle. **(a)** Overview of the three phases of the Calvin cycle. The figure tracks the carbon atoms in the molecules in the cycle. **(b)** Reactions and enzymes of the Calvin cycle (the enzymes are printed in rust). Reaction 1 first produces an unstable, six-carbon intermediate (not shown), which splits almost immediately into two molecules of 3PGA, the substance detected by the labeling experiments as the first product of the light-independent reactions.

O_2
CO_2
Light
ATP
NADPH
Light
reactions
Calvin
cycle
ADP
$NADP^+$
H_2O
Sugars
Chloroplast
stroma

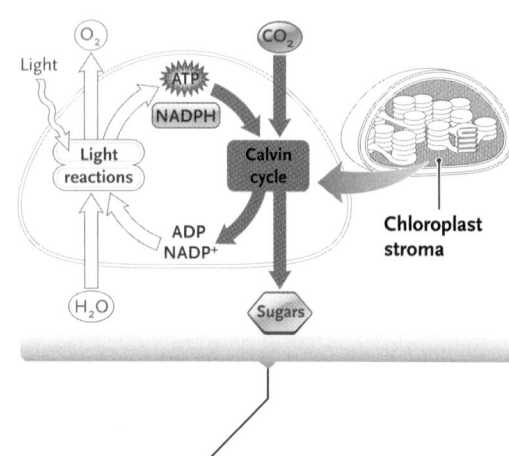

b. Reactions and enzymes of the Calvin Cycle

1 Ribulose 1,5-bisphosphate (RuBP) reacts with CO_2, producing two molecules of 3PGA. This reaction converts CO_2 into organic form.
2 A phosphate group from ATP is added to each of the two 3PGA molecules, producing two molecules of 1,3-bisphosphoglycerate. The reaction raises the energy content of the products to a level high enough to enter the next reaction of the cycle.
3 The two molecules of 1,3-bisphosphoglycerate are reduced by electrons carried by NADPH. One phosphate is removed from each reactant at the same time, yielding two molecules of G3P.
4 Some of the G3P produced by reaction 3 enters a series of reactions yielding ribulose 5-phosphate.
5 Another phosphate is transferred from ATP to ribulose 5-phosphate to produce RuBP. This reaction regenerates the RuBP used in reaction 1, and the cycle is ready to turn again.
H_2C—O—P
C=O
HCOH
HCOH
H_2C—O—P
Ribulose
1,5-bisphos-
phate (RuBP)
CO_2
RuBP
carboxylase
H_2C—O—P
HCOH
COO^-
3-Phospho-
glycerate
(3PGA)
3-Phospho-
glycerate kinase
ATP
ADP
H_2C—O—P
HCOH
HC—O—P
O
1,3-Bisphospho-
glycerate
NADPH
$NADP^+$
1,3-Bisphospho-
glycerate kinase
P_i
H_2C—O—P
HCOH
CH
O
Glyceraldehyde-
3-phosphate
(G3P)
Calvin Cycle
Some net G3P
To reactions synthesizing sugars and other organic compounds
Complex
reactions
regenerating
ribulose
5-phosphate
H_2COH
C=O
HCOH
HCOH
H_2C—O—P
Ribulose
5-phosphate
ATP
ADP
Ribulose
5-phosphate
kinase

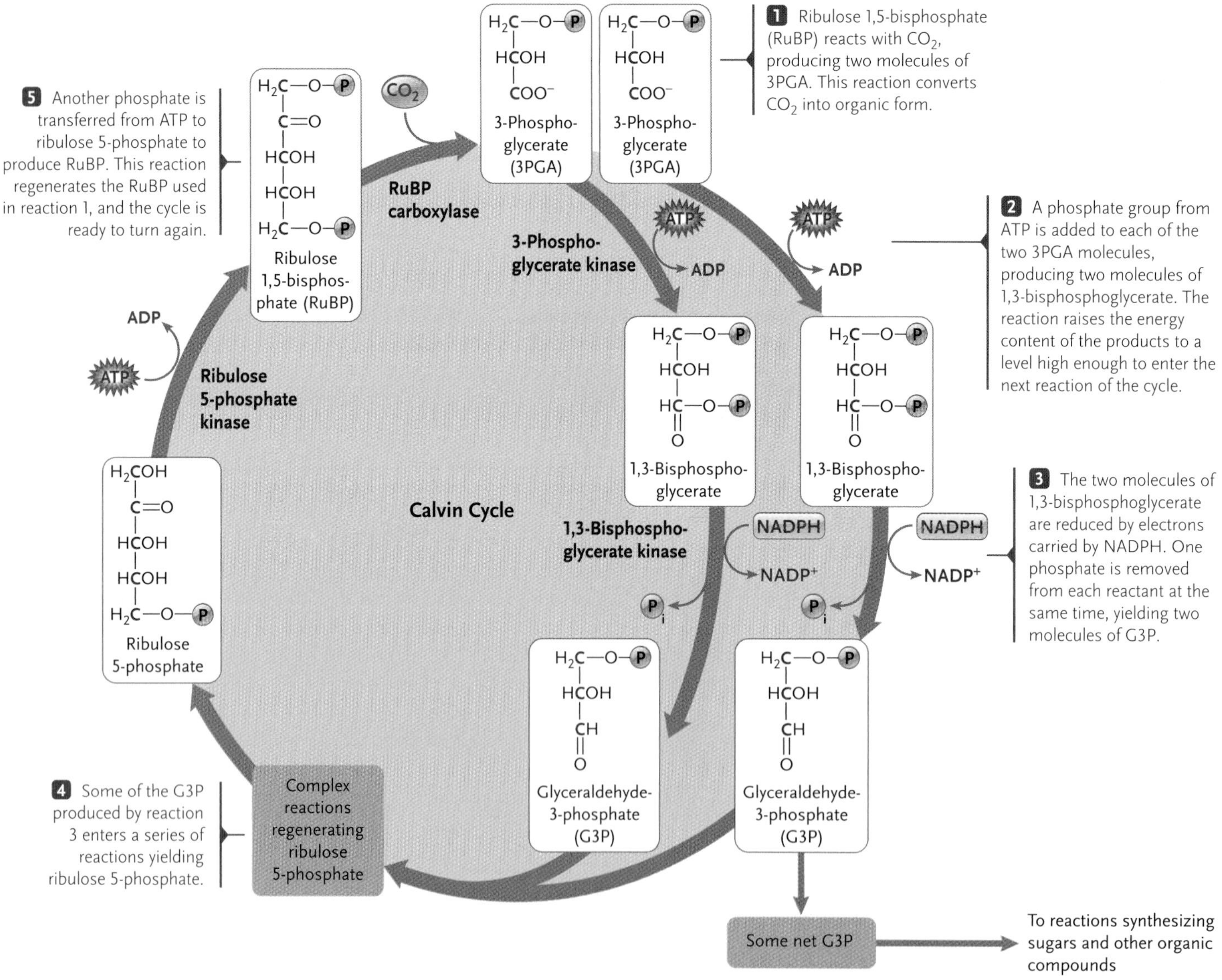

Small but Pushy

We noted that all the active sites of rubisco appear to be on the large polypeptide subunit of the enzyme. Even so, 99% of the enzyme's catalytic activity is lost if the small subunit is removed. What does the small subunit do?

Betsy A. Reed and F. Robert Tabita of The Ohio State University set out to answer this question using molecular techniques. They hypothesized that the structure of a specific region of the small subunit was critical to its function. To test this hypothesis, the investigators used DNA cloning techniques (described in Chapter 18) to produce five versions of the small subunit, each with a different amino acid substituted for the normal one at five different positions in the protein, and examined the effects of the substitutions on enzyme activity. One of the modified small subunits, which had glutamine substituted for arginine at position 88

in the small subunit amino acid sequence, was unable to assemble with the large subunit to form a complete enzyme complex, showing that the arginine in position 88 is essential for normal enzyme assembly.

The four remaining versions of the small subunit assembled normally with large subunits. Each complete rubisco complex was placed in a test tube system containing RuBP and other factors required for the initial reaction of the Calvin cycle. The altered versions of the enzyme were all able to recognize and bind their substrate—RuBP, CO_2, or O_2—as ably as the normal enzyme. Therefore, these four alterations induced in the small subunit had no effect on the specificity of the enzyme.

The investigators next checked the rates at which the enzymes catalyzed CO_2 fixation. Three of the four altered

enzymes ran the first reaction of the Calvin cycle at only 35% of the rate of the normal enzyme. The most active worked only about half as fast as the normal enzyme. In other words, the small subunit has a very significant effect on the enzyme's rate of catalysis. The effect is critically important when considered in the context of the comparatively slow reaction rate of the normal enzyme. The enzyme's multiple form—eight copies of each subunit, massed together, all doing the same thing—and the very large amount of the enzyme packed into leaves compensate for the slow rate. Evidently, the small subunit evolved as yet another way to compensate for the enzyme's slow action, by pushing the large subunit to do its job faster. It may do so by altering the three-dimensional folding of the large subunit into patterns that increase its catalytic rate.

cules in chloroplasts that the enzyme may make up 50% or more of the total protein of plant leaves. As such, it is also the world's most abundant protein, estimated to total some 40 million tons worldwide, equivalent to about 10 kg for every human.

Rubisco has essentially the same overall structure in almost all photosynthetic organisms: eight copies each of a large and a small polypeptide, joined together in a 16-subunit structure. The large subunit contains all of the known binding sites for substrates, including CO_2 and RuBP. Although the small subunit has no active sites, it is still essential for efficient operation of the enzyme. *Insights from the Molecular Revolution* describes a recent effort to determine the molecular functions of the small subunit.

Rubisco is also the key regulatory site of the Calvin cycle. The enzyme is stimulated by both NADPH and ATP; as long as these substances are available from the light-dependent reactions, the enzyme is active and the light-independent reactions proceed. During the daytime, when sunlight powers the light-dependent reactions, the abundant NADPH and ATP supplies keep the Calvin cycle running; in darkness, when NADPH and ATP become unavailable, the enzyme is inhibited and the Calvin cycle slows or stops. Similar controls based on the avail-

ability of ATP and NADPH also regulate the enzymes that catalyze other reactions of the Calvin cycle, including reactions 2 and 3 in Figure 9.13b.

G3P Is the Starting Point for Synthesis of Many Other Organic Molecules

The net G3P formed in the Calvin cycle is the starting point for production of a wide variety of organic molecules. More complex carbohydrates such as glucose and other monosaccharides are made from G3P by reactions that, in effect, reverse the first half of glycolysis. Once produced, the monosaccharides enter biochemical pathways that make disaccharides such as sucrose, polysaccharides such as starches and cellulose, and other complex carbohydrates of cell walls. Other pathways manufacture amino acids, fatty acids and lipids, proteins, and nucleic acids. The reactions forming these products occur both within chloroplasts and in the surrounding cytosol and nucleus.

Sucrose, a disaccharide consisting of glucose linked to fructose, is the main form in which the products of photosynthesis circulate from cell to cell in higher plants. Organic nutrients are stored in most higher plants as sucrose or starch, or as a combination of the two in proportions that depend on the plant species. Sugar cane and sugar beets, which contain stored sucrose in high concentrations, are the main sources of the sucrose we use as table sugar.

9.4 Photorespiration and the C_4 Cycle

Oxygen can compete with carbon dioxide for the active site of rubisco. When oxygen binds to the active site, rubisco acts as an *oxygenase* instead of a carboxylase. As an oxygenase, it catalyzes a reaction in which O_2 instead of CO_2 is added to RuBP. The products of the reaction are toxic and cannot be used by plants for synthesis of carbohydrates. Instead, the products are eliminated by pathways that *release* CO_2. Because O_2 is taken up by rubisco's oxygenase activity, and CO_2 is released at later steps, the entire process is known as **photorespiration.**

Photorespiration reduces the efficiency of energy use in photosynthesis and impairs the growth of many plants, including some of the crop plants that provide food for our population. However, many plants have evolved ways of dealing with photorespiration, including a preliminary reaction series known as the **C_4 cycle,** which allows CO_2 to be fixed by a different carboxylase

that is unaffected by high oxygen concentrations. This adaptation is combined with other adaptations that restrict rubisco's carboxylase activity to conditions where oxygen concentration remains low.

The Oxygenase Activity of Rubisco Leads to the Formation of a Toxic Molecule

Figure 9.14 shows the result of the oxygenase activity of rubisco. First, the reaction converts RuBP into one molecule of 3PGA and one molecule of a two-carbon substance, phosphoglycolate. No carbon is fixed during this reaction, and energy must then be used to salvage the carbons from phosphoglycolate. The pathway for the latter process begins with the removal of the phosphate group from phosphoglycolate, producing *glycolate,* a toxic substance that is eliminated by oxidation inside microbodies (microbodies are discussed in Section 5.3). The products of this oxidation enter reaction pathways that yield CO_2. Thus, as an overall pathway, photorespiration uses O_2 and releases CO_2.

The balance of the carboxylase and oxygenase activities of rubisco depends on the relative concentrations of O_2 and CO_2 inside leaves and other structures carrying out photosynthesis. As O_2 concentration rises and CO_2 concentration falls, the oxygenase activity of rubisco increases proportionately.

Why does rubisco have the oxygenase activity? One possibility is that the enzyme evolved before the water-splitting reaction of photosynthesis appeared, at a time when the atmosphere was rich in CO_2 and low in O_2. Under these conditions, there would be no selection against the oxygenase activity of the enzyme.

Elevated Temperatures Increase the Level of Photorespiration in Many Plants

Oxygen concentration rises in leaves, and CO_2 concentration falls, when photosynthesis proceeds at high rates, as it does on hot, sunny days. Other physiological responses of plants to hot weather also tend to tip the O_2/CO_2 balance in the direction of O_2. As photosynthesis proceeds during the day, plants open their stomata to release the O_2 made in photosynthesis and to let in CO_2. However, opening the stomata also releases water, leading to dehydration of the plants during hot weather. As the stomata close in response to the water loss, O_2 builds up and CO_2 concentration falls in the leaves, increasing the oxygenase activity of RuBP carboxylase and the rate of photorespiration.

Unfortunately, many economically important crop plants are among those seriously impaired by high photorespiration rates at elevated temperatures—rice, barley, wheat, soybeans, tomatoes, and potatoes, to name a few. The detrimental effects of photorespiration on these plants can be estimated by growing them at elevated temperatures in hothouses contain-

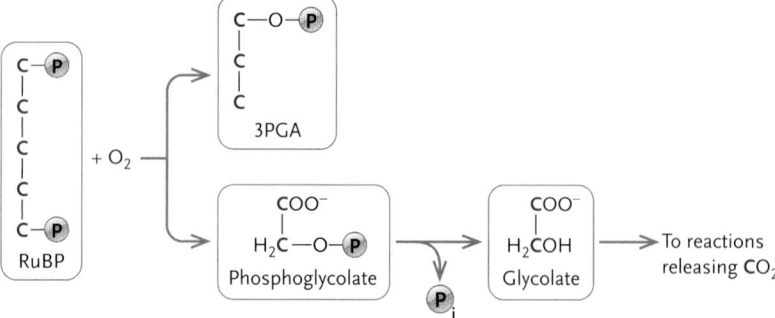

Figure 9.14
Photorespiration, an alternative pathway for rubisco in which, in the presence of oxygen, the oxygenase activity of the enzyme produces glycolate. Glycolate, a toxic product, is eliminated by reactions that convert carbon back to inorganic form as CO_2.

ing CO_2 in high concentrations. Under these conditions, which curtail photorespiration, some of the crops grow as much as five times faster (as measured by dry weight) than they do at the CO_2 concentrations of the atmosphere.

The C_4 Cycle Circumvents Photorespiration by Using a Carboxylase That Has No Oxygenase Activity

In the C_4 cycle **(Figure 9.15),** CO_2 initially combines with a three-carbon molecule, *phosphoenolpyruvate (PEP),* producing the four-carbon intermediate oxaloacetate. The reaction is catalyzed by the critical enzyme of the C_4 cycle, *PEP carboxylase.* The C_4 cycle gets its name because its first product is a four-carbon molecule rather than a three-carbon molecule as in the Calvin cycle (the Calvin cycle is often called the C_3 cycle to make this distinction). Oxaloacetate is then reduced to *malate,* a four-carbon acid, by electrons transferred from NADPH.

With some variations in intermediates and products, the C_4 cycle takes place in several groups of plants, including important cereal crops such as corn.

The C_4 cycle runs when O_2 concentrations are high. PEP carboxylase has no activity as an oxygenase, and is therefore unaffected when O_2 concentrations are high in leaves. Later, at a location or time when O_2 concentrations are low, the malate produced by the C_4 cycle is oxidized to a three-carbon product, pyruvate, with release of CO_2:

$$\text{malate} + \text{NADP}^+ \rightarrow \text{pyruvate} + \text{NADPH} + CO_2$$

The CO_2 then enters the Calvin cycle for fixation by rubisco into G3P and other products of the light-independent reactions. Because O_2 concentrations are low and CO_2 concentrations are high, the oxygenase activity of rubisco is limited, and the Calvin cycle proceeds normally.

The pyruvate returns to the C_4 cycle, where it is converted to PEP at the expense of one molecule of

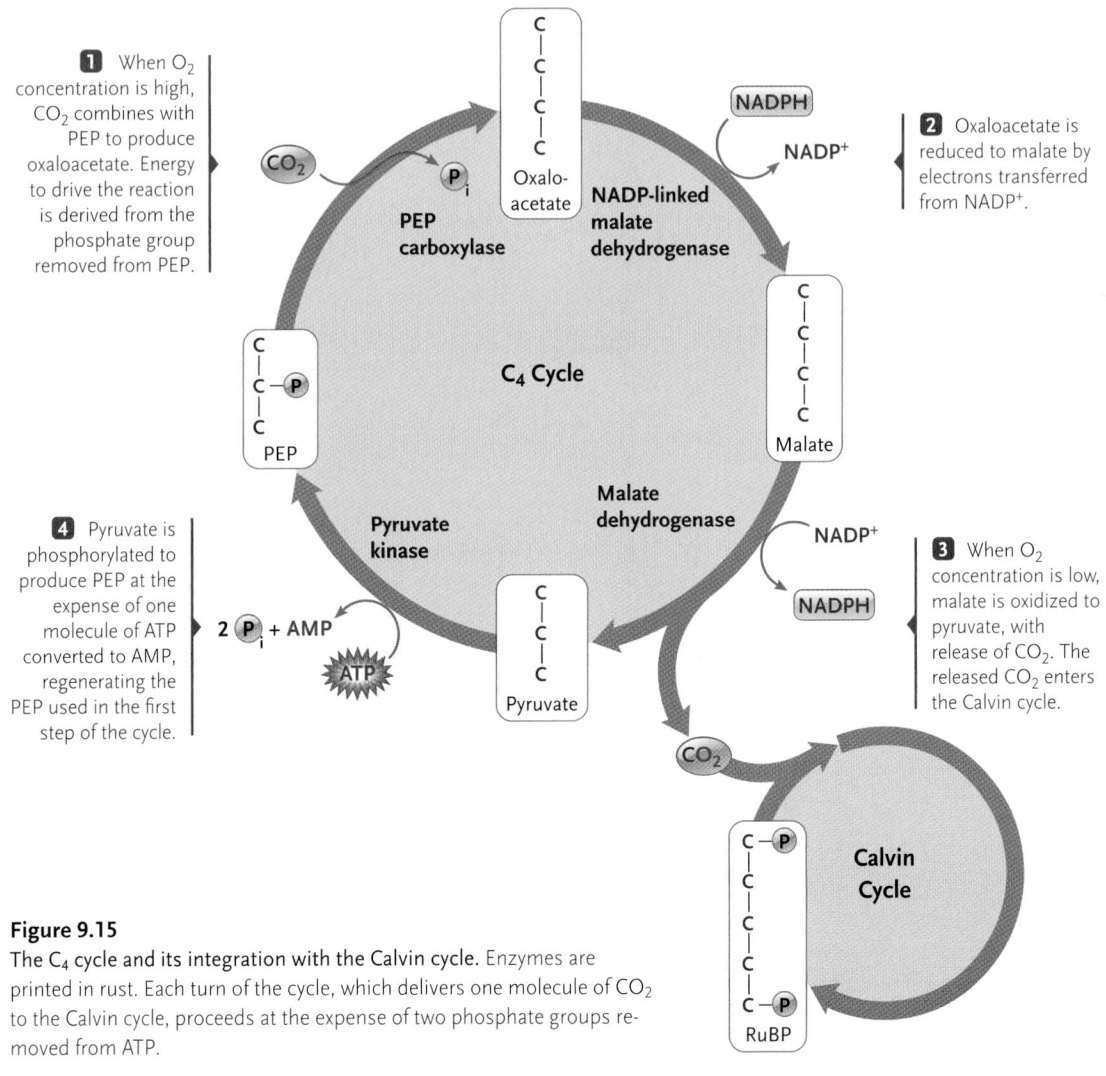

Figure 9.15

The C_4 cycle and its integration with the Calvin cycle. Enzymes are printed in rust. Each turn of the cycle, which delivers one molecule of CO_2 to the Calvin cycle, proceeds at the expense of two phosphate groups removed from ATP.

1 When O_2 concentration is high, CO_2 combines with PEP to produce oxaloacetate. Energy to drive the reaction is derived from the phosphate group removed from PEP.

2 Oxaloacetate is reduced to malate by electrons transferred from NADP$^+$.

3 When O_2 concentration is low, malate is oxidized to pyruvate, with release of CO_2. The released CO_2 enters the Calvin cycle.

4 Pyruvate is phosphorylated to produce PEP at the expense of one molecule of ATP converted to AMP, regenerating the PEP used in the first step of the cycle.

ATP converted to AMP. Because two phosphates are removed from ATP for each turn of the C$_4$ cycle, making each molecule of glucose by the combined activities of the C$_4$ and Calvin pathways requires an additional 12 ATP.

In spite of the extra penalty paid in ATP, the ability to bypass photorespiration gives plants using the C$_4$/Calvin cycle combination an advantage in warmer climates over plants with only the Calvin cycle. The crossover point lies at about 30°C. Above this temperature, C$_4$ plants become significantly more efficient than Calvin-limited plants; below 30°C, the additional ATP used by C$_4$ plants makes Calvin-limited species more efficient in spite of photorespiration.

The 30°C crossover point gives C$_4$ plants an advantage in the tropics and in temperate regions with high summer temperatures, such as the southern and central United States. In colder areas, Calvin plants have the advantage. For example, in Florida 80% of all native species are C$_4$ plants, compared with 0% in Manitoba, Canada.

The C$_4$ pathway occurs in at least 16 different families of flowering plants (angiosperms; discussed in Chapter 27). Some of the families are only distantly related, suggesting that the C$_4$ pathway may have developed independently several times in the evolution of higher plants.

Some Plants Circumvent Photorespiration by Running the C$_4$ and Calvin Cycles in Different Locations

Some C$_4$ plants run the Calvin and C$_4$ cycles at the same time, but in locations with differing CO$_2$ and O$_2$ concentrations. In corn, the C$_4$ cycle occurs in *mesophyll* cells, which lie close to the surface of leaves and stems, where O$_2$ is abundant **(Figures 9.16a and b)**. The malate product of the C$_4$ cycle diffuses from the meso-

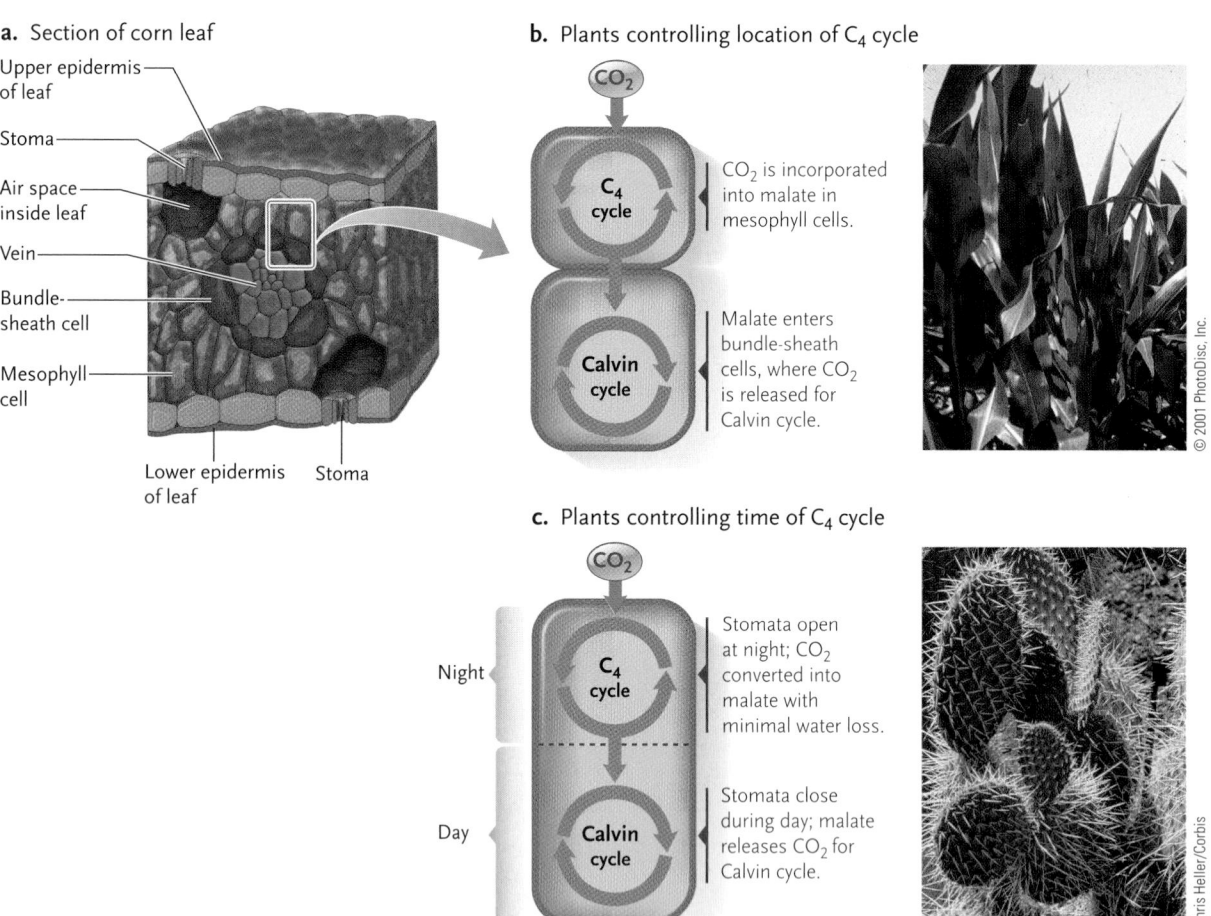

a. Section of corn leaf

Upper epidermis of leaf
Stoma
Air space inside leaf
Vein
Bundle-sheath cell
Mesophyll cell
Lower epidermis of leaf   Stoma

b. Plants controlling location of C$_4$ cycle

CO$_2$

C$_4$ cycle — CO$_2$ is incorporated into malate in mesophyll cells.

Calvin cycle — Malate enters bundle-sheath cells, where CO$_2$ is released for Calvin cycle.

© 2001 PhotoDisc, Inc.

c. Plants controlling time of C$_4$ cycle

CO$_2$

Night — C$_4$ cycle — Stomata open at night; CO$_2$ converted into malate with minimal water loss.

Day — Calvin cycle — Stomata close during day; malate releases CO$_2$ for Calvin cycle.

Chris Heller/Corbis

Figure 9.16
Coordination of the C$_4$ and Calvin cycles to minimize photorespiration. **(a)** and **(b)** Some C$_4$ plants separate the two cycles into different locations internally, as in the corn leaf shown in this diagram. The mesophyll cells (lighter green), which are closer to the leaf surfaces, carry out the C$_4$ cycle in a relatively O$_2$-rich environment. The bundle-sheath cells (darker green), which are cut off from O$_2$ by the surrounding layer of mesophyll cells, carry out the Calvin cycle. **(c)** Other C$_4$ plants carry out the two cycles at different times, as in the beavertail cactus *(Opuntia basilaris)* in the photo.

Photosynthesis is considered by many to be the most important biological process on Earth. In particular, directly or indirectly (through herbivorous animals), photosynthesis provides all of our food requirements. Research on photosynthesis therefore is of high importance and is likely to have significant benefit for humankind. For example, a complete understanding of the chemistry of photosynthesis, the regulation of the process, and the genes that encode the components of the process could be applicable to other endeavors of human interest, such as solar energy conversion and the development of therapeutic drugs.

From research on agricultural crops, we have learned that photosynthesis is not a very efficient process. Estimates are that only 1% to 2% of the solar energy that strikes the planet's surface is converted to new photosynthetic products. Research is being done to learn enough about photosynthesis so that crop plants can be engineered to be more efficient. An area of particular relevance here is photorespiration, which reduces the efficiency of energy use in photosynthesis. Hopefully, research will give us a better understanding of the biochemical control of photorespiration and provide clues about breeding new, more energy-efficient plants.

Let us consider two specific avenues of research.

How is the efficiency of photosynthesis regulated?

The laboratory of David Kramer at Washington State University is interested in the energetics and control of photosynthesis, the electron transfer reactions, the coupling of electron transfer reactions to ATP synthesis, and photosynthesis in extreme environments.

As you have learned, energy conversion by the chloroplast involves the capture of light energy and the channeling of that energy through an electron transfer system with the eventual synthesis of NADPH and ATP. At high concentrations, many of the intermediates produced in this energy conversion can potentially destroy the photosynthetic apparatus, a phenomenon called photoinhibition. To prevent such damage, the effi-

ciency of some of the photosystem components is reduced by the release of some of the energy as heat. Increased heat lowers the efficiency of photosynthesis, however. Evidence from a range of studies indicates that the balance between protection against photoinhibition and photosynthetic efficiency is important in enabling plants to acclimate to environmental changes. Kramer's group is doing research to develop an understanding of the structure and function of ATP synthase and the cytochrome complex, and the effects of these components on the proton-motive force, which is known to play a pivotal role in balancing photoinhibition and photosynthetic efficiency. The results will illuminate how the specific mechanisms of photosynthesis determine plant growth and survival. In addition, the technology developed as part of the research may lead to applications in plant breeding and farming, providing farmers with a means to assess the physiological states of the plants they are growing and, therefore, to modify the conditions for optimal growth.

How are chloroplast thylakoid membrane-protein complexes assembled?

Research by Andrew Webber's group at Arizona State University is directed to understanding the formation of chloroplast thylakoid membrane-protein complexes. Those complexes are key to the process of photosynthesis, yet their assembly is not understood. Using molecular biology and biochemistry techniques, Webber's group is studying how the synthesis of chloroplast proteins, some of which are encoded by genes in the chloroplast and others of which are encoded by genes in the nucleus, is coordinated and regulated. The researchers are also using molecular techniques to change specific amino acids in the chloroplast proteins with the aim of elucidating how those amino acids are involved in assembly and functioning of the complexes. The results will add more detailed knowledge about the structure and function of components that are key to the process of photosynthesis.

Peter J. Russell

phyll cells to *bundle sheath cells,* located in deeper tissues where O_2 is less abundant. In these cells, in which the Calvin cycle operates, the malate enters chloroplasts and is converted to pyruvate and CO_2. Because O_2 concentration is low, and CO_2 concentration is high because of its release by malate breakdown, the oxygenase activity of rubisco is inhibited and carboxylase activity is promoted. The pyruvate produced by malate oxidation returns to the mesophyll cells to enter another turn of the C_4 cycle.

Several tropical and temperate crop plants in addition to corn, including sugar cane, sorghum, and some pasture grasses, use the C_4 cycle to control the location of initial CO_2 fixation in leaf cells. Unfortunately, many highly successful weed pests, such as Bermuda grass and crabgrass, also use the same adaptation to compete successfully with lawn grasses and crops in warm climates.

Other Plants Control Photorespiration by Running the C₄ and Calvin Cycles at Different Times

Instead of running the Calvin and C_4 cycles simultaneously in different locations, some C_4 plants run the cycles at different times to circumvent photorespiration. The plants in this group include many with thick, succulent leaves or stems such as the cactus shown in **Figure 9.16c.** These plants are known collectively as **CAM plants,** named for *crassulacean acid metabolism,* from the Crassulaceae family in which the CAM adaptation was first observed.

CAM plants typically live in regions that are hot and dry during the day and cool at night. Their fleshy leaves or stems have a low surface-to-volume ratio, and their stomata are reduced in number. Further, the stomata open only at night, when they release O_2 that accumulates from photosynthesis during the day and allow CO_2 to enter the leaves. The entering CO_2 is fixed by the C_4 pathway into malate, which accumulates throughout the night and is stored in large cell vacuoles.

Daylight initiates the second phase of the strategy. As the sun comes up and the temperature rises, the stomata close, reducing water loss and cutting off the exchange of gases with the atmosphere. Malate diffuses from cell vacuoles into the cytosol, where it is oxidized to pyruvate, and CO_2 is released in high concentration. The high CO_2 concentration favors the carboxylase activity of rubisco carboxylase, allowing the Calvin cycle to proceed at maximum efficiency with little loss of organic carbon from photorespiration. The pyruvate produced by malate breakdown accumulates during the day; as night falls, it enters the C_4 reactions converting it back to malate. During the night, oxygen is released by the plants, and more CO_2 enters.

Reduction of water loss by closure of the stomata during the hot daylight hours has the added benefit of making CAM plants highly resistant to dehydration.

As a result, CAM species can tolerate extreme daytime heat and dryness.

In this chapter, you have seen how photosynthesis supplies the organic molecules used as fuels by almost all the organisms of the world. It is a story of electron flow: electrons, pushed to high energy levels by the absorption of light energy, are added to CO_2, which is fixed into carbohydrates and other fuel molecules. The high-energy electrons are then removed from the fuel molecules by the oxidative reactions of cellular respiration, which use the released energy to power the activities of life. Among the most significant of these activities are cell growth and division, the subjects of the next chapter.

STUDY BREAK

1. When does photorespiration occur? What are the reactions of photorespiration, and what are the energetic consequences of the process?
2. How do C_4 plants circumvent photorespiration?

Review

Go to **ThomsonNOW** at www.thomsonedu.com/login to access quizzing, animations, exercises, articles, and personalized homework help.

9.1 Photosynthesis: An Overview

- In photosynthesis, plants, algae, and photosynthetic prokaryotes use the energy of sunlight to drive synthesis of organic molecules from simple inorganic raw materials. The organic molecules are used by the photosynthesizers themselves as fuels; they also form the primary energy source for heterotrophs.

- The two overall stages of photosynthesis are the light-dependent and light-independent reactions. In eukaryotes, both stages take place inside chloroplasts (Figures 9.2 and 9.3).

- Photosynthesizers use the energy of sunlight to push electrons to elevated energy levels. In eukaryotes and many prokaryotes, water is split as the source of the electrons for this process, and oxygen is released to the environment as a by-product.

- The high-energy electrons provide an indirect energy source for ATP synthesis and also for CO_2 fixation, in which CO_2 is fixed into organic substances by the addition of both electrons and protons.

Animation: Photosynthesis overview

Animation: Sites of photosynthesis

9.2 The Light-Dependent Reactions of Photosynthesis

- In the light-dependent reactions of photosynthesis, light is converted to chemical energy when electrons, excited by absorption of light in a pigment molecule, are passed from the pigment to a stable orbital in a primary acceptor molecule (Figure 9.5).

- Chlorophylls and carotenoids, the photon-absorbing pigments in eukaryotes and cyanobacteria, together absorb light energy at

a range of wavelengths, enabling a wide spectrum of light to be used in photosynthesis (Figures 9.6 and 9.7).

- In organisms that split water as their electron source, the pigments are organized with proteins into two photosystems. Specialized forms of chlorophyll *a* pass excited electrons to primary acceptor molecules in the photosystems (Figure 9.8).

- Electrons obtained from splitting water are used for the synthesis of NADPH and ATP. In the noncyclic electron flow pathway, electrons first flow through photsytem II, becoming excited there to a higher energy level, and then pass through an electron transfer system to photosystem I releasing energy that is used to create an H^+ gradient across the membrane. The gradient is used by ATP synthase to drive synthesis of ATP. The net products of the light-dependent reactions are ATP, NADPH, and oxygen (Figures 9.9 and 9.10).

- Electrons can also flow cyclically around photosystem I and the electron transfer system, building the H^+ concentration and allowing extra ATP to be produced, but no NADPH (Figure 9.11).

Interaction: Wavelengths of light

Animation: Noncyclic pathway of electron flow

9.3 The Light-Independent Reactions of Photosynthesis

- In the light-independent reactions of photosynthesis, CO_2 is reduced and converted into organic substances by the addition of electrons and hydrogen carried by the NADPH produced in the light-dependent reactions. ATP, also derived from the light-dependent reactions, provides additional energy. The key enzyme of the light-independent reactions is rubisco (RuBP carboxylase/oxygenase), which catalyzes the reaction that combines CO_2 into organic compounds (Figure 9.13).

- In the process, NADPH is oxidized to $NADP^+$, and ATP is hydrolyzed to ADP and phosphate. These products of the light-independent reactions cycle back as inputs to the light-dependent reactions.
- The Calvin cycle produces surplus molecules of G3P, which are the starting point for synthesis of glucose, sucrose, starches, and other organic molecules. The light-independent reactions take place in the chloroplast stroma in eukaryotes and in the cytoplasm of photosynthetic bacteria.

Animation: Calvin cycle

9.4 Photorespiration and the C_4 Cycle

- When oxygen concentrations are high relative to CO_2 concentrations, rubisco acts as an oxygenase, catalyzing the combination of RuBP with O_2 rather than CO_2 and forming toxic products that cannot be used in photosynthesis. The toxic products are eliminated by reactions that release carbon as CO_2, greatly reducing the efficiency of photosynthesis. The entire process is called photorespiration because it uses oxygen and releases CO_2 (Figure 9.14).
- Some plants have evolved the C_4 pathway, a supplemental system that bypasses the oxygenase activity of rubisco. In the pathway, initial fixation of CO_2 is catalyzed by a carboxylase that has no oxygenase activity, in specific locations or at times within the plant when oxygen is overabundant. In later steps, the CO_2 is released in relatively oxygen-free regions or times for final fixation in the reactions using RuBP in the Calvin cycle (Figures 9.15 and 9.16).

Animation: C_3-C_4 comparison

Questions

Self-Test Questions

1. An organism exists for long periods by using only CO_2 and H_2O. It could be classified as a (an):
   a. herbivore.
   b. carnivore.
   c. decomposer.
   d. autotroph.
   e. heterotroph.

2. During the light-dependent reactions:
   a. CO_2 is fixed.
   b. NADPH and ATP are synthesized using electrons derived from splitting water.
   c. glucose is synthesized.
   d. water is split and the electrons generated are used for glucose synthesis.
   e. photosystem I is unlinked from photosystem II.

3. Which of the following is a correct step in the light-dependent reactions of the Z system?
   a. Light is absorbed at P700, and electrons flow through a pathway to the NADPH acceptor.
   b. Electrons flow from photosystem II to water.
   c. $NADP^+$ is oxidized to NADPH as it accepts electrons.
   d. Water is degraded to activate P680.
   e. Electrons pass through a thylakoid membrane to create energy to pump H^+ through the cytochrome complex.

4. The light-dependent reactions of photosynthesis resemble aerobic respiration as both:
   a. synthesize NADPH.
   b. synthesize NADH.
   c. require electron transfer systems to synthesize ATP.
   d. require oxygen as the final electron acceptor.
   e. have the same initial energy source.

5. The molecules that link the light-dependent and light-independent reactions are:
   a. ADP and H_2O.
   b. RuBP and CO_2.
   c. cytochromes and water.
   d. G3P and RuBP.
   e. ATP and NADPH.

6. You bite into a spinach leaf. Which one of the following is true?
   a. You are getting 50% of the protein in the leaf in the form of ribulose 1,5-bisphosphate carboxylase.
   b. The major pigment you are ingesting is a carotenoid.
   c. The water in the leaf is a product of the light-independent reactions.
   d. Any energy from the leaf you can use directly is in the form of ATP.
   e. The spinach most likely was grown in an area with a low CO_2 concentration.

7. The molecule produced by the light-dependent reactions that is used for the synthesis of glucose and other organic molecules is:
   a. ADP.
   b. G3P.
   c. CO_2.
   d. $NADP^+$.
   e. NADPH.

8. Which of the following statements about the C_4 cycle is *incorrect*?
   a. CO_2 initially combines with PEP.
   b. PEP carboxylase catalyzes a reaction to produce oxaloacetate.
   c. Oxaloacetate transfers electrons from NADPH and is reduced to malate.
   d. Less ATP is used to run the C_4 cycle than the C_3 cycle.
   e. The cycle runs when O_2 concentration is high.

9. In one turn of the Calvin cycle, one molecule of CO_2 generates:
   a. 6 ATP.
   b. 6 NADH.
   c. 6 ATP and 6 NADPH.
   d. one (CH_2O) unit of carbohydrate.
   e. one molecule of glucose.

10. All of the following are adaptations that assist C_4 plants in surviving in hot dry regions *except:*
    a. closing stomata.
    b. using crassulacean acid metabolism.
    c. increasing their rate of photorespiration.
    d. running cycles at different times.
    e. running cycles at different positions in the cell.

Questions for Discussion

1. Suppose a garden in your neighborhood is filled with red, white, and blue petunias. Explain the floral colors in terms of which wavelengths of light are absorbed and reflected by the petals.

2. About 200 years ago, Jan Baptista van Helmont tried to determine the source of raw materials for plant growth. To do so, he planted a young tree weighing 5 pounds in a barrel filled with 200 pounds of soil. He watered the tree regularly. After 5 years, he again weighed the tree and the soil. At that time the tree weighed 169 pounds, 3 ounces, and the soil weighed

199 pounds, 14 ounces. Because the tree's weight had increased so much, and the soil's weight had remained about the same, he concluded that the tree gained weight as a result of the water he had added to the barrel. Criticize his conclusion in terms of the information you have learned from this chapter.

3. Like other accessory pigments, the carotenoids extend the range of wavelengths absorbed in photosynthesis. They also protect plants from a potentially lethal process known as *photooxidation*. This process begins when excitation energy in chlorophylls drives the conversion of oxygen into free radicals, substances that can damage organic compounds and kill cells. When plants that cannot produce carotenoids are grown in light, they bleach white and die. Given this observation, what molecules in the plants are likely to be destroyed by photooxidation?

4. What molecules would you have to provide a plant, theoretically speaking, for it to make glucose in the dark?

Experimental Analysis

Space travelers of the future land on a planet in a distant galaxy, where they find populations of a carbon-based life form. The beings on this planet are of a vibrantly purple color. The travelers sus-pect that the beings secure the energy necessary for survival by a process similar to photosynthesis on Earth. How might they go about testing this conclusion?

Evolution Link

If global warming raises the temperature of our climate signifi-cantly, will C_3 plants or C_4 plants be favored by natural selection? How will global warming change the geographical distributions of plants?

How Would You Vote?

The oxygen in Earth's atmosphere is a sure indicator that photo-synthetic organisms flourish here. New technologies will allow astronomers in search of life to measure the oxygen content of the atmosphere of planets too far away for us to visit. Should public funds be used to continue this research? Go to www .thomsonedu.com/login to investigate both sides of the issue and then vote.

A cell in mitosis (fluorescence micrograph). The spindle (red) is separating copies of the cell's chromosomes (green) prior to cell division.

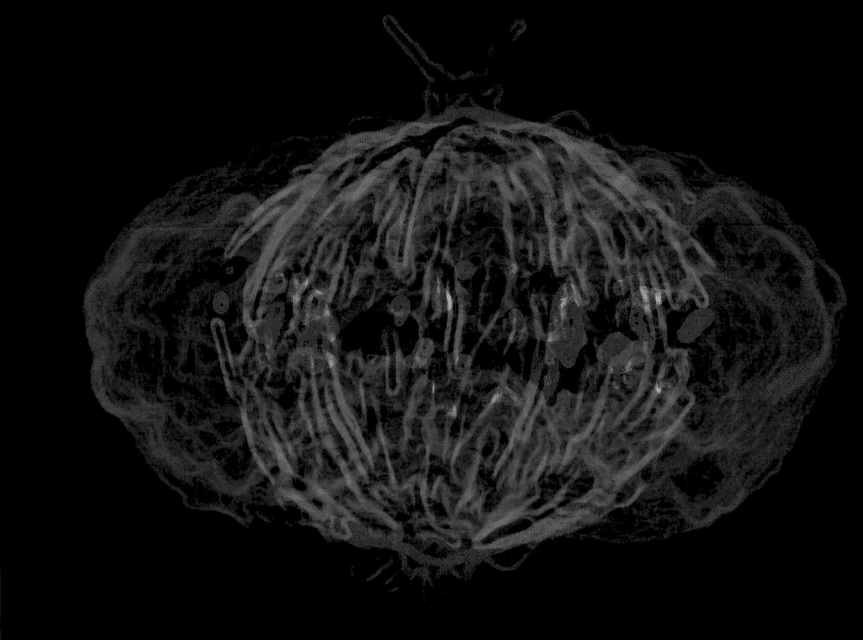

Dr. Paul Andrews, University of Dundee/Science Photo Library/Photo Researchers, Inc.

10 Cell Division and Mitosis

WHY IT MATTERS

The first rays of the sun dance over the wild Alagnak River of the Alaskan tundra. This September morning, life is both beginning and ending in the clear, cold waters. By the thousands, mature silver salmon have returned from the open ocean to spawn in their native freshwater stream. The salmon rest briefly in quiet eddies, then continue upstream **(Figure 10.1).** They are tinged with red, the color of spawning.

A female salmon pauses, then hollows out a shallow nest in the gravel riverbed. Now scores of translucent pink eggs emerge from her body (see Figure 10.1, inset). Within moments, a male salmon appears and sheds a cloud of sperm over the eggs. Trout and other predators will consume most of the eggs; but a few fertilized eggs will survive and give rise to a new generation of salmon.

The female lingers for some hours, but depleted of eggs and with vital organs failing, she soon dies and floats to the surface. A bald eagle loses no time in retrieving her carcass and consuming it on the riverbank. Yet, her remains speak of a remarkable journey. That female silver salmon started life as a pea-sized egg that was fertilized in the Alagnak's gravel riverbed. She hatched in the

Chris Huss

Figure 10.1

The end of one generation of silver salmon (*Oncorhynchus kisutch*) and the beginning of the next in the Alagnak River in Alaska. The inset shows eggs being laid by a female salmon.

stream, fed, and grew for a time, then migrated to the sea; within 3 years in the ocean, she became a fully grown adult salmon, fashioned from billions of cells. Early in her development, some of her cells were destined for reproduction, and in time, they gave rise to eggs that, after her return to the stream of her birth, were laid as part of an ongoing story of birth and reproduction.

For humans, as for the silver salmon and all other organisms, reproduction depends on the capacity of individual cells to grow and then to divide. Starting with a fertilized egg in your mother's body, a single cell divided into two, the two into four, and so on, until billions of cells were growing, developing along genetically determined pathways, and dividing further to produce the tissues and organs. Cell divisions still continue in many parts of the body. For example, constant cell divisions produce enough cells to replace the lining of the small intestine every 5 days; more than 2 million cells divide *each second* to maintain the supply of red blood cells. Cell divisions also underlie the development of egg or sperm cells in your body. All human cell divisions proceed almost without error despite the complexities of the mechanism.

The high accuracy of eukaryotic cell division depends on three elegantly interrelated systems. One system is DNA replication, which duplicates a DNA molecule into two copies with almost perfect fidelity. The second system is a mechanical system of microtubules, which divides the DNA copies precisely between the daughter cells. The third mechanism is an elaborate system of molecular controls that regulates when and where division occurs and corrects random mistakes. This chapter focuses on the mechanical and regulatory systems of cell division.

10.1 The Cycle of Cell Growth and Division: An Overview

As a prelude to dividing, most eukaryotic cells enter a period of growth, in which they synthesize proteins, lipids, and carbohydrates and at one stage replicate the nuclear DNA. After the growth period, the nuclei divide and, usually, *cytokinesis* (*cyto* = cell, derived from "hollow vessel"; *kinesis* = movement)—the division of the cytoplasm—follows, partitioning nuclei to daughter cells. Each daughter nucleus contains one copy of the replicated DNA. The sequence of events—a period of growth followed by nuclear division and cytokinesis—is known as the **cell cycle**.

The Products of Mitosis Are Genetic Duplicates of the Dividing Cell

In eukaryotic cell cycles, nuclear division after the growth period occurs by one of two mechanisms: *mitosis* or *meiosis*. **Mitosis** divides the replicated DNA equally and with great precision, producing daughter nuclei that are exact genetic copies of the parental nucleus. Cytokinesis segregates the daughter nuclei into separate cells. This version of the cell cycle—growth and mitosis followed by cytokinesis—is the mechanism by which multicellular eukaryotes increase into size and maintain their body mass. It is also the mechanism by which many single-celled eukaryotes such as yeast and protozoa reproduce. Another cell division process, meiosis, produces daughter nuclei that differ genetically from the parental nuclei entering the process. Meiosis occurs as part of the developmental changes that produce gametes in animals and spores in plants and many fungi.

This chapter concentrates on mitosis; meiosis and its role in generating genetic diversity are covered in Chapter 11. How prokaryotic organisms grow and divide also is explored in this chapter. We begin our discussion with **chromosomes**, the nuclear units of genetic information divided and distributed by mitotic cell division.

Chromosomes Are the Genetic Units Divided by Mitosis

In all eukaryotes, the hereditary information of the nucleus is distributed among individual, linear DNA molecules. These DNA molecules are combined with proteins, which stabilize the DNA molecules, maintain their structure, and control the activity of individual genes, the segments of DNA that code for proteins. Each linear DNA molecule, with its associated proteins, is known as a *chromosome* (*chroma* = color, referring to the strong colors the chromosomes of dividing cells take on when stained with dyes used to

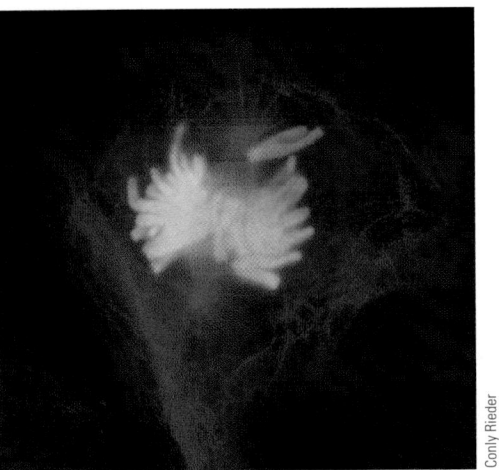

Figure 10.2
Eukaryotic chromosomes (blue) in a dividing animal cell.

prepare cells for light microscopy, and *soma* = body; **Figure 10.2**).

Many eukaryotes have two copies of each type of chromosome in their nuclei, so their chromosome complement is said to be **diploid**, or *2n*. For example, humans have 23 pairs of chromosomes for a diploid number of 46 chromosomes. Other eukaryotes, mostly microorganisms, have only one copy of each type of chromosome in their nuclei, so their chromosome complement is said to be **haploid**, or *n*. For example, yeast is a haploid organism with 16 different chromosomes. Still others, such as many plant species, have three, four, or even more complete sets of chromosomes in each cell. The number of chromosome sets is called the **ploidy** of a cell or species.

During replication, each chromosome is duplicated into two exact copies called **sister chromatids**. Mitosis separates the sister chromatids and places one in each of the two daughter nuclei produced by the division. *As a result of this precise division, each daughter nucleus receives exactly the same number and types of chromosomes and contains the same genetic information as the parent cell entering the division.* The equal distribution of daughter chromosomes to each of the two cells that result from cell division is **chromosome segregation**.

The precision of chromosome replication and segregation in the mitotic cell cycle underlies the growth of all multicellular eukaryotes. Each person's development from a fertilized egg, through billions of mitotic divisions, reflects the precision of mitotic division.

STUDY BREAK

Compare the DNA content of daughter cells with that of the parent cell.

10.2 The Mitotic Cell Cycle

Growth and division of both diploid and haploid cells occurs by the mitotic cell cycle. The first stage of the mitotic cell cycle is **interphase**. During this stage, the cell grows and replicates its DNA before undergoing mitosis (also called *M phase*) and cytokinesis (**Figure 10.3**). Internal regulatory controls trigger each phase, ensuring that the processes of one phase are completed successfully before the next phase can begin. In multicellular eukaryotes, the internal controls are modified by external signal molecules such as hormones, which coordinate the division of individual cells with the overall developmental and metabolic processes of the organism.

Interphase Extends from the End of One Mitosis to the Beginning of the Next Mitosis

Interphase begins as a daughter cell from a previous division cycle enters an initial period of cytoplasmic growth. During this initial growth stage, called the **G_1 phase** of the cell cycle, the cell makes proteins and other types of cellular molecules but not nuclear DNA (the G in G_1 stands for *gap*, referring to the absence of DNA synthesis). Then, if the cell is going to divide,

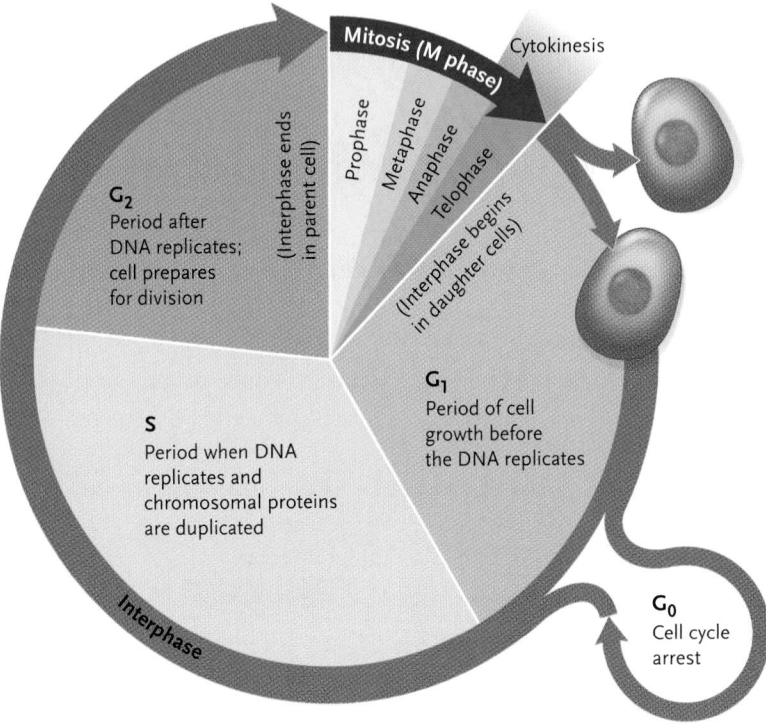

Figure 10.3
The cell cycle. The length of G_1 varies, but for a given cell type, the timing of S, G_2, and mitosis is usually relatively uniform. Cytokinesis (segment at 2 o'clock) usually begins while mitosis is in progress and reaches completion as mitosis ends. Cells in a state of division arrest are considered to enter a side loop or shunt from G_1 called G_0.

DNA replication begins, initiating the **S phase** of the cell cycle (S stands for *synthesis,* meaning DNA synthesis).

During the S phase, the cell duplicates the chromosomal proteins, as well as the DNA, and continues the synthesis of other cellular molecules. As the S phase is completed, the cell enters the **G₂ phase** of the cell cycle (G_2 refers to the second gap during which there is no DNA synthesis). During G_2, the cell continues to synthesize proteins, including those required for mitosis, and the cell continues to grow. At the end of G_2, which marks the end of interphase, mitosis begins. During all the steps of interphase, the chromosomes are in their extended form, making them invisible under a light microscope.

Usually, G_1 is the only phase of the cell cycle that varies in length. The other phases are typically uniform in length within a species. Thus, whether cells divide rapidly or slowly primarily depends on the length of G_1. Once DNA replication begins, most mammalian cells take about 10 to 12 hours to proceed through the S phase, about 4 to 6 hours to go through G_2, and about 1 hour or less to complete mitosis.

G_1 is also the stage in which many cell types stop dividing. This state of division arrest is often designated the **G₀ phase** (see Figure 10.3). For example, in humans, most cells of the nervous system stop dividing once they are fully mature.

The events of interphase are an important focus of research, particularly the regulatory controls for the transition from the G_1 phase to the S phase, and with it, the commitment to cell division. Understanding the molecular events that regulate the G_1/S phase transition is important because one of the hallmarks of cancer is loss of the normal control of that transition.

After Interphase, Mitosis Proceeds in Five Stages

Once it begins, mitosis proceeds continuously, without significant pauses or breaks. However, for convenience in study, biologists separate mitosis into five sequential stages: *prophase* (*pro* = before), *prometaphase* (*meta* = between), *metaphase, anaphase* (*ana* = back), and *telophase* (*telo* = end). Mitosis in an animal cell and a plant cell is shown in **Figures 10.4** and **10.5,** respectively. The entire process takes from 1 to 4 hours in most eukaryotes.

Prophase. During **prophase,** the duplicated chromosomes within the nucleus *condense* from the greatly

Figure 10.4

The stages of mitosis. Light micrographs show mitosis in an animal cell (whitefish embryo). Diagrams show mitosis in an animal cell with two pairs of chromosomes.

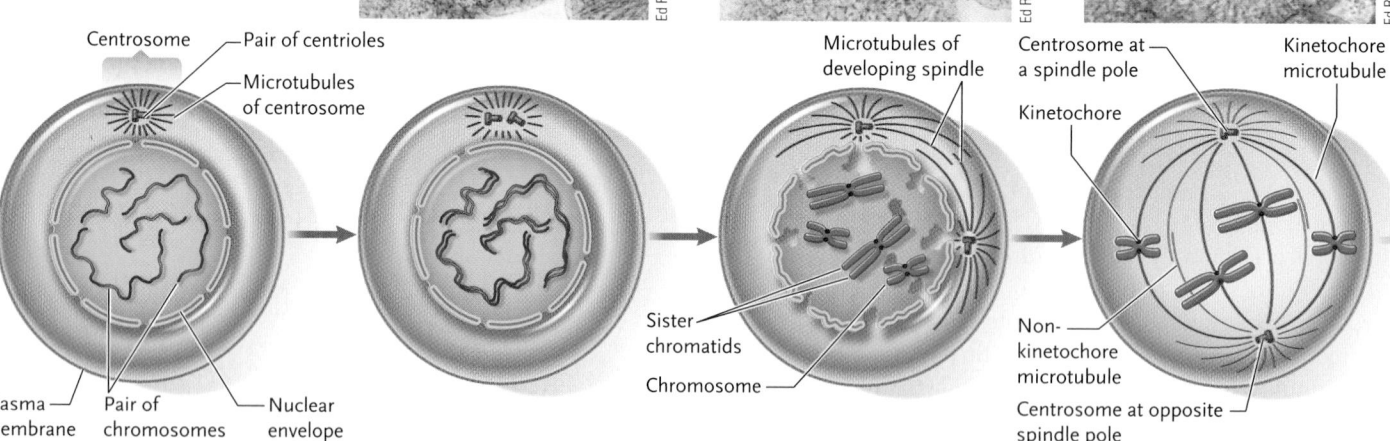

Interphase

Mitosis

Ed Reschke

Ed Reschke

Ed Reschke

Centrosome — Pair of centrioles

— Microtubules of centrosome

Plasma membrane — Pair of chromosomes — Nuclear envelope

Sister chromatids

Chromosome —

Microtubules of developing spindle

Centrosome at a spindle pole

Kinetochore

Kinetochore microtubule

Non-kinetochore microtubule

Centrosome at opposite spindle pole

G₁ of interphase	G₂ of interphase	Prophase	Prometaphase
The chromosomes are unreplicated and extend throughout the nucleus. For simplicity we show only two pairs of chromosomes. One of each pair was inherited from one parent, and the other was inherited from the other parent.	After replication during the S phase of interphase, each chromosome is double at all points and now consists of two sister chromatids. The centrioles within the centrosome have also doubled into pairs.	The chromosomes condense into threads that become visible under the light microscope. Each chromosome is double as a result of replication. The centrosome has divided into two parts, which are generating the spindle as they separate.	The nuclear envelope has disappeared and the spindle enters the former nuclear area. Microtubules from opposite spindle poles attach to the two kinetochores of each chromosome.

extended state typical of interphase into compact, rod-like structures. As they condense, the chromosomes appear as thin threads under the light microscope. (The word *mitosis* [*mitos* = thread] is derived from this threadlike appearance.) At this point, each chromosome is a double structure made up of two identical sister chromatids. While condensation is in progress, the nucleolus becomes smaller and eventually disappears in most species. The disappearance reflects a shutdown of all types of RNA synthesis, including the ribosomal RNA made in the nucleolus.

Why is condensation necessary? Each diploid human cell, although on average only 40 to 50 nm in diameter, contains *2 meters* of DNA distributed among 23 pairs of chromosomes. Condensation during prophase packs these long DNA molecules into units small enough to be divided successfully during mitosis.

In the cytoplasm, the mitotic **spindle** (**Figure 10.6;** see also Figure 10.11), the structure that actually separates chromatids, begins to form between the two centrosomes as they start migrating toward the opposite ends of the cell, where they will form the **spindle poles.** The spindle develops as two bundles of microtubules that radiate from the two spindle poles.

Prometaphase. At the end of prophase, the nuclear envelope breaks down, heralding the beginning of **prometaphase.** The developing spindle now enters the former nuclear area. Bundles of spindle microtubules grow from centrosomes at the *opposite spindle poles* toward the center of the cell. By this time, a complex of several proteins, a **kinetochore,** has formed on each chromatid at the **centromere,** a region named because it lies centrally in many chromosomes and because it forms a segment that is often narrower than the rest of the chromosome. *Kinetochore microtubules* bind to the kinetochores. These connections determine the outcome of mitosis, because they attach the sister chromatids of each chromosome to microtubules leading to the opposite spindle poles (see Figure 10.6). Nonkinetochore microtubules overlap those from the opposite spindle pole.

Metaphase. During **metaphase,** the spindle reaches its final form and the spindle microtubules move the chromosomes into alignment at the spindle midpoint, also called the *metaphase plate.* The chromosomes complete their condensation in this stage. The pattern of condensation gives each chromosome a characteristic shape, determined by the location of the centromere

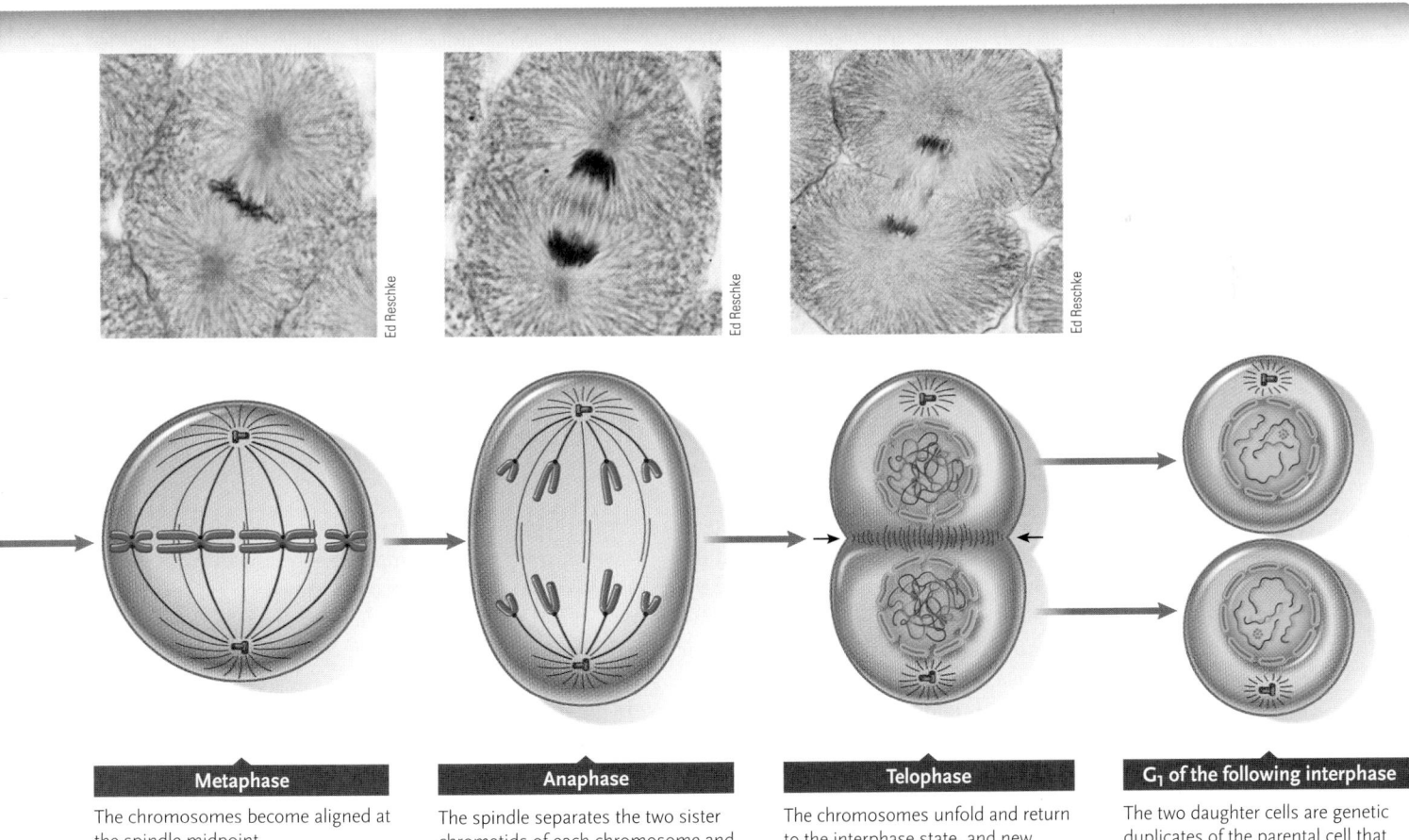

Metaphase	Anaphase	Telophase	G₁ of the following interphase
The chromosomes become aligned at the spindle midpoint.	The spindle separates the two sister chromatids of each chromosome and moves them to opposite spindle poles.	The chromosomes unfold and return to the interphase state, and new nuclear envelopes form around the daughter nuclei. The cytoplasm is beginning to divide by furrowing at the points marked by arrows.	The two daughter cells are genetic duplicates of the parental cell that entered mitotic division.

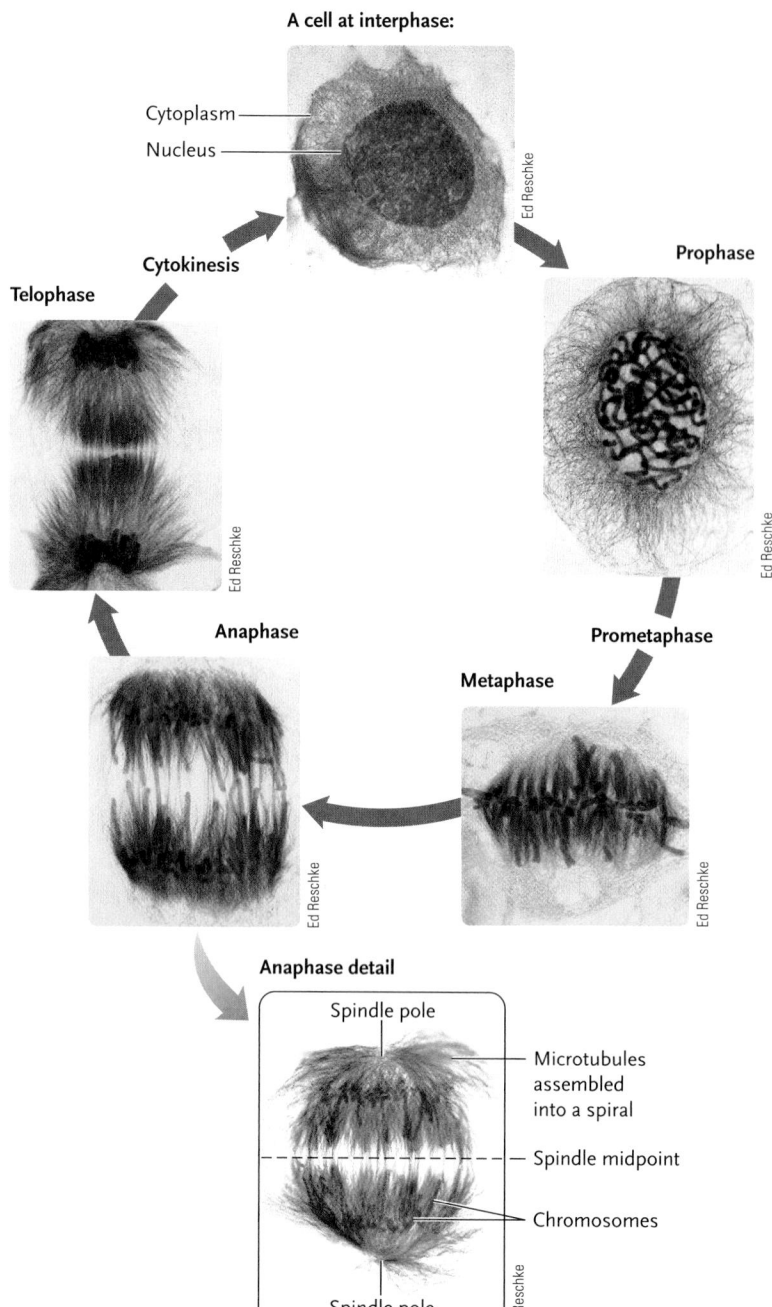

A cell at interphase:

Cytoplasm

Nucleus

Cytokinesis

Telophase

Prophase

Anaphase

Prometaphase

Metaphase

Anaphase detail

Spindle pole

Microtubules
assembled
into a spiral

Spindle midpoint

Chromosomes

Spindle pole

Figure 10.5
Mitosis in the
blood lily
Haemanthus. The
chromosomes are
stained blue; the
spindle microtu-
bules are stained
red.

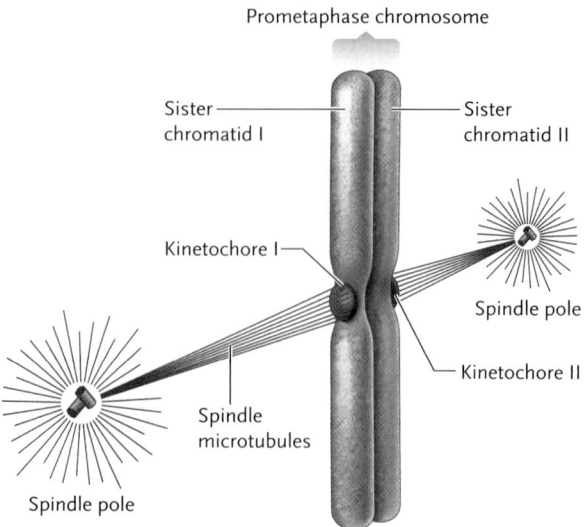

Prometaphase chromosome

Sister
chromatid I

Sister
chromatid II

Kinetochore I

Spindle pole

Kinetochore II

Spindle
microtubules

Spindle pole

Figure 10.6
Spindle connections made by chromosomes at mitotic prometaphase. The two kinetochores of the chromosome connect to
opposite spindle poles, ensuring that the chromatids are separated and moved to opposite spindle poles during anaphase.

be seen at the centromeres, where tension developed
by the spindle pulls the kinetochores toward opposite
poles. The movement continues until the separated
chromatids, now called *daughter chromosomes,* have
reached the two poles. At this point, chromosome segregation has been completed.

Telophase. During **telophase,** the spindle disassembles
and the chromosomes at each spindle pole decondense
and return to the extended state typical of interphase.
As decondensation proceeds, the nucleolus reappears,
RNA transcription resumes, and a new nuclear envelope forms around the chromosomes at each pole producing the two daughter nuclei. At this point, nuclear
division is complete, and the cell has two nuclei.

Cytokinesis Completes Cell Division by Dividing the Cytoplasm between Daughter Cells

Cytokinesis, the division of the cytoplasm, usually follows the nuclear division stage of mitosis and produces
two daughter cells, each containing one of the daughter nuclei. In most cells, cytokinesis begins during
telophase or even late anaphase. By the time cytokinesis is completed, the daughter nuclei have progressed
to the interphase stage and entered the G_1 phase of the
next cell cycle.

Cytokinesis proceeds by different pathways in the
various kingdoms of eukaryotic organisms. In animals,
protists, and many fungi, a groove, the **furrow,** girdles
the cell and gradually deepens until it cuts the cytoplasm into two parts. In plants, a new cell wall, called
the **cell plate,** forms between the daughter nuclei and
grows laterally until it divides the cytoplasm. In both

and the length and thickness of the arms that extend
from the centromere. The shapes and sizes of all the
chromosomes at metaphase form the **karyotype** of the
species. In many cases, the karyotype is so distinctive
that a species can be identified from this characteristic
alone. How human chromosomes are prepared for
analysis as a karyotype is shown in **Figure 10.7.**

Once the chromosomes are assembled at the spindle midpoint, with the sister chromatids of each chromosome attached to microtubules leading to opposite
spindle poles, metaphase is complete.

Anaphase. During **anaphase,** the spindle separates
sister chromatids and pulls them to opposite spindle
poles. The first signs of chromosome movement can

Figure 10.7 Research Method

Preparing a Human Karyotype

PURPOSE: A karyotype is a display of chromosomes of an organism arranged in pairs. A normal karyotype has a characteristic appearance for each species. Examination of the karyotype of the chromosomes from a particular individual indicates whether the individual has a normal set of chromosomes or whether there are abnormalities in number or appearance of individual chromosomes, and also indicates the species.

PROTOCOL:

1. Add sample (for example, blood) to culture medium that has stimulator for growth and division of white blood cells. Incubate at 37°C. Add colchicine, which causes spindle to disassemble, to arrest mitosis at metaphase.

2. Stain the cells so that the chromosomes are distinguished. Some stains produce chromosome-specific banding patterns, as shown in the photograph below.

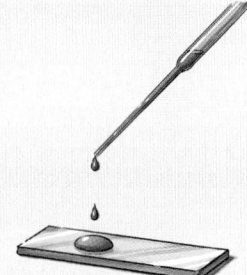

3. View the stained cells under a microscope equipped with a digital imaging system and take a digital photograph. A computer processes the photograph to arrange the chromosomes in pairs and number them according to size and shape.

Pair of homologous chromosomes

Pair of sister chromatids closely aligned side-by-side

1 2 3 4 5
6 7 8 9 10 11 12
13 14 15 16 17 18
19 20 21 22
XX

© Leonard Lessin/Peter Arnold, Inc.

Peter Arnold, Inc.

INTERPRETING THE RESULTS: The karyotype is evaluated with respect to the scientific question being asked. For example, it may identify a particular species, or it may indicate whether or not the chromosome set of a human (fetus, child, or adult) is normal or aberrant.

---

cases, the plane of cytoplasmic division is determined by the layer of microtubules that persist at the former spindle midpoint.

Furrowing. In furrowing, the layer of microtubules that remains at the former spindle midpoint expands laterally until it stretches entirely across the dividing cell **(Figure 10.8)**. As the layer develops, a band of microfilaments forms just inside the plasma membrane, forming a belt that follows the inside boundary of the cell in the plane of the microtubule layer (microfilaments are discussed in Section 5.3). Powered by motor proteins, the microfilaments slide together, tightening the band and constricting the cell. The constriction forms a groove—the furrow—in the plasma membrane. The furrow gradually deepens, much like the tightening of a drawstring, until the daughter cells are completely separated. The cy-

toplasmic division separates the daughter nuclei into the two cells and, at the same time, distributes the organelles and other structures (which also have doubled) approximately equally between the cells.

Cell Plate Formation. In cell plate formation, the layer of microtubules that persists at the former spindle midpoint serves as an organizing site for vesicles produced by the endoplasmic reticulum (ER) and Golgi complex **(Figure 10.9)**. As the vesicles collect, the layer expands until it spreads entirely across the dividing cell. During this expansion, the vesicles fuse together and their contents assemble into a new cell wall, the cell plate, that stretches completely across the former spindle midpoint. The junction separates the cytoplasm and its organelles into two parts and isolates the daughter nuclei in separate cells. The plasma membranes that line the two sur-

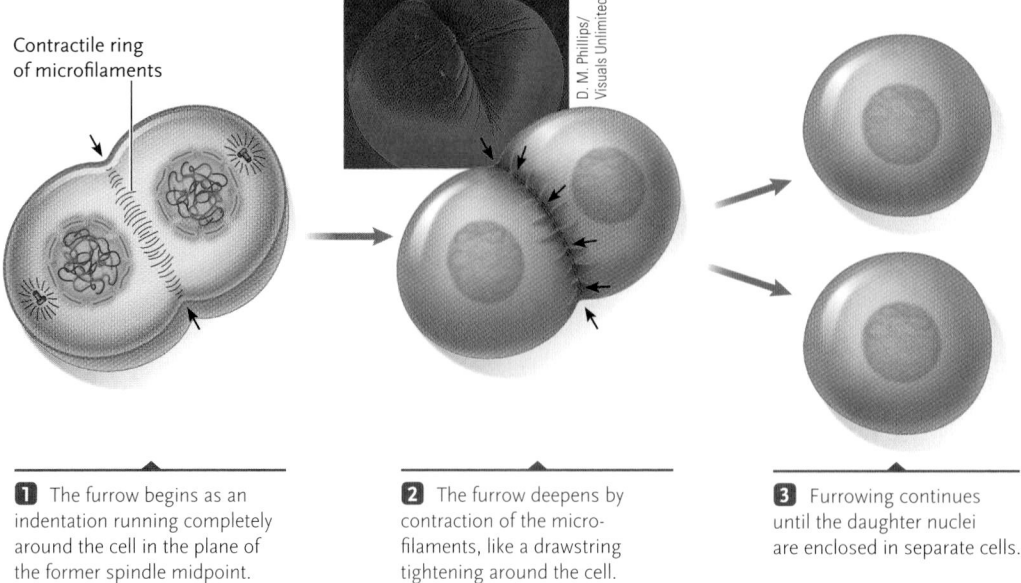

Contractile ring of microfilaments

D. M. Phillips/ Visuals Unlimited

1 The furrow begins as an indentation running completely around the cell in the plane of the former spindle midpoint.

2 The furrow deepens by contraction of the micro-filaments, like a drawstring tightening around the cell.

3 Furrowing continues until the daughter nuclei are enclosed in separate cells.

Figure 10.8
Cytokinesis by furrowing. The micrograph shows a furrow developing in the first division of a fertilized egg cell.

faces of the cell plate are derived from the vesicle membranes.

Microscopic pores, lined with plasma membrane, remain open in the cell plate. These openings, called *plasmodesmata* (singular, *plasmodesma*), form membrane-lined channels that directly connect the cytoplasm of the daughter cells. Molecules and ions that flow through the channels create direct avenues of communication between the daughter cells (see Section 5.4).

The Mitotic Cell Cycle Is Significant for Both Development and Reproduction

The mitotic cycle of interphase, nuclear division, and cytokinesis accounts for the growth of multicellular eukaryotes from single initial cells, such as a fertilized egg, to fully developed adults. Mitosis also serves as a method of reproduction called **vegetative** or **asexual reproduction**, which occurs in many kinds of plants and protists and in some animals. In asexual reproduction, daughter cells produced by mitotic cell division are released from the parent and grow separately by further mitosis into complete individuals. For example, asexual reproduction occurs when a single-celled protozoan such as an amoeba divides by mitosis to produce two separate individuals or when a leaf cutting is used to generate an entire new plant.

A group of cells produced by mitotic division of a single cell is known as a *clone*. Except for chance DNA mutations, all the cells of a clone are genetically identical because they are produced by mitosis. A clone may consist either of two or more individual cells or two or more entire multicellular organisms. (The *Focus on Research* explains how cells are grown as clones for biological experimentation.)

Figure 10.9
Cytokinesis by cell plate formation.

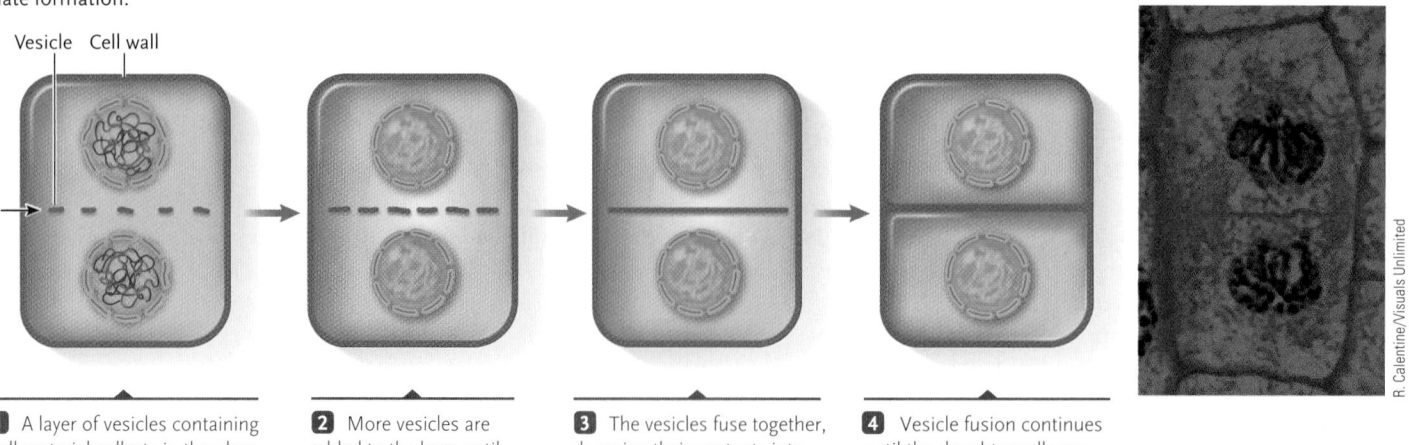

Vesicle   Cell wall

R. Calentine/Visuals Unlimited

1 A layer of vesicles containing wall material collects in the plane of the former spindle midpoint (arrow).

2 More vesicles are added to the layer until it extends across the cell.

3 The vesicles fuse together, dumping their contents into a gradually expanding wall between the daughter cells.

4 Vesicle fusion continues until the daughter cells are separated by a continuous new wall, the cell plate.

Basic Research: Growing Cell Clones in Culture

How can investigators safely test whether a particular substance is toxic to human cells or whether it can cure or cause cancer? One widely used approach is to work with **cell cultures**—living cells grown in laboratory vessels. Many types of prokaryotic and eukaryotic cells can be grown in this way.

When cell cultures are started from single cells, they form **clones**: barring mutations, all the individuals descending from the original cell are genetically identical. Clones are ideal for experiments in genetics, biochemistry, molecular biology, and medicine because the cells lack genetic differences that could affect the experimental results.

Microorganisms such as yeasts and many bacteria are easy to grow in laboratory cultures. For example, the human intestinal bacterium *Escherichia coli* can be grown in solutions (growth media) that contain only an organic carbon source such as glucose, a nitrogen source, and inorganic salts. Under optimal conditions, the cycle of cell growth and division of *E. coli* cells takes 20 minutes. As a result, large numbers of cells are produced in a short time. The cells may be

grown in liquid suspensions or on the surface of a solid growth medium such as an agar gel (agar is a polysaccharide extracted from an alga). Many thousands of bacterial strains are used in a wide variety of experimental studies.

Many types of plant cells can also be cultured as clones in specific growth media. With the addition of plant growth hormones, complete plants can often be grown from single cultured cells. Growing plants from cultured cells is particularly valuable in genetic engineering, in which genes introduced into cultured cells can be tracked in fully developed plants. Plants that have been engineered successfully can then be grown simply by planting their seeds.

Animal cells vary in what is needed to culture them. For many types, the culture medium must contain essential amino acids—that is, the amino acids that the cells cannot make for themselves. In addition, mammalian cells require specific growth factors provided by adding blood serum, the fluid part of the blood left after red and white blood cells are removed.

Even with added serum, many types of normal mammalian cells can-

not be grown in long-term cultures. Eventually, the cells stop dividing and die. By contrast, tumor cells often form cultures that grow and divide indefinitely.

The first successful culturing of cancer cells was performed in 1951 in the laboratory of George and Margaret Gey (Johns Hopkins University, Baltimore, MD). Gey and Gey's cultures of normal cells died after a few weeks, but the researchers achieved success with a culture of tumor cells from a cancer patient. The cells in culture continued to grow and divide; in fact, descendants of those cells are still being cultured and used for research today. The cells were given the code name *HeLa*, from the first two letters of the patient's first and last names—Henrietta Lacks. Unfortunately, the tumor cells in Henrietta's body also continued to grow, and she died within 2 months of her cancer diagnosis.

Other types of human cells have since been grown successfully in culture, derived either from tumor cells or normal cells that have been "immortalized" by inducing genetic changes that transformed them into tumorlike cells.

Mitosis Varies in Detail But Always Produces Duplicate Nuclei

Although variations occur in the details of mitosis, particularly among protists, fungi, and primitive plants, its function is to duplicate nuclei each with the same set of chromosomes as the nucleus of the parent cell. The process is the same no matter what the chromosome number of the cell is. That is, the number of chromosome sets does not affect the outcome of mitosis because each chromosome attaches individually to spindle microtubules and proceeds independently through the division process.

STUDY BREAK

1. What is the order of the stages of mitosis?
2. What is the importance of centromeres to mitosis?

3. Colchicine, an alkaloid extracted from plants, prevents the formation of spindle microtubules. What would happen if a cell enters mitosis when colchicine is present?

10.3 Formation and Action of the Mitotic Spindle

The mitotic spindle is central to both mitosis and cytokinesis. The spindle is made up of microtubules and their motor proteins, and its activities depend on their changing patterns of organization during the cell cycle.

Microtubules form a major part of the interphase cytoskeleton of eukaryotic cells. (Section 5.3 outlines the patterns of microtubule organization in the cytoskeleton.) As mitosis approaches, the microtubules disassemble from their interphase arrangement and

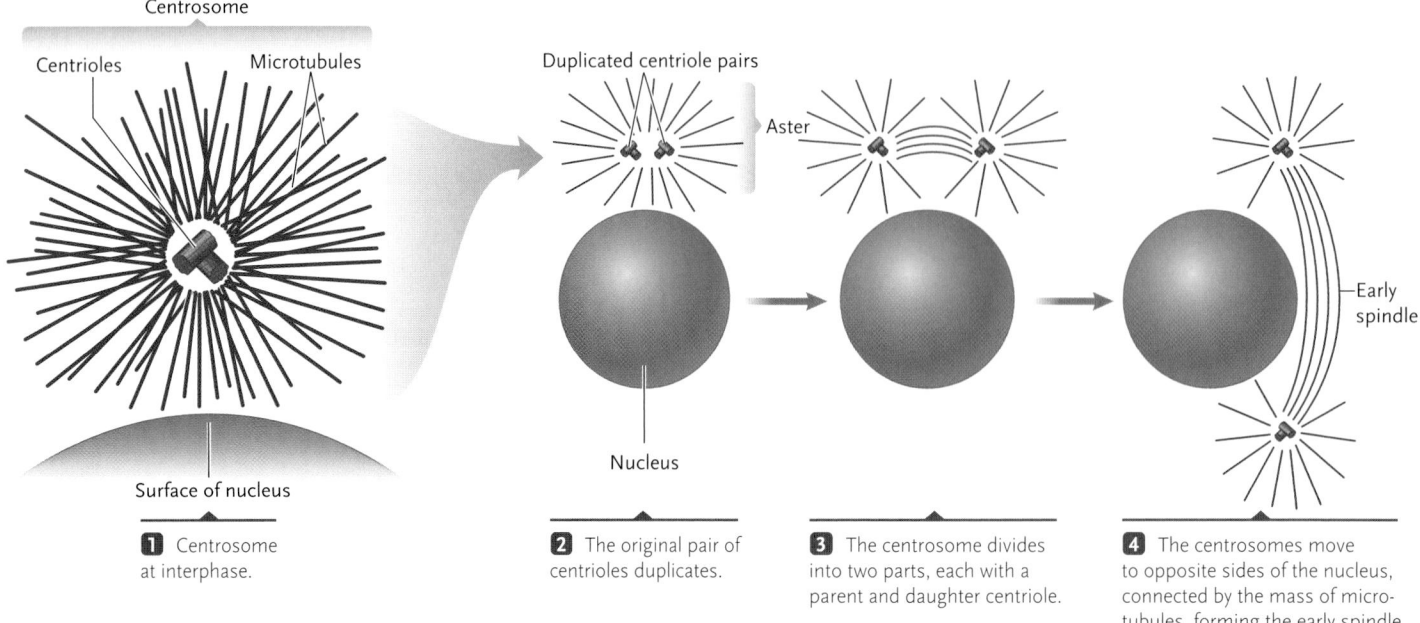

1 Centrosome at interphase.

2 The original pair of centrioles duplicates.

3 The centrosome divides into two parts, each with a parent and daughter centriole.

4 The centrosomes move to opposite sides of the nucleus, connected by the mass of microtubules, forming the early spindle.

Figure 10.10
The centrosome and its role in spindle formation.

reorganize into the spindle, which grows until it fills almost the entire cell. This reorganization follows one of two pathways in different organisms, depending on the presence or absence of a *centrosome* during interphase. However, once organized, the basic function of the spindle is the same, regardless of whether a centrosome is present.

Animals and Plants Form Spindles in Different Ways

Animal cells and many protists have a **centrosome**, a site near the nucleus from which microtubules radiate outward in all directions (**Figure 10.10**, step 1). The centrosome organizes the microtubule cytoskeleton during interphase and positions many of the cytoplasmic organelles (see Section 5.3). In fact, the centrosome is the main **microtubule organizing center (MTOC)** of the cell. The centrosome contains a pair of **centrioles**, usually arranged at right angles to each other. The radiating microtubules of the centrosome surround the centrioles. These microtubules, rather than the centrioles, generate the spindle. That is, if experimenters remove the centrioles, the spindle still forms by essentially the same pattern.

At the time that DNA replicates during the S phase of the cell cycle, the centrioles within the centrosome also duplicate, producing two pairs of centrioles (Figure 10.10, step 2). As *prophase* begins in the M phase, the centrosome separates into two parts, each containing one "old" and one "new" centriole—one centriole of the original pair and its copy (step 3). The duplicated centrosomes, with the centrioles inside them, continue to separate until they reach opposite ends of the nucleus (step 4). As they move apart, the microtubules between the centrosomes lengthen and increase in number.

By *late prophase,* when the centrosomes are fully separated, the microtubules that extend between them form a large mass around one side of the nucleus called the *early spindle.* When the nuclear envelope subsequently breaks down at the end of prophase, the spindle moves into the region formerly occupied by the

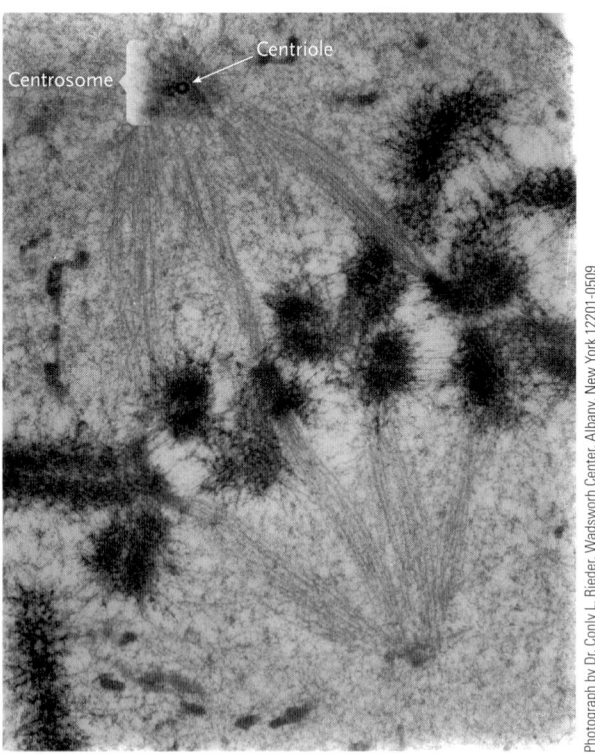

Figure 10.11
A fully developed spindle in a mammalian cell. Only microtubules connected to chromosomes have been caught in the plane of this section. One of the centrioles is visible in cross section in the centrosome at the top of the micrograph. Original magnification ×14,000.

nucleus and continues growing until it fills the cytoplasm. The microtubules that extend from the centrosomes also grow in length and extent, producing radiating arrays called **asters** that appear starlike under the light microscope.

By dividing the duplicated centrioles, the spindle ensures that when the cytoplasm divides during cytokinesis, the daughter cells each receive a pair of centrioles and that centrioles are maintained in the cell line. In the cell and its descendents, centrioles carry out their primary function: they generate flagella or cilia, the whiplike extensions that provide cell motility, at one or more stages of the life cycle of a species (see Section 5.5).

No centrosome or centrioles are present in angiosperms (flowering plants) or in most gymnosperms, such as conifers. Instead, the spindle forms from microtubules that assemble in all directions from multiple MTOCs surrounding the entire nucleus (see Figure 10.5). When the nuclear envelope breaks down at the end of prophase, the spindle moves into the former nuclear region, as in animals.

Mitotic Spindles Move Chromosomes by a Combination of Two Mechanisms

When fully formed at metaphase, the spindle may contain from hundreds to many thousands of microtubules, depending on the species **(Figure 10.11)**. In almost all eukaryotes, these microtubules are divided into two groups. *Kinetochore microtubules* connect the chromosomes to the spindle poles **(Figure 10.12a)**. *Nonkinetochore microtubules* extend between the spindle poles without connecting to chromosomes; at the spindle midpoint, these microtubules from one pole overlap with microtubules from the opposite pole **(Figure 10.12b)**. The separation of the chromosomes at anaphase results from a combination of separate but coordinated movements produced by the two types of microtubules.

In kinetochore microtubule–based movement, the motor proteins in the kinetochores of the chromosomes "walk" along the kinetochore microtubules, pulling the chromosomes with them until they reach the poles **(Figure 10.13)**. The kinetochore microtubules disassemble as the kinetochores pass along them; thus, the microtubules become shorter as the movement progresses (see Figure 10.12a). The movement is similar to a locomotive traveling over a railroad track, except that the track is disassembled as the locomotive passes by.

In nonkinetochore microtubule–based movement, the entire spindle is lengthened, pushing the poles farther apart (see Figure 10.12b). The pushing movement is produced by microtubules sliding over one another in the zone of overlap, powered by proteins acting as microtubule motors. In many species, the nonkinetochore microtubules also push the poles apart by growing in length as they slide.

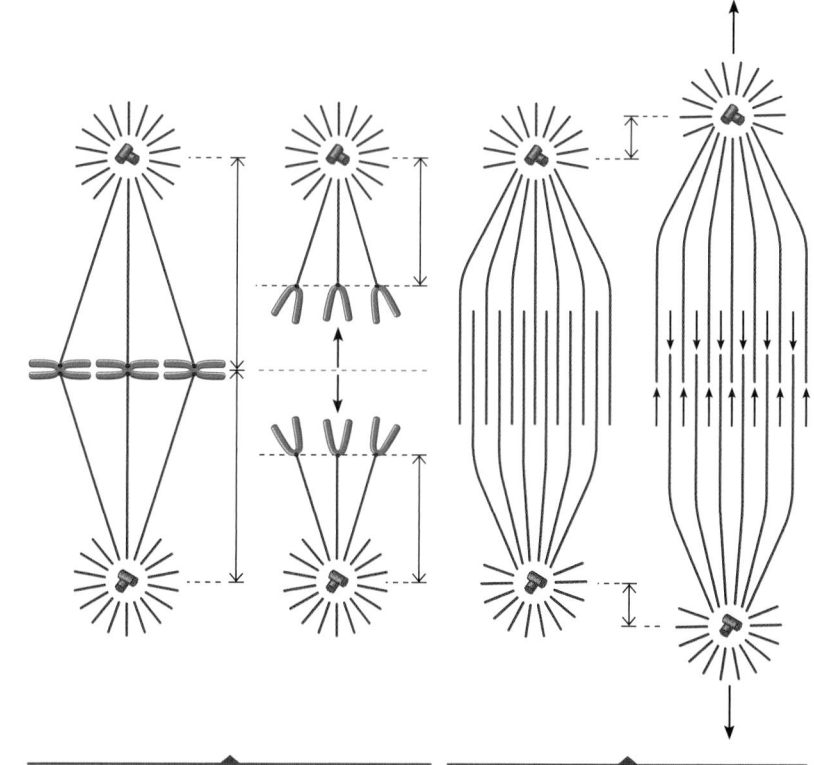

a. The kinetochore microtubules connected to the kinetochores of the chromosomes become shorter, lessening the distance from the chromosomes to the poles.

b. Sliding of the nonkinetochore microtubules in the zone of overlap at the spindle midpoint pushes poles farther apart and increases the total length of the spindle.

Figure 10.12
The two microtubule-based movements of the anaphase spindle.

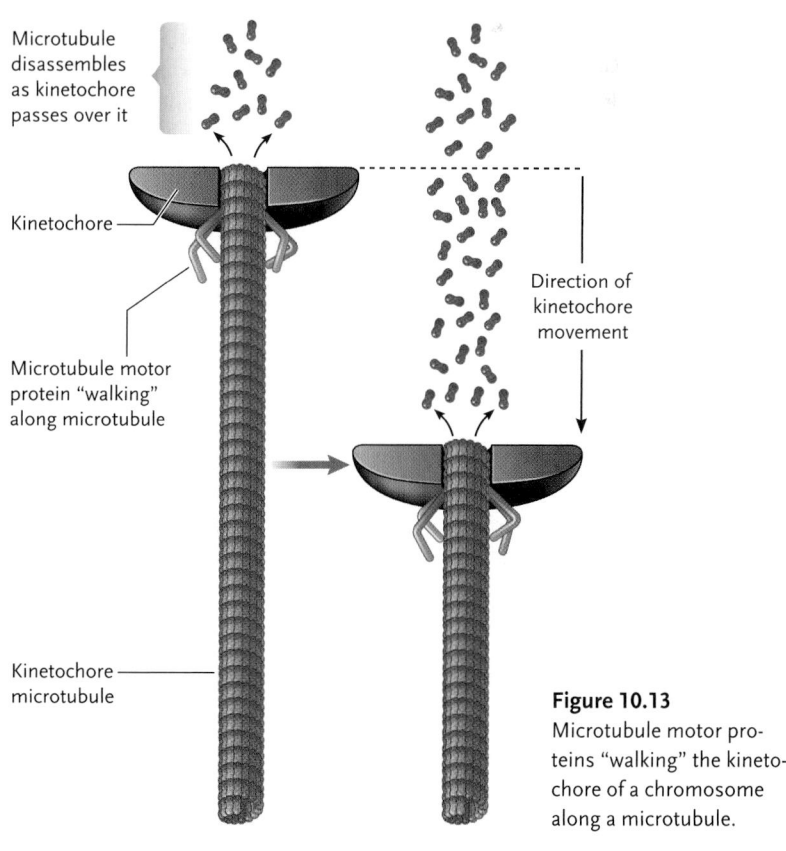

Microtubule disassembles as kinetochore passes over it

Kinetochore

Direction of kinetochore movement

Microtubule motor protein "walking" along microtubule

Kinetochore microtubule

Figure 10.13
Microtubule motor proteins "walking" the kinetochore of a chromosome along a microtubule.

Figure 10.14 Experimental Research

How Do Chromosomes Move during Anaphase of Mitosis?

HYPOTHESIS: One hypothesis was that kinetochore microtubules move during anaphase, pulling chromosomes to the poles. An alternative hypothesis was that chromosomes move by sliding over or along kinetochore microtubules.

EXPERIMENT: G. J. Gorbsky and his colleagues made regions of the kinetochore microtubules visibly distinct to test the hypotheses.

1. Kinetochore microtubules were combined with a dye molecule that bleaches when it is exposed to light.

2. The region of the spindle between the kinetochores and the poles was exposed to a microscopic beam of light that bleached a narrow stripe across the microtubules. The bleached region could be seen with a light microscope and analyzed as anaphase proceeded.

RESULTS: The bleached region remained at the same distance from the pole as the chromosomes moved toward the pole.

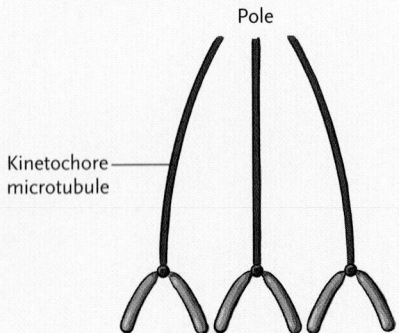

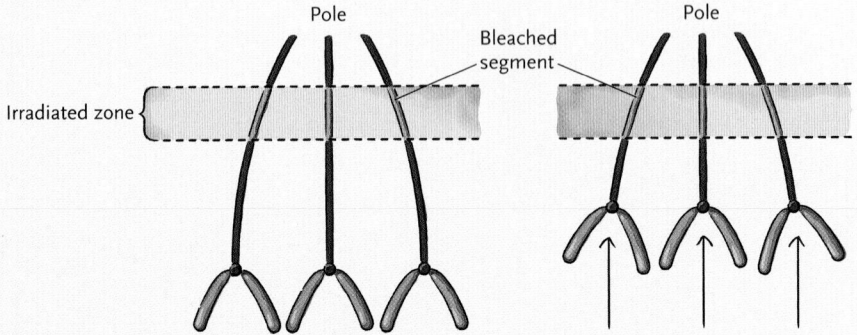

CONCLUSION: The results support the hypothesis that chromosomes move by sliding over or along kinetochore microtubules.

Researchers discovered kinetochore-based movement by tagging kinetochore microtubules at points with a microscopic beam of ultraviolet light, producing bleached sites that could be seen in the light microscope **(Figure 10.14)**. As the chromosomes were pulled to the spindle poles, the bleached sites stayed in the same place. This result showed that the kinetochore microtubules do not move with respect to the poles during the anaphase movement; it is the basis for our understanding of chromosomes' movement along the microtubules.

STUDY BREAK

1. How does spindle formation differ in animals and plants?
2. How do mitotic spindles move chromosomes?

10.4 Cell Cycle Regulation

We have noted that a number of internal and external regulatory mechanisms control the mitotic cell cycle. As part of the internal controls, the cell cycle has built-in **checkpoints** to prevent critical phases from beginning until the previous phases are completed. Hormones, growth factors, and other external controls coordinate the cell cycle with the needs of an organism by stimulating or inhibiting division. Some key research contributing to our understanding of cell cycle regulation, particularly defining the genes involved and their protein products, was done using yeast. The *Focus on Research: Model Research Organisms* describes yeast and its role in research in more detail.

Cyclins and Cyclin-Dependent Kinases Are the Internal Controls That Directly Regulate Cell Division

A major factor that regulates cell division is the complex of a protein called **cyclin** with an enzyme called **cyclin-dependent kinase (CDK)**. The CDK is a *protein kinase*, which adds phosphate groups to target proteins. The activity of the CDK directly affects the cell cycle, whereas cyclin turns CDK "on" or "off." R. Timothy Hunt, Imperial Cancer Research Fund, London, UK, received a Nobel Prize in 2001 for discovering cyclins.

FOCUS ON RESEARCH

Model Research Organisms: The Yeast *Saccharomyces cerevisiae*

Saccharomyces cerevisiae, commonly known as baker's yeast or brewer's yeast, was probably the first microorganism to have been grown and kept in cultures—a beer-brewing vessel is basically a *Saccharomyces* culture. Favorite strains of baker's and brewer's yeast have been kept in continuous cultures for centuries. The yeast has also been widely used in scientific research; its microscopic size and relatively short generation time make it easy and inexpensive to culture in large numbers in the laboratory.

The cells growing in *Saccharomyces* cultures are haploids. If the culture conditions are kept at optimal levels (which requires only a source of a fermentable sugar such as glucose, a nitrogen source, and minerals), the cells reproduce asexually by budding. *Saccharomyces* has two mating types (that is, sexes). If two yeast cells of different mating types contact one another, they fuse—mate—producing a diploid cell. Diploid cells can also reproduce asexually by budding. Diploid yeast can be induced to reproduce sexually by adjusting cultures to less favorable conditions, such as a reduced nitrogen supply. The cells then undergo meiosis, producing haploid spores. When

conditions again become favorable, the spores germinate into haploid cells, which reproduce asexually.

Sexual reproduction allows *Saccharomyces* to be used for genetic crosses. Its large number of offspring makes it possible to detect relatively rare genetic events, as can be done with bacteria. Genetic studies with *Saccharomyces* led to the discovery of some of the genes that control the eukaryotic cell cycle, including those for the entry into DNA replication and both mitotic and meiotic cell division. (Leland Hartwell, Fred Hutchinson Cancer Research Center, Seattle, Washington, and Paul Nurse, Imperial Cancer Research Fund, London, UK, received a Nobel Prize for their work in this area.) Many of these genes, after their first discovery in yeast cells, were found to have counterparts in animals and plants. Genetic studies with *Saccharomyces* were also the first to show the genes carried in the DNA of mitochondria and their patterns of inheritance. The complete DNA sequence of *S. cerevisiae*, which includes more than 12 million base pairs that encode about 6000 genes, was the first eukaryotic genome to be obtained.

Another advantage of yeast for genetic studies is that plasmids, extrachromosomal segments of DNA, have been produced that can be used for introducing genes into yeast cells. Using the plasmids, researchers can alter essentially any of the yeast genes experimentally to test their functions and can introduce genes or DNA samples from other organisms for testing or cloning. These genetic engineering studies have demonstrated that many mammalian genes can replace yeast genes when introduced into the fungi, confirming their close relationships, even though mammals and fungi are separated by millions of years of evolution.

Saccharomyces has been so important to genetic studies in eukaryotes that it is often called the eukaryotic *E. coli*. Research with another yeast, *Schizosaccharomyces pombe*, has been similarly productive, particularly in studies of genes that control the cell cycle.

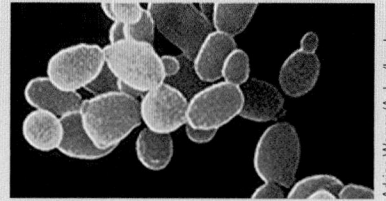

Adrian Warren/Ardea/London

CDK enzymes are "cyclin-dependent" because they are active only when combined with a cyclin molecule. The levels of the CDKs are the same throughout the cell cycle, but the levels of the cyclins fluctuate, reaching amounts capable of activating CDKs only at particular points of the cell cycle. The name *cyclin* reflects these cyclic changes in its concentration.

Several different cyclin/CDK combinations regulate cell cycle transitions at checkpoints. For example, the two cyclin/CDK combinations that control the cell cycle at the G_1-to-S and the G_2-to-M checkpoints are shown in **Figure 10.15.** At the G_1-to-S checkpoint, cyclin E has reached a concentration high enough to form a complex with CDK2 and activate it. The CDK2 of the complex then phosphorylates a number of cell cycle control proteins, which trigger the cell to make the transition into the S phase. After the transition is made, the level of cyclin E decreases by degradation of the

protein. CDK2 then becomes activated again only when cyclin E levels are high at the next G_1-to-S checkpoint.

Similar events occur at the G_2-to-M checkpoint. Cyclin B reaches a sufficient level to complex with CDK1. When the cell is ready to enter mitosis, the CDK1 of the complex phosphorylates a number of target proteins that move the cell from the G_2-to-M phase and promote the stages of mitosis. The activity of the CDK1/cyclin B complex is highest during metaphase; during anaphase, the cyclin B component is degraded and the transition from mitosis to G_1 soon occurs.

Internal Checkpoints Stop the Cell Cycle if Stages Are Incomplete

The cyclin/CDK combinations directly control the cell cycle, but other factors within the cell act as indirect controls by altering the activity of the cyclin/CDK com-

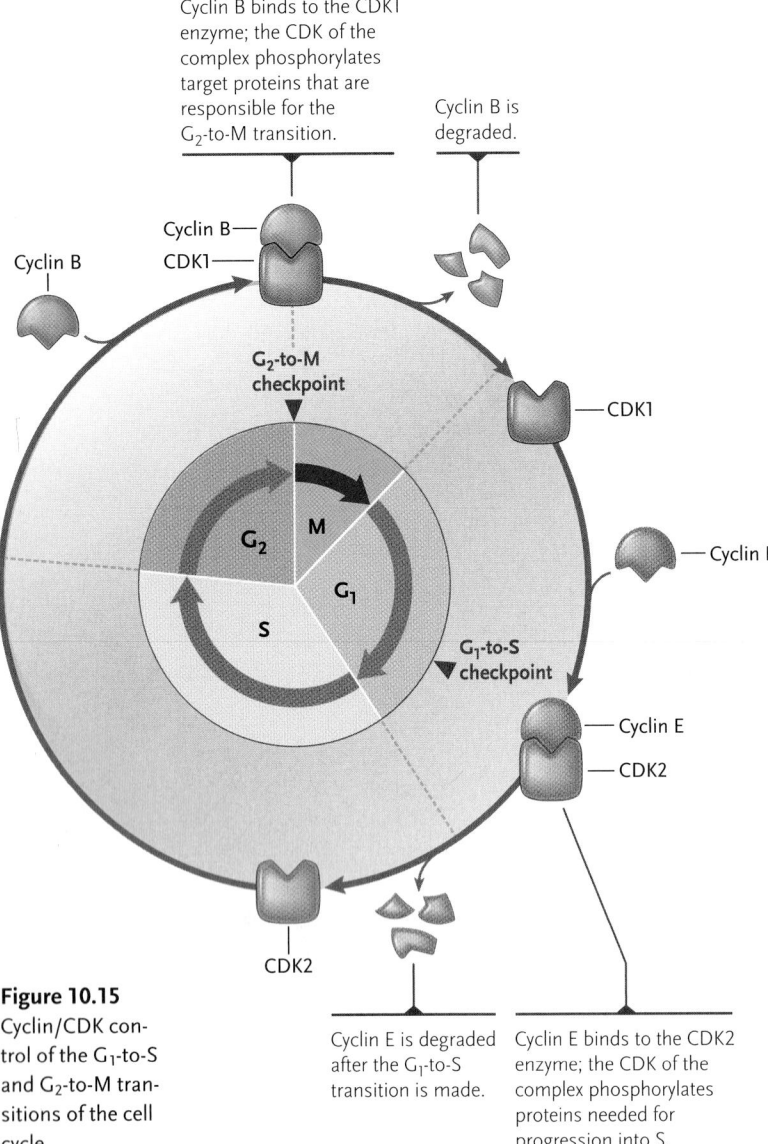

Cyclin B binds to the CDK1 enzyme; the CDK of the complex phosphorylates target proteins that are responsible for the G₂-to-M transition.

Cyclin B is degraded.

Cyclin B

Cyclin B—
CDK1—

G₂-to-M checkpoint

CDK1

M

G₂

Cyclin E

G₁

S

Cyclin E

G₁-to-S checkpoint

CDK2

CDK2

Cyclin E is degraded after the G₁-to-S transition is made.

Cyclin E binds to the CDK2 enzyme; the CDK of the complex phosphorylates proteins needed for progression into S.

Figure 10.15
Cyclin/CDK control of the G₁-to-S and G₂-to-M transitions of the cell cycle.

plexes. Among the most critical controls are the checkpoints that keep the cycle from progressing to the next phase unless the actions of a previous phase are successfully completed. At each key checkpoint, regulatory events block the cyclin/CDK complex from triggering the associated cell cycle transition until the cell is ready and able to undergo that transition.

For example, as we just described, the cyclin B/CDK1 complex stimulates the cell to enter the M phase. Until the cell is ready to enter mitosis, phosphorylation of a site on CDK1 in the complex by another kinase keeps the CDK inactive. When the cell is ready, a phosphatase removes the inhibitory phosphate, the CDK becomes active, and the cell is moved into mitosis. Control at checkpoints is exerted in many types of circumstances. For instance, if not all of the DNA is replicated during the S phase, the cell slows its progress during G₂ to allow more time for replication to be completed. Similarly, if radiation or chemicals damage DNA, inhibitory events at checkpoints around

the cell cycle will slow the cycle to give the cell time to potentially repair the damage.

External Controls Coordinate the Mitotic Cell Cycle of Individual Cells with the Overall Activities of the Organism

The internal controls that regulate the cell cycle are modified by signal molecules that originate from outside the dividing cells. In animals, these signal molecules include the peptide hormones and similar proteins called *growth factors.*

Many of the external factors bind to receptors at the cell surface, which respond by triggering reactions inside the cell. These reactions often include steps that add inhibiting or stimulating phosphate groups to the cyclin/CDK complexes, particularly to the CDKs. The reactions triggered by the activated receptor may also directly affect the same proteins regulated by the cyclin/CDK complexes. The overall effect is to speed, slow, or stop the progress of cell division, depending on the particular hormone or growth factor and the internal pathway that is stimulated. Some growth factors are even able to break the arrest of cells shunted into the G₀ stage and return them to active division. (Hormones, growth factors, and other signal molecules are part of the cell communication system, as discussed in Chapter 7.)

Cell-surface receptors in animals also recognize contact with other cells or with molecules of the extracellular matrix (see Section 5.5). The contact triggers internal reaction pathways that inhibit division by arresting the cell cycle, usually in the G₁ phase. The response, called **contact inhibition**, stabilizes cell growth in fully developed organs and tissues. As long as the cells of most tissues are in contact with one another or the extracellular matrix, they are shunted into the G₀ phase and prevented from dividing. If the contacts are broken, the freed cells often enter rounds of division.

Contact inhibition is easily observed in cultured mammalian cells grown on a glass or plastic surface. In such cultures, division proceeds until all the cells are in contact with their neighbors in a continuous, unbroken, single layer. At this point, division stops. If a researcher then scrapes some of the cells from the surface, cells at the edges of the "wound" are released from inhibition and divide until they form a continuous layer and all the cells are again in contact with their neighbors.

Cell Cycle Controls Are Lost in Cancer

Cancer occurs when cells lose the normal controls that determine when and how often they will divide. Cancer cells divide continuously and uncontrollably, producing a rapidly growing mass called a *tumor* **(Figure 10.16).** Cancer cells also typically lose their adhesions to other cells and often become actively mobile. As a result, in a process called *metastasis,* they tend to break loose from an original tumor, spread throughout the body,

Herpesviruses and Uncontrolled Cell Division

Almost all of us harbor one or more herpesviruses as more or less permanent residents in our cells. Fortunately, most of the herpesviruses are relatively benign—one group is responsible for the bothersome but nonlethal oral and genital ulcers known commonly as cold sores or "herpes." But another virus, *herpesvirus 8,* has been implicated as the cause of two kinds of cancer: Kaposi's sarcoma and lymphomas of the body cavity. How does herpesvirus 8 cause the uncontrolled cell division characteristic of malignant tumors? To answer this question, investigators in London and at the Friedrich-Alexander University in Germany decided to examine the effects of herpesvirus 8 on the primary transition point that leads to cell division, the change from G_1 to S. The investigators focused on how the virus might interfere with regulatory mechanisms that regulate the rate of cell division.

One way that cells slow their rate of division is to use regulatory proteins that inhibit cyclin D/CDK complexes. Cylin D combined with either CDK4 or CDK6 contributes to the G_1/S transition and thus stimulates cell division. Three important regulatory proteins, p16, p21, and p27, can slow cell division by binding to cyclin D/CDK complexes. These proteins prevent normal cells from becoming transformed into cancer cells (*p* stands for protein; the number indicates the molecular weight in thousands). Because these proteins have that ability, they are called *tumor suppressor proteins.*

The investigators knew that the DNA of herpesvirus 8 encodes a protein that acts as a cyclin. Could this viral cyclin, *K-cyclin,* be the means by which the herpesvirus bypasses normal controls and triggers the rapid cell division characteristic of cancer? To answer this question, researchers first inserted the DNA coding for K-cyclin into a benign virus. When they infected cultured human cells with this virus, the virus produced K-cyclin, which bound to the human CDK6. These K-cyclin/CDK6 complexes stimulated the initiation of the S phase much faster than the normal cyclin D/CDK6 complexes. In addition, the tumor suppressor proteins that normally regulate cell division by binding to cyclin D/CDK6 complexes were unable to bind to the K-cyclin complexes, resulting in uncontrolled division.

Thus, herpesvirus 8 has evolved as a mechanism that overrides normal cellular controls and triggers cell division. At some point in its evolution, the virus may have picked up a copy of a cyclin gene, which through mutation and selection became the K-cyclin that is unaffected by the inhibitors. Researchers hope that their findings may lead to treatments that can switch off K-cyclin and stop the virally induced tumor growth.

---

and grow into new tumors in other body regions. Metastasis is promoted by changes that defeat contact inhibition and alter the cell-surface molecules that link cells together or to the extracellular matrix.

Enlarging tumors damage surrounding normal tissues by compressing them and interfering with blood supply and nerve function. Tumors may also break through barriers such as the outer skin, internal cell layers, or the gut wall. The breakthroughs cause bleeding, open the body to infection by microorganisms, and destroy the separation of body compartments necessary for normal function. Both compression and breakthroughs can cause pain that, in advanced cases, may become extreme. As tumors increase in mass, the actively growing and dividing cancer cells may deprive normal cells of their required nutrients, leading to generally impaired body functions, muscular weakness, fatigue, and weight loss.

Cancer cells typically have a number of genes of different types with functions that have been altered in some way to promote uncontrolled cell division or metastasis. One type of altered gene, called an **oncogene**, has a normal, unaltered counterpart in nontumor cells in the same organism. Some of the genes that become oncogenes encode components of the cyclin/CDK system that

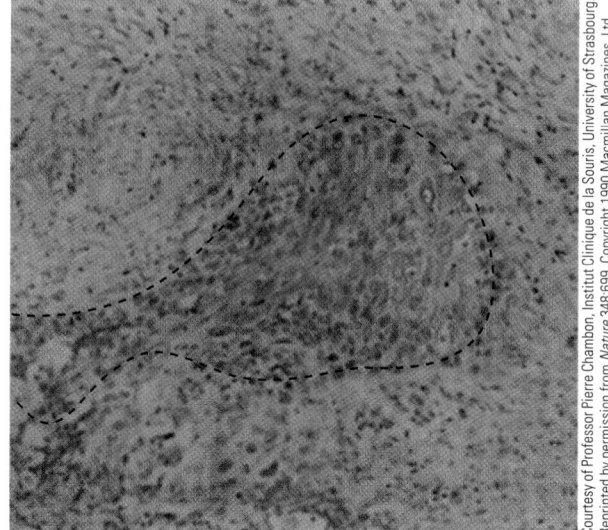

Figure 10.16
A mass of tumor cells (dashed line) embedded in normal tissue. As is typical, the tumor cells appear more densely packed because they have less cytoplasmic volume than normal cells. Original magnification ×270.

regulates cell division; others encode proteins that regulate gene activity, form cell surface receptors, or make up elements of the systems controlled by the receptors.

For example, one oncogene encodes a faulty surface receptor that is constantly active, even without binding an extracellular signal molecule. As a result, the internal reaction pathways triggered by the receptor, which induce cell division, are continually turned on. Another oncogene encodes a faulty cyclin that constantly activates its CDK and triggers DNA replication and the rest of the cell cycle. (*Insights from the Molecular Revolution* describes an experiment testing the effects of a viral system that induces cancer by overriding normal controls of the cyclin/CDK system.) Cancer, oncogenes, and the alterations that convert normal genes to oncogenes are discussed in further detail in Chapter 16.

The overview of the mitotic cell cycle and its regulation presented in this chapter only hints at the complexity of cell growth and division. The greatest wonder of the processes is that, despite the complexity of the events, the cell cycle functions almost without error in every multicellular organism. For example, the 2 million per second mitotic divisions that produce red blood cells proceed with few or no detectable errors throughout the lifetimes of most humans.

STUDY BREAK

1. Why is a CDK not active throughout the entire cell cycle?
2. How do cyclin/CDK complexes typically trigger transitions in the cell cycle?
3. What is an oncogene? How might an oncogene affect the cell cycle?

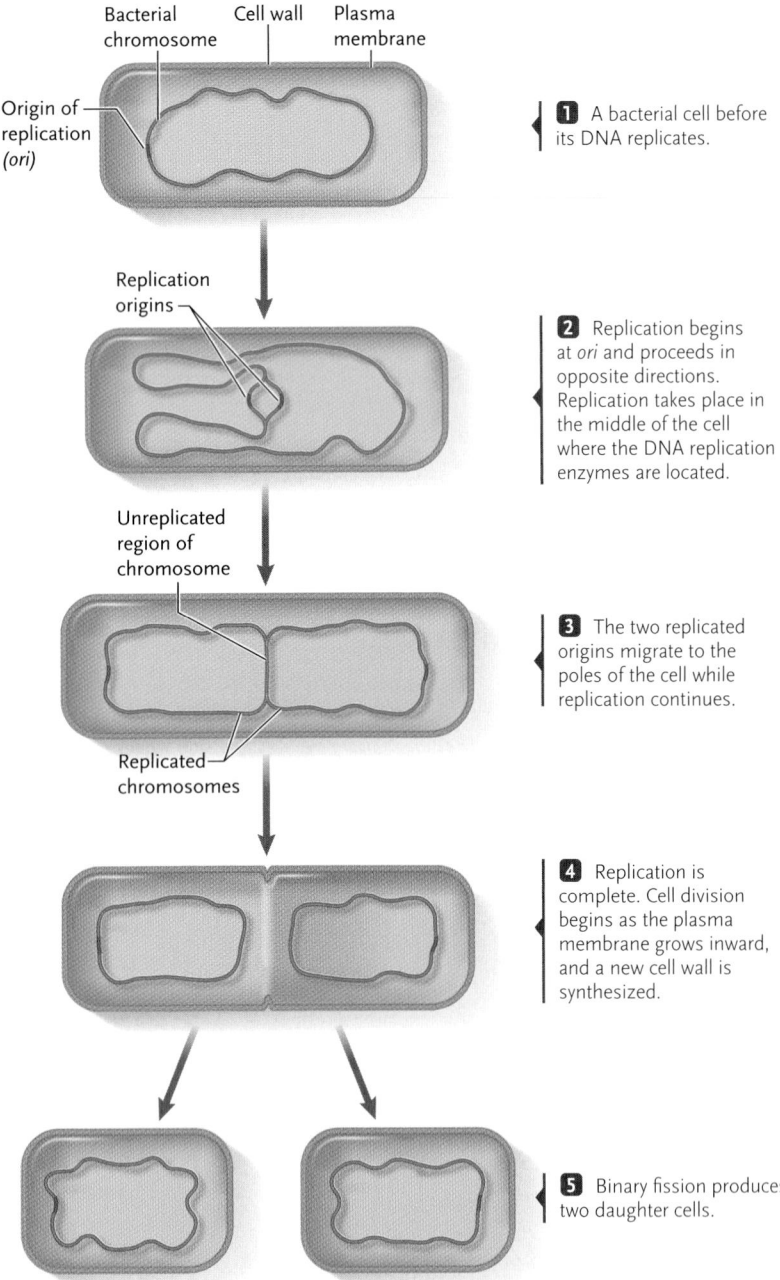

Figure 10.17
Model for the segregation of replicated bacterial chromosomes to daughter cells.

Bacterial chromosome   Cell wall   Plasma membrane

Origin of replication (*ori*)

1 A bacterial cell before its DNA replicates.

Replication origins

2 Replication begins at *ori* and proceeds in opposite directions. Replication takes place in the middle of the cell where the DNA replication enzymes are located.

Unreplicated region of chromosome

3 The two replicated origins migrate to the poles of the cell while replication continues.

Replicated chromosomes

4 Replication is complete. Cell division begins as the plasma membrane grows inward, and a new cell wall is synthesized.

5 Binary fission produces two daughter cells.

10.5 Cell Division in Prokaryotes

Prokaryotes undergo a cycle of cytoplasmic growth, DNA replication, and cell division, producing two daughter cells from an original parent cell. The entire mechanism of prokaryotic cell division is called **binary fission**—that is, splitting or dividing into two parts. There is no known prokaryotic equivalent of mitosis, nor are there prokaryotic equivalents of microtubules, a spindle apparatus, or the cyclin/CDK control proteins.

Replication Occupies Most of the Cell Cycle in Rapidly Dividing Prokaryotic Cells

In most prokaryotes, hereditary information is encoded in a single, circular DNA molecule known as a **bacterial chromosome (Figure 10.17,** step 1). In prokaryotic cells dividing at the maximum rate, DNA replication occupies most of the period between cytoplasmic divisions. As soon as replication is complete, the cytoplasm divides to complete the cell cycle. For example, in *Escherichia coli* cells, which divide every 20 minutes, DNA replication occupies all but 1 minute of the entire division cycle.

Replicated Chromosomes Are Distributed Actively to the Halves of the Prokaryotic Cell

In the 1960s, François Jacob of The Pasteur Institute, Paris, France, proposed a model for the segregation of bacterial chromosomes to the daughter cells in which the two chromosomes attach to the plasma membrane near the middle of the cell and separate as a new plasma membrane is added between the two sites during cell elongation. The essence of this model is that chromosome separation is passive. However, current research indicates that bacterial chromosomes rapidly separate in an active way that is linked to DNA replication events and is independent of cell elongation. The new model is shown in Figure 10.17.

Replication of a bacterial chromosome commences at a specific region called the **origin of replication** (ori). The ori is in the middle of the cell where the enzymes for DNA replication are located. Once the ori has been duplicated, the two origins migrate toward the two ends (poles) of the cell as replication continues for the rest of the chromosome. This active movement distributes the two replicated chromosomes to the two ends of the cell. How this movement occurs is unknown.

Next, cytoplasmic division in prokaryotes occurs through an inward growth of the plasma membrane, along which new cell wall material is assembled to cut the cell into two parts (Figure 10.17, step 5). The new wall divides the replicated DNA molecules and the cytoplasmic structures and molecules equally between the daughter cells.

Mitosis Has Evolved from Binary Fission

The prokaryotic mechanism works effectively because most prokaryotic cells have only a single chromosome; so, if a daughter cell receives at least one copy of the chromosome, its genetic information is complete. By contrast, the genetic information of eukaryotes is divided among several to many chromosomes, with each chromosome containing a much greater length of DNA than a bacterial chromosome. If a daughter cell fails to receive a copy of even one chromosome, the effects are usually lethal. The evolution of mitosis solved the mechanical problems associated with distributing long DNA molecules without breakage. Mitosis provided the level of precision required to ensure that each daughter cell receives a complete complement of the chromosomes, together with a complete copy of the genetic information for the parental cell.

Scientists believe that the ancestral division process was binary fission and that mitosis evolved from that process. Variations in the mitotic apparatus in modern-day organisms illuminate possible intermediates in this evolutionary pathway. For example, in

UNANSWERED QUESTIONS

Disrupted or defective control of cell growth and division can lead to diseases such as cancer. Complex, interacting molecular networks within the cell fine tune the division of each cell in both unicellular and multicellular organisms. Identifying the genes and proteins involved in these networks is crucial both for a complete understanding of cell growth and division and for developing models for diseases caused by cell cycle defects. Many researchers are working in this area worldwide.

How are transitions between phases of the cell cycle regulated?

Research in many labs has shown that transitions are important control points for progression through the cell cycle. If a cell in G_1 phase has damaged DNA, for instance, the cell pauses to repair the DNA before entering S phase, to ensure that any mutations are not passed on to progeny cells. More specifically, one researcher, Raymond Deshaies of Caltech, is using mammalian cells in tissue culture and the yeast *Saccharomyces cerevisiae* as model organisms to characterize the molecular events involved in two cell cycle transitions, G_1 to S and mitosis to G_1. At the G_1-to-S transition, DNA replication enzymes become active. At the mitosis-to-G_1 transition, the mitotic spindle breaks down, allowing the cell to return to G_1 phase. In yeast, both of these transitions are controlled by a specific cellular process for breaking down proteins. Deshaies's research group is performing experiments to determine how the proteins involved in the G_1-to-S and mitosis-to-G_1 transitions work at the molecular and cellular levels and how their activities are controlled.

Can targeting p53 be an effective anticancer therapy?

The factors that trigger uncontrolled cell division to convert normal cells into cancer cells is one of the most relevant areas of current research to humankind in general. The hope is that by working out the molecular events that regulate the cell cycle, the factors that cause cancer can be identified and the progression to cancer can be reversed or at least controlled, thus stopping tumor growth. Investigations of tumor suppressor proteins have developed some of the most promising leads. In their normal role in cells, these factors stop or slow cell division. For example, p53, so named for its molecular weight of 53,000 daltons, is one of the factors that normally prevent cells from progressing from S to G_2 and mitosis when DNA is damaged in some way. Investigators have found that loss of function of p53 is involved in more than 50% of all human cancers. The mutant proteins build up to abnormally high levels in cancer cells.

Clearly, p53 represents a major target for developing new anticancer therapies. Rainer Brachmann at the University of California, Irvine, who is doing research in this direction, has studied many common p53 mutant proteins found in cancers. He has found that by changing particular amino acids at certain positions in the p53 mutant proteins, tumor suppressor function is restored. His lab group is determining the structures of a number of the p53 mutant proteins by X-ray crystallography, and they hope to be able to design small molecules that can stabilize the mutant proteins and perhaps be an effective anticancer therapy.

Peter J. Russell

many primitive eukaryotes, such as dinoflagellates (a type of single-celled alga), the nuclear envelope remains intact during mitosis, and the chromosomes bind to the inner membrane of the nuclear membrane. When the nucleus divides, the chromosomes are segregated.

A more advanced form of the mitotic apparatus is seen in yeasts and diatoms (another type of single-celled alga). In these organisms, the mitotic spindle forms and chromosomes segregate to daughter nuclei without the disassembly and reassembly of the nuclear envelope. Currently, scientists think that the types of mitosis seen in yeasts and diatoms, as well as in animals and higher plants, evolved separately from a common ancestral type. Mitotic cell division, the subject of this chapter, produces two cells that have the same genetic information as the parental cell entering division. In the next chapter, you will learn about meiosis, a specialized form of cell division that produces gametes, which have half the number of chromosomes as that present in diploid cells.

STUDY BREAK

1. How do prokaryotes divide?
2. What processes involved in eukaryotic cell division are absent from prokaryotic cell division?

Review

Go to **ThomsonNOW** at www.thomsonedu.com/login to access quizzing, animations, exercises, articles, and personalized homework help.

10.1 The Cycle of Cell Growth and Division: An Overview

- In mitotic cell division, DNA replication is followed by the equal separation—that is, segregation—of the replicated DNA molecules and their delivery to daughter cells. The process ensures that the two cell products of a division have the same genetic information as the parent cell entering division.

- Mitosis is the basis for growth and maintenance of body mass in multicelled eukaryotes, and for the reproduction of many single-celled eukaryotes.

- The DNA of eukaryotic cells is divided among individual, linear chromosomes located in the cell nucleus.

- DNA replication and duplication of chromosomal proteins converts each chromosome into two exact copies known as sister chromatids.

10.2 The Mitotic Cell Cycle

- Mitosis and interphase constitute the mitotic cell cycle. Mitosis occurs in five stages. In prophase (stage 1), the chromosomes condense into short rods and the spindle forms in the cytoplasm (Figures 10.3 and 10.4).

- In prometaphase (stage 2), the nuclear envelope breaks down, the spindle enters the former nuclear area, and the sister chromatids of each chromosome make connections to opposite spindle poles. Each chromatid has a kinetochore that attaches to spindle microtubules (Figures 10.3 and 10.6).

- In metaphase (stage 3), the spindle is fully formed and the chromosomes, moved by the spindle microtubules, become aligned at the metaphase plate (Figures 10.3 and 10.4).

- In anaphase (stage 4), the spindle separates the sister chromatids and moves them to opposite spindle poles. At this point, chromosome segregation is complete (Figures 10.3 and 10.4).

- In telophase (stage 5), the chromosomes decondense and return to the extended state typical of interphase and a new nuclear envelope forms around the chromosomes (Figures 10.3 and 10.4).

- Cytokinesis, the division of the cytoplasm, completes cell division by producing two daughter cells, each containing a daughter nucleus produced by mitosis (Figures 10.3 and 10.4).

- Cytokinesis in animal cells proceeds by furrowing, in which a band of microfilaments just under the plasma membrane contracts, gradually separating the cytoplasm into two parts (Figure 10.8).

- In plant cytokinesis, cell wall material is deposited along the plane of the former spindle midpoint; the deposition continues until a continuous new wall, the cell plate, separates the daughter cells (Figure 10.9).

Animation: The cell cycle

Animation: Mitosis step-by-step

Animation: Cytoplasmic division

10.3 Formation and Action of the Mitotic Spindle

- In animal cells, the centrosome divides and the two parts move apart. As they do so, the microtubules of the spindle form between them. In plant cells with no centrosome, the spindle microtubules assemble around the nucleus (Figure 10.10).

- In the spindle, kinetochore microtubules run from the poles to the kinetochores of the chromosomes, and nonkinetochore microtubules run from the poles to a zone of overlap at the spindle midpoint without connecting to the chromosomes (Figure 10.12).

- During anaphase, the kinetochores move along the kinetochore microtubules, pulling the chromosomes to the poles. The nonkinetochore microtubules slide over each other, pushing the poles farther apart (Figures 10.12 and 10.13).

Animation: Mechanisms for chromosome movement

10.4 Cell Cycle Regulation

- The cell cycle is controlled directly by complexes of cyclin and a cyclin-dependent protein kinase (CDK). CDK is activated when combined with a cyclin and then adds phosphate groups to target proteins, activating them. The activated proteins trigger the cell to progress to the next cell cycle stage. Each major stage of the cell cycle begins with activation of one or more cyclin/CDK complexes and ends with deactivation of the complexes by breakdown of the cyclins (Figure 10.15).

- Important internal controls create checkpoints to ensure that the reactions of one stage are complete before the cycle proceeds to the next stage.

- External controls are based primarily on surface receptors that recognize and bind signals such as peptide hormones and growth factors, surface groups on other cells, or molecules of the extracellular matrix. The binding triggers internal reactions that speed, slow, or stop cell division.
- In cancer, control of cell division is lost, and cells divide continuously and uncontrollably, forming a rapidly growing mass of cells that interferes with body functions. Cancer cells also break loose from their original tumor (metastasize) to form additional tumors in other parts of the body.

Animation: Cancer and metastasis

10.5 Cell Division in Prokaryotes

- Replication begins at the origin of replication of the bacterial chromosome in reactions catalyzed by enzymes located in the middle of the cell. Once the origin of replication is duplicated, the two origins migrate to the two ends of the cells. Division of the cytoplasm then occurs through a partition of cell wall material that grows inward until the cell is separated into two parts (Figure 10.17).

Questions

Self-Test Questions

1. During the cell cycle, the DNA mass of a cell:
   a. decreases during G_1.
   b. decreases during metaphase.
   c. increases during the S phase.
   d. increases during G_2.
   e. decreases during interphase.

2. A protein, p21, inhibits CDKs. The earliest effect of p21 on the cell cycle would be to stop the cell cycle at:
   a. early G_1.   d. G_2.
   b. late G_1.   e. the mitotic prophase.
   c. the S phase.

3. Which of the following is *not* characteristic of eukaryotic cell division?
   a. a system of internal molecular controls
   b. DNA replication
   c. external growth factors
   d. microtubular organizing center
   e. G_0 stage

4. The major microtubule organizing center of the animal cell is:
   a. chromosomes, composed of chromatids.
   b. the centrosome, composed of centrioles.
   c. the chromatin, composed of chromatids.
   d. chromosomes, composed of centromere.
   e. centrioles, composed of centrosome.

5. The chromatids separate into chromosomes:
   a. during prophase.
   b. going from prophase to metaphase.
   c. going from anaphase to telophase.
   d. going from metaphase to anaphase.
   e. going from telophase to interphase.

6. Which of the following statements about mitosis is *incorrect*?
   a. Microtubules from the spindle poles attach to the kinetochores on the chromosomes.
   b. In anaphase, the spindle separates sister chromatids and pulls them apart.
   c. In metaphase, spindle microtubules align the chromosomes at the spindle midpoint.
   d. Cytokinesis describes the movement of chromosomes.
   e. Both the animal cell furrow and the plant cell plate form at their former spindle midpoint.

7. Mitomycin C is an anticancer drug that stops cell division by inserting itself into the strands of DNA and binding them together. This action is thought to have its major effect at:
   a. late G_1, early S phases.   d. metaphase.
   b. late G_2.   e. anaphase.
   c. prophase.

8. Which of the following statements about cell cycle regulators is *incorrect*?
   a. Cyclin is synthesized during the S phase.
   b. Cyclin and CDKs are at the highest level during G_1.
   c. CDKs combine with cyclin to phosphorylate target proteins.
   d. CDKs combine with cyclin to move the cycle into mitosis.
   e. During anaphase of mitosis, cyclin is degraded, allowing mitosis to end.

9. Which of the following is *not* characteristic of cancer cells?
   a. metastasis
   b. contact inhibition
   c. avoidance of the G_0 stage
   d. oncogene overactivation of cyclin
   e. extra growth factor receptors

10. In bacteria:
   a. several chromosomes undergo mitosis.
   b. binary fission produces four daughter cells.
   c. replication begins at the origin, and the DNA strands separate.
   d. the plasma membrane plays an important role in separating the duplicated chromosomes into the two daughter cells.
   e. the daughter cells receive different genetic information from the parent cell.

Questions for Discussion

1. You have a means of measuring the amount of DNA in a single cell. You first measure the amount of DNA during G_1. At what points during the remainder of the cell cycle would you expect the amount of DNA per cell to change?

2. A cell has 38 chromosomes. After mitosis and cell division, one daughter cell has 39 chromosomes and the other has 37. What might have caused these abnormal chromosome numbers? What effects do you suppose this might have on cell function? Why?

3. Taxol (Bristol-Myers Squibb, New York), a substance derived from Pacific yew *(Taxus brevifolia)*, is effective in the treatment of breast and ovarian cancers. It works by stabilizing microtubules, thereby preventing them from disassembling. Why would this activity slow or stop the growth of cancer cells?

4. A cell has 24 chromosomes at G_1 of interphase. How many chromosomes would you expect it to have at G_2 of interphase? At metaphase of mitosis? At telophase of mitosis?

Experimental Analysis

Many chemicals in the food we eat potentially have effects on cancer cells. Chocolate, for example, contains a number of flavonoid compounds, which act as natural antioxidants. Design an experiment to determine whether any of the flavonoids in chocolate inhibit the cell cycle of breast cancer cells growing in culture.

Evolution Link

The genes and proteins involved in cell cycle regulation in prokaryotes and eukaryotes are very different. However, both types of organisms use similar molecular regulatory reactions to coordinate DNA synthesis with cell division. What does this observation mean from an evolutionary perspective?

How Would You Vote?

It is illegal to sell your organs, but you can sell your cells, including eggs, sperm, and blood cells. HeLa cells are still being sold all over the world by cell culture firms. Should the family of Henrietta Lacks share in the profits? Go to www.thomsonedu.com/login to investigate both sides of the issue and then vote.

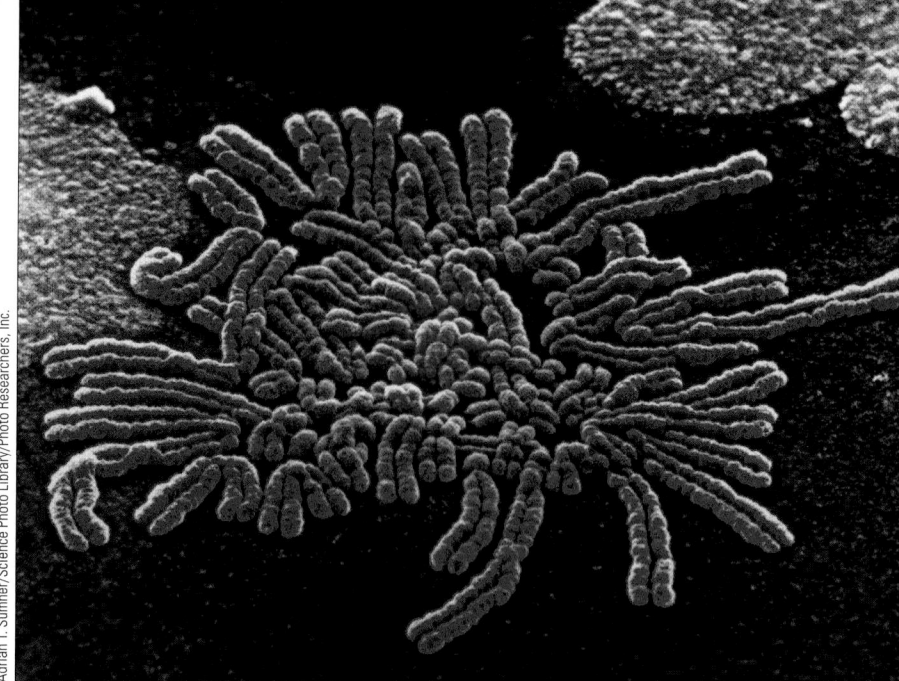

Chromosomes aligned during metaphase of the first division of meiosis, the process that produces gametes such as eggs and sperm (colorized SEM).

Adrian T. Sumner/Science Photo Library/Photo Researchers, Inc.

11 Meiosis: The Cellular Basis of Sexual Reproduction

WHY IT MATTERS

A couple clearly shows mutual interest. First, he caresses her with one arm, then another—then another, another, and another. She reciprocates. This interaction goes on for hours; a hug here, a squeeze there. At the climactic moment, the male reaches deftly under his mantle and removes a packet of sperm, which he inserts under the mantle of the female. For every one of his sperm that successfully performs its function, a fertilized egg can develop into a new octopus.

For the octopus, sex is an occasional event, preceded by a courtship ritual that involves intermingled tentacles. For another marine animal, the slipper limpet, sex is a lifelong group activity. Slipper limpets are relatives of snails. Like many other animals, a slipper limpet passes through a free-living immature stage before it becomes a sexually mature adult. When the time comes for an immature limpet to transform into an adult, it settles onto a rock or other firm surface. If the limpet settles by itself, it develops into a female. If instead it settles on top of a female, it develops into a male. If another slipper limpet settles down on that male, it, too, becomes a male. Adult slipper limpets almost always live in such piles, with the one on the bottom always being a female. All the male limpets continually contribute sperm that fertilize

221

eggs shed by the female. If the one female dies, the surviving male at the bottom of the pile changes into a female and reproduction continues.

These octopuses and slipper limpets are engaged in forms of **sexual reproduction**, the production of offspring through union of male and female **gametes**—for example, eggs and sperm cells in animals. Sexual reproduction depends on **meiosis**, a specialized process of cell division that produces gametes. Meiosis reduces the number of chromosomes, producing gametes with half the number of chromosomes present in the **somatic cells** (body cells) of a species. The derivation of the word *meiosis* (*meioun* = to diminish) reflects this reduction. At **fertilization**, the nuclei of an egg and sperm cell fuse, producing a cell called the **zygote**, in which the chromosome number typical of the species is restored. Without the halving of chromosome number by the meiotic divisions, fertilization would double the number of chromosomes in each subsequent generation.

Both meiosis and fertilization also mix genetic information into new combinations; thus, none of the offspring of a mating pair is likely to be genetically identical. By contrast, asexual reproduction generates genetically identical offspring because they are the products of mitotic divisions (asexual reproduction is discussed in Chapter 10). Sexual reproduction generates the variability that is the basis of most inherited differences among individual sexually reproducing organisms. This variability is the source of raw material for the process of evolution.

The halving of the chromosome number and mixing of genetic information into new combinations—both by meiosis—and the restoration of the chromosome number by fertilization, are the biological foundations of sexual reproduction. Intermingled tentacles in octopuses, communal sex among limpets, and the courting and mating rituals of humans, are nothing more or less than variations of the means for accomplishing fertilization.

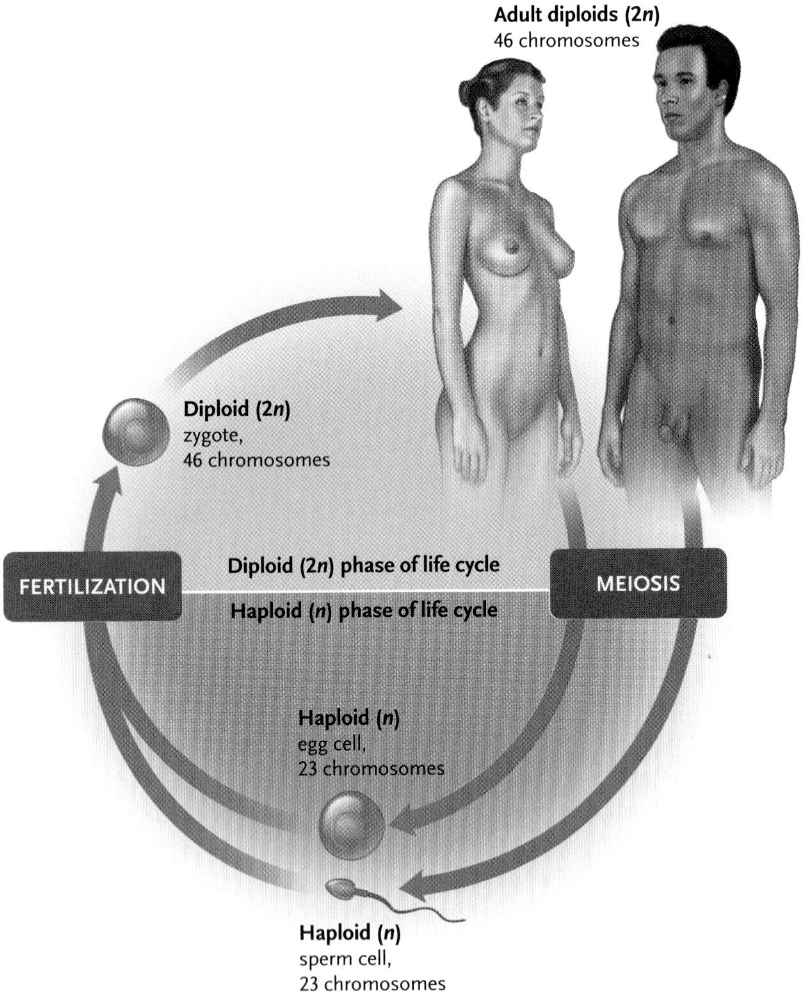

11.1 The Mechanisms of Meiosis

Meiosis occurs only in eukaryotes that reproduce sexually and only in organisms that are at least diploid—that is, organisms that have at least two representatives of each chromosome.

Meiosis Is Based on the Interactions and Distribution of Homologous Chromosome Pairs

To follow the steps of meiosis, you must understand clearly the significance of the chromosome pairs in diploid organisms. As discussed in Section 10.1, the two representatives of each chromosome in a diploid cell constitute a *homologous pair*—they have the same genes, arranged in the same order in the DNA of the chromosomes. One chromosome of each homologous pair, the **paternal chromosome**, is derived from the male parent of the organism, and the other chromosome, the **maternal chromosome**, is derived from its female parent.

Although the two chromosomes of a homologous pair contain the same genes, arranged in the same order, different versions of the genes, called **alleles**, may be present in either member of the pair. The alleles of a gene have different DNA sequences and encode distinct versions of the same protein, which may have different structures, different biochemistry, or both.

For example, humans normally have 46 chromosomes in their cells, which make up 23 homologous pairs (see Figure 10.7). However, each individual (except for identical twins, identical triplets, and so forth) has a unique combination of the alleles in the two chromosomes of each homologous pair. The distinct set of alleles, arising from the mixing mechanisms of meiosis and fertilization, gives each individual his or her unique combination of inherited traits, including such attributes

Adult diploids (2n)
46 chromosomes

Diploid (2n)
zygote,
46 chromosomes

FERTILIZATION

Diploid (2n) phase of life cycle

Haploid (n) phase of life cycle

MEIOSIS

Haploid (n)
egg cell,
23 chromosomes

Haploid (n)
sperm cell,
23 chromosomes

Figure 11.1
The cycle of meiosis and fertilization. Meiosis reduces the chromosome number from the diploid level of two representatives of each chromosome to the haploid level of one representative of each chromosome. Fertilization restores the chromosome number to the diploid level.

as height, hair and eye color, susceptibility to certain diseases, and even aspects of personality and intelligence.

Meiosis separates the homologous pairs, thereby reducing the diploid or 2*n* number of chromosomes to the **haploid** or *n* number (**Figure 11.1**). Each gamete produced by meiosis receives only one member of each homologous pair. For example, a human egg or sperm cell contains 23 chromosomes, one of each pair. When the egg and sperm combine in sexual reproduction to produce the zygote—the first cell of the new individual—the diploid number of 46 chromosomes (23 pairs) is regenerated. The processes of DNA replication and mitotic cell division ensure that this diploid number is maintained in the body cells as the zygote develops.

The Meiotic Cell Cycle Produces Four Genetically Different Daughter Cells with Half the Parental Number of Chromosomes

Meiosis is a two-part (**meiosis I** and **meiosis II**) process of cell division in sexually reproducing organisms. In meiosis, the duplicated chromosomes in the parental cell are distributed to four daughter cells, each of which, therefore, has half the number chromosomes as does the parental cell. By contrast, in mitosis, each chromosome duplication is followed by a division. Consequently, the chromosome number remains constant from one cell generation to the next.

The meiotic cell cycle begins with a premeiotic interphase in which DNA replicates and the chromosomal proteins are duplicated. (This interphase passes through G_1, S, and G_2 stages as does a premitotic interphase.) As in a premitotic interphase, the two resulting copies are the identical *sister chromatids* of each chromosome (**Figure 11.2**). Following premeiotic interphase, cells enter the two meiotic divisions, meiosis I and meiosis II. Homologous chromosomes pair during meiosis I and undergo an exchange of chromosome segments; this process is called *recombination*. The homologous chromosomes separate after recombination as the cell continues through the first division. Completion of meiosis I produces two cells, each with the haploid number of chromosomes, with each chromosome still consisting of two chromatids. During the second meiotic division, meiosis II, the sister chromatids separate and segregate into different cells. A total of four cells, each with the haploid

number of chromosomes, is the result of the two meiotic divisions.

For convenience, biologists separate each meiotic division into the same key stages as mitosis: *prophase, prometaphase, metaphase, anaphase,* and *telophase*. The stages are identified as belonging to the two divisions, meiosis I and meiosis II, by a *I* or *II*, as in *prophase I* and *prophase II*. A brief interphase called **interkinesis** separates the two meiotic divisions, *but no DNA replication occurs during interkinesis.*

Prophase I. At the beginning of prophase I, the replicated chromosomes, each consisting of two sister chromatids, begin to fold and condense into threadlike structures in the nucleus (**Figure 11.3**, step 1). The two

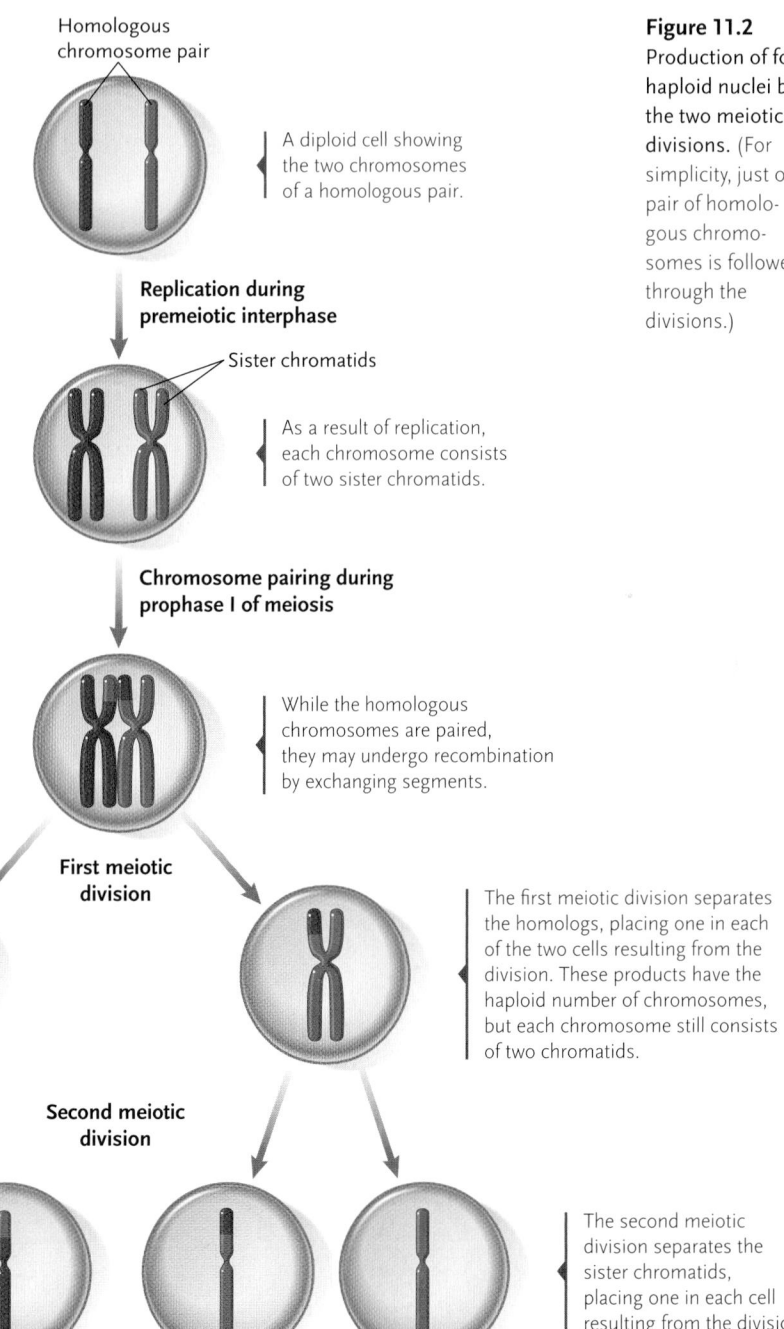

Figure 11.2
Production of four haploid nuclei by the two meiotic divisions. (For simplicity, just one pair of homologous chromosomes is followed through the divisions.)

Homologous chromosome pair

A diploid cell showing the two chromosomes of a homologous pair.

Replication during premeiotic interphase

Sister chromatids

As a result of replication, each chromosome consists of two sister chromatids.

Chromosome pairing during prophase I of meiosis

While the homologous chromosomes are paired, they may undergo recombination by exchanging segments.

First meiotic division

The first meiotic division separates the homologs, placing one in each of the two cells resulting from the division. These products have the haploid number of chromosomes, but each chromosome still consists of two chromatids.

Second meiotic division

The second meiotic division separates the sister chromatids, placing one in each cell resulting from the division.

First meiotic division

Prophase I

Plasma membrane

Duplicated centrioles

Nuclear envelope

Tetrad

Homologous chromosomes

Two sister chromatids

Condensation of chromosomes

1 At the beginning of prophase I the chromosomes begin to condense into threadlike structures. Each consists of two sister chromatids, as a result of DNA replication during premeiotic interphase. The chromosomes of two homologous pairs, one long and one short, are shown.

Synapsis

2 Homologous chromosomes come together and pair.

Recombination

3 While they are paired, the chromatids of homologous chromosomes undergo recombination by exchanging segments. The enlarged circle shows a site under-going recombination (arrow).

Prometaphase I

4 In prometaphase I, the nuclear envelope breaks down, and the spindle moves into the former nuclear area. Kinetochore microtubules connect to the chromosomes—kinetochore microtubules from one pole attach to both sister kinetochores of one duplicated chromosome, and kinetochore microtubules from the other pole attach to both sister kinetochores of the other duplicated chromosome.

Second meiotic division

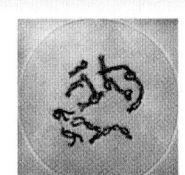

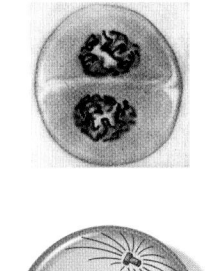

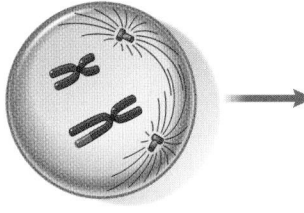

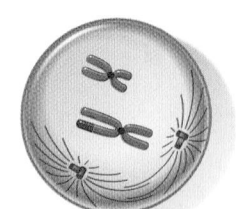

Figure 11.3
The meiotic divisions. The sequence is shown as it would occur in a male animal. (Two homologous pairs of chromosomes are shown.) (Micrographs with thanks to the John Innes Foundation Trustees.)

Prophase II

8 The chromosomes condense and a spindle forms.

Prometaphase II

9 The nuclear envelope breaks down, the spindle enters the former nuclear area, and kinetochore microtubules from the opposite spindle poles attach to the kinetochores of each chromosome.

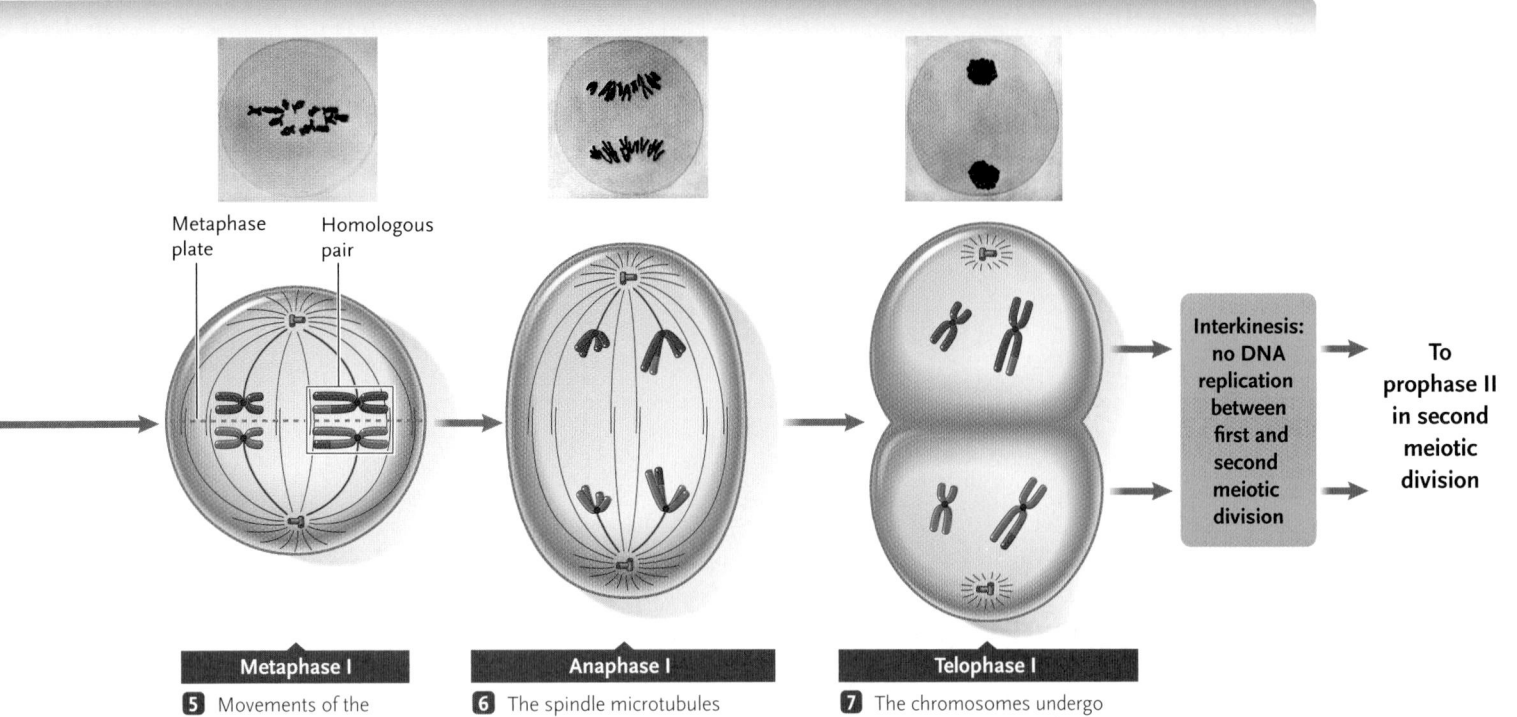

Metaphase I

5 Movements of the spindle microtubules align the tetrads in the equatorial plane–metaphase plate–between the two spindle poles.

Anaphase I

6 The spindle microtubules separate the two chromosomes of each homologous pair and move them to opposite spindle poles. The poles now contain the haploid number of chromosomes. However, each chromosome at the poles still contains two chromatids.

Telophase I

7 The chromosomes undergo little or no change except for limited decondensation or unfolding in some species. The spindle of the first meiotic division disassembles, and two new spindles form for the second division.

Interkinesis: no DNA replication between first and second meiotic division

To prophase II in second meiotic division

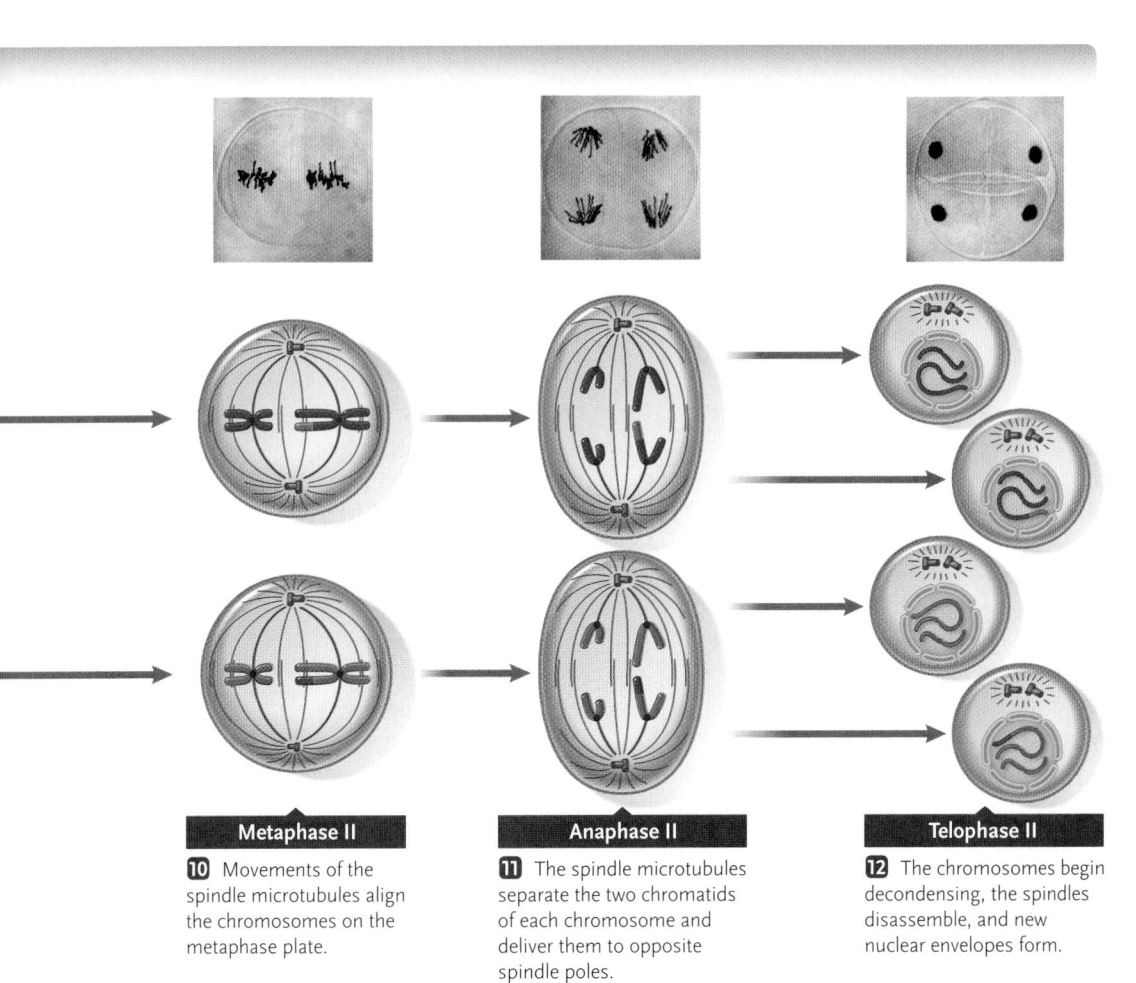

Metaphase II

10 Movements of the spindle microtubules align the chromosomes on the metaphase plate.

Anaphase II

11 The spindle microtubules separate the two chromatids of each chromosome and deliver them to opposite spindle poles.

Telophase II

12 The chromosomes begin decondensing, the spindles disassemble, and new nuclear envelopes form.

INSIGHTS FROM THE MOLECULAR REVOLUTION

Fertile Fields in the Human Y Chromosome

Part of the human Y chromosome contains active genes and pairs with homologous regions of the X chromosome during meiosis. The remainder of the Y, which does not pair with the X, was once thought to be an inert region containing no functional genes. However, researchers have identified several genes in this region. One gene is *SRY*, which governs formation of the testes in developing male embryos. Other genes include a recently discovered cluster of eight "housekeeping" genes—that is, genes that encode proteins with basic functions, such as protein synthesis, that occur in all cells. Housekeeping genes are well known on the X chromosome and other, non-sex chromosomes, but it was a surprise to find them in the nonpairing region of the Y.

This unexpected discovery led two investigators at the Massachusetts Institute of Technology, Bruce T. Lahn and David C. Page, to question whether other genes might be in the nonpairing region of the Y. One clue that more genes might exist there is that deletions in this region often lead to infertility and testicular tumors in male individuals, suggesting that the Y chromosome contains sequences that are vital to male fertility.

To focus on possible unknown genes, Lahn and Page first used genetic analysis techniques to eliminate noncoding regions of the Y chromosome, which do not contain genes, as well as genes that were already known. A small set of chromosome fragments potentially containing previously unidentified genes remained. Molecular analysis of these fragments showed the presence of 12 genes, all of which are in the nonpairing region of the Y. Lahn and Page used a computer program to compare the amino acid sequences encoded by the 12 genes with data banks of known protein sequences. The comparison showed that five of the genes are housekeeping genes—for example, one gene encodes a ribosomal protein.

Each of these five genes is also on the X chromosome.

The proteins encoded in the seven remaining genes showed no clear relationships to any known proteins. However, most of them contain combinations of amino acids that are characteristic of proteins that bind to DNA or RNA sequences or to chromatin (the DNA-protein fibers of chromosomes). These characteristics suggest that the proteins encoded in these genes may regulate genes or stabilize DNA, RNA, or chromatin; thus, that they may have a role in the developmental processes that produce viable sperm cells.

Lahn and Page's research shows that regions of the Y chromosome once thought to be genetic wastelands actually contain functional genes, including some that may be required for normal male fertility. Identification of the functions of these genes may lead to treatments for male infertility resulting from defects in their functions.

---

chromosomes of each homologous pair then come together and line up side-by-side in a zipperlike way; this process is called **pairing** or **synapsis** (step 2). The fully paired homologs are called **tetrads**, referring to the fact that each homologous pair consists of four chromatids. No equivalent of chromosome pairing exists in mitosis.

While they are paired, the chromatids of homologous chromosomes physically exchange segments (step 3). This physical exchange, called **recombination**, is the step that mixes the alleles of the homologous chromosomes into new combinations and contributes to the generation of variability in sexual reproduction (see later in this chapter for details of the recombination mechanism).

As prophase I finishes, a spindle forms in the cytoplasm by the same basic mechanisms described in Section 10.3.

Prometaphase I. In prometaphase I, the nuclear envelope breaks down and the spindle enters the former nuclear area (Figure 11.3, step 4). The two chromosomes of each pair attach to kinetochore microtubules leading to opposite spindle poles. That is, both sister chromatids of one homolog attach to microtubules leading to one spindle pole, whereas both sister chromatids of the other homolog attach to microtubules leading to the opposite pole.

Metaphase I and Anaphase I. At metaphase I, movements of the spindle microtubules have aligned the tetrads on the equatorial plane—the *metaphase plate*—between the two spindle poles (Figure 11.3, step 5). Then, the two chromosomes of each homologous pair separate and move to opposite spindle poles as the spindle microtubules contract during anaphase I (step 6). The movement segregates homologous pairs, delivering a haploid set of chromosomes to each pole of the spindle. However, all the chromosomes at the poles are still double structures composed of two sister chromatids.

Rarely, chromosome segregation fails. For example, both chromosomes of a homologous pair may connect to the same spindle pole in anaphase I. In the resulting **nondisjunction**, as it is called, the spindle fails to separate the homologous chromosomes of the tetrad. As a result, one pole receives both chromosomes of the homologous pair, whereas the other pole has no copies of that chromosome. Zygotes that receive an extra chromosome because of nondisjunction have three copies of one chromosome instead of two. In humans, most zygotes of this kind do not result in live births. One exception is Down syndrome, which results from three copies of chromosome 21 instead of the normal two copies. Down syndrome involves characteristic alterations in body and facial structure, mental retardation, and significantly reduced fertility (see Chapter 13 for a more detailed discussion of Down syndrome.)

Telophase I and Interkinesis. Telophase I is a brief, transitory stage in which there is little or no change in the chromosomes (Figure 11.3, step 7). New nuclear envelopes form in some species but not in others. Telophase I is followed by an interkinesis in which the single spindle of the first meiotic division disassembles and the microtubules reassemble into two new spindles for the second division.

Prophase II, Prometaphase II, and Metaphase II. The second meiotic division, meiosis II, is similar to a mitotic division. During prophase II, the chromosomes condense and a spindle forms (Figure 11.3, step 8). During prometaphase II, the nuclear envelope breaks down, the spindle enters the former nuclear area, and spindle microtubules leading to opposite spindle poles attach to the two kinetochores of each chromosome (step 9). At metaphase II, movements of the spindle microtubules have aligned the chromosomes on the metaphase plate (step 10).

Anaphase II and Telophase II. Anaphase II begins as the spindles separate the two chromatids of each chromosome and pull them to opposite spindle poles (Figure 11.3, step 11). At the completion of anaphase II, the chromatids—now called chromosomes—have been segregated to the two poles. During telophase II, the chromatids decondense to the extended interphase state, the spindles disassemble, and new nuclear envelopes form around the masses of chromatin (step 12). The result is four haploid cells, each with a nucleus containing half the number of chromosomes present in a G_1 nucleus of the same species.

Sex Chromosomes in Meiosis. In many eukaryotes, including most animals, one or more pairs of chromosomes, called the **sex chromosomes,** are different in male and female individuals of the same species. For example, in humans, the cells of females contain a pair of sex chromosomes called the *XX* pair (the sex chromosomes are visible in Figure 10.7). Male humans contain a pair of sex chromosomes that consist of one X chromosome and a smaller chromosome called the *Y chromosome.* The two X chromosomes in females are fully homologous, whereas the male X and Y chromosomes are homologous only through a short region. However, an X chromosome from the mother is able to pair up with either an X or a Y from the father and follow the same pathways through the meiotic divisions as the other chromosome pairs.

As a result of meiosis, a gamete formed in females may receive either member of the X pair. A gamete formed in males receives either an X or a Y chromosome. (*Insights from the Molecular Revolution* outlines studies that identify a key sex-determining gene on the mammalian Y chromosome.)

The sequence of steps in the two meiotic divisions accomplishes the major outcomes of meiosis: the generation of genetic variability and the reduction of chromosome number. (**Figure 11.4** reviews the two meiotic divisions and compares them with the single division of mitosis.)

Study Break

1. How does the outcome of meiosis differ from that of mitosis?
2. What is recombination, and in what stage of meiosis does it occur?
3. Which of the two meiotic divisions is similar to a mitotic division?

11.2 Mechanisms That Generate Genetic Variability

The generation of genetic variability is a prime evolutionary advantage of sexual reproduction. The variability increases the chance that at least some offspring will be successful in surviving and reproducing in changing environments.

The variability produced by sexual reproduction is apparent all around us, particularly in the human population. Except for identical twins (or identical triplets, identical quadruplets, and so forth), no two humans look alike, act alike, or have identical biochemical and physiological characteristics, even if they are members of the same immediate family. Other species that reproduce sexually show equivalent variability arising from meiosis.

During meiosis and fertilization, genetic variability arises from three sources: (1) recombination, (2) the differing combinations of maternal and paternal chromosomes segregated to the poles during anaphase I, and (3) the particular sets of male and female gametes that unite in fertilization. The three mechanisms, working together, produce so much total variability that no two gametes produced by the same or different individuals and no two zygotes produced by union of the gametes are likely to have the same genetic makeup. Each of these sources of variability is discussed in further detail in the following sections.

Variability Generated by Recombination Depends on Chromosome Pairing and Physical Exchanges between Homologous Chromatids

Recombination, the key genetic event of prophase I, starts when homologous chromosomes pair (**Figure 11.5,** step 1). As the homologous chromosomes pair, they are held together tightly by a protein framework called the **synaptonemal complex (Figure 11.6).** Supported by this framework, regions of homologous chromatids exchange segments, producing new combinations of alleles (Figure 11.5, step 2). The exchange process is very precise and involves the breakage and rejoining of DNA molecules by enzymes. Each recombination event involves two of the four chromatids; the other two chromatids are not involved. When meiosis is completed, each of the four nuclei produced by the meiotic divisions receives one of the four chromatids (Figure 11.5, step 3); two receive unchanged chromatids, and two receive chromatids that have new combinations of alleles due to recombination. When the exchange is complete toward the end of prophase I, the synaptonemal complex disassembles and disappears.

The sites where recombination occurs can be seen later in prophase I, when increased condensation of the chromosomes thickens the chromosomes enough to make them visible under the light microscope (see Figure 11.3, steps 3 and 4). The sites, called **crossovers** or **chiasmata** (singular, *chiasma* = crosspiece), clearly show that two of the four chromatids have exchanged segments. Because of the shape produced, the recombination process is also called **crossing over.**

Recombination takes place largely at random, at almost any position along the chromosome arms, between any two of the four chromatids of a homologous pair. One or more additional recombination events may occur in the same chromosome pair and involve the same or different chromatids exchanging segments in the first event. In most species, recombination occurs at two or three sites in each set of paired chromosomes.

Random Segregation of Maternal and Paternal Chromosomes Is the Second Major Source of Genetic Variability in Meiosis

Random segregation of chromosomes of maternal and paternal origin accounts for the second major source of genetic variability in meiosis. Recall that metaphase I is the stage of meiosis in which the homologous pairs of chromosomes attach to the spindle poles. The maternal and paternal chromosomes of each pair typically carry different alleles of many of the genes on that chromosome. For each homologous pair, one chromosome makes spindle connections leading to one pole and the other chromosome connects to the opposite pole. In making these connections, all the maternal chromosomes may connect to one pole and all the paternal chromosomes may connect to the opposite pole. Or, as is most likely, any random combination of connections between these possibilities may be made. As a result, any combination of chromosomes of maternal and paternal origin may be segregated to the spindle poles **(Figure 11.7).** The second meiotic division segregates these random combinations of chromosomes to gamete nuclei.

The number of possible combinations depends on the number of chromosome pairs in a species. For example, the 23 chromosome pairs of humans allow 2^{23} different combinations of maternal and paternal chromosomes to be delivered to the poles, producing potentially 8,388,608 genetically different gametes from this source of variability alone.

Figure 11.4
Comparison of key steps in meiosis and mitosis. Both diagrams use an animal cell as an example. Maternal chromosomes are shown in red; paternal chromosomes are shown in blue.

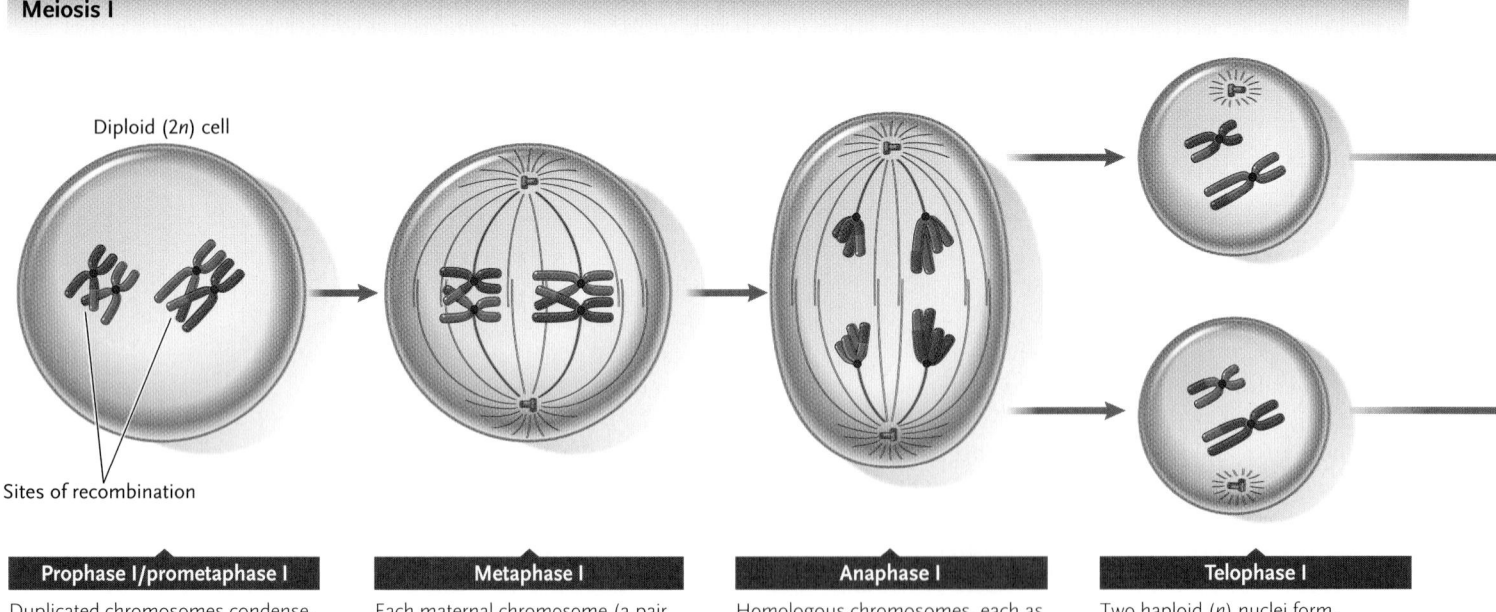

Meiosis I

| **Prophase I/prometaphase I** | **Metaphase I** | **Anaphase I** | **Telophase I** |

Duplicated chromosomes condense. Homologous chromosomes pair and exchange segments by recombination. Chromosomes attach to spindle in homologous pairs.

Each maternal chromosome (a pair of sister chromatids) and its paternal homolog align randomly at the spindle midpoint.

Homologous chromosomes, each as a pair of sister chromatids, separate and move to opposite poles.

Two haploid (*n*) nuclei form.

Random Joining of Male and Female Gametes in Fertilization Adds Additional Variability

The male and female gametes produced by meiosis are genetically diverse. Which two gametes join in fertilization is a matter of chance. This chance union of gametes amplifies the variability of sexual reproduction. Considering just the variability from random separation of homologous chromosomes and that from fertilization, the possibility that two children of the same parents could receive the same combination of maternal and paternal chromosomes is 1 chance out of $(2^{23})^2$ or 1 in 70,368,744,000,000 (~70 trillion), a number that far exceeds the number of people in the entire human population. The further variability introduced by recombination makes it practically impossible for humans and most other sexually reproducing organisms to produce genetically identical gametes or offspring. The only exception is identical twins (or identical triplets, identical quadruplets, and so forth), which arise not from the combination of identical gametes during fertilization but from mitotic division of a single fertilized egg into separate cells that give rise to genetically identical individuals.

Mitosis

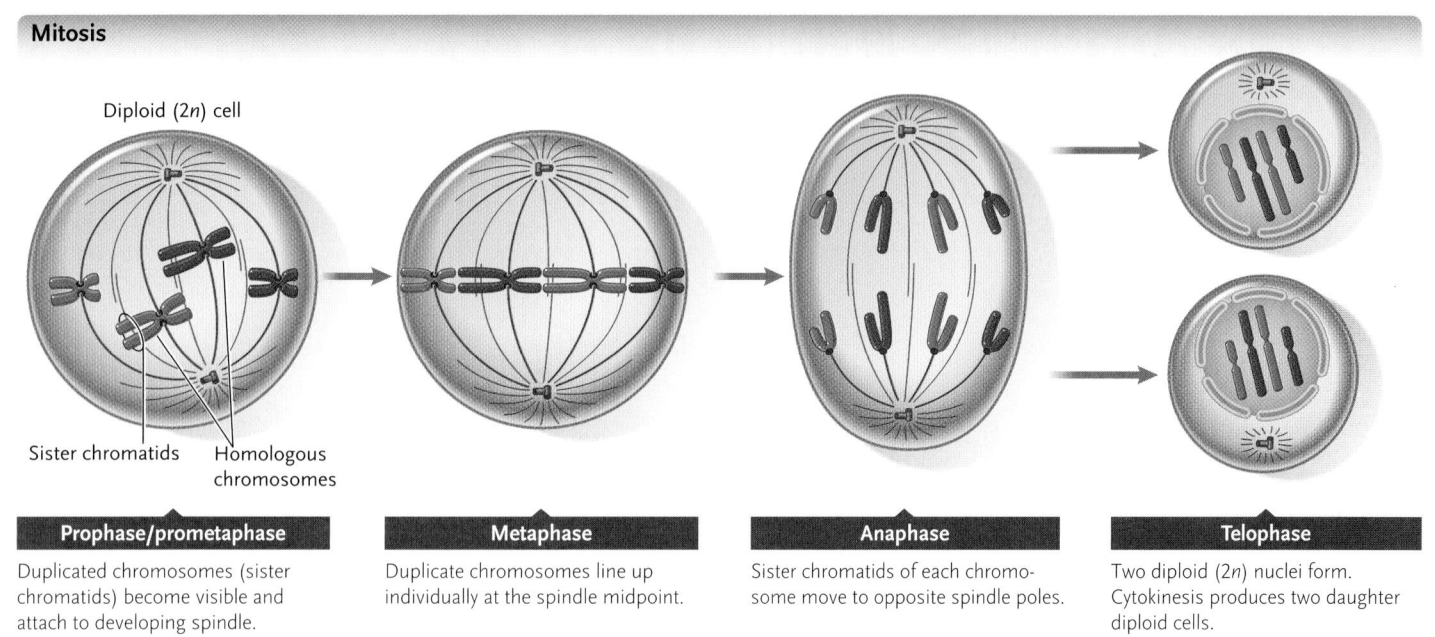

Diploid (2n) cell

Sister chromatids   Homologous chromosomes

Prophase/prometaphase
Duplicated chromosomes (sister chromatids) become visible and attach to developing spindle.

Metaphase
Duplicate chromosomes line up individually at the spindle midpoint.

Anaphase
Sister chromatids of each chromosome move to opposite spindle poles.

Telophase
Two diploid (2n) nuclei form. Cytokinesis produces two daughter diploid cells.

Meiosis II

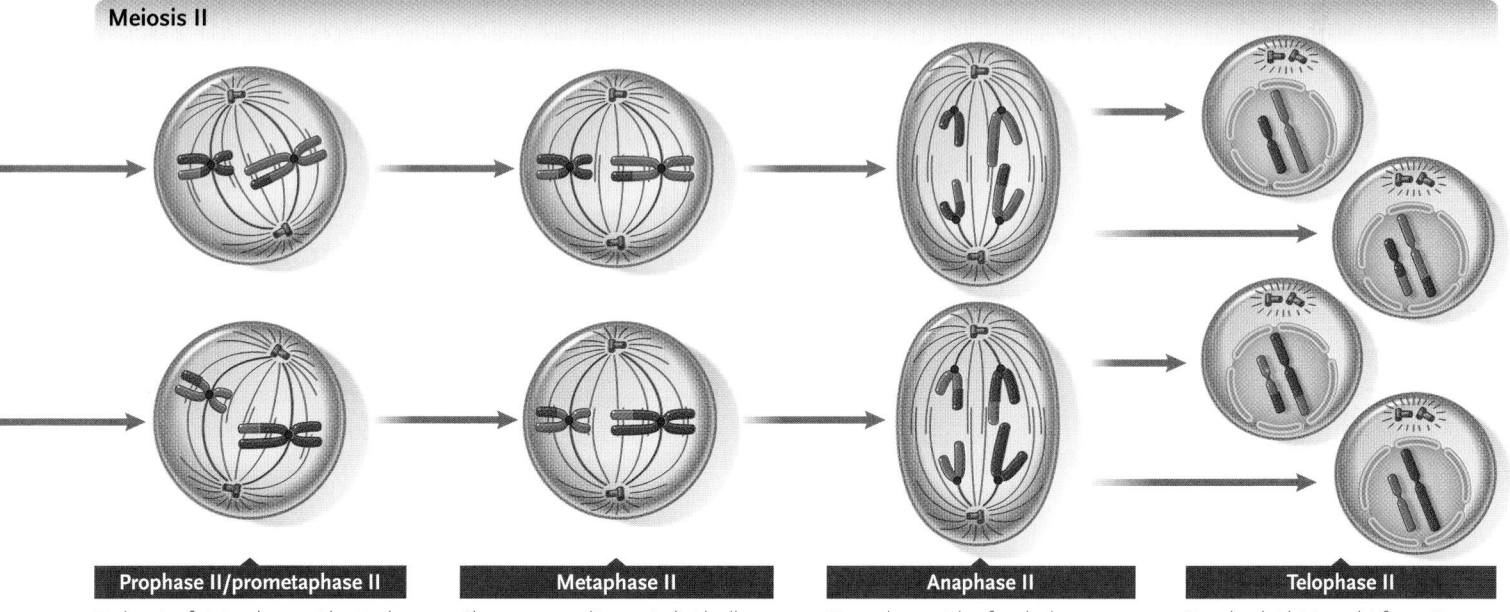

Prophase II/prometaphase II
Each pair of sister chromatids attaches to newly formed spindle.

Metaphase II
Chromosomes line up individually at the spindle midpoint.

Anaphase II
Sister chromatids of each chromosome move to opposite poles.

Telophase II
Four haploid (n) nuclei form. Cytokinesis produces four haploid cells.

Figure 11.5

Effects of the exchanges between chromatids that accomplish genetic recombination. The letters indicate two alleles, *A* and *a*, of one gene, and two alleles, *B* and *b*, of another gene. The parents have these alleles in the combinations *A-B* and *a-b*; as a result of the recombination, two of the chromatids—the recombinants—have the new combinations *a-B* and *A-b*.

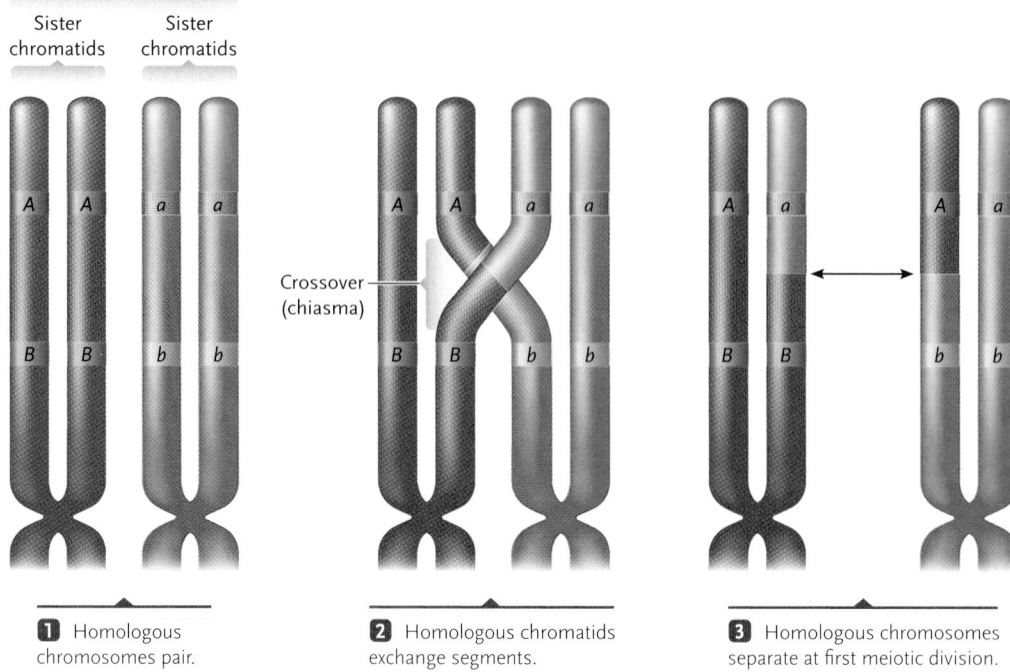

Homologous pair

Sister chromatids | Sister chromatids

Crossover (chiasma)

1 Homologous chromosomes pair.

2 Homologous chromatids exchange segments.

3 Homologous chromosomes separate at first meiotic division.

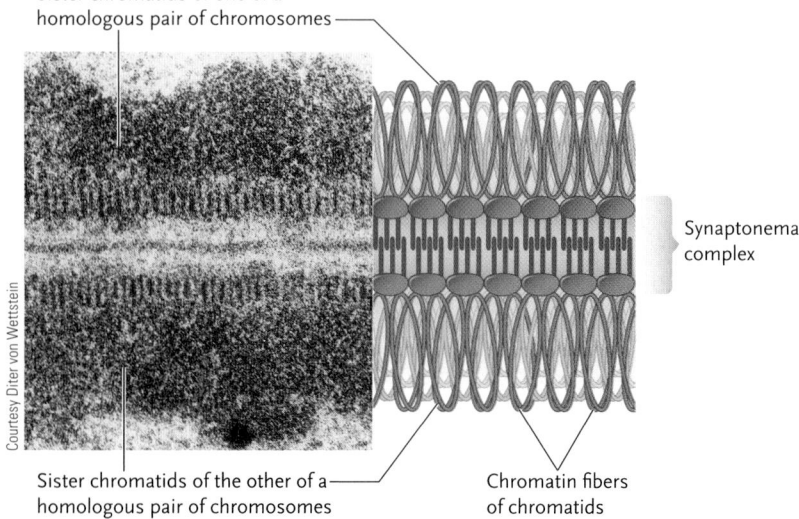

Courtesy Diter von Wettstein

Sister chromatids of one of a homologous pair of chromosomes

Synaptonemal complex

Sister chromatids of the other of a homologous pair of chromosomes

Chromatin fibers of chromatids

Figure 11.6

The synaptonemal complex as seen in a meiotic cell of the fungus *Neotiella*. The relationship of the complex to the chromatin fibers of the paired chromosomes is shown.

STUDY BREAK

1. What are the three ways in which sexual reproduction generates genetic variability?
2. Consider an animal with six pairs of chromosomes; one set of six chromosomes is from this animal's male parent, and the homologous set of six chromosomes is from this animal's female parent. When this animal produces gametes, what proportion of these gametes will have chromosomes, all of which originate from the animal's female parent?

11.3 The Time and Place of Meiosis in Organismal Life Cycles

The time and place at which meiosis occurs follows one of three major patterns in the life cycles of eukaryotes **(Figure 11.8)**. The differences reflect the portions of the life cycle spent in the haploid and diploid phases and whether mitotic divisions intervene between meiosis and the formation of gametes.

In Animals, the Diploid Phase Is Dominant, the Haploid Phase Is Reduced, and Meiosis Is Followed Directly by Gamete Formation

Animals follow the pattern (see Figure 11.8a) in which the diploid phase dominates the life cycle, the haploid phase is reduced, and meiosis is followed directly by gamete formation. In male animals, each of the four nuclei produced by meiosis is enclosed in a separate cell by cytoplasmic divisions, and each of the four cells differentiates into a functional sperm cell. In female animals, only one of the four nuclei becomes functional as an egg cell nucleus.

Fertilization restores the diploid phase of the life cycle. Thus, animals are haploids only as sperm or eggs, and no mitotic divisions occur during the haploid phase of the life cycle.

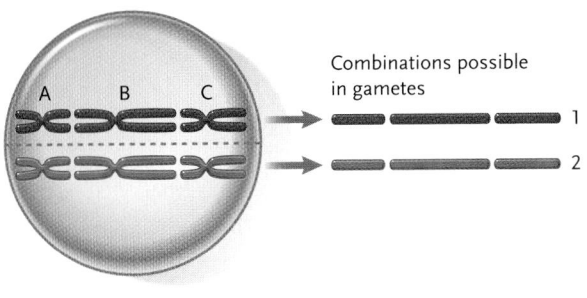

Combinations possible in gametes

1
2

or

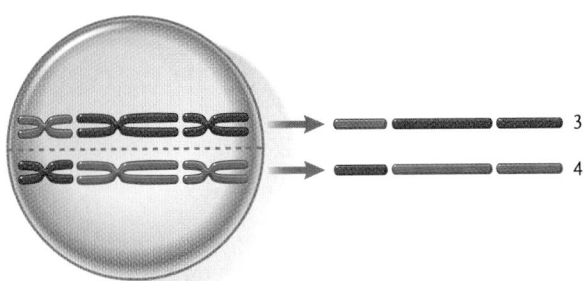

3
4

or

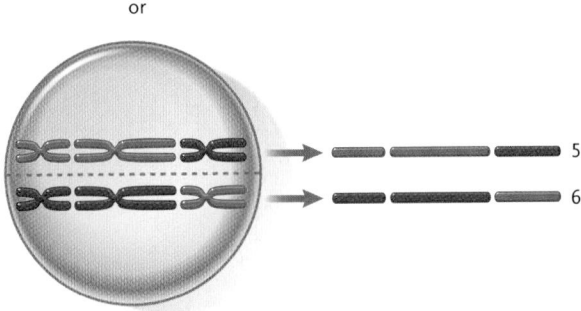

5
6

or

7
8

Figure 11.7

Possible outcomes of the random spindle connections of three pairs of chromosomes at metaphase I of meiosis. The three types of chromosomes are labeled A, B, and C. Maternal chromosomes are red; paternal chromosomes are blue. There are four possible patterns of connections, giving eight possible combinations of maternal and paternal chromosomes in gametes (labeled 1–8).

In Most Plants and Fungi, Generations Alternate between Haploid and Diploid Phases That Are Both Multicellular

Most plants and some algae and fungi follow the life cycle pattern shown in Figure 11.8b. These organisms alternate between haploid and diploid generations in which, depending on the organism, either generation may dominate the life cycle, and mitotic divisions oc-

a. Animal life cycles

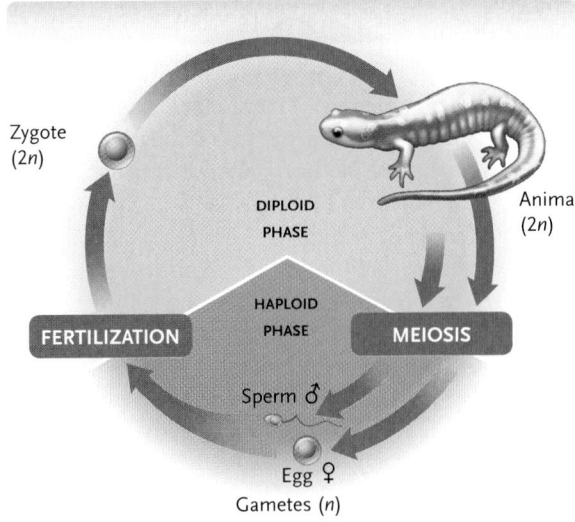

b. All plants and some fungi and algae (fern shown; relative length of the two phases varies widely in plants)

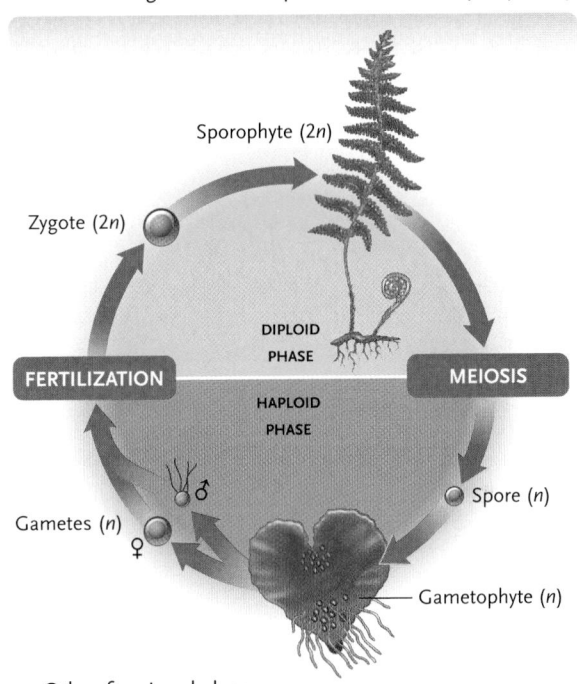

c. Other fungi and algae

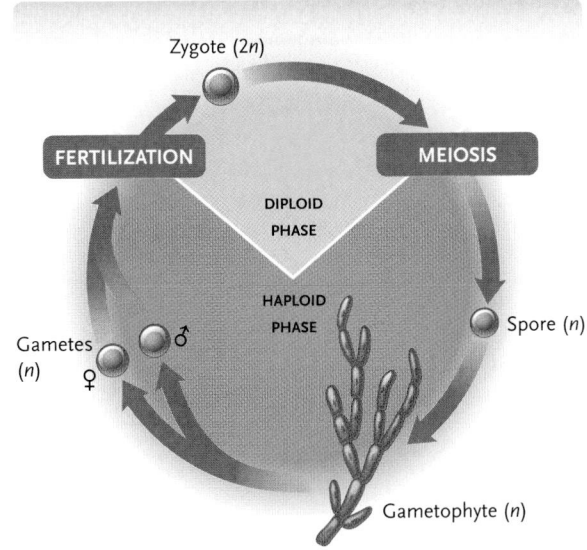

Figure 11.8

Variations in the time and place of meiosis in eukaryotes. The diploid phase of the life cycles is shaded in green; the haploid phase is shaded in yellow. n = haploid number of chromosomes; $2n$ = diploid number. **(a)** Meiosis in animal life cycles. **(b)** Meiosis in most plants and some fungi and algae. **(c)** Meiosis in other fungi and algae.

cur in both phases. In these organisms, fertilization produces the diploid generation, in which the individuals are called **sporophytes** (*spora* = seed; *phyta* = plant). After the sporophytes grow to maturity by mitotic divisions, some of their cells undergo meiosis, producing haploid, genetically different, reproductive cells called **spores**. The spores are not gametes; they germinate and grow directly by mitotic divisions into a generation of haploid individuals called **gametophytes** (*gameta* = gamete). At maturity, the nuclei of some cells in gametophytes develop into egg or sperm nuclei. All the egg or sperm nuclei produced by a particular gametophyte are genetically identical because they arise through mitosis; meiosis does not occur in gametophytes. Fusion of a haploid egg and sperm nucleus produces a diploid zygote nucleus that divides by mitosis to produce the diploid sporophyte generation again.

In many plants, including most bushes, shrubs, trees, and flowers, the diploid sporophyte generation is the most visible part of the plant. The gametophyte generation is reduced to an almost microscopic stage that develops in the reproductive parts of the sporophytes—in flowering plants, in the structures of the flower. The female gametophyte remains in the flower; the male gametophyte is released from flowers as microscopic pollen grains. When pollen contacts a flower of the same species, it releases a haploid nucleus that fertilizes a haploid egg cell of a female gametophyte in the flower. The resulting cell reproduces by mitosis to form a sporophyte.

In Some Fungi and Other Organisms, the Haploid Phase Is Dominant and the Diploid Phase Is Reduced to a Single Cell

The life cycle of some fungi and algae follows the third life cycle pattern (see Figure 11.8c). In these organisms, the diploid phase is limited to a single cell, the zygote, produced by fertilization. Immediately after fertilization, the diploid zygote undergoes meiosis to produce the haploid phase. Mitotic divisions occur only in the haploid phase.

During fertilization, two haploid gametes, usually designated simply as positive (+) or negative (−) because they are similar in structure, fuse to form a diploid nucleus. This nucleus immediately enters meiosis, producing four haploid cells. These cells develop directly or after one or more mitotic divisions into haploid spores. These spores germinate to produce haploid individuals, the gametophytes, which grow or in-

---

UNANSWERED QUESTIONS

The overall mechanism and outcomes of meiosis have been known for a long time, since the turn of the twentieth century. However, despite the fundamental importance of meiosis in sexual reproduction, the biochemical, genetic, and molecular mechanisms of meiosis are poorly understood. For example, how do homologous chromosomes recognize their appropriate pairing partners? How do they become aligned in a configuration that allows the formation of crossovers? How is the number of crossover events regulated to ensure that each chromosome pair will have a crossover? Developing a deeper understanding of the molecular mechanisms that regulate meiosis is highly important in human biology, because mis-segregation of chromosomes during meiosis I is a major cause of birth defects and the leading cause of miscarriages.

What molecular interactions initiate the chromosome pairing process, and what forces bring the chromosomes together?

Before pairing begins in prophase I, the chromosomes of homologous pairs are separated widely in the nucleus. During pairing, the chromosomes move together. What molecular interactions initiate the pairing process, and what forces bring the chromosomes together? Abby Dernburg at Lawrence Berkeley National Laboratory and her collaborators are studying these questions using as a model organism, the nematode *Caenorhabditis elegans*. She and her colleagues have found that there are regions called pairing centers near one end of each chromosome. During meiosis, the pairing centers stabilize a previously unrecognized intermediate in the pairing process, and they promote synapsis. However, intermediates can form, even when two chromosomes are not a perfect match and do not recombine. Their working model to explain this occurrence is that the pairing centers hold a pair of chromosomes together long enough for the quality of the molecular match to be assessed, and those with a good match then proceed to synapsis. How the molecular match is assessed is unknown and is the subject of current investigations. The researchers are also searching for and characterizing specific genes that are involved in pairing center function.

What are the genetic and molecular controls of meiosis?

Many unanswered questions surround the regulatory controls that switch cells from mitosis and that control the many individual steps of the division process. Both the yeast *Saccaromyces cerevisiae* and the fruit fly, *Drosophila melanogaster,* have been used widely in this research. Researchers have discovered a large number of mutant genes—well over a hundred in the two species combined—with effects in meiosis. Many of the genes identified are related to genes in humans. Many questions surround the control and operation of these genes and the identity and functions of the proteins they encode. For instance, Michael Lichten at the National Cancer Institute, National Institutes of Health, is using yeast in a project to describe the molecular steps of meiotic recombination from start to finish, including the chromosomal structure changes that occur when homologous chromosomes pair and recombine. For example, one of Lichten's research projects is using genomics scale techniques to study gene expression changes related to chromosomal structure during meiosis. That is, he is surveying all the genes in the genome to determine which ones increase or decrease gene expression during chromosome condensation, recombination, and segregation during meiosis.

Peter J. Russell

---

crease in number by mitotic divisions. Eventually positive and negative gametes are formed in these individuals by differentiation of some of the cells produced by the mitotic divisions. Because the gametes are produced by mitosis, all the gametes of an individual are genetically identical.

In this chapter, we have seen that meiosis has three outcomes that are vital to sexual reproduction. Meiosis reduces the chromosomes to the haploid number so that the chromosome number does not double at fertilization. Through recombination and random separation of maternal and paternal chromosomes, meiosis produces genetic variability in gametes; further variability is provided by the random combination of gametes in fertilization. The next chapter shows how the outcomes of meiosis and fertilization underlie the inheritance of traits in sexually reproducing organisms.

STUDY BREAK

How does the place of meiosis differ in the life cycles of animals and most plants?

Review

11.1 The Mechanisms of Meiosis

- The major cellular processes that underlie sexual reproduction are the halving of chromosome number by meiosis and restoration of the number by fertilization. Meiosis and fertilization also produce new combinations of genetic information (Figure 11.1).

- Meiosis occurs only in eukaryotes that reproduce sexually and only in organisms that are at least diploid—that is, organisms that have at least two representatives of each chromosome.

- DNA replicates and the chromosomal proteins are duplicated during the premeiotic interphase, producing two copies, the sister chromatids, of each chromosome.

- During prophase I of the first meiotic division (meiosis I), the replicated chromosomes condense and come together and pair. While they are paired, the chromatids of homologous chromosomes undergo recombination by exchanging segments. While these events are in progress, the spindle forms in the cytoplasm (Figures 11.2 and 11.3).

- During prometaphase I, the nuclear envelope breaks down, the spindle enters the former nuclear area, and kinetochore microtubules leading to opposite spindle poles attach to one kinetochore of each pair of sister chromatids of homologous chromosomes (Figure 11.3).

- At metaphase I, spindle microtubule movements have aligned the tetrads on the metaphase plate, the equatorial plane between the two spindle poles. The connections of kinetochore microtubules to opposite poles ensure that the homologous pairs separate and move to opposite spindle poles during anaphase I, reducing the chromosome number to the haploid value. Each chromosome at the poles still contains two chromatids.

- Telophase I and interkinesis are brief and transitory stages; no DNA replication occurs during interkinesis. During these stages, the single spindle of the first meiotic division disassembles and the microtubules reassemble into two new spindles for the second division.

- During prophase II, the chromosomes condense and a spindle forms. During prometaphase II, the nuclear envelope breaks down, the spindle enters the former nuclear area, and spindle microtubules leading to opposite spindle poles attach to the two kinetochores of each chromosome. At metaphase II, the chromosomes become aligned on the metaphase plate. The connections of kinetochore microtubules to opposite spindle poles ensure that during anaphase II, the chromatids of each chromosome are separated and segregate to those opposite spindle poles.

- During telophase II, the chromosomes decondense to their extended interphase state, the spindles disassemble, and new nu-

clear envelopes form. The result is four haploid cells, each containing half the number of chromosomes present in a G_1 nucleus of the same species.

Animation: Gamete-producing organs

Animation: Meiosis step-by-step

Animation: Meiosis I and II

Animation: Comparing mitosis and meiosis

11.2 Mechanisms That Generate Variability

- Recombination is the first source of the genetic variability produced by meiosis (Figures 11.5 and 11.6). During recombination, chromatids generate new combinations of alleles by physically exchanging segments. The exchange process involves precise breakage and joining of DNA molecules. It is catalyzed by enzymes and occurs while the homologous chromosomes are held together tightly by the synaptonemal complex.

- The crossovers visible between the chromosomes at late prophase I reflect the exchange of chromatid segments that occurred during the molecular steps of genetic recombination.

- The random segregation of homologous chromosomes is the second source of genetic variability produced by meiosis. The homologous pairs separate at anaphase I of meiosis, segregating random combinations of maternal and paternal chromosomes to the spindle poles (Figure 11.7).

- Random joining of male and female gametes in fertilization is the third source of genetic variability.

Animation: Crossing over

Animation: Random alignment

11.3 Time and Place of Meiosis in Organismal Life Cycles

- The time and place of meiosis follow one of three major pathways in the life cycles of eukaryotes, which reflect the portions of the life cycle spent in the haploid and diploid phases and whether mitotic divisions intervene between meiosis and the formation of gametes (Figure 11.8).

- In animals, the diploid phase dominates the life cycle; mitotic divisions occur only in this phase. Meiosis in the diploid phase gives rise to products that develop directly into egg and sperm cells without undergoing mitosis (Figure 11.8a).

- In most plants and fungi, the life cycle alternates between haploid and diploid generations that both grow by mitotic divisions. Fertilization produces the diploid sporophyte generation; after growth by mitotic divisions, some cells of the sporophyte undergo meiosis and produce haploid spores. The spores germinate and grow by mitotic divisions into the gametophyte generation. After growth of the gametophyte, cells develop directly into egg or sperm nuclei, which fuse in fertilization to produce the diploid sporophyte generation again (Figure 11.8b).

- In some fungi and protists, meiosis occurs immediately after fertilization, producing a haploid phase, which dominates the life cycle; mitosis occurs only in the haploid phase. At some point in the life cycle, haploid cells differentiate directly into gametes, which fuse together as pairs to produce the brief diploid phase (Figure 11.8c).

Animation: Generalized life cycles

Questions

Self-Test Questions

1. The diploid number of this individual is 6.

This figure represents:
a. mitotic metaphase.
b. meiotic metaphase I.
c. meiotic metaphase II.
d. a gamete.
e. six nonhomologous chromosomes.

2. Which of the following is *not* associated with meiosis?
a. daughter cells identical to the parent cell
b. variety in resulting cells
c. chromosome number halved in resulting cells
d. four daughter cells arising from one parent cell
e. 23 chromosomes in the human egg or sperm

3. Chiasmata:
a. form during metaphase II of meiosis.
b. occur between two nonhomologous chromosomes.
c. represent chromosomes independently assorting.
d. are sites of DNA exchange between homologous chromatids.
e. ensure the resulting cells are identical to the parent cell.

4. If $2n$ is four, the number of possible combinations in the resulting gametes is:
a. 1. b. 2. c. 4. d. 8. e. 16.

5. The number of human chromosomes in a cell in prophase I of meiosis is ___ and in telophase II is ___.
a. 92; 46   d. 23; 16
b. 46; 23   e. 4; 2
c. 23; 23

6. In meiosis:
a. homologous chromosomes pair up at prophase II.
b. chromosomes separate from their homologous partners at anaphase I.
c. the centromeres split at anaphase I.
d. a female gamete has two X chromosomes.
e. reduction of chromosome number occurs in meiosis II.

7. The DNA content in a diploid cell in G_2 is X. If that cell goes into meiosis at its metaphase II, the DNA content would be:
a. 0.1X. b. 0.5X. c. X. d. 2X. e. 4X.

8. Metaphase in mitosis is similar to what stage in meiosis?
a. prophase I   d. metaphase II
b. prophase II   e. crossing over
c. metaphase I

9. In the human gamete:
a. there must be one chromosome of each type, except for the sex chromosomes, where both an X and a Y chromosome are present.
b. a chromosome must be represented from each parent.
c. there must be an unequal mixture of chromosomes from both parents.
d. there must be representation of chromosomes from only one parent.
e. there is the possibility of 2^{46} different combinations of maternal and paternal chromosomes.

10. In plants, the adult diploid individuals are called:
a. spores.   d. gametophytes.
b. sporophytes.   e. zygotes.
c. gametes.

Questions for Discussion

1. You have a technique that allows you to measure the amount of DNA in a cell nucleus. You establish the amount of DNA in a sperm cell of an organism as your baseline. Which multiple of this amount would you expect to find in a nucleus of this organism at G_2 of premeiotic interphase? At telophase I of meiosis? During interkinesis? At telophase II of meiosis?

2. One of the human chromosome pairs carries a gene that influences eye color. In an individual human, one chromosome of this pair has an allele of this gene that contributes to the formation of blue eyes. The other chromosome of the pair has an allele that contributes to brown eye color (other genes also influence eye color in humans). After meiosis in the cells of this individual, what fraction of the nuclei will carry the allele that contributes to blue eyes? To brown eyes?

3. Mutations are changes in DNA sequence that can create new alleles. In which cells of an individual, somatic or meiotic cells, would mutations be of greatest significance to that individual? What about to the species to which the individual belongs?

Experimental Analysis

Design experiments to determine whether a new pesticide on the market adversely affects egg production and fertilization in frogs.

Evolution Link

Explain aspects of the processes of mitosis and meiosis that would lead you to conclude that they are evolutionarily related processes. Do you think that mitosis evolved from meiosis, or did the opposite occur? Explain your conclusion.

How Would You Vote?

Japanese researchers have successfully created a "fatherless" mouse that contains the genetic material from the eggs of two females. The mouse is healthy and fully fertile. Do you think researchers should be allowed to try the same process with human eggs? Go to www.thomsonedu.com/login to investigate both sides of the issue and then vote.

Mice, showing genetic variation in coat color.

Carolyn A. McKeone/Science Photo Library/Photo Researchers, Inc.

12 Mendel, Genes, and Inheritance

WHY IT MATTERS

Parties and champagne were among the last things on Ernest Irons's mind on New Year's Eve, 1904. Irons, a medical intern, was examining a blood specimen from a new patient and was sketching what he saw through his microscope—peculiarly elongated red blood cells (**Figure 12.1**). He and his supervisor, James Herrick, had never seen anything like them. The shape of the cells was reminiscent of a sickle, a cutting tool with a crescent-shaped blade.

The patient had complained of weakness, dizziness, shortness of breath, and pain. His father and two sisters had died from mysterious ailments that had damaged their lungs or kidneys. Did those deceased family members also have sickle-shaped red cells in their blood? Was there a connection between the abnormal cells and the ailments? How did the cells become sickled?

The medical problems that baffled Irons and Herrick killed their patient when he was only 32 years old. The patient's symptoms were characteristic of a genetic disorder now called *sickle-cell disease*. This disease develops when a person has received two copies of a gene (one from each parent) that codes for an altered subunit of hemoglobin, the oxygen-transporting protein in red blood cells. When oxygen sup-

a.

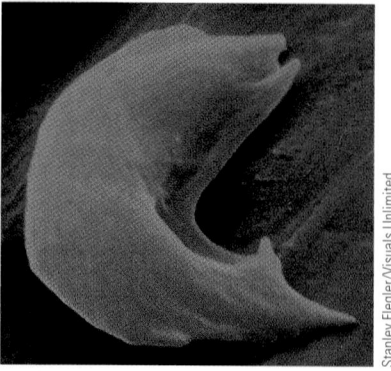

b.

Stanley Flegler/Visuals Unlimited

Figure 12.1
Red blood cell shape in sickle-cell disease. **(a)** A normal red blood cell. **(b)** A sickled red blood cell.

plies are low, the altered hemoglobin forms long, fibrous, crystal-like structures that push red blood cells into the sickle shape. The altered protein differs from the normal protein by just a single amino acid.

The sickled red blood cells are too elongated and inflexible to pass through the capillaries, the smallest vessels in the circulatory system. As a result, the cells block the capillaries. The surrounding tissues become starved for oxygen and saturated with metabolic wastes, causing the symptoms experienced by Irons and Herrick's patient. The problem worsens as oxygen concentration falls in tissues and more red blood cells are pushed into the sickled form. (You will learn more about sickle-cell disease in this chapter and in Chapter 13.)

Researchers have studied sickle-cell disease in great detail at both the molecular and the clinical levels. You may find it curious, though, that our understanding of sickle-cell disease—and all other heritable traits—actually began with studies of pea plants in a monastery garden.

Fifty years before Ernest Irons sketched sickled red blood cells, a scholarly monk named Gregor Mendel **(Figure 12.2)** used garden peas to study patterns of

Moravian Museum, Brno

Figure 12.2
Gregor Mendel (1822–1884), the founder of genetics.

inheritance. To test his hypotheses about inheritance, Mendel bred generation after generation of pea plants and carefully observed the patterns by which parents transmit traits to their offspring. Through his experiments and observations, Mendel discovered the fundamental rules that govern inheritance. His discoveries and conclusions founded the science of genetics and still have the power to explain many of the puzzling and sometimes devastating aspects of inheritance that continue to occupy our attention.

12.1 The Beginnings of Genetics: Mendel's Garden Peas

Until about 1900, scientists and the general public believed in the **blending theory of inheritance**, which suggested that hereditary traits blend evenly in offspring through mixing of the parents' blood, much like the effect of mixing coffee and cream. Even today, many people assume that parental characteristics such as skin color, body size, and facial features blend evenly in their offspring, with the traits of the children appearing about halfway between those of their parents. Yet if blending takes place, why don't extremes, such as very tall and very short individuals, gradually disappear over generations as repeated blending takes place? Also, why do children with blue eyes keep turning up among the offspring of brown-eyed parents?

Gregor Mendel's experiments with garden peas, performed in the 1860s, provided the first answers to these questions and many more. Mendel was an Augustinian monk who lived in a monastery in Brünn, now part of the Czech Republic. But he had an unusual education for a monk in the mid-nineteenth century. He had studied mathematics, chemistry, zoology, and botany at the University of Vienna under some of the foremost scientists of his day. He had also been reared on a farm and was well aware of agricultural principles and their application. He kept abreast of breeding experiments published in scientific journals. Mendel also won several awards for developing improved varieties of fruits and vegetables.

In his work with peas, Mendel studied a variety of heritable characteristics called **characters**, such as flower color or seed shape. A variation in a character, such as purple or white flower color, is called a **trait**. Mendel established that characters are passed to offspring in the form of discrete hereditary factors, which now are known as genes. Mendel observed that, rather than blending evenly, many parental traits appear unchanged in offspring, whereas others disappear in one generation to reappear unchanged in the next. Although Mendel did not know it, the inheritance patterns he observed are the result of the segregation of chromosomes, on which the genes are located, to gametes in meiosis (see Chapter 11). Mendel's methods illustrate, perhaps as well as any experiments in the history of science, how rigorous scientific work is conducted: through observation, making hypotheses, and testing the hypotheses with experiments.

Mendel Chose True-Breeding Garden Peas for His Experiments

Mendel chose the garden pea (*Pisum sativum;* **Figure 12.3**) for his research because the plant could be grown easily in the monastery garden, without elaborate equipment. As in other flowering plants, gametes are produced in structures of the flowers (see Figure 12.3). The male gametes are sperm nuclei contained in the pollen, which is produced in the *anthers* of the flower. The female gametes are egg cells, produced in the *carpel* of the flowers. Normally, pea plants **self-fertilize** (also known as **self-pollinate**, or more simply, *self*): sperm nuclei in pollen produced by anthers fertilize egg cells housed in the carpel of the same flower. However, for his experiments, Mendel prevented self-fertilization simply by cutting off the anthers. Pollen to fertilize these flowers must then come from a different plant. This technique is called **cross-pollination**, or more simply, a *cross*. This technique allowed Mendel to test the effects of mating pea plants of different parental types.

To begin his experiments, Mendel chose pea plants that were known to be **true-breeding** (also called *pure-breeding*); that is, when self-fertilized, or more simply, *selfed,* they passed traits without change from one generation to the next.

Mendel First Worked with Single-Character Crosses

Flower color was among the seven characters Mendel selected for study; one true-breeding variety of peas had purple flowers, and the other true-breeding variety had white flowers (see Figure 12.3). Would these traits blend evenly if plants with purple flowers were cross-pollinated with plants with white flowers?

To answer this question, Mendel took pollen from the anthers of plants with purple flowers and placed it in the flowers of white-flowered plants. He placed the pollen on the *stigma,* the part of the carpel that receives pollen in flowers (see Figure 12.3). He also performed the reciprocal experiment by placing pollen from white-flowered plants on the stigmas of purple-flowered plants. Seeds were the result of the crosses; each seed contains a zygote, or embryo, that will develop into a new pea plant. The plants that develop from the seeds produced by the cross—the first generation of offspring from the cross—are the F_1 **generation** (F stands for *filial; filius* = son). The plants used in the initial cross are called the parental or **P generation.** The plants that grew from the F_1 seeds all formed purple flowers, as if the trait for white flowers had disappeared. The flowers showed no evidence of blending.

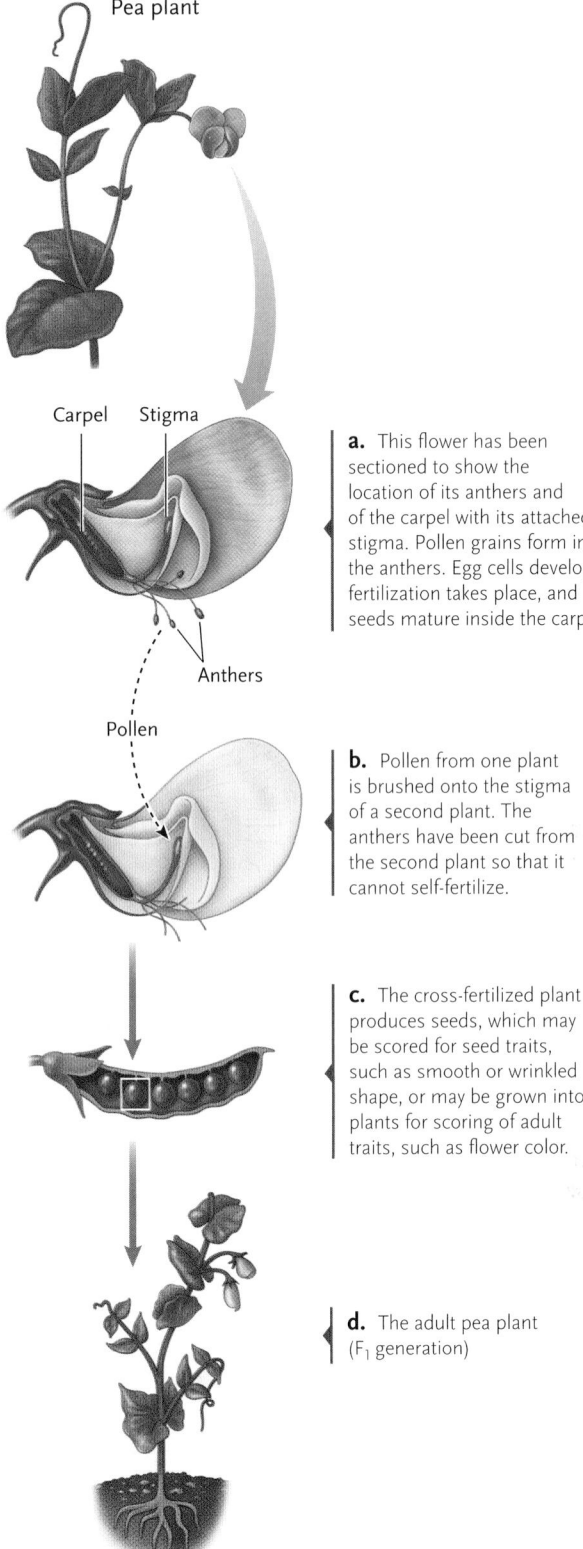

a. This flower has been sectioned to show the location of its anthers and of the carpel with its attached stigma. Pollen grains form in the anthers. Egg cells develop, fertilization takes place, and seeds mature inside the carpel.

b. Pollen from one plant is brushed onto the stigma of a second plant. The anthers have been cut from the second plant so that it cannot self-fertilize.

c. The cross-fertilized plant produces seeds, which may be scored for seed traits, such as smooth or wrinkled shape, or may be grown into plants for scoring of adult traits, such as flower color.

d. The adult pea plant (F_1 generation)

Figure 12.3
The garden pea (*Pisum sativum*), the focus of Mendel's experiments.

Mendel then allowed the purple-flowered F_1 plants to self, producing seeds that represented the F_2 **generation.** When he planted the F_2 seeds produced by this cross, the white-flowered trait reappeared: both purple-flowered and white-flowered plants were produced. Mendel counted 705 plants with purple flowers and 224 with white flowers, in a ratio that he noted was close to 3 : 1, or about 75% purple-flowered plants and 25% white-flowered plants.

Mendel made similar crosses that involved six other characters with pairs of traits **(Figure 12.4)**; for example, the character of seed color has the traits yellow and green. In all cases, he observed a uniform F_1 generation, in which only one of the two traits was present. In the F_2 generation, the missing trait reappeared, and both traits were present among the offspring. Moreover, the trait present in the F_1 generation was present in a definite, predictable proportion among the offspring.

Mendel's Single-Character Crosses Led Him to Propose the Principle of Segregation

Using his knowledge of mathematics, Mendel developed a set of hypotheses to explain the results of his crosses. His first hypothesis was: *The adult plants carry a* pair *of factors that govern the inheritance of each character.* He correctly deduced that for each character, an organism inherits one factor from each parent.

In modern terminology, Mendel's factors are called *genes,* which are located on chromosomes; the different versions of a gene, producing different traits of a character, are **alleles** of the gene (see Section 11.1). Although Mendel did not use the modern terms *genes* and *alleles,* we use them in this chapter in our description of Mendel's work. Thus, there are two alleles of the gene that govern flower color in garden peas: one allele for purple flower color and the other allele for white flower color. Organisms with two copies of each gene are now known as diploids (see Section 11.1); the two alleles of a gene in a diploid individual may be identical or different.

How can the disappearance of one of the traits, such as white flowers, in the F_1 generation and its reappearance in the F_2 generation be explained? Mendel deduced that the trait that had seemed to "disappear" in the F_1 generation actually was present but was masked in some way by the "stronger" allele. Mendel called the masking effect **dominance**. Accordingly, Mendel's second hypothesis stated: *If an individual's pair of genes consists of different alleles, one allele is dominant over the other.* This hypothesis assumes that one allele is **dominant** and the other allele is **recessive**. When a dominant allele for a trait is paired with a recessive allele for the same trait, the dominant allele is

Figure 12.4
Mendel's crosses with seven different characters in peas, including his results and the calculated ratios of offspring.

Character	Traits crossed	F_1	F_2		Ratio
Seed shape	round × wrinkled	All round	5474 round	1850 wrinkled	2.96:1
Seed color	yellow × green	All yellow	6022 yellow	2001 green	3.01:1
Pod shape	inflated × constricted	All inflated	882 inflated	299 constricted	2.95:1
Pod color	green × yellow	All green	428 green	152 yellow	2.82:1
Flower color	purple × white	All purple	705 purple	224 white	3.15:1
Flower position	axial (along stems) × terminal (at tips)	All axial	651 axial	207 terminal	3.14:1
Stem length	tall × dwarf	All tall	787 tall	277 dwarf	2.84:1

expressed. By contrast, a recessive allele is expressed only when two copies of the allele are present. For example, for flower color in Mendel's experiments, the allele for purple flowers was dominant and the allele for white flowers was recessive.

As a third hypothesis, Mendel proposed: *The pairs of alleles that control a character* **segregate** *(separate) as gametes are formed; half the gametes carry one allele, and the other half carry the other allele.* This hypothesis is now known as Mendel's **Principle of Segregation**. During fertilization, fusion of the haploid maternal and paternal gametes produces a diploid nucleus called the *zygote nucleus.* The zygote nucleus receives one allele for the character from the male gamete and one allele for the same character from the female gamete, reuniting the pairs.

Mendel's three hypotheses explained the results of the crosses **(Figure 12.5)**. Both alleles of the gene that governs flower color in the original, true-breeding parent plant with purple flowers are the same. The symbol P is used here to designate this allele, with the capital letter indicating that it is dominant, which gives this true-breeding parent the PP combination of alleles. Such an individual is called a **homozygote** (*homo* = same) and is said to be **homozygous** for the P allele. In other words, the individual has two copies of the same allele of the flower color gene. Therefore, when the individual produces gametes and the paired alleles separate, all the gametes of this individual will receive a P allele (see the left side heading in Figure 12.5a).

In the original true-breeding parent with white flowers, both alleles of the gene are also the same. The symbol p is used here to designate this allele, with the lowercase letter indicating that it is recessive, which gives this true-breeding plant the homozygous pp combination of alleles. These alleles also separate during gamete formation, producing gametes with one p allele (see the top heading in Figure 12.5a). (Mendel originated the practice of using uppercase and lowercase letters to designate dominant and recessive alleles.)

All the F_1 plants produced by crossing purple-flowered and white-flowered plants—the cross $PP \times pp$—receive the same combination of alleles, Pp (see the cell in Figure 12.5a). An individual of this type, with two different alleles of a gene, is called a **heterozygote** (*hetero* = different) and is said to be **heterozygous** for the trait. Because P is dominant over p, all the Pp plants have purple flowers, even though they also carry the allele for white flowers. An F_1 heterozygote produced from a cross that involves a single character is called a **monohybrid** (*mono* = one; *hybrid* = an offspring of parents with different traits).

According to Mendel's hypotheses, all the Pp plants in the F_1 generation produce two kinds of gametes. Because the heterozygous Pp pair separates during gamete formation, half of the gametes receive the P allele and half receive the p allele. Figure 12.5b shows how these gametes can combine during selfing of F_1 plants. Gener-

ally, a cross between two individuals that are each heterozygous for the same pair of alleles— $Pp \times Pp$ here—is called a **monohybrid cross**. The gametes are entered in both the rows and columns in Figure 12.5b; the cells show the possible combinations. Combining two gametes that both carry the P allele produces a PP F_2 plant; combining P from one parent and p from the other produces a Pp plant; and combining p from both F_1 parents produces a pp F_2 plant. The homozygous PP and heterozygous Pp plants in the F_2 generation have purple flowers, the dominant trait; the homozygous pp offspring have white flowers, the recessive trait.

Mendel's hypotheses explain how individuals may differ genetically but still look the same. The PP and Pp plants, although genetically different, both have purple flowers. In modern terminology, **genotype** refers to the *genetic constitution of an organism,* and **phenotype** (Greek *phainein* = to show) refers to its *outward appearance.* In this case, the two different genotypes PP and Pp produce the same purple-flower phenotype.

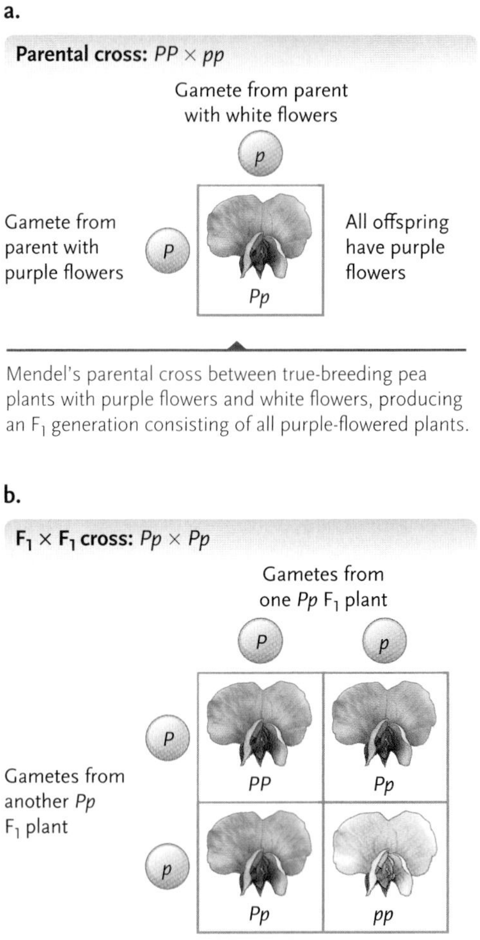

Thus, the results of Mendel's crosses support his three hypotheses:

1. The genes that govern genetic characters occur in pairs in individuals.
2. If different alleles are present in an individual's pair of genes, one allele is dominant over the other.
3. The two alleles of a gene segregate and enter gametes singly.

Mendel Could Predict Both Classes and Proportions of Offspring from His Hypotheses

Mendel could predict both classes and proportions of offspring from his hypotheses. To understand how Mendel's hypotheses allowed him to predict the proportions of offspring resulting from a genetic cross,

a.

Parental cross: $PP \times pp$

Mendel's parental cross between true-breeding pea plants with purple flowers and white flowers, producing an F_1 generation consisting of all purple-flowered plants.

b.

$F_1 \times F_1$ **cross:** $Pp \times Pp$

Mendel's cross between F_1 plants with purple flowers, producing an F_2 generation consisting of $3/4$ purple-flowered and $1/4$ white-flowered plants.

Figure 12.5
The principle of segregation in Mendel's crosses studying the inheritance of flower color in garden peas.

let's review the mathematical rules that govern **probability**—that is, the possibility that an outcome will occur if it is a matter of chance, as in the random fertilization of an egg by a sperm cell that contains one allele or another.

In the mathematics of probability, the likelihood of an outcome is predicted on a scale of 0 to 1. An outcome that is certain to occur has a probability of 1, and an outcome that cannot possibly happen has a probability of 0. If two different outcomes are equally likely, as in getting heads or tails in flipping a coin, we determine the probability of one of the outcomes by dividing that outcome by the total number of possible outcomes. For obtaining heads in flipping a coin, the probability is 1 divided by 2, or 1/2. The probabilities of all the possible outcomes, when added together, must equal 1. Thus, a coin flip has only two possible outcomes, heads or tails, each with a probability of 1/2; the sum of these probabilities is: $1/2 + 1/2 = 1$.

The Product Rule in Probability. What is the chance of flipping two heads in succession? Because the outcome of one flip has no effect on the next one, the two successive flips are independent. When two or more events are independent, the probability that they will occur in succession is calculated using the **product rule**—their individual probabilities are multiplied. That is, the probability that events A and B *both* will occur equals the probability of event A *multiplied* by the probability of event B. For example, the probability of getting heads on the first flip is 1/2; the probability of heads on the second flip is also 1/2 **(Figure 12.6)**. Because the events are independent, the probability of getting two heads in a row is $1/2 \times 1/2 = 1/4$.

Applying the same principles, the probability of getting two tails is also $1/2 \times 1/2 = 1/4$ (see Figure 12.6). Similarly, because the sex of one child has no effect on the sex of the next child in a family, the probability of having four girls in a row is the product of their individual probabilities (very close to 1/2 for each birth): $1/2 \times 1/2 \times 1/2 \times 1/2 = 1/16$.

The Sum Rule in Probability. Another relationship, the **sum rule**, applies when there are two or more different ways of obtaining the same outcome; that is, the probability that *either* event A *or* event B will occur equals the probability of event A *plus* the probability of event B. Returning to the coin toss example, the probability of getting a head and a tail in two tosses can be determined. We could toss the coin twice and get a head, then a tail. The probability that this will occur is 1/2 for the head $\times$ 1/2 for the tail = 1/4 (see Figure 12.6). However, we could toss the coin twice and get first a tail, then a head. The probability that this will occur is also 1/4 (see Figure 12.6). Both of these outcomes must be considered together, because both give a head and a tail. That is, there are two ways of obtaining the same outcome. Therefore, for the probability of tossing and head and a tail, we sum the individual probabilities to get the final probability: here, $1/4 + 1/4 = 1/2$.

Probability in Mendel's Crosses. The same rules of probability just discussed apply to Mendel's crosses. For example, in the crosses that involve the purple-flowered and white-flowered traits, half of the gametes of the F₁ generation contain the *P* allele of the gene and half contain the *p* allele (see Figure 12.5b). To produce a *PP* zygote, two *P* gametes must combine. The probability of selecting a *P* gamete from one F₁ parent is 1/2, and the probability of selecting a *P* gamete from the other F₁ parent is also 1/2. Therefore, the probability of producing a *PP* zygote from this monohybrid cross is $1/2 \times 1/2 = 1/4$. That is, by the product rule, one-fourth of the offspring of the F₁ cross *Pp* $\times$ *Pp* are expected to be *PP*, which have purple flowers **(Figure 12.7a)**. By the same line of reasoning, one-fourth of the F₂ offspring are expected to be *pp*, which have white flowers **(Figure 12.7b)**.

What about the production of *Pp* offspring? The cross *Pp* $\times$ *Pp* can produce *Pp* in two different ways. A *P* gamete from the first parent can combine with a *p* gamete from the second parent *(Pp)*, or a *p* gamete from the first parent can combine with a *P* gamete from the second parent *(pP)* **(Figure 12.7c)**. Because there are two different ways to get the same outcome, we apply the sum rule to obtain the combined probability. Each of the ways to get *Pp* has an individual probability of 1/4; when we add these individual probabilities, we have $1/4 + 1/4 = 1/2$. Therefore, half of the offspring are expected to be *Pp*, which have purple flowers. We could get the same result from the requirement that all of the individual probabilities must

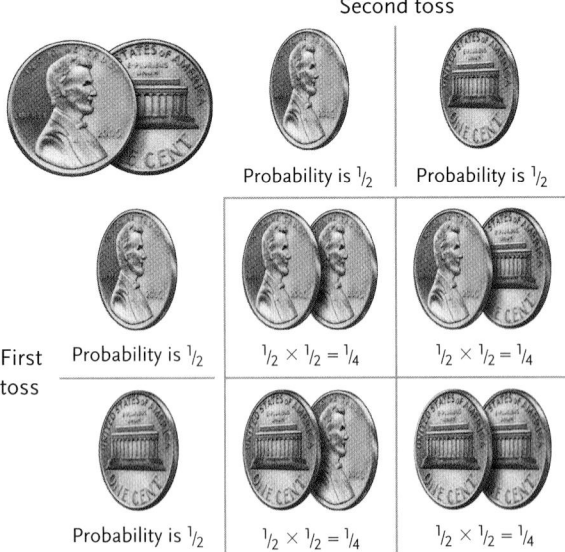

Figure 12.6

Rules of probability. For each coin toss, the probability of a head is 1/2; the probability of a tail is also 1/2. Because the outcome of the first toss is independent of the outcome of the second, the combined probabilities of the outcomes of successive tosses are calculated by multiplying their individual probabilities according to the product rule.

Second toss

Probability is ½ | Probability is ½

First toss — Probability is ½ | $½ \times ½ = ¼$ | $½ \times ½ = ¼$

Probability is ½ | $½ \times ½ = ¼$ | $½ \times ½ = ¼$

add up to 1. If the probability of *PP* is 1/4 and the probability of *pp* is 1/4, then the probability of the remaining possibility, *Pp*, must be 1/2, because the total of the individual probabilities must add up to 1: 1/4 + 1/4 + 1/2 = 1.

What if we want to know the probability of obtaining purple flowers in the cross *Pp* × *Pp*? In this case, the rule of addition applies, because there are two ways to get purple flowers: genotypes *PP* and *Pp*. Adding the individual probabilities of these combinations, 1/4 *PP* + 1/2 *Pp*, gives a total of 3/4, indicating that three-fourths of the F_2 offspring are expected to have purple flowers. Because the total probabilities must add up to 1, the remaining one-fourth of the offspring are expected to have white flowers (1/4 *pp*). These proportions give the ratio 3 : 1, which is close to the ratio Mendel obtained in his cross.

What we have just stepped through in describing Figure 12.7 is the **Punnett square** method for determining the genotypes of offspring and their expected proportions. To use the Punnett square, write the probability of obtaining gametes with each type of allele from one parent at the top of the diagram and write the chance of obtaining each type of allele from the other parent on the left side. Then fill in the cells by combining the alleles from the top and from the left and multiply their individual probabilities.

Mendel Used a Testcross to Check the Validity of His Hypotheses

Mendel realized that he could assess the validity of his hypotheses by determining whether they could be used successfully to *predict* the outcome of a cross of a different type than he had tried so far. Accordingly, he crossed an F_1 plant with purple flowers, assumed to have the heterozygous genotype *Pp*, with a true-breeding white-flowered plant, with the homozygous genotype *pp*. In this cross, *Pp* × *pp*, all the gametes of the *pp* plant contain a single *p* allele. Therefore, the probability that a gamete from this parent contains *p* is 1. The gamete and its probability of 1 are entered as the row heading of the Punnett square in **Figure 12.8.** The *Pp* parent produces two types of gametes, half that contain the *P* allele and half that contain the *p* allele. These values, 1/2 *P* and 1/2 *p*, are entered as the column headings. Filling in the possible combinations in the cells gives the two expected classes, *Pp* and *pp*, both with a probability of 1/2. Thus, half the offspring of this cross are expected to have purple flowers and half are expected to have white flowers; the ratio is 1 : 1. Mendel's actual results in this cross were 85 purple-flowered plants and 81 white-flowered plants, which closely approach the expected 1 : 1 ratio. Mendel also made the same type of cross with all the other traits used in his study, including those traits affecting seed shape, seed color, and plant height, and found the same 1 : 1 ratio.

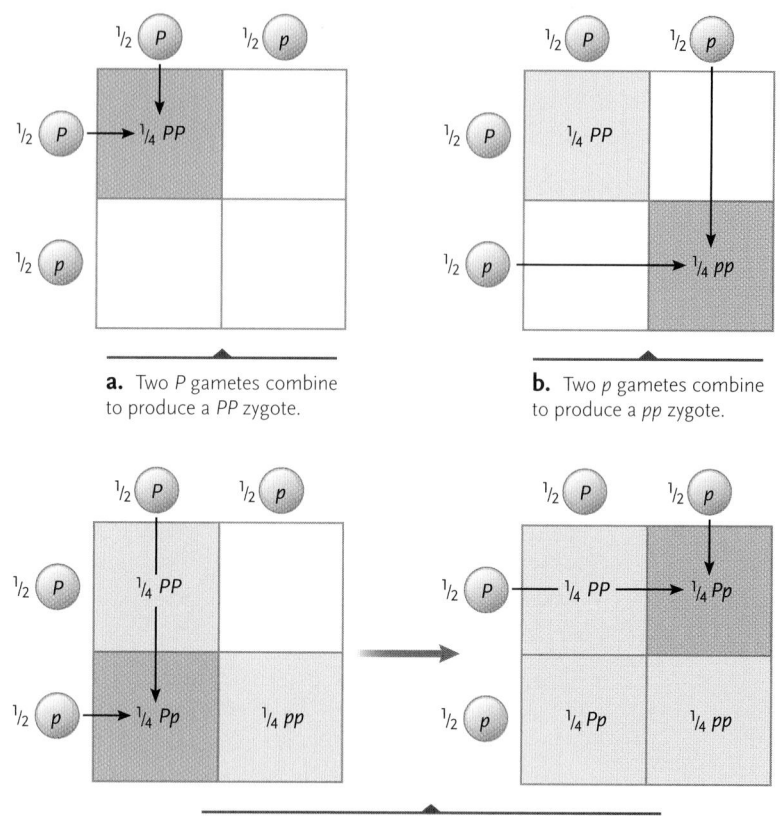

a. Two *P* gametes combine to produce a *PP* zygote.

b. Two *p* gametes combine to produce a *pp* zygote.

c. A *P* and a *p* gamete combine to produce a *Pp* zygote in two squares, for a total of 1/4 *Pp* + 1/4 *Pp* = 1/2 *Pp*.

Figure 12.7

Punnett square method for predicting offspring and their ratios in genetic crosses. The example is the F_1 × F_1 cross of purple-flowered plants from Figure 12.5. Each cell shows the genotype and proportion of one type of zygote.

a.

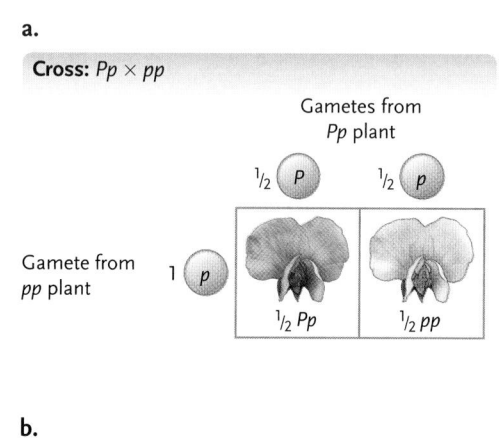

Cross: *Pp* × *pp*

Gametes from *Pp* plant

1/2 *P*   1/2 *p*

Gamete from *pp* plant   1 *p*

1/2 *Pp*   1/2 *pp*

b.

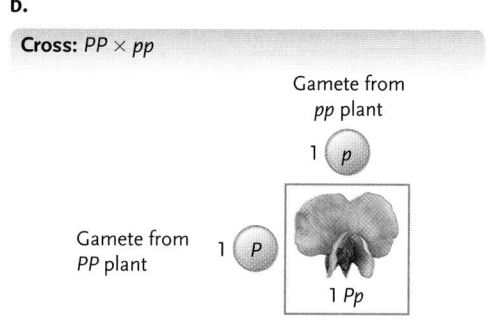

Cross: *PP* × *pp*

Gamete from *pp* plant

1 *p*

Gamete from *PP* plant   1 *P*

1 *Pp*

Figure 12.8

Method of predicting the outcome of genetic crosses.

A cross between an individual with the dominant phenotype and a homozygous recessive individual, such as the one described, is called a **testcross.** Geneticists use a testcross as a standard test to determine whether an individual with a dominant trait is a heterozygote or a homozygote, because these cannot be distinguished phenotypically. If the offspring of the testcross are of two types, with half displaying the dominant trait and half the recessive trait, then the individual in question must be a heterozygote (see Figure 12.8). If all the offspring display the dominant trait, the individual in question must be a homozygote. For example, the cross $PP \times pp$ gives all Pp progeny, which show the dominant purple phenotype (see Figure 12.8).

Obviously, the testcross method cannot be used for humans. However, it can be used in reverse, by noting the traits present in families over several generations and working backward to deduce whether a parent must have been a homozygote or a heterozygote (see also Chapter 13).

Mendel Tested the Independence of Different Genes in Crosses

Mendel next asked what happens in crosses when more than one character is involved. Would the alleles of different characters be inherited independently, or would they interact to alter their expected proportions in offspring?

To answer these questions, Mendel crossed parental stocks that had differences in two of the hereditary characters he was studying: seed shape and seed color. His single-character crosses had shown each was controlled by a pair of alleles. For seed shape, the RR or Rr genotypes produce round seeds and the rr genotype produces wrinkled seeds. For seed color, yellow is dominant. The homozygous YY or heterozygous Yy genotypes produce yellow seeds; the homozygous yy genotype produces green seeds.

Mendel crossed plants that bred true for the production of round and yellow seeds *(RR YY)* with plants that bred true for the production of wrinkled and green seeds *(rr yy)* **(Figure 12.9).** The cross, $RR\ YY \times rr\ yy$, yielded an F_1 generation that consisted of all round yellow seeds, with the genotype $Rr\ Yy$. A zygote produced from a cross that involves two characters is called a **dihybrid** (di = two).

Mendel then planted the F_1 seeds, grew the plants to maturity, and selfed them; that is, he crossed the F_1 plants to themselves. A cross between two individuals that are heterozygous for two pairs of alleles—here, $Rr\ Yy \times Rr\ Yy$—is called a **dihybrid cross** (see Figure 12.9). The seeds produced by these plants, representing the F_2 generation, included 315 round yellow seeds, 101 wrinkled yellow seeds, 103 round green seeds, and 32 wrinkled green seeds. Mendel noted that these numbers were close to a 9:3:3:1 ratio (3:1 for round:wrinkled, and 3:1 for yellow:green).

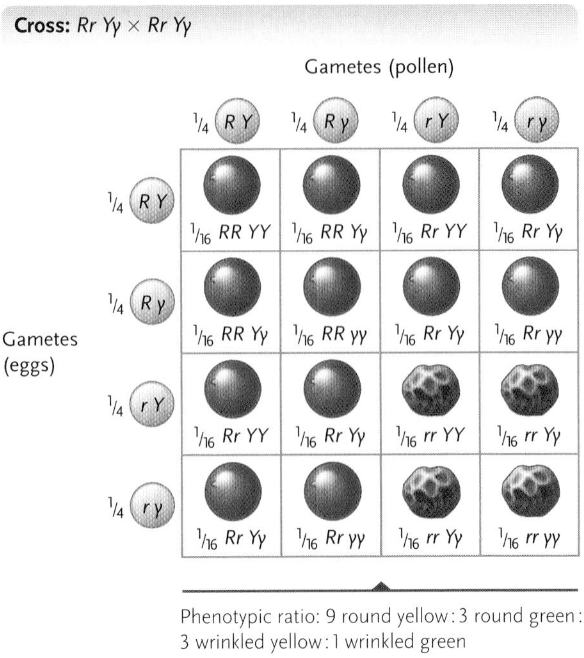

Cross: $Rr\ Yy \times Rr\ Yy$

Phenotypic ratio: 9 round yellow : 3 round green : 3 wrinkled yellow : 1 wrinkled green

Figure 12.9
The principle of independent assortment in Mendel's crosses involving two hereditary characters in garden peas, seed shape, and seed color.

This 9:3:3:1 ratio was consistent with Mendel's previous findings if he added one further hypothesis: *The alleles of the genes that govern the two characters segregate independently during formation of gametes.* That is, the allele for seed shape that the gamete receives (R or r) has no influence on which allele for seed color it receives (Y or y) and vice versa. The two events are completely independent. Mendel termed this assumption **independent assortment**; it is now known as Mendel's **Principle of Independent Assortment.**

To understand the effect of independent assortment in the cross, assume that the $RR\ YY$ parent produces only $R\ Y$ gametes and the $rr\ yy$ parent produces only $r\ y$ gametes. In the F_1 generation, all possible combinations of these gametes produce only one genotype, $Rr\ Yy$, in the offspring. As observed, all the F_1 will be round yellow seeds.

If the alleles that control seed shape and seed color assort independently in gamete formation, each F_1 plant grown from the seeds would produce four types of gametes. The R allele for seed shape can be delivered independently to a gamete with either the Y or y allele for seed color, and similarly, the r allele can be delivered to a gamete with either the Y or y allele. Thus, the independent assortment of genes from the $Rr\ Yy$ parents is expected to produce four types of gametes with equal probability: 1/4 $R\ Y$, 1/4 $R\ y$, 1/4 $r\ Y$, and 1/4 $r\ y$. These gametes and their probabilities are entered as the row and column headings of the Punnett square in Figure 12.9.

Filling in the cells of the diagram (see Figure 12.9) gives 16 combinations of alleles, all with an equal probability of 1 in every 16 offspring. Of these, the genotypes *RR YY, RR Yy, Rr YY,* and *Rr Yy* all have the same phenotype: round yellow seeds. These combinations occur in 9 of the 16 cells in the diagram, giving a total probability of 9/16. The genotypes *rr YY* and *rr Yy,* which produce the wrinkled yellow seeds, are found in three cells, giving a probability of 3/16 for this phenotype. Similarly, the genotypes *RR yy* and *Rr yy,* which yield round green seeds, occur in three cells, giving a probability of 3/16. Finally, the genotype *rr yy,* which produces wrinkled green seeds, is found in only one cell and therefore has a probability of 1/16.

These probabilities of round yellow seeds, wrinkled yellow seeds, round green seeds, and wrinkled green seeds, in a 9:3:3:1 ratio, closely approximate the actual results of 315:101:108:32 obtained by Mendel. Thus, Mendel's first three hypotheses, with the added hypothesis of independent assortment, explain the observed results of his dihybrid cross. Mendel's testcrosses completely confirmed his hypotheses; for example, the testcross *Rr Yy* × *rr yy* produced 55 round yellow seeds, 51 round green seeds, 49 wrinkled yellow seeds, and 53 wrinkled green seeds. This distribution corresponds well with the expected 1:1:1:1 ratio in the offspring. (Try to set up a Punnett square for this cross and predict the expected classes of offspring and their frequencies.)

Mendel's first three hypotheses provided a coherent explanation of the pattern of inheritance for alternate traits of the same character, such as purple and white for flower color. His fourth hypothesis, independent assortment, addressed the inheritance of traits for different characters, such as seed shape, seed color, and flower color, and showed that, instead of being inherited together, the traits of different characters were distributed independently to offspring.

Mendel's Research Founded the Field of Genetics

Mendel's techniques and conclusions were so advanced for his time that their significance was not immediately appreciated. Mendel's success was based partly on a good choice of experimental organism. He was also lucky. The characters he chose all segregate independently; that is, none of them is physically near each other on the chromosomes, a condition that would have given ratios other than 9:3:3:1, showing that they do not assort independently. Mendel's findings anticipated in detail the patterns by which genes and chromosomes determine inheritance. Yet, when Mendel first reported his findings, during the nineteenth century, the structure and function of chromosomes and the patterns by which they are separated and distributed to gametes were unknown; meiosis remained to be discovered. In addition, his use of mathematical analysis was a new and radical departure from the usual biological techniques of his day.

Mendel reported his results to a small group of fellow intellectuals in Brünn and presented his results in 1866 in a natural history journal published in the city. His article received little notice outside of Brünn, and those who read it were unable to appreciate the significance of his findings. His work was overlooked until the early 1900s, when three investigators—Hugo de Vries in Holland, Carl Correns in Germany, and Erich von Tschermak in Austria—independently performed a series of breeding experiments similar to Mendel's and reached the same conclusions. These investigators, in searching through previously published scientific articles, discovered to their surprise Mendel's article about his experiments conducted 34 years earlier. Each gave credit to Mendel's discoveries, and the quality and far-reaching implications of his work were at last realized. Mendel died in 1884, 16 years before the rediscovery of his experiments and conclusions, and thus he never received the recognition that he so richly deserved during his lifetime.

Mendel was unable to relate the behavior of his "factors" (genes) to cell structures because the critical information he required was not obtained until later, through the discovery of meiosis during the 1890s. The next section describes how a genetics student familiar with meiosis was able to make the connection between Mendel's factors and chromosomes.

Sutton's Chromosome Theory of Inheritance Related Mendel's Genes to Chromosomes

By the time Mendel's results were rediscovered in the early 1900s, critical information from studies of meiosis was available. It was not long before a genetics student, Walter Sutton, recognized the similarities between the inheritance of the genes discovered by Mendel and the behavior of chromosomes in meiosis and fertilization **(Figure 12.10)**.

In a historic article published in 1903, Sutton, then a graduate student at Columbia University in New York, drew all the necessary parallels between genes and chromosomes:

- Chromosomes occur in pairs in sexually reproducing, diploid organisms, as do the alleles of each gene.
- The chromosomes of each pair are separated and delivered singly to gametes, as are the alleles of a gene.
- The separation of any pair of chromosomes in meiosis and gamete formation is independent of the separation of other pairs (see Figure 12.10), as in the independent assortment of the alleles of different genes in Mendel's dihybrid crosses.
- Finally, one member of each chromosome pair is derived in fertilization from the male parent, and the other member is derived from the female parent, in an exact parallel with the two alleles of a gene.

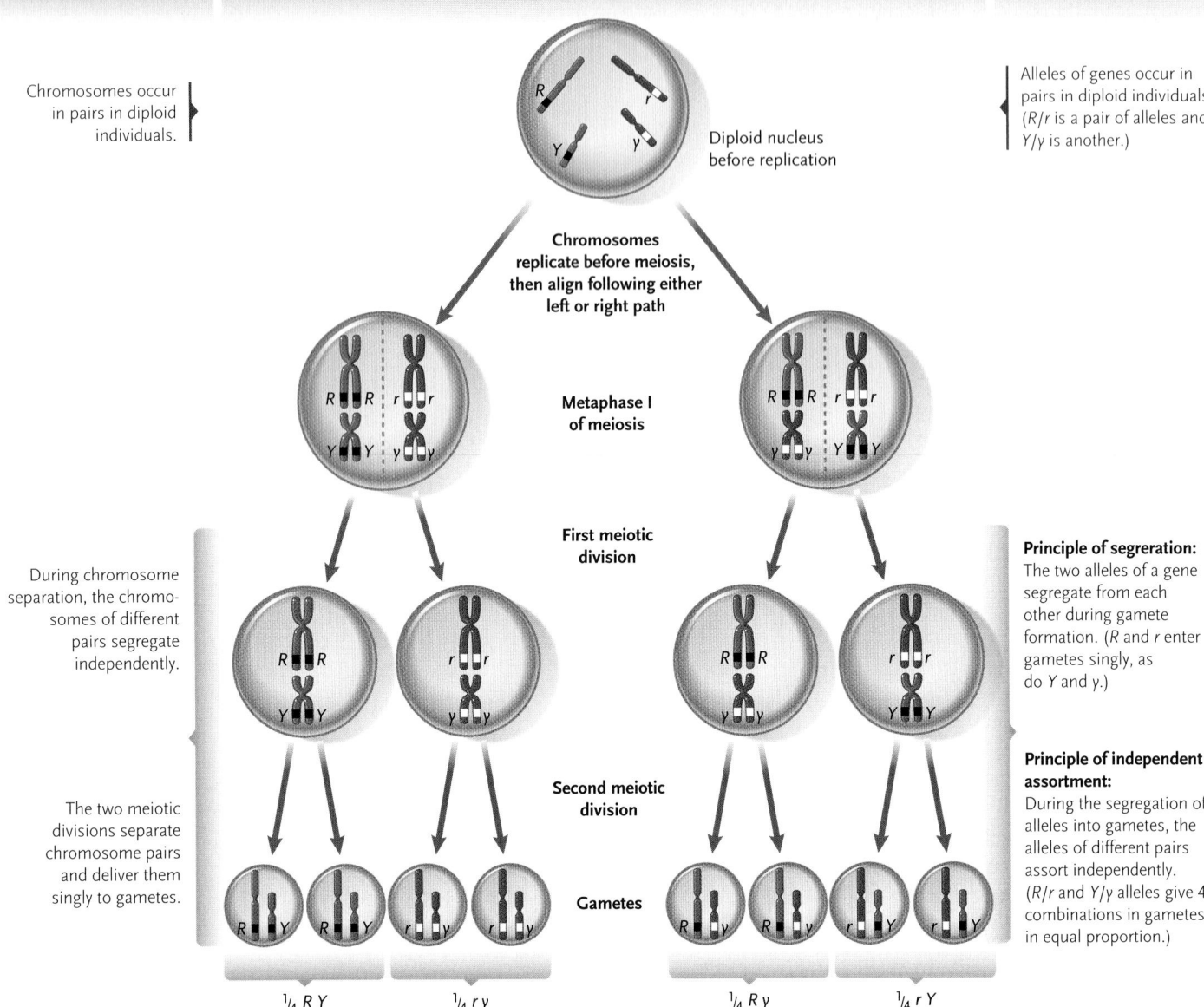

Behavior of chromosomes in meiosis

Chromosomes occur in pairs in diploid individuals.

During chromosome separation, the chromosomes of different pairs segregate independently.

The two meiotic divisions separate chromosome pairs and deliver them singly to gametes.

Meiosis in male or female diploid parent

Diploid nucleus before replication

Chromosomes replicate before meiosis, then align following either left or right path

Metaphase I of meiosis

First meiotic division

Second meiotic division

Gametes

¼ R Y   ¼ r y   ¼ R y   ¼ r Y

Behavior of genes and alleles in meiosis and their correspondence to Mendel's principles

Alleles of genes occur in pairs in diploid individuals (*R/r* is a pair of alleles and *Y/y* is another.)

Principle of segregation:
The two alleles of a gene segregate from each other during gamete formation. (*R* and *r* enter gametes singly, as do *Y* and *y*.)

Principle of independent assortment:
During the segregation of alleles into gametes, the alleles of different pairs assort independently. (*R/r* and *Y/y* alleles give 4 combinations in gametes in equal proportion.)

Figure 12.10

The parallels between the behavior of chromosomes and genes and alleles in meiosis. The gametes show four different combinations of alleles produced by independent segregation of chromosome pairs.

From this total coincidence in behavior, Sutton correctly concluded that genes and their alleles are carried on the chromosomes, a conclusion known today as the **chromosome theory of inheritance.**

The exact parallel between the principles set forth by Mendel, and the behavior of chromosomes and genes during meiosis, is shown in Figure 12.10 for an *Rr Yy* diploid. For a cross of *Rr Yy* × *Rr Yy*, when the gametes fuse randomly, the progeny will show a phenotypic ratio of 9:3:3:1. This mechanism explains the same ratio of gametes and progeny as the *Rr Yy* × *Ry Yy* cross in Figure 12.9.

The particular site on a chromosome at which a gene is located is called the **locus** (plural, *loci*) of the gene. The locus is a particular DNA sequence that encodes (typically) a protein responsible for the phenotype controlled by the gene. A locus for a gene with two al-

leles, *A* and *a*, on a homologous pair of chromosomes is shown in **Figure 12.11.** At the molecular level, different alleles consist of small differences in the DNA sequence of a gene, which may result in functional differences in the protein encoded by the gene. These differences are detected as distinct phenotypes in the offspring of a cross. *Insights from the Molecular Revolution* describes a molecular study that uncovered the mechanisms that control height in pea plants, one of the seven characteristics originally examined by Mendel.

All the genetics research conducted since the early 1900s has confirmed Mendel's basic hypotheses about inheritance. This research has shown that Mendel's conclusions apply to all types of organisms, from yeast and fruit flies to humans, and has led to the rapidly growing field of human genetics. In humans, a number of easily seen traits show inheritance patterns that

Why Mendel's Dwarf Pea Plants Were So Short

Two independent research teams worked out the molecular basis for one of the seven characters Mendel studied—dwarfing, which is governed by stem length in garden peas. The investigators, including Diane Lester and her colleagues at the University of Tasmania in Australia and David Martin and his coworkers at Oregon State University, were interested in learning the molecular differences in the alleles of the gene that produced tall or dwarf plants. The dominant *T* allele (*T* = tall) of the gene produces plants of normal height; the recessive *t* allele produces dwarf plants with short stems. How can a single gene control the overall height of a plant?

Lester's team discovered that the gene codes for an enzyme that carries out a preliminary step in the synthesis of the plant hormone gibberellin, which, among other effects, causes the stems of plants to elongate. Martin's group cloned the gene and determined its complete DNA sequence. (Cloning techniques and DNA sequencing are described in Sections 18.1 and 18.3.) The sequence showed that the *T* and *t* alleles of the gene encode two versions of the enzyme that catalyzes gibberellin synthesis, which differ by only a single amino acid. Lester's group found that the faulty enzyme encoded by the *t* allele carries out its step (addition of a hydroxyl group to a precursor) much more slowly than the enzyme encoded by the normal *T* allele. As a result, plants with the *t* allele have only about 5% as much gibberellin in their stems as *T* plants. The reduced gibberellin levels limit stem elongation, producing the dwarf plants.

Thus, the methods of molecular biology allowed contemporary researchers to study a gene first discovered in the mid-nineteenth century. The findings leave little doubt that a change in a single amino acid leads to the dwarf phenotype Mendel observed in his monastery garden.

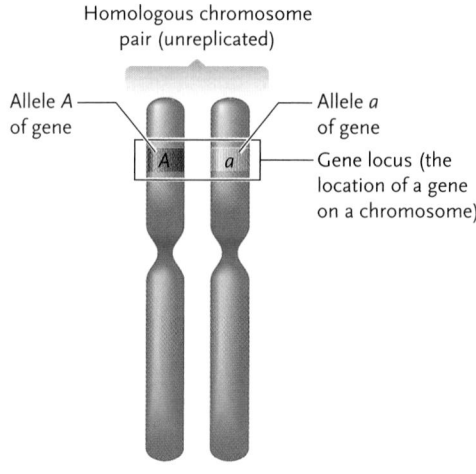

Figure 12.11
A locus, the site occupied by a gene on a pair of homologous chromosomes. Two alleles, *A* and *a*, of the gene are present at this locus in the homologous pair. These alleles have differences in the DNA sequence of the gene.

follow Mendelian principles **(Figure 12.12)**; for example, albinism, the lack of normal skin color, is recessive to normal skin color, and normally separated fingers are recessive to fingers with webs between them. Similarly, achondroplasia, the most frequent form of short-limb dwarfism, is a dominant trait that involves abnormal bone growth. Many human disorders that cannot be seen easily also show simple inheritance patterns. For instance, cystic fibrosis, in which a defect in the membrane transport of chloride ions leads to pulmonary and digestive dysfunctions and eventually death, is a recessive trait.

The post-Mendel research has demonstrated additional patterns of inheritance (see the next section)

that were not anticipated by Mendel and, in some circumstances, require modifications or additions to his hypotheses.

STUDY BREAK

1. Two pairs of traits are segregating in a cross. Two parents produce 156 progeny that fall into 4 phenotypes. The numbers of offspring in the 4 phenotypes are 89, 31, 28, and 8. What are the genotypes of the two parents?
2. If instead, the four phenotypes in question 1 occur in approximately equal numbers, what are the genotypes of the parents? What is this kind of cross called?

12.2 Later Modifications and Additions to Mendel's Hypotheses

The rediscovery of Mendel's research in the early 1900s produced an immediate burst of interest in genetics. The research that followed greatly expanded our understanding of genes and their inheritance. The discovery that the alleles of many genes are neither fully dominant nor fully recessive was among these new findings. Some alleles show incomplete dominance, in which recessive alleles do have some effect on the phenotype of heterozygotes. Other alleles are codominant; that is, they have different and approximately equal effects in heterozygotes.

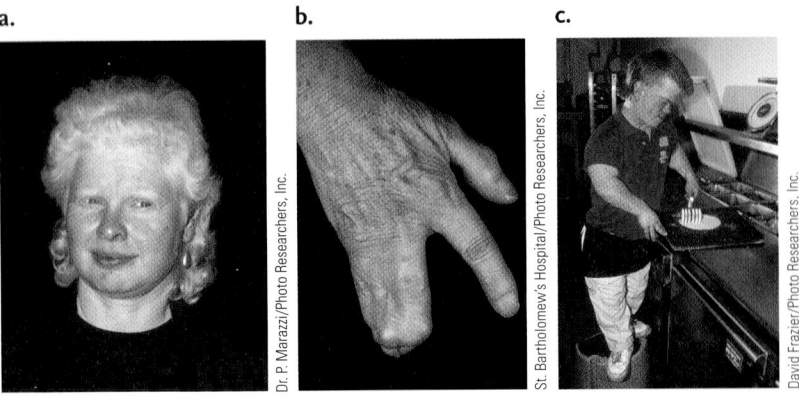

a. **b.** **c.**

Figure 12.12
Human traits showing inheritance patterns that follow Mendelian principles. **(a)** Lack of normal skin color (albinism). **(b)** Webbed fingers. **(c)** Achondroplasia, or short-limbed dwarfism.

Further research also demonstrated that more than two alleles of a gene may be present among all the members of a population. This condition, called multiple alleles, is still consistent with Mendel's conclusions because each sexually reproducing, diploid individual in a population has only two alleles of each gene—a pair—which are inherited and passed on according to Mendel's principles.

Geneticists also found that the activity of one gene can influence the activity of a different gene, a phenomenon called epistasis. Furthermore, some characters are explained by polygenic inheritance, in which several different genes each contribute to the phenotype. In addition, alterations in a single gene sometimes affect more than one phenotype in an organism; this phenomenon is called pleiotropy. The following sections discuss each of these so-called extensions of Mendel's fundamental principles.

In Incomplete Dominance, Dominant Alleles Do Not Completely Mask Recessive Alleles

Incomplete dominance occurs when the effects of recessive alleles can be detected to some extent in heterozygotes. Flower color in snapdragons shows incomplete dominance **(Figure 12.13)**. If true-breeding, red-flowered and white-flowered snapdragon plants are crossed, all the F_1 offspring have pink flowers (see Figure 12.13). The pink color might make it appear that the pure red and white colors have blended out and disappeared—mixing red and white makes pink—until two F_1 plants are crossed. The cross demonstrates that the red and white traits both reappear in the F_2 generation, which has red, pink, and white flowers in numbers approximating a $1:2:1$ ratio.

This outcome can be explained by incomplete dominance between a C^R allele for red color and a C^W

allele for white color. When one allele is not completely dominant to the other, we use a superscript to signify the character. In this case, C signifies the character for flower color and the superscripts indicate the alleles (R for red and W for white). Therefore, the initial cross is $C^R C^R$ (red) $\times$ $C^W C^W$ (white), which produces $C^R C^W$ F_1 (pink) plants. The C^R allele encodes an enzyme that produces a red pigment, but two alleles ($C^R C^R$) are necessary to produce enough of the active form of the enzyme to produce fully red flowers. The enzyme is completely inactive in $C^W C^W$ plants, which produce colorless flowers that appear white because of the scattering of light by cell walls and other structures. With their single C^R allele, the $C^R C^W$ heterozygotes of the F_1 generation can produce only enough pigment to give the flowers a pink color. When the pink $C^R C^W$ F_1 plants are crossed, the fully red and white colors reappear, together with the pink color, in the F_2 generation, in a ratio of $1/4$ $C^R C^R$ (red), $1/2$ $C^R C^W$ (pink), and $1/4$ $C^W C^W$ (white). This ratio is exactly the same as the ratio of genotypes produced from a cross of two heterozygotes in Mendel's experiments (for example, see Figure 12.7).

Some human disorders show incomplete dominance. For example, sickle-cell disease (see the introduction to this chapter) is characterized by an alteration in the hemoglobin molecule that changes the shape of red blood cells when oxygen levels are low. An individual with sickle-cell disease is homozygous for a recessive allele that encodes a defective form of one of the polypeptides of the hemoglobin molecule. Individuals heterozygous for that recessive allele and the normal allele have a condition known as *sickle-cell trait*, which is a milder form of the disease because the individuals still produce normal polypeptides from the normal allele.

Familial hypercholesterolemia is another example of incomplete dominance. The gene involved encodes the low-density lipoprotein (LDL) receptor, a cell membrane protein responsible for removing excess cholesterol from the blood (see Section 6.5). Individuals with familial hypercholesterolemia are homozygous for a defective LDL receptor gene, produce no LDL receptors, and have a severe form of the disease. These individuals have six times the normal level of cholesterol in the blood and therefore are very prone to atherosclerosis (hardening of the arteries). Many individuals with familial hypercholesterolemia have heart attacks as children. Heterozygous individuals have half the normal number of receptors, which results in a milder form of the disease. Their symptoms are twice the normal blood cholesterol level, an unusually high risk of atherosclerosis, and a high risk of heart attacks before age 35.

Many alleles that appear to be completely dominant are actually incomplete in their effects when analyzed at the biochemical or molecular level. For example, for pigments that produce fur or flower colors, biochemical studies often show that even though heterozygotes may produce enough pigment to make them look the same externally as homozygous dominants, a difference in the amount of pigment is measurable at the biochemical level. Thus, whether dominance between alleles is complete or incomplete often depends on the level at which the effects of the alleles are examined.

A similar situation occurs in humans who carry the recessive allele that causes Tay–Sachs disease. Children who are homozygous for the recessive allele do not have a functional version of an enzyme that breaks down gangliosides, a type of membrane lipid. As a result, gangliosides accumulate in the brain, leading to mental impairment and eventually to death. Heterozygotes are without symptoms of the disease, even though they have one copy of the recessive allele. However, at the biochemical level, reduced breakdown of gangliosides can be detected in heterozygotes, evidently due to a reduced quantity of the active enzyme.

In Codominance, the Effects of Different Alleles Are Equally Detectable in Heterozygotes

Codominance occurs when alleles have approximately equal effects in individuals, making the alleles equally detectable in heterozygotes. The inheritance of the human blood types, M, MN, and N, is an example of codominance. These are different blood types from the familiar blood types of the ABO blood group. The L^M and L^N alleles of the MN blood group gene that control this character encode different forms of a glycoprotein molecule located on the surface of red blood cells. If the genotype is $L^M L^M$, only the M form of the glycoprotein is present and the blood type is M; if it is $L^N L^N$, only the N form is present and the blood type is N. In heterozygotes with the $L^M L^N$ genotype, both glycoprotein types are present and can be detected, producing the blood type MN. Because each genotype has a different phenotype, the inheritance pattern for the MN blood group alleles is generally the same as for incompletely dominant alleles.

The MN blood types do not affect blood transfusions and have relatively little medical importance. However, they have been invaluable in tracing human evolution and prehistoric migrations, and they are frequently used in initial tests to determine the paternity of a child. Among their primary advantages in research and paternity determination is that the genotype of all individuals, including heterozygotes, can be detected directly—and inexpensively—from their phenotype, with no requirement for further genetic tests or analysis.

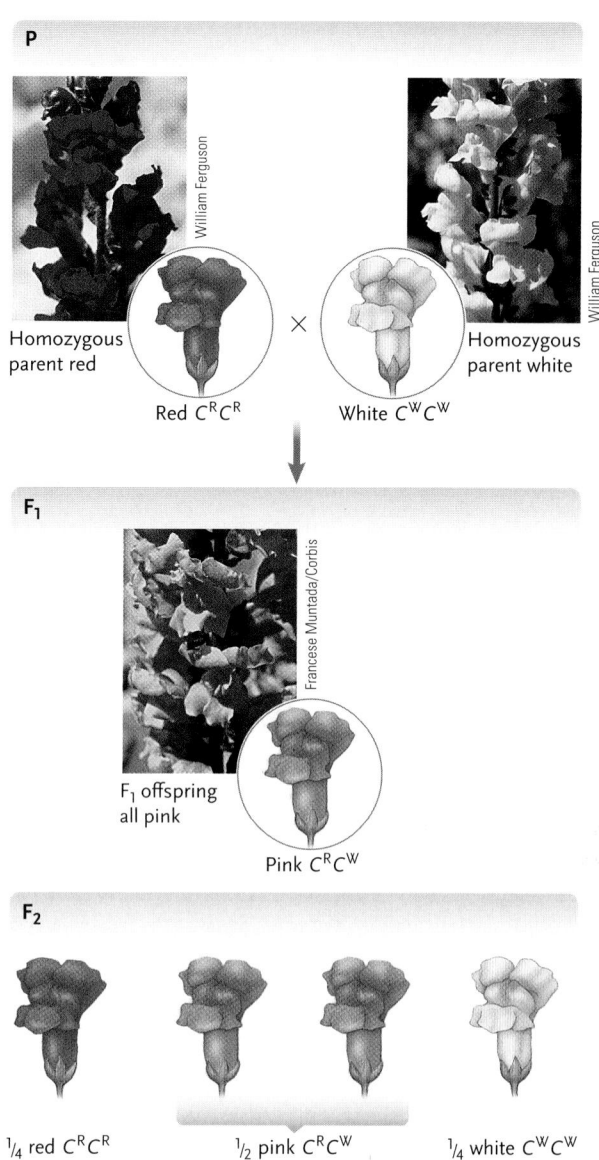

Figure 12.13
Incomplete dominance in the inheritance of flower color in snapdragons.

In Multiple Alleles, More Than Two Alleles of a Gene Are Present in a Population

One of Mendel's major and most fundamental assumptions was that alleles occur in pairs in individuals; in the pairs, the alleles may be the same or different. After the rediscovery of Mendel's principles, it soon became apparent that although alleles do indeed occur in pairs in individuals, **multiple alleles** (more than two different alleles of a gene) may be present if all the individuals of a population are taken into account. For example, for a gene B, there could be the normal allele, B, and several alleles with alterations in the gene named, for example, b_1, b_2, b_3, and so on. Some individuals in a population may have the B and b_1 alleles of a gene; others, the b_2 and b_3 alleles; still others, the b_3 and b_5 alleles; and so on, for all possible combinations. Thus, although any one individual can

Figure 12.14

Multiple alleles. Multiple alleles consist of small differences in the DNA sequence of a gene at one or more points, which result in detectable differences in the structure of the protein encoded by the gene. The *B* allele is the normal allele, which encodes a protein with normal function. The three *b* alleles each have alterations of the normal protein-coding DNA sequence that may adversely affect the function of that protein.

have only two alleles of the gene, there are more than two alleles in the population as a whole. Genes may certainly occur in many more than the four alleles of the example; for instance, one of the genes that plays a part in the acceptance or rejection of organ transplants in humans has more than 200 different alleles.

The multiple alleles of a gene each contain differences at one or more points in their DNA sequences **(Figure 12.14),** which cause detectable alterations in the structure and function of proteins encoded by the alleles. Multiple alleles present no real difficulty in genetic analysis because each diploid individual still has only two of the alleles, allowing gametes to be predicted and traced through crosses by the usual methods.

Human ABO Blood Group. The human *ABO* blood group provides another interesting example of multiple alleles, in a system that also exhibits both dominance and codominance. The ABO blood group was discovered in 1901 by Karl Landsteiner, an Austrian biochemist who was investigating the sometimes fatal outcome of attempts to transfer whole blood from one person to another. Landsteiner found that only certain combinations of four blood types, designated A, B, AB, and O, can be mixed safely in transfusions **(Table 12.1).**

Landsteiner determined that, in the wrong combinations, red blood cells from one blood type are agglutinated or clumped by an agent in the serum of another type (the serum is the fluid in which the blood cells are suspended). The clumping was later found to depend on

the action of an antibody in the blood serum. (Antibodies, protein molecules that interact with specific substances called antigens, are discussed in Chapter 43.)

The antigens responsible for the blood types of the ABO blood group are the carbohydrate parts of glycoproteins located on the surfaces of red blood cells (unrelated to the glycoprotein carbohydrates responsible for the blood types of the MN blood group). People with type A blood have *antigen A* on their red blood cells, and people with type B blood have *antigen B* on their red blood cells. At the same time, people with type A blood have antibodies against antigen B, and people with type B blood have antibodies against antigen A. People with type O blood have neither antigen A nor antigen B on their red blood cells, but they have antibodies against both of these antigens. People with type AB blood have neither anti-A nor anti-B antibodies, but they have both the A and B antigens, and their red blood cells are clumped by antibodies in the blood of all the other groups.

The four blood types—A, B, AB, and O—are produced by different combinations of multiple (three) alleles of a single gene *I* **(Figure 12.15).** The three alleles, designated I^A, I^B, and *i*, produce the following blood types:

$I^A I^A$ = type A blood	$I^B I^B$ = type B blood
$I^A i$ = type A blood	$I^B i$ = type B blood
$I^A I^B$ = type AB blood	ii = type O blood

In addition, I^A and I^B are codominant alleles that are each dominant to the *i* allele.

In Epistasis, Genes Interact, with the Activity of One Gene Influencing the Activity of Another Gene

The genetic characters discussed so far in this chapter, such as flower color, seed shape, and the blood types of the ABO group, are all produced by the alleles of single genes, with each gene functioning on its own. This is not the case for every gene. In **epistasis** (*epi* = on or over; *stasis* = standing or stopping), genes interact, with one or more alleles of a gene at one locus inhibiting or masking the effects of one or more alleles of a gene at a different locus. The result of epistasis is that some expected phenotypes do not appear among offspring.

Labrador retrievers (Labs) may have black, chocolate brown, or yellow fur **(Figure 12.16).** The different colors result from variations in the amount and distribution in hairs of a brownish black pigment called melanin. One gene, coding for an enzyme involved in melanin production, determines how much melanin is produced. The dominant *B* allele of this gene produces black fur color in *BB* or *Bb* Labs; less pigment is produced in *bb* dogs, which are chocolate brown. How-

Table 12.1 | **Blood Types of the Human ABO Blood Group**

Blood Type	Antigens	Antibodies	Blood Types Accepted in a Transfusion
A	A	Anti-B	A or O
B	B	Anti-A	B or O
AB	A and B	None	A, B, AB, or O
O	None	Anti-A, anti-B	O

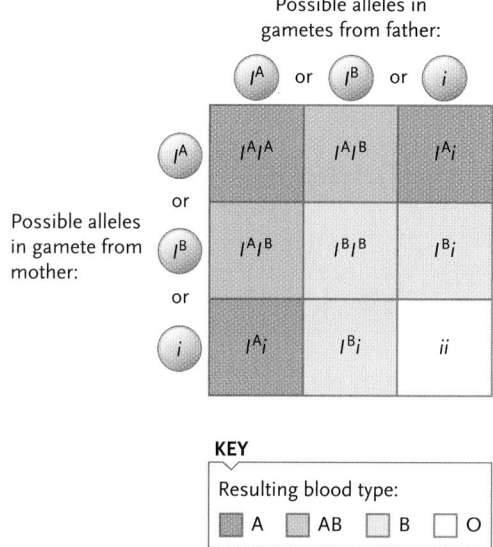

Figure 12.15
Inheritance of the blood types of the human ABO blood group.

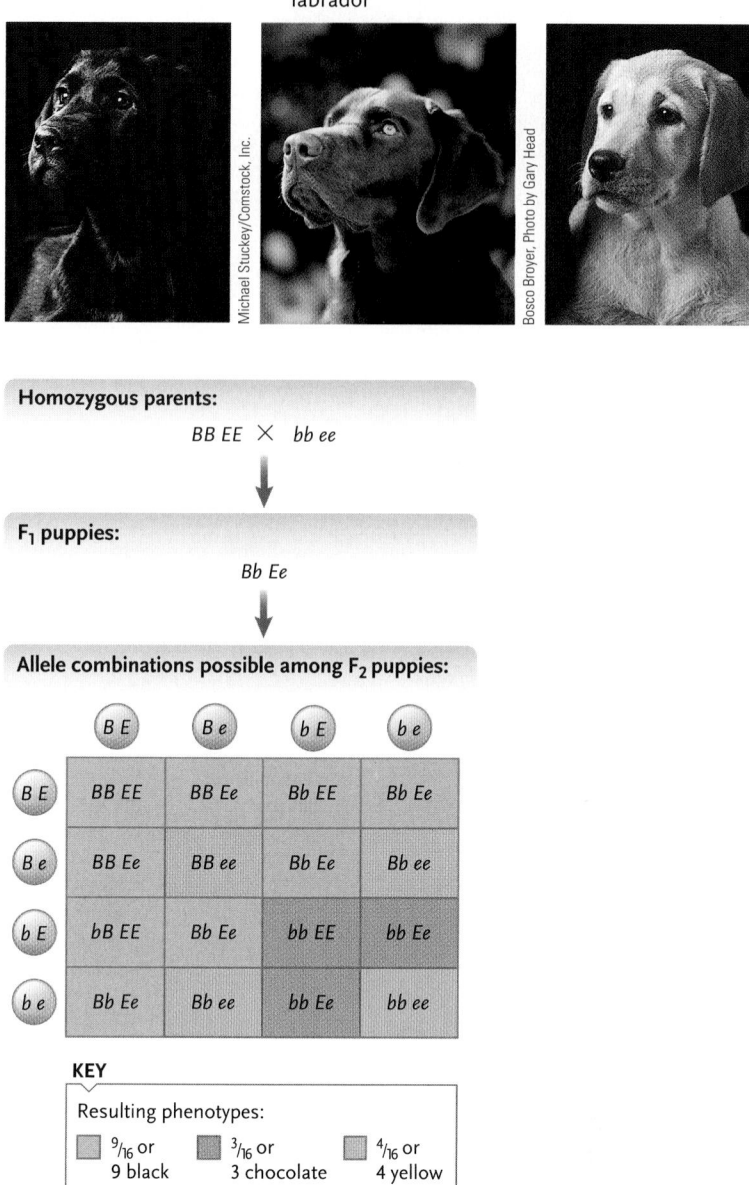

a. Black labrador   **b.** Chocolate brown labrador   **c.** Yellow labrador

Homozygous parents:

$BB\ EE\ \times\ bb\ ee$

F₁ puppies:

$Bb\ Ee$

Allele combinations possible among F₂ puppies:

	B E	B e	b E	b e
B E	BB EE	BB Ee	Bb EE	Bb Ee
B e	BB Ee	BB ee	Bb Ee	Bb ee
b E	bB EE	Bb Ee	bb EE	bb Ee
b e	Bb Ee	Bb ee	bb Ee	bb ee

KEY

Resulting phenotypes:
- ⁹/₁₆ or 9 black
- ³/₁₆ or 3 chocolate
- ⁴/₁₆ or 4 yellow

Figure 12.16
An example of epistasis: the inheritance of coat color in Labrador retrievers.

ever, another gene at a different locus determines whether the black or chocolate color appears at all, by controlling the deposition of pigment in hairs. A dominant allele *E* of this second gene permits pigment deposition, so that the black color in *BB* or *Bb* individuals, or the chocolate color in *bb* individuals, actually appears in the fur. Pigment deposition is almost completely blocked in homozygous recessive *ee* individuals, so the fur lacks melanin and has a yellow color whether the genotype for the *B* gene is *BB*, *Bb*, or *bb*. Thus, the *E* gene is epistatic to the *B* gene (that is, *E* and *B* interact).

Epistasis by the *E* gene eliminates some of the expected classes from crosses among Labs. Rather than two separate classes, as would be expected from a dihybrid cross without epistasis, the *BB ee*, *Bb ee*, *bB ee*, and *bb ee* genotypes produce a single yellow phenotype, giving the distribution: 9/16 black, 3/16 chocolate, and 4/16 yellow. That is, the ratio is 9 : 3 : 4 instead of the expected 9 : 3 : 3 : 1 ratio. Many other dihybrid crosses that involve epistatic interactions produce distributions that differ from the expected 9 : 3 : 3 : 1 ratio.

In human biology, researchers believe that gene interactions and epistasis are common. Current thinking is that epistasis is an important factor in determining an individual's susceptibility to common human diseases. That is, the different degrees of susceptibility are the result of different gene interactions in the individuals. A specific example is insulin resistance, a disorder in which muscle, fat, and liver cells do not use insulin correctly, with the result that glucose and insulin levels become high in the blood. This disorder is believed to be determined by several genes often interacting with one another.

In Polygenic Inheritance, a Character Is Controlled by the Common Effects of Several Genes

Some characters follow a pattern of inheritance in which there is a more or less even gradation of types, forming a continuous distribution, rather than "on" or "off" (discontinuous) effects such as the production of purple or white flowers in pea plants. For example, in the human population, people range from short to tall, in a continuous distribution of gradations in height between limits of about 4 and 7 feet. Typically, a continuous distribution of this type is the result of **polygenic inheritance**, in which several to many differ-

a. Students at Brigham Young University, arranged according to height

b. Actual distribution of individuals in the photo according to height

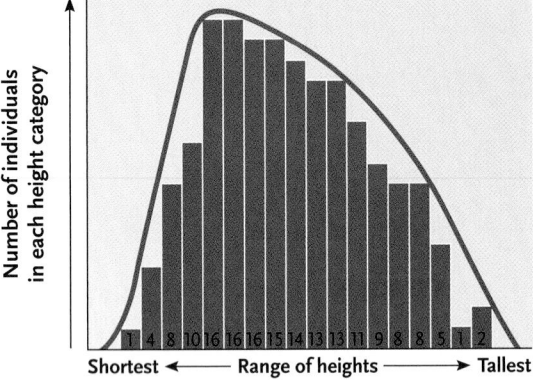

Number of individuals in each height category

1 4 8 10 16 16 16 15 14 13 13 11 9 8 8 5 1 2

Shortest ← Range of heights → Tallest

c. Idealized bell-shaped curve for a population that displays continuous variation in a trait

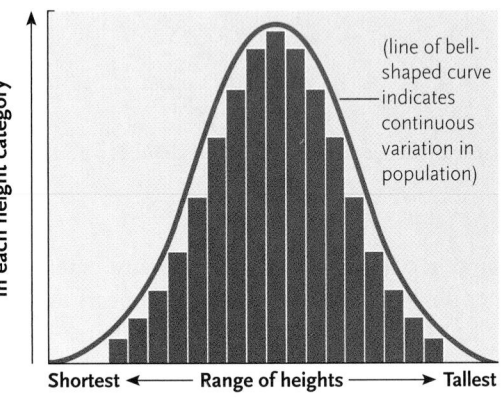

Number of individuals in each height category

(line of bell-shaped curve indicates continuous variation in population)

Shortest ← Range of heights → Tallest

If the sample in the photo included more individuals, the distribution would more closely approach this ideal.

Figure 12.17
Continuous variation in height due to polygenic inheritance.

ent genes contribute to the same character. Other characters that undertake a similar continuous distribution include skin color and body weight in humans, ear length in corn, seed color in wheat, and color spotting in mice. These characters are also known as *quantitative traits*.

Polygenic inheritance can be detected by defining classes of a variation, such as human body height of 60 inches in one class, 61 inches in the next class, 62 inches in the next class, and so on. The number of individuals in each class is then plotted as a graph. If the plot produces a bell-shaped curve, with fewer individuals at the extremes and the greatest numbers clustered around the midpoint, it is a good indication that the trait is quantitative **(Figure 12.17)**.

Polygenic inheritance is often modified by the environment. For example, height in humans is not the result of genetics alone. Poor nutrition during infancy and childhood is one environmental factor that can limit growth and prevent individuals from reaching the height expected from genetic inheritance; good nutrition can have the opposite effect. Thus, the average young adult in Japan today is several inches taller than

the average adult in the 1930s, when nutrition was poorer. Similarly, individuals who live in cloudy, northern or southern climates usually have lighter skin color than individuals with the same genotype who live in sunny climates.

At first glance, the effects of polygenic inheritance might appear to support the idea that characteristics of parents are blended in their offspring. Commonly, people believe that the children in a family with one tall and one short parent will be of intermediate height. Although the children of such parents are most likely to be of intermediate height, careful genetic analysis of many such families shows that their offspring actually range over a continuum from short to tall, forming a typical bell-shaped curve. Careful analysis of the inheritance of skin color produces the same result: Although the skin color of children is most often intermediate between that of the parents, a typical bell-shaped distribution is obtained in which some children at the extremes are lighter or darker than either parent. Thus, genetic analysis does not support the idea of blending or even mixing of parental traits in polygenic characteristics such as body size or skin color.

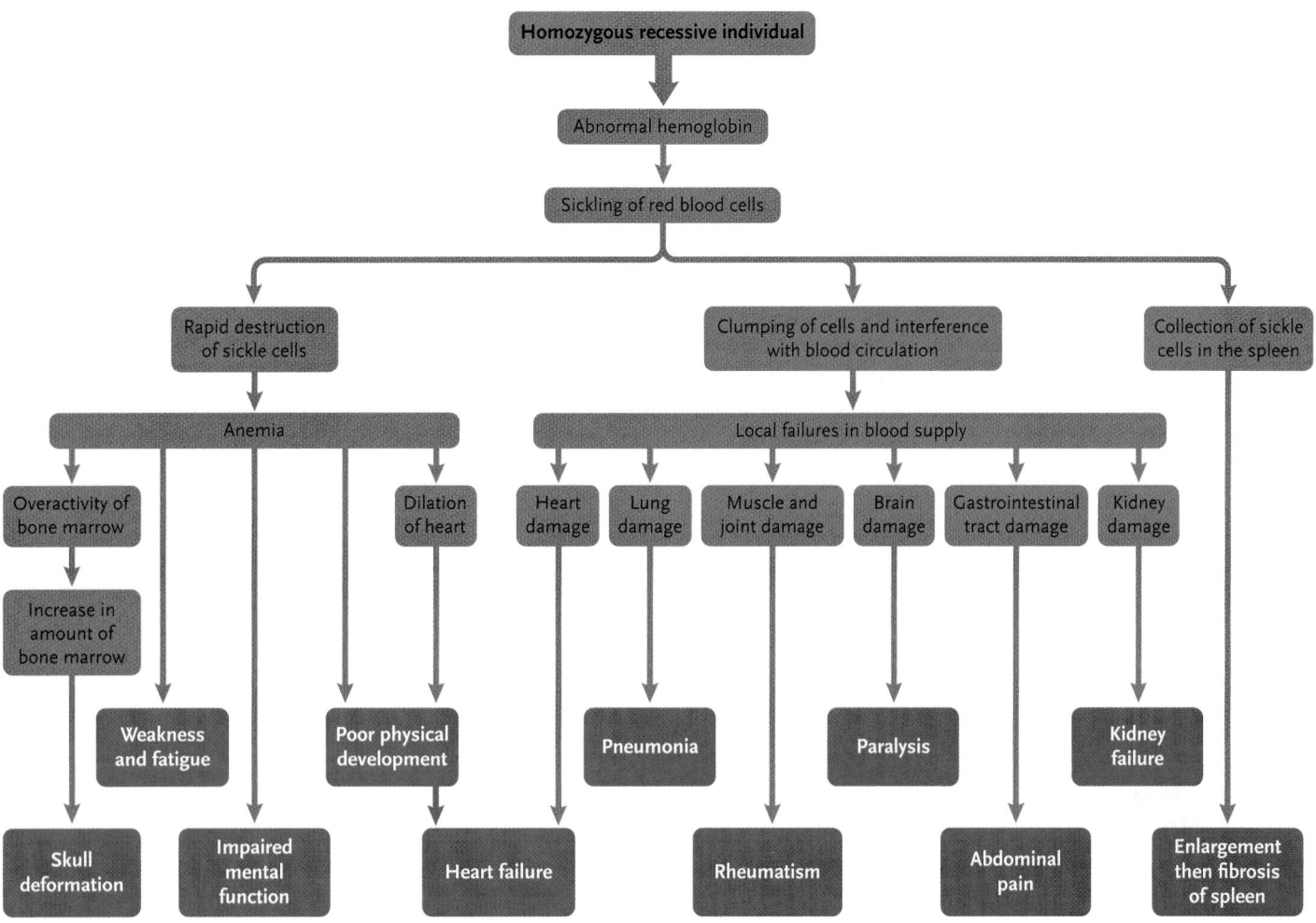

In Pleiotropy, Two or More Characters Are Affected by a Single Gene

In **pleiotropy**, single genes affect more than one character of an organism. For example, sickle-cell disease (see earlier discussion) is caused by a recessive allele of a single gene that affects hemoglobin structure and function. However, the altered hemoglobin, the primary phenotypic change of the sickle-cell mutation, leads to blood vessel blockage, which can damage many tissues and organs in the body and affect many body functions, producing such wide-ranging symptoms as fatigue, abdominal pain, heart failure, paralysis, and pneumonia **(Figure 12.18)**. Physicians recognize these wide-ranging pleiotropic effects as symptoms of sickle-cell disease.

The next chapter describes additional patterns of inheritance that were not anticipated by Mendel, including the effects of recombination during meiosis. These additional patterns also extend, rather than contradict, Mendel's fundamental principles.

Figure 12.18
Pleiotropy, as demonstrated by the wide-ranging, multiple effects of the single mutant allele responsible for sickle-cell disease.

STUDY BREAK

1. Palomino horses have a golden coat color, with a white mane and tail. Palominos do not breed true. Instead, there is a 50% chance that a foal will be a Palomino. What is the explanation?
2. A true-breeding rabbit with *agouti* (mottled, grayish brown) fur crossed with a true-breeding rabbit with *chinchilla* (silver) fur produces all agouti offspring. A true-breeding *chinchilla* rabbit crossed with a true-breeding *Himalayan* rabbit (white fur with pigmented nose, ears, tail, and legs) produces all *chinchilla* offspring. A true-breeding *Himalayan* rabbit crossed with a true-breeding *albino* rabbit produces all *Himalayan* offspring. Explain the inheritance of the fur colors.

The determination of genetic principles by Mendel and later geneticists involved crosses of plants and animals with visible traits, that is, phenotypes that could be seen by visual examination. Examples are smooth and wrinkled seeds of garden peas and red and white eyes of fruit flies. Until recently, it was impossible to determine the biochemical or molecular basis for traits. Even now, we do not know the molecular basis for most of the traits mentioned in this chapter, or for many others. For example, the molecular reason for the dwarf (short stem) phenotype of Mendel's peas was determined only as recently as the 1990s. Similar research is ongoing to determine the molecular bases for other visible genetic traits in a wide variety of organisms, including humans.

Is epistasis involved in human diseases?

Current research involves molecular investigations into epistasis, which is increasingly being recognized as an important phenomenon in common human diseases. Consequently, research is focused on identifying and understanding key epistatic relationships to help in the treatment and cure of such diseases.

For example, Nicholas Katsanis's research group at Johns Hopkins University is studying Bardet–Biedl syndrome, a genetic disorder characterized by obesity and learning defects. The severity of the symptoms varies dramatically among patients. Katsanis's group has shown that epistasis is involved in the development of Bardet–Biedl syndrome. These researchers have identified a DNA sequence that interacts with other molecules known to be altered in patients with Bardet–Biedl syndrome. For example, they showed in particular families that a more severe form of the disease occurs in individuals who have both a mutation in the DNA sequence and mutations in known Bardet–Biedl syndrome genes. However, mutations in the DNA sequence alone do not cause the illness. Experiments with zebrafish as a model organism have provided more evidence that the DNA sequence has an epistatic effect on other Bardet–Biedl syndrome mutations. This research opens the way to developing a better understanding of disorders that involve gene interactions.

Is polygenic inheritance involved in human disorders?

Polygenic inheritance involves a number of genes contributing to the same phenotypic traits, such as height of an individual or weight of a pumpkin. The traits are known as quantitative traits, and the genes responsible are called quantitative trait loci, or QTLs. Researchers are making great strides in identifying and characterizing the QTL genes involved with the goal of obtaining a molecular understanding of the traits. Such traits include a number of genetic disorders in humans, such as alcoholism, arthritis, behavioral and psychiatric disorders, cancer, diabetes, obesity, hypertension, and sensitivity to drugs.

For example, Xiasong Wang of The Jackson Laboratory is studying the genetics of atherosclerosis, the basis for coronary artery disease. The major risk factors for atherosclerosis are mainly genetically determined, notably high blood levels of low-density lipoproteins (LDLs), and low levels of high-density lipoproteins (HDLs). The level of the HDL is the more important of the two, because high levels of HDLs are known to be protective against cardiovascular disease. Wang's group has identified 21 mouse and 27 human atherosclerosis QTLs, and 37 mouse and 30 human HDL-regulating QTLs. Currently, these researchers are using genomic and bioinformatics approaches to determine the nature of the QTL genes that are common to both organisms.

Peter J. Russell

Review

Go to **ThomsonNOW** at www.thomsonedu.com/login to access quizzing, animations, exercises, articles, and personalized homework help.

12.1 The Beginnings of Genetics: Mendel's Garden Peas

- Mendel was successful in his research because of his good choice of experimental organism, which had clearly defined characters, such as flower color or seed shape, and because he analyzed his results quantitatively (Figures 12.3 and 12.4).

- Mendel showed that traits are passed from parents to offspring as hereditary factors (now called genes and alleles) in predictable ratios and combinations, disproving the notion of blended inheritance (Figure 12.5).

- Mendel realized that his results with crosses that involve single characters (monohybrid crosses) could be explained if three hypotheses were true: (1) The genes that govern genetic characters occur in pairs in individuals. (2) If different alleles of a gene are present in the pair of an individual, one allele is dominant over the other. (3) The two alleles of a gene segregate and enter gametes singly (Figures 12.5 and 12.7).

- Mendel confirmed his hypotheses by a testcross between an F_1 heterozygote with a homozygous recessive parent. This type of testcross is still used to determine whether an individual is homozygous or heterozygous for a dominant allele (Figure 12.8).

- To explain the results of his crosses with individuals showing differences in two characters—dihybrid crosses—Mendel added a fourth hypothesis: The alleles of the genes that govern the two characters segregate independently during formation of gametes (Figure 12.9).

- Walter Sutton was the first person to note the similarities between the inheritance of genes and the behavior of chromosomes in meiosis and fertilization. These parallels made it obvious that genes and alleles are carried on the chromosomes. Sutton's parallels are called the chromosome theory of inheritance (Figure 12.10).

- A locus is the site occupied by a gene on a chromosome (Figure 12.11).

Animation: Crossing garden pea plants

Animation: Genetic terms

Animation: Monohybrid cross

Animation: F_2 ratios interaction

Practice: Testcross

Animation: Dihybrid cross

Animation: Crossover review

12.2 Later Modifications and Additions to Mendel's Hypotheses

- In incomplete dominance, some or all alleles of a gene are neither completely dominant nor recessive. In such cases, the phenotype of heterozygotes with different alleles of the gene can be distinguished from that of either homozygote (Figure 12.13).

- In codominance, different alleles of a gene have approximately equal effects in heterozygotes, also allowing heterozygotes to be distinguished from either homozygote.

- Many genes may have multiple alleles if all the individuals in a population are taken into account. However, any diploid individual in a population has only two alleles of these genes, which are inherited and passed on according to Mendel's principles (Figures 12.14 and 12.15).

- In epistasis, genes interact, with one or more alleles of one locus inhibiting or masking the effects of one or more alleles at a different locus. The result is that some expected phenotypes do not appear among offspring (Figure 12.16).

- In polygenic inheritance, genes at several to many different loci interact to control the same character, producing a more or less continuous variation in the character from one extreme to another. Plotting the distribution of such characters among individuals typically produces a bell-shaped curve (Figure 12.17).

- In pleiotropy, one gene affects more than one character of an organism (Figure 12.18).

Animation: Comb shape in chickens

Animation: Codominance: ABO blood types

Animation: Incomplete dominance

Animation: Coat color in Labrador retrievers

Animation: Pleiotropic effects of Marfan syndrome

Animation: Coat color in the Himalayan rabbit

Interaction: Continuous variation in height

Questions

Self-Test Questions

1. The dominant C allele of a gene that controls color in corn produces kernels with color; plants homozygous for a recessive c allele of this gene have colorless or white kernels. What kinds of gametes, and in what proportions, would be produced by the plants in the following crosses? What seed color, and in what proportions, would be expected in the offspring of the crosses?
   a. $CC \times Cc$ b. $Cc \times Cc$ c. $Cc \times cc$

2. In peas, the allele T produces tall plants and the allele t produces dwarf plants. The T allele is dominant to t. If a tall plant is crossed with a dwarf, the offspring are distributed about equally between tall and dwarf plants. What are the genotypes of the parents?

3. The ability of humans to taste the bitter chemical phenylthiocarbamide (PTC) is a genetic trait. People with at least one copy of the normal, dominant allele of the PTC gene can taste PTC; those who are homozygous for a mutant, recessive allele cannot taste it. Could two parents able to taste PTC have a nontaster child? Could nontaster parents have a child able to taste PTC? A pair of taster parents, both of whom had one parent able to taste PTC and one nontaster parent, are expecting their first child. What is the chance that the child will be able to taste PTC? Unable to taste PTC? Suppose the first child is a nontaster. What is the chance that their second child will also be unable to taste PTC?

4. One gene has the alleles A and a; another gene has the alleles B and b. For each of the following genotypes, what types of gametes will be produced, and in what proportions, if the two gene pairs assort independently?
   a. $AA\ BB$ c. $Aa\ bb$
   b. $Aa\ BB$ d. $Aa\ Bb$

5. What genotypes, and in what frequencies, will be present in the offspring from the following matings?
   a. $AA\ BB \times aa\ BB$ c. $Aa\ Bb \times aa\ bb$
   b. $Aa\ BB \times AA\ Bb$ d. $Aa\ Bb \times Aa\ Bb$

6. In addition to the two genes in problem 4, assume you now study a third independently assorting gene that has the alleles C and c. For each of the following genotypes, indicate what types of gametes will be produced:
   a. $AA\ BB\ CC$ c. $Aa\ BB\ Cc$
   b. $Aa\ BB\ cc$ d. $Aa\ Bb\ Cc$

7. A man is homozygous dominant for alleles at 10 different genes that assort independently. How many genotypically different types of sperm cells can he produce? A woman is homozygous recessive for the alleles of 8 of these 10 genes, but she is heterozygous for the other 2 genes. How many genotypically different types of eggs can she produce? What hypothesis can you suggest to describe the relationship between the number of different possible gametes and the number of heterozygous and homozygous genes that are present?

8. In guinea pigs, an allele for rough fur (R) is dominant over an allele for smooth fur (r); an allele for black coat (B) is dominant over that for white (b). You have an animal with rough, black fur. What cross would you use to determine whether the animal is homozygous for these traits? What phenotype would you expect in the offspring if the animal is homozygous?

9. You cross a lima bean plant from a variety that breeds true for green pods with another lima bean from a variety that breeds true for yellow pods. You note that all the F_1 plants have green pods. These green-pod F_1 plants, when crossed, yield 675 plants with green pods and 217 with yellow pods. How many genes probably control pod color in this experiment? Give the alleles letter designations. Which is dominant?

10. Some recessive alleles have such a detrimental effect that they are lethal when present in both chromosomes of a pair. Homozygous recessives cannot survive and die at some point during embryonic development. Suppose that the allele r is lethal in the homozygous rr condition. What genotypic ratios would you expect among the living offspring of the following crosses?
   a. $RR \times Rr$
   b. $Rr \times Rr$

11. In garden peas, the genotypes GG or Gg produce green pods and gg produces yellow pods; TT or Tt plants are tall and tt plants are dwarfed; RR or Rr produce round seeds and rr produces wrinkled seeds. If a plant of a true-breeding, tall variety with green pods and round seeds is crossed with a plant of a true-breeding, dwarf variety with yellow pods and wrinkled seeds, what phenotypes are expected, and in what ratios, in the F_1 generation? What phenotypes, and in what ratios, are expected if F_1 individuals are crossed?

12. In chickens, feathered legs are produced by a dominant allele *F*. Another allele *f* of the same gene produces featherless legs. The dominant allele *P* of a gene at a different locus produces pea combs; a recessive allele *p* of this gene causes single combs. A breeder makes the following crosses with birds 1, 2, 3, and 4; all parents have feathered legs and pea combs:

Cross	Offspring
1 × 2	all feathered, pea comb
1 × 3	3/4 feathered; 1/4 featherless, all pea comb
1 × 4	9/16 feathered, pea comb; 3/16 featherless, pea comb; 3/16 feathered, single comb; 1/16 featherless, single comb

What are the genotypes of the four birds?

13. A mixup in a hospital ward caused a mother with O and MN blood types to think that a baby given to her really belonged to someone else. Tests in the hospital showed that the doubting mother was able to taste PTC (see problem 3). The baby given to her had O and MN blood types and had no reaction when the bitter PTC chemical was placed on its tongue. The mother had four other children with the following blood types and tasting abilities for PTC:
   a. Type A and MN blood, taster
   b. Type B and N blood, nontaster
   c. Type A and M blood, taster
   d. Type A and N blood, taster

   Without knowing the father's blood types and tasting ability, can you determine whether the child is really hers? (Assume that all her children have the same father.)

14. In cats, the genotype *AA* produces tabby fur color; *Aa* is also a tabby, and *aa* is black. Another gene at a different locus is epistatic to the gene for fur color. When present in its dominant *W* form (*WW* or *Ww*), this gene blocks the formation of fur color and all the offspring are white; *ww* individuals develop normal fur color. What fur colors, and in what proportions, would you expect from the cross *Aa Ww* × *Aa Ww*?

15. Having malformed hands with shortened fingers is a dominant trait controlled by a single gene; people who are homozygous for the recessive allele have normal hands and fingers. Having woolly hair is a dominant trait controlled by a different gene; homozygous recessive individuals have normal, nonwoolly hair. Suppose a woman with normal hands and nonwoolly hair marries a man who has malformed hands and woolly hair. Their first child has normal hands and nonwoolly hair. What are the genotypes of the mother, the father, and the child? If this couple has a second child, what is the probability that it will have normal hands and woolly hair?

Questions for Discussion

1. The eyes of brown-eyed people are not alike, but rather vary considerably in shade and pattern. What do you think causes these differences?

2. Explain how individuals of an organism that are phenotypically alike can produce different ratios of progeny phenotypes.

3. ABO blood type tests can be used to exclude paternity. Suppose a defendant who is the alleged father of a child takes a blood type test and the results do not exclude him as the father. Do the results indicate that he is the father? What arguments could a lawyer make based on the test results to exclude the defendant from being the father? (Assume the tests were performed correctly.)

Experimental Analysis

Imagine that you are a breeder of Labrador retriever dogs. Labs can be black, chocolate brown, or yellow. Suppose that a yellow Lab is donated to you and you need to know its genotype. You have a range of dogs with known genotypes. What cross would you make to determine the genotype of the donated dog? Explain how the resulting puppies show you the Lab's genotype.

Evolution Link

How could an epistatic interaction shelter a harmful allele from the action of natural selection?

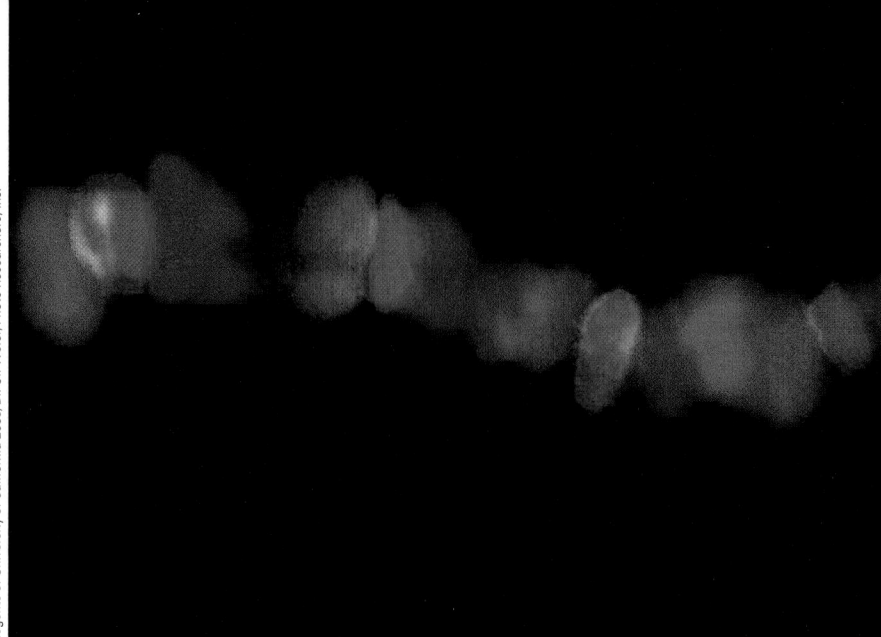

Fluorescent probes bound to specific sequences along human chromosome 10 (light micrograph). New ways of mapping chromosome structure yield insights into the inheritance of normal and abnormal traits.

Regents of University of California 2005/Dr. Uli Weier/Photo Researchers, Inc.

13 Genes, Chromosomes, and Human Genetics

WHY IT MATTERS

Imagine being 10 years old and trapped in a body that each day becomes more shriveled, frail, and old. You are just tall enough to peer over the top of the kitchen counter, and you weigh less than 35 pounds. Already you are bald, and you probably have only a few more years to live. But if you are like Mickey Hayes or Fransie Geringer **(Figure 13.1),** you still have not lost your courage or your childlike curiosity about life. Like them, you still play, laugh, and celebrate birthdays.

Progeria, the premature aging that afflicts Mickey and Fransie, is caused by a genetic error that occurs in only 1 of every 8 million human births. The error is perpetuated each time cells of the embryo—then of the child—duplicate their chromosomes and divide. The outcome of that rare mistake is an acceleration of aging and a greatly reduced life expectancy.

Progeria affects both boys and girls. Usually, symptoms begin to appear before the age of 2 years. The rate of body growth declines to abnormally low levels. Skin becomes thinner, muscles become flaccid, and limb bones start to degenerate. Children with progeria never reach puberty, and most die in their early teens from a stroke or heart

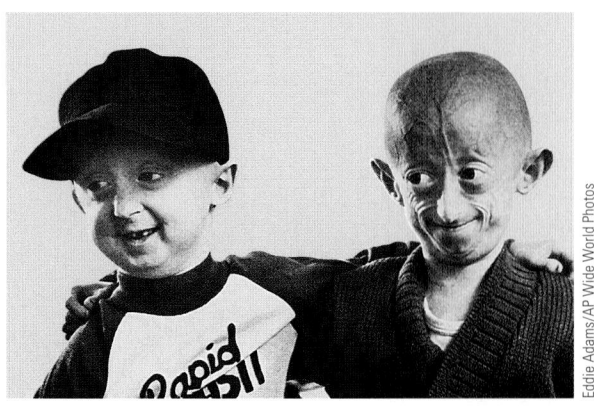

Figure 13.1
Two boys, both younger than 10, who have progeria, a genetic disorder characterized by accelerated aging and extremely reduced life expectancy.

attack brought on by hardening of the arteries, a condition typical of advanced age.

The plight of Mickey and Fransie provides a telling and tragic example of the dramatic effects that gene defects can have on living organisms. We are the products of our genes, and the characteristics of each individual, from humans to pine trees to protozoa, depend on the combination of genes, alleles, and chromosomes inherited from its parents, as well as on environmental effects. This chapter delves deeply into genes and the role of chromosomes in inheritance.

13.1 Genetic Linkage and Recombination

In his historic experiments, Gregor Mendel carried out crosses with seven different characters in garden peas, controlled by seven different genes. He found that each of the genes assorted independently of the others in the formation of gametes. If Mendel had extended his study to additional characters, he would have found exceptions to this principle. This should not be surprising, because an organism has far more genes than chromosomes. Conceptually, then, chromosomes contain many genes, with each gene at a particular locus. Genes located on different chromosomes assort independently in gamete formation because the two chromosomes behave independently of one another during meiosis. Genes located on the same chromosome may be inherited together in genetic crosses—that is, not assort independently—because the chromosome is inherited as a single physical entity in meiosis. Genes on the same chromosome are known as **linked genes**, and the phenomenon is called **linkage.**

The Principles of Linkage and Recombination Were Determined with *Drosophila*

In the early part of the twentieth century, Thomas Hunt Morgan and his coworkers at Columbia University were using the fruit fly, *Drosophila melanogaster,* as a

model organism to investigate Mendel's principles in animals. (*Focus on Research* describes the development and use of *Drosophila* in research.) In 1911, Morgan crossed a true-breeding fruit fly with normal red eyes and normal wing length, genotype $pr^+pr^+\ vg^+vg^+$, with a true-breeding fly with the recessive traits of purple eyes and vestigial (that is, short and crumpled) wings, genotype *prpr vgvg*, to analyze the segregation of the two traits (**Figure 13.2**, step 1).

This gene symbolism is new to us. Morgan devised this symbolism, and it is commonly used, much more so than the *A/a* system we have used until now. In this system, the plus (+) symbol indicates a wild-type—normal—allele of a gene. Typically, but not always, a wild-type allele is the most common allele found in a population. In most instances, the wild-type allele is dominant to mutant alleles, but there are exceptions. The letter is chosen based on the phenotype of the organism that expresses the mutant allele, for example, *pr* for *purple* eyes. Thus, we refer to the gene as the *purple* or *pr* gene; the dominant wild-type allele of the gene, pr^+, gives the wild-type red eye color.

The F_1 (first-generation) offspring of Morgan's cross were all $pr^+pr\ vg^+vg$, and because of the dominance of the wild-type alleles, they had red eyes and normal wings (see Figure 13.2, step 2). Next, Morgan testcrossed F_1 females to males with purple eyes and vestigial wings, genotype *prpr vgvg*. The testcross was used here because you can follow the meiotic events in just one parent; that is, all of the gametes from the other parent carry recessive alleles for the genes in the cross (see Section 12.1). Based on Mendel's principle of independent assortment (see Section 12.1), there should be four classes of phenotypes in the offspring, in the approximate 1:1:1:1 ratio of red eyes, normal wings : purple, vestigial : red, vestigial : purple, normal. But Morgan did not observe this result (step 3); instead, of the 2839 progeny flies, 1339 were red-normal and 1195 were purple-vestigial. These phenotypes are identical to the two original parental flies (the parents of the female parent used in the testcross); therefore, they are called **parental** phenotypes. The remaining progeny flies consisted of 151 red-vestigial and 154 purple-normal. Because these two classes have phenotypes with different combinations of traits from those of the original parents, they are called **recombinant** phenotypes. If the genes had shown independent assortment, there would have been 25% of each of the four classes, or 50% parental and 50% recombinant phenotypes. In numbers, there would have been 710 (approximately) of each of the 4 classes.

How could the low frequency of recombinant phenotypes be explained? Morgan hypothesized that the two genes are linked genetically—physically associated on the same chromosome. That is, *pr* and *vg* are linked genes. He further hypothesized that the behavior of these linked genes in the testcross is explained by what he called *chromosome recombination,* a process in which

Figure 13.2 Experimental Research

Evidence for Gene Linkage

QUESTION: Do the purple-eye vestigial-wing genes of *Drosophila* assort independently?

EXPERIMENT: Morgan crossed true-breeding wild-type flies with red eyes and normal wings with purple-eyed, vestigial-winged flies. The F_1 dihybrids were all wild-type in phenotype. Next he crossed the F_1 dihybrid flies with purple-eyed, vestigial-winged flies (this is a testcross) and analyzed the phenotypes of the progeny.

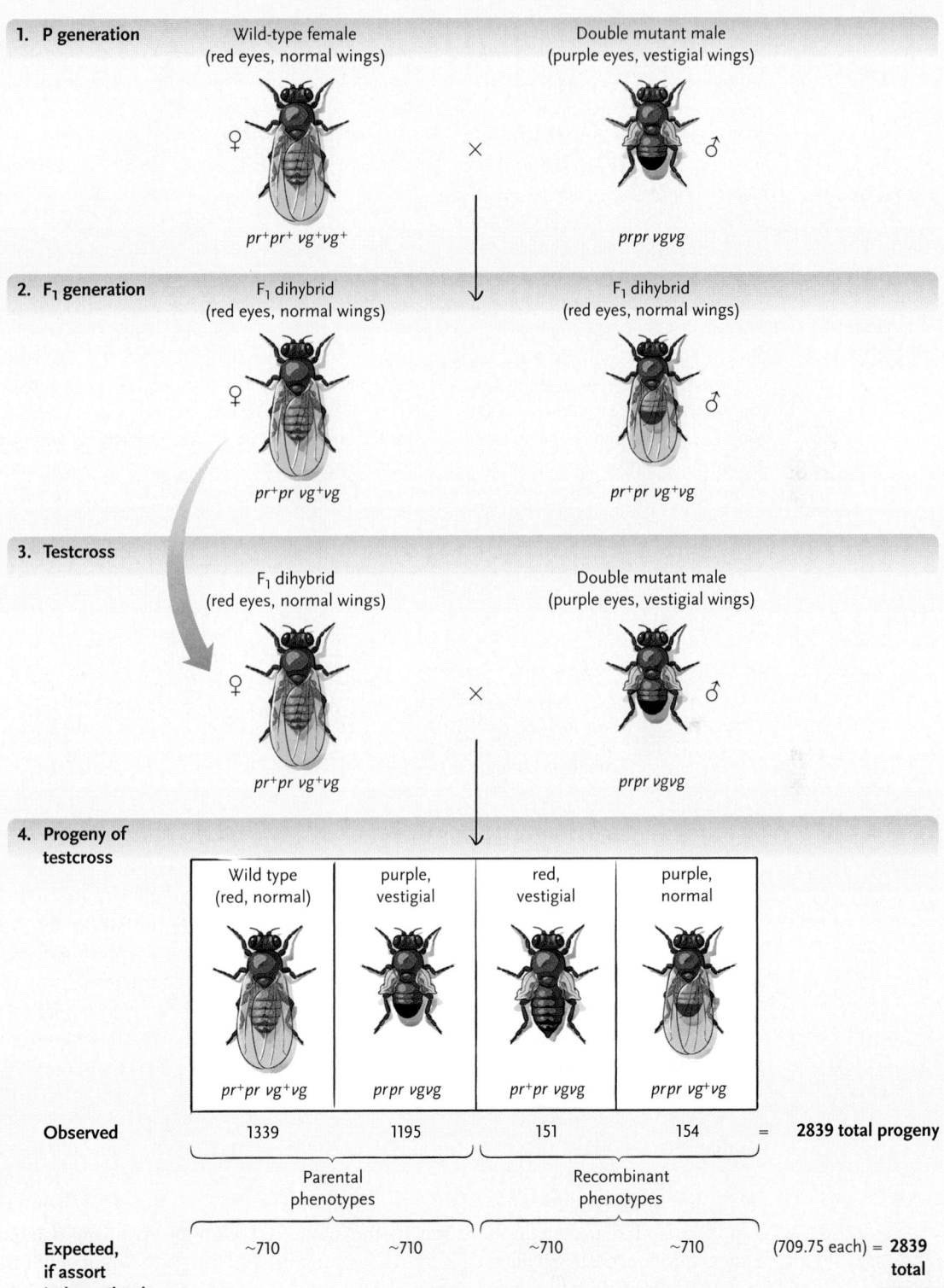

1. **P generation**

Wild-type female (red eyes, normal wings) × Double mutant male (purple eyes, vestigial wings)

$pr^+pr^+ vg^+vg^+$ $prpr vgvg$

2. **F_1 generation**

F_1 dihybrid (red eyes, normal wings) F_1 dihybrid (red eyes, normal wings)

$pr^+pr vg^+vg$ $pr^+pr vg^+vg$

3. **Testcross**

F_1 dihybrid (red eyes, normal wings) × Double mutant male (purple eyes, vestigial wings)

$pr^+pr vg^+vg$ $prpr vgvg$

4. **Progeny of testcross**

Wild type (red, normal)	purple, vestigial	red, vestigial	purple, normal
$pr^+pr vg^+vg$	$prpr vgvg$	$pr^+pr vgvg$	$prpr vg^+vg$

	Wild type (red, normal)	purple, vestigial	red, vestigial	purple, normal	
Observed	1339	1195	151	154	= **2839 total progeny**
	Parental phenotypes		Recombinant phenotypes		
Expected, if assort independently	~710	~710	~710	~710	(709.75 each) = **2839 total progeny**

RESULTS: 2534 of the testcross progeny flies had parental phenotypes, wild-type and purple, vestigial, while 305 of the progeny had recombinant phenotypes of red, vestigial and purple, normal. If the genes assorted independently, the expectation is a 1:1:1:1 ratio for testcross progeny: approximately 1420 of both parental and recombinant progeny.

CONCLUSION: The purple-eye and vestigial-wing genes do not assort independently. The simplest alternative is that the two genes are linked on the same chromosome.

Model Research Organisms: The Marvelous Fruit Fly, *Drosophila melanogaster*

Herman Eisenbeiss/Photo Researchers, Inc.

The unobtrusive little fruit fly that appears seemingly from nowhere when rotting fruit or a fermented beverage is around is one of the mainstays of genetic research. It was first described in 1830 by C. F. Fallén, who named it *Drosophila*, meaning "dew lover." The species identifier became *melanogaster*, which means "black belly."

The great geneticist Thomas Hunt Morgan began to culture *D. melanogaster* in 1909 in the famous "Fly Room" at Columbia University. Many important discoveries in genetics were made in the Fly Room, including sex-linked genes and sex linkage and the first chromosome map. The subsequent development of methods to induce mutations in *Drosophila* led, through studies of the mutants produced, to many other discoveries that collectively established or confirmed essentially all the major principles and conclusions of eukaryotic genetics.

One reason for the success of *D. melanogaster* as a subject for genetics research is the ease of culturing it.

It is grown usually at 25°C in small milk bottles stopped with a cotton or plastic foam wad and filled about one-third of the way with a fermenting medium that contains water, corn meal, agar, molasses, and yeast. The several hundred eggs laid by each adult female hatch rapidly and progress through larval and pupal stages to produce adult flies in about 10 days, which are ready to breed within 10 to 12 hours.

Males and females can be identified easily with the unaided eye. Many types of mutations produce morphologic differences, such as changes in eye color, wing shape, or the numbers and shapes of bristles, that can be seen with the unaided eye or under a low-power binocular microscope. The salivary gland cells of the fly larvae also have giant chromosomes. The chromosomes are so large that differences can be observed directly with the light microscope.

The availability of a wide range of mutants, comprehensive linkage maps of each of its chromosomes, and the ability to manipulate genes readily by molecular techniques made the fruit fly one of the model organisms for genome sequencing in the Human Genome Project. The sequencing of *Drosophila*'s genome was completed in 2001; there are approximately 14,000 genes in its 165 million base-pair genome. (A database of the *Drosophila* genome is available at http://flybase.bio.indiana.edu). Importantly, the relationship between fruit fly and human genes is close, to the point that many human disease genes have counterparts in the fruit fly genome. This similarity enables the fly genes to be studied as models of human disease genes in efforts to understand better the functions of those genes and how alterations in them lead to disease.

The analysis of fruit fly embryonic development has also contributed significantly to the understanding of development in humans. For example, experiments on mutants that affect fly development have provided insights into the genetic basis of many human birth defects.

Lastly, *Drosophila* molecular biologists are known for their sense of humor in naming genes based on the mutant phenotypes they identify. Their names contrast markedly from the serious naming of yeast and human genes. For instance, the *tinman* mutant of fruit flies lacks a heart in the embryo, *cheap date* mutants are especially sensitive to alcohol, and *indy* (I'm not dead yet) mutants have a doubled life span (named after a character in *Monty Python and the Holy Grail* who, when being taken to be buried, protests "I'm not dead yet.").

---

two homologous chromosomes exchange segments (crossover) with each other during meiosis (see Figure 11.6). Furthermore, he proposed that the frequency of this recombination is a function of the distance between linked genes. The nearer two genes are, the greater the chance they will be inherited together (resulting in parental phenotypes) and the lower the chance that recombinant phenotypes will be produced. The farther apart two genes are, the lower the chance that they will be inherited together and the greater the chance that recombinant phenotypes will be produced. These brilliant and far-reaching hypotheses were typical of Morgan, who founded genetics research in the United States, developed *Drosophila* as a research organism, and made discoveries that were almost as significant to the development of genetics as those of Mendel.

We show how this applies to the purple-vestigial cross in **Figure 13.3**, which presents the alleles of the genes with cartoons of the chromosomes themselves. This figure allows us to follow pictorially the consequences of crossing-over during meiosis in the production of gametes.

The pr^+pr vg^+vg F_1 dihybrid parent produces four types of gametes (see Figure 13.3). The two parental gametes, pr^+ vg^+ and pr vg, are generated by simple segregation of the chromosomes during meiosis without any crossing-over (recombination) between the genes. The two recombinant gametes, pr^+ vg and pr vg^+, result from crossing-over between the homologous chromatids when they are paired in prophase I of meiosis (see Section 11.1 and Figure 11.4). The offspring of the cross are produced by fusion of each of these four gametes with a pr vg gamete produced by the $prpr$ $vgvg$ male parent. The parental or recombinant phenotypes of the offspring directly reflect the genotypes of the gametes of the

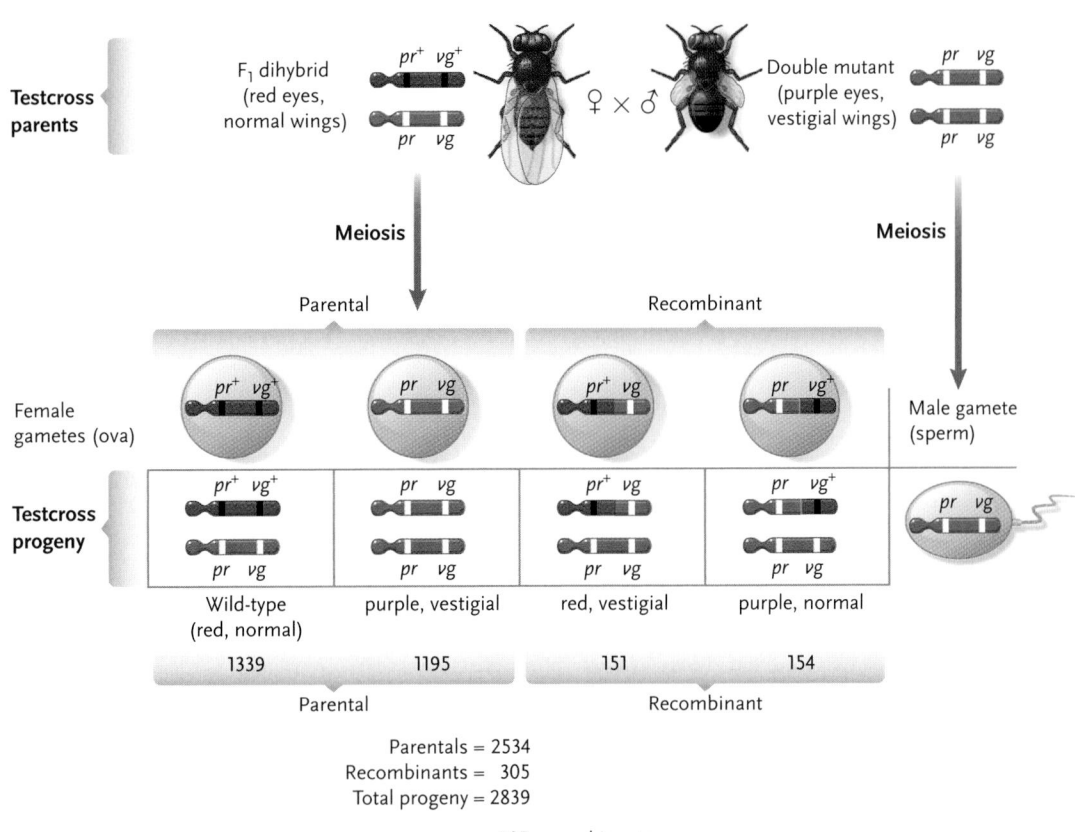

Figure 13.3
Recombination, the result of crossing-over between homologous chromosomes. The testcross of Figure 13.2 is redrawn here showing the two linked genes on chromosomes. The two parental homologs are colored differently to allow us to follow them during the cross. The parental phenotypes in the testcross progeny are generated by segregation of the parental chromosomes, whereas the recombinant phenotypes are generated by crossing-over between the two linked genes.

dihybrid parent. **Genetic recombination** is the process by which the combinations of alleles for different genes in two parental individuals become shuffled into new combinations in offspring individuals.

To determine the distance between the two genes on the chromosome, we calculate the **recombination frequency**, the percentage of testcross progeny that are recombinants. For this testcross, the recombination frequency is 10.7% (see Figure 13.3).

Recombination Frequency Can Be Used to Map Chromosomes

The recombination frequency of 10.7% for the *pr* and *vg* genes of *Drosophila* means that 10.7% of the gametes originating from the $pr^+pr\ vg^+vg$ parent contained recombined chromosomes. That recombination frequency is characteristic for those two genes. In other crosses that involve linked genes, Morgan found that the recombination frequency was characteristic of the two genes involved, varying from less than 1% to 50% (see the next section).

From these observations, Alfred Sturtevant, then an undergraduate at Columbia University working with Morgan, realized that the variations in recombination frequencies could be used as a means of mapping genes on chromosomes. Sturtevant himself later recalled his lightbulb moment:

> I suddenly realized that the variations in the strength of linkage already attributed by Morgan to difference in the

spatial separation of the gene offered the possibility of determining sequence in the linear dimensions of a chromosome. I went home and spent most of the night (to the neglect of my undergraduate homework) in producing the first chromosome map.

Sturtevant's revelation was that the recombination frequency observed between any two linked genes reflects the distance between them on their chromosome. The greater this distance, the greater the chance that a crossover can form between the genes and the greater the recombination frequency.

Therefore, recombination frequencies can be used to make a **linkage map** of a chromosome showing the relative locations of genes. For example, assume that the three genes *a*, *b*, and *c* are carried together on the same chromosome. Crosses reveal a 9.6% recombination frequency for *a* and *b*, an 8% recombination frequency for *a* and *c*, and a 2% recombination frequency for *b* and *c*. These frequencies allow the genes to be arranged in only one sequence on the chromosomes as follows:

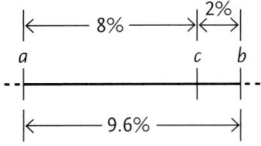

You will note that the *a-b* recombination frequency does not exactly equal the sum of the *a-c* and *c-b* recombination frequencies. This is because genes farther

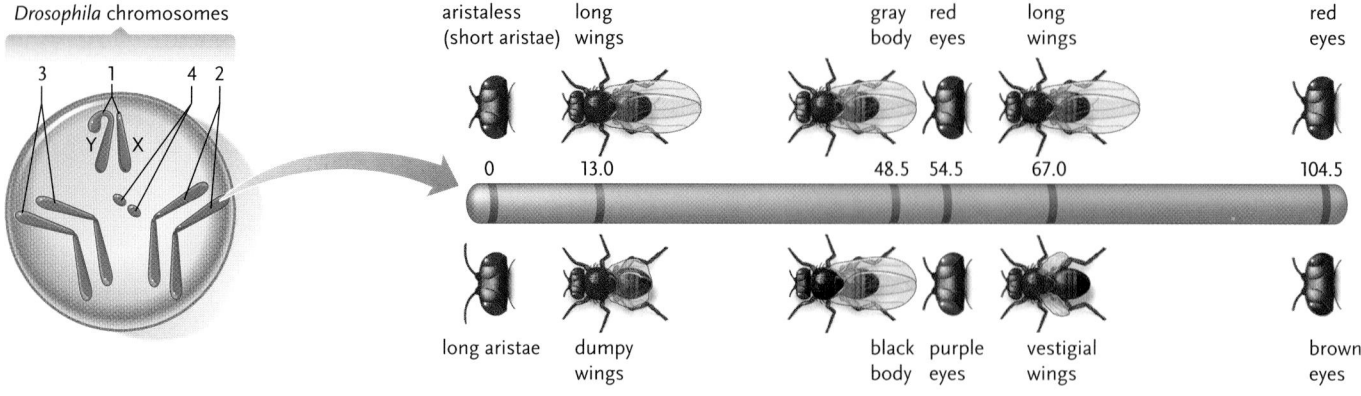

Figure 13.4

Relative map locations of several genes on chromosome 2 of *Drosophila*, as determined by recombination frequencies. For each gene, the diagram shows the normal or "wild-type" phenotype on the top and the mutant phenotype on the bottom. Mutant alleles at two different locations alter wing structure, one producing the dumpy wing and the other the vestigial wing phenotypes; the normal allele at these locations results in normal long-wing structure. Mutant alleles at two different locations also alter eye color.

apart on a chromosome are more likely to have more than one crossover occur between them. Whereas a single crossover between two genes gives recombinants, a double crossover (two single crossovers occurring in the same meiosis) between two genes gives parentals. You can see this simply by drawing single and double crossovers between two genes on a piece of paper. In our example, double crossovers that occur between *a* and *b* have slightly decreased the recombination frequency between these two genes.

Using this method, Sturtevant created the first linkage map showing the arrangement of six genes on the *Drosophila* X chromosome. (A partial linkage map of a *Drosophila* chromosome is shown in **Figure 13.4**.)

Since the time of Morgan, many *Drosophila* genes and those of other eukaryotic organisms widely used for genetic research, including *Neurospora* (a fungus), yeast, maize (corn), and the mouse, have been mapped using the same approach. Recombination frequencies, together with the results of other techniques, have also been used to create linkage maps of the locations of genes in the DNA of prokaryotes such as the human intestinal bacterium *Escherichia coli* (see Chapter 17).

The unit of a linkage map, called a **map unit** (abbreviated mu), is equivalent to a recombination frequency of 1%. The map unit is also called the **centimorgan** in honor of Morgan's discoveries of linkage and recombination. Map units are not absolute physical distances in micrometers or nanometers; rather, they are *relative,* showing the positions of genes with respect to each other. One of the reasons that the units are relative and not absolute distances is that the frequency of crossing-over varies to some extent from one position to another on chromosomes.

In recent years, the linkage maps of a number of species have been supplemented by DNA sequencing of whole genomes, which shows the precise physical locations of genes in the chromosomes.

Widely Separated Linked Genes Assort Independently

Genes can be so widely separated on a chromosome that recombination is likely to occur at some point between them in every cell undergoing meiosis. When this is the case, no linkage is detected and the genes assort independently. In other words, even though the alleles of the genes are carried on the same chromosome, the approximate 1:1:1:1 ratio of phenotypes is seen in the offspring of a dihybrid × double mutant testcross. That is, 50% of the progeny are parentals and 50% are recombinants. Linkage between such widely separated genes can still be detected, however, by testing their linkage to one or more genes that lie between them. For example, the genes *a* and *c* in **Figure 13.5** are located so far apart that they assort independently and show no linkage. However, crosses show that *a* and *b* are 23 map units apart (recombination frequency of 23%), and crosses that show *b* and *c* are 34 map units apart. Therefore, *a* and *c* must also be linked and carried on the same chromosome at 23 + 34 = 57 map units apart. Obviously, we could not see a recombination frequency of 57% in testcross progeny because the maximum frequency of recombinants is 50%.

We now know that some of the genes Mendel studied assort independently even though they are on the same chromosome. For example, the genes for flower color and seed color are located on the same chromosome pair, but they are so far apart that the frequent

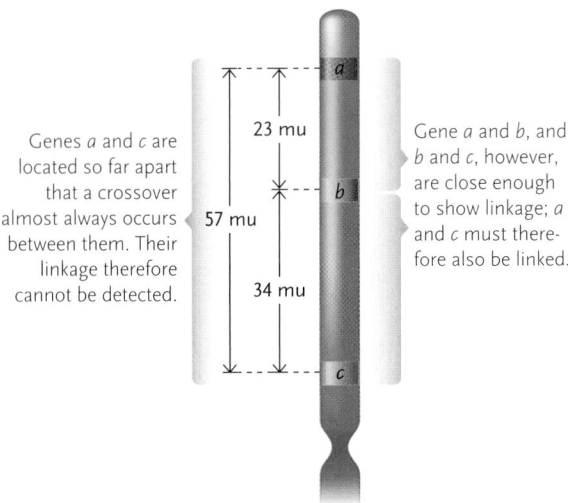

Figure 13.5

Genes far apart on the same chromosome. Genes *a* and *c* are far apart and will not show linkage, suggesting they are on different chromosomes. However, linkage between such genes can be established by noting their linkage to another gene or genes located between them—gene *b* here.

Genes *a* and *c* are located so far apart that a crossover almost always occurs between them. Their linkage therefore cannot be detected.

23 mu

57 mu

34 mu

Gene *a* and *b*, and *b* and *c*, however, are close enough to show linkage; *a* and *c* must therefore also be linked.

recombination between them makes them appear to be unlinked.

STUDY BREAK

You want to determine whether genes *a* and *b* are linked. What cross would you use and why? How would this cross tell you if they are linked?

13.2 Sex-Linked Genes

In many organisms, one or more pairs of chromosomes are different in males and females (see Section 11.1). Genes located on these chromosomes, the *sex chromosomes,* are called **sex-linked genes;** they are inherited differently in males and females. (Note that the word *linked* in *sex-linked gene* means that the gene is on a sex chromosome, whereas the use of the term *linked* when considering two or more genes means that the genes are on the same chromosome, not necessarily a sex chromosome.) Chromosomes other than the sex chromosomes are called **autosomes;** genes on these chromosomes have the same patterns of inheritance in both sexes. In humans, chromosomes 1 to 22 are the autosomes.

Females Are XX and Males Are XY in Both Humans and Fruit Flies

In most species with sex chromosomes, females have two copies of a chromosome known as the **X chromosome,** forming a homologous XX pair, whereas males have only one X chromosome. An-

other chromosome, the **Y chromosome,** occurs in males but not in females, giving males an XY combination; the XX-XY human chromosome complement is shown in Figure 10.7.

Each normal gamete produced by an XX female carries an X chromosome. Half the gametes produced by an XY male carry an X chromosome and half carry a Y. When a sperm cell carrying an X chromosome fertilizes an X-bearing egg cell, the new individual develops into an XX female. Conversely, when a sperm cell carrying a Y chromosome fertilizes an X-bearing egg cell, the combination produces an XY male **(Figure 13.6).** The Punnett square shows that fertilization is expected to produce females and males with an equal frequency of 1/2. This expectation is closely matched in the human and *Drosophila* populations.

Other sex chromosome arrangements occur, as in some insects with XX females and XO males (the O means there is no Y chromosome). In birds, butterflies, and some reptiles, the situation is reversed: males have a homologous pair of sex chromosomes (ZZ instead of XX), and females have the equivalent of an XY combination (ZW).

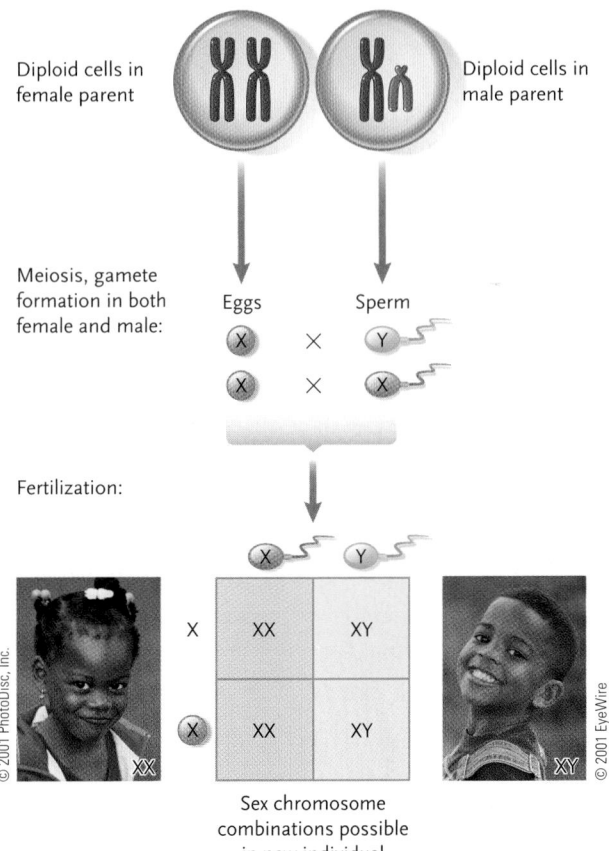

Figure 13.6

Sex chromosomes and the chromosomal basis of sex determination in humans. Females have two X chromosomes and produce gametes (eggs), all of which have the X sex chromosome. Males have one X and one Y chromosome and produce gametes, half with an X and half with a Y chromosome. Males transmit their Y chromosome to their sons, but not to their daughters. Males receive their X chromosome only from their mother.

Human Sex Determination Depends on the Y Chromosome

The human X chromosome carries about 2350 genes. Although some of these genes are associated with sexual traits, such as differing distributions of body fat in males and females, most are concerned with nonsexual traits such as the ability to perceive color, metabolize certain sugars, or form blood clots when tissues are injured. Human sex determination depends on the Y chromosome, which contains the *SRY* gene (for *sex-determining region of the Y*) that switches development toward maleness at an early point in embryonic development.

For the first month or so of embryonic development in humans and other mammals, the rudimentary structures that give rise to reproductive organs and tissues are the same in XX or XY embryos. After 6 to 8 weeks, the *SRY* gene becomes active in XY embryos, producing a protein that regulates the expression of other genes, thereby stimulating part of these structures to develop as testes. As a part of stimulation by hormones secreted in the developing testes and elsewhere, tissues degenerate that would otherwise develop into female structures such as the vagina and oviducts. The remaining structures develop into the penis and scrotum. In XX embryos, which do not have a copy of the *SRY* gene, development proceeds toward female reproductive structures. The rudimentary male structures degenerate in XX embryos because the hormones released by the developing testes in XY embryos are not present. Further details of the *SRY* gene and its role in human sex determination are presented in Chapter 48 (specifically, see *Insights from the Molecular Revolution* in that chapter).

Sex-Linked Genes Were First Discovered in *Drosophila*

The different sets of sex chromosomes in males and females affect the inheritance of the alleles on these chromosomes in a distinct pattern known as *sex linkage*.

Two features of the XX-XY arrangement cause sex linkage. One is that alleles carried on the X chromosome occur in two copies in females but in only one copy in males. The second feature is that alleles carried on the Y chromosome are present in males but not females.

Morgan discovered sex-linked genes and their pattern of sex linkage in 1910. It started when he found a male fly in his stocks with white eyes instead of the normal red eyes **(Figure 13.7)**. He crossed the white-eyed male with a true-breeding female with red eyes and observed that all the F_1 flies had red eyes **(Figure 13.8a)**. He concluded that the white-eye trait was recessive. Next, he allowed the F_1 flies to interbreed. Based on Mendel's principles, he expected that both male and female F_2 flies would show a 3 : 1 ratio of red-eyed flies to white-eyed flies. Morgan was surprised to find that all the F_2 females had red eyes, and half of the F_2 males had red eyes and half had white eyes **(Figure 13.8b)**.

Morgan hypothesized that the alleles segregating in the cross were of a gene located on the X chromosome—now termed a *sex-linked gene*. The white-eyed male parent in the cross had the genotype X^wY: an X chromosome with a white (X^w) allele, and no other allele of that gene on the Y. The red-eyed female parent in the cross had the genotype $X^{w+}X^{w+}$: each X chromosome carries the dominant normal allele for red eyes, X^{w+}.

We can follow the alleles in this cross (see Figure 13.8a). The F_1 flies of a cross $X^{w+}X^{w+} \times X^wY$ are produced as follows. The X chromosome of the males comes from their mother; therefore, their genotype is $X^{w+}Y$, and their phenotype is red eyes. The females receive one X from each parent; therefore, their genotype is $X^{w+}X^w$, and their phenotype is red eyes due to the dominance of the X^{w+} allele.

In the F_2 generation, the females receive an X^{w+} allele from the male F_1 parent and either an X^{w+} or X^w allele from the female F_1 parent; these genotypes result in red eyes (see Figure 13.8a). The males receive their one X chromosome from the female F_1 parent, which has the genotype $X^{w+}X^w$. Therefore, F_2 males are half $X^{w+}Y$ (red eyes) and half X^wY (white eyes).

Morgan also made a *reciprocal cross* of the one just described; that is, the phenotypes were switched between the sexes. The reciprocal cross here was a white-eyed female (X^wX^w) with a red-eyed male ($X^{w+}Y$) (see Figure 13.8b). The F_1 males all had white eyes because they received the X^w-bearing chromosome from the female parent; thus, their genotype is X^wY. The F_1 females have red eyes; therefore, they are all heterozygous $X^{w+}X^w$. This result is clearly different from the cross in Figure 13.8a.

In the F_2 generation of this second cross, both male and female flies showed a 1 : 1 ratio of red eyes to white eyes (see Figure 13.8b). Again, this result differs markedly from that of the cross in Figure 13.8a.

In summary, Morgan's work showed that there is a distinctive pattern in the phenotypic ratios for recip-

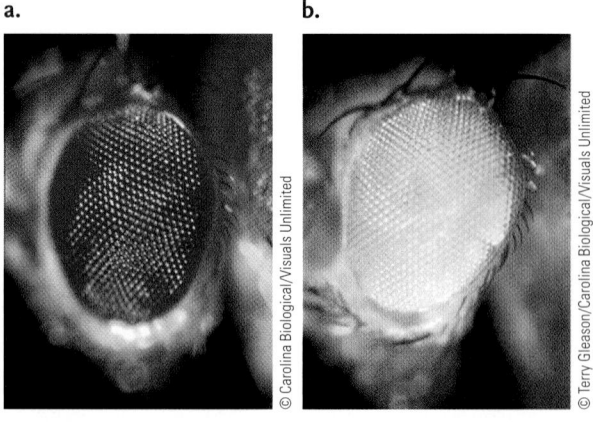

a.   **b.**

© Carolina Biological/Visuals Unlimited

© Terry Gleason/Carolina Biological/Visuals Unlimited

Figure 13.7

Eye color phenotypes in *Drosophila*. **(a)** Normal, red wild-type eye color. **(b)** Mutant white eye color caused by a recessive allele of a sex-linked gene carried on the X chromosome.

Figure 13.8 Experimental Research

Evidence for Sex-Linked Genes

QUESTION: How is the white-eye gene of *Drosophila* inherited?

EXPERIMENT: Morgan crossed a white-eyed male *Drosophila* with a true-breeding female with red eyes and then crossed the F₁ flies to produce the F₂ generation. He also performed the reciprocal cross in which the phenotypes were switched in the parental flies—true-breeding white-eyed female × red-eyed male.

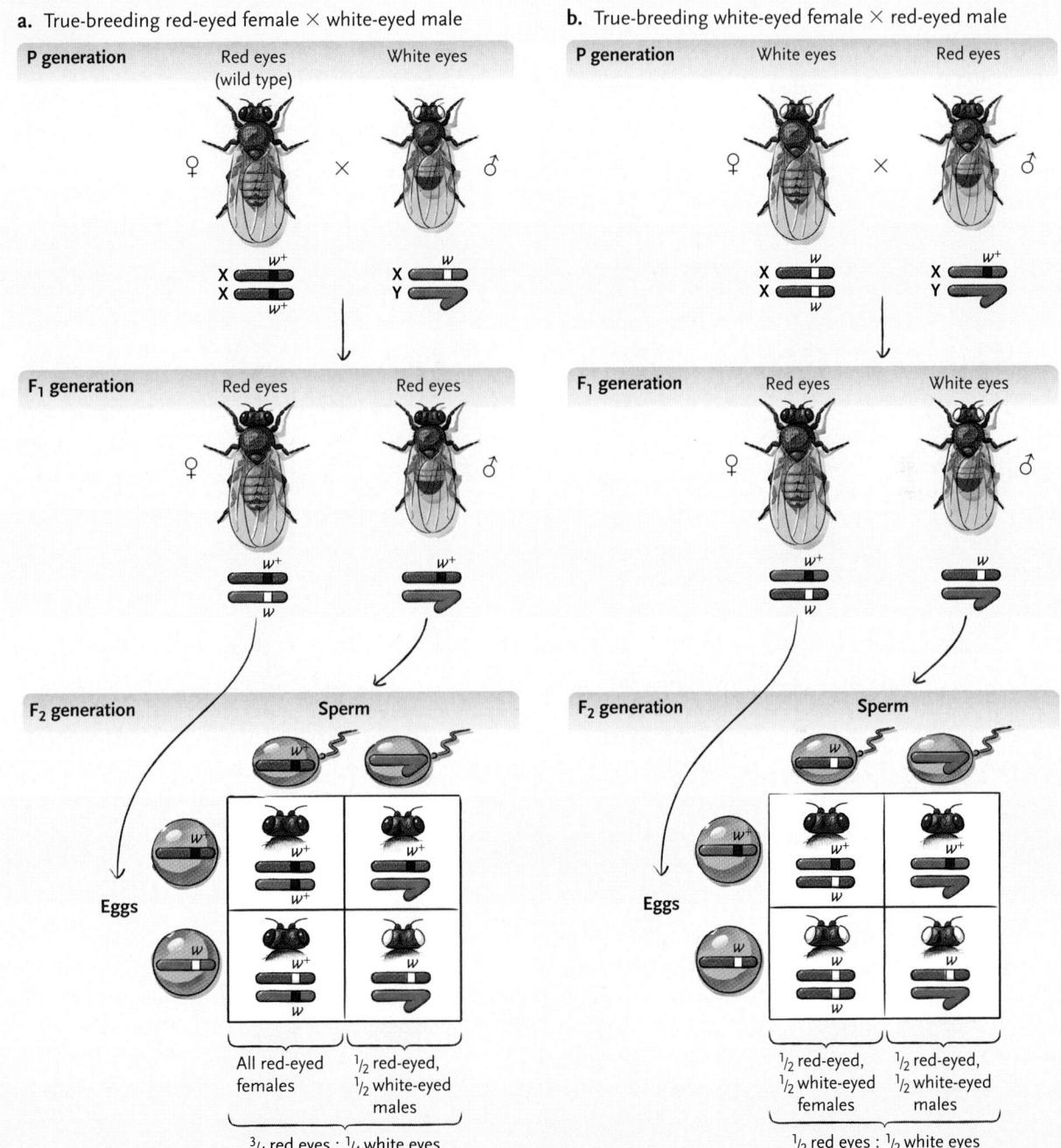

a. True-breeding red-eyed female × white-eyed male

b. True-breeding white-eyed female × red-eyed male

RESULTS: Differences were seen in both the F₁ and F₂ generations for the red ♀ × white ♂ and white ♀ × red ♂ crosses.

CONCLUSION: The segregation pattern for the white-eye trait showed that the white-eye gene is a sex-linked gene located on the X chromosome.

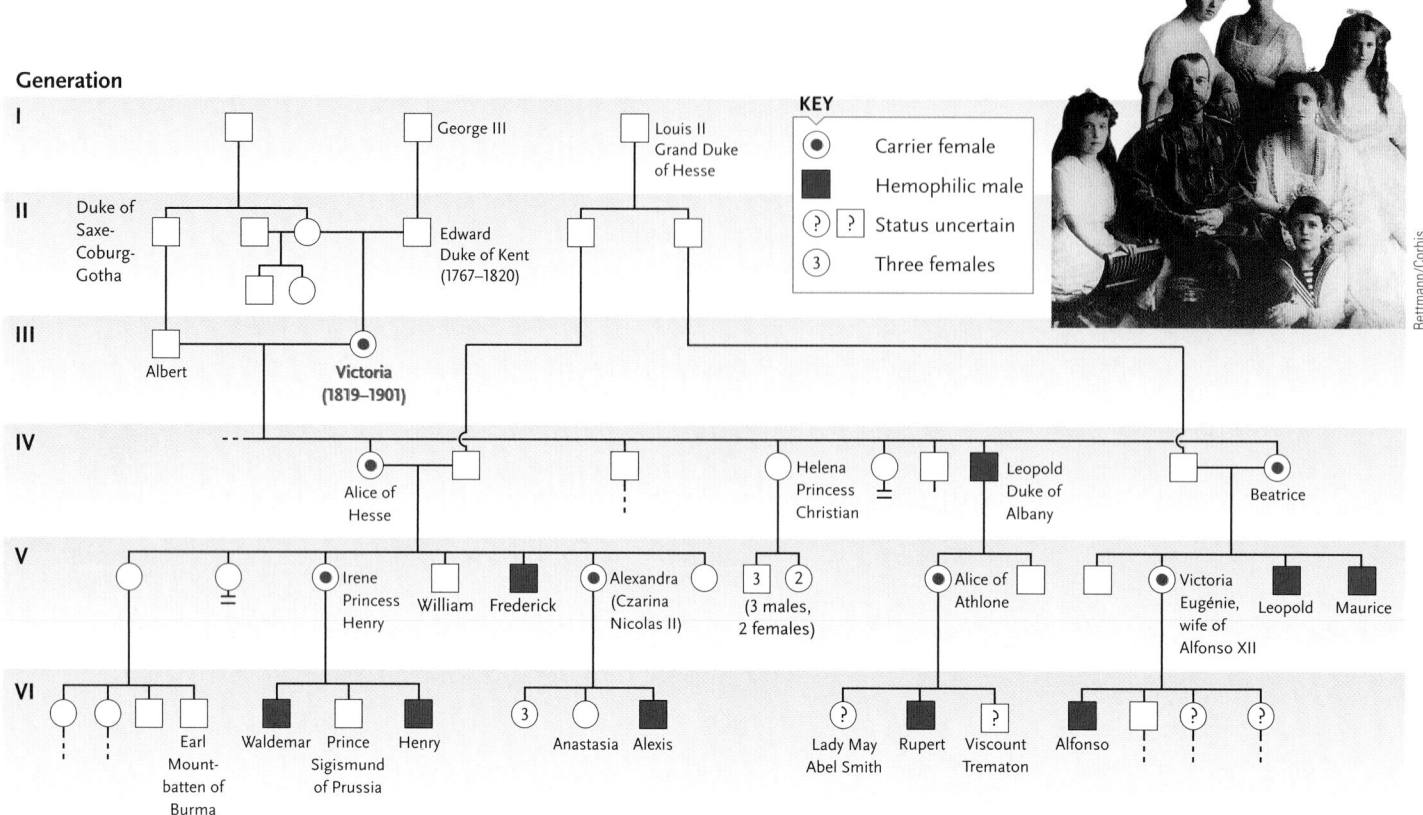

Generation

KEY
- Carrier female
- Hemophilic male
- Status uncertain
- Three females

Figure 13.9
Inheritance of hemophilia in descendants of Queen Victoria of England. The photograph shows the Russian royal family in which the son, Crown Prince Alexis, had hemophilia. His mother was a carrier of the mutated gene.

rocal crosses in which the gene involved is on the X chromosome. A key indicator of this sex linkage is when all male offspring of a cross between a true-breeding mutant female and a wild-type male have the mutant phenotype. As we have seen, this occurs because a male receives his X chromosome from his female parent.

Sex-Linked Genes in Humans Are Inherited as They Are in *Drosophila*

For obvious reasons, experimental genetic crosses cannot be conducted with humans. However, a similar analysis can be made by interviewing and testing living members of a family and reconstructing the genotypes and phenotypes of past generations from family records. The results are summarized in a chart called a **pedigree**, which shows all parents and offspring for as many generations as possible, the sex of individuals in the different generations, and the presence or absence of the trait of interest. Females are designated by a circle and males by a square; a solid circle or square indicates the presence of the trait.

In humans, as in fruit flies, sex-linked recessive traits appear more frequently among males than females because males need to receive only one copy of the allele on the X chromosome inherited from their mothers to develop the trait. Females must receive two

copies of the recessive allele, one from each parent, to develop the trait. Two examples of human sex-linked traits are red–green color blindness, a recessive trait in which the affected individual is unable to distinguish between the colors red and green because of a defect in light-sensing cells in the retina, and hemophilia, a recessive trait in which affected individuals have a defect in blood clotting.

Hemophiliacs—people with hemophilia—are "bleeders"; that is, if they are injured, they bleed uncontrollably because a protein required for forming blood clots is not produced in functional form. Males are bleeders if they receive an X chromosome that carries the recessive allele. The disease also develops in females with the recessive allele on both of their X chromosomes—a rare combination. Although affected persons, with luck and good care, can reach maturity, their lives are tightly circumscribed by the necessity to avoid injury of any kind. Even internal bleeding from slight bruises can be fatal. The disease, which affects about 1 in 7000 males, can be treated by injection of the required clotting molecules.

Hemophilia has had effects reaching far beyond individuals who inherit the disease. The most famous cases occurred in the royal families of Europe descended from Queen Victoria of England **(Figure 13.9)**. The disease was not recorded in Queen Victoria's ancestors, so the recessive allele for the trait probably appeared as a spontaneous mutation in the queen or one of her parents. Queen Victoria was heterozygous for the recessive hemophilia allele; that is, she was a **carrier**, meaning that she carried the mutant allele and could pass it on to her offspring but she did not have symptoms of the disease. A carrier is indicated in a pedigree by a male or female symbol with a central dot.

Note in Queen Victoria's pedigree in Figure 13.9 that Leopold, Duke of Albany, had hemophilia, as did his grandson, Rupert, Viscount Trematon. The trait alternates from generation to generation in males because a father does not pass his X chromosome to his sons; the X chromosome received by a male always comes from his mother. The appearance of a trait in the males of alternate generations therefore indicates that the allele under study is recessive and carried on the X chromosome.

At one time, 18 of Queen Victoria's 69 descendants were affected males or female carriers. Because so many sons of European royalty were affected, the trait influenced the course of history. In Russia, Crown Prince Alexis was one of Victoria's hemophiliac descendants. His affliction drew together his parents, Czar Nicholas II and Czarina Alexandra (a granddaughter of Victoria and a carrier), and the hypnotic monk Rasputin, who manipulated the family to his advantage by convincing them that only he could control the boy's bleeding. The situation helped trigger the Russian Revolution of 1917, which ended the Russian monarchy and led to the establishment of a Communist government in the former Soviet Union, a significant event in twentieth century history.

Hemophilia affected only sons in the royal lines but could have affected daughters if a hemophiliac son had married a carrier female. Because the disease is rare in the human population as a whole, the chance of such a mating is so low that only a few unions of this type have been recorded.

Inactivation of One X Chromosome Evens out Gene Effects in Mammalian Females

Although mammalian females have twice as many X chromosomes as males, the effects of most genes carried on the X chromosome in females is equalized in the male and female offspring of placental mammals by a *dosage compensation mechanism* that inactivates one of the two X chromosomes in most body cells of female mammals.

As a result of the equalizing mechanism, the activity of most genes carried on the X chromosome is essentially the same in males and females. The inactivation occurs by a condensation process that folds and packs the chromatin of one of the two X chromosomes into a tightly coiled state similar to the condensed state of chromosomes during cell division. The inactive, condensed X chromosome can be seen at one side of the nucleus in cells of females as a dense mass of chromatin called the **Barr body**.

The inactivation occurs during embryonic development. Which of the two X chromosomes becomes inactive in a particular embryonic cell line is a random event. But once one of the X chromosomes is inactivated in a cell, that same X is inactivated in all descendants of the cell. Thus, within one female, one of the X chromosomes is active in particular cells and inactive in others and vice versa.

If the two X chromosomes carry different alleles of a gene, one allele will be active in cell lines in which one X chromosome is active, and the other allele will be active in cell lines in which the other X chromosome is active. For many sex-linked alleles, such as the recessive allele that causes hemophilia, random inactivation of either X chromosome has little overall whole-body effect in heterozygous females because the dominant allele is active in enough of the critical cells to produce a normal phenotype. However, for some genes, the inactivation of either X chromosome in heterozygotes produces recognizably different effects in distinct regions of the body.

For example, the orange and black patches of fur in calico cats result from inactivation of one of the two X chromosomes in regions of the skin of heterozygous females **(Figure 13.10)**. Males, which get only one of the two alleles, normally have either black or orange fur. Similarly, in humans, females who are heterozygous for an allele on the X chromosome that blocks development of sweat glands may have a patchy distribution

Ulrike Schanz/Animals, Animals

Figure 13.10

A female cat with the calico color pattern in which patches of orange and black fur are produced by random inactivation of one of the two X chromosomes. Two genes control the black and orange colors: the *O* gene on the X chromosome is for orange fur color, and the *B* gene on an autosome is for black fur color. A calico cat has the genotype *Oo BB* (or *Oo Bb*). An orange patch results when the X chromosome carrying the mutant *o* allele is inactivated. In this case, the *O* gene masks the expression of the *B* gene and orange fur is produced. (This example is of epistasis; see Section 12.2.) A black patch results when the X chromosome carrying the *O* allele is inactivated. In this case, the mutant *o* allele cannot mask *B* gene expression and black fur results. The white patches result from interactions with a different, autosomal gene that entirely blocks pigment deposition in the fur.

of skin areas with and without the glands. Females with the patchy distribution are not seriously affected and may be unaware of the condition.

As we have seen, the discovery of genetic linkage, recombination, and sex-linked genes led to the elaboration and expansion of Mendel's principles of inheritance. Next, we examine what happens when patterns of inheritance are modified by changes in the chromosomes.

STUDY BREAK

You have a true-breeding strain of miniature-winged fruit flies, where this wing trait is recessive to the normal long wings. How would you show whether the miniature wing trait is sex-linked or autosomal?

13.3 Chromosomal Alterations That Affect Inheritance

Although chromosomes are relatively stable structures, they are sometimes altered by breaks in the DNA, which can be generated by agents such as radiation or certain chemicals or by enzymes encoded in some infecting viruses. The broken chromosome fragments may be lost, or they may reattach to the same or different chromosomes. The resulting changes in chromosome structure may have genetic consequences if alleles are eliminated, mixed in new combinations, duplicated, or placed in new locations by the alterations in cell lines that lead to the formation of gametes.

Genetic changes may also occur through changes in chromosome number, including addition or loss of one or more chromosomes or even entire sets of chromosomes. Both chromosomal alterations and changes in chromosome number can be a source of disease and disability, as well as a source of variability during evolution.

Deletions, Duplications, Translocations, and Inversions Are the Most Common Chromosomal Alterations

Chromosomal alterations after breakages occur in four major forms (Figure 13.11):

- A **deletion** occurs if a broken segment is lost from a chromosome.
- A **duplication** occurs if a segment is broken from one chromosome and inserted into its homolog. In the receiving homolog, the alleles in the inserted fragment are added to the ones already there.
- A **translocation** occurs if a broken segment is attached to a different, nonhomologous chromosome.
- An **inversion** occurs if a broken segment reattaches to the same chromosome from which it was lost, but in reversed orientation, so that the order of genes is reversed.

To be inherited, chromosomal alterations must occur or be included in cells of the germ line leading to development of eggs or sperm.

Deletions and Duplications. A deletion (see Figure 13.11a) may cause severe problems if the missing segment contains genes that are essential for normal development or cellular functions. For example, one deletion from human chromosome 5 typically leads to severe mental retardation and a malformed larynx. The cries of an affected infant sound more like a meow than a human cry. Hence the name of the disorder, *cri-du-chat* (meaning "cat's cry").

A duplication (see Figure 13.11b) may have effects that vary from harmful to beneficial, depending on the genes and alleles contained in the duplicated region. Although most duplications are likely to be detrimental, some have been important sources of evolutionary change. That is, because there are duplicate genes, one copy can mutate into new forms without seriously affecting the basic functions of the organism. For example, mammals have genes that encode several types of

Figure 13.11
Chromosome deletion, duplication, translocation (a reciprocal translocation is shown), and inversion.

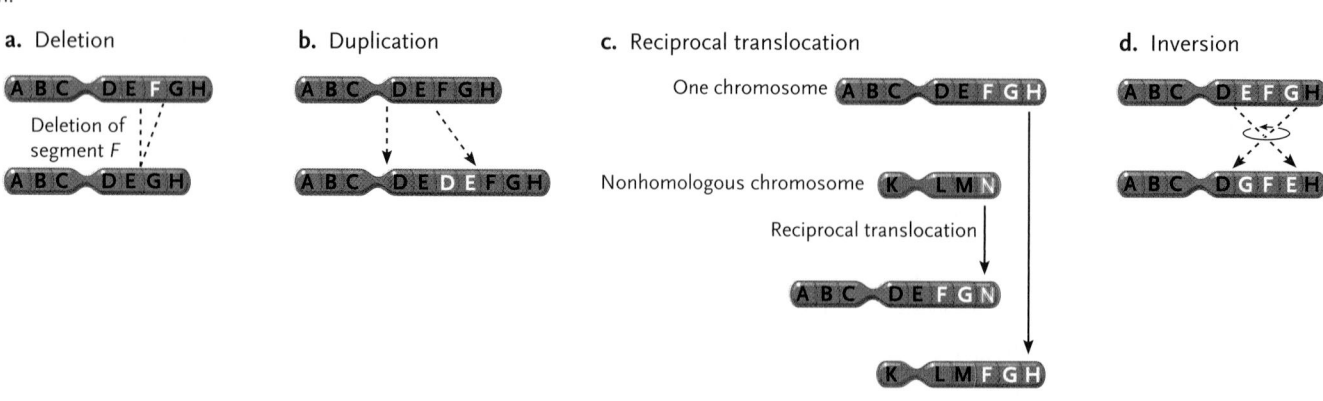

a. Deletion

A B C D E F G H

Deletion of segment *F*

A B C D E G H

b. Duplication

A B C D E F G H

A B C D E D E F G H

c. Reciprocal translocation

One chromosome A B C D E F G H

Nonhomologous chromosome K L M N

Reciprocal translocation

A B C D E F G N

K L M F G H

d. Inversion

A B C D E F G H

A B C D G F E H

hemoglobin that are not present in vertebrates, such as sharks, which evolved earlier; the additional hemoglobin genes of mammals are believed to have appeared through duplications, followed by mutations in the duplicates that created new and beneficial forms of hemoglobin as further evolution took place. Duplications sometimes arise during recombination in meiosis, if crossing-over occurs unequally, so that a segment is deleted from one chromosome of a homologous pair and inserted in the other.

Translocations and Inversions. In a translocation, a segment breaks from one chromosome and attaches to another, nonhomologous chromosome. In many cases, a translocation is reciprocal, meaning that two nonhomologous chromosomes exchange segments (see Figure 13.11c). Reciprocal translocations resemble genetic recombination, except that the two chromosomes involved in the exchange do not contain the same genes.

For example, a particular cancer of the human immune system, Burkitt lymphoma, is caused by a translocation that moves a segment of human chromosome 8 to the end of chromosome 14. The break does not interrupt any genes required for normal cell function. The translocated segment contains genes that control cell division. These genes are precisely regulated at their normal location but are near the control regions of highly active genes in the new location, causing them to be overactive and leading to uncontrolled cell division and the development of a cancer.

In an inversion, a chromosome segment breaks and then reattaches to the same chromosome, but in reverse order (see Figure 13.11d). Inversions have essentially the same effects as translocations—genes may be broken internally by the inversion, with loss of function, or they may be transferred intact to a new location within the same chromosome, producing effects that range from beneficial to harmful.

Inversions and translocations have been important factors in the evolution of plants and some animals, including insects and primates. For example, five of the chromosome pairs of humans show evidence of translocations and inversions that are not present in one of our nearest primate relatives, gorillas, and therefore must have occurred after the gorilla and human evolutionary lineages split.

The Number of Entire Chromosomes May Also Change

At times, whole, single chromosomes are lost or gained from cells entering or undergoing meiosis, resulting in a change of chromosome number. Most often, these changes occur through **nondisjunction**—the failure of homologous pairs to separate during the first meiotic division or of chromatids to separate during the second

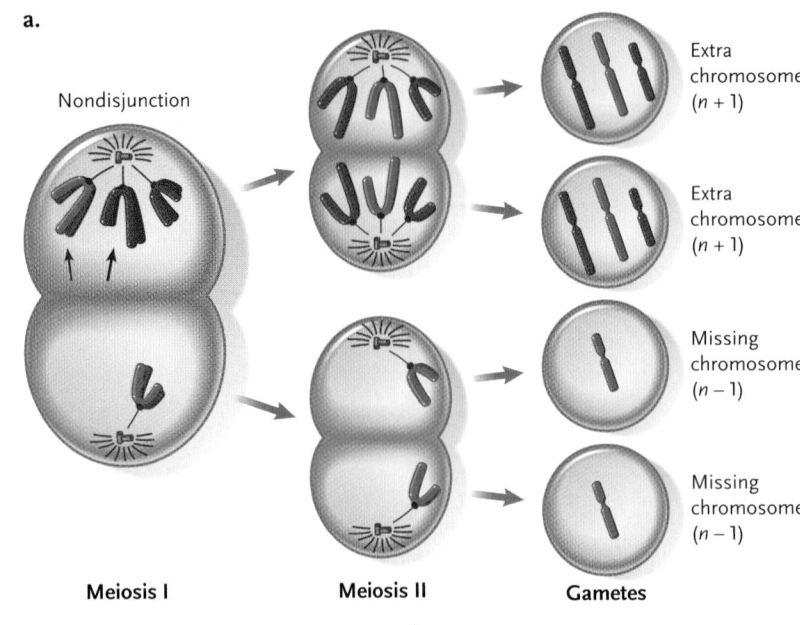

a.

Nondisjunction during the first meiotic division causes both chromosomes of one pair to be delivered to the same pole of the spindle. The nondisjunction produces two gametes with an extra chromosome and two with a missing chromosome.

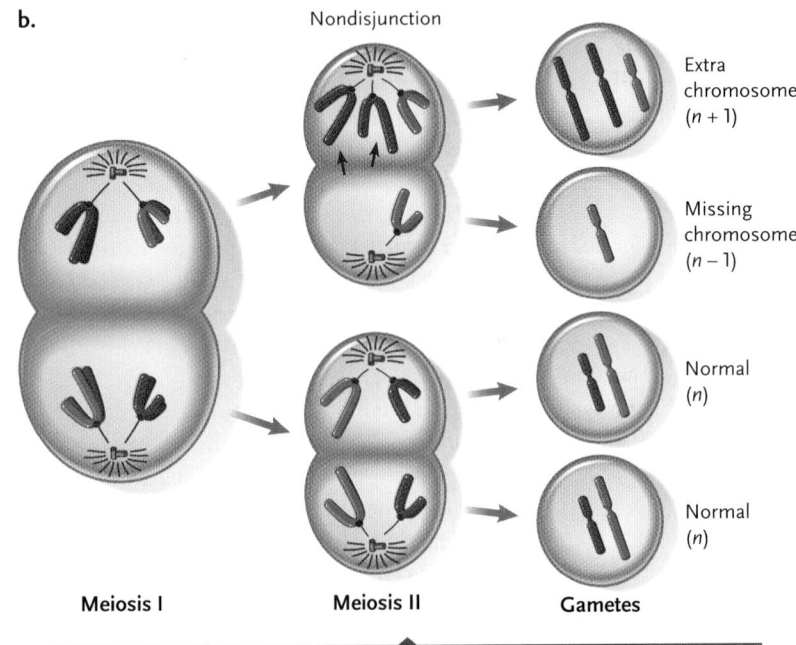

b.

Nondisjunction during the second meiotic division produces two normal gametes, one gamete with an extra chromosome and one gamete with a missing chromosome.

Figure 13.12
Nondisjunction during **(a)** the first meiotic division and **(b)** the second meiotic division.

meiotic division **(Figure 13.12)**. As a result, gametes are produced that lack one or more chromosomes or contain extra copies of the chromosomes. Fertilization by these gametes produces an individual with extra or missing chromosomes. Such individuals are called **aneuploids**, whereas individuals with a normal set of chromosomes are called **euploids.**

Changes in chromosome number can also occur through duplication of entire sets, meaning individuals may receive one or more extra copies of the entire

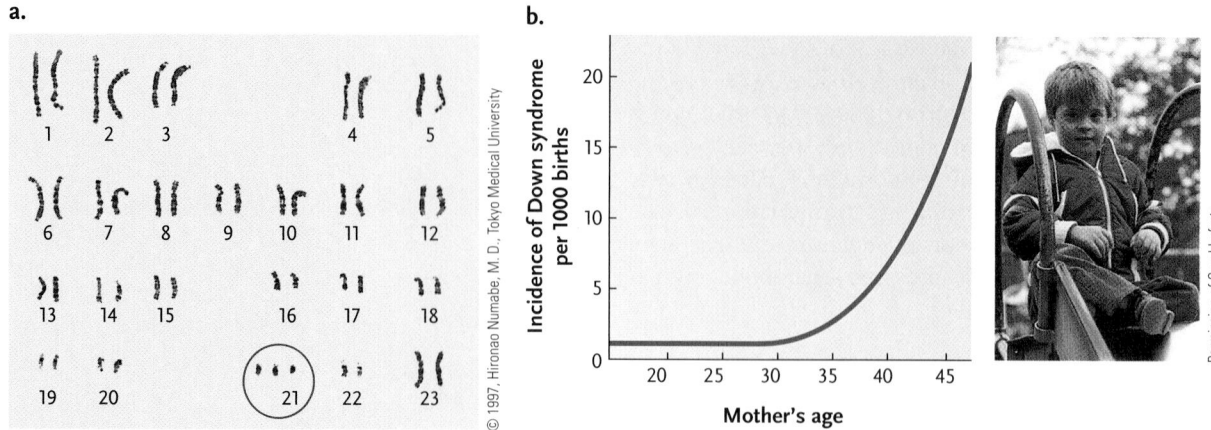

Figure 13.13

Down syndrome. **(a)** The chromosomes of a human female with Down syndrome showing three copies of chromosome 21 (circled in red). **(b)** The increase in the incidence of Down syndrome with increasing age of the mother, from a study conducted in Victoria, Australia, between 1942 and 1957.

haploid complement of chromosomes. Such individuals are called **polyploids**. *Triploids* have three copies of each chromosome instead of two; *tetraploids* have four copies of each chromosome. Multiples higher than tetraploids also occur.

Aneuploids. The effects of addition or loss of whole chromosomes vary depending on the chromosome and the species. In animals, aneuploidy of autosomes usually produces debilitating or lethal developmental abnormalities. These abnormalities also occur in humans; addition or loss of an autosomal chromosome causes embryos to develop so abnormally that they are aborted naturally. For reasons that are not understood, aneuploidy is as much as 10 times more frequent in humans than in other mammals. Of human embryos that have been miscarried and examined, about 70% are aneuploids.

In some cases, autosomal aneuploids survive. This is the case with humans who receive an extra copy of chromosome 21—one of the smallest chromosomes **(Figure 13.13a).** Many of these individuals survive until young adulthood. The condition produced by the extra chromosome, called *Down syndrome* or *trisomy 21,* is characterized by short stature and moderate to severe mental retardation. About 40% of individuals with Down syndrome have heart defects, and skeletal development is slower than normal. Most do not mature sexually and remain sterile. However, with attentive care and special training, individuals with Down syndrome can participate with reasonable success in many activities.

Down syndrome arises from nondisjunction of chromosome 21 during the meiotic divisions, primarily in women (about 5% of nondisjunctions that lead to Down syndrome occur in men). The nondisjunction occurs more frequently as women age, increasing the chance that a child may be born with the syndrome **(Figure 13.13b).** In all, 1 in every 1000 children is born with Down syndrome in the United States, making it one of the most common serious human genetic defects.

Aneuploidy of sex chromosomes can also arise by nondisjunction during meiosis **(Figure 13.14** and **Table 13.1).** Unlike autosomal aneuploidy, which usually has drastic effects on survival, altered numbers of X and Y chromosomes are often tolerated, producing indi-

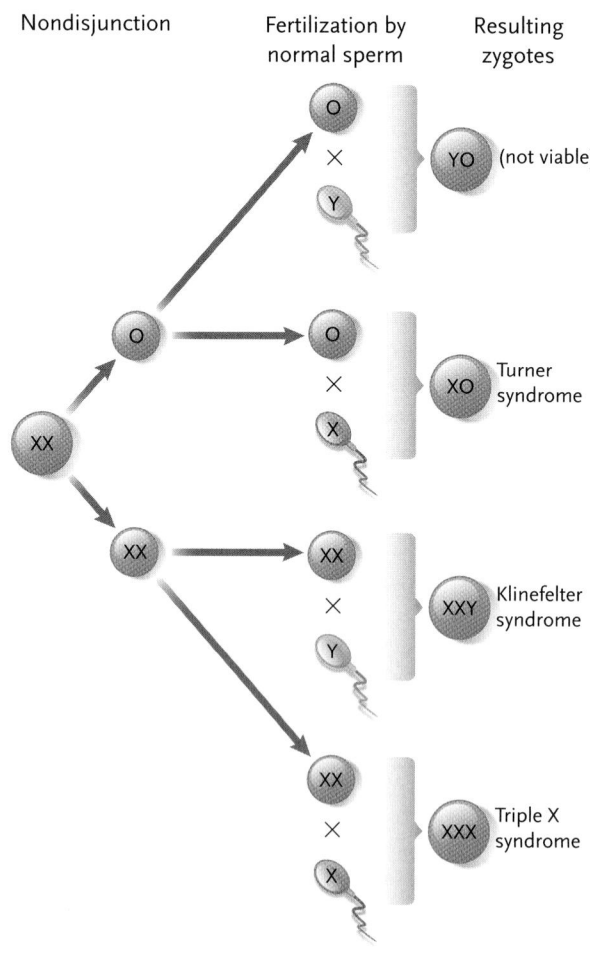

Figure 13.14

Some abnormal combinations of sex chromosomes resulting from nondisjunction of X chromosomes in females.

viduals who progress through embryonic development and grow to adulthood. In the case of multiple X chromosomes, the X-chromosome inactivation mechanism converts all but one of the X chromosomes to a Barr body, so the dosage of active X-chromosome genes is the same as in normal XX females and XY males. However, X chromosomes are not inactivated until about 15 to 16 days after fertilization. Expression of the extra X chromosome genes early in development results in some deterious effects.

Because sexual development in humans is pushed toward male or female reproductive organs primarily by the presence or absence of the Y chromosome, people with a Y chromosome are externally malelike, no matter how many X chromosomes are present. If no Y chromosome is present, X chromosomes in various numbers give rise to femalelike individuals. (Table 13.1 lists the effects of some alterations in sex chromosome number.) Similar abnormal combinations of sex chromosomes also occur in other animals, including *Drosophila*, with varying effects on viability.

Polyploids. Polyploidy often originates from failure of the spindle to function normally during mitosis in cell lines leading to germ-line cells. In these divisions, the spindle fails to separate the duplicated chromosomes, which are incorporated into a single nucleus with twice the usual number of chromosomes. Eventually, meiosis takes place and produces gametes with two copies of each chromosome instead of one. Fusion of one such gamete with a normal haploid gamete produces a triploid, and fusion of two such gametes produces a tetraploid.

The effects of polyploidy vary widely between plants and animals. In plants, polyploids are often hardier and more successful in growth and reproduction than the diploid plants from which they were derived. As a result, polyploidy is common and has been an important source of variability in plant evolution. About half of all flowering plant species are polyploids, including important crop plants such as wheat and other cereals, cotton, strawberries, and bananas.

By contrast, among animals, polyploidy is uncommon because it usually has lethal effects during embryonic development. For example, in humans, all but about 1% of polyploids die before birth, and the few who are born die within a month. The lethality is probably due to disturbance of animal developmental pathways, which are typically much more complex than those of plants.

We now turn to a description of the effects of altered alleles on human health and development.

STUDY BREAK

What mechanisms are responsible for (a) duplication of a chromosome segment, (b) generation of a Down syndrome individual, (c) a chromosome translocation, and (d) polyploidy?

Table 13.1	Effects of Unusual Combinations of Sex Chromosomes in Humans	
Combination of Sex Chromosomes	Approximate Frequency	Effects
XO	1 in 5000 births	Turner syndrome: females with underdeveloped ovaries; sterile; intelligence and external genitalia are normal; typically, individuals are short in stature with underdeveloped breasts
XXY	1 in 2000 births	Klinefelter syndrome: male external genitalia with very small and underdeveloped testes; sterile; intelligence usually normal; sparse body hair and some development of the breasts; similar characteristics in XXXY and XXXXY individuals
XYY	1 in 1000 births	XYY syndrome: apparently normal males but often taller than average
XXX	1 in 1000 births	Triple-X syndrome: apparently normal female with normal or slightly retarded mental function

13.4 Human Genetics and Genetic Counseling

We have already noted a number of human genetic traits and conditions caused by mutant alleles or chromosomal alterations (see **Table 13.2** for a more detailed list). All these traits are of interest as examples of patterns of inheritance that amplify and extend Mendel's basic principles. Those with harmful effects are also important because of their impact on human life and society.

In Autosomal Recessive Inheritance, Heterozygotes Are Carriers and Homozygous Recessives Are Affected by the Trait

Sickle-cell anemia and cystic fibrosis are examples of human traits caused by recessive alleles on autosomes. Many other human genetic traits follow a similar pattern of inheritance (see Table 13.2). These traits are passed on according to the pattern known as **autosomal recessive inheritance**, in which individuals who are homozygous for the dominant allele are free of symptoms and are not carriers; heterozygotes are usually symptom free but are carriers. People who are homozygous for the recessive allele show the trait.

For sickle-cell anemia, between 10% and 15% of African Americans in the United States are carriers—that is, they have sickle-cell trait (see Section 12.2). Although carriers make enough normal hemoglobin through the activity of the dominant allele to be essentially unaffected, the mutant, sickle-cell form of the hemoglobin molecule is also present in their red blood

Table 13.2	Examples of Human Genetic Traits
Trait	**Adverse Health Effects**
Autosomal Recessive Inheritance	
Albinism	Absence of pigmentation (melanin)
Attached ear lobes	None
Cystic fibrosis	Excess mucus in lungs and digestive cavities
Sickle-cell disease	Severe tissue and organ damage
Galactosemia	Brain, liver, and eye damage
Phenylketonuria	Mental retardation
Tay–Sachs disease	Mental retardation, death
Autosomal Dominant Inheritance	
Free ear lobes	None
Achondroplasia	Defective cartilage formation that causes dwarfism
Early balding in males	None
Campodactyly	Rigid, bent little fingers
Curly hair	None
Huntington disease	Progressive, irreversible degeneration of nervous system
Syndactyly	Webbing between fingers
Polydactyly	Extra digits
Brachydactyly	Short digits
Progeria	Premature aging
X-Linked Inheritance	
Hemophilia A	Deficient blood-clotting
Red–green color blindness	Inability to distinguish red from green
Testicular feminizing syndrome	Absence of male organs, sterility
Changes in Chromosome Structure	
Cri-du-chat	Mental retardation, malformed larynx
Changes in Chromosome Number	
Down syndrome	Mental retardation, heart defects

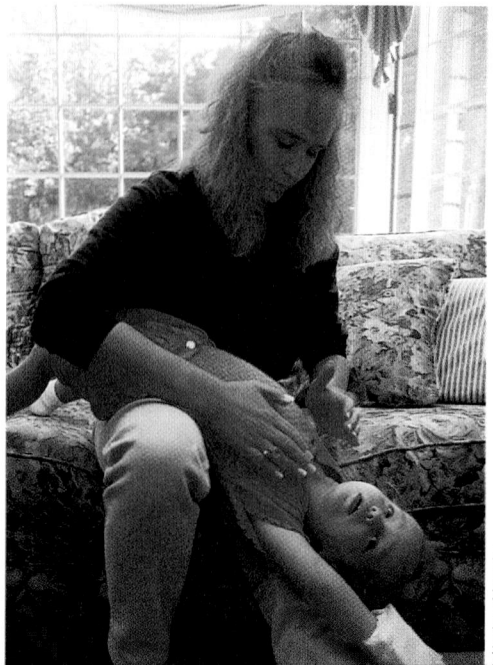

© Abraham Menashe

Figure 13.15

A child affected by cystic fibrosis. Daily chest thumps, back thumps, and repositioning dislodge thick mucus that collects in airways to the lungs.

cells. Carriers can be identified by a simple test for the mutant hemoglobin. In countries where malaria is common, including several countries in Africa, carriers are less susceptible to contraction of the disease, which helps explain the increased proportions of the recessive allele among races that originated in malarial areas.

Cystic fibrosis (CF), one of the most common genetic disorders among persons of Northern European descent, is another autosomal recessive trait **(Figure 13.15).** About 1 in every 25 people from this line of descent is an unaffected carrier with one copy of the recessive allele, and about 1 in 2500 is homozygous for the recessive allele. The homozygous recessives have an altered membrane transport protein that results in excess Cl^- (chloride ions) to the extracellular fluids. Through pathways that are not completely understood, the alteration in chloride transport causes thick, sticky mucus to collect in airways of the lungs, in the ducts of glands such as the pancreas, and in the digestive tract. The accumulated mucus impairs body functions and, in the lungs, promotes pneumonia and other infections. With current management procedures, the life expectancy for a person with cystic fibrosis is about 40 years.

Another autosomal recessive disease, *phenylketonuria* (PKU), appears in about 1 of every 15,000 births. Affected individuals cannot produce an enzyme that converts the amino acid phenylalanine to another amino acid, tyrosine. As a result, phenylalanine builds up in the blood and is converted in the body into other products, including phenylpyruvate. Elevations in both phenylalanine and phenylpyruvate damage brain tissue and can lead to mental retardation. If diagnosed early enough, an affected infant can be placed on a phenylalanine-restricted diet, which can prevent the PKU symptoms. Most U.S. hospitals routinely test newborn infants to detect the disorder.

In Autosomal Dominant Inheritance, Only Homozygous Recessives Are Unaffected

Some human traits follow a pattern of **autosomal dominant inheritance** (see Table 13.2). In this case, the allele that causes the trait is dominant, and people who

Achondroplastic Dwarfing by a Single Amino Acid Change

Researchers recently found that the gene responsible for achondroplastic dwarfing is on chromosome 4. The same chromosome is known to carry a gene encoding a receptor that binds human growth hormone (receptors are proteins in the plasma membrane; when they bind a hormone, they undergo a change in their three-dimensional structure that triggers a response inside the cell). An intriguing question was whether the achondroplastic gene is a mutant form of the gene that encodes the growth hormone receptor. Arnold Munnich and his colleagues at the Hospital of Children's Diseases in Paris, France, performed molecular experiments designed to answer this question.

The receptor binds the *fibroblast growth factor (FGF)*, a growth hormone that stimulates a wide range of mammalian cells to grow and divide. The many receptors that bind this hormone form a family called the *fibroblast growth factor receptors (FGFRs)*. The gene that encodes the

FGFR on chromosome 4 is active in chondrocytes—cells that form cartilage and bone.

Munnich and his colleagues isolated the gene that encodes the FGFR and obtained its DNA sequence. They found two versions of the gene's sequence with a single difference—one version had an adenine-thymine (A-T) base pair and the other had a guanine-cytosine (G-C) base pair at the same position in the DNA sequence. The change substitutes arginine for glycine at one position in the amino acid sequence of the encoded protein. These amino acids have very different chemical properties. The substitution occurs in a segment of the protein that extends across the membrane, connecting a hormone-binding site outside the cell with a site inside the cell that triggers the internal response.

The investigators then looked for the A-T–to–G-C substitution in the mutant form of the gene on chromosome 4 that causes achondroplastic dwarfing. The substitution was present

in copies of the gene isolated from 6 families of achondroplastic dwarfs, but absent in 120 people who lack the trait. This result supported the hypothesis that a mutation in the gene that codes for FGFR protein is responsible for achondroplastic dwarfing.

How does the single amino acid substitution cause dwarfing? The cause is not known exactly. The change may inhibit the transmission of the signal triggered by a hormone binding to the receptor on the outer membrane. Evidently, if the signal is not transmitted, the chondrocyte cells fail to respond to the hormone by growing and dividing, interfering with elongation of the limb bones and producing the short arms and legs characteristic of achondroplastic dwarfs.

Identification of the gene responsible for achondroplastic dwarfing opens the future to finding a cure for the condition, possibly through genetic engineering of infants or young children who carry the mutation.

---

are either homozygous or heterozygous for the dominant allele are affected. Individuals homozygous for the recessive normal allele are unaffected.

Achondroplasia, a type of dwarfing that occurs in about 1 in 10,000 people, is caused by an autosomal dominant allele of a gene on chromosome 4. Of individuals with the dominant allele, only heterozygotes survive embryonic development; homozygous dominants are usually stillborn. When limb bones develop in heterozygous children, cartilage formation is defective, leading to disproportionately short arms and legs. The trunk and head, however, are of normal size. Affected adults are usually not much more than 4 feet tall. Achondroplastic dwarfs are of normal intelligence, are fertile, and can have children. The gene responsible for this trait has been identified (see *Insights from the Molecular Revolution*).

Males Are More Likely to Be Affected by X-Linked Recessive Traits

Red–green color blindness and hemophilia have already been presented as examples of human traits that demonstrate **X-linked recessive inheritance**, that is, traits due to inheritance of recessive alleles carried on the X chro-

mosome. Another X-linked recessive human disease trait is Duchenne muscular dystrophy. In affected individuals, muscle tissue begins to degenerate late in childhood; by the onset of puberty, most individuals with this disease are unable to walk. Muscular weakness progresses, with later involvement of the heart muscle; the average life expectancy for individuals with Duchenne muscular dystrophy is 25 years. (*Insights from the Molecular Revolution* in Chapter 41 discusses molecular research that could lead to a cure for the disease.)

Human Genetic Disorders Can Be Predicted, and Many Can Be Treated

Of all newborns, between 1% and 3% have mutant alleles that encode defective forms of proteins required for normal functions. Possibly 1% have pronounced difficulties due to a chromosomal rearrangement or other aberration. Of all patients in children's hospitals, 10% to 25% are treated for problems arising from inherited disorders. Several approaches, which include genetic counseling, prenatal diagnosis, and genetic screening, can reduce the number of children born with genetic diseases.

Genetic counseling allows prospective parents to assess the possibility that they might have an affected child. For example, parents may seek counseling if they, a close relative, or one of their existing children has a genetic disorder. Genetic counseling begins with identification of parental genotypes through family pedigrees or direct testing for an altered protein or DNA sequence. With this information in hand, counselors can often predict the chances of having a child with the trait in question. Couples can then make an informed decision about whether to have a child.

Genetic counseling is often combined with techniques of **prenatal diagnosis**, in which cells derived from a developing embryo or its surrounding tissues or fluids are tested for the presence of mutant alleles or chromosomal alterations. In **amniocentesis**, cells are obtained from the amniotic fluid—the watery fluid surrounding the embryo in the mother's uterus. In **chorionic villus sampling**, cells are obtained from portions of the placenta that develop from tissues of the embryo. More than 100 genetic disorders can now be detected by these tests. If prenatal diagnosis detects a serious genetic defect, the prospective parents can reach an informed decision about whether to continue the pregnancy, including religious and moral considerations, as well as genetic and medical advice.

Once a child is born, inherited disorders are identified by **genetic screening**, in which biochemical or molecular tests for disorders are routinely applied to children and adults or to newborn infants in hospitals. The tests can detect inherited disorders early enough to start any available preventive measures before symptoms develop. We have noted that most hospitals in the United States now test all newborns for PKU, making it possible to use dietary restrictions to prevent symptoms of the disorder from developing. As a result, it is becoming less common to see individuals debilitated by PKU.

In addition to the characters and traits described so far in this chapter, some patterns of inheritance depend on genes located not in the cell nucleus, but in mitochondria or chloroplasts in the cytoplasm, as discussed in the following section.

STUDY BREAK

1. A man has Simpson syndrome, an addiction to a certain television show. His wife does not have this syndrome. This couple has four children, two boys and two girls. One of the boys and one of the girls has this syndrome; the other children are normal. Can Simpson syndrome be an autosomal recessive trait? A sex-linked recessive trait?

2. In another family, a female child has wiggly ears, whereas her brother does not. Both parents are normal. Can the wiggly ear trait be an autosomal recessive trait? A sex-linked recessive trait?

13.5 Nontraditional Patterns of Inheritance

We consider two examples of nontraditional patterns of inheritance in this section. In **cytoplasmic inheritance**, the pattern of inheritance follows that of genes in the cytoplasmic organelles, mitochondria or chloroplasts. In **genomic imprinting**, the expression of a nuclear gene is based on whether an individual organism inherits the gene from the male or female parent.

Cytoplasmic Inheritance Follows the Pattern of Inheritance of Mitochondria or Chloroplasts

Chapter 5 noted that both chloroplasts and mitochondria contain DNA. Organelle DNA contains genes and alleles that, like nuclear genes, are also subject to being mutated. Mutant genes in some cases result in altered phenotypes, but the inheritance pattern of these mutant genes is fundamentally different from that of mutant genes carried on chromosomes in the nucleus. The two major differences are as follows: (1) ratios typical of Mendelian segregation are not found because genes are not segregating by meiosis, and (2) genes usually show uniparental inheritance from generation to generation. In *uniparental inheritance,* all progeny (both males and females) have the phenotype of only one of the parents. For most multicellular eukaryotes, the mother's phenotype is expressed, a phenomenon called *maternal inheritance.* Maternal inheritance occurs because the amount of cytoplasm in the female gamete usually far exceeds that in the male gamete. Hence, a zygote receives most of its cytoplasm, including mitochondria and (in plants) chloroplasts, from the female parent and little from the male parent.

The first example of cytoplasmic inheritance of a mutant trait was found in 1909 by the German scientist Carl Correns, one of the geneticists who rediscovered Mendel's principles. Correns made his discovery through his genetic studies of a plant, *Mirabilis* (the four-o'clock), using mutant plants that had a variegated pattern of green and white **(Figure 13.16)**. In the white segments, chloroplasts are colorless instead of green. Correns fertilized flowers in a green region of the plant with pollen from a white region and vice versa. He discovered that the phenotype of the progeny seedlings showed maternal inheritance. That is, it was always that of the female segment the pollen fertilized: white for the white ♀ × green ♂ cross, and green for the green ♀ × white ♂ cross.

Many examples of maternal inheritance of mutant traits involving the chloroplast are now known. In each case, the trait results from a mutation of one of the genes in the chloroplast genome. Mutant traits that show maternal inheritance have also been character-

Figure 13.16
A four-o'clock *(Mirabilis)* plant with a variegated (patchy) distribution of green and white segments.

ized in many eukaryotic species, including in plants, animals, protists, and fungi. Similar to the mutant traits of chloroplasts, each mutant trait results from an alteration of a gene in the mitochondrial genome.

In humans, several inherited diseases have been traced to mutations in mitochondrial genes **(Table 13.3)**. Recall that the mitochondrion plays a critical role in synthesizing ATP, the energy source for many cellular reactions. The mutations producing the diseases in Table 13.3 are in mitochondrial genes that encode components of the ATP-generating system of the organelle. The resulting mitochondrial defects are especially destructive to the organ systems most dependent on mitochondrial reactions for energy: the central nervous system, skeletal and cardiac muscle, the liver, and the kidneys. These inherited diseases show maternal inheritance.

In Genomic Imprinting, the Allele Inherited from One of the Parents Is Expressed While the Other Allele Is Silent

Genomic imprinting is a phenomenon in which the expression of an allele of a gene is determined by the parent that contributed it. In some cases, the paternally derived allele is expressed; in others, the maternally derived allele is expressed. The silent allele—that which is not expressed—is called the *imprinted allele*. The imprinted allele is not inactivated by mutation. Rather, it is silenced by chemical modification (methylation) of certain bases in its sequence.

As an example, Prader-Willi syndrome (PWS) and Angelman syndrome (AS) in humans are each caused by genomic imprinting of a particular gene on a chromosome inherited from one parent, coincident with deletion of the same gene on the homologous chromosome inherited from the other parent. The syndromes

differ with respect to the gene imprinted. Both PWS and AS occur in about 1 in 15,000 births and are characterized by serious developmental, mental, and behavioral problems. PWS individuals are compulsive overeaters (leading to obesity), have short stature, have small hands and feet, and show mild to moderate mental retardation. AS individuals are hyperactive, are unable to speak, have seizures, show severe mental retardation, and display a happy disposition with bursts of laughter.

How is genomic imprinting responsible for these two syndromes? PWS is caused when an individual has a normal maternally derived chromosome 15 and a paternally derived chromosome 15 with a deletion of a small region of several genes that includes the PWS gene. The PWS gene is imprinted, and therefore silenced, on maternally derived chromosomes. As a result, when there is no PWS gene on the paternally derived chromosome, there is no PWS gene activity and PWS results. Similarly, AS is caused when an individual has a normal paternally derived chromosome 15 and a maternally derived chromosome 15 with a deletion of the same region; that region also includes the AS gene, the normal function of which is also required for normal development. In this case, genomic imprinting silences the AS gene on the paternally derived chromosome, and because there is no AS gene on the maternally derived chromosome, there is no AS gene activity and AS syndrome develops.

The mechanism of imprinting involves the modification of the DNA in the region that controls the expression of a gene by the addition of methyl ($—CH_3$) groups to cytosine nucleotides. The methylation of the control region of a gene prevents it from being expressed. (Note that there are a few instances of methylation-activating genes.) The regulation of gene expression by methylation of DNA is discussed further in Chapter 16. Genomic imprinting occurs in the

Table 13.3	Some Human Diseases Caused by Mutations in Mitochondrial Genes
Disease	**Symptoms**
Kearns–Sayre syndrome	May include muscle weakness, mental deficiencies, abnormal heartbeat, short stature
Leber hereditary optic neuropathy	Vision loss from degeneration of the optic nerve, abnormal heartbeat
Mitochondrial myopathy and encephalomyopathy	May include seizures, strokelike episodes, hearing loss, progressive dementia, abnormal heartbeat, short stature
Myoclonic epilepsy	Vision and hearing loss, uncoordinated movement, jerking of limbs, progressive dementia, heart defects

gametes where the allele destined to be inactive in the new embryo after fertilization—either the father's or the mother's depending on the gene—is methylated. That methylated (silenced) state of the gene is passed on as the cells grow and divide to produce the somatic (body) cells of the organism.

A number of cancers are associated with the failure to imprint genes. For instance, the mammalian *Igf2* (insulin growth factor 2) gene encodes a growth factor, a molecule that stimulates cells to grow and divide. *Igf2* is an imprinted gene, with the paternally derived allele on and the maternally derived allele off. In some cases, the imprinting mechanism for this gene does not work, resulting in both alleles of *Igf2* being active, a phenomenon known as **loss of imprinting**.

The resulting double dose of the growth factor disrupts the cell division cycle, contributing to uncontrolled growth and cancer.

In this chapter, you have learned about genes and the role of chromosomes in inheritance. In the next chapter, you will learn about the molecular structure and function of the genetic material and about the molecular mechanism by which DNA is replicated.

STUDY BREAK

What key feature or features would suggest to you that a mutant trait shows cytoplasmic inheritance?

UNANSWERED QUESTIONS

Does recombination protect against cancer?

In this chapter, we learned that genetic recombination during meiosis in germ-line cells is a mechanism for the exchange of genetic information between chromosomes. Researchers have discovered that genetic recombination also occurs in somatic cells and is a vital mechanism for the repair of damaged or broken chromosomes. If recombination is defective, unrepaired chromosome breaks or gene translocations can have serious consequences to the cell and even lead to cancer. For instance, the *BRCA1* and *BRCA2* genes, which predispose patients to breast cancer, have recombination and repair defects that lead to genome instability. Researchers are using tissue culture cells derived from patients with these genes to find out exactly what goes wrong when recombination is inadequate and how recombination may act to maintain genome stability and provide protection against cancer and other diseases.

How did the *SRY* gene on the Y chromosome evolve?

The key gene on the Y chromosome, which determines the sex of a mammal, is the *SRY* gene. How did this gene evolve? Research is turning up information that it probably evolved from a brain-determining gene on the X chromosome. How did that occur? What other genes are involved in the sex-determining pathway, and how do they interact to coordinate the important developmental events in sex determination? And what genes on the Y chromosome affect male fertility? Researchers are attempting to answer these questions.

How are X chromosomes counted in the mammalian X-chromosome inactivation system?

There is a difference in the number of X chromosomes in female and male mammals. The extra dosage of genes in females is compensated for by the X-chromosome inactivation mechanism (see earlier discussion). Remarkably, this mechanism operates no matter how many X chromosomes are present. That is, in an XXXY cell, all but one of the X chromosomes will be inactivated. How does the cell count the number of X chromosomes? Some information about this exists, but the full molecular details need to be unraveled. Jeannie Lee, a Howard Hughes Medical Institute (HHMI) investigator at Harvard Medical School, is conducting research toward this end.

Peter J. Russell

Review

Go to **Thomson**NOW™ at www.thomsonedu.com/login to access quizzing, animations, exercises, articles, and personalized homework help.

13.1 Genetic Linkage and Recombination

- Genes, consisting of sequences of nucleotides in DNA, are arranged linearly in chromosomes.
- Genes carried on the same chromosome are linked together in their transmission from parent to offspring. Linked genes are inherited in patterns similar to those of single genes, except for changes in the linkage due to recombination (Figure 13.2).
- In recombination, alleles linked on the same chromosome are mixed into new combinations by exchange of segments be-

tween the chromosomes of a homologous pair. The exchanges occur while homologous chromosomes during prophase I of meiosis.

- The amount of recombination between any two genes located on the same chromosome pair reflects the distance between them on the chromosome. The greater this distance, the greater the chance that chromatids will exchange segments at points between the genes and the greater the recombination frequency.
- The relationship between separation and recombination frequencies is used to produce chromosome maps in which genes are assigned relative locations with respect to each other (Figure 13.4).

13.2 Sex-Linked Genes

- Sex linkage is a pattern of inheritance produced by genes carried on sex chromosomes: chromosomes that differ in males and females. In humans and fruit flies, which have XX females and XY males, most sex-linked genes are carried on the X chromosome.

- Since males have only one X chromosome, they need to receive only one copy of a recessive allele from their mothers to develop the trait. Females must receive two copies of the recessive allele, one from each parent, to develop the trait (Figures 13.6–13.8).

- In mammals, inactivation of one of the two X chromosomes in cells of the female makes the dosage of X-linked genes the same in males and females (Figure 13.10).

Animation: Human sex determination

Animation: Morgan's reciprocal crosses

13.3 Chromosomal Alterations That Affect Inheritance

- Inheritance is influenced by processes that delete, duplicate, or invert segments within chromosomes, or translocate segments between chromosomes (Figure 13.11).

- Chromosomes also change in number by addition or removal of individual chromosomes or entire sets. Changes in single chromosomes usually occur through nondisjunction, in which homologous pairs fail to separate during meiosis I, or sister chromatids fail to separate during meiosis II. As a result, one set of gametes receives an extra copy of a chromosome and the other set is deprived of the chromosome.

- Polyploids have one or more extra copies of the entire chromosome set. Polyploids usually arise when the spindle fails to function during mitosis in cell lines leading to gamete formation, producing gametes that contain double the number of chromosomes typical for the species (Figures 13.12–13.14).

Animation: Karyotype preparation

Animation: Duplication

Animation: Deletion

Animation: Inversion

Animation: Translocation

13.4 Human Genetics and Genetic Counseling

- Three modes of inheritance are most significant in human heredity: autosomal recessive, autosomal dominant, and X-linked recessive inheritance.

- In autosomal recessive inheritance, males or females carry a recessive allele on an autosome. Heterozygotes are carriers that are usually unaffected, but homozygous individuals show symptoms of the trait.

- In autosomal dominant inheritance, a dominant gene is carried on an autosome. Individuals that are homozygous or heterozygous for the trait show symptoms of the trait; homozygous recessives are normal.

- In X-linked recessive inheritance, a recessive allele for the trait is carried on the X chromosome. Male individuals with the recessive allele on their X chromosome or female individuals with the recessive allele on both X chromosomes show symptoms of the trait. Heterozygous females are carriers but usually show no symptoms of the trait.

- Genetic counseling, based on identification of parental genotypes by constructing family pedigrees and prenatal diagnosis, allow prospective parents to reach an informed decision about whether to have a child or continue a pregnancy.

Animation: Autosomal dominant inheritance

Animation: Autosomal recessive inheritance

Animation: X-linked inheritance

Animation: Pedigree diagrams

Animation: Amniocentesis

13.5 Nontraditional Patterns of Inheritance

- Cytoplasmic inheritance depends on genes carried on DNA in mitochondria or chloroplasts. Cytoplasmic inheritance follows the maternal line: it parallels the inheritance of the cytoplasm in fertilization, in which most or all of the cytoplasm of the zygote originates from the egg cell (Figure 13.16).

- Genomic imprinting is a phenomenon in which the expression of an allele of a gene is determined by the parent that contributed it. In some cases, the allele inherited from the father is expressed; in others, the allele from the mother is expressed. Commonly, the silencing of the other allele is the result of methylation of the region adjacent to the gene that is responsible for controlling the expression of that gene.

Questions

Problems

1. In humans, red–green color blindness is an X-linked recessive trait. If a man with normal vision and a color-blind woman have a son, what is the chance that the son will be color-blind? What is the chance that a daughter will be color-blind?

2. The following pedigree shows the pattern of inheritance of red–green color blindness in a family. Females are shown as circles and males as squares; the squares or circles of individuals affected by the trait are filled in black.

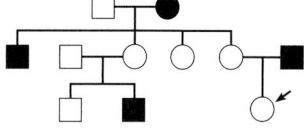

What is the chance that a son of the third-generation female indicated by the arrow will be color-blind if the father is a normal man? If the father is color-blind?

3. Individuals affected by a condition known as polydactyly have extra fingers or toes. The following pedigree shows the pattern of inheritance of this trait in one family:

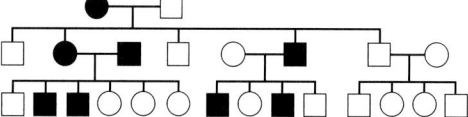

From the pedigree, can you tell if polydactyly comes from a dominant or recessive allele? Is the trait sex-linked? As far as you can determine, what is the genotype of each person in the pedigree with respect to the trait?

4. A number of genes carried on the same chromosome are tested and show the following crossover frequencies. What is their sequence in the map of the chromosome?

Genes	Crossover Frequencies between Them
C and *A*	7%
B and *D*	3%
B and *A*	4%
C and *D*	6%
C and *B*	3%

5. In *Drosophila*, two genes, one for body color and one for eye color, are carried on the same chromosome. The wild-type gray body color is dominant to black body color, and wild-type red eyes are dominant to purple eyes. You make a cross between a fly with gray body and red eyes and a fly with black body and purple eyes. Among the offspring, about half have gray bodies and red eyes and half have black bodies and purple eyes. A small percentage have (a) black bodies and red eyes or (b) gray bodies and purple eyes. What alleles are carried together on the chromosomes in each of the flies used in the cross? What alleles are carried together on the chromosomes of the F₁ flies with black bodies and red eyes, and those with gray bodies and purple eyes?

6. Another gene in *Drosophila* determines wing length. The dominant wild-type allele of this gene produces long wings; a recessive allele produces vestigial (short) wings. A female that is true-breeding for red eyes and long wings is mated with a male that has purple eyes and vestigial wings. F₁ females are then crossed with purple-eyed, vestigial-winged males. From this second cross, a total of 600 offspring are obtained with the following combinations of traits:

    252 with red eyes and long wings
    276 with purple eyes and vestigial wings
    42 with red eyes and vestigial wings
    30 with purple eyes and long wings

Are the genes linked, unlinked, or sex-linked? If they are linked, how many map units separate them on the chromosome?

Drosophila with vestigial wings

7. One human gene, which is suspected to be carried on the Y chromosome, controls the length of hair on men's ears. One allele produces nonhairy ears, and another produces hairy ears. If a man with hairy ears has sons, what percentage will also have hairy ears? What percentage of his daughters will have hairy ears?

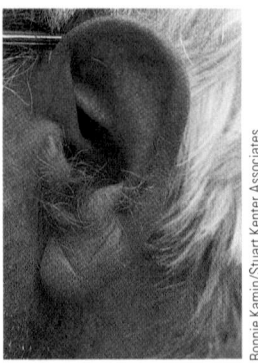

Male with hairy ears

8. You conduct a cross in *Drosophila* that produces only half as many male as female offspring. What might you suspect as a cause?

Questions for Discussion

1. Can a linkage map be made for a haploid organism that reproduces sexually?

2. Crossing-over does not occur between any pair of homologous chromosomes during meiosis in male *Drosophila*. From what you have learned about meiosis and crossing-over, propose one hypothesis for why this might be the case.

3. Even though X inactivation occurs in XXY (Klinefelter syndrome) humans, they do not have the same phenotype as normal XY males. Similarly, even though X inactivation occurs in XX individuals, they do not have the same phenotype as XO (Turner syndrome) humans. Why might this be the case?

4. All mammals have evolved from a common ancestor. However, the chromosome number varies among mammals. By what mechanism might this have occurred?

Experimental Analysis

Assume that genes *a, b, c, d, e,* and *f* are linked. Explain how you would construct a linkage map that shows the order of these six genes and the map units between them.

Evolution Link

How would the effects of natural selection differ on alleles that cause diseases fatal in childhood (such as progeria) and those that cause diseases that shorten life expectancy to 40 or 50 years (such as cystic fibrosis)?

How Would You Vote?

Advances in genetics have led to our ability to detect mutant genes that cause medical disorders in human embryos and fetuses. Should society encourage women to give birth only if their child will not develop severe medical problems? How severe? Go to www.thomsonedu.com/login to investigate both sides of the issue and then vote.

A digital model of DNA (based on data generated by X-ray crystallography).

STUDY PLAN

14 DNA Structure, Replication, and Organization

WHY IT MATTERS

One might have wondered, in the spring of 1868, why Johann Friedrich Miescher, a Swiss physician and physiological chemist, was collecting pus cells from discarded bandages. His intentions were purely scientific: Miescher wanted to study the chemical composition of the cell nucleus. He used pus cells because much of the volume of the white blood cells in pus is occupied by the nucleus. From the nuclei of these cells, Miescher extracted large quantities of an acidic substance with a high phosphorus content. He called the unusual substance "nuclein." His discovery is at the root of the development of our molecular understanding of life: nuclein is now known by its modern name, **deoxyribonucleic acid**, or **DNA**, the molecule that is the genetic material of all living organisms.

At the time of Miescher's discovery, scientists knew nothing about the molecular basis of heredity and very little about genetics. Although Mendel had already published the results of his genetic experiments with garden peas, the significance of his findings was not widely known or appreciated. It was not known which chemical substance in cells actually carries the instructions for reproducing parental traits in offspring. Not until 1952, more than 80 years after

Figure 14.1
James D. Watson and Francis H. C. Crick demonstrating their 1953 model for DNA structure, which revolutionized the biological sciences.

A. C. Barrington Brown © 1968 J. D. Watson

Miescher's discovery, did scientists fully recognize that the hereditary molecule was DNA.

After DNA was established as the hereditary molecule, the focus of research changed to the three-dimensional structure of DNA. Among the scientists striving to work out the structure were James D. Watson, a young American postdoctoral student at Cambridge University in England, and the Englishman Francis H. C. Crick, then a graduate student at Cambridge University. Using chemical and physical information about DNA, in particular Rosalind Franklin's analysis of the arrangement of atoms in DNA, the two investigators assembled molecular models from pieces of cardboard and bits of wire. Eventually they constructed a model for DNA that fit all the known data **(Figure 14.1).** Their discovery was of momentous importance in biology. The model enabled scientists to understand key processes in cells for the first time in terms of the structure and interaction of molecules. For example, the model immediately made it possible to understand how genetic information is stored in the structure of DNA and how DNA replicates. Unquestionably, the discovery launched a molecular revolution within biology, making it possible for the first time to relate the genetic traits of living organisms to a universal molecular code present in the DNA of every cell. In addition, Watson and Crick's discovery opened the way for numerous advances in fields such as medicine, forensics, pharmacology, and agriculture, and eventually gave rise to the current rapid growth of the biotechnology industry.

14.1 Establishing DNA as the Hereditary Molecule

In the first half of the twentieth century, many scientists believed that proteins were the most likely candidates for the hereditary molecules because they ap-peared to offer greater opportunities for information coding than did nucleic acids. That is, proteins contain 20 types of amino acids, whereas nucleic acids have only 4 different nitrogenous bases available for coding. Other scientists believed that nucleic acids were the hereditary molecules. In this section, we describe the experiments showing that DNA, and not protein, is the genetic material.

Experiments Began When Griffith Found a Substance That Could Genetically Transform Pneumonia Bacteria

In 1928, Frederick Griffith, a British medical officer, observed an interesting phenomenon in his experiments with the bacterium *Streptococcus pneumoniae*, which causes a severe form of pneumonia in mammals. Griffith was trying a make a vaccine to prevent pneumonia infections in the epidemics that occurred after World War I. He used two strains of the bacterium in his attempts. The smooth strain—*S*—has a polysaccharide capsule surrounding each cell and forms colonies that appear smooth and glossy when grown on a culture plate. When he injected the *S* strain into mice, it was virulent (highly infective, or pathogenic), causing pneumonia and killing the mice in a day or two **(Figure 14.2,** step 1).The rough strain—*R*—does not have a polysaccharide capsule and forms colonies with a non-shiny, rough appearance. When Griffith injected the *R* strain into mice, it was avirulent (not infective, or nonpathogenic); the mice lived (step 2). Evidently the capsule was responsible for the virulence of the *S* strain.

If Griffith killed the *S* bacteria by heating before injecting them into the mice, the mice remained healthy (step 3). However, quite unexpectedly, Griffith found that if he injected living *R* bacteria along with the heat-killed *S* bacteria, many of the mice died (step 4). Also, he was able to isolate living *S* bacteria with polysaccharide capsules from the infected mice. In some way, living *R* bacteria had acquired the ability to make the polysaccharide capsule from the dead *S* bacteria, and they had changed—transformed—into virulent *S* cells. The transformed bacteria were altered permanently; the smooth, infective trait was stably inherited by the descendants of the transformed bacteria. Griffith called the conversion of *R* bacteria to *S* bacteria *transformation* and called the agent responsible the *transforming principle*. What was the nature of the molecule responsible for the transformation? The most likely candidates were proteins or nucleic acids.

Avery and His Coworkers Identified DNA as the Molecule That Transforms Rough *Streptococcus* to the Infective Form

In the 1940s, Oswald Avery, a physician and medical researcher at the Hospital at Rockefeller Institute for Medical Research, and his coworkers Colin MacLeod

Figure 14.2 Experimental Research

Griffith's Experiment with Infective and Noninfective Strains of Streptococcus pneumoniae

QUESTION: What is the nature of the genetic material?

EXPERIMENT: Frederick Griffith studied the conversion of a nonvirulent (noninfective) R form of the bacterium *Streptococcus pneumoniae* to a virulent (infective) S form. The S form has a capsule surrounding the cell, giving colonies of it on a laboratory dish a smooth, shiny appearance. The R form has no capsule, so the colonies have a rough, nonshiny appearance. Griffith injected the bacteria into mice and determined how the mice were infected.

1. Mice injected with live, infective S cells (control to show effect of S cells)

2. Mice injected with live, noninfective R cells (control to show effect of R cells)

3. Mice injected with heat-killed S cells (control to show effect of dead S cells)

4. Mice injected with heat-killed S cells plus live R cells

 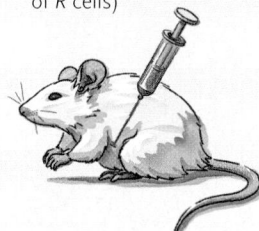 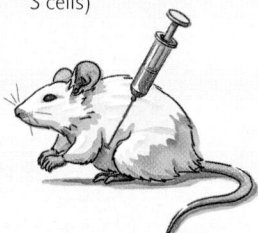 

RESULT: Mice die. Live, infective S cells in their blood; shows that S cells are virulent.

RESULT: Mice live. No live R cells in their blood; shows that R cells are nonvirulent.

RESULT: Mice live. No live S cells in their blood; shows that live S cells are necessary to be virulent to mice.

RESULT: Mice die. Live S cells in their blood; shows that living R cells can be converted to virulent S cells with some factor from dead S cells.

CONCLUSION: Griffith concluded that some molecules released when S cells were killed could change living R cells genetically to the virulent S form. He called the molecule the *transforming principle* and the process of genetic change *transformation*.

---

and Maclyn McCarty performed an experiment designed to identify the chemical nature of the transforming principle that can change R *Streptococcus* bacteria into the S infective form. Rather than working with mice, they attempted to reproduce the transformation using bacteria growing in culture tubes. They used heat to kill virulent S bacteria and then treated the macromolecules extracted from the cells in turn with enzymes that break down each of the three main candidate molecules for the hereditary material—protein; DNA; or the other nucleic acid, RNA. When they destroyed proteins or RNA, the researchers saw no effect; the extract of S bacteria still transformed R bacteria into virulent S bacteria—the cells had polysaccharide capsules and produced smooth colonies on culture plates. When they destroyed DNA, however, no transformation occurred—no smooth colonies were seen on culture plates.

In 1944, Avery and his colleagues published their discovery that the transforming principle was DNA. At the time, many scientists firmly believed that the genetic material was protein. So, although their findings were clearly revolutionary, Avery and his colleagues presented their conclusions in the paper cautiously,

offering several interpretations of their results. Some scientists accepted their results almost immediately. However, those who believed that the genetic material was protein argued that it was possible not all protein was destroyed by their enzyme treatments and, as contaminants in their DNA transformation reaction, these remaining proteins were in fact responsible for the transformation. Further experiments were needed to convince all scientists that DNA is the hereditary molecule.

Hershey and Chase Found the Final Evidence Establishing DNA as the Hereditary Molecule

A final series of experiments conducted in 1952 by bacteriologist Alfred D. Hershey and his laboratory assistant Martha Chase at the Cold Spring Harbor Laboratory removed any remaining doubts that DNA is the hereditary molecule. Hershey and Chase studied the infection of the bacterium *Escherichia coli* by bacteriophage T2. *E. coli* is a bacterium normally found in the intestines of mammals. **Bacteriophages** (or simply **phages**; see Chapter 17) are viruses that infect bacte-

Figure 14.3 Experimental Research

The Hershey and Chase Experiment Demonstrating That DNA Is the Hereditary Molecule

QUESTION: Is DNA or protein the genetic material?

EXPERIMENT: Hershey and Chase performed a definitive experiment to show whether DNA or protein is the genetic material. They used phage T2 for their experiment; it consists only of DNA and protein.

1. They infected *E. coli* growing in the presence of radioactive ^{32}P or ^{35}S with phage T2. The progeny phages were either labeled in their DNA with ^{32}P or in their protein with ^{35}S.

2. Fresh *E. coli* cells were infected with the radioactively labeled phages.

3. After infecting the bacteria, the cells were mixed in a blender to remove the phage coats from the cell surface. The components were analyzed for radioactivity.

4. Progeny phages analyzed for radioactivity.

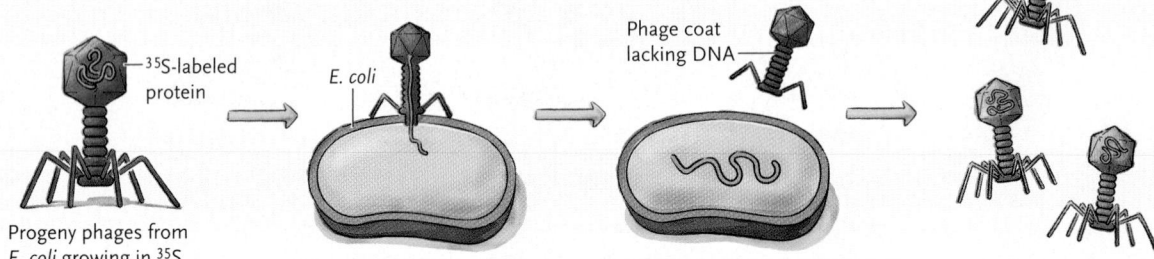

^{35}S-labeled protein

E. coli

Phage coat lacking DNA

Progeny phages from *E. coli* growing in ^{35}S

RESULT: No radioactivity within cell; ^{35}S in phage coat

RESULT: No radioactivity in progeny phages

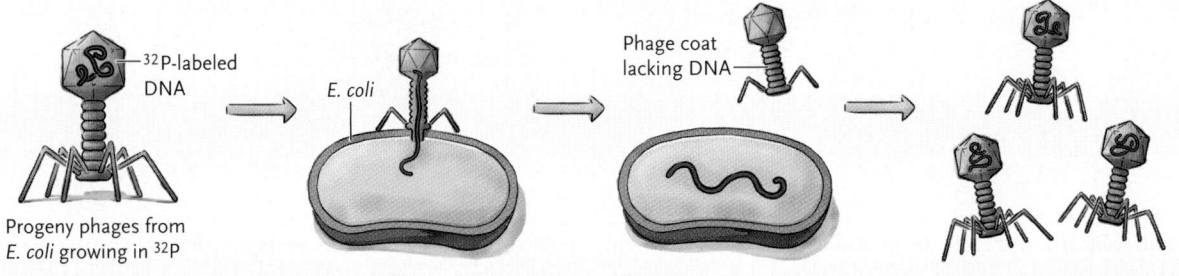

^{32}P-labeled DNA

E. coli

Phage coat lacking DNA

Progeny phages from *E. coli* growing in ^{32}P

RESULT: ^{32}P within cell; not in phage coat

RESULT: ^{32}P in progeny phages

CONCLUSION: ^{32}P, the radioisotope used to label DNA, was found within phage-infected cells and in progeny phages, indicating that DNA is the genetic material. ^{35}S, the radioisotope used to label proteins, was found in phage coats after infection, but was not found in the infected cell or in progeny phages, showing that protein is not the genetic material.

---

ria. A **virus** is an infectious agent that contains either DNA or RNA surrounded by a protein coat. Viruses cannot reproduce except in a host cell. When a virus infects a cell, it can use the cell's resources to produce more virus particles.

The phage life cycle begins when a phage attaches to the surface of a bacterium. For phages such as T2, the infected cell quickly stops producing its own molecules and instead starts making progeny phages. After about 100 to 200 phages are assembled inside the bacterial cell, a viral enzyme breaks down the cell wall,

killing the cell and releasing the new phages. The whole life cycle takes approximately 90 minutes.

The T2 phage that Hershey and Chase studied consists of only a core of DNA surrounded by proteins. Therefore, one of these molecules must be the genetic material that enters the bacterial cell and directs the infective cycle within. But which one? Hershey and Chase prepared two batches of phages, one with the protein tagged with a radioactive label and the other with the DNA tagged with a radioactive label. To obtain labeled phages, they added T2 to *E. coli* growing in the presence of either the radioactive isotope of sulfur (^{35}S) or the radioactive isotope of phosphorus (^{32}P) (**Figure 14.3,** step 1). The progeny phages produced in the ^{35}S medium had labeled proteins and unlabeled DNA because sulfur is a component of proteins but not of DNA. The phages produced in the ^{32}P medium had labeled DNA and unlabeled proteins because phosphorus is a component of DNA but not of proteins.

Hershey and Chase then infected separate cultures of *E. coli* with the two types of labeled phages (step 2). After a short period to allow the genetic material to enter the bacterial cell, they mixed the bacteria in a kitchen blender. They reasoned that only the genetic material was injected into the bacterial cell, leaving the rest of the phage outside. By mixing the cells in a blender, they could shear off the phage parts that did not enter the bacteria and collect them separately for analysis.

When they infected the bacteria with phages that contained labeled protein coats, they found no radioactivity in the bacterial cells but could easily measure it in the material removed by the blender (step 3, top). They also found no radioactivity in the progeny phages (step 4, top). However, if the infecting phages contained radioactive DNA, they found radioactivity inside the infected bacteria but none in the phage coats removed by the blender (step 3, bottom). In addition, radioactivity *was* seen in the progeny phages (step 4, bottom). The results were unequivocal: the genetic material of the phage was DNA, not protein.

When taken together, the experiments of Griffith, Avery and his coworkers, and Hershey and Chase established that DNA, not proteins, carries genetic information. Their research also established the term *transformation*, which is still used in molecular biology. **Transformation** is the conversion of a cell's hereditary type by the uptake of DNA released by the breakdown of another cell, as in the Griffith and Avery experiments. Having identified DNA as the hereditary molecule, scientists turned next to determine its structure.

Study Break

Imagine that ^{35}S labeled *both* protein and DNA, whereas ^{32}P labeled only DNA. How would Hershey and Chase's results have been different?

14.2 DNA Structure

The experiments that established DNA as the hereditary molecule were followed by a highly competitive scientific race to discover the structure of DNA. The race ended in 1953, when Watson and Crick elucidated the structure of DNA, ushering in a new era of molecular biology.

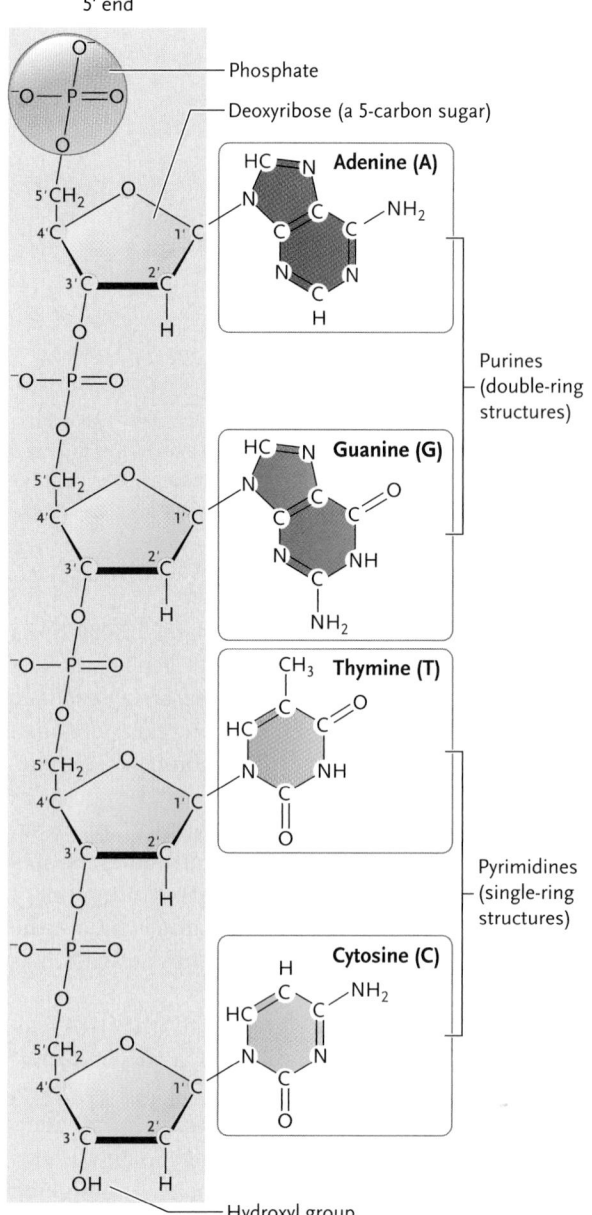

Figure 14.4

The four nucleotide subunits of DNA, linked into a polynucleotide chain. The sugar–phosphate backbone of the chain is highlighted in gray. The connection between adjacent deoxyribose sugars is a phosphodiester bond. The polynucleotide chain has polarity; at one end, the 5′ end, a phosphate group is bound to the 5′ carbon of a deoxyribose sugar, whereas at the other end, the 3′ end, a hydroxyl group is bound to the 3′ carbon of a deoxyribose sugar.

Watson and Crick Brought Together Information from Several Sources to Work Out DNA Structure

Before Watson and Crick began their research, other investigators had established that DNA contains four different nucleotides. Each nucleotide consists of the five-carbon sugar *deoxyribose* (carbon atoms on deoxyribose are numbered with primes from 1′ to 5′), a phosphate group, and one of the four nitrogenous bases—adenine (A), guanine (G), thymine (T), or cytosine (C) (**Figure 14.4**). Two of the bases, **adenine** and

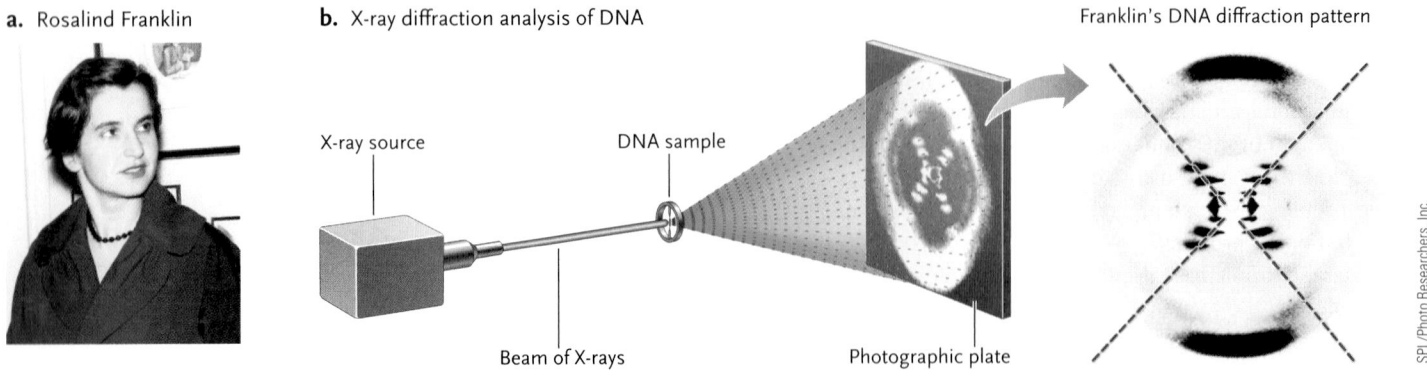

a. Rosalind Franklin

b. X-ray diffraction analysis of DNA

X-ray source

DNA sample

Beam of X-rays

Photographic plate

Franklin's DNA diffraction pattern

SPL/Photo Researchers, Inc.

Figure 14.5
X-ray diffraction analysis of DNA. **(a)** Rosalind Franklin. **(b)** The X-ray diffraction method to study DNA and the diffraction pattern Rosalind Franklin obtained. The X-shaped pattern of spots (dashed lines) was correctly interpreted by Franklin to indicate that DNA has a helical structure similar to a spiral staircase.

guanine, are *purines,* nitrogenous bases built from a pair of fused rings of carbon and nitrogen atoms. The other two bases, **thymine** and **cytosine**, are *pyrimidines,* built from a single carbon ring. An organic chemist, Erwin Chargaff, measured the amounts of nitrogenous bases in DNA and discovered that they occur in definite ratios. He observed that the amount of purines equals the amount of pyrimidines, but more specifically, the amount of adenine equals the amount of thymine, and the amount of guanine equals the amount of cytosine; these relationships are known as *Chargaff's rules.*

Researchers had also determined that DNA contains nucleotides joined to form a *polynucleotide chain.* In a polynucleotide chain, the deoxyribose sugars are linked by phosphate groups in an alternating sugar–phosphate–sugar–phosphate pattern, forming a **sugar–phosphate backbone** (highlighted in gray in Figure 14.4). Each phosphate group is a "bridge" between the 3′ carbon of one sugar and the 5′ carbon of the next sugar; the entire linkage, including the bridging phosphate group, is called a *phosphodiester bond.*

The polynucleotide chain of DNA has polarity—directionality. That is, the two ends of the chain are not the same: at one end, a phosphate group is bound to the 5′ carbon of a deoxyribose sugar, whereas at the other end, a hydroxyl group is bonded to the 3′ carbon of a deoxyribose sugar (see Figure 14.4). Consequently, the two ends are called the **5′ end** and **3′ end**, respectively.

Those were the known facts when Watson and Crick began their collaboration in the early 1950s. However, the number of polynucleotide chains in a DNA molecule and the manner in which they fold or twist in DNA were unknown. Watson and Crick themselves did not conduct experiments to study the structure of DNA; instead, they used the research data of others for their analysis, relying heavily on

data gathered by physicist Maurice H. F. Wilkins and research associate Rosalind Franklin **(Figure 14.5a),** at King's College, London. These researchers were using X-ray diffraction to study the structure of DNA **(Figure 14.5b).** In **X-ray diffraction,** an X-ray beam is directed at a molecule in the form of a regular solid, ideally in the form of a crystal. Within the crystal, regularly arranged rows and banks of atoms bend and reflect the X-rays into smaller beams that exit the crystal at definite angles determined by the arrangement of atoms in the crystal. If a photographic film is placed behind the crystal, the exiting beams produce a pattern of exposed spots. From that pattern, researchers can deduce the positions of the atoms in the crystal.

Wilkins and Franklin did not have DNA crystals with which to work, but they were able to obtain X-ray diffraction patterns from DNA molecules that had been pulled out into a fiber (see Figure 14.5). The patterns indicated that the DNA molecules within the fiber were cylindrical and about 2 nm in diameter. Separations between the spots showed that major patterns of atoms repeat at intervals of 0.34 and 3.4 nm within the DNA. Franklin interpreted an X-shaped distribution of spots in the diffraction pattern (see dashed lines in Figure 14.5) to mean that DNA has a helical structure.

The New Model Proposed That Two Polynucleotide Chains Wind into a DNA Double Helix

Watson and Crick constructed scale models of the four DNA nucleotides and fitted them together in different ways until they arrived at an arrangement that satisfied both Wilkins' and Franklin's X-ray data and Chargaff's chemical analysis. Watson and Crick's trials led them to a double-stranded model for DNA structure in which two polynucleotide chains twist around each other in a right-handed way, like a double-spiral staircase **(Figure 14.6).** They were the first to propose the famous double-helix model for DNA.

In the **double-helix model** the two sugar–phosphate backbones are separated from each other by a regular distance. The bases extend into and fill this central space. A purine and a pyrimidine, if paired together, are exactly wide enough to fill the space between the backbone chains in the double helix. However, a purine–purine base pair is too wide to fit the space exactly, and a pyrimidine–pyrimidine pair is too narrow. From Chargaff's data, Watson and Crick proposed that the purine–pyrimidine base pairs in DNA are A-T and G-C pairs. That is, wherever an A occurs in one strand, a T must be opposite it in the other strand; wherever a G occurs in one strand, a C must be opposite it. This feature of DNA is called **complementary base pairing,** and one strand is said to be *complementary* to the other. The base pairs, which fit together like pieces of a jigsaw puzzle, are stabilized by hydrogen bonds—two between A and T and three between G and C (see Figures 14.6 and 3.30; hydrogen bonds are discussed in Section 2.3). The hydrogen bonds between the paired bases, repeated along the double helix, hold the two strands together in the helix.

The base pairs lie in flat planes almost perpendicular to the long axis of the DNA molecule. In this state, each base pair occupies a length of 0.34 nm along the long axis of the double helix (see Figure 14.6). This spacing accounts for the repeating 0.34-nm pattern noted in the X-ray diffraction patterns. The larger 3.4-nm repeat pattern was interpreted to mean that each full turn of the double helix takes up 3.4 nm along the length of the molecule and therefore 10 base pairs are packed into a full turn.

Watson and Crick also realized that the two strands of a double helix fit together in a stable chemical way only if they are **antiparallel,** that is, only if they run in opposite directions (see Figure 14.6, arrows). In other words, the *3′ end* of one strand is opposite the *5′ end* of the other strand. This antiparallel arrangement is highly significant for the process of replication, which is discussed in the next section.

As hereditary material, DNA must faithfully store and transmit genetic information for the entire life cycle of an organism. Watson and Crick recognized that this information is coded into the DNA by the particular sequence of the four nucleotides. Although only four different kinds of nucleotides exist, combining them in groups allows an essentially infinite number of different sequences to be "written," just as the 26 letters of the alphabet can be combined in groups to write a virtually unlimited number of words. Chapter 15 shows how taking the four nucleotides in groups of three forms enough words to spell out the structure of any conceivable protein.

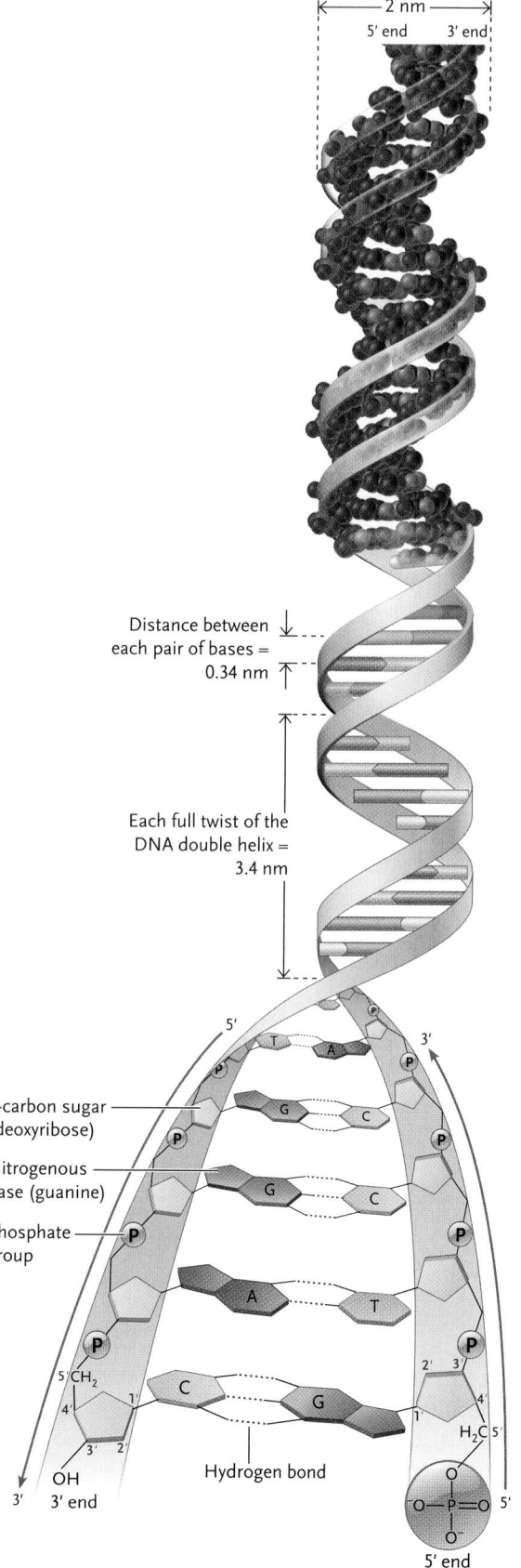

Figure 14.6

DNA double helix. Arrows and labeling of the ends show that the two polynucleotide chains of the double helix are antiparallel—that is, they have opposite polarity in that they run in opposite directions. In the space-filling model at the top, the spaces occupied by atoms are indicated by spheres. There are 10 base pairs per turn of the helix; only 8 base pairs are visible because the other 2 are obscured where the backbones pass over each other.

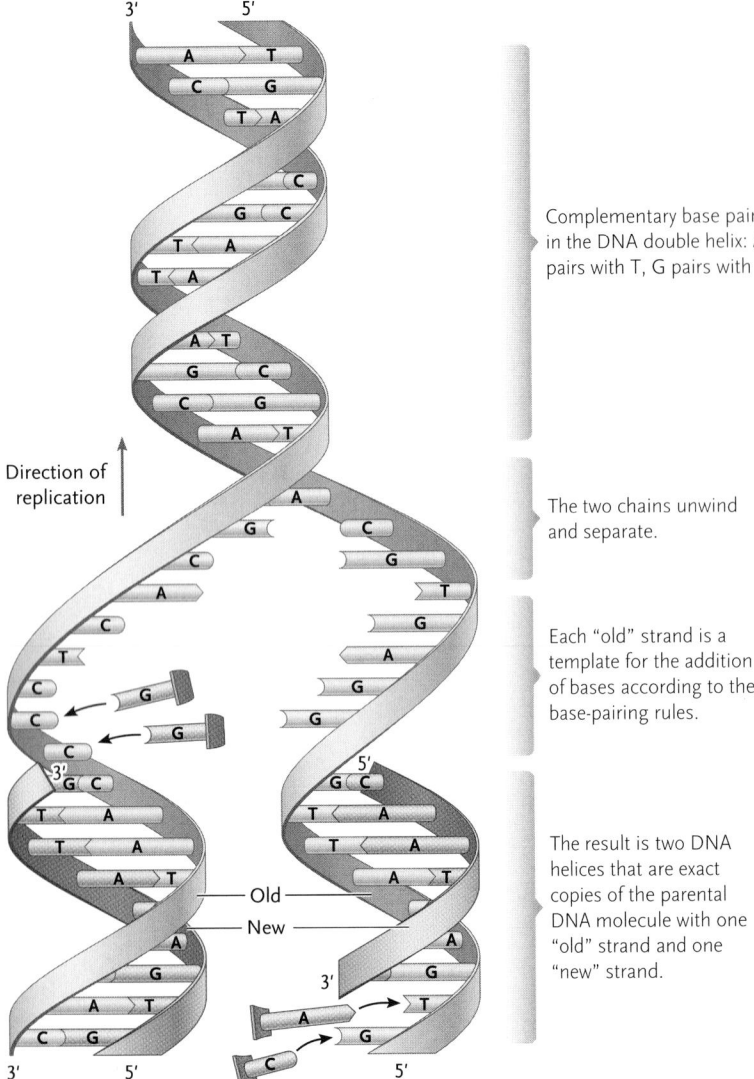

Figure 14.7
Watson and Crick's model for DNA replication. The original DNA molecule is shown in gray. A new polynucleotide chain (red) is assembled on each original chain as the two chains unwind. The template and complementary copy chains remain wound together when replication is complete, producing molecules that are half old and half new. The model is known as the semiconservative model for DNA replication.

Direction of replication

Complementary base pairing in the DNA double helix: A pairs with T, G pairs with C.

The two chains unwind and separate.

Each "old" strand is a template for the addition of bases according to the base-pairing rules.

The result is two DNA helices that are exact copies of the parental DNA molecule with one "old" strand and one "new" strand.

Old
New

Watson and Crick announced their model for DNA structure in a brief but monumental paper published in the journal *Nature* in 1953. Watson and Crick shared a Nobel Prize with Wilkins in 1962 for their discovery of the molecular structure of DNA. Rosalind Franklin might have been a candidate for a Nobel Prize had she not died of cancer at age 38 in 1958. (The Nobel Prize is given only to living investigators.) Unquestionably, Watson and Crick's discovery of DNA structure opened the way to molecular studies of genetics and heredity, leading to our modern understanding of gene structure and action at the molecular level.

14.3 DNA Replication

Once they had discovered the structure of DNA, Watson and Crick realized immediately that complementary base pairing between the two strands could explain how DNA replicates **(Figure 14.7)**. They imagined that, for replication, the hydrogen bonds between the two strands break, and the two strands unwind and separate. Each strand then acts as a template for the synthesis of its partner strand. When replication is complete, there are two double helices, each of which has one strand derived from the parental DNA molecule base paired with a newly synthesized strand. Most important, each of the two new double helices has the identical base-pair sequence as the parental DNA molecule.

The model of replication Watson and Crick proposed is termed **semiconservative replication (Figure 14.8a)**. Other scientists proposed two other models for replication. In the *conservative replication model,* the two strands of the original molecule serve as templates for the two strands of a new DNA molecule, then rewind into an all "old" molecule **(Figure 14.8b)**. After the two complementary copies separate from their templates, they wind together into an all "new" molecule. In the *dispersive replication model,* neither parental strand is conserved and both chains of each replicated molecule contain old and new segments **(Figure 14.8c)**.

Meselson and Stahl Showed That DNA Replication Is Semiconservative

A definitive experiment published in 1958 by Matthew Meselson and Franklin Stahl of the California Institute of Technology demonstrated that DNA replication is semiconservative **(Figure 14.9)**. In their experiment, Meselson and Stahl had to be able to distinguish parental DNA strands from newly synthesized DNA. To do this they used a nonradioactive "heavy" nitrogen isotope to tag the parental DNA strands. The heavy isotope, ^{15}N, has one more neutron in its nucleus than the

a. Semiconservative replication   **b. Conservative replication**   **c. Dispersive replication**

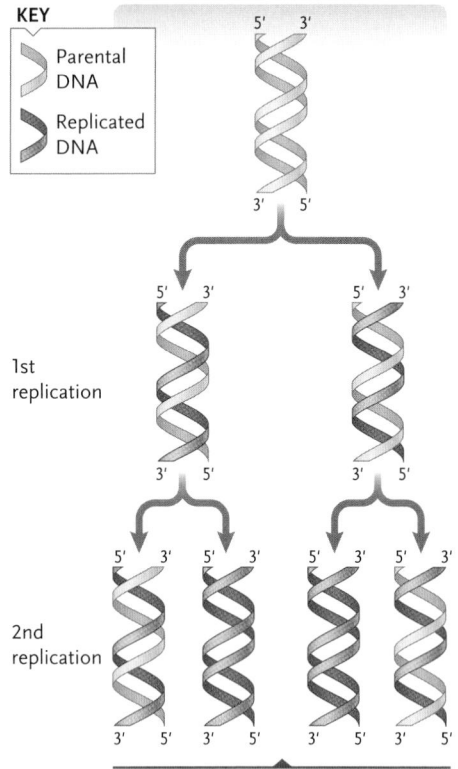
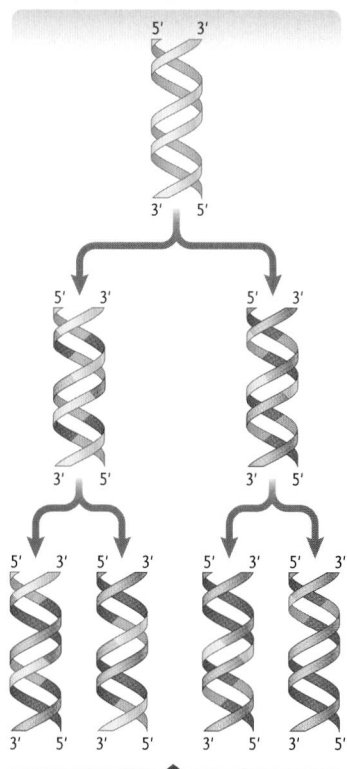

KEY

Parental DNA

Replicated DNA

1st replication

2nd replication

The two parental strands of DNA unwind, and each is a template for synthesis of a new strand. After replication has occurred, each double helix has one old strand paired with one new strand. This model was the one proposed by Watson and Crick themselves.

The parental strands of DNA unwind, and each is a template for synthesis of a new strand. After replication has occurred, the parental strands pair up again. Therefore, the two resulting double helices consist of one with two old strands, and the other with two new strands.

The original double helix splits into double-stranded segments onto which new double-stranded segments form. These newly formed sections somehow assemble into two double helices, both of which are a mixture of the original double-stranded DNA interspersed with new double-stranded DNA.

Figure 14.8
Semiconservative **(a)**, conservative **(b)**, and dispersive **(c)** models for DNA replication.

normal ^{14}N isotope. Molecules containing ^{15}N are measurably heavier (denser) than molecules of the same type containing ^{14}N.

As the first step in their experiment, Meselson and Stahl grew the bacterium *E. coli* in a culture medium containing the heavy ^{15}N isotope (Figure 14.9, step 1). The heavy isotope incorporated into the nitrogenous bases of DNA, resulting in all the DNA being labeled with ^{15}N. Then they transferred the bacteria to a culture medium containing the light ^{14}N isotope (step 2). All new DNA synthesized after the transfer contained the light isotope. Just before the transfer to the medium with the ^{14}N isotope, and after each round of replication following the transfer, they took a sample of the cells and extracted the DNA (step 3).

Meselson and Stahl then mixed the DNA samples with cesium chloride (CsCl) and centrifuged the mixture at very high speed (step 4). During the centrifugation, the CsCl forms a density gradient and DNA molecules move to a position in the gradient where their density matches that of the CsCl. Therefore, DNA of different densities is separated into bands, with the densest DNA settling closer to the bottom of the tube. Figure

14.9 "Result" shows the outcome of these experiments, and "Conclusions" shows why the results were compatible with only the semiconservative replication model.

DNA Polymerases Are the Primary Enzymes of DNA Replication

During replication, complementary nucleotide chains are assembled from individual nucleotides by enzymes known as **DNA polymerases.** More than one kind of DNA polymerase is required for DNA replication in both eukaryotes and prokaryotes. *Nucleoside triphosphates* are substrates for the polymerization reaction catalyzed by DNA polymerases **(Figure 14.10).** A nucleoside triphosphate is a nitrogenous base linked to a sugar, which is linked, in turn, to a chain of three phosphate groups. You have encountered a nucleoside triphosphate before, namely the ATP produced in cellular respiration (see Chapter 8). The nucleoside triphosphates used in DNA replication differ from ATP by having the sugar deoxyribose rather than the sugar ribose. Because four different bases are found in DNA—adenine (A), guanine (G), cytosine (C), and

Figure 14.9 Experimental Research

The Meselson and Stahl Experiment Demonstrating the Semiconservative Model to Be Correct

QUESTION: Does DNA replicate semiconservatively?

EXPERIMENT: Matthew Meselson and Franklin Stahl proved that the semiconservative model of DNA replication is correct and that the conservative and dispersive models are incorrect.

1. Bacteria grown in ^{15}N (heavy) medium. All DNA is heavy.

2. Bacteria transferred to ^{14}N (light) medium and allowed to grow and divide for several generations. All new DNA is light.

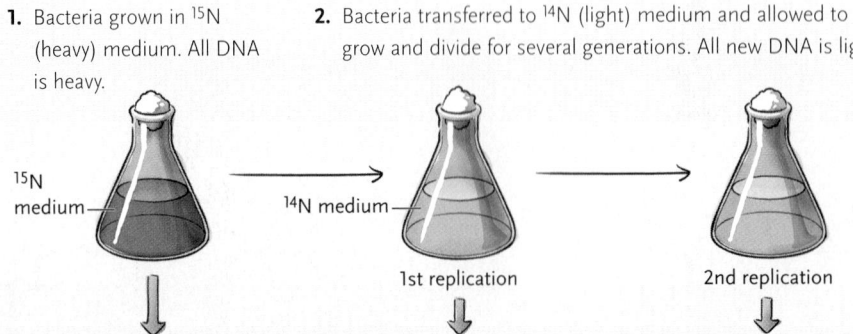

3. DNA extracted from bacteria cultured in ^{15}N medium and after each generation in ^{14}N medium.

4. DNA mixed with cesium chloride (CsCl) and centrifuged at very high speed for about 48 hours.

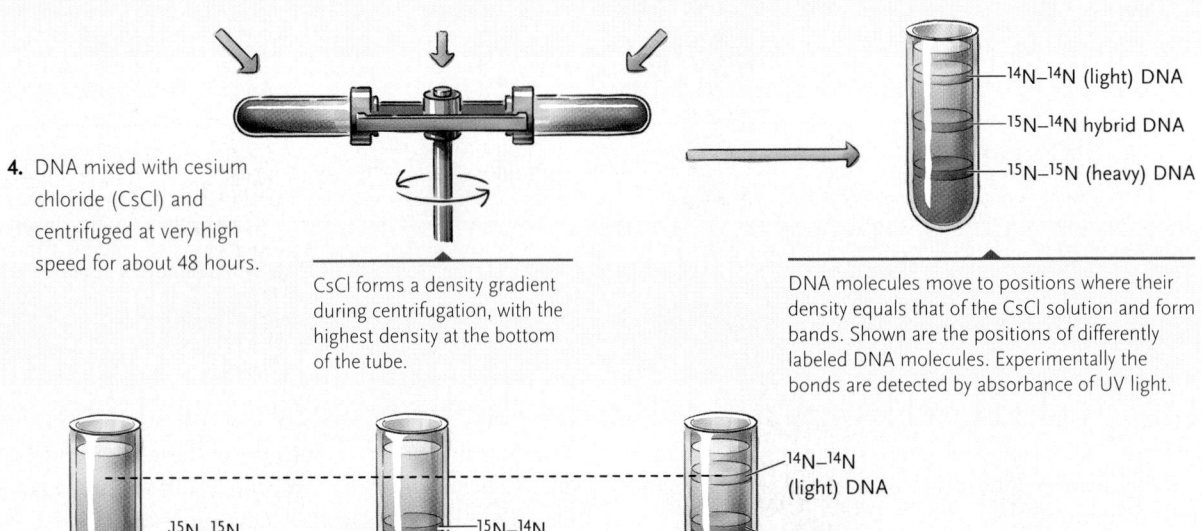

CsCl forms a density gradient during centrifugation, with the highest density at the bottom of the tube.

DNA molecules move to positions where their density equals that of the CsCl solution and form bands. Shown are the positions of differently labeled DNA molecules. Experimentally the bonds are detected by absorbance of UV light.

RESULT: Meselson and Stahl obtained the following results:

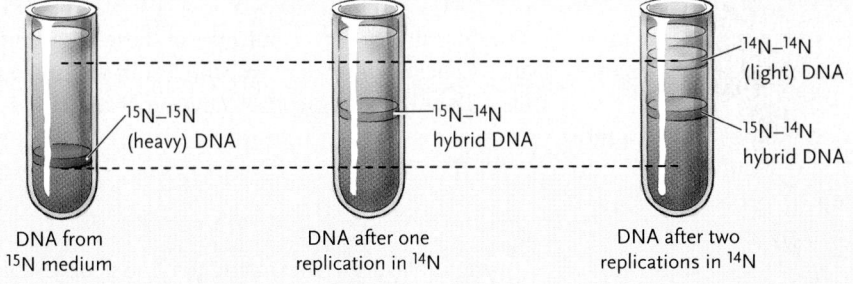

CONCLUSIONS: The predicted DNA banding patterns for the three DNA replication models were:

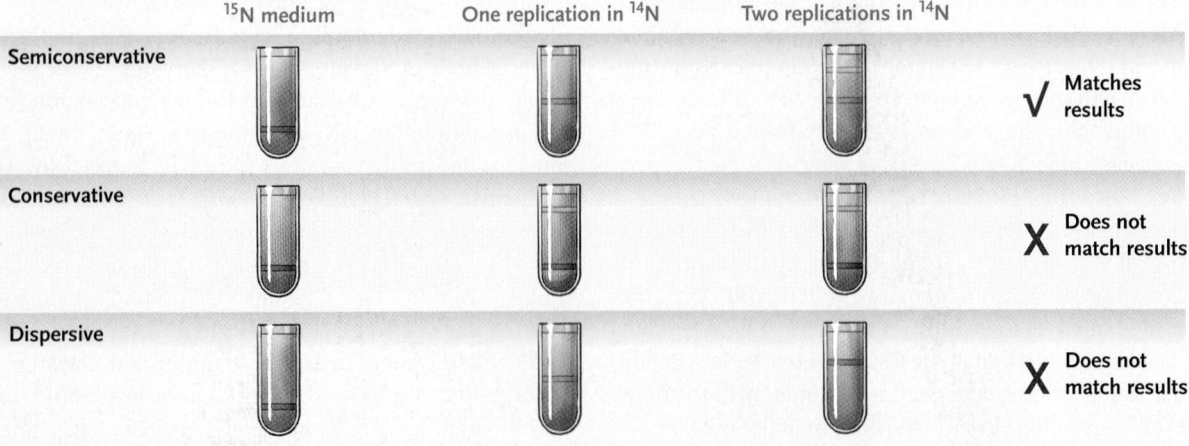

The results support the semiconservative model.

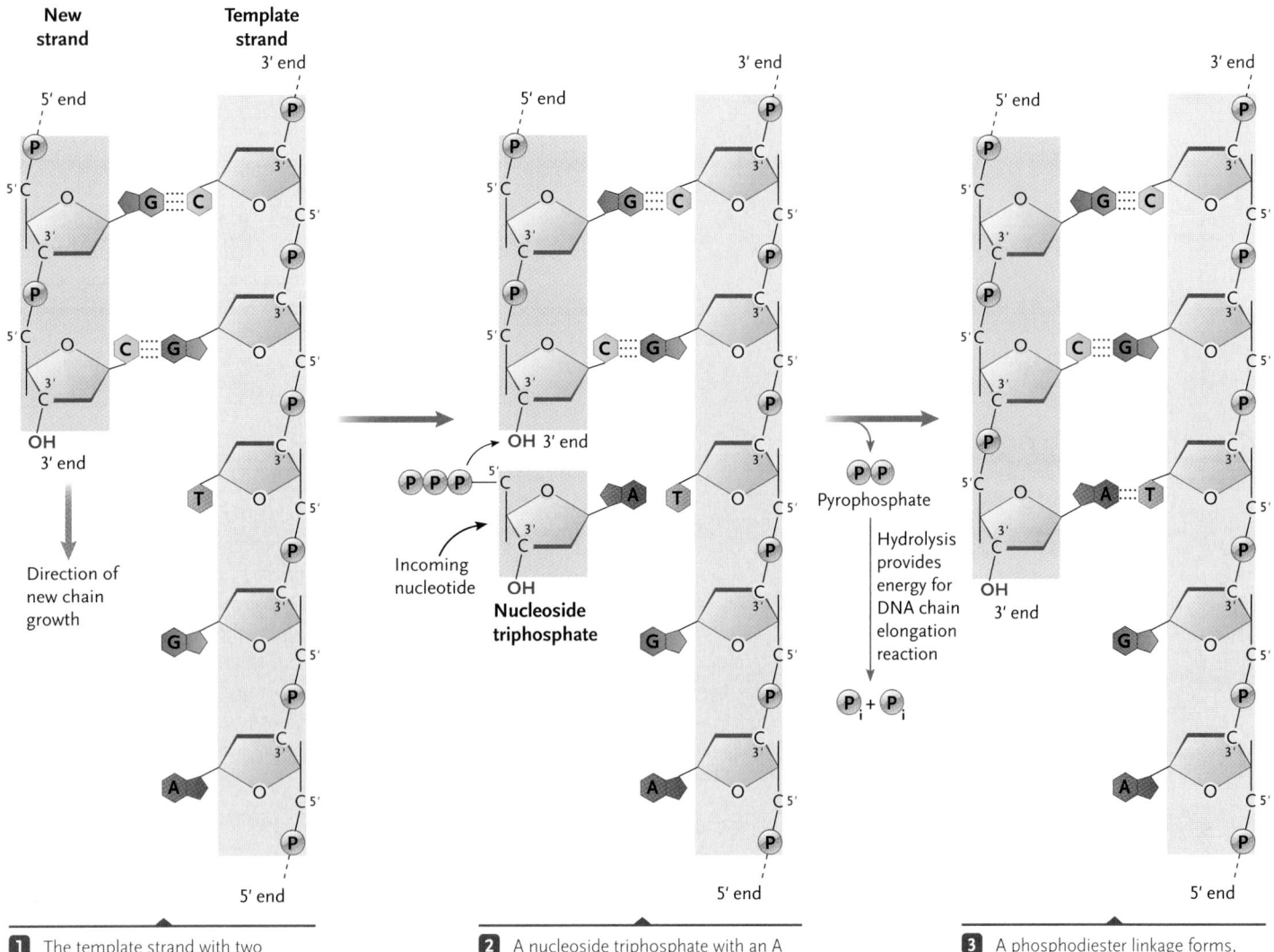

1 The template strand with two nucleotides of the new strand assembled.

2 A nucleoside triphosphate with an A base forms a complementary base pair with the next nucleotide of the template strand.

3 A phosphodiester linkage forms, linking the newly added nucleotide to the end of the primer, lengthening the strand by one.

Figure 14.10
Reactions assembling a complementary chain in the 5′→3′ direction on a template DNA strand, showing the phosphodiester linkage created when the DNA polymerase enzyme adds each nucleotide to the chain.

thymine (T)—four different nucleoside triphosphates are used for DNA replication. By analogy with the ATP naming, the nucleoside triphosphates for DNA replication are given the short names dATP, dGTP, dCTP, and dTTP, where the "d" stands for "deoxyribose."

Figure 14.10 presents a section of a DNA polynucleotide chain being replicated to show how DNA polymerase catalyzes the assembly of a new DNA strand that is complementary to the template strand. To understand Figure 14.10, remember that the carbons in the deoxyriboses of nucleotides are numbered with primes. Each DNA strand has two distinct ends: the 5′ end has an exposed phosphate group attached to the 5′ carbon of the sugar, and the 3′ end has an exposed hydroxyl group attached to the 3′ carbon of the sugar. As we learned earlier, because of the antiparallel nature of the DNA double helix, the 5′ end of one strand is opposite the 3′ end of the other.

Part of a template strand with two nucleotides of a new strand hydrogen bonded to it by complementary base pairing is shown in step 1 of Figure 14.10. One of the characteristics of DNA polymerase is that it can add a nucleotide *only to the 3′ end of an existing nucleotide chain*. The next template nucleotide has a T base. This means the DNA polymerase will bind a nucleoside triphosphate with an A base (dATP) from the surrounding solution (step 2). The enzyme then catalyzes the formation of the phosphodiester bond involving the 3′−OH group at the end of the existing chain and the innermost of the three phosphate groups of the dATP, releasing the other two phosphates as a pyrophosphate molecule (step 3). Hydrolysis of the bond between the two phosphates provides the energy for the formation of the new bond.

The DNA polymerase then moves to the next base on the DNA template, shown as guanine in step 3, binds a dCTP, and, using the reaction just described, catalyzes the formation of a phosphodiester bond, inserting the C nucleotide to the growing new strand. The process then continues, adding complementary nucleotides one by one to the growing DNA strand.

As a new DNA strand is assembled, a 3′−OH group is always exposed at its "newest" end; the "old-

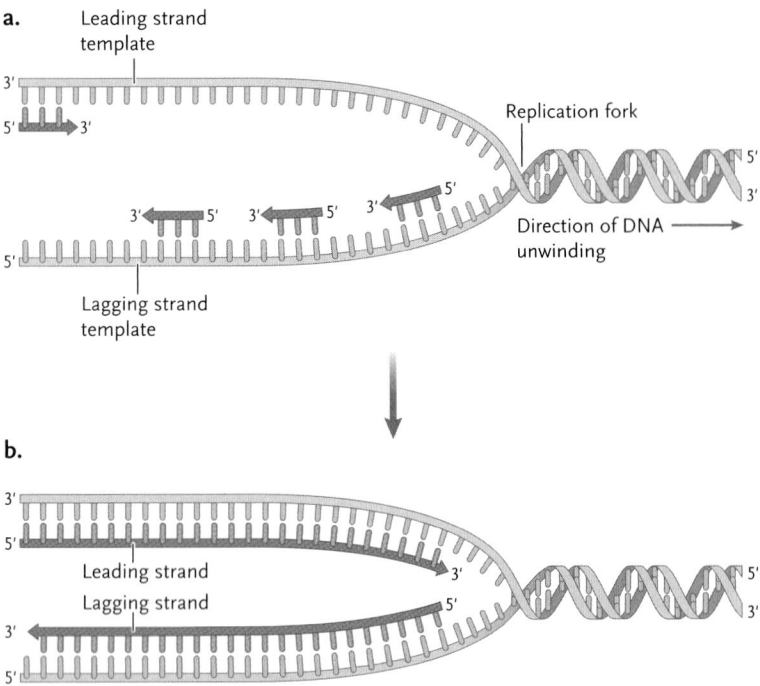

a.

Leading strand
template

3'
5' 3'

Replication fork

5'
3'

Direction of DNA ⟶
unwinding

5' 3' 5' 3' 5' 3' 5'

5'

Lagging strand
template

b.

3'
5'

Leading strand

3'

5'

Lagging strand

5'

3'

3'
5'

Figure 14.11

How antiparallel template strands are replicated at a fork. The template strand presented to DNA polymerase in the "wrong" 5'→3' direction—the strand on the bottom in **(a)**—is copied in short lengths that run opposite to the direction of fork movement. The short lengths are then linked into a continuous chain **(b).** The overall effect is synthesis of both strands in the direction of fork movement.

est" end of the new chain has an exposed 5' triphosphate. DNA polymerases are therefore said to assemble nucleotide chains in the 5'→3' direction. Because of the antiparallel nature of DNA, the template strand is "read" in the 3'→5' direction for this new synthesis.

The key molecular events of DNA replication described in this section are:

1. The two strands of the DNA molecule unwind for replication to occur.
2. Nucleotides are added only to an existing chain.
3. The overall direction of new synthesis is in the 5'→3' direction, which is a direction antiparallel to that of the template strand.
4. Nucleotides enter into a newly synthesized chain according to the A-T and G-C complementary base-pairing rules.

The following sections describe how enzymes and other proteins conduct these molecular events.

Helicases Unwind DNA to Expose Template Strands for New DNA Synthesis

For replication to be semiconservative, the two strands of the parental DNA molecule must unwind and separate to expose template strands for new DNA synthesis

during the replication process. The unwinding produces a Y-shaped structure called a **replication fork**, which consists of the two unwound template strands transitioning to double-helical DNA. An enzyme, **DNA helicase**, catalyzes the unwinding, which exposes both strands for the next steps in replication. The helicase uses the energy of ATP hydrolysis to unwind the DNA helix. The exposed single-stranded segments of DNA become coated with **single-stranded binding proteins**, which stabilize the DNA for the replication process. These proteins are displaced as the replication enzymes make the new polynucleotide chain on the template strands.

Let us consider a possible consequence of the unwinding of DNA by helicases. If the DNA is circular, as is the case for the genomes of most bacteria, unwinding the DNA will eventually cause the still-wound DNA ahead of the unwinding to become knotted. You can visualize this by making a small circular double helix with a pair of shoelaces. Now pick a place and pull apart the laces. You will see that the more you pull, the more the laces become overtwisted and strained on the other side of the circle. In the cell, the overtwisting and strain of DNA ahead of the replication fork during replication is avoided by the action of enzymes known as **topoisomerases**, which remove the overtwisting as it forms.

RNA Primers Provide the Starting Point for DNA Polymerase to Begin Synthesizing a New DNA Chain

DNA polymerases can add nucleotides only to the 3' end of an existing strand. How, then, can a new strand begin, since there is no existing strand in place? The answer lies in a short nucleotide chain called a **primer**, made of RNA instead of DNA. The primer, assembled by the enzyme **primase**, is laid down as the first series of nucleotides in a new DNA strand. RNA primers are removed and replaced with DNA later in replication.

One New DNA Strand Is Synthesized Continuously; the Other, Discontinuously

DNA polymerases assemble a new DNA strand on a template strand in the 5'→3' direction. Because the two strands of a DNA molecule are antiparallel, only one of the template strands runs in a direction that allows DNA polymerase to make a 5'→3' complementary copy in the direction of unwinding. That is, on this template strand (top strand in **Figure 14.11**), the new DNA strand is synthesized continuously in the direction of unwinding of the double helix. However, the other template strand (bottom strand in Figure 14.11) runs in the opposite direction; this means DNA polymerase has to copy it in the direction opposite to the unwinding.

How is the new DNA strand made in the opposite direction to the unwinding? The polymerases make this strand in short lengths that are actually synthesized in the direction opposite to that of DNA unwinding (see Figure 14.11). The short lengths produced by this **discontinuous replication** are then covalently linked into a continuous polynucleotide chain. The short lengths are called *Okazaki fragments,* in honor of Reiji Okazaki, the Japanese scientist who first detected them. The new DNA strand assembled in the direction of DNA unwinding is called the **leading strand** of DNA replication; the strand assembled discontinuously in the opposite direction is called the **lagging strand.** The template strand for the leading strand is the *leading strand template,* and the template strand for the lagging strand is the *lagging strand template.*

Multiple Enzymes Coordinate Their Activities in DNA Replication

Helicase, primase, and DNA polymerase coordinate their activities with additional enzymes to replicate DNA. In the first step of the process, a helicase unwinds the template DNA to produce a replication fork (**Figure 14.12,** step 1). Just behind the site of unwinding, primases lay down short RNA primers about 10 nucleotides in length. The primers are assembled in the $5' \rightarrow 3'$ direction on both template chains—in the direction of unwinding on one chain and in the opposite direction on the other.

DNA polymerase then adds DNA nucleotides to the RNA primers (step 2). Helicase continues to unwind the DNA. Leading strand synthesis continues in the direction of unwinding, whereas on the lagging strand template, primase creates a new RNA primer and DNA polymerase adds DNA nucleotides to the new primer (step 3). When this second fragment reaches the primer of the first fragment, the DNA polymerase leaves and a different type of DNA polymerase binds. This polymerase removes the RNA primer on the first fragment, replacing the RNA nucleotides with DNA nucleotides (step 4). At this point, the two newly synthesized fragments are not covalently joined—they have a "nick" between them (see step 4). Another enzyme, **DNA ligase** (*ligare* = to tie), closes the nick, joining the two fragments into one larger fragment (step 5). The replication process continues in the same way until the entire DNA molecule is copied (step 6). **Table 14.1** summarizes the activities of the major enzymes replicating DNA.

The entire replication mechanism, including the activities of the helicase, primase, DNA polymerases, DNA ligase, and other proteins involved in the process, advances at a rate of about 500 to 1000 nucleotides per second in prokaryotes and at a rate of about 50 to 100 per second in eukaryotes. The entire process is so rapid that the RNA primers and gaps left by discontinuous synthesis persist for only seconds or fractions of a second. Consequently, the replication enzymes operate only at the replication fork. A short distance behind the fork, the new DNA chains are fully continuous and wound with their template strands into complete DNA double helices. Each helix consists of one "old" and one "new" polynucleotide chain.

Researchers identified the enzymes that replicate DNA through experiments with a variety of prokaryotes and eukaryotes and with viruses that infect both types of cells. Experiments with the bacterium *E. coli* have provided the most complete information about DNA replication, particularly in the laboratory of Arthur Kornberg at Stanford University. Kornberg received a Nobel Prize in 1959 for his discovery of the mechanism for DNA synthesis.

Telomerases Solve a Specialized Replication Problem at the Ends of Linear DNA Molecules

The priming mechanism outlined in Figure 14.12 leaves one major problem unsolved for linear chromosomes, such as those in eukaryotes. Think about replication that occurs at one end of a chromosome (**Figure 14.13,** step 1). To begin the new strand, an RNA primer is laid down opposite the end of the template strand and then a DNA polymerase adds new DNA nucleotides from the end of this primer (step 2). Once replication is under way, this first primer is removed, leaving a gap at the beginning (5′) end of the new strand (as in step 3). In a similar way, a gap is produced at the 5′ end of the new strand made starting at the other end of the chromosome. Therefore, when these new, now shortened DNA strands are used as a template for the next round of DNA replication, the new chromosome will be shorter. Indeed, when most somatic cells go through the cell cycle, the chromosomes shorten with each division. Deletion of genes by such shortening would have serious, eventually lethal, consequences for the cell.

In most chromosomes, however, the genes are protected by a buffer of noncoding DNA. That is, at the ends of each eukaryotic chromosome are telomeres (*telo* = end; *mere* = segment). **Telomeres** are short sequences repeated hundreds to thousands of times, which do not code for proteins. In humans, the repeated sequence, the *telomere repeat,* is 5′-TTAGGG-3′ on the leading template strand. With each replication, a fraction of the telomere repeats is lost but the genes are unaffected. The buffering fails only when the entire telomere is lost.

The enzyme **telomerase** can stop the shortening by adding telomere repeats to the chromosome ends.

Figure 14.12 **Steps in DNA replication, including the activities of the helicase, primase, DNA polymerases, and DNA ligase taking part in the process.** Primer synthesis, removal, gap filling, and nick sealing occur primarily in the lagging strand. The drawings simplify the process. In reality, the enzymes assemble at the fork, replicating both strands from that position as the template strands fold and pass through the assembly.

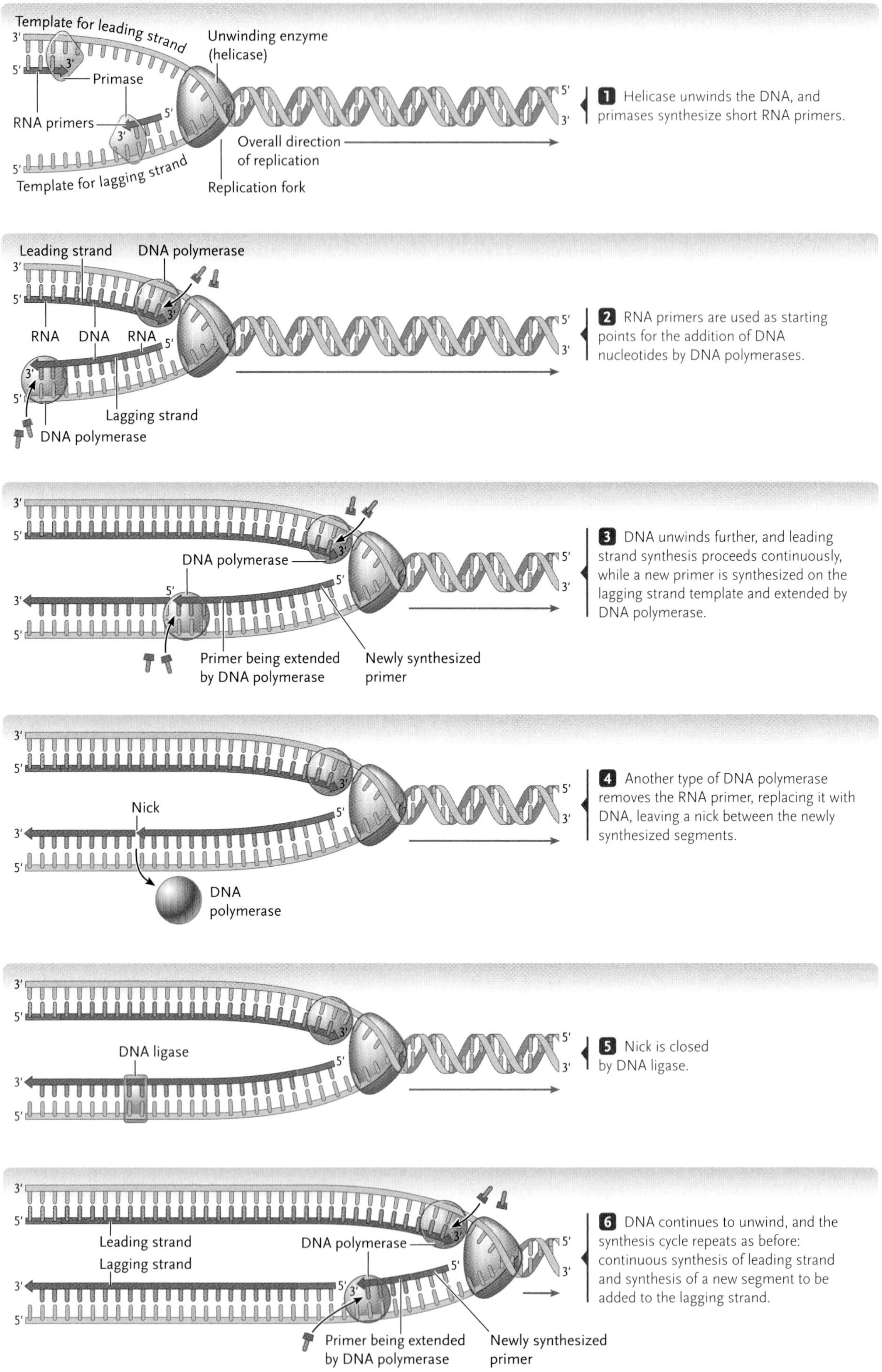

Template for leading strand
Unwinding enzyme (helicase)
Primase
RNA primers
Overall direction of replication
Template for lagging strand
Replication fork
1 Helicase unwinds the DNA, and primases synthesize short RNA primers.
Leading strand
DNA polymerase
RNA
DNA
RNA
Lagging strand
DNA polymerase
2 RNA primers are used as starting points for the addition of DNA nucleotides by DNA polymerases.
DNA polymerase
Primer being extended by DNA polymerase
Newly synthesized primer
3 DNA unwinds further, and leading strand synthesis proceeds continuously, while a new primer is synthesized on the lagging strand template and extended by DNA polymerase.
Nick
DNA polymerase
4 Another type of DNA polymerase removes the RNA primer, replacing it with DNA, leaving a nick between the newly synthesized segments.
DNA ligase
5 Nick is closed by DNA ligase.
Leading strand
Lagging strand
DNA polymerase
Primer being extended by DNA polymerase
Newly synthesized primer
6 DNA continues to unwind, and the synthesis cycle repeats as before: continuous synthesis of leading strand and synthesis of a new segment to be added to the lagging strand.

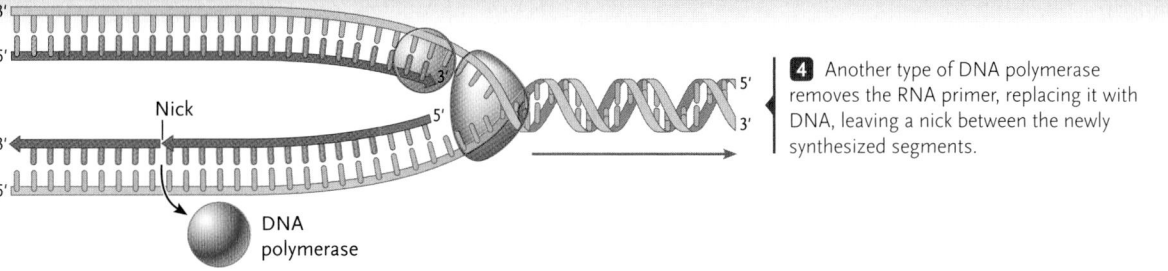

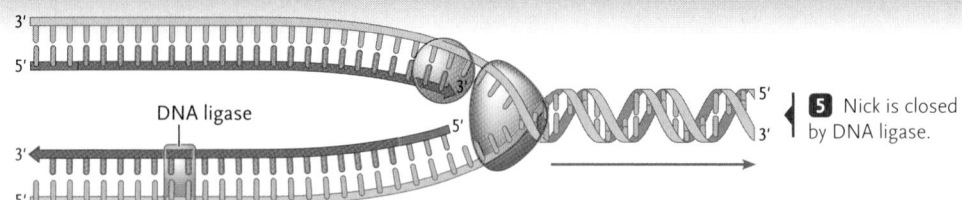

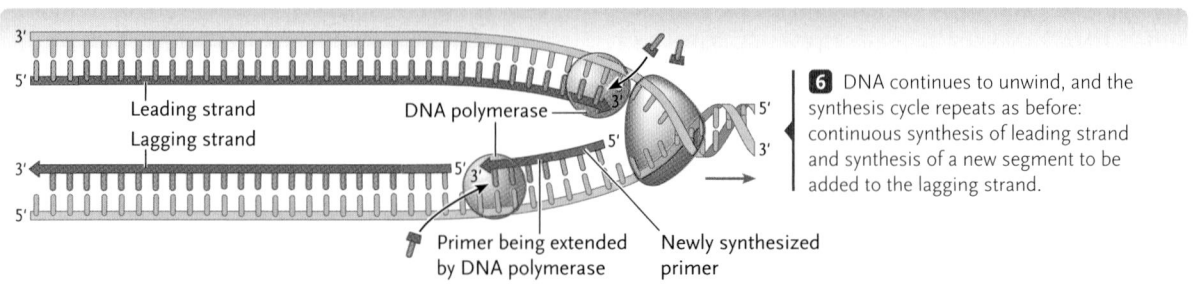

Table 14.1	Major Enzymes of DNA Replication
Enzyme	Activity
Helicase	Unwinds DNA helix
Single-stranded binding proteins	Stabilize DNA in single-chain form
Primase	Assembles RNA primers
DNA polymerases	Assemble DNA chains on primers; replace primers while simultaneously replacing primer nucleotides with DNA nucleotides
DNA ligase	Seals nicks left after RNA primers replaced with DNA
Topoisomerases	Relieve overtwisting and strain of DNA ahead of replication fork (in circular DNA)

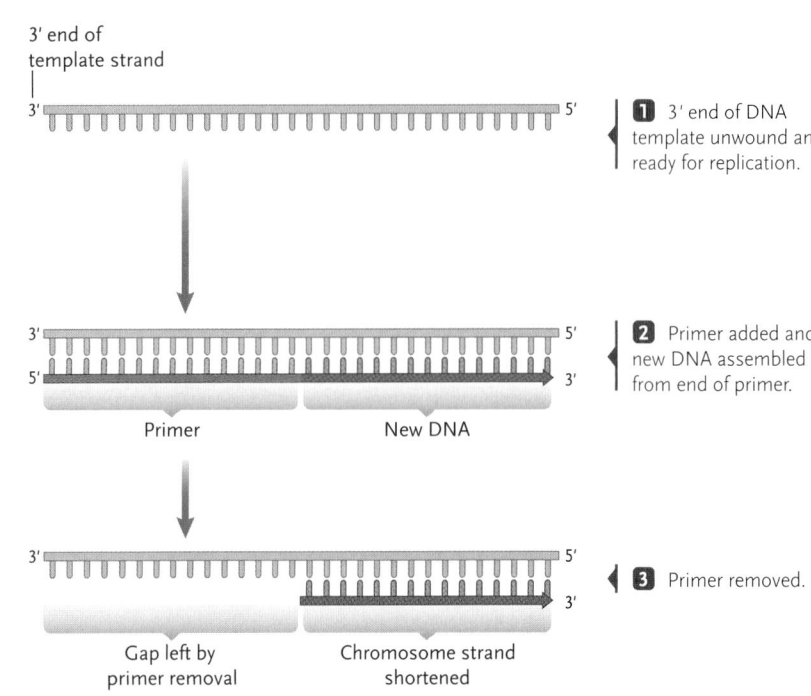

Figure 14.13

How a gap is left by primer removal at the 5′ end of a replicating linear DNA molecule.

Discovered in 1985 by Elizabeth H. Blackburn and her graduate student Carol W. Greider at the University of California, Berkeley, telomerase adds additional telomere repeats to the end of the *template strand* before DNA replication begins (**Figure 14.14,** step 1; compare with Figure 14.13, step 1). After the addition, the primer of the leading strand is laid down, using the newly added telomere repeats as the template. A DNA polymerase then extends the new DNA strand as usual (Figure 14.14, step 2). The primer is removed, which still leaves an unfilled gap at the beginning of the leading chain (step 3). However, the chromosome has not shortened because it now has the extra telomere repeats added by the telomerase. The telomerase enzyme, which appears to be present in all eukaryotes, is an unusual enzyme that consists of protein subunits complexed with RNA; the RNA part is the template for making the extra telomere repeats.

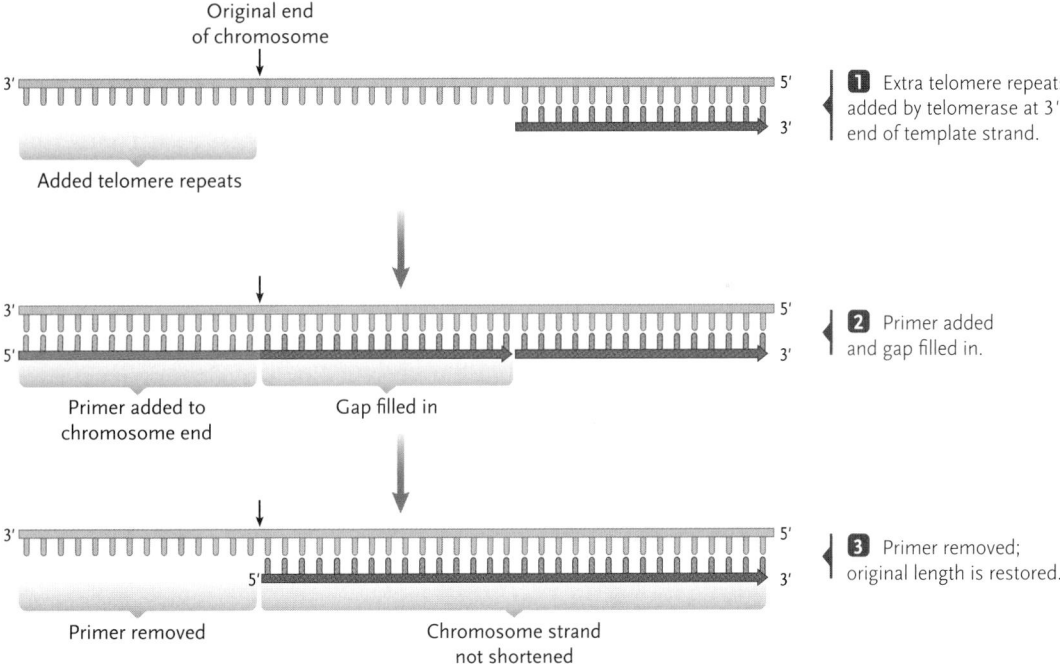

Figure 14.14

How telomere repeats added to eukaryotic chromosomes prevent chromosome shortening.

Telomerase is active in some cells but not in others. In particular, telomerase is active in sperm and eggs, which is necessary to maintain chromosome length from generation to generation. It is also active in the rapidly dividing cells of the early embryo. However, telomerase becomes inactive after a number of divisions, meaning subsequent telomeres shorten as the cells continue to divide. As a result, a cell is capable of only a certain number of mitotic divisions before it stops dividing and dies. Could telomere shortening, then, contribute to the aging process in multicellular animals? Telomere shortening has indeed been linked to the aging process, but it is unknown whether it contributes to or is a result of aging.

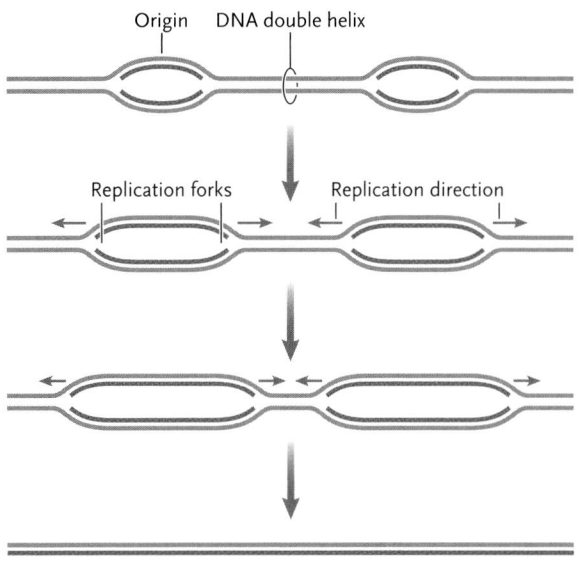

Figure 14.15
Replication from
multiple origins in
the chromosomes
of eukaryotes.

meet along the chromosomes to produce fully replicated DNA molecules.

Normally, a replication origin is activated only once during the S phase of a eukaryotic cell cycle, so no portion of the DNA is replicated more than once. *Insights from the Molecular Revolution* describes abnormal extra DNA replication that underlies a common cause of mental retardation in humans.

STUDY BREAK

1. What is the importance of complementary base pairing to DNA replication?
2. Why is a primer needed for DNA replication? How is the primer made?
3. Two DNA polymerases are used in DNA replication. What are their roles?
4. Why are telomeres important?

Some observations have made it difficult to draw firm conclusions on this issue. For example, humans, a long-lived species, have telomeres that are much shorter than mice, which live just a few years. Clearly telomeres alone do not determine the life span of an organism.

An unexpected link between telomerases and cancer was found when investigators discovered that more than 90% of cancer cells have fully active telomerase enzymes, regardless of the type of body cell from which they are derived. Evidently, as body cells develop into cancer cells, their telomerases are reactivated, preserving chromosome length during the rapid divisions characteristic of cancer. A positive side of this discovery is that it may lead to an effective cancer treatment, if a means can be found to switch off the telomerases in tumor cells. The chromosomes in the rapidly dividing cancer cells would then eventually shorten to the length at which they break down, leading to cell death and elimination of the tumor.

DNA Replication Begins at Replication Origins

Replication begins at sites called **replication origins.** Hundreds of replication origins may be present in the long chromosomes of eukaryotes. The origins are recognized by proteins that bind to the DNA and stimulate helicases to start the unwinding, followed by primer synthesis and DNA replication. In most cases, replication proceeds from both sides of a replication origin, producing two replication forks that move in opposite directions **(Figure 14.15).** (This means that the leading strands and lagging strands are reversed on the two sides.) The forks eventually

14.4 Mechanisms That Correct Replication Errors

DNA polymerases make very few errors as they assemble new nucleotide chains. Most of the mistakes that do occur, called **base-pair mismatches,** are corrected, either by a proofreading mechanism carried out during replication by the DNA polymerases themselves or by a DNA repair mechanism that corrects mismatched base pairs after replication is complete.

Proofreading Depends on the Ability of DNA Polymerases to Reverse and Remove Mismatched Bases

The **proofreading mechanism,** first proposed in 1972 by Arthur Kornberg and Douglas L. Brutlag of Stanford University, depends on the ability of DNA polymerases to back up and remove mispaired nucleotides from a DNA strand. Only when the most recently added base is correctly paired with its complementary base on the template strand can the DNA polymerases continue to add nucleotides to a growing chain. The correct pairs allow the fully stabilizing hydrogen bonds to form **(Figure 14.16,** step 1). If a newly added nucleotide is mismatched (step 2), the DNA polymerase reverses, using a built-in deoxyribonuclease to remove the newly added incorrect nucleotide (step 3). The enzyme resumes working forward, now inserting the correct nucleotide (step 4).

Several experiments have confirmed that the major DNA polymerases of replication can actually proofread their work. For example, when the primary DNA polymerase that replicates DNA in bacteria is intact, with its reverse activity working, its overall error rate is astonishingly low—only about 1 mispair survives in the DNA for every 1 million nucleotides assembled in the test tube. If the proofreading activity of the enzyme is experimentally inhibited, the error rate increases to about 1 mistake for every 1000 to 10,000 nucleotides assembled. Experiments with eukaryotes have yielded similar results.

DNA Repair Corrects Errors That Escape Proofreading

Any base-pair mismatches that remain after proofreading face still another round of correction by **DNA repair mechanisms**. These **mismatch repair** mechanisms increase the accuracy of DNA replication well beyond the one-in-a million errors that persist after proofreading. As noted earlier, the "correct" A-T and G-C base pairs fit together like pieces of a jigsaw puzzle, and their dimensions separate the sugar–phosphate backbone chains by a constant distance. Mispaired bases are too large or small to maintain the correct separation, and they cannot form the hydrogen bonds characteristic of the normal base pairs. As a result, base mismatches distort the structure of the DNA helix. These distortions provide recognition sites for the enzymes catalyzing mismatch repair.

The repair enzymes move along newly replicated DNA molecules, "scanning" the DNA for distortions in the newly synthesized nucleotide chain. If the enzymes encounter a distortion, they remove a portion of the new chain, including the mismatched nucleotides (**Figure 14.17,** step 1). The gap left by the removal (step 2) is then filled by a DNA polymerase, using the template strand as a guide (step 3). The repair is completed by a DNA ligase, which seals the nucleotide chain into a continuous DNA molecule (step 4).

The same repair mechanisms also detect and correct alterations in DNA caused by the damaging effects of chemicals and radiation, including the ultraviolet light in sunlight. Some idea of the importance of the repair mechanisms comes from the unfortunate plight of individuals with *xeroderma pigmentosum,* a hereditary disorder in which the repair mechanism is faulty. Because of the effects of unrepaired alterations in their DNA, skin cancer can develop quickly in these individuals if they are exposed to sunlight.

Very few replication errors remain in DNA after proofreading and DNA repair. The errors that persist, although extremely rare, are a primary source of **mutations,** differences in DNA sequence that appear and remain in the replicated copies. When a mutation occurs in a gene, it can alter the property of the pro-

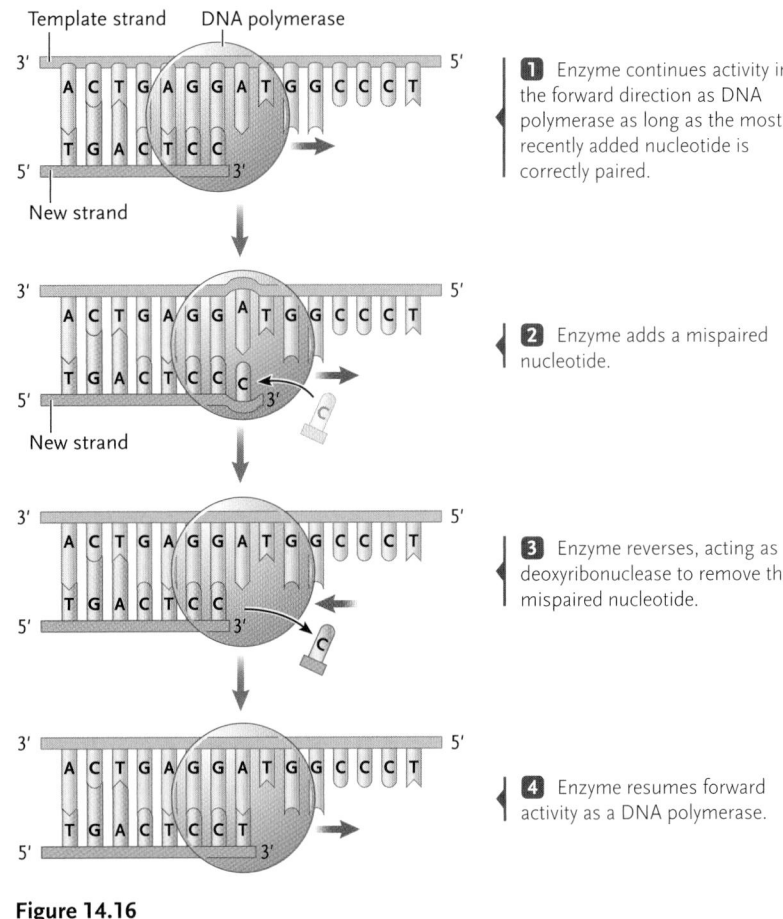

1 Enzyme continues activity in the forward direction as DNA polymerase as long as the most recently added nucleotide is correctly paired.

2 Enzyme adds a mispaired nucleotide.

3 Enzyme reverses, acting as a deoxyribonuclease to remove the mispaired nucleotide.

4 Enzyme resumes forward activity as a DNA polymerase.

Figure 14.16
Proofreading by a DNA polymerase.

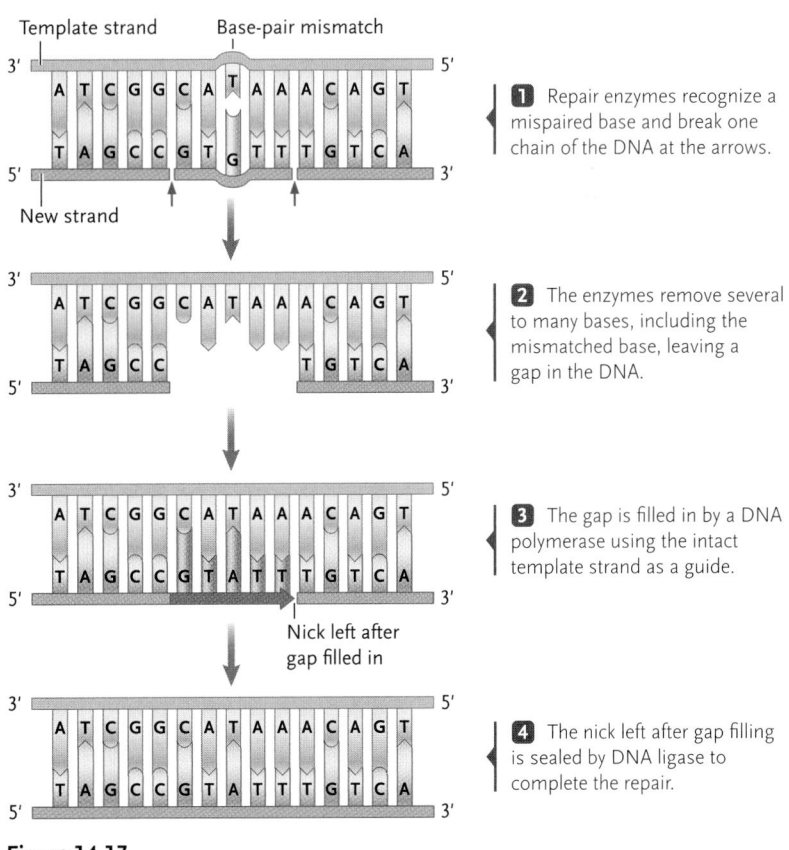

1 Repair enzymes recognize a mispaired base and break one chain of the DNA at the arrows.

2 The enzymes remove several to many bases, including the mismatched base, leaving a gap in the DNA.

3 The gap is filled in by a DNA polymerase using the intact template strand as a guide.

4 The nick left after gap filling is sealed by DNA ligase to complete the repair.

Figure 14.17
Repair of mismatched bases in replicated DNA.

A Fragile Connection between DNA Replication and Mental Retardation

One of the most common sources of inherited mental retardation in humans results from breaks that occur in a narrow, constricted region near one end of the X chromosome (see figure). Because the region breaks easily when cultured cells divide, the associated disabilities are called the *fragile X syndrome*. In addition to mental retardation, affected individuals may have an unusually long face and protruding ears; affected males may have oversized testes.

The disorder affects males more frequently than females, as is typical with X-linked traits—about 1 in 1500 male and 1 in 2500 female births are affected by the fragile X syndrome. However, the inheritance pattern of the syndrome also has some unusual characteristics. The disease is passed from a grandfather through his daughter to his grandchild. The grandfather and daughter have apparently normal X chromosomes, although the daughters sometimes have symptoms; however, abnormal X chromosomes and symptoms appear with high frequency in the grandchildren.

Geneticists were baffled by this unusual pattern of inheritance until a partial explanation was supplied by findings in the laboratories of Grant R. Sutherland of Adelaide Children's Hospital in Australia and others. The investigators examined DNA from individuals with fragile X syndrome using "probes"—short, artificially synthesized DNA sequences that are complementary to, and can pair with, DNA sequences that are of interest. They found that probes containing C and G nucleotides in high proportions

paired most strongly with DNA in the fragile X region. Sequencing of DNA that paired with those probes showed that the region contains many repeats of the three-nucleotide sequence CCG.

Interestingly, the number of CCG repeats varies in the different groups: from 6 to 50 copies in people without the syndrome, 50 to 200 copies in people with mild or no symptoms who transmit the syndrome, and about 230 to 1000 copies in seriously affected people. Somehow the number of CCG repeats increases, which initiates the serious disease symptoms.

The increase in copies occurs by overreplication of the CCG sequence, which begins to occur when the number of copies exceeds about 50. As more CCG copies are added, the region becomes increasingly unstable and the tendency for overreplication also increases. This feature of the process explains why symptoms of the disease often become worse in successive generations.

Scientists still do not understand what causes the overreplication or

how it causes fragile X syndrome. However, the increase in CCG copies appears to turn off a nearby gene called *FMR-1*, which is necessary for normal mental development. The increase also inhibits other genes located elsewhere that have the same effect. One hypothesis takes note of the fact that methyl groups are added to cytosines as part of the controls that turn off large blocks of genes in mammals. According to this idea, the extra cytosines of the added CCG groups provide many additional methylation sites, leading to inactivation of the genes near the fragile X region.

The probe that pairs with CCG groups can be used to estimate the number of copies of the sequence in people without the syndrome who may have increased numbers of the sequence in the fragile X region. If the number is elevated above 50 repeats, these individuals can be counseled about the possibility that they could transmit the disease to their offspring several generations down the line.

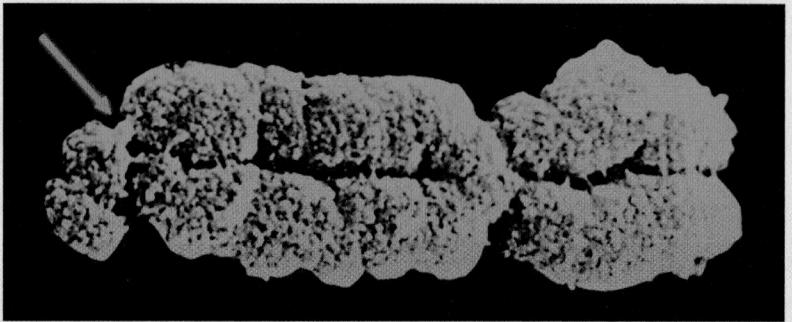

The constricted region *(arrow)* in the human X chromosome associated with fragile X syndrome. The chromosome is double because it has been duplicated in preparation for cell division.

tein encoded by the gene, which, in turn, may alter how the organism functions. Hence, mutations are highly important to the evolutionary process because they are the ultimate source of the variability in off-spring acted on by natural selection.

We now turn from DNA replication and error correction to the arrangements of DNA in eukaryotic and prokaryotic cells. These arrangements organize super-structures that fit the long DNA molecules into the

microscopic dimensions of cells and also contribute to the regulation of DNA activity.

STUDY BREAK

Why is a proofreading mechanism important for DNA replication, and what are the mechanisms that correct errors?

14.5 DNA Organization in Eukaryotes and Prokaryotes

Enzymatic proteins are the essential catalysts of every step in DNA replication. In addition, numerous proteins of other types organize the DNA in both eukaryotes and prokaryotes and control its function.

In eukaryotes, two major types of proteins, the histone and nonhistone proteins, are associated with DNA structure and regulation in the nucleus. These proteins are known collectively as the **chromosomal proteins** of eukaryotes. The complex of DNA and its associated proteins, termed **chromatin**, is the structural building block of a chromosome.

By comparison, the single DNA molecule of a prokaryotic cell is more simply organized and has fewer associated proteins. However, prokaryotic DNA is still associated with two classes of proteins with functions similar to those of the eukaryotic histones and nonhistones: one class that organizes the DNA structurally and one that regulates gene activity. We begin this section with the major DNA-associated proteins of eukaryotes.

Histones Pack Eukaryotic DNA at Successive Levels of Organization

The **histones** are a class of small, positively charged (basic) proteins that are complexed with DNA in the chromosomes of eukaryotes. (Most other cellular proteins are larger and are neutral or negatively charged.) The histones link to DNA by an attraction between their positive charges and the negatively charged phosphate groups of the DNA.

Five types of histones exist in most eukaryotic cells: H1, H2A, H2B, H3, and H4. The amino acid sequences of these proteins are highly similar among eukaryotes, suggesting that they perform the same functions in all eukaryotic organisms.

One function of histones is to pack DNA molecules into the narrow confines of the cell nucleus. For example, each human cell nucleus contains 2 meters of DNA. Combination with the histones compacts this length so much that it fits into nuclei that are only about 10 μm in diameter. Another function is the regulation of DNA activity.

Histones and DNA Packing. The histones pack DNA at several levels of chromatin structure. In the most fundamental structure, called a **nucleosome**, two molecules each of H2A, H2B, H3, and H4 combine to form a beadlike, eight-protein **nucleosome core particle** around which DNA winds for almost two turns **(Figure 14.18).** A short segment of DNA, the **linker**, extends between one nucleosome and the next. Under the electron microscope, this structure looks like beads on a string. The diameter of the beads (the nucleosomes) gives this structure its name—the **10-nm chromatin fiber** (see Figure 14.18).

Each nucleosome and linker includes about 200 base pairs of DNA. Nucleosomes compact DNA by a factor of about 7; that is, a length of DNA becomes about 7 times shorter when it is wrapped into nucleosomes.

Histones and Chromatin Fibers. The fifth histone, H1, brings about the next level of chromatin packing. One H1 molecule binds both to the nucleosome at

Figure 14.18
Levels of organization in eukaryotic chromatin and chromosomes.

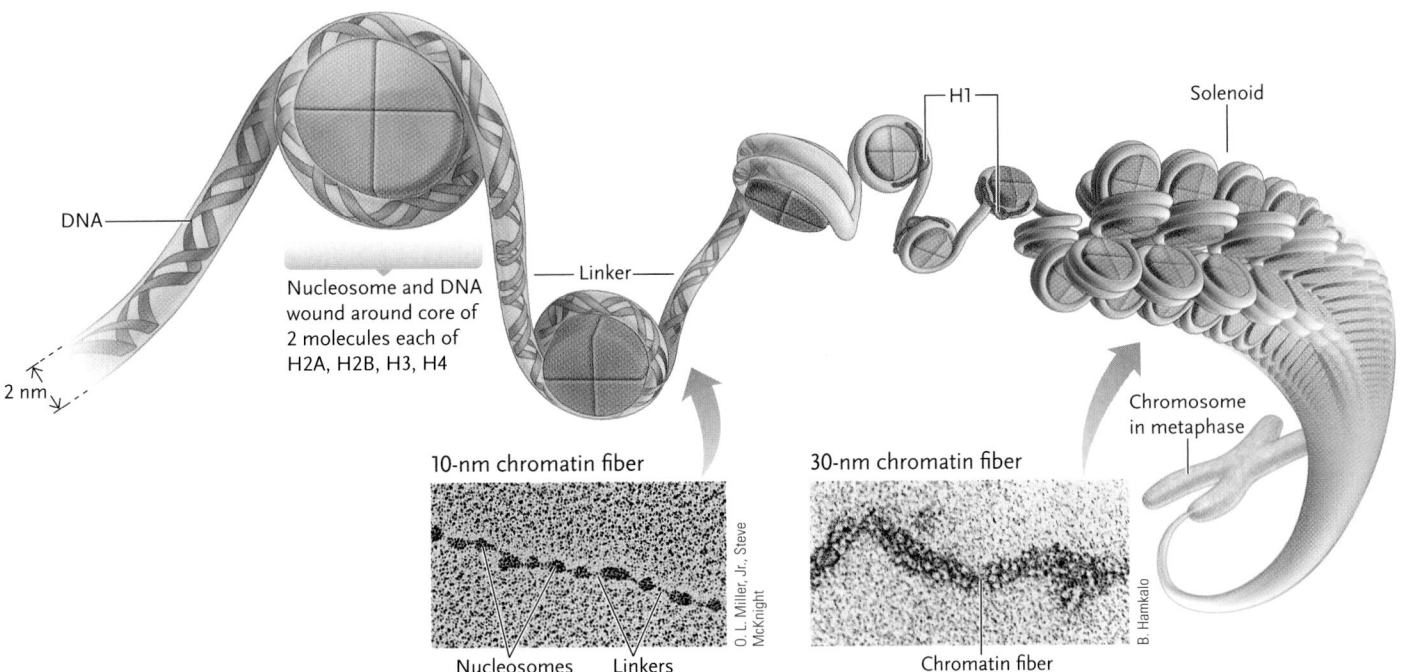

DNA

2 nm

Nucleosome and DNA wound around core of 2 molecules each of H2A, H2B, H3, H4

Linker

H1

Solenoid

Chromosome in metaphase

10-nm chromatin fiber

Nucleosomes   Linkers

O. L. Miller, Jr., Steve McKnight

30-nm chromatin fiber

Chromatin fiber

B. Hamkalo

the point where the DNA enters and leaves the core particle and to the linker DNA. This binding causes the nucleosomes to package into a coiled structure 30 nm in diameter, called the **30-nm chromatin fiber** or **solenoid**, with about six nucleosomes per turn (see Figure 14.18).

The arrangement of DNA in nucleosomes and solenoids compacts the DNA and probably also protects it from chemical and mechanical damage. In the test tube, DNA wound into nucleosomes and chromatin fibers is much more resistant to attack by deoxyribonuclease (a DNA-digesting enzyme) than when it is not bound to histone proteins. Therefore, DNA must unwind almost entirely from solenoids and nucleosomes when it becomes active. When genes become active, however, their DNA becomes almost as susceptible to attack as naked DNA in the test tube.

Packing at Still Higher Levels: Euchromatin and Heterochromatin. In interphase nuclei, chromatin fibers are loosely packed in some regions and densely packed in others. The loosely packed regions are known as **euchromatin** (*eu* = true, regular, or typical), and the densely packed regions are called **heterochromatin** (*hetero* = different). Chromatin fibers also fold and pack into the thick, rodlike chromosomes visible during mitosis and meiosis. Some experiments indicate that links formed between H1 histone molecules contribute to the packing of chromatin fibers, both into heterochromatin and into the chromosomes visible during nuclear division (see discussion in Section 10.2 as well as the more detailed discussion in Section 15.2). However, the exact mechanism for the more complex folding and packing is not known.

Several experiments indicate that heterochromatin represents large blocks of genes that have been turned off and placed in a compact storage form. For example, recall the process of X-chromosome inactivation in mammalian females (see Section 13.2). As one of the two X chromosomes becomes inactive in cells early in development, it packs down into a block of heterochromatin called the *Barr body*, which is large enough to see under the light microscope. These findings support the idea that, in addition to organizing nuclear DNA, histones play a role in regulating gene activity.

Many Nonhistone Proteins Have Key Roles in the Regulation of Gene Expression

Nonhistone proteins are loosely defined as all the proteins associated with DNA that are not histones. Nonhistones vary widely in structure; most are negatively charged or neutral, but some are positively charged. They range in size from polypeptides smaller than histones to some of the largest cellular proteins.

Many nonhistone proteins help control the expression of individual genes. (The regulation of gene expression is the subject of Chapter 16.) For example, expression of a gene requires that the enzymes and proteins for that process be able to access the gene in the chromatin. If a gene is packed into heterochromatin, it is unavailable for activation. If the gene is in the more-extended euchromatin, it is more accessible. Many nonhistone proteins affect gene accessibility by modifying histones to change how the histones associate with DNA in chromatin, either loosening or tightening the association. Other nonhistone proteins are regulatory proteins that activate or repress the expression of a gene. Yet others are components of the enzyme–protein complexes that are needed for the expression of any gene.

DNA Is Organized More Simply in Prokaryotes Than in Eukaryotes

Several features of DNA organization in prokaryotes differ fundamentally from eukaryotic DNA. In contrast to the linear DNA in eukaryotes, the primary DNA molecule of most prokaryotic cells is circular, with only one copy per cell. In parallel with eukaryotic terminology, the DNA molecule is called a **bacterial chromosome.** The chromosome of the best-known bacterium, *E. coli,* includes about 1460 μm of DNA, which is equivalent to 4.6 million base pairs. There are exceptions: some bacteria have two or more different chromosomes in the cell, and some bacterial chromosomes are linear.

Figure 14.19
Replication from a single origin in the DNA circle of prokaryotes.

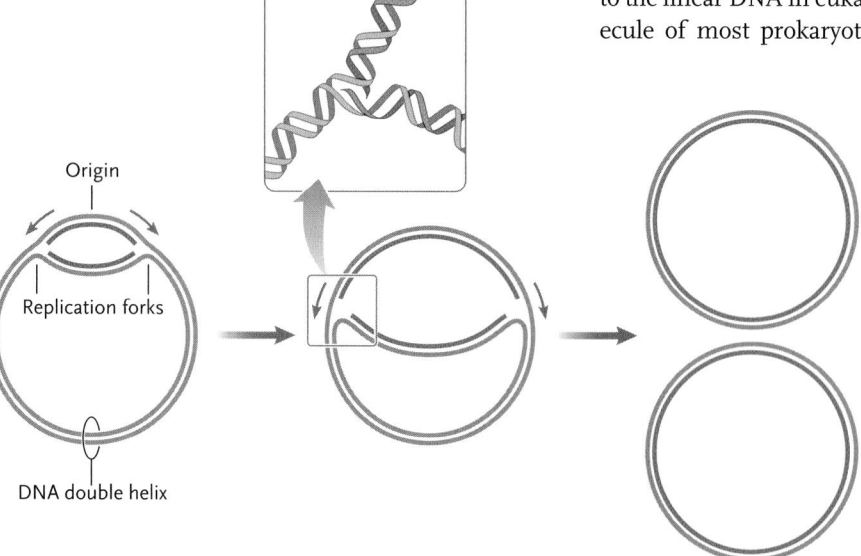

Origin

Replication forks

DNA double helix

Replication begins from a single origin in the DNA circle, forming two forks that travel around the circle in opposite directions. Eventually, the forks meet at the opposite side from the origin to complete replication **(Figure 14.19)**.

Inside prokaryotic cells, the DNA circle is packed and folded into an irregularly shaped mass called the **nucleoid** (shown in Figure 5.7). The DNA of the nucleoid is suspended directly in the cytoplasm with no surrounding membrane.

Many prokaryotic cells also contain other DNA molecules, called **plasmids**, in addition to the main chromosome of the nucleoid. Most plasmids are circular, although some are linear. Plasmids have replication origins and are duplicated and distributed to daughter cells together with the bacterial chromosome during cell division.

Although bacterial DNA is not organized into nucleosomes, there are positively charged proteins that combine with bacterial DNA. Some of these proteins help organize the DNA into loops, thereby providing some compaction of the molecule. Bacterial DNA also combines with many types of genetic regulatory proteins that have functions similar to those of the nonhistone proteins of eukaryotes (see Chapter 16).

Unanswered Questions

In this chapter, we learned about the structure and replication of DNA, the key role of telomeres in maintaining the ends of chromosomes, and the packaging of DNA into chromosomes in eukaryotes. Research is ongoing in all of these areas, both to understand basic cellular processes and in applied areas such as understanding and treating human diseases and infections.

Does DNA polymerase affect cellular aging?

As cells grow and divide, they eventually become senescent and then die. Understanding cellular aging is critical to understanding the aging process of an organism and, therefore, to developing treatments for age-related diseases.

Senescing cells eventually fail to initiate DNA synthesis and are therefore unable to transition from the G_1 to the S phase of the cell cycle (see Chapter 10). This transition requires the expression of a particular set of genes, some of which encode proteins involved in DNA replication. Loss of regulation of expression of these genes is hypothesized to contribute to cellular aging. In particular, research has shown that the activity and function of a DNA polymerase involved in the initiation of Okazaki fragments on the lagging strand template may be a crucial determinant of aging. Research is ongoing to test this hypothesis, focusing on correlating changes in the activity of the enzyme with cellular aging and the interactions of the enzyme with other replication proteins.

Can DNA replication be an effective target for antiherpes drugs?

Certain antiviral drugs target DNA replication enzymes involved in viral reproduction inside cells. For example, acyclovir, first developed in the 1970s, is the standard drug for treating various herpes simplex viruses (HSV). The genome of HSV is DNA, and a virus-encoded DNA polymerase replicates the viral DNA. Not only can HSV produce sores on the mouth and genitals, it can also kill people who have compromised immune systems, including newborns, elderly individuals, and AIDS patients. Acyclovir is a chemical analog of a DNA nucleoside; that is, it resembles a natural DNA nucleoside (deoxyribose sugar + base) in structure. Acyclovir is taken into cells. If the cell is infected with HSV, a virus-encoded enzyme efficiently converts the acyclovir to acyclovir monophosphate, which cellular enzymes then convert to acyclovir triphosphate. The triphosphate derivative is used by the HSV-encoded DNA polymerase during viral DNA replication, but incorporation of the analog blocks further DNA synthesis. In short, acyclovir blocks viral DNA replication by the viral DNA polymerase. Fortunately, acyclovir does not inhibit cellular DNA polymerase at the concentrations that are effective against the viral enzyme. Furthermore, the efficient conversion of acyclovir to acyclovir triphosphate depends on the presence of a viral enzyme, which means that acyclovir has little effect on cells not infected with HSV.

Other viral enzymes are the target of a recently developed new class of antiherpes drugs. Drugs in this class target the HSV-encoded helicase and primase enzymes used for viral DNA replication; these two proteins work as a complex at the replication fork. The new drugs have been shown to be highly effective in blocking HSV replication in tissue culture cells, as well as reducing the death rate of virus-infected mice. Moreover, they are more potent than acyclovir. Another benefit of these drugs is that they are effective when given 65 hours after mice were infected, whereas acyclovir must be given within a few hours after infection. Further research is needed to determine whether the drugs will be effective in humans, and at what safe dose. However, research in this area requires significant amounts of money, but pharmaceutical companies appear reluctant to invest in such projects because of the large profits that can be made with acyclovir.

More broadly, the results of the experiments indicate that targeting aspects of replication of viral genomes can be an effective mode of treatment.

What is the role of telomerase in cell division?

The University of California, San Francisco, research group of Elizabeth Blackburn, one of the discoverers of telomerase, is investigating the function of telomeres and telomerase in yeast and human cancer cells. Blackburn has demonstrated that some mutants of the RNA molecule in telomerase cause telomere shortening and cellular aging (senescence) in the protozoan *Tetrahymena*. In addition, she found that chemical inhibitors of telomerase cause human telomeres to shorten. Other mutations of the telomere length regulation system cause telomeres to become too long, triggering degradation of the telomeres and potentially causing the cells to stop mitosis at anaphase. Continuing research is directed at determining the precise roles of telomeres in the cell division process. The research findings are also being used as a platform for developing an anticancer strategy, using cells from human breast, prostate, and bladder cancers.

Peter J. Russell

With this description of prokaryotic DNA organization, our survey of DNA structure and its replication and organization is complete. The next chapter revisits the same structures and discusses how they function in the expression of information encoded in the DNA.

Review

14.1 Establishing DNA as the Hereditary Molecule

- Griffith found that a substance derived from killed infective pneumonia bacteria could transform noninfective living pneumonia bacteria to the infective type (Figure 14.2).

- Avery and his coworkers showed that DNA, and not protein or RNA, was the molecule responsible for transforming pneumonia bacteria into the infective form.

- Hershey and Chase showed that the DNA of a phage, not the protein, enters bacterial cells to direct the life cycle of the virus. Taken together, the experiments of Griffith, Avery and his coworkers, and Hershey and Chase established that DNA is the hereditary molecule (Figure 14.3).

Animation: Griffith's experiment

Animation: The Hershey and Chase experiments

14.2 DNA Structure

- Watson and Crick discovered that a DNA molecule consists of two polynucleotide chains twisted around each other into a right-handed double helix. Each nucleotide of the chains consists of deoxyribose, a phosphate group, and either adenine, thymine, guanine, or cytosine. The deoxyribose sugars are linked by phosphate groups to form an alternating sugar–phosphate backbone. The two strands are held together by adenine–thymine and guanine–cytosine base pairs. Each full turn of the double helix involves 10 base pairs (Figures 14.4 and 14.6).

- The two strands of the DNA double helix are antiparallel.

Animation: The nucleotides of DNA

Animation: The DNA double helix

Practice: Constructing DNA

14.3 DNA Replication

- DNA is duplicated by semiconservative replication, in which the two strands of a parental DNA molecule unwind and each serves as a template for the synthesis of a complementary copy (Figures 14.7–14.9).

- DNA replication is catalyzed by several enzymes. Helicase unwinds the DNA; primase synthesizes an RNA primer used as a starting point for nucleotide assembly by DNA polymerases. DNA polymerases assemble nucleotides into a chain one at a time, in a sequence complementary to the sequence of bases in the template strand. After a DNA polymerase removes the primers and fills in the resulting gaps, DNA ligase closes the remaining single-chain nicks (Figures 14.10 and 14.12).

- As the DNA helix unwinds, only one template strand runs in a direction allowing the new DNA strand to be made continuously in the direction of unwinding. The other template strand is copied in short lengths that run in the direction opposite to unwinding. The short lengths produced by this discontinuous replication are then linked into a continuous strand (Figures 14.11 and 14.12).

- The ends of eukaryotic chromosomes consist of telomeres, short sequences repeated hundreds to thousands of times. These repeats provide a buffer against chromosome shortening during replication. Although most somatic cells show this chromosome shortening, some cell types do not because they have a telomerase enzyme that adds telomere repeats to the chromosome ends (Figures 14.13 and 14.14).

- DNA synthesis begins at sites that act as replication origins and proceeds from the origins as two replication forks moving in opposite directions (Figure 14.15).

Animation: Overview of DNA replication and base pairing

Animation: DNA replication in detail

14.4 Mechanisms That Correct Replication Errors

- In proofreading, the DNA polymerase reverses and removes the most recently added base if it is mispaired as a result of a replication error. The enzyme then resumes DNA synthesis in the forward direction (Figure 14.16).

- In DNA mismatch repair, enzymes recognize distorted regions caused by mispaired base pairs and remove a section of DNA that includes the mispaired base from the newly synthesized nucleotide chain. A DNA polymerase then resynthesizes the section correctly, using the original template chain as a guide (Figure 14.17).

14.5 DNA Organization in Eukaryotes and Prokaryotes

- Eukaryotic chromosomes consist of DNA complexed with histone and nonhistone proteins.

- In eukaryotic chromosomes, DNA is wrapped around a nucleosome consisting of two molecules each of histones H2A, H2B, H3, and H4. Linker DNA connects adjacent nucleosomes. The binding of histone H1 causes the nucleosomes to package into a coiled structure called a solenoid (Figure 14.18).

- Chromatin is distributed between euchromatin, a loosely packed region in which genes are active in RNA transcription, and heterochromatin, densely packed masses in which the genes are inactive. Chromatin also folds and packs to form thick, rodlike chromosomes during nuclear division.

- Nonhistone proteins help control the expression of individual genes.

- The bacterial chromosome is a closed, circular molecule of DNA; it is packed into the nucleoid region of the cell. Replication begins from a single origin and proceeds in both directions. Many bacteria also contain plasmids, which replicate independently of the host chromosome (Figure 14.19).

- Bacterial DNA is organized into loops through interaction with proteins. Other proteins similar to eukaryotic nonhistones regulate gene activity in prokaryotes.

  Animation: Chromosome structural organization

Questions

Self-Test Questions

1. Working on the Amazon River, a biologist isolated DNA from two unknown organisms, P and Q. He discovered that the adenine content of P was 15% and the cytosine content of Q was 42%. This means that:
   a. the amount of guanine in P is 15%.
   b. the amount of guanine and cytosine combined in P is 70%.
   c. the amount of adenine in Q is 42%.
   d. the amount of thymine in Q is 21%.
   e. it takes more energy to unwind the DNA of P than the DNA of Q.

2. The Hershey and Chase experiment showed that viral:
   a. ^{35}S entered bacterial cells.
   b. ^{32}P remained outside of bacterial cells.
   c. protein entered bacterial cells.
   d. DNA entered bacterial cells.
   e. DNA mutated in bacterial cells.

3. Pyrimidines include:
   a. cytosine and thymine.
   b. adenine, cytosine, and guanine.
   c. adenine and thymine.
   d. cytosine and guanine.
   e. adenine and guanine.

4. Which of the following statements about DNA replication is *false*?
   a. Synthesis of the new DNA strand is from 3′ to 5′.
   b. Synthesis of the new DNA strand is from 5′ to 3′.
   c. DNA unwinds, primase adds RNA primer, and DNA polymerases synthesize the new strand and remove the RNA primer.
   d. Many initiation points exist in each eukaryotic chromosome.
   e. Okazaki fragments are synthesized in the opposite direction from the direction in which the replication fork moves.

5. Which of the following statements about DNA is *false*?
   a. Phosphate is linked to the 5′ and 3′ carbons of adjacent deoxyribose molecules.
   b. DNA is bidirectional in its synthesis.
   c. Each side of the helix is antiparallel to the other.
   d. The binding of adenine to thymine is through three hydrogen bonds.
   e. Avery identified DNA as the transforming factor in crosses between smooth and rough bacteria.

6. In the Meselson and Stahl experiment, the DNA in the parental generation was all ^{15}N^{15}N, and after one round of replication, the DNA was all ^{15}N^{14}N. What DNAs were seen after three rounds of replication, and in what ratio were they found?
   a. one ^{15}N^{14}N : one ^{14}N:^{14}N
   b. one ^{15}N^{14}N : two ^{14}N:^{14}N
   c. one ^{15}N^{14}N : three ^{14}N:^{14}N
   d. one ^{15}N^{14}N : four ^{14}N:^{14}N
   e. one ^{15}N^{14}N : seven ^{14}N:^{14}N

7. During replication, DNA is synthesized in a 5′→3′ direction. This implies that:
   a. the template is read in a 5′→3′ direction.
   b. successive nucleotides are added to the 3′–OH end of the newly forming chain.

   c. because both strands are replicated nearly simultaneously, replication must be continuous on both.
   d. ligase unwinds DNA in a 5′→3′ direction.
   e. primase acts on the 3′ end of the replicating strand.

8. Telomerase:
   a. is active in cancer cells.
   b. is more active in adult than embryonic cells.
   c. complexes with the ribosome to form telomeres.
   d. acts on unique genes called telomeres.
   e. shortens the ends of chromosomes.

9. Mismatch repair is the ability:
   a. to seal Okazaki fragments with ligase into a continual DNA strand.
   b. of primase to remove the RNA primer and replace it with the correct DNA.
   c. of some enzymes to sense the insertion of an incorrect nucleotide, remove it, and use a DNA polymerase to insert the correct one.
   d. to correct mispaired chromosomes in prophase I of meiosis.
   e. to remove worn-out DNA by telomerase and replace it with newly synthesized nucleotides.

10. Prokaryotic DNA:
    a. is surrounded by densely packed histones.
    b. has many sites for the initiation of DNA replication.
    c. has both strands synthesized in the same direction.
    d. is packaged as euchromatin and heterochromatin.
    e. is packaged as a large circular chromosome.

Questions for Discussion

1. Chargaff's data suggested that adenine pairs with thymine and guanine pairs with cytosine. What other data available to Watson and Crick suggested that adenine–guanine and cytosine–thymine pairs normally do not form?

2. Eukaryotic chromosomes can be labeled by exposing cells to radioactive thymidine during the S phase of interphase. If cells are exposed to radioactive thymidine during the S phase, would you expect both or only one of the sister chromatids of a duplicated chromosome to be labeled at metaphase of the following mitosis (see Section 10.2)?

3. If the cells in question 2 finish division and then enter another round of DNA replication in a medium that has been washed free of radioactive label, would you expect both or only one of the sister chromatids of a duplicated chromosome to be labeled at metaphase of the following mitosis?

4. During replication, an error uncorrected by proofreading or mismatch repair produces a DNA molecule with a base mismatch at the indicated position:

   AATTCCGACTCCTATGG
   TTAAGGTTGAGGATACC
               ↑

   The mismatch results in a mutation. This DNA molecule is received by one of the two daughter cells produced by mitosis. In the next round of replication and division, the mutation appears in only one of the two daughter cells. Develop a hypothesis to explain this observation.

5. Strains of bacteria that are resistant to an antibiotic sometimes appear spontaneously among other bacteria of the same type that are killed by the antibiotic. In view of the information in this chapter about DNA replication, what might account for the appearance of this resistance?

Experimental Analysis

Design an experiment using radioactive isotopes to show that the process of bacterial transformation involves DNA and not protein.

Evolution Link

The amino acid sequences of the DNA polymerases found in bacteria show little similarity to those of the DNA polymerases found in eukaryotes and in archaeans. By contrast, the amino acid sequences of the DNA polymerases of eukaryotes and archaeans show a high degree of similarity. Interpret these observations from an evolutionary point of view.

Transcription of a eukaryotic gene to produce messenger RNA (mRNA), a type of RNA that acts as a template for protein synthesis. The DNA of the gene unwinds from the nucleosome (left side) and is copied by an RNA polymerase (center) into mRNA (exiting the top).

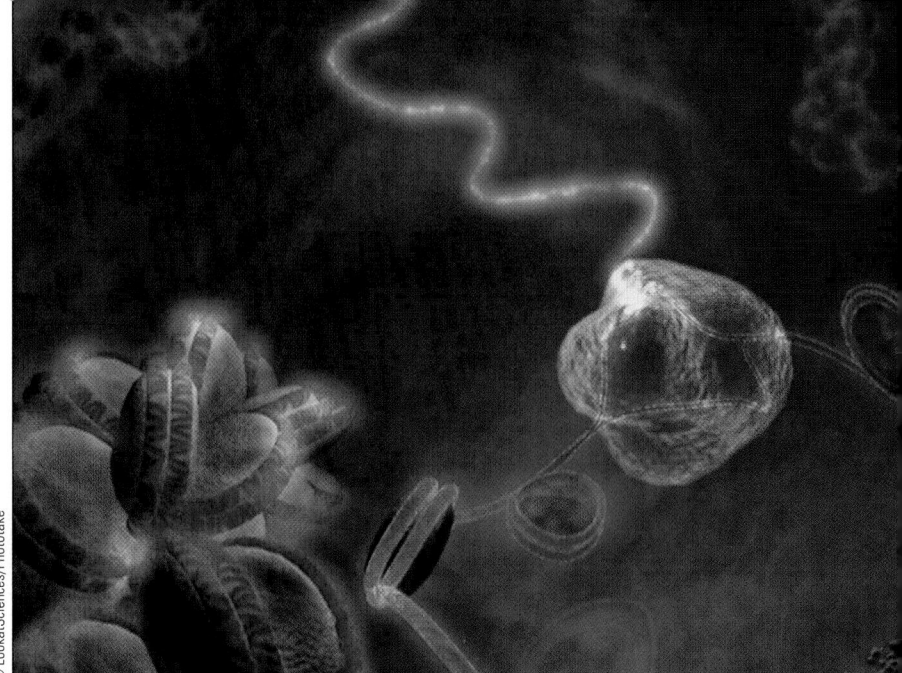

© LookatSciences/Phototake

15 From DNA to Protein

WHY IT MATTERS

The marine mussel *Mytilus* **(Figure 15.1)** lives in one of the most demanding environments on Earth—it clings permanently to rocks pounded by surf day in and day out, constantly in danger of being dashed to pieces or torn loose by foraging predators. The mussel is remarkably resistant to disturbance, however; if you try to pry one loose you will find how difficult it is to tear the tough, elastic fibers that hold it fast, or even to cut them with a knife.

The fibers holding mussels to the rocks are a complex of proteins secreted by the muscular foot of the animal. The proteins, which include *keratin* (an intermediate filament protein; discussed in Section 5.3) form a tough, adhesive material called *byssus*.

Byssus is a premier underwater adhesives. It fascinates biochemists, adhesive manufacturers, dentists, and surgeons looking for better ways to hold repaired body parts together. Genetic engineers have inserted segments of mussel DNA into yeast cells, which reproduce in large numbers and serve as "factories" translating the mussel genes into byssus and other proteins. With byssus produced in this way, investigators are learning how to use or imitate the mussel glue for human needs. This exciting work, like the mussel's own byssus-building,

Figure 15.1
The marine mussel *Mytilus* and its natural habitat.

starts with one of life's universal truths: *every protein is assembled on ribosomes according to instructions that are copied from DNA.*

In this chapter we trace the reactions by which proteins are made, beginning with the instructions encoded in DNA and leading through RNA to the sequence of amino acids in a protein. Many enzymes and other proteins are players as well as products in this story, as are several kinds of RNA and the cell's protein-making machines, the ribosomes. The same basic steps produce the proteins of all organisms. Our discussion begins with an overview of the entire process, starting with DNA and ending with a finished protein.

15.1 The Connection between DNA, RNA, and Protein

You have learned that genes encode proteins. In this section you will learn how that connection was discovered. This section also presents an overview of the molecular steps from gene to protein: transcription and translation.

Proteins Are Specified by Genes

How do scientists know that genes encode—specify the amino acid sequence of—proteins? Two key pieces of research involving defects in metabolism proved this connection unequivocally. The first began in 1896 with Archibald Garrod, an English physician. He studied *alkaptonuria*, a human disease that does little harm but is easily detected: the patient's urine turns black in air. Garrod and an English geneticist, William Bateson, studied families of patients with the disease and concluded that it is an inherited trait. Garrod also found that people with alkaptonuria excrete a particular chemical in their urine. It is this chemical that turns

black in air. Garrod deduced that normal people are able to metabolize the chemical, whereas people with alkaptonuria cannot. In 1908 Garrod concluded that the disease was an *inborn error of metabolism*. He did not know it at the time, but alkaptonuria results from a change in a gene that encodes an enzyme that metabolizes a key chemical. The altered gene causes a defect in the function of the enzyme, which leads to the phenotype of the disease. Garrod's work was the first evidence of a specific relationship between genes and metabolism.

In the second piece of research, George Beadle and Edward Tatum, working at Stanford University in the 1940s with the orange bread mold *Neurospora crassa,* obtained results showing a direct relationship between genes and enzymes. Beadle and Tatum chose *Neurospora* for their work because it is a haploid fungus, with simple nutritional needs. That is, wild-type *Neurospora*—the form of the mold found in nature—grows readily on a minimal medium (MM) consisting of a number of inorganic salts, sucrose, and a vitamin. The researchers reasoned that the fungus uses the simple chemicals in MM to synthesize all of the more complex molecules needed for growth and reproduction, including amino acids for proteins and nucleotides for DNA and RNA.

Beadle and Tatum exposed spores of wild-type *Neurospora* to X-rays. An X-ray is a *mutagen,* an agent that causes mutations. They found that some of the treated spores would not germinate and grow on MM unless they supplemented the medium with additional nutrients, such as amino acids or vitamins. Mutant strains that are unable to grow on MM are called *auxotrophs* (*auxo* = increased; *troph* = eater) or *nutritional mutants.* Beadle and Tatum hypothesized that each auxotrophic strain had a defect in a gene that codes for an enzyme needed to synthesize a particular nutrient. The wild-type strain could make the nutrient for itself from raw materials in the MM, but the mutant strain could grow only if the researchers supplied the nutrient. By testing each mutant strain on MM with a single added nutrient, they discovered what specific nutrient the strain needed in order to grow and, therefore, generally what gene defect it had. For example, a mutant that requires the addition of the amino acid arginine to grow has a defect in a gene for an enzyme involved in the synthesis of arginine. Such arginine auxotrophs are known as *arg* mutants. The assembly of arginine from raw materials is a multistep "assembly-line" process with a different enzyme catalyzing each step. Each *arg* mutant differs in the enzyme that is defective and therefore in the step of the assembly pathway that is blocked.

Beadle and Tatum studied four *arg* mutants—*argE, argF, argG,* and *argH*—to determine the metabolic defect each had; that is, where in the arginine synthesis pathway each was blocked. They took samples from each culture and tested whether the samples could grow on MM, or on MM supplemented either with ornithine, citrulline, argininosuccinate—three compounds known to be involved in the synthesis of arginine—or with arginine itself **(Figure 15.2).** None of the four mutants grew on MM, but all grew on MM + arginine. Each of the *arg* mutants showed a different pattern of growth on the supplemented MM (see Figure 15.2). Beadle and Tatum deduced that the biosynthesis of arginine occurs in a number of steps, with each step controlled by a gene that encodes the enzyme for the step (see Figure 15.2, Conclusion). For example, the *argH* mutant grows on MM + arginine, but not on MM + any of the other three compounds; this means that the mutant is blocked at the last step in the pathway that produces arginine. Similarly, the *argG* mutant grows on MM + arginine or argininosuccinate, but not on MM + any of the other supplements; this means that *argG* is blocked in the pathway before argininosuccinate is made (see Figure 15.2, Conclusion). With similar analysis, the researchers deduced the whole pathway from precursor to arginine and showed which gene encoded the enzyme that carried out each step. In sum, Beadle and Tatum had shown the direct relationship between genes and enzymes, which they put forward as the **one gene–one enzyme hypothesis.** Their experiment was a keystone in the development of molecular biology. As a result of their work, they were awarded a Nobel Prize in 1958.

As you learned in Chapter 3, enzymes are just one form of proteins, the amino acid-containing macromolecules that carry out many vital functions in living organisms. A functional protein consists of one or more subunits, called *polypeptides.* The protein hemoglobin, for instance, is made up of four polypeptides, two each of an α subunit and a β subunit. Hemoglobin's ability to transport oxygen is a functional property belonging only to the complete protein, and not to any of the polypeptides individually. Since a different gene encodes each distinct polypeptide, two different genes are needed to encode the hemoglobin protein: one for the α polypeptide and one for the β polypeptide. Due to the fact that some proteins consist of more than one polypeptide, and not all proteins are enzymes, Beadle and Tatum's hypothesis is now restated as the **one gene–one polypeptide hypothesis.** It is important to keep in mind the distinction between a protein, the functional molecule, and a polypeptide, the molecule encoded by a gene, as we discuss transcription and translation in the rest of this chapter.

The Pathway from Gene to Polypeptide Involves Transcription and Translation

The pathway from gene to polypeptide has two major steps, *transcription* and *translation.* **Transcription** is the mechanism by which the information encoded in DNA is made into a complementary RNA copy. It is called transcription because the information in one nucleic acid type is transferred to another nucleic acid type. **Translation** is the use of the information encoded in the RNA to assemble amino acids into a polypeptide. It is called translation because the information in a nucleic acid, in the form of nucleotides, is converted into a different kind of molecule—amino acids. In 1956, Francis Crick gave the name *central dogma* to the flow of information from DNA to RNA to protein.

In transcription, the enzyme RNA polymerase copies the DNA sequence of a gene into an RNA sequence. The process is similar to DNA replication, except that only one of the two DNA strands—the **template strand**—is copied into an RNA strand, and only part of the DNA sequence of the genome is copied in any cell at any given time. A gene encoding a polypeptide is a *protein-coding gene,* and the RNA transcribed from it is called **messenger RNA (mRNA).**

In translation, an mRNA associates with a *ribosome,* a particle on which amino acids are linked into polypeptide chains. As the ribosome moves along the mRNA, the amino acids specified by the mRNA are joined one by one to form the polypeptide encoded by the gene.

The processes of transcription and translation are similar in prokaryotes and eukaryotes **(Figure 15.3).** One key difference is that in eukaryotes, RNA polymerase makes a precursor-mRNA (pre-mRNA) in the nucleus that is processed to produce the functional mRNA. That mRNA exits the nucleus and is translated in the cytoplasm.

The Genetic Code Is Written in Three-Letter Words Using a Four-Letter Alphabet

Conceptually, the transcription of DNA into RNA is straightforward. The DNA "alphabet" consists of the four letters A, T, G, and C, representing the four DNA

Figure 15.2 Experimental Research

Relationship between Genes and Enzymes

QUESTION: Do genes specify enzymes?

EXPERIMENT: Test *arg* mutants of the orange bread mold *Neurospora crassa* for growth on MM (minimal medium), MM + ornithine, MM + citrulline, MM + argininosuccinate, and MM + arginine. *Arg* mutants are unable to synthesize the amino acid arginine, which is essential for growth.

RESULTS:

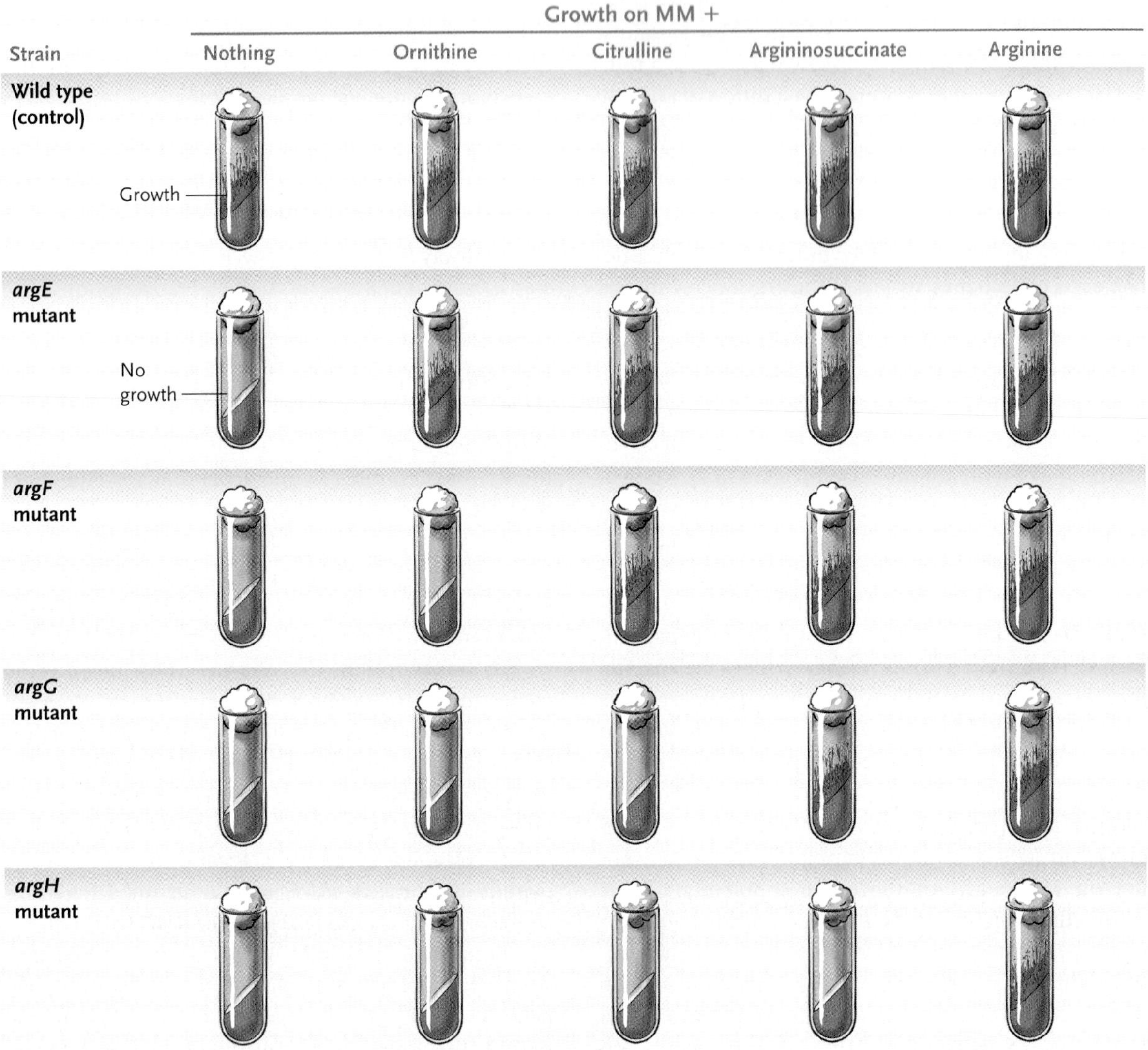

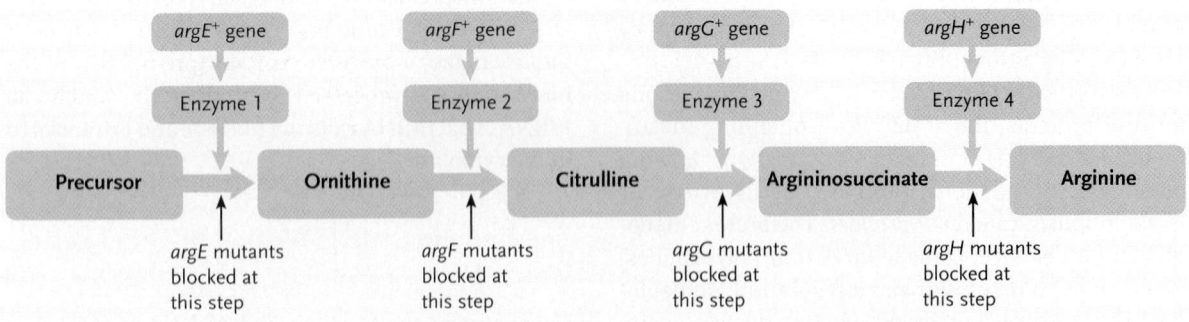

Growth on MM +

Strain	Nothing	Ornithine	Citrulline	Argininosuccinate	Arginine
Wild type (control)	Growth				
argE mutant	No growth				
argF mutant					
argG mutant					
argH mutant					

CONCLUSION: Arginine is synthesized in a biochemical pathway. Each step of the pathway is catalyzed by an enzyme, and each enzyme is encoded by a gene:

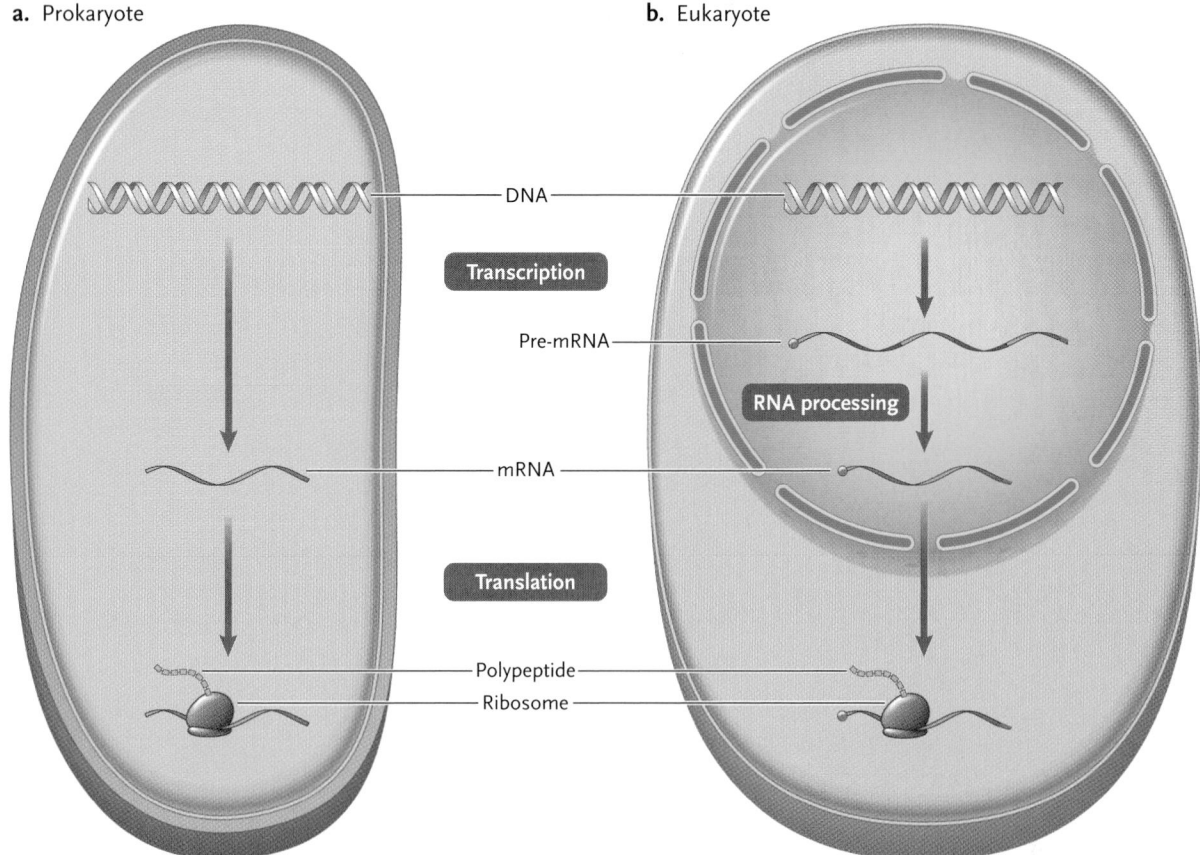

a. Prokaryote

b. Eukaryote

DNA

Transcription

Pre-mRNA

RNA processing

mRNA

Translation

Polypeptide

Ribosome

Figure 15.3

Transcription and translation in **(a)** prokaryotes and **(b)** eukaryotes. In prokaryotes RNA polymerase synthesizes an mRNA molecule that is ready for translation on ribosomes. In eukaryotes, RNA polymerase synthesizes a precursor-mRNA (pre-mRNA molecule) that has extra segments that are removed by RNA processing to produce a translatable mRNA. That mRNA exits the nucleus through a nuclear pore and is translated on ribosomes in the cytoplasm.

nucleotide bases: adenine, thymine, guanine, and cytosine. The RNA "alphabet" consists of the four letters A, U, G, and C, representing the four RNA bases: adenine, uracil, guanine, and cytosine. In other words, the nucleic acids share three of the four bases but differ in the other one; T in DNA is equivalent to U in RNA. But while there are four RNA bases, there are 20 amino acids. How is nucleotide information in an mRNA translated into the amino acid sequence of a polypeptide?

Breaking the Genetic Code. The nucleotide information that specifies the amino acid sequence of a polypeptide is called the **genetic code.** Scientists realized that the four bases in an mRNA (A, U, G, C) would have to be used in combinations of at least three to provide the capacity to code for 20 different amino acids. One- and two-letter words were eliminated because if the code used one-letter words, only four different amino acids could be specified (that is, 4^1); if two-letter words

were used, only 16 different amino acids could be specified (that is, 4^2). But if the code used three-letter words, 64 different amino acids could be specified (that is, 4^3), more than enough to specify 20 amino acids. We know now that the genetic code is a three-letter code; each three-letter word (triplet) of the code is called a **codon.** **Figure 15.4** illustrates the relationship between codons in a gene, codons in an mRNA, and the amino acid sequence of a polypeptide. The three-letter codons in DNA are first transcribed into complementary three-letter RNA codons. The process is similar to DNA replication except that in mRNA, the complement to adenine (A) in the template strand is uracil (U) instead of thymine (T) as in DNA replication.

How do the RNA codons correspond to the amino acids? The identity of most of the codons was established in 1964 by Marshall Nirenberg and Philip Leder of the National Institutes of Health (NIH). These researchers found that short, artificial mRNAs of codon length—three nucleotides—could bind to ribosomes in a test tube and cause a single transfer RNA (tRNA), with its linked amino acid, to bind to the ribosome. (As you will learn in Section 15.4, tRNAs are a special class of RNA molecules that bring amino acids to the ribosome for assembly into the polypeptide chain.) Nirenberg and Leder then made 64 of the short mRNAs, each consisting of a different, single codon. They added the mRNAs, one at a time, to a test tube containing ribosomes and all the different tRNAs, each linked to its own amino

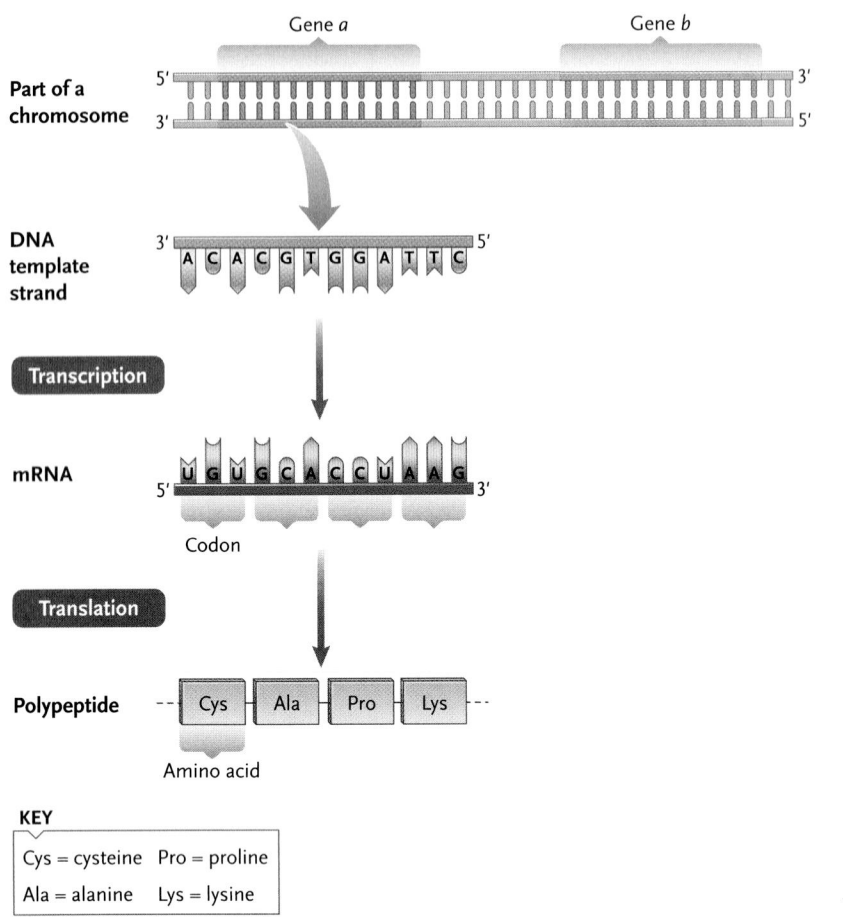

Part of a chromosome

Gene *a*   Gene *b*

5′ ——————————————————————— 3′
3′ ——————————————————————— 5′

DNA template strand

3′ A C A C G T G G A T T C 5′

Transcription

mRNA

5′ U G U G C A C C U A A G 3′

Codon

Translation

Polypeptide

Cys — Ala — Pro — Lys

Amino acid

KEY

Cys = cysteine   Pro = proline
Ala = alanine   Lys = lysine

Figure 15.4
Relationship between a gene, codons in an mRNA, and the amino acid sequence of a polypeptide.

Another approach, carried out in 1966 by H. Gobind Khorana and his coworkers at Massachusetts Institute of Technology, used long, artificial mRNA molecules containing only one nucleotide repeated continuously, or different nucleotides in repeating patterns. Each artificial mRNA was added to ribosomes in a test tube, and the sequence of amino acids in the polypeptide chain made by the ribosomes was analyzed. For example, an artificial mRNA containing only uracil nucleotides in the sequence UUUUUU . . . resulted in a polypeptide containing only the amino acid phenylalanine: UUU must be the codon for phenylalanine. Khorana's approach, combined with the results of Nirenberg and Leder's experiments, identified the coding assignments of all the codons. Nirenberg and Khorana received a Nobel Prize in 1968 for their research in solving the nucleic acid code.

Features of the Genetic Code. **Figure 15.5** shows the genetic code of the 64 possible codons. By convention, scientists write the codons in the 5′→3′ direction, and as they appear in mRNAs, in which U substitutes for the T of DNA. Of the 64 codons, 61 specify amino acids. These are known as **sense codons.** One of these codons, AUG, specifying the amino acid methionine, is the first codon read in an mRNA in translation in both prokaryotes and eukaryotes. In that position, AUG is called a **start codon** or **initiator codon.** The three codons that do not specify amino acids—UAA, UAG, and UGA—are **stop codons** (also called **nonsense codons** and **termination codons**) that act as "periods" indicating the end of a polypeptide-encoding sentence. When a ribosome reaches one of the stop codons, poly-

acid. The idea was that each single-codon mRNA would link to the tRNA in the mixture that carried the amino acid corresponding to the codon. The experiment worked for 50 of the 64 codons, allowing those codons to be assigned to amino acids definitively.

Figure 15.5
The genetic code, written in the form in which the codons appear in mRNA. The AUG initiator codon, which codes for methionine, is shown in green; the three terminator codons are boxed in red.

Second base of codon

First base of codon	U	C	A	G	Third base of codon
U	UUU UUC — Phe UUA UUG — Leu	UCU UCC UCA UCG — Ser	UAU UAC — Tyr UAA UAG	UGU UGC — Cys UGA UGG — Trp	U C A G
C	CUU CUC CUA CUG — Leu	CCU CCC CCA CCG — Pro	CAU CAC — His CAA CAG — Gln	CGU CGC CGA CGG — Arg	U C A G
A	AUU AUC — Ile AUA AUG — Met	ACU ACC ACA ACG — Thr	AAU AAC — Asn AAA AAG — Lys	AGU AGC — Ser AGA AGG — Arg	U C A G
G	GUU GUC GUA GUG — Val	GCU GCC GCA GCG — Ala	GAU GAC — Asp GAA GAG — Glu	GGU GGC GGA GGG — Gly	U C A G

KEY

Ala = alanine
Arg = arginine
Asn = asparagine
Asp = aspartic acid
Cys = cysteine
Gln = glutamine
Glu = glutamic acid
Gly = glycine
His = histidine
Ile = isoleucine
Leu = leucine
Lys = lysine
Met = methionine
Phe = phenylalanine
Pro = proline
Ser = serine
Thr = threonine
Trp = tryptophan
Tyr = tyrosine
Val = valine

peptide synthesis stops and the new polypeptide chain is released from the ribosome.

Only two amino acids, methionine and tryptophan, are specified by a single codon. All the rest are represented by more than one codon, some by as many as six. In other words, there are many *synonyms* in the nucleic acid code, a feature known as **degeneracy** (also called *redundancy*). For example, UGU and UGC both specify cysteine, and CCU, CCC, CCA, and CCG all specify proline.

Another feature of the genetic code is that it is **commaless**; that is, the words of the nucleic acid code are sequential, with no indicators such as commas or spaces to mark the end of one codon and the beginning of the next. The code can be read correctly only by starting at the right place—at the first base of the first three-letter codon at the beginning of a coded message—and reading three nucleotides at a time from this beginning codon. In other words, there is only one correct reading frame for each mRNA. For example, if you read the message SADMOMHASMOPCUTOFFBOYTOT three letters at a time, starting with the first letter of the first "codon," you would find that a mother reluctantly had her small child's hair cut. However, if you start incorrectly at the second letter of the first codon, you read the gibberish message ADM OMH ASM OPC UTO FFB OYT OT.

The code is also **universal.** With a few exceptions, the same codons specify the same amino acids in all living organisms, and also in viruses. The universality of the nucleic acid code indicates that it was established in its present form very early in the evolution of life and has remained virtually unchanged through billions of years of evolutionary history. (The evolution of life and the genetic code are discussed further in Chapter 24.) Minor exceptions to the universality of the genetic code have been found in a few organisms including a yeast, some protozoans, a prokaryote, and in the genetic systems of mitochondria and chloroplasts.

STUDY BREAK

1. On the basis of their work with auxotrophic mutants of the fungus, *Neurospora crassa,* Beadle and Tatum proposed the one gene–one enzyme hypothesis. Why is it now known as the one gene–one polypeptide hypothesis?
2. If the codon were five bases long, how many different codons would exist in the genetic code?

15.2 Transcription: DNA-Directed RNA Synthesis

Transcription is the process by which information coded in DNA is transferred to a complementary RNA copy. The process of RNA transcription is similar to DNA replication (**Figure 15.6**). However, there are some important differences between transcription and replication. In transcription:

- Only one of the two DNA nucleotide strands acts as a template for synthesis of a complementary copy, instead of both as in replication.
- Only a relatively small part of a DNA molecule—the sequence encoding a single gene—serves as a template, rather than all of both strands as in DNA replication.
- **RNA polymerases** catalyze the assembly of nucleotides into an RNA strand, rather than the DNA polymerases that catalyze replication.
- The RNA molecules resulting from transcription are single polynucleotide chains, not double ones as in DNA replication.
- Where adenine appears in the DNA template chain, a uracil is matched to it in the RNA transcript instead of thymine as in DNA replication (see Figure 15.4 and Figure 15.6).

Transcription is similar in prokaryotes and eukaryotes. Throughout this section, we will point out the important differences between prokaryote and eukaryote processes.

RNA Polymerases Work Like DNA Polymerases, but Require No Primer

Transcription begins as RNA polymerase binds to the DNA and unwinds it near the beginning of a gene (Figure 15.6, step 1). Unlike DNA polymerases, RNA polymerases can start the complementary copy with no need for a primer already in place (primers for DNA replication are discussed in Section 14.3). Like DNA, RNA is made in the 5′→3′ direction using the 3′→5′ DNA strand as template (step 2). Thus, we refer to the beginning of the RNA strand as the *5′ end,* and the other end as the *3′ end.* The RNA polymerase continues adding nucleotides one at a time until the gene is transcribed completely. At this point the newly synthesized RNA molecule and the enzyme are released from the DNA template (step 3).

Specific Sequences of Nucleotides in the DNA Indicate Where Transcription of a Gene Begins and Ends

An organism's genome contains a large number of genes. For example, scientists analyzing the human genome sequence believe between 20,000 and 25,000 protein-coding genes are needed to make a human. Transcription is the process whereby particular genes are expressed in any given cell at a given time. Some of those genes are protein-coding genes which encode mRNAs that are translated; others are non–protein-coding genes which encode RNAs that are

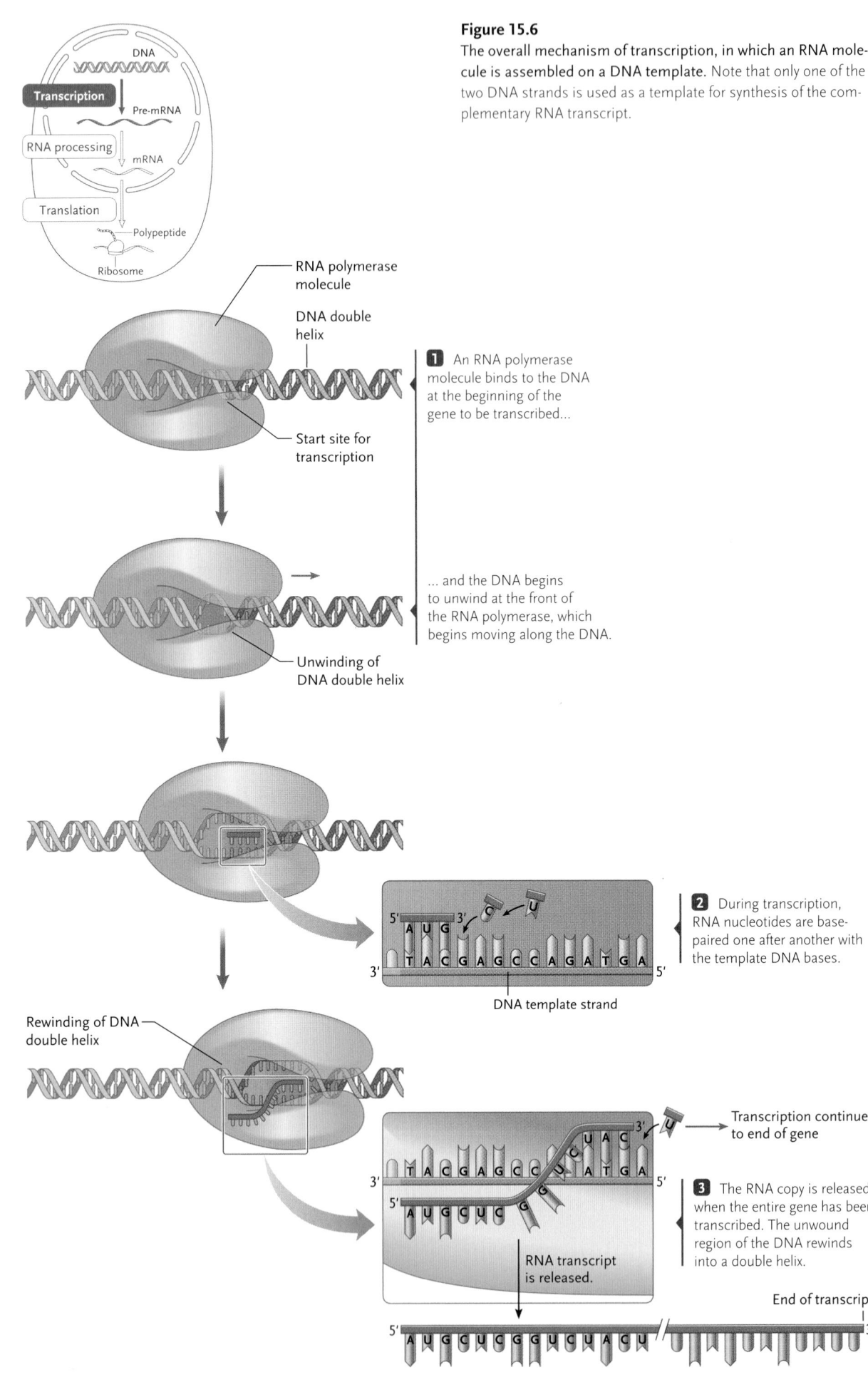

Figure 15.6
The overall mechanism of transcription, in which an RNA molecule is assembled on a DNA template. Note that only one of the two DNA strands is used as a template for synthesis of the complementary RNA transcript.

DNA

Transcription

Pre-mRNA

RNA processing

mRNA

Translation

Polypeptide

Ribosome

RNA polymerase molecule

DNA double helix

Start site for transcription

1 An RNA polymerase molecule binds to the DNA at the beginning of the gene to be transcribed...

... and the DNA begins to unwind at the front of the RNA polymerase, which begins moving along the DNA.

Unwinding of DNA double helix

2 During transcription, RNA nucleotides are base-paired one after another with the template DNA bases.

DNA template strand

Rewinding of DNA double helix

Transcription continues to end of gene

3 The RNA copy is released when the entire gene has been transcribed. The unwound region of the DNA rewinds into a double helix.

RNA transcript is released.

End of transcript

not translated, such as ribosomal RNAs (rRNAs) and transfer RNAs (tRNAs) The following sections describe the basic steps of transcription for a protein-coding gene.

Organization of a Gene and the Steps of Transcription. Let us first outline the structure of a gene and how it is transcribed into an RNA **(Figure 15.7).** At one end of a gene is a control sequence called a **promoter** (Figure 15.7, step 1). The part of the gene that is copied into RNA is called the **transcription unit.** To *initiate* transcription, RNA polymerase binds to the promoter, unwinds the DNA in that region, and starts synthesizing a new RNA molecule at the *transcription start point* (step 1). As RNA polymerase moves along the DNA, unwinding it at the forward end of the enzyme, the RNA molecule *elongates* as new nucleotides are added one by one (step 2). The newly synthesized portion of the RNA molecule winds temporarily with the template strand of the DNA into a hybrid RNA-DNA double helix. Beyond this short region the growing RNA strand unwinds from the DNA and extends from the RNA polymerase as a single nucleotide chain. At the back end of the RNA polymerase, the DNA double helix reforms. Elongation of the RNA chain continues until the end of the transcription unit, at which point RNA synthesis *terminates* and the completed RNA transcript and RNA polymerase are released from the DNA (step 3).

Once an RNA polymerase molecule has started transcription and progressed past the beginning of a gene, another molecule of RNA polymerase may start transcribing as soon as there is room at the promoter. In most genes this process continues until there are many RNA polymerase molecules spaced closely along a gene, each making an RNA transcript.

The Promoter of Protein-Coding Genes and Transcription Initiation. The promoter specifies where in the DNA transcription begins. In prokaryotes, the promoters are immediately upstream of the site where transcription initiates. RNA polymerase itself recognizes key DNA sequences in the promoter, binds to the promotor, and begins transcription of the mRNA. The same RNA polymerase also transcribes the tRNA and rRNA genes in prokaryotes; those genes have promoters highly similar in sequence to those of protein-coding genes.

In eukaryotes, RNA polymerase II transcribes protein-coding genes. RNA polymerases I and III transcribe genes for non–protein-coding RNAs, such as rRNAs in the ribosomes and tRNAs. Like prokaryotic promoters, the promoters of eukaryotic protein-coding genes are immediately upstream of the transcription start point, but they are typically more complex than those in prokaryotes. For example, a key element of the promoter of most eukaryotic protein-coding genes, the **TATA box,** plays an important role in transcription ini-

tiation. Other sequences further upstream of the gene are important for regulating the rate of transcription (discussed in Chapter 16).

Transcription Termination. In prokaryotes, specific DNA sequences called **terminators** signal the end of transcription of the gene. A protein binds to the terminator, triggering the termination of transcription and the release of the RNA and RNA polymerase from the template. Eukaryotic DNA has no equivalent sequences. Instead, the 3′ end of the mRNA is specified by a very different process, which will be discussed in the next section.

Study Break

1. If the DNA template strand has the sequence 3′-CAAATTGGCTTATTACCGGATG-5′, what would be the sequence of an RNA transcribed from it?
2. What is the role of the promoter in transcription?

15.3 Production of mRNAs in Eukaryotes

Both prokaryotic and eukaryotic mRNAs contain regions that code for proteins, along with noncoding regions that play key roles in the process of protein synthesis. In prokaryotic mRNAs, the coding region is flanked by untranslated ends, the 5′ untranslated region (5′ UTR) and a 3′ untranslated region (3′ UTR). The same elements are present in eukaryotic mRNAs along with additional noncoding elements. The synthesis of mRNA in eukaryotes is the focus of this section. Roger Kornberg of Stanford University received a Nobel Prize in 2006 for describing the molecular structure of the eukaryotic transcription apparatus and how it acts in transcription.

Eukaryotic Protein-Coding Genes Are Transcribed into Precursor-mRNAs That Are Modified in the Nucleus

A eukaryotic protein-coding gene is typically transcribed into a **precursor-mRNA (pre-mRNA)** that must be processed in the nucleus to produce the translatable mRNA **(Figure 15.8;** and see Figure 15.3). The mRNA exits the nucleus and is translated in the cytoplasm.

Modifications of Pre-mRNA and mRNA Ends. At the 5′ end of the pre-mRNA is the **5′ cap,** consisting of a guanine-containing nucleotide that is reversed so that its 3′-OH group faces the beginning rather than the end of the molecule. A *capping enzyme* adds the 5′ cap to the pre-mRNA soon after RNA polymerase II begins

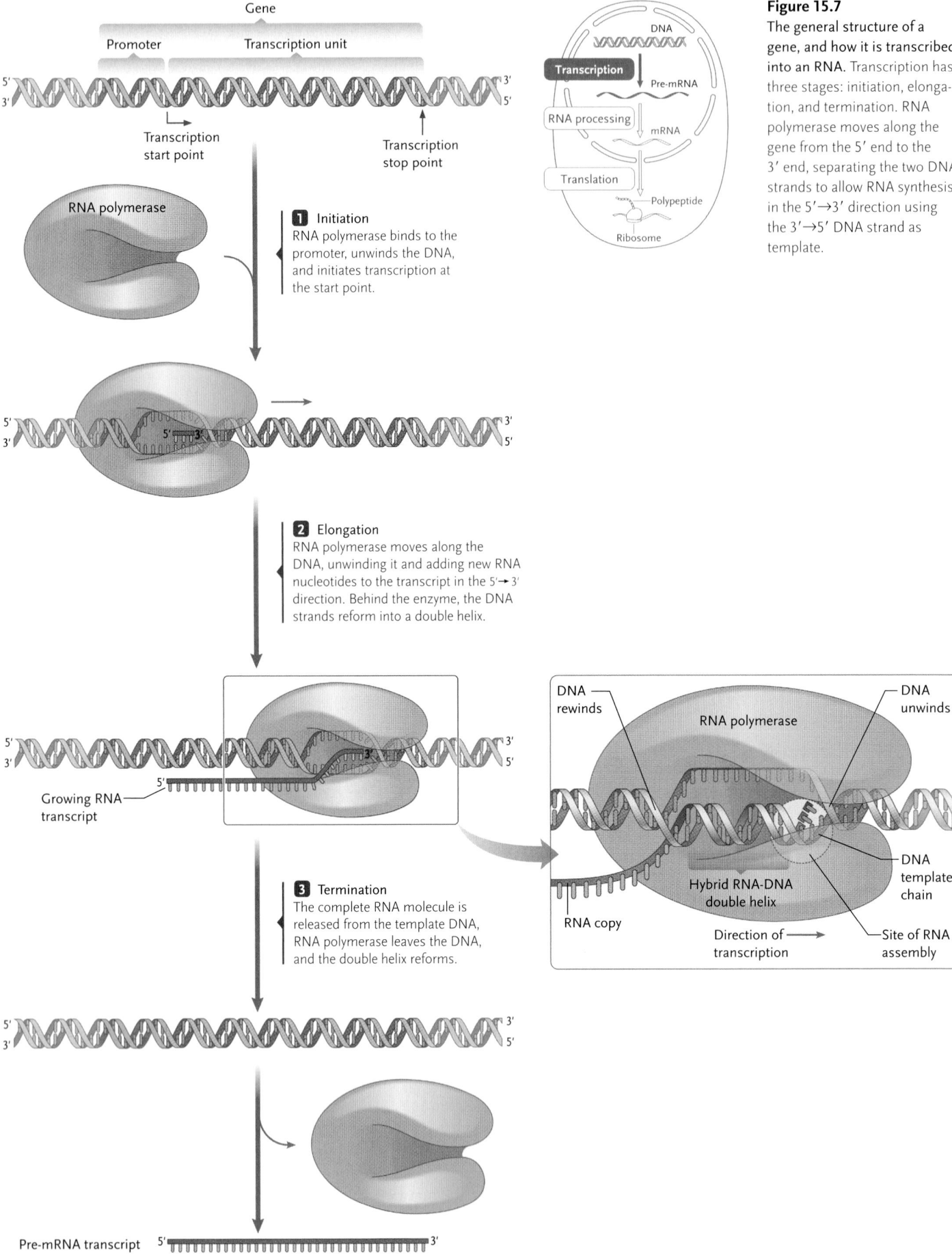

Figure 15.7
The general structure of a gene, and how it is transcribed into an RNA. Transcription has three stages: initiation, elongation, and termination. RNA polymerase moves along the gene from the 5' end to the 3' end, separating the two DNA strands to allow RNA synthesis in the 5'→3' direction using the 3'→5' DNA strand as template.

Gene

Promoter   Transcription unit

Transcription start point

Transcription stop point

RNA polymerase

1 Initiation
RNA polymerase binds to the promoter, unwinds the DNA, and initiates transcription at the start point.

2 Elongation
RNA polymerase moves along the DNA, unwinding it and adding new RNA nucleotides to the transcript in the 5'→3' direction. Behind the enzyme, the DNA strands reform into a double helix.

Growing RNA transcript

3 Termination
The complete RNA molecule is released from the template DNA, RNA polymerase leaves the DNA, and the double helix reforms.

DNA rewinds

RNA polymerase

DNA unwinds

Hybrid RNA-DNA double helix

DNA template chain

RNA copy

Direction of transcription

Site of RNA assembly

Pre-mRNA transcript

DNA

Transcription

Pre-mRNA

RNA processing

mRNA

Translation

Polypeptide

Ribosome

transcription. The cap, which is connected to the rest of the chain by three phosphate groups, remains when pre-mRNA is processed to mRNA. The cap is the site where ribosomes attach to mRNAs at the start of translation.

The termination of transcription of a eukaryotic protein-coding gene is different from that of a prokaryotic gene in that there is no terminator sequence at the end of the gene in the DNA that signals RNA polymerase to stop transcription. Instead, at the 3′ end of the gene is a sequence that is transcribed into the pre-mRNA. Proteins bind to this *polyadenylation signal,* and cleave the pre-mRNA at that point. Then, the enzyme *poly(A) polymerase* adds a chain of 50 to 250 adenine nucleotides, one nucleotide at a time, to that 3′ end of the pre-mRNA. This string of A nucleotides, called the **poly(A) tail**, enables the mRNA produced from the pre-mRNA to be translated efficiently, and protects it from attack by RNA-digesting enzymes in the cytoplasm. If it is removed experimentally, the mRNAs are quickly degraded inside cells.

Sequences Interrupting the Protein-Coding Sequence.

The transcription unit of a eukaryotic protein-coding gene—the RNA-coding sequence—also contains non–protein-coding sequences called **introns** that interrupt the protein-coding sequence (shown in Figure 15.8). The introns are transcribed into pre-mRNAs, but removed from pre-mRNAs during processing in the nucleus, so that the coded messages in the finished mRNAs are read continuously, without interruptions. The amino acid–coding sequences that are retained in finished mRNAs are called **exons**. The mechanisms by which introns originated in genes is a mystery.

Introns were discovered by several methods, including direct comparisons between the nucleotide sequences of mature mRNAs and either pre-mRNAs or the genes encoding them. The majority of known eukaryotic genes contain at least one intron; some contain more than 60. The original discoverers of introns, Richard Roberts of New England Biolabs and Phillip Sharp of Massachusetts Institute of Technology, received a Nobel Prize in 1993 for their findings.

Introns Are Removed During Pre-mRNA Processing to Produce the Translatable mRNA

A process called **mRNA splicing**, which occurs in the nucleus, removes introns from pre-mRNAs and joins exons together **(Figure 15.9)**. mRNA splicing takes place in a **spliceosome**, a complex formed between the pre-mRNA and a handful of **small ribonucleoprotein particles**. A ribonucleoprotein particle is a complex of RNA and proteins. The small ribonucleoprotein parti-

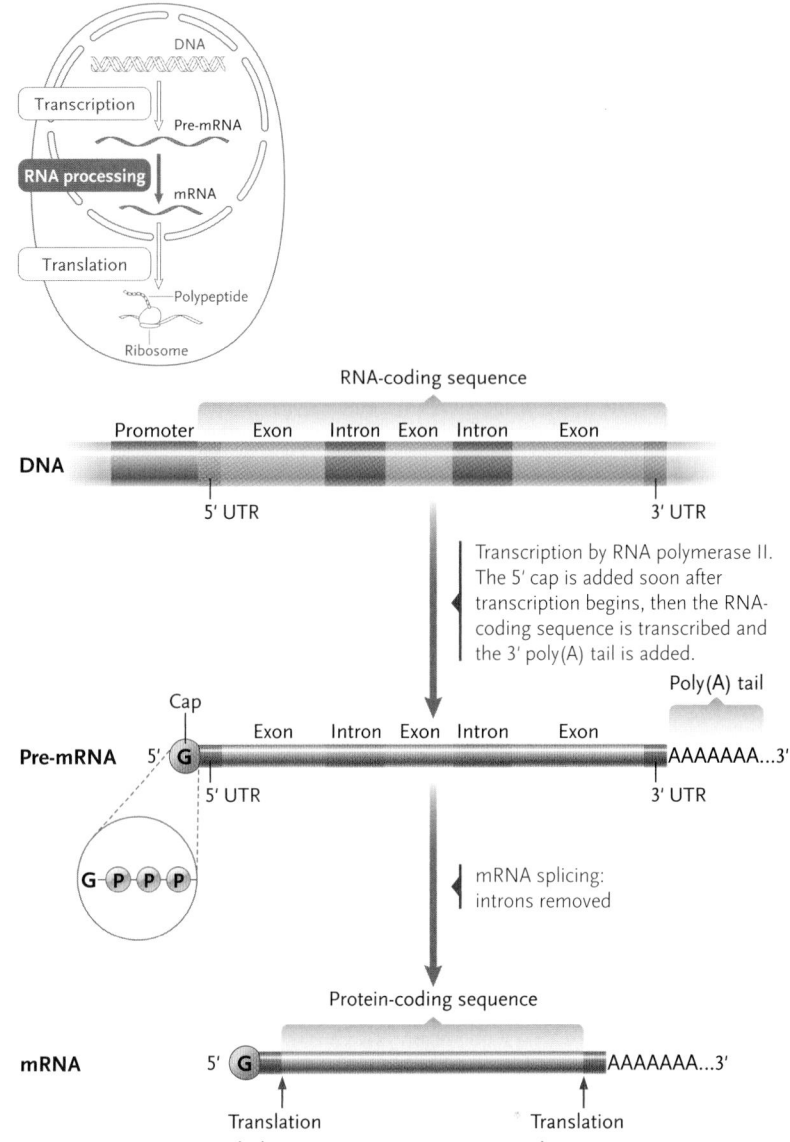

Figure 15.8
Relationship between a eukaryotic protein-coding gene, the pre-mRNA transcribed from it, and the mRNA processed from the pre-mRNA.

cles involved in mRNA splicing are located in the nucleus; each consists of a relatively short *small nuclear RNA* (snRNA) bound to a number of proteins. The particles are therefore known as snRNPs, pronounced "snurps."

The snRNPs bind in a particular order to an intron in the pre-mRNA. The first snRNPs are those with snRNAs that recognize and pair with sequences at the junctions of the intron with the adjacent exons. The complex then recruits the other snRNPs, producing a larger complex that loops out the intron and brings the two exon ends close together. At this point the active spliceosome has been formed. The spliceosome cleaves the pre-mRNA at the junction between the 3′ end of the exon (exon 1 in Figure 15.9) and the 5′ end of the intron, and the intron loops back to bond with itself near its 3′ end. The spliceosome then cleaves the pre-mRNA at the junction between the 3′ end of the intron and exon 2, releasing the intron

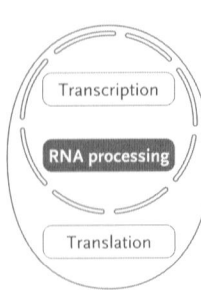

and joining together the two exons (exon 1 and exon 2 in Figure 15.9). Because of the shape of the released intron, it is called a *lariat structure*. Enzymes degrade the intron and the snRNPs are released and used in other mRNA splicing reactions.

The cutting and splicing are so exact that not a single base of an intron is retained in the finished mRNA, nor is a single base removed from the exons. Without this precision, removing introns would change the reading frame of the coding portion of the mRNA, producing gibberish from the point of a mistake onward.

Introns Contribute to Protein Variability

Introns seem wasteful in terms of the energy and raw materials required to replicate and transcribe them, and the elaborate cellular machinery required to remove them during pre-mRNA processing. Why are they present in mRNA-encoding genes? Among a number of possibilities, introns may provide advantages by increasing the coding capacity of existing genes through a process called *alternative splicing* and by generating new proteins through a process called *exon shuffling*.

Alternative Splicing. Many pre-mRNAs are processed by reactions that join exons in different combinations to produce different mRNAs from a single gene. The mechanism, called **alternative splicing**, greatly increases the number and variety of proteins encoded in the cell nucleus without increasing the size of the genome. For example, geneticists estimate that three-quarters of all human pre-mRNAs are subjected to alternative splicing. In each case the different mRNAs produced from the "parent" pre-mRNA are translated to produce a family of related proteins with various combinations of amino acid sequences derived from the exons. Each protein in the family, then, will vary to a degree in its function. Alternative splicing helps us understand why humans have only about 25,000 genes. As a result of the alternative splicing process, the number of proteins produced far exceeds the number of genes, and it is proteins that direct an organism's functions.

Figure 15.10 shows an example of alternative splicing that occurs in mammals, including humans. The pre-mRNA transcript of the α-tropomyosin gene is alternatively spliced in various ways in different tissues—smooth muscle (for example, muscles of the intestine and bladder), skeletal muscle (for example, biceps, glutes), fibroblast (connective tissue cell that makes collagen), liver, and brain—to produce different forms of the α-tropomyosin protein that are functionally optimized for each tissue type. **Figure 15.10** shows the alternative splicing of the α-tropomyosin pre-mRNA to the

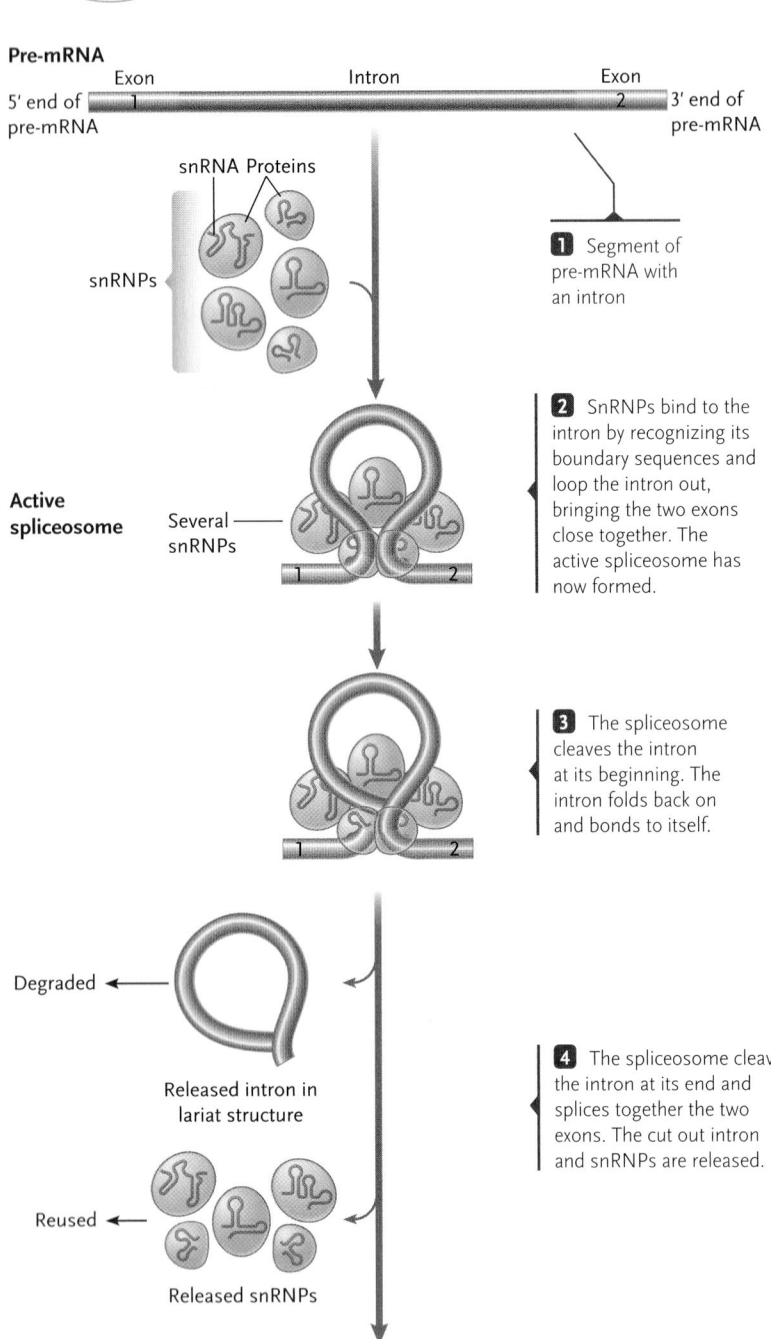

Pre-mRNA

1 Segment of pre-mRNA with an intron

2 SnRNPs bind to the intron by recognizing its boundary sequences and loop the intron out, bringing the two exons close together. The active spliceosome has now formed.

3 The spliceosome cleaves the intron at its beginning. The intron folds back on and bonds to itself.

4 The spliceosome cleaves the intron at its end and splices together the two exons. The cut out intron and snRNPs are released.

Figure 15.9
mRNA splicing—the removal from pre-mRNA of introns and joining of exons in the spliceosome.

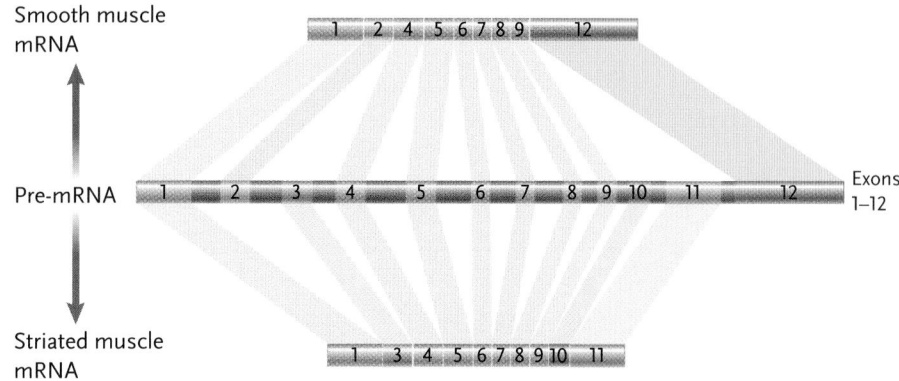

Figure 15.10

Alternative splicing of the α-tropomyosin pre-mRNA to distinct mRNA forms found in smooth muscle and striated muscle. All of the introns are removed in both mRNA splicing pathways, but exons 3, 10, and 11 are also removed to produce the smooth muscle mRNA, and exons 2 and 12 are also removed to produce the striated muscle mRNA.

mRNAs found in smooth muscle and striated muscle. Exons 2 and 12 are exclusive to the smooth muscle mRNA, while exons 3, 10, and 11 are exclusive to the striated muscle mRNA.

Exon Shuffling. Another advantage provided by introns may come from the fact that intron-exon junctions often fall at points dividing major functional regions in encoded proteins, as they do in the genes for antibody proteins, hemoglobin blood proteins, and the peptide hormone insulin. The functional divisions may have allowed new proteins to evolve by **exon shuffling**, a process by which existing protein regions or domains, already selected for their functions by the evolutionary process, are mixed into novel combinations to create new proteins. Evolution of new proteins by this mechanism would produce changes much more quickly and efficiently than by alterations in individual amino acids at random points. The process resembles automobile design, in which new models are produced by combining proven parts and substructures of previous models, rather than starting with an entirely new design each year.

15.4 Translation: mRNA-Directed Polypeptide Synthesis

Translation is the assembly of amino acids into polypeptides. In prokaryotes, translation takes place throughout the cell, while in eukaryotes it takes place mostly in the cytoplasm, although, as you will see, a few specialized genes are transcribed and translated in mitochondria and chloroplasts.

Figure 15.11 summarizes the translation process. For prokaryotes, the mRNA produced by transcription is immediately available for translation. For eukaryotes, the mRNA produced by splicing of the pre-mRNA first exits the nucleus, and then is translated in the cytoplasm. In translation, the mRNA associates with a ribosome, and tRNAs, another type of RNA, bring amino acids to the complex to be joined one by one into the polypeptide chain. The sequence of amino acids in the polypeptide chain is determined by the sequence of codons in the mRNA, while the ribosome is simply a facilitator of the translation process.

In this section we start by discussing the key players in the process, the tRNAs and ribosomes, and then walk through the translation process from a start codon to a stop codon.

tRNAs Are Small, Highly Specialized RNAs That Bring Amino Acids to the Ribosome

Transfer RNAs (tRNAs) bring amino acids to the ribosome for addition to the polypeptide chain.

tRNA Structure. tRNAs are small RNAs, about 75 to 90 nucleotides long (mRNAs are typically hundreds of nucleotides long), with a highly distinctive structure that accomplishes their role in translation (**Figure 15.12**). All tRNAs can wind into four double-helical segments, forming in two dimensions what is known as the *cloverleaf* pattern. At the tip of one of the double-helical segments is the **anticodon**, the three-nucleotide segment that pairs with a codon in mRNAs. Opposite the anticodon, at the other end of the cloverleaf, is a double-helical segment that links to the amino acid corresponding to the anticodon. For example, a tRNA that is linked to serine (Ser) pairs with the codon 5'-AGU-3' (see Figure 15.11). The anticodon of the tRNA that pairs with this codon is 3'-UCA-5'. (The an-

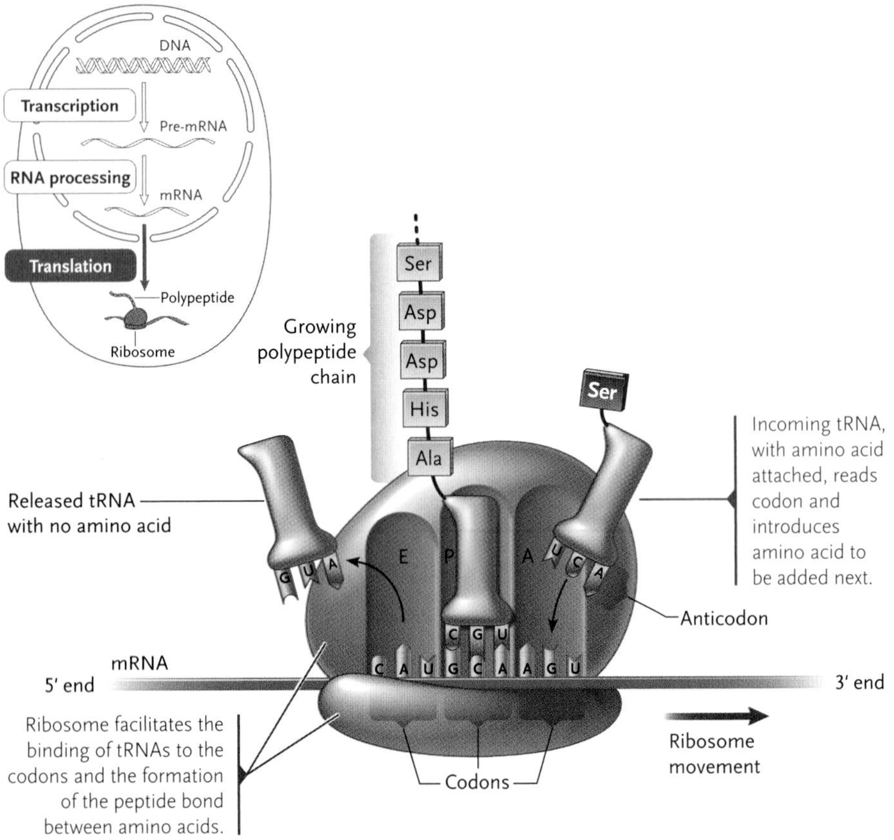

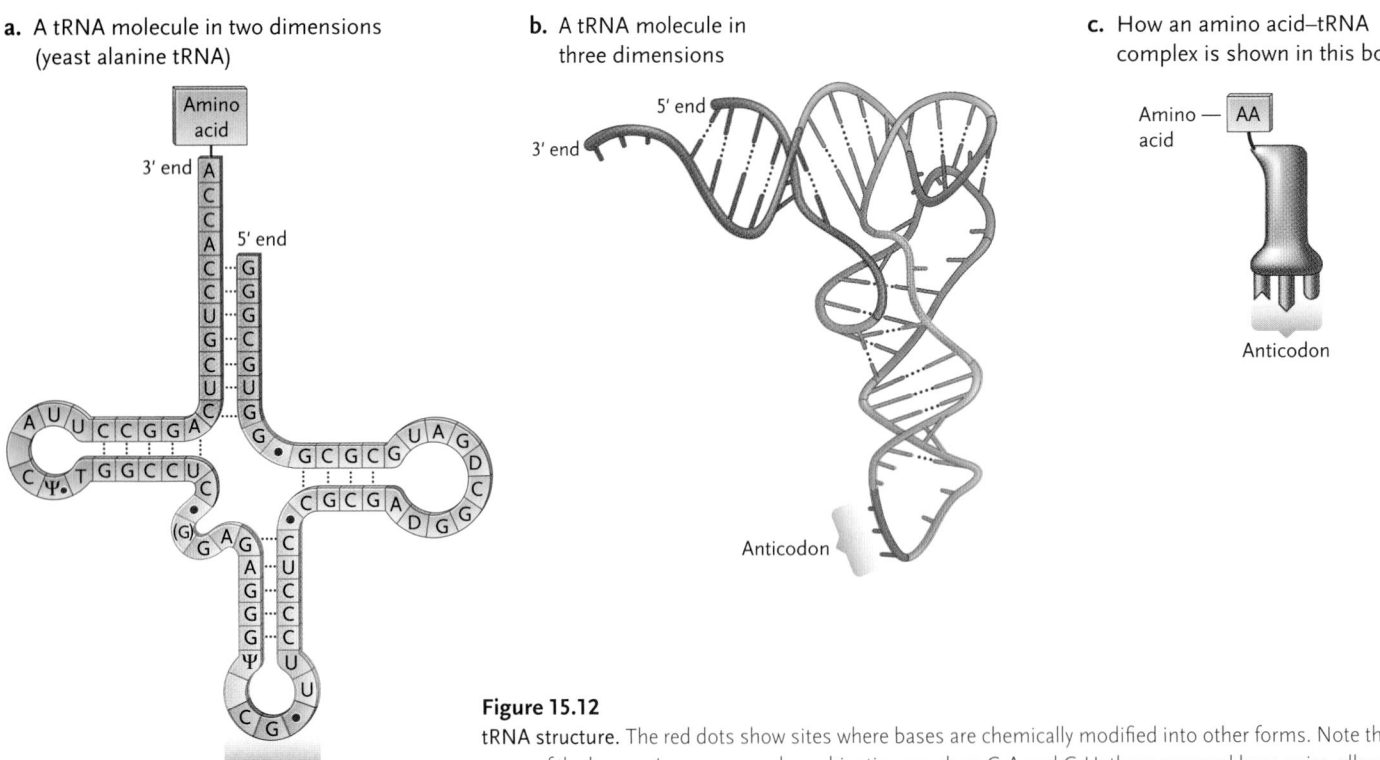

Figure 15.11

An overview of translation, in which ribosomes assemble amino acids into a polypeptide chain. The figure shows a ribosome in the process of translation. A tRNA molecule with an amino acid bound to it is entering the ribosome on the right. The anticodon on the tRNA will pair with the codon in the mRNA. Its amino acid will then be added to the growing polypeptide that is currently attached to the tRNA in the middle of the ribosome.

a. A tRNA molecule in two dimensions (yeast alanine tRNA)

b. A tRNA molecule in three dimensions

c. How an amino acid–tRNA complex is shown in this book

Figure 15.12

tRNA structure. The red dots show sites where bases are chemically modified into other forms. Note that some of the base pairs are unusual combinations such as G-A and G-U; these unusual base pairs, allowed by the greater flexibility of short RNA chains, are common in tRNAs.

ticodon and codon pair in an antiparallel manner, as do the strands in DNA. We will write anticodons in the $3' \rightarrow 5'$ direction to make it easy to see how they pair with codons.)

The tRNA cloverleaf folds in three dimensions into the L-shaped structure shown in Figure 15.12b. The anticodon and the segment binding the amino acid are located at the opposite tips of the L.

We learned earlier that 61 of the 64 codons of the genetic code specify an amino acid. Does this mean that 61 different tRNAs read the sense codons? The answer is no. Francis Crick's **wobble hypothesis** states that the complete set of 61 sense codons can be read by fewer than 61 distinct tRNAs because of particular pairing properties of the bases in the anticodons. That is, the pairing of the anticodon with the first two nucleotides of the codon is always precise, but the anticodon has more flexibility in pairing with the third nucleotide of the codon. In many cases the same tRNA anticodon can read codons that have either U or C in the third position; for example, a tRNA carrying phenylalanine can read both codons UUU and UUC. Similarly the same tRNA anticodon can read two codons that have A or G in the third position; for example, a tRNA carrying glutamine can read both codons CAA and CAG.

Addition of Amino Acids to Their Corresponding tRNAs.
The correct amino acid must be present on a tRNA if translation is to be accurate. The process of adding an amino acid to a tRNA is called **aminoacylation** (literally, the addition of an amino acid) or **charging** (because the process adds free energy as the amino acid-tRNA combinations are formed).

The finished product of charging, a tRNA linked to its "correct" amino acid, is called an **aminoacyl-tRNA**. Twenty different enzymes called **aminoacyl-tRNA synthetases**—one synthetase for each of the 20 amino acids—catalyze aminoacylation **(Figure 15.13)**. First, a molecule of ATP and the amino acid (AA) bind to the enzyme, and the enzyme links the two, with the release of two phosphate groups (Figure 15.13, step 1):

$$AA + ATP \rightarrow AA\text{-}AMP + 2\,P_i$$

Much of the energy released by the breakdown is retained in the aminoacyl-AMP molecule. Next, the correct tRNA binds to the enzyme (step 2). Third, the enzyme transfers the amino acid from the AA-AMP to the tRNA to form the aminoacyl-tRNA (step 3):

$$AA\text{-}AMP + tRNA \rightarrow AA\text{-}tRNA + AMP$$

Finally, the charged tRNA is released from the enzyme; the enzyme can then perform other activation reactions (step 4). The aminoacyl-tRNA also retains much of the energy released by ATP breakdown. This energy eventually drives the formation of the peptide bond linking amino acids during translation.

With the tRNAs attached to their corresponding amino acids, our attention moves to the ribosome,

where the amino acids are removed from their tRNAs and linked into polypeptide chains.

Ribosomes Are rRNA-Protein Complexes That Work as Automated Protein Assembly Machines

Ribosomes are ribonucleoprotein particles that carry out protein synthesis by translating mRNA into chains of amino acids. Like some automated machines, such as those forming complicated metal parts by a series of machining steps, ribosomes use an information tape—an mRNA molecule—as the directions required to accomplish a task. For ribosomes, the task is joining amino acids in ordered sequences to make a polypeptide chain.

In prokaryotes, ribosomes carry out their assembly functions throughout the cell. In eukaryotes, ribosomes function in the cytoplasm, either suspended freely in the cytoplasmic solution, or attached to the membranes of the endoplasmic reticulum (ER), the system of tubular or flattened sacs in the cytoplasm (discussed in Section 5.5).

A finished ribosome is made up of two parts of dissimilar size, called the *large* and *small ribosomal subunits* **(Figure 15.14)**. Each subunit is a combination of **ribosomal RNA (rRNA)** and ribosomal proteins.

Prokaryotic and eukaryotic ribosomes are similar in structure and function. However, the differences in their molecular structure, particularly in the ribosomal proteins, give them distinguishable properties. For example, the antibiotics streptomycin and erythromycin are effective antibacterial agents because they inhibit the function of the bacterial ribosome, but not the eukaryotic ribosome. Streptomycin and erythromycin affect translation activities in the small and large ribosomal subunit, respectively.

In translation, the mRNA moves through a groove in the ribosome. The ribosome also has binding sites where tRNAs interact with the mRNA (see Figure 15.14 and refer also to Figure 15.11). The **A site** (aminoacyl site) is where the incoming aminoacyl-tRNA carrying the next amino acid to be added to the polypeptide chain binds to the mRNA. The **P site** (peptidyl site) is where the tRNA carrying the growing polypeptide chain is bound. The **E site** (exit site) is where an exiting tRNA binds prior to release from the ribosome. You will learn more about these functional sites as we discuss the stages of translation.

Translation Initiation Brings the Ribosomal Subunits, an mRNA, and the First Aminoacyl-tRNA Together

Translation is similar in prokaryotes and eukaryotes. We will present translation from a eukaryotic perspective, and indicate how it differs in prokaryotes.

Figure 15.13
Aminoacylation (aka charging): the addition of an amino acid to a tRNA.

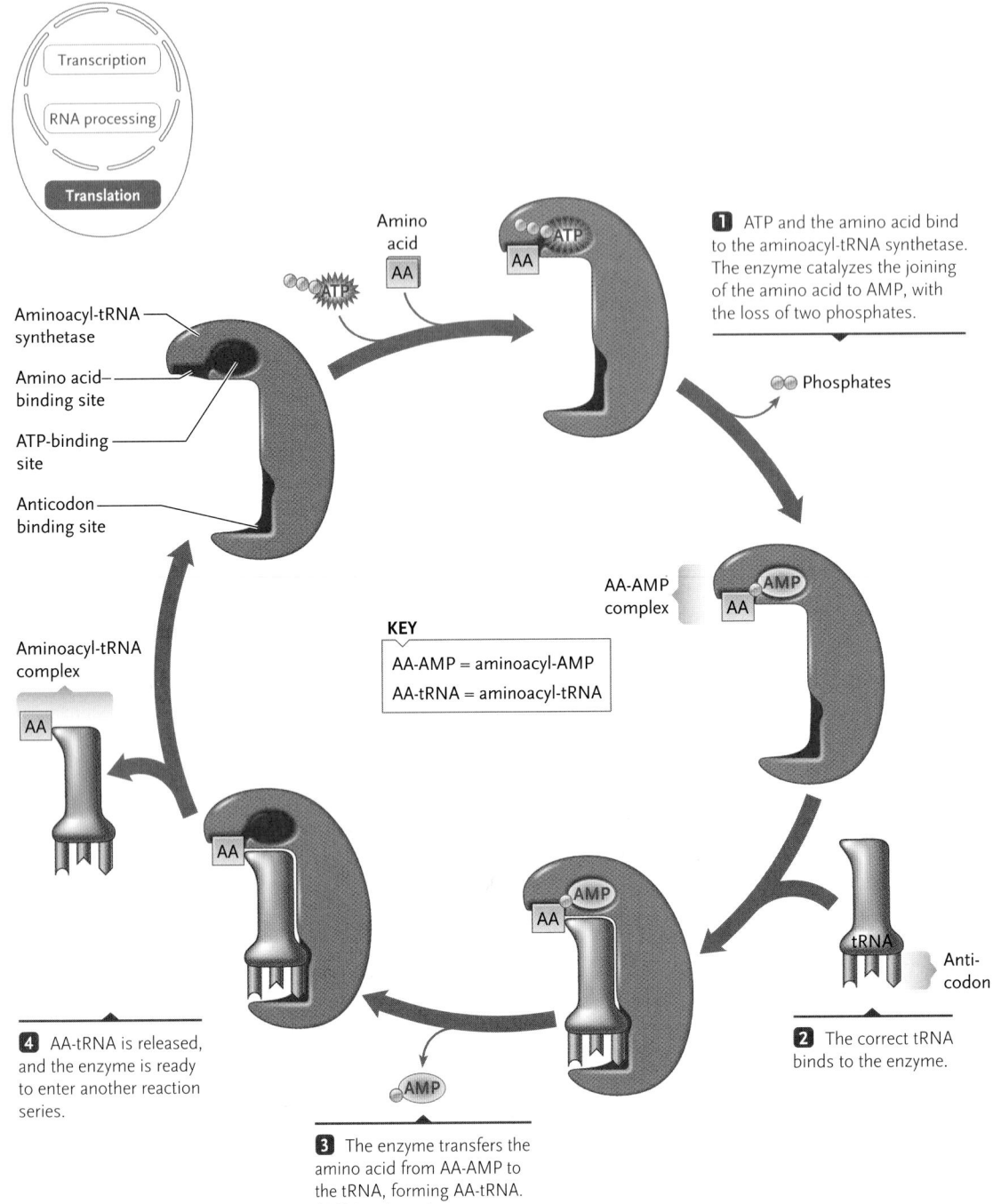

Transcription

RNA processing

Translation

Amino acid

AA

1 ATP and the amino acid bind to the aminoacyl-tRNA synthetase. The enzyme catalyzes the joining of the amino acid to AMP, with the loss of two phosphates.

Phosphates

Aminoacyl-tRNA synthetase

Amino acid–binding site

ATP-binding site

Anticodon binding site

AA-AMP complex

KEY

AA-AMP = aminoacyl-AMP

AA-tRNA = aminoacyl-tRNA

Aminoacyl-tRNA complex

AA

tRNA

Anti-codon

2 The correct tRNA binds to the enzyme.

4 AA-tRNA is released, and the enzyme is ready to enter another reaction series.

AMP

3 The enzyme transfers the amino acid from AA-AMP to the tRNA, forming AA-tRNA. AMP is released.

There are three major stages of translation: initiation, elongation, and termination. During initiation the translation components assemble on the start codon of the mRNA. In elongation the assembled complex reads the string of codons in the mRNA one at a time while joining the specified amino acids into the polypeptide. Termination completes the translation process when the complex disassembles after the last amino acid of the polypeptide specified by the mRNA has been added to the polypeptide.

In translation initiation, a large and a small ribosomal subunit associates with an mRNA molecule and the first aminoacyl-tRNA of the new protein chain binds to the AUG start codon **(Figure 15.15)**. The aminoacyl-tRNA used for initiation is a specialized **initiator tRNA** with an anticodon to the methionine-specifying AUG start codon. Each step in translation initiation is aided by proteins called *initiation factors*.

In the first step of the initiation process, the initiator methionine-tRNA (Met-tRNA—anticodon 3′-UAC-5′) with a molecule of GTP bound to it forms a complex with the small ribosomal subunit (Figure 15.15, step 1). The complex binds to the mRNA at the 5′ cap and then moves along the mRNA—a process called *scanning*—until it

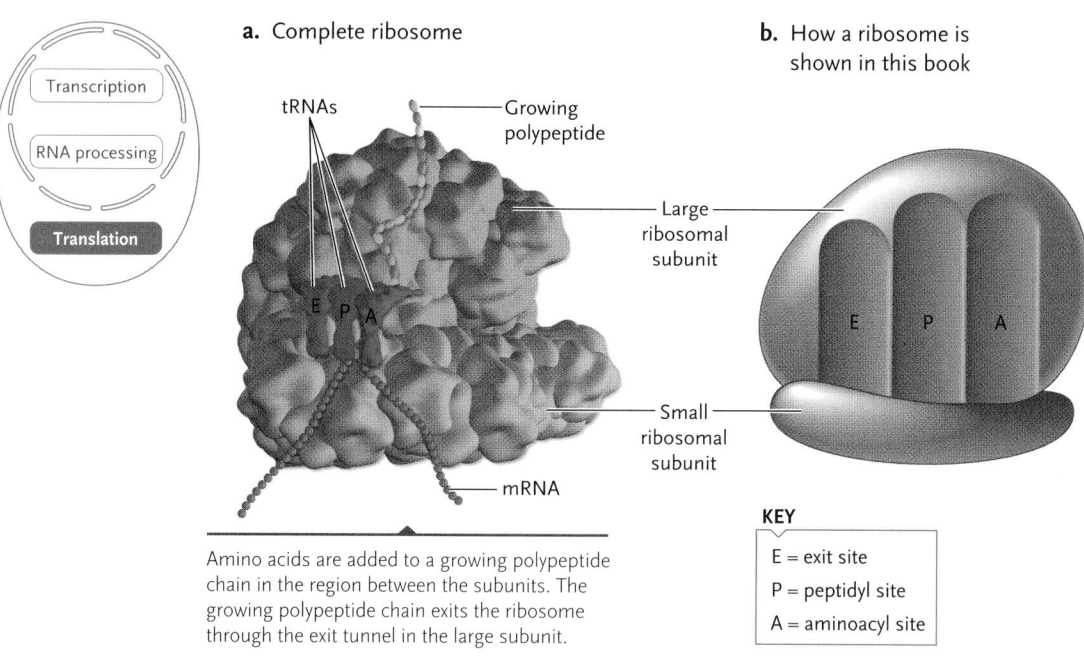

a. Complete ribosome

Transcription
RNA processing
Translation

tRNAs
Growing polypeptide
E P A
Large ribosomal subunit
Small ribosomal subunit
mRNA

Amino acids are added to a growing polypeptide chain in the region between the subunits. The growing polypeptide chain exits the ribosome through the exit tunnel in the large subunit.

b. How a ribosome is shown in this book

E P A

KEY

E = exit site
P = peptidyl site
A = aminoacyl site

Figure 15.14
Ribosome structure.
(a) Computer model of a ribosome in the process of translation. **(b)** The ribosome as we will show it during translation. (a: Michael W. Davidson/ Molecular Expressions, Florida State Research Foundation.)

reaches the first AUG codon (step 2). This is the start codon and it is recognized by the anticodon of the Met-tRNA. The large ribosomal subunit then binds, completing the ribosome (step 3). The initiation factors are released when GTP is hydrolyzed to GDP + phosphate. At the end of initiation, the initiator Met-tRNA is in the P site.

In prokaryotes, translation initiation is different: Rather than scanning from the 5′ end of the mRNA, the small ribosomal subunit, the initiator Met-tRNA, GTP, and protein initiator factors bind directly to the region of the mRNA with the AUG start codon. A **ribosome binding site** just upstream of the start codon directs the small ribosomal subunit in this initiation step. The large ribosomal subunit then binds to the small subunit to complete the ribosome.

After the initiator tRNA pairs with the AUG initiator codon, the subsequent stages of translation simply read the nucleotide bases three at a time on the mRNA. The initiator tRNA-AUG pairing thus establishes the correct *reading frame*—the series of codons for the polypeptide encoded by the mRNA.

Polypeptide Chains Grow during the Elongation Stage of Translation

The central reactions of translation take place in the elongation stage, which adds amino acids one at a time to a growing polypeptide chain. The individual steps of elongation depend on the binding properties of the P, A, and E sites of the ribosome. Protein *elongation factors* aid the elongation events.

The P site, with one exception, can bind only to a **peptidyl-tRNA**—a tRNA linked to a growing polypeptide chain containing two or more amino acids. The

exception is the initiator tRNA, which is recognized by the P site as a peptidyl-tRNA even though it carries only a single amino acid, methionine. The A site can bind only to an aminoacyl-tRNA. The tRNA that previously was in the P site binds to the E site and then leaves the ribosome.

How the P, A, and E sites are used through the elongation cycle is shown in **Figure 15.16.** The cycle begins at the point when an initiator tRNA with its attached methionine is bound to the P site, and the A site is empty. First, an aminoacyl-tRNA with an appropriate anticodon binds to the codon in the A site of the ribosome; GTP is hydrolyzed in this step (Figure 15.16, step 1).

Next, the amino acid (here, the initiator methionine) is cleaved from the tRNA in the P site and forms a peptide bond with the amino acid on the tRNA in the A site (step 2). **Peptidyl transferase** catalyzes this reaction. Researchers were surprised to discover that this enzyme is not a protein but a part of an rRNA of the large ribosomal subunit. An RNA molecule that catalyzes a reaction like a protein enzyme does is called a *catalytic RNA* or a **ribozyme** (*ribo*nucleic acid en*zyme*).

At the close of the reaction, the (now) polypeptide chain is attached to the tRNA in the A site and an "empty" tRNA remains at the P site. Next, the ribosome moves—translocates—along the mRNA to the next codon, using energy from GTP hydrolysis (step 3). The two tRNAs remain bound to their respective codons, so this step positions the just-formed peptidyl-tRNA in the P site, and generates a new vacant A site. The empty tRNA that was in the P site is now in the E site, from where it is released from the ribosome (step 4). With the A site empty and a peptidyl tRNA in the P site, the ribosome repeats the elongation cycle. In subsequent

1 Met-tRNA with GTP bound to it and the small ribosomal subunit form a complex.

2 The complex binds to the 5′ cap of the mRNA and scans along it until it reaches AUG start codon.

3 The large ribosomal subunit binds and GTP is hydrolyzed, completing initiation.

turns of the cycle, the growing polypeptide on the tRNA in the P site is transferred to the amino acid on the A site tRNA. (*Insights from the Molecular Revolution* describes a technique for investigating the structure of the P and A sites.)

Elongation is highly similar in prokaryotes and eukaryotes, with no significant conceptual differences. During each elongation cycle, which turns at the rate of about one to three times per second in eukaryotes and 15 to 20 times per second in prokaryotes, the ribo-

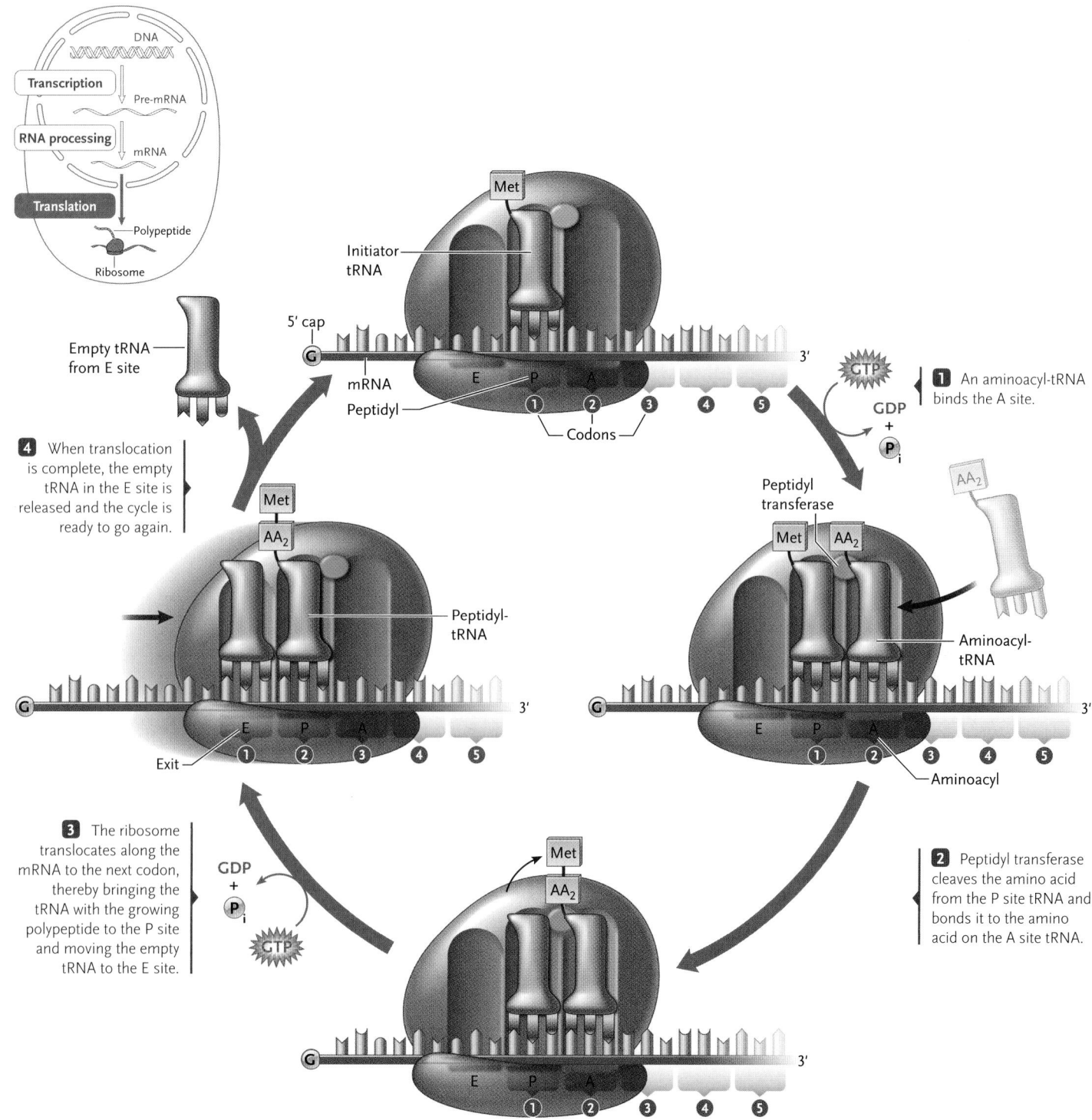

Figure 15.16
The steps in the elongation stage of translation. (For simplicity, protein elongation factors and GTP are not shown in the figure.)

1 An aminoacyl-tRNA binds the A site.

2 Peptidyl transferase cleaves the amino acid from the P site tRNA and bonds it to the amino acid on the A site tRNA.

3 The ribosome translocates along the mRNA to the next codon, thereby bringing the tRNA with the growing polypeptide to the P site and moving the empty tRNA to the E site.

4 When translocation is complete, the empty tRNA in the E site is released and the cycle is ready to go again.

some advances by one codon along the mRNA and adds one amino acid to the growing polypeptide chain. The growing polypeptide chain extends from the ribosome through the exit tunnel (see Figure 15.14) as elongation continues.

Termination Releases a Completed Polypeptide from the Ribosome

Translation switches from the elongation to the termination stage when the A site of a ribosome arrives at one of the UAA, UAG, or UGA stop codons on the mRNA (**Figure 15.17,** step 1). When a stop codon appears at the A site, a **release factor** (RF; also called a **termination factor**) binds at this site instead of an aminoacyl-tRNA (step 2). In response, the polypeptide chain is released from the tRNA at the P site as usual (step 3). However, because no amino acid is present at the A site, the freed polypeptide chain is released from the ribosome (step 4). At the same time, the ribosomal subunits separate and detach from the mRNA. The empty tRNA and the release factor are also released. Termination is highly similar in prokaryotes and eukaryotes.

Measuring Ribosomes with a Molecular Ruler

One of the major problems in the research on the three-dimensional structure of ribosomes has been unraveling the structure and arrangement of the rRNA molecules of ribosomes. It is important to know this because the rRNAs play important roles in translation; the peptidyl transferase, for example, is an enzyme activity of the large rRNA of the large ribosomal subunit. The problem of determining rRNA positions is difficult because the rRNAs twist and fold throughout the large or small ribosomal subunits. Which part of each rRNA type is located in active sites of the ribosome, such as the E, P, and A sites and the peptidyl transferase site, and what are the functions of the rRNA segments that wind through these locations? An ingenious molecular approach, devel-

oped by Harry F. Noller and his coworkers at the University of California at Santa Cruz, provides a method for answering some of these questions. The technique provides a *molecular scissors* that leaves an identifying snip at points in the rRNA molecules of the ribosome.

Noller and his colleagues prepared artificial tRNAs that extend in a straight line, with an anticodon at one end and a sequence at the other end that mimics the part of a tRNA that binds to an amino acid. Instead of an amino acid, Noller's team attached an Fe^{2+} atom to the acceptor end of the structure (see figure). The iron atom can break the sugar-phosphate backbone of an RNA molecule. Thus the artificial tRNAs can work as a molecular scissors, with one end able to bind to an mRNA in either

the A, P, or E site, and the other end able to make an identifying snip in the part of an rRNA molecule that is located in its vicinity. And, the scissors were made in different, precisely calculated lengths, running in regular increments from 4 to 33 base pairs. The artificial tRNAs would thus act as "molecular rulers" that could measure distances between the anticodon-binding region and parts of the rRNA molecules in ribosomes.

The molecular scissors were reacted with ribosomes, one length at a time, and then the rRNAs were extracted to see where the snips occurred. To locate the sites of the snips, the researchers sorted the extracted rRNAs by gel electrophoresis, a method that is capable of separating RNA molecules that differ in

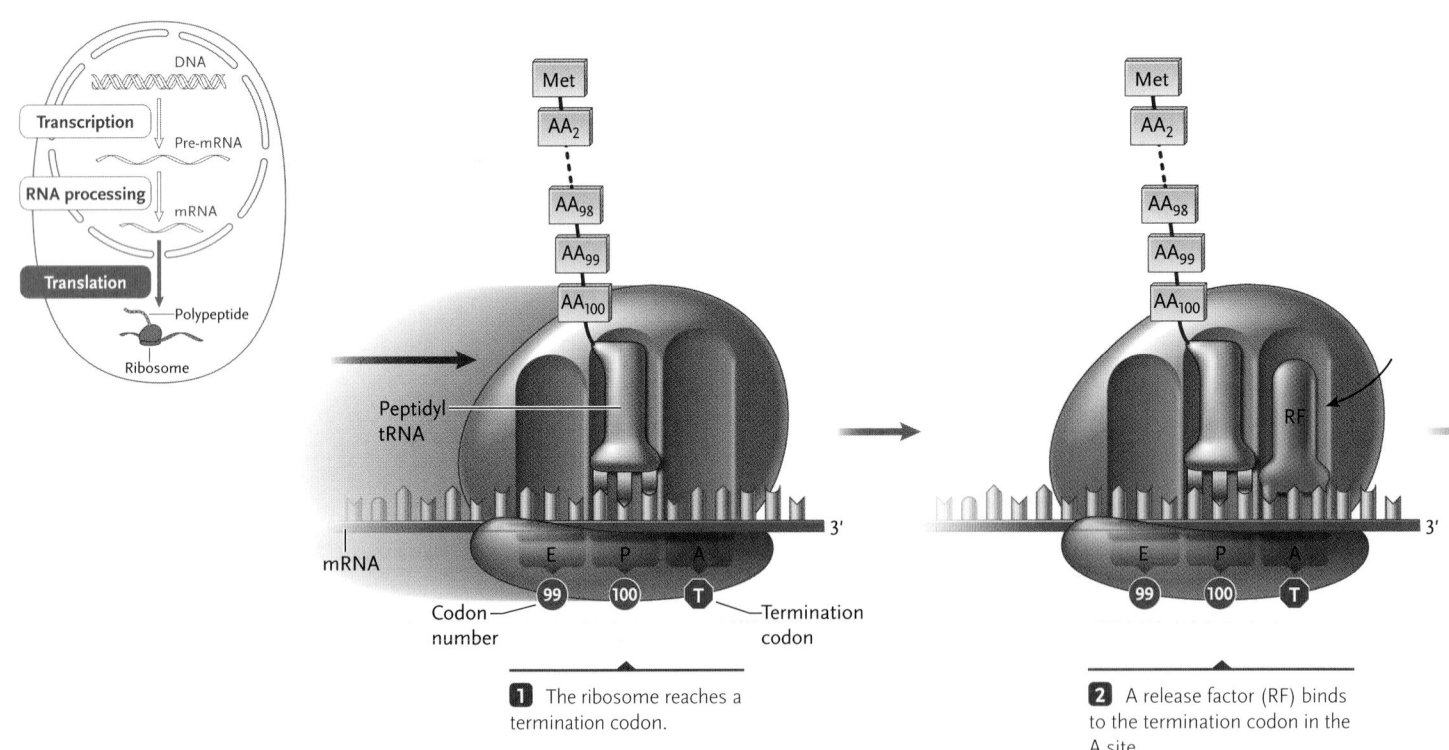

1 The ribosome reaches a termination codon.

2 A release factor (RF) binds to the termination codon in the A site.

Figure 15.17
The steps in the termination stage of translation.

length by only a single base. (Gel electrophoresis is described in Figure 18.7.) By comparing the lengths of the snipped rRNA molecules with those of intact molecules, the investigators could locate the exact points in the sequences at which the RNAs were cut. This information allowed them to map the segments of the rRNAs forming parts of the E, P, and A sites.

The rulers were bound to bacterial ribosomes, which contain three types of rRNAs: 5S; 23S, which forms part of the large subunit; and 16S, which forms part of the small subunit. (The S values are a measure of molecular size, with the larger numbers indicating larger size.) When the rulers attached to the A site, the snipping end of the artificial tRNAs made cuts in both 23S rRNA and 16S rRNA, showing that both the large and small ribosomal subunits cooperate to form the A site.

When bound to the P site, the molecular rulers made cuts in all three rRNA types. These results show that, as in the A site, both the large and small subunits cooperate to form the P site and that all three rRNA types evidently form parts of the P site. The locations of the cuts also gave clues about the three-dimensional structure of the 16S and other rRNAs in the A and P sites. This method may allow the three-dimensional structure of the E, P, and A sites to be reconstructed as well as identify the rRNA types forming the sites. The next step is to try to work out the functions of these segments in polypeptide assembly.

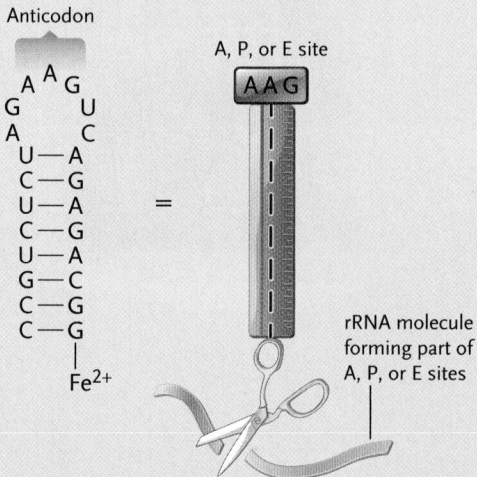

Molecular "ruler"

A molecular ruler used to measure the arrangement of tRNAs and rRNAs in ribosomes. By varying the length of the ruler, rRNA molecules located at different points within the A, P, and E sites can be marked by cuts.

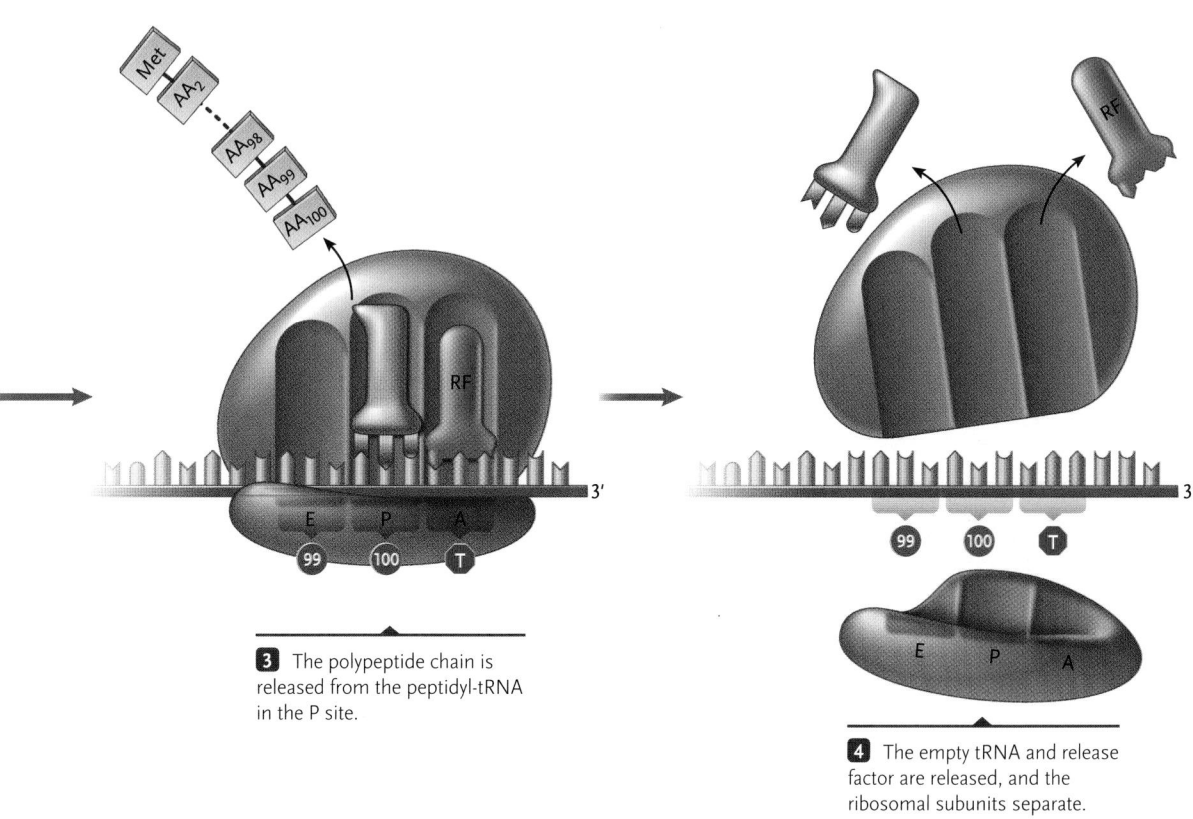

3 The polypeptide chain is released from the peptidyl-tRNA in the P site.

4 The empty tRNA and release factor are released, and the ribosomal subunits separate.

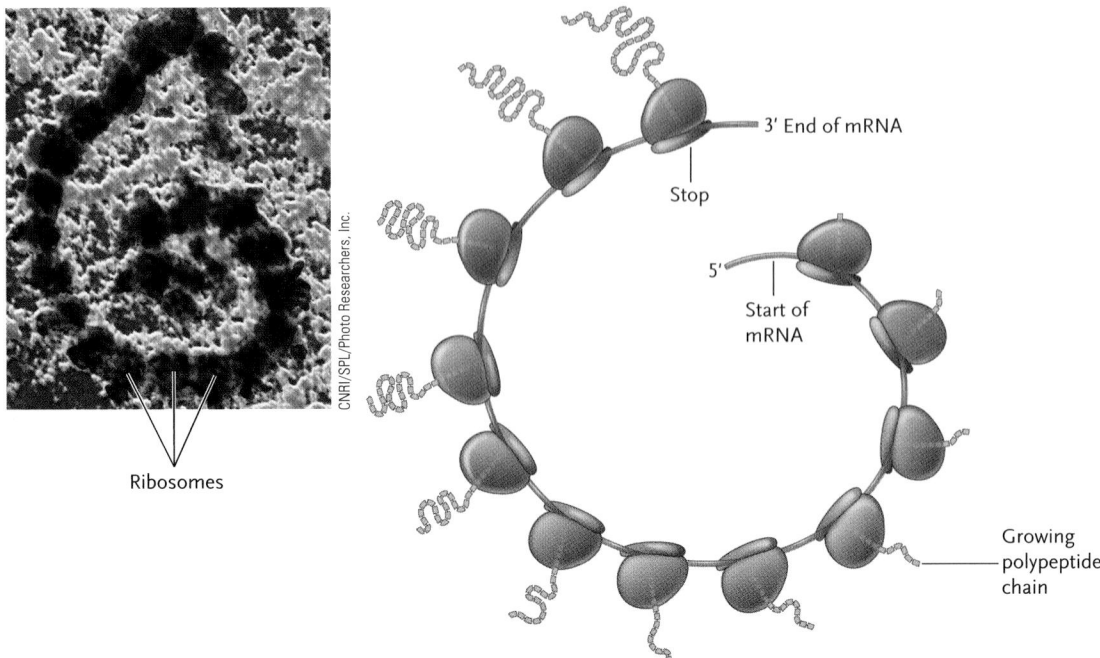

Figure 15.18
Polysomes, consisting of a series of ribosomes "reading" the same mRNA.

Labels on figure: 3' End of mRNA; Stop; 5'; Start of mRNA; Growing polypeptide chain; Ribosomes

Credit: CNRI/SPL/Photo Researchers, Inc.

Multiple Ribosomes Simultaneously Translate a Single mRNA

Once the first ribosome has begun translation, another one can assemble with an initiator tRNA as soon as there is room on the mRNA. Ribosomes continue to attach as translation continues and become spaced along the mRNA like beads on a string. The entire structure of an mRNA molecule and the multiple ribosomes attached to it is known as a **polysome** (a contraction of *polyribosome;* **Figure 15.18**). The multiple ribosomes greatly increase the overall rate of polypeptide synthesis from a single mRNA. The total number of ribosomes in a polysome depends on the length of the coding region of its mRNA molecule, ranging from a minimum of one or two ribosomes on the smallest mRNAs to as many as 100 on the longest mRNAs.

In prokaryotes, because of the absence of a nuclear envelope, transcription and translation typically are coupled. That is, as soon as the 5' end of a new mRNA emerges from the RNA polymerase, ribosomal subunits attach and initiate translation **(Figure 15.19)**. In essence the polysome forms while the mRNA is still being made. By the time the mRNA is completely transcribed, it is covered with ribosomes from end to end, each assembling a copy of the encoded polypeptide.

Newly Synthesized Polypeptides Are Processed and Folded into Finished Form

Most eukaryotic proteins are in an inactive, unfinished form when ribosomes release them. Processing reactions that convert the new proteins into finished form include the removal of amino acids from the ends or interior of the polypeptide chain and the addition of larger organic groups, including carbohydrate or lipid structures.

Proteins fold into their final three-dimensional shapes as the processing reactions take place. For many proteins, helper proteins called *chaperones* or *chaperonins* assist the folding process by combining with the folding protein, promoting "correct" three-dimensional structures, and inhibiting incorrect ones (see Section 3.5 and Figure 3.24).

In some cases the same initial polypeptide may be processed by alternative pathways that produce different mature polypeptides, usually by removing different, long stretches of amino acids from the interior of the polypeptide chain. Alternative processing is another mechanism that increases the number of polypeptides encoded by a single gene.

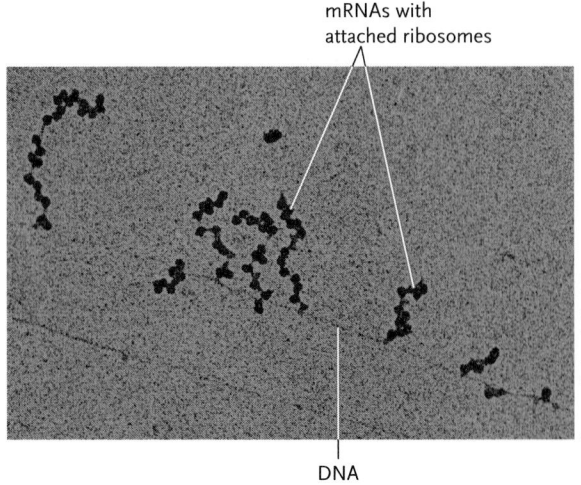

Figure 15.19
Simultaneous transcription and translation in progress in an electron microscope preparation extracted from *E. coli.* ×57,000.

Labels on figure: mRNAs with attached ribosomes; DNA

Credit: Courtesy Barbara A. Hamkalo

Other proteins are processed into an initial, inactive form that is later activated at a particular time or location by removal of a covering segment of the amino acid chain. The digestive enzyme pepsin, for example, is made by processing reactions within cells lining the stomach into an inactive form called *pepsinogen*. When the cells secrete pepsinogen into the stomach, the high acidity of that organ triggers removal of a segment of amino acids from one end of the protein's amino acid chain; the removal converts the enzyme into the active form in which it rapidly breaks proteins in food particles into shorter pieces. The initial production of the protein as inactive pepsinogen protects the cells that make it from having their proteins degraded by the enzyme.

Finished Proteins Contain Sorting Signals That Direct Them to Cellular Locations

Proteins are found in all parts of the eukaryotic cell, including the cytosol, the nucleus, the plasma membrane, and the membranes or interior of various organelles; they are also transported to the cell exterior. How are newly synthesized proteins directed to these locations? Proteins that remain in the cytosol, such as microtubule proteins or the enzymes of glycolysis, have no signals. They are made on ribosomes called *free ribosomes*, which are suspended in the cytosol, and they enter the cytosol as they are made.

For all other proteins, a system of "zip codes," in the form of amino acid sequences that form *sorting*

signals in the proteins, direct the proteins to their cellular locations, or out of the cell. The signals are coded in the DNA, transcribed into mRNAs, and "printed" in proteins as they are made. The signals, first discovered by Günter Blobel, Peter Walter, and their coworkers at Rockefeller University, are recognized and bound by receptors in the locations to which the proteins are addressed. Blobel received a Nobel Prize in 1999 for his work with the mechanism sorting proteins in cells.

One major signal pathway sorts proteins to the endoplasmic reticulum **(Figure 15.20)**. In these proteins, a short segment of amino acids called the **signal peptide** (also called a **signal sequence**) is in the very first part of the polypeptide chain to be made. When the signal peptide emerges from the ribosome, a protein-RNA complex called the **signal recognition particle (SRP)** binds to it and temporarily blocks further translation (Figure 15.20, step 1). Next, the SRP binds a protein in the ER membrane called the **SRP receptor**; this step "docks" the ribosome on the ER membrane (step 2). (The docked ribosomes give the dotted appearance to the rough ER; see Section 5.3.) The ribosome now can continue protein synthesis, and the growing polypeptide is pushed through the ER membrane into the rough ER lumen (see step 2) where an enzyme, *signal peptidase,* removes the signal sequence (step 3). Synthesis of the polypeptide continues until it is complete (step 4). Depending on other built-in signals, the polypeptide may move to any part of the ER-based system: the ER itself, the Golgi complex, the plasma membrane,

Figure 15.20
The signal mechanism directing proteins to the ER. The figure shows several ribosomes at different stages of translation of the mRNA.

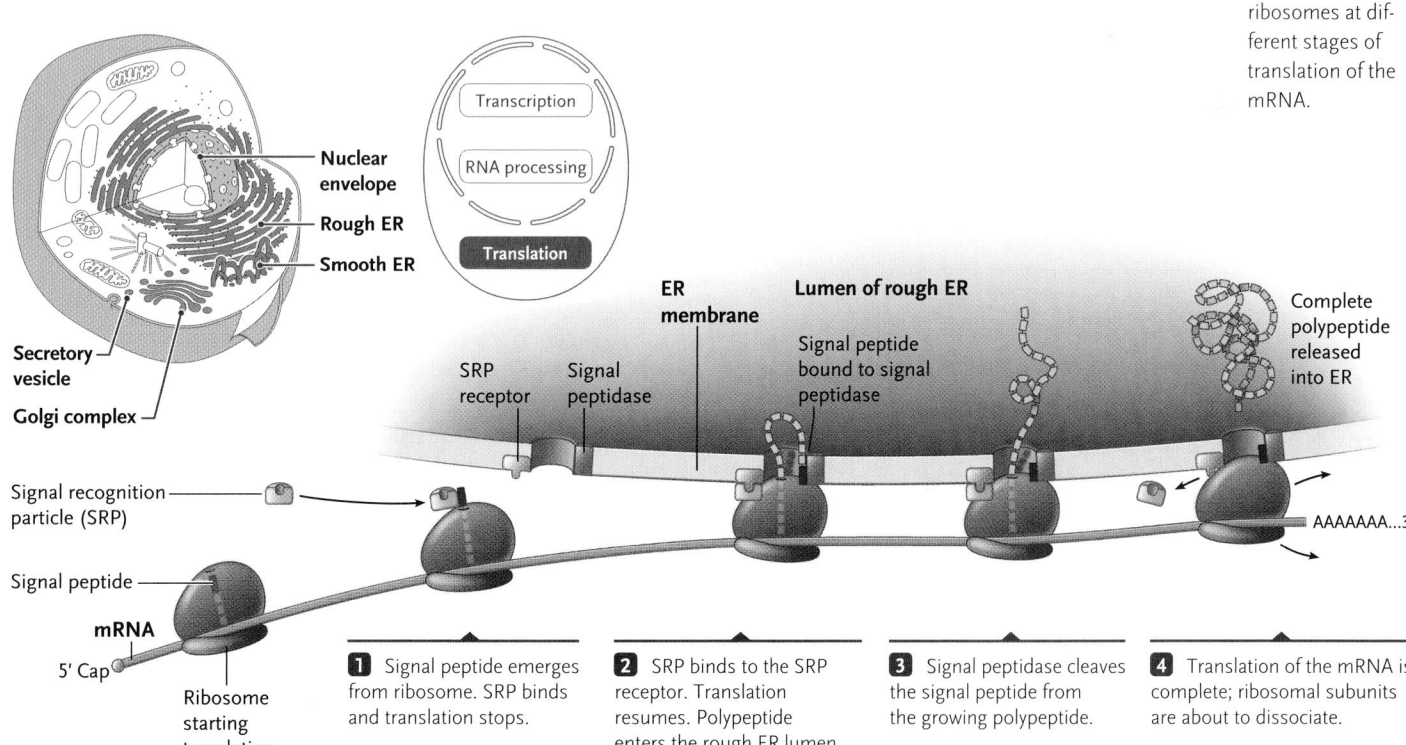

1 Signal peptide emerges from ribosome. SRP binds and translation stops.

2 SRP binds to the SRP receptor. Translation resumes. Polypeptide enters the rough ER lumen and binds to signal peptidase.

3 Signal peptidase cleaves the signal peptide from the growing polypeptide.

4 Translation of the mRNA is complete; ribosomal subunits are about to dissociate.

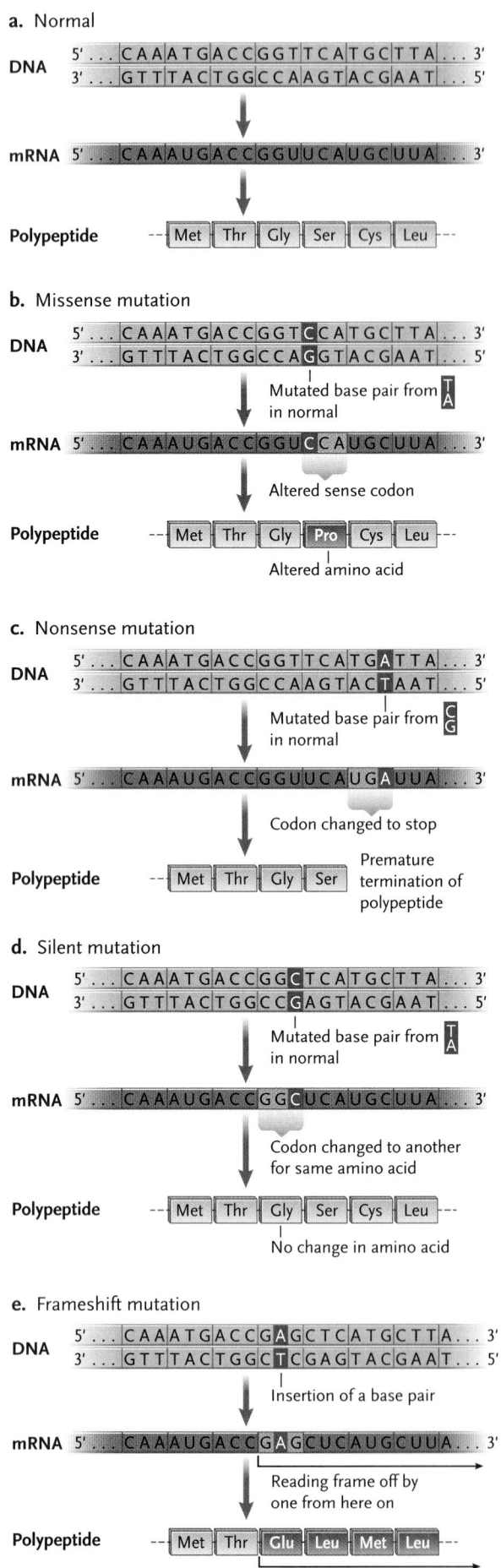

Figure 15.21
Effects of mutations in protein-coding genes on the amino acid sequence of the encoded polypeptide.

the nuclear envelope, secretory vesicles, or via secretory vesicles to the cell exterior (these destinations are shown in the inset to Figure 15.20).

Nuclear proteins include a signal that is bound by receptors in the pore complexes of the nuclear envelope (shown in Figure 5.11). Once bound, they are pushed through the pore complex into the nuclear interior, in a process that requires ATP energy. For these proteins, the signal remains because the proteins need to enter the nucleus each time the nuclear envelope breaks down and reforms during the cell division cycle.

Many proteins that are to become part of organelles such as mitochondria, chloroplasts, or microbodies are also made on free ribosomes. However, these proteins have signals that are bound by receptors in the organelle membranes, targeting them for entry into the organelles. Further signals on the proteins direct them to the different membranes or compartments inside the organelles.

The sorting system, in all its remarkable complexity, routes newly synthesized proteins and gets them to their final destinations in the eukaryotic cell. Without it, cells would wind up as a jumble of proteins floating about in the cytoplasm, with none of the spatial organization that makes cellular life possible.

The same basic system of sorting signals distributes proteins throughout prokaryotic cells, indicating that this mechanism probably evolved with the first cells. In prokaryotes, signals similar to the ER-directing signals of eukaryotes direct newly synthesized bacterial proteins to the plasma membrane (bacteria do not have ER membranes); further information built into the proteins keeps them in the plasma membrane or allows them to enter the cell wall or to be secreted outside the cell. Proteins without sorting signals remain in the cytoplasmic solution.

Interestingly, the bacterial and eukaryotic routing signals are interchangeable. That is, a bacterial signal peptide grafted to a polypeptide made in a eukaryotic cell routes the molecule to the ER membrane, and a eukaryotic ER-directing signal peptide grafted to a polypeptide made in a bacterial cell directs the molecule to the plasma membrane. The interchangeability of the bacterial and eukaryotic signal peptides indicates that the sorting mechanism appeared early in the evolution of cellular life.

Base-Pair Mutations Can Affect Protein Structure and Function

Mutations are changes to the genetic material. How do mutations affect protein structure and function? **Base-pair substitution mutations** are particular mutations involving changes to individual base pairs in the genetic material. If a base-pair substitution mutation occurs in the protein-coding portion of a gene, it can affect the structure and function of the protein.

That is, a base-pair change will change a base in a codon.

Let us consider a theoretical stretch of DNA encoding a string of amino acids in a polypeptide to see the possible consequences of base-pair substitution mutations in the coding region of a gene. The normal (unmutated) DNA and amino acid sequences are shown in **Figure 15.21a.** If the codon is altered to specify a different amino acid, then the protein will have a different amino acid sequence. A mutation such as this is called a **missense mutation**, because the amino acid that is placed in the polypeptide is not the normal one **(Figure 15.21b).** Whether the function of a polypeptide is altered significantly depends on the amino acid change that occurs. Individuals homozygous for a missense mutation in the gene for one of the two polypeptide types found in the oxygen-carrying protein hemoglobin **(Figure 15.22)** have the genetic disease sickle-cell disease, described in Chapter 12 (pp. 235–236). Many other human genetic diseases are caused by homozygous missense mutations.

A second type of base-pair substitution mutation is a **nonsense mutation (Figure 15.21c).** In this case the base-pair change in the DNA results in a change from a sense (amino acid–coding) codon to a nonsense (termination) codon in the mRNA. Translation of an mRNA containing a nonsense mutation results in a shorter-than-normal polypeptide and, in many

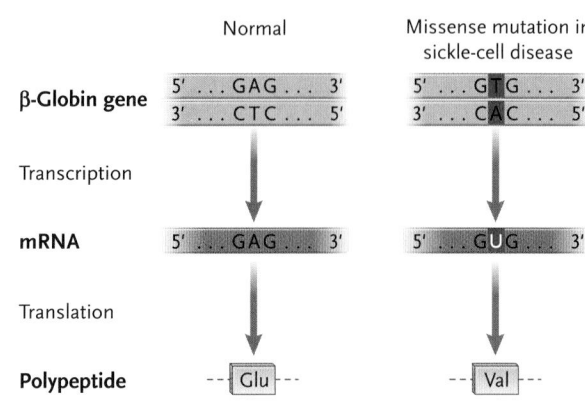

Figure 15.22
Missense mutation in a gene for one of the two polypeptides of hemoglobin that is the cause of sickle-cell disease.

cases, this polypeptide will be only partially functional at best.

Because of the degeneracy of the genetic code, some base-pair substitution mutations do not alter the amino acid specified by the gene because the changed codon specifies the same amino acid as in the normal polypeptide. Such mutations are known as **silent mutations (Figure 15.21d).**

If a single base pair is deleted or inserted in the coding region of a gene, the reading frame of the resulting mRNA is altered. That is, after that point, the ribosome reads codons that are not the same as for the normal mRNA, producing a different amino acid sequence in the polypeptide from then on. This type of

---

UNANSWERED QUESTIONS

What are the structures and specific functions of the proteins involved in transcription?

Even though molecular geneticists have a good understanding of the initiation, elongation, and termination stages of transcription, they have a lot to learn about the precise molecular events involved. Researchers are actively exploring the molecular structures of RNA polymerase, the transcription factors, and other proteins, and are developing models for how they interact to initiate transcription and how that interaction is controlled. For example, in yeast, the transcription machinery assembled at a promoter includes nearly 100 polypeptides! What does each of these components do? How are they assembled and positioned correctly on the template? Among the researchers studying this topic is Richard Ebright, a Howard Hughes Medical Institute investigator at Rutgers University.

How is alternative splicing regulated?

The discovery that a majority of human genes show alternative splicing was very surprising. How is alternative splicing regulated? This is not a peculiarly human question, because alternative splicing is prevalent in all eukaryotes. Not only will this research provide basic information about the mechanism of mRNA splicing, but it is likely also to have significant implications for human medicine. For example, the same disease gene often produces variable symptoms in different individuals. Is this because of effects on the regulation of alternative splicing

resulting in different defective proteins with different effects? If so, how does this altered regulation occur, and what regulates it? The answers to these questions would open the door to the development of drugs to treat the disease symptoms.

How does the ribosome work?

The ribosome is a large complex of rRNAs and proteins. Research in recent years resulted in the unexpected conclusion that the rRNAs are functionally important in protein synthesis, rather than simply being a scaffold on which the ribosomal proteins hang. Peptidyl transferase activity is the property of rRNA, for instance, as we discussed in this chapter. Current research is directed at understanding how the three-dimensional structure of the ribosome gives it its functional properties. For example, how does the ribosome translocate along the mRNA? This process must involve changes in the three-dimensional shape of the ribosome. What are the ribosome's moving parts? A surprising result about this from Harry Noller's lab at the University of California, Santa Cruz (see *Insights from the Molecular Revolution* for other research from this lab), indicates that translocation is a property of the ribosome itself and not of the protein factors and GTP. The molecular basis of translocation is being intensely researched in light of this new knowledge.

Peter J. Russell

mutation is called a **frameshift mutation (Figure 15.21e,** insertion mutation shown); the resulting polypeptide often is nonfunctional because of the significantly altered amino acid sequence.

Both transcription and translation are steps in the process of gene expression, the realization of the gene's coded information in the makeup and activities of a cell. However, the flow of information is not one way; organisms and cells also exert control over how their genes are expressed, as you will see in the next chapter.

STUDY BREAK

1. How does translation initiation occur in eukaryotes versus prokaryotes?
2. Distinguish between the P, A, and E sites of the ribosome.
3. How are proteins directed to different parts of a eukaryotic cell?

Review

Go to ThomsonNOW™ at www.thomsonedu.com/login to access quizzing, animations, exercises, articles, and personalized homework help.

15.1 The Connection between DNA, RNA, and Protein

- In their genetic experiments with *Neurospora crassa,* Beadle and Tatum found a direct correspondence between gene mutations and alterations of enzymes. Their one gene–one enzyme hypothesis is now restated as the one gene–one polypeptide hypothesis (Figure 15.2).

- The pathway from genes to proteins involves transcription then translation. In transcription, a sequence of nucleotides in DNA is copied into a complementary sequence in an RNA molecule. In translation, the sequence of nucleotides in an mRNA molecule specifies an amino acid sequence in a polypeptide (Figure 15.3).

- The genetic code is a triplet code. AUG at the beginning of a coded message establishes a reading frame for reading the codons three nucleotides at a time. The code is redundant: most of the amino acids are specified by more than one codon (Figures 15.4 and 15.5).

- The genetic code is essentially universal.

  Animation: Uracil–thymine comparison

  Animation: Protein synthesis summary

  Practice: The major differences between prokaryotic and eukaryotic protein synthesis

15.2 Transcription: DNA-Directed RNA Synthesis

- Transcription is the process by which information coded in DNA is transferred to a complementary RNA copy (Figure 15.6).

- Transcription begins when an RNA polymerase binds to a promoter sequence in the DNA and starts synthesizing an RNA molecule. The enzyme then adds RNA nucleotides in sequence according to the DNA template. At the end of the transcribed sequence, the enzyme and the completed RNA transcript release from the DNA template. The mechanism of termination is different in eukaryotes and prokaryotes (Figure 15.7).

  Animation: Gene transcription details

15.3 Production of mRNAs in Eukaryotes

- A gene encoding an mRNA molecule includes the promoter, which is recognized by the regulatory proteins and transcription factors that promote DNA unwinding and the initiation of transcription by an RNA polymerase. Transcription in eukaryotes produces a pre-mRNA molecule that consists of a 5′ cap, the 5′ untranslated region, interspersed exons (amino acid-coding segments) and introns, the 3′ untranslated region, and the 3′ poly(A) tail. All are copied from DNA except the 5′ cap and poly(A) tail, which are added during transcription (Figure 15.8).

- Introns in pre-mRNAs are removed to produce functional mRNAs by splicing. snRNPs bind to the introns, loop them out of the pre-mRNA, clip the intron at each exon boundary, and join the adjacent exons together (Figure 15.9).

- Many pre-mRNAs are subjected to alternative splicing, a process that joins exons in different combinations to produce different mRNAs encoded by the same gene. Translation of each mRNA produced in this way generates a protein with different function (Figure 15.10).

  Animation: Pre-mRNA transcript processing

15.4 Translation: mRNA-Directed Polypeptide Synthesis

- Translation is the assembly of amino acids into polypeptides. Translation occurs on ribosomes. The P, A, and E sites of the ribosome are used for the stepwise addition of amino acids to the polypeptide as directed by the mRNA (Figures 15.11 and 15.14).

- Amino acids are brought to the ribosome attached to specific tRNAs. Amino acids are linked to their corresponding tRNAs by aminoacyl-tRNA synthetases. By matching amino acids with tRNAs, the reactions also provide the ultimate basis for the accuracy of translation (Figures 15.12 and 15.13).

- Translation proceeds through the stages of initiation, elongation, and termination. In initiation, a ribosome assembles with an mRNA molecule and an initiator methionine-tRNA. In elongation, amino acids linked to tRNAs add one at a time to the growing polypeptide chain. In termination, the new polypeptide is released from the ribosome and the ribosomal subunits separate from the mRNA (Figures 15.15–15.17).

- After they are synthesized on ribosomes, polypeptides are converted into finished form by processing reactions, which include removal of one or more amino acids from the protein chains, addition of organic groups, and folding guided by chaperones.

- Proteins are distributed in cells by means of signals spelled out by amino acid sequences (Figure 15.20).

- Base-pair substitution mutations alter the mRNA and can lead to changes in the amino acid sequence of the encoded polypeptide. A missense mutation changes one sense codon to one that specifies a different amino acid, a nonsense mutation changes a sense codon to a stop codon, and a silent mutation changes one sense codon to another sense codon that specifies the same amino acid. A base-pair insertion or deletion is a frameshift mutation that alters the reading frame beyond the point of the mutation, leading to a different amino acid sequence from then on in the polypeptide (Figures 15.21 and 15.22).

Animation: Structure of a ribosome

Animation: Translation

Animation: Base-pair substitution

Animation: Frameshift mutation

Questions

Self-Test Questions

1. Which statement about the following pathway is false?

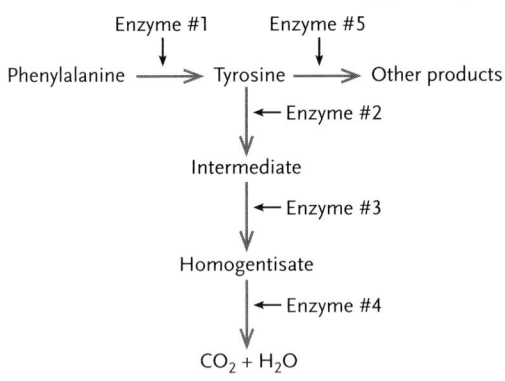

a. A mutation for enzyme #1 causes phenylalanine to build up.
b. A mutation for enzyme #2 prevents tyrosine from being synthesized.
c. A mutation at enzyme #3 prevents homogentistate from being synthesized.
d. A mutation for enzyme #2 could hide a mutation in enzyme #4.
e. Each step in a pathway such as this is catalyzed by an enzyme, which is coded by a gene.

2. Eukaryotic mRNA:
a. uses snRNPs to cut out introns and seal together translatable exons.
b. uses a spliceosome mechanism made of DNA to recognize consensus regions to cut and splice.
c. has a guanine cap on its 3′ end and a poly(A) tail on its 5′ end.
d. is composed of adenine, thymine, guanine, and cytosine.
e. codes the guanine cap and poly(A) tail from the DNA template.

3. A segment strand of DNA has a base sequence of 5′-GCATTAGAC-3′. What would be the sequence of an RNA molecule complementary to that sequence?
a. 5′-GUCTAATGC-3′   d. 5′-GUCUAAUGC-3′
b. 5′-GCAUUAGAC-3′   e. 5′-CGUAAUCUG-3′
c. 5′-CGTAATCTG-3′

4. Which of the following statements about the initiation phase of translation is false?
a. An initiation factor allows 5′ mRNA to attach to the small ribosomal subunit.
b. Initiation factors complex with GTP to help Met-tRNA and AUG pair.
c. mRNA attaches first to the small ribosomal subunit.
d. GTP is synthesized.
e. 3′-UAC-5′ on the tRNA binds 5′-AUG-3′ on mRNA.

5. Which of the following statements about aminoacylation is false?
a. It precedes translation.
b. It occurs in the ribosome.
c. It requires ATP to bind an aminoacyl-tRNA synthetase.

d. It joins the correct amino acid to a specific tRNA based on the tRNA's anticodon.
e. It uses three binding sites on aminoacyl-tRNA synthetase.

6. Translation is in progress, with methionine bound to a tRNA in the P site, and a phenylalanine bound to a tRNA in the A site. The order of the next steps in the elongation cycle is:
a. the ribosome translocates → a new aminoacyl-tRNA enters the A site → peptidyl transferase catalyzes a peptide bond between the two amino acids → empty tRNA is released from the ribosome.
b. peptidyl transferase catalyzes a peptide bond between the two amino acids → a new aminoacyl-tRNA enters the A site → empty tRNA is released from the ribosome → the ribosome translocates.
c. peptidyl transferase catalyzes a peptide bond between the two amino acids → empty tRNA is released from the ribosome → a new aminoacyl-tRNA enters the A site → the ribosome translocates.
d. peptidyl transferase catalyzes a peptide bond between the two amino acids → the ribosome translocates → empty tRNA is released from the ribosome → a new aminoacyl-tRNA enters the A site.
e. the ribosome translocates → peptidyl transferase catalyzes a peptide bond between the two amino acids → empty tRNA is released from the ribosome → a new aminoacyl-tRNA enters the A site.

7. Which of the following statements is false?
a. GTP is an energy source during various stages of translation.
b. In the ribosome, peptidyl transferase catalyses peptide bond formation between amino acids.
c. When the mRNA code UAA reaches the ribosome, there is no tRNA to bind to it.
d. A long polypeptide is cut off the tRNA in the A site so its Met amino acid links to the amino acid in the P site.
e. Forty-two amino acids of a protein are encoded by 126 nucleotides of the mRNA.

8. Which item binds to SRP receptor and to the signal sequence to guide a newly synthesized protein to be secreted to its proper "channel"?
a. ribosome
b. signal recognition particle
c. endoplasmic reticulum
d. signal peptidase
e. receptor protein

9. A part of an mRNA molecule with the sequence 5′-UGC GCA-3′ is being translated by a ribosome. The following activated tRNA molecules are available. Two of them can correctly bind the mRNA so that a dipeptide can form.

tRNA Anticodon	Amino Acid
3′-GGC-5′	Proline
3′-CGU-5′	Alanine
3′-UGC-5′	Threonine
3′-CCG-5′	Glycine
3′-ACG-5′	Cysteine
3′-CGG-5′	Alanine

a. cysteine-alanine d. alanine-alanine
b. proline-cysteine e. threonine-glycine
c. glycine-cysteine

10. A missense mutation cannot be:
   a. the code for the sickle-cell gene.
   b. caused by a frameshift.
   c. the deletion of a base in a coding sequence.
   d. the addition of two bases in a coding sequence.
   e. the same as a silent mutation.

Questions for Discussion

1. Which do you think are more important to the accuracy by which amino acids are linked into proteins: nucleic acids or enzymatic proteins? Why?

2. A mutation appears that alters an anticodon in a tRNA from AAU to AUU. What effect will this change have on protein synthesis in cells carrying this mutation?

3. The normal form of a gene contains the nucleotide sequence:

   5'-ATGCCCGCCTTTGCTACTTGGTAG-3'
   3'-TACGGGCGGAAACGATGAACCATC-5'

   When this gene is transcribed, the result is the following mRNA molecule:

   5'-AUGCCCGCCUUUGCUACUUGGUAG-3'

   In a mutated form of the gene, two extra base pairs (underlined) are inserted:

   5'-ATGCCCGCCT<u>AA</u>TTGCTACTTGGTAG-3'
   3'-TACGGGCGGA<u>TT</u>AACGATGAACCATC-5'

   What effect will this particular mutation have on the structure of the protein encoded in the gene?

4. A geneticist is attempting to isolate mutations in the genes for four enzymes acting in a metabolic pathway in the bacterium *Escherichia coli*. The end product *E* of the pathway is absolutely essential for life:

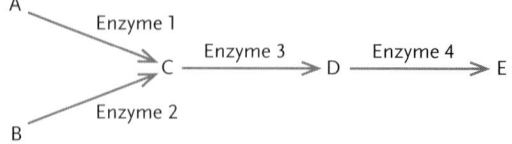

   The geneticist has been able to isolate mutations in the genes for enzymes 1 and 2, but not for enzymes 3 and 4. Develop a hypothesis to explain why.

Experimental Analysis

How could you show experimentally that the genetic code is universal; namely, that it is the same in bacteria as it is in eukaryotes such as fungi, plants, and animals?

Evolution Link Question

How might the process of alternative splicing and exon shuffling affect the rate at which new proteins evolve?

How Would You Vote?

Ricin, a molecule that inactivates ribosomes, is difficult to disperse through the air and is unlikely to be used in a large-scale terrorist attack. However, ricin powder did turn up in a Senate office building. Scientists are working to develop a vaccine against ricin. If mass immunizations were to be offered, would you sign up to be vaccinated? Go to www.thomsonedu.com/login to investigate both sides of the issue and then vote.

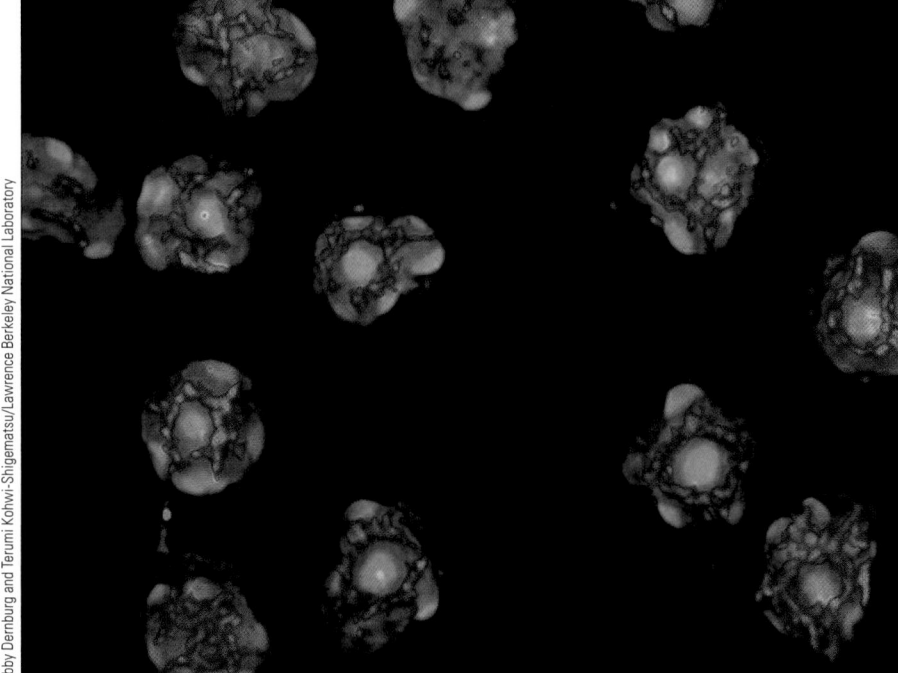

Chromatin remodeling proteins (gold) binding to chromatin (blue). Chromatin remodeling, a change in chromosome structure in the region of a gene, is a key step in the activation of genes in eukaryotes.

16 Control of Gene Expression

WHY IT MATTERS

A human egg cell is almost completely inactive metabolically when it is released from the ovary. It remains quiescent as it begins its travel down a fallopian tube leading from the ovary to the uterus, carried along by movements of cilia lining the walls of the tube **(Figure 16.1).** It is here, in the fallopian tube, that egg and sperm cells meet and embryonic development begins. Within seconds after the cells unite, the fertilized egg breaks its quiescent state and begins a series of divisions that continues as it moves through the fallopian tube and enters the uterus. Subsequent divisions produce specialized cells that *differentiate* into the distinct types tailored for specific functions in the body, such as muscle cells and cells of the nervous system.

All the nucleated cells of the developing embryo retain the same set of genes. The structural and functional differences in the cell types are determined not by the presence or absence of certain genes but rather through differences in *gene activity*. Some genes, known as housekeeping genes, are active in almost all cells; other genes are turned on or off ("expressed" or "not expressed") depending on the cell type. Each differentiated cell is characterized by genes that are active in only that cell type. For example, all mammalian cells carry

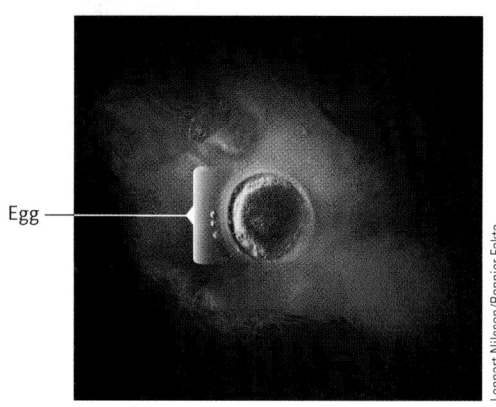

Egg

Figure 16.1

A human egg at the time of its release from the ovary. The outer layer appearing light blue in color is a coat of poly-saccharides and glycoproteins that surrounds the egg. Within the egg, genes and regulatory proteins are poised to enter the pathways initiating embryonic development.

the genes for hemoglobin, but these genes are active only in cells that give rise to red blood cells. Specific regulatory events that take place only in red blood cells activate the hemoglobin genes in those cells. Those genes are not activated in other cell types.

The fundamental mechanisms that control gene activity are common to all multicellular eukaryotes. Even single-celled eukaryotes and prokaryotes have systems that turn genes on or off when required. With few exceptions, however, prokaryotic systems are limited almost exclusively to short-term responses to environmental changes; eukaryotic cells exhibit both short-term response and long-term differentiation.

The processes that directly control gene activity are known collectively as **transcriptional regulation.** Transcriptional regulation, the fundamental level of control, determines which genes are transcribed into mRNA. Additional controls fine-tune regulation by affecting the processing of mRNA *(posttranscriptional regulation)*, its translation into proteins *(translational regulation)*, and the life span and activity of the proteins themselves *(posttranslational regulation)*.

These levels of regulation ultimately affect more than proteins, because among the proteins are enzymes that determine the types and kinds of all other molecules made in the developing cell. So, effectively, these regulatory mechanisms tailor the production of all cellular molecules. The entire spectrum of controls constitutes an exquisitely sensitive mechanism regulating when, where, and what kinds and numbers of cellular molecules are produced.

In this chapter we examine the mechanisms of transcriptional regulation and its fine-tuning by additional controls at the posttranscriptional, translational, and posttranslational levels. Our discussion begins with bacterial systems, where researchers first discovered a mechanism for transcriptional regulation, and then moves to eukaryotic systems where the regulation of gene activity is more complicated. How genes regulate development is discussed in Chapter 48.

16.1 Regulation of Gene Expression in Prokaryotes

Transcription and translation are closely regulated in prokaryotes in ways that reflect prokaryotic life histories. Prokaryotes are relatively simple, single-celled organisms with generations that come and go in a matter of minutes. Rather than the complex patterns of long-term cell differentiation and development typical of multicellular eukaryotes, prokaryotic cells typically undergo rapid and reversible alterations in biochemical pathways that allow them to adapt quickly to changes in their environment.

The human intestinal bacterium *Escherichia coli,* for example, can metabolize a wide range of nutrients including lactose (milk sugar). When lactose is present, *E. coli* makes enzymes for metabolizing the sugar, but it does not make those enzymes when lactose is absent. The versatile and responsive control system allows the bacterium to use the nutrients available in the surrounding medium with maximum efficiency.

The Operon Is the Unit of Transcription in Prokaryotes

When the environment in which a bacterium lives changes, some metabolic processes are stopped and others are started. Typically, this involves turning off the genes for the metabolic processes not needed and turning on the genes for the new metabolic processes. For each metabolic process, there are a few to many genes involved, and the regulation of those genes must be coordinated. For example, three genes encode enzymes for the metabolism of lactose by *E. coli*. In the absence of lactose, the three genes are not expressed, while in the presence of lactose, the genes are expressed. That is, the control of these genes is at the transcription level.

In 1961, François Jacob and Jacques Monod of the Pasteur Institute in Paris proposed the *operon model* for the control of the expression of genes for lactose metabolism in *E. coli*. Subsequently, the *operon model* has been shown to be widely applicable to the regulation of gene expression in bacteria and their viruses. Jacob and Monod received the Nobel Prize in 1965 for their discovery and explanation of bacterial operons and their regulation by repressors.

An **operon** is a cluster of prokaryotic genes and the DNA sequences involved in their regulation. One of those DNA sequences is the **promoter**, which is the site to which RNA polymerase binds as it starts transcription of the gene. Each operon, which can contain several to many genes, is transcribed as a unit from the promoter into a single mRNA, and as a result the mRNA contains codes for several proteins. The cluster of genes transcribed into a single mRNA is called a **transcription unit.** A ribosome translates the mRNA from one end to the other, sequentially making each protein encoded in the mRNA. Typically, the proteins encoded by an operon catalyze steps in the same function, such as enzymes acting in sequence in a biochemical pathway.

The other DNA regulatory sequence in the operon is the **operator,** a short segment to which a regulatory protein binds. The regulatory protein is encoded by a gene separate from the operon the protein controls. Some operons are controlled by a regulatory protein termed a **repressor,** which, when active, prevents the operon genes

Lennart Nilsson/Bonnier Fakta

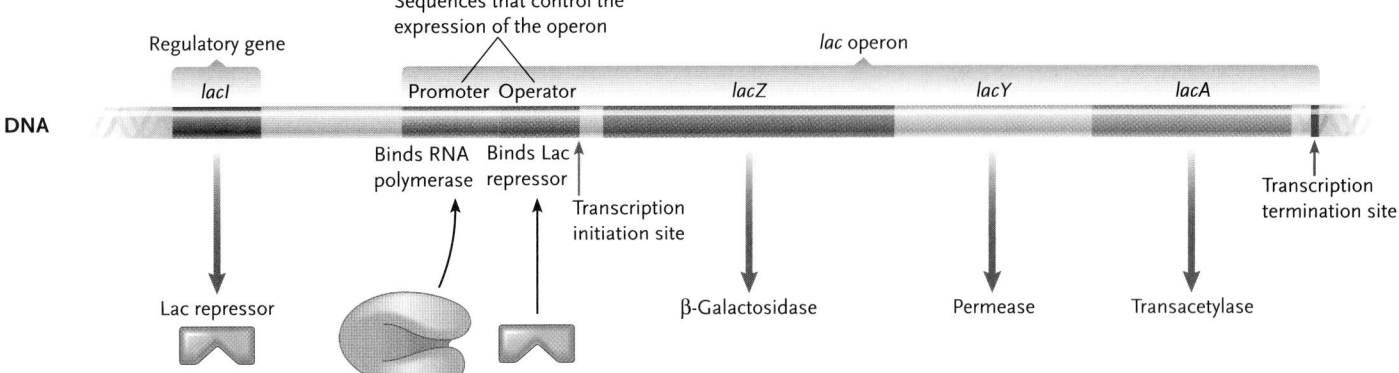

Figure 16.2

The *E. coli lac* operon. The *lacZ, lacY,* and *lacA* genes encode the enzymes taking part in lactose metabolism. The separate regulatory gene, *lacI,* encodes the Lac repressor, which plays a pivotal role in the control of the operon. The promoter binds RNA polymerase, and the operator binds activated Lac repressor. The transcription unit, which extends from the transcription initiation site to the transcription termination site, contains the genes.

from being expressed. Other operons are controlled by a regulatory protein termed an *activator,* which, when active, turns on the expression of the genes.

Many operons are controlled by more than one regulatory mechanism, and a number of the repressors or activators control more than one operon. The result is a complex network of superimposed controls that provides total regulation of transcription and allows almost instantaneous responses to changing environmental conditions.

The *lac* Operon for Lactose Metabolism Is Transcribed When an Inducer Inactivates a Repressor

Jacob and Monod researched the genetic control of lactose metabolism in *E. coli.* Lactose is a sugar that, when metabolized, provides energy for the cell. Jacob and Monod used genetic and biochemical approaches to study the genetic control of lactose metabolism in *E. coli.* Their genetic studies showed that three genes are involved: *lacZ, lacY,* and *lacA,* for lactose metabolism **(Figure 16.2)**. These three genes are adjacent to one another on the chromosome in the order Z-Y-A. The genes are transcribed as a unit into a single mRNA starting with the *lacZ* gene; the promoter for the transcription unit is upstream of *lacZ.* The *lacZ* gene encodes the enzyme β-galactosidase, which catalyzes the conversion of the disaccharide sugar, lactose, into the monosaccharide sugars, glucose and galactose. These sugars are then metabolized by other enzymes, producing energy for the cell. The *lacY* gene encodes a permease enzyme that transports lactose actively into the cell, and the *lacA* gene encodes a transacetylase enzyme, the function of which is unknown.

Jacob and Monod called the cluster of genes and adjacent sequences that control their expression the *lac* operon (see Figure 16.2). They coined the name *operon*

from a key DNA sequence they discovered for regulating transcription of the operon—the **operator.** The operator was named because it controls the operation of the genes adjacent to it. For the *lac* operon, the operator is a short DNA sequence between the promoter and the *lacZ* gene.

The two investigators found that the *lac* operon was controlled by a regulatory protein that they termed the *Lac repressor.* The Lac repressor is encoded by the regulatory gene *lacI,* which is nearby but separate from the *lac* operon (see Figure 16.2), and is synthesized in active form. When lactose is absent from the medium, active Lac repressor binds to the operator, thereby blocking the RNA polymerase from binding to the promoter; as a result, transcription cannot occur **(Figure 16.3a)**. Actually, the repressor occasionally falls off, allowing transcription to occur—but at a very slow rate, leading to just a few molecules of each encoded enzyme in the cell.

When lactose is added to the medium, the *lac* operon is turned on and all three enzymes are synthesized rapidly **(Figure 16.3b)**. How does this occur? Lactose enters the cell and the β-galactosidase molecules already present convert some of it to *allolactose,* an isomer of lactose. Allolactose is an **inducer** for the *lac* operon—the isomer turns on the three genes in the operon. Allolactose does this by binding to the Lac repressor, inactivating it by altering its shape so that it no longer can bind to the operator. With the repressor out of the way, RNA polymerase then is able to bind to the promoter, and it transcribes the three genes. The *lac* operon is called an **inducible operon** because an inducer molecule increases its expression.

When the lactose is used up from the medium, the regulatory system again switches the *lac* operon off. That is, the absence of lactose means that there are no allolactose inducer molecules to inactivate the repressor; the again-active repressor binds to the operator, blocking transcription of the operon. The controls are aided by

a. Lactose absent from medium

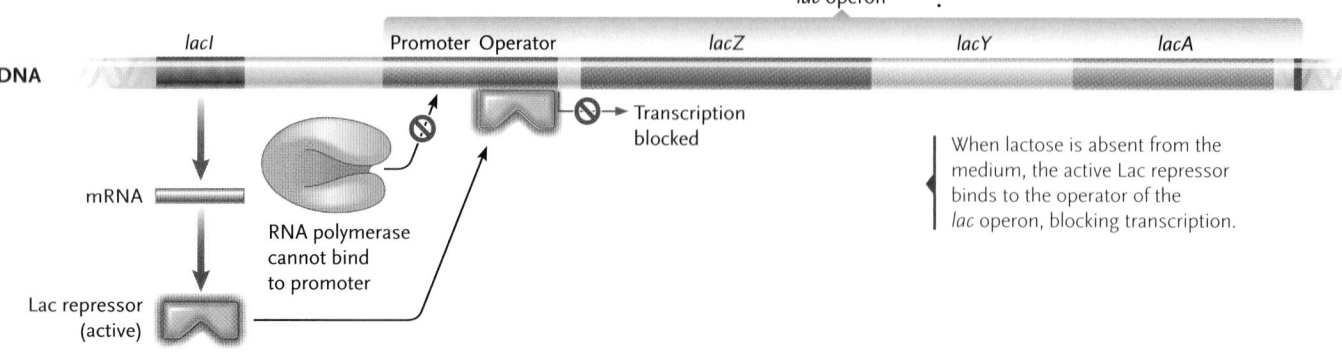

When lactose is absent from the medium, the active Lac repressor binds to the operator of the *lac* operon, blocking transcription.

b. Lactose present in medium

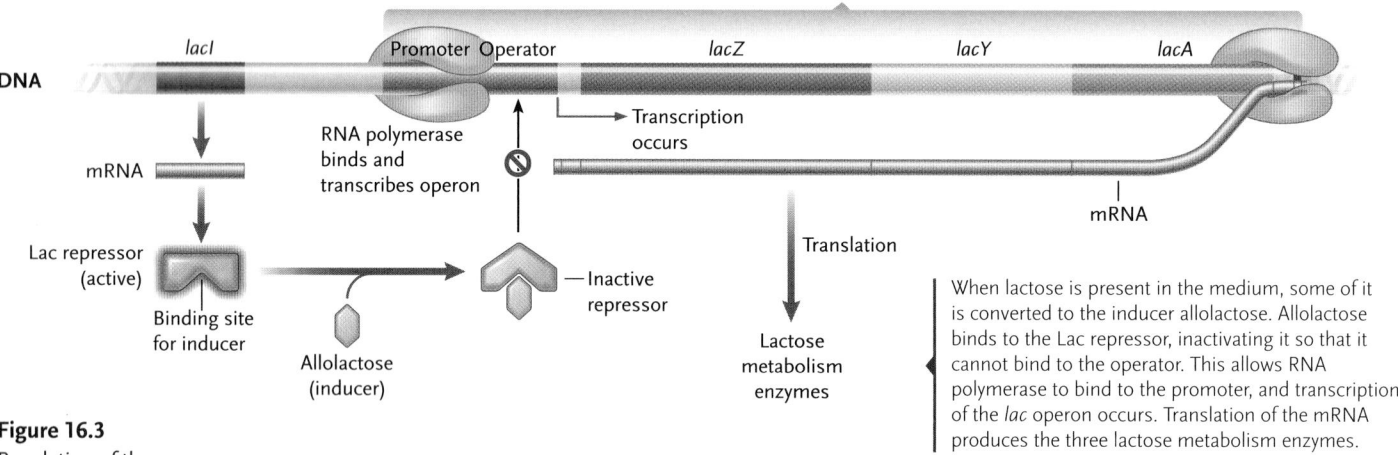

When lactose is present in the medium, some of it is converted to the inducer allolactose. Allolactose binds to the Lac repressor, inactivating it so that it cannot bind to the operator. This allows RNA polymerase to bind to the promoter, and transcription of the *lac* operon occurs. Translation of the mRNA produces the three lactose metabolism enzymes.

Figure 16.3
Regulation of the inducible *lac* operon by the Lac repressor in the absence **(a)** and presence **(b)** of lactose.

the fact that bacterial mRNAs are very short-lived, about 3 minutes on the average. This quick turnover permits the cytoplasm to be cleared quickly of the mRNAs transcribed from an operon. The enzymes themselves also have short lifetimes, and are quickly degraded.

Transcription of the *trp* Operon Genes for Tryptophan Biosynthesis Is Repressed When Tryptophan Activates a Repressor

Tryptophan is an amino acid that is used in the synthesis of proteins. If tryptophan is absent from the medium, *E. coli* must make tryptophan so that it can synthesize its proteins. If tryptophan is present in the medium, then the cell will use that source of the amino acid rather than making its own.

Tryptophan biosynthesis also involves an operon, the *trp* operon **(Figure 16.4).** The five genes in this operon, *trpA–trpE*, encode the enzymes for the steps in the tryptophan biosynthesis pathway. Upstream of the *trpE* gene are the operon's promoter and operator sequences. Expression of the *trp* operon is controlled by the Trp repressor, a regulatory protein encoded by the *trpR* gene, which is located elsewhere in the genome (not nearby as was the case for the repressor gene for

the *lac* operon). In contrast to the Lac repressor, though, the Trp repressor is synthesized in an inactive form in which it cannot bind to the operator.

When tryptophan is absent from the medium and must be made by the cell, the *trp* operon genes are expressed (see Figure 16.4a). This is the default state: since the Trp repressor is inactive and cannot bind to the operator, RNA polymerase can bind to the promoter and transcribe the operon. The resulting mRNA is translated to produce the five tryptophan biosynthetic enzymes that catalyze the reactions for tryptophan synthesis.

If tryptophan is present in the medium, there is no need for the cell to make tryptophan, so the *trp* operon is shut off (see Figure 16.4b). This occurs by a mechanism in which the tryptophan entering the cell binds to the Trp repressor and activates it. The active Trp repressor then binds to the operator of the *trp* operon and blocks RNA polymerase from binding to the promoter—the operon cannot be transcribed.

For the *trp* operon, then, the presence of tryptophan represses the expression of the tryptophan biosynthesis genes; hence, this operon is an example of a **repressible operon.** Here, tryptophan acts as a **corepressor,** a regulatory molecule that combines with a repressor to activate it and shut off the operon.

a. Tryptophan absent from medium

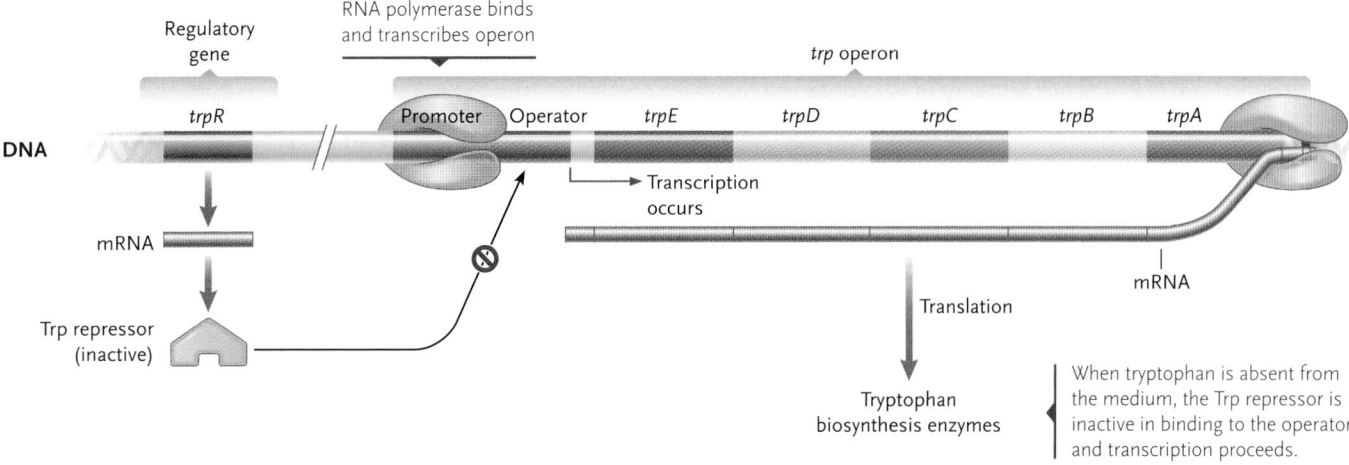

b. Tryptophan present in medium

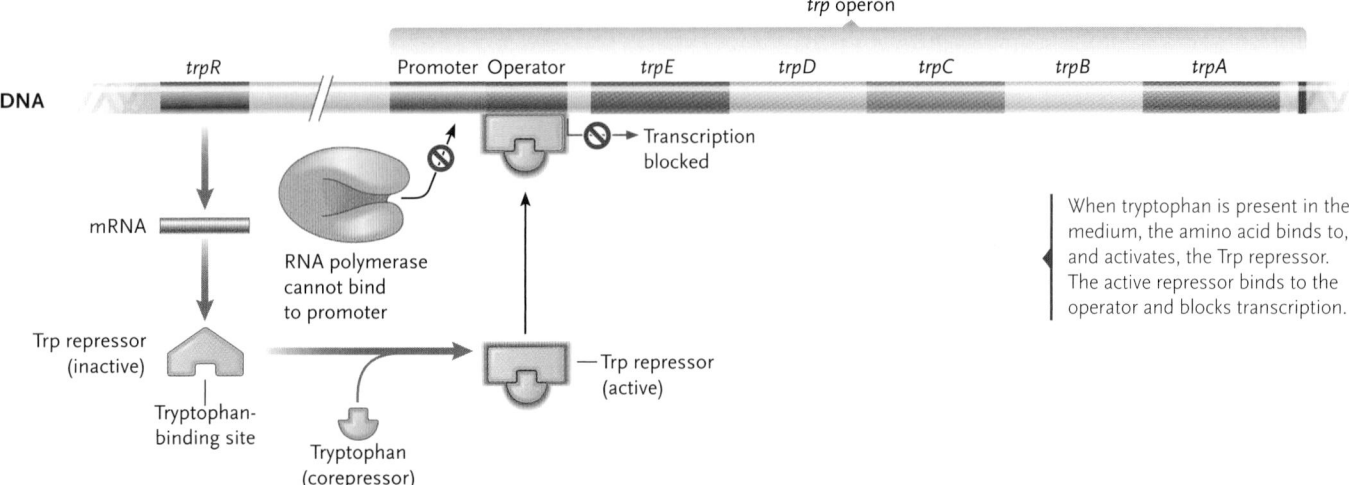

Figure 16.4
Regulation of the repressible *trp* operon by the Trp repressor in the absence **(a)** and presence **(b)** of tryptophan.

To compare and contrast the two operons we have discussed: (1) In the *lac* operon, the repressor is synthesized in an active form. When the inducer (allolactose) is present, it binds to the repressor and inactivates it. The operon is then transcribed. (2) In the *trp* operon, the repressor is synthesized in an inactive form. When the corepressor (tryptophan) is present, it binds to the repressor and activates it. The active repressor blocks transcription of the operon.

Inducible and repressible operons illustrate two types of *negative gene regulation* because both are regulated by a repressor that turns off gene expression when it is in active form. Genes are expressed only when the repressor is in inactive form.

Transcription of the *lac* Operon Is Also Controlled by a Positive Regulatory System

Several years after Jacob and Monod proposed their operon model for the lactose metabolism genes, researchers found a *positive gene regulation* system that also regulates the *lac* operon. This system ensures that the *lac* operon is transcribed if lactose is provided as an energy source, but not if glucose is present in addition to lactose. This is because glucose is a more efficient source of energy than is lactose. Glucose can be used directly in the glycolysis pathway to produce energy for the cell (see Chapter 8). Lactose, on the other hand, must first be converted into glucose and galactose, and the galactose then converted into glucose. These conversions require energy from the cell. Thus the cell gains more net energy by metabolizing glucose than by metabolizing lactose, or for that matter any other sugar.

Figure 16.5a shows the positive gene regulation system working when lactose is present and glucose is absent in the growth medium. In essence, we are adding to the model shown earlier in Figure 16.3b. Lactose is metabolized to the inducer, allolactose, which binds to and inactivates the Lac repressor. RNA polymerase is then recruited to the promoter by active *CAP (catabolite activator protein)* at the *CAP site*, a DNA sequence immediately upstream of the pro-

a. Lactose present; glucose low or absent

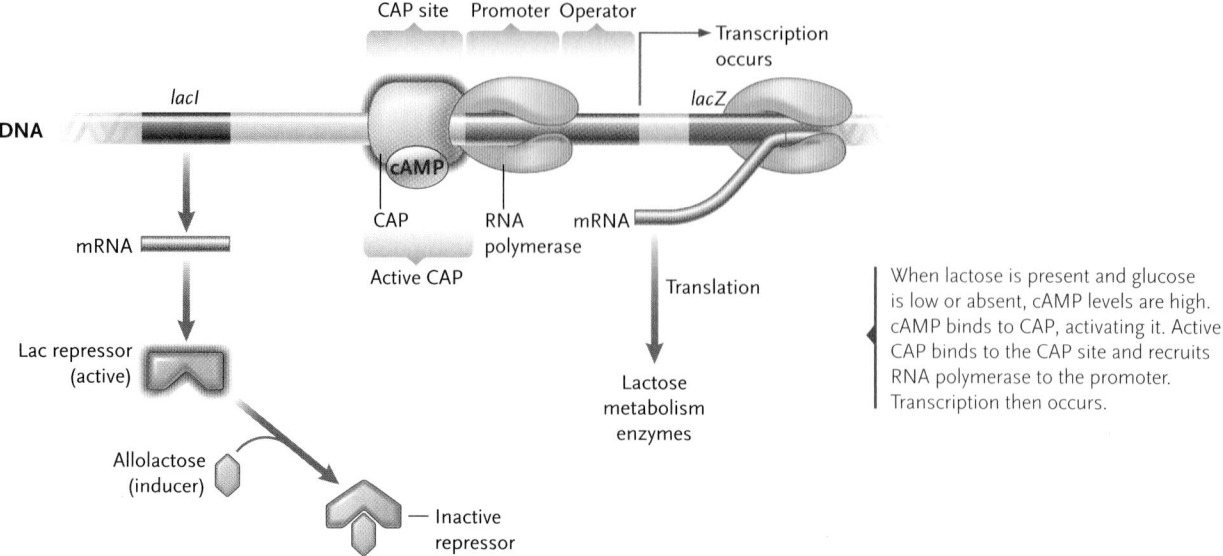

When lactose is present and glucose is low or absent, cAMP levels are high. cAMP binds to CAP, activating it. Active CAP binds to the CAP site and recruits RNA polymerase to the promoter. Transcription then occurs.

b. Lactose present; glucose present

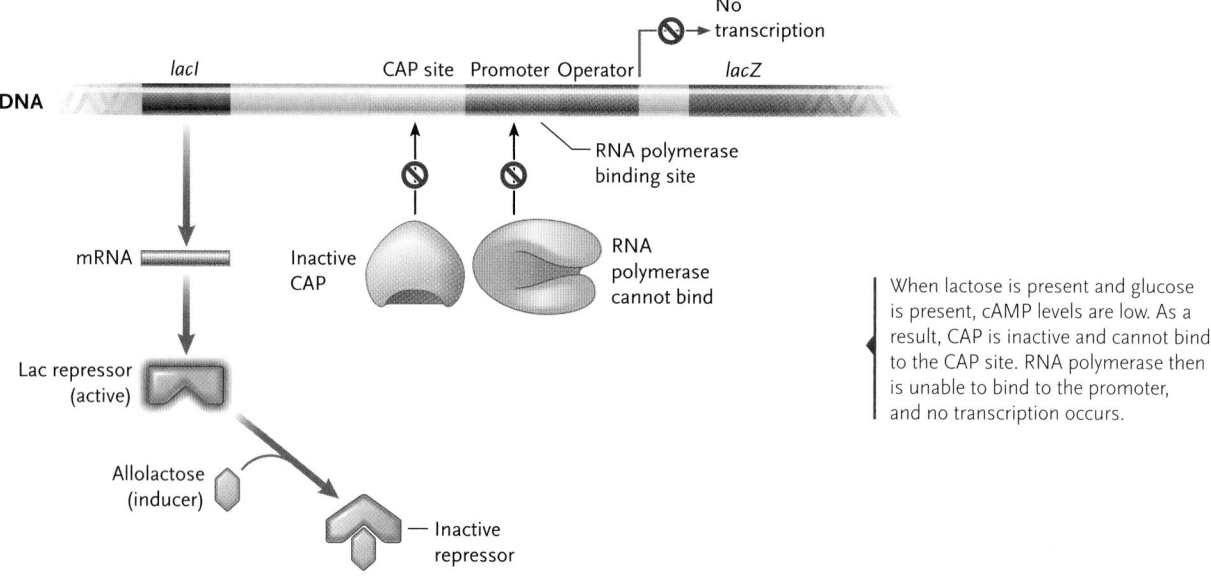

When lactose is present and glucose is present, cAMP levels are low. As a result, CAP is inactive and cannot bind to the CAP site. RNA polymerase then is unable to bind to the promoter, and no transcription occurs.

Figure 16.5
Positive regulation of the *lac* operon by the CAP activator. Other operons involved in the metabolism of various sugars are regulated in the same way.

moter. CAP is an **activator,** a regulatory protein that stimulates gene expression. It is synthesized in *inactive* form and is activated when cAMP (cyclic AMP) binds to it (cAMP is a nucleotide that plays a role in regulating cellular processes in both prokaryotes in eukaryotes; see Section 7.4). When glucose is absent from the medium, cAMP is abundant in the cell, so CAP is active under these conditions and can bind to the CAP site.

If both lactose and glucose are present in the medium, the *lac* operon is not transcribed **(Figure 16.5b).** Metabolism of the incoming glucose triggers a series of events leading to inactivation of adenylyl cyclase, the enzyme that catalyzes the synthesis of cAMP from ATP.

The level of cAMP drops drastically, reaching a point when it is too low to activate CAP. Without active CAP bound to the CAP site, RNA polymerase is unable to bind to the promoter, and the operon cannot be transcribed. In short, gene expression cannot be activated under these conditions. When the glucose is depleted from the medium, the bacteria then shift to metabolizing the lactose in the medium. Inactivation of adenylyl cyclase is reversed, cAMP levels rise again, and CAP is activated. The events of Figure 16.5a then occur.

The same positive gene regulation system using CAP and cAMP regulates a large number of other operons that control the metabolism of many sugars. In each case, the system functions so that glucose, if it is present in the growth medium, is metabolized first.

In sum, regulation of gene expression in prokaryotes occurs primarily at the transcription level. There are also some examples of regulation at the translation level. For example, some proteins can bind to the mRNAs that produce them and modulate their translation. This serves as a feedback mechanism to fine-tune the amounts of the proteins in the cell. In the remainder of the chapter we discuss the regulation of gene expression in eukaryotes. You will see that regulation occurs at several points between the gene and the protein, and that regulatory mechanisms are more complex than those in prokaryotes.

STUDY BREAK

1. Suppose the *lacI* gene is mutated so that the Lac repressor is not made. How does this mutation affect the regulation of the *lac* operon?
2. Answer the equivalent question for the *trp* operon: How would a mutation that prevents the Trp repressor from being made affect the regulation of the *trp* operon?

16.2 Regulation of Transcription in Eukaryotes

As you just learned, gene expression in prokaryotes is commonly regulated at the transcription level with genes organized in functional units called operons. The molecular mechanisms in operon function are a simple means of coordinating synthesis of proteins with related functions. In eukaryotes, the coordinated synthesis of proteins with related functions also occurs, but the genes involved usually are scattered around the genomes; that is, they are not organized into operons. Nonetheless, like operons, individual eukaryotic genes also consist of protein-coding sequences and adjacent regulatory sequences.

There are two general categories of eukaryotic gene regulation. Short-term regulation involves regulatory events in which gene sets are quickly turned on or off in response to changes in environmental or physiological conditions in the cell's or organism's environment. This type of regulation is most similar to prokaryotic gene regulation. Long-term gene regulation involves regulatory events required for an organism to develop and differentiate. Long-term gene regulation occurs in multicellular eukaryotes and not in simpler, unicellular eukaryotes. The mechanisms we discuss in this and the next section are applicable to both short-term and long-term regulation. The specific molecules and genes involved are different and, of course, so is the outcome to the cell or organism.

In Eukaryotes, Regulation of Gene Expression Occurs at Several Levels

The regulation of gene expression is more complicated in eukaryotes than in prokaryotes because eukaryotic cells are more complex, because the nuclear DNA is organized with histones into chromatin, and because multicellular eukaryotes produce large numbers and types of cells. Further, the eukaryotic nuclear envelope separates the processes of transcription and translation, whereas in prokaryotes translation can start on an mRNA that is still being made. Consequently, gene expression in eukaryotes is regulated at more levels. That is, there is transcriptional regulation, posttranscriptional regulation, translational regulation, and posttranslational regulation **(Figure 16.6)**. The most important of these is transcriptional regulation.

Chromatin Structure Plays an Important Role in Whether a Gene Is Active or Inactive

Eukaryotic DNA is organized into chromatin by combination with histone proteins (discussed in Section 14.5). Recall that DNA is wrapped around a core of two molecules each of histones H2A, H2B, H3, and H4, forming the nucleosome (see Figure 14.18). Higher levels of chromatin organization occur when histone H1 links adjacent nucleosomes.

Genes in regions of the DNA that are tightly wound around histones in chromatin are inactive, because their promoters are not accessible to the proteins that initiate transcription. Activating a gene involves changing the state of the chromatin so that the proteins that initiate transcription can bind to their promoters, a process called **chromatin remodeling**. In one type of chromatin remodeling, an activator binds to a regulatory sequence upstream of the gene's promoter and recruits a *remodeling complex,* a protein complex that displaces a nucleosome from the chromatin, exposing the promoter **(Figure 16.7)**. In a second type of chromatin remodeling, an activator binds to a regulatory sequence upstream of the gene's promoter, and recruits an enzyme that acetylates (adds acetyl groups: $CH_3CO—$) to histones in the nucleosome where the promoter is located. Acetylation causes the histones to loosen their association with DNA, and the promoter becomes accessible. This type of remodeling is reversed by deacetylation enzymes that remove the acetyl groups from the histones. Many activators use both of these chromatin remodeling mechanisms to regulate gene activity.

Regulation of Transcription Initiation Involves the Effects of Proteins Binding to a Gene's Promoter and Regulatory Sites

Chromatin remodeling is a crucial initial event in facilitating gene expression. Remodeling opens the way for transcription initiation to occur. Transcription ini-

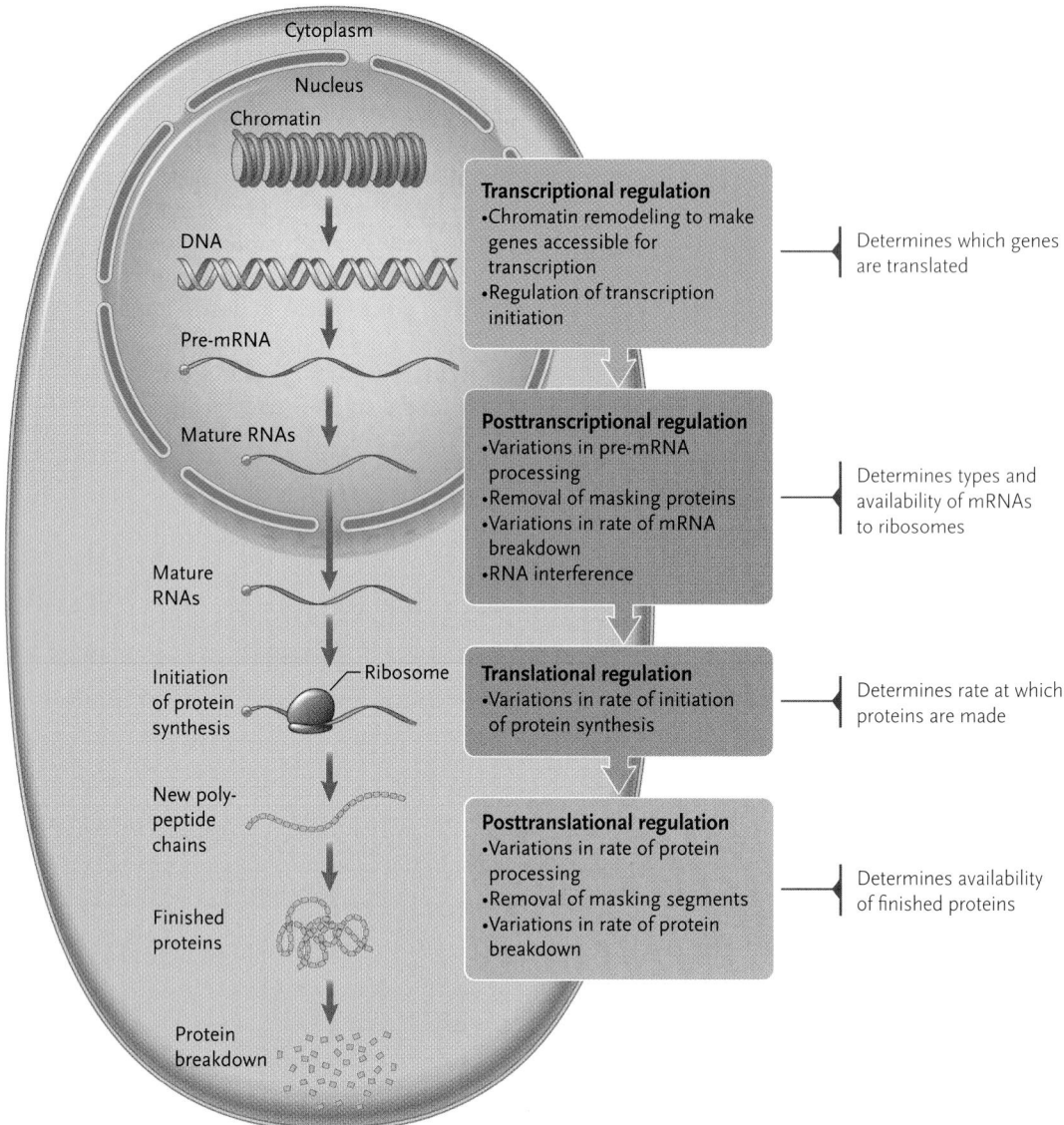

Figure 16.6
Steps in transcriptional, posttranscriptional, translational, and posttranslational regulation of gene expression in eukaryotes.

tiation is the most important level at which the regulation of gene expression takes place.

Organization of a Eukaryotic Protein-Coding Gene.
Figure 16.8 shows a eukaryotic gene, emphasizing the regulatory sites involved in its expression. Immediately upstream of the transcription unit is the promoter, a short region often containing the TATA box. The TATA box plays an important role in transcription initiation. RNA polymerase II itself cannot recognize the promoter sequence. Instead, proteins called transcription factors recognize and bind to the TATA box and then recruit the polymerase. Once the RNA polymerase II–transcription factor complex forms, the polymerase unwinds the DNA and transcription begins. Adjacent to the promoter, further upstream, is the **promoter proximal region**, which contains regulatory sequences called *promoter proximal elements*. Promoter proximal elements are part of a regulatory system for increasing the rate of transcription. More distant from the begin-

ning of the gene is the **enhancer**, which contains regulatory sequences that determine whether the gene is transcribed at its maximum possible rate.

Activation of Transcription. To initiate transcription, proteins called **general transcription factors** (also called *basal transcription factors*) bind to the promoter in the area of the TATA box **(Figure 16.9)**. These factors recruit the enzyme RNA polymerase II, which alone cannot bind to the promoter, and orient the enzyme to start transcription at the correct place. The combination of general transcription factors with RNA polymerase II is the **transcription initiation complex**. On its own, this complex brings about only a low rate of transcription initiation, which leads to just a few mRNA transcripts.

Activators—regulatory proteins that control the expression of one or more genes—bind to the promoter proximal elements to increase the rate of transcription. When bound, activators interact directly with

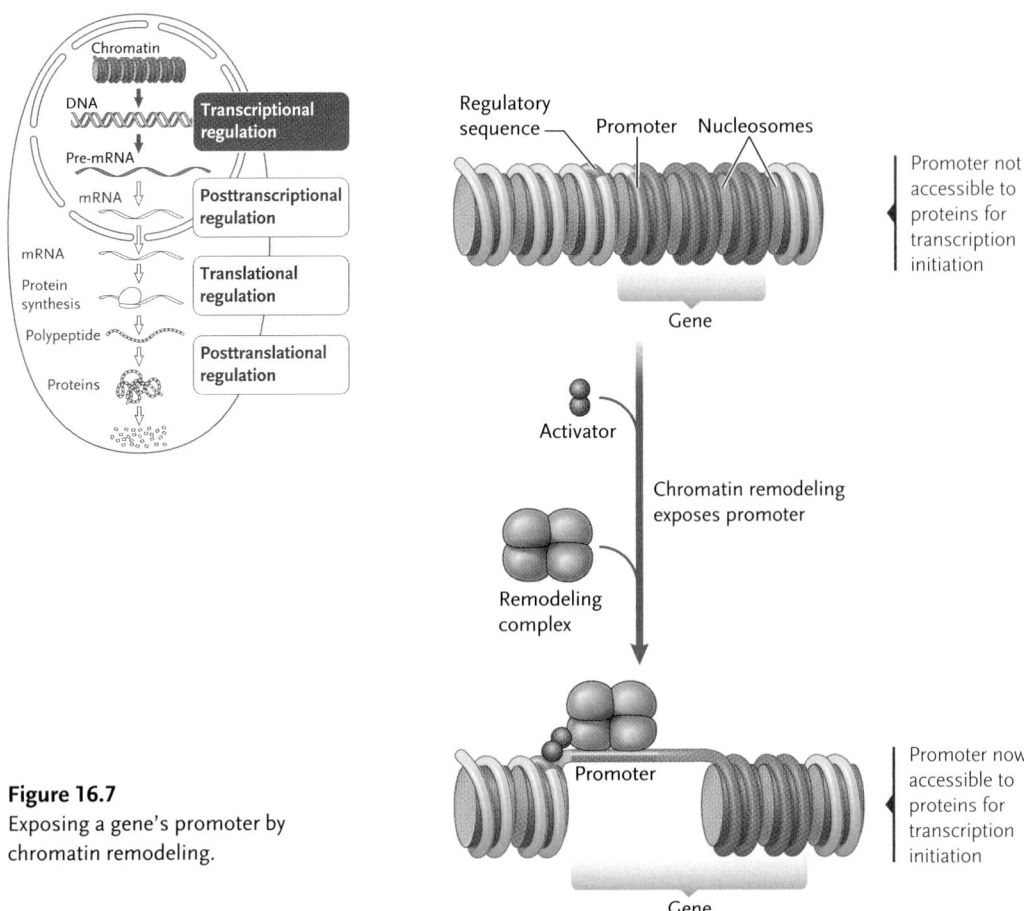

Figure 16.7
Exposing a gene's promoter by chromatin remodeling.

Figure 16.8
Organization of a eukaryotic gene. The transcription unit is the segment that is transcribed into the pre-mRNA; it contains the 5′ UTR (untranslated region), exons, introns, and 3′ UTR. Immediately upstream of the transcription unit is the promoter, which often contains the TATA box. Adjacent to the promoter and further upstream of the transcription unit is the promoter proximal region, which contains regulatory sequences called promoter proximal elements. More distant from the gene is the enhancer, which contains regulatory sequences that control the rate of transcription of the gene.

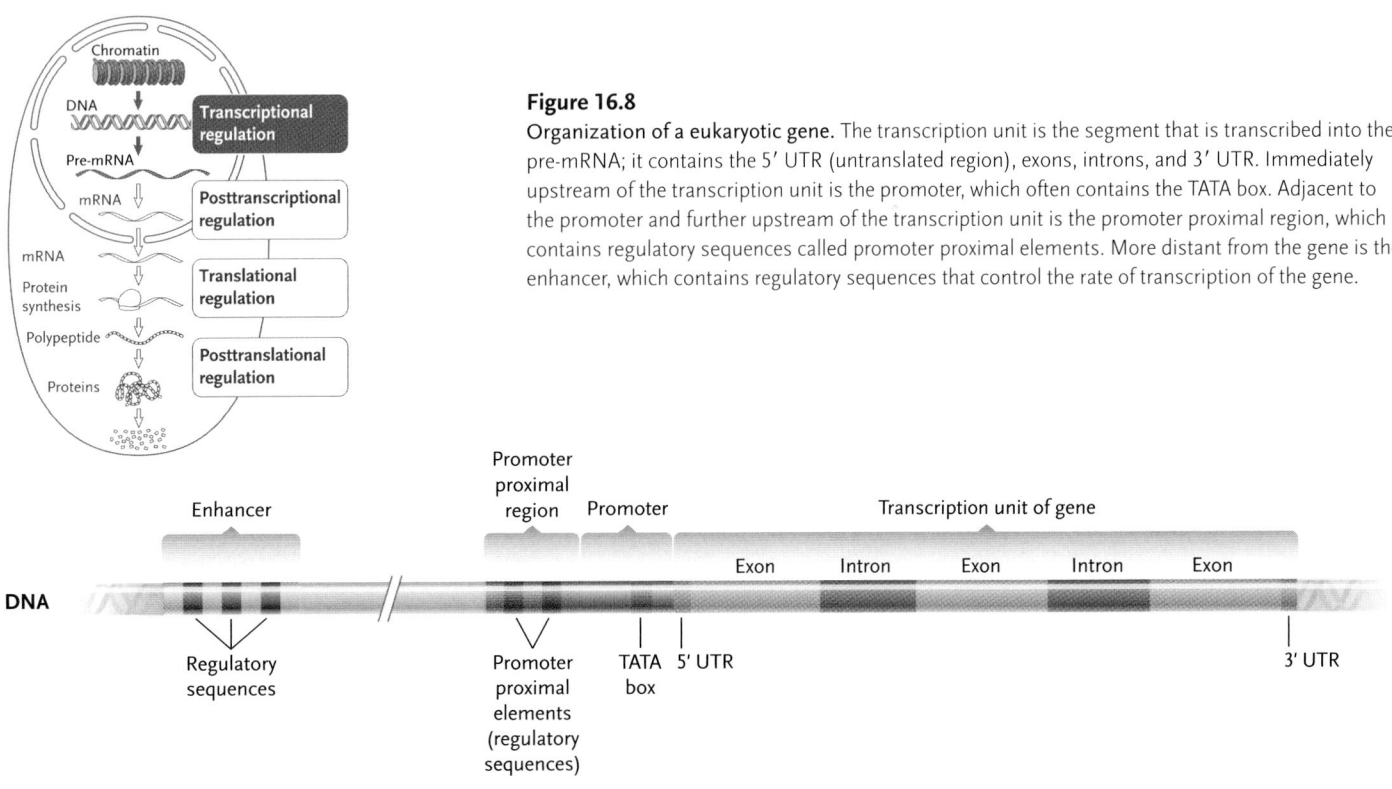

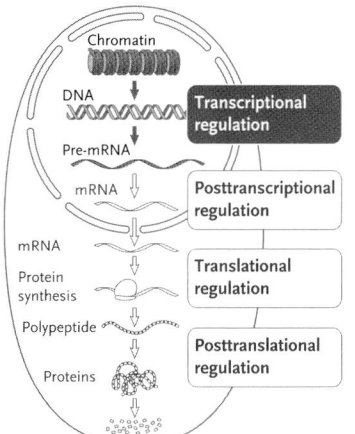

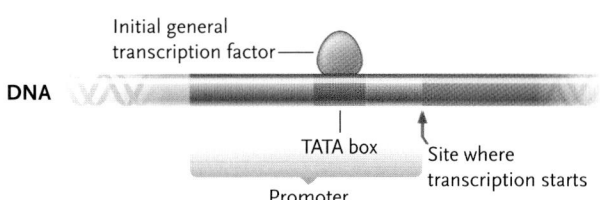

DNA

Initial general
transcription factor

TATA box

Site where
transcription starts

Promoter

1 The first general transcription factor recognizes and binds to the TATA box of a protein-coding gene's promoter.

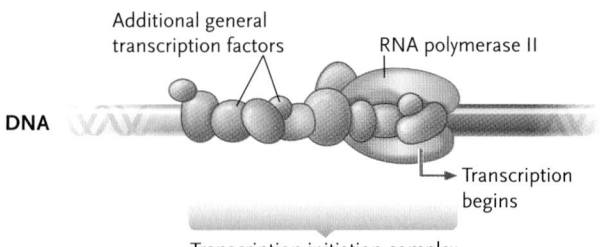

DNA

Additional general
transcription factors

RNA polymerase II

Transcription
begins

Transcription initiation complex

2 Additional general transcription factors and then RNA polymerase add to the complex, and then transcription begins.

the general transcription factors to stimulate transcription initiation, so many more transcripts are synthesized in a given time. Housekeeping genes—genes that are expressed in all cell types for basic cellular functions such as glucose metabolism—have promoter proximal elements that are recognized by activators present in all cell types. By contrast, genes expressed only in particular cell types or at particular times have promoter proximal elements that are recognized by activators found only in those cell types or at those times. To turn this around, the particular set of activators present within a cell at a given time is responsible for determining which genes in that cell are expressed to a significant level.

Events at the enhancer determine whether a gene is transcribed at its maximal rate **(Figure 16.10)**. Particular activators bind to the regulatory sequences within the enhancer. A **coactivator** (also called a *mediator*), a large multiprotein complex, forms a bridge between the activators at the enhancer and the proteins at the promoter and promoter proximal region, and causes the DNA to loop around on itself. The interactions between the coactivator, the proteins at the promoter, and

the RNA polymerase stimulate transcription to its maximal rate.

Repression of Transcription. In some genes, repressors oppose the effect of activators, thereby blocking or reducing the rate of transcription. The final rate of transcription then depends upon the "battle" between the activation signal and the repression signal.

Repressors in eukaryotes work in various ways. Some repressors bind to the same regulatory sequence to which activators bind (often in the enhancer), thereby preventing activators from binding to that site. Other repressors bind to their own specific site in the DNA near where the activator binds and interact with the activator so that it cannot interact with the coactivator. Yet other repressors recruit histone deacetylation enzymes that modify histones, leading to chromatin compaction and making a gene's promoter inaccessible to the transcription machinery.

Combinatorial Gene Regulation. Let us review the key elements of regulation of transcription of a protein-coding gene. General transcription factors bind to cer-

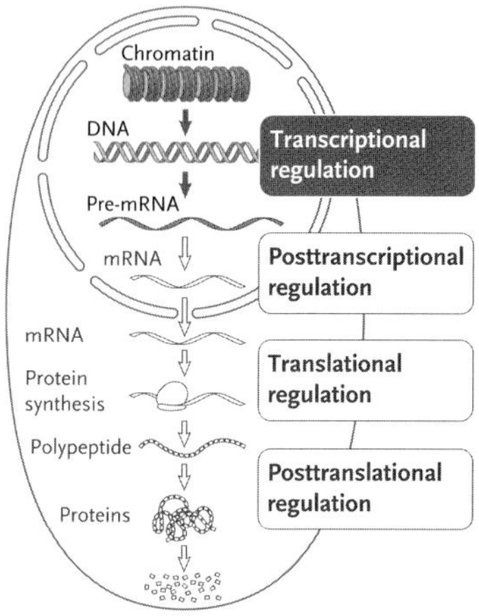

Figure 16.10
Interactions between activators at the enhancer, a coactivator, and general transcription factors at the promoter lead to maximal transcription of the gene.

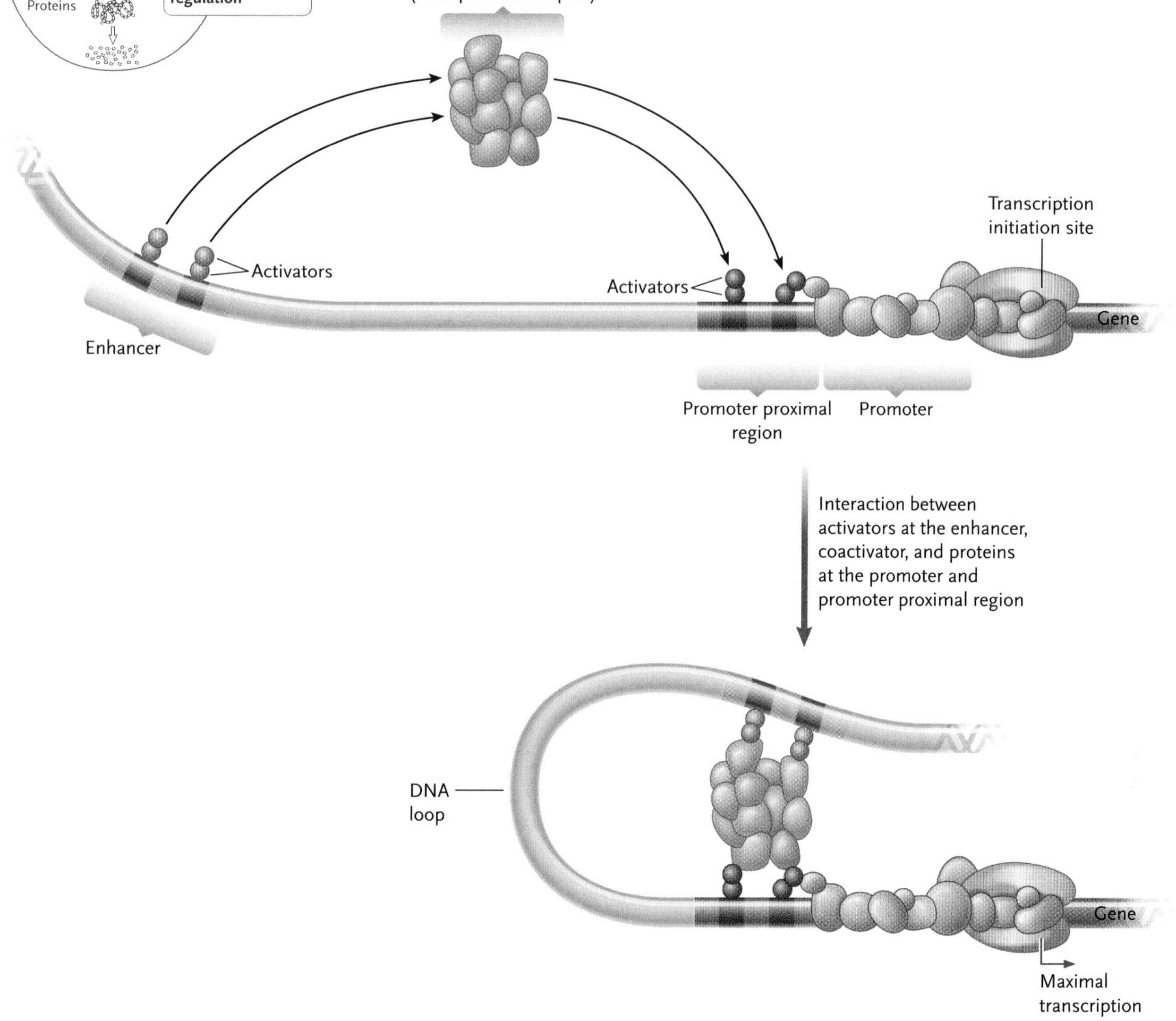

tain promoter sequences such as the TATA box and recruit RNA polymerase II; this results in a basal level of transcription. Specific activators bind to promoter proximal elements and stimulate the rate of transcription initiation. Activators also bind to the enhancer to give maximal transcription of the gene.

How are these events coordinated in regulating gene expression? Characteristic of any given gene is the number and types of promoter proximal elements. In some genes there may be only one regulatory element, but genes under complex regulatory control have many regulatory elements. Similarly, the number

and types of regulatory sequences in the enhancer is specific for each gene.

Both promoter proximal regions and enhancers are important in regulating the transcription of a gene. Each different regulatory sequence in those two regions binds a specific regulatory protein. Since some regulatory proteins are activators and others are repressors, the overall effect of regulatory sequences on transcription depends on the particular proteins that bind to them. If activators bind to both the regulatory sequences in the promoter proximal region and to the enhancer, transcription is activated maximally, mean-

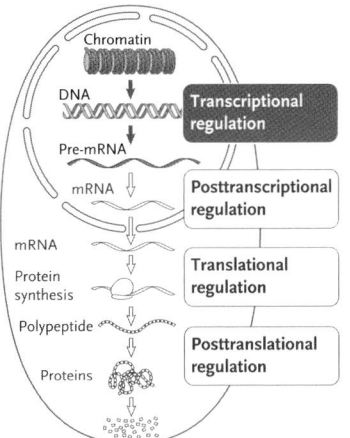

Figure 16.11
Combinatorial gene regulation. A relatively small number of regulatory proteins control transcription of all protein-coding genes. Different combinations of activators bind to enhancer regulatory sequences to control the rate of transcription of each gene.

a. A unique combination of activators controls gene *A*.

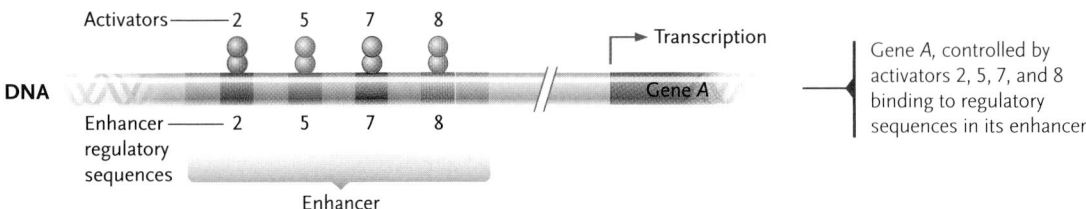

b. A different combination of activators controls gene *B*.

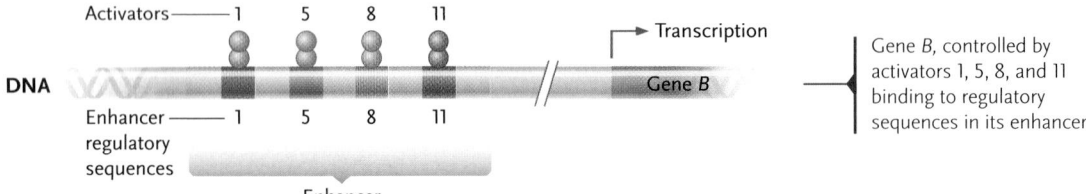

ing a high rate of transcription and therefore the production of a high level of the mRNA encoded by the gene. But, if a repressor binds to the enhancer and an activator binds to the promoter proximal element, the amount of gene expression depends upon the relative strengths of those two regulatory proteins. For example, if the repressor is strong, gene expression, in terms of the rate of transcription and the consequent level of the mRNA encoded by the gene, will be reduced.

A relatively small number of regulatory proteins (activators and repressors) control transcription of all protein-coding genes. By combining a few regulatory proteins in particular ways, the transcription of a wide array of genes can be controlled, and a large number of cell types can be specified. The process is called **combinatorial gene regulation**. Let us consider a theoretical example of two genes, each with activators already bound to the respective promoter proximal elements **(Figure 16.11)**. Maximal transcription of gene *A* requires activators 2, 5, 7, and 8 binding to their regulatory sequences in the enhancer, whereas maximal transcription of gene *B* requires activators 1, 5, 8, and

11 binding to its enhancer. That is, both genes require activators 5 and 8 for full activation in combination with different other activators.

This operating principle solves a basic dilemma in gene regulation—if each gene were regulated by a single, distinct protein, the number of genes encoding regulatory proteins would have to equal the number of genes to be regulated. Regulating the regulators would require another set of genes of equal number, and so on until the coding capacity of any chromosome set, no matter how large, would be exhausted. But because different genes require different combinations of regulatory proteins, the number of genes encoding regulatory proteins can be much lower than the number of genes they control.

Coordinated Regulation of Transcription of Genes with Related Functions. In the discussion of prokaryotic operons, you learned that genes with related function are often clustered *and* they are transcribed from one promoter onto a single mRNA. That mRNA is translated from one end to the other to produce the several proteins encoded by the genes. There are no operons in eukaryotes, yet the transcription of genes with related function is coordinately controlled. How is this accomplished?

The answer is that all genes that are coordinately regulated have the same regulatory sequences associated with them. Therefore, with one signal, the transcription of all of the genes can be controlled simultaneously. Let us consider an example of this: the control of gene expression by steroid hormones in mammals. A **hormone** is a molecule produced by one tissue and transported via the bloodstream to another specific tissue to alter its physiological activity. A **steroid** is a type of lipid derived from cholesterol (see Section 3.4). Examples of steroid hormones are testosterone and glucocorticoid. Testosterone regulates the expression of a large number of genes associated with the maintenance of primary and secondary male characteristics. Glucocorticoid, among other actions, regulates the expression of genes involved in the maintenance of the concentration of glucose and other fuel molecules in the blood.

A steroid hormone acts on specific target tissues in the body because only cells in those tissues have *steroid hormone receptors* in their cytoplasm that recognize and bind the hormone (see Section 7.5). The steroid hormone moves through the plasma membrane into the cytoplasm and the receptor binds to it **(Figure 16.12)**. The hormone-receptor complex then enters the nucleus and binds to specific regulatory sequences adjacent to the genes whose expression is controlled by the hormone. This binding activates transcription of those genes, and proteins encoded by the genes are made rapidly.

All genes regulated by a specific steroid hormone have the same DNA sequence to which the hormone-receptor complex binds. This sequence is called a **steroid hormone response element.** For example, all genes controlled by glucocorticoid have a glucocorticoid response element associated with them. Therefore, the release of glucocorticoid into the bloodstream coordinately activates the transcription of genes with that response element.

Methylation of DNA Can Control Gene Transcription

DNA methylation, in which a methyl group (—CH$_3$) is added enzymatically to cytosine bases in the DNA, can regulate transcription of a gene. Specifically, methylation of cytosines in promoters inhibits transcription and turns the genes off, a phenomenon called **silencing.**

For example, genes encoding the blood protein hemoglobin are highly methylated and inactive in most vertebrate body cells. In the cell lines giving rise to red blood cells, however, enzymes remove the methyl groups from the hemoglobin genes, which are then transcribed.

DNA methylation in some cases silences large blocks of genes, or even chromosomes. For example, in body cells of female placental mammals, including humans, one of the two X chromosomes packs tightly

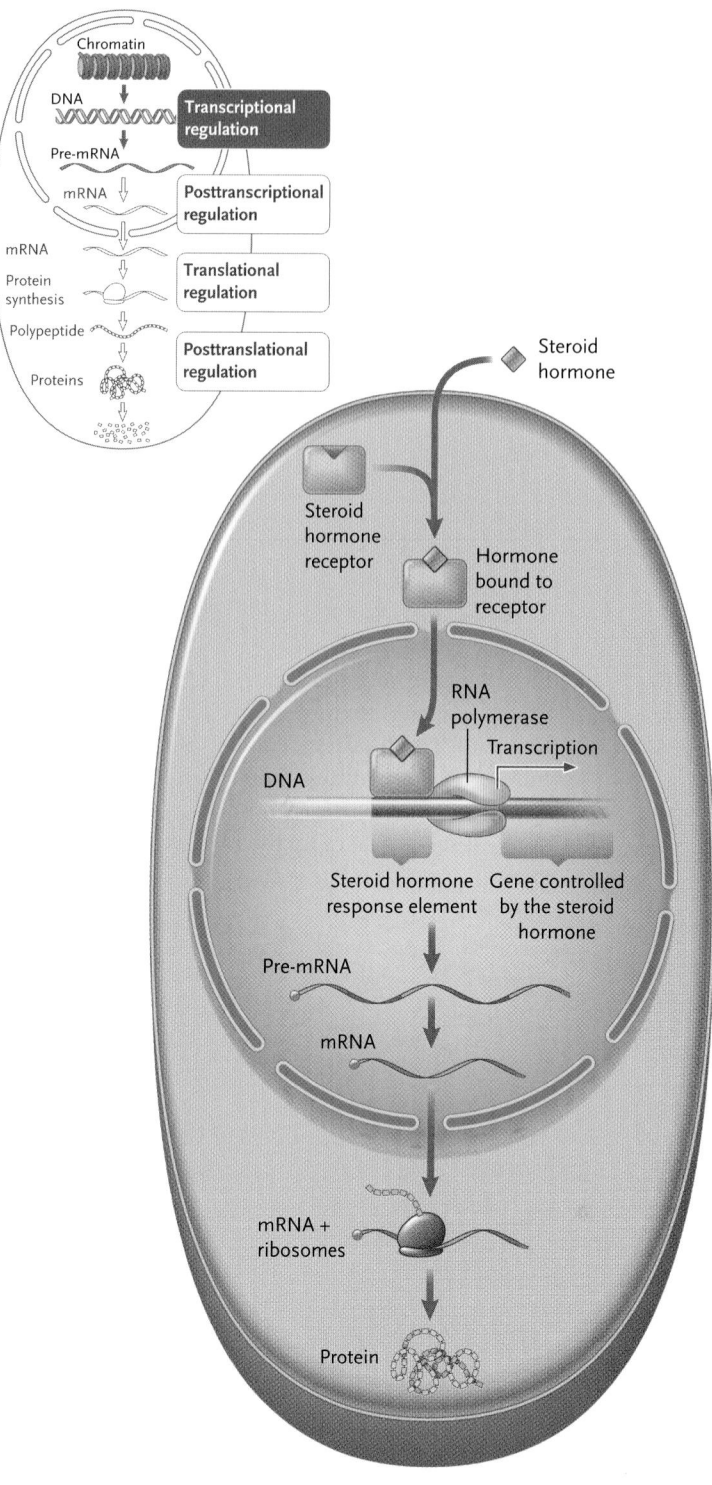

Figure 16.12

Steroid hormone regulation of gene expression. A steroid hormone enters the cell and forms a complex in the cytoplasm with a steroid hormone receptor that is specific to the hormone. Steroid hormone-receptor complexes migrate to the nucleus, bind to the steroid hormone response element next to each gene they control (one such gene is shown in the figure), and affect transcription of those genes.

into a mass known as a Barr body, in which essentially all the genes of the X chromosome are turned off. As part of this general inactivation, which also includes chromatin modifications, cytosines in the DNA become methylated.

DNA methylation underlies **genomic imprinting**, in which methylation permanently silences transcription of either the inherited maternal or paternal allele of a particular gene (see Section 13.5). The methylation occurs during gametogenesis in a parent. An inherited methylated allele is not expressed—it is silenced. That allele is known as the *imprinted allele*. The expression of the gene involved therefore depends upon expression of the nonimprinted allele inherited from the other parent. The methylation of the parental allele is maintained as the DNA is replicated, so that the silenced allele remains inactive in progeny cells. Some examples of genomic imprinting were presented in Section 13.5. In one of those examples, the mammalian *Igf2* (insulin growth factor 2) gene is inherited with the paternally derived allele nonmethylated and, therefore, active, and with the maternally derived allele methylated and, therefore, silenced.

Once mRNAs are transcribed from active genes, further regulation occurs at each of the major steps in the pathway from genes to proteins: during pre-mRNA processing and the movement of finished mRNAs to the cytoplasm (posttranslational regulation), during protein synthesis (translational regulation), and after translation is complete (posttranslational regulation). The next section takes up the regulatory mechanisms operating at each of these steps.

STUDY BREAK

1. What is the role of histones in gene expression? How does acetylation of the histones affect gene expression?
2. What are the roles of general transcription factors, activators, and coactivators in transcription of a protein-coding gene?

16.3 Posttranscriptional, Translational, and Posttranslational Regulation

Transcriptional regulation determines which genes are copied into mRNAs. This basic level of regulation is fine-tuned by posttranscriptional, translational, and posttranslational controls, the subjects of this section (refer again to Figure 16.6).

Posttranscriptional Regulation Controls mRNA Availability

Posttranscriptional regulation regulates translation by controlling the availability of mRNAs to ribosomes. The controls work by several mechanisms, including changes in pre-mRNA processing and the rate at which mRNAs are degraded.

Variations in Pre-mRNA Processing. In Chapter 15 we noted that mRNAs are transcribed initially as pre-mRNA molecules. These pre-mRNAs are processed to produce the finished mRNAs, which then enter protein synthesis. Variations in pre-mRNA processing can regulate *which* proteins are made in cells. As described in Section 15.3, pre-mRNAs can processed by *alternative splicing*. Alternative splicing produces different mRNAs from the same pre-mRNA by removing different combinations of exons (the amino acid–coding segments) along with the introns (the noncoding spacers). The resulting mRNAs are translated to produce a family of related proteins with various combinations of amino acid sequences derived from the exons. Alternative splicing itself is under regulatory control. Regulatory proteins specific to the type of cell control which exons are removed from pre-mRNA molecules by binding to regulatory sequences within those molecules. The outcome of alternative splicing is that appropriate proteins within a family are synthesized in cell types or tissues in which they are optimally functional. Perhaps three-quarters of human genes are alternatively spliced at the pre-mRNA level.

Posttranscriptional Control by Masking Proteins. Some posttranscriptional controls operate by means of "masking" proteins that bind to mRNAs and make them unavailable for protein synthesis. These controls are important in many animal eggs, in which they keep mRNAs in an inactive form until the egg has been fertilized and embryonic development is under way. When an mRNA is to become active, other factors—other proteins, made as part of the developmental pathway—remove the masking proteins and allow the mRNA to enter protein synthesis.

Variations in the Rate of mRNA Breakdown. The rate at which eukaryotic mRNAs break down can also be controlled posttranscriptionally. The mechanism involves a regulatory molecule, such as a steroid hormone, directly or indirectly affecting the mRNA breakdown steps, either slowing or increasing the rate of those steps. For example, in the mammary gland of the rat, the mRNA for casein (a milk protein) has a half-life of about 5 hours (meaning that it takes 5 hours for half of the mRNA present at a given time to break down). The half-life of casein mRNA changes to about 92 hours if the peptide hormone prolactin is present. Prolactin is synthesized in the brain and in other tissues, including the breast. The most important effect of prolactin is to stimulate the mammary glands to produce milk (that is, it stimulates lactation). During milk production, a large amount of casein must be synthesized, and this is accomplished in part by radically decreasing the rate of breakdown of the casein mRNA.

Nucleotide sequences in the 5′ UTR (untranslated region; see Section 15.3) appear also to be important in determining mRNA half life. If the 5′ UTR is transferred experimentally from one mRNA to another, the

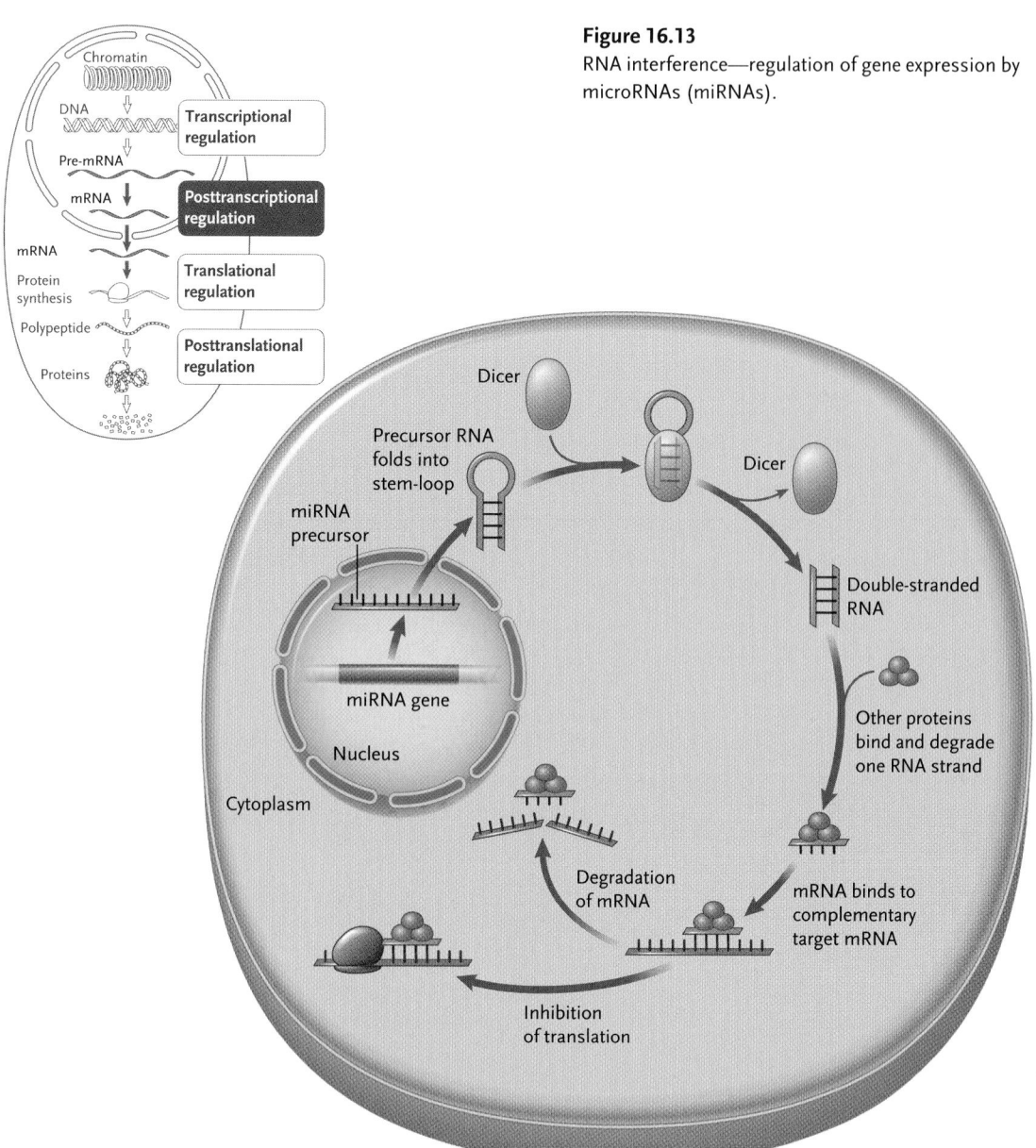

Figure 16.13

RNA interference—regulation of gene expression by microRNAs (miRNAs).

half-life of the receiving mRNA becomes the same as that of the donor mRNA. The controlling sequences in the 5′ UTR of an mRNA might be recognized by proteins that regulate its stability.

Regulation of Gene Expression by Small RNAs. The relatively recent discovery of *micro-RNAs* (miRNAs) has revolutionized our understanding of gene control. miRNAs are small, single-stranded RNAs found in organisms as diverse as worms, flies, plants, and mammals, where they regulate important processes such as development, growth, and behavior. What are miRNAs and how do they work?

Each miRNAs is encoded by a non-protein-coding gene. Transcription of the gene produces an RNA that is the precursor to the miRNA **(Figure 16.13).** The precursor RNA folds and base-pairs with itself, forming a stem-loop structure. An enzyme named Dicer cuts the stem-loop to produce a double-stranded RNA,

about 21–22 base pairs long. A protein complex then binds to the double-stranded RNA and degrades one of the two RNA strands, leaving a small single-stranded RNA—the miRNA. Still bound to the protein complex, the miRNA binds to any mRNA that has a complementary sequence. Gene expression is then silenced in one of two ways: either the proteins in the complex cleave the mRNA where the miRNA is bound to it, or the double-stranded segment formed between the miRNA and the mRNA blocks ribosomes from translating the mRNA.

Researchers think that there are 120 genes for miRNAs in worms and 250 genes in humans. Many of these miRNAs are expressed in developmentally regulated patterns. The targets of the miRNA's action are often mRNAs for regulatory proteins that control the development of the organism.

The phenomenon of silencing a gene posttranscriptionally by a small, single-stranded RNA that is complementary to part of an mRNA is termed **RNA interference (RNAi)**. miRNAs are one class of single-stranded RNAs that cause RNAi; another class is known as **small interfering RNA (siRNA)**. Whereas miRNA is produced from RNA that is encoded in the cell's genome, siRNA is produced from double-stranded RNA that is *not* encoded by nuclear genes. For example, the life cycle and replication of many viruses involves a double-stranded RNA stage. Viral double-stranded RNA enters the RNAi process as described for miRNAs: double-stranded RNA is cut by Dicer into short double-stranded RNA molecules, and then a protein complex binds to the molecules and degrades one of the RNA strands to produce siRNA. The protein complex is the same one that acts on the double-stranded RNA precursors of miRNAs. In the RNAi process, siRNA acts exactly like microRNA—mRNAs complementary to the siRNA are targeted and either they are degraded or their translation is blocked. In our viral example, the targeted mRNAs would be mRNAs for proteins needed for viral genome replication and the production of new virus particles.

Any gene can be silenced experimentally by RNAi. To silence a gene, researchers introduce into the cell a double-stranded RNA that can be processed by Dicer and the protein complex into an siRNA complementary to the mRNA transcribed from that gene. Indeed, RNAi has become a powerful new technique for silencing specific

genes experimentally in a variety of organisms. Andrew Fire of the Massachusetts Institute of Technology and Craig Mello of Harvard University received a Nobel Prize in 2006 for their discovery of RNA interference.

Translational Regulation Controls the Rate of Protein Synthesis

At the next regulatory level, translational regulation controls the rate at which mRNAs are used in protein synthesis. Translational regulation occurs in essentially all cell types and species. For example, translational regulation is involved in cell cycle control in all eukaryotes and in many processes during development in multicellular eukaryotes, such as red blood cell differentiation in animals. Significantly, many viruses exploit translational regulation to control their infection of cells and to shut off the host cell's own genes.

Let us consider the general role of translational regulation in animal development. During early development of most animals, little transcription occurs. The changes in protein synthesis patterns seen in developing cell types and tissues instead derive from the activation, repression, or degradation of maternal mRNAs, the mRNAs that were in the mother's egg before fertilization. One important mechanism for translational regulation involves adjusting the length of the poly(A) tail of the mRNA. (Recall from Section 15.3 that the poly(A) tail—a string of adenine-containing nucleotides—is added to the 3′ end of pre-mRNA and is retained on the mRNA produced from the pre-mRNA after introns are removed.) That is, enzymes can change the length of the poly(A) tail on an mRNA in the cytoplasm in either direction: by shortening it or lengthening it. Increases in poly(A) tail length result in increased translation; decreases in length result in decreased translation. For example, during embryogenesis (the formation of the embryo) of the fruit fly, *Drosophila,* key proteins are synthesized when the poly(A) tails on the mRNAs for those proteins are lengthened in a regulated way. Evidence for this came from experiments in which poly(A) tail lengthening was blocked; the result was that embryogenesis was inhibited. But, while researchers know that the length of poly(A) tails is regulated in the cytoplasm, how this process occurs is not completely understood.

Posttranslational Regulation Controls the Availability of Functional Proteins

Posttranslational regulation controls the availability of functional proteins mainly in three ways: chemical modification, processing, and degradation. Chemical modification involves the addition or removal of chemical groups, which reversibly alters the activity of the protein. For example, you saw in Section 7.2 how the addition of phosphate groups to proteins involved in signal transduction pathways either stimulates or in-

Figure 16.14
Protein degradation by ubiquitin addition and enzymatic digestion within a proteasome.

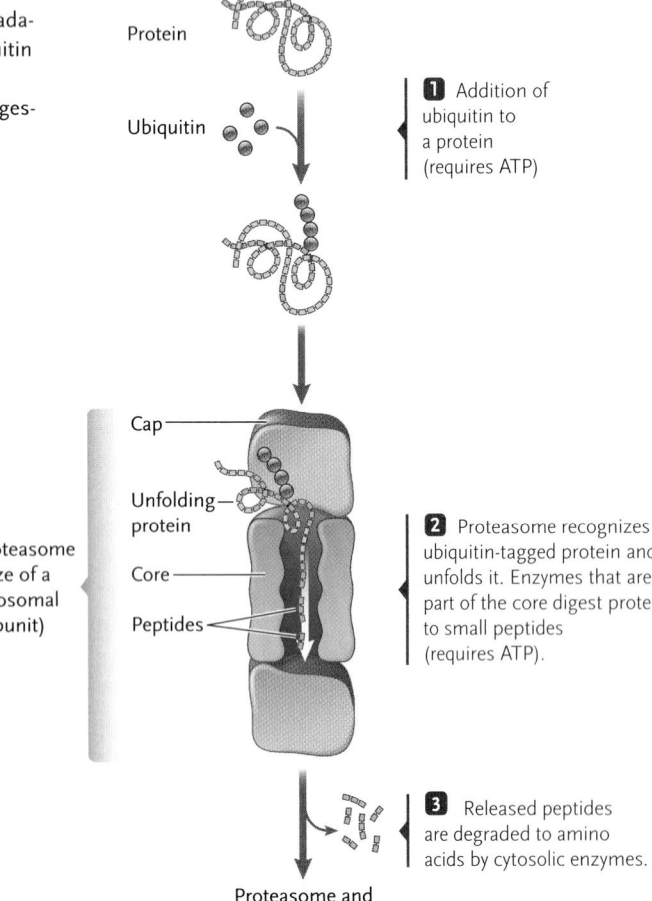

Protein

Ubiquitin

1 Addition of ubiquitin to a protein (requires ATP)

Cap

Unfolding protein

Proteasome (size of a ribosomal subunit)

Core

Peptides

2 Proteasome recognizes ubiquitin-tagged protein and unfolds it. Enzymes that are part of the core digest protein to small peptides (requires ATP).

3 Released peptides are degraded to amino acids by cytosolic enzymes.

Proteasome and ubiquitin are recycled.

hibits the activity of those proteins. Further, in Section 10.4 you learned how the addition of phosphate groups to target proteins plays a crucial role in regulating how a cell progresses through the cell division cycle. And, in Section 16.2 you saw how acetylation of histones altered the properties of the nucleosome, loosening its association with DNA in chromatin.

In processing, proteins are synthesized as inactive precursors, which are converted to an active form under regulatory control. For example, you saw in Section 15.4 that the digestive enzyme pepsin is synthesized as pepsinogen, an inactive precursor that activates by removal of a segment of amino acids. Similarly, the glucose-regulating hormone insulin is synthesized as a precursor called proinsulin; processing of the precursor removes a central segment but leaves the insulin molecule, which consists of two polypeptide chains linked by disulfide bridges.

The rate of degradation of proteins is also under regulatory control. Some proteins in eukaryotic cells last for the lifetime of the individual, while others persist only for minutes. Proteins with relatively short cellular lives include many of the proteins regulating transcription. Typically, these short-lived proteins are marked for breakdown by enzymes that attach a "doom tag" consisting of a small protein called *ubiquitin* (**Figure 16.14,** step 1). The protein is given this name because it is indeed ubiquitous—present in almost the same form in essentially all eukaryotes. The ubiquitin tag labels the doomed proteins so that they are recognized and attacked by a *proteasome*, a large cytoplasmic complex of a number of different proteins (step 2). The proteasome unfolds the protein, and protein-digesting enzymes within the core digest the protein into small peptides. The peptides are released from the proteasome and cytosolic enzymes further digest the peptides into individual amino acids, which are recycled for use in protein synthesis or oxidized as an energy source (step 3). The ubiquitin protein and proteasome are also recycled. Aaron Ciechanover and Avram Herhsko, both of the Israel Institute of Technology, Haifa, Israel, and Irwin Rose of the University of California, Irvine, received a Nobel Prize in 2004 for the discovery of ubiquitin-mediated protein degradation.

Control of protein breakdown is the last of the opportunities for control of gene expression. We will now look at cancer, a disease in which control of gene expression goes awry.

Study Break

1. How does a microRNA silence gene expression?
2. If the poly(A) tail on a mRNA was removed, what would likely be the effect on the translation of that mRNA?

16.4 The Loss of Regulatory Controls in Cancer

The cell division cycle of all eukaryotic cells from single-celled microorganisms to cells that are components of multicellular organisms is controlled by genes. The types of genes exerting this control are basically the same in terms of functions in all eukaryotes. Mutations in these genes can disrupt normal cell growth and division. The effects of such mutations are more significant and profound in complex multicellular organisms, particularly mammals. For example, occasionally, dividing and differentiating cells deviate from their normal genetic program and give rise to tissue masses called *tumors*. In other words, the cells lose their normal regulatory controls and revert partially or completely to an embryonic developmental state, in a process called *dedifferentiation*. If the deviant cells stay together in a single mass, the tumor is said to be *benign*. Benign tumors usually are not life threatening, and their surgical removal generally results in a complete cure.

If the cells of a tumor invade and disrupt surrounding tissues, the tumor is said to be *malignant* and is called a cancer (**Figure 16.15** shows a cancer cell). Sometimes, cells from malignant tumors break off and move through the blood system or lymphatic system, forming new tumors at other locations in the body. The spreading of a malignant tumor is called *metastasis* (meaning "change of state"). Malignant tumors can result in debilitation and death in various ways, including damage to critical organs, metabolic problems, hemorrhage, and secondary malignancies. In some cases, malignant tumors can be eliminated from the body by surgery or destroyed by chemicals *(chemotherapy)* or radiation.

Cancer cell

White blood cells

Lennart Nilsson, © Boehringer Ingelheim International GmbH

Figure 16.15
A scanning electron micrograph of a cancer cell surrounded by several white blood cells.

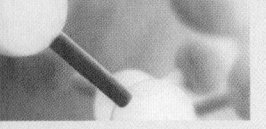

A Viral Tax on Transcriptional Regulation

The human *T-cell leukemia virus (HTLV)* causes a virulent form of cancer by triggering rapid, uncontrolled division of white blood cells. It does so by speeding up a pathway that triggers division at normal rates in uninfected cells. The pathway is a G-protein coupled receptor-response pathway involving cyclic AMP (cAMP) as a second messenger (see Section 7.4).

Normally, the pathway is triggered by cAMP released when an infection takes place. In the pathway, a specific activator called CREB (CRE-binding protein) is activated by phosphorylation and binds to CRE (cAMP-response element), in the enhancers for a group of genes that controls cell division. The binding turns on the genes and leads to the rapid division of white blood cells characteristic of the immune response.

HTLV takes advantage of the pathway by means of a short sequence CGTCA, in its DNA that mimics part of the human CRE enhancer sequence TGACGTCA. When the host cell's CREB is activated, it binds avidly to the enhancer sequence in the virus, turns on the viral genes, and leads to reproduction of the virus. Unfortunately for someone infected with HTLV, CREB in an infected cell also binds more avidly to the enhancers of cell division genes, leading to the uncontrolled division of white cells and, hence, to leukemia.

How HTLV produced these effects on cell division was puzzling because, in a test tube, CREB binds only weakly to the viral mimic of the CRE enhancer. An explanation was provided by Susanne Wagner and Michael R. Green at the University of Massachusetts Medical Center in Worcester, who found that HTLV uses one of its own proteins, called *Tax*, to get around the problem of weak binding. When Tax is present, CREB binds strongly to the viral CRE imitation. Tax also greatly increases the ability of CREB to bind to the normal CRE enhancer, leading to rapid and uncontrolled growth of infected white cells and to leukemia.

How does the Tax protein accomplish this feat? The CREB activator interacts with DNA as a dimer (a pair of molecules that together form the functional form). Wagner and Green found that the Tax protein greatly increases the ability of CREB to form dimers and thus to bind to the viral CRE enhancer. Evidently, Tax acts as a sort of molecular "safety pin" that holds the dimer together with either the viral imitation or the cell's CRE enhancer.

The Tax protein thus compensates for the imperfection of the viral sequence that imitates CRE by greatly increasing the ability of CREB to form dimers and bind to it. Although Tax provides a major advantage to HTLV, it may also open a chink in its armor. If a means can be found to interfere with Tax or its synthesis, it may be possible to stop the uncontrolled growth of white cells, and thus the leukemia caused by the virus.

Most Cancers Are Caused by Genes That Have Lost Their Normal Controls

All the characteristics of cancer cells—dedifferentiation, uncontrolled division, and metastasis—reflect changes in gene activity. Many of the genes that become altered encode proteins that control the cell division cycle of normal cells. That is, healthy cells grow and divide only when the balance of stimulatory and inhibitory signals received from outside the cell favors cell division. A cancer cell, by contrast, does not respond properly to the usual signals and divides without the usual constraints.

Two main types of genes commonly show altered activities as cells become cancer cells. One class is the **proto-oncogenes** (*oncos* = bulk or mass), genes in normal cells that encode various kinds of proteins that stimulate cell division. In cancer cells, the proto-oncogenes are altered to become **oncogenes**, genes that stimulate the cell to progress to the cancerous state. Among the mechanisms that can convert proto-oncogenes to oncogenes:

- Mutations in a gene's promoter or other control sequences may disrupt normal regulatory controls, making the gene abnormally active. The mutations can occur spontaneously or be induced by radiation or by particular chemicals.
- Mutations in the coding segment of the gene may produce an altered form of the encoded protein that is abnormally active.
- Translocation, in which a segment of a chromosome breaks off and attaches to a different chromosome (discussed in Section 13.3), may move a gene that controls cell division to a new location near the promoter or enhancer sequence of a highly active gene, making the cell division gene overactive.
- Infecting viruses may introduce genes to regions in the chromosomes where the expression of the genes disrupts cell cycle control or alters regulatory proteins to turn genes on. (*Insights from the Molecular Revolution* describes a virus that causes a blood cancer by altering a transcription factor.)

For example, translocation may affect *MYC*, a proto-oncogene controlling cell division. The activity of *MYC* is normally tightly regulated. However, *MYC* lies in a chromosome region that often breaks, causing a translocation that places *MYC* near the enhancer and promoter of a highly active antibody gene. The placement makes *MYC* continuously active, converting it into an oncogene that triggers rapid and uncontrolled cell division.

Several proto-oncogenes encode cell surface receptors that bind extracellular signal molecules such as peptide hormones or growth factors. In general, the oncogene forms of these receptors are continually activated, whether they are bound to the external signal molecule or not. As a result, the internal pathways they trigger, including those that cause cells to divide, are also continually active.

Another key group of proto-oncogenes encodes enzymes forming parts of the internal reaction pathways triggered by surface receptors (see Chapter 7). Most important are genes encoding the protein kinases, which regulate the activity of other proteins by adding phosphate groups to them. Some of the proteins phosphorylated by the protein kinases directly take part in gene regulation or initiation of cell division; others form parts of the cellular response pathways linked to surface receptors. The protein kinases encoded by the oncogene forms of the genes are continually active, constantly phosphorylating the control protein so that cell division continues at high and uncontrolled rates.

The other main class of genes that shows altered activity in cancer cells is the **tumor-suppressor genes**, which, in normal cells, encode proteins that inhibit cell division. Both alleles of a tumor suppressor gene must be inactivated for inhibitory activity to be lost in cancer cells. The best known of these genes is *TP53*, so called because its encoded protein, p53, has a molecular weight of 53,000 daltons. Among other activities, normal p53 stops cell division by combining with and inhibiting cyclin-dependent protein kinases that trigger entry into critical stages of DNA replication and mitosis (discussed in Section 10.4). Without the normal form of the p53 protein, the cyclin-dependent protein kinases are continually active in triggering cell division. Inactive *TP53* genes are found in many types of cancers.

Cancer Develops Gradually by Multiple Steps

Cancer rarely develops by alteration of a single proto-oncogene to an oncogene, or inactivation of a single tumor-suppressor gene. Instead, in almost all cancers, successive alterations in several to many genes gradually accumulate to tilt normal cells to cancer cells. This gradual mechanism is called the *multistep progression of cancer* (**Figure 16.16**). The gradual nature of the process explains why smokers, for example, may not develop cancer until years after the first mutations caused by chemicals in tobacco smoke may occur, soon after smoking begins. It also offers some hope to those who quit smoking, for stopping the exposure to the carcinogenic smoke may halt multistep progression before it reaches its deadly conclusion in cancer.

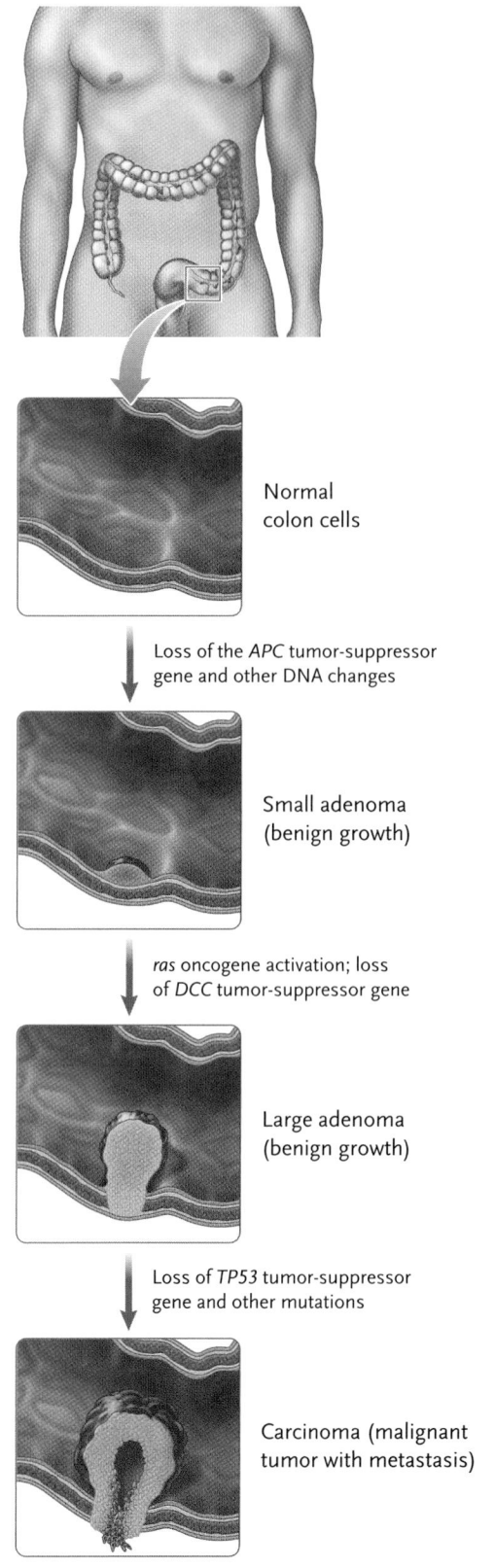

Normal colon cells

Loss of the *APC* tumor-suppressor gene and other DNA changes

Small adenoma (benign growth)

ras oncogene activation; loss of *DCC* tumor-suppressor gene

Large adenoma (benign growth)

Loss of *TP53* tumor-suppressor gene and other mutations

Carcinoma (malignant tumor with metastasis)

Figure 16.16
A multistep model for the development of a type of colorectal cancer.

How are specific patterns of gene expression generated and maintained in a developing eukaryotic organism?

You learned in this chapter that chromatin remodeling is necessary to "open the door" for the transcription machinery to assemble at a promoter. However, researchers do not understand completely how the chromatin remodeling complexes are targeted to particular genes and regulated to give specific patterns of gene expression in a eukaryotic cell, or throughout the development of multicellular eukaryotic organism to maturity, and then through subsequent life. Since mutations in genes encoding chromatin remodeling components are directly linked to human cancers, understanding such basic functions of the complexes is a highly important goal.

Can RNA interference silence disease?

RNA interference (RNAi) is the process in which a small RNA in a complex with several proteins interferes with the expression of an mRNA, either by cleaving it or by blocking its translation. Potentially, RNAi therapy will have many clinical applications, including the treatment of cancers by targeting out-of-control oncogenes. Let us consider two RNAi therapies that are in the works.

Macular degeneration treatment. Some human genetic diseases such as macular degeneration, the leading cause of blindness among those age 55 and older in the United States, are characterized by an overabundance of a protein called VEGF (vascular endothelial growth factor), which promotes blood vessel growth. In macular degeneration patients, the overabundance of VEGF leads to excess blood vessels behind the retina. These blood vessels leak, leading to clouded and often complete loss of vision. Researchers are investigating whether RNAi could be an effective therapy for such diseases. In fact, two biotechnology companies have recently starting testing an RNAi therapy for macular degeneration, targeting the expression of the gene for VEGF. Practically speaking, such research involves first understanding the expression of the disease gene and then working out the way to deliver the small interfering RNA to the diseased cells to eliminate the gene product or decrease its level.

Anti-viral treatment. Similarly, important research is being done to see if RNAi can be an effective therapy for viral infections. Viral targets being investigated by research groups include HIV (the virus that causes AIDS), and hepatitis B and hepatitis C (viruses that cause liver disease). This research involves developing an effective small interfering RNA that can block expression of a vital viral gene, and then perfecting a system to deliver it to patients. In fact, geneticists Anton McCaffrey and Mark Kay of Stanford University have had success in using RNAi to control hepatitis C in laboratory mice. However, the method used to introduce the RNAi is not feasible with humans and, hence, other delivery approaches are being explored.

Peter J. Russell

The ravages of cancer, probably more than any other example, bring home the critical extent to which humans and all other multicellular organisms depend on the mechanisms controlling gene expression to develop and live normally. In a sense, the most amazing thing about these control mechanisms is that, in spite of their complexity, they operate without failures throughout most of the lives of all eukaryotes.

In the next chapter, you will learn about the molecular genetics of bacteria and their phages and about DNA sequences in prokaryotic and eukaryotic genomes that have the ability to move to different chromosomal locations.

STUDY BREAK

1. What is the normal function of a tumor-suppressor gene? How do mutations in tumor-suppressor genes contribute to the onset of cancer?
2. What is the normal function of a proto-oncogene? How can mutations in proto-oncogenes contribute to the onset of cancer?

Review

Go to **ThomsonNOW** at www.thomsonedu.com/login to access quizzing, animations, exercises, articles, and personalized homework help.

16.1 Regulation of Gene Expression in Prokaryotes

- Transcriptional control in prokaryotes involves short-term changes that turn specific genes on or off in response to changes in environmental conditions. The changes in gene activity are controlled by regulatory proteins that recognize operators of operons (Figure 16.2).

- Regulatory proteins may be repressors, which slow the rate of transcription of operons, or activators, which increase the rate of transcription.

- Some repressors are made in an active form, in which they bind to the operator of an operon and inhibit its transcription. Combination with an inducer blocks the activity of the repressor and allows the operon to be transcribed (Figure 16.3).

- Other repressors are made in an inactive form, in which they are unable to inhibit transcription of an operon unless they combine with a corepressor (Figure 16.4).

- Activators typically are made in inactive form, in which they cannot bind to their binding site next to an operon. Combining with another molecule, often a nucleotide, converts the activator into the form in which it binds with its binding site and recruits RNA polymerase, thereby stimulating transcription of the operon (Figure 16.5).

Animation: The lactose operon

Animation: Negative control of the lactose operon

16.2 Regulation of Transcription in Eukaryotes

- Operons are not found in eukaryotes. Instead, genes that encode proteins with related functions typically are scattered through the genome, while being regulated in a coordinated manner.

- Two general types of gene regulation occur in eukaryotes. Short-term regulation involves relatively rapid changes in gene expression in response to changes in environmental or physiological conditions. Long-term regulation involves changes in gene expression associated with the development and differentiation of an organism.

- Gene expression in eukaryotes is regulated at the transcriptional level (where most regulation occurs) and at posttranscriptional, translational, and posttranslational levels (Figure 16.6).

- Transcriptionally active genes have a looser chromatin structure than transcriptionally inactive genes. The change in chromatin structure that accompanies the activation of transcription of a gene involves chromatin remodeling—specific histone modifications—particularly in the region of a gene's promoter (Figure 16.7).

- Regulation of transcription initiation involves proteins binding to a gene's promoter and regulatory sites. At the promoter, general transcription factors bind and recruit RNA polymerase II, giving a very low level of transcription. Activator proteins bind to promoter proximal elements and increase the rate of transcription. Other activators bind to the enhancer and, through interaction with a coactivator, which binds also to the proteins at the promoter, greatly stimulate the rate of transcription (Figures 16.8–16.10).

- The overall control of transcription of a gene depends on the particular regulatory proteins that bind to promoter proximal elements and enhancers. The regulatory proteins are cell-type specific and may be activators or repressors. This gene regulation is achieved by a relatively low number of regulatory proteins, acting in various combinations (Figure 16.11).

- The coordinate expression of genes with related functions is achieved by each of the related genes having the same regulatory sequences associated with them.

- Sections of chromosomes or whole chromosomes can be inactivated by DNA methylation, a phenomenon called silencing. DNA methylation is also involved in genomic imprinting, in which transcription of either the inherited maternal or paternal allele of a gene is inhibited permanently.

Animation: Controls of eukaryotic gene expression

Animation: X-chromosome inactivation

16.3 Posttranscriptional, Translational, and Posttranslational Regulation

- Posttranscriptional, translational, and posttranslational controls operate primarily to regulate the quantities of proteins synthesized in cells (Figure 16.6).

- Posttranscriptional controls regulate pre-mRNA processing, mRNA availability for translation, and the rate at which mRNAs are degraded. In alternative splicing, different mRNAs are derived from the same pre-mRNA. In another process, small single-stranded RNAs complexed with proteins bind to mRNAs that have complementary sequences, and either the mRNA is cleaved or translation is blocked (Figure 16.13).

- Translational regulation controls the rate at which mRNAs are used by ribosomes in protein synthesis.

- Posttranslational controls regulate the availability of functional proteins. Mechanisms of regulation include the alteration of protein activity by chemical modification, protein activation by processing of inactive precursors, and affecting the rate of degradation of a protein.

16.4 The Loss of Regulatory Controls in Cancer

- In cancer, cells partially or completely dedifferentiate, divide rapidly and uncontrollably, and break loose to form additional tumors in other parts of the body.

- Proto-oncogenes and tumor-suppressor genes typically are altered in cancer cells. Proto-oncogenes encode proteins that stimulate cell division. Their altered forms, oncogenes, are abnormally active. Tumor-suppressor genes in their normal form encode proteins that inhibit cell division. Mutated forms of these genes lose this inhibitory activity.

- Most cancers develop by multistep progression involving the successive alteration of several to many genes (Figure 16.16).

Questions

Self-Test Questions

1. The control of the delivery of mRNA to the cytoplasm is an example of:
   a. translational regulation.
   b. posttranslational regulation.
   c. transcriptional regulation.
   d. posttranscriptional regulation.
   e. deoxyribonucleic regulation.

2. For the *E. coli lac* operon, when glucose is absent and lactose is added:
   a. allolactose binds to the operator.
   b. the *lac* gene cannot make Lac repressor protein.
   c. allolactose binds the Lac repressor protein to remove it from the operator.

   d. the genes *lacZ, lacY,* and *lacA* are turned off.
   e. β-galactosidase decreases in the cell.

3. For the *E. coli lac* operon, when lactose is present:
   a. and glucose is absent, cAMP binds and activates catabolic activator protein (CAP).
   b. and glucose is absent, the level of cAMP decreases.
   c. activated CAP binds the repressor protein to remove it from the operator gene.
   d. the cell prefers lactose over glucose.
   e. RNA polymerase cannot bind to the promoter.

4. For the *trp* operon:
   a. tryptophan is an inducer.
   b. when end-product tryptophan binds to the Trp repressor, it stops transcription of the tryptophan biosynthesis genes.
   c. Trp repressor is synthesized in active form.

d. low levels of tryptophan bind to the *trp* operator and block transcription of the tryptophan biosynthesis genes.
e. high levels of tryptophan activate RNA polymerase and induce transcription.

5. Chromatin remodeling activates gene expression when it:
   a. allows proteins initiating transcription to disengage from the promoter.
   b. winds genes tightly around histones.
   c. deacetylates histones.
   d. inserts nucleosomes into chromatin.
   e. recruits a protein complex that displaces nucleosome from the promoter.

6. Which statement about activation of transcription is *not* correct?
   a. A transcription factor binds the TATA box.
   b. A coactivator called a mediator forms a bridge between the promoter and the gene to be transcribed.
   c. Transcription factors bind the promoter and RNA polymerase.
   d. Activators bind to the enhancer region on DNA.
   e. RNA is transcribed downstream from the promoter region.

7. Which of the following statements does not support the idea of combinatorial gene regulation?
   a. Promoter proximal regions and enhancers regulate transcription of genes.
   b. A few regulatory genes can control a large number of transcribable genes.
   c. If repressor binding to enhancer is strong, gene expression is reduced.
   d. Genes requiring complex regulation have a single regulatory element.
   e. The number and types of regulatory sequences in the enhancer vary with each gene.

8. Normal ears in a certain mammal are perky; mutants have droopy ears. In males of these mammals, the gene encoding perky ears is transcribed only from the female parent. This is because the gene from the male parent is silenced by methylation. If the maternal gene is mutated:
   a. male offspring have droopy ears.
   b. male offspring have perky ears.
   c. male offspring have one droopy ear and one perky ear.
   d. the genetic mechanism is called alternative splicing.
   e. this is an example of posttranscriptional regulation.

9. Which of the following statements does not describe microRNA?
   a. MicroRNA is encoded by non-protein-coding genes.
   b. MicroRNA has a precursor that is folded and then cut by a Dicer enzyme.
   c. MicroRNA is an example of a molecule that induces RNA interference or gene silencing.
   d. MicroRNA is synthesized in vitro but not in vivo.
   e. MicroRNA has a similar function to that of small interfering RNAs.

10. Which of the following is not a characteristic of cancer cells?
    a. proto-oncogenes converting to active oncogenes
    b. the position of the *MYC* gene near a repressor gene
    c. the mutation of the *TP53* gene
    d. their stepwise developmental stages
    e. amplification of growth factors and growth factor receptors

Questions for Discussion

1. In a mutant strain of *E. coli*, the CAP protein is unable to combine with its target region of the *lac* operon. How would you expect the mutation to affect transcription when cells of this strain are subjected to the following conditions?
   a. lactose and glucose are both available
   b. lactose is available but glucose is not
   c. both lactose and glucose are unavailable

2. Duchenne muscular dystrophy, an inherited genetic disorder, affects boys almost exclusively. Early in childhood, muscle tissue begins to break down in affected individuals, who typically die in their teens or early twenties as a result of respiratory failure. Muscle samples from women who carry the mutation reveal some regions of degenerating muscle tissue adjacent to other regions that are normal. Develop a hypothesis explaining these observations.

3. Eukaryotic transcription is generally controlled by binding of regulatory proteins to DNA sequences rather than by modification of RNA polymerases. Develop a hypothesis explaining why this is so.

Evolution Link

Fruit flies homozygous for a mutation in the tumor-suppressor gene *HIPPO* develop tumors in every organ. Expression of the human gene *MST2* in flies homozygous for *HIPPO* show greatly reduced or no tumors. What does this result suggest about the evolution of tumor suppressor genes in animals?

Experimental Analysis

Design an experiment using rats as the model organism to test the hypothesis that human chorionic gonadotrophin (hCG), a hormone produced during pregnancy, leads to a significant protection against breast cancer.

How Would You Vote?

Some females at high risk of developing breast cancer opt for prophylactic mastectomy, the surgical removal of one or both breasts even before cancer develops. Many of them would never have developed cancer. Should the surgery be restricted to cancer treatment? Go to www.thomsonedu.com/login to investigate both sides of the issue and then vote.

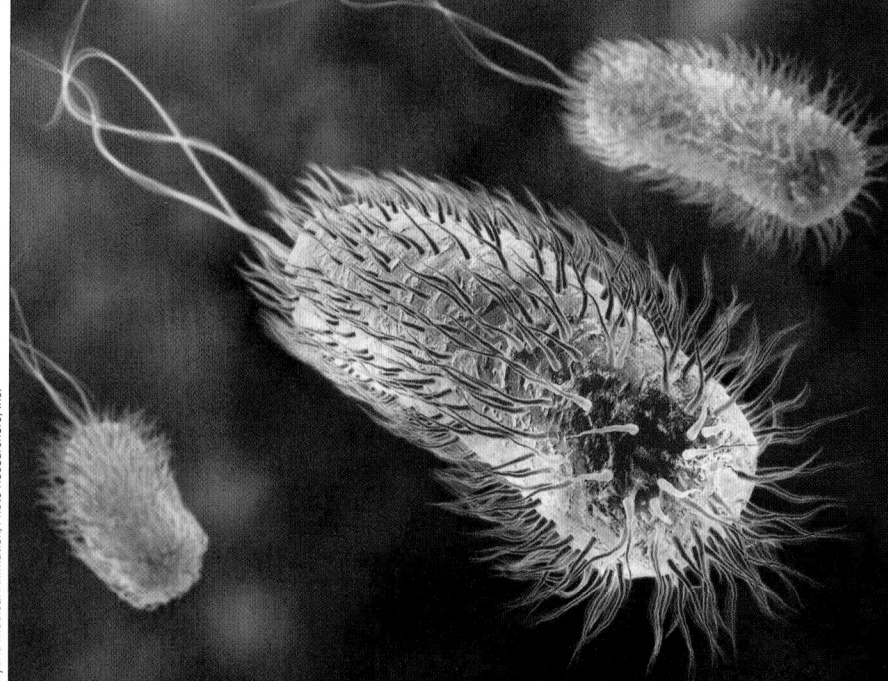

Escherichia coli, a model research organism for several types of biological studies, including bacterial genetics (computer rendering).

17 Bacterial and Viral Genetics

WHY IT MATTERS

In 1885, a Viennese pediatrician, Theodor Escherich, discovered a bacterium that caused severe diarrhea in infants. He named it *Bacterium coli*. Surprisingly, however, researchers discovered that *Bacterium coli* was also present in healthy infants, and was a normal inhabitant of the human intestine. Only certain strains cause human diseases. Further, researchers in the twentieth century found that if they mixed together bacteria of different strains, the organisms produced some progeny with a mixture of traits from more than one strain—evidence that bacteria could have some kind of sexual reproduction.

As an organism that is readily available and easy to grow, the bacterium has been of central interest to scientists since its first discovery. Renamed *Escherichia coli* in honor of Escherich, the bacterium brings several distinct advantages to scientific investigation. It can be grown quickly in huge numbers in nutrient solutions that are simple to prepare. And, it is infected with a group of viruses called **bacteriophages (phages** for short) that have been just as valuable to scientists as *E. coli* itself because these phages can also be grown by the billions in cultures of the bacterium. The rapid generation times

FOCUS ON RESEARCH

Model Research Organisms: *Escherichia coli*

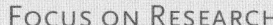

We probably know more about *E. coli* than any other organism. For example, microbiologists have deciphered the complete DNA sequence of the genome of a standard laboratory strain of *E. coli*, including the sequence of the approximately 4400 genes in its genome. The functions of about one third of these genes are still unidentified, however.

E. coli got its start in laboratory research because of the ease with which it can be grown in cultures. Because *E. coli* cells can divide about every 20 minutes under optimal conditions, a clone of 1 billion cells can be grown in a matter of hours, in only 10 mL of culture medium. The same amount of medium can accommodate as many as 10 billion cells before the growth rate begins to slow. *E. coli* strains can be grown in the laboratory with minimal equipment, requiring little more than culture vessels in an incubator held at 37°C.

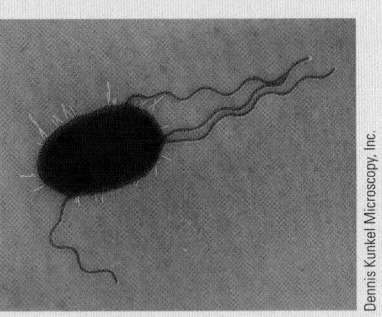

Dennis Kunkel Microscopy, Inc.

The major advantage of *E. coli* for scientific research in the early days it was used, however, can be summed up in a single word: sex. When Joshua Lederberg and Edward Tatum discovered that *E. coli* can enter into a form of sexual reproduction, they and other scientists realized they could carry out genetic crosses with the bacterium, producing genetic recombinants that could indicate the relative positions of genes on the chromosome. Knowing these relative positions, Lederberg, Tatum, and other scientists were able to generate a genetic map of the *E. coli* chromosome. The map showed that genes with related functions are clustered together, a fact that had significant implications for the regulation of expression of those genes. For example, François Jacob and Jacques Monod's work with the genes for lactose metabolism led to the pioneering operon model (described in Section 16.1). In their work, at The Pasteur Institute in Paris, they used conjugation to map the genes and generated partial diploids to help understand the details of the regulation of transcription of those genes.

The development of *E. coli* as a model organism for studying gene organization and the regulation of gene expression led to the field of molecular genetics. The study of naturally occurring plasmids in *E. coli* and of enzymes that

cut DNA at specific sequences eventually resulted in techniques for combining DNA from different sources, such as inserting a gene from an organism into a plasmid. Today *E. coli* is used for creating plasmids that contain inserted genes or other sequences, and for amplifying (cloning) them once they are made.

In essence, the biotechnology industry has its foundation in molecular genetics studies of *E. coli*, and large-scale *E. coli* cultures are widely used as "factories" for production of desired proteins. For example, the human insulin hormone, required for treatment of certain forms of diabetes, can be produced by *E. coli* factories. (Chapter 18 explains more about cloning and other types of DNA manipulation.)

Laboratory strains of *E. coli* are harmless to humans. Similarly, the natural *E. coli* cells in the colon of humans and other mammals are usually harmless. There are pathogenic strains of *E. coli*, though; sometimes they make the news when humans eating food that contains a pathogenic strain develop disease symptoms, such as diarrhea and fever. The genomes of several pathogenic *E. coli* strains have been sequenced, and it is notable that they have more genes than the lab strain, or the strain in the human colon. The extra genes include the genes that make the bacterium pathogenic.

of *E. coli,* its phages, and their numerous offspring make them especially valuable to geneticists. Geneticists have used them to analyze genetic crosses and their outcomes much more quickly than they can with eukaryotes. They can also detect genetic events that occur only once within millions of offspring. The characteristics of these rare events helped scientists to work out the structure, activity, and recombination of genes at the molecular level.

Because *E. coli* can be cultured in completely defined chemical solutions—solutions in which the identity and amount of each chemical is known—it is particularly useful for biochemical investigations. What goes in, what comes out, and what biochemical changes occur inside can be detected and closely followed in normal and mutant bacteria, and in bacteria infected by viruses. These biochemical studies have added immeasurably to the definition of genes and their activi-

ties, and have identified many biochemical pathways and the enzymes catalyzing them. (*Focus on Research* tells more about *E. coli*'s advantages as a model laboratory organism.)

After research with bacteria and their viruses showed the way, biologists successfully applied the same techniques to eukaryotes such as *Neurospora* and *Aspergillus,* fungi with short generation times that can also be grown and analyzed biochemically in large numbers. Molecular studies are now easy to carry out with a wide variety of eukaryotes, including yeast, the fruit fly *Drosophila,* and the plant *Arabidopsis.* The results of this research showed that the molecular characteristics of genes discovered in prokaryotes apply to eukaryotes as well.

This chapter outlines the basic findings of molecular genetics in bacteria and their phages. It also describes *transposable elements*—sequences that can move from place to place in bacterial DNA—and compares the bacterial transposable elements with those of eukaryotes. We begin our discussion with the sex life of bacteria.

17.1 Gene Transfer and Genetic Recombination in Bacteria

In the first half of the twentieth century, foundational genetic experiments with eukaryotes revealed the processes of genetic recombination during sexual reproduction and led to the construction of genetic maps of chromosomes for a number of organisms (see Chapters 12 and 13). Bacteria became the subject of genetics research in the middle of the twentieth century. A key early question was whether gene transfer and genetic recombination can occur in bacteria even though these organisms do not reproduce sexually by meiosis. For particular bacteria, the answer to the question was yes—genes can be transferred from one bacterium to another by several different mechanisms, and the newly introduced DNA can recombine with DNA already present. Such genetic recombination performs the same function as it does in eukaryotes: it generates genetic variability through the exchange of alleles between homologous regions of DNA molecules from two different individuals.

E. coli is one of the bacteria in which genetic recombination occurs. By the 1940s geneticists knew that E. coli and many other bacteria could be grown in a **minimal medium** containing water, an organic carbon source such as glucose, and a selection of inorganic salts including one, such as ammonium chloride, that provides nitrogen. The growth medium can be in liquid form or in the form of a gel made by adding agar to the liquid medium. (Agar is a polysaccharide material, indigestible by most bacteria, that is extracted from algae.)

Since it is not practical to study a single bacterium for most experiments, researchers soon developed techniques for starting bacterial cultures from a single cell, generating cultures with a large number of genetically identical cells. Cultures of this type are called **clones.** To start bacterial clones, the scientist spreads a drop of a bacterial culture over a sterile agar gel in a culture dish. The culture is diluted enough to ensure that cells will be widely separated on the agar surface. Each individual cell divides many times to produce a separate colony that is a clone of the cell. Cells can be removed from a clone and introduced into liquid cultures or spread on agar and grown in essentially any quantity.

Genetic Recombination Occurs in *E. coli*

In 1946, two scientists at Yale University, Joshua Lederberg and Edward L. Tatum, set out to determine if genetic recombination occurs in bacteria, using *E. coli* as their experimental organism. In essence, they were testing whether bacteria had a sexual reproduction process. As a first step, they induced genetic mutations in *E. coli* bacteria by exposing the cells to mutagens such as X-rays or ultraviolet light. After the exposure, Lederberg and Tatum found that some bacteria had become auxotrophs, mutant strains that could not grow on minimal medium (see Section 15.1). One mutant strain could grow only if the vitamin biotin and the amino acid methionine were added to the culture medium; evidently the genes that encoded the enzymes required to make these substances had mutated in this strain. A second mutant strain did not need biotin or methionine in its growth medium, but could grow only if the amino acids leucine and threonine were added along with the vitamin thiamine. These two genetic strains of *E. coli* were represented in genetic shorthand as:

Strain 1: bio^- met^- leu^+ thr^+ thi^+

Strain 2: bio^+ met^+ leu^- thr^- thi^-

In this shorthand, *bio* refers to the gene that governs a cell's ability to synthesize biotin from inorganic precursors. The designation bio^+ indicates that the allele is normal; bio^- represents the mutant allele, which produces cells that cannot make biotin for themselves. Similarly, met^+, met^-, leu^+, leu^-, thr^+, thr^-, and thi^+, thi^- are the respective normal and mutant alleles for methionine, leucine, threonine, and thiamine synthesis.

Lederberg and Tatum mixed about 100 million cells of the two mutant strains together and placed them on a minimal medium **(Figure 17.1).** None of the cells were expected to be able to grow on the minimal medium unless some form of recombination between DNA molecules from the two parental types produced the new combination with normal alleles for each of the five genes:

bio^+ met^+ leu^+ thr^+ thi^+

Several hundred colonies grew on the minimal medium, indicating that genetic recombination had actually taken place in the bacteria. Lederberg and Tatum eliminated the possibility that the colonies arose from chance mutations back to normal alleles by placing hundreds of millions of cells of strains 1 or 2 separately on the surface of a minimal medium: no colonies grew.

Bacterial Conjugation Brings DNA of Two Cells Together, Allowing Recombination to Occur

Lederberg and Tatum's results led to a major question: How did DNA molecules with different alleles get together to undergo genetic recombination? Recombina-

Figure 17.1 Experimental Research

Genetic Recombination in Bacteria

QUESTION: Does genetic recombination occur in bacteria?

EXPERIMENT: Lederberg and Tatum tested whether genetic recombination occurred between two mutant strains of *E. coli*. Mutant strain 1 required biotin and methionine to grow, but not leucine, threonine, or thiamine—its genotype was $bio^- \ met^- \ leu^+ \ thr^+ \ thi^+$. Mutant strain 2 required leucine, threonine, and thiamine to grow, but not biotin or methionine—its genotype was $bio^+ \ met^+ \ leu^- \ thr^- \ thi^-$.

Lederberg and Tatum mixed together large numbers of the two mutant strains and plated them on minimal medium, which lacked any of the nutrients the strains needed for growth. As controls, they also plated large numbers of the two mutant strains individually on minimal medium.

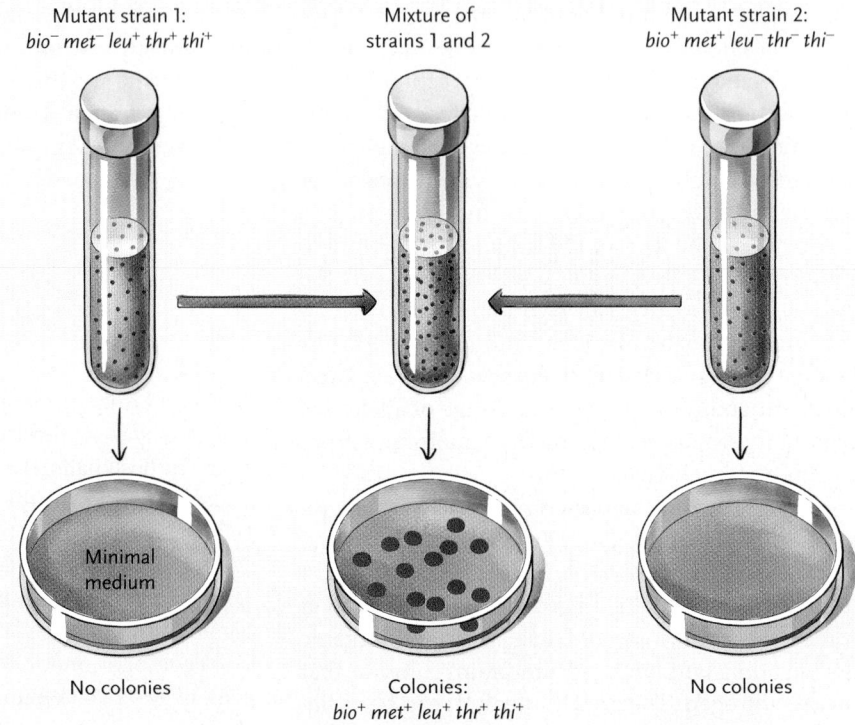

Mutant strain 1:
$bio^- \ met^- \ leu^+ \ thr^+ \ thi^+$

Mixture of strains 1 and 2

Mutant strain 2:
$bio^+ \ met^+ \ leu^- \ thr^- \ thi^-$

Minimal medium

No colonies

Colonies:
$bio^+ \ met^+ \ leu^+ \ thr^+ \ thi^+$

No colonies

RESULTS: No colonies grew on the control plates, indicating that the mutant alleles in the two strains did not mutate back to normal alleles, which would have produced growth on minimal medium. However, many colonies grew on plates spread with a mixture of mutant strain 1 and mutant strain 2.

CONCLUSION: In order to grow on minimal medium, the bacteria must have the genotype $bio^+ \ met^+ \ leu^+ \ thr^+ \ thi^+$. Lederberg and Tatum reasoned that the colonies on the plate must have resulted from genetic recombination between mutant strains 1 and 2.

---

tion in eukaryotes occurs in diploid cells, by an exchange of segments between the chromatids of homologous chromosome pairs (discussed in Section 13.1). Bacteria typically have a single, circular chromosome—they are haploid organisms (see Section 14.5). At first, bacterial cells were thought to fuse together, producing the prokaryotic equivalent of a diploid zygote. However, it was later established that instead of fusing, bacterial cells *conjugate*: they contact each other, initially becoming connected by a long tubular structure called a *sex pilus* **(Figure 17.2a),** and then forming a cytoplasmic bridge that connects two cells **(Figure 17.2b).** During **conjugation,**

a copy of part of the DNA of one cell, the *donor* (the bristly cell in Figure 17.2a), moves through the cytoplasmic bridge into the other cell, the *recipient*. Once donor DNA enters the recipient, it pairs with the homologous region of the recipient cell's DNA, and genetic recombination can occur. Through this unidirectional transfer of DNA, bacterial conjugation thus accomplishes a prokaryotic form of sexual reproduction.

The F Factor and Conjugation. The donor bacterial cell in a pairing initiates conjugation. The ability to conjugate depends on the presence within a donor cell of a

a. Attachment by sex pilus

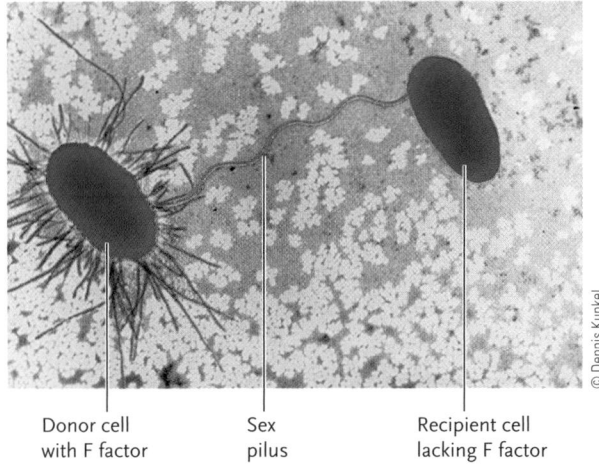

Donor cell with F factor

Sex pilus

Recipient cell lacking F factor

© Dennis Kunkel

b. Cytoplasmic bridge formed

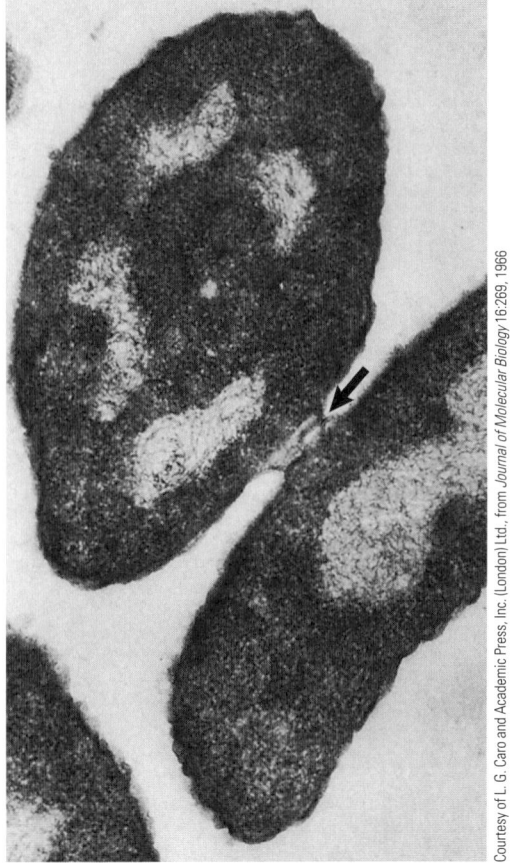

Courtesy of L. G. Caro and Academic Press, Inc. (London) Ltd., from *Journal of Molecular Biology* 16:269, 1966

Figure 17.2
Conjugating *E. coli* cells. **(a)** Initial attachment of two cells by the sex pilus. **(b)** A cytoplasmic bridge (arrow) has formed between the cells, through which DNA moves from one cell to the other.

a. Bacterial DNA released from cell

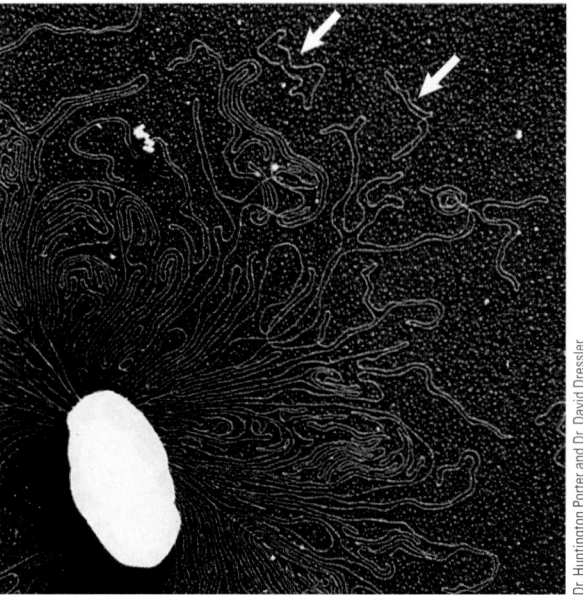

Dr. Huntington Porter and Dr. David Dressler

b. Plasmid

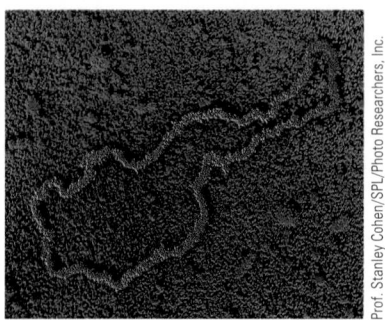

Prof. Stanley Cohen/SPL/Photo Researchers, Inc.

Figure 17.3
Electron micrographs of DNA released from a disrupted bacterial cell. **(a)** Plasmids (arrows) near the mass of chromosomal DNA. **(b)** A single plasmid at higher magnification (colorized).

plasmid called the **F factor** (F = fertility). Plasmids are small circles of DNA that occur in bacteria in addition to the main circular chromosomal DNA molecule **(Figure 17.3).** Plasmids contain several to many genes and a replication origin that permits them to be duplicated and passed on during bacterial division. Donor cells in conjugation are called **F⁺ cells** because they contain the F factor. They are able to mate with recipi-

ent cells but not with other donor cells. Recipient cells lack the F factor and, hence, are called **F⁻ cells.**

The F factor carries about 20 or so genes. Several of the genes encode proteins of the **sex pilus**, also called the **F pilus** (plural, *pili*). The sex pilus is a long, tubular structure on the cell surface that allows an F⁺ donor cell to attach to a F⁻ recipient (see Figure 17.2a). Once attached, the cells form a cytoplasmic bridge and conjugate (**Figure 17.4a,** step 1, and see Figure 17.2b). During conjugation, the F plasmid replicates using a special type of DNA replication. When the two strands of the plasmid DNA separate during replication, one of the strands is transferred from the F⁺ cell through the cytoplasmic bridge to the F⁻ cell (Figure 17.4a, step 2). In the recipient cell, synthesis of the complementary strand to the entering DNA strand occurs (Figure 17.4a, step 3). When the entire F factor strand has entered and its complementary strand has been synthesized, the F factor circularizes into a complete F factor, changing the cell to F⁺ (Figure 17.4a, step 4). No chromosomal DNA is transferred between cells in this

a. Transfer of the F factor

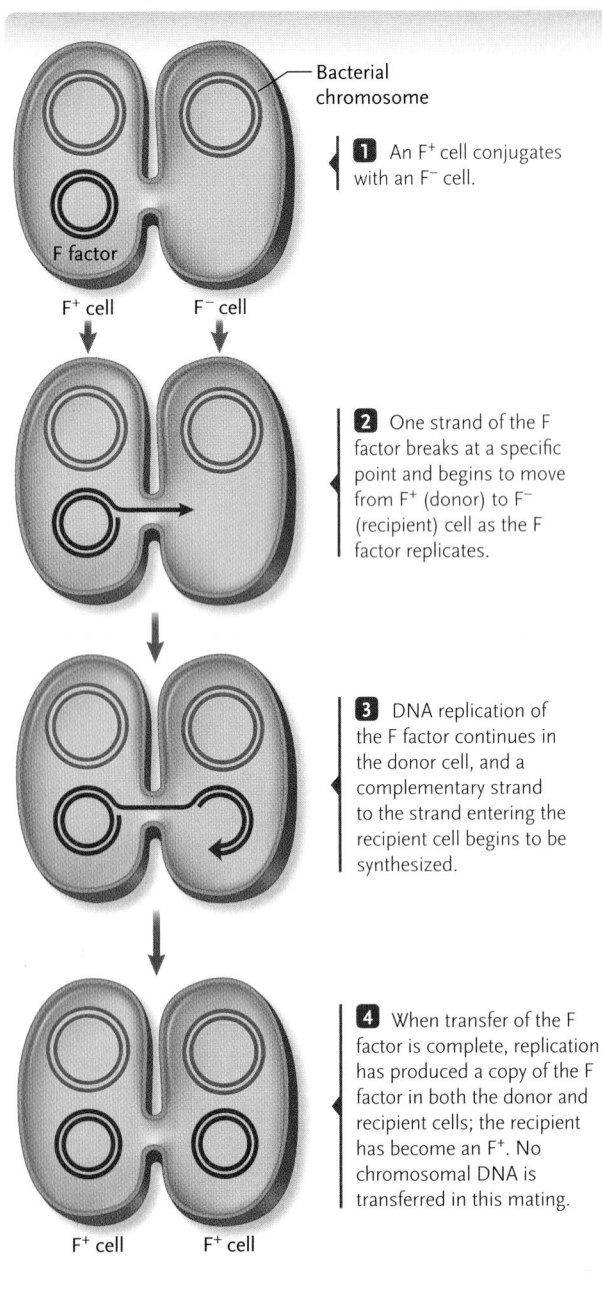

Bacterial chromosome

1 An F$^+$ cell conjugates with an F$^-$ cell.

F factor

F$^+$ cell   F$^-$ cell

2 One strand of the F factor breaks at a specific point and begins to move from F$^+$ (donor) to F$^-$ (recipient) cell as the F factor replicates.

3 DNA replication of the F factor continues in the donor cell, and a complementary strand to the strand entering the recipient cell begins to be synthesized.

4 When transfer of the F factor is complete, replication has produced a copy of the F factor in both the donor and recipient cells; the recipient has become an F$^+$. No chromosomal DNA is transferred in this mating.

F$^+$ cell   F$^+$ cell

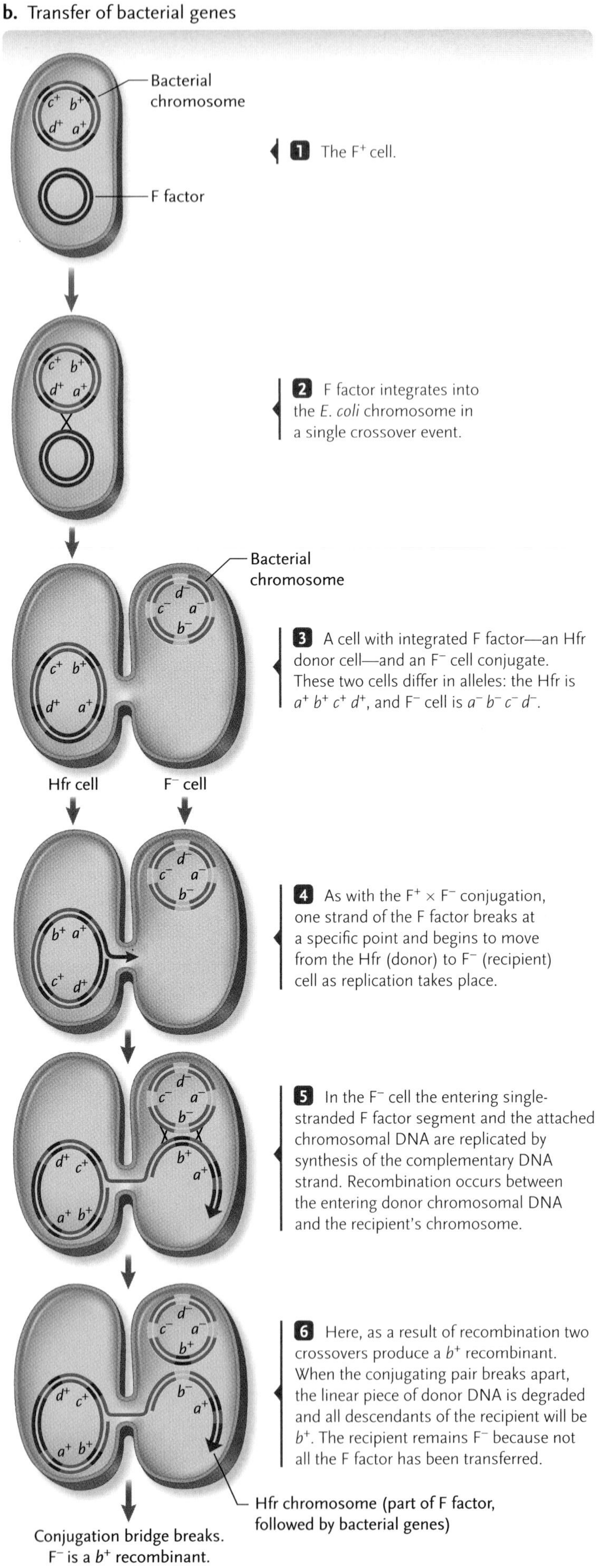

b. Transfer of bacterial genes

Bacterial chromosome

c^+ b^+
d^+ a^+

1 The F$^+$ cell.

F factor

c^+ b^+
d^+ a^+

2 F factor integrates into the E. coli chromosome in a single crossover event.

Bacterial chromosome

d^-
c^- a^-
b^-

c^+ b^+
d^+ a^+

3 A cell with integrated F factor—an Hfr donor cell—and an F$^-$ cell conjugate. These two cells differ in alleles: the Hfr is a^+ b^+ c^+ d^+, and F$^-$ cell is a^- b^- c^- d^-.

Hfr cell   F$^-$ cell

d^-
c^- a^-
b^-

b^+ a^+
c^+ d^+

4 As with the F$^+$ × F$^-$ conjugation, one strand of the F factor breaks at a specific point and begins to move from the Hfr (donor) to F$^-$ (recipient) cell as replication takes place.

c^- d^-
a^-
b^-
b^+
a^+

d^+ c^+
a^+ b^+

5 In the F$^-$ cell the entering single-stranded F factor segment and the attached chromosomal DNA are replicated by synthesis of the complementary DNA strand. Recombination occurs between the entering donor chromosomal DNA and the recipient's chromosome.

c^- d^-
a^-
b^+

d^+ c^+
a^+ b^+

b^-
a^+

6 Here, as a result of recombination two crossovers produce a b^+ recombinant. When the conjugating pair breaks apart, the linear piece of donor DNA is degraded and all descendants of the recipient will be b^+. The recipient remains F$^-$ because not all the F factor has been transferred.

Hfr chromosome (part of F factor, followed by bacterial genes)

Conjugation bridge breaks. F$^-$ is a b^+ recombinant.

Figure 17.4

Transfer of genetic material during conjugation between *E. coli* cells. **(a)** Transfer of the F factor during conjugation between F$^+$ and F$^-$ cells. **(b)** Transfer of bacterial genes and production of recombinants during conjugation between Hfr and F$^-$ cells.

process, however, so no genetic recombination results from $F^+ \times F^-$ mating.

Hfr Cells and Genetic Recombination.

How does genetic recombination of bacterial genes occur as a result of conjugation if no chromosomal DNA transfers when an F factor is transferred in a mating? The answer is that in some F^+ cells (**Figure 17.4b,** step 1), the F factor integrates into the bacterial chromosome by crossing-over (Figure 17.4b, step 2), which produces a donor that can transfer genes on the bacterial chromosome to a recipient. These special donor cells are known as **Hfr cells** (Hfr = high frequency recombination). Because the F factor genes are still active when the plasmid is integrated into the bacterial chromosome, an Hfr cell can conjugate with an F^- cell. Figure 17.4b, step 3, shows an Hfr $\times$ F^- mating where the two cell types differ in alleles. DNA replication begins in the middle of the integrated F factor, and a segment of the F factor moves through the conjugation bridge into the recipient (Figure 17.4b, step 4), bringing the chromosomal DNA behind it (Figure 17.4b, step 5). Synthesis of the complementary DNA strand to the entering DNA from the donor occurs in the recipient cell. The conjugation bridge between the mating cells soon breaks, but DNA transfer usually continues for long enough so that at least some genes of the donor cell follow the F segment into the recipient cell. The recipient therefore becomes a **partial diploid** for the donor chromosomal DNA segment that goes through the conjugation bridge.

For our example, the recipient cell in Figure 17.4b, step 5, has become $a^+ b^+/a^- b^-$. The recipient's DNA and the homologous DNA fragment from the donor can pair and recombine. Genetic recombination occurs by a double crossover event exchanging donor gene(s) with recipient gene(s) using essentially the same mechanisms as in eukaryotes (discussed in Section 13.1)—Figure 17.4b, step 6, shows the generation of an b^+ recombinant. In other pairs in the mating population, the a^+ gene could recombine with the homologous recipient gene, or both a^+ and b^+ genes could recombine. The genetic recombinants observed in the Lederberg and Tatum experiment described earlier were produced in this same general way. Since conjugation usually breaks off long before the second part of the F plasmid has been transferred (it would be the last DNA piece transferred), the recipient cell remains F^-.

Recombinants produced during conjugation can be detected only if the alleles of the genes in the DNA transferred from the donor differ from those in the recipient's chromosome. Following recombination, the bacterial DNA replicates and the cell divides normally, producing a cell line with the new combination. Any remnants of the DNA fragment that originally entered the cell are degraded as division proceeds and do not contribute further to genetic recombination or cell heredity.

Mapping Genes by Conjugation.

Genetic recombination by conjugation was discovered by two scientists, François Jacob (the same scientist who proposed the operon model for the regulation of gene expression in bacteria; see Section 16.1) and Elie L. Wollman, at the Pasteur Institute in Paris. They began their experiments by mating Hfr and F^- cells that differed in a number of alleles. At regular intervals after conjugation commenced, they removed some of the cells and agitated them in a blender to break apart mating cells. They then cultured the separated cells and analyzed them for recombinants. They found that the longer they allowed cells to conjugate before separation, the greater the number of donor genes that entered the recipient and produced recombinants. From this result, Jacob and Wollman concluded that during conjugation, the Hfr cell slowly injects a copy of its DNA into the F^- cell. Full transfer of an entire DNA molecule to an F^- cell would take about 90 to 100 minutes. In nature, however, the entire DNA molecule is rarely transferred because the cytoplasmic bridge between conjugating cells is fragile and easily broken by random molecular motions before transfer is complete.

The pattern of gene transfer from Hfr to F^- cells was used to map the *E. coli* chromosome. The F factor integrates into one of a few possible fixed positions around the circular *E. coli* DNA. As a result, the genes of the bacterial DNA follow the F factor segment into the recipient cell in a definite order, with the gene immediately behind the F factor segment entering first and the next genes following. In the theoretical example shown in Figure 17.4b, donor genes will enter in the order $a^+–b^+–c^+–d^+$. By breaking off conjugation at gradually increasing times, investigators allowed longer and longer pieces of DNA to enter the recipient cell, carrying more and more genes from the donor cell (detected by the appearance of recombinants). By noting the order and time at which genes were transferred, investigators were able to map and assign the relative positions of most genes in the *E. coli* chromosome. The resulting genetic map has distances between genes in units of minutes. To this day, the genetic map of *E. coli* shows map distances as minutes, reflecting the mapping of genes by conjugation.

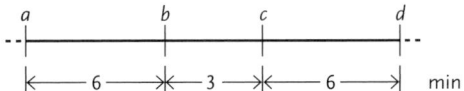

The genetic maps from *E. coli* conjugation experiments indicated that the genes are arranged in a circle, reflecting the circular form of the *E. coli* chromosome. More recently, direct sequencing of the *E. coli* genome has confirmed the results obtained by genetic mapping.

In addition to the F plasmid, bacteria also contain other types of plasmids. Some plasmids contain genes that provide resistance to unfavorable conditions, such as exposure to antibiotics (plasmids providing such resistance genes are called **R plasmids**). The competitive advantage provided by the genes in some plasmids may account for the wide distribution of plasmids of all kinds in prokaryotic cells.

Transformation and Transduction Provide Additional Sources of DNA for Recombination

The discovery of conjugation and genetic recombination in *E. coli* showed that genetic recombination is not restricted to eukaryotes. Further discoveries demonstrated that DNA can transfer from one bacterial cell to another by two additional mechanisms, *transformation* and *transduction*. Like conjugation, these mechanisms transfer DNA in one direction, and create partial diploids in which recombination can occur between alleles in the homologous DNA regions.

Transformation. In **transformation**, bacteria take up pieces of DNA that are released as other cells disintegrate. Fred Griffith, a medical officer in the British Ministry of Health, London, discovered this mechanism in 1928, when he found that a noninfective form of the bacterium *Streptococcus pneumoniae*, unable to cause pneumonia in mice, could be transformed to the infective form if it was exposed to heat-killed cells of an infective strain. The key difference between the strains is a polysaccharide capsule around the infective strain that is absent in the noninfective strain because of a genetic difference between the strains. In 1944, Oswald Avery and his colleagues at New York University found that the substance derived from the killed infective cells, the substance capable of transforming noninfective bacteria to the infective form, was DNA (discussed in Section 14.1).

Subsequently, geneticists established that in the transformation of *Streptococcus,* the linear DNA fragments taken up from disrupted infective cells recombine with the chromosomal DNA of the noninfective cells by double crossovers, much in the same way as genetic recombination takes place in conjugation. The recombination introduces the normal allele for capsule formation into the DNA of the noninfective cells; expression of that normal allele generates a capsule around the cell and its descendants, making them infective.

Only some species of bacteria can take up DNA from the surrounding medium by natural mechanisms. Such bacteria typically have a DNA-binding protein on the outer surface of the cell wall. When DNA from the cell's surroundings binds to the protein, a deoxyribonuclease enzyme breaks the DNA into short pieces that pass through the cell wall and plasma membrane into the cytosol. The entering DNA can then recombine with the recipient cell's chromosome if it contains homologous regions.

E. coli cells do not normally take up DNA from their surroundings. However, they can be induced to take up DNA by *artificial transformation*. One way this

is accomplished is to expose *E. coli* cells to calcium ions and the DNA of interest, incubate them on ice, and then give them a quick heat shock. This treatment alters the plasma membrane so that DNA can penetrate and enter. The entering DNA undergoes recombination if it contains regions that are homologous to part of the chromosomal DNA.

Another technique for artificial transformation, called *electroporation*, exposes cells briefly to rapid pulses of an electrical current. The electrical shock alters the plasma membrane so that DNA can enter. The method works well with many of the bacterial species that are unable to take up DNA on their own, and also with many types of eukaryotic cells.

Artificial transformation is often used to insert plasmids containing DNA sequences of interest into *E. coli* cells as a part of cloning techniques. After the cells are transformed, clones of the cells are grown in large numbers to increase the quantity of the inserted DNA to the amounts necessary for sequencing or genetic engineering. (DNA cloning and genetic engineering are discussed further in Chapter 18.)

Transduction. In **transduction**, DNA is transferred to recipient bacterial cells by an infecting phage (see Section 14.1). When new phages assemble in an infected bacterial cell, they sometimes incorporate fragments of the host cell DNA along with or instead of the viral DNA. After the phages are released from the host cell, they may attach to another cell and inject the bacterial DNA (and the viral DNA if it is present) into that cell. The introduction of this DNA, as in conjugation and transformation, makes the recipient cell a partial diploid and allows recombination to take place. Joshua Lederberg and his graduate student, Norton Zinder, then at the University of Wisconsin at Madison, discovered transduction in 1952 in experiments with the bacterium *Salmonella typhimurium* and phage P22. Lederberg received a Nobel Prize in 1958 for his discovery of conjugation and transduction in bacteria.

Replica Plating Allows Genetic Recombinants to Be Identified and Counted

How do researchers identify and count genetic recombinants in conjugation, transformation, or transduction experiments? Joshua Lederberg and Esther Lederberg developed a now widely applied technique called **replica plating** for doing this. In replica plating, a plate of solid growth medium with colonies on it—the master plate— is pressed gently onto sterile velveteen **(Figure 17.5).** This transfers some of each colony to the velveteen in the same pattern as the colonies were on the plate. The velveteen is then pressed gently onto new plates of solid growth medium—the replica plates—thereby transferring some of the cells from each original colony to those plates. The new plates each have been inoculated with a "replica" of the original set of colonies on the starting

Figure 17.5 Research Method

Replica Plating

PURPOSE: Replica plating is used to identify different strains with respect to their growth requirements in a heterogeneous mixture of strains.

PROTOCOL:

1. Press sterile velveteen gently onto a plate of solid growth medium with colonies on it. Some of each colony transfers to the velveteen in the same pattern as the colonies on the plate. In the example, a mixture of colonies of normal and auxotrophic strains are on a plate of complete medium.

2. Press the velveteen gently onto a new plate—the replica plate—to transfer some of each strain. In the example, the replica plate contains minimal medium. Incubate to allow colonies to grow and compare the pattern of colonies on the replica plate with that on the master plate.

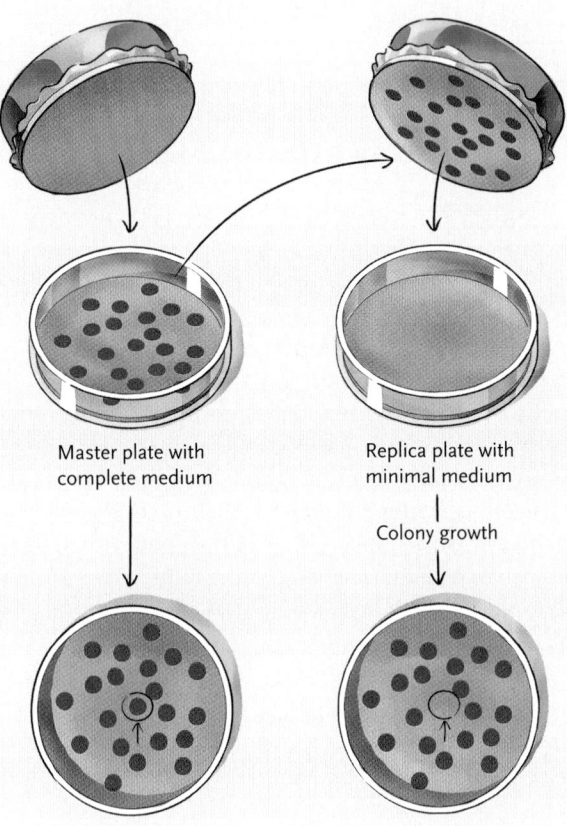

Master plate with complete medium

Replica plate with minimal medium

Colony growth

INTERPRETING THE RESULTS: A colony present on the master plate but not on the replica plate indicates that strain requires some growth substance missing from the minimal medium in order to grow. In other words, the strain is an auxotroph. In actual experiments, the compositions of the master plate and replica plate media are chosen to be appropriate for the goals of the experiment.

plate. The new plates are incubated to allow new colonies to grow. Therefore, by replica plating from the original plate to plates containing growth media with different compositions, an investigator can determine the mutations a strain carries. That is, the compositions of the growth media on the new plates are adjusted to promote the growth of colonies with particular characteristics. **Complete medium** is a full complement of nutrient substances, including amino acids and other chemicals that normal strains can make for themselves; in a minimal medium, normal cells will grow but auxotrophic mutants will not, because they are unable to make one or more of the missing substances.

Figure 17.5 shows the identification of auxotrophic mutants of *E. coli* by replica plating colonies that were grown in complete medium onto minimal medium. All strains grow on the complete medium, but auxotrophic mutants will not grow on the minimal me-

dium. Therefore, an investigator compares the colony patterns on the complete medium (original plate) and the minimal medium to determine missing colonies on the minimal medium plate. The corresponding colonies on the original plate are taken and studied further. In an actual experiment, the compositions of the media are appropriate for the goals of the experiment. For example, to identify a *met*⁺ recombinant in a conjugation experiment, the starting plate contains methionine and the colonies are replica plated to a plate lacking methionine. Comparison of the colony patterns on the two plates identifies *met*⁺ recombinants because they grow on the plate lacking methionine, whereas *met*⁻ parentals do not.

While the phages mentioned in this section infect only bacteria, viruses infect all living organisms. Many viruses have become important subjects for research into, among other things, the molecular nature of recombination and the genetic control of virus infection. Viruses and viral recombination are the subjects of the next section.

STUDY BREAK

Describe the properties of F⁺, F⁻, and Hfr cells of *E. coli*.

17.2 Viruses and Viral Recombination

As agents of transduction, the phages that infect bacteria are important tools in research on bacterial genetics. The same viruses are also important for studying *viral* recombination and genetics. Viruses can undergo genetic recombination when the DNA of two viruses, carrying different alleles of one or more genes, infect a single cell. Using phages, researchers study viral genetics with the same molecular and biochemical techniques used to investigate their bacterial hosts.

Viruses in the Free Form Consist of a Nucleic Acid Core Surrounded by a Protein Coat

Viruses in the free form, outside their host cells, consist of a nucleic acid **core** surrounded by a protective protein **coat. Figure 17.6** shows a phage and an animal virus. Some animal viruses have an additional layer—the envelope—derived from the plasma membrane of the host cell. Viruses are carried about passively by random molecular movements and perform none of the metabolic activities of life. However, once they or their nucleic acid genome enters a host cell, viruses typically subvert the host's cellular machinery for the replication of the viral nucleic acid and the synthesis of viral proteins. Viruses of different kinds infect bacterial, plant, and animal cells. (Viruses are described further in Chapter 25.)

The viral nucleic acid molecules—DNA in some viruses, RNA in others—may contain from a few to around a hundred genes. All viruses have genes encoding at least their coat proteins and the enzymes required for nucleic acid replication. Many viruses also have genes encoding recognition proteins that become implanted in the coat surface. These coat proteins recognize and bind to the host cell, promoting entry of the viral particle or its nucleic acid core into that cell.

E. coli's Bacteriophages Are Widely Used in Genetic Research

Several bacteriophages that infect *E. coli* are used by scientists studying both bacterial and viral genetics. These include **virulent bacteriophages**, which kill their host cells during each cycle of infection, and **temperate bacteriophages**, which may enter an inactive phase in which the host cell replicates and passes on the bacteriophage DNA for generations before the phage becomes active and kills the host.

E. coli's Virulent Bacteriophages. Among the virulent bacteriophages infecting *E. coli*, the **T-even bacteriophages** T2, T4, and T6 have been most valuable in genetic studies (shown in Figure 17.6a). The coat of these phages is divided into a *head* and a *tail*. Packed into the head is a single linear molecule of DNA. The tail, assembled from several different proteins, has recognition proteins at its tip that can bind to the surface of the host cell.

As the first step in a cycle of infection, a T-even phage collides randomly with the surface of an *E. coli* cell and the tail attaches to the cell wall (**Figure 17.7**, step 1). The tail then contracts and injects the phage's DNA through the cell wall and plasma membrane and into the cytoplasm (step 2). The coat proteins remain outside. Throughout its life cycle within the bacterial cell, the phage uses host cell machinery to express its genes. One of the proteins produced early in infection is an enzyme that breaks down the bacterial chromosome. Also early in infection, the phage gene for a DNA polymerase that replicates the phage chromosome is expressed, and replication of the phage DNA begins; eventually 100 to 200 new viral DNA molecules are made (step 3). Later in infection, the host cell machinery transcribes the other phage genes and translates their mRNAs, producing the viral coat proteins (step 4). As the head and tail proteins assemble, the replicated viral DNA packs into the heads (step 5).

a. Bacteriophages **b.** Animal virus

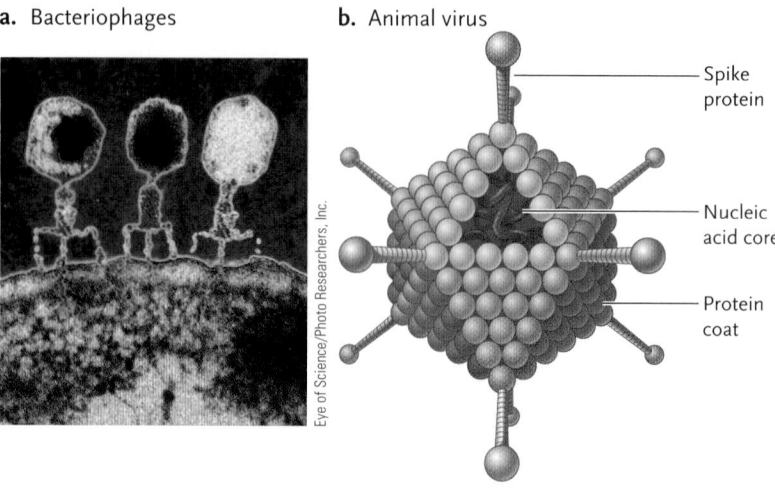

Spike protein

Nucleic acid core

Protein coat

Eye of Science/Photo Researchers, Inc.

Figure 17.6

Viruses. **(a)** Bacteriophages injecting their DNA into *E. coli*. **(b)** An animal virus. A portion of the protein coat has been cut away to show the nucleic acid core. The spike proteins are recognition proteins that allow the viral particle to bind to the surface of a host cell.

When viral assembly is complete, a final enzyme encoded in the viral DNA ruptures the cell. The rupture releases viral particles into the surrounding medium; these progeny phages can infect *E. coli* cells they encounter (step 6). This whole series of events from infection of one cell through the release of progeny phages from broken open, or lysed, cells is called the **lytic cycle.**

For some virulent phages (although not T-even phages), fragments of the host DNA may be included in the heads as the viral particles assemble, providing the basis for transduction of bacterial genes during the next cycle of infection. Because genes are randomly incorporated from essentially any DNA fragments, gene transfer by this mechanism is termed **generalized transduction.**

A Scientist's Favorite Temperate *E. coli* Bacteriophage, lambda (λ). The infective cycle of bacteriophage *lambda,* an *E. coli* bacteriophage much used in research, is typical of temperate phages. Phage λ infects *E. coli* in much the same way as the T-even phages do. The phage injects its linear DNA chromosome into the bacterium (**Figure 17.8,** step 1). Once inside, the linear chromosome forms a circle, and then follows one of two paths. Sophisticated molecular switches govern which path is followed at the time of infection.

One path is the lytic cycle, which is like the lytic cycles of virulent phages. The lytic cycle starts with steps 1 and 2 (infection), then goes directly to steps 7 through 9 (production and release of progeny virus), and back to step 1.

The second and more common path is the **lysogenic cycle.** This cycle begins when the circular lambda chromosome integrates into the host cell's DNA by crossing-over (Figure 17.8, step 3). The DNA of a temperate phage typically inserts at one or possibly a few specific sites in the bacterial chromosome through the action of a phage-encoded enzyme that recognizes certain sequences in the host DNA. In the case of lambda, there is one integration site in the *E. coli* chromosome. Once integrated, the lambda genes are mostly inactive and, therefore, no phage components are made. As a consequence, the phage does not affect its host cell and its descendants. The viral genome is known as a **prophage** while it is inserted in the host cell DNA. In the integrated state, the viral DNA is replicated and passed on in division along with the host cell DNA (steps 4 and 5).

In response to certain environmental signals, such as UV irradiation, the lambda phage in an infected cell becomes active and enters the lytic cycle. Genes that were inactive in the prophage are now transcribed. Among the first viral proteins synthesized in response to the environmental signal are enzymes that excise the lambda chromosome from the host chromosome (step 6). Excision occurs by a crossing-over process that reverses the integration step. The result is a circular

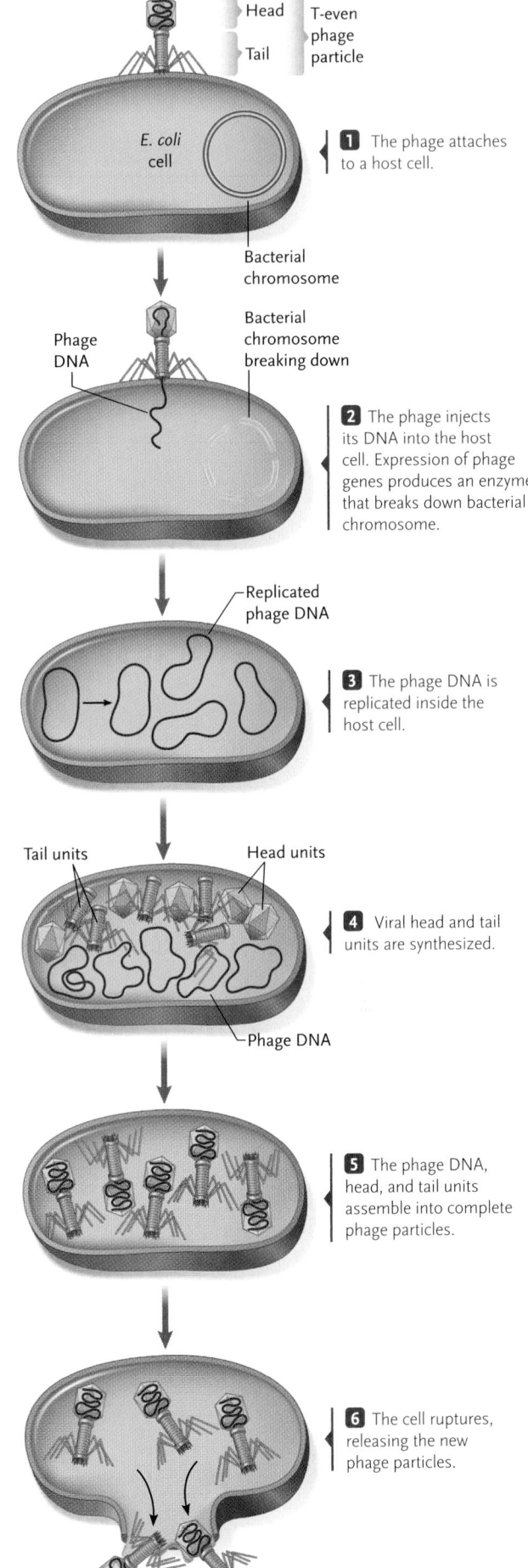

Figure 17.7
The infective cycle of a T-even bacteriophage, an example of a virulent phage.

① The phage attaches to a host cell.

② The phage injects its DNA into the host cell. Expression of phage genes produces an enzyme that breaks down bacterial chromosome.

③ The phage DNA is replicated inside the host cell.

④ Viral head and tail units are synthesized.

⑤ The phage DNA, head, and tail units assemble into complete phage particles.

⑥ The cell ruptures, releasing the new phage particles.

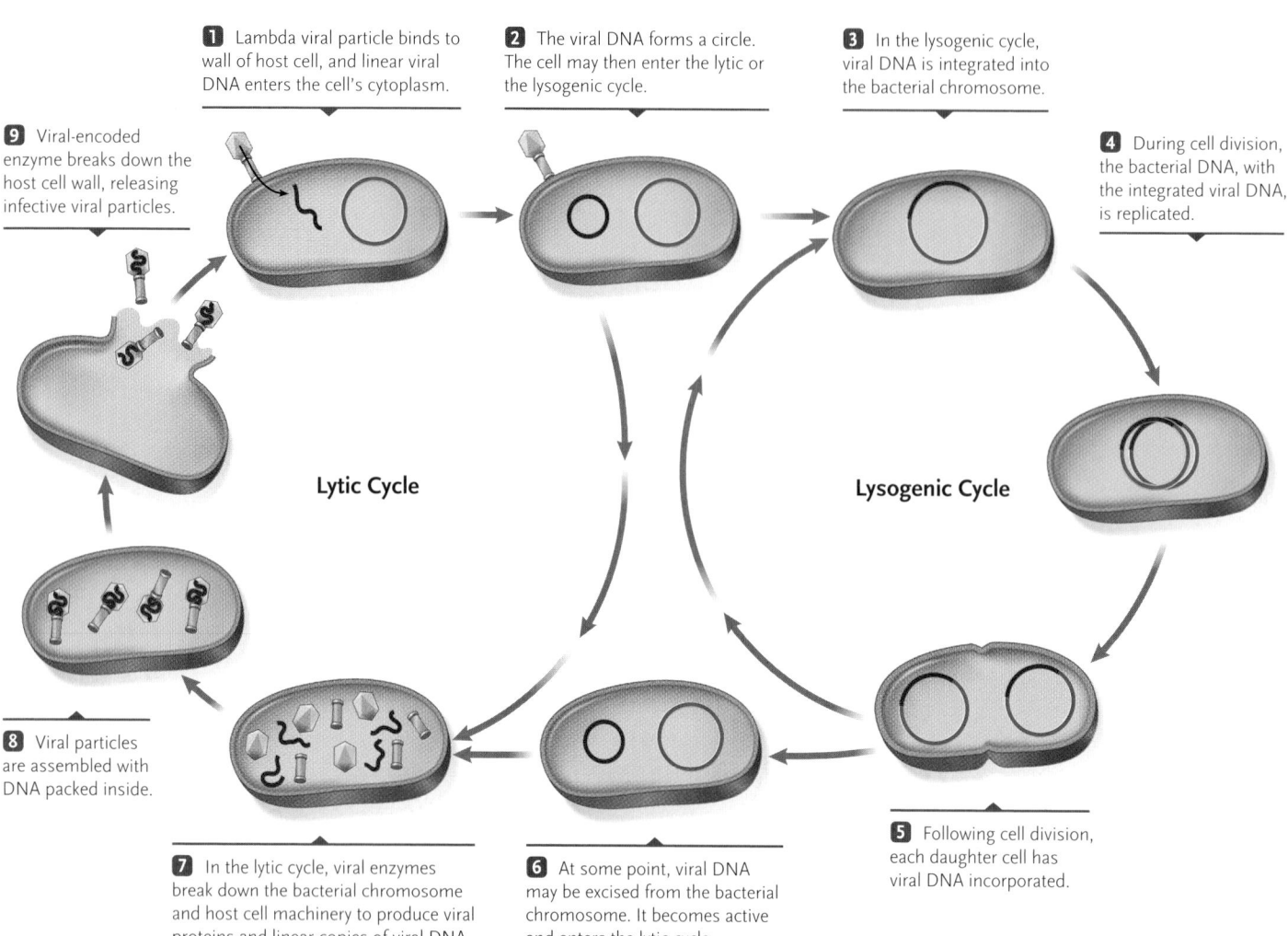

1 Lambda viral particle binds to wall of host cell, and linear viral DNA enters the cell's cytoplasm.

2 The viral DNA forms a circle. The cell may then enter the lytic or the lysogenic cycle.

3 In the lysogenic cycle, viral DNA is integrated into the bacterial chromosome.

9 Viral-encoded enzyme breaks down the host cell wall, releasing infective viral particles.

4 During cell division, the bacterial DNA, with the integrated viral DNA, is replicated.

Lytic Cycle

Lysogenic Cycle

8 Viral particles are assembled with DNA packed inside.

5 Following cell division, each daughter cell has viral DNA incorporated.

7 In the lytic cycle, viral enzymes break down the bacterial chromosome and host cell machinery to produce viral proteins and linear copies of viral DNA.

6 At some point, viral DNA may be excised from the bacterial chromosome. It becomes active and enters the lytic cycle.

Figure 17.8
The infective cycle of lambda, an example of a temperate phage, which can go through the lytic cycle or the lysogenic cycle.

lambda chromosome. Replication of that chromosome produces many copies of linear lambda chromosomes. Expression of genes on those chromosomes generates coat proteins, which assemble with the chromosomes to produce the viral particles (steps 7 and 8). This active stage culminates in rupture of the host cell with the release of infective viral particles (step 9), and the beginning of a new cycle (step 1).

At times, the excision of the lambda chromosome from the *E. coli* DNA is not precise, resulting in the inclusion of one or more host cell genes. These genes are replicated with the viral DNA and packed into the coats, and may be carried to a new host cell in the next cycle of infection. Clearly only genes that are adjacent to the integration site(s) of a temperate phage can be cut out with the viral DNA, included in phage particles during the lytic stage, and undergo transduction. Accordingly, this mechanism of gene transfer is termed **specialized transduction.**

We have seen several examples of how bacteria and viruses generate genetic variability through exchange of genes between organisms. We now turn to a source of genetic variability that involves gene transfer within the genome of a single organism.

STUDY BREAK

What is the difference between a virulent phage and a temperate phage?

17.3 Transposable Elements

All organisms contain particular segments of DNA that can move from one place to another within a cell's genome. The movable sequences are called *transposable genetic elements,* or more simply, **transposable elements (TEs).** They have also been called "jumping genes."

The movement of TEs, called *transposition,* involves a type of genetic recombination process. However, the location in the DNA where the TE moves to—the *target site*—is not homologous with the TE. In this respect, transposition differs from the genetic recombination process in bacteria conjugation, transformation, and transduction and in eukaryote meiosis, which involves crossing-over between homologous DNA molecules.

Transposition of a TE occurs at a low frequency. Depending on the TE, transposition occurs in one of two ways: (1) a cut-and-paste process, in which the TE leaves its original location and transposes to a new location **(Figure 17.9a)**; or (2) a copy-and-paste process, in which a copy of a TE transposes to a new location, leaving the original TE behind **(Figure 17.9b).** For most TEs, transposition starts with contact between the TE and the target site. This also means that TEs do not exist free of the DNA in which they are integrated; hence, their popular name of "jumping genes" is actually inaccurate.

TEs are important because of the genetic changes they cause. For example, they produce mutations by transposing into genes and knocking out their functions, and they increase or decrease gene expression by transposing into regulatory sequences of genes. As such, TEs are an important source of genetic variability.

Insertion Sequence Elements and Transposons Are the Two Major Types of Bacterial Transposable Elements

Bacterial TEs were discovered in the 1960s. They have been shown to move from site to site within the bacterial chromosome, between the bacterial chromosome and plasmids, and between plasmids. The frequency of transposition is low but constant for a given TE. Some bacterial TEs insert randomly, at any point in the DNA, while others recognize certain sequences as "hot spots" for insertion and insert preferentially at these locations.

The two major types of bacterial TEs are *insertion sequences* (*IS*) and *transposons* **(Figure 17.10). Insertion sequences** are the simplest TEs. They are relatively small and contain only genes for their transposition, notably the gene for **transposase**, an enzyme that catalyzes some of the reactions inserting or removing

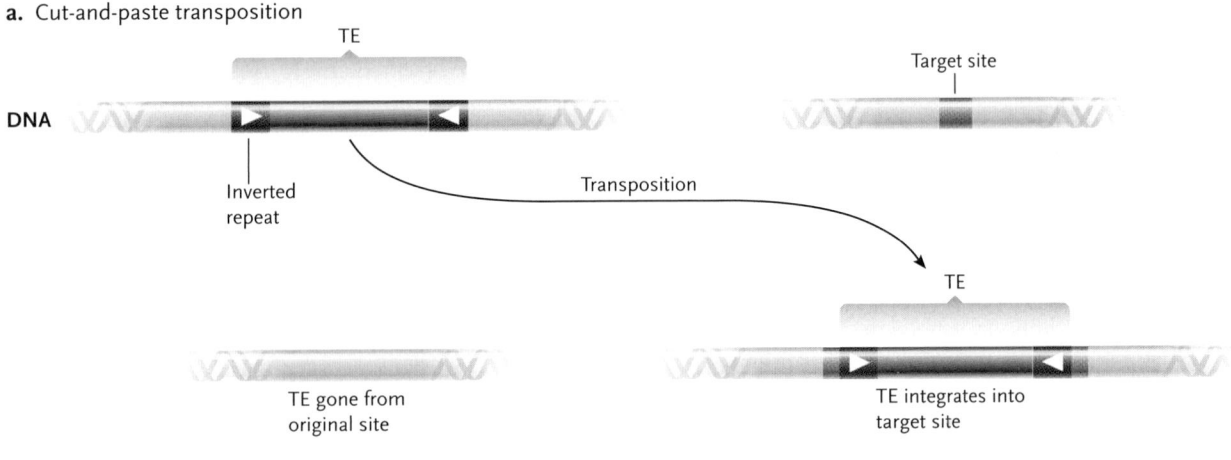

a. Cut-and-paste transposition

b. Copy-and-paste transposition

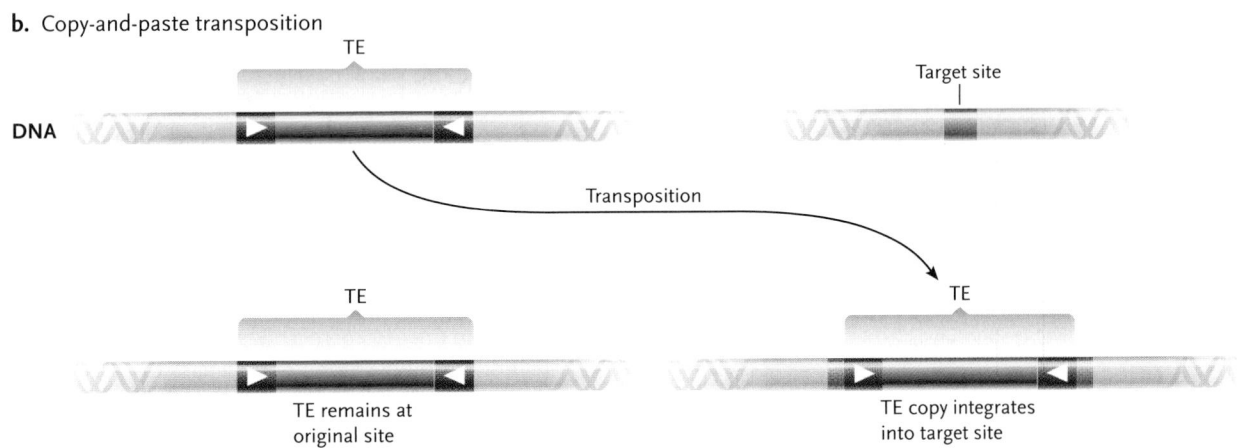

Figure 17.9

Two transposition processes for transposable elements. **(a)** Cut-and-paste transposition, in which the TE leaves one location in the DNA and moves to a new location. **(b)** Copy-and-paste transposition, in which a copy of the TE moves to a new location, leaving the original TE behind.

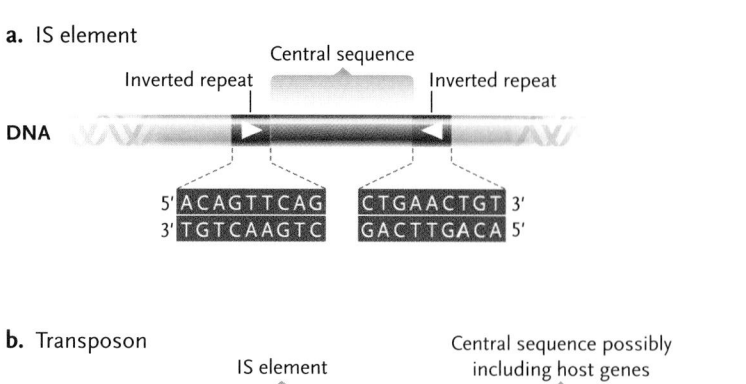

a. IS element

Central sequence

Inverted repeat   Inverted repeat

DNA

5' ACAGTTCAG CTGAACTGT 3'
3' TGTCAAGTC GACTTGACA 5'

b. Transposon

IS element   Central sequence possibly   IS element
                    including host genes

DNA

Figure 17.10

Types of bacterial transposable elements. **(a)** Insertion sequence. **(b)** Transposon.

the TE from the DNA (see Figure 17.10a). At the two ends of an IS is a short *inverted repeat* sequence—the same DNA sequence running in opposite directions (shown by directional arrows in the figure). The inverted repeat sequences enable the transposase enzyme to identify the ends of the TE when it catalyzes transposition.

The second type of bacterial TE, called a **transposon**, has an inverted repeat sequence at each end enclosing a central region with one or more genes. In a number of bacterial transposons, the inverted repeat sequences are insertion sequences, which provide the transposase for movement of the element (see Figure 17.10b). Bacterial transposons without IS ends have short inverted repeat end sequences, and a transposase gene is within the central region. Additional gene(s) in the central region typically are for antibiotic resistance; they originated from the main bacterial DNA circle or from plasmids. These non-IS genes included in transposons are carried along

as the TEs move from place to place within and between species.

Many antibiotics, such as penicillin, erythromycin, tetracycline, ampicillin, and streptomycin, that were once successful in curing bacterial infections have lost much of their effectiveness because of resistance genes carried in transposons. Movements of the transposons, particularly to plasmids that have been transferred by conjugation within and between bacterial species, greatly increase the spread of genes providing antibiotic resistance. Resistance genes have made many bacterial diseases difficult or impossible to treat with standard antibiotics. (Chapter 25 discusses bacterial resistance further.)

Transposable Elements Were First Discovered in Eukaryotes

TEs were first discovered in a eukaryote, maize (corn), in the 1940s by Barbara McClintock, a geneticist working with corn at the Cold Spring Harbor Laboratory in New York. McClintock noted that some mutations affecting kernel and leaf color appeared and disappeared rapidly under certain conditions. Mapping the alleles by linkage studies produced a surprising result—the map positions changed frequently, indicating that the alleles could move from place to place in the corn chromosomes. Some of the movements were so frequent that changes in their effects could be noticed at different times in a single developing kernel **(Figure 17.11)**.

When McClintock first reported her results, her findings were regarded as an isolated curiosity, possibly applying only to corn. This was because the then-prevailing opinion among geneticists was that genes are fixed in the chromosomes and do not move to other locations. Her conclusions were widely accepted only after TEs were detected and characterized in bacteria in the 1960s. By the 1970s, further examples of TEs were discovered in other eukaryotes, including yeast and mammals. McClintock was awarded a Nobel Prize in 1983 for her pioneering work, after these discoveries

Nik Kleinberg

P. J. Maugham

Figure 17.11

Barbara McClintock and corn kernels showing different color patterns due to the movement of transposable elements. As TEs move into or out of genes controlling pigment production in developing kernels, the ability of cells and their descendants to produce the dark pigment is destroyed or restored. The result is random patterns of pigmented and colorless (yellow) segments in individual kernels.

confirmed her early findings that TEs are probably universally distributed among both prokaryotes and eukaryotes.

Eukaryotic Transposable Elements Are Classified as Transposons or Retrotransposons

Eukaryotic TEs fall into two major classes, *transposons* and *retrotransposons,* distinguished by the way the TE sequence moves from place to place in the DNA. However, eukaryotes have no TEs resembling insertion sequences. Researchers detect both classes of eukaryotic TEs through DNA sequencing or through their effects on genes at or near their sites of insertion.

Eukaryotic transposons are similar to bacterial transposons in their general structure and in the way they transpose. A gene for transposase is in the central region of the transposon, and most have inverted repeat sequences at their ends. Depending on the transposon, transposition is by the cut-and-paste or copy-and-paste mechanism (see Figure 17.9).

Members of the other class of eukaryotic TEs, the **retrotransposons,** transpose by a copy-and-paste mechanism but, unlike the other TEs we have discussed, their transposition occurs via an intermediate RNA copy of the TE **(Figure 17.12).** First, the retrotransposon, which is a DNA element integrated into the chromosomal DNA, is transcribed into a complementary RNA copy. Next, an enzyme called **reverse transcriptase,** which is encoded by one of the genes

of the retrotransposon, uses the RNA as a template to make a DNA copy of the retrotransposon. The DNA copy is then inserted into the DNA at a new location, leaving the original in place. Some retrotransposons are bounded by sequences that are directly repeated rather than in inverted form; others have no repeated sequences at their ends.

Cellular genes may become incorporated into the central region of either a transposon or a retrotransposon and travel with it as it moves to a new location. The trapped genes may become continuously active through the effects of regulatory sequences in the TE. The trapped genes may also become abnormally active if moved in a TE to the vicinity of an enhancer or promoter of an intensely transcribed cellular gene. Certain forms of cancer have been linked to the TE-instigated abnormal activation of genes that regulate cell division.

Once TEs are inserted in the chromosomes, they become more or less permanent residents, duplicated and passed on during cell division along with the rest of the DNA. TEs inserted into the DNA of reproductive cells that produce gametes may be inherited, thereby becoming a permanent part of the genetic material of a species. *Insights from the Molecular Revolution* tells about a transposon that has become established in *Drosophila melanogaster* in recent times.

Long-standing TEs are subject to mutation along with other sequences in the DNA. Such mutations may accumulate in a TE, gradually altering it into a nonmobile, residual sequence in the DNA. The DNA of many

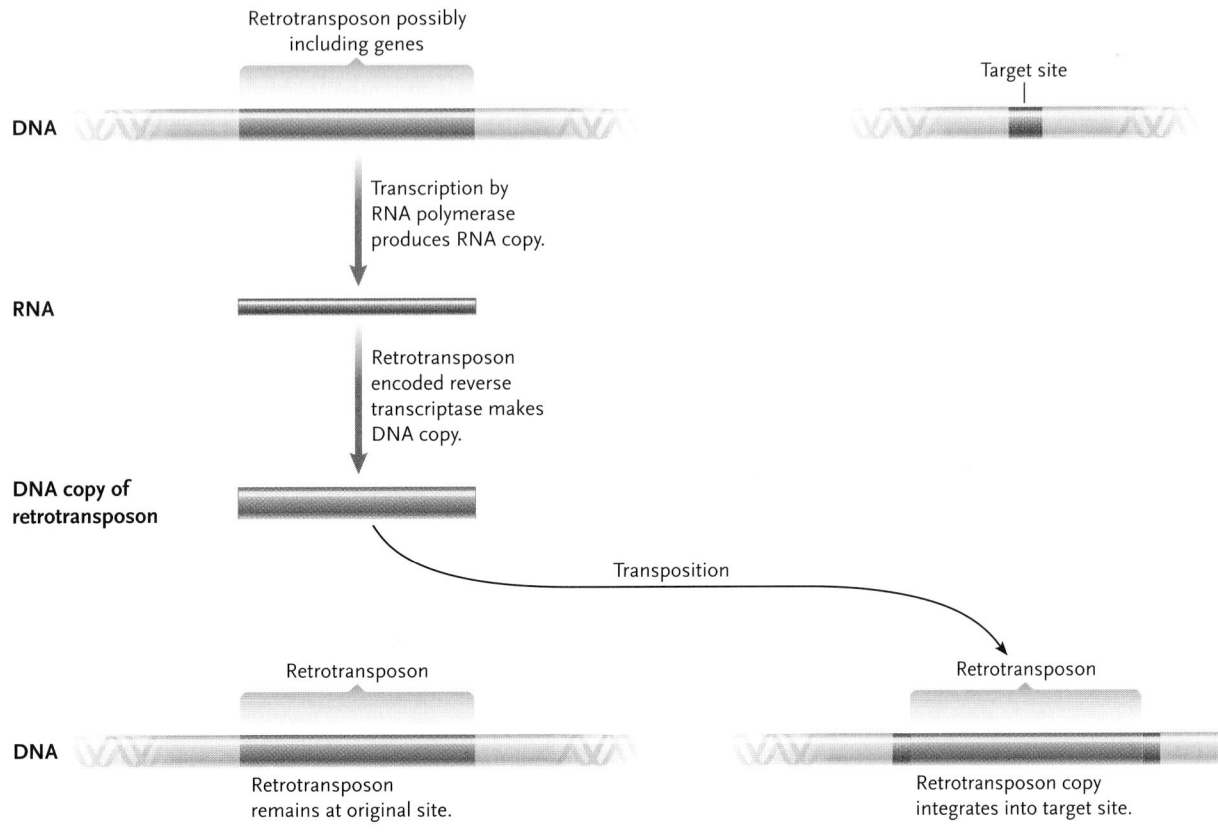

Figure 17.12
Transposition of a eukaryotic retrotransposon to a new location by means of an intermediate RNA copy.

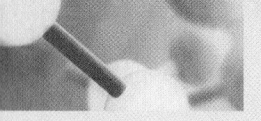

Genes That Jump a Mite Too Far

The fruit fly *Drosophila melanogaster* is cultured in genetic laboratories all over the world. Many of the flies in laboratory cultures are home to a transposon known as a *P element*. Strains of flies that have this transposon in their DNA are known as *P strains;* strains of flies without the element are *M strains.*

A curious feature is the absence of P elements in *D. melanogaster* laboratory cultures that have been maintained in total isolation since 1950. The worldwide distribution of P elements among wild *D. melanogaster* populations is also uneven, as if the elements have not had time to become established in wild fruit flies in all parts of the world. These characteristics make it likely that P elements invaded wild *D. melanogaster* populations recently, probably within the last 30 to 50 years.

Margaret G. Kidwell of the University of Arizona and Marilyn A. Houck, now at Texas Tech University, hypothe-

sized that the P element first appeared in *D. melanogaster* by "jumping" from another *Drosophila* species, *D. willistoni,* in which the element is universally distributed. Even though the two species do not interbreed, the P elements in *D. willistoni* and *D. melanogaster* are almost identical. The alternative hypothesis—that P elements were inherited from a common ancestor and were present in both species when *D. melanogaster* and *D. willistoni* separated millions of years ago—predicts that the elements would likely have mutated into significantly different forms in the two species, and clearly this is not the case.

How might P elements have moved from *D. willistoni* to *D. melanogaster?* Kidwell and Houck point to a mite, *Protolaelaps regalis,* as the possible vehicle. (Mites are small relatives of spiders.) The mite feeds on fruit flies and their eggs. Houck noticed that the mouth parts of the mite resemble the microscopic needles used by investiga-

tors to transfer DNA from cell to cell. This resemblance gave her the idea that the mites, which feed on both *D. willistoni* and *D. melanogaster,* transferred DNA containing the P elements from one species to the other. After having fed on a *D. willistoni* fly, a mite with DNA containing P elements on its mouth parts may have fed on a *D. melanogaster* egg, at the same time injecting the P element DNA into the egg. If the egg survived, and the P element DNA became integrated into the egg's DNA, an adult produced from the egg once it is fertilized would be a P-strain *D. melanogaster.*

In support of their idea, the investigators have found mites with P element DNA on their mouth parts and in their digestive tracts after feeding on P-strain flies. If the DNA transfer actually occurred in this way, the interspecies transfer also opens the possibility that evolution may occur through genes introduced by this mechanism of natural genetic engineering.

---

eukaryotes, including humans, contains many nonfunctional TEs likely created in this way.

Retroviruses Are Similar to Retrotransposons

RNA to DNA reverse transcription is associated with certain viruses as well as with retrotransposons. A **retrovirus** is a virus with an RNA genome that repli-

cates via a DNA intermediate. When a retrovirus infects a host cell, a reverse transcriptase carried in the viral particle is released and copies the single-stranded RNA genome into a double-stranded DNA copy. The viral DNA is then inserted into the host DNA, where it is replicated and passed to progeny cells during cell division. Similar to the prophage of temperate bacteriophages, the inserted viral DNA is known as a provirus **(Figure 17.13).**

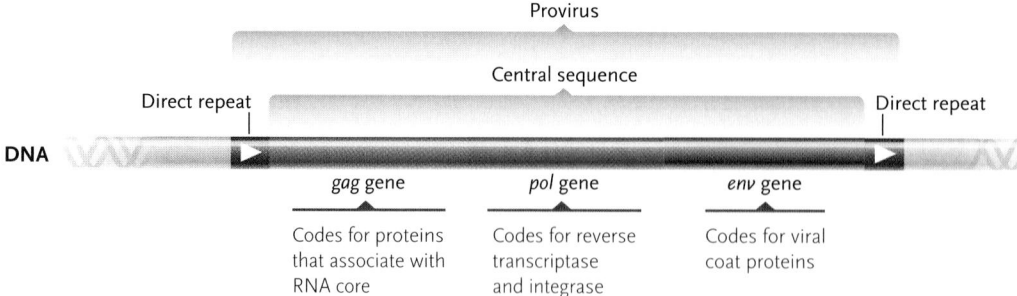

Figure 17.13

A mammalian retrovirus in the provirus form in which it is inserted into chromosomal DNA. The direct repeats at either end contain sequences capable of acting as enhancer, promoter, and termination signals for transcription. The central sequence contains genes coding for proteins, concentrated in the *gag, pol,* and *env* regions. The provirus of HIV, the virus that causes AIDS, takes this form.

Retroviruses are found in a wide range of organisms, with most so far identified in vertebrates. You, as well as most other humans and mammals, probably contain from one to as many as 100 or more retroviruses in your genome as proviruses. Many of these retroviruses do not ever produce viral particles. However, they may sometimes cause genetic disturbances of various kinds, including alterations of gene activity or DNA rearrangements such as deletions and translocations, some of which may be harmful to the host. The AIDS retrovirus, called *HIV* (for *Human Immunodeficiency Virus*), does produce viral particles. When HIV infects a human, its primary effect is to interfere with white blood cells of the immune system (HIV and AIDS are discussed further in Chapter 43).

Some retroviruses, such as the *avian sarcoma virus*, have been linked to cancer (in this case in chickens). Many of these cancer-causing retroviruses have picked up a host gene that triggers entry of cells into uncontrolled DNA replication and cell division. When included in a retrovirus, the gene comes under the influence of the highly active retroviral promoter, which makes the gene continually active and leads to the uncontrolled cell division characteristic of cancer. In other words, the host gene has become an oncogene, a gene that promotes the development of cancer by stimulating cell division (see Section 16.4). Usually, the host gene replaces one or more retrovirus genes, making the virus unable to produce viral particles.

Retroviruses also activate genes related to cell division by moving them to the vicinity of an active host cell promoter or enhancer, or by delivering an enhancer or active promoter to the vicinity of a host cell gene. In either case, the result may be uncontrolled cell division. Harmless retroviruses are being developed as a means to introduce genes into mammalian and other animal cells for genetic engineering (discussed further in Chapter 18).

The close similarities between retrotransposons and retroviruses have led to the proposal that retroviruses may have evolved from retrotransposons. This might have occurred through the chance enclosure of the RNA intermediates of retrotransposons by the coat proteins of infecting viruses, giving the retrotransposons the ability to escape from an original host and move to new individuals. The reverse process, the evolution of retrotransposons from retroviruses, is thought to be less likely because eukaryotic cells contain several additional types of retrotransposons that are not related to retroviruses except that they also move through formation of an RNA intermediate. For example, one of these non-retroviral retroposons, called *Alu*, is

---

UNANSWERED QUESTIONS

How is movement of transposable elements regulated?

You learned in this chapter that transposable elements (TEs) are mobile genetic elements in genomes. TEs are found in most living organisms—unlike viruses, they are permanent residents. Research with a variety of TEs and organisms has shown that there is a delicate balance between the transposons and the host genome. By studying the types, numbers, and genetic locations of TEs in genomes, researchers can learn about the evolution of the mechanisms the elements use to move and about the regulation of that movement.

A TE may cause a mutation if it moves into the coding region of a gene or if it inactivates its promoter. However, in normal cells, TE movement around the genome is actively restricted, and therefore, mutational damage is low. Researchers are finding that this negative regulation of TE movement can involve factors and mechanisms provided by the host cell and/or by gene regulatory limitations specific to the transposon. Through continued research, scientists hope to get a much better understanding of the molecular interactions between host cell and TE that bring about this relatively peaceful relationship, and of how these interactions developed during evolution.

Is there an adaptive function for a retrotransposon?

Amazingly, over 70% of the barley genome consists of retrotransposons. You would expect, therefore, that any changes in the number of these retrotransposons would have a significant impact on genome size. An interesting research project has shown a nearly threefold variation in the copy number of one barley retrotransposon, BARE-1, over a 300-meter span of a particular canyon in Israel. The highest copy numbers were found in plants in drier areas higher in the canyon. The researchers hypothesize that this might be because a stress-response regulatory sequence in the BARE-1 promoter activates the retrotransposon to cause transposition of the element in these plants. The movement leaves the original element in place, so the effect is to increase the number of BARE-1 elements. The scientists propose that the increase in retrotransposon number may have adaptive value for the plants growing under dry conditions where their growth rates are relatively slow. Possibly the increase in retrotransposon number, and its effect on overall genome size, produces larger cell sizes, which would offset the slower growth rates. The model is an interesting one, and is being pursued through further research with this system, and with others involving expanding retrotransposon numbers.

Peter J. Russell

clearly derived evolutionarily from an RNA molecule that occurs widely in eukaryotes as part of the signal-recognition particle (the signal-recognition particle, described in Section 15.4, is an RNA-protein complex that helps attach ribosomes to the endoplasmic reticulum during protein synthesis). In total, retrotransposons and retroviruses of all types occupy some 40% of the human genome.

The genetic elements discussed in this chapter, particularly plasmids and retroviruses, often act as natural genetic engineers by moving genes between species. The next chapter describes how human genetic engineers manipulate and clone DNA and how they analyze genomes at the DNA level.

STUDY BREAK

Among eukaryotic transposable elements, how do transposons, retrotransposons, and retroviruses differ?

Review

Go to ThomsonNOW™ at www.thomsonedu.com/login to access quizzing, animations, exercises, articles, and personalized homework help.

17.1 Gene Transfer and Genetic Recombination in Bacteria

- The rapid generation times and numerous offspring of bacteria and viruses make it possible to trace genetic crosses and their outcomes much more quickly than in eukaryotes. These characteristics make it possible to detect rare genetic events. The results of these crosses show that recombination may occur within the boundaries of a gene as well as between genes.

- Recombination occurs in both bacteria and eukaryotes by exchange of segments between homologous DNA molecules. In bacteria, the DNA of the bacterial chromosome may recombine with DNA brought into the cell from outside.

- Three primary mechanisms bring DNA into bacterial cells from the outside: conjugation, transformation, and transduction.

- In conjugation, which is the basis of bacterial sexual reproduction, two bacterial cells form a cytoplasmic bridge, and part or all of the DNA of one cell moves into the other through the bridge. The donated DNA can then recombine with homologous sequences of the recipient cell's DNA (Figures 17.1, 17.2, and 17.4).

- *E. coli* bacteria that are able to act as DNA donors in conjugation have an F plasmid, making them F⁺; recipients have no F plasmid and are F⁻. In Hfr strains of *E. coli*, the F plasmid is within the main chromosome. As a result, genes of the main chromosome are often transferred into F⁻ cells along with a portion of the F-plasmid DNA. Researchers have mapped genes on the *E. coli* chromosome by noting the order in which they are transferred from Hfr to F⁻ cells during conjugation (Figure 17.4).

- In transformation, intact cells absorb pieces of DNA released from cells that have disintegrated. The entering DNA fragments can recombine with the recipient cell's DNA.

- In transduction, DNA is transferred from one cell to another by an infecting virus.

  Practice: Distinguishing between the three major processes:
  conjugation, transformation, and transduction

17.2 Viruses and Viral Recombination

- When in the free form, viruses consist of a nucleic acid core, either DNA or RNA, surrounded by a protein coat (Figure 17.6).

- The cycle of viral infection begins when the nucleic acid molecule of a virus is introduced into a host cell and replicated. Viral coat proteins are made and assembled with the DNA into new viral particles (Figure 17.7).

- Virulent phages kill the host cell by releasing an enzyme that ruptures the plasma membrane and cell wall and releases the new viral particles (Figure 17.7).

- Temperate phages do not always kill their host cell. They may enter the lytic cycle, in which the viral DNA becomes active, exits the host DNA, and begins replication, or a lysogenic cycle. In the lysogenic cycle, the phage's DNA is integrated into the host cell's DNA and may remain for many generations. At some point, the virus may enter the lytic cycle and begin replication. After production of viral coats, the DNA is assembled into new viral particles, which are released as the cell ruptures (Figure 17.8).

- During a cycle of viral infection with particular phages, one or more fragments of host cell DNA may be incorporated into viral particles. As an infected cell breaks down, it releases the viral particles containing host cell DNA. These particles, which form the basis of bacterial transduction, may infect a second cell and introduce the bacterial DNA segment into the new host, where it may recombine with the host DNA.

17.3 Transposable Elements

- Both prokaryotes and eukaryotes contain TEs (transposable elements)—DNA sequences that can move from place to place in the DNA. The TEs may move from one location in the DNA to another, or generate duplicated copies that insert in new locations while leaving the "parent" copy in its original location (Figure 17.9).

- Genes of the host cell DNA may become incorporated into a TE and may be carried with it to a new location. There, the genes may become abnormally active when placed near sequences that control the activity of genes within the TE, or near the control elements of active host genes.

- Eukaryotic TEs occur as transposons, which release from one location in the DNA and insert at a different site, or as retrotransposons, which move by making an RNA copy, which is then replicated into a DNA copy that is inserted at a new location. The "parent" copy remains at the original location. Like retrotransposons, retroviruses integrate into chromosomal DNA by making a DNA copy of their RNA genome. Retroviruses may have evolved from retrotransposons (Figures 17.11–17.13).

- TE-instigated abnormal activation of genes regulating cell division has been linked to the development of some forms of cancer in humans and other complex animals.

Questions

Self-Test Questions

1. When studying the differences in the genes of bacteria, researchers:
   a. do not grow bacteria on minimal medium as the medium lacks needed nutrients.
   b. use a bacterial clone, which is a group of cells from different bacteria of varying genetic makeup.
   c. use bacteria diploid for their full genome because they can grow on minimal medium.
   d. can study only one genetic trait in a single recombinant event.
   e. can measure the passage of genes between cells during conjugation, transduction, and transformation.

2. If crossovers occurred between two bacterial genomes as shown in the figure, the result would be:

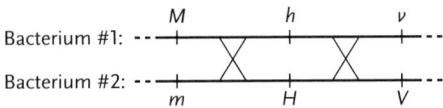

   a. MHv and mhV.
   b. MHV and mhv.
   c. Mhv and mHv.
   d. MHV and mhV.
   e. mhv and MhV.

3. In conjugation, when a bacterial F factor is transferred:
   a. the donor cell becomes F^-.
   b. the recipient cell becomes F^+.
   c. the recipient cell becomes F^-.
   d. the donor cell turns into a recipient cell.
   e. viral DNA integrates into the recipient DNA.

4. Which of the following is *not* correct for bacterial conjugation?
   a. Both Hfr and F^+ bacteria have the ability to code for a sex pilus.
   b. After an F^- cell has conjugated with an F^+, its plasmid holds the F^+ factor.
   c. The recipient cell usually becomes Hfr following conjugation.
   d. In an Hfr $\times$ F^- mating, DNA of the main chromosome moves to a recipient cell.
   e. Genes on the F factor encode proteins of the sex pilus.

5. Which of the following is *not* correct for bacterial transformation?
   a. Artificial transformation is used in cloning procedures.
   b. Avery was able to transform live noninfective bacteria with DNA from dead infective bacteria.
   c. The cell wall and plasma membrane must be penetrated for transformation to proceed.
   d. A virus is required for the process.
   e. Electroporation is a form of artificial transformation used to introduce DNA into cells.

6. Transduction:
   a. may allow recombination of newly introduced DNA with host cell DNA.
   b. is the movement of DNA from one bacterial cell to another by means of a plasmid.
   c. can cause the DNA of the donor to change but not the DNA of the recipient.
   d. is the movement of viral DNA but not bacterial DNA into a recipient bacterium.
   e. requires a physical contact between two bacterium.

7. Viruses:
   a. have a protein core.
   b. have a nucleic acid coat.
   c. that infect bacteria are called bacteriophages.
   d. were probably the first forms of life on Earth.
   e. if temperate, kill host cells.

8. A virus in its lysogenic cycle:
   a. is lysing the host cell.
   b. is transducing a bacterial cell.
   c. is assembling viral particles for cell rupture.
   d. is damaging the host cell.
   e. has its genome integrated in host DNA.

9. Which of the following is *not* correct about transposable elements?
   a. They can be recognized by their ends of inverted transposable elements.
   b. They have an internal portion that can be transcribed.
   c. They encode a transposase enzyme.
   d. They have no harmful effects on cell function.
   e. They move by a cut-and-paste or copy-and-paste mechanism.

10. Which is *not* correct about retroviruses?
    a. They are RNA viruses.
    b. They are believed to be the source of retrotransposons.
    c. They encode an enzyme for their insertion into host cell DNA.
    d. They encode single-stranded viral DNA from viral RNA.
    e. They encode a reverse transcriptase enzyme for RNA to DNA synthesis.

Questions for Discussion

1. You set up an experiment like the one carried out by Lederberg and Tatum, mixing millions of *E. coli* of two strains with the following genetic constitutions:

   Strain 1: bio^- met^- thr^+ leu^+

   Strain 2: bio^+ met^+ thr^- leu^-

   Among the bacteria obtained after mixing, you find some cells that do not require threonine, leucine, or biotin to grow, but still need methionine. How might you explain this result?

2. As a control for their experiments with bacterial recombination, Lederberg and Tatum placed cells of either "parental" strain 1 or 2 on the surface of a minimal medium. If you set up this control and a few scattered colonies showed up, what might you propose as an explanation? How could you test your explanation?

3. Experimental systems have been developed in which transposable elements can be induced to move under the control of a researcher. Following the induced transposition of a yeast TE element, two mutants were identified with altered activities of enzyme X. One of the mutants lacked enzyme activity completely, while the other had five times as much enzyme activity as normal cells did. Both mutants were found to have the TE inserted into the gene for enzyme X. Propose hypotheses for how the two different mutant phenotypes were produced.

Experimental Analysis

You have a culture of Hfr *E. coli* cells that cannot make biotin for themselves. To this culture you add some wild-type *E. coli* cells that have been heat killed, and then subject the culture to electroporation. After the addition, you find some cells that can grow on minimal medium. How could you establish whether the wild-type *bio*+ allele was inserted in a plasmid or the chromosomal DNA of the Hfr cells?

Evolution Link

Are viruses evolutionarily derived from complex organisms that can reproduce themselves, or are they remnants of precellular "life"? Argue your case.

Protein microarray, a key tool of proteomics, the study of the complete set of proteins that can be expressed by an organism's genome. Each colored dot is a protein, with a specific color for each protein being studied.

Pasteka/SPL/Photo Researchers, Inc.

18 DNA Technologies and Genomics

WHY IT MATTERS

In early October 1994, 32-year-old Shirley Duguay, a mother of five, disappeared from her home on Prince Edward Island, Canada. Within a few days, her car was found abandoned; bloodstains inside matched her blood type. Several months later the Royal Canadian Mounted Police (RCMP) found Duguay's body in a shallow grave. Among the chief suspects in the murder was her estranged common-law husband, Douglas Beamish, who was living nearby with his parents.

While searching for Duguay, the RCMP discovered a plastic bag containing a man's leather jacket with the victim's blood on it. Beamish's friends and family acknowledged that Beamish had a similar jacket, but none could or would positively identify it. In the lining of the jacket investigators found 27 white hairs, which forensic scientists identified as cat hairs. The RCMP remembered that Beamish's parents had a white cat named Snowball **(Figure 18.1).** Could they prove that the cat hair in the jacket was Snowball's?

A Mountie investigator used the Internet to find two experts on cat genomes, Marilyn Menotti-Raymond and Stephen J. O'Brien of the Laboratory of Genome Diversity at the U.S. National Cancer Institute. Menotti-Raymond and O'Brien analyzed DNA from the root of one of

371

Figure 18.1
Snowball, the key
to a murder case.
(© Marilyn Menotti-
Raymond, The Na-
tional Cancer Institute–
Frederick)

the cat hairs taken from the jacket and from a blood sample taken from Snowball. They then used a technique called the polymerase chain reaction (PCR) to amplify 10 specific regions of the cat genome, each of which varies among cats in the number of copies of a short (two-nucleotide) repeated sequence. They found that the hair and blood samples matched perfectly, providing strong evidence that the hair came from Snowball.

Beamish was tried for the murder of Shirley Duguay. The evidence presented by Menotti-Raymond and O'Brien helped convict him, and he was sentenced to 18 years in prison.

The researchers' analysis of the cat hair and blood is an example of DNA fingerprinting. (Though human DNA fingerprinting evidence is now extensively used in court cases, the Beamish case was the first to admit nonhuman animal DNA fingerprinting data as evidence.) DNA fingerprinting is an application of **DNA technologies**, techniques to isolate, purify, analyze, and manipulate DNA sequences. Scientists use DNA technologies both for basic research into the biology of organisms and for applied research. The use of DNA technologies to alter genes for practical purposes is called **genetic engineering.**

Genetic engineering is the latest addition to the broad area known as *biotechnology,* which is any technique applied to biological systems or living organisms to make or modify products or processes for a specific purpose. Thus, biotechnology includes manipulations that do not involve DNA technologies, such as the use of yeast to brew beer and bake bread and the use of bacteria to make yogurt and cheese.

In this chapter, we focus on how biologists isolate genes and manipulate them for basic and applied research. You will learn about the basic DNA technologies and their applications to research in biology, to genetic engineering, and to the analysis of genomes. You will also learn about some of the risks and controversies surrounding genetic engineering and about some of the scientific, social, and ethical questions related to its application.

We begin our discussion with a description of methods used to obtain genes in large quantities, an essential step for their analysis or manipulation.

18.1 DNA Cloning

Remember from Chapter 17 that a *clone* is a line of genetically identical cells or individuals derived from a single ancestor. DNA cloning is a method for pro-

ducing many copies of a piece of DNA, such as a gene of interest; that is, a gene that a researcher wants to study or manipulate. Scientists clone DNA for many reasons. For example, a researcher might be interested in how a particular human gene functions. Each human cell contains only two copies of most genes, amounting to a very small fraction of the total amount of DNA in a diploid cell. In its natural state in the genome, then, the gene is extremely difficult to study. However, through DNA cloning, a researcher can produce a sample large enough for scientific experimentation.

Cloned genes are used in basic research to find out about their biological functions. For example, researchers can determine the DNA sequence of a cloned gene, giving them the ultimate information about its structure. Also, by manipulating the gene and inducing mutations in it, they can gain information about its function and about how its expression is regulated. Cloned genes can be expressed in bacteria, and the proteins encoded by the cloned genes can be produced in quantity and purified. Those proteins can be used in basic research or, in the case of genes encoding proteins of pharmaceutical or clinical importance, they can be used in applied research.

An overview of one common method for cloning a gene of interest from a genome is shown in **Figure 18.2;** the method uses bacteria (commonly, *Escherichia coli*) and plasmids, the small circular DNA molecules that replicate separately from the bacterial chromosome (see Section 17.1). The researcher extracts DNA containing a gene of interest from cells and cuts it into fragments. The fragments are inserted into plasmids producing *recombinant DNA molecules*—**recombinant DNA** is DNA from two or more different sources joined together. The recombinant plasmids are introduced into bacteria; each bacterium receives a different plasmid. The bacterium continues growing and dividing, and as it does, the plasmid continues to replicate. It is through replication of the plasmid that amplification of the piece of DNA inserted into the plasmid occurs. The final step, then, is to identify the bacteria containing the plasmid with the gene of interest and isolate it for further study.

Bacterial Enzymes Called Restriction Endonucleases Form the Basis of DNA Cloning

The key to DNA cloning is the specific joining of two DNA molecules from different sources, such as a genomic DNA fragment and a bacterial plasmid (see Figure 18.2). This specific joining of DNA is made possible, in part, by bacterial enzymes called **restriction endonucleases** (also called **restriction enzymes**), discovered in the late 1960s. Restriction enzymes recognize short, specific DNA sequences called *restriction sites,* typically four to eight base pairs long, and cut the

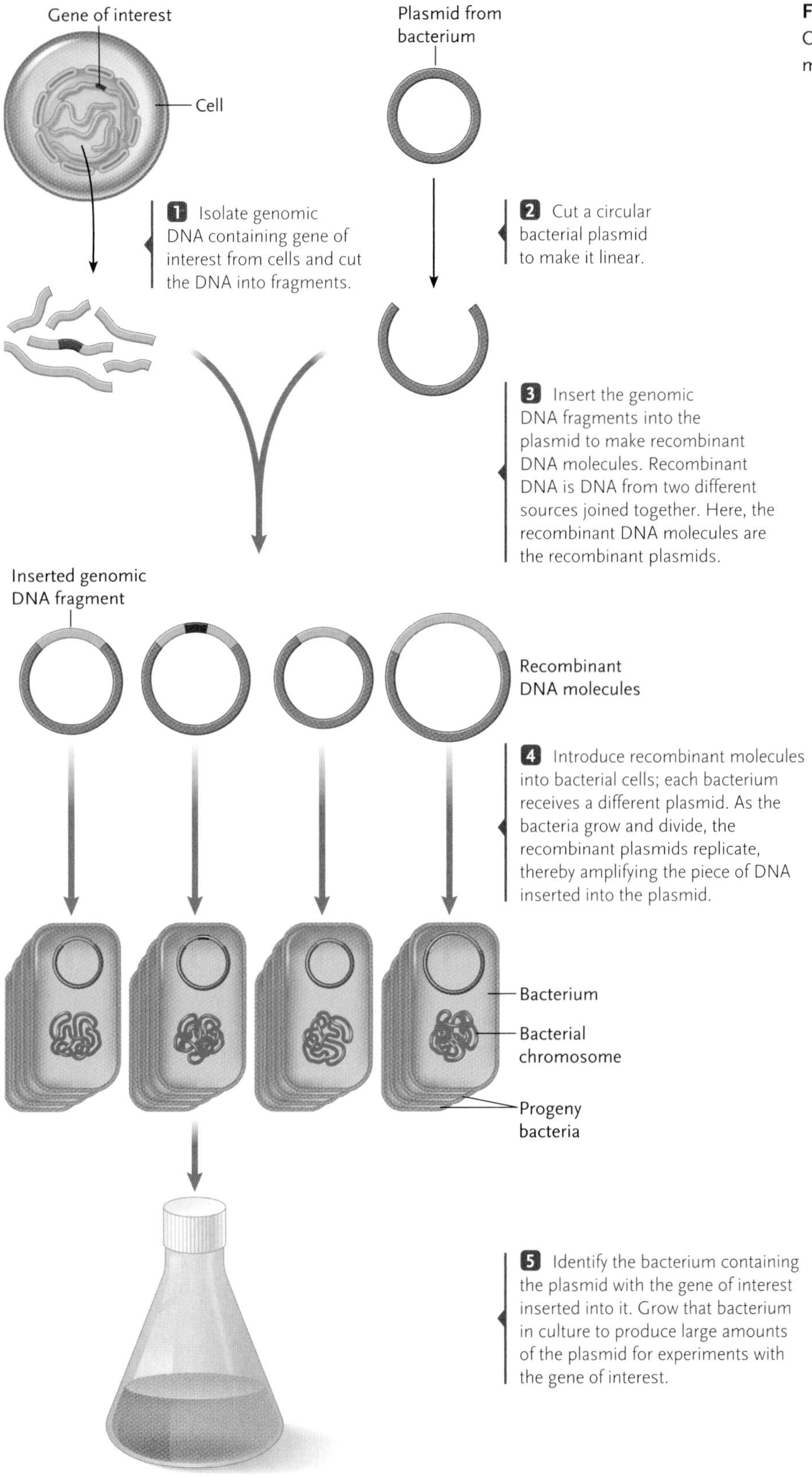

Gene of interest

Cell

Plasmid from bacterium

Figure 18.2
Overview of cloning DNA fragments in a bacterial plasmid.

1 Isolate genomic DNA containing gene of interest from cells and cut the DNA into fragments.

2 Cut a circular bacterial plasmid to make it linear.

3 Insert the genomic DNA fragments into the plasmid to make recombinant DNA molecules. Recombinant DNA is DNA from two different sources joined together. Here, the recombinant DNA molecules are the recombinant plasmids.

Inserted genomic DNA fragment

Recombinant DNA molecules

4 Introduce recombinant molecules into bacterial cells; each bacterium receives a different plasmid. As the bacteria grow and divide, the recombinant plasmids replicate, thereby amplifying the piece of DNA inserted into the plasmid.

Bacterium

Bacterial chromosome

Progeny bacteria

5 Identify the bacterium containing the plasmid with the gene of interest inserted into it. Grow that bacterium in culture to produce large amounts of the plasmid for experiments with the gene of interest.

DNA at specific locations within those sequences. The DNA fragments produced by cutting a long DNA molecule with a restriction enzyme are known as **restriction fragments.**

The "restriction" in the name of the enzymes refers to their normal role inside bacteria, in which the enzymes defend against viral attack by breaking down (restricting) the DNA molecules of infecting viruses. The bacterium protects the restriction sites in its own DNA from cutting by modifying bases in those sites enzymatically, thereby blocking the action of its restriction enzyme.

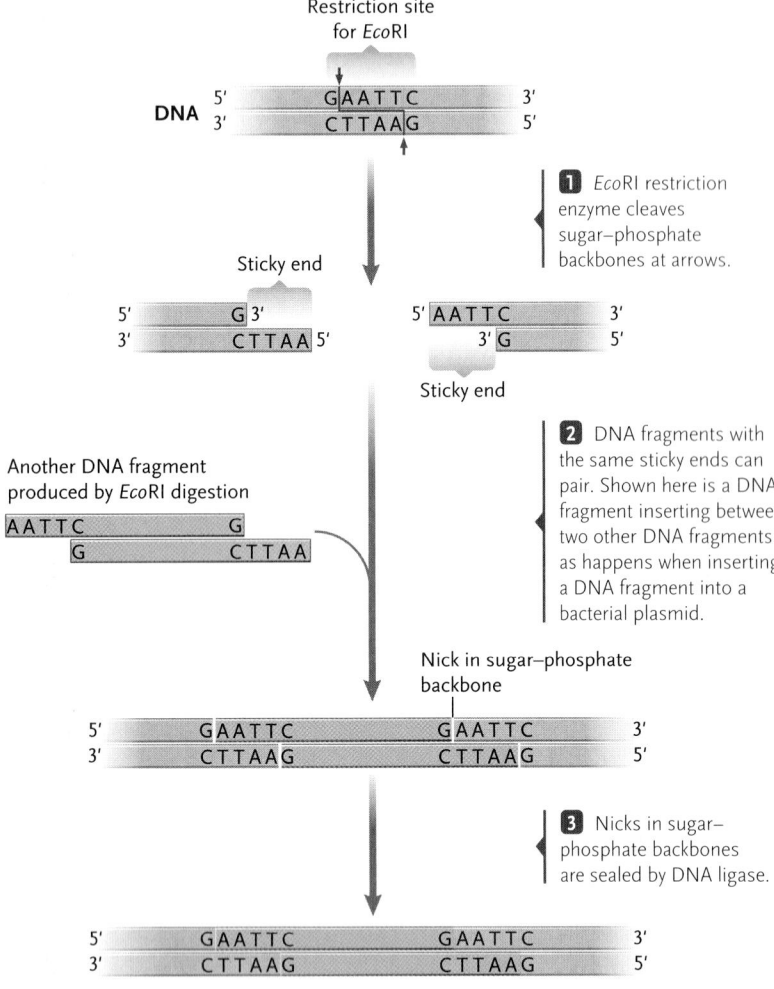

Figure 18.3
The restriction site for the restriction enzyme *Eco*RI, and the generation of a recombinant DNA molecule by complementary base pairing of DNA fragments produced by digestion with the same restriction enzyme.

Hundreds of different restriction enzymes have been identified, each one cutting DNA at a specific restriction site. As illustrated by the restriction site of *Eco*RI **(Figure 18.3)**, most restriction sites are symmetrical in that the sequence of nucleotides read in the 5′→3′ direction on one strand is same as the sequence read in the 5′→3′ direction on the complementary strand. The restriction enzymes most used in cloning—such as *Eco*RI—cleave the sugar–phosphate backbones of DNA to produce DNA fragments with single-stranded ends (step 1). The ends are called **sticky ends** because the short single-stranded regions can form hydrogen bonds with complementary sticky ends on any other DNA molecules cut with the same enzyme. For example, step 2 shows the insertion of a DNA molecule with sticky ends produced by *Eco*RI between two other DNA molecules with the same sticky ends. The pairings leave nicks in the sugar–phosphate backbones of the DNA strands that are sealed by *DNA ligase,* an enzyme that has the same function in DNA replication (step 3; see Section 14.3).

The result is DNA from two different sources joined together—a recombinant DNA molecule.

Bacterial Plasmids Illustrate the Use of Restriction Enzymes in Cloning

The bacterial plasmids used for cloning are examples of cloning vectors—DNA molecules into which a DNA fragment can be inserted to form a recombinant DNA molecule for cloning. Bacterial plasmid cloning vectors are not naturally in bacteria; they are plasmids modified to have special features. Commonly, plasmid cloning vectors are engineered to contain two genes that are useful in the final steps of a cloning experiment for sorting bacteria that have recombinant plasmids from those that do not. The *amp*^R gene encodes an enzyme that breaks down the antibiotic ampicillin; when the plasmid is introduced into *E. coli* and the *amp*^R gene is expressed, the bacteria become resistant to ampicillin. The *lacZ*^+ gene encodes β-galactosidase (recall the *lac* operon from Section 16.1), which hydrolyzes the sugar lactose, as well as a number of synthetic substrates. Restriction sites are located within the *lacZ*^+ gene, but do not alter the gene's function. For a given cloning experiment, one of these restriction sites is chosen.

Cloning a Gene of Interest. **Figure 18.4** expands on the overview of Figure 18.2 to show the steps used to clone a gene of interest using a plasmid cloning vector and restriction enzymes. Genomic DNA isolated from the organism in which the gene is found is cut with a restriction enzyme, and a plasmid cloning vector is cut within the *lacZ*^+ gene with the same restriction enzyme (steps 1 and 2). Mixing the DNA fragments and cut plasmid together with DNA ligase produces various joined molecules as the sticky ends pair and the enzyme seals them together. Some of these molecules are recombinant plasmids consisting of, in each case, a DNA fragment inserted into the plasmid cloning vector; others are nonrecombinant plasmids resulting from the cut plasmid resealed into a circle without an inserted fragment (step 3). In addition, ligase joins together pieces of genomic DNA with no plasmid involved. Only the recombinant plasmids are important in the cloning of the gene of interest; we sort out the other two undesired molecules in later steps.

Next, the DNA molecules are transformed—introduced—into ampicillin-sensitive, *lacZ*^− *E. coli* (which cannot make β-galactosidase), and the transformed bacteria are spread on a plate of agar growth medium containing ampicillin and the β-galactosidase substrate X-gal (steps 4 and 5). (Section 17.1 describes techniques for transformation of DNA into bacteria.) Only bacteria with a plasmid can grow and form colonies because expression of the plasmid's *amp*^R gene makes the bacteria resistant to ampicillin (see Figure 18.4 results). Within each cell of a colony, the plasmids have replicated until a hundred or so are present.

Figure 18.4 Research Method

Cloning a Gene of Interest in a Plasmid Cloning Vector

PURPOSE: Cloning a gene produces many copies of a gene of interest that can be used, for example, to determine the DNA sequence of the gene, to manipulate the gene in basic research experiments, to understand its function, and to produce the protein encoded by the gene.

PROTOCOL:

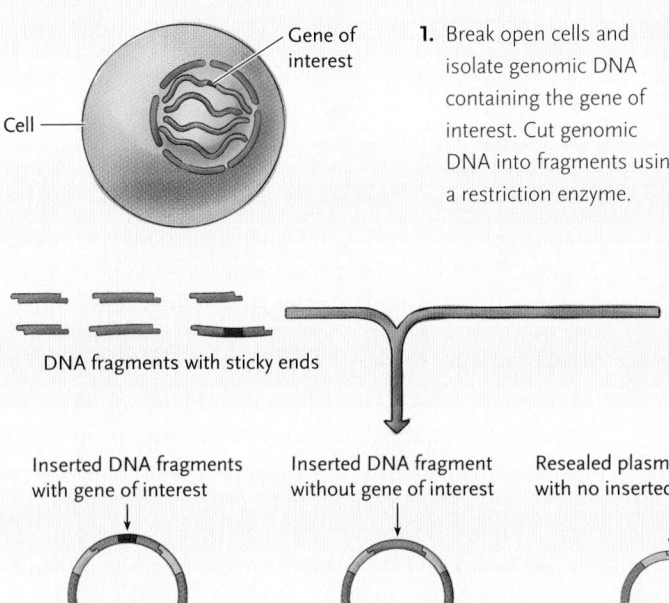

1. Break open cells and isolate genomic DNA containing the gene of interest. Cut genomic DNA into fragments using a restriction enzyme.

Gene of interest

Cell

DNA fragments with sticky ends

Restriction site

amp^R gene

lacZ^+ gene

Plasmid cloning vector

2. Cut a circular plasmid cloning vector with the same restriction enzyme to make it linear. The restriction site for the enzyme is within the *lacZ*^+ gene.

Cut plasmid cloning vectors with a restriction enzyme to produce sticky ends

3. Combine the cut genomic DNA fragments with the cut plasmid. DNA molecules will join by base pairing of their sticky ends, and DNA ligase is added to seal them toegether. The result is a mixture of recombinant and nonrecombinant plasmids.

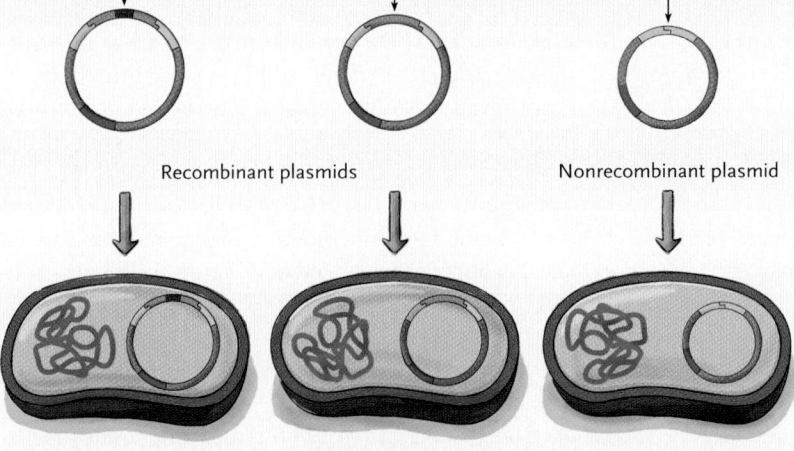

Inserted DNA fragments with gene of interest

Inserted DNA fragment without gene of interest

Resealed plasmid cloning vector with no inserted DNA fragment

Recombinant plasmids

Nonrecombinant plasmid

4. Transform the plasmids into *E. coli*. In this step, some bacteria will take up a plasmid while others will not.

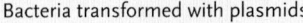

Bacteria transformed with plasmids

Bacteria not transformed with a plasmid

Selection:
Transformed bacteria grow on medium containing ampicillin because of *amp*^R gene on plasmid.

Untransformed bacterium cannot grow on medium containing ampicillin.

Screening:
Blue colony contains bacteria with a nonrecombinant plasmid; that is, the *lacZ*^+ gene is intact.

White colony contains bacteria with a recombinant plasmid, that is, the vector with an inserted DNA fragment. Once the white colony with the gene of interest is identified, it can be grown in culture to produce large quantities of the plasmid.

Plate containing ampicillin and X-gal

5. Spread the bacterial cells on a plate of growth medium containing ampicillin and X-gal, and incubate until colonies appear.

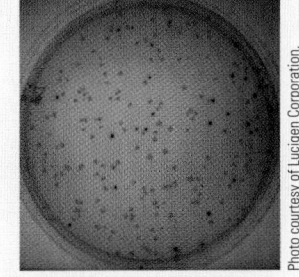

INTERPRETING THE RESULTS: Cloning and blue-white screening produce clearly identifiable white colonies containing recombinant plasmids. Most of the white colonies will contain plasmids that do not contain the gene of interest. Further screening will be done to identify the particular white colony that contains a plasmid with the gene of interest (Figure 18.5). Once identified, the colony can be cultured to produce large quantities of the plasmid for analysis or manipulation of the gene.

The X-gal in the medium distinguishes between bacteria that have been transformed with recombinant plasmids and nonrecombinant plasmids by *blue-white screening* (see Figure 18.4 results). If a colony produces β-galactosidase, it converts X-gal to a blue product and the colony turns blue, but if a colony does not produce the enzyme, X-gal is unchanged and the colony remains white. Colonies containing nonrecombinant plasmids have an intact *lacZ*+ gene, produce the enzyme, and turn blue. Colonies containing recombinant plasmids are white because those plasmids each contain a DNA fragment inserted into the *lacZ*+ gene, so they do not produce the enzyme. The white colonies are examined to find the one containing a recombinant plasmid with the gene of interest.

Two researchers, Paul Berg and Stanley Cohen, were prime movers in the development of DNA cloning techniques using restriction enzymes and bacterial plasmids. Berg and Cohen received a Nobel Prize in 1980 for their research, which pushed DNA technology to the forefront of biological investigations.

Identifying the Clone Containing the Gene of Interest.
How is a clone containing the gene of interest identified amongst the population of clones? The gene of interest has a unique DNA sequence, which is the basis for one commonly used identification technique. In this technique, called **DNA hybridization**, the gene of interest is identified in the set of clones when it base pairs with a short, single-stranded complementary DNA or RNA molecule called a *nucleic acid probe* (**Figure 18.5**). The probe is typically labeled with a radioactive or a nonradioactive tag, so investigators can detect it. In our example, if we know the sequence of part of the gene of interest, we can use that information to design and synthesize a nucleic acid probe. Or, we can take advantage of DNA sequence similarities of evolutionarily related organisms. For instance, we could make a probe for a human gene based on the sequence of an equivalent mouse gene. Once a colony containing plasmids with the gene of interest has been identified, that colony can be used to produce large quantities of the cloned gene.

DNA Libraries Contain Collections of Cloned DNA Fragments

As you have seen, the starting point for cloning a gene of interest is a large set of plasmid clones carrying fragments representing all of the DNA of an organism's genome. A collection of clones that contains a copy of every DNA sequence in a genome is called a **genomic library**. A genomic library can be made using plasmid cloning vectors or any other kind of cloning vector. The number of clones in a genomic library increases with the size of the genome. For example, a yeast genomic library of plasmid clones consists of hundreds of plasmids, whereas a human genomic library of plasmid clones consists of thousands of plasmids.

A genomic library is a resource containing all of the DNA of an organism cut into pieces. Just as for a book library, where you can search through the same set of books on various occasions to find different passages of interest, you can search through the same genomic library on various occasions to find and isolate different genes or other DNA sequences.

Researchers also commonly use another kind of DNA library that is made starting with mRNA molecules isolated from a cell. To convert single-stranded mRNA to double-stranded DNA for cloning (RNA cannot be cloned) first they use the enzyme *reverse transcriptase* (made by retroviruses) to make a single-stranded DNA that is complementary to the mRNA. Then they degrade the mRNA strand with an enzyme, and use DNA polymerase to make a second DNA strand that is complementary to the first. The result is a **complementary DNA (cDNA)**. After adding restriction sites to each end, they insert the cDNA into a cloning vector as described for the genomic library. The entire collection of cloned cDNAs made from the mRNAs isolated from a cell is a **cDNA library.**

Not all genes are active in every cell. Therefore, a cDNA library is limited in that it includes copies of only the genes that were active in the cells used as the starting point for creation of the library. This limitation can be an advantage, however, in identifying genes active in one cell type and not another. cDNA libraries are useful, therefore, for providing clues to the changes in gene activity that are responsible for cell differentiation and specialization. An ingenious method for comparing the cDNAs libraries produced by different cell types—the DNA chip—is described later in this chapter.

cDNA libraries provide a critical advantage to genetic engineers who wish to insert eukaryotic genes into bacteria, particularly when the bacteria are to be used as "factories" for making the protein encoded in the gene. The genes in eukaryotic nuclear DNA typically contain many *introns*, spacer sequences that interrupt the amino acid-coding sequence of a gene (see Section 15.3). Because bacterial DNA does not contain introns, bacteria are not equipped to process eukaryotic genes correctly. However, the cDNA copy of a eukaryotic mRNA already has the introns removed, so bacteria can transcribe and translate it accurately to make eukaryotic proteins.

Genomic and cDNA libraries both depend on cloning in a living cell to produce multiple copies of the DNA of interest. Next we look at a highly automated method of making copies of a targeted piece of DNA in a genome.

The Polymerase Chain Reaction (PCR) Amplifies DNA in Vitro

Producing multiple DNA copies by cloning requires a series of techniques and considerable time. A much more rapid process, **polymerase chain reaction (PCR)**,

Figure 18.5 Research Method

DNA Hybridization to Identify a DNA Sequence of Interest

PROTOCOL:

1. Prepare master plates of white colonies detected in the blue-white screening step of Figure 18.4. These colonies contain bacteria with recombinant plasmids. Hundreds or thousands of colonies can be screened for the gene of interest by using many master plates.

2. Lay a special filter paper on the plate to pick up some cells from each colony. This produces a replica of the colony pattern on the filter.

3. Treat the filter to break open the cells and to denature the released DNA to single strands. The single-stranded DNA sticks to the filter in the same position as the colony from which it was derived.

4. Add a labeled single-stranded probe (DNA or RNA) for the gene of interest and incubate. The label can be radioactive or nonradioactive. If a recombinant plasmid's inserted DNA fragment is complementary to the probe, the two will hybridize, that is, form base pairs. Wash off excess labeled probe.

5. Detect the hybridization event by looking for the labeled tag on the probe. If the probe was radioactively labeled, place the filter against photographic film. The decaying radioactive compound exposes the film, giving a dark spot when the film is developed. Correlate the position of any dark spot on the film to the original colony pattern on the master plate. Isolate the colony and use it to produce large quantities of the gene of interest.

PURPOSE: Hybridization with a specific DNA probe allows researchers to detect a specific DNA sequence, such as a gene, within a population of DNA molecules. Here, DNA hybridization is used to screen a collection of bacterial colonies to identify those containing a recombinant plasmid with a gene of interest.

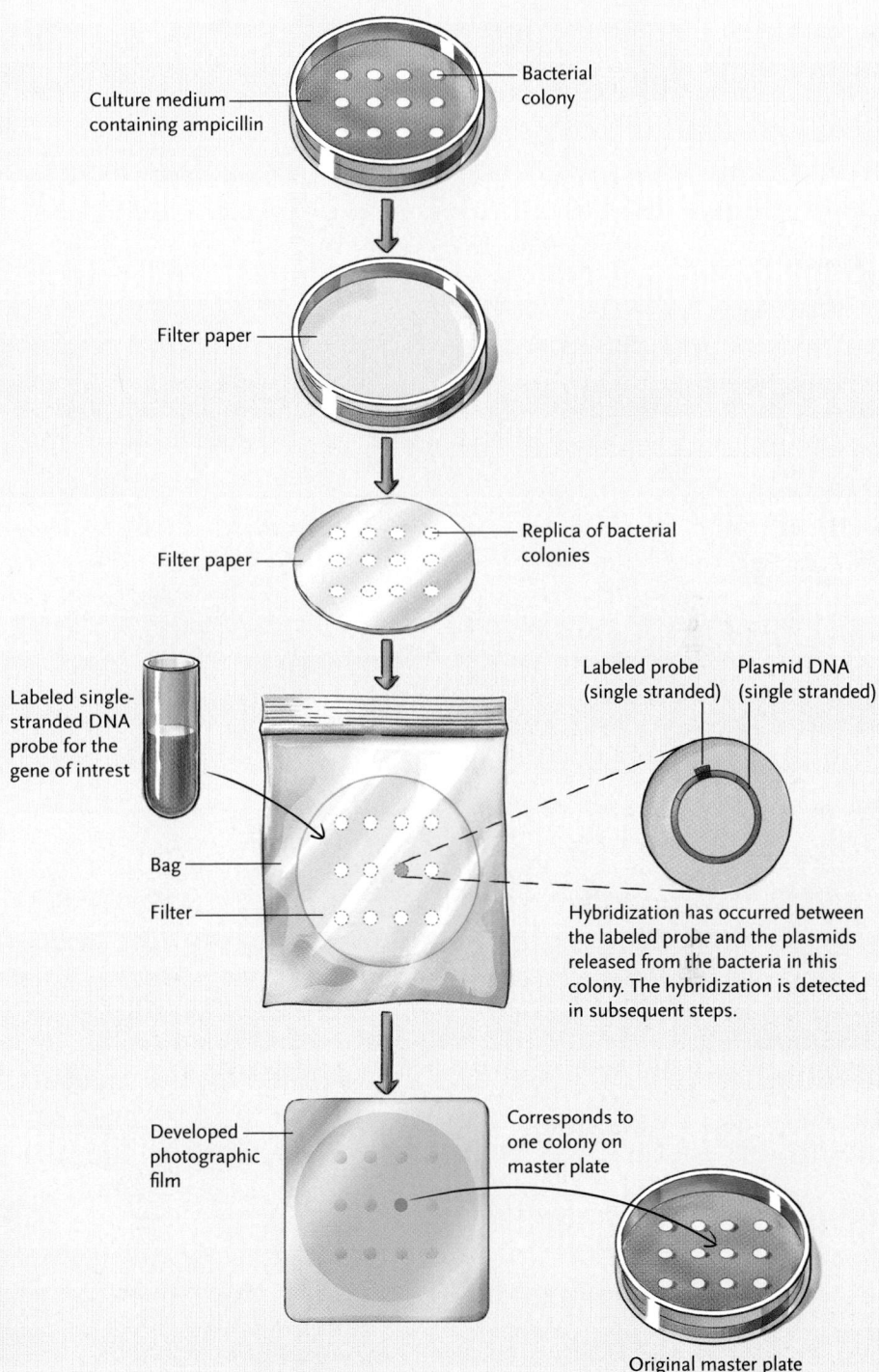

INTERPRETING THE RESULTS: DNA hybridization with a labeled probe enables a researcher to home in on a sequence of interest. If the probe is for a particular gene, it allows the specific identification of a colony containing bacteria with recombinant plasmids carrying that gene. The specificity of the method depends directly on the probe used. The same collection of bacterial clones can be used again and again to search for recombinant plasmids carrying different genes or different plasmids of interest simply by changing the probe used in the experiment.

Figure 18.6 Research Method

The Polymerase Chain Reaction (PCR)

PURPOSE: To amplify—produce large numbers of copies of—a target DNA sequence in the test tube without cloning.

PROTOCOL: A polymerase chain reaction mixture has four key elements: **(1)** the DNA with the target sequence to be amplified; **(2)** a pair of DNA primers, one complementary to one end of the target sequence and the other complementary to the other end of the target sequence; **(3)** the four nucleoside triphosphate precursors for DNA synthesis (dATP, dTTP, dGTP, and dCTP); and **(4)** DNA polymerase. Since PCR uses high temperatures for some of the steps, a heat-stable DNA polymerase is used, typically one isolated from a microorganism that grows in a high-temperature area such as a thermal pool or near a deep-sea vent.

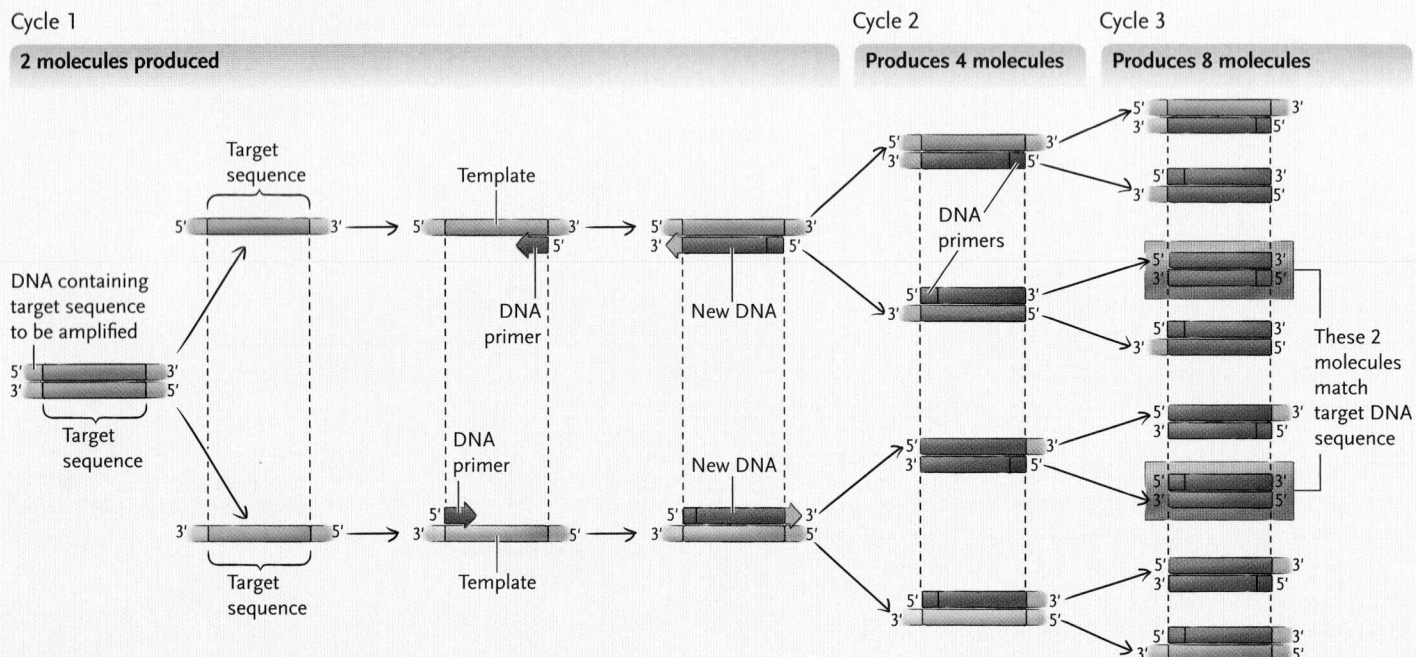

1. Denaturation: Heat DNA containing target sequence to denature it to single strands.

2. Annealing: Cool the mixture to allow the two primers to anneal to their complementary sequences at the two ends of the target sequence.

3. Heat to the optimal temperature for DNA polymerase to extend the primers, using the four nucleoside triphosphate precursors to make complementary copies of the two template strands. This completes cycle 1 of PCR; the end result is two molecules.

4. Repeat the same steps of denaturation, annealing of primers, and extension in cycle 2, producing a total of four molecules.

5. Repeat the same steps in cycle 3, producing a total of eight molecules. Two of the eight match the exact length of the target DNA sequence (highlighted in yellow).

INTERPRETING THE RESULTS: After three cycles, PCR produces a pair of molecules matching the target sequence. Subsequent cycles amplify these molecules to the point where they outnumber all other molecules in the reaction by many orders of magnitude.

produces an extremely large number of copies of a specific DNA sequence from a DNA mixture without having to clone the sequence in a host organism. The process is called *amplification* because it increases the amount of DNA to the point where it can be analyzed or manipulated easily. Developed in 1983 by Kary B. Mullis and F. Faloona at Cetus Corporation (Emeryville, CA), PCR has become one of the most important tools in modern molecular biology, finding wide application in all areas of biology. Mullis received a Nobel Prize in 1993 for his role in the development of PCR.

How PCR is performed is shown in **Figure 18.6.** PCR essentially is DNA replication, but a special case in which a DNA polymerase replicates just a portion of a DNA molecule rather than the whole molecule. PCR takes advantage of a characteristic common to all DNA polymerases: these enzymes add nucleotides only to the end of an existing chain called the *primer* (see Section 14.3). For replication to take place, a primer therefore must be in place, base-paired to the template chain at which replication is to begin. By cycling 20 to 30 times through a series of steps, PCR amplifies the target sequence, producing millions of copies.

Since the primers used in PCR are designed to bracket only the sequence of interest, the cycles replicate only this sequence from a mixture of essentially any DNA molecules. Thus PCR not only finds the "needle in the haystack" among all the sequences in a mixture, but also makes millions of copies of the "needle"—the DNA sequence of interest. Usually no further purification of the amplified sequence is necessary.

The characteristics of PCR allow extremely small DNA samples to be amplified to concentrations high enough for analysis. PCR is used, for example, to produce enough DNA for analysis from the root of a single human hair, or from a small amount of blood, semen, or saliva, such as the traces left at the scene of a crime. It is also used to extract and multiply DNA sequences from skeletal remains; ancient sources such as mammoths, Neanderthals, and Egyptian mummies; and, in rare cases, from amber-entombed fossils, fossil bones, and fossil plant remains.

A successful outcome of PCR is shown by analyzing a sample of the amplified DNA using **agarose gel electrophoresis** to see if the copies are the same length as the target **(Figure 18.7).** Gel electrophoresis is a technique by which DNA, RNA, or protein molecules are separated in a gel subjected to an electric field. The type of gel and the conditions used vary with the experiment, but in each case the gel functions as a molecular sieve to separate the macromolecules based on size, electrical charge, or other properties. For separating large DNA molecules, such as those typically produced by PCR, a gel made of agarose, a natural molecule isolated from seaweed, is used because of its large pore size.

For PCR experiments, the size of the amplified DNA is determined by comparing the position of the DNA band with the positions of DNA fragments of known size separated on the gel at the same time. If that size matches the predicted size for the target DNA, PCR is deemed successful. In some cases, such as DNA from ancient sources, a size prediction may not be possible; in this case, agarose gel electrophoresis analysis simply indicates whether there was DNA in the sample that could be amplified.

The advantages of PCR have made it the technique of choice for researchers, law enforcement agencies, and forensic specialists whose primary interest is in the amplification of specific DNA fragments up to a practical maximum of a few thousand base pairs. Cloning remains the technique of choice for amplification of longer fragments. The major limitation of PCR relates to the primers. In order to design a primer for PCR, the researcher must first have sequence information about the target DNA. By contrast, cloning can be used to amplify DNA of unknown sequence.

STUDY BREAK

1. What features do restriction enzymes have in common? How do they differ?
2. Plasmid cloning vectors are one type of cloning vector that can be used with *E. coli* as a host organism. What features of a plasmid cloning vector make it useful for constructing and cloning recombinant DNA molecules?
3. What is a cDNA library, and from what cellular material is it derived? How does a cDNA library differ from a genomic library?
4. What information and materials are needed to amplify a region of DNA using PCR?

18.2 Applications of DNA Technologies

The ability to clone pieces of DNA—genes, especially—and to amplify specific segments of DNA by PCR revolutionized biology. These and other DNA technologies are now used for research in all areas of biology, including cloning genes to determine their structure, function, and regulation of expression; manipulating genes to determine how their products function in cellular or developmental processes; and identifying differences in DNA sequences among individuals in ecological studies. The same DNA technologies also have practical applications, including medical and forensic detection, modification of animals and plants, and the manufacture of commercial products. In this section, case studies provide examples of how the techniques are used to answer questions and solve problems.

DNA Technologies Are Used in Molecular Testing for Many Human Genetic Diseases

Many human genetic diseases are caused by defects in enzymes or other proteins that result from mutations at the DNA level. Once scientists have identified the specific mutations responsible for human genetic diseases, they can often use DNA technologies to develop molecular tests for those diseases. One example is sickle-cell disease (see *Why It Matters* in Chapter 12,

Figure 18.7 Research Method

Separation of DNA Fragments by Agarose Gel Electrophoresis

PURPOSE: Gel electrophoresis separates DNA molecules, RNA molecules, or proteins according to their sizes, electrical charges, or other properties through a gel in an electric field. Different gel types and conditions are used for different molecules and types of applications. A common gel for separating large DNA fragments is made of agarose.

PROTOCOL:

1. Prepare a gel consisting of a thin slab of agarose and place it in a gel box in between two electrodes. The gel has wells for placing the DNA samples to be analyzed. Add buffer to cover the gel.

2. Load DNA sample solutions, such as PCR products, into wells of the gel, alongside a well loaded with marker DNA fragments of known sizes.

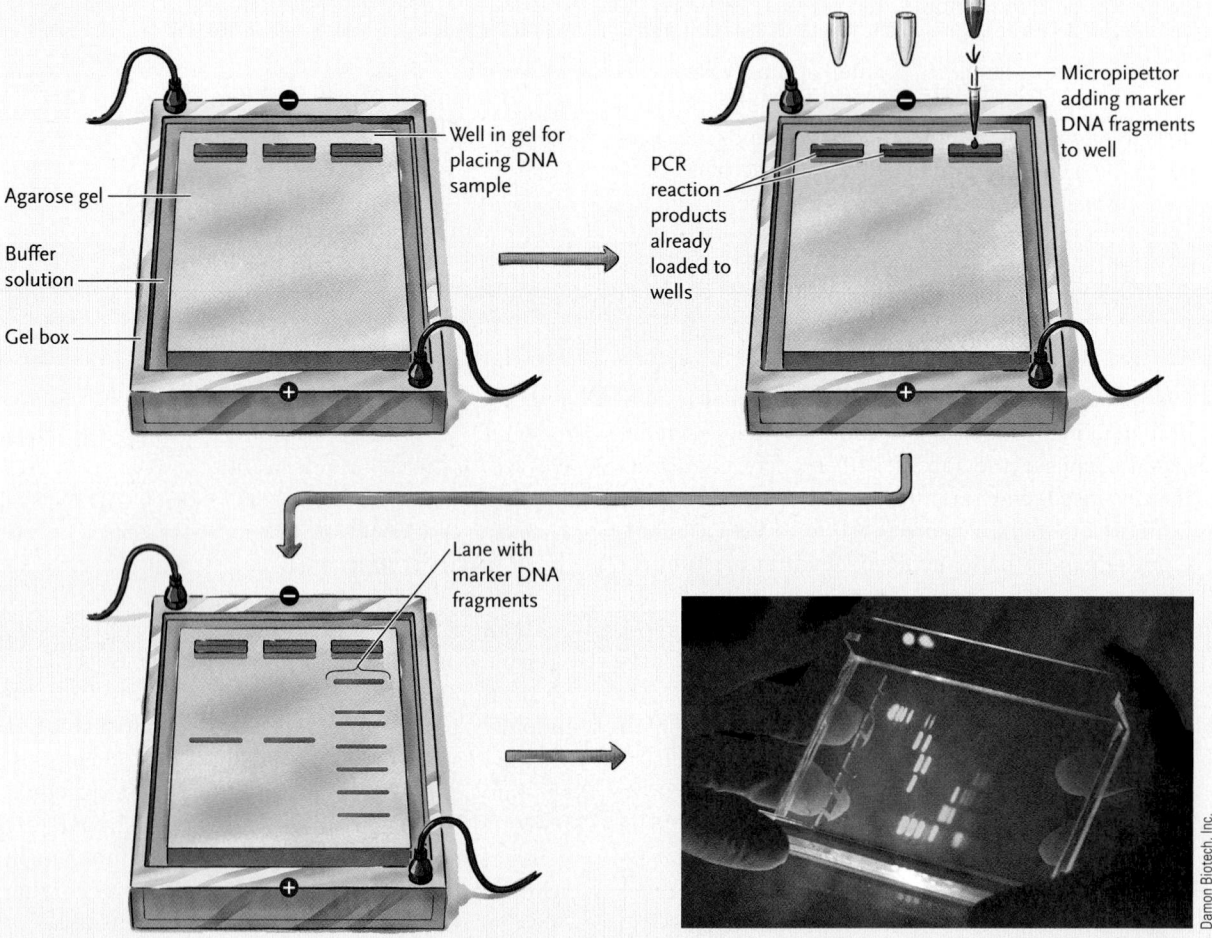

Well in gel for placing DNA sample

Agarose gel

Buffer solution

Gel box

Micropipettor adding marker DNA fragments to well

PCR reaction products already loaded to wells

Lane with marker DNA fragments

Damon Biotech, Inc.

3. Apply an electric current to the gel; DNA fragments are negatively charged, so they migrate to the positive pole. Shorter DNA fragments migrate faster than longer DNA fragments. At the completion of the separation, DNA fragments of the same length have formed bands in the gel. At this point, the bands are invisible.

4. Stain the gel with a dye that binds to DNA. The dye fluoresces under UV light, enabling the DNA bands to be seen and photographed. An actual gel showing separated DNA bands stained and visualized this way is shown.

INTERPRETING THE RESULTS: Agarose gel electrophoresis separates DNA fragments according to their length. The lengths of the DNA fragments being analyzed are determined by measuring their migration distances and comparing those distances to a calibration curve of the migration distances of the marker bands, which have known length. For PCR, agarose gel electrophoresis shows whether DNA of the correct length was amplified. For restriction enzyme digests, this technique shows whether fragments are produced as expected.

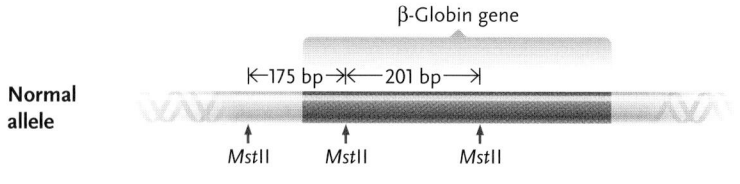

Normal allele

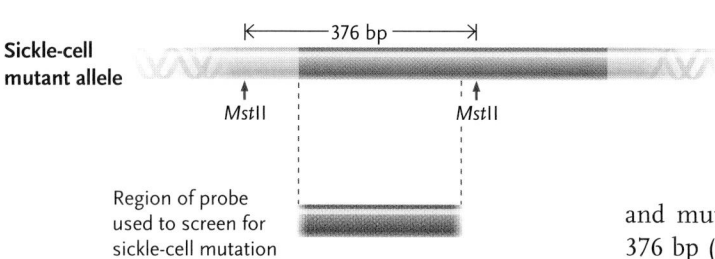

Sickle-cell mutant allele

Region of probe used to screen for sickle-cell mutation

Figure 18.8
Restriction site differences between the normal and sickle-cell mutant alleles of the β-globin gene. The figure shows a DNA segment that can be used as a probe to identify these alleles in subsequent analysis (see Figure 18.9).

Section 12.2, and Section 13.4). People with this disease are homozygous for a DNA mutation that affects hemoglobin, the oxygen carrying molecule of the blood. Hemoglobin consists of two copies each of the α-globin and β-globin polypeptides. The mutation, which is in the β-globin gene, alters one amino acid in the polypeptide. As a consequence, the function of hemoglobin is significantly impaired in individuals homozygous for the mutation (who have sickle-cell anemia), and mildly impaired in individuals heterozygous for the mutation (who have sickle-cell trait).

The sickle-cell mutation changes a restriction site in the DNA **(Figure 18.8)**. Three restriction sites for *Mst*II are associated with the normal β-globin gene, two within the coding sequence of the gene and one upstream of the gene. The sickle-cell mutation eliminates the middle site of the three. Cutting the β-globin gene with *Mst*II produces two DNA fragments from the normal gene and one fragment from the mutated gene (see Figure 18.8). Restriction-enzyme-generated DNA fragments of different lengths from the same region of the genome such as in this example are known as **restriction fragment length polymorphisms** (RFLPs, pronounced "riff-lips").

RFLPs typically are analyzed using **Southern blot analysis** (named after its inventor, researcher Edward Southern) **(Figure 18.9)**. In this technique, genomic DNA is digested with a restriction enzyme, and the DNA fragments are separated using agarose gel electrophoresis. The fragments are then transferred— blotted—to a filter paper, and a labeled probe is used to identify a DNA sequence of interest from among the many thousands of fragments on the filter paper.

Analyzing DNA for the sickle-cell mutation by *Mst*II digestion and Southern blot analysis is straightforward (see Figure 18.9). An individual with sickle-cell disease will have one DNA band of 376 bp detected by the probe (lane A), a healthy individual will have two DNA bands of 175 and 201 bp (lane B), and an individual with sickle-cell trait (heterozygous for normal

and mutant alleles) will have three DNA bands of 376 bp (mutant allele), and 201 and 175 bp (normal allele) (lane C). The same probe detects all three RFLP fragments by binding to all or part of the sequence.

Restriction enzyme digestion and Southern blot analysis may be used to test for a number of other human genetic diseases, including phenylketonuria and Duchenne muscular dystrophy. In some cases, restriction enzyme digestion is combined with PCR for a quicker, easier analysis. The gene or region of the gene with the restriction enzyme variation is first amplified using PCR, and the amplified DNA is then cut with the diagnostic restriction enzyme. Amplification produces enough DNA so that separation by size on an agarose gel produces clearly visible bands, positioned according to fragment length. Researchers can then determine whether the fragment lengths match a normal or abnormal RFLP pattern. This method eliminates the need for a probe or for Southern blotting.

DNA Fingerprinting Is Used to Identify Human Individuals as well as Individuals of Other Species

Just as each human has a unique set of fingerprints, each also has unique combinations and variations of DNA sequences (with the exception of identical twins) known as *DNA fingerprints*. **DNA fingerprinting** is a technique used to distinguish between individuals of the same species using DNA samples. Invented by Sir Alec Jeffreys in 1985, DNA fingerprinting has become a mainstream technique for distinguishing human individuals, notably in forensics and paternity testing. And, as *Why It Matters* indicated, the technique is applicable to all kinds of organisms. We focus on humans in the following discussion.

DNA Fingerprinting Principles. In DNA fingerprinting, scientists use molecular techniques, most typically PCR, to analyze DNA variations at various loci in the genome. In the United States, 13 loci in noncoding regions of the genome are the standards for PCR analysis. Each locus is an example of a *short tandem repeat* (STR) sequence, meaning that it has a short sequence of DNA repeated

Figure 18.9 Research Method

Southern Blot Analysis

PURPOSE: The Southern blot technique allows researchers to identify DNA fragments of interest after separating DNA fragments on a gel. One application is to compare different samples of genomic DNA cut with a restriction enzyme to detect specific restriction fragment length polymorphisms. Here the technique is used to distinguish between individuals with sickle-cell disease, individuals with sickle-cell trait, and normal individuals.

PROTOCOL:

1. Isolate genomic DNA and digest with a restriction enzyme. Here, genomic DNA is isolated from three individuals: A, sickle-cell disease (homozygous for the sickle-cell mutant allele); B, normal (homozygous for the normal allele); and C, sickle-cell trait (heterozygote for sickle-cell mutant allele). Digest the DNA with *Mst*II.

2. Separate the DNA fragments by agarose gel electrophoresis. The thousands of differently sized DNA fragments generated results in a smear of DNA down the length of each lane in the gel, which can be seen after staining the DNA. (Gel electrophoresis and gel staining are shown in Figure 18.7).

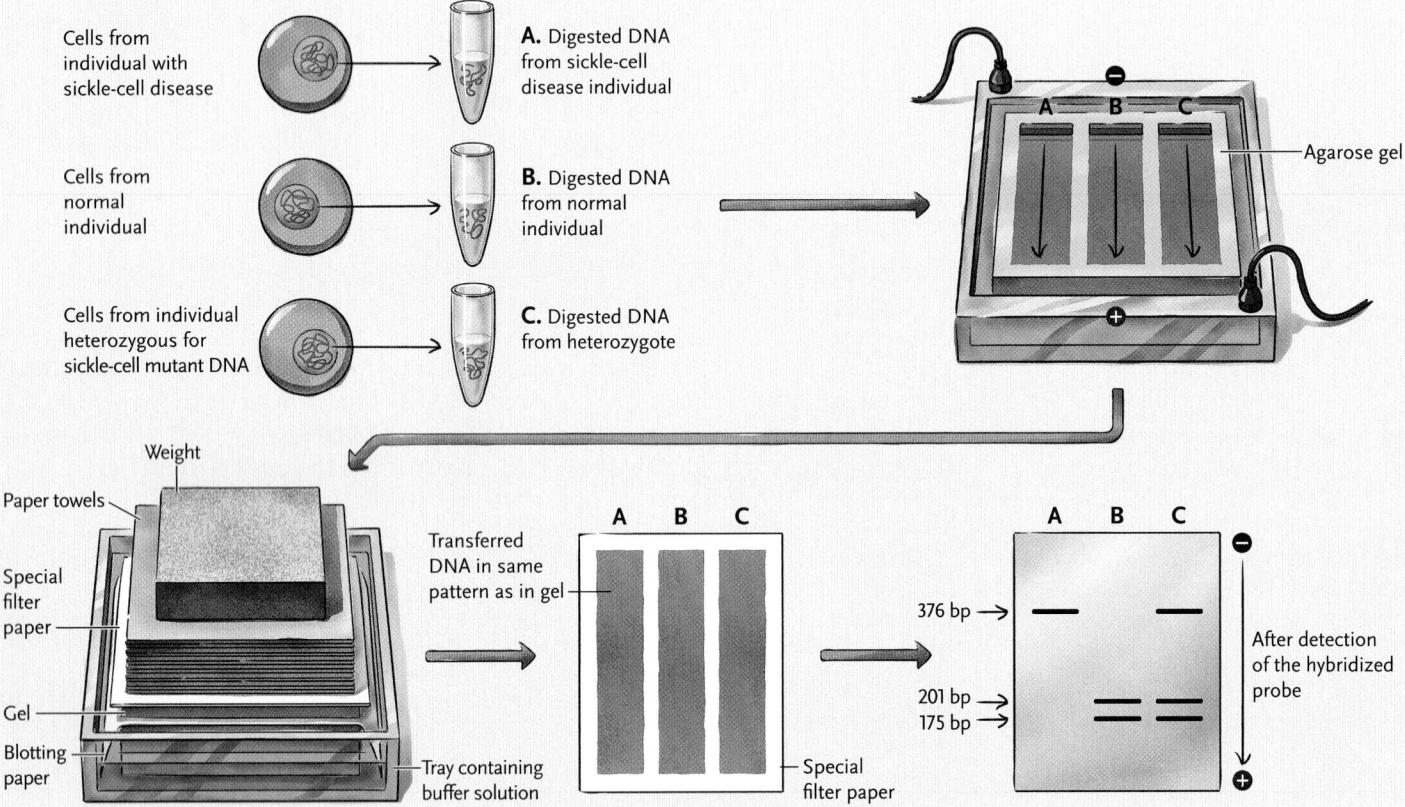

3. Hybridization with a labeled DNA probe to identify DNA fragments of interest cannot be done directly with an agarose gel. Edward Southern devised a method to transfer the DNA fragments from a gel to a special filter paper. First, treat the gel with a solution to denature the DNA to single strands. Next, place the gel on a piece of blotting paper with ends of the paper in the buffer solution and place the special filter paper on top of the gel. Capillary action wicks the buffer solution in the tray up the blotting paper, through the gel and special filter paper, and into the weighted stack of paper towels on top of the gel. The movement of the solution transfers—blots—the single-stranded DNA fragments to the filter paper, where they stick. The pattern of DNA fragments is the same as it was in the gel.

4. To home in on a particular region of the genome, use DNA hybridization with a labeled probe. That is, incubate a labeled, single-stranded probe with the filter and, after washing off excess probe, detect hybridization of the probe with DNA fragments on the filter. For a radioactive probe, the filter is placed against photographic film, which, after development will show a band or bands where the probe hybridized. In this experiment, the probe is a cloned piece of DNA from the area shown in Figure 18.8 that can bind to all three of the *Mst*II fragments of interest.

INTERPRETING THE RESULTS: The hybridization result indicates that the probe has identified a very specific DNA fragment or fragments in the digested genomic DNA. The RFLPs for the β-globin gene can be seen in Figure 18.8. DNA from the sickle-cell disease individual cut with *Mst*II results in a single band of 376 bp detected by the probe, while DNA from the normal individual results in two bands of 201, and 175 bp. DNA from a sickle-cell trait heterozygote results in three bands of 376 bp (from the sickle-cell mutant allele), and 201 and 175 bp (the latter two from the normal allele). This type of analysis in general is useful for distinguishing normal and mutant alleles of genes where the mutation involved alters a restriction site.

a. Alleles at an STR locus

b. DNA fingerprint analysis of the STR locus by PCR

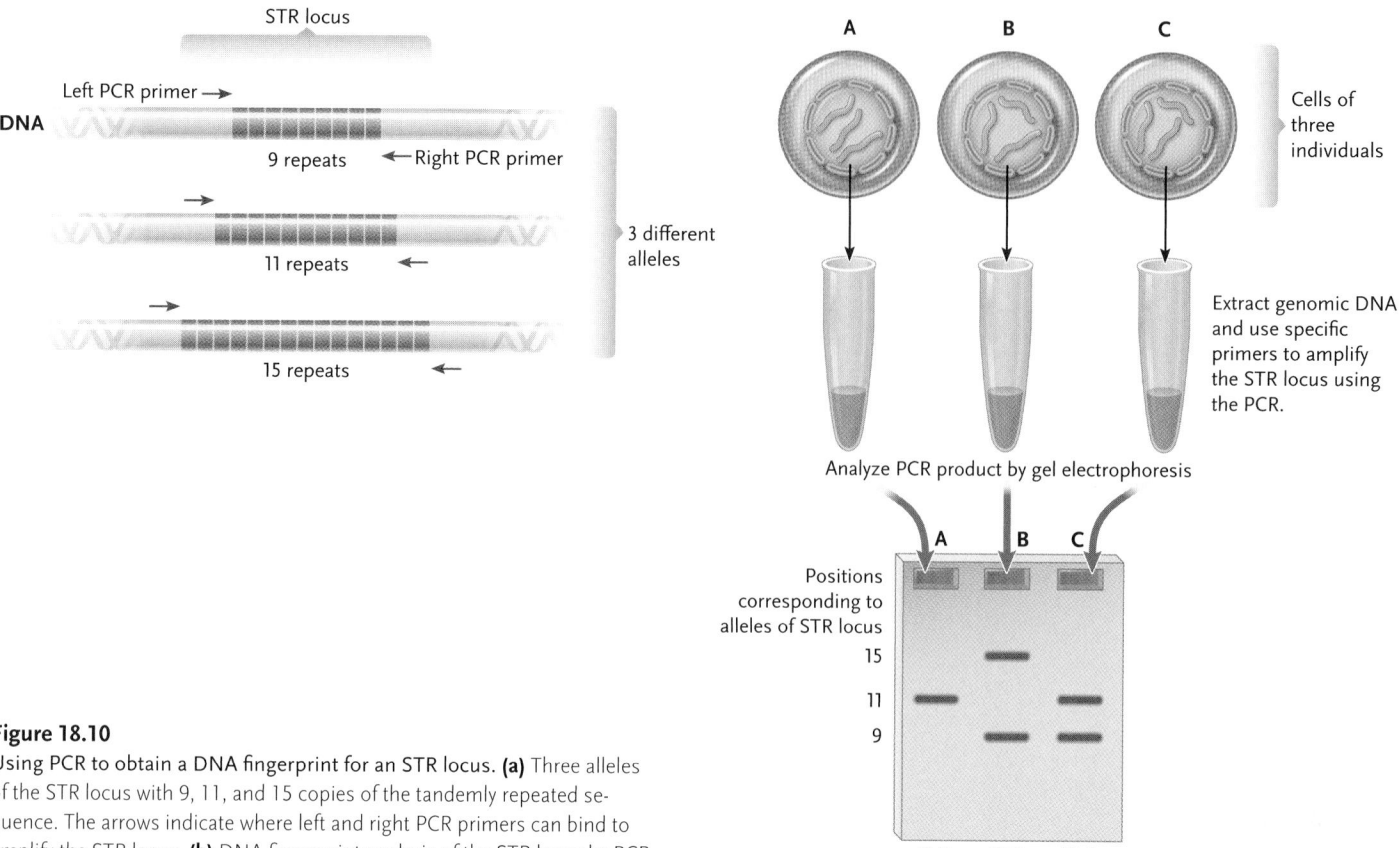

Figure 18.10

Using PCR to obtain a DNA fingerprint for an STR locus. **(a)** Three alleles of the STR locus with 9, 11, and 15 copies of the tandemly repeated sequence. The arrows indicate where left and right PCR primers can bind to amplify the STR locus. **(b)** DNA fingerprint analysis of the STR locus by PCR.

in series, with each repeat about 3 to 5 bp. Each locus has a different repeated sequence, and the number of repeats varies among individuals in a population. For example, one STR locus has the sequence AGAT repeated between 8 and 20 times. As a further source of variation, a given individual is either homozygous or heterozygous for an STR allele; perhaps you are homozygous for the eleven-repeat allele, or heterozygous for a nine-repeat allele and a fifteen-repeat allele. Likely your DNA fingerprint for this locus is different from most of the others in your class. Because each individual has an essentially unique combination of alleles (identical twins are the exception), analysis of multiple STR loci can discriminate between DNA of different individuals.

Figure 18.10 illustrates how PCR is used to obtain a DNA fingerprint for a theoretical STR locus with three alleles of 9, 11, and 15 tandem repeats (Figure 18.10a). Using primers that flank the STR locus, the locus is amplified from genomic DNA using PCR, and the PCR products are analyzed by gel electrophoresis (Figure 18.10b). The number of bands on the gel and the sizes of the DNA in the bands show the STR alleles that were amplified. One band indicates that the individual was homozygous for an STR allele with a particular number of repeats, while two bands indicates the individual is heterozygous for two STR alleles with different numbers of repeats. In the result shown in Figure 18.10b, the A individual is homozygous for a 11-repeat allele

(designated 11,11), B is heterozygous for a 15-repeat allele and a 9-repeat allele (15,9), and C is heterozygous for the 11-repeat allele and the 9-repeat allele (11,9).

DNA Fingerprinting in Forensics. DNA fingerprints are routinely used to identify criminals or eliminate innocent persons as suspects in legal proceedings. For example, a DNA fingerprint prepared from a hair found at the scene of a crime, or from a semen sample, might be compared with the DNA fingerprint of a suspect to link the suspect with the crime. Or, a DNA fingerprint of blood found on a suspect's clothing or possessions might be compared with the DNA fingerprint of a victim. Typically, the evidence is presented in terms of the probability that the particular DNA sample could have come from a random individual. Hence the media report probability values, such as one in several million, or in several billion, that a person other than the accused could have left his or her DNA at the crime scene.

Although courts initially met with legal challenges to the admissibility of DNA fingerprints, experience has shown that they are highly dependable as a line of evidence if DNA samples are collected and prepared with care and if a sufficient number of polymorphic loci are examined. There is always concern, though, about the possibility of contamination of the sample with DNA from another source during the path from crime scene to forensic lab analysis. Moreover, in some cases criminals themselves have planted fake DNA samples at crime scenes to confuse the investigation.

There are many examples of the use of DNA fingerprinting to identify a criminal. For example, in a case in England, the DNA fingerprints of more than 4000 men were made during an investigation of the rape and murder of two teenage girls. The results led to the release of a man wrongly imprisoned for the crimes and to the confession and conviction of the actual killer. And the application of DNA fingerprinting techniques to stored forensic samples has led to the release of a number of persons wrongly convicted for rape or murder.

DNA Fingerprinting in Testing Paternity and Establishing Ancestry. DNA fingerprints are also widely used as evidence of paternity because parents and their children share common alleles in their DNA fingerprints. That is, each child receives one allele of each locus from one parent and the other allele from the other parent. A comparison of DNA fingerprints for a number of loci can prove almost infallibly whether a child has been fathered or mothered by a given person. DNA fingerprints have also been used for other investigations, such as confirming that remains discovered in a remote region of Russia were actually those of Czar Nicholas II and members of his family, murdered in 1918 during the Russian revolution.

DNA fingerprinting is also widely used in studies of other organisms, including other animals, plants, and bacteria. Examples include testing for pathogenic *E. coli* in food sources such as hamburger meat, investigating cases of wildlife poaching, detecting genetically modified organisms among living organisms or in food, and comparing the DNA of ancient organisms with present-day descendants.

Genetic Engineering Uses DNA Technologies to Alter the Genes of a Cell or Organism

We have seen the many ways scientists use DNA technologies to ask, and answer, questions that were once completely inaccessible. Genetic engineering goes beyond gathering information; it is the use of DNA technologies to modify genes of a cell or organism. The goals of genetic engineering include using prokaryotes, fungi, animals, and plants as factories for the production of proteins needed in medicine and scientific research; correcting hereditary disorders; and improving animals and crop plants of agricultural importance. In many of these areas genetic engineering has already been spectacularly successful. The successes and potential benefits of genetic engineering, however, are tempered by ethical and social concerns about its use, along with the fear that the methods may produce toxic or damaging foods, or release dangerous and uncontrollable organisms to the environment.

Genetic engineering uses DNA technologies of the kind discussed already in this chapter. DNA—perhaps a modified gene—is introduced into target cells of an organism. Organisms that have undergone a gene transfer are called **transgenic**, meaning that they have been modified to contain genetic information—the *transgene*—from an external source.

The following sections discuss examples of applications of genetic engineering to bacteria, animals, and plants, and assess major controversies arising from these projects.

Genetic Engineering of Bacteria to Produce Proteins. Transgenic bacteria have been made, for example, to make proteins for medical applications, break down toxic wastes such as oil spills, produce industrial chemicals such as alcohols, and process minerals. *E. coli* has been the organism of choice for many of these applications of DNA technologies.

Using *E. coli* to make a protein from a foreign source is conceptually straightforward. First, the gene for the protein is cloned from the appropriate organism. Then the gene is inserted in a special type of bacterial plasmid called an *expression vector,* which has a bacterial promoter adjacent to the restriction site used for inserting the gene. The resulting recombinant plasmid is transformed into *E. coli,* which transcribes the gene and translates the resulting mRNA to make the desired protein. The protein is either extracted from the bacterial cells and purified or, if the protein is secreted, it is purified from the culture medium.

For example, *E. coli* bacteria have been genetically engineered to make the human hormone insulin; the commercial product is called humulin. Insulin is required by persons with some forms of diabetes. Humulin is a perfect copy of the human insulin hormone. Many other proteins, including human growth hormone to treat human growth disorders, tissue plasminogen activator to dissolve blood clots that cause heart attacks, and a vaccine against hoof-and-mouth disease of cattle (a highly contagious and sometimes fatal viral disease), have been developed for commercial production in bacteria by similar methods.

Although they offer many benefits, genetically engineered bacteria pose the risk that they may be released accidentally into the environment where any adverse effects are currently unknown. Scientists minimize the danger of accidental release by growing the bacteria in laboratories that follow appropriate biosafety protocols. In addition, the bacterial strains used typically are genetically modified so that they will not survive outside of the growth media used in the laboratory.

Genetic Engineering of Animals. Many animals, including fruit flies, fish, mice, pigs, sheep, goats, and cows, have been altered successfully by genetic engineering. There are many purposes for these alterations, including basic research, correcting genetic disorders in humans and other mammals, and producing pharmaceutically important proteins.

Genetic Engineering Methods for Animals. Several methods are used to introduce a gene of interest into animal cells. The gene may be introduced into *germ-line cells,* which develop into sperm or eggs and thus enable the introduced gene to be passed from generation to generation. Or, the gene may be introduced into *somatic* (body) *cells,* differentiated cells that are not part of lines producing sperm or eggs, in which case the gene is not transmitted from generation to generation.

Germ-line cells of embryos are often used as targets for introducing genes, particularly in mammals **(Figure 18.11).** The treated cells are then cultured in quantity and reintroduced into early embryos. If the technique is successful, some of the introduced cells become founders of cell lines that develop into eggs or sperm with the desired genetic information integrated into their DNA. Individuals produced by crosses using the engineered eggs and sperm then contain the introduced sequences in all their cells. Several genes have been introduced into the germ lines of mice by this approach, resulting in permanent, heritable changes in the engineered individuals.

A related technique involves introducing desired genes into *stem cells,* which are capable of differentiating into almost any adult cell type and tissue. Stem cells that have taken up the gene are then injected into an early embryo, where they differentiate into a variety of tissues along with cells of the embryo itself, including sperm and egg cells. Males and females are then bred, leading to offspring that are either homozygotes, containing two copies of the introduced gene, or heterozygotes, containing one introduced gene and one gene that was native to the embryo receiving the engineered stem cells.

Introduction of genes into stem cells has been performed mostly in mice. One of the highly useful results is the production of a "knockout mouse," a homozygous recessive that receives two copies of a gene altered to a nonfunctional state, and thus has no functional copies. The effect of the missing gene on the knockout mouse is a clue to the normal function of the gene. In some cases, knockout mice are used to model human genetic diseases.

For introducing genes into somatic cells, typically somatic cells are removed from the body, cultured, and then transformed with DNA containing the transgene. The modified cells are then reintroduced into the body where the transgene functions. Because germ cells and their products are not involved, the transgene remains in the individual and is not passed to offspring.

Gene Therapy: Correcting Genetic Disorders. The path to **gene therapy**—correcting genetic disorders—in humans began with experiments using mice. In 1982, Richard Palmiter at the University of Washington,

Figure 18.11 Research Method

Introduction of Genes into Mouse Embryos Using Embryonic Germ-Line Cells

PURPOSE: To make a transgenic animal that can transmit the transgene to offspring. The embryonic germ-line cells that receive the transgene develop into the reproductive cells of the animal.

PROTOCOL:

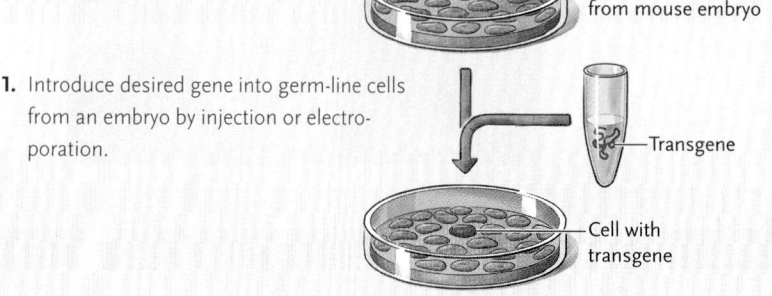

Germ-line cells derived from mouse embryo

1. Introduce desired gene into germ-line cells from an embryo by injection or electroporation.

Transgene

Cell with transgene

2. Clone cell that has the incorporated transgene to produce a pure culture of transgenic cells.

Pure population of transgenic cells

3. Inject transgenic cells into early-stage embryos (called a blastocyst).

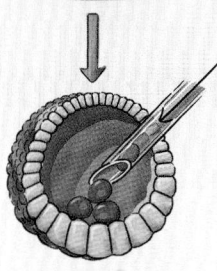

4. Implant embryos into surrogate (foster) mothers.

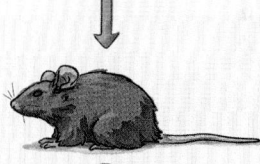

5. Allow embryos to grow to maturity and be born.

6. Interbreed the progeny mice.

Mice have transgenic cells in body regions including germ line

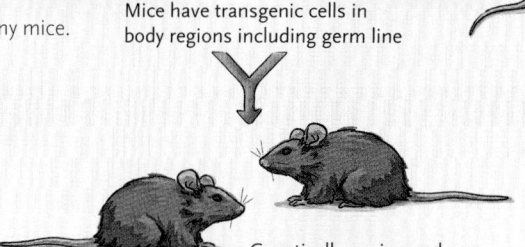

Genetically engineered offspring—all cells transgenic

INTERPRETING THE RESULTS: The result of the breeding is some offspring in which all cells are transgenic—a genetically engineered animal has been produced.

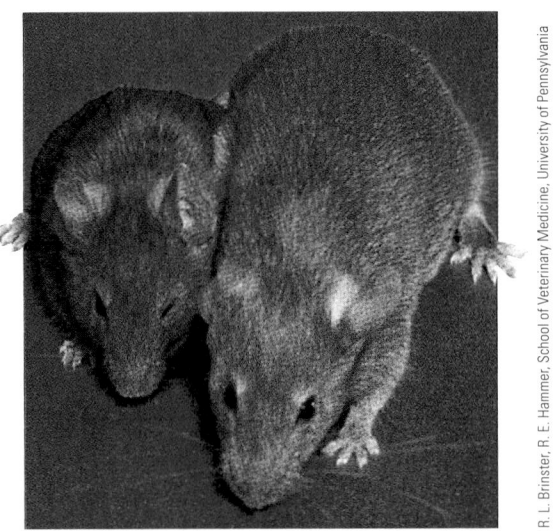

R. L. Brinster, R. E. Hammer, School of Veterinary Medicine, University of Pennsylvania

Figure 18.12

A genetically engineered giant mouse (right) produced by the introduction of a rat growth hormone gene into the animal. A mouse of normal size is on the left.

Ralph Brinster of the University of Pennsylvania, and their colleagues injected a growth hormone gene from rats into fertilized mouse eggs and implanted the eggs into a surrogate mother. She gave birth to some normal-sized mouse pups that grew faster than normal and became about twice the size of their normal litter mates. These *giant mice* (**Figure 18.12**) attracted extensive media attention from around the world.

Palmiter and Brinster next attempted to cure a genetic disorder by gene therapy. In this experiment, they were able to correct a genetic growth hormone deficiency that produces dwarf mice. They introduced a normal copy of the growth hormone gene into fertilized eggs taken from mutant dwarf mice and implanted them into a surrogate mother. The transgenic mouse pups grew to slightly larger than normal, demonstrating that the genetic defect in those mice had been corrected.

This sort of experiment, in which a gene is introduced into germ-line cells of an animal to correct a genetic disorder, is **germ-line gene therapy.** For ethical reasons, germ-line gene therapy is not permitted with humans. Instead, humans are treated with **somatic gene therapy,** in which genes are introduced into somatic cells (as described in the previous section).

The first successful use of somatic gene therapy with a human subject who had a genetic disorder was carried out in the 1990s by W. French Anderson and his colleagues at the National Institutes of Health. The subject was a young girl with *adenosine deaminase deficiency (ADA)*. Without the adenosine deaminase enzyme, white blood cells cannot mature (see Chapter 43); without normally functioning white blood cells, the body's immune response is so deficient that most children with ADA die of infections before reaching puberty. The researchers successfully introduced a functional ADA gene into mature white blood cells isolated from the patient. Those cells were reintroduced into the girl, and expression of the ADA gene provided a temporary cure for her ADA deficiency. The cure was not perma-

nent because mature white blood cells, produced by differentiation of stem cells in the bone marrow, are nondividing cells with a finite life time. Therefore, the somatic gene therapy procedure has to be repeated every few months. Indeed, the subject of this example still receives periodic gene therapy to maintain the necessary levels of the ADA enzyme in her blood. In addition, she receives direct doses of the normal enzyme.

Successful somatic gene therapy has also been achieved for sickle-cell disease. In December 1998, a 13-year-old boy's bone marrow cells were replaced with stem cells from the umbilical cord of an unrelated infant. The hope was that the stem cells would produce healthy bone marrow cells, the source of blood cells. The procedure worked, and the patient has been declared cured of the disease.

However, despite enormous efforts, human somatic gene therapy has not been the panacea people expected. Relatively little progress has been made since the first gene therapy clinical trial for ADA deficiency described, and, in fact, there have been major setbacks. In 1999, for example, a teenage patient in a somatic gene therapy trial died as a result of a severe immune response to the viral vector being used to introduce a normal gene to correct his genetic deficiency. Furthermore, some children in gene therapy trials involving the use of retrovirus vectors to introduce genes into blood stem cells have developed a leukemia-like condition. In short, somatic gene therapy is not yet an effective treatment for human genetic disease, even though the approach has been successful in a number of cases to correct models of human genetic disorders in experimental mammals. Although no commercial human gene therapy product has been approved for use, research and clinical trials continue as scientists try to circumvent the difficulties.

Turning Domestic Animals into Protein Factories. Another successful application of genetic engineering turns animals into pharmaceutical factories for the production of proteins required to treat human diseases or other medical conditions. Most of these *pharming* projects, as they are called, engineer the animals to produce the desired proteins in milk, making the production, extraction, and purification of the proteins harmless to the animals.

One of the first successful applications of this approach was carried out with sheep engineered to produce a protein required for normal blood clotting in humans. The protein, called a *clotting factor,* is deficient in persons with one form of hemophilia, who require frequent injections of the factor to avoid bleeding to death from even minor injuries. Using DNA cloning techniques, researchers joined the gene encoding the normal form of the clotting factor to the promoter sequences of the β-lactoglobin gene, which encodes a protein secreted in milk, and introduced it into fertilized eggs. Those cells were implanted into a surrogate mother, and the transgenic sheep born were allowed to mature. The β-lactoglobin promoter controlling the clotting factor gene became activated in mammary gland cells of females, resulting in the production of clotting factor. The clotting factor was then secreted into the milk. Production in the milk is harmless to the sheep and yields the protein in a form that can easily be obtained and purified.

Other similar projects are under development to produce particular proteins in transgenic mammals. These include a protein to treat cystic fibrosis, collagen to correct scars and wrinkles, human milk proteins to be added to infant formulas, and normal hemoglobin for use as an additive to blood transfusions.

Producing Animal Clones. Making transgenic mammals is expensive and inefficient. And, because only one copy of the transgene typically becomes incorporated into the treated cell, not all progeny of a transgenic animal inherit that gene. Scientists reasoned that an alternative to breeding a valuable transgenic mammal to produce progeny with the transgene would be to clone the mammal. Each clone would be identical to the original, including the expression of the transgene. That this is possible was shown in 1997 when two Scottish scientists, Ian Wilmut and Keith H. S. Campbell of the Roslin Institute, Edinburgh, announced that that they had successfully cloned a sheep from a single somatic cell derived from an adult sheep **(Figure 18.13)**—the first mammalian clone made. For their experiment, the researchers fused a diploid cell derived from the mammary gland of a 6-year-old adult sheep with an unfertilized egg cell from which the nucleus had been removed. Signals from the egg cytoplasm triggered DNA replication and cell division, producing a cluster of cells derived from the mammary gland cell. The cluster was implanted into the uterus of an adult female sheep, where it developed into an embryo that grew to full term and was delivered as an apparently normal lamb, named Dolly. Their cloning success rate, though, was very low—Dolly represents less than 0.4% of the transgenic cells they made. Dolly developed to sexual maturity and produced four normal offspring. She was euthanized at age 6 after contracting a fatal, virus-induced lung disease that her cloners believe was unrelated to the cloning.

After the successful cloning experiment producing Dolly, many additional mammals have been cloned, including mice, goats, pigs, monkeys, rabbits, dogs, a male calf appropriately named Gene, and a domestic cat called *CC* (for *Copy Cat*).

Cloning farm animals has been so successful that several commercial enterprises now provide cloned copies of champion animals. One example is a clone of an American Holstein cow, Zita, who was the U.S. national champion milk producer for many years. Animal breeders estimate that there are now more than 100 cloned animals on American farms, and breeders plan to produce entire herds if government approval is granted.

Figure 18.13
Dolly, the cloned sheep.

The cloning of domestic animals has its drawbacks. Many cloning attempts fail, leading to the death of the transplanted embryos. Cloned animals often suffer from health defects from conditions such as birth defects and poor lung development. Genes may be lost during the cloning process or may be expressed abnormally in the cloned animal. For example, molecular studies have shown that the expression of perhaps hundreds of genes in the genomes of clones is regulated abnormally.

Genetic Engineering of Plants. Genetic engineering of plants has led to increased resistance to pests and disease; greater tolerance to heat, drought, and salinity; greater crop yields; faster growth; and resistance to herbicides. Another aim is to produce seeds with higher levels of amino acids. The essential amino acid lysine, for example, is present only in limited quantities in cereal grains such as wheat, rice, oats, barley, and corn; the seeds of legumes such as beans, peas, lentils, soybeans, and peanuts are deficient in the essential amino acids methionine or cysteine. Increasing the amounts of the deficient amino acids in plant seeds by genetic engineering would greatly improve the diet of domestic animals and human populations that rely on seeds as a primary food source. Efforts are also under way to increase the content of vitamins and minerals in crop plants.

Other possibilities for plant genetic engineering include plant pharming to produce pharmaceutical products. Plants are ideal for this purpose, because they are primary producers at the bottom rung of the food chain and can be grown in huge numbers with maximum conservation of the sun's energy captured in photosynthesis.

Some plants, such as *Arabidopsis*, tobacco, potato, cabbage, and carrot, have special advantages for genetic engineering because individual cells can be removed from an adult, altered by the introduction of a desired gene, and then grown in cultures into a multicellular mass of cloned cells called a *callus*. Subsequently, roots,

Figure 18.14

A crown gall tumor on the trunk of a California pepper tree. The tumor, stimulated by genes introduced from the bacterium *Rhizobium radiobacter*, is the bulbous, irregular growth extending from the trunk.

Stephen Wolfe, *Molecular and Cellular Biology*

stems, and leaves develop in the callus, forming a young plant that can then be grown in containers or fields by the usual methods. In the plant, each cell contains the introduced gene. The gametes produced by the transgenic plants can then be used in crosses to produce offspring, some of which will have the transgene, as in the similar experiments with animals.

Methods Used to Insert Genes into Plants. Genes are inserted into plant cells by several techniques. A commonly used method takes advantage of a natural process that causes crown gall disease, which is characterized by bulbous, irregular growths—tumors, essentially—that can develop at wound sites on the trunks and limbs of deciduous trees **(Figure 18.14)**. Crown gall disease is caused by the bacterium *Rhizobium radiobacter* (formerly *Agrobacterium tumefaciens*, recently reclassified on the basis of genome analysis). This bacterium contains a large, circular plasmid called the **Ti (tumor inducing) plasmid.** The interaction between the bacterium and a plant cell it infects stimulates the excision of a segment of the Ti plasmid called *T DNA* (for transforming DNA), which then integrates into the plant cell's genome. Genes on the T DNA are then expressed; the products stimulate the transformed cell to grow and divide and therefore to produce a tumor. The tumors provide essential nutrients for the bacterium. The Ti plasmid is used as a vector for making transgenic plants, in much the same way as bacterial plasmids are used as vectors to introduce genes into bacteria **(Figure 18.15)**.

Successful Plant Genetic Engineering Projects. An early visual demonstration of the successful use of genetic engineering techniques to produce a transgenic plant is the glowing tobacco plant **(Figure 18.16)**. The transgenic plant contained luciferase, the gene for the firefly enzyme. When the plant was soaked in the substrate for the enzyme, it became luminescent.

The most widespread application of genetic engineering of plants involves the production of transgenic crops. Thousands of such crops have been developed and field tested, and many have been approved for commercial use. If you examine the processed plant-based foods at a national supermarket chain, you will likely find that at least two-thirds contain transgenic plants.

In many cases, plants are modified to make them resistant to insect pests, viruses, or herbicides. Crops modified for insect resistance include corn, cotton, and potatoes. The most common approach to making plants resistant to insects is to introduce the gene from the bacterium *Bacillus thuringiensis* that encodes the *Bt* toxin, an organic pesticide. This toxin has been used in powder form to kill insects in agriculture for many years, and now transgenic plants making their own *Bt* toxin are resistant to specific groups of insects that feed on them. Millions of acres of crop plants planted in the United States, amounting to about 70% of the nation's agricultural acreage, are now *Bt*-engineered varieties.

Virus infections cause enormous crop losses worldwide. Transgenic crops that are virus-resistant would be highly valuable to the agricultural community. There is some promise in this area. By some unknown process, transgenic plants expressing certain viral proteins become resistant to infections by whole viruses that contain those same proteins. Two virus-resistant genetically modified crops made so far are papaya and squash.

Several crops have also been engineered to become resistant to herbicides. For example, *glyphosate* (commonly known by its brand name, Roundup) is a highly potent herbicide that is widely used in weed control. The herbicide works by inhibiting a particular enzyme in the chloroplast. Unfortunately, it also kills crops. But transgenic crops have been made in which a bacterial form of the chloroplast enzyme has been added to the plants. The bacterial-derived enzyme is not affected by Roundup, and farmers that use these herbicide-resistant crops can spray fields of crops to kill weeds without killing the crops. Now most of the corn, soybean, and cotton plants grown in the United States and many other countries are the genetically engineered, glyphosate-resistant ("Roundup-ready") varieties.

Crop plants are also being engineered to alter their nutritional qualities. For example, a strain of rice plants has been produced with seeds rich in β-carotene, a precursor of vitamin A, as well as iron **(Figure 18.17)**. The new rice, which is given a yellow or golden color by the carotene, may provide improved nutrition for the billions of people that depend on rice as a diet staple. In particular, the rice may help improve the nutrition of children younger than age 5 in southeast Asia, 70% of whom suffer from impaired vision because of vitamin A deficiency. *Insights from the Molecular Revolution* describes an experiment in which rice plants were genetically engineered to develop resistance to a damaging bacterial blight.

Plant pharming is also an active area both in university research labs and at biotechnology companies. Plant pharming involves the engineering of transgenic plants to produce medically valuable products. The approach is one described earlier: the gene for the product is cloned into a cloning vector adjacent to a promoter, in this case one active in plants, and the recombinant DNA molecule is introduced into plants. Products under development include vaccines for various bacterial and viral diseases, protease inhibitors to treat or prevent virus infections, collagen to treat scars and wrinkles, and aprotinin to reduce bleeding and clotting during heart surgery.

In contrast to animal genetic engineering, genetically altered plants have been widely developed and appear to be here to stay as mainstays of agriculture. But, as the next section discusses, both animal and plant genetic engineering have not proceeded without concerns.

DNA Technologies and Genetic Engineering Are a Subject of Public Concern

When recombinant DNA technology was developed in the early 1970s, researchers quickly recognized that in addition to the anticipated many benefits, there might be deleterious outcomes. One key concern at the time was that a bacterium carrying a recombinant DNA molecule might escape into the environment. Perhaps it could transfer that molecule to other bacteria and produce new, potentially harmful, strains. To address these concerns, the U.S. scientists who developed the technology drew up safety guidelines for recombinant DNA research in the United States. Adopted by the National Institutes of Health (NIH), the guidelines listed the precautions to be used in the laboratory when constructing recombinant DNA molecules and included the design and use of host organisms that could survive only in growth media in the laboratory. Since that time countless thousands of experiments involving recombinant DNA molecules have been done in laboratories around the world. Those experiments have shown that recombinant DNA manipulations can be done safely. Over time, therefore, the recombinant DNA guidelines have become more relaxed. Nonetheless, stringent regulations still exist for certain areas of recombinant DNA research that pose significant risk, such as cloning genes from highly pathogenic bacteria or viruses, or gene therapy experiments. In essence, as the risk increases, the research facility must increase its security and it must obtain more levels of approval by peer scientist groups.

Guidelines for genetic engineering also extend to research in several areas that have been the subject of public concern and debate. While the public is concerned little about genetically engineered microorganisms, for example those cleaning up oil spills and hazardous chemicals, it is concerned about possible problems with **genetically modified organisms (GMOs)** used as food. A GMO is a transgenic organism; the majority of GMOs are crop plants. Issues are the safety of GMO-containing food and the possible adverse effects

Figure 18.15 Research Method

Using the Ti Plasmid of Rhizobium radiobacter *to Produce Transgenic Plants*

PURPOSE: To make transgenic plants. This technique is one way to introduce a transgene into a plant for genetic engineering purposes.

PROTOCOL:

1. Isolate the Ti plasmid from *Rhizobium radiobacter*. The plasmid contains a segment called T DNA (T = transforming), which induces tumors in plants.

2. Digest the Ti plasmid with a restriction enzyme that cuts within the T DNA segment. Mix with a gene of interest on a DNA fragment that was produced by digesting with the same enzyme. Use DNA ligase to join the two DNA molecules together to produce a recombinant plasmid.

3. Transform the recombinant Ti plasmid into a disarmed *Rhizobium radiobacter* that cannot induce tumors, and use the transformed bacterium to infect cells in plant fragments in a test tube. In infected cells, the T DNA with the inserted gene of interest excises from the Ti plasmid and integrates into the plant cell genome.

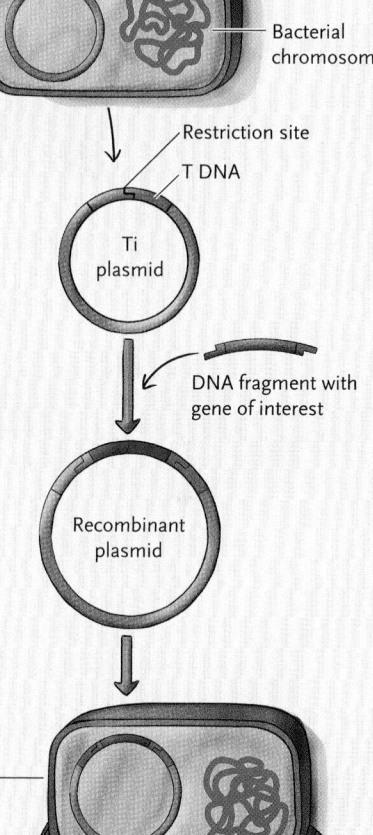

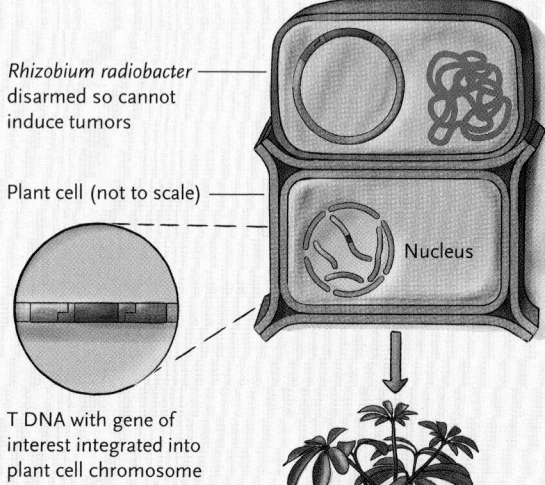

4. Culture the transgenic plant fragments to regenerate whole plants.

INTERPRETING THE RESULTS: The plant has been genetically engineered to contain a new gene. The transgenic plant will express a new trait based on that gene, perhaps resistance to an herbicide or the production of an insect toxin according to the goal of the experiment.

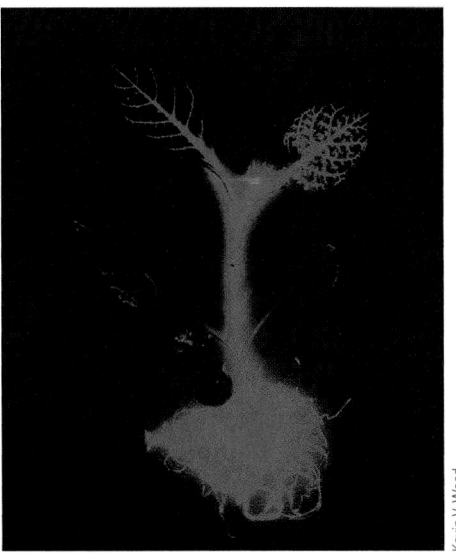

Figure 18.16
A genetically engineered tobacco plant, made capable of luminescence by the introduction of a firefly gene coding for the enzyme luciferase.

Kevin V. Wood

of the GMOs to the environment, such as by interbreeding with natural species or by harming beneficial insect species. For example, could introduced genes providing herbicide or insect resistance move from crop plants into related weed species through cross pollination, producing "super weeds" that might be difficult or impossible to control? And *Bt*-expressing corn was originally thought to have adverse effects on monarch butterflies who fed on the pollen. The most recent of a series of independent studies investigating this possibility has indicated that the risk to the butterflies is extremely low.

More broadly, different countries have reacted to GMOs in different ways. In the United States, transgenic crops are widely planted and harvested. Before commercialization, such GMOs are evaluated for potential risk by appropriate government regulatory agencies, including the NIH, Food and Drug Administration (FDA), Department of Agriculture, and Environmental Protection Agency (EPA). Typically, opposition to GMOs has come from particular activist and consumer groups.

Political opposition to GMOs has been greater in Europe, dampening the use of transgenic crop plants in the fields and GMOs in food. In 1999, the European Union (EU) imposed a 6-year moratorium on all GMOs, leading to a bitter dispute with the United States, Canada, and Argentina, the leading growers of transgenic crops. More recently, the EU has revised the GMO regulations in all member states. Basically, the EU has decided that using genetic engineering in agriculture and food production is permissible provided the GMO or food containing it is safe for humans, animals, and the environment. All use of GMOs in the field or in food requires authorization following a careful review process.

On a global level, an international agreement, the **Cartagena Protocol on Biosafety,** "promotes biosafety by establishing practical rules and procedures for the safe transfer [between countries], handling and use of GMOs." Separate procedures have been set up for GMOs that are to be introduced into the environment and those that are to be used as food or feed or for processing. To date, 132 countries have ratified the Protocol. (As of November 2006, the United States is not one of them.)

In sum, the use of DNA technologies in biotechnology has the potential for tremendous benefits to humankind. Such experimentation is not without risk, and so for each experiment, researchers must assess that risk and make a judgment about whether to proceed and, if so, how to do so safely. Furthermore, agreed-upon guidelines and protocols should ensure a level of biosafety for researchers, consumers, politicians, and governments.

We now turn to the analysis of whole genomes.

STUDY BREAK

1. What are the principles of DNA fingerprinting?
2. What is a transgenic organism?
3. What is the difference between using germ-line cells and somatic cells for gene therapy?

18.3 Genome Analysis

The development of DNA technologies for analyzing genes and gene expression revolutionized experimental biology. DNA sequencing techniques (described in this section) have made it possible to analyze the sequences of cloned genes and genes amplified by PCR. Having the complete sequence of a gene aids researchers tremendously in unraveling how the gene functions. But a gene is only part of a genome. Researchers want to know about the organization of genes in a complete genome, and how genes work together in networks to control life. Of particular interest, of course, is the human genome. The complete sequencing of the approximately 3 billion base-pair human genome—the

Regular rice

Genetically engineered golden rice containing β-carotene

Dr. Jorge Mayer, Golden Rice Project

Figure 18.17
Rice genetically engineered to contain β-carotene.

Insights from the Molecular Revolution

Engineering Rice for Blight Resistance

Rice is common in the diet of the entire human population; for one-third of humanity, more than 2 billion people, it is the primary nutrient source. Worldwide, some 146 million hectares are planted with rice, producing 560 million tons of the grain annually.

This major human staple is threatened by a rice blight caused by the bacterium *Xanthomonas oryzae*. Rice plants infected by the bacterium turn yellow and wilt; in many Asian and African rice fields, as much as half of the crop is lost to the blight. Although some wild forms of rice, which are not usable as crop plants, have a natural resistance to the *Xanthomonas* blight, none of the cultivated varieties is resistant. Attempts to develop resistant strains by crossbreeding crop rice with wild varieties have produced only one resistant type that, unfortunately, is essentially unusable as a food source.

At best, crossbreeding plants to develop resistance can take many years to produce results. As a shortcut, Pamela Ronald and her colleagues at the University of California at Davis decided to try genetic engineering. They were aided by the results of genetic experiments in which a gene, *Xa21*, was found that confers resistance to *Xanthomonas* in a wild rice.

Ronald and her coworkers set out to clone the DNA of the *Xa21* gene. Their first step was to locate it in wild rice chromosomes, using traditional genetic crosses to set up a linkage map. The studies showed that *Xa21* is located near several known genes, including one that was so close that it rarely recombined with *Xa21* in genetic crosses.

The DNA of the known gene was used as a probe to find nearby sequences in a genomic library of the resistant wild rice genome. The probe paired with 16 DNA fragments in the library. To determine whether any of the fragments included the *Xa21* gene, the researchers introduced each fragment individually into rice crop plants that were susceptible to *Xanthomonas* infection, using a gene gun. Out of 1500 plants that incorporated the fragments, 50 plants, all containing the same 9600-base-pair fragment, proved to be resistant to the blight. These plants were the first rice crop plants to be engineered successfully to resist *Xanthomonas* infections.

The 9600-base-pair fragment was then broken into subfragments that were cloned and tested individually for their ability to confer resistance. This technique allowed the researchers to identify and isolate the specific part of the fragment containing the resistance gene. Sequencing the gene revealed that it encodes a typical plasma membrane receptor protein, which in some way triggers an internal cellular response on exposure to *Xanthomonas* or its molecular components. The response alters cell structure or biochemistry to inhibit growth of the bacterium.

Ronald and her coworkers have introduced the *Xa21* gene successfully into three varieties of rice that are widely grown as crops in Asia and Africa. These genetically engineered plants will be field-tested to see if they have the yield, taste, and resistance needed to make them successful crop plants. Ronald has also sent copies of the *Xa21* gene to investigators in Europe, Africa, Asia, and other locations in the United States, so they can experiment with introducing blight resistance into local varieties. If these efforts are successful, the genetically engineered rice plants promise to greatly increase the world output of this economically vital crop.

Human Genome Project (HGP)—began in 1990. The task was completed in 2003 by an international consortium of researchers and by a private company, Celera Genomics. As part of the official HGP, for purposes of comparison the genomes of several important model organisms commonly used in genetic studies were sequenced: *E. coli* (representing prokaryotes), the yeast *Saccharomyces cerevisiae* (representing single-celled eukaryotes), *Drosophila melanogaster* and *Caenorhabditis elegans* (the fruit fly and nematode worm, respectively, representing multicellular animals of moderate genome complexity), and *Mus musculus* (the mouse, representing a mammal of genome complexity comparable to that of humans). In addition, the sequences of the genomes of many organisms beyond this list, including plants, have been completed or are in progress at this time. What researchers are learning from analyzing complete genomes is of enormous importance to our understanding of biology and the evolution of organisms.

DNA Sequencing Techniques Are Based on DNA Replication

DNA sequencing is the key technology for genome sequencing projects. DNA sequencing is also used on a smaller scale, for example, in determining the sequence of individual genes that have been cloned or amplified by PCR.

DNA sequencing was first developed in the late 1970s by Allan M. Maxam, a graduate student, and his mentor, Walter Gilbert of Harvard University; within a few years, another investigator, Frederick Sanger of Cambridge University, designed the method that is most used today. Gilbert and Sanger were awarded a Nobel Prize in 1980.

The Sanger method is based on the properties of nucleotides known as *dideoxyribonucleotides*—the method, therefore, is also called *dideoxy sequencing* **(Figure 18.18).** Dideoxyribonucleotides have a single —H bound to the 3′ carbon of the deoxyribose sugar

Figure 18.18 Research Method

Dideoxy (Sanger) Method for Sequencing DNA

PROTOCOL:

1. A dideoxy sequencing reaction has the following components: the fragment of DNA to be sequenced (denatured to single strands); a DNA primer that will bind to the 3′ end of the sequence to be determined; a mixture of the four deoxyribonucleotide precursors for DNA synthesis; and a mixture of the four dideoxyribonucleotides (dd) precursors, each labeled with a different fluorescent molecule, and DNA polymerase to catalyze the DNA synthesis reaction.

2. Synthesis of the new DNA strand is in the 5′→3′ direction starting at the 3′ end of the primer. New synthesis continues until a dideoxyribonucleotide is incorporated into the DNA instead of a normal deoxyribonucleotide. For a large population of template DNA strands, the dideoxy sequencing reaction produces a series of new strands, with lengths from one on up. At the 3′ end of each new strand is the labeled dideoxyribonucleotide that terminated the synthesis.

3. The labeled strands produced by the reaction are separated by gel electrophoresis. The principle of separation is the same as for agarose gel electrophoresis described in Figure 18.9. But here it is necessary to discriminate between DNA strands that differ in length by one nucleotide, which agarose gels cannot do. In this case, therefore, a gel made of polyacrylamide is prepared in a capillary tube for separating the DNA fragments. As the bands of DNA fragments move near the bottom of the tube, a laser beam shining through the gel excites the fluorescent labels on each DNA fragment. The fluorescence is registered by a detector with the wavelength of the fluorescence indicating which of the four dideoxyribonucleotides is at the end of the fragment in each case.

PURPOSE: Obtain the sequence of a piece of DNA, such as in gene sequencing or genome sequencing. The method is shown here with an automated sequencing system.

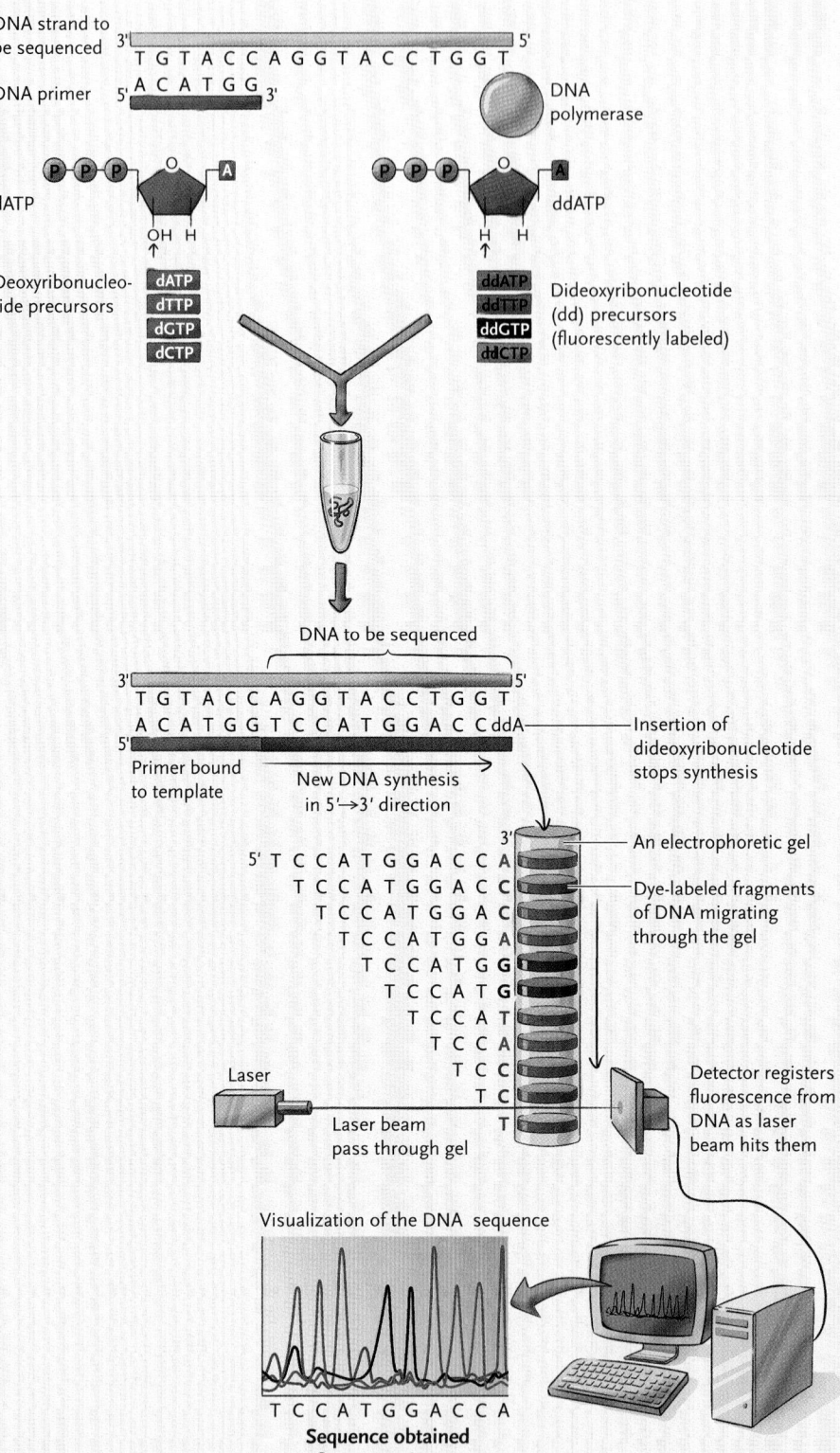

INTERPRETING THE RESULTS: The data from the laser system are sent to a computer that interprets which of the four possible fluorescent labels is at the end of each DNA strand. The results show the colors of the labels as the DNA bands passed the detector. They may be seen on the computer screen or in printouts. The sequence of the newly synthesized DNA, which is complementary to the template strand, is read from left (5′) to right (3′). (The sequence shown here begins after the primer.)

instead of the —OH normally appearing at this position in deoxyribonucleotides. DNA polymerases, the replication enzymes, recognize the dideoxyribonucleotides and place them in the DNA just as they do the normal deoxyribonucleotides. However, because a dideoxyribonucleotide has no 3'-OH group available for addition of the next base, replication of a nucleotide chain stops when one of these nucleotides is added to a growing nucleotide chain. (Remember from Section 14.3 that a 3'-OH group must be present at the growing end of a nucleotide chain for the next nucleotide to be added during DNA replication.) In a dideoxy sequencing reaction, researchers use a mixture of dideoxyribonucleotides and normal nucleotides, so that chain termination will occur randomly at each position where a particular nucleotide appears in the population of DNA molecules being replicated. Each chain-termination event generates a newly synthesized DNA strand that ends with the dideoxyribonucleotide; hence, for this particular strand, the base at the 3' end is known and, because of base-pairing rules, the base on the template strand being sequenced is deduced. Once they know the base at the end of each terminated DNA strand, researchers can work out the complete sequence of the template DNA strand.

The dideoxy sequencing method can be used with any pure piece of DNA, such as a cloned DNA fragment or a fragment amplified by PCR. An unambiguous sequence of about 500 to 750 nucleotides can be obtained from each sequencing experiment.

Structural Genomics Determines the Complete DNA Sequence of Genomes

Genome analysis consists of two main areas: *structural genomics* and *functional genomics*. **Structural genomics** is the actual sequencing of genomes and the analysis of the nucleotide sequences to locate genes and other functionally important sequences within the genome. **Functional genomics** is the study of the functions of genes and of other parts of the genome. In the case of genes, this includes developing an understanding of the regulation of their expression, the proteins they encode, and the role played by the proteins in the organism's metabolic processes.

The most widely used method for sequencing a genome is the *whole-genome shotgun method* (**Figure 18.19**). In this method, the entire genome is broken into thousands to millions of random, overlapping fragments, and each fragment is cloned and sequenced. The genome sequence then is assembled by computer on the basis of the sequence overlaps between fragments.

The first genome sequence reported, that of the bacterium *Haemophilus influenzae*, was determined using the whole-genome shotgun method by J. Craig Venter and his associates at Celera Genomics (the developers of the method). Originally it was thought that

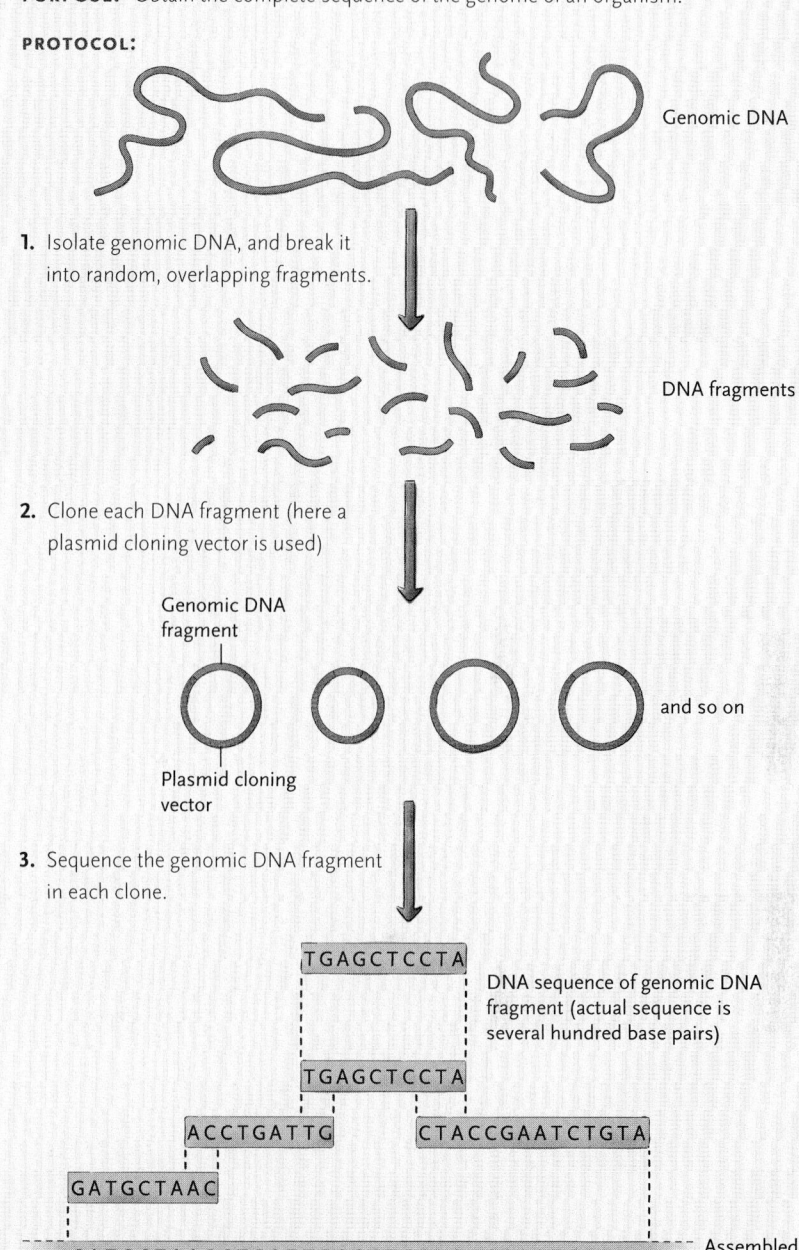

Figure 18.19 Research Method

Whole-Genome Shotgun Sequencing

PURPOSE: Obtain the complete sequence of the genome of an organism.

PROTOCOL:

Genomic DNA

1. Isolate genomic DNA, and break it into random, overlapping fragments.

DNA fragments

2. Clone each DNA fragment (here a plasmid cloning vector is used)

Genomic DNA fragment

Plasmid cloning vector

and so on

3. Sequence the genomic DNA fragment in each clone.

TGAGCTCCTA

DNA sequence of genomic DNA fragment (actual sequence is several hundred base pairs)

TGAGCTCCTA

ACCTGATTG   CTACCGAATCTGTA

GATGCTAAC

GATGCTAACCTGATTGAGCTCCTACCGAATCTGTA   Assembled sequence

4. Enter the DNA sequences of the fragments into a computer, and use the computer to assemble overlapping sequences into the continuous sequences of each chromosome of the organism. This technique is analogous to taking 10 copies of a book that have been torn randomly into smaller sets of a few pages each and, by matching overlapping pages of the leaflets, assembling a complete copy of the book with the pages in the correct order.

INTERPRETING THE RESULTS: The method generates the complete sequence of the genome of an organism.

the much larger genomes of eukaryotes would be too difficult to sequence using this method. But improvements in sequencing technologies and in the computer algorithms used to identify overlapping sequences have made it easier to assemble the segment sequences into the sequence of a whole genome. Whole genome shotgun sequencing is now the method of choice for sequencing essentially any genome.

Functional Genomics Focuses on the Functions of Genes and Other Parts of the Genome

The genomes of a large number of viruses and more than 180 organisms have been sequenced, and those of more species are continually being added to the total. Among those already sequenced are the cytomegalovirus, bacteria including *E. coli,* various archaean species, and eukaryotes including the brewer's yeast *Saccharomyces cerevisiae,* the protozoan *Plasmodium falciparium* (the malarial parasite), the roundworm *Caenorhabditis elegans,* the plants *Arabidopsis thaliana* and rice, the fruit fly *Drosophila melanogaster,* the chicken, the mouse, the rat, the dog, the chimpanzee, and human.

Analysis of Genome Sequences. The complete genome sequence for an organism is basically a very long string of letters, which means little without further analysis. Discovering the functions of genes and other parts of the genome is one important goal of this analysis. Most research is focused on the genes because they control the functions of cells and, therefore, of organisms. Functional genomics relies on laboratory experiments by molecular biologists and sophisticated computer analyses by researchers in the rapidly growing field of **bioinformatics,** which fuses biology with mathematics and computer science. Bioinformatics is used, for example, to find genes within a genomic sequence, align sequences in databases to determine the degree of matching, predict the structure and function of gene products, and postulate evolutionary relationships for sequences.

Protein-coding genes are of particular interest in genome analysis. Once a genome sequence is determined, researchers use computer algorithms to search both strands of the sequence for these genes. They identify possible protein-coding genes by searching for open reading frames (ORFs), that is, a start codon (ATG, at the DNA level) in frame (separated by a multiple of three nucleotides) with one of the stop codons (TAG, TAA, or TGA at the DNA level). This process is easy for prokaryotic genomes, because the genes have no introns. In eukaryotic protein-coding genes, which typically have introns, more sophisticated algorithms are used to try to identify the junctions between exons and introns in scanning for open reading frames.

Each open reading frame found by computer analysis of a genome can be "translated" by computer to give the amino acid sequence of the protein it could encode. Researchers may then be able to assign a function to the open reading frame by performing a *sequence similarity search,* a computer-based comparison of a DNA or amino acid sequence with databases of sequences of known genes or proteins. That is, if an open reading frame or its protein product resemble those of a previously sequenced gene, the two genes are related in an evolutionary sense and are likely to have similar functions.

Many new features of genetic organization have been discovered, or previous conclusions reinforced, through the findings of genome sequencing. One of the more surprising discoveries is that the eukaryotic genomes sequenced to date contain large numbers of previously unknown genes, many more than scientists expected to find. In *Caenorhabditis,* for example, 12,000 of the 19,000 genes are of unknown function. Identifying these genes and their functions is one of the major challenges of contemporary molecular genetics.

Another revelation is the degree to which different organisms, some of them widely separated in evolutionary origins, contain similar genes. For example, even though the yeast *Saccharomyces* is a fungus separated from our species by millions of years of evolutionary history, about 2300 of its approximately 6000 genes are related to those of mammals, including many genes that control progress through the cell cycle. The similarities are so close that the yeast and human versions of many genes can be interchanged with little or no effect on cell functions in either organism.

The sequences also confirm that eukaryotic genomes contain large numbers of noncoding sequences, most of them in the form of repeated sequences of various lengths and numbers. Most of these sequences, which make up from about 25% to 50% of the total genomic DNA in different eukaryotic species, have no determined function at this point in time.

Features of the Human Genome Sequence. The human genome sequence consists of 3.2 *billion* base pairs. Until the genome was sequenced, researchers expected that human cells might contain as many as 100,000 different protein-coding genes. The best current estimate is 20,000 to 25,000 protein-coding genes. However, although the number of protein-coding genes is unexpectedly small, the total number of different proteins produced in humans is much greater and probably approaches the 100,000 figure originally proposed for genes. The additional proteins arise through such processes as alternative splicing during mRNA processing (see Section 15.3) and differences in protein processing (discussed further in the following).

All the protein-coding sequences occupy less than 2% of the human genome. Introns—the noncoding spacers in genes—occupy another 24% of the genome. The rest of the DNA, almost three-quarters of the genome, occupies the spaces between genes. Some of this intergenic DNA is functional and includes regulatory sequences such as promoters and enhancers, but much of it, more than 50% of the total genome, consists of repeated sequences that have no known function.

Completing the human genome sequence is only the beginning of human genomics. The next steps are to determine the functions of the unknown genes and of the sequence elements in intergenic regions. This *data mining,* as it is called, may answer fundamental questions about genome organization and the mechanisms controlling genes in development and cell differentiation. Genes related to human health and disease, including cancer, are of particular interest. The analysis of these disease-related genes may suggest methods to predict individual susceptibility to diseases and may possibly lead to means for their diagnosis and treatment.

On an even larger scale, the human genome is being compared with the genomes of other species to determine the molecular basis of differences in anatomy, physiology, and developmental patterns between species. Ultimately, species comparisons may reveal the mutational changes underlying the evolution of our species and many others. This area of genomics is known as *comparative genomics.*

There are bioethics issues concerning the human genome. To address those issues, the U.S. Department of Energy and the NIH have funded studies of the ethical, legal, and social issues surrounding the availability of genetic information from human genome research. Among the questions being looked at: Who should have access to personal genetic information, and how should it be used? To what extent should genetic information be private and confidential? How will genetic tests be evaluated and regulated? How can people be informed sufficiently about the genetic information from genomic analysis so that they can make informed personal medical choices? Does a set of genes predispose a person's behavior, and can the person control that behavior?

Studying Differential Gene Activity in Entire Genomes with DNA Microarrays. As a part of genome research, investigators are interested in comparing which genes are active in different cell types of humans and other organisms, and tracking the changes in total gene activity in the same cell types as development progresses. In some cases, the researcher wants to know whether or not particular genes are being expressed, and in other cases how the level of expression varies in different circumstances. This research has been revolutionized by a technique using **DNA microarrays.** The microarrays are also called **DNA chips** for short because the techniques used to "print" the arrays resemble those used to lay out electronic circuits on a computer

chip. The surface of a DNA chip is divided into a microscopic grid of about 60,000 spaces. On each space of the grid, a computerized system deposits a microscopic spot containing about 10,000,000 copies of a DNA probe about 20 nucleotides long.

Studies of gene activity using DNA microarrays involves comparing gene expression under a defined experimental condition with expression under a reference (control) condition. For instance, DNA microarrays can be used to answer basic biological questions, such as: How does gene expression change when a cell goes from a resting state (reference condition) to a dividing state (experimental condition); that is, how is gene expression different in different stages of development? DNA microarrays can also be used to address many questions of medical significance, such as: How are genes differentially expressed in normal cells and cells of various cancers? In these experiments, investigators might focus on which genes are active and inactive under the two conditions, or on how the levels of expression of genes change under the two conditions.

Figure 18.20 shows how a DNA microarray is used to compare gene expression in normal cells and in cancer cells in humans. mRNAs are isolated from each cell type, and cDNAs are made from them, incorporating different fluorescent labels: green for one, red for the other. The two cDNAs are mixed and added to the DNA chip, where they hybridize with any complementary probes. A laser locates and quantifies the green and red fluorescence, enabling a researcher to see which genes are expressed in the cells and, for those that are expressed, to quantify differences in gene expression between the two cell types (see Interpreting the Results in Figure 18.20). The results can help researchers understand how the cancer develops and progresses.

DNA microarrays are also used to screen individuals for particular mutations. To detect mutations, the probes spotted onto the chip include probes for the normal sequence of the genes of interest along with probes for sequences of all known mutations. A fluorescent spot at a site on the chip printed with a probe for a given mutation immediately shows the presence of the mutation in the individual. Such a test is currently used to screen patients for whether they carry any one of a number of mutations of the *breast cancer 1 (BRCA1)* gene known to be associated with the possible development of breast cancer.

Studying the Array of Expressed Proteins Is the Next Level of Genomic Analysis

Genome research also includes analysis of the proteins encoded by a genome, for proteins are largely responsible for cell function and, therefore, for all the functions of an organism. The term **proteome** has been coined to refer to the complete set of proteins that can be expressed by an organism's genome. A *cellular proteome* is a subset of those proteins, the collection of

Figure 18.20 Research Method

DNA Microarray Analysis of Gene Expression Levels

PURPOSE: DNA microarrays can be used in various experiments, including comparing the levels of gene expression in two different tissues, as illustrated here. The power of the technique is that the entire set of genes in a genome can be analyzed simultaneously.

PROTOCOL:

Normal cells (reference)

Cancer cells (experimental)

mRNA

cDNA

Each spot has a different probe

1. Isolate mRNAs from a control cell type (here, normal cells) and an experimental cell type (here, cancer cells).

2. Prepare cDNA libraries from each mRNA sample. For the normal cell (control) library use nucleotides with a green fluorescent label, and for the cancer cell (experimental) library use nucleotides with a red fluorescent label.

3. Denature the cDNAs to single strands, mix them, and pump them across the surface of a DNA microarray containing a set of single-stranded probes representing every protein-coding gene in the human genome. The probes are spotted on the surface, with each spot containing a probe for a different gene. Allow the labeled cDNAs to hybridize with the gene probes on the surface of the chip, and then wash excess cDNAs off.

4. Locate and quantify the fluorescence of the labels on the hybridized cDNAs with a laser detection system.

Gene expressed in both cell types

Gene expressed in normal cells only

Colored spots are where labeled cDNAs have hybridized

Gene expressed in cancer cells only

Actual DNA microarray result

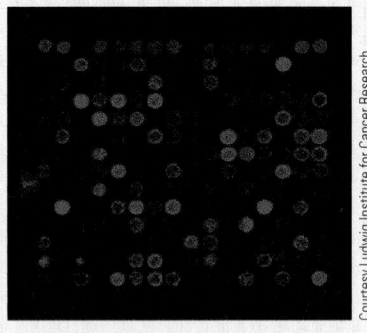

Courtesy Ludwig Institute for Cancer Research

INTERPRETING THE RESULTS: The colored spots on the microarray indicate where the labeled cDNAs have bound to the gene probes attached to the chip and, therefore, which genes were active in normal and/or cancer cells. Moreover, we can quantify the gene expression in the two cell types by the color detected. A purely green spot indicates the gene was active in the normal cell, but not in the cancer cell. A purely red spot indicates the gene was active in the cancer cell, but not in the normal cell. A yellow spot indicates the gene was equally active in the two cell types, and other colors tell us the relative levels of gene expression in the two cell types. For this particular experiment, we would be able to see how many genes have altered expression in the cancer cells, and exactly how their expression was changed.

proteins found in a particular cell type under a particular set of environmental conditions.

The study of the proteome is the field of **proteomics.** The number of possible proteins encoded by the genome is larger than the number of protein-coding genes in the genome, at least in eukaryotes. In eukaryotes, alternative splicing of gene transcripts and variation in protein processing means that expression of a gene may yield more than one protein product. Therefore, proteomics is a more challenging area of research than is genomics.

The two major immediate goals of proteomics are to determine (1) the number and structure of proteins in the proteome and (2) the functional interactions between the proteins. The interactions are particularly important because they help us understand how proteins work together to determine the phenotype of the cell. For instance, if a particular set of interacting proteins characterized a lung tumor cell, then drugs could be developed that specifically target the interactions.

What are the tools of proteomics? For many years it has been possible to separate and identify proteins by gel electrophoresis (using polyacrylamide to make the gels, the same material used for separating DNA fragments in DNA sequencing) or mass spectrometry. However, to study an entire cellular proteome, many more proteins must be analyzed simultaneously than is possible with either of those techniques. A big step in that direction is the development of **protein microarrays (protein chips)**, which are similar in concept to DNA microarrays. For example, one type of protein microarray involves binding antibodies prepared against different proteins to different locations on the protein chip. An antibody for a foreign substance such

UNANSWERED QUESTIONS

Can pharmaceuticals be customized to the individual?

One exciting avenue of investigation arising from genomics research is pharmacogenomics. The word is a combination of "pharmacology" and "genomics" and means the study of how the genome of an individual affects the body's responses.

Currently, pharmaceuticals are dispensed under the assumption that all humans are basically the same. Of course, the dose may be adjusted to the weight or age of the individual. However, not everyone is the same, as we know from the fact that every drug provokes an allergic reaction in some people. Indeed, a variety of factors affect a person's response to medicines, including the genome (the expression of the particular array of genes in tissues targeted by the pharmaceutical), diet, health, age, and the environment.

The goal is for the pharmaceutical to be as effective as it can be, with the best available drug therapy given to a patient right from the beginning. Pharmacogenomics offers the promise of customizing drugs by adapting to them to key cellular proteomes of the patient. Conceptually drugs could be chemically matched to the proteins, enzymes, and RNA molecules that are associated with diseases. Such drugs therefore would be much more targeted to specific diseases than present-day drugs are, the result being that the therapeutic benefits will be maximized while side effects will be minimized. And drug dosages will be tailored to an individual's genetic makeup (as manifested in the proteome) so as to take into account how and at what rate a person metabolizes a drug.

There are very few examples of pharmacogenomics in action in the clinic at the moment. Many research groups are currently working at the lab bench to develop applications.

Can gene expression at the genomic level be used to identify specific cancers?

DNA microarrays can be used to analyze the differences in gene expression between two cell or tissue types (see Figure 18.20). In 1999, Eric Lander's research group at Whitehead Institute, MIT, was the first to show that gene expression arrays can distinguish between two types of cancer. The cancers, acute myeloid leukemia (AML) and acute lymphoblastic leukemia (ALL), are very similar in clinical symptoms, but because the treatments are different the diagnosis must be precise. Before Lander's discovery, the cancers could be distinguished only through a series of expensive tests that took precious time. Lander used DNA microarray analysis based on 6800 human genes and accurately distinguished the two types.

DNA microarray analysis has now been developed to identify a number of different types of cancers based on gene expression profiles. Undoubtedly many more cancers will be added to the list as time goes on.

Can genome-wide approaches produce tools to combat pathogens?

Animals are susceptible to attack by a wide range of pathogenic agents, including virulent bacteria and viruses. Humans are vaccinated against some of the important viral pathogens, an action that has saved millions of lives. Still, we are invaded. In healthy individuals, the immune system is activated to combat the invasion. Immunocompromised individuals, however, have weakened immune systems; they risk serious illness or death from infection with pathogens that healthy individuals combat easily. Pathogens are also highly relevant to our lives in another way: agricultural animals and plants are vital for maintaining our food supply, yet they also may be or may become susceptible to pathogen invasion.

Biodefense proteomics is an area of research with the goal of using genome-wide approaches to produce tools to combat pathogens. It is anticipated that the research will discover targets for the next generation of vaccines, therapeutics, and diagnostics. The proteomics approaches being used include characterizing the proteomes of pathogens (for example, for the pathogen itself if it is cellular, or for an infected cell if the pathogen is viral); to determine the mechanisms of microbial pathogenesis; and to develop an understanding of animal immune responses and nonimmune mediate responses that are triggered by microbial pathogens.

Peter J. Russell

as a protein is generated by the immune system of an animal that has been injected with that substance. The antibody is isolated from the blood of that animal and can be used to bind specifically to the protein in experiments. Proteins are isolated from cells and labeled, and then pumped over the surface of the protein microarray. Each labeled protein binds to the antibody for that protein. After washing off excess proteins, the protein microarray is analyzed much as for DNA microarrays to determine where the proteins bound and to quantify that binding. With this technique, a researcher can quantify proteins in different cell types and different tissues. Researchers can also compare proteins under different conditions, such as during differentiation, or with and without a particular disease condition, or with and without a particular drug treatment. In the future, we can expect protein arrays to become routine for studying cellular proteomes.

Systems Biology Is the Study of the Interactions between All the Components of an Organism

Traditional biology research focuses on identifying and studying the functions of individual genes, proteins, and cells. Although such research has provided an enormous body of knowledge—this textbook being an example—it provides only a limited insight into how a whole organism functions at the cellular and molecular levels. For instance, studying separately the individual components of a bicycle does not tell you what the whole bicycle is or what it does.

Systems biology is an area of biology that seeks to overcome the limitations of traditional biology approaches by studying the organism as a whole to unravel the integrated and interacting network of genes, proteins, and biochemical reactions responsible for

life. That is, systems biologists work from the premise that those interactions are responsible for an organism's form and function. Present-day research in systems biology has been stimulated by the development of techniques for genomic and proteomic analysis and by the data from those analyses.

Systems biologists use genomics and proteomics techniques, such as those discussed in this section along with information from other sources. They typically obtain very complex data and use sophisticated quantitative analysis to generate models for the interactions within an organism.

Systems biologists study organisms of many kinds. Some focus on humans and have the ambitious goal of transforming the practice of medicine. The vision is to define the interactions between all the components that affect the health of an individual human. It may then be possible to predict more accurately than is currently possible whether a person will develop particular diseases and to personalize treatments for those diseases.

Unquestionably DNA technologies and genomics research have resulted in remarkable achievements so far, probably touching every person on the planet directly or indirectly. They, and proteomics research, also have enormous potential to lead to major advances and achievements in the future.

STUDY BREAK

1. How are possible protein-coding genes identified in a genome sequence of a bacterium? Of a mammal?
2. How would you determine how a steroid hormone affects gene expression in human tissue culture cells?

Review

Go to **ThomsonNOW**™ at www.thomsonedu.com/login to access quizzing, animations, exercises, articles, and personalized homework help.

18.1 DNA Cloning

- Producing multiple copies of genes by cloning is a common first step for studying the structure and function of genes or for manipulating genes. Cloning involves cutting genomic DNA and a cloning vector with the same restriction enzyme, joining the fragments to produce recombinant plasmids, and introducing those plasmids into a living cell such as a bacterium, where replication of the plasmid takes place (Figures 18.2–18.4).

- A clone containing a gene of interest may be identified among a population of clones by using DNA hybridization with a labeled nucleic acid probe (Figure 18.5).

- A genomic library is a collection of clones that contains a copy of every DNA sequence in the genome. A cDNA (complemen-

tary DNA) library is the entire collection of cloned cDNAs made from the mRNAs isolated from a cell. A cDNA library contains only sequences from the genes that are active in the cell when the mRNAs are isolated.

- PCR amplifies a specific target sequence in DNA, such as a gene, defined by a pair of primers. PCR increases DNA quantities by successive cycles of denaturing the template DNA, annealing the primers, and extending the primers in a DNA synthesis reaction catalyzed by DNA polymerase; with each cycle, the amount of DNA doubles (Figure 18.6).

Animation: Base pairing of DNA fragments

Animation: Formation of recombinant DNA

Animation: Restriction enzymes

Animation: How to make cDNA

Animation: Use of a radioactive probe

Animation: Polymerase chain reaction (PCR)

Animation: Automated DNA sequencing

18.2 Applications of DNA Technologies

- Recombinant DNA and PCR techniques are used in DNA molecular testing for human genetic disease mutations. One approach exploits restriction site differences between normal and mutant alleles of a gene that create restriction fragment length polymorphisms (RFLPs) detectable by DNA hybridization with a labeled nucleic acid probe (Figures 18.8 and 18.9).

- Human DNA fingerprints are produced from a number of loci in the genome characterized by tandemly repeated sequences that vary in number in all individuals (except identical twins). To produce a fingerprint, the PCR is used to amplify the region of genomic DNA for each locus, and the lengths of the PCR products indicate the alleles an individual has for the repeated sequences at each locus. DNA fingerprints are widely used to establish paternity, ancestry, or criminal guilt (Figure 18.10).

- Genetic engineering is the introduction of new genes or genetic information to alter the genetic makeup of humans, other animals, plants, and microorganisms such as bacteria and yeast. Genetic engineering primarily aims to correct hereditary defects, improve domestic animals and crop plants, and provide proteins for medicine, research, and other applications. (Figures 18.11, 18.12, and 18.15).

- Genetic engineering has enormous potential for research and applications in medicine, agriculture, and industry. Potential risks include unintended damage to living organisms or the environment.

Animation: DNA fingerprinting

Animation: Gene transfer using a Ti plasmid

Animation: Transferring genes into plants

18.3 Genome Analysis

- Genome analysis consists of two main areas: structural genomics, the sequencing of genomes and the identification of the genes the sequences contain, and functional genomics, the study of the function of genes and other parts of the genome.

- Sequencing a genome involves a replication reaction with a DNA template, a DNA primer, the four normal deoxyribonucleotides, and a mixture of four dideoxyribonucleotides, each labeled with a different fluorescent tag, and DNA polymerase. Replication stops at any place in the sequence in which a dideoxyribonucleotide is substituted for the normal deoxyribonucleotide. The lengths of the terminated DNA chains and the label on them indicate the overall sequence of the DNA chain being sequenced (Figure 18.18).

- The whole-genome shotgun method of sequencing a genome involves breaking up the entire genome into random, overlapping fragments, cloning each fragment, determining the sequence of the fragment in each clone, and using computer algorithms to assemble overlapping sequences into the sequence of the complete genome (Figure 18.19).

- Once a gene is sequenced, the sequence of the protein encoded in a prokaryotic gene can be deduced by reading the coding portion of the gene three nucleotides at a time, starting at the AUG codon that indicates the beginning of a coding sequence.

- Complete genome sequences have been obtained for many viruses, a large number of prokaryotes, and many eukaryotes, including the human. The sequences have revealed that all eukaryotes share related gene sequences, and they have also revealed a significant proportion of genes whose functions are not presently known.

- Having the complete genome of an organism makes it possible to study the expression of all of the genes in the genome simultaneously, including comparing gene expression in two different cell types. The DNA microarray (or DNA chip) is typically used for the comparison; this technique can provide information about which genes are active in the two cell types as well as relative levels of expression of those genes (Figure 18.20).

- Proteomics is the study of the complete set of proteins in an organism or in a particular cell type. Protein numbers, protein structure, and protein interactions are all topics of proteomics.

- Systems biology combines data derived from genomics, proteomics, and other sources of information. Using sophisticated quantitative analysis, it seeks to model the total array of interactions responsible for an organism's form and function.

Questions

Self-Test Questions

1. Using cDNA is associated with which of the following?
   a. Introns can be identified and sequenced by this method.
   b. It measures both active and inactive DNA.
   c. Promoter regions can be identified by this method.
   d. One can identify start and stop regions by this method.
   e. One can identify active mRNA and make a complementary DNA sequence to the mRNA.

2. Restriction endonucleases, ligases, plasmids, viral or yeast vectors, electrophoretic gels, and a bacterial gene resistant to an antibiotic are all required for:
   a. dideoxyribonucleotide analysis.
   b. PCR.
   c. DNA cloning.
   d. DNA fingerprinting.
   e. DNA sequencing.

3. The PCR technique is distinguished from other processes discussed in this chapter by the use of:
   a. primers.
   b. DNA.
   c. RNA.
   d. Taq polymerase.
   e. the four nucleoside triphosphates.

4. Restriction fragment length polymorphisms:
   a. are produced by reaction with restriction endonucleases and are detected by Southern blot analysis.
   b. are of the same length for mutant and normal β-globin alleles.
   c. determine the sequence of bases in a DNA fragment.
   d. have in their middle short fragments of DNA that are palindromic.
   e. are used as vectors.

5. DNA fingerprinting:
   a. compares one stretch of the same DNA between two or more people.
   b. measures different lengths of DNA from many repeating noncoding regions for comparison between two or more people.
   c. requires the largest DNA lengths to run the greatest distance on a gel.
   d. requires amplification after the gels are run.
   e. can easily differentiate DNA between identical twins.

6. Dolly, a sheep, was an example of reproductive (germ line) cloning. Required to perform this process was:
   a. implantation of uterine cells from one strain into the mammary gland of another.
   b. the fusion of the mammary cell from one strain with an enucleated egg of another strain.
   c. the fusion of an egg from one strain with the egg of a different strain.
   d. the fusion of an embryonic diploid cell with an adult haploid cell.
   e. the fusion of two nucleated mammary cells from two different strains.

7. All of the following are true for somatic cell gene therapy *except:*
   a. White blood cells can be used.
   b. Somatic cells are cultured, and the desired DNA is introduced into them.
   c. Cells with the introduced DNA are returned to the body.
   d. The technique is still very experimental.
   e. The inserted genes are passed on to the offspring.

8. The sequence of the human genome:
   a. was obtained by sequencing overlapping DNA fragments.
   b. revealed far more genes than expected.
   c. revealed 3 trillion base pairs.
   d. used techniques not applicable to mapping other species.
   e. revealed 250,000 protein-coding genes.

9. Sanger's DNA sequencing technique:
   a. uses dideoxyribonucleotides to make new full-length strands of DNA.
   b. is based on cellular transcription.
   c. requires an RNA template, RNA primer, RNA polymerase, reverse transcriptase, and the dideoxyribonucleotides, ddATP, ddUTP, ddCTP, and ddGTP.
   d. places the RNA template to be sequenced on a gel and then adds the other ingredients from (c).
   e. is based on DNA replication.

10. A microarray could be used to:
    a. sequence DNA from several chromosomes in one individual.
    b. synthesize multiple copies of DNA from several sources.
    c. propagate human germ-line cells for cloning.
    d. compare coding DNA from a patient's normal lung cells with coding DNA from his cancerous lung cells.
    e. determine proteins that are expressed under certain environmental conditions.

Questions for Discussion

1. Do you think that genetic engineering is worth the risk? Who do you think should decide whether genetic engineering experiments and projects should be carried out: scientists, judges, politicians?

2. Do you think that human germ-line cells should be modified by genetic engineering to cure birth defects? To increase intelligence or beauty?

3. Write a paragraph supporting genetic engineering, and one arguing against it. Which argument carries more weight, in your opinion?

4. What should juries know to interpret DNA evidence? Why might juries sometimes ignore DNA evidence?

5. A forensic scientist obtained a small DNA sample from a crime scene. In order to examine the sample, he increased its quantity by the polymerase chain reaction. He estimated that there were 50,000 copies of the DNA in his original sample. Derive a simple formula and calculate the number of copies he will have after 15 cycles of the PCR.

6. A market puts out a bin of tomatoes that have outstanding color, flavor, and texture. A sign posted above them identifies them as genetically engineered produce. Most shoppers pick unmodified tomatoes in an adjacent bin, even though they are pale, mealy, and nearly tasteless. Which tomatoes would you pick? Why?

Experimental Analysis

Suppose a biotechnology company has developed a GMO, a transgenic plant that expresses *Bt* toxin. The company sells its seeds to a farmer under the condition that the farmer may plant the seed, but not collect seed from the plants that grow and use it to produce crops in the subsequent season. The seeds are expensive, and the farmer buys seeds from the company only once. How could the company show experimentally that the farmer has violated the agreement and is using seeds collected from the first crop to grow the next crop?

Evolution Link

Search for the words "comparative genomics" on the Internet to answer this question: How can complete genome sequences provide more accurate information about the evolution of species than sequences of one or a few genes?

How Would You Vote?

Nutritional labeling is required on all packaged food in the United States, but genetically modified food products may be sold without labeling. Should food distributors be required to label all products made from genetically modified plants or livestock? Go to www.thomsonedu.com/login to investigate both sides of the issue and then vote.

A replica of H.M.S. *Beagle*, the ship that carried Charles Darwin on his round-the-world journey of discovery.

Christopher Railing

19 Development of Evolutionary Thought

WHY IT MATTERS

On June 18, 1858, Charles Darwin received the shock of his life. Alfred Russel Wallace, a young naturalist working in the Asian tropics, had solicited Darwin's opinion of a short manuscript about how species change through time. Darwin quickly realized that Wallace had independently described a mechanism for biological evolution that was nearly identical to the one he had been studying for more than 20 years but had not yet described in print.

Like researchers today, scientists in the nineteenth century had to publish their work quickly to establish the "priority" on which scientific reputations are made. Darwin's friend and colleague, the geologist Charles Lyell, had encouraged him to publish a preliminary essay on evolution 2 years before Wallace's letter arrived. But Darwin procrastinated, and because Wallace was the first to prepare his work for publication, Darwin feared that history would credit the younger man with these new ideas. Despite his anxiety, Darwin forwarded Wallace's manuscript to Lyell, who passed it along to the botanist Joseph Hooker. Lyell and Hooker engineered a solution that gave credit to both men **(Figure 19.1)**. On July 1, 1858, papers by Darwin and Wallace were presented to the Linnaean Society of London, a prestigious scientific organization.

Charles Darwin

Alfred Russel Wallace

Figure 19.1

Pioneers of evolutionary theory. Charles Darwin (1809–1882) and Alfred Russel Wallace (1823–1913) independently discovered the mechanism of natural selection.

Darwin worked feverishly after this harrowing experience, and his now-famous book, *On the Origin of Species by Means of Natural Selection,* was published on November 24, 1859. The first printing of 1250 copies sold out in one day. Today, we honor Darwin for developing the seminal idea about how biological evolution occurs and for the vast documentation that he accumulated over decades of study.

In *The Origin,* Darwin proposed that natural mechanisms produce and transform the diversity of life on Earth. His concept of evolution still forms the unifying intellectual paradigm within which all biological research is undertaken. Even when researchers do not address explicitly evolutionary questions, their observations, theories, hypotheses, and experiments are formulated with the implicit knowledge that all forms of life are related and have evolved from ancestral forms.

Biological evolution occurs in populations when specific *processes* cause the genomes of organisms to differ from those of their ancestors. These genetic changes, and the phenotypic modifications they cause, are the *products* of evolution. By studying the products of evolution, biologists strive to understand the processes that cause evolutionary change.

The theory of evolution is so widely accepted that most people cannot think about the biological world in any other way. But the biological changes implied by Darwin's ideas and by modern evolutionary theory had not been included in earlier worldviews.

19.1 Recognition of Evolutionary Change

The historical development of evolutionary theory is a fascinating tale of scientists struggling to reconcile evidence of change with a prevailing philosophy that change was impossible in a perfectly created universe.

Europeans Integrated Ideas from Ancient Greek Philosophy into Christian Doctrine

The Greek philosopher Aristotle (384–322 B.C.) was a keen observer of nature, and he is generally considered the first student of **natural history,** the branch of biology that examines the form and variety of organisms in their natural environments. Aristotle believed that both inanimate objects and living species had fixed characteristics. Careful study of their differences and similarities enabled him to create a ladder-like classification of nature from simplest to most complex forms: minerals ranked below plants, plants below animals, animals below humans, and humans below the gods of the spiritual realm.

By the fourteenth century, Europeans had merged Aristotle's classification with the biblical account of creation: all of the different kinds of organisms had been specially created by God, species could never change or become extinct, and new species could never arise. Biological research became dominated by **natural theology,** which sought to name and catalog all of God's creation. Careful study of each species would identify its position and purpose in the *Scala Naturae,* or Great Chain of Being, as Aristotle's ladder of life was called. In the eighteenth century, the Swedish botanist Carolus Linnaeus (1707–1778), who developed the science of **taxonomy,** the branch of biology that classifies organisms (see Chapter 23), undertook this important work *ad majorem Dei gloriam* ("for the greater glory of God").

Scholars also used a literal interpretation of scripture to date the time of creation precisely. By tabulating the human generations described in the Bible, they determined that the creation had occurred around 4000 B.C., making Earth a bit less than 6000 years old. Thus, Earth hardly seemed old enough for much change to have taken place.

Scientists Slowly Became Aware of Change in the Natural World

Modern science came of age in the fifteenth through eighteenth centuries. The English philosopher and statesman Sir Francis Bacon (1561–1626) established the importance of observation, experimentation, and inductive reasoning. Other scientists, notably Nicolaus Copernicus (1473–1543), Galileo Galilei (1564–1642), René Descartes (1596–1650), and Sir Isaac Newton (1643–1727), proposed mechanistic theories to explain physical events. In addition, three new disciplines—biogeography, comparative morphology, and geology—promoted a growing awareness of change.

Questions about Biogeography. As long as naturalists encountered organisms only from Europe and surrounding lands, the task of understanding the *Scala Naturae* was manageable. But global explorations in the fifteenth through seventeenth centuries provided

Ostrich (*Struthio camelus*)
of Africa

Rhea (*Rhea americana*)
of South America

Emu (*Dromaius novaehollandiae*)
of Australia

Wolfgang Kaehler/Corbis

Kenneth W. Fink/Photo Researchers, Inc.

Dave Watts/A. N. T. Photo Library

Figure 19.2
Large, flightless birds. Three large bird species with greatly reduced wings occupy similar habitats in geographically separated regions.

naturalists with thousands of unknown plants and animals from Asia, sub-Saharan Africa, the Pacific Islands, and the Americas. Although some were similar to European species, others were new and very strange.

Studies of the world distribution of plants and animals, now called **biogeography**, raised puzzling questions. Was there no limit to the number of species created by God? Where did all these species fit in the *Scala Naturae*? If all species had been created in the Garden of Eden, why were the species found in Africa or Asia different from those found in Europe? Why was each species found only in certain places and not others **(Figure 19.2)?**

Questions about Comparative Morphology. When biologists began to compare the **morphology** (anatomical structure) of organisms, they discovered interesting similarities and differences. For example, the front legs of pigs, the flippers of dolphins, and the wings of bats differ markedly in size, shape, and function **(Figure 19.3).** But these appendages have similar locations in the animals' bodies; all are constructed of bones, muscles, and skin; and all develop similarly in the animals' embryos. If these limbs were specially created for different means of locomotion, why didn't the Creator use different materials and structures for walking, swimming, and flying?

Natural theologians answered that some general body plans were perfect, and there was no need to invent a new plan for every species. But a French scientist, George-Louis Leclerc (1707–1788), le Comte (Count) de Buffon, was still puzzled by the existence of body parts with no apparent function. For example, he noted that the feet of pigs and some other mammals have two toes that never touch the ground (see Figure 19.3). If each species is anatomically perfect for its particular way of life, why do useless structures exist?

Buffon proposed that some animals must have *changed* since their creation; he suggested that **vestigial structures,** the useless parts we observe today, must have functioned in ancestral organisms. Buffon offered no explanation of how functional structures became vestigial, but he clearly recognized

that some species were "conceived by Nature and produced by Time."

Questions about Fossils. By the mid-eighteenth century, geologists were mapping the **stratification,** or horizontal layering, of sedimentary rocks beneath the soil surface (see Figure 22.3). Different layers held different kinds of **fossils** (*fossilis* = dug up). Relatively small and simple fossils appeared in the deepest layers. Fossils in the layers above them were more complex. Those in the uppermost layers often resembled living organisms. Moreover, fossils found in any particular layer were often similar, even if they were collected from geographically separated sites. What were these fossils, and why did they vary more from one layer of

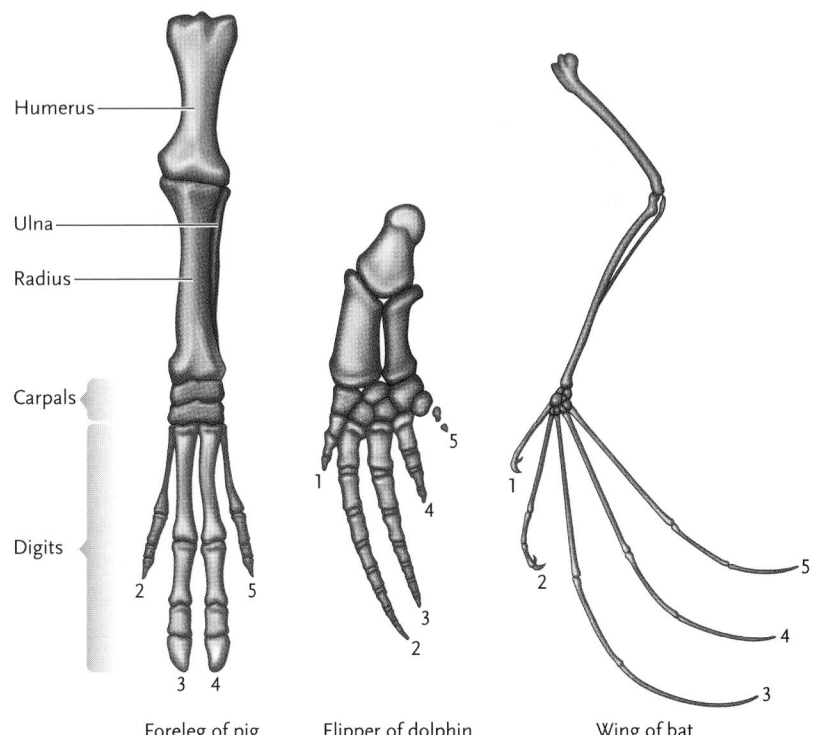

Humerus

Ulna

Radius

Carpals

Digits

5

1

4

5

1

2

2

5

2

3

4

2

3

3 4

3

Foreleg of pig   Flipper of dolphin   Wing of bat

Figure 19.3
Mammalian forelimbs and locomotion. Pigs use their legs to walk or run, dolphins use their flippers to swim, and bats use their wings to fly. Homologous bones are pictured in the same color, and digits (fingers) are numbered; pigs have lost the first digit over evolutionary time.

rock to another than from one geographical region to another?

Some scientists suggested that fossils were the remains of extinct organisms, but natural theology did not allow extinction. Thomas Jefferson, the third president of the United States and an amateur fossil hunter, thought that fossils were the remains of species that were now extremely rare; he believed that nature could not have "permitted any one race of her animals to become extinct" or "formed any link in her great works so weak [as] to be broken." He even asked Lewis and Clark to keep an eye out for giant ground sloths, now known to be extinct, during their exploration of the Pacific Northwest.

Georges Cuvier (1769–1832), a French zoologist and a founder of comparative morphology, as well as **paleobiology** (the study of ancient organisms), realized that the layers of fossils represented organisms that had lived at successive times in the past. He suggested that the abrupt changes between geological strata marked dramatic shifts in ancient environments. Cuvier and his followers developed the theory of **catastrophism**, reasoning that each layer of fossils represented the remains of organisms that had died in a local catastrophe, such as a flood. Somewhat different species then recolonized the area, and when another catastrophe struck, they formed a different set of fossils in the next higher layer.

Lamarck Developed an Early Theory of Biological Evolution

A contemporary of Cuvier and a student of Buffon, Jean Baptiste de Lamarck (1744–1829) proposed the first comprehensive theory of biological evolution based on specific mechanisms. He proposed that a metaphysical "perfecting principle" caused organisms to become better suited to their environments. Simple organisms evolved into more complex ones, moving up the ladder of life; microscopic organisms were replaced at the bottom by spontaneous generation.

Lamarck theorized that two mechanisms fostered evolutionary change. According to his *principle of use and disuse,* body parts grow in proportion to how much they are used, as anyone who "pumps iron" well knows. Conversely, structures that are not often used get weaker and shrink, such as the muscles of an arm immobilized in a cast. According to his second principle, the *inheritance of acquired characteristics,* changes that an animal acquires during its lifetime are inherited by its offspring. Thus, Lamarck argued

that long-legged wading birds, such as herons **(Figure 19.4),** are descended from short-legged ancestors that stretched their legs to stay dry while feeding in shallow water. Their offspring inherited slightly longer legs, and after many generations, their legs became extremely long.

Today, we know that Lamarck's proposed mechanisms do not cause evolutionary change. Although muscles do grow larger through continued use, most structures do not respond in the way Lamarck predicted. Moreover, structural changes acquired during an organism's lifetime are not inherited by the next generation. Even in his own day, Lamarck's ideas were not widely accepted.

Despite the shortcomings of his theory, Lamarck made four tremendously important contributions to the development of an evolutionary worldview. First, he proposed that all species change through time. Second, he recognized that new characteristics are passed from one generation to the next. Third, he suggested that organisms change in response to their environments. And fourth, he hypothesized the existence of specific mechanisms that caused evolutionary change. The first three of these ideas became cornerstones of Darwin's evolutionary theory. Perhaps Lamarck's most important contribution was that he fostered discussion. By the mid-nineteenth century, most educated Europeans were talking about evolutionary change, whether they believed in it or not.

Geologists Recognized That Earth Had Changed over Time

In 1795, the Scottish geologist James Hutton (1726–1797) argued that slow and continuous physical processes, *acting over long periods of time,* produced Earth's major geological features; for example, the movement of water in a river slowly erodes the land and deposits sediments near the mouth of the river. Given enough time, erosion creates deep canyons, and sedimentation creates thick topsoil on flood plains. Hutton's **gradualism,** the view that Earth changed *slowly* over its history, contrasted sharply with Cuvier's catastrophism.

The English geologist Charles Lyell (1797–1875) championed and extended Hutton's ideas in an influential series of books, *Principles of Geology.* Lyell argued that the geological processes that sculpted Earth's surface over long periods of time—such as volcanic eruptions, earthquakes, erosion, and the formation and movement of glaciers—are exactly the same as the processes observed today. This concept, **uniformitarianism,** undermined any remaining notions of an unchanging Earth. Also, because geological processes proceed very slowly, it must have taken millions of years, not just a few thousand, to mold the landscape into its current configuration.

Figure 19.4
A great blue heron *(Ardea herodias).* Like many other wading birds, herons have long, stiltlike legs.

Rich Kirchner/Foto Natura/Photo Researchers, Inc.

STUDY BREAK

1. Why did the existence of vestigial structures make Buffon question the idea that living systems never changed?
2. What were Lamarck's contributions to an evolutionary worldview?
3. How do the concepts of gradualism and uniformitarianism in geology undermine the belief that Earth is only about 6000 years old?

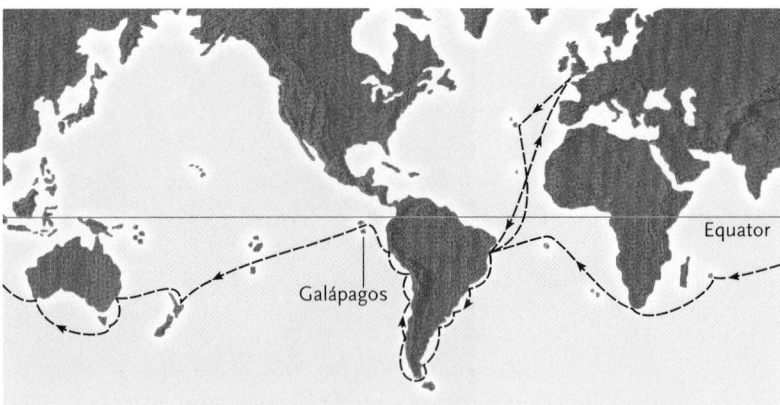

Figure 19.5
Darwin's voyage.
H.M.S. *Beagle* circumnavigated the globe between 1831 and 1836.

19.2 Darwin's Journeys

In 1831, in the midst of this intellectual ferment, young Charles Darwin wondered what to do with his life. Raised in a wealthy English household, he had always collected shells and studied the habits of insects and birds; he preferred hunting and fishing to classical studies. Despite lackluster performance as a student, Darwin was expected to continue the family tradition of practicing medicine. But he abandoned medical studies after 2 years. Instead, he followed his interest in natural history over the objections of his father, who reputedly told him, "You care for nothing but shooting, dogs, and rat-catching and you will be a disgrace to yourself and all of your family."

At the suggestion of his father, Darwin studied for a career as a clergyman, earning a degree at Cambridge University. There, he found a mentor in the Reverend John Henslow, a leading botanist, who arranged for Darwin to travel as the captain's dining companion aboard H.M.S. *Beagle,* a naval surveying ship. Darwin thus embarked on a sea voyage and an intellectual journey that altered the foundations of modern thought.

Darwin Saw the World on the Voyage of the *Beagle*

The *Beagle* sailed westward to map the coastline of South America and then circumnavigated the globe **(Figure 19.5).** When the ship's naturalist quit his post midjourney, Darwin replaced him in an unofficial capacity. For nearly 5 years Darwin toured the world, and because he suffered from seasickness, he seized every chance to go ashore. He collected plants and animals in Brazilian rain forests and fossils in Patagonia. He hiked the grasslands of the pampas and climbed the Andes in Chile. Armed with Henslow's parting gift, the first volume of Lyell's *Principles of Geology,* Darwin was primed to apply gradualism and uniformitarianism to the living world.

What Darwin Saw. When he began his travels, Darwin had no clue that biological evolution had produced the mind-boggling variety of species that he would en-

counter. Three broad sets of observations later helped him unravel the mystery of evolutionary change.

First, while exploring along the coast of Argentina, Darwin discovered fossils that often resembled organisms that inhabit the same region today. For example, despite an enormous size difference, living armadillos and fossilized glyptodonts had similar body armor, but they were unlike any other species known to science **(Figure 19.6).** If both species had been created at the same time and both were found in South America, why didn't glyptodonts live alongside armadillos? Darwin later wondered whether armadillos might be the living descendants of the now-extinct glyptodonts.

Second, Darwin observed that the animals he encountered in different South American habitats clearly resembled each other but differed from species that occupied similar habitats in Europe. For example, he

Figure 19.6
Ancestors and descendants. An extinct glyptodont (top) probably weighed 300 to 400 times as much as its living descendant, a nine-banded armadillo *(Dasypus novemcinctus).*

a. South American nutria

b. European beaver

Figure 19.7

Morphologic differences in species from different continents. Darwin noted that **(a)** South American nutria *(Myocastor coypus)* and **(b)** European beavers *(Castor fiber)* differ in appearance, even though both species are aquatic rodents that feed on vegetation. Notice that nutria have long, round tails, whereas beavers have short, flat tails.

a. The Galápagos

b. Galápagos tortoise

c. Marine iguana

d. Blue-footed booby

• Darwin

• Wolf

Pinta

Marchena   Genovesa

Santiago   Equator

Bartolomé

Seymour
Rábida   Baltra

Fernandina   Pinzón

Santa Cruz

Santa Fe

Tortuga   San Cristóbal

Isabela

Española

Floreana

Figure 19.8

The Galápagos. **(a)** Volcanic eruptions created the Galápagos archipelago (located 1000 km west of Ecuador) between 3 and 5 million years ago. **(b)** The islands were named for the giant tortoises found there (in Spanish, *galápa* means tortoise); this tortoise *(Geochelone elephantopus)* is native to Isla Santa Cruz. **(c)** Marine iguanas *(Amblyrhynchus cristatus)* dive into the Pacific Ocean to feed on algae. **(d)** A male blue-footed booby *(Sula nebouxii)* engages in a courtship display.

a. *Certhidea olivacea* **b.** *Geospiza scandens* **c.** *Geospiza magnirostris* **d.** *Camarhynchus pallidus*

Figure 19.9

Bill shape and food habits. The 13 finch species that inhabit the Galápagos are descended from a common ancestor, a seed-eating ground finch that migrated to the islands from South America. **(a)** *Certhidea olivacea* uses its slender bill to probe for insects in vegetation. **(b)** *Geospiza scandens* has a medium-sized bill suitable for eating cactus flowers and fruit. **(c)** *Geospiza magnirostris* uses its thick, strong bill to crush cactus seeds. **(d)** *Camarhynchus pallidus* uses its bill to hammer at bark and to hold cactus spines, with which it probes for wood-boring insects, such as termites.

noted that nutria *(Myocastor coypus),* a semiaquatic rodent in South America, bore a closer resemblance to rodent species from the mountains or grasslands of that continent than it did to the European beaver *(Castor fiber),* another semiaquatic rodent that had once been common in England **(Figure 19.7).** Why did animals from markedly different South American environments resemble each other, and why were animals that lived in similar environments on separate continents different? Darwin later understood that animals in South America resembled each other because they had inherited their similarities from a common ancestor.

Third, Darwin observed fascinating patterns in the distributions of species on the Galápagos **(Figure 19.8).** There he found strange and wonderful creatures, including giant tortoises and lizards that dove into the sea to feed on algae. Darwin quickly noted that the animals on different islands varied slightly in form. Indeed, experienced sailors could easily identify a tortoise's island of origin by the shape of its shell. Moreover, many species resembled those on the distant South American mainland. Why did so many different organisms occupy one small island cluster, and why did these species resemble others from the nearest continent? Darwin later hypothesized that the plants and animals of the Galápagos were descended from South American ancestors, and that each species had changed after being isolated on a particular island.

Darwin's Reflections after His Voyage. The *Beagle* returned to England in 1836, and Darwin began his first notebook on the *Transmutation of Species* the fol-

lowing year. He realized that changes in species over time provided the only plausible explanation for his observations.

A diverse group of finches from the Galápagos **(Figure 19.9)** provided the single greatest spark for Darwin's work. He had noticed great variability in the shapes of their bills, but he had incorrectly assumed that birds on different islands belonged to the same species. Thus, he had not recorded the island where he had captured each specimen. Luckily, the *Beagle*'s captain, Robert Fitzroy, had more thoroughly documented his own collection, allowing Darwin to study the relationships and geographical distributions of a dozen species. As Darwin reviewed his data, he began to focus on two aspects of a general problem. Why were the finches on a particular island slightly different from those on nearby islands, and how did all these different species arise?

Darwin Used Common Knowledge and Several Inferences to Develop His Theory

With a substantial inheritance and burdened by chronic illness, Darwin led a reclusive life as he embarked on an intellectual journey every bit as exciting as his voyage on the *Beagle* (see *Focus on Research*). His lifetime goal was to accumulate evidence of evolutionary change and identify the mechanism that caused it.

Selective Breeding and Heredity. Having grown up in the country, Darwin was well aware that "like begets like"; that is, offspring frequently resemble their parents. Plant and animal breeders had applied this basic truth of inheritance for thousands of years. By

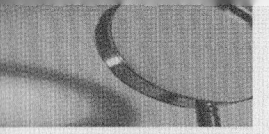

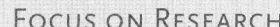

Basic Research: Charles Darwin's Life as a Scientist

Darwin's observations during the voyage of H.M.S. *Beagle* convinced him that species change through time, and that natural processes produced Earth's biodiversity. He spent the rest of his life gathering data to support his ideas and unravel the workings of natural selection.

Shortly after the *Beagle* returned to England in 1836, Darwin began his first notebook on the "transmutation of species." But he put his study of evolution aside while he wrote up the geological and biological research that he had undertaken during the voyage. This task took him 10 years to complete—twice as long as the journey itself. The results of these efforts were numerous articles and several books, including the now famous *Journal of the Voyage of the Beagle*, published in 1839.

After preparing a sketch of his ideas about evolution in 1844, Darwin continued to write up his observations from the voyage. But he had trouble classifying one species of barnacle, a small marine invertebrate, which he had collected in Chile. For the next 8 years he studied barnacles, examining more than 10,000 specimens and revising the entire classification of these animals. His colleagues saw this study as a strange diversion from his work on evolution, but Darwin's detailed examination of barnacle anatomy sharpened his observational skills and provided a test case in which he could apply his ideas about descent with modification to a large and diverse group of organisms. He published four volumes about barnacles in 1854.

While studying barnacles, Darwin continued to think about "the species question." He kept notebooks about variation in plants and animals, focusing on variation that was amplified by selective breeding. He was a tireless collector of facts, which he sought from every possible source. He badgered dog breeders, horse farmers, and horticulturists with long lists of questions about their work. His enthusiasm was infectious, and workers throughout the world supplied him with data and specimens. Darwin was also an eager and skilled experimentalist, and he took up pigeon breeding, marveling at the huge variety of morphological traits that he and other breeders could produce. In the late 1850s, a communication from another naturalist, Alfred Russel Wallace, forced him to finally complete *The Origin*, which revolutionized the study of biology.

Even after *The Origin* was published, Darwin continued to gather facts and write about evolution, working almost up to the day he died in 1882 at age 74. He published a detailed analysis of how earthworms improve the soil *(The Formation of Vegetable Mould through the Action of Worms)* and wrote books on several botanical topics, among them plants that eat animals *(Insectivorous Plants)*, pollination and fertilization systems *(Fertilisation in Orchids* and *The Effects of Self- and Cross-Fertilisation)*, and the tendency of plants to grow toward sunlight *(The Power of Movement in Plants)*. Darwin's work always had an evolutionary focus, however, and he produced several revisions of *The Origin*, as well as books on artificial selection *(Variation of Animals and Plants under Domestication)*, human ancestry *(The Descent of Man)*, and animal behavior *(The Expression of the Emotions in Men and Animals)*.

William Perlman/Star Ledger/Corbis

Darwin's study. Darwin undertook most of his life's work in this room at Down House. He hesitated to discard old papers and specimens, believing that he would find a use for them as soon as they were carried away in the trash.

selectively breeding individuals with favorable characteristics, they enhanced those traits in future generations.

Farmers use selective breeding to improve domesticated plants and animals. If one cow produces more milk than any other, the farmer selectively breeds her (rather than others), hoping that her offspring will also be good milk producers. Although the mechanism of heredity was not yet understood, this principle had been applied countless times to produce bigger beets, plumper pigs, and fancier pigeons (see Figure 1.10). Darwin was well aware of this process, which he called **artificial selection**, but he puzzled over how it could operate in nature. (*Insights from the Molecular Revolu-*

INSIGHTS FROM THE MOLECULAR REVOLUTION

Artificial Selection in the Test Tube

From Darwin's time until very recently, artificial selection was the province of plant and animal breeders, who chose individuals with desired traits to be the parents of the next generation. Now the laborious and time-consuming techniques of the breeders have been bypassed by rapid artificial selection experiments on DNA and protein molecules in the test tube.

One example of artificial selection in the test tube was provided by John J. Toole and his colleagues at Gilead Sciences in Foster City, California. They were interested in developing DNA molecules that could interfere with blood clotting by binding to thrombin, a blood protein that forms a major part of blood clots. The DNA could be used to treat people who are in danger of developing blood clots that might clog arteries in the heart, brain, or other critical organs. Nucleic acid molecules would be particularly useful as anticlotting agents because, unlike the proteins now used for this purpose, they rarely induce an immune reaction in the person being treated.

The investigators began their experiments by using a commercially available apparatus to make short, artificial DNA molecules of random sequence. They ran the apparatus long enough to produce more than 10^{13} (10 trillion!) different DNA sequences, and then made multiple copies of the sequences using the polymerase chain reaction (PCR; see Section 18.1). To select for DNA molecules that could bind to thrombin, they poured the entire DNA preparation through a column that contained thrombin molecules attached to glass beads. Only a few sequences among the trillions, about 0.01% of the total DNA sample, were able to bind strongly to thrombin. The researchers used PCR to multiply the sequences they had captured, generating 10 trillion "progeny" molecules. These progeny DNA molecules were poured through another column that contained thrombin molecules attached to glass beads. This time, a larger percentage of the molecules bound strongly to the thrombin molecules. These strongly binding DNA molecules were then used as the "parents" to generate another 10 trillion progeny. After five repetitions of the total process, producing five generations of DNA molecules, 40% of the DNA molecules in the preparation could recognize and bind strongly to thrombin.

The final products of the artificial selection were tested for their ability to interfere with the activity of thrombin in the blood clotting reaction. These experiments were successful; the anti-thrombin DNA molecules are being tested in monkeys and baboons, in which they appear to work effectively as anticlotting agents.

Toole and his team thus mimicked the evolutionary process on the molecular scale. Their experimental process selected DNA molecules that could bind to thrombin from the many random nucleotide sequences available in the test tube. The sequences that survived the selection test produced the greatest number of progeny molecules in the next generation. The same selection pressure, exerted over five generations of progeny molecules, greatly increased the percentage that could bind strongly to the protein. As a result, the DNA population evolved in the test tube from one with little or no ability to bind thrombin to one with high ability.

This approach is being used in many laboratories to develop DNA and RNA molecules with desired functions. By starting with DNA molecules that encode enzymes, researchers hope to select biological catalysts that can speed chemical reactions with scientific, medical, or industrial purposes.

---

tion describes how modern researchers apply artificial selection to molecules in a test tube.)

The Struggle for Existence. Darwin had a revelation about how selective breeding could occur naturally when he read the famous publication by Thomas Malthus, *Essay on the Principles of Population*. Malthus, an English clergyman and economist, was worried about the fate of the nation's poor. England's population was growing much faster than its agricultural capacity, and with individuals competing for limited food resources, some would inevitably starve.

Darwin applied Malthus's argument to organisms in nature. Species typically produce many more offspring than are needed to replace the parent generation, yet the world is not overrun with sunflowers, tortoises, or bears. Darwin even calculated that, if its reproduction went unchecked, a single pair of elephants, the slowest breeding animal known, would leave roughly 19 million descendants after only 750 years. Happily for us (and all other species that might get underfoot), the world is not so crowded with elephants. Instead, some members of every population survive and reproduce, whereas others die without reproducing.

Darwin's Inferences. Darwin's discovery of a mechanism for evolutionary change required him to infer the nature of a process that no one had envisioned, much less documented **(Table 19.1).** First, individuals within populations vary in size, form, color, behavior, and other characteristics. Second, many of these variations are hereditary. What if variations in hereditary traits enabled some individuals to survive and reproduce more than others? Organisms with favorable traits would leave many offspring, whereas those that lacked favorable traits would die leaving

Table 19.1 — Darwin's Observations and Inferences about Evolution by Means of Natural Selection

Observations	Inferences	
Most organisms produce more than one or two offspring.	Individuals within a population compete for limited resources.	A population's characteristics will change over the generations as advantageous, heritable characteristics become more common.
Populations do not increase in size indefinitely.		
Food and other resources are limited for most populations.		
Individuals within populations exhibit variability in many characteristics.	Hereditary characteristics may allow some individuals to survive longer and reproduce more than others.	
Many variations have a genetic basis that is inherited by subsequent generations.		

few, if any, descendants. Thus, favorable hereditary traits would become more common in the next generation. If the next generation was subjected to the same process of selection, the traits would be even more common in the third generation. Because this process is analogous to artificial selection, Darwin called it **natural selection.**

As an evolutionary mechanism, natural selection favors **adaptive traits**, genetically based characteristics that make organisms more likely to survive and reproduce. And by favoring individuals that are well adapted to the environments in which they live, natural selection causes species to change through time. As shown in Figure 19.9, each species of Galápagos finch has a distinctive bill. Variations in bill size and shape make some birds better adapted for crushing seeds and others for capturing insects. Imagine an island where large seeds were the only food available; individuals with a stout bill would be more likely to survive and reproduce than would birds with slender bills. These favored individuals would pass the genes that produce stout bills to their descendants, and after many generations, their bills might resemble those of *Geospiza magnirostris* (see Figure 19.9c). Natural selection also changes nonmorphologic characteristics of populations; for example, insect populations that are exposed to insecticides develop resistance to these toxic chemicals over time (see Figure 19.11).

Darwin realized that natural selection could also account for striking differences between populations and, given enough time, for the production of new species. For example, suppose that small insects were the only food available to finches on a different island. Birds with long thin bills might be favored by natural selection, and the population of finches might eventually possess a bill shaped like that of *Certhidea olivacea* (see Figure 19.9a). If we apply parallel reasoning to the many characteristics that affect survival and reproduction, natural selection would cause the populations to become more different over time, a process called **evolutionary divergence.**

Darwin's Theory Revolutionized the Way We Think about the Living World

It would be hard to overestimate the impact of Darwin's theory on Western thought. In *The Origin,* Darwin proposed a logical mechanism for evolutionary change and provided enough supporting evidence to convince the educated public.

Darwin argued that all the organisms that have ever lived arose through **descent with modification,** the evolutionary alteration and diversification of ancestral species. He envisioned this pattern of descent as a tree growing through time **(Figure 19.10).** The base of the

Present

Time

Origin of life

Figure 19.10
The tree of life. Darwin envisioned the history of life as a tree. Branching points represent the origins of new lineages; branches that do not reach the top represent extinct groups.

trunk represents the ancestor of all organisms. Branching points above it represent the evolutionary divergence of ancestors into their descendants. Each limb represents a body plan suitable for a particular way of life; smaller branches represent more narrowly defined groups of organisms; and the uppermost twigs represent living species.

Darwin proposed natural selection as the mechanism that drives evolutionary change. In fact, most of *The Origin* was an explanation of how natural selection acted on the variability within groups of organisms, preserving favorable traits and eliminating unfavorable ones.

Four characteristics distinguish Darwin's theory from earlier explanations of biological diversity and adaptive traits:

1. Darwin provided purely physical, rather than spiritual, explanations about the origins of biological diversity.
2. Darwin recognized that evolutionary change occurs in groups of organisms, rather than in individuals: some members of a group survive and reproduce more successfully than others.
3. Darwin described evolution as a multistage process: variations arise within groups, natural selection eliminates unsuccessful variations, and the next generation inherits successful variations.
4. Like Lamarck, Darwin understood that evolution occurs because some organisms function better than others *in a particular environment.*

What is most amazing about Darwin's intellectual achievement is that he knew nothing about Mendelian genetics (see Chapter 12). Thus, he had no clear idea of how variation arose or how it was passed from one generation to the next.

Evolution was a popular topic in Victorian England, and Darwin's theory was both praised and ridiculed. Although he had not speculated about the evolution of humans in *The Origin,* many readers were quick to extrapolate Darwin's ideas to our own species. Needless to say, certain influential Victorians were not amused by the suggestion that humans and apes share a common ancestry.

Nevertheless, Darwin's painstaking logic and careful documentation convinced most readers that evolution really does take place. Thomas Huxley, so staunch an advocate that he was known as "Darwin's bulldog," summed up the reaction of many when he quipped that the theory was so obvious, once articulated, that he was surprised he had not thought of it himself. Darwin's vision of common ancestry quickly became the intellectual framework for nearly all biological research. Many readers, however, did not readily accept the mechanism of natural selection. The major stumbling block was that Darwin had not provided any plausible theory of heredity.

STUDY BREAK

1. What observations that Darwin made on his round-the-world voyage influenced his later thoughts about evolution?
2. How did Darwin's understanding of artificial selection enable him to envision the process of natural selection?
3. What were the four great intellectual triumphs of Darwin's theory?

19.3 Evolutionary Biology since Darwin

Although Gregor Mendel published his work on genetics in 1866, it was not well known in England until 1900. At that time, scientists perceived a fundamental conflict between Darwin's and Mendel's theories. One problem was that Darwin had used complex characteristics, such as the structure of bird bills, to illustrate how natural selection worked. We now know that at least several genes often control such traits. By contrast, Mendel had studied simpler characteristics, such as the height of pea plants (see Chapter 12). A single gene often controls simple traits, which is one reason Mendel could interpret his experimental results so clearly. Biologists had a hard time applying Mendel's straightforward experimental results to Darwin's complex examples.

A second problem arose because Darwin believed that biological evolution occurred gradually over many generations. However, early twentieth-century geneticists, focusing on simple traits such as those Mendel had studied, sometimes observed very rapid and dramatic changes in certain characteristics. A widely accepted theory, *mutationism* suggested that evolution occurred in spurts, induced by the chance appearance of "hopeful monsters," rather than by gradual change.

The Modern Synthesis Created a Unified Theory of Evolution

In the 1910s and 1920s, geneticists and mathematicians forged a critical link between Darwinism and Mendelism. The new discipline, **population genetics**, recognized the importance of genetic variation as the raw material of evolution. Population geneticists constructed mathematical models, which applied equally well to simple and complex traits, to predict how natural selection and other processes influence a population's genetics.

In the 1930s and 1940s, a unified theory of evolution, the **modern synthesis,** interpreted data from biogeography, comparative morphology, comparative

Figure 19.11 Experimental Research

How Exposure to Insecticide Fosters the Evolution of Insecticide Resistance

QUESTION: Does exposure to insecticide foster the evolution of insecticide resistance in insect populations?

EXPERIMENT: Researchers studied samples of wild mosquitoes *(Anopheles culicifacies)* captured at a small village in India, where public health officials frequently sprayed the insecticide dichloro-diphenyl-trichloroethane (DDT) to control these pests. For each test, the researchers exposed samples of mosquitoes to a 4% concentration of DDT for 1 hour and then measured the percentage that died during the next 24 hours. Tests were repeated 12 months and 16 months after the first experiment.

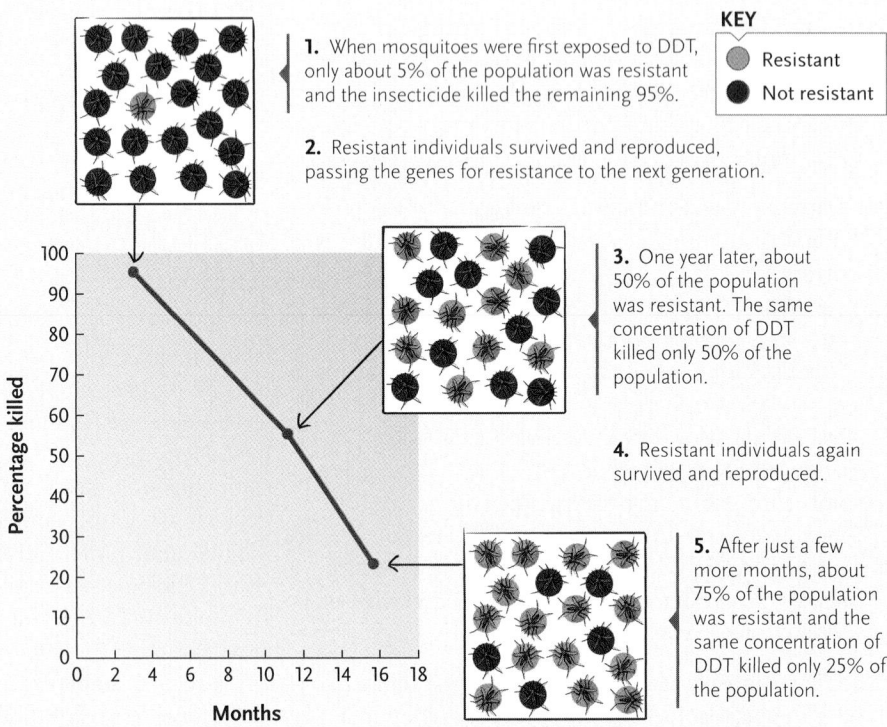

KEY
- ◒ Resistant
- ● Not resistant

1. When mosquitoes were first exposed to DDT, only about 5% of the population was resistant and the insecticide killed the remaining 95%.

2. Resistant individuals survived and reproduced, passing the genes for resistance to the next generation.

3. One year later, about 50% of the population was resistant. The same concentration of DDT killed only 50% of the population.

4. Resistant individuals again survived and reproduced.

5. After just a few more months, about 75% of the population was resistant and the same concentration of DDT killed only 25% of the population.

RESULTS: Over the course of the experiment, smaller and smaller percentages of the mosquitoes died after their exposure to the test concentration of the insecticide.

CONCLUSION: The indiscriminate use of DDT established natural selection that favored DDT-resistant individuals. Exposure to DDT therefore fostered the evolution of an adaptive resistance to DDT in the mosquito population.

---

embryology, paleontology, and taxonomy within an evolutionary framework. The authors of the modern synthesis focused on evolutionary change within populations, and although they considered natural selection the primary mechanism of evolution, they acknowledged the importance of other processes (see Chapter 20). Proponents of the modern synthesis also embraced Darwin's idea of gradualism and deemphasized the significance of mutations that changed traits suddenly and dramatically.

The modern synthesis also tried to link the two levels of evolutionary change that Darwin had identified: microevolution and macroevolution. **Microevolution** describes the small-scale genetic changes that populations undergo, often in response to shifting environmental circumstances; a small evolutionary shift in the size of the bill of a finch species is an example of microevolution. **Macroevolution** describes large-scale patterns in the history of life, such as the appearance and then relatively sudden disappearance of gigantic dinosaurs. According to the modern synthesis, macroevolution results from the gradual accumulation of microevolutionary changes, but researchers are just beginning to unravel the genetic mechanisms that establish a relationship between these two levels of evolutionary change (see Chapter 22).

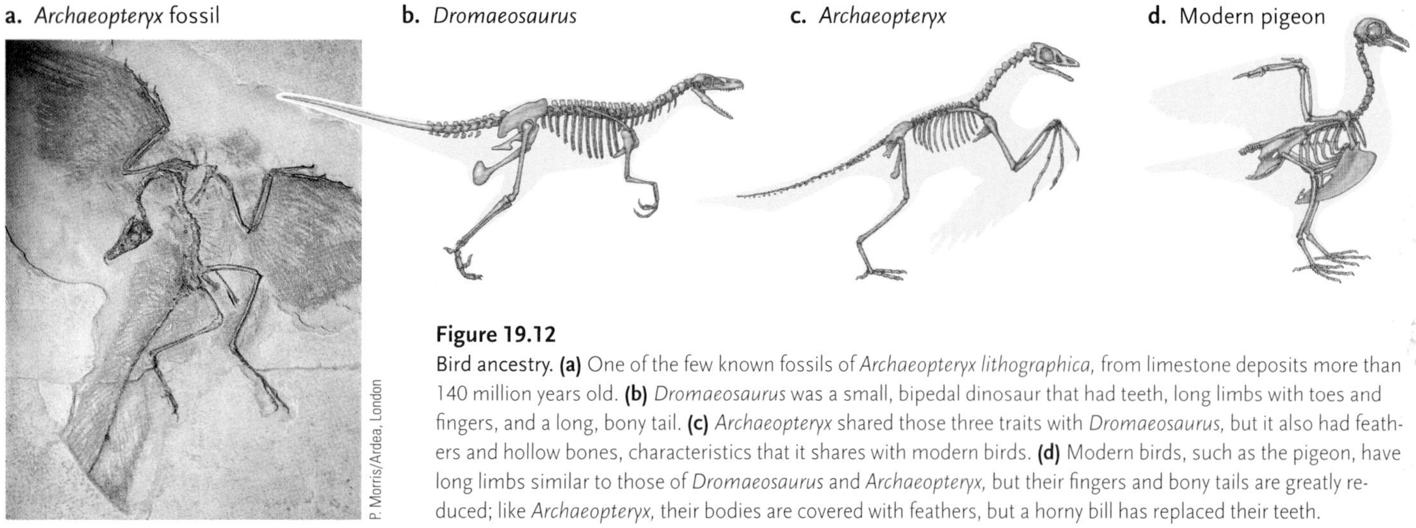

a. *Archaeopteryx* fossil   **b.** *Dromaeosaurus*   **c.** *Archaeopteryx*   **d.** Modern pigeon

Figure 19.12

Bird ancestry. **(a)** One of the few known fossils of *Archaeopteryx lithographica*, from limestone deposits more than 140 million years old. **(b)** *Dromaeosaurus* was a small, bipedal dinosaur that had teeth, long limbs with toes and fingers, and a long, bony tail. **(c)** *Archaeopteryx* shared those three traits with *Dromaeosaurus*, but it also had feathers and hollow bones, characteristics that it shares with modern birds. **(d)** Modern birds, such as the pigeon, have long limbs similar to those of *Dromaeosaurus* and *Archaeopteryx*, but their fingers and bony tails are greatly reduced; like *Archaeopteryx*, their bodies are covered with feathers, but a horny bill has replaced their teeth.

Research in Many Fields Has Provided Evidence of Evolutionary Change

During the past 100 years, scientists have assembled a huge and compelling body of evidence from many biological disciplines indicating that biological evolution is a fact of life on Earth.

Adaptation by Natural Selection. Biologists interpret the products of natural selection as evolutionary adaptations. For example, the wings of birds, which have been modified by evolutionary processes over millions of years, have an obvious function that helps these animals survive and reproduce. Sometimes, however, natural selection operates on a short time scale, as illustrated by the development of pesticide resistance in insects. When we first use a new pesticide, a low concentration often kills a large percentage of the pests. However, just by chance, a few insects may have genetic characteristics that confer resistance to the poison. The surviving individuals produce offspring, many of which inherit the resistance. As a result, a given concentration of the poison kills a smaller percentage of insects in the next generation; therefore, over time, the entire population may become highly resistant **(Figure 19.11)**.

The Fossil Record. Because evolution results from the modification of existing species, Darwin's theory proposes that all species that have ever lived are genetically related. The fossil record documents such continuity, providing clear evidence of ongoing change in many **biological lineages**, evolutionary sequences of ancestral organisms and their descendants (see Chapter 22). For example, the evolution of modern birds can be traced from a dinosaur ancestor through fossils such as *Archaeopteryx lithographica* **(Figure 19.12)**. This species, discovered only 2 years after *The Origin* was published, resembled both dinosaurs and birds. Like small carnivorous dinosaurs, *Archaeopteryx* walked on its hind legs and had teeth, claws on its forelimbs, and a long, bony tail. Like modern birds, it had hollow bones, an enlarged sternum, and feathers that covered its body.

Historical Biogeography. Analyses of **historical biogeography,** the study of the geographical distributions of plants and animals in relation to their evolutionary history, are generally consistent with Darwin's theory of evolution. Species on oceanic islands often closely resemble species on the nearest mainland, suggesting that the island and mainland species share a common ancestry. Moreover, species on a continental land mass are clearly related to one another and are often distinct from those on other continents. For example, monkeys in South America have long, prehensile tails and broad noses, traits that they inherited from a shared South American ancestor. By contrast, monkeys in Africa and Asia evolved from a different common ancestor in the Old World, and their shorter tails and narrower noses distinguish them from their American cousins.

Comparative Morphology. Other evidence of evolution comes from **comparative morphology,** analyses of the structure of living and extinct organisms. Such analyses are based on the comparison of **homologous traits,** characteristics that are similar in two species because they inherited the genetic basis of the trait from their common ancestor. For example, the forelimbs of all four-legged vertebrates are homologous because they evolved from a common ancestor with a forelimb composed of the same component parts (see Figure 19.3, which shows homologous bones in the same color). Even though the shapes of the bones are different in pigs, dolphins, and bats, similarities in the three limbs are apparent. The differences in structural details arose over evolutionary time, allowing pigs to walk, dolphins to swim, and bats to fly. The arms of humans and the wings of birds are also constructed of comparable elements, suggesting that they, too, share a common ancestor with the three species illustrated.

Comparative Embryology. The early embryos of different species within a major group of organisms are often strikingly similar. For example, certain components of the circulatory system emerge in all vertebrate embryos at corresponding stages of development **(Figure 19.13).** In addition, the early embryos of humans and other four-limbed vertebrates possess gill pouches (similar to those in adult fishes) and a tiny tail. These embryonic similarities indicate that fishes, amphibians, reptiles, birds, and mammals all evolved from a common ancestor. Additional genetic instructions have also evolved, causing their adult morphology to diverge.

Comparative Molecular Biology. The genes and proteins of different species also contain information about evolutionary relationships. The very existence of a common genetic code is powerful evidence for the relatedness of all forms of life. Moreover, some genes and their protein products are present in most living organisms, an observation that is most easily explained by the hypothesis of common ancestry. For example, cytochrome *c,* a protein involved in cellular respiration (see Section 8.4), is found within the mitochondria of

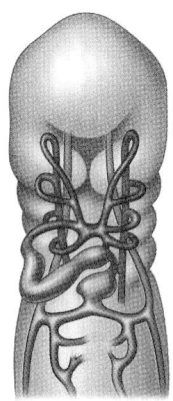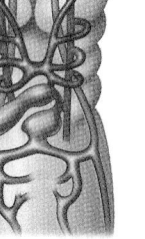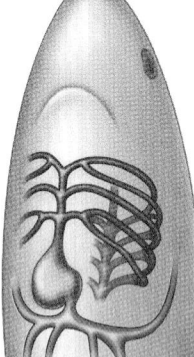

Human embryo   Adult shark

Figure 19.13
Embryologic clues to evolutionary history. Related species often show similar patterns of embryonic development. The aortic arches (red), a two-chambered heart (orange), and a set of veins (blue) in an early human embryo are also present in the embryos of other vertebrates. These structures persist into adulthood in some fishes, such as sharks.

all eukaryotic organisms. Evolutionary processes have modified the gene that codes for this protein, establishing variations in its amino acid sequence among different groups of organisms. Closely related species—for example, humans and their fellow primates, chimpanzees and rhesus monkeys—exhibit few differences in the amino acid sequence; more distantly related organisms, such as humans and yeast, exhibit many differences **(Figure 19.14).**

Some People Misinterpret the Theory of Evolution

The theory of evolution has always been a contentious subject because it challenges deeply held traditional views of how living organisms originated. Many of Darwin's contemporaries were dismayed by the suggestion that all organisms share a common ancestry. Some people even misinterpreted this assertion as "humans evolved from chimpanzees or gorillas." But the theory of evolution makes no such claims. Instead, it suggests that humans and apes are descended from an apelike common ancestor (see Section 30.13). In other words, an ancient population of organisms left descendants, which now include the living species of apes, as well as our own species. Moreover, the theory recognizes that evolution is an ongoing process: humans and apes have been evolving up until this very moment and will continue to evolve for as long as their descendants persist.

Early in the twentieth century, some scientists embraced the notion of **orthogenesis,** or progressive, goal-oriented evolution. This idea, derived from the *Scala Naturae,* suggests that evolution produces new species with the goal of improvement "in mind." We now know that evolution proceeds as an ongoing process of dynamic adjustment, not toward any fixed goal. Natural selection preserves the genes of organisms that function well in particular environments, but it cannot predict future environmental change. Imagine a population of plants with genes that affect how well they function under wet versus dry conditions. After a 5-year drought, the population would include mostly dry-adapted plants. If a series of wet years follows the drought, these plants will be poorly adapted to the altered conditions. The process that favored drought-adapted plants operated under the prevailing dry conditions, not in anticipation of how conditions might change in the future.

Evolution is the core theory of modern biology because its explanatory power touches on every aspect of the living world. And the application of molecular techniques to the study of evolutionary biology has greatly enhanced our knowledge. Despite some common misunderstandings about what the theory predicts, the study of evolution is alive and

Figure 19.14 Observational Research

How Differences in Amino Acid Sequences among Species Reflect Their Evolutionary Relationships

HYPOTHESIS: The genetic instructions coding for proteins are more similar in closely related species than they are in more distantly related species.

PREDICTION: The amino acid sequences for a particular protein will be more similar in closely related species than in more distantly related species.

OBSERVATIONAL METHODS: Researchers gathered the amino acid sequences for the protein cytochrome *c* from a variety of organisms and compared them with the 104 amino acid sequence of this protein in humans.

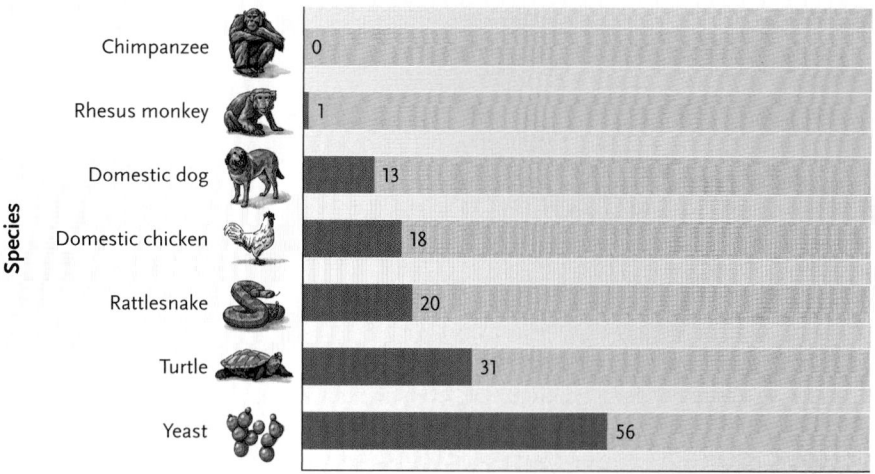

Number of amino acids that differ from the human sequence

RESULTS: Species that are closely related to humans, such as chimpanzees and rhesus monkeys, have amino acid sequences that are identical or nearly identical to the sequence in humans. More distantly related species, such as turtles and yeasts, exhibit sequences that are quite different from the sequence in humans.

CONCLUSION: Closely related species have very similar amino acid sequences in their proteins, reflecting similarities in their genetic makeup. More distantly related species exhibit substantial differences in amino acid sequences, reflecting the genetic divergence among them.

well. In fact, in late 2005, *Science* magazine, a prestigious scientific journal devoted to all of the natural sciences, declared "Evolution in Action" as the breakthrough of the year. The editorial staff cited exciting recent discoveries about genetic differences among organisms ranging from bacteria to humans, mechanisms that promote species formation, and the regulatory genes that may bridge the gap between microevolution and macroevolution.

In the remaining chapters of this unit you will discover how contemporary evolutionary theory explains changes at every level of biological organization from adaptive modifications within populations (see Chapter 20), to the development of new species (see Chapter 21), to the history of life (see Chapter 22), and the classification of all organisms on Earth (see Chapter 23).

STUDY BREAK

1. What two problems slowed the acceptance of Darwin's theory among scientists?
2. What is the difference between microevolution and macroevolution?
3. What types of data provide evidence that evolution has adapted organisms to their environments and promoted the diversification of species?

What determines whether a species adapts to a changing environment or becomes extinct?

Natural selection has produced marvelous adaptations in every species on Earth, and we know that evolutionary adaptation to certain environmental changes has allowed many species to persist. But we also know that more than 99% of the species that have ever lived became extinct, evidently because they failed to adapt to changes in climate, natural competitors or enemies, or other environmental factors. But what kinds of genetic variation are required for adaptation, and what kinds of characteristics must evolve to allow survival? This is a critical question today, because human activities are changing environments so rapidly and drastically that many species face the threat of extinction. Can aquatic species adapt to various kinds of water pollution? Can animals and plants that lived in prairies adapt to different habitats, now that most prairies have been destroyed? Can Arctic species adapt to changes in climate as human production of carbon dioxide increases Earth's average temperature faster than ever before?

Is adaptation by natural selection responsible for most of the genetic differences between species?

New genetic variations sometimes become more common within populations or species because the proteins for which they code are advantageous and preserved by natural selection. But biologists who study molecular evolution have discovered that a large part of the genome in most organisms (about 98% of the human genome, for example) does not code for proteins and therefore appears to have no function. If this observation is generally correct, why do the noncoding parts of genomes exist? Are evolutionary changes in noncoding regions and the differences in noncoding sequences among species adaptive? For example, only about 1% of the DNA base pairs differ between human and chimpanzee genomes—but this amounts to about 34 million base-pair differences altogether, at least 60,000 of which alter the amino acid sequences of proteins. How can we determine which of these differences are adaptive and which differences underlie the unique characteristics of humans?

How do pathways of embryonic development evolve?

The characteristics of adult organisms are the product of developmental events, starting with the fertilized egg, that include growth in size, changes in the shape of various body parts, and the differentiation of cell types. These processes are largely controlled by genes, with input from the environment. Although biologists are beginning to learn how the genetic foundations of developmental processes evolve, many questions remain. For example, how do genetic changes induce differences in the branching patterns of antlers among species of deer, or differences in the length of the tails of monkeys and apes (including humans), or differences in the number and size of scales among species of lizards? We know that the proteins forming the lens of the eye are actually enzymes that play different roles in other cells, and that they have been "recruited" to form the lens, but what mechanisms induce them to assume this new role? And why do different enzymes form the lens in eyes of birds and mammals? Evolutionary developmental biology, which is discussed in Chapter 22, is one of the most active, exciting fields in biology at this time.

 Douglas J. Futuyma is Distinguished Professor in the Department of Evolution and Ecology at Stony Brook University. His research interests focus on speciation and the evolution of ecological interactions among species, and in particular on insect–plant interactions. Learn more about his work at http://life.bio.sunysb.edu/ee/people/futuyindex.html.

Review

Go to **ThomsonNOW** at www.thomsonedu.com/login to access quizzing, animations, exercises, articles, and personalized homework help.

19.1 Recognition of Evolutionary Change

- Ancient Greek philosophers classified the natural world, ranking inanimate objects and living organisms from simple to complex.

- Natural theologians, who merged Greek philosophy with the biblical account of creation, believed that all species were specially created and perfectly adapted. Existing species could not change or become extinct, and new species could not arise. Studies in biogeography, comparative morphology, and paleontology led scientists to wonder whether species might change through time (Figures 19.2 and 19.3).

- Lamarck developed the first comprehensive theory of biological evolution; he proposed that species evolved into more complex forms that functioned better in their environments. He hypothesized that structures in an organism changed when they were used, and that those changes were inherited by the organism's offspring. Experiments have refuted Lamarck's proposed mechanisms.

- Two geologists, Hutton and Lyell, recognized that major features on Earth were created by the long-term action of the very slow geological processes that scientists observe today. Their insights suggested that Earth was much older than natural theologians had supposed.

19.2 Darwin's Journeys

- Darwin's observations during his voyage on the *Beagle* provided much of the data and inspiration for the development of his theory of evolution (Figures 19.5–19.8).

- Darwin based the theory of evolution by means of natural selection on three inferences: (1) individuals within a population compete for limited resources, (2) hereditary characteristics allow some individuals to survive longer and reproduce more than others, and (3) a population's characteristics change over time as advantageous heritable characteristics become more common (Table 19.1).

- Darwin also proposed that the accumulation of differences fostered by natural selection could cause populations to diverge over time. Such evolutionary divergence can lead to the production of new species, which can, in turn, give rise to new evolutionary lineages (Figures 19.9 and 19.10).

Animation: The Galápagos
Animation: Finches of the Galpágos

19.3 Evolutionary Biology since Darwin

- Scientists working in population genetics developed theories of evolutionary change by integrating Darwin's ideas with Mendel's research on genetics.

- In the 1930s and 1940s, the modern synthesis provided a unified view of evolution that drew on studies from many biological disciplines. It emphasized evolution within populations, the central role of variation in the evolutionary process, and the gradualism of evolutionary change.

- Studies of adaptation, the fossil record, historical biogeography, comparative morphology, comparative embryology, and comparative molecular biology provide compelling evidence of evolutionary change (Figures 19.11–19.14).

- Evolutionary biology is an active field of study, and the application of molecular techniques is yielding new answers to old questions.

Questions

Self-Test Questions

1. Which of the following statements about evolutionary studies is *not* true?
   a. Biologists study the products of evolution to understand the processes causing it.
   b. Biologists design molecular experiments to examine evolutionary processes operating over short time periods.
   c. Biologists study the inheritance of characteristics that a parent acquired during its lifetime.
   d. Biologists study variation in homologous structures among related organisms.
   e. Biologists examine why a huge variety of species may inhabit a small island cluster.

2. Which of the following ideas is *not* included in Darwin's theory?
   a. All organisms that have ever existed arose through evolutionary modifications of ancestral species.
   b. The great variety of species alive today resulted from the diversification of ancestral species.
   c. Natural selection drives some evolutionary change.
   d. Natural selection preserves favorable traits.
   e. Natural selection eliminates adaptive traits.

3. The father of taxonomy is:
   a. Charles Darwin.
   b. Charles Lyell.
   c. Alfred Wallace.
   d. Carolus Linnaeus.
   e. Jean Baptiste de Lamarck.

4. The wings of birds, the legs of pigs, and the flippers of whales provide an example of:
   a. vestigial structures.
   b. homologous structures.
   c. acquired characteristics.
   d. artificial selection.
   e. uniformitarianism.

5. Which of the following statements is *not* compatible with Darwin's theory?
   a. All organisms have arisen by descent with modification.
   b. Evolution has altered and diversified ancestral species.
   c. Evolution occurs in individuals rather than in groups.
   d. Natural selection eliminates unsuccessful variations.
   e. Evolution occurs because some individuals function better than others in a particular environment.

6. Which of the following does *not* contribute to the study of evolution?
   a. population genetics
   b. inheritance of acquired characteristics
   c. the fossil record
   d. DNA sequencing
   e. comparative morphology

7. Which of the following could be an example of microevolution?
   a. a slight change in a bird population's color due to a small genetic change in the population
   b. large differences between fossils found near the ground surface and those found in deep rock layers
   c. the sudden disappearance of an entire genus
   d. the direct evolutionary link between living primates and humans
   e. a flood that drowns all members of a population

8. Which of the following ideas proposed by Lamarck was *not* included in Darwin's theory?
   a. Organisms change in response to their environments.
   b. Changes that an organism acquires during its lifetime are passed to its offspring.
   c. All species change with time.
   d. Genetic changes may be passed from one generation to the next.
   e. Specific mechanisms cause evolutionary change.

9. Medical advances now allow many people who suffer from genetic diseases to survive and reproduce. These advances:
   a. refute Darwin's theory.
   b. support Lamarck's theory.
   c. disprove descent with modification.
   d. reduce the effects of natural selection.
   e. eliminate adaptive traits.

10. The belief that evolution is progressive or goal-oriented is called:
    a. gradualism.
    b. uniformitarianism.
    c. taxonomy.
    d. orthogenesis.
    e. the modern synthesis.

Questions for Discussion

1. Explain why the characteristics we see in living organisms adapt them to the environments in which their ancestors lived rather than to the environments in which they live today.

2. Imagine a population of mice that includes both brown and black individuals. They live in a habitat with brown soil, where predatory hawks can see black mice more easily than they can see brown ones. Design a study that would allow you to determine whether the brown mice are better adapted to this environment than black mice.

3. Find examples from popular publications or advertisements for consumer products that misrepresent the theory of biological evolution. Explain how the theory is misrepresented.

Experimental Analysis

Design an experiment to test Lamarck's hypothesis that characteristics acquired during an organism's lifetime are inherited by their offspring. (You may wish to review the components of a well-designed experiment in Chapter 1 before formulating your answer.) Can you think of examples of acquired characteristics that are *not* inherited by offspring?

Evolution Link

Identify three discoveries or inventions that have changed how humans are affected by natural selection. Describe in detail how each discovery influences survival or reproduction in our species.

How Would You Vote?

A large asteroid could obliterate civilization and much of Earth's biodiversity. Should nations around the world contribute to locating and tracking asteroids? Go to www.thomsonedu.com/login to investigate both sides of the issue and then vote.

Phenotypic variation. The frog *Dendrobates pumilio* exhibits dramatic color variation in populations that inhabit the Bocas del Toro Islands, Panama.

© Mark Moffett/Foto Natura/Minden Pictures

20 Microevolution: Genetic Changes within Populations

WHY IT MATTERS

On November 28, 1942, at the height of American involvement in World War II, a disastrous fire killed more than 400 people in Boston's Cocoanut Grove nightclub. Many more would have died later but for a new experimental drug, penicillin. A product of *Penicillium* mold, penicillin fought the usually fatal infections of *Staphylococcus aureus,* a bacterium that enters the body through damaged skin. Penicillin was the first antibiotic drug based on a naturally occurring substance that kills bacteria.

Until the disaster at the Cocoanut Grove, the production and use of penicillin had been a closely guarded military secret. But after its public debut, the pharmaceutical industry hailed penicillin as a wonder drug, promoting its use for the treatment of the many diseases caused by infectious microorganisms. Penicillin became widely available as an over-the-counter remedy, and Americans dosed themselves with it, hoping to cure all sorts of ills **(Figure 20.1)**. But in 1945, Alexander Fleming, the scientist who discovered penicillin, predicted that some bacteria could survive low doses, and that the offspring of those germs would be more resistant to its effects. In 1946—just 4 years after penicillin's use in Boston—14% of the *Staphylococcus* strains

Figure 20.1

Selling penicillin. This ad, from a 1944 issue of *Life* magazine, credits penicillin with saving the lives of wounded soldiers.

microorganisms—along with the genes that confer antibiotic resistance—become more common in later generations. In other words, bacterial strains adapt to antibiotics through the evolutionary process of selection. Our use of antibiotics is comparable to artificial selection by plant and animal breeders (see Chapter 19), but when we use antibiotics, we inadvertently select for the success of organisms that we are trying to eradicate.

The evolution of antibiotic resistance in bacteria is an example of **microevolution**, which is a heritable change in the genetics of a population. A **population** of organisms includes all the individuals of a single species that live together in the same place and time. Today, when scientists study microevolution, they analyze variation—the differences between individuals—in natural populations and determine how and why these variations are inherited. Darwin recognized the importance of heritable variation within populations; he also realized that natural selection can change the pattern of variation in a population from one generation to the next. Scientists have since learned that microevolutionary change results from several processes, not just natural selection, and that sometimes these processes counteract each other.

In this chapter, we first examine the extensive variation that exists within natural populations. We then take a detailed look at the most important processes that alter genetic variation within populations, causing microevolutionary change. Finally, we consider how microevolution can fine-tune the functioning of populations within their environments.

isolated from patients in a London hospital were resistant. By 1950, more than half the strains were resistant.

Scientists and physicians have discovered numerous antibiotics since the 1940s, and many strains of bacteria have developed resistance to these drugs. In fact, according to the Centers for Disease Control and Prevention, between 30,000 and 40,000 Americans die each year from infection by antibiotic-resistant bacteria.

How do bacteria become resistant to antibiotics? The genomes of bacteria—like those of all other organisms—vary among individuals, and some bacteria have genetic traits that allow them to withstand attack by antibiotics. When we administer antibiotics to an infected patient, we create an environment favoring bacteria that are even slightly resistant to the drug. The surviving bacteria reproduce, and resistant

20.1 Variation in Natural Populations

In some species, individuals vary dramatically in appearance; but in most species, the members of a population look pretty much alike **(Figure 20.2)**. Even those that look alike, such as the *Cerion* snails on the right in Figure 20.2, are not identical, however. With a scale and ruler, you could detect differences in their weight as well as in

a. European garden snails

b. Bahaman land snails

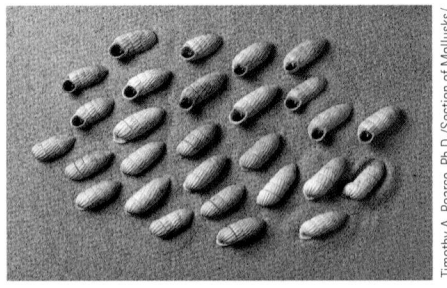

Figure 20.2

Phenotypic variation. **(a)** Shells of the European garden snail *(Cepaea nemoralis)* from a population in Scotland vary considerably in appearance. **(b)** By contrast, shells of *Cerion christophei* from a population in the Bahamas look very similar.

the length and diameter of their shells. With suitable techniques, you could also document variations in their individual biochemistry, physiology, internal anatomy, and behavior. All of these are examples of **phenotypic variation**, differences in appearance or function that are passed from generation to generation.

Evolutionary Biologists Describe and Quantify Phenotypic Variation

Darwin's theory recognized the importance of heritable phenotypic variation, and today, microevolutionary studies often begin by assessing phenotypic variation within populations. Most characters exhibit **quantitative variation**: individuals differ in small, incremental ways. If you weighed everyone in your biology class, for example, you would see that weight varies almost continuously from your lightest to your heaviest classmate. Humans also exhibit quantitative variation in the length of their toes, the number of hairs on their heads, and their height, as discussed in Chapter 12.

We usually display data on quantitative variation in a bar graph or, if the sample is large enough, as a curve **(Figure 20.3)**. The width of the curve is proportional to the variability—the amount of variation—among individuals, and the *mean* describes the average value of the character. As you will see shortly, natural selection often changes the mean value of a character or its variability within populations.

Other characters, like those Mendel studied (see Section 12.1), exhibit **qualitative variation**: they exist in two or more discrete states, and intermediate forms are often absent. Snow geese, for example, have *either* blue *or* white feathers **(Figure 20.4)**. The existence of discrete variants of a character is called a **polymorphism** (*poly* = many; *morphos* = form); we describe such traits as *polymorphic*. The *Cepaea nemoralis* snail shells in Figure 20.2a are polymorphic in background color, number of stripes, and color of stripes. Biochemical polymorphisms, like the human A, B, AB, and O blood groups (described in Section 12.2), are also common.

We describe phenotypic polymorphisms quantitatively by calculating the percentage or *frequency* of each trait. For example, if you counted 123 blue snow geese and 369 white ones in a population of 492 geese, the frequency of the blue phenotype would be 123/492 or 0.25, and the frequency of the white phenotype would be 369/492 or 0.75.

Phenotypic Variation Can Have Genetic and Environmental Causes

Phenotypic variation within populations may be caused by genetic differences between individuals, by differences in the environmental factors that individuals experience, or by an interaction between genetics and the environment. As a result, genetic and pheno-

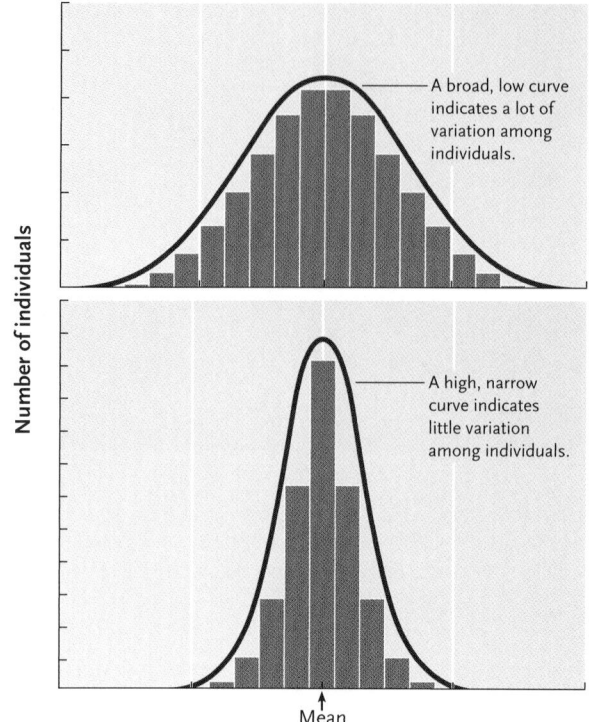

Figure 20.3

Quantitative variation. Many traits vary continuously among members of a population, and a bar graph of the data often approximates a bell-shaped curve. The mean defines the average value of the trait in the population, and the width of the curve is proportional to the variability among individuals.

typic variations may not be perfectly correlated. Under some circumstances, organisms with different genotypes exhibit the same phenotype. For example, the black coloration of some rock pocket mice from Arizona is caused by certain mutations in the *Mc1r* gene (see Section 1.2); but black mice from New Mexico do not share those mutations—that is, they have different genotypes—even though they exhibit the same phenotype. On the other hand, organisms with the same genotype sometimes exhibit different phenotypes. For

Figure 20.4

Qualitative variation. Individual snow geese *(Chen caerulescens)* are either blue or white. Although both colors are present in many populations, geese tend to associate with others of the same color.

Figure 20.5

Environmental effects on phenotype. Soil acidity affects the expression of the gene controlling flower color in the common garden plant *Hydrangea macrophylla*. When grown in acid soil, it produces deep blue flowers. In neutral or alkaline soil, its flowers are bright pink.

example, the acidity of soil influences flower color in some plants **(Figure 20.5).**

Knowing whether phenotypic variation is caused by genetic differences, environmental factors, or an interaction of the two is important because *only genetically based variation is subject to evolutionary change.* Moreover, knowing the causes of phenotypic variation has important practical applications. Suppose, for example, that one field of wheat produced more grain than another. If a difference in the availability of nutrients or water caused the difference in yield, a farmer might choose to fertilize or irrigate the less productive field. But if the difference in productivity resulted from genetic differences between plants in the two fields, a farmer might plant only the more productive genotype. Because environmental factors can influence the expression of genes, an organism's phenotype is frequently the product of an interaction between its genotype and its environment. In our hypothetical example, the farmer may maximize yield by fertilizing and irrigating the better genotype of wheat.

How can we determine whether phenotypic variation is caused by environmental factors or by genetic differences? We can test for an environmental cause experimentally by changing one environmental variable and measuring the effects on genetically similar subjects. You can try this yourself by growing some cuttings from an ivy plant in shade and other cuttings from the same plant in full sun. Although they all have the same genotype, the cuttings grown in sun will produce smaller leaves and shorter stems.

Breeding experiments can demonstrate the genetic basis of phenotypic variation. For example, Mendel inferred the genetic basis of qualitative traits, such as flower color in peas, by crossing plants with different phenotypes. Moreover, traits that vary quantitatively will respond to artificial selection only if the variation has some genetic basis. For example, re-

searchers observed that individual house mice (*Mus domesticus*) differ in activity levels, as measured by how much they use an exercise wheel and how fast they run. John G. Swallow, Patrick A. Carter, and Theodore Garland, Jr., then at the University of Wisconsin at Madison, used artificial selection to produce lines of mice that exhibit increased wheel-running behavior, demonstrating that the observed differences in these two aspects of activity level have a genetic basis **(Figure 20.6).**

Breeding experiments are not always practical, however, particularly for organisms with long generation times. Ethical concerns also render these techniques unthinkable for humans. Instead, researchers sometimes study the inheritance of particular traits by analyzing genealogical pedigrees, as discussed in Section 13.2, but this approach often provides poor results for analyses of complex traits.

Several Processes Generate Genetic Variation

Genetic variation, the raw material molded by microevolutionary processes, has two potential sources: the production of new alleles and the rearrangement of existing alleles. Most new alleles probably arise from small scale mutations in DNA (described later in this chapter). The rearrangement of existing alleles into new combinations can result from larger scale changes in chromosome structure or number and from several forms of genetic recombination, including crossing over between homologous chromosomes during meiosis, the independent assortment of non-homologous chromosomes during meiosis, and random fertilizations between genetically different sperm and eggs.

The shuffling of *existing* alleles into new combinations can produce an extraordinary number of novel genotypes and phenotypes in the next generation. By one estimate, more than 10^{600} combinations of alleles are possible in human gametes, yet there are fewer than 10^{10} humans alive today. So unless you have an identical twin, it is extremely unlikely that another person with your genotype has ever lived or ever will.

Populations Often Contain Substantial Genetic Variation

How much genetic variation actually exists within populations? In the 1960s, evolutionary biologists began to use gel electrophoresis (see Figure 18.7) to identify biochemical polymorphisms in diverse organisms. This technique separates two or more forms of a given protein if they differ significantly in shape, mass, or net electrical charge. The identification of a protein polymorphism allows researchers to infer genetic variation at the locus coding for that protein.

Figure 20.6 Experimental Research

Using Artificial Selection to Demonstrate That Activity Level in Mice Has a Genetic Basis

QUESTION: Do observed differences in activity level among house mice have a genetic basis?

EXPERIMENT: Swallow, Carter, and Garland knew that a phenotypic character responds to artificial selection only if it has a genetic, rather than an environmental, basis. In an experiment with house mice *(Mus domesticus)*, they selected for the phenotypic character of increased wheel-running activity. In four experimental lines, they bred those mice that ran the most. Four other lines, in which breeders were selected at random with respect to activity level, served as controls.

RESULTS: After 10 generations of artificial selection, mice in the experimental lines ran longer distances and ran faster than mice in the control lines. Thus, artificial selection on wheel-running activity in house mice increased **(a)** the distance that mice run per day and **(b)** their average speed. The data illustrate responses of females in four experimental lines and four control lines. Males showed similar responses.

a. Distance run

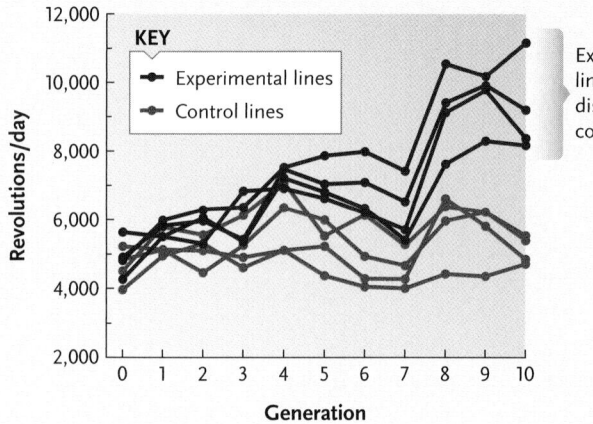

b. Average speed

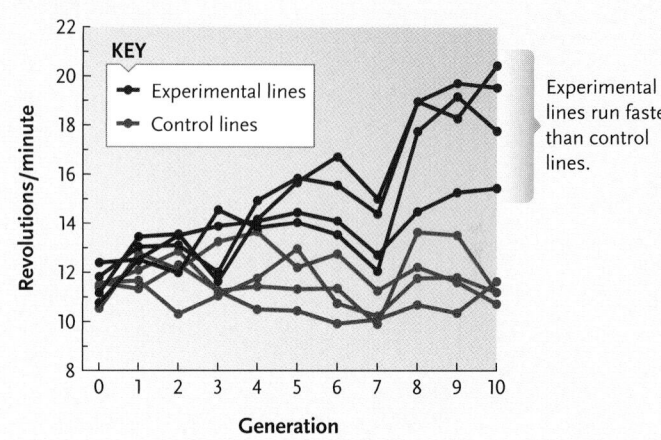

CONCLUSION: Because two measures of activity level responded to artificial selection, researchers concluded that variation in this behavioral character has a genetic basis.

Researchers discovered much more genetic variation than anyone had imagined. For example, nearly half the loci surveyed in many populations of plants and invertebrates are polymorphic. Moreover, gel electrophoresis actually underestimates genetic variation because it doesn't detect different amino acid substitutions if the proteins for which they code migrate at the same rate.

Advances in molecular biology now allow scientists to survey genetic variation directly, and researchers have accumulated an astounding knowledge of the structure of DNA and its nucleotide sequences. In general, studies of chromosomal and mitochondrial DNA suggest that every locus exhibits some variability in its nucleotide sequence. The variability is apparent in comparisons of individuals from a single population, populations of one species, and related species. However, some variations detected in the protein-coding regions of DNA may not affect phenotypes because, as explained on page 426, they do not change the amino acid sequences of the proteins for which the genes code.

STUDY BREAK

1. If a population of skunks includes some individuals with stripes and others with spots, would you describe the variation as quantitative or qualitative?
2. In the experiment on house mice described in Figure 20.6, how did researchers demonstrate that variations in activity level had a genetic basis?
3. What factors contribute to phenotypic variation in a population?

20.2 Population Genetics

To predict how certain factors may influence genetic variation, population geneticists first describe the genetic structure of a population. They then create hypotheses, which they formalize in mathematical models, to describe how evolutionary processes may change the genetic structure under specified conditions. Finally, researchers test the predictions of these models to evaluate the ideas about evolution that are embodied within them.

All Populations Have a Genetic Structure

Populations are made up of individuals, each with its own genotype. In diploid organisms, which have pairs of homologous chromosomes, an individual's genotype includes two alleles at every gene locus. The sum of all alleles at all gene loci in all individuals is called the population's **gene pool.**

To describe the structure of a gene pool, scientists first identify the genotypes in a representative sample and calculate **genotype frequencies,** the percentages of individuals possessing each genotype. Knowing that each diploid organism has two alleles (either two copies of the same allele or two different alleles) at each gene locus, a scientist can then calculate **allele frequencies,** the relative abundances of the different alleles. For a locus with two alleles, scientists use the symbol p to identify the frequency of one allele, and q the frequency of the other.

The calculation of genotype and allele frequencies for the two alleles at the gene locus governing flower color in snapdragons (genus *Antirrhinum*) is straightforward **(Table 20.1).** This locus is easy to study because it exhibits incomplete dominance (see Section 12.2). Individuals that are homozygous for the C^R allele ($C^R C^R$) have red flowers; those homozygous for the C^W allele ($C^W C^W$) have white flowers; and heterozygotes ($C^R C^W$) have pink flowers. Genotype frequencies represent how the C^R and C^W alleles are distributed among individuals. In this example, examination of the plants reveals that 45% of individuals have the $C^R C^R$ genotype, 50% have the heterozygous $C^R C^W$

genotype, and the remaining 5% have the $C^W C^W$ genotype. Allele frequencies represent the commonness or rarity of each allele in the gene pool. As calculated in the table, 70% of the alleles in the population are C^R and 30% are C^W. Remember that for a gene locus with two alleles, there are three genotype frequencies, but only two allele frequencies (p and q). The sum of the three genotype frequencies must equal 1; so must the sum of the two allele frequencies.

The Hardy-Weinberg Principle Is a Null Model That Defines How Evolution Does Not Occur

When designing experiments, scientists often use control treatments to evaluate the effect of a particular factor: the control tells us what we would see if the experimental treatment had no effect. As you may recall from the hypothetical example presented in Chapter 1 (see Figure 1.14), to determine whether fertilizer has an effect on plant growth, you must compare the growth of fertilized plants (the experimental treatment) to the growth of plants that received no fertilizer (the control treatment). However, in studies that use observational rather than experimental data, there is often no suitable control. In such cases, investigators develop conceptual models, called **null models,** which predict what they would see if a particular factor had no effect. Null models serve as theoretical reference points against which observations can be evaluated.

Early in the twentieth century, geneticists were puzzled by the persistence of recessive traits because they assumed that natural selection replaced recessive or rare alleles with dominant or common ones. An English mathematician, G. H. Hardy, and a German physician, Wilhelm Weinberg, tackled this problem independently in 1908. Their analysis, now known as the

| Table 20.1 | Calculation of Genotype Frequencies and Allele Frequencies for the Snapdragon Flower Color Locus |

Because each diploid individual has two alleles at each gene locus, the entire sample of 1000 individuals has a total of 2000 alleles at the C locus.

Flower Color Phenotype	Genotype	Number of Individuals	Genotype Frequency[1]	Total Number of C^R Alleles[2]	Total Number of C^W Alleles[2]
Red	$C^R C^R$	450	450/1000 = 0.45	2 × 450 = 900	0 × 450 = 0
Pink	$C^R C^W$	500	500/1000 = 0.50	1 × 500 = 500	1 × 500 = 500
White	$C^W C^W$	50	50/1000 = 0.05	0 × 50 = 0	2 × 50 = 100
	Total	1000	0.45 + 0.50 + 0.05 = 1.0	1400	600

Calculate allele frequencies using the total of 1400 + 600 = 2000 alleles in the sample:

$$p = \text{frequency of } C^R \text{ allele} = 1400/2000 = 0.7$$
$$q = \text{frequency of } C^W \text{ allele} + 600/2000 = 0.3$$
$$p + q = 0.7 + 0.3 = 1.0$$

[1]Genotype frequency = the number of individuals possessing a particular genotype divided by the total number of individuals in the sample.
[2]Total number of C^R or C^W alleles = the number of C^R or C^W alleles present in one individual with a particular genotype multiplied by the number of individuals with that genotype.

Hardy-Weinberg principle, specifies the conditions under which a population of diploid organisms achieves **genetic equilibrium**, the point at which neither allele frequencies nor genotype frequencies change in succeeding generations. Their work also showed that dominant alleles need not replace recessive ones, and that the shuffling of genes in sexual reproduction does not in itself cause the gene pool to change.

The Hardy-Weinberg principle is a mathematical model that describes how genotype frequencies are established in sexually reproducing organisms. According to this model, genetic equilibrium is possible only if *all* of the following conditions are met:

1. No mutations are occurring.
2. The population is closed to migration from other populations.
3. The population is infinite in size.
4. All genotypes in the population survive and reproduce equally well.
5. Individuals in the population mate randomly with respect to genotypes.

If the conditions of the model are met, the allele frequencies of the population will never change, and the genotype frequencies will stop changing after one generation. In short, under these restrictive conditions, microevolution will *not* occur (see *Focus on Research*). The Hardy-Weinberg principle is thus a null model that serves as a reference point for evaluating the circumstances under which evolution *may* occur.

If a population's genotype frequencies do not match the predictions of this model or if its allele frequencies change over time, microevolution may be occurring. Determining which of the model's conditions are not met is a first step in understanding how and why the gene pool is changing. Natural populations never fully meet all five requirements simultaneously, but they often come pretty close.

STUDY BREAK

1. What is the difference between the genotype frequencies and the allele frequencies in a population?
2. Why is the Hardy-Weinberg principle considered a null model of evolution?
3. If the conditions of the Hardy-Weinberg principle are met, when will genotype frequencies stop changing?

20.3 The Agents of Microevolution

A population's allele frequencies will change over time if conditions of the Hardy-Weinberg model are violated. The processes that foster microevolutionary

Table 20.2 Agents of Microevolutionary Change

Agent	Definition	Effect on Genetic Variation
Mutation	A heritable change in DNA	Introduces new genetic variation into population
Gene flow	Change in allele frequencies as individuals join a population and reproduce	May introduce genetic variation from another population
Genetic drift	Random changes in allele frequencies caused by chance events	Reduces genetic variation, especially in small populations; can eliminate alleles
Natural selection	Differential survivorship or reproduction of individuals with different genotypes	One allele can replace another or allelic variation can be preserved
Nonrandom mating	Choice of mates based on their phenotypes and genotypes	Does not directly affect allele frequencies, but usually prevents genetic equilibrium

change—which include mutation, gene flow, genetic drift, natural selection, and nonrandom mating—are summarized in **Table 20.2**.

Mutations Create New Genetic Variations

A **mutation** is a spontaneous and heritable change in DNA. Mutations are rare events; during any particular breeding season, between one gamete in 100,000 and one in 1 million will include a new mutation at a particular gene locus. New mutations are so infrequent, in fact, that they exert little or no immediate effect on allele frequencies in most populations. But over evolutionary time scales, their numbers are significant—mutations have been accumulating in biological lineages for billions of years. And because it is a mechanism through which entirely new genetic variations arise, *mutation is a major source of heritable variation*.

For most animals, only mutations in the germ line (the cell lineage that produces gametes) are heritable; mutations in other cell lineages have no direct effect on the next generation. In plants, however, mutations may occur in meristem cells, which eventually produce flowers as well as nonreproductive structures (see Chapter 31); in such cases, a mutation may be passed to the next generation and ultimately influence the gene pool.

Deleterious mutations alter an individual's structure, function, or behavior in harmful ways. In mammals, for example, a protein called collagen is an essential component of most extracellular structures. Several simple mutations in humans cause forms of Ehlers-Danlos syndrome, a disruption of collagen synthesis that may result in loose skin, weak joints, or sudden death from the rupture of major blood vessels, the colon, or the uterus.

By definition, *lethal mutations* cause the death of organisms carrying them. If a lethal allele is dominant, both homozygous and heterozygous carriers suffer

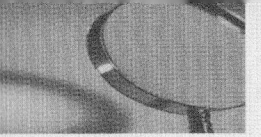

FOCUS ON RESEARCH

Basic Research: Using the Hardy-Weinberg Principle

To see how the Hardy-Weinberg principle can be applied, we will analyze the snapdragon flower color locus, using the hypothetical population of 1000 plants described in Table 20.1. This locus includes two alleles—C^R (with its frequency designated as p) and C^W

(with its frequency designated as q)—and three genotypes—homozygous C^RC^R, heterozygous C^RC^W, and homozygous C^WC^W. Table 20.1 lists the number of plants with each genotype: 450 have red flowers (C^RC^R), 500 have pink flowers (C^RC^W), and 50 have white flowers

(C^WC^W). It also shows the calculation of both the genotype frequencies ($C^RC^R = 0.45$, $C^RC^W = 0.50$, and $C^WC^W = 0.05$) and the allele frequencies ($p = 0.7$ and $q = 0.3$) for the population.

Let's assume for simplicity that each individual produces only two gametes and that both gametes contribute to the production of offspring. This assumption is unrealistic, of course, but it meets the Hardy-Weinberg requirement that all individuals in the population contribute equally to the next generation. In each parent, the two alleles segregate and end up in different gametes:

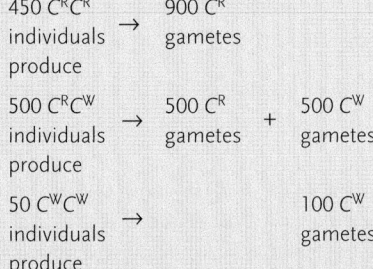

450 C^RC^R individuals produce	$\rightarrow$	900 C^R gametes	
500 C^RC^W individuals produce	$\rightarrow$	500 C^R gametes	+ 500 C^W gametes
50 C^WC^W individuals produce	$\rightarrow$	100 C^W gametes	

You can readily see that 1400 of the 2000 total gametes carry the C^R allele and 600 carry the C^W allele. The frequency of C^R gametes is 1400/2000 or 0.7, which is equal to p; the frequency of C^W gametes is 600/2000 or 0.3, which is equal to q. Thus, the allele frequencies in the gametes are exactly the same as the allele frequencies in the parent generation—it could not be

Sperm

C^R frequency
$p = 0.7$
C^R

C^W frequency
$q = 0.3$
C^W

Eggs

C^R frequency
$p = 0.7$
C^R

C^RC^R offspring
frequency = $p^2 = 0.49$

C^WC^R offspring
frequency = $pq = 0.21$

C^W frequency
$q = 0.3$
C^W

C^RC^W offspring
frequency = $pq = 0.21$

C^WC^W offspring
frequency = $q^2 = 0.09$

from its effects; if recessive, it affects only homozygous recessive individuals. A lethal mutation that causes death before the individual reproduces is eliminated from the population.

Neutral mutations are neither harmful nor helpful. Recall from Section 15.1 that in the construction of a polypeptide chain, a particular amino acid can be specified by several different codons. As a result, some DNA sequence changes—especially certain changes at the third nucleotide of the codon—do not alter the amino acid sequence. Not surprisingly, mutations at the third position appear to persist longer in populations than those at the first two positions. Other mutations may change an organism's phenotype without influencing its survival and reproduction. A neutral mutation might even be beneficial later if the environment changes.

Sometimes a change in DNA produces an *advantageous mutation,* which confers some benefit on an individual that carries it. However slight the advantage, natural selection may preserve the new allele and even increase its frequency over time. Once the mutation has been passed to a new generation, other agents of microevolution determine its long-term fate.

Gene Flow Introduces Novel Genetic Variants into Populations

Organisms or their gametes (for example, pollen) sometimes move from one population to another. If the immigrants reproduce, they may introduce novel alleles into the population they have joined. This phenomenon, called **gene flow**, violates the Hardy-Weinberg requirement that populations must be closed to migration.

otherwise because each gamete carries one allele at each locus.

Now assume that these gametes, both sperm and eggs, encounter each other at random. In other words, individuals reproduce without regard to the genotype of a potential mate. We can visualize the process of random mating in the mating table on the left.

We can also describe the consequences of random mating—$(p + q)$ sperm fertilizing $(p + q)$ eggs—with an equation that predicts the genotype frequencies in the offspring generation:

$$(p + q) \times (p + q) = p^2 + 2pq + q^2$$

If the population is at genetic equilibrium for this locus, p^2 is the predicted frequency of the C^RC^R genotype, $2pq$ the predicted frequency of the C^RC^W genotype, and q^2 the predicted frequency of the C^WC^W genotype. Using the gamete frequencies determined above, we can calculate the predicted genotype frequencies in the next generation:

frequency of $C^RC^R =$
$$p^2 = (0.7 \times 0.7) = 0.49$$

frequency of $C^RC^W =$
$$2pq = 2(0.7 \times 0.3) = 0.42$$

frequency of $C^WC^W =$
$$q^2 = (0.3 \times 0.3) = 0.09$$

Notice that the predicted genotype frequencies in the offspring generation have changed from those in the parent generation: the frequency of heterozygous individuals has decreased, and the frequencies of both types of homozygous individuals have increased. This result occurred because the starting population was *not already* in equilibrium at this gene locus. In other words, the distribution of parent genotypes did not conform to the predicted $p^2 + 2pq + q^2$ distribution.

The 2000 gametes in our hypothetical population produced 1000 offspring. Using the genotype frequencies we just calculated, we can predict how many offspring will carry each genotype:

490 red (C^RC^R)
420 pink (C^RC^W)
90 white (C^WC^W)

In a real study, we would examine the offspring to see how well their numbers match these predictions.

What about the allele frequencies in the offspring? The Hardy-Weinberg principle predicts that they did not change. Let's calculate them and see. Using the method shown in Table 20.1 and the prime symbol (′) to indicate offspring allele frequencies:

$$p' = ([2 \times 490] + 420)/2000 =$$
$$1400/2000 = 0.7$$

$$q' = ([2 \times 90] + 420)/2000 =$$
$$600/2000 = 0.3$$

You can see from this calculation that the allele frequencies did not change from one generation to the next, even though the alleles were rearranged to produce different proportions of the three genotypes. Thus, the population is now at genetic equilibrium for the flower color locus; neither the genotype frequencies nor the allele frequencies will change in succeeding generations as long as the population meets the conditions specified in the Hardy-Weinberg model.

To verify this, you can calculate the allele frequencies of the gametes for this offspring generation and predict the genotype frequencies and allele frequencies for a third generation. You could continue calculating until you ran out of either paper or patience, but these frequencies will not change.

Researchers use calculations like these to determine whether an actual population is near its predicted genetic equilibrium for one or more gene loci. When they discover that a population is not at equilibrium, they infer that microevolution is occurring and can investigate the factors that might be responsible.

Gene flow is common in some animal species. For example, young male baboons typically move from one local population to another after experiencing aggressive behavior by older males. And many marine invertebrates disperse long distances as larvae carried by ocean currents.

Dispersal agents, such as pollen-carrying wind or seed-carrying animals, are responsible for gene flow in most plant populations. For example, blue jays foster gene flow among populations of oaks by carrying acorns from nut-bearing trees to their winter caches, which may be as much as a mile away **(Figure 20.7).** Transported acorns that go uneaten may germinate and contribute to the gene pool of a neighboring oak population.

Documenting gene flow among populations is not always easy, particularly if it occurs infrequently. Researchers can use phenotypic or genetic markers to

Figure 20.7
Gene flow. Blue jays *(Cyanocitta cristata)* serve as agents of gene flow for oaks (genus *Quercus*) when they carry acorns from one oak population to another. An uneaten acorn may germinate and contribute to the gene pool of the population into which it was carried.

identify immigrants in a population, but they must also demonstrate that immigrants reproduced, thereby contributing to the gene pool of their adopted population. In the San Francisco Bay area, for example, Bay checkerspot butterflies *(Euphydryas editha bayensis)* rarely move from one population to another because they are poor fliers (see Figure 53.16). When adult females do change populations, it is often late in the breeding season, and their offspring have virtually no chance of finding enough food to mature. Thus, many immigrant females do not foster gene flow because they do not contribute to the gene pool of the population they join.

The evolutionary importance of gene flow depends upon the degree of genetic differentiation between populations and the rate of gene flow between them. If two gene pools are very different, a little gene flow may increase genetic variability within the population that receives immigrants, and it will make the two populations more similar. But if populations are already genetically similar, even lots of gene flow will have little effect.

Genetic Drift Reduces Genetic Variability within Populations

Chance events sometimes cause the allele frequencies in a population to change unpredictably. This phenomenon, known as **genetic drift**, has especially dramatic effects on small populations, which clearly violate the Hardy-Weinberg assumption of infinite population size.

A simple analogy clarifies why genetic drift is more pronounced in small populations than in large ones. When individuals reproduce, male and female gametes often pair up randomly, as though the allele in any particular sperm or ovum was determined by a coin toss. Imagine that "heads" specifies the R allele and "tails" specifies the r allele. If the two alleles are equally common (that is, their frequencies, p and q, are both equal to 0.5), heads should be as likely an outcome as tails. But if you toss the coin 20 or 30 times to simulate random mating in a small population, you won't often see a 50-50 ratio of heads and tails. Sometimes heads will predominate and sometimes tails will—just by chance. Tossing the coin 500 times to simulate random mating in a somewhat larger population is more likely to produce a 50-50 ratio of heads and tails. And if you tossed the coin 5000 times, you would get even closer to a 50-50 ratio.

Chance deviations from expected results—which cause genetic drift—occur whenever organisms engage in sexual reproduction, simply because their population sizes are not infinitely large. But genetic drift is particularly common in small populations because only a few individuals contribute to the gene pool and because any given allele is present in very few individuals.

Genetic drift generally leads to the loss of alleles and reduced genetic variability. Two general circumstances, population bottlenecks and founder effects, often foster genetic drift.

Population Bottlenecks. On occasion, a stressful factor such as disease, starvation, or drought kills a great many individuals and eliminates some alleles from a population, producing a **population bottleneck.** This cause of genetic drift greatly reduces genetic variation even if the population numbers later rebound.

In the late nineteenth century, for example, hunters nearly wiped out northern elephant seals *(Mirounga angustirostris)* along the Pacific coast of North America **(Figure 20.8).** Since the 1880s, when the species received protected status, the population has increased to more than 30,000, all descended from a group of about 20 survivors. Today the population exhibits no variation in 24 proteins studied by gel electrophoresis. This low level of genetic variation, which is unique among seal species, is consistent with the hypothesis that genetic drift eliminated many alleles when the population experienced the bottleneck.

Founder Effect. When a few individuals colonize a distant locality and start a new population, they carry only a small sample of the parent population's genetic variation. By chance, some alleles may be totally missing from the new population, whereas other alleles that were rare "back home" might occur at relatively high frequencies. This change in the gene pool is called the **founder effect.**

The human medical literature provides some of the best-documented examples of the founder effect. The Old Order Amish, an essentially closed religious community in Lancaster County, Pennsylvania, have an exceptionally high incidence of Ellis–van Creveld syndrome, a genetic disorder caused by a recessive allele. In the homozygous state, the allele produces dwarfism, shortened limbs, and polydactyly (extra fin-

Figure 20.8

Population bottleneck. Northern elephant seals *(Mirounga angustirostris)* at the Año Nuevo State Reserve in California are descended from a population that was decimated by hunting late in the nineteenth century. In this photo, two large bulls fight to control a harem of females.

Genetic Variation Preserved in Humpback Whales

For centuries, hunters slaughtered humpback whales (*Megaptera novaeangliae*) for their meat and oil. By 1966, when an international agreement limited whale hunting, the worldwide population of humpbacks had been reduced to fewer than 5000 individuals. These survivors were distributed among three distinct populations in the North Atlantic, North Pacific, and Southern oceans. Since the hunting agreement was imposed, the populations have recovered to include more than 20,000 individuals.

The derivation of present-day humpback populations from the relatively small number surviving in 1966 is of concern because the population bottleneck may have reduced genetic variability. Such a loss could have adverse effects on the surviving population's reproductive capacity, resistance to disease, and ability to survive unfavorable environmental changes.

How serious was the bottleneck for the surviving humpback whales? A large group of researchers working in Hawaii, the continental United States, Australia, South Africa, Canada, Mexico, and the Dominican Republic set out to answer this question, using molecular techniques to measure the amount of genetic variability in the surviving whale populations.

The researchers chose mitochondrial DNA (mtDNA) for their measurements because it is small, it is easily extracted and identified, and almost all of its variability comes from chance mutations that occur at a steady rate rather than from genetic recombination (see Section 13.5). Except for the few changes produced by mutations since the population bottleneck (which can be estimated from the mutation rate and subtracted from the total), the variability of mtDNA should be the amount remaining from the population that existed before the bottleneck.

Using biopsy darts, the researchers obtained small skin samples from 90 humpback whales distributed among the three oceanic populations. They extracted the mtDNA from the skin samples and amplified it using the polymerase chain reaction (see Figure 18.6). They then isolated a 463-base-pair segment containing the promoters and replication origin for mtDNA, along with spacer sequences. The DNA base sequence was determined for each sample.

The researchers were surprised to find that the mtDNA sequence variation was relatively high in most of their sample, between 76% and 82% of the average variation found in all animal species studied to date. However, a subpopulation of the north Pacific population living near Hawaii showed low genetic variability; in fact, no variability at all was detected in the mtDNA segment of this subpopulation. Why the Hawaiian humpbacks have no variability in the mtDNA segment examined is unclear. One possibility is that this subpopulation originated recently, perhaps during the twentieth century. Information supporting this idea comes from whaling records, which list no sightings or catches of humpbacks in the Hawaiian region during the nineteenth century. Furthermore, the native Hawaiian people have no legends or words describing whales of the humpback type (baleen whales). Perhaps the subpopulation was started by a few whales with the same genetic make-up in the mtDNA region, providing an example of the founder effect.

With the exception of this Hawaiian subpopulation, humpback whales appear to have retained genetic variability comparable to other animals. This retention of variability in the face of near extinction may result from the whales' relatively long generation time. Because they have a potential life span of about 50 years, some individuals that survived the period of commercial hunting are still alive today. The researchers suggest that enough of these long-lived individuals survived to provide a reservoir of variability from the old populations.

These results indicate that the hunting ban came in time to prevent a significant loss of genetic variability in humpback whales. Hopefully, the same is true of other whale species that were hunted nearly to extinction.

---

gers). Genetic analysis suggests that, although this syndrome affects less than 1% of the Amish in Lancaster County, as many as 13% may be heterozygous carriers of the allele. All of the individuals exhibiting the syndrome are descended from one couple who helped found the community in the mid-1700s.

Conservation Implications. Genetic drift has important implications for conservation biology. By definition, endangered species experience severe population bottlenecks, which result in the loss of genetic variability. Moreover, the small number of individuals available for captive breeding programs may not fully represent a species' genetic diversity. Without such variation, no matter how large a population may become in the future, it will be less resistant to diseases or less able to cope with environmental change.

For example, scientists believe that an environmental catastrophe produced a population bottleneck in the African cheetah *(Acinonyx jubatus)* 10,000 years ago. Cheetahs today are remarkably uniform in genetic make-up. Their populations are highly susceptible to diseases; they also have a high proportion of sperm cell abnormalities and a reduced reproductive capacity. Thus, limited genetic variation, as well as small numbers, threatens populations of endangered species. *Insights from the Molecular Revolution* describes techniques used to determine whether hunting has had the same effect on humpback whales.

Natural Selection Shapes Genetic Variability by Favoring Some Traits over Others

The Hardy-Weinberg model requires all genotypes in a population to survive and reproduce equally well. But as you know from Section 19.2, heritable traits enable some individuals to survive better and reproduce more than others. **Natural selection** is the process by which such traits become more common in subsequent generations. Thus, natural selection violates a requirement of the Hardy-Weinberg equilibrium.

Although natural selection can change allele frequencies, *it is the phenotype of an individual organism,* *rather than any particular allele, that is successful or not.* When individuals survive and reproduce, their alleles—both favorable and unfavorable—are passed to the next generation. Of course, an organism with harmful or lethal dominant alleles will probably die before reproducing, and all the alleles it carries will share that unhappy fate, even those that are advantageous.

To evaluate reproductive success, evolutionary biologists consider **relative fitness,** the number of surviving offspring that an individual produces compared with the number left by others in the population. Thus, a particular allele will increase in frequency in the next generation if individuals carrying that allele leave *more*

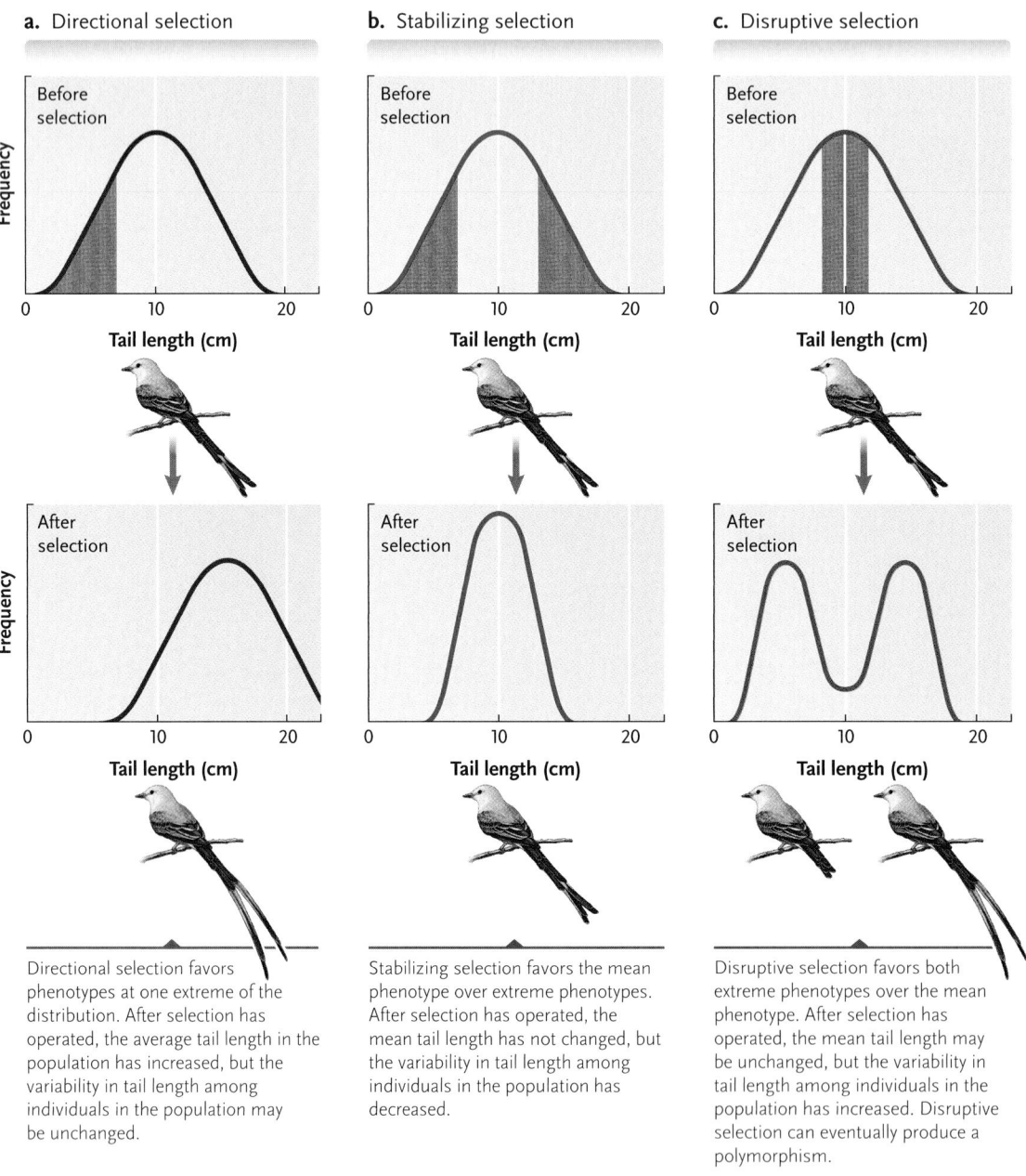

a. Directional selection

b. Stabilizing selection

c. Disruptive selection

Directional selection favors phenotypes at one extreme of the distribution. After selection has operated, the average tail length in the population has increased, but the variability in tail length among individuals in the population may be unchanged.

Stabilizing selection favors the mean phenotype over extreme phenotypes. After selection has operated, the mean tail length has not changed, but the variability in tail length among individuals in the population has decreased.

Disruptive selection favors both extreme phenotypes over the mean phenotype. After selection has operated, the mean tail length may be unchanged, but the variability in tail length among individuals in the population has increased. Disruptive selection can eventually produce a polymorphism.

Figure 20.9

Three modes of natural selection. This hypothetical example uses tail length of birds as the quantitative trait subject to selection. The yellow shading in the top graphs indicates phenotypes that natural selection does *not* favor. Notice that the area under each curve is constant because each curve presents the frequencies of all phenotypes in the population. When stabilizing selection **(b)** reduces variability in the trait, the curve becomes higher and narrower.

offspring than individuals carrying other alleles. Differences in the *relative* success of individuals are the essence of natural selection.

Natural selection tests fitness differences at nearly every stage of the life cycle. One plant may be fitter than others in the population because its seeds survive colder conditions, because the arrangement of its leaves captures sunlight more efficiently, or because its flowers are more attractive to pollinators. However, natural selection exerts little or no effect on traits that appear during an individual's postreproductive life. For example, Huntington disease, a dominant-allele disorder that first strikes humans after the age of 40, is not subject to strong selection. Carriers of the disease-causing allele reproduce before the onset of the condition, passing it to the next generation.

Biologists measure the effects of natural selection on phenotypic variation by recording changes in the mean and variability of characters over time (see Figure 20.3). Three modes of natural selection have been identified: directional selection, stabilizing selection, and disruptive selection (Figure 20.9).

Directional Selection. Traits undergo **directional selection** when individuals near one end of the phenotypic spectrum have the highest relative fitness. Directional selection shifts a trait away from the existing mean and toward the favored extreme (see Figure 20.9a). After selection, the trait's mean value is higher or lower than before.

Directional selection is extremely common. For example, predatory fish promote directional selection for larger body size in guppies when they selectively feed on the smallest individuals in a guppy population (see *Focus on Research* in Chapter 49). And most cases of artificial selection, including the experiment on the activity levels of house mice, are directional, aimed at increasing or decreasing specific phenotypic traits. Humans routinely use directional selection to produce domestic animals and crops with desired characteristics, such as the small size of chihuahuas and the intense "bite" of chili peppers.

Stabilizing Selection. Traits undergo **stabilizing selection** when individuals expressing intermediate phenotypes have the highest relative fitness (see Figure 20.9b). By eliminating phenotypic extremes, stabilizing selection reduces genetic and phenotypic variation and increases the frequency of intermediate phenotypes. Stabilizing selection is probably the most common mode of natural selection, affecting many familiar traits. For example, very small and very large human newborns are less likely to survive than those born at an intermediate weight (Figure 20.10).

Warren G. Abrahamson and Arthur E. Weis of Bucknell University have shown that opposing forces of directional selection can sometimes produce an overall pattern of stabilizing selection (Figure 20.11).

Figure 20.10 Observational Research

Evidence for Stabilizing Selection in Humans

HYPOTHESIS: Human birth weight has been adjusted by natural selection.

NULL HYPOTHESIS: Natural selection has not affected human birth weight.

METHOD: Two noted human geneticists, Luigi Cavalli-Sforza and Sir Walter Bodmer of Stanford University, collected data on the variability in human birth weight, a character exhibiting quantitative variation, and on the mortality rates of babies born at different weights. The researchers then searched for a relationship between birth weight and mortality rate by plotting both data sets on the same graph. A lack of correlation between birth weight and mortality rate would support the null hypothesis.

RESULTS: When plotted together on the same graph, the bar graph (birth weight) and the curve (mortality rate) illustrate that the mean birth weight is very close to the optimum birth weight (the weight at which mortality is lowest). The two data sets also show that few babies are born at the very low and very high weights associated with high mortality.

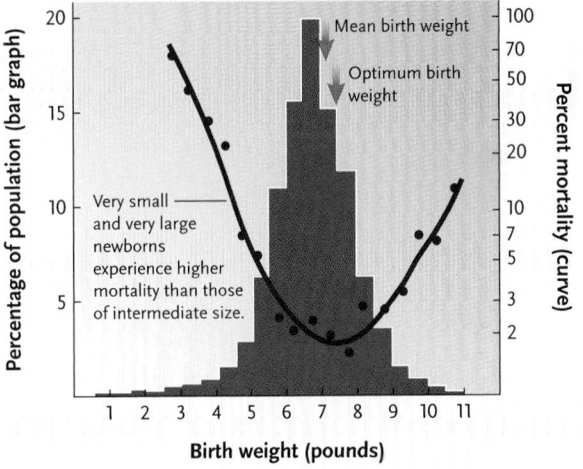

CONCLUSION: The shapes and positions of the birth weight bar graph and the mortality rate curve suggest that stabilizing selection has adjusted human birth weight to an average of 7 to 8 pounds.

The gallmaking fly *(Eurosta solidaginis)* is a small insect that feeds on the tall goldenrod plant *(Solidago altissima)*. When a fly larva hatches from its egg, it bores into a goldenrod stem, and the plant responds by producing a spherical growth deformity called a gall. The larva feeds on plant tissues inside the gall. Galls vary dramatically in size; genetic experiments indicate that gall size is a heritable trait of the fly, although plant genotype also has an effect.

Fly larvae inside galls are subjected to two opposing patterns of directional selection. On one hand, a tiny wasp *(Eurytoma gigantea)* parasitizes gallmaking flies by laying eggs in fly larvae inside their galls. After hatching, the young wasps feed on the fly larvae, killing them in the process. However, adult wasps are

Figure 20.11 Observational Research

How Opposing Forces of Directional Selection Produce Stabilizing Selection

HYPOTHESIS: The size of galls made by larvae of the gallmaking fly *(Eurosta solidaginis)* is governed by conflicting selection pressures established by parasitic wasps and predatory birds.

PREDICTION: Gallmaking flies that produce galls of intermediate size will be more likely to survive than those that make either small galls or large galls.

METHOD: Abrahamson and his colleagues surveyed galls made by the larvae of the gallmaking fly in Pennsylvania. They measured the diameters of the galls they encountered, and, for those galls in which the larvae had died, they determined whether they had been killed by **(a)** a parasitic wasp *(Eurytoma gigantea)* or **(b)** a predatory bird, such as the downy woodpecker *(Dendrocopus pubescens)*.

a. *Eurytoma gigantea*, a parasitic wasp

Forrest W. Buchanan/Visuals Unlimited

b. *Dendrocopus pubescens*, a predatory bird

Gregory K. Scott/Photo Researchers, Inc.

c.

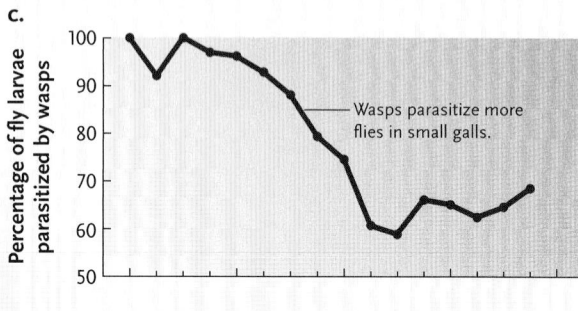

Wasps parasitize more flies in small galls.

d.

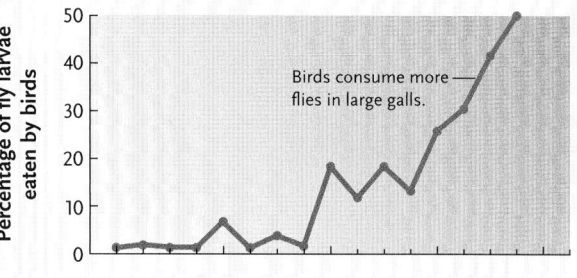

Birds consume more flies in large galls.

e.

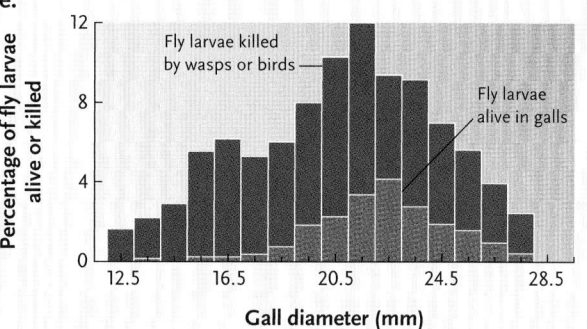

Fly larvae killed by wasps or birds

Fly larvae alive in galls

RESULTS: Tiny wasps are more likely to parasitize gall-making fly larvae inside small galls **(c)**, fostering directional selection in favor of large galls. By contrast, birds usually feed on fly larvae inside large galls **(d)**, fostering directional selection in favor of small galls. These opposing patterns of directional selection create stabilizing selection for the size of galls that the fly larvae make **(e).**

CONCLUSION: Because wasps preferentially parasitize fly larvae in small galls and birds preferentially eat fly larvae in large galls, the opposing forces of directional selection establish an overall pattern of stabilizing selection in favor of medium-sized galls.

---

so small that they cannot easily penetrate the thick walls of a large gall; they generally lay eggs in fly larvae occupying small galls. Thus, wasps establish directional selection favoring flies that produce large galls, which are less likely to be parasitized. On the other hand, several bird species open galls to feed on mature fly larvae; these predators preferentially open large galls, fostering directional selection in favor of small galls.

In about one-third of the populations surveyed in central Pennsylvania, wasps and birds attacked galls with equal frequency, and flies producing galls of intermediate size had the highest survival rate. The smallest and largest galls—as well as the genetic pre-

Geospiza conirostris

Birds with long bills open cactus fruits to feed on the fleshy pulp.

Birds with intermediate bills may be favored during nondrought years when many types of food are available.

Birds with deep bills strip bark from trees to locate insects.

Figure 20.12
Disruptive selection. Cactus finches *(Geospiza conirostris)* on Genovesa exhibit extreme variability in the size and shape of their bills.

disposition to make very small or very large galls—were eliminated from the population.

Disruptive Selection. Traits undergo **disruptive selection** when extreme phenotypes have higher relative fitness than intermediate phenotypes (see Figure 20.9c). Thus, alleles producing extreme phenotypes become more common, promoting polymorphism. Under natural conditions, disruptive selection is much less common than directional selection and stabilizing selection.

Peter Grant of Princeton University, the world's expert on the ecology and evolution of the Galápagos finches, has analyzed a likely case of disruptive selection on the size and shape of the bill in a population of cactus finches *(Geospiza conirostris)* on the island of Genovesa. During normal weather cycles the finches feed on ripe cactus fruits, seeds, and exposed insects. During drought years, when food is scarce, they also search for insects by stripping bark from the branches of bushes and trees.

During the long drought of 1977, about 70% of the cactus finches on Genovesa died; the survivors exhibited unusually high variability in their bills **(Figure 20.12).** Grant suggested that this morphological variability allowed birds to specialize on particular foods. Birds that stripped bark from branches to look for insects had particularly deep bills, and birds that opened cactus fruits to feed on the fleshy interior had especially long bills. Thus, birds with extreme bill phenotypes appeared to feed efficiently on specific resources, establishing disruptive selection on the size and shape of their bills. The selection may be particularly strong when drought limits the variety and overall availability of food. However, intermediate bill morphologies may be favored during nondrought years when insects and small seeds are abundant.

Sexual Selection Often Exaggerates Showy Structures in Males

Darwin hypothesized that a special process, which he called **sexual selection,** has fostered the evolution of showy structures—such as brightly colored feathers, long tails, or impressive antlers—as well as elaborate courtship behavior in the males of many animal spe-

cies. Sexual selection encompasses two related processes. As the result of *intersexual selection* (that is, selection based on the interactions between males and females), males produce these otherwise useless structures simply because females find them irresistibly attractive. Under *intrasexual selection* (that is, selection based on the interactions between members of the same sex), males use their large body size, antlers, or tusks to intimidate, injure, or kill rival males. In many species, sexual selection is the most probable cause of **sexual dimorphism,** differences in the size or appearance of males and females.

Like directional selection, sexual selection pushes phenotypes toward one extreme. But the products of sexual selection are sometimes bizarre—such as the ridiculously long tail feathers of male African widowbirds. How could evolutionary processes favor the production of such costly structures? Malte Andersson of the University of Gothenburg, Sweden, conducted a field experiment to determine whether the long tail feathers were the product of either intersexual selection or intrasexual selection **(Figure 20.13).** Male widowbirds compete vigorously for favored patches of habitat in which they court females. After surveying the behavior of birds under natural conditions, Andersson lengthened the tails of some males, shortened those of others, and left some males essentially unaltered to serve as controls. His results suggest that females are more strongly attracted to males with long tails than to males with short tails, but that tail length had no effect on a male's ability to compete with other males for space in the habitat. Thus, the long tail of the African widowbird is a product of intersexual selection, not intrasexual selection. Behavioral aspects of sexual selection are described further in Chapter 55.

Nonrandom Mating Can Influence Genotype Frequencies

The Hardy-Weinberg model requires individuals to select mates randomly with respect to their genotypes. This requirement is, in fact, often met; humans, for example, generally marry one another in total ignorance of their genotypes for digestive enzymes or blood types.

Nevertheless, many organisms mate nonrandomly, selecting a mate with a particular phenotype

Figure 20.13 Experimental Research

Sexual Selection in Action

QUESTION: Is the long tail of the male long-tailed widowbird *(Euplectes progne)* the product of intrasexual selection, intersexual selection, or both?

EXPERIMENT: Andersson counted the number of females that associated with individual male widowbirds in the grasslands of Kenya. He then shortened the tails of some individuals by cutting the feathers, lengthened the tails of others by gluing feather extensions to their tails, and left a third group essentially unaltered as a control. One month later, he again counted the number of females associating with each male and compared the results from the three groups.

RESULTS: Males with experimentally lengthened tails attracted more than twice as many mates as males in the control group, and males with experimentally shortened tails attracted fewer. Andersson observed no differences in the ability of altered males and control group males to maintain their display areas.

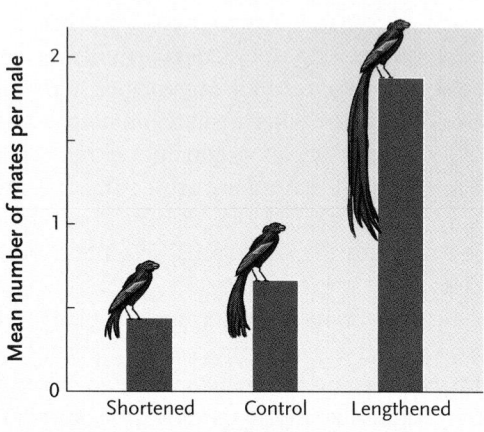

© 2008 Josef Hlasak

CONCLUSION: Female widowbirds clearly prefer males with experimentally lengthened tails to those with normal tails or experimentally shortened tails. Tail length had no obvious effect on the interactions between males. Thus, the long tail of male widowbirds is the product of intersexual selection.

and underlying genotype. Snow geese, for example, usually select mates of their own color, and a tall woman is more likely to marry a tall man than a short man. If no one phenotype is preferred by all potential mates, nonrandom mating does not establish selection for one phenotype over another. But because individuals with similar genetically based phenotypes mate with each other, the next generation will contain fewer heterozygous offspring than the Hardy-Weinberg model predicts.

Inbreeding is a special form of nonrandom mating in which individuals that are genetically related mate with each other. Self-fertilization in plants (see Chapter 34) and a few animals (see Chapter 47) is an extreme example of inbreeding because offspring are produced from the gametes of a single parent. However, other organisms that live in small, relatively closed populations often mate with related individuals. Because relatives often carry the same alleles, inbreeding generally increases the frequency of homozygous genotypes and decreases the frequency of heterozygotes. Thus, recessive phenotypes are often expressed.

For example, the high incidence of Ellis–van Creveld syndrome among the Old Order Amish population, mentioned earlier, is caused by inbreeding. Although the founder effect originally established the disease-causing allele in this population, inbreeding increases the likelihood that it will be expressed. Most human societies discourage matings between genetically close relatives, thereby reducing inbreeding and the production of recessive homozygotes.

STUDY BREAK

1. Which agents of microevolution tend to increase genetic variation within populations, and which ones tend to decrease it?
2. Which mode of natural selection increases the representation of the average phenotype in a population?
3. In what way is sexual selection like directional selection?

20.4 Maintaining Genetic and Phenotypic Variation

Evolutionary biologists continue to discover extraordinary amounts of genetic and phenotypic variation in most natural populations. How can so much variation persist in the face of stabilizing selection and genetic drift?

Diploidy Can Hide Recessive Alleles from the Action of Natural Selection

The diploid condition reduces the effectiveness of natural selection in eliminating harmful recessive alleles from a population. Although such alleles are disadvantageous in the homozygous state, they may have little or no effect on heterozygotes. Thus, recessive alleles can be protected from natural selection by the phenotypic expression of the dominant allele.

In most cases, the masking of recessive alleles in heterozygotes makes it almost impossible to eliminate them completely through selective breeding. Experimentally, we can prevent homozygous recessive organisms from mating. But, as the frequency of a recessive allele decreases, an increasing proportion of its remaining copies is "hidden" in heterozygotes **(Table 20.3)**. Thus, the diploid state preserves recessive alleles at low frequencies, at least in large populations. In small populations, a combination of natural selection and genetic drift can eliminate harmful recessive alleles.

Natural Selection Can Maintain Balanced Polymorphisms

A **balanced polymorphism** is one in which two or more phenotypes are maintained in fairly stable proportions over many generations. Natural selection preserves balanced polymorphisms when heterozygotes have higher relative fitness, when different alleles are favored in different environments, and when the rarity of a phenotype provides an advantage.

Heterozygote Advantage. A balanced polymorphism can be maintained by **heterozygote advantage**, when heterozygotes for a particular locus have higher relative fitness than either homozygote. The best-documented example of heterozygote advantage is the maintenance of the *HbS* (sickle) allele, which codes for a defective form of hemoglobin in humans. As you learned in Chapter 12, hemoglobin is an oxygen-transporting molecule in red blood cells. The hemoglobin produced by the *HbS* allele differs from normal hemoglobin (coded by the *HbA* allele) by just one amino acid. In *HbS/HbS* homozygotes, the faulty hemoglobin forms long fibrous chains under low oxygen conditions, causing red blood cells to assume a sickle shape (as shown

Table 20.3 | **Masking of Recessive Alleles in Diploid Organisms**

When a recessive allele is common in a population (top), most copies of the allele are present in homozygotes. But when the allele is rare (bottom), most copies of it exist in heterozygotes. Thus, rare alleles that are completely recessive are protected from the action of natural selection because they are masked by dominant alleles in heterozygous individuals.

Frequency of Allele *a*	Genotype Frequencies*			% of Allele *a* Copies in	
	AA	Aa	aa	Aa	aa
0.99	0.0001	0.0198	0.9801	1	99
0.90	0.0100	0.1800	0.8100	10	90
0.75	0.0625	0.3750	0.5625	25	75
0.50	0.2500	0.5000	0.2500	50	50
0.25	0.5625	0.3750	0.0625	75	25
0.10	0.8100	0.1800	0.0100	90	10
0.01	0.9801	0.0198	0.0001	99	1

*Population is assumed to be in genetic equilibrium.

in Figure 12.1). Homozygous *HbS/HbS* individuals often die of sickle-cell disease before reproducing, yet in tropical and subtropical Africa, *HbS/HbA* heterozygotes make up nearly 25% of many populations.

Why is the harmful allele maintained at such high frequency? It turns out that sickle-cell disease is most common in regions where malarial parasites infect red blood cells in humans **(Figure 20.14)**. When heterozygous *HbA/HbS* individuals contract malaria, their infected red blood cells assume the same sickle shape as those of homozygous *HbS/HbS* individuals. The sickled cells lose potassium, killing the parasites, which limits their spread within the infected individual. Heterozygous individuals often survive malaria because the parasites do not multiply quickly inside them; their immune systems can effectively fight the infection; and they retain a large population of uninfected red blood cells. Homozygous *HbA/HbA* individuals are also subject to malarial infection, but because their infected cells do not sickle, the parasites multiply rapidly, causing a severe infection with a high mortality rate.

Therefore, *HbA/HbS* heterozygotes have greater resistance to malaria and are more likely to survive severe infections in areas where malaria is prevalent. Natural selection preserves the *HbS* allele in these populations because heterozygotes in malaria-prone areas have higher relative fitness than homozygotes for the normal *HbA* allele.

Selection in Varying Environments. Genetic variability can also be maintained within a population when different alleles are favored in different places or at different times. For example, the shells of European garden

a. Distribution of *HbS* allele

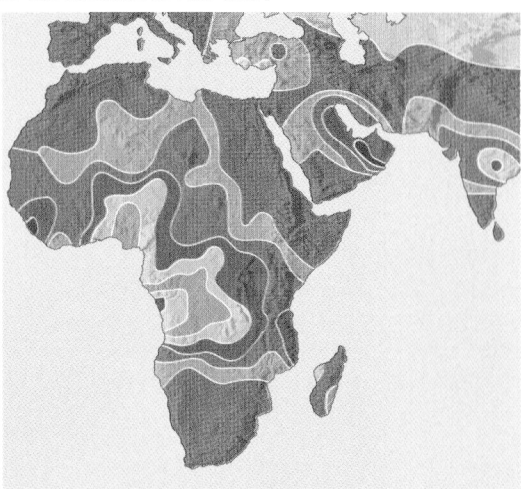

b. Distribution of malarial parasite

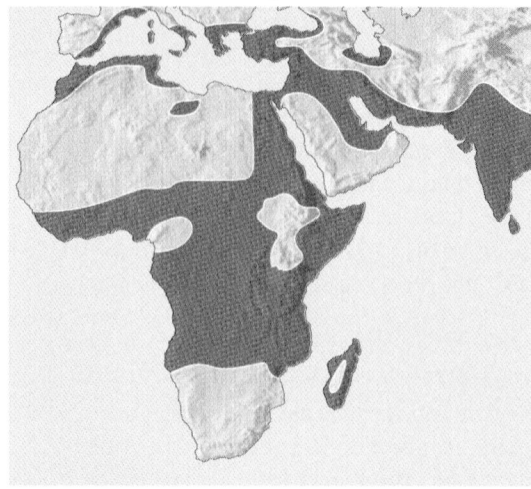

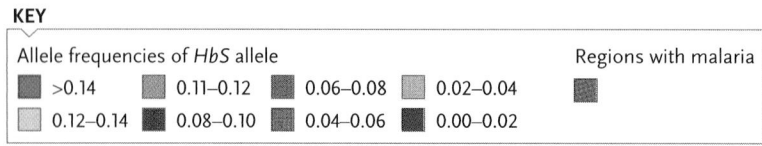

KEY

Allele frequencies of *HbS* allele				Regions with malaria
>0.14	0.11–0.12	0.06–0.08	0.02–0.04	
0.12–0.14	0.08–0.10	0.04–0.06	0.00–0.02	

Figure 20.14

Heterozygote advantage. The distribution of the *HbS* allele **(a)**, which causes sickle-cell disease in homozygotes, roughly matches the distribution of the malarial parasite *Plasmodium falciparum* **(b)** in southern Europe, Africa, the Middle East, and India. Gene flow among human populations has carried the *HbS* allele to some malaria-free regions.

snails range in color from nearly white to pink, yellow, or brown, and may be patterned by one to five stripes of varying color (see Figure 20.2a). This polymorphism, which is relatively stable through time, is controlled by several gene loci. The variability in color and in striping pattern can be partially explained by selection for camouflage in different habitats.

Predation by song thrushes *(Turdus ericetorum)* is a major agent of selection on the color and pattern of these snails in England. When a thrush finds a snail, it smacks it against a rock to break the shell. The bird eats the snail, but leaves the shell near its "anvil." Researchers used the broken shells near an anvil to compare the phenotypes of captured snails to a random sample of the entire snail population. Their analyses indicated that thrushes are visual predators, usually capturing snails that are easy to find. Thus, well-camouflaged snails survive, and the alleles that specify their phenotypes increase in frequency.

The success of camouflage varies with habitat, however; local subpopulations of the snail, which occupy different habitats, often differ markedly in shell color and pattern. The predators eliminate the most conspicuous individuals in each habitat; thus, natural selection differs from place to place **(Figure 20.15).** In woods where the ground is covered with dead leaves, snails with unstriped pink or brown shells predominate. In hedges and fields, where the vegetation in-

cludes thin stems and grass, snails with striped yellow shells are the most common. In populations that span several habitats, selection preserves different alleles in different places, thus maintaining variability in the population as a whole.

Frequency-Dependent Selection. Sometimes genetic variability is maintained in a population simply because rare phenotypes—whatever they happen to be—have higher relative fitness than more common phenotypes. The rare phenotype will increase in frequency until it becomes so common that it loses its advantage. Such phenomena are examples of **frequency-dependent selection** because the selective advantage enjoyed by a particular phenotype depends on its frequency in the population.

Predator-prey interactions can establish frequency-dependent selection because predators often focus their attention on the most common types of prey (see Chapter 50). For example, the aquatic insects called water boatmen occur in three different shades of brown. When all three shades are available at moderate frequencies, fish preferentially feed on the darkest individuals, which are the least camouflaged. But if any one phenotype is very common, fish will learn to focus their attention on that phenotype (see Chapter 54), consuming it in disproportionately large numbers **(Figure 20.16).**

Some Genetic Variations May Be Selectively Neutral

Many biologists believe that some genetic variations are neither preserved nor eliminated by natural selection. According to the **neutral variation hypothesis**, some of the genetic variation at loci coding for enzymes and other soluble proteins is **selectively neutral**. Even if various alleles code for slightly different amino acid sequences in proteins, the different forms of the proteins may function equally well. In those cases, natural selection would not favor some alleles over others.

Biologists who support the neutral variation hypothesis do not question the role of natural selection in producing complex anatomical structures or useful biochemical traits. They also recognize that selection reduces the frequency of harmful alleles. But they argue that we should not simply assume that every genetic variant that persists in a population has been preserved by natural selection. In practice, it is often very difficult to test the natural variation hypothesis because the fitness effects of different alleles are often subtle and vary with small changes in the environment.

The neutral variation hypothesis helps to explain why we see different levels of genetic variation in different populations. It proposes that genetic variation is directly proportional to a population's size and the length of time over which variations have accumulated. Small populations experience fewer mutations than large populations simply because they include fewer replicating genomes. Small populations also lose rare alleles more readily through genetic drift. Thus, small populations should exhibit less genetic variation than large ones, and a population, like the northern elephant seals, that has experienced a recent population bottleneck should exhibit an exceptionally low level of genetic variation. These predictions of the neutral variation hypothesis are generally supported by empirical data.

STUDY BREAK

1. How does the diploid condition protect harmful recessive alleles from natural selection?
2. What is a balanced polymorphism?
3. Why is the allele that causes sickle-cell disease very rare in human populations that are native to northern Europe?

20.5 Adaptation and Evolutionary Constraints

Although natural selection preserves alleles that confer high relative fitness on the individuals that carry them, researchers are cautious about interpreting the benefits that particular traits may provide.

Figure 20.15 Observational Research

Habitat Variation in Color and Striping Patterns of European Garden Snails

HYPOTHESIS: Genetically based variations in the shell color and striping patterns of the European garden snail *(Cepaea nemoralis)* differ substantially from one type of vegetation to another because birds and other visual predators establish strong selection for camouflage in local populations.

PREDICTION: Snails with plain, dark-colored shells will be most abundant in woodland habitats, but snails with striped, light-colored shells will be most abundant in hedges and fields.

METHOD: Two British researchers, A. J. Cain and P. M. Shepard, surveyed the distribution of color and striping patterns of snails in many local populations. They plotted the data on a graph showing the percentage of snails with yellow shells versus the percentage of snails with striped shells, noting the vegetation type where each local population lived.

RESULTS: The shell color and striping patterns of snails living in a particular vegetation type tend to be clustered on the graph, reflecting phenotypic differences that enable the snails to be camouflaged in different habitats. Thus, the alleles that control these characters vary from one local population to another.

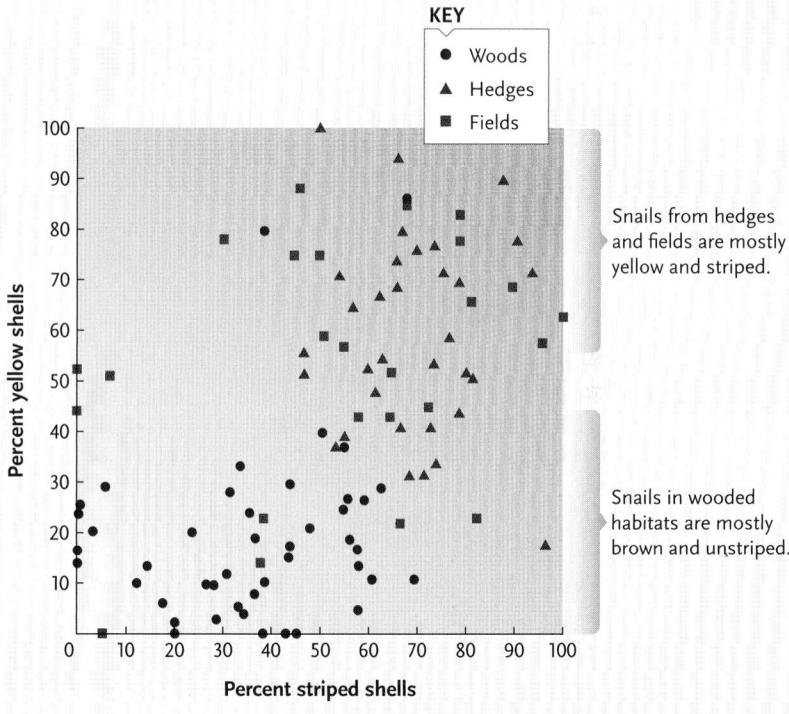

CONCLUSION: Variations in the color and striping patterns on the shells of European garden snails allow most snails to be camouflaged in whatever habitat they occupy. Because these traits are genetically based, the frequencies of the alleles that control them also differ among snails living in different vegetation types. Natural selection therefore favors different alleles in different local populations, maintaining genetic variability in populations that span several vegetation types.

Figure 20.16 Experimental Research

Demonstration of Frequency-Dependent Selection

QUESTION: How does the frequency of a prey type influence the likelihood that it will be captured by predators?

EXPERIMENT: Water boatmen *(Sigara distincta)* occur in three color forms, which vary in the effectiveness of their camouflage. Researchers offered different proportions of the three color forms to predatory fishes in the laboratory and recorded how many of each form were eaten.

RESULTS: When all three phenotypes were available, predatory fishes consumed a disproportionately large number of the most common form, thereby reducing its frequency in the population.

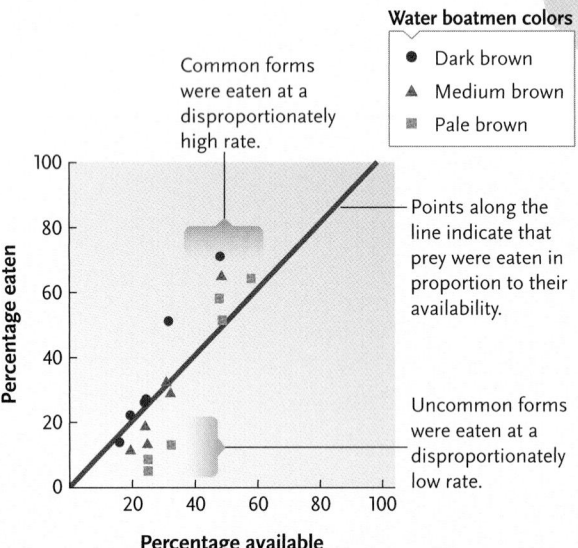

Water boatmen colors
- ● Dark brown
- ▲ Medium brown
- ■ Pale brown

Common forms were eaten at a disproportionately high rate.

Points along the line indicate that prey were eaten in proportion to their availability.

Uncommon forms were eaten at a disproportionately low rate.

y-axis: Percentage eaten (0, 20, 40, 60, 80, 100)
x-axis: Percentage available (20, 40, 60, 80, 100)

CONCLUSION: Predators tend to feed disproportionately on whatever form of their prey is most abundant, thereby reducing its frequency in the prey population.

Scientists Construct Hypotheses about the Evolution of Adaptive Traits

An **adaptive trait** is any product of natural selection that increases the relative fitness of an organism in its environment. **Adaptation** is the accumulation of adaptive traits over time, and this book describes many examples. The change in the oxygen-binding capacity of hemoglobin in response to carbon dioxide concentration, the water-retaining structures and special photosynthetic pathways of desert plants, and the warning coloration of poisonous animals can all be interpreted as adaptive traits.

In fact, we can concoct an adaptive explanation for almost any characteristic we observe in nature. But such explanations are just fanciful stories unless they are framed as testable hypotheses about the relative fitness of different phenotypes and genotypes. Unfor-

tunately, evolutionary biologists cannot always conduct straightforward experiments because they sometimes study traits that do not vary much within a population or species. In such cases, they may compare variations of a trait in closely related species living in different environments. For example, one can test how the traits of desert plants are adaptive by comparing them to traits in related species from moister habitats.

When biologists try to unravel how and why a particular characteristic evolved, they must also remember that a trait they observe today may have had a different function in the past. For example, the structure of the shoulder joint in birds allows them to move their wings first upward and backward and then downward and forward during flapping flight. But analyses of the fossil record reveal that this adaptation, which is essential for flight, did not originate in birds: some predatory nonflying dinosaurs, including the ancestors of birds, had a similarly constructed shoulder joint. Researchers hypothesize that these fast-running predators may have struck at prey with a flapping motion similar to that used by modern birds. Thus, the structure of the shoulder may have first evolved as an adaptation for capturing prey, and only later proved useful for flapping flight. This hypothesis—however plausible it may be—cannot be tested by direct experimentation because the nonflying ancestors of bird have been extinct for millions of years. Instead, evolutionary biologists must use anatomical studies of birds and their ancestors as well as theoretical models about the mechanics of movement to challenge and refine the hypothesis.

Finally, although evolution has produced all the characteristics of organisms, not all are necessarily adaptive. Some traits may be the products of chance events and genetic drift. Others are produced by alleles that were selected for unrelated reasons (see Section 12.2). And still other characteristics result from the action of basic physical laws. For example, the seeds of many plants fall to the ground when they mature, reflecting the inevitable effect of gravity.

Several Factors Constrain Adaptive Evolution

When we analyze the structure and function of an organism, we often marvel at how well adapted it is to its environment and mode of life. However, the adaptive traits of most organisms are compromises produced by competing selection pressures. Sea turtles, for example, must lay their eggs on beaches because their embryos cannot acquire oxygen under water. Although flippers allow females to crawl to nesting sites on beaches, they are not ideally suited for terrestrial locomotion. Their structure reflects their primary function in underwater locomotion.

Moreover, no organism can be perfectly adapted to its environment because environments change over

What are the evolutionary forces affecting molecular variation within populations?

This question may sound like a simple restatement of the entire chapter you have just read, but it is one of the *fundamental* questions in population genetics today—and we have only begun to scratch its surface. The Hardy-Weinberg principle provides a useful null hypothesis, but since we know that evolution happens routinely, that null hypothesis is very frequently rejected. Recent studies have attempted to address this question using theoretical models, extensive DNA sequence data, and detailed measures of recombination rate.

Recombination generates new variation, and, most importantly, it causes the evolutionary forces acting on some genes to become independent of forces acting on other genes. Let's imagine that genes *A* and *B* are on the same chromosome, as shown in this depiction of chromosomes sampled from different individuals within a population:

Gene *B* has two alleles (*B* and *b*), but they have no phenotypic effect, and natural selection does not act on them. Suppose that a new advantageous allele at gene *A* (designated *a*) arises in one chromosome. If there is no recombination between genes *A* and *B*, then as allele *a* spreads in the population by selection, so too will allele *b*, even though there was no selection directly favoring the *b* allele. This effect of selection on nearby genes is called a *selective sweep*. By contrast, if genes *A* and *B* frequently recombine, then allele *a* may not remain associated with allele *b*. Under frequent recombination, the spread of allele *a* may have little or no effect on gene *B*: sometimes *a* will be associated with *b*, but at other times *a* will be associated with *B*.

In the 1990s, evolutionary geneticists were greatly excited by several studies that identified a strong and positive relationship between the recombination rate between particular genes and the amount of genetic variation within those genes. In other words, genes that experienced a lot of recombination also exhibited a great deal of variability. This relationship is consistent with the hypothesis that natural selection often occurs throughout the genome—new advantageous alleles arise frequently, and the impact of their "sweeps" is proportional to their recombination rates. This relationship between recombination and genetic variation was first documented in *Drosophila* (fruit flies) by Chip

Aquadro and his team at Cornell University, but it has since been demonstrated in humans and various plants. Hence, this pattern appears to be very general.

However, our initial interpretation may be too simplistic. Brian Charlesworth, then at the University of Chicago, suggested that the observed pattern may result from the frequent appearance of detrimental mutations that eliminate variation in regions of low recombination—called *background selection*—rather than from sweeps associated with the spread of advantageous alleles. Given that detrimental mutations arise far more frequently than advantageous ones, background selection surely explains some of this general pattern, and perhaps much of it.

An alternative hypothesis that may explain the relationship between recombination rate and genetic variation suggests that recombination rate and the level of genetic variation may be mechanistically connected. A direct connection may operate if recombination itself induces mutations, resulting in higher mutation rates in regions of high recombination. Alternatively, the connection may be indirect: recombination rate is known to be related to the base composition in specific regions of the genome, and base composition is known to influence mutation rates. In 2006, Chris Spencer and his colleagues at Oxford University examined the impact of recombination rates on patterns of nucleotide variation at a very fine scale across the human genome. They found that recombination rates had very local effects on variation, an observation that is consistent with the alternative hypothesis of a mechanistic connection between recombination and mutation rate; their results are not consistent with explanations involving natural selection.

Although biologists first thought that the observed relationship between recombination rate and genetic variation had solved questions about the evolutionary forces that affect molecular variation, this observation has become a puzzle in and of itself. Many of us continue to address this question, now using whole-genome sequences and theoretical and empirical tools for estimating recombination rates. We know that the "final answer" will be that all of the processes described above contribute to this relationship, but knowing their specific contributions will help us understand how, how much, and what kinds of natural selection shape variation within genomes.

Mohamed Noor is an associate professor of biology at Duke University. His research interests include speciation and evolutionary genetics, and recombination. To learn more about his research go to http://www.biology.duke.edu/noorlab/Noorlab.html.

Dr. Noor was a PhD student with Dr. Jerry Coyne, who contributed the Unanswered Questions for Chapter 21.

time. When selection occurs in a population, it preserves alleles that are successful under the prevailing environmental conditions. Thus, each generation is adapted to the environmental conditions under which its parents lived. If the environment changes from one generation to the next, adaptation will always lag behind.

Another constraint on the evolution of adaptive traits is historical. Natural selection is not an engineer that designs new organisms from scratch. Instead, it acts on new mutations and existing genetic variation. Because new mutations are fairly rare, natural selection works primarily with alleles that have been present for many generations. Thus, adaptive changes in the morphology of an organism are almost inevitably based on small modifications of existing structures. The bipedal (two-footed) posture of humans, for example, evolved from the quadrupedal (four-footed) posture of our ancestors. Natural selection did not produce an entirely new skeletal design to accompany this radical behavioral shift. Instead, existing characteristics of the spinal column and the musculature of the legs and back were modified, albeit imperfectly, for an upright stance.

The agents of evolution cause microevolutionary changes in the gene pools of populations. In the next chapter, we examine how microevolution in different populations can cause their gene pools to diverge. The extent of genetic divergence is sometimes sufficient to cause the populations to evolve into different species.

STUDY BREAK

1. How can a biologist test whether a trait is adaptive?
2. Why are most organisms adapted to the environments in which their parents lived?

Review

Go to **ThomsonNOW™** at www.thomsonedu.com/login to access quizzing, animations, exercises, articles, and personalized homework help.

20.1 Variation in Natural Populations

- Phenotypic traits exhibit either quantitative or qualitative variation within populations of all organisms (Figures 20.2 and 20.3).

- Genetic variation, environmental factors, or an interaction between the two cause phenotypic variation within populations. Only genetically based phenotypic variation is heritable and subject to evolutionary change.

- Genetic variation arises within populations largely through mutation and genetic recombination. Artificial selection experiments and analyses of protein and DNA sequences reveal that most populations include significant genetic variation (Figure 20.6).

20.2 Population Genetics

- All the alleles in a population comprise its gene pool, which can be described in terms of allele frequencies and genotype frequencies.

- The Hardy-Weinberg principle of genetic equilibrium is a null model that describes the conditions under which microevolution will not occur: mutations do not occur; populations are closed to migration; populations are infinitely large; natural selection does not operate; and individuals select mates at random. Microevolution, a change in allele frequencies through time, occurs in populations when the restrictive requirements of the model are not met.

Animation: How to find out if a population is evolving

20.3 The Agents of Microevolution

- Several processes cause microevolution in populations. Mutation introduces completely new genetic variation. Gene flow carries novel genetic variation into a population through the arrival and reproduction of immigrants. Genetic drift causes random changes in allele frequencies, especially in small populations. Natural selection occurs when the genotypes of some individuals enable them to survive and reproduce more than others. Nonrandom mating within a population can cause its genotype frequencies to depart from the predictions of the Hardy-Weinberg equilibrium.

- Natural selection alters phenotypic variation in one of three ways (Figure 20.9). Directional selection increases or decreases the mean value of a trait, shifting it toward a phenotypic extreme. Stabilizing selection increases the frequency of the mean phenotype and reduces variability in the trait (Figure 20.10). Disruptive selection increases the frequencies of extreme phenotypes and decreases the frequency of intermediate phenotypes (Figure 20.12).

- Sexual selection promotes the evolution of exaggerated structures and behaviors (Figure 20.13).

- Although nonrandom mating does not change allele frequencies, it can affect genotype frequencies, producing more homozygotes and fewer heterozygotes than the Hardy-Weinberg model predicts.

Animation: Directional selection

Animation: Change in moth population

Animation: Stabilizing selection

Animation: Disruptive selection

Animation: Disruptive selection among African finches

Animation: Simulation of genetic drift

20.4 Maintaining Genetic and Phenotypic Variation

- Diploidy can maintain genetic variation in a population if alleles coding for recessive traits are not expressed in heterozygotes and are thus hidden from natural selection.

- Polymorphisms are maintained in populations when heterozygotes have higher relative fitness than both homozygotes (Figure 20.14), when natural selection occurs in variable environments (Figure 20.15), or when the relative fitness of a phenotype varies with its frequency in the population (Figure 20.16).

- Some biologists believe that many genetic variations are selectively neutral, conferring neither advantages nor disadvantages on the individuals that carry them. The neutral variation hypothesis explains why large populations and those that have not experienced a recent population bottleneck exhibit the highest levels of genetic variation.

Animation: Distribution of sickle-cell trait

Animation: Life cycle of *Plasmodium*

20.5 Adaptation and Evolutionary Constraints

- Adaptive traits increase the relative fitness of individuals carrying them. Adaptive explanations of traits must be framed as testable hypotheses.

- Natural selection cannot result in perfectly adapted organisms because most adaptive traits represent compromises among conflicting needs; because most environments are constantly changing; and because natural selection can affect only existing genetic variation.

Animation: Adaptation to what?

Questions

Self-Test Questions

1. Which of the following represents an example of qualitative phenotypic variation?
   a. the lengths of people's toes
   b. the body sizes of pigeons
   c. human ABO blood groups
   d. the birth weights of humans
   e. the number of leaves on oak trees

2. A population of mice is at Hardy-Weinberg equilibrium at a gene locus that controls fur color. The locus has two alleles, M and m. A genetic analysis of one population reveals that 60% of its gametes carry the M allele. What percentage of mice contains both the M and m alleles?
   a. 60%   d. 36%
   b. 48%   e. 16%
   c. 40%

3. If the genotype frequencies in a population are 0.60 AA, 0.20 Aa, and 0.20 aa, and if the requirements of the Hardy-Weinberg principle apply, the genotype frequencies in the offspring generation will be:
   a. 0.60 AA, 0.20 Aa, 0.20 aa.
   b. 0.36 AA, 0.60 Aa, 0.04 aa.
   c. 0.49 AA, 0.42 Aa, 0.09 aa.
   d. 0.70 AA, 0.00 Aa, 0.30 aa.
   e. 0.64 AA, 0.32 Aa, 0.04 aa.

4. The reason spontaneous mutations do not have an immediate effect on allele frequencies in a large population is that:
   a. mutations are random events, and mutations may be either beneficial or harmful.
   b. mutations usually occur in males and have little effect on eggs.
   c. many mutations exert their effects after an organism has stopped reproducing.
   d. mutations are so rare that mutated alleles are greatly outnumbered by nonmutated alleles.
   e. most mutations do not change the amino acid sequence of a protein.

5. The phenomenon in which chance events cause unpredictable changes in allele frequencies is called:
   a. gene flow.
   b. genetic drift.
   c. inbreeding.
   d. balanced polymorphism.
   e. stabilizing selection.

6. An Eastern European immigrant carrying the allele for Tay Sachs disease settled in a small village on the St. Lawrence River. Many generations later, the frequency of the allele in that village is statistically higher than it is in the immigrant's homeland. The high frequency of the allele in the village probably provides an example of:
   a. natural selection.
   b. the concept of relative fitness.
   c. the Hardy-Weinberg genetic equilibrium.
   d. phenotypic variation.
   e. the founder effect.

7. If a storm kills many small sparrows in a population, but only a few medium-sized and large ones, which type of selection is probably operating?
   a. directional selection
   b. stabilizing selection
   c. disruptive selection
   d. intersexual selection
   e. intrasexual selection

8. Which of the following phenomena explains why the allele for sickle-cell hemoglobin is common in some tropical and subtropical areas where the malaria parasite is prevalent?
   a. balanced polymorphism
   b. heterozygote advantage
   c. sexual dimorphism
   d. neutral selection
   e. stabilizing selection

9. The neutral variation hypothesis proposes that:
   a. complex structures in most organisms have not been fostered by natural selection.
   b. most mutations have a strongly harmful effect.
   c. some mutations are not affected by natural selection.
   d. natural selection cannot counteract the action of gene flow.
   e. large populations are subject to stronger natural selection than small populations.

10. Phenotypic characteristics that increase the fitness of individuals are called:
    a. mutations.
    b. founder effects.
    c. heterozygote advantages.
    d. adaptive traits.
    e. polymorphisms.

Questions for Discussion

1. Most large commercial farms routinely administer antibiotics to farm animals to prevent the rapid spread of diseases through a flock or herd. Explain why you think that this practice is either wise or unwise.

2. Many human diseases are caused by recessive alleles that are not expressed in heterozygotes. Explain why it is almost impossible to eliminate such genetic traits from human populations.

3. Using two types of beans to represent two alleles at the same gene locus, design an exercise to illustrate how population size affects genetic drift.

4. In what ways are the effects of sexual selection, disruptive selection, and nonrandom mating different? How are they similar?

Experimental Analysis

Design an experiment to test the hypothesis that the differences in size among adult guppies are determined by the amount of food they eat rather than by genetic factors.

Evolution Link

Captive breeding programs for endangered species often have access to a limited supply of animals for a breeding stock. As a result, their offspring are at risk of being highly inbred. Why and how might zoological gardens and conservation organizations avoid or minimize inbreeding?

How Would You Vote?

The symptoms of Huntington disease and some other genetically based diseases in humans appear only after the carriers of the disease-causing allele have already reproduced. As a result, they pass the alleles to their offspring and the disease persists in the population. Do you think that all people should be screened for disease-causing alleles and that carriers of such alleles should be discouraged or even prevented from having children? Go to www.thomsonedu.com/login to investigate both sides of the issue and then vote.

Two closely related species of parrot, the scarlet macaw *(Ara chloroptera)* and the blue and yellow macaw *(Ara arauna)*, perching together in the Amazon jungle of Peru.

21 Speciation

WHY IT MATTERS

In 1927, nearly 100 years after Darwin boarded the *Beagle,* a young German naturalist named Ernst Mayr embarked on his own journey, to the highlands of New Guinea. He was searching for rare "birds of paradise," no trace of which had been seen in Europe since plume hunters had returned years before with ornate and colorful feathers that were used to decorate ladies' hats **(Figure 21.1).** On his trek through the remote Arfak Mountains, Mayr identified 137 bird species (including many birds of paradise) based on differences in their size, plumage, color, and other external characteristics.

To Mayr's surprise, the native Papuans—who were untrained in the ways of Western science, but who hunted these birds for food and feathers—had their own names for 136 of the 137 species he had identified. The close match between the two lists confirmed Mayr's belief that the *species* is a fundamental level of organization in nature. Each species has a unique combination of genes underlying its distinctive appearance and habits. Thus, people who observe them closely—whether indigenous hunters or Western scientists—can often distinguish one species from another.

443

Figure 21.1

Birds of paradise. A male Count Raggi's bird of paradise *(Paradisaea raggiana)* has clearly attracted the attention of a female (the smaller, less colorful bird) with his showy plumage and conspicuous display. There are 43 known bird of paradise species, 35 of them found only on the island of New Guinea.

Mayr also discovered some remarkable patterns in the geographical distributions of the bird species in New Guinea. For example, each mountain range he explored was home to some species that lived nowhere else. Closely related species often lived on different mountaintops, separated by deep valleys of unsuitable habitat. In 1942, Mayr published the book *Systematics and the Origin of Species,* in which he described the role of geography in the evolution of new species; the book quickly became a cornerstone of the modern synthesis (which was outlined in Section 19.3).

What mechanisms produce distinct species? As you discovered in Chapter 20, microevolutionary processes alter the pattern and extent of genetic and phenotypic variation within populations. When these processes differ between populations, the populations will diverge, and they may eventually become so different that we recognize them as distinct species. Although Darwin's famous book was titled *On the Origin of Species,* he didn't dwell on the question of *how* new species arise. But the concept of **speciation**—the process of species formation—was implicit in his insight that similar species often share inherited characteristics and a common ancestry. Darwin also recognized that "descent with modification" had generated the amazing diversity of organisms on Earth.

Today evolutionary biologists view speciation as a *process,* a series of events that occur through time. However, they usually study the *products* of speciation, species that are alive today. Because they can rarely witness the process of speciation from start to finish,

scientists make inferences about it by studying organisms in various stages of species formation. In this chapter, we consider four major topics: how biologists define and recognize species; how species maintain their genetic identity; how the geographical distributions of organisms influence speciation; and how different genetic mechanisms produce new species.

21.1 What Is a Species?

Like the hunters of the Arfak Mountains, most of us recognize the different species that we encounter every day. We can distinguish a cat from a dog and sunflowers from roses. The concept of species is based on our perception that Earth's biological diversity is packaged in discrete, recognizable units, and not as a continuum of forms grading into one another. As evolutionary scientists learn more about the causes of microevolution, they refine our understanding of what a species really is.

The Morphological Species Concept Is a Practical Way to Identify Species

Biologists often describe new species on the basis of visible anatomical characteristics, a process that dates back to Linnaeus' classification of organisms in the eighteenth century (described in Chapter 23). This approach is based on the **morphological species concept,** the idea that all individuals of a species share measurable traits that distinguish them from individuals of other species.

The morphological species concept has many practical applications. For example, paleobiologists use morphological criteria to identify the species of fossilized organisms (see Chapter 22). And because we can observe the external traits of organisms in nature, field guides to plants and animals list diagnostic (that is, distinguishing) physical characters that allow us to recognize them **(Figure 21.2).**

Nevertheless, relying exclusively on a morphological approach can present problems. Consider the variation in the shells of *Cepaea nemoralis* (shown earlier in Figure 20.2). How could anyone imagine that so variable a collection of shells represents just one species

Yellow-throated warbler   Myrtle warbler

Figure 21.2

Diagnostic characters. Yellow-throated warblers *(Dendroica dominica)* and myrtle warblers *(Dendroica coronata)* can be distinguished by the color of feathers on the throat and rump.

of snail? Moreover, morphology does not help us distinguish some closely related species that are nearly identical in appearance. Finally, morphological species definitions tell us little about the evolutionary processes that produce new species.

The Biological and Phylogenetic Species Concepts Derive from Evolutionary Theory

The **biological species concept** emphasizes the dynamic nature of species. Ernst Mayr defined biological species as "groups of . . . interbreeding natural populations that are reproductively isolated from [do not produce fertile offspring with] other such groups." The concept is based on reproductive criteria and is easy to apply, at least in principle: if the members of two populations interbreed and produce fertile offspring *under natural conditions,* they belong to the same species; their fertile offspring will, in turn, produce the next generation of that species. If two populations do not interbreed in nature, or fail to produce fertile offspring when they do, they belong to different species.

The biological species concept defines species in terms of population genetics and evolutionary theory. The first half of Mayr's definition notes the genetic *cohesiveness* of species: populations of the same species experience gene flow, which mixes their genetic material. Thus, we can think of a species as one large gene pool, which may be subdivided into local populations.

The second part of the biological species concept emphasizes the genetic *distinctness* of each species. Because populations of different species are reproductively isolated, they cannot exchange genetic information. In fact, the process of speciation is frequently defined as the evolution of reproductive isolation between populations.

The biological species concept also explains why individuals of a species generally look alike: members of the same gene pool share genetic traits that determine their appearance. Individuals of different species generally do not resemble one another as closely because they share fewer genetic characteristics. In practice, biologists often use similarities or differences in morphological traits as convenient markers of genetic similarity or reproductive isolation.

However, the biological species concept does not apply to the many forms of life that reproduce asexually, including most bacteria; some protists, fungi, and plants; and a few animals. In these species, individuals don't interbreed, so it is pointless to ask whether different populations do. Similarly, we cannot use the biological species concept to study extinct organisms, because we have little or no data on their reproductive habits. These species must all be defined using morphological or biochemical criteria. Yet, despite its limitations, the biological species concept currently provides the best evolutionary definition of a sexually reproducing species.

Recognizing the limitations of the biological species concept, some researchers have proposed a **phylogenetic species concept.** Using both morphological and genetic sequence data, scientists first reconstruct the evolutionary tree for the populations of interest. They then define a phylogenetic species as a cluster of populations—the tiniest twigs on the tree—that emerge from the same small branch. Thus, a phylogenetic species comprises populations that share a recent evolutionary history. We will consider this approach for defining species as well as more inclusive evolutionary groups in Chapter 23.

Many Species Exhibit Substantial Geographical Variation

Populations change in response to shifting environments, and separate populations of a species frequently differ both genetically and phenotypically. Neighboring populations often have shared characteristics because they live in similar environments, exchange individuals, and experience comparable patterns of natural selection. Widely separated populations, by contrast, may live under different conditions and experience different patterns of selection; because gene flow is less likely to occur between distant populations, their gene pools and phenotypes often differ.

When geographically separated populations of a species exhibit dramatic, easily recognized phenotypic variation, biologists may identify them as different **subspecies (Figure 21.3),** which are local variants of a species. Individuals from different subspecies usually interbreed where their geographical distributions

Figure 21.3
Subspecies. Five subspecies of rat snake (Elaphe obsoleta) in eastern North America differ in color and in the presence or absence of stripes or blotches.

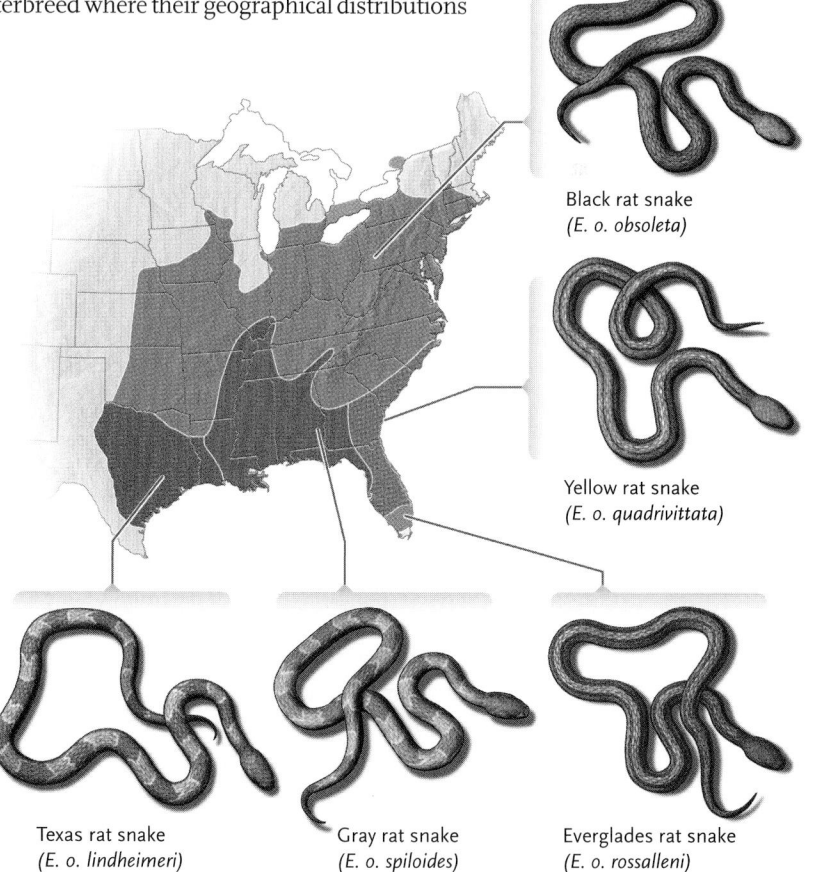

Black rat snake
(E. o. obsoleta)

Yellow rat snake
(E. o. quadrivittata)

Texas rat snake
(E. o. lindheimeri)

Gray rat snake
(E. o. spiloides)

Everglades rat snake
(E. o. rossalleni)

Figure 21.4

Ring species. Six of the seven subspecies of the salamander *Ensatina eschscholtzii* are distributed in a ring around California's Central Valley. Subspecies often interbreed where their geographical distributions overlap. However, the two subspecies that nearly close the ring in the south (marked with an arrow), the Monterey salamander and the yellow-blotched salamander, rarely interbreed.

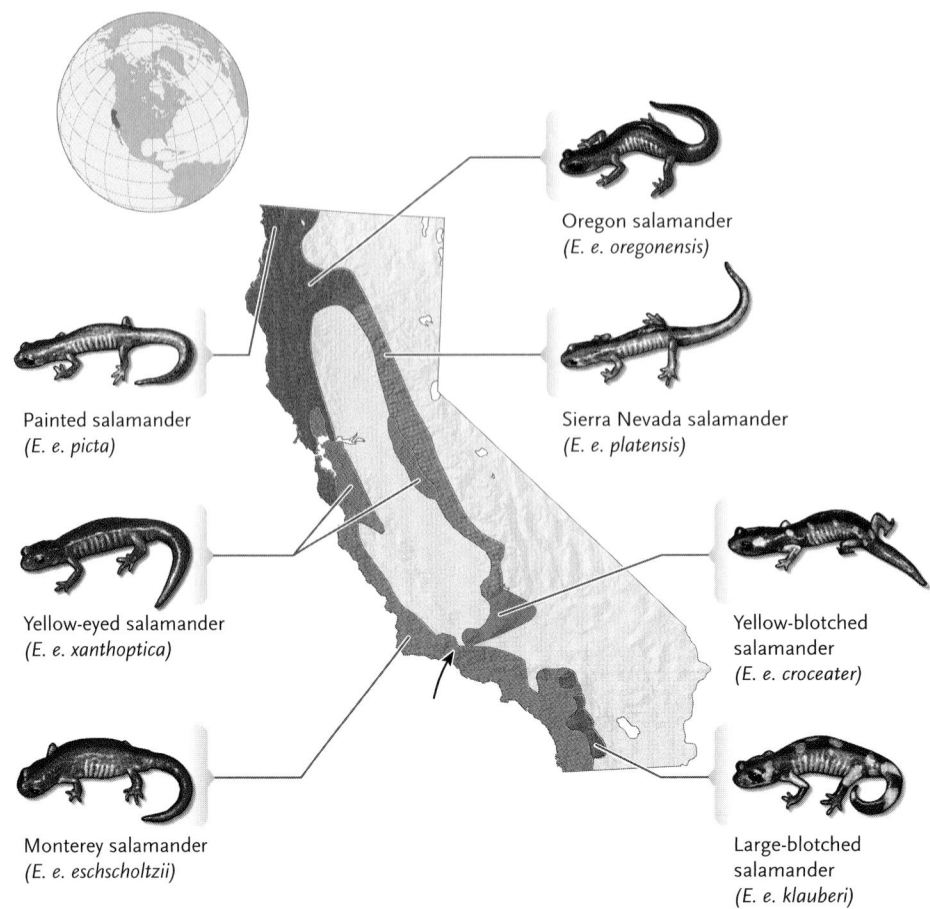

Oregon salamander
(*E. e. oregonensis*)

Painted salamander
(*E. e. picta*)

Sierra Nevada salamander
(*E. e. platensis*)

Yellow-eyed salamander
(*E. e. xanthoptica*)

Yellow-blotched salamander
(*E. e. croceater*)

Monterey salamander
(*E. e. eschscholtzii*)

Large-blotched salamander
(*E. e. klauberi*)

meet, and their offspring often exhibit intermediate phenotypes. Biologists sometimes use the word "race" as shorthand for the term "subspecies."

Various patterns of geographical variation have provided great insight into the speciation process. Two of the best-studied patterns are *ring species* and *clinal variation.*

Ring Species. Some plant and animal species have a ring-shaped geographical distribution that surrounds uninhabitable terrain. Adjacent populations of these so-called **ring species** can exchange genetic material directly, but gene flow between distant populations occurs only through the intermediary populations.

The lungless salamander *Ensatina eschscholtzii,* an example of a ring species, is widely distributed in the coastal mountains and the Sierra Nevada of California, but it cannot survive in the hot, dry Central Valley **(Figure 21.4).** Seven subspecies differ in biochemical traits, color, size, and ecology. Individuals from adjacent subspecies often interbreed where their geographical distributions overlap, and intermediate phenotypes are fairly common. But at the southern end of the Central Valley, adjacent subspecies rarely interbreed. Apparently, they have differentiated to such an extent that they can no longer exchange genetic material directly.

Are the southernmost populations of this salamander subspecies or different species? A biologist who saw *only* the southern populations, which coexist without interbreeding, might define them as separate species. However, they still have the potential to exchange genetic material through the intervening populations that form the ring. Hence, biologists recognize these populations as belonging to the same species. Most likely, the southern subspecies are in an intermediate stage of species formation.

Clinal Variation. When a species is distributed over a large, environmentally diverse area, some traits may exhibit a **cline,** a pattern of smooth variation along a geographical gradient. Clinal variation usually results from gene flow between adjacent populations that are each adapting to slightly different conditions. For example, many birds and mammals in the northern hemisphere show clinal variation in body size **(Figure 21.5)** and the relative length of their appendages: in general, populations living in colder environments have larger bodies and shorter appendages, a pattern that is usually interpreted as a mechanism to conserve heat (see Chapter 46). If a cline extends over a large geographical gradient, populations at the opposite ends may be very different.

Despite the geographical variation that many species exhibit, most closely related species are genetically and morphologically different from each other. In the next section, we consider the mechanisms that maintain the genetic distinctness of closely related species by preventing their gene pools from mixing.

STUDY BREAK

1. How does the morphological species concept differ from the biological species concept?
2. What is clinal variation?

21.2 Maintaining Reproductive Isolation

Reproductive isolation is central to the biological species concept. A **reproductive isolating mechanism** is a biological characteristic that prevents the gene pools of two species from mixing. Biologists classify reproductive isolating mechanisms into two categories (summarized in **Table 21.1**): **prezygotic isolating mechanisms** exert their effects before the production of a zygote, or fertilized egg, and **postzygotic isolating mechanisms** operate after zygote formation. These isolating mechanisms are not mutually exclusive; two or more of them may operate simultaneously.

Prezygotic Isolating Mechanisms Prevent the Production of Hybrid Individuals

Biologists have identified five mechanisms that can prevent interspecific (between species) matings or fertilizations, and thus prevent the production of hybrid (mixed species) offspring. These five prezygotic mechanisms are *ecological, temporal, behavioral, mechanical,* and *gametic isolation.*

Species living in the same geographical region may experience **ecological isolation** if they live in different habitats. For example, lions and tigers were both common in India until the mid-nineteenth century, when hunters virtually exterminated the Asian lions. However, because lions live in open grasslands and tigers in dense forests, the two species did not encounter one another and did not interbreed. Lion-tiger hybrids are sometimes born in captivity, but do not occur under natural conditions.

Species living in the same habitat can experience **temporal isolation** if they mate at different times of day or different times of year. For example, the fruit flies *Drosophila persimilis* and *Drosophila pseudo-obscura* overlap extensively in their geographical distributions, but they do not interbreed, in part because *D. persimilis* mates in the morning and *D. pseudo-obscura* in the

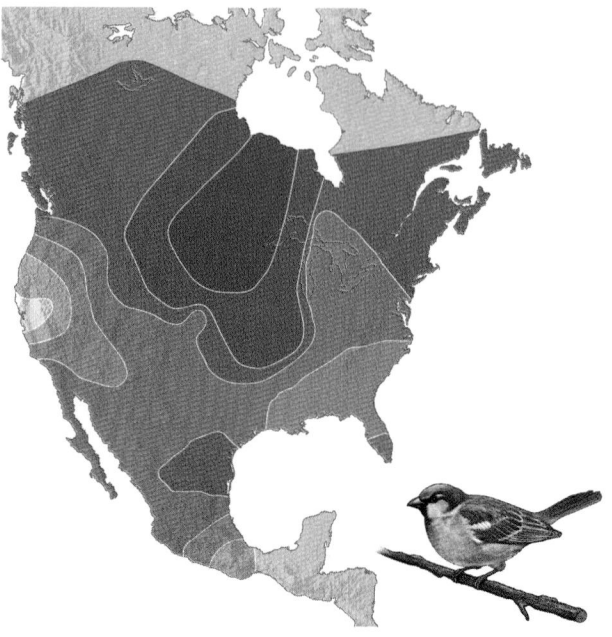

Figure 21.5

Clinal variation. House sparrows *(Passer domesticus)* exhibit clinal variation in overall body size, which was summarized from measurements of 16 skeletal features. Darker shading indicates larger size.

Table 21.1	Reproductive Isolating Mechanisms	
Timing Relative to Fertilization	**Mechanism**	**Mode of Action**
Prezygotic ("premating") mechanisms	Ecological isolation	Species live in different habitats
	Temporal isolation	Species breed at different times
	Behavioral isolation	Species cannot communicate
	Mechanical isolation	Species cannot physically mate
	Gametic isolation	Species have nonmatching receptors on gametes
Postzygotic ("postmating") mechanisms	Hybrid inviability	Hybrid offspring do not complete development
	Hybrid sterility	Hybrid offspring cannot produce gametes
	Hybrid breakdown	Hybrid offspring have reduced survival or fertility

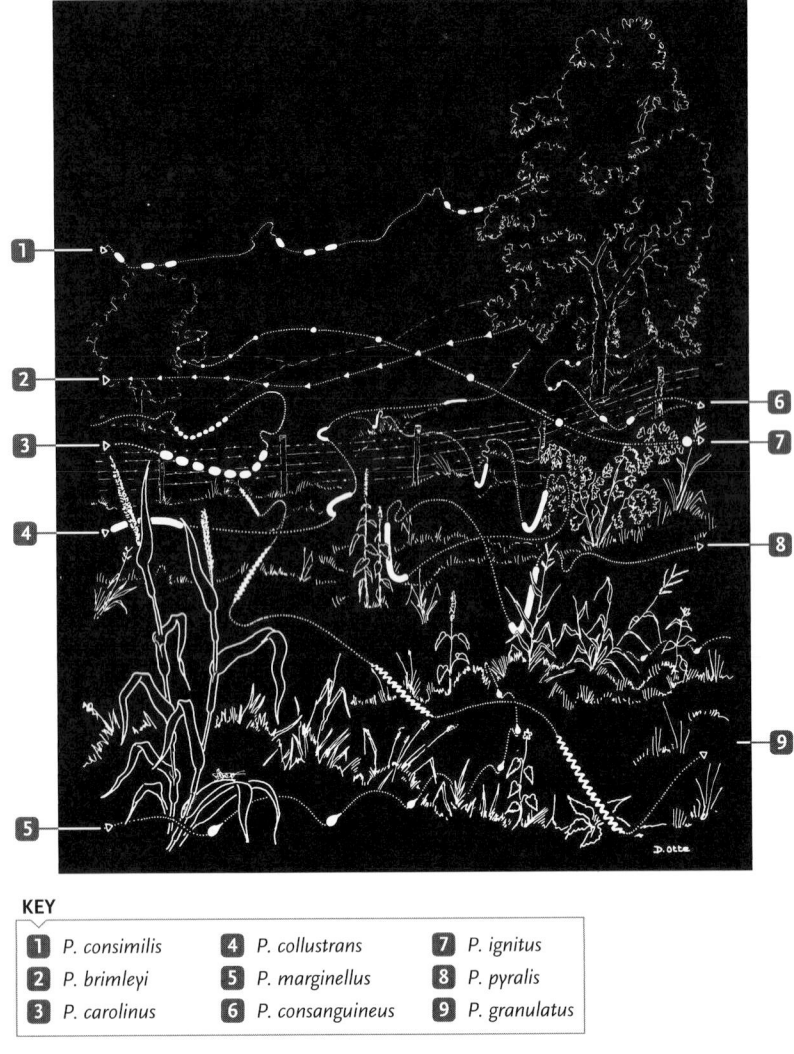

KEY

1	*P. consimilis*	**4**	*P. collustrans*	**7**	*P. ignitus*
2	*P. brimleyi*	**5**	*P. marginellus*	**8**	*P. pyralis*
3	*P. carolinus*	**6**	*P. consanguineus*	**9**	*P. granulatus*

Figure 21.6

Behavioral reproductive isolation. Male fireflies (*Photinus* species) use bioluminescent signals to attract potential mates. The different flight paths and flashing patterns of males in nine North American species are represented here. Females respond only to the display given by males of their own species.

(Courtesy of James E. Lloyd. Miscellaneous Publications of the Museum of Zoology of the University of Michigan, 130:1–195, 1966.)

afternoon. Two species of pine in California are reproductively isolated where their geographical distributions overlap: even though both rely on the wind to carry male gametes (pollen grains) to female gametes (ova) in other cones, *Pinus radiata* releases pollen in February and *Pinus muricata* releases pollen in April.

Many animals rely on specific signals, which often differ dramatically between species, to identify the species of a potential mate. **Behavioral isolation** results when the signals used by one species are not recognized by another. For example, female birds rely on the song, color, and displays of males to identify members of their own species. Similarly, female fireflies identify males by their flashing patterns **(Figure 21.6)**. These behaviors (collectively called *courtship displays*) are often so complicated that signals sent by one species are like a foreign language that another species simply does not understand.

Mate choice by females and sexual selection (discussed in Section 20.3) generally drive the evolution of mate recognition signals. Females often spend substantial energy in reproduction, and choosing an appropriate mate—that is, a male of her own species—is critically important for the production of successful young. By contrast, a female that mates with a male from a different species is unlikely to leave any surviving offspring at all. Over time, the number of males with recognizable traits, as well as the number of females able to recognize the traits, increases in the population.

Differences in the structure of reproductive organs or other body parts—**mechanical isolation**—may prevent individuals of different species from interbreeding. In particular, many plants have anatomical features that allow only certain pollinators, usually particular bird or insect species, to collect and distribute pollen (see Chapter 27). For example, the flowers and nectar of two native California plants, the monkey-flowers *Mimulus lewisii* and *Mimulus cardinalis,* attract different animal pollinators **(Figure 21.7)**. *Mimulus lewisii* is pollinated by bumblebees. It has shallow pink flowers with broad petals that provide a landing platform for the bees. Bright yellow streaks on the petals serve as "nectar guides," directing bumblebees to the short nectar tube and reproductive parts, which are located among the petals. Bees enter the flowers to drink their concentrated nectar, and they pick up and deliver pollen as they brush against the reproductive parts of the flowers. *Mimulus cardinalis,* by contrast, is pollinated by hummingbirds. It has long red flowers with no yellow streaks, and the reproductive parts extend above the petals. The red color attracts hummingbirds but lies outside the color range detected by bumblebees. The nectar of *M. cardinalis* is more dilute than that of *M. lewisii* but is produced in much greater quantity, making it easier for hummingbirds to

Mimulus lewisii   *Mimulus cardinalis*

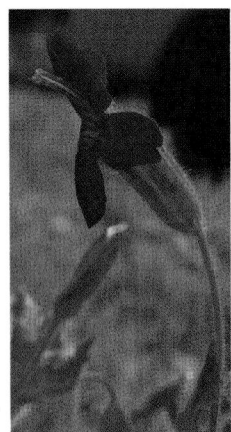

Figure 21.7

Mechanical reproductive isolation. Because of differences in floral structure, two species of monkey-flower attract different animal pollinators. *Mimulus lewisii* attracts bumblebees and *Mimulus cardinalis* attracts hummingbirds.

ingest. When a hummingbird visits *M. cardinalis* flowers, it pushes its long bill down the nectar tube, and its forehead touches the reproductive parts, picking up and delivering pollen. Recent research has demonstrated that where the two monkey-flower species grow side-by-side, animal pollinators restrict their visits to either one species or the other 98% of the time, providing nearly complete reproductive isolation.

Even when individuals of different species mate, **gametic isolation,** an incompatibility between the sperm of one species and the eggs of another, may prevent fertilization. Many marine invertebrates release gametes into the environment for external fertilization. The sperm and eggs of each species recognize one another's complementary surface proteins (see Chapter 47), but the surface proteins on the gametes of different species don't match. In animals with internal fertilization, sperm of one species may not survive and function within the reproductive tract of another. Interspecific matings between some *Drosophila* species, for example, induce a reaction in the female's reproductive tract that blocks "foreign" sperm from reaching eggs. Parallel physiological incompatibilities between a pollen tube and a stigma prevent interspecific fertilization in some plants.

Postzygotic Isolating Mechanisms Reduce the Success of Hybrid Individuals

If prezygotic isolating mechanisms between two closely related species are incomplete or ineffective, sperm from one species sometimes fertilizes an egg of the other species. In such cases the two species will be reproductively isolated if their offspring, called interspecific (between species) hybrids, have lower fitness than those produced by intraspecific (within species) matings. Three postzygotic isolating mechanisms—*hybrid inviability, hybrid sterility,* and *hybrid breakdown*—can reduce the fitness of hybrid individuals.

Many genes govern the complex processes that transform a zygote into a mature organism. Hybrid individuals have two sets of developmental instructions, one from each parent species, which may not interact properly for the successful completion of embryonic development. As a result, hybrid organisms frequently die as embryos or at an early age, a phenomenon called **hybrid inviability.** For example, domestic sheep and goats can mate and fertilize one another's ova, but the hybrid embryos always die before coming to term, presumably because the developmental programs of the two parent species are incompatible.

Although some hybrids between closely related species develop into healthy and vigorous adults, they may not produce functional gametes. This **hybrid sterility** often results when the parent species differ in the number or structure of their chromosomes, which cannot pair properly during meiosis. Such hybrids have zero fitness because they leave no descendants. The most familiar example is a mule, the product of mating be-

Figure 21.8

Interspecific hybrids. Horses and zebroids (hybrid offspring of horses and zebras) run in a mixed herd. Zebroids are usually sterile.

Jen and Des Bartlett/Bruce Coleman USA

tween a female horse ($2n = 64$) and a male donkey ($2n = 62$). Zebroids, the offspring of matings between horses and zebras, are also usually sterile **(Figure 21.8).**

Some first-generation hybrids (F_1; see Section 12.1) are healthy and fully fertile. They can breed with other hybrids and with both parental species. However, the second generation (F_2), produced by matings between F_1 hybrids, or between F_1 hybrids and either parental species, may exhibit reduced survival or fertility, a phenomenon known as **hybrid breakdown.** For example, experimental crosses between *Drosophila* species may produce functional hybrids, but their offspring experience a high rate of chromosomal abnormalities and harmful types of genetic recombination. Thus, reproductive isolation is maintained between the species because there is little long-term mixing of their gene pools.

STUDY BREAK

1. What is the difference between prezygotic and postzygotic isolating mechanisms?
2. When a male duck of one species performed a courtship display to a female of another species, she interpreted his behavior as aggressive rather than amorous. What type of reproductive isolating mechanism does this scenario illustrate?

21.3 The Geography of Speciation

As Ernst Mayr recognized, geography has a huge impact on whether gene pools have the opportunity to mix. Biologists define three modes of speciation, based on the geographical relationship of populations as they become

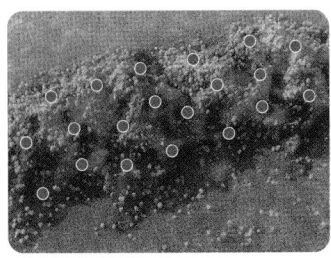

1 At first, a population is distributed over a large geographical area.

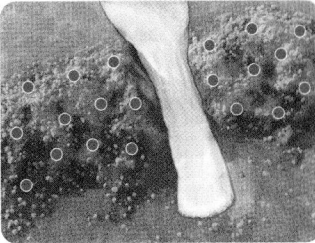

2 A geographical change, such as the advance of a narrow glacier, separates the original population, creating a barrier to gene flow.

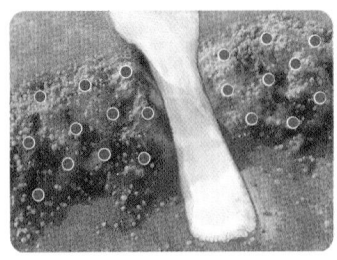

3 In the absence of gene flow, the separated populations evolve independently and diverge into different species.

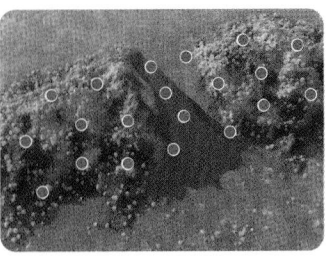

4 When the glacier later melts, allowing individuals of the two species to come into secondary contact, they do not interbreed.

Figure 21.9
The model of allopatric speciation and secondary contact.

reproductively isolated: *allopatric speciation* (*allo* = different; *patria* = homeland), *parapatric speciation* (*para* = beside), and *sympatric speciation* (*sym* = together).

Allopatric Speciation Occurs between Geographically Separated Populations

Allopatric speciation may take place when a physical barrier subdivides a large population or when a small population becomes separated from a species' main geographical distribution. Probably the most common mode of speciation in large animals, allopatric speciation occurs in two stages. First, two populations become *geographically* separated, preventing gene flow between them. Then, as the populations experience distinct mutations as well as different patterns of natural selection and genetic drift, they may accumulate genetic differences that isolate them *reproductively*.

Geographical separation sometimes occurs when a barrier divides a large population into two or more units **(Figure 21.9)**. For example, hurricanes may create new channels that divide low coastal islands and the populations inhabiting them. Uplifting mountains or landmasses as well as advancing glaciers can also pro-

duce barriers that subdivide populations. The uplift of the Isthmus of Panama, caused by movements of Earth's crust about five million years ago (see the *Focus on Research* in Chapter 22), separated a once-continuous shallow sea into the eastern tropical Pacific Ocean and the western tropical Atlantic Ocean. Populations of marine organisms were subdivided by this event, and pairs of closely related species now live on either side of this divide **(Figure 21.10)**.

In other cases, small populations may become isolated at the edge of a species' geographical distribution. Such peripheral populations often differ genetically from the central population because they are adapted to somewhat different environments. Once a small population is isolated, genetic drift and natural selection as well as limited gene flow from the parent population foster further genetic differentiation. In time, the accumulated genetic differences may lead to reproductive isolation.

Populations on oceanic islands represent extreme examples of this phenomenon. Founder effects, an example of genetic drift (see Section 20.3), make the populations genetically distinct. And on oceanic archipelagos, such as the Galápagos and Hawaiian islands, individuals from one island may colonize nearby islands, found-

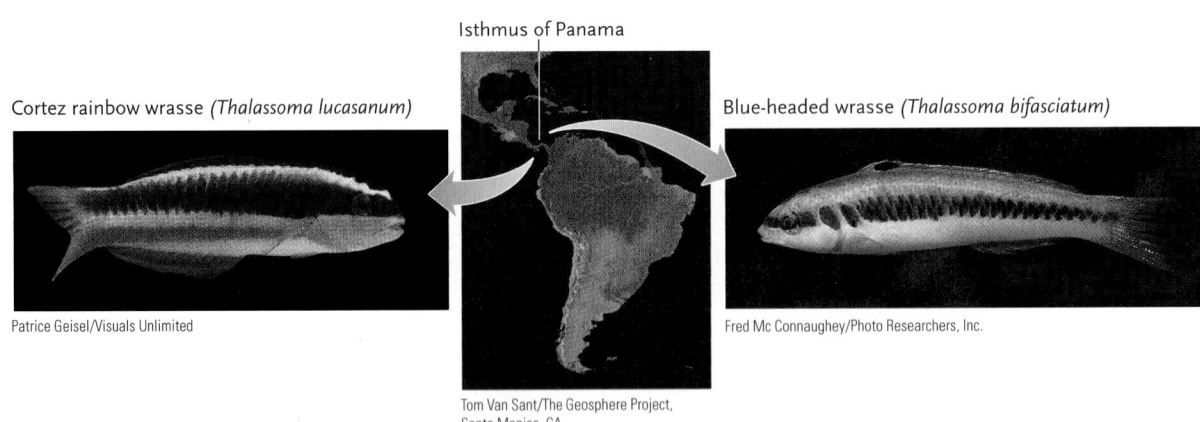

Isthmus of Panama

Cortez rainbow wrasse *(Thalassoma lucasanum)*

Blue-headed wrasse *(Thalassoma bifasciatum)*

Patrice Geisel/Visuals Unlimited

Fred Mc Connaughey/Photo Researchers, Inc.

Tom Van Sant/The Geosphere Project, Santa Monica, CA

Figure 21.10
Geographical separation. The uplift of the Isthmus of Panama divided an ancestral wrasse population. The Cortez rainbow wrasse now occupies the eastern Pacific Ocean, and the blue-headed wrasse now occupies the western Atlantic Ocean.

ing populations that differentiate into distinct species. Each island may experience multiple invasions, and the process may be repeated many times within the archipelago, leading to the evolution of a **species cluster**, a group of closely related species recently descended from a common ancestor **(Figure 21.11)**. The nearly 800 species of fruit flies on the Hawaiian Islands, described in *Focus on Research,* form several species clusters.

Sometimes, allopatric populations reestablish contact when a geographical barrier is eliminated or breached (see Figure 21.9, step 4). This *secondary contact* provides a test of whether or not the populations have diverged into separate species. If their gene pools did not differentiate much during geographical separation, the populations will interbreed and merge. But if the populations have differentiated enough to be reproductively isolated, they have become separate species.

During the early stages of secondary contact, prezygotic reproductive isolation may be incomplete. Some members of each population may mate with individuals from the other, producing viable, fertile offspring, in areas called **hybrid zones**. Although some hybrid zones have persisted for hundreds or thousands of years **(Figure 21.12)**, they are generally narrow, and ecological or geographical factors maintain the separation of the gene pools for the majority of individuals in both species.

If hybrid offspring have lower fitness than those produced within each population, natural selection will favor individuals that mate only with members of their own population. Recent studies of *Drosophila* suggest that this phenomenon, called **reinforcement**, enhances reproductive isolation that had begun to develop while the populations were geographically separated. Thus, natural selection may promote the evolution of prezygotic isolating mechanisms.

Parapatric Speciation May Occur between Adjacent Populations

Sometimes a single species is distributed across a discontinuity in environmental conditions, such as a major change in soil type. Although organisms on one side of the discontinuity may interbreed freely with those on the other side, natural selection may favor different alleles on either side, limiting gene flow. In such cases, **parapatric speciation**—speciation arising between adjacent populations—may occur if hybrid offspring have low relative fitness.

Some strains of bent grass *(Agrostis tenuis),* a common pasture plant in Great Britain, have the physiological ability to grow on mine tailings where the soil is heavily polluted by copper or other metals. Plants of the copper-tolerant strains grow well on polluted soils, but plants of the pasture strain do not. Conversely, copper-tolerant plants don't survive as well as pasture plants on unpolluted soils. These strains often grow within a few meters of each other where polluted and unpolluted soils form an intricate mosaic. Because

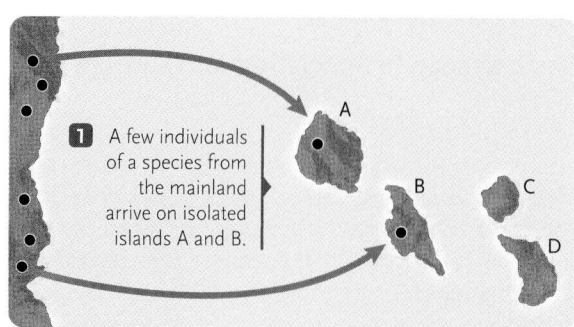

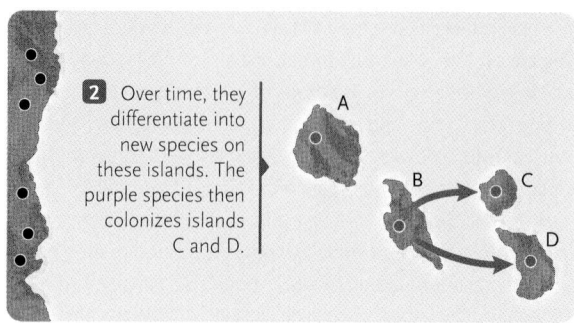

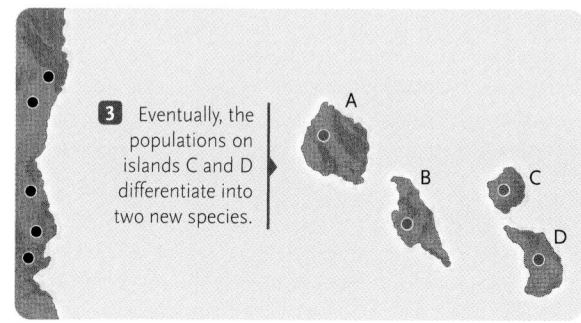

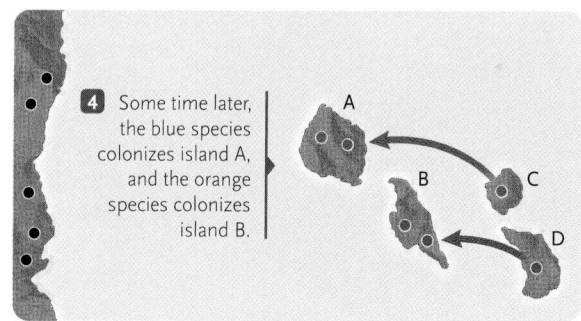

Figure 21.11

Evolution of a species cluster on an archipelago. Letters identify four islands in a hypothetical archipelago, and colored dots represent different species. The ancestor of all the species is represented by black dots on the mainland. At the end of the process, islands A and B are each occupied by two species, and islands C and D are each occupied by one species, all of which evolved on the islands.

bent grass is wind-pollinated, pollen is readily transferred from one strain to another.

Thomas McNeilly and Janis Antonovics of University College of North Wales crossed these strains in the laboratory and determined that they are still fully interfertile. However, copper-tolerant plants flower about one week earlier than nearby pasture plants, which promotes prezygotic (temporal) isola-

Basic Research: Speciation in Hawaiian Fruit Flies

After Darwin published his analyses of island species, evolutionary biologists realized that oceanic archipelagos provide "natural laboratories" for studies of speciation. The islands of the Hawaiian archipelago have been geographically isolated throughout their history, lying at least 3200 km (1900 miles) from the nearest continents or other islands **(Figure a).** They were built by undersea volcanic eruptions over hundreds of thousands of years and emerged from the sea from northwest to southeast: Kauai is at least 5 million years old, and Hawaii, the "Big Island," is less than 1 million years old. Individual islands differ in maximum elevation and include a wide range of habitats, from dry zones of sparse vegetation to wet tracts of lush forest.

Resident species must have arrived from distant mainland localities or evolved on the islands from colonizing ancestors. The islands' isolation, differ-

ent ages, and geographical and ecological complexity provide environmental conditions that foster repeated interisland colonizations followed by allopatric speciation events. Thus, it is not surprising that species clusters have evolved in several groups of organisms (including flowering plants, insects, and birds).

Nearly 800 species of fruit flies have been identified on the archipelago, and most species live on only one island. Biologists used many characters to identify the different fruit fly species, including external and internal anatomy, cell structure, chromosome structure, ecology, and mating behavior. Their data suggest that the vast majority of native Hawaiian species arose from one ancestral species that colonized the archipelago long ago, probably from eastern Asia. After repeated speciation events, the fruit flies of the Hawaiian Islands represent more than 25% of all known fruit fly species.

Hampton Carson, now of the University of Hawaii, has spearheaded studies on the evolutionary relationships of Hawaiian fruit flies. He and his colleagues have gathered data on hundreds of fly species—a daunting task. Most species are sexually dimorphic. Although the females of different species may be similar in appearance, the males of even closely related species differ in virtually every aspect of their external anatomy: body size, head shape, and the structure of their eyes, antennae, mouthparts, bristles, legs, and wings. Their mating behavior and choice of mating sites also vary dramatically.

Nevertheless, closely related species on different islands occupy comparable habitats and associate with related plant species. Carson suggests that speciation in these flies resulted from the evolution of different genetically determined *mating systems,* the behaviors and sexual characteristics that males display when seeking a mate. The mating systems serve as prezygotic isolating mechanisms.

The 100 or more species of "picture-wing" *Drosophila,* relatively large flies with patterns on their wings, illustrate the evolution of a species cluster. Carson and his colleagues used similarities and differences in the banding patterns on

the flies' giant salivary chromosomes (described in the *Focus on Research* in Chapter 13), to trace the evolutionary origin of species on the younger islands by identifying their closest relatives on the older islands. Their analysis of 26 species on Hawaii, the youngest island, suggests that flies from the older islands colonized Hawaii at least 19 different times, and each founder population evolved into a new species there. Additional species apparently evolved when lava flows on Hawaii subdivided existing populations.

Among the picture-wing fruit flies, some interspecies matings result in hybrid sterility or hybrid breakdown. But for the majority of species, prezygotic reproductive isolation is maintained by differences in their mating systems. For example, *Drosophila silvestris* and *Drosophila heteroneura,* which produce healthy and fertile hybrids in the laboratory, have similar geographical distributions; however, differences in courtship behavior and in the shape of the males' heads, a characteristic that females use to recognize males of their own species **(Figure b),** keep these two species reproductively isolated. In nature, they hybridize only in one small geographical area.

The work of Carson and his colleagues suggests that most speciation in Hawaiian *Drosophila* has resulted from founder effects. When a fertile female—or a small group of males and females—moves to a new island, this founding population responds to novel selection pressures in its new environment. Sexual selection then exaggerates distinctive morphological and behavioral characteristics, maintaining the population's reproductive isolation from its new neighbors. The tremendous variety of Hawaiian fruit flies has undoubtedly been produced by repeated colonizations of newer islands by flies from older islands and by the back-colonization of older islands by newly evolved species. Thus, they represent what evolutionary biologists describe as an *adaptive radiation,* a cluster of closely related species that are ecologically different (as described further in Chapter 22).

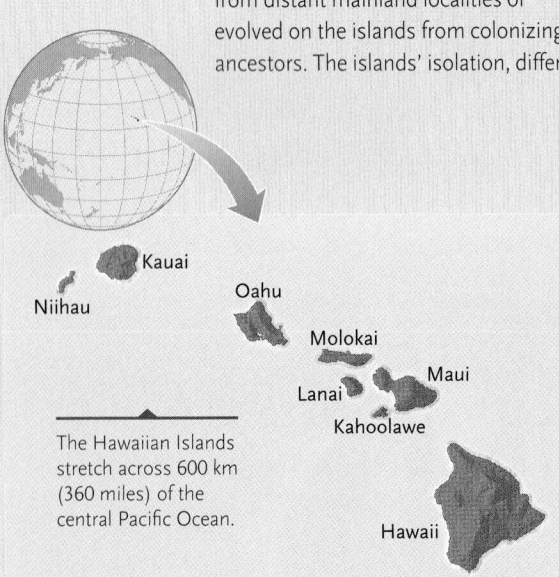

Figure a
The Hawaiian Islands

The Hawaiian Islands stretch across 600 km (360 miles) of the central Pacific Ocean.

Drosophila heteroneura

Drosophila silvestris

Kenneth Y. Kaneshiro, University of Hawaii

Kenneth Y. Kaneshiro, University of Hawaii

Figure b
Two *Drosophila* species in which the males' head shapes differ.

Bullock's oriole

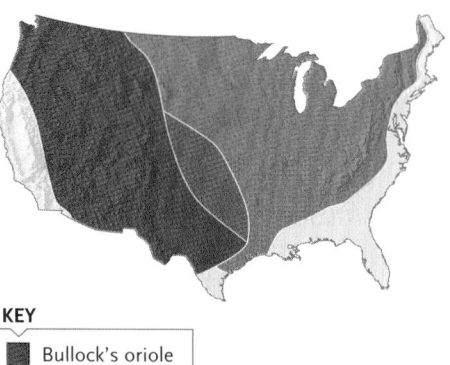

Baltimore oriole

KEY
- Bullock's oriole
- Hybrid zone
- Baltimore oriole

Figure 21.12

Hybrid zones. Males of the Baltimore oriole *(Icterus galbula)* and Bullock's oriole *(Icterus bullockii)* differ in color and courtship song. The populations have maintained a hybrid zone for hundreds of years, and once were considered subspecies of the same species. The American Ornithologists' Union recognized them as separate species in 1997. They now hybridize less frequently than they once did, leading some researchers to suggest that their reproductive isolation evolved recently.

tion of the two strains **(Figure 21.13)**. If the flowering times become further separated, the two strains may attain complete reproductive isolation and become separate species.

Some biologists argue that the places where parapatric populations of bent grass interbreed are really hybrid zones where allopatric populations have established secondary contact. Unfortunately, there is no way to determine whether the hybridizing populations were parapatric or allopatric in the past. Thus, a thorough evaluation of the parapatric speciation hypothesis must await the development of techniques that enable biologists to distinguish clearly between the products of allopatric and parapatric speciation.

Sympatric Speciation Occurs within One Continuously Distributed Population

In **sympatric speciation,** reproductive isolation evolves between distinct subgroups that arise within one population. Models of sympatric speciation do not require that the populations be either geographically or environmentally separated as their gene pools diverge. We examine below general models of sympatric speciation in animals and plants; the genetic basis of sympatric speciation is one of the topics we consider in the next section.

Insects that feed on just one or two plant species are among the animals most likely to evolve by sympatric speciation. These insects generally carry out most important life cycle activities on or near their "host" plants. Adults mate on the host plant; females lay their eggs on it; and larvae feed on the host plant's tissues, eventually developing into adults, which initiate another round of the life cycle. Host plant choice is

genetically determined in many insect species. In others, individuals associate with the host plant species they ate as larvae.

Theoretically, a genetic mutation could suddenly change some insects' choice of host plant. Mutant individuals would shift their life cycle activities to the new host, and then interact primarily with others preferring the same new host, an example of ecological isolation. These individuals would collectively form a separate subpopulation, called a **host race.** Reproductive isolation could evolve between different host races if the individuals of each host race are more likely to mate with members of their own host race than with members of another. Some biologists criticize this model, however, because it assumes that the genes controlling two traits, the insects' host plant choice and their mating preferences, change simultaneously. Moreover, host plant choice is controlled by multiple gene loci in some insect species, and it is clearly influenced by prior experience in others.

The apple maggot *(Rhagoletis pomonella)* is the most thoroughly studied example of possible sympatric speciation in animals **(Figure 21.14)**. This fly's natural host plant in eastern North America is the hawthorn *(Crataegus* species), but at least two host races have appeared in little more than 100 years. The larvae of a new host race were first discovered feeding on apples in New York state in the 1860s. In the 1960s, a cherry-feeding host race appeared in Wisconsin.

Recent research has shown that variations at just a few gene loci underlie differences in the feeding preferences of *Rhagoletis* host races; other genetic differences cause the host races to develop at different rates. Moreover, adults of the three races mate during different summer months. Nevertheless, individuals

Figure 21.13 Observational Research

Evidence for Reproductive Isolation in Bent Grass

QUESTION: Do adjacent populations of bent grass *(Agrostis tenuis)* living on different soil types exhibit any signs of reproductive isolation?

HYPOTHESIS: McNeilly and Antonovics hypothesized that adjacent populations of bent grass flowered at slightly different times, which could foster prezygotic reproductive isolation between them.

METHODS: On a late summer day in 1965, the researchers compared the flowers of bent grass growing on polluted soil at a copper mine with those of plants growing on unpolluted soil in a nearby pasture. A meter-wide stretch of polluted pasture (indicated by cross-hatching) formed a boundary between the two populations. Researchers assigned a score to every flower, with immature flowers scored as 3 and mature flowers as 4.

RESULTS: On the day that they were surveyed, flowers of the copper-tolerant plants had higher scores, indicating that they were more mature—and thus would complete pollination earlier—than the flowers of the pasture plants.

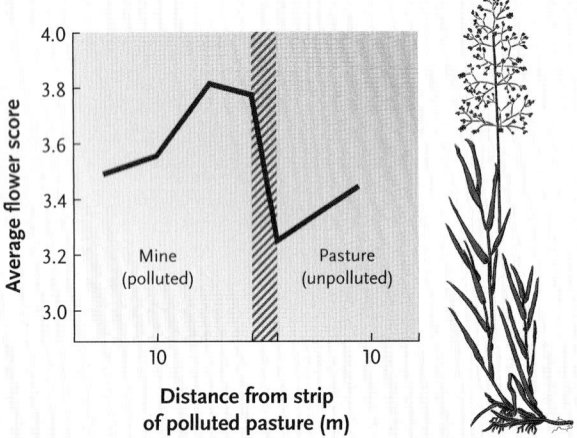

CONCLUSION: Because adjacent populations of bent grass flower at slightly different times, temporal reproductive isolation may be developing between them.

Figure 21.14

Sympatric speciation in animals. Male and female apple maggots *(Rhagoletis pomonella)* court on a hawthorn leaf. The female will later lay her eggs on the fruit, and the offspring will feed, mate, and lay their eggs on hawthorns as well.

ing with individuals of the parent species. Nearly half of all flowering plant species are polyploid, including many important crops and ornamental species. The genetic mechanisms that produce polyploid individuals in plant populations are well understood; we describe them in detail as part of a larger discussion of the genetics of speciation.

STUDY BREAK

1. What are the two stages required for allopatric speciation?
2. What factor appears to promote parapatric speciation in bent grass?
3. Why might insects from different host races be unlikely to mate with each other?

21.4 Genetic Mechanisms of Speciation

What genetic changes lead to reproductive isolation between populations, and how do these changes arise? In this section we examine three genetic mechanisms that can lead to reproductive isolation: *genetic divergence* between allopatric populations, *polyploidy* in sympatric populations, and *chromosome alterations,* which occur independently of the geographical distributions of populations.

Genetic Divergence in Allopatric Populations Can Lead to Speciation

In the absence of gene flow, geographically separated populations inevitably accumulate genetic differences. Most postzygotic isolating mechanisms probably develop as accidental by-products of mutation, genetic

show no particular preference for mates of their own host race, at least under simplified laboratory conditions. Thus, although behavioral isolation has not developed between races, ecological and temporal isolation may separate adults in nature. Researchers are still not certain that the different host races are reproductively isolated under natural conditions.

Sympatric speciation often occurs in plants through a genetic phenomenon, **polyploidy**, in which an individual receives one or more *extra* copies of the entire haploid complement of chromosomes (see Section 13.3). As we explain in the next section, polyploidy can lead to speciation because these large-scale genetic changes may prevent polyploid individuals from breeding

Monkey-Flower Speciation

Reproductive isolation is the primary criterion that biologists use to distinguish species. A molecular study by H. D. Bradshaw and his coworkers at the University of Washington indicates that the amount of genetic change required to establish reproductive isolation, and thus new species, may be surprisingly small in some cases.

These scientists studied two monkey-flower species, *Mimulus lewisii* and *Mimulus cardinalis,* that experience mechanical reproductive isolation because differences in flower structure keep bumblebees or hummingbirds from carrying pollen from one species to the other (see Figure 21.7). Although these species do not hybridize in nature, they are easily crossed in the laboratory and produce fertile hybrids. The F_2 offspring of the laboratory crosses have flowers with various forms intermediate between the parental *lewisii* and *cardinalis* types, indicating that several gene loci control the traits separating the species. But how many?

Relatively little is known about the genetics of the two monkey-flower species, so a direct genetic analysis of their hereditary differences was impractical. Instead, the investigators studied 153 randomly chosen DNA sequences distributed throughout the haploid number of eight chromosomes in the two species. They correlated the distribution of these sequences with the distribution of flower traits in 93 plants of the F_2 generation. Some of the DNA sequences segregated so closely with a particular trait, such as yellow pigment, that they are almost certainly located near that trait in the chromosomes. Because the sequences can pair with complementary DNA in the chromosomes, the investigators used them as "probes" to find the sites in the chromosomes from which they originated. From the close linkage of the sequences to the traits, the investigators could estimate the positions and approximate number of genes that establish reproductive isolation.

Their results indicate that reproductive isolation of *M. lewisii* and *M. cardinalis* results from differences in eight floral traits—the amount of (1) anthocyanin pigments and (2) carotenoid pigments in petals; (3) flower width; (4) petal width; (5) nectar volume; (6) nectar concentration; and the lengths of the stalks supporting the (7) male and (8) female reproductive parts. Although the investigators could not directly determine the number of genes controlling each trait, the characteristics of the traits, their locations at eight sites on six of the chromosomes, and their pattern of inheritance make it most likely that each trait is controlled by a single gene, giving a likely minimum of eight genes. Thus mutations in as few as eight genes may have established reproductive isolation and speciation in the monkey-flowers.

This research was the first in which random differences in DNA sequences were used to answer the fundamental evolutionary question of how much genetic change is needed to produce a new species.

---

drift, and natural selection. Note, however, that natural selection cannot promote the evolution of reproductive isolating mechanisms between *allopatric* populations directly: individuals in such populations do not encounter one another and therefore have no opportunity to produce hybrid offspring. And if there are no hybrid offspring, natural selection cannot select against the matings that would have produced them. Nevertheless, natural selection may sometimes foster adaptive changes that create postzygotic reproductive isolation between populations when they later reestablish contact. And, if postzygotic isolating mechanisms reduce the fitness of hybrid offspring, natural selection can reinforce the evolution of prezygotic isolating mechanisms.

How much genetic divergence is necessary for speciation to occur? To understand the genetic basis of speciation in closely related species, researchers first identify the specific causes of reproductive isolation. They then use standard techniques of genetic analysis along with new molecular approaches such as gene mapping and sequencing to analyze the genetic mechanisms that establish reproductive isolation. As explained in *Insights from the Molecular Revolution,* these techniques now allow researchers to determine the minimum number of genes responsible for reproductive isolation in particular pairs of species.

In cases of postzygotic reproductive isolation, mutations in at least a few gene loci establish reproductive isolation. For example, if two common aquarium fishes, swordtails *(Xiphophorus helleri)* and platys *(Xiphophorus maculatus),* mate, two genes induce the development of lethal tumors in their hybrid offspring. When hybrid sterility is the primary cause of reproductive isolation between *Drosophila* species, at least 5 gene loci are responsible. About 55 gene loci contribute to postzygotic reproductive isolation between the toads *Bombina bombina* and *Bombina variegata.*

In cases of prezygotic reproductive isolation, some mechanisms have a surprisingly simple genetic basis. For example, a single mutation reverses the direction of coiling in the shells of some snail species. Snails with shells that coil in opposite directions cannot approach each other closely enough to mate, making reproduction between them mechanically impossible.

Many traits that now function as prezygotic isolating mechanisms may originally have evolved in response to sexual selection (described in Section 20.3).

Figure 21.15

Sexual selection and prezygotic isolation. In closely related species, such as mallard ducks *(Anas platyrhynchos)* and pintails *(Anas acuta)*, males have much more distinctive coloration than females, a sure sign of sexual selection at work.

Mallards

Pintails

This evolutionary process exaggerates showy structures and courtship behaviors in males, the traits that females use to identify appropriate mates. When two species encounter one another on secondary contact, these traits may also prevent interspecific mating. For example, many closely related duck species exhibit dramatic variation in the appearance of males, but not females **(Figure 21.15),** an almost certain sign of sexual selection. Yet these species hybridize readily in captivity, producing offspring that are both viable and fertile. Speciation in these birds probably resulted from geographical isolation and sexual selection without significant genetic divergence: only a few morphological and behavioral characters are responsible for their reproductive isolation. Thus, sometimes the evolution of reproductive isolation may not require much genetic change at all.

Polyploidy Is a Common Mechanism of Sympatric Speciation in Plants

Polyploidy is common among plants, and it may be an important factor in the evolution of some fish, amphibian, and reptile species. Polyploid individuals can arise from chromosome duplications within a single species (autopolyploidy) or through hybridization of different species (allopolyploidy).

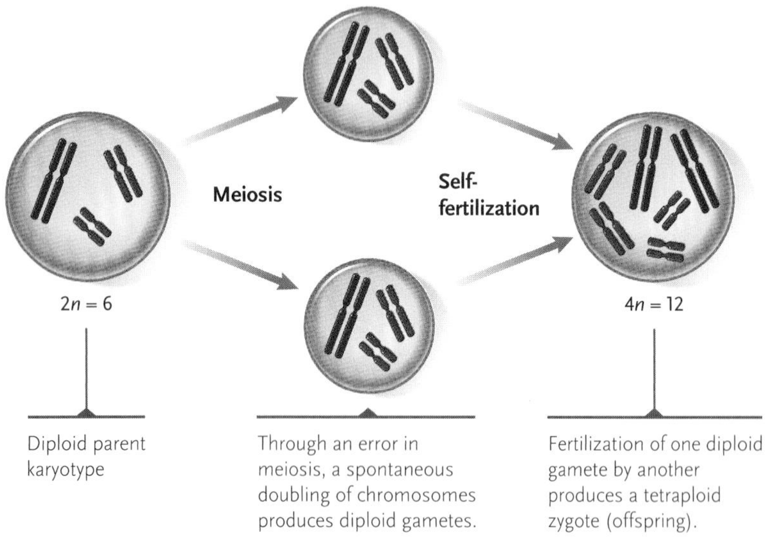

Autopolyploidy. In **autopolyploidy (Figure 21.16),** a diploid (2*n*) individual may produce, for example, tetraploid (4*n*) offspring, each of which has four complete chromosome sets. Autopolyploidy often results when gametes, through an error in either mitosis or meiosis, spontaneously receive the same number of chromosomes as a somatic cell. Such gametes are called **unreduced gametes** because their chromosome number has not been reduced compared with that of somatic cells.

Diploid pollen can fertilize the diploid ovules of a self-fertilizing individual, or it may fertilize diploid ovules on another plant with unreduced gametes. The resulting tetraploid offspring can reproduce either by self-pollination or by breeding with other tetraploid individuals. However, a tetraploid plant cannot produce fertile offspring by hybridizing with its diploid parents. The fusion of a diploid gamete with a normal haploid gamete produces a triploid (3*n*) offspring, which is usually sterile because its odd number of chromosomes cannot segregate properly during meiosis. Thus, the tetraploid is reproductively isolated from the original diploid population. Many species of grasses, shrubs, and ornamental plants, including violets, chrysanthemums, and nasturtiums, are autopolyploids, having anywhere from four to 20 complete chromosome sets.

Allopolyploidy. In **allopolyploidy (Figure 21.17),** two closely related species hybridize and subsequently form polyploid offspring. Hybrid offspring are sterile if the two parent species have diverged enough that their chromosomes do not pair properly during meiosis. However, if the hybrid's chromosome number is doubled, the chromosome complement of the gametes is also doubled, producing homologous chromosomes that *can* pair during meiosis. The hybrid can then produce polyploid gametes and, through self-fertilization or fertilization with other doubled hybrids, establish a population of a new polyploid species. Compared with speciation by genetic divergence, speciation by allopolyploidy is extremely rapid, causing a new species to arise in one generation without geographical isolation.

Meiosis

Self-fertilization

2*n* = 6

4*n* = 12

Diploid parent karyotype

Through an error in meiosis, a spontaneous doubling of chromosomes produces diploid gametes.

Fertilization of one diploid gamete by another produces a tetraploid zygote (offspring).

Figure 21.16

Speciation by autopolyploidy in plants. A spontaneous doubling of chromosomes during meiosis produces diploid gametes. If the plant fertilizes itself, a tetraploid zygote will be produced.

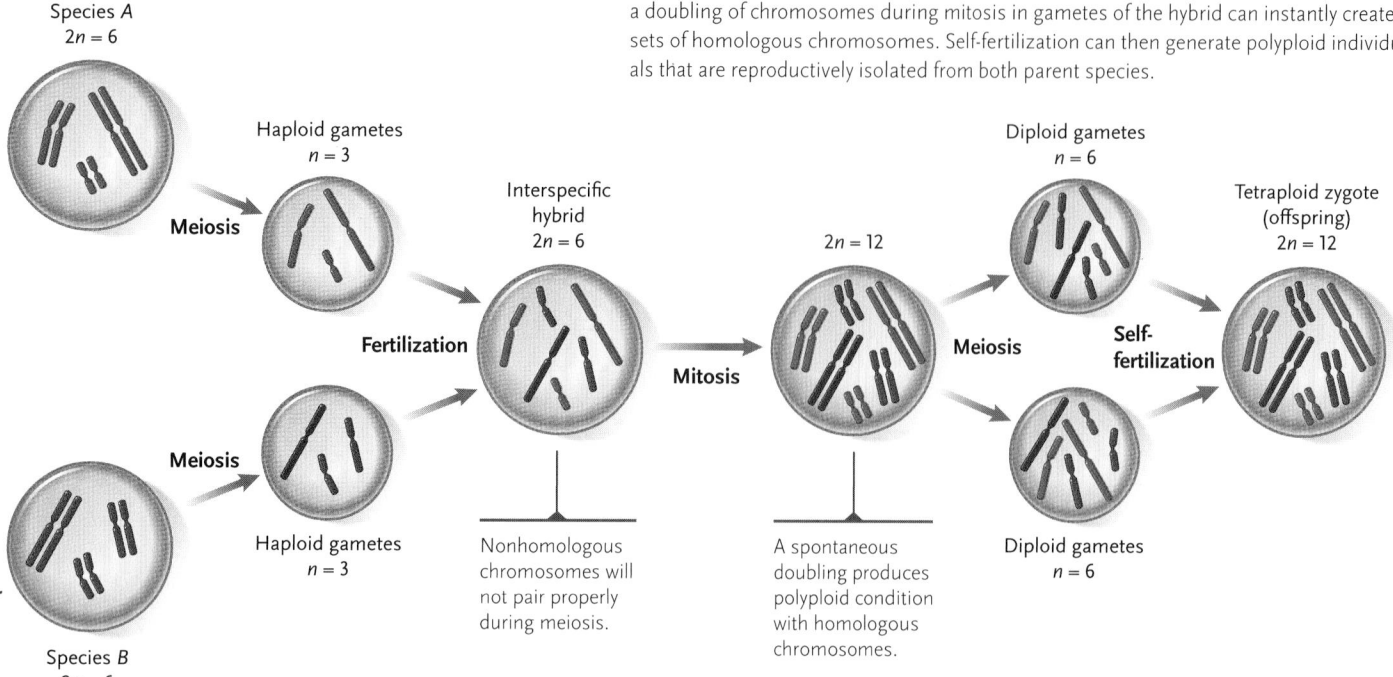

Figure 21.17

Speciation by allopolyploidy in plants. A hybrid mating between two species followed by a doubling of chromosomes during mitosis in gametes of the hybrid can instantly create sets of homologous chromosomes. Self-fertilization can then generate polyploid individuals that are reproductively isolated from both parent species.

Species *A*
2*n* = 6

Meiosis

Haploid gametes
n = 3

Fertilization

Interspecific hybrid
2*n* = 6

Meiosis

Haploid gametes
n = 3

Species *B*
2*n* = 6

Nonhomologous chromosomes will not pair properly during meiosis.

Mitosis

2*n* = 12

A spontaneous doubling produces polyploid condition with homologous chromosomes.

Diploid gametes
n = 6

Diploid gametes
n = 6

Meiosis

Self-fertilization

Tetraploid zygote (offspring)
2*n* = 12

Even when sterile, polyploids are often robust, growing larger than either parent species. For that reason, both autopolyploids and allopolyploids have been important to agriculture. For example, the wheat used to make flour (*Triticum aestivum*) has six sets of chromosomes **(Figure 21.18).** Other polyploid crop plants include plantains (cooking bananas), coffee, cotton, potatoes, sugarcane, and tobacco.

Plant breeders often try to increase the probability of forming an allopolyploid by using chemicals that

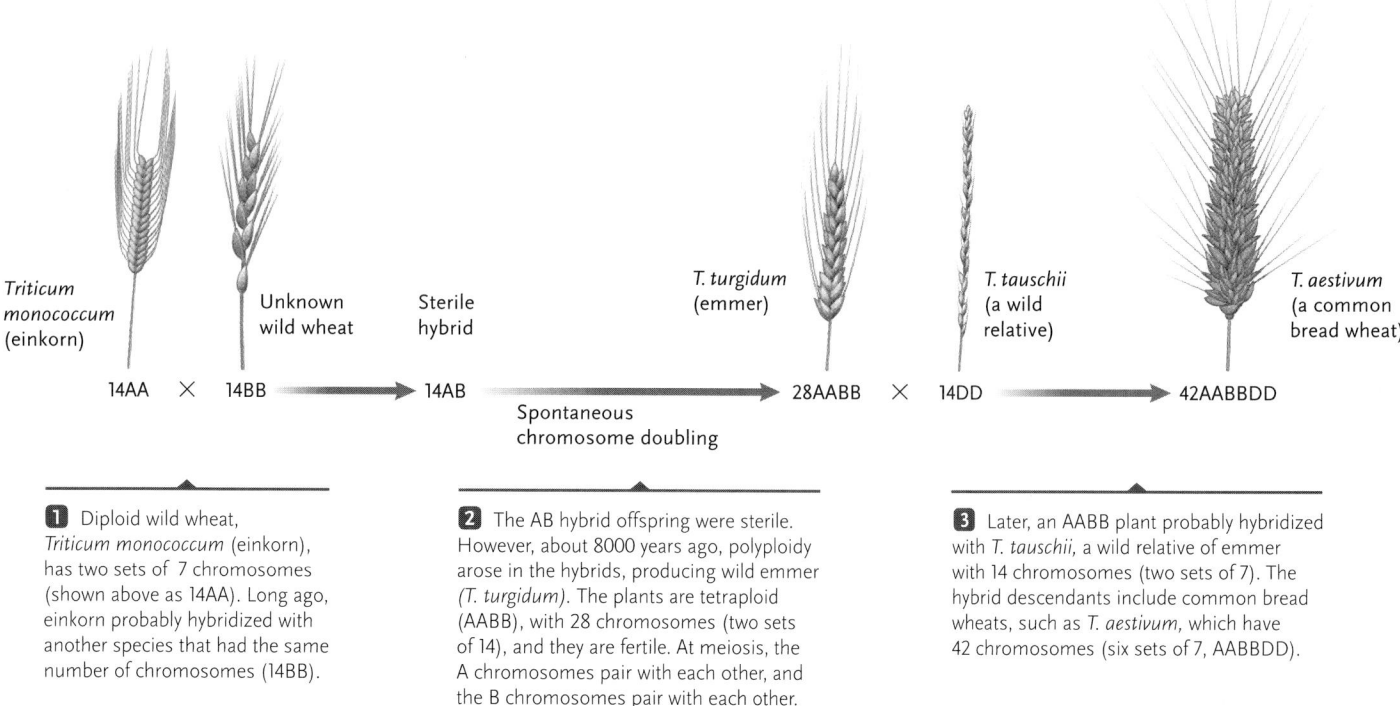

Triticum monococcum (einkorn)

Unknown wild wheat

Sterile hybrid

T. turgidum (emmer)

T. tauschii (a wild relative)

T. aestivum (a common bread wheat)

14AA ✕ 14BB ⟶ 14AB ⟶ 28AABB ✕ 14DD ⟶ 42AABBDD

Spontaneous chromosome doubling

1 Diploid wild wheat, *Triticum monococcum* (einkorn), has two sets of 7 chromosomes (shown above as 14AA). Long ago, einkorn probably hybridized with another species that had the same number of chromosomes (14BB).

2 The AB hybrid offspring were sterile. However, about 8000 years ago, polyploidy arose in the hybrids, producing wild emmer (*T. turgidum*). The plants are tetraploid (AABB), with 28 chromosomes (two sets of 14), and they are fertile. At meiosis, the A chromosomes pair with each other, and the B chromosomes pair with each other.

3 Later, an AABB plant probably hybridized with *T. tauschii*, a wild relative of emmer with 14 chromosomes (two sets of 7). The hybrid descendants include common bread wheats, such as *T. aestivum*, which have 42 chromosomes (six sets of 7, AABBDD).

Figure 21.18

The evolution of wheat *(Triticum).* Cultivated wheat grains more than 11,000 years old have been found in the Eastern Mediterranean region. Researchers believe that speciation in wheat occurred through hybridization and polyploidy.

Figure 21.19 Observational Research

Chromosomal Similarities and Differences among the Great Apes

QUESTION: Does chromosome structure differ between humans and their closest relatives among the great apes?

HYPOTHESIS: Yunis and Prakash hypothesized that chromosome structure would differ markedly between humans and their close relatives among the apes: chimpanzees, gorillas, and orangutans.

METHODS: The researchers used Giemsa stain to visualize the banding patterns on metaphase chromosome preparations from humans, chimpanzees, gorillas, and orangutans. By matching the banding patterns on the chromosomes, the researchers verified that they were comparing the same segments of the genomes in the four species. They then searched for similarities and differences in the structure of the chromosomes.

RESULTS: The analysis of human chromosome 2 reveals that it was produced by the fusion of two smaller chromosomes that are still present in the other three species. Although the position of the centromere in human chromosome 2 matches that of the centromere in one of the chimpanzee chromosomes, in gorillas and orangutans it falls within an inverted segment of the chromosome.

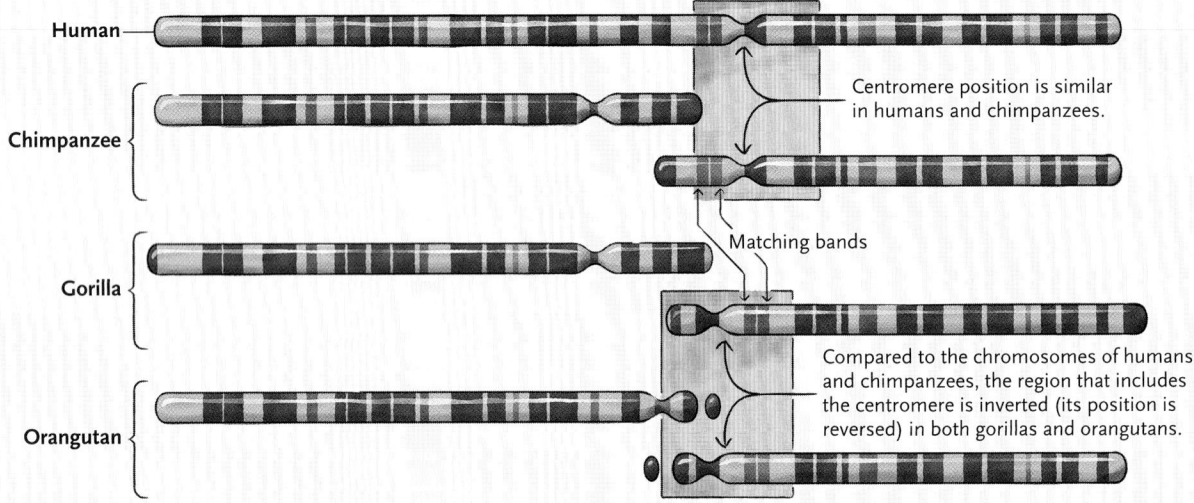

Human

Chimpanzee

Gorilla

Orangutan

Centromere position is similar in humans and chimpanzees.

Matching bands

Compared to the chromosomes of humans and chimpanzees, the region that includes the centromere is inverted (its position is reversed) in both gorillas and orangutans.

CONCLUSION: Differences in chromosome structure between humans and both gorillas and orangutans are more pronounced than they are between humans and chimpanzees. Structural differences in the chromosomes of these four species may contribute to their reproductive isolation.

foster nondisjunction of chromosomes during mitosis. In the first such experiment, undertaken in the 1920s, scientists crossed a radish and a cabbage, hoping to develop a plant with both edible roots and leaves. Instead, the new species, *Raphanobrassica,* combined the least desirable characteristics of each parent, growing a cabbagelike root and radishlike leaves. Recent experiments have been more successful. For example, plant scientists have produced an allopolyploid grain, triticale, that has the disease-resistance of its rye parent and the high productivity of its wheat parent.

Chromosome Alterations Can Foster Speciation

Other changes in chromosome structure or number may also foster speciation. Closely related species often have a substantial number of chromosome differences between them, including inversions, translocations, deletions, and duplications (described in Section 13.3). These differences may foster postzygotic isolation.

In 1982, Jorge J. Yunis and Om Prakash of the University of Minnesota Medical School compared the chromosome structures of humans and their closest relatives among the apes—chimpanzees, gorillas, and orangutans—by examining the *banding patterns* in stained chromosome preparations. In all species, banding patterns vary from one chromosome segment to another. When researchers find identical banding patterns in chromosome segments from two or more related species, they know that they are examining comparable portions of the species' genomes. Thus, the banding patterns allow scientists to identify specific chromosome segments and compare their positions in the chromosomes of different species.

Nearly all of the 1000 bands that Yunis and Prakash identified are present in humans and in the three ape species. However, the banding patterns revealed that whole sections of chromosomes have been rearranged over evolutionary time **(Figure 21.19)**. For example, humans have a diploid chromosome complement of 46 chromosomes, whereas chimpanzees, gorillas, and orangutans have 48. The difference can be traced to the fusion (that is, the joining together) of two ancestral

chromosomes into chromosome 2 of humans; the ancestral chromosomes are separate in the other three species.

Moreover, banding patterns suggest that the position of the centromere in human chromosome 2 closely matches that of a centromere in one of the chimpanzee chromosomes, reflecting their close evolutionary relationship. But this centromere falls within an inverted region of the chromosome in gorillas and orangutans, reflecting their evolutionary divergence from chimpanzees and humans. (Recall from Section 13.3 that an inverted chromosome segment has a reversed orientation, so the order of genes on it is reversed relative to the order in a segment that is not inverted.) Nevertheless, humans and chimps differ from each other in centromeric inversions in six other chromosomes.

How might such chromosome rearrangements promote speciation? In a paper published in 2003, Arcadi Navarro of the Universitat Pompeu Fabra in Spain and Nick H. Barton of the University of Edin-

UNANSWERED QUESTIONS

Do asexual organisms form species?

As you learned in this chapter, the biological species concept applies only to sexually reproducing organisms because only those organisms can evolve barriers to gene flow (asexual organisms reproduce more or less clonally). Nevertheless, research is starting to show that organisms whose reproduction is almost entirely asexual, such as bacteria, seem to form distinct and discrete clusters in nature. (These clusters could be considered "species.") That is, bacteria and other asexual forms may be as distinct as the species of birds described by Ernst Mayr in New Guinea. Workers are now studying the many species of bacteria in nature (only a small number of which have been discovered) to see if they indeed fall into distinct groups. If they do, then scientists will need a special theory, independent of reproductive isolation, to explain this distinctness. Scientists are now working on theories of whether the existence of discrete ecological niches in nature might explain the possible discreteness of asexual "species."

How often does speciation occur allopatrically versus sympatrically or parapatrically?

Scientists do not know how often speciation occurs between populations that are completely isolated geographically (allopatric speciation) compared with how often it occurs in populations that exchange genes (parapatric or sympatric speciation). The relative frequency of these modes of speciation in nature is an active area of research. The ongoing work includes studies on small isolated islands: if an invading species divides into two or more species in this situation, it probably did so sympatrically or parapatrically, since geographical isolation of populations in small islands is unlikely. In addition, biologists are reconstructing the evolutionary history of speciation using molecular tools and correlating this history with the species' geographical distributions. If

this line of research were to show, for example, that the most closely related pairs of species always had geographically isolated distributions, it would imply that speciation was usually allopatric. These lines of research should eventually answer the controversial question of the relative frequency of various forms of speciation.

What are the genetic changes underlying speciation?

Biologists know a great deal about the types of reproductive isolation that prevent gene flow between species, but almost nothing about their genetic bases. Which genes control the difference between flower shape in monkey-flower species? Which genes lead to inviability and sterility of *Drosophila* hybrids? Which genes cause species of ducks to preferentially mate with members of their own species over members of other species? Do the genetic changes that lead to reproductive isolation tend to occur repeatedly at the same genes in a group of organisms, or at different genes? Do the changes occur mostly in protein-coding regions of genes, or in the noncoding regions that control the production of proteins? Were the changes produced by natural selection or by genetic drift? Biologists are now isolating "speciation genes" and sequencing their DNA. With only a handful of such genes known, and all of these causing hybrid sterility or inviability, there will undoubtedly be a lot to learn about the genetics of speciation in the next decade.

Jerry Coyne conducts research on speciation and teaches at the University of Chicago. To learn more about his research go to http://pondside.uchicago.edu/ecol-evol/faculty/coyne_j.html.

burgh in Scotland compared the rates of evolution in protein-coding genes that lie within rearranged chromosome segments of humans and chimpanzees to those in genes outside the rearranged segments. They discovered that proteins evolved more than twice as quickly in the rearranged chromosome segments. Navarro and Barton reasoned that because chromosome rearrangements inhibit chromosome pairing and recombination during meiosis, new genetic variations favored by natural selection would be conserved within the rearranged segments. These variations accumulate over time, contributing to genetic divergence between populations with the rearrangement and those without it. Thus, chromosome rearrangements can be a trigger for speciation: once a chromosome rearrangement becomes established within a population, that population will diverge more rapidly from populations lacking the rearrangement. The genetic divergence eventually causes reproductive isolation.

In the next chapter we consider the effects of speciation over vast spans of time as we examine paleobiology and patterns of macroevolution.

STUDY BREAK

1. How can natural selection promote reproductive isolation in allopatric populations?
2. What group of organisms has frequently undergone speciation by polyploidy?

Review

Go to **ThomsonNOW**™ at www.thomsonedu.com/login to access quizzing, animations, exercises, articles, and personalized homework help.

21.1 What Is a Species?

- In practice, most biologists describe, identify, and recognize species on the basis of morphological characteristics that serve as indicators of their genetic similarity to or divergence from other species (Figure 21.2).

- The biological species concept defines species as groups of interbreeding populations that are reproductively isolated from populations of other species in nature. A biological species thus represents a gene pool within which genetic material is potentially shared among populations. The biological species concept cannot be applied to organisms that reproduce only asexually, to those that are extinct, or to geographically separated populations. The phylogenetic species concept defines a species as a group of populations with a recently shared evolutionary history.

- Most species exhibit geographical variation of phenotypic and genetic traits. When marked geographical variation in phenotypes is discontinuous, biologists sometimes name subspecies (Figure 21.3). In ring species, populations are distributed in a ring around unsuitable habitat (Figure 21.4). Many species exhibit clinal variation of characteristics, which change smoothly over a geographical gradient (Figure 21.5).

Animation: Morphological differences within a species

21.2 Maintaining Reproductive Isolation

- Reproductive isolating mechanisms are characteristics that prevent two species from interbreeding.

- Prezygotic isolating mechanisms either prevent individuals of different species from mating or prevent fertilization between their gametes. Prezygotic isolation occurs because species live in different habitats, breed at different times, use different courtship behavior (Figure 21.6), or differ anatomically (Figure 21.7). Prezygotic isolation can also result from genetic and physiological incompatibilities between male and female gametes.

- Postzygotic isolating mechanisms reduce the fitness of interspecific hybrids through hybrid inviability, hybrid sterility (Figure 21.8), or hybrid breakdown.

Animation: Reproductive isolating mechanisms

Animation: Temporal isolation among cicadas

21.3 The Geography of Speciation

- The model of allopatric speciation proposes that speciation results from divergent evolution in geographically separated populations (Figures 21.9–21.11). If allopatric populations accumulate enough genetic differences, they will be reproductively isolated upon secondary contact. Nevertheless, some species hybridize over small areas of secondary contact (Figure 21.12).

- The model of parapatric speciation suggests that reproductive isolation can evolve between parts of a population that occupy opposite sides of an environmental discontinuity (Figure 21.13).

- A model of sympatric speciation in insects suggests that reproductive isolation may evolve between host races that rarely contact one another under natural conditions (Figure 21.14). Sympatric speciation commonly occurs in flowering plants by allopolyploidy.

Animation: Models of speciation

Animation: Allopatric speciation on an archipelago

Animation: Sympatric speciation in wheat

21.4 Genetic Mechanisms of Speciation

- Allopatric populations inevitably accumulate genetic differences, some of which contribute to their reproductive isolation. Reproductive isolating mechanisms evolve as by-products of genetic changes that occur during divergence. Prezygotic isolating mechanisms may evolve in populations experiencing secondary contact (Figure 21.15).

- We cannot yet generalize about how many gene loci participate in the process of speciation, but at least several gene loci are usually involved.

- Speciation by polyploidy in flowering plants involves the duplication of an entire chromosome complement through nondisjunction of chromosomes during meiosis or mitosis. Polyploids can arise among the offspring of a single species (autopolyploidy; Figure 21.16) or, more commonly, after hybridization between closely related species (allopolyploidy; Figures 21.17 and 21.18).

- Chromosome alterations can promote speciation by fostering the genetic divergence of, and reproductive isolation between, populations with different numbers of chromosomes or different chromosome structure (Figure 21.19).

Questions

Self-Test Questions

1. The biological species concept defines species on the basis of:
   a. reproductive characteristics.
   b. biochemical characteristics.
   c. morphological characteristics.
   d. behavioral characteristics.
   e. all of the above

2. Biologists can apply the biological species concept *only* to species that:
   a. reproduce asexually.
   b. lived in the past.
   c. are allopatric to each other.
   d. hybridize in captivity.
   e. reproduce sexually.

3. A characteristic that exhibits smooth changes in populations distributed along a geographical gradient is called a:
   a. ring species.
   b. subspecies.
   c. cline.
   d. hybrid breakdown.
   e. subspecies.

4. If two species of holly (genus *Ilex*) flower during different months, their gene pools may be kept separate by:
   a. mechanical isolation.
   b. ecological isolation.
   c. gametic isolation.
   d. temporal isolation.
   e. behavioral isolation.

5. Prezygotic isolating mechanisms:
   a. reduce the fitness of hybrid offspring.
   b. generally prevent individuals of different species from producing zygotes.
   c. are found only in animals.
   d. are found only in plants.
   e. are observed only in organisms that reproduce asexually.

6. In the model of allopatric speciation, the geographical separation of two populations:
   a. is sufficient for speciation to occur.
   b. occurs only after speciation is complete.
   c. allows gene flow between them.
   d. reduces the relative fitness of hybrid offspring.
   e. inhibits gene flow between them.

7. Adjacent populations that produce hybrid offspring with low relative fitness may be undergoing:
   a. clinal isolation.
   b. parapatric speciation.
   c. allopatric speciation.
   d. sympatric speciation.
   e. geographical isolation.

8. An animal breeder, attempting to cross a llama with an alpaca for finer wool, found that the hybrid offspring rarely lived more than a few weeks. This outcome probably resulted from:
   a. genetic drift.
   b. prezygotic reproductive isolation.
   c. postzygotic reproductive isolation.
   d. sympatric speciation.
   e. polyploidy.

9. Which of the following could be an example of allopolyploidy?
   a. One parent has 32 chromosomes, the other has 10, and their offspring have 42.
   b. Gametes and somatic cells have the same number of chromosomes.
   c. Chromosome number increases by one in a gamete and in the offspring it produces.
   d. Chromosome number decreases by one in a gamete and in the offspring it produces.
   e. Chromosome number in the offspring is exactly half of what it is in the parents.

10. Which of the following genetic characteristics is shared by humans and chimpanzees?
    a. They have the same number of chromosomes.
    b. The position of the centromere on human chromosome 2 matches the position of a centromere on a chimpanzee chromosome.
    c. A fusion of ancestral chromosomes formed chromosome 2.
    d. Centromeres on all of their chromosomes fall within inverted chromosome segments.
    e. all of the above

Questions for Discussion

1. All domestic dogs are classified as members of the species *Canis familiaris*. But it is hard to imagine how a tiny Chihuahua could breed with a gigantic Great Dane. Do you think that artificial selection for different breeds of dogs will eventually create different dog species?

2. Human populations often differ dramatically in external morphological characteristics. On what basis are all human populations classified as a single species?

3. If intermediate populations in a ring species go extinct, eliminating the possibility of gene flow between populations at the two ends of the ring, would you now identify those remaining populations as full species? Explain your answer.

Experimental Analysis

Design an experiment to test whether populations of birds on different islands belong to the same species.

Evolution Link

How do human activities (such as destruction of natural habitats, diversion of rivers, and the construction of buildings) influence the chances that new species of plants and animals will evolve in the future? Frame your answer in terms of the geographical and genetic factors that foster speciation.

How Would You Vote?

Often, when a species is at the brink of extinction, some individuals are captured and brought to zoos for captive breeding programs. Some people object to this practice. They say that keeping a species alive in a zoo is a distraction from more meaningful conservation efforts, and captive animals seldom are successfully restored to the wild. Do you support captive breeding of highly endangered species? Go to www.thomsonedu.com/login to investigate both sides of the issue and then vote.

Fossil of a dragonfly *(Cordulagomphus tuberculatus)* from the Cretaceous period, discovered in Ceara Province, Brazil.

22 Paleobiology and Macroevolution

WHY IT MATTERS

In January 1796, Georges Cuvier surprised his audience at the National Institute of Sciences and Arts in Paris by suggesting that fossils were the remains of species that no longer lived on Earth. Natural historians had long recognized the organic origin of fossils, but they did not believe that any creature could become extinct. They thought that the species preserved as fossils still lived in remote and inaccessible places.

Cuvier realized that he could not use the abundant fossils of small marine animals to demonstrate the reality of extinction: these species might still live in the deep sea or other unexplored regions. However, he reasoned that the world was already so well explored that scientists were unlikely to discover any new large terrestrial mammals. Thus, if he could show that fossilized mammals were different from living mammals, he could logically conclude that the fossilized species were truly extinct.

Now credited as the founder of comparative morphology, Cuvier thought that animals were essentially like machines. Each anatomical structure was a crucial part of a perfectly integrated whole. For example, a carnivore requires limbs to pursue prey, claws to catch it, teeth

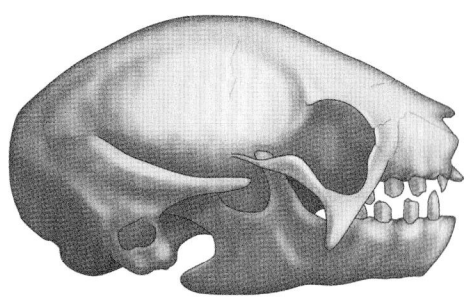

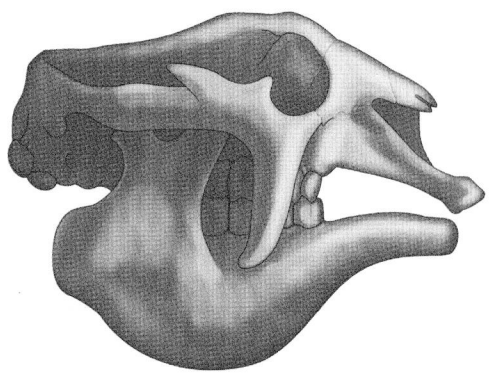

Figure 22.1

Comparing living organisms to fossils. Georges Cuvier compared the skull of a living sloth (top) to a fossilized skull from Paraguay (bottom). The fossilized skull has been reduced in size to facilitate the comparison.

to tear its flesh, and internal organs to digest meat. Thus, from the study of a few critical parts, a knowledgeable anatomist could make reasonable inferences about an animal's overall structure.

Cuvier is also recognized as the founder of paleobiology because he used the anatomy of living species to analyze fossils, which are rarely complete. Paleobiologists often use their knowledge of comparative morphology to make inferences about missing parts. Thus, when asked to analyze a large fossilized skull from Paraguay, Cuvier compared it to specimens in the museum and declared it to be a sloth **(Figure 22.1)**. But living sloths are small, whereas this specimen was gigantic, so Cuvier concluded that it was extinct. If such a large species were still living, naturalists would surely have discovered it while exploring South America.

Cuvier studied fossils of other large mammals, especially elephants and rhinoceroses. In every case, he demonstrated that fossilized species were anatomically different from living species. And because no one had seen living examples of the fossilized species, Cuvier concluded that they must be extinct. In 1812, he produced a multivolume treatise in which he acknowledged Earth's great age and documented the appearance and disappearance of species over time. He even noted that fossils lying near the ground surface more closely resembled living species than did

those that were deeply buried. Despite these extraordinary insights, Cuvier never embraced the concept of evolution. If all anatomical features of an animal's body were perfectly integrated, as he believed, how could any part change without upsetting that delicate functional balance?

Cuvier was an early student of macroevolution, the large-scale changes in morphology and diversity that characterize the 3.8-billion-year history of life. Macroevolution has occurred over so vast a span of time and space that the evidence for it is fundamentally different from that for microevolution and speciation. In this chapter we consider what paleobiology and the new field of evolutionary developmental biology tell us about macroevolutionary patterns.

22.1 The Fossil Record

Paleobiologists discover, describe, and name new fossil species and analyze the morphology and ecology of extinct organisms. Because fossils provide physical evidence of life in the past, they are our primary sources of data about the evolutionary history of many organisms.

Fossils Form When Organisms Are Buried by Sediments or Preserved in Oxygen-Poor Environments

Most fossils form in sedimentary rocks. Rain and runoff constantly erode the land, carrying fine particles of rock and soil downstream to a swamp, a lake, or the sea. Particles settle to the bottom as sediments, forming successive layers over millions of years. The weight of newer sediments compresses the older layers beneath them into a solid-matter matrix: sand into sandstone and silt or mud into shale. Fossils form within the layers when the remains of organisms are buried in the accumulating sediments.

The process of fossilization is a race against time because the soft remains of organisms are quickly consumed by scavengers or decomposed by microorganisms. Thus, fossils usually preserve the details of hard structures, such as the bones, teeth, and shells of animals and the wood, leaves, and pollen of plants. During fossilization, dissolved minerals replace some parts molecule by molecule, leaving a fossil made of stone **(Figure 22.2a)**; other fossils form as molds, casts, or impressions in material that is later transformed into solid rock **(Figure 22.2b)**.

In some environments, the near absence of oxygen prevents decomposition, and even soft-bodied organisms are preserved. Some insects, plants, and tiny lizards and frogs are embedded in amber, the fossilized resin of coniferous trees **(Figure 22.2c)**. Other organisms are preserved in glacial ice, coal, tar pits, or the highly acidic water of peat bogs **(Figure**

22.2d). Sometimes organisms are so well preserved that researchers can examine their internal anatomy, cell structure, and food in their digestive tracts. Biologists have even analyzed samples of DNA from a 40-million-year-old magnolia leaf.

a. Petrified wood

b. An invertebrate

c. Insects in amber

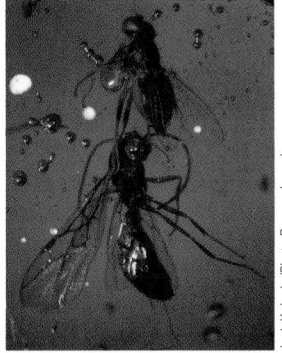

d. Mammoth in permafrost

Figure 22.2

Fossils. **(a)** Petrified wood, from the Petrified Forest National Park in Arizona, formed when minerals replaced the wood of dead trees molecule by molecule. **(b)** The soft tissues of an invertebrate (genus *Dickinsonia*) from the Proterozoic era were preserved as an impression in very fine sediments. **(c)** This 30-million-year-old fly (above) and wasp were trapped in the oozing resin of a coniferous tree and are now encased in amber. **(d)** A frozen baby mammoth (genus *Mammonteus*) that lived about 40,000 years ago was discovered embedded in Siberian permafrost in 1989.

The Fossil Record Provides an Incomplete Portrait of Life in the Past

The 300,000 described fossil species represent less than 1% of all the species that have ever lived. Several factors make the fossil record incomplete. First, soft-bodied organisms do not fossilize as easily as species with hard body parts. Moreover, we are unlikely to find the fossilized remains of species that were rare and locally distributed. Finally, fossils rarely form in habitats where sediments do not accumulate, such as mountain forests. The most common fossils are those of hard-bodied, widespread, and abundant organisms that lived in swamps or shallow seas, where sedimentation is ongoing.

Most fossils are composed of stone, but they don't last forever. Many are deformed by pressure from overlying rocks or destroyed by geological disturbances like volcanic eruptions and earthquakes. Once they are exposed on Earth's surface, where scientists are most likely to find them, rain and wind cause them to erode. Because the effects of these destructive processes are additive, old fossils are much less common than those formed more recently.

Scientists Assign Relative and Absolute Dates to Geological Strata and the Fossils They Contain

The sediments found in any one place form distinctive strata (layers) that differ in color, mineral composition, particle size, and thickness **(Figure 22.3).** If they have not been disturbed, the strata are arranged in the order in which they formed, with the youngest layers on top. However, strata are sometimes uplifted, warped, or even inverted by geological processes.

Geologists of the early nineteenth century deduced that the fossils discovered in a particular sedimentary stratum, no matter where it is found, represent organisms that lived and died at roughly the same time in the past. Because each stratum formed at a specific time, the sequence of fossils in the lowest (oldest) to the highest (newest) strata reveals their *relative ages*. Geologists used the sequence of strata and their distinctive fossil assemblages to establish the geological time scale **(Table 22.1).**

Although the geological time scale provides a relative dating system for sedimentary

Figure 22.3

Geological strata in the Grand Canyon. Millions of years of sedimentation in an old ocean basin produced layers of rock that differ in color and particle size. Tectonic forces later lifted the land above sea level, and the flow of the Colorado River carved this natural wonder.

Table 22.1 The Geological Time Scale and Major Evolutionary Events

Eons (Duration drawn to scale)	Eon	Era	Period	Epoch	Millions of Years Ago	Major Evolutionary Events
Cenozoic / Mesozoic / Paleozoic (Phanerozoic) ; Proterozoic	Phanerozoic	Cenozoic	Quaternary	Holocene	0.01	Origin of humans; major glaciations
				Pleistocene	1.7	Origin of ape-like human ancestors
			Tertiary	Pliocene	5.2	Angiosperms and mammals further diversify and dominate terrestrial habitats
				Miocene	23	Divergence of primates; origin of apes
				Oligocene	33.4	Angiosperms and insects diversify; modern orders of mammals differentiate
				Eocene	55	Grasslands and deciduous woodlands spread; modern birds and mammals diversify; continents approach current positions
				Paleocene	65	Many lineages diversify: angiosperms, insects, marine invertebrates, fishes, dinosaurs; asteroid impact causes mass extinction at end of period, eliminating dinosaurs and many other groups
		Mesozoic	Cretaceous		144	Gymnosperms abundant in terrestrial habitats; first angiosperms; modern fishes diversify; dinosaurs diversify and dominate terrestrial habitats; frogs, salamanders, lizards, and birds appear; continents continue to separate
			Jurassic		206	Predatory fishes and reptiles dominate oceans; gymnosperms dominate terrestrial habitats; radiation of dinosaurs; origin of mammals; Pangaea starts to break up; mass extinction at end of period
			Triassic		251	

Eon	Era	Period	Millions of years ago	Major events
Phanerozoic (continued)	Paleozoic	Permian	290	Insects, amphibians, and reptiles abundant and diverse in swamp forests; some reptiles colonize oceans; fishes colonize freshwater habitats; continents coalesce into Pangaea, causing glaciation and decline in sea level; mass extinction at end of period eliminates 85% of species
		Carboniferous	354	Vascular plants form large swamp forests; first seed plants and flying insects; amphibians diversify; first reptiles appear
		Devonian	417	Terrestrial vascular plants diversify; fungi and invertebrates colonize land; first insects appear; first amphibians colonize land; major glaciation at end of period causes mass extinction, mostly of marine life
		Silurian	443	Jawless fishes diversify; first jawed fishes; first vascular plants on land
		Ordovician	490	Major radiations of marine invertebrates and fishes; major glaciation at end of period causes mass extinction of marine life
		Cambrian	543	Diverse radiation of modern animal phyla (Cambrian explosion); simple marine communities
Proterozoic			2500	High concentration of oxygen in atmosphere; origin of aerobic metabolism; origin of eukaryotic cells; evolution and diversification of protists, fungi, soft-bodied animals
Archaean			3800	Evolution of prokaryotes, including anaerobic bacteria and photosynthetic bacteria; oxygen starts to accumulate in atmosphere
			4600	Formation of Earth at start of era; Earth's crust, atmosphere, and oceans form; origin of life at end of era

Archaean

Figure 22.4 Research Method

Radiometric Dating

PURPOSE: Radiometric dating allows researchers to estimate the absolute age of a rock sample or fossil.

PROTOCOL:

1. Knowing the approximate age of a rock or fossil, select a radioisotope that has an appropriate half-life. Because different radioisotopes have half-lives ranging from seconds to billions of years, it is usually possible to choose one that brackets the estimated age of the sample under study. For example, if you think that your fossil is more than 10 million years old, you might use uranium-235. The half-life of ^{235}U, which decays into the lead isotope ^{207}Pb, is about 700 million years. Or if you think that your fossil is less than 70,000 years old, you might select carbon-14. The half-life of ^{14}C, which decays into the nitrogen isotope ^{14}N, is 5730 years.

Radioisotopes Commonly Used in Radiometric Dating

Radioisotope (Unstable)	More Stable Breakdown Product	Half-Life (Years)	Useful Range (Years)
Samarium-147 →	Neodymium-143	106 billion	>100 million
Rubidium-87 →	Strontium-87	48 billion	>10 million
Thorium-232 →	Lead-208	14 billion	>10 million
Uranium-238 →	Lead-206	4.5 billion	>10 million
Uranium-235 →	Lead-207	700 million	>10 million
Potassium-40 →	Argon-40	1.25 billion	>100,000
Carbon-14 →	Nitrogen-14	5730	<70,000

2. Prepare a sample of the material and measure the quantities of the parent radioisotope and its more stable breakdown product.

INTERPRETING THE RESULTS: Compare the relative quantities of the parent radioisotope and its breakdown product (or some other stable isotope) to determine what percentage of the original parent radioisotope remains in the sample. Then use a graph of radioactive decay for that isotope to determine how many half-lives have passed since the sample formed.

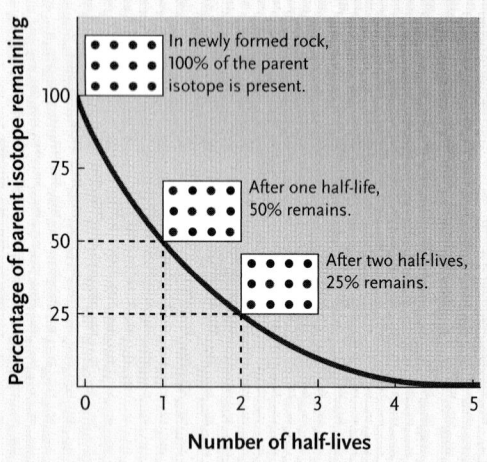

In newly formed rock, 100% of the parent isotope is present.

After one half-life, 50% remains.

After two half-lives, 25% remains.

Knowing the number of half-lives that have passed allows you to estimate the age of the sample.

A living mollusk absorbed trace amounts of ^{14}C, a rare radioisotope of carbon, and large amounts of ^{12}C, which is the more stable and common isotope of carbon.

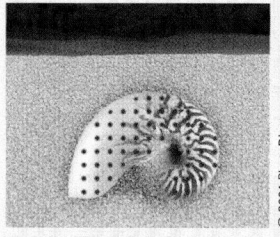

When the mollusk died, it was buried in sand and fossilized. From the moment of its death, the ratio of ^{14}C to ^{12}C began to decline through radioactive decay. Because the half-life of ^{14}C is 5730 years, half of the original ^{14}C was eliminated from the fossil in 5730 years and half of what remained was eliminated in another 5730 years.

After the fossil was discovered, a scientist determined that its ^{14}C to ^{12}C ratio was one-eighth of the ^{14}C to ^{12}C ratio in living organisms. Thus, radioactive decay had proceeded for three half-lives—or about 17,000 years—since the mollusk's death.

---

strata, it does not tell us how old the rocks and fossils actually are. But many rocks contain radioisotopes, which, from the moment they form, begin to break down into other, more stable elements. The breakdown proceeds at a steady rate that is unaffected by chemical reactions or environmental conditions such as temperature or pressure. Using a technique called **radiometric dating**, scientists can estimate the age of a rock by noting how much of an unstable "parent" isotope has decayed to another form. By measuring the relative amounts of the parent radioisotope and its breakdown products and comparing this ratio with the isotope's **half-life**—the time it takes for half of a given

amount of radioisotope to decay—researchers can estimate the *absolute age* of the rock (**Figure 22.4**). Table 22.1 presents these age estimates along with the major geological and evolutionary events of each period.

Radiometric dating works best with volcanic rocks, which form when lava cools and solidifies. But most fossils are found in sedimentary rocks. To date sedimentary fossils, scientists determine the age of volcanic rocks from the same strata. Using this method, investigators have linked fossils to deposits that are hundreds of millions of years old.

Fossils that still contain organic matter, such as the remains of bones or wood, can be dated directly by measuring their content of the radioactive carbon isotope ^{14}C, which decays to ^{14}N. Living organisms absorb traces of ^{14}C and large quantities of ^{12}C, a stable carbon isotope, from the environment and incorporate them into biological molecules. As long as an organism is still alive, its ^{14}C content remains constant because any ^{14}C that decays is replaced by the uptake of other ^{14}C atoms. But as soon as the organism dies, no further replacement occurs and ^{14}C begins its steady radioactive decay. Scientists use the ratio of ^{14}C to ^{12}C present in a fossil to determine its age, as explained in Figure 22.4.

To develop a feeling for geological time, imagine the 4.5-billion-year history of Earth scaled onto an annual calendar; each day represents a little over 12 million years. The planet was formed on January 1. Animal life originated in mid-November, dinosaurs lived between December 14 and December 26, and the primate ancestors of modern humans appeared during the last 4 hours of December 31.

Fossils Provide Abundant Information about Life in the Past

Imperfect as it is, the fossil record provides our only direct information about life in the past. Fossilized skeletons, shells, stems, leaves, and flowers tell us about the size and appearance of ancient animals and plants. The fossil record also allows scientists to see how structures were modified as they became adapted for specialized uses (see Figure 19.3). Moreover, fossils chronicle the proliferation and extinction of evolutionary lineages and provide data on their past geographical distributions.

Fossils can also provide indirect data about behavior, physiology, and ecology. For example, the fossilized footprints of some dinosaurs suggest that adults surrounded their young when the group moved, perhaps to protect them from predators. Complex scrolls of bone in the nasal passages of early mammals suggest that they had a well-developed sense of smell, and fossilized teeth and dung provide data about the diets of extinct animals. The study of fossilized pollen allows paleobiologists to reconstruct the vegetation and climate of ancient sites. The changing arrays of fossils that document biological evolution partly reflect large-scale shifts in Earth's physical environments, a topic that we explore in the next section.

STUDY BREAK

1. What biological materials are the most likely to fossilize?
2. Why does the fossil record provide an incomplete portrait of life in the past?
3. What sort of information can paleobiologists discern from the fossil record?

22.2 Earth History, Biogeography, and Convergent Evolution

Organisms interact constantly with their environments. Some of these interactions have caused fundamental changes in Earth's physical environment, such as the development of an oxidizing atmosphere (see Chapter 24). In this section we consider other aspects of Earth's history and their profound effects on living systems.

Geological Processes Have Often Changed Earth's Physical Environment

Long-term shifts in geography and climate—as well as brief but catastrophic events—have significantly altered environments on Earth. Major geological and climatic shifts occur because the planet's crust is in motion.

According to the theory of **plate tectonics**, Earth's crust is broken into irregularly shaped plates of rock that float on its semisolid mantle **(Figure 22.5)**. Currents in the mantle cause the plates—and the continents embedded in them—to move, a phenomenon called **continental drift**. About 250 million years ago, Earth's landmasses coalesced into a single supercontinent called Pangaea; continental drift later separated Pangaea into a northern continent, Laurasia, and a southern continent, Gondwana. Laurasia and Gondwana subsequently broke into the continents we know today **(Figure 22.6)**.

The drifting continents induced global changes in Earth's climate. For example, the movement of continents toward the poles encouraged the formation of glaciers, which caused temperature and rainfall to decrease worldwide. As a result of complex continental movements, Earth's average temperature has fluctuated widely. During one geologically recent cold spell (about 20,000 years ago), the northern polar ice cap extended into southern Indiana and Ohio.

Unpredictable events have also changed physical environments on Earth. Massive volcanic eruptions and asteroid impacts have occasionally altered the planet's atmosphere and climate drastically. These cataclysmic events have sometimes caused many forms of life to disappear over relatively short periods of geological time.

Historical Biogeography Explains the Broad Geographical Distributions of Organisms

More than a century after Darwin published his observations, the theory of plate tectonics refocused attention on biogeography. Historical biogeographers try to explain how organisms acquired their geographical distributions over evolutionary time.

Continuous and Disjunct Distributions. Many species have a **continuous distribution:** they live in suitable habitats throughout a geographical area. For example, herring gulls *(Larus argentatus)* live along the coastlines of all northern continents. Continuous distributions usually require no special historical explanation.

Other groups exhibit **disjunct distributions,** in which closely related species live in widely separated locations. For example, magnolia trees *(Magnolia* species)

a. Earth's crustal plates

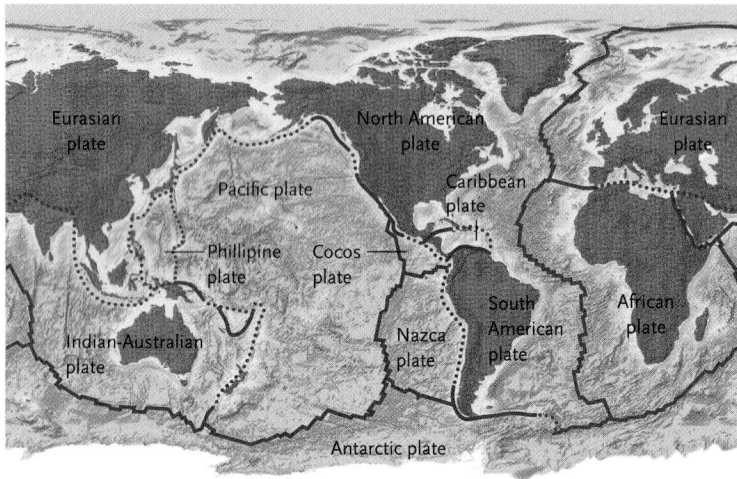

KEY

—— Oceanic ridge ······ Oceanic trench

b. Model of plate tectonics

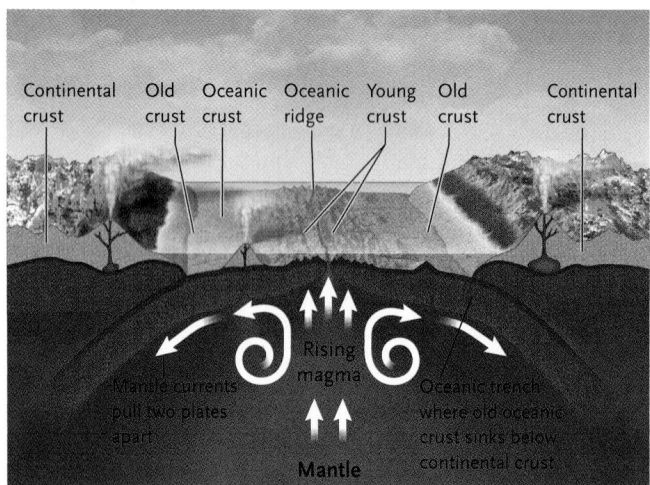

Figure 22.5

Plate tectonics. **(a)** Earth's crust is broken into large, rigid plates. New crust is added at oceanic ridges, and old crust is recycled into the mantle at oceanic trenches. **(b)** Oceanic ridges form where pressure in the mantle forces magma (molten rock) through fissures in the sea floor. Mantle currents pull the plates apart on either side of the ridge, forcing the sea floor to move laterally away from the ridge. This phenomenon, seafloor spreading, is widening the Atlantic Ocean about 3 cm per year. Oceanic trenches form where plates collide. The heavier oceanic crust sinks below the lighter continental crust, and it is recycled into the mantle, a process called subduction. The highest mountain ranges (including the Rockies, Himalayas, Alps, and Andes) formed where subduction uplifted continental crust. Earthquakes and volcanoes are also common near trenches.

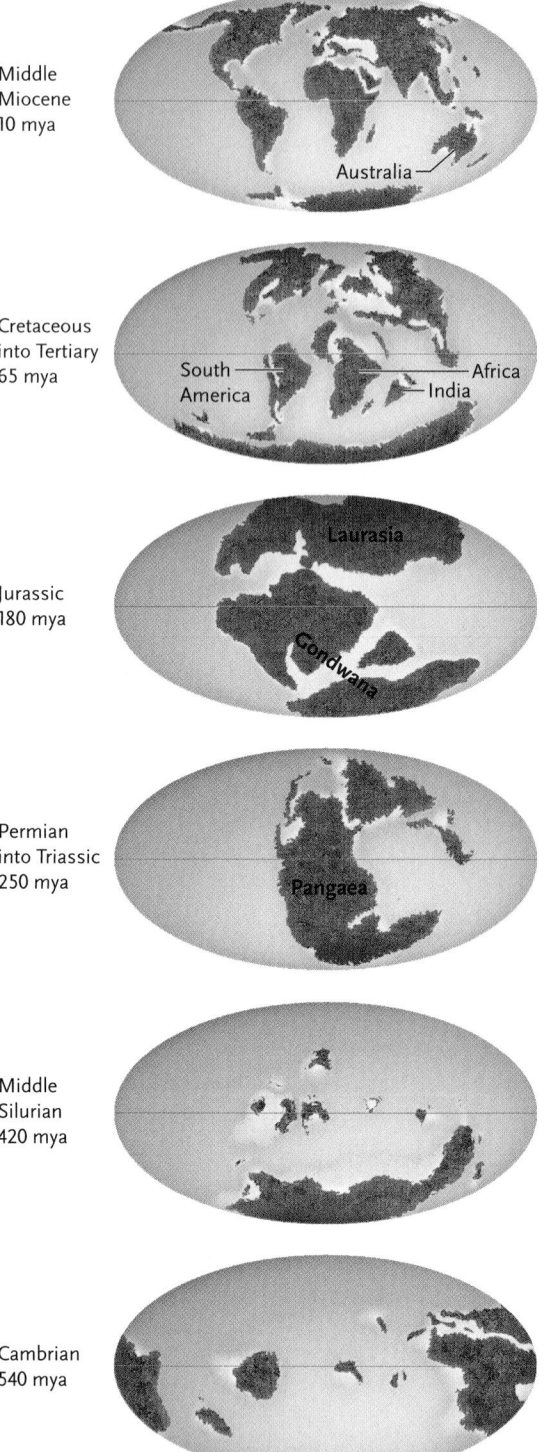

Figure 22.6

History of long-term changes in the positions of continents. Earth's many landmasses coalesced during the Permian period, forming the supercontinent Pangaea. About 180 million years ago (mya), Pangaea separated into Gondwana and Laurasia. Then Gondwana began to break apart. Africa and India pulled away first, opening the South Atlantic and Indian Oceans. Australia separated from Antarctica about 55 million years ago and slowly drifted northward. South America separated from Antarctica shortly thereafter. Laurasia remained nearly intact until 43 million years ago when North America and Greenland together separated from Europe and Asia. Movement of the continents also changed the shapes and the sizes of the oceans.

occur in parts of North, Central, and South America as well as in China and Southeast Asia, but nowhere in between.

Two phenomena—dispersal and vicariance—create disjunct distributions. **Dispersal** is the movement of organisms away from their place of origin; it can produce a disjunct distribution if a new population becomes established on the far side of a geographical barrier. **Vicariance** is the fragmentation of a continuous geographical distribution by external factors. Over the course of evolutionary history, dispersal and vicariance have together influenced the geographical distributions of organisms on a very grand scale (see *Focus on Research*).

Biogeographical Realms. For species that were widespread in the Mesozoic era, Pangaea's breakup was a powerful vicariant experience. The subsequent geographical isolation of continents fostered the evolution of distinctive regional **biotas** (all organisms living in a region). Alfred Russel Wallace used the biotas to define six **biogeographical realms**, which we still recognize today **(Figure 22.7)**.

The Australian and Neotropical realms, which have been geographically isolated since the Mesozoic, contain many **endemic species** (those that occur nowhere else on Earth). The Australian realm, in particular, has had no complete land connection to any other continent for approximately 55 million years. As a result, Australia's mammalian fauna (all the mammals living in the region) is unique, made up almost entirely of endemic marsupials. Few native placental mammals occur in Australia because the placental lineage arose elsewhere after Australia had become isolated.

The biotas of the Nearctic and Palearctic realms are, by contrast, fairly similar. North America and Eurasia were frequently connected by land bridges; eastern North America was attached to Western Europe until the breakup of Laurasia 43 million years ago, and northwestern North America had periodic contact with northeastern Asia over the Bering land bridge during much of the past 60 million years.

Convergent Evolution Produces Similar Adaptations in Distantly Related Organisms

Distantly related species living in different biogeographical realms are sometimes very similar in appearance. For example, the overall form of cactuses in the Americas is almost identical to that of spurges in Africa **(Figure 22.8)**. But these lineages arose independently long after those continents had separated; thus, cactuses and spurges did not inherit their similarities from a shared ancestor. Their overall resemblance is the product of **convergent evolution,** the evolution of similar adaptations in distantly related organisms that occupy similar environments.

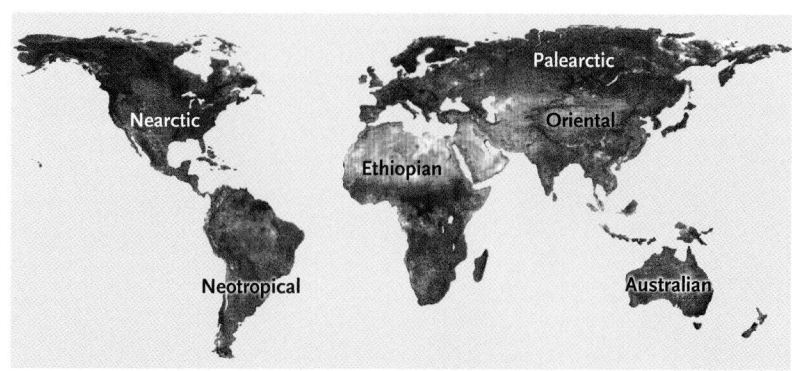

Figure 22.7
Wallace's biogeographical realms. Each realm contains a distinctive biota.

Convergent evolution also creates similarities in distantly related animals that use the same mode of locomotion. Some marine fishes, birds, and mammals have torpedo-shaped bodies and appendages modified for swimming in strong ocean currents **(Figure 22.9)**. Even entire faunas can develop convergent morphologies. For example, the marsupial mammals of Australia and placental mammals of North America—groups that arose long after the breakup of Pangaea—include many pairs of morphologically convergent species that also occupy similar habitats and feed on similar foods. To understand convergent evolution as well as most other macroevolutionary patterns, biologists must analyze the evolutionary history of individual lineages, a sometimes-controversial activity that we consider next.

STUDY BREAK

1. How did the process of continental drift affect the geographical distributions of organisms?
2. Why do some distantly related species that live in different biogeographical realms sometimes resemble each other?

a. Cactus    **b.** Spurge

Figure 22.8
Convergent evolution in plants. **(a)** *Echinocereus* and other North American cactuses (family Cactaceae) are strikingly similar to **(b)** *Euphorbia* and other African spurges (family Euphorbiaceae). Convergent evolution adapted both groups to desert environments with thick, water-storing stems, spiny structures that discourage animals from feeding on them, CAM photosynthesis (see Section 9.4), and stomata that open only at night.

Basic Research: The Great American Interchange

a. Jurassic Period

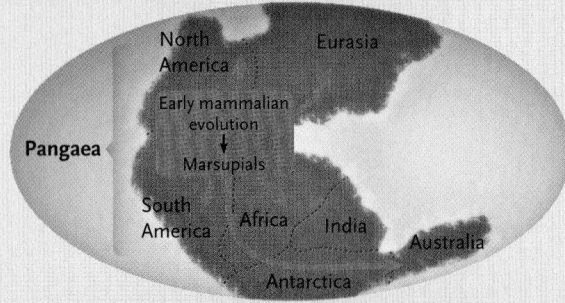

b. Cretaceous Period

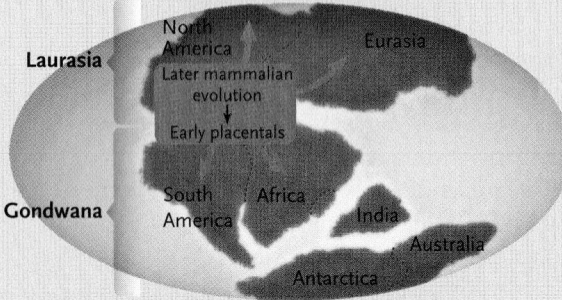

c. Miocene Epoch

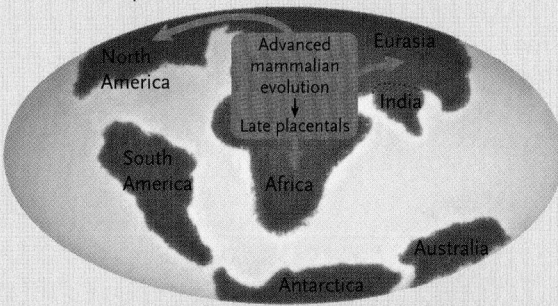

d. Pliocene Epoch

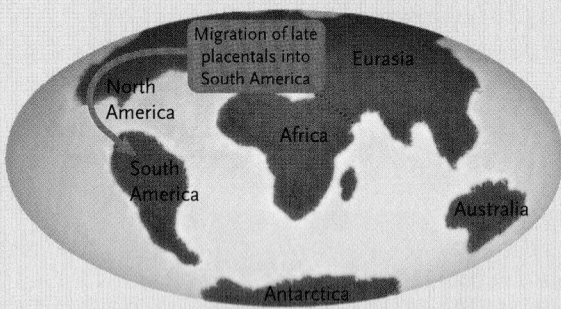

Figures a–d

Dispersal and vicariance changed the geographical distributions of marsupial and placental mammals.

Paleobiologists reconstruct the biogeographical history of a lineage by dating its fossils and mapping their geographical distributions at specific times in the past. The complex evolutionary history of mammals, especially in North and South America, illustrates the effects of dispersal and vicariance.

Mammals first arose in western Pangaea (now part of North America), where they diverged into several evolutionary lineages. The earliest marsupials (whose young complete development in a pouch on the mother's belly) dispersed to the future Eurasia, Africa, South America, Antarctica, and Australia during the Jurassic period **(Figure a)**. Somewhat later, but before the continents completely separated during the Cretaceous period, the earliest placentals (whose young complete development within the mother's uterus) also dispersed from North America into Eurasia, Africa, and South America **(Figure b)**.

The breakup of Pangaea did not destroy all of these dispersal paths immediately. Persistent land connections allowed many organisms to migrate freely between North America and Eurasia throughout the Cretaceous period. Modern placentals further diversified in Eurasia during the Miocene epoch, and these new forms quickly dispersed back into North America **(Figure c)**. As a result, the mammalian faunas of these continents have always been very similar.

By contrast, Australia and South America experienced substantial geographical isolation, particularly after the breakup of Gondwana. In South America, a distinctive mammalian fauna evolved from the marsupials and early placentals that had arrived during the Mesozoic. Small marsupials fed primarily on insects and other invertebrates, but larger species, including a marsupial saber-toothed cat, ate other vertebrates. The early South

American placentals gave rise to many large ungulates (hoofed herbivorous mammals) as well as to armadillos, sloths, and anteaters, some of which still live in South America today.

Periodic dispersal events slowly added to South America's mammal fauna. For example, rodents and primates first arrived about 25 million years ago, during the Oligocene epoch. They probably came from Africa by island-hopping across the slowly widening South Atlantic. These rodents eventually gave rise to guinea pigs and their relatives, and these primates to all the living New World monkeys. South America then began to drift northwest toward North America. By the late Miocene epoch, 6 million to 8 million years ago, North American rodents and raccoons were able to disperse into South America across the narrow water gap. By about 3 million years ago, in the Pliocene epoch, the Panamanian land bridge was established between North and South America, allowing mammals to migrate in both directions **(Figure d)**.

A group of paleobiologists led by Larry G. Marshall of the Field Museum of Natural History in Chicago and S. David Webb of the Florida State Museum and the University of Florida have determined that about 10% of the mammal species on each continent dispersed across the land bridge to the other side. But North America—with its greater size and long-standing connections to Eurasia—had a greater variety of mammals than South America. Thus, more different types of mammals moved from north to south than in the opposite direction. Dispersal was so extensive during the Pliocene that paleobiologists describe these movements as the Great American Interchange.

The dispersal of so many northern mammals into South America fundamentally changed its ecological communities. Carnivorous cats, dogs, and weasels and herbivorous camels, deer, elephants, horses, rabbits, and tapirs swept into the continent. Many new arrivals were wildly successful, apparently because they ate resources that native South American mammals were not using. Moreover, the northern immigrants had high rates of speciation, producing numerous descendant species.

As climates periodically cooled during the Pleistocene epoch, many mammals became extinct on both continents. Descendants of the northern immigrants have fared well over the long term: about half the mammals in South America today—including all the cats, llamas, tapirs, and many rodents—are the descendants of northern ancestors. Most South American species that moved north were not as successful. Perhaps they could not adapt to physical conditions in the north, especially during the Pleistocene glaciations; or perhaps they could not prevail in competition with mammals that were already there. Today, relatively few mammals of southern origin persist in North America. Armadillos, monkeys, and anteaters are restricted to the southernmost parts. Only one opossum species (*Didelphis virginiana*, **Figure e**) and one porcupine species (*Erethizon dorsatum*) have moved further north.

Figure e

Opposums (*Didelphis virginiana*) are among the few mammals of South American origin that are successful in North America.

a. Shark

b. Penguin

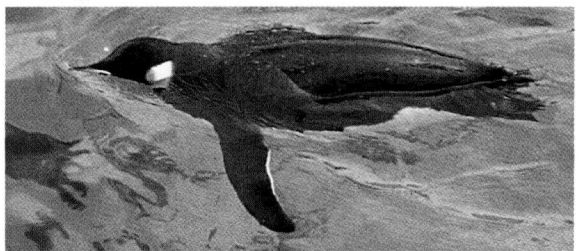

c. Porpoise

Figure 22.9

Convergent evolution in marine vertebrates. Convergent evolution produced similar body forms and appendages in distantly related marine predators: **(a)** sharks, which are cartilaginous fishes; **(b)** penguins, which are birds; and **(c)** porpoises, which are mammals. The resemblances are superficial, however. The tails of sharks are vertical, whereas those of penguins and porpoises are horizontal. Penguins also lack fins along their backs.

22.3 Interpreting Evolutionary Lineages

As newly discovered fossils demand the reinterpretation of old hypotheses, biologists constantly refine their ideas about the history of life. The evolution of horses is a case in point.

Modern Horses Are Living Representatives of a Once-Diverse Lineage

The earliest known ancestors of modern horses were first identified by Othniel C. Marsh of Yale University just a year after Darwin published *On the Origin of Species*. These early horses, *Hyracotherium,* stood 25 to 50 cm high and weighed no more than 20 kg. Their toes (four on the front feet and three on the hind) were each capped with a tiny hoof, but the animals walked on soft pads as dogs do today. Their faces were short, their teeth were small, and they browsed on soft leaves in woodland habitats.

In 1879, Marsh published his analysis of 60 million years of horse family history. He described the evolution of this group of mammals as a sequence of stages from the tiny *Hyracotherium* through intermediates represented by *Mesohippus, Merychippus,* and *Pliohippus* to the modern *Equus* **(Figure 22.10a).** (Each of these names refers to a genus, a group of closely related species.) Marsh inferred a pattern of descent characterized by gradual, directional evolution in several skeletal features. Changes in the legs and feet allowed horses to run more quickly, and changes in the face and teeth accompanied a switch in diet from soft leaves to tough grasses.

a. Marsh's reconstruction of horse evolution

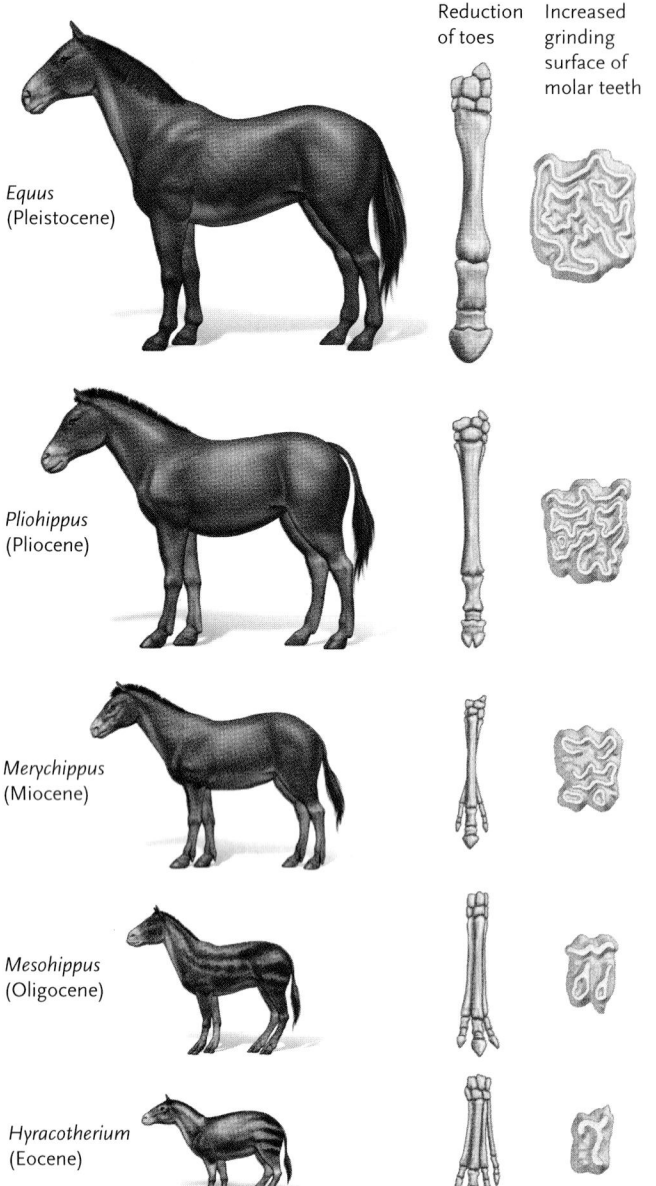

Reduction of toes

Increased grinding surface of molar teeth

Equus (Pleistocene)

Pliohippus (Pliocene)

Merychippus (Miocene)

Mesohippus (Oligocene)

Hyracotherium (Eocene)

b. Modern reconstruction of horse evolution

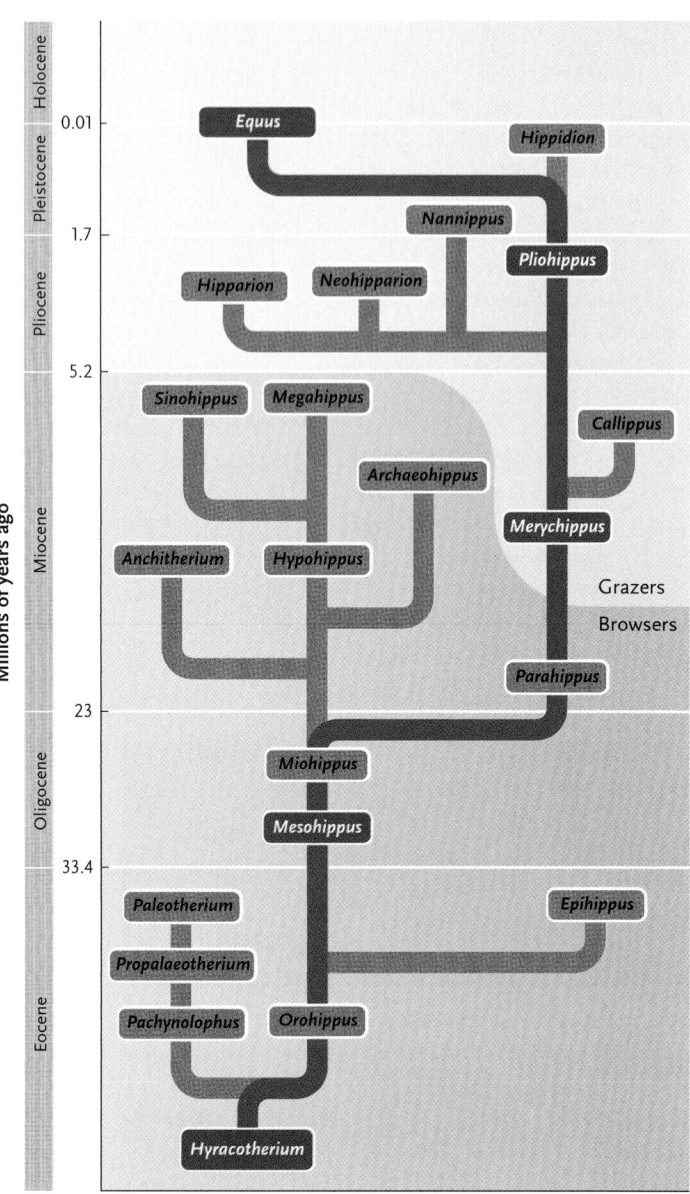

c. Changes in body size of horse species over time

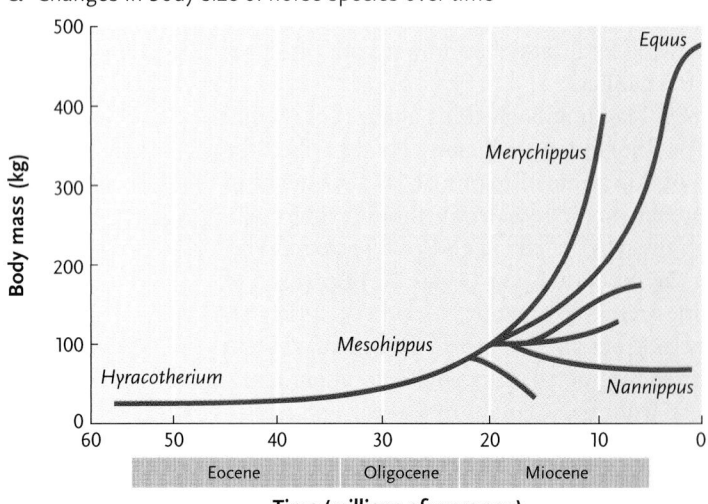

Figure 22.10

Evolution of horses. **(a)** Marsh depicted the evolution of horses as a linear pattern of descent characterized by an increase in body size, a reduction in the number of toes, increased fusion of the bones in the lower leg, elongation of the face, and an increase in the size of grinding teeth at the back of the mouth. **(b)** Recent studies revealed that the horse family includes numerous evolutionary branches with variable morphology. The horses in Marsh's analysis are highlighted in green. **(c)** Although many branches of the lineage evolved a larger body size, some remained as small as the earliest horses.

The fossil record for horses is superb, and we now have fossils of more than 100 extinct species from five continents. These data reveal a macroevolutionary history very different from Marsh's interpretation. *Hyracotherium* was not gradually transformed into *Equus* along a linear track. Instead, the evolutionary tree for horses was highly branched **(Figure 22.10b),** and *Hyracotherium*'s descendants differed in size, number of toes, tooth structure, and other traits. Although many branches of this lineage lived in the Miocene and Pliocene epochs, all but one are now extinct. The species of the genus *Equus* living today (horses, donkeys, and zebras) are the surviving tips of that one branch.

When we study extinct organisms, we tend to focus on traits that characterize modern species. Marsh, for example, assumed that the differences between *Hyracotherium* and *Equus* were typical of the changes that characterized the group's evolutionary history. But not all fossil horses were larger **(Figure 22.10c),** had fewer toes, or were better adapted to feed on grass than their ancestors. And if a branch other than *Equus* had survived, Marsh's description of trends in horse evolution would have been very different. All evolutionary lineages have extinct branches, and any attempt to trace a linear evolutionary path—as Marsh did for horses and many people do for humans—imposes artificial order on an inherently disorderly history.

Evolutionary Biologists Debate the Mode and Tempo of Macroevolution

What evolutionary processes produce the numerous branches of a lineage such as the horse lineage, and over what time scale does a lineage evolve?

Modes of Evolutionary Change. The species that paleobiologists identify may arise by one of two modes, or processes of change, called anagenesis and cladogenesis. **Anagenesis** refers to the accumulation of changes in a lineage as it adapts to changing environments. If morphological changes are large, we may give the organisms different names at different times in their history. One might say, for example, that Species A from the late Mesozoic era had evolved into Species B from the middle Cenozoic era **(Figure 22.11a).** Anagenesis does not increase the number of species—it is the evolutionary transformation of an existing species rather than the production of new ones.

Cladogenesis refers to the evolution of two or more descendant species from a common ancestor. If the fossilized remains of the descendants are distinct, paleobiologists will recognize them as different species **(Figure 22.11b).** Cladogenesis does increase the number of species on Earth.

Tempo of Morphological Change. Macroevolutionists have developed two alternative hypotheses to describe the tempo, or timing, of morphological change.

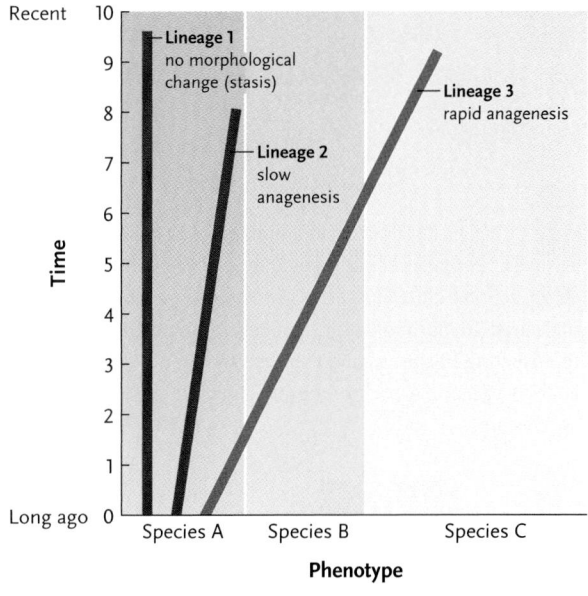

a. Anagenesis

Three lineages begin with the same phenotype, identified as fossil Species A, at time 0. The rate of evolutionary change is shown by the angle of the line for each lineage: lineage 1 undergoes no change over time, lineage 2 changes slowly, and lineage 3 changes so rapidly that its phenotype shifts far to the right in the graph. Paleobiologists might assign different names to the fossils of lineage 3 at different times in its history—Species A at time 1, Species B at times 2 through 6, and Species C at times 7 through 9—even though no additional species evolved. By contrast, fossils of lineages 1 and 2 change so little over time that they would be identified as Species A throughout their evolutionary history.

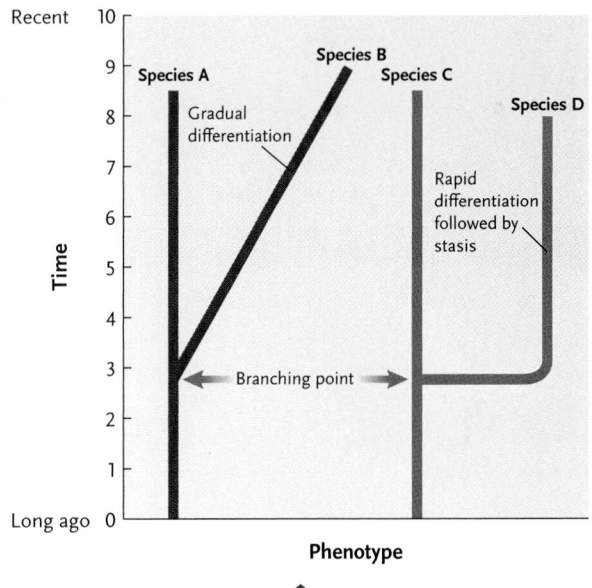

b. Cladogenesis

Each branching point represents a new line of descent. The branching may show either gradual (left) or rapid (right) morphological differentiation from the parent species.

Figure 22.11
Patterns of evolution. In these hypothetical examples, the vertical axis represents geological time and the horizontal axis represents variation in phenotypic traits.

Figure 22.12 Observational Research

Evidence Supporting the Punctuated Equilibrium Hypothesis

HYPOTHESIS: The punctuated equilibrium hypothesis states that most morphological change within evolutionary lineages appears during periods of rapid speciation.

PREDICTION: The fossil record will reveal that most species experienced relatively little morphological change for long periods of time, but that new, morphologically distinctive species arose suddenly.

METHOD: Cheetham examined numerous fossilized samples of populations of a small marine invertebrate, the ectoproct *Metrarabdotos,* from the Dominican Republic. He measured 46 morphological characters in populations representing 18 species and then used a complex statistical analysis to summarize how morphologically different the species are.

RESULTS: The morphology of most *Metrarabdotos* species changed very little over millions of years, but new species, morphologically very different from their ancestors, often appeared suddenly in the fossil record. Each dot in the graph represents a sample of a population from the fossil record. The horizontal axis reflects the overall morphological difference between samples of one species over time or between samples of ancestral species and descendant species. The dashed lines represent gaps in the fossil record for this genus.

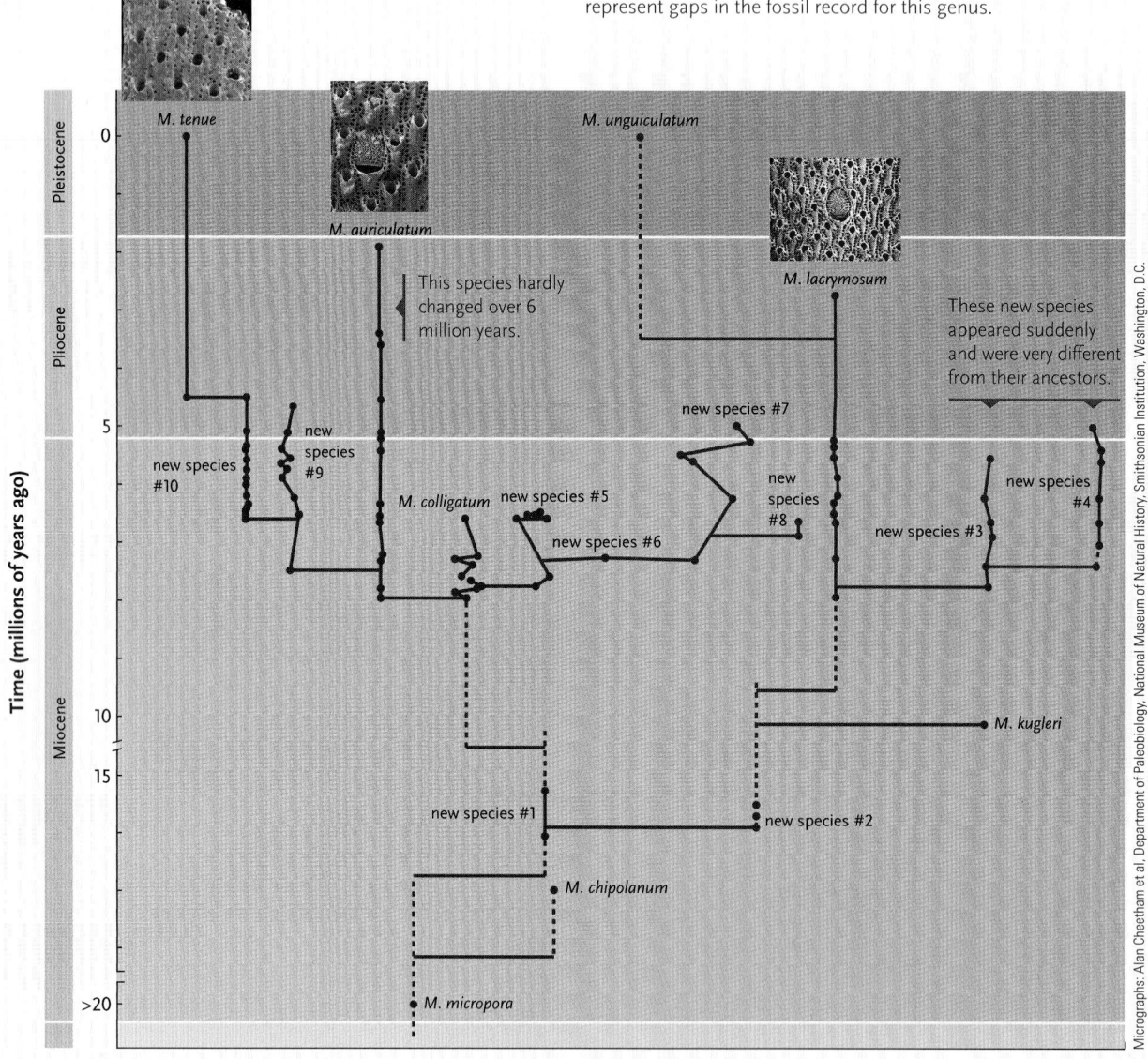

Micrographs: Alan Cheetham et al, Department of Paleobiology, National Museum of Natural History, Smithsonian Institution, Washington, D.C.

Overall morphological difference

CONCLUSION: The *Metrarabdotos* lineage exhibits a pattern of morphological evolution that is consistent with the predictions of the punctuated equilibrium hypothesis.

According to the **gradualist hypothesis,** large changes result from the slow, continuous accumulation of small changes over time. If this hypothesis is correct, we might expect to find a series of transitional fossils that document gradual evolution. In fact, we rarely find evidence of perfectly gradual change in any lineage. Most species appear suddenly in a particular stratum, persist for some time with little change, and then disappear from the fossil record. Then another species with different traits suddenly appears in the next higher stratum.

In the early 1970s, Niles Eldredge of the American Museum of Natural History and Stephen Jay Gould of Harvard University published an explanation for the absence of transitional forms, or "missing links." Their **punctuated equilibrium hypothesis** suggested that speciation usually occurs in isolated populations at the edge of a species' geographical distribution. Such populations experience substantial genetic drift and distinctive patterns of natural selection (as described in Section 21.3). According to this hypothesis, morphological variations arise rapidly during cladogenesis. Thus, most species exhibit long periods of morphological equilibrium or stasis (little change in form), punctuated by brief periods of cladogenesis and rapid morphological evolution. If this hypothesis is correct, transitional forms live only for short periods of geological time in small, localized populations—the very conditions that discourage broad representation in the fossil record. Darwin himself used this line of reasoning to explain puzzling gaps in the fossil record: new species appear as fossils only after they become abundant and widespread and begin a period of morphological stasis.

Some evolutionists point to flaws in the punctuated equilibrium hypothesis. First, rapid morphological evolution frequently occurs without cladogenesis. For example, in North America, variations in the body size of house sparrows evolved within 100 years without the appearance of new sparrow species (see Figure 21.5). Furthermore, geographical variation in most widespread species (see Section 21.1) provides compelling evidence of morphological evolution without speciation.

Second, critics challenge the hypothesis' definition of rapid morphological change, particularly given our inability to resolve time precisely in the fossil record. To a paleobiologist with a geological perspective, "instantaneous" events occur over tens or hundreds of thousands of years. But to a population geneticist, those time scales may encompass thousands of generations, ample time for gradual microevolutionary change.

Third, examples of evolutionary stasis may not be as static as they appear. Alternating periods of directional selection that favor opposite patterns of change could produce the appearance of stasis. For example, if natural selection favored slight increases in body size for 2000 years and then favored slight decreases for the next 2000 years, paleobiologists would probably detect no change in body size at all.

The fossil record provides some support for both hypotheses. A punctuated pattern is evident in the evolutionary history of *Metrarabdotos,* a genus of ectoprocts from the Caribbean Sea. Ectoprocts are small colonial animals that build hard skeletons (see Figure 29.15a), the details of which are well preserved in fossils. Alan Cheetham of the Smithsonian Institution measured 46 morphological traits in fossils of 18 *Metrarabdotos* species. He then used a summary statistic to describe the morphological difference between populations of a single species over time and between ancestral species and their descendants. His results indicate that most species did not change much over millions of years, but new species, which were morphologically different from their ancestors, often appeared quite suddenly **(Figure 22.12).**

By contrast, a study of Ordovician trilobites supports the gradualist hypothesis of evolution. The number of "ribs" in their tail region changed continuously over 3 million years. The change was so gradual that a sample from any given stratum was almost always intermediate between samples from the strata just above and below it. The changes in rib number probably evolved without cladogenesis **(Figure 22.13).**

The punctuationalist and gradualist hypotheses represent extremes on a continuum of possible macroevolutionary patterns. The mode and tempo of evolution vary among lineages, and both viewpoints are validated by data on some organisms but not others. Although some biologists still question the punctuated equilibrium hypothesis, its publication rekindled interest in paleobiology and macroevolution, inspiring much new research. Some of the most interesting results have focused on morphological changes within lineages and on long-term changes in the number of living species.

STUDY BREAK

1. Did the horse lineage undergo a steady increase in body size over its evolutionary history?
2. How do the predictions of the gradualist and the punctuationalist hypotheses differ?

22.4 Macroevolutionary Trends in Morphology

Some evolutionary lineages exhibit trends toward larger size and greater morphological complexity, and others are marked by the development of novel structures.

Figure 22.13 Observational Research

Evidence Supporting the Gradualist Hypothesis

HYPOTHESIS: The gradualist hypothesis states that most morphological change within evolutionary lineages results from the accumulation of small, incremental changes over long periods of time.

PREDICTION: The fossil record will reveal that the morphology of fossils from a given stratum will be intermediate between those of fossils from the strata immediately below and above it.

METHOD: Peter R. Sheldon of Trinity College, Dublin, Ireland, counted the number of "ribs" in the tail region of the exoskeletons of approximately 15,000 trilobite fossils from central Wales, United Kingdom. The fossils had formed over a span of about 3 million years during the Ordovician period. Sheldon plotted the mean number of ribs found in successive samples of each lineage.

RESULTS: Sheldon's data reveal gradual changes in the mean number of "ribs" in these animals with no evidence of speciation.

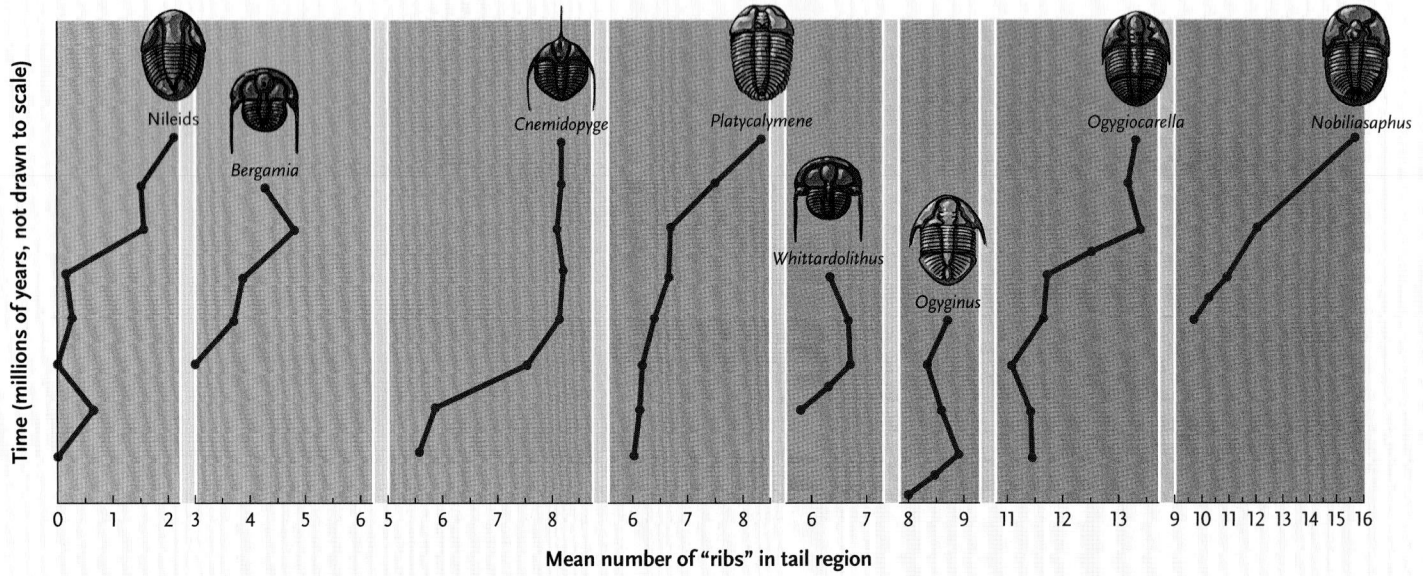

CONCLUSION: Morphological changes in Ordovician trilobites from central Wales are consistent with the predictions of the gradualist hypothesis.

The Body Size of Organisms Has Generally Increased over Time

Body size affects most aspects of an organism's physiology and ecology. When we look at the entire history of life, organisms have generally become larger over time. The earliest organisms were tiny, as most still are today. But the change from replicating molecules to acellular, unicellular, and finally multicellular organization must have demanded an increase in body size.

Within evolutionary lineages, increases in body size are not universal, but they are common. The nineteenth-century paleobiologist Edward Drinker Cope first noted this trend toward larger body size, now known as *Cope's Rule*, in vertebrates. Although Cope's Rule also applies to some invertebrate and plant lineages, no one has conducted a truly broad survey to test the generality of the hypothesis. Insects, for example, are a major exception to Cope's Rule. Most insects have remained small since their appearance in the Devonian, probably because of the constraints imposed by an external skeleton (see Section 29.7).

We can readily imagine why natural selection may sometimes favor larger size. Large organisms maintain a more constant internal environment than small ones. They may also have access to a wider range of resources, harvest resources more efficiently, and be less likely to be captured by predators. Moreover, larger females may produce more young, and larger males may have greater access to mates. Unfortunately, we cannot test such hypotheses about extinct life forms directly. We can only analyze past events with an understanding of how natural selection affects organisms living today.

In the 1970s, Steven Stanley of Johns Hopkins University proposed an explanation for how macroevolutionary trends may develop. He suggested that certain traits might make some species more likely to undergo speciation than others. This mechanism, called **species selection**, is analogous to natural selection. In natural selection, the evolutionary success of an individual is measured by the number of its surviving offspring. In species selection, the evolutionary success of a species is measured by the number of its descendant species. Thus, the traits of species that frequently undergo cladogenesis become more common, establishing a trend within a lineage. For example, if large species leave more descendant species than small ones do, the number of large species will increase faster than the number of small species. As a result, the average size of species in the lineage will increase over time. Stanley's hypothesis has not been widely tested.

Morphological Complexity Has Also Generally Increased over Time

In general, the evolutionary increase in size has been accompanied by an increase in morphological complexity. Among contemporary organisms, for example, species with large body size have a greater variety of cell types than do species with small body size. We can probably assume that new cell types arose when larger organisms first evolved.

However, under some circumstances, natural selection has simplified traits. The single toe and fused leg bones of modern horses are stronger, but mechanically less complex, than the ancestral structures in *Hyracotherium*. Similarly, snakes, which evolved from lizards with well-developed legs, have lost their limbs entirely. These changes, which increase the efficiency of locomotion, represent decreases in morphological complexity.

Several Phenomena Trigger the Evolution of Morphological Novelties

Sometimes a trait that is adaptive in one context turns out to be advantageous in another. Natural selection may then modify the trait to enhance its new function. Such **preadaptations** are just lucky accidents; they never evolve *in anticipation* of future evolutionary needs.

John Ostrom of Yale University described how some carnivorous dinosaurs, the immediate ancestors of *Archaeopteryx* and modern birds, were preadapted for flight (see Figure 19.12). These small, agile creatures were bipedal with lightweight hollow bones and long forelimbs to capture prey; some even had rudimentary feathers that may have retained body heat. But all these traits evolved because they were useful adaptations in highly active and mobile predators, not because they would someday allow flight.

The morphology of individuals sometimes changes over time because of **allometric growth** (*allo* = different;

metro = measure), the differential growth of body parts. In humans, for example, the relative sizes of different body parts change because human heads, torsos, and limbs grow at different rates **(Figure 22.14a)**.

Allometric growth can also create morphological differences in closely related species. For example, the skulls of chimpanzees and humans are similar in newborns, but markedly different in adults **(Figure 22.14b)**. Some regions of the chimp skull grow much faster than others, while the proportions of the human skull

Figure 22.14
Examples of allometric growth.

a. Allometric growth in humans

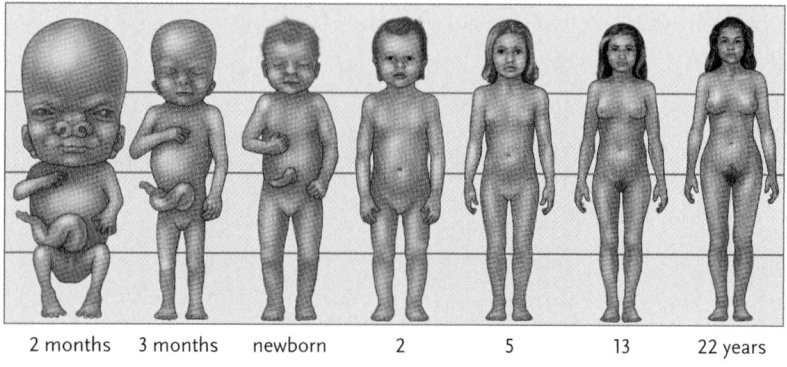

2 months   3 months   newborn   2   5   13   22 years

Humans exhibit allometric growth from prenatal development until adulthood. Our heads grow more slowly than other body parts; our legs grow faster.

b. Differential growth in the skulls of chimpanzees and humans

Changes in chimpanzee skull

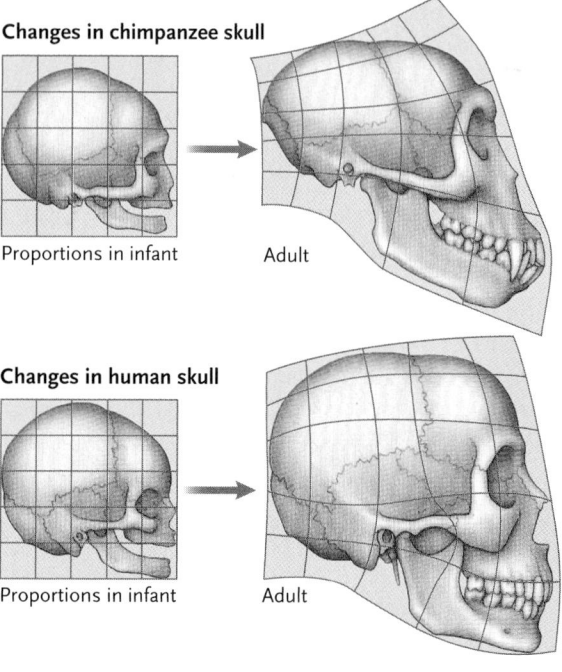

Proportions in infant   Adult

Changes in human skull

Proportions in infant   Adult

Although the skulls of newborn humans and chimpanzees are remarkably similar, differential patterns of growth make them diverge during development. Imagine that the skulls are painted on a blue rubber sheet marked with a grid. Stretching the sheet deforms the grid in particular ways, mimicking the differential growth of various parts of the skull.

Figure 22.15

Paedomorphosis in salamanders. Some small-mouthed salamanders *(Ambystoma talpoideum)* undergo metamorphosis, losing their gills and developing lungs (left). Others are paedomorphic: they retain juvenile morphological characteristics, such as gills, after attaining sexual maturity (right).

David Scott/SERL

STUDY BREAK

1. How have the sizes of organisms changed since life first appeared?
2. What processes can trigger the evolution of morphological novelties?

22.5 Macroevolutionary Trends in Biodiversity

The number of species living on Earth—its overall **biodiversity**—changes over time as a result of both adaptive radiation and extinction.

Adaptive Radiations Are Clusters of Related Species with Diverse Ecological Adaptations

In some lineages, rapid speciation produces a cluster of closely related species that occupy different habitats or consume different foods; we describe such a lineage as an **adaptive radiation.** The Galápagos finches **(Figure 22.17)** and the Hawaiian fruit flies described in Chapter 21 are examples of adaptive radiations.

Adaptive radiation usually occurs after an ancestral species moves into an unfilled **adaptive zone**, a general way of life. Browsing on soft leaves in the forest is the adaptive zone that early horses occupied, and grazing on grass in open habitats is the adaptive zone that horses occupy today. Feeding on plastic in landfills might become an adaptive zone in the future if some organism develops the ability to digest that now-abundant resource.

An organism may move into a new adaptive zone after the chance evolution of a key morphological innovation that allows it to use the environment in a unique way. For example, the dehydration-resistant eggs of early reptiles enabled them to complete their life cycle on land, opening terrestrial habitats to them. Similarly, the evolution of flowers that attract insect pollinators was a key innovation in the history of flowering plants.

An adaptive zone may also open up after the demise of a successful group. Mammals, for example, were relatively inconspicuous during their first 150 million years on Earth, presumably because dinosaurs dominated terrestrial habitats. But after dinosaurs declined in the late Mesozoic era, mammals underwent an explosive adaptive radiation. Today they are the dominant vertebrates in many terrestrial habitats.

Extinctions Have Been Common in the History of Life

Increased biodiversity is counteracted by **extinction,** the death of the last individual in a species or the last species in a lineage. Paleobiologists recognize two dis-

change much less. Differences in the adult skulls may simply reflect changes in one or a few genes that regulate the pattern of growth.

Changes in the timing of developmental events, called **heterochrony** *(hetero = different; chronos = time)*, also cause the morphology of closely related species to differ. **Paedomorphosis** *(paedo = child; morpho = form)*, the development of reproductive capability in an organism with juvenile characteristics, is a common form of heterochrony.

Many salamanders, for example, undergo metamorphosis from an aquatic juvenile into a morphologically distinct terrestrial adult. However, populations of several species are paedomorphic—they grow to adult size and become reproductively mature without changing to the adult form **(Figure 22.15)**. The evolutionary change causing these differences may be surprisingly simple. In amphibians, including salamanders, the hormone thyroxine induces metamorphosis (see Chapter 40). Paedomorphosis could result from a mutation that either reduces thyroxine production or limits the responsiveness of some developmental processes to thyroxine concentration.

Changes in developmental rates also influence the morphology of plants **(Figure 22.16)**. The flower of a larkspur species, *Delphinium decorum*, includes a ring of petals that guide bees to its nectar tube and structures on which bees can perch. By contrast, *Delphinium nudicaule*, a more recently evolved species, has tight flowers that attract hummingbird pollinators, which can hover in front of the flowers. Slower development in *D. nudicaule* flowers causes the structural difference: a mature flower in the descendant species resembles an unopened (juvenile) flower of the ancestral species.

Novel morphological structures, such as the wings of birds, often appear suddenly in the fossil record. How do novel features evolve? Scientists have identified several mechanisms including preadaptation, allometric growth, and heterochrony. We describe new research about the genetic basis of some morphological innovations in the last section of this chapter.

tinct patterns of extinction in the fossil record, back-ground extinction and mass extinction.

Species and lineages have been going extinct since life first appeared. We should expect species to disappear at some low rate, the **background extinction rate;** as environments change, poorly adapted organisms will not survive and reproduce. In all likelihood, more than 99.9% of the species that have ever lived are now extinct. David Raup of the University of Chicago has suggested that, on average, as many as 10% of species go extinct every million years and that more than 50% go extinct every 100 million years. Thus, the history of life has been characterized by an ongoing turnover of species.

On at least five occasions, extinction rates rose well above the background rate. During these **mass extinctions,** large numbers of species and lineages died out over relatively short periods of geological time (**Figure 22.18**). The Permian extinction was the most severe: more than 85% of the species alive at that time—including all trilobites, many amphibians, and the trees of the coal swamp forests—disappeared forever. During the last mass extinction, at the end of the Cretaceous, half the species on Earth, including most dinosaurs, became extinct. A sixth mass extinction, potentially the largest of all, may be occurring now as a result of human degradation of the environment (see Chapter 53).

Different factors were responsible for the five mass extinctions. Some were probably caused by tectonic activity and associated changes in climate. For example, the Ordovician extinction occurred after Gondwana moved toward the South Pole, triggering a glaciation that cooled the world's climate and lowered sea levels. The Permian extinction coincided with a major glaciation and a decline in sea level induced by the formation of Pangaea.

Many researchers believe that an asteroid impact caused the Cretaceous mass extinction. The resulting dust cloud may have blocked the sunlight necessary for photosynthesis, setting up a chain reaction of extinctions that began with microscopic marine organisms. Geological evidence supports this hypothesis. Rocks dating to the end of the Cretaceous period (65 million years ago) contain a highly concentrated layer of iridium, a metal that is rare on Earth but common in asteroids. The impact from an iridium-laden asteroid only 10 km in diameter could have caused an explosion equivalent to that of a billion tons of TNT, scattering iridium dust around the world. Geologists have identified the Chicxulub crater, 180 km in diameter, on the edge of Mexico's Yucatán peninsula as the likely site of the impact.

Although scientists agree that an asteroid struck Earth at that time, many question its precise relationship to the mass extinction. Dinosaurs had begun their decline at least 8 million years earlier, but many persisted for at least 40,000 years after the impact. More-

Figure 22.16 Observational Research

Paedomorphosis in Delphinium *Flowers*

HYPOTHESIS: The narrow tubular shape of the flowers of *Delphinium nudicaule*, which are pollinated by hummingbirds, is the product of paedomorphosis, the retention of juvenile characteristics in a reproductive adult.

PREDICTION: The flowers of *D. nudicaule* grow more slowly and mature at an earlier stage of development than those of *Delphinium decorum*, a species with broad, open flowers that are pollinated by bees.

D. decorum   *D. nudicaule*

METHOD: Edward O. Guerrant of the University of California at Berkeley measured 42 bud and flower characteristics in *D. nudicaule* and *D. decorum* as their flowers developed and used the number of days since the completion of meiosis in pollen grains as a measure of flower maturity. He then used a complex statistical analysis to compare the characteristics of the buds and flowers of both species.

RESULTS: The mature flowers of *D. nudicaule* resemble the buds of both species more closely than they resemble the flowers of *D. decorum*. Although the time required for maturation of the reproductive structures is similar in the two species, the rate of petal growth (measured as petal blade length) is slower in *D. nudicaule*. As a result, the mature flowers of *D. nudicaule* do not open as widely as those of *D. decorum*. Because of these morphological differences, bees can pollinate flowers of *D. decorum*, but they can't land on the flowers of *D. nudicaule*, which are instead pollinated by hummingbirds.

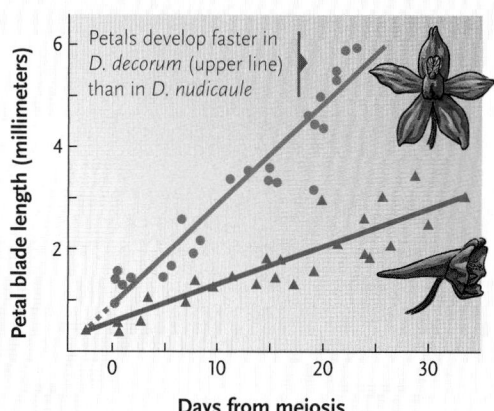

CONCLUSION: The narrower and more tubular shape of *D. nudicaule* flowers, which mature at an earlier stage of development than *D. decorum* flowers, is the product of paedomorphosis.

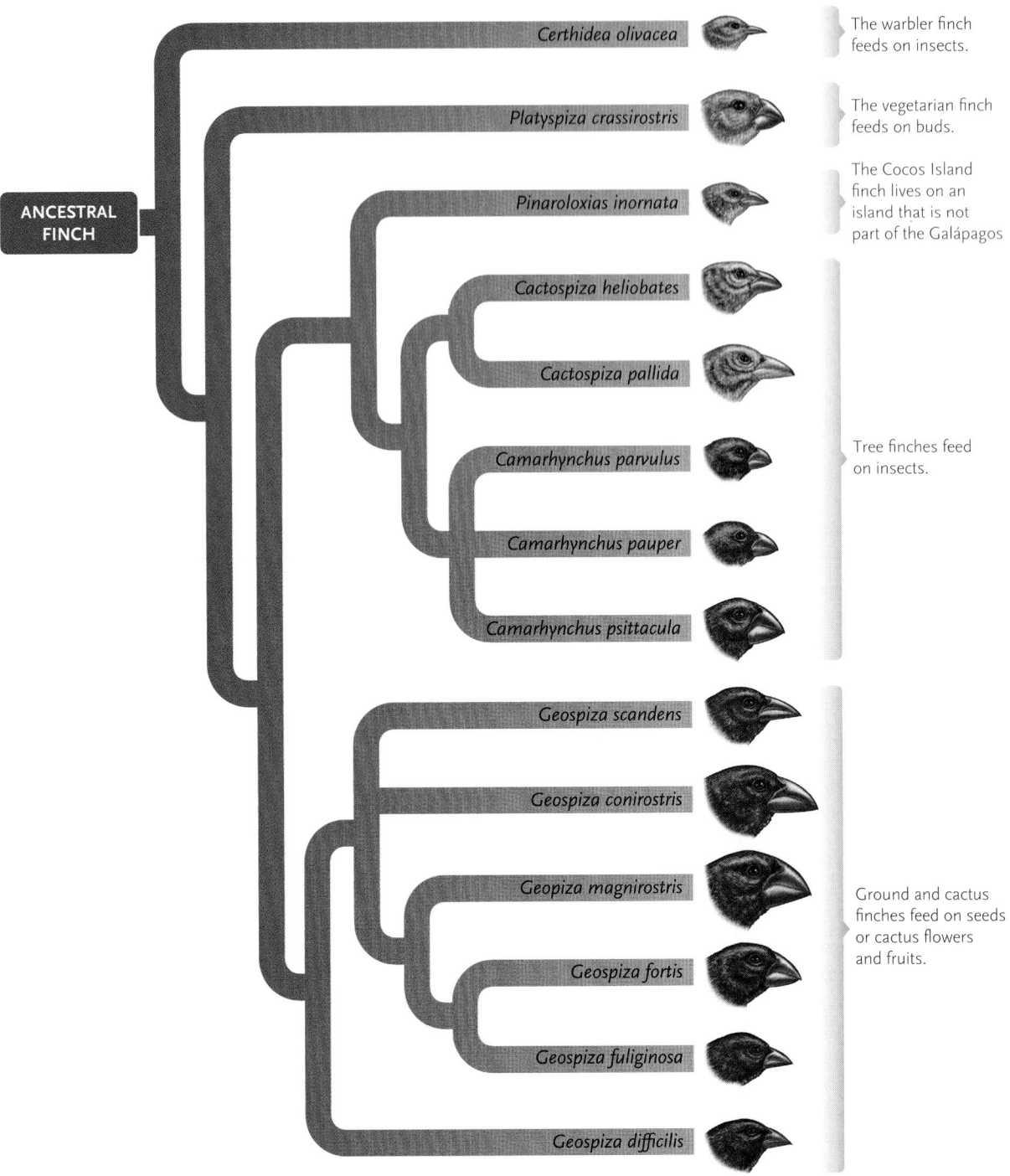

Certhidea olivacea — The warbler finch feeds on insects.

Platyspiza crassirostris — The vegetarian finch feeds on buds.

Pinaroloxias inornata — The Cocos Island finch lives on an island that is not part of the Galápagos

Cactospiza heliobates

Cactospiza pallida

Camarhynchus parvulus

Camarhynchus pauper

Camarhynchus psittacula

Tree finches feed on insects.

Geospiza scandens

Geospiza conirostris

Geopiza magnirostris

Geospiza fortis

Geospiza fuliginosa

Geospiza difficilis

Ground and cactus finches feed on seeds or cactus flowers and fruits.

ANCESTRAL FINCH

Figure 22.17
Adaptive radiation. The 14 species of Galápagos finches are descended from one ancestral species.

over, other groups of organisms did not suddenly disappear, as one would expect after a global calamity. Instead, the Cretaceous extinction took place over tens of thousands of years.

Biodiversity Has Increased Repeatedly over Evolutionary History

Although mass extinctions temporarily reduce biodiversity, they also create evolutionary opportunities. Some species survive because they have highly adaptive traits, large population sizes, or widespread distributions. And some surviving species undergo adaptive

radiation, filling adaptive zones that mass extinctions made available.

Sometimes the success of one lineage comes at the expense of another. Although the diversity of terrestrial vascular plants has increased almost continuously since the Devonian period, this trend includes booms and busts in several lineages **(Figure 22.19)**. Ferns and conifers recovered rapidly after the Permian extinction, maintaining their diversity until the end of the Mesozoic era. However, angiosperms, which arose and diversified in the late Jurassic and early Cretaceous periods, may have hastened the decline of these groups by replacing them in many environments.

The superb fossil record left by certain marine animals reveals three major periods of adaptive radiation **(Figure 22.20).** The first occurred during the Cambrian, more than 500 million years ago, when all animal phyla, the major categories of animal life, arose. Most of these phyla became extinct, and a second wave of radiations established the dominant Paleozoic fauna during the Ordovician period. A third evolutionary fauna emerged in the Triassic period, right after the great Permian extinction; it produced the immediate ancestors of modern marine animals. The diversity of marine animals has increased consistently since the early Triassic, in large measure because of continental drift. As continents and shallow seas became increasingly isolated, regional biotas diversified independently of one another, increasing worldwide biodiversity.

Historical increases in biodiversity can also be attributed to the evolution of ecological interactions. For example, the number of plant species found *within* fossil assemblages has increased over time, suggesting the evolution of mechanisms that allow more species to coexist. In addition, insects diversified dramatically in the Cretaceous period, possibly because the angiosperms created a new adaptive zone for them. New insect species then provided a novel set of pollinators that may have stimulated the radiation of angiosperms. Such long-term evolutionary interactions between ecologically intertwined lineages have played an important role in structuring ecological communities, which are described more fully in Chapter 50.

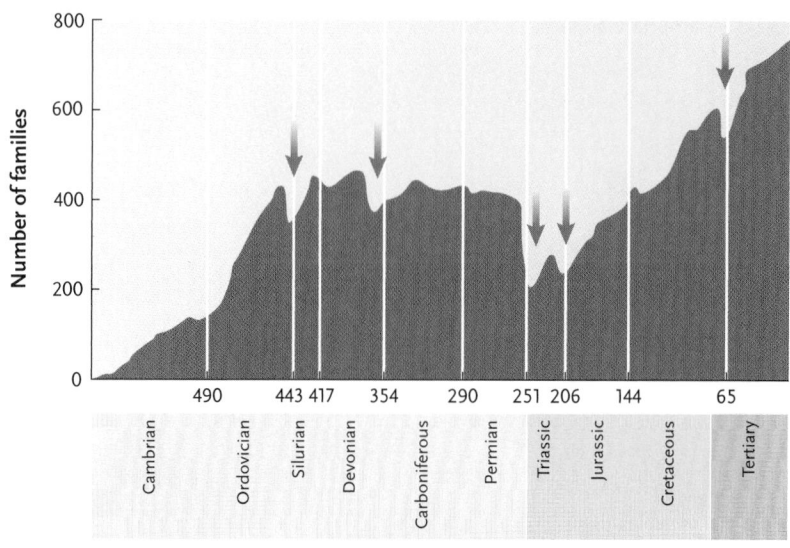

Figure 22.18

Mass extinctions. Biodiversity, indicated by the height of the dark blue area in the graph, was temporarily reduced by at least five mass extinctions (arrows) during the history of life. The data presented in this graph record the family-level diversity of marine animals. A family is a group of genera descended from a common ancestor.

Study Break

1. What factors might allow a population of organisms to occupy a new adaptive zone?
2. Did the mass extinction at the end of the Cretaceous period occur quickly or over a long period of time?
3. When did the first major adaptive radiation of animals occur?

22.6 Evolutionary Developmental Biology

Historically, evolutionary biologists compared the embryos of different species to study their evolutionary history (see Figure 19.13), but they often worked independently from scientists studying the embryonic development of organisms. As a result, evolutionary biologists were unable to construct a coherent picture of the specific developmental mechanisms that contributed to morphological innovations. Since the late 1980s, however, advances in molecular genetics have allowed scientists to explore the genomes of organisms in great detail, fostering a new approach to these stud-

ies. **Evolutionary developmental biology**—evo-devo, for short—asks how evolutionary changes in the genes regulating embryonic development can lead to changes in body shape and form.

The study of the genetics of embryonic development helps us understand macroevolutionary trends

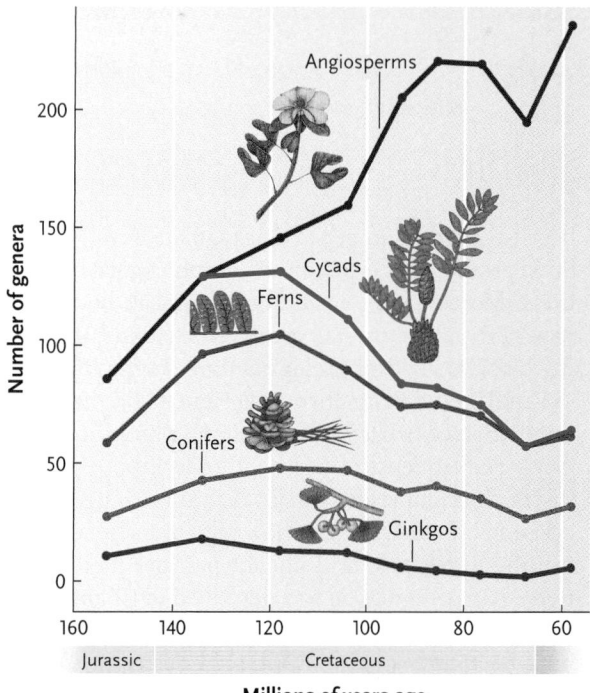

Figure 22.19

History of vascular plant diversity. The diversity of angiosperms increased during the Mesozoic era as the diversity of other groups declined.

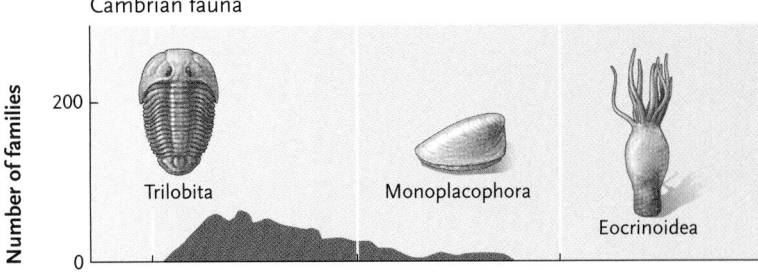

Cambrian fauna

Trilobita

Monoplacophora

Eocrinoidea

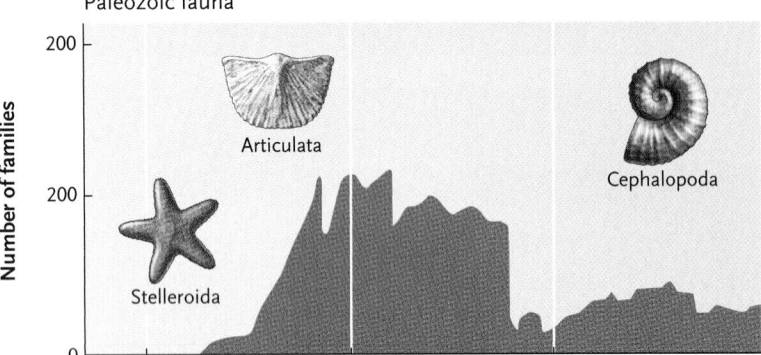

Paleozoic fauna

Articulata

Cephalopoda

Stelleroida

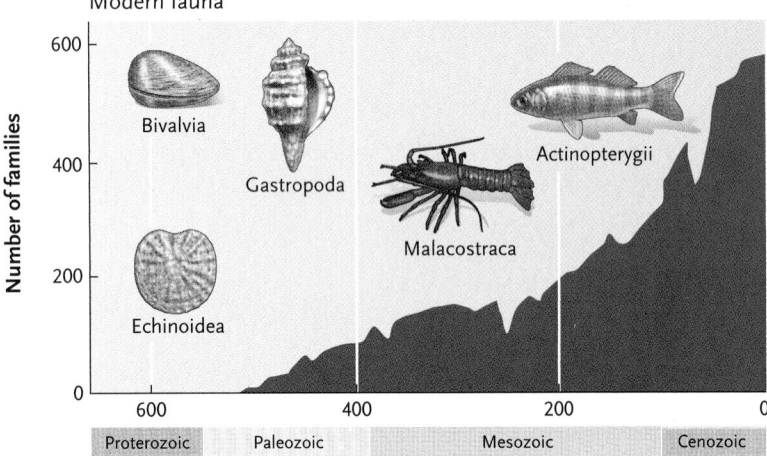

Modern fauna

Bivalvia

Gastropoda

Actinopterygii

Malacostraca

Echinoidea

| Proterozoic | Paleozoic | Mesozoic | Cenozoic |

Millions of years ago

Figure 22.20

History of marine animal diversity. Marine animals have undergone three major radiations. Few remnants of the Cambrian fauna remain alive today.

because changes in genes that regulate development often promote the evolution of morphological innovations. Moreover, the resulting changes in body plan have sometimes fostered adaptive radiations, increasing biodiversity over geological time. In the life cycle of a multicellular organism (see Figure 1.7), the many different body parts of the adult develop in a highly controlled sequence of steps that is specified by genetic instructions in the single cell of a fertilized egg. Developmental biologists study how regulatory genes control the development of phenotypes and their variations. (Gene regulation was described in Chapter 16.) When these genes code for transcription factors that bind regulatory sites on DNA, either activating or repressing the expression of other genes that contribute to an organism's form, they are called **homeotic genes** (described further in Chapters 34 and 48). In this section we describe a few intriguing discoveries about the genetic mechanisms that underlie some macroevolutionary trends in animals.

Most Animals Share the Same Genetic Tool Kit That Regulates Their Development

Comparisons of genome sequence data reveal that most animals, regardless of their complexity or position in the tree of life, share a set of several hundred homeotic genes that control their development. Collectively, these genes have been dubbed the "genetic tool kit," because they govern the basic design of the body plan by controlling the activity of thousands of other genes. Some of the tool-kit genes must be at least 500 million years old, because all living animals inherited them from a common ancestor alive at that time. Some of the same tool-kit genes are also present in plants, fungi, and prokaryotes, suggesting that those genes may date back to the earliest forms of life.

Structurally, tool-kit genes do not differ much among the animals that possess them, and they generally play the same role in development for all species. For example, genes in the *Hox* family control the overall body plan of animals. All *Hox* genes include a 180-nucleotide sequence called a **homeobox,** which codes for a **homeodomain,** part of a protein that functions as a transcription factor. When bound to a regulatory site on a strand of DNA, the homeodomain either activates or represses a downstream gene involved in development.

Among other functions, *Hox* genes specify where appendages—wings in flies and legs in mice—will develop on the animal's body. They do so by producing transcription factors that activate the genes specifying wings or legs in the body regions where these appendages typically grow. The different *Hox* genes, which are expressed at different positions along the head-to-tail axis of a developing embryo, are arranged on a chromosome in the same sequence in which they are expressed in the body. Remarkably, the *Hox* genes and their relative positions on chromosomes have been conserved by evolution; nearly identical genes are found in animals as different as fruit flies and mice **(Figure 22.21).** Genes with comparable functions control aspects of development in plants (see Chapter 34).

Another example of a highly conserved and widely distributed tool-kit gene, the *Pax-6* gene, triggers the formation of light-sensing organs as diverse as the eye spots in flatworms, the compound eyes of insects and other arthropods, and the camera eyes of vertebrates (see Chapter 39). Like the *Hox* genes, *Pax-6* also contains a homeobox, indicating that the protein for which it codes either activates or represses gene transcription. The proteins coded by *Pax-6* in different animals are so similar that when researchers genetically engineered fruit fly larvae to express the *Pax-6* gene taken from a squid or a mouse, the flies responded by developing eyes. The induced eyes were, however, fruit fly eyes—not squid eyes or mouse eyes. Thus, *Pax-6* triggers activity in the genes that carry the specific instructions for making an eye typical of the species. Apparently, the ancient genetic sequence for *Pax-6,* the master regulatory gene for eye development, has been conserved over the hundreds of millions of years since the common ancestor of squids, fruit flies, and mice lived.

Evolutionary Changes in Developmental Switches May Account for Much Evolutionary Change

If most animals share the same tool-kit genes, how has evolution produced different body plans among species? What makes a squid, a fruit fly, and a mouse different? Researchers in evo-devo have proposed that morphological differences among species arise when mutations alter the effects of developmental regulatory genes. As you will discover in Chapter 48, the developmental programs of animals involve complex networks of many interacting genes. Varying combinations of tool-kit genes may be expressed at different times and in different body regions. According to this hypothesis, the several hundred tool-kit genes encode proteins that work either as activators or repressors in a multitude of possible combinations. Thus, they can generate an unimaginably large number of different gene expression patterns, each with the potential to alter morphology.

Sean Carroll of the Howard Hughes Medical Institute and the University of Wisconsin at Madison has described the regulatory sites that transcription factors can bind as *switches,* like those we use to turn lights on or off. When a combination of transcription factors turns on a regulatory switch, a gene further downstream is activated. When transcription factors turn off a regulatory switch, a downstream gene is inactivated.

Although all the cells in an animal contain exactly the same set of genes, the differential expression of genes in different body regions and at different times during embryonic development causes different structures to be made. Allometric growth can result from evolutionary changes in developmental switches that cause certain body parts to grow larger or faster than others. Similarly, heterochrony can be explained as an evolutionary change in the switches that either delays the development of adult characteristics or speeds up the development of reproductive maturity.

If Carroll's hypothesis is correct, morphological novelties arise when evolutionary changes in developmental switches alter the expression patterns of *existing* genes. This view contrasts markedly with the explanation proposed in the modern synthesis (the unified theory of evolution described in Chapter 19), that most morphological novelties arise as mutations slowly accumulate in genes that carry the blueprints for building particular structures. According to the modern synthe-

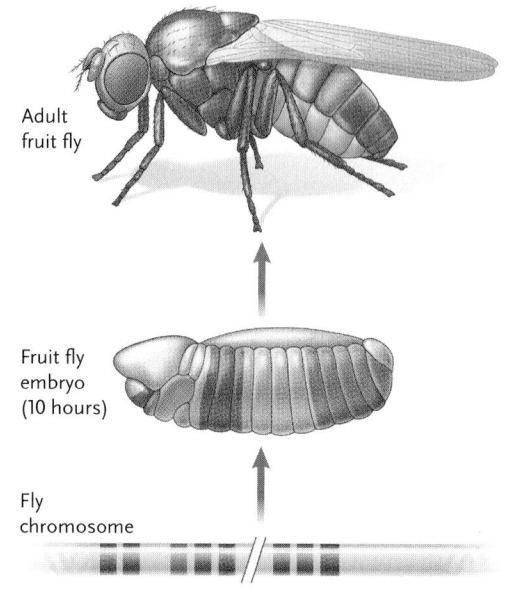
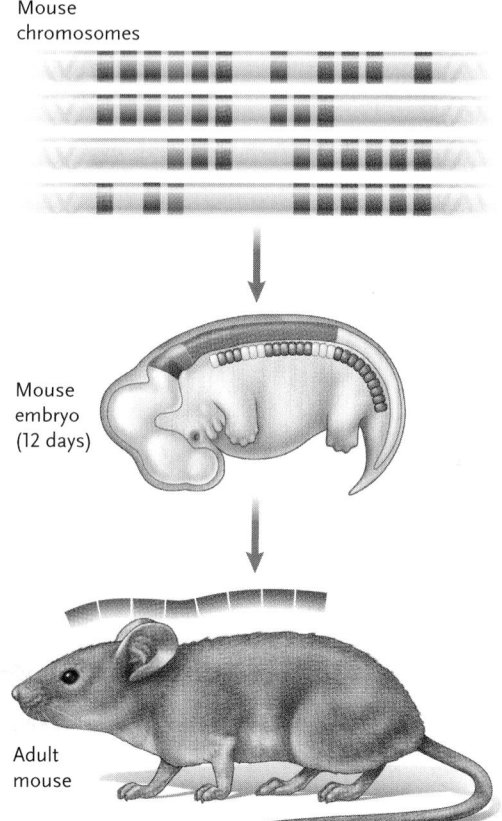

Figure 22.21

Hox genes. The linear sequence of *Hox* genes on chromosomes and the expression of *Hox* genes in different body regions have been conserved by evolution. Each color-coded band on the chromosomes in the illustration represents a different gene in the *Hox* family of genes. Fruit flies have one set of *Hox* genes, which are arranged on a single chromosome in the same order that they are expressed in the fruit fly embryo. Like all mammals, mice have four sets of *Hox* genes, arranged on four chromosomes that are expressed in mouse embryos in the same order as the *Hox* genes in fruit flies. The illustrations of the adult fruit fly and mouse show the adult body regions that are influenced by the expression of *Hox* genes in their embryos.

Fancy Footwork from Fins to Fingers

Both fishes and tetrapods (four-footed animals) have two pairs of appendages, one anterior and one posterior.

a. Fishes

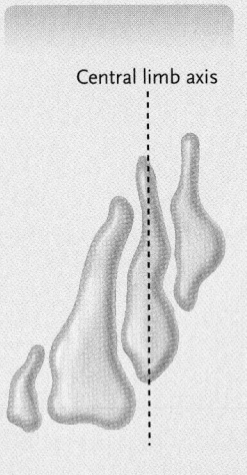

Central limb axis

Bones in the fin of a fish develop from centers of cartilage formation along a central axis (dashed line).

b. Tetrapods

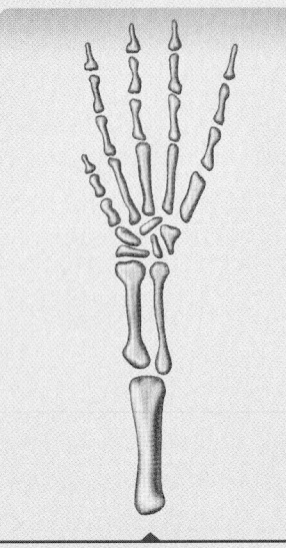

Bones in the limb and digits of a tetrapod also develop from centers of cartilage formation in the central axis.

c. Fishes   d. Tetrapods

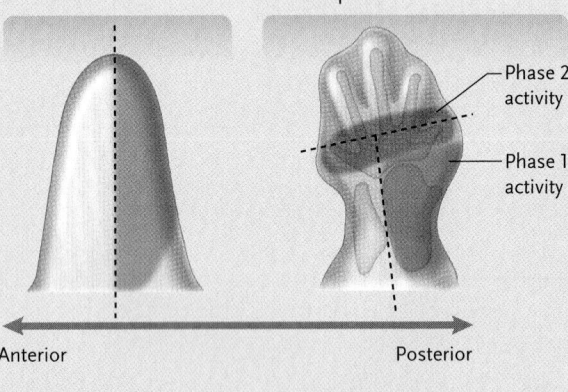

Phase 2 activity

Phase 1 activity

Anterior   Posterior

During development of the fin in fishes, *HoxD* genes become active in cells posterior to the central axis of the fin (shown in blue).

During development of the limb and digits in tetrapods, *HoxD* genes first become active in cells posterior to the central axis of the limb (blue). Later, these genes are active in a band of cells perpendicular to the central axis of the limb (green).

They develop similarly in both groups during early embryonic stages. They start out as buds of mesoderm—the middle of the three primary embryonic tissue layers—and thicken by increased cell division. As the buds elongate, cartilage is deposited at localized centers, and bones of the appendages later form in these centers.

In fishes, the bones develop along a central axis from the base to the tip of the limb **(Figure a)**. In tetrapods, the centers of cartilage formation generate the long bones of the limb and the five digits of the foot **(Figure b)**. In humans, the digits are the thumb and fingers of the hand or the toes of the foot.

For some time, evolutionists wondered whether the digits of tetrapods were modifications of the bones radiating from the central limb axis in fishes or novel evolutionary structures. Molecular research by Paolo Sordino, Frank van der Hoeven, and Denis Duboule at the University of Geneva in Switzerland indicates that animal digits are a morphological novelty.

In all animals with paired anterior and posterior appendages, groups of homeobox genes control their development. Sordino and his colleagues compared the activity of a group called the *HoxD* genes in the zebrafish *Danio rerio* (a common aquarium fish and a model research organism) with the previously known *HoxD* patterns in birds and rodents.

To begin their work, the researchers used the DNA of a rodent *HoxD* gene as a probe to search for similar genes in fragmented zebrafish DNA. The probe paired with fragments of zebrafish DNA that, when cloned and sequenced, proved to include three *HoxD* genes—*HoxD-11*, *HoxD-12*, and *HoxD-13*—arranged in the same order

as in rodents. They tested the activity of the *HoxD* genes in developing zebrafish limbs by using a nucleic acid probe that could pair with mRNA products of the genes. The probe was linked to a blue dye molecule, so that cells in which a particular *HoxD* gene was active would appear blue in the light microscope. The investigators found that in zebrafish, the *HoxD* genes became active in cells along the posterior side of the central axis **(Figure c)**. As fin development neared completion, activity of the *HoxD* genes dropped off.

Using the same techniques to study tetrapods, the investigators found that the *HoxD* genes were activated in two distinct phases **(Figure d)**. In phase 1, gene activity was restricted to the posterior half of the limb, as it is in zebrafish; this period of activity corresponds to development of the long limb bones. Later, in phase 2, the *HoxD* genes became active in a band of cells perpendicular to the central axis; the cartilage centers that form the bones of the digits develop in this anterior-posterior band. Sordino and his colleagues found no equivalent band of activity in zebrafish; the *HoxD* gene activity remained restricted to a single phase along the posterior half of the fin.

Thus, the phase of *HoxD* gene activity corresponding to development of the digits is a separate pattern unique to tetrapods; it therefore appears to be a morphological novelty. If this is the case, fishes probably have no bones homologous to the five digits of the hand and foot, which were added as new structures during the evolutionary events that split the ancestors of fishes from those of four-footed animals.

sis, the accumulated mutations eventually create *new* genes that specify the creation of new structures.

Although Carroll's hypothesis argues that changes in genes regulating development cause most morphological change, proponents of evo-devo recognize that mutations in developmental regulatory genes and their effects on morphology are subject to the action of the

same microevolutionary processes—natural selection, genetic drift, and gene flow—that influence the frequencies of genotypes and phenotypes in populations. Thus, every morphological change induced by a mutation in a homeotic gene or in a developmental switch is tested by the success or failure of the individual that carries it.

Numerous studies have shown that changes in the expression of homeotic genes can have dramatic effects on morphology. *Insights from the Molecular Revolution* explains how a change in the number and expression of *Hox* genes produced a striking alteration in the structure of vertebrate appendages.

In another example, researchers have determined how an adaptive morphological change in a small fish, the three-spined stickleback *(Gasterosteus aculeatus),* results from the deactivation of a homeotic gene. The freshwater stickleback populations in North American lakes are the descendants of marine ancestors that colonized the lakes after the retreat of glaciers between 10,000 and 20,000 years ago. Marine sticklebacks have bony armor along their sides and prominent spines; lake-dwelling sticklebacks have greatly reduced armor and, in many populations, lack spines on their pelvic fins **(Figure 22.22).**

Natural selection has apparently fostered these morphological differences in response to the dominant predators in each habitat. In marine environments, long spines prevent some predatory fishes from

UNANSWERED QUESTIONS

Does morphological evolution always proceed gradually or can it occur in great leaps and bounds?

As you read in this chapter, biologists disagree about whether evolutionary changes in morphology can occur very rapidly. Although biologists have proposed various hypotheses to explain the abrupt changes that we sometimes find in the fossil record, evo-devo studies provide insight into one mechanism for how dramatic changes can arise: the spatial redeployment of homeotic genes. *Homeosis* is defined as the complete replacement of one type of organ with another. In one famous example, a *Hox* gene mutation in *Drosophila* replaces the antennae with legs. If such a mutation were to occur in nature, the organism would probably not have a selective advantage. But what if it did? These kinds of mutant phenotypes first inspired Richard Goldschmidt to develop his idea of the "hopeful monster" early in the twentieth century. Stephen Jay Gould later revised and updated this idea in the context of his punctuated equilibrium hypothesis about the tempo and mode of evolution. The hypothesis suggests that if, on very rare occasions, truly dramatic morphological changes provide a selective advantage, they may lead to the rapid formation of a new species based on only a few genetic differences. What types of organisms are the most likely to exhibit homeotic change in an evolutionary context? The best candidates are those with highly modular bodies made up of repeating units—like the segments of an insect or the bones in the spine of a vertebrate. Such animals often express different organ identities in different modular units—such as the antennae, claws, and legs of a lobster—and these identities may be redeployed to different positions along the body axis. Plants are among the most modular organisms on Earth. They produce serially repeated structures—a leaf, a bud at the base of the leaf, and a stem—to generate their bodies. Many exciting and promising questions in plant evo-devo relate to how evolutionary homeosis may have generated rapid change in plant morphology.

Has homeosis contributed to the appearance and diversification of the angiosperms?

The sudden appearance of flowering plants, the angiosperms, in the fossil record so puzzled Charles Darwin that he dubbed their evolution an "abominable mystery." How did the gymnosperms, which always bear their male and female reproductive structures separately, give rise to the hermaphroditic (that is, bearing both male and female structures) flower? Our current understanding of the genetics of floral developmental provides a simple solution to this puzzle: the genetic program controlling floral organ identity is homeotic. Thus, it is possible for very simple genetic changes to transform an entirely male set of reproductive organs into a combination of male and female parts. Such models have been outlined by Günter Theissen at the Friedrich-Schiller-Universität in Germany as well as David Baum and Lena Hileman at the University of Wisconsin and University of Kansas, respectively. In addition to fostering the origin of the angiosperms, homeosis may have played a role in the group's diversification. Commonly observed shifts in the morphology of sepals and petals or in the number of stamens (male reproductive structures) are suggestive of homeotic changes. These examples are more suitable for experimental verification than the question on the origin of the angiosperms is, because they are much more recent occurrences. Although these hypotheses are very attractive, they remain to be confirmed through molecular genetic analyses.

How have new floral organ identity programs evolved?

The homeotic scenarios described here involve spatial shifts in the expression of preexisting identity programs, such as a stamen developing where there was previously a petal. But what about cases where a whole new type of floral organ appears? Across the angiosperms there are many examples of flowers that have more than the four most common types of organs—sepals, petals, stamens, and carpels, described in Chapter 34—which suggests that new organ identity programs must have evolved. In such instances, do the new organs evolve through modification of preexisting identity programs, or are entirely new gene pathways recruited? Studies using a new model plant for genetics, *Aquilegia*—commonly known as columbine—suggest that a fifth type of floral organ has evolved through modification of the stamen identity program. Many questions about how this process actually occurred remain unanswered. Was the derivation of the new program achieved through just a few genetic changes, or many? Did it involve changes in regulatory gene function, or only shifts in gene expression? There is much more to learn about how completely new types of floral organs have evolved.

Elena M. Kramer is the John L. Loeb Associate Professor of Biology at Harvard University, where she studies the evolution of the genetic mechanisms controlling floral development. To learn more about Dr. Kramer's research go to http://www.oeb.harvard.edu/faculty/kramer/index.htm.

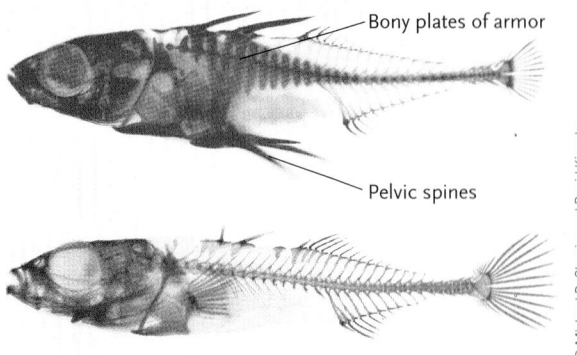

Figure 22.22

Sticklebacks. Marine populations (top) of three-spined sticklebacks (*Gasterosteus aculeatus*) have prominent bony plates along their sides and large spines on their dorsal and pelvic fins. Many freshwater populations of the same species (bottom) lack the bony plates and spines. Pelvic spines do not develop in the freshwater fishes because they do not express the *Pitx1* gene in their fin buds during embryonic development. The skeletons of these specimens, each about 8 cm long, were dyed bright red.

Bony plates of armor

Pelvic spines

© Michael D. Shapiro and David Kingsley

glands and sensory organs in the head. In long-spined marine sticklebacks, *Pitx1* is expressed in the embryonic buds from which pelvic fins develop, promoting the development of spines. But *Pitx1* is not expressed in the fin buds of the freshwater sticklebacks; hence, pelvic spines do not develop. However, freshwater sticklebacks have not *lost* the *Pitx1* gene; it is still expressed elsewhere in the fishes' bodies. Apparently, a mutation somehow blocks its expression in the developing pelvic fin, thereby blocking the production of pelvic spines.

In the next chapter, we examine how biologists explore the evolutionary relationships among the many species they encounter and how they organize that information into a useful framework for researchers in every biological discipline. In the following unit, we will revisit evo-devo and examine some recent discoveries about how changes in homeotic genes have diversified body plans in the major evolutionary groups of organisms.

swallowing sticklebacks. But long spines are a liability in lakes, where voracious dragonfly larvae grab hold of sticklebacks by their spines and then devour them; freshwater sticklebacks that lack spines are more likely to escape from their clutches.

The presence or absence of spines on the pelvic fins of these fishes is governed by the expression of the gene *Pitx1*. Pelvic spines are part of the pelvic fin skeleton, the fishes' equivalent of a hind limb. In fact, *Pitx1* also contributes to the development of hind limbs in four-legged vertebrates as well as certain

Study Break

1. What evidence suggests that many developmental control genes have been conserved by evolution?
2. What genetic factor is apparently responsible for the presence or absence of spines in stickleback fish?

Review

Go to **ThomsonNOW**™ at www.thomsonedu.com/login to access quizzing, animations, exercises, articles, and personalized homework help.

22.1 The Fossil Record

- Fossils are the parts of organisms preserved in sedimentary rocks or in oxygen-poor environments (Figure 22.2).
- The fossil record is incomplete because few organisms fossilize completely, because some organisms are more likely to fossilize than others, and because natural processes destroy many fossils.
- Fossils provide a relative dating system, the geological time scale, for the strata in which they occur (Figure 22.3). Radiometric dating techniques establish the absolute age of rocks and fossils (Figure 22.4 and Table 22.1).
- The fossil record provides data on changes in morphology, biogeography, ecology, and behavior of organisms; some fossils also contain biological molecules.

Animation: Radioisotope decay

Animation: Radiometric dating

Animation: Geologic time scale

22.2 Earth History, Biogeography, and Convergent Evolution

- Earth's crust is composed of plates of solid rock that float on a semisolid mantle (Figure 22.5). New crust is constantly generated and old crust is recycled, and currents in the mantle cause the continents to move over geological time (Figure 22.6). Continental movements cause variations in patterns of glaciation, sea level, and climate. Asteroid impacts and volcanic eruptions have also influenced the environment.
- Disjunct distributions of species are produced by dispersal and vicariance. Dispersal results in a disjunct distribution when a new population is established on the far side of a barrier. Vicariance results in a disjunct distribution when external factors such as continental drift fragment the landscape.
- Continental drift has created six major biogeographical realms, each with a characteristic biota (Figure 22.7).
- Convergent evolution produces similar adaptations in distantly related species that live in similar environments (Figures 22.8 and 22.9).

Animation: Plate margins

Animation: Geologic forces

22.3 Interpreting Evolutionary Lineages

- The horse lineage is complex and highly branched. It includes species of various sizes and diverse morphological adaptations (Figure 22.10).

- Anagenesis and cladogenesis produce morphological change (Figure 22.11). Anagenesis is the accumulation of many small changes in a species over long periods of time. Cladogenesis is the evolutionary division of an ancestral species into multiple descendant species. Only cladogenesis increases the number of species living at a particular time.

- The gradualist hypothesis of evolution suggests that major morphological changes result from the accumulation of small changes over long periods of time. The punctuated equilibrium hypothesis suggests that most morphological evolution occurs during short periods of cladogenesis. Both patterns occur in nature (Figures 22.12 and 22.13).

Animation: Evolutionary tree diagrams

22.4 Macroevolutionary Trends in Morphology

- In many lineages, body size has increased over evolutionary history. Evolutionary trends may be produced by species selection if certain traits are associated with high rates of speciation.

- In a general way, morphological complexity has also increased over evolutionary time, although certain morphological features have become simplified in some lineages.

- A preadaptation is a trait that turns out to be useful in a new environmental context even before natural selection refines its form. Morphological novelties can arise from evolutionary changes in the relative growth rates of body parts (Figure 22.14), the timing of developmental events (Figures 22.15 and 22.16),

or changes in homeobox genes that control developmental processes.

Animation: Morphological divergence

22.5 Macroevolutionary Trends in Biodiversity

- Adaptive radiation produces morphological diversity within lineages (Figure 22.17).

- Extinction decreases species diversity. Mass extinctions have occurred at least five times in the history of life (Figure 22.18). Tectonic activity, climatic change, and asteroid strikes are probable causes of mass extinctions.

- Biodiversity has increased since life first evolved, partly in response to increased geographical separation of the continents and partly because complex interactions evolve among existing species (Figures 22.19 and 22.20).

22.6 Evolutionary Developmental Biology

- Evolutionary developmental biology—evo-devo—examines how evolutionary changes in genes that regulate embryonic development can foster changes in body shape and form.

- Most organisms share an ancient tool kit of several hundred genes that regulate the expression of thousands of genes involved in development (Figure 22.21). The tool-kit genes produce transcription factors that collectively activate or repress genes in a complex developmental network. *Hox* genes control aspects of the overall body plan of animals, and the *Pax-6* gene triggers the development of light-sensing organs.

- Evolutionary changes in developmental switches may account for many morphological changes. The differential expression of the *Pitx1* gene in sticklebacks determines whether or not a fish grows pelvic spines (Figure 22.22).

Questions

Self-Test Questions

1. The fossil record:
   a. provides direct evidence about life in the past.
   b. supports the punctuated equilibrium hypothesis, but not the gradualist hypothesis.
   c. provides abundant data about rare species with local distributions.
   d. is equally good for all organisms that ever lived.
   e. provides no evidence about the physiology or behavior of ancient organisms.

2. The absolute age of a geological stratum is determined by:
   a. the thickness of its rocks.
   b. the particle size in its rocks.
   c. the types of fossils found within it.
   d. anagenetic analysis.
   e. radiometric dating techniques.

3. The observation that fossils of *Premedosaurus* are found only in Argentina and Northern Europe provides an example of:
   a. a continuous distribution.
   b. a disjunct distribution.
   c. species selection.
   d. allometry.
   e. gradualism.

4. The evolutionary history of horses demonstrates that:
   a. modern horses are the direct, lineal descendants of the earliest horses.
   b. the leg bones of modern horses are more complex than those of the earliest horses.

   c. horses have always had specialized teeth that allow them to feed on tough grasses.
   d. horses diversified greatly, but only a few types survived to the present.
   e. the first horses lived in open, grassy habitats.

5. The punctuated equilibrium hypothesis:
   a. recognizes that morphological evolution may occur slowly or quickly.
   b. suggests that major morphological novelties can arise by anagenesis.
   c. may help explain why there are so many "missing links" in the fossil record.
   d. suggests that the fossil record is usually complete.
   e. links mass extinctions to the impact of asteroids striking Earth.

6. Macroevolutionary trends in body size could be caused by:
   a. plate tectonics.   d. heterochrony.
   b. paedomorphosis.   e. convergent evolution.
   c. species selection.

7. The differential growth of body parts is called:
   a. allometry.   d. cladogenesis.
   b. paedomorphosis.   e. preadaptation.
   c. heterochrony.

8. Preadaptations are traits that:
   a. prepare some organisms for future environmental changes.
   b. appear in lineages as a result of an adaptive radiation.
   c. evolve in anticipation of a species' future needs.

d. are useful in new situations before natural selection changes them.

e. occur in animals, but not in plants.

9. Adaptive radiations often follow mass extinctions because:

    a. mass extinctions limit the impact of species selection.

    b. mass extinctions foster allometry and heterochrony.

    c. mass extinctions decimate all forms of life on Earth.

    d. species that undergo frequent cladogenesis survive mass extinctions.

    e. extinctions open adaptive zones that had been previously occupied.

10. Homeotic genes are defined as genes that:

    a. bind directly to a regulatory site on DNA.

    b. code for transcription factors activating or repressing genes that influence an organism's form.

    c. determine whether or not a morphological innovation leads to an adaptive radiation.

    d. have been inherited from an ancient ancestor by nearly all forms of life.

    e. help biologists differentiate between plants and animals.

Questions for Discussion

1. Many millions of years from now, continental drift may obliterate the Pacific Ocean, pushing North America into physical contact with Asia. What effects might these events have on the organisms living at that time?

2. The species selection hypothesis measures evolutionary success in terms of the number of descendant species that a given species produces. Should our species, *Homo sapiens*, be considered successful under this definition?

3. Extinctions are common in the history of life. Why are biologists alarmed by the current wave of extinctions caused by human activity?

Experimental Analysis

Design a study to determine whether the wings of birds, bats, and insects and their ability to fly are the products of convergent evolution.

Evolution Link

The geological evolution of Earth has had an obvious effect on biological evolution. Consider the reverse: How has the evolution of different organisms, such as photosynthetic microorganisms or humans, changed the physical environment on Earth?

How Would You Vote?

Scientifically important fossils are sometimes found on privately owned land, creating disputes about who owns the fossils and how they should be used. For example, ownership of "Sue," the largest *Tyrannosaurus* fossil ever discovered, had to be settled in a court of law. Although the fossil was unearthed on a privately owned ranch on a Sioux Indian reservation, the land was held in trust by the U.S. government. The government argued that the fossil was public property, but the court eventually decided that the rancher owned the fossil. He could keep it or dispose of it however he chose. He sold it at auction for more than $8 million, the highest price ever paid for a fossil. A group of corporate sponsors raised the funds to buy the fossil on behalf of the Field Museum in Chicago, where it is now on public display. Do you think that scientifically important specimens should be the property of any one individual, or should they belong to the government, a museum, or some other research institution? Go to www.thomsonedu.com/login to investigate both sides of the issue and then vote.

A new plant species from Idaho. Sacajawea's bitterroot (*Lewisia sacajaweana*) was formally described in 2006. It is named in honor of Sacajawea, the Native American woman who guided Lewis and Clark in their exploration of the Pacific Northwest in the early 1800s.

STUDY PLAN

23 Systematic Biology: Phylogeny and Classification

WHY IT MATTERS

Mention the word "malaria," and people envision the tropics: explorers wander through the jungle in pith helmets and sleep under mosquito netting; clouds of insects hover nearby, ready to infect them with *Plasmodium,* the parasite that causes this disease. You may be surprised to learn, however, that less than 100 years ago, malaria was also a serious threat in the southeastern United States and much of western Europe.

Scientists puzzled over the cause of malaria for thousands of years. Hippocrates, a Greek physician who worked in the fifth century B.C., knew that people who lived near malodorous marshes often suffered from fevers and swollen spleens. Indeed, the name malaria is derived from the Latin for "bad air." By 1900, scientists had established that mosquitoes, *Plasmodium*'s intermediate hosts, transmit the parasite to humans. Mosquitoes breed in standing water, and anyone living nearby is likely to suffer their bites.

Until the 1920s, scientists thought that the mosquito species *Anopheles maculipennis* carried malaria in Europe. But some areas with huge mosquito populations had little human malaria, whereas other areas had relatively few mosquitoes and a high incidence of the disease.

Then, a French researcher reported variation in the mosquitoes, and Dutch scientists identified two forms of the "species," only one of which seemed to carry malaria. The breakthrough came in 1924, when a retired public health inspector in Italy discovered that individual mosquitoes—all thought to be the same species—produced eggs with one of six distinctive surface patterns **(Figure 23.1).**

Further research revealed that the name *Anopheles maculipennis* had been applied to six separate mosquito species. Although the adults of these species are very similar, their eggs are clearly different. The species are reproductively isolated from each other, and they differ ecologically: some breed in brackish coastal marshes, others in freshwater inland marshes, and still others in slow-moving streams. Only some of these species have a preference for human blood, and researchers eventually determined that only three of them routinely transmit malaria to humans.

These discoveries explained why the geographical distributions of mosquitoes and malaria did not always match. And government agencies could finally fight malaria by eradicating the disease-carrying species. Health workers drained marshes to prevent mosquitoes from breeding. They applied insecticides to kill mosquito larvae or introduced *Gambusia,* the mosquito

fish, which eats them. These targeted control programs were very successful.

The eradication of malaria in Europe owes a debt to **systematics,** the branch of biology that studies the diversity of life and its evolutionary relationships. Systematic biologists—systematists for short—identify, describe, name, and classify organisms, organizing their observations within a framework that reflects evolutionary relationships. In this chapter we first describe the goals of systematics and the traditional classification scheme that has been used for more than 200 years. Next we consider some of the evidence that systematists use and how that evidence must be interpreted to infer evolutionary relationships. Finally, we consider the analytical methods that contemporary systematists embrace.

23.1 Systematic Biology: An Overview

By organizing information about the biological world, systematics facilitates research in all fields of biology.

The Twin Goals of Systematics Are Reconstruction of Evolutionary History and Classification of Species

The science of systematics has two major goals. One is to reconstruct the **phylogeny,** the evolutionary history, of a group of organisms. Phylogenies are illustrated in **phylogenetic trees,** formal hypotheses that identify likely relationships among species. Like all hypotheses, they are revised as scientists gather new data.

The second goal of systematics is **taxonomy,** the identification and naming of species and their placement in a classification. A **classification** is an arrangement of organisms into hierarchical groups that reflect their relatedness. Most systematists want classifications to mirror phylogenetic history and, thus, the pattern of branching evolution.

Systematics Provides Essential Information for All of the Biological Sciences

Systematics is sometimes maligned as "stamp collecting" by those who think that systematists just collect, describe, and maintain specimens. In fact, systematists study the patterns of phenotypic and genetic variation discussed in Chapters 20 and 21. Thus, their work enhances our understanding of microevolution, speciation, adaptive radiation, and extinction. While studying these phenomena, systematists also prepare guidebooks to biodiversity.

The ability to identify species is also crucial for controlling agricultural pests and agents of disease, such as malaria-carrying mosquitoes. Systematics also helps us to identify endangered species, manage wild-

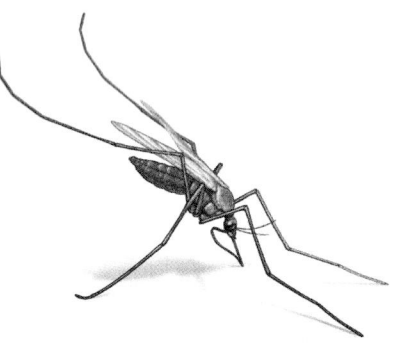

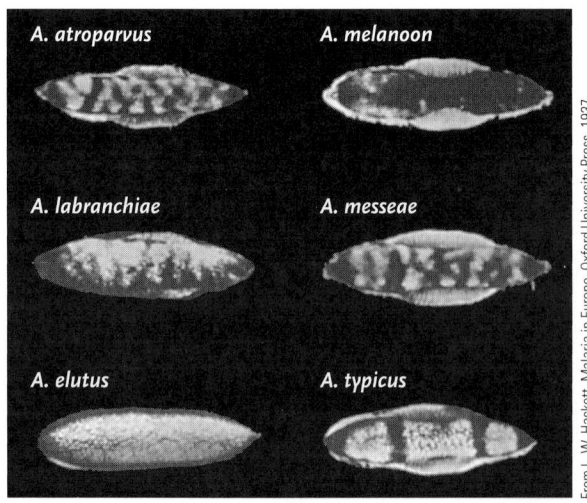

A. atroparvus   A. melanoon

A. labranchiae   A. messeae

A. elutus   A. typicus

From L. W. Hackett, Malaria in Europe. Oxford University Press, 1937

Figure 23.1

Eggs of European mosquitoes. Differences in surface patterns on the eggs of *Anopheles* mosquitoes in Europe helped researchers identify six separate species. The adults of all six species look remarkably alike. An adult *Anopheles atroparvus* is illustrated.

life effectively, and choose wild plants and animals for selective breeding and genetic engineering projects.

Data collected and organized by systematists also allows biologists to select appropriate organisms for their work. Most biological experiments are first conducted with individuals of a single species, because each species is a closed genetic system that may respond uniquely to experimental conditions. If a researcher inadvertently used two species, and these species responded differently, the mixed results probably wouldn't make much sense.

Finally, accurate phylogenetic trees are essential components of the comparative method, which biologists use to analyze evolutionary processes. Without a good phylogenetic hypothesis, we could not distinguish similarities inherited from a common ancestor from those that evolved independently in response to similar environments. For example, if biologists did not know the ancestry of sharks, penguins, and porpoises, they could not determine that their similarities were produced by convergent evolution (see Figure 22.9).

STUDY BREAK

1. What is the difference between a phylogenetic tree and a classification?
2. How does work in systematics allow biologists to select appropriate organisms for research?

23.2 The Linnaean System of Taxonomy

The practice of naming and classifying organisms originated with the Swedish naturalist Carl von Linné (1707–1778), better known by his Latinized name, Carolus Linnaeus. A professor at the University of Uppsala, Linnaeus sent ill-prepared students around the world to gather specimens, losing perhaps a third of his followers to the rigors of their expeditions. Although not a commendable student adviser, Linnaeus developed the basic system of naming and classifying organisms still in use today.

Linnaeus Developed the System of Binomial Nomenclature

Linnaeus invented the system of **binomial nomenclature,** in which species are assigned a Latinized two-part name, or **binomial.** The first part identifies a group of species with similar morphology, called a **genus** (plural, *genera*). The second part is the **specific epithet,** or species name.

A combination of the generic name and the specific epithet provides a unique name for every species. For example, *Ursus maritimus* is the polar bear and

Ursus arctos is the brown bear. By convention, the first letter of a generic name is always capitalized; the specific epithet is never capitalized; and the entire binomial is italicized. In addition, the specific epithet is never used without the full or abbreviated generic name preceding it because the same specific epithet is often given to species in different genera. For instance, *Ursus americanus* is the American black bear, *Homarus americanus* is the Atlantic lobster, and *Bufo americanus* is the American toad. If you were to order just "*americanus*" for dinner, you might be dismayed when your plate arrived—unless you have an adventurous palate!

Nonscientists often use different common names to identify a species. For example, *Bothrops asper,* a poisonous snake native to Central and South America, is called *barba amarilla* (meaning "yellow beard") in some places and *cola blanca* (meaning "white tail") in others; biologists have recorded about 50 local names for this species. Adding to the confusion, the same common name is sometimes used for several different species. Binomials, however, allow people everywhere to discuss organisms unambiguously.

Many binomials are descriptive of the organism or its habitat. *Asparagus horridus,* for example, is a spiny plant. Other species, such as the South American bird *Rhea darwinii,* are named for notable biologists. The naming of newly discovered species follows a formal process of publishing a description of the species in a scientific journal. International commissions meet periodically to settle disputes about scientific names.

The Taxonomic Hierarchy Organizes Huge Amounts of Systematic Data

Linnaeus described and named thousands of species on the basis of their similarities and differences. Keeping track of so many species was no easy task, so he devised a **taxonomic hierarchy** for arranging organisms into ever more inclusive categories **(Figure 23.2).** A **family** is a group of genera that closely resemble one another. Similar families are grouped into **orders,** similar orders into **classes,** similar classes into **phyla** (singular, *phylum*), and similar phyla into **kingdoms.** Finally, all life on Earth is classified into three **domains,** described in Section 1.3. The organisms included within any category of the taxonomic hierarchy compose a **taxon** (plural, *taxa*). Woodpeckers, for example, are a taxon (Picidae) at the family level, and pine trees are a taxon *(Pinus)* at the genus level.

Linnaeus did not believe in evolution. His goals were to illuminate the details of God's creation and to devise a practical way for naturalists to keep track of their discoveries. Nevertheless, the taxonomic hierarchy he defined was easily applied to Darwin's concept of branching evolution, which is itself a hierarchical phenomenon. As we discussed in the preceding two chapters, ancestral species give rise to descendant species through repeated branching of a lineage. Organ-

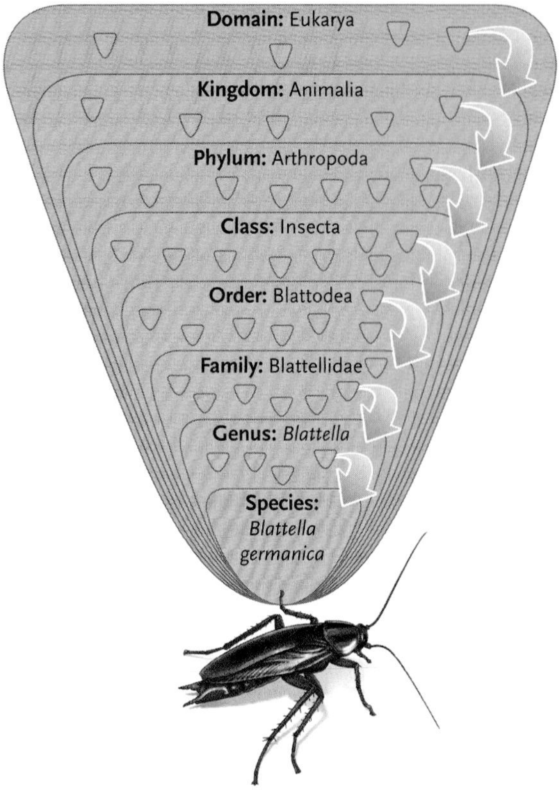

Figure 23.2
The Linnaean hierarchy of classification. The classification of a common household pest, the German cockroach *(Blattella germanica)*, illustrates the nested hierarchy that Linnaeus developed. The German cockroach is one of many closely related species classified together in the genus *Blattella*, which is in turn one of nine genera in the family Blattellidae. Six distinctive cockroach families compose the order Blattodea, one of about 30 orders grouped into the class Insecta. The phylum Arthropoda contains about a dozen classes of animals, including insects, horseshoe crabs, spiders, crabs, and centipedes. Arthropoda is one of approximately 30 phyla, each representing a major lineage and body plan, within the kingdom Animalia. The classification of animal diversity is described in detail in Chapters 29 and 30.

isms in the same genus generally share a fairly recent common ancestor, whereas those assigned only to the same class or phylum share a common ancestor from the more distant past.

Study Break

1. How does the system of binomial nomenclature minimize ambiguity in the naming and identification of species?
2. Which taxonomic category is immediately above family? Which is immediately below it?

23.3 Organismal Traits as Systematic Characters

Systematists compare organisms and then group species that share certain characteristics. Linnaeus focused on external anatomy. For example, he defined birds as a class of oviparous (egg-laying) animals with feathered bodies, two wings, two feet, and a bony beak. No other animals possess all these characteristics, which distinguish birds from "quadrupeds" (his term for mammals), "amphibians" (among which he included reptiles), fishes, insects, and "worms."

For roughly 200 years, systematists building on Linnaeus' work relied on a variety of organismal traits to analyze evolutionary relationships and classify organisms: chromosomal anatomy; details of physiological functioning; the morphology of subcellular structures, cells, organ systems, and whole organisms; and patterns of behavior. Today, systematists often focus on the molecular sequences of nucleic acids and proteins (see Section 23.6). Here we consider two commonly studied organismal characteristics: *morphological traits* and *behavioral traits.*

Morphological Characters Provide Abundant Clues to Evolutionary Relationships

Morphological differences often reflect genetic differences between organisms (see Section 20.1), and they are easy to measure in preserved or living specimens. Moreover, morphological characteristics are often clearly preserved in the fossil record, allowing the comparison of living species with their extinct relatives.

Useful morphological traits vary from group to group. In flowering plants, the details of flower anatomy often reveal common ancestry. Among vertebrates, the presence or absence of scales, feathers, and fur as well as the structure of the skull help scientists to reconstruct the evolutionary history of major groups. Sometimes systematists use obscure characters of unknown function. But differences in the number of scales on the back of a lizard or in the curvature of a vein in the wing of a bee may be good indicators of the genetic differentiation that accompanied or followed speciation—even if we do not know *why* these differences evolved.

Sometimes we rely on characteristics found only in the earliest stages of an organism's life cycle to provide evidence of evolutionary relationships. As described in Chapter 30, analyses of the embryos of vertebrates reveal that they are rather closely related to sea cucumbers, sea stars, and sea urchins and even more closely related to a group of nearly shapeless marine invertebrates called sea squirts or tunicates.

Behavioral Characters Offer Additional Data When Species Are Not Morphologically Distinct

Sometimes external morphology cannot be used to differentiate species. For example, two species of treefrog (*Hyla versicolor* and *Hyla chrysoscelis*) commonly occur together in forests of the central and eastern United

States. Both species have bumpy skin and adhesive pads on their toes that enable them to climb vegetation. They also have gray backs, white bellies, yellowish-orange coloration on their thighs, and large white spots below their eyes. The frogs are so similar that even experts cannot easily tell them apart.

How do we know that these frogs represent two species? During the breeding season, males of each species use a distinctive mating call to attract females **(Figure 23.3)**. The difference in calls is a prezygotic reproductive isolating mechanism that prevents females from mating with males of a different species (see Section 21.2). Prezygotic isolating mechanisms are excellent systematic characters because they are often the traits that animals themselves use to recognize members of their own species. The two frog species also differ in chromosome number—*Hyla chrysoscelis* is diploid and *Hyla versicolor* is tetraploid—which is a postzygotic isolating mechanism.

Hyla versicolor   *Hyla chrysoscelis*

Figure 23.3

Look-alike frog species. The frogs *Hyla versicolor* and *Hyla chrysoscelis* are so similar in appearance that one photo can depict both species. Male mating calls, visualized in sound spectrograms for the two species, are very different. The spectrograms, which depict call frequency on the vertical axis and time on the horizontal axis, show that *H. chrysoscelis* has a faster trill rate.

(Sound spectrograms from The Amphibians and Reptiles of Missouri, by T. R. Johnson © 1987 by the Conservation Commission of the State of Missouri. Reprinted by permission.)

Study Break

1. Why are morphological traits often helpful in tracing the evolutionary relationships within a group of organisms?
2. Why are prezygotic isolating mechanisms useful characters for systematic studies of animals?

23.4 Evaluating Systematic Characters

With a wealth of traits available for analysis, systematists use several guidelines to select characters for study. In this section we examine the most important of these principles.

Characters Must Be Independent Markers of Underlying Genetic Similarity and Differentiation

Ideally, systematists would create phylogenetic hypotheses and classifications by analyzing the genetic changes that caused speciation and differentiation. But in many cases they have had to rely on phenotypic traits as indicators of genetic similarity or divergence. Thus, systematists study traits in which phenotypic variation reflects genetic differences; they exclude differences caused by environmental variation (see Section 20.1).

Characters must also be genetically *independent,* reflecting different parts of the organisms' genomes. This precaution is necessary because different organismal characters can have the same genetic basis—and we want to use each genetic variation only once in an analysis. For example, tropical *Anolis* lizards climb

trees using small adhesive pads on the underside of their toes. The number of pads varies from species to species, and researchers have used the number of pads on the fourth toe of the left hind foot as a systematic character. They do not also use the number of pads on the fourth toe of the *right* hind foot as a separate character, because the same genes almost certainly control the number of pads on the toes of both feet.

Only Homologous Characters Provide Data about Evolutionary Relationships

A basic premise of systematic analyses is that phenotypic similarities between organisms reflect their underlying genetic similarities. As you may recall from Figure 19.3, species that are morphologically similar have often inherited the genetic basis of their resemblance from a common ancestor. Similarities that result from shared ancestry, such as the four limbs of all tetrapod vertebrates, are called **homologies** (or homologous characters). *Systematic analyses rely on the comparison of homologous characters as indicators of common ancestry and genetic relatedness.*

Even though homologous structures were inherited from a common ancestor, they may differ greatly among species, especially if their function has changed. For example, the stapes, a bone in the middle ear of tetrapod vertebrates, evolved from—and is therefore homologous to—the hyomandibula, a bone that supported the jaw joint of early fishes. The ancestral function of the bone is retained in some modern fishes, but its structure, position, and function are different in tetrapods **(Figure 23.4)**.

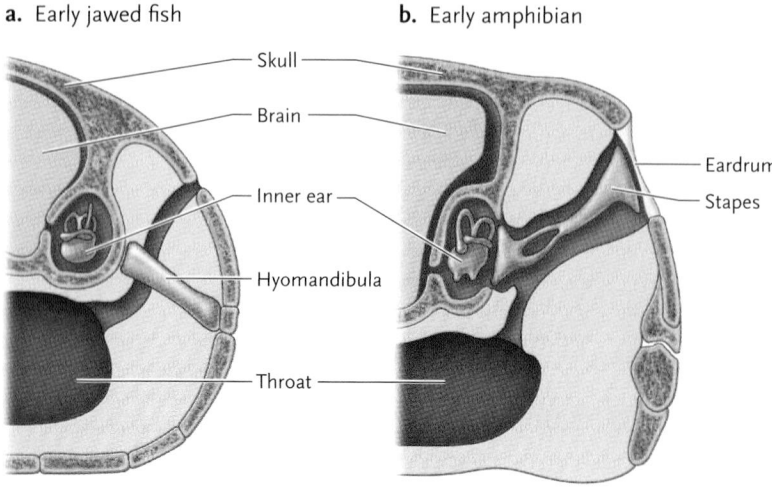

a. Early jawed fish

- Skull
- Brain
- Inner ear
- Hyomandibula
- Throat

b. Early amphibian

- Eardrum
- Stapes

Figure 23.4

Homologous bones, different structure and function. The hyomandibula, which braced the jaw joint against the skull in early jawed fishes **(a)**, is homologous to the stapes, which transmits sound to the inner ear in the four-legged vertebrates, exemplified here by an early amphibian **(b).** Both diagrams show a cross section through the head just behind the jaw joint.

As you know from the discussion of convergent evolution in Section 22.2, organisms that are not closely related sometimes bear a striking resemblance to one another, especially when they live in similar environments. Phenotypic similarities that evolved independently in different lineages are called **homoplasies** (or homoplasious characters). Some biologists use the terms *analogies* or *analogous characters* for homoplasious characters that serve a similar function in different species. *Systematists exclude homoplasies from their analyses, because homoplasies provide no information about shared ancestry or genetic relatedness.*

If homoplasies are similar and homologies are sometimes different, how can we tell them apart? First, homologous structures are similar in anatomical detail and in their relationship to surrounding structures. For example, the bones within the wings of birds and bats are homologous **(Figure 23.5).** Both wings include the same basic structural elements with similar spatial relationships to each other and to the bones that attach the wing to the rest of the skeleton. However, the large flat surfaces of their wings are homoplasious, the products of convergent evolution. The bird's wing is made of feathers, whereas the bat's wing is formed of skin.

Second, homologous characters emerge from comparable embryonic structures and grow in similar ways during development. Systematists have put great stock in embryological indications of homology on the assumption that evolution has conserved the pattern of embryonic development in related organisms. Indeed, recent discoveries in evolutionary development biology (described in Section 22.6 and explored further in Chapters 29 and 30) have revealed that the genetic controls of developmental pathways are very similar across a wide variety of organisms.

Systematists Focus Attention on Derived Versions of Characters

In all evolutionary lineages, some characteristics evolve slowly and others evolve rapidly, a phenomenon called **mosaic evolution.** Because mosaic evolution is pervasive, every species displays a mixture of **ancestral characters** (old forms of traits) and **derived characters** (new forms of traits). Derived characters provide the most useful information about evolutionary relationships because once a derived character becomes established, it is usually present in all of that species' descendants. Thus, unless they are lost or replaced by newer characters over evolutionary time, derived characters serve as markers for entire evolutionary lineages.

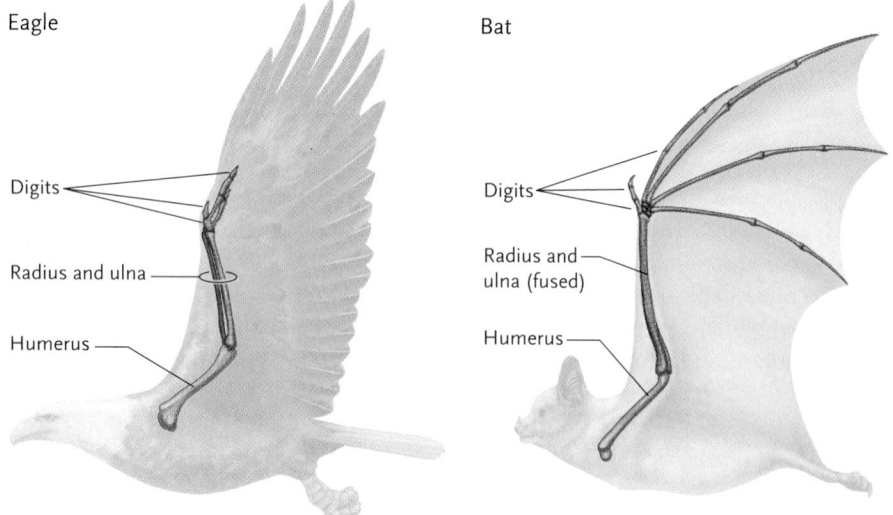

Eagle

- Digits
- Radius and ulna
- Humerus

Bat

- Digits
- Radius and ulna (fused)
- Humerus

Figure 23.5

Assessing homology. The wing skeletons of birds and bats are homologous structures with the same basic elements. However, the flat wing surfaces are homoplasious structures.

a. Caddis fly

b. Orange palm dart butterfly

c. Monarch butterfly

Figure 23.6

Outgroup comparison. Most adult insects, like the **(a)** caddis fly (family Limnephilidae) and the **(b)** orange palm dart butterfly *(Cephrenes auglades)*, have six walking legs. This comparison of butterflies with other insects suggests that the four walking legs of the **(c)** monarch butterfly *(Danaus plexippus)* represents the derived character state.

Systematists score characters as either ancestral or derived only when comparing them among organisms. Thus, any particular character is derived *only in relation to* what occurs in other organisms—either an older version of the same character or, in the case of an entirely new trait, the absence of it altogether. For example, most species of animals lack a vertebral column and the other components of an internal skeleton. However, one animal lineage—the vertebrates, including fishes, amphibians, reptiles, birds, and mammals—has those structures. Thus, when systematists compare vertebrates to all of the animals that lack a vertebral column, they score the absence of a vertebral column as the ancestral condition and the presence of a vertebral column as derived.

How can systematists distinguish between ancestral and derived characters? In other words, how can they determine the direction in which a character has evolved? The fossil record, if it is detailed enough, can provide unambiguous information. For example, biologists are confident that the presence of a vertebral column is a derived character because fossils of the earliest animals lack that structure.

For some traits, researchers use embryological evidence. Derived characters often appear later during embryonic development as modifications of an ancestral developmental plan. Recall, for example, that the early embryos of mammals first develop fishlike features in their circulatory systems (as shown in Figure 19.13) and only later develop the characteristic adult morphology. This developmental sequence suggests that the two-chambered linear hearts of fishes are ancestral, and that the four-chambered, double-loop hearts of mammals are derived.

Systematists frequently use a technique called **outgroup comparison** to identify ancestral and derived characters by comparing the group under study to more distantly related species that are not otherwise included in the analysis. Most modern butterflies, for example, have six walking legs, but species in two families have four walking legs and two small, nonwalking

legs. Which is the ancestral character state, and which is derived? Outgroup comparison with other insects, most of which have six walking legs as adults, suggests that six walking legs is ancestral and four is derived **(Figure 23.6).**

STUDY BREAK

1. Why do systematists use homologous characters in their phylogenetic analyses?
2. What is outgroup comparison?

23.5 Phylogenetic Inference and Classification

After exploring two guiding principles of research in systematics, we describe how systematists use their analyses of organismal characters to reconstruct phylogenetic histories and create classifications.

Many Systematic Studies Rely on the Principles of Monophyly and Parsimony

Phylogenetic trees portray the evolutionary diversification of lineages as a hierarchy that reflects the branching pattern of evolution. Each branch represents the descendants of a single ancestral species. When converting the phylogenetic tree into a classification, systematists use the **principle of monophyly;** that is, they try to define **monophyletic taxa,** each of which contains a single ancestral species and all of its descendants **(Figure 23.7).** By contrast, **polyphyletic taxa**—which systematists never intentionally define—would include species from separate evolutionary lineages. A taxon that included convergent species, such as sharks, penguins, and dolphins, would be polyphyletic. **Paraphyletic taxa** each contain an ancestor and some, but not all, of

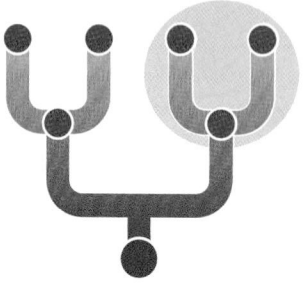

Monophyletic taxon

A monophyletic taxon includes
an ancestral species and all of
its descendants.

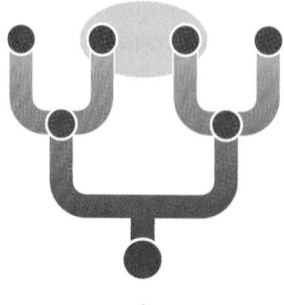

Polyphyletic taxon

A polyphyletic taxon includes
species from different
evolutionary lineages.

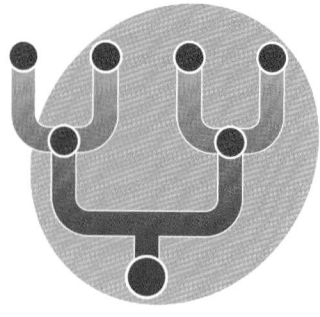

Paraphyletic taxon

A paraphyletic taxon includes an
ancestral species and only some
of its descendants.

Figure 23.7
Defining taxa in a classification. Systematists can create different classifications from the same phylogenetic tree by identifying different groups of species as a single taxon (shaded).

its descendants. For example, the traditional taxon Reptilia is paraphyletic, as described in the next section. These distinctions are crucial when making classifications.

Many systematists also strive to create *parsimonious* phylogenetic hypotheses, which means that they include the fewest possible evolutionary changes to account for the diversity within a lineage. According to the **principle of parsimony,** any particular evolutionary change is an unlikely event; therefore it is extremely unlikely that the same change evolved twice in one lineage. For example, phylogenetic trees place all birds on a single branch, implying that feathered wings evolved once in their common ancestor. This hypothesis is more parsimonious than one proposing that feathered wings evolved independently in two or more vertebrate lineages.

Traditional Evolutionary Systematics Was Based on Linnaeus' Methods

For a century after Darwin published *On the Origin of Species,* most systematists followed Linnaeus' practice of using phenotypic similarities and differences to infer evolutionary relationships. This approach, called **traditional evolutionary systematics,** groups together species that share both ancestral and derived characters. For example, mammals are defined by their internal skeleton, vertebral column, and four limbs—all ancestral characters among the tetrapod vertebrates—as well as hair, mammary glands, and a four-chambered heart—all of which are derived characters.

The classifications produced by traditional systematics reflect both evolutionary branching and morphological divergence **(Figure 23.8a).** For example, among the tetrapod vertebrates, the amphibian and mammalian lineages each diverged early, followed shortly thereafter by the turtle lineage. The remaining organisms then diverged into two groups: lepidosaurs gave rise to lizards and snakes, and archosaurs gave rise to crocodilians, dinosaurs, and birds. Thus, although

crocodilians outwardly resemble lizards, they share a more recent common ancestor with birds. Birds differ from crocodilians because birds experienced substantial morphological change when they emerged as a distinct group.

Even though the phylogenetic tree shows six living groups, the traditional classification recognizes only four classes of tetrapod vertebrates: Amphibia, Mammalia, Reptilia, and Aves (birds). These groups are given equal ranking because each represents a distinctive body plan and way of life. The class Reptilia, however, is clearly a paraphyletic taxon: it includes *some* of the descendants of the common ancestor labeled A in Figure 23.8a, namely turtles, lizards, snakes, and crocodilians; but it omits birds, and thus does not include *all* descendants.

Traditional evolutionary systematists justify this definition of the Reptilia because it includes morphologically similar animals with close evolutionary relationships. Crocodilians are classified with lizards, snakes, and turtles because they share a distant common ancestry and are covered with dry, scaly skin. Traditional systematists also argue that the key innovations initiating the adaptive radiation of birds—wings, feathers, high metabolic rates, and flight—represent such extreme divergence from the ancestral morphology that birds merit recognition as a separate class.

Cladistics Uses Shared Derived Characters to Trace Evolutionary History

In the 1950s and 1960s, some researchers criticized classifications that were based on two distinct phenomena, branching evolution and morphological divergence, as inherently unclear. After all, how can we tell *why* two groups are classified in the same higher taxon? They may have shared a recent common ancestor, as did lizards and snakes. Alternatively, they may have retained similar ancestral characteristics after being separated on different branches of a phylogenetic tree, as is the case for lizards and crocodilians.

a. Traditional phylogenetic tree with classification

b. Cladogram with classification

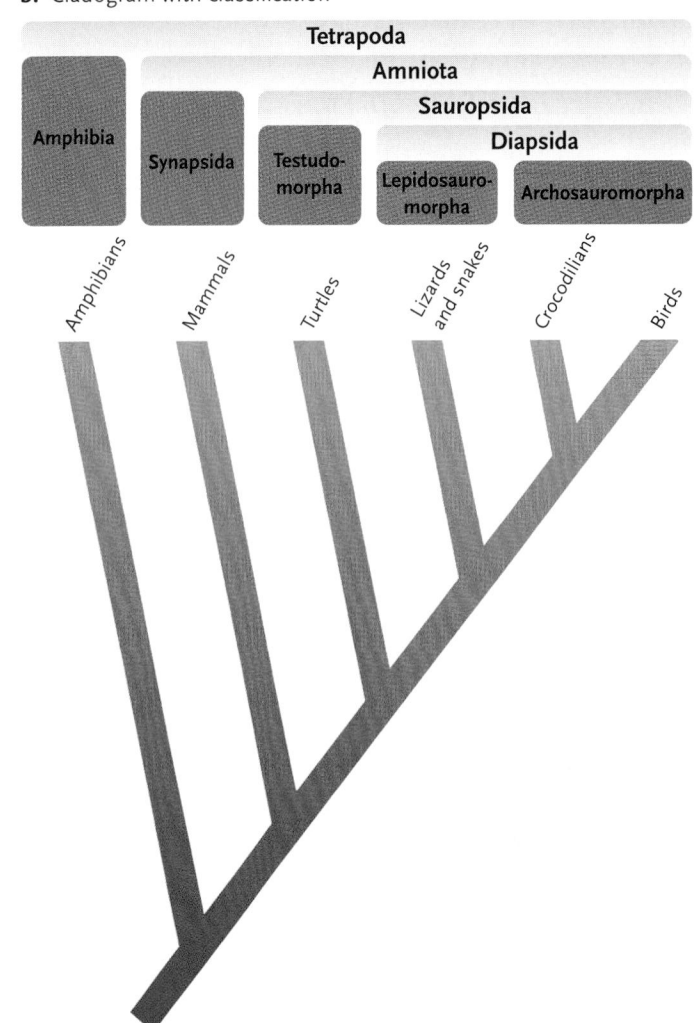

| Class Amphibia | Class Mammalia | Class Reptilia | Class Aves |

Amphibians
Mammals
Turtles
Lizards and snakes
Crocodilians
Birds

LEPIDOSAURS
ARCHOSAURS

A ——— Common ancestor of reptiles and birds

Figure 23.8
Phylogenetic trees and classifications for tetrapod vertebrates. **(a)** Traditional and **(b)** cladistic approaches produce different phylogenetic trees and classifications. Classifications are presented above the trees.

To avoid such confusion, many systematists quickly followed the philosophical and analytical lead of Willi Hennig, a German entomologist, who published an influential book, *Phylogenetic Systematics,* in 1966. Hennig and his followers argued that classifications should be based solely on evolutionary relationships. **Cladistics,** as this approach is known, produces phylogenetic hypotheses and classifications that reflect only the branching pattern of evolution; it ignores morphological divergence altogether.

Cladists group together only species that *share derived characters.* For example, cladists argue that mammals form a monophyletic lineage—a **clade**—because they possess a unique set of derived characters: hair, mammary glands, reduction of bones in the lower jaw, and a four-chambered heart. The ancestral characters found in mammals—internal skeleton, vertebral column, and four legs—do not distinguish them from other tetrapod vertebrates, so these traits are excluded from the analysis.

The phylogenetic trees produced by cladists, called **cladograms,** thus illustrate the hypothesized sequence of evolutionary branchings, with a hypothetical ancestor at each branching point **(Figure 23.8b).** They portray strictly monophyletic groups and are usually constructed using the principle of parsimony. Once a researcher identifies derived, homologous characters, constructing a cladogram is straightforward **(Figure 23.9).**

The classifications produced by cladistic analysis often differ radically from those of traditional evolutionary systematics (compare the two parts of Figure 23.8). Pairs of higher taxa are defined directly from the two-way branching pattern of the cladogram. Thus, the clade Tetrapoda (the traditional amphibians, reptiles, birds, and mammals) is divided into two taxa, the Amphibia (tetrapods that do not have an amnion, as discussed in Section 30.3) and the Amniota (tetrapods that have an amnion). The Amniota is subdivided into two taxa on the basis of skull morphology and other characteristics: Synapsida (mammals) and Sauropsida (turtles, lizards, snakes, crocodilians, and birds). The Sauropsida is further divided into the Testudomorpha (turtles) and the Diapsida (lizards and snakes, crocodilians, and birds). Finally, the Diapsida is subdivided

Figure 23.9 Research Method

Constructing a Cladogram

PURPOSE: Systematists construct cladograms to visualize hypothesized evolutionary relationships by grouping together organisms that share derived characters. The cladogram also illustrates where derived characters first evolved.

PROTOCOL:

1. *Select the organisms to study.* To demonstrate the method, we develop a cladogram for the nine groups of living vertebrates: lampreys, sharks (and their relatives), bony fishes (and their relatives), amphibians (frogs and salamanders), turtles, lizards (including snakes), crocodilians (including alligators), birds, and mammals. We also include marine animals called lancelets (phylum Chordata, subphylum Cephalochordata), which are closely related to vertebrates (see Chapter 30). Lancelets are the outgroup in our analysis.

2. *Choose the characters on which the cladogram will be based.* Our simplified example is based on the presence or absence of 10 characters: (1) vertebral column, (2) jaws, (3) swim bladder or lungs, (4) paired limbs (with one bone connecting each limb to the body), (5) extraembryonic membranes (such as the amnion), (6) mammary glands, (7) dry, scaly skin somewhere on the body, (8) two openings on each side near the back of the skull, (9) one opening on each side of the skull in front of the eye, and (10) feathers.

3. *Score the characters as either ancestral or derived in each group.* As the outgroup, lancelets possess the ancestral character; any deviation from the lancelet pattern is derived. Because lancelets lack all of the characters in our analysis, the presence of each character is the derived condition. We tabulate data on the distribution of ancestral (−) and derived (+) characters, listing lancelets first and the other organisms in alphabetical order.

	Vertebrae	Jaws	Swim bladder or lungs	Paired limbs	Extraembryonic membranes	Mammary glands	Dry, scaly skin	Two openings at back of skull	One opening in front of eye	Feathers
Lancelets	−	−	−	−	−	−	−	−	−	−
Amphibians	+	+	+	+	−	−	−	−	−	−
Birds	+	+	+	+	+	−	+	+	+	+
Bony fishes	+	+	+	−	−	−	−	−	−	−
Crocodilians	+	+	+	+	+	−	+	+	+	−
Lampreys	+	−	+	+	−	−	−	−	−	−
Lizards	+	+	+	+	+	−	+	+	−	−
Mammals	+	+	+	+	+	+	−	−	−	−
Sharks	+	+	−	−	−	−	−	−	−	−
Turtles	+	+	+	+	−	−	+	−	−	−

4. *Construct the cladogram from information in the table, grouping organisms that share derived characters.* All groups except lancelets have vertebrae. Thus, we group organisms that share this derived character on the right-hand branch, identifying them as a monophyletic lineage. Lancelets are on their own branch to the left, indicating that they lack vertebrae.

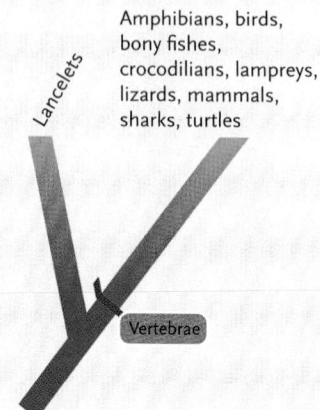

All of the remaining organisms except lampreys have jaws. (Lancelets also lack jaws, but we have already separated them out, and do not consider them further.) Place all groups with jaws, a derived character, on the right-hand branch. Lampreys are separated out to the left, because they lack jaws. Again, the branch on the right represents a monophyletic lineage.

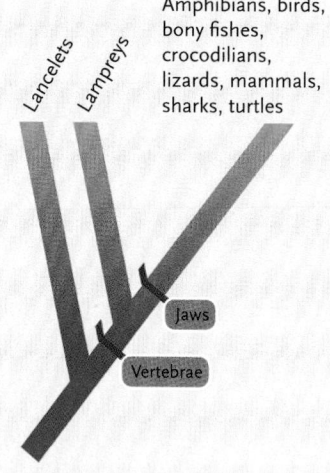

into two more recently evolved taxa—the Lepidosauromorpha (lizards and snakes) and the Archosauromorpha (crocodilians and birds). Thus, a strictly cladistic classification exactly parallels the pattern of branching evolution that produced the organisms included in the classification. These parallels are the essence and strength of the cladistic method.

Most biologists now use the cladistic approach because of its evolutionary focus, clear goals, and precise methods. In fact, some systematists advocate abandoning the Linnaean hierarchy for classifying and naming organisms. They propose using a strictly cladistic system, called **PhyloCode**, that identifies and names clades instead of pigeonholing or-

5. *Construct the rest of the cladogram using the same step-by-step procedure to separate the remaining groups.* In our completed cladogram, seven groups share a swim bladder or lungs; six share paired limbs; and five have extraembryonic membranes during development. Some groups are distinguished by the unique presence of a derived character, such as feathers in birds.

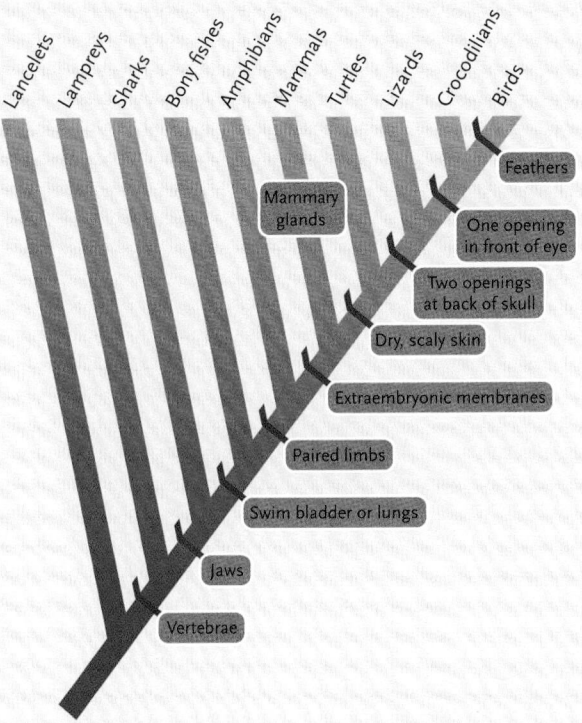

INTERPRETING THE RESULTS: Although cladograms provide information about evolutionary relationships, the common ancestors represented by the branch points are often hypothetical. You can tell from the cladogram, however, that birds are more closely related to lizards than they are to turtles. Follow the branches of the cladogram from birds and lizards back to their intersection, or node. Next, trace the branches of birds and turtles to their node. You can see that the bird–turtle node is closer to the bottom of the cladogram than the bird–lizard node. Nodes that are closer to the bottom of the cladogram indicate a more distant common ancestry than those closer to the top.

ganisms into the familiar taxonomic levels embraced by more traditional systematists. Although traditional evolutionary systematics has guided many people's understanding of biological diversity, we use a cladistic approach to describe evolutionary lineages and taxa in the Biodiversity unit that follows this chapter.

STUDY BREAK

1. How does a monophyletic taxon differ from a polyphyletic taxon?
2. Why is the traditionally defined group Reptilia a paraphyletic taxon?
3. What characteristics are used to group organisms in a cladistic analysis?

23.6 Molecular Phylogenetics

Most systematists now conduct phylogenetic analyses using molecular characters, such as the nucleotide base sequences of DNA and RNA or the amino acid sequences of the proteins for which they code. Because DNA is inherited, shared changes in molecular sequences—insertions, deletions, or substitutions—provide clues to the evolutionary relationships of organisms. Technological advances have automated many of the necessary laboratory techniques, and analytical software makes it easy to compare new data to information filed in data banks accessible over the Internet.

Molecular Characters Have Both Advantages and Disadvantages over Organismal Characters

Molecular sequences have certain practical advantages over organismal characters. First, they provide abundant data: every amino acid in a protein and every base in a nucleic acid can serve as a separate, independent character for analysis. Moreover, because many genes have been conserved by evolution, molecular sequences can be compared between distantly related organisms that share no organismal characteristics. Molecular characters can also be used to study closely related species with only minor morphological differences. Finally, many proteins and nucleic acids are not directly affected by the developmental or environmental factors that cause nongenetic morphological variations such as those described in Section 20.1.

Molecular characters have certain drawbacks, however. For example, only four alternative character states (the four nucleotide bases) exist at each position in a DNA or RNA sequence and only 20 alternative character states (the 20 amino acids) at each position in a protein. (You may want to review Sections 14.2 and 15.1 on the structure of these molecules.) And if two species have the same nucleotide base substitution at a given position in a DNA segment, their similarity may well have evolved independently. As a result, systematists often find it difficult to verify that molecular similarities were inherited from a common ancestor.

For organismal characters, biologists can establish that similarities are homologous by analyzing the characters' embryonic development or details of their func-

Whales with Cow Cousins?

More than 50 million years ago, whales evolved from terrestrial mammals into streamlined creatures, spectacularly adapted to life in the sea. But which mammals were their ancestors? Using morphological comparisons of living and fossil species, evolutionists had hypothesized that modern cetaceans—whales, dolphins, and porpoises—evolved from wolflike mammals called mesonychians. However, recent work by molecular biologists suggests that cetaceans are part of a lineage that includes an ungulate ancestor of cows and hippopotamuses.

Several molecular studies support this surprising conclusion. Mitsuru Shimamura and his colleagues at the Tokyo Institute of Technology and other Japanese institutions examined the distribution of transposable elements (TEs) in whales and ungulates. TEs are sequences that move to new locations in DNA (see Section 17.3). The TEs that the researchers studied in whales move by making RNA copies of themselves; the RNA copies then act as templates for making DNA copies, which are inserted in new locations. The mechanism leaves the original copy still in place in the DNA.

The TEs studied by the Shimamura team are called SINEs (for *Short INter-*spersed *Elements*). These elements, which occur only in mammals, are particularly useful for evolutionary studies because the pattern by which they duplicate and move to new locations is unique in each evolutionary lineage. If SINEs occur at the same sites in the nuclear DNA of several species, those species are likely to be members of the same lineage.

To begin their work, Shimamura and his coworkers isolated two types of SINEs from whales, which they designated CHR-1 and CHR-2. They found that the DNA of these SINEs could pair with sequences in the nuclear DNA of hippos, cows, and other ruminants, but not with sequences of pigs and camels. This result showed that the CHR-1 and CHR-2 SINEs are present in whales, cows, and hippos but not in pigs and camels.

The researchers then used similar techniques to work out the locations of the SINEs in the DNA, with particular focus on SINEs that may have inserted into known protein-encoding genes. SINEs can insert into genes without serious damage if they do so in introns, the surplus segments that are transcribed but spliced out of the messenger RNA copy of the gene (see Section 15.3). To find genes containing the SINEs, the researchers added probes—labeled DNA sequences that could pair with CHR-1 and CHR-2—to DNA preparations containing all the genes of the species under study. They also searched through electronic databanks of known gene sequences of the species, looking for genes with introns containing either of the two SINEs.

The probes and computer searches produced seven "hits" among protein-encoding genes. Three CHR-1 insertions were found at the same locations in genes of cetaceans, ruminants, and hippos, but were absent from these locations in camels and pigs. The results indicate that the SINEs inserted at these locations in a common ancestor of cetaceans, ruminants, and hippos after camels and pigs had split off as a separate group (see **figure**). Additionally, some other SINEs evidently inserted later, after an evolutionary split had separated the ruminants and cetaceans. Two CHR-1 insertions were found in ruminants but not in cetaceans, hippos, camels, or pigs; two CHR-2 insertions were found only in cetaceans. These data enabled the investigators to construct the phylogenetic tree shown in the figure; the gene loci within which they found CHR-1 and CHR-2 insertions are labeled on the branches of the tree. Molecular studies testing the distribution of other DNA sequences, including mitochondrial DNA, support the close relationships between whales and cows suggested by the Shimamura experiments.

Some evolutionists contested the conclusions from molecular studies because they considered the database too limited and because morphological studies supported other hypotheses. Pigs, ruminants, camels, and hippos share a mobile heel joint that is different from the nonmobile joint in all other mammals. With their greatly reduced hind limbs, modern whales have no heel joint; but a land-living fossil believed to be an ancestor of whales has a nonmobile heel joint. Further, the teeth of pigs, ruminants, camels, and hippos are different from those of cetaceans. These morphological characters support a traditional classification in which ruminants, pigs, camels, and hippos form one lineage, and cetaceans a separate one. However, in 2001, Philip D. Gingerich of the University of Michigan and his colleagues in Pakistan reported the discovery of two ancient whale fossils, both of which had mobile heel joints. These new findings provide strong evidence in support of the conclusion that whales are closely related to ruminants and hippos.

Cetaceans | Ruminants | Hippos | Pigs | Camels

Pm 52
Pm 72

pgha 3
C21-352

aaa 228
aaa 792
Gm 5

Gene loci that include CHR-1 and CHR-2 insertions

tion. But molecular characters have no embryonic development, and biologists still do not understand the functional significance of most molecular differences. Despite these disadvantages, molecular characters represent the genome directly, and researchers use them with great success in phylogenetic analyses. *Insights from the Molecular Revolution* describes an example using sequences called transposable elements.

Variations in the Rates at Which Molecules Evolve Govern the Molecules Chosen for Phylogenetic Analyses

Although molecular phylogenetics is based on the observation that many molecules have been conserved by evolution, different adaptive changes and neutral mutations accumulate in separate lineages from the moment they first diverge. Mutations in some types of DNA appear to arise at a relatively constant rate. Thus, differences in the DNA sequences of two species can serve as a **molecular clock**, indexing their time of divergence. Large differences imply divergence in the distant past, whereas small differences suggest a more recent common ancestor.

Because mosaic evolution exists at the molecular level, different molecules exhibit individual rates of change, and every molecule is an independent clock, ticking at its own rate. Researchers study different molecules to track evolutionary divergences that occurred over different time scales. For example, mitochondrial DNA (mtDNA) evolves relatively quickly; it is useful for dating evolutionary divergences that occurred within the last few million years. Studies of mtDNA have illuminated aspects of the evolutionary history of humans, as described in Section 30.13. By contrast, chloroplast DNA (cpDNA) and genes that encode ribosomal RNA evolve much more slowly, providing information about divergences that date back hundreds of millions of years.

To synchronize molecular clocks, some researchers study DNA sequences that are not parts of protein-encoding genes. Because they don't affect protein structure, mutations in these sequences are probably not often eliminated by natural selection. Thus, the sequence differences between species in noncoding regions probably result from mutation alone and therefore reflect the ticking of the molecular clock more directly. Some researchers also calibrate molecular clocks to the fossil record, so that actual times of divergence can be predicted from molecular data with a fair degree of certainty.

The Analysis of Molecular Characters Requires Specialized Approaches

Molecular phylogenetics relies on the same basic logic that underlies analyses based on organismal characters: species that diverged recently from a common ancestor should share many similarities in their molecular sequences, whereas more distantly related species should exhibit fewer similarities. Nevertheless, the practice of molecular phylogenetics is based on a set of distinctive methods.

Determining the Molecular Sequence. After selecting a protein molecule or appropriate segment of a nucleic acid for analysis, systematists determine the exact sequence of amino acids (in the case of proteins) or nucleotide bases (in the case of DNA or RNA) that compose the molecule.

Amino acid sequencing allows systematists to compare the primary structure of protein molecules directly. As you may recall from Chapter 15, the amino acid sequence of a protein is determined by the sequence of nucleotide bases in the gene encoding that protein. When two species exhibit similar amino acid sequences for the same protein, systematists infer their genetic similarity and evolutionary relationship. For example, researchers have used sequence data from the protein cytochrome *c* to construct a phylogenetic tree for organisms as different as slime molds, vascular plants, and humans **(Figure 23.10)**.

Most systematic studies are now based, at least in part, on DNA sequencing data, which provide a detailed view of the genetic material that evolutionary processes change. The polymerase chain reaction (PCR) makes it easy for researchers to produce numerous copies of specific segments of DNA for comparison (see Section 18.1). This technique allows scientists to sequence minute quantities of DNA taken from dried or preserved specimens in museums and even from some fossils.

Aligning Molecular Sequences. Before comparing molecular sequences from different organisms, systematists must ensure that the homologous sequences being compared are properly "aligned." In other words, they must be certain that they are comparing nucleotide bases or amino acids at exactly the same positions in the nucleic acid or protein molecule. This crucial step is necessary because mutations often change the length of a DNA sequence and the relative locations of specific positions through the insertion or deletion of base pairs (see Section 15.4). Such mutations make sequence comparisons more difficult; but, by determining where such insertions or deletions have occurred, systematists can match up the positions of—in other words, *align*—the nucleotides for comparison. Although alignments can be done "by eye" in many cases, most systematists use computer programs to accomplish this task. **Figure 23.11** provides a simplified example of this step in the process.

Constructing Phylogenetic Trees. Once the molecules are aligned, a systematist can compare the nucleotide base or amino acid sequences to determine whether

Figure 23.10 Observational Research

Using Amino Acid Sequences to Construct a Phylogenetic Tree

HYPOTHESIS: Because the amino acid sequences of proteins change over evolutionary time, sequence differences between organisms should reflect their evolutionary relationships.

PREDICTION: Closely related species will exhibit similar amino acid sequences, whereas more distantly related species will exhibit greater differences in their amino acid sequences.

METHOD: Researchers determined the amino acid sequence of cytochrome *c*, a protein in the electron transport system that has been conserved by evolution, using samples from a wide variety of eukaryotic species classified in four kingdoms. They compared the data derived from the different species and used the sequences to construct a phylogenetic tree.

RESULTS: The amino acid sequence of cytochrome *c* is surprisingly similar in distantly related organisms that diverged from a common ancestor hundreds of millions of years ago. Gold shading marks the amino acids that are identical in the sequences for yeast (top row), wheat (middle row), and human (bottom row). Abbreviations for the amino acids listed below are derived from those in Figure 3.15.

The phylogenetic tree based on similarities and differences in cytochrome *c* sequences is remarkably consistent with trees based on organismal characters. The vertical axis gives the approximate time of each evolutionary branching, estimated from the amino acid sequence data.

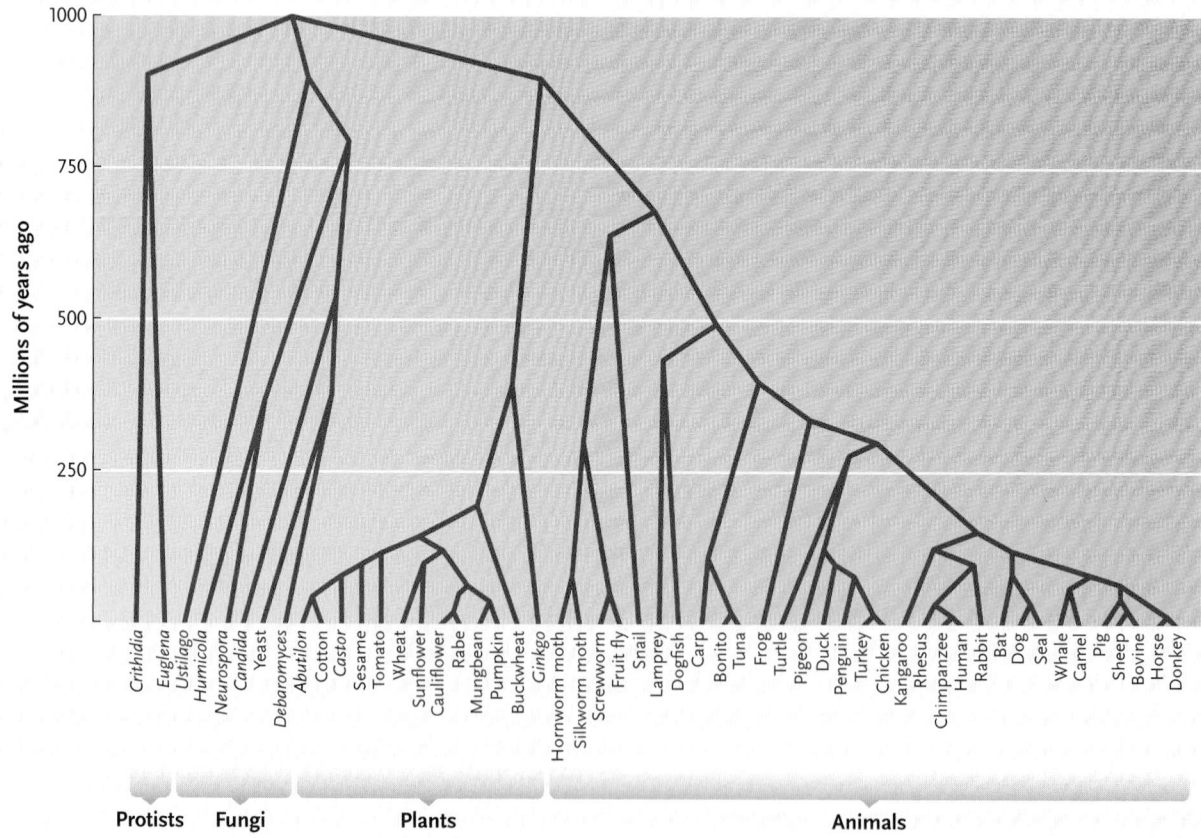

CONCLUSION: Amino acid sequence data can be used to construct phylogenetic trees for species that share essentially no organismal characteristics.

mutations or other processes have produced evolutionary changes in the sequences. The similarities and differences can then be used to reconstruct the phylogenetic tree. Every phylogenetic tree is a hypothesis about evolutionary relationships, and different assumptions can yield multiple alternative trees for any data set. Indeed, systematists have developed several approaches for comparing molecular sequences and constructing trees.

For DNA sequences, the simplest approach is to count the number of similarities and differences between every pair of organisms being compared. Systematists use such data to estimate the *genetic distances* between species and to construct a phylogenetic tree by grouping together those organisms that exhibit the smallest genetic distances. However, this approach reconstructs phylogenies with both ancestral and derived characters, the same criticism that was leveled against traditional evolutionary systematics.

An alternative approach for converting molecular sequence data into a phylogenetic tree follows a cladistic method, using the principle of parsimony, which requires the identification of ancestral and derived character states. In other words, systematists must determine, for each position in the sequence, which nucleotide base is ancestral and which is derived. As is the case for organismal characters, the analysis of homologous sequences in a designated outgroup can provide that information. Under the parsimony approach, a computer program then tests all possible phylogenetic trees and identifies the one that accounts for the diversity of organisms in the group with the fewest evolutionary changes in molecular sequences.

In recent years, researchers have faulted the parsimony approach because identical changes in nucleotides often arise independently. To avoid this problem, systematists have begun using a series of sophisticated statistical techniques collectively called *maximum likelihood methods*. This approach reconstructs phylogenetic history from molecular sequence data by making assumptions about variations in the rate at which different segments of DNA evolve. These statistical models can take into account variations in the rates of evolution between genes or between species as well as changes in evolutionary rates over time. Maximum likelihood programs construct numerous alternative phylogenetic trees and estimate how likely it is that each tree represents the true evolutionary history. Systematists then accept the phylogenetic tree that is most likely to be true—until more data are available.

Molecular Phylogenetics Has Clarified Many Evolutionary Relationships

As you will see in the next unit, molecular phylogenetics has enabled systematists to resolve some longstanding disputes about evolutionary relationships.

Figure 23.11 Research Method

Aligning DNA Sequences

PURPOSE: The insertion or deletion of base pairs often changes the length of a DNA sequence and the relative locations of specific positions along its length. Systematists must therefore "align" the sequences that they are comparing. This procedure ensures that the nucleotide bases being compared are at exactly the same positions in the nucleic acid molecules. By determining where insertions or deletions have occurred, systematists can match up the positions of—in other words, *align*—the nucleotides for comparison. In this hypothetical example, imagine that the DNA segments were obtained from three different species. A comparable procedure is used to align the amino acid sequences of proteins.

PROTOCOL:

1. Before alignment, three DNA segments differ in length and exhibit nucleotide differences in many positions.

Segment A AATTGACCTTCTAAGTGTAAT
Segment B AATTGAGCCTTCTAAGTCTAAT
Segment C AATTGATTCTAAGTGTAAT

2. The computer program detects similar sequences in parts of the three segments.

Segment A AATTGACCTTCTAAGTGTAAT
Segment B AATTGAGCCTTCTAAGTCTAAT
Segment C AATTGATTCTAAGTGTAAT

3. The three segments are aligned under the hypotheses that segment B included a one-nucleotide insertion and segment C had experienced a two-nucleotide deletion.

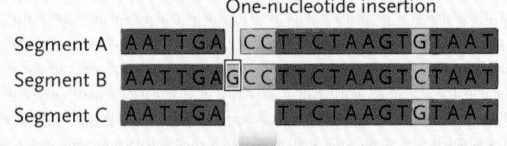

One-nucleotide insertion

Segment A AATTGA CCTTCTAAGTGTAAT
Segment B AATTGAG CCTTCTAAGTCTAAT
Segment C AATTGA   TTCTAAGTGTAAT

Two-nucleotide deletion

INTERPRETING THE RESULTS: After alignment, the sequences can be compared at every position. In addition to the one-nucleotide insertion in segment B and the two-nucleotide deletion in segment C, the comparison reveals one nucleotide substitution in segment B.

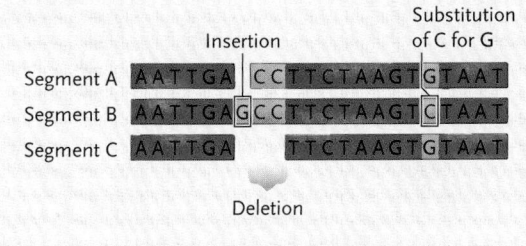

Insertion   Substitution of C for G

Segment A AATTGA CCTTCTAAGTGTAAT
Segment B AATTGAG CCTTCTAAGTCTAAT
Segment C AATTGA   TTCTAAGTGTAAT

Deletion

As one example, analyses of morphological data had produced conflicting hypotheses about the origin and relationships of flowering plants. In 1999, four teams of researchers, analyzing different parts of flowering plant genomes, independently identified

Figure 23.12

The ancestral flowering plant. DNA sequencing studies identified *Amborella trichopoda* as a living representative of the earliest group of flowering plants.

Amborella branch

Thomas J. Lemieux, University of Colorado

Amborella flower

Sandra Floyd, University of Colorado

Amborella trichopoda, a bush native to the South Pacific island of New Caledonia, as a living representative of the most ancient group of flowering plants yet discovered **(Figure 23.12)**. The first team to publish their results, Sarah Mathews and Michael Donoghue of Harvard University, studied phytochrome genes (*PHYA* and *PHYC*) that had duplicated early in the evolutionary history of this group. Other researchers, who studied chloroplast, mitochondrial, and ribosomal sequences, obtained similar results, providing strong support for this phylogenetic hypothesis.

On a very grand scale, molecular phylogenetics has revolutionized our view of the entire tree of life. The first efforts to create a phylogenetic tree for all forms of life were based on morphological analyses. However, these analyses did not resolve branches of the tree containing prokaryotes, which lack significant structural variability, or the relationships of those branches to eukaryotes.

In the 1960s and early 1970s, biologists organized living systems into five kingdoms. All prokaryotes were grouped into the kingdom Monera. The eukaryotic organisms were grouped into four kingdoms: Fungi, Plantae, Animalia, and Protista. The Protista was always recognized as a polyphyletic "grab bag" of unicellular or acellular eukaryotic organisms. Unfortunately, phylogenetic analyses based on morphology were unable to sort these organisms into distinct evolutionary lineages.

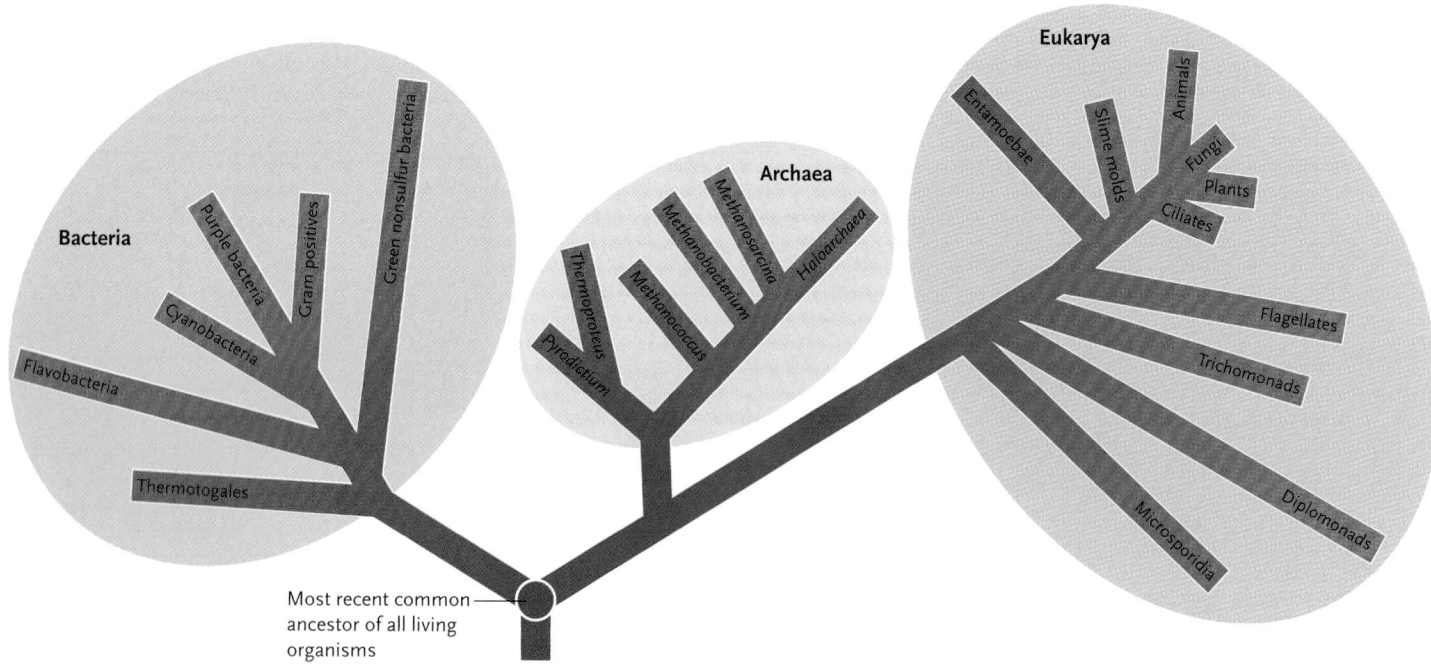

Figure 23.13

Three domains: the tree of life. Carl R. Woese's analysis of rRNA sequences suggests that all living organisms can be classified into one of three domains: Bacteria, Archaea, and Eukarya.

In the 1970s, biologists realized that molecular phylogenetics provides an alternative approach. They simply needed to identify and analyze molecules that have been conserved by evolution over billions of years. Carl R. Woese, a microbiologist at the University of Illinois at Urbana-Champaign, identified the small subunit of ribosomal RNA as a suitable molecule for analysis. Ribosomes, the structures that translate messenger RNA molecules into proteins (see Section 15.1), are remarkably similar in all forms of life. They are apparently so essential to cellular processes that the genes specifying ribosomal structure exhibit similarities in their nucleotide sequences in organisms from bacteria to humans. Thus, it is possible to sequence these genes and align them for analysis.

The phylogenetic tree based on rRNA sequences divides living organisms into three primary lineages called domains: Bacteria, Archaea, and Eukarya **(Figure 23.13).** According to this hypothesis, two domains, Bacteria and Archaea, consist of prokaryotic organisms, and one, Eukarya, consists of eukaryotes. Bacteria in-

UNANSWERED QUESTIONS

Should we abandon the traditional Linnaean hierarchy in favor of a more evolutionary classification?

Diligent Kindly Professors Cannot Often Fail Good Students—or some equally silly mnemonic device for remembering the Linnaean taxonomic hierarchy—is all that many students recall about systematics. Even if they remember the underlying rank names—is G for "group" or "genus"?—they often forget that Linnaeus conceived his system of classification more than a century before Darwin articulated his theory of evolution, which revolutionized our understanding of biological diversity. In the roughly 150 years since Darwin published *On the Origin of Species*, systematists have sought not only to categorize life's diversity but, more importantly, to understand its origins. The broad relevance of studies in systematics has become increasingly clear as biologists have discovered that systematic principles are as important to tracing the emergence and spread of avian flu as they are to distinguishing a duck from a dove.

As we approach the sesquicentennial of Darwin's theory, its impact becomes increasingly revolutionary. Perhaps the most striking recent example is a call for the complete abandonment of the Linnaean taxonomic hierarchy. Although biologists thought they had reconciled the perspectives of Darwin and Linnaeus, a growing minority of systematists now believe that any effort to catalog and categorize life's diversity must be explicitly phylogenetic and free of the arbitrary ranks that Linnaeus invented. This movement, which has been codified in the PhyloCode initiative, is fueled largely by newly available molecular data, vastly improved phylogenetic methodologies, and increasingly fast computers. These advances offer the potential to reconstruct accurate and fully resolved phylogenetic trees at a scale never before possible. For the first time, biologists see real progress in accurately reconstructing the entire tree of life. Although we are still far from achieving this goal, every day millions of new, phylogenetically informative DNA fragments are being sequenced and analyzed by thousands of computers running around the clock.

Although PhyloCode's synthesis of taxonomy and evolutionary systematics may be long overdue, this attempted coup is not without controversy. For example, some systematists contend that such a radical revision of our taxonomic system will introduce confusion and in-stability in the naming of species. Even the revolution's adherents recognize that we still face many challenging limitations to the synthesis between taxonomic practice and Darwinian principles. Nowhere is this more evident than in the definition of species.

During Linnaeus' time, species were viewed as immutable natural types created by God. Darwin, however, formulated his theory on the principle that species change over time. Although the truth of this basic hypothesis is no longer a subject of debate, its practical implications for delimiting species boundaries and understanding how new species form are among the most exciting areas of study in modern systematics. Most practicing systematists view species as real (that is, biologically meaningful) categories, but the criteria for recognizing species vary dramatically among systematists working on different types of organisms (plants versus animals, or organisms that reproduce asexually versus those that reproduce sexually). Biologists are now beginning to use new molecular tools to address the challenge of understanding the origin of new species. Using these tools and sophisticated genetic experiments, evolutionary biologists are beginning to probe the precise genetic basis of species. Over the past decade a small number of "speciation genes" have been identified; more such discoveries are sure to follow in the coming years. Although many of these studies have been restricted to model research organisms, such as fruit flies, the new tools offered by the fields of genomics and bioinformatics offer the potential to address similar questions in an increasingly broad array of organisms.

Simply put, the systematics of today is not that of your grandparents. Given the enormous challenge involved in categorizing and understanding the origin and evolutionary relationships of millions of species, many additional changes are on the horizon. For the next generation of systematists, however, a better mnemonic to remember may be "Keep Probing Charles' *Origin* For Good Systematics."

Rich Glor conducts research on the evolution of *Anolis* lizards at the University of Rochester. To learn more about his research, go to http://www.lacertilia.com.

cludes well-known microorganisms, and Archaea includes microorganisms that live in physiologically harsh environments, such as hot springs or very salty habitats. Eukarya includes the familiar animals, plants, and fungi, as well as the many lineages formerly included among the Protista. The next unit of this book is devoted to detailed analyses of the biology and evolutionary relationships between and within these three domains.

STUDY BREAK

1. What are three advantages of using molecular characters in phylogenetic analyses?
2. How can molecular sequence data be used as a molecular clock?
3. Why was a phylogenetic analysis of prokaryotes based on molecular sequence data more successful than the analysis based on morphological data?

Review

Go to **ThomsonNOW**™ at www.thomsonedu.com/login to access quizzing, animations, exercises, articles, and personalized homework help.

23.1 Systematic Biology: An Overview

- Systematic biology has two goals: the reconstruction of evolutionary history and the naming and classification of organisms. Phylogenetic trees and classifications are hypotheses about the relationships of organisms.
- By providing a guide to biological diversity, systematics allows biologists to identify species for research, for the control of harmful organisms, and for conservation (Figure 23.1).

23.2 The Linnaean System of Taxonomy

- Linnaeus invented a system of binomial nomenclature in which each species receives a unique two-part name.
- Species are organized into a taxonomic hierarchy (Figure 23.2), which reflects the pattern of branching evolution. Species classified in the same genus or family have a more recent common ancestor than species classified only in the same class or phylum.

Animation: Classification systems

23.3 Organismal Traits as Systematic Characters

- Systematists have always studied organismal characters, such as morphology, chromosome structure and number, physiology, and behavior.
- Morphological traits often allow the reconstruction of a group's phylogeny, that is, its evolutionary history.
- Behavioral characters are useful for understanding the relationships of animals that are not morphologically different (Figure 23.3).

Animation: Evolutionary tree for plants

23.4 Evaluating Systematic Characters

- Systematists study characters that are genetically independent, reflecting different parts of the organisms' genomes.
- Most systematists use homologous characters that reflect genetic similarities and differences among species (Figures 23.4 and 23.5).
- Because characters evolve at different rates, systematists select traits that evolved at a rate consistent with the timing of branching evolution.

- Systematists base their analyses on derived versions of homologous traits (Figure 23.6).

23.5 Phylogenetic Inference and Classification

- Phylogenetic trees and classifications include only monophyletic taxa, each of which contains a single ancestral species and all of its descendants (Figure 23.7). Many systematists create parsimonious phylogenies, which include the fewest possible evolutionary changes to account for the diversity within a lineage.
- Traditional evolutionary systematics emphasizes branching evolution and morphological divergence. Using both ancestral and derived characters, this approach sometimes creates classifications with paraphyletic taxa, which include an ancestor and some, but not all, of its descendants (Figure 23.8a).
- Cladistics emphasizes only evolutionary branching to define monophyletic taxa (Figure 23.8b). Cladists create phylogenetic hypotheses and classifications by grouping organisms that share derived characters (Figure 23.9).

Animation: Constructing a cladogram

Animation: Interpreting a cladogram

Animation: Current evolutionary tree

23.6 Molecular Phylogenetics

- Contemporary systematists use the structure of proteins and nucleic acids in their analyses. Molecular characters provide abundant data and can be compared among many morphologically distinct forms of life, but because molecular similarities in different species may have evolved independently, systematists cannot always verify that they were inherited from a common ancestor.
- Molecular characters may act as molecular clocks, providing data that allows researchers to determine the times when lineages first diverged (Figure 23.10).
- The use of molecular characters in phylogenetic studies requires the sequencing and alignment of molecules (Figure 23.11). Several methods, including genetic distances, parsimony, and maximum likelihood, have been proposed for the construction of phylogenetic trees.
- Molecular phylogenetics has clarified relationships among the flowering plants (Figure 23.12) and provided insights into the evolutionary relationships of all organisms (Figure 23.13).

Animation: Cytochrome *c* comparison

Questions

Self-Test Questions

1. The evolutionary history of a group of organisms is called its:
   - a. classification.
   - b. taxonomy.
   - c. phylogeny.
   - d. domain.
   - e. outgroup.

2. In the Linnaean hierarchy, the organisms classified within the same taxonomic category are called:
   - a. a phylum.
   - b. a taxon.
   - c. a genus.
   - d. a binomial.
   - e. an epithet.

3. When systematists study morphological or behavioral traits to reconstruct the evolutionary history of a group of animals, they assume that:
   - a. similarities and differences in phenotypic characters reflect underlying genetic similarities and differences.
   - b. the animals use exactly the same traits to identify appropriate mates.
   - c. differences in these traits caused speciation in the past.
   - d. the adaptive value of these traits can be explained.
   - e. variations in these traits are produced by environmental effects during development.

4. Which statement best describes the concept of mosaic evolution?
   - a. Some phenotypic variation is caused by environmental factors.
   - b. Homologous characters are those that different organisms inherit from a common ancestor.
   - c. Different organismal traits may reflect the same part of an organism's genome.
   - d. Some characters evolve more quickly than others.
   - e. The fossil record provides clues about the ancestral versions of characters.

5. Which of the following pairs of structures are homoplasious?
   - a. the wing skeleton of a bird and the wing skeleton of a bat
   - b. the wing of a bird and the wing of a fly
   - c. the eye of a fish and the eye of a human
   - d. the bones in the foot of a duck and the bones in the foot of a chicken
   - e. the adhesive toe pads on the right hind foot of an *Anolis* lizard and those on the left hind foot

6. Which of the following does *not* help systematists determine which version of a morphological character is ancestral and which is derived?
   - a. outgroup comparison
   - b. patterns of embryonic development
   - c. studies of the fossil record
   - d. studies of the character in more related species
   - e. dating of the character by molecular clocks

7. In a cladistic analysis, a systematist groups together organisms that share:
   - a. derived homologous traits.
   - b. derived homoplasious traits.
   - c. ancestral homologous traits.
   - d. ancestral homoplasious traits.
   - e. all of the above.

8. A monophyletic taxon is one that contains:
   - a. an ancestor and all of its descendants.
   - b. an ancestor and some of its descendants.
   - c. organisms from different evolutionary lineages.
   - d. an ancestor and those descendants that still resemble it.
   - e. organisms that resemble each other because they live in similar environments.

9. Which of the following is *not* an advantage of using molecular characters in a systematic analysis?
   - a. Molecular characters provide abundant data.
   - b. Systematists can compare molecules among species that are morphologically very similar.
   - c. Systematists can compare molecules among species that share few morphological characters.
   - d. Amino acid sequences in proteins are generally not influenced by environmental factors.
   - e. Systematists can easily determine whether base substitutions in the DNA of two species are homologous.

10. To construct a cladogram by applying the principles of parsimony to molecular sequence data, one would:
    - a. start by making assumptions about variations in the rates at which different DNA segments evolve.
    - b. group together organisms that share the largest number of ancestral sequences.
    - c. group together organisms that share derived sequences, matching the groups to those defined by morphological characters.
    - d. group together organisms that share derived sequences, minimizing the number of hypothesized evolutionary changes.
    - e. identify derived sequences by studying the embryology of the organisms.

Questions for Discussion

1. Systematists use both amino acid sequences and DNA sequences to determine evolutionary relationships. Think about the genetic code (Section 15.1), and explain why phylogenetic hypotheses based on DNA sequences may be more accurate than those based on amino acid sequences.

2. Traditional evolutionary systematists identify the Reptilia as one class of vertebrates, even though we know that this taxon is paraphyletic. Describe the advantages and disadvantages of defining paraphyletic taxa in a classification.

3. The following table provides information about the distribution of ancestral and derived states for six systematic characters (labeled 1 through 6) in five species (labeled A through E). A "d" means that the species has the derived form of the character, and an "a" means that it has the ancestral form. Construct a cladogram for the five species using the principle of parsimony; in other words, assume that each derived character evolved only once in this group of organisms. Mark the branches of the cladogram to show where each character changed from the ancestral to the derived state.

Species	Character					
	1	2	3	4	5	6
A	a	a	a	a	a	a
B	d	a	a	a	a	d
C	d	d	d	a	a	a
D	d	d	d	a	d	a
E	d	d	a	d	a	a

4. Imagine that you are a systematist studying a group of little-known flowering plants. You discover that the phylogenetic tree based on flower morphology differs dramatically from the phylogenetic tree based on DNA sequences. How would you try to resolve the discrepancy? Which tree would you believe is more accurate?

5. Create an imaginary phylogenetic tree for an ancestral species and its 10 descendants. Circle a monophyletic group, a polyphyletic group, and a paraphyletic group on the tree. Explain why the groups you identify match the definitions of the three types of groups.

Experimental Analysis

Imagine that you are trying to determine the evolutionary relationships among six groups of animals that look very much alike because they have few measurable morphological characters. What data would you collect to reconstruct their phylogenetic history?

Evolution Link

How do the two models of macroevolution (gradualist versus punctuated equilibrium) relate to the philosophies of phylogenetic inference espoused by traditional evolutionary systematists and cladists? You may want to review material in Section 22.3 before answering this question.

Black smoker hydrothermal vents on the ocean floor. Many scientists support the theory that life developed near hydrothermal vents, where superheated, mineral-rich water is found.

Dr. Ken Macdonald/SPL/Photo Researchers, Inc.

24 The Origin of Life

WHY IT MATTERS

In 1927, Belgian priest and astronomer George Lemaître proposed the Big Bang Theory, which is now the dominant scientific theory about the origin of the universe. According to this theory, an incomprehensibly vast explosion about 14 billion years ago produced the matter and energy of our universe. Most of the matter was initially distributed in clouds of gas and dust; some of these clouds still exist today **(Figure 24.1)**. As the universe expanded, gravitational attraction caused the dust clouds to condense in some regions into more concentrated collections of matter. In our small corner of the early universe, the dust clouds condensed into the sun and its surrounding planets, including Earth.

Earth is estimated to have formed approximately 4.6 billion years ago, when it condensed out of cosmic dust and began its long transition into the environment we know today. There is no record of the time when life first formed, but microscopic deposits resembling bacteria have been found in Australia, in rocks laid down as sediments about 3.5 billion years ago during the Archaean era (inset to Figure 24.1). If these deposits are actually fossil prokaryotes, then life may have appeared during the first billion years or so of Earth's existence.

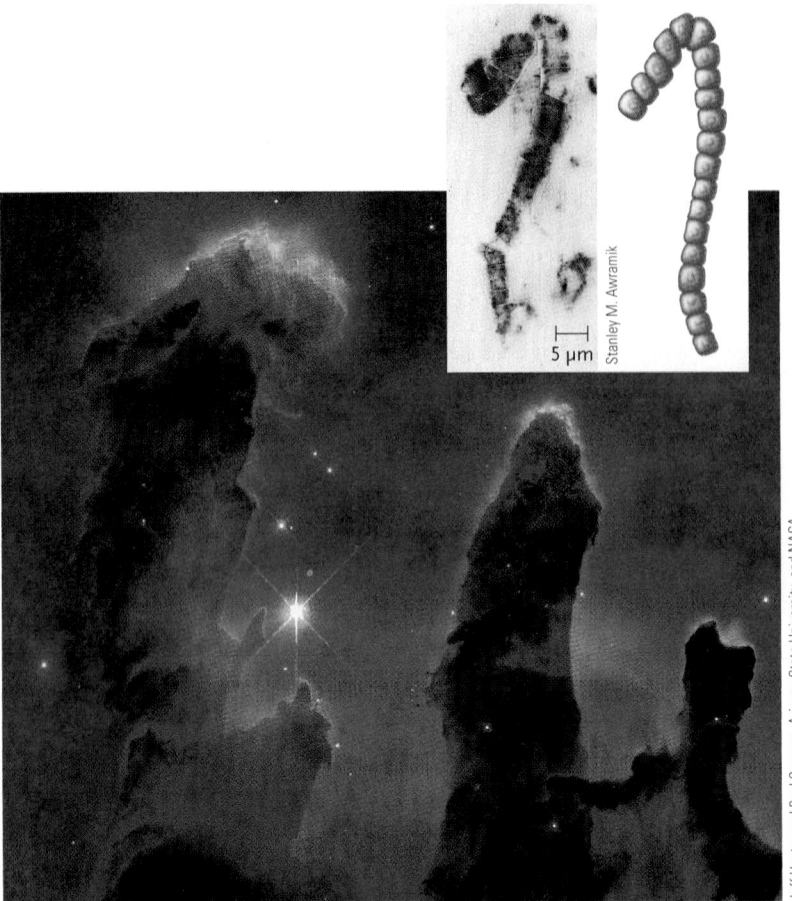

Figure 24.1
The Eagle nebula, a cloud of gas and dust particles some 7000 light years from Earth. Gas is condensing and forming stars, and perhaps planets, in this nebula. The inset shows structures that are believed to be a strand of fossil prokaryote cells in a rock sample 3.5 billion years old.

Figure 24.2 outlines the key events in the early evolution of life, which we will examine in this chapter. The earliest events are uncertain, but probably include the formation of organic molecules and the development of **protocells,** primitive cell-like structures that have some of the properties of life and that might have been the precursors of cells. Prokaryotic cells arose during the first billion years or so after the formation of Earth, and about 500 million years later some of them developed the capacity to perform photosynthesis, which released oxygen into the atmosphere. The oxygen-enriched environment was probably essential to the development of the first eukaryotic cells, which may have occurred as long as 2.2 billion years ago.

24.1 The Formation of Molecules Necessary for Life

All present-day living cells are complex; they have (1) a boundary membrane separating the cell interior from the exterior; (2) one or more nucleic acid coding molecules located in a nuclear region (a nucleus in eukaryotes and a nucleoid region in prokaryotes); (3) a

system using the coded information to make proteins and, through them, other biological molecules; and (4) a metabolic system providing energy for these activities. Because these systems are so complex, it is highly unlikely that living cells appeared suddenly from nonliving matter. Rather, there must have been a transition from nonliving to living matter.

No fossils or other records exist to inform us about this transition, but much evidence supports the idea that life did emerge from the nonliving world. Living organisms are composed entirely of elements common in the nonliving, physical world on Earth and throughout the universe. Moreover, all of the reactions that sustain life are elaborations of those in the physical world. Most scientists study the origin of life by assuming that it originated from nonliving matter on Earth, through chemical and physical processes no different from those operating today. Hypotheses made under these assumptions are testable to the extent that the chemical and physical processes can be duplicated in the laboratory.

But some scientists have not ruled out an extraterrestrial origin of life. Analysis of meteorites has shown that they contain some organic molecules characteristic of living organisms. Could a living cell or organism have arrived in such a way? Most scientists believe it is unlikely that a cell or an organism could have survived a long journey in space, even if protected from radiation, or that it could have survived intense heating while traveling through Earth's atmosphere and the actual impact with Earth. However, other scientists argue that conditions inside some meteorites might have been less extreme and allowed "life" to continue. At this point the hypothesis that life arrived on Earth by interplanetary transport cannot be ruled out. Nonetheless, even if a living organism arrived from space and

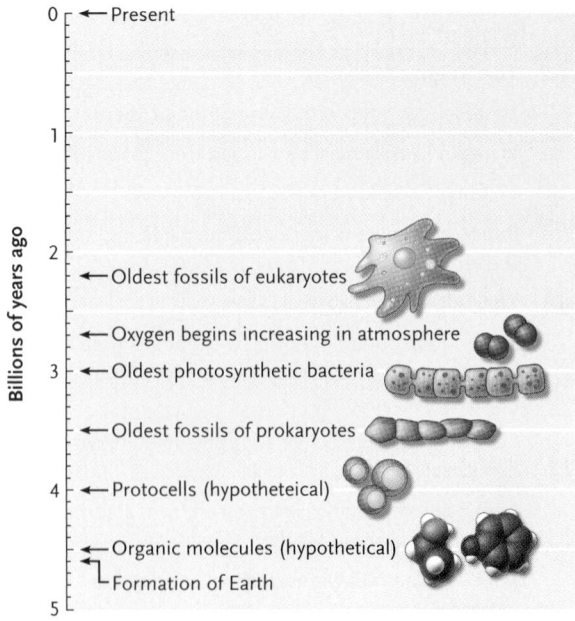

Figure 24.2
A timeline for the evolution of cellular life.

spawned a population on this planet, life would still have had to arise from nonliving matter in a similar way on the organism's home planet.

Conditions on Primordial Earth Led to the Formation of Organic Molecules

As we noted in the introduction, astronomers estimate that our solar system condensed from an interstellar dust cloud some 4.6 billion years ago. Intense heat and pressure generated in the central region of the cloud by the condensation set off a thermonuclear reaction that established the star of our solar system, the sun. The remainder of the spiraling dust and gas condensed into the planets and other bodies orbiting the sun.

Gravitational compression caused internal temperatures in our planet to rise to 1000° to 3000°C, causing its matter to melt and stratify into layers. Metallic elements sank to the core and lighter substances, such as silicates, carbides, and sulfides of the metallic elements, floated to the surface **(Figure 24.3)**. As the planet radiated away some of its heat, the surface layers cooled and solidified into the rocks of the crust. Earth's gravitational pull was strong enough to hold an atmosphere around the planet, derived partly from the original dust cloud and partly from gases released from the planet's interior as it cooled.

Primordial Earth met several basic conditions necessary for life to begin. Although its gravitational pull was strong enough to retain an atmosphere, it was not strong enough to compress the atmospheric gases into liquid form. Earth's distance from the sun was such that, on average, sunlight warmed the surface enough to keep much of the liquid water (much of which may have come from icy objects from the main asteroid belt colliding with Earth) from freezing, but not enough to boil the water. This allowed liquid water to accumulate in rivers, lakes, and seas. *Liquid water is essential for the chemistry of biological systems* (see Chapter 2).

Evaporation of water at the surface would have contributed water vapor to the atmosphere. Besides water vapor, the primordial atmosphere probably contained hydrogen and nitrogen molecules. Erupting volcanoes probably released large quantities of hydrogen sulfide, carbon dioxide, and carbon monoxide. Any molecular oxygen would have reacted with elements of the crust and atmosphere to form oxides. Spontaneous reactions of hydrogen, nitrogen, and carbon would have produced ammonia (NH_3) and methane (CH_4).

As Earth's surface cooled, natural sources of energy caused chemical bonds to break and reform, leading to the formation of organic molecules. In addition to sunlight and electrical discharges during storms, radioactivity from atomic decay and heat from volcanoes, geysers, and hydrothermal (hot water) vents in the sea floor all acted on the primordial atmosphere and crust—as they still do today. As many as a half-billion years may have passed before the concentrations of organic mol-

Figure 24.3
An artist's depiction of Earth during its early cooling stage, still too hot to support life.

ecules reached levels where their interactions formed more complex organic substances. We now consider the current thinking about how simple molecules were converted into the key molecules of life.

The Oparin-Haldane Hypothesis Initiated Scientific Investigations into the Origin of Life

Scientific efforts to explain the origin of life began with a major hypothesis proposed independently in the 1920s by two investigators, Aleksandr I. Oparin, a Russian plant biochemist at Moscow State University in Russia, and J. B. S. Haldane, a Scottish geneticist and evolutionary biologist at Cambridge University in England. Their hypothesis rested on the critical assumption that Earth's primordial atmosphere was radically different from today's atmosphere. They proposed that, rather than being an oxygen-rich (oxidizing) atmosphere as it is now, the early atmosphere was composed of substances such as hydrogen (H_2), methane (CH_4), ammonia (NH_3), and water, which are *fully reduced*—they contain the maximum possible number of electrons and hydrogens (see Section 8.1). These substances, they concluded, would have given the primordial atmosphere a *reducing* character; it contained an abundance of electrons and hydrogens available for reduction reactions, which could create organic molecules from inorganic elements and compounds. Energy to drive the reductions, according to the hypothesis, came from solar energy and other natural sources such as the electrical energy of lightning in atmospheric storms.

The absence of oxygen in the primitive atmosphere is essential to the Oparin-Haldane hypothesis. Oxygen can reverse reductions by removing electrons and hydrogens from organic molecules (see Section 8.1). In other words, if oxygen was present, the newly formed molecules would have been broken down quickly by oxidation.

Oparin and Haldane proposed that reductions occurring on the primordial Earth produced great quantities of organic molecules. The molecules accumulated because the two main routes by which such substances break down today, chemical attack by oxygen and decay by microorganisms, could not take place. According to Oparin and Haldane's hypothesis, the organic substances would have became so concentrated that the oceans and other bodies of water resembled a "prebiotic soup."

Oparin and Haldane assumed that these highly concentrated organic molecules would tend to aggregate in random combinations and that, by chance, some of the combinations were able to carry out one or more primitive reactions characteristic of life, such as increasing in mass by adding new materials. Later, scientists reasoned that these combinations were able to compete successfully against less efficient combinations for space and materials in the organic soup. As a result, they persisted and became more numerous.

Chemistry Simulation Experiments Support the Oparin-Haldane Hypothesis

In the 1950s, new discoveries in chemistry provided direct support for the most basic proposals of Oparin and Haldane's hypothesis. In 1953, Stanley L. Miller, a graduate student in Harold Urey's laboratory at the University of Chicago, tested the hypothesis by creating a laboratory simulation of conditions Oparin and Haldane believed existed on early Earth. Miller placed components of a reducing atmosphere—hydrogen, methane, ammonia, and water vapor—in a closed apparatus and exposed the gases to an energy source in the form of continuously sparking electrodes (**Figure 24.4**). Water vapor was added to the "atmosphere" by boiling water in one part of the apparatus, and it was

removed by cooling and condensation in another part. After running the apparatus for only a week, Miller found a large assortment of organic compounds in the water, including urea, amino acids, and lactic, formic, and acetic acids. In fact, as much as 15% of the carbon was now in the form of organic compounds. Two percent of the carbon was in the form of amino acids, which form easily under sufficiently reducing conditions. The significance of the finding at the time was enormous: amino acids, which are essential to cellular life, could be made under the conditions scientists believed existed on early Earth.

Other chemicals have been tested in the Miller-Urey apparatus. For example, hydrogen cyanide (HCN) and formaldehyde (CH_2O) were considered likely to have been among the earliest substances formed in the primitive atmosphere. When HCN and CH_2O molecules were added to the simulated primitive atmosphere in Miller's apparatus, all the building blocks of complex biological molecules were produced. Among the products were amino acids; fatty acids; the purine and pyrimidine building blocks of nucleic acids; sugars such as glyceraldehyde, ribose, deoxyribose, glucose, fructose, mannose, and xylose; and phospholipids, which form the lipid bilayers of biological membranes.

The synthesis of complex biological molecules in a reducing atmosphere in the Miller-Urey experiment supported the Oparin-Haldane hypothesis. However, it is only a conjecture that a reducing atmosphere was present at the time key organic molecules were formed on early Earth. Indeed, current thinking is that early Earth's atmosphere was not reducing but that it contained large amounts of oxidants such as CO_2 and N_2. In such an oxidizing atmosphere, any organic molecules generated spontaneously in the environment would be oxidized quickly back to inorganic forms by combination with the oxygen in the atmosphere. This is supported experimentally: running the Miller-Urey experiment in the presence of oxygen results in essentially no organic molecules. Moreover, amino acids cannot be produced in such an atmosphere, making the origin of life impossible.

In addition, the Miller-Urey experiment required the input of a large amount of energy. In the experiment, energy was provided continuously, but in the atmosphere of early Earth it would have been delivered, at best, intermittently from lightning storms. Scientists think that amino acids and other organic compounds may well have formed under these conditions, but not in the amounts seen in the laboratory experiment.

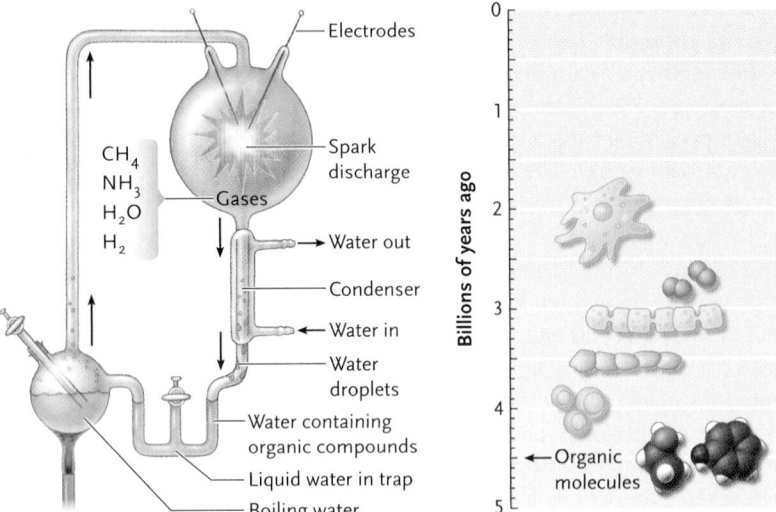

Figure 24.4

The Miller-Urey apparatus demonstrating that organic molecules can be synthesized spontaneously under conditions simulating the primordial Earth. Operation for 1 week converted 15% of the carbon in the "atmosphere" inside the apparatus into a surprising variety of organic compounds.

(Redrawn from an original courtesy of S. L. Miller. Copyright 1955 by the American Chemical Society.)

Scientists Have New Theories about the Sites for the Origin of Life

If organic compounds were not generated in a reducing atmosphere, how else could they have arisen? Scientists have developed a number of theories. All of them assume the presence of liquid water, which is a reasonable assumption. Remember that water is essential for the chemistry of biological systems (see Chapter 2). Two of the more reasonable theories are described here.

One current theory for the origin of life, which has significant support among scientists, is that life developed near hydrothermal vents in the sea floor. Many such vents exist in today's oceans, emitting bursts of mineral-rich water superheated to up to 400°C by submarine volcanoes. Scientists exploring hydrothermal vents find complex ecosystems associated with them.

Life might have originated near oceanic hydrothermal vents because reducing conditions existed there along with an abundance of the chemicals essential for life. Even now, there are high levels of hydrogen gas, methane, and ammonia around the vents. Indeed, based on simulation experiments, scientists believe that hydrothermal vents could have produced a lot more organic material than that generated in the Miller-Urey experiment. However, if life did evolve near hydrothermal vents, we would expect many present hydrothermal-vent life forms to be ancient. This is not the case: in most cases these organisms are closely related to modern non-vent organisms. Critics of the hydrothermal-vent origin of life theory also argue that the temperature at the vents is too high to permit the origin of life. The critics argue that, at the high temperature found at vents, the organic molecules are too unstable and would be destroyed as soon as they form. Supporters of the theory counter that the necessary organic molecules for life are formed not at the vent itself, but somewhere in the gradient between the hot water at the vent and the near-freezing water surrounding the vent.

Scientists debate whether organic molecules could be produced in the temperature gradient in the amounts needed. Recently, Koichiro Matsuno and his colleagues at Nagaoka University of Technology in Japan assembled an artificial system simulating the environment of ocean bottom hydrothermal vents, and added the feature of cycling materials between heat and cold. This feature accommodated the possibility that chemical products made near the vents were quenched in the surrounding colder water and then reentered the vent area where they could undergo further reactions. Their experiments demonstrated that amino acids are formed and that they can polymerize into short polypeptides under these conditions. They argue that the amounts are sufficient to form complicated molecules.

Another theory is that some organic compounds had an extraterrestrial origin. Interestingly, many of the compounds made in the Miller-Urey experiment exist in outer space. For example, a meteorite that fell on Murchison, Australia, in 1969 contained more than 90 amino acids, only 19 of which are found on Earth. Since amino acids appear to be able to survive in outer space, they could potentially have been present when Earth was formed. And perhaps other organic compounds arrived by meteor or comet impact.

STUDY BREAK

1. Why is the issue of the reducing nature of early Earth's atmosphere key to the origin of molecules necessary for life?
2. How do the theories about the sites for the origin of life differ?

24.2 The Origin of Cells

Whether organic molecules originated in the atmosphere, in hydrothermal vents, or in outer space, they still do not qualify as life. In this section, we discuss the key stage in the origin of life, the formation of the first cells.

Protocells Formed with Some of the Properties of Life

How did organic building blocks such as amino acids assemble into macromolecules such as proteins and nucleic acids? To answer this question, researchers have proposed and tested several processes. One process is the concentration of subunits by the evaporation of water. Another is *dehydration synthesis (condensations),* in which subunits assemble into larger molecules through removal of the elements of a molecule of water (see Section 3.1). Experiments with these processes under simulated conditions showed that both evaporation and condensation reactions can produce polypeptide chains from amino acids, polysaccharides from glucose and other monosaccharides, and nucleotides and nucleic acids from nitrogenous bases, ribose, and phosphates.

Scientists reason that spontaneous condensations and other reactions produced significant quantities of all the major biological molecules over the hundreds of millions of years following the initial formation of Earth. They hypothesize that the accumulation of organic matter set up the conditions necessary for the next stage, the chance assembly of molecules into aggregations that became membrane-bound to form primitive protocells. Protocells are key to the origin of life, because life depends upon reactions occurring in a controlled and sequestered environment, the cell. Researchers have proposed several mechanisms for the assembly of organic molecules into aggregates, each of which has been successfully duplicated in laboratory experiments simulating primordial conditions. Two of those mechanisms are absorption into clays and lipid bilayer assembly.

Absorption into Clays. Could clays have provided an ideal environment for molecular aggregation and interaction on the primitive Earth? Clays consist of very thin layers of minerals separated by layers of water only a few nanometers thick. The layered structure readily absorbs ions and organic molecules and promotes their interactions, including condensations and other assembly reactions. Clays can also store potential energy, and therefore could have channeled some of the energy into reactions taking place inside them.

Several experiments have supported these proposals. For example, Noam Lahav at the Hebrew University in Israel and Sherwood Chang of NASA's Ames Research Center added amino acids to clays and exposed the mixtures to water-content changes and fluctuating temperatures, as they might be in a tidal flat. After several cycles of the fluctuating conditions, polypeptides were detected in the clays. Other researchers found that RNA nucleotides linked to phosphate chains could combine into RNA-like molecules in clays. Accumulation of these and other macromolecules in the clays could have provided an environment in which they could react to carry out the first reactions of life.

However, even if molecules became organized in clay and some of the reactions of life commenced, it is not clear how a lipid bilayer membrane could have formed around them. Such a membrane is necessary to organize the molecules into protocells, the presumed precursors of cells. (The biological importance of lipid bilayers and membranes are discussed in Sections 2.4, 3.4, and 6.1.)

Lipid Bilayer Assembly. In the 1950s, R. J. Goldacre at Chester Beatty Research Institute, London, hypothesized that protocells could have formed starting with lipid bilayers that had assembled spontaneously. In the 1970s, David W. Deamer at the University of California at Davis and other investigators tested this hypothesis, finding that phospholipids and some other types of lipid molecules could form under simulated conditions. The phospholipids self-assembled readily into bilayers when suspended in water (see Section 6.1). Often, the bilayers rounded up into stable, closed vesicles consisting of a continuous-boundary "membrane" surrounding an inner space **(Figure 24.5).**

Further tests showed that the bilayers formed in these experiments have many properties of living membranes. For example, they can incorporate proteins onto their surfaces or into the hydrophobic membrane interior, and they form vesicles that can trap other substances in the fluid enclosed by the membrane. Potentially, on early Earth, the concentration of organic molecules in such vesicles could have stimulated their growth and eventual fragmentation into smaller vesicles, providing a primitive form of reproduction. These mechanisms of aggregation, as well as others, may have worked separately or together to form protocells.

Living Cells May Have Developed from Protocells

Eventually the chemical reactions taking place in the primitive protocells became organized enough to make the transition to living cells. Of the several critical events necessary for this transition, we will look closely at two: the development of pathways that captured and harnessed the energy required to drive molecular synthesis and the development of a system for the storage, replication, and translation of information for protein synthesis. Remember that proteins are the catalysts for most cellular reactions.

Development of Energy-Harnessing Reaction Pathways. Oxidation-reduction reactions (see Section 8.1) were probably among the initial energy-releasing reactions of the primitive protocells. In an oxidation, electrons are removed from a substance; the removal releases free energy that can be used to drive synthesis and other reactions. In a reduction, electrons are added to a substance; the added electrons provide energy that can contribute to the formation of complex molecules from simpler building blocks.

At first the electrons removed in an oxidation would have been transferred directly to the substances being reduced, in a one-step process. However, the greater efficiency of stepwise energy release would have favored development of intermediate carriers and opened the way for primitive electron transfer systems. Evolved from those primitive systems are the present-day electron transfer systems of mitochondria and chloroplasts (see Sections 8.4 and 9.2).

As part of the energy-harnessing reactions, ATP became established as the coupling agent that links energy-releasing reactions to those requiring energy. ATP may first have entered protocells as one of many organic molecules absorbed from the primitive environment. Initially, it was probably simply hydrolyzed into ADP and phosphate as an energy source. Later, as protocells developed, some of the free energy released

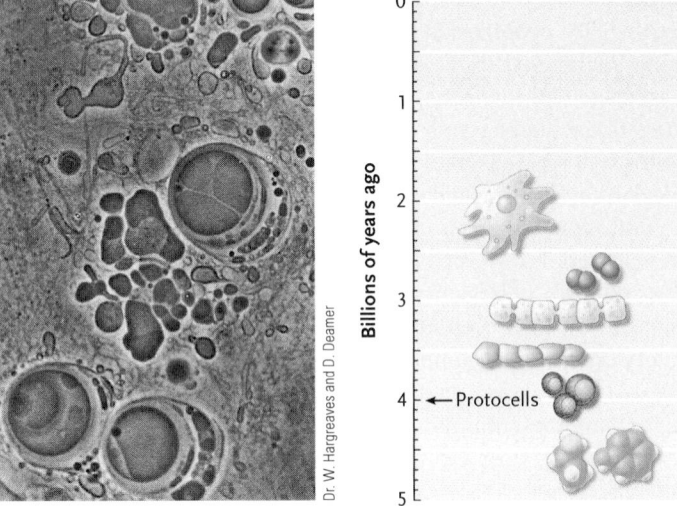

Figure 24.5
An electron micrograph of vesicles of various sizes and shapes assembled from phospholipids synthesized under simulated primordial conditions. When the vesicles are more highly magnified than in this micrograph, their walls can be seen to consist of a lipid bilayer.

Dr. W. Hargreaves and D. Deamer

Billions of years ago

0

1

2

3

4 ← Protocells

5

during electron transfer was probably used to synthesize ATP directly from ADP and inorganic phosphate. Because of the efficiency and versatility of energy transfer by ATP, it gradually became the primary substance connecting energy-releasing and energy-requiring reactions in early cells.

Origin of the Information System. A system that could store, reproduce, and translate the information required for protein synthesis was a second critical event for the transition from protocells to living cells. How the information system developed is crucial to the understanding of the origin of life.

In contemporary organisms, information flows from DNA to RNA to protein. This nucleic acid–based information system depends mostly on enzymatic proteins for replication, transcription, and translation of the nucleic acids. However, the specificity of enzymatic proteins depends on their amino acid sequences, which are determined by the sequences of nucleotides in nucleic acids. Thus, proteins depend on nucleic acids for their structure, and nucleic acids depend on proteins to catalyze their activities. How could one have appeared before the other? Scientists believe the information system developed in stages, although the order of the steps is a subject of debate. There are two main hypotheses: the RNA-first hypothesis and the protein-first hypothesis.

The *RNA-first hypothesis* states that the first genes and enzymes were RNA molecules. That is, *ribozymes*—RNA molecules capable of catalyzing biochemical reactions—may have functioned both as informational molecules and as catalysts in protocells, without requiring protein enzymes for catalytic reactions (ribozymes are discussed in Section 4.6). Thus, a self-catalyzed "RNA world" may have been the first step in the development of an information system.

Ribozymes may have originally developed by the chance assembly of RNA nucleotides taking part in oxidative and other metabolic reactions in protocells (RNA nucleotides such as ATP, NAD, and coenzyme A form important parts of many metabolic pathways, including glycolysis, respiration, and photosynthesis; see Chapters 8 and 9). The RNA molecules then developed the capacity to replicate themselves and other RNA molecules. That is, these RNA molecules acted both as templates—like mRNA—and as catalysts—like ribosomal RNA (see Section 15.4). Then ribozymes could replicate ribozymes, with no need for protein enzymes. Such self-replicating systems may have provided the basis of an RNA-based informational system, and founded the RNA world. *Insights from the Molecular Revolution* describes an experiment in which ribozymes that can replicate RNA were generated in a test tube.

In the RNA world, DNA would have developed as a subsequent step. At first, DNA nucleotides may have been produced by random removal of an oxygen atom from the ribose subunits of the RNA nucleotides. At some point, the DNA nucleotides paired with the RNA informational molecules, and were assembled into complementary copies of the RNA sequences. Some modern day viruses carry out this RNA-to-DNA reaction using the enzyme reverse transcriptase (see Section 18.1). Once the DNA copies were made, selection may have favored DNA as the informational storage molecule because it has greater chemical stability and can be assembled into much longer coding sequences than RNA. RNA was left to function at intermediate steps between the stored information in DNA and protein synthesis, as it still does today.

As the RNA-based information system evolved, some RNAs may have acted as tRNA-like molecules, linking to amino acids and pairing with the RNA informational molecules. These associations could have led to the assembly of polypeptides of ordered sequence—the development of an RNA genetic code. When DNA took over information storage from RNA, the code would have been transferred to DNA.

Modern analysis of the ribosome, the organelle responsible for translation of mRNA (see Section 15.4) has shown that the enzyme that catalyzes the formation of a peptide bond between amino acids is a property of one of the RNA molecules of the ribosome. This finding supports the proposal that, in addition to replicating themselves, RNA molecules also generated the first proteins.

The second hypothesis, the *protein-first hypothesis,* states that proteins were the first informational molecules to arise. Then, once complex enzymes developed within protocells, nucleic acids—both DNA and RNA—were assembled enzymatically from small molecules, and replication and transcription processes developed.

Of course, we have no way of knowing exactly how life originated. Sifting through the various models and theories we can perhaps agree that there were some basic steps: (1) the abiotic (nonliving) synthesis of organic molecules such as amino acids; (2) the assembly of complex organic molecules from simple molecules, including protein or RNA or both; and (3) the aggregation of complex organic molecules inside membrane-bound protocells. Once the information system had developed in the protocells, and the protocells could divide, they had become true living cells. The advent of living cells marked the beginning of biological evolution, which depends on cells that can reproduce and pass on information to their descendants.

Prokaryotic Cells Were the First Living Cells

The change to biological evolution set the stage for the appearance of all the features of cellular life. One of these features was a nuclear region that contained the DNA of the coding system and the mechanisms replicating the DNA and transcribing it into RNA. Another feature was a cytoplasmic region containing ribosomes and the enzymes required to translate RNA informa-

Replicating the RNA World

The discovery of ribozymes led to the proposal that an RNA world was the first step in the evolution of a molecular information system that could store, reproduce, and translate the information required for protein synthesis. In an RNA world, RNA molecules would have to act both as templates for their own replication and as catalysts to carry out the replication. The catalytic ability of RNA molecules has been amply demonstrated, but could they carry out RNA replication?

Wendy K. Johnston and her coworkers at the Massachusetts Institute of Technology decided to answer this question by using a ribozyme (a catalytic RNA) as the starting point for developing an RNA molecule that could replicate itself, as might have happened during the evolution of cellular life. The ribozyme they chose is an RNA ligase, which can catalyze one of the most fundamental reactions of replication, linking together short chains of nucleotides.

To achieve their feat, Johnston and her coworkers used a technique that accomplishes molecular evolution in a test tube (the RNA ligase used to start their experiments was the product of an earlier test-tube experiment). They assembled a reaction mixture containing the RNA ligase with an added a 76-nucleotide RNA strand of random sequence to serve as a template for self-replication. They then generated 1×10^{15} versions (a quintillion!) of the ligase with different sequences concentrated in the added strand. To the mutated versions in a test tube they added RNA nucleoside triphosphates (NTPs), an RNA template chain, and an RNA primer, with the RNA primer linked covalently to the ribozyme (see **figure**).

In the initial run, the template was only two bases longer than the primer, so to be successful a ribozyme had only to add two nucleotides to the primer. To detect the successful ribozymes, the investigators used RNA nucleoside triphosphates that were tagged with a chemical label. Any ribozymes that added the nucleotides to the primer would become labeled and thus be identifiable among the unsuccessful ribozymes in the test tube.

After the first round of selection, the investigators selected the labeled ribozyme variants, which had added nucleotides to the primer, and multiplied them using PCR (the polymerase chain reaction; see Section 18.1). They then added all the elements to the test tube for another round of replication and selection. This cycle of replication and selection was repeated through 18 successive rounds. As part of the process, additional mutations were induced in the ribozymes after round 10, and the selection pressure was increased by several methods. One was to make the template longer in successive rounds, so that the ribozymes had to add more nucleotides to the primer to be successful. Another was to alter the sequence of the template chain, so that the ribozymes had to be able to copy a template of any sequence to be successful. Also, the investigators shortened the time allowed for replication in successive runs, so that ribozymes had to work faster to be successful.

By the 18th round of replication, the selection process had produced a ribozyme that could replicate an RNA template 14 nucleotides longer than the primer. The template could be of any sequence. In addition, the template did not have to be covalently linked to the ribozyme for replication to occur.

To check on the accuracy of replication, the investigators gave the 18th-round ribozyme a template chain that was 11 nucleotides longer than the primer and then sequenced 100 of the complementary chains produced by the ribozyme. Of the replication products, 89 of 100 were precise complementary copies, all matched exactly to the template. In the remaining 11 products, only 12 base mismatches were found, slightly more than one base mismatch per copy.

Thus, the selected ribozyme was able to work as an RNA polymerase, faithfully replicating an RNA template into a complementary copy and thereby meeting a major requirement for an RNA world. The research continues, with further test-tube selection experiments designed to increase the accuracy of replication, the length of the template, and the rate of replication. These are small steps compared to the enormous task involved in the evolution of a full-fledged information system, but it is likely that life evolved in the same pattern, through the accumulation of small changes over hundreds of millions of years of molecular trial and error.

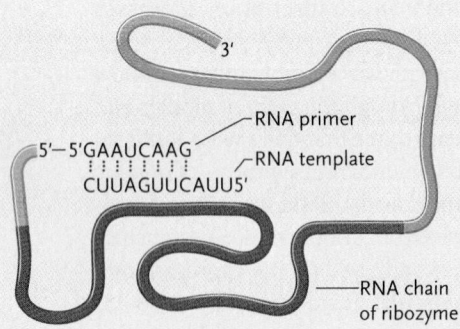

The general arrangement of sequences in the ribozymes used by Johnston and her coworkers. The blue and green portions are sequences added to provide raw materials for test-tube evolution. The template shown is three nucleotides longer than the primer and would require a ribozyme to add two nucleotides to the primer to be selected as successful (the last nucleotide in the template cannot be copied).

tion into sequences of amino acids in proteins. The cytoplasm also contained an oxidative system supplying chemical energy for protein synthesis and assembly of other required molecules. A mechanism of cell division also evolved, allowing replicated DNA to be distributed equally between daughter cells. All these systems were enclosed by a membrane controlling the flow of molecules and ions in and out of the cell. The stages leading to this level may have taken more than a billion years, occupying the period from Earth's formation 4.6 billion years ago to the earliest known prokaryotic fossils, dated as 3.5 billion years old.

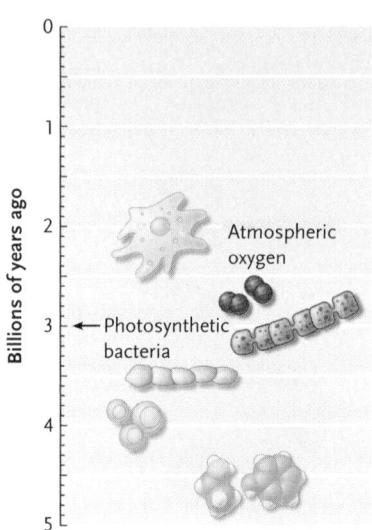

Bill Bachmann/Photo Researchers, Inc.

Figure 24.6
Stromatolites exposed at low tide in Western Australia's Shark Bay. These mounds, which consist of mineral deposits made by photosynthetic cyanobacteria, are about 2000 years old; they are highly similar in structure to fossil stromatolites that formed more than 3.0 billion years ago. As a result of photosynthesis by cyanobacteria, oxygen began to accumulate in the atmosphere.

Subsequent Events Increased the Oxidizing Nature of the Atmosphere

According to Richard E. Dickerson of UCLA and others, the earliest form of photosynthesis evolved about 3.5 billion years ago in the early prokaryotes. This form of photosynthesis probably used electron donors such as hydrogen sulfide (H_2S) that do not release oxygen. However, at some point, an enzymatic system evolved that could use the most abundant molecule of the environment, water (H_2O), as the electron donor for photosynthesis. This reaction split water into protons, electrons, and oxygen, which was released into the atmosphere.

The oxygen released by the water-splitting reaction accumulated in the atmosphere and set the stage for the development of electron transfer systems using oxygen as the final electron acceptor. These transfer systems arose when some cells developed cytochromes that could deliver low-energy electrons to oxygen (see Section 8.4). These cells were able to tap the greatest possible amount of energy from the electrons before releasing them from electron transfer, making the cells highly successful in their environment.

When might water-splitting photosynthesizers have appeared? A possible answer to this question has been found in rock formations laid down at least 3 billion years ago. These rocks contain **stromatolites**, fossils of ancient prokaryotes (cyanobacteria) that carried out photosynthesis by the water-splitting reaction **(Figure 24.6)**. Thus, oxygen-producing bacteria were present at least 3 billion years ago and perhaps evolved soon after the first prokaryotes appeared. Scientists believe that it may have taken another billion years for oxygen to accumulate to significant quantities in the atmosphere.

These major events established the preconditions for the evolution of eukaryotic cells. The next section traces this evolution, which was pivotal to the later evolution of large-scale multicellularity and the plants, animals, and the other organisms of the domain Eukarya.

STUDY BREAK

Several mechanisms have been proposed for the assembly of organic molecules into protocells. Why is the model involving a lipid bilayer membrane a particularly attractive one?

24.3 The Origins of Eukaryotic Cells

Present-day eukaryotic cells have several interrelated characteristics that distinguish them from prokaryotes: (1) the separation of DNA and cytoplasm by a nuclear envelope; (2) the presence in the cytoplasm of membrane-bound compartments with specialized metabolic and synthetic functions—mitochondria, chloroplasts, the endoplasmic reticulum (ER), and the Golgi complex, among others; and (3) highly specialized motor (contractile) proteins that move cells and internal cell parts. In this section we discuss how eukaryotes most probably evolved from associations of prokaryotes.

The Endosymbiont Hypothesis Proposes that Mitochondria and Chloroplasts Evolved from Ingested Prokaryotes

The **endosymbiont hypothesis**, put forward by Lynn Margulis at the University of Massachusetts, Amherst, proposes that the membranous organelles of eukaryotic cells, the mitochondria and chloroplasts, may each have originated from symbiotic (mutually advantageous) relationships between two prokaryotic cells **(Figure 24.7)**.

Mitochondria began to develop when photosynthetic and nonphotosynthetic prokaryotes coexisted in an oxygen-rich atmosphere. The nonphotosynthetic prokaryotes fed themselves by ingesting organic molecules from their environment. These prokaryotes included both anaerobes, unable to use oxygen as the final acceptor for electron transfer, and aerobes, fully

Figure 24.7

The endosymbiont hypothesis. Mitochondria and chloroplasts of eukaryotic cells are thought to have originated from various bacteria that lived as endosymbionts within other cells.

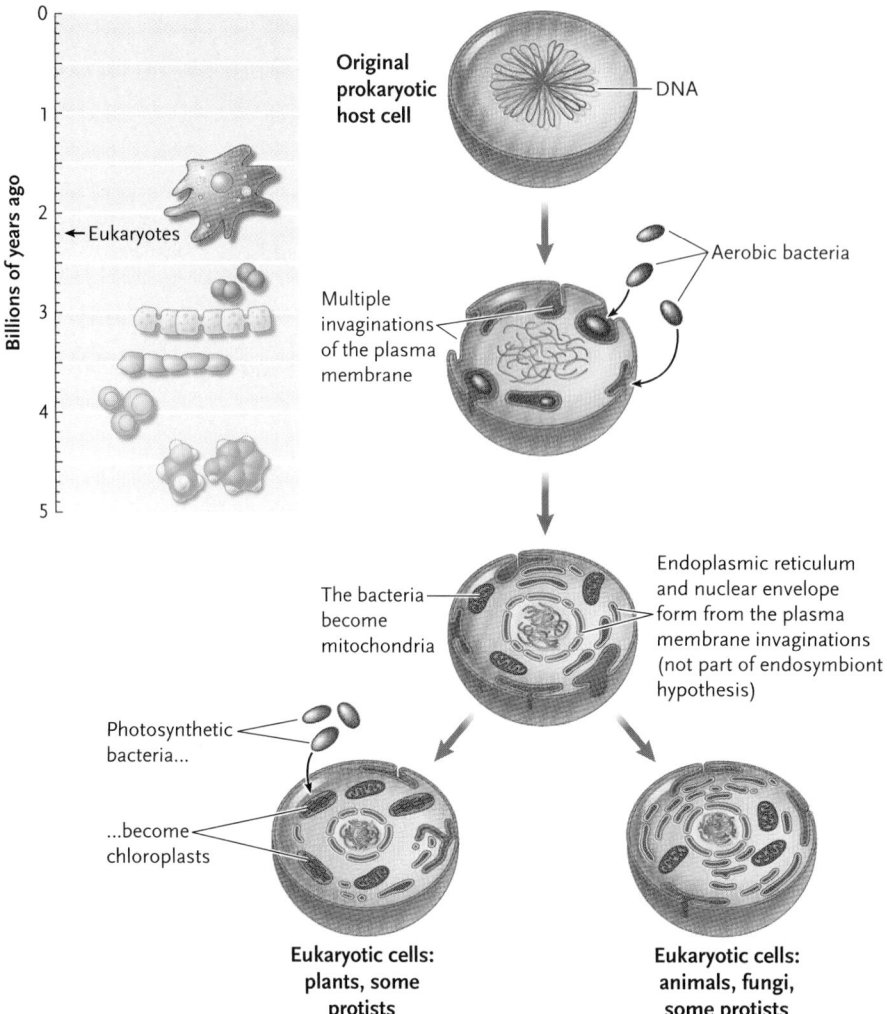

Billions of years ago

← Eukaryotes

Original prokaryotic host cell — DNA

Aerobic bacteria

Multiple invaginations of the plasma membrane

The bacteria become mitochondria

Endoplasmic reticulum and nuclear envelope form from the plasma membrane invaginations (not part of endosymbiont hypothesis)

Photosynthetic bacteria...

...become chloroplasts

Eukaryotic cells: plants, some protists

Eukaryotic cells: animals, fungi, some protists

capable of using oxygen. Only the aerobes could fully exploit the energy stored in organic molecules, but predatory anaerobes could capture that energy by eating aerobic cells. These anaerobic prokaryotes had become efficient predators, and lived by ingesting other cells. Among the ingested cells were some aerobic prokaryotes; instead of being digested, some of them persisted in the cytoplasm of the predators and continued to respire aerobically in their new location. They had become *endosymbionts,* organisms that live symbiotically within a host cell. The cytoplasm of the host anaerobe, formerly limited to the use of organic molecules as final electron acceptors, was now home to an aerobe capable of carrying out the much more efficient transfer of electrons to oxygen.

As a part of the transition to a true eukaryotic cell, the cell also evolved to acquire other membranous structures, the major ones being the nuclear envelope, the ER, and the Golgi complex. Endocytosis, the process of infolding of the plasma membrane (see Figure 5.14), is believed to be responsible for the evolution of these structures. (These events are not part of the endosymbiont hypothesis.) Researchers believe that, in cell lines leading from prokaryotes to eukaryotes, pockets of the plasma membrane formed during endocyto-

sis may have extended inward and surrounded the nuclear region. Some of these membranes fused around the DNA, forming the nuclear envelope and, hence, the nucleus. The remaining membranes formed vesicles in the cytoplasm that gave rise to the ER and the Golgi complex **(Figure 24.8).**

Next, according to the endosymbiont hypothesis, many functions duplicated in the aerobic endosymbiont were taken over by the host cell. As part of this transfer of function, most of the genes of the aerobe moved to the cell nucleus and became integrated into the host cell's DNA. At the same time, the host anaerobe became dependent for its survival on the respiratory capacity of the symbiotic aerobe. The ingested aerobe presumably benefited as well, because the host cell brought in large quantities of food molecules to be oxidized. This gradual process of mutual adaptation culminated in transformation of the cytoplasmic aerobes into mitochondria. The first eukaryotic cells had appeared, the ancestors of all modern-day eukaryotes.

The endosymbiont hypothesis proposes that a similar mechanism led to the appearance of the membrane-bound plastids (the general term for chloroplasts and related organelles, both photosynthetic and nonphotosynthetic) some time after mitochondria evolved. Plastids originated when aerobic cells that had mitochondria, but were unable to carry out photosynthesis, ingested photosynthetic prokaryotes resembling present-day cyanobacteria (see Figure 24.7). These photosynthetic prokaryotes gradually changed into plastids by evolutionary processes similar to those that produced mitochondria. The cells with both plastids and mitochondria founded the cell lines that gave rise to the modern eukaryotic algae and plants.

Several Lines of Evidence Support the Endosymbiont Hypothesis

Researchers reasoned that if the endosymbiont hypothesis is correct, then both mitochondria and plastids would have structures and biochemical reactions more like those of prokaryotes than those of eukaryotes. This has been shown to be the case. For example, both organelles typically contain circular DNA molecules that closely resemble prokaryotic DNA, and code for rRNAs and ribosomes that resemble prokaryotic forms.

Another line of evidence supports a key assumption of the endosymbiont hypothesis by showing that engulfed cells or organelles can survive in the cytoplasm of the ingesting cell. Among animals, no less than 150 living genera, distributed among 11 phyla, include species that contain eukaryotic algae or cyanobacteria as residents in the cytoplasm of their cells. For example, larvae of the marine snail *Elysia* initially contain no chloroplasts, but after they begin feeding on algae, chloroplasts from the algal cells are taken up into the cells lining the gut. When the larvae develop into adult snails, the chloroplasts continue to carry out photosynthesis in their new location and produce carbohydrates that are used by the snails. The uptake of functional chloroplasts has also been observed among the Protoctista (the protists; see Chapter 26); **Figure 24.9** shows a protist with chloroplasts that closely resemble cyanobacteria.

How long did it take for evolutionary mechanisms to produce fully eukaryotic cells? The oldest known fossil eukaryotes are 2.2 billion years old. If prokaryotic cells first evolved some 3.5 billion years ago, it took up to 1.3 billion years for eukaryotic cells to evolve from prokaryotes (see Figure 24.2). If so, this long interval probably reflects the complexity of the adaptations leading from prokaryotic to eukaryotic cells. Of course, it is possible that eukaryotic cells evolved more quickly, and we have yet to find the evidence.

Eukaryotic Cells May Have Evolved from a Common Ancestral Line Shared with Archaeans

The system of classification that has gained acceptance among biologists, and the one used in this book, groups all living organisms into three domains. One domain, the Eukarya, contains the eukaryotes. The second domain, the Bacteria, includes one of two groups of prokaryotes, the bacteria, which consists of both photosynthesizing and nonphotosynthesizing species. The third domain, the Archaea, contains the other group of prokaryotes, many of which inhabit extreme environments, including highly saline environments and hot springs.

There is little question that the three domains originated from a common ancestral cell line, because all share common fundamental characteristics—they all use the same genetic code, for example, and DNA and RNA molecules carry out the same basic functions in transcription and translation. However, the events leading from this common ancestry to the three domains of life remain unclear. The most difficult questions surround the role of the archaeans in both bacterial and eukaryotic evolution.

Archaeans have some features that are typical of bacteria, including a genome organized into a single, circular DNA molecule that is suspended in a nuclear region of the cytoplasm with no surrounding nuclear envelope. There are no membrane-bound organelles in the cytoplasm equivalent to mitochondria, chloroplasts, the ER, or the Golgi complex. However, the archaeans also have some features that are typically eukaryotic. One is the presence of interrupting, noncoding sequences called introns (see Section 15.3) in their genes;

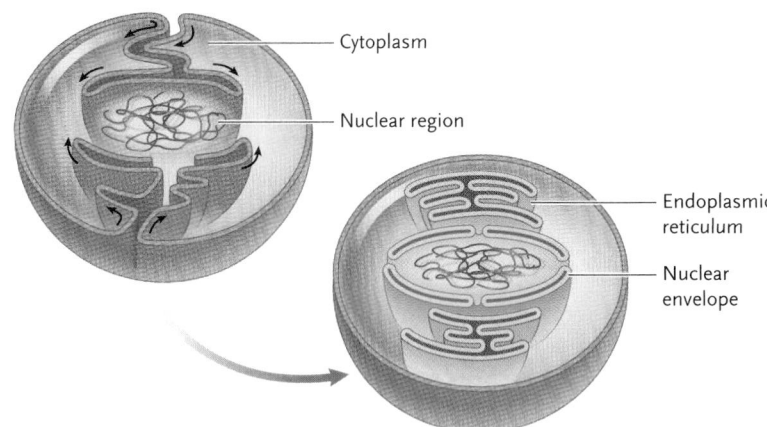

Figure 24.8
A hypothetical route for formation of the nuclear envelope and endoplasmic reticulum, through segments of the plasma membrane that were brought into the cytoplasm by endocytosis.

Cytoplasm

Nuclear region

Endoplasmic reticulum

Nuclear envelope

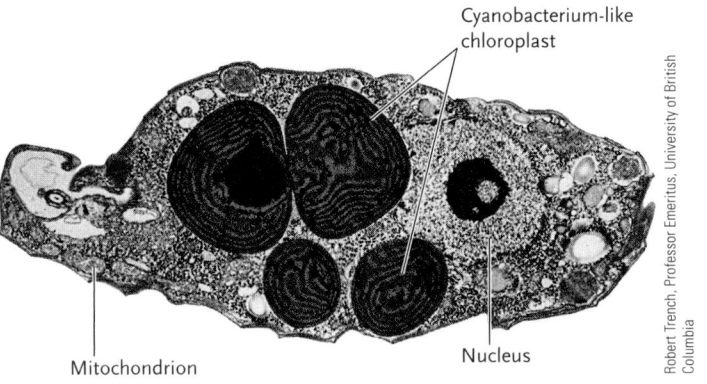

Cyanobacterium-like chloroplast

Mitochondrion

Nucleus

Robert Trench, Professor Emeritus, University of British Columbia

Figure 24.9
Cyanophora paradoxa, a protist with chloroplasts that closely resemble cyanobacteria without cell walls.

in some archaean genes that have counterparts in eukaryotes, the introns occur in exactly the same positions. By contrast, introns are rare or nonexistent in bacteria. The archaeans also have some characteristics that are unique to their domain, including features of gene and rRNA sequences, and features of cell wall and plasma membrane structure that are found nowhere else among living organisms. The characteristics shared by archaeans and eukaryotes suggest that their roots may lie in a common ancestral line that split off from the line leading to bacteria. At some point, this ancestral line split into the lines leading to archaeans and eukaryotes.

Multicellular Eukaryotes Probably Evolved in Colonies of Cells

The first eukaryotes were unicellular. They are the ancestors of the present-day diversity of unicellular eukaryotes. Multicellular eukaryotes evolved from unicellular eukaryotes and then diverged to produce the present-day multicellular eukaryotes. Molecular clock analysis indicates the first multicellular eukaryote likely arose between 800 and 1000 million years ago, while the first fossil records (of small algae) are from 600 to 800 million years ago.

According to the prevalent theory, multicellular eukaryotes arose by the congregation of cells of the same species into a colony. The ability to act in a coordinated way, probably increased the capacity of colonies to adapt to changes in the environment. Subsequently, differentiation of cells into various specialized cell types with distinct functions produced organisms with a wider range of capabilities and adaptability. Cell differentiation in a colony would have required cell signals that affected gene expression. That is, because each cell in the colony has the same genome, the development of specific func-

---

UNANSWERED QUESTIONS

What was the first polymer of life?

As discussed in this chapter, many researchers hypothesize that RNA was the first polymer of life because it both self-replicates and can catalyze chemical reactions. In addition, it is neatly connected with contemporary life, which is based on nucleic acids and proteins. There are several problems with this hypothesis, however. One of them is that the synthesis of RNA building blocks, nucleotides, and in particular their ribose fragment, is quite difficult under primordial conditions. To circumvent this difficulty, several researchers have proposed that other genetic polymers, whose monomers are simpler to synthesize, might have preceded RNA. For example, Albert Eschenmoser of the Swiss Federal Institute of Technology in Zurich, Switzerland, replaced ribose with the sugar pyranose, and Peter Nielsen from the University of Copenhagen, Denmark, synthesized a polymer with a peptide-like backbone. These polymers are stable and capable of self-replication. Another popular proposal is that the initial complement of nucleic acid bases was different from A, U, G, and C. One reason for this proposal is poor stability of cytosine in water. Although we have no evidence that transitional polymers were present on early Earth, it is important to realize that alternatives to nucleic acids exist and might be used by life elsewhere.

The protein-first hypothesis is currently not in favor with scientists, even though these polymers are excellent catalysts of chemical reactions and their building blocks, amino acids, existed on prebiotic Earth. This is because there is no known mechanism for proteins to self-replicate. Some researchers speculate that a limited replication of proteins is possible. An alternative hypothesis, supported by computer simulations, is that replication of individual polymers was not necessary at the origin of life and, instead, the reproduction of protein functions in a population was initially sufficient. Currently, neither view has much experimental support, but as we learn more about the structure and functions of small proteins major surprises might be in store.

Can we recreate protocells in a laboratory?

As you read this chapter, you must have noticed that our knowledge about the origin of life is still incomplete. But do we know enough to test our understanding by building in a laboratory a simple life capable of self-reproduction and Darwinian evolution? Several groups of scientists are attempting to do just that. Conceptually, the simplest design is "the minimal RNA cell" proposed by Jack Szostak from Harvard Medical School. It consists of only two ribozymes encapsulated in a membrane-bound structure. One of them is capable of copying both ribozymes; the other catalyzes the synthesis of the membrane-forming molecules from their precursors. In principle, such a system could self-reproduce and undergo evolution through mutations of the ribozymes. However, the apparent simplicity of this construct is somewhat deceiving—no actual ribozymes that function together in this way are currently known. An international team of scientists is attempting to build a simple cell using a set of already existing components, as originally proposed by Steen Rasmussen from Los Alamos National Laboratory and Liaohai Chen from Argonne National Laboratory. This cell would differ from everything we know, however, and would therefore represent an example of "alien life." Craig Venter and several other researchers have taken yet another approach. Starting with a simple, contemporary microorganism as a template, they are trying to delete nonessential genes or substitute natural or synthetic genes that are smaller in size. So far, each of these strategies has encountered a surprising number of conceptual and technical difficulties, and none has been successful. This shows that synthesizing life is more complex that one would expect. If any of these efforts eventually succeeds, it will open the doors not only to many new investigations on the origin of life on Earth but also to the exploration of alternative forms of life and to applications of artificial cells in biotechnology and medicine.

Andrew Pohorille heads the NASA Center for Computational Astrobiology and Fundamental Biology at NASA's Ames Research Center. He is also professor of Chemistry and Pharmaceutical Chemistry at the University of California, San Francisco. For his work on the origin of life he was awarded the 2002 NASA Exceptional Scientific Achievement Medal.

tions (phenotypes) would require intracellular signals that would change the program of gene regulation. Over time, as genomes evolved, the division of function among cells led to the evolution of the tissues and organ systems of complex eukaryotes.

Multicellularity evolved several times in early eukaryotes, producing a number of lineages of algae as well as the ancestors of present day fungi, plants, and animals.

Life May Have Been the Inevitable Consequence of the Physical Conditions of the Primitive Earth

The events outlined in this chapter, leading from Earth's origin to the appearance of eukaryotic cells, may seem improbable. But, as scientist and author George Wald of Harvard University put it, given the total time span of these events, more than 3.5 billion years, "the impossible becomes possible, the possible probable, and the probable virtually certain. One has only to wait; time itself performs the miracles."

Some researchers go a step further and maintain that the evolution of life on our planet was an inevitable outcome of the initial physical and chemical conditions established by Earth's origin, among them a reducing atmosphere (at least in some locations), a size that generates moderate gravitational forces, and a distance from the Sun that results in average surface temperatures between the freezing and boiling points of water. Given the same conditions and sufficient time, according to these scientists, it is inevitable that life has evolved or is evolving now on other planets in the universe.

The chapters to follow in this unit trace the course of evolution and its products after eukaryotic cells were added to the prokaryotes already on Earth. Among prokaryotes, evolution established two major groups, the Bacteria and Archaea; among eukaryotes, further evolution established the protists, fungi, plants, and animals. The survey begins in the next chapter with a description of present-day Bacteria and Archaea, and of the viruses that infect prokaryotes and eukaryotes.

STUDY BREAK

Summarize the key points of the theory of endosymbiont origins for mitochondria and chloroplasts.

Review

Go to ThomsonNOW™ at www.thomsonedu.com/login to access quizzing, animations, exercises, articles, and personalized homework help.

24.1 The Formation of Molecules Necessary for Life

- Living cells are characterized by a boundary membrane, one or more nucleic acid coding molecules in a nuclear region, a system for using the coded information to make proteins, and a metabolic system providing energy for those activities.

- Oparin and Haldane independently hypothesized that life arose de novo under the conditions they thought prevailed on the primitive Earth, including a reducing atmosphere that lacked oxygen. Reduction reactions, fueled by natural energy acting on the primitive atmosphere, produced organic molecules. Random aggregations of these molecules were able to carry out primitive reactions characteristic of life that gradually became more complex until life appeared. Chemistry simulation experiments support the hypothesis that organic molecules would form under these conditions (Figure 24.4).

- Present thinking is that early Earth's atmosphere was not reducing, but in fact contained significant amounts of oxidants. This has caused skepticism about Oparin and Haldane's hypothesis. One new theory proposes that life developed near hydrothermal vents in the sea floor.

Animation: Miller's reaction chamber experiment

Animation: Milestones in the history of life

24.2 The Origin of Cells

- Organic molecules produced in early Earth's environment by chance formed aggregates that became membrane-bound in protocells, primitive cell-like structures with some of the properties of life. Protocells may have been the precursors of cells (Figure 24.5).

- Next, living cells may have developed from protocells by the development of several critical components, notably energy-harnessing pathways, and a system based on nucleic acids that could store and pass on the information required to make proteins.

- Subsequently, fully cellular life evolved, with a nuclear region containing DNA and the mechanisms for copying its information into RNA messages; a cytoplasmic region containing systems for utilizing energy and systems for translating RNA messages into proteins; a membrane separating the cell from its surroundings; and a reproductive system duplicating the informational molecules and dividing them among daughter cells.

- The first living cells were prokaryotes. Eventually, some early cells developed the capacity to carry out photosynthesis using water as an electron donor; the oxygen produced as a byproduct accumulated and the oxidizing character of Earth's atmosphere increased. From this time on organic molecules produced in the environment were quickly broken down by oxidation, and life could arise only from preexisting life, as in today's world.

24.3 The Origins of Eukaryotic Cells

- According to the endosymbiont hypothesis, mitochondria developed from ingested prokaryotes that were capable of using oxygen as final electron acceptor; chloroplasts developed from ingested cyanobacteria (Figure 24.7).

- Eukaryotic structures such as the ER, Golgi complex, and nuclear envelope appeared through infoldings of the plasma membrane as a part of endocytosis (Figure 24.8).

- Multicellular eukaryotes probably evolved by differentiation of cells of the same species that had congregated into colonies. Multicellularity evolved several times, producing lineages of several algae and ancestors of fungi, plants, and animals.

Animation: Eukaryotic evolution

Questions

Self-Test Questions

1. Earth was formed _____ years ago, whereas the oldest known living cell formed about _____ years ago.
   a. 400×10^3; 3.6×10^6
   b. 4.6×10^9; 1.0×10^9
   c. 3.8×10^9; 4.6×10^7
   d. 4.6×10^9; 3.5×10^9
   e. 2.0×10^9; 600×10^6

2. Which of the following is *not* a characteristic of all living organisms?
   a. They replicate genetic information and convert the information into proteins.
   b. They pass genetic information between generations.
   c. They get energy from molecules in a controlled fashion.
   d. They use external energy to drive internal reactions requiring energy.
   e. They use mitochondria to transform energy for their cells' needs.

3. The greatest leap in evolution is from:
   a. nonlife to prokaryotes.
   b. prokaryotes to one-celled eukaryotes.
   c. ancient archaeans to modern archaeans.
   d. one-celled eukaryotes to fungi.
   e. one-celled eukaryotes to insects.

4. According to the Oparin-Haldane hypothesis, the atmosphere when life began was believed to be composed primarily of:
   a. H_2O, N_2, and CO_2.
   b. H_2, H_2O, NH_3, and CH_4.
   c. H_2O, N_2, O_2, and CO_2.
   d. O_2 and no H_2.
   e. H_2 only.

5. The Miller-Urey experiment:
   a. was based on the belief the atmosphere was oxidizing.
   b. was able to synthesize amino acids and macromolecules from reduced gases.
   c. did not require much energy or a continuous energy source to keep synthesizing.
   d. did not require water to produce organic molecules.
   e. used free oxygen as a reactant.

6. An unknown organism was found in a park. It was one-celled, had no nuclear membrane around its DNA, and contained no mitochondria and no chloroplasts. It belongs to the group:
   a. eukaryotes.
   b. vertebrates or plants.
   c. bacteria or archaea.
   d. plants or fungi.
   e. fungi.

7. Hydrothermal vents are theorized as sources for the origin of life because:
   a. the temperature of the water around them supports most life.
   b. most organic molecules undergo dehydration synthesis at high temperatures.
   c. the amino acids degrade in the colder water soon after synthesis.
   d. water is needed by living things.
   e. reducing conditions with needed molecules surround them.

8. The proposed first macromolecule for the beginning of life is:
   a. DNA to code the cell's activities.
   b. protein to be used in cell functions.
   c. ribozymes to act as information and catalytic molecules.
   d. H_2O as needed by all living things.
   e. chlorophyll for photosynthesis.

9. As part of the evolution of eukaryotic cell, endocytosis, the process of infolding of the plasma membrane, led to the formation of:
   a. chromosomes.
   b. the cell wall.
   c. ribosomes.
   d. the nuclear envelope.
   e. microtubules.

10. Which of the following is *not* part of the evidence supporting the theory of endosymbiosis: Both mitochondria and plastids:
   a. are each the size of many bacterial cells.
   b. have structures and biochemical reactions more like prokaryotes than eukaryotes.
   c. code mRNA, rRNA, and tRNA similar to prokaryotes.
   d. contain circular DNA.
   e. have DNA similar to nuclear DNA.

Questions for Discussion

1. What evidence supports the idea that life originated through inanimate chemical processes?

2. Explain, in terms of hydrophilic and hydrophobic interactions, how protocells might have formed in water from aggregations of lipids, proteins, and nucleic acids.

3. What conditions would likely be necessary for a planet located elsewhere in the universe to evolve life similar to that on Earth?

4. Most scientists agree that life on Earth can arise only from preexisting life, but also that life could have originated spontaneously on the primordial Earth. Can you reconcile these seemingly contradictory statements?

Experimental Analysis

Suppose you discover a hot springs-fed pool on a remote mountain never before explored by humans. In the pool you find a cellular life form that appears to be prokaryotic. What experiments would you do to distinguish between the alternative hypotheses that this organism evolved on Earth from ancestral prokaryotes or is descended from a life form that arrived at that location in a meteorite?

Evolution Link

In the evolution unit, you learned how changes in the environment can foster evolutionary changes in biological systems. How have changing biological systems influenced the evolution of changes in Earth's physical environment?

How Would You Vote?

Private companies make millions of dollars selling an enzyme first isolated from cells in Yellowstone National Park. Should the federal government let private companies bioprospect within the boundaries of national parks, as long as it shares in the profits from any discoveries? Go to www.thomsonedu.com/login to investigate both sides of the issue and then vote.

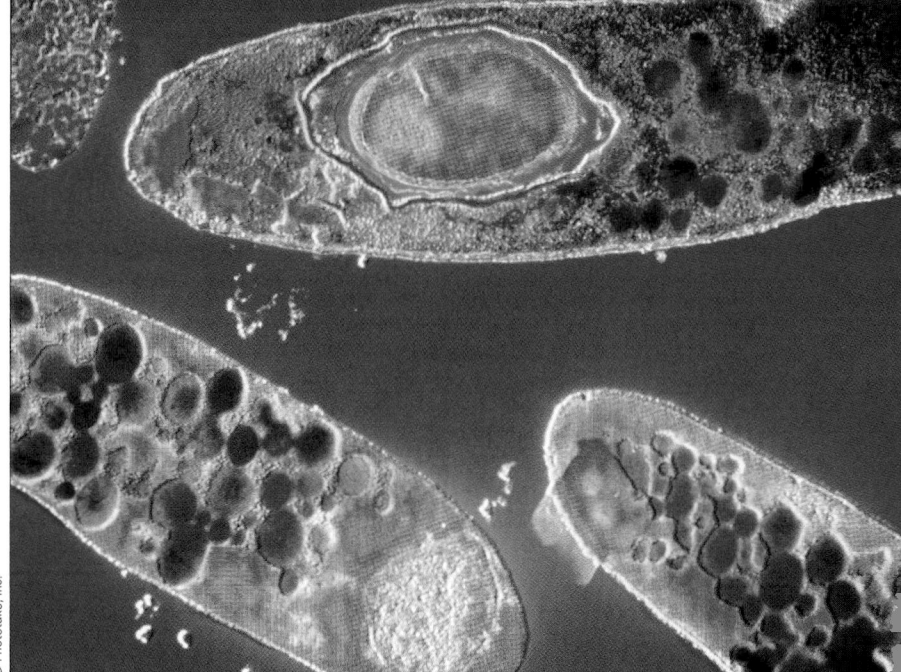

The bacterium *Clostridium butyricum*, one of the *Clostridium* species that produces the toxin botulin (colorized TEM).

© Phototake, Inc.

25 Prokaryotes and Viruses

WHY IT MATTERS

You wait in line with anticipation at a fast-food restaurant, biding your time until you reach the counter and get your hamburger. Somewhere in the back of your mind may be the worry that the hamburger will contain bacteria that could make you sick or even cost you your life. The hamburger you receive will be well done, almost to the crispy stage, because of that fear. Not too many years ago, people were sickened, and a few even died, because their fast-food hamburgers were contaminated by a pathogenic strain of the bacterium *Escherichia coli*, the normally harmless bacteria that inhabit our intestinal tract. Since then, fast-food restaurants have cooked their hamburgers well beyond the point required to kill any lurking *E. coli* or other pathogenic bacteria.

The bacterium *E. coli* is a prokaryote, an organism lacking a true nucleus. Prokaryotes, the main topic of this chapter, are the smallest organisms of the world **(Figure 25.1)**. Few species are more than 1 to 2 μm long; from 500 to 1000 of them would fit side by side across the dot above this letter "i."

Prokaryotes are small, but their total collective mass (their *biomass*) on Earth may be greater than that of all plant life. They colonize

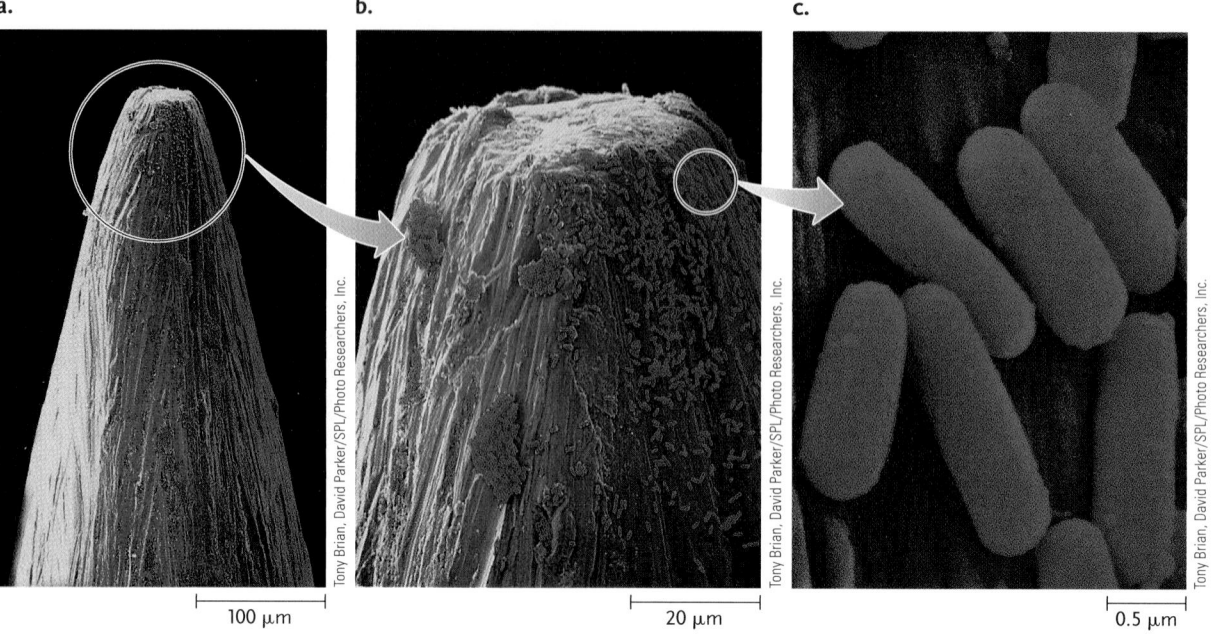

a.   b.   c.

100 μm   20 μm   0.5 μm

Figure 25.1

Bacillus bacteria on the point of a pin. Cells magnified **(a)** 70 times, **(b)** 350 times, and **(c)** 14,000 times.

every niche on Earth that supports life, meaning that they are found essentially everywhere. Huge numbers of bacteria inhabit surfaces and cavities of the human body, including the skin, the mouth and nasal passages, the large intestine, and the vagina. Collectively, the bacteria in and on the human body outnumber all the cells in the body.

Biologists classify prokaryotes into two of the three domains of life, the **Archaea** and the **Bacteria** (the third domain, the **Eukarya**, includes all eukaryotes). Bacteria are the prokaryotic organisms most familiar to us, including many types responsible for diseases of humans and other animals and many other types found in a wide variety of ecosystems. Many of the Archaea (*archaios* = ancient) live under conditions so extreme, including high salinity, acidity, or temperature, that their environments cannot be tolerated by other organisms, including bacteria.

As a group, prokaryotes have a wide range of metabolic capabilities. Their metabolic activities are crucial for maintenance of the biosphere. In particular, prokaryotes are the key players in the life-sustaining recycling of the elements carbon, nitrogen, and oxygen, and this recycling is necessary to sustain life. For example, prokaryotes are involved in breaking down organic material in dead plants and animals, releasing carbon dioxide that is used for plant growth. Prokaryotes are also the only living source of nitrogen, an element essential for all life. And a significant amount of the oxygen in the atmosphere originates from bacterial photosynthesis. An illustration of prokaryotes' importance is Biosphere 2, an attempt by scientists to build a completely closed ecosystem in Arizona. The attempt failed, in part because the researchers did not have a complete enough understanding of the activities of the microorganisms in the soil. Through respiration by soil microorganisms, the oxygen level in the Biosphere

structure decreased to lower-than-expected levels and the ecosystem ceased to be self-sustaining. This small-scale example illustrates the essential role of prokaryotes in enabling life of all forms to exist.

Prokaryotes also have a great impact on the lives of humans. Among other things, they are important for the production of certain foods, they carry out chemical reactions that are of importance in industry, they are used for the production of pharmaceutical products, they cause diseases, and they are used for bioremediation of polluted sites.

Viruses, the other subject of this chapter, are also extremely important in the biosphere. Smaller still than prokaryotes, viruses are present in the environment in even greater numbers than bacteria. In some aquatic habitats, viruses that infect bacteria alone exist at concentrations approaching 100 million per milliliter! Viruses are classified separately from the three domains of life because they are considered to be nonliving. However, viruses of one kind or another can infect the cells of just about every kind of living organism.

25.1 Prokaryotic Structure and Function

Prokaryotes show great diversity in their ability to colonize areas that can sustain life. Their cells are small, but relatively complex in organization. For instance, although they do not have a membrane-bound nucleus or organelles, their DNA and some proteins are localized in particular places. They vary in how their cell membrane is protected, and some species have specialized surface structures that protect them from their environment or that enable them to move actively. Prokaryotes also show great diversity in the

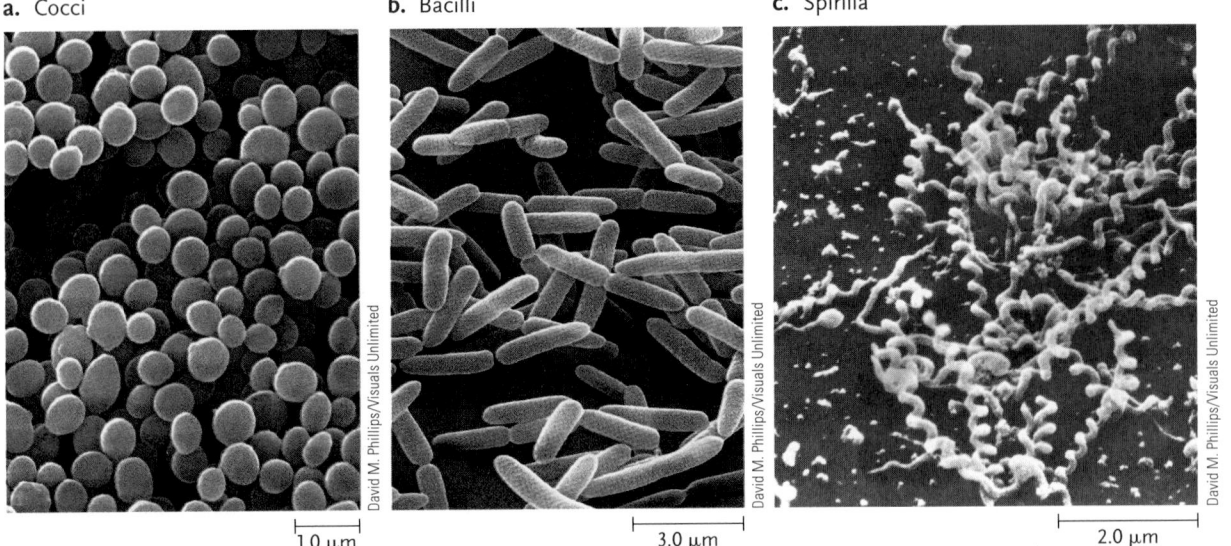

a. Cocci   **b.** Bacilli   **c.** Spirilla

1.0 μm   3.0 μm   2.0 μm

Figure 25.2

Common shapes among prokaryotes. **(a)** Scanning electron microscope (SEM) image of *Micrococcus*, a coccus bacterium; **(b)** SEM of *Salmonella*, a bacillus bacterium; **(c)** SEM of *Spiroplasma*, a spiral prokaryote of the spirillum type.

ways they obtain energy and in their metabolic activities.

The diversity of prokaryotes has arisen through rapid adaptation to their environments as a result of evolution by natural selection. Genetic variability in prokaryotic populations, the basis for this rapid adaptation, derives largely from mutation, and to a lesser degree from transfer of genes between organisms by transformation, transduction, and conjugation (see Chapter 17). Since prokaryotes have much shorter generation times than eukaryotes, and small genomes (roughly 1000 times smaller than an average eukaryote), prokaryotes have roughly 1000 times more mutations per gene, per unit time, per individual than is the case for eukaryotes. Further, prokaryotes typically have much larger population sizes than eukaryotes, contributing to their greater genetic variability. In short, prokaryotes have an enormous capacity to adapt and this has been key to their evolutionary success.

Prokaryotes Are Simple in Structure Compared with Eukaryotic Cells

Prokaryote cells examined under an electron microscope typically reveal little more than a cell wall and plasma membrane surrounding a cytoplasm with DNA concentrated in one region and ribosomes scattered throughout. They have no cytoplasmic organelles equivalent to the mitochondria, chloroplasts, endoplasmic reticulum, or Golgi complex of eukaryotic cells. With few exceptions, the reactions carried out by these organelles in eukaryotes are distributed between the cytoplasmic solution and the plasma membrane in prokaryotes.

Three shapes are common among prokaryotes: spherical, rodlike, and spiral **(Figure 25.2).** The spherical prokaryotes are **cocci** (singular, *coccus* = berry). Cylindrical or rod-shaped prokaryotes are **bacilli** (singular, *bacillus* = small staff or rod). The spiral prokaryotes are the **vibrios** (*vibrare* = to vibrate), which are curved and commalike, and the **spirilla** (singular, spirillum), which are twisted helically like a corkscrew. Among the prokaryotes of all structural types are some that live singly and others that link into chains or aggregates of cells.

Internal Structures. The genome of most prokaryotes consists of a single, circular DNA molecule called the *prokaryotic chromosome*. There are exceptions: a few bacterial species, for example the causative agent of Lyme disease (*Borrelia borgdorfri*), have a linear chromosome. Genome sequencing projects have shown that the range of genome sizes among bacteria and archaeans is about 20-fold, with the smallest genome, that of *Mycoplasma genitalium*, being about 580,000 bp. In all prokaryotes, the chromosome is packed into an area of the cell called the **nucleoid**. There is no nucleolus in the nucleoid, and it has no boundary membranes equivalent to the nuclear envelope of eukaryotes **(Figure 25.3).**

Besides the DNA of the nucleoid, many prokaryotes also contain small circles of DNA called **plasmids**, distributed in the cytoplasm. The plasmids, which often contain genes with functions that supplement those in the nucleoid, contain a replication origin that allows them to replicate along with the nucleoid DNA and be passed on during cell division (see Section 14.5).

Prokaryotic ribosomes are smaller than eukaryotic ribosomes and contain fewer proteins and RNA molecules. Archaeal ribosomes resemble those of bacteria

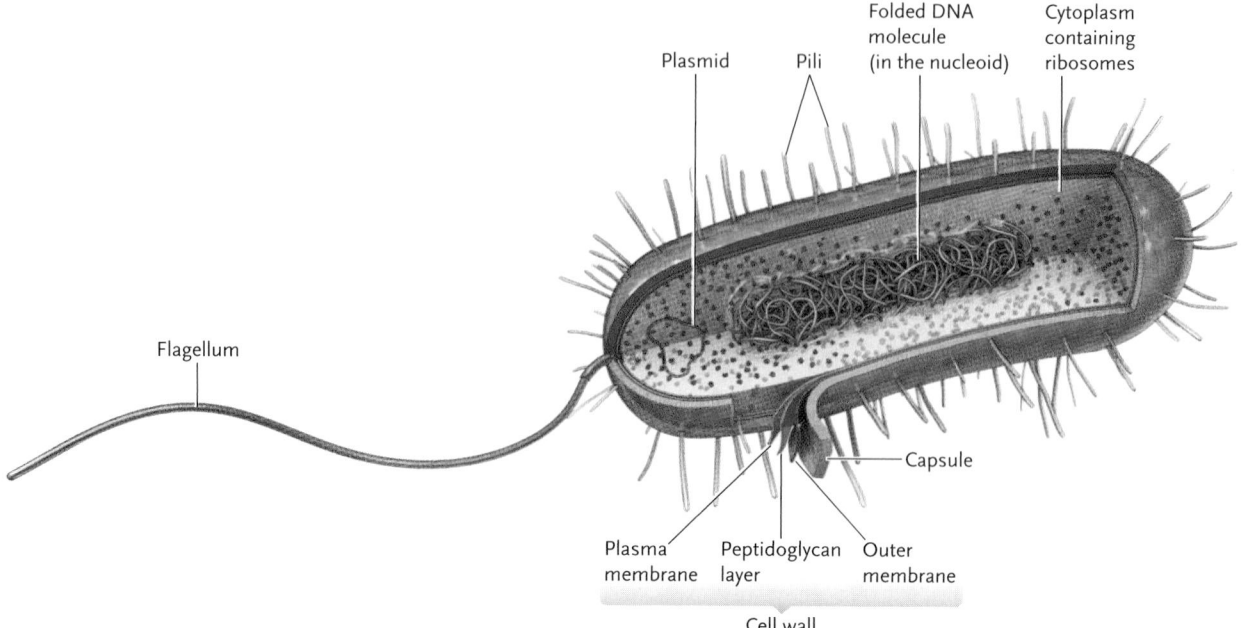

Plasmid Pili Folded DNA molecule (in the nucleoid) Cytoplasm containing ribosomes

Flagellum

Capsule

Plasma membrane Peptidoglycan layer Outer membrane

Cell wall

Figure 25.3
The structures of a bacterial cell.

in size, but differ in structure. Scientists have demonstrated that, with some differences in detail, bacterial ribosomes carry out protein synthesis by the same mechanisms as those of eukaryotes (see Section 15.4). Interestingly, protein synthesis in archaeans is a combination of bacterial and eukaryotic processes, with some unique archaeal features. As a result, antibiotics that stop bacterial infections by targeting ribosome activity do not stop protein synthesis of archaeans.

Some prokaryotes are capable of photosynthesis. These microorganisms have membranous structures corresponding to those that carry out photosynthesis in plants, but they are organized differently.

The cytoplasm of many prokaryotes also contains storage granules holding glycogen, lipids, phosphates, or other materials. The stored material is used as an energy reserve or a source of building blocks for synthetic reactions.

Prokaryotic Cell Walls. All prokaryotic cells are bounded by a plasma membrane. This membrane must withstand both high intracellular osmotic pressures and the action of natural chemicals in the environment that have detergent properties. Most prokaryotes have one or more layers of materials coating the plasma membrane that provide the necessary protection.

Bacteria typically are surrounded by a cell wall that lies outside the plasma membrane. The primary structural molecules of bacterial cell walls are **peptidoglycans**, polymeric substances formed from a polysaccharide backbone tied together by short polypeptides. The peptidoglycans vary in chemical structure among different bacterial species.

Differences in bacterial cell wall composition are important clinically. In 1882, Hans Christian Gram, a Danish physician, developed a staining method to dis-

tinguish in bodily fluids two types of bacteria, each of which could cause pneumonia. In this **Gram stain technique,** an investigator treats bacteria with the dye crystal violet and then with iodine, which fixes the dye to the cell wall. Next the bacteria are washed with alcohol, and then treated with a second strain, either fuchsin or safranin. Bacteria that appear purple after these steps have retained the crystal violet stain; they are **Gram-positive.** Bacteria that appear pink after these steps have lost the crystal violet stain in the alcohol wash and are stained pink with the second dye; they are **Gram-negative.** (Gram-positive cells also react with the second dye, but the stain does not affect the color imparted by the crystal violet.)

The staining difference reflects differences in the cell walls of the bacteria **(Figure 25.4).** The cell wall of Gram-positive bacteria consists of a thick peptidoglycan layer (see Figure 25.4a). In contrast, the cell wall of Gram-negative bacteria consists of a thin layer of peptidoglycans (see Figure 25.4b). Outside of the thin cell wall is an additional boundary membrane, called the **outer membrane,** which covers the peptidoglycan layer. The outer membrane contains **lipopolysaccharides,** assembled from lipid and polysaccharide subunits found nowhere else in nature. The outer membrane protects Gram-negative bacteria from potentially damaging substances in their environment. For example, the outer membrane of *E. coli* protects it from the detergent effects of bile released into the intestinal tract, which otherwise would lyse (break open) the bacterium and kill it.

Rapidly distinguishing between Gram-positive and Gram-negative bacteria is important for determining the first line of treatment for bacterial-caused human diseases. Most pathogenic bacteria are Gram-negative species; their outer membrane protects them against the body's defense systems and blocks the en-

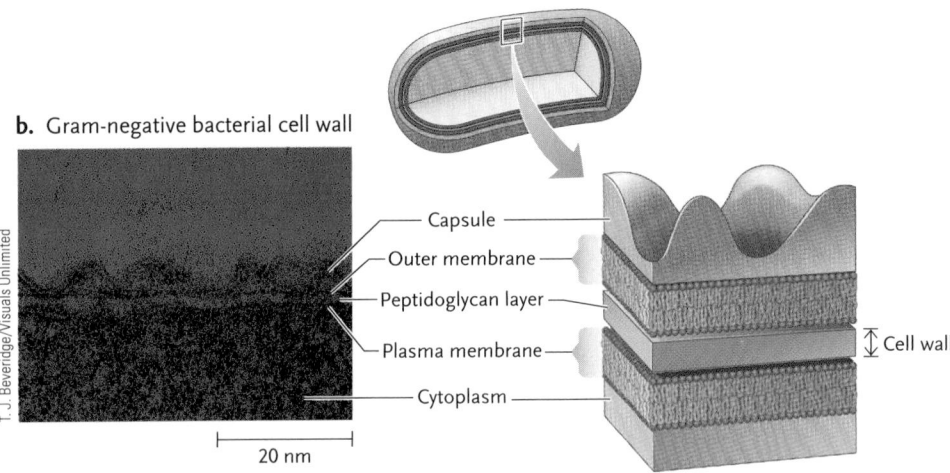

a. Gram-positive bacterial cell wall

T. J. Beveridge/Visuals Unlimited

20 nm

Peptidoglycan layer

Plasma membrane

Cytoplasm

Capsule
may be
present

Cell
wall

b. Gram-negative bacterial cell wall

T. J. Beveridge/Visuals Unlimited

20 nm

Capsule

Outer membrane

Peptidoglycan layer

Plasma membrane

Cytoplasm

Cell wall

Figure 25.4
Cell wall structure in Gram-positive and Gram-negative bacteria. **(a)** The thick cell wall in Gram-positive bacteria. **(b)** The thin cell wall of Gram-negative bacteria.

try of drugs such as antibiotics. For example, the antibiotic penicillin blocks new bacterial cell wall formation by inhibiting peptidoglycan crosslinking. The weakened cell wall soon leads to the death of the bacterium. Penicillin is effective against Gram-positive pathogens, but it is less effective against Gram-negative pathogens because their outer membrane inhibits entry of the antibiotic.

Many Gram-positive and Gram-negative bacteria are surrounded by a slime coat typically composed of polysaccharides. When the slime is attached to the cells, it is a **capsule (Figure 25.5),** and when it is loosely associated with the cells, it is a **slime layer,** although there is no sharp distinction between the two. Depending on the species, the capsule ranges from a layer that is thinner than the cell wall to many times thicker than the entire cell. Slime typically is essential for survival of the bacteria in natural environments. For example, the slime helps protect the cells from desiccation and antibiotics.

In many bacteria, the capsule prevents bacterial viruses and molecules such as enzymes, antibiotics, and antibodies from reaching the cell surface. In many pathogenic bacteria, the presence or absence of the protective capsule differentiates infective from noninfective forms. For example, normal *Streptococcus pneumoniae* bacteria are capsulated and are virulent, caus-

ing severe pneumonia in humans and other mammals. Mutant *S. pneumoniae* without capsules are nonvirulent and can easily be eliminated by the body's immune system if they are injected into mice or other animals (see Section 14.1).

Flagella and Pili. Many bacteria and archaeans can move actively through liquids and across wet surfaces. The most common mechanism for movement involves

Figure 25.5
The capsule surrounding the cell wall of *Rhizobium,* a Gram-negative soil bacterium.

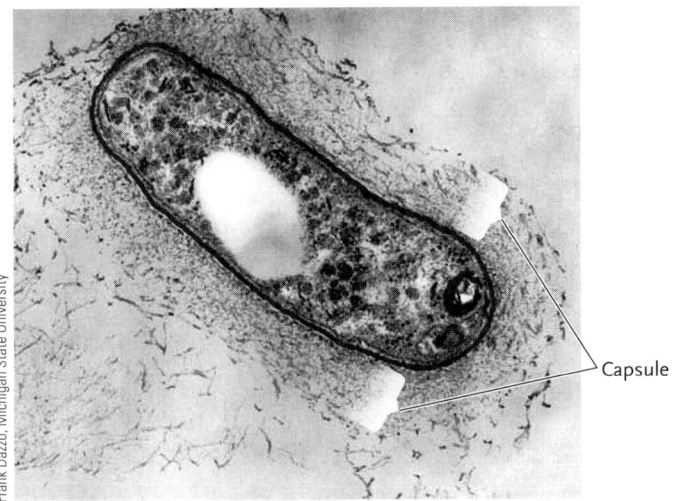

Frank Dazzo, Michigan State University

Capsule

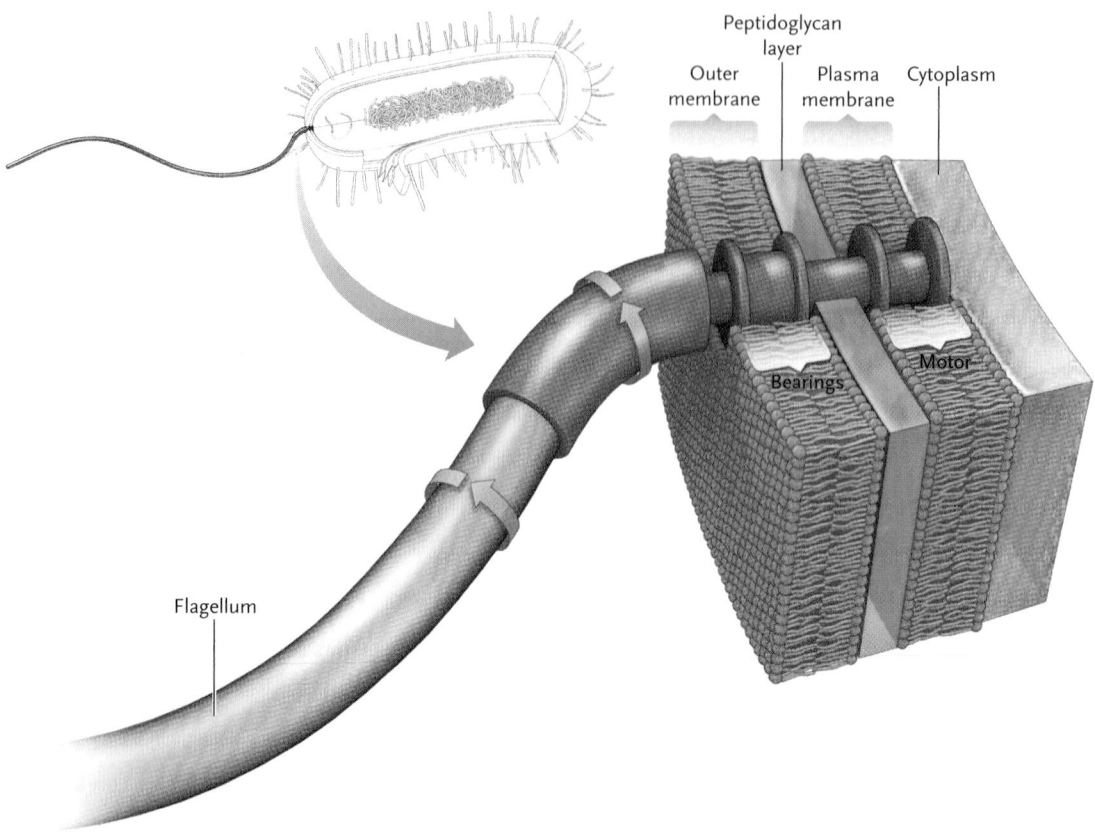

Peptidoglycan layer

Outer membrane

Plasma membrane

Cytoplasm

Motor

Bearings

Flagellum

Figure 25.6
A flagellum of a Gram-negative bacterium. A proton (H⁺) gradient drives the motor, which rotates the flagellum in a counterclockwise direction.

the action of **flagella** (singular, flagellum, meaning whip) extending from the cell wall **(Figure 25.6)**. These flagella are much smaller and simpler than the flagella of eukaryotic cells and contain no microtubules (eukaryotic flagella are discussed in Section 5.3).

Bacterial flagella consist of a helical fiber of protein that rotates in a socket in the cell wall and plasma membrane, much like the propeller of a boat. The rotation, produced by what is essentially a tiny electric motor, pushes the cell through liquid. The motor is powered by a gradient of hydrogen or sodium ions, which flow through it as positive charges, creating an electrical repulsion that makes the flagellum rotate.

Archaeal flagella are analogous, not homologous, to bacterial flagella. That is, they carry out the same function, but the genes for the two types of flagellar systems are different.

Some bacteria and archaeans have rigid shafts of protein called **pili** (singular, pilus) extending from their cell walls **(Figure 25.7)**. Among bacteria, pili are characteristic primarily of Gram-negative bacteria; relatively few Gram-positive bacteria produce these structures. A recognition protein at the tip of a pilus allows bacterial cells to adhere to other cells. One type, called *sex pili*, allows bacterial cells to adhere to each other as a prelude to conjugation, a primitive form of sexual reproduction (see Section 17.1). Other types help bacteria to bind to animal cells. For example, *Neisseria gonorrhoeae*, the Gram-negative bacterium

that causes gonorrhea, has pili that allow it to attach to cells of the throat, eye, urogenital tract, or rectum in humans.

In sum, prokaryotes are simpler and less structurally diverse than eukaryotic cells. However, bacteria are much more diverse metabolically, as we will now explore.

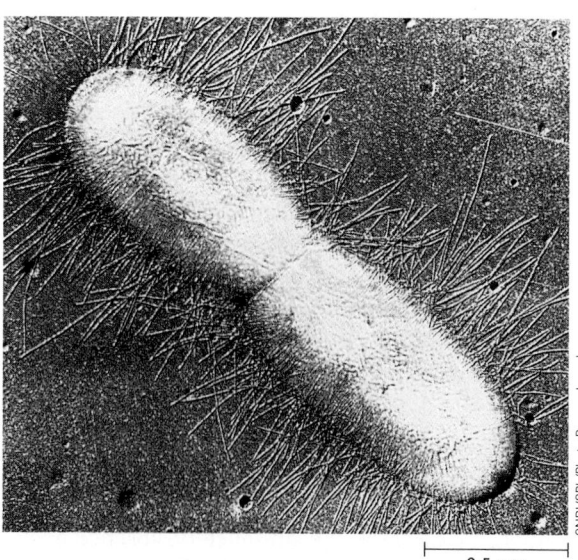

0.5 μm

CNRI/SPL/Photo Researchers, Inc.

Figure 25.7
Pili extending from the surface of a dividing *E. coli* bacterium.

Prokaryotes Have the Greatest Metabolic Diversity of All Living Organisms

All organisms take in carbon and energy in some form, but prokaryotes show the greatest diversity in their modes of securing these resources **(Figure 25.8)**. Some prokaryotes are **autotrophs** (*auto* = self; *troph* = nourishment), meaning that they, like plants, obtain carbon from an inorganic molecule, CO_2. (Note that, while CO_2 contains a carbon atom, oxides containing carbon are considered inorganic molecules.) Others are **heterotrophs,** meaning that they, like humans and other animals, obtain carbon from organic molecules. Bacterial heterotrophs obtain carbon from the organic molecules of living hosts, or from organic molecules in the products, wastes, or remains of dead organisms.

Prokaryotes are also divided according to the source of the energy they use to drive biological activities. **Chemotrophs** (*chemo* = chemical; *troph* = nourishment) obtain energy by oxidizing inorganic or organic substances, while **phototrophs** obtain energy from light. Combining carbon source and energy gives us the following four types (see Figure 25.8):

1. **Chemoautotrophs:** Prokaryotic chemoautotrophs obtain energy by oxidizing inorganic substances such as hydrogen, iron, sulfur, ammonia, nitrites, and nitrates and use CO_2 as their carbon source. They use the electrons they remove in the oxidations to make organic molecules by reducing CO_2 or to provide energy for ATP synthesis (using an electron transfer system embedded in the plasma membrane). Chemoautotrophs occur widely among the prokaryotes, including many bacteria and most archaeans, but are not found among eukaryotes.
2. **Chemoheterotrophs:** Prokaryotic chemoheterotrophs oxidize organic molecules as their energy source and obtain carbon in organic form. They include most of the bacteria that cause disease in humans, domestic animals, and plants and many bacteria responsible for decomposing matter. They are the largest prokaryotic group in terms of numbers of species.
3. **Photoautotrophs:** Photoautotrophs are photosynthetic organisms that use light as their energy source and CO_2 as their carbon source. They include several groups of bacteria, for example, the *cyanobacteria,* the *green sulfur bacteria,* and the *purple sulfur bacteria,* as well as plants and many protists. The cyanobacteria use water as their source of electrons for reducing CO_2, while the two types of sulfur bacteria use sulfur or sulfur compounds.
4. **Photoheterotrophs:** Photoheterotrophs use light as their ultimate energy source but obtain carbon in organic form rather than as CO_2. Photoheterotrophs are limited to two groups of bacteria, the *green* and *purple nonsulfur bacteria.* "Nonsulfur"

Carbon source		Energy source	
		Oxidation of molecules*	Light
CO_2		**CHEMOAUTOTROPH** Found in some bacteria and archaeans; not found in eukaryotes	**PHOTOAUTOTROPH** Found in some photosynthetic bacteria, in some protists, and in plants
Organic molecules		**CHEMOHETEROTROPH** Include some bacteria and archaeans, and also in protists, fungi, animals, and plants	**PHOTOHETEROTROPH** Found in some photosynthetic bacteria

*Inorganic molecules for chemoautotrophs and organic molecules for chemoheterotrophs.

Figure 25.8

Modes of nutrition among Bacteria and Archaea. All four modes of nutrition occur in the Bacteria with chemoheterotrophs as the most common type; among the Archaea, chemoautotrophs are most common, while others are chemoheterotrophs.

indicates they are unable to oxidize sulfur or other inorganic substances as an ultimate source of electrons for reductions; instead, they use a variety of substrates, including H_2, alcohols, or organic acids.

Prokaryotes Differ in Whether Oxygen Can Be Used in Their Metabolism

Prokaryotes also differ in how their metabolic systems function with respect to oxygen (see Chapter 8). **Aerobes** require oxygen for cellular respiration (in other words, oxygen is the final electron acceptor for that process); **obligate aerobes** cannot grow without oxygen. **Anaerobes** do not require oxygen to live. **Obligate anaerobes** are poisoned by oxygen, and survive either by fermentation, in which organic molecules are the final electron acceptors (see Section 8.5), or by a form of respiration in which inorganic molecules such as nitrate ions (NO_3^-) or sulfate ions (SO_4^{2-}) are used as final electron acceptors. **Facultative anaerobes** use O_2 when it is present, but under anaerobic conditions, they live by fermentation.

Prokaryotes Fix and Metabolize Nitrogen

Nitrogen is a component of amino acids and nucleotides and, hence, is of vital importance for the cell. Prokaryotes are able to metabolize nitrogen in many forms. For example, a number of bacteria and archaeans are able to reduce atmospheric nitrogen (N_2, the major component of Earth's atmosphere) to ammonia (NH_3), a process called **nitrogen fixation.** The ammonia is quickly ionized to ammonium (NH_4^+), which the cell then uses in biosynthetic pathways to produce nitrogen-containing molecules such as amino acids and nucleic acids. Nitrogen fixation is an exclusively prokaryotic process and is the only means of replenishing the nitrogen sources used by most microorganisms and by all plants and animals. In other words, all organisms use nitrogen fixed by bacteria. Examples of nitrogen-fixing bacteria are some of the cyanobacteria and *Azotobacter* among free-living bacteria and *Rhizobium* among bacteria that are symbiotic with plants (see Chapter 33).

Not all bacteria convert fixed nitrogen directly into organic molecules. Some bacteria carry out **nitrification**, the conversion of ammonium (NH_4^+) to nitrate (NO_3^-). This is carried out in two steps by two types of *nitrifying bacteria*. One type of nitrifying bacteria converts ammonium to nitrite (NO_2^-) (for example, *Nitrosomonas*), while the other converts nitrite to nitrate (for example, *Nitrobacter*). Because of this specialization, both types of nitrifying bacteria are usually present in soils and water, with some converting ammonium to nitrite and others using that nitrite to produce nitrate. The nitrate can be used by plants and fungi to incorporate nitrogen into organic molecules. Animals obtain nitrogen in organic form by eating other organisms.

In sum, nitrification makes nitrogen available to many other organisms, including plants and animals and bacteria that cannot metabolize ammonia. You will learn more about nitrogen metabolism in connection with the nitrogen cycle (see Chapter 51). The metabolic versatility of the prokaryotes is one factor that accounts for their abundance and persistence on the planet; another factor is their impressive reproductive capacity.

Prokaryotes Reproduce Asexually or, Rarely, by a Form of Sexual Reproduction

In prokaryotes, asexual reproduction is the normal mode of reproduction. In this process, a parent cell divides by binary fission into two daughter cells that are exact genetic copies of the parent (see Figure 10.18).

Conjugation, in which two parent cells join or "mate," occurs in some bacterial and archaeal species. Conjugation depends upon genes carried by a plasmid that replicates separately from the prokaryotic chromosome. Usually only the plasmid is passed on during conjugation, but in some bacteria, the plasmid integrates into the chromosome of the host so that host genes transfer from one parent (donor) to the other (recipient). Genetic recombination then occurs, thereby achieving a prokaryotic form of sexual reproduction. The recombinant cell divides to produce daughter cells that differ in genetic information from either parent. (Conjugation and the transfer of host genes between bacterial cells is described in Section 17.1.)

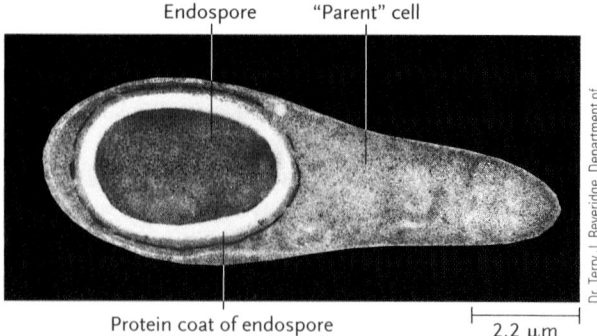

Endospore   "Parent" cell

Protein coat of endospore   2.2 μm

Figure 25.9

A developing endospore of the bacterium *Clostridium tetani*, a dangerous pathogen that causes tetanus.

A small number of bacteria can produce an **endospore**, so-called because it develops *within* the cell (**Figure 25.9**). The endospore, which typically develops when environmental conditions become unfavorable, is metabolically inactive and highly resistant to heat, desiccation, and attack by enzymes or other chemical agents. When an endospore forms, binary fission cuts the parent cell into parts of unequal size. The larger cell then envelops the smaller one and surrounds it with a tough, chemically resistant protein coat; the smaller cell develops into the endospore. Rupture of the larger cell releases the endospore to the environment. If environmental conditions become favorable for growth, the spore germinates: it becomes permeable, water enters the cell, its surface coat breaks, and the cell is released in a metabolically active form.

No one is certain how long endospores can survive. There are claims that endospores survive for thousands or millions of years, but the data are controversial.

In Nature, Bacteria May Live in Communities Attached to a Surface

Researchers grow prokaryotes as individuals in liquid cultures or as isolated colonies on solid media. The results from studies using pure cultures have been crucial in developing an understanding of, among many other things, the nature of the genetic material, DNA replication, gene expression, and gene regulation. But, since pure cultures are extremely rare in nature, some of the information learned from them may not apply to populations of prokaryotes in nature.

Researchers have discovered that, in nature, prokaryotes may live in communities where they interact in a variety of ways. The communities may consist of one or more species of bacteria, or archaeans, or both bacteria and archaeans. Eukaryotic microorganisms may also be in the communities. One important type of prokaryotic community is known as a **biofilm**, which consists of a complex aggregation of microorganisms attached to a surface. Benefits of biofilm formation to prokaryotes include adherence of the organisms to hospitable surfaces, the transfer of genes between species, and living off the products of other organisms in the biofilm. Biofilms form on any surface with sufficient water and nutrients for prokaryotes to grow. For instance, they may be found on lake surfaces, on rocks in freshwater or marine environments (making them slippery), surrounding plant roots and root hairs, and on animal tissues such as intestinal mucosa and teeth (human dental plaque is a biofilm).

Biofilms have practical consequences for humans, both beneficial and detrimental. On the beneficial side, for example, biofilms on solid supports are used in sewage treatment plants for processing organic matter before the water is discharged, and they can be effective in bioremediation (biological clean-up) of toxic organic molecules contaminating the groundwater. On the detrimental side, however, biofilms can be harmful to human health. For example, biofilms adhere to many kinds of surgical equipment and supplies, including catheters and synthetic implants such as pacemakers and artificial joints. When pathogenic bacteria are involved, infections occur. Those infections are difficult to treat, because pathogenic bacteria in a biofilm are up to 1000 times more resistant to antibiotics than are the same bacteria in liquid cultures. Other examples of medical conditions resulting from activities of biofilms include middle-ear infections, bacterial endocarditis (an infection of the heart's inner lining or the heart valves), and Legionnaire's disease (an acute respiratory infection caused by breathing in pieces of biofilms containing the pathogenic bacterium *Legionnella*).

How does a biofilm form? Imagine a surface, living or environmental, over which water containing nutrients is flowing **(Figure 25.10)**. The surface rapidly becomes coated with polymeric organic molecules from the liquid, such as polysaccharides or glycoproteins. Once the surface is conditioned with organic molecules,

free bacteria attach in a reversible manner in a matter of seconds (see Figure 25.10, step 1). If the bacteria remain attached, the association may become irreversible (step 2), at which point the bacteria grow and divide on the surface (step 3). Next, the physiology of the bacteria changes and the cells begin to secrete *extracellular polymer substances* (EPS), a slimy, gluelike substance similar to the molecules found in bacterial slime layers. EPS extends between cells in the mixture, forming a matrix that binds cells to each other and anchors the complex to the surface, thereby establishing the biofilm (step 4). The slime layer entraps a variety of materials, such as dead cells and insoluble minerals. Over time, other organisms are attracted to and join the biofilm; depending on the environment, these may include other bacterial species, algae, fungi, or protozoa, producing diverse microbial communities (step 5).

Genomic and proteomic studies have shown that the changes in prokaryote physiology accompanying the formation of a biofilm result from marked changes in the prokaryote's gene expression pattern. In effect, the prokaryote becomes a significantly different organism. This change has large implications when pathogenic bacteria are involved, for example, because most research on the control of those bacteria is done with liquid cultures. The challenge now is to devise new treatment strategies for biofilm-caused diseases. If we can gain a better understanding of the genetic changes involved in the transition from free-floating to biofilm state, then perhaps we can devise treatments that will switch the bacteria back to the free-living state, where they are more susceptible to antibiotics.

In sum, we must recognize that rather than living as individuals as once was thought, prokaryotes typically live in communities in nature. Much remains to be learned about how bacteria form a biofilm, how the change in gene expression during the transition is regulated, and how they interact.

In the next two sections, we describe the major groups of prokaryotes.

Figure 25.10
Steps in the formation of a biofilm.

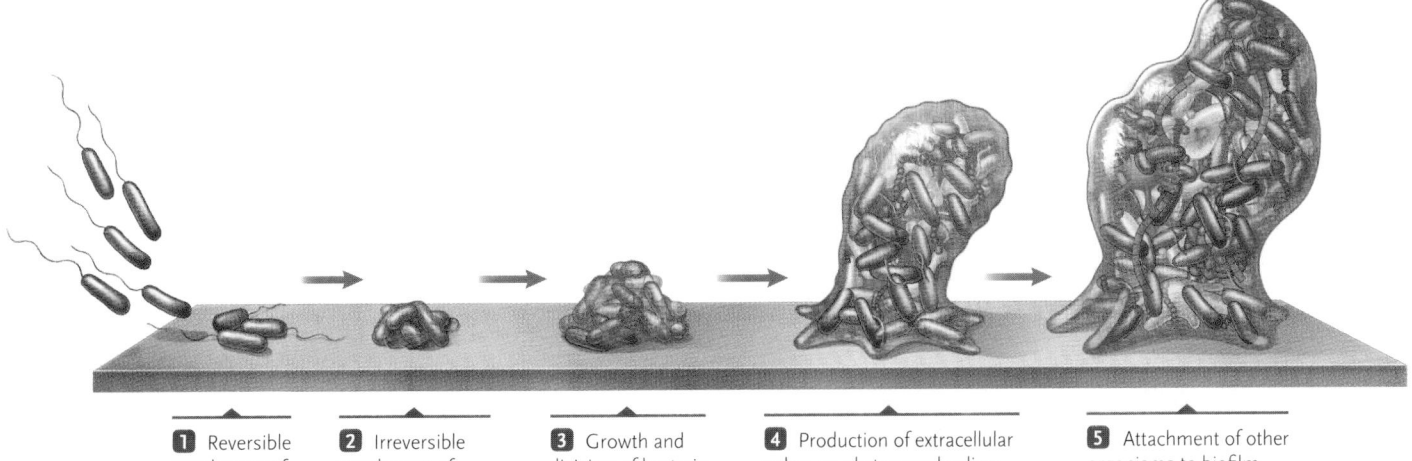

1 Reversible attachment of bacteria (sec)

2 Irreversible attachment of bacteria (sec–min)

3 Growth and division of bacteria (hr–days)

4 Production of extracellular polymer substances, leading to biofilm formation (hr–days)

5 Attachment of other organisms to biofilm (days–months)

1. What distinguishes a prokaryotic cell from a eukaryotic cell?
2. What is the difference between a chemoheterotroph and a photoautotroph?
3. What is the difference between an obligate anaerobe and a facultative anaerobe?
4. What is the difference between nitrogen fixation and nitrification? Why are nitrogen-fixing prokaryotes important?
5. What is a biofilm? Give an example of a biofilm that is beneficial to humans and one that is harmful.

25.2 The Domain Bacteria

Prokaryote classification has been revolutionized by molecular techniques that allow researchers to obtain and compare bacterial DNA, RNA, and protein sequences as tests of relatedness and evolutionary origin. Ribosomal RNA (rRNA) sequences, which are present in all organisms, have been most widely used in the evolutionary studies of prokaryotes. Under the assumption that mutations causing sequence changes occur at constant rates, researchers use the degree of sequence divergence to estimate how much time has passed since any two species shared the same ancestor (see Section 23.6). The sequencing studies thus provide a means to trace the evolutionary origins of prokaryotes and to place them in taxonomic groups. In this way, prokaryotes have been classified into the domains Bacteria and Archaea (the Eukarya is the third domain of life) **(Figure 25.11)**. Researchers have identified several evolutionary branches within each prokaryote domain. In the future, full genomic sequences will

likely be compared to refine this taxonomic classification. We discuss the major groups of the domain Bacteria in this section and of the domain Archaea in the next section.

Molecular Studies Reveal More Than a Dozen Evolutionary Branches in the Bacteria

Sequencing studies reveal that bacteria have more than 12 distinct and separate evolutionary branches, variously called kingdoms, subkingdoms, phyla, or divisions. Although all these groups are of significance to science, medicine, and the human economy, we restrict our discussion to six that are particularly important—the proteobacteria, the green bacteria, the cyanobacteria, the Gram-positive bacteria, the spirochetes, and the chlamydias (see Figure 25.11).

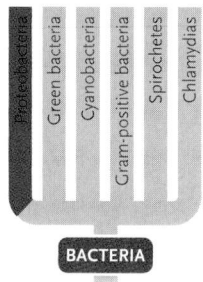

Proteobacteria: The Purple Bacteria and Their Relatives. The proteobacteria are a highly diverse group of Gram-negative bacteria that scientists hypothesize derive from a purple, photosynthesizing evolutionary ancestor. Many present-day species retain those characteristics, carrying out photosynthesis as either photoautotrophs (the purple sulfur bacteria) or photoheterotrophs (the purple nonsulfur bacteria). "Purple" refers to the color given to the cells by their photosynthetic pigment, a type of chlorophyll distinct from that of plants. Proteobacteria carry out a type of photosynthesis that does not use water as an electron donor and does not release oxygen as a by-product of photosynthesis.

Other present-day proteobacteria are chemoheterotrophs that are thought to have evolved as an evolutionary branch following the loss of photosynthetic capabilities in an early proteobacterium. The evolutionary ancestors of mitochondria are considered likely to have been ancient nonphotosynthetic proteobacteria.

Among the chemoheterotrophs classified with the proteobacteria are bacteria that cause human diseases such as bubonic plague, Legionnaire's disease, gonorrhea, and various forms of gastroenteritis and dysentery; bacterial plant pathogens that cause rots, scabs, and wilts; and the colon-inhabiting *E. coli* (shown dividing in Figure 25.7). The proteobacteria also include both free-living and symbiotic nitrogen-fixing bacteria.

Among the more unusual nonphotosynthetic proteobacteria are the myxobacteria, which form colonies held together by the slime they produce. Enzymes secreted by the colonies digest "prey"—other bacteria, primarily—that become stuck in the slime. When environmental conditions become unfavorable, as when soil nutrients or water are depleted, myxobacteria form a *fruiting body* **(Figure 25.12)**, which contains clusters of

Figure 25.11
An abbreviated phylogenetic tree of prokaryotes.

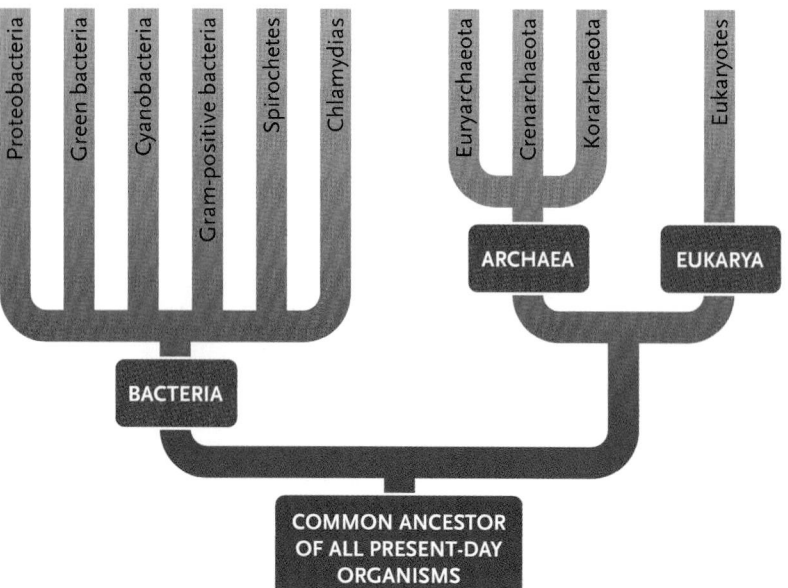

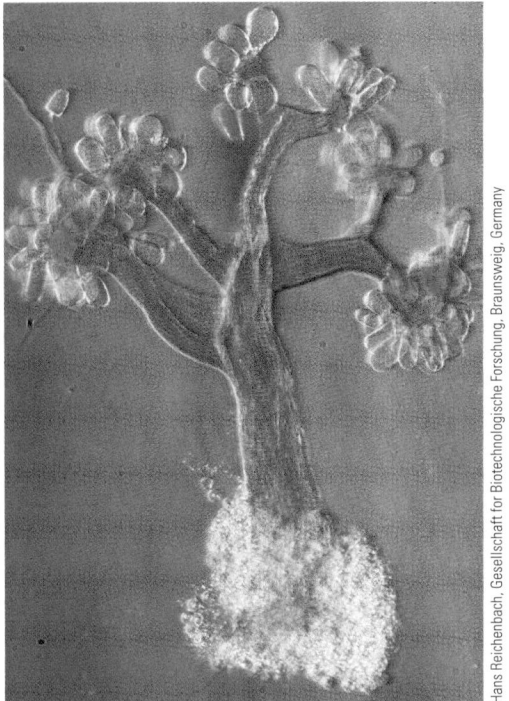

Figure 25.12

The fruiting body of *Chondromyces crocatus*, a myxobacterium. Cells of this species collect together to form the fruiting body.

spores. When the fruiting body bursts, the spores disperse and form new colonies.

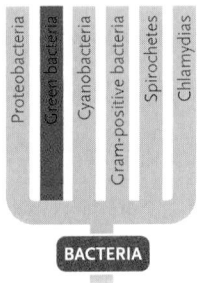

Green Bacteria. The green bacteria are a diverse group of Gram-negative photosynthesizers with photosynthetic pigments that give the cells a green color. The pigments are a form of chlorophyll distinct from the chlorophyll of plants. Like the purple bacteria, they do not release oxygen as a byproduct of photosynthesis. Green bacteria occur in two subgroups: green sulfur bacteria, which are photoautotrophs, and green nonsulfur bacteria, which are photoheterotrophs. The green sulfur bacteria are fairly closely related to the Archaea and are usually found in hot springs. The green nonsulfur bacteria are found typically in marine and high-salt environments.

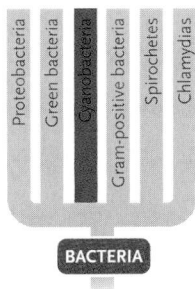

Cyanobacteria. The cyanobacteria **(Figure 25.13)** are Gram-negative photoautotrophs that have a blue-green color and carry out photosynthesis by the same pathways as eukaryotic algae and plants, using the same chlorophyll as in plants as their primary photosynthetic pigment. They release oxygen as a by-product of photosyn-

thesis. The first appearance of oxygen in quantity in Earth's atmosphere depended on the activities of ancient cyanobacteria.

The direct ancestors of present-day cyanobacteria were the first organisms to use the water-splitting reactions of photosynthesis. As such, they were critical to the appearance of oxygen in the atmosphere, which allowed the evolutionary development of aerobic organisms. Chloroplasts probably evolved from early cyanobacteria that were incorporated into the cytoplasm of primitive eukaryotes, which eventually gave rise to the algae and higher plants (see Section 24.3). Besides releasing oxygen, present-day cyanobacteria help fix nitrogen into organic compounds in aquatic habitats and in lichens, which are symbiotic organisms consisting of a cyanobacterium with a filamentous fungus (see Chapter 28).

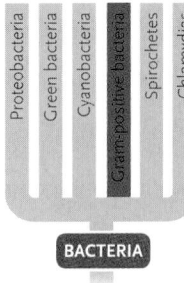

Gram-Positive Bacteria. The large group of Gram-positive bacteria contains many species that live primarily as chemoheterotrophs. One species, *Bacillus subtilis*, is studied by biochemists and geneticists almost as extensively as is *E. coli*. A number of Gram-positive bacteria cause human diseases, including *Bacillus an-*

a.

b.

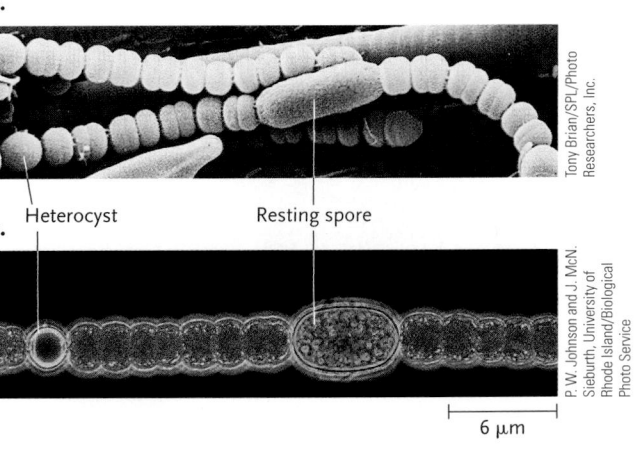

c.

Heterocyst Resting spore

6 μm

Figure 25.13

Cyanobacteria. **(a)** A population of cyanobacteria covering the surface of a pond. **(b)** and **(c)** Chains of cyanobacterial cells. Some cells in the chains form spores. The heterocyst is a specialized cell that fixes nitrogen.

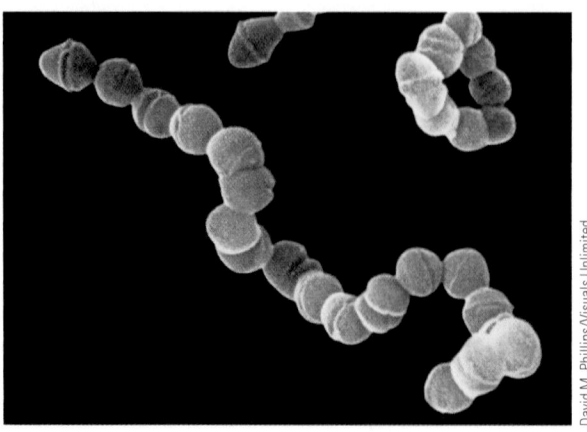

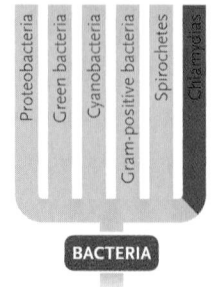

Figure 25.14
Streptococcus bacteria forming the long chains of cells typical of many species in this genus.

thracis, which causes anthrax and has been much in the news as a possible terrorist weapon; *Staphylococcus,* which causes some forms of food poisoning, skin infections such as pimples and boils, toxic shock syndrome, pneumonia, and meningitis; and *Streptococcus* **(Figure 25.14),** which causes strep throat, some forms of pneumonia, scarlet fever, and kidney infections. Nevertheless, some Gram-positive bacteria are beneficial; *Lactobacillus,* for example, carries out the lactic acid fermentation used in the production of pickles, sauerkraut, and yogurt. One unusual group of bacteria, the mycoplasmas, is placed among the Gram-positive bacteria by molecular studies even though they are Gram-negative. Their staining reaction reflects that they are naked cells that secondarily lost their cell walls in evolution. Some mycoplasmas, with diameters from 0.1 to 0.2 μm, are the smallest known cells.

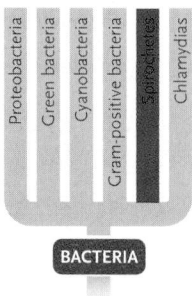

Spirochetes. The spirochetes are Gram-negative bacteria with helically spiraled bodies and an unusual form of movement in which bacterial flagella, embedded in the cytoplasm, cause the entire cell to twist in a corkscrew pattern. Their corkscrew movements enable them to move in viscous environments such as mud and sewage, where they are common. Two spirochetes, *Treponema denticola* and *Treponema vincentii,* are more or less

harmless inhabitants of the human mouth; another species, *Treponema pallidum,* is the cause of syphilis. Other pathogenic spirochetes cause relapsing fever and Lyme disease. Beneficial spirochetes in termite intestines aid in the digestion of plant fiber.

Chlamydias. The chlamydias are structurally unusual among bacteria because, although they are Gram-negative and have cell walls with a membrane outside of them, they lack peptidoglycans. All the known chlamydias are intracellular parasites that cause various diseases in animals. One bacterium of this group, *Chlamydia trachomatis* **(Figure 25.15),** is responsible for one of the most common sexually transmitted infections of the urinary and reproductive tracts of humans. The same bacterium causes trachoma, an infection of the cornea that is the leading cause of preventable blindness in humans.

Bacteria Cause Diseases by Several Mechanisms

As you have just learned, some bacteria cause diseases while others are beneficial. Here we focus on pathogenic bacteria.

Bacteria vary in the pathways by which they cause diseases. A number of bacterial lineages produce **exotoxins,** toxic proteins that leak from or are secreted from the bacterium and interfere with the biochemical processes of body cells in various ways. For example, the exotoxin of the Gram-positive bacterium *Clostridium botulinum* is found as a contaminant of poorly preserved foods, and causes botulism. The botulism exotoxin is one of the most poisonous substances known: a few nanograms can cause illness, and a few hundred grams could kill every human on Earth. It acts by interfering with the transmission of nerve impulses. The muscle paralysis produced by the exotoxin can be fatal if the muscles that control breathing are affected. Interestingly, the botulism exotoxin, with the brand name Botox, is used in low doses for the cosmetic removal of wrinkles, and in the treatment of migraine headaches, involuntary contraction of the eye muscles, and some other medical conditions.

Some other bacteria cause disease through **endotoxins.** Endotoxins are not released by living cells as exotoxins are; instead, they are lipopolysaccharides released from the outer membrane that surrounds cell walls when bacteria die and lyse. Endotoxins are natural components of the outer membrane of all Gram-negative bacteria, which include *E. coli, Salmonella,* and *Shigella.* These lipopolysaccharides cause disease by overstimulating the host's immune system, often triggering inflammation. Endotoxin release has different effects, depending on the bacterial species and the site

Figure 25.15
Cells of *Chlamydia trachomatis* inside a human cell. This bacterium is a major infectious cause of human eye and genital disease.

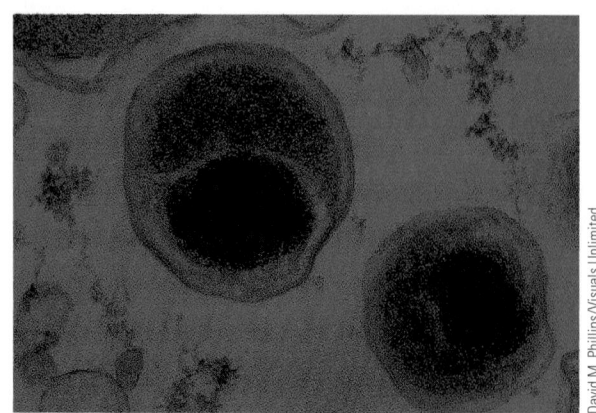

of infection, that include typhoid or other fevers, diarrhea, and, in severe cases, organ failure and death. For example, *Salmonella typhi,* the cause of typhoid, enter the human intestines and penetrate the intestinal wall, eventually ending up in the lymph nodes. There they multiply, and some of the cells die and lyse, releasing endotoxins into the bloodstream. This both triggers the host's immune response and causes blood poisoning, a serious medical condition in which the circulatory system becomes dysfunctional. If the infection is not successfully treated, the condition can progress to multiple organ system failure and, eventually, death.

Some bacteria release **exoenzymes,** enzymatic proteins that digest plasma membranes and cause cells of the infected host to rupture and die. Exoenzymes may also digest extracellular materials such as collagen, causing connective tissue diseases. Some exoenzymes attack red or white blood cells, leading to anemias, impairment of the immune response, or interference with blood clotting. Among the bacteria that release exoenzymes are *Streptococcus, Staphylococcus,* and *Clostridium.* Necrotizing fasciitis (flesh-eating disease), the spectacularly destructive and rapid degeneration of subcutaneous tissues in the skin, is caused by an exoenzyme released by *Streptococcus* and some other bacteria.

Some of the ill effects of bacteria have little to do with exotoxins, endotoxins, or exoenzymes, but are caused purely by the body's responses to infection. The severe pneumonia caused by *Streptococcus pneumoniae,* for example, results from massive accumulation of fluid and white blood cells in the lungs in response to the infection. The white blood cells have little effect on the bacteria, however, because of the bacterial cell's protective capsule. As the fluid, white blood cells, and bacteria continue to accumulate, they block air passages in the lungs and severely impair breathing.

Pathogenic Bacteria Commonly Develop Resistance to Antibiotics

Antibiotics are routinely used to treat bacterial infections. These substances, produced as defensive molecules by some bacteria and fungi, or by chemical synthesis, kill or inhibit the growth of other microbial species. For example, streptomycins, produced by soil bacteria, block protein synthesis in their targets. Penicillins, produced by fungi, prevent formation of covalent bonds that hold bacterial cell walls together, weakening the wall and causing the cells to rupture.

Many pathogenic bacteria develop resistance to antibiotics through mutations that allow them to break down the drugs or otherwise counteract their effects (see *Why It Matters,* Chapter 20). Resistance is also acquired through genes carried on plasmids, picked up by conjugation or on DNA brought into pathogens by other pathways such as transformation and transduction (see Section 17.1). Taking antibiotics routinely in

mild doses, or failing to complete a prescribed dosage, contribute to the development of resistance by selecting strains that can survive in the presence of the drug. Overprescription of antibiotics for colds and other virus-caused diseases can also promote bacterial resistance. That is, viruses are unaffected by antibiotics, but the presence of antibiotics in the system can lead to resistance as just described. Antibacterial agents that may promote resistance are also commonly included in such commercial products as soaps, detergents, and deodorants. Resistance is a form of evolutionary adaptation; antibiotics alter the bacterium's environment, conferring a reproductive advantage on those strains best adapted to the altered conditions.

The development of resistant strains has made tuberculosis, cholera, typhoid fever, gonorrhea, "staph," and other diseases caused by bacteria difficult to treat with antibiotics. For example, as recently as 1988, drug-resistant strains of *Streptococcus pneumoniae,* which causes pneumonia, meningitis, and middle-ear infections, were practically unheard of in the United States. Now, resistant strains of *S. pneumoniae* are common and increasingly difficult to treat.

In this section, you have seen that bacteria thrive in nearly every habitat on Earth, including the human body. However, some members of the second prokaryotic domain, the Archaea, the subject of the next section, live in habitats that are too forbidding even for the bacteria.

STUDY BREAK

1. What methodologies have been used to classify prokaryotes?
2. What were the likely characteristics of the evolutionary ancestor of present-day proteobacteria?
3. What are the differences between the way photosynthesis is carried out by photosynthetic Proteobacteria and by cyanobacteria?
4. What is an exotoxin, an endotoxin, and an exoenzyme, and how do they differ with respect to how they cause disease?

25.3 The Domain Archaea

Archaea were first discovered in 1977, and scientists believed they were bacteria. However, research showed that they have some eukaryotic features, some bacterial features, and some features that are unique to the group (also discussed in Section 24.3; **Table 25.1** compares the characteristics of Bacteria, Archaea, and Eukarya). Based on research by Carl Woese and his colleagues that compared their DNA and rRNA sequences with those of other organisms, Archaea were

Table 25.1

Table 25.1   Characteristics of the Bacteria, Archaea, and Eukarya

Characteristic	Bacteria	Archaea	Eukarya
DNA arrangement	Single, circular in most, but some linear and/or multiple	Single, circular	Multiple linear molecules
Chromosomal proteins	Prokaryotic histone-like proteins	Five eukaryotic histones	Five eukaryotic histones
Genes arranged in operons	Yes	Yes	No
Nuclear envelope	No	No	Yes
Mitochondria	No	No	Yes
Chloroplasts	No	No	Yes
Peptidoglycans in cell wall	Present	Present but modified, or absent	Absent
Membrane lipids	Unbranched; linked by ester linkages	Branched; linked by ether linkages	Unbranched; linked by ester linkages
RNA polymerase	One type	Multiple types	Multiple types
Ribosomal proteins	Prokaryotic	Some prokaryotic, some eukaryotic	Eukaryotic
First amino acid placed in proteins	Formylmethionine	Methionine	Methionine
Aminoacyl-tRNA synthetases	Prokaryotic	Eukaryotic	Eukaryotic
Cell division proteins	Prokaryotic	Prokaryotic	Eukaryotic
Proteins of energy metabolism	Prokaryotic	Prokaryotic	Eukaryotic
Sensitivity to chloramphenicol and streptomycin	Yes	No	No

subsequently classified as a separate domain of life. (*Insights from the Molecular Revolution* describes the research that first revealed the complete DNA sequence of an archaean.) Scientists use sequencing studies and the archeans' unique characteristics to identify the organisms in this group.

Archaea Have Some Unique Characteristics

The first-studied Archaea were found in extreme environments, such as hot springs, hydrothermal vents on the ocean floor, and salt lakes (**Figure 25.16**). For that reason, these prokaryotes were called *extremophiles*

("extreme lovers"). Subsequently archaeans have also been found living in normal environments; like bacteria, these are *mesophiles*.

Many Archaea are chemoautotrophs that obtain energy by oxidizing inorganic substances, while others are chemoheterotrophs that oxidize organic molecules. No known member of the Archaea has been shown to be pathogenic.

The cell structure of archaeans is basically prokaryotic. Among their unique characteristics are certain features of the plasma membrane and cell wall. The lipid molecules in archaean plasma membranes have a chemical bond between the hydrocarbon chains and

Figure 25.16
Typically extreme archaean habitats.
(a) Highly saline water in Great Salt Lake, Utah, colored red-purple by Archaea.
(b) Hot, sulfur-rich water in Emerald Pool, Yellowstone National Park, colored brightly by the oxidative activity of archaeans, which converts H_2S to elemental sulfur.

a.

Barry Rokeach

b.

© Alan L. Detrick/Science Source/Photo Researchers, Inc.

INSIGHTS FROM THE MOLECULAR REVOLUTION

Extreme but Still in Between

In 1996 Carol J. Bult, Carl R. Woese, J. Craig Venter, and 37 other scientists at the Institute for Genomic Research obtained the complete DNA sequence of the archaean *Methanococcus jannaschii*. It was the first archaean genome to be sequenced. The results were obtained by sequencing randomly chosen overlapping DNA fragments from a DNA library until the entire genome was completed (the whole-genome shotgun approach, described in Section 18.3). Comparisons of the *Methanococcus* sequence with bacterial sequences and that of a eukaryote, the brewer's yeast *Saccharomyces cerevisiae*, give strong support to the proposal that the Archaea are a separate domain of living organisms.

Many archaeans have a lifestyle clearly different from those of the bac-

teria and eukaryotes, and *Methanococcus* is no exception. It was first discovered by the deep-sea submarine *Alvin* in a hot-water vent at a depth of more than 2600 m. It can live at temperatures as high as 94°C, only a few degrees less than the temperature of boiling water, and can tolerate pressure as high as 200 times the pressure of air at sea level.

The *Methanococcus* main genome, which includes 1,664,976 base pairs, was found to contain 1682 protein-encoding sequences. Two plasmids also contain protein-encoding genes, one plasmid with 44 genes and the other with 12. Of the total of 1738 protein-encoding sequences, only 38%—less than half—could be given probable identities based on sequence similarities with those of genes coding

for known proteins in other organisms. Some of the sequences were similar to proteins of bacteria, and some to those of eukaryotes. Among the eukaryote-like genes are those encoding all five of the histone chromosomal proteins typical of eukaryotes and eukaryotic forms of the enzymes carrying out DNA replication and RNA transcription.

Other identified genes encode proteins unique to the Archaea, such as those encoding some enzymes and other proteins of the pathway reducing CO_2 to methane. Many other proteins with no known counterparts in the Bacteria or Eukarya are among the unidentified 62%, demonstrating the unique character of the Archaea and providing a rich lode of new proteins for mining by molecular biologists and other scientists.

---

glycerol unlike that in the plasma membranes of all other organisms. The difference is significant because the exceptional linkage is more resistant to disruption, making the plasma membranes of the Archaea more tolerant of the extreme environmental conditions under which many of these organisms live.

The cell walls of some archaeans are assembled from molecules related to the peptidoglycans, but with different molecular components and bonding structure. Others have walls assembled from proteins or polysaccharides instead of peptidoglycans. The cell walls of archaeans are as resistant to physical disruption as the plasma membrane is; some archaeans can be boiled in strong detergents without disruption. Different archaeans stain as either Gram-positive or Gram-negative.

Molecular Studies Reveal Three Evolutionary Branches in the Archaea

Based on differences between the rRNA coding sequences in their genomes, the domain Archaea is divided into three groups (see Figure 25.11). Two major groups, the **Euryarchaeota** and the **Crenarchaeota**, contain archaeans that have been cultured and examined in the laboratory. The third group, the **Korarchaeota**, has been recognized solely on the basis of rRNA coding sequences in DNA taken from environmental samples. A fourth group, the **Nanoarchaeota**, was proposed based on rRNA sequence analysis of a thermophilic archaean found in a symbiotic relationship with an-

other thermophilic archaean. Genome sequence comparisons have now shown that the Nanoarchaeota are most probably a subgroup of the Euryarchaeota.

Euryarchaeota. The Euryarchaeota are found in different extreme environments. They include methogens, which produce methane; extreme halophiles, which live in high concentrations of salt; and some extreme thermophiles, which live under high-temperature conditions.

Methanogens (methane generators) live in reducing environments **(Figure 25.17)**, and represent about one half of all known species of archaeans. All known methanogens belong to the Euryarchaeota. Examples are *Methanococcus* and *Methanobacterium*. Methanogens are obligate anaerobes,

Figure 25.17
A colony of the methanogenic archaean *Methanosarcina*, which lives in the sulfurous, waterlogged soils of marshes and swamps.

meaning they are killed by oxygen. They are found in the anoxic (oxygen-lacking) sediments of swamps, lakes, marshes, and sewage works, as well as in more moderate environments, such as the rumen of cattle, sheep, and camels; the large intestine of dogs and humans; and the hindguts of insects such as termites and cockroaches. Methanogens generate energy by converting at least ten different substrates such as carbon dioxide and hydrogen gas, methanol, or acetate into methane gas (CH_4), which is released into the atmosphere. A single species may use two or three substrates, for example converting carbon dioxide and hydrogen into methane and water.

Halophiles are salt-loving organisms. Extreme halophilic Archaea live in highly saline (salty) environments such as the Great Salt Lake or the Dead Sea, and on foods preserved by salting. Moreover, they require a high concentration of salt to live: they need a minimum NaCl concentration of about 1.5 M (about 9% solution), and can live in a fully saturated solution (5.5 M, or 32%). All known extreme halophilic Archaea belong to the Euryarchaeota. Most are aerobic chemoheterotrophs; they obtain energy from sugars, alcohols, and amino acids using pathways similar to those of bacteria. Examples are *Halobacterium* and *Natrosobacterium*. *Halobacterium,* like a number of extreme halophiles, uses light as a secondary energy source supplementing the oxidations that are its primary source of energy.

Extreme thermophiles live in extremely hot environments. Extreme thermophilic Archaea live in thermal areas such as ocean floor hydrothermal vents and hot springs such as those in Yellowstone National Park. Their optimal temperature range for growth is 70° to 95°C, approaching the boiling point of water. By comparison, no eukaryotic organism is known to live at a temperature higher than 60°C. Some extreme thermophiles are members of the Euryarchaeota. Some of them, such as *Pyrophilus,* are obligate anaerobes, while others, such as *Thermoplasma,* are facultative anaerobes that grow on a variety of organic compounds.

Crenarchaeota. The group Crenarchaeota contains most of the extreme thermophiles. Their optimal temperature range of 75° to 105°C is higher than that of the Euryarchaeota. Most are unable to grow at temperatures below 70°C. The most thermophilic member of the group, *Pyrobolus,* grows optimally at 106°C, but dies below 90°C. It can also grow at 113°C and survive an hour of autoclaving at 121°C! *Pyrobolus* lives in ocean floor hydrothermal vents where the pressure makes it possible to have temperatures above 100°C, the boiling point of water on Earth's surface.

Also within this group are **psychrophiles** ("cold loving"), organisms that grow optimally in cold temperatures in the range −10 to 20°C. These organisms are found mostly in the Antarctic and Arctic oceans, which are frozen most of the year, and in the intense cold at ocean depths.

Mesophilic members of the Crenarchaeota comprise a large part of plankton in cool, marine waters where they are food sources for other marine organisms. As yet, no individual species of these archaeans has been isolated and characterized.

Crenarchaeota archaeans exhibit a wide range of metabolism with regard to oxygen, including obligate anaerobes, facultative anaerobes, and aerobes.

Korarchaeota. The group Korarchaeota has been recognized solely on the basis of analyzing rRNA sequences in DNA obtained from marine and terrestrial hydrothermal environments, such as the Obsidian Pool at Yellowstone National Park. To date, no members of this group have been isolated and cultivated in the lab, and nothing is known about their physiology. They are the oldest lineage in the domain Archaea according to molecular data.

Thermophilic archaeans are important commercially. For example, enzymes from some species are used in basic and applied research, such as the thermostable DNA polymerase used in the polymerase chain reaction (PCR; see Chapter 18). Other enzymes from thermophilic archaeans are being tested for addition to detergents, where it is hoped that they will be active under high temperatures and acidic pH.

From the highly varied prokaryotes, we now turn to the viruses, which occur in most environments in even greater numbers than bacteria and archaeans. The next section also discusses prions and viroids, infective agents that are even simpler and smaller than viruses.

STUDY BREAK

1. What distinguishes members of the Archaea from members of the Bacteria and Eukarya?
2. How does a methanogen obtain its energy? In which group or groups of Archaea are methanogens found?
3. Where do extreme halophilic archaeans live? How do they obtain energy? In which group or groups of Archaea are the extreme halophiles found?
4. What are extreme thermophiles and psychrophiles?

25.4 Viruses, Viroids, and Prions

A **virus** (Latin for poison) is a biological particle that can infect the cells of a living organism. Viral infections usually have detrimental effects on their hosts.

The study of viruses is called *virology*, and researchers studying viruses are known as *virologists*.

All viruses contain a nucleic acid molecule (the genome), surrounded by a layer of protein called the **coat** or **capsid**. The complete virus particle is also called a **virion**. Viruses are considered nonliving primarily because they have no metabolic system of their own to provide energy for their life cycles; instead, they are dependent upon the host cells they infect for that function. That is, expression of virus genes in an infected host cell directs that cell to use its own machinery to duplicate the virus. However, their genome contains all the information required to convert host cells to the duplication of viruses of the same type. Although they are considered to be nonliving material, viruses are classified by the International Committee on Taxonomy of Viruses into orders, families, genera, and species using several criteria, including size and structure, type and number of nucleic acid molecules, method of replication of the nucleic acid molecules inside host cells, host range, and infective cycle. More than 4000 species of viruses have been classified into more than 80 families according to these criteria.

One or more kinds of viruses probably infect all living organisms. Usually a virus infects only a single species or a few closely related species. (A virus may even infect only one organ system, or a single tissue or cell type in its host.) However, some viruses are able to infect unrelated species, either naturally or after mutating. For example, some humans have contracted bird flu from being infected with the natural bird flu virus as a result of contact with virus-infected birds. At least 65 deaths of people in Asia have been attributed to bird flu. The bird flu virus has the potential to mutate to give efficient human-to-human transmission, raising significant concern about the possibility of a worldwide epidemic of bird flu virus infections of humans, with the possibility of millions of deaths. Of the viral families, 21 include viruses that cause human diseases. Viruses also cause diseases of wild and domestic animals; plant viruses cause annual losses of millions of tons of crops, especially cereals, potatoes, sugar beets, and sugar cane. (**Table 25.2** lists some important families that infect animals.)

The effects of viruses on the organisms they infect range from undetectable, through merely bothersome, to seriously debilitating or lethal. For instance, some viral infections of humans, such as those causing cold sores, chicken pox, and the common cold, are usually little more than a nuisance to healthy adults. Others, including AIDS, encephalitis, yellow fever, and smallpox, are among the most severe and deadly human diseases. While most viruses have detrimental effects, some may be considered beneficial. One of the primary reasons why bacteria do not completely overrun the planet is that they are destroyed in incredibly huge numbers by viruses known as **bacteriophages**, or **phages** for short (*phagein* = to eat; see Chapters 14 and 17 for the use of phages in important discoveries in

molecular biology and bacterial genetics). Viruses also provide a natural means to control some insect pests.

Viral Structure Is Reduced to the Minimum Necessary to Transmit Nucleic Acid Molecules from One Host Cell to Another

The nucleic acid genome of a virus, depending on the viral type, may be either DNA or RNA, in either double- or single-stranded form. The nucleic acid molecule contains genes encoding proteins of the viral coat, and often also enzymes required to duplicate the genome. The simplest viral nucleic acid molecules contain only a few genes, but those of the most complex viruses may contain a hundred or more.

Some viruses have coats assembled from protein molecules of a single type; more complex viruses have coats made up of several different proteins—in some, 50 or more, including the recognition proteins that bind to host cells. The particles of some viruses also

Table 25.2 | **Major Animal Viruses**

Viral Family	Envelope	Nucleic Acid	Diseases
Adenoviruses	No	ds DNA	Respiratory infections, tumors
Flaviviruses	Yes	ss RNA	Yellow fever, dengue, hepatitis C
Hepadnaviruses	Yes	ds DNA	Hepatitis B
Herpesviruses	Yes	ds DNA	
H. simplex I			Oral herpes, cold sores
H. simplex II			Genital herpes
Varicella-zoster			Chicken pox, shingles
Orthomyxovirus	Yes	ss RNA	Influenza
Papovaviruses	No	ds DNA	Benign and malignant warts
Paramyxoviruses	Yes	ss RNA	Measles, mumps, pneumonia
Picornaviruses	No	ss RNA	
Enteroviruses			Polio, hemorrhagic eye disease, gastroenteritis
Rhinoviruses			Common cold
Hepatitis A virus			Hepatitis A
Apthovirus			Foot-and-mouth disease in livestock
Poxviruses	Yes	ds DNA	Smallpox, cowpox
Retroviruses	Yes	ss RNA	
HTLV I, II			T-cell leukemia
HIV			AIDS
Rhabdoviruses	Yes	ss RNA	Rabies, other animal diseases

ss = single-stranded; ds = double-stranded.

contain the DNA or RNA polymerase enzymes required for viral nucleic acid replication and an enzyme that attacks cell walls or membranes.

Most viruses take one of two basic structural forms, helical or polyhedral. In **helical viruses** the protein subunits assemble in a rodlike spiral around the genome **(Figure 25.18a).** A number of viruses that infect plant cells are helical. In **polyhedral viruses** the coat proteins form triangular units that fit together like the parts of a geodesic sphere **(Figure 25.18b).** The polyhedral viruses include forms that infect animals, plants, and bacteria. In some polyhedral viruses, protein spikes that provide host cell recognition extend from the corners where the facets fit together. Some viruses, the **enveloped viruses,** are covered by a surface membrane derived from their host cells; both enveloped helical and enveloped polyhedral viruses are known **(Figure 25.18c).** For example, HIV (for *h*uman *i*mmunodeficiency *v*irus), the virus that causes AIDS, is an enveloped polyhedral virus. Protein spikes extend through the membrane, giving the particle its recognition and adhesion functions.

A number of bacteriophages with DNA genomes, such as T2 (see Section 14.1), have a **tail** attached at one side of a polyhedral **head,** forming what is known as a **complex virus (Figure 25.18d).** The genome is packed into the head; the tail is made up of proteins forming a collar, sheath, baseplate, and tail fibers. The tail has recognition proteins at its tip and, once attached to a host cell, functions as a sort of syringe that injects the DNA genome into the cell.

Viruses Infect Bacterial, Animal, and Plant Cells by Similar Pathways

Free viral particles move by random molecular motions until they contact the surface of a host cell. For infection to occur, the virus or the viral genome must then enter the cell. Inside the cell, typically the viral genes are expressed, leading to replication of the viral genome and assembly of progeny viruses. The viruses are then released from the host cell, a process that often ruptures the host cell, killing it.

Infection of Bacterial Cells. Bacteriophages vary as to whether they have a DNA or an RNA genome, and whether that nucleic acid is double-stranded or single-stranded. A DNA bacteriophage such as the virulent phage T2 (see Figure 25.18d) infects a bacterial cell and goes through the *lytic cycle*, in which the host cell is killed in each cycle of infection (described in Section 17.2 and Figure 17.9). In brief review, the lytic cycle of phage T2 is as follows: After the phage attaches to a host cell, an enzyme present in the baseplate of the viral coat, *lysozyme,* digests a hole in the cell wall through which the DNA of the phage enters the bacterium while the proteins of the viral coat remain outside. Once inside the bacterium, expression of phage genes directs the replication of the phage DNA, synthesis of phage proteins, and assembly of progeny phage particles. Next, the phage directs synthesis of a phage-encoded lysozyme enzyme that lyses the bacterial cell wall, causing the cell to rupture and releasing the progeny phages to the surroundings where they can infect other bacteria.

Some bacteriophages alternate between a lytic cycle and a *lysogenic cycle*, in which the viral DNA inserts into the host cell DNA and production of new viral particles is delayed (see Section 17.2). During the lysogenic cycle, the integrated viral DNA, known as the *prophage,* remains partially or completely inactive, but is replicated and passed on with the host DNA to all descendants of the infected cell. In response to certain environmental signals, the prophage loops out of the chromosome and the lytic cycle of the phage proceeds.

Infection of Animal Cells. Viruses infecting animal cells follow a similar pattern except that both the viral coat and genome, which is DNA or RNA depending

Figure 25.18
Viral structure.
The tobacco mosaic virus in **(a)** assembles from more than 2000 identical protein subunits.

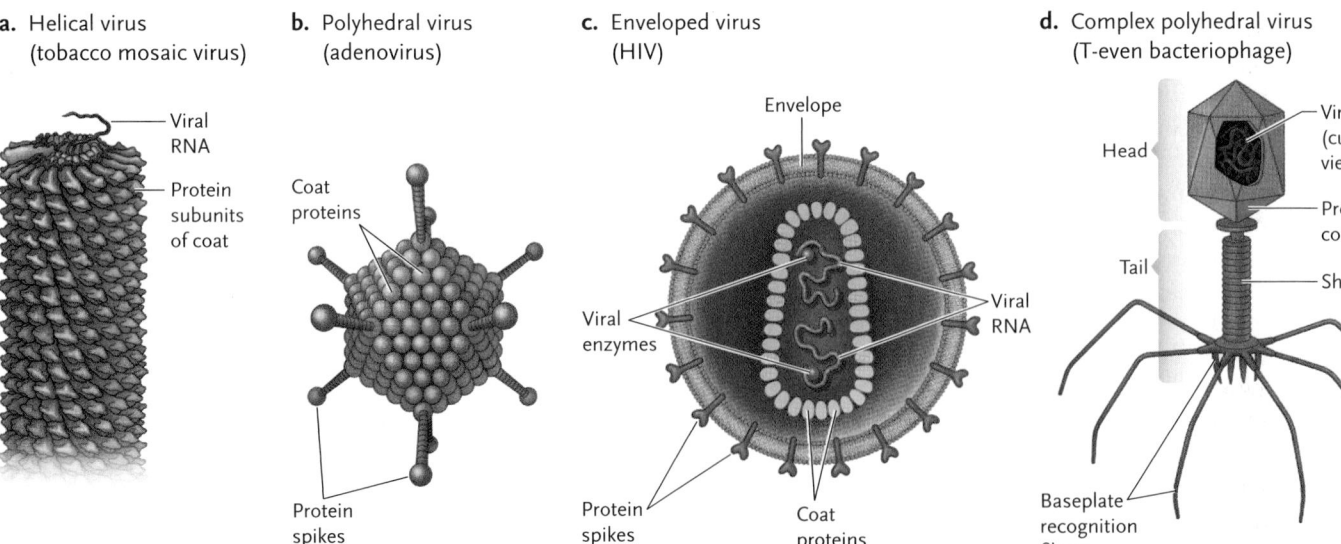

a. Helical virus
(tobacco mosaic virus)

Viral RNA

Protein subunits of coat

b. Polyhedral virus
(adenovirus)

Coat proteins

Protein spikes

c. Enveloped virus
(HIV)

Envelope

Viral enzymes

Viral RNA

Protein spikes

Coat proteins

d. Complex polyhedral virus
(T-even bacteriophage)

Viral DNA (cutaway view)

Head

Protein coat

Tail

Sheath

Baseplate recognition fibers

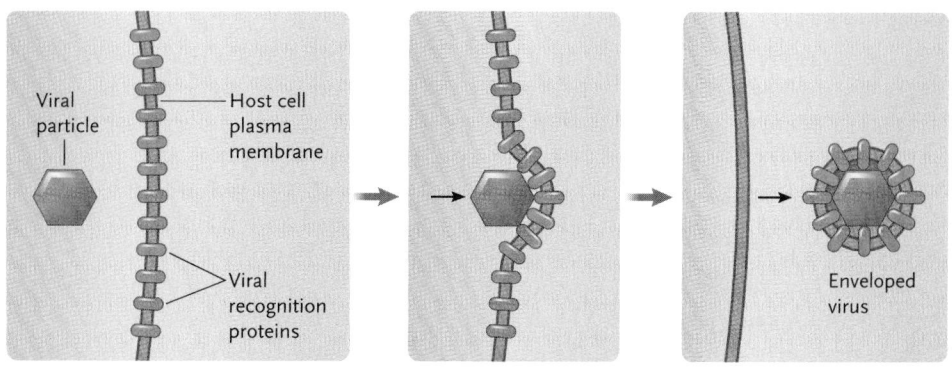

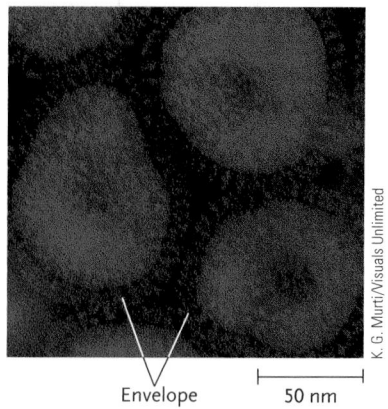

Envelope   50 nm

K. G. Murti/Visuals Unlimited

Figure 25.19
How enveloped viruses acquire their envelope. The micrograph shows the influenza virus with its envelope. Note the recognition proteins studding the envelope.

on the viral type, enter a host cell. Depending on the virus, removal of the coat to release the genome occurs during or after cell entry; the envelope does not enter the cell.

Viruses without an envelope, such as adenovirus (DNA genome) and poliovirus (RNA genome), bind by their recognition proteins to the plasma membrane and are then taken into the host cell by endocytosis. The virus coat and genome of enveloped viruses, such as herpesviruses and pox viruses (DNA genome), and HIV and influenza virus (RNA genome), enter the host cell by fusion of their envelope with the host cell plasma membrane.

Once inside the host cell, the genome directs the synthesis of additional viral particles by basically the same pathways as bacterial viruses. Newly completed viruses that do not acquire an envelope are released by rupture of the cell's plasma membrane, typically killing the cell. In contrast, most enveloped viruses receive their envelope as they pass through the plasma membrane, usually without breaking the membrane **(Figure 25.19)**. This pattern of viral release typically does not injure the host cell.

The vast majority of animal virus infections are asymptomatic; pathogenesis—the causation of disease—is of no value to the virus. However, there are a number of pathogenic viruses, and they cause diseases in a variety of ways. Some viruses, for instance, cause cell death when progeny viruses are released from the cell. This can lead to massive cell death, destroying vital tissues such as nervous tissue or white or red blood cells, or causing lesions such as ulcers in skin and mucous membranes. Some other viruses release cellular molecules when infected cells break down, which can induce fever and inflammation. Yet other viruses alter gene function when they insert into the host cell DNA, leading to cancer and other abnormalities.

Some animal viruses enter a **latent phase** in which the virus remains in the cell in an inactive form: the viral nucleic acid is present in the cytoplasm or nuclear DNA, but no complete viral particles or viral release can be detected. (The latent phase is similar to the lysogenic cycle that is part of the life cycle of some bacteriophages.) At some point, the latent phase may end

as the viral DNA is replicated in quantity, coat proteins are made, and completed viral particles are released from the cell. The herpesviruses causing oral and genital ulcers in humans remains in a latent phase of this type in the cytoplasm of some body cells for the life of the individual. At times, particularly during periods of metabolic stress, the virus becomes active in some cells, directing viral replication and causing ulcers to form as cells break down during viral release.

Infection of Plant Cells. Plant viruses may be rodlike or polyhedral; although most include RNA as their nucleic acid, some contain DNA. None of the known plant viruses have envelopes. Plant viruses enter cells through mechanical injuries to leaves and stems or through transmission from plant to plant by biting and feeding insects such as leaf hoppers and aphids, by nematode worms, and by pollen during fertilization. Plant viruses can also be transmitted from generation to generation in seeds. Once inside a cell, plant viruses replicate in the same patterns as animal viruses. Within plants, virus particles pass from infected to healthy cells through plasmodesmata, the openings in cell walls that directly connect plant cells, and through the vascular system.

Plant viruses are generally named and classified by the type of plant they infect and their most visible effects. *Tomato bushy stunt virus,* for example, causes dwarfing and overgrowth of leaves and stems of tomato plants, and *tobacco mosaic virus* causes a mosaic-like pattern of spots on leaves of tobacco plants. Most species of crop plants can be infected by at least one destructive virus.

The tobacco mosaic virus was the first virus to be isolated, crystallized, disassembled, and reassembled in the test tube, and the first viral structure to be established in full molecular detail (see Figure 25.18a).

Viral Infections Are Typically Difficult to Treat

Viral infections are unaffected by the antibiotics and other treatment methods used for bacterial infections. As a result, many viral infections are allowed to run their course, with treatment limited to relieving the symptoms while the natural immune defenses of the patient attack the virus. Some viruses, however, cause serious and sometimes deadly symptoms upon infection and, consequently, researchers have spent considerable effort to develop antiviral drugs to treat them. Many of these drugs fight the virus directly by targeting a stage of the viral life cycle; they include amantidine (inhibits hepatitis B and hepatitis C virus entry into cells), acyclovir (analog of nucleosides [*analog* means it is chemically similar] that inhibits replication of the genomes of herpesviruses), and zanamivir (inhibits release of influenza virus particles from cells).

The influenza virus illustrates the difficulties inherent in treating viral diseases. The influenza type A virus (see Figure 25.19) causes flu epidemics that sweep over the world each year. It has many unusual features that tend to keep it a step ahead of efforts to counteract its infections. One is the genome of the virus, which consists of eight separate pieces of RNA. When two different influenza viruses infect the same individual, the pieces can assemble in random combinations derived from either parent virus. The new combinations can change the protein coat of the virus, making it unrecognizable to antibodies developed against either parent virus. Antibodies are highly specific protein molecules produced by the immune system that recognize and bind to foreign proteins originating from a pathogen (see Chapter 43). The invisibility to antibodies means that new virus strains can infect people who have already had the flu or who have had flu shots that stimulate the formation of antibodies effective only against the earlier strains of the virus. Random mutations in the RNA genome of the virus add to the variations in the coat proteins that make previously formed antibodies ineffective.

Luckily, most flu infections, although debilitating, are not dangerous, except for individuals who are very young or very old or who have compromised immune systems. However, some flu epidemics have been devastatingly lethal. The worst recorded example is the epidemic of 1918. A strain of influenza virus known as the Spanish flu infected approximately 20% of the world's 1.8 billion people, killing about 50 million.

The exact type of virus responsible for this deadly epidemic was finally determined in 2005, when researchers led by Jeffrey Taubenberger at the U.S. Armed Forces Institute of Pathology reconstructed the genome of the virus and produced infectious, pathogenic viruses in the laboratory. The team worked mainly with tissue from a 1918 flu victim found in permafrost in Alaska. Using modern DNA technology (see Chapter 18), they pieced together the sequences of the virus's eight genes and characterized their protein products. They also transformed clones of the genes into animal cells and were able to produce complete virus particles. These newly reconstructed 1918 viruses are about 50 times more virulent than modern-day human influenza viruses; they kill a higher percentage of mice and kill them much more quickly, for instance. (All of these experiments were done with appropriate approval and under highly controlled experimental conditions.) By studying the 1918 virus genome and its pathogenicity, the researchers are learning how highly virulent viruses can be produced. What they have learned so far is that the 1918 virus had mutations in polymerase genes for replicating the viral genome in host cells, likely making this strain capable of replicating more efficiently. Some of the mutations are similar to those found in bird flu viruses, including the one causing human deaths in Asia. The scientists believe that the 1918 flu virus likely arose directly from a bird flu virus and not from an assembly of RNA genome segments from a bird flu virus and a human flu virus in the same cell. If this is true, the concern about a devastating bird flu epidemic in the near future is well founded.

Other human viruses are also considered to have evolved from a virus that previously infected other animals. HIV is one of these; until the second half of the twentieth century, infections of this virus were apparently restricted almost entirely to African monkeys. Now the virus infects nearly 40 million people worldwide, with the greatest concentration of infected individuals in sub-Saharan Africa.

Viruses May Have Evolved from Fragments of Cellular DNA or RNA

Where did viruses come from? Because viruses can duplicate only by infecting a host cell, they probably evolved after cells appeared. They may represent fragments of DNA molecules that once formed part of the genetic material of living cells, or an RNA copy of such a fragment. In some way, the fragments became surrounded by a protective layer of protein with recognition functions and escaped from their parent cells. As viruses evolved, the information encoded in the core of the virus became reduced to a set of directions for producing more viral particles of the same kind.

Viroids and Prions Are Infective Agents Even Simpler in Structure than Viruses

Viroids, first discovered in 1971, are plant pathogens that consist of strands or circles of RNA, smaller than any viral DNA or RNA molecule, that have no protein coat. Some of the infective RNAs acting as viroids contain fewer than 300 nucleotides. Infection by viroids can rapidly destroy entire fields of citrus, potatoes, tomatoes, coconut palms, and other crop plants.

The manner in which viroids cause disease remains ill defined. In fact, researchers believe that there is more than one mechanism. Some recent research has defined one pathway to disease in which viroid RNA activates a protein kinase (an enzyme that adds phosphate groups to proteins) in plants. This process leads to a reduction in protein synthesis and protein activity, and disease symptoms result.

Prions, named by Stanley Prusiner of the University of California San Francisco in 1982 as a loose acronym for *proteinaceous infectious particles*, are the only known infectious agents that do not include a nucleic acid molecule.

Prions have been identified as the causal agents of certain diseases that degenerate the nervous system in mammals. One of these diseases is *scrapie*, a brain disease that causes sheep to rub against fences, rocks, or trees until they scrape off most of their wool. Another prion-based disease is bovine spongiform encephalopathy (BSE), also called *mad cow disease* **(Figure 25.20).** The disease produces spongy holes and deposits of proteinaceous material in brain tissue. In 1996, 150,000 cattle in Great Britain died from an outbreak of BSE, which was traced to cattle feed containing ground-up tissues of sheep that had died of scrapie. Humans are subject to a fatal prion infection called *Creutzfeldt-Jakob disease (CJD)*. The symptoms of CJD include rapid mental deterioration, loss of vision and speech, and paralysis; autopsies show spongy holes and deposits in brain tissue similar to those of cattle with BSE. Classic CJD occurs as a result of the spontaneous transformation of normal proteins into prion proteins. Fewer than 300 cases a year occur in the United States. Variant CJD

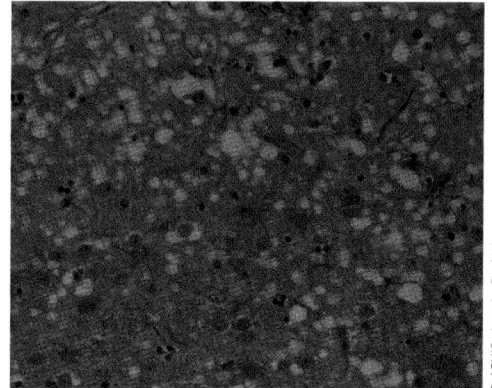

Figure 25.20
Bovine spongiform encephalopathy (BSE). The light-colored patches in this section from a brain damaged by BSE are areas where tissue has been destroyed.

Unanswered Questions

Do viruses infect archaeans?

Viruses of bacteria, and of the many types of eukaryotes, are well defined morphologically and molecularly. Do viruses infect members of the Archaea? If so, do these viruses resemble known viruses?

Mark Young's research group at Montana State University has focused on characterizing viruses from extreme thermophilic archaeans belonging to Crenoarchaeota. The researchers have discovered a number of viruses in archaeans from Yellowstone National Park acidic thermal areas. The morphology and molecular features of some of these viruses are novel and unrelated to those of any other known viruses. Young's group has sequenced the genomes of several of these new viruses, and their results indicate that the genes they carry have little or no similarity to known genes. Another archaean virus from the same area has a morphology also found in viruses of Bacteria and Eukarya. This result is of evolutionary significance because it suggests that the structure of the virus particle existed before the separation of each domain. The long-term goal of the research is to determine the mechanisms by which the viruses replicate in their extremely hot environment and to use them as tools for characterizing the special mechanisms the organisms use to survive at high temperatures. The research will also contribute to our understanding of the role viruses played in evolution.

How can West Nile virus be controlled?

West Nile virus is typically spread by mosquitoes. Usually, a mosquito becomes a carrier after biting a bird infected with the virus, and it then transmits the virus to other birds. Infected mosquitoes can also transmit the virus to humans and a number of other hosts, such as horses.

West Nile virus first entered the United States in 1999, and a number of humans have been infected. Humans infected with West Nile virus usually have mild symptoms such as fever, headache, body aches, rash, and swollen lymph glands. In some infected individuals, though, the virus enters the brain, where it can cause meningitis (inflammation of the lining of the brain and spinal cord) or encephalitis (inflammation of the brain), both of which can be fatal.

Researchers are trying to understand the infection cycle of the virus and how the virus causes disease. Specific research questions include how the virus replicates in the host and how the virus spreads through the body. Answers to these questions should aid efforts to develop effective vaccines and drugs to prevent and treat this disease. (At present, there is no vaccine for humans; one is available for horses.)

How do prion proteins move within the brain?

The brain-wasting diseases caused by prions are not well understood, despite much research. We know that prion proteins invade nerve cells and ultimately lead to fatal degeneration of the nervous system. To understand disease progression, scientists have investigated how prion proteins move through the nervous system. Using labeled-protein techniques, researchers have tracked infectious prion proteins from sites of infection up to the brain. Recently, the research groups of Bryon Caughey at the Rocky Mountain Laboratories in Montana, and Marco Prado at the University of Minas Gerais, Brazil, followed prion proteins as they invaded mouse brain cells growing in tissue culture. One exciting observation in these experiments was that prion proteins moved through the wirelike projections of the nerve cells to points of contact with other cells. Perhaps in a living organism, the prion proteins would be able to cross into the adjacent cell. The results are heralded as a significant step toward developing therapies to stop the spread of brain-wasting diseases by blocking the pathways by which prion proteins invade cells, replicate, move within the cell, and invade adjacent cells.

Peter J. Russell

is a form of the disease caused by eating nervous tissue containing meat or meat products from cattle with BSE. Another prion-based disease of humans, *kuru,* originally spread among cannibals in New Guinea, who became infected by eating raw human brain during ritual feasts following the death of an individual.

For several decades, scientists had hypothesized that a slow virus—a disease-causing virus with a long incubation period and gradual onset of pathogenicity—was responsible for these diseases. Prusiner was the first to hypothesize that infectious proteins were responsible. The research community mostly rejected this hypothesis out of hand because they held to the dogma that infectious agents required genes in the form of DNA or RNA to cause disease. Prusiner obtained experimental data supporting his hypothesis, and showed that prions are proteins normally made in the cell that misfold and cause other proteins of the same type to misfold, thereby "replicating" structural information from one molecule to the next. Typically, the misfolded prion proteins aggregate, whereas the normal proteins do not. If a misfolded prion protein is transferred from one animal to another, infection occurs; the transferred prions cause the recipient's proteins to misfold and eventually symptoms of the neurodegenerative disease characteristic of the prion will develop. Proteins with prion behavior are also found

naturally in yeast and other fungi; no diseases are associated with these prions. Prusiner received a Nobel Prize in 1997 for his discovery of prions.

The diseases caused by prions share symptoms that include loss of motor control, dementia, and eventually death. Progression of the disease is slow but there is no present cure. Under the microscope, aggregates of misfolded proteins called amyloid fibers are seen in brain tissues; the accumulation of these proteins in the brain is the likely cause of the brain damage in animals with prion diseases. The normal forms of the prion proteins are found on the surface of many types of cells, including brain cells. However, scientists do not know the function of the protein's normal form.

We began this chapter with prokaryotes, the simplest living organisms, and we end with still simpler entities, viruses, viroids, and prions, which are derived from living organisms and retain only some of the properties of life. In the next six chapters we turn to life at its most complex: the eukaryotic kingdoms of protists, plants, fungi, and animals.

STUDY BREAK

Distinguish between a virus, a viroid, and a prion.

Review

Go to **ThomsonNOW**™ at www.thomsonedu.com/login to access quizzing, animations, exercises, articles, and personalized homework help.

25.1 Prokaryotic Structure and Function

- Three shapes are common in prokaryotes: spherical, rodlike, and spiral (Figure 25.2).
- Prokaryotic genomes typically consist of a single, circular DNA molecule packaged into the nucleoid. Many prokaryotic species also contain plasmids, which replicate independently of the main DNA (Figure 25.3).
- Gram-positive bacteria have a cell wall consisting of a thick peptidoglycan layer. Gram-negative bacteria have a thin peptidoglycan layer. The thin cell wall is surrounded by an outer membrane (Figure 25.4).
- A polysaccharide capsule or slime layer surrounds many bacteria. This sticky, slimy layer both protects the bacteria and helps them adhere to surfaces (Figure 25.5).
- Some prokaryotes have flagella, corkscrew-shaped protein fibers that rotate like propellers, and pili, protein shafts that help bacterial cells adhere to each other or to eukaryotic cells (Figure 25.7).
- Prokaryotes show great diversity in their modes of obtaining energy and carbon. Chemoautotrophs obtain energy by oxidizing inorganic substrates and use carbon dioxide as their carbon source. Chemoheterotrophs obtain both energy and carbon from organic molecules. Photoautotrophs are photosynthetic organisms that use light as a source of energy and carbon dioxide as their carbon source. Photoheterotrophs use light as a source of energy and obtain their carbon from organic molecules (Figure 25.8).

- Some prokaryotes are capable of nitrogen fixation, the conversion of atmospheric nitrogen to ammonia; others are responsible for nitrification, the two-step conversion of ammonium to nitrate.
- Prokaryotes normally reproduce asexually by binary fission. Some prokaryotes are capable of conjugation, in which part of the DNA of one cell is transferred to another cell.
- In nature, prokaryotes may live in an interacting community, such as a biofilm. Biofilms have both detrimental and beneficial consequences; they can harm human health, but they also can be effective in, for example, bioremediation (Figure 25.10).

Animation: Prokaryotic body plan

Animation: Gram staining

Animation: Prokaryotic fission

Animation: Prokaryotic conjunction

25.2 The Domain Bacteria

- Bacteria are divided into more than a dozen evolutionary branches (Figure 25.11).
- The proteobacteria are Gram-negative bacteria that include purple sulfur (photoautotrophic) and nonsulfur (photoheterotrophic) photosynthetic species, and nonphotosynthetic species. Free-living proteobacteria include the spore-forming myxobacteria and species that fix nitrogen (Figure 25.12).
- The green bacteria are Gram-negative and include sulfur (photoautotrophic) and nonsulfur (photoheterotrophic) photosynthetic bacteria.

- The cyanobacteria are Gram-negative photoautotrophs that carry out photosynthesis and release oxygen as a by-product (Figure 25.13).
- The Gram-positive bacteria are primarily chemoheterotrophs that include many pathogenic species (Figure 25.14).
- The spirochetes are spiral-shaped bacteria that are propelled by twisting movements produced by the rotation of flagella.
- Chlamydias are Gram-negative intracellular parasites that cause various diseases in animals. They have cell walls with an outer membrane, but they lack peptidoglycans (Figure 25.15).
- Bacteria cause disease through exotoxins, endotoxins, and exoenzymes.
- Pathogenic bacteria may develop resistance to antibiotics through mutation of their own genes, or by acquiring resistance genes from other bacteria or plasmids.

Animation: Examples of Eubacteria

25.3 The Domain Archaea

- The Archaea have some features that are like those of bacteria, others that are eukaryotic, and some that are uniquely archaean (Table 25.1).
- The archaean plasma membrane contains unusual lipid molecules. The cell walls of archaeans consist of distinct molecules similar to peptidoglycans, or of protein or polysaccharide molecules.

- The Archaea are classified into three groups. The Euryarchaeota include the methanogens, the extreme halophiles, and some extreme thermophiles. The Crenarchaeota contain most of the archaean extreme thermophiles, as well as psychrophiles and mesophiles. Obligate anaerobes, facultative anaerobes, and aerobes are found among the Crenarchaeota. The Korarchaeota are recognized only on the basis of sequences in DNA samples.

25.4 Viruses, Viroids, and Prions

- Viruses are nonliving infective agents. A free virus particle consists of a nucleic acid genome enclosed in a protein coat. Recognition proteins enabling the virus to attach to host cells extend from the surface of infectious viruses (Figure 25.18).
- Viruses reproduce by entering a host cell and directing the cellular machinery to make new particles of the same kind.
- Viruses are unaffected by antibiotics and most other treatment methods; hence, infections caused by them are difficult to treat.
- Viroids, which infect crop plants, consist only of a very small, single-stranded RNA molecule. Prions, which cause brain diseases in some animals, are infectious proteins with no associated nucleic acid. Prions are misfolded versions of normal cellular proteins that can induce other normal proteins to misfold.

Animation: Body plans of viruses

Animation: Bacteriophage multiplication cycles

Questions

Self-Test Questions

1. A urologist identifies cells in a man's urethra as bacterial. Which of the following descriptions applies to the cells?
   a. They have sex pili, which give them motility.
   b. They have flagella, which allow them to remain in one position in the urethral tube.
   c. They are covered by a capsule, which enables them to multiply quickly.
   d. They are covered by pili, which keep them attached to the urethral walls.
   e. They contain a peptidoglycan cell wall, which gives them buoyancy to float in the fluids of the urethra.

2. A bacterium that uses nitrites as its only energy source was found in a deep salt mine. It is a:
   a. chemoautotroph.   d. heterotroph.
   b. parasite.   e. photoheterotroph.
   c. photoautotroph.

3. The _____ are all oxygen-producing photoautotrophs.
   a. spirochetes   d. Gram-positive bacteria
   b. chlamydias   e. proteobacteria
   c. cyanobacteria

4. At the health center, a fecal sample was taken from a feverish student. Organisms with corkscrew-like flagella and no endomembranes but with cell walls were isolated as the cause for the illness. These organisms belong to the group:
   a. protists with nuclei.
   b. bacteria with ribosomes.
   c. fungi with endoplasmic reticulum.
   d. plants with chloroplasts
   e. Archaea with Golgi bodies.

5. Which of the following is *not* a property of an endospore?
   a. resistant to boiling—must be autoclaved to be killed
   b. metabolically inactive
   c. can survive millions of years
   d. provides a method to preserve bacterial DNA under harsh conditions
   e. is a means that bacterial cells use to multiply

6. Each bacterial cell is traditionally thought to act independently of others. An exception to this is:
   a. biofilm aggregates.
   b. photosynthesis.
   c. peptidoglycan layering.
   d. toxin release.
   e. facultative anaerobic metabolism.

7. Penicillin, an antibiotic, inhibits the formation of cross-links between sugar groups in peptidoglycan. Bacteria treated with penicillin should be:
   a. aerobic.   d. Gram-positive.
   b. anaerobic.   e. flagellated.
   c. Gram-negative.

8. The best choice when using/prescribing antibiotics is to:
   a. increase the dosage when the original amount does not work.
   b. determine the kind of bacterium causing the problem.
   c. stop taking the antibiotic when you feel better but the prescription has not run out.
   d. ask the doctor to prescribe a drug as a precaution for an infection you do not have.
   e. choose soaps that are labeled "antibacterial."

9. When a virus enters the lysogenic stage:
   a. the viral DNA is replicated outside the host cell.
   b. it enters the host cell and kills it immediately.
   c. it enters the host cell, picks up host DNA, and leaves the cell unharmed.
   d. it sits on the host cell plasma membrane with which it covers itself and then leaves the cell.
   e. The viral DNA integrates into the host genome.

10. An infectious material is isolated from a nerve cell. It contains protein with amino acid sequences identical to the host protein but no nucleic acids. It belongs to the group:
    a. prions.   d. viroids.
    b. Archaea.   e. sporeformers.
    c. toxin producers.

Questions for Discussion

1. The digestive tract of newborn chicks is free of bacteria until they eat food that has been exposed to the feces of adult chickens. The ingested bacteria establishes a population in the digestive tract that is beneficial for the digestion of food. However, if *Salmonella* are present in the adult feces, this bacterium, which can be pathogenic for humans who ingest it, may become established in the digestive tracts of the chicks. To eliminate the possibility that *Salmonella* might become established, should farmers feed newborn chicks a mixture of harmless known bacteria from a lab culture, or a mixture of unknown fecal bacteria from healthy adult chickens? Design an experiment to answer this question.

2. Investigators in Australia found that mats of pond scum formed by the bacterium *Botyrococcus braunii* decayed into a substance resembling crude oil when the ponds dried up. Formulate a hypothesis explaining how this process may have contributed to Earth's oil deposits.

3. What rules would you suggest to prevent the spread of mad cow disease (BSE)?

Experimental Analysis

Suppose you isolate a previously unknown virus that has caused infection in humans. Describe how you would show experimentally to what virus genus and species this new virus is most closely related.

Evolution Link

Prion diseases cause similar fatal brain degeneration in a large number of animals, including human, baboon, chimpanzee, mule deer, cow, sheep, pig, golden hamster, rat, mouse, and rabbit. Can you make any evolutionary hypotheses based on this observation? How might you determine the evolutionary relationships of prion proteins?

How Would You Vote?

Eliminating mosquitoes is the best defense against West Nile virus. Many local agencies are spraying pesticides wherever mosquitoes are likely to breed. Some people fear ecological disruptions and bad effects on health and say spraying will never eliminate all mosquitoes anyway. Would you support a spraying program in your community? Go to www.thomsonedu.com/login to investigate both sides of the issue and then vote.

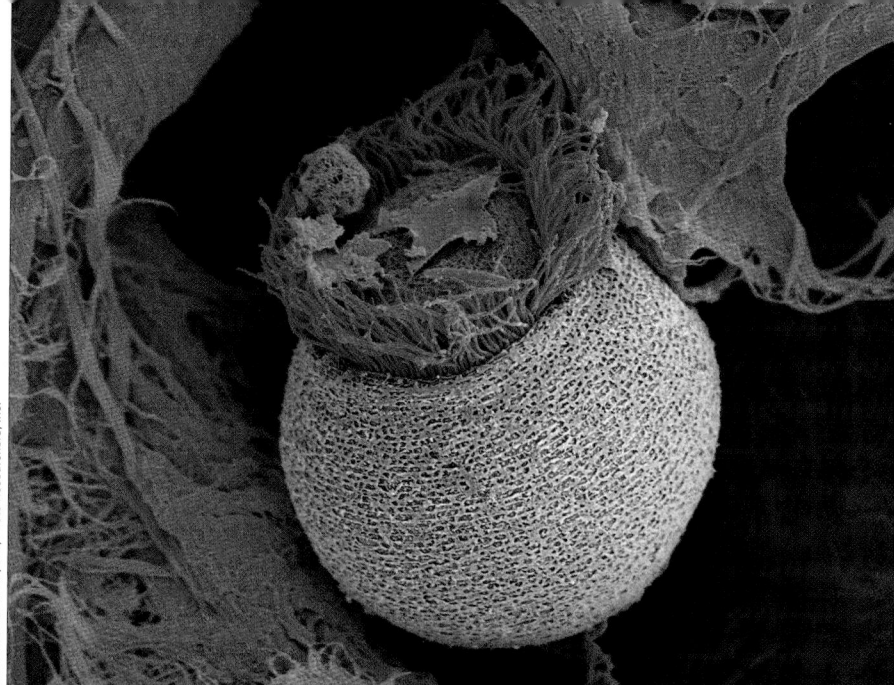

Steve Gschmeissner/SPL/Photo Researchers, Inc.

A ciliated protozoan, a type of protist (colorized SEM). This protozoan lives in water, feeding on bacteria and decaying organic matter.

26 Protists

WHY IT MATTERS

Go for a swim just about anywhere in the natural world and you will share the water with multitudes of diverse organisms called protists. Like their most ancient ancestors, almost all of these eukaryotic species are aquatic. Structurally, single-celled protists are the simplest of all eukaryotes. Although most are microscopic in size, many have had or have a significant impact on the world. For example, the protist *Phytophthora infestans,* a water mold also referred to as a downy mildew, infects valuable crop plants such as potatoes. Pototoes were the main food staple in Europe in the nineteenth century, and *P. infestans* destroyed potato crops, causing potato famines that spread across Europe; millions died in these famines. In Ireland, for instance, the growing seasons between 1845 and 1860 were cool and damp. Year after year, *P. infestans* spores spread along thin films of water on the plants. Late blight, a rotting of plant parts, became epidemic. One-third of the Irish population starved to death, died of typhoid fever (a secondary effect), or fled to other countries.

Today, related species threaten forests in the United States, Europe, and Australia. For example, when conditions favored its growth, *Phytophthora ramorum,* started an epidemic of sudden oak death in California, during which tens of thousands of oak trees have died. As the name suggests, infected trees die rapidly. The first sign of infection is a dark red-to-black sap oozing from the bark surface. The pathogen has now jumped to madrones, redwoods, and certain other trees and shrubs. Cascading ecological changes resulting from tree death caused by this pathogen will reduce sources of food and shelter for forest species.

Protists are the subject of this chapter. Also known as *protoctists*, they are members of the kingdom **Protoctista**. Protists are the results of the varied early branching of eukaryotic evolution. They are abundant on Earth and play key ecological, economic, and medical roles in the world's biological communities.

We begin this chapter with a discussion of the identity of the protists. As our discussion will show, the members of the kingdom Protoctista are so diverse that they are best defined as what they are not—that is, by contrasting them with other kingdoms.

26.1 What Is a Protist?

Protists are easily the most varied of all Earth's creatures. **Figure 26.1** shows a number of protists, illustrating their great diversity. Protists include both microscopic single-celled and large multicellular organisms. They may inhabit aquatic environments, moist soils, or the bodies of animals and other organisms, and they may live as predators, photosynthesizers, parasites, or decomposers. The extreme diversity of the group has made the protists so difficult to classify that their status as a kingdom remains highly unsettled.

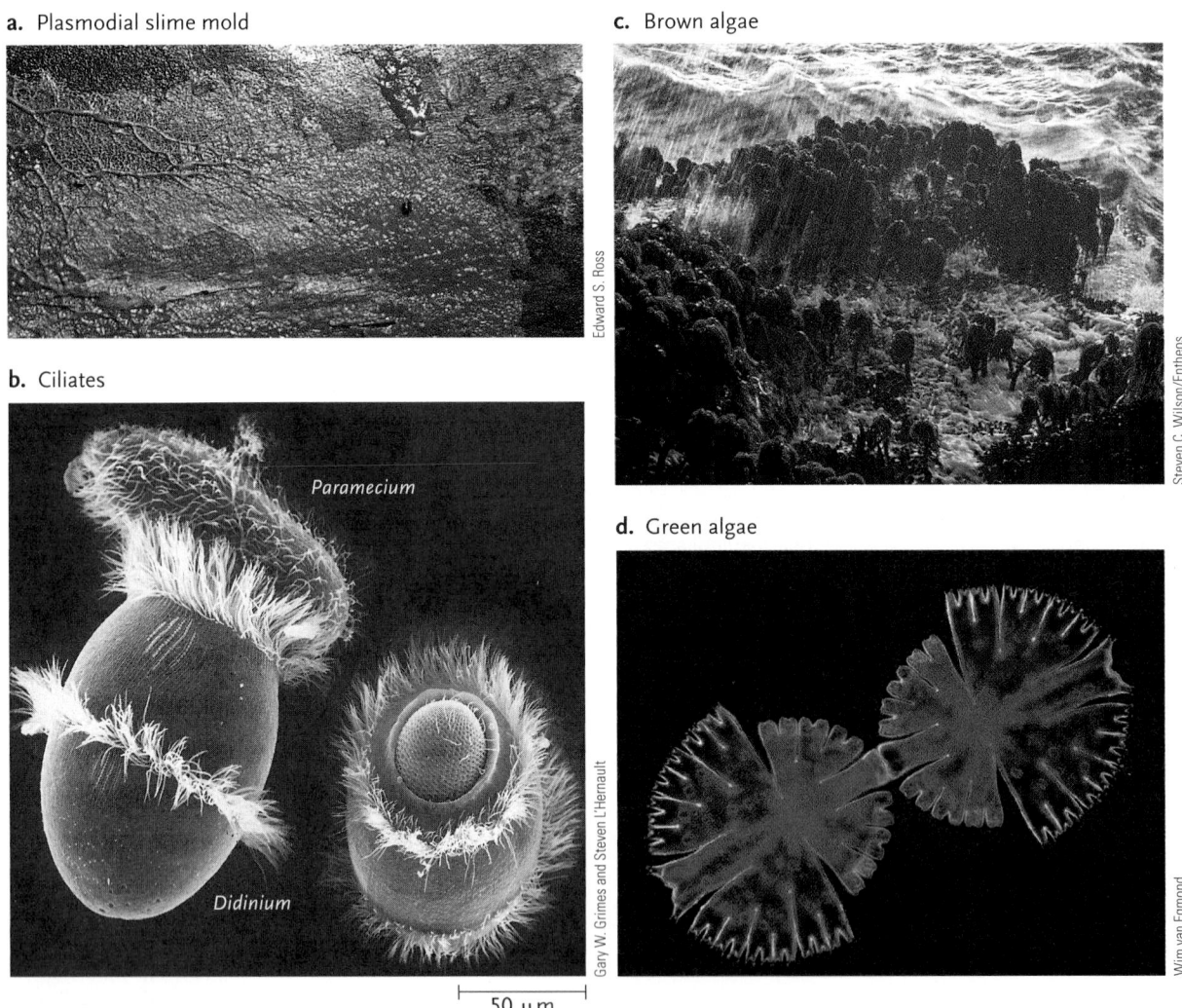

a. Plasmodial slime mold

b. Ciliates

Paramecium

Didinium

50 μm

c. Brown algae

d. Green algae

Figure 26.1

A sampling of protist diversity. **(a)** *Physarum*, a plasmodial slime mold (yellow shape, lower part of figure) migrating over a rotting log. **(b)** *Didinium*, a ciliate, consuming another ciliate, *Paramecium*. **(c)** *Postelsia palmaeformis* (the sea palm), a brown alga, thriving in the surf pounding a California coast. **(d)** *Micrasterias*, a single-celled green alga, here shown dividing in two.

Protists Are Most Easily Classified by What They Are Not

The one reasonable certainty about protist classification is that the organisms lumped together in the kingdom Protoctista are not prokaryotes, fungi, plants, or animals.

Because protists are eukaryotes, the boundary between them and prokaryotes is clear and obvious. Unlike prokaryotes, protists have a true nucleus, with multiple, linear chromosomes. In addition to cytoplasmic organelles—including mitochondria (in most but not all species), endoplasmic reticulum, Golgi complex, and chloroplasts (in some species)—protists and other eukaryotes have microtubules and microfilaments, which provide motility and cytoskeletal support. They reproduce asexually by mitosis or sexually by meiosis and union of sperm and egg cells, rather than by binary fission as do prokaryotes.

The phylogenetic relationship between protists and other eukaryotes is more complex **(Figure 26.2)**. From its beginning, the eukaryotic family tree branched out in many directions. All of the organisms in the eukaryotic lineages consist of protists except for three groups, the animals, land plants, and fungi, which arose from protist ancestors. Although some protists have features that resemble those of the fungi, plants, or animals, several characteristics are distinctive. For instance, cell wall components in protists differ from those of the fungi (molds and yeasts, for example). In contrast to land plants, protists lack highly differentiated structures equivalent to true roots, stems, and leaves; they also lack the protective structures that encase developing embryos in plants. Protists are distinguished from animals by their lack of highly differentiated structures such as limbs and a heart, and by the absence of features such as nerve cells, complex developmental stages, and an internal digestive tract. Pro-

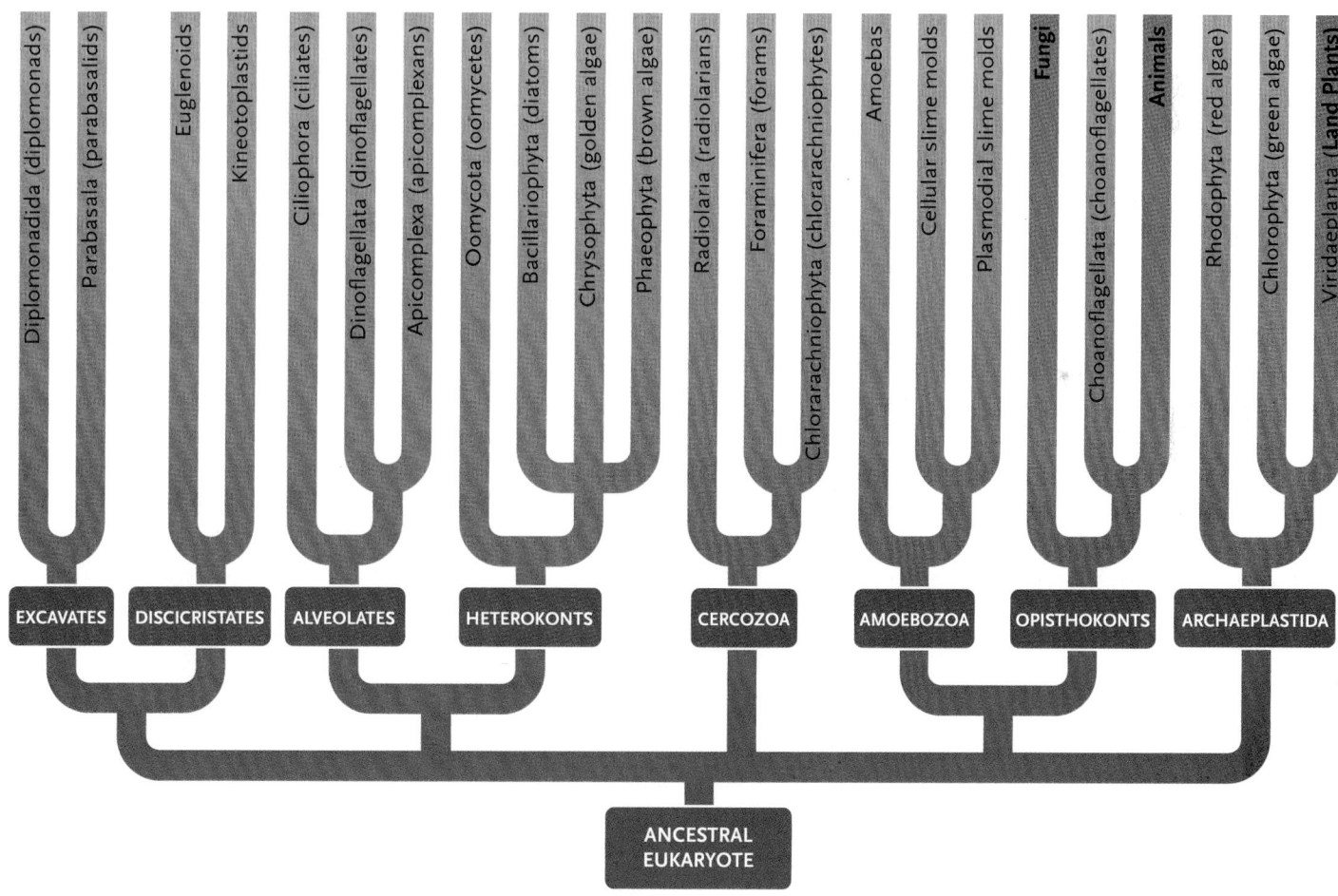

Figure 26.2

The phylogenetic relationship between the evolutionary groups within the kingdom Protoctista and the other eukaryotes. The Archaeplastida include the land plants of the kingdom Plantae (boxed), and the Opisthokonts include the animals of the kingdom Animalia and the fungi of the kingdom Fungi (boxed). The tree was constructed based on a consensus of molecular and ultrastructural data.

tists also lack collagen, the characteristic extracellular support protein of animals. Thus, the kingdom Protoctista is a catchall group that includes all the eukaryotes that are not fungi, plants, or animals.

Until recently, the protists were classified into phyla according to criteria such as body form, modes of nutrition and movement, and forms of meiosis and mitosis. However, comparisons of nucleic acid and amino acid sequences, now considered the most informative method for determining the evolutionary relationships of protists, show that most of the organisms previously grouped together do not share a common lineage. Further, many organisms within the phyla are no more closely related to each other than they are to the fungi, plants, or animals.

Given the extreme diversity of the protists, some evolutionists maintain that the kingdom Protoctista is actually a collection of many kingdoms—as many as 30, depending on differing evaluations of the lineages indicated by the sequence comparisons. Evolutionary lineages within the kingdoms are variously described as clades, subkingdoms, or phyla, and the existing schemes are constantly revised as new information is obtained.

For simplicity, we retain the Protoctista as a single kingdom in this book, with the understanding that it is a collection of largely unrelated organisms placed together for convenience. We will refer to the major evolutionary clusterings indicated by molecular and structural comparisons as groups (see Figure 26.2).

Figure 26.3
A ciliate, *Paramecium*, showing the cytoplasmic structures typical of many protists.
(Top: Frieder Sauer/ Bruce Coleman Ltd.; bottom: Redrawn from V. & J. Pearse and M. & R. Buchsbaum, Living Invertebrates, The Boxwood Press, 1987.)

Protist Diversity Is Reflected in Their Metabolism, Reproduction, Structure, and Habitat

As you might expect from the broad range of organisms included in the kingdom, protists are highly diverse in metabolism, reproduction, structure, and habitat.

Metabolism. Almost all protists are aerobic organisms that live either as heterotrophs—by obtaining their organic molecules from other organisms—or as autotrophs—by producing organic molecules for themselves by photosynthesis. Among the heterotrophs, some protists obtain organic molecules by directly ingesting part or all of other organisms and digesting them internally. Others absorb organic molecules from their environment. A few protists can live as either heterotrophs or autotrophs.

Reproduction. Reproduction may be asexual by mitosis or sexual by meiotic cell division and formation of gametes. In protists that reproduce by both mitosis and meiosis, the two modes of cell division are combined into a **life cycle** that is highly distinctive among the different protist groups.

Structure. Many protists live as single cells or as **colonies** in which individual cells show little or no differentiation and are potentially independent. Within colonies, individuals use cell signaling to cooperate on tasks such as feeding or movement. Some protists are large multicellular organisms, in which cells are differentiated and completely interdependent. For example, seaweeds are multicellular marine protists that include the largest and most differentiated organisms of the group; their structures include a hodlfast to secure the organism to the rocks, leaflike fronds, and, in some cases, an air bladder for flotation. The giant kelp of coastal waters rival forest trees in size.

Some single-celled and colonial protists have complex intracellular structures, some found nowhere else among living organisms **(Figure 26.3).** For example, many freshwater protists have a mechanism to maintain water balance in and out of the cell to prevent lysis. Excess water entering cells by osmosis (see Section 6.3) is handled using a specialized cytoplasmic organelle, the **contractile vacuole.** The contractile vacuole gradually fills with water; when it reaches maximum size it moves to the plasma membrane and forcibly contracts, expelling the water to the outside through a pore in the membrane. Many protists also have **food vacuoles** that digest prey or other organic material engulfed by the cells. Enzymes secreted into the food vacuoles digest the organic molecules; any remaining undigested matter is expelled to the outside by a mechanism similar to the expulsion of water by contractile vacuoles.

The cells of some protists are supported by an external cell wall, or by an internal or external shell built up

Vacuoles   Contractile vacuoles

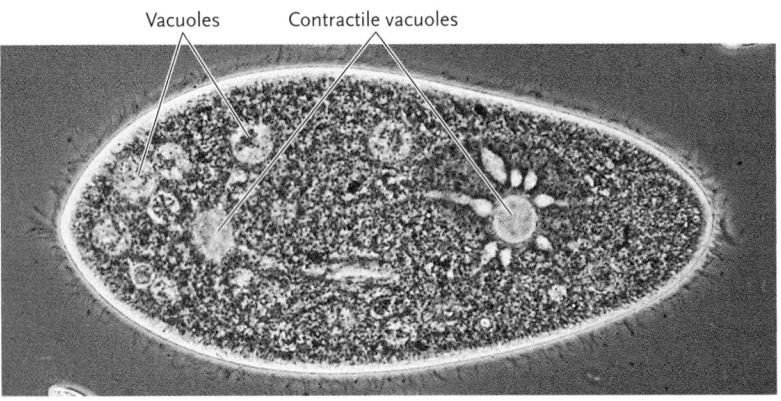

20 μm

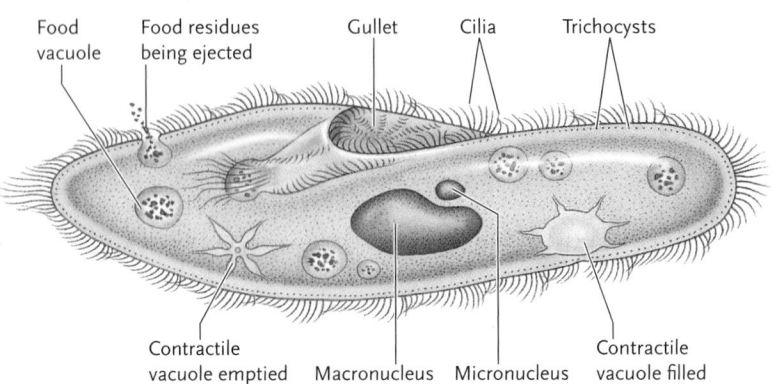

Food vacuole   Food residues being ejected   Gullet   Cilia   Trichocysts

Contractile vacuole emptied   Macronucleus   Micronucleus   Contractile vacuole filled

from organic or mineral matter; in some, the shell takes on highly elaborate forms. Other protists have a **pellicle**, a layer of supportive protein fibers located inside the cell, just under the plasma membrane, providing strength and flexibility instead of a cell wall (see Figure 26.5).

Almost all protists have structures providing motility at some time during their life cycle. Some move by amoeboid motion, in which the cell extends one or more lobes of cytoplasm called **pseudopodia** ("false feet"; see Figure 26.15). The rest of the cytoplasm and the nucleus then flow into a pseudopodium, completing the movement. Other protists move by the beating of flagella or cilia (see Section 5.3; cilia are essentially the same as flagella, except that cilia are often shorter and occur in greater numbers on a cell). In some protists, cilia are arranged in complex patterns, with an equally complex network of microtubules and other cytoskeletal fibers supporting the cilia under the plasma membrane. Among the protists are the most complex single cells known because of the wide variety of cytoplasmic structures they have.

Habitat. Protists live in aqueous habitats, including aquatic or moist terrestrial locations such as oceans, freshwater lakes, ponds, streams, and moist soils, and within host organisms. In bodies of water, small photosynthetic protists collectively make up the **phytoplankton** (*phytos* = plant; *planktos* = drifting), the abundant organisms that capture the energy of sunlight in nearly all aquatic habitats. These photosynthetic protists provide organic substances and oxygen for heterotrophic bacteria and protists and for the small crustaceans and animal larvae that are the primary constituents of **zooplankton** (*zoe* = life, usually meaning animal life); although protists are not animals, biologists often include them among the zooplankton. The phytoplankton and the larger multicellular protists forming seaweeds collectively account for about half of the total organic matter produced by photosynthesis.

In the moist soils of terrestrial environments, protists play important roles among the detritus feeders that recycle matter from organic back to inorganic form. In their roles in phytoplankton, zooplankton, and as detritus feeders, protists are enormously important in the world ecosystem.

Protists that live in host organisms are parasites, obtaining nutrients from the host. Indeed, many of the parasites that have significant effects on human health are protists, causing diseases such as malaria, sleeping sickness, and giardiasis.

Study Break

What distinguishes protists from prokaryotes? What distinguishes them from fungi, plants, and animals?

26.2 The Protist Groups

This section considers the biological features of each of the groups of protists included in Figure 26.2. This taxonomic tree represents a current consensus, based both on molecular data, such as comparative genomics, and on fine structures that have a distinctive form in a particular group.

The Excavates Lack Mitochondria

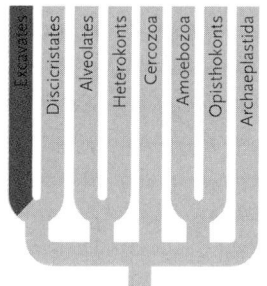

All members of the Excavates are single-celled animal parasites that lack mitochondria and move by means of flagella; most have a hollow (excavated) ventral feeding groove. Because they lack mitochondria, they are limited to glycolysis as an ATP source. However, the nuclei of Excavates contain genes derived from mitochondria, meaning that the ancestors of these protists probably had mitochondria. They may have lost their mitochondria as an adaptation to the parasitic way of life, in which oxygen is in short supply. We consider two groups here, the Diplomonadida and the Parabasala.

Diplomonadida. Diplomonad cells have two nuclei and move by means of multiple freely beating flagella. In addition to lacking mitochondria, they also lack a clearly defined endoplasmic reticulum and Golgi complex. The best-known representative of the group, *Giardia lamblia* **(Figure 26.4a)**, infects the mammalian intestinal tract, inducing severe diarrhea and abdominal cramps. *Giardia* is spread by contamination of water with feces, in which resistant cysts of the protist can be present in large numbers. So many streams and lakes in wilderness areas of the United States have become contaminated with *Giardia* cysts that hikers must boil water from these sources before drinking it, or pass it through filters able to remove particles as small as 1 μm. Treating water with chemicals such as chlorine or iodine does not kill the cysts.

Parabasala. In addition to freely beating flagella, species among the Parabasala have a sort of fin called an **undulating membrane,** formed by a flagellum buried in a fold of the cytoplasm. The buried flagellum allows parabasalans to move through thick and viscous fluids. Among the Parabasala are the trichomonads, including *Trichomonas vaginalis* **(Figure 26.4b),** a worldwide nuisance responsible for infections of the urinary and reproductive tracts in both men and women. The infective trichomonad is passed from person to person primarily, but not exclusively, by sexual intercourse. It lives in the vagina in women and in the urethra of both sexes. The

a. *Giardia lamblia*

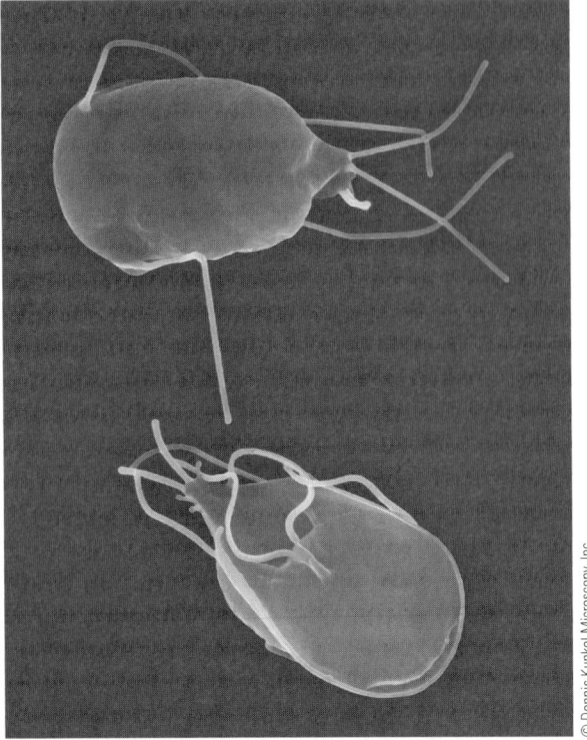

b. *Trichomonas vaginalis*

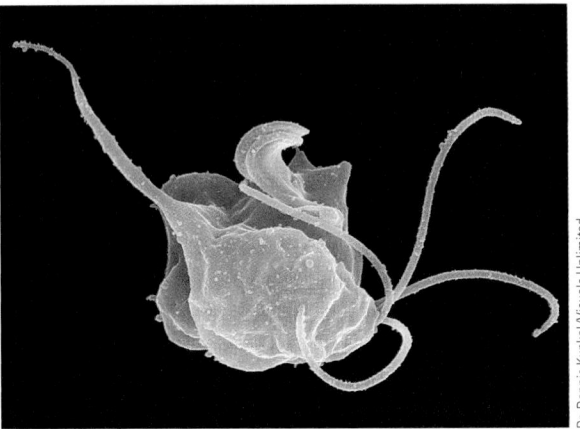

Figure 26.4
The Diplomonadida and Parabasala of the Excavates. **(a)** A diplomonad, *Giardia lamblia,* that causes intestinal disturbances. **(b)** A parabasalid, *Trichomonas vaginalis,* that causes a sexually transmitted disease, trichomoniasis.

most are photosynthetic, some are facultative heterotrophs; some can even alternate between photosynthesis and life as a heterotroph. The best-known members of this group are the euglenoids, with *Euglena gracilis* **(Figure 26.5)** as the best-known species. Another group, the parasitic kinetoplastids, includes some organisms responsible for human diseases. A commonly used nontaxonomic name for protists of the Discicristates is *protozoa* ("first animal"), referring to their similarity to animals with respect to ingesting food and moving by themselves.

Euglenoids. With the exception of a few marine species, the euglenoids inhabit freshwater ponds, streams, and lakes. Most are autotrophs that carry out photosynthesis by the same mechanisms as plants, using the same photosynthetic pigments, including chlorophylls *a* and *b* and β-carotene. Many of the photosynthetic euglenoids, including *E. gracilis,* can also live as heterotrophs by absorbing organic molecules through the plasma membrane. Some euglenoids lack chloroplasts and live entirely as heterotrophs.

Euglena gracilis and other euglenoids have a profusion of cytoplasmic organelles, including a contractile vacuole and, in photosynthetic species, chloroplasts (see Figure 26.5). Rather than an external cell wall, the euglenoids have a spirally grooved pellicle formed from transparent, protein-rich material. Most of the photosynthetic euglenoids, including *E. gracilis,* have an *eyespot* containing carotenoid pigment granules in association with a light-sensitive structure. The eyespot is part of a sensory mechanism that stimulates cells to swim toward moderately bright light or away from intensely bright light so that the organism is in light conditions for optimal photosynthetic activity. The cells swim by whiplike movements of flagella that extend from a pocketlike depression at one end of the cell. Most have two flagella, one rudimentary and short, the other long.

Kinetoplastids. The kinetoplastids are a group of nonphotosynthetic, heterotrophic cells that live as animal parasites **(Figure 26.6)**. Their name reflects the structure of the single mitochondrion in a cell of this group, which contains a large DNA-protein deposit called a *kinetoplast.* Most kinetoplastids have a leading and a trailing flagellum, which are used for movement. In some cases, the trailing flagellum is attached to the side of the cell, forming an undulating membrane that is often used to enable the organism to glide along or attach to surfaces.

The kinetoplastids include the trypanosomes, responsible for several diseases afflicting millions of humans in tropical regions. *Trypanosoma brucei* (see Figure 26.6) causes African sleeping sickness, transmitted from one host to another by bites of the tsetse fly. Early symptoms include fever, headaches, rashes, and anemia. Untreated, the disease damages the cen-

infection is usually symptomless in men, but in women *T. vaginalis* can cause severe inflammation and irritation of the vagina and vulva. It is easily cured by drugs.

The Discicristates Include the Euglenoids and Kinetoplastids, Which Are Motile Protists

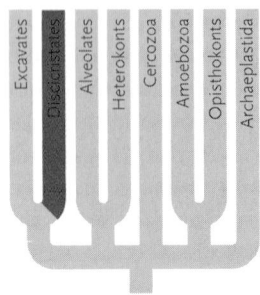

The **Discicristates** are named for their disc-shaped mitochondrial cristae (inner mitochondrial membranes). The group includes about 1800 species, almost all single-celled, highly motile cells that swim by means of flagella. While

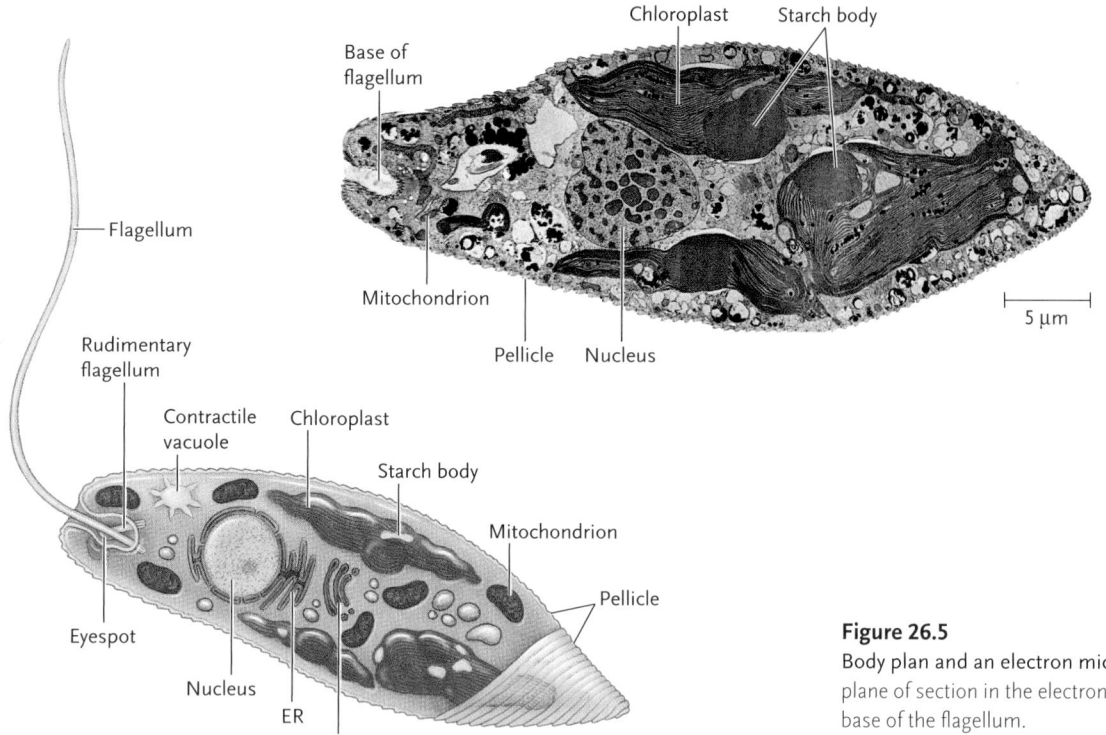

Figure 26.5
Body plan and an electron micrograph of *Euglena gracilis*. The plane of section in the electron micrograph has cut off all but the base of the flagellum.

(Micrograph: P. L. Walne and J. H. Arnott, Planta, 77:325–354, 1967.)

tral nervous system, leading to a sleeplike coma and eventual death. The disease has proved difficult to control because the same trypanosome infects wild mammals, providing an inexhaustible reservoir for the parasite. Other trypanosomes, also transmitted by insects, cause Chagas disease in the southwestern United States and Central and South America, and leishmaniasis in the tropics. Humans with Chagas disease have an enlarged liver and spleen and may experience severe brain and heart damage; people with leishmaniasis have skin sores and ulcers that may become very deep and disfiguring, particularly to the face.

The Alveolates Have Complex Cytoplasmic Structures and Use Flagella or Cilia to Move

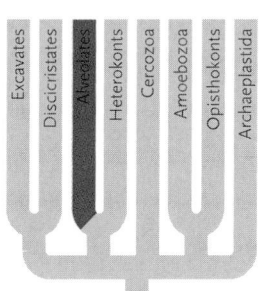

The Alveolates are so called because they have small, membrane-bound vesicles called *alveoli* (*alvus* = belly) in a layer just under the plasma membrane. The Alveolates include two motile, primarily free-living groups, the Ciliophora and Dinoflagellata, and a nonmotile, parasitic group, the Apicomplexa.

Ciliophora: The Ciliates. The Ciliophora—the ciliates—includes nearly 10,000 known species of primarily single-celled but highly complex heterotrophic organisms that swim by means of cilia (see Figures 26.1b and 26.3). Ciliates were among the first organisms observed in the seventeenth century by the pioneering microscopist Anton van Leeuwenhoek. Essentially any sample of pond water or bottom mud contains a wealth of these creatures.

The organisms in the Ciliophora have many highly developed organelles, including a mouthlike gullet lined with cilia; structures that exude mucins, toxins, or other defensive and offensive materials from the cell surface; contractile vacuoles; and complex systems of

Figure 26.6
Trypanosoma brucei, the parasitic kinetoplastid that causes African sleeping sickness.

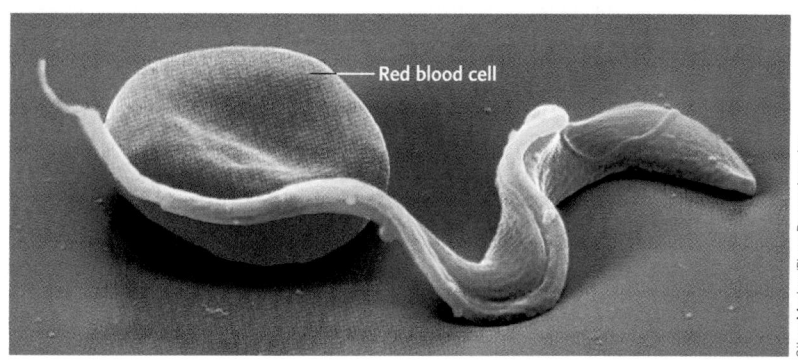

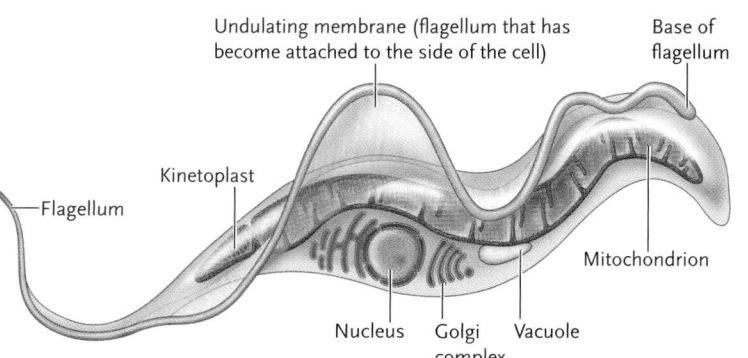

food vacuoles. A pellicle reinforces cell shape. A complex cytoskeletal network of microtubules and other fibers anchors the cilia just below the pellicle and coordinates the ciliary beating. The cilia can stop and reverse their beating in synchrony, allowing ciliates to stop, back up, and turn if they encounter negative stimuli. Evidence that the cytoskeletal network organizes ciliary beating comes from microsurgical experiments in which a segment of the body surface was cut out and reinserted in the opposite direction. The cilia in the reversed segment beat in the opposite direction to those on the rest of the organism.

The ciliates are the only eukaryotes that have two types of nuclei in each cell: one or more small nuclei called *micronuclei*, and a single larger *macronucleus* (see Figure 26.3b). A **micronucleus** is a diploid nucleus that contains a complete complement of genes. It functions primarily in cellular reproduction, which may be asexual or sexual. The number of micronuclei present depends on the species. The **macronucleus** develops from a micronucleus, but loses all genes except those required for basic "housekeeping" functions of the cell and for ribosomal RNAs. These DNA sequences are duplicated many times, greatly increasing its capacity to transcribe the mRNAs needed for these functions, and the rRNAs needed to make ribosomes.

In asexual reproduction by mitosis, both types of nuclei replicate their DNA, divide, and are passed on to daughter cells. In sexual reproduction of *Paramecium*, for example, two cells **conjugate** by first forming a cytoplasmic bridge **(Figure 26.7).** Next, the micronucleus in each cell undergoes meiosis, producing four haploid micronuclei. In a series of steps, three of the four micronuclei in each cell degenerate, and the macronucleus also begins degenerating. The remaining micronucleus divides by mitosis, and one of the two micronuclei in each cell then passes through the cytoplasmic bridge into the other cell (step 5). The two haploid micronuclei in each cell now fuse to form a diploid micronucleus, with pairs of homologous chromosomes, one from each of the original parents. The two cells then separate. Through a further series of divisions, the micronucleus in each cell gives rise to two micronuclei and two macronuclei. Finally, each cell divides to produce two daughter cells, completing sexual reproduction.

Ciliates abound in freshwater and marine habitats, where they feed voraciously on bacteria, algae, and each other. *Paramecium* is a typical member of the group (see Figure 26.3). Its rows of cilia drive it through its watery habitat, rotating the cell on its long axis while it moves forward, or backs and turns. The cilia also sweep water laden with prey and food particles into the gullet, where food vacuoles form. The ciliate digests food in the vacuoles and eliminates indigestible material through an anal pore. Contractile vacuoles with elaborate, raylike extensions remove excess water from the cytoplasm and expel it to the outside. When under attack or otherwise stressed, *Paramecium* discharges many dartlike protein threads from surface organelles called **trichocysts.**

Some ciliates live individually while others are colonial. Certain ciliates are animal parasites; others live and reproduce in their hosts as mutually beneficial symbionts. (*Symbiosis* is the interaction between two organisms living together in close association, sometimes one inside another.) A compartment of the stomach of cattle and other grazing animals contains large numbers of symbiotic ciliates that digest the cellulose in their host's plant diet. The animals then digest the excess ciliates.

One ciliate, *Balantidium coli*, is a human intestinal parasite that causes diarrhea, with stools typically containing blood and pus. It is passed on when humans eat food contaminated by the feces of animals infected by *Balantidium*, particularly pigs. Less than 1% of the human population is infected worldwide.

Dinoflagellata: The Dinoflagellates. Of over 4000 known dinoflagellate species, most are single-celled organisms in marine phytoplankton. They live as heterotrophs or autotrophs; many can carry out both modes of nutrition. Some contain algae as symbionts. Typically, they have a shell formed from cellulose plates **(Figure 26.8).** The beating of flagella, which fit into grooves in the plates, makes dinoflagellates spin like a top (*dinos* = spinning).

The cytoplasmic structures of dinoflagellates include mitochondria, chloroplasts in photosynthetic species, and other internal membrane systems characteristic of eukaryotes. The photosynthetic dinoflagellates contain chlorophylls *a* and *c* along with accessory pigments that make them golden-brown or brown; algal symbionts give some a green, blue, or red color.

Their abundance in phytoplankton makes dinoflagellates a major primary producer of ocean ecosystems. Some species live as symbionts in the tissues of other marine organisms such as jellyfish, sea anemones, corals, and mollusks. For example, dinoflagellates in coral use the coral's carbon dioxide and nitrogenous waste, while supplying 90% of the coral's nutrition. The vast numbers of dinoflagellates living as photosynthetic symbionts in tropical coral reefs allow the reefs to reach massive size; without the dinoflagellates many coral species would die.

Some dinoflagellates are **bioluminescent**—they glow or release a flash of light, particularly when disturbed. The production of light depends on the enzyme *luciferase* and its substrate *luciferin*, in forms similar to the system that produces light in fireflies. Dinoflagellate fluorescence can make the sea glow in the wake of a boat at night and coat nocturnal surfers and swimmers with a ghostly light.

At times dinoflagellate populations grow to such large numbers that they color the seas red, orange, or brown. The resulting **red tides** are common in spring and summer months along the warmer coasts of the world, including all the U.S. coasts. Some red-tide dinoflagellates produce a toxin that interferes with nerve

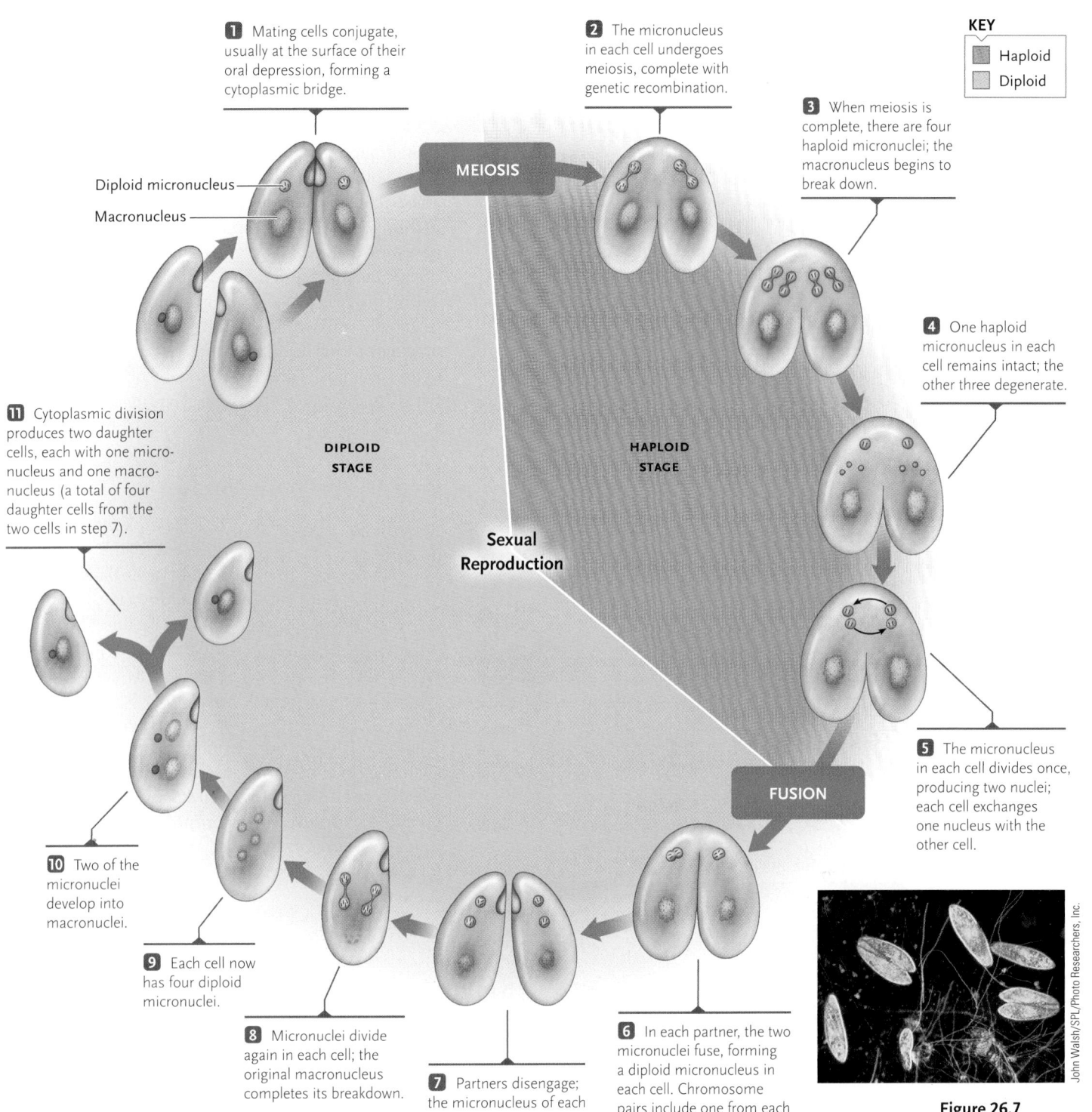

1 Mating cells conjugate, usually at the surface of their oral depression, forming a cytoplasmic bridge.

2 The micronucleus in each cell undergoes meiosis, complete with genetic recombination.

3 When meiosis is complete, there are four haploid micronuclei; the macronucleus begins to break down.

4 One haploid micronucleus in each cell remains intact; the other three degenerate.

5 The micronucleus in each cell divides once, producing two nuclei; each cell exchanges one nucleus with the other cell.

6 In each partner, the two micronuclei fuse, forming a diploid micronucleus in each cell. Chromosome pairs include one from each of the original parents.

7 Partners disengage; the micronucleus of each divides mitotically.

8 Micronuclei divide again in each cell; the original macronucleus completes its breakdown.

9 Each cell now has four diploid micronuclei.

10 Two of the micronuclei develop into macronuclei.

11 Cytoplasmic division produces two daughter cells, each with one micronucleus and one macronucleus (a total of four daughter cells from the two cells in step 7).

Diploid micronucleus

Macronucleus

KEY
Haploid
Diploid

MEIOSIS

FUSION

DIPLOID STAGE

HAPLOID STAGE

Sexual Reproduction

John Walsh/SPL/Photo Researchers, Inc.

Figure 26.7
Sexual reproduction by conjugation in a ciliate, *Paramecium*.

function in animals that ingest these protists. Fish that feed on plankton, and birds that feed on the fish, may be killed in huge numbers by the toxin. Dinoflagellate toxin does not noticeably affect clams, oysters, and other mollusks, but it becomes concentrated in their tissues. Eating the tainted mollusks can cause respiratory failure and death for humans and other animals. The toxin is especially deadly for mammals because it paralyzes the diaphragm and other muscles required for breathing.

Apicomplexa. The apicomplexans are all nonmotile parasites of animals. They absorb nutrients through their plasma membranes rather than by engulfing food particles, and they lack food vacuoles. They get their name from the *apical complex,* a group of organelles at one end of the cell that functions in attachment and invasion of host cells.

Typically, apicomplexan life cycles involve both asexual and sexual reproduction. All the apicomplexans, which includes almost 4000 known species, pro-

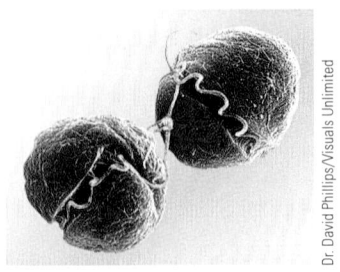

Figure 26.8
Karenia brevis, a toxin-producing dinoflagellate.

duce infective sporelike stages called *sporozoites*. The sporozoites reproduce asexually in cells they infect, eventually bursting them, which releases the progeny to infect new cells. At some point they generate specialized cells that form gametes; fusion of gametes produces resistant cells known as *cysts*. Usually, a host is infected by ingesting cysts, which divide to produce sporozoites. This basic life cycle pattern varies considerably among the apicomplexans, and many of these organisms use more than one host species for different stages of their life cycle.

One apicomplexan genus, *Plasmodium,* is responsible for malaria, one of the most widespread and debilitating diseases of humans. The disease is transmitted by the bite of 60 different species of mosquitoes, all members of the genus *Anopheles*. Although the disease is now rare in the United States, *Anopheles* mosquitoes are common enough to spread malaria if *Plasmodium* is introduced by travelers from other countries. The infective cycle of *Plasmodium,* described in *Focus on Research,* is representative of the complex life cycles of apicomplexans.

Another organism in this group, *Toxoplasma,* has a sexual phase of its life cycle in cats and asexual phases in humans, cattle, pigs, and other animals. Cysts of the parasite in the feces of infected cats are spread in household and garden dust. Humans ingesting or inhaling the cysts develop toxoplasmosis, a disease that is usually mild in adults but can cause severe brain damage or even death to a fetus. Because of the danger of toxoplasmosis, pregnant women should avoid emptying litter boxes or otherwise cleaning up after a cat.

The Heterokonts Include the Largest Protists, the Brown Algae

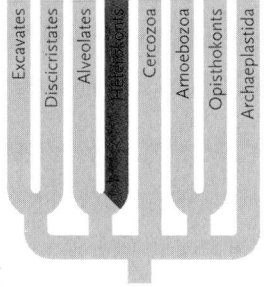

The Heterokonts (*hetero* = different; *kontos* = pole, referring to the flagellum) are named for their two different flagella: one with hollow tripartite hairs that give the flagellum a "hairy" appearance and a second one that is plain. The flagella occur only on reproductive cells such as eggs and sperm, except in the golden algae, in which cells are flagellated in all stages. The heterokonts include the Oomycota (water molds, white rusts, and mildews—formerly classified as fungi), Bacillariophyta (diatoms), Chrysophyta (golden algae), and Phaeophyta (brown algae).

Oomycota: Water Molds, White Rusts, and Downy Mildews. The Oomycota (**Figure 26.9**) are funguslike heterokonts that lack chloroplasts and live as heterotrophs. Like fungi, they secrete enzymes that digest the complex molecules of surrounding dead or alive organic matter into simpler substances small enough to be absorbed into their cells. The water molds live almost exclusively in freshwater lakes and streams or moist terrestrial habitats; the white rusts and downy mildews are parasites of plants. Oomycota may reproduce asexually or sexually.

Like fungi, many Oomycota grow as microscopic, nonmotile filaments called **hyphae** (singular, hypha), which form a network called a **mycelium (Figure 26.10)**. Other features, however, set the Oomycota apart from the fungi; chief among them are differences in nucleotide sequence, which clearly indicate close evolutionary relationships to the heterokonts rather than to the fungi. Further, nuclei in hyphae are diploid in the Oomycota, rather than haploid as in the fungi, and repro-

a. Water mold

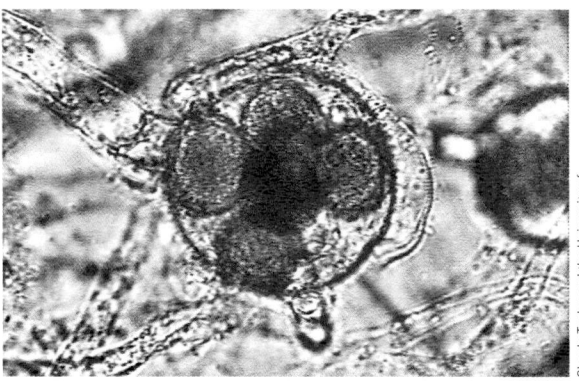

b. Water mold infecting fish

c. Downy mildew

Figure 26.9
Oomycota. **(a)** The water mold *Saprolegnia parasitica*. **(b)** *S. parasitica* growing as cottony white fibers on the tail of an aquarium fish. **(c)** A downy mildew, *Plasmopara viticola*, growing on grapes. At times it has nearly destroyed vineyards in Europe and North America.

Applied Research: Malaria and the *Plasmodium* Life Cycle

Although malaria is uncommon in the United States, it is a major epidemic in many other parts of the world. From 300 million to 500 million people become infected with malaria each year in tropical regions, including Africa, India, southeast Asia, the Middle East, Oceania, and Central and South America. Of these, about 2 million die each year, twice as many as from AIDS worldwide. It is particularly deadly for children younger than 6. In many countries where malaria is common, people are often infected repeatedly, with new infections occurring alongside preexisting infections.

Four different species of the apicomplexan genus *Plasmodium* cause malaria. In the life cycle of the parasites (see **figure**), sporozoites develop in the female *Anopheles* mosquito, which transmits them by its bite to human or bird hosts. The infecting parasites divide repeatedly in their hosts, initially in liver cells and then in red blood cells. Their growth causes red blood cells to rupture in regular cycles every 48 or 72 hours, depending on the *Plasmodium* species. The ruptured red blood cells clog vessels and release the parasite's metabolic wastes, causing cycles of chills and fever.

The victim's immune system is ineffective because, during most of the infective cycle, the parasite is inside body cells and thus "hidden" from antibodies. Further, *Plasmodium* regularly changes its surface molecules, continually producing new forms that are not recognized by antibodies developed against a previous form. In this way, the parasite keeps one step ahead of the immune system, often making malarial infections essentially permanent.

Travelers in countries with high rates of malaria are advised to use antimalarial drugs such as chloroquine, quinine, or quinidine as a preventative.

However, many *Plasmodium* strains in Africa, India, and southeast Asia have developed resistance to the drugs. Vaccines have proved difficult to develop; because vaccines work by inducing the production of antibodies that recognize surface groups on the parasites, they are defeated by the same mechanisms the parasite uses inside the body to keep one step ahead of the immune reaction.

While in a malarial region, travelers should avoid exposure to mosquitoes by remaining indoors from dusk until dawn and sleeping inside mosquito nets treated with insect repellent. When out of doors, travelers should wear clothes that expose as little skin as possible and are thick enough to prevent mosquitoes from biting through the cloth. An insect repellent containing DEET should be spread on any skin that is exposed.

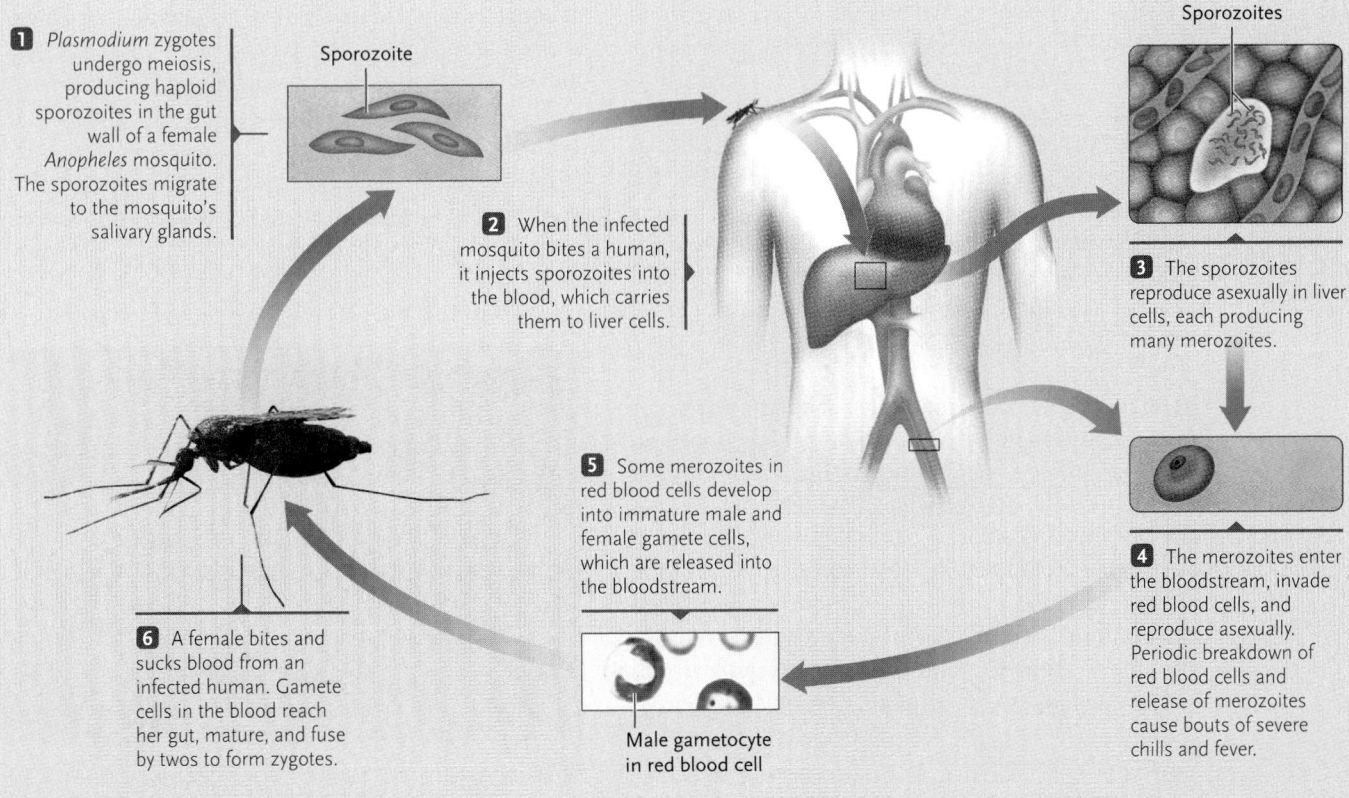

1 *Plasmodium* zygotes undergo meiosis, producing haploid sporozoites in the gut wall of a female *Anopheles* mosquito. The sporozoites migrate to the mosquito's salivary glands.

Sporozoite

2 When the infected mosquito bites a human, it injects sporozoites into the blood, which carries them to liver cells.

Sporozoites

3 The sporozoites reproduce asexually in liver cells, each producing many merozoites.

4 The merozoites enter the bloodstream, invade red blood cells, and reproduce asexually. Periodic breakdown of red blood cells and release of merozoites cause bouts of severe chills and fever.

5 Some merozoites in red blood cells develop into immature male and female gamete cells, which are released into the bloodstream.

Male gametocyte in red blood cell

6 A female bites and sucks blood from an infected human. Gamete cells in the blood reach her gut, mature, and fuse by twos to form zygotes.

Life cycle of a *Plasmodium* species that causes malaria.
(Photo: Sinclair Stammers/Photo Researchers, Inc.; micrograph: Steven L'Hernault.)

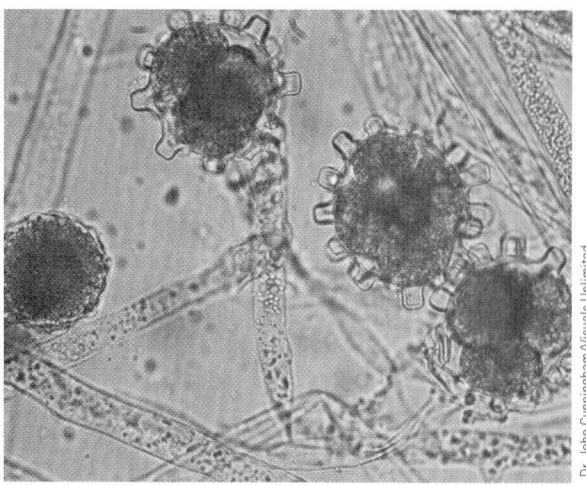

Figure 26.10
The funguslike body form of the Oomycota, consisting of filaments called hyphae, which grow into a network called a mycelium.

Dr. John Cunningham/Visuals Unlimited

ductive cells are flagellated and motile; fungi have no motile stages. Finally, the cell walls of most Oomycota contain cellulose (see Figure 3.7c); fungal cell walls instead contain a different polysaccharide, chitin (see Figure 3.7d).

Most water molds are key decomposers of both aquatic and moist terrestrial habitats. Dead animal or plant material immersed in water commonly becomes coated with cottony water molds. Other water molds parasitize living aquatic animals, such as the mold growing on the fish shown in Figure 26.9a. The white rusts and downy mildews are parasites of land plants (see Figure 26.9c).

Some water molds have had drastic effects on human history. *P. infestans,* a water mold that causes rotting of potato and tomato plants, was responsible for the Irish potato famine of 1845 to 1860. In this famine more than a million people, a third of Ireland's population, starved to death. Many of the survivors migrated in large numbers to other countries, including the United States and Canada.

Bacillariophyta: Diatoms. The Bacillariophyta, or diatoms, are single-celled organisms that are covered by a glassy silica shell, which is intricately formed and beautiful in many species. The two halves of the shell fit together like the top and bottom of a candy box **(Figure 26.11).** Substances move to and from the plasma membrane through elaborately patterned perforations in the shell. Although flagella are present only in gametes, many diatoms move by an unusual mechanism in which a secretion released through grooves in the shell propels them in a gliding motion.

Diatoms are autotrophs that carry out photosynthesis by pathways similar to those of plants. The pri-

mary photosynthetic organisms of marine plankton, they fix more carbon into organic material than any other planktonic organism. They are also abundant in freshwater habitats as both phytoplankton and bottom-dwelling species. Although most diatoms are free living, some are symbionts inside other marine protists. One diatom, *Pseudonitzschia,* produces a toxic amino acid that can accumulate in shellfish. The amino acid, which acts as a nerve poison, causes amnesic shellfish poisoning when ingested by humans; the poisoning can be fatal.

Asexual reproduction in diatoms occurs by mitosis followed by a form of cytoplasmic division in which each daughter cell receives either the top or bottom half of the parent shell. The daughter cell then secretes the missing half, which becomes the smaller, inside shell of the box. The daughter cell receiving the larger top half grows to the same size as the parent shell, but the cell receiving the smaller bottom half is limited to the size of this shell. As asexual divisions continue, the cells receiving bottom halves become progressively smaller. Very small diatoms may switch to a sexual mode of reproduction; they enter meiosis and produce flagellated gametes, which lose their shells and fuse in pairs to form a zygote. The zygote grows to normal size before secreting a completely new shell with full-size top and bottom halves.

The shells of diatoms are common in fossil deposits. In fact, more diatoms are known as fossils than as living species—some 35,000 extinct species have been described as compared with 7000 living species. For about 180 million years the shells of diatoms have been accumulating into thick layers of sediment at the bottom of lakes and seas. Since diatoms store food as oil, fossil diatoms may be a source of oil in many oil deposits.

Grinding the fossilized shells into a fine powder produces *diatomaceous earth,* which is used in abra-

Jan Hinsch/SPL/Photo Researchers, Inc.

Figure 26.11
Diatom shells. Depending on the species, the shells are either radially or bilaterally symmetrical, as seen in this sample.

a. Golden alga

b. Brown alga, *Macrocystis*

c. Brown alga, *Postelsia palmaeformis*

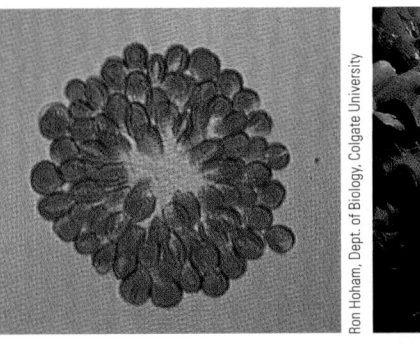

Figure 26.12

Golden and brown algae. **(a)** A microscopic, swimming colony of *Synura*, a golden alga. Each cell bears two flagellae, which are not visible in this light micrograph. **(b)** A brown alga, *Macrocystis*. Note the whitish gas bladders that keep the blades floating. **(c)** The holdfast, stemlike stipes, and leaflike blades, as seen in another brown alga, the sea palm *Postelsia palmaeformis*.

sives and filters, as an insulating material, and as a pesticide. Diatomaceous earth kills crawling insects by abrading their exoskeleton, causing them to dehydrate and die. Insect larvae are killed in the same way. Insects also die when they eat the powder but larger animals, including humans, are unaffected by it.

Chrysophyta: Golden Algae. Most golden algae **(Figure 26.12a)** are colonial forms in which each cell of the colony bears a pair of flagella. The golden algae have glassy shells, but in the form of plates or scales rather than in the candy-box form of the diatoms.

Nearly all chrysophytes are autotrophs and carry out photosynthesis using pathways similar to those of plants. Their color is due to a brownish carotenoid pigment, fucoxanthin, which masks the green color of the chlorophylls. Golden algae are important in freshwater habitats and in "nanoplankton," a community of marine phytoplankton composed of huge numbers of extremely small cells. During the spring and fall, "blooms" of golden algae can give a fishy taste and brownish color to the water.

Phaeophyta: Brown Algae. The brown algae (*phaios* = brown) are photosynthetic autotrophs that range from microscopic forms to giant kelps reaching lengths of 50 m or more (Figure 26.1c and **Figure 26.12b** and **c**). Their color is also due to fucoxanthin. Their cell walls contain cellulose and a mucilaginous polysaccharide, alginic acid.

Nearly all of the 1500 known phaeophyte species inhabit temperate or cool coastal marine waters. The kelps form vast underwater forests; fragments of these algae litter the beaches in coastal regions where they grow. Great masses of another brown alga, *Sargassum*, float in an area of the mid-Atlantic Ocean called the Sargasso Sea, which covers millions of

square kilometers between the Azores and the Bahamas.

Kelps are the largest and most complex of all protists. Their tissues are differentiated into leaflike *blades,* stalklike *stipes,* and rootlike *holdfasts* that anchor them to the bottom. Hollow, gas-filled bladders give buoyancy to the stipes and blades and help keep them upright. The stalks of some kelps contain tubelike vessels, similar to the vascular elements of plants, which rapidly distribute dissolved sugars and other products of photosynthesis throughout the body of the alga.

Life cycles among the brown algae are typically complex and in many species consist of alternating haploid and diploid generations **(Figure 26.13).** The large structures that we recognize as kelps and other brown seaweeds are diploid **sporophytes,** so called because they give rise to haploid spores by meiosis. The spores, which are flagellated swimming cells, germinate and divide by mitosis to form an independent, haploid **gametophyte** generation. The gametophytes give rise to haploid gametes, the egg and sperm cells. Most brown algal gametophytes are multicellular structures only a few centimeters in diameter. Cells in the gametophyte, produced by mitosis, differentiate to form flagellated, swimming sperm cells or nonmotile eggs. Fusion of a sperm and an egg cell gives rise to a diploid zygote, which grows by mitotic divisions into the sporophyte generation. Other variations occur in smaller brown algae, including some life cycles in which the sporophytes and gametophytes are the same size and some in which the gametophyte is larger than the sporophyte.

The alginic acid in brown algal cell walls, called **algin** when extracted, is an essentially tasteless and nontoxic substance used to thicken such diverse prod-

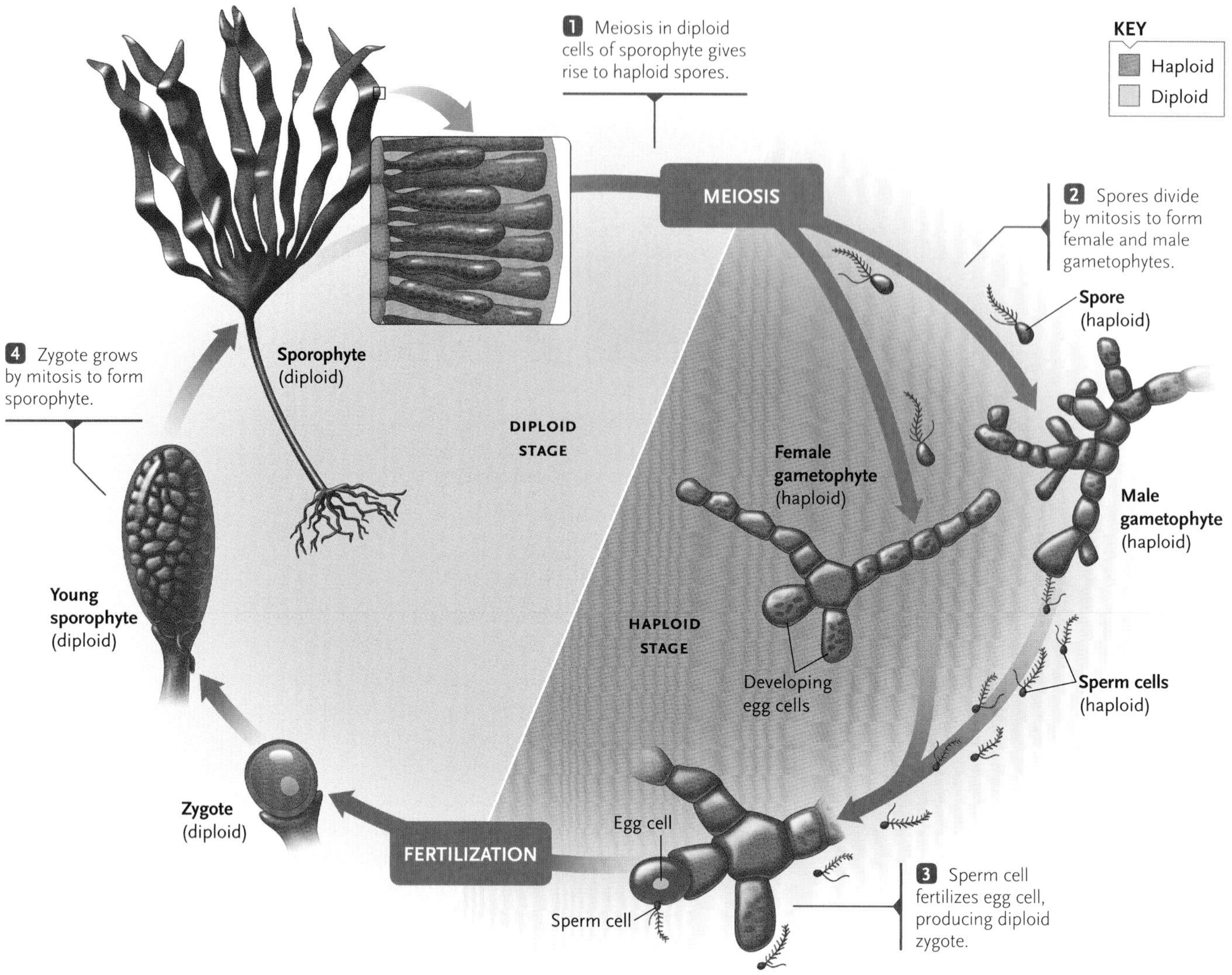

Figure 26.13

The life cycle of the brown alga *Laminaria*, which alternates between a diploid sporophyte stage and a haploid gametophyte stage.

1 Meiosis in diploid cells of sporophyte gives rise to haploid spores.

KEY
- Haploid
- Diploid

MEIOSIS

2 Spores divide by mitosis to form female and male gametophytes.

Spore (haploid)

4 Zygote grows by mitosis to form sporophyte.

Sporophyte (diploid)

DIPLOID STAGE

Female gametophyte (haploid)

Male gametophyte (haploid)

Young sporophyte (diploid)

HAPLOID STAGE

Developing egg cells

Sperm cells (haploid)

Zygote (diploid)

Egg cell

FERTILIZATION

Sperm cell

3 Sperm cell fertilizes egg cell, producing diploid zygote.

---

ucts as ice cream, pudding, salad dressing, jellybeans, cosmetics, paper, and floor polish. Brown algae are also harvested as food crops and fertilizers.

The Cercozoa Are Amoebas with Filamentous Pseudopods

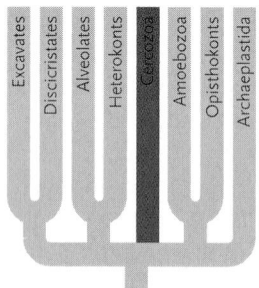

Amoeba is a descriptive term for a single-celled protist that moves by means of temporary cellular projections called pseudopods. Several major groups of protists contain amoebas, which are similar in form but are not all closely related. The amoebas classified in cercozoa produce stiff, filamentous pseudopodia, and many produce hard outer shells, also called *tests*. We consider here two heterotrophic groups of cercozoan amoebae, the Radiolaria and the Foramin-

ifera, and a third, photosynthesizing group, the Chlorarachniophyta.

Radiolaria. Radiolarians are distinguished by axopods, slender, raylike strands of cytoplasm supported internally by long bundles of microtubules. They engulf prey organisms that stick to the axopods and digest them in food vacuoles.

Radiolarians live in marine environments. They secrete a glassy internal skeleton from which the axopods project (**Figure 26.14a** and **b**). Just outside the skeleton, the cytoplasm is crowded with frothy vacuoles and lipid droplets, which provide buoyancy.

The skeletons of dead radiolarians sink to the bottom and become part of the sediment. Over time, they harden into sedimentary rocks that form an important part of the geological record.

Foraminifera: Forams. Foraminifera, or forams, live in marine environments. Their shells consist of organic matter reinforced by calcium carbonate (**Figure**

a. Radiolarian skeletons

Wim van Egmond

b. Living foram

Courtesy of Allen W. H. Be and David A. Caron

c. Foram shells

John Clegg/Ardea, London

d. Foram body plan

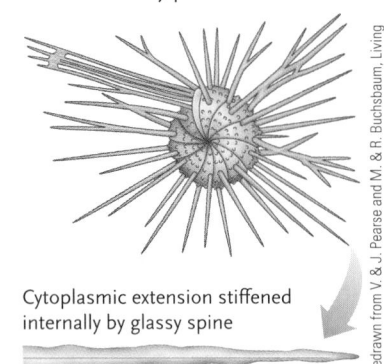

Cytoplasmic extension stiffened internally by glassy spine

Redrawn from V. & J. Pearse and M. & R. Buchsbaum, Living Invertebrates, The Boxwood Press, 1987.

Figure 26.14

Radiolarians and forams. **(a)** The internal skeletons of two radiolarian species, possibly *Pterocorys* and *Stylosphaera*. Bundles of microtubules support the cytoplasmic extensions of the radiolarians. **(b)** A living foram, showing the cytoplasmic strands extending from its shell. **(c)** Empty foram shells. **(d)** The body plan of a foram. Needlelike, glassy spines support the cytoplasmic extensions of the forams.

26.14c–d). Most foram shells are chambered, spiral structures that, although microscopic, resemble those of mollusks. Forams are identified and classified primarily by the form of the shell; about 250,000 species are known. Some species are planktonic, but they are most abundant on sandy bottoms and attached to rocks along the coasts. Their name comes from the perforations in their shells (*foramen* = little hole), through which extend long, slender strands of cytoplasm supported internally by a network of needlelike spines. The forams engulf prey that adhere to the strands and conduct them through the holes in the shell into the central cytoplasm, where they are digested in food vacuoles. Some forams have algal symbionts that carry out photosynthesis, allowing them to live as both heterotrophs and autotrophs.

Marine sediments are typically packed with the shells of dead forams. The sediments may be hundreds of feet thick; the White Cliffs of Dover in England are composed primarily of the shells of ancient forams. Most of the world's deposits of limestone and marble contain foram shells; the great pyramids and many other monuments of ancient Egypt are built from blocks cut from fossil foram deposits. Because distinct species lived during different geological periods, they are widely used to establish the age of sedimentary rocks containing their shells. They, along with radiolarian species, are also used as indicators by oil prospectors because layers of forams often overlie oil deposits.

Chlorarachniophyta. Chloroarachniophytes are green, photosynthetic amoebas that also engulf food. They contain chlorophylls *a* and *b,* but they are phylogenetically distinct from other chlorophyll *b*–containing eukaryotes. Many filamentous pseudopodia extend from the cell surface.

The Amoebozoa Includes Most Amoebas and Two Types of Slime Molds

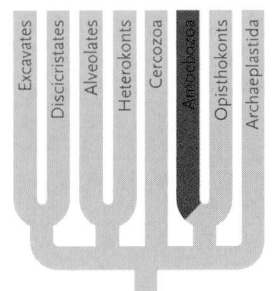

The Amoebozoa includes most of the amoebas (others are in the Cercozoa) as well as the cellular and plasmodial slime molds. All members of this group use pseudopods for locomotion and feeding for all or part of their life cycles.

Amoebas. Amoebas of the Amoebozoa are single-celled organisms that are abundant in marine and freshwater environments and in the soil. They use pseudopods for locomotion and feeding. The pseudopods extend and retract at any point on their body surface and are unsupported by any internal cellular organization. This type of pseudopod—called a *lobose* ("lobelike") *pseudopod*—distinguishes these amoebas from those in the Cercozoa, which have stiff, supported pseudopods. As a result of their pseudopod activity, and the ability to flatten or round up, these amoebas have no fixed body shape. A number of species are parasites, but most species feed on algae, bacteria, other protists, and bits of organic matter. The ingested matter is enclosed in food vacuoles and digested by enzymes secreted into the vacuoles. Any undigested matter is expelled to the outside by fusion of the vacuole with the plasma membrane. Their reproduction is entirely asexual, through mitotic divisions.

The most-studied amoebozoan is *Amoeba proteus* **(Figure 26.15).** Its natural habitat is in freshwater ponds and streams. Another member, *Acanthamoeba,* which lives in the soil, is widely used as a source of actin and

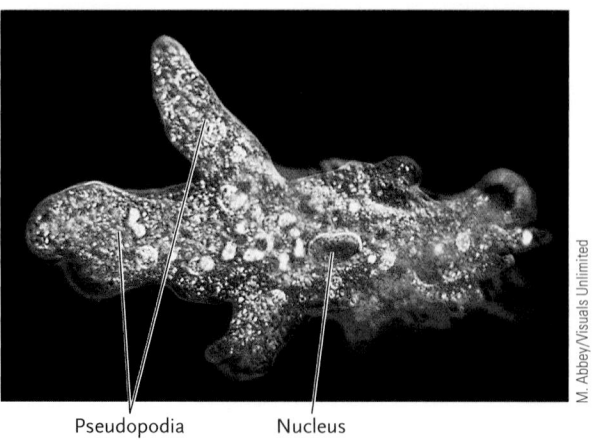

Pseudopodia   Nucleus

Figure 26.15
Amoeba proteus of the Amoebozoa is perhaps the most familiar protist of all.

myosin for scientific studies of amoeboid motion and cytoplasmic streaming.

The parasitic amoebas include some 45 species that infect the human digestive tract, one in the mouth and the rest in the intestine. One of the intestinal parasites, *Entamoeba histolytica,* causes amoebic dysentery. Cysts of this amoeba contaminate water supplies and soil in regions with inadequate sewage treatment. When ingested, a cyst breaks open to release an amoeba that feeds and divides rapidly in the digestive tract. Enzymes released by the amoebas destroy cells lining the intestine, producing the ulcerations, painful cramps, and debilitating diarrhea characteristic of the disease. Amoebic dysentery afflicts millions of people worldwide; in less-developed countries, it is a leading cause of death among infants and small children. Other parasitic amoebas cause less severe digestive upsets.

Slime Molds. Slime molds are heterotrophic protists that, at some stage of their life cycle, exist as individuals that move by amoeboid motion but the remainder of the time exist in more complex forms. They live on moist, rotting plant material such as decaying leaves and bark. The cells engulf particles of dead organic matter, and also bacteria, yeasts, and other microorganisms, and digest them internally. At one stage of their life cycles, they differentiate into a funguslike, stalked structure called a **fruiting body,** which forms spores by either asexual or sexual reproduction. Some species are brightly colored in hues of yellow, green, red, orange, brown, violet, or blue. The two major evolutionary lineages of slime molds, the cellular slime molds and the plasmodial slime molds, differ in cellular organization.

The Cellular Slime Molds. **Cellular slime molds** exist primarily as individual cells, either separately or as a coordinated mass. Among the 70 or so species of cellular slime molds, *Dictyostelium discoideum* is best known; its genome sequence was reported in May 2005. Its life cycle begins when a haploid spore lands in a suitably moist environment containing decaying organic matter **(Figure 26.16).** The spore germinates into an amoeboid cell that grows and divides mitotically into separate haploid cells as long as the food source lasts. When the food supply dwindles, some of the cells release a chemical signal (cyclic AMP; see Section 7.4) in pulses; in response, the amoebas move together and form a sausage-shaped mass that crawls in coordinated fashion like a slug. Some "slugs," although not much more than a millimeter in length, contain more than 100,000 individual cells. At some point the slug stops moving and differentiates into a stalked fruiting body, with cell walls reinforced by cellulose. When mature, the head of the fruiting body bursts, releasing spores that are carried by wind, water, or animals to new locations. Because the cells forming the slug and fruiting body are all products of mitosis, this pattern of reproduction is asexual.

Cellular slime molds also reproduce sexually by a pattern in which two haploid cells fuse to form a diploid zygote (also shown in Fig. 26.16) that enters a dormant stage. Eventually, the zygote undergoes meiosis, producing four haploid cells that may multiply inside the spore by mitosis. When conditions are favorable, the spore wall breaks down, releasing the cells. These grow and divide into separate amoeboid cells.

The Plasmodial Slime Molds. **Plasmodial slime molds** exist primarily as a large composite mass, the **plasmodium,** in which individual nuclei are suspended in a common cytoplasm surrounded by a single plasma membrane. (This is not to be confused with *Plasmodium,* the genus of apicomplexans that causes malaria.) There are about 500 known species of plasmodial slime molds. The main phase of the life cycle, the plasmodium (see Figure 26.1a), flows and feeds as a single huge amoeba—a single cell that contains thousands to millions or even billions of diploid nuclei surrounded by a single plasma membrane. Typically, a plasmodium, which may range in size from a few centimeters to more than a meter in diameter, moves in thick, branching strands connected by thin sheets. The movements occur by cytoplasmic streaming, driven by actin microfilaments and myosin (see Section 5.3). You may have seen one of these slimy masses crossing a lawn, moving over a mat of dead leaves, climbing a tree, or even in the movies—a slime mold in effect stars as a monster in the science fiction movie *The Blob.*

At some point, often in response to unfavorable environmental conditions, fruiting bodies form at sites on the plasmodium. At the tips of the fruiting bodies, nuclei become enclosed in separate cells, each surrounded by its own plasma membrane and cell wall. Depending on the species, either chitin or cellulose may reinforce the walls. These cells undergo meiosis, forming haploid, resistant spores that are released from the fruiting bodies and carried about by water or wind. If they reach a favorable environment, the spores germinate to form flagellated or unflagellated gametes, depending on the species, that fuse to form a diploid

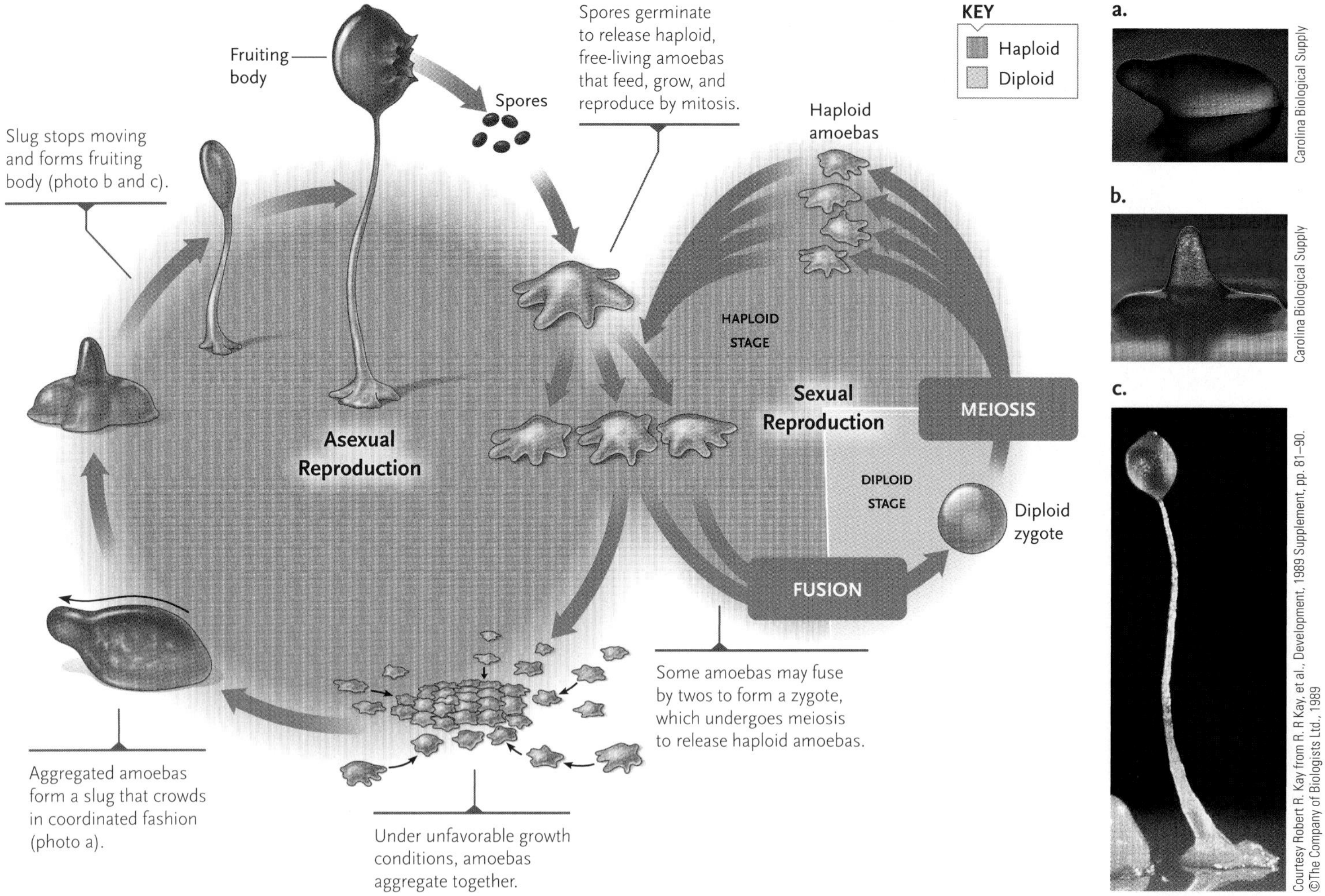

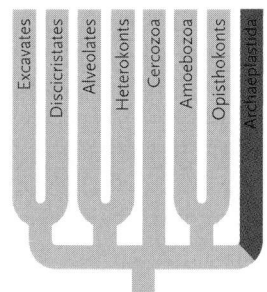

Fruiting body

Spores

Slug stops moving and forms fruiting body (photo b and c).

Spores germinate to release haploid, free-living amoebas that feed, grow, and reproduce by mitosis.

Haploid amoebas

Asexual Reproduction

HAPLOID STAGE

Sexual Reproduction

MEIOSIS

DIPLOID STAGE

Diploid zygote

FUSION

Some amoebas may fuse by twos to form a zygote, which undergoes meiosis to release haploid amoebas.

Aggregated amoebas form a slug that crowds in coordinated fashion (photo a).

Under unfavorable growth conditions, amoebas aggregate together.

KEY
- Haploid
- Diploid

a.

b.

c.

Carolina Biological Supply

Carolina Biological Supply

Courtesy Robert R. Kay from R. R Kay, et al., Development, 1989 Supplement, pp. 81–90. ©The Company of Biologists Ltd., 1989

Figure 26.16
Life cycle of the cellular slime mold *Dictyostelium discoideum*. The light micrographs show **(a)** a migrating slug, **(b)** an early stage in fruiting body formation, and **(c)** a mature fruiting body.

zygote. The zygote nucleus then divides repeatedly without an accompanying division of the cytoplasm, forming many diploid nuclei suspended in the common cytoplasm of a new plasmodium.

The Slime Molds in Science. Both the cellular and plasmodial slime molds, particularly *Dictyostelium* (cellular) and *Physarum* (plasmodial; see Figure 26.1a), have been of great interest to scientists because of their ability to differentiate into fruiting bodies with stalks and spore-bearing structures. This differentiation is much simpler than the complex developmental pathways of other eukaryotes, providing a unique opportunity to study cell differentiation at its most fundamental level. One such study, examining the role of cyclic AMP in differentiation, is described in *Insights from the Molecular Revolution*.

The plasmodial slime molds are particularly useful in this kind of research because they become large enough to provide ample material for biochemical and molecular analyses. Actin and myosin extracted from *Physarum polycephalum,* for example, have been much used in studies of actin-based motility. A further advantage of plasmodial slime molds is that the many nuclei of a plasmodium usually replicate and pass through mitosis in synchrony, making them useful in research tracking the changes that take place in the cell cycle.

The Archaeplastida Include the Red and Green Algae, and Land Plants

Excavates | Discicristates | Alveolates | Heterokonts | Cercozoa | Amoebozoa | Opisthokonts | Archaeplastida

The Archaeplastida consist of the red and green algae, which are protists, and the land plants (the *viridaeplantae,* or "true plants"), which comprise the kingdom Plantae. These three groups have a common evolutionary origin, and they are all photosynthesizers. Here we describe the two types of algae; we discuss land plants in Chapter 27.

Rhodophyta: The Red Algae. Nearly all the 4000 known species of red algae, which are also known as the Rhodophyta (*rhodon* = rose), are small marine seaweeds **(Figure 26.17).** Fewer than 200 species are found in freshwater lakes and streams or in soils. Most red algae grow

Getting the Slime Mold Act Together

Development of differentiated structures can be followed at its simplest level in slime molds. In the cellular slime mold *Dictyostelium discoidium*, the aggregation of individual cells leading to differentiation begins when unfavorable living conditions induce some cells to secrete cyclic AMP (cAMP). Other *Dictyostelium* cells move toward the regions of highest cAMP concentration and aggregate into the slug stage. Further pulses of cAMP trigger differentiation into a stalk and spores.

Within the aggregating cells, the cAMP activates a cAMP-dependent protein kinase (PKA; see Section 7.4). The PKA, which is active only when cAMP is present, adds phosphate groups to target proteins in the cells. The target proteins, activated or deactivated by addition of the phosphate groups, trigger cellular developmental processes that lead to slug formation and differentiation of the stalk and spores.

These observations prompt several questions about development in *Dictyostelium*. Which is more important to the process, cAMP or the PKA? Is the PKA the only enzyme activated by the cAMP signal, or are other cAMP-activated pathways also essential to cell differentiation in the slime mold?

Adam Kuspa and his graduate student Bin Wang at Baylor College of Medicine in Houston, Texas, set out to answer these questions. They were aided by the availability of a mutant strain of *Dictyostelium* that lacks a normal form of *adenylyl cyclase*, the enzyme that converts ATP into cAMP.

Kuspa and Wang constructed an artificial gene by linking the promoter of an actin gene to the protein-encoding portion of a gene for the PKA. They chose the actin promoter because it is highly active and would induce essentially continuous transcription of the gene to which it is attached. The enzyme encoded in the artificial PKA gene was a modified form that does not require cAMP to be active, making it a *cAMP-independent* protein kinase. The researchers induced the mutant cells to take up the artificial gene by exposing them to Ca^{2+} ions (see Section 17.1). Once inside the cells, the actin promoter resulted in transcription of the artificial PKA gene, raising internal PKA concentration to levels about 1.6 times the amount in normal cAMP-activated cells.

Kuspa and Wang found that the cells with the artificial PKA gene aggregated into slugs when their cultures were deprived of food (in this case, bacteria). Moreover, the slugs differentiated normally into fruiting bodies. Tests for cAMP failed to detect the signal molecule, indicating that activated PKA by itself can trigger all the steps in the developmental pathway. Thus the requirement for cAMP in normal slime mold development is primarily or exclusively to stimulate the PKA. And, because development can proceed with active PKA alone, this protein kinase is probably more central to the growth and differentiation processes of the slime mold than is cAMP. Further, it appears from the results that no essential developmental pathways other than those involving PKA are triggered by cAMP.

These results are of more than passing interest because both cAMP and cAMP-dependent protein kinases are also active in animal development and intercellular signaling, including that of humans and other mammals. They also show that in *Dictyostelium*, a single molecule, the PKA normally activated by cAMP, can trigger all stages of development and differentiation.

---

attached to sandy or rocky substrates, but a few occur as plankton. Although most are free-living autotrophs, some are parasites that attach to other algae or plants.

Red algae are typically multicellular organisms, with plantlike bodies composed of interwoven fila-

ments. The base of the body is differentiated into a holdfast, which anchors it to the bottom or other solid substrate, and into stalks with leaflike plates. Their cell walls contain cellulose and mucilaginous pectins that give them a slippery texture. In some species, the walls

Figure 26.17
Red algae.
(a) *Antithamnion plumula*, showing the filamentous and branched body form most common among red algae. **(b)** A sheetlike red alga growing on a tropical reef.

a. Filamentous red alga

Wim van Egmond

b. Sheetlike red alga

Douglas Faulkner/Sally Faulkner Collection

are hardened with stonelike deposits of calcium carbonate. Many of the red algae with stony cell walls resemble corals and occur with corals in reefs and banks.

Although most red algae are reddish in color, some are greenish purple or black. The color differences are produced by accessory pigments, mainly *phycobilins,* which mask the green color of their chlorophylls. The phycobilins are unusual photosynthetic pigments with structures related to the ring structure of hemoglobin. The accessory pigments of some red algae make them highly efficient in absorbing the shorter wavelengths of light that penetrate to the ocean depths, allowing them to grow at deeper levels than any other algae. Some red algae live at depths to 260 m if the water is clear enough to transmit light to these levels.

Red algae have complex reproductive cycles involving alternation between diploid sporophytes and haploid gametophytes. No flagellated cells occur in the red algae; instead, gametes are released into the water to be brought together by random collisions in currents.

Extracts containing the mucilaginous pectins of red algal cell walls are widely used in industry and science. Extracted **agar** is used as a moisture-preserving, inert agent in cosmetics and baked goods, as a setting agent for jellies and desserts, and as a culture medium in the laboratory. **Carrageenan,** extracted from the red alga *Eucheuma,* is used to thicken and stabilize paints, dairy products such as pudding and ice cream, and many other creams and emulsions.

Some red algae are harvested as food in Japan and China. *Porphyra,* one of these harvested algae, is used in sushi bars as the *nori* wrapped around fish and rice. Different *Porphyra* species have different flavors; all are nutritious.

Chlorophyta: The Green Algae. The green algae or Chlorophyta (*chloros* = green) are autotrophs that carry out photosynthesis using the same pigments as plants. They include single-celled, colonial, and multicellular species (**Figure 26.18;** see also Figure 26.1d). Most green algae are microscopic, but some range upward to the size of small seaweeds. Although the multicellular green algae have bodies that are filamentous, tubular, or leaflike, there is relatively little cellular differentiation as compared with the brown algae. However, the most complex green algae, such as the sea lettuce *Ulva* (see Figure 26.18c), have tissues differentiated into a leaflike body and a holdfast.

a. Single-celled green alga

c. Multicellular green alga

Linda Sims/Visuals Unlimited

b. Colonial green alga

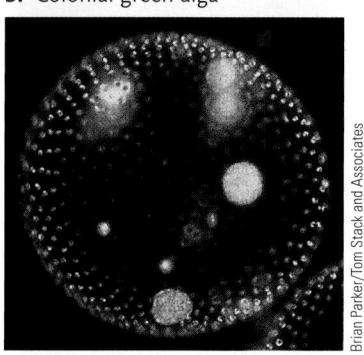

Brian Parker/Tom Stack and Associates

Manfrage Kage/Peter Arnold, Inc.

Figure 26.18

Green algae. **(a)** A single-celled green alga, *Acetabularia,* which grows in marine environments. Each individual in the cluster is a large single cell with a rootlike base, stalk, and cap. **(b)** A colonial green alga, *Volvox.* Each green dot in the spherical wall of the colony is a potentially independent, flagellated cell. Daughter colonies can be seen within the parent colony. **(c)** A multicellular green alga, *Ulva,* common to shallow seas around the world.

Figure 26.19
The life cycle of
the green alga
Ulothrix, in which
the haploid stage
is multicellular
and the diploid
stage is a single
cell, the zygote.
"+" and "−" are
morphologically
identical mating
types ("sexes") of
the alga.

With at least 16,000 species, green algae show more diversity than any other algal group. Most live in freshwater aquatic habitats, but some are marine, or live on rocks and soil surfaces, on tree bark, or even on snow. The green, slimy mat that grows profusely in stagnant pools and ponds, for example, consists of filaments of a green alga. A few species live as symbionts in other protists or in fungi and animals. Lichens (see Figure 28.14) are the primary example of a symbiotic relationship between green algae and fungi. Many animal phyla, including some marine snails and sea anemones, contain green algal chloroplasts, or entire green algae, as symbionts in their cells.

Life cycles among the green algae are as diverse as their body forms. Many can reproduce either sexually or asexually, and some alternate between haploid and diploid generations. Gametes in different species may be undifferentiated flagellated cells, or differentiated as a flagellated sperm cell and a nonmotile egg cell. Most common is a life cycle with a multicellular haploid phase and a single-celled diploid phase (Figure 26.19).

Among all the algae, the nucleic acid sequences of green algae are most closely related to those of land plants. In addition, as we have noted, green algae use the same photosynthetic pigments as plants, including chlorophylls *a* and *b*, and have the same complement of carotenoid accessory pigments. In some green algae, the thylakoid membranes within chloroplasts are arranged into stacks resembling the grana of plant chloroplasts (see Section 5.4). As storage reserves,

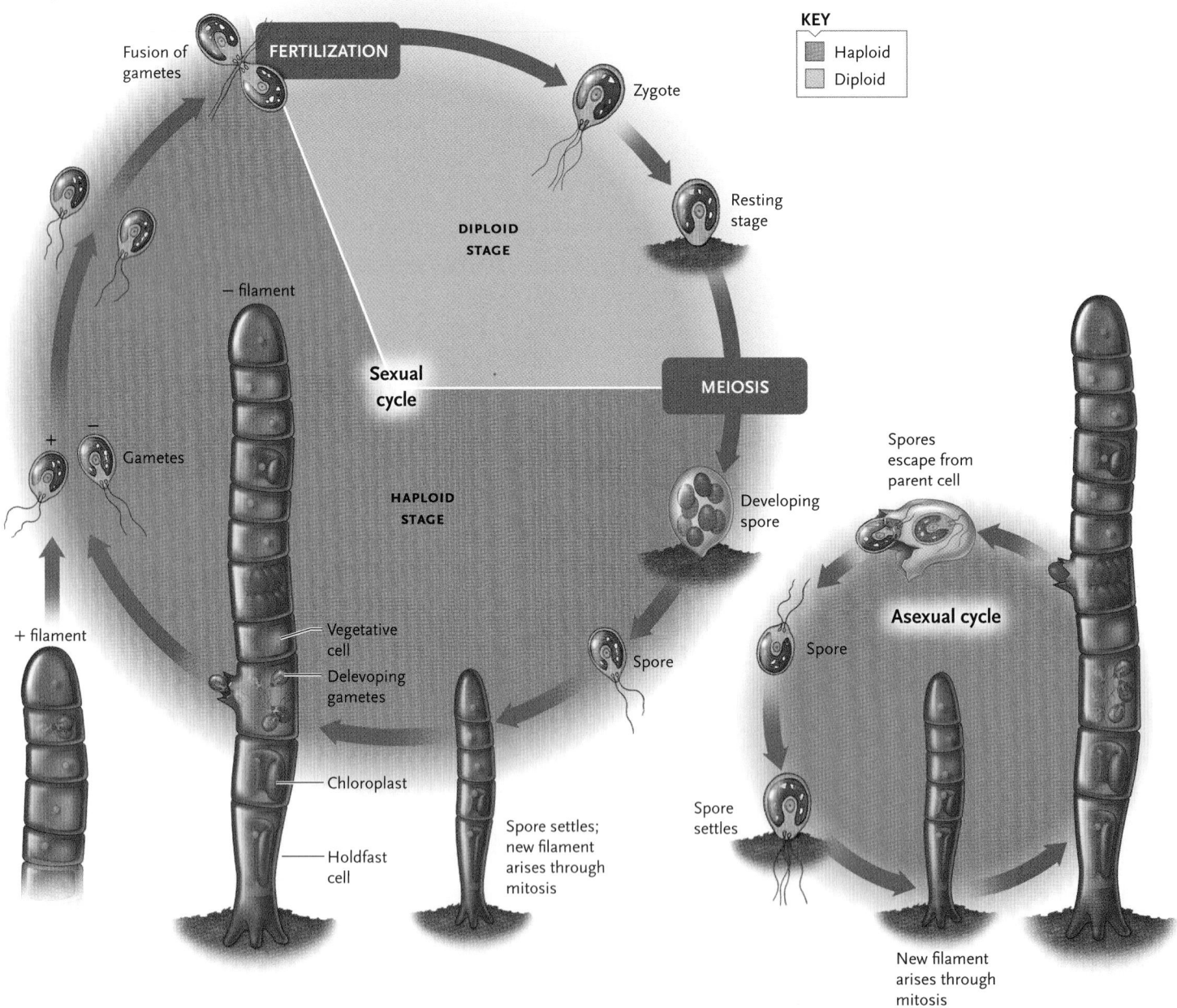

green algae contain starches of the same types as plants, and the cell walls of some green algal species contain cellulose, pectins, and other polysaccharides like those of plants. On the basis of these similarities, many biologists propose that some ancient green algae gave rise to the evolutionary ancestors of modern-day plants.

What green alga might have been the ancestor of modern land plants? Many biologists consider a group known as the **charophytes** to be most similar to the algal ancestors of land plants. These organisms, including *Chara* **(Figure 26.20)**, *Spirogyra, Nitella,* and *Coleochaete,* live in freshwater ponds and lakes. Their ribosomal RNA and chloroplast DNA sequences are more closely related to plant sequences than those of any other green alga. Further, the new cell wall separating daughter cells in charophytes is formed through development of a cell plate, by a mechanism closely similar to that of plants (see Section 10.2). The body form is distinctly plantlike, with a stemlike axis upon which whorls of leaflike blades occur at intervals.

The Opisthokonts Include the Choanoflagellates, Which May Be the Ancestors of Animals

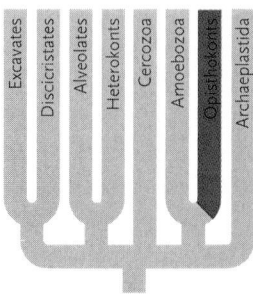

Opisthokonts (*opistho* = posterior) are a broad group of eukaryotes that includes the choanoflagellates, protists thought to be the ancestors of fungi and animals. A single posterior flagellum is found at some stage in the life cycle of these organisms; sperm in animals is an example.

Choanoflagellata (*choanos* = collar) are named for a collar of closely packed microvilli that surrounds the single flagellum by which these protists move and take in food **(Figure 26.21)**. The collar resembles an upside-down lampshade. There are about 150 species of choanoflagellates. They live in fresh and marine waters. Some species are mobile, with the flagellum pushing the cells along, as is the case with animal sperm, in contrast to most flagellates, which are pulled by their flagella. Most choanoflagellates, though, are *sessile;* that is, attached via a stalk to a surface. A number of species are colonial with a cluster of cells on a single stalk.

Choanoflagellates have the same basic structure as choanocytes (collar cells) of sponges, and they are similar to collared cells that act as excretory organs in organisms such a the flatworms and rotifers. These morphological similarities, as well as molecular sequence comparison data, indicate that a choanoflagel-

Reproductive structures

Dr. John Clayton, National Institute of Water and Atmospheric Research, New Zealand

Figure 26.20
The charophyte *Chara*, representative of a group of green algae that may have given rise to the plant kingdom.

late type of protist is likely to have been the ancestor of animals and, of course, of present-day choanoflagellates.

In Several Protist Groups, Plastids Evolved from Endosymbionts

We have encountered chloroplasts in a number of eukaryotic organisms in this chapter: red algae, green algae, land plants, euglenoids, dinoflagellates, heterokonts, and chlorarachniophytes. How did these chloroplasts evolve?

In Section 24.3 we discussed the endosymbiont hypothesis for the origin of eukaryotes. In brief, an anaerobic prokaryote ingested an aerobic prokaryote, which survived as an endosymbiont (see Figure 24.7). Over time, the endosymbiont became an organelle, the mitochondrion, which was incapable of free living, and the result was a true eukaryotic cell. Cells of animals,

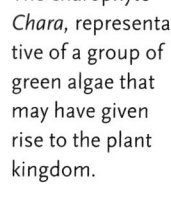

Figure 26.21
A choanoflagellate.

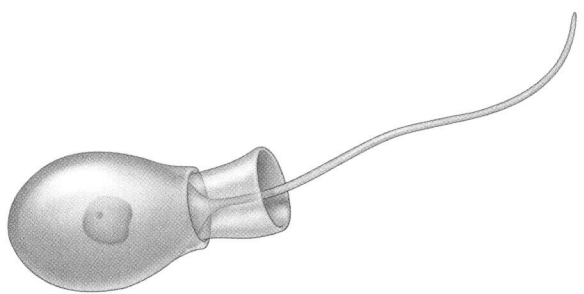

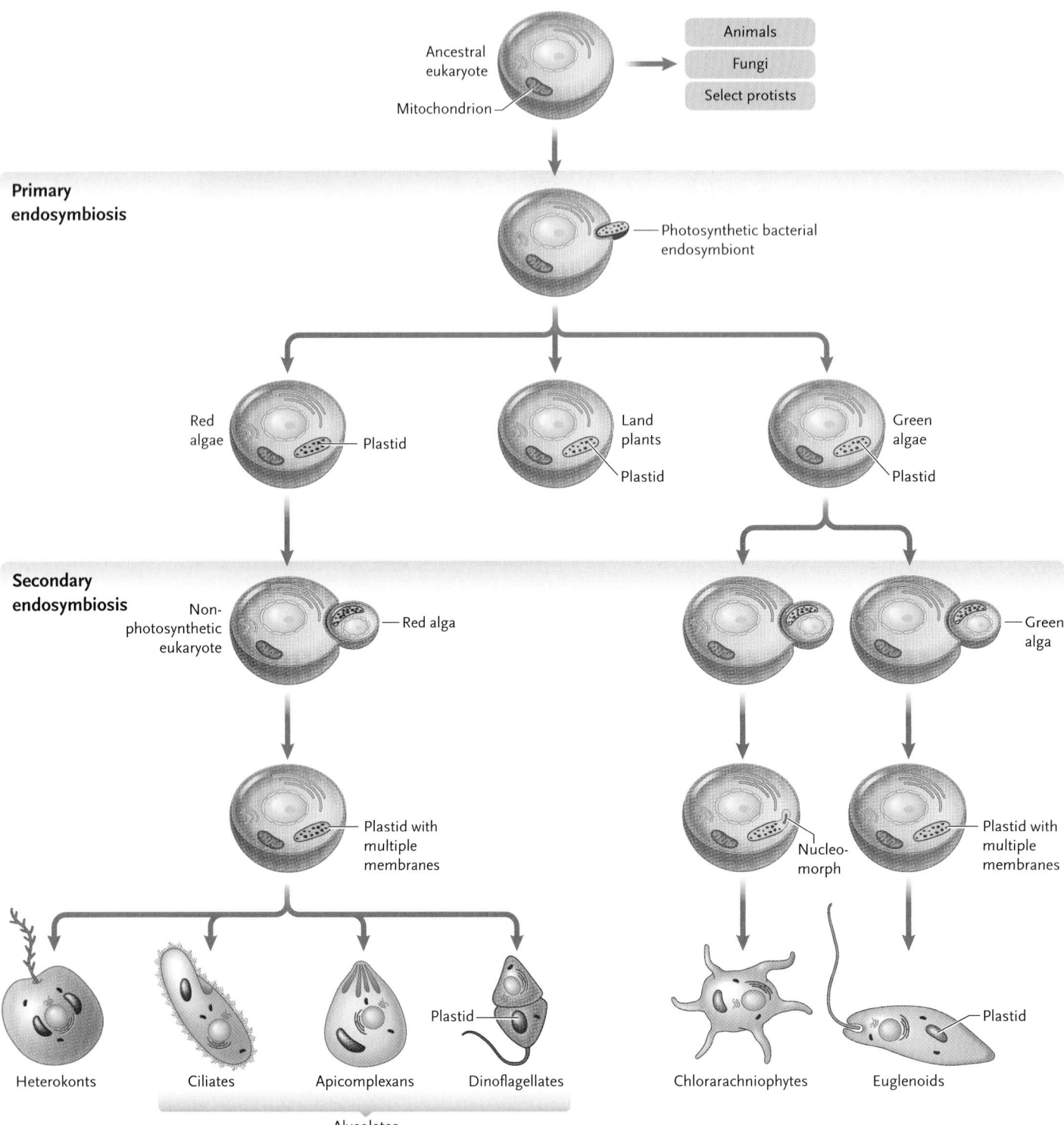

Figure 26.22
The origin and distribution of plastids among the eukaryotes by primary and secondary endosymbiosis.

fungi, and some protists derive from this ancestral eukaryote. The addition of plastids (the general term for chloroplasts and related organelles) through further endosymbiotic events produced the cells of all photosynthetic eukaryotes, including land plants, algae, and some other protists.

Figure 26.22 presents a model for the origin of plastids in eukaryotes through two major endosymbiosis events. First, in a single **primary endosymbiosis** event

perhaps 600 million years ago, a eukaryotic cell engulfed a photosynthetic cyanobacterium (a photosynthetic prokaryote, remember). In some such cells, the cyanobacterium was not digested, but instead formed a symbiotic relationship with the engulfing host cell—it become an endosymbiont. Over time the symbiont lost genes no longer required for independent existence, and most of the remaining genes migrated from the prokaryotic genome to the host's nuclear genome.

The symbiont had become an organelle—a chloroplast. All plastids subsequently evolved from this original chloroplast. Evidence for the single origin of plastids comes from a variety of sequence comparisons, including recent sequencing of the genomes of key protists, a red alga and a diatom.

The first photosynthesizing eukaryote was essentially an ancestral single-celled alga. The chloroplasts of the Archaeplastida—the red algae, green algae, and land plants—result from evolutionary divergence of this organism. Their chloroplasts, which originate from primary endosymbiosis, have two membranes, one from the plasma membrane of the engulfing eu-

karyote and the other from the plasma membrane of the cyanobacterium.

At least three **secondary endosymbiosis** events led to the plastids in other protists (see Figure 26.22). In each case, a nonphotosynthetic eukaryote engulfed a photosynthetic eukaryote, and new evolutionary lineages were produced. In one of these events, a red alga ancestor was engulfed and became an endosymbiont. In models accepted by a number of scientists, the transfer of functions that occurred over evolutionary time led to the chloroplasts of the heterokonts and the dinoflagellates. And, from the same photosynthetic ancestor, loss of chloroplast functions occurred

UNANSWERED QUESTIONS

What was the first eukaryote?

Since prokaryotes precede eukaryotes in the fossil record, we assume that eukaryotes arose after prokaryotes. The first eukaryote would have been some sort of protist—a single-celled organism with a nucleus and some rudimentary organelles, perhaps even a half-tamed mitochondrion. One approach to identifying which of the surviving protists is the most ancient has been to infer evolutionary trees from gene sequence data. To determine the earliest branching eukaryote, these trees need to include the prokaryotes. But herein lies the problem—prokaryotes are very distant, evolutionarily speaking, from even the simplest eukaryotes, and the mathematical models used to construct evolutionary trees are not yet up to the job. Initially, these models suggested that some protist parasites, like the excavates *Giardia* and *Trichomonas*, might be the most ancient eukaryotes, and this idea fit nicely with the fact that these protists lacked mitochondria. Indeed, for a time it was thought that the excavates might actually have diverged from the eukaryotic branch of life before the establishment of mitochondria. Nowadays, we know that *Giardia* and *Trichomonas* did initially have mitochondria. The latest research shows that they even have a tiny relic of the mitochondrion, though exactly what it does in these oxygen-shunning parasites remains to be figured out. Thus, trees depicting *Giardia* and *Trichomonas* at the base of the great expansion of eukaryotic life must be viewed with some caution—these protists might be the surviving representatives of the earliest cells with a nucleus, but they might not be. We simply need better methods for identifying just what the first eukaryotes were like.

How many times did plastids arise by endosymbioses?

For many years researchers thought that the green algae, plants, and red algae were the only organisms to have primary endosymbiosis-derived plastids. However, a second, independent primary endosymbiosis has been recently discovered in which a shelled amoeba has captured and partially domesticated a cyanobacterium. This organism, known as *Paulinella*, is a vital window into the process by which autotrophic eukaryotes first arose some 600 million years. *Paulinella* has tamed the cyanobacterium sufficiently to have it divide and segregate in coordination with host cell division, but the endosymbiont

is still very much a cyanobacterium and has undergone little of the modification and streamlining we see in the red or green algal plastids.

After a primary endosymbiosis was established, the second chapter in plastid acquisition could take place. Secondary endosymbiosis involves a eukaryotic host engulfing and retaining a eukaryotic alga. Essentially, secondary endosymbiosis can convert a heterotrophic organism into an autotroph by hijacking a photosynthetic cell and putting it to work as a solar-powered food factory. Secondary endosymbiosis results in plastids with three or four membranes, and we know that it occurred at least three times—once for the euglenoids, once for the chlorarachniophytes, and once for the chromalveolates (a proposed grouping of heterokonts and alveolates). We can even tell what kind of endosymbiont was involved by the biochemistry and genetic makeup of the plastid: a green alga for euglenoids and chlorarachniophytes, and a red alga for chromalveolates. The number of secondary endosymbioses is hotly debated, largely because not all protistologists support the existence of chromalveolates. Some contend that there were multiple, independent enslavements of different red algae to produce the dinoflagellates, heterokonts, and apicomplexans. Understanding these events is crucial to confirming or refuting the proposed chromalveolate "supergroup."

A nice example of secondary endosymbiosis-in-action was recently discovered by Japanese scientists who found a flagellate, *Hatena*, with a green algal endosymbiont. *Hatena* hasn't yet assumed control of endosymbiont division and has to get new symbionts each time it divides, so it appears to be at a very early stage in establishing a relationship. We also want to know how secondary endosymbioses proceed because they have been a major driver in eukaryotic evolution. The heterokonts, for instance, are the most important ocean phytoplankton and are key to ocean productivity and global carbon cycling. Knowing exactly how they got to be autotrophs in the first place is fundamental to understanding the world we live in.

 Dr. Geoff McFadden is a professor of botany at the University of Melbourne. He studies the early evolution of eukaryotes, especially the origin and evolution of plastids and mitochondria. You can learn more about his research by visiting http://homepage.mac.com/fad1/McFaddenLab.html.

in the lineage of the Apicomplexa, which have a remnant plastid. In an independent event, a nonphotosynthetic eukaryote engulfed a green alga ancestor. Subsequent evolution in this case produced the euglenoids. In a different event, a similar endosymbiosis involving a green alga led to the chlorarachniophytes. In these protists, the chloroplast is contained still within the remnants of the original symbiont cell, with a vestige of the original nucleus (the nucleomorph) also present.

Note that secondary endosymbiosis has produced plastids with additional membranes acquired from the new host, or series of hosts. For example, euglenoids have plastids with three membranes, while chlorarachniophytes have plastids with four membranes (see Figure 26.22). Sequencing the genomes of the chlorarachniophyte's nucleus, chloroplast, and vestigial nucleus is providing interesting information about the early endosymbiosis event that generated these organisms.

In sum, the protists are a highly diverse and ecologically important group of organisms. Their complex evolutionary relationships, which have long been the subject of contention, are now being revised as new information is discovered, including more complete genome sequences. A deeper understanding of protists is also contributing to a better understanding of their recent descendents, the fungi, plants, and animals. We turn to these descendents in the next four chapters, beginning with the fungi.

Study Break

1. What is the evidence that the Excavates, which lack mitochondria, derive from ancestors that had mitochondria rather than from ancestors that were in lineages that never contained mitochondria?
2. In primary endosymbiosis, a nonphotosynthetic eukaryotic cell engulfed a photosynthetic cyanobacterium. How many membranes surround the chloroplast that evolved?

Review

26.1 What Is a Protist?

- Protists are eukaryotes that differ from fungi in having motile stages in their life cycles and distinct cell wall molecules. Unlike plants, they lack true roots, stems, and leaves. Unlike animals, protists lack collagen, nerve cells, and an internal digestive tract, and they lack complex developmental stages (Figures 26.1 and 26.2).

- Protists are aerobic organisms that live as autotrophs or heterotrophs, or by a combination of both nutritional modes. Some are parasites or symbionts living in or among the cells of other organisms.

- Protists live in aquatic or moist terrestrial habitats, or as parasites within animals as single-celled, colonial, or multicellular organisms, and range in size from microscopic to some of Earth's largest organisms.

- Reproduction may be asexual by mitotic cell divisions, or sexual by meiosis and union of gametes in fertilization.

- Many protists have specialized cell structures including contractile vacuoles, food vacuoles, eyespots, and a pellicle, cell wall, or shell. Most are able to move by means of flagella, cilia, or pseudopodia (Figure 26.3).

26.2 The Protist Groups

- The Excavates, exemplified by the Diplomonadida and Parabasala are flagellated, single cells that lack mitochondria (Figure 26.4).

- The Discicristates are almost all single-celled, autotrophic or heterotrophic (some are both), motile protists that swim using flagella. The free-living, photosynthetic forms—the euglenoids—typically have complex cytoplasmic structures, including eyespots (Figures 26.5 and 26.6).

- Alveolates include the ciliates, dinoflagellates, and apicomplexans. The ciliates swim using cilia and have complex cytoplasmic structures and two types of nuclei, the micronucleus and macronucleus. The dinoflagellates swim using flagella and are primarily marine organisms; some are photosynthetic. The apicomplexans are nonmotile parasites of animals (Figures 26.7 and 26.8).

- Heterokonts include the funguslike Oomycota, which live as saprophytes or parasites, and three photosynthetic groups, the diatoms, golden algae, and brown algae. For most heterokonts, flagella occur only on reproductive cells. Many Oomycota grow as masses of microscopic hyphal filaments and secrete enzymes that digest organic matter in their surroundings. Diatoms are single-celled organisms covered by a glassy silica shell; golden algae are colonial forms; brown algae are primarily multicellular marine forms that include large seaweeds with extensive cell differentiation (Figures 26.9–26.13).

- Cercozoa are amoebas with filamentous pseudopods supported by internal cellular structures. Many produce hard outer shells. Radiolara (radiolarians) are primarily marine organisms that secrete a glassy internal skeleton. They feed by engulfing prey that adhere to their axopods. Foraminifera (forams) are marine, single-celled organisms that form chambered, spiral shells containing calcium. They engulf prey that adhere to the strands of cytoplasm extending from their shells. Chlorarachniophytes engulf food using their pseudopodia (Figure 26.14).

- The Amoebozoa includes most amoebas and two heterotrophic slime molds, cellular (which move as individual cells) and plasmodial (which move as large masses of nuclei sharing a common cytoplasm). The amoebas in this group are heterotrophs abundant in marine and freshwater environments and in the soil. They move by extending pseudopodia (Figures 26.15 and 26.16).

- The Archaeplastida include the red and green algae, as well as the land plants that comprise the kingdom Plantae. The red algae are typically multicellular, primarily photosynthetic organisms of marine environments, with plantlike bodies composed of interwoven filaments. They have complex life cycles including alternation of generations, with no flagellated cells at any stage. The green algae are single-celled, colonial, and multicellular species that live primarily in freshwater habitats and carry out photosynthesis by mechanisms like those of plants; all produce flagellated gametes (Figures 26.17–26.20).

- The Opisthokonts are a broad group of eukaryotes that includes the choanoflagellates. These protists are characterized by a collar of microvilli surrounding a single flagellum. A choanoflagellate type of protist is considered likely to have been the ancestor of animals (Figure 26.21).

- Several groups of protists, as well as land plants, contain chloroplasts. Present-day chloroplasts and other plastids result from endosymbiosis events that took place millions of years ago: In a primary endosymbiosis event, a eukaryotic cell engulfed a cyanobacterium, which became an endosymbiont. Over time, the symbiont became an organelle, the chloroplast. This first photosynthesizing organism was a green alga. Evolutionary divergence produced the red algae, green algae, and land plants. By

secondary endosymbiosis, in which a nonphotosynthetic eukaryote engulfed a photosynthetic eukaryote, the various photosynthetic protists were produced (Figure 26.22).

Animation: Body plan of Euglena

Animation: Paramecium body plan

Animation: Ciliate conjugation

Animation: Apicomplexan life cycle

Animation: Red alga life cycle

Animation: Green alga life cycle

Animation: Amoeboid motion

Animation: Cellular slime mold life cycle

Questions

Self-Test Questions

1. Protists are characterized by:
   a. division by binary fission.
   b. multicellular structures.
   c. complex digestive systems.
   d. peptidoglycan cell walls.
   e. organelles and reproduction by meiosis/mitosis.

2. Which of the following is *not* found among the protist groups?
   a. life cycles
   b. contractile vacuoles
   c. pellicles
   d. collagen
   e. pseudopodia

3. Freely beating flagella buried in a fold of cytoplasm moving through viscous fluids of humans and commonly found as an infective agent in U.S. college health centers describes a member of:
   a. Ciliophora.
   b. Discicristates.
   c. Diplomonadida.
   d. Parabasala.
   e. Alveolates.

4. When *Paramecium* conjugate:
   a. cytoplasmic division produces four daughter cells, each having two micronuclei and two macronuclei.
   b. one haploid micronucleus in each cell remains intact; the other three degenerate. The micronucleus of each cell divides once, producing two nuclei, and each cell exchanges one nucleus with the other cell. In each partner the two micronuclei fuse, forming a diploid zygote micronucleus in each cell.
   c. and the partners disengage, the micronucleus of each divides meiotically. Macronuclei divide again in each cell and the original micronucleus breaks down. Each cell has two haploid micronuclei; one of the macronuclei develops into a micronucleus.
   d. the mating cells join together at opposite sites of their oral depression.
   e. the micronucleus in each cell undergoes mitosis. When mitosis is complete there are four diploid macronuclei; the micronucleus then breaks down.

5. The protist group Diplomonadida is characterized by:
   a. a mouthlike gullet and hairlike surface. *Paramecium* is an example.
   b. flagella and a lack of mitochondria. *Giardia* is an example.

   c. nonmotility, parasitism, and sporelike infective stages. *Toxoplasma* is an example.
   d. switching between autotrophic and heterotrophic life styles. *Euglena* is an example.
   e. large protein deposits. Movement is by two flagella, which are part of an undulating membrane. *Trypanosoma* is an example.

6. The greatest contributors to protist fossil deposits are:
   a. Oomycota.
   b. Chrysophyta.
   c. Bacillariophyta.
   d. Sporophyta.
   e. Alveolates.

7. The group with the distinguishing characteristic of gas-filled bladders and a cell wall composed of alginic acid is:
   a. Chrysophyta.
   b. Phaeophyta.
   c. Oomycota.
   d. Bacillariophyta.
   e. none of the preceding.

8. *Plasmodium* is transmitted to humans by the bite of a mosquito *(Anopheles)* and engages in a life cycle with infective spores, gametes, and cysts. This infective protist belongs to the group:
   a. Apicomplexa.
   b. Heterokonts.
   c. Dinoflagellata.
   d. Oomycota.
   e. Ciliophora.

9. Tripping on a rotten log, a hunter notices a mucus-looking mass moving slowly toward brightly colored fruiting bodies. The organisms in the mass are:
   a. amoebas in the group Cercozoa.
   b. slime molds.
   c. red algae.
   d. green algae.
   e. charophytes.

10. The latest stage for evolving the double membrane seen in modern day algal chloroplasts is thought to be the combining of:
    a. two ancestral nonphotosynthetic prokaryotes.
    b. two ancestral photosynthetic prokaryotes.
    c. a nonphotosynthetic eukaryote with a photosynthetic eukaryote.
    d. a photosynthetic prokaryote with a nonphotosynthetic eukaryote.
    e. mitochondria with an already established plastid.

Questions for Discussion

1. You decide to vacation in a developing country where sanitation practices and standards of personal hygiene are inadequate. Considering the information about protists covered in this chapter, what would you consider safe to drink in that country? What treatments could make the water safe to drink? What kinds of foods might be best avoided? What kinds of preparation might make foods safe to eat?

2. The overreproduction of dinoflagellates, producing red tides, is sometimes caused by fertilizer runoff into coastal waters. The red tides kill countless aquatic species, birds, and other wildlife. Would you consider drastic cutbacks in the use of fertilizers as a means to lessen the red tides? Why?

Experimental Analysis

Design an experiment to demonstrate whether the flagellated protist *Euglena* is phototropic, that is, is attracted to and moves toward light. Also propose a follow-up experiment (on the assumption of a positive result) to determine the wavelength range and light intensity range sufficient to cause phototropic movement.

Evolution Link

Use the Internet to research why studies of a molecular sensor, receptor tyrosine kinase (see Section 7.3), supports the hypothesis that a choanoflagellate type of protist is the ancestor of animals. Summarize your findings.

How Would You Vote?

The pathogen that causes sudden oak death has already infected 26 kinds of plants in California and Oregon. Some infected species are commonly sold as nursery stock. Should the states that are free of this pathogen be allowed to prohibit shipping of all plants from the states that are affected? Go to www.thomsonedu.com/login to investigate both sides of the issue and then vote.

A temperate forest with representatives of three major groups of land plants—mosses (bryophytes), conifers (gymnosperms), and flowering plants (angiosperms).

Animals, Animals—Earth Scenes

27 Plants

WHY IT MATTERS

Ages ago, along the edges of the ancient supercontinent Laurentia, the only sound was the rhythmic muffled crash of waves breaking in the distance. There were no birds or other animals, no plants with leaves rustling in the breeze. In the preceding eons, oxygen-producing photosynthetic cells had come into being and had gradually changed the atmosphere. Solar radiation had converted much of the oxygen into a dense ozone layer—a shield against lethal doses of ultraviolet radiation, which had kept early organisms below the water's surface. Now, they could populate the land.

Cyanobacteria were probably the first to adapt to intertidal zones and then to spread into shallow, coastal streams. Later, green algae and fungi made the same journey. Seven to eight hundred million years ago, green algae living near the water's edge, or perhaps in a moist terrestrial environment, became the ancestors of modern plants. Several lines of evidence indicate that these algae were charophytes, a group discussed in Chapter 26. Today the **Kingdom Plantae** encompasses more than 300,000 living species, organized in this textbook into 10 phyla. These modern plants range from mosses, horsetails, and ferns to conifers and flowering plants **(Figure 27.1).** Most

a. Mosses growing on rocks

b. A ponderosa pine

Craig Wood/Visuals Unlimited

c. An orchid

© Craig Alikas/www.orchidworks.com

Robert Potts, California Academy of Sciences

Figure 27.1

Representatives of the Kingdom Plantae. **(a)** Mosses growing on rocks. Mosses evolved relatively soon after plants made the transition to land. **(b)** A ponderosa pine, *Pinus ponderosa*. This species and other conifers belonging to the phylum Coniferophyta represent the gymnosperms. **(c)** An orchid, *Cattalya rojo*, a showy example of a flowering plant.

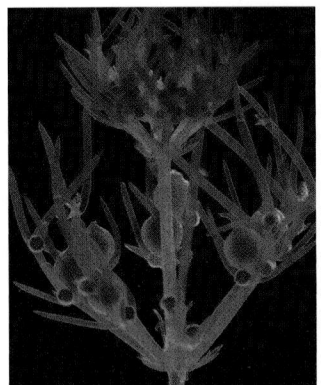

Courtesy Microbial Culture Collection, National Institute for Environmental Studies, Japan

Figure 27.2

Chara, a stonewort. This representative of the charophyte lineage is known commonly as muskweed because of its skunky odor.

plants living today are terrestrial, and nearly all plants are multicellular autotrophs that use sunlight energy, water, carbon dioxide, and dissolved minerals to produce their own food. Together with photosynthetic bacteria and protists, plant tissues provide the nutritional foundation for nearly all communities of life. Humans also use plants as sources of medicinal drugs, wood for building, fibers used in paper and clothing, and a wealth of other products.

While the ancestors of land plants were making the transition to a fully terrestrial life, some remarkable adaptive changes unfolded. Eons of natural selection sorted out solutions to fundamental problems, among them avoiding desiccation, physically supporting the plant body in air, obtaining nutrients from soil, and reproducing sexually in environments where water would not be available for dispersal of eggs and sperm. With time, plants evolved features that not only addressed these problems but also provided access to a wide range of terrestrial environments. Those ecological opportunities opened the way for a dramatic radiation of varied plant species—and for the survival of plant-dependent organisms such as ourselves.

27.1 The Transition to Life on Land

Land plants and green algae share several fundamental traits: they have cellulose in their cell walls, they store energy captured during photosynthesis as starch, and their light-absorbing pigments include both chlorophyll *a* and chlorophyll *b*. Like other green algae, the charophyte lineage that produced the ancestor of land plants arose in water and has aquatic descendants today **(Figure 27.2)**. Yet because terrestrial environments pose very different challenges than aquatic environments, evolution in land plants produced a range of adaptations crucial to survival on dry land.

The algal ancestors of plants probably invaded land between 425 and 490 million years ago (mya). We say "probably" because the fossil record is sketchy in pinpointing when the first truly terrestrial plants appeared. A British and Arab research team working in the Middle East found fossilized tissue and spores from what appears to be a land plant in rocks dated to 475 mya. If the remains indeed are from a plant, they represent the earliest known plant fossils. Even in more recent deposits the most common finds of possible plant parts are microscopic bits and pieces. Obvious leaves, stems, roots, and reproductive parts seldom occur together, or if they do, it can be difficult to determine if the fossilized bits all belong to the same individual. Whole plants are extremely rare. Adding to the challenge, some chemical and structural adaptations to life on land arose independently in several plant lineages. Consequently, a fossil may have some but not all the features of modern land plants, leaving the puzzled paleobotanist to guess whether a given specimen was aquatic, terrestrial, or a transitional form. Despite these problems, botanists have been able to gain insight into several innovations and overall trends in plant evolution.

Early Biochemical and Structural Adaptations Enhanced Plant Survival on Land

To survive on land, plants had to have protection against drying out, a demand that had not been a problem for algae in their aquatic habitats. The earliest land plants may have benefited from an inherited ability to make **sporopollenin**, a resistant polymer that surrounds the zygotes of modern charophytes. In land plants, sporopollenin is a major component of the thick wall that protects reproductive spores from drying and other damage. Some of the first land plants also evolved an outer waxy layer called a **cuticle**, which slows water loss, helping to prevent desiccation **(Figure 27.3a)**. Another multifaceted adaptation was the presence of **stomata**, tiny passageways through the cuticle-covered surfaces **(Figure 27.3b)**. Stomata (singular, *stoma; stoma* = mouth), which can open and close, became the main route for plants to take up carbon di-

a. Cuticle on the surface of a leaf

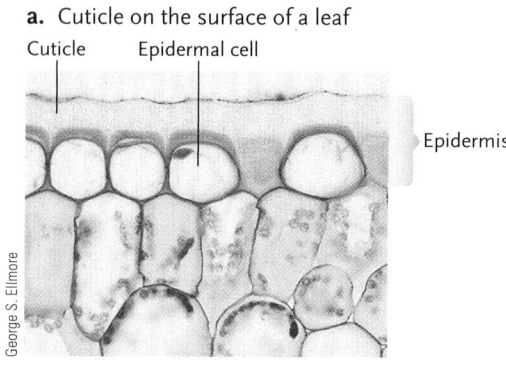

Cuticle   Epidermal cell

Epidermis

George S. Ellmore

b. Stomata

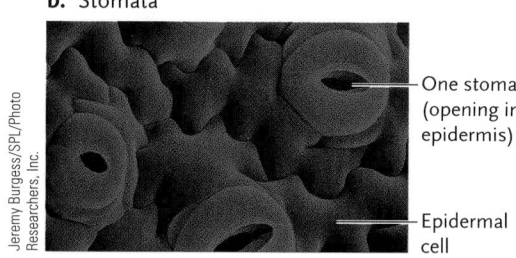

One stoma
(opening in
epidermis)

Epidermal
cell

Jeremy Burgess/SPL/Photo Researchers, Inc.

Figure 27.3

Land plant adaptations for limiting water loss. **(a)** A waxy cuticle, which covers the epidermis of land plants and helps reduce water loss. **(b)** Surface view of stomata in the epidermis (surface layer of cells) of a leaf. Stomata allow carbon dioxide and water to enter plant tissues and oxygen to leave.

port vessels, and the **vascular plants**, or **tracheophytes**. This split correlates with the appearance of several fundamental adaptations in the vascular plant lineage **(Table 27.1)**. Transport vessels, which we describe shortly, was one adaptation. Another was **lignin**, a tough, rather inert polymer that strengthens the secondary walls of various plant cells and thus helps vascular plants to grow taller and stay erect on land, giving photosynthetic tissues better access to sunlight. Another was the **apical meristem**, a region of unspecialized dividing cells near the tips of shoots and roots. Descendants of such unspecialized cells differentiate and form all mature plant tissues. Meristem tissue is the foundation for a vascular plant's extensively branching stem parts, and is a central topic of Chapter 31.

Other land plant adaptations were related to the demands of reproduction in a dry environment. As described in more detail shortly, they included multicellular chambers that protect developing gametes, and a dependent, multicellular embryo that is sheltered inside tissues of a parent plant. Botanists use the term **embryophyte** (*phyton* = plant) as a synonym for land plants because all land plants produce an embryo during their reproductive cycle.

Vascular Tissue Was an Innovation for Transporting Substances within a Large Plant Body

The Latin *vas* means duct or vessel, and vascular plants have specialized tissues made up of cells arranged in lignified, tubelike structures that branch throughout the plant body, conducting water and solutes. One type

oxide and control water loss by evaporation. The next unit describes these tissue specializations more fully.

By about 470 million years ago, land plants had split into two major groups, the **nonvascular plants**, or **bryophytes**, such as mosses, which lack internal trans-

Table 27.1	Trends in Plant Evolution				
Traits derived from algal ancestor: cell walls with cellulose, energy stored in starch, two forms of chlorophyll (*a* and *b*); possibly, sporopollenin in spore wall					
Bryophytes	**Ferns and Their Relatives**	**Gymnosperms**	**Angiosperms**		**Functions in Land Plants**
Cuticle ———————————————————————————→					Protection against water loss, pathogens
Stomata ———————————————————————————→					Regulation of water loss and gas exchange (CO_2 in, O_2 out)
Nonvascular ——→	Vascular ———————————————————→				Internal tubes that transport water, nutrients
	Lignin ———————————————————————→				Mechanical support for vertical growth
	Apical meristem ——————————————→				Branching shoot system
	Roots, stems, leaves ————————————→				Enhanced uptake, transport of nutrients and enhanced photosynthesis
Haploid phase dominant ——→	Diploid phase dominant ——————————→				Genetic diversity
One spore type (homospory) ——→	Two spore types (heterospory) ———————→				Promotion of genetic diversity
Motile gametes ————————————→		Nonmotile gametes ———————→			Protection of gametes within parent body
Seedless ———————————————→		Seeds ———————————→			Protection of embryo

Figure 27.4

Fossil of one of the earliest vascular plants, *Cooksonia*, which dates to about 420 mya. Cooksonia was small and, as this image shows, its stems lacked leaves and probably were less than 3 cm long. The cup-shaped structures at the top of the stems produced reproductive spores.

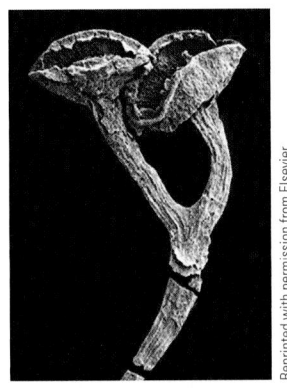

Reprinted with permission from Elsevier

of vascular tissue, called **xylem**, distributes water and dissolved mineral ions through plant parts. Another vascular tissue, **phloem**, distributes sugars manufactured during photosynthesis. Chapter 32 explains how xylem and phloem perform these key internal transport functions.

Ferns, conifers, and flowering plants—most of the plants you are familiar with—are vascular plants. Supported by lignin and with a well-developed vascular system, the body of a plant can grow large. Extreme examples are the giant redwood trees of the northern California coast, some of which are more than 300 feet tall. By contrast, nonvascular plants lack lignin, and have very simple internal transport systems, or none at all. As a result, modern nonvascular plants generally are small, as are the examples you will read about shortly.

Root and Shoot Systems Were Adaptations for Nutrition and Support

The body of a bryophyte is not differentiated into true roots and stems—structures that are fundamental adaptations for absorbing nutrients from soil and for support of an erect plant body. The evolution of sturdy stems—the basis of an aerial *shoot system*—went hand in hand with the capacity to synthesize lignin. To become large, land plants would also require a means of anchoring aerial parts in the soil, as well as effective strategies for obtaining soil nutrients. **Roots**—anchoring structures that also absorb water and nutrients—were the eventual solution to these problems. The earliest fossils showing clear evidence of roots are from vascular plants, although the exact timing of this change is uncertain. The first unquestioned fossils of a vascular plant, a small plant called *Cooksonia* **(Figure 27.4),** were found in deposits that date to about 420 mya. *Cooksonia* fossils have been unearthed in various locales but, frustratingly, none has ever included the lower portion of the plant—only its leafless, branching upper stems. *Cooksonia* probably was supported physically only by a **rhizome**—a horizontal, modified stem that can penetrate a substrate and anchor the plant. At some point, however, ancestral forms of vascular plants did come to have true roots. Ultimately, vascular plants developed specialized **root systems**, which generally consist of underground, cylindrical absorptive structures with a large surface area that favors the rapid uptake of soil water and dissolved mineral ions.

Above ground, the simple stems of early land plants also became more specialized, evolving into **shoot systems** in vascular plants. Shoot systems have stems and leaves that arise from apical meristems and that function in the absorption of light energy from the sun and carbon dioxide from the air. Stems grew larger and branched extensively after the evolution of lignin. The mechanical strength of lignified tissues almost certainly provided plants with several adaptive advantages. For instance, a strong, internal scaffold could support upright stems bearing leaves and other photosynthetic structures—and so help increase the surface area for intercepting sunlight. Also, reproductive structures borne on aerial stems might serve as platforms for more efficient launching of spores from the parent plant.

Structures we think of as "leaves" arose several times during plant evolution. In general, leaves represent modifications of stems, and **Figure 27.5** illustrates the basic steps of two main evolutionary pathways. In at least one early group of plants, the club mosses described in Section 27.3, leaflike parts evolved as outgrowths of the plant's main vertical axis (see Figure 27.5a). In other groups, leaves arose when small, neighboring stem branches became joined by thin, weblike tissue containing cells that had chloroplasts (see Figure 27.5b).

a. Leaf development as an offshoot of the main vertical axis

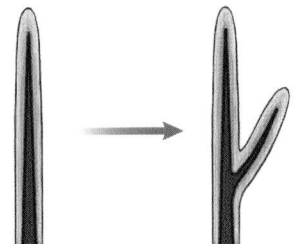

b. Development of leaves in a branching pattern

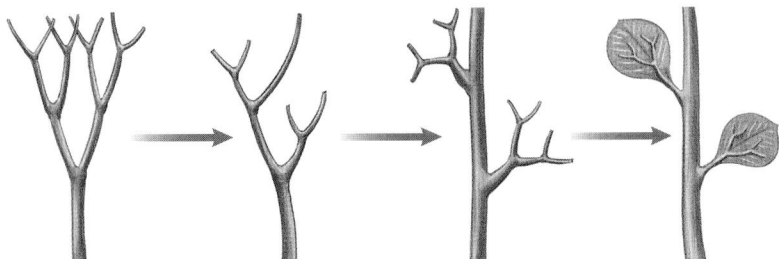

Figure 27.5

Evolution of leaves. **(a)** One type of early leaflike structure may have evolved as offshoots of the plant's main vertical axis; there was only one vein (transport vessel) in each leaf. Today, the seedless vascular plants known as lycophytes (club mosses) have this type of leaf. **(b)** In other groups of seedless vascular plants, leaves arose in a series of steps that began when the main stem evolved a branching growth pattern. Small side branches then fanned out and photosynthetic tissue filled the space between them, becoming the leaf blade. With time the small branches became modified into veins.

In the Plant Life Cycle, the Diploid Phase Became Dominant

As early plants moved into drier habitats, their life cycles also were modified considerably. You may recall that in sexually reproducing organisms, meiosis in diploid cells produces haploid (*n*) reproductive cells (see Chapter 11). These cells may be gametes—sperm or eggs—or they may be **spores**, which can give rise to a new haploid individual asexually, without mating.

As noted in Chapter 26, in green algae the haploid phase that starts at meiosis is usually the greater part of the life cycle, and the haploid alga spends much of its life producing and releasing gametes into the surrounding water. A much shorter diploid (2*n*) phase starts when gametes fuse at fertilization. Plants also cycle between haploid and diploid life phases, a phenomenon called **alternation of generations (Figure 27.6).** The diploid generation produces haploid spores and is called a **sporophyte** ("spore grower"). The haploid generation produces gametes and is called a **gametophyte** ("gamete grower"). As plants evolved on land, the haploid gametophyte phase became physically smaller and less complex and had a shorter life span while just the opposite occurred with the diploid sporophyte phase. In mosses and other nonvascular plants the sporophyte is a little larger and long-lived than in green algae, and in vascular plants the sporophyte clearly is larger and more complex and lives much longer than the gametophyte **(Figure 27.7).** When you look at a pine tree, for example, you see a large, long-lived sporophyte.

The sporophyte generation begins after fertilization, when the resulting zygote grows mitotically into a multicellular, diploid organism. Its body will eventually develop capsules called **sporangia** ("spore chambers"; singular, *sporangium*), which produce spores. Many botanists hypothesize that the trend toward "diploid dominance" in vascular plants reflects the advantages conferred by genetic diversity in land environments, where the supply of water and nutrients is inconsistent. Whereas haploid organisms are genetically identical to the parent, in a changeable environment the new combinations of parental alleles in a diploid organism may provide the genetic basis for adaptations to varying circumstances.

The haploid phase of the plant life cycle begins in the reproductive parts of the sporophyte. There, meiosis produces haploid spores in the sporangia. The spores then divide by mitosis and give rise to multicellular haploid gametophytes. A gametophyte's function is to nourish and protect the forthcoming generation. Unlike nonvascular plants, most groups of vascular plants retain spores and gametophytes until environmental conditions favor fertilization.

Some Vascular Plants Evolved Separate Male and Female Gametophytes

As already noted, during sexual reproduction in plants, meiosis produces spores. When a plant makes only one type of spore it is said to be **homosporous** ("same spore"). A gametophyte that develops from such a spore is bisexual—it can produce both sperm and eggs.

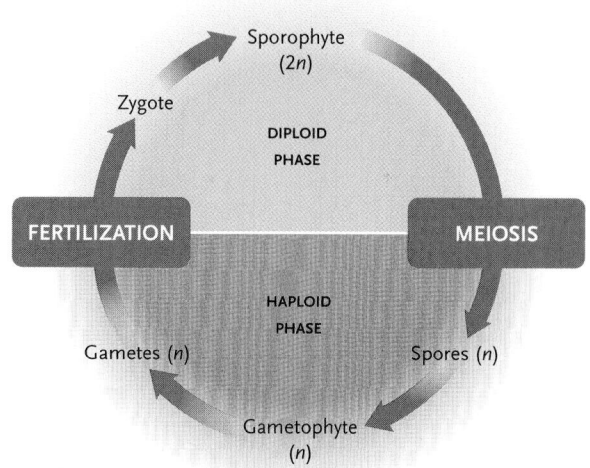

The sperm have flagella and are motile, for they must swim through liquid water in order to encounter female gametes. Other vascular plants, including gymnosperms and angiosperms, are **heterosporous.** They produce two types of spores in two different types of sporangia, and those spores develop into small, sexually different gametophytes. The smaller spore type develops into a male gametophyte—a *pollen grain.* The larger one develops into a female gametophyte, in which eggs form and fertilization occurs. The pollen grains of most vascular plants produce nonmotile sperm and also the structures required to deliver them to the egg.

Figure 27.6
Overview of the alternation of generations, the basic pattern of the plant life cycle.
The relative dominance of haploid and diploid phases is different for different plant groups (compare with Figure 27.7).

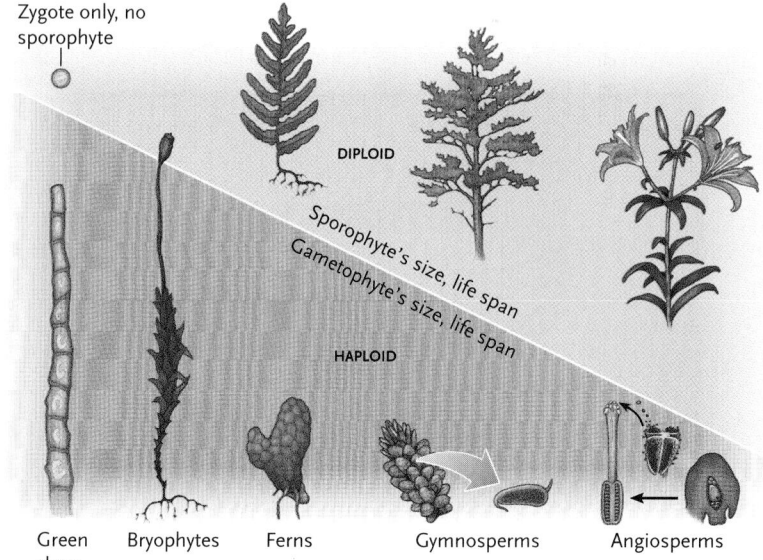

Figure 27.7
Evolutionary trend from dominance of the gametophyte (haploid) generation to dominance of the sporophyte (diploid) generation, represented here by existing species ranging from a green alga *(Ulothrix)* to a flowering plant. This trend developed as early plants were colonizing habitats on land. In general, the sporophytes of vascular plants are larger and more complex than those of bryophytes, and their gametophytes are reduced in size and complexity. In this diagram the fern represents seedless vascular plants.

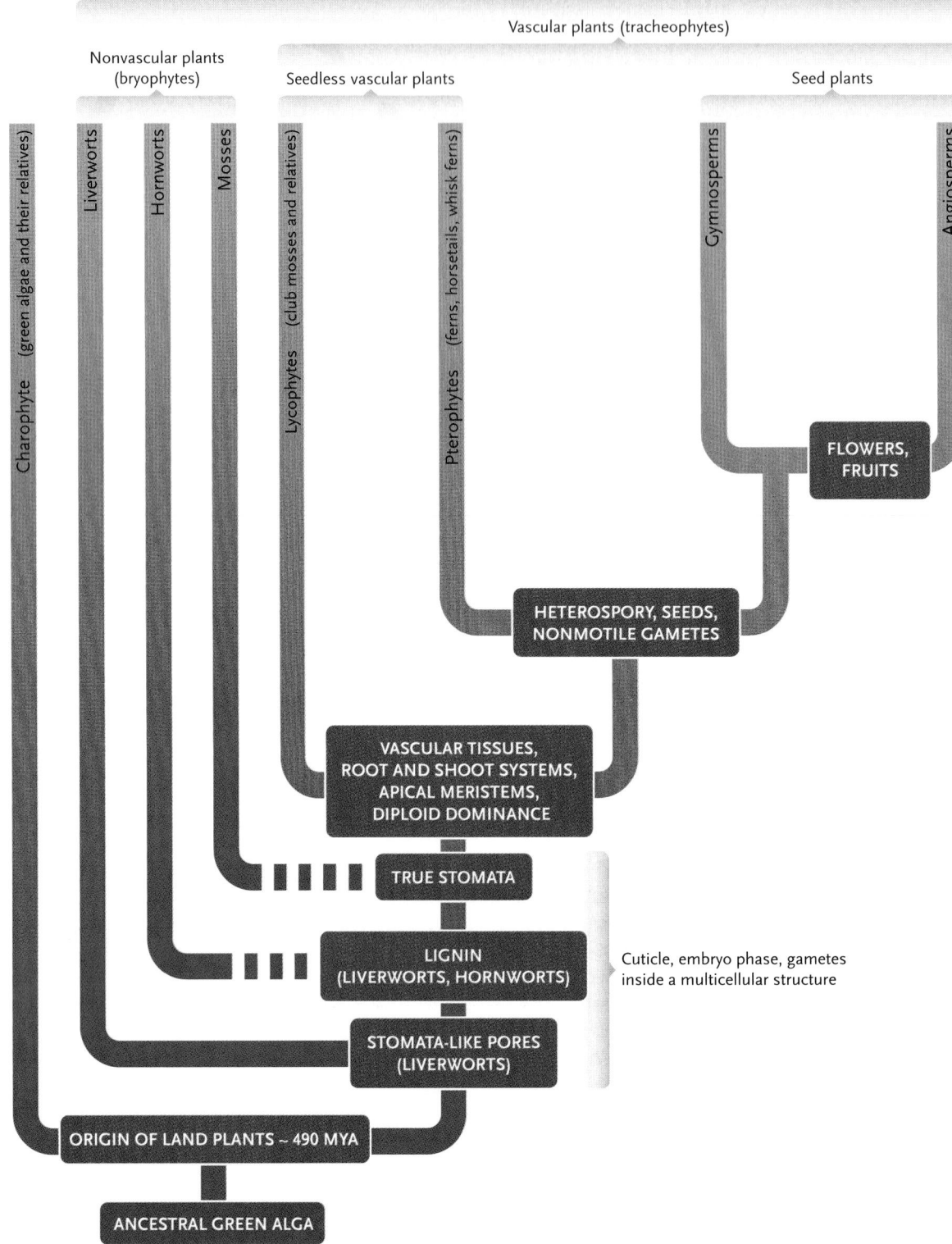

Figure 27.8
Overview of the possible phylogenetic relationships between major groups of land plants. Plant systematists do not agree on the relative place of bryophytes in this evolutionary history, hence the dashed lines. This diagram provides only a general picture of the points in land plant evolution where major adaptations took hold. For example, heterospory and seeds are shown as adaptations common to all seed plants, but some living fern species also are heterosporous. Fossil evidence indicates that certain ancient lycophytes and horsetails also produced two types of spores and some had seeds as well. Cycads and ginkgoes are unlike other gymnosperms in that they have motile sperm.

As you will read in a later section, the evolution of pollen grains and pollination helped spark the rapid diversification of plants in the Devonian period, 408–360 mya. During this time another innovation, the seed, contributed to this diversification. In fact, so many new fossils appear in Devonian rocks that paleobotanists—scientists who specialize in the study of fossil plants—have thus far been unable to determine which fossil lineages gave rise to the modern plant phyla. Clearly, however, as each major lineage came into being, its characteristic adaptations included major modifications of existing structures and functions **(Figure 27.8)**. The next sections fill out this general picture, beginning with the plants that most clearly resemble the plant kingdom's algal ancestors.

STUDY BREAK

1. How did plant adaptations such as a root system, a shoot system, and a vascular system collectively influence the transition to terrestrial life?
2. Describe the difference between homospory and heterospory, and explain how heterospory paved the way for other reproductive adaptations in land plants.

27.2 Bryophytes: Nonvascular Land Plants

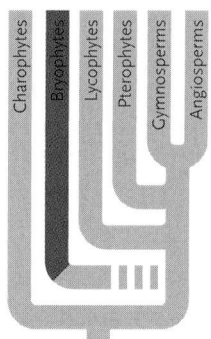

The **bryophytes** (*bryon* = moss)—liverworts, hornworts, and mosses—have a curious combination of traits that allow them to bridge aquatic and land environments. Because bryophytes lack a well-developed system for conducting water, it is not surprising that they commonly grow on wet sites along creek banks or on rocks just above running water; in bogs, swamps, or the dense shade of damp forests; and on moist tree trunks or rooftops. Some species are **epiphytes** (*epi* = upon)—they grow independently (that is, not as a parasite) on another organism and in a host of other moist places, ranging from the splash zone just above high tide on rocky shores, to edges of snowbanks, to coastal salt marshes.

In general, bryophytes are strikingly algalike. They produce flagellated sperm that must swim through water to reach eggs, and as noted they do not have a complex vascular system (although some have a primitive type of conducting tissue). Bryophytes have parts that are rootlike, stemlike, and leaflike. However, the "roots" are rhizoids (slender rootlike structures), and bryophyte "stems" and "leaves" did not evolve from the same structures as vascular plant stems and leaves did. (Said another way, stems and leaves are not homologous in the two groups.) Also, as already mentioned, bryophyte tissues do not contain lignin. The absence of this strengthening material and the lack of internal pipelines for efficient nutrient transport partly account for bryophytes' small size—typically less than 20 cm long—and for their tendency to grow sprawled along surfaces instead of upright.

In other ways, bryophytes are clearly adapted to land. Along with their leaflike, stemlike, and fibrous, rootlike organs, sporophytes of some species have a water-conserving cuticle and stomata. Like most plants, bryophytes also have both sexual and asexual reproductive modes. And as is true of all plants, the life cycle has both gametophyte (*n*) and sporophyte (*2n*) phases, though the sporophyte is tiny and lives only a short time. **Figure 27.9** shows the green, leafy gametophyte of a moss plant, with miniscule diploid sporophytes attached to it by slender stalks. Bryophyte gametophytes produce gametes sheltered within a layer of protective cells called a **gametangium** (plural, *gametangia*). The gametangia in which bryophyte eggs form are flask-shaped structures called **archegonia** (*archi* = first, *gonos* = seed). Flagellated sperm form in rounded gametangia called **antheridia** (*antheros* = flowerlike). The sperm swim through a film of water to the archegonia and fertilize eggs. Each fertilized egg gives rise to a diploid embryo sporophyte, which stays attached to the gametophyte, produces spores—and the cycle repeats.

Despite these similarities with more complex plants, bryophytes are unique in several ways. Unlike in vascular species, the gametophyte is much larger than the sporophyte and obtains its nutrition independently of the sporophyte body. In fact, the comparatively tiny sporophyte remains attached to the gametophyte and depends on the gametophyte for much of its nutrition.

Because of bryophytes' mix of characteristics, their position in plant evolution is still an open question. The basic bryophyte body plan may be similar to the ancestral condition from which higher plants evolved, but it is also possible that bryophytes represent structurally simplified vascular plant lineages that evolved after vascular plants had already appeared. In another view, they are a side shoot of evolution, completely separate from the path that led to vascular plants. The fossil record provides little help in resolving the issue, because the first undisputed bryophyte fossils appear in late-Devonian rocks 350 mya, after vascular plants were already on the scene. (Fossil remains that may resemble liverworts, however, have recently been discovered in rocks that are 50 to 100 million years older.)

Despite questions raised by recent fossil finds, most current molecular, biochemical, cellular, and

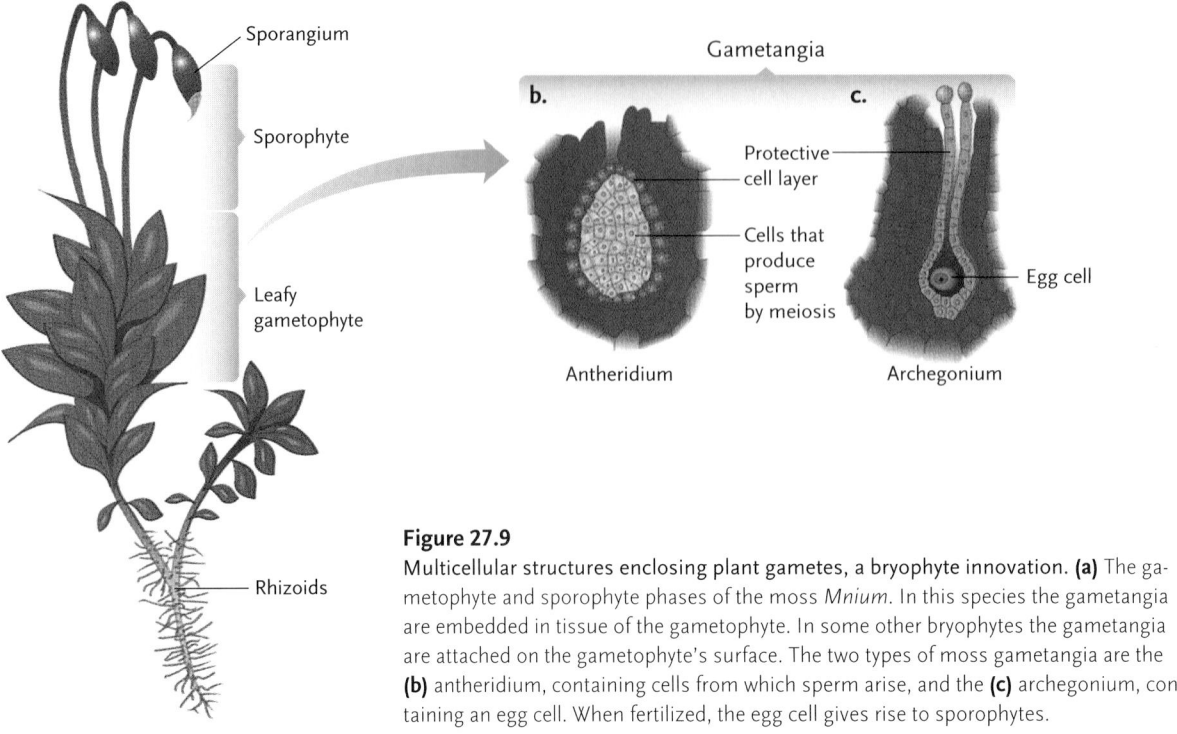

a. Moss gametophyte with attached sporophytes

Sporangium

Sporophyte

Leafy
gametophyte

Rhizoids

Gametangia

b.

Protective
cell layer

Cells that
produce
sperm
by meiosis

Antheridium

c.

Egg cell

Archegonium

Figure 27.9

Multicellular structures enclosing plant gametes, a bryophyte innovation. **(a)** The gametophyte and sporophyte phases of the moss *Mnium*. In this species the gametangia are embedded in tissue of the gametophyte. In some other bryophytes the gametangia are attached on the gametophyte's surface. The two types of moss gametangia are the **(b)** antheridium, containing cells from which sperm arise, and the **(c)** archegonium, containing an egg cell. When fertilized, the egg cell gives rise to sporophytes.

morphological evidence supports the view that bryophytes are not a monophyletic group. Instead, the various bryophytes evolved as separate lineages, in parallel with vascular plants. The relationships are far from resolved, however. For example, molecular evidence can be interpreted to mean that liverworts diverged early on from the lineage that led to all other land plants, with hornworts diverging later and mosses later still. Until new discoveries and interpretive work clarify this picture, the classification scheme in this chapter places liverworts, hornworts, and mosses in separate phyla. Our survey of nonvascular plants begins with the liverworts and hornworts, the simplest of the group, and concludes with mosses—plants that not only are more familiar to most of us, but whose structure and physiology more closely resemble that of vascular plants.

Liverworts May Have Been the First Land Plants

Liverworts make up the phylum **Hepatophyta**, and early herbalists thought that these small plants were shaped like the lobes of the human liver (*hepat* = liver; *wort* = herb). The resemblance might be a little vague to modern eyes: many of the 6000 species of liverworts consist of a flat, branching, ribbonlike plate of tissue closely pressed against damp soil. This simple body, called a **thallus** (plural, *thalli*) is the gametophyte generation. Threadlike rhizoids anchor the gametophytes to their substrate. About two-thirds of liverwort species have leaflike structures and some have stemlike parts.

None have true stomata, the openings that regulate gas exchange in most other land plants, although some species do have pores that open and close. Mitochondrial gene sequence data show that liverworts lack a few features (three introns) that are present in other bryophytes and in vascular plants. Taken together with liverwort morphology, this finding has led many researchers to conclude that liverworts were probably the first land plants.

In species of the liverwort genus *Marchantia* **(Figure 27.10)**, male and female gametophytes are separate plants. Male plants produce antheridia and female plants produce archegonia on specialized stalked organs (see Figure 27.10a–b). The motile sperm released from an antheridium swim through surface water, and some eventually encounter an egg inside an archegonium of a female gametophyte. After fertilization, a small, diploid sporophyte develops inside the archegonium, matures there, and produces haploid spores by meiosis. During meiosis, *Marchantia* sex chromosomes segregate, so some spores have the male genotype and others the female genotype. As in other liverworts, the spores develop inside jacketed sporangia that split open to release the spores. The capsules contain elongated cells twisted into a corkscrew shape. When certain regions of the cell wall absorb water and swell, the "corkscrews" rapidly unwind, helping to eject spores to the outside. A spore that is carried by air currents to a suitable location germinates and gives rise to a haploid gametophyte, which is either male or female. *Marchantia* also can reproduce asexually by way of **gemmae** (*gem* = bud), small cell masses that form in cuplike growths on a thallus (see Figure 27.10c). Gemmae can grow into new thalli when rainwater splashes them out of the cups and onto an appropriately moist substrate.

Hornworts Have Both Plantlike and Algalike Features

Roughly 100 species of hornworts make up the phylum **Anthocerophyta.** Many of them have cell features in common with green algae, including the presence in each cell of a single large chloroplast that contains algalike protein bodies called pyrenoids. No other group of land plants has this feature, and some biologists have speculated that the distinction of "first land plant" should be assigned not to liverworts but to hornworts instead. Like some liverworts, a hornwort gametophyte has a flat thallus, but the sporangium of the sporophyte phase is long and pointed, like a horn **(Figure 27.11).** The sporangia split into two or three ribbonlike sections when they release spores. Sexual reproduction occurs in basically the same way as in liverworts: free-swimming sperm fertilize eggs, which give rise to the sporophytes. Hornworts sometimes reproduce asexually by fragmentation as pieces of a thallus break off, form rhizoids, and develop into new individuals.

Mosses Most Closely Resemble Vascular Plants

Chances are that you have seen, touched, or sat upon at least several of the approximately 10,000 species of mosses, and the use of the name **Bryophyta** for this phylum underscores the fact that mosses are the best-known bryophytes. They also are structurally and functionally most similar to the vascular plants we will consider in following sections. Their spores, produced by the tens of millions in sporangia, give rise to thread-like, haploid gametophytes that grow into the familiar moss plants, which often form tufts or carpets of vegetation on the surface of rocks, soil, or bark.

The moss life cycle, diagrammed in **Figure 27.12,** begins when a haploid (*n*) spore lands on a wet soil surface. After the spore germinates it elongates and branches into a filamentous web of tissue called a **protonema** ("first thread"), which can become dense enough to color the surface of soil, rocks, or bark visibly green. After several weeks of growth, the bud-like cell masses on a protonema develop into leafy, green gametophytes anchored by rhizoids. A single protonema can be extremely prolific, producing bud after bud—and in this way giving rise to a dense clone of genetically identical gametophytes. Leafy mosses also may reproduce asexually by gemmae produced at the surface of rhizoids as well as on above-ground parts.

When a leafy moss is sexually mature, gametangia develop on its gametophytes and gametes form in

a. Male plant

Male gametophyte

b. Female plant

Female gametophyte

c. Asexual reproduction
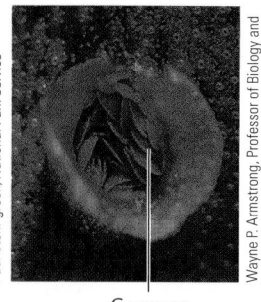
Gemmae

Figure 27.10

The bryophyte *Marchantia*, the only liverwort to produce **(a)** male and **(b)** female gametophytes on separate plants. *Marchantia* also reproduces asexually by way of **(c)** gemmae, multicellular vegetative bodies that develop in tiny cups on the plant body. Gemmae can grow into new plants when splashing raindrops transport them to suitable sites.

them. In some moss genera, plants are unisexual and produce male *or* female gametangia—antheridia at the tips of male gametophytes and archegonia at the tips of female gametophytes. In other genera, plants are bisexual and produce both antheridia and archegonia. Propelled by a pair of flagella, sperm released from antheridia swim through a film of dew or rainwater and down a channel in the neck of the archegonium, attracted by a chemical gradient secreted by each egg. Fertilization produces the new sporophyte generation inside the archegonium, in the form of diploid zygotes that develop into small, mature sporophytes, each consisting of a sporangium on a stalk. Moss sporophytes may eventually develop chloroplasts and nourish themselves photosynthetically, but initially they depend on the gametophytes for food. And even after a moss sporophyte begins photosynthesis, it still must obtain water, carbohydrates, and some other nutrients from the gametophyte.

Certain moss gametophytes are structurally complex, with features similar to those of higher plants. For example, some species have a central strand of primitive conducting tissue. One kind of tissue is made up of elongated, thin-walled, dead and empty cells that

Figure 27.11
The hornwort *Anthoceros.* The base of each long, slender sporophyte is embedded in the flattened, leafy gametophyte.

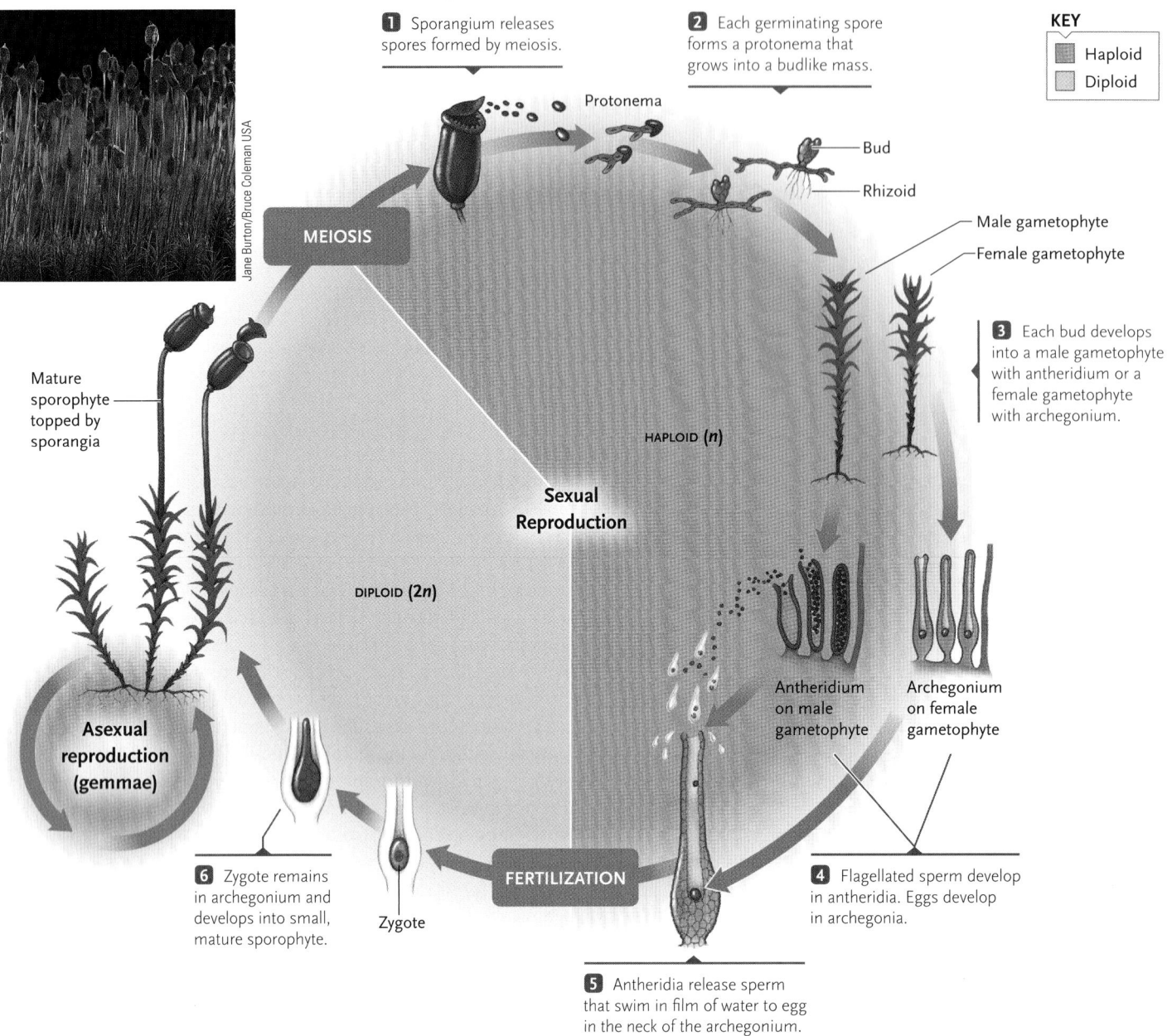

1 Sporangium releases spores formed by meiosis.

2 Each germinating spore forms a protonema that grows into a budlike mass.

KEY

Haploid
Diploid

Jane Burton/Bruce Coleman USA

Protonema

Bud
Rhizoid

Male gametophyte
Female gametophyte

MEIOSIS

3 Each bud develops into a male gametophyte with antheridium or a female gametophyte with archegonium.

Mature sporophyte topped by sporangia

HAPLOID (*n*)

Sexual Reproduction

DIPLOID (*2n*)

Antheridium on male gametophyte

Archegonium on female gametophyte

Asexual reproduction (gemmae)

6 Zygote remains in archegonium and develops into small, mature sporophyte.

Zygote

FERTILIZATION

4 Flagellated sperm develop in antheridia. Eggs develop in archegonia.

5 Antheridia release sperm that swim in film of water to egg in the neck of the archegonium.

Figure 27.12
Life cycle of a moss, *Polytrichum*.

conduct water. These specialized cells, called hydroids, have oblique end walls that sometimes are partly dissolved or perforated with pores. Experiments with dyes show that water moves through them, as it does in similar xylemlike arrangements in vascular plants (see Chapters 31 and 32). In a few mosses, the water-conducting cells are surrounded by sugar-conducting tissue resembling the phloem of vascular plants.

Mosses and other bryophytes are important both ecologically and economically. As colonizers of bare land, their small bodies trap particles of organic and inorganic matter, helping to build soil on bare rock and stabilizing soil surfaces with a biological crust in harsh places like coastal dunes, inland deserts, and embankments created by road construction. Some hornworts harbor mutualistic nitrogen-fixing cyanobacteria, and so increase the amount of nitrogen available to other plants. In arctic tundras, bryophytes constitute as

much as half the biomass, and they are crucial components of the food web that supports animals in that ecosystem. People have long used *Sphagnum* and other absorbent "peat" mosses (which typically grow in bogs) for everything from diapering babies and filtering whiskey to increasing the water-holding capacity of garden soil. Peat moss also has found use as a fuel; each day the Rhode generating station in Ireland, one of several in that nation, burns 2000 metric tons of peat to produce electricity.

In the next section we turn to the vascular plants, which have specialized tissues that can transport water, minerals, and sugars. Without the capacity to move these substances efficiently throughout the plant body, large sporophytes could not have survived on land. Unlike bryophytes, modern vascular plants are monophyletic—all groups are descended from a common ancestor.

27.3 Seedless Vascular Plants

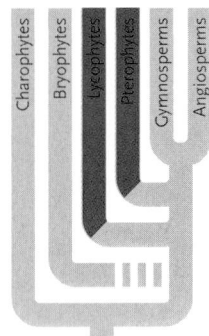

The first vascular plants, which did not "package" their embryos inside protective seeds, were the dominant plants on Earth for almost 200 million years, until seed plants became abundant. The fossil record shows that seedless vascular plants were well established by the late Silurian, some 428 mya, and they flourished until the end of the Carboniferous, about 250 mya. Some living seedless vascular plants have certain bryophyte-like traits, whereas others have some characteristics of seed plants. On one hand, like bryophytes, seedless vascular plants reproduce sexually by releasing spores, and they have swimming sperm that require free water to reach eggs. On the other hand, as in seed plants, the sporophyte of a seedless vascular plant separates from the gametophyte at a certain point in its development and has well-developed vascular tissues (xylem and phloem). Also, the sporophyte is the larger, longer-lived stage of the life cycle and the gametophytes are very small. Some bryophytes even lack chlorophyll. **Table 27.2** summarizes these characteristics and gives an overview of seedless vascular plant features within the larger context of modern plant phyla.

Seedless vascular plants once encompassed a huge number of diverse species of trees, shrubs, and herbs. In the late Paleozoic era, they were Earth's dominant vegetation. Some lineages have endured to the present, but collectively these survivors total fewer than 14,000 species. The taxonomic relationships between various lines are still under active investigation, and comparisons of gene sequences from the genomes in plastids, cell nuclei, and sometimes mitochondria are revealing previously unsuspected links between some of them. In this book we assign seedless vascular plants to two phyla, the Lycophyta (club mosses and their close relatives) and the Pterophyta (ferns, whisk ferns, and horsetails).

Early Seedless Vascular Plants Flourished in Moist Environments

The extinct plant genus *Cooksonia* (see Figure 27.4) probably was one of the earliest ancestors of modern seedless vascular plants. Like other members of its extinct phylum (Rhyniophyta) *Cooksonia* was small, rootless, and leafless, but its simple stems had a central core of xylem, an arrangement seen in many existing vascular plants. Mudflats and swamps of the damp Devonian period were dominated by plants like *Cooksonia* and *Rhynia* **(Figure 27.13)**. While these and other now-extinct phyla came and went, ancestral forms of both modern phyla of seedless vascular plants appeared. In botanical terms, the earliest seedless vascular plants were "herbs"—that is, they did not have woody, lignified tissue. By the start of the Carboniferous period, however, the small herbaceous Devonian plants had given rise to larger shrubby species and to trees with some woody tissue, bark, roots, leaves, and even seeds.

Carboniferous forests were swampy places dominated by members of the phylum **Lycophyta**, and fascinating fossil specimens of this group have been unearthed in North America and Europe. One example is *Lepidodendron,* which had broad, straplike leaves and sporangia near the ends of the branches **(Figure 27.14a)**. It also had xylem and several other types of tissues that are typical of all modern vascular plants (although probably not in the same proportions as seen today). Like trees growing in modern year-round tropical climates, the fossils do not exhibit growth rings. This observation implies that the continents of Europe and North America lay along the equator during the Carboniferous period. Also abundant at the time were representatives of the phylum **Pterophyta**, including ferns such as *Medullosa* and giants such as *Calamites*—huge horsetails that could have a trunk diameter of 30 cm. The sturdy, upright stems were attached to a system of rhizomes—horizontal underground stems. Ferns populated the forest understory. Some early seed plants also were present, including now-extinct fernlike plants, called seed ferns, which bore seeds at the tips of leaves **(Figure 27.14b)**.

Lepidodendron and *Calamites* dominated lush swamp forests in a subtropical climate. After leaves, branches, and old trees fell to the ground, they became buried in anaerobic sediments. Over geologic time, these buried remains became compressed and fossilized, and today they form much of the world's coal reserves. This is why coal is called a "fossil fuel," and the Carboniferous period is called the Coal Age. Characterized by a moist climate over much of the planet, and by the dominance of seedless vascular plants, the Carboniferous period continued for 150 million years, ending when climate patterns changed during the Paleozoic era.

Most modern seedless vascular plants are ferns, and like their ancestors they also are confined largely to wet or humid environments because they require external water for reproduction. Except for whisk ferns, their gametophytes have no vascular tissues for water transport, and male gametes must swim through water to reach eggs. The few vascular seedless plants that are

Table 27.2 Plant Phyla and Major Characteristics

Phylum	Common Name	Number of Species	Common General Characteristics
Bryophytes: Nonvascular plants. Gametophyte dominant, free water required for fertilization, cuticle and stomata present in some.			
Hepatophyta	Liverworts	6000	Leafy or simple flattened thallus, rhizoids; spores in capsules. Moist, humid habitats.
Anthocerophyta	Hornworts	100	Simple flattened thallus, rhizoids; hornlike sporangia. Moist, humid habitats.
Bryophyta	Mosses	10,000	Feathery or cushiony thallus, some have hydroids; spores in capsules. Moist, humid habitats; colonizes bare rock, soil, or bark.
Seedless vascular plants: Sporophyte dominant, free water required for fertilization, cuticle and stomata present.			
Lycophyta	Club mosses	1000	Simple leaves, true roots; most species have sporangia on sporophylls. Mostly wet or shady habitats.
Pterophyta	Ferns, whisk ferns, horsetails	13,000	*Ferns:* Finely divided leaves, woody stems in tree ferns; sporangia in sori. Habitats from wet to arid. *Whisk ferns:* Branching stem from rhizomes; sporangia on stem scales. Tropical to subtropical habitats. *Horsetails:* Hollow photosynthetic stem, scalelike leaves, sporangia in strobili. Swamps, disturbed habitats.
Gymnosperms: Vascular plants with "naked" seeds. Sporophyte dominant, fertilization by pollination, cuticle and stomata present.			
Cycadophyta	Cycads	185	Shrubby or treelike with palmlike leaves, pithy stems; male and female strobili on separate plants. Widespread distribution.
Ginkgophyta	Ginkgo	1	Woody-stemmed tree, deciduous fan-shaped leaves. Male, female structures on separate plants. Temperate areas of China.
Gnetophyta	Gnetophytes	70	Shrubs or woody vines; one has strappy leaves. Male and female strobili on separate plants. Limited to deserts, tropics.
Coniferophyta	Conifers	550	Mostly evergreen, woody trees and shrubs with needlelike or scalelike leaves; male and female cones usually on same plant.
Angiosperms: Plants with flowers and seeds protected inside fruits. Sporophyte dominant, fertilization by pollination, cuticle and stomata present. Major groups: Monocots, eudicots.			
Anthophyta	Flowering plants	268,500+ (including magnoliids, other basal angiosperms)	Wood and herbaceous plants. Nearly all land habitats, some aquatic.
Monocots	Grasses, palms, lilies, orchids, and others	60,000	Pollen grains have a single groove; one cotyledon. Parallel-veined leaves common.
Eudicots	Most fruit trees, roses, cabbages, melons, beans, potatoes, and others	200,000	Pollen grains have three grooves. Most species have two cotyledons; net-veined leaves common.

adapted to dry environments such as deserts can reproduce sexually only when adequate water is available, as during seasonal rains.

Modern Lycophytes Are Small and Have Simple Vascular Tissues

Lycophytes such as club mosses were highly diverse 350 mya, when some tree-sized forms inhabited lush swamp forests. Today, however, such giants are no more. The most familiar of the 1000 or so living species of lycophytes are club mosses, including members of genera such as *Lycopodium* and *Selaginella,* which grow on forest floors **(Figure 27.15a)**. Other groups include the spike mosses and quillworts. The sporophyte of a club moss has upright or horizontal stems that contain a small amount of xylem and bear small green leaves and roots—both of which have vascular tissue. Sporangia are clustered at the bases of specialized leaves, called **sporophylls,** that occur near stem tips. A cluster of sporophylls forms a **cone** or **strobilus** (plural, *strobili*). In some species the sporangia release haploid spores produced by meiosis **(Figure 27.15b)**. If a spore eventually germinates (which can occur even several years after it is released), it forms a free-living gametophyte, but one that differs markedly from the sporophyte. Ranging in size from nearly invisible to several centimeters, the gametophyte easily becomes buried under decompos-

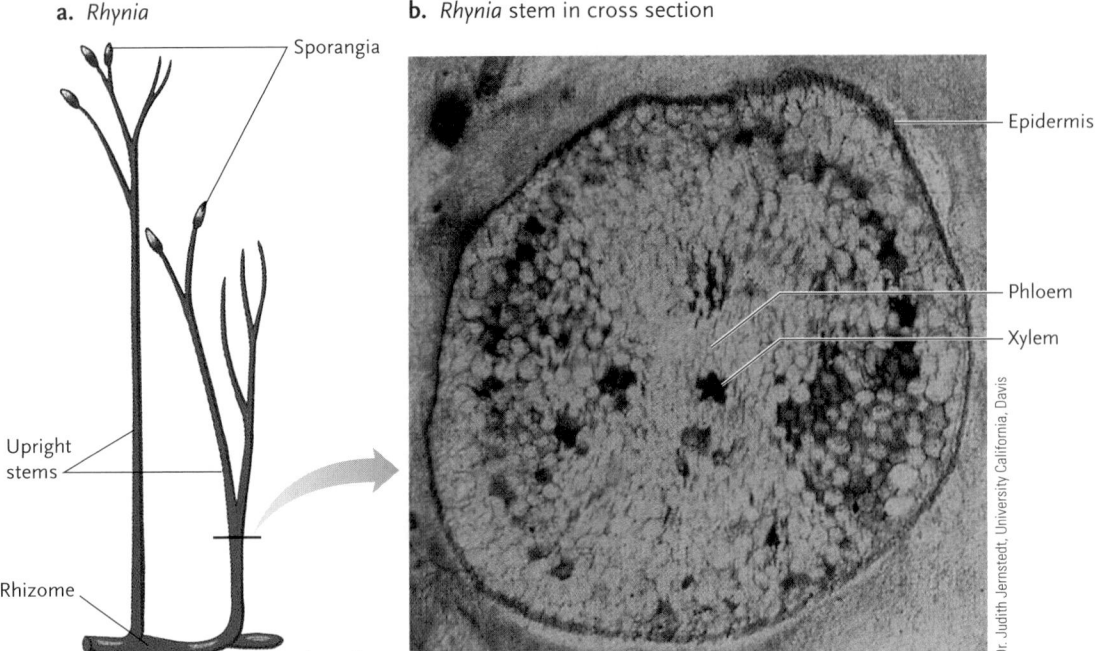

a. *Rhynia*

Sporangia

Upright stems

Rhizome

Rhizoids

b. *Rhynia* stem in cross section

Epidermis

Phloem

Xylem

Dr. Judith Jernstedt, University California, Davis

Figure 27.13
Rhynia, an early seedless vascular plant. **(a)** Fossil-based reconstruction of the entire plant, about 30 cm tall. **(b)** Cross section of the stem, approximately 3 mm in diameter. This fossil was embedded in chert approximately 400 million years ago. Still visible in it are traces of the transport tissues xylem and phloem, along with other specialized tissues.

Figure 27.14
Reconstruction of the lycophyte tree (*Lepidodendron*) and its environment. **(a)** Fossil evidence suggests that *Lepidodendron* grew to be about 35 m tall with a trunk 1 m in diameter. **(b)** Artist's depiction of a Coal Age forest.

ing plant litter. There rhizoids attach it to its substrate. It cannot photosynthesize, and instead obtains nutrients by way of mycorrhizae. Although all species of *Lycopodium* are homosporous—that is, one bisexual gametophyte produces both eggs and sperm—those of other genera are heterosporous. Regardless, as with ancestral lycophytes, the sperm require water in which they can swim to the eggs. After fertilization, the life cycle comes full circle as the zygote develops into a diploid embryo that grows into a sporophyte.

a. The lycophyte tree (*Lepidodendron*)

Stem of a giant lycophyte (*Lepidodendron*)

b. Artist's depiction of a Coal Age forest

Seed fern (*Medullosa*); probably related to the progymnosperms, which may have been among the earliest seed-bearing plants

Stem of a giant horsetail (*Calamites*)

Field Museum of Natural History, Chicago

Figure 27.15
Lycophytes. **(a)** *Lycopodium* sporophyte, showing the conelike strobili in which spores are produced. **(b)** A fossilized lycophyte spore bearing a characteristic Y-shaped mark (arrow) called a trilete scar.

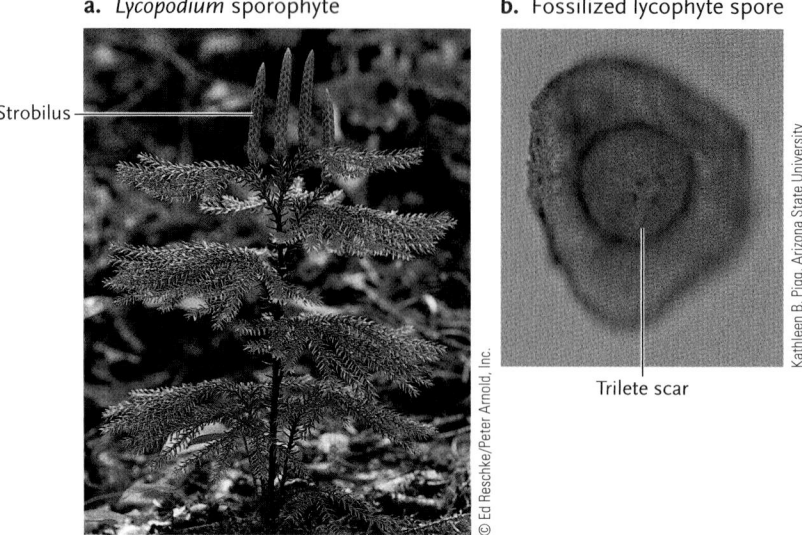

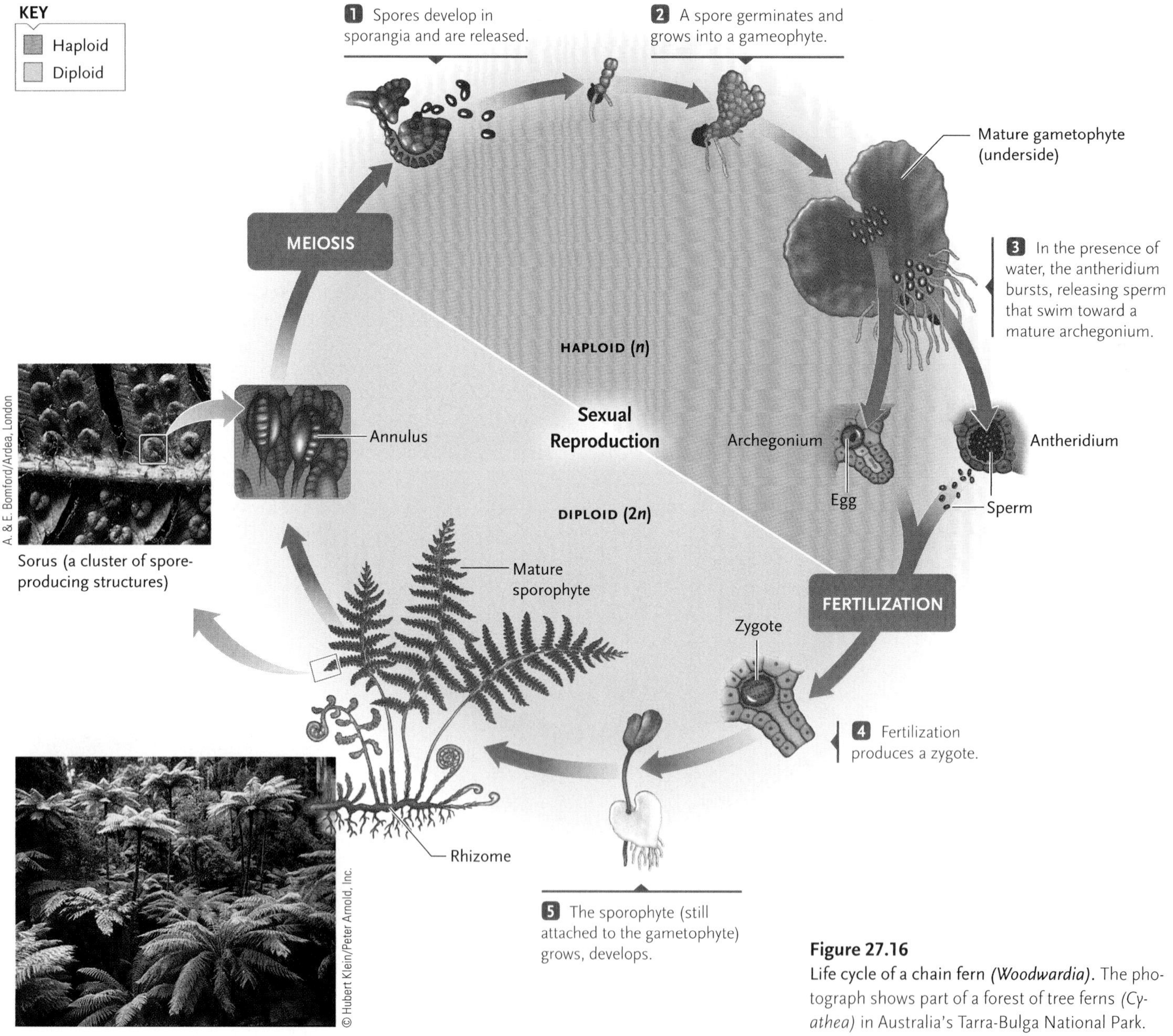

Figure 27.16
Life cycle of a chain fern *(Woodwardia)*. The photograph shows part of a forest of tree ferns *(Cyathea)* in Australia's Tarra-Bulga National Park.

Ferns, Whisk Ferns, Horsetails, and Their Relatives Make Up the Diverse Phylum Pterophyta

Second only to the flowering plants, the phylum Pterophyta (*pteron* = wing) contains a large and diverse group of vascular plants—the 13,000 or so species of ferns, whisk ferns, and horsetails. Most ferns, including some that are poplar houseplants, are native to tropical and temperate regions. Some floating species are less than 1 cm across, while some tropical tree ferns grow to 25 m tall. Other species are adapted to life in arctic and alpine tundras, salty mangrove swamps, and semi-arid deserts.

Complex Anatomical Features in Ferns. The familiar plant body of a fern is the sporophyte phase **(Figure 27.16).** It produces an above-ground clump of fern leaves, called fronds. Often finely divided and feather-like, and containing multiple strands of vascular tissue, fronds are the most complex leaves of the plant kingdom. Young fronds are tightly coiled, and as they emerge above the soil these "fiddleheads" (so named because they resemble the scrolled pegheads of violins) unroll and expand. Before they unfurl, fiddleheads may be gathered by people who relish them as a gastronomic treat (albeit with care—the fiddleheads of some species contain a carcinogen). Leaves of some species last for only a single growing season, while in others they grow for several years. A typical frond has a well-developed epidermis with chloroplasts in the epidermal cells and stomata on the lower surface.

Except for tropical tree ferns, the stems of most ferns are underground rhizomes. The stem's vascular system is organized into a complex, interconnecting network of bundles, each having a central core of xylem surrounded by phloem. Roots descend along the length of the rhizomes. A rhizome can live for centuries, growing at its tip and extending outward horizontally through the soil, sometimes over a considerable area. In most ferns, the fronds arise from nodes positioned along the rhizome. A **node** is the point on a stem where one or more leaves are attached.

A fern sporophyte produces sporangia on the lower surface or margin of some leaves. Often, several sporangia are clustered into a rust-colored **sorus** (plural, *sori*). Sori may be exposed or they may be protected with a flap of tissue. Each sporangium is a delicate case, shaped rather like an old-fashioned pocket watch and covered by a layer of epidermal cells. In the layer, a row of thick-walled cells called the **annulus** ("ring") nearly encircles the sporangium.

Inside the sporangium, haploid spores arise by meiosis. Meanwhile, the sporangium slowly dries out, and as it does so the annulus steadily contracts. Eventually the force of the contracting annulus rips open the sporangium, which snaps back on itself, flinging out the mature spores. In this way fern spores can be dis-persed up to 2 m away from the parent plant. Wind may carry them much farther: on board the *Beagle,* Charles Darwin collected fern spores hundreds of miles from shore.

A germinating spore develops into a gametophyte, which is typically a small, heart-shaped plant anchored to the soil by rhizoids. Both antheridia and archegonia are present on the lower surface of each gametophyte, where moisture is trapped. Inside an antheridium is a globular packet of haploid cells, each of which develops into a helical sperm with many flagella. When water is present, the antheridium bursts, releasing the sperm. If mature archegonia are nearby, the sperm swim toward them, drawn by a chemical attractant that diffuses from the neck of the archegonium, which is open when free water is present.

After a sperm fertilizes an egg, the diploid zygote begins dividing and developing into an embryo, which at this stage obtains nutrients from the gametophyte. In a short time, however, the embryo develops into a young sporophyte that is larger than the gametophyte and has its own green leaf and a root system. The sporophyte now is nutritionally independent and the parent gametophyte degenerates and dies.

Features of Early Vascular Plants in Whisk Ferns. The whisk ferns and their relatives are represented by two genera, *Psilotum* (pronounced si-lo′-tum) and *Tmesipteris* (may-sip′-ter-is), with only about 10 species in all. These rather uncommon plants grow in tropical and subtropical regions, often as epiphytes. In the United States the range for *Psilotum* species **(Figure 27.17)** includes Hawaii, Gulf Coast states such as Florida and Louisiana, and parts of the West.

The sporophytes of whisk ferns are up to 60 cm tall and resemble the extinct *Cooksonia* and *Rhynia*. Like those early vascular plants, they lack true roots and leaves. Instead, small leaflike scales adorn an upright, green, branching stem, which arises from a horizontal rhizome system anchored by rhizoids. The absorptive rhizoids have mycorrhizal fungi associated with them, which provide enhanced access to some nutrients.

A whisk fern's stem is structurally and functionally multifaceted. The stem's epidermal cells carry out photosynthesis, while its core has the transport tissues xylem and phloem and other anatomical features of more complex vascular plants. Sporangia rest atop some of the stem scales. Inside them, meiotic divisions of specialized cells produce haploid spores.

Horsetails, Possibly the Most Ancient Living Plant Species. The ancient relatives of modern-day horsetails included treelike forms taller than a two-story building. Only fifteen species in a single genus,

Figure 27.17
Sporophytes of a whisk fern *(Psilotum)*, a seedless vascular plant. Three-lobed sporangia occur at the ends of stubby branchlets; inside the sporangia, meiosis gives rise to haploid spores.

Kingsley R. Stern

a. Sporophyte stem

b. Sporangia

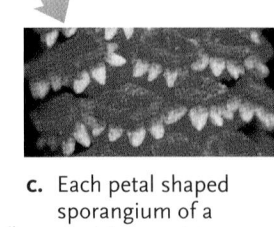

Strobilus, an aggregation of sporangia at the tip of the horsetail sporophyte

William Ferguson

W. H. Hodges

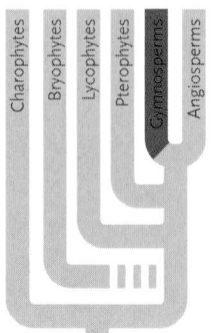

Kratz/Zefa

c. Each petal shaped sporangium of a strobilus contains spores that formed by meiosis.

Figure 27.18
A species of *Equisetum*, the horsetails. **(a)** Vegetative stem. **(b)** Strobili, which bear sporangia. **(c)** Close-up of sporangia and associated structures on a strobilus.

Equisetum, have survived to the present **(Figure 27.18)**. Horsetails grow in moist soil along streams and in disturbed habitats, such as roadsides and beds of railroad tracks. Their sporophytes typically have underground rhizomes and roots that anchor the rhizome to the soil. The scalelike leaves are arranged in whorls about a photosynthetic stem that is stiff and gritty because horsetails accumulate silica in their tissues. American pioneers used them to scrub out pots and pans, hence their other common name, "scouring rushes."

Equisetum sporangia are borne in strobili on highly specialized stem structures quite different from the sporophylls of club mosses. In most horsetails the strobili occur on ordinary vegetative shoots, but in a few species they occur only on special fertile shoots. Each stalked spore-bearing structure in a strobilus resembles an umbrella and is attached at right angles to a main axis. Haploid spores develop in sporangia attached near the edge of the "umbrella's" underside, and air currents disperse them. They must germinate within a few days to produce gametophytes, which are free-living plants about the size of a small pea.

STUDY BREAK

1. Compare and contrast the lycophyte and bryophyte life cycles with respect to the sizes and longevity of gametophyte and sporophyte phases.
2. In ferns, whisk ferns, and horsetails, what kinds of structures fulfill the roles of roots and leaves?
3. How does the life cycle of a horsetail differ from that of a fern?

27.4 Gymnosperms: The First Seed Plants

Charophytes Bryophytes Lycophytes Pterophytes Gymnosperms Angiosperms

Gymnosperms are the conifers and their relatives. The earliest fossils identified as gymnosperms are found in Devonian rocks. By the Carboniferous, when nonvascular plants were dominant, many lines of gymnosperms had also evolved, and the first true conifers appeared. These radiated during the Permian period; the Mesozoic era that followed, 248 to 65 mya, was the age not only of the dinosaurs but of the gymnosperms as well.

The evolution of gymnosperms involved sweeping changes in plant structures related to reproduction. As a prelude to our survey of modern gymnosperms, we begin by considering some of these innovations, which opened new adaptive options for land plants.

Major Reproductive Adaptations Occurred as Gymnosperms Evolved

The word *gymnosperm* is derived from the Greek *gymnos*, meaning naked, and *sperma*, meaning seed. The evolution of gymnosperms included important reproductive adaptations—pollen and pollination, the ovule, and the seed. The fossil record has not revealed the sequence in which these changes arose, but all of them contributed to the radiation of gymnosperms into land environments. **Figure 27.19** shows an artist's rendering of *Archaeopteris*, which may have been one of the first true trees. Called a *progymnosperm*, it belonged to an evolutionary line that is thought to have given rise to modern seed plants.

Pollen and Ovules: Shelter for Spores. Unlike bryophytes and seedless vascular plants, gymnosperm sporophytes do not disperse their spores. The sporophyte produces haploid spores by meiosis, but it retains these spores inside reproductive structures where they give rise to multicellular haploid gametophytes. As noted briefly earlier, sperm arise inside a **pollen grain**, a male gametophyte that typically has walls reinforced with the polymer sporopollenin. All but a few gymnosperms have nonmotile sperm. Usually, two of these nonswimming sperm develop inside each pollen grain—very different from the flagellated, swimming sperm of algae and plants that do not produce seeds.

An **ovule** is a structure in a sporophyte in which a female gametophyte develops, complete with an egg. Physically connected to the sporophyte and surrounded by the ovule's protective layers, a female gametophyte no longer faces the same risks of predation or environmental assault that can threaten a free-living gametophyte.

Figure 27.19
Fossil-based reconstruction of *Archaeopteris*, a large Devonian progymnosperm. It could grow 25 m tall and may have been a seed-forming ancestor of modern gymnosperms.

Pollination is the transfer of pollen to female reproductive parts via air currents or on the bodies of animal pollinators. Pollen and pollination were enormously important adaptations for gymnosperms, because the shift to nonswimming sperm along with a means for delivering them to female gametes meant that reproduction no longer required liquid water. The only gymnosperms that have retained swimming sperm are the cycads and ginkgoes, described shortly, which have relatively few living species and are restricted to just a few native habitats.

Seeds: Protecting and Nourishing Plant Embryos. A **seed** is the structure that forms when an ovule matures after a pollen grain reaches it and a sperm fertilizes the egg. It consists of three basic parts: (1) the embryo sporophyte, (2) tissues around it containing carbohydrates, proteins, and lipids that nourish the embryo until it becomes established as a plantlet with leaves and roots, and (3) a tough, protective outer seed coat **(Figure 27.20).** This complex structure makes seeds ideal packages for sheltering an embryo from drought, cold, or other adverse conditions. As a result, seed-making plants enjoy a tremendous survival advantage over species that simply release spores to the environment. Encased in a seed, the embryo also can be transported far from its parent, as when ocean currents carry coconut seeds ("coconuts" protected in large, buoyant fruits) hundreds of kilometers across the sea. As discussed in Chapter 34, some plant embryos housed in seeds can remain dormant for months or years before environmental conditions finally prompt them to germinate and grow.

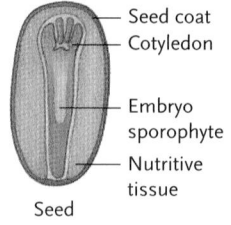

Figure 27.20
Generalized view of a seed—in this case, the seed of a pine, a gymnosperm.

Modern Gymnosperms Include Conifers and a Few Other Groups

Today there are about 800 gymnosperm species. The sporophytes of nearly all are large trees or shrubs, although a few are woody vines. The most widespread and familiar gymnosperms are the conifers (Coniferophyta). Others are the cycads (Cycadophyta), ginkgoes (Ginkgophyta), and gnetophytes (Gnetophyta).

Economically, gymnosperms, particularly conifers, are vital to human societies. They are sources of lumber, paper pulp, turpentine, and resins, among other products. They also have huge ecological importance. Their habitats range from tropical forests to deserts, but gymnosperms are most dominant in the cool-temperate zones of the northern and southern hemispheres. They flourish in poor soils where flowering plants don't compete as well. In North America, for example, gymnosperm forests cover more than one-third of the continent's landmass—although in some areas, logging has significantly reduced the once-lush forest cover. Our survey of gymnosperms begins, however, with the cycads, ginkgoes, and gnetophytes—the latter two groups remnants of lineages that have all but vanished from the modern scene.

Cycads Are Restricted to Warmer Climates

During the Mesozoic era, the **Cycadophyta** (*kykas* = palm), or cycads, flourished along with the dinosaurs. About 185 species have survived to the present, but they are confined to the tropics and subtropics.

At first glance, you might mistake a cycad for a small palm tree **(Figure 27.21).** Some cycads have massive, cone-shaped strobili (clusters of sporophylls) that bear either pollen or ovules. Air currents or crawling insects transfer pollen from male plants to the developing gametophyte on female plants. Poisonous alkaloids that may help deter insect predators occur in various cycad tissues. In tropical Asia, some people consume cycad seeds and flour made from cycad trunks, but only after the toxic compounds have been rinsed away. Much in demand from fanciers of unusual plants, cycads in some countries are uprooted and sold in what amounts to a black-market trade—greatly diminishing their numbers in the wild.

Figure 27.21
The cycad *Zamia*. Note the large, terminal female cone and fernlike leaves.

Ginkgoes Are Limited to a Single Living Species

The phylum **Ginkgophyta** has only one living species, the ginkgo (or maiden-hair) tree *(Ginkgo biloba)*, which grows wild today only in warm-temperate forests of central China. Ginkgo trees are large, diffusely branching trees with characteristic fan-shaped leaves **(Figure 27.22)** that turn a brilliant yellow in autumn. Nursery-propagated male trees often are planted in cities because they are resistant to insects, disease, and air pollutants. The female trees are equally pollution-resistant, but gardeners shy away from them—their fleshy fruits produce a notoriously foul odor.

Gnetophytes Include Simple Seed Plants with Intriguing Features

The phylum Gnetophyta contains three genera—*Gnetum, Ephedra,* and *Welwitschia*—that together include about 70 species. Moist, tropical regions are home to about 30 species of *Gnetum,* which includes both trees and leathery leafed vines (lianas). About 35 species of *Ephedra* grow in desert regions of the world **(Figure 27.23a–c)**.

Of all the gymnosperms, *Welwitschia* is the most bizarre. This seed-producing plant grows in the hot deserts of south and west Africa. The bulk of the plant is a deep-reaching taproot. The only exposed part is a woody disk-shaped stem that bears cone-shaped strobili and leaves. The plant never produces more than two strap-shaped leaves, which split lengthwise repeatedly as the plant grows older, producing a rather scraggly pile **(Figure 27.23d)**.

Although gnetophytes are structurally and functionally simpler than most other seed plants, recent studies of sexual reproduction mechanisms in *Gnetum* and *Ephedra* species uncovered a two-step process of fertilization—which is a hallmark of angiosperms, the most advanced seed plants. This discovery raised some provocative evolutionary questions, even leading some investigators to propose that ancient gnetophytes may have given rise to flowering plants. Complicating this picture, however, are molecular findings, such as those arrived at by a research team at the Academia Sinica in

a. Ginkgo tree

b. Fossil and modern gingko leaves

c. Male cone

d. Ginkgo seeds

Figure 27.22
Ginkgo biloba. **(a)** A ginkgo tree. **(b)** Fossilized ginkgo leaf compared with a leaf from a living tree. The fossil formed at the Cretaceous–Tertiary boundary. Even though 65 million years have passed, the leaf structure has not changed much. **(c)** Pollen-bearing cones and **(d)** fleshy-coated seeds of the *Ginkgo*.

Taiwan, People's Republic of China. When the team compared 65 nuclear rRNA sequences from ferns, gymnosperms, and angiosperms, their analysis supported the hypothesis that cycads and ginkgoes represent the earliest gymnosperm lineage, with a divergent lineage of gnetophytes and conifers arising later. The team found no molecular evidence for a link between the Gnetophyta and angiosperms.

Conifers Are the Most Common Gymnosperms

About 80% of all living gymnosperm species are members of one phylum, the **Coniferophyta**, or conifers ("cone-bearers"). Conifer trees and shrubs are longer-lived, and anatomically and morphologically more complex, than any sporophyte phase we have discussed so far. Characteristically, they form woody reproductive cones, and most of the 550 conifer species are woody trees or shrubs with needlelike or scalelike leaves, which are anatomically adapted to aridity. For instance, needles have a thick cuticle, sunken stomata, and a fibrous epidermis, all traits that reduce the loss of water vapor.

Most conifers are evergreens. That is, although they shed old leaves, often in autumn, they retain enough leaves so that they still look "green," unlike deciduous species like maples, which shed *all* their leaves as winter approaches. Familiar conifer examples are the pines, spruces, firs, hemlocks, junipers, cypresses, and redwoods. Like other seed plants, conifers are heterosporous, producing pollen in clusters of small strobili and eggs in larger, woody ones. Both of these structures are often referred to as cones. Seeds develop on the shelflike scales of the female cones.

Pines and many other gymnosperms produce resins, a mix of organic compounds that are by-products of metabolism. Resin accumulates and flows in long resin ducts through the wood, inhibiting the activity of wood-boring insects and certain microbes. Pine resin extracts are the raw material of turpentine and (minus the volatile terpenes) the sticky rosin used to treat violin bows.

We know a great deal about the pine life cycle **(Figure 27.24),** so it is a convenient model for gymnosperms. All but 1 of the 93 pine species are trees (*Pinus mugo,* native to high elevations in Europe, is a shrub). The male cones (strobili) are relatively small and delicate, only about 1 cm long, and are borne on the lower branches. Each one consists of many small scales, which are specialized leaves (called sporophylls) attached to the cone's axis in a spiral. Two sporangia develop on the underside of each scale. Inside the sporangia, spore "mother cells" called microsporocytes undergo meiosis and give rise to haploid **microspores.** Each microspore then undergoes mitosis to develop into a winged pollen grain—an immature male gametophyte. At this stage the pollen grain consists of four

a. *Ephedra* plant

b. *Ephedra* male cone

William Ferguson

c. *Ephedra* female cone

Edward S. Ross

Robert & Linda Mitchell Photography

d. *Welwitschia* plant with female cones

© Fletcher and Baylis/Photo Researchers, Inc.

Figure 27.23

Gnetophytes. **(a)** Sporophyte of *Ephedra*, with close-ups of **(b)** its pollen-bearing cones and **(c)** a seed-bearing cone, which develop on separate plants. **(d)** Sporophyte of *Welwitschia mirabilis*, with seed-bearing cones.

cells, two that will degenerate and two that will function later in reproduction.

Young female cones develop higher in the tree, at the tips of upper branches. The cone scales bear ovules. Inside each ovule is a spore mother cell called a megasporocyte. Unlike microsporocytes, the megasporocyte in an ovule undergoes meiosis only when conditions are right and produces four haploid spores called **megaspores.** Only one megaspore survives, however, and it develops slowly, becoming a mature female gametophyte only when pollination is underway. In a pine, the process takes well over a year. The mature female gametophyte is a small oval mass of cells with several archegonia at one end, each containing an egg.

Each spring, air currents lift vast numbers of pollen grains off male cones—by some estimates, billions may be released from a single pine tree. The extravagant numbers assure that at least some pollen grains will land on female cones. The process is not as random as it might seem: studies have shown that the contours of female cones create air currents that

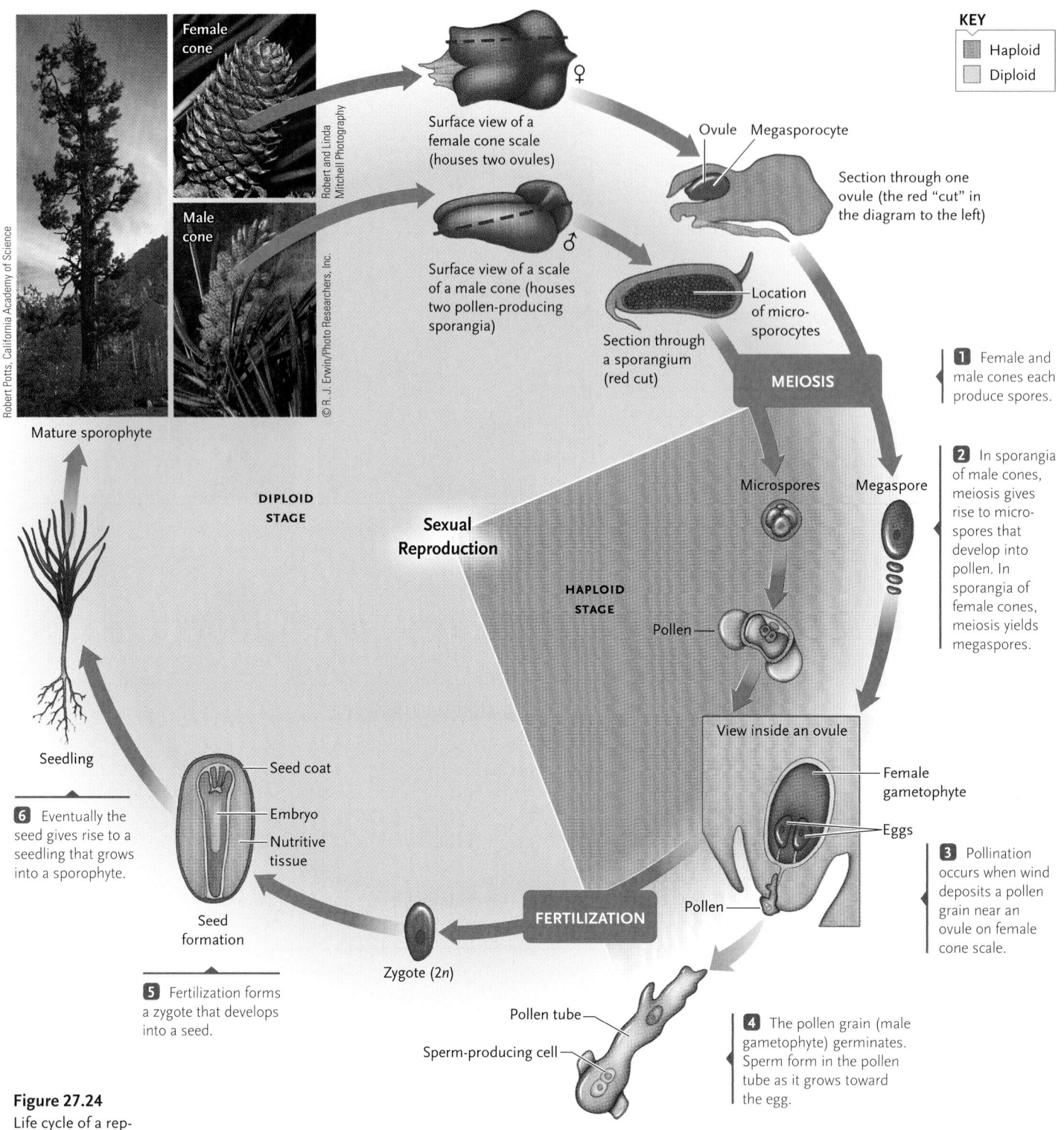

Figure 27.24
Life cycle of a representative conifer, a ponderosa pine *(Pinus ponderosa).* Pines are the dominant conifers in the Northern Hemisphere and their large sporophytes provide a heavily exploited source of wood.

Labels within the figure:

Female cone

Male cone

Robert Potts, California Academy of Science

Robert and Linda Mitchell Photography

© R. J. Erwin/Photo Researchers, Inc.

Surface view of a female cone scale (houses two ovules) ♀

Surface view of a scale of a male cone (houses two pollen-producing sporangia) ♂

Section through a sporangium (red cut)

Ovule Megasporocyte

Section through one ovule (the red "cut" in the diagram to the left)

Location of micro-sporocytes

KEY
- Haploid
- Diploid

MEIOSIS

1 Female and male cones each produce spores.

2 In sporangia of male cones, meiosis gives rise to microspores that develop into pollen. In sporangia of female cones, meiosis yields megaspores.

Microspores Megaspore

Pollen

Mature sporophyte

DIPLOID STAGE

Sexual Reproduction

HAPLOID STAGE

Seedling

View inside an ovule

Female gametophyte

Eggs

6 Eventually the seed gives rise to a seedling that grows into a sporophyte.

Seed coat

Embryo

Nutritive tissue

Pollen

3 Pollination occurs when wind deposits a pollen grain near an ovule on female cone scale.

Seed formation

FERTILIZATION

Zygote (2n)

Pollen tube

Sperm-producing cell

5 Fertilization forms a zygote that develops into a seed.

4 The pollen grain (male gametophyte) germinates. Sperm form in the pollen tube as it grows toward the egg.

can favor the "delivery" of pollen grains near the cone scales. After pollination, the pollen grain develops into a *pollen tube* that grows toward the female spore mother cell. As it does, sperm form in the tube and stimulate maturation of the female gametophyte and the production of eggs. When a pollen tube reaches an egg, the stage is set for fertilization, the formation of a zygote, and early development of the plant embryo. Often, fertilization occurs months to a year after pollination. Once an embryo forms, a pine seed—which, recall, includes the embryo, female gametophyte tissue, and seed coat—eventually is shed from the cone. The seed coat protects the embryo from drying out, and the female gametophyte tissue serves as its food reserve. This tissue makes up the bulk of a "pine nut."

STUDY BREAK

1. What are the four major reproductive adaptations that evolved in gymnosperms?
2. What are the basic parts of a seed, and how is each one adaptive?
3. Describe some features that make conifers structurally more complex than other gymnosperms.

27.5 Angiosperms: Flowering Plants

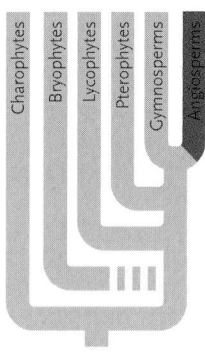

Of all plant phyla, the flowering plants, or **angiosperms**, are the most successful today. At least 260,000 species are known (**Figure 27.25** shows a few examples), and botanists regularly discover new ones in previously unexplored regions of the tropics. The word angiosperm is derived from the Greek *angeion* (meaning a case or vessel) and *sperma* (seed). The "vessel" refers to the modified leaf, called a *carpel,* that surrounds and protects the ovules and later, the seeds of angiosperms. Carpels are **flowers**, reproductive structures that are a key defining feature of angiosperms. Another defining feature is the **fruit**—botanically speaking, a structure that surrounds the angiosperm embryo and aids seed dispersal.

In addition to having flowers and fruits, angiosperms are the most ecologically diverse plants on Earth, growing on dry land and in wetlands, fresh water, and the seas. They range in size from tiny duckweeds about 1 mm long to towering *Eucalyptus* trees more than 100 m tall. Most are free-living photosynthesizers. Others lack chloroplasts and feed on nonliving organic matter or are parasites that feed on living host organisms.

The Fossil Record Provides Little Information about the Origin of Flowering Plants

The evolutionary origin of angiosperms has confounded plant biologists for well over a hundred years. Charles Darwin called it the "abominable mystery," because

a. Flowering plants in a desert

b. Alpine angiosperms

c. Triticale, a grass

d. A parasitic angiosperm

Figure 27.25

Flowering plants. Diverse photosynthetic species are adapted to nearly all environments, ranging from **(a)** deserts to **(b)** snowlines of high mountains. **(c)** Triticale, a hybrid grain derived from parental stocks of wheat *(Triticum)* and rye *(Secale)*, is one example of the various grasses utilized by humans. **(d)** The parasitic flowering plant Indian pipe *(Monotropa uniflora)* having no chlorophyll of its own, obtains food by associating with mycorrhizae, which are in turn associated with the roots of photosynthetic plants.

Archaefructus sinensis

© David Dilcher, Florida Museum of Natural History/Paleobotany Laboratory

Sketch of *Archaefructus sinensis*

Figure 27.26
Fossil of *Archaefructus sinensis*, thought to have been an early flowering plant. The sketch shows what this small, possibly aquatic plant may have looked like.

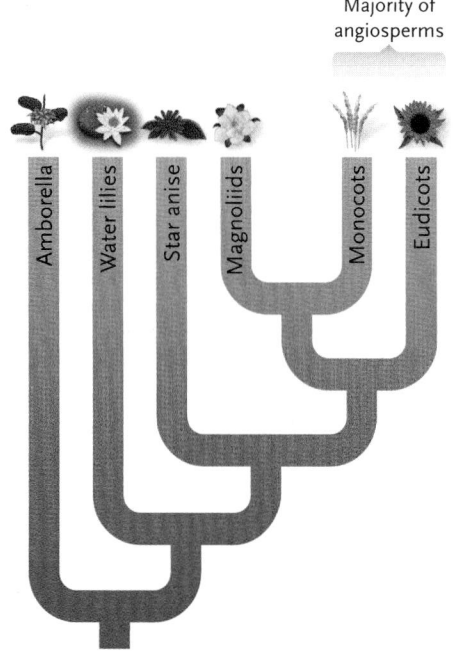

Majority of angiosperms

Amborella · Water lilies · Star anise · Magnoliids · Monocots · Eudicots

Figure 27.27
A hypothetical phylogenetic tree for flowering plants.

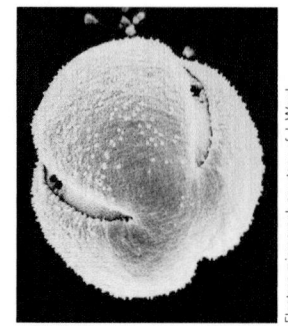

Electron micrograph courtesy of J. Ward

Figure 27.28
Eudicot pollen grain. Eudicot pollen grains have three slitlike grooves, only two of which are visible here. Pollen made by all other seed plants, including monocots, have just one groove.

or seedless vascular plants. As with gymnosperms, attempts to reconstruct the earliest flowering plant lineages from morphological, developmental, and biochemical characteristics have produced several conflicting classifications and family trees. Some paleobotanists hypothesize that flowering plants arose in the Jurassic period; others propose they evolved in the Triassic from now-extinct gymnosperms or from seed ferns.

As the Mesozoic era ended and the modern Cenozoic era began, great extinctions occurred among both plant and animal kingdoms. Gymnosperms declined, and dinosaurs disappeared. Flowering plants, mammals, and social insects flourished, radiating into new environments. Today we live in what has been called "the age of flowering plants."

Angiosperms Are Subdivided into Several Clades, Including Monocots and Eudicots

Angiosperms are assigned to the phylum **Anthophyta**, a name that derives from the Greek *anthos*, meaning flower. **Figure 27.27** shows one current model of major clades within the phylum. The great majority of angiosperms are classified either as monocots or eudicots. **Monocots** are distinguished by the morphology of their embryos, which have a single seed leaf called a cotyledon ("cuplike hollow"). **Eudicots** ("true dicots"), which generally have two cotyledons, are set apart from other angiosperms by the structure of their pollen grains, which have three grooves **(Figure 27.28)**. By contrast, the pollen of monocots and all other seed plants (including more than 8500 species once lumped with eudicots under the term "dicots") have only a single groove. Paleobotanists use this clear structural difference not only to help establish the general type of plant that produced fossil pollen, but also what types of plants were present in fossil deposits of a particular age or geographic location.

While most angiosperms can fairly easily be categorized as either monocots or eudicots, figuring out the appropriate classification for other angiosperms is an ongoing challenge and an extremely active area of plant research. The diagram in Figure 27.27 reflects a synthesis of evidence from both morphological and molecular studies, an approach examined in this chapter's *Insights from the Molecular Revolution*. Along with eudicots and monocots, botanists currently recognize four other clades **(Figure 27.29)**. The **magnoliids**, a group that includes magnolias (see Figure 27.29a), laurels, and avocados, are more closely related to monocots than to eudicots. Some researchers also place plants that are the sources of spices such as peppercorns, nutmeg, and cinnamon in the magnoliid clade. The other three clades are considered to be **basal angiosperms** representing the earliest branches of the flowering plant lineage. They include the star anise group (see Figure 27.29b), water lilies (see Figure

flowering plants appear suddenly in the fossil record, without a fossil sequence that links them to any other plant groups. The oldest well-documented fossil specimens date back 125 million years. Discovered in China, these remarkable fossils show complex and strikingly modern-looking plants that have leaves, stems, fruits, and seeds **(Figure 27.26)**. Two species have been unearthed and have been assigned to the genus *Archaefructus*, representing a newly discovered, extinct angiosperm group.

The fossil record has yet to reveal obvious transitional organisms between flowering plants and either gymnosperms

The Powerful Genetic Toolkit for Studying Plant Evolution

Unlike animals and most other eukaryotic organisms, plants have three distinct sets of genes—in the cell nucleus, in mitochondria, and in chloroplasts. Chloroplast DNA, or cpDNA, has been especially useful for evolutionary studies, particularly the chloroplast *rbcL* gene. Mutations of the gene have occurred slowly, at about one-fourth to one-fifth the rate of genes in the nucleus. As a result, the DNA sequences of *rbcL* genes of different species diverge less than those of most other plant genes. Further, there are no introns—noncoding sequences—interrupting the coding sequence of the *rbcL* gene. Researchers can compare *rbcL* DNA from different species base by base, with no need to subtract introns. At the same time, the *rbcL* genes of different species are different enough to allow researchers to assemble evolutionary trees based on the degree of sequence variation.

Studies using cpDNA have helped fuel several fundamental shifts in our understanding of branch points in plant evolution. For example, together with gene sequence data from nuclear DNA, analysis of *rbcL* genes provided the molecular foundation for the now widely accepted view that charophyte green algae were the evolutionary forerunners of land plants. Similarly, in the late 1990s an international research team led by Yin-Long Qiu at the University of Massachusetts at Amherst correlated the loss of introns from two mitochondrial genes with the hypothesis that the first land plants were liverworts. Qiu and his colleagues carried out a genetic survey of more than 350 land plants representing all major lineages. They discovered that the noncoding sequences were present in all other bryophytes and all major lines of vascular plants, but were absent in liverworts, green algae, and all other eukaryotes. The findings are supported by analysis of *rbcL* sequences in various plant groups. Data from cpDNA and mtDNA analyses also underlies the hypothesis that, as land plants evolved, the ancient relatives of club mosses (lycophytes) were the forerunners of other vascular plants. Clearly, these varied molecular tools, and cpDNA in particular, are helping plant scientists explore evolutionary relationships across the whole spectrum of the Kingdom Plantae.

27.29c), and an intriguing ancient line represented by a single shrub, *Amborella trichopoda* (see Figure 27.29d). Found only in cloud forests of the South Pacific island of New Caledonia, *Amborella*'s small white flowers and vascular system are structurally simpler than those of other angiosperms, and its female gametophyte differs as well. These morphological differences and a comparison of the nucleotide sequences of genes encoding the two angiosperm phytochromes (photoreceptors discussed in Chapter 35) suggest that *Amborella* is the closest living relative of the first flowering plants.

Figure 27.30a gives some idea of the variety of living monocots, which include grasses, palms, lilies, and orchids. The world's major crop plants (wheat, corn, rice, rye, sugarcane, and barley) are domesticated grasses, and all are monocots. There are at least 60,000 species of monocots, including 10,000 grasses and 20,000 orchids. Eudicots are even more diverse, with nearly 200,000 species **(Figure 27.30b)**. They include flowering shrubs and trees, most nonwoody (herbaceous) plants, and cacti. **Figure 27.31** shows the life cycle of a lily, a monocot. The life cycle of a typical eudicot is described in detail in the next unit, which focuses on the structure and function of flowering plants.

Many Factors Contributed to the Adaptive Success of Angiosperms

At this writing, molecular studies place the origin of flowering plants at least 140 mya. It took only about 40 million years—a short span in geological time—for angiosperms to eclipse gymnosperms as the prevailing form of plant life on land (see Figure 22.19). Several

Figure 27.29
Representatives of basal angiosperm clades. **(a)** Southern magnolia (*Magnolia grandiflora*), a magnoliid. **(b)** Star anise (*Illicium floridanum*). **(c)** Sacred lotus (*Nelumbo nucifera*), a water lily. **(d)** *Amborella trichopoda*.

a. Southern magnolia (*Magnolia grandiflora*), a magnoliid

D. Harms/Peter Arnold, Inc.

b. Star anise (*Illicium floridanum*)

Rob & Ann Simpson/Visuals Unlimited

c. Sacred lotus (*Nelumbo nucifera*), a water lily

© Gregory C. Dimijian/Photo Researchers, Inc.

d. *Amborella*

© Sangtae Kim, University of Florida

a. Representative monocots

Wheat *(Triticum)*

Tulips *(Tulipa)*

Eastern prairie fringed orchid
(Platanthera leucophaea)

b. Representative eudicots

Rose *(Rosa)*

Yellow bush lupine
(Lupinus arboreus)

Cherry *(Prunus)*

Claret cup cactus
(Echinocereus triglochidratus)

Figure 27.30
Examples of monocots and eudicots.

factors fueled this adaptive success. As with other seed plants, the large, diploid sporophyte phase dominates a flowering plant's life cycle, and the sporophyte retains and nourishes the much smaller gametophytes. But flowering plants also show some evolutionary innovations not seen in gymnosperms.

More Efficient Transport of Water and Nutrients. Where gymnosperms have only one type of water-conducting cell (tracheids), angiosperms have an additional, more specialized type (called vessel elements). As a result, an angiosperm's xylem vessels move water more rapidly from roots to shoot parts. Also, modifications in angiosperm phloem tissue allow it to more efficiently transport sugars produced in photosynthesis through the plant body.

Enhanced Nutrition and Physical Protection for Embryos. Other changes in angiosperms made it more likely that reproduction would succeed. For example, a two-step *double fertilization* process in the seeds of flowering plants gives rise to both an embryo and a unique nutritive tissue (called endosperm) that nour-

ishes the embryonic sporophyte. The ovule containing a female gametophyte is enclosed within an **ovary,** which develops from a carpel and shelters the ovule against desiccation and against attack by herbivores or pathogens. In turn, ovaries develop into the fruits that house angiosperm seeds. As noted earlier, a fruit not only protects seeds, but helps disperse them—for instance, when an animal eats a fruit, seeds may pass through the animal's gut none the worse for the journey and be released in a new location in the animal's feces. Above all, angiosperms have flowers, the unique reproductive organs that you will read much more about in the next unit.

Angiosperms Coevolved with Animal Pollinators

The evolutionary success of angiosperms correlates not only with the adaptations just described, but also with efficient mechanisms of transferring pollen to female reproductive parts. While a conifer depends on air currents to disperse its pollen, angiosperms coevolved with pollinators—insects, bats, birds, and other animals that

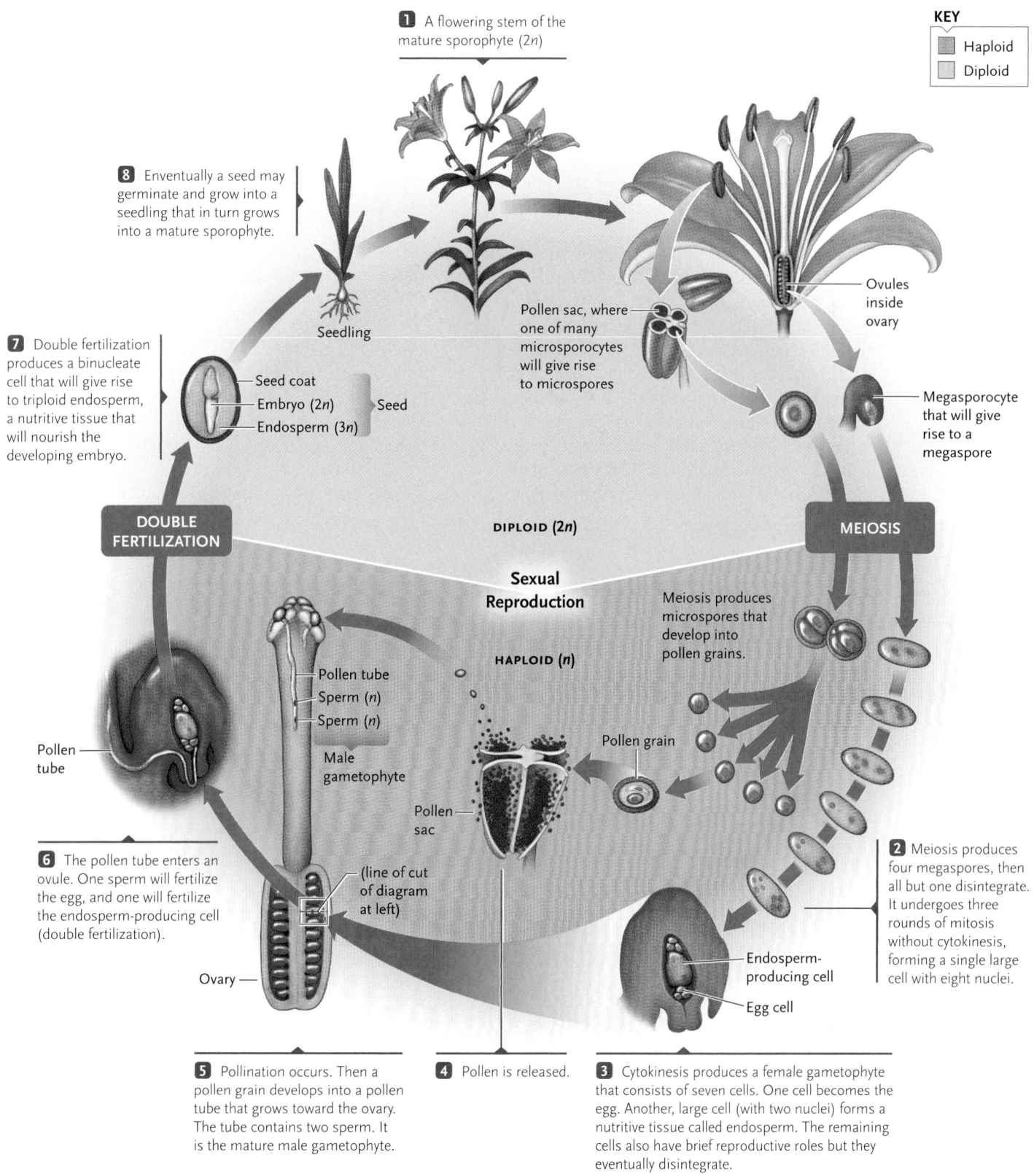

1 A flowering stem of the mature sporophyte (2*n*)

KEY
Haploid
Diploid

8 Enventually a seed may germinate and grow into a seedling that in turn grows into a mature sporophyte.

Seedling

Pollen sac, where one of many microsporocytes will give rise to microspores

Ovules inside ovary

Megasporocyte that will give rise to a megaspore

7 Double fertilization produces a binucleate cell that will give rise to triploid endosperm, a nutritive tissue that will nourish the developing embryo.

Seed coat
Embryo (2*n*)   Seed
Endosperm (3*n*)

DOUBLE FERTILIZATION

DIPLOID (2*n*)

MEIOSIS

Meiosis produces microspores that develop into pollen grains.

Sexual Reproduction

HAPLOID (*n*)

Pollen tube
Sperm (*n*)
Sperm (*n*)

Male gametophyte

Pollen grain

Pollen tube

Pollen sac

2 Meiosis produces four megaspores, then all but one disintegrate. It undergoes three rounds of mitosis without cytokinesis, forming a single large cell with eight nuclei.

6 The pollen tube enters an ovule. One sperm will fertilize the egg, and one will fertilize the endosperm-producing cell (double fertilization).

(line of cut of diagram at left)

Endosperm-producing cell

Egg cell

Ovary

5 Pollination occurs. Then a pollen grain develops into a pollen tube that grows toward the ovary. The tube contains two sperm. It is the mature male gametophyte.

4 Pollen is released.

3 Cytokinesis produces a female gametophyte that consists of seven cells. One cell becomes the egg. Another, large cell (with two nuclei) forms a nutritive tissue called endosperm. The remaining cells also have brief reproductive roles but they eventually disintegrate.

Figure 27.31
Life cycle of a flowering plant, the monocot *Lilium*. Double fertilization is a notable feature of the cycle. The male gametophyte delivers two sperm to an ovule. One sperm fertilizes the egg, forming the embryo, and the other fertilizes the endosperm-producing cell, which nourishes the embryo.

withdraw pollen from male floral structures (often while obtaining nectar) and inadvertently transfer it to female reproductive parts. **Coevolution** occurs when two or more species interact closely in the same ecological setting. A heritable change in one species affects selection pressure operating between them, and the other species evolves as well. Over time, plants that came to have distinctive flowers, scents, and sugary nectar coevolved with animals that could take advantage of the rich food source.

In general, a flower's reproductive parts are positioned so that visiting pollinators will brush against them. In addition, many floral features correlate with specific pollinators. For example, reproductive parts may be located above nectar-filled floral tubes the same length as the feeding structure of a preferred pollinator. Nectar-sipping bats **(Figure 27.32a)** and moths forage by night. They pollinate intensely sweet-smelling flowers with white or pale petals that are more visible than colored petals in the dark. Long, thin mouthparts of moths and butterflies reach nectar in narrow floral tubes or flora spurs. The Madagascar hawkmoth uncoils a mouthpart the same length—an astonishing 22 cm—as a narrow floral spur of an orchid it pollinates, *Angraecum sesquipedale* **(Figure 27.32b).** Red and yellow flowers attract birds **(Figure 27.32c),** which have good daytime vision but a poor sense of smell. Hence bird-pollinated plants do not squander metabolic resources to make fragrances. By contrast, flowers of species that are pollinated by beetles or flies may smell like rotten meat, dung, or decaying matter. Daisies and other fragrant flowers with distinctive patterns, shapes, and red or orange components attract butterflies, which forage by day.

Bees see ultraviolet light and visit flowers with sweet odors and parts that appear to humans as yellow, blue, or purple **(Figure 27.32d).** Produced by pigments that absorb ultraviolet light, the colors form patterns called "nectar guides" that attract bees—which may pick up or "drop off" pollen during the visit. Here, as in our other examples, flowers contribute to the reproductive success of plants that bear them.

a. Bat pollinating a giant saguaro

b. Hawk moth pollinating an orchid

c. Hummingbird visiting a hibiscus flower

d. Bee-attracting pattern of a marsh marigold

Visible light   UV light

Figure 27.32

Coevolution of flowering plants and animal pollinators. The colors and configurations of some flowers, and the production of nectar or odors, have coevolved with specific animal pollinators. **(a)** At night, nectar-feeding bats sip nectar from flowers of the giant saguaro (*Carnegia gigantea*), transferring pollen from flower to flower in the process. **(b)** The hawkmoth *Xanthopan morgani praedicta* has a proboscis long enough to reach nectar at the base of the equally long floral spur of the orchid *Angraecum sesquipedale*. **(c)** A Bahama woodstar hummingbird (*Calliphlox evelynae*) sipping nectar from a hibiscus blossom *(Hibiscus)*. The long narrow bill of hummingbirds coevolved with long, narrow floral tubes. **(d)** Under ultraviolet light, the bee-attracting pattern of a gold-petaled marsh marigold becomes visible to human eyes.

Current Research Focuses on Genes Underlying Transitions in Plant Traits

Improvements in the ability of plant scientists to manipulate, analyze, and compare modern plant genomes, coupled with advances in the analysis of fossil plants, are having a profound impact on our understanding of the evolution of flowering plants. A case in point is research on the gene *LFY*, which encodes the regulatory protein LEAFY (Chapter 16 discusses regulatory proteins in detail). The LEAFY protein typically controls expression of several genes by binding to the genes' control sequences. All land plants carry the *LFY* gene, but its effects on phenotype vary markedly in different plant groups. In mosses, which arose almost 400 million years ago, the LEAFY protein regulates growth throughout the plant. In ferns and gymnosperms, which arose later, LEAFY controls growth in a subset of tissues. In angiosperms, LEAFY regulates gene expression only in the particular type of meristem

UNANSWERED QUESTIONS

Where did flowering plants come from?

Flowers are a unique feature of the angiosperms, yet botanists still understand little of their evolutionary origin. When flowering plants appear in the Cretaceous fossil record, they appear suddenly and diversify immediately, a situation Darwin famously referred to as an "abominable mystery." What did the first angiosperms and the first flower look like? And where did they arise?

As described in this chapter, recent molecular analyses have converged on *Amborella trichopoda* as the living representative of the most ancient lineage in the angiosperm family tree. This research has shed light on many questions. For example, *Amborella* flowers have some features considered evolutionarily primitive, such as petals and sepals that are not distinctly different in form. This observation supports the hypothesis that two other types of flower parts, the calyx and corolla, arose later in angiosperm evolution. But *Amborella* also has some features thought to have evolved much more recently, such as single-sex flowers that have either male or female reproductive parts (but never both). Should we be surprised to find both primitive and advanced traits in this ancient lineage? Not at all. *Amborella* has existed on Earth for millions of years and its flowers may have evolved new features over that time.

The puzzle of where angiosperms came from and what the first flowering plants looked like has not been solved by fossil studies, either. This chapter discusses the fossil species *Archaefructus*, which dates from the Jurassic and is thought to be the oldest known fossil flower. It consists of an elongated axis with what its discoverers described as stamens (male reproductive structures) toward the base and carpels (female reproductive structures) toward the apex, and no sepals and petals. This elongated flower is unlike the flowers of any modern angiosperm, and its structure suggests that the earliest flowers may have been very different from what we see today. However, some paleobotanists have reinterpreted the *Archefructus* "flower" as an inflorescence (a flower cluster), with male flowers at the base and female flowers toward the apex. In addition, radiometric dating places *Archaefructus* in the early-mid Cretaceous, a period from which other early angiosperm fossils are known. Thus, *Archaefructus* may not be the oldest flower, and the fossil specimen may represent a cluster of flowers instead of a single flower. This debate continues.

Botanists also disagree about the ancestors of angiosperms. Some gnetophytes—gymnosperms that include *Welwitschia* and *Ephedra* species (refer to Figure 27.23)—have features similar to angiosperms. Botanists long speculated that the two groups were closely related, with a common ancestor that had flowerlike features. However, recent analyses based on DNA sequence data suggest that gnetophytes are not closely related to angiosperms after all. There are also fossil gymnosperm taxa with features that might be forerunners of carpels or other flower parts, but paleobotanists disagree on these interpretations as well. Thus examinations of fossils and extant species have yet to resolve key questions about the evolution of angiosperms.

What, then, can molecular data tell us? Studies of the genetic mechanisms that guide the development of flower parts have provided a framework for understanding how genes control flower formation (a topic of Chapter 34). This research has also given us insight into what kinds of molecular changes may have led to the evolution of flowers. For instance, certain genes that encode transcription factors required for the formation of reproductive organs in flowers are found also in gymnosperms. This finding is not surprising, because gymnosperms also form male and female reproductive structures; the most logical hypothesis is that angiosperms retained the gymnosperm developmental program for these organs. Yet genes for other transcription factors active in flower formation are *not* found in gymnosperms. We know that transcription factors may turn on and off entire developmental pathways, such as those that cause undifferentiated tissue (called meristem tissue) to form a flower. One hypothesis is that in an ancient gymnosperm ancestor, duplications in a particular gene family gave rise to genes that in turn accumulated mutations allowing them to perform new functions that resulted in the formation of the first flowers.

As much insight as these molecular studies give us into events that might have resulted in the evolution of flowers, they have not brought us any closer to understanding the fundamental question of where angiosperms arose. Additional fossil data may help provide the answer, but it is also possible that the earliest angiosperms, or their direct ancestors, lived in habitats where fossils do not readily form. Additional molecular data may deepen our understanding of how changes in genes produced the first flower. But molecular data based on contemporary species will not help decipher what the first angiosperm and the first flower looked like. Thus, it is possible that the abominable mystery will live on.

Amy Litt is Director of Plant Genomics and Cullman Curator at the New York Botanical Garden, where she also earned her Ph.D. Her main interests lie in the evolution of plant form and how changes in gene function during the course of plant evolution have produced novel plant forms and functions—particularly new flower and fruit morphologies. Learn more about her work at http://sciweb.nybg.org/science2/Profile_106.asp.

tissue that gives rise to flowers (a topic of Chapter 34). Curious about the evolutionary shift from a general to a specific effect, Alexis Maizel and his team at the Max Planck Institute for Developmental Biology in Germany compared *LFY* sequences and their corresponding proteins in fourteen species, including a moss, ferns, gymnosperms, and the angiosperms *Arabidopsis* (thale cress) and snapdragon. Remarkably, they discovered that the evolutionary honing of the effects of the LEAFY protein correlated with only a handful of changes in the base sequence of the *LFY* gene. Each change affected how—or if—the LEAFY protein regulated the expression of a given gene. Over time, LEAFY took on its highly specific, crucial role in angiosperms, helping to direct the developmental events that produce flowers.

Today some of the most exciting research in all of biology involves studies exploring the connections between genetic changes and key evolutionary transitions in plant form and functioning. As the genes of many more plant species are sequenced and correlated with evidence from comparative morphology and the fossil record, we can expect a steady stream of new insights about the evolutionary journey of all major plant lineages. In Chapter 28 a very different group of organisms, the fungi, takes center stage. Although many fungal species seem superficially plantlike, biologists today are avidly exploring evolutionary links between fungi and animals.

STUDY BREAK

1. How has the relative lack of fossil early angiosperms affected our understanding of this group?
2. Describe two basic features that distinguish monocots from eudicots, and give some examples of species in each clade.
3. List at least three adaptations that have contributed to the evolutionary success of angiosperms as a group.

Review

Go to ThomsonNOW™ at www.thomsonedu.com/login to access quizzing, animations, exercises, articles, and personalized homework help.

27.1 The Transition to Life on Land

- Plants are thought to have evolved from charophyte green algae between 425 and 490 million years ago (Figure 27.2).
- Adaptations to terrestrial life in early land plants include an outer cuticle that helps prevent desiccation, lignified tissues, spores protected by a wall containing sporopollenin, multicellular chambers that protect developing gametes, and an embryo sheltered inside a parent plant (Figure 27.3).
- Other key evolutionary trends among land plants included the development of vascular tissues, root systems, and shoot systems, including lignified stems and leaves equipped with stomata; a shift from dominance by a long-lived, larger haploid gametophyte to dominance of a long-lived, larger diploid sporophyte, and a shift from homospory to heterospory with separate male and female gametophytes (Figures 27.5–27.7).
- Male gametophytes (pollen) became specialized for dispersal without liquid water, and female gametophytes became specialized for enclosing embryo sporophytes in seeds.

Animation: Milestones in plant evolution

Animation: Haploid to diploid dominance

Animation: Evolutionary tree for plants

Animation: The importance of alternation of generations

27.2 Bryophytes: Nonvascular Land Plants

- Existing nonvascular land plants, or bryophytes, include the liverworts (Hepatophyta), hornworts (Anthocerophyta), and mosses (Bryophyta). Liverworts may have been the first land plants.

- Bryophytes produce flagellated sperm that swim through free water to reach eggs. They lack a vascular system; true roots, stems, and leaves; and lignified tissue. A larger, dominant gametophyte (*n*) phase alternates with a small, fleeting sporophyte (*2n*) phase. Spores develop inside jacketed sporangia (Figures 27.9 and 27.12).

Animation: Moss life cycle

Animation: *Marchantia*, a liverwort

27.3 Seedless Vascular Plants

- Existing seedless vascular land plants include the lycophytes (club mosses), whisk ferns, horsetails, and ferns. Like bryophytes, they release spores and have swimming sperm. Unlike bryophytes, they have well-developed vascular tissues. The sporophyte is the larger, longer-lived stage of the life cycle and develops independently of the small gametophyte.
- Club mosses (Lycophyta) have sporangia clustered at the bases of specialized leaves called sporophylls. Each sporophylls cluster forms a strobilus (cone). Haploid spores dispersed from the sporangia germinate to form small, free-living gametophytes. Ferns, whisk ferns, and horsetails (Pterophyta) have a similar life cycle. Horsetail sporophytes typically have underground stems (rhizomes) anchored to the soil by roots.
- Ferns are the largest and most diverse group of seedless vascular plants. Most species do not have aboveground stems, only leaves that arise from nodes along an underground rhizome. Fern leaves typically have well-developed stomata, and the vascular system consists of bundles, each with xylem surrounded by phloem. Sporangia on the lower surface of sporophylls (fronds) release spores that develop into gametophytes. Sexual reproduction produces a much larger, long-lived sporophyte (Figure 27.16).

Animation: Seedless vascular plants

Animation: Fern life cycle

27.4 Gymnosperms: The First Seed Plants

- Gymnosperms (conifers and their relatives), together with angiosperms (flowering plants), are the seed-bearing vascular plants. Reproductive innovations include pollination, the ovule, and the seed. An ovule is a sporangium containing a female gametophyte, so the female gametophyte is attached to and protected by the sporophyte. The smaller spore type produces a male gametophyte. Since pollination takes place via air currents or animal pollinators, plants fertilized by pollination do not require liquid water to reproduce. A seed forms when an ovule matures following fertilization; in gymnosperms, its main function is to protect and help disperse the embryonic sporophyte (Figure 27.24).

- During the Mesozoic, gymnosperms were the dominant land plants. Today conifers are the primary vegetation of forests at higher latitudes and elevations and have important economic uses as sources of lumber, resins, and other products.

Animation: *Pinus* cones

Animation: Pine life cycle

27.5 Angiosperms: Flowering Plants

- Angiosperms (Anthophyta) have dominated the land for more than 100 million years and currently are the most diverse plant group. There are two main angiosperm clades: monocots and eudicots. Other clades are represented by magnolias and their relatives (magnoliids), water lilies, the star anise group, and *Amborella*, a single species thought to be the most basal living angiosperm (Figures 27.29 and 27.30).

- The angiosperm vascular system moves water from roots to shoots more efficiently than in gymnosperms, and the phloem tissue moves sugars more efficiently through the plant body. Reproductive adaptations include a protective ovary around the ovule, endosperm, fruits that aid seed dispersal, the complex organs called flowers, and the coevolution of flower characteristics with the structural and/or physiological characteristics of animal pollinators (Figures 27.31 and 27.32).

Animation: Flower parts

Animation: Monocot life cycle

Questions

Self-Test Questions

1. Which of the following is *not* an evolutionary trend among plants?
   a. developing vascular tissues
   b. becoming seedless
   c. having a dominant diploid generation
   d. producing nonmotile gametes
   e. producing two types of spores

2. As plants made the evolutionary transition to a terrestrial existence, they benefited from adaptations that:
   a. increased the motility of their gametes on dry land.
   b. flattened the plant body to expose it to the sun.
   c. reduced the number and distribution of roots to prevent drying.
   d. provided mechanisms for gaining access to nutrients in soil.
   e. allowed stems and leaves to absorb water from the atmosphere.

3. Land plants no longer required water as a medium for reproduction with the evolution of:
   a. fruits and roots.
   b. flowers and leaves.
   c. cell walls and rhizoids.
   d. lignified stems.
   e. seeds and pollen.

4. Which is the correct matching of phylum and plant group?
   a. Anthophyta: pines
   b. Bryophyta: gnetophytes
   c. Coniferophyta: angiosperms
   d. Hepatophyta: cycads
   e. Pterophyta: horsetails

5. A homeowner noticed moss growing between bricks on his patio. Closer examination revealed tiny brown stalks with cuplike tops emerging from green leaflets. These brown structures were:
   a. the sporophyte generation.
   b. the gametophyte generation.
   c. elongated haploid reproductive cells.
   d. archegonia.
   e. antheridia.

6. Horsetails are most closely related to:
   a. mosses and whisk ferns.
   b. liverworts and hornworts.
   c. cycads and ginkgos.
   d. club mosses and ferns.
   e. gnetophytes and gymnosperms.

7. Which feature(s) do ferns share with all other land plants?
   a. sporophyte and gametophyte life cycle stages
   b. gametophytes supported by a thallus
   c. dispersal of spores from a sorus
   d. asexual reproduction by way of gemmae
   e. water uptake by means of rhizoids

8. The evolution of true roots is first seen in:
   a. liverworts.
   b. seedless vascular plants.
   c. mosses.
   d. flowering plants.
   e. conifers.

9. Based solely on numbers of species, the most successful plants today are:
   a. angiosperms.
   b. ferns.
   c. gymnosperms.
   d. mosses.
   e. the bryophytes as a group.

10. Angiosperms and gymnosperms share the following characteristic(s):
    a. pollination by means of water.
    b. seeds protected within an ovary.
    c. embryonic cotyledons.
    d. a dominant sporophyte generation.
    e. a seasonal loss of all leaves.

Questions for Discussion

1. Suggest adjustments in the angiosperm life cycle that would better suit plants to some future world where environments were generally hotter and more arid. Do the same for a colder and wetter environment.

2. Working in the field, you discover a fossil of a previously undescribed plant species. The specimen is small and may not be complete; the parts you have do not include any floral organs. What sorts of observations would you need in order to classify the fossil as a seedless vascular plant with reasonable accuracy? What evidence would you need in order to distinguish between a fossil lycopod and a fern?

3. Modern humans emerged about 100,000 years ago. How accurate is it to state that our species has lived in the Age of Wood? Explain.

4. Compare the size, anatomical complexity, and degree of independence of a moss gametophyte, a fern gametophyte, a Douglas fir female gametophyte, and a dogwood female gametophyte. Which one is the most protected from the external environment? Which trends in plant evolution does your work on this question bring to mind?

Experimental Analysis

You are studying mechanisms that control the development of flowers, and your research to date has focused on eudicots, which tend to have showier flowers than monocots. A colleague has suggested that you broaden your analysis to include representative basal angiosperms. Outline the rationale for this expanded approach and indicate which additional species or group(s) you plan to include. Discuss the type(s) of data you plan to gather and why you feel the information will make your study more complete.

Evolution Link

Plant evolutionary biologist Spencer C. H. Barrett has written that the reproductive organs of angiosperms are more varied than the equivalent structures of any other group of organisms. Which angiosperm organs was Barrett talking about? Explain why you agree or disagree with his view.

How Would You Vote?

Demand for paper is a big factor in deforestation. However, using recycled paper can add to the cost of a product. Are you willing to pay more for papers, books, and magazines that are printed on recycled paper? Go to www.thomsonedu.com/login to investigate both sides of the issue and then vote.

The mushroom-forming fungus *Inocybe fastigiata*, a forest-dwelling species that commonly lives in close association with conifers and hardwood trees.

Fritz Polking/Peter Arnold, Inc.

28 Fungi

WHY IT MATTERS

In a forest, decay is everywhere—rotting leaves, moldering branches, perhaps the disintegrating carcass of an insect or a small mammal. Each year in most terrestrial ecosystems, an astounding amount of organic matter is produced, cast off, broken down, and its elements gradually recycled. This recycling has a huge impact on world ecosystems; for example, each year it returns at least 85 billion tons of carbon, in the form of carbon dioxide, to the atmosphere. Chief among the recyclers are the curious organisms of the **Kingdom Fungi**—about 60,000 described species of molds, mushroom-forming fungi, yeasts, and their relatives **(Figure 28.1),** and an estimated 1.6 million more that are yet to be described.

Fungi are eukaryotes, most are multicellular, and all are heterotrophs, obtaining their nutrients by breaking down organic molecules that other organisms have synthesized. Molecular evidence suggests that fungi were present on land at least 500 million years ago, and possibly much earlier. In the course of the intervening millennia, evolution equipped fungi with a remarkable ability to break down organic matter, ranging from living and dead organisms and animal wastes to your groceries, clothing, paper and wood, even photographic

605

Sulfur shelf fungus, *Polyporus*

Big laughing mushroom, *Gymnopilus*

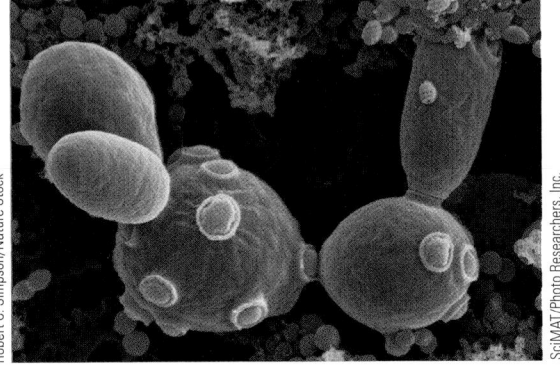

Baker's yeast cells, *Saccharomyces cerevisiae*

Figure 28.1
Examples of fungi that hint at the rich diversity within the Kingdom Fungi.

film. Along with heterotrophic bacteria, they have become Earth's premier decomposers.

Fungi collectively also are the single greatest cause of plant diseases, and a host of species cause disease in humans and other animals. Some even produce carcinogenic toxins. On the other hand, 90% of plants obtain needed minerals by way of a symbiotic relationship with a fungus. Humans have harnessed the metabolic activities of certain fungi to obtain substances ranging from flavorful cheeses and wine to therapeutic drugs such as penicillin and the immunosuppressant cyclosporin. And, as you know from previous chapters, species such as the yeast *Saccharomyces cerevisiae* and the mold *Neurospora crassa* have long been pivotal model organisms in studies of DNA structure and function, and the yeast has also been important in the development of genetic engineering methods.

Despite their profound impact on ecosystems and other life forms, most of us have only a passing acquaintance with the fungi—perhaps limited to the mushrooms on our pizza or the invisible but annoying types that cause skin infections like athlete's foot. This chapter provides you with an introduction to mycology, the study of fungi (*mykes* = mushroom; *logos* = knowledge). We begin with general characteristics of this kingdom and then discuss its major divisions.

28.1 General Characteristics of Fungi

We begin our survey of fungi by examining how fungi differ from other forms of life, how fungi obtain nutrients, and the adaptations for reproduction and growth that enable fungi to spread far and wide through the environment.

Fungi May Be Single-Celled or Multicellular

Two basic body forms, single-celled and multicellular, emerged as the lineages of fungi evolved. Some fungi are single cells, a form called **yeast**, while others exist in a multicellular form made up of threadlike filaments. Still others alternate between yeast and multicellular forms at different stages of the life cycle. Whether a fungus is single-celled or multicellular, a rigid wall usually surrounds the plasma membrane of its cells. Generally the polysaccharide **chitin** provides this rigidity, the same function it serves in the external skeletons of insects and other arthropods.

In a multicellular fungus, exploiting food sources is the province of a cottony mesh of tiny filaments that branch repeatedly as they grow over or into organic matter. Each filament is a **hypha** (*hyphe* = web; plural, hyphae); the combined mass of hyphae is a **mycelium** (plural, mycelia). Hyphae generally are tube-shaped **(Figure 28.2).** In most multicellular fungi the hyphae are partitioned by cross walls called **septa** (*saeptum* = partition; singular, septum). The septa create cell-like compartments that contain organelles. However, in one group, the zygomycetes described shortly, most hyphae are *aseptate*—they lack cross walls—although septa do arise to separate reproductive structures from the rest of the hypha.

The unusual features of fungal hyphae have led many mycologists to question whether "multicellular" is really an accurate description for most fungal architecture. For instance, depending on the species, hyphal cells may have more than one nucleus, and septa have pores that permit nuclei and other organelles to move between hyphal cells. These passages also allow cytoplasm to extend from one hyphal cell to the next, throughout the whole mycelium. By a mechanism called **cytoplasmic streaming**, cytoplasm containing nutrients can flow unimpeded through the hyphae, from food-absorbing parts of the fungal body to other, nonabsorptive parts such as reproductive structures.

A multicellular fungus grows larger as its hyphae elongate and branch. Each hypha elongates at its tip as new wall polymers (delivered by vesicles) are incorporated and additional cytoplasm, including organelles, is synthesized. A hypha branches a few micrometers behind its tip, and as the new hyphae elongate, then branch themselves, an extensive mycelium can form quickly. Each forming branch fills with cytoplasm that includes new nuclei produced by mitosis. Although the rapid branching of hyphae is what allows multicellular fungi to grow aggressively—sometimes increasing in mass many times over within a few days—researchers have only recently gained the tools to explore the mechanisms that underlie this phenomenon. Studies spurred by the sequencing of the genome of *Neurospora crassa* suggest that multiple steps involving a variety of genes and their interacting protein products determine where and when a new branch arises. Given that the rapid growth of fungal mycelia has such a tremendous impact in nature, fungal diseases, and many other areas, this topic is a central focus of much mycological research.

Beyond their role in nutrient transport, aggregations of hyphae are the structural foundation for all other parts that arise as a multicellular fungus develops. For example, in many fungi a subset of hyphae interweave tightly, becoming prominent reproductive structures (sometimes called fruiting bodies). Grocery store mushrooms are examples. But while a mushroom or some analogous structure may be the most conspicuous part of a given fungus, it usually represents only a small fraction of the organism's total mass. The rest penetrates the food source the fungus is slowly digesting. In some fungi, modified hyphae called **rhizoids** anchor the fungus to its substrate. Most fungi

a. Multicellular fungus

— Mycelium —

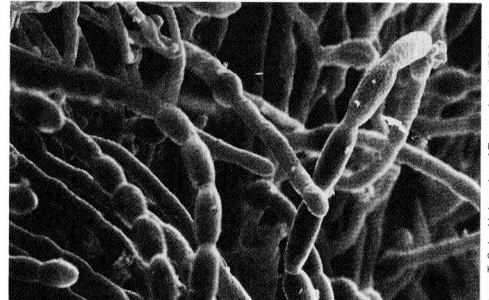

b. Fungal hyphae

Garry T. Cole, University of Texas, Austin/BPS

Figure 28.2
Fungal mycelia. **(a)** Sketch of the mycelium of a mushroom-forming fungus, which consists of branching septate hyphae. **(b)** Micrograph of fungal hyphae.

that parasitize living plants produce hyphal branches called **haustoria** (*haustor* = drinker) that penetrate the walls of a host plant's cells and channel nutrients back to the fungal body.

Fungi Obtain Nutrients by Extracellular Digestion and Absorption

Some major challenges have shaped the adaptations by which fungi obtain nutrients. As heterotrophs, fungi must secure nutrients by breaking down organic substances formed by other organisms. Nearly all fungi are terrestrial, but unlike other land-dwelling heterotrophs (such as animals), fungi are not mobile. They also lack mouths or appendages for seizing, handling, and dismantling food items. Instead, fungi have a very different suite of adaptations for obtaining nutrients.

To begin with, most species of fungi can synthesize nearly all their required nutrients from a few raw materials, including water, some minerals and vitamins (especially B vitamins), and a sugar or some other organic carbon source. For many species, carbohydrates in dead organic matter are the carbon sources, and fungi with this mode of nutrition are called **saprobes** (*sapros* = rotten). Other fungi are parasites, which extract carbohydrates from tissues of a living host, harming it in the process. Parasitic fungi include those responsible for many devastating plant diseases, such as wheat rust and Dutch elm disease. Still other fungi are nourished by plants with which they have a mutually beneficial symbiotic association.

Regardless of their nutritional mode, all fungi gain the raw materials required to build and maintain their cells by absorption from the environment. Fungi can absorb many small molecules directly from their surroundings, and gain access to other nutrients through extracellular digestion. In this process, a fungus releases enzymes that digest nearby organic matter, breaking down larger molecules into absorbable fragments. Fungal species differ in the particular digestive enzymes they synthesize, so a substrate that is a suitable food source for one species may be unavailable to another. Although there are exceptions, fungi typically thrive only in moist environments where they can directly absorb water, dissolved ions, simple sugars, amino acids, and other small molecules. When some of a mycelium's hyphal filaments contact a source of food, growth is channeled in the direction of the food source. Nutrients are absorbed only at the porous tips of hyphae; small atoms and molecules pass readily through these tips, and then transport mechanisms move them through the underlying plasma membrane.

Large organic molecules, such as the carbohydrate cellulose (see Section 3.3), *cannot* directly enter any part of a fungus. To use such substances as a food source, a fungus must secrete hydrolytic enzymes that break down the large molecules into smaller, absorbable subunits. Depending on the size of the subunit, further digestion may occur inside cells.

With their adaptations for efficient extracellular digestion, fungi are masters of the decay so vital to terrestrial ecosystems. For instance, in a single autumn one elm tree can shed 400 pounds of withered leaves; and in a tropical forest, a year's worth of debris may total 60 tons per acre. Without the metabolic activities of saprobic fungi and other decomposers such as bacteria, natural communities would rapidly become buried in their own detritus. As fungi digest dead tissues of other organisms, they also make a major contribution to the recycling of chemical elements those tissues contain. For instance, over time the degradation of organic compounds by saprobic fungi helps return key nutrients such as nitrogen and phosphorus to ecosystems. But the prime example of this recycling virtuosity involves carbon. The respiring cells of fungi and other decomposers give off carbon dioxide, liberating carbon that would otherwise remain locked in the tissues of dead organisms. Each year this activity recycles a vast amount of carbon to plants, the primary producers of nearly all ecosystems on Earth.

All Fungi Reproduce by Way of Spores, but Other Aspects of Reproduction Vary

Biologists have observed a striking number of reproductive variations in fungi, differences that are part of what makes them fascinating to study. As you will learn in the next section, fungi have traditionally been classified on the basis of their reproductive characteristics, although today evidence from molecular analysis also plays a prominent role.

Overall, most fungi have the capacity to reproduce both sexually and asexually. Although no single diagram can depict all the variations, **Figure 28.3** gives an overview of the life cycle stages that mycologists have observed in several groups of fungi. The figure illustrates two general points. First, the life cycle of multicellular fungi typically involves a diploid stage (2n), a haploid stage (n), and a dikaryotic ("two nuclei") stage in which the fungus forms hyphae (and a mycelium) that are n + n—neither strictly haploid *nor* diploid. Depending on the type of fungus, this stage may be long lasting or extremely brief, and it is described more fully later in this section. Second, all fungi, whether they are multicellular or in a single-celled, yeast form, can reproduce via **fungal spores.** The spores are microscopic, and in all but one group they are not motile—that is, they are not propelled by flagella. Each spore is a walled single cell or multicellular structure that is dispersed from the parent body, often via wind or water. The spores of single-celled fungi form inside the parent cell, then escape when the wall breaks open. In multicellular fungi, spores arise in or on specialized hyphal structures and may develop thick walls that help them withstand cold or drying out after they are released.

Reproduction by way of spores is one of the crucial fungal adaptations. Most fungi are opportunists, obtaining energy by exploiting food sources that occur unpredictably in the environment. Having lightweight spores that are easily disseminated by air or water increases opportunities for finding food. And releasing vast numbers of spores, as some fungi do, improves the odds that at least a few spores will germinate and produce a new individual.

In nature generally, opportunistic organisms are adapted to reach new food sources quickly and utilize them rapidly. Fungi that are adept at degrading simple sugars and starches often are among the first decomposers to exploit a new source of food. They meet with keen competition from each other and from other decomposers. However, once fungal spores encounter potential food and favorable conditions, they can quickly develop into new individuals that simultaneously feed and rapidly make more spores.

Many opportunistic fungi develop rapidly, growing and reproducing before the food source is depleted. A common trade-off for speed, however, is small, even microscopic body size. Larger species of fungi are often adapted to move in later, exploiting food sources such as cellulose and lignin (a complex polymer in the walls of many plant cells), which their predecessors may have lacked the enzymatic machinery to digest efficiently. Some of these fungi may produce huge mycelia (and reproductive "fruiting bodies" such as mushrooms) by extracting nutrients from dead trees that

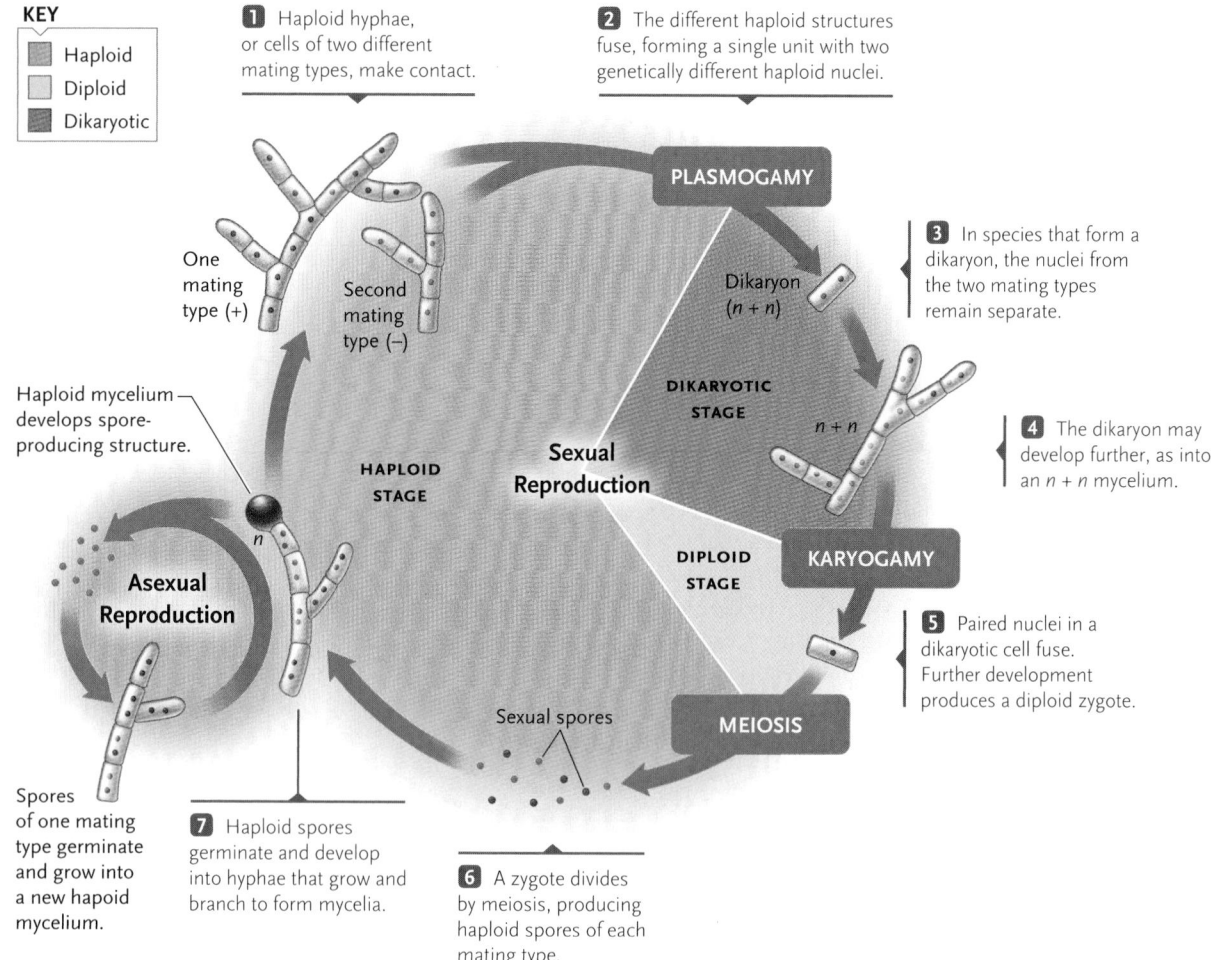

KEY

▦	Haploid
▦	Diploid
▦	Dikaryotic

1 Haploid hyphae, or cells of two different mating types, make contact.

2 The different haploid structures fuse, forming a single unit with two genetically different haploid nuclei.

One mating type (+)

Second mating type (−)

PLASMOGAMY

3 In species that form a dikaryon, the nuclei from the two mating types remain separate.

Dikaryon (*n* + *n*)

DIKARYOTIC STAGE

Haploid mycelium develops spore-producing structure.

HAPLOID STAGE

Sexual Reproduction

n + *n*

4 The dikaryon may develop further, as into an *n* + *n* mycelium.

n

DIPLOID STAGE

KARYOGAMY

Asexual Reproduction

5 Paired nuclei in a dikaryotic cell fuse. Further development produces a diploid zygote.

Sexual spores

MEIOSIS

Spores of one mating type germinate and grow into a new haploid mycelium.

7 Haploid spores germinate and develop into hyphae that grow and branch to form mycelia.

6 A zygote divides by meiosis, producing haploid spores of each mating type.

Figure 28.3

Generalized life cycle for many fungi. Overall, fungi are diploid for only a short time. The duration of the dikaryon stage varies considerably, being lengthy for some species and extremely brief in others. Some types of fungi reproduce only asexually while in others shifts in environmental factors, such as the availability of key nutrients, can trigger a shift from asexual to sexual reproduction or vice versa. For still others sexual reproduction is the norm.

contain enough organic material to sustain an extended period of growth.

Features of Asexual Reproduction in Fungi. When a fungus produces spores asexually (see Figure 28.3), it may disperse billions of them into the environment. Some fungi (including many yeasts) also can reproduce asexually by budding or fission, or, in multicellular types, when fragments of hyphae break away from the mycelium and grow into separate individuals. In still others, environmental factors may determine whether the fungus produces hyphal fragments *or* asexual spores. These asexual reproductive strategies all result in new individuals that are essentially clones of the parent fungus. They can be viewed as another adaptation for speed, because the alternative—sexual reproduction—requires the presence of a suitable partner and generally involves several more steps.

The asexual stage of many multicellular fungi—including the pale gray fuzz you might see on berries or bread—is often called a **mold**. The term can be con-

fusing if you are attempting to keep track of taxonomic groupings; for example, the water molds and slime molds described in Chapter 26 are protists, although they were grouped with fungi until additional research revealed their true evolutionary standing. The mold visible on an overripe raspberry is actually a mycelium with aerial structures bearing sacs of haploid spores at their tips.

Features of Sexual Reproduction in Fungi. Although asexual reproduction is the norm, quite a few fungi shift to sexual reproduction when environmental conditions (such as a lack of nitrogen) or other influences dictate. As you may remember from Chapter 11, in sexual reproduction two haploid cells unite, and in most species fertilization—the fusion of two gamete nuclei to form a diploid zygote nucleus—soon follows. In fungi, however, the partners in sexual union can be two hyphae, two gametes, or other types of cells; the particular combination depends on the species involved. And in sharp contrast to other life forms, many days, months, or even years may pass between the time fertilization gets underway and when it is completed.

During the initial sexual stage, called **plasmogamy** (*plasma* = a formed thing; *gamos* = union), the cytoplasms of two genetically different partners fuse. The resulting new cell, a **dikaryon** (*di* = two; *karyon* = nucleus), contains two haploid nuclei, one from each parent. A dikaryon itself is not haploid (the condition of having one set of chromosomes) because it contains two nuclei. But neither is it diploid, because the nuclei are not fused. So, to be precise, we say that a dikaryon has an *n* + *n* nuclear condition.

Plasmogamy can occur when hyphal cells of two different **mating type**, termed plus (+) and minus (−), fuse, a process that occurs in most fungi. The uniting hyphae belong to mycelia of different individuals of the same species that happen to grow near one another. The fusion of different mating types ensures genetic diversity in new individuals.

Once a dikaryon forms, the amount of time that elapses before the next stage begins depends on the type of fungus, as described in the next section. Sooner or later, however, a second phase of fertilization unfolds: The nuclei in the dikaryotic cell fuse to make a 2*n* zygote nucleus. This process is called **karyogamy** ("nuclear union"); in fungi that form mushrooms, it occurs in the tips of hyphae that end in the gills, which you may be able to see if you look closely at the underside of a mushroom cap (see Figure 28.1). In animals, a zygote is the first cell of a new individual, but in the world of fungi the zygote has a different fate. After it forms, meiosis converts the zygote nucleus into four haploid (*n*) nuclei. Those nuclei are packaged into haploid "sexual spores," which vary genetically from each parent. Then the spores are released to spread throughout the environment.

To sum up, in fungi both asexual and sexual spores are haploid, and both can germinate into haploid individuals. However, asexual spores are genetically identical products of asexual reproduction, while sexual spores are genetically varied products of sexual reproduction. We turn now to current ideas on the evolutionary history of fungi, and a survey of the major taxonomic groups in this kingdom.

STUDY BREAK

1. What features distinguish the two basic fungal body forms?
2. What is a fungal spore, and how does it function in reproduction?
3. Fungi reproduce sexually or asexually, but for many species the life cycle includes an unusual stage not seen in other organisms. What is this genetic condition, and what is its role in the life cycle?

28.2 Major Groups of Fungi

The evolutionary origins and lineages of fungi have been obscure ever since the first mycologists began puzzling over the characteristics of this group. With the advent of molecular techniques for research, these topics have become extremely active and exciting areas of biological research that may shed light on fundamental events in the evolution of all eukaryotes. Not surprisingly when so much new information is coming to light, mycologists hold a wide range of views on how different groups arose and may be related. Even so, there is wide agreement on five phyla of fungi, known formally as the Chytridiomycota, Zygomycota, Glomeromycota, Ascomycota, and Basidiomycota **(Figure 28.4)**.

In a sixth group, termed conidial fungi, asexual reproduction produces spores called conidia. "Conidial" is not a true taxonomic classification, however. Rather, it serves as a holding station for fungal species that have not yet been assigned to one of the five phyla because no sexual reproductive phase has been observed. This is another instance in which the name for a fungal group can be confusing, because numerous species belonging to the Zygomycota, Ascomycota, and Basidiomycota also form conidia as part of their asexual reproductive cycle.

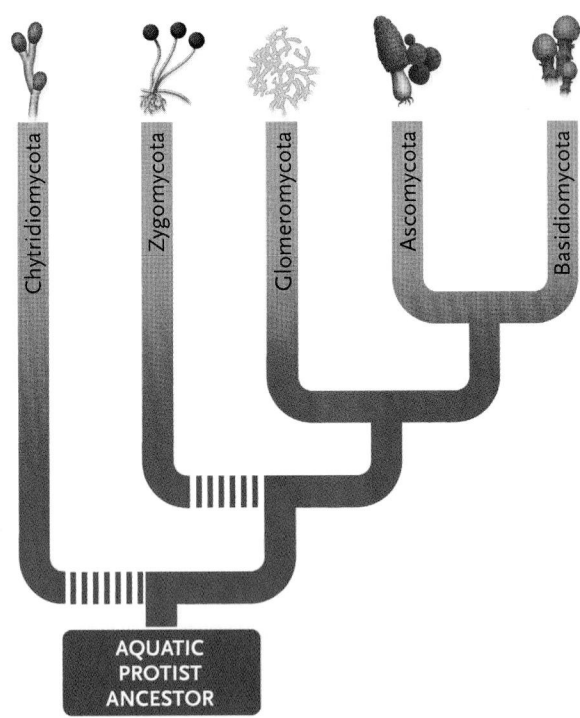

Figure 28.4

A phylogeny of fungi. This scheme represents a widely accepted view of the general relationships between major groups of fungi, but it may well be revised as new molecular findings provide more information. The dashed lines indicate that two groups, the chytrids and zygomycetes, are probably paraphyletic—they include subgroups that are not all descended from a single ancestor.

There Was Probably a Fungus among Us

The relationships of the fungi to protists, plants, and animals are buried so far back in evolutionary history that they have proved difficult to reconstruct. On the basis of morphological comparisons, for a long time taxonomists classified fungi as more closely related to the protists or plants than to animals. However, an investigation of ribosomal RNA (rRNA) sequences led to the conclusion that fungi and animals are more closely related to each other than either group is to protists or plants.

Patricia O. Wainwright, Gregory Hinkle, Mitchell L. Sogin, and Shawn K. Stickel of Rutgers University and the Woods Hole Marine Biology Laboratory carried out the analysis by comparing sequences of 18S rRNA, an rRNA molecule that forms part of the small ribosomal subunit in eukaryotes (see Section 15.4). The investigators began their work by sequencing the 18S rRNA molecules of species among the sponges, ctenophores, and cnidarians (see Chapter 29), which had never been sequenced before. These sequences were then compared with the 18S rRNA sequences of fungi, plants, and several protists, including protozoans and algae, which had been previously obtained by others. For the comparisons, the investigators used a computer program that sorts the rRNA sequences into related groups under the assumption that species with the greatest similarities in 18S rRNA sequence are most closely related. The sequence information was entered into the program in several different combinations; each time the analysis came up with the same family tree (see **figure**).

The family tree placed animals as the branch most closely related to fungi, and indicates that the two groups share a common ancestor not shared with any of the other groups. Other investigators have cited similarities in biochemical pathways in fungi and animals, such as pathways that make the amino acid hydroxyproline, the protein ferritin (which combines with iron atoms), and the polysaccharide chitin, which is a primary constituent of both fungal cell walls and arthropod exoskeletons. Studies of fungi called chytrids (p. 612) also are providing provocative insights on this topic.

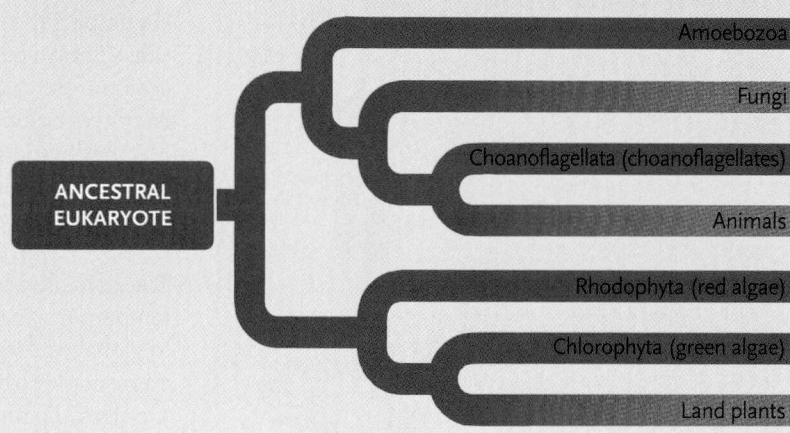

Finally, we will briefly consider a particular odd group of single-celled parasites called **microsporidia.** Based on genetic studies, many mycologists believe they make up a possible sixth phylum within the Kingdom Fungi.

ganism that does not fossilize well. Although traces of what may be fossil fungi exist in rock formations nearly 1 billion years old, the oldest fossils that we can confidently assign to the modern Kingdom Fungi appear in rock strata laid down about 500 million years ago.

Fungi Were Present on Earth by at Least 500 Million Years Ago

Many fungi look plantlike, and for many years fungi were classified as plants. As biologists learned more about the distinctive characteristics of fungi, however, it became clear that fungi merited a separate kingdom. The discovery of chitin in fungal cells, and recent comparisons of DNA and RNA sequences, all indicate that fungi and animals are more closely related to each other than they are to other eukaryotes (see *Insights from the Molecular Revolution*). Analysis of the sequences of several genes suggests that the lineages leading to animals and fungi may have diverged around 965 million years ago. Whenever the split developed, phylogenetic studies indicate that fungi first arose from a single-celled, flagellated protist—the sort of or-

Once They Appeared, Fungi Radiated into at Least Five Major Lineages

Most likely, the first fungi were aquatic. When other kinds of organisms began to colonize land, they may well have brought fungi along with them. For example, researchers have discovered what appear to be mycorrhizae—symbiotic associations of a fungus and a plant—in fossils of the some of the earliest known land plants (see Chapter 27). The final section of this chapter examines the nature of mycorrhizae more fully.

Over time, fungi diverged into the strikingly diverse lineages that we consider in the rest of this section **(Table 28.1).** As the lineages diversified, different adaptations associated with reproduction arose. For this reason, mycologists traditionally assigned fungi to

Table 28.1 | Summary of Fungal Phyla

Phylum	Body Type	Key Feature	
Chytridiomycota (chytrids)	One to several cells	Motile spores propelled by flagella; usually asexual	
Zygomycota (zygomycetes)	Hyphal	Sexual stage in which a resistant zygospore forms for later germination	
Glomeromycota (glomeromycetes)	Hyphal	Hyphae associated with plant roots, forming arbuscular mycorrhizae	
Ascomycota (ascomycetes)	Hyphal	Sexual spores produced in sacs called asci	
Basidiomycota (basidiomycetes)	Hyphal	Sexual spores (basidiospores) form in basidia of a prominent fruiting body (basidiocarp)	

phyla according to the type of structure that houses the final stages of sexual reproduction and releases sexual spores. These features can still be useful indicators of the phylogenetic standing of a fungus, although now the powerful tools of molecular analysis are bringing many revisions to our understanding of the evolutionary journey of fungi. Our survey begins with chytrids, which probably most closely resemble the fungal kingdom's most ancient ancestors.

Chytrids Produce Motile Spores That Have Flagella

The phylum Chytridiomycota includes about a thousand species, referred to simply as chytrids. Chytrids are the only fungi that produce motile spores, which swim by way of flagella. Nearly all chytrids are microscopic **(Figure 28.5a)**, and mycologists have recently begun paying significant attention to them, in part because their characteristics strongly suggest that the group arose near the beginning of fungal evolution. Another reason for research interest is the discovery that the chytrid *Batrachochytrium dendrobatis* is responsible for a disease epidemic that recently has wiped out an estimated two-thirds of the species of harlequin frogs *(Atelopus)* of the American tropics **(Figure 28.5b)**. The epidemic has correlated with the rising average temperature in the frogs' habitats, an increase credited to global warming. Studies show that the warmer environment provides optimal growing temperatures for the chytrid pathogen.

Most chytrids are aquatic, although a few live as saprobes in soil, feeding on decaying plant and animal matter; as parasites on insects, plants, and some animals or even as symbiotic partners in the gut of cattle

a. *Chytriomyces hyalinus*

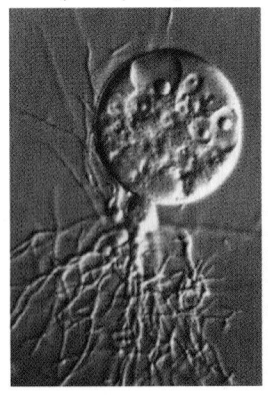

b. Chytridiomycosis in a frog

Skin surface

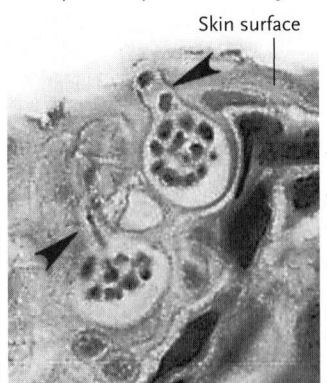

c. Harlequin frog

Figure 28.5

Chytrids. **(a)** *Chytriomyces hyalinus*, one of the few chytrids that reproduces sexually. **(b)** Chytridiomycosis, a fungal infection, shown here in the skin of a frog. The two arrows point to flask-shaped cells of the parasitic chytrid *Batrachochytrium dendrobatis*, which has devastated populations of harlequin frogs **(c)**.

and some other herbivores. Wherever a chytrid lives, reproduction requires at least a film of water through which the swimming spores can move.

A chytrid may advance though its entire life cycle within a matter of days, and for most species, much of this brief lifetime is spent in asexual reproduction. Although individuals initially exist as a vegetative (nonreproductive) phase, the fungus soon shifts into a reproductive mode. First, one or more spore-forming chambers called **sporangia** (*angeion* = vessel; singular, sporangium) develop, each containing one or more haploid nuclei. More developmental steps package the nuclei one by one in flagella-bearing spores. The spores are released to the environment through a pore or tube, and each swims briefly until it comes to rest on a substrate and a tough cyst forms around it. Under proper conditions, it will soon germinate and launch the life cycle anew.

A few chytrids reproduce sexually. Mycologists have observed a remarkable variety of sexual modes, but in all of them spores of different mating types unite. Karyogamy directly follows plasmogamy to produce a 2*n* zygote. This cell may form a mycelium that gives rise to sporangia, or it may directly give rise to either asexual or sexual spores.

Zygomycetes Form Zygospores for Sexual Reproduction

The phylum Zygomycota—fungi that reproduce sexually by way of structures called *zygospores*—contains fewer than a thousand species. What zygomycetes lack in numbers, however, they make up for in impact on other organisms. Many zygomycetes are saprobes that live in soil, feeding on plant detritus. There, their metabolic activities release mineral nutrients in forms that plant roots can take up. Some zygomycetes are parasites of insects (and even other zygomycetes), and some wreak havoc on human food supplies, spoiling stored grains, bread, fruits, and vegetables such as sweet potatoes. Others, however, have become major partners in commercial enterprises, where they are used in manufacturing products that range from industrial pigments to pharmaceuticals.

Most zygomycetes have aseptate hyphae, a feature that distinguishes them from the other multicellular fungi. Like other fungi, however, zygomycetes usually reproduce asexually, as shown at the lower left in **Figure 28.6.** When a haploid spore lands on a favorable substrate, it germinates and gives rise to a branching mycelium. Some of the hyphae grow upward, and saclike, thin-walled sporangia form at the tips of these aerial hyphae. Inside the sporangia the asexual cycle comes full circle as new haploid spores arise through mitosis and are released.

The black bread mold, *Rhizopus stolonifer,* may produce so many charcoal-colored sporangia **(Figure 28.7a)** that moldy bread looks black. The spores released are lightweight, dry, and readily wafted away by air currents. In fact, winds have dispersed *R. stolonifer* spores just about everywhere on Earth, including the Arctic. Another zygomycete, *Pilobolus* **(Figure 28.7b),** forcefully spews its sporangia away from the dung in which it grows. A grazing animal may eat a sporangium on a blade of grass; the spores then pass through the animal's gut unharmed and begin the life cycle again in a new dung pile.

Mycelia of many zygomycetes may occur in either the + or − mating type, and the nuclei of the different mating types are equivalent to gametes. Each strain secretes steroidlike hormones that can stimulate the development of sexual structures in the complementary strain and cause sexual hyphae to grow toward each other. When + and − hyphae come into close proximity, a septum forms behind the tip of each hypha, producing a terminal **gametangium** that contains several haploid nuclei (see Figure 28.6). When the gametangia of the two strains make contact, cellular enzymes digest the wall between them, yielding a single large, thin-walled cell that contains many nuclei from both parents. In other words, plasmogamy has occurred, and this new cell is a dikaryon. Gradually a second, inner wall forms, thickens, and hardens. This structure, with the multinucleate cell inside it, is a **zygospore**, the structure that gives this fungal group its scientific name. It becomes dormant and sometimes stays dormant for months or years.

Karyogamy follows plasmogamy, but the timing varies in different groups of zygomycetes. The exact trigger is unknown, but eventually the diploid zygospore ends its dormancy. The cell undergoes meiosis and produces a stalked sporangium (see Figure 28.6, step 5). The sporangium contains haploid spores of each mating type, which are released to the outside world. When a spore later germinates, it produces either a + or a − mycelium, and the sexual cycle can continue.

Zygomycetes that have aseptate hyphae are structurally simpler than the species in most other fungal groups. Although septa wall off the reproductive structures, in effect the branching mycelium of each fungus is a single, huge, multinucleate cell—the same body structure as found in some algae and certain protists. Because such zygomycetes have numerous nuclei in a common cytoplasm, these fungi are said to be **coenocytic,** which means "contained in a shared vessel." By contrast, in other fungal groups septa at least partially divide the hyphae into individual cells, which typically contain two or more nuclei.

Presumably, having hyphae that lack septa confers some selective advantages. One benefit may be that without septa to impede the flow, nutrients can move freely from the absorptive hyphal tips to other hyphae where reproductive parts develop. Hence the fungus may be able to reproduce faster.

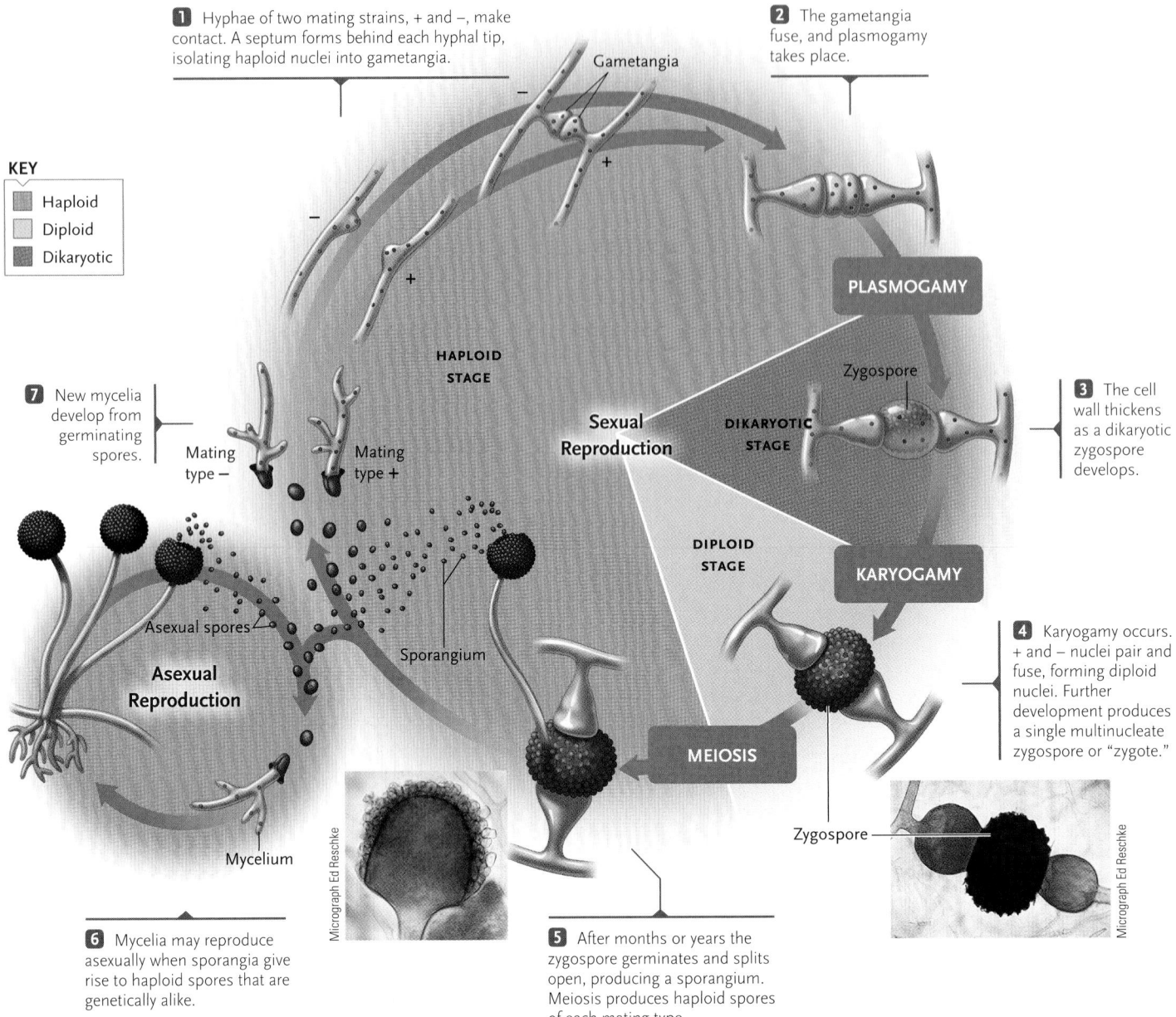

1 Hyphae of two mating strains, + and −, make contact. A septum forms behind each hyphal tip, isolating haploid nuclei into gametangia.

Gametangia

2 The gametangia fuse, and plasmogamy takes place.

PLASMOGAMY

KEY

	Haploid
	Diploid
	Dikaryotic

HAPLOID STAGE

Zygospore

Sexual Reproduction

DIKARYOTIC STAGE

3 The cell wall thickens as a dikaryotic zygospore develops.

7 New mycelia develop from germinating spores.

Mating type −

Mating type +

DIPLOID STAGE

KARYOGAMY

Asexual spores

Asexual Reproduction

Sporangium

4 Karyogamy occurs. + and − nuclei pair and fuse, forming diploid nuclei. Further development produces a single multinucleate zygospore or "zygote."

MEIOSIS

Zygospore

Micrograph Ed Reschke

Mycelium

Micrograph Ed Reschke

6 Mycelia may reproduce asexually when sporangia give rise to haploid spores that are genetically alike.

5 After months or years the zygospore germinates and splits open, producing a sporangium. Meiosis produces haploid spores of each mating type.

Figure 28.6
Life cycle of the bread mold *Rhizopus stolonifer,* a zygomycete. Asexual reproduction is common, but different mating types (+ and −) also reproduce sexually. In both cases, haploid spores form and give rise to new mycelia.

In zygomycetes, aggregations of "cooperating" hyphae may form body structures specialized for certain functions. However, such structures are more common in the three groups of more complex fungi that we consider next.

Glomeromycetes Form Spores at the Ends of Hyphae

The 160 known members of the phylum Glomeromycota are all specialized to form the associations called mycorrhizae with plant roots. It would be hard to overestimate their ecological impact, for Glomeromycota collectively make up roughly half of the fungi in soil and form mycorrhizae with an estimated 80% to 90% of all land plants. Virtually all glomeromycetes reproduce asexually, by way of spores that form at the tips of hyphae. The hyphae also secrete enzymes that

allow them to enter plant roots, where their tips branch into treelike clusters. As you will read in the next section, the clusters, called arbuscules, nourish the fungus by taking up sugars from the plant and in return supply the plant roots with a steady supply of dissolved minerals from the surrounding soil.

Ascomycetes, the Sac Fungi, Produce Sexual Spores in Saclike Asci

The phylum Ascomycota includes more than 30,000 species that produce reproductive structures called *asci* **(Figure 28.8).** A few ascomycetes prey upon various agricultural insect pests and thus have potential for use as "biological pesticides." Many more are destructive plant pathogens, including *Venturia inaequalis,* the fungus responsible for apple scab, and *Ophiostoma ulmi,* which causes Dutch elm disease. Several ascomycetes can be serious pathogens of humans. For example, *Claviceps purpurea*, a parasite on rye and other grains, causes ergotism, a disease marked by vomiting, hallucinations, convulsions, and in severe cases, gangrene and even death. Other ascomycetes cause nuisance infections such as athlete's foot and ringworm. Strains of *Aspergillus* grow in damp grain or peanuts; their metabolic wastes, known as aflatoxins, can cause cancer in humans who eat the poisoned food over an extended period. A few ascomycetes even show trapping behavior, ensnaring small worms that they then digest **(Figure 28.9a).** Yet some ascomycetes are valuable to humans: one species, the orange bread mold *Neurospora crassa*, has been important in genetic research, including the elucidation of the one gene–one enzyme hypothesis (see Section 15.1). And certain species of *Penicillium* **(Figure 28.9b)** are the source of the penicillin family of antibiotics, while others produce the aroma and distinctive flavors of Camembert and Roquefort cheeses. This multifaceted division also includes gourmet delicacies such as truffles *(Tuber melanosporum)* and the succulent true morel *Morchella esculenta*.

Although yeasts and filamentous fungi with a yeast stage in the life cycle occur in all fungal groups except chytrids, many of the best-known yeasts are ascomycetes. The yeast *Candida albicans* **(Figure 28.10)** infects mucous membranes, especially of the mouth (where it causes a disorder called thrush) and the vagina. *Saccharomyces cerevisiae,* which produces the ethanol in alcoholic beverages and the carbon dioxide

a. Sporangia of *Rhizopus stolonifer*

b. Sporangia (dark sacs) of *Pilobolus*

500 μm

Figure 28.7

Two of the numerous strategies for spore dispersal by zygomycetes. **(a)** The sporangia of *Rhizopus stolonifer,* shown here on a slice of bread, release powdery spores that are easily dispersed by air currents. **(b)** In *Pilobolus,* the spores are contained in a sporangium (the dark sac) at the end of a stalked structure. When incoming rays of sunlight strike a light-sensitive portion of the stalk, turgor pressure (pressure against a cell wall due to the movement of water into the cell) inside a vacuole in the swollen portion becomes so great that the entire sporangium may be ejected outward as far as 2 m—a remarkable feat, given that the stalk is only 5 to 10 mm tall.

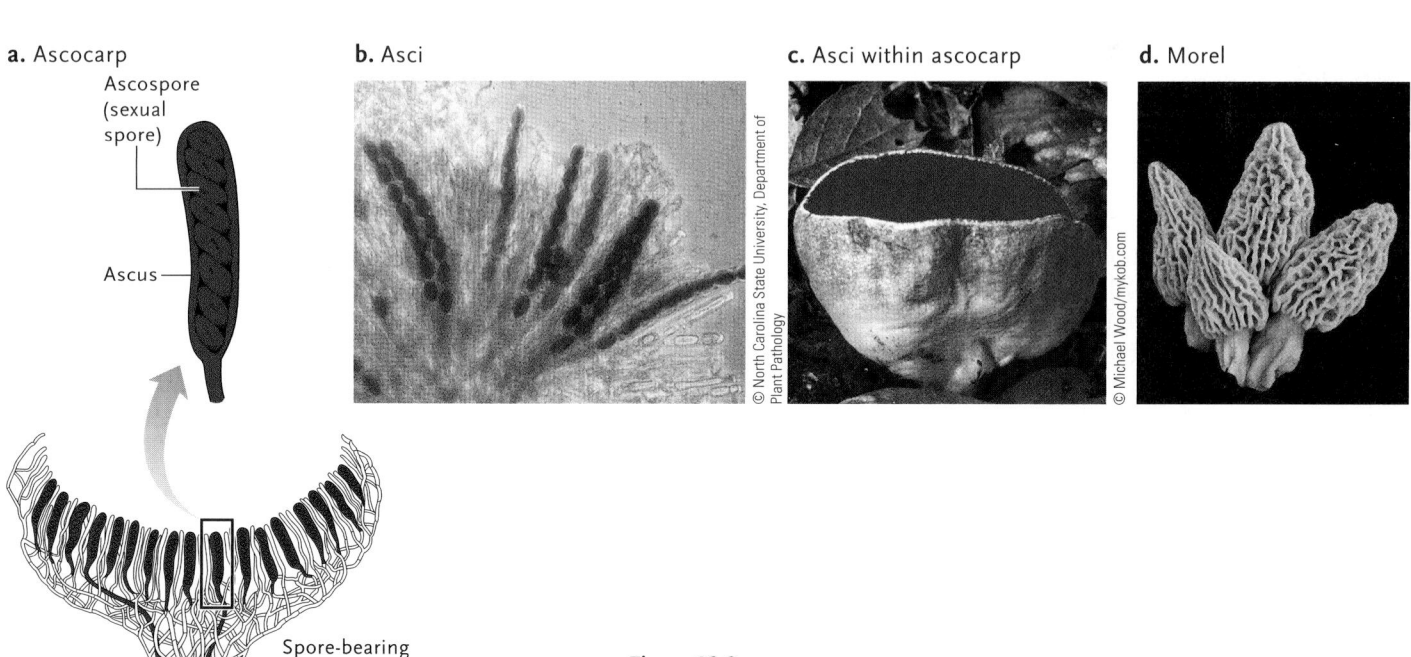

a. Ascocarp
Ascospore (sexual spore)
Ascus
Spore-bearing hypha of this ascocarp

b. Asci

c. Asci within ascocarp

d. Morel

Figure 28.8

A few of the ascomycetes, or sac fungi. The examples shown are multicellular species that form mushrooms as reproductive structures. **(a)** A cup-shaped ascocarp, composed of tightly interwoven hyphae. The spore-producing asci occur inside the cup. **(b)** Asci on the inner surface of an ascocarp. **(c)** Scarlet cup fungus (*Sarcoscypha*). **(d)** A true morel (*Morchella esculenta*), a prized edible fungus.

a. A penicillium species   **b.** A trapping ascomycete

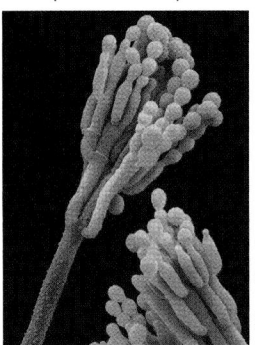

N. Allin and G. L. Barron

© Dennis Kunkel Microscopy, Inc.

Figure 28.9

Other ascomycete representatives. **(a)** *Eupenicillium*. Notice the rows of conidia (asexual spores) atop the structures that produce them. **(b)** Hyphae of *Arthrobotrys dactyloides*, a trapping ascomycete, form nooselike rings. When the fungus is stimulated by the presence of a prey organism, rapid changes in ion concentrations draw water into the hypha by osmosis. The increased turgor pressure shrinks the "hole" in the noose and captures this nematode. The hypha then releases digestive enzymes that break down the worm's tissues.

Yeast cells

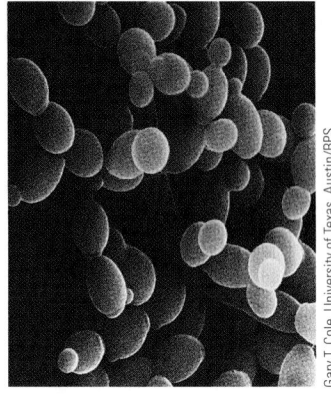

Gary T. Cole, University of Texas, Austin/BPS

Figure 28.10

Candida albicans, cause of yeast infections of the mouth and vagina.

that leavens bread, has also been a model organism for genetic research. By one estimate it has been the subject of more genetic experiments than any other eukaryotic microorganism. Yeasts commonly reproduce asexually by fission or budding from the parent cell, but many also can reproduce sexually after the fusion of two cells of different mating types (analogous to the mating types described earlier). Many ascomycete yeasts are found naturally in the nectar of flowers and on fruits and leaves. At least 1500 species have been described, and mycologists suspect that thousands more are yet to be identified.

Tens of thousands of ascomycetes, however, are not yeasts. They are multicellular, with tissues built up from septate hyphae. Although septa do slow the flow of nutrients (which, recall, can cross septa through pores), they also confer advantages. For example, septa present barriers to the loss of cytoplasm if a hypha is torn or punctured, whereas in an aseptate zygomycete, fluid pressure may force out a significant amount of cytoplasm before a breach can be sealed by congealing cytoplasm. In ways that are not well understood, septa can also limit the damage from toxins that are secreted by competing fungi.

As with zygomycetes, certain hyphae in ascomycetes are specialized for asexual reproduction. Instead of making spores inside sporangia, however, many ascomycetes produce asexual spores called **conidia** ("dust"; singular, conidium). In some of the species, the conidia form in chains that elongate from modified hyphal branches called **conidiophores**. In other ascomycetes, the conidia may pinch off from the hyphae in a series

of "bubbles," a bit like a string of detachable beads. Either way, an ascomycete can form and release spores much more quickly than a zygomycete can. Each newly formed conidium contains a haploid nucleus and some of the parent hypha's cytoplasm. Conidia and conidiophores of some ascomycete species are visible as the white powdery mildew that attacks grapes, roses, grasses, and the leaves of squash plants.

Ascomycetes can also reproduce sexually, and are commonly termed sac fungi because the meiotic divisions that generate haploid sexual spores occur in saclike cells called **asci** (*askos* = bladder; singular, ascus). In *Neurospora crassa* **(Figure 28.11)** and other complex ascomycetes, reproductive bodies called **ascocarps** bear or contain the asci. Some ascocarps resemble globes, others flasks or open dishes. An ascocarp begins to develop when two haploid mycelia of + and − mating types fuse (step 1). Plasmogamy then takes place, with the details differing from species to species. (In some species, hormonal signals cause the tip of one hypha to enlarge and form a "female" reproductive organ called an ascogonium, while the other hyphal tip develops into a "male" antheridium.) Paired nuclei, one from each mating type, migrate into the hyphae. During plasmogamy, the fused sexual structures give rise to dikaryotic hyphae, which develop inside the ascocarp. Asci form at the hyphal tips. Inside them, karyogamy takes place, producing a diploid zygote nucleus. It divides by meiosis, producing four haploid nuclei. In yeasts and some other ascomycetes cell division stops at this point, but in *N. crassa* and in many other species a round of mitosis ensues and results in eight nuclei. Regardless, the nuclei, other organelles, and a portion of cytoplasm then are incorporated into ascospores that may germinate on a suitable substrate and continue the life cycle.

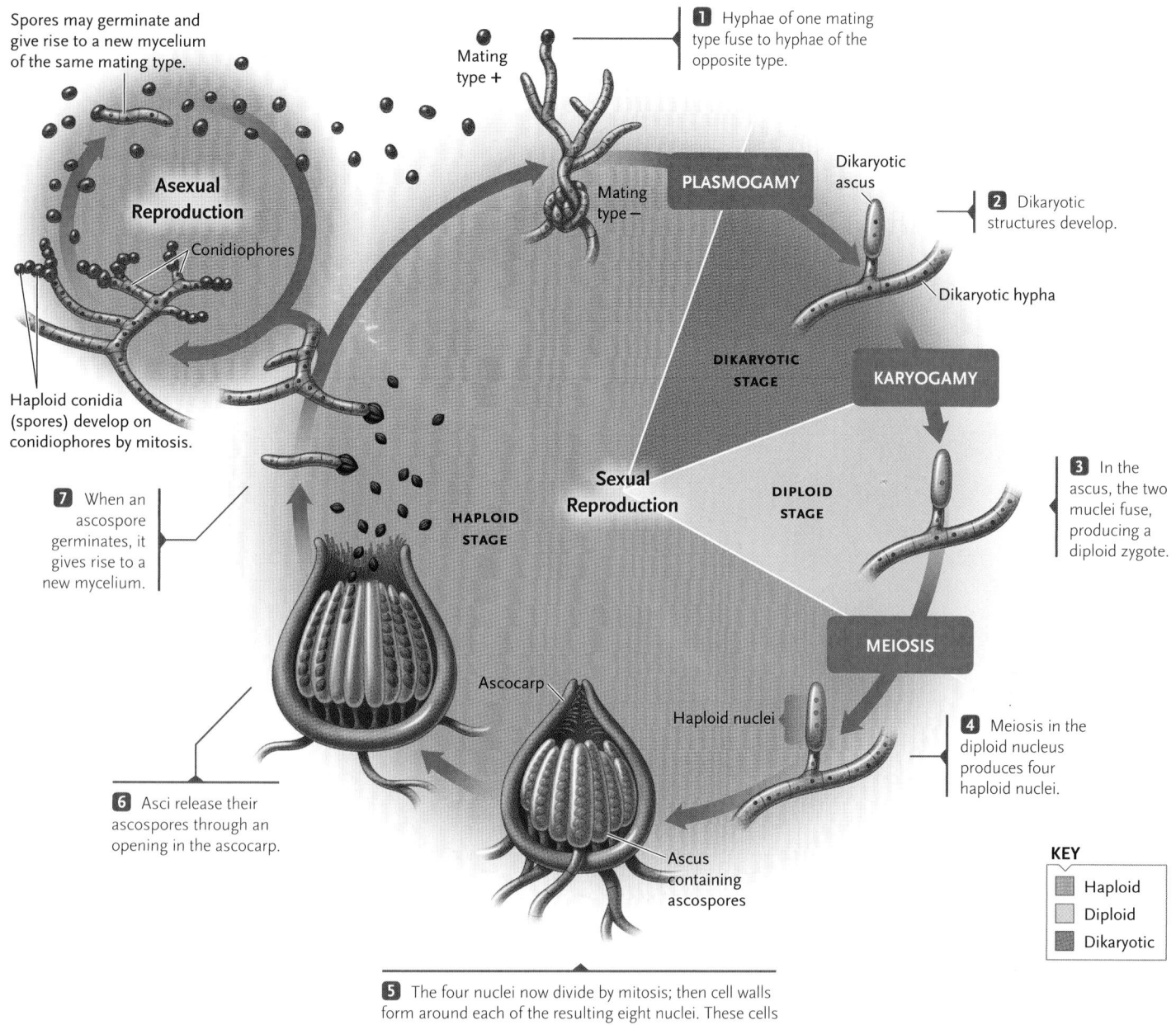

Spores may germinate and give rise to a new mycelium of the same mating type.

Asexual Reproduction

Conidiophores

Haploid conidia (spores) develop on conidiophores by mitosis.

Mating type +

Mating type −

1 Hyphae of one mating type fuse to hyphae of the opposite type.

PLASMOGAMY

Dikaryotic ascus

2 Dikaryotic structures develop.

Dikaryotic hypha

DIKARYOTIC STAGE

KARYOGAMY

Sexual Reproduction

DIPLOID STAGE

3 In the ascus, the two nuclei fuse, producing a diploid zygote.

HAPLOID STAGE

MEIOSIS

Haploid nuclei

4 Meiosis in the diploid nucleus produces four haploid nuclei.

Ascocarp

Ascus containing ascospores

7 When an ascospore germinates, it gives rise to a new mycelium.

6 Asci release their ascospores through an opening in the ascocarp.

5 The four nuclei now divide by mitosis; then cell walls form around each of the resulting eight nuclei. These cells are ascospores. Asci develop inside an ascocarp, which began to form soon after sexual reproduction began.

KEY

	Haploid
	Diploid
	Dikaryotic

Figure 28.11
Life cycle of the ascomycete *Neurospora crassa.*

Basidiomycetes, the Club Fungi, Form Sexual Spores in Club-Shaped Basidia

The 25,000 or so species of fungi in the phylum Basidiomycota include the mushroom-forming species, shelf fungi, coral fungi, bird's nest fungi, stinkhorns, smuts, rusts, and puffballs **(Figure 28.12)**. The common name for this group is club fungi, so named because the spore-producing cells, called **basidia** (meaning base or foundation), usually are club shaped. Some species have enzymes for digesting cellulose and lignin and are important decomposers of woody plant debris. A surprising number of basidiomycetes, including the prized edible oyster mushrooms *(Pleurotus ostreatus)*, also can trap and consume bacteria and small animals such as rotifers and nematodes by secreting paralyzing toxins or gluey substances that immobilize the prey. This adaptation gives the fungus access to a rich source of molecular nitrogen, an essential nutrient that often is scarce in terrestrial habitats.

Many basidiomycetes take part in vital mutualistic associations with the roots of forest trees, as discussed later in this chapter. Others, the rusts and smuts, are parasites that cause serious diseases in wheat, rice, and other plants. Still others produce millions of dollars' worth of the reproductive structures commonly called mushrooms.

Amanita muscaria, the fly agaric mushroom (see Figure 28.12d), has been used as a fly poison, from

a. Coral fungus

b. Shelf fungus

c. White-egg bird's nest fungus

d. Fly agaric mushroom

e. Scarlet hood

Figure 28.12

Representative basidiomycetes, or club fungi. **(a)** The light red coral fungus *Ramaria*. **(b)** The shelf fungus *Polyporus*. **(c)** The white-egg bird's nest fungus *Crucibulum laeve*. Each tiny "egg" contains spores. Raindrops splashing into the "nest" can cause "eggs" to be ejected, thereby spreading spores into the surrounding environment. **(d)** The fly agaric mushroom *Amanita muscaria*, which causes hallucinations. **(e)** The scarlet hood *Hygrophorus*.

which it gets its common name. Due to its hallucinogenic effects, *A. muscaria* also is used in the religious rituals of ancient societies in Central America, Russia, and India. Other species of this genus, including the death cap mushroom *Amanita phalloides,* produce deadly toxins. The *A. phalloides* toxin, called α-amanitin, halts gene transcription, and hence protein synthesis, by inhibiting the activity of RNA polymerase. Within 8 to 24 hours of ingesting as little as 5 mg of the toxin, vomiting and diarrhea begin. Later, kidney and liver cells start to degenerate; without intensive medical care, death can follow within a few days.

A few basidiomycetes generally reproduce only by asexual means, by budding or shedding a fragment of a hypha. One is *Cryptococcus neoformans,* which causes a form of meningitis in humans. In general, however, basidiomycetes do reproduce sexually, producing large numbers of haploid sexual spores. **Figure 28.13** shows the life cycle of a typical basidiomycete.

Basidia typically develop on a **basidiocarp**, which is the reproductive body of the fungus. A basidiocarp consists of tight clusters of hyphae; the feeding mycelium is buried in the soil or decaying wood. The shelflike bracket fungi visible on trees are basidiocarps, and about 10,000 species of club fungi produce the basidiocarps we call mushrooms. Each is a short-lived reproductive body consisting of a stalk and a cap. Basidia develop on "gills," which are the sheets of tissue on the underside of the cap. The basidia undergo meiosis to produce microscopic, haploid **basidiospores** (Figure 28.13, inset) that disperse throughout the environment.

When a basidiospore lands on a suitable food source, it germinates and gives rise to a haploid mycelium. Two compatible mating types growing near each other may undergo plasmogamy. The resulting mycelium is dikaryotic, its cells containing one nucleus from each mating type. The dikaryotic stage of a basidiomycete is the feeding mycelium that can grow for years—a major departure from an ascomycete's short-lived dikaryotic stage. Accordingly, a basidiomycete has many more opportunities for producing sexual spores, and the mycelium can give rise to reproductive bodies many times.

After an extensive mycelium develops, and when environmental conditions such as moisture are favorable, basidiocarps grow from the mycelium and develop basidia. At first, each basidium in the mushroom or other reproductive body is dikaryotic, but then the two

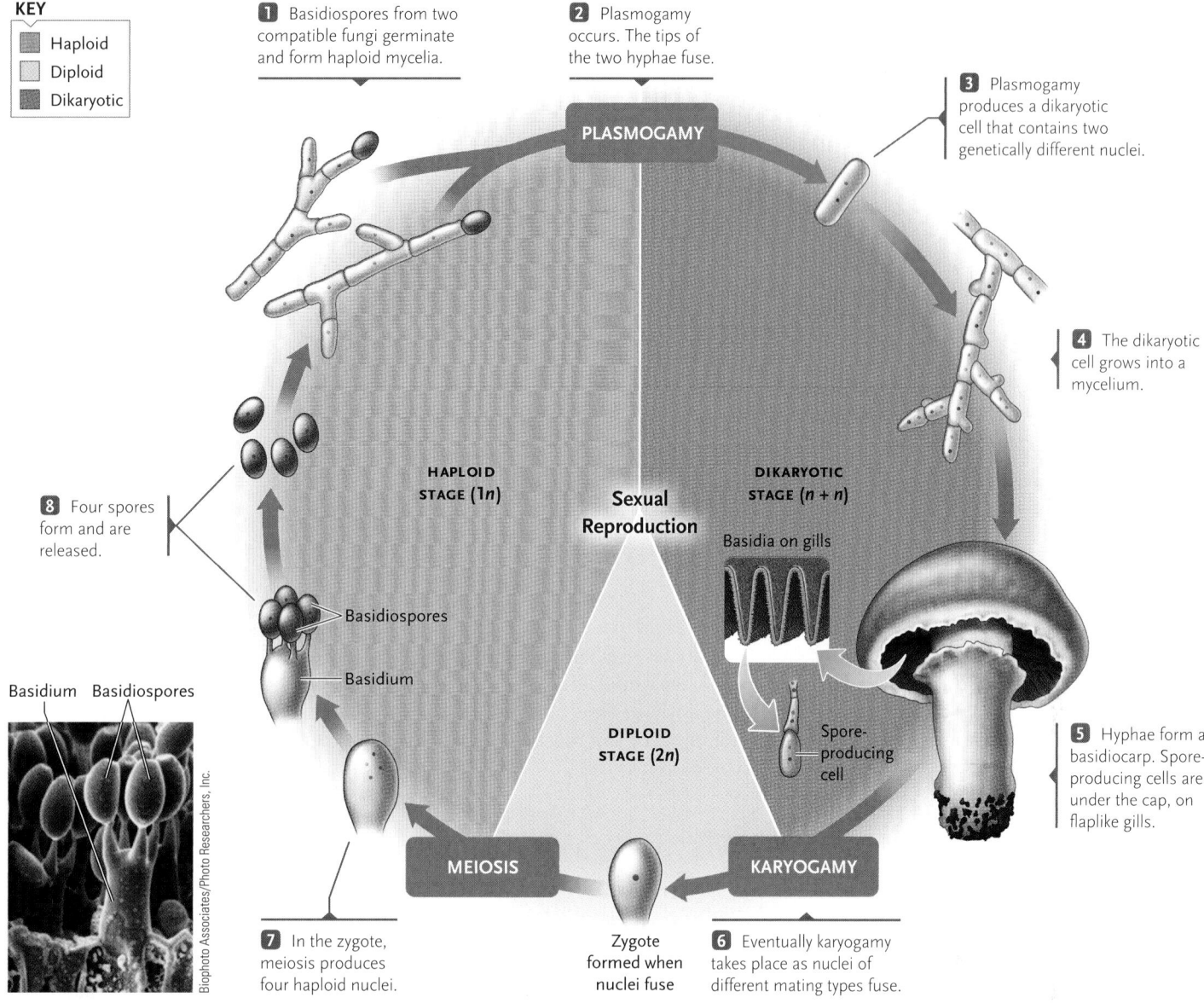

Figure 28.13

Generalized life cycle of the basidiomycete *Agaricus bisporus*, a species known commonly as the button mushroom. During the dikaryotic stage, cells contain two genetically different nuclei, shown here in different colors. Inset: Micrograph showing basidia and basidiospores.

nuclei undergo karyogamy, fusing to form a diploid zygote nucleus. The zygote exists only briefly; meiosis soon produces haploid basidiospores, which are wafted away from the basidium by air currents. Basidia can produce huge numbers of spores—for many species, estimates run as high as 100 million spores *per hour* during reproductive periods, day after day.

Squirrels and many other small animals may eat mushrooms almost as soon as they appear, but in some species the underlying mycelium can live for many years. For example, U.S. Forest Service scientists have found that the mycelium of a single individual of *Armillaria ostoyae* covers an area equivalent to 1665 football fields in an eastern Oregon forest. By one estimate, it measures an average of 1 m deep and nearly 6000 m across, making it perhaps one of the largest organisms

on Earth. As such a mycelium grows, specialized mechanisms during cell division maintain the dikaryotic condition and the paired nuclei in each hyphal cell.

Conidial Fungi Are Species for Which No Sexual Phase Is Known

As noted earlier, fungi generally are classified on the basis of their structures for sexual reproduction. When a sexual phase is absent or has not yet been detected, the fungal species is said to be anamorphic ("no related form") and is lumped into a convenience grouping, the conidial fungi (recall that conidia are asexual spores). This classification is the equivalent of "unidentified." Other names for this grouping are "imperfect fungi" and deuteromycetes.

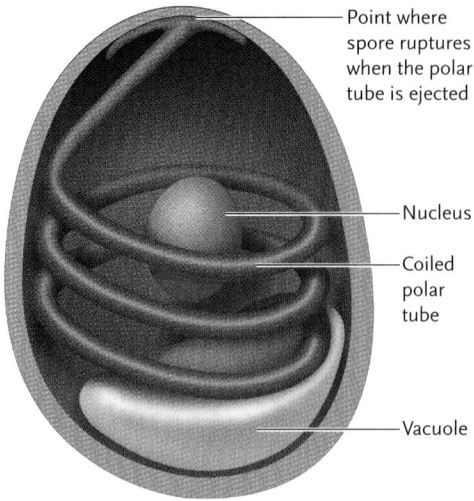

Figure 28.14

Structure of microsporidia. When a spore germinates, its vacuole expands and forces the coiled "polar tube" outward and into a nearby, soon-to-be host cell. The nucleus and cytoplasm of the parasite enter the host through the tube, launching developmental steps that lead to the development of more microsporidia inside the host.

Labels in figure:
- Point where spore ruptures when the polar tube is ejected
- Nucleus
- Coiled polar tube
- Vacuole

When researchers discover a sexual phase for a conidial fungus, or when molecular studies establish a clear relationship to a sexual species, the conidial fungus is reassigned to the appropriate phylum. Thus far, some have been classified as basidiomycetes, but most conidial fungi have turned out to be ascomycetes.

Microsporidia Are Single-Celled Sporelike Parasites

There are more than 1200 species of the single-celled parasites called **microsporidia**. They are known to infect insects including honeybees and grasshoppers, and vertebrates including fish and humans—especially individuals with compromised immune systems such as people with AIDS. Microsporidia are rather mysterious organisms. Physically they resemble spores **(Figure 28.14)**, but they lack mitochondria and have several other puzzling characteristics. Molecular studies suggest that they are related to zygomycetes, and some researchers have proposed that the group may have lost many typical fungal features as it evolved a highly specialized parasitic lifestyle.

STUDY BREAK

1. Name the five phyla of the Kingdom Fungi and describe the reproductive adaptations that distinguish each one.
2. In terms of structure, which are the simplest fungal groups? The most complex?
3. Describe some ways, positive or negative, that members of each fungal phylum interact with other life forms.

28.3 Fungal Associations

Many fungi are partners in mutually beneficial interactions with photosynthetic organisms, and these associations play major roles in the functioning of ecosystems. A **symbiosis** is a state such as parasitism or mutualism in which two or more species live together in close association. Chapter 50 discusses general features of symbiotic associations more fully; here we are interested in some examples of the symbioses fungi form with photosynthetic partners—cyanobacteria, green algae, and plants.

A Lichen Is an Association between a Fungus and a Photosynthetic Partner

You may be familiar with one type of lichen, the leathery patches of various colors growing on certain rocks. Technically, a **lichen** is a single vegetative body that is the result of an association between a fungus and a photosynthetic partner. The fungal partner in a lichen, called the **mycobiont**, usually makes up only about 10% of the whole. The other 90% is the photosynthetic partner, called the **photobiont.** Most frequently, these are green algae of the genus *Trebouxia* or cyanobacteria of the genus *Nostoc*. Thousands of ascomycetes and a few basidiomycetes form this kind of symbiosis, but only about 100 photosynthetic species serve as photobionts.

Lichens often live in harsh, dry microenvironments, including on bare rock and wind-whipped tree trunks. Yet lichens have vital ecological roles and important human uses. Lichens secrete acid that eats away at rock, breaking it down and converting it to soil that can support larger plants. Some paleobiologists have suggested that lichens may have been some of the earliest land organisms, covering bare rocks during the Ordovician period (roughly 500 million to 425 million years ago). In this scenario, millennia of decaying lichens would have created the first soils in which the earliest land plants could grow. Today, lichens continue to enhance the survival of other life forms. For instance, in arctic tundra, where plants are scarce, reindeer and musk oxen can survive by eating lichens. Insects, slugs, and some other invertebrates also consume lichens, and they are nest-building materials for many birds and small mammals. People have derived dyes from lichens; they are even a component of garam masala, an ingredient in Indian cuisine. Some environmental chemists monitor air pollution by observing lichens, most of which cannot grow in heavily polluted air (see *Focus on Research*).

Because lichens are composite organisms, it may seem odd to talk of lichen "species." Biologists do give lichens binomial names, however, based on the characteristics of the mycobiont. More than 13,500 lichens are recognized, each one a unique combination of a

Applied Research: Lichens as Monitors of Air Pollution's Biological Damage

Lichens have become reliable pollution-monitoring devices all over the world—in some cases, replacing costly electronic monitoring stations. Different species are vulnerable to specific pollutants. For example, *Ramalina* lichens are damaged by nitrate and fluoride salts. Elevated levels of sulfur dioxide (a major component of acid rain) cause old man's beard *(Usnea trichodea)* to shrivel and die, but strongly promote the growth of a crusty European lichen, *Lecanora conizaeoides*. The sensitivity of yellow *Evernia* lichens to SO$_2$ enabled the scientist who discovered its damage at remote Isle Royale in Michigan to point the finger northward to coal-burning furnaces at Thunder Bay, Canada. Conversely, healthy lichens on damaged trees of Germany's Black Forest lifted suspicion from French

coal-burning power plants and allowed investigators to identify the true source of the tree damage: nitrogen oxides from automobile exhausts. The

result was Germany's first auto emission standards, which went into effect in the 1990s.

Usnea (old man's beard), a pendent (hanging) lichen.

Mark Mattock/Planet Earth Pictures

particular species of fungus and one or more species of photobiont. The relationship often begins when a fungal mycelium contacts a free-living cyanobacterium, algal cell, or both. The fungus parasitizes the photosynthetic host cell, sometimes killing it. If the host cell can survive, however, it multiplies in association with the fungal hyphae. The result is a tough, pliable body called a **thallus**, which can take a variety of forms **(Figure 28.15a)**. Short, specialized hyphae penetrate algal cells of the thallus, which become the fungus's sole source of nutrients. Often, the mycobiont of a lichen absorbs up to 80% of the carbohydrates the photobiont produces.

Benefits for the photobiont are less clear-cut, in part because the drain on nutrients hampers its growth and reproduction. In one view, many and possibly most lichens are parasitic symbioses in which the photobiont does not receive equal benefit. On the other hand, it is relatively rare to find a lichen's photobiont species living independently in the same conditions under which the lichen survives, whereas as part of a lichen it may eke out an enduring existence; some lichens have been dated as being more than 4000 years old! Studies have also revealed that at least some green algae do clearly benefit from the relationship. Such algae are sensitive to desiccation and intense ultraviolet radiation. Sheltered by a lichen's fungal tissues, a green alga can thrive in locales where alone it would perish. Clearly, we still have quite a bit

to learn about the physiological interactions between lichen partners.

As you might expect with such a communal life form, reproduction has its quirky aspects. In lichens that involve an ascomycete, the fungus produces ascospores that are dispersed by the wind. The spores germinate to form hyphae that may colonize new photosynthetic cells and so establish new symbioses. A lichen itself can also reproduce in at least two ways. In some types, a section of the thallus detaches and grows into a new lichen. In about one-third of lichens, specialized regions of the thallus give rise asexually to reproductive cell clusters called **soredia** (*soros* = heap; singular, soredium). Each cluster includes both algal and hyphal cells **(Figure 28.15b)**. As the lichen grows, the soredia detach and are dispersed by water, wind, or passing animals.

Mycorrhizae Are Symbiotic Associations of Fungi and Plant Roots

A **mycorrhiza** ("fungus-root") is a mutualistic symbiosis in which fungal hyphae associate intimately with plant roots. Mycorrhizae greatly enhance the plant's ability to extract various nutrients, especially phosphorus and nitrogen, from soil (see Chapter 33).

In **endomycorrhizae**, the fungal hyphae penetrate the cells of the root. This kind of association occurs on the roots of nearly all flowering plants, and in most

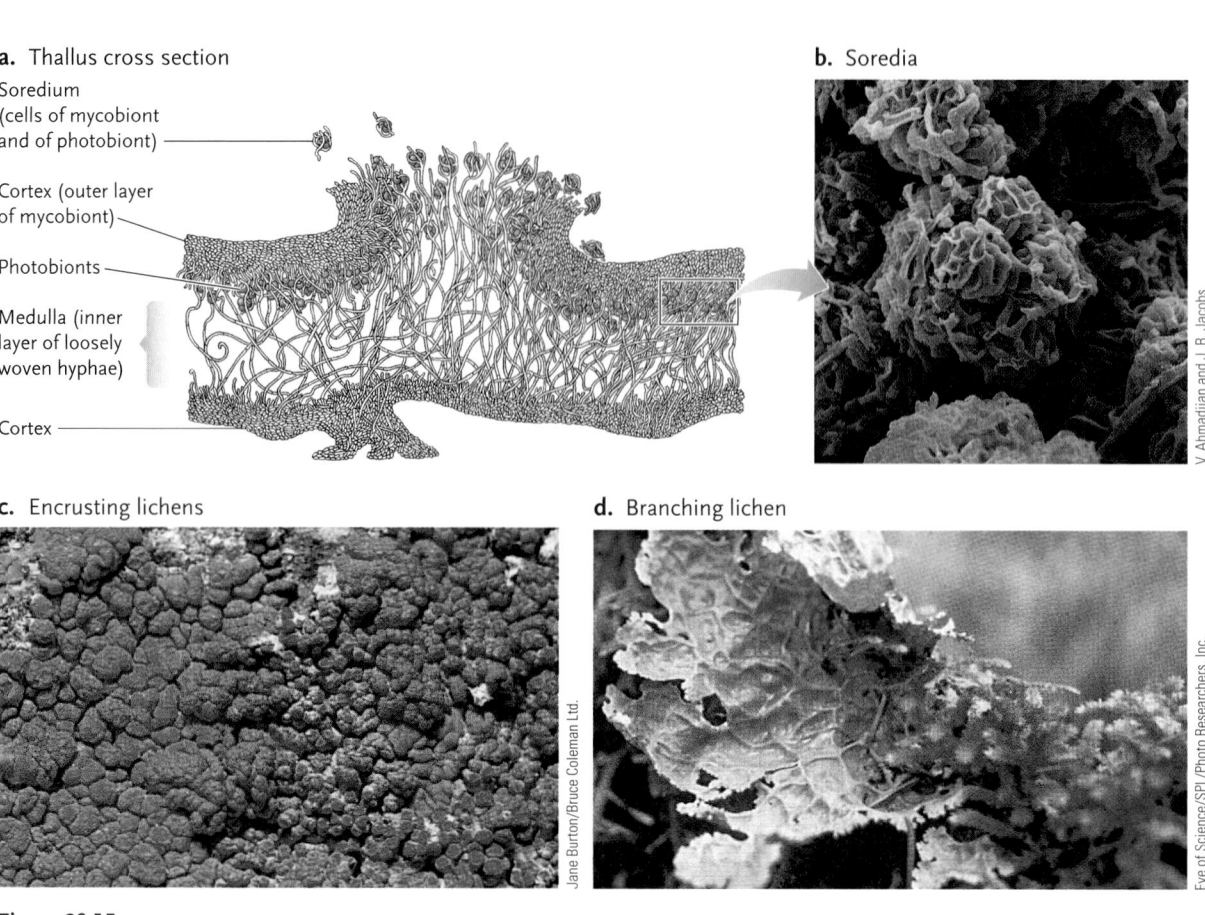

a. Thallus cross section

Soredium
(cells of mycobiont
and of photobiont)

Cortex (outer layer
of mycobiont)

Photobionts

Medulla (inner
layer of loosely
woven hyphae)

Cortex

b. Soredia

V. Ahmadjian and J. B. Jacobs

c. Encrusting lichens

Jane Burton/Bruce Coleman Ltd.

d. Branching lichen

Eye of Science/SPL/Photo Researchers, Inc.

Figure 28.15

Lichens. **(a)** Sketch of a cross section through the thallus of the lichen *Lobaria verrucosa*. The soredia **(b)**, which contain both hyphae and algal cells, are a type of dispersal fragment by which lichens reproduce asexually. **(c)** Encrusting lichens. **(d)** Erect, branching lichen, *Cladonia rangiferina*.

cases a glomeromycete is the fungal partner. The tree-like, branched hyphae of endomycorrhizae are called arbuscules **(Figure 28.16)**, and glomeromycetes are sometimes referred to as arbuscular fungi.

Basidiomycetes are the usual fungal partners in **ectomycorrhizae (Figure 28.17),** in which hyphal tips grow between and around the young roots of trees and shrubs but never enter the root cells. Ectomycorrhizal associations—often several of them—are very common with trees. For instance, the extensive root system of a mature pine may be studded with ectomycorrhizae involving dozens of fungal species. The musky-flavored truffles *(Tuber melanosporum)* prized by gourmets are ascomycetes that form ectomycorrhizal associations with oak trees (genus *Quercus*).

Orchids are partners in a unique mycorrhizal relationship. The fungal partner, usually a basidiomycete, lives inside the orchid's tissues and provides the plant with a variety of nutrients. In fact, seeds of wild orchids germinate, and seedlings survive, only when such mycorrhizae are present.

In general, mycorrhizae represent a "win-win" situation for the partners. The fungal hyphae absorb carbohydrates synthesized by the plant, along with some amino acids and perhaps growth factors as well. The growing plant in turn absorbs mineral ions made accessible to it by the fungus. Collectively, the fungal hyphae have a tremendous surface area for absorbing mineral ions from a large volume of the surrounding soil. Dissolved mineral ions accumulate in the hyphae when they are plentiful in the soil, and are released to the plant when they are scarce. This service is a survival boon to a great many plants, especially species that cannot readily absorb mineral ions, particularly phosphorus **(Figure 28.18)**. For plants that inhabit soils poor in mineral ions, such as in tropical rain forests, mycorrhizal associations are crucial for survival. Likewise, in temperate forests, species of spruce, oak, pine, and some other trees die unless mycorrhizal fungi are present. Plants that live in dry habitats often rely on specialized mycorrhizal hyphae that serve as conduits for water into the root. Like lichens, mycorrhizae are highly vulnerable to damage from pollutants, especially acid rain.

Mycorrhizae have a long evolutionary history. Fossils show that endomycorrhizae were common among ancient land plants, and some biologists have speculated they might have been key for enhancing

Figure 28.16
Endomycorrhizae. **(a)** In this instance, the roots of leeks are growing in association with the glomeromycete *Glomus versiforme* (longitudinal section). Notice the arbuscules that have formed as the fungal hyphae branched after entering the leek root **(b)**.

a. Leek root with endomycorrhizae (black)

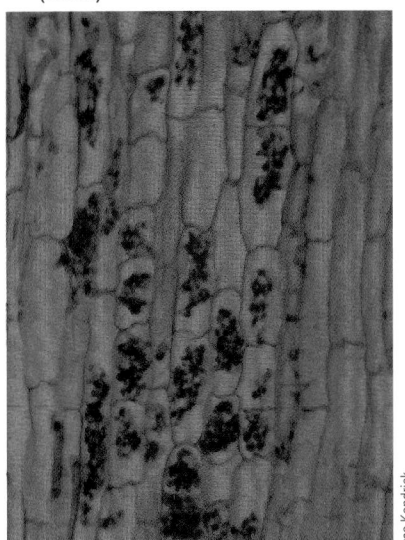

Bryce Kendrick

b. Arbuscule

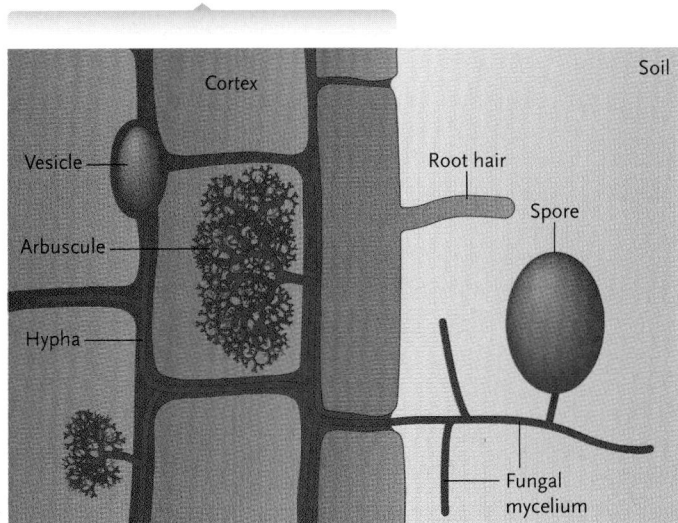

Root

Soil

Cortex

Vesicle

Arbuscule

Hypha

Root hair

Spore

Fungal mycelium

a. Lodgepole pine

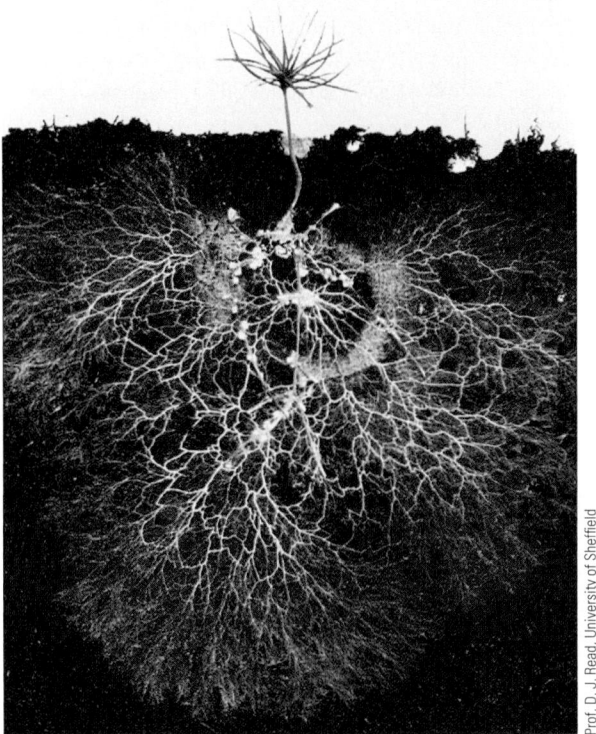

Prof. D. J. Read, University of Sheffield

b. Mycorrhiza

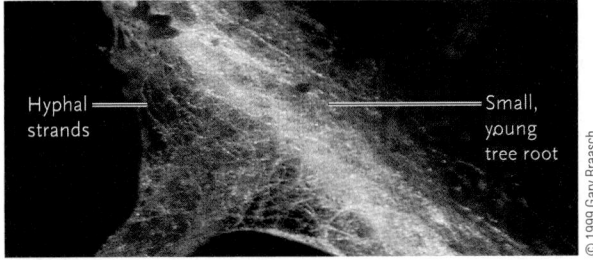

Hyphal strands

Small, young tree root

© 1999 Gary Braasch

Figure 28.17
Ectomycorrhizae. **(a)** Lodgepole pine, *Pinus contorta*, seedling, longitudinal section. Notice the extent of the mycorrhiza compared with the above-ground portion of the seedling, which is only about 4 cm tall. **(b)** Mycorrhiza of a hemlock tree.

Figure 28.18
Effect of mycorrhizal fungi on plant growth. The six-month-old juniper seedlings on the left were grown in sterilized low-phosphorus soil inoculated with a mycorrhizal fungus. The seedlings on the right were grown under the same conditions but without the fungus.

F. B. Reeves

the transport of water and minerals to the plants. In that scenario, endomycorrhizae may have played a crucial role in allowing plants to make the transition to life on land.

STUDY BREAK

1. Explain what a lichen is, and how each partner contributes to the whole.
2. Describe the biological and ecological roles of mycorrhizae.
3. How do endomycorrhizae and ectomycorrhizae differ?

How do plant pathogenic fungi invade plants?

Many species of fungi are pathogenic to plants. Of particular interest to humans are the pathogenic fungi that invade crop plants. In general, to invade a plant the pathogenic fungus must first break down any form of natural resistance that the plant has, and then establish an infection. Moreover, each pathogenic fungus has specificity—it invades only a particular set of plants, not all plants. For a number of pathogenic fungi, scientists are beginning to gain an understanding of the cellular and molecular events involved in invasion and the spread of the infection through the plant. A complete understanding of these processes will open the way to developing approaches that protect crop plants from fungal invasion, or at least reduce the extent of damage to the plants.

One example of the research being done in this area concerns the ascomycete fungus *Cochliobolus carbonum*. This fungus is pathogenic to maize, causing leaf blight (early drying of the leaves) and ear rot disease. *C. carbonum* secretes a toxin called HC-toxin to infect maize hosts. Guri Johal of Purdue University is studying the infection process, in particular investigating the molecular mechanisms by which HC-toxin leads to fungal colonization of maize tissues. Currently, little is known about those mechanisms.

Another example of research with pathogenic fungi concerns the ascomycete *Magnaporthe grisea*, the fungus that causes rice blast (lesions on leaves and other parts of the plant). The genome of this fungus has been sequenced, making possible the use of genomic/proteomic tools and approaches for studying pathogenesis. Dan Ebbole at Texas A&M University is using those tools and approaches to analyze proteins secreted by *M. grisea* with the aim of understanding their roles in the interaction of the pathogen with rice plants. Specifically, Ebbole and his group are looking at 300 proteins that, based on analysis of the genome, are predicted to be secreted. They produce tagged versions of the proteins by expressing the genes for them in fungal cultures. Then they test the purified proteins directly on plants one by one to see if any elicits a specific response by the host plant. They anticipate that this approach will serve as a screen to identify proteins that play significant roles in the pathogen–plant interaction. Those proteins will then be analyzed more completely, with the objective of developing cellular and molecular models for pathogenesis.

What are the interactions between all the molecular components of a fungus?

As you learned in Section 18.3, the study of the interactions between all of the molecular components of a cell or organism is systems biology. Over the years, significant advances have been made toward a molecular understanding of many processes in fungi, particularly in model fungi such as the yeast *Saccharomyces cerevisiae* and the mold *Neurospora crassa*. In addition, genome sequences have been obtained for a number of fungi, including the two species just mentioned as well as some pathogenic species. For a number of fungi, then, researchers are poised for systems biology studies. To that end, scientists from around the world have established the Yeast Systems Biology Network (YSBN) to coordinate research efforts in the systems biology of *S. cerevisiae*. The researchers argue that this yeast is a particularly appropriate model system for a concentrated effort to obtain a systems-level understanding of biological processes. Indeed, yeast has been a model system for eukaryotic cell structure and function, and for a number of aspects of fungal biology (see Chapter 10's *Focus on Research*). It was also one of the original model eukaryotes chosen for genome sequencing in the Human Genome Project (see Chapter 18).

Peter J. Russell

Review

Go to **ThomsonNOW** at www.thomsonedu.com/login to access quizzing, animations, exercises, articles, and personalized homework help.

28.1 General Characteristics of Fungi

- Fungi are key decomposers contributing to the recycling of carbon and some other nutrients. They occur as single-celled yeasts or multicellular filamentous organisms.

- The fungal mycelium consists of filamentous hyphae that grow throughout the substrate the fungus feeds upon (Figure 28.2). A wall containing chitin surrounds the plasma membrane, and in most species septa partition the hyphae into cell-like compartments. Pores in septa permit cytoplasm and organelles to move between hyphal cells. Aggregations of hyphae form all other tissues and organs of a multicellular fungus.

- Fungi gain nutrients by extracellular digestion and absorption. Saprobic species feed on nonliving organic matter. Parasitic types obtain nutrients from tissues of living organisms. Many fungi are partners in symbiotic relationships with plants.

- All fungi may reproduce via spores generated either asexually or sexually (Figure 28.3). Some types also may reproduce asexually by budding or fragmentation of the parent body. Sexual reproduction usually has two stages. First, in plasmogamy, the cytoplasms of two haploid cells fuse to become a dikaryon containing a haploid nucleus from each parent. Later, in karyogamy, the nuclei fuse and form a diploid zygote. Meiosis then generates haploid spores.

Animation: Mycelium

28.2 Major Groups of Fungi

- The main phyla of fungi are the Chytridiomycota (which have motile spores), Zygomycota (zygospore-forming fungi), Glomeromycota, Ascomycota (sac fungi), and Basidiomycota (club fungi) (Figure 28.4). The phyla traditionally have been distinguished mainly on the basis of the structures that arise as part of sexual reproduction. When a sexual phase cannot be detected or is absent from the life cycle, the specimen is assigned to an informal grouping, the conidial fungi.

- Chytrids usually are microscopic. They are the only fungi that produce motile, flagellated spores. Many are parasites (Figure 28.5).

- Zygomycetes have aseptate hyphae and are coenocytic, with many nuclei in a common cytoplasm. They sometimes reproduce sexually by way of hyphae that occur in + and − mating types; haploid nuclei in the hyphae function as gametes. Further development produces the zygospore, which may go dormant for a time. When the zygospore breaks dormancy it produces a stalked sporangium containing haploid spores of each mating type, which are released (Figures 28.6 and 28.7).

- Glomerulomycetes form a distinct type of endomycorrhizae in association with plant roots. They reproduce asexually, by way of spores that form at the tips of hyphae.

- Most ascomycetes are multicellular (Figure 28.9). In asexual reproduction, chains of haploid asexual spores called conidia elongate or pinch off from the tips of conidiophores (modified aerial hyphae; Figure 28.10). In sexual reproduction, haploid sexual spores called ascospores arise in saclike cells called asci. In the most complex species, reproductive bodies called ascocarps bear or contain the asci. Ascospores can give rise to a new haploid mycelium (Figures 28.8 and 28.11).

- Most basidiomycete species reproduce only sexually. Club-shaped basidia develop on a basidiocarp and bear sexual spores on their surface. When dispersed, these basidiospores may germinate and give rise to a haploid mycelium (Figure 28.13).

- Microsporidia are single-celled sporelike parasites of arthropods, fish, and humans (Figure 28.14).

Animation: Zygomycete life cycle

Animation: Sac fungi

Animation: Club fungus life cycle

28.3 Fungal Associations

- Many ascomycetes and a few basidiomycetes enter into symbioses with cyanobacteria or green algae to produce the communal life form called a lichen, which has a spongy body called a thallus. The algal cells supply the lichen's carbohydrates, most of which are absorbed by the fungus. In some lichens a section of the thallus may detach and grow into a new individual. In others, specialized regions of the thallus give rise asexually to reproductive soredia that include both algal and hyphal cells (Figure 28.15).

- In the symbiosis called a mycorrhiza, fungal hyphae make mineral ions and sometimes water available to the roots of a plant partner. The fungus in turn absorbs carbohydrates, amino acids, and possibly other growth-enhancing substances from the plant (Figures 28.16–28.18). In endomycorrhizae, the fungal hyphae (usually of a glomeromycete) penetrate the cells of the root. With ectomycorrhizae, hyphal tips grow close to young roots but do not enter roots cells; the usual fungal partner is a basidiomycete.

Animation: Lichens

Animation: Mycorrhiza

Questions

Self-Test Questions

1. Which of the following attributes best exemplifies a filamentous saprobic fungus?
   a. reproduction by spores on week-old bread
   b. metabolic by-products that make bread rise
   c. extracellular digestion of tissues in a fallen log
   d. extracellular digestion of a living leaf's cellulose with hydrolytic enzymes
   e. aggressive expansion of the fungal mycelium into the tissues of a living elm tree

2. Which of the following events is/are *not* part of a typical fungal life cycle involving asexual reproduction?
   a. formation of a dikaryon
   b. hyphae developing into a mycelium
   c. formation of a diploid zygote
   d. plasmogamy, which occurs when hyphae fuse at their tips
   e. production and release of large numbers of spores

3. A trait common to all fungi is:
   a. reproduction via spores.
   b. parasitism.
   c. septate hyphae.
   d. a dikaryotic phase inside a zygospore.
   e. plasmogamy after an antheridium and ascogonium come into contact.

4. The chief characteristic used to classify fungi into the major fungal phyla is:
   a. nutritional dependence on nonliving organic matter.
   b. recycling of nutrients in terrestrial ecosystems.
   c. adaptations for obtaining water.
   d. features of reproduction.
   e. cell wall metabolism.

5. At lunch George ate a mushroom, some truffles, a little Camembert cheese, and a bit of moldy bread. Which of the following groups was *not* represented in the meal?
   a. Basidiomycota
   b. Ascomycota
   c. conidial fungi
   d. chytrids
   e. Zygomycota

6. Which of the following fungal reproductive structures is diploid?
   a. basidiocarps
   b. ascospores
   c. conidia
   d. gametangia
   e. zygospores

7. A mushroom is:
   a. the food-absorbing region of an ascomycete.
   b. the food-absorbing region of a basidiomycete.
   c. a reproductive structure formed only by basidiomycetes.
   d. a specialized form of mycelium not constructed of hyphae.
   e. a collection of saclike cells called asci.

8. A zygomycete is characterized by:
   a. aseptate hyphae.
   b. mostly sexual reproduction.
   c. absence of + and − mating types.
   d. the tendency to form mycorrhizal associations with plant roots.
   e. a life cycle in which karyogamy does not occur.

9. Which best describes a lichen?
   a. It is a fungus that breaks down rock to provide nutrients for an alga.
   b. It colonizes bare rocks and slowly degrades them to small particles.
   c. It spends part of the life cycle as a mycobiont and part as a fungus.
   d. It is an association between a basidiomycete and an ascomycete.
   e. It is an association between a photobiont and a fungus.

10. In a college greenhouse a new employee observes fuzzy mycorrhizae in the roots of all the plants. Destroying no part of the plants, he carefully removes the mycorrhizae. The most immediate result of this "cleaning" is that the plants cannot:
    a. carry out photosynthesis.
    b. absorb water through their roots.
    c. transport water up their stems.
    d. extract phosphorus and nitrogen from water.
    e. store carbohydrates in their roots.

Questions for Discussion

1. A mycologist wants to classify a specimen that appears to be a new species of fungus. To begin the classification process, what kinds of information on body structures and/or functions must the researcher obtain in order to assign the fungus to one of the major fungal groups?

2. In a natural setting—a pile of horse manure in a field, for example—the sequence in which various fungi appear illustrates ecological succession, the replacement of one species by another in a community (see Chapter 50). The earliest fungi are the most efficient opportunists, for they can form and disperse spores most rapidly. In what order would you expect representatives from each division of fungi to appear on the manure pile? Why?

3. As the text noted, conifers, orchids, and some other types of plants cannot grow properly if their roots do not form associations with fungi, which provide the plant with minerals such as phosphate and in return receive carbohydrates and other nutrients synthesized by the plant. In some instances, however, the plant receives proportionately more nutrients than the fungus does. Even so, biologists still consider this to be a mycorrhizal association. Explain why you agree or disagree.

4. Humans are fundamentally diploid organisms. Explain how this state of affairs compares with the fungal life cycle, then compare the two general life cycles in light of the two groups' overall reproductive strategies.

Experimental Analysis

Experiments on the orange bread mold *Neurospora crassa,* an ascomycete, were pivotal in elucidating the concept that each gene encodes a single enzyme. As *N. crassa* ascospores arise through meiosis and then mitosis in an ascus, each ascospore occupies a particular position in the final string of eight spores the ascus contains:

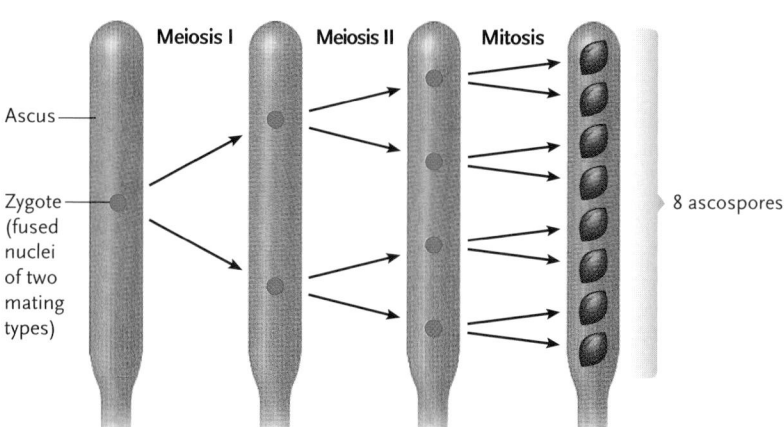

This quirk of ascospore development was extremely useful to early geneticists, because it vastly simplified the task of figuring out which alleles ended up in particular ascospores following meiosis. Recalling genetics topics discussed in Chapter 11, why was the analysis easier?

Evolution Link

The hypothesis that fungi are more closely related to animals than to plants has receive support from studies of fungus genomes. For instance, scientists have documented striking similarities in the structure of many fungal and human genes—similarities that may be especially important in medicine. One mycologist, John Taylor of the University of California at Berkeley, suggests that a close biochemical relationship between fungi and animals may explain why fungal infections are typically so resistant to treatment, and why it has proven rather difficult to develop drugs that kill fungi without damaging their human or other animal hosts. About 100 fungal genomes have been or soon will be sequenced, including genomes of several medically important species. If you are a researcher working to develop new antifungal drugs, how could you make use of this growing genetic understanding? Using Web resources, can you find examples of antifungal drugs that exploit biochemical differences between animals and fungi?

How Would You Vote?

The disappearance of lichens and soil fungi may be an early indication that coal-fired power plants are emitting pollutants that also can endanger human health. Controlling emissions raises the cost of energy for consumers. Should pollution standards for these power plants be tightened? Go to www.thomsonedu.com/login to investigate both sides of the issue and then vote.

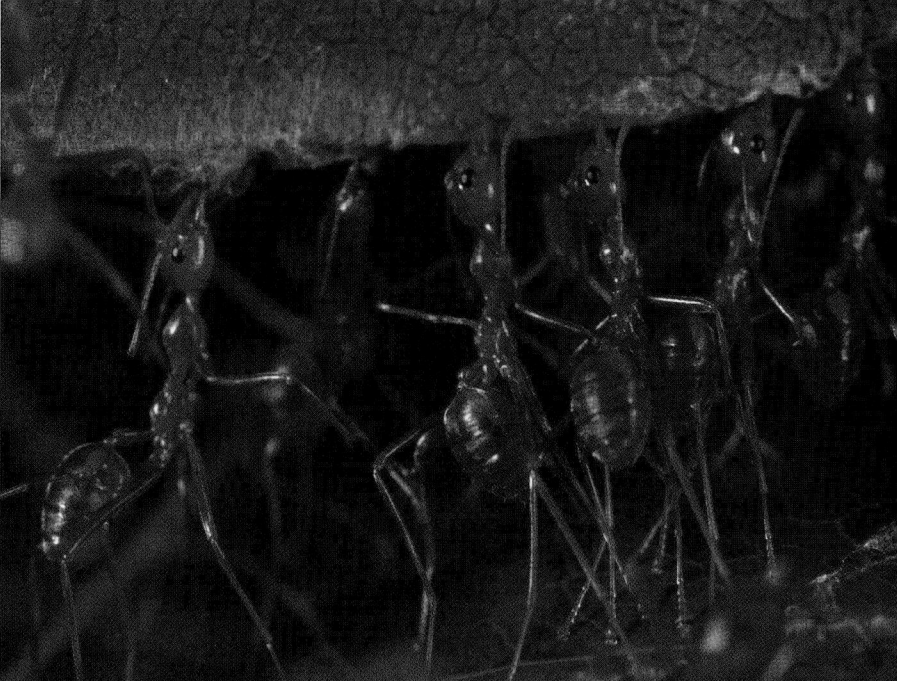

Weaver ants *(Oecophylla longinoda)* carry a leaf to repair their nest in Papua New Guinea.

Mark Moffett/Minden Pictures

29 Animal Phylogeny, Acoelomates, and Protostomes

WHY IT MATTERS

In 1909, a lucky fossil hunter named Charles Wolcott tripped over a rock on a mountain path in British Columbia, Canada. Under the force of his hammer, the rock split apart, revealing the discovery of a lifetime. Wolcott and other workers soon found fossils of more than 120 species of previously undescribed animals from the Cambrian period. These creatures had lived on the muddy sediments of a shallow ocean basin. About 530 million years ago, an underwater avalanche buried them in a rain of silt that was eventually compacted into finely stratified shale. Over millions of years, the shale was uplifted by tectonic activity and incorporated into the mountains of western Canada. It is now known as the Burgess Shale formation.

Some animals in the Burgess Shale were truly bizarre **(Figure 29.1).** For example, *Opabinia* was about as long as a tube of lipstick; it had five eyes on its head and a grasping organ that it may have used to capture prey. No living animals even remotely resemble *Opabinia*. The smaller *Hallucigenia* sported seven pairs of large spines on one side and seven pairs of soft organs on the other. Recent research suggests that *Hallucigenia* may belong in the phylum Onychophora, described in Section 29.7. Nevertheless, most species of the Burgess

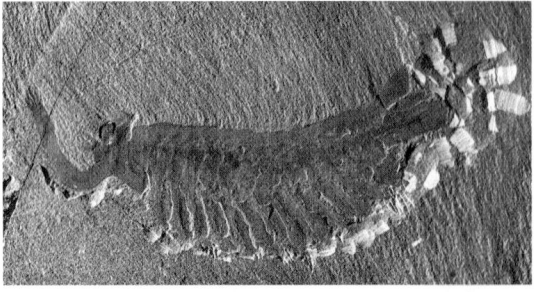
Opabinia

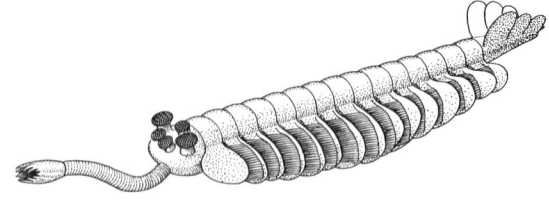

Hallucigenia

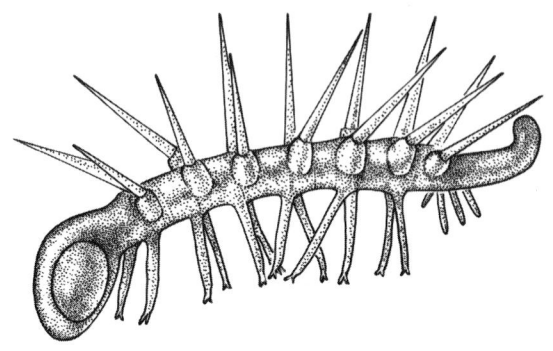

Figure 29.1
Animals of the Burgess Shale. *Opabinia* had five eyes and a grasping organ on its head. *Hallucigenia* had seven pairs of spines and soft protuberances.
(Images: Dr. Chip Clark, National Museum of Natural History, Smithsonian Institution.)

Shale left no descendants that are still alive today. Thus, this remarkable assemblage of fossils provides a glimpse of some evolutionary novelties that—whether through the action of natural selection or just plain bad luck—were ultimately unsuccessful.

Other animal lineages have shown much greater longevity. Zoologists have described nearly 2 million living species in the kingdom **Animalia**, about five times as many as in all the other kingdoms combined. The familiar **vertebrates**, animals with a backbone, encompass only a small fraction (about 47,000 species) of the total. The overwhelming majority of animals fall within the descriptive grouping of **invertebrates**, animals without a backbone.

The remarkable evolutionary diversification of animals resulted from their ability to consume other organisms as food and, for most groups, their ability to move from one place to another. Today animals are important consumers in nearly every environment on Earth. Their diversification has been accompanied by the evolution of specialized tissues and organ systems as well as complex behaviors.

In this chapter, we introduce the general characteristics of animals and a phylogenetic hypothesis about their evolutionary history and classification. We also survey some of the major invertebrate phyla; a *phylum* is an ancient monophyletic lineage with a distinctive body plan. In Chapter 30 we examine the deuterostome lineage, which includes the vertebrates and their nearest invertebrate relatives.

29.1 What Is an Animal?

Biologists recognize the Kingdom Animalia as a monophyletic group that is easily distinguished from the other kingdoms.

All Animals Share Certain Structural and Behavioral Characteristics

Animals are eukaryotic, multicellular organisms. Their cells lack cell walls, a trait that differentiates them from plants and fungi. The individual cells of most animals are similar in size, so that very large animals like elephants have many more cells than small ones like fleas. In large animals, most cells are far from the body surface, but specialized tissues and organ systems deliver nutrients and oxygen to them and carry wastes away.

All animals are **heterotrophs**: they acquire energy and nutrients by eating other organisms. Food is ingested (eaten) and then digested (broken down) and absorbed by specialized tissues. Animals use oxygen to metabolize the food they eat through the biochemical pathways of aerobic respiration, and most store excess energy as glycogen, oil, or fat.

All animals are **motile**—able to move from place to place—at some time in their lives. They travel through the environment to find food or shelter and to interact with other animals. Most familiar animals are motile as adults. However, in some species, such as mussels and barnacles, only the young are motile; they eventually settle down as **sessile**—unable to move from one place to another—adults. The advantages of motility have fostered the evolution of locomotor structures, including fins, legs, and wings. And in many animals, locomotion results from the action of muscles, specialized contractile tissues that move individual body parts. Most animals also have sensory and nervous systems that allow them to receive, process, and respond to information about the environment.

Animals reproduce either asexually or sexually; in many groups they switch from one mode to the other. Sexually reproducing species produce short-lived, haploid **gametes** (eggs and sperm), which fuse to form diploid **zygotes** (fertilized eggs). Animal life cycles generally include a period of development during which mitosis transforms the zygote into a multicelled **embryo**, which develops into a sexually immature juvenile or a free-living **larva**, which becomes a sexually mature adult. Larvae often differ markedly from adults, and they may occupy different habitats and consume different foods.

The Animal Lineage Probably Arose from a Colonial Choanoflagellate Ancestor

An overwhelming body of morphological and molecular evidence indicates that all animal phyla had a common ancestor. For example, all animals share similarities in their cell-to-cell junctions and the molecules in their extracellular matrices (see Section 5.5) as well as similarities in the structure of their ribosomal RNAs.

Most biologists agree that the common ancestor of all animals was probably a colonial, flagellated protist that lived at least 700 million years ago, during the Precambrian era. It may have resembled the minute, sessile choanoflagellates that live in both freshwater and marine habitats today (see Figure 26.21). In 1874 the German embryologist Ernst Haeckel proposed a colonial, flagellated ancestor, suggesting that it was a hollow, ball-shaped organism with unspecialized cells. According to his hypothesis, its cells became specialized for particular functions, and a developmental re-

organization produced a double-layered, sac-within-a-sac body plan **(Figure 29.2)**. As you will see in Chapter 48, the embryonic development of many living animals roughly parallels this hypothetical evolutionary transformation.

STUDY BREAK

1. What characteristics distinguish animals from plants?
2. How does the ability of animals to move through the environment relate to their acquisition of nutrients and energy?

29.2 Key Innovations in Animal Evolution

Once established, the animal lineage diversified quickly into an amazing array of body plans. Before the development of molecular sequencing techniques, biologists used several key morphological innovations to unravel the evolutionary relationships of the major animal groups.

Tissues and Tissue Layers Appeared Early in Animal Evolution

The presence or absence of **tissues**, groups of cells that share a common structure and function, divides the animal kingdom into two distinct branches. One branch, the sponges, or Parazoa (*para* = alongside; *zoon* = animal), lacks tissues. All other animals, collectively grouped in the Eumetazoa (*eu* = true; *meta* = later), have tissues.

During the development of eumetazoans, embryonic tissues form as either two or three concentric **primary cell layers**. The innermost layer, the **endoderm**, eventually develops into the lining of the gut (digestive system) and, in some animals, respiratory organs. The outermost layer, the **ectoderm**, forms the external cover-

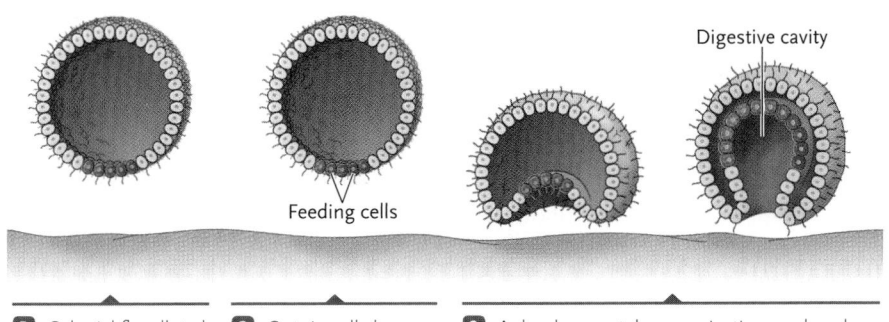

Figure 29.2

Animal origins. Many biologists believe that animals arose from a colonial, flagellated protist in which cells became specialized for specific functions and a developmental reorganization produced two cell layers. The cell movements illustrated here are similar to those that occur during the development of many animals, as described in Chapter 48.

1 Colonial flagellated protist with unspecialized cells

2 Certain cells became specialized for feeding and other functions.

3 A developmental reorganization produced a two-layered animal with a sac-within-a-sac body plan.

ing and nervous system. Between the two, the **mesoderm** forms the muscles of the body wall and most other structures between the gut and the external covering. Some simple animals have a **diploblastic** body plan that includes only two layers, endoderm and ectoderm. However, most animals are **triploblastic**, having all three primary cell layers.

Most Animals Exhibit either Radial or Bilateral Symmetry

The most obvious feature of an animal's body plan is its shape **(Figure 29.3).** Most animals are **symmetrical**; in other words, their bodies can be divided by a plane into mirror-image halves. By contrast, most sponges have irregular shapes and are therefore **asymmetrical**.

Most eumetazoans exhibit one of two body symmetry patterns. The Radiata includes two phyla, Cnidaria (hydras, jellyfishes, and sea anemones) and Ctenophora (comb jellies), which have **radial symmetry**. Their body parts are arranged regularly around a central axis, like the spokes on a wheel. Thus, any cut down the long axis of a hydra divides it into matching halves. Radially symmetrical animals are usually sessile or slow moving and receive sensory input from all directions.

All other eumetazoan phyla fall within the Bilateria, animals that have **bilateral symmetry**. In other words, only a cut along the midline from head to tail divides them into left and right sides that are essentially mirror images of each other. Bilaterally symmetrical animals also have **anterior** (front) and **posterior** (back) ends as well as **dorsal** (upper) and **ventral** (lower) surfaces. As these animals move through the environment, the anterior end encounters food, shelter, or enemies first. Thus, in bilaterally symmetrical animals, natural selection also favored **cephalization**, the development of an anterior head where sensory organs and nervous system tissue are concentrated.

Many Animals Have Body Cavities That Surround Their Internal Organs

The body plans of many bilaterally symmetrical animals include a body cavity that separates the gut from the muscles of the body wall **(Figure 29.4).** Acoelomate animals (*a* = not; *koilos* = hollow), such as flatworms (phylum Platyhelminthes), do not have such a cavity; a continuous mass of tissue, derived largely from mesoderm, packs the region between the gut and the body wall (see Figure 29.4a). **Pseudocoelomate** animals (*pseudo* = false), including the roundworms (phylum Nematoda) and wheel animals (phylum Rotifera), have a **pseudocoelom**, a fluid- or organ-filled space between the gut and the muscles of the body wall (see Figure 29.4b). Internal organs lie within the pseudocoelom and are bathed by its fluid. **Coelomate** animals have a true **coelom**, a fluid-filled body cavity completely lined by the **peritoneum**, a thin tissue derived from mesoderm (see Figure 29.4c). Membranous extensions of the inner and outer layers of the peritoneum, the **mesenteries**, surround the internal organs and suspend them within the coelom.

Biologists describe the body plan of pseudocoelomate and coelomate animals as a "tube within a tube"; the digestive system forms the inner tube, the body wall forms the outer tube, and the body cavity lies between them. The body cavity separates internal organs from the body wall, allowing them to function independently of whole-body movements. The fluid within the cavity also protects delicate organs from mechanical damage. And, because the volume of the body cavity is fixed, the incompressible fluid within it serves as a **hydrostatic skeleton**, which provides support; in some animals muscle contractions can shift the fluid, changing the animals' shape and allowing them to move from place to place (see Section 41.2).

Developmental Patterns Mark a Major Divergence in Animal Ancestry

Embryological and molecular evidence suggests that bilaterally symmetrical animals are divided into two lineages: the protostomes, which includes most phyla of invertebrates, and the deuterostomes, which includes the vertebrates and their nearest invertebrate relatives. Protostomes and deuterostomes differ in several developmental characteristics **(Figure 29.5).**

Shortly after fertilization, an egg undergoes a series of mitotic divisions called **cleavage** (see Section 48.1). The first two cell divisions divide a zygote as you might slice an apple, cutting it into four wedges from top to bottom. In many protostomes, subsequent cell divisions produce daughter cells that lie *between* the pairs of cells below them; this pattern is called **spiral cleavage** (left side of Figure 29.5a). In deuterostomes, by contrast, subsequent cell divisions produce a mass

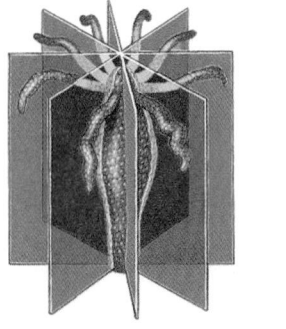
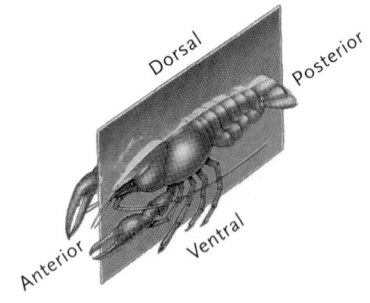
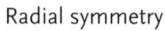

Radial symmetry   Bilateral symmetry

Figure 29.3

Patterns of body symmetry. Most animals have either radial or bilateral symmetry.

a. In acoelomate animals, no body cavity separates the gut and body wall.

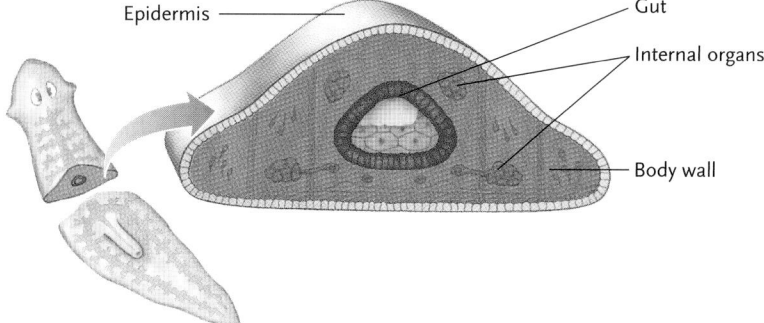

Epidermis

Gut

Internal organs

Body wall

Figure 29.4
Body plans of triploblastic animals.

b. In pseudocoelomate animals, the pseudocoelom forms between the gut (a derivative of endoderm) and the body wall (a derivative of mesoderm).

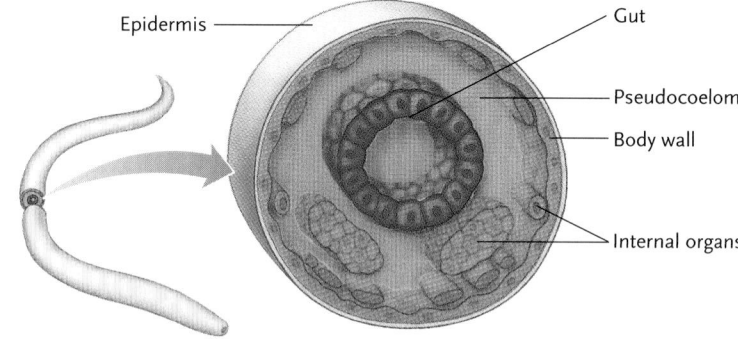

Epidermis

Gut

Pseudocoelom

Body wall

Internal organs

c. In coelomate animals, the coelom is completely lined by peritoneum (a derivative of mesoderm).

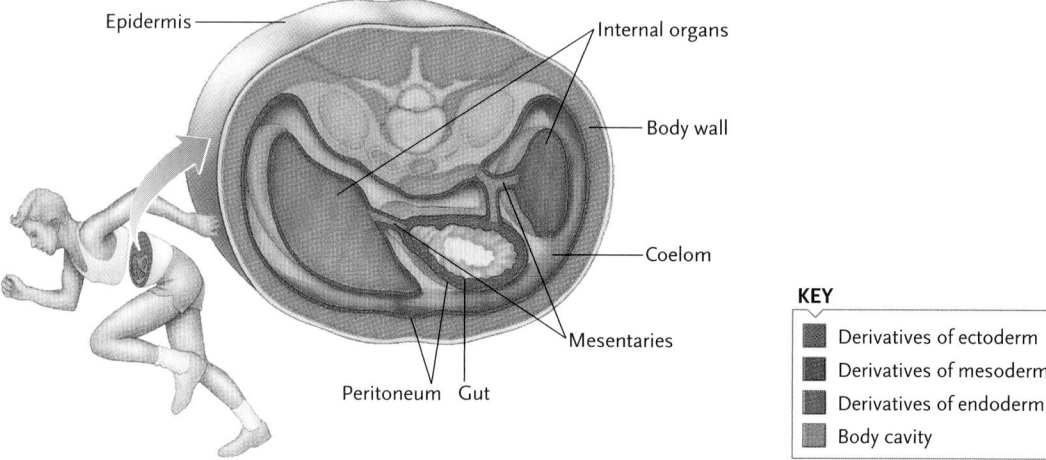

Epidermis

Internal organs

Body wall

Coelom

Mesentaries

Peritoneum   Gut

KEY

▪	Derivatives of ectoderm
▪	Derivatives of mesoderm
▪	Derivatives of endoderm
▪	Body cavity

of cells that are stacked directly above and below one another; this pattern is called **radial cleavage** (right side of Figure 29.5a).

Protostomes and deuterostomes often differ in the timing of important developmental events. During cleavage, certain genes are activated at specific times, determining a cell's developmental path and ultimate fate. Many protostomes undergo **determinate cleavage:** each cell's developmental path is determined as the cell is produced. Thus, one cell isolated from a two- or four-cell protostome embryo cannot develop into a functional embryo or larva. By contrast, many deuterostomes have **indeterminate cleavage:** the developmental fates of cells are determined later. A cell isolated from a four-cell deuterostome embryo will develop into a functional, although smaller than usual, embryo or larva. Like other deuterostomes, humans have indeterminate cleavage; thus, the cells produced by the first few cleavage divisions sometimes separate and develop into identical twins.

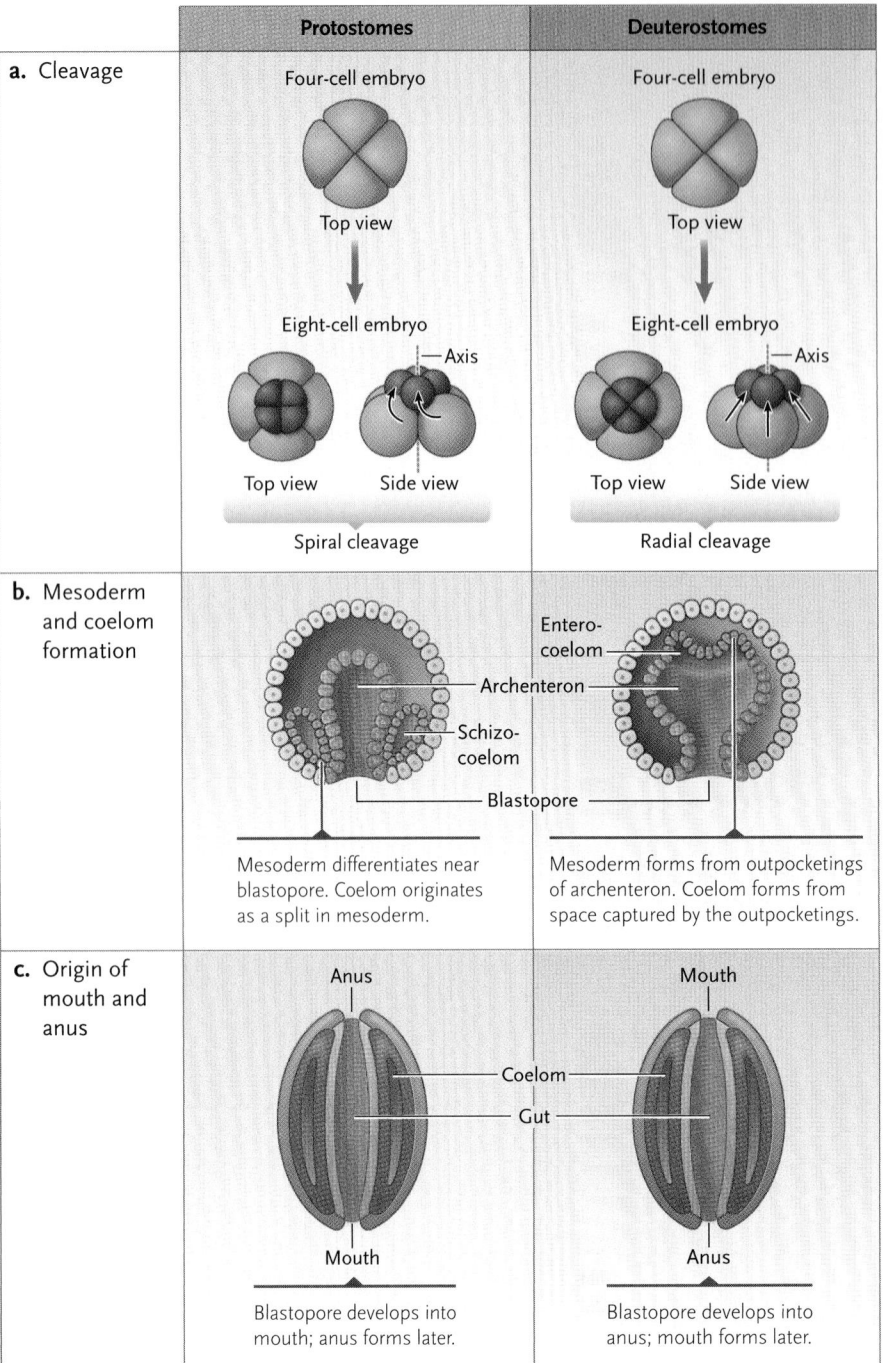

KEY

- Derivatives of ectoderm
- Derivatives of mesoderm
- Derivatives of endoderm
- Body cavity

	Protostomes	Deuterostomes
a. Cleavage	Four-cell embryo — Top view Eight-cell embryo — Top view / Side view — Axis Spiral cleavage	Four-cell embryo — Top view Eight-cell embryo — Top view / Side view — Axis Radial cleavage
b. Mesoderm and coelom formation	Archenteron, Schizocoelom, Blastopore Mesoderm differentiates near blastopore. Coelom originates as a split in mesoderm.	Enterocoelom, Archenteron, Blastopore Mesoderm forms from outpocketings of archenteron. Coelom forms from space captured by the outpocketings.
c. Origin of mouth and anus	Anus, Coelom, Gut, Mouth Blastopore develops into mouth; anus forms later.	Mouth, Coelom, Gut, Anus Blastopore develops into anus; mouth forms later.

Figure 29.5

Protostomes and deuterostomes. The two lineages of coelomate animals differ in **(a)** cleavage patterns, **(b)** the origin of mesoderm and the coelom, and **(c)** the polarity of the digestive system.

As development proceeds, an opening on the surface of the embryo connects the developing gut, called the **archenteron**, to the outside environment. This initial opening is called the **blastopore** (see Figure 29.5b). Later in development, a second opening at the opposite end of the embryo transforms the pouchlike gut into a digestive tube (see Figure 29.5c). In protostomes (*protos* = first; *stoma* = mouth), the blastopore develops into the mouth, and the second opening forms the anus. In some deuterostomes (*deuteros* = second), the blastopore develops into the anus, and the second opening becomes the mouth.

Protostomes and deuterostomes also differ in the origin of mesoderm and the coelom (see Figure 29.5b). In most protostomes, mesoderm originates from a few specific cells near the blastopore. As the mesoderm grows and develops, it splits into inner and outer layers. The space between the layers is called a **schizocoelom** (*schizo* = split). In deuterostomes, mesoderm often forms from outpocketings of the archenteron. The space pinched off by the outpocketings is called an **enterocoelom** (*enteron* = intestine).

Several other characteristics also differ in adult protostomes and deuterostomes. For example, the central nervous system of protostomes is generally positioned on the ventral side of the body, and their brain surrounds the opening of the digestive tract. By contrast, the nervous system and brain of deuterostomes lie on the dorsal side of the body.

Segmentation Divides the Bodies of Some Animals into Repeating Units

Some phyla in both the protostome and deuterostome lineages exhibit varying degrees of **segmentation**, the production of body parts as repeating units. During development, segmentation first arises in the mesoderm, the middle tissue layer that produces most of the body's bulk. In humans and other vertebrates, we see evidence of segmentation in the vertebral column (backbone), ribs, and associated muscles, such as the "six-pack abs" that sit-ups accentuate. In some animals, segmentation is also reflected in structures derived from the endoderm and ectoderm. For example, the ringlike pattern on an earthworm or a caterpillar matches the underlying segments.

Segmentation provides several advantages. In markedly segmented animals, such as earthworms and their relatives, each segment may include a complete set of important organs, including respiratory surfaces and parts of the nervous, circulatory, and excretory systems. Thus, a segmented animal may survive damage to the organs in one segment, because those in other segments perform the same functions. Segmentation also improves control over movement, especially in wormlike animals. Each segment has its own set of muscles, which can act independently of those in other segments. Thus, an earthworm can move its anterior end to the left while it swings its posterior end to the right. The segmented backbone and body wall musculature of vertebrates allow greater flexibility of movement than would unsegmented structures.

STUDY BREAK

1. What is a tissue, and what three primary tissue layers are present in the embryos of most animals?
2. What type of body symmetry do humans have?
3. What is the functional significance of the coelom?
4. What are some advantages of having a segmented body?

29.3 An Overview of Animal Phylogeny and Classification

For many years, biologists used the morphological innovations and embryological patterns described earlier to trace the phylogenetic history of animals. These efforts were sometimes hampered by the difficulty of identifying homologous structures in different phyla

and by morphological data that led to contradictory interpretations. Recently, biologists have used molecular sequence data to reanalyze animal relationships. Although biologists now recognize nearly 40 animal phyla, we focus primarily on the phyla that include substantial numbers of species.

Molecular Analyses Have Refined Our Understanding of Animal Phylogeny

Molecular analyses of animal relationships are often based on nucleotide sequences in small subunit ribosomal RNA and mitochondrial DNA (see Chapter 15). Recent analyses of *Hox* gene sequences provide similar results. (*Hox* genes are described in Sections 22.6 and 48.6.) These molecular analyses are still reasonably new, and they include studies of relatively few genes. Thus, the phylogenetic tree based on molecular sequences **(Figure 29.6)** represents a working hypothesis; its details will likely change as researchers accumulate more data.

The phylogenetic tree based upon molecular characters includes the major lineages that biologists had defined using the morphological innovations and embryological characters just described. For example, molecular data confirm the distinctions between the Parazoa and the Eumetazoa and between the Radiata and the Bilateria. They also confirm the separation of the deuterostome phyla from all others within the Bilateria.

However, the molecular phylogeny groups many other phyla—including the acoelomate animals, pseudocoelomate animals, protostomes, and a few others—into one taxon, the Protostomia. This group is, in turn, subdivided into two major lineages, the Lophotrochozoa and the Ecdysozoa, groups that were not previously recognized. The name Lophotrochozoa (*lophos* = crest; *trochos* = wheel) refers to both the lophophore, a feeding structure found in three phyla (illustrated in Figure 29.15), and the trochophore, a type of larva found in annelids and mollusks (illustrated in Figure 29.23). The name Ecdysozoa (*ekdysis* = escape) refers to the cuticle or external skeleton that these species secrete and periodically molt (or "escape from") when they experience a growth spurt or begin a different stage of the life cycle (illustrated in Figure 29.34); the molting process is called **ecdysis**.

The Molecular Phylogeny Reveals Surprising Patterns in the Evolution of Key Morphological Innovations

Phylogenetic trees contain explicit hypotheses about evolutionary change, and the molecular phylogeny has forced biologists to reevaluate the evolution of several important morphological innovations. For example, traditional phylogenies based upon morphology and embryology usually inferred that the absence of a body cavity, the acoelomate condition, was ancestral and that the

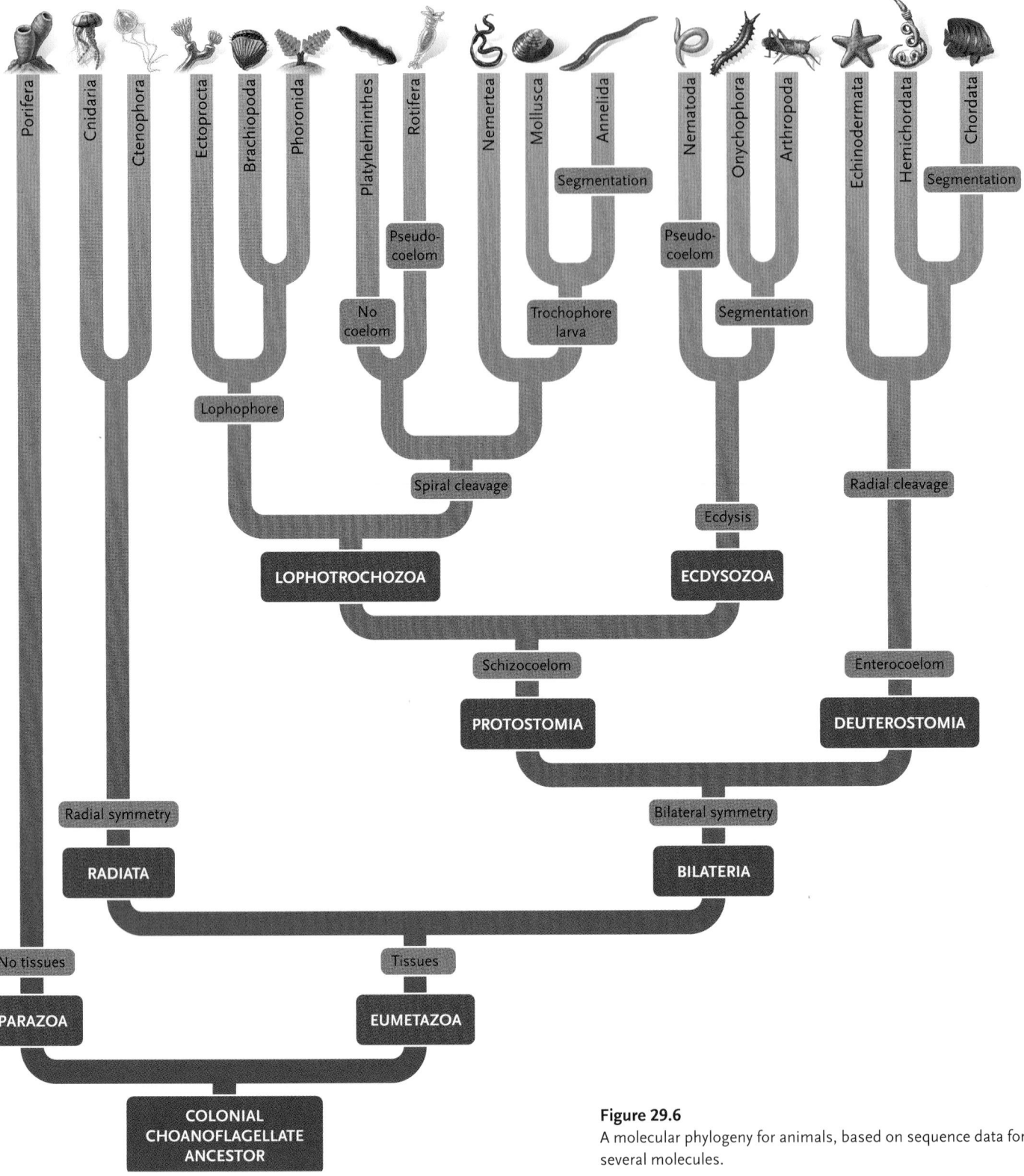

Figure 29.6

A molecular phylogeny for animals, based on sequence data for several molecules.

presence of a body cavity, the pseudocoelomate or coelomate condition, was derived. But the molecular tree provides a very different view. Because most protostome phyla have a schizocoelom, the molecular tree suggests that this trait is ancestral within the lineage, having evolved in the common ancestor of the lineage. If that hypothesis is correct, then the acoelomate condition of flatworms (phylum Platyhelminthes) represents the evolutionary *loss* of the schizocoelom, *not* an ancestral condition. Similarly, the molecular tree hypothesizes that the pseudocoelom evolved independently in rotifers (Lophotrochozoa, phylum Rotifera) and in roundworms (Ecdysozoa, phylum Nematoda) as modifications of the ancestral schizocoelom. Thus, according to the molecular tree, the pseudocoelomate condition of these organisms is the product of convergent evolution.

Traditional phylogenies also suggested that the segmented body plan of several protostome phyla was inherited from a segmented common ancestor and that segmentation arose independently in the chordates by convergent evolution. The molecular tree, by contrast, suggests that segmentation evolved independently in *three* lineages—segmented worms (Lophotrochozoa, phylum Annelida), arthropods and velvet worms (Ecdysozoa, phyla Arthropoda and Onychophora), and chordates (Deuterostomia, phylum Chordata)—rather than in just two lineages.

Despite these surprising findings about morphological evolution, most biologists now embrace the hypothesis provided by molecular sequence studies. In the future, new data will undoubtedly foster active discussions, heated disputes, and revisions of the phylogeny. Students may be understandably frustrated by the lack of consensus among experts and an ever-changing phylogeny for animals. But these disputes highlight the uniqueness of science as a way of knowing the natural world through the process of collecting evidence and rigorously challenging accepted hypotheses. The phylogenetic tree based on molecular sequence data is truly revolutionary, and the dust has yet to settle on the disagreements that these new analyses have provoked.

STUDY BREAK

1. Which major groupings of animals defined on the basis of morphological characters have been confirmed by molecular sequence studies?
2. What type of body cavity is ancestral within the Protostomes?

29.4 Animals without Tissues: Parazoa

The Parazoa is a lineage that includes just one group of animals, the sponges.

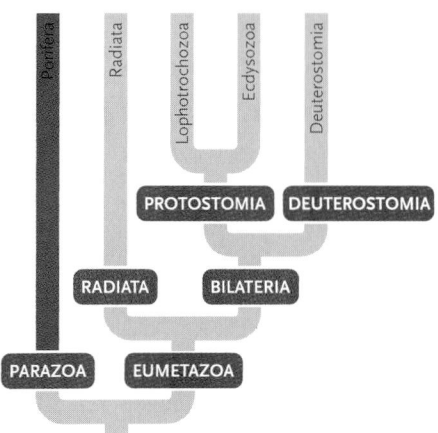

Sponges Have Simple Body Plans and Lack Tissues

Sponges, phylum Porifera (meaning "pore bearers"), lack true tissues: during development, their cells do not form the layers typical of other phyla. Mature sponges are sessile, and their shapes are less fixed than those of other animals, because mobile cells allow them to change shape in response to local conditions **(Figure 29.7)**. Sponges have been abundant since the Cambrian, especially in shallow coastal areas. Most of the 8000 living species are marine. Mature sponges range in size from 1 cm to 2 m.

Sponges have simple body plans **(Figure 29.8)**. Flattened cells form an outer layer, the **pinacoderm.** The inner surface of saclike sponges is lined by collar cells, called **choanocytes,** each equipped with a beating flagellum and a surrounding "collar" of modified microvilli. (Choanocytes resemble the cells of choanoflagellates, the hypothesized ancestor of all animals.) Amoeboid cells wander through the gelatinous **mesohyl** between the two layers; they secrete a supporting skeleton of a fibrous protein and *spicules,* small needlelike structures of calcium carbonate or silica (see Figure 29.8d). The natural sponges that we use are the fibrous remains of the bath sponge *(Spongia),* which lacks mineralized parts.

A sponge's body is an elaborate system for filtering food particles from the surrounding water. Water flows through pores in the pinacoderm into a central chamber, the **spongocoel,** and then out of the sponge through one or more openings called **oscula** (singular, *osculum*). The beating flagellae of the choanocytes maintain a constant flow, and contractile pore cells *(porocytes)* adjust the flow rate. Even a small sponge may filter as much as 20 liters of water per day. The choanocytes capture suspended parti-

Figure 29.7
Asymmetry in sponges. The shapes of sponges vary with their habitats. Those that occupy calm waters, such as this stinker vase sponge *(Ircinia campana),* may be lobed, tubular, cuplike, or vaselike.

Marty Snyderman/Planet Earth Pictures

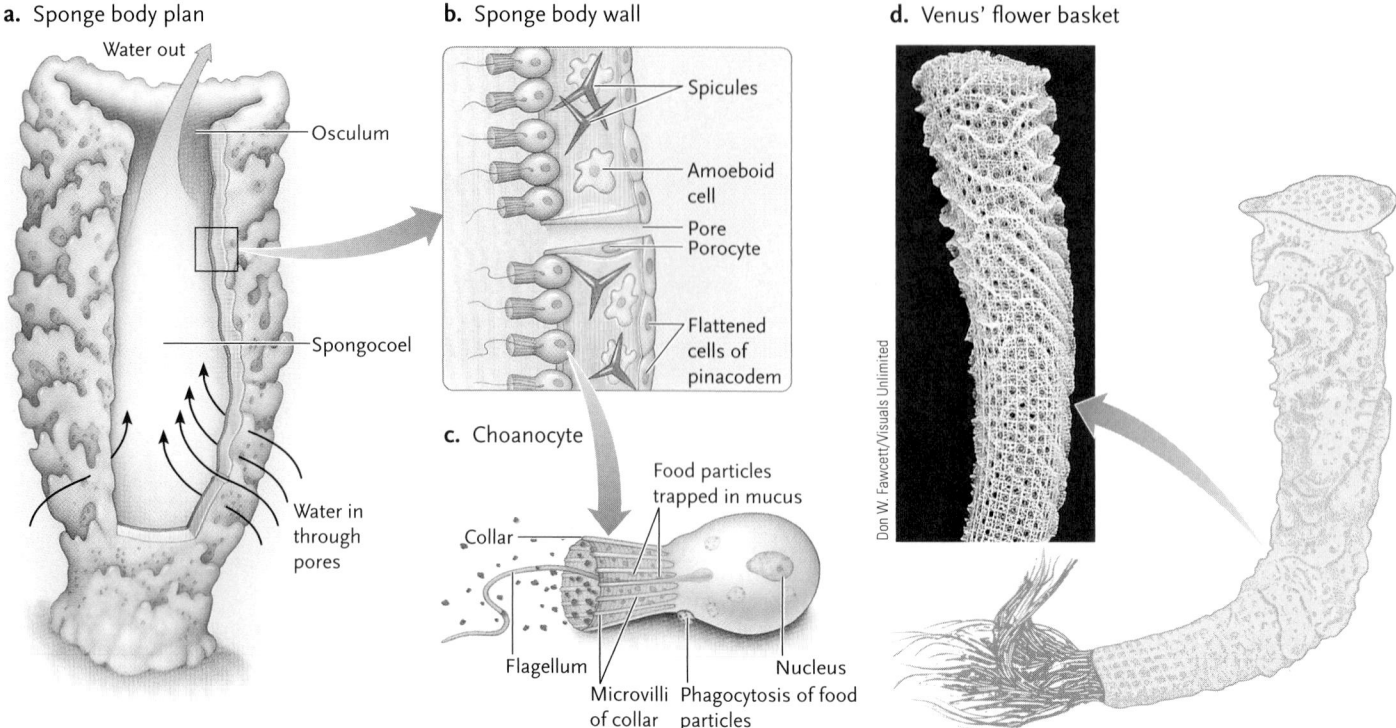

a. Sponge body plan

Water out

Osculum

Spongocoel

Water in through pores

b. Sponge body wall

Spicules

Amoeboid cell

Pore
Porocyte

Flattened cells of pinacoderm

c. Choanocyte

Food particles trapped in mucus

Collar

Flagellum

Microvilli of collar

Phagocytosis of food particles

Nucleus

d. Venus' flower basket

Don W. Fawcett/Visuals Unlimited

Figure 29.8

The body plan of sponges. Most sponges have **(a)** simple body plans and **(b)** relatively few cell types. **(c)** Beating flagella on the choanocytes create a flow of water through incurrent pores, into the spongocoel, and out through the osculum. **(d)** Venus' flower basket (*Euplectella* species), a marine sponge, has spicules of silica fused into a rigid framework.

cles and microorganisms from the water and pass this food to mobile amoeboid cells, which carry nutrients to cells of the pinacoderm.

Most sponges are **hermaphroditic**: individuals produce both sperm and eggs. Sperm are released into the spongocoel and then out into the environment through oscula; eggs remain in the mesohyl, where sperm from other sponges, drawn in with water, fertilize them. Zygotes develop into flagellated larvae that are expelled to fend for themselves. Surviving larvae attach to substrates and undergo **metamorphosis** (a reorganization of form) into sessile adults. Some sponges also reproduce asexually; small fragments break off an adult and grow into new sponges. Many species also produce *gemmules,* clusters of cells with a resistant covering that allows them to survive unfavorable conditions; gemmules germinate into new sponges when conditions improve.

STUDY BREAK

1. What type of body symmetry do sponges exhibit?
2. How does a sponge gather food from its environment?

29.5 Eumetazoans with Radial Symmetry

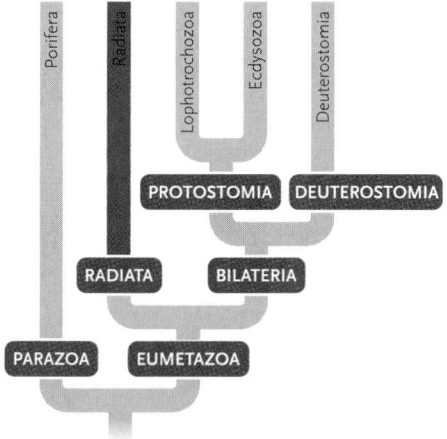

Unlike sponges, eumetazoans have true tissues, which develop from distinct layers in the embryo. Working together, the cells of a tissue perform complex functions beyond the capacity of individual cells. For example, nerve tissue transmits information rapidly through an animal's body, and epithelial tissue forms barriers that surround the body and line body cavities. In this section, we describe eumetazoans with radial symmetry, which enables them to sense stimuli from all directions, an effective adaptation for life in open water.

Two phyla of soft-bodied organisms, Cnidaria and Ctenophora, have radial symmetry and tissues, but they lack organ systems and a coelom. Species in both phyla possess a **gastrovascular cavity** that serves both digestive and circulatory functions. It has a single opening, the mouth. Gas exchange and excretion occur by diffusion because no cell is far from a body surface.

The radiate phyla have a diploblastic body plan with only two tissue layers, the inner *gastrodermis* (an endoderm derivative) and the outer *epidermis* (an ectoderm derivative). Most species also possess a gelatinous *mesoglea* (*mesos* = middle; *glia* = glue) between the two layers. The mesoglea contains widely dispersed fibrous and amoeboid cells.

Cnidarians Use Nematocysts to Stun or Kill Prey

Nearly all of the 8900 species in the phylum Cnidaria (*knide* = stinging nettle, a plant with irritating surface hairs) live in the sea **(Figure 29.9)**. Their body plan is organized around a saclike gastrovascular cavity; the mouth is ringed with tentacles, which push food into it. Cnidarians may be vase-shaped, upward-pointing **polyps** or bell-shaped, downward-pointing **medusae** (see Figure 29.9a). Most polyps attach to a substrate at the *aboral* (opposite the mouth) end; medusae are unattached and float.

Cnidarians are the simplest animals that exhibit a division of labor among specialized tissues (see Figure 29.9b, c). (Sponges have specialized cells, but no tissues.) The gastrodermis includes gland cells and phagocytic nutritive cells. Gland cells secrete enzymes for the extracellular digestion of food, which is then engulfed by nutritive cells and exposed to intracellular digestion. The epidermis includes nerve cells, sensory cells, contractile cells, and cells specialized for prey capture. A layer of acellular mesoglea separates the gastrodermis from the epidermis.

Cnidarians prey on crustaceans, fishes, and other animals. The epidermis includes unique cells, **cnidocytes**, each armed with a stinging **nematocyst** **(Figure 29.10)**. The nematocyst is an encapsulated, coiled thread that is fired at prey or predators, sometimes releasing a toxin through its tip. Discharge of nematocysts may be triggered by touch, vibrations, or chemical stimuli. The toxin can paralyze small prey by disrupting nerve cell membranes. The painful stings of some jellyfishes and corals result from the discharge of nematocysts.

Cnidarians engage in directed movements by contracting specialized cells in the epidermis. In medusae, the mesogleal jelly serves as a deformable skeleton against which contractile cells act. Rapid contractions narrow the bell, forcing out jets of water that propel the animal. Polyps use their water-filled gastrovascular cavity as a hydrostatic skeleton. When some cells contract, fluid within the chamber is shunted about,

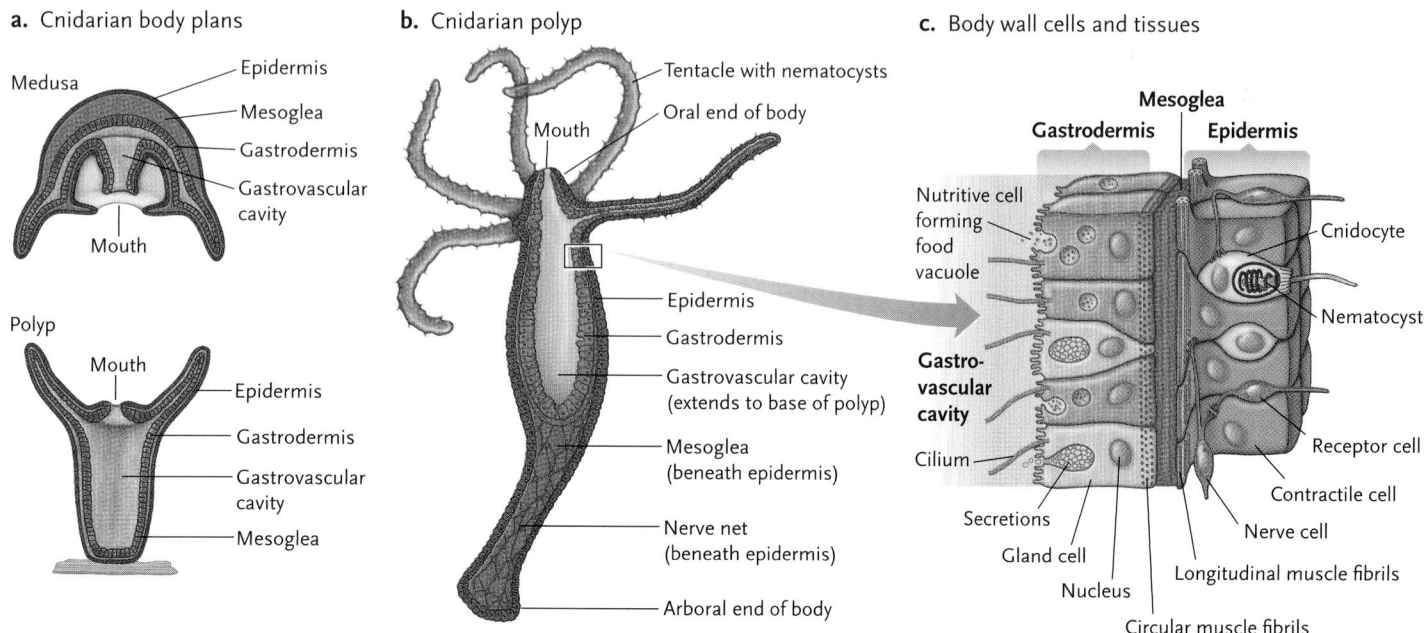

Figure 29.9

The cnidarian body plan. **(a)** Cnidarians exist as either polyps or medusae. **(b)** The body of both forms is organized around a gastrovascular cavity, which extends all the way to the aboral end of the animal. **(c)** The two tissue layers in the body wall, the gastrodermis and the epidermis, include a variety of cell types.

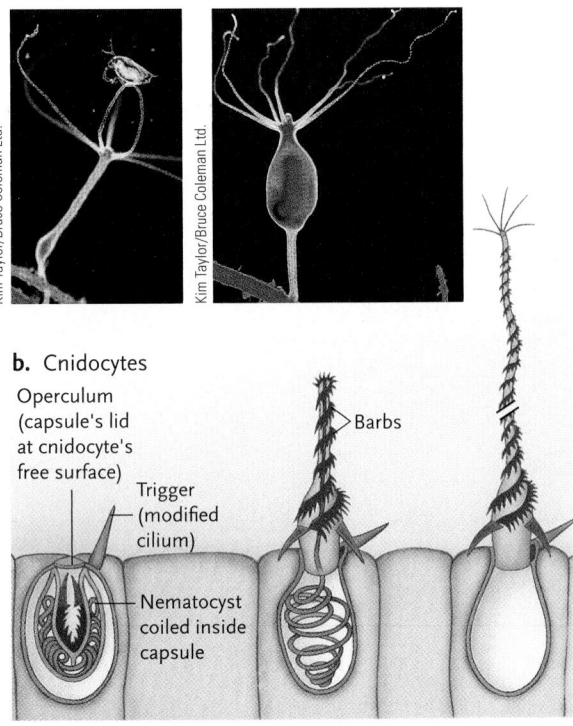

a. *Hydra* consuming a crustacean

Kim Taylor/Bruce Coleman Ltd.

Kim Taylor/Bruce Coleman Ltd.

b. Cnidocytes

Operculum
(capsule's lid
at cnidocyte's
free surface)

Trigger
(modified
cilium)

Nematocyst
coiled inside
capsule

Barbs

Figure 29.10

Predation by cnidarians. **(a)** A polyp of a freshwater *Hydra* captures a small crustacean with its tentacles and swallows it whole. **(b)** Cnidocytes, special cells on the tentacles, encapsulate nematocysts, which are discharged at prey.

changing the body's shape and moving it in a particular direction.

The **nerve net**, which threads through both tissue layers, is a simple nervous system that coordinates responses to stimuli but has no central control organ or brain. Impulses initiated by sensory cells are transmitted in all directions from the site of stimulation.

Many cnidarians exist in only the polyp or the medusa form, but some have a life cycle that alternates between them **(Figure 29.11)**. In the latter type, the polyp often produces new individuals asexually from buds that break free of the parent (see Figure 47.2). The medusa is often the sexual stage, producing sperm and eggs, which are released into the water. The four lineages of Cnidaria differ in the form that predominates in the life cycle.

Hydrozoa. Most of the 2700 species in the Hydrozoa have both polyp and medusa stages in their life cycles (see Figure 29.11). The polyps form sessile colonies that develop asexually from one individual. A colony can include thousands of polyps, which may be specialized for feeding, defense, or reproduction. They share food through their connected gastrovascular cavities.

Unlike most hydrozoans, freshwater species of *Hydra* (see Figure 29.10a) live as solitary polyps that attach temporarily to rocks, twigs, and leaves. Under favorable conditions hydras reproduce by budding. Under adverse conditions they produce eggs and sperm; the zygotes, which are encapsulated in a protective coating, develop and grow when conditions improve.

Scyphozoa. The medusa stage predominates in the 200 species of the Scyphozoa **(Figure 29.12a)**, or jellyfishes. They range from 2 cm to more than 2 m in diameter. Nerve cells near the margin of the bell control their tentacles and coordinate the rhythmic activity of contractile cells, which move the animal. Specialized sensory cells are clustered at the edge of the bell: **statocysts** sense gravity and **ocelli** are sensitive to light. Scyphozoan medusae are either male or female; they release gametes into the water where fertilization takes place.

Cubozoa. The 20 known species of box jellyfish, the Cubozoa **(Figure 29.12b)**, exist primarily as cube-shaped medusas only a few centimeters tall; the largest species grows to 25 cm in height. Nematocyst-rich tentacles grow in clusters from the four corners of the boxlike medusa, and groups of light receptors and image-forming eyes occur on the four sides of the bell. Unlike the true jellyfish, cubozoans are active swimmers. They feed on small fishes and invertebrates, immobilizing their prey with one of the deadliest toxins produced by animals. Cubozoans live in tropical and subtropical coastal waters, where they sometimes pose a serious threat to swimmers.

Anthozoa. The Anthozoa includes 6000 species of corals and sea anemones **(Figure 29.13)**. Anthozoans exist only as polyps, and often reproduce by budding or fission; most also reproduce sexually. Corals are always sessile and usually colonial. Most species build calcium carbonate skeletons, which sometimes accumulate into gigantic underwater reefs. The energy needs of many corals are partly fulfilled by the photosynthetic activity of symbiotic protists that live within the corals' cells. For this reason, corals are restricted to shallow water where sunlight can penetrate. Sea anemones, by contrast, are soft-bodied, solitary polyps, ranging from 1 cm to 10 cm in diameter. They occupy shallow coastal waters. Most species are essentially sessile, but some move by crawling slowly or by using the gastrovascular cavity as a hydrostatic skeleton.

Ctenophores Use Tentacles to Feed on Microscopic Plankton

Like the cnidarians, the 100 species of comb jellies in the phylum Ctenophora (*kteis* = comb; *-phoros* = bearing) have radial symmetry, mesoglea, and feeding tentacles. However, they differ from cnidarians in signifi-

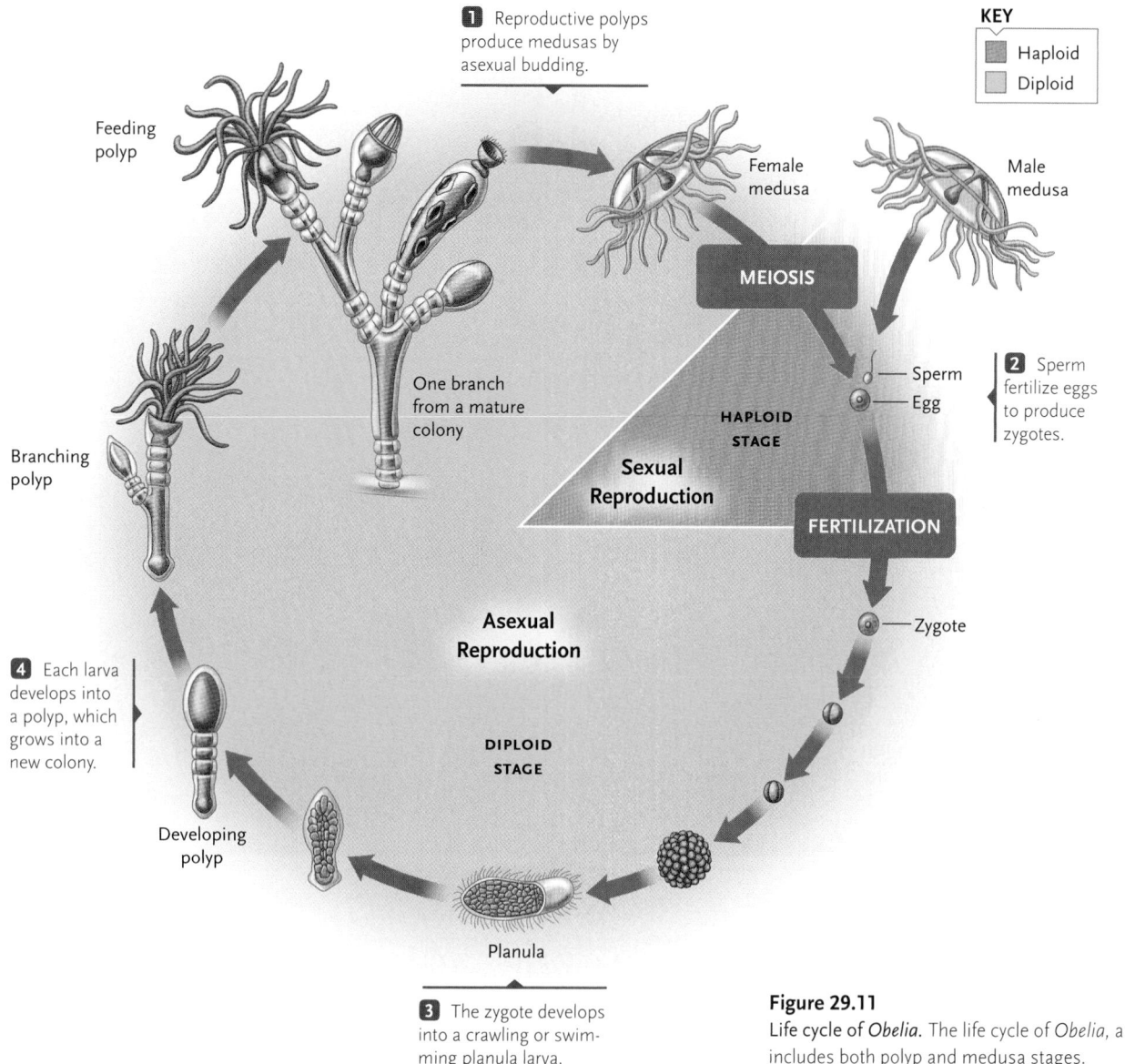

1 Reproductive polyps produce medusas by asexual budding.

Feeding polyp

One branch from a mature colony

Branching polyp

KEY
- Haploid
- Diploid

Female medusa

Male medusa

MEIOSIS

Sperm

Egg

2 Sperm fertilize eggs to produce zygotes.

HAPLOID STAGE

Sexual Reproduction

FERTILIZATION

Zygote

Asexual Reproduction

4 Each larva develops into a polyp, which grows into a new colony.

DIPLOID STAGE

Developing polyp

Planula

3 The zygote develops into a crawling or swimming planula larva.

Figure 29.11

Life cycle of *Obelia*. The life cycle of *Obelia*, a colonial hydrozoan, includes both polyp and medusa stages.

cant ways: they lack nematocysts; they expel some waste through anal pores opposite the mouth; and certain tissues appear to be of mesodermal origin. These transparent, and often luminescent, animals range in size from a few millimeters to a few meters **(Figure 29.14).** They live primarily in coastal regions of the oceans.

Ctenophores move by beating cilia arranged on eight longitudinal plates that resemble combs. They are the largest animals to use cilia for locomotion, but they are feeble swimmers. Nerve cells connected to the cilia coordinate the animals' movements, and a gravity-sensing statocyst helps them maintain an upright position. They capture microscopic plankton with their two tentacles, which have specialized cells that discharge sticky filaments; the food-laden tenta-

cles are drawn across the mouth. Ctenophores are hermaphroditic, producing gametes in cells that line the gastrovascular cavity. Eggs and sperm are expelled through the mouth or from special pores, and fertilization occurs in the open water.

STUDY BREAK

1. How do cnidarians capture, consume, and digest their prey?
2. Which group of cnidarians has only a polyp stage in its life cycle?
3. What do ctenophores eat, and how do they collect their food?

a. Scyphozoan

b. Cubozoan

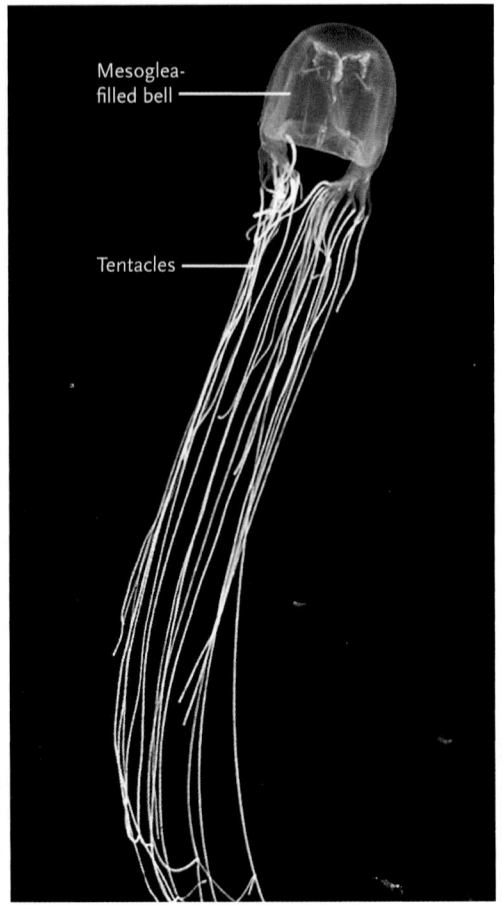

Mesoglea-filled bell

Tentacles

© Michael Durham/Minden Pictures

Courtesy of Dr. William H. Hamner

Figure 29.12

Scyphozoans and cubozoans. (a) Most scyphozoans, like the sea nettle *(Chrysaora quinquecirrha)*, live as floating medusae. Their tentacles trap prey, and the long oral arms transfer it to the mouth on the underside of the bell. **(b)** Cubozoans, like the sea wasp *(Chironex fleckeri)*, are strong swimmers that actively pursue small fishes and invertebrates.

a. Coral

b. Sea anemone escape behavior

Tentacle of one polyp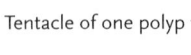

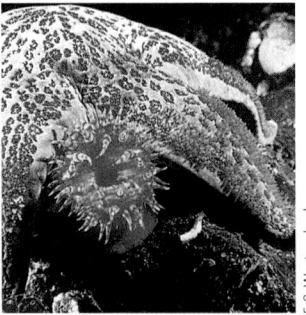

F. S. Westmorland

F. S. Westmorland

F. S. Westmorland

Christian DellaCorte

Interconnected skeletons of polyps of a colonial coral

Figure 29.13

Anthozoans. (a) Many corals, like the staghorn coral *(Acropora cervicornis)*, are colonial, and their polyps build a hard skeleton of calcium carbonate. The skeletons accumulate to form coral reefs in shallow tropical waters (see Figure 52.28b). **(b)** A white-spotted sea anemone *(Urticina lofotensis)* detaches from its substrate to escape from a predatory sea star.

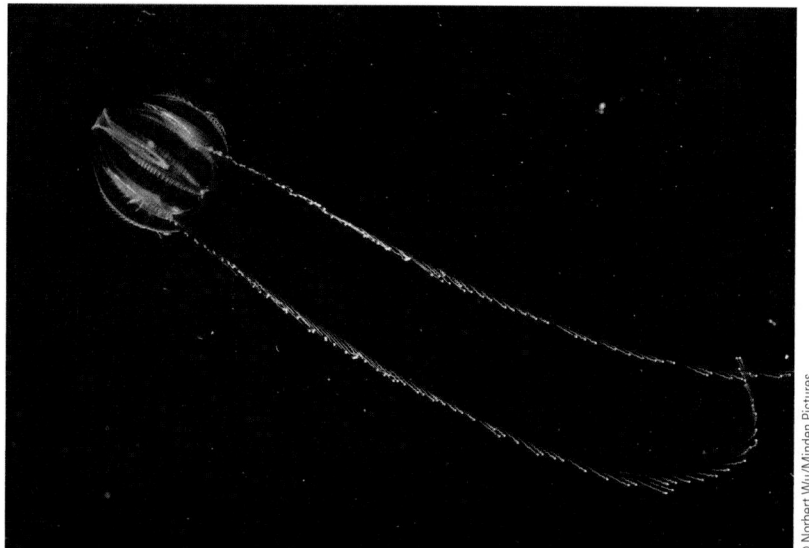

Figure 29.14
Ctenophores. The comb jelly *Mertensia ovum* collects microscopic prey on its two long sticky tentacles and then wipes the food-laden tentacles across its mouth.

© Norbert Wu/Minden Pictures

29.6 Lophotrochozoan Protostomes

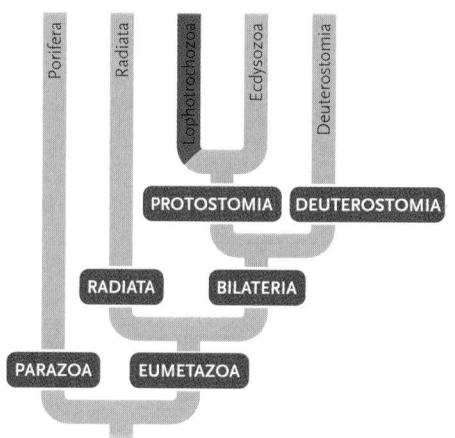

The remaining organisms described in this chapter fall within the group Bilateria: they have bilateral symmetry and a greater variety of tissues than do the Radiata. Bilaterians also have organ systems, structures that include two or more tissue types organized to perform specific functions; most of them also possess a coelom or pseudocoelom. With bilateral symmetry and sensory organs concentrated at the anterior end of the body, most bilaterians engage in highly directed, often rapid movements in pursuit of food or mates or to escape danger. And their complex organ systems accomplish tasks more efficiently than simple tissues. For example, animals that have a tubular digestive system surrounded by a space (the coelom or pseudocoelom) use muscular contractions of the digestive system to move ingested food past specialized epithelia that break it down and absorb the breakdown products.

Molecular analyses group eight of the phyla that we consider into the Lophotrochozoa, one of the two main protostome lineages (see Figure 29.6).

The Lophophorate Phyla Share a Distinctive Feeding Structure

Three small groups of aquatic, coelomate animals—the phyla Ectoprocta, Brachiopoda, and Phoronida—possess a **lophophore**, a circular or U-shaped fold with one or two rows of hollow, ciliated tentacles surrounding the mouth **(Figure 29.15)**. Molecular sequence data as well as the lophophore suggest that these phyla share a common ancestry.

a. Ectoprocta *(Plumatella repens)*

b. Brachiopoda *(Terebraulina septentrionalis)*

© Andrew J. Martinez/Photo Researchers, Inc.

c. Phoronida *(Phoronopsis californica)*

© blickwinkel/Hecker/Alamy

© Lawrence Naylor/Photo Researchers, Inc.

Figure 29.15
Lophophorate animals. Although the lophophorate animals differ markedly in appearance, they all use a lophophore—the feathery structures in the photos—to acquire food.

The lophophore, which looks like a crown of tentacles at the anterior end of the animal, serves as a site for gas exchange and waste elimination as well as for food capture. Most lophophorates are sessile suspension-feeders as adults: movement of cilia on the tentacles brings food-laden water toward the lophophore, the tentacles capture small organisms and debris, and the cilia transport them to the mouth. The lophophorates have a complete digestive system, which is U-shaped in most species, with the anus lying outside the ring of tentacles.

Phylum Ectoprocta. The Ectoprocta (sometimes called Bryozoa) are tiny colonial animals that mainly occupy marine habitats (see Figure 29.15a). They secrete a hard covering over their soft bodies and feed by extending the lophophore through a hole. Each colony, which may include more than a million individuals, is produced asexually by a single animal. Ectoproct colonies are permanently attached to solid substrates, where they form encrusting mats, bushy upright growths, or jellylike blobs. Nearly 4000 living species are known.

Phylum Brachiopoda. The Brachiopoda, or lampshells, have two calcified shells that develop on the animal's dorsal and ventral sides (see Figure 29.15b). Most species attach to substrates with a stalk that protrudes through one of the shells. The lophophore is held within the two shells, and the animal feeds by opening its shell and drawing water over its tentacles. Although only 250 species of brachiopods live today, more than 30,000 extinct species are known as fossils, mostly from Paleozoic seas.

Phylum Phoronida. The 18 or so species of phoronid worms vary in length from a few millimeters to 25 cm (see Figure 29.15c). They usually build tubes in soft ocean sediments or on hard substrates, and feed by protruding the lophophore from the top of the tube. The animal can withdraw into the tube when disturbed. Phoronida reproduce both sexually and by budding.

Flatworms Have Digestive, Excretory, Nervous, and Reproductive Systems, but Lack a Coelom

The 13,000 flatworm species in the phylum Platyhelminthes (*platys* = flat; *helmis* = worm) live in aquatic and moist terrestrial habitats. Like cnidarians, flatworms can swim or float in water, but they are also able to crawl over surfaces. They range from less than 1 mm to more than 20 m in length, but most are just a few millimeters thick. Free-living species eat live prey or decomposing carcasses, whereas parasitic species consume the tissues of living hosts.

Like the radiate phyla, flatworms are acoelomate, but they have a complex structural organization that

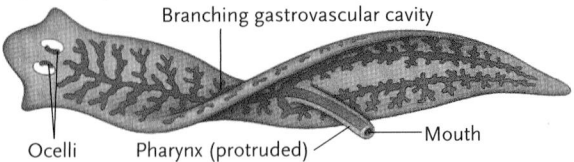
Digestive system

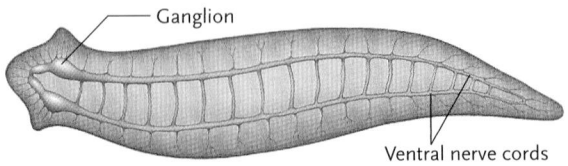
Nervous system

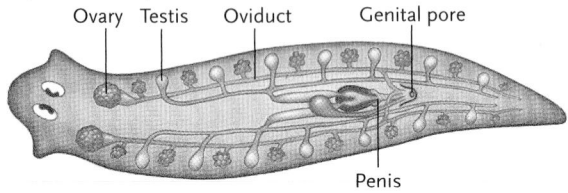
Reproductive system

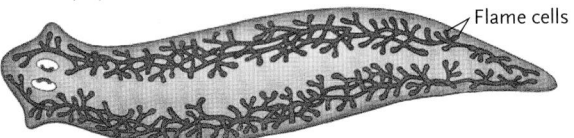
Excretory system

Figure 29.16

Flatworms. The phylum Platyhelminthes, exemplified by a freshwater planarian, have well-developed digestive, excretory, nervous, and reproductive systems. Because flatworms are acoelomate, their organ systems are embedded in a solid mass of tissue between the gut and the epidermis.

reflects their triploblastic construction **(Figure 29.16)**. Endoderm lines the digestive cavity with cells specialized for the chemical breakdown and absorption of ingested food. Mesoderm, the middle tissue layer, produces muscles and reproductive organs. Ectoderm produces a ciliated epidermis, the nervous system, and the *flame cell system,* a simple excretory system; flame cells regulate the concentrations of salts and water within body fluids, allowing free-living flatworms to live in freshwater habitats. Flatworms do not have circulatory or respiratory systems, but, because all cells of their dorsoventrally (top-to-bottom) flattened bodies are near an interior or exterior surface, diffusion supplies them with nutrients and oxygen.

The flatworm nervous system includes two or more longitudinal ventral nerve cords interconnected by numerous smaller nerve fibers, like rungs on a ladder. An anterior **ganglion**, a concentration of nervous system tissue that serves as a primitive brain, integrates their behavior. Most free-living species have *ocelli*, or eye spots, that distinguish light from dark and tiny chemoreceptor organs that sense chemical cues.

Figure 29.17

Turbellaria. A few turbellarians, such as *Pseudoceros dimidiatus*, are colorful marine worms.

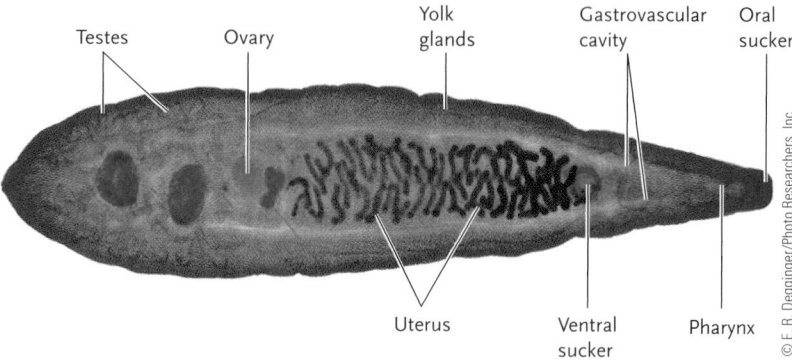

Figure 29.18

Trematoda. The hermaphroditic Chinese liver fluke *(Opisthorchis sinensis)* uses a well-developed reproductive system to produce thousands of eggs.

The phylum Platyhelminthes includes four lineages, defined largely by their anatomical adaptations to free-living or parasitic habits.

Turbellaria. Most free-living flatworms (Turbellaria) live in the sea **(Figure 29.17),** but the familiar planarians and a few others live in fresh water or on land. Turbellarians swim by undulating the body wall musculature, or they crawl across surfaces by using muscles and cilia to glide on mucous trails produced by the ventral epidermis.

The gastrovascular cavity in free-living flatworms is similar to that in cnidarians. Food is ingested and wastes eliminated through a single opening, the mouth, located on the ventral surface. Most turbellarians also acquire food with a muscular **pharynx** that connects the mouth to the digestive cavity (see Figure 29.16). Chemicals secreted into the saclike cavity digest the food into particles; then cells throughout the gastrovascular surface engulf the particles and subject them to intracellular digestion. In some species, the digestive cavity is highly branched, increasing the surface area for digestion and absorption.

Nearly all turbellarians are hermaphroditic, with complex reproductive systems (see Figure 29.16). When they mate, each partner functions simultaneously as a male and a female. Many free-living species also reproduce asexually by simply separating the anterior half of the animal from the posterior half. Both halves subsequently regenerate the missing parts.

Trematoda and Monogenoidea. Flukes (Trematoda and Monogenoidea) are parasites that obtain nutrients from host tissues **(Figure 29.18).** Most adult trematodes are **endoparasites,** living in the gut, liver, lungs, bladder, or blood vessels of vertebrates. Monogenes are **ectoparasites,** attaching to the gills or skin of aquatic vertebrates. Flukes are structurally specialized for a parasitic existence. They use suckers or hooks to attach to hosts, and a tough outer covering protects them from chemical attack. They produce large numbers of eggs that can readily infect new hosts. Monogene flukes usually have simple life cycles with a single host species. Trematodes, by contrast, have complex life cycles and multiple hosts. Humans suffer potentially fatal infections by many flukes, as discussed in *Focus on Applied Research.*

Cestoda. Tapeworms (Cestoda) develop, grow, and reproduce within the intestines of vertebrates **(Figure 29.19).** Through evolution, they have lost their mouths

a. Tapeworm

b. Scolex

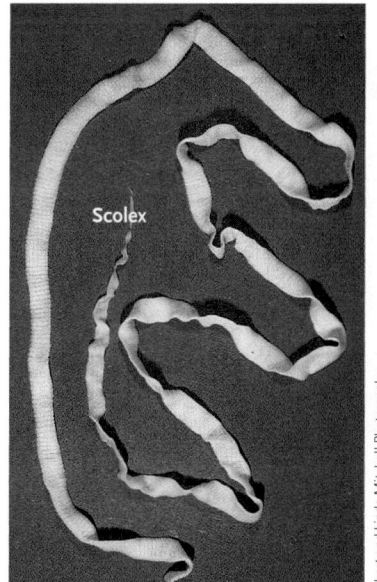

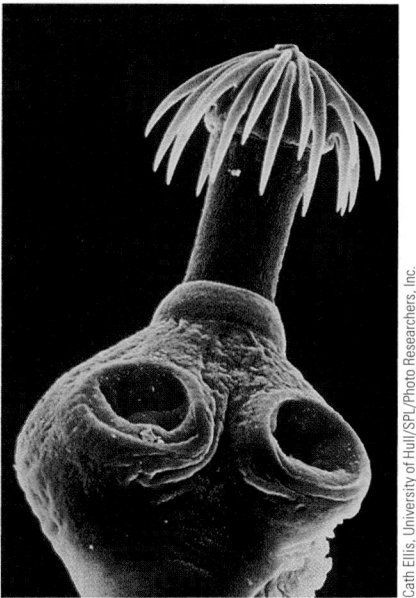

Figure 29.19

Cestoda. **(a)** Tapeworms have long bodies comprised of a series of proglottids that each produce thousands of fertilized eggs. **(b)** The anterior end is a scolex with hooks and suckers that attach to the host's intestinal wall.

Applied Research: A Rogue's Gallery of Parasitic Worms

Many parasitic flatworms (phylum Platyhelminthes) and roundworms (phylum Nematoda) call the human body home, frequently causing disfiguring or life-threatening infections. The effort to control or eliminate these infections often begins when parasitologists or public health researchers study the worms' life cycles and ecology. This approach is successful because the worms often have more than one host: humans may be the *primary host,* harboring the sexually mature stage of the parasites' life cycles, but other animals serve as *intermediate hosts* to the larval stages. If researchers can learn the details of a parasite's life cycle, they can identify ways to cut it short before the parasite infects a human host.

More than 200 million people in tropical and subtropical regions suffer from *schistosomiasis,* a disease caused by three species of flatworms called blood flukes (Trematoda). Japanese blood flukes (*Schistosoma japonicum*) mature and mate in blood vessels of the human intestine **(Figure a).** Sharp spines on their eggs rupture the blood vessels, releasing the eggs into the lumen of the gut, from which they are passed with feces. When the infected human waste enters standing fresh water, the eggs hatch into ciliated larvae, which burrow into certain aquatic snails. The larvae feed on the snail's tissues and reproduce asexually. Their tailed offspring leave the snail and, when they contact human skin, bore inward to a blood vessel. They eventually reach the intestine, where they complete their complex life cycle and produce fertilized eggs. Infected humans mount an immune response against flukes, but it is always a losing battle. Severe infections cause coughs, rashes, pain, and eventually diarrhea, anemia, and permanent damage to the intestines, liver, spleen, bladder, and kidneys. Death often results. Drug therapy can reduce the symptoms, but schistosomiasis is most common in less-developed countries with limited access to medical care. Research has demonstrated that the disease is best controlled by proper sanitation and the elimination of snails that serve as intermediate hosts.

Tapeworms (Cestoda), another group of flatworms, rely primarily on vertebrate hosts. Fishes, hogs, or cattle become intermediate hosts by inadvertently eating tapeworm eggs **(Figure b).** When the host's digestive enzymes dissolve the protective covering on the eggs, the newly freed larvae bore through the intestinal wall and travel through the bloodstream to muscles or other tissues, where they *encyst* (produce a protective covering and enter a resting stage). Humans and other carnivores ingest living cysts when they eat the undercooked flesh of these animals. The scolex of the tapeworm larva then attaches to the primary host's intestinal wall, and the worm begins to grow. When mature, its proglottids produce huge numbers of eggs, which are released with feces to begin the next generation. Tapeworm infection can result in malnutrition of the host, because tapeworms consume much of the nutrients that the host ingests. They can also grow large enough to cause intestinal blockage. A full understanding of the tapeworm life cycle suggests that careful inspection and adequate cooking of meat can prevent infection.

The roundworm *Trichinella spiralis* causes the painful and sometimes fatal symptoms of *trichinosis.* Adult trichinas breed in the small intestine of their hosts, including hogs and some game animals. Female worms release juveniles, which burrow into blood vessels and travel to various organs, where they become encysted **(Figure c).** Humans become infected when they consume insufficiently cooked pork or other meat that contains encysted larvae. Once in the human digestive tract, the encysted worms complete their development, mate, and produce larvae. Most of the

Figure a
Life cycle of the Japanese blood fluke *(Schistosoma japonicum).*

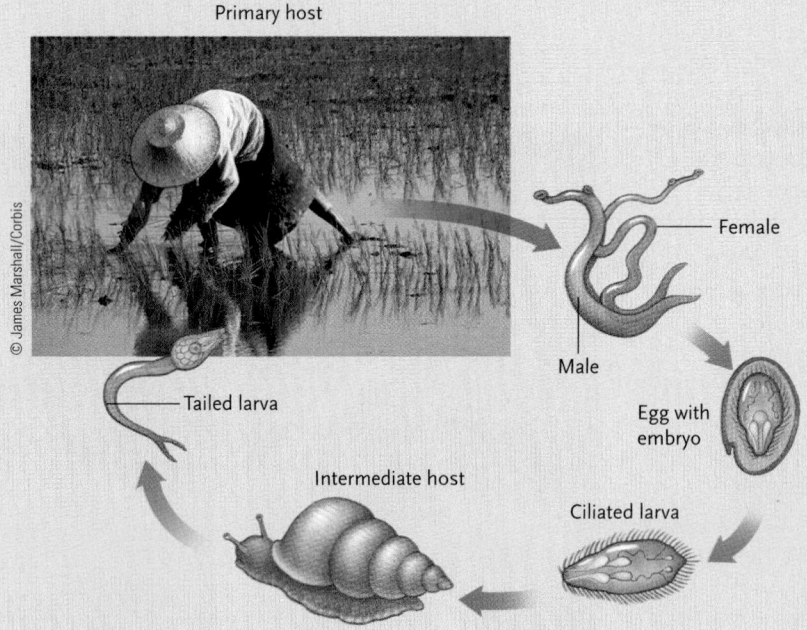

Primary host

Female

Male

Tailed larva

Egg with embryo

Intermediate host

Ciliated larva

© James Marshall/Corbis

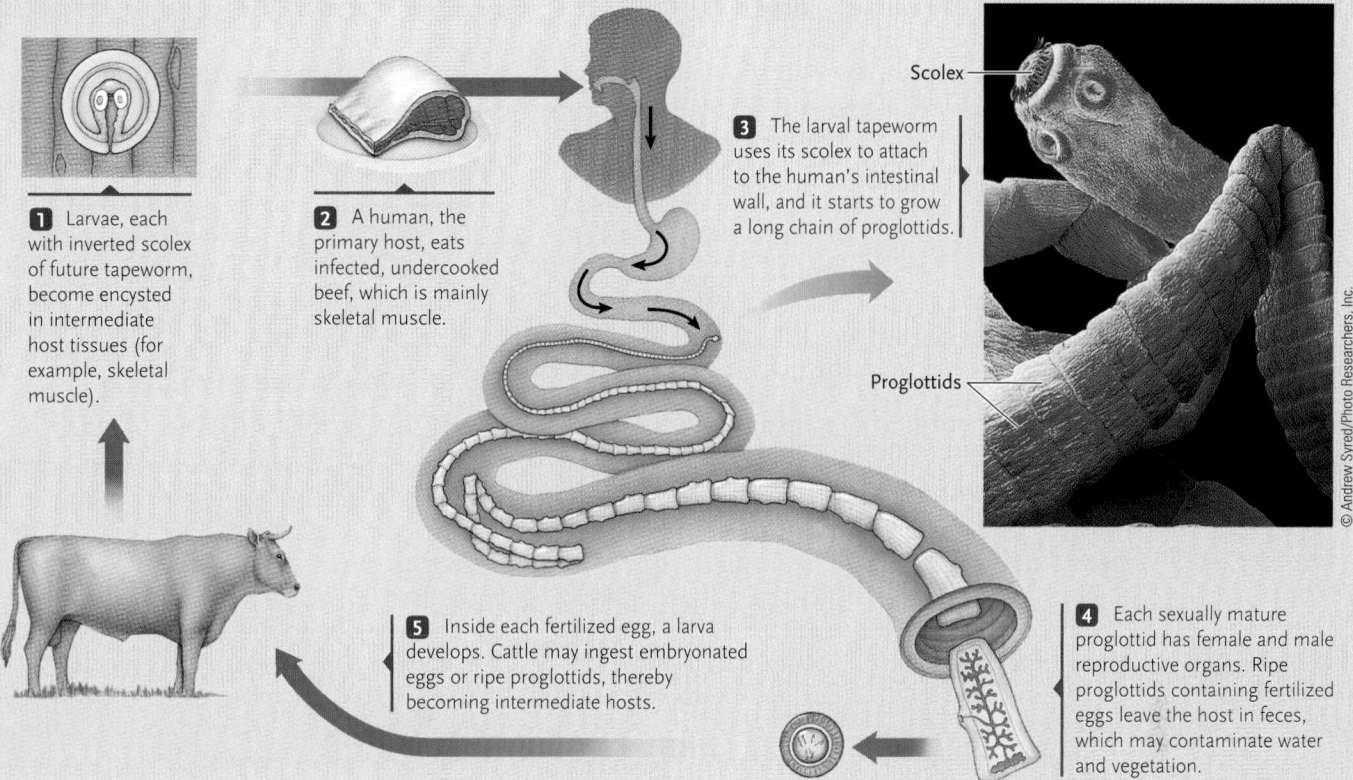

1 Larvae, each with inverted scolex of future tapeworm, become encysted in intermediate host tissues (for example, skeletal muscle).

2 A human, the primary host, eats infected, undercooked beef, which is mainly skeletal muscle.

3 The larval tapeworm uses its scolex to attach to the human's intestinal wall, and it starts to grow a long chain of proglottids.

Scolex

Proglottids

© Andrew Syred/Photo Researchers, Inc.

4 Each sexually mature proglottid has female and male reproductive organs. Ripe proglottids containing fertilized eggs leave the host in feces, which may contaminate water and vegetation.

5 Inside each fertilized egg, a larva develops. Cattle may ingest embryonated eggs or ripe proglottids, thereby becoming intermediate hosts.

Figure b
Life cycle of the beef tapeworm.

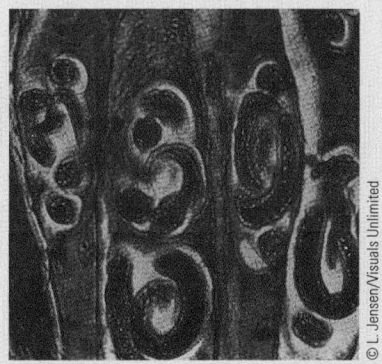

Figure c
Trichinella spiralis juveniles in muscle tissue.

© L. Jensen/Visuals Unlimited

awful symptoms of trichinosis, including severe pain, high fever, and debilitating anemia, are produced by the migration of millions of larvae throughout the primary host's tissues. Once begun, the infection is difficult or impossible to control, but, as with tapeworms, thorough cooking of meat can prevent infection.

Filarial worms (*Wuchereria* and *Brugia*) cause another debilitating nematode infection. These large roundworms (up to 10 cm long) live in the lymphatic system, where they obstruct the normal flow of lymphatic fluid to the bloodstream. Female worms release first-stage larvae, which are acquired by mosquitoes, the intermediate host, when they feed on human blood. The larvae develop into second-stage larvae in mosquitoes, and when a mosquito bites another human, it may transmit those larvae to a new host. If victims experience severe filarial worm infection, their lymphatic vessels can be so obstructed that surrounding tissues swell grotesquely, a condition know as *elephantiasis* (**Figure d**). Public health programs that reduce or eliminate mosquito populations lower the incidence of this disease.

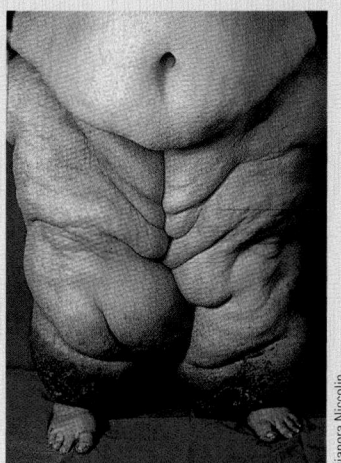

Dianora Niccolin

Figure d
Elephantiasis caused by *Wuchereria bancrofti*.

and digestive systems and absorb nutrients through their body wall. Tapeworms have a specialized structure, the *scolex,* with hooks and suckers that attach to the host's intestine; like the flukes, they also have a protective covering resistant to digestive enzymes.

Most of a tapeworm's body, which may be up to 20 m long, consists of a series of identical structures, *proglottids,* that contain little more than male and female reproductive systems. New proglottids are generated near the scolex; older proglottids, each carrying as many as 80,000 fertilized eggs, break off from the tapeworm's posterior end and leave the host's body in feces. As described in *Focus on Applied Research,* tapeworms are important parasites on humans, pets, and livestock.

Rotifers Are Tiny Pseudocoelomates with a Jawlike Feeding Apparatus

Most of the 1800 species in the phylum Rotifera (*rota* = wheel; *ferre* = to carry) live in fresh water **(Figure 29.20).** All are microscopic—about the size of a ciliate protist—but they have well-developed digestive, reproductive, excretory, and nervous systems as well as a pseudocoelom. In some habitats, rotifers make up a large part of the zooplankton, tiny animals that float in open water.

Rotifers use coordinated movements of cilia, arranged in a wheel-like *corona* around the head, to propel themselves in the environment. Cilia also bring food-laden water to their mouths. Ingested microorganisms are conveyed to the *mastax,* a toothed grinding organ, and then passed to the stomach and intestine. Rotifers have a **complete digestive tract:** food enters through the mouth, and undigested waste is voided through the anus.

The life history patterns of some rotifers are adapted to the ever-changing environments in small bodies of water. During most months, rotifer populations include only females that reproduce by **parthenogenesis,** a form of asexual reproduction in which unfertilized eggs develop into diploid females (see Section 47.1). When environmental conditions deteriorate, females produce eggs that develop into haploid males. The males fertilize haploid eggs to produce diploid female zygotes. The fertilized eggs have durable shells and food reserves to survive drying or freezing.

Ribbon Worms Use a Proboscis to Capture Food

The 650 species of ribbon worms or proboscis worms (phylum Nemertea) vary from less than 1 cm to 30 m in length **(Figure 29.21).** Most species are marine, but a few occupy moist terrestrial habitats. Although the often brightly colored ribbon worms superficially resemble free-living flatworms, their body plans are more complex. First, they possess both a mouth and an anus; thus, they have a complete digestive tract. Second, nemerteans have a circulatory system in which fluid flows through **circulatory vessels** that carry nutrients and oxygen to tissues and remove wastes (see Section 42.1). Finally, they have a muscular, mucus-covered proboscis, a tube that can be everted (turned inside out) through the mouth or a separate pore to capture prey. The proboscis is housed within a chamber, the *rhynchocoel,* which is unique to this phylum.

Figure 29.20
Phylum Rotifera. **(a)** Despite their small size, rotifers such as *Philodina roseola* have complex body plans and organ systems. **(b)** This rotifer, another *Philodina* species, is laying eggs.

a. Rotifer body plan

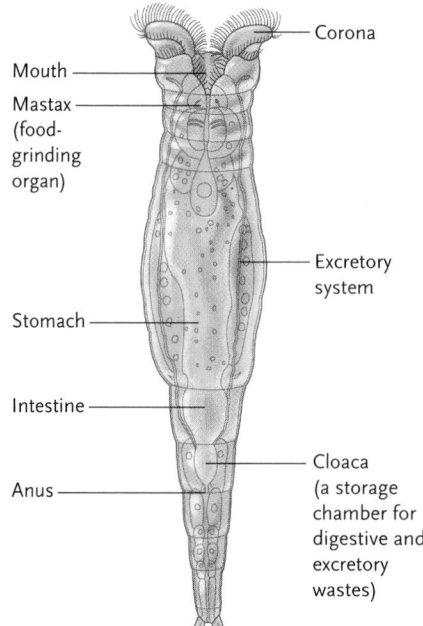

- Corona
- Mouth
- Mastax (food-grinding organ)
- Excretory system
- Stomach
- Intestine
- Anus
- Cloaca (a storage chamber for digestive and excretory wastes)

b. Rotifer laying eggs

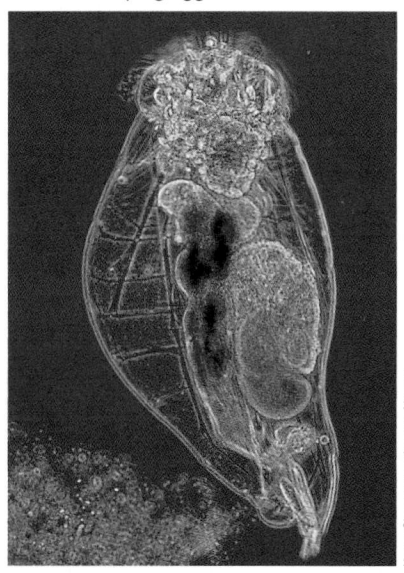

Herve Chaumeton/Agence Nature

a. Ribbon worm

b. Ribbon worm anatomy

Proboscis pore · Proboscis · Rhynchocoel

Mouth · Intestine · Proboscis retractor muscle · Anus

Everted proboscis

Kjell B. Sandved

Figure 29.21
Phylum Nemertea. **(a)** The flattened, elongated bodies of ribbon worms, such as this species in the genus *Lineus*, are often brightly colored. **(b)** Ribbon worms have a complete digestive system and a specialized cavity, the rhynchocoel, that houses a protrusible proboscis.

Mollusks Have a Muscular Foot and a Mantle That Secretes a Shell or Aids in Locomotion

Most of the 100,000 species in the phylum Mollusca (*mollis* = soft)—including clams, snails, octopuses, and their relatives—are marine. However, many clams and snails occupy freshwater habitats, and some snails live on land. Mollusks vary in length from clams less than 1 mm to the giant squids, which can exceed 18 m.

In mollusks, the body is divided into three regions: the visceral mass, the head–foot, and the mantle **(Figure 29.22).** The **visceral mass** contains the digestive, excretory, reproductive systems, and heart. The muscular **head–foot** often provides the major means of locomotion. In the more active groups, the head region is well

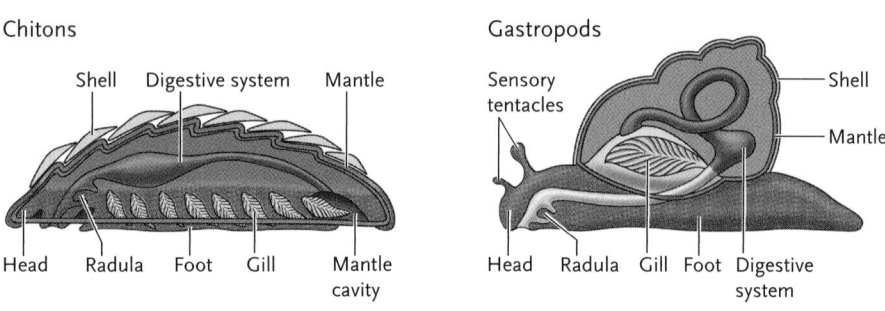

Chitons

Shell · Digestive system · Mantle

Head · Radula · Foot · Gill · Mantle cavity

Gastropods

Sensory tentacles · Shell · Mantle

Head · Radula · Gill · Foot · Digestive system

Figure 29.22
Molluskan body plans. The bilaterally symmetrical body plans of mollusks include a muscular head–foot, a visceral mass, and a mantle.

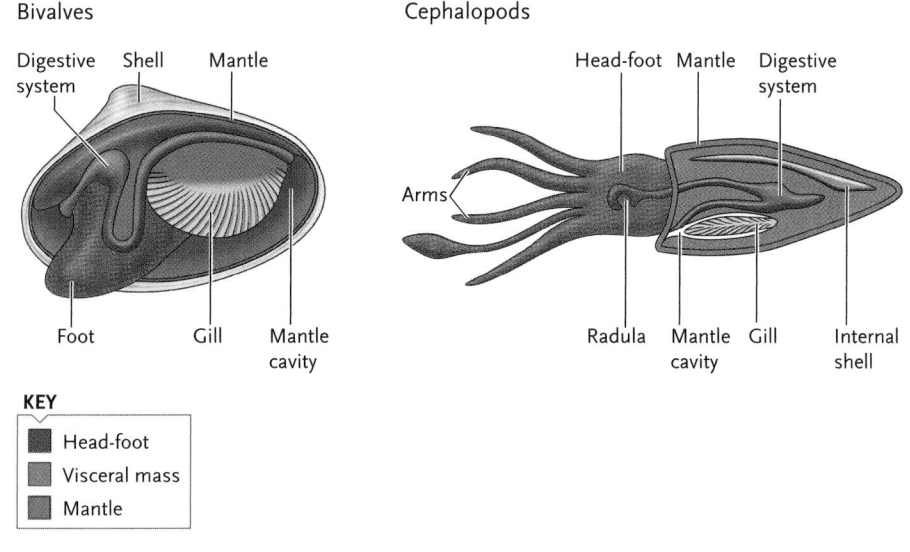

Bivalves

Digestive system · Shell · Mantle

Foot · Gill · Mantle cavity

Cephalopods

Head-foot · Mantle · Digestive system

Arms

Radula · Mantle cavity · Gill · Internal shell

KEY
- ■ Head-foot
- ■ Visceral mass
- ■ Mantle

Figure 29.23

Trochophore larva. At the conclusion of their embryological development, both mollusks and annelids typically pass through a trochophore stage. The top-shaped trochophore larva has a band of cilia just anterior to its mouth.

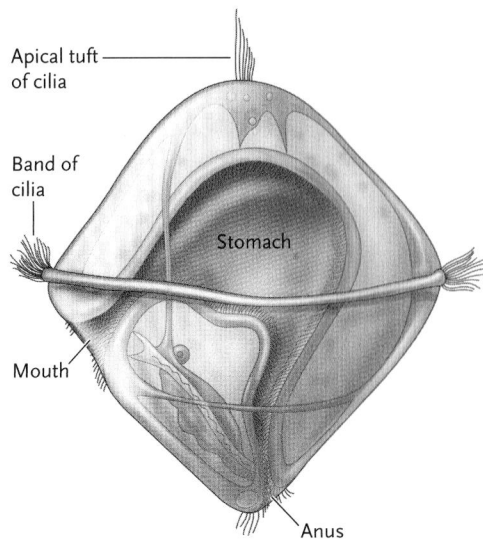

Apical tuft of cilia

Band of cilia

Stomach

Mouth

Anus

defined and carries sensory organs and a brain. The mouth often includes a toothed **radula,** which scrapes food into small particles or drills through the shells of prey. Many mollusks are covered by a protective shell of calcium carbonate secreted by the **mantle,** one or two folds of the body wall that often enclose the visceral mass. The mantle also defines a space, the *mantle cavity,* that houses the *gills,* delicate respiratory structures with enormous surface area. In most mollusks, cilia on the mantle and gills generate a steady flow of water into the mantle cavity.

The large size of mollusks—as well as their possession of a true coelom—requires a circulatory system to maintain cells that are far from the body surface. Most mollusks have an **open circulatory system** in which **hemolymph,** a bloodlike fluid, leaves the circulatory vessels and bathes tissues directly (see Figure 42.3a). Hemolymph pools in spaces called *sinuses,* and then drains into vessels that carry it back to the heart.

The sexes are usually separate, although many snails are hermaphroditic. Fertilization may be internal or external. The zygotes of marine species often develop into free-swimming, ciliated **trochophore** larvae **(Figure 29.23),** typical of both this phylum and the phylum Annelida, which we describe next. In some

mollusks, the trochophore develops into a second larval stage, called a *veliger,* before metamorphosing into an adult. Some snails as well as octopuses and squids have direct development: embryos develop into miniature replicas of the adults.

Mollusca includes eight lineages. In the following sections, we examine the four that are most commonly encountered.

Polyplacophora. The 600 species of chitons (Polyplacophora: *poly* = many; *plax* = flat surface) are sedentary mollusks that graze on algae along rocky marine coasts. The oval, bilaterally symmetrical body has a dorsal shell divided into eight plates that allow it to conform to irregularly shaped surfaces **(Figure 29.24).** When a chiton is disturbed or exposed to strong wave action, the muscles of its broad foot maintain a tenacious grip, and the mantle's edge functions like a suction cup to hold fast to the substrate.

Gastropoda. Snails and slugs (Gastropoda: *gaster* = belly; *pod* = foot) are the largest molluskan group, numbering 40,000 species **(Figure 29.25).** Aquatic species use gills to acquire oxygen, but in terrestrial species a modified mantle cavity functions as an air-breathing lung. Gastropods feed on algae, vascular plants, or animal prey. Some are scavengers, and a few are parasites.

The visceral mass of most snails is housed in a coiled or cone-shaped shell that is balanced above the rest of the body, much as you balance a backpack full of books (see Figure 29.25a, b). Most shelled snails undergo **torsion,** a curious realignment of body parts that is independent of shell coiling. Muscle contractions and differential growth twist the visceral mass and mantle 180° relative to the head–foot. This rearrangement moves the mantle cavity forward so that the head can be withdrawn into the shell in times of danger. But it also brings the gills, anus, and excretory openings above the head—a potentially messy configuration, were it not for cilia that sweep away wastes.

Some gastropods, including terrestrial slugs and colorful nudibranchs (sea slugs), are shell-less, a condition that leaves them somewhat vulnerable to predators (see Figure 29.25c). Some nudibranchs consume cnidarians and then transfer undischarged nematocysts to projections on their dorsal surface, where these "borrowed" stinging capsules provide protection.

The nervous and sensory systems of gastropods are well developed. Tentacles on the head include chemical and touch receptors; the eyes detect changes in light intensity but don't form images.

Bivalvia. The 8000 species of clams, scallops, oysters, and mussels (Bivalvia: *bi* = two; *valva* = folding door) are restricted to aquatic habitats. They are enclosed within a pair of shells, hinged together dorsally by a ligament **(Figure 29.26).** Contraction of the ligament

Figure 29.24
Polyplacophora. Chitons live on rocky shores, where they use their foot and mantle to grip rocks and other hard substrates. This bristled chiton (*Mopalia ciliata*) lives in Monterey Bay, California.

Jeff Foott/Tom Stack & Associates

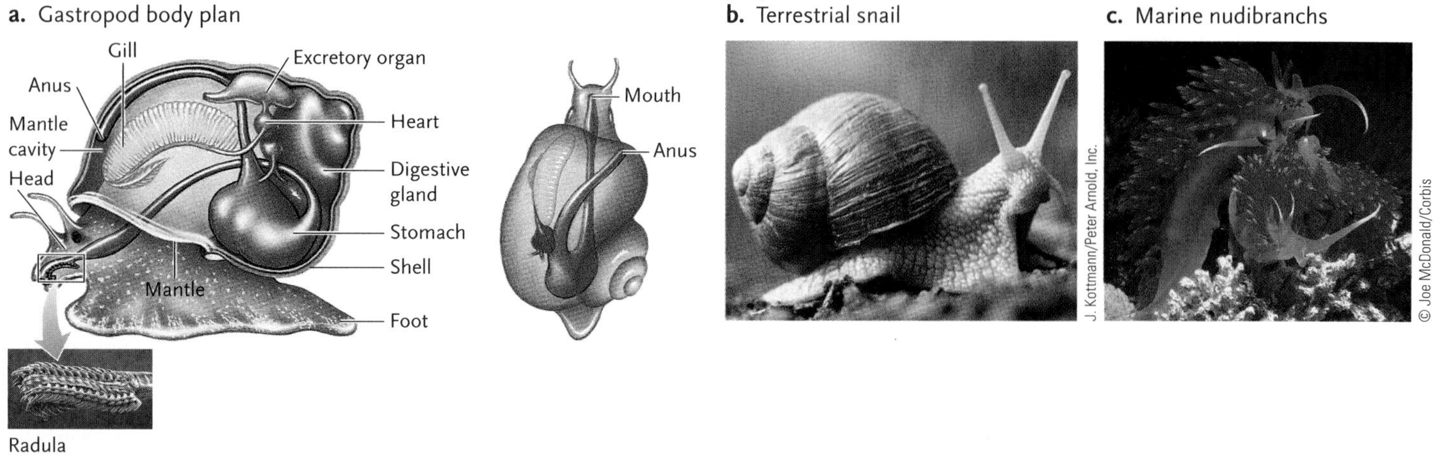

a. Gastropod body plan

Gill

Anus

Mantle cavity

Head

Excretory organ

Heart

Digestive gland

Stomach

Shell

Foot

Mantle

Mouth

Anus

Radula

b. Terrestrial snail

c. Marine nudibranchs

Figure 29.25

Gastropoda. **(a)** Most gastropods have a coiled shell that houses the visceral mass. A developmental process called torsion causes the digestive and excretory systems to eliminate wastes into the mantle cavity, near the animal's head. **(b)** The edible snail *(Helix pomatia)* is a terrestrial gastropod. **(c)** Nudibranchs, like these Spanish shawl nudibranchs *(Flabellina iodinea)*, are shell-less marine snails. (Micrograph: Danielle C. Zacherl with John McNulty.)

opens the shell by pulling the two sides apart, and contraction of one or two **adductor muscles** closes it by pulling them together (see Figure 29.26a). Although some bivalves are tiny, giant clams of the South Pacific can be more than 1 m across and weigh 225 kg.

Adult mussels and oysters are sessile and permanently attached to hard substrates, but many clams are mobile and use their muscular foot to burrow in sand or mud. Some bivalves, such as young scallops, swim by rhythmically clapping their valves together, forcing a current of water out of the mantle cavity (see Figure 29.26b). The "scallops" that we eat are their well-developed adductor muscles.

Bivalves have a reduced head, and they lack a radula. Part of the mantle forms two tubes called *siphons* (see Figure 29.26c). Beating of cilia on the gills and

mantle carry water into the mantle cavity through the incurrent siphon and out through the excurrent siphon. Incurrent water carries dissolved oxygen and food particles to the gills, where oxygen is absorbed. Mucus strands on the gills trap the food, which is then transported by cilia to *palps,* where final sorting takes place; acceptable bits are carried to the mouth. The excurrent water carries away metabolic wastes and feces.

Despite their sedentary existence, bivalves have moderately well developed nervous systems; sensory organs that detect chemicals, touch, and light; and statocysts to sense their orientation. When they encounter pollutants, many bivalves stop pumping water and close their shells. When confronted by a predator, some burrow into sediments or swim away.

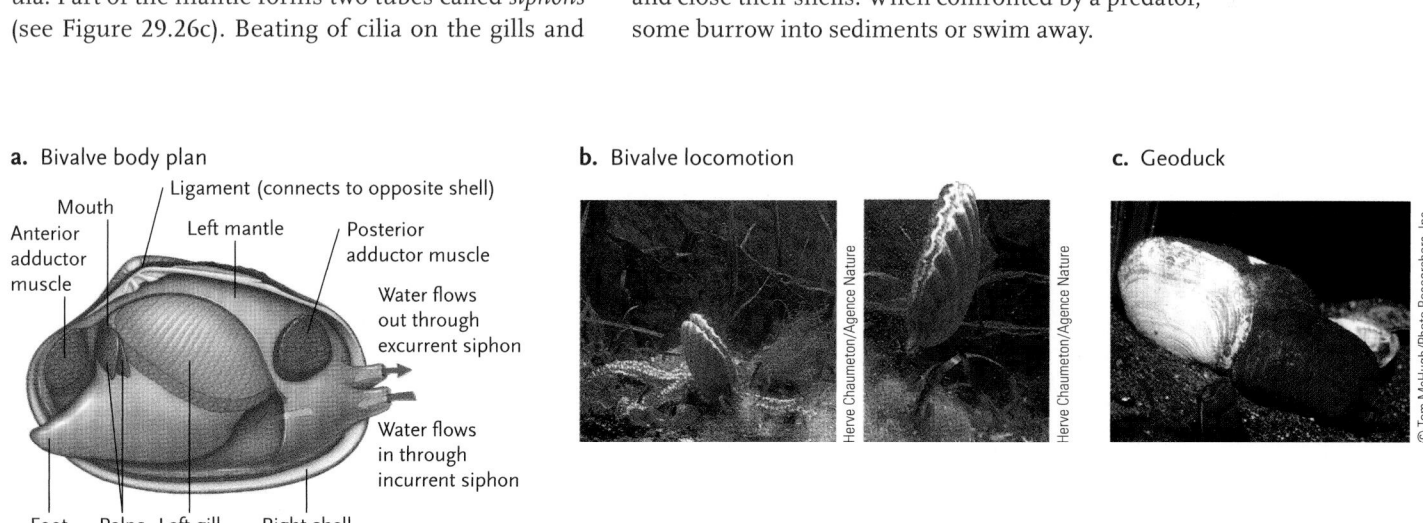

a. Bivalve body plan

Mouth

Anterior adductor muscle

Ligament (connects to opposite shell)

Left mantle

Posterior adductor muscle

Water flows out through excurrent siphon

Water flows in through incurrent siphon

Foot   Palps   Left gill   Right shell

b. Bivalve locomotion

c. Geoduck

Figure 29.26

Bivalvia. **(a)** Bivalves are enclosed in a hinged two-part shell. Part of the mantle forms a pair of water-transporting siphons. **(b)** When threatened by a predator (in this case a sea star), some scallops clap their shells together rapidly, propelling the animal away from danger. **(c)** The geoduck *(Panope generosa)* is a clam with enormous muscular siphons.

Cephalopoda. The 600 species of octopuses, squids, and nautiluses (Cephalopoda: *kephale* = head) are active marine predators, including the fastest and most intelligent invertebrates **(Figure 29.27)**. They vary in length from a few centimeters to 18 m. Giant squids, the largest invertebrates known, may be the source of "sea monster" stories.

The cephalopod body has a fused head and foot (see Figure 29.27d). The head comprises the mouth and eyes. The ancestral "foot" forms a set of arms and tentacles, equipped with suction pads, adhesive structures, or hooks. Cephalopods use these structures to capture prey and a pair of beaklike jaws to bite or crush it. Venomous secretions often speed the captive's death. Some species use their radula to drill through the shells of other mollusks.

Cephalopods have a highly modified shell. Octopuses have no remnant of a shell at all. In squids and cuttlefishes, it is reduced to a stiff internal support. Only the chambered nautilus and its relatives retain an external shell.

Squids are the most mobile cephalopods, moving rapidly by a kind of jet propulsion. When muscles in the mantle relax, water is drawn into the mantle cavity. When they contract, a jet of water is squeezed out through a funnel-shaped excurrent siphon. By manipulating the position of the mantle and siphon, the animal can control the rate and direction of its locomotion. When threatened, a squid can make a speedy escape. Many species simultaneously release a dark fluid, commonly called "ink," that obscures their direction of movement. Being highly active, cephalopods need lots of oxygen, and they alone among the mollusks have a **closed circulatory system:** hemolymph is confined within the walls of hearts and blood vessels, providing increased pressure to the vascular fluid. Moreover, accessory hearts speed the flow of hemolymph through the gills, enhancing the uptake of oxygen and release of carbon dioxide.

Compared with other mollusks, cephalopods have large and complex brains. Giant nerve fibers connect the brain with the muscles of the mantle, enabling quick responses to food or danger. Their image-forming eyes are similar to those of vertebrates. Cephalopods are also highly intelligent. Octopuses, for example, learn to recognize objects with distinctive shapes or colors, and they can be trained to approach or avoid them.

Cephalopods have separate sexes and elaborate courtship rituals. Males store sperm within the mantle cavity and use a specialized tentacle to transfer packets of sperm into the female's mantle cavity, where fertilization occurs. Fertilized eggs, wrapped in a protective jelly, are attached to objects in the environment. The young hatch with an adult body form.

Annelids Exhibit a Serial Division of the Body Wall and Some Organ Systems

The 15,000 species of segmented worms (phylum Annelida: *anellus* = ring) occupy marine, freshwater, and moist terrestrial habitats. Terrestrial annelids eat organic debris; aquatic species consume algae, microscopic organisms, detritus, or other animals. They range from a few millimeters to several meters in length.

The annelid body is highly segmented: the body wall muscles and some organs—including respiratory surfaces, parts of the nervous, circulatory, and excretory systems, and the coelom itself—are divided into similar repeating units **(Figure 29.28)**. Body segments are separated by transverse partitions called **septa.** The digestive system and major blood vessels are not segmented and run the length of the animal.

The body wall muscles of annelids have both circular and longitudinal layers. Alternate contractions of these muscle groups allow annelids to make directed movements, using the pressure of the fluid in the coe-

a. Squid **b.** Octopus **c.** Chambered nautilus

Eye Eye

d. Internal anatomy of squid

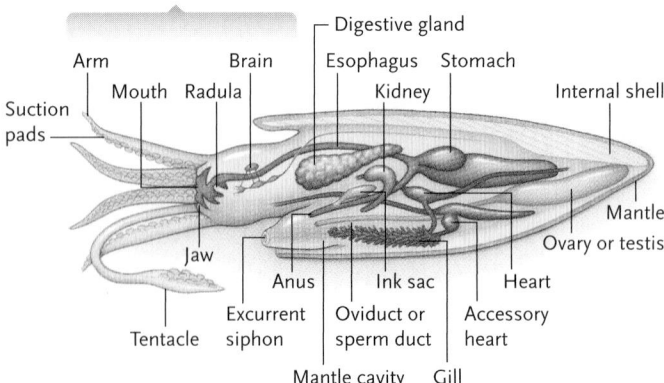

Figure 29.27
Cephalopoda. **(a)** Squids, such as *Dosidicus gigas*, and **(b)** octopuses, such as *Octopus vulgaris*, are the most familiar cephalopods. **(c)** The chambered nautilus (*Nautilus macromphalus*) and its relatives retain an external shell. **(d)** Like other cephalopods, the squid body includes a fused head and foot; most organ systems are enclosed by the mantle.

a. Digestive system

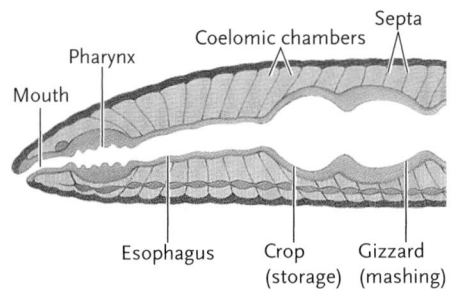

Mouth
Pharynx
Coelomic chambers
Septa
Esophagus
Crop (storage)
Gizzard (mashing)

b. Circulatory system

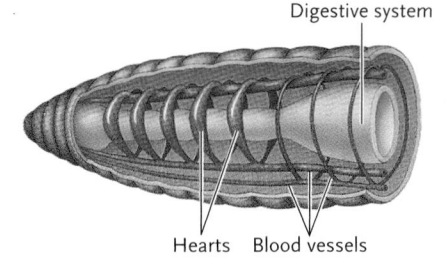

Digestive system
Hearts
Blood vessels

c. Nervous system

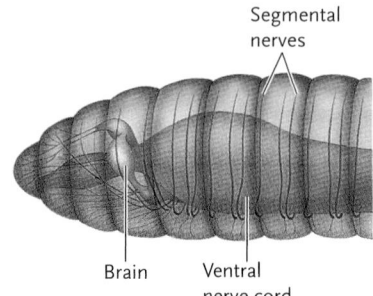

Segmental nerves
Brain
Ventral nerve cord

d. Excretory system (metanephridium)

Bladderlike storage region of excretory organ
Excretory organ's thin loop reabsorbs some solutes, returns them to blood

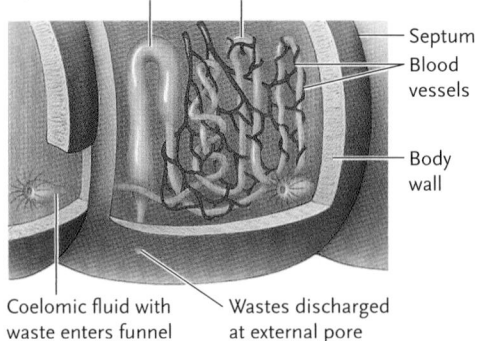

Septum
Blood vessels
Body wall
Coelomic fluid with waste enters funnel
Wastes discharged at external pore

e. Cross section

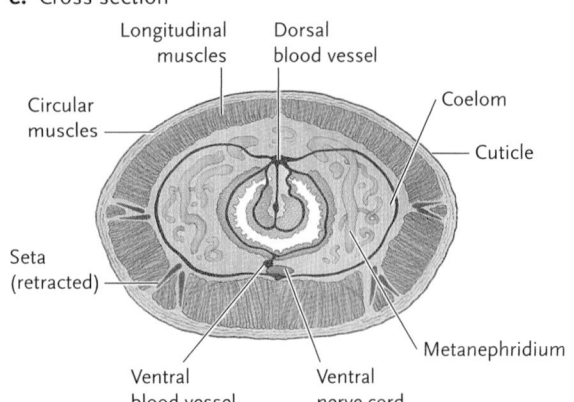

Longitudinal muscles
Dorsal blood vessel
Circular muscles
Coelom
Cuticle
Seta (retracted)
Ventral blood vessel
Ventral nerve cord
Metanephridium

Figure 29.28
Segmentation in the phylum Annelida. Although the digestive system **(a)**, the longitudinal blood vessels **(b)**, and the ventral nerve cord **(c)** are not segmented, the coelom **(a)**, blood vessels **(b)**, nerves **(c)**, and excretory organs **(d)** appear as repeating structures in most segments. The body musculature **(e)** includes both circular and longitudinal layers that allow these animals to use the coelomic chambers as a hydrostatic skeleton.

lom as a hydrostatic skeleton. All annelids except leeches also have chitin-reinforced bristles, called **setae** (sometimes written *chaetae*), which protrude outward from the body wall. Setae anchor the worm against the substrate, providing traction.

Annelids have a complete digestive system and a closed circulatory system. However, they lack a discrete respiratory system; oxygen and carbon dioxide diffuse through the skin. The excretory system is composed of paired **metanephridia**, which usually occur in all body segments posterior to the head. The nervous system is highly developed, with ganglia (local control centers) in every segment, a simple brain in the head, and sensory organs that detect chemicals, moisture, light, and touch.

Most freshwater and terrestrial annelids are hermaphroditic, and worms exchange sperm when they mate. Newly hatched worms have an adult morphology. Some terrestrial annelids also reproduce asexually by fragmenting and regenerating missing parts. Marine annelids usually have separate sexes, and release gametes into the sea for fertilization. The zygotes develop into trochophore larvae that add segments, gradually assuming an adult form.

Annelida includes three lineages, each of which is largely restricted to one environment.

Polychaeta. The 10,000 species of bristle worms (Polychaeta: *chaite* = bristles) are primarily marine **(Figure 29.29).** Many live under rocks or in tubes constructed from mucus, calcium carbonate secretions, grains of sand, and small shell fragments. Their setae project from well-developed **parapodia** (singular, *parapodium* = closely resembling a foot), fleshy lateral extensions of the body wall used for locomotion and gas exchange. Sense organs are concentrated on a well-developed head.

Crawling or swimming polychaetes are often predatory; they use sharp jaws in a protrusible muscular pharynx to grab small invertebrate prey. Other species graze on algae or scavenge organic matter. A few tube dwellers draw food-laden water into the tube by beating their parapodia; most others collect food by extending feathery, ciliated, mucus-coated tentacles.

Oligochaeta. Most of the 3500 species of oligochaete worms (Oligochaeta: *oligos* = few) are terrestrial **(Figure 29.30),** but they are restricted to moist habitats because

a. Fan worm

b. Polychaete feeding structures

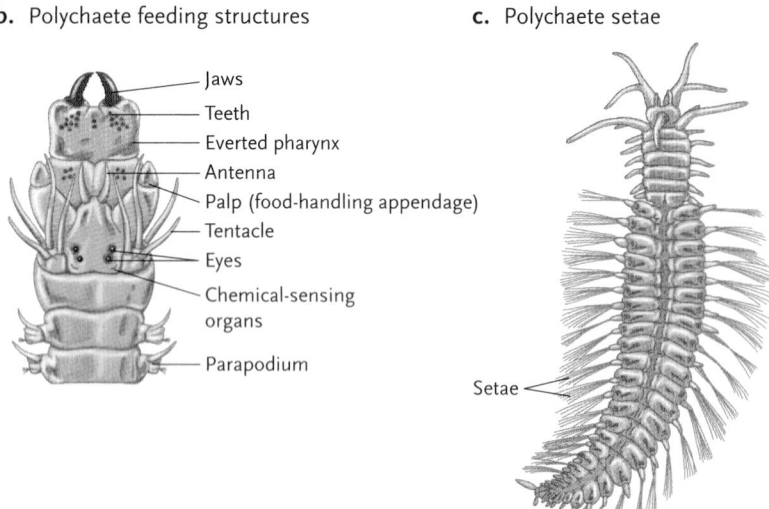

Jaws
Teeth
Everted pharynx
Antenna
Palp (food-handling appendage)
Tentacle
Eyes
Chemical-sensing organs
Parapodium

c. Polychaete setae

Setae

Figure 29.29
Polychaeta. **(a)** The tube-dwelling fan worm (*Sabella melanostigma*) has mucus-covered tentacles that trap small food particles. **(b)** Some polychaetes, like *Nereis*, actively seek food; when they encounter a suitable tidbit, they evert their pharynx, exposing sharp jaws that grab the prey and pull it into the digestive system. **(c)** Many marine polychaetes (such as *Proceraea cornuta*, shown here) have numerous setae, which they use for locomotion.

Figure 29.30
Oligochaeta. Earthworms (*Lumbricus terrestris*) generally move across the ground surface at night.

they quickly dehydrate in dry air or soil. They range in length from a few millimeters to more than 3 m. Terrestrial oligochaetes, the earthworms, are nocturnal, spending their days in burrows that they excavate. Aquatic species live in mud or detritus at the bottom of lakes, rivers, and estuaries.

Earthworms are scavengers on decomposing organic matter. In his book *The Formation of Vegetable Mould through the Action of Worms,* Darwin noted that earthworms can ingest their own weight in soil every day. He calculated that a typical population of 16,000 worms per hectare (10,000 sq m) consumes more than 20 tons of soil in a year. This impressive activity aerates soil and makes nutrients available to plants by mixing

the subsoil with the topsoil. Earthworms have complex organ systems (see Figure 29.28), and they sense light and touch at both ends of the body. In addition, they have moisture receptors, an important adaptation in organisms that must stay wet to allow gas exchange across the skin.

Hirudinea. The 500 species of leeches (Hirudinea: *hirudo* = leech) are mostly freshwater parasites. They have dorsoventrally flattened, tapered bodies with a sucker at each end. Although the body wall is segmented, the coelom is reduced and not partitioned. Many leeches are ectoparasites of vertebrates, but some attack small invertebrate prey.

Parasitic leeches feed on the blood of their hosts. Most attach to the host with the posterior sucker, and then use their sharp jaws to make a small, often painless, triangular incision. A sucking apparatus draws blood from the prey, while a special secretion, *hirudin*, maintains the flow by preventing the host's blood from coagulating. Leeches have a highly branched gut that allows them to consume huge blood meals **(Figure 29.31).** For centuries, doctors used medicinal leeches (*Hirudo medicinalis*) to "bleed" patients; today, surgeons use them to drain excess fluid from tissues after recon-

Leech before feeding

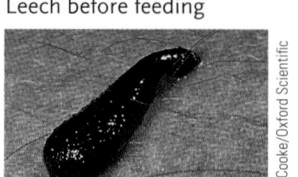

Leech after feeding

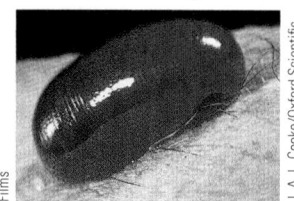

Figure 29.31
Hirudinea. Parasitic leeches consume huge blood meals, as shown by these before and after photos of a medicinal leech (*Hirudo medicinalis*). Because suitable hosts are often hard to locate, gorging allows a leech to take advantage of any host it finds.

structive surgery, reducing swelling until the patient's blood vessels regenerate and resume this function.

STUDY BREAK

1. What organs systems are present in free-living flatworms (Turbellaria)? Which of these organ systems is absent in tapeworms (Cestoda)?
2. What characteristic reveals the close evolutionary relationship of ectoprocts, brachiopods, and phoronid worms?
3. What anatomical structures and physiological systems allow squids and other cephalopods to be much more active than other types of mollusks?
4. Which organs systems exhibit segmentation in most annelid worms?

29.7 Ecdysozoan Protostomes

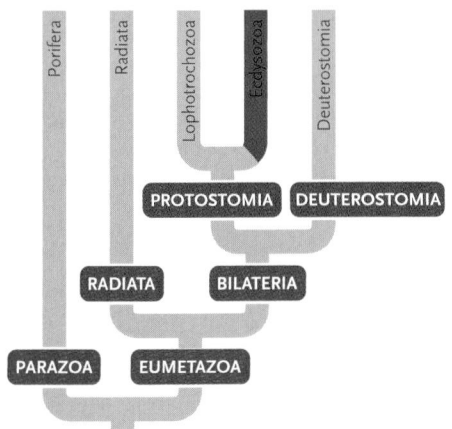

The three phyla in the protostome group Ecdysozoa all have an external covering that they shed periodically. The outer covering protects these animals from harsh environmental conditions, and it helps parasitic species resist host defenses. Although many of these animals live in either aquatic or moist terrestrial habitats, a tough exoskeleton allows one group, the insects, to thrive on dry land and in the air.

Nematodes Are Unsegmented Worms Covered by a Flexible Cuticle

Roundworms (phylum Nematoda: *nema* = thread) are perhaps the most abundant animals on Earth. A cupful of rich soil, a dead earthworm, or a rotting fruit may contain thousands of them. Although 80,000 species have been described, experts estimate that more than half a million exist. Many nematodes are almost microscopically small, but some species are a meter or more long. They occupy nearly every freshwater, ma-

rine, and terrestrial habitat on Earth, consuming detritus, microorganisms, plants, or animals.

The roundworm body is cylindrical and usually tapered at both ends **(Figure 29.32)**. None of the cells have cilia or flagella. Roundworms are covered in a tough but flexible, water-resistant **cuticle**, which is replaced by the underlying epidermis as the worm grows. The cuticle prevents the animal from dehydrating in dry environments, and, in parasitic species, it resists attack by acids and enzymes in a host's digestive system. Beneath the cuticle and epidermis, a layer of longitudinal muscles extends the length of the body. Nematodes use alternating muscle contractions to push against the substrate and propel themselves forward, usually with a thrashing motion.

The adults of one soil-dwelling species, *Caenorhabditis elegans,* are transparent and contain fewer than 1000 cells. As *Focus on Model Research Organisms* explains, biologists have studied this worm inside and out.

Nematodes reproduce sexually, and the sexes are separate in most species. In some, internal fertilization produces many thousands of fertile eggs per day. The eggs of many species can remain dormant if environmental conditions are unsuitable.

Because of their great numbers, nematodes have enormous ecological, agricultural, and medical significance. Free-living species are responsible for decomposition and nutrient recycling in many habitats. Parasitic nematodes attack the roots of plants, causing tremendous crop damage. Roundworms also parasitize animals, including humans. Although some, like pinworms *(Enterobius),* are more of a nuisance than a danger, others, like trichinas or filarial worms, can cause serious disease, disfigurement, or even death (see *Focus on Applied Research* on pages 644–645). More than 1 billion people worldwide suffer from debilitating and life-threatening nematode infections.

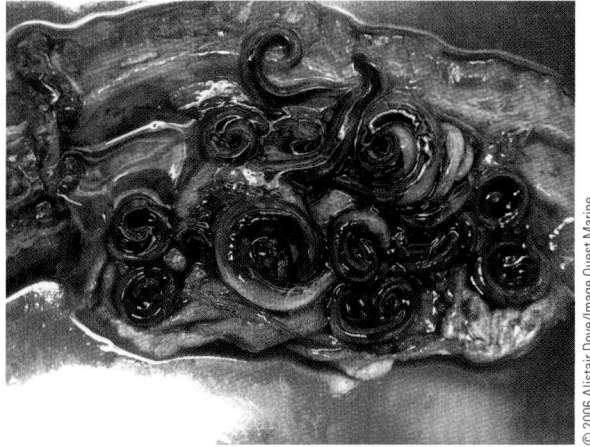

Figure 29.32

Phylum Nematoda. Some roundworms, like these *Anguillicola crassus* in the swim bladder of an eel, are parasites of plants or animals. Others are important consumers of dead organisms in most ecosystems.

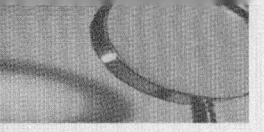

Model Research Organisms: *Caenorhabditis elegans*

Researchers studying the tiny, free-living nematode *Caenorhabditis elegans* have made many recent advances in molecular genetics, animal development, and neurobiology. It is so popular as a model research organisms that most workers simply refer to it as "the worm."

Several attributes make *C. elegans* a model research organism. It has an adult size of about 1 mm and thrives on cultures of *E. coli* or other bacteria; thus, thousands can be raised in a culture dish. It is hermaphroditic and often self-fertilizing, which allows researchers to maintain pure genetic strains. It completes its life cycle from egg to reproductive adult within 3 days at room temperature. Furthermore, stock cultures can be kept alive indefinitely by freezing them in liquid nitrogen or in an ultra-cold freezer set to −80°C. Researchers can therefore store new mutants for later research without having to clean, feed, and maintain active cultures.

Best of all, the worm is anatomically simple (see **figure**); an adult contains just 959 cells (excluding the gonads). Having a fixed cell number is relatively uncommon among animals, and developmental biologists have made good use of this trait. The eggs, juveniles, and adults of the worm are completely transparent, and researchers can observe cell divisions and cell movements in living animals with straightforward microscopy techniques. There is no need to kill, fix, and stain specimens for study. And virtually every cell in the worm's body is accessible for manipulation by laser microsurgery, microinjection, and similar approaches.

The genome of *C. elegans*, which was sequenced in 1998, is also simple, consisting of 100 million base pairs organized into roughly 17,000 genes on six pairs of chromosomes. The genome, which is about the same size as one human chromosome, specifies the amino acid sequences of about 10,000 protein molecules, far fewer than are found in more complex animals.

The worm entered the biological limelight in 1965 when Sidney Brenner of the Medical Research Council's Laboratory of Molecular Biology in Cambridge, England, identified it as an ideal organism for research on the genetic control of development. By 1983 numerous researchers had collectively identified the cell lineage for every one of its 959 cells. We now know the exact patterns of cell divisions, cell migrations, and programmed cell death that generate an adult from a fertilized egg.

Research on *C. elegans* has generated interesting results, some of which contradict old assumptions about animal development. For example, researchers once believed that a particular tissue always arises from the same embryonic germ layer, but in *C. elegans* some muscle cells do not develop from mesoderm and some nervous system cells do not arise from ectoderm. Moreover, the bilaterally symmetrical body of the adult worm does not arise from symmetrical events; matching cells on the left and right sides of the worm sometimes arise from different developmental pathways. Finally, researchers were surprised by the important role of cell death in this organism. Developmental events in the embryo produce 671 cells, but 113 of them are programmed to die before the larva hatches from the egg.

Studies of the worm's nervous system have been equally fruitful. Researchers have mapped the development of all 302 neurons and their 7000 connections. Now they are working to identify the molecules that function in sensory recognition and cell-to-cell signaling. Molecular analyses reveal that some of the proteins that carry messages between nerve cells in the worm are similar to those found in vertebrates.

The knowledge gained from research on *C. elegans* is highly relevant to studies of larger and more complex organisms, including vertebrates. Recent research demonstrates some striking similarities among nematodes, fruit flies, and mice in the genetic control of development, in some of the proteins that govern important events like cell death, and in the molecular signals used for cell-to-cell communication. Using a relatively simple model like *C. elegans*, researchers can answer research questions more quickly and more efficiently than they could if they studied larger and more complex animals.

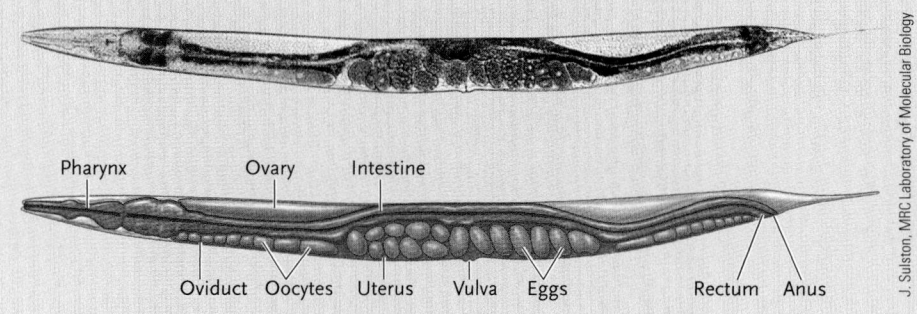

Pharynx   Ovary   Intestine

Oviduct   Oocytes   Uterus   Vulva   Eggs   Rectum   Anus

J. Sulston, MRC Laboratory of Molecular Biology

Velvet Worms Have Segmented Bodies and Numerous Unjointed Legs

The 65 living species of velvet worms (phylum Onychophora: *onyx* = claw) live under stones, logs, and forest litter in the tropics and in moist temperate habitats in the southern hemisphere. They range in size from 15 mm to 15 cm and feed on small invertebrates.

Onychophorans have a flexible cuticle, segmented bodies, and numerous pairs of unjointed legs **(Figure 29.33)**. Like the annelids, they have pairs of excretory organs in most segments. But unlike annelids, they

Figure 29.33

Phylum Onychophora. Members of the small phylum Onychophora, such as species in the genus *Dnycophor*, have segmented bodies and unjointed appendages.

have an open circulatory system, a specialized respiratory system, relatively large brains, jaws, and tiny claws on their feet. Many produce live young, which, in some species, are nourished within a uterus. Fossil evidence indicates that the onychophoran body plan has not changed much over the last 500 million years.

Arthropods Are Segmented Animals with a Hard Exoskeleton and Jointed Appendages

The more than 1 million known species of arthropods (phylum Arthropoda: *arthron* = joint) include more than half the animal species on Earth—and only a fraction of the living arthropods have been described. This huge lineage includes insects, spiders, crustaceans, millipedes, centipedes, the extinct trilobites, and their relatives.

Arthropods have a segmented body encased in a rigid **exoskeleton.** This external covering is made of chitin, a mix of polysaccharide fibers glued together with glycoproteins, as well as waxes and lipids that block the passage of water. In some marine groups, such as crabs and lobsters, it is hardened with calcium carbonate. The exoskeleton probably first evolved in marine species, providing protection against predators. In terrestrial habitats, it provides support against gravity and protection from dehydration, contributing to the success of insects in even the driest places on Earth. The exoskeleton is especially thin and flexible at the joints between body segments and in the appendages. Contractions of muscles attached to the inside of the exoskeleton move individual body parts like levers, allowing highly coordinated movements and patterns of locomotion that are more precise than those in soft-bodied animals with hydrostatic skeletons.

Although the exoskeleton has obvious advantages, it is nonexpandable and therefore could limit growth of the animal. But, like other Ecdysozoa, arthropods grow and periodically develop a soft, new exoskeleton beneath the old one, which they shed in the complex process of ecdysis **(Figure 29.34).** After shedding the old exoskeleton, aquatic species swell with water and terrestrial species swell with air before the new one hardens. They are especially vulnerable to predators at these times. "Soft-shelled" crabs, prized as food in many countries, are ones that have recently molted.

As arthropods evolved, body segments became fused in various ways, reducing their overall number. Each region, along with its highly modified paired appendages, is specialized, but the structure and function vary greatly among groups. In insects (see Figure 29.43), which have three body regions, the **head** includes a brain, sensory structures, and some sort of feeding apparatus. The segments of the **thorax** bear walking legs and, in some insects, wings. The **abdomen** includes much of the digestive system and sometimes part of the reproductive system.

The coelom of arthropods is greatly reduced, but another cavity, the *hemocoel,* is filled with bloodlike hemolymph. The heart pumps the hemolymph through an open circulatory system, bathing tissues directly.

Arthropods are active animals and require substantial quantities of oxygen. Different groups have distinctive mechanisms for gas exchange, because oxygen cannot cross the impermeable exoskeleton. Marine and freshwater species, like crabs and lobsters, rely on diffusion across gills. The terrestrial groups—

Figure 29.34

Ecdysis in insects. Like all other arthropods, this cicada (*Graptopsaltsia nigrofusca*) sheds its old exoskeleton as it grows and when it undergoes metamorphosis into a winged adult.

Unscrambling the Arthropods

This book follows a classification that divides arthropods into five subphyla—trilobites (now extinct), chelicerates, crustaceans, myriapods, and hexapods. In the past, biologists grouped hexapods and myriapods together (**figure, left**) because they share certain morphological characteristics, which may indicate a common ancestry: unbranched appendages, a tracheal system for gas exchange, Malpighian tubules for excretion, and one pair of antennae. Some biologists have argued, however, that these traits may have been produced by convergent evolution. Furthermore, the comparison of other morphological traits suggests that hexapods are more closely related to crustaceans than to myriapods. For example, both hexapods and crustaceans have jawlike mandibles on the fourth head segment, similar compound eyes, comparable development of the nervous system, and similarities in the structure of thoracic appendages.

A molecular study by Jeffry L. Boore and his colleagues at the University of Michigan lends support to the hypothesis that hexapods may indeed be more closely related to crustaceans than to myriapods. For their study, Boore and his coworkers compared the arrangement of genes in the mitochondrial DNA of representative arthropod species. Mitochondrial DNA (mtDNA) was chosen for study because it has been subject to many random rearrangements during the evolutionary history of the arthropods, producing wide variations in the order and placement of the 36 or 37 genes typically present. As long as the rearrangements do not disrupt gene function, they may have little or no effect on fitness; therefore, they may not be affected by natural selection. As Boore and his colleagues noted, the large number of possible rearrangements of

the mtDNA genes makes it unlikely that the same gene order would appear by chance in any two groups; thus, groups with the same arrangement are likely to share a common ancestor in which the rearrangement first appeared.

The investigators sequenced the mtDNA of a chelicerate, two crustaceans, and a myriapod (listed in the figure caption). They also sequenced the mtDNAs of a mollusk and an annelid, animals that are only distantly related to arthropods, to provide outgroup comparisons. They compared these sequences with those in 14 other invertebrates, including four hexapods and one crustacean, already obtained by other workers.

The locations of two genes encoding leucine-carrying tRNAs provide the strongest clue to the evolutionary relationships of these organisms. These genes are positioned next to each other in the mtDNA of the chelicerate, the mollusk, and the annelid. This arrangement appears to reflect the ancient gene order present in the common ancestor of annelids, mollusks, and arthropods. However, in the hexapods and the crustaceans studied, one of the leucine tRNA genes is located at a different position, between two genes

coding for electron transfer proteins. Moreover, this rearrangement does not appear in the myriapod examined. It is extremely unlikely that the same translocation of the leucine tRNA gene occurred independently in hexapods and crustaceans. Instead, this gene rearrangement probably occurred in an ancestor shared by hexapods and crustaceans after this ancestor diverged from the lineage that includes the chelicerates and myriapods.

The shared gene arrangement in insects and crustaceans and the absence of a matching arrangement in myriapods suggests that hexapods and crustaceans are closely related, and that hexapods and myriapods are not, in spite of their morphological similarities (**figure, right**). Other investigators, who compared the sequences of ribosomal RNA genes in a variety of arthropod species, reached the same conclusions independently.

Because the molecular research studied a relatively few characteristics, it cannot yet be accepted as the definitive answer to questions about arthropod lineages. However, the results give strong support to the idea that hexapods and crustaceans are more closely related than the traditional family tree suggested.

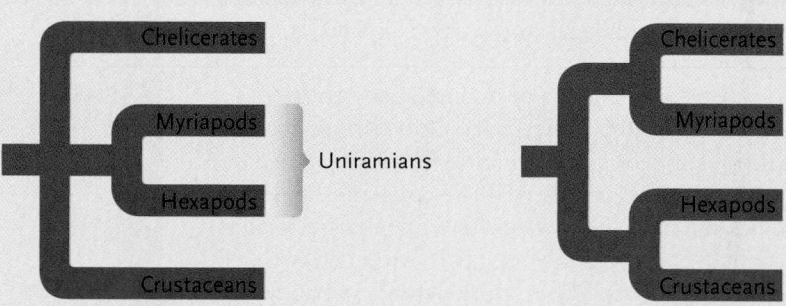

The traditional phylogenetic tree for arthropods (left) differs from that suggested by the research described in this box. The DNA sequences used to construct the new phylogenetic tree (right) were obtained from the chelicerate *Limulus*, the myriapod *Thyrophygus*, the hexapod *Drosophila*, and the crustaceans *Daphnia* and *Homarus*.

insects and spiders—have developed unique and specialized respiratory systems.

High levels of activity also require intricate sensory structures. Many arthropods are equipped with a highly organized central nervous system, touch recep-

tors, chemical sensors, **compound eyes** that include multiple image-forming units, and in some, hearing organs.

Arthropod systematics is an active area of research, and scientists are currently using molecular, morpho-

Figure 29.35
Subphylum Trilobita. Trilobites, like *Olenellus gilberti*, bore many pairs of relatively undifferentiated appendages.

logical, and developmental data to reexamine relationships within this immense group. As *Insights from the Molecular Revolution* explains, hypotheses about the phylogeny and classification of arthropods are in a state of flux. Some researchers even argue for splitting them into four or more phyla. We follow the traditional definition of five *subphyla*, partly because this classification adequately reflects arthropod diversity and partly because no alternative hypothesis has been widely adopted by experts.

Subphylum Trilobita. The trilobites (subphylum Trilobita: *tri* = three; *lobos* = lobe), now extinct, were among the most numerous animals in shallow Paleozoic seas. They disappeared in the Permian mass extinction, but the cause of their demise is unknown. Most trilobites were ovoid, dorsoventrally flattened, and heavily armored, with two deep longitudinal grooves that divided the body into the one median and two lateral lobes for which the group is named **(Figure 29.35)**. Their segmented bodies were organized into a head, which included a pair of sensory **antennae** (chemosensory organs) and two compound eyes, and a thorax and an abdomen, both of which bore pairs of walking legs.

The position of trilobites in the fossil record indicates that they were among the earliest arthropods. Thus, biologists are confident that their three body regions and numerous unspecialized appendages—one pair per segment—represent ancestral traits within the phylum. As you will learn as you read about the other four subphyla, the subsequent evolution of the different arthropod groups included dramatic remodeling of the major body regions as well as modifications of the ancestral, unspecialized paired appendages into structures specialized for different functions.

Subphylum Chelicerata. In spiders, ticks, mites, scorpions, and horseshoe crabs (subphylum Chelicerata: *chela* = claw; *keras* = horn), the first pair of appendages, the **chelicerae**, are fanglike structures used for biting prey. The second pair of appendages, the *pedipalps*, serve as grasping organs, sensory organs, or walking legs. All chelicerates have two major body regions, the **cephalothorax** (a fused head and thorax) and the abdomen. The group originated in shallow Paleozoic seas, but most living species are terrestrial. They vary in size from less than a millimeter to 20 cm; all are predators or parasites.

The 60,000 species of spiders, scorpions, mites, and ticks (Arachnida) represent the vast majority of chelicerates **(Figure 29.36)**. Arachnids have four pairs

a. Wolf spider

b. Spider anatomy

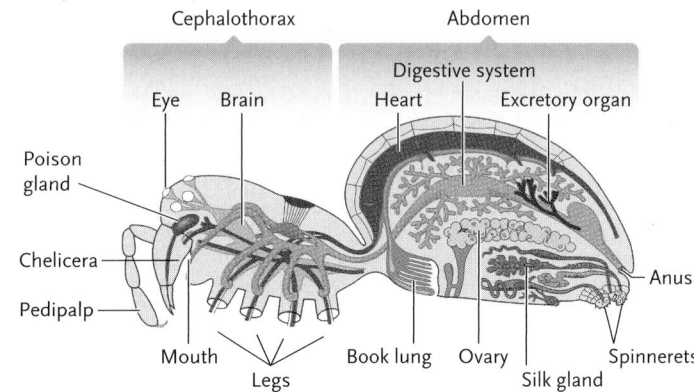

c. Scorpion

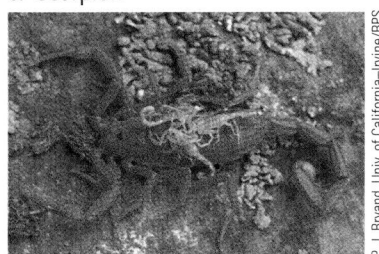

d. House dust mite

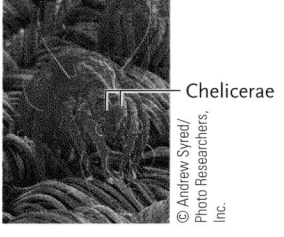

Chelicerae

Figure 29.36
Subphylum Chelicerata, subgroup Arachnida. **(a)** The wolf spider (*Lycosa* species) is harmless to humans. **(b)** The arachnid body plan includes a cephalothorax and abdomen. **(c)** Scorpions have a stinger at the tip of the segmented abdomen. Many, like *Centruroides sculpuratus*, protect their eggs and young. **(d)** House dust mites (*Dermatophagoides pteronyssinus*), shown in a scanning electron micrograph, feed on microscopic debris.

of walking legs on the cephalothorax and highly modified chelicerae and pedipalps. In some spiders, males use their pedipalps to transfer packets of sperm to females. Scorpions use them to shred food and to grasp one another during courtship. Many predatory arachnids have excellent vision, provided by simple eyes on the cephalothorax. Scorpions and some spiders also have unique pocketlike respiratory organs called **book lungs,** derived from abdominal appendages.

Spiders, like most other arachnids, subsist on a liquid diet. They use their chelicerae to inject paralyzing poisons and digestive enzymes into prey and then suck up the partly digested tissues. Many spiders are economically important predators, helping to control insect pests. Only a few are a threat to humans. The toxin of a black widow *(Latrodectus mactans)* causes paralysis, and the toxin of the brown recluse *(Loxosceles reclusa)* destroys tissues around the site of the bite.

Although many spiders hunt actively, others capture prey on silken threads secreted by **spinnerets,** which are modified abdominal appendages. Some species weave the threads into complex, netlike webs. The silk is secreted as a liquid, but quickly hardens on contact with air. Spiders also use silk to make nests, to protect their egg masses, as a safety line when moving through the environment, and to wrap prey for later consumption.

Most mites are tiny, but they have a big impact. Some are serious agricultural pests that feed on the sap of plants. Others cause mange (patchy hair loss) or painful and itchy welts on animals. House dust mites cause allergic reactions in many people. Ticks, which are generally larger than mites, are blood-feeding ectoparasites that often transmit pathogens, such as those causing Rocky Mountain spotted fever and Lyme disease.

The subphylum Chelicerata also includes five species of horseshoe crabs (Merostomata), an ancient lineage that has not changed much during its 350 million year history **(Figure 29.37).** Horseshoe crabs are carnivorous bottom feeders in shallow coastal waters. Beneath their characteristic shell, they have one pair

Figure 29.37
Marine chelicerates. Horseshoe crabs, like *Limulus polyphemus,* are included in the Merostomata.

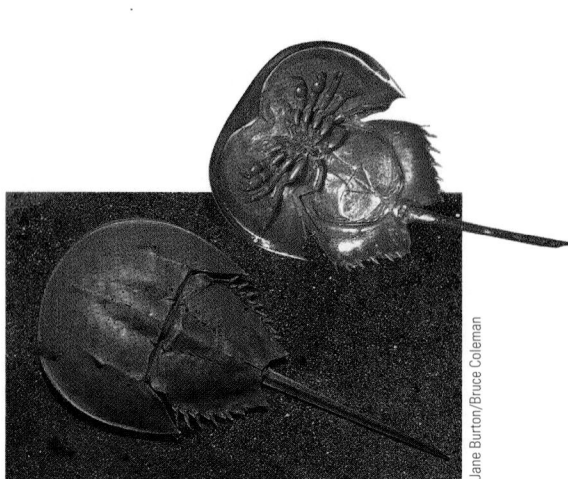

Jane Burton/Bruce Coleman

of chelicerae, a pair of pedipalps, four pairs of walking legs, and a set of paperlike gills, derived from ancestral walking legs.

Subphylum Crustacea. The 35,000 species of shrimps, lobsters, crabs, and their relatives (subphylum Crustacea, meaning "encrusted") represent a lineage that emerged more than 500 million years ago. They are abundant in marine and freshwater habitats. A few species, such as sowbugs and pillbugs, live in moist, sheltered terrestrial environments. In many crustaceans two, or even all three, of the arthropod body regions—head, thorax, and abdomen—are fused; a fused cephalothorax and a separate abdomen is an especially common pattern. The edible "tail" of a lobster or crayfish is actually a highly muscularized abdomen. In some, the exoskeleton includes a **carapace,** a protective covering that extends backward from the head. Crustaceans vary in size from water fleas less than 1 mm long to lobsters that can grow to 60 cm in length and weigh as much as 20 kg.

Crustaceans generally have five characteristic pairs of appendages on the head **(Figure 29.38).** Most have two pairs of sensory antennae and three pairs of mouthparts. The latter include one pair of *mandibles,* which move laterally to bite and chew, and two pairs of *maxillae,* which hold and manipulate food. Numerous paired appendages posterior to the mouthparts vary among groups.

Most crustaceans are active animals that exhibit complex patterns of movement. Their activities are coordinated by elaborate sensory and nervous systems, including chemical and touch receptors in the antennae, compound eyes, statocysts on the head, and sensory hairs embedded in the exoskeleton throughout the body. The nervous system is similar to that in annelids, but the ganglia are larger and more complex, allowing a finer level of motor control. High levels of activity require substantial oxygen, and larger species have complex, feathery gills tucked beneath the carapace. Activity also produces abundant metabolic wastes that are excreted by diffusion across the gills or, in larger species, by glands located in the head.

The sexes are typically separate, and courtship rituals are often complex. Eggs are usually brooded on the surface of the female's body or beneath the carapace. Many have free-swimming larvae that, after undergoing a series of molts, gradually assume an adult form.

The subphylum includes so many different body plans that it is usually divided into six major groups with numerous subgroups. The crabs, lobsters, and shrimps (Decapoda, meaning "10 feet," a subgroup of the Malacostraca) number more than 10,000 species. The vast majority of decapods are marine, but a few shrimps, crabs, and crayfishes occupy freshwater habitats. Some crabs also live in moist terrestrial habitats, where they scavenge dead vegetation, clearing the forest floor of debris.

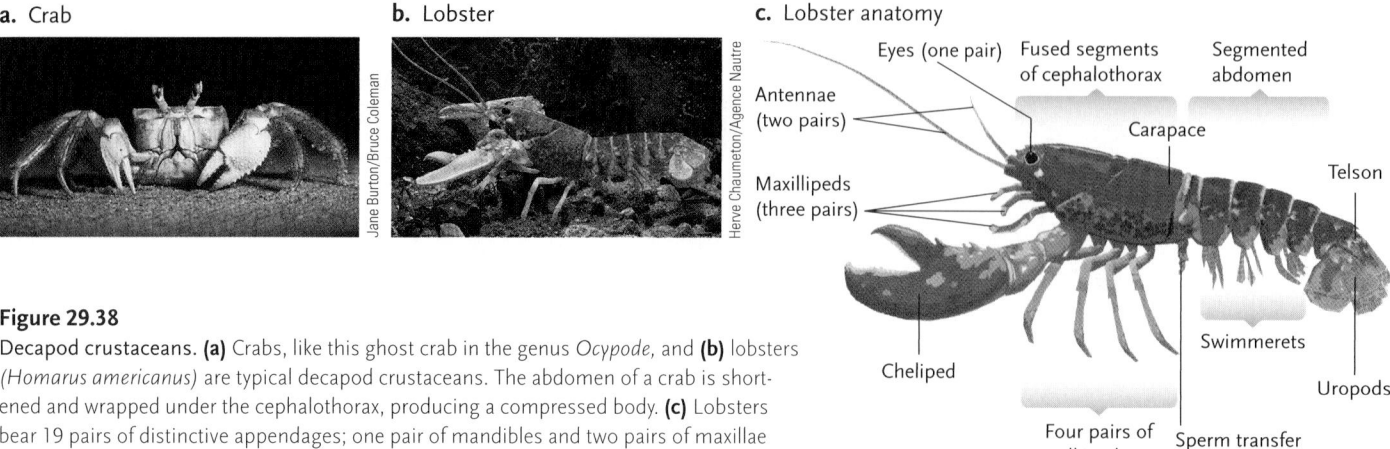

a. Crab

b. Lobster

c. Lobster anatomy

Eyes (one pair)
Fused segments of cephalothorax
Segmented abdomen
Antennae (two pairs)
Carapace
Maxillipeds (three pairs)
Telson
Cheliped
Swimmerets
Uropods
Four pairs of walking legs
Sperm transfer appendage

Figure 29.38
Decapod crustaceans. **(a)** Crabs, like this ghost crab in the genus *Ocypode,* and **(b)** lobsters *(Homarus americanus)* are typical decapod crustaceans. The abdomen of a crab is shortened and wrapped under the cephalothorax, producing a compressed body. **(c)** Lobsters bear 19 pairs of distinctive appendages; one pair of mandibles and two pairs of maxillae are not illustrated in this lateral view.

All decapods exhibit extreme specialization of their appendages. In the American lobster, for example, each of the 19 pairs of appendages is different (see Figure 29.38c). Behind the antennae, mandibles, and maxillae, the thoracic segments have three pairs of *maxillipeds,* which shred food and pass it up to the mouth, a pair of large *chelipeds* (pinching claws), and four pairs of walking legs. The abdominal appendages include a pair specialized for sperm transfer (in males only), *swimmerets* for locomotion and for brooding eggs, and *uropods,* a pair of appendages that, combined with the *telson,* the tip of the abdomen, form a fan-shaped tail. If any appendage is damaged, the animal can autotomize (drop) it and begin growing a new one before its next molt.

Representatives of several crustacean groups—fairy shrimps, amphipods, water fleas, krill, ostracods, and copepods **(Figure 29.39)**—live as plankton in the upper waters of oceans and lakes. Most are only a few millimeters long, but are present in huge numbers. They feed on microscopic algae or detritus and are themselves food for larger invertebrates, fishes, and some suspension-feeding marine mammals like the baleen whales. Planktonic crustaceans are among the most abundant animals on Earth.

Adult barnacles (Cirripedia, meaning "hairy footed," a subgroup of the Maxillopoda) are sessile, marine crustaceans that live within a strong, calcified cup-shaped shell **(Figure 29.40).** Their free-swimming larvae attach permanently to substrates—rocks, wooden pilings, the hulls of ships, the shells of mollusks, even the skin of whales—and secrete the shell, which is actually a modified exoskeleton. To feed, barnacles open the shell and extend six pairs of feathery legs. The beating legs capture microscopic plankton and transfer it the mouth. Unlike most crustaceans, barnacles are hermaphroditic.

Subphylum Myriapoda. The 3000 species of centipedes (Chilopoda) and 10,000 species of millipedes (Diplopoda) are classified together in the subphylum Myriap-

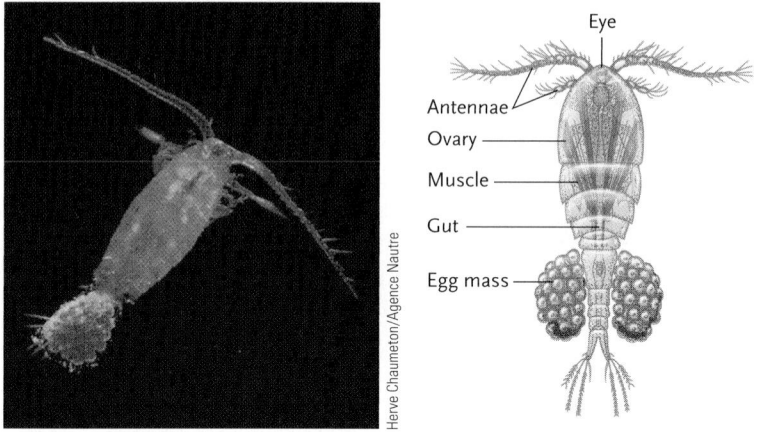

Eye
Antennae
Ovary
Muscle
Gut
Egg mass

Figure 29.39
Copepods. Tiny crustaceans, like these copepods (*Calanus* species on the left, *Cyclops* species on the right), occur by the billions in freshwater and marine plankton.

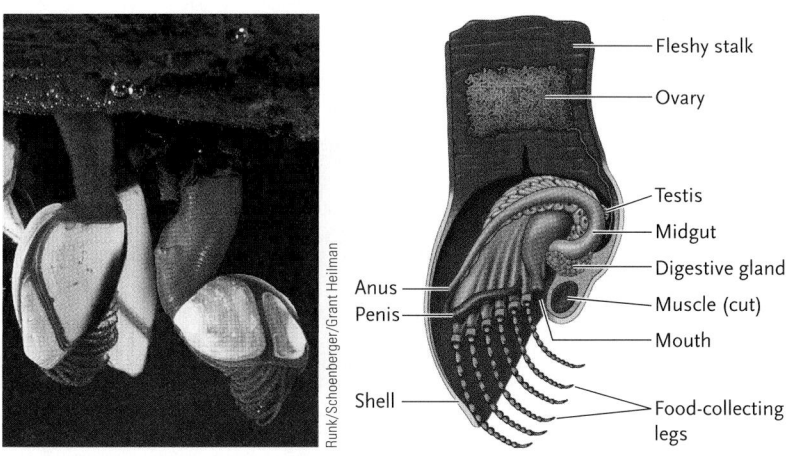

Fleshy stalk
Ovary
Testis
Midgut
Digestive gland
Anus
Muscle (cut)
Penis
Mouth
Shell
Food-collecting legs

Figure 29.40
Barnacles. Gooseneck barnacles *(Lepas anatifera)* attach to the underside of floating debris. Like other barnacles, they open their shells and extend their feathery legs to collect particulate food from seawater.

a.

b.

Figure 29.41

Millipedes and centipedes. **(a)** Millipedes, like *Spirobolus* species, feed on living and decaying vegetation. They have two pairs of walking legs on most segments. **(b)** Like all centipedes, this Southeast Asian species *(Scolopendra subspinipes)*, shown feeding on a small frog, is a voracious predator. Centipedes have one pair of walking legs per segment.

Figure 29.42

Insect diversity. Insects are grouped into about 30 orders, 8 of which are illustrated here.

oda (meaning "countless feet"). Myriapods have two body regions, a head and a segmented trunk **(Figure 29.41)**. The head bears one pair of antennae, and the trunk bears one (centipedes) or two (millipedes) pairs of walking legs on most of its many segments. Myriapods are terrestrial, and many species live under rocks or dead leaves. Centipedes are fast and voracious predators, using powerful toxins to kill their prey; they generally feed on invertebrates, but some eat small vertebrates. The bite of some species is harmful to humans. Although most species are less than 10 cm long, some grow to 25 cm. The millipedes are slow but powerful

a. Silverfish (Thysanura, *Ctenolepisma longicaudata*) are primitive wingless insects.

b. Dragonflies, like the flame skimmer (Odonata, *Libellula saturata*), have aquatic larvae that are active predators; adults capture other insects in mid-air.

c. Male praying mantids (Mantodea, *Mantis religiosa*) are often eaten by the larger females during or immediately after mating.

d. This rhinoceros beetle (Coleoptera, *Dynastes granti*) is one of more than 250,000 beetle species that have been described.

e. Fleas (Siphonoptera, *Hystrichopsylla dippiei*) have strong legs with an elastic ligament that allows these parasites to jump on and off their animal hosts.

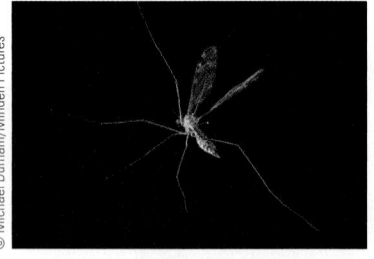

f. Crane flies (Diptera, *Tipula* species) look like giant mosquitoes, but their mouthparts are not useful for biting other animals; the adults of most species live only a few days and do not feed at all.

g. The luna moth (Lepidoptera, *Actias luna*), like other butterflies and moths, has wings that are covered with colorful microscopic scales.

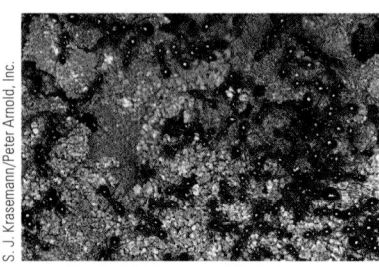

h. Like many other ant species, fire ants (Hymenoptera, *Solenopsis invicta*) live in large cooperative colonies. Fire ants—named for their painful sting—were introduced into southeastern North America, where they are now serious pests.

herbivores or scavengers. The largest species attain a length of nearly 30 cm. Although they lack a poisonous bite, they curl into a ball and exude noxious liquids when disturbed.

Subphylum Hexapoda. In terms of sheer numbers, diversity, and the range of habitats they occupy, the 1,000,000 or more species of insects and their closest relatives (subphylum Hexapoda, meaning "six feet") are the most successful animals on Earth. They were among the first animals to colonize terrestrial habitats, where most species still live. The oldest insect fossils date from the Devonian, 380 million years ago. Insects have one pair of antennae on the head, a pair of mandibles for feeding, and unbranched appendages. Insects are generally small, ranging from 0.1 mm to 30 cm in length. The group is divided into about 30 subgroups **(Figure 29.42).** The insect body plan always includes a head, thorax, and abdomen **(Figure 29.43).** The head is equipped with multiple mouthparts, a pair of compound eyes, and one pair of sensory antennae. The thorax has three pairs of walking legs and often one or two pairs of wings. Insects are the only invertebrates capable of flight. Their wings, which are made of lightweight but durable sheets of chitin and sclerotin, arise embryonically from the body wall; unlike the wings of birds and bats, insect wings are not derived from ancestral appendages.

Studies in evolutionary developmental biology have begun to unravel the genetic changes that fostered certain aspects of the insect body plan. For example, the *Distal-less* gene (*Dll* for short) is a highly conserved tool-kit gene (see Section 22.6) that triggers the development of appendages in all sorts of animals—the legs of chickens, the fins of fishes, the parapodia of polychaete worms, and the diverse appendages of arthropods. All arthropods also have a gene called *Ultrabithorax* (*Ubx* for short). It is one of the *Hox* genes that control the overall body plans of animals. In insects, the *Ubx* gene contains a unique mutation, not found in other arthropods, that causes the protein for which it codes to repress *Dll*, thereby preventing the formation of appendages wherever *Ubx* is expressed. And because insects express *Ubx* in their abdomen, they do not grow abdominal appendages. All other arthropods, which have the ancestral, nonrepressing form of the *Ubx* gene, have appendages in the posterior region of their body. Thus, one mutation in a *Hox*-family gene has fostered the evolution of a highly distinctive morphological trait in insects—having legs on the thorax, but not on the abdomen.

Insects exchange gases through a specialized **tracheal system,** a branching network of tubes that carry oxygen from small openings in the exoskeleton to tissues throughout the body. Insects excrete nitrogenous wastes through specialized **Malpighian tubules** that transport wastes to the digestive system for dis-

External anatomy of a grasshopper

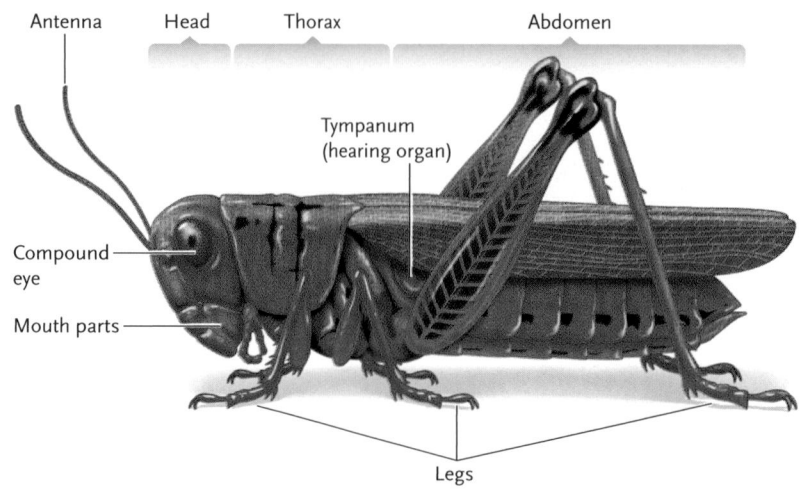

Internal anatomy of a female grasshopper

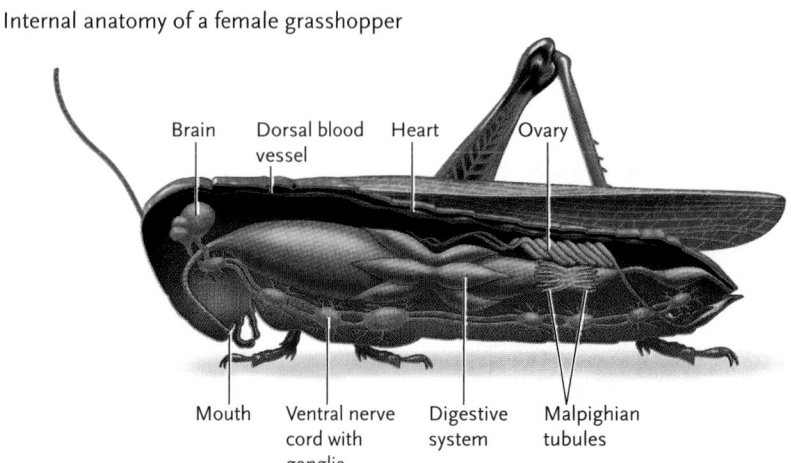

Figure 29.43
The insect body plan. Insects have distinct head, thorax, and abdomen. Of all the internal organ systems, only the dorsal blood vessel, ventral nerve cord, and some muscles are strongly segmented.

posal with feces. Both of these organ systems are unique among animals. Insect sensory systems are diverse and complex. Besides image-forming compound eyes, many insects have light-sensing ocelli on their heads. Many also have hairs, sensitive to touch, on their antennae, legs, and other regions of the body. Chemical receptors are particularly common on the legs, allowing the identification of food. And many groups of insects have hearing organs to detect predators and potential mates. The familiar chirping of crickets, for example, is a mating call emitted by males that may repel other males and attract females.

As a group, insects feed in every conceivable way and on most other organisms **(Figure 29.44).** Species that eat plants, such as grasshoppers, have a pair of rigid mandibles, which chew food before it is ingested. Behind the mandibles is a pair of maxillae, which may also aid in food acquisition. Insects also have inflexible upper and lower lips, the *labrum* and *labium,* respectively. A tonguelike structure just dorsal to the labium in chewing insects houses the openings of salivary glands. But evolution has modified this ancestral mandibulate pattern in numerous ways. In some biting

Figure 29.44

Specialized insect mouthparts. The **(a)** ancestral chewing mouthparts have been modified by evolution, allowing different insects to **(b)** sponge up food, **(c)** drink nectar, and **(d)** pierce skin to drink blood.

a. Grasshopper

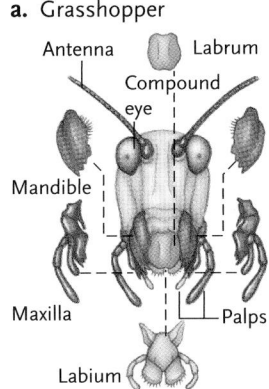
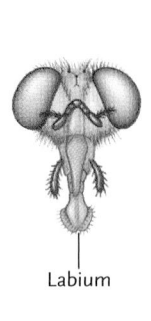

Antenna · Labrum
Compound eye
Mandible
Maxilla
Maxilla · Palps
Labium
Labium

b. Housefly

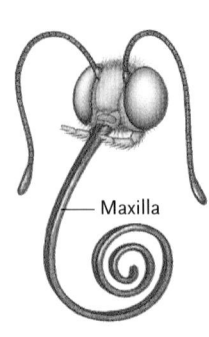

Labium

c. Butterfly

Maxilla

d. Mosquito

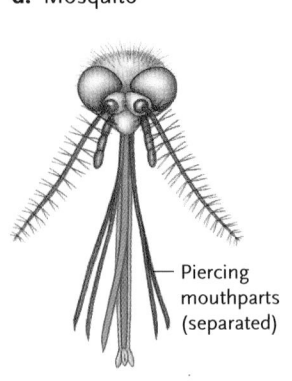

Piercing mouthparts (separated)

flies, like mosquitoes and blackflies, the mouthparts have evolved into piercing structures. In butterflies and moths the mouthparts include a long proboscis to drink nectar. And in houseflies, the mouthparts are adapted for sopping up liquid food.

After it hatches from an egg, an insect passes through a series of developmental stages called *instars*. Several hormones control development and ecdysis, which marks the passage from one instar to the next. Insects exhibit one of three basic patterns of postem-

Unanswered Questions

What are the evolutionary relationships among the invertebrate lineages?

If we step back from our vertebrate and terrestrial biases, invertebrates are the dominant form of life on Earth. Many students of natural history are overwhelmed by the sheer numbers of organisms that one encounters, particularly in the aquatic environment, and the challenge of categorizing them taxonomically. Sorting them out in terms of their ecological roles is equally daunting. The problem we face is that invertebrates do it all—they are predators, herbivores, parasites, detritivores, and the primary symbiotic organisms on Earth. This adaptability of form and function is a fascinating hallmark of the animal way of life, but it poses a legion of questions, many still unanswered. Recent advances in genetic technology have advanced our understanding, but there is much left to do.

In the past two decades, our ability to compare genetic information from various invertebrate groups has led to a remarkable reshuffling of the long-established categories used to classify these organisms. The first categories to be eliminated were groups based on superficial phenotypic resemblances, such as the "pseudocoelomates," which had plagued student understanding of diversity. Today, we have a much deeper knowledge of the evolutionary relationships of these organisms. Perhaps the most exciting discovery is that much of the diversity we see is not the product of slow changes in protein-coding gene sequences, but rather the result of variations in the timing and location of the expression of genes that affect development. Evo-devo, the melding of evolutionary and developmental biology—mostly made possible by intensive studies of two model invertebrates, *Caenorhabditis elegans* and *Drosophila melanogaster*—has revealed that changes in the expression of relatively simple sets of genes have brought about the myriad forms of life we see among the invertebrates. As systematists incorporate these new discoveries in their analyses, our understanding of the evolutionary relationships among the invertebrates will surely change.

What is the genetic basis of the diversity of form and function observed among invertebrates?

Because of advances in genetics research, we are on the cusp of being able to answer some fundamental questions. How does an animal's body develop either radial or bilateral symmetry? How does an organism develop a head with a concentration of nervous system tissue, and how do all the exquisite sensory systems associated with a big brain develop? From where do the respiratory pigments, which increase the capacity of the hemolymph or blood to carry oxygen, increasing an organism's capacity for activity, arise? How does the immune system develop, and what genetic change fosters the quantum leap from a nonspecific defense system to one that responds specifically to foreign invaders? More practically, are there genetic switches that we can manipulate? Can we make blood-feeding invertebrates, such as mosquitoes, or parasitic species, like tapeworms or filarial worms, innocuous? Is it possible to use genetic engineering to reduce the ability of the mosquito *Anopheles gambiae* (the "deadliest organism on Earth") to transmit malaria? Will it be possible to forestall or reverse the global decline of coral reefs, one of the richest habitats on Earth?

The answers to these and many more basic and applied questions lie within a deep knowledge and understanding of the invertebrates and the roles they play on our planet. If one looks at invertebrates as dynamic systems, as rich sources of clues to life on Earth, the questions they pose easily provide a lifetime of investigation and reward.

William S. Irby is an associate professor in the Department of Biology at Georgia Southern University in Statesboro. His research focuses on the ecology and evolution of blood-feeding behavior in mosquitoes. To learn more about his work, go to http://www.bio.georgiasouthern.edu/amain/fac-list.html.

a. No metamorphosis

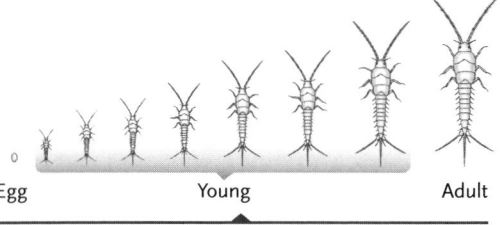

Egg   Young   Adult

Some wingless insects, like silverfish (order Thysanura), do not undergo a dramatic change in form as they grow.

b. Incomplete metamorphosis

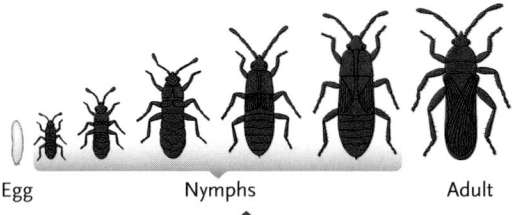

Egg   Nymphs   Adult

Other insects, such as true bugs (order Hemiptera), have incomplete metamorphosis; they develop from nymphs into adults with relatively minor changes in form.

c. Complete metamorphosis

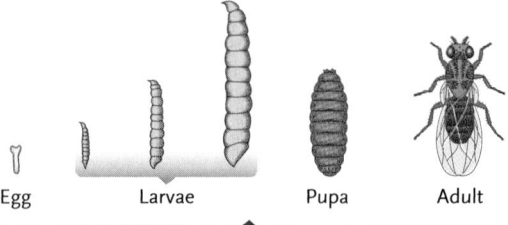

Egg   Larvae   Pupa   Adult

Fruit flies (order Diptera) and many other insects have complete metamorphosis; they undergo a total reorganization of their internal and external anatomy when they pass through the pupal stage of the life cycle.

Figure 29.45
Patterns of postembryonic development in insects.

erpillars, grubs, or maggots) are often worm-shaped, with chewing mouthparts. They grow and molt several times, retaining their larval morphology. Before they transform into sexually mature adults, they spend a period of time as a sessile **pupa**. During this stage, the larval tissues are drastically reorganized. The adult that emerges is so different from the larva that it is often hard to believe that they are of the same species. Moths, butterflies, beetles, and flies are examples of insects with complete metamorphosis. Their larval stages specialize in feeding and growth, whereas the adults are adapted for dispersal and reproduction. In some species, the adults never feed, relying on the energy stores accumulated during the larval stage.

The 240-million-year-history of insects has been characterized by innovations in morphology, life cycle patterns, locomotion, feeding, and habitat use. Their well-developed nervous systems govern exceptionally complex patterns of behavior, including parental care, a habit that reaches its zenith in the colonial social insects, the ants, bees, and wasps (see Chapter 55). The factors that contribute to the insects' success also make them our most aggressive competitors. They destroy vegetable crops, stored food, wool, paper, and timber. They feed on blood from humans and domesticated animals, sometimes transmitting disease-causing pathogens as they do so. Nevertheless, insects are essential members of terrestrial ecological communities. Many species pollinate flowering plants, including important crops. Many others attack or parasitize species that are harmful to human activities. And most insects are a primary source of food for other animals. Some even make useful products like honey, beeswax, and silk.

In the next chapter we consider the lineage of deuterostomes, which includes the vertebrates and their closest invertebrate relatives.

bryonic development **(Figure 29.45)**. Primitive, wingless species simply grow and shed their exoskeleton without undergoing major changes in morphology. Other species undergo **incomplete metamorphosis**; they hatch from the egg as a *nymph,* which lacks functional wings. In many species, such as grasshoppers (order Orthoptera), the nymphs resemble the adults. In other insects, such as dragonflies (order Odonata), the aquatic nymphs are morphologically very different from the adults.

Most insects undergo **complete metamorphosis:** the larva that hatches from the egg differs greatly from the adult. Larvae and adults often occupy different habitats and consume different food. The larvae (cat-

STUDY BREAK

1. What part of a parasitic nematode's anatomy protects it from the digestive enzymes of its host?
2. If an arthropod's rigid exoskeleton cannot be expanded, how does the animal grow?
3. How do the number of body regions and the appendages on the head differ among the four subphyla of living arthropods?
4. How do the life stages differ between insects that have incomplete metamorphosis and those that have complete metamorphosis?

Review

Go to ThomsonNOW™ at www.thomsonedu.com/login to access quizzing, animations, exercises, articles, and personalized homework help.

29.1 What Is an Animal?

- Animals are eukaryotic, multicellular heterotrophs that are motile at some time in their lives.
- Animals probably arose from a colonial flagellated ancestor during the Precambrian era (Figure 29.2).

29.2 Key Innovations in Animal Evolution

- All animals except sponges have tissues, organized into either two or three tissue layers.
- Although some animals exhibit radial symmetry, most exhibit bilateral symmetry (Figure 29.3).
- Acoelomate animals have no body cavity. Pseudocoelomate animals have a body cavity between the derivatives of endoderm and mesoderm. Coelomate animals have a body cavity that is entirely lined by derivatives of mesoderm (Figure 29.4).
- Two lineages of animals differ in developmental patterns (Figure 29.5). Most protostomes exhibit spiral, determinate cleavage, and their coelom forms from a split in a solid mass of mesoderm. Most deuterostomes have radial, indeterminate cleavage, and the coelom usually forms within outpocketings of the primitive gut.
- Four animal phyla exhibit segmentation.

Animation: Types of body symmetry

Animation: Types of body cavities

Animation: Developmental differences between protostomes and deuterostomes

29.3 An Overview of Animal Phylogeny and Classification

- Analyses of molecular sequence data have refined our view of animal evolutionary history (Figure 29.6). The molecular phylogeny recognizes some major lineages that had been identified on the basis of morphological and embryological characters. Sponges are grouped in the Parazoa. All other lineages are grouped in the Eumetazoa. Among the Eumetazoa, the Radiata includes animals with two tissue layers and radial symmetry, and the Bilateria includes animals with three tissue layers and bilateral symmetry.
- Bilateria is further subdivided into Protostomia and Deuterostomia. The new phylogeny divides the Protostomia into the Lophotrochozoa and the Ecdysozoa.
- The molecular phylogeny suggests that ancestral protostomes had a coelom and that segmentation arose independently in three lineages.

29.4 Animals without Tissues: Parazoa

- Sponges (phylum Porifera) are asymmetrical animals with limited integration of cells in their bodies (Figure 29.7).
- The body of many sponges is a water-filtering system (Figure 29.8). Flagellated choanocytes draw water into the body and capture particulate food.

Animation: Body plan of a sponge

29.5 Eumetazoans with Radial Symmetry

- The two major radiate phyla, Cnidaria and Ctenophora, have two well-developed tissue layers with a gelatinous mesoglea between (Figure 29.9). They lack organ systems. All are aquatic.
- Cnidarians capture prey with tentacles and stinging nematocysts (Figures 29.10, 29.12, and 29.13). Their life cycles may include polyps, medusae, or both (Figure 29.11).
- Ctenophores use long tentacles to capture particulate food and use rows of cilia for locomotion (Figure 29.14).

Animation: Cnidarian body plans

Animation: Nematocyst action

Animation: Cnidarian life cycle

29.6 Lophotrochozoan Protostomes

- The taxon Lophotrochozoa includes eight phyla.
- Three small phyla (Ectoprocta, Brachiopoda, and Phoronida) use a lophophore to feed on particulate matter (Figure 29.15).
- Free-living flatworm species (phylum Platyhelminthes) have well-developed digestive, excretory, reproductive, and nervous systems (Figures 29.16 and 29.17). Parasitic species attach to their animal hosts with suckers or hooks (Figures 29.18 and 29.19).
- The rotifers (phylum Rotifera) are tiny and abundant inhabitants of freshwater and marine ecosystems (Figure 29.20). Movements of cilia in the corona control their locomotion and bring food to their mouths.
- The ribbon worms (phylum Nemertea) are elongate and often colorful animals with a proboscis housed in a rhynchocoel (Figure 29.21).
- Mollusks (phylum Mollusca) have fleshy bodies that are often enclosed in a hard shell. The molluskan body plan includes a head–foot, visceral mass, and mantle (Figures 29.22, 29.24–29.27).
- Segmented worms (phylum Annelida) generally exhibit segmentation of the coelom and of the muscular, circulatory, excretory, respiratory, and nervous systems. They use the coelom as a hydrostatic skeleton for locomotion (Figures 29.28–29.31).

Animation: Planarian organ systems

Animation: Blood fluke life cycle

Animation: Tapeworm life cycle

Animation: Earthworm body plan

Animation: Molluscan groups

Animation: Snail body plan

Animation: Torsion in gastropods

Animation: Clam body plan

Animation: Cuttlefish body plan

29.7 Ecdysozoan Protostomes

- The taxon Ecdysozoa includes three phyla that periodically shed their cuticle or exoskeleton.
- Roundworms (phylum Nematoda) feed on decaying organic matter or parasitize plants or animals (Figure 29.32). They move by contracting longitudinal muscles of the body wall.

- The velvet worms (phylum Onychophora) have segmented bodies and unjointed legs (Figure 29.33). Some species bear live young, which develop in a uterus.
- The segmented bodies of the arthropods (phylum Arthropoda) have specialized appendages for feeding, locomotion, or reproduction. Arthropods shed their exoskeleton as they grow or enter a new stage of the life cycle (Figure 29.34). They have an open circulatory system, a complex nervous system, and, in some groups, highly specialized respiratory and excretory systems.
- Arthropods are divided into five subphyla. The extinct trilobites (subphylum Trilobita), with three-lobed bodies and relatively undifferentiated appendages, were abundant in Paleozoic seas (Figure 29.35). Chelicerates have a cephalothorax and abdomen; appendages on the head serve as pincers or fangs and pedipalps

(Figures 29.36 and 29.37). Crustaceans have a carapace that covers the cephalothorax as well as highly modified appendages, including antennae and mandibles (Figures 29.38–29.40). Myriapods have a head and an elongate, segmented trunk (Figure 29.41). Hexapods have three body regions, three pairs of walking legs on the thorax, and three pairs of feeding appendages on the head (Figures 29.42–29.44). Insects exhibit three patterns of postembryonic development (Figure 29.45).

Animation: Roundworm body plan

Animation: Crab life cycle

Animation: Chelicerates

Animation: Insect head parts

Animation: Insect development

Questions

Self-Test Questions

1. Which of the following characteristics is *not* typical of most animals?
   a. heterotrophic
   b. sessile
   c. bilaterally symmetrical
   d. multicellular
   e. motile at some stage of life cycle

2. A body cavity that separates the digestive system from the body wall but is *not* completely lined with mesoderm is called a:
   a. schizocoelom.   d. pseudocoelom.
   b. mesentery.   e. hydrostatic skeleton.
   c. peritoneum.

3. Protostomes and deuterostomes typically differ in:
   a. their patterns of body symmetry.
   b. the number of germ layers during development.
   c. their cleavage patterns.
   d. the size of their sperm.
   e. the size of their digestive systems.

4. The nematocysts of cnidarians are used primarily for:
   a. capturing prey.
   b. detecting light and dark.
   c. courtship.
   d. sensing chemicals.
   e. gas exchange.

5. Which organ system is absent in flatworms (phylum Platyhelminthes)?
   a. nervous system
   b. reproductive system
   c. circulatory system
   d. digestive system
   e. excretory system

6. Which part of a mollusk secretes the shell?
   a. visceral mass   d. head–foot
   b. radula   e. mantle
   c. trochophore

7. What is the major morphological innovation seen in annelid worms?
   a. a complete digestive system
   b. image-forming eyes
   c. a respiratory system
   d. an open circulatory system
   e. body segmentation

8. Which phylum includes the most abundant animals in soil?
   a. Nematoda   d. Annelida
   b. Rotifera   e. Brachiopoda
   c. Mollusca

9. Which body region of an insect bears the walking legs?
   a. head   d. thorax
   b. carapace   e. trunk
   c. abdomen

10. Ecdysis refers to a process in which:
    a. bivalves use siphons to pass water across their gills.
    b. arthropods shed their old exoskeletons.
    c. cnidarians build skeletons of calcium carbonate.
    d. rotifers produce unfertilized eggs.
    e. squids escape from predators in a cloud of ink.

Questions for Discussion

1. Many invertebrate species are hermaphroditic. What selective advantages might this characteristic offer? In what kinds of environments might it be most useful?

2. People who eat raw clams and oysters harvested from sewage-polluted waters often develop mild to severe gastrointestinal infections. These mollusks are suspension feeders. Develop a hypothesis about why people who eat them raw may be at risk.

3. On a voyage to the ocean bottom, a biologist discovers a worm that appears to be new to science. What characteristics of this animal should the biologist examine to determine whether or not she has discovered a previously undescribed phylum?

4. The phylogenetic tree and classification based on molecular sequence data suggests that segmentation evolved independently in Lophotrochozoa (phylum Annelida), Ecdysozoa (phyla Onychophora and Arthropoda), and Deuterostomia (phylum Chordata). What morphological evidence would you try to collect to confirm that segmentation is not homologous in these three groups?

5. What are the relative advantages and disadvantages of radially symmetrical and bilaterally symmetrical body plans?

Experimental Analysis

Design an experiment to test the hypothesis that the cuticle of parasitic nematodes protects them from the acids and enzymes present in the digestive systems of their hosts. Your design must include both experimental and control treatments.

Evolution Link

Many insects have a larval stage that is morphologically different from the adult and that feeds on different foods. What selection pressures may have fostered the evolution of a life cycle with such distinctive life stages? Your answer should address the different biological activities that characterize each life cycle stage.

How Would You Vote?

Cone snails are diverse, but most kinds have a limited geographic range, which makes them highly vulnerable to extinction. We do not know how many are harvested, because no one monitors the trade. Should the United States push to extend regulations on trade in endangered species to cover any species captured from the wild? Go to www.thomsonedu.com/login to investigate both sides of the issue and then vote.

Snow monkeys *(Macaca fuscata)*. These snow monkeys, which have the northernmost distribution of any nonhuman primate, are soaking in a hot spring in Japan.

© Ingo Arndt/Foto Natura/Minden Pictures

Continued on next page

30 Deuterostomes: Vertebrates and Their Closest Relatives

WHY IT MATTERS

In 1798, naturalists at the British Museum skeptically probed a curious specimen that had been sent from Australia. The furry creature—about the size of a housecat—had webbed front feet, a ducklike bill, and a flat, paddlelike tail **(Figure 30.1)**. The scientists eagerly searched for evidence that a prankster had stitched together parts from wildly different animals, but they found no signs of trickery and soon accepted the duck-billed platypus *(Ornithorhynchus anatinus)* as a genuine zoological novelty.

Further study has revealed that the platypus is even stranger than those scientists could have imagined. Like other mammals, the platypus is covered with fur, and females produce milk that the offspring lick off the fur on their mother's belly. But like turtles and birds, a platypus has no teeth, and it reproduces by laying eggs instead of giving birth to its offspring. And like turtles, birds, lizards, snakes, and crocodilians, it has a cloaca, a multipurpose chamber through which it releases feces, urine, and eggs. Scientists had never before seen such a weird combination of traits, and they didn't quite know what to make of them.

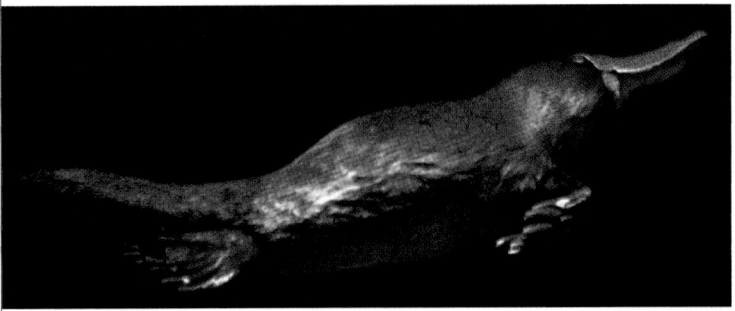

Figure 30.1

A puzzling animal. Because of its strange mixture of traits, the platypus (*Ornithorhynchus anatinus*) amazed the first European zoologists who saw it.

Studies of the platypus under natural conditions have helped biologists make sense of its characteristics. The platypus inhabits streams and lagoons in Australia and Tasmania. It rests in streamside burrows during the day, but at night it slips into the water to hunt for invertebrates. Its dense fur keeps its body warm and dry under water, and its tail serves both as a rudder and as a storehouse for energy-rich fat. It uses its bill to scoop up food and the horny pads that line its jaws to grind up prey. While underwater, the platypus clamps shut its eyes, ears, and nostrils, relying on roughly 800,000 sensory receptors in its bill to detect the movements and weak electrical discharges of nearby prey.

The platypus, with its strange combination of characteristics, illustrates the remarkable diversity of adaptations that enable vertebrates—animals with backbones—to occupy nearly every habitat on Earth. Despite the platypus's mixed characteristics, biologists eventually classified it as a member of the mammal lineage because, like all other mammals, it has hair on its body and produces milk to nourish its offspring. Today biologists know that it is one of just a few remaining survivors of an early lineage of egg-laying mammals.

In this chapter, we survey the Deuterostomia, a monophyletic lineage of animals that dates to the Paleozoic. The deuterostomes are defined by features of early embryological development and molecular sequence data (see Chapter 29). There are three living phyla of deuterostomes; we briefly consider two phyla of invertebrate deuterostomes before focusing on the Phylum Chordata, which includes a few thousand species of invertebrates as well as nearly 50,000 living species of vertebrates.

30.1 Invertebrate Deuterostomes

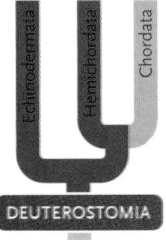

Deuterostome body plans have been so modified by evolution that a casual observer would not readily group the two phyla of invertebrate deuterostomes—Echinodermata and Hemichordata—together with the Phylum Chordata. However, embryological and molecular analyses agree that all three are indeed closely related.

Echinoderms Have Secondary Radial Symmetry and an Internal Skeleton

The phylum Echinodermata (*echinos* = spiny; *derma* = skin) includes 6600 species of sea stars, sea urchins, sea cucumbers, brittle stars, and sea lilies. These slow moving or sessile, bottom-dwelling animals are important herbivores and predators in shallow coastal waters and, paradoxically, the ocean depths. The phylum was diverse in the Paleozoic, but only a remnant of that fauna remains. Living species vary in size from less than 1 cm to more than 50 cm in diameter.

Echinoderms develop from a bilaterally symmetrical, free-swimming larva. As a larva develops, it assumes a secondary radial symmetry, often organized around five rays or "arms" **(Figure 30.2)**. Many echinoderms have an *oral surface,* with the mouth facing the substrate, and an *aboral surface* facing in the opposite direction. Virtually all echinoderms have an internal skeleton made of calcium-stiffened *ossicles* that develop from mesoderm. In some groups, fused ossicles form a rigid container called a *test.* In most, spines or bumps project from the ossicles.

The internal anatomy of echinoderms is unique among animals **(Figure 30.3)**. They have a well-defined coelom and a complete digestive system (see Figure 30.3a), but no excretory or respiratory systems, and most have only a minimal circulatory system. In many, gases are exchanged and metabolic wastes eliminated through projections of the epidermis and peritoneum near the base of the spines. Given their radial symmetry, there is no head or central brain; the nervous system is organized around nerve cords that encircle the mouth and branch into the rays. Sensory cells are abundant in the skin.

Echinoderms move using a unique system of fluid-filled canals, the *water vascular system* (see Figure 30.3b). In a sea star, for example, water enters the system through the *madreporite,* a sievelike plate on the aboral surface. A short tube connects it to the *ring canal,* which surrounds the esophagus. The ring canal branches into five *radial canals* that extend into the arms. Each radial canal is connected to numerous *tube feet* that protrude through holes in the plates. Each tube foot has a mucus-covered, suckerlike tip and a small muscular bulb, the *ampulla,* that lies inside the body. When an ampulla contracts, fluid is forced into the tube foot, causing it to lengthen and attach to the substrate (see Figure 30.3c). The tube foot then contracts, pulling the animal along. As the tube foot shortens, water is forced back into the ampulla, and the tube foot releases its grip on the substrate. The tube foot can then take another step forward, reattaching to the substrate. Although each tube foot has limited strength, the coordinated action of hundreds or even thousands of them is so strong that they can hold an echinoderm to a substrate even against strong wave action.

Echinoderms have separate sexes, and most reproduce by releasing gametes into the water. Radial cleavage is so clearly apparent in the transparent eggs of some sea urchins that they are commonly used for demonstrations of cleavage in introductory biology laboratories. A few echinoderms also reproduce asexually by splitting in half and regenerating the missing parts; some can regenerate body parts lost to predators.

Echinoderms are divided into six groups, one of which, the sea daisies (Concentricycloidea) was discovered only in 1986. These small, medusa-shaped animals occupy sunken, waterlogged wood in the deep sea. In the following sections, we describe the five other groups, which are more diverse and better known.

Asteroidea. The 1500 species of sea stars (Asteroidea, from *asteroeides* = starlike) live from rocky shorelines to depths of 10,000 m. The body consists of a central disk surrounded by 5 to 20 radiating "arms" (see Figure 30.2a), with the mouth centered on the oral surface. The ossicles of the endoskeleton are not fused, permitting flexibility of the arms and disk. Small pincers, **pedicellariae,** at the base of short spines remove debris that falls onto the animal's surface (see Figure 30.3c). Many sea stars feed on invertebrates and small fishes. Species that consume bivalve mollusks pry apart the two valves using their tube feet and slip their everted stomachs between the bivalve's shells. The stomach secretes digestive enzymes that dissolve the mollusk's tissues. Some sea stars are destructive predators of corals, endangering many reefs.

Ophiuroidea. The 2000 species of brittle stars and basket stars (Ophiuroidea, from *ophioneos* = snakelike) occupy roughly the same range of habitats as sea stars. Their bodies have a well-defined central disk and slender, elongate arms that are sometimes branched (see Figure 30.2b). Ophiuroids can crawl fairly swiftly across substrates by moving their arms in coordinated fashion. As their common name implies, the arms are delicate and easily broken, an adaptation that allows them to escape from predators with only minor losses. Brittle stars feed on small prey, suspended plankton, or detritus that they extract from muddy deposits.

Echinoidea. The 950 species of sea urchins and sand dollars (Echinoidea, "having spines") lack arms altogether (see Figure 30.2c). Their ossicles are fused into solid tests, which provide excellent protection but restrict flexibility. The test is spherical in sea urchins and flattened in sand dollars. Five rows of tube feet, used primarily for locomotion, emerge through pores in the test. Most echinoids have movable spines, some with poison glands; a jab from certain tropical species can cause severe pain and inflammation to a careless swimmer. Echinoids graze on algae and other organisms that cling to marine surfaces. In the center of an

Figure 30.2
Echinoderm diversity. Echinoderms exhibit secondary radial symmetry, usually organized as five rays around an oral-aboral axis.

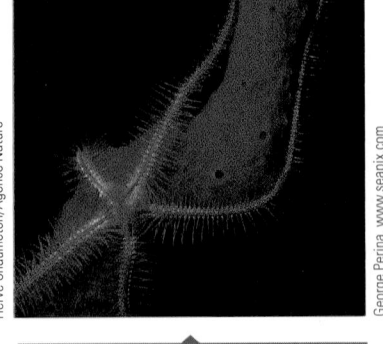

a. Asteroidea: This sea star *(Fromia milleporella)* lives in the intertidal zone.

b. Ophiuroidea: A brittle star *(Ophiothrix swensonii)* perches on a coral branch.

c. Echinoidea: A sea urchin *(Strongylocentrotus purpuratus)* grazes on algae.

d. Holothuroidea: A sea cucumber *(Cucumaria miniata)* extends its tentacles, which are modified tube feet, to trap particulate food.

e. Crinoidea: A feather star *(Himerometra robustipinna)* feeds by catching small particles with its numerous arms.

a. Internal anatomy

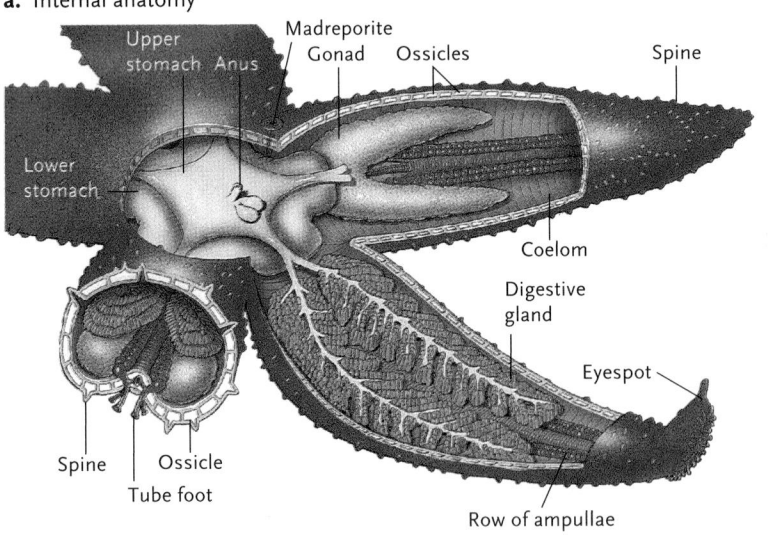

Upper stomach · Anus · Madreporite · Gonad · Ossicles · Spine · Coelom · Digestive gland · Eyespot · Lower stomach · Spine · Ossicle · Tube foot · Row of ampullae

b. Water vascular system

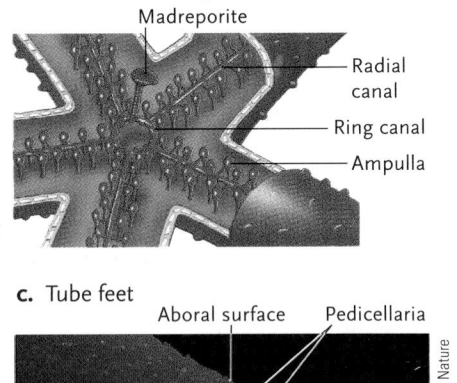

Madreporite · Radial canal · Ring canal · Ampulla

c. Tube feet

Aboral surface · Pedicellaria · Oral surface · Tube foot

Figure 30.3
Internal anatomy of a sea star. **(a)** The coelom is well developed in echinoderms, as illustrated by this cutaway diagram of a sea star. **(b)** The water vascular system, unique in the animal kingdom, operates the tube feet **(c)**, which are responsible for locomotion. Note the pedicellariae on the upper surface of the sea star's arm **(c)**.

urchin's oral surface is a five-part nipping jaw that is controlled by powerful muscles. Some species damage kelp beds, disrupting the habitat of young lobsters and other crustaceans. But echinoid ovaries are a delicacy in many countries, making these animals a prized natural resource.

Holothuroidea. The 1500 species of sea cucumbers (Holothuroidea, from *holothourion* = water polyp) are elongate animals that lie on their sides on the ocean bottom (see Figure 30.2d). Although they have five rows of tube feet, their endoskeleton is reduced to widely separated microscopic plates. The body, which is elongated along the oral-aboral axis, is soft and fleshy, with a tough, leathery covering. Modified tube feet form a ring of tentacles around the mouth, which points to the side or upward. Some species secrete a mucous net that traps plankton or other food particles. The net and tentacles are inserted into the mouth where the net and trapped food are ingested. Other species extract food from bottom sediments. Many sea cucumbers exchange gases through an extensively branched *respiratory tree* that arises from the rectum, the part of the digestive system just inside the anus at the aboral end of the animal. A well-developed circulatory system distributes oxygen and nutrients to tissues throughout the body.

Crinoidea. The 600 living species of sea lilies and feather stars (Crinoidea, from *krinon* = lily) are the surviving remnants of a fauna that was diverse and abundant 500 million years ago (see Figure 30.2e). Most species occupy marine waters of medium depth. The central disk and mouth point upward rather than toward the substrate. Between five and several hundred branched arms surround the disk; new arms are added as a crinoid grows larger. The branches of the arms are covered with tiny mucus-coated tube feet, which trap suspended microscopic organisms. The sessile sea lilies have the central disk attached to a flexible stalk that can reach a meter in length. Adult feather stars can swim or crawl weakly, attaching temporarily to substrates.

Acorn Worms Use Gill Slits and a Pharynx to Acquire Food and Oxygen

The 80 species of acorn worms (phylum Hemichordata, from *hemi* = half and *chorda* referring to the phylum Chordata) are sedentary marine animals that live in U-shaped tubes or burrows in coastal sand or mud. Their soft bodies, which range from 2 cm to 2 m in length, are organized into an anterior proboscis, a tentacled collar, and an elongate trunk **(Figure 30.4)**. They use their muscular, mucus-coated proboscis to construct burrows and trap food particles. Acorn worms also have pairs of **gill slits** in the pharynx, the part of the digestive system just posterior to the mouth. Beating cilia create a flow of water, which enters the phar-

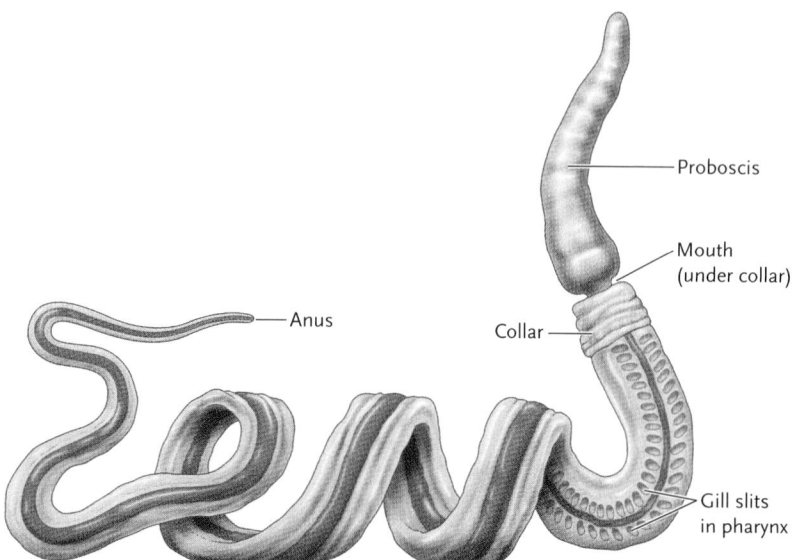

Figure 30.4
Phylum Hemichordata. Acorn worms draw food- and oxygen-laden water in through the mouth and expel it through gill slits in the anterior region of the trunk.

ynx through the mouth and exits through the gill slits. As water passes through, suspended food is trapped and shunted into the digestive system, and gases are exchanged across the partitions between gill slits. The coupling of feeding and respiration—as well as a dorsal nerve cord—reflects the close evolutionary relationship between hemichordates and chordates, the phylum that we consider next.

STUDY BREAK

1. What organ system is unique to echinoderms, and what is its function?
2. How does a perforated pharynx enable hemichordates to acquire food and oxygen from seawater?

30.2 Overview of the Phylum Chordata

The phylum Chordata contains three subphyla: two lineages of invertebrates, Urochordata and Cephalochordata, and a diverse lineage of vertebrates, Vertebrata.

Key Morphological Innovations Distinguish Chordates from Other Deuterostome Phyla

Chordates are distinguished from other deuterostomes by a set of key morphological innovations: a *notochord, segmental muscles in the body wall and tail, a dorsal hollow nerve chord,* and a *perforated pharynx* **(Figure 30.5)**. These structures foster higher levels of activity, unique modes of aquatic locomotion, and more efficient feeding and oxygen acquisition.

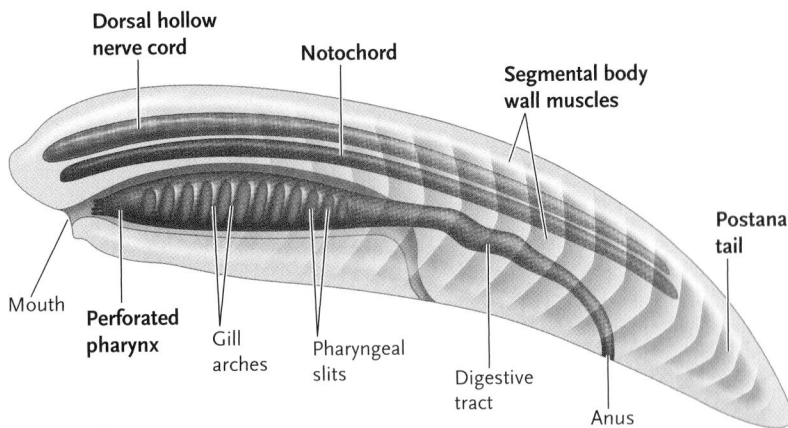

Figure 30.5

Diagnostic chordate characters. Chordates have a notochord; a muscular postanal tail; a segmental body wall and tail muscles; a dorsal, hollow nerve cord; and a perforated pharynx.

Segmental Body Wall and Tail Muscles. Chordates evolved in water, and they swim by contracting segmentally arranged blocks of muscles in the body wall and tail. The chordate tail, which is posterior to the anus, provides much of the propulsion in aquatic species. Segmentation allows each muscle block to contract independently; waves of contractions pass down one side of the animal and then down the other, sweeping the body and tail back and forth in a smooth and continuous movement.

Dorsal Hollow Nerve Cord. The central nervous system of chordates is a hollow nerve cord on the dorsal side of the animal (see Section 48.3). By contrast, most nonchordate invertebrates have solid nerve cords on the ventral side. In vertebrates, an anterior enlargement of the nerve cord forms a brain; in invertebrates, an anterior concentration of nervous system tissue is usually described as a *ganglion*.

Notochord. Early in chordate development, mesoderm that is dorsal to the developing digestive system forms a **notochord** (*noton* = the back; *chorda* = string). This flexible rod, constructed of fluid-filled cells surrounded by tough connective tissue, supports the embryo from head to tail. The notochord forms the skeleton of invertebrate chordates. Their body wall muscles are anchored to the notochord, and when these muscles contract, the notochord bends, but it does not shorten. As a result, the chordate body swings left and right during locomotion, propelling the animal forward; unlike annelids and other nonchordate invertebrates, the chordate body does not shorten when the animal is moving. Remnants of the notochord persist as gelatinous disks in the backbones of adult vertebrates.

Perforated Pharynx. The chordate pharynx, the part of the digestive system just posterior to the mouth, typically contains perforations or slits during some stage of the life cycle. These paired openings originated as exit holes for water that carried particulate food into the mouth, allowing chordates to gather large quantities of food. Invertebrate chordates also collect oxygen and release carbon dioxide across the walls of the pharynx. In fishes, **gill arches**, the supporting structures between the slits in the pharynx, are often sites of gas exchange, allowing animals to extract oxygen efficiently from the water. Invertebrate chordates and fishes retain a perforated pharynx throughout their lives. In most air-breathing vertebrates, the slits are present only during embryonic development and in larvae.

a. Larval tunicate

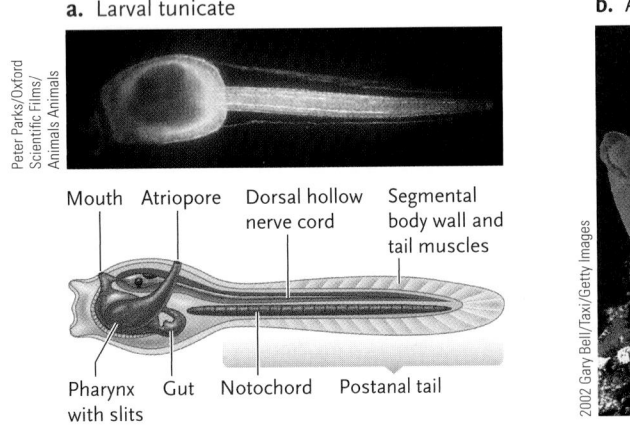

b. Adult tunicates

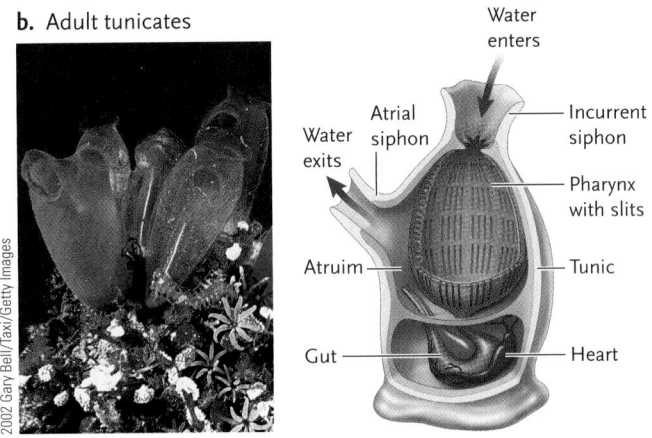

Figure 30.6

Urochordates. **(a)** A tadpolelike tunicate larva metamorphoses into **(b)** a sessile adult; shown here is *Rhopalaea crassa*. After a larva attaches to a substrate at its anterior end, the tail, notochord, and most of the nervous system are recycled to form new tissues. Slits in the pharynx multiply, the mouth becomes the incurrent siphon, and the atriopore becomes the atrial siphon.

Invertebrate Chordates Are Small, Marine Suspension Feeders

Two subphyla of invertebrate chordates exhibit the basic chordate body plan in its simplest form.

Subphylum Urochordata. The 2500 species of tunicates, sometimes called sea squirts (subphylum Urochordata, from *oura* = tail), float in surface waters or attach to substrates in shallow marine habitats. The sessile adults of many species secrete a gelatinous or leathery "tunic" around their bodies and squirt water through a siphon when disturbed; adults grow to several centimeters **(Figure 30.6)**. In the most common group of sea squirts (Ascidiacea), the swimming larvae possess the defining chordate features. Larvae eventually attach to substrates and transform into sessile adults. During metamorphosis, they lose most traces of the notochord, dorsal nerve cord, and tail, and their basketlike pharynx enlarges. In adults, beating cilia pull water into the pharynx through an **incurrent siphon.** A mucous net traps particulate food, which is carried, with the mucus, to the gut. Water passes through the pharyngeal slits, enters a chamber called the **atrium,** and is expelled—along with digestive wastes and carbon dioxide—through the **atrial siphon.** Oxygen is absorbed across the walls of the pharynx.

Subphylum Cephalochordata. The 28 lancelet species (subphylum Cephalochordata, from *kephale* = head) occupy warm, shallow marine habitats where they lie mostly buried in sand **(Figure 30.7)**. Although generally sedentary, they have well-developed body wall muscles and a prominent notochord. Most species are included in the genus *Branchiostoma* (formerly *Amphioxus*). Lancelet bodies, which are 5 to 10 cm long, are pointed at both ends like the double-edged surgical tools for which they are named. Adults have light receptors on the head as well as chemical sense organs on tentacles that grow from the **oral hood.** Lancelets use cilia to draw food-laden water through hundreds of pharyngeal slits; water flows into the atrium and is expelled through the **atriopore.** Most gas exchange occurs across the skin.

Vertebrates Possess Several Unique Tissues, Including Bone and Neural Crest

The most distinctive anatomical characteristic of the subphylum Vertebrata is an internal skeleton that provides structural support for muscles and protection for the nervous system and other organs. The skeleton and the muscles attached to it enable most vertebrates to move rapidly through the environment. A vertebrate's skeleton is composed of many separate, bony elements. Indeed, vertebrates are the only animals that have **bone,** a connective tissue in which living cells secrete the mineralized matrix that surrounds them (see Figure 36.5d). The **vertebral column,** made up of individual **vertebrae,** surrounds and protects the dorsal nerve cord, and a bony **cranium** surrounds the brain. The cranium, vertebral column, ribs, and sternum (breastbone) make up the **axial skeleton.** Most vertebrates also have a **pectoral girdle** anteriorly and a **pelvic girdle** posteriorly that attach bones in the fins or limbs to the axial skeleton. Bones of the two girdles and the appendages constitute the **appendicular skeleton.** One vertebrate lineage, Chondrichthyes, has lost its bone over evolutionary time; its skeleton is made of cartilage, a dense but flexible connective tissue that is often a developmental precursor of bone (see Section 36.2).

Vertebrates also possess a unique cell type, **neural crest,** which is distinct from endoderm, mesoderm, and ectoderm. Neural crest cells arise next to the developing nervous system, but later migrate throughout a vertebrate's body. They ultimately contribute to many uniquely vertebrate structures, including parts of the cranium, teeth, sensory organs, cranial nerves, and the medulla (that is, the interior part) of the adrenal glands.

Finally, the brains of vertebrates are much larger and more complex than those of invertebrate chordates.

Figure 30.7 Cephalochordates. **(a)** The unpigmented skin of adult lancelets reveals their segmental body wall muscles. A cutaway view **(b)** illustrates their internal anatomy.

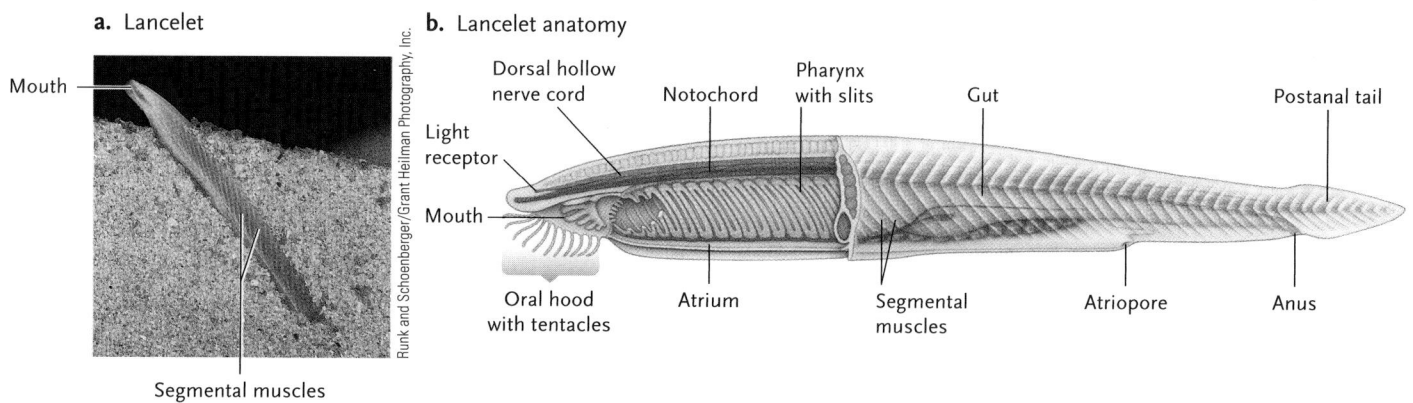

a. Lancelet

Mouth

Segmental muscles

Runk and Schoenberger/Grant Heilman Photography, Inc.

b. Lancelet anatomy

Dorsal hollow nerve cord

Light receptor

Mouth

Oral hood with tentacles

Notochord

Atrium

Pharynx with slits

Segmental muscles

Gut

Atriopore

Postanal tail

Anus

Moreover, the vertebrate brain is divided into three regions—the forebrain, midbrain, and hindbrain—each of which governs distinct nervous system functions (see Section 38.1).

STUDY BREAK

1. On a field trip to a lake, a college student captures a worm-shaped animal with segmental body wall muscles. While examining the specimen in laboratory the following day, she determines that the main nerve cord runs along the ventral side of the animal. Is this animal a chordate?
2. What structures distinguish vertebrates from invertebrate chordates?

30.3 The Origin and Diversification of Vertebrates

Biologists use embryological, molecular, and fossil evidence to trace the origin of vertebrates and to chronicle their evolutionary diversification.

Vertebrates Probably Arose from a Cephalochordate-Like Ancestor through the Duplication of Genes That Regulate Development

Molecular sequence studies suggest that vertebrates are more closely related to cephalochordates than to urochordates. The evolution of vertebrates from a cephalochordate-like ancestor was marked by the emergence of neural crest, bone, and other typically vertebrate traits. What genetic changes were responsible for these remarkable developments? Biologists now hypothesize that an increase in the number of homeotic—structure determining—genes may have made the development of more complex anatomy possible. (Homeotic genes are described further in Sections 22.6 and 48.6.)

In animals, one group of homeotic genes, the *Hox* genes, influences the three-dimensional shape of the animal and the locations of important structures—such as eyes, wings, and legs—particularly along the head to tail axis of the body. *Hox* genes are arranged on the chromosomes in a particular order, forming what biologists call the *Hox* gene complex (see Section 22.6). Each gene in the complex governs the development of particular structures. Animal groups with the simplest structure,

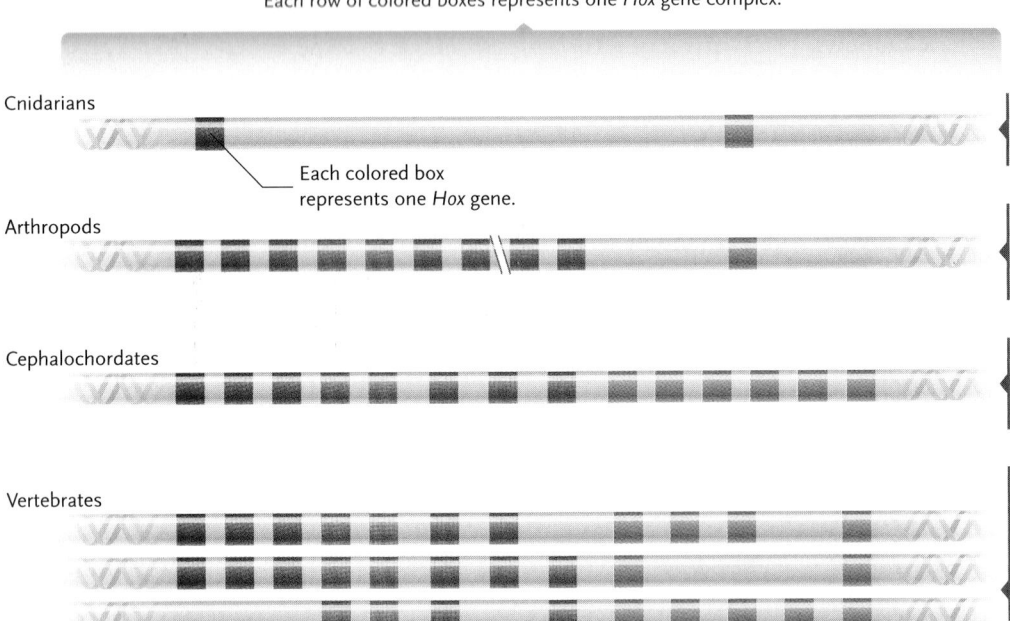

Each row of colored boxes represents one *Hox* gene complex.

Cnidarians

Each colored box represents one *Hox* gene.

Arthropods

Cephalochordates

Vertebrates

a. Invertebrates with simple anatomy, such as cnidarians, have a single *Hox* gene complex that includes just a few *Hox* genes.

b. Invertebrates with more complicated anatomy, such as arthropods, have a single *Hox* gene complex, but with a larger number of *Hox* genes.

c. Invertebrate chordates, such as cephalochordates, also have a single *Hox* gene complex, but with even more *Hox* genes than are found in nonchordate invertebrates.

d. Vertebrates, such as the laboratory mouse, have numerous *Hox* genes, arranged in two to seven *Hox* gene complexes. The additional *Hox* gene complexes are products of wholesale duplications of the ancestral *Hox* gene complex. The additional copies of *Hox* genes specify the development of uniquely vertebrate characteristics, such as the cranium, vertebral column, and neural crest cells.

Figure 30.8

Hox genes and the evolution of vertebrates. The *Hox* genes in different animals appear to be homologous, indicated here by their color and position in the complex. Vertebrates have many more individual *Hox* genes than most invertebrates do, and the entire *Hox* gene complex was duplicated in the vertebrate lineage.

such as cnidarians, have two *Hox* genes. Those with more complex anatomy, such as insects, have 10. Chordates have as many as 13 or 14. Thus, lineages with many *Hox* genes generally have more complex anatomy than do those with fewer *Hox* genes.

Molecular analyses also reveal that the entire *Hox* gene complex was duplicated several times in the evolution of vertebrates, producing multiple copies of all its genes **(Figure 30.8)**. The cephalochordate *Branchiostoma* has just one *Hox* gene complex, but the most primitive living vertebrates, the jawless hagfishes described later, have two. All vertebrates that possess jaws, a derived characteristic, have at least four sets, and some fishes have as many as seven. Evolutionary developmental biologists hypothesize that the duplication of *Hox* genes and other tool-kit genes allowed the evolution of new structures: while the original copies of these genes maintained their ancestral functions, the duplicate copies assumed *new* functions, directing the development of novel structures, such as the vertebral column and jaws.

Early Vertebrates Diversified into Numerous Lineages with Distinctive Adaptations

The oldest known vertebrate fossils were discovered in the late 1990s, when scientists in China described several species from the early Cambrian period, about 550 million years ago. Both *Myllokunmingia* and *Haikouichthys* were fish-shaped animals about 3 cm long **(Figure 30.9)**. In both species the brain was surrounded by a cranium, which, in these cases, was formed of fibrous connective tissue or cartilage. They also had segmental body wall muscles and fairly well-developed fins, but neither shows any evidence of bone.

The early vertebrates gave rise to numerous descendants, which varied greatly in anatomy, physiology, and ecology. New feeding mechanisms and locomotor structures were often crucial to their success. Today, vertebrates occupy nearly every habitat and feed on virtually all other organisms. Here we briefly introduce the major vertebrate lineages **(Figure 30.10)**.

Although biologists use four key morphological innovations—a cranium, vertebrae, bone, and neural crest cells—to identify vertebrates, these structures did not arise all at once. Instead, they appeared somewhat independently of one another as new groups arose. Some researchers and textbooks present a phylogeny and classification that places the "vertebrates" (animals that have vertebrae) within a larger lineage called the "craniates" (animals that have a cranium). But only one small group, the hagfishes (Myxinoidea, described later), has a cranium but no vertebrae, and some recent molecular analyses do not support its separation from the other vertebrates. Thus, for the sake of simplicity, we describe organisms that possessed any of the four key innovations as "vertebrates."

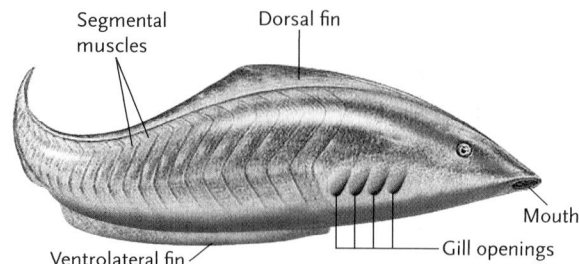

Figure 30.9
An early vertebrate. *Myllokunmingia*, one of the earliest vertebrates yet discovered, had no bones; it was about 3 cm long.

Several groups of early jawless vertebrates are described as *agnathans* (*a* = not; *gnathos* = jawed), but they do not form a monophyletic group. Although most became extinct in the Paleozoic era, two ancient lineages, Myxinoidea and Petromyzontoidea, still live today. All other vertebrates possess moveable jaws; they are members of the monophyletic lineage **Gnathostomata** ("jawed mouth"). The first jawed fishes, the Acanthodii and Placodermi, are now extinct, but several other lineages of jawed fishes are still abundant in aquatic habitats: the Chondrichthyes includes fishes with cartilaginous skeletons, such as sharks and skates; the Actinopterygians and Sarcopterygians comprise the bony fishes, which have bony endoskeletons. All jawless vertebrates and jawed fishes are restricted to aquatic habitats, and they use gills to extract oxygen from the water that surrounds them.

The Gnathostomata also includes the monophyletic lineage **Tetrapoda** (*tetra* = four; *pod* = foot); most tetrapods use four limbs for locomotion. Many tetrapods are semiterrestrial or terrestrial, although some, like sea turtles and porpoises, have secondarily returned to aquatic habitats. Adult tetrapods generally use air-breathing lungs for gas exchange. Within the Tetrapoda, one lineage, the amphibians, includes animals, such as frogs and salamanders, that typically need standing water to complete their life cycles. Another lineage, the **Amniota**, comprises animals with specialized eggs that can develop on land. Shortly after their appearance, the amniotes diversified into three lineages: one is ancestral to living mammals; another to the living turtles; and a third to lizards, snakes, alligators, and birds. We consider the detailed evolutionary history of the amniotes in Section 30.7.

STUDY BREAK

1. How do the *Hox* genes of vertebrates differ from those of cephalochordates?
2. Which of the taxonomic groups Amniota, Gnathostomata, and Tetrapoda includes the largest number of species? Which includes the fewest?

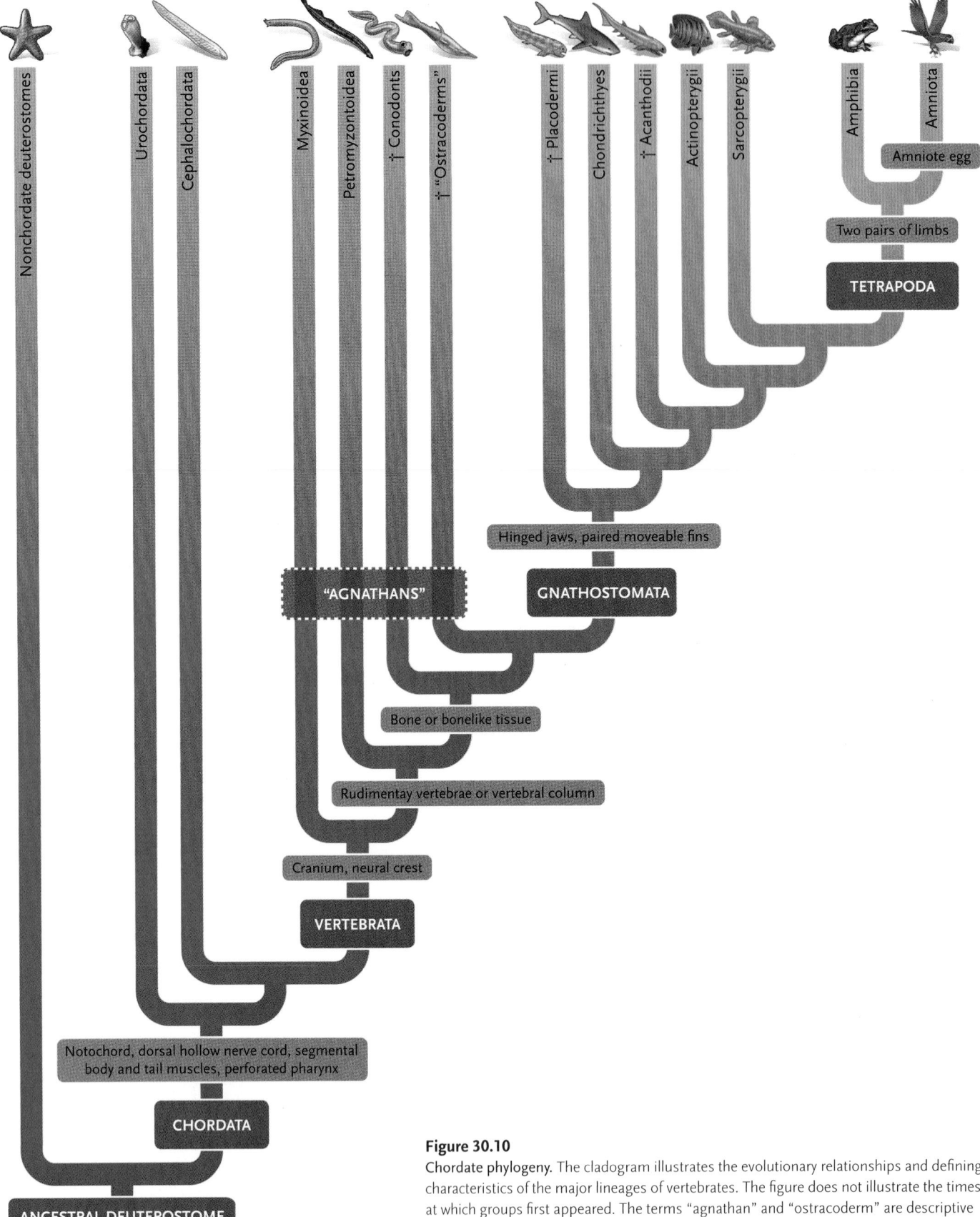

Figure 30.10

Chordate phylogeny. The cladogram illustrates the evolutionary relationships and defining characteristics of the major lineages of vertebrates. The figure does not illustrate the times at which groups first appeared. The terms "agnathan" and "ostracoderm" are descriptive and do not identify monophyletic groups. Extinct groups are marked with a dagger.

The following labels appear on the cladogram:

- Nonchordate deuterostomes
- Urochordata
- Cephalochordata
- Myxinoidea
- Petromyzontoidea
- † Conodonts
- † "Ostracoderms"
- † Placodermi
- Chondrichthyes
- † Acanthodii
- Actinopterygii
- Sarcopterygii
- Amphibia
- Amniota

- Amniote egg
- Two pairs of limbs
- **TETRAPODA**
- Hinged jaws, paired moveable fins
- "AGNATHANS"
- **GNATHOSTOMATA**
- Bone or bonelike tissue
- Rudimentay vertebrae or vertebral column
- Cranium, neural crest
- **VERTEBRATA**
- Notochord, dorsal hollow nerve cord, segmental body and tail muscles, perforated pharynx
- **CHORDATA**
- **ANCESTRAL DEUTEROSTOME**

30.4 Agnathans: Hagfishes and Lampreys, Conodonts and Ostracoderms

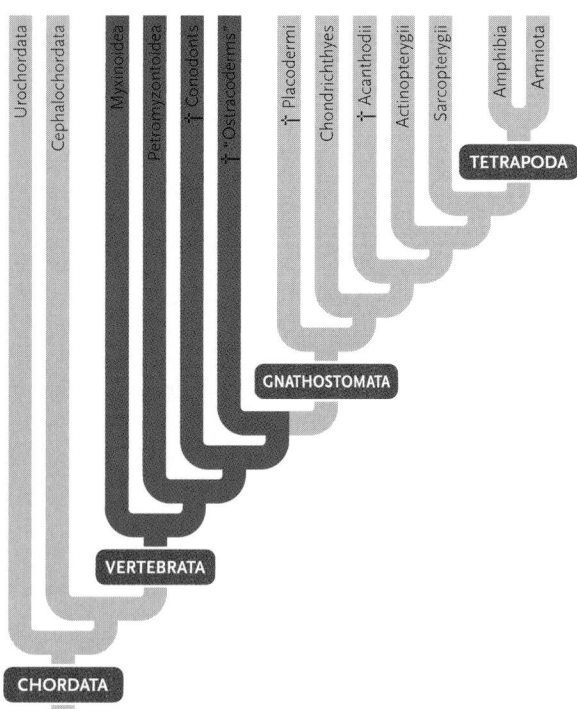

Lacking jaws, most of the earliest vertebrates used a muscular pharynx to suck edible tidbits into their mouths. The two living groups of agnathans as well as species that flourished in the Paleozoic vary greatly in size and shape as well as in the number of vertebrate characters they possess.

Hagfishes and Lampreys Are the Living Descendants of Ancient Agnathan Lineages

Two apparently separate lineages of jawless vertebrates, hagfishes (Myxinoidea) and lampreys (Petromyzontoidea), still live today. Both have skeletons composed entirely of cartilage. Although scientists have found no fossilized hagfishes or lampreys from the early Paleozoic era, the absence of jaws and bone in their living descendants suggests that their lineages arose early in vertebrate history, before the evolution of bone. Hagfishes and lampreys have a well-developed notochord, but no true vertebrae or paired fins, and their skin has no scales. Individuals grow to a maximum length of about 1 m **(Figure 30.11).**

The axial skeleton of the 60 living species of hagfishes includes only a cranium and a notochord; it has no specialized structures surrounding the dorsal nerve cord. Some biologists do not even include hagfishes among the Vertebrata, because they lack any sign of vertebrae. Hagfishes are marine scavengers that burrow in sediments on continental shelves. They feed on inverte-

brate prey and on dead or dying fishes. In response to predators, they secrete an immense quantity of sticky, noxious slime; when no longer threatened, a hagfish ties itself into a knot and wipes the slime from its body. Hagfish life cycles are simple and lack a larval stage.

The 40 or so living species of lampreys have traces of an axial skeleton. Their notochord is surrounded by dorsally pointing cartilages that partially cover the nerve cord; many biologists suspect that this arrangement may reflect an early stage in the evolution of the vertebral column. Most lamprey species are parasitic as adults. They have a circular mouth surrounded by a sucking disk with which they attach to a fish or other vertebrate host; they feed on a host's body fluids after rasping through its skin. In most species, sexually mature adults migrate from the ocean or a lake to the headwaters of a stream, where they lay eggs and then die. Their suspension-feeding larvae, which resemble adult cephalochordates, burrow into mud and develop for as long as seven years before undergoing metamorphosis and migrating to the sea or a lake to live as parasitic adults.

Conodonts and Ostracoderms Were Early Jawless Vertebrates with Bony Structures

Mysterious bonelike fossils, most less than 1 mm long, have long been known in oceanic rocks dating from the early Paleozoic era through the early Mesozoic era. Called **conodont** ("cone tooth") elements, these abun-

a. Living jawless fishes

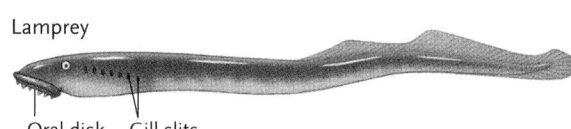

b. Mouth of a lamprey

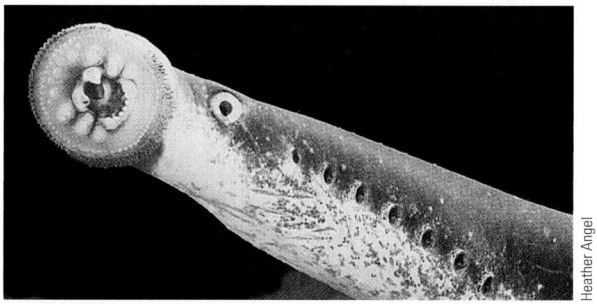

Figure 30.11

Living agnathans. **(a)** Two groups of jawless fishes, hagfishes and lampreys, survive today. **(b)** Lampreys use a toothed oral disk to attach to a host and feed on its blood and soft tissues.

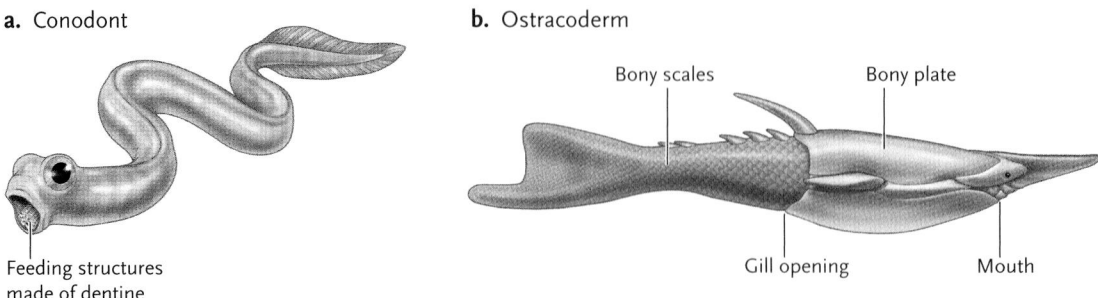

a. Conodont

Feeding structures
made of dentine

b. Ostracoderm

Bony scales

Bony plate

Gill opening

Mouth

Figure 30.12

Extinct agnathans. **(a)** Conodonts were elongate, soft-bodied animals with bonelike feeding structures in the mouth and pharynx. **(b)** *Pteraspis*, an ostracoderm, had large bony plates on its head and small bony scales on the rest of its body; it was about 6 cm long.

dant fossils were once described as the support structures of marine algae or the feeding structures of ancient invertebrates. However, recent analyses of their mineral composition reveal that they were made of dentine, a bonelike component of vertebrate teeth. In the 1980s and 1990s, several research teams discovered fossils of intact conodont animals with these elements in place.

Conodonts were elongate, soft-bodied animals; most were 3 to 10 cm long. They had a notochord, cranium, segmental body wall muscles, and large, moveable eyes **(Figure 30.12a).** The conodont elements at the front of the mouth were forward pointing, hook-shaped structures that apparently functioned to collect food; those in the pharynx were stouter, suitable for crushing items that had been consumed. Paleontologists now classify conodonts as vertebrates—the earliest vertebrates with bonelike structures.

An assortment of jawless fishes, representing several evolutionary lineages and collectively called **ostracoderms** (*ostrakon* = shell), were abundant from the Ordovician through the Devonian periods **(Figure 30.12b).** Like their invertebrate chordate ancestors, ostracoderms used their pharynx to extract small food particles from mud and water. However, the ostracoderms' muscular pharynx enabled them to *suck* mud and water into their mouths, providing a much stronger flow than the cilia-driven currents of invertebrate chordates. The greater flow rate allowed ostracoderms to collect food more rapidly. It also supported a larger body size: although most were much smaller, some ostracoderms reached a length of 2 m.

The skin of ostracoderms was heavily armored with plates and scales formed of bone. Although some ostracoderms had paired lateral extensions of their bony armor, they could not move them the way living fishes move their paired fins. Ostracoderms lacked a true vertebral column, but they had rudimentary support structures surrounding the nerve cord. They also had other distinctly vertebrate-like characteristics. For example, imprints in the head shields indicate that their brains had the three regions—forebrain, mid-

brain, and hindbrain—typical of all later vertebrates (see Section 38.1).

STUDY BREAK

1. What characteristics of the living hagfishes and lampreys suggest that their lineages arose very early in vertebrate evolution?
2. What traits in conodonts and ostracoderms are derived relative to those in hagfishes and lampreys?

30.5 Jawed Fishes

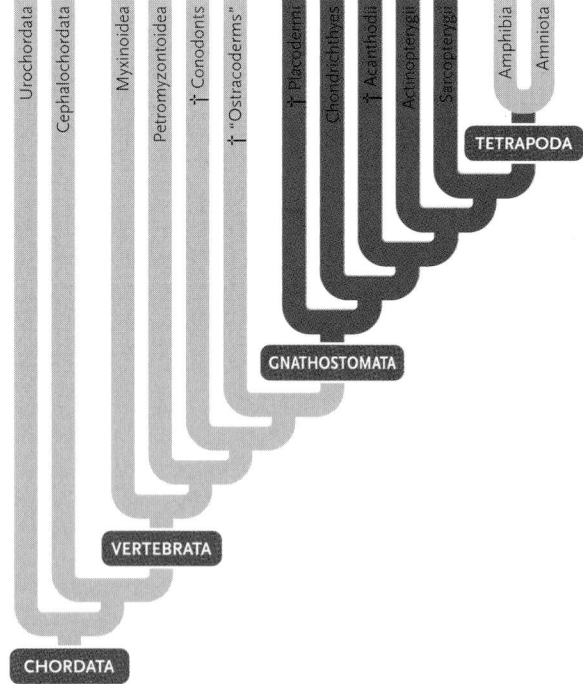

The first gnathostomes were jawed fishes. Key derived traits made their feeding and locomotion more efficient than those of their ancestors.

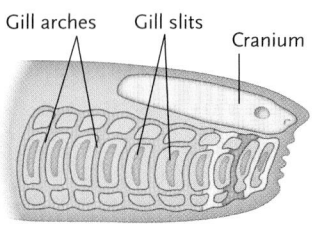

a. Jaws evolved from gill arches in the pharynx of jawless fishes.

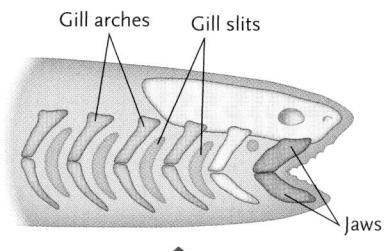

b. In early jawed fishes, the upper jaw was firmly attached to the cranium.

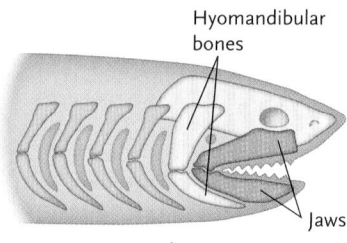

c. In later jawed fishes, the jaws were supported by the hyomandibular bones, which were derived from a second pair of gill arches.

Figure 30.13
The evolution of jaws.

Jawed Fishes First Appeared in the Paleozoic Era

The renowned anatomist and paleontologist Alfred Sherwood Romer of Harvard University described the evolution of jaws as "perhaps the greatest of all advances in vertebrate history." Hinged jaws allow vertebrates to grasp, kill, shred, and crush large food items. Some species also use their jaws for defense, for grooming, to construct nests, and to transport their young.

The Origin of Jaws and Fins. Embryological evidence suggests that jaws evolved from paired gill arches in the pharynx of a jawless ancestor **(Figure 30.13).** One pair of ancestral gill arches formed bones in the upper and lower jaws, while a second pair was transformed into the hyomandibular bones that braced the jaws against the cranium. Nerves and muscles of the ancestral suspension-feeding pharynx control and move the jaws.

Innovative locomotor mechanisms have often appeared at roughly the same time as innovative feeding mechanisms in the vertebrate lineage, and many early jawed fishes also had fins. The earliest fins were folds of skin and moveable spines that stabilized locomotion and deterred predators. Moveable fins appeared independently in several lineages, and by the Devonian period, most fishes had unpaired (dorsal, anal, and caudal) and paired (pectoral and pelvic) fins **(Figure 30.14).**

Early Jawed Fishes. In two early lineages of jawed fishes, spiny sharks and placoderms, the upper jaw was firmly attached to the cranium (see Figure 30.13b); their inflexible mouths simply snapped open and shut **(Figure 30.15).** Both groups also show evidence of an internal skeleton.

Spiny sharks (Acanthodii, from *akantha* = thorn), which persisted from the late Ordovician through the Permian periods, were less than 20 cm long. Their small, light scales; streamlined bodies; well-developed eyes; large jaws; and numerous teeth suggest that they were fast swimmers and efficient predators. Most had a row of ventral spines and fins with internal skeletal support on each side of the body. Acanthodian anatomy suggests that they are closely related to the bony fishes alive today.

Placoderms (Placodermi, from *plax* = flat surface) appeared in the Silurian and diversified in the Devonian and Carboniferous periods, but they left no direct descendants. Some, like *Dunkleosteus*, reached a length of 10 m. Their bodies were covered with large, heavy plates of bone anteriorly and smaller scales posteriorly. Their jaws had sharp cutting edges, but not separate teeth, and their paired fins had internal skeletons and powerful muscles.

a. Spiny shark

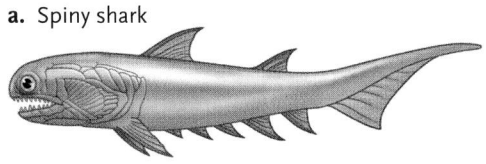

b. Placoderm

Figure 30.15
Early gnathostomes. **(a)** *Climatius*, an acanthodian, was small, reaching a total length of about 8 cm. **(b)** The placoderm *Dunkleosteus* was gigantic, sometimes growing to 10 m in length. Some acanthodians had teeth on their jaws, but placoderms had only sharp, cutting edges.

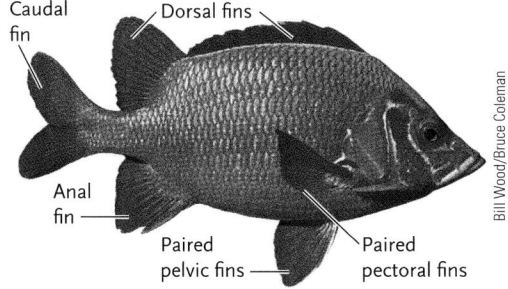

Bill Wood/Bruce Coleman

Figure 30.14
Fish fins. Most fishes have both paired and unpaired fins.

Chondrichthyes Includes Fishes with Cartilaginous Endoskeletons

The 850 living species in the Chondrichthyes (*chondros* = cartilage; *ichthys* = fish) have skeletons composed entirely of cartilage, which is much lighter than bone. The absence of bone in Chondrichthyes is a derived trait, however, because all earlier fishes had bony armor or bony endoskeletons.

Most living chondrichthyans are grouped in the Elasmobranchii, which includes the skates, rays, and sharks; nearly all are marine predators **(Figure 30.16).** Skates and rays are dorsoventrally flattened (see Figure 30.16a). They swim by undulating their enlarged pectoral fins. Most are bottom dwellers that often lie partly buried in sand. They feed on hard-shelled invertebrates, which they crush with massive, flattened teeth. The largest species, the manta ray *(Manta birostris),* which measures 6 m across, feeds on plankton in the open ocean. Some rays have electric organs that stun prey with as much as 200 volts.

Sharks (see Figure 30.16b) are among the ocean's dominant predators. Flexible fins, lightweight skeletons, streamlined bodies, and the absence of heavy body armor allow most sharks to pursue prey rapidly. Their livers often contain **squalene,** an oil that is lighter than water, which increases their buoyancy. The great white shark *(Carcharodon carcharias)* is the largest predatory species, attaining a length of 10 m. The whale shark *(Rhincodon typus),* which grows to 18 m, is the largest fish; it feeds on plankton.

Elasmobranchs—including sharks, skates, and rays—exhibit remarkable adaptations for acquiring and processing food. Their teeth develop in whorls under the fleshy parts of the mouth. New teeth migrate forward as old, worn teeth break free. In many sharks, the upper jaw is loosely attached to the cranium, and it swings down during feeding. As the jaws open, the mouth spreads widely, sucking in large, hard-to-digest chunks of prey, which are swallowed intact. Although the elasmobranch digestive system is short, it includes a corkscrew-shaped **spiral valve,** which slows the passage of material and increases the surface area available for digestion and absorption.

Elasmobranchs also have well-developed sensory systems. In addition to vision and olfaction, they use **electroreceptors** to detect weak electric currents produced by other animals. And their **lateral-line system,** a row of tiny sensors in canals along both sides of the body, detects vibrations in water (see Figure 39.4).

Chondrichthyans exhibit numerous reproductive specializations. Males have a pair of organs, the **claspers,** on the pelvic fins, which help transfer sperm into the female's reproductive tract. Fertilization occurs internally. In many species, females produce yolky eggs with tough leathery shells (see Figure 30.16c).

a. Manta ray

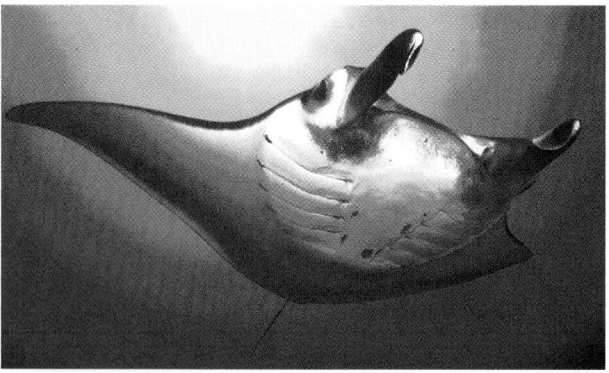

b. Galápagos shark

c. Swell shark egg case

Figure 30.16
Chondrichthyes. **(a)** Skates and rays, like the manta ray *(Manta birostris),* as well as **(b)** sharks, like the Galápagos shark *(Carcharhinus galapagensis),* are grouped in the Elasmobranchii. **(c)** Many shark egg cases, like that of the swell shark *(Cephaloscylium ventricosum),* include a large yolk that nourishes the developing embryo.

Others retain the eggs within the oviduct until the young hatch. A few species nourish young within a uterus.

The Actinopterygii and Sarcopterygii Are Fishes with Bony Endoskeletons

In terms of diversity and sheer numbers, the fishes with bony endoskeletons—a cranium, vertebral column with ribs, and bones supporting their moveable fins—are the most successful of all vertebrates. The endoskeleton provides lightweight support, particularly compared with the heavy bony armor of ostracoderms and placoderms, and enhances their locomotor efficiency. Bony fishes first appeared in the Silurian period and rapidly diversified into two lineages. The ray-finned fishes (Actinopterygii, from *aktis* = ray and *pteron* = wing) have fins that are supported by thin and flexible bony rays. The fleshy-finned fishes (Sarcopterygii, from *sarco* = flesh) have fins that are supported by muscles and an internal bony skeleton. Ray-finned fishes have always been more diverse, and they vastly outnumber the fleshy-finned fishes today. The 21,000 living species of bony fishes occupy nearly every aquatic habitat and represent more than 95% of living fish species. Adults range from 1 cm to more than 6 m in length.

Bony fishes have numerous adaptations that increase their swimming efficiency. In many modern ray-finned fishes, a gas-filled **swim bladder** serves as a hydrostatic organ that increases buoyancy (see Figure 30.18a). The swim bladder is derived from an ancestral air-breathing lung that allowed early actinopterygians to gulp air, supplementing their gill respiration in aquatic habitats where dissolved oxygen concentration was low. The scales of most bony fishes are small, smooth, and lightweight. And their bodies are covered with a protective coat of mucus, which retards bacterial growth and smoothes the flow of water.

Actinopterygii. The most primitive living actinopterygians, sturgeons and paddlefishes, have mostly cartilaginous skeletons **(Figure 30.17a)**. These large fishes live in rivers and lakes of the northern hemisphere. Sturgeons feed on detritus and invertebrates; paddlefish consume plankton. Gars and bowfins are remnants of a more recent radiation **(Figure 30.17b)**. They occur only in the eastern half of North America, where they feed on fishes and other prey. Gars are protected from predators by a heavy coat of bony scales.

Teleosts, the latest radiation of Actinopterygii, are the most diverse, successful, and familiar bony fishes. Evolution has produced a wide range of body forms **(Figure 30.18)**. Teleosts have an internal skeleton made almost entirely of bone. On either side of the head, a flap of the body wall, the **operculum**, covers a chamber that houses the gills. Sensory systems generally include large eyes, a lateral-line system, sound receptors, chemoreceptive nostrils, and taste buds. Variations in jaw structure allow different teleosts to consume plankton, seaweed, invertebrates, or other vertebrates.

Teleosts exhibit remarkable feeding and locomotor adaptations. When some teleosts open their mouths, bones at the front of the jaws swing forward to create a circular opening. Folds of skin extend backward, forming a tube through which they suck food (see Figure 30.18f). Many also have symmetrical caudal fins, posterior to the vertebral column, which provide power for locomotion. And their pectoral fins lie high on the sides of the body, providing fine control over swimming. Some species use their pectoral fins for acquiring food, for courtship, and for care of eggs and young. Some teleosts even use them for crawling on land or gliding in air.

Most marine species produce small eggs that hatch into planktonic larvae. Eggs of freshwater teleosts are generally larger and hatch into tiny versions of the adults. Parents often care for their eggs and young, fanning oxygen-rich water over them, removing fungal growths, and protecting them from predators. Some freshwater species, such as guppies, give birth to live young.

Sarcopterygii. Two groups of fleshy-finned fishes (Sarcopterygii), lobe-finned fishes and lungfishes, are now represented by only eight living species **(Figure 30.19)**.

a. Lake sturgeon

b. Long-nosed gar

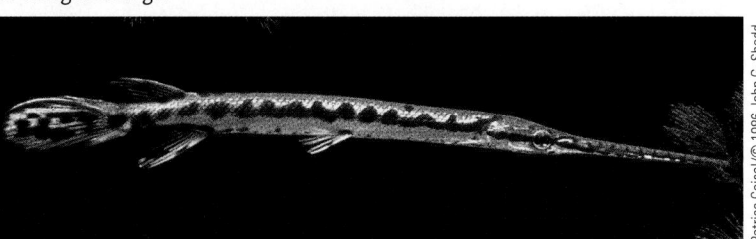

Figure 30.17

Primitive actinopterygians. **(a)** A lake sturgeon *(Accipenser fulvescens)* and **(b)** a long-nosed gar *(Lepisosteus osseus)* are living representatives of early actinopterygian radiations.

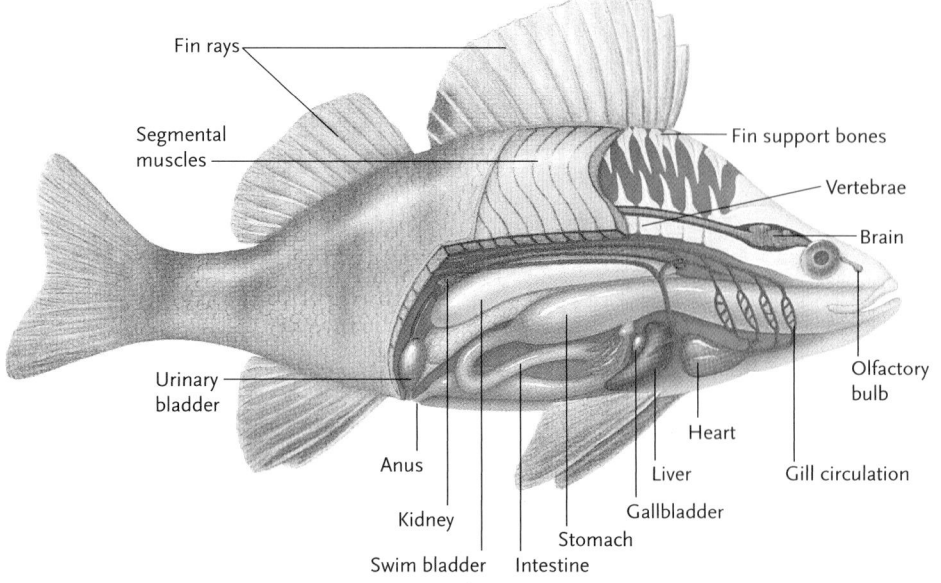

Fin rays

Segmental muscles

Fin support bones

Vertebrae

Brain

Urinary bladder

Anus

Kidney

Swim bladder

Intestine

Stomach

Gallbladder

Liver

Heart

Gill circulation

Olfactory bulb

a. Teleost internal anatomy

b. Sea horses, like the northern sea horse *(Hippocampus hudsonius)*, use a prehensile tail to hold on to substrates; they are weak swimmers.

Digital Vision/Getty Images, Inc.

c. The long, flexible body of a spotted moray eel *(Gymnothorax moringa)* can wiggle through the nooks and crannies of a reef.

Kit Kittle/Corbis

d. Flatfishes, like this European flounder *(Platichthys flesus)*, lie on one side and leap at passing prey.

F. Graner/Peter Arnold, Inc.

Operculum

e. Open ocean predators, like the yellowfin tuna *(Thunnus albacares)*, have strong, torpedo-shaped bodies and powerful caudal fins.

Brandon Cole/Visuals Unlimited

f. Kissing Gouramis *(Helostoma temmincki)* extend their jaws into a tube that sucks food into the mouth.

Arthur W. Ambler/Photo Researchers, Inc.

Figure 30.18

Teleost diversity. Although all teleosts share similar internal features, their diverse shapes adapt them to different diets and types of swimming.

a. Coelacanth

b. Australian lungfish

Figure 30.19

Sarcopterygians. **(a)** The coelacanth *(Latimeria chalumnae)* is one of two living species of lobe-finned fishes. **(b)** The Australian lungfish *(Neoceratodus forsteri)* is one of only six living lungfish species.

Although lobe-finned fishes were once thought to have been extinct for 65 million years, a living coelacanth *(Latimeria chalumnae)* was discovered in 1938 near the Comoros Islands, off the southeastern coast of Africa. We now know that a population of this meter-long fish lives at depths of 70 to 600 m, feeding on fishes and squid. Remarkably, a second population of coelacanths was discovered in 1998, when a specimen was found in an Indonesian fish market, 10,000 km east of the Comoros population. Based on analyses of its DNA, it is a distinct species *(Latimeria menadoensis)*.

Lungfishes have changed relatively little over the last 200 million years. Six living species are distributed on southern continents. The Australian lungfishes, which live in rivers and pools, use their lungs to supplement gill respiration when dissolved oxygen concentration is low. The South American and African species, which live in swamps, use their lungs to collect oxygen during the annual dry season, which they spend encased in a mucus-lined burrow in the dry mud. When the rains begin, water fills the burrow and the fishes awaken from dormancy.

STUDY BREAK

1. What characteristics of sharks and rays make them more efficient predators than the acanthodians or placoderms?
2. How do the air bladder and fins of ray-finned bony fishes increase their locomotor abilities?
3. How do the lungs of lungfishes allow them to survive in stressful environments?

30.6 Early Tetrapods and Modern Amphibians

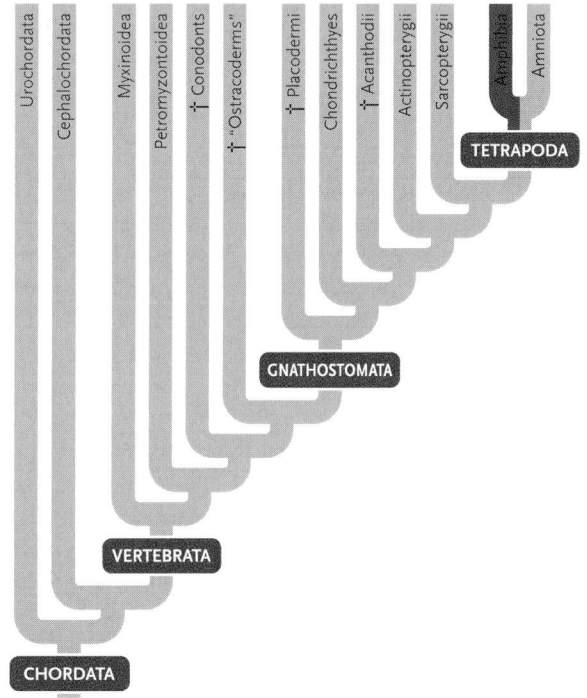

The fossil record suggests that tetrapods arose from a group of fleshy-finned fishes, the *osteolepiforms*, in the late Devonian period. Osteolepiforms and early tetrapods shared several derived characteristics: both had curious infoldings of their tooth surfaces, a trait with unknown function; and the shapes and positions of bones on the dorsal side of their crania and in their appendages were similar **(Figure 30.20)**.

a. *Eusthenopteron*, an osteolepiform fish

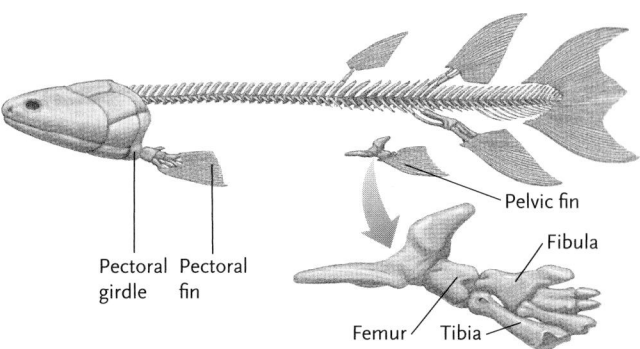

Pectoral girdle • Pectoral fin • Pelvic fin • Fibula • Femur • Tibia

b. *Ichthyostega*, an early tetrapod

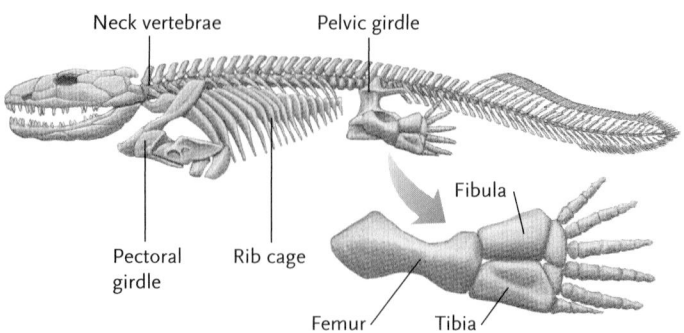

Neck vertebrae • Pelvic girdle • Fibula • Pectoral girdle • Rib cage • Femur • Tibia

Figure 30.20

Evolution of tetrapod limbs. The limb skeleton of osteolepiform fishes, such as **(a)** *Eusthenopteron*, is homologous to that of early tetrapods, such as **(b)** *Ichthyostega*. Although *Ichthyostega* retained many fishlike characteristics, its pectoral girdle was completely freed from the cranium and it had a heavy rib cage. Fossils of its forefoot have not yet been discovered.

Key Adaptations Facilitated the Transition to Land

Fishes are not adapted to live on land, and the first tetrapods faced serious environmental challenges. First, because air is less dense than water, it provides less support for an animal's body. Second, animals exposed to air inevitably lose body water by evaporation. Third, the sensory systems of fishes, which work well under water, do not function well on land. However, swampy late Devonian habitats also offered distinct advantages. Land plants, worms, and arthropods provided abundant food; oxygen was more readily available in air than in water; and no predators lived in these new habitats.

In some ways, osteolepiforms were preadapted for terrestrial life (see Figure 30.20a). Most had strong, stout fins that enabled them to crawl on the muddy bottom of shallow pools, and their vertebral column included crescent-shaped bones that provided good support. They had nostrils leading to sensory pits that housed olfactory (smell) receptors. And they almost certainly had lungs to augment gill respiration in the swampy, oxygen-poor waters where they lived.

The earliest tetrapod for which we have nearly complete skeletal data is the semiterrestrial, meter-long *Ichthyostega* (see Figure 30.20b). Compared with its fleshy-finned ancestors, *Ichthyostega* had a stronger vertebral column, sturdier girdles and appendages, a rib cage that protected its internal organs (including lungs), and a neck. Fishes have no neck: the pectoral girdle is fused to the cranium. But several vertebrae separated these structures in *Ichthyostega*, allowing it to move its head to scan the environment and to capture food. However, *Ichthyostega* retained a fishlike lateral-line system, caudal fin, and scaly covering on its body.

Life on land also required changes in sensory systems. In fishes, for example, the body wall picks up sound vibrations and transfers them to sensory receptors directly. But sound waves are harder to detect in air. Early tetrapods developed a **tympanum,** a specialized membrane on either side of the head that is vibrated by airborne sounds. The tympanum connects to the **stapes,** a bone that is homologous to the hyomandibula, which had supported the jaws of fishes (see Figure 23.4). The stapes, in turn, transfers vibrations to the sensory cells of an inner ear.

Modern Amphibians Are Very Different from Their Paleozoic Ancestors

Most of the more than 6000 species of living amphibians—including frogs, salamanders, and caecelians—are small, and their skeletons contain fewer bones than those of Paleozoic tetrapods like *Ichthyostega*. All living amphibians are carnivorous as adults, but the aquatic larvae of some species are herbivores.

Most living amphibians have a thin, scaleless skin, well supplied with blood vessels, that is a major site of gas exchange. Because gases must enter the body across a thin layer of water, the skin of most amphibians must remain moist, restricting them to aquatic or wet terrestrial habitats. Adults of some species also acquire oxygen through saclike lungs. The evolution of lungs was accompanied by modifications of the heart and circulatory system that increase the efficiency with which oxygen is delivered to body tissues (see Section 42.1).

The life cycles of many amphibians (*amphi* = both; *bios* = life) include both larval and adult stages. Eggs are laid and fertilized in water, where they hatch into larvae, such as the tadpoles of frogs, that eventu-

a. A frog

b. A salamander

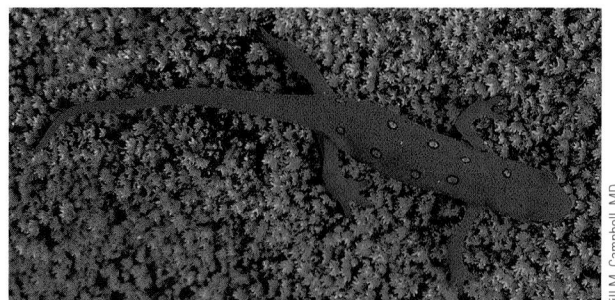

c. A caecelian

Figure 30.21

Living amphibians. **(a)** Anurans, like the northern leopard frog *(Rana pipiens)*, have compact bodies and long hind legs. **(b)** Urodeles, such as the red-spotted newt *(Notophthalmus viridescens)*, have an elongate body and four legs. **(c)** Caecelians, like *Caecelia nigricans* from Colombia, are legless burrowing animals.

ally metamorphose into adults (see Figure 40.9). Although the larvae of most species are aquatic, adults may be aquatic, amphibious, or terrestrial. Some salamanders are paedomorphic: the larval stage attains sexual maturity without changing its form or moving to land. By contrast, some frogs and salamanders reproduce on land, skipping the larval stage altogether. But even though they are terrestrial breeders, their eggs dry out quickly unless they are laid in moist places.

Modern amphibians are represented by three lineages **(Figure 30.21)**. Populations of practically all amphibians have declined rapidly in recent years, probably because of exposure to acid rain, high levels of ultraviolet B radiation, or parasitic infections.

Anura. The 3700 species of frogs and toads (Anura, from *an* = without; *oura* = tail) have short, compact bodies, and adults lack tails. Their elongate hind legs and webbed feet allow them to hop on land or swim. A few species are adapted to dry habitats, withstanding periods of drought by encasing themselves in mucous cocoons.

Urodela. Most of the 400 species of salamanders (Urodela, from *oura* = tail; *delos* = visible) have an elongate, tailed body and four legs. They walk by alternately contracting muscles on either side of the body much the way fishes swim. Species in the most diverse group, the lungless salamanders, are fully terrestrial throughout their lives, using their skin and the lining of the throat for gas exchange.

Gymnophiona. The 200 species of caecelians (Gymnophiona, from *gymnos* = naked; *ophioneos* = snakelike) are legless burrowing animals with wormlike bodies. They occupy tropical habitats throughout the world. Unlike other modern amphibians, caecelians have small bony scales embedded in their skin. Fertilization is internal, and females give birth to live young.

STUDY BREAK

1. For the first tetrapods, what were the advantages and disadvantages of moving onto the land?
2. What parts of the life cycle in most modern amphibians are dependent on water or very moist habitats?

30.7 The Origin and Mesozoic Radiations of Amniotes

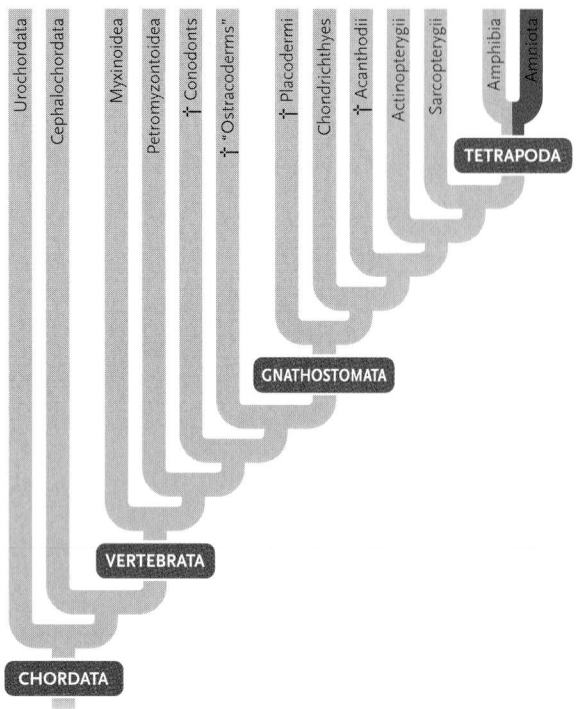

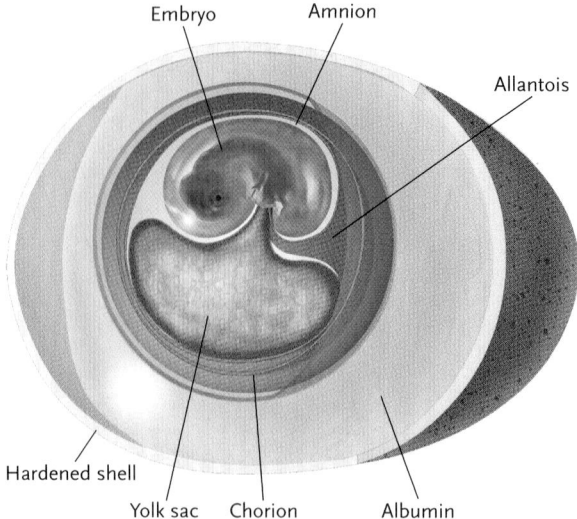

Figure 30.22

The amniote egg. A water-retaining egg with four specialized membranes (the amnion, allantois, chorion, and yolk sac) and a hard or leathery shell allowed amniotes and their descendants to reproduce in dry environments.

The amniote lineage arose during the Carboniferous period, when seed plants and insects, which served as excellent food resources, began to occupy higher ground. The lineage is named for the amnion, a fluid-filled sac that surrounds the embryo during development.

Key Adaptations Allow Amniotes to Live a Fully Terrestrial Life

Although the fossil record includes abundant skeletons of early amniotes, it provides little direct information about their soft body parts and physiology. For amniotes living today, three key adaptations allow them to live in dry habitats, freeing them from a dependency on moist surroundings and standing water. First, they have a tough, dry skin. Its cells are filled with keratin and lipids, which are relatively impermeable to water. Thus, amniotes do not dehydrate in air as quickly as amphibians do.

Second, many amniotes produce an **amniote egg**, which can survive and develop on dry land. The eggs of modern reptiles and birds have four specialized membranes and a hard or leathery shell perforated by microscopic pores **(Figure 30.22)**. The membranes protect the developing embryo and facilitate gas exchange and excretion; the shell mediates the exchange of air and water between the egg and its environment. The egg also includes generous supplies of **yolk**, the embryo's main energy source, and **albumin**, a source of nutrients and water. Compared with those of amphib-

ians, amniote eggs are large; and lacking a larval stage, the young hatch as miniature versions of the adult. By contrast to reptiles and birds, the eggs of virtually all mammals lack a shell; embryos, with the same four membranes, implant in the wall of the mother's uterus and receive nutrients and oxygen directly from her.

Third, some amniotes produce uric acid as a waste product of nitrogen metabolism (see Chapter 46). By contrast, fishes and amphibians produce ammonium ions or urea, toxic materials that require lots of water to flush them from body tissues. Because uric acid is less toxic than these other compounds, it can be excreted as a semisolid paste, conserving body water.

Amniotes Diversified into Three Main Lineages

Based on the abundance and diversity of their fossils, amniotes were extremely successful; they quickly replaced many nonamniote species in terrestrial habitats. During the Carboniferous and Permian periods, amniotes produced three radiations: synapsids, anapsids, and diapsids **(Figure 30.23)**. Differences in skull structure—specifically, the number of bony arches in the temporal region of the skull—distinguish the three groups. In those animals that have temporal arches, the openings between the arches provide space for

Figure 30.23

Amniote ancestry. The early amniotes gave rise to three lineages (anapsids, synapsids, and diapsids) and numerous descendants. The lineages are distinguished by the number of bony arches in the temporal region of the skull (indicated on the small icons).

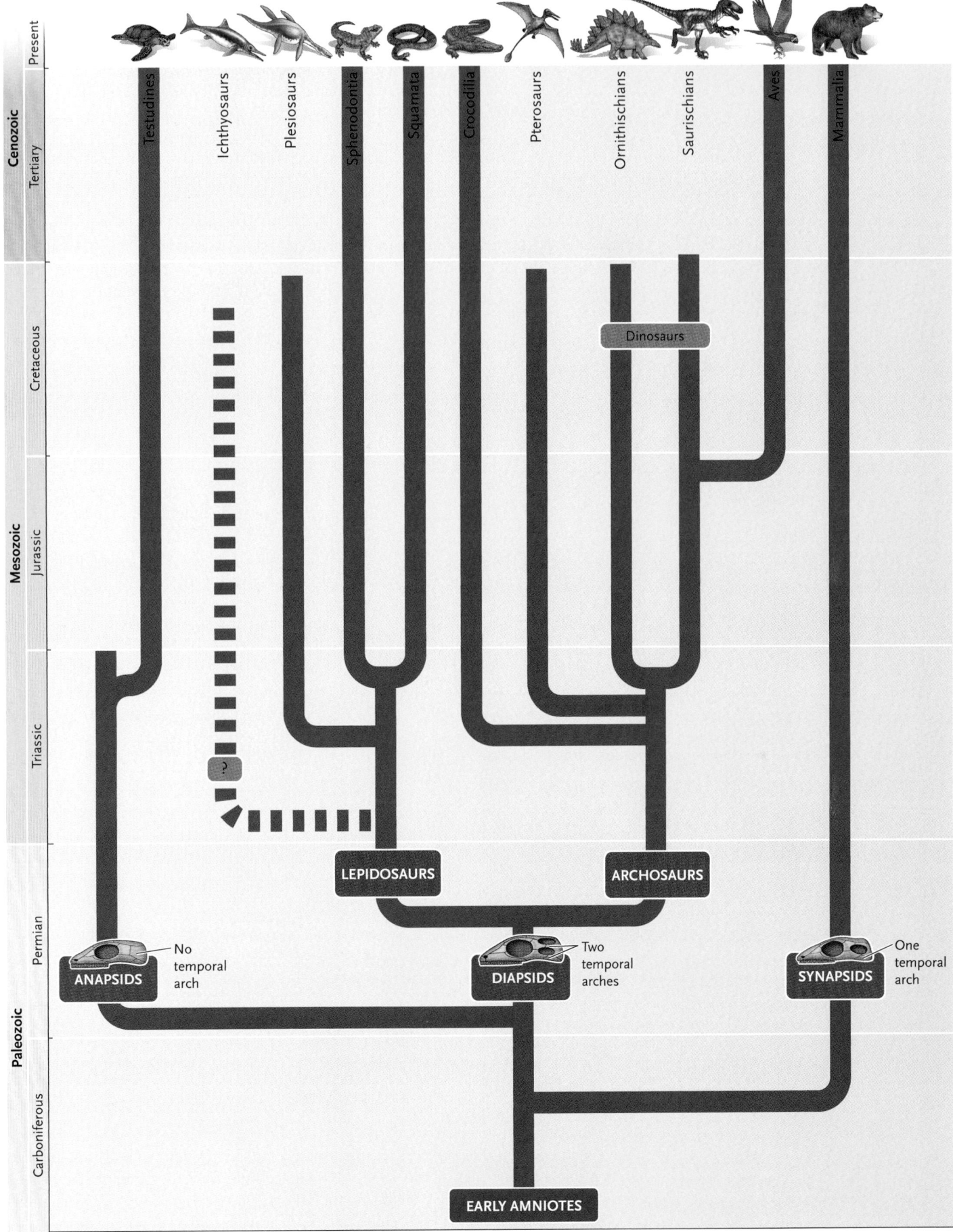

large and powerful jaw muscles to bunch up when they contract.

Synapsida. The first offshoot from the ancestral amniotes was a group of small terrestrial predators, the **synapsids** (*syn* = with; *apsis* = arch), which had one temporal arch on each side of the head. Synapsids emerged late in the Permian period; mammals are their living descendants.

Anapsida. A second lineage to emerge was the **anapsids** ("no arch"), which had no temporal arches and no spaces on the sides of the skull. Many biologists believe that turtles are living representatives of this group.

Diapsida. Most Mesozoic amniotes belong to the third lineage, **diapsids** ("two arches"), which had two temporal arches. Their living descendants include lizards and snakes, crocodilians, and birds.

Diapsids Diversified Wildly during the Mesozoic Era

The early diapsids differentiated into two lineages, the **Archosauromorpha** (*archos* = ruler; *saurus* = lizard; *morphe* = form) and the **Lepidosauromorpha** (*lepis* = scale), which differed in many skeletal characteristics. The archosauromorphs (commonly called archosaurs), or "ruling reptiles," include crocodilians, pterosaurs, and dinosaurs. Crocodilians, which first appeared during the Triassic period, have bony armor and a laterally flattened tail that propels them through water. Pterosaurs, now extinct, were flying predators of the Jurassic and Cretaceous periods. Their wings, which spanned as much as 13 m, were composed of thin sheets of skin attached to the sides of the body and supported by an elongate finger. Small pterosaurs may have been active fliers, but large ones probably soared on air currents as vultures do today.

Two lineages of dinosaurs, "lizard-hipped" saurischians and "bird-hipped" ornithischians proliferated in the Triassic and Jurassic periods. As their names imply, they differed in the anatomy of their pelvic girdles. The saurischian lineage included bipedal carnivores and quadrupedal herbivores. Most carnivorous saurischians were swift runners. Their forelimbs, however, were often ridiculously short. *Tyrannosaurus*, which was 15 m long and stood 6 m high, is the most familiar, but most species were much smaller. One group of small carnivorous saurischians was ancestral to birds. By the Cretaceous period, some herbivorous saurischians had also attained gigantic size, and many had long, flexible necks. For example, *Apatosaurus* (previously known as *Brontosaurus*) was 25 m long and may have weighed 50,000 kg.

The largely herbivorous ornithischian dinosaurs had enormous, chunky bodies. This lineage included the armored or plated dinosaurs (*Ankylosaurus* and *Stegosaurus*), the duck-billed dinosaurs (*Hadrosaurus*), horned dinosaurs (*Styracosaurus*), and some with remarkably thick skulls (*Pachycephalosaurus*). The ornithischians were most abundant in the Jurassic and Cretaceous periods.

The second major lineage of diapsids was the lepidosauromorphs (commonly called lepidosaurs), a diverse group that included both marine and terrestrial animals. Plesiosaurs were marine, fish-eating creatures that used long, paddlelike limbs to row through the water. Ichthyosaurs were porpoiselike animals with laterally flattened tails. They were so highly specialized for marine life that they could not venture onto land, even to reproduce. Instead, they gave birth to live young, as porpoises and whales do today. A third important group within this lineage is the squamates, which includes the living lizards and snakes.

30.8 Testudines: Turtles

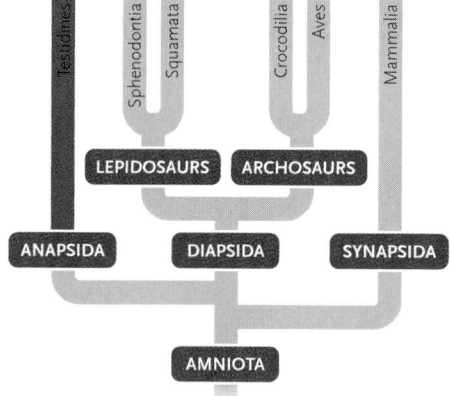

Turtles Have Bodies Encased in a Bony Shell

The turtle body plan, largely defined by a bony, boxlike shell, has changed little since the group first appeared during the Triassic period **(Figure 30.24).** The shell includes a dorsal **carapace** and a ventral **plastron.** A turtle's ribs are fused to the inside of the carapace, and, in contrast to other tetrapods, the pectoral and pelvic girdles lie within the ribcage. Large keratinized scales cover the bony plates that form the shell.

The 250 living species of turtles occupy terrestrial, freshwater, and marine habitats. They range from 8 cm to 2 m in length. All species lack teeth, but they use a keratinized beak and powerful jaw muscles to feed on plants or animal prey. When threatened, most species retract into their shells. Many species are now highly endangered because adults are hunted for meat, their eggs are consumed by humans and other predators, and their young are collected for the pet trade.

STUDY BREAK

How does the overall structure of turtles distinguish them from other amniotes?

30.9 Living Nonfeathered Diapsids: Sphenodontids, Squamates, and Crocodilians

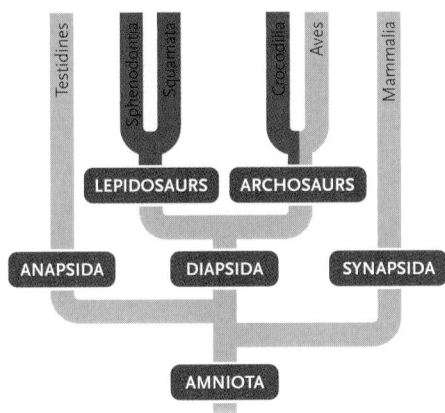

Biologists once grouped living turtles, lizards and snakes, and crocodilians in the class Reptilia, because they all have a dry, scaly skin, produce amniote eggs, and have low metabolic rates and variable body temperatures. But as you read in Section 23.5, these animals do not represent a monophyletic lineage. Turtles are probably not closely related to the other groups. Moreover, the "class Reptilia" excludes birds, which, like crocodilians, are part of the archosaur lineage. In this section, we describe all of the living diapsids except birds, which we consider separately in Section 30.10 because of their unique derived traits and their conspicuous evolutionary success.

Living Sphenodontids Are Remnants of a Diverse Mesozoic Lineage

The tuatara *(Sphenodon punctatus)* is one of two living representatives of the sphenodontids *(sphen = wedge; odont = tooth)*, a diverse Mesozoic lineage **(Figure 30.25a)**. These lizardlike animals survive on a few is-

a. The turtle skeleton

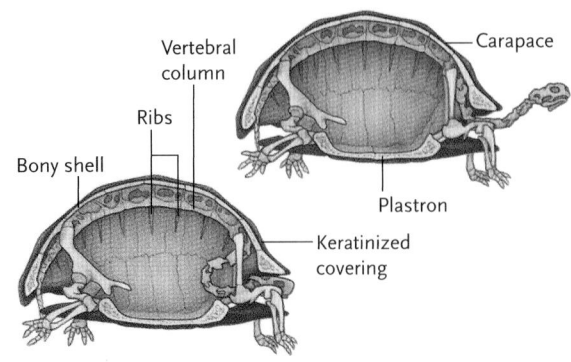

b. An aquatic turtle

Paul J. Fusco/Photo Researchers, Inc.

Figure 30.24

Testudines. **(a)** Most turtles can withdraw their heads and legs into a bony shell. **(b)** Aquatic turtles, like the Eastern painted turtle *(Chrysemys picta)*, often bask in the sun to warm up. The sunlight may also help eliminate parasites that cling to the turtle's skin.

lands off the coast of New Zealand. Adults are about 60 cm long. They live in dense colonies, where males and females defend small territories using vocal and visual displays. They often share underground burrows with seabirds, feeding mainly on invertebrates and small vertebrates. They are primarily nocturnal and maintain low body temperatures during periods of activity.

Squamates—Lizards and Snakes—Are Covered by Overlapping, Keratinized Scales

The skin of lizards and snakes (Squamata, from *squama* = scale) is composed of overlapping, keratinized scales that protect against dehydration. Squamates periodically shed their skin as they grow, much the way arthropods shed their old exoskeletons (see Section 29.7). Most squamates regulate their body temperature behaviorally: they are active only when weather conditions are favorable, and they shuttle between sunny and shady places when they need to warm up or cool down (see Section 46.7).

Most of the 3700 lizard species are less than 15 cm long **(Figure 30.25b)**. However, the Komodo dragon *(Varanus komodoensis)* grows to nearly 3 m. Lizards occupy a wide range of habitats, but they are especially

a. Sphenodontia includes the tuatara *(Sphenodon punctatus)* and one other species.

Pete & Judy Morrin/Ardea London

b. Basilisk lizards *(Basiliscus basiliscus)* escape from predators by running across the surface of streams.

© Stephen Dalton/Photo Researchers, Inc.

Venom gland

Hollow fang

Andrew Dennis/A.N.T. Photo Library

c. A western diamondback rattlesnake *(Crotalus atrox)* of the American southwest bares its fangs with which it injects a powerful toxin into prey.

Mary Ann McDonald/Corbis

d. Crocodilia includes semiaquatic predators, like this resting African Nile crocodile *(Crocodylus niloticus)*, that frequently bask in the sun.

Figure 30.25
Living nonfeathered diapsids.

common in deserts and the tropics; one species *(Zootoca vivipara)* occurs within the Arctic Circle. Most lizards feed on insects, although some eat leaves or meat. The diverse tropical genus *Anolis* has become a frequent subject of research, as described in *Focus on Research.*

The 2300 species of snakes evolved from a lineage of lizards that lost their legs over evolutionary time **(Figure 30.25c)**. Streamlined bodies make snakes efficient burrowers or climbers. Many subterranean species are only 10 or 15 cm long, but the giant constrictors may grow to 10 m. Unlike lizards, all snakes are predators that swallow prey whole. Snakes have smaller skull bones than their lizard ancestors did, and the bones are connected to each other by elastic ligaments that stretch remarkably, allowing some snakes to swallow food that is larger than their head. Snakes also have well-developed sensory systems for detecting prey. The flicking tongue carries airborne molecules to sensory receptors in the roof of the mouth. Most snakes can detect vibrations on the ground, and some, like rattlesnakes, have heat-sensing organs (see Figure 39.22). Many snakes kill by constriction, which suffocates

prey, and several groups produce toxins that immobilize, kill, and partially digest it.

Crocodilians Are Semiaquatic, Predatory Archosaurs

The 21 species of alligators and crocodiles (Crocodilia, from *crocodilus* = crocodile), along with the birds, are the remnants of the once-diverse archosaur lineage **(Figure 30.25d)**. The largest species, the Australian saltwater crocodile *(Crocodylus porosus)*, grows to 7 m. Crocodilians are aquatic predators that consume other vertebrates. Striking anatomical adaptations distinguish them from living lepidosaurs, including a four-chambered heart that is homologous to the heart in birds.

American alligators *(Alligator mississippiensis)* exhibit strong maternal behavior, which also reflects their relationship to birds. Females guard their nests ferociously and free their offspring from the nest after they hatch. Young stay close to the mother for about a year, feeding on scraps that fall from her mouth and

Model Research Organisms: *Anolis* Lizards of the Caribbean

The lizard genus *Anolis* has been a model system for studies in ecology and evolutionary biology since the 1960s, when Ernest E. Williams of Harvard University's Museum of Comparative Zoology first began studying it. With more than 400 known species—and new ones being described all the time—*Anolis* is one of the most diverse vertebrate genera. Most anoles are less than 10 cm long, not including the tail, and many occur at high densities, making it easy to collect lots of data in a relatively short time. Male anoles defend territories, and their displays make them conspicuous even in dense forests.

Anolis species are widely distributed in South America and Central America, but nearly 40% occupy Caribbean islands. The number of species on an island is generally proportional to the island's size. Cuba, the largest island, has more than 50 species, whereas small islands have just one or two.

Studies by Williams and others suggest that the anoles on some large islands are the products of independent adaptive radiations. Eight of the 10 *Anolis* species now found on Puerto Rico probably evolved on that island from a common ancestor. Similarly, the seven *Anolis* species on Jamaica shared a common ancestor, which was different from the ancestor of the Puerto Rican species. The anole faunas on Cuba and Hispaniola are the products of several independent radiations on each island.

Williams discovered that these independent radiations had produced similar-looking species on different islands. He developed the concept of the *ecomorph*, a group of species that have similar morphological, behavioral, and ecological characteristics even though they are not closely related within the genus. Williams named the ecomorphs after the vegetation that they commonly used (see **figure**). For example, grass anoles are small, slender species that usually perch on low, thin vegetation. Trunk-ground anoles have chunky bodies and large heads, and they perch low on tree trunks, frequently jumping to the ground to feed. Although the grass anoles or the trunk-ground anoles on different islands are similar in many ways, they are not closely related to each other. Their resemblances are the products of convergent evolution.

Ecomorphs exist because evolutionary processes have accentuated the morphological differences among species that occupy different types of vegetation. Jon Losos of Harvard University has demonstrated that trunk-ground anoles, which have relatively long legs and tails, can run faster on wide surfaces and jump farther than species with relatively short legs. And in nature the trunk-ground anoles run and jump more frequently than the other ecomorphs do.

Different ecomorphs on an island use different parts of their habitats by choosing different perch sites (grass, tree trunks, rocks). When two or more species of the same ecomorph inhabit the same island, they occupy habitats with different temperature and shade conditions (see the figure). For example, in Puerto Rico, one species of trunk-ground anole (*Anolis gundlachi*) occupies cool, shady uplands; another (*Anolis cristatellus*) lives in warm, fairly open lowland habitats; and a third species (*Anolis cooki*) lives in desert habitats. Other species in Puerto Rico exhibit similar differences in their distributions. These differences in geographical distribution and habitat use presumably allow the different species to avoid competition with each other and gain access to the resources they need to survive and reproduce.

Evolutionary processes have also fostered physiological differences that reinforce the ecological separation established by the lizards' use of different habitats. For example, *A. cristatellus* maintains higher body temperatures than *A. gundlachi,* and neither is physiologically adapted to the environment of the other: *A. cristatellus* dies in the high altitude forests where *A. gundlachi* thrives, while *A. gundlachi* suffers heat stress at body temperatures that are typical for *A. cristatellus*.

Researchers throughout the Americas continue to explore the ecology and evolution of anoles. Some unravel their biogeography and systematic relationships; others focus on the ecology of populations and communities; still others study their social behavior or sensory physiology. With so many species distributed across hundreds of Caribbean islands, the lizard genus *Anolis* provides fertile ground for testing hypotheses about nearly every aspect of vertebrate biology.

A. cooki

Manuel Leal, Duke University

A. poncensis

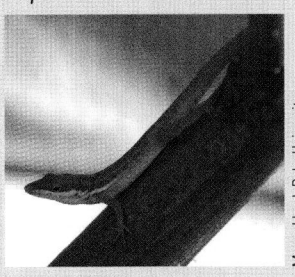

Manuel Leal, Duke University

Sunny lowlands | Shaded uplands

A. cuvieri

A. occultus

Desert habitats

A. stratulus

A. evermanni

A. cristatellus

A. gundlachi

A. cooki

A. pulchellus

A. krugi

A. poncensis

Manuel Leal, Duke University

A. gundlachi

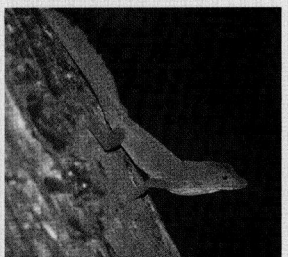

Manuel Leal, Duke University

A. krugi

Manuel Leal, Duke University

living under her watchful protection. Most alligator and crocodile species are highly endangered. Their habitats have been disrupted by human activities, and they have been hunted for meat and leather. Protection efforts have been extremely successful, however. American alligators, for example, recently recovered from the brink of extinction.

STUDY BREAK

1. In addition to losing their legs over evolutionary time, how do snakes differ from their lizard ancestors?
2. What anatomical and behavioral characteristics of crocodilians demonstrate their relatively close relationship to birds?

30.10 Aves: Birds

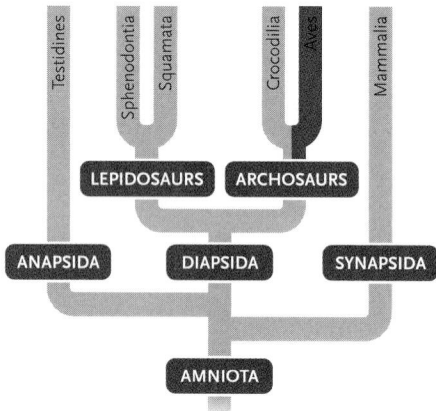

Birds (Aves) appeared in the Jurassic period as descendants of carnivorous, bipedal dinosaurs. Thus, they are full-fledged members of the archosaur lineage. Their evolutionary relationship to dinosaurs is evident in their skeletal anatomy, the scales on their legs and feet, and their posture when walking. However, powered flight gave birds access to new adaptive zones, setting the stage for their astounding evolutionary success **(Figure 30.26a).**

Key Adaptations Reduce Body Weight and Provide Power for Flight

The skeletons of birds are both lightweight and strong **(Figure 30.26b).** For example, the endoskeleton of the frigate bird, which has a 1.5 m wingspan, weighs little more than 100 g, far less than its feathers. Most birds have hollow limb bones with small supporting struts that crisscross the internal cavities. Evolution has also reduced the number of separate bony elements in the wings, skull, and vertebral column (especially the tail),

making the skeleton light and rigid. And all modern birds lack teeth, which are dense and heavy; they acquire food with a lightweight, keratinized bill. Many species have a long, flexible neck, which allows them to use their bills for feeding, grooming, nest-building, and social interactions.

The bones associated with flight are generally large. The forelimb and forefoot are elongate, forming the structural support for the wing. And most modern birds possess a **keeled sternum** (breastbone) to which massive flight muscles attach **(Figure 30.26c).** Not all birds are strong fliers, however; ostriches and other bipedal runners have strong, muscular legs but small wings and flight muscles (see Figure 19.2).

Like the skeleton, soft internal organs are modified in ways that reduce weight. Most birds lack a urinary bladder; uric acid paste is eliminated with digestive wastes. Females have only one ovary and never carry more than one mature egg; eggs are laid as soon as they are shelled.

All birds also possess **feathers (Figure 30.26d),** sturdy, lightweight structures derived from scales in the skin of their ancestors. Each feather has numerous barbs and barbules with tiny hooks and grooves that maintain the feathers' structure, even during vigorous activity. Flight feathers on the wings provide lift; contour feathers streamline the surface of the body; and down feathers form an insulating cover close to the skin. Worn feathers are replaced once or twice each year.

Other adaptations for flight allow birds to harness the energy needed to power their flight muscles. Their metabolic rates are eight to ten times higher than those of other comparably sized diapsids, and they process energy-rich food rapidly. A complex and efficient respiratory system (see Figure 44.7) and four-chambered heart (see Figure 42.5d) enable them to consume and distribute oxygen efficiently. As a consequence of high rates of metabolic heat production, birds maintain a high and constant body temperature (see Section 46.8).

Flying Birds Were Abundant by the Cretaceous Period

Although the earliest known bird, the pigeon-sized *Archaeopteryx,* had feathers, its skeleton was essentially that of a small dinosaur (see Figure 19.12). It had digits and claws on the forelimbs, teeth on its jaws, many bones in its wings and vertebral column, and only a poorly developed sternum. How could flight evolve in so unbirdlike an animal? Biologists hypothesize that *Archaeopteryx* ran after prey, using its feathered wings like fly swatters. Larger wings would have provided extra lift when they jumped at prey, and gradual evolutionary modifications of the wing bones and muscles could have led to powered flight.

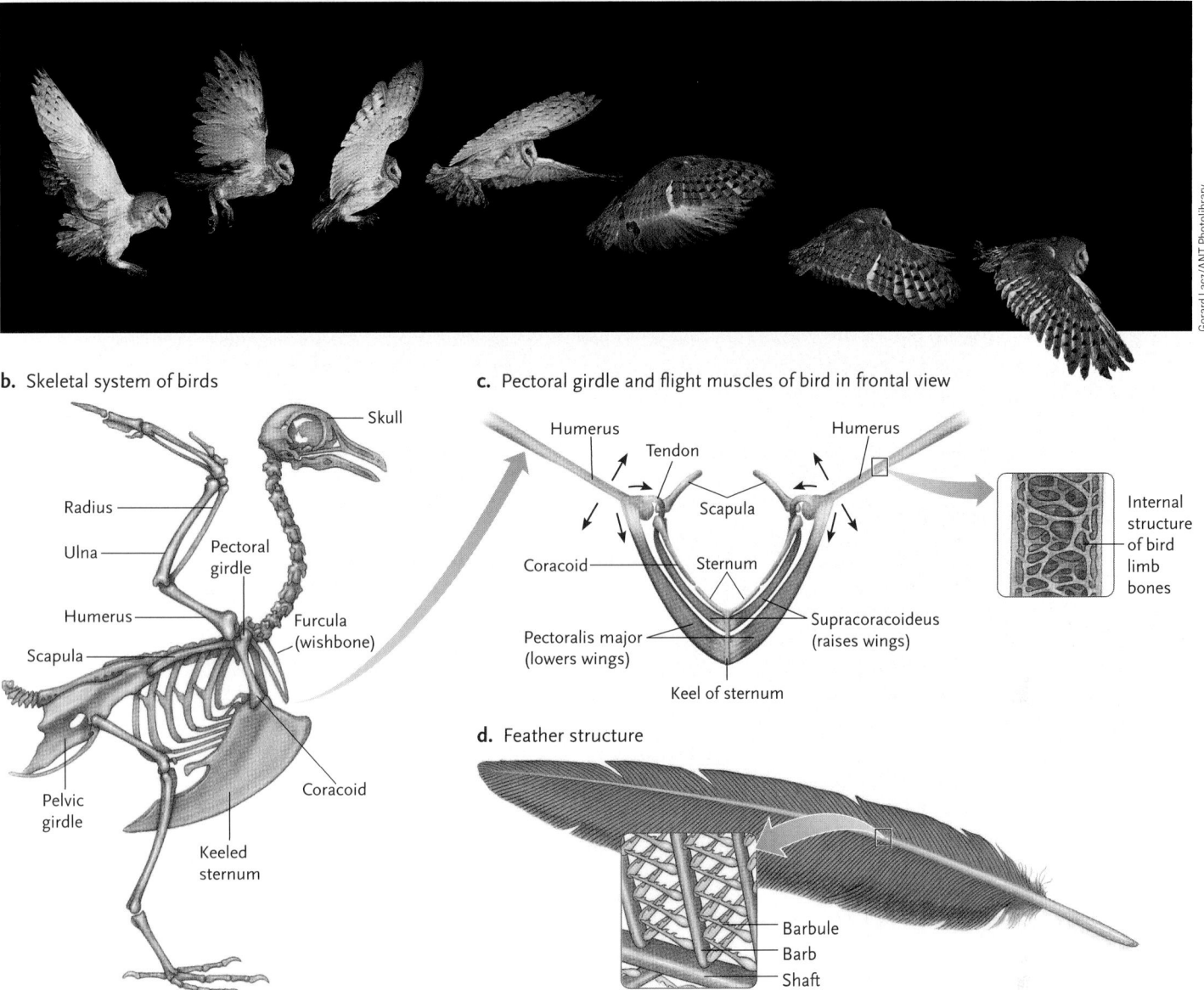

a. Wing movements of an owl during flight

Gerard Lacz/ANT Photolibrary

b. Skeletal system of birds

Skull
Radius
Ulna
Pectoral girdle
Humerus
Scapula
Furcula (wishbone)
Pelvic girdle
Coracoid
Keeled sternum

c. Pectoral girdle and flight muscles of bird in frontal view

Humerus
Tendon
Humerus
Scapula
Coracoid
Sternum
Pectoralis major (lowers wings)
Supracoracoideus (raises wings)
Keel of sternum
Internal structure of bird limb bones

d. Feather structure

Barbule
Barb
Shaft

Figure 30.26

Adaptations for flight in birds. **(a)** The flapping movements of a bird's wing provide *thrust* for forward movement and *lift* to counteract gravity. **(b)** The bird skeleton includes a boxlike trunk, short tail, long neck, lightweight skull and beak, and well-developed limbs. In large birds, limb bones are hollow. **(c)** Two sets of flight muscles attach to a keeled sternum; one set raises the wings, and the other lowers it. **(d)** Flexible feathers form an airfoil on the wing surface.

Crow-sized birds with full flight capability appeared by the early Cretaceous period. They had a keeled sternum and other modern skeletal features. The modern groups of wading birds and seabirds first appear in late Cretaceous rocks; fossils of other modern groups are found in slightly later deposits. Woodpeckers, perching birds, birds of prey, pigeons, swifts, the flightless ratites, penguins, and some other groups were all present by the end of the Oligocene; birds continued to diversify through the Miocene (see Table 22.1).

Modern Birds Vary in Morphology, Diet, Habits, and Patterns of Flight

The 9000 living bird species show extraordinary ecological specializations, but they share the same overall body plan. Living birds are traditionally classified into nearly 30 groups **(Figure 30.27).** They vary in size from the bee hummingbird *(Mellisuga helenae)* of Cuba, which weighs little more than 1 g, to the ostrich *(Struthio camelus),* which can weigh as much as 150 kg.

a. The Laysan albatross (Procellariiformes, *Phoebastria immutabilis*) has the long thin wings typical of birds that fly great distances.

b. The roseate spoonbill (Ciconiformes, *Ajaia ajaja*) uses its bill to strain food particles from water.

c. The bald eagle (Falconiformes, *Haliaeetus leucocephalus*) uses its sharp bill and talons to capture and tear apart prey.

d. A European nightjar (Caprimulgiformes, *Caprimulgus europaeus*) uses its wide mouth to capture flying insects.

e. A Bahama woodstar hummingbird (Apodiformes, *Calliphlox evelynae*) hovers before a hibiscus blossom to drink nectar from the base of the flower.

f. The chestnut-backed chickadee (Passeriformes, *Parus rufescens*) uses its thin bill to probe for insects in dense vegetation.

Figure 30.27
Bird diversity.

The structure of the bill usually reflects a bird's diet. Seed and nut eaters, such as finches and parrots, have deep, stout bills that crack hard shells. Carnivorous hawks and carrion-eating vultures have sharp beaks to rip flesh, and nectar-feeding hummingbirds have slender bills to reach into flowers. The bills of ducks are modified to extract particulate matter from water, and many perching birds have slender bills to feed on insects.

Birds also differ in the structure of their feet and wings. Predators have large, strong talons (claws), whereas ducks and other swimming birds have webbed feet that serve as paddles. Long-distance fliers like albatrosses have narrow wings; those that hover at flowers, such as hummingbirds, have wide ones. The wings of some species, like penguins, are so specialized for swimming that they are incapable of aerial flight.

All birds have well-developed sensory and nervous systems, and their brains are proportionately larger than those of other diapsids of comparable size. Large eyes provide sharp vision, and most species also have good hearing, which nocturnal hunters like owls use to locate prey. Some vultures and other species have a good sense of smell, which they use to find food. Migrating birds use polarized light, changes in air pressure, and Earth's magnetic field for orientation.

Most birds exhibit complex social behavior, including courtship, territoriality, and parental care. Many species communicate with vocalizations and visual displays to challenge other individuals or attract mates. Most raise their young in a nest, using body heat to incubate eggs. The nest may be a simple depression on a gravely beach, a cup woven from twigs and grasses, or a feather-lined hole in a tree.

Many bird species embark on a semiannual long-distance migration (see Section 55.1). The golden plover *(Pluvialis dominica)*, for example, migrates 20,000 km twice each year. Migrations are a response to seasonal changes in climate. Birds travel toward the tropics as winter approaches; in spring, they return to high latitudes using seasonally available food sources.

STUDY BREAK

1. What specific adaptations allow birds to fly?
2. How do the structures of a bird's bill, wings, and feet reflect its dietary and habitat specializations?

30.11 Mammalia: Monotremes, Marsupials, and Placentals

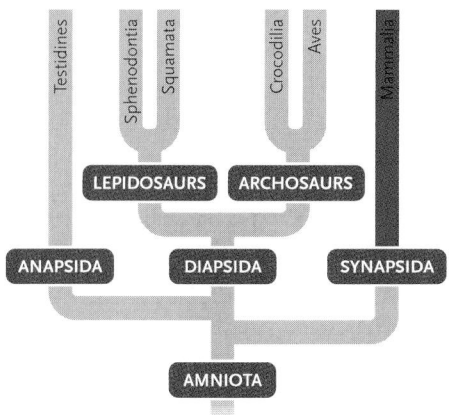

The synapsid lineage, which includes the living mammals, was the first group of amniotes to diversify broadly on land. Indeed, during the late Paleozoic era, medium- to large-sized synapsids were the most abundant predators in terrestrial habitats. One particularly successful and persistent branch of the synapsid lineage, the *therapsids,* exhibited many mammal-like characteristics in their legs, skulls, jaws, and teeth. And by the end of the Triassic period, the earliest mammals—most of them no bigger than a rat—had emerged from therapsid ancestors. Several mammalian lineages coexisted with dinosaurs and other diapsids throughout the Mesozoic, but paleontologists hypothesize that most Mesozoic mammals were active only at night to avoid predatory dinosaurs, which were active during the day. Two mammalian lineages, the egg-laying Prototheria (or monotremes) and the live-bearing Theria (marsupials and placentals), survived the mass extinction that eliminated most dinosaurs at the end of the Mesozoic. The Theria diversified into the mammalian groups that are most familiar today.

Mammals Exhibit Key Adaptations in Anatomy, Physiology, and Behavior

Four sets of key adaptations fostered the success of mammals.

High Metabolic Rate and Body Temperature. Like birds, mammals have high metabolic rates that release enough energy to maintain high activity levels and enough heat to maintain high body temperatures (see Section 46.8). An outer covering of fur and a layer of subcutaneous fat help retain body heat. Using metabolic heat to stay warm requires lots of oxygen, and mammals have a muscular organ, the diaphragm, that fills their lungs with air (see Figure 44.8). Four-chambered hearts and complex circulatory systems deliver oxygen to active tissues (see Figure 42.5d).

Specializations of the Teeth and Jaws. Mammals also have anatomical features that allow them to feed efficiently. Ancestrally, mammals have four types of teeth (see Figure 45.17): flattened **incisors** nip and cut food; pointed **canines** pierce and kill prey; and two sets of cheek teeth, **premolars** and **molars,** grind and crush food. Moreover, teeth in the upper and lower jaws occlude (that is, fit together) tightly as the mouth is closed; thus, mammals can use their large jaw muscles to chew food thoroughly.

Parental Care. Mammals provide more parental care to their young than any other animals. In most species, young complete development within a female's uterus, deriving nourishment through the **placenta,** a specialized organ that mediates the delivery of oxygen and nutrients (see Section 47.2). Females also have **mammary glands,** specialized structures that produce energy-rich milk, a watery mixture of fats, sugars, proteins, vitamins, and minerals. This perfectly balanced diet is the sole source of nutrients for newborn offspring.

Complex Brains. Finally, mammals have larger brains than other tetrapods of equivalent body size; the difference lies primarily in the **cortex,** the part of the forebrain responsible for information processing and learning (see Figure 38.6). Extensive postnatal care provides opportunities for offspring to learn from older individuals. Thus, mammalian behavior is strongly influenced by past experience as well as by genetically programmed instincts.

The Major Groups of Modern Mammals Differ in Their Reproductive Adaptations

Biologists recognize a primary distinction between two lineages of modern mammals: the egg-laying Prototheria (*protos* = first; *therion* = wild beast), or monotremes, and the live-bearing Theria. The Theria, in turn, diversified into two sublineages, the Metatheria (*meta* = between), or marsupials, and the Eutheria (*eu* = true), or placentals, which also differ in their reproductive adaptations.

Monotremes. The three living species of monotremes (Prototheria), which are limited to the Australian region, reproduce with a leathery-shelled egg **(Figure 30.28).** Newly hatched young lap up milk secreted by modified sweat glands on the mother's belly. The duck-billed platypus *(Ornithorhynchus anatinus)* lives in burrows along riverbanks and feeds on aquatic invertebrates. The two species of echidnas, or spiny anteaters (*Tachyglossus aculeatus* and *Zaglossus bruijni*), feed on ants or termites.

Marsupials. The 240 species of marsupials (Metatheria) have short gestation: the young are nourished through a placenta very briefly—sometimes only for

a. Short-nosed echidna

b. Duck-billed platypus

Figure 30.28
Monotremes. **(a)** The short-nosed echidna *(Tachyglossus aculeatus)* is terrestrial. **(b)** The duck-billed platypus *(Ornithorhynchus anatinus)* raises its young in a streamside burrow.

Figure 30.29
Marsupials. An Eastern gray kangaroo *(Macropus giganteus)* carries her "joey" in her pouch.

way between North and South America (see *Focus on Research* in Chapter 22).

Placentals. The 4000 species of placental mammals (Eutheria) are the dominant mammals today. They complete embryonic development in the mother's uterus, nourished through a placenta until they reach a fairly advanced stage of development. Some species, like humans, are helpless at birth, but others, such as horses, are quickly mobile.

Biologists divide eutherians into about 18 groups, only eight of which contain more than 50 living species **(Figure 30.30)**. Rodents (Rodentia) make up about 45% of eutherian species, and bats (Chiroptera) comprise another 22%. Our own group, Primates, is represented by fewer than 170 living species (less than 5% of all mammalian species), many of which are highly endangered. Researchers still do not agree on the details of eutherian evolution. *Insights from the Molecular Revolution* describes the use of molecular techniques to resolve one question about their relationships.

Some eutherians have highly specialized locomotor structures. Whales and dolphins (Cetacea) and manatees and dugongs (Sirenia) are descended from terrestrial ancestors, but their appendages do not function on land, and they are now restricted to aquatic habitats. By contrast, seals and walruses (Carnivora) feed under water but rest and breed on land. Bats (Chiroptera) use wings for powered flight.

Although early mammals appear to have been insectivorous, the diets of modern eutherians are diverse. Odd-toed ungulates *(ungula* = hoof) like horses and rhinoceroses (Perissodactyla), even-toed ungulates like cows and camels (Artiodactyla), and rabbits and hares (Lagomorpha) feed on vegetation. Carnivores (Carnivora) consume other animals. Most insectivores (Insectivora) and bats eat insects, but some feed on flowers,

8 to 10 days—before birth. Newborns use their forelimbs to drag themselves across the mother's belly fur and enter her abdominal pouch, the **marsupium**, where they complete development attached to a teat. Marsupials are the dominant native mammals of Australia and a minor component of the South American fauna **(Figure 30.29)**; only one species, the opossum *(Didelphis virginiana)*, occurs in North America. South America once had a diverse marsupial fauna, but it declined after the Isthmus of Panama bridged the sea-

a. The capybara (Rodentia, *Hydrochoerus hydrochaeris*), the largest rodent, feeds on vegetation in South American wetlands.

b. Most bats, like the Yuma Myotis (Chiroptera, *Myotis yumanensis*), are nocturnal predators on insects.

c. Walruses (Carnivora, *Obodenus rosmarus*) feed primarily on marine invertebrates in frigid arctic waters.

d. The black rhinoceros (Perissodactyla, *Diceros bicornis*) feeds on grass in sub-Saharan Africa.

e. Arabian camels (Artiodactyla, *Camelus dromedarius*) use enlarged foot pads to cross hot desert sands.

f. Antillean manatees (Sirenia, *Trichechus manatus*) are herbivores that live in warm coastal marshes and rivers from Florida to northern South America.

Figure 30.30
Eutherian diversity.

fruit, and nectar. Many whales and dolphins prey on fishes and other animals, but some eat plankton. And some groups, including rodents and primates, feed opportunistically on both plant and animal matter.

STUDY BREAK

1. During the Mesozoic era, why were most mammals active only at night?
2. Which key adaptations in mammals allow them to be active under many types of environmental conditions?
3. On what basis are the major groups of living mammals distinguished?

30.12 Nonhuman Primates

We now focus our attention on Primates, the mammalian lineage that includes humans, apes, monkeys, and their close relatives. The first Primates appeared early in the Eocene epoch, about 55 million years ago, in forested habitats in North America, Europe, Asia, and North Africa.

Key Derived Traits Enabled Primates to Become Arboreal, Diurnal, and Highly Social

Several derived traits allow primates to be arboreal (to live in trees rather than on the ground). For example, most primates have a more erect posture than

The Guinea Pig Is Not a Rat

Using the Linnaean system of taxonomy, the Rodentia has traditionally included more than 1800 species distributed among 29 families, including squirrels, rats and mice, guinea pigs, and porcupines. Their placement in the same order implies that they have a common evolutionary ancestor not shared by any other groups within the mammals. Biologists commonly accepted this interpretation until a molecular study compared the amino acid sequences of 15 proteins encoded in the nuclear DNA of various rodents. The comparisons revealed differences suggesting that guinea pigs should be placed in a separate order. Since then, further molecular evaluations of nuclear genes have produced contradictory results, with some studies supporting the traditional classification of guinea pigs as rodents and others placing them outside the Rodentia.

A cooperative study by Anna Maria D'Erchia and her colleagues at universities and institutes in Italy and Sweden now adds molecular weight to the conclusion that guinea pigs belong in an order of their own. The research team used mitochondrial DNA (mtDNA) sequences because they are easy to isolate and purify, and typically undergo many random mutations and rearrangements that have no apparent effect on gene function. Thus the changes observed in mtDNA are expected to reflect more faithfully the ticking of the molecular clock that tracks the time course of evolutionary events.

For their study, the researchers sequenced mtDNA of the guinea pig and another mammal considered by some biologists to be closely related to rodents, rabbits (Lagomorpha). Other workers had previously sequenced the mtDNAs of 14 other mammals in eight orders, including primates (Primates), seals (Carnivora), cows (Artiodactyla), whales (Cetacea), horses (Perissodactyla), mice and rats (Rodentia), hedgehogs (Insectivora), and opossums (Marsupialia).

The researchers evaluated these sequences with three different statistical programs that use similarities and differences in mtDNA sequences to construct evolutionary trees. They also conducted analyses using nuclear DNA, including separate evaluations of the entire nuclear genome, the protein-encoding sequences, and the DNA encoding ribosomal RNA. Significantly, all the methods produced essentially the same family tree (shown in **Figure a**).

The tree places guinea pigs in a group of their own, sharing a more recent common ancestor with all of the mammalian orders examined except those represented by rodents, hedgehogs, and opossums. The lineage that includes guinea pigs and most other mammals shared a common ancestor with the lineage leading to rodents at a point further back in evolutionary time. And the lineage that includes rodents, guinea pigs, and most other mammals split off from the ancestors of hedgehogs and opossums even earlier. Thus, guinea pigs merit placement in a separate group from rodents. The results for rabbits also indicate that they are more closely related to other mammals than they are to mice and rats. Incidentally, the tree also supports the conclusion from other molecular studies that cows and whales are more closely related to each other than to other mammals (see *Insights from the Molecular Revolution* in Chapter 23).

Figure a
Molecular phylogeny for guinea pigs and other mammals. The numbers next to the branches in the tree indicate the probability that the branch is correct, on a scale of 0 to 100.

other mammals, and they have flexible hip and shoulder joints, which allow a variety of locomotor activities. They can grasp objects with their hands and feet, because they have nails, not claws, on their fingers and toes; their fingertips are well endowed with sensory nerves that enhance the sense of touch. Unlike other mammals, most primates have an opposable big toe, which can touch the tips of other digits and the sole of the foot; many species also have an opposable thumb.

Most primates are diurnal (active during daylight hours), and, unlike most mammals, they rely more on vision than on their sense of smell. Thus, they generally have short snouts and small olfactory lobes of the brain. Most species have forward-facing eyes with overlapping fields of vision, providing excellent depth perception, which comes in handy when moving through trees. Many species have color vision.

Primate brains—especially the regions that integrate information—are large and complex. As a result, they have an exceptional capacity to learn. Most species live in social groups; thus, young primates, which mature slowly, can interact with and learn from their elders and peers during an extended period of parental care. Females give birth to only one or two young at a time, allowing them to devote substantial attention to each offspring.

Living Primates Include Two Major Lineages

Primatologists recognize two lineages within the Primates **(Figure 30.31)**, the Strepsirhini (*streptos* = twisted or turned, *rhin* = nose) and the Haplorhini (*haploos* = single or simple).

Strepsirhini. The 36 living species of Strepsirhini—lemurs, lorises, and galagos—possess many ancestral morphological traits, including moist, fleshy noses and eyes that are positioned somewhat laterally on their heads **(Figure 30.32)**. Strepsirhines generally have short gestation periods and rapid maturation. Today, 22 lemur species survive on Madagascar, a large island off the east coast of Africa; they are ecologically diverse and range in size from 40 g to 7 kg; some lemurs are arboreal, whereas others spend substantial time on the ground. The 12 species of lorises and galagos occupy tropical forest and subtropical woodlands in Africa, India, and Southeast Asia; they are all arboreal and nocturnal.

Haplorhini. Most species in the Haplorhini—the familiar monkeys and apes—have many derived primate characteristics, including compact, dry noses, and forward-facing eyes.

However, five species of tarsiers, which are restricted to tropical forests on the islands of Southeast Asia, exhibit several ancestral traits: small body size

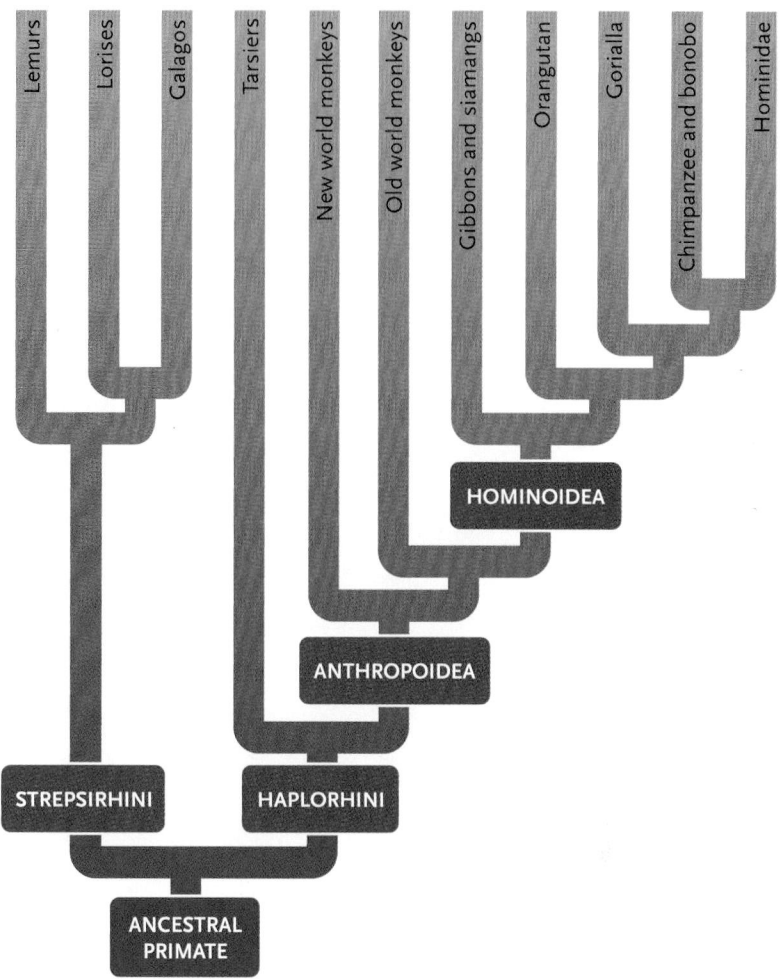

Figure 30.31
Primate phylogeny. A phylogenetic tree for the Primates illustrates the two main lineages: Strepsirhini and Haplorhini. Note that chimpanzees are the closest living relatives of humans.

(about 100 g), large eyes and ears, and two grooming claws on each foot **(Figure 30.33)**. But they share the derived characteristics of dry noses and forward-facing eyes with the other haplorhines; and DNA sequence data link them to the monkeys and apes and not to the strepsirhines described earlier.

The 130 or so species of monkeys, 13 species of apes, and humans constitute the monophyletic haplorhine lineage **Anthropoidea**, which probably arose in Africa; fossils of a diverse and abundant radiation of forest-dwelling anthropoids, dating from the late Eocene epoch, have been discovered in northern Egypt. Continental drift then established long-term geographical and evolutionary separation of anthropoids in the New World and Old World **(Figure 30.34)**.

By the middle of the Oligocene epoch, about 30 million years ago, the ancestors of the New World monkeys had arrived in South America and begun to diversify there. They probably rafted across the

Cagan Sekercioglu/Visuals Unlimited

Figure 30.32
Strepsirhines. The ring-tailed lemur *(Lemur catta)* of Madagascar has ancestral primate characteristics, such as a long snout and wet nose.

Larry Burrows/Aspect Photolibrary

Figure 30.33
Tarsiers. Tarsiers *(Tarsius bancanus)* are classified as haplorhines, but they retain many ancestral characteristics.

a. Spider monkey

b. Hamadryas baboon

Joseph Van Os/Getty Images, Inc.

Figure 30.34
New World and Old World monkeys. **(a)** Many New World monkeys, such as the spider monkey *(Ateles geoffroyi)*, have prehensile tails, which they use as a fifth limb. Old World monkeys lack prehensile tails, and many, such as **(b)** the Hamadryas baboon *(Papio hamadryas)*, are largely terrestrial.

Atlantic, which was narrower at that time, on trees or other storm debris. New World monkeys now live in Central and South America (see Figure 30.34a). They range in size from tiny marmosets and tamarins (350 g) to hefty howler monkeys (10 kg). Most are exclusively arboreal and diurnal. The larger species may hang below branches by their arms, and some use a prehensile (grasping) tail as a fifth limb.

Anthropoids diversified most spectacularly in the Old World, however, eventually giving rise to two lineages—one ancestral to Old World monkeys and the other to apes and humans. Although many people assume that the apes are descended from Old World monkeys, the fossil record contradicts that impression. The earliest hominoid (ape) fossils date to the early Miocene, roughly 23 million years ago, but the oldest known Old World monkeys appeared several million years later.

Old World monkeys, which occupy habitats ranging from tropical rain forests to deserts in Africa and Asia, may grow as large as 35 kg (see Figure 30.34b). Many species are sexually dimorphic; in other words, males and females attain different adult sizes (see Section 20.3). Arboreal species use all four limbs for locomotion, but none has a prehensile tail. Some species, such as baboons, often walk or run on the ground.

Within the anthropoid lineage, the **Hominoidea** ("humanlike") is a monophyletic group that includes apes and humans. The climate of the early Miocene was wetter than it is today, and eastern Africa, where many early hominoid fossils are found, was covered with extensive forests. A climate shift in the middle Miocene, around 14 million years ago, converted dense forests into woodlands and grasslands. Hominoids probably adopted a more terrestrial existence and shifted their diets. Miocene hominoids ranged in size from 4 kg to 80 kg. They occupied both forest and open woodland habitats; some were probably ground dwelling.

Although hominoids are closely related to Old World monkeys, several characteristics distinguish them. Apes lack a tail, and great apes (orangutans, gorillas, chimpanzees, and bonobos) are much larger than monkeys. Moreover, the posterior region of the vertebral column is shorter and more stable in apes. Apes also show more complex behavior.

The gibbons and siamangs, which live in tropical forests in Southeast Asia, are the smallest of the apes, ranging in weight from 6 to 11 kg. With extremely long arms and strong shoulders, they hang below branches by their arms and swing themselves forward, a pattern of locomotion called **brachiation (Figure 30.35a)**. The much larger orangutan *(Pongo pygmaeus)*, now restricted to forested areas on the islands of Borneo and Sumatra, can grow to 90 kg. Orangutans use both hands and feet to climb trees; they sometimes venture onto the ground on all fours.

Gorillas *(Gorilla gorilla)*, which are currently restricted to two large central African forests, are the largest of the living primates. Males can weigh 180 kg; females are about half that size. Because of their size, gorillas spend most of their time on the ground. They often use "knuckle-walking" locomotion, leaning forward and supporting part of their weight on the backs of their hands. Gorillas are almost exclusively vegetarian.

Chimpanzees *(Pan troglodytes)* are also forest dwellers, weighing up to 45 kg **(Figure 30.35b)**. Like gorillas, they spend most of their waking hours on the ground; they often knuckle-walk, but sometimes adopt a **bipedal** (two-legged) stance and swagger short distances. Groups of related males form loosely defined communities of up to 50 individuals, which may cooperate in hunts and foraging. Bonobos *(Pan paniscus)*, sometimes called pygmy chimpanzees, are restricted to a small area in central Africa. Somewhat smaller than chimps, they have longer legs and smaller heads.

The Primates also includes humans *(Homo sapiens)*, which occupy virtually all terrestrial habitats. Humans have adaptations that allow an upright posture and bipedal locomotion. They are ground-dwelling animals with extremely broad diets and complex social behavior.

a. Black-handed gibbon

b. Chimpanzee

Art Wolfe/Photo Researchers, Inc.

Kenneth Garrett/National Geographic Image Collection

Figure 30.35

Apes. **(a)** Small-bodied apes, such as the black-handed gibbon *(Hylobates agilis)* are agile brachiators that swing through the trees with ease. **(b)** Among the large-bodied apes, chimpanzees *(Pan troglodytes)* have opposable thumbs and big toes.

Figure 30.36

Adaptations for bipedal locomotion. Differences in the posture, skeleton, and muscles of monkeys, great apes, and humans illustrate the anatomical changes that accompanied upright, bipedal locomotion. Evolutionary changes in the spine, pelvis, hip, knee, ankle, and foot were accompanied by changes in the sizes of leg muscles and their points of attachment to the bones they move.

Monkey   Ape   Human

STUDY BREAK

1. What characteristics of primates allow them to spend a great deal of time in trees?
2. What is the lowest taxonomic group that includes monkeys, apes, and humans? What is the lowest taxonomic group that includes only apes and human?
3. Which species of ape spend most of the time on the ground?

30.13 The Evolution of Humans

Genetic analyses of living hominoid species indicate that African hominoids diverged into several lineages between 10 million and 5 million years ago; one lineage, the **hominids**, includes modern humans and our bipedal ancestors.

Hominids First Walked Upright in East Africa about 6 Million Years Ago

Upright posture and bipedal locomotion are key adaptations that distinguish hominids from apes. Researchers identify early hominid fossils from features of the skull, spine, pelvis, knees, ankles, and feet that make bipedal locomotion possible **(Figure 30.36)**. As a conse-

quence of bipedal locomotion, the hands were no longer used for locomotor functions, allowing them to become specialized for other activities, such as tool use. Evolutionary refinements in grasping ability allow hominids to hold objects tightly with a *power grip* or manipulate them precisely with a *precision grip* **(Figure 30.37)**. Hominids also developed larger brains.

Paleontologists have uncovered fossil of numerous hominids that lived in East Africa and South Africa from roughly 6 million to 1 million years ago **(Figure 30.38)**. In 2000, researchers found 13 fossils of *Orrorin tugenensis* ("first man" in a local African language), a species that lived about 6 million years ago in East Af-

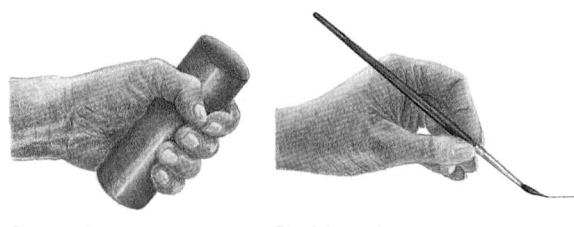

Power grip   Precision grip

Figure 30.37

Power grip versus precision grip. Hominids grasp objects in two distinct ways. **(a)** The power grip allows us to grasp an object firmly. **(b)** The precision grip allows us to manipulate objects with fine movements.

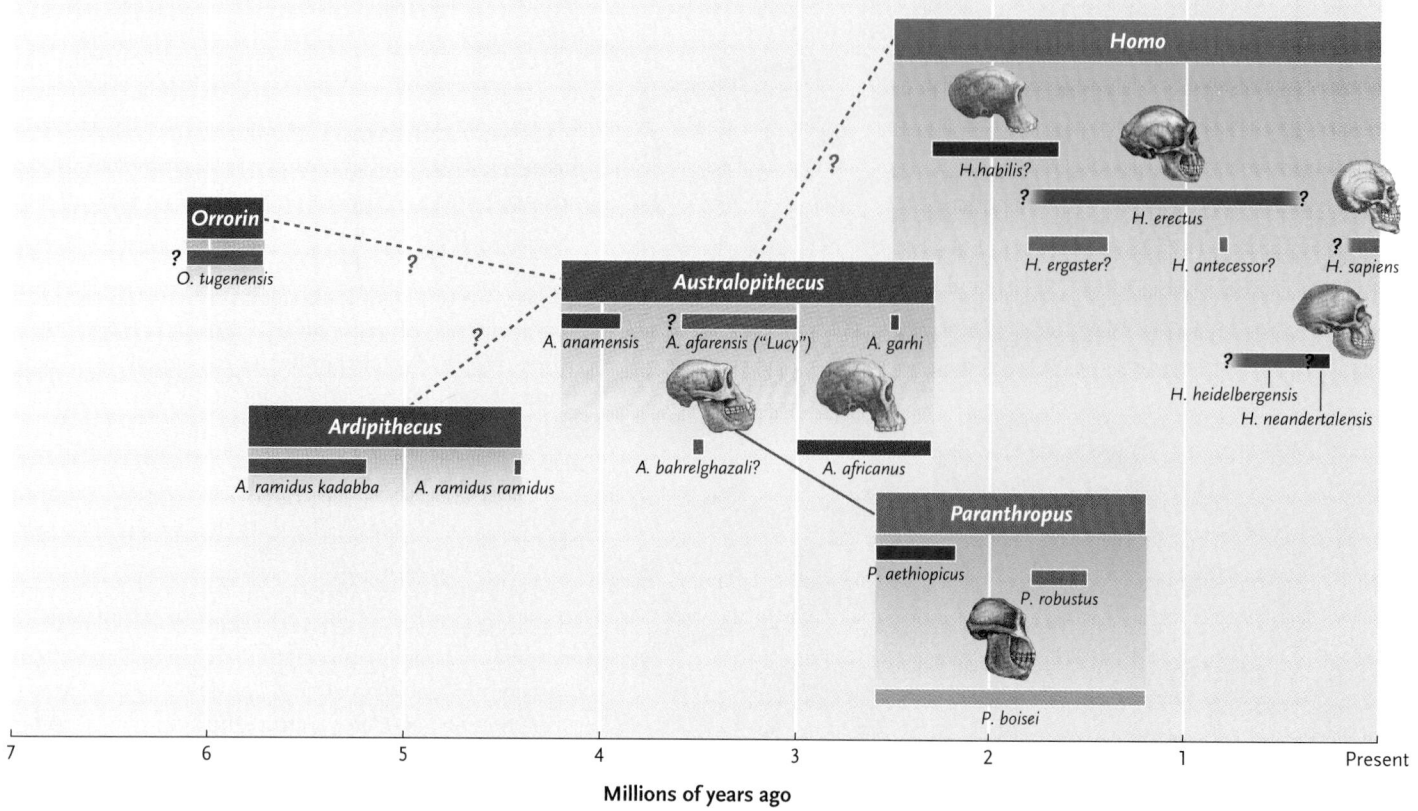

Figure 30.38

Hominid time line. Several species of hominids lived simultaneously at sites in eastern and southern Africa. The time line for each species and genus reflects the ages of known fossils. The numerous question marks indicate researchers' uncertainty about the classification, the ages of fossils, and the evolutionary relationships among the species. Some of the skulls pictured are reconstructed from fragmentary fossils.

rican forests. However, these remains are fragmentary, and experts are still evaluating them. The best-studied early hominid fossils, the remains of 50 individuals discovered in the East African Rift Valley, date from about 5 million years ago. Named *Ardipithecus ramidus,* these hominids stood 120 cm tall and had apelike teeth. Other *Ardipithecus* fossils, recently discovered at a different site, appear to be much older (5.8 million years) and show evidence of bipedal locomotion.

Hominid fossils from 4.2 million to 1.2 million years ago are known from many sites in East, Central, and South Africa. They are currently assigned to the genera *Australopithecus* (*australis* = southern; *pithekos* = ape) and *Paranthropus* (*para* = beside; and *anthropos* = human being). With their large faces, protruding jaws, and small skulls and brains, most of these hominids had an apelike appearance (see Figure 30.38). *Australopithecus anamensis,* which lived in East Africa around 4 million years ago, is the oldest known species. It had thick enamel covering on its teeth, a derived hominid characteristic; the structure of a fossilized leg bone suggests that it was bipedal.

Specimens of more than 60 individuals of *Australopithecus afarensis* have been found in northern Ethiopia, including about 40% of a female's skeleton, named "Lucy" by its discoverers **(Figure 30.39).** *A. afarensis* lived 3.5 million to 3 million years ago, but it retained several ancestral characteristics. For example, it had moderately large and pointed canine teeth, and a relatively small brain. Males and females were 150 cm and 120 cm tall, respectively. Skeletal analyses suggest that *A. afarensis* was fully bipedal, a conclusion supported by fossilized footprints preserved in a layer of volcanic ash.

Other species of *Australopithecus* and *Paranthropus* lived in East Africa or South Africa between 3.7 million and 1 million years ago. Adult males ranged from 40 to 50 kg in weight and from 130 to 150 cm in height; females were smaller. Most species had deep jaws and large molars. Several species had a crest of bone along the midline of the skull, providing a large surface for the attachment of jaw muscles. These anatomical features suggest that they fed on hard food, such as nuts, seeds, and other vegetable products. One species,

a. "Lucy"   **b.** Australopithecine footprints

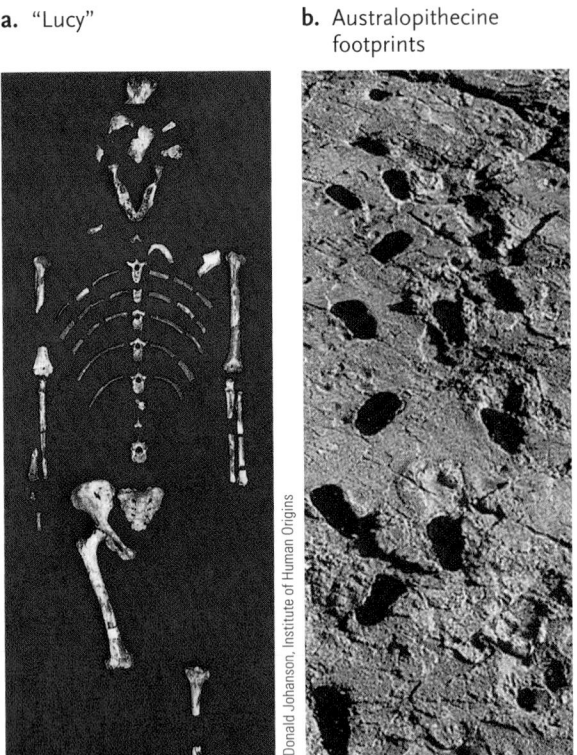

Dr. Donald Johanson, Institute of Human Origins

Louise M. Robbins

Figure 30.39

Australopithecines. **(a)** Researchers named the most complete fossil of *Australopithecus afarensis* "Lucy." **(b)** Mary Leakey discovered australopithecine footprints, made in soft, damp volcanic ash about 3.7 million years ago. The footprints indicate that australopithecines were fully bipedal.

Australopithecus africanus, known only from South Africa, had small jaws and teeth, indicating that it probably consumed a softer diet. The phylogenetic relationships of the species classified as *Australopithecus* and *Paranthropus*—and their exact relationships to later hominids—are not yet fully understood. But most scientists agree that *Australopithecus* was ancestral to humans, which are classified in the genus *Homo*.

Homo habilis Was Probably the First Hominid to Manufacture Stone Tools

Pliocene fossils of the earliest humans, which may have included several species, are fragmentary. They are also widely distributed in space and time, complicating analyses of their relationships. For the sake of simplicity, we describe them as belonging to one species, *Homo habilis* ("handy man").

From 2.3 million to 1.7 million years ago, *H. habilis* occupied the woodlands and savannas of eastern and southern Africa, sharing these habitats with various species of *Paranthropus*. The two genera are easy to tell apart because the brains of *H. habilis* were at least 20% larger, and they had larger incisors and smaller molars than their hominid cousins. Their diet included hard-shelled nuts and seeds as well as soft fruits, tubers, leaves, and insects. They may also have hunted small prey or scavenged carcasses left by large predators.

Researchers have found numerous tools dating to the time of *H. habilis,* but they are not sure which species used them. Many of the hominid species alive at the time probably cracked marrowbones with rocks or scraped flesh from bones with sharp stones. Paleoanthropologist Louis Leakey was the first to discover evidence of tool *making* at East Africa's Olduvai Gorge, which cuts through a great sequence of sedimentary rock layers. The oldest tools at this site are crudely chipped pebbles, which were probably manufactured by *H. habilis.*

Homo erectus Dispersed from Africa to Other Continents

Early in the Pleistocene epoch, about 1.8 million years ago, new species of humans appeared in East Africa. Most fossils are fragmentary. For convenience, we describe them all as *Homo erectus* ("upright man"), recognizing that they probably represent several species. One nearly complete skeleton suggests that *H. erectus* was taller than its ancestors, had a much larger brain, a thicker skull, and protruding brow ridges **(Figure 30.40)**.

H. erectus made fairly sophisticated tools, including the hand axe (see Figure 30.40b), which they apparently used to cut food and other materials, to scrape meat from bones, and to dig for roots. *H. erectus* probably fed on both plants and animals; they may have hunted and scavenged

a. *Homo erectus*   **b.** Hand axe

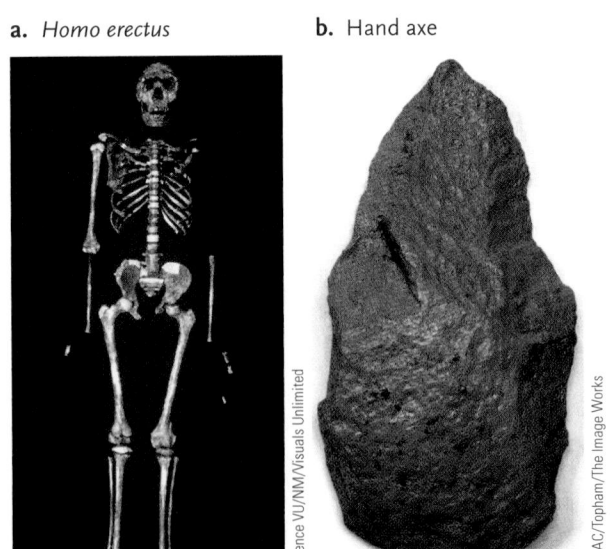

Science VU/NM/Visuals Unlimited

AAAC/Topham/The Image Works

Figure 30.40

Homo erectus. **(a)** A nearly complete skeleton of *Homo erectus* was discovered in Kenya. **(b)** Hand axes are frequently found at *H. erectus* sites.

animal prey. Archaeological data points to their use of fire to process food and to keep themselves warm.

The pressure of growing populations apparently forced groups of *H. erectus* out of Africa about 1.5 million years ago. They dispersed northward from East Africa into both northwestern Africa and Eurasia. Some moved eastward through Asia as far as the island of Java. Recent discoveries in Spain indicate that *H. erectus* also occupied parts of Western Europe.

Modern Humans Are the Only Surviving Descendants of *Homo erectus*

Judging from its geographical distribution, *Homo erectus* was successful in many environments. It produced several descendant species, of which modern humans (*Homo sapiens*, "wise man") are the only survivors.

Fossils from Africa, Asia, and Europe indicate that archaic humans, the now-extinct descendants of *H. erectus,* first appeared at least 400,000 years ago. They generally had a larger brain, rounder skull, and smaller molars than *H. erectus.*

Neanderthals. The Neanderthals *(Homo neanderthalensis),* who occupied Europe and western Asia from 150,000 to 28,000 years ago, are the best-known archaic humans. Compared with modern humans, they had a heavier build, more-pronounced brow ridges, and slightly larger brains (see Figure 30.38). Neanderthals were culturally and technologically sophisticated. They made complex tools, including wooden spears, stone axes, and flint scrapers and knives. At some sites they built shelters of stones, branches, and animal hides, and they routinely used fire. They were successful hunters and probably consumed nuts, berries, fishes, and bird eggs. Some groups buried their dead, and they may have had rudimentary speech.

Researchers once classified Neanderthals as a subspecies of *H. sapiens,* but most scientists now believe that they were a separate species. In 1997 two teams of researchers, Matthias Kring and Svante Pääbo of the University of Munich and Anne Stone and Mark Stoneking of Pennsylvania State University, independently analyzed short segments of mitochondrial DNA (mtDNA) extracted from the fossilized arm bone of a Neanderthal. Unlike nuclear DNA, which individuals inherit from both parents, only mothers pass mtDNA to offspring. It does not undergo genetic recombination (see Section 13.5), and it has a high mutation rate, making it useful for phylogenetic analyses. Many scientists believe that mutation rates in mtDNA are fairly constant, allowing this molecule to serve as a molecular clock (see Section 23.6). Comparing the Neanderthal sequence with mtDNA from 986 living humans, the researchers discovered three times more differences between the Neanderthals and modern humans than between pairs of modern humans in their sample. These results suggest that Neanderthals and modern humans are different species that diverged from a common ancestor 690,000 to 550,000 years ago—hundreds of thousands of years before modern humans appeared.

Modern Humans. Modern humans *(Homo sapiens)* differ from Neanderthals and other archaic humans in having a slighter build, less-protruding brow ridges, and a more prominent chin. The earliest fossils of modern humans found in Africa and Asia are 150,000 years old; those from the Middle East are 100,000 years old. Fossils from about 20,000 years ago are known from Western Europe, the most famous being those of the Cro-Magnon deposits in southern France. The widespread appearance of modern humans roughly coincided with the demise of Neanderthals in Western Europe and the Middle East 40,000 to 28,000 years ago. Although the two species apparently coexisted in some regions for thousands of years, we have little concrete evidence that they interacted.

One Origin or Many? *Homo erectus* apparently left Africa in waves between 1.5 million and 500,000 years ago. But when and where did modern humans first arise? Researchers use fossils and genetic data from contemporary human populations to address two competing hypotheses about this question.

According to the **African Emergence Hypothesis,** a population of *H. erectus* gave rise to several descendant species between 1.5 million and 0.5 million years ago. The early descendants, archaic humans, left Africa and established populations in the Middle East, Asia, and Europe. Some time later, 200,000 to 100,000 years ago, *H. sapiens* arose in Africa. These modern humans also migrated into Europe and Asia, and eventually drove the archaic humans to extinction. Thus, the African Emergence Hypothesis suggests that all modern humans are descended from a fairly recent African ancestor.

According to the **Multiregional Hypothesis,** populations of *H. erectus* and archaic humans had spread through much of Europe and Asia by 0.5 million years ago. Modern humans then evolved from archaic humans in many regions simultaneously. Although these geographically separated populations may have experienced some evolutionary differentiation (see Section 21.3), gene flow between them prevented reproductive isolation and maintained them as a single, but variable, species, *H. sapiens.*

Paleontological data do not clearly support either hypothesis. Some scientists argue that human remains with a mixture of archaic and modern characteristics confirm the Multiregional Hypothesis. In late 1998, for example, researchers in Portugal discovered a fossilized child that had been buried only 24,000 years ago, when only modern humans are thought to have occupied Europe. This fossil shows a surprising mix of Neanderthal and modern human traits, possibly indicating that the two groups interbred. On the other hand, recent finds in the Mideast indicate that Neanderthals and modern humans coexisted without interbreeding for 50,000 years. Thus, Neanderthals could not have been the ancestors of those modern humans.

Scientists also use DNA sequences from modern humans to evaluate the two hypotheses. In 1987, Rebecca Cann, Mark Stoneking, and Allan Wilson of the University of California at Berkeley and their colleagues published an analysis of mtDNA sequences from more than 100 ethnically diverse humans on four continents. They found that contemporary African populations contain the greatest variation in mtDNA. One explanation for this observation is that neutral mutations have been accumulating in African

UNANSWERED QUESTIONS

What causes the evolution of diversity?

In this chapter, you have read about the extensive diversity—in size, shape, color, structure, lifestyle, and habitats—of the vertebrate animals. Their mechanisms for maintaining themselves—including behaviors such as moving, feeding, reproducing, hiding, fighting, and sleeping—are equally varied. But nearly all vertebrates share some fundamental features of life. Recent research shows us that this generality is especially true of early development, but it is also apparent in many of the basic homeostatic mechanisms that vertebrates share—features of digestion and metabolism, respiration, and other characteristics regulated by the products of genetic networks and cascades. Biologists had long thought that once we understood the genetics of a variety of species, we would also understand the basis for their evolution and the maintenance of diversity. But now that a number of animal genomes have been sequenced completely, we have actually learned more about the genetic information that is shared by all animal species than we have about the genes that promote diversification. One of the great unanswered questions in biology, therefore, is how diversity arises and how it is maintained, given that so much genetic information is shared among even distantly related species. Equally important is why the same kinds of features evolved almost identically time after time in many unrelated lineages.

The ever-increasing body of genetic information is opening the "black box" of why there are so many species, and how they came to be. For example, we've long known that limbs evolved from fins, based on fossil evidence (which continues to accumulate, providing additional supportive evidence). But we haven't known what the *mechanisms* for forming a fin or a limb are, and how selection works to modify such structures. Now, with our knowledge of *Hox* and other tool-kit genes that control embryonic development, we understand how fins and limbs are formed. We can also experimentally manipulate development and perform selection experiments to test possible pathways of evolution—the "how they came to be."

Why do we see the recurrent, independent evolution of common themes?

A phenomenon that we often see, but don't yet fully understand, is why certain themes—such as body elongation, limblessness, and tooth modification—recur among distantly related vertebrates. The evolution of viviparity, live-bearing reproduction, is one such theme. In the vast majority of animals, reproduction occurs when a female lays her eggs in water and a male sprays sperm over them; typically both parents then abandon the fertilized eggs. But some species in many separate vertebrate lineages have evolved forms of viviparity. Cartilaginous and bony fishes exhibit diverse modes of embryonic nutrition, and in one group, the sea horses, it is the males that become pregnant. Amphibians also exhibit diverse patterns of viviparity: some have pregnant fathers and others have mothers that brood embryos in the skin of their backs, in their stomachs, or in their oviducts. And many squamate reptiles grow placentas similar to those of mammals. All mammals except monotremes are live-bearers with maternal nutrition. In fact, among the living vertebrate groups, birds are the only group that has not evolved the live-bearing habit. Can you think of some reasons why?

In some fishes, amphibians, and squamates, the mother supplies all the nutrition for the developing young through her investment in the egg; but in other species, females resorb their yolks and provide nutrients directly to offspring that are born as fully metamorphosed juveniles. Given that viviparity has evolved independently in approximately 200 vertebrate groups, it is not surprising that we see such a variety of reproductive patterns. Biologists are just beginning to understand the evolution of viviparity and to determine which patterns are shared among different evolutionary lineages and which are not. It appears that the hormonal basis for viviparity is similar in many groups, although the timing, the receptors involved, and the physiological responses vary. Researchers are now identifying candidate genes in the hope of unraveling the genetic networks that have fostered the different modes of viviparity. Ecological studies are revealing the interactions of potential selection regimes that influence reproductive modes. However, we still have much to learn about how genetics, development, physiology, and ecology interact in the evolution of diversity and common recurring themes.

 Marvalee H. Wake is a professor of the graduate school in the Department of Integrative Biology at the University of California at Berkeley. She studies vertebrate evolutionary morphology, development, and reproductive biology, with the goal of understanding evolutionary patterns and processes. To learn more about her research, go to http://ib.berkeley.edu/research/interests/research_profile.php?person=236.

Basic parts of the shoot system of an apple tree (*Malus domestica*), including leaves, stems, and vividly colored fruits.

© Mark Bolton/Corbis

31 The Plant Body

WHY IT MATTERS

Food, fibers for clothing, wood and other materials for construction, paper and inks, dozens of pharmaceuticals—these and many other essentials of modern human life derive from the parts of plants. In fact, members of the genus *Homo* have been depending on plant parts for their entire history. Fossil teeth discovered in the East African Rift Valley indicate that our early ancestors' diet likely included hard-shelled nuts, dry seeds, soft fruits, and leaves. By about 11,000 years ago humans were domesticating seed plants to provide stable food supplies. Directly or indirectly, leaves, stems, roots, flowers, seeds, and fruits of plants are the basic sources of energy for Earth's human inhabitants and all other animals as well **(Figures 31.1)**.

As you saw in Chapter 27, plants that made the transition from aquatic to terrestrial life did so only as adaptations in form and function helped solve problems posed by the terrestrial environment. These evolutionary adaptations included a shoot system that helps support leaves and other body parts in air, a root system that anchors the plant in soil and provides access to soil nutrients and water, tissues for internal transport of nutrients, and specializations for preventing water loss.

a. Wheat

b. Antelope feeding on leaves

c. Cedar waxwing consuming berries

Figure 31.1

Examples of plant parts that provide food for animals. **(a)** Mechanized harvesting of *Triticum* seeds, commonly known as wheat grains. **(b)** A pronghorn *(Antilocarpa americana)*, which consumes leaves and grasses. **(c)** Cedar waxwing *(Bombycilla cedrorum)* feeding on plump berries of *Sorbus americana*, the American mountain ash.

This unit surveys the structure and functioning of plants—their morphology, anatomy, and physiology. A plant's *morphology* is its external form, such as the shape of its leaves, and *anatomy* is the structure and arrangement of its internal parts. Plant *physiology* refers to the mechanisms by which the plant's body functions in its environment. Our focus is the plant phylum Anthophyta—angiosperms, or flowering plants—in terms of distribution and sheer numbers of species, the most successful plants on Earth.

31.1 Plant Structure and Growth: An Overview

Plants are photosynthetic autotrophs—"self-feeding" photosynthesizers that need sunlight and the carbon dioxide available in air as well as the water available in soil. In addition, many plants require nutrients that are usually available only in soil, and their aboveground parts may need the physical support of structures anchored in the ground. The evolutionary response to these challenges produced a plant body consisting of two closely linked but quite different components—a photosynthetic *shoot system* extending upward into the air and a nonphotosynthetic *root system* extending downward into the soil. Each system consists of various **organs**—body structures that contain two or more types of tissues and that have a definite form and function. Plant organs include leaves, stems, and roots, among others. A **tissue** is a group of cells and intercellular substances that function together in one or more specialized tasks.

In All Plant Tissues the Cells Share Some General Features

All plant cells share certain features, regardless of the tissue in which they reside. New plant cells develop a primary cell wall around the **protoplast**, the botanical term for the cell's cytoplasm, organelles, and plasma membrane. The primary wall contains cellulose, an insoluble polysaccharide made up of glucose subunits that is embedded in a matrix of other polysaccharides called hemicelluloses. This combination helps make the wall rigid but flexible. Pectin, another polysaccharide, is abundant in the primary wall and in the middle lamella, the layer between the primary walls of neighboring cells that helps bind cells together in tissues. (Plant pectin is often used to congeal jams and jellies.) As a young plant grows, the protoplast of many types of plant cells deposits additional cellulose and other materials inside the primary wall, forming a strong secondary cell wall (see Figure 5.25).

As in animals, all of a plant's cells have the same genes in their nuclei. As each cell matures and *differentiates* (becomes specialized for a particular function), specific genes become active. For the most part, fully differentiated animal cells perform their functions while alive, but some types of plant cells die after differentiating, and their protoplasts disappear. The walls that remain, however, serve key functions, particularly in vascular tissue.

The secondary cell walls of some plants contain lignin, a water-insoluble, inert polymer. **Lignification,** the deposition of lignin in cell walls, anchors the cellulose fibers in the walls, making them stronger and more rigid, and protects the other wall components from physical or chemical damage. Because water cannot penetrate and soften lignified cell walls, lignification also creates a waterproof barrier around the wall's cellulose strands. Many biologists believe that the evolution of large vascular plants became possible when certain cells developed biochemical pathways leading to lignification and could therefore become organized into watertight conducting channels.

Substances pass from one lignified cell to another through various routes. Solutes such as amino acids and sugars move in the plasmodesmata linking adjacent cells (see Figure 5.25). Water moves from cell to cell across *pits,* narrow regions where the secondary wall is absent and the primary wall is thinner and more porous than elsewhere.

Shoot and Root Systems Perform Different but Integrated Functions

A flowering plant's **shoot system** typically consists of stems, leaves, buds, and—during part of the plant's life cycle—reproductive organs known as flowers **(Figure 31.2).** A stem with its attached leaves and buds is a *vegetative* (nonreproductive) shoot; a bud eventually gives rise to an extension of the shoot or to a new, branching shoot. A *reproductive* shoot produces flowers, which later develop fruits containing seeds.

The shoot system is highly adapted for photosynthesis. Leaves greatly increase a plant's surface area and thus its exposure to light. Stems are frameworks for upright growth, which favorably position leaves for light exposure and flowers for pollination. Some parts of the shoot system also store carbohydrates manufactured during photosynthesis.

The **root system** usually grows belowground. It anchors the plant, and sometimes structurally supports its upright parts. It also absorbs water and dissolved minerals from soil and stores carbohydrates. Adaptations in the structure and function of plant cells and tissues were an integral part of the evolution of shoots and roots. For example, vascular tissues specialized to serve as internal pipelines conduct water, minerals, and organic substances throughout the plant. The root hairs sketched in Figure 31.2 are surface cells specialized for absorbing water from soil.

Meristems Produce New Tissues Throughout a Plant's Life

As you know from experience, animals generally grow to a certain size, and then their growth slows dramatically or stops. This pattern is called **determinate growth.** In contrast, a plant can grow throughout its life, a pattern called **indeterminate growth.** Individual plant parts such as leaves, flowers, and fruits exhibit determinate growth, but every plant also has self-perpetuating embryonic tissue at the tips of shoots and roots. Under the influence of plant hormones,

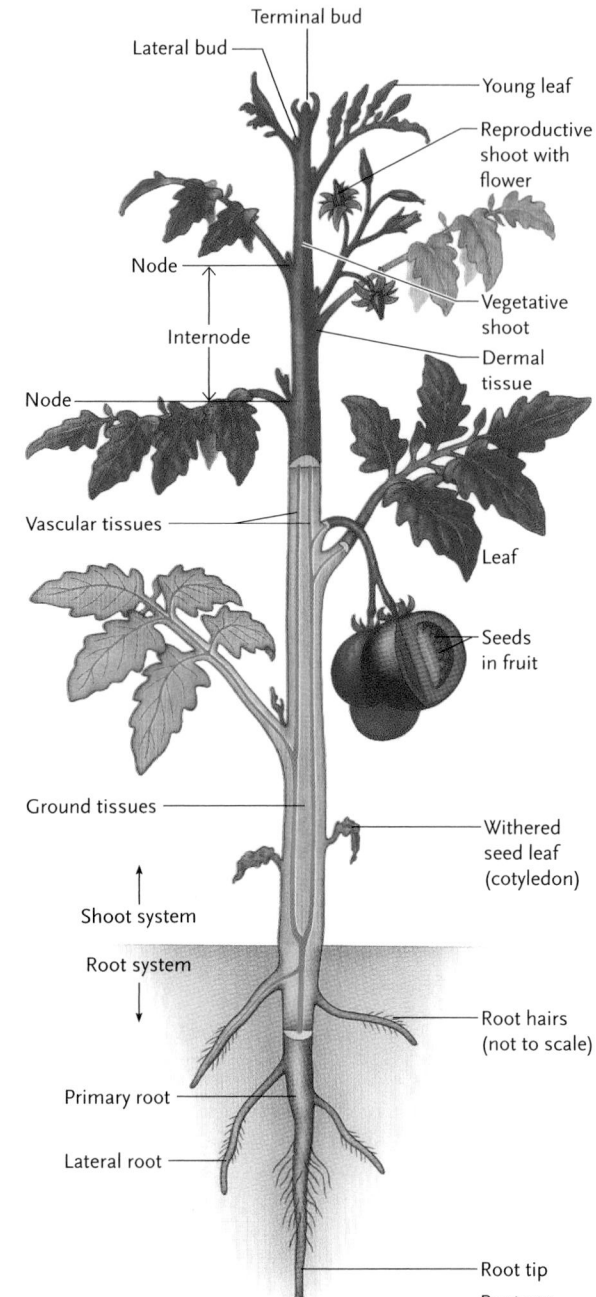

Figure 31.2
Body plan for the commercially grown tomato plant *Solanum lycopersicum,* a typical angiosperm. Vascular tissues (purple) conduct water, dissolved minerals, and organic substances. They thread through ground tissues, which make up most of the plant body. Dermal tissues (epidermis, in this case) cover the surfaces of the root and shoot systems.

these **meristems** (*merizein* = to divide) produce new tissues more or less continuously while a plant is alive. A capacity for indeterminate growth gives plants a great deal of flexibility—or what biologists often call *plasticity*—in their possible responses to changes in environmental factors such as light, temperature, water, and nutrients. This plasticity has major adaptive benefits for an organism that cannot move about, as most animals can. For example, if external factors change the direction of incoming light for photosyn-

a. Plants increase in length by cell divisions in apical meristems and by elongation of the daughter cells.

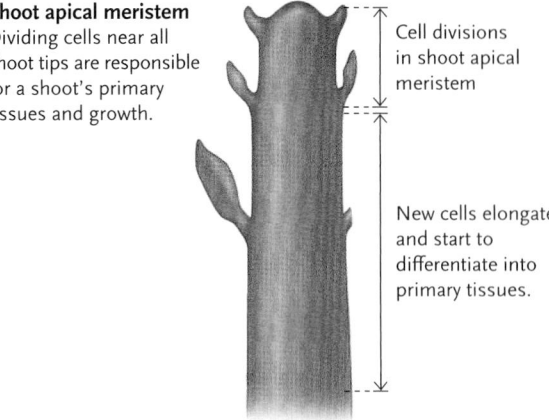

Shoot apical meristem
Dividing cells near all shoot tips are responsible for a shoot's primary tissues and growth.

Cell divisions in shoot apical meristem

New cells elongate and start to differentiate into primary tissues.

Root apical meristem
Dividing cells near all root tips are responsible for a root's primary tissues and growth.

New cells elongate and start to differentiate into primary tissues.

Cell divisions in root apical meristem

b. Some plants increase in girth by way of cell divisions in lateral meristems.

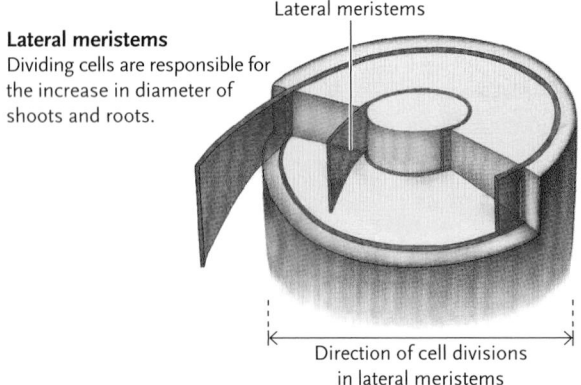

Lateral meristems

Lateral meristems
Dividing cells are responsible for the increase in diameter of shoots and roots.

Direction of cell divisions in lateral meristems

Figure 31.3
Approximate locations of types of meristems that are responsible for increases in the length and diameter of the shoots and roots of a vascular plant.

thesis, stems can "shift gears" and grow in that direction. Likewise, roots can grow outward toward water. These and other plant movements, called tropisms, are a major topic of Chapter 35.

As you know, animals grow mainly by mitosis, which increases the number of body cells. Plants, however, grow by two mechanisms—an increase in the number of cells by mitotic cell division in the meri-

stems, *and* an increase in the size of individual cells. In regions adjacent to the meristems in the tips of shoots and roots, the daughter cells rapidly increase in size—especially in length—for some time after they are produced. In contrast, when animal cells divide mitotically the daughter cells usually increase in size only a little.

Meristems Are Responsible for Growth in Both Height and Girth

Some plants have only one kind of meristem while others have two **(Figure 31.3)**. All vascular plants have **apical meristems,** clumps of self-perpetuating tissue at the tips of their buds, stems, and roots (see Figure 31.3a). Tissues that develop from apical meristems are called **primary tissues** and make up the **primary plant body.** Growth of the primary plant body is called **primary growth.**

Some species of plants—grasses and dandelions, for example—show only primary growth, which occurs at the tips of roots and shoots. Others, particularly plants that have a woody body, show **secondary growth,** which originates at self-perpetuating cylinders of tissue called **lateral meristems.** Secondary growth increases the diameter of older roots and stems (see Figure 31.3b). The tissues that develop from lateral meristems, called **secondary tissues,** make up the woody **secondary plant body** we see in trees and shrubs.

Primary and secondary growth can go on simultaneously in a single plant, with primary growth increasing the length of shoot parts and secondary growth adding girth. Each spring, for example, a maple tree undergoes primary growth at each of its root and shoot tips, while secondary growth increases the diameter of its older woody parts. Plant hormones govern these growth processes and other key events that are described in Chapter 35.

Monocots and Eudicots Are the Two General Structural Forms of Flowering Plants

As noted in Chapter 27, several broad categories of body architecture arose as flowering plants evolved. The two major ones are the **monocot** and **eudicot** lineages. Grasses, daylilies, irises, cattails, and palms are examples of monocots. Eudicots include nearly all familiar angiosperm trees and shrubs, as well as many nonwoody (herbaceous) plants. Examples are maples, willows, oaks, cacti, roses, poppies, sunflowers, and garden beans and peas.

Monocots and eudicots, recall, differ in the number of *cotyledons*—the seed leaves associated with plant embryos. Monocot seeds have one cotyledon and eudicot seeds have two. Although monocots and eudicots have similar types of tissues, their body structures differ in distinctive ways **(Table 31.1)**. As we discuss the morphology of flowering plants, we will refer frequently to these structural differences.

Flowering Plants Can Be Grouped according to Type of Growth and Lifespan

In evolutionary terms, the distinction between monocot and eudicot flowering plants is most important structurally and developmentally. Yet botanists sometimes use other criteria to distinguish between flowering plants—for example, by whether they have secondary growth. Most monocots and some eudicots are *herbaceous* plants, showing little or no secondary growth during their life cycle. In contrast, many eudicots (and all gymnosperms) are *woody* plants, which do have secondary growth.

We can also distinguish plants by lifespan. **Annuals** are herbaceous plants in which the life cycle is completed in one growing season. With minimal or no secondary growth, annuals typically have only apical meristems. Examples are marigolds (a eudicot) and corn (a monocot). **Biennials** complete the life cycle in two growing seasons, and limited secondary growth occurs in some species. Roots, stems, and leaves form in the first season, then the plant flowers, forms seeds, and dies in the second. Examples are carrots and celery (eudicots). In **perennials**, vegetative growth and reproduction continue year after year. Many perennials, such as trees, shrubs, and some vines, have secondary tissues, although others, such as irises and daffodils, do not.

STUDY BREAK

1. Compare and contrast the components and functions of a land plant's shoot and root systems.
2. Explain what meristem tissue is, and name and describe the functions of the basic types of meristems.

31.2 The Three Plant Tissue Systems

Plants develop three tissue systems that provide the foundation for the various plant organs. The **ground tissue system**, which makes up most of the plant body, functions in metabolism, storage, and support. The **vascular tissue system** consists of various tubes that transport water and nutrients throughout the plant. The tubes are organized in bundles that are dispersed through the ground tissues. The **dermal tissue system** serves as a skinlike protective covering for the plant body. Figure 31.2 shows the general location of each system in the shoot and root.

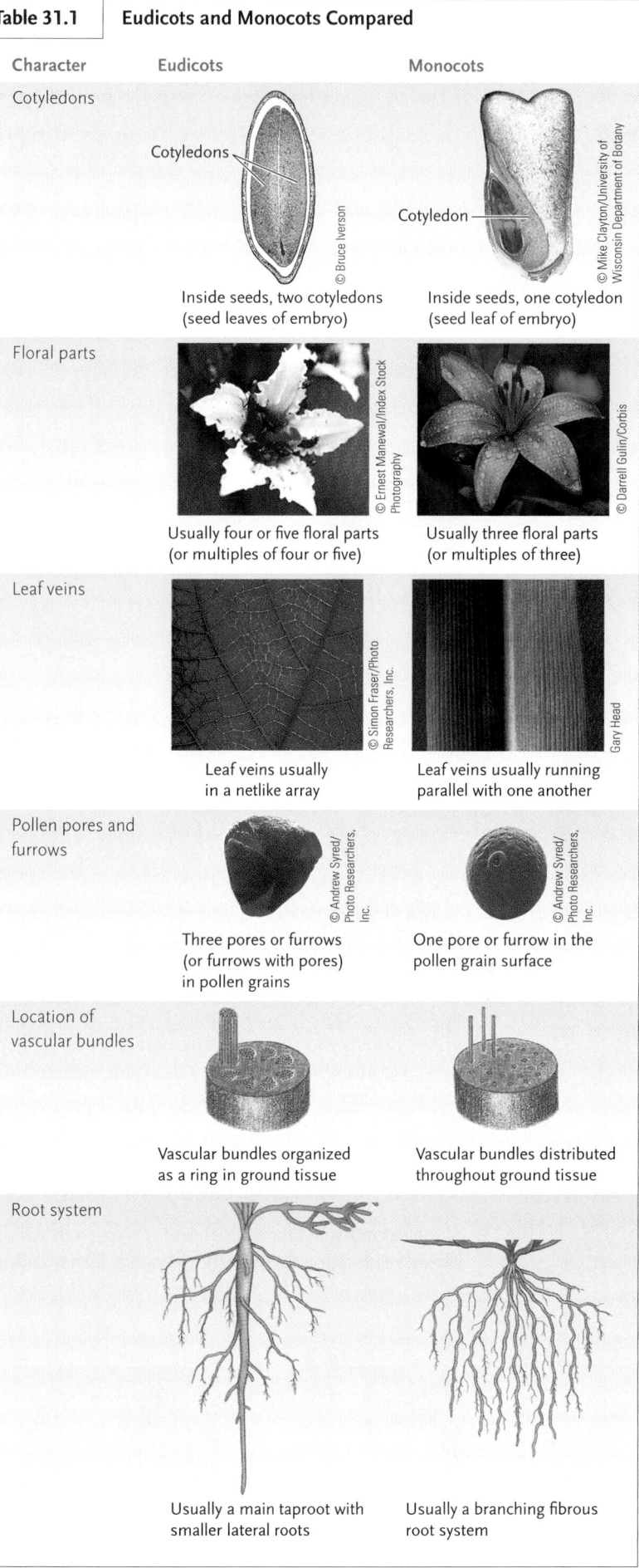

Table 31.1 **Eudicots and Monocots Compared**

Character	Eudicots	Monocots
Cotyledons	Inside seeds, two cotyledons (seed leaves of embryo)	Inside seeds, one cotyledon (seed leaf of embryo)
Floral parts	Usually four or five floral parts (or multiples of four or five)	Usually three floral parts (or multiples of three)
Leaf veins	Leaf veins usually in a netlike array	Leaf veins usually running parallel with one another
Pollen pores and furrows	Three pores or furrows (or furrows with pores) in pollen grains	One pore or furrow in the pollen grain surface
Location of vascular bundles	Vascular bundles organized as a ring in ground tissue	Vascular bundles distributed throughout ground tissue
Root system	Usually a main taproot with smaller lateral roots	Usually a branching fibrous root system

Table 31.2 — Summary of Flowering Plant Tissues and Their Components

Tissue System	Name of Tissue	Cell Types in Tissue	Tissue Function
Ground tissue	Parenchyma	Parenchyma cells	Photosynthesis, respiration, storage, secretion
	Collenchyma	Collenchyma cells	Flexible strength for growing plant parts
	Sclerenchyma	Fibers or sclereids	Rigid support, deterring herbivores
Vascular tissue	Xylem	Conducting cells (tracheids, vessel members); parenchyma cells; sclerenchyma cells	Transport of water and dissolved minerals
	Phloem	Conducting cells (sieve tube members); parenchyma cells; sclerenchyma cells	Sugar transport
Dermal tissue	Epidermis	Undifferentiated cells; guard cells and other specialized cells	Control of gas exchange, water loss; protection
	Periderm	Cork; cork cambium; secondary cortex	Protection

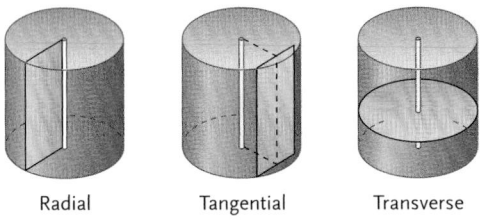

Radial   Tangential   Transverse

Figure 31.4

Terms that identify how tissue specimens are cut from a plant. Along the radius of a stem or root, longitudinal cuts give radial sections. Cuts at right angles to a root or stem radius give tangential sections. Cuts perpendicular to the long axis of a stem or root give transverse sections (cross sections).

Figure 31.5

Locations of ground, vascular, and dermal tissues in one kind of plant stem, transverse section. Ground tissues are simple tissues while vascular and dermal tissues are complex, containing various types of specialized cells. (Micrograph: James D. Mauseth, Plant Anatomy, Benjamin Cummings, 1988.)

Each tissue system includes several types of tissue, and each tissue is made up of cells with specializations for different functions **(Table 31.2)**. *Simple* tissues have only one type of cell. Other tissues are *complex,* with organized arrays of two or more types of cells. **Figure 31.4** will help you interpret micrographs of plant tissues, beginning with the tissues in a transverse section of a stem shown in **Figure 31.5**.

Ground Tissues Are All Structurally Simple, but They Exhibit Important Differences

Plants have three types of ground tissue, each with a distinct structure and function—*parenchyma, collenchyma,* and *sclerenchyma* **(Figure 31.6)**. Each type is

structurally simple, being composed mainly of one kind of cell. In a very real sense the cells in ground tissues are the "worker bees" of plants, carrying out photosynthesis, storing carbohydrates, providing mechanical support for the plant body, and performing other basic functions. Each kind of cell has a distinctive wall structure, and some have variations in the protoplast as well.

Parenchyma: Soft Primary Tissues. Parenchyma (*para* = around, *chein* = fill in, or pour) makes up the bulk of the soft, moist primary growth of roots, stems, leaves, flowers, and fruits. Most parenchyma cells have only a thin primary wall and so are pliable and permeable to water. Often the cells are spherical or many-sided, although they also can be elongated like a sausage, as in Figure 31.6a. Parenchyma cells sometimes have air spaces between them, especially in leaves (see Figure 31.17). The air spaces may be sizeable in the stems and leaves of aquatic plants, such as water lilies. This adaptation facilitates the movement of oxygen from aerial leaves and stems to submerged parts of the plant, and it also helps the leaves float upward toward the light.

Parenchyma cells may be specialized for tasks as varied as storage, secretion, and photosynthesis. They can occur both as part of parenchyma tissue and as individual cells in other tissues. In many plant species, modified parenchyma cells are specialized for short-distance transport of solutes. Such cells are common in tissues in which water and solutes must be rapidly moved from cell to cell—for example, in vascular tissues and in tissues that secrete nectar. Parenchyma cells usually remain alive and,

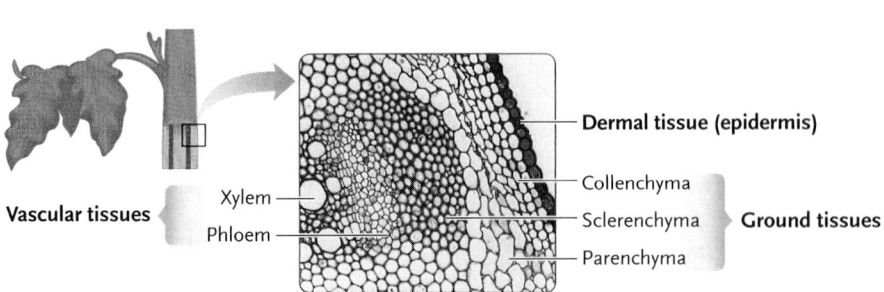

Vascular tissues { Xylem, Phloem }

Dermal tissue (epidermis)

Collenchyma
Sclerenchyma } **Ground tissues**
Parenchyma

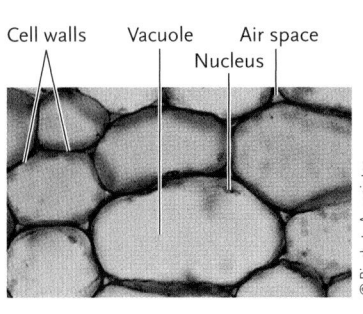

Cell walls   Vacuole   Air space
        Nucleus

a. Parenchyma tissues consist of soft, living cells specialized for storage, other functions.

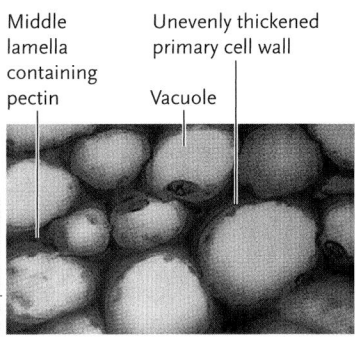

Middle lamella containing pectin   Unevenly thickened primary cell wall

Vacuole

b. Collenchyma tissues provide flexible support.

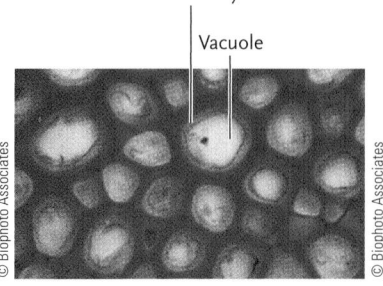

Thick secondary wall

Vacuole

c. Sclerenchyma tissues provide rigid support and protection.

Figure 31.6

Examples of ground tissues from the stem of a sunflower plant (*Helianthus annuus*).

© Biophoto Associates

---

when mature, retain the capacity to divide; in fact, their mitotic divisions often heal wounds in plant parts.

Collenchyma: Flexible Support. The "strings" in celery are examples of the flexible ground tissue called **collenchyma** (see Figure 31.6b), which helps strengthen plant parts that are still elongating (*kolla* = glue). Collenchyma cells are typically elongated, and collectively they often form strands or a sheathlike cylinder under the dermal tissue of growing shoot regions and the stalklike petioles that attach leaves to stems.

The primary walls of collenchyma cells are built of alternating layers of cellulose and pectin. These walls thicken and stretch as the cell enlarges. Mature collenchyma cells are alive and metabolically active, and they continue to synthesize cellulose and pectin layers as the plant grows.

Sclerenchyma: Rigid Support and Protection. Mature plant parts gain additional mechanical support and protection from **sclerenchyma** (*skleros* = hard). The cells of this ground tissue develop thick secondary walls (see Figure 31.6c), which commonly are lignified and perforated by pits through which water can pass. After mature sclerenchyma cells become encased in lignin, they die because their protoplasts can no longer exchange gases, nutrients, and other materials with the environment. The walls, however, continue to provide protection and support.

The two types of sclerenchyma cells—*sclereids* and *fibers*—differ in their shape and arrangement. Sheet-like arrays of rigid **sclereids** form a protective coat around seeds; examples are the hard casings of a coconut shell or peach pit. Sclereids come in a range of shapes; the gritty texture of a pear comes from the roughly cube-shaped sclereids scattered through its flesh **(Figure 31.7a).** The long, tapered cells called **fibers (Figure 31.7b)** resist stretching, but are more pliable than sclereids. Fibers in the stems of flax plants are massed in parallel; they can flex and twist without stretching and are used to manufacture rope, paper, and linen cloth.

Vascular Tissues Are Specialized for Conducting Fluids

Vascular tissues are complex tissues composed of specialized conducting cells, parenchyma cells, and fibers. *Xylem* and *phloem,* the two kinds of vascular tissues in flowering plants, are organized into bundles of interconnected cells that extend throughout the plant.

Xylem: Transporting Water and Minerals. Xylem (*xylon* = wood) conducts water and dissolved minerals absorbed from the soil upward from a plant's roots to the shoot. As you read in Chapter 27, xylem was a key early adaptation allowing plants to make the transition to life on land. Xylem contains two types of conducting cells: *tracheids* and *vessel members.* Both develop thick, lignified secondary cell walls and die at maturity. The empty cell walls of abutting cells serve as pipelines for water and minerals.

Tracheids are elongated, with tapered, overlapping ends **(Figure 31.8a).** In plants adapted to drier soil conditions, they have strong secondary walls that keep them from collapsing when less water is present. As in sclerenchyma, water can move from cell to cell through

a. Sclereids   **b.** Fibers

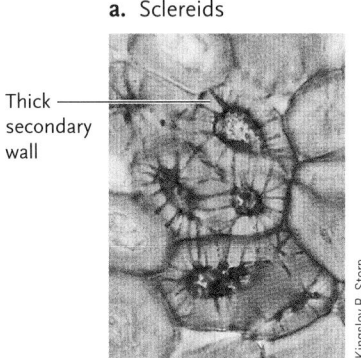

Thick secondary wall

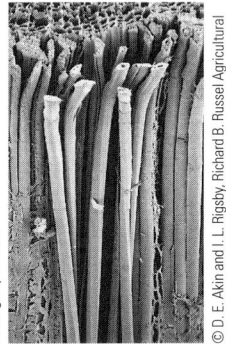

© Kingsley R. Stern

© D. E. Akin and I. L. Rigsby, Richard B. Russel Agricultural Research Service, U. S. Department of Agriculture, Athens, Georgia

Figure 31.7

Examples of sclerenchyma cells. **(a)** From the flesh of a pear (*Pyrus*), one type of sclereid: stone cells, each with a thick, lignified wall. **(b)** Strong fibers from stems of a flax plant (*Linum*).

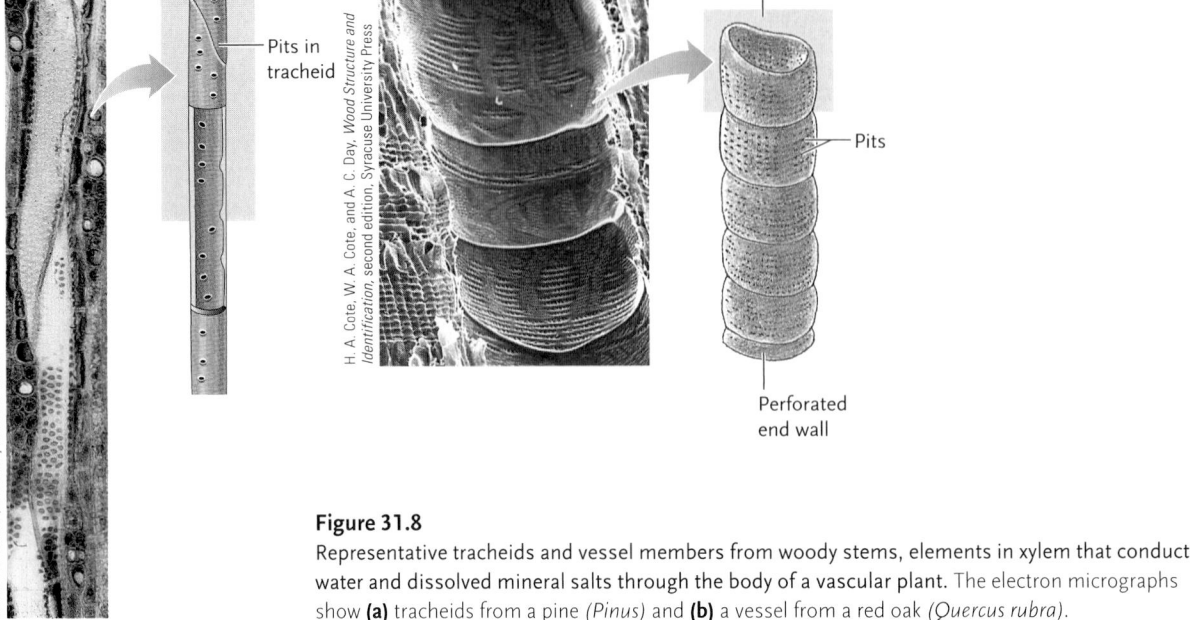

a. Tracheids, tangential section

Pits in tracheid

b. Part of a vessel

One vessel member

Pits

Perforated end wall

Alison W. Roberts, University of Rhode Island

H. A. Cote, W. A. Cote, and A. C. Day, *Wood Structure and Identification*, second edition, Syracuse University Press

Figure 31.8

Representative tracheids and vessel members from woody stems, elements in xylem that conduct water and dissolved mineral salts through the body of a vascular plant. The electron micrographs show **(a)** tracheids from a pine *(Pinus)* and **(b)** a vessel from a red oak *(Quercus rubra)*.

pits. Usually, a pit in one cell is opposite a pit of an adjacent cell, so water seeps laterally from tracheid to tracheid.

Vessel members (or vessel elements) are shorter cells joined end to end in tubelike columns called vessels **(Figure 31.8b). Vessels** are typically several centimeters long, and in some vines and trees they may be many meters long. Like tracheids, vessel members have pits; however, they also have another adaptation that greatly enhances water flow. As vessel members mature, enzymes break down portions of their end walls, producing perforations. Some vessel members have a single, large perforation, so that the end is completely open (see Figure 31.8b). Others have a cluster of small, round perforations, or ladderlike bars, extending across the open end (see Figure 31.8). The predictability of the perforation patterns suggests that this process is under precise genetic control.

Fossil evidence shows that the forerunners of modern plant species relied solely on tracheids for water transport, and today ferns and most gymnosperms still have only tracheids. Nearly all angiosperms and a few other types of plants have both tracheids and vessel members, however, which confers an adaptive advantage. Flowing water sometimes incorporates air bubbles, which represent a potentially lethal threat to the plant. Water can flow rapidly through vessel members that are linked end to end, but the open channel cannot prevent air bubbles from forming and possibly blocking the flow through the whole vessel. By contrast, even though water moves more slowly in tracheids, the pit membranes are impermeable to air bubbles, and a bubble that forms in

one tracheid stays there; water continues to move between other tracheids.

Conducting cells that form after the surrounding tissue has reached its maximum size have complete secondary walls, with pits or perforations. In growing plants, however, only a partial secondary wall forms in cells of xylem, so the cell can elongate as the tissue it services grows. At maturity, tracheids and vessel members die as genetic cues cause their protoplasts to degenerate and lignin to be deposited in cell walls.

Parenchyma cells in xylem participate in the transport of minerals through vessel members and tracheids. Sclerenchyma fibers function like steel cables in concrete, helping keep the tissue fairly rigid and lending structural support to the plant.

Phloem: Transporting Sugars and Other Solutes. The vascular tissue **phloem** (*phloios* = tree bark) transports solutes, notably the sugars made in photosynthesis, throughout the plant body. The main conducting cells of phloem are **sieve tube members (Figure 31.9),** which connect end to end, forming a **sieve tube.** As the name implies, their end walls, called sieve plates, are studded with pores. In flowering plants the phloem is strengthened by fibers and sclereids.

Immature sieve tube members contain the usual plant organelles. Over time, however, the cell nucleus and internal membranes in plastids break down, mitochondria shrink, and the cytoplasm is reduced to a thin layer lining the interior surface of the cell wall. Even without a nucleus, the cell lives up to several years in most plants, and much longer in some trees.

a. Sieve-tube members

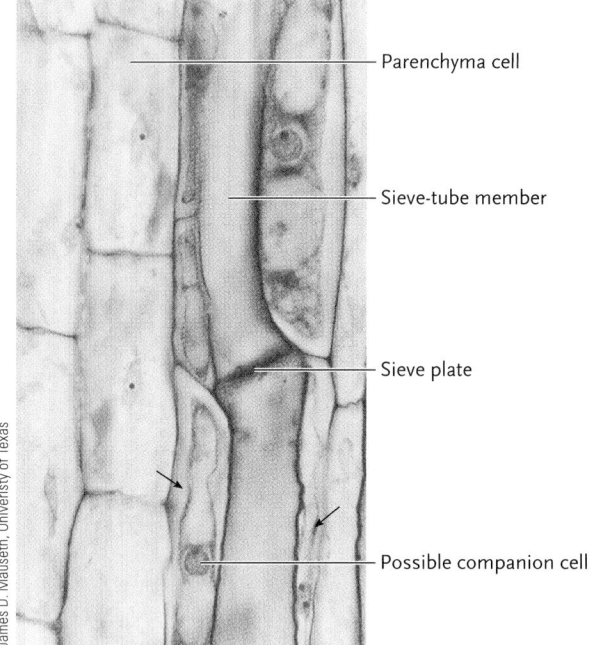

Parenchyma cell

Sieve-tube member

Sieve plate

Possible companion cell

James D. Mauseth, University of Texas

b. Sieve plate

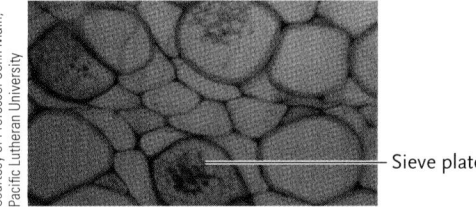

Sieve plate

Courtesy of Professor John Main, Pacific Lutheran University

Figure 31.9

Structure of sieve tube members. **(a)** Micrograph showing sieve tube members in longitudinal section. The arrows point to cells that may be companion cells. Long tubes of sieve tube members conduct sugars and other organic compounds. **(b)** Sieve plate in a cell in phloem, cross section.

In many flowering plants, specialized parenchyma cells known as **companion cells** are connected to mature sieve tube members by plasmodesmata. Unlike sieve tube members, companion cells retain their nucleus when mature. They assist sieve tube members both with the uptake of sugars and with the unloading of sugars in tissues engaged in food storage or growth. They may also help regulate the metabolism of mature sieve tube members. We return to the functions of phloem cells in Chapter 32.

Dermal Tissues Protect Plant Surfaces

A complex tissue called **epidermis** covers the primary plant body in a single continuous layer **(Figure 31.10a)** or sometimes in multiple layers of tightly packed cells. The external surface of epidermal cell walls is coated with waxes that are embedded in cutin, a network of chemically linked fats. Epidermal cells secrete this coating, or **cuticle**, which resists water loss and helps fend off attacks by microbes. A cuticle coats all plant parts except the very tips of the shoot and most absorptive parts of roots; other root regions have an extremely thin cuticle.

Most epidermal cells are relatively unspecialized, but some are modified in ways that represent important adaptations for plants. Young stems, leaves, flower parts, and even some roots have pairs of crescent-shaped **guard cells (Figure 31.10b)**. Unlike other cells of the epidermis, guard cells contain chloroplasts and so can carry out photosynthesis. The pore between a pair of guard cells is termed a **stoma** (plural, *stomata*). Water vapor, carbon dioxide, and oxygen move across the epidermis through the stomata, which open and close by way of mechanisms we consider in Chapter 32.

With their exact spacing and vital role in regulating the exchange of gases between a plant and its environment, stomata have captured the interest of

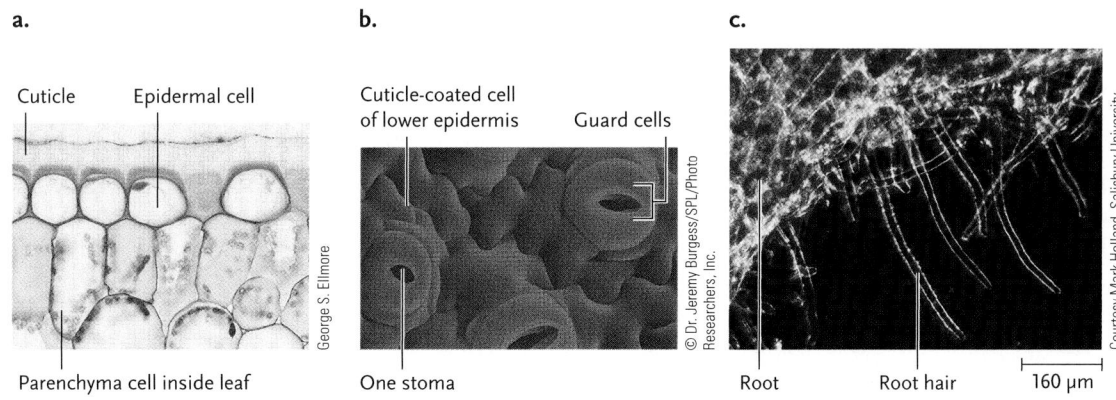

a.

Cuticle   Epidermal cell

Parenchyma cell inside leaf

George S. Ellmore

b.

Cuticle-coated cell of lower epidermis   Guard cells

One stoma

© Dr. Jeremy Burgess,/SPL/Photo Researchers, Inc.

c.

Root   Root hair   160 μm

Courtesy Mark Holland, Salisbury University

Figure 31.10

Structure and examples of epidermal tissue. **(a)** Cross section of leaf epidermis from a bush lily *(Clivia miniata)*. **(b)** Scanning electron micrograph of a leaf surface, showing cuticle-covered epidermal cells and stomata. **(c)** Root hairs, an epidermal specialization.

Shaping up Flower Color

Different pigments in flowers produce different colors, but are pigments the whole story? A molecular study of flower color in snapdragons (*Antirrhinum majus*) provided a surprising answer. Kenichi Noda of the Nippon Oil Company in Japan and Beverly J. Glover and her colleagues at the John Innes Institute in England were interested in a mutant snapdragon called *mixta*, which produces pale red flower petals with a dull, flat surface rather than the deep red, velvety petals of wild-type plants **(Figure a)**.

Through a series of steps, the investigators isolated and cloned the *mixta* gene. Sequencing the gene revealed close similarities to a regulatory gene that activates genes in some other plants. The similarities suggested that the normal snapdragon gene also codes for a regulatory protein that produces normal flower color. When the transposable element inserts in the gene, the regulatory protein is lost.

How does the regulatory protein govern flower color? At first Noda

Figure a
Wild-type snapdragon, which has flowers with deep red petals.

and his colleagues thought it regulated production of anthocyanin, a pigment that gives flowers a red color, and that loss of the protein in mutant plants hampered anthocyanin production. They discarded this hypothesis when both the wild-type and *mixta* plants were found to have normal levels of anthocyanin. However, microscopic examination of flower petals revealed that wild-type and *mixta* epidermal cells are shaped differently. Normal plants have conical epidermal cells, with the tip of the cone pointing outward and giving the petals a velvety appearance. Epidermal cells of *mixta* mutants have a flat, irregular surface that produces a dull appearance. **Figure b** shows the surface of a variegated flower petal, which has both conical and flat cells. This structural difference suggested that in wild-type petals, the cone tips act as prisms that make the red pigment

Flat cells

Conical cells

Figure b
Scanning electron micrograph of the surface of a variegated snapdragon petal, showing conical cells in the bright red colored areas and flat cells in the pale colored areas. The genetic events that produce variegated petals were clues that helped the research team identify the *mixta* gene.

clearly visible, while the irregular surface of the mutant cells scatters light and masks the pigment color.

As a test, the research team removed the cell walls from the epidermal petal cells, a step that eliminated differences in cell shape. Both the normal and mutant cells had the same intense, red color. On this basis, the researchers proposed that the regulatory protein encoded in the normal *mixta* gene activates other genes whose protein products in some way produce the conical cell shape.

geneticists probing the molecular underpinnings of plant development. Working with *Arabidopsis thaliana* (thale cress) plants, researchers have identified an enzyme—encoded by the gene *YDA*—that appears to ultimately control where and how many stomata form. In mutant plants with a defective enzyme, the epidermis is blanketed with stomata packed side by side. The plants often die early in development or are stunted and appear fuzzy—hence the enzyme's name, YODA, recalling the short, hairy Star Wars character. In nonmutated wild-type plants, unequal divisions of precursor cells produce one smaller and one larger daughter cell, and the smaller one gives rise to the two guard cells of a stoma. (The larger cell either divides again or becomes an underlying epidermal cell.) YODA comes into play when a series of precursor reactions phosphorylate it. The activated enzyme then triggers a cascade of reactions that, by some as-yet-unknown mechanism, either promote or restrict these asymmetric divisions.

Other epidermal specializations are the single-celled or multicellular outgrowths collectively called **trichomes,** which give the stems or leaves of some plants a hairy appearance. Some trichomes exude sugars that attract insect pollinators. Leaf trichomes of *Urtica,* the stinging nettle, provide protection by injecting an irritating toxin into the skin of animals that brush against the plant or try to eat it. **Root hairs,** which develop as extensions in the outer wall of root epidermal cells **(Figure 31.10c),** are also trichomes. Root hairs absorb much of a plant's water and minerals from the soil.

The epidermal cells of flower petals (which are modified leaves) synthesize pigments that are partly responsible for a blossom's colors. However, molecular studies have revealed that flower colors and their intensity or brightness also depend on the shape of the epidermal cells, as described in *Insights from the Molecular Revolution*.

STUDY BREAK

1. Describe the defining features, cellular components, and functions of the ground tissue system.
2. What are the functions of xylem and phloem?
3. What are the cellular components and functions of the dermal tissue system?

31.3 Primary Shoot Systems

A young flowering plant's shoot system consists of the main stem, leaves, and buds as well as flowers and fruits. Chapter 34 looks more closely at flowers and fruits; here we focus on the growth and organization of stems, buds, and leaves of the primary shoot system.

Stems Are Adapted to Provide Support, Routes for Vascular Tissues, Storage, and New Growth

Stems are structurally adapted for four main functions. First, they provide mechanical support, generally along a vertical (upright) axis, for body parts involved in growth, photosynthesis, and reproduction. These parts include meristematic tissues, leaves, and flowers. Second, they house the vascular tissues (xylem and phloem), which transport products of photosynthesis, water and dissolved minerals, hormones, and other substances throughout the plant. Third, they often are modified to store water and food. And finally, buds and specific stem regions contain meristematic tissue that gives rise to new cells of the shoot.

The Modular Organization of a Stem. A plant stem develops in a pattern that divides the stem into modules, each consisting of a *node* and an *internode*. A **node** is a place on the stem where one or more leaves are attached; the area between two nodes is thus an **internode**. The upper angle between the stem and an attached leaf is an **axil**. New primary growth occurs in buds—a **terminal bud** at the apex of the main shoot, and **lateral buds**, which produce branches (lateral shoots), in the leaf axils. Meristematic tissue in buds gives rise to leaves, flowers, or both **(Figure 31.11)**.

In eudicots, most growth in a stem's length occurs directly below the apical meristem, as internode cells divide and elongate. Internode cells nearest the apex are most active, so the most visible new growth occurs at the ends of stems. In grasses and some other monocots, by contrast, the upper cells of an internode stop dividing as the internode elongates, and cell divisions are limited to a meristematic region at the base of the internode. The stems of bamboo and other grasses elongate as the internodes are "pushed up" by the growth of such meristems. This adaptation allows grasses to grow back readily after grazing by herbivores (or being chopped off by a lawnmower), because the meristem is not removed.

Terminal buds release a hormone that inhibits the growth of nearby lateral buds, a phenomenon called **apical dominance.** Gardeners who want a bushier plant can stimulate lateral bud growth by periodically cutting off the terminal bud. The flow of hormone signals then dwindles to a level low enough that lateral buds begin to grow. In nature, apical dominance is an adaptation that directs the plant's resources into growing up toward the light.

a. Location of nodes and buds

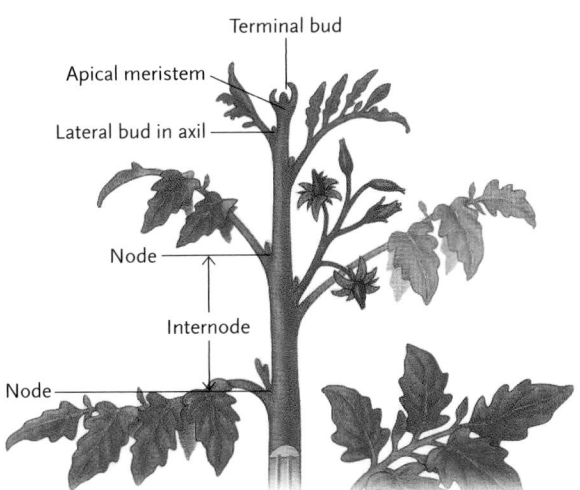

Terminal bud
Apical meristem
Lateral bud in axil
Node
Internode
Node

b. Leaves at a terminal bud

Jakub Jasinski/Visuals Unlimited

Figure 31.11
Modular structure of a stem. **(a)** The arrangement of nodes and buds on a plant stem. **(b)** Formation of leaves at a terminal bud of a dogwood (genus *Cornus*).

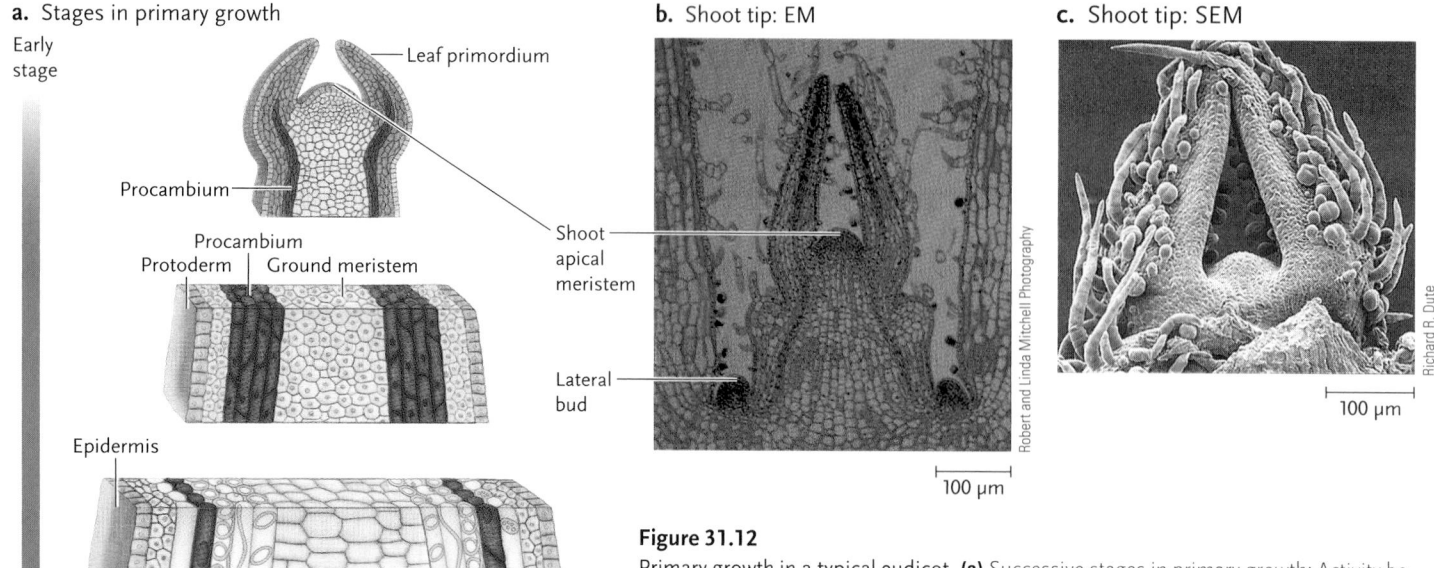

a. Stages in primary growth

Early stage

Leaf primordium

Procambium

Procambium
Protoderm | Ground meristem

Shoot apical meristem

Lateral bud

Epidermis

Later stage

Cortex | Pith | Primary phloem
Procambium | Primary xylem

b. Shoot tip: EM

Robert and Linda Mitchell Photography

100 μm

c. Shoot tip: SEM

Richard R. Dute

100 μm

Figure 31.12

Primary growth in a typical eudicot. **(a)** Successive stages in primary growth: Activity begins at the shoot apical meristem and continues at the primary meristems derived from it. Notice the progressive differentiation of most of the tissue regions. **(b)** Light micrograph of a *Solenostemon* shoot tip, cut longitudinally through its center. **(c)** Scanning electron micrograph of its surface.

Primary Growth and Structure of a Stem. Primary growth, the cell divisions and enlargement that produce the primary plant body, begins in the shoot and root apical meristems. The sequence of events is similar in roots and shoots; it is shown for a eudicot shoot in **Figure 31.12.**

The shoot apical meristem is a dome-shaped mass of cells. When one cell divides, one of its daughter cells becomes an **initial,** a cell that remains as part of the meristem. The other daughter cell becomes a **derivative.** The derivative typically divides once or twice and then enters on the path to differentiation. When initials divide, they replenish the supply of derivatives in the meristem.

As derivatives differentiate, they give rise to three **primary meristems:** *protoderm, procambium,* and *ground meristem* (see Figure 31.12). These primary meristems are relatively unspecialized tissues with cells that differentiate in turn into specialized cells and tissues. In eudicots, the primary meristems are also responsible for elongation of the plant body.

How do the genetically identical cells of an apical meristem give rise to three types of primary meristem cells, and ultimately to all the specialized cells of the plant? *Focus on Research* describes some experiments that are probing the genetic mechanisms underlying meristem activity.

Each primary meristem occupies a different position in the shoot tip, as shown in Figure 31.12a. Outermost is **protoderm,** a meristem that will produce the stem's epidermis. While protoderm cells divide and the resulting derivatives are maturing, the shoot tip continues to grow. Eventually, the protoderm cells differentiate into specific types of epidermal cells, including

guard cells and trichomes. Some monocots, such as palms, have a primary thickening meristem just under the protoderm; this tissue contributes to both lateral growth and elongation of the stem.

Inward from the protoderm is the **ground meristem,** which will give rise to ground tissue, most of it parenchyma. **Procambium,** which produces the primary vascular tissues, is sandwiched between ground meristem layers. Procambial cells are long and thin, and their spatial orientation foreshadows the future function of the tissues they produce. In most plants, inner procambial cells give rise to xylem and outer procambial cells to phloem. In plants with secondary growth, a thin region of procambium between the primary xylem and phloem remains undifferentiated. Later on it will give rise to the lateral meristems.

The developing vascular tissues become organized into **vascular bundles,** multistranded cords of primary xylem and phloem that are wrapped in sclerenchyma and thread lengthwise through the parenchyma. In the stems and roots of most eudicots and some conifers, the vascular bundles form a **stele** (Greek *stele* = pillar), also known as a *vascular cylinder,* that vertically divides the column of ground tissue into an outer **cortex** and an inner **pith (Figure 31.13a).** Both cortex and pith consist mainly of parenchyma; in some plant species the pith parenchyma stores starch reserves. In the stems of most monocots, vascular bundles are dispersed through the ground tissue **(Figure 31.13b),** so separate cortical and pith regions do not form. In some monocots, including bamboo, the pith breaks down, leaving the stem with a hollow core. The hollow stems of certain hard-walled bamboo species are used to make bamboo flutes.

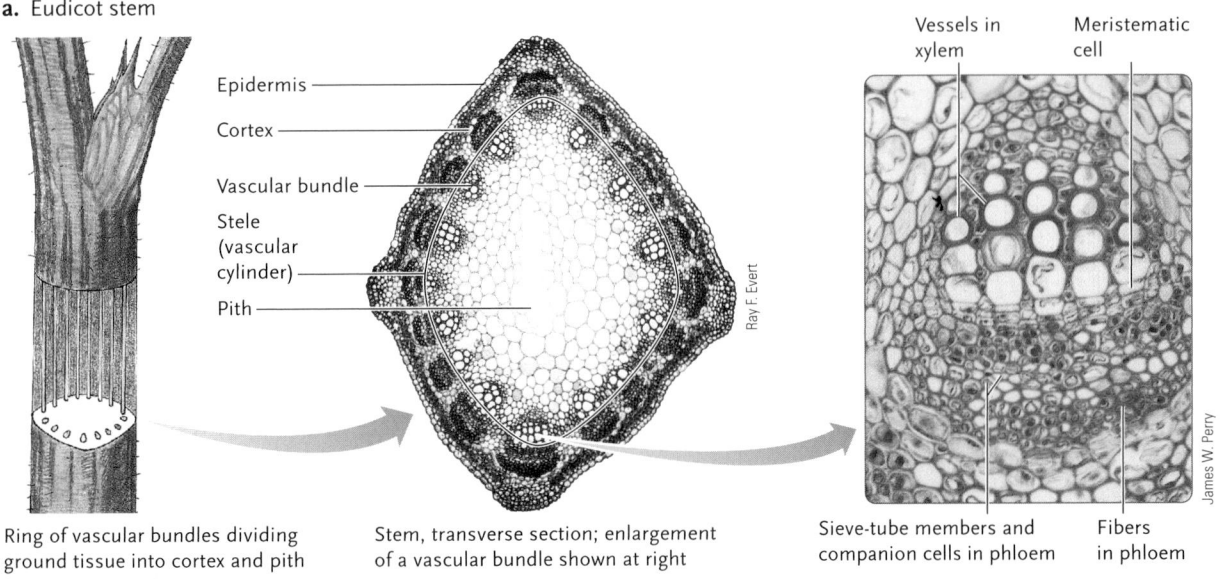

a. Eudicot stem

Epidermis

Cortex

Vascular bundle

Stele (vascular cylinder)

Pith

Ray F. Evert

Ring of vascular bundles dividing ground tissue into cortex and pith

Stem, transverse section; enlargement of a vascular bundle shown at right

Vessels in xylem

Meristematic cell

James W. Perry

Sieve-tube members and companion cells in phloem

Fibers in phloem

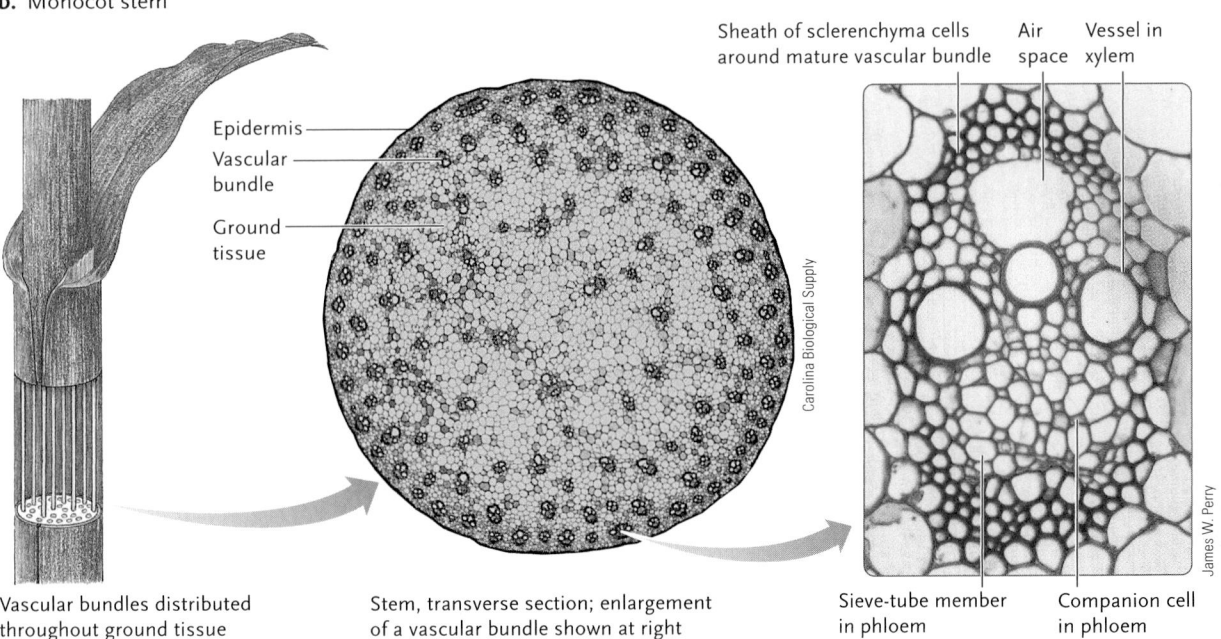

b. Monocot stem

Epidermis

Vascular bundle

Ground tissue

Carolina Biological Supply

Vascular bundles distributed throughout ground tissue

Stem, transverse section; enlargement of a vascular bundle shown at right

Sheath of sclerenchyma cells around mature vascular bundle

Air space

Vessel in xylem

James W. Perry

Sieve-tube member in phloem

Companion cell in phloem

Figure 31.13

Organization of cells and tissues inside the stem of a eudicot and a monocot. **(a)** Part of a stem from alfalfa *(Medicago),* a eudicot. In many species of eudicots and conifers, the vascular bundles develop in a more or less ringlike array in the ground tissue system, as shown here. **(b)** Part of a stem from corn *(Zea mays),* a monocot. In most monocots and some herbaceous eudicots, vascular bundles are scattered through the ground tissue, as shown here.

As leaves and buds appear along a stem, some vascular bundles in the stem branch off into these developing tissues. The arrangement of vascular bundles in a plant ultimately depends on the number of branch points to leaves and buds and on the number and distribution of leaves.

Stem Modifications. Evolution has produced a range of stem specializations, including structures modified for reproduction, food storage, or both **(Figure 31.14).** An onion or a garlic head is a *bulb,* a modified shoot that consists of a bud with fleshy leaves. *Tubers* are stem regions enlarged by the presence of starch-storing parenchyma cells; examples of plants that form tubers are the potato and the cassava (the source of tapioca). The "eyes" of a potato are buds at nodes of the modified stem, and the regions between eyes are internodes. Many grasses, such as Bermuda grass, and some weeds are difficult to eradicate because they have *rhizomes*—long underground stems that can

Basic Research: Homeobox Genes: How the Meristem Gives Its Marching Orders

How do descendents of some dividing cells in a shoot apical meristem (SAM) "know" to become stem tissues, while others embark on the developmental path that produces leaves or other shoot parts? Although the full answer to this question is not yet known, research teams at several laboratories around the world have found evidence of a genetic mechanism in plant meristem cells that appears to guide the process.

Working with SAM tissue from maize (*Zea mays*, generally known in North America as corn), investigators have identified more than a dozen regulatory genes whose protein products activate groups of other genes in differentiating cells. Some genes that act in this way to guide development along a particular path are called *homeotic genes*, because they contain a nucleotide sequence called the homeobox. The homeobox (see Chapter

48) binds to a specific promoter region shared by all of the genes that a homeotic gene controls. Interaction with a homeobox sequence turns the affected genes on or off. Homeobox genes were first discovered in studies of how legs, antennae, and other structures develop in *Drosophila*, the common fruit fly.

Researcher Sarah Hake of the Plant Gene Expression Center (U.S. Department of Agriculture) was curious about the action of a homeotic gene in maize known as *knotted-1 (KN-1)*. Normally the *KN-1* gene is expressed in apical meristems, where it maintains the meristem in an undifferentiated state. When a mutated form, *kn-1*, is expressed, however, the mutation causes abnormal knobby growths on leaves—hence the gene's name. Hake's research helped establish that *KN-1* defines developmental pathways that unfold in meristems. For example,

when Hake cloned the *KN-1* gene and inserted it into tobacco leaf cells, the cells *de*differentiated and began acting like meristem cells. As they divided, they produced lines that could differentiate into leaves and stems.

Subsequent studies of *KN-1* in species as diverse as sunflowers and garden peas have led to the identification of the family of what are now called knotted-1-like genes, all of which encode regulatory proteins that influence developmental pathways. As in maize, some are typically expressed in SAM tissue. In sunflower, tomato, and perhaps other species, knotted-1-like genes also appear to be expressed in differentiated plant parts including leaves, flowers, stems, and even roots. The early work on SAM tissue and homeobox genes in maize has blossomed into a wide-ranging investigation of the molecular signals that shape plant architecture.

---

extend as much as half a meter deep into the soil and rapidly produce new shoots when existing ones are pulled out. The pungent, starchy "root" of ginger is a rhizome also. Crocuses and some other ornamental plants develop elongated, fleshy underground stems called *corms*, another starch-storage adaptation. Tubers, rhizomes, and corms all have meristematic tissue at nodes from which new plants can be propagated—a vegetative (asexual) reproductive mode. Other plants, including the strawberry, reproduce vegetatively via slender stems called *stolons*, which grow along the soil surface. New plants arise at nodes along the stolon.

Leaves Carry Out Photosynthesis and Gas Exchange

Each spring a mature maple tree heralds the new season by unfurling roughly 100,000 leaves. Some other tree species produce leaves by the millions. For these

---

a. Onion bulb   **b.** Potato tuber   **c.** Ginger rhizome   **d.** Crocus corm   **e.** Strawberry stolons

Figure 31.14

A selection of modified stems. **(a)** The fleshy bulbs of onions *(Allium cepa)* are modified shoots in which the plant stores starch. **(b)** A potato *(Solanum tuberosum)*, a tuber. **(c)** Ginger "root," the pungent, starchy rhizome of the ginger plant *(Zingiber officinale)*. **(d)** Crocus plants (genus *Crocus*) typically grow from a corm. **(e)** A strawberry plant *(Fragaria ananassa)* and stolon.

and most other plants, leaves are the main organs of photosynthesis and gas exchange.

Leaf Morphology and Anatomy. In both eudicots and monocots, the leaf **blade** provides a large surface area for absorbing sunlight and carbon dioxide **(Figure 31.15a)**. Studies show that in general, leaves of flowering plants are oriented on the stem axis so that they can capture the maximum amount of sunlight; the stems and leaves of some plants follow the sun's movement during the course of a day by changing position (this phenomenon is described in Chapter 35).

Many eudicot leaves, such as those of maples, have a broad, flat blade attached to the stem by a stalklike **petiole**. Depending on the species, the petiole can be long, short, or in between. A celery stalk is a fleshy petiole. Unless a petiole is very short, it holds a leaf away from the stem and helps prevent individual leaves from shading one another. In many plant species petioles allow leaves to move in the breeze. This helps circulate air around the leaf, replenishing the supply of carbon dioxide for photosynthesis. In most monocot leaves, such as those of rye grass or corn, the blade is longer and narrower and its base simply forms a sheath around the stem.

Leaf Modifications. Leaf forms are based on two basic patterns: simple leaves, which have a single blade **(Figure 31.15b)**, and compound leaves, in which the leaf blade is divided into smaller leaflets **(Figure 31.15c)**. As with other plant parts, there is huge variety in the morphology of leaves. For instance, leaf edges or margins may be smooth, toothed, or lobed. Some leaves are modified as spines **(Figure 31.16a)**, while others have trichomes that take the form of hairs or hooks—all possibly adaptations for defense against herbivores. Leaves or parts of leaves also may be modified into tendrils, like those of the sweet pea **(Figure 31.16b)**, or other structures. Epidermal cells on the leaves of the saltbush *Atriplex spongiosa* form balloonlike structures **(Figure 31.16c)** that contain concentrated Na^+ and Cl^- taken up from the salty soil. Eventually, the salt-filled epidermal cells burst or fall off the leaf, releasing the salt to the outside. This adaptation helps control the salt concentration in the plant's tissues—another example of the link between structure, function, and the environment in which a plant lives.

Leaf Primary Growth and Internal Structure. In both angiosperms and gymnosperms, leaves develop on the sides of the shoot apical meristem. Initially, meristem cells near the apex divide and their derivatives elongate. The resulting bulge enlarges into a thin, rudimentary leaf, or **leaf primordium** (see Figure 31.12). As the plant grows and internodes elongate, the leaves that form from leaf primordia become spaced at intervals along the length of the stem or its branches.

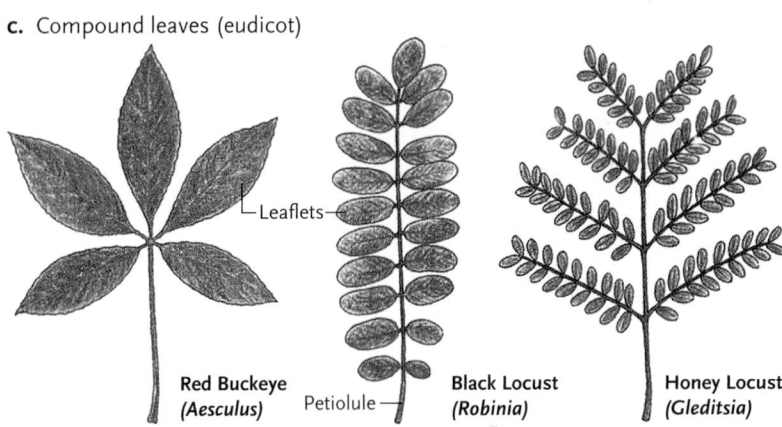

a. Common forms of eudicot and monocot leaves

Petiole
Axillary bud
Blade
Node
Blade
Sheath
Stem
Node

Eudicot   **Monocot**

b. Simple leaves (eudicot)

Poplar (Populus)   **Oak (Quercus)**   **Maple (Acer)**

c. Compound leaves (eudicot)

Leaflets

Red Buckeye (Aesculus)   Petiolule   **Black Locust (Robinia)**   **Honey Locust (Gleditsia)**

Figure 31.15
Leaf forms. **(a)** Common forms of eudicot and monocot leaves. **(b)** Examples of simple eudicot leaves. **(c)** Examples of compound eudicot leaves.

Leaf tissues typically form several layers **(Figure 31.17)**. Uppermost is epidermis, with cuticle covering its outer surface. Just beneath the epidermis is **mesophyll** (*mesos* = middle; *phyllon* = leaf), ground tissue composed of loosely packed parenchyma cells that contain chloroplasts. The leaves of many plants, especially eudicots, contain two layers of mesophyll. *Palisade mesophyll* cells contain more chloroplasts and are arranged in compact columns with smaller air spaces between them, typically toward the upper leaf surface. *Spongy mesophyll,* which tends to be located toward the underside of a leaf, consists of irregularly arranged cells with a conspicuous network of air spaces that gives it a spongy appearance. Air spaces between mesophyll cells enhance the uptake of carbon dioxide and release of oxygen during photosynthesis and account for 15% to 50% of a leaf's volume. Mesophyll also contains collenchyma and sclerenchyma cells, which support the photosynthetic cells.

Below the mesophyll is another cuticle-covered epidermal layer. Except in grasses and a few other

a. Cactus spines

b. Tendrils

Joseph Devenney/Getty Images Inc.

Maxine Adcock/Sciene Photo Library/Photo Researchers, Inc.

c. Salt bladders, a form of trichome

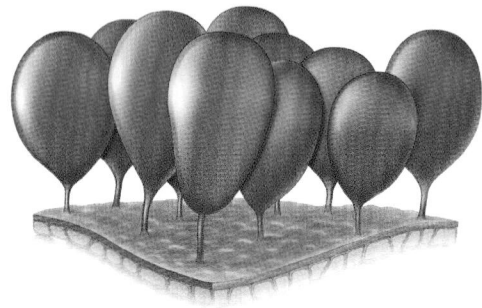

Figure 31.16

A few adaptations of leaves. **(a)** Spines on a barrel cactus *(Ferro-cactus covillei)* thwart browsing herbivores and limit the surface area from which water is lost in the plant's arid environment. **(b)** The tendrils of a sweet pea *(Lathyrus odoratus)* help to support the climbing plant's stem. **(c)** SEM of salt bladders on the leaf of a saltbush plant *(Atriplex spongiosa)*. The "bladders" are trichomes, specialized outgrowths of the leaf epidermis in which excess salt from the plant's tissue fluid accumulates. The salt-laden trichomes eventually burst or slough off.

a. Typical stucture of an angiosperm leaf

b. Fine structure of a bean leaf *(Phaseolus)*

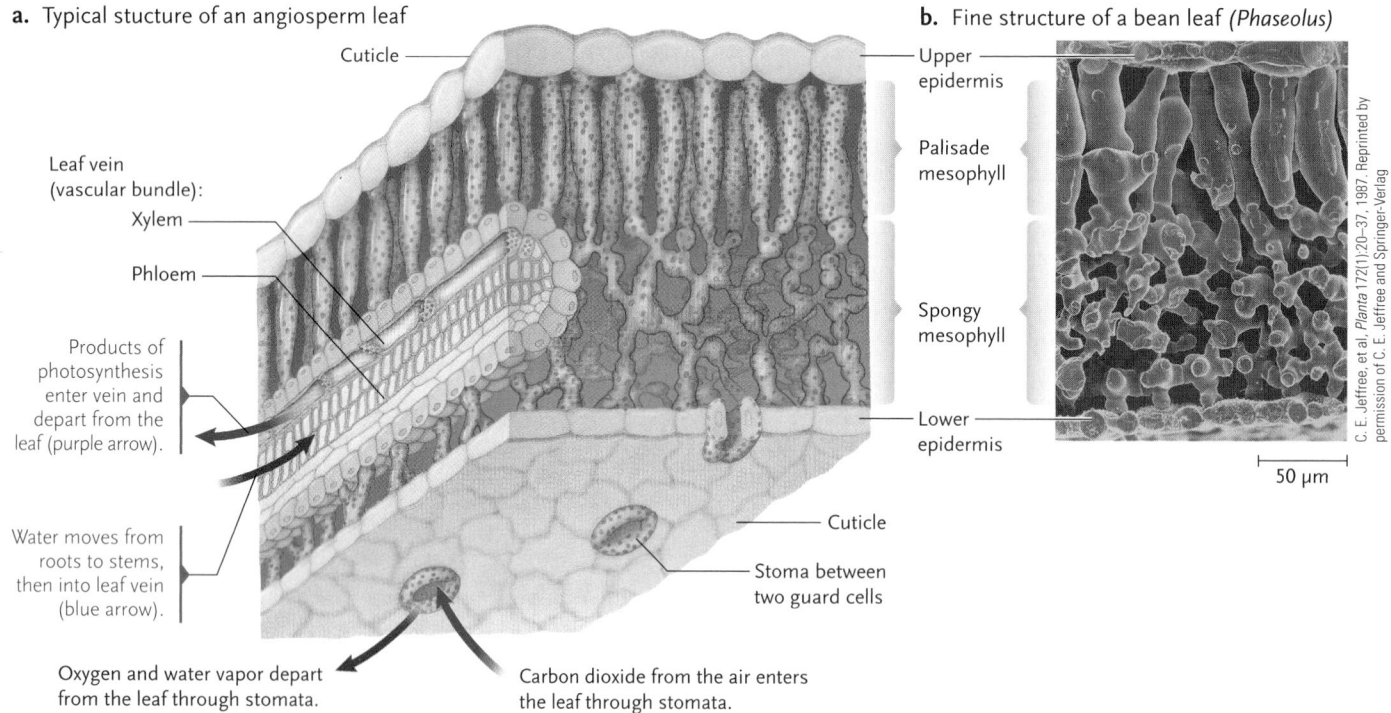

Cuticle

Leaf vein (vascular bundle):

Xylem

Phloem

Products of photosynthesis enter vein and depart from the leaf (purple arrow).

Water moves from roots to stems, then into leaf vein (blue arrow).

Oxygen and water vapor depart from the leaf through stomata.

Carbon dioxide from the air enters the leaf through stomata.

Upper epidermis

Palisade mesophyll

Spongy mesophyll

Lower epidermis

Cuticle

Stoma between two guard cells

C. E. Jeffree, et al. Planta 172(1):20–37, 1987. Reprinted by permission of C. E. Jeffree and Springer-Verlag

50 μm

Figure 31.17

Internal structure of a leaf. **(a)** Diagram of typical leaf structure for many kinds of flowering plants. **(b)** Scanning electron micrograph of tissue from the leaf of a kidney bean plant *(Phaseolus)*, transverse section. Notice the compact organization of epidermal cells. See Figure 31.10b for a scanning electron micrograph of stomata.

plants, this layer contains most of the stomata through which water vapor exits the leaf and gas exchange occurs. For example, the upper surface of an apple leaf has no stomata, while a square centimeter of the lower surface has more than 20,000. A square centimeter of the upper epidermis of a tomato leaf has about 1200 stomata, whereas the same area of the lower epidermis has 13,000. The positioning of stomata on the side of the leaf that faces away from the sun may be an adaptation limiting water loss by evaporation through stomatal openings.

Vascular bundles form a lacy network of **veins** throughout the leaf. Eudicot leaves typically have a branching vein pattern; in monocot leaves, veins tend to run in parallel arrays.

In temperate regions, most leaves are temporary structures. In deciduous (*deciduus* = falling off, shedding) species such as birches and maples, hormonal signals cause the leaves to drop from the stem as days shorten in autumn. Other temperate species, such as camellias or hollies, as well as conifers, also drop leaves, but they appear "evergreen" because the leaves may persist for several years and do not all drop at the same time.

Plant Shoots May Have Juvenile and Adult Forms

Leaf shape and other shoot characteristics can mirror the progress of a long-lived plant through its life cycle. Plants that live many years may spend part of their lives in a juvenile phase, then shift to a mature or adult phase. The differences between juveniles and adults often are reflected in leaf size and shape, in the arrangement of leaves on the stem, or in a change from vegetative growth to a reproductive stage—or sometimes all three. For example, oak saplings (genus *Quercus*) have fewer leaves than mature oaks do, but the leaves are considerably larger—an adaptation that probably provides saplings with increased leaf surface area for taking in carbon (in carbon dioxide). Young English ivy plants *(Hedera helix)* grow as vines, have leaves with multiple lobes arranged on the stem in an alternating pattern, and do not flower **(Figure 31.18a)**. By contrast, mature English ivy is a flowering shrub with oval leaves that arise on the stem in a spiral pattern **(Figure 31.18b)**. A magnolia tree *(Magnolia grandiflora)* doesn't flower until its juvenile phase ends, which can be 20 years or more from the time the *Magnolia* seed sprouts. Most woody plants must attain a certain size before their meristem tissue can respond to the hormonal signals that govern flower development, a topic we consider in Chapter 35.

Phase changes provide more examples of the plasticity that characterizes plant development. They almost certainly are associated with changes in the expression of genes that control the development of stem nodes, leaf and flower buds, and other basic aspects of plant growth.

STUDY BREAK

1. Describe the functions of stems and stem structure, and list the basic steps in primary growth of stems.
2. Explain the general function of leaves and how leaf anatomy supports this role in eudicots and monocots.
3. Describe the steps in primary growth of a leaf and the structures that result from the process.
4. Describe two examples of the life phases of long-lived plant species.

31.4 Root Systems

Plants must absorb enough water and dissolved minerals to sustain growth and routine cellular maintenance, a task that requires a tremendous root surface. In one study, measurements of the root system of a rye plant *(Secale cereale)* that have been growing for only 4 months may have a surface area of more than 700 m²—about 130 times greater than the surface area of its shoot system. The roots of carrots, sugar beets, and most other plants also store nutrients produced in photosynthesis, some to be used by root cells and some to be transported later to cells of the shoot. As a root system penetrates downward and spreads out, it also anchors the aboveground parts.

Figure 31.18
Age-related phase changes in English ivy *(Hedera helix)*. **(a)** The juvenile, vine-type growth habit. **(b)** The mature shrub.

a. Young English ivy

Thomas L. Rost

b. Mature English ivy

Thomas L. Rost

Taproot and Fibrous Root Systems Are Specialized for Particular Functions

Most eudicots have a **taproot system**—a single main root, or taproot, that is adapted for storage and smaller branching roots called **lateral roots (Figure 31.19a).** As the main root grows downward, its diameter increases, and the lateral roots emerge along the length of its older, differentiated regions. The youngest lateral roots are near the root tip. Carrots and dandelions have a taproot system, as do pines and many other conifers. A pine's taproot system can penetrate 6 m or more into the soil.

Grasses and many other monocots develop a **fibrous root system** in which several main roots branch to form a dense mass of smaller roots **(Figure 31.19b).** Fibrous root systems are adapted to absorb water and nutrients from the upper layers of soil, and tend to spread out laterally from the base of the stem. Fibrous roots are important ecologically because dense root networks help hold topsoil in place and prevent erosion. During the 1930s, drought, overgrazing by livestock, and intensive farming in the North American Midwest destroyed hundreds of thousands of acres of native prairie grasses, contributing to soil erosion on a massive scale. Swirling clouds of soil particles prompted journalists to name the area the Dust Bowl.

In some plants, **adventitious roots** arise from the stem of the young plant. "Adventitious" refers to any structure arising at an unusual location, such as roots that grow from stems or leaves. Adventitious roots and their branchings all are about the same length and diameter. Those of English ivy and some other climbing plants produce a gluelike substance (from trichomes) that allows them to cling to vertical surfaces. The *prop roots* of a corn plant are adventitious roots that develop from the shoot node nearest the soil surface; they both support the plant and absorb water and nutrients. Mangroves and other trees that grow in marshy habitats often have huge prop roots, which develop from branches as well as from the main stem **(Figure 31.19c).**

Root Structure Is Specialized for Underground Growth

Like shoots, roots have distinct anatomical parts, each with a specific function. In most plants, primary growth of roots begins when an embryonic root (called a *radicle*) emerges from a germinating seed and its meristems become active. **Figure 31.20** shows the structure of a root tip. Notice that the root apical meristem terminates in a dome-shaped cell mass, the **root cap.** The meristem produces the cap, which in turn surrounds and protects the meristem as the root elongates through the soil. Certain cells in the cap respond to gravity, which guides the root tip downward. Cap cells also secrete a polysaccharide-rich substance that lubricates the tip and eases the growing root's passage through the soil. Outer root cap cells are continually abraded off and replaced by new cells at the cap's base.

Zones of Primary Growth in Roots. Primary growth takes place in successive stages, beginning at the root tip and progressing upward. Just inside the root cap

Figure 31.19
Types of roots.
(a) Taproot system of a California poppy *(Eschscholzia californica).*
(b) Fibrous root system of a grass plant. **(c)** The prop roots of red mangrove trees *(Rhizophora),* examples of adventitious roots.

a. Taproot system

b. Fibrous root system

c. Adventitious roots

© Beth Davidow/Visuals Unlimited

some roots have a small clump of apical meristem cells called the **quiescent center.** Unlike other meristematic cells, cells of the quiescent center divide very slowly unless the root cap or the apical meristem is injured; then they become active and can regenerate the damaged part. The quiescent center also may include cells that synthesize plant hormones controlling root development.

The root apical meristem and the actively dividing cells behind it form the **zone of cell division.** As in the stem, cells of the apical meristem segregate into three primary meristems. Cells in the center of the root tip become the procambium; those just outside the procambium become ground meristem; and those on the periphery of the apical meristem become protoderm.

The zone of cell division merges into the **zone of elongation.** Most of the increase in a root's length comes about here as cells become longer as their vacuoles fill with water. This "hydraulic" elongation pushes the root cap and apical meristem through the soil as much as several centimeters a day.

Above the zone of elongation, cells do not increase in length but they may differentiate further and take on specialized roles in the **zone of maturation.** For instance, epidermal cells in this zone give rise to root hairs, and the procambium, ground meristem, and protoderm complete their differentiation in this region.

Tissues of the Root System. Coupled with primary growth of the shoot, primary root growth produces a unified system of vascular pipelines extending from root tip to shoot tip. The root procambium produces cells that mature into the root's xylem and phloem **(Figure 31.21).** Ground meristem gives rise to the root's cortex, its ground tissue of starch-storing parenchyma cells that surround the stele. In eudicots, the stele runs through the center of the root (see Figure 31.21a). In corn and some other monocots, the stele forms a ring that divides the ground tissue into cortex and pith (see Figure 31.21b).

The cortex contains air spaces that allow oxygen to reach all of the living root cells. Numerous plasmodesmata connect the cytoplasm of adjacent cells of the cortex. In many flowering plants, the outer root cortex cells give rise to an **exodermis,** a thin band of cells that, among other functions, may limit water losses from roots and help regulate the absorption of ions. The innermost layer of the root cortex is the **endodermis,** a thin, selectively permeable barrier that helps control the movement of water and dissolved minerals into the stele. We look in more detail at the roles of exodermis and endodermis in Chapter 32.

Between the stele and the endodermis is the **pericycle,** consisting of one or more layers of parenchyma cells that can still function as meristem. The pericycle gives rise to lateral roots **(Figure 31.22).** In response to chemical growth regulators, **root primordia**

a.

Endodermis
Pericycle
Cortex
Epidermis
Xylem
Phloem
Stele

Fully grown root hair

Zone of maturation
The tissue systems complete their differentiation and begin to take on their specialized roles. Root hairs begin to form.

Zone of elongation
Most cells stop dividing but increase in length. The primary meristems begin to differentiate into tissue systems; the phloem matures and the xylem starts to form.

b.

Zone of cell division
Rapidly dividing cells of the root apical meristem segregate into three primary meristems.

Quiescent center

Root cap

John Limbaugh/Ripon Microslides, Inc.

100 μm

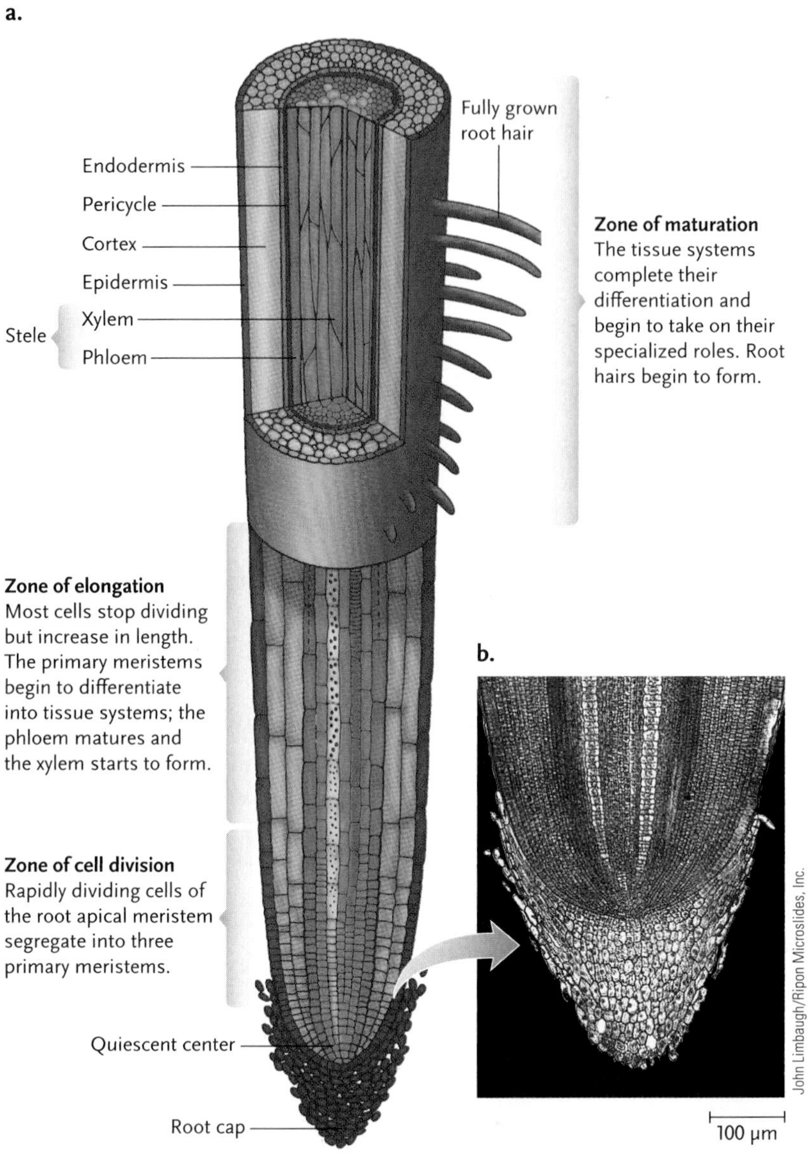

Figure 31.20
Tissues and zones of primary growth in a root tip.
(a) Generalized root tip, longitudinal section. **(b)** Micrograph of a corn root tip, longitudinal section.

(rudimentary roots) arise at specific sites in the pericycle. Gradually, the lateral roots emerge and grow out through the cortex and epidermis, aided by enzymes released by the root primordium that help break down the intervening cells. The distribution and frequency of lateral root formation partly control the overall shape of the root system—and the extent of the soil area it can penetrate.

In some cells in the developing root epidermis the outer surface becomes extended into root hairs (see Figure 31.20). Root hairs can be more than a centimeter long and can form in less than a day. Collectively, the thousands or millions of them on a plant's roots greatly increase the plant's absorptive surface. Root hair structure supports this essential function. Each hair is a slender tube with thin walls made sticky on their surface by a coating of pectin. Soil particles tend to adhere to the wall, providing an intimate association between the hair and the surrounding earth, thus facilitating the uptake of water molecules and

a. Eudicot root

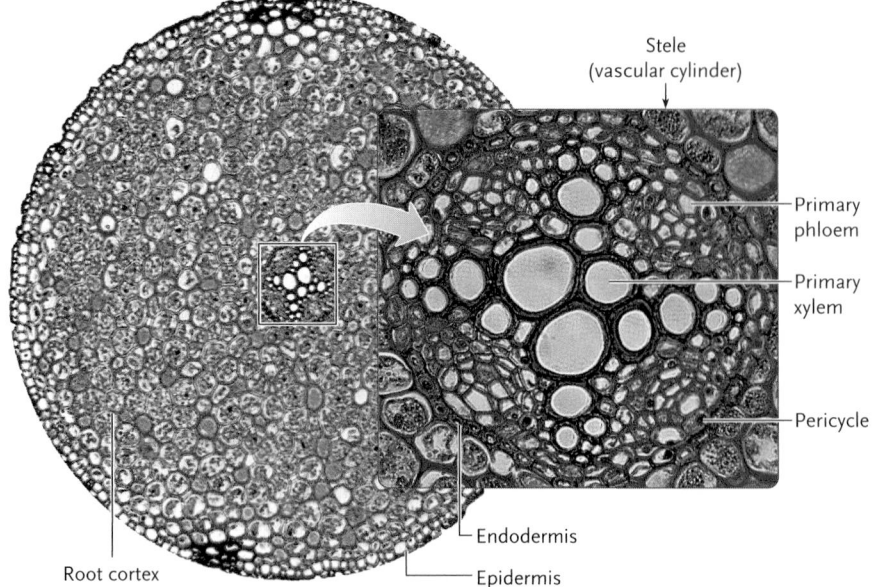

Stele (vascular cylinder)

Primary phloem

Primary xylem

Pericycle

Endodermis

Epidermis

Root cortex

b. Monocot root

Pith

Root cortex

Epidermis

Stele

Primary xylem

Primary phloem

Figure 31.21
Stele structure in eudicot and monocot roots compared. **(a)** A young root of the buttercup *Ranunculus*, a eudicot. The close-up shows details of the stele. **(b)** Root of a corn plant *(Zea mays)*, a monocot. Notice how the stele divides the ground tissue into cortex and pith. Both roots are shown in transverse section.
(a: Chuck Brown; b: Carolina Biological Supply.)

mineral ions from soil. When plants are transplanted, rough handling can tear off much of the fragile absorptive surface. Unable to take up enough water and minerals, the transplant may die before new root hairs can form.

STUDY BREAK

1. Compare the two general types of root systems.
2. Describe the zones of primary growth in roots.
3. Describe the various tissues that arise in a root system and their functions.

31.5 Secondary Growth

All plants undergo primary growth of the root and stem. In addition, some plants have secondary growth processes that add girth to roots and stems over two or more growing seasons. In plant species that have secondary growth, older stems and roots become more massive and woody through the activity of two types of lateral meristems, or *cambia* (singular, cambium). One of these meristems, the **vascular cambium**, produces

secondary xylem and phloem. The other, called the **cork cambium**, produces **cork**, a secondary epidermis that is one element of the multilayered structure known as bark. In cells of these tissues, mitosis is periodically reactivated. Hence secondary growth permits woody plants to grow taller and live longer than herbaceous plants.

Vascular Cambium Gives Rise to Secondary Growth in Stems

Recall that after the stem of a woody plant completes its primary growth, each vascular bundle contains a layer of undifferentiated cells between the primary xylem and the primary phloem. These cells, along with parenchyma cells between the bundles, eventually give rise to a cylinder of vascular cambium that wraps around the xylem and pith of the stem **(Figure 31.23)**. Vascular cambium consists of two types of cells—*fusiform initials* and *ray initials*—that have different shapes and functions (see Figure 31.23b). Secondary growth takes place as these cells divide. Initials divide at right angles to the stem surface, so their descendants add girth to the stem instead of length. **Fusiform initials**, which are derived from cambium inside the vascular bundles, give rise to secondary xylem and phloem cells. Secondary xylem forms on the inner face of the vascular cambium, and secondary phloem forms on the outer face. **Ray initials** are derived from the parenchyma cells between vascular bundles. As they divide, their descendants form spokelike *rays*. These

Figure 31.22
Micrographs showing the formation of a lateral root from the pericycle of a willow tree *(Salix)*. These micrographs show transverse sections. (All images: © Omnikron/Photo Researches, Inc.)

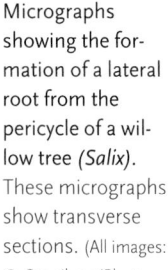 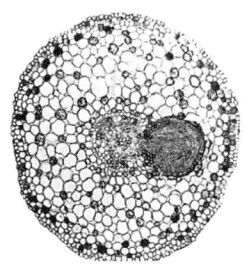 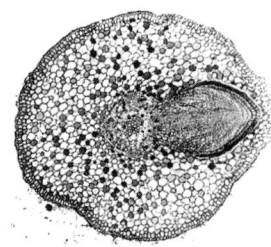 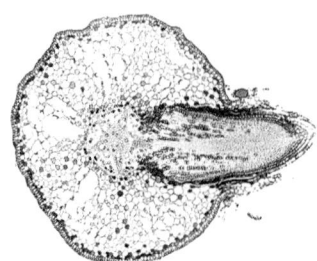

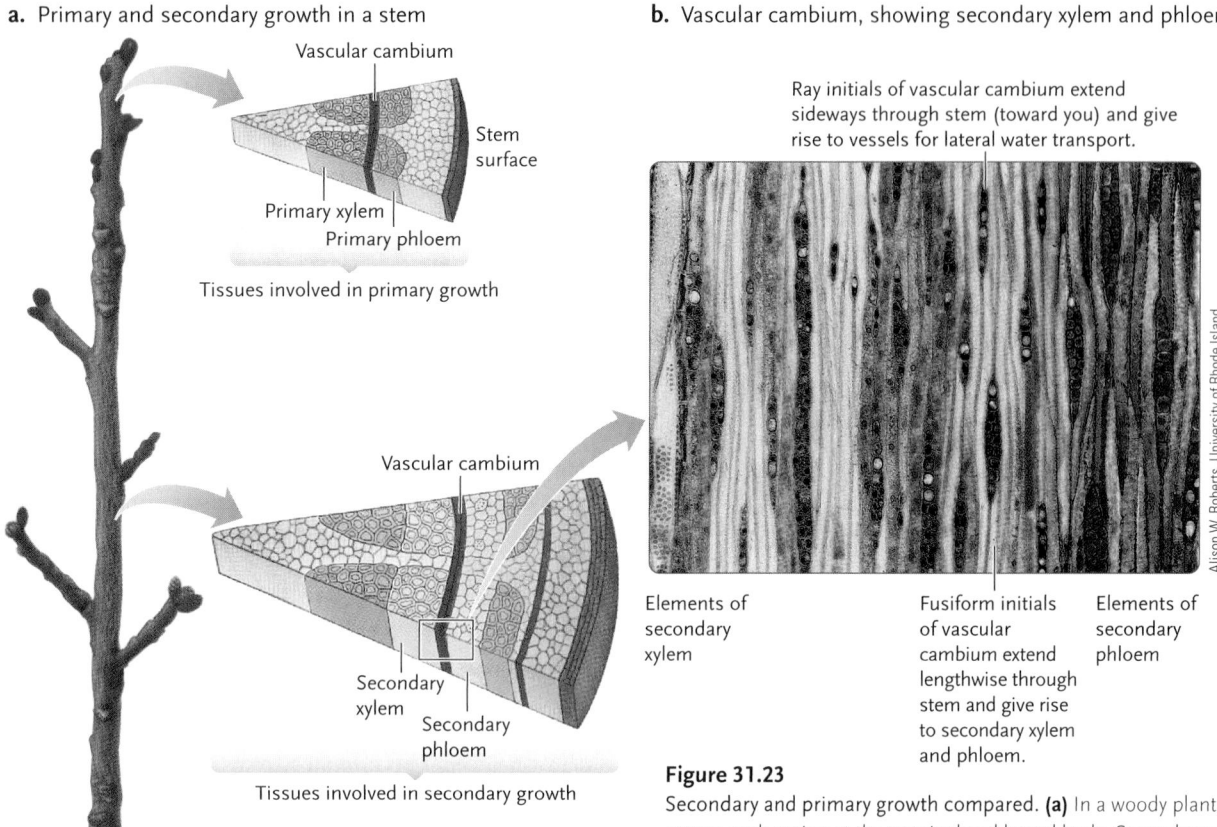

a. Primary and secondary growth in a stem

Vascular cambium

Stem surface

Primary xylem

Primary phloem

Tissues involved in primary growth

Vascular cambium

Secondary xylem

Secondary phloem

Tissues involved in secondary growth

b. Vascular cambium, showing secondary xylem and phloem

Ray initials of vascular cambium extend sideways through stem (toward you) and give rise to vessels for lateral water transport.

Alison W. Roberts, University of Rhode Island

Elements of secondary xylem

Fusiform initials of vascular cambium extend lengthwise through stem and give rise to secondary xylem and phloem.

Elements of secondary phloem

Figure 31.23
Secondary and primary growth compared. **(a)** In a woody plant, primary growth resumes each spring at the terminal and lateral buds. Secondary growth resumes at the vascular cambium inside the stem. **(b)** Fusiform initials and ray initials of the vascular cambium of a walnut tree *(Juglans)*, tangential section.

horizontal channels carry water sideways through the stem, in a radial pattern that resembles a sliced pie. While xylem and phloem mainly conduct fluid lengthwise in the stem, rays ensure that water and solutes also move laterally as stems thicken.

With time, the mass of secondary xylem inside the ring of vascular cambium increases, forming the hard tissue known as **wood.** Outside the vascular cambium, secondary phloem cells also are added each year **(Figure 31.24).** (The primary phloem cells, which have thin walls, are destroyed as they are pushed outward by secondary growth.) As a stem increases in diameter, the growing mass of new tissue eventually causes the cortex, and the secondary phloem beyond

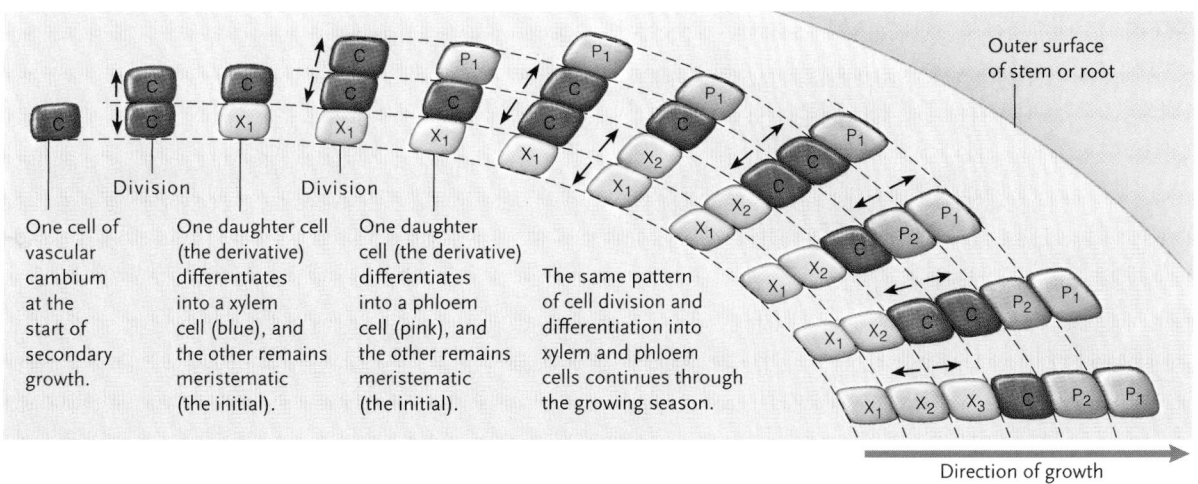

Outer surface of stem or root

Division

Division

One cell of vascular cambium at the start of secondary growth.

One daughter cell (the derivative) differentiates into a xylem cell (blue), and the other remains meristematic (the initial).

One daughter cell (the derivative) differentiates into a phloem cell (pink), and the other remains meristematic (the initial).

The same pattern of cell division and differentiation into xylem and phloem cells continues through the growing season.

Direction of growth

Figure 31.24
Relationship between the vascular cambium and its derivative cells (secondary xylem and phloem). The drawing shows stem growth through successive seasons. Notice how the ongoing divisions displace the cambial cells, moving them steadily outward even as the core of xylem increases the stem or root thickness.

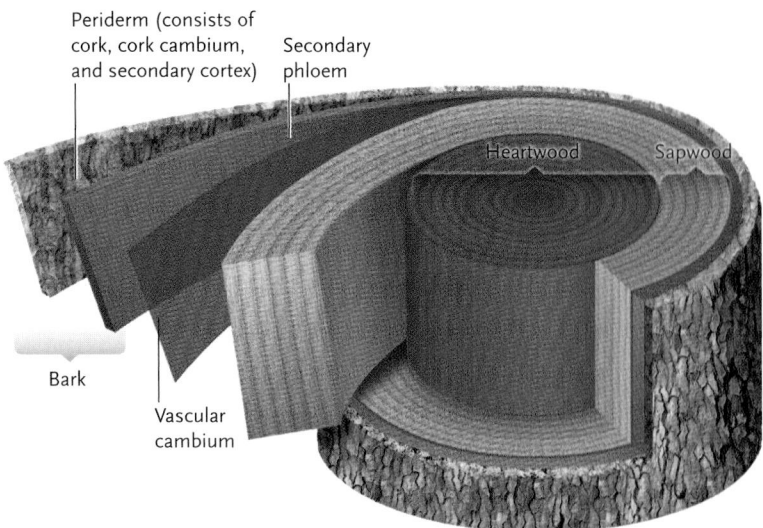

Figure 31.25
Structure of a woody stem showing extensive secondary growth. Heartwood, the mature tree's core, has no living cells. Sapwood, the cylindrical zone of xylem between the heartwood and vascular cambium, contains some living parenchyma cells among the nonliving vessels and tracheids. Everything outside the vascular cambium is bark. Everything inside it is wood.

it, to rupture. Parts of the cortex split away and carry epidermis with them. Cork cambium—produced early in the stem's secondary development by meristem cells in the cortex, epidermis, or secondary phloem—replaces the lost epidermis with cork. The cork cambium produces cork to the outside and secondary cortex to the inside.

Bark encompasses all the living and nonliving tissues between the vascular cambium and the stem surface. It includes the secondary phloem and the **periderm** (*peri* = surrounding; *derma* = skin), the outermost portion of bark that consists of cork, cork

cambium, and secondary cortex **(Figure 31.25)**. Girdling a tree by removing a belt of bark around the trunk is lethal because it destroys the secondary phloem layer, and so nutrients from photosynthesis in leaves cannot reach the tree's roots. Natural corks used to seal bottles are manufactured from the especially thick outer bark of the cork oak, *Quercus suber*. Tubular openings called *lenticels* develop in the periderm. They function a bit like snorkels, permitting exchanges of oxygen and carbon dioxide between the living tissues and the outside air.

As a tree ages, changes also unfold in the appearance and function of the wood itself. In the center of its older stems and roots is **heartwood**, dry tissue that no longer transports water and solutes and is a storage depot for some defensive compounds. In time, these substances—including resins, oils, gums, and tannins—clog and fill in the oldest xylem pipelines. Typically they darken heartwood, strengthen it, and make it more aromatic and resistant to decay. **Sapwood** is secondary growth located between heartwood and the vascular cambium. Compared with heartwood, it is wet and not as strong.

In temperate climates, trees produce secondary xylem seasonally, with larger-diameter cells produced in spring and smaller-diameter cells in summer. This "spring wood" and "summer wood" reflect light differently, and it is possible to identify them as alternating light and dark bands. The alternating bands represent annual growth layers, or "tree rings" **(Figure 31.26)**.

Secondary Growth Can Also Occur in Roots

The roots of grasses, palms, and other monocots are almost always the product of primary growth alone. In plants with roots that have secondary growth, the continuous ring of vascular cambium develops differently than it does in stems. When their primary growth is complete, these roots have a layer of residual procambium between the xylem and phloem of the stele **(Figure 31.27,** step 1). The vascular cambium arises in part from this residual cambium, and in part from the pericycle (step 2). Eventually, the cambial tissues arising from the procambium and those arising from the pericycle merge into a complete cylinder of vascular cambium (step 3). The vascular cambium functions in roots as it does in stems, giving rise to secondary xylem to the inside and secondary phloem to the outside. As secondary xylem accumulates, older roots can become extremely thick and woody. Their ongoing secondary growth is powerful enough to break through concrete sidewalks and even dislodge the foundations of homes.

Cork cambium also forms in roots, where it is produced by the pericycle. In many woody eudicots and in all gymnosperms, most of the root epidermis and cortex fall away, and the surface consists entirely of periderm (step 4).

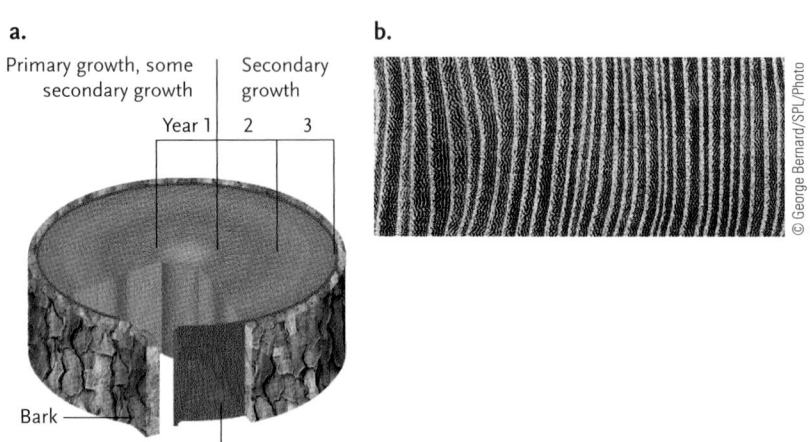

Figure 31.26
Secondary growth and tree ring formation. **(a)** Radial cut through a woody stem that has three annual rings, corresponding to secondary growth in years 2 through 4. **(b)** Tree rings in an elm (*Ulmas*). Each ring corresponds to one growing season. Differences in the widths of tree rings correspond to shifts in climate, including the availability of water.

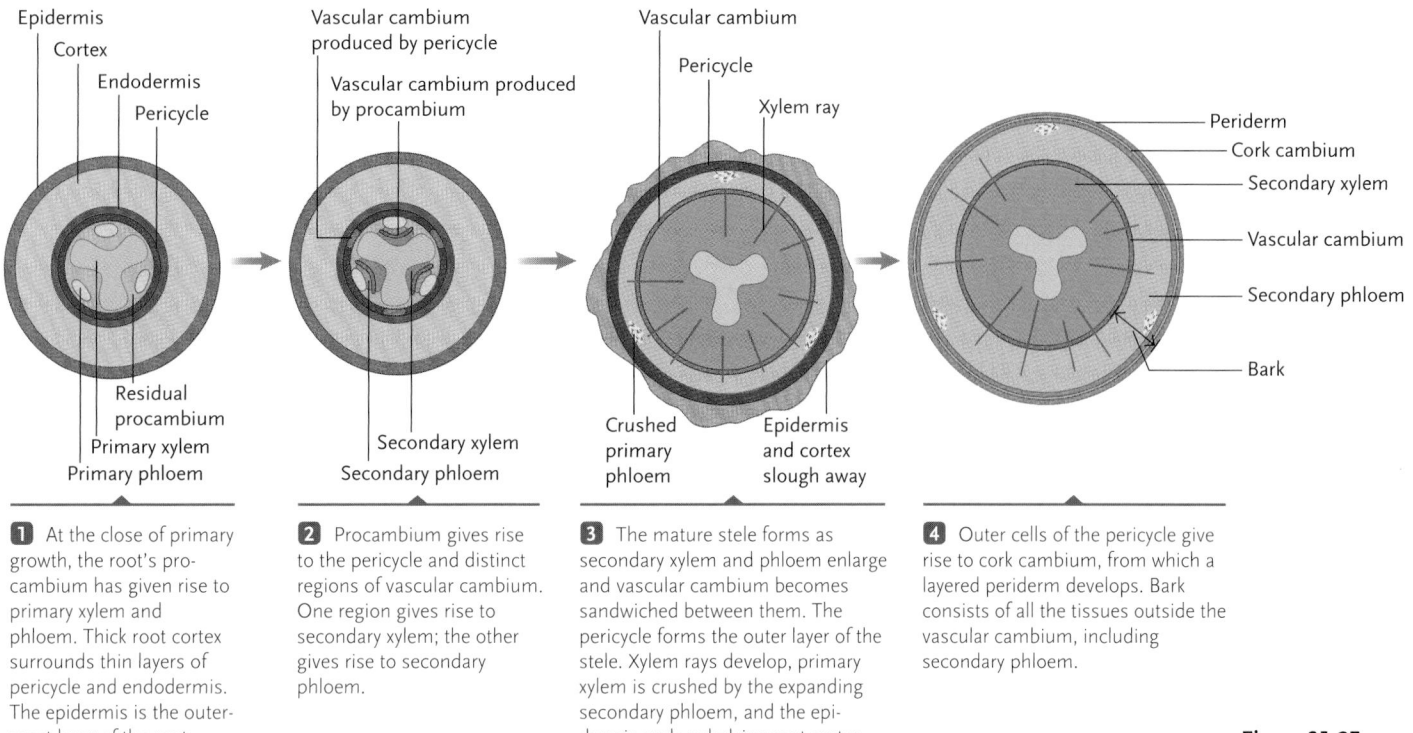

Epidermis Cortex Endodermis Pericycle Residual procambium Primary xylem Primary phloem	Vascular cambium produced by pericycle Vascular cambium produced by procambium Secondary xylem Secondary phloem	Vascular cambium Pericycle Xylem ray Crushed primary phloem Epidermis and cortex slough away	Periderm Cork cambium Secondary xylem Vascular cambium Secondary phloem Bark

1 At the close of primary growth, the root's pro-cambium has given rise to primary xylem and phloem. Thick root cortex surrounds thin layers of pericycle and endodermis. The epidermis is the outer-most layer of the root.

2 Procambium gives rise to the pericycle and distinct regions of vascular cambium. One region gives rise to secondary xylem; the other gives rise to secondary phloem.

3 The mature stele forms as secondary xylem and phloem enlarge and vascular cambium becomes sandwiched between them. The pericycle forms the outer layer of the stele. Xylem rays develop, primary xylem is crushed by the expanding secondary phloem, and the epidermis and underlying root cortex begin to slough away.

4 Outer cells of the pericycle give rise to cork cambium, from which a layered periderm develops. Bark consists of all the tissues outside the vascular cambium, including secondary phloem.

Figure 31.27
Secondary growth in the root of one type of woody plant.

Secondary Growth Is an Adaptive Response

Plants, like all living organisms, compete for resources, and woody stems and roots confer some advantages. Plants with taller stems or wider canopies that defy the pull of gravity can intercept more of the light energy from the sun. With a greater energy supply for photosynthesis, they have the metabolic means to increase their root and shoot systems, and thus are better able to acquire resources—and ultimately to reproduce successfully.

In every stage of a plant's growth cycle, growth maintains a balance between the shoot system and root system. Leaves and other photosynthetic parts of the shoot must supply root cells with enough sugars to support their metabolism, and roots must provide shoot structures with water and minerals. As long as a plant is growing, this balance is maintained, even as the complexity of the root and shoot systems increases, whether the plant lives only a few months or—like some bristlecone pines—for 6000 years.

STUDY BREAK

1. Explain the nature of secondary growth and where it typically occurs in plants.
2. Describe the components of vascular cambium and their roles in secondary growth in stems, including the development of tissues such as bark, cork, and wood.
3. Compare secondary growth in stems and in roots.

Are plants developmental procrastinators?

It is well established that plants can survive physical insults and exposure to a wide range of environmental fluctuations. What biological resources of plants make them so resilient given their lifestyle constraints? What is the source of phenotypic plasticity that allows a plant's body form to change in response to changes in its habitat? Perhaps the answer lies in the ability of plants to put off making developmental decisions in response to environmental shifts.

Unlike most animal cells, plant cells are pluripotent, retaining their developmental flexibility. Thus, they can behave as stem cells capable of proliferating and producing new structures and even new individuals. Furthermore, many types of plant cells will readily transdifferentiate and assume a new cellular identity even after reaching developmental maturity. In other words, few developmental decisions appear to be final, and many can be tailored to the environmental constraints imposed on the plant. Some plant species appear to have this flexibility even during more global developmental events, such as switching from vegetative growth to reproductive growth or flowering. Why is that the case? Recent findings suggest that by "leaving all options open," plants can quickly adapt to environmental changes and produce progeny, which is the ultimate biological goal for all living organisms.

Research has documented that shifts in environmental context activate genetic changes underlying plants' developmental and phenotypic plasticity. For example, at the whole-organism level some plants, such as *Impatiens balsamina*, can switch between making leaves and making flowers if relative day length changes. In fact, these plants can make leaves that are partial flowers, or flowers that are partial leaves, if light conditions are alternated between short days and long days. Nicholas Battey at the University of Reading in England and his colleagues, who have investigated this phenomenon for many years, have demonstrated that a genetic basis exists for this ability to change body form in response to changing environmental cues. Furthermore, Battey's group suggests that among flowering plants a genetic continuum exists from species that require constant reminders to initiate flowering to species that only require a single signal. For perennial plants such as trees, developmental reprogramming is essential because it allows them to orchestrate seasonally appropriate formation and growth of different organs from the same meristem. Currently, a major effort is under way to understand the genetic basis of developmental evolution as well as how genetic variation may influence phenotypic plasticity. Since plasticity appears to be closely associated with environmental factors, one approach is to study the natural variants of a species from different geographical origins.

Plants respond to environmental variation both spatially and temporally. Perhaps a sort of biological global positioning system (GPS) exists that provides developmentally relevant information in time and space, which the plant translates into a variety of responses. In some species the GPS may be on all the time, while in other species it may only operate at certain times of year, or it may only function once during the plant's life time. What might these genetic GPS devices be? How would we test this idea? Have candidate genes already been identified that might be components within the GPS? Is there a link between a plant's GPS and the genetic basis for its ability to procrastinate developmentally? In short, the answer to all these questions appears to be "maybe," and in all likelihood the full answer will be a complex one. Research conducted by Christopher Cullis at Case Western Reserve University shows that environmentally induced changes in the physical features of flax plants *(Linum usitatissimum)* are accompanied by changes in the entire genome of affected plants, and some of these genetic alterations are heritable. These findings are particularly striking because they demonstrate that in the short term plants can respond to environmental fluctuations not only by altering their developmental output (body form and phenotype) but also by "revising" their genomes. That the very blueprint of life, DNA, is also imbued with significant plasticity is particularly exciting and opens a new realm of inquiry into the mechanisms by which plants may respond to environmental challenges. In some biological contexts being a procrastinator can be advantageous.

Marianne Hopkins is a postdoctoral fellow in the Biology Department at the University of Waterloo in Waterloo, Canada. Her expertise lies in plant genetics and plant molecular biology.

Susan Lolle is an associate professor of biology at the University of Waterloo in Waterloo, Canada. Her research interests include plant development, genetics, and genome biology. To learn more go to http://www.biology.uwaterloo.ca.

Review

Go to **ThomsonNOW**™ at www.thomsonedu.com/login to access quizzing, animations, exercises, articles, and personalized homework help.

31.1 Plant Structure and Growth: An Overview

- The vascular plant body consists of an aboveground shoot system with stems, leaves, and flowers, and an underground root system (Figure 31.2).
- Meristems (Figure 31.3) give rise to the plant body and are responsible for a plant's lifelong growth. Each meristem cell divides to produce an initial, which functions as meristem, and a derivative, which may differentiate into a specialized body cell.

- Primary growth of roots and shoots originates at apical meristems at root and shoot tips. Some plants show secondary growth as lateral meristems increase the diameter of stems and roots.
- The two major classes of flowering plants (angiosperms) are monocots and eudicots (Table 31.1).

31.2 The Three Plant Tissue Systems

- Growing plant cells form secondary walls outside the primary walls. Maturing cells become specialized for specific functions, with some functions accomplished by walls of dead cells.

- Plants have three tissue systems (Figure 31.5). Ground tissues make up most of the plant body, vascular tissues serve in transport, and dermal tissue forms a protective cover.
- Of the three types of ground tissues, parenchyma is active in photosynthesis, storage, and other tasks (Figure 31.6); collenchyma and sclerenchyma provide mechanical support.
- Xylem and phloem are the plant vascular tissues. Xylem conducts water and solutes and consists of conducting cells called tracheids and vessel members (Figure 31.8). Phloem, the food-conducting tissue, contains living cells (sieve tube members) joined end to end in sieve tubes (Figure 31.9).
- The dermal tissue, epidermis (Figure 31.10) is coated with a waxy cuticle that restricts water loss. Water vapor and other gases enter and leave the plant through pores called stomata, which are flanked by specialized epidermal cells called guard cells. Epidermal specializations also include trichomes, such as root hairs.

Animation: Tissue systems of a tomato plant

Animation: Apical meristems

Animation: Shoot differentiation

31.3 Primary Shoot Systems

- The primary shoot system consists of the main stem, leaves, and buds, plus any attached flowers and fruits. Stems provide mechanical support, house vascular tissues, and may store food and fluid.
- Stems are organized into modular segments. Nodes are points where leaves and buds are attached, and internodes fall between nodes (Figure 31.11). The terminal bud at a shoot tip consists of shoot apical meristem (Figure 31.12). Lateral buds occur at intervals along the stem. Meristem tissue in buds gives rise to leaves, flowers, or both.
- Derivatives of the apical meristem produce three primary meristems: protoderm makes the stem's epidermis, procambium gives rise to primary xylem and phloem, and ground meristem gives rise to ground tissue.
- Vascular tissues are organized into vascular bundles, with phloem surrounding xylem in each bundle (Figure 31.13).
- Monocot and eudicot leaves have blades of different forms, all providing a large surface area for absorbing sunlight and carbon dioxide (Figure 31.15). Leaf modifications are adaptive responses to environmental selection pressures (Figure 31.16). Leaf characteristics such as shape or arrangement may change over the life cycle of a long-lived plant (Figure 31.18).

Animation: Ground tissues

Animation: Vascular tissues

Animation: Monocot and dicot leaves

Animation: Simple and compound leaves

Animation: Leaf organization

31.4 Root Systems

- Roots absorb water and dissolved minerals and conduct them to aerial plant parts; they anchor and sometimes support the plant and often store food. Root morphologies include taproot systems, fibrous root systems, and adventitious roots (Figure 31.19).
- During primary growth of a root, the primary meristem and actively dividing cells make up the zone of cell division, which merges into the zone of elongation. Past the zone of elongation, cells may differentiate and perform specialized roles in the zone of cell maturation (Figure 31.20).
- A root's vascular tissues (xylem and phloem) usually are arranged as a central stele (Figure 31.21). Parenchyma around the stele forms the root cortex. The root endodermis also wraps around the stele. Inside it is the pericycle, containing parenchyma that can function as meristem. It gives rise to root primordia from which lateral roots emerge (Figure 31.22). Root hairs greatly increase the surface available for absorbing water and solutes.

Animation: Root organization

Animation: Root cross section

Animation: Root systems

31.5 Secondary Growth

- In plants with secondary growth, older stems and roots become more massive and woody via the activity of vascular cambium and cork cambium.
- Vascular cambium consists of two types of cells: fusiform initials, which generate secondary xylem and phloem, and ray initials, which produce horizontal water transport channels called xylem rays (Figures 31.23 and 31.24). Secondary growth takes place as these cells divide.
- Cork cambium gives rise to cork, which replaces epidermis lost when stems increase in diameter. Together, cork cambium and cork make up the periderm (Figure 31.25), the outer portion of bark.
- In root secondary growth, a thin layer of procambium cells between the xylem and phloem differentiates into vascular cambium (Figure 31.27), which gives rise to secondary xylem and phloem. The pericycle produces root cork cambium.

Animation: Secondary growth

Animation: Secondary growth in a root

Animation: Growth in a walnut twig

Animation: Layers in a woody stem

Animation: Annual rings

Questions

Self-Test Questions

1. With respect to growth, plants differ from animals in that:
   a. plant growth involves only an increase in the total number of the organism's cells.
   b. plant cells remain roughly the same size after cell division, whereas animal cells increase in size after they form.
   c. all plants form woody tissues during growth.
   d. plants have indeterminate growth; animals have determinate growth.
   e. plants can grow only when young; animals grow for many years.

2. Identify the correct pairing of a plant tissue and its function.
   a. epidermis: rigid support
   b. xylem: sugar transport
   c. parenchyma: photosynthesis, respiration
   d. phloem: water and mineral transport
   e. periderm: control of gas exchange

3. Identify the correct pairing of a structure and its component(s).
   a. epidermis: companion cells
   b. phloem: sieve tube members
   c. sclerenchyma: nonlignified cell walls

d. secondary cell wall: cuticle
e. parenchyma: sclereids

4. Which of the following is *not* part of a stem?
   a. petiole
   b. pith
   c. xylem
   d. procambium
   e. ground meristem

5. Which of the following would be absent in a eudicot leaf?
   a. spongy mesophyll
   b. palisade mesophyll
   c. pericycle
   d. vascular bundles
   e. stoma

6. A student left a carrot in her refrigerator. Three weeks later she noticed slender white fibers growing from its surface. They were not a fungus. Instead they represented:
   a. lateral roots on a taproot.
   b. adventitious roots.
   c. root hairs on a fibrous root.
   d. root hairs on a lateral root.
   e. young prop roots.

7. Which of the following is *not* a structure that results from secondary plant growth?
   a. periderm
   b. sapwood
   c. cork
   d. pith
   e. heartwood

8. Which characteristic do monocots and eudicots share?
   a. the position of the vascular bundles
   b. the pattern of leaf veins
   c. the number of grooves in the pollen grains
   d. the number of cotyledons
   e. the formation of flowers

9. A student forgets to water his plant and the leaves start to droop. The structures first affected by water loss and now not functioning are the:
   a. sieve tubes.
   b. sclereids and fibers.
   c. vessel members and tracheids.
   d. companion cells.
   e. guard cells and stoma.

10. The greatest mitotic activity in a root takes place in the:
    a. zone of maturation.
    b. zone of cell division.
    c. zone of elongation.
    d. root cap.
    e. endodermis.

Questions for Discussion

1. Leaves are modified in diverse ways. Cactus leaves, for example, are transformed into spines. Cacti are adapted to arid habitats in which relatively few other plant species grow. What kinds of selection pressures may have operated to favor the evolution of spinelike cactus leaves?

2. While camping in a national forest you notice a "Do Not Litter" sign nailed onto the trunk of a mature fir tree about 7 feet off the ground. When you return 5 years later, will the sign be at the same height, or will the tree's growth have raised it higher?

3. Peaches, cherries, and other fruits with pits are produced only on secondary branches that are 1 year old. To renew the fruiting wood on a peach tree, how often would you prune it? Where on a branch would you make the cut, and why?

4. African violets and some other flowering plants are propagated commercially using leaf cuttings. Initially, a leaf detached from a parent plant is placed in a growth medium. In time, adventitious shoots and roots develop from the leaf blade, producing a new plant. Which cells in the original leaf tissue are the most likely to give rise to the new structures? What property of the cells makes this propagation method possible?

Experimental Analysis

The sticky cinquefoil (*Potentilla glandulosa*) is a small, deciduous plant with bright yellow flowers that lives in throughout the American West, and its leaf phenotype can vary depending on environmental conditions. Inland, where there are dramatic seasonal temperature swings and unpredictable droughts, plants shed their large "summer leaves" in autumn when the temperature begins to drop. In the spring new leaves are smaller and develop in a compact rosette. This phenotype persists for several months and is thought to be an adaptation that makes the plants less vulnerable to drought (because less water evaporates from reduced leaf surfaces). By contrast, the leaves of *P. glandulosa* plants growing in a coastal climate are always large. In their habitat, seasonal temperature swings are not as great and the annual cycle of winter rain and summer drought is highly predictable. Suppose you decide to explore the hypothesis that the coastal population is genetically capable of exhibiting the same seasonal shift in leaf morphology as the inland plants. Would you need access to a greenhouse where you can control variables, or would it be just as easy to do experiments in the wild? Explain your reasoning and outline your experimental design—including the variable or variables you will test in the first experiment.

Evolution Link

About 90 million years ago flowering plants began their rapid (in a geologic timeframe) rise to dominance in the modern Kingdom Plantae. The first angiosperms may originally have been small, treelike plants in tropical regions, but at some point they began diversifying rapidly into other habitats where early gymnosperms flourished. In the 1990s South African botanist William Bond proposed the "slow seedling" hypothesis to help explain this evolutionary change, and botanists continue to explore and refine it. The hypothesis proposes that angiosperms were able to encroach on and eventually dominate many habitats where ancient gymnosperms lived in part because flowering species increasingly evolved adaptations that made them fast-growing herbaceous plants. Gymnosperms, by contrast grow more slowly. Based on your reading in this chapter and Chapter 27, what are some structural and biochemical features of gymnosperms (such as conifers) that might result in slower growth, putting them at a competitive disadvantage in this scenario?

How Would You Vote?

Large-scale farms and large cities compete for clean, fresh water, which is becoming scarcer as human population growth skyrockets. Should cities restrict urban growth to reduce conflicts over water supplies? Go to www.thomsonedu.com/login to investigate both sides of the issue and then vote.

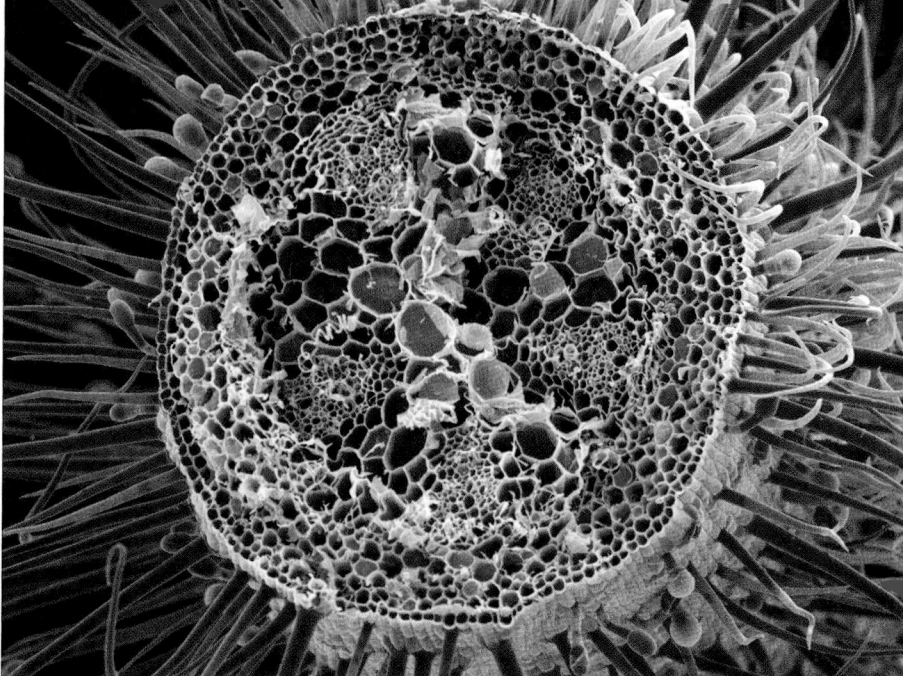

Cross section of the stem of a geranium *(Pelargonium)* showing parenchyma (pink) wrapping around vessels that transport water and nutrients in plants. In this false-color SEM, large-diameter vessels (xylem) that carry water and minerals appear whitish and bundles of smaller vessels (phloem), which transport sugars, appear pale green.

32 Transport in Plants

WHY IT MATTERS

The coast redwood, *Sequoia sempervirens* **(Figure 32.1),** takes life to extremes. Redwood trees can live for more than 2000 years, and they can grow taller than any other organism on Earth. The tallest known specimen, located in Redwood National Park in California, soars 115.5 m, roughly 379 ft, from the dank forest floor. Botanists who have studied these giants estimate that such massive plants consume thousands of liters of water each day to survive. And that water—with its cargo of dissolved nutrients—must be transported the great distances between roots and leaves.

At first, movement of fluids and solutes 100 m or more from a mature redwood's roots to its leafy crown may seem to challenge the laws of physics. Raising water that high above ground in a pipe requires a powerful mechanical pump at the base and substantial energy to counteract the pull of gravity. You also require a pump—your heart—to move fluid over a vertical distance of less than 3 m. Yet a redwood tree has no pump. As you'll learn in this chapter, the evolutionary adaptations that move water and solutes throughout the plant body can move large volumes over great distances by harnessing the cumulative effects of seemingly weak interactions such as cohesion

Figure 32.1
Redwoods *(Sequoia sempervirens)* such as this tree growing in coastal California have reached recorded heights of over 100 m during life spans of more than 2000 years. Such extremely tall trees exemplify the ability of plants to move water and solutes from roots to shoots over amazingly long distances.

those for long-distance transport. Short-distance transport mechanisms move substances into and between cells across membranes, and also to and from vascular tissues. For example, water, oxygen, and minerals enter roots by crossing the cell membranes of root hairs **(Figure 32.2a)**, and nutrients such as carbohydrates from photosynthesis cross plasma membranes to nourish cells of the plant body. Similarly, water and other substances move short distances to and from a plant's xylem and phloem, which are arranged in vascular bundles **(Figure 32.2b)**. Long-distance transport mechanisms move substances between roots and shoot parts **(Figure 32.2c)**. Thus water and dissolved minerals travel in the xylem from roots to other plant parts, and products of photosynthesis move in the phloem from the leaves and stems into roots and other structures. Carbon dioxide for photosynthesis enters photosynthetic tissues in the shoot.

We consider transport processes in the xylem and phloem later in this chapter. For the moment our focus is on mechanisms that move water and solutes into and out of specific cells in roots, leaves, and stems. Keep in mind that the plant cell wall does not prevent solutes from moving into plant cells. Most solutes can cross the wall by way of the plasmodesmata that connect adjacent cells (see Section 5.4).

Both Passive and Active Mechanisms Move Substances into and out of Plant Cells

Recall from Chapter 6 that in all cells there are two general mechanisms for transporting water and solutes across the plasma membrane into and between cells. In **passive transport**, substances move down a concentration gradient or, if the substance is an ion, down an electrochemical gradient. **Active transport** requires the cell to expend energy in moving substances *against* a gradient, usually by hydrolysis of ATP.

True to its name, simple diffusion is the simplest form of passive transport: as described in Section 6.2, oxygen, carbon dioxide, water, and some other small molecules can readily diffuse across cell plasma membranes, following a concentration gradient. By contrast, in all other types of membrane transport, ions and some larger molecules cross cell membranes assisted by carriers collectively called **transport proteins**, which are embedded in the membrane.

Passive transport of substances down an electrochemical gradient is called *facilitated diffusion* because the transport protein involved "facilitates" the process in some way. Transport proteins called *channel proteins* are configured to form a pore in the plasma membrane. Those called *carrier proteins* change shape in a way that releases the substance to the other side of the membrane.

In active transport, membrane transport proteins use energy to move substances against a concentration gradient or an electrochemical gradient. As you may

and evaporation. Overall, plant transport mechanisms solve a fundamental biological problem—the need to acquire materials from the environment and distribute them throughout the plant body.

Our discussion begins with a brief review of the principles of water and solute movement in plants, a topic introduced in Chapter 6. Then we examine how those principles apply to the movement of water and solutes into and through a plant's vascular pipelines.

32.1 Principles of Water and Solute Movement in Plants

In plants, as in all organisms, the movement of water and solutes begins at the level of individual cells and relies on mechanisms such as osmosis and the operation of transport proteins in the plasma membrane. Once water and nutrients enter a plant's specialized transport systems—the vascular tissues xylem and phloem—other mechanisms carry them between various regions of the plant body in response to changing demands for those substances. Ultimately, these movements of materials result from the integrated activities of the individual cells, tissues, and organs of a single, smoothly functioning organism—the whole plant.

Plant transport mechanisms fall into two general categories—those for short-distance transport and

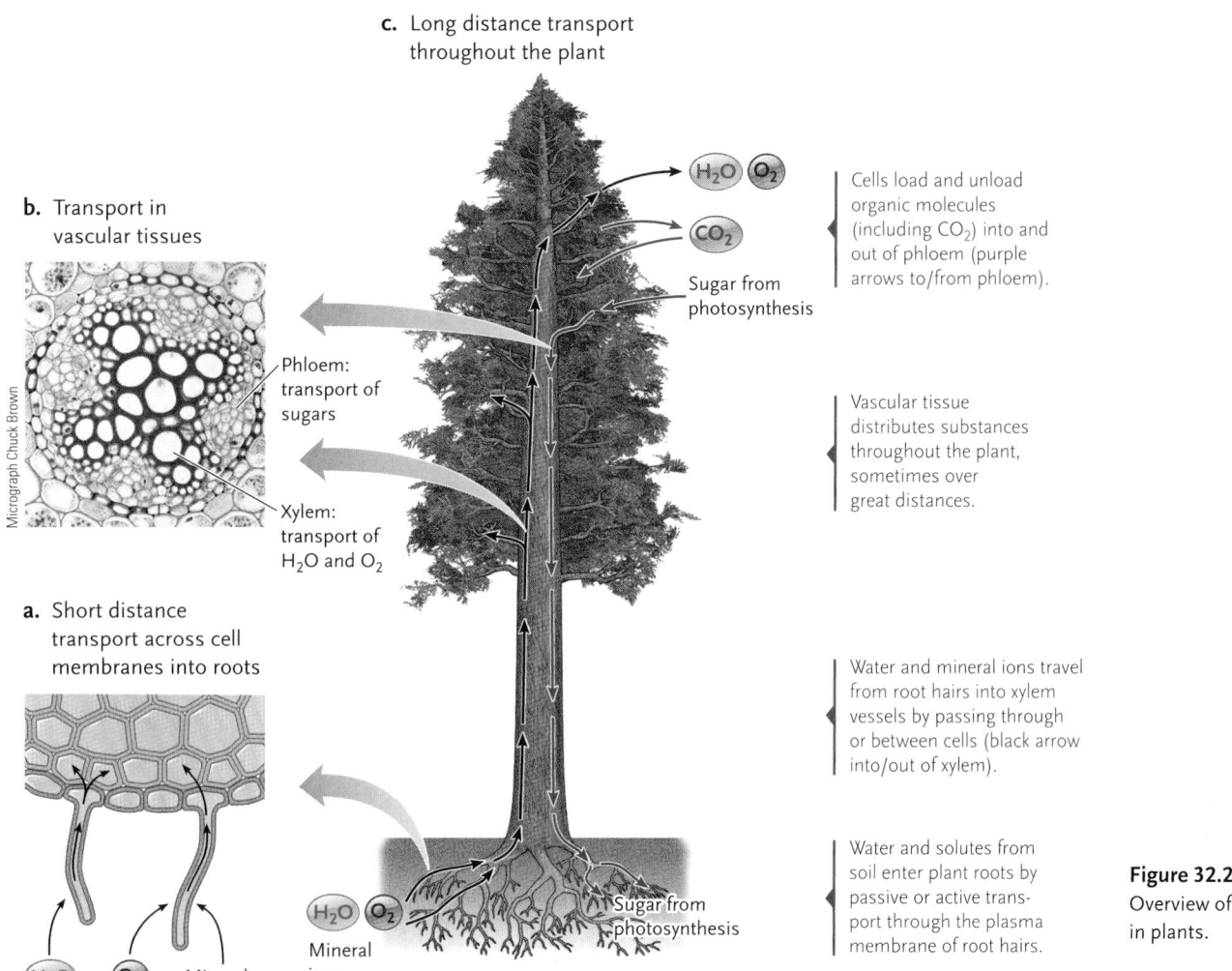

c. Long distance transport throughout the plant

b. Transport in vascular tissues

Micrograph Chuck Brown

Phloem: transport of sugars

Xylem: transport of H_2O and O_2

a. Short distance transport across cell membranes into roots

H_2O O_2 Minerals

H_2O O_2

Mineral ions

Sugar from photosynthesis

H_2O O_2

CO_2

Sugar from photosynthesis

Cells load and unload organic molecules (including CO_2) into and out of phloem (purple arrows to/from phloem).

Vascular tissue distributes substances throughout the plant, sometimes over great distances.

Water and mineral ions travel from root hairs into xylem vessels by passing through or between cells (black arrow into/out of xylem).

Water and solutes from soil enter plant roots by passive or active transport through the plasma membrane of root hairs.

Figure 32.2
Overview of transport routes in plants.

recall from Section 6.4, an electrochemical gradient exists across cell membranes when the concentrations of various ions differ inside or outside the cell. The differences in ion concentration result in a difference in electrical charge across the plasma membrane. In plant cells the cytoplasm is slightly more negative than the fluid outside the cell. This charge difference is measured as an electrical voltage called the **membrane potential.** The word "potential" refers to the fact that the movement of ions across a membrane is a potential source of energy—that is, such ion movements can perform cellular work.

ATP provides the energy for active transport of substances into and out of plant cells. Hydrogen ions (protons), which tend to be more concentrated outside the cell than in the negatively charged cytoplasm, play a central role in the process. First, a proton pump pushes H^+ across the plasma membrane against its electrochemical gradient, from the inside to the outside of the cell **(Figure 32.3a).** As protons accumulate outside the cell, the electrochemical gradient becomes steeper and significant potential energy is available. Crucial solutes such as cations (positively charged ions) often are more concentrated in the extracellular fluid. One result of the increased charge difference cre-

ated by proton pumping is that cations move into the cell through their membrane channels **(Figure 32.3b).** These cations include mineral ions that have essential roles in plant cell metabolism.

The H^+ gradient also powers *secondary active transport,* a process in which a concentration gradient of an ion is used as the energy source for active transport of another substance. The two secondary mechanisms—*symport* and *antiport*—actively transport ions, sugars, and amino acids into and out of plant cells against their concentration gradient. In **symport,** the potential energy released as H^+ follows its gradient into the cell is coupled to the simultaneous uptake of another ion or molecule **(Figure 32.3c).** In this way, plant cells can take up metabolically important anions such as nitrate (NO_3^-) and potassium (K^+). Nearly all organic substances that enter plant cells move in by symport as well.

In **antiport,** the energy released as H^+ diffuses into the cell powers the active transport of a second molecule, such as Ca^{2+}, in the opposite direction, *out of* the cell **(Figure 32.3d).** One of antiport's key functions is to remove excess Na^+, which readily moves into plant cells by facilitated diffusion through channel proteins. If the Na^+ were not eliminated, it would quickly build up to toxic levels.

a. H⁺ pumped against its electrochemical gradient

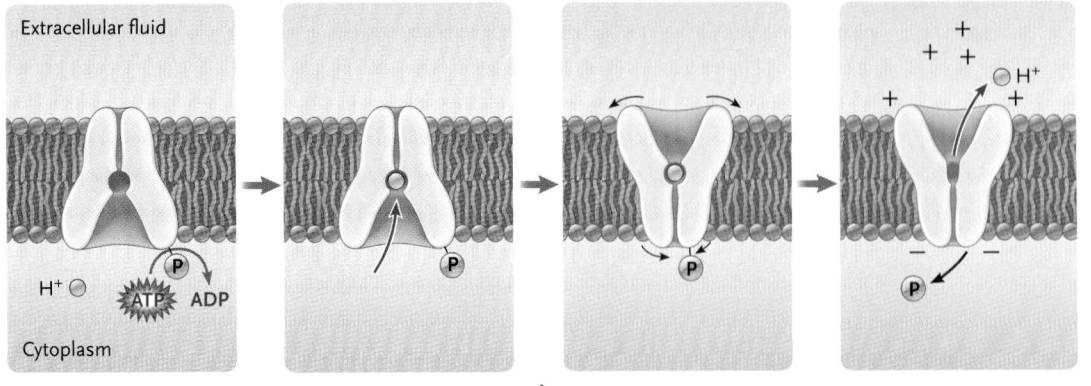

b. Uptake of cations

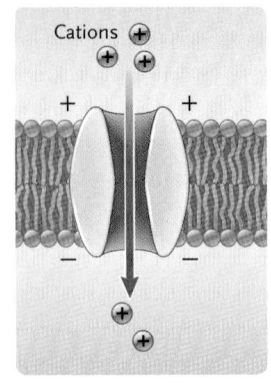

ATP energy pumps hydrogen ions (H⁺) out of the cytoplasm, creating an H⁺ gradient.

The concentration of H⁺ becomes higher outside the membrane than inside. Inward diffusion of H⁺ in response to the gradient becomes a source of energy for transporting other ions and neutral molecules such as sugar into the plant cell.

Some cations, such as NH_4^+, enter the cell through selective channel proteins, following the electrochemical gradient created by H⁺ pumping.

c. Symport

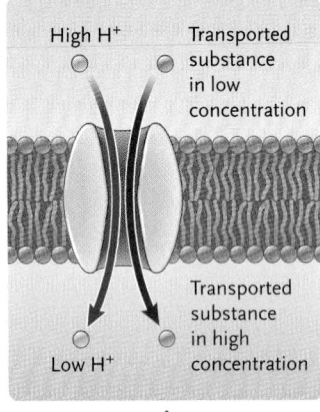

d. Antiport

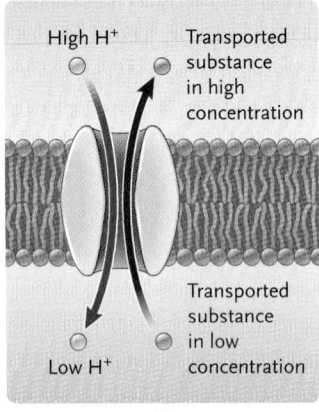

Figure 32.3
Ion transport across the plasma membrane.

In symport, the inward diffusion of H⁺ is coupled with the simultaneous active transport of another substance into the cell.

In antiport, H⁺ moving into the cell powers the movement of another solute in the opposite direction.

Both passive and active transport are selective transport mechanisms that transport specific substances. Two factors govern this specificity. One is the size of the interior channel, which allows only molecules in a particular size range to pass through. The other factor is the distribution of charges along the inside of the channel. A channel that permits cations such as Na⁺ to pass through easily may completely bar anions such as Cl⁻ and vice versa.

Relatively speaking, only small amounts of mineral ions and other solutes move into and out of plant cells. As we see next, H_2O is another matter. Throughout a plant's life large volumes of water enter and exit its cells and tissues by way of osmosis.

Osmosis Governs Water Movement in Plants

One of the most important aspects of plant physiology is how water moves into and through plant cells and tissues. Inside a plant's tubelike vascular tissues, large amounts of water or any other fluid travel by **bulk flow**—the group movement of molecules in response to a difference in pressure between two locations, like water in a closed plumbing system gushing from an open faucet. For example, the dilute solution of water and ions that flows in the xylem, called **xylem sap**, moves by bulk flow from roots to shoot parts. The solution is pulled upward through the plant body in a process that relies on the cohesion of water molecules and that we will consider more fully later in this chapter. Individual cells, however, gain and lose water by **osmosis**, the passive movement of water across a selectively permeable membrane in response to solute concentration gradients, a pressure gradient, or both (see Section 6.3). The driving force for osmosis is energy stored in the water itself. This potential energy, called **water potential**, is symbolized by the Greek letter ψ. By convention, pure water has a ψ value of zero. Two factors that strongly influence this value in living plants are the presence of solutes and physical pressure.

The effect of dissolved solutes on water's tendency to move across a membrane is called *solute potential*, symbolized by ψ_S. In practical terms, water potential is higher where there are more water molecules in a solution relative to the number of solute molecules. Likewise, the water potential is *lower* in a solution with relatively more solutes. The relationship between water potential and solute potential is vital to understanding transport phenomena in plants because water tends to move by osmosis from regions where water potential is higher to regions where it is lower. Solutes are usually more concentrated inside plant cells than in the fluid surrounding them. This means the water potential is higher outside plant cells than inside them, so water tends to enter the cells by osmosis. This in fact is the mechanism that draws soil water into a plant's roots.

The osmotic movement of water into a plant cell cannot continue indefinitely, however, because eventually physical pressure counterbalances it. The wall around a plant cell strictly limits how much the cell can expand. Accordingly, as water moves into plant cells, the pressure inside them increases. This pressure, called **turgor pressure**, rises until it is high enough to prevent more water from entering a cell by osmosis. In effect, when osmotic water movement stops, turgor pressure has increased the water potential inside the cell until it equals the potential of the water outside the cell. The physical pressure required to halt osmotic water movement across a membrane is termed a solution's *pressure potential* and is symbolized as ψ_P.

By convention, plant physiologists measure water potential in units of pressure called **megapascals** (MPa). They use standard atmospheric pressure as a baseline, assigning it a value of zero. Accordingly, the water potential of pure water at standard atmospheric pressure is expressed as 0 MPa. This notation can be used to describe the changing effects under different conditions of solute potential and pressure potential **(Figure 32.4)**. Adding pressure increases the MPa while adding solutes reduces it (because the relative concentration of water is lower), and water will flow from a solution of higher MPa to a solution of lower MPa. With these principles in mind, consider now how they operate in living plant cells.

Recall from Section 5.4 that a large **central vacuole** occupies most of the volume of a mature plant cell. The central vacuole, which is surrounded by a vacuolar membrane, or **tonoplast**, contains a dilute solution of sugars, proteins, other organic molecules, and salts. The cell cytoplasm is confined to a thin layer between the tonoplast and the plasma membrane. A major role of the central vacuole is to maintain turgor pressure in

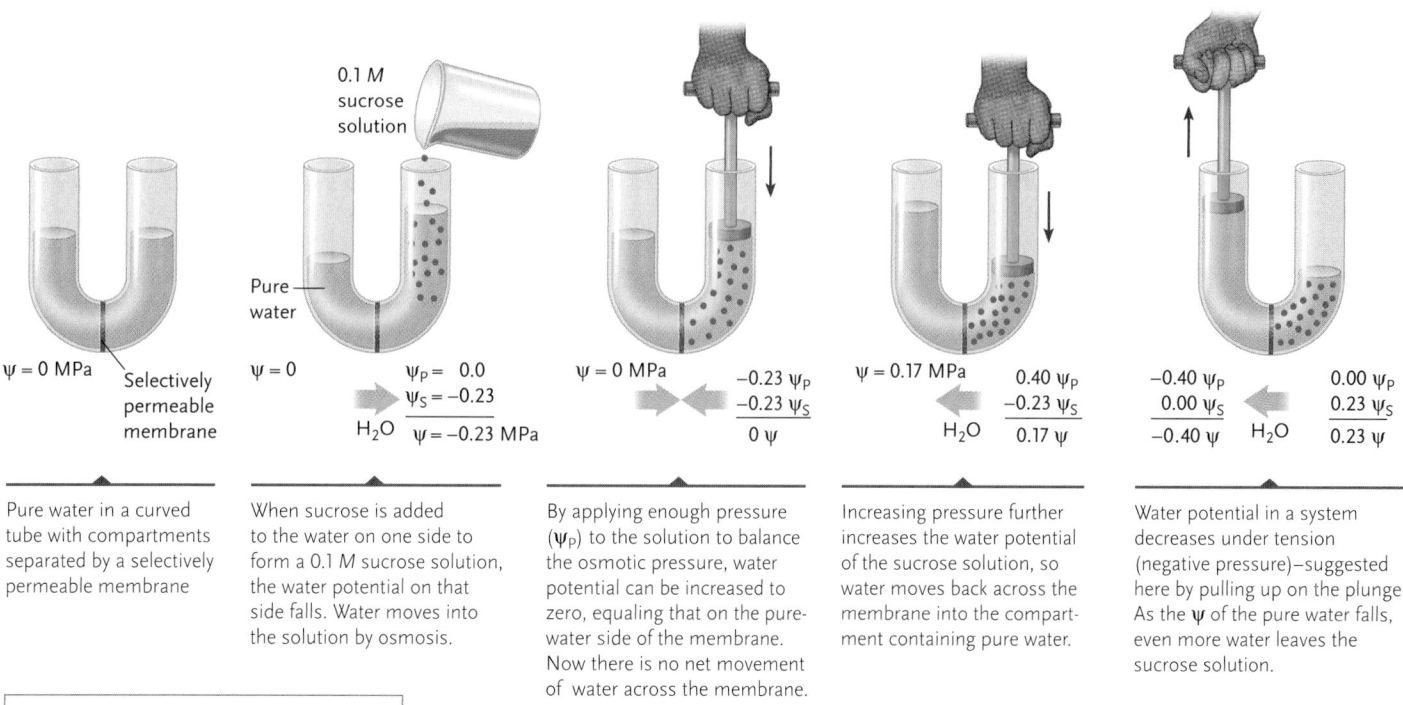

Plant physiologists assign a value of 0 MPa to the water potential (ψ) of pure water in an open container under normal atmospheric pressure and temperature.

Figure 32.4

The relationship between osmosis and water potential. If the water potential is higher on one side of a selectively permeable membrane, water will cross the membrane to the area of lower water potential. This diagram shows pure water on one side of a selectively permeable membrane and a simple sucrose solution on the other side. In an organism, however, the selectively permeable membranes of cells are rarely if ever in contact with pure water.

A Plant Water Channel Gives Oocytes a Drink

Water moves into or out of the central vacuole of plant cells to compensate for gains or losses of water in the surrounding cytoplasm. Does this water simply diffuse through the lipid part of the tonoplast, or does it move through an aquaporin? Christophe Maurel and his colleagues at the University of California, San Diego, sought to answer this question. They were encouraged by the discovery of aquaporins in the plasma membranes of red blood cells and by the fact that a closely related protein called TIP (tonoplastintrinsic protein) occurs in the tonoplast.

To find out whether TIP functions as the water channel of tonoplasts, the team began by isolating the gene that encodes TIP in *Arabidopsis thaliana* plants. For the later experiments they selected animal cells (oocytes of the frog *Xenopus laevis*) to ensure no other proteins made in plant cells could affect the outcome.

Next they cloned the coding sequence of the TIP gene, inserting it into a bacterial plasmid cloning vector. The vector contained a promoter that allowed in vitro transcription of a cloned coding sequence. In addition, the research team had added to it DNA sequences for 5′ and 3′ UTRs (untranslated regions; see Section 15.3) that function in the processing of the coding sequence in the mRNA transcripts of a *Xenopus* gene. The TIP coding sequence was inserted between the DNA for the UTRs, which ensured that the *Xenopus* oocytes could efficiently translate mRNAs transcribed from the TIP sequence clone. That is, the test-tube transcription of the engineered TIP clone produced mRNAs in a form that could readily be translated into TIP proteins inside *Xenopus* cells.

The test-tube TIP mRNA molecules were then injected into mature *Xenopus* oocytes, which are normally only slightly permeable to water. After 2 to 3 days in an isotonic medium, the oocytes were transferred to a hypotonic medium. Thy swelled and ruptured within 6 minutes. Control oocytes that were not injected with the TIP mRNA, or that were injected only with distilled water, swelled only slightly during the same interval and did not burst when placed in a hypotonic medium. These results supported the conclusion that the TIP protein forms an aquaporin when inserted into a membrane. In this system, the TIP proteins inserted into the oocyte plasma membrane, since animal cells do not have tonoplasts. In its normal location in the tonoplast of plant cells, TIP evidently allows water to move readily in either direction, compensating for water movement between the thin layer of cytoplasm and the extracellular space.

---

the cell. Many solutes that enter a plant cell are actively transported from the cytoplasm into the central vacuole through channels in the tonoplast. As the solutes accumulate, water follows by osmosis.

The plant cell's relatively small amount of cytoplasm must compensate fairly quickly for water gains or losses caused by changes in osmotic flow. If the medium around a plant cell becomes hypertonic (has a high solute concentration), for example, water flows rapidly out of the cell. Water from the central vacuole replaces it, entering the cytoplasm through water-conducting channel proteins called **aquaporins**. Experiments that identified this channel are the topic of this chapter's *Insights from the Molecular Revolution*.

The water mechanics we have been discussing have major implications for land plants. For instance, the drooping of leaves and stems called **wilting** occurs when environmental conditions cause a plant to lose more water than it gains. Conditions that lead to wilting include dry soil, in which case the water potential in the soil falls below that in the plant. Then the turgor pressure inside the cells falls, and the protoplast shrinks away from the cell wall **(Figure 32.5a).** By contrast, as long as the ψ of soil is higher than that in root epidermal cells, water will follow the ψ gradient and enter root cells, making them turgid, or firm **(Figure 32.5b).** As we see in the next section, water and solutes entering roots may move through the plant body by several routes.

STUDY BREAK

1. Explain the role(s) of a gradient of protons in moving substances across a plant cell's plasma membrane.
2. How do symport and antiport differ? Give examples of key substances each mechanism transports.
3. What is "water potential," and why is it important with respect to plant cells?

32.2 Transport in Roots

Soil around roots provides a plant's water and minerals, but roots don't simply "soak up" these essential substances. Instead, water and minerals that enter roots first travel laterally through the root cortex to the root xylem. Only then do they begin their journey upward to stems, leaves, and other tissues.

Water Travels to the Root Xylem by Three Pathways

Soil water always enters a root through the root epidermis. Once inside a root, however, water may take one of three routes into the root xylem, traveling either through living cells or in nonliving areas of the root **(Figure 32.6).**

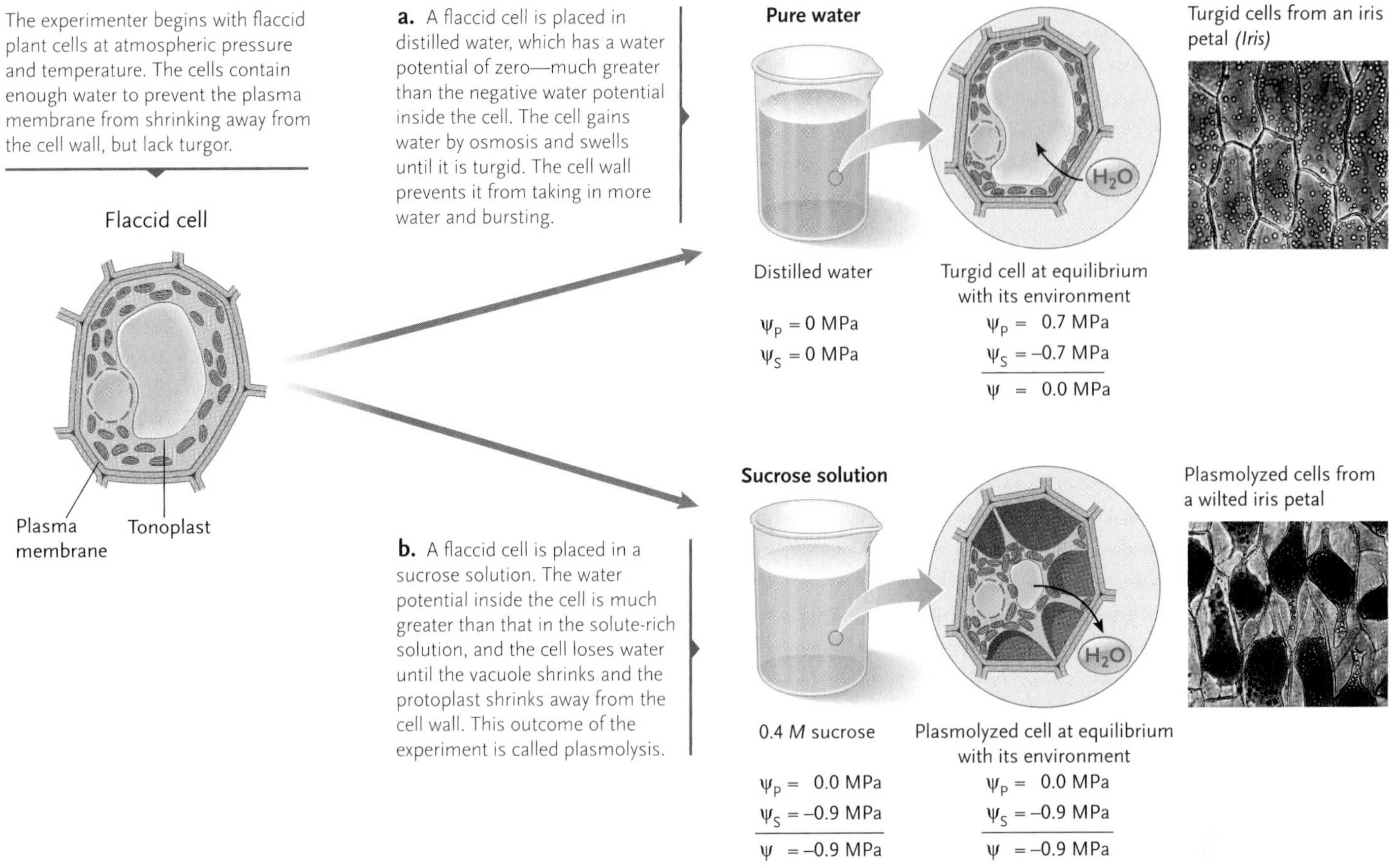

The experimenter begins with flaccid plant cells at atmospheric pressure and temperature. The cells contain enough water to prevent the plasma membrane from shrinking away from the cell wall, but lack turgor.

Flaccid cell

Plasma membrane   Tonoplast

a. A flaccid cell is placed in distilled water, which has a water potential of zero—much greater than the negative water potential inside the cell. The cell gains water by osmosis and swells until it is turgid. The cell wall prevents it from taking in more water and bursting.

Pure water

Distilled water

$\psi_P = 0$ MPa
$\psi_S = 0$ MPa

Turgid cell at equilibrium with its environment

$\psi_P = 0.7$ MPa
$\underline{\psi_S = -0.7}$ MPa
$\psi = 0.0$ MPa

Turgid cells from an iris petal *(Iris)*

b. A flaccid cell is placed in a sucrose solution. The water potential inside the cell is much greater than that in the solute-rich solution, and the cell loses water until the vacuole shrinks and the protoplast shrinks away from the cell wall. This outcome of the experiment is called plasmolysis.

Sucrose solution

0.4 *M* sucrose

$\psi_P = 0.0$ MPa
$\underline{\psi_S = -0.9}$ MPa
$\psi = -0.9$ MPa

Plasmolyzed cell at equilibrium with its environment

$\psi_P = 0.0$ MPa
$\underline{\psi_S = -0.9}$ MPa
$\psi = -0.9$ MPa

Plasmolyzed cells from a wilted iris petal

Figure 32.5

An experiment to test the effects of different osmotic environments on plant cells. Notice that in both **(a)** and **(b)** the final condition is the same: the water potential of the plant cell and its environment become equal. (Micrographs: © Claude Nuridsany and Marie Perennou/Science Photo Library/Photo Researchers, Inc.)

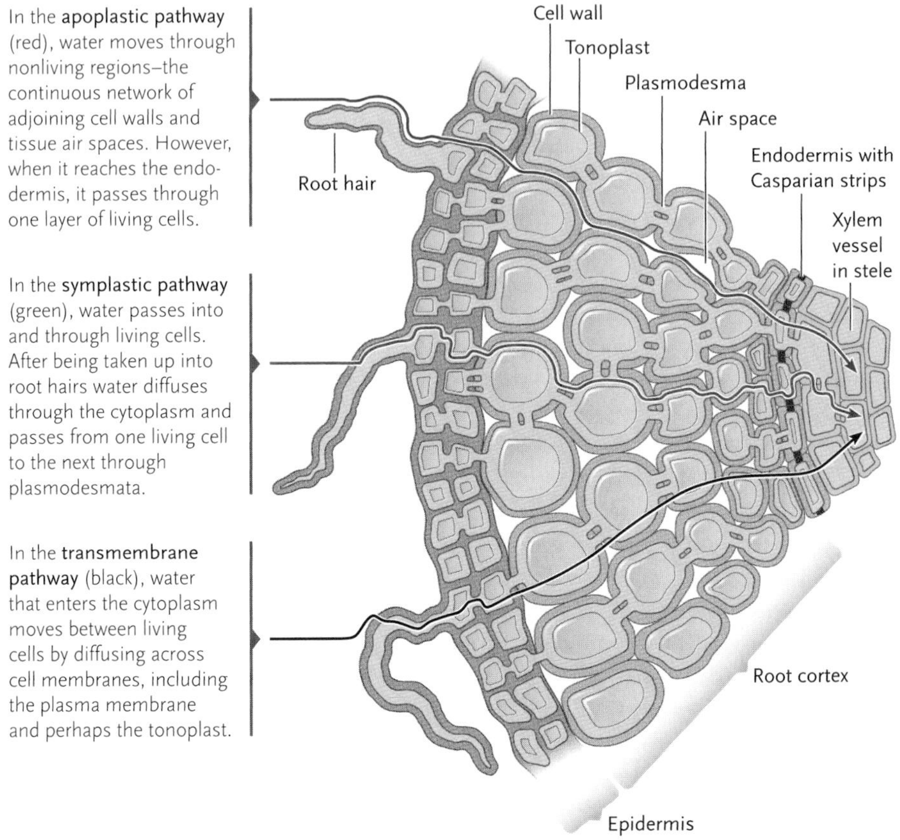

In the **apoplastic pathway** (red), water moves through nonliving regions—the continuous network of adjoining cell walls and tissue air spaces. However, when it reaches the endodermis, it passes through one layer of living cells.

In the **symplastic pathway** (green), water passes into and through living cells. After being taken up into root hairs water diffuses through the cytoplasm and passes from one living cell to the next through plasmodesmata.

In the **transmembrane pathway** (black), water that enters the cytoplasm moves between living cells by diffusing across cell membranes, including the plasma membrane and perhaps the tonoplast.

Cell wall
Tonoplast
Plasmodesma
Air space
Endodermis with Casparian strips
Xylem vessel in stele
Root hair
Root cortex
Epidermis

Figure 32.6

Pathways for the movement of water into roots. Ions also enter roots via these three pathways, but must be actively transported into cells when they reach the Casparian strips of the endodermis. In this way, only certain solutes in soil water are allowed to enter the stele.

Nonliving regions of a plant such as the continuous network of adjoining cell walls and air spaces in root tissue are called the *apoplast*. Thus water follows an **apoplastic pathway** when it moves through the apoplast of roots, a route that does not cross cell membranes. Botanists refer to a plant's living parts as the *symplast,* and water moving through roots in the **symplastic pathway** moves from cell to cell through the open channels of plasmodesmata. Water also can enter root cells across the cell plasma membranes, a **transmembrane pathway**. Water crosses the tonoplast of the central vacuole in this way as well.

When water enters a root, some diffuses into epidermal cells, entering the symplast. But a great deal of the water taken up by plant roots moves into the apoplast, moving along through cell walls and intercellular spaces. This apoplastic water (and any solutes dissolved in it) travels rapidly inward until it encounters the endodermis, the sheetlike single layer of cells that separates the root cortex from the stele. Cells in the root cortex generally have air spaces between them (which helps aerate the tissue), but endodermal cells are tightly packed **(Figure 32.7a)**. Each one also has a beltlike **Casparian strip** in its radial and transverse walls, positioned somewhat like a ribbon of packing tape around a rectangular package

(Figure 32.7b–c). The strip is impregnated with suberin, a waxy substance impermeable to water. Thus the Casparian strip blocks the apoplastic pathway at the endodermis, preventing water and solutes in the apoplast from automatically passing on into the stele. Instead, if molecules are to move into the stele, they must detour across the plasma membranes of endodermal cells, entering the cells (and the symplast) where the wall is not blanketed by a Casparian strip **(Figure 32.7d)**. From there water and solutes can pass through plasmodesmata to cells in the outer layer of the stele (the pericycle), then on into the xylem.

Although water molecules can easily cross an endodermal cell's plasma membrane, the semipermeable membrane allows only a subset of the solutes in soil water to cross. Undesirable solutes may be barred, while desirable ones may move into the cell by facilitated diffusion or active transport. Conversely, the endodermis prevents needed substances in the xylem from leaking out, back into the root cortex. In this way the endodermis provides important control over which substances enter and leave a plant's vascular tissue. The roots of most flowering plants also have a second layer of cells with Casparian strips just inside the root

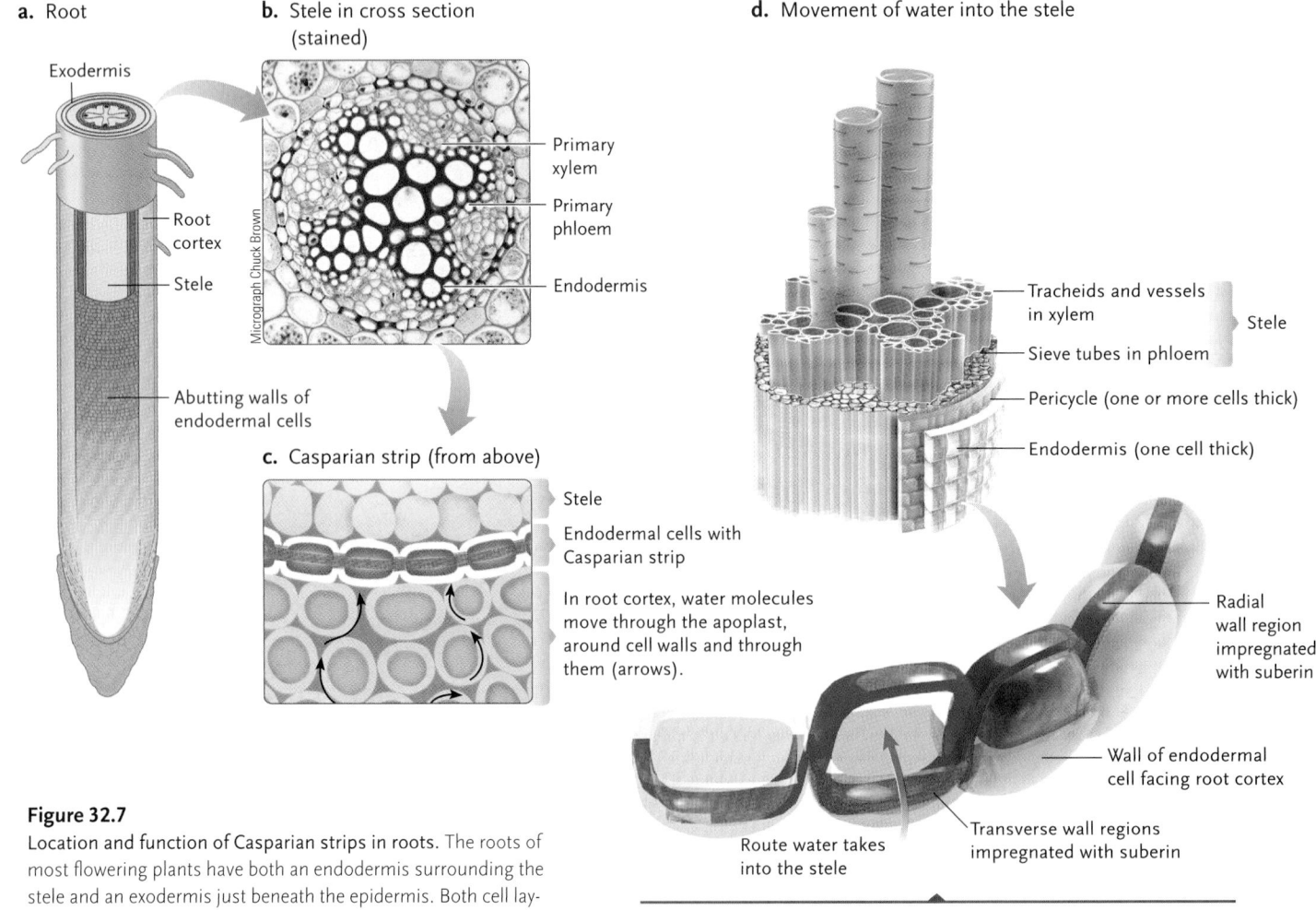

a. Root

Exodermis
Root cortex
Stele
Abutting walls of endodermal cells

b. Stele in cross section (stained)

Micrograph Chuck Brown

Primary xylem
Primary phloem
Endodermis

c. Casparian strip (from above)

Stele
Endodermal cells with Casparian strip
In root cortex, water molecules move through the apoplast, around cell walls and through them (arrows).

d. Movement of water into the stele

Tracheids and vessels in xylem
Stele
Sieve tubes in phloem
Pericycle (one or more cells thick)
Endodermis (one cell thick)
Radial wall region impregnated with suberin
Wall of endodermal cell facing root cortex
Transverse wall regions impregnated with suberin
Route water takes into the stele

Waxy, water-impervious Casparian strip (gold) in abutting walls of endodermal cells that control water and nutrient uptake

Figure 32.7

Location and function of Casparian strips in roots. The roots of most flowering plants have both an endodermis surrounding the stele and an exodermis just beneath the epidermis. Both cell layers have an impervious Casparian strip that helps to control the uptake of water and dissolved nutrients.

epidermis. This layer, the exodermis, functions like the endodermis.

Roots Take Up Ions by Active Transport

Mineral ions in soil water also enter roots through the epidermis. Some enter the apoplast along with water, but most ions important for plant nutrition tend to be much more concentrated in roots than in the surrounding soil, so they cannot follow a concentration gradient into root epidermal cells. Instead the epidermal cells actively transport ions inward—that is, ions enter the symplast immediately. They travel to the xylem via the symplastic or transmembrane pathways. Other ions can still move inward following the apoplastic pathway until they reach the Casparian strip of the endodermis. If they are to contribute to the plant's nutrition, however, they must be actively transported from the exodermis into cells of the root cortex and, as just described, from the endodermis into the stele. In short, mechanisms that control which solutes will be absorbed by root cells ultimately determine which solutes will be distributed through the plant.

Once an ion reaches the stele, it diffuses from cell to cell until it is "loaded" into the xylem. Experiments to determine whether the loading is passive (by diffusion) or active have been inconclusive, so the details of this final step are not entirely clear. Because the xylem's conducting elements are not living, water and ions in effect reenter the apoplastic pathway when they reach either tracheids or vessels. Once in the xylem, water can move laterally to and from tissues or travel upward in the conducting elements. Minerals are distributed to living cells and taken up by active transport. The following section examines how this "distribution of the wealth" takes place.

STUDY BREAK

1. Explain two key differences in how the apoplastic and symplastic pathways route substances laterally in roots.
2. How does an ion enter a root hair and then move to the xylem?

32.3 Transport of Water and Minerals in the Xylem

We return now to the question that opened this chapter: How does the solution of water and minerals called xylem sap move—100 m or more in the tallest trees—from roots to stems, then into leaves? Xylem sap is mostly water, and we know that it moves upward by bulk flow through the tracheids and vessels in xylem. Yet because mature xylem cells are dead, they cannot expend energy to move water into and through the plant shoot. Instead, the driving force for the upward movement of xylem sap from root to shoot is sunlight, which causes water to evaporate from leaves and other aerial parts of land plants. Experiments show that only a small fraction of the water in xylem sap is used in a plant's growth and metabolism. The rest evaporates into the air in a phenomenon called **transpiration.** As described next, transpiration drives the ascent of sap.

The Mechanical Properties of Water Have Key Roles in Its Transport

Chapter 2 introduced several biologically important mechanical properties of water. Two of them interest us here. First, water molecules are strongly *cohesive:* they tend to form hydrogen bonds with one another. Second, water molecules are *adhesive:* they form hydrogen bonds with molecules of other substances, including the carbohydrates in plant cell walls. Water's cohesive and adhesive forces jointly pull water molecules into exceedingly small spaces, such as crevices in cell walls or narrow tubes such as xylem vessels in roots, stems, and leaves. In 1914, plant physiologist Henry Dixon explained the ascent of sap in terms of the relationship between transpiration and water's mechanical properties. His model of xylem transport is now called the **cohesion–tension mechanism of water transport (Figure 32.8).**

According to the cohesion–tension model, water transport begins as water evaporates from the walls of mesophyll cells inside leaves and into the intercellular spaces. This water vapor escapes by transpiration through open stomata, the minute passageways in the leaf surface. As water molecules exit the leaf, they are replaced by others from the mesophyll cell cytoplasm. The water loss gradually reduces the water potential in a transpiring cell below the water potential in the leaf xylem. Now, water from the xylem in the leaf veins follows the gradient into cells, replacing the water lost in transpiration.

In the xylem, water molecules are confined in narrow, tubular xylem cells. The water molecules form a long chain, like a string of weak magnets, held together by hydrogen bonds between individual molecules. When a water molecule moves out of a leaf vein into the mesophyll, its hydrogen bonds with the next molecule in line stretch but don't break. The stretching creates *tension*—a negative pressure gradient—in the column. Adhesion of the water column to xylem vessel walls adds to the tension. Under continuous tension from above, the entire column of water molecules in xylem is drawn upward, in a fashion somewhat analogous to the way water moves up through a drinking straw. Botanists refer to this root-to-shoot flow as the *transpiration stream.*

Transpiration continues regardless of whether evaporating water is replenished by water rapidly taken

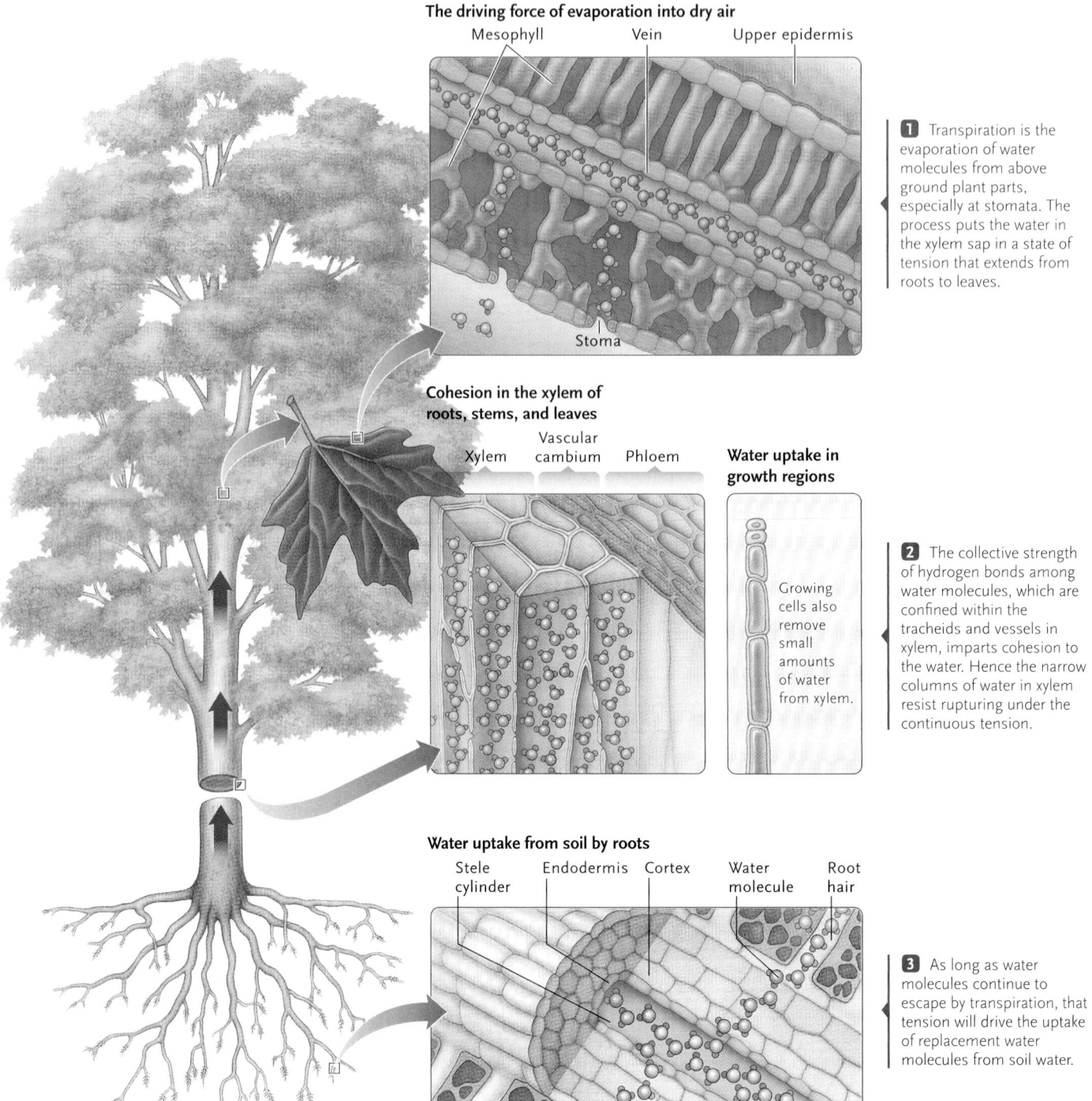

The driving force of evaporation into dry air

Mesophyll Vein Upper epidermis

Stoma

1 Transpiration is the evaporation of water molecules from above ground plant parts, especially at stomata. The process puts the water in the xylem sap in a state of tension that extends from roots to leaves.

Cohesion in the xylem of roots, stems, and leaves

Xylem Vascular cambium Phloem

Water uptake in growth regions

Growing cells also remove small amounts of water from xylem.

2 The collective strength of hydrogen bonds among water molecules, which are confined within the tracheids and vessels in xylem, imparts cohesion to the water. Hence the narrow columns of water in xylem resist rupturing under the continuous tension.

Water uptake from soil by roots

Stele cylinder Endodermis Cortex Water molecule Root hair

3 As long as water molecules continue to escape by transpiration, that tension will drive the uptake of replacement water molecules from soil water.

Figure 32.8
Cohesion–tension mechanism of water transport. Transpiration, the evaporation of water from shoot parts, creates tension on the water in xylem sap. This tension, which extends from root to leaf, pulls upward columns of water molecules that are hydrogen-bonded to one another.

up from the soil. Wilting is visible evidence that the water-potential gradient between soil and a plant's shoot parts has shifted. Remember that as soil dries out, the remaining water molecules are held ever more tightly by the soil particles. In effect, the action of soil particles reduces the water potential in the soil surrounding plant roots, and as this happens the roots take up water more slowly. However, because the water

that evaporates from the plant's leaves is no longer being fully replaced, the leaves wilt as turgor pressure drops. Reducing the water potential in soil by adding solutes such as NaCl and other salts can have the same wilting effect. When the water potential in the soil finally equals that in leaf cells, a gradient no longer exists. Then movement of water from the soil into roots and up to the leaves comes to a halt.

Leaf Anatomy Contributes to Cohesion–Tension Forces

Leaf anatomy is key to the processes that move water upward into plants. To begin with, as much as two-thirds of a leaf's volume consists of air spaces—thus there is a large internal surface area for evaporation. Leaves also may have thousands to millions of stomata, through which water vapor can escape. Both these factors increase transpiration. Also, every square centimeter of a leaf contains thousands of tiny xylem veins, so that most leaf cells lie within half a millimeter of a vein. This close proximity readily supplies water to cells and the spaces between them, from which the water can readily evaporate.

As water evaporates from a leaf, surface tension at the interface between the water film and the air in the leaf space translates into negative pressure that draws water from the leaf veins. This tension is multiplied many times over in all of the leaves and xylem veins of a plant. It increases further as the plant's metabolically active cells take up xylem sap.

In the Tallest Trees, the Cohesion–Tension Mechanism May Reach Its Physical Limit

A variety of experiments have tested the premises of the cohesion–tension model, and thus far the data strongly support it. For example, the model predicts that xylem sap will begin to move upward at the top of a tree early in the day when water begins to evaporate from leaves. Experiments with several different tree species have confirmed that this is the case. The experiments also showed that sap transport peaks at midday when evaporation is greatest, then tapers off in the evening as evaporative water loss slows.

Other experiments have probed the relationship between xylem transport and tree height. One team of researchers studied eight of the tallest living redwoods, including one that towers nearly 113 m above the forest floor. When the scientists measured the maximum tension exerted in the xylem sap in twigs at the tops of the trees, they discovered that it approached the known physical limit at which the bonds between water molecules in a column of water in a conifer's xylem will rupture. Based on this finding and other evidence, the team has predicted that the maximum height for a healthy redwood tree is 122 to 130 m. Therefore it is possible that the tallest redwoods alive today may grow taller still.

Root Pressure Contributes to Upward Water Movement in Some Plants

The cohesion–tension mechanism accounts for upward water movement in tall trees. In some nonwoody plant species, however—lawn grasses, for instance—a positive pressure can develop in roots and force xy-

Figure 32.9
Guttation, caused by root pressure. The drops of water appear at the endings of xylem veins along the leaf edges of a strawberry plant (*Fragaria*).

lem sap upward. This **root pressure** operates under conditions that reduce transpiration, such as high humidity or low light. In fact, the mechanism that produces root pressure often operates at night, when solar-powered transpiration slows or stops. Then, active transport of ions into the stele sets up a water potential gradient across the endodermis. Because the Casparian strip of the endodermis tends to prevent ions from moving back into the root cortex, the water potential difference becomes quite large. It can move enough water and dissolved solutes into the xylem to produce a relatively high positive pressure. Although not sufficient to force water to the top of a very tall plant, in some smaller plant species root pressure is strong enough to force water out of leaf openings, in a process called **guttation (Figure 32.9).** Pushed up and out of vein endings by root pressure, tiny droplets of water that look like dew in the early morning emerge from modified stomata at the margins of leaves.

Stomata Regulate the Loss of Water by Transpiration

Three environmental conditions have major effects on the rate of transpiration: relative humidity, air temperature, and air movement. The most important is relative humidity, which is a measure of the amount of water vapor in air. The less water vapor in the air, the more evaporates from leaves (because the water potential is higher in the leaves than in the dry air). The air temperature at the leaf surface also speeds evaporation as it rises. Although evaporation does cool the leaf somewhat, the amount of water lost can double for each 10°C rise in air temperature. Air movement at the leaf surface carries water vapor away from the surface and so makes a steeper gradient. Together these factors explain why on extremely hot, dry, breezy days, the leaves of certain plants must completely replace their water each hour.

a. Open stoma   **b.** Closed stoma

Guard cell   Guard cell

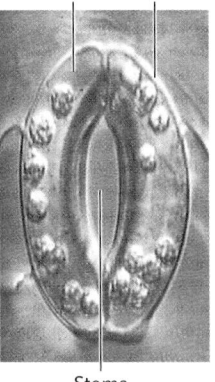
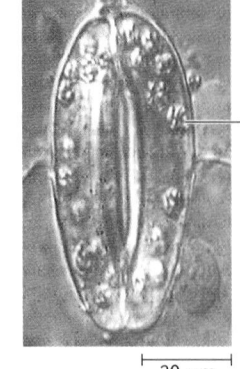

Chloroplast (guard cells are the only epidermal cells that have these organelles)

Stoma

20 μm

Figure 32.10

Guard cells and stomatal action. **(a)** An open stoma. Water entered collapsed guard cells, which swelled under turgor pressure and moved apart, thus forming the stoma in the needlelike leaf of the rock needlebush (*Hakea gibbosa*). **(b)** A closed stoma. Water exited the swollen guard cells, which collapsed against each other and closed the stoma.

Even when conditions are not so drastic, more than 90% of the water moving into a leaf can be lost through transpiration. About 2% of the water remaining in the leaf is used in photosynthesis and other activities. These measurements emphasize the need for controls over transpiration, for if water loss exceeds water uptake by roots, the resulting dehydration of plant tissues interferes with normal functioning, and the plant may wilt and die.

The cuticle-covered epidermis of leaves and stems reduces the rate of water loss from aboveground plant parts, but it also limits the rate at which CO_2 for photosynthesis can diffuse into the leaf. The functioning

a. Open stomata, with potassium mostly in guard cells

b. Closed stomata, with potassium mostly in epidermal cells

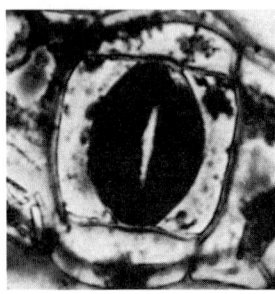
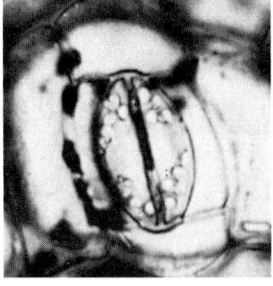

T. A. Masefield

T. A. Masefield

Figure 32.11

Evidence for potassium accumulation in stomatal guard cells undergoing expansion. Strips from the leaf epidermis of a dayflower (*Commelina communis*) were immersed in a solution containing a stain that binds preferentially with potassium ions. **(a)** In leaf samples with open stomata, most of the potassium was concentrated in the guard cells. **(b)** In leaf samples with closed stomata, little potassium was in guard cells; most was present in adjacent epidermal cells.

of stomata also affects a plant's water balance. When stomata are open, carbon dioxide can be absorbed, but unless the relative humidity of external air is 100%, water always moves out. However, plants have evolved adaptations that balance water loss with CO_2 uptake. This "transpiration–photosynthesis compromise" involves the regulation of transpiration and gas exchange by opening and closing stomata as environmental conditions change.

Opening and Closing of Stomata. Two guard cells flank each stomatal opening **(Figure 32.10)**. Their elastic walls are reinforced by cellulose microfibrils that wrap around the walls like a series of belts. The inward-facing walls are thicker and less elastic than the outer walls.

The opening and closing of stomata are good examples of a symport mechanism (see Figure 32.3c). Stomata open when potassium ions (K^+) flow into the guard cells through ion channels. As a first step, an active transport pump in the plasma membrane begins pumping H^+ ions out of the guard cells. Recall from Section 32.1 that H^+ pumped out of the cell can then follow its concentration gradient back into the cell. This inward flow of H^+ powers the active transport of K^+ into the guard cell. As a result, the K^+ concentration in turgid guard cells may be four to eight times higher than that in flaccid (limp) guard cells **(Figure 32.11)**. Water follows inward by osmosis. As turgor pressure builds, the thick inner wall does not expand much, but the outer walls of each guard cell expand lengthwise, so the two cells bend away from each other and create a stoma ("mouth") between them. Stomata close when the H^+ active transport protein stops pumping. K^+ flows passively out of the guard cells, and water follows by osmosis. When the water content of the guard cells dwindles, turgor pressure drops. The guard cells collapse against each other, closing the stomata.

In most plants, stomata open at first light, stay open during daylight, and close at night. Experiments have shown that guard cells respond to a number of environmental and chemical signals, any of which can induce the ion flows that open and close stomata. These signals include light, CO_2 concentration in the air spaces inside leaves, and the amount of water available to the plant.

Light and CO_2 Concentration. Light induces stomata to open through stimulation of blue-light receptors, probably located in the plasma membrane of guard cells. When stimulated, the receptors start the chain of events leading to stomatal opening by triggering activity of the H^+ pumps. Also, as photosynthesis begins in response to light, CO_2 concentration drops in the leaf air spaces as chloroplasts use the gas in carbohydrate production. In some way, this drop in CO_2

concentration sets off the series of events increasing the flow of K^+ into guard cells and furthers stomatal opening. The effects of reduced CO_2 concentration have been tested by placing plants in the dark in air containing no CO_2. Even in the absence of light, as the CO_2 concentration falls in leaves, guard cells swell and the stomata open.

Normally, when the sun goes down, a plant's demand for CO_2 drops as photosynthesis comes to a halt. Yet aerobic respiration continues to produce CO_2, which accumulates in leaves. As CO_2 concentration rises, and the blue-light wavelengths that activated the H^+ pumps wane, K^+ is lost from the guard cells and they collapse, closing the stomata. Thus, at night transpiration is reduced and water is conserved.

Water Stress. As long as water is readily available to a plant's roots, the stomata remain open during daylight. However, if water loss stresses a plant, the stomata close or open only slightly, regardless of light intensity or CO_2 concentration. Some simple but elegant experiments have shown that the stress-related closing of stomata depends on a hormone, abscisic acid (ABA), that is released by roots when water is unavailable. Test plants were suspended in containers so that only one-half the root system received water. Even though the roots with access to water could absorb enough water to satisfy the needs of all the plants' leaves, the stomata still closed. Tissue analysis revealed that water-stressed roots rapidly synthesize ABA. Transported through the xylem, this hormone stimulates K^+ loss by guard cells, and water moves out of the cell by osmosis—so the stomata close **(Figure 32.12)**. Mesophyll cells also take up ABA from the xylem and release it, with the same effects on stomata, when their turgor pressure falls due to excessive water loss. ABA can also cause stomata to close when the hormone is added experimentally to leaves.

The Biological Clock. Besides responding to light, CO_2 concentration, and water stress, stomata apparently open and close on a regular daily schedule imposed by a biological clock. Even when plants are placed in continuous darkness, their stomata open and close (for a time) in a cycle that roughly matches the day/night cycle of Earth. Such *circadian rhythms* (*circa* = around; *dies* = day) are also common in animals, and several, including wake/sleep cycles in mammals, are known to be controlled by hormones—a topic pursued in Chapter 40.

In Dry Climates, Plants Exhibit Various Adaptations for Conserving Water

Many plants have other evolutionary adaptations that conserve water, including modifications in structure or physiology **(Figure 32.13)**. Oleanders, for example,

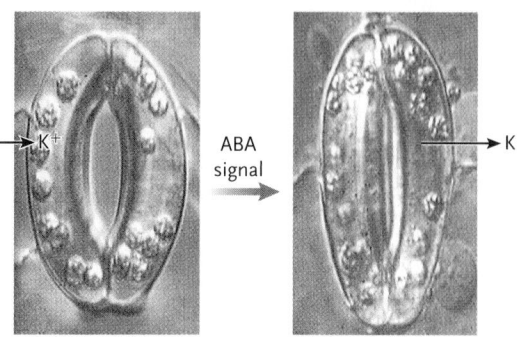

a. Stoma is open; water has moved in. **b.** Stoma is closed; water has moved out.

Figure 32.12

Hormonal control of stomatal closing. **(a)** When a stoma is open, high solute concentrations in the cytoplasm of both guard cells have raised the turgor pressure, keeping the cells swollen open. **(b)** In a water-stressed plant, the hormone abscisic acid (ABA) binds to receptors on the guard cell plasma membrane. Binding activates a signal transduction pathway that lowers solute concentrations inside the cells, which lowers the turgor pressure—so the stoma closes.

have stomata on the underside of the leaf at the bottom of pitlike invaginations of the leaf epidermis lined by hairlike trichomes (see Figure 32.13b). Sunken stomata are less exposed to drying breezes, and trichomes help retain water vapor at the pore opening, so that water evaporates from the leaf much more slowly.

The leaves of *xerophytes*—plants adapted to hot, dry environments in which water stress can be severe—have a thickened cuticle that gives them a leathery feel and provides enhanced protection against evaporative water loss. An example is mesquite *(Prosopis)*. In still other plants that inhabit arid landscapes, such as cacti, stems are thick, leaflike pads covered by sharp spines that actually are modified leaves (see Figure 32.13c). These structural alterations reduce the surface area for transpiration.

One intriguing variation on water-conservation mechanisms occurs in CAM plants, including cacti, orchids, and most succulents. As discussed in Section 9.4, **crassulacean acid metabolism** (CAM) is a biochemical variation of photosynthesis that was discovered in a member of the family Crassulaceae. CAM plants generally have fewer stomata than other types of plants, and their stomata follow a reversed schedule. They are closed during the day when temperatures are higher and the relative humidity is lower, and open at night. At night, the plant temporarily fixes carbon dioxide by converting it to malate, an organic acid. In the daytime, the CO_2 is released from malate and diffuses into chloroplasts, so photosynthesis takes place even though a CAM plant's stomata are closed. This adaptation prevents heavy evaporative water losses during the heat of the day.

a. Oleanders

BIOS Matt Alexander/Peter Arnold, Inc.

b. Oleander leaf

Cuticle
Multilayer epidermis

Recessed stoma

Thomas L. Rost

c. Spines (modified leaves) on a cactus stem

Fritz Polking/Visuals Unlimited

d. CAM plant

Fritz Polking/Visuals Unlimited

Figure 32.13

Some adaptations that enable plants to survive water stress. **(a)** Oleanders *(Nerium oleander)* are adapted to arid conditions. **(b)** As shown in the micrograph, oleander leaves have recessed stomata on their lower surface and a multilayer epidermis covered by a thick cuticle on the upper surface. **(c)** Like many other cacti, the leaves of the Graham dog cactus *(Opuntia grahamii)* are modified into spines that protrude from the underlying stem. Transpiration and photosynthesis occur in the green stems, such as the oval stem in this photograph. **(d)** *Sedum*, a CAM plant, in which the stomata open only at night.

STUDY BREAK

1. Explain the key steps in the cohesion–tension mechanism of water transport in a plant.
2. How and when do stomata open and close? In what ways is their functioning important to a plant's ability to manage water loss?

32.4 Transport of Organic Substances in the Phloem

A plant's phloem is another major long-distance transport system, and a superhighway at that: it carries huge amounts of carbohydrates, lesser but vital amounts of amino acids, fatty acids, and other organic compounds, and still other essential substances such as hormones. And unlike the xylem's unidirectional upward flow, the phloem transports substances throughout the plant to wherever they are used or stored. Organic compounds and water in the sieve tubes of phloem are under pressure and driven by concentration gradients.

Organic Compounds Are Stored and Transported in Different Forms

Plants synthesize various kinds of organic compounds, including large amounts of carbohydrates that are stored mainly as starch. Yet regardless of where in a plant a particular compound is destined to be used or stored, starch, protein, and fat molecules cannot leave the cells in which they are formed because all are too large to cross cell membranes. They also may be too insoluble in water to be transported to other regions of the plant body. Consequently, in leaves and other plant parts, specific reactions convert organic compounds to transportable forms. For example, hydrolysis of starch liberates glucose units, which combine with fructose to form sucrose—the main form in which sugars are transported through the phloem of most plants. Proteins are broken down into amino acids, and lipids converted into fatty acids. These forms are also better able to cross cell membranes by passive or active mechanisms.

Organic Solutes Move by Translocation

In plants, the long-distance transport of substances is called **translocation.** Botanists most often use this term to refer to the distribution of sucrose and other organic compounds by phloem, and they understand the mechanism best in flowering plants. The phloem of flowering plants contains interconnecting sieve tubes formed by living sieve tube member cells (see Figure 31.9). Sieve tubes lie end to end within vascular bundles, and they extend through all parts of the plant. Water and organic compounds, collectively called **phloem sap,** flow rapidly through large pores on the sieve tubes' end walls—another example of a structural adaptation that suits a particular function.

Phloem Sap Moves from Source to Sink under Pressure

Over the decades, plant physiologists have proposed several mechanisms of translocation, but it was the tiny aphid, an insect that annoys gardeners, that helped demonstrate that organic compounds flow under pressure in the phloem. An aphid attacks plant leaves and stems, forcing its needlelike stylet (a mouthpart) into sieve tubes to obtain the dissolved sugars and other nutrients inside. Numerous experiments with aphids have

Figure 32.14 Experimental Research

Translocation Pressure

HYPOTHESIS: High pressure forces phloem sap to flow through sieve tubes from a source to a sink.

EXPERIMENT: In the late 1970s, John Wright and Donald Fisher at the University of Georgia devised an experiment to directly measure the turgor pressure in sieve tubes of weeping willow saplings *(Salix babylonica)* under nondestructive conditions, using aphids that feed on *S. babylonica* in the wild. Weeping willow saplings were grown in a greenhouse under natural conditions of light and moisture. Aphids were placed on the trees and allowed to begin feeding by inserting their stylets into sieve tubes in the normal fashion. After being anesthetized by exposure to high concentrations of carbon dioxide, the aphids' bodies were cut away and only their stylets were left embedded in the sieve tubes. A tiny pressure-measuring device called a micromanometer then was glued over the end of each stylet. The micromanometer registered the volume and pressure of phloem sap as it was exuded from the stylet over time periods ranging from 30 to 90 minutes.

a. Aphid releasing honeydew

b. Micrograph of aphid stylet in sieve tube

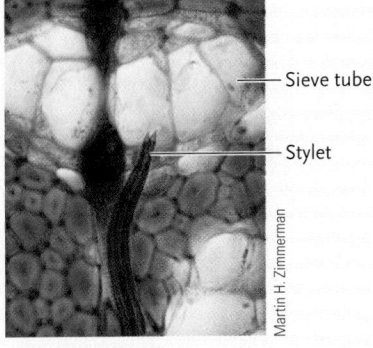

Sieve tube

Stylet

Martin Zimmerman, Science, 1961, 133:73–79, © AAAS

Martin H. Zimmerman

RESULTS: In nearly all cases, a high volume of pressurized sap flowed through the severed stylets into the micromanometer during the test periods.

CONCLUSION: The evidence supports pressure flow as the mechanism that moves phloem sap through sieve tubes.

Other experiments have confirmed that both turgor pressure and the concentration of sucrose are highest in sieve tubes closest to the sap source. Phloem sap also moves most rapidly closest to the source, where pressure is highest.

shown that in most plant species, sucrose is the main carbohydrate being translocated through the phloem. Studies also verify that the contents of sieve tubes are under high pressure, often five times as much as in an automobile tire. **Figure 32.14** explains a simple and innovative experiment that provided direct confirmation that phloem sap flows under pressure. When a live aphid feeds on phloem sap, this pressure forces the fluid through the aphid's gut and (minus nutrients absorbed) out its anus as "honeydew." If you park your car

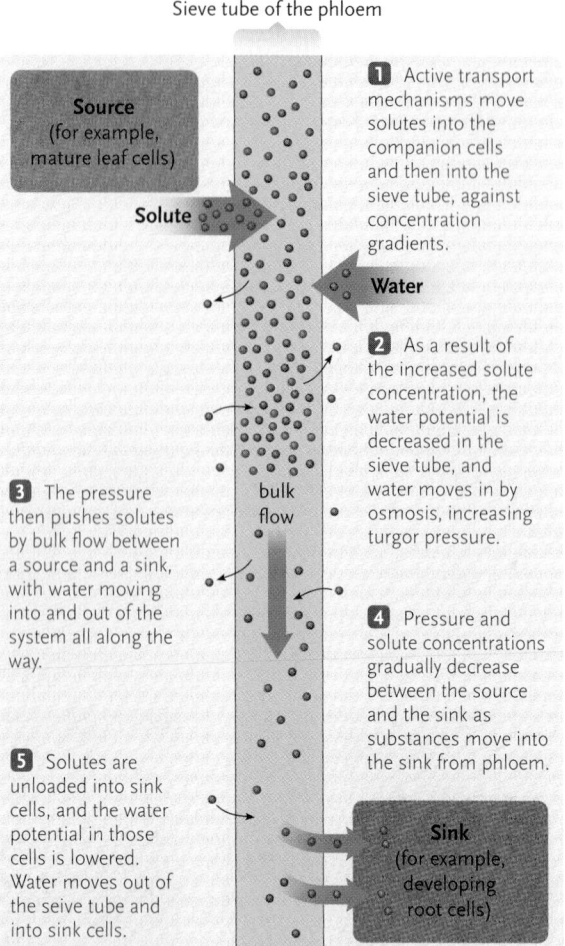

Sieve tube of the phloem

Source
(for example,
mature leaf cells)

Solute

Water

1 Active transport mechanisms move solutes into the companion cells and then into the sieve tube, against concentration gradients.

2 As a result of the increased solute concentration, the water potential is decreased in the sieve tube, and water moves in by osmosis, increasing turgor pressure.

bulk flow

3 The pressure then pushes solutes by bulk flow between a source and a sink, with water moving into and out of the system all along the way.

4 Pressure and solute concentrations gradually decrease between the source and the sink as substances move into the sink from phloem.

5 Solutes are unloaded into sink cells, and the water potential in those cells is lowered. Water moves out of the seive tube and into sink cells.

Sink
(for example, developing root cells)

Figure 32.15
Summary of the pressure flow mechanism in the phloem of flowering plants. Organic solutes are loaded into sieve tubes at a source, such as a leaf, and move by bulk flow toward a sink, such as roots or rapidly growing stem parts.

under a tree being attacked by aphids, it might get spattered with sticky honeydew droplets, thanks to the high fluid pressure in the tree's phloem.

A great deal of what botanists know about the transport of phloem sap has come from studies of sucrose transport in flowering plants. A fundamental discovery is that in flowering plants sucrose-laden phloem sap flows from a starting location, called the *source,* to another site, called the *sink,* along gradients of decreasing solute concentration and pressure. A **source** is any region of the plant where organic substances are being loaded into the phloem's sieve tube system. A **sink** is any region where organic substances are being unloaded from the sieve tube system and used or stored. What causes sucrose and other solutes produced in leaf mesophyll to flow from a source to a sink? In flowering plants, the **pressure flow mechanism** builds up at the source end of a sieve tube system and pushes those solutes by bulk flow toward a sink, where they are removed. **Figure 32.15** summarizes this mechanism.

The site of photosynthesis in mature leaves is an example of a source. Another example is a tulip bulb. In spring, stored food is mobilized for transport upward to growing plant parts, but after the plants bloom, the bulb becomes a sink as sugars manufactured in the tulip plant's leaves are translocated into it for storage. Young leaves, roots, and developing fruits generally start out as sinks, only to become sources when the season changes or the plant enters a new developmental phase. In general, sinks receive organic compounds from sources closest to them. Hence, the lower leaves on a rose bush may supply sucrose to roots, while leaves farther up the shoot supply the shoot tip.

Most substances carried in phloem are loaded into sieve tube members by active transport **(Figure 32.16a)**. Sucrose will be our example here. In leaves, sucrose formed inside mesophyll cells is exported and eventually reaches the apoplast (adjoining cell walls and air spaces) next to a small phloem vein. Here, it is actively pumped into companion cells by symport (see Figure 32.3), in which H^+ ions moves into the cell through the same carrier that takes up the sugar molecules. From the companion cells, most sucrose crosses into the living sieve tube members through plasmodesmata. Some sucrose also is loaded into sieve tube members by symport.

In some plants, companion cells become modified into **transfer cells** that facilitate the short-distance transport of organic solutes from the apoplast into the symplast. Transfer cells generally form when large amounts of solutes must be loaded or unloaded into the phloem, and they shunt substances through plasmodesmata to sieve tube members. As a transfer cell is forming, parts of the cell wall grow inward like pleats. This structural feature increases the surface area across which solutes can be taken up. The underlying plasma membrane, packed with transport proteins, then expands to cover the ingrowths. Transfer cells also enhance solute transport between living cells in the xylem, and they occur in glandlike tissues that secrete nectar. Botanists have discovered transfer cells in species from every taxonomic group in the plant kingdom, as well as in fungi and algae. In part because they arise from differentiated cells (instead of from meristem cells like other plant types of plant cells), researchers are working to define the molecular mechanisms that trigger their development.

When sucrose is loaded into sieve tubes its concentration rises inside the tubes. Thus the water potential falls, and water flows into the sieve tubes by osmosis. In fact, the phloem typically carries a great deal of water. As water enters sieve tubes, turgor pressure in the tubes increases, and the sucrose-rich fluid moves by bulk flow into the increasingly larger sieve tubes of larger veins. Eventually, the fluid is pushed out of the leaf into the stem and toward a sink **(Figure 32.16b)**. When sucrose is unloaded at the sink, water in the tube "follows solutes," moving by osmosis into the surrounding cells **(Figure 32.16c)**. Ultimately, the water enters the xylem and is recirculated.

Sieve tubes are mostly passive conduits for translocation. The system works because companion cells

a. Loading at a source

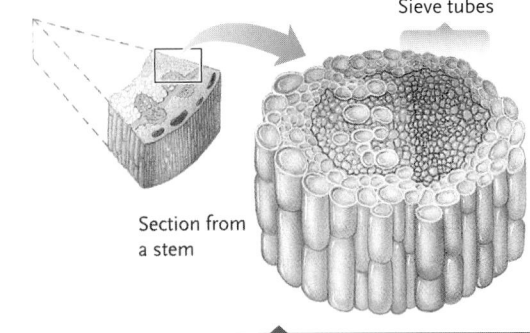

Upper epidermis

Photosynthetic cell

Sieve tube in phloem

Companion cell

Section from a leaf

Lower epidermis

Photosynthetic cells in leaves are a common source of carbohydrates that must be distributed through a plant. Small, soluble forms of these compounds move from the cells into phloem (in a leaf vein).

Figure 32.16
Translocation in the tissues of *Sonchus,* commonly called sow thistle. Research on *Sonchus* provided experimental evidence for the pressure flow mechanism.

b. Translocation along a distribution path

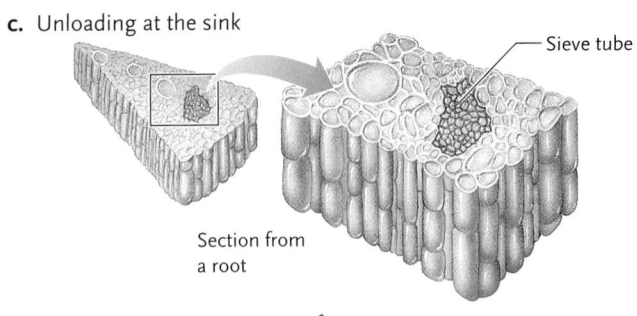

Sieve tubes

Section from a stem

Fluid pressure is greatest inside sieve tubes at the source. It pushes the solute-rich fluid to a sink, which is any region where cells are growing or storing food. There, the pressure is lower because cells are withdrawing solutes from the tubes and water follows the solutes.

c. Unloading at the sink

Sieve tube

Section from a root

Solutes are unloaded from sieve tubes into cells at the sink; water follows. Translocation continues as long as solute concentration gradients and a pressure gradient exist between the source and the sink.

Unanswered Questions

What are plasmodesmata made of, and exactly how do they function?

Plasmodesmata, the cytoplasmic channels through plant cell walls, connect plant cells to each other. Yet two fundamental questions about plasmodesmata remain unanswered: Exactly how do plasmodesmata function, and what are their structural components?

As described in this chapter and in Chapter 5, botanists have long assumed that nutrients, water, and small molecules that serve as growth regulators move through plasmodesmata, which form part of the symplastic pathway in plant tissues. Recent studies have demonstrated that larger molecules, including viruses and important proteins involved in plant growth and development, also move from cell to cell through plasmodesmata. For example, Patricia Zambryski and K. M. Crawford at the University of California at Berkeley reported that proteins, including transcription factors, travel via plasmodesmata from the cell that produces the proteins to adjacent cells where the factors promote or inhibit the expression of particular genes.

While the normal functions of plasmodesmata in plant growth and development still are not well understood, ongoing research by Zambryski and other plant scientists has begun to shed light on the workings of these vital channels. For instance, a variety of studies of the processes by which viruses spread through plant tissues have revealed that plasmodesmata are not simply static, open channels. Instead they are dynamic structures with the capacity to close, reopen, widen, and narrow. This capacity for structural change is not triggered by viral infection: rather, it seems that viruses simply take over the plant's natural mechanism for moving molecules from one cell to another.

Plasmodesmata were first observed using electron microscopy several decades ago, and they appear to be lined with proteins as well as membranes. Multiple biochemical approaches have failed to identify the proteins, probably because of the difficulty of purifying proteins that are associated with both a membrane and the cell wall. Genetic screens to identify plasmodesmata proteins, as well as the genes that regulate the functioning of plasmodesmata, are currently under way and may finally reveal details of plasmodesmata structure. As our understanding of the architecture of plasmodesmata and how they function grows, so will insights into the mechanisms of plant development, how plants interact with viral pathogens, and other questions as well.

Beverly McMillan

supply most of the energy that loads sucrose and other solutes at the source, and because solutes are removed at their sinks. As sucrose enters a sink, for example, its concentration in sieve tubes decreases, with a corresponding decrease in pressure. Thus for sucrose and other solutes transported in the phloem, there is always a gradient of concentration from source to sink—and a pressure gradient that keeps the solute moving along.

As noted previously, phloem sap moving through a plant carries a wide variety of substances, including hormones, amino acids, organic acids, and agricultural chemicals. The phloem also transports organic nitrogen compounds and mineral ions that are removed from dying leaves and stored for reuse in root tissue.

The transport functions of xylem and phloem are closely integrated with phenomena discussed later in this unit—reproduction and embryonic development, and the hormone-based regulation of plant growth.

STUDY BREAK

1. Compare and contrast translocation and transpiration.
2. Using sucrose as your example, summarize how a substance moves from a source into sieve tubes and then is unloaded at a sink. What is this mechanism called, and why?

Review

Go to **ThomsonNOW** at www.thomsonedu.com/login to access quizzing, animations, exercises, articles, and personalized homework help.

32.1 Principles of Water and Solute Movement in Plants

- Plants have mechanisms for moving water and solutes (1) into and out of cells, (2) laterally from cell to cell, and (3) long-distance from the root to shoot or vice versa (Figure 32.2).

- Both passive and active mechanisms move substances into and out of plant cells. Solutes generally are transported by carriers (facilitated diffusion), either passively down a concentration or electrochemical gradient (in the case of ions), or actively against a gradient, which requires cellular energy. An H^+ gradient creates the membrane potential that drives the cross-membrane transport of many ions or molecules (Figure 32.3).

- Most organic substances enter plant cells by symport, in which the energy of the H^+ gradient is coupled with uptake of a different solute. Some substances cross the plant cell membrane by antiport, in which energy of the H^+ gradient powers movement of a second solute out of cells.

- Water crosses plant cell membranes by osmosis, which is driven by water potential (ψ). Water tends to move osmotically from regions where water potential is higher to regions where it is lower.

- Water potential reflects a balance between turgor pressure and solute potential. Water potential is measured in megapascals (MPa) (Figures 32.4 and 32.5).

- Water and solutes also move into and out of the cell's central vacuole, transported from the cytoplasm across the tonoplast. Aquaporins across the tonoplast enhance water movement. Water in the central vacuole is vital for maintaining turgor pressure inside a plant cell.

- Bulk flow of fluid occurs when pressure at one point in a system changes with respect to another point in the system.

32.2 Transport in Roots

- Water and mineral ions entering roots travel laterally through the root cortex to the root xylem, following one or more of three major routes: the apoplastic pathway, the symplastic pathway, and the transmembrane pathway (Figure 32.6).

- In the apoplastic pathway, water diffuses into roots between the walls of root epidermal cells. By contrast, water and solutes absorbed by roots can enter either the symplastic or transmembrane pathway, both of which pass through cells.

- Casparian strips form a barrier that forces water and solutes in the apoplastic pathway to pass through cells in order to enter the stele. When an ion reaches the stele, it diffuses from cell to cell to reach the xylem (Figure 32.7). Roots of many flowering plants have a second layer of cells with Casparian strips (exodermis) just inside the root epidermis.

  Animation: Water absorption

  Animation: Root functioning

32.3 Transport of Water and Minerals in the Xylem

- In the conducting cells of xylem, tension generated by transpiration extends down from leaves to roots. By the cohesion–tension mechanism of water transport, water molecules are pulled upward by tension created as water exits a plant's leaves (Figure 32.8).

- In tall trees, negative pressure generated in the shoot drives bulk flow of xylem sap. In some plants, notably herbaceous species, positive pressure sometimes develops in roots and can force xylem sap upward (Figure 32.9).

- Transpiration and carbon dioxide uptake occur mostly through stomata. Environmental factors such as relative humidity, air temperature, and air movement at the leaf surface affect the transpiration rate.

- Most plants lose water and take up carbon dioxide during the day, when stomata are open. At night, when stomata close, plants conserve water and the inward movement of carbon dioxide falls.

- Stomata open in response to falling levels of carbon dioxide in leaves and also to incoming light wavelengths that activate photoreceptors in guard cells.

- Activation of photoreceptors triggers active transport of K^+ into guard cells. Simultaneous entry of anions such as Cl^- and synthesis of negatively charged organic acids increase the solute concentration, lowering the water potential so that water enters by osmosis. As turgor pressure builds, guard cells swell and draw apart, producing the stomatal opening (Figure 32.10).

- Guard cells close when light wavelengths used for photosynthesis wane. The stomata of water-stressed plants close regardless of light or CO_2 needs, possibly under the influence of the plant hormone ABA. The leaves of species native to arid environments typically have adaptations (such as an especially thick cuticle) that enhance the plant's ability to conserve water (Figures 32.12 and 32.13).

Animation: Stomata

Animation: Transpiration

Animation: Interdependent processes

32.4 Transport of Organic Substances in the Phloem

- In flowering plants, phloem sap is translocated in sieve tube members. Differences in pressure between source and sink regions drive the flow. Sources include mature leaves; sinks include growing tissues and storage regions (such as the tubers of a potato) (Figures 32.14 and 32.15).

- In leaves, the sugar sucrose is actively transported into companion cells adjacent to sieve tube members, then loaded into the sieve tubes through plasmodesmata.

- In some plants, transfer cells take up materials and pass them to sieve tube members. Transfer cells in xylem enhance the transport of solutes between tissues.

- As the sucrose concentration increases in the sieve tubes, water potential decreases. The resulting influx of water causes pressure to build up inside the sieve tubes, so the sucrose-laden fluid flows in bulk toward the sink, where sucrose and water are unloaded and distributed among surrounding cells and tissues (Figure 32.16).

Questions

Self-Test Questions

1. Antiport transport mechanisms:
   a. move dissolved materials by osmosis.
   b. transport molecules in the opposite direction of H^+ transported by proton pumps.
   c. transport molecules in the same direction as H^+ is pumped.
   d. are not affected by the size of molecules to be transported.
   e. are not affected by the charge of molecules to be transported.

2. All the following have roles in transporting materials between plant cells except:
   a. the stele.
   b. symport.
   c. the cell membrane.
   d. stomata.
   e. transport proteins.

3. Turgor pressure is best expressed as the:
   a. movement of water into a cell by osmosis.
   b. driving force for osmotic movement of water (ψ).
   c. group movement of large numbers of molecules due to a difference in pressure between two locations.
   d. equivalent of water potential.
   e. pressure exerted by fluid inside a plant cell against the cell wall.

4. Water potential is:
   a. the driving force for the osmotic movement of water into plant cells.
   b. higher in a solution that has more solute molecules relative to water molecules.
   c. a measure of the physical pressure required to halt osmotic water movement across a membrane.
   d. a measure of the combined effects of a solution's pressure potential and its solute potential.
   e. the functional equivalent of turgor pressure.

5. To regulate the flow of water and minerals in the root, the:
   a. Casparian strip of endodermal cells blocks the apoplastic pathway, forcing water and solutes to cross cell plasma membranes in order to pass into the stele.
   b. apoplastic pathway is expanded, allowing a greater variety of substances to move into the stele.
   c. symplastic pathway is modified in ways that make plasma membranes of root cortex cells more permeable to water and solutes.
   d. symplastic pathway shuts down entirely so that substances can move only through the apoplast.

   e. transmembrane pathway augments transport via the apoplast, shunting substances around cells.

6. An indoor gardener leaving for vacation completely wraps a potted plant with clear plastic. Temperature and light are left at low intensities. The effect of this strategy is to:
   a. halt photosynthesis.
   b. reduce transpiration.
   c. cause guard cells to shrink and stomata to open.
   d. destroy cohesion of water molecules in the xylem.
   e. increase evaporation from leaf mesophyll cells.

7. Stomata open when:
   a. water has moved out of the leaf by osmosis.
   b. K^+ flows out of guard cells.
   c. turgor pressure in the guard cells lessens.
   d. the H^+ active transport protein stops pumping.
   e. outward flow of H^+ sets up a concentration gradient that moves K^+ in via symport.

8. A factor that contributes to the movement of water up a plant stem is:
   a. active transport of water into the root hairs.
   b. an increase in the water potential in the leaf's mesophyll layer.
   c. cohesion of water molecules in stem and leaf xylem.
   d. evaporation of water molecules from the walls of cells in root epidermis and cortex and in the stele.
   e. absorption of raindrops on a leaf's epidermis.

9. In translocation of sucrose-rich phloem sap:
   a. the sap flows toward a source as pressure builds up at a sink.
   b. crassulacean acid metabolism reduces the rate of photosynthesis.
   c. companion cells use energy to load solutes at a source and the solutes then follow their concentration gradients to sinks.
   d. sucrose diffuses into companion cells while H^+ simultaneously leaves the cells by a different route.
   e. companion cells pump sucrose into sieve tube members.

10. In Vermont in early spring, miles of leafless maple trees have buckets hanging from "spigots" tapped into them to capture the fluid raw material for making maple syrup. This fluid flows into the buckets because:
   a. the tap drains phloem sap stored in the heartwood.
   b. phloem sap is moving from its source in maple tree roots to its sink in the developing leaf buds.

c. phloem sap is moving from where it was synthesized to the closest sink.
d. bulk flow results as phloem sap is actively transported from smaller to larger veins.
e. phloem sap is diverted into the tap from transfer cells.

Questions for Discussion

1. Many popular houseplants are native to tropical rain forests. Among other characteristics, many nonwoody species have extraordinarily broad-bladed leaves, some so ample that indigenous people use them as umbrellas. What environmental conditions might make a broad leaf adaptive in tropical regions, and why?

2. Insects such as aphids that prey on plants by feeding on phloem sap generally attack only young shoot parts. Other than the relative ease of piercing less mature tissues, suggest a reason why it may be more adaptive for these animals to focus their feeding effort on younger leaves and stems.

3. So-called systemic insecticides often are mixed with water and applied to the soil in which a plant grows. The chemicals are effective against sucking insects no matter which plant tissue the insects attack, but often don't work as well against chewing insects. Propose a reason for this difference.

4. Concerns about global warming and the greenhouse effect (see Chapter 51) center on rising levels of greenhouse gases, including atmospheric carbon dioxide. Plants use CO_2 for photosynthesis, and laboratory studies suggest that increased CO_2 levels could cause a rise in photosynthetic activity. However, as one environmentalist noted, "What plants do in environmental chambers may not happen in nature, where there are many other interacting variables." Strictly from the standpoint of physiological effects, what are some possible ramifications of a rapid doubling of atmospheric CO_2 on plants in temperate environments? In arid environments?

Experimental Analysis

In an experiment designed to explore possible links between ion uptake by roots and loading of ions into the xylem, a length of root was suspended through an impermeable barrier that separated two compartments—the root tip in one compartment and the cut end of the root in the other. Initially the solutions in the two compartments were identical, except that a known quantity of a radioactive tracer (representing an ion) was added to the one in which the root tip was suspended. The experimenters could then measure the relationship between ion uptake in the root and loading of the ion into the root xylem under different chemical conditions (such as the addition of a hormone or protein synthesis inhibitor). The research has provided evidence that ion uptake in the root is independent of loading of the ion into xylem. How does the experimental design support this kind of testing?

Evolution Link

A variety of structural features of land plants reflect the conflicting demands for conserving water and taking in carbon dioxide for photosynthesis. Identify at least four fundamental structural adaptations that help resolve this dilemma and explain how each one contributes to a land plant's survival.

How Would You Vote?

Phytoremediation using genetically engineered plants can increase the efficiency with which a contaminated site is cleaned up. Do you support planting genetically engineered plants for such projects? Go to www.thomsonedu.com/login to investigate both sides of the issue and then vote.

Lush azaleas *(Rhododendron)* and a stately Southern live oak *(Quercus virginiana)* draped with the unusual flowering plant called Spanish moss *(Tillandsia usneoides)*. The roots of shrubs, trees, and most other plants take up water and minerals from soil, but Spanish moss is an epiphyte—it lives independently on other plants and obtains nutrients via absorptive hairs on its leaves and stems.

33 Plant Nutrition

WHY IT MATTERS

Tropical rainforests are remarkable for many reasons, but for biologists the key one may be that they are the most biologically diverse ecosystems on Earth. In addition to containing countless thousands of species of animals, fungi, protists, and prokaryotes, these amazingly lush domains are dense with broadleaved, evergreen trees, some of which soar 40 or 50 m skyward. With rain a near-daily event, it may not seem surprising that the trees' foliage is a luxuriant deep green **(Figure 33.1).** Yet tropical rainforests are demanding places for plants to survive, in large part because the soil is chronically deficient in nutrients, the chemical elements necessary for plant metabolism. This nutrient scarcity is a direct outcome of the incessant rain and the high acidity of tropical rainforest soil. There is ample moisture in the upper layer of soil, but in acid soil mineral nutrients vital to plant metabolism, such as potassium, calcium, magnesium, and phosphorus, are subject to **leaching**—being washed into deeper soil levels that are not as accessible to plant roots. In addition, in the warm, moist environment of a tropical rainforest, bacteria and fungi speedily decompose fallen leaves and other organic remains. Just as rapidly, established trees and vines

Figure 33.1
A lush tropical rain forest growing in Southeast Asia.

take up any nutrients these decomposers have released, leaving few or none to enrich the soil. As falling rain dissolves some atmospheric CO_2, it creates carbonic acid—a type of "acid rain" that exacerbates the leaching problem even more.

Such poor soil and the near perpetual twilight at the forest floor make it extremely difficult for small shrubs and herbaceous plants to survive there. Nearly all such plants climb upward as vines using the tree trunks for mechanical support, or they live attached to the upper branches of taller species, where they can absorb needed minerals from falling dust or from the surfaces of other plants. These intricate adaptations to their particular environment allow the plants to secure energy and raw materials and to utilize both for growth and development.

Tropical rainforests are not unique in posing nutritional challenges for plants. In fact, plants rarely have ready access to a full complement of necessary resources. In a rainforest, the carbon, hydrogen, and oxygen plants need for photosynthesis are relatively easy to come by: plants there usually get enough carbon from the CO_2 in air, and their roots can take up enough water to gain the necessary hydrogen and oxygen. But soils in other environments are frequently dry, making water a limited resource, and almost nowhere in nature do soils hold lavish amounts of dissolved minerals such as nitrogen, calcium, and others that are vital for a plant's survival. In response to the challenge of obtaining nutrients, plants have evolved the range of structural and physiological adaptations that we consider in this chapter.

33.1 Plant Nutritional Requirements

No organism grows normally when deprived of a chemical element essential for its metabolism. In the latter half of the nineteenth century, plant physiologists exploited rapid advances in chemistry to probe both the chemical composition of plants and the essential nutrients plants need to survive. Because plants require some nutrients in only trace amounts, in recent times researchers have brought to bear sophisticated methods in their studies of plant nutrition.

Plants Require Macronutrients and Micronutrients for Their Metabolism

By weight, the tissues of most plants are more than 90% water. Early researchers could obtain a rough idea of the composition of a plant's dry weight by burning the plant and then analyzing the ash. This method typically yielded a long list of elements, but the results were flawed. Chemical reactions during burning can dissipate quantities of some important elements, such as nitrogen. Also, plants take up a variety of ions that they don't use; depending on the minerals present in the soil where a plant grows, a plant's tissues can contain nonnutritive elements such as gold, lead, arsenic, and uranium.

Studying Plant Nutrition Using Hydroponics. In 1860, German plant physiologist Julius von Sachs pioneered an experimental method for identifying the minerals

Figure 33.2 Research Method

Hydroponic Culture

PURPOSE: In studies of plant nutritional requirements, using hydroponic culture allows a researcher to manipulate and precisely define the types and amounts of specific nutrients that are available to test plants.

PROTOCOL: In a typical hydroponic apparatus, many plants are grown in a single solution containing pure water and a defined mix of mineral nutrients. The solution is replaced or refreshed as needed and is aerated with a bubbling system:

a. Basic components of a hydroponic apparatus

Plant support

Nutrient solution

Air pumped into bubbling system

b. Procedure for identifying elements essential for proper plant nutrition

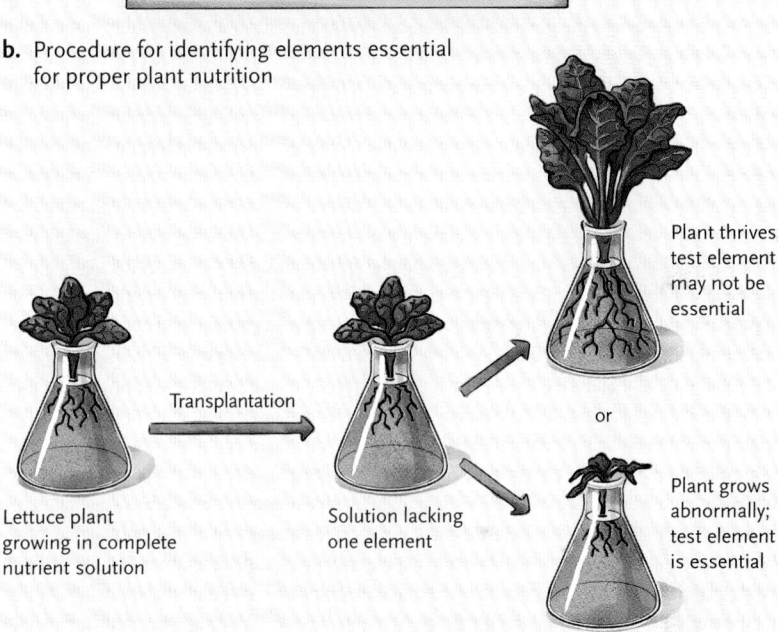

A "complete" solution contains all the known and suspected essential plant nutrients. An "incomplete" solution contains all but one of the same nutrients, in the same amounts. For experiments, researchers first grow plants in a complete solution, then transplant some of the plants to an incomplete solution.

Lettuce plant growing in complete nutrient solution

Transplantation

Solution lacking one element

or

Plant thrives; test element may not be essential

Plant grows abnormally; test element is essential

INTERPRETING THE RESULTS: Normal growth of test plants suggests that the missing nutrient is not essential, while abnormal growth is evidence that the missing nutrient may be essential.

---

absorbed into plant tissues that are essential for plant growth. Sachs carefully measured amounts of compounds containing specific minerals and mixed them in different combinations with pure water. He then grew plants in the solutions, a method called now **hydroponic culture** (*hydro* = water; *ponos* = work). By eliminating one element at a time and observing the results, Sachs deduced a list of six essential plant nutrients, in descending order of the amount required: nitrogen, potassium, calcium, magnesium, phosphorus, and sulfur.

Sachs's innovative research paved the way for decades of increasingly sophisticated studies of plant nutrition. In the spirit of his work, one basic experimental method involves growing a plant in a solution containing a complete spectrum of known and possible essential nutrients **(Figure 33.2a)**. The healthy plant is then transferred to a solution that is identical, except that it lacks one element having an unknown nutritional role **(Figure 33.2b)**. Abnormal growth of the plant in this solution is evidence that the missing element is essential. If the plant grows normally, the missing element may not be essential; however, only further experimentation can confirm this hypothesis.

In a typical modern hydroponic apparatus, the nutrient solution is refreshed regularly, and air is bubbled into it to supply oxygen to the roots. Without sufficient oxygen for respiration, the plants' roots do not absorb nutrients efficiently. (The same effect occurs in poorly aerated soil.) Variations of this technique are used on a commercial scale to grow some vegetables, such as lettuce and tomatoes.

Essential Macronutrients and Micronutrients. Hydroponics research has revealed that plants generally require 17 essential elements **(Table 33.1).** By definition, an **essential element** is necessary for normal growth and reproduction, cannot be functionally replaced by a different element, and has one or more roles in plant metabolism. With enough sunlight and the 17 essential elements, plants can synthesize all the compounds they need.

Nine of the essential elements are **macronutrients**, meaning that plants incorporate relatively large amounts

Table 33.1	Essential Plant Nutrients and Their Functions		
Element	Commonly Absorbed Forms	Some Known Functions	Some Deficiency Symptoms
Macronutrients			
Carbon*	CO_2	Raw materials for photosynthesis	Rarely deficient
Hydrogen*	H_2O		No symptoms; available from water
Oxygen*	O_2, H_2O, CO_2		No symptoms; available from water and CO_2
Nitrogen	NO_3^-, NH_4^+	Component of proteins, nucleic acids, coenzymes, chlorophylls	Stunted growth; light-green older leaves; older leaves yellow and die (chlorosis)
Phosphorus	$H_2PO_4^-$, HPO_4^{2+}	Component of nucleic acids, phospholipids, ATP, several coenzymes	Purplish veins; stunted growth; fewer seeds, fruits
Potassium	K^+	Activation of enzymes; key role in maintaining water-solute balance and so influences osmosis	Reduced growth; curled, mottled, or spotted older leaves; burned leaf edges; weakened plant
Calcium	Ca^{2+}	Roles in formation and maintenance of cell walls and in membrane permeability; enzyme cofactor	Leaves deformed; terminal buds die; poor root growth
Sulfur	SO_4^{2-}	Component of most proteins, coenzyme A	Light-green or yellowed leaves; reduced growth
Magnesium	Mg^{2+}	Component of chlorophyll; activation of enzymes	Chlorosis; drooping leaves
Micronutrients			
Chlorine	Cl^-	Role in root and shoot growth, and in photosynthesis	Wilting; chlorosis; some leaves die (deficiency not seen in nature)
Iron	Fe^{2+}, Fe^{3+}	Roles in chlorophyll synthesis, electron transport; component of cytochrome	Chlorosis; yellow and green striping in grasses
Boron	H_3BO_3	Roles in germination, flowering, fruiting, cell division, nitrogen metabolism	Terminal buds, lateral branches die; leaves thicken, curl, and become brittle
Manganese	Mn^{2+}	Role in chlorophyll synthesis; coenzyme action	Dark veins, but leaves whiten and fall off
Zinc	Zn^{2+}	Role in formation of auxin, chloroplasts, and starch; enzyme component	Chlorosis; mottled or bronzed leaves; abnormal roots
Copper	Cu^+, Cu^{2+}	Component of several enzymes	Chlorosis; dead spots in leaves; stunted growth
Molybdenum	MoO_4^{2-}	Component of enzyme used in nitrogen metabolism	Pale green, rolled or cupped leaves
Nickel	Ni^{2+}	Component of enzyme required to break down urea generated during nitrogen metabolism	Dead spots on leaf tips (deficiency not seen in nature)

*Carbon, hydrogen, and nitrogen are the nonmineral plant nutrients. All others are minerals.

of them into their tissues. Three of these elements—carbon, hydrogen, and oxygen—account for about 96% of a plant's dry mass. Together, these three elements are the key components of lipids and of carbohydrates such as cellulose; with the addition of nitrogen, they form the basic building blocks of proteins and nucleic acids. Plants also use phosphorus in constructing nucleic acids, ATP, and phospholipids, and they use potassium for functions ranging from enzyme activation to mechanisms that control the opening and closing of stomata. Rounding out the list of macronutrients are calcium, sulfur, and magnesium. Carbon, hydrogen, and oxygen come from the air and water, and are the only plant nutrients that are not considered to be minerals. The other six macronutrients are minerals, inorganic substances available to plants through the soil as ions dissolved in water. Most minerals that serve as nutrients in plants are derived from the weathering of rocks and inorganic particles in the Earth's crust.

The other elements essential to plants are also minerals, and are classed as **micronutrients** because plants require them only in trace amounts. Nevertheless, they are just as vital as macronutrients to a plant's health and survival. For example, 5 metric tons of potatoes contain roughly the amount of copper in a single (copper-plated) penny—yet without it, potato plants are sickly and do not produce normal tubers.

Chlorine was identified as a micronutrient nearly a century after Sachs's experiments. The researchers who discovered its role performed hydroponic culture experiments in a California laboratory near the Pacific Ocean, where the air, like coastal air everywhere, contains sodium chloride. The investigators found that their test plants could obtain tiny but sufficient quantities of chlorine from the air, as well as from sweat (which also contains NaCl) on the researchers' own hands. Great care had to be taken to exclude chlorine from the test plants' growing environment in order to prove that it was essential.

In some cases, plant seeds contain enough of certain trace minerals to sustain the adult plant. For example, nickel (Ni^{2+}) is a component of urease, the enzyme required to hydrolyze urea. Urea is a toxic by-product of the breakdown of nitrogenous compounds, and it will kill cells if it accumulates. In the late 1980s investigators found that barley seeds contain enough nickel to sustain two complete generations of barley plants. Plants grown in the absence of nickel did not begin to show signs of nickel deficiency until the third generation.

Besides the 17 essential elements, some species of plants may require additional micronutrients. Experiments suggest that many, perhaps most, plants adapted to hot, dry conditions require sodium; many plants that photosynthesize by the C_4 pathway (see Section 9.4) appear to be in this group. A few plant species require selenium, which is also an essential micronutrient for animals. Horsetails (*Equisetum*) require silicon, and some grasses (such as wheat) may also need it. Scientists continue to discover additional micronutrients for specific plant groups.

Both micronutrients and macronutrients play vital roles in plant metabolism. Many function as cofactors or coenzymes in protein synthesis, starch synthesis, photosynthesis, and aerobic respiration. As you read in Section 32.1, some also have a role in creating solute concentration gradients across plasma membranes, which are responsible for the osmotic movement of water.

Nutrient Deficiencies Cause Abnormalities in Plant Structure and Function

Plants differ in the quantity of each nutrient they require—the amount of an essential element that is adequate for one plant species may be insufficient for another. Lettuce and other leafy plants require more nitrogen and magnesium than do other plant types, for example, and alfalfa requires significantly more potassium than do lawn grasses. An adequate amount of an essential element for one plant may even be harmful to another. For example, the amount of boron required for normal growth of sugar beets is toxic for soybeans. For these reasons, the nutrient content of soils is an important factor in determining which plants will grow well in a given location.

Plants that are deficient in one or more of the essential elements develop characteristic symptoms (Table 33.1 lists some observable symptoms of nutrient deficiencies). The symptoms give some indication of the metabolic roles the missing elements play. Deficiency symptoms typically include stunted growth, abnormal leaf color, dead spots on leaves, or abnormally formed stems **(Figure 33.3).** For instance, iron is a component of the cytochromes upon which the cellular electron transfer system depends, and it plays a role in reactions that synthesize chlorophyll. Iron deficiency causes **chlorosis**, a yellowing of plant tissues that results from a lack of chlorophyll (see Figure 33.3b). Because ionic iron (Fe^{3+}) is relatively insoluble in water, gardeners often fertilize plants with a soluble iron compound called chelated iron to stave off or cure chlorosis. Similarly, because magnesium is a necessary component of chlorophyll, a plant deficient in this element has fewer chloroplasts than normal in its leaves and other photosynthetic parts. It appears paler green than normal, and its growth is stunted because of reduced photosynthesis (see Figure 33.3c).

Plants that lack adequate nitrogen may also become chlorotic (see Figure 33.3d), with older leaves yellowing first because the nitrogen is preferentially shunted to younger, actively growing plant parts. This adaptation is not surprising, given nitrogen's central role in the synthesis of amino acids, chlorophylls, and other compounds vital to plant metabolism. With some other mineral deficiencies, young leaves are the first to

a. Plant grown using a complete growth solution

b. Iron deficiency

c. Magnesium deficiency

d. Nitrogen deficiency

e. Phosphorus deficiency

f. Potassium deficiency

g. Sulfur deficiency

h. Calcium deficiency

Figure 33.3

Leaves and stems of tomato plants showing visual symptoms of seven different mineral deficiencies. The plants were grown in the laboratory, where the experimenter could control which nutrients were available.

(Photos by E. Epstein, University of California, Davis.)

show symptoms. These kinds of observations underscore the point that plants utilize different nutrients in specific, often metabolically complex ways.

Soils are more likely to be deficient in nitrogen, phosphorus, potassium, or some other essential mineral than to contain too much, and farmers and gardeners typically add nutrients to suit the types of plants they wish to cultivate. They may observe the deficiency symptoms of plants grown in their locale or have soil tested in a laboratory, then choose a fertilizer with the appropriate balance of nutrients to compensate for the deficiencies. Packages of commercial fertilizers use a numerical shorthand (for example, 15-30-15) to indicate the percentages of nitrogen, phosphorus, and potassium they contain.

STUDY BREAK

1. What are the two main categories of the essential elements plants need? Give several examples of each.
2. Do all plants require the same basic nutrients in the same amounts? Explain.

33.2 Soil

Soil anchors plant roots and is the main source of the inorganic nutrients plants require. It also is the source of water for most plants, and of oxygen for respiration in root cells. The physical texture of soil determines whether root systems have access to sufficient water and dissolved oxygen. These characteristics reinforce the conclusion that the physical and chemical properties of soils in different habitats have a major impact on the ability of plant species to grow, survive, and reproduce there.

The Components of a Soil and the Size of the Particles Determine Its Properties

Soil is a complex mix of mineral particles, chemical compounds, ions, decomposing organic matter, air, water, and assorted living organisms. Most soils develop from the physical or chemical weathering of rock (which also liberates mineral ions). The different kinds of soil particles range in size from sand (2.0–0.02 mm) to silt (0.02–0.002 mm) and clay (diameter less than 0.002 mm). These mineral particles usually are mixed with various organic components, including **humus**—decomposing parts of plants and animals, animal droppings, and other organic matter. Dry humus has a loose, crumbly texture. It can absorb a great deal of water and thus contributes to the capacity of soil to hold water. Organic molecules in humus are reservoirs of nutrients, including nitrogen, phosphorus, and sulfur, that are vital to living plants.

The relative proportions of the different sizes of mineral particles give soil its basic texture—gritty if the soil is largely sand, smooth if silt predominates, and dense and heavy if clay is the major component. A soil's texture in turn helps determine the number and volume of pores—air spaces—that it contains. The relative amounts of sand, silt, and clay determine whether a soil is sticky when wet, with few air spaces (mostly clay), or dries quickly and may wash or blow away (mostly sand). Clay soils are more than 30% clay, while sandy soils contain less than 20% clay or silt.

The piles of bagged humus for sale at garden centers each spring reflect the fact that the amount of humus in a soil also affects plant growth. Its plentiful organic material feeds decomposers whose metabolic activities in turn release minerals that plant roots can take up, but that is not its only value in soil. Humus helps retain soil water and, with its loose texture, helps aerate soil as well. Well-aerated soils containing roughly equal proportions of humus, sand, silt, and clay are **loams**, and they are the soils in which most plants do best.

Soil also contains living organisms. Trillions of bacteria, hundreds of millions of fungi, and several million nematodes—not to mention earthworms and insects—are present in every square meter of fertile soil. Together with the roots of living plants, these organisms have a major influence on the composition and characteristics of soil. Bacteria and fungi decompose organic matter; burrowing creatures such as earthworms aerate the soil; and when plant roots die they contribute their organic matter to the soil.

As soils develop naturally, they tend to take on a characteristic vertical profile, with a series of layers or **horizons (Figure 33.4).** Each horizon has a distinct texture and composition that varies with soil type. The top layer of surface litter—twigs and leaves, animal dung, fungi, and similar organic matter—is accordingly called the *O horizon*. The most fertile soil layer, called **topsoil**, occurs just below and forms the *A horizon*. This layer may be less than a centimeter deep on steep slopes to more than a meter deep in grasslands. It consists of humus mixed with mineral particles and usually is fairly loose; it is here that the roots of most herbaceous plants are located. Below the topsoil is the **subsoil** or *B horizon,* a layer of larger soil particles containing relatively little organic matter. Mineral ions, including those that serve as nutrients in plants, tend to accumulate in the B horizon, and mature tree roots generally extend into this layer. Under it is the *C horizon,* a layer of mineral particles and rock fragments that extends down to bedrock.

Regions where the topsoil is deep and rich in humus are ideal for agriculture; the vast grasslands of the North American Midwest and Ukraine are prime examples. Without soil management and intensive irrigation, crops

William Ferguson

O horizon
Fallen leaves and other organic material littering the surface of mineral soil

A horizon
Topsoil, which contains some percentage of decomposed organic material and which is of variable depth; here it extends about 30 cm below the soil surface

B horizon
Subsoil; larger soil particles than the A horizon, not much organic material, but greater accumulation of minerals; here it extends about 60 cm below the A horizon

C horizon
No organic material, but partially weathered fragments and grains of rock from which soil forms; extends to underlying bedrock

Bedrock

Figure 33.4
A representative profile of soil horizons.

usually cannot be grown in deserts due to the lack of rainfall and low humus in the soil. Nor can agriculture flourish for long on land cleared of a tropical rainforest, due to the soil leaching and lack of nutrients described in the chapter introduction.

The Characteristics of Soil Affect Root–Soil Interactions

In different regions, and even in different parts of a local area, the proportions of the types of soil particles can differ dramatically, with corresponding variations in the soil's suitability for plant growth.

Plants have evolved adaptations to many otherwise inhospitable soil environments, as you will see in the following section. First, however, we consider the general ways in which soil composition influences the ability of plant roots to obtain water and minerals.

Water Availability. As water flows into and through soil, gravity pulls much of the water down through the spaces between soil particles into deeper soil layers. This available water is part of the **soil solution (Figure 33.5),** a combination of water and dissolved substances that coats soil particles and partially fills pore spaces. The solution develops through ionic interactions between water molecules and soil particles. Clay particles and the organic components in soil (especially proteins) often bear negatively charged ions on their surfaces. The negative charges attract the polar water molecules, which form hydrogen bonds with the soil particles (see Section 2.4).

Unless a soil is irrigated, the amount of water in the soil solution depends largely on the amount and pattern of precipitation (rain or snow) in a region. How much of this water is actually available to plants depends on the soil's composition—the size of the air

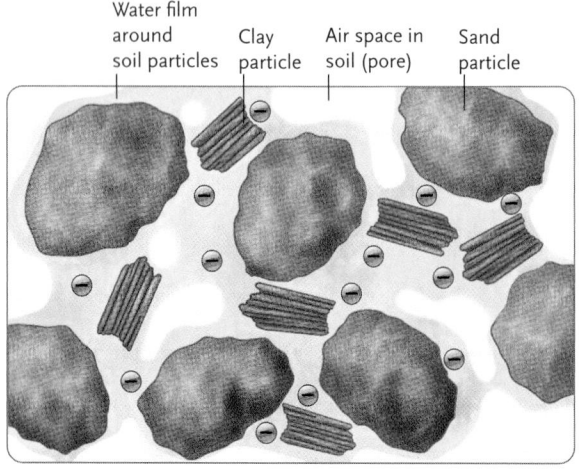

Figure 33.5

Location of the soil solution. Negatively charged ions on the surfaces of soil particles attract water molecules, which coat the particles and fill spaces between them (blue). Hydrogen bonds between water and soil components counteract the pull of gravity and help hold some water in the soil spaces.

spaces in which water can accumulate and the proportions of water-attracting particles of clay and organic matter. By volume, soil is about one-half solid particles and one-half air space.

The size of the particles in a given soil has a major effect on how well plants will grow there. Sandy soil has relatively large air spaces, so water drains rapidly below the top two soil horizons where most plant roots are located. Soils rich in clay or humus are often high in water content, but in the case of clay, ample water is not necessarily an advantage for plants. Whereas a humus-rich soil contains lots of air spaces, the closely layered particles in clay allow few air spaces—and what spaces there are tend to hold tightly the water that enters them. The lack of air spaces in clay soils also severely limits supplies of oxygen available to roots for cellular respiration, and the plant's metabolic activity suffers. Thus, few plants can grow well in clay soils, even when water content is high. (Overwatered houseplants die because their roots are similarly "smothered" by water.) Plants do not fare much better in drier clay-rich soils, because roots cannot extract the existing water and cannot easily penetrate the densely packed clay. These characteristics explain why good agricultural soils tend to be sandy or silty loams, which contain a mix of humus and coarse and fine particles.

As you learned in Chapter 31, root hairs are specialized extensions of root epidermal cells; they directly contact the soil solution and allow roots to absorb water (and dissolved ions). The soil solution usually contains fewer dissolved solutes than does the water in the cells of plant roots. Accordingly, water tends to diffuse from wet soil into the roots, following the osmotic gradient (see Section 32.2). As roots extract water from the surrounding soil, however, the remaining water molecules are held to the negatively charged clay surfaces with ever-increasing force. Plants start to wilt when the forces that draw water into their root cells equal those holding water in soil. Under these conditions, water no longer diffuses into roots, but it continues to evaporate from leaves and to be used in photosynthesis. Plants that survive in deserts or in salty soils have adaptations that permit their roots to absorb water even when osmotic conditions in soil do not favor water movement into the plant.

Mineral Availability. Some mineral nutrients enter plant roots as cations (positively charged ions) and some as anions (negatively charged ions). Although both cations and anions may be present in soil solutions, they are not equally available to plants.

Cations such as magnesium (Mg^{2+}), calcium (Ca^{2+}), and potassium (K^+) cannot easily enter roots because they are attracted by the net negative charges on the surfaces of soil particles. To varying degrees, they become reversibly bound to negative ions on the surfaces. Attraction in this form is called *adsorption.* The cations are made available to plant roots through

cation exchange, a mechanism in which one cation, usually H^+, replaces a soil cation (**Figure 33.6**). There are two main sources for the hydrogen ions. Respiring root cells release carbon dioxide, which dissolves in the soil solution, yielding carbonic acid (H_2CO_3). Subsequent reactions ionize H_2CO_3 to produce bicarbonate (HCO_3^-) and hydrogen ions (H^+). Reactions involving organic acids inside roots also produce H^+, which is excreted. As H^+ enters the soil solution, it displaces adsorbed mineral cations attached to clay and humus, freeing them to move into roots. Other types of cations may also participate in this type of exchange, as shown in Figure 33.6.

By contrast, anions in the soil solution, such as nitrate (NO_3^-), sulfate (SO_4^{2-}), and phosphate (PO_4^-), are only weakly bound to soil particles, and so they generally move fairly freely into root hairs. However, because they are so weakly bound compared with cations, anions are more subject to loss from soil by leaching.

The pH of soil also affects the availability of some mineral ions. Soil pH is a function of the balance between cation exchange and other processes that raise or lower the concentration of H^+ in soil. As noted earlier, in areas that receive heavy rainfall, soils tend to become acidic (that is, they have a pH of less than 7). This acidification occurs in part because moisture promotes the rapid decay of organic material in humus; as the material decomposes, the organic acids it contains are released. Acid precipitation, which results from the release of sulfur and nitrogen oxides into the air, also contributes to soil acidification. By contrast, the soil in arid regions, where precipitation is low, often is alkaline (the pH is greater than 7).

Although most plants are not directly sensitive to soil pH, chemical reactions in very acid (pH < 5.5) or very alkaline (pH > 9.5) soils can have a major impact on whether plant roots take up various mineral cations. For example, experiments have demonstrated that in the presence of OH^- in alkaline soil, calcium and phosphate ions react to form insoluble calcium phosphates. The phosphate captured in these compounds is as unavailable to roots as if it were completely absent from the soil.

For a soil to sustain plant life over long periods, the mineral ions that plants take up must be replenished naturally or artificially. Over the long run, some mineral nutrients enter the soil from the ongoing weathering of rocks and smaller bits of minerals. In the shorter run, minerals, carbon, and some other nutrients are returned to the soil by the decomposition of organisms and their parts or wastes. Other inputs occur when airborne compounds, such as sulfur in volcanic and industrial emissions, become dissolved in rain and fall to earth. Still others, including compounds of nitrogen and phosphorus, may enter soil in fertilizers.

Although the use of commercial fertilizers maintains high crop yields, agricultural chemicals do not

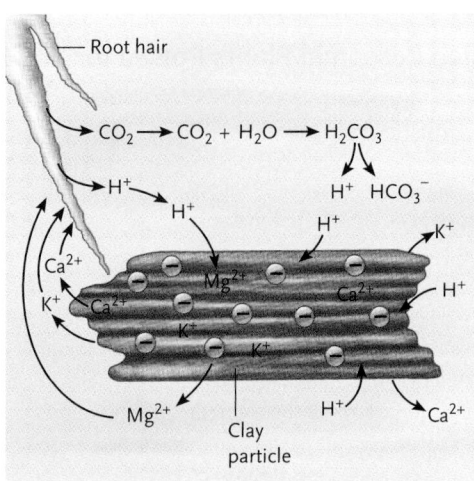

Figure 33.6

Cation exchange on the surface of a clay particle. When cations come into contact with the negatively charged surface of the particle, they become adsorbed. As one type of cation, such as H^+, becomes adsorbed, other ions are liberated and can be taken up by plant roots.

add humus to the soil. Their use can also cause serious problems, as when nitrogen-rich runoff from agricultural fields promotes the serious overgrowth of algae in lakes and bays. In many parts of the world, industrial pollutants such as cadmium, lead, and mercury are increasingly serious soil contaminants. The use of plants to remove such materials from soil, called *phytoremediation*, is the topic of this chapter's *Focus on Research*.

STUDY BREAK

1. Why is humus an important component of fertile soil?
2. How does the composition of a soil affect a plant's ability to take up water?
3. What factors affect a plant's ability to absorb minerals from the soil?

33.3 Obtaining and Absorbing Nutrients

Soil managed for agriculture can be plowed, precisely irrigated, and chemically adjusted to provide air, water, and nutrients in optimal quantities for a particular crop. By contrast, in natural habitats there are wide variations in soil minerals, humus, pH, the presence of other organisms, and other factors that influence the availability of essential elements. Although adequate carbon, hydrogen, and oxygen are typically available, other essential elements may not be as abundant. In particular, nitrogen, phosphorus, and potassium are often relatively scarce. The evolutionary solutions to

FOCUS ON RESEARCH

Applied Research: Plants Poised for Environmental Cleanup

For several decades researchers have been searching for efficient modes of *phytoremediation*—the use of plants to remove pollutants from the environment. A high-profile target is the highly toxic organic compound methylmercury (MeHg). This substance is present in coastal soils and wetlands contaminated by industrial wastes that contain an ionic form of the element mercury called Hg (II). Bacteria in contaminated sediments metabolize Hg(II) and generate MeHg as a metabolic by-product. Once MeHg forms, it enters the food web and eventually becomes concentrated in tissues of fishes and other animals. In humans MeHg can lead to degeneration of the nervous system and is the cause of most cases of mercury poisoning due to consuming contaminated fish.

In the 1990s a team of scientists including Scott Bizily and Richard Meagher at the University of Georgia decided to try to modify plants genetically so that they could detoxify mercury-contaminated soil and wetlands. It was already known that bacteria in contaminated sediments possess two genes, *merA* and *merB*, which encode enzymes that convert MeHg into elemental mercury (Hg)—a relatively inert substance that is much less dangerous to organisms. Both these bacterial mercury-resistance genes had already been cloned by others. After modifying the cloned genes so that they could be expressed in plants, the team used a vector (the bacterium *Rhizobium radiobacter*) to introduce each gene into several different sets of *Arabidopsis thaliana* plants (thale cress). They eventually obtained three groups of transgenic plants: some that were *merA* only, some that were *merB* only, and some that were *merA* and *merB*. In a series of experiments, seeds from each group were grown (along with wild-type controls) in five different growth media—one containing no mercury and the other four containing increasing concentrations of methylmercury. Wild-type and *merA* seeds germinated and grew only in the mercury-free growth medium. The *merB* seedlings fared somewhat better: they germinated and grew briefly even at the highest concentrations of MeHg, but soon became chlorotic and died. By contrast, seeds with the *merA/merB* genotype not only germinated, but the resulting seedlings grew into robust plants with healthy root and shoot systems. In later tests *merA/merB* plants were grown in chambers in which the chemical composition of the air was monitored. This study revealed that the doubly transgenic plants also were transpiring large amounts of Hg. The implication of these findings was clear: *A. thaliana* plants having both *merA* and *merB* genes were able to take up the toxic methylmercury with no ill effects and convert it to a harmless form. Meagher and his colleagues now are experimenting with ways of increasing the efficiency of phytoremediating enzymes when plant cells express *merA* and *merB*. They also are studying the mechanisms by which ionic mercury taken up by roots may be transported via the xylem to leaves and other shoot parts. The goal is to engineer plants that accumulate large quantities of mercury in aboveground tissues that can be harvested, leaving the living plant to continue its "work" of detoxifying a contaminated landscape.

---

these challenges include an array of adaptations in the structure and functioning of plant roots.

Root Systems Allow Plants to Locate and Absorb Essential Nutrients

Immobile organisms such as plants must locate nutrients in their immediate environment, and for plants the adaptive solution to this problem is an extensive root system. Roots make up 20% to 50% of the dry weight of many plants, and even more in species growing where water or nutrients are especially scarce, such as arctic tundra. As long as a plant lives, its root system continues to grow, branching out through the surrounding soil. Roots don't necessarily grow *deeper* as a root system branches out, however. In arid regions, a shallow-but-broad root system may be better positioned to take up water from occasional rains that may never penetrate below the first few inches of soil.

A root system grows most extensively in soil where water and mineral ions are abundant. As described in Section 31.4, roots take up ions in the regions just above the root tips. Over successive growing seasons, long-lived plants such as trees can develop millions, even billions, of root tips, each one a potential absorption site.

Root hairs, the diminutive absorptive structures shown in Figure 31.10c, are another significant adaptation for the uptake of mineral ions and water. In a plant such as a mature red oak *(Quercus rubra)*, which has a vast root system, the total number of root hairs is astronomical. Even in young plants, root hairs greatly increase the root surface area available for absorbing water and ions.

Recall from Chapter 32 that plant cell membranes also have ion-specific transport proteins by which they selectively absorb ions from soil. For example, from studies of plants such as *Arabidopsis thaliana,* a weed that has become a key model organism for plant research, we know that transport channels for potassium ions (K^+) are embedded in the cell membranes of root cortical cells. Such ion transporters absorb more or less of a particular ion depending on chemical conditions in the surrounding soil.

Getting to the Roots of Plant Nutrition

One way that mycorrhizal fungi benefit their host plants is by increasing their phosphate uptake from soils. How do the fungi accomplish this beneficial process? A molecular answer to this question came from Maria J. Harrison and Marianne L. van Buuren at the Samuel Roberts Noble Foundation in Ardmore, Oklahoma, who were able to identify a gene in one of these fungi that encodes a phosphate transport protein.

Harrison and van Buuren began with a cDNA library prepared from a plant that had been colonized by the mycorrhizal fungus *Glomus versiforme*. They then probed the plant-derived cDNA with a gene that encodes a phosphate transporter in yeast. The goal was to determine whether any cDNA sequences were similar to the yeast gene. (Recall from Section 18.1 that a cDNA library is a cloned collection of DNA sequences derived from mRNAs isolated from a cell. Hence it represents sequences that encode proteins.) The transport protein encoded by the yeast gene is embedded in the plasma membrane, where it uses an H$^+$ gradient as an energy source to move phosphate ions into yeast cells by active transport (see Section 6.4).

When mixed with the plant-derived cDNA sequences, the probe did indeed pair with one of the cDNA sequences. Subsequent sequencing of the segment revealed that the cDNA coded for a protein with a structure typical of many eukaryotic and prokaryotic membrane transport proteins.

To eliminate the possibility that the probe was identifying a plant cDNA in the library rather than one from the mycorrhizal fungus, the investigators next used the identified cDNA to probe a preparation containing all the DNA of a plant that had not been colonized by *Glomus*. No pairing occurred with any of the plant DNA fragments, confirming that the cDNA represented a gene came from the fungus. Additional experiments supported this finding.

Harrison and van Buuren carried their investigation further to see whether the fungal gene actually encoded a phosphate transport protein. For this set of experiments, the investigators used a yeast mutant with a nonfunctional phosphate transporter. Because these mutant yeast cells cannot readily take in phosphate, they grow very slowly, even in a culture medium containing a high concentration of phosphate ions.

The researchers added the *Glomus* gene to the mutants under conditions that increased the likelihood that the yeast cells would take and incorporate the DNA. In response, the yeast cells began to grow rapidly and normally, indicating that they could now synthesize a functional phosphate transporter. When radioactive phosphate ions were added to the culture, the cells rapidly became labeled, confirming that they were taking up phosphate ions at a much greater rate than untreated mutants.

Harrison and van Buuren's study was the first to reveal the molecular basis of phosphate transport by the mycorrhizal fungi. More recent studies with potato plants *(Solanum tuberosum)* have identified a gene encoding a phosphate transporter protein that is expressed in parts of potato roots where mycorrhizae form. These lines of research may lead to methods for reducing the amount of phosphate fertilizers added to crop plants by identifying mycorrhizal fungi providing the most efficient phosphate uptake—or by engineering crop plants with an enhanced capacity to take in this essential nutrient.

Mycorrhizae, symbiotic associations between a fungus and the roots of a plant (see Section 28.3) also promote the uptake of water and ions—especially phosphate and nitrogen—in most species of plants. As shown in Figure 28.17, the fungal partner in the association often grows as a network of hyphal filaments around and beyond the plant's roots. Collectively, the hyphae provide a tremendous surface area for absorbing ions from a large volume of soil. As with plant roots, transport proteins shepherd ions into hyphae. Researchers have recently verified experimentally that hyphal transport proteins are encoded by the DNA of the fungus, not that of the plant (as described in *Insights from the Molecular Revolution*). Some of the plant's sugars and nitrogenous compounds nourish the fungus, and as the root grows, it uses some of the minerals that the fungus has secured. In other types of mycorrhizae, the fungus actually lives inside cells of the root cortex. Orchids, for example, depend on this type of mutualistic association. And, as will be described shortly, some other plants gain access to nitrogen by way of mutually beneficial associations with bacteria.

Nutrients Move into and through the Plant Body by Several Routes

Most mineral ions enter plant roots passively along with the water in which they are dissolved. Some enter root cells immediately. Others travel in solution *between* cells until they meet the endodermis sheathing the root's stele (see Figure 32.6). At the endodermis, the ions are actively transported into the endodermal cells and then into the xylem for transport throughout the plant.

Inside cells, most mineral ions enter vacuoles or the cell cytoplasm, where they become available for metabolic reactions. Some nutrients, such as nitrogen-containing ions, move in phloem from site to site in the plant, as dictated by growth and seasonal needs. In plants that shed their leaves in autumn, before the leaves age and fall significant amounts of nitrogen, phosphorus, potassium, and magnesium move out of them and into twigs and branches. This adaptation conserves the nutrients, which will be used in new growth the next season. Likewise, in late summer, mineral ions move to the roots and lower stem tissues of perennial range grasses that typically die back during the winter. These activities are regulated by hormonal signals, which are the topic of Chapter 35.

Plants Depend on Bacterial Metabolism to Provide Them with Usable Nitrogen

A lack of nitrogen is the single most common limit to plant growth. Air contains plenty of gaseous nitrogen—almost 80% by volume—but plants lack the enzyme necessary to break apart the three covalent bonds in each N_2 molecule ($N \equiv N$). Some nitrogen from the atmosphere reaches the soil in the form of nitrate, NO_3^-, and ammonium ion, NH_4^+. Plants can absorb both these inorganic nitrogen compounds, but usually there is not nearly enough of them to meet plants' ongoing needs.

Nitrogen also enters the soil in organic compounds as dead organisms and animal wastes decompose. For example, dried blood is about 12% nitrogen by weight and chicken manure is about 5% nitrogen, but the nitrogen is bound up in complex organic molecules such as proteins, and in that form it is unavailable to plants. Instead, the main natural processes that replenish soil nitrogen and convert it to absorbable form are carried out by bacteria. These processes are described later and summarized in **Figure 33.7**. They are part of the *nitrogen cycle,* the global movement of nitrogen in its various chemical forms from the environment to organisms and back to the environment, which is described in Chapter 51.

Production and Assimilation of Ammonium and Nitrate. The incorporation of atmospheric nitrogen into compounds that plants can take up is called **nitrogen fixation.** Metabolic pathways of *nitrogen-fixing bacteria* living in the soil or in mutualistic association with plant roots add hydrogen to atmospheric N_2, producing two molecules of NH_3 (ammonia) and one H_2 for each N_2 molecule. The process requires a substantial input of ATP and is catalyzed by the enzyme nitrogenase. In a final step, H_2O and NH_3 react, forming NH_4^+ (ammonium) and OH^-.

Another bacterial process, called **ammonification,** also produces NH_4^+ when soil bacteria known as *ammonifying bacteria* break down decaying organic matter. In this way, nitrogen already incorporated into plants and other organisms is recycled.

Although plants use NH_4^+ to synthesize organic compounds, most plants absorb nitrogen in the form of nitrate, NO_3^-. Nitrate is produced in soil by **nitrification,**

Figure 33.7
How plants obtain nitrogen from soil. Many commercial nitrogen fertilizers are in the chemical form of nitrate, which plant roots readily take up, or in the form of ammonium, which nitrifying bacteria convert to nitrate.

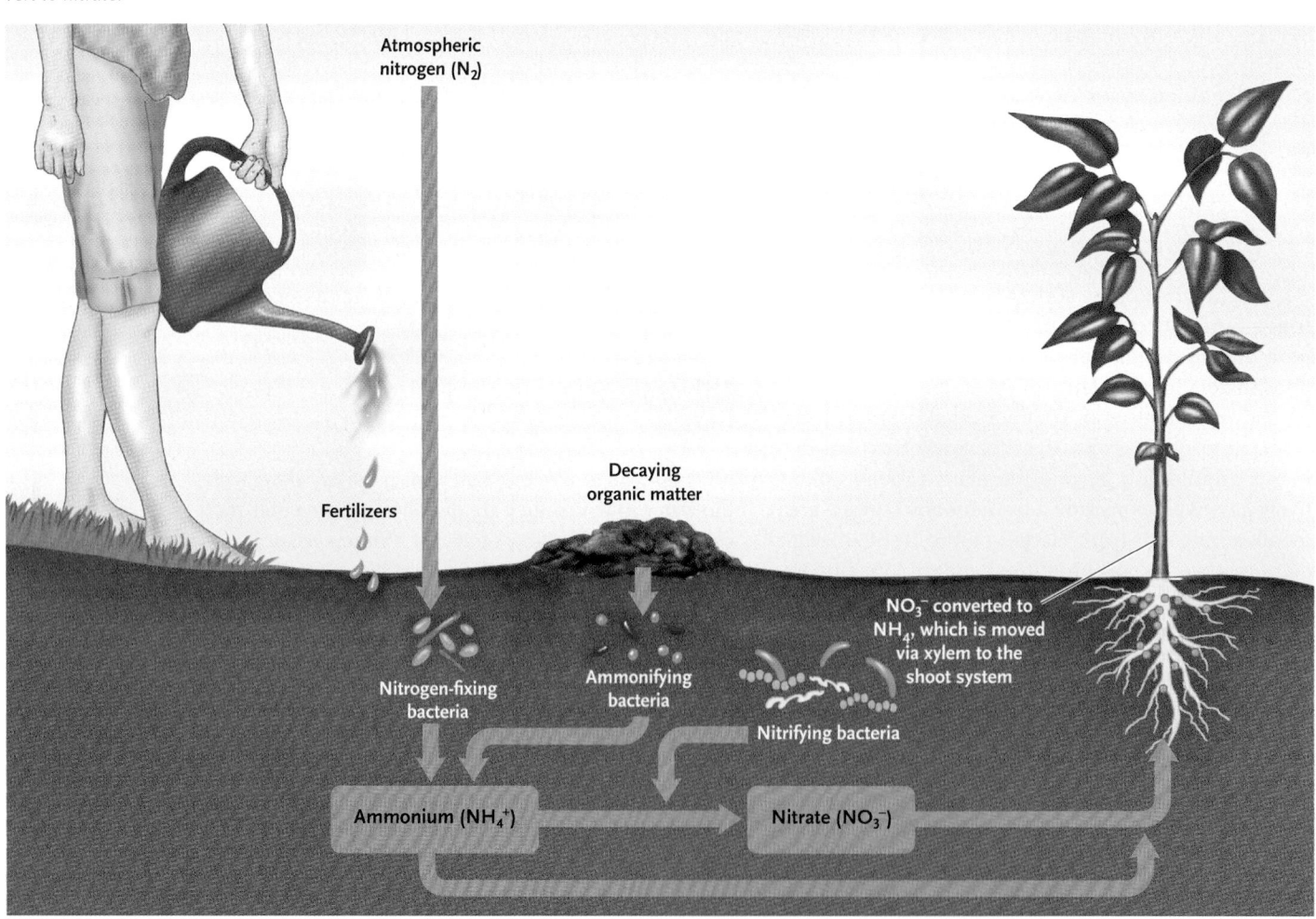

Atmospheric nitrogen (N_2)

Fertilizers

Decaying organic matter

NO_3^- converted to NH_4^+, which is moved via xylem to the shoot system

Nitrogen-fixing bacteria

Ammonifying bacteria

Nitrifying bacteria

Ammonium (NH_4^+)

Nitrate (NO_3^-)

a. Root nodules

b. Field experiment with soybeans *(Glycine max)* and *Rhizobium*

c. Bacteroids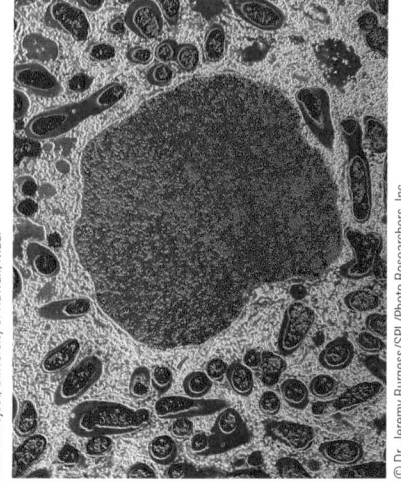

Root nodule

Figure 33.8

The beneficial effect of root nodules. **(a)** Root nodules on a soybean plant *(Glycine max)*. **(b)** Soybean plants growing in nitrogen-poor soil. The plants on the right were inoculated with *Rhizobium* cells and developed root nodules. **(c)** False-color transmission electron micrograph showing membrane-bound bacteroids (red) in a root nodule cell. Membranes that enclose the bacteroids appear blue. The large yellow-green structure is the cell's nucleus.

in which NH_4^+ is oxidized to NO_3^-. Soils generally teem with *nitrifying bacteria,* which carry out this process. Because of ongoing nitrification, nitrate is far more abundant than ammonium in most soils. Usually, the only soils from which plant roots take up ammonium directly are highly acidic, such as in bogs, where the low pH is toxic to nitrifying bacteria.

Nitrogen Assimilation. Once inside root cells, absorbed NO_3^- is converted by a multistep process back to NH_4^+. In this form, nitrogen is rapidly used to synthesize organic molecules, mainly amino acids. These molecules pass into the xylem, which transports them throughout the plant. In some plants, the nitrogen-rich precursors travel in xylem to leaves, where different organic molecules are synthesized. Those molecules travel to other plant cells in the phloem.

Nitrogen Fixation in Plant–Bacteria Associations. Although some nitrogen-fixing bacteria live free in the soil (see Figure 33.7), by far the largest percentage of nitrogen is fixed by species of *Rhizobium* and *Bradyrhizobium,* which form mutualistic associations with the roots of plants in the legume family. The host plant supplies organic molecules that the bacteria use for cellular respiration, and the bacteria supply NH_4^+ that the plant uses to produce proteins and other nitrogenous molecules. In legumes (peas, beans, clover, and alfalfa), the nitrogen-fixing bacteria reside in **root nodules**, localized swellings on roots **(Figure 33.8)**. Farmers may exploit root nodules to increase soil nitrogen by rotating crops (for example, planting soybeans and corn in alternating years). When the legume crop is harvested, the root nodules

and other tissues remaining in the soil enrich its nitrogen content.

Decades of research have revealed the details of how this remarkable relationship unfolds. Usually, a single species of nitrogen-fixing bacteria colonizes a single legume species, drawn to the plant's roots by chemical attractants—primarily compounds called flavonoids—that the roots secrete. Through a sequence of exchanged molecular signals, bacteria are able to penetrate a root hair and form a colony inside the root cortex.

An association between a soybean plant *(Glycine max)* and *Bradyrhizobium japonicum* illustrates the process. In response to a specific flavonoid released by soybean roots, bacterial genes called *nod* genes (for *nodule*) begin to be expressed **(Figure 33.9a)**. Products of the *nod* gene cause the tip of the root hair to curl toward the bacteria and trigger the release of bacterial enzymes that break down the root hair cell wall **(Figure 33.9b)**. As bacteria enter the cell and multiply, the plasma membrane forms a tube called an **infection thread** that extends into the root cortex, allowing the bacteria to invade cortex cells **(Figure 33.9c)**. The enclosed bacteria, now called **bacteroids**, enlarge and become immobile. Stimulated by still other *nod* gene products, cells of the root cortex begin to divide. This region of proliferating cortex cells forms the root nodule **(Figure 33.9d)**. Typically, each cell in a root nodule contains several thousand bacteroids; the plant takes up some of the nitrogen fixed by the bacteroids, and the bacteroids utilize some compounds produced by the plant.

Inside bacteroids, N_2 is reduced to NH_4^+ (ammonium) using ATP produced by cellular respiration. The process is catalyzed by nitrogenase. Ammonium

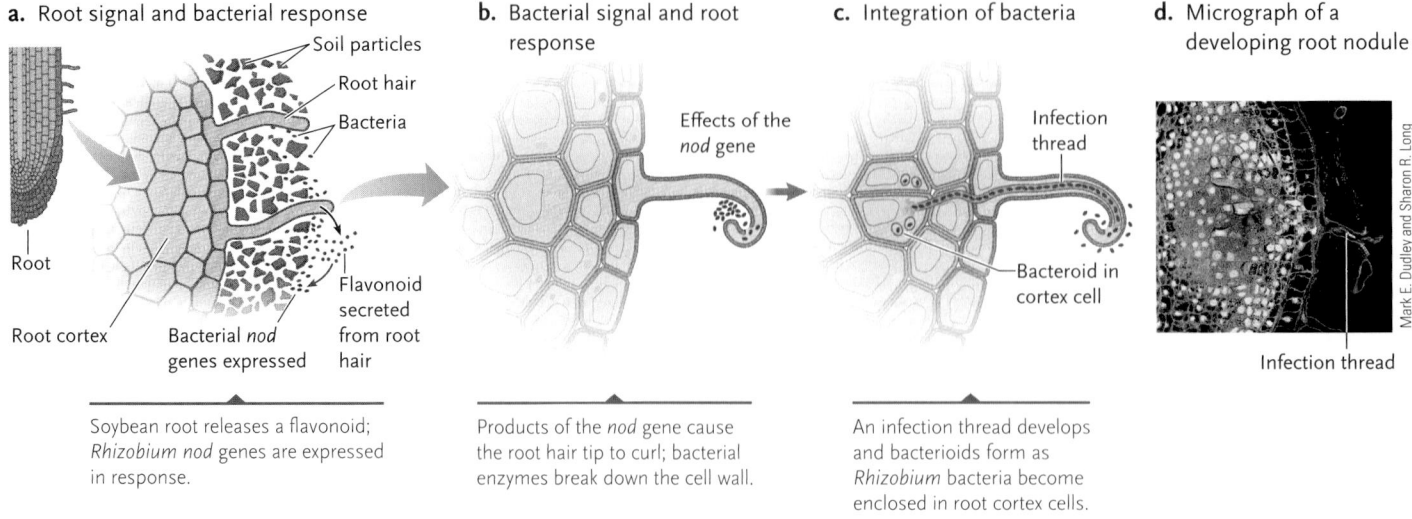

a. Root signal and bacterial response

Soil particles
Root hair
Bacteria

Root

Root cortex

Bacterial *nod* genes expressed

Flavonoid secreted from root hair

Soybean root releases a flavonoid; *Rhizobium nod* genes are expressed in response.

b. Bacterial signal and root response

Effects of the *nod* gene

Products of the *nod* gene cause the root hair tip to curl; bacterial enzymes break down the cell wall.

c. Integration of bacteria

Infection thread

Bacteroid in cortex cell

An infection thread develops and bacteroids form as *Rhizobium* bacteria become enclosed in root cortex cells.

d. Micrograph of a developing root nodule

Mark E. Dudley and Sharon R. Long

Infection thread

Figure 33.9
Root nodule formation in legumes, which interact mutualistically with the nitrogen-fixing bacteria *Rhizobium* and *Bradyrhizobium*.

is highly toxic to cells if it accumulates, however. Thus, NH_4^+ is moved out of bacteroids into the surrounding nodule cells immediately and converted to other compounds, such as the amino acids glutamine and asparagine.

One factor encoded by the bacterial *nod* genes stimulates plant nodule cells to produce a protein called **leghemoglobin** ("legume hemoglobin"). Like the hemoglobin of animal red blood cells, leghemoglobin contains a reddish, iron-containing heme group that

Unanswered Questions

Is "networking" the key to success for plants in some environments?

Key factors that influence plants' ability to take root, grow, and thrive—notably the availability of water and mineral nutrients—are belowground, in the form of mycorrhizae. As you have read in this chapter, field studies and traditional laboratory analyses established that the symbiotic associations between mycorrhizal fungi and plant roots are crucial elements in the survival of the vast majority of vascular plants. Many researchers, including Peter Kennedy and his colleagues at the University of California at Berkeley, also have wondered about possible broader impacts of mycorrhizal associations, such as the extent of their role, if any, in determining the diversity of plant species in different ecological settings and in determining the particular combinations of species that occur. Now Kennedy and others are harnessing molecular tools to shed light on new kinds of questions about interactions among plants and mycorrhizal fungi.

Researchers' ability to define and amplify fungal DNA sequences has revealed the existence of common mycorrhizal networks (CMNs), in which roots of individual plants of the same or different species all form mycorrhizae with the same individual fungus. This discovery has raised several questions: Do mineral ions or other resources pass between plants in a CMN? Several studies indicate that the answer is yes, but much more research is needed to refine scientific

understanding of these interchanges. Does a CMN moderate the effects of competition among plants of different species? Does formation of a CMN improve the survival chances of seedlings, and so help shape the distribution of specific plant species in a given area? Kennedy and his coworkers are exploring these and other questions with respect to CMNs involving two tree species that grow in mixed forests near San Francisco, California—the coast Douglas fir *(Pseudotsuga menziesii)*, a gymnosperm, and the tanbark oak *(Lithocarpus densiflora)*, an angiosperm.

Research efforts by Kennedy and others are examining competition among different species of mycorrhizal fungi, which differ markedly in their resistance to drought and their capacity to take up nutrients. Among other objectives, these studies aim to determine if, or to what extent, the ability of a given plant species to withstand water stress or to gain access to soil nutrients depends on the particular species of fungus with which it forms mycorrhizae. And do the benefits of mycorrhizae increase or decline as environmental conditions change? Answers to such questions will add a new dimension to our understanding of plant nutrition, as well as to our appreciation of what has been called "possibly the most important form of symbiosis in nature."

Beverly McMillan

trate directly absorbed by roots is reduced to ammonium, which then is converted to nontoxic forms.

- In many plant species, root cells synthesize amino acids and other organic nitrogenous compounds, and these molecules are transported in xylem throughout the plant. In some plants, the nitrogen-rich precursors travel in xylem to leaves, where different organic molecules are synthesized. Those molecules then travel to other plant cells in phloem.

- A few plant species have evolved alternative mechanisms for obtaining some or all of their nutrients (Figure 33.10). So-called

carnivorous plants typically produce insect-attracting secretions that contain enzymes which digest the animal's tissues. The plant then absorbs the released nutrients.

- Some plant species parasitize other plants. The parasite may or may not contain chlorophyll and carry out photosynthesis; species that do not photosynthesize obtain all of their nutrition from the host. Epiphytes grow on other plants but obtain nutrients independently.

Animation: Uptake of nutrients by plants

Questions

Self-Test Questions

1. Which best describes a micronutrient?
   a. It makes up 96% of the plant's dry mass.
   b. It cannot be replaced artificially.
   c. It is early on the periodic chart compared with macronutrients.
   d. It is required in large amounts during sunlight hours.
   e. It is an essential element.

2. Nutrient runoff from fertilizing lush lawns often causes "algal blooms" in nearby lakes, making swimming impossible. The fertilizer components most likely to have caused the blooms are:
   a. iron, magnesium, and nitrogen.
   b. nitrogen, phosphorus, and sulfur.
   c. nitrogen, potassium, and phosphorus.
   d. selenium, magnesium, and potassium.
   e. nitrogen, magnesium, and nickel.

3. Which of the following is/are not among the ideal soil conditions for growing crops?
   a. extremely large air spaces
   b. sandy or silty loam
   c. blend of sand and clay
   d. thick top soil
   e. less than 5% humus

4. Which of the following processes contributes to the uptake of mineral ions by plant roots?
   a. chlorosis
   b. osmosis
   c. anion exchange
   d. cation exchange
   e. growth of root hairs

5. Which of the following does not influence soil pH?
   a. rainfall
   b. hydroponic growth
   c. release of sulfur and nitrogen oxides into the air
   d. decomposition of organisms
   e. weathering of rock

6. Which of the following is a process that helps plants utilize nitrogen?
   a. nitrogen-fixing bacteria synthesizing nitrate
   b. ammonifying bacteria using ammonium to produce nitrate
   c. nitrifying bacteria converting NH_4^+ to NO_3^-
   d. the absorption of NH_4^+ by root hairs
   e. the absorption of atmospheric N_2 into the xylem

7. The nod genes in the bacteria in soybean nodules allow the bacteria to fix nitrogen. Which of the following is not a step in this process?
   a. The products of nod genes cause cells of the root cortex to divide and become the root nodule in which bacteroids fix nitrogen for the plant.

   b. In the cortex cells bacteria enlarge and become immobile, forming bacteroids.
   c. Bacteria enter the root hair cell and multiply, causing the cell plasma membrane to form an infection thread that extends into the root cortex.
   d. Roots release flavonoid, which turns on the expression of bacterial nod genes. Products of nod genes cause the tip of the root hair to curl toward the bacteria.
   e. Root hairs trigger release of bacterial enzymes that break down root hair cell walls.
   f. All of the above are steps in the process.

8. Carnivorous plants are deficient in:
   a. oxygen.   d. nitrogen.
   b. phosphorus.   e. carbon.
   c. potassium.

9. Haustorial roots are characteristic of plants that are:
   a. parasites.   d. leghemoglobin users.
   b. epiphytes.   e. carnivorous.
   c. nitrate fixers.

10. Identify the correct match of a nutrient with its function.
    a. chlorine: component of several enzymes
    b. potassium: component of nucleic acids
    c. phosphorus: component of most proteins
    d. manganese: role in shoot and root growth
    e. calcium: maintenance of cell walls and membrane permeability

Questions for Discussion

1. If you want to study factors that affect plant nutrition in nature, what would be the advantages and disadvantages of using a hydroponic culture method?

2. Gardeners often add a humus-rich "soil conditioner" to garden plots before they plant. Adding the conditioner helps aerate the soil, and the decomposing organic materials in humus provide nutrients. If the plot is for annual plants, it often must be reconditioned year after year, even though the gardener faithfully pulls weeds, fertilizes seedlings, applies chemicals to curtail disease-causing soil microbes, and immediately tosses out the mature plants (along with any plant debris) when they have finished bearing. Suggest some reasons why reconditioning is necessary in this scenario, and some strategies that could help limit the need for it.

3. One effect of acid rain is to dissolve rock, liberating minerals into soil. Accordingly, can a case be made that acid rain confers environmental benefits as well as doing harm? What are some other factors, especially with regard to plant adaptations for gaining nutrients, that bear on this question?

4. Using Table 33.1 as a guide, describe some of the known roles of nitrogen, phosphorus, and potassium in plant function. What are some of the signs that a plant suffers a deficiency in those elements?

Experimental Analysis

A plant in your garden is undersized and develops chlorotic leaves even though you fertilize it with a mixture that contains nitrogen, potassium, and phosphorus. After determining that the plant receives enough sunlight for photosynthesis, you next decide to test whether its mineral nutrition is adequate. What specific hypothesis will your experiment test? How will your experimental design test the hypothesis?

Evolution Link

This chapter's *Focus on Research* discusses phytoremediation, the use of plants to remove environmental pollutants such as heavy metals. Some plant species are "hyperaccumulaters" that take up arsenic and other metallic contaminants and sequester such toxins in shoot parts. How might this activity confer a selective advantage?

The reproductive structures of an ornamental poppy *(Papaver rhoeas)*. Male gametophytes, which produce pollen, surround the female gametophyte, which produces eggs and is the site of fertilization and seed development (photographer's close-up).

© Ted Kinsman/SPL/Photo Researchers, Inc.

34 Reproduction and Development in Flowering Plants

WHY IT MATTERS

Seeds of a small flowering tree, *Theobroma cacao,* produce the raw material that modern confectioners turn into chocolate. The tree evolved in the undergrowth of tropical rain forests in Central America, where it was domesticated by the Maya and Aztec peoples. Today cacao trees flourish on vast plantations in the tropical lowlands of Central and South America, the West Indies, West Africa, and New Guinea. Unlike most angiosperms, which produce flowers at the tips of floral shoots, *T. cacao* flowers grow directly from buds on the tree trunk. The flowers are pollinated by insects, primarily midges of the genus *Forcipomyia*. Pollination is the first step toward fertilization of the eggs, and within about 6 months, large, heavy fruits develop from them **(Figure 34.1)**. Each podlike fruit contains from 20 to 60 seeds— the cacao "beans" that chocolate manufacturers process into cocoa, chocolate, and other commercial products.

As in other flowering plants, cacao seeds result from sexual reproduction. Angiosperms have elaborate reproductive systems— housed in flowers—that produce, protect, and nourish sperm, eggs, and developing embryos. As with cacao, the flowers of many species also serve as invitations to animal pollinators, which function in

775

Figure 34.1
Flowers and fruits growing from the trunk of a cacao tree *(Theobroma cacao)*, in Central America. Each fruit is the mature ovary of a *T. cacao* flower.

bringing sperm and egg together. Once a new individual forms and begins to grow, finely regulated gene interactions guide the development of flowers and other plant parts. Under certain circumstances, many plants—including cacao—also reproduce asexually, so that individuals of the new generation are clones, genetically identical to their parents.

Sexual reproduction dominates the life cycle of flowering plants, however, and it will be our main focus in the first three sections of this chapter. We then consider asexual reproduction and conclude with a discussion of early plant development. Using methods of molecular biology and a variety of model organisms, plant biologists are beginning to elucidate some of the mechanisms by which plant developmental pathways unfold.

34.1 Overview of Flowering Plant Reproduction

In the living world, sexual reproduction occurs when male and female haploid gametes unite to create a fertilized egg. This fertilized egg—the diploid zygote—then embarks on a developmental course of mitotic cell divisions, cell enlargement, and cell differentiation. In flowering plants, subsequent steps result in distinctive haploid and diploid forms of the individual.

Diploid and Haploid Generations Arise in the Angiosperm Life Cycle

Once an angiosperm zygote has formed, the developmental sequence generates an embryo enclosed within a seed. In a seed, early versions of the basic plant tissue systems are already in place, so the embryo technically is already a **sporophyte**—the diploid, spore-producing body of a plant (see Section 27.1).

When most people look at a cherry tree or a rosebush, what they think of as "the plant" is the sporophyte **(Figure 34.2)**.

At some point during one or more seasons of an angiosperm sporophyte's growth and development, one or more of its vegetative shoots undergo changes in structure and function and become *floral shoots*—that is, reproductive shoots that will give rise to a flower or inflorescence (a group of flowers on the same floral shoot). Certain cells in the flowers divide by meiosis. Unlike in animals, however, meiosis in plants does not yield gametes directly. Instead, meiosis gives rise to haploid **spores,** walled cells that develop by mitosis into multicellular haploid **gametophytes.** The gametophytes produce haploid sex cells, the gametes, again by mitosis. Male gametophytes produce sperm cells, the male gametes of flowering plants; female gametophytes produce eggs. This division of a life cycle into a diploid, spore-producing generation and a haploid, gamete-producing one is called **alternation of generations** (a phenomenon described more fully in Chapter 27).

In virtually all plants, the gametophyte and sporophyte are strikingly different from one another in both function and structure. For instance, in bryophytes (mosses and liverworts) the gametophyte is usually larger than the sporophyte; the sporophyte grows out of the gametophyte and is nourished by it (see Section 27.2). In ferns, which are seedless vascular plants, the gametophyte is much smaller than the sporophyte and is free-living for much of its lifespan; in most fern species the gametophyte nourishes itself by photosynthesis. In angiosperms and other seed plants, gametophytes are small structures that are retained *inside* the sporophyte for all or part of their lives. The female gametophyte of a flowering plant usually consists of only seven cells that are embedded in floral tissues, as you will read shortly. Male gametophytes are released into the environment as pollen grains, so small that they

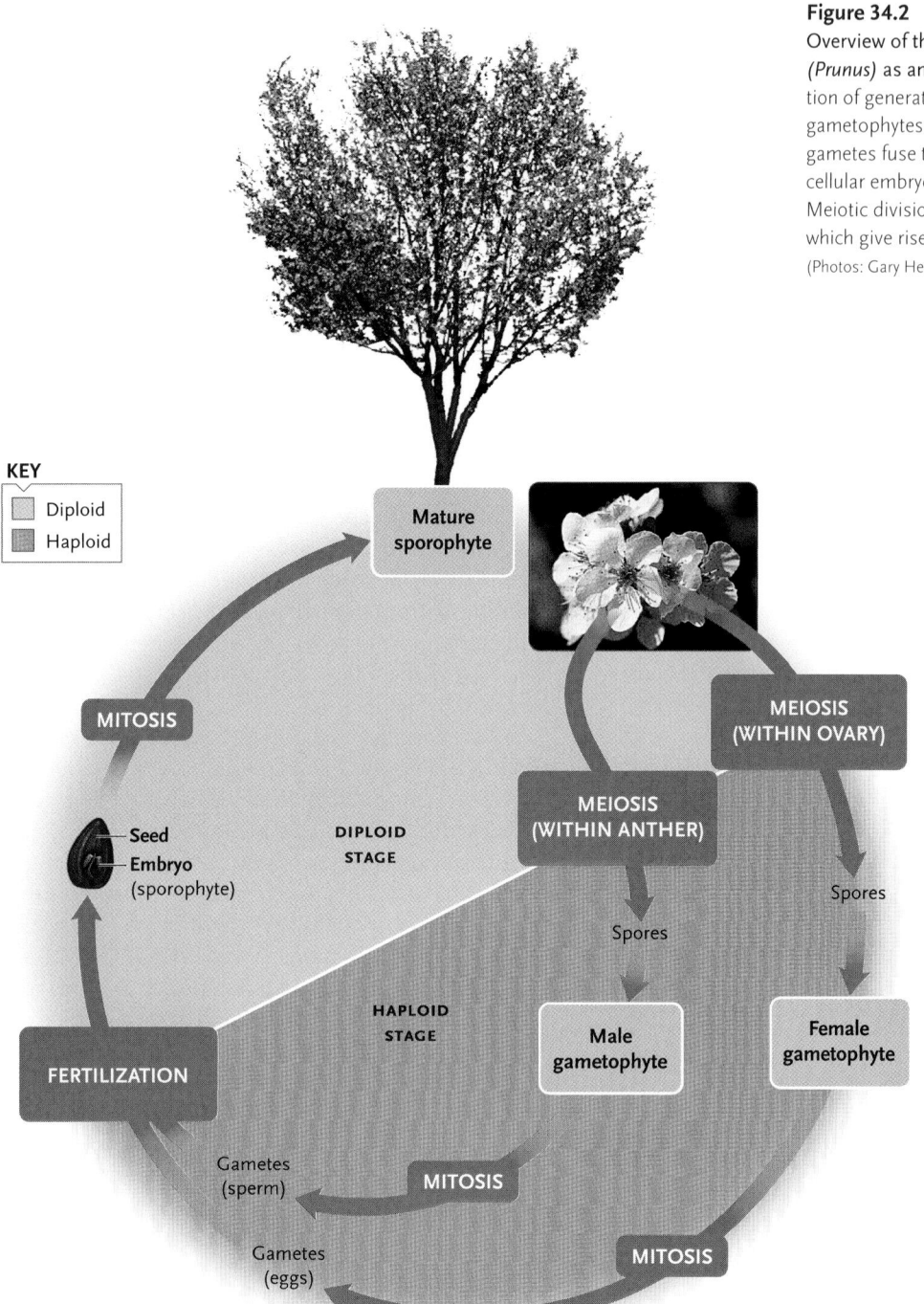

Figure 34.2
Overview of the flowering plant life cycle, using the cherry (*Prunus*) as an example. This type of reproductive cycle, alternation of generations, has a haploid phase in which multicellular gametophytes produce gametes and a diploid phase in which two gametes fuse to form a zygote. This zygote develops into a multicellular embryo within a seed and then into a mature sporophyte. Meiotic divisions in the flower of the sporophyte produce spores, which give rise to new gametophytes.
(Photos: Gary Head.)

KEY
- ☐ Diploid
- ▨ Haploid

Mature sporophyte

MITOSIS

MEIOSIS (WITHIN OVARY)

MEIOSIS (WITHIN ANTHER)

Seed

Embryo (sporophyte)

DIPLOID STAGE

Spores

Spores

HAPLOID STAGE

FERTILIZATION

Male gametophyte

Female gametophyte

Gametes (sperm)

MITOSIS

Gametes (eggs)

MITOSIS

are measured in micrometers. The pollen grain matures when it reaches a compatible ovule, resulting in fertilization and production of a new generation of seeds.

Sporophytes may also reproduce asexually. For instance, strawberry plants send out horizontal stolons, and new roots and shoots develop at each node along the stems. Short underground stems of onions and lilies put out buds that grow into new plants. In summer and fall, Bermuda grass produces new plants at nodes along its subterranean rhizomes. Asexual reproduction also can be induced artificially. Whole orchards of genetically identical fruit trees have been grown from cuttings or buds of a single parent tree.

We turn now to our consideration of sexual reproduction in angiosperms, beginning with the crucial step in which flowers develop.

STUDY BREAK

1. What are the two "alternating generations" of plants?
2. How do these two life phases differ in structure and function?

34.2 The Formation of Flowers and Gametes

Flowering marks a developmental shift for an angiosperm. Biochemical signals—triggered in part by environmental cues such as day length and temperature—travel to the apical meristem of a shoot and set in motion changes in the activity of cells there. Instead of continuing vegetative growth, the shoot is modified into a floral shoot that will give rise to floral organs.

In Angiosperms, Flowers Contain the Organs for Sexual Reproduction

A flower develops from the end of the floral shoot, called the **receptacle.** Cells in the receptacle differentiate to produce up to four types of concentric tissue regions called *whorls.* The arrangement and number of whorl types varies in different species; **Figure 34.3** shows a typical example in which a flower has one of each of the four whorls. The two outer whorls consist of nonfertile, vegetative structures. The outermost whorl (whorl 1), the **calyx**, is made up of leaflike **sepals.** The calyx is usually green, and, early in the flower's development, it encloses all the other parts, as in an unopened rose bud. The next whorl, the **corolla**, includes the **petals.** Corollas are the "showy" parts of flowers; they have distinctive colors, patterning, and shapes, and these features often function in attracting bees and other animal pollinators.

A flower's two inner whorls are specialized for making gametes. Inside the corolla is the whorl of **stamens** (whorl 3), in which male gametophytes form. In almost all living flowering plant species, a stamen consists of a slender **filament** (stalk) capped by a bilobed **anther.** Each anther contains four **pollen sacs**, in which pollen develops.

The innermost whorl (whorl 4) consists of one or more **carpels**, in which female gametophytes form. The lower part of a carpel is the **ovary.** Inside it is one or more **ovules**, in which an egg develops and fertilization takes place. A seed is a mature ovule. In many flowers that have more than one carpel, the carpels fuse into a single, common ovary containing multiple ovules. Typically, the carpel's slender **style** widens at its upper end, terminating in the **stigma**, which serves as a landing platform for pollen. Fused carpels may share a single stigma and style, or each may retain separate ones. The name angiosperm ("seed vessel") refers to the carpel.

Some species have so-called **complete flowers**, in which all four whorls are present. In other species, flowers lack one or more of the whorls, and thus botanists describe them as **incomplete flowers (Figure 34.4).** Botanists also distinguish flowers on the basis of the sexual parts they contain. Most angiosperms produce **perfect flowers**, which have both kinds of sexual parts—that is, both stamens and carpels. **Imperfect flowers** are a type of incomplete flower that has stamens or carpels, but not both. (Notice that all imperfect flowers are also incomplete.) Species with imperfect flowers are further divided according to whether individual plants produce both sexual types of flowers, or only one. In **monoecious** ("one house") species, such as oaks, each plant has some "male" flowers with only stamens and some "female" flowers with only carpels. In **dioecious** ("two houses") species, such as willows, a given plant produces flowers having only stamens or only carpels. With this basic angiosperm reproductive anatomy in mind, we now turn to the processes by which male and female gametes come into being.

Pollen Grains Arise from Microspores in Anthers

Most of a flowering plant's reproductive life cycle, from production of sperm and eggs to production of a mature seed, takes place within its flowers. **Figure 34.5** shows this cycle as it unfolds in a perfect flower. The spores that give rise to male gametophytes are produced in a flower bud's anthers (see Figure 34.5, left). The pollen sacs inside each anther hold diploid microsporocytes (or *microspore mother cells*); each microsporocyte undergoes meiosis and eventually produces four small haploid **microspores.** Like most plant cells, the

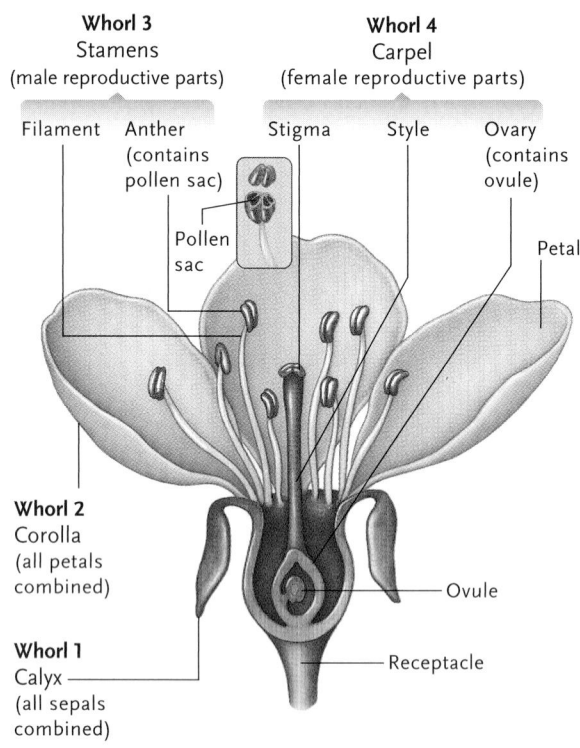

Figure 34.3

Structure of a cherry *(Prunus)* flower, with the four whorls indicated. Like the flowers of many angiosperms, it has a single carpel and several stamens. The anthers of the stamen produce haploid pollen. The stigma of the carpel receives the pollen, and the ovule inside the ovary contains the haploid eggs.

a. Complete flower of an apple tree *(Malus)*

Janet Jones

b. Incomplete flower of a
Hubbard squash *(Cucurbita)*

Karlene V. Schwartz

Figure 34.4

Examples of complete and incomplete flowers.
(a) Apple flowers *(Malus)* are complete. Each has
many stamens, carpels, and petals, along with petal-
like sepals. **(b)** As with other plants in the pumpkin
family, the flowers of this Hubbard squash *(Cucurbita
maxima)* are both incomplete and imperfect because
each has either stamens or carpels, but never both.

microspores are walled, and inside its wall each micro-spore divides again, this time by mitosis. The result is an immature, haploid male gametophyte—a **pollen grain.**

Of the two nuclei produced by the mitotic division of a microspore, one again divides. After this second round of mitosis the male gametophyte consists of three cells—two sperm cells plus a third cell that controls the development of a **pollen tube.** When pollen lands on a stigma, this tube grows through the tissues of a carpel and carries the sperm cells to the ovary. A mature male gametophyte consists of the pollen tube and sperm cells—the male gametes.

The walls of pollen grains are hardened by the decay-resistant polymer *sporopollenin,* and are tough enough to protect the male gametophyte during the somewhat precarious journey from anther to stigma. These walls are so distinctive that the family to which a plant belongs usually can be identified from pollen alone—based on the size and wall sculpturing of the grains, as well as the number of pores in the wall **(Figure 34.6).** Because they withstand decay, pollen grains fossilize well and can provide revealing clues about the evolution of seed plants and the ecological communities that lived in the past.

Eggs and Other Cells of Female Gametophytes Arise from Megaspores

Meanwhile, in the ovary of a flower, one or more dome-shaped masses form on the inner wall. Each mass becomes an ovule (see Figure 34.5, right), which, if all goes well, develops into a seed. Only one ovule forms in the carpel of some flowers, such as the cherry. Dozens, hundreds, or thousands may form in the carpels of other flowers, such as those of a bell pepper plant *(Capsicum annuum).* At one end, the ovule has a small opening, called the **micropyle.**

Inside the cell mass, a diploid megasporocyte (or *megaspore mother cell*) divides by meiosis, forming four haploid **megaspores.** In most plants, three of these megaspores disintegrate. The remaining megaspore enlarges and develops into the female gametophyte in a sequence of steps tracked in Figure 34.5.

First, three rounds of mitosis occur *without* cytoplasmic division; the result is a single cell with eight nuclei arranged in two groups of four. Next, one nucleus in each group migrates to the center of the cell; these two **polar nuclei** ("polar" because they migrate from opposite ends of the cell) may fuse or remain separate. The cytoplasm then divides, and a cell wall forms around the two polar nuclei, forming a single large *central cell.* A wall also forms around each of the other nuclei. Three of these walled nuclei become *antipodal cells,* which eventually disintegrate. Three others form a cluster (called the "egg apparatus") near the micropyle; one of them is an **egg cell** that may eventually be fertilized. The other two, called *synergids,* will have a role in pollination. The eventual result of all these events is an **embryo sac** containing seven cells and eight nuclei. This embryo sac is the female gametophyte.

In about a third of flowering plants, biologists have observed variations in the events that produce a female gametophyte. In lilies, for example, changes in the sequence of cell divisions produce several cells with triploid nuclei (see Figure 27.31). The egg cell is not involved, however, so such differences do not affect reproduction. They may have roles in the development and functioning of other embryonic tissues.

As the male and female gametophytes complete their maturation, the stage is set for fertilization and the development of a new individual.

STUDY BREAK

1. What is the biological role of flowers, and what fundamental physiological change must occur before an angiosperm can produce a flower?
2. Explain the steps leading to the formation of a mature male gametophyte, beginning with microsporocytes in a flower's anthers. Which structures are diploid and which haploid?
3. Trace the development of a female gametophyte, beginning with the megasporocyte in an ovule of a flower's ovary. Which structures are diploid and which haploid?

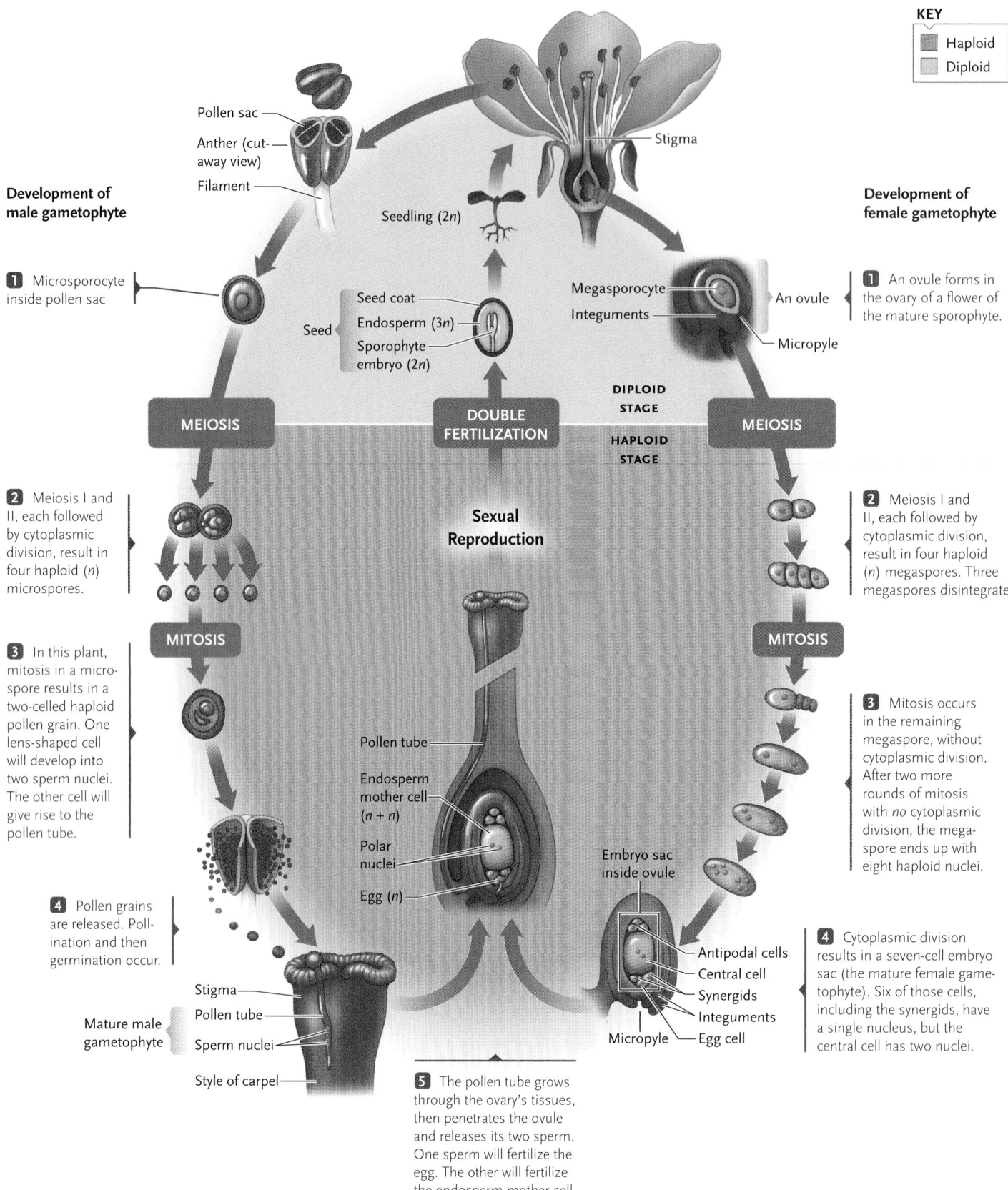

KEY

■ Haploid
□ Diploid

Development of male gametophyte

Pollen sac

Anther (cut-away view)

Filament

Stigma

Seedling (2n)

1 Microsporocyte inside pollen sac

Seed coat

Endosperm (3n)

Seed

Sporophyte embryo (2n)

Megasporocyte

Integuments

An ovule

Micropyle

Development of female gametophyte

1 An ovule forms in the ovary of a flower of the mature sporophyte.

MEIOSIS

DOUBLE FERTILIZATION

DIPLOID STAGE

HAPLOID STAGE

MEIOSIS

2 Meiosis I and II, each followed by cytoplasmic division, result in four haploid (n) microspores.

Sexual Reproduction

2 Meiosis I and II, each followed by cytoplasmic division, result in four haploid (n) megaspores. Three megaspores disintegrate.

3 In this plant, mitosis in a microspore results in a two-celled haploid pollen grain. One lens-shaped cell will develop into two sperm nuclei. The other cell will give rise to the pollen tube.

MITOSIS

MITOSIS

3 Mitosis occurs in the remaining megaspore, without cytoplasmic division. After two more rounds of mitosis with *no* cytoplasmic division, the megaspore ends up with eight haploid nuclei.

Pollen tube

Endosperm mother cell (n + n)

Polar nuclei

Egg (n)

Embryo sac inside ovule

4 Pollen grains are released. Pollination and then germination occur.

Stigma

Pollen tube

Sperm nuclei

Style of carpel

Mature male gametophyte

Antipodal cells

Central cell

Synergids

Integuments

Micropyle   Egg cell

4 Cytoplasmic division results in a seven-cell embryo sac (the mature female gametophyte). Six of those cells, including the synergids, have a single nucleus, but the central cell has two nuclei.

5 The pollen tube grows through the ovary's tissues, then penetrates the ovule and releases its two sperm. One sperm will fertilize the egg. The other will fertilize the endosperm mother cell.

Figure 34.5

Life cycle of cherry *(Prunus)*, a eudicot. Pollen grains develop in pollen sacs within the anthers. An embryo sac forms in the single ovule within the cherry flower's ovary, and an egg forms within the embryo sac. When the pollen grains are released and contact the stigma, double fertilization occurs. An embryo sporophyte and nutritive endosperm develop and become encased in a seed coat.

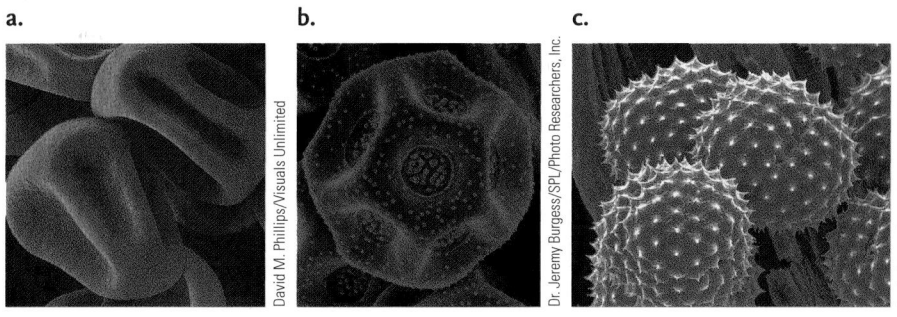

a.

David M. Phillips/Visuals Unlimited

b.

Dr. Jeremy Burgess/SPL/Photo Researchers, Inc.

c.

David Scharf/Peter Arnold, Inc.

Figure 34.6

Some examples of pollen grain diversity. Scanning electron micrographs of pollen grains from **(a)** a grass, **(b)** chickweed (*Stellaria*), and **(c)** ragweed (*Ambrosia*) plants.

34.3 Pollination, Fertilization, and Germination

The process by which plants produce seeds—which have the potential to give rise to new individuals—begins with *pollination,* when pollen grains make contact with the stigma of a flower. Air or water currents, birds, bats, insects, or other agents make the transfer. (Section 27.5 discussed the complex relationship between some flowering plants and their animal pollinators.)

Pollination is the first in a series of events leading to *fertilization,* the fusion of an egg and sperm inside the flower's ovary. The resulting embryo and its ovule mature into a seed housing a young sporophyte, and when the seed *germinates,* or sprouts, the sporophyte begins to grow.

Pollination Requires Compatible Pollen and Female Tissues

Even after pollen reaches a stigma, in most cases pollination and fertilization can take place only if the pollen and stigma are compatible. For example, if pollen from one species lands on a stigma from another, chemical incompatibilities usually prevent pollen tubes from developing.

Even when the sperm-bearing pollen and a stigma are from the same species, pollination may not lead to fertilization unless the pollen and stigma belong to genetically distinct individuals. For instance, when pollen from a given plant lands on that plant's own stigma, a pollen tube may begin to develop, but stop before reaching the embryo sac. This **self-incompatibility** is a biochemical recognition and rejection process that prevents self-fertilization, and it apparently results from interactions between proteins encoded by *S* (self) genes.

Research has shown that *S* genes usually have multiple alleles—in some species there may be hundreds—and a common type of incompatibility occurs when pollen and stigma carry an identical *S* allele. The result is a biochemical signal that prevents proper formation of the pollen tube **(Figure 34.7).** For example, studies on plants of the mustard family have revealed that pollen contacting an incompatible stigma produces a protein that prevents the stigma from hydrating the relatively dry pollen grain, an essential step if the pollen tube is to grow. A wide range of self-incompatibility responses has been discovered, however. In cacao, for instance, when incompatible pollen contacts a stigma, a pollen tube grows normally but a hormonal response soon causes the flower to drop off the plant, preventing fertilization.

Self-incompatibility prevents inbreeding and promotes genetic variation, which is the raw material for

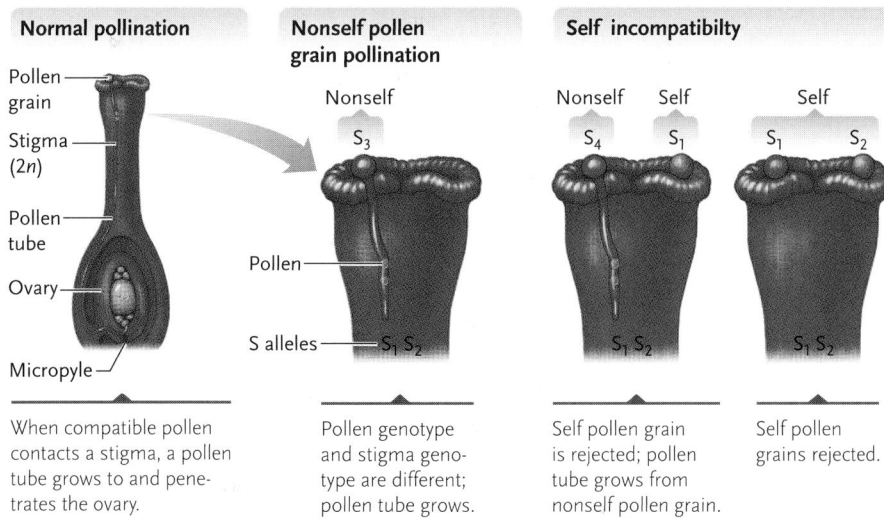

Normal pollination

Pollen grain
Stigma (2*n*)
Pollen tube
Ovary
Micropyle

When compatible pollen contacts a stigma, a pollen tube grows to and penetrates the ovary.

Nonself pollen grain pollination

Nonself
S_3
Pollen
S alleles — $S_1 S_2$

Pollen genotype and stigma genotype are different; pollen tube grows.

Self incompatibilty

Nonself   Self
S_4 S_1
$S_1 S_2$

Self pollen grain is rejected; pollen tube grows from nonself pollen grain.

Self
S_1 S_2
$S_1 S_2$

Self pollen grains rejected.

Figure 34.7

Self-incompatibility. When a pollen grain has an *S* allele that matches one in the stigma (which is diploid), the result is a biochemical response that prevents fertilization—in this illustration, by preventing the growth of a pollen tube.

natural selection and adaptation. Even so, many flowering plants do self-pollinate, either partly or exclusively, because that mode, too, has benefits in some circumstances. (Mendel's peas are a classic example.) For instance, "selfing" may help preserve adaptive traits in a population. It also reduces or eliminates a plant's reliance on wind, water, or animals for pollination, and thus ensures that seeds will form when conditions for cross-pollination are unfavorable, as when pollinators or potential mates are scarce.

Double Fertilization Occurs in Flowering Plants

If a pollen grain lands on a compatible stigma, it absorbs moisture and germinates a pollen tube, which burrows through the stigma and style toward an ovule. Chemical cues from the two synergid cells lying close to the egg cell help guide the pollen tube toward its destination. Before or during these events, the pollen grain's haploid sperm-producing cell divides by mitosis, forming two haploid sperm. When the pollen tube reaches the ovule, it enters through the micropyle and an opening forms in its tip. By this time one synergid has begun to die (an example of programmed cell death), and the two sperm are released into the disintegrating cell's cytoplasm. Experiments suggest that elements of the synergid's cytoskeleton guide the sperm onward, one to the egg cell and the other to the central cell.

Next there occurs a remarkable sequence of events called **double fertilization**, which has been observed only in flowering plants and (in a somewhat different version) in the gnetophyte *Ephedra* (see Section 27.4). Typically, one sperm nucleus fuses with the egg to form a diploid (2n) zygote. The other sperm nucleus fuses with the central cell, forming a cell with a triploid (3n) nucleus. Tissues derived from that 3n cell are called **endosperm** ("inside the seed"). They nourish the embryo and, in monocots, the seedling, until its leaves form and photosynthesis has begun.

Embryo-nourishing endosperm forms only in flowering plants, and its evolution coincided with a

Figure 34.8

Stages in the embryonic development of shepherd's purse *(Capsella bursa-pastoris)*, a eudicot. Figure 34.16 looks in more detail at the development of early plant embryos. The micrographs are not to the same scale.

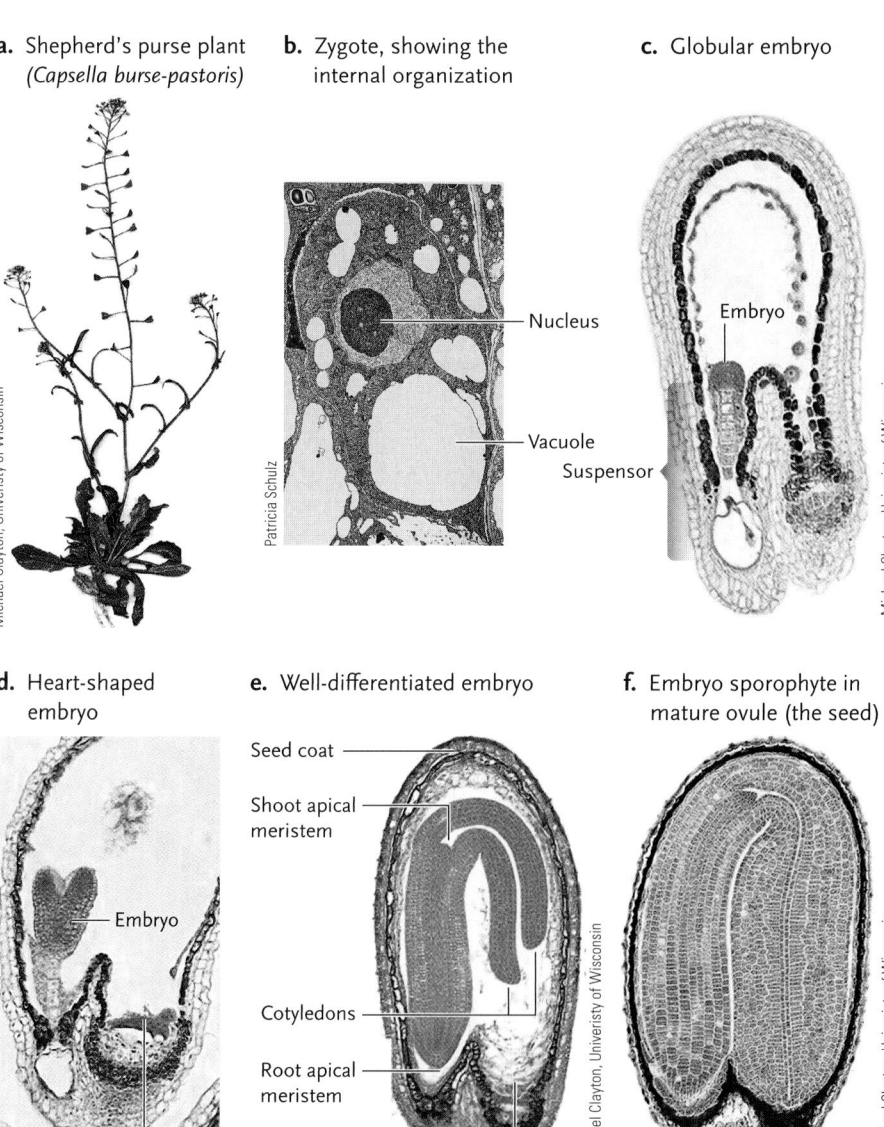

a. Shepherd's purse plant *(Capsella burse-pastoris)*

Michael Clayton, University of Wisconsin

b. Zygote, showing the internal organization

Nucleus

Vacuole

Patricia Schulz

c. Globular embryo

Embryo

Suspensor

Michael Clayton, Univeristy of Wisconsin

d. Heart-shaped embryo

Embryo

Dr. Charles Good, Ohio State University–Lima

e. Well-differentiated embryo

Seed coat

Shoot apical meristem

Cotyledons

Root apical meristem

Endosperm

Michael Clayton, Univeristy of Wisconsin

f. Embryo sporophyte in mature ovule (the seed)

Michael Clayton, University of Wisconsin

reduction in the size of the female gametophyte. In other land plants, such as gymnosperms and ferns, the gametophyte itself contains enough stored food to nourish the embryonic sporophytes.

The Embryonic Sporophyte Develops inside a Seed

When the zygote first forms, it starts to develop and elongate even before mitosis begins. For example, in shepherd's purse *(Capsella),* shown in **Figure 34.8,** most of the organelles in the zygote, including the nucleus, become situated in the top half of the cell, while a vacuole takes up most of the lower half (see Figure 34.8b). The first round of mitosis divides the zygote into an upper *apical cell* and a lower *basal cell.* The apical cell then gives rise to the multicellular embryo, while most descendants of the basal cell form a simple row of cells, the **suspensor,** which transfers nutrients from the parent plant to the embryo (see Figure 34.8c).

The first apical cell divisions produce a globe-shaped structure attached to the suspensor. As they continue to grow, embryos of *Capsella* and other eudicots become heart-shaped (see Figure 34.8d); each lobe of the "heart" is a developing cotyledon (seed leaf), which provides nutrients for growing tissues. Typically,

the two cotyledons absorb much of the nutrient-storing endosperm and become plump and fleshy. For instance, mature seeds of a sunflower *(Helianthus annuus)* have no endosperm at all. In some eudicots, however, the cotyledons remain as slender structures; they produce enzymes that digest the seed's ample endosperm and transfer the liberated nutrients to the seedling. Monocots have one, large cotyledon; in many monocot species, especially grasses such as corn and rice, the cotyledon absorbs the endosperm after germination, when the embryo inside the seed begins to grow.

By the time the ovule is mature—that is, a fully developed seed—it has become encased by a protective **seed coat.** Inside the seed, the sheltered embryo has a lengthwise axis with a root apical meristem at one end and a shoot apical meristem at the other (see Figure 34.8e, f).

Figure 34.9a and **Figure 34.9b** illustrate the structure of the seeds of two eudicots, the kidney bean *(Phaseolus vulgaris)* and the castor bean *(Ricinus communis).* The kidney bean has broad, fleshy cotyledons and the castor bean much thinner ones, but in other ways the embryos are quite similar. The **radicle,** or embryonic root, is located near the micropyle, where the pollen tube entered the ovule prior to fertilization. The radicle

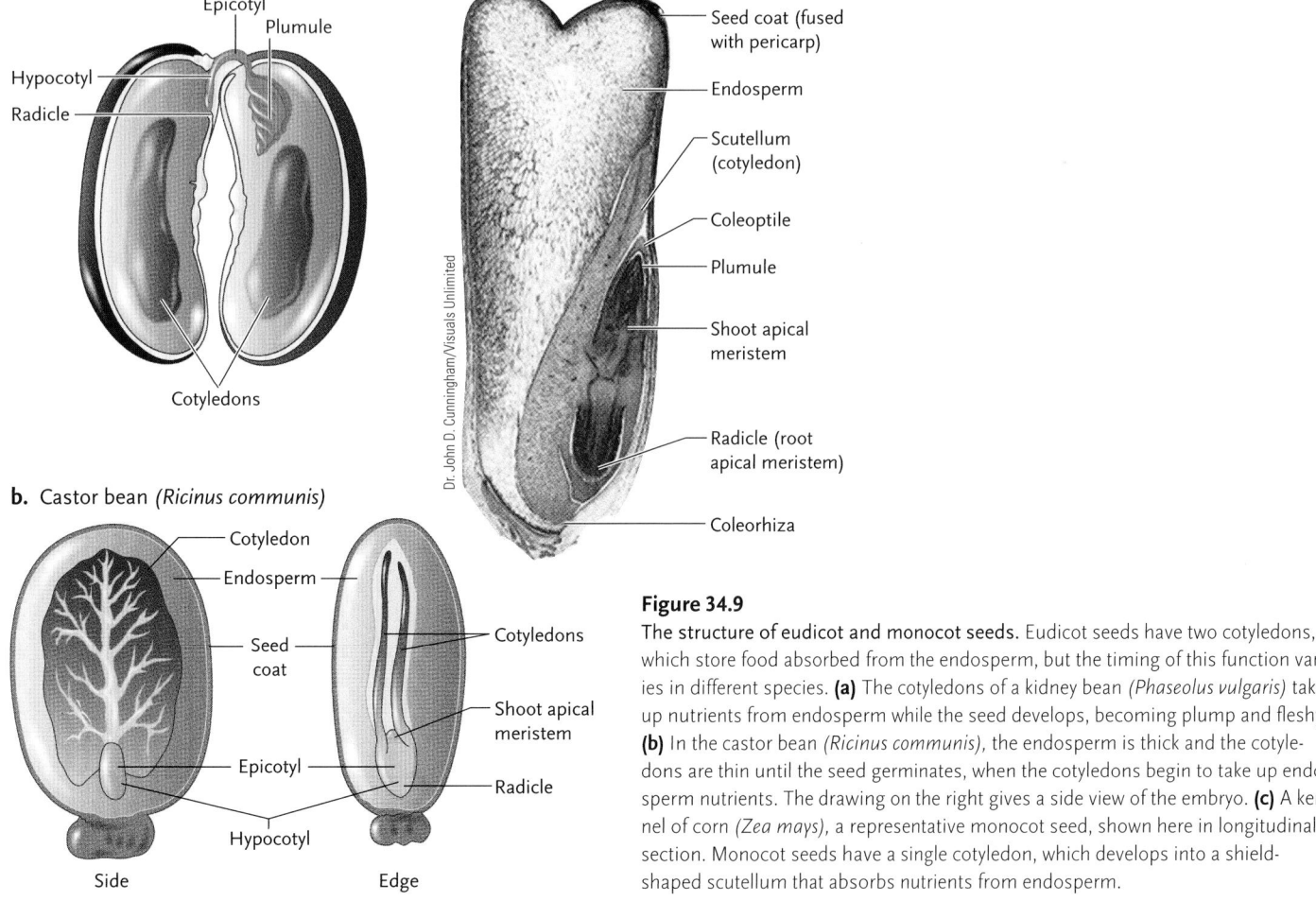

Figure 34.9

The structure of eudicot and monocot seeds. Eudicot seeds have two cotyledons, which store food absorbed from the endosperm, but the timing of this function varies in different species. **(a)** The cotyledons of a kidney bean *(Phaseolus vulgaris)* take up nutrients from endosperm while the seed develops, becoming plump and fleshy. **(b)** In the castor bean *(Ricinus communis),* the endosperm is thick and the cotyledons are thin until the seed germinates, when the cotyledons begin to take up endosperm nutrients. The drawing on the right gives a side view of the embryo. **(c)** A kernel of corn *(Zea mays),* a representative monocot seed, shown here in longitudinal section. Monocot seeds have a single cotyledon, which develops into a shield-shaped scutellum that absorbs nutrients from endosperm.

attaches to the cotyledon at a region of cells called the **hypocotyl.** Beyond the hypocotyl is the **epicotyl,** which has the shoot apical meristem at its tip and which often bears a cluster of tiny foliage leaves, the **plumule.** At germination, when the root and shoot first elongate and emerge from the seed, the cotyledons are positioned at the first stem node with the epicotyl above them and the hypocotyl below them.

The embryos of monocots such as corn differ structurally from those of eudicots in several ways **(Figure 34.9c).** They have only one very large cotyledon, called a **scutellum.** In addition, the root and shoot apical meristems of monocots are blanketed by protective tissues. The shoot apical meristem and plumule are covered by a **coleoptile,** a sheath of cells that protects them during upward growth through the soil. A similar covering, the **coleorhiza,** sheathes the radicle until it breaks out of the seed coat and enters the soil as the primary root. The actual embryo of a corn plant is buried deep within the corn "kernel," which technically is called a *grain.* Most of the moist interior of a fresh corn grain is endosperm; the single cotyledon forms a plump, shield-shaped mass that absorbs nutrients from the endosperm.

Fruits Protect Seeds and Aid Seed Dispersal

Most angiosperm seeds are housed inside fruits, which provide protection and often aid seed dispersal. A **fruit** is a matured or ripened ovary. Usually, fruits begin to develop after a flower's ovule or ovules are fertilized by pollen, and the start of ovule growth after pollination is called "fruit set." The fruit wall, called the **pericarp,** develops from the ovary wall and can have several layers. Hormones in pollen grains provide the initial stimulus that turns on the genetic machinery leading to fruit development; additional signals come from hormones produced by the developing seeds.

Fruits are extremely diverse, and biologists classify them into types based on combinations of structural features. A major defining feature is the nature of the pericarp, which may be fleshy (as in peaches) or dry (as in a hazelnut). A fruit also is classified according to the number of ovaries or flowers from which it develops. **Simple fruits,** such as peaches, tomatoes, and the cacao fruits pictured in Figure 34.1, develop from a single ovary, and in many of them at least one layer of the pericarp is fleshy and juicy. Other simple fruits, including grains and nuts, have a thin, dry pericarp, which may be fused to the seed coat. The garden pea *(Pisum sativa)* is a simple fruit, the peas being the seeds and the surrounding shell the pericarp. **Aggregate fruits** are formed from several ovaries in a single flower. Examples are raspberries and strawberries, which develop from clusters of ovaries. Strawberries also qualify as *accessory* fruits, in which floral parts in addition to the ovary become incorporated as the fruit develops. For instance, anatomically, the fleshy part of a straw-

berry is an expanded receptacle (the end of the floral shoot) and the strawberry fruits are the tiny, dry nubbins (called *achenes*) you see embedded in the fleshy tissue of each berry. **Multiple fruits** develop from several ovaries in multiple flowers. For example, a pineapple is a multiple fruit that develops from the enlarged ovaries of several flowers clustered together in an inflorescence. **Figure 34.10** shows examples of some different types of fruits.

Fruits have two functions: they protect seeds, and they aid seed dispersal in specific environments. For example, the shell of a sunflower seed is a pericarp that protects the seeds within. A pea pod is a pericarp that in nature splits open to disperse the seeds (peas) inside. Maple fruits have winglike extensions for dispersal (see Figure 34.10e). When the fruit drops, the wings cause it to spin sideways and also can carry it away on a breeze. This aerodynamic property propels maple seeds to new locations, where they will not have to compete with the parent tree for water and minerals. Fruits also may have hooks, spines, hairs, or sticky surfaces, and they are ferried to new locations when they adhere to feathers, fur, or blue jeans of animals that brush against them. Fleshy fruits such as blueberries and cherries are nutritious food for many animals, and their seeds are adapted for surviving digestive enzymes in the animal gut. The enzymes remove just enough of the hard seed coats to increase the chance of successful germination when the seeds are expelled from the animal's body in feces.

Seed Germination Continues the Life Cycle

A mature seed is essentially dehydrated. On average, only about 10% of its weight is water—too little for cell expansion or metabolism. After a seed is dispersed and germinates, the embryo inside it becomes hydrated and resumes growth. Ideally, a seed germinates when external conditions favor the survival of the embryo and growth of the new sporophyte. This timing is important, for once germination is underway the embryo loses the protection of the seed coat and other structures that surround it. Overall, the amount of soil moisture and oxygen, the temperature, day length, and other environmental factors influence when germination takes place.

In some species, the life cycle may include a period of seed **dormancy** (*dormire* = to sleep), in which biological activity is suspended. Botanists have described a striking array of variations in the conditions required for dormant seeds to germinate. For instance, seeds may require minimum periods of daylight or darkness, repeated soaking, mechanical abrasion, or exposure to certain enzymes, the high heat of a fire, or a freeze–thaw cycle before they finally break dormancy. In some desert plants, hormones in the seed coat inhibit growth of a seedling until heavy rains flush them away. This adaptation prevents seeds from germinating unless there is

a. Peach (*Prunus*), a simple fruit **b.** Raspberry (*Rubus*), an aggregate fruit **c.** Strawberry (*Fragaria*), an accessory fruit

Fleshy pericarp

Fruit wall

d. Pineapple (*Ananus comosus*), a multiple fruit **e.** Maple (*Acer*) fruit

One of many individual fruits

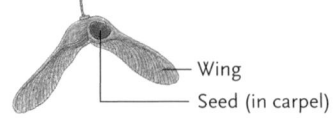

Wing

Seed (in carpel)

Figure 34.10

Fruits. **(a)** Peach, a fleshy simple fruit. **(b)** Raspberry (*Rubus*), an aggregate fruit. **(c)** Strawberry (*Fragaria ananassa*), an accessory fruit that is also an aggregate fruit. **(d)** Pineapple (*Ananas comosus*), a multiple fruit. **(e)** Winged fruits of maple (*Acer*).

enough water in the soil to support growth of the plant through the flowering and seed production stages before the soil dries once again. Many desert plants—and plants in harsh environments such as alpine tundra—cycle from germination to growth, flowering, and seed development in the space of a few weeks, and their offspring remain dormant as seeds until conditions once again favor germination and growth.

Seeds of some species appear to remain viable for amazing lengths of time. Thousand-year-old lotus seeds (*Nelumbo lutea*) discovered in a dry lake bed have germinated trouble-free. And in one startling case, seeds of arctic lupine (*Lupinus arcticus*) were discovered in the 10,000-year-old frozen entrails of a lemming. When they were thawed, they readily germinated as well.

Germination begins with **imbibition**, in which water molecules move into the seed, attracted to hydrophilic groups of stored proteins. As water enters, the seed swells, the coat ruptures, and the radicle begins

its downward growth into the soil. Within this general framework, however, there are many variations among plants.

Once the seed coat splits, water and oxygen move more easily into the seed. Metabolism switches into high gear as cells divide and elongate to produce the seedling. Stable enzymes that were synthesized before dormancy become active; other enzymes are produced as the genes encoding them begin to be expressed. Among other roles, the increased gene activity and enzyme production mobilize the seed's food reserves in cotyledons or endosperm. Nutrients released by the enzymes sustain the rapidly developing seedling until its root and shoot systems are established.

The events of seed germination have been studied extensively in cereal grains, and **Figure 34.11** illustrates them in barley. Notice that the seed's endosperm is separated from the pericarp by a thin layer of cells called the **aleurone.** As a hydrating seed imbibes water, the embryo produces a *gibberellin*, a hormone that

Figure 34.11
How food reserves are mobilized in a germinated seed of barley *(Hordeum vulgare),* a monocot.

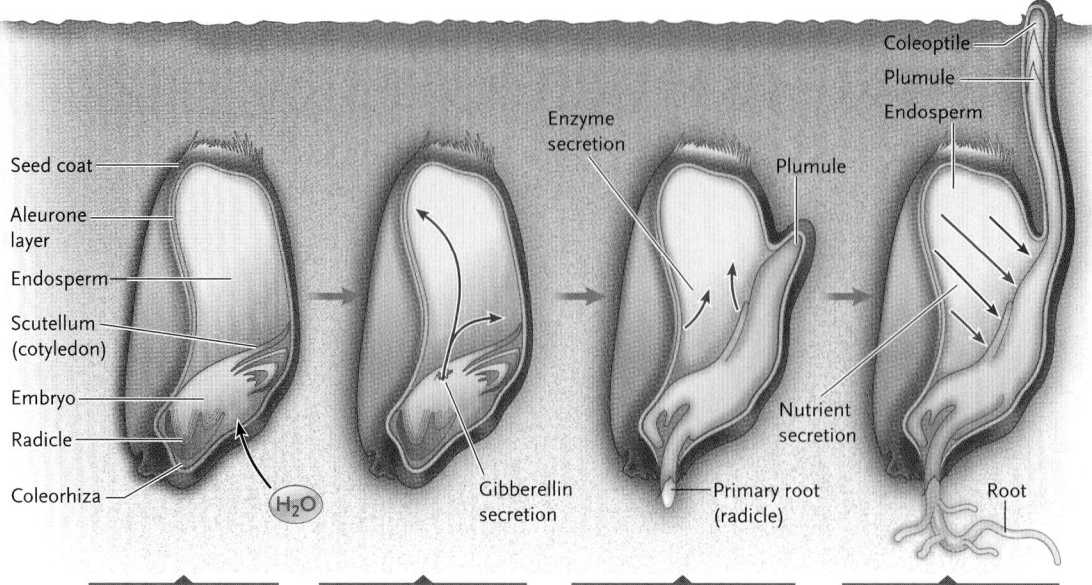

a. Germination begins when the embryo imbibes water.

b. The hydrated embryo produces a gibberellin, a hormone that acts on the cells of the aleurone and the scutellum, activating genes that encode hydrolytic enzymes.

c. The enzymes are secreted into the endosperm, where they break down endosperm cell walls and macromolecules, releasing nutrients.

d. The growing embryo absorbs nutrients released from the endosperm.

Figure 34.12
Stages in the development of a representative eudicot, the kidney bean *(Phaseolus vulgaris).*

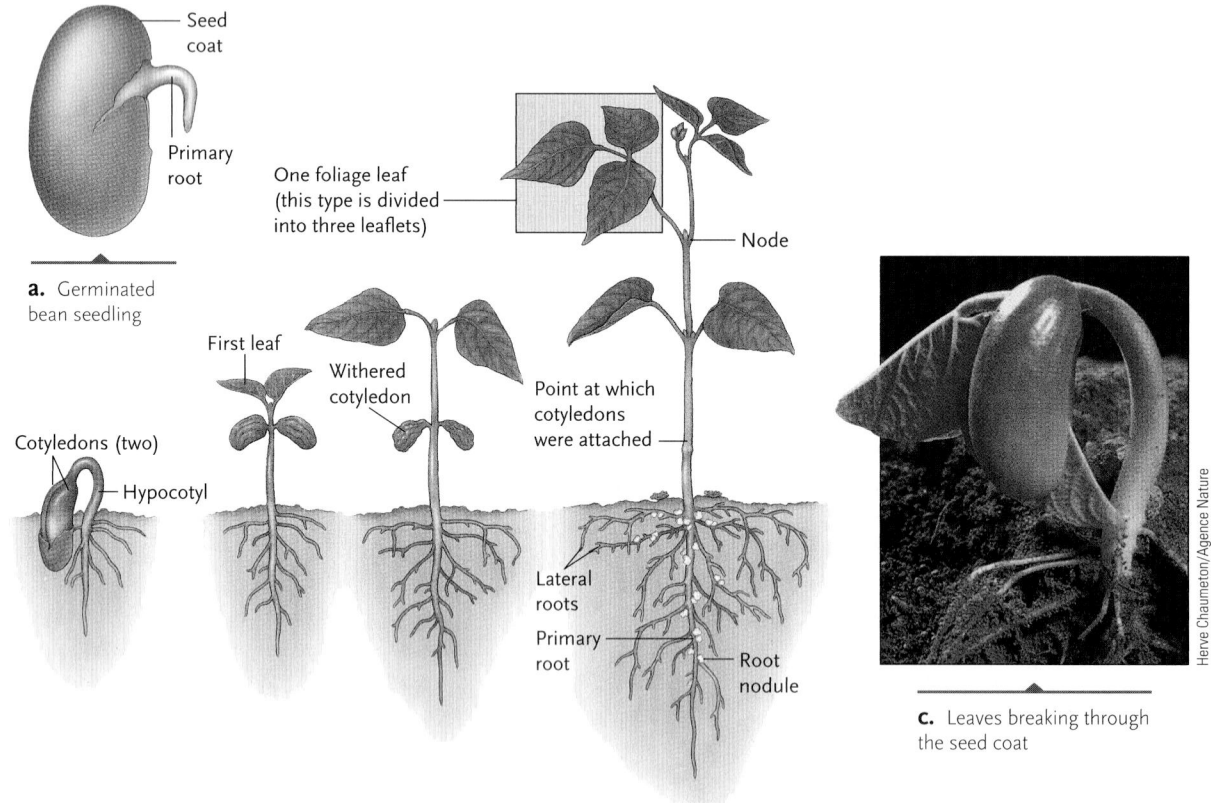

a. Germinated bean seedling

c. Leaves breaking through the seed coat

b. Food-storing cotyledons are lifted above the soil surface when cells of the hypocotyl elongate. The hypocotyl becomes hook-shaped and forces a channel through the soil as it grows. At the soil surface, the hook straightens in response to light. For several days, cells of the cotyledons carry out photosynthesis; then the cotyledons wither and drop off. Photosynthesis is taken over by the first leaves that develop along the stem and later by foliage leaves.

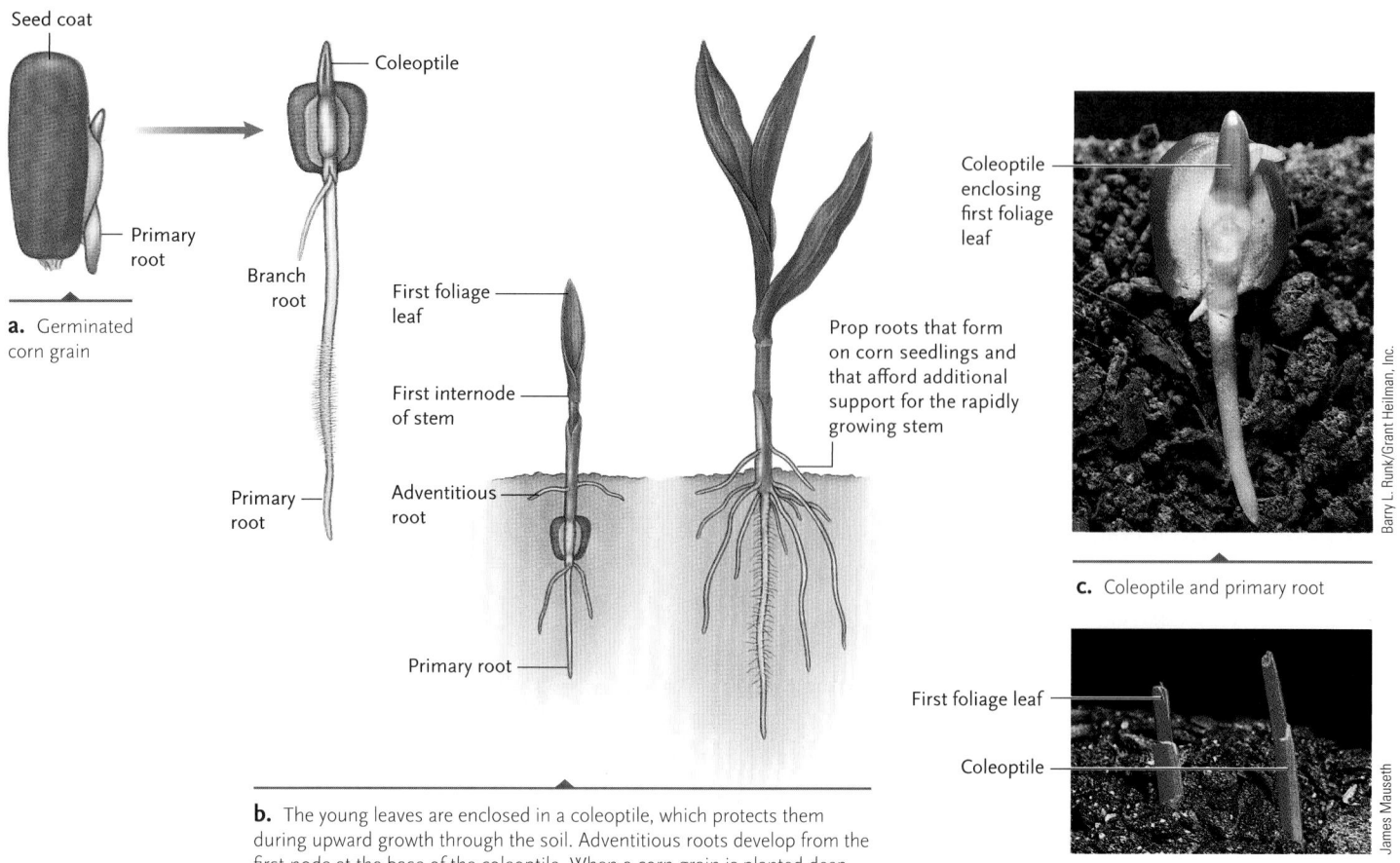

a. Germinated corn grain

b. The young leaves are enclosed in a coleoptile, which protects them during upward growth through the soil. Adventitious roots develop from the first node at the base of the coleoptile. When a corn grain is planted deep, the first internode elongates, separating the primary and adventitious roots. When a grain is planted close to the soil surface, light inhibits elongation of the first internode and the primary and adventitious roots look as if they originate in the same region of the stem.

c. Coleoptile and primary root

Barry L. Runk/Grant Heilman, Inc.

d. Coleoptile and first foliage leaf of two seedlings breaking through the soil surface

James Mauseth

Figure 34.13
Stages in the development of a representative monocot, the corn plant *(Zea mays)*.

stimulates aleurone cells to manufacture and secrete hydrolytic enzymes. Some of these enzymes digest components of endosperm cell walls; others digest proteins, nucleic acids, and starch of the endosperm, releasing nutrient molecules for use by cells of the young root and shoot. Although it is clear that nutrient reserves are also mobilized by metabolic activity in eudicots and in gymnosperms, the details of the process are not well understood.

Inside a germinating seed, embryonic root cells are generally the first to divide and elongate, giving rise to the radicle. When the radicle emerges from the seed coat as the primary root, germination is complete. **Figure 34.12** and **Figure 34.13** depict the stages of early development in a kidney bean, a eudicot, and in corn, a monocot. As the young plant grows, its development continues to be influenced by interactions of hormones and environmental factors, as you will read in Chapter 35.

Most plants give rise to large numbers of seeds because, in the wild, only a tiny fraction of seeds survive, germinate, and eventually grow into another mature plant. Also, flowers, seeds, and fruits represent major investments of plant resources. Asexual reproduction, discussed next, is a more "economical" means by which many plants can propagate themselves.

STUDY BREAK

1. Explain the sequence of events in a flowering plant that begins with formation of a pollen tube and culminates with the formation of a diploid zygote and the $3n$ cell that will give rise to endosperm in a seed.
2. Early angiosperm embryos undergo a series of general changes as a seed matures. Summarize this sequence, then describe the structural differences that develop in the seeds of monocots and eudicots.
3. Germination begins when a seed imbibes water. What are the next key biochemical and developmental events that bring an angiosperm's life cycle full circle?

34.4 Asexual Reproduction of Flowering Plants

As noted in Chapter 31, nodes in the stolons of strawberries and the rhizomes of Bermuda grass each can give rise to new individuals. So can "suckers" that sprout from the roots of blackberry bushes and "eyes" in the tubers of potatoes. All these examples involve asexual or **vegetative reproduction** from a nonreproductive plant part, usually a bit of meristematic tissue in a bud on the root or stem. All of them produce offspring that are clones of the parent. Vegetative reproduction relies on an intriguing property of plants—namely, that many fully differentiated plant cells are **totipotent** ("all powerful"). That is, they have the genetic potential to develop into a whole, fully functional plant. Under appropriate conditions, a totipotent cell can *dedifferentiate*: it returns to an unspecialized embryonic state, and the genetic program that guides the development of a new individual is turned on.

Vegetative Reproduction Is Common in Nature

Various plant species have developed different mechanisms for reproducing asexually. In the type of vegetative reproduction called **fragmentation,** cells in a piece of the parent plant dedifferentiate and then can regenerate missing plant parts. Many gardeners have discovered to their frustration that a chunk of dandelion root left in the soil can rapidly grow into a new dandelion plant in this way.

When a leaf falls or is torn away from a jade plant (*Crassula* species), a new plant can develop from meristematic tissue in the detached leaf adjacent to the wound surface. In the "mother of thousands" plant, *Kalanchoe daigremontiana,* meristematic tissue in notches along the leaf margin gives rise to tiny plantlets **(Figure 34.14)** that eventually fall to the ground, where they can sprout roots and grow to maturity.

Some flowering plants, including some citrus species and the grass variety known as Kentucky blue grass (*Poa pratensis*), can reproduce asexually through a mechanism called **apomixis.** Typically, a diploid embryo develops from an unfertilized egg or from diploid

Figure 34.14
Kalanchoe daigremontiana, the mother-of-thousands plant. Each tiny plant growing from the leaf margin can become a new, independent adult plant.

Ed Reschke/Peter Arnold, Inc.

cells in the ovule tissue around the embryo sac. The resulting seed is said to contain a **somatic embryo,** which is genetically identical to the parent.

In wild plant species, most types of asexual reproduction result in offspring located near the parent. These clonal populations lack the variability provided by sexual reproduction, variation that enhances the odds for survival when environmental conditions change. Yet asexual reproduction offers an advantage in some situations. It usually requires less energy than producing complex reproductive structures such as seeds and showy flowers to attract pollinators. Moreover, clones are likely to be well suited to the environment in which the parent grows.

Many Commercial Growers and Gardeners Use Artificial Vegetative Reproduction

For centuries, gardeners and farmers have used asexual plant propagation to grow particular crops and trees and some ornamental plants. They routinely use *cuttings,* pieces of stems or leaves, to generate new plants; placed in water or moist soil, a cutting may sprout roots within days or a few weeks. Trees and wine grapes often are propagated by grafting a bud or branch from a plant with desirable fruit traits—the *scion*—and joining it to a root or stem from a plant with useful root traits—the *stock*. A grafted plant usually produces flowers and fruit identical to those of the scion's parent plant. The scion of a grafted wine grape variety may be chosen for the quality of its fruit and the stock for its hardy, disease-resistant root system. Vegetative propagation can also be used to grow plants from single cells. Rose bushes and fruit trees from nurseries and commercially important fruits and vegetables such as Bartlett pears, McIntosh apples, Thompson seedless grapes, and asparagus come from plants produced vegetatively in tissue culture conditions that cause their cells to dedifferentiate to an embryonic stage.

Vegetative Propagation in Tissue Culture. In groundbreaking experiments in the 1950s, Frederick C. Steward explored the totipotency of plant cells. Together with his coworkers at Cornell University, Steward propagated whole carrot plants in the laboratory by culturing carrot root phloem. Later researchers confirmed that almost any plant cell that has a nucleus and lacks a secondary cell wall may be totipotent.

The method of plant tissue culture Steward pioneered is simple in its general outlines **(Figure 34.15).** Bits of tissue are excised from a plant and grown in a nutrient medium. The procedure disrupts normal interactions between cells in the tissue sample, and the cells dedifferentiate and form an unorganized, white cell mass called a **callus.** When cultured with nutrients and growth hormones, some cells of the callus regain totipotency and develop into plantlets with roots and shoots.

Steward's work laid the foundation for large-scale commercial applications of plant tissue culture, as well as for a whole new field of research on *somatic embryogenesis* in plants. Single cells derived from a callus generated from shoot meristem are placed in a medium containing nutrients and hormones that promote cell differentiation. With some species, totipotent cells in the sample eventually give rise to diploid somatic embryos that can be packaged with nutrients and hormones in artificial "seeds" (see Figure 34.15). Endowed with the same traits as their parent, crop plants grown from somatic embryos are genetically uniform.

However, mutations often occur in the DNA of somatic embryos derived from callus culture. Screening techniques can identify such *somaclonal* mutants with desirable traits—for example, resistance to a disease that attacks wild-type plants of the same species. In plants that are infected with viruses, callus cultures can be restricted to virus-free cells and thus generate virus-free clones. Tissue culture propagation can then produce hundreds or thousands of identical plants from a single specimen. This technique, called **somaclonal selection**, is now a staple tool in efforts to improve major food crops, such as corn, wheat, rice, and soybeans. It is also being used to rapidly increase stocks of hybrid orchids, lilies, and other valued ornamental plants. The yellow and orange tomatoes that have become common in produce markets are the fruits of plants developed by somaclonal selection.

Research on tobacco and some other species has shown that plants can also be regenerated by **protoplast fusion.** In this method, the walls of living cells in solution are first digested away by enzymes, leaving the protoplasts. Then the protoplasts are induced to fuse, either by applying an electric current, a laser beam, or chemical additives to the solution, which briefly "loosen" the plasma membranes. The resulting cell (now 4*n*, or tetraploid) is transferred to a solid nutrient medium and allowed to develop into a callus; then individual callus cells are stimulated to develop into embryos. If the fused protoplasts come from somatic cells of a single species, the embryos often grow into fertile plants. It has proven more difficult to grow fertile hybrids from fused protoplasts of different species and genera, probably because there are species-specific signals that govern key physiological functions. Even so, this method has produced the pomato, a cross between a potato and a tomato.

Regardless of how it comes into being, an embryonic sporophyte changes significantly as it begins the developmental journey toward maturity, when it will be capable of reproducing. Next we explore what researchers are learning about these developmental changes.

STUDY BREAK

1. Describe three modes of asexual reproduction that occur in flowering plants.
2. What is totipotency, and how do methods of tissue culture exploit this property of plant cells?

34.5 Early Development of Plant Form and Function

Unlike animals, plants have specialized body parts such as leaves and flowers that may arise from meristems throughout an individual's life—sometimes for thousands of years. Accordingly, in plants the biological role of embryonic development is not to generate the tissues and organs of the adult, but to establish a basic body plan—the root–shoot axis and the radial, "outside-to-inside" organization of epidermal, ground, and vascular tissues (see Section 31.1)—and the precursors of the primary meristems. Though they may sound simple, these fundamentals and the stages beyond them all require an intricately orchestrated sequence of molecular events that plant scientists are defining through sophisticated experimentation.

One of the most fruitful approaches has been the study of plants with natural or induced gene mutations that block or otherwise affect steps in development—and accordingly lend insight into the developmental roles of the normal, wild-type versions of those abnormal genes. Some of these genes are **homeotic genes**, regulatory genes in the genome of an organism that encode transcription factors. The transcription factors are proteins that control the expression of other genes, which in turn direct events in development (see *Focus on Research* in Chapter 31). While researchers work with various species to probe the genetic underpinnings of early plant development, the thale cress (*Arabidopsis thaliana*) has become a favorite model organism for plant genetic research (see *Focus on Research*).

Within Hours, an Early Plant Embryo's Basic Body Plan Is Established

The entire *Arabidopsis* genome has been sequenced, providing a powerful molecular "database" for determining how various genes contribute to shaping the plant body. Experimenters' ability to trace the expression of specific genes has shed considerable light on how the root–shoot axis is set and how the three basic plant tissue systems arise.

The Root–Shoot Axis. Shortly after fertilization gives rise to an *Arabidopsis* zygote, the single cell divides. Electron microscopy shows that, as with the *Capsella*

Figure 34.15 Research Method

Plant Cell Culture

PURPOSE: To grow in the laboratory plants that have the same genetic makeup as a parent plant.

PROTOCOL:

1. Typically, bits of somatic tissue are excised, often from root and shoot tips or meristems, because these parts tend to be free of viruses. The excised tissue is cultured in a nutrient medium, under strictly controlled environmental conditions.

2. Within a few days, cells in the excised tissue dedifferentiate and form an unorganized tissue mass called a callus.

3. Individual callus cells can be separated out and cultured in a medium containing growth hormones.

4. Totipotent cells eventually give rise to plantlets with roots and shoots.

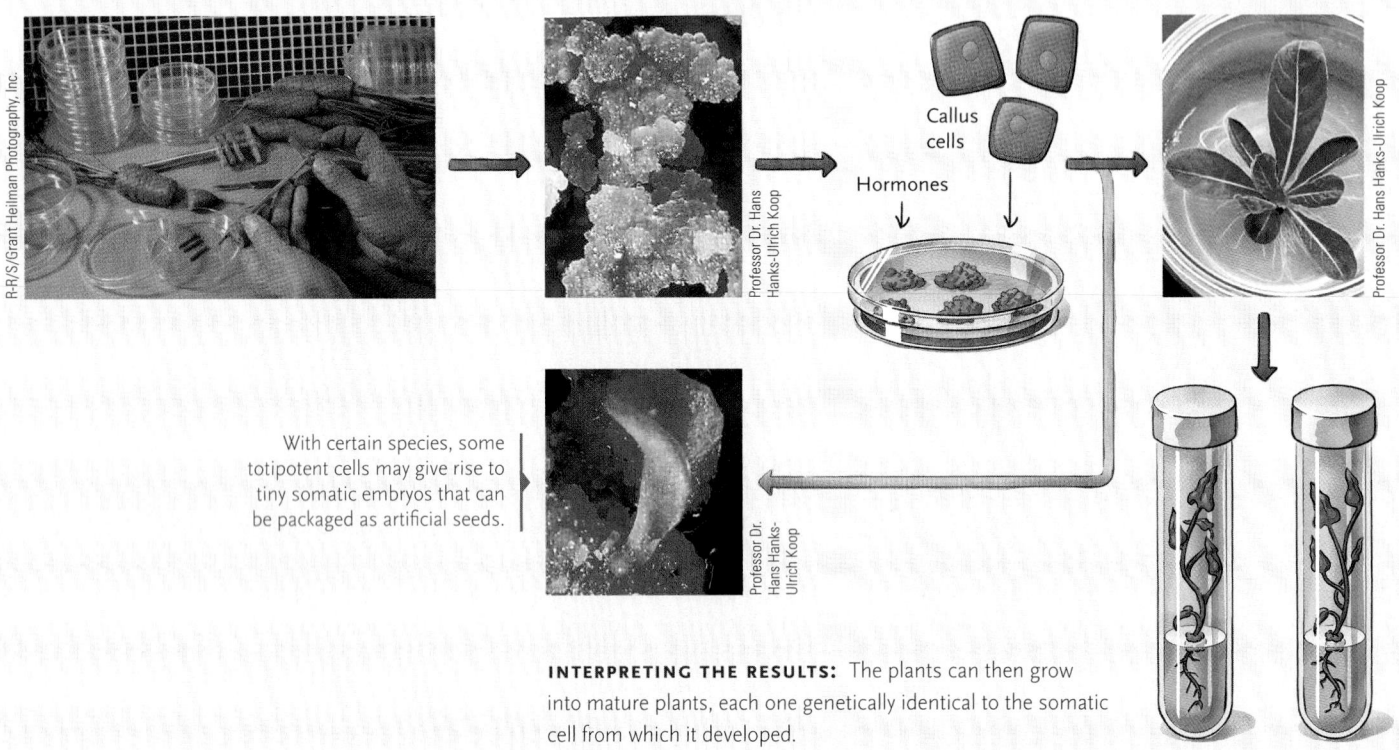

Callus cells

Hormones

With certain species, some totipotent cells may give rise to tiny somatic embryos that can be packaged as artificial seeds.

R-R/S/Grant Heilman Photography, Inc.

Professor Dr. Hans Hanks-Ulrich Koop

Professor Dr. Hans Hanks-Ulrich Koop

Professor Dr. Hans Hanks-Ulrich Koop

INTERPRETING THE RESULTS: The plants can then grow into mature plants, each one genetically identical to the somatic cell from which it developed.

zygote described earlier, this first round of mitosis produces a small apical cell and a larger basal cell **(Figure 34.16a).** The apical cell receives the lion's share of the cytoplasm, while the basal cell receives the zygote's large vacuole and less cytoplasm. Researchers have confirmed that this asymmetrical division of the zygote results in the daughter cells receiving different mixes of mRNAs—the gene transcripts that will be translated into proteins.

Translation of differing mRNAs produces proteins that include several transcription factors, and it marks the genetic threshold of the separation of the plant body into root and shoot regions. As transcription factors trigger the expression of differing genes, distinct biochemical pathways unfold in sequence in the two cells. For example, a basal cell initially exports a signaling molecule (a hormone of the auxin family, discussed in Chapter 35) to the apical cell, and this sets in motion steps leading to the development of the various embryonic shoot features. Later, gene expression

and the flow of chemical signals shift in ways that promote the development of specific structures from the basal cell, including portions of the root apical meristem.

Several of the genes that influence root–shoot polarity have been identified, and when any of them is disrupted, the result can be a serious defect. For example, when an embryo receives two copies of a mutant gene called *gnom,* the embryo doesn't develop distinct root and shoot regions. Instead it remains a lumpy blob **(Figure 34.16b).**

Radial Organization of Tissue Layers. A day or so after an *Arabidopsis* egg cell is fertilized, the embryo consists of eight cells. Even at this early stage both the root–shoot axis and the beginnings of tissue systems are present. When an embryo reaches the so-called torpedo stage, cells representing all three basic tissue systems are in place **(Figure 34.16c).** Again working with mutant plants, investigators have identified

Model Research Organisms: *Arabidopsis thaliana*

For plant geneticists, the little white-flowered thale cress, *Arabidopsis thaliana,* has attributes that make it a prime subject for genetic research. A tiny member of the mustard family, *Arabidopsis* is revealing answers to some of the biggest questions in plant development and physiology.

Each plant grows only a few centimeters tall, so little laboratory space is required to house a large population. As long as *Arabidopsis* is provided with damp soil containing basic nutrients, it grows easily and rapidly in artificial light. Like Mendel's peas, *Arabidopsis* is self-compatible and self-fertilizing, and the flowers of a single plant can yield thousands of seeds per mating. Seeds grow to mature plants in just over a month and then flower and reproduce themselves in another 3 to 4 weeks. This permits investigators to perform desired genetic crosses and obtain large numbers of offspring having known, desired genotypes with

relative ease. Individual *Arabidopsis* cells also grow well in culture.

The *Arabidopsis* genome was the first complete plant genome to be sequenced; at this writing researchers have identified approximately 28,000 genes arranged on five pairs of chromosomes. The genome contains relatively little repetitive DNA, so it is fairly easy to isolate *Arabidopsis* genes, which can then be cloned using genetic engineering techniques. Cloned genes are inserted into bacterial plasmids and the recombinant plasmids transferred to the bacterial species *Agrobacterium tumefaciens*, which readily infects *Arabidopsis* cells. Amplified by the bacteria, the genes and their protein products can be sequenced or studied in other ways.

Typically, researchers use chemical mutagens or recombinant bacteria to introduce changes in the *Arabidopsis* genome. These mutants have become powerful tools for exploring molecular

and cellular mechanisms that operate in plant development—for example, elucidation of the homeotic genes responsible for flower development described in this chapter. *Arabidopsis* mutants are also being used to probe fundamental questions such as how plant cells respond to gravity and the role of pigments called phytochromes in plant responses to light.

An ambitious, multinational research effort called the 2010 Project aims to determine the functions of all *Arabidopsis* genes by 2010. The Arabidopsis Information Resource (TAIR) recently estimated the percentages of *A. thaliana* genes in different functional categories **(Figure a).** The goal of Project 2010 is to create a comprehensive genetic portrait of a flowering plant—how each gene affects the functioning of not only individual cells but the plant as a whole.

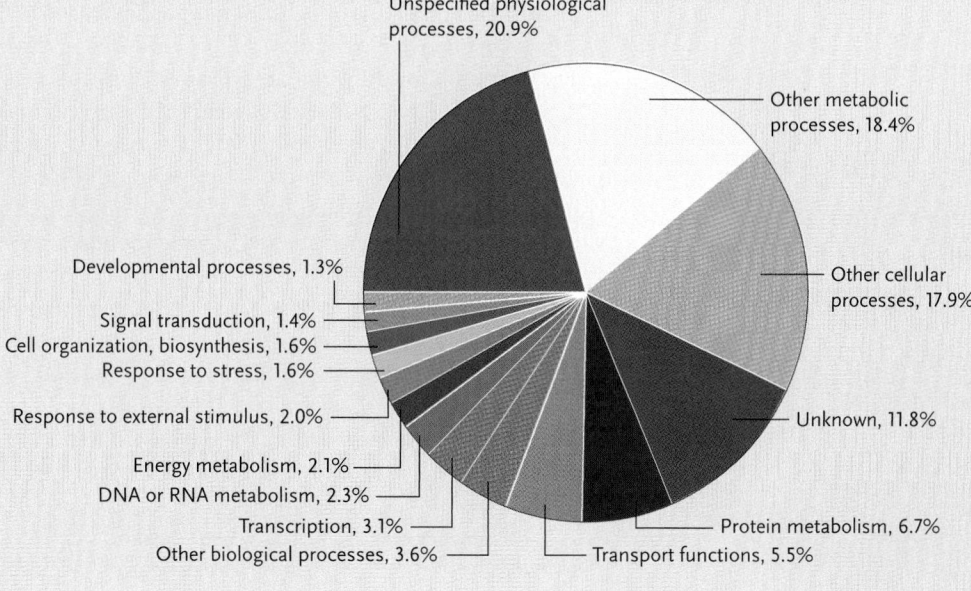

Unspecified physiological processes, 20.9%

Other metabolic processes, 18.4%

Other cellular processes, 17.9%

Developmental processes, 1.3%

Signal transduction, 1.4%
Cell organization, biosynthesis, 1.6%
Response to stress, 1.6%

Response to external stimulus, 2.0%

Energy metabolism, 2.1%
DNA or RNA metabolism, 2.3%

Transcription, 3.1%
Other biological processes, 3.6%

Unknown, 11.8%

Protein metabolism, 6.7%

Transport functions, 5.5%

Figure a
The percentages of *A. thaliana* genes that influence different functional categories.
(Courtesy of the Arabidopsis Information Resource, 2005.)

several *Arabidopsis* genes that help govern early development of tissue systems. For example, a gene called *SCR* encodes a protein that apparently regulates mitotic divisions that produce the first cells of a developing root's cortex and endoderm tissue layers (Figure 31.20 shows the locations of these tissues in a mature root). The roots of a mutant *scr* seedling contain cells with jumbled characteristics of both tissue layers.

No matter what the species, nearly all new plant embryos have the general body plan we have been discussing. As development proceeds, cells at different sites become specialized in prescribed ways as a particular set of genes is expressed in each type of cell—a process known as *differentiation*. Differentiated cells in turn are the foundation of specialized tissues and organs, which come about through processes we consider next.

a. Developing embryos

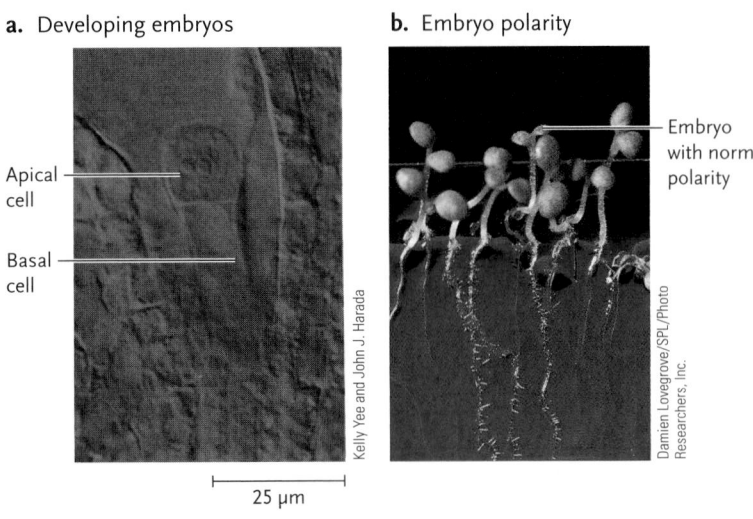

Apical cell

Basal cell

Kelly Yee and John J. Harada

b. Embryo polarity

Embryo with normal polarity

Damien Lovegrove/SPL/Photo Researchers, Inc.

25 µm

c. Beginnings of plant tissue systems

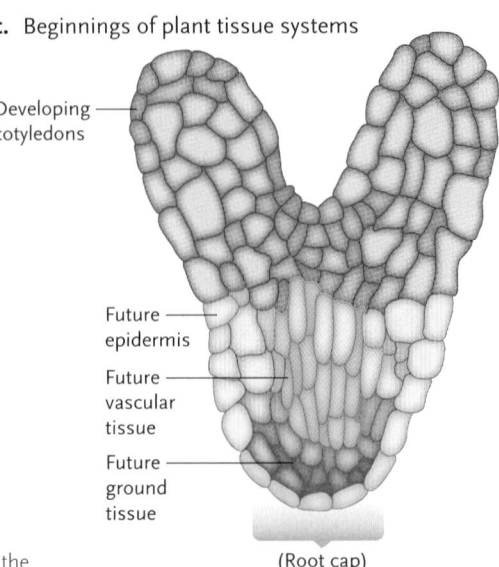

Developing cotyledons

Future epidermis

Future vascular tissue

Future ground tissue

(Root cap)

Figure 34.16
Stages in the development of the basic body plan of a plant embryo. **(a)** After a zygote forms, the first round of cell division produces an embryo with an apical cell that contains much of the zygote's cytoplasm, and a larger basal cell that receives the vacuole and less cytoplasm. The division allots different transcription factors to each cell and establishes the plant's root–shoot axis. **(b)** Normal *A. thaliana* seedlings, in which the root–shoot polarity has become established. **(c)** The approximate locations of early embryonic cells that are the forerunners of epidermal, ground, and vascular tissue systems, respectively.

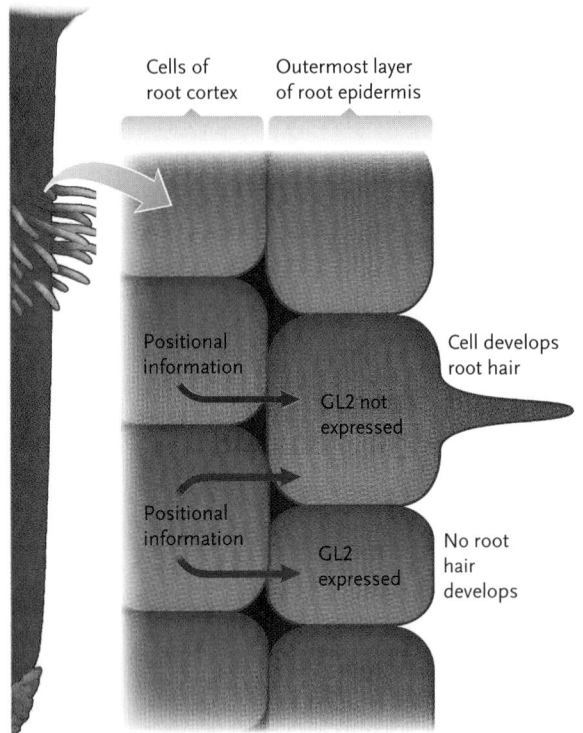

Cells of root cortex

Outermost layer of root epidermis

Positional information

GL2 not expressed

Cell develops root hair

Positional information

GL2 expressed

No root hair develops

Figure 34.17
One model of how positional information influences the development of root hairs. In this model, the only epidermal cells that develop root hairs are those whose inner wall is in contact with two root cortex cells. Such positioning gives rise to signals that block the expression of the GL2 (GLABRA2) gene. When GL2 is expressed, a root epidermal cell will not develop a root hair.

Key Developmental Cues Are Based on a Cell's Position

Although many of the specifics of development differ in animals and plants, one fundamental holds true for both: Normal development produces ordered spatial arrangements of differentiated tissues. Examples in plants include root and shoot apical meristems at opposite ends of the root–shoot axis, the cotyledons that divide the shoot into an upper epicotyl and a lower hypocotyl, and the nested layers of vascular, ground, and epidermal tissue systems. Developmental biologists call this progressive ordering of parts **pattern formation**, and a wealth of research has shown that it is guided by the position of cells relative to one another. Such *positional information* helps establish a cell's developmental fate: that is, it provides cues that "tell" cells where they are in the developing embryo and thus lay the groundwork for an appropriate genetic response.

Numerous researchers have explored how cells in a developing plant or plant part receive and respond to positional information. Experiments have demonstrated, for example, that only certain cells in the epidermis of an embryonic root will give rise to root hairs, the type of trichomes that take up water and minerals from soil (see Section 31.2). These specialized root epidermal cells all share the same position with respect to the underlying root cortex—each abuts two cortical cells. By contrast, no root hair extension will develop from an epidermal cell that lines up against only one cortical cell. **Figure 34.17** diagrams one model of what happens next. In this scenario, one or more

Trichomes: Window on Development in a Single Plant Cell

The delicate plant cell extensions called trichomes are helping to illuminate developmental processes that go on in a single plant cell as it differentiates—that is, as it acquires its ultimate specialized structure and function. In *Arabidopsis* each of these minute protuberances consists of a single cell with a branching tripartite pattern **(Figure a).**

A curious feature of trichomes is that as one differentiates, increases in size, and extends branches in different directions, its chromosomes—and the cell's DNA—and duplicated several times over without mitosis (a process

called endoreduplication). As a result, the cell has multiple copies of chromosomes. Experiments that isolate the effects of different mutants have helped confirm that the amount of DNA in the cell strongly influences the cell's structure, and that several genes interact to determine it. One of these genes is called *TRY* (for *TRIPTYCHON*); when it is mutated the affected plant's trichomes have a double complement of DNA and develop five branches **(Figure b).** But genes that regulate the cell cycle are only part of the story. Experiments with other mutants show that several other genes also help produce

the characteristic three-pronged trichome branching. For example, when a gene called *TUBULIN FOLDING CO-FACTOR C (TFCC)* is affected, the normal organization of microtubules in mutant *tfcc* trichomes is disrupted and the resulting trichome has just two short branches, resembling the oar handles of a canoe. When yet another gene, *STICHEL*, is mutated, *sti* mutants don't develop any branches at all **(Figure c).** The underlying reason for this phenotype is not yet understood.

The examples described here underscore how complex molecular interactions affecting multiple aspects of a cell's functioning ultimately shape a cell's form and function. Because the genes that operate in trichomes are also involved in the development of other types of plant cells, understanding their effects and interactions promises to shed light on processes that operate to generate differentiated cells throughout the plant body.

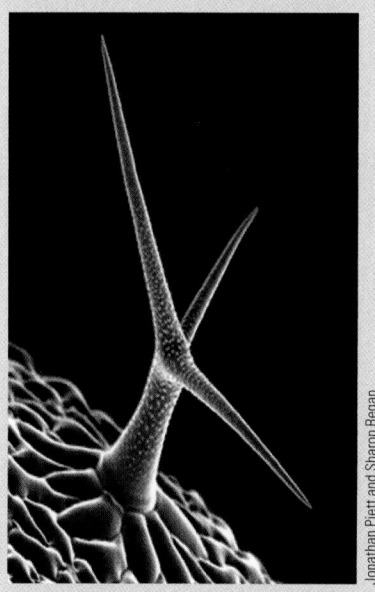

Figure a
Normal trichome from the epidermis of a leaf of *Arabidopsis thaliana*.

Jonathan Piett and Sharon Regan

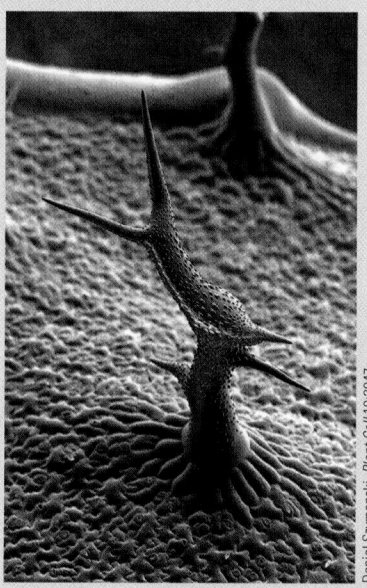

Figure b
Five-pronged trichome from a *try* mutant.

Daniel Szymanski, *Plant Cell* 10:2047

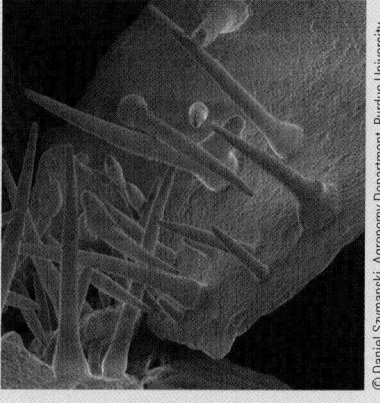

Figure c
The unbranched trichomes of an *sti* mutant.

© Daniel Szymanski; Agronomy Department, Purdue University

chemical signals may cross from cortical to epidermal cells by way of plasmodesmata. When an epidermal cell receives signals from a single cortex cell, a series of genes are expressed in a cascade of effects that culminate in the expression of a gene called GL2 (or GLABRA2). The product of GL2 blocks the formation of root hairs. If, on the other hand, an epidermal cell aligns with two cortex cells, it receives signals from both and the cascade of gene effects blocks expression of GL2—and a root hair develops. *Insights from the Molecular Revolution* gives more examples of ways that trichomes such as root hairs have become popular experimental models for studying the differentiation of plant cells.

Morphogenesis Shapes the Plant Body

As a plant embryo grows and tissues of differentiated cells form, the stage is set for different body regions to develop characteristic shapes and structures that correlate with their function. This process, called **morphogenesis**, shapes the new shoot and root parts produced by dividing cells in meristems. In animals, morphogenesis involves localized cell division and growth, as well as migration of cells and entire tissues from one site to another (see Chapter 48). Plant cells, however, are enclosed within thick walls and usually cannot move. Thus morphogenesis in plants relies on mechanisms that don't require mobility. One of these

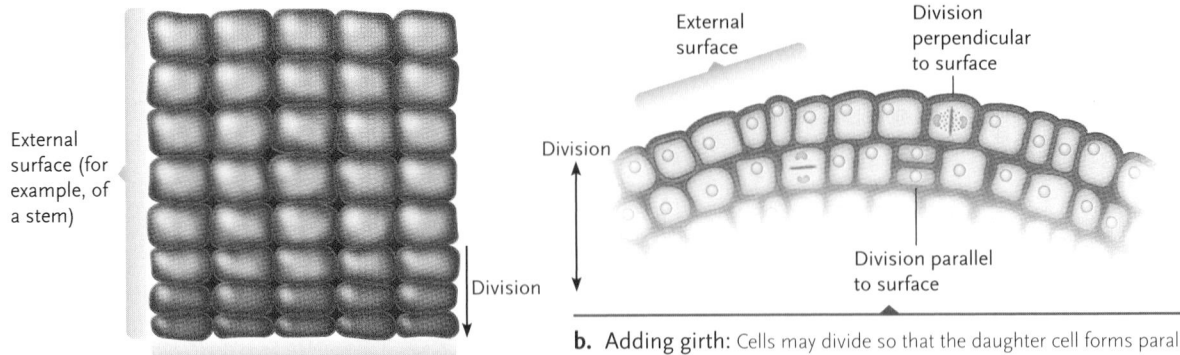

External surface (for example, of a stem)

New cells formed by division in a transverse plane

Division

a. Adding length: Division in a transverse plane parallel to the root-shoot axis produces daughter cells stacked like bricks.

External surface

Division perpendicular to surface

Division

Division parallel to surface

b. Adding girth: Cells may divide so that the daughter cell forms parallel to the plants nearest surface or so that the new cell wall forms perpendicular to the nearest surface.

Figure 34.18
Plant cell division in different planes. The external surface nearest the dividing cell is the reference point for establishing division planes.

mechanisms is **oriented cell division**, which establishes the overall shape of a plant organ, and another is **cell expansion**, which enlarges the cells in specific directions in a developing organ.

Oriented Cell Division. As described in Chapter 31, roots and stems grow lengthwise as the division and expansion of cells produce columns of cells parallel to the root–shoot axis. The cell divisions occur in a transverse plane—that is, new cell walls, and then the cell plate, form so that the cells become stacked one atop the other like wooden blocks **(Figure 34.18a).** A plant adds girth—increases in circumference—by way of cell divisions in other planes. For instance, new cell walls

Figure 34.19
How the plane of cell division is determined in a plant cell. This series of micrographs shows events in onion (*Allium cepa*) root tip cells. The arrows mark the eventual location of the new cell wall.
(All: S. M. Wick, *J Cell Biol*, 89:685, 1987, Rockefeller University Press.)

Preprophase band

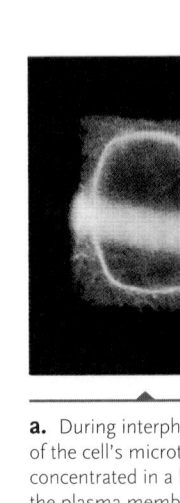

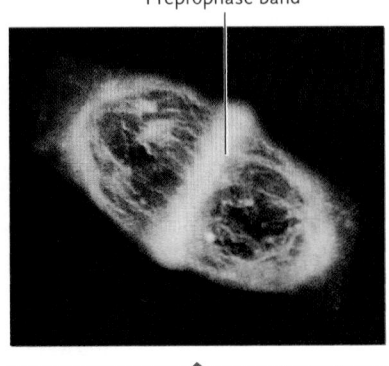

a. During interphase, most of the cell's microtubules are concentrated in a layer under the plasma membrane.

b. In preprophase of mitosis, the microtubules break their links to the plasma membrane and assemble into a thick preprophase band.

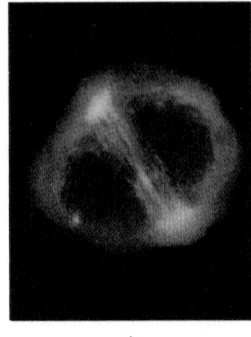
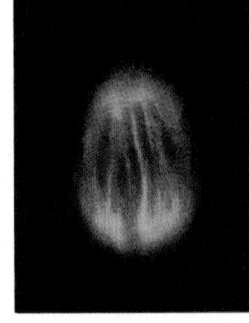
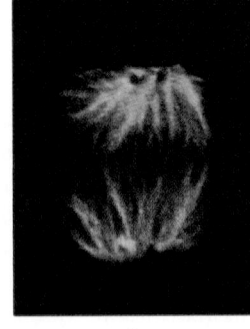
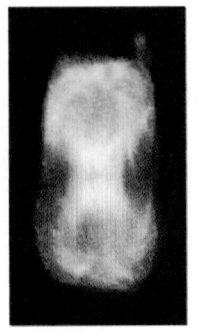
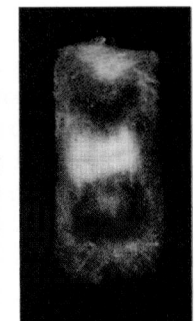

c. As cell division begins, the preprophase band disassembles and microtubules organize into a spindle.

d. After division is complete, the spindle disassembles, leaving a layer of microtubules in the same plane as the preprophase band. The new cell wall and cell plate form there.

may form parallel to the nearest plant surface, or perpendicular both to the nearest surface and to the transverse plane (**Figure 34.18b**).

You may recall from Chapter 10 that the cell plate forms during the cytokinesis phase of mitosis; it establishes the plane of the middle lamella that will eventually separate the parent and daughter cells. The capacity of dividing plant cells to synthesize a new cell plate in a different plane from the old one underlies morphogenesis in nearly all plant groups. In meristematic tissue, changes in the plane of cell division establish the direction in which structures such as lateral roots, branches, and leaf and flower buds will grow, and so gives the plant body its overall form.

Figure 34.19 shows how the plane of a cell plate is established. While a plant cell destined to divide is still in interphase, hours before mitosis begins, the cell nucleus migrates to a particular location in the cell. The nucleus becomes surrounded by a layer of microtubules and microfilaments that radiate outward from it. Where the layer contacts the cell wall, a belt of microtubules and microfilaments called the *preprophase band* forms briefly, encircling the cell cytoplasm. The band usually disappears as mitosis gets underway in the cell, but its position marks the site where the cell plate forms during cytokinesis. Remnants of microfilaments may guide the edges of the developing cell plate into the proper position against the cell wall.

Cell Expansion. Once a cell has divided, the daughter cells expand to mature size. Yet plant cells are encased in a primary wall of nonliving material. Botanists are beginning to learn how the cell wall expands to accommodate the enlarging cell within.

Primary cell walls consist of a loose mesh of cellulose microfibrils embedded in a gel-like matrix. As plant cells mature, they may elongate to as much as 100 times their embryonic lengths. During this elongation, the cellulose meshwork is first loosened and then stretched. Turgor pressure supplies the force for stretching. The exact mechanism that loosens the wall structure is not known, although experiments indicate that it depends on a dramatic drop in pH. Some researchers suggest that an auxin in the cell cytoplasm may stimulate a plasma membrane proton pump that moves H^+ into the cell wall (see Section 6.4). The acidic wall conditions may activate hydrolytic enzymes that break bonds between wall components, or they may promote loosening in some other way.

During expansion, enzyme complexes in the cell's plasma membrane synthesize new cellulose microfibrils from glucose in the cytoplasm. When each microfibril is fully formed, it is bound in place in the growing wall by pectins and other wall components.

The direction of cell expansion depends on the orientation of the newly formed cellulose microfibrils (**Figure 34.20**). If the microfibrils are randomly oriented,

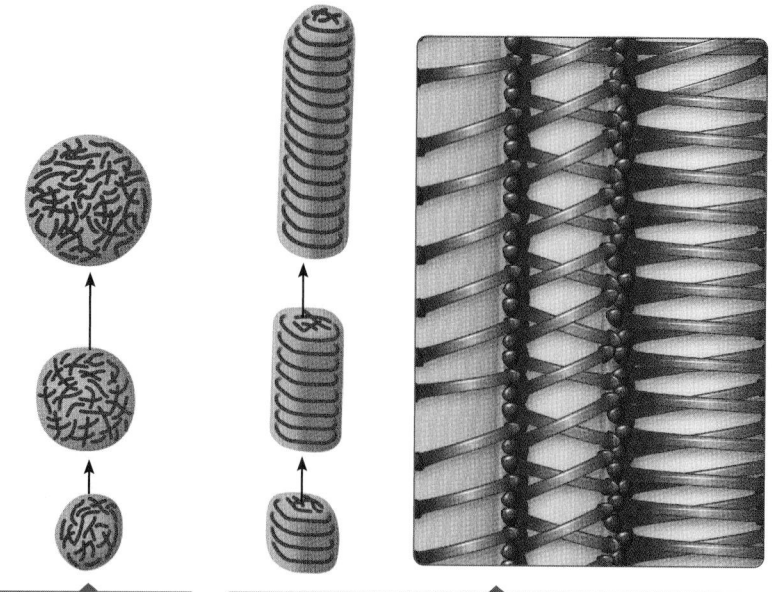

a. When microfibrils are oriented at random, the primary wall is elastic all over, so the cell can grow in all directions.

b. When microfibrils are oriented transversely, the cell can grow only longitudinally.

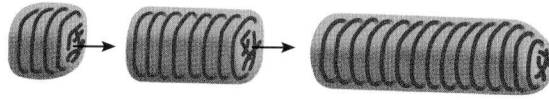

c. When microfibrils are oriented longitudinally, the cell can grow only laterally.

Figure 34.20
Cell expansion and the orientation of cellulose microfibrils. In each cell, microtubules inherited from the parent cell are already oriented in prescribed patterns that govern how cellulose microfibrils will be oriented in the cell wall. Their orientation in turn governs the direction in which a cell can expand.

the cell expands equally in all directions. If they are oriented at right angles to the cell's long axis, the cell expands lengthwise. And if new fibrils are deposited parallel to the long axis of the cell, the cell expands laterally.

Patterns of Cell Division during Early Growth. Like the first mitotic division in an *Arabidopsis* zygote, it's not uncommon for cell divisions in a growing plant to be asymmetrical, so that one daughter cell ends up with more cytoplasm than the other. The unequal distribution of cytoplasm means that the daughter cells differ in their composition and structure, and the differences affect how they interact with their neighbors during growth, even though all cells carry the same genes. Their cytoplasmic differences and interactions with one another trigger selective gene expression. Such events seal the developmental fate of particular cell lineages. Their descendant cells divide in prescribed planes and expand in set directions, producing plant parts with diverse shapes and functions.

Figure 34.21 Experimental Research

Probing the Roles of Floral Organ Identity Genes

QUESTION: What are the genetic mechanisms that govern the formation of the parts of a flower?

EXPERIMENTS: Meyerowitz and his colleagues grew *Arabidopsis thaliana* plants having mutated, inactivated versions of the genes suspected of controlling the proper development of floral organs. They compared the types and arrangements of floral organs in the test plants with the organs present in normal, wild-type *A. thaliana* flowers.

a. Normal *A. thaliana*

Carpels (whorl 4)
Stamen (whorl 3)
Petal (whorl 2)
Sepal (whorl 1)

Jose Luis Riechmann

Whorls in transverse section

Carpels
Stamens
Petals
Sepals

Normal arrangement of organs: carpels in whorl 4, stamens in whorl 3, petals in whorl 2, and sepals in whorl 1

RESULTS: At least three classes of homeotic genes (A, B, and C) regulate different aspects of normal floral organ development.

b. When mutation inactivates the *APETALA2* gene, class A genes are not expressed.

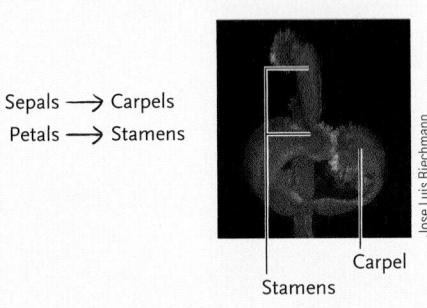

Jose Luis Riechmann

Sepals ⟶ Carpels
Petals ⟶ Stamens

Carpel
Stamens

Stamens replace petals and carpels replace sepals. Organ identity in the other whorls does not change.

c. When mutation inactivates the *APETALA3* or *PISTILLATA* gene, class B genes are not expressed.

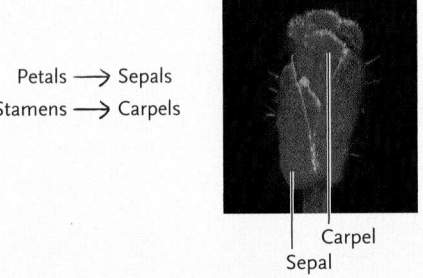

Jose Luis Riechmann

Petals ⟶ Sepals
Stamens ⟶ Carpels

Carpel
Sepal

Carpels replace stamens and sepals replace petals. Organ identity in the other whorls does not change.

d. When mutation inactivates the *AGAMOUS* gene, class C genes are not expressed.

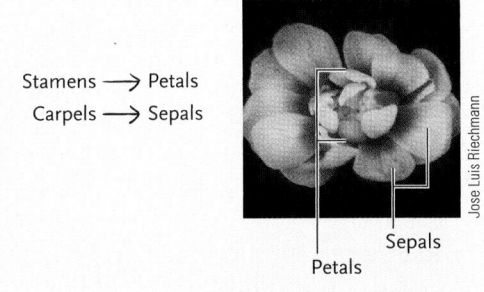

Jose Luis Riechmann

Stamens ⟶ Petals
Carpels ⟶ Sepals

Sepals
Petals

No carpels; instead petals develop in whorl 3 and a version of a floral meristem develops where whorl 4 would normally be. It gives rise to extra petals and sepals. Organ identity in other whorls does not change.

Further experiments with double mutants (inactivation of both A and B, A and C, and B and C) all produce abnormal flowers. For example, only carpels develop in mutants having only an active C class gene, and only sepals develop in plants having only an active B class gene.

e. Overlapping activity fields of floral organ identity genes

Carpels			C	E	Whorl 4
Stamens		B	C	E	Whorl 3
Petals	A	B		E	Whorl 2
Sepals	A				Whorl 1

CONCLUSIONS: In *A. thaliana*, A, B, and C activity genes, expressed alone or in pairs, underlie the development of a normal pattern of floral organs. The fields of activity overlap. In addition, A and C activity apparently counteract each other, and if one is absent the other can spread beyond the whorls where it normally appears. Subsequent research revealed that a fourth E class of gene activity is required for proper expression of other organ identity genes.

Regulatory Genes Guide the Development of Floral Organs

Research with several plant species has shed light on the genetic mechanisms that govern the formation of the parts of a flower. For example, experiments with *Arabidopsis* carried out by Elliot Meyerowitz and his colleagues at the California Institute of Technology showed that *floral organ homeotic genes* regulate the development of the sepals, petals, stamens, and carpels in flowers.

The Meyerowitz team studied plants with various mutations in floral organs. By observing the effects of specific mutations on the structure of *Arabidopsis* flowers, the investigators eventually identified three classes of homeotic gene activity—which they named A, B, and C—that regulate different aspects of normal flower development. Subsequent studies by other scientists identified an E class of gene activity that appears to be an essential partner in the functioning of A, B, and C class genes. The effects of the genes overlap, so that A, B, and C class genes are expressed in two whorls, and E class genes in three: Class A genes are expressed in whorls 1 and 2 (sepals and petals), class B genes in whorls 2 and 3 (petals and stamens), class C genes in whorls 3 and 4 (stamens and carpels), and class E genes in whorls 2, 3, and 4 **(Figure 34.21)**.

Abnormal floral patterns such as those in Figure 34.21b–d show how mutations in the floral organ homeotic genes can affect the identity of flower parts. For example, a mutation that deactivates the A-class gene *APETALA2* produces a flower with carpels and stamens in whorl 1 (see Figure 34.21b). Another intriguing finding is that the A and C activity classes normally oppose each other. When no A gene is expressed, C activity spreads into whorls where the A usually occurs, and vice versa. Subsequent studies have examined many other floral homeotic genes, as well as the genes that control the various gene classes.

As the genes governing flower development are isolated, they can be cloned and their nucleotide sequences defined and manipulated. Such cloned genes already are of keen interest in plant genetic engineering, because food grains such as wheat and many other vital agricultural commodities come directly or indirectly from flowers.

Leaves Arise from Leaf Primordia in a Closely Regulated Sequence

A mature leaf may have many millions of differentiated cells organized into tissues such as epidermis and mesophyll. As described in Chapter 31, leaves develop from leaf primordia that arise just behind the tips of shoot apical meristems **(Figure 34.22)**.

Clonal analysis has opened a window on many aspects of plant development, including how leaf primordia originate and give rise to leaves. In this method, the investigator cultures meristematic tissue that contains a mutated embryonic cell having a readily observable trait, such as the absence of normal pigment. (In the laboratory, this kind of mutation can be induced by chemicals or radiation.) The unusual trait then serves as a marker that identifies the mutant cell's clonal descendants, making it possible to map the growing structure. Researchers have used clonal analysis to study leaf development in garden peas, tomatoes, grasses, and tobacco, among others.

Like flowers, leaves arise through a developmental program that begins with gene-regulated activity in meristematic tissue. Hormones or other signals may arrive at target cells via the stem's vascular tissue, activating genes that regulate development. Studies show that small phloem vessels first penetrate a young leaf primordium almost immediately after it begins to bulge out from the underlying meristematic tissue, and xylem soon follows. The early phloem connections are especially vital to the leaf's survival, because the leaf does not begin photosynthesis until it attains one-third of its mature size.

A growing primordium becomes cone-shaped, wider at its base than at its tip. At a certain point, mitosis speeds up in cells along the flanks of the lengthening

a. Leaf primordia of *Coleus*

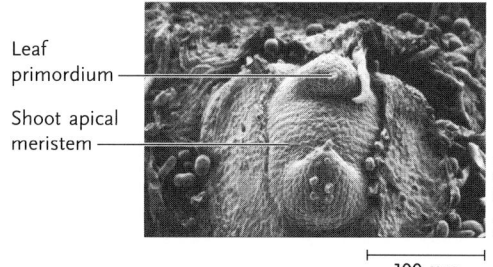

Leaf primordium

Shoot apical meristem

100 μm

b. Early stages of leaf development

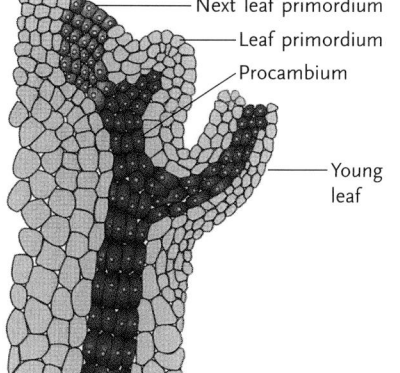

Next leaf primordium

Leaf primordium

Procambium

Young leaf

Roland R. Dute

Figure 34.22

Early stages in leaf development. **(a)** Leaf primordia at the shoot tip of *Coleus*, a genus that includes several popular house plants. **(b)** Diagram showing leaf primordia in different stages of development. See also Figure 34.12a, which shows the progression of leaves that form during the early growth and development of a eudicot.

cone. In eudicots, the rapid cell divisions occur perpendicular to the surface and produce the leaf blade. In monocot grasses, which have long, narrow leaves, vertical "files" of cells develop as cells in the meristem at the base of the cone divide in a plane parallel to the surface. The ultimate shape of a leaf depends in part on variations in the plane and rate of these cell divisions.

Cells at the leaf tip, and those of xylem and phloem that service it, are the oldest in the leaf, and it is here that photosynthesis begins. Commonly, leaf tip cells also are the first to stop dividing. By the time a leaf has expanded to its mature size, all mitosis has ended and the leaf is a fully functional photosynthetic organ.

In nature, genes that govern leaf and flower development switch on or off in response to changing environmental conditions. Their signals determine the course of a plant's vegetative growth throughout its life. In many perennials, new leaves begin to develop inside buds in autumn, then become dormant until the following spring, when external conditions favor further growth. Environmental cues stimulate the gene-guided production of hormones that travel through the plant in xylem and phloem, triggering renewed leaf growth and expansion. Leaves and other shoot parts also age, wither, and fall away from the plant as hormonal signals change. The far-reaching effects of plant hormones on growth and development are the subject of Chapter 35.

STUDY BREAK

1. What is a homeotic gene? Give at least two examples of plant tissues such genes might govern in a species such as *Arabidopsis*.
2. What are the two basic mechanisms of morphogenesis in plants? Describe the patterns of cell division by which a plant part (a) grows longer and (b) adds girth.
3. Summarize the gene-guided developmental program that gives rise to a leaf.

UNANSWERED QUESTIONS

What are the signaling events that mediate pollen-tube guidance during compatible pollination?

As you learned in this chapter, after pollen lands on the surface of the carpel, it forms a pollen tube, which then invades the carpel, migrates past several different cell types, and enters the micropyle to fertilize an egg and central cell. Recent work has shown that pollen-tube migration is mediated by a series of cell–cell interactions such as attraction, repulsion, and adhesion. However, most of these conclusions have been based on analyzing static images and fixed tissues, and it has been hypothesized that many more subtle, dynamic interactions between the pollen tube and female cells exist to ensure compatible pollination. It has been shown, for example, that pollen tubes gain the competence to successfully find the ovules only when they grow on the pistil tissues, a process that is functionally analogous to the transformation that mammalian spermatozoa undergo after residence for a finite amount of time in the female reproductive tract. It was also recently demonstrated that once a pollen tube penetrates the ovule, additional pollen tubes are prevented from gaining access into the targeted ovule, a process that is functionally analogous to the prevention of polyspermy by a fertilized mammalian egg. These novel signaling events are spurring researchers to take a closer look at pollen-tube guidance to ovules. Only when such signaling events are described can appropriate efforts be taken to identify the cues that mediate these events (see next question).

What kinds of approaches are being taken to identify novel signaling events? Researchers usually first identify a mutant plant that is defective in any of the pollen-tube guidance steps that are essential for successful fertilization. Subsequently, they analyze the defects to learn more about how this process normally happens. Other researchers develop microscopy-based real-time assays to directly observe pollen-tube behavior with a variety of female tissues.

What are the chemical cues from female tissues that facilitate compatible pollination?

As you learned in this chapter, if a pollen tube lands on a compatible stigma, chemical cues produced by the female tissue then guide the pollen tube from the stigma to the embryo sac of an ovule. Recent research has revealed that cues produced by both sporophyte and gametophyte are essential for proper guidance of pollen tubes to ovules. Despite these advances, the identities of these cues remain unknown. What are the hurdles that have hampered efforts to uncover pollen-tube navigation cues of even known signaling events described above? First, the pistil tissue within which the pollen tube elongates to the ovule is comprised of several types of tissue, including stigma, style, and transmitting tract, and these tissues are not readily accessible. Second, analyzing the dynamic responses of pollen tubes is difficult given that pollen-tube navigation occurs well within opaque pistils. Third, it appears that multiple, stage-specific, short-range, and readily labile signals produced in minute quantities mediate pollen-tube guidance. However, recent development of global approaches that are highly sensitive and assays that directly monitor pollen-tube elongation offer hope that guidance cues will be uncovered sooner rather than later.

Ravi Palanivelu is an assistant professor in the Department of Plant Sciences at the University of Arizona. His current research focuses on the isolation and characterization of pollen-tube guidance signals during *Arabidopsis* reproduction, with the long-term goal of understanding the molecular basis of how cells communicate with each other. To learn more about Dr. Palanivelu's research, go to http://www.ag.arizona.edu/research/ravilab.

Review

Go to **ThomsonNOW** at www.thomsonedu.com/login to access quizzing, animations, exercises, articles, and personalized homework help.

34.1 Overview of Flowering Plant Reproduction

- In most flowering plant life cycles, a multicellular diploid sporophyte (spore-producing plant) stage alternates with a multicellular haploid gametophyte (gamete-producing plant) stage. The sporophyte develops roots, stems, leaves, and, at some point, flowers. The separation of a life cycle into diploid and haploid stages is called alternation of generations (Figure 34.2).

Animation: Flowering plant life cycle

34.2 The Formation of Flowers and Gametes

- A flower develops at the tip of a floral shoot. It can have up to four whorls supported by the receptacle. The calyx and corolla consist of the sepals and petals, respectively. The third whorl consists of stamens, and carpels make up the innermost whorl (Figure 34.3).

- The anther of a stamen contains sacs where pollen grains develop. If compatible pollen lands on the carpel's stigma, which contains an ovary in which eggs develop, fertilization takes place.

- A complete flower has both male and female reproductive parts. In monoecious species each plant has both types of flowers; in dioecious species the "male" and "female" flowers are on different plants (Figure 34.4).

- In pollen sacs, meiosis produces haploid microspores. Mitosis inside each microspore produces a pollen grain, an immature male gametophyte. One of its cells develops into two sperm cells, the male gametes of flowering plants. Another cell produces the pollen tube (Figures 34.5 and 34.6).

- An ovule forms inside a carpel, on the wall of the ovary. Development in the ovule produces a female gametophyte—the embryo sac with egg cell.

- In the ovule, meiosis produces four haploid megaspores. Usually all but one disintegrate. The remaining megaspore undergoes mitosis three times without cytokinesis, producing eight nuclei in a single large cell. Two of these (polar nuclei) migrate to the center of the cell. When cytokinesis occurs, cell walls form around the nuclei, with the two polar nuclei enclosed in a single wall. The result is the seven-celled embryo sac, one cell of which is the haploid egg. The cell containing polar nuclei will help give rise to endosperm.

Animation: Floral structure and function

Animation: Flower parts

Animation: Microspores to pollen

Animation: Apple fruit structure

34.3 Pollination, Fertilization, and Germination

- Upon pollination, the pollen grain resumes growth. A pollen tube develops, and mitosis of the male gametophyte's sperm-producing cell produces two sperm nuclei (Figure 34.7).

- In double fertilization, one sperm nucleus fuses with one egg nucleus to form a diploid ($2n$) zygote. The other sperm nucleus and the two polar nuclei of the remaining cell also fuse, forming a cell that will give rise to triploid ($3n$) endosperm in the seed (Figure 34.8).

- After the endosperm forms, the ovule expands, and the embryonic sporophyte develops. A mature ovule is a seed. Inside the seed, the embryo has a lengthwise axis with a root apical meristem at one end and a shoot apical meristem at the other.

- Eudicot embryos have two cotyledons. The embryonic shoot consists of an upper epicotyl and a lower hypocotyl; also present is an embryonic root, the radicle. The single cotyledon of a monocot forms a scutellum that absorbs nutrients from endosperm. Apical meristems of a monocot embryo are protected by a coleoptile over the shoot tip and a coleorhiza over the radicle (Figure 34.9).

- A fruit is a matured or ripened ovary. Fruits protect seeds and disperse them by animals, wind, or water.

- Fruits are simple, aggregate, or multiple, depending on the number of flowers or ovaries from which they develop. Fruits also vary in the characteristics of their pericarp, which surrounds the seed (Figure 34.10).

- The seeds of most plants remain dormant until external conditions such as moisture, temperature, and day length favor the survival of the embryo and the development of a new sporophyte (Figures 34.11–34.13).

Animation: Bee-attracting flower pattern

34.4 Asexual Reproduction of Flowering Plants

- Many flowering plants also reproduce asexually, as when new plants arise by mitosis at nodes or buds along modified stems of the parent plant. New plants also may arise by vegetative propagation (Figure 34.14).

- Tissue culture methods for developing new plants from a parent plant's somatic (nonreproductive) cells include somatic embryogenesis and protoplast fusion (Figure 34.15).

Animation: Eudicot life cycle

Animation: Eudicot seed development

34.5 Early Development of Plant Form and Function

- In plants that reproduce sexually, development starts at fertilization. Early on, a new embryo acquires its root–shoot axis, and cells in different regions begin to become specialized for particular functions (Figure 34.16). In morphogenesis, body regions develop characteristic shapes and structures that correlate with their function (Figure 34.17).

- Dividing plant cells can synthesize a new cell plate in a different plane from the old one. Such changes establish the direction in which structures such as lateral roots, branches, and leaf and flower buds grow (Figures 34.18 and 34.19).

- Chemical signals that help guide morphogenesis appear to act on certain cells in meristematic tissue, activating homeotic genes that ultimately regulate cell division and differentiation (Figures 34.20 and 34.21).

Animation: ABC model for flowering

Questions

Self-Test Questions

1. An angiosperm life cycle includes:
   a. meiosis within the male gametophyte to produce sperm.
   b. meiosis within the female gametophyte to produce eggs.
   c. meiosis within the ovary to produce megaspores.
   d. fertilization to produce microspores.
   e. fertilization to produce megaspores.

2. In a flower:
   a. the ovary contains the ovule.
   b. the stamens support the petals.
   c. the anther contains the megaspores.
   d. the carpel includes the sepals.
   e. the corolla includes the receptacle.

3. Double fertilization in a flower means:
   a. six sperm fertilize two groups of three eggs each.
   b. one sperm fertilizes the egg; a second sperm fertilizes the $2n$ mother cell.
   c. one microspore becomes a pollen grain; the other microspore becomes a sperm-producing cell.
   d. one pollen grain can make sperm nuclei and a pollen tube.
   e. one sperm can fertilize two endosperm mother cells.

4. A seed is best described as a (an):
   a. epicotyl.
   b. endosperm.
   c. ovary.
   d. mature spore.
   e. mature ovule.

5. The primary root develops from the embryonic:
   a. epicotyl.
   b. hypocotyl.
   c. coleoptile.
   d. radicle.
   e. plumule.

6. Which of the following is *not* a step in the germination of a monocot seed?
   a. Enzymes secreted into the endosperm digest the endosperm cell wall and macromolecules.
   b. The embryo imbibes water and then produces gibberellin.
   c. The embryo absorbs nutrients released from the endosperm.
   d. Endosperm develops as a food reserve.
   e. Gibberellin acts on the cells of the aleurone and scutellum to encode hydrolytic enzymes.

7. A student cuts off a leaflet from a plant and places it in a glass of water. Within a week roots appear on the base of the cutting. A month later she places the growing cutting into soil and it grows to the full size of the "parent" plant. This is an example of:
   a. parthenocarpy.
   b. fragmentation.
   c. grafting.
   d. vegetative reproduction.
   e. tissue culture propagation.

8. Which of the following is *not* an example of pattern formation in developing plants?
   a. an epidermal cell receiving developmental signals from a cortical cell
   b. the loosening of the cell wall to allow the elongation of selected cells to reach mature size
   c. regulation by homeotic genes of the position of different flower parts
   d. oriented cell division that establishes the shape of an organ
   e. cell expansion that directs specific cells to undergo mitosis at a given time and place

9. During the development of a leaf:
   a. mitotic cell divisions occur on planes specific to different plant groups.
   b. xylem vessels are the first to penetrate the leaf primordium.
   c. the growing leaf primordium becomes wider at its base than at its tip.
   d. the leaf primordium bulges from the region behind the shoot apical meristem.
   e. All of the above occur during leaf development.

10. In spring a lone walnut tree in your backyard develops attractive white flowers, and by the end of summer roughly half the flowers have given rise to the shelled fruits we know as walnuts. Walnut trees are self-pollinating. Assuming that pollination was 100% efficient in the case of your tree, which of the following statements best describes your tree's reproductive parts?
    a. Its flowers are in the botanical category of "perfect" flowers.
    b. The tree is monoecious.
    c. The tree is dioecious.
    d. The tree has imperfect, monoecious flowers.
    e. a and b together provide the best description of the tree's flowers.

Questions for Discussion

1. A plant physiologist has succeeded in cloning a gene for pest resistance into petunia cells. How can she use tissue culture to propagate a large number of petunia plants having the gene?

2. A large tree may have tens of thousands of shoot tips, and the cells in each tip can differ genetically from cells in other tips, sometimes substantially. Propose a hypothesis to explain this finding, and speculate about how it might be beneficial to the plant. How might this sort of natural variation be useful to human society?

3. Grocery stores separate displays of fruits and vegetables according to typical uses for these plant foods. For instance, bell peppers, cucumbers, tomatoes, and eggplants are in the vegetable section, while apples, pears, and peaches are displayed with other fruits. How does this practice relate to the biological definition of a fruit?

Experimental Analysis

The developmental genetics of flowers are of keen interest in plant biotechnology, especially with regard to food plants such as wheat and rice. Outline a research program for a crop species that would exploit the genetics of flower development, including the effects of homeotic genes, to engineer a more productive variety.

Evolution Link

Botanists estimate that half or more of angiosperm species may be polyploids that arose initially through hybridization. *Polyploidy*—having more than a diploid set of the parental chromosomes—can result from nondisjunction of homologous chromosomes during meiosis, or when cytokinesis fails to occur in a dividing cell. *Hybridization* is the successful mating of individuals from two different species. Such an interspecific hybrid is likely to be sterile because it has uneven numbers of parental chromosomes, or because the chromosomes are too different to pair during meiosis. A sterile hybrid may reproduce asexually, however, and if by chance its offspring should become polyploid, that plant will be fertile because the original set of chromosomes will have homologs that can pair normally during meiosis. Explain why both the hybrid parent and fertile polyploid offspring may be considered a new species, and describe at least two ways in which this route to speciation differs from speciation in the animal kingdom.

How Would You Vote?

Microencapsulated pesticides are easy to apply and effective for long periods. But they are about the size of pollen grains and are a tempting but toxic threat to certain pollinators. Should we restrict their use? Go to www.thomsonedu.com/login to investigate both sides of the issue and then vote online.

© Garry Black/Masterfile

Sunflower plants *(Helianthus)* with flower heads oriented toward the sun's rays—an example of a plant response to the environment.

35 Control of Plant Growth and Development

WHY IT MATTERS

In the early 1920s, a researcher in Japan, Eiichi Kurosawa, was studying a rice plant disease that the Japanese called *bakanae*—the "foolish seedling" disease. Stems of rice seedlings that had become infected with the fungus *Gibberella fujikuroi* elongated twice as much as uninfected plants. The lanky stems were weak and eventually toppled over before the plants could produce seeds. Kurosawa discovered that extracts of the fungus also could trigger the disease. Eventually, other investigators purified the fungus's disease-causing substance, naming it gibberellin (GA).

Botanists today recognize more than 100 chemically different gibberellins, the largest class of plant hormones. Gibberellins have been isolated from fungi and from flowering plants, and may exist in other plant groups as well. Like other hormones, gibberellins are intercellular signaling molecules. In flowering plants, gibberellins have major, predictable effects **(Figure 35.1),** beginning with seed germination.

Basic aspects of plant growth and development are adaptations that promote the survival of organisms that cannot move through their environment. These adaptations range from the triggers for seed germination to the development of a particular body form, the shift

Figure 35.1
Effects of the hormone gibberellin on stem growth of rice plants (Oryza).

from a vegetative phase to a reproductive one, and the timed death of flowers, leaves, and other parts. Although many of the details remain elusive or disputed, ample evidence exists that an elaborate system of molecular signals regulates many of these phenomena. We know, for example, that plant hormones alter patterns of growth, cell metabolism, and morphogenesis in response to changing environmental rhythms, including seasonal changes in day length and temperature and the daily rhythms of light and dark. They also adjust those patterns in response to environmental conditions, such as the amount of sunlight or shade, moisture, soil nutrients, and other factors. Some hormones govern growth responses to directional stimuli, such as light, gravity, or the presence of nearby structures. Often, hormonal effects involve changes in gene expression, although sometimes other mechanisms are at work.

We begin by surveying the different groups of plant hormones and other signaling molecules, and then turn our attention to the remarkable diversity of responses to both internal and environmental signals.

35.1 Plant Hormones

In plants, a **hormone** (*horman* = to stimulate) is a signaling molecule that regulates or helps coordinate some aspect of the plant's growth, metabolism, or development. Plant hormones act in response to two general types of cues: Internal chemical conditions related to growth and development, and circum-

stances in the external environment that affect plant growth, such as light and the availability of water. Some plant hormones are transported from the tissue that produces them to another plant part, while others exert their effects in the tissue where they are synthesized.

All plant hormones share certain characteristics. They are rather small organic molecules, and all are active in extremely low concentrations. Another shared feature is specificity: each one affects a given tissue in a particular way. Hormones that have effects outside the tissue where they are produced typically are transported to their target sites in vascular tissues, or they diffuse from one plant part to another. Within these general parameters, however, plant hormones vary greatly in their effects. Some stimulate one or more aspects of the plant's growth or development, whereas others have an inhibiting influence. Adding to the potential for confusion, a given hormone can have different effects in different tissues, and the effects also can differ depending on a target tissue's stage of development. And as researchers have increasingly discovered, many physiological responses result from the interaction of two or more hormones.

Biologists recognize at least seven major classes of plant hormones **(Table 35.1)**: auxins, gibberellins, cytokinins, ethylene, brassinosteroids, abscisic acid (ABA), and jasmonates. Recent discoveries have added other hormonelike signaling agents to this list of established plant hormones. We now consider each major class of plant hormones and discuss some of the newly discovered signaling molecules as well.

Auxins Promote Growth

Auxins are synthesized primarily in the shoot apical meristem and young stems and leaves. Their main effects are to stimulate plant growth by promoting cell elongation in stems and coleoptiles, and by governing growth responses to light and gravity. Our focus here is indoleacetic acid (IAA), the most important natural auxin. Botanists often use the general term "auxin" to refer to IAA, a practice we follow here.

Experiments Leading to the Discovery of Auxins. Auxins were the first plant hormones identified. The path to their discovery began in the late nineteenth century in the library of Charles Darwin's home in the English countryside (see *Focus on Research* in Chapter 19). Among his many interests, Darwin was fascinated by plant **tropisms**—movements such as the bending of a houseplant toward light. This growth response, triggered by exposure to a directional light source, is an example of a **phototropism**.

Working with his son Francis, Darwin explored phototropisms by germinating seeds of two species of grasses, oat *(Avena sativa)* and canary grass *(Phalaris canariensis)*, in pots on the sill of a sunny window. Re-

Table 35.1   Major Plant Hormones and Signaling Molecules

Hormone/Signaling Compound	Where Synthesized	Tissues Affected	Effects
Auxins	Apical meristems, developing leaves and embryos	Growing tissues, buds, roots, leaves, fruits, vascular tissues	Promote growth and elongation of stems; promote formation of lateral roots and dormancy in lateral buds; promote fruit development; inhibit leaf abscission; orient plants with respect to light, gravity
Gibberellins	Root and shoot tips, young leaves, developing embryos	Stems, developing seeds	Promote cell divisions and growth and elongation of stems; promote seed germination and bolting
Cytokinins	Mainly in root tips	Shoot apical meristems, leaves, buds	Promote cell division; inhibit senescence of leaves; coordinate growth of roots and shoots (with auxin)
Ethylene	Shoot tips, roots, leaf nodes, flowers, fruits	Seeds, buds, seedlings, mature leaves, flowers, fruits	Regulates elongation and division of cells in seedling stems, roots; in mature plants regulates senescence and abscission of leaves, flowers, and fruits
Brassinosteroids	Young seeds; shoots and leaves	Mainly shoot tips, developing embryos	Stimulate cell division and elongation, differentiation of vascular tissue
Abscisic acid	Leaves	Buds, seeds, stomata	Promotes responses to environmental stress, including inhibiting growth/promoting dormancy; stimulates stomata to close in water-stressed plants
Jasmonates	Roots, seeds, probably other tissues	Various tissues, including damaged ones	In defense responses, promote transcription of genes encoding protease inhibitors; possible role in plant responses to nutrient deficiencies
Oligosaccharins	Cell walls	Damaged tissues; possibly active in most plant cells	Promote synthesis of phytoalexins in injured plants; may also have a role in regulating growth
Systemin	Damaged tissues	Damaged tissues	To date known only in tomato; roles in defense, including triggering jasmonate-induced chemical defenses
Salicylic acid	Damaged tissues	Many plant parts	Triggers synthesis of pathogenesis-related (PR) proteins, other general defenses

call from Chapter 34 that the shoot apical meristem and plumule of grass seedlings are sheathed by a protective coleoptile—a structure that is extremely sensitive to light. Darwin did not know this detail, but he observed that as the emerging shoots grew, within a few days they bent toward the light. He hypothesized that the tip of the shoot somehow detected light and communicated that information to the coleoptile. Darwin tested this idea in several ways **(Figure 35.2)** and concluded that when seedlings are illuminated from the side, "some influence is transmitted from the upper to the lower part, causing them to bend."

The Darwins' observations spawned decades of studies—a body of work that illustrates how scientific understanding typically advances step-by-step, as one set of experimental findings stimulates new research. First, scientists in Denmark and Poland showed that the bending of a shoot toward a light source was caused by something that could move through agar (a jellylike culture material derived from certain red algae) but not through a sheet of the mineral mica. This finding prompted experiments establishing that indeed the stimulus was a chemical produced in the shoot tip. Soon afterward, in 1926, experiments by the Dutch plant physiologist Frits Went confirmed that

the growth-promoting chemical diffuses downward from the shoot tip to the stem below **(Figure 35.3).** Using oat seeds, Went first sliced the tips from young shoots that had been grown under normal light conditions. He then placed the tips on agar blocks and left them there long enough for diffusible substances to move into the agar. Meanwhile, the decapitated stems stopped growing, but growth quickly resumed in seedlings that Went "capped" with the agar blocks (see Figure 35.3a). Clearly, a growth-promoting substance in the excised shoot tips had diffused into the agar, and from there into the seedling stems. Went also attached an agar block to one side of a decapitated shoot tip; when the shoot began growing again it bent away from the agar (see Figure 35.3b). Importantly, Went performed his experiments in total darkness, to avoid any "contamination" of his results by the possible effects of light.

Went did not determine the mechanism— differential elongation of cells on the shaded side of a shoot—by which the growth promoter controlled phototropism. However, he did develop a test that correlated specific amounts of the substance, later named auxin (*auxein* = to increase), with particular growth effects. This careful groundwork culminated several

Figure 35.2 Experimental Research

The Darwins' Experiments on Phototropism

QUESTION: Why does a plant stem bend toward the light?

EXPERIMENT 1: The Darwins observed that the first shoot of an emerging grass seedling, which is sheathed by a coleoptile, bends toward sunlight shining through a window. They removed the shoot tip from a seedling and illuminated one side of the seedling.

Original observation

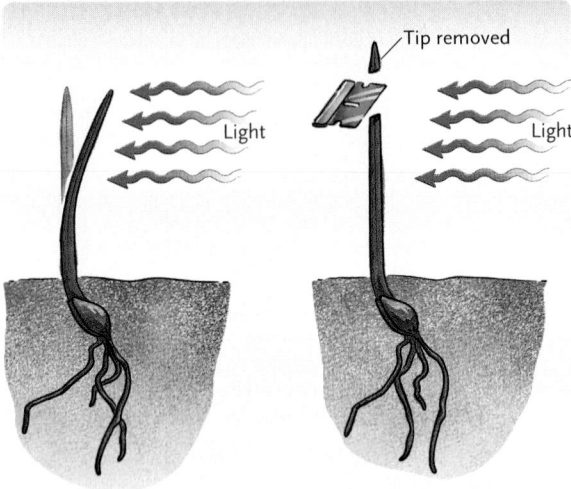

Light

Tip removed

Light

RESULT: The seedling neither grew nor bent.

EXPERIMENT 2: The Darwins divided seedlings into two groups. They covered the shoot tips of one group with an opaque cap and the shoot tips of the other group with a translucent cap. All the seedlings were illuminated from the same side.

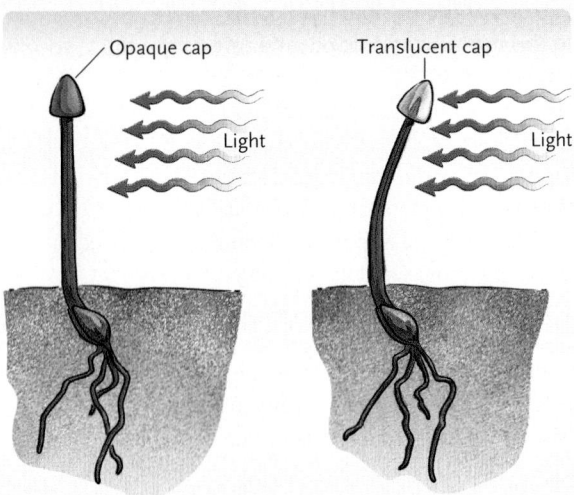

Opaque cap

Translucent cap

Light

Light

RESULT: The seedlings with opaque caps grew but did not bend. Those with translucent caps both grew *and* bent toward the light.

CONCLUSION: When seedlings are illuminated from one side, an unknown factor transmitted from a seedling's tip to the tissue below causes it to bend toward the light.

years later when other researchers identified auxin as indoleacetic acid (IAA).

Effects of Auxins. As already noted, auxin stimulates aspects of plant growth and development. In fact, recent studies of plant development have revealed that auxin is one of the first chemical signals to help shape the plant body. When the zygote first divides, forming an embryo that consists of a basal cell and an apical cell (see Section 34.3), auxin exported by the basal cell to the apical cell helps guide the development of the various features of the embryonic shoot. As the embryo develops further, IAA is produced mainly by the leaf primordium of the young shoot (see Figure 34.22). While the developing shoot is underground, IAA is actively transported downward, stimulating the primary growth of the stem and root **(Figure 35.4).** Once an elongating shoot breaks through the soil surface, its tip is exposed to sunlight, and the first leaves unfurl and begin photosynthesis. Shortly thereafter the leaf tip stops producing IAA and that task is assumed first by cells at the leaf edges, then by cells at base of the young leaf. Even so, as Section 35.3 discusses more fully, IAA continues to influence a plant's responses to light and plays a role in plant growth responses to gravity as well. IAA also stimulates cell division in the vascular cambium and promotes the formation of secondary xylem, as well as the formation of new root apical meristems, including lateral meristems. Not all of auxin's effects promote growth, however. IAA also maintains apical dominance, which inhibits growth of lateral meristems on shoots and restricts the formation of branches (see Section 31.3). Hence, auxin is a signal that the shoot apical meristem is present and active.

Commercial orchardists spray synthetic IAA on fruit trees because it promotes uniform flowering and helps set the fruit; it also helps prevent premature fruit drop. These effects mean that all the fruit may be picked at the same time, with considerable savings in labor costs.

Some synthetic auxins are used as herbicides, essentially stimulating a target plant to "grow itself to death." An **herbicide** is any compound that, at proper concentration, kills plants. Some herbicides are selective, killing one class of plants and not others. The most widely used herbicide in the world is the synthetic auxin 2,4-D (2,4-dichlorophenoxyacetic acid). This chemical is used extensively to prevent broadleaf weeds (which are eudicots) from growing in fields of cereal crops such as corn (which are monocots). By an unknown mechanism, 2,4-D causes an abnormal burst of growth in which eudicot stems elongate more than 10 times faster than normal—much faster than the plant can support metabolically.

Auxin Transport. To exert their far-reaching effects on plant tissues, auxins must travel away from their main synthesis sites in shoot meristems and young leaves.

a. The procedure showing that IAA promotes elongation of cells below the shoot tip

b. The procedure showing that cells in contact with IAA grow faster than those farther away

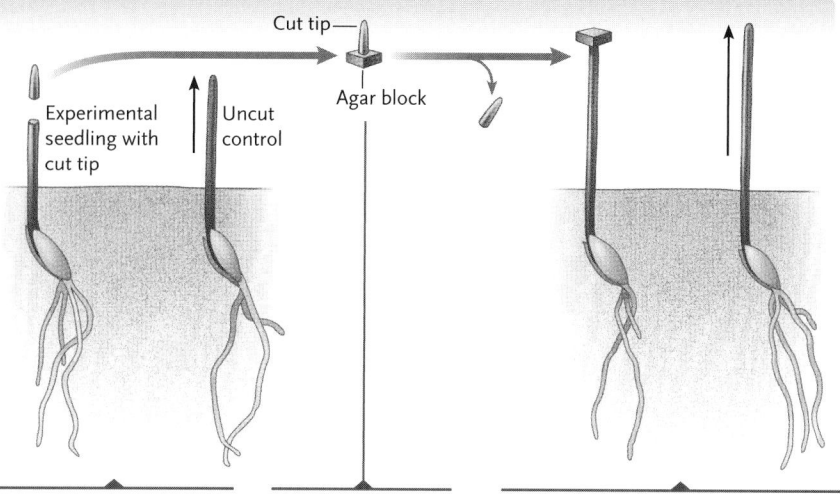
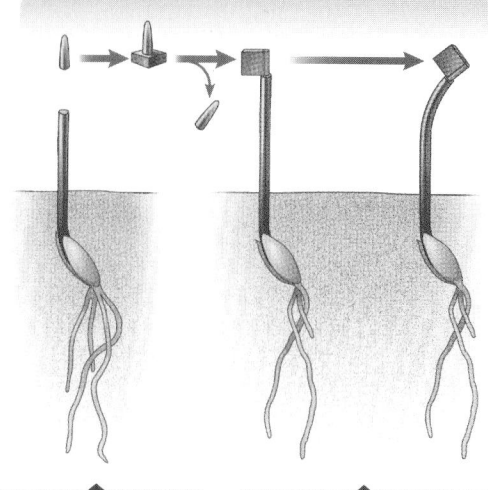

1 After Went cut off the tip of an oat seedling, the shoot stopped elongating, while a control seedling with an intact tip continued to grow.

2 He placed the excised tip on an agar block for 1–4 hours. During that time, IAA diffused into the agar block from the cut tip.

3 Went then placed the agar block containing auxin on another detipped oat shoot, and the shoot resumed elongation, growing about as rapidly as in a control seedling with an intact shoot tip.

1 Went removed the tip of a seedling and placed it on an agar block.

2 He placed the agar block containing auxin on one side of the shoot tip. Auxin moved into the shoot tip on that side, causing it to bend away from the hormone.

Figure 35.3
Two experiments by Frits Went demonstrating the effect of IAA on an oat coleoptile. Went carried out the experiments in darkness to prevent effects of light from skewing the results.

Yet xylem and phloem sap usually do not contain auxins. Moreover, experiments have shown that while IAA moves through plant tissues slowly—roughly 1 cm/hr—this rate is 10 times faster than could be explained by simple diffusion. How, then, is auxin transported?

Plant physiologists adapted the agar block method pioneered by Went to trace the direction and rate of auxin movements in different kinds of tissues. A research team led by Winslow Briggs at Stanford University determined that the shaded side of a shoot tip contains more IAA than the illuminated side. Hypothesizing that light causes IAA to move laterally from the illuminated to the shaded side of a shoot tip, the team then inserted a vertical barrier (a thin slice of mica) between the shaded and illuminated sides of a shoot tip. IAA could not cross the barrier, and when the shoot tip was illuminated it did not bend. In addition, the concentrations of IAA in the two sides of the shoot tip remained about the same. When the barrier was shortened so that the separated sides of the tip again touched, the IAA concentration in the shaded area increased significantly, and the tip *did* bend. The study confirmed that IAA initially moves laterally in the shoot tip, from the illuminated side to the shaded side, where it triggers the elongation of cells and curving of the tip toward light. Subsequent research showed that IAA then moves downward in a shoot by way of a top-to-bottom mechanism called **polar transport**. That is, IAA in a coleoptile or shoot tip travels from the apex of the tissue to its base, such as from the tip of a developing leaf to the stem. **Figure 35.5** outlines the experimental method that demonstrated polar transport. When IAA reaches roots, it moves toward the root tip.

Inside a stem, IAA appears to be transported via parenchyma cells adjacent to vascular bundles. IAA again moves by polar transport as it travels through and between cells: It enters at one end by diffusing passively through cell walls and exits at the opposite end by active transport across the plasma membrane. The mostly widely accepted explanation for polar IAA transport from cell to cell proposes different mechanisms for moving IAA into and out of plant cells. In this model, IAA enters cells as the result of a high outside/low inside hydrogen ion (H^+) concentration gradient produced by the H^+ pumps in the plasma membrane of all plant cells **(Figure 35.6)**. The movement of H^+ ions out of the cytoplasm into the cell wall also produces an electrochemical gradient. In a neutral pH environment, IAA bears a negative charge (IAA^-), but in acidic surroundings, such as in a cell wall into which H^+ has been pumped, IAA^- reacts with H^+ to form an uncharged molecule, IAAH. The uncharged molecules may then dif-

Figure 35.4
The effect of auxin treatment on a gardenia *(Gardenia)* cutting. Four weeks after an auxin was applied to the base of the cutting on the left, its stem and roots have elongated, but the number of leaves is unchanged. The plant cutting on the right was not treated.

Treated with auxin   Untreated

Figure 35.5 Experimental Research

Evidence for the Polar Transport of Auxin in Plant Tissues

QUESTION: Can IAA move both upward and downward in a plant shoot?

EXPERIMENT 1: As a preliminary step, the researchers excised sections of shoot tips from grass seedlings and placed them between blocks of agar containing different concentrations of IAA labeled with radioactive carbon-14. Labeling the IAA allowed them to easily track its movements. The researchers positioned an upright shoot tip section on an untreated agar block, and then placed a second, "donor" agar block containing labeled IAA atop the section.

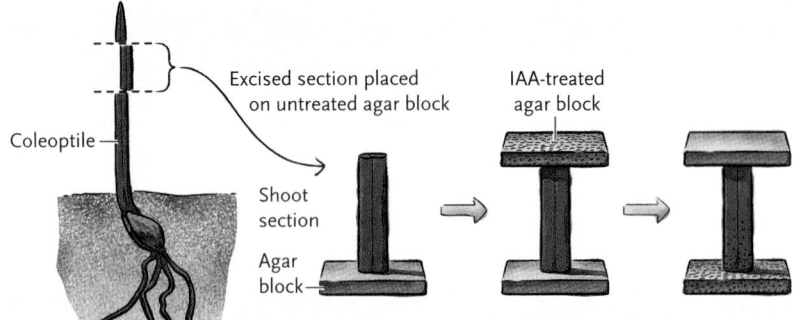

Coleoptile

Excised section placed on untreated agar block

IAA-treated agar block

Shoot section

Agar block

RESULT: IAA traveled from the upper block to the lower one, indicating that the hormone moved downward through the vertical shoot tip.

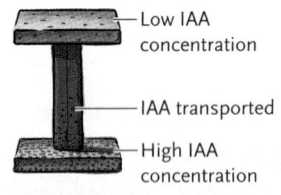

Low IAA concentration

IAA transported

High IAA concentration

EXPERIMENT 2: The researchers positioned an upright shoot tip section on an agar block containing a high concentration of labeled IAA and placed a donor agar block containing a lower concentration of IAA on top.

RESULT: Even against its concentration gradient, IAA was transported from the upper block to the lower one.

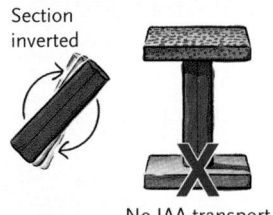

Section inverted

No IAA transport through inverted shoot section

EXPERIMENT 3: The shoot tip from step 2 was inverted (reversing its normal orientation), and the same procedure was performed.

RESULT: No IAA was transported downward.

CONCLUSION: In plant shoot tips, IAA is transported in only one direction, from the shoot tip downward to plant parts below.

---

fuse across the plasma membrane through membrane transporters called AUX1 proteins (after the gene that encodes them in *Arabidopsis thaliana* plants), or they may enter via cotransport with H^+—or perhaps they move by both means.

A different mechanism moves IAA out of the cell at the opposite pole. Once IAAH reaches the electrically neutral cytoplasm at the apical end of the cell, it dissociates into H^+ and IAA^-. Then, the hormone crosses the cell and diffuses out of it by way of transporters called PIN proteins, which tend to be clustered at the cell's basal end. When the IAA^- diffuses through the transport proteins into the acidic cell wall, it reacts

again with H^+, and the process continues with the next cell in line.

There is increasing evidence that auxin also may travel rapidly through plants in the phloem. As this work continues, researchers will undoubtedly gain a clearer understanding of how plants distribute this crucial hormone to their growing parts.

Possible Mechanisms of IAA Action. Ever since auxin was discovered, researchers have actively sought to understand how IAA stimulates cell elongation. You learned in Section 34.5 that as a plant cell elongates, the cellulose meshwork of the cell wall is first loosened

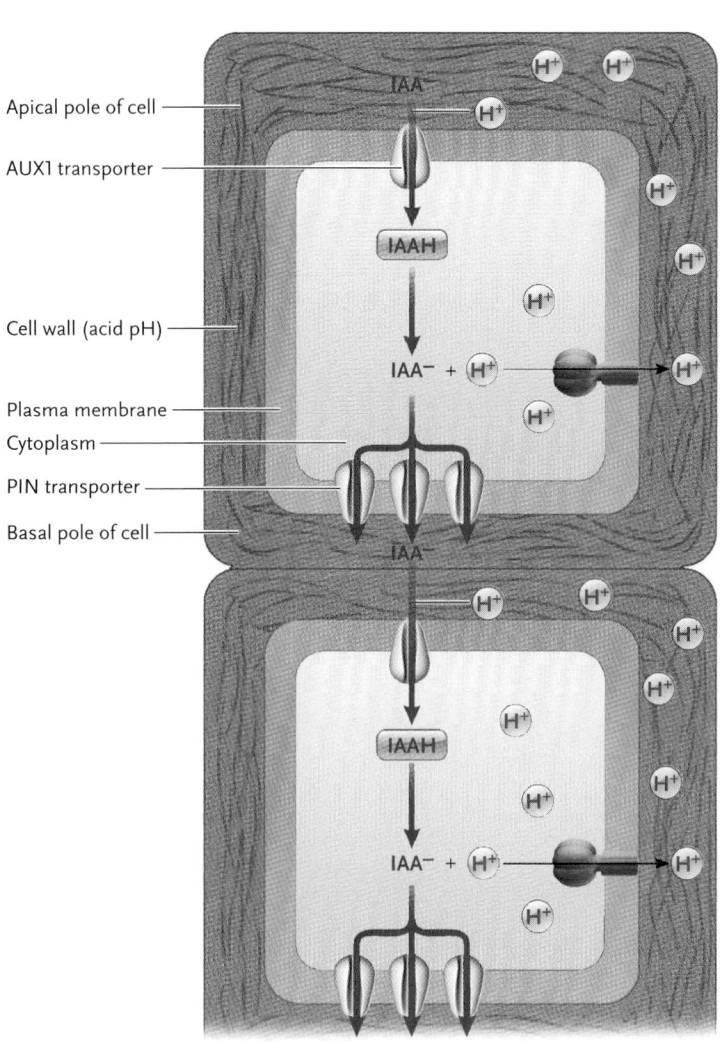

Apical pole of cell

AUX1 transporter

Cell wall (acid pH)

Plasma membrane

Cytoplasm

PIN transporter

Basal pole of cell

IAA

IAAH⁺

IAA⁻ + H⁺

1 As auxin (IAA⁻) diffuses through the cell wall, the acid pH makes H⁺ bind to it. The resulting nonionized form is IAAH.

2 AUX1 transports IAAH into the cell cytoplasm.

3 In the less acidic cytoplasm, auxin gives up H⁺ and reverts to its ionized form.

4 As H⁺ is pumped out of the cell, the acidity of the wall increases.

5 Auxin moves out passively through PIN transporters.

6 These steps are repeated in each adjoining parenchyma cell. Thus the auxin transport shows polarity, from auxin's source in a shoot tip and leaves, downward toward the base of the stem.

Figure 35.6
A model for polar auxin transport. A plasma membrane H⁺ pump maintains gradients of pH and electrical charge across the membrane, moving H⁺ out of the cell using energy from ATP hydrolysis. These gradients are key to transporting IAA from the apical region to the basal region of a cell in a column. Following the gradients, at the basal end of a cell IAA diffuses through transport proteins into the cell wall, then (as IAAH) into the next cell in line.

and then stretched by turgor pressure. Several hormones, and auxin especially, apparently increase the plasticity (irreversible stretching) of the cell wall. Two major hypotheses seek to explain this effect, and both may be correct.

Plant cell walls grow much faster in an acidic environment—that is, when the pH is less than 7. The **acid-growth hypothesis** suggests that auxin causes cells to secrete acid (H⁺) into the cell wall by stimulating the plasma membrane H⁺ pumps to move hydrogen ions from the cell interior into the cell wall; the increased acidity activates proteins called *expansins*, which penetrate the cell wall and disrupt bonds between cellulose microfibrils in the wall **(Figure 35.7)**. In the laboratory, it is easy to measure an increase in the rate at which coleoptiles or stem tissues release acid when they are treated with IAA. Activation of the plasma membrane H⁺ pump also produces a membrane potential that pulls K⁺ and other cations into the cell; the resulting osmotic gradient draws water into the cell, increasing turgor pressure and helping to stretch the "loosened" cell walls.

A second hypothesis, which also is supported by experimental evidence, suggests that auxin triggers the expression of genes encoding enzymes that play roles in the synthesis of new wall components. Plant cells exposed to IAA don't show increased growth if they are treated with a chemical that inhibits protein synthesis. However, researchers have identified mRNAs that rapidly increase in concentration within 10 to 20 minutes after stem sections have been treated with auxin, although they still do not know exactly which proteins these mRNAs encode.

Gibberellins Also Stimulate Growth, Including the Elongation of Stems

Gibberellins stimulate several aspects of plant growth. Perhaps most apparent to humans is their ability to promote the lengthening of plant stems by stimulating both cell division and cell elongation. Synthesized in shoot and root tips and young leaves, gibberellins, like auxin, modify the properties of plant cell walls in ways that promote expansion (although the gibberellin mechanism does not involve acidification of the cell wall). It may be that the two hormones both affect expansins, or are functionally linked in some other way yet to be discovered. Gibberellins have other known effects as well, such as helping to break the dormancy of seeds and buds.

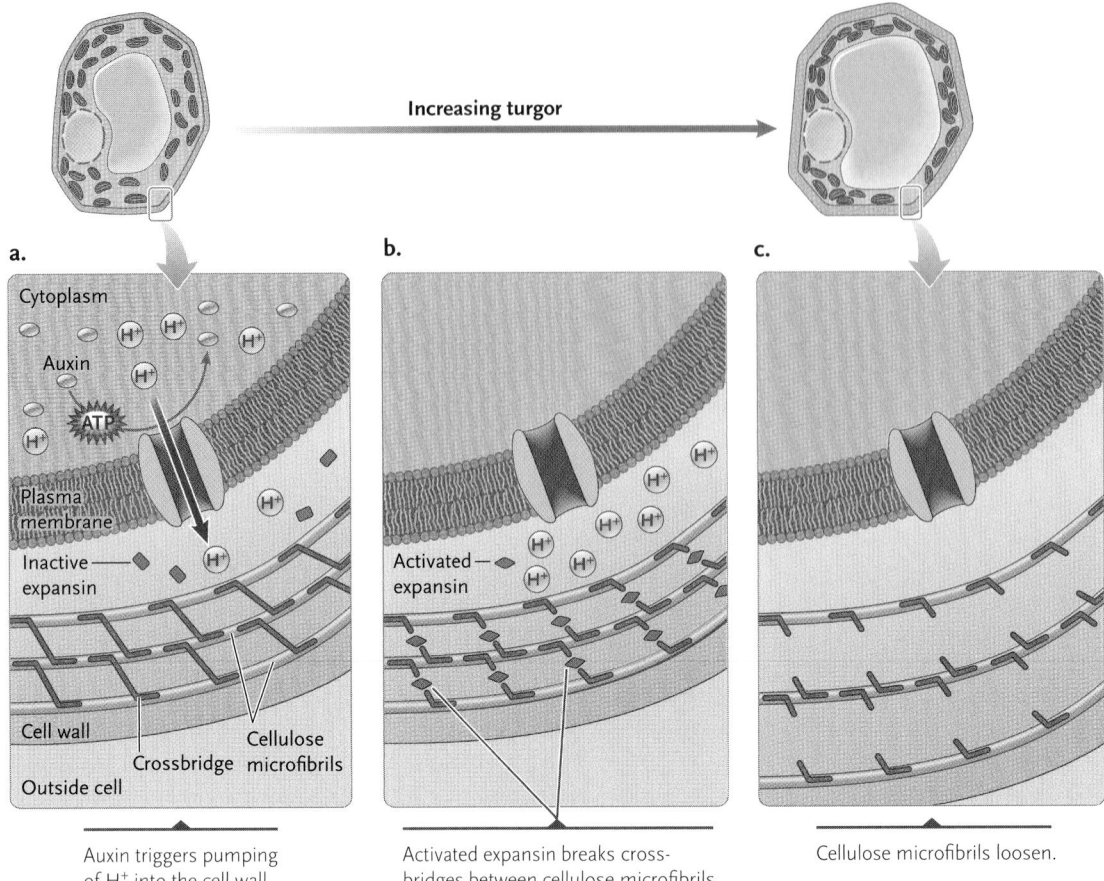

Increasing turgor

a.

Cytoplasm

Auxin

ATP

Plasma membrane

Inactive expansin

Cell wall

Crossbridge

Cellulose microfibrils

Outside cell

Auxin triggers pumping of H⁺ into the cell wall.

b.

Activated expansin

Activated expansin breaks crossbridges between cellulose microfibrils.

c.

Cellulose microfibrils loosen.

Figure 35.7

How auxin may regulate expansion of plant cells. According to the acid-growth hypothesis, plant cells secrete acid (H⁺) when auxin stimulates the plasma membrane H⁺ pumps to move hydrogen ions into the cell wall; the increased acidity activates enzymes called *expansins*, which disrupt bonds between cellulose microfibrils in the wall. As a result, the wall becomes extensible and the cell can expand.

Of the 100-plus compounds of the gibberellin family, relatively few are biologically active as hormones. The others are inactive forms or serve as precursors to active forms.

Gibberellins are active in eudicots as well as in a few monocots. In most plant species that have been analyzed, the main controller of stem elongation is the gibberellin called GA_1. Normally, GA_1 is synthesized in small amounts in young leaves and transported throughout the plant in the phloem. When GA_1 synthesis goes awry, the outcome is a dramatic change in the plant's stature. For example, experiments with a dwarf variety of peas *(Pisum sativum)* and some other species show that these plants and their taller relatives differ at a single gene locus. Normal plants make an enzyme required for gibberellin synthesis; dwarf plants of the same species lack the enzyme, and their internodes elongate very little.

Another stark demonstration of the effect gibberellins can have on internode growth is **bolting**, growth of a floral stalk in plants that form vegetative rosettes, such as cabbages *(Brassica oleracea)* and iceberg lettuce *(Lactuca sativa)*. In a rosette plant, stem internodes are

so short that the leaves appear to arise from a single node. When these plants flower, however, the stem elongates rapidly and flowers develop on the new stem parts. An experimenter can trigger exaggerated bolting by spraying a plant with gibberellin **(Figure 35.8)**. In nature, external cues such as increasing day length or warming after a cold snap stimulate gibberellin synthesis, and bolting occurs soon afterward. This observation supports the hypothesis that in rosette plants and possibly some others, gibberellins switch on internode lengthening when environmental conditions favor a shift from vegetative growth to reproductive growth.

Other experiments using gibberellins have turned up a striking number of additional roles for this hormone family. For example, a gibberellin helps stimulate buds and seeds to break dormancy and resume growth in the spring. Research on barley embryos showed that gibberellin provides signals during germination that lead to the enzymatic breakdown of endosperm, releasing nutrients that nourish the developing seedling (see Section 34.3). In monoecious species, which have flowers of both sexual types on the same plant, applications of gibberellin seem to encourage proportionately more "male" flowers to develop. As a result, there may be more pollen available to pollinate "female" flowers and, eventually, more fruit produced. A gibberellin used by commercial grape growers promotes fruit set and lengthens the stems on which fruits develop, allowing space for individual grapes to grow larger. One result is fruit with greater consumer appeal **(Figure 35.9)**.

Cytokinins Enhance Growth and Retard Aging

Cytokinins play a major role in stimulating cell division (hence the name, which refers to cytokinesis). These hormones were first discovered during experiments designed to define the nutrient media required for plant tissue culture. Researchers found that in addition to a carbon source such as sucrose or glucose, minerals, and certain vitamins, cells in culture also required two other substances. One was auxin, which promoted the elongation of plant cells but did not stimulate the cells to divide. The other substance could be coconut milk, which is actually liquid endosperm, or it could be DNA that had been degraded into smaller molecules by boiling. When either was added to a culture medium along with an auxin, the cultured cells would begin dividing and grow normally.

We now know that the active ingredients in both boiled DNA and endosperm are cytokinins, which have a chemical structure similar to that of the nucleic acid base adenine. The most abundant natural cytokinin is zeatin, so-called because it was first isolated from the endosperm of young corn seeds *(Zea mays)*. In endosperm, zeatin probably promotes the burst of cell division that takes place as a fruit matures. As you might expect, cytokinins also are abundant in the rapidly dividing meristem tissues of root and shoot tips. Cytokinins occur not only in flowering plants but also in many conifers, mosses, and ferns. They are also synthesized by many soil-dwelling bacteria and fungi and may be crucial to the growth of mycorrhizae, which help nourish thousands of plant species (see Section 33.3). Conversely, *Agrobacterium* and other microbes that cause plant tumors carry genes that regulate the production of cytokinins.

Cytokinins are synthesized largely (although not only) in root tips and apparently are transported through the plant in xylem sap. Besides promoting cell division, they have a range of effects on plant metabolism and development, probably by regulating protein synthesis. For example, cytokinins promote expansion of young leaves (as leaf cells expand), cause chloroplasts to mature, and retard leaf aging. Another cytokinin effect—coordinating the growth of roots and shoots, in concert with auxin—underscores the point that plant hormones often work together to evoke a particular response. Investigators culturing

tobacco tissues found that the relative amounts of auxin and a cytokinin strongly influenced not only growth, but also development **(Figure 35.10)**. When the auxin-to-cytokinin ratio is about 10:1, the growing tissue did not differentiate but instead remained as a loose mass of cells, or *callus*. When the relative auxin concentration was increased slightly, the callus produced roots. When the relative concentration of the cytokinin was increased, chloroplasts in the callus cells matured, the callus became green and more compact, and it produced shoots. In nature, the interaction of a cytokinin and auxin may produce the typical balanced growth of roots and shoots, with each region providing the other with key nutrients.

Natural cytokinins can prolong the life of stored vegetables. Similar synthetic compounds are already widely used to prolong the shelf life of lettuces and mushrooms and to keep cut flowers fresh.

Ethylene Regulates a Range of Responses, Including Senescence

Most parts of a plant can produce **ethylene**, which is present in fruits, flowers, seeds, leaves, and roots. In different species it helps regulate a wide variety of plant physiological responses, including dormancy of seeds and buds, seedling growth, stem elongation, the ripening of fruit, and the eventual separation of fruits, leaves, and flowers from the plant body. Ethylene is an unusual hormone, in part because it is structurally simple (see Table 35.1) and in part because it is a gas at normal temperature and pressure.

Before a bean or pea seedling emerges from the soil, ethylene simultaneously slows elongation of the stem and stimulates cell divisions that increase stem girth.

Figure 35.8 A dramatic example of bolting in cabbage *(Brassica oleracea)*, a plant commonly grown as a winter vegetable. The rosette form (left) reflects the plant's growth habit when days are short (and nights are long). Gibberellin was applied to the plants at the right, triggering the rapid stem elongation and subsequent flowering, characteristic of bolting.

Two untreated cabbages (controls)   Cabbages treated with gibberellins

Sylvan H. Wittwer/Visuals Unlimited

Figure 35.9 Effect of gibberellin on seedless grapes *(Vitis vinifera)*. The grapes on the right developed on vines that were treated with a gibberellin.

NORMAL   G.A. TREATED

Sylvan Wittwer/Visuals Unlimited

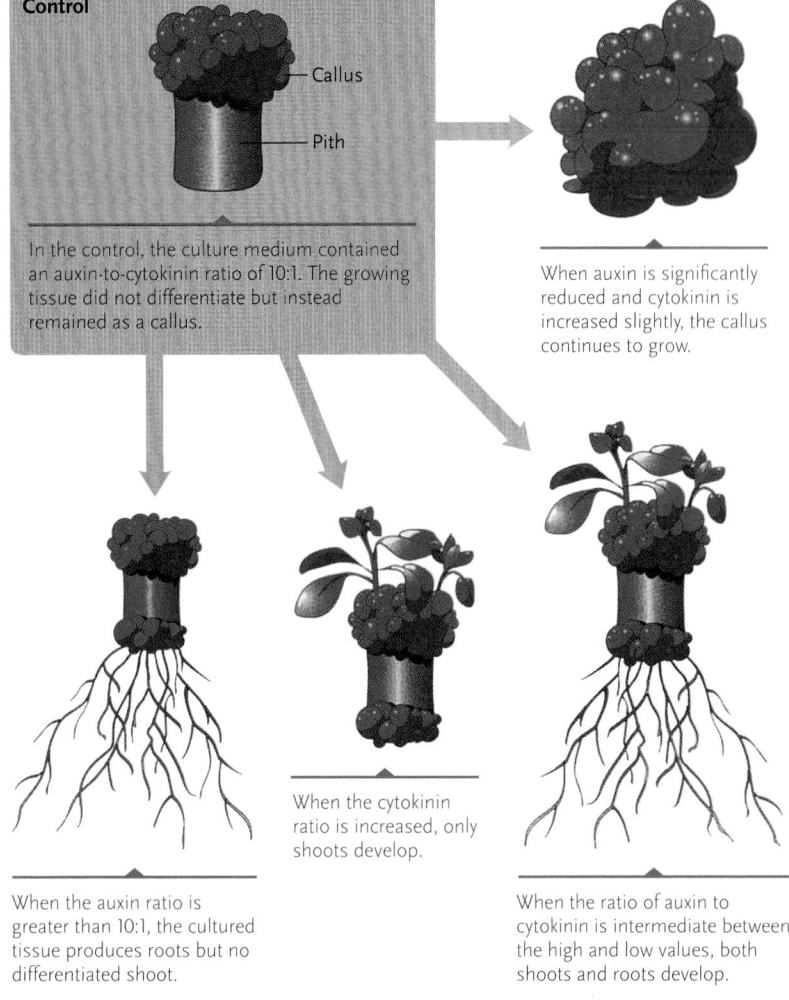

Control

Callus

Pith

In the control, the culture medium contained an auxin-to-cytokinin ratio of 10:1. The growing tissue did not differentiate but instead remained as a callus.

When auxin is significantly reduced and cytokinin is increased slightly, the callus continues to grow.

When the cytokinin ratio is increased, only shoots develop.

When the auxin ratio is greater than 10:1, the cultured tissue produces roots but no differentiated shoot.

When the ratio of auxin to cytokinin is intermediate between the high and low values, both shoots and roots develop.

Figure 35.10
Effects of varying ratios of auxin and cytokinin on tobacco tissues *(Nicotiana tabacum)* grown in culture. The method starts with a block of stem pith, essentially a core of ground tissue removed from the center of a stem. The callus growing on the pith is a disorganized mass of undifferentiated cells.

These alterations push the curved hypocotyl through the soil and into the air (see Figure 34.12). Such ethylene-induced horizontal growth also can help a growing seedling "find its way" into the air if the seed happens to germinate under a pebble or some other barrier.

Ethylene also governs the biologically complex process of aging, or **senescence**, in plants. Senescence is a closely controlled process of deterioration that leads to the death of plant cells. In autumn the leaves of deciduous trees senesce, often turning yellow or red as chlorophyll and proteins break down, allowing other pigments to become more noticeable. Ethylene triggers the expression of genes leading to the synthesis of chlorophyllases and proteases, enzymes that launch the breakdown process. In many plants, senescence is associated with **abscission**, the dropping of flowers, fruits, and leaves in response to environmental signals. In this process, ethylene apparently stimulates the activity of enzymes that digest cell walls in an abscission zone—a localized region at the base of the petiole. The petiole detaches from the stem at that point **(Figure 35.11)**.

For some species, the funneling of nutrients into reproductive parts may be a cue for senescence of leaves, stems, and roots. When the drain of nutrients is halted by removing each newly emerging flower or seed pod, a plant's leaves and stems stay green and vigorous much longer **(Figure 35.12)**. Gardeners routinely remove flower buds from many plants to maintain vegetative growth. Senescence requires other cues, however. For instance, when a cocklebur is induced to flower under winterlike conditions, its leaves turn yellow regardless of whether the nutrient-demanding young flowers are left on or pinched off. It is as if a "death signal" forms that leads to flowering and senescence when there are fewer hours of daylight (typical of winter days). This observation underscores the general theme that many plant responses to the environment involve the interaction of multiple molecular signals.

Fruit ripening is a special case of senescence. Although the precise mechanisms are not well understood, ripening begins when a fruit starts to synthesize ethylene. The ripening process may involve the conversion of starch or organic acids to sugars, the softening of cell walls, or the rupturing of the cell membrane and loss of cell fluid. The same kinds of events occur in wounded plant tissues, which also synthesize ethylene.

Ethylene from an outside source can stimulate senescence responses, including ripening, when it binds to specific protein receptors on plant cells. The ancient Chinese observed that they could induce picked fruit to ripen faster by burning incense; later, it was found that the incense smoke contains ethylene. Today ethylene gas is widely used to ripen tomatoes, pineapples, bananas, honeydew melons, mangoes, papayas, and other fruit that has been picked and shipped while still green. Ripening fruit itself gives off ethylene, which is why placing a ripe banana in a closed sack of unripe peaches (or some other green fruit) often can cause the fruit to ripen. Oranges and other citrus fruits may be exposed to ethylene to brighten the color of their rind. Conversely, limiting fruit exposure to ethylene can delay ripening. Apples will keep for months without rotting if they are exposed to a chemical that inhibits ethylene production or if they are stored in an environment that inhibits the hormone's effects—including low atmospheric pressure and a high concentration of CO_2, which may bind ethylene receptors.

Brassinosteroids Regulate Plant Growth Responses

The dozens of steroid hormones classed as **brassinosteroids** all appear to be vital for normal growth in plants, for they stimulate cell division and elongation in a wide range of plant cell types. Confirmed as plant hormones in the 1980s, brassinosteroids now are the subject of intense research on their sources and effects. While brassinosteroids have been detected in a wide variety of plant tissues and organs, the highest concentrations are found in shoot tips and in developing seeds and embryos—all examples of young, actively developing parts. In laboratory studies, the hormones have different effects depending on the tissue where they are active. They have promoted cell elongation, differentiation of vascular tissue, and elongation of a pollen tube after a flower is pollinated. By contrast, they inhibit the elongation of roots. First isolated from pollen of a plant in the mustard family, *Brassica napus* (a type of canola), in nature brassinosteroids seem to regulate the expression of genes associated with a plant's growth responses to light. This role was underscored by the outcomes of experiments using mutant *Arabidopsis* plants that were homozygous for a defective genes called *bri1* (for brassinosteroid-insensitive receptor) **(Figure 35.13)**; the results provided convincing evidence that brassinosteroids mediate growth responses to light.

Abscisic Acid Suppresses Growth and Influences Responses to Environmental Stress

Plant scientists ascribe a variety of effects to the hormone **abscisic acid** (ABA), many of which represent evolutionary adaptations to environmental challenges. Plants apparently synthesize ABA from carotenoid pigments inside plastids in leaves and possibly other plant parts. Several ABA receptors have been identified, and in general, we can group its effects into changes in gene expression that result in long-term inhibition of growth, and rapid, short-term physiological changes that are responses to immediate stresses, such as a lack of water, in a plant's surroundings. As its name suggests, at one time ABA was thought to play a major role in abscission. As already described, however, abscission is largely the domain of ethylene.

Suppressing Growth in Buds and Seeds. Operating as a counterpoint to growth-stimulating hormones like gibberellins, ABA inhibits growth in response to environmental cues, such as seasonal changes in temperature and light. This growth suppression can last for many months or even years. For example, one of ABA's major growth-inhibiting effects is apparent in perennial plants, in which the hormone promotes dormancy in leaf buds—an important adaptive advantage in places where winter cold can damage young leaves. If ABA is applied to a growing leaf bud, the bud's normal development stops, and instead protective *bud scales*—modified, nonphotosynthetic leaves that are small, dry, and tough—form around the apical meristem and insulate it from the elements **(Figure 35.14)**. After the scales develop, most cell metabolic activity shuts down and the leaf bud becomes dormant.

In some plants that produce fleshy fruits, such as apples and cherries, abscisic acid is associated with the dormancy of seeds as well. As the seed develops,

N.R. Lersten

Abscission zone at base of leaf where it joins the stem

Figure 35.11
Abscission zone in a maple *(Acer)*. This longitudinal section at the left is through the base of the petiole of a leaf.

ABA accumulates in the seed coat, and the embryo does not germinate even if it becomes hydrated. Before such a seed can germinate, it usually will require a long period of cool, wet conditions, which stimulate the breakdown of ABA. The buildup of ABA in developing seeds does more than simply inhibit development, however. As early development draws to a close, ABA stimulates the transcription of certain genes, and large amounts of their protein products are synthesized. These proteins are thought to store nitrogen and other nutrients that the embryo will use when it eventually does germinate. ABA and related growth inhibitors are often applied to plants slated to be shipped to plant nurseries. Dormant plants suffer less shipping damage, and the effects of the inhibitors can be reversed by applying a gibberellin.

Responses to Environmental Stress. ABA also triggers plant responses to various environmental stresses, including cold snaps, high soil salinity, and drought. A

Larry D. Nooden

Control plant (pods not removed)   Experimental plant (pods removed)

Figure 35.12
Experimental results showing that the removal of seed pods from a soybean plant *(Glycine max)* delays its senescence.

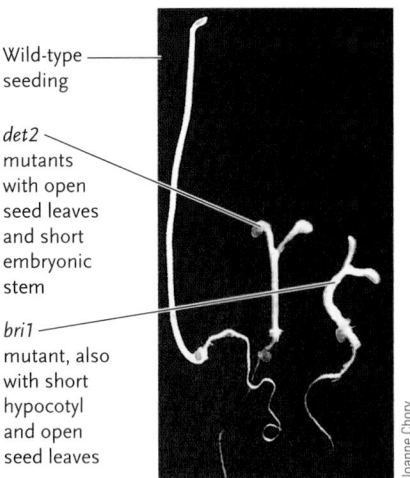

Wild-type seeding

det2 mutants with open seed leaves and short embryonic stem

bri1 mutant, also with short hypocotyl and open seed leaves

Joanne Chory

Figure 35.13
Experimental evidence that brassinosteroids can mediate a plant's responses to light by regulating gene expression. In *Arabidopsis*, wild-type seedlings synthesize a protein (encoded by the *DET2* gene) that prevents leaves from developing *(seedling at left)* until photosynthesis is possible, after the seedling breaks out of the dark environment of soil. When the gene is defective, a mutant *det2* plant *(center)* will develop a short hypocotyl (embryonic stem) and open seed leaves (cotyledons) even when there is no light for photosynthesis. Experiments with *bri1* mutants, which lack functioning receptors for a brassinosteroid, resulted in a similar phenotype *(right)*. These findings supported the hypothesis that a brassinosteroid is necessary for normal expression of the *DET2* gene.

great deal of research has focused on how ABA influences plant responses to a lack of water. When a plant is water-stressed, ABA helps prevent excessive water loss by stimulating stomata to close. As described in Section 32.3, flowering plants depend heavily on the proper functioning of stomata. When a lack of water leads to wilting, mesophyll cells in wilted leaves rapidly synthesize and secrete ABA. The hormone diffuses to guard cells, where an ABA receptor binds it. Binding stimulates the release of K^+ and water from the guard cells, and within minutes the stomata close.

Once bound to its receptor, ABA may exert its effects through a cascade of signals that includes phosphorylated proteins. Experiments have shown that an *Arabidopsis* mutant unable to respond to ABA lacks an enzyme that removes phosphate groups from certain proteins. This condition suggests that cleaving phosphates is one step in the ABA response. *Insights from the Molecular Revolution* highlights recent research filling in other steps in the ABA-induced response pathway.

Jasmonates and Oligosaccharins Regulate Growth and Have Roles in Defense

In recent years, studies of plant growth and development have helped define the roles—or revealed the existence—of several other hormonelike compounds in plants. Like the well-established plant hormones just described, these substances are organic molecules and only tiny amounts are required to alter some aspect of a plant's functioning. Some have long been known to exist in plants, but the extent of their signaling roles has only recently become better understood. This group includes **jasmonates** (JA), a family of about 20 compounds derived from fatty acids. Experiments with *Ara-*

bidopsis and other plants have revealed numerous genes that respond to JA, including genes that help regulate root growth and seed germination. JA also appears to help plants "manage" stresses due to deficiencies of certain nutrients (such as K^+). The JA family is best known, however, as part of the plant arsenal to limit damage by pathogens and predators, the topic of the following section.

Some other substances also are drawing keen interest from plant scientists, but because their signaling roles are still poorly understood they are not widely accepted as confirmed plant hormones. A case in point involves the complex carbohydrates that are structural elements in the cell walls of plants and some fungi. Several years ago, researchers observed that in some plants, some of these oligosaccharides could serve as signaling molecules. Such compounds were named **oligosaccharins**, and one of their known roles is to defend the plant against pathogens. In addition, oligosaccharins have been proposed as growth regulators that adjust the growth and differentiation of plant cells, possibly by modulating the influences of growth-promoting hormones such as auxin. At this writing, researchers in many laboratories are pursuing a deeper understanding of this curious subset of plant signaling molecules.

STUDY BREAK

1. Which plant hormones promote growth and which inhibit it?
2. Give examples of how some hormones have both promoting and inhibiting effects in different parts of the plant at different times of the life cycle.

35.2 Plant Chemical Defenses

Plants don't have immune systems like those that have evolved in animals (the subject of Chapter 43). Even so, over the millennia, in higher plants virtually constant exposure to predation by herbivores and the onslaught

Figure 35.14
Bud scales, here on the bud of a perennial cornflower (*Centaurea montana*).

Amanda Darcy/Getty Images Inc.

Stressing Out in Plants and People

Unlike people, plants cannot move to more favorable locations when an environmental stress threatens. Instead, to survive stresses plants adjust their responses to environmental factors such as temperature and the availability of water. Recent molecular work shows that responses to stress imposed by drought and cold involve some of the same chemical steps in plants and humans, indicating an ancient link to a common evolutionary ancestor. The research may also point the way to genetic engineering strategies to modify major crop plants for earlier and better responses to stress.

Many plant stress responses are triggered by the hormone abscisic acid (ABA). Although individual steps in the response pathway are unclear, it is known that calcium ions increase in concentration in the cytoplasm when plant cells are exposed to ABA. Soon after the rise in Ca^{2+}, genes are activated that compensate for the stressful situation.

Nam-Hai Chua and his colleagues at the Rockefeller University and the University of Minnesota were interested in piecing together the molecular steps in the plant pathway. One substance they thought might be involved is *cyclic ADP-ribose (cADPR)*, a signaling molecule that was first implicated in calcium release pathways in animal cells.

The Chua team began by injecting two plant genes, *rd29A* and *kin2*, into tomato cells. The two genes are activated by ABA as part of the stress response. Each of the injected genes was linked to an unrelated marker gene that would also be turned on if the gene became active. When ABA was injected into stressed tomato plants grown from the injected cells, the markers were activated, indicating that *rd29A* and *kin2* were turned on by ABA.

The next step was to inject Ca^{2+} and to note whether the injected genes were activated. The injection activated the *rd29A* and *kin2* genes, confirming the role of calcium ions in the pathway. A chemical called EGTA that removes Ca^{2+} from the cytoplasm cancelled the gene activation, as expected if calcium is part of the response pathway.

Then, the investigators injected cADPR to see if it activated *rd29A* and *kin2*. This result was also positive; cADPR had the same effect as either ABA or Ca^{2+}. EGTA blocked the positive response to cADPR, indicating that cADPR lies between ABA and calcium release in the signal pathway.

Another experiment determined whether protein phosphorylation might be part of the pathway. To accomplish this, the investigators injected an inhibitor of protein kinases, the enzymes that phosphorylate proteins as a part of

many cellular response pathways (see Section 7.2). After the inhibitor was added, injecting ABA, cADPR, or Ca^{2+} failed to activate *rd29A* and *kin2*, indicating that protein phosphorylation occupies a critical step following these elements in the pathway.

From these results the Chua team was able to reconstruct a major part of the pathway:

ABA → ? → cADPR →
$\quad$ Ca^{2+} → protein kinases →
$\quad\quad$? → stress gene activation

The question marks indicate one or more unknown steps. Most significant is the specific ABA receptor that carries out the first step in the response pathway. A recently identified ABA receptor that is involved in the events that cause stomata to close (among other effects) may be this "missing link."

In addition to possible benefits for agriculture, the Chua team's research may also shed light on signal pathways in animals in which cADPR plays a part, including one that adjusts the heartbeat and another that regulates insulin release in response to elevated blood glucose. Thus Chua's work may help fill in steps in both plant and animal responses, in pathways inherited from a common ancestor predating both plants and people.

---

of pathogens have resulted in a striking array of chemical defenses that ward off or reduce damage to plant tissues from infectious bacteria, fungi, worms, or plant-eating insects **(Table 35.2)**. You will discover in this section that as with the defensive strategies of animals, plant defenses include both general responses to any type of attack and specific responses to particular threats. Some get underway almost as soon as an attack begins, while others help promote the plant's long-term survival. And more often than not, multiple chemicals interact as the response unfolds.

Jasmonates and Other Compounds Interact in a General Response to Wounds

When an insect begins feeding on a leaf or some other plant part, the plant may respond to the resulting wound by launching what in effect is a cascade of

chemical responses. These complex signaling pathways often rely on interactions among jasmonates, ethylene, or some other plant hormone. As the pathway unfolds it triggers expression of genes leading to chemical and physical defenses at the wound site. For example, in some plants jasmonate induces a response leading to the synthesis of protease inhibitors, which disrupt an insect's capacity to digest proteins in the plant tissue. The protein deficiency in turn hampers the insect's growth and functioning.

A plant's capacity to recognize and respond to the physical damage of a wound apparently has been a strong selection pressure during plant evolution. When a plant is wounded experimentally, numerous defensive chemicals can be detected in its tissues in relatively short order. One of these, **salicylic acid**, or **SA** (a compound similar to aspirin, which is acetylsalicylic acid), seems to have multiple roles in plant defenses,

Table 35.2 | **Summary of Plant Chemical Defenses**

Type of Defense	Effects
General Defenses	
Jasmonate (JA) responses to wounds/injury by pathogens; pathways often include other hormones such as ethylene	Synthesis of defensive chemicals such as protease inhibitors
Hypersensitive response to infectious pathogens (e.g., fungi, bacteria)	Physically isolates infection site by surrounding it with dead cells
PR (pathogenesis-related) proteins	Enzymes, other proteins that degrade cell walls of pathogens
Salicylic acid (SA)	Mobilized during other responses and independently; induces the synthesis of PR proteins, operates in systemic acquired resistance
Systemin (in tomato)	Triggers JA response
Secondary metabolites	
Phytoalexins	Antibiotic
Oligosaccharins	Trigger synthesis of phytoalexins
Systemic acquired resistance (SAR)	Long-lasting protection against some pathogens; components include SA and PR proteins that accumulate in healthy tissues
Specific Defenses	
Gene-for-gene recognition of chemical features of specific pathogens (by binding with receptors coded by R genes)	Triggers defensive response (e.g., hypersensitive response, PR proteins) against pathogens
Other	
Heat-shock responses (encoded by heat-shock genes)	Synthesis of chaperone proteins that reversibly bind other plant proteins and prevent denaturing due to heat stress
"Antifreeze" proteins	In some species, stabilize cell proteins under freezing conditions

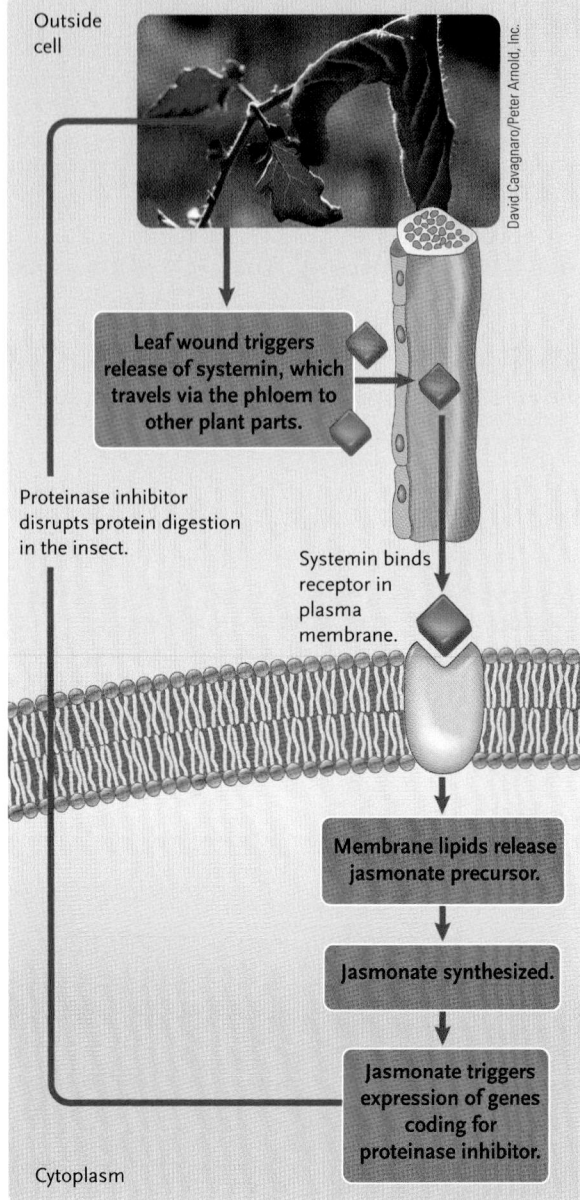

Figure 35.15

The systemin response to wounding. When a plant is wounded, it responds by releasing the protein hormone systemin. Transported through the phloem to other plant parts, in receptive cells systemin sets in motion a sequence of reactions that lead to the expression of genes encoding protease inhibitors—substances that can seriously disrupt an insect predator's capacity to digest protein.

including interacting with jasmonates in signaling cascades.

Researchers are regularly discovering new variations of hormone-induced wound responses in plants. For example, experiments have elucidated some of the steps in an unusual pathway that thus far is known only in tomato *(Lycopersicum esculentum)* and a few other plant species. As diagrammed in **Figure 35.15,** the wounded plant rapidly synthesizes **systemin**, the first peptide hormone to be discovered in plants. (Various animal hormones are peptides, a topic covered in Chapter 40.) Systemin enters the phloem and is transported throughout the plant. Although various details of the signaling pathway have yet to be worked out, when receptive cells bind systemin, their plasma membranes release a lipid that is the chemical precursor of jasmonate. Next jasmonate is synthesized, and it in turn sets in motion the expression of genes that encode

protease inhibitors, which protect the plant against attack, even in parts remote from the original wound.

The Hypersensitive Response and PR Proteins Are Other General Defenses

Often, a plant that becomes infected by pathogenic bacteria or fungi counters the attack by way of a **hypersensitive response**—a defense that physically cordons off an infection site by surrounding it with dead cells. Initially, cells near the site respond by producing a burst of highly reactive oxygen-containing compounds (such as hydrogen peroxide, H_2O_2) that can break down nucleic acids, inactivate enzymes, or have other toxic effects on cells. The burst is catalyzed by enzymes in the plant cell's plasma membrane. It may begin the process of killing cells close to the attack site and, as the response advances, programmed cell death may also come into play. In short order, the "sacrificed" dead cells wall off the infected area from the rest of the plant. Thus denied an ongoing supply of nutrients, the invading pathogen dies. A common sign of a successful hypersensitive response is a dead spot surrounded by healthy tissue **(Figure 35.16)**.

While the hypersensitive response is underway, salicylic acid triggers other defensive responses by an infected plant. One of its effects is to induce the synthesis of **pathogenesis-related proteins,** or **PR proteins.** Some PR proteins are hydrolytic enzymes that break down components of a pathogen's cell wall. Examples are chitinases that dismantle the chitin in the cell walls of fungi and so kill the cells. In some cases, plant cell receptors also detect the presence of fragments of the disintegrating wall and set in motion additional defense responses.

Secondary Metabolites Defend against Pathogens and Herbivores

Many plants counter bacteria and fungi by making **phytoalexins,** biochemicals of various types that function as antibiotics. When an infectious agent breaches a plant part, genes encoding phytoalexins begin to be transcribed in the affected tissue. For instance, when a fungus begins to invade plant tissues, the enzymes it secretes may trigger the release of oligosaccharins. In addition to their roles as growth regulators (described in Section 35.1), these substances also can promote the production of phytoalexins, which have toxic effects on a variety of fungi. Plant tissues may also synthesize phytoalexins in response to attacks by viruses.

Phytoalexins are among many *secondary metabolites* produced by plants. Such substances are termed "secondary" because they are not routinely synthesized in all plant cells as part of basic metabolism. A wide range of plant species deploy secondary metabolites as defenses against feeding herbivores. Examples are alkaloids such as caffeine, cocaine, and the poison strychnine (in seeds of the *nux vomica* tree, *Strychnos nux-vomica*), tannins such as those in oak acorns, and various terpenes. The terpene family includes insect-repelling substances in conifer resins and cotton, and essential oils produced by sage and basil plants. Because these terpenes are volatile—they easily diffuse out of the plant into the surrounding air—they also can provide indirect defense to a plant. Released from the wounds created by a munching insect, they attract other insects that prey on the herbivore. Chapter 50

looks in detail at the interactions between plants and herbivores.

Gene-for-Gene Recognition Allows Rapid Responses to Specific Threats

One of the most interesting questions with respect to plant defenses is how plants first sense that an attack is underway. In some instances plants apparently can detect an attack by a specific predator through a mechanism called **gene-for-gene recognition.** This term refers to a matchup between the products of dominant alleles of two types of genes: a so-called **R gene** (for "resistance") in a plant, and an **Avr gene** (for "avirulence") in a particular pathogen. Thousands of R genes have been identified in a wide range of plant species. Dominant R alleles confer enhanced resistance to plant pathogens including bacteria, fungi, and nematode worms that attack roots.

The basic mechanism of gene-for-gene recognition is simple: The dominant R allele encodes a receptor in plasma membranes of a plant's cells, and the dominant pathogen Avr allele encodes a molecule that can bind the receptor. "Avirulence" implies "not virulent," and binding of the Avr gene product triggers an immediate defense response in the plant. Trigger molecules run the gamut from proteins to lipids to carbohydrates that have been secreted by the pathogen or released from its surface **(Figure 35.17).** Experiments have demonstrated a rapid-fire sequence of early biochemical changes that follow binding of the Avr-encoded molecule; these include changes in ion concentrations inside and outside plant cells and the production of biologically active oxygen compounds that heralds the hypersensitive response. In fact, of the instances of gene-for-gene recognition plant scientists have observed thus far, most trigger the hypersensitive response and the ensuing synthesis of PR proteins, with their antibiotic effects.

Systemic Acquired Resistance Can Provide Long-Term Protection

The defensive response to a microbial invasion may spread throughout a plant, so that the plant's healthy tissues become less vulnerable to infection. This phenomenon is called **systemic acquired resistance,** and experiments using *Arabidopsis* plants have shed light on how it comes about **(Figure 35.18).** In a key early step, salicylic acid builds up in the affected tissues. By some route, probably through the phloem, the SA passes

Figure 35.16 Evidence of the hypersensitive response. The dead spots on these leaves of a strawberry plant *(Fragaria)* are sites where a pathogen invaded, triggering the defensive destruction of the surrounding cells.

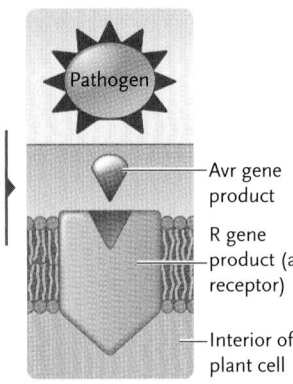

Required precondition
A plant has a dominant R gene encoding a receptor that can bind the product of a specfic pathogen dominant Avr gene.

1 When the R-encoded receptor binds its matching Avr product, the binding triggers signaling pathways, leading to various defense responses in the plant.

2 Fluxes of ions and enzyme activity at the plasma membrane contribute to the hypersensitive response. Soon PR proteins, phytoalexins, and salicyic acid (SA) are synthesized. The PR proteins and phytoalexins combat pathogens directly. SA promotes systemic acquired resistance.

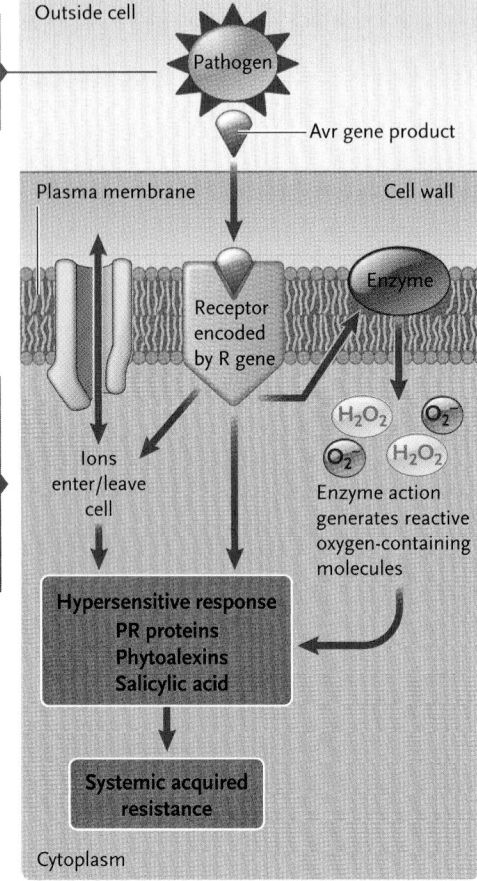

Figure 35.17
Model of how gene-for-gene resistance may operate. For resistance to develop, the plant must have a dominant R gene and the pathogen must have a corresponding dominant Avr gene. Products of such "matching" genes can interact physically, rather like the lock-and-key mechanism of an enzyme and its substrate. Most R genes encode receptors at the plasma membranes of plant cells. As diagrammed in step 1, when one of these receptors binds an Avr gene's product, the initial result may be changes in the movements of specific ions into or out of the cell and the activation of membrane enzymes that catalyze the formation of highly reactive oxygen-containing molecules. Such events help launch other signaling pathways that lead to a variety of defensive responses, including the hypersensitive response (step 2).

from the infected organ to newly forming organs such as leaves, which begin to synthesize PR proteins—again, providing the plant with a "home-grown" antimicrobial arsenal. How does the SA exert this effect? It seems that when enough SA accumulates in a plant

cell's cytoplasm, a regulatory protein called NPR-1 (for *n*onexpressor of *p*athogenesis-*r*elated genes) moves from the cytoplasm into the cell nucleus. There it interacts with factors that promote the transcription of genes encoding PR proteins.

In addition to synthesizing SA that will be transported to other tissues by a plant's vascular system, the damaged leaf also synthesizes a chemically similar compound, methyl salicylate. This substance is volatile, and researchers speculate that it may serve as an airborne "harm" signal, promoting defense responses in the plant that synthesized it and possibly in nearby plants as well.

Extremes of Heat and Cold Also Elicit Protective Chemical Responses

Plant cells also contain **heat-shock proteins (HSPs)**, a type of chaperone protein (see Section 3.5) found in cells of many species. HSPs bind and stabilize other proteins, including enzymes, that might otherwise stop functioning if they were to become denatured by rising temperature. Plant cells may rapidly synthesize HSPs in response to a sudden temperature rise. For example, experiments with cells and seedlings of soybean *(Glycine max)* showed that when the temperature rose 10°–15°C, in less than five minutes mRNA transcripts coding for as many as 50 different HSPs were present in cells. When the temperature returns to a normal range, HSPs release bound proteins, which can then resume their usual functions. Further studies have revealed that heat-shock proteins help protect plant cells subjected to other environmental stresses as well, including drought, salinity, and cold.

Like extreme heat, freezing can also be lethal to plants. If ice crystals form in cells they can literally tear the cell apart. In many cold-resistant species, dormancy (discussed in Section 35.4) is the long-term strategy for dealing with cold, but in the short term, such as an unseasonable cold snap, some species also undergo a rapid shift in gene expression that equips cold-stressed cells with so-called antifreeze proteins. Like heat-shock proteins, these molecules are thought to help maintain the structural integrity of other cell proteins.

STUDY BREAK

1. Which plant chemical defenses are general responses to attack, and which are specific to a particular pathogen?
2. Why is salicylic acid considered to be a general systemic response to damage?
3. How is the hypersensitive response integrated with other chemical defenses?

35.3 Plant Responses to the Environment: Movements

Although a plant cannot move from place to place as external conditions change, plants do alter the orientation of their body parts in response to environmental stimuli. As noted earlier in the chapter, growth toward or away from a unidirectional stimulus, such as light or gravity, is called a tropism. Tropic movement involves permanent changes in the plant body because cells in particular areas or organs grow differentially in response to the stimulus. Plant physiologists do not fully understand how tropisms occur, but they are fascinating examples of the complex abilities of plants to adjust to their environment. This section will also touch upon two other kinds of movements—developmental responses to physical contact, and changes in the position of plant parts that are not related to the location of the stimulus.

Phototropisms Are Responses to Light

Light is a key environmental stimulus for many kinds of organisms. Phototropisms, which we have already discussed in the section on auxins, are growth responses to a directional light source. As the Darwins discovered, if light is more intense on one side of a stem, the stem may curve toward the light **(Figure 35.19a)**. Phototropic movements are extremely adaptive for photosynthesizing organisms because they help maximize the exposure of photosynthetic tissues to sunlight.

How do auxins influence phototropic movements? In a coleoptile that is illuminated from one side, IAA moves by polar transport into the cells on the shaded

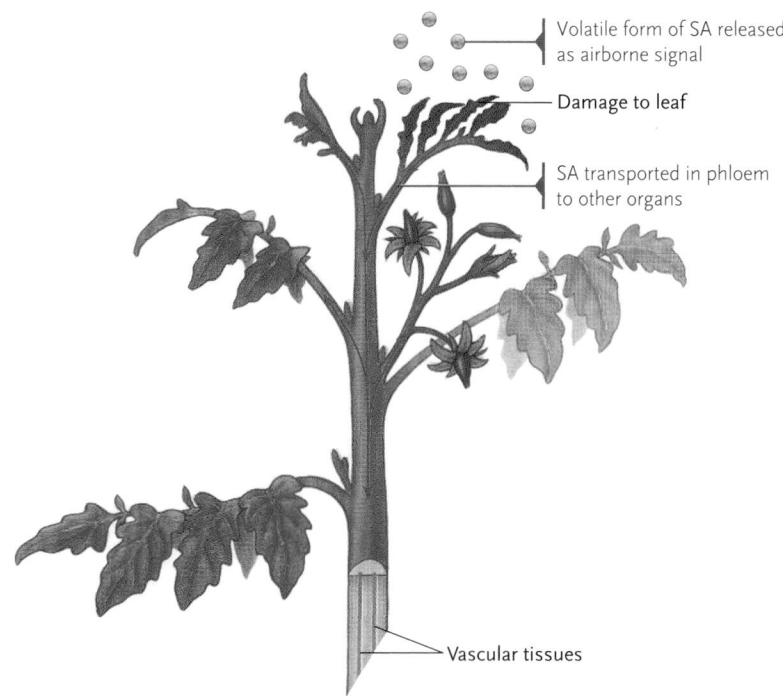

Volatile form of SA released as airborne signal

Damage to leaf

SA transported in phloem to other organs

Vascular tissues

Figure 35.18

A proposed mechanism for systemic acquired resistance. When a plant successfully fends off a pathogen, the defensive chemical salicylic acid (SA) is transported in the phloem to other plant parts, where it may help protect against another attack by stimulating the synthesis of PR proteins. In addition, the plant synthesizes and releases a slightly different, more volatile form of SA. This chemical may serve as an airborne signal to other parts of the plant as well as to neighboring plants.

side **(Figure 35.19b–d).** Phototropic bending occurs because cells on the shaded side elongate more rapidly than do cells on the illuminated side.

The main stimulus for phototropism is light of blue wavelengths. Experiments on corn coleoptiles have shown that a large, yellow pigment molecule

Cathlyn Melloan/Stone/Getty Images

a. Seedlings bend toward light.

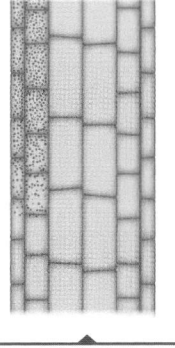

b. Rays from the sun strike one side of a shoot tip.

c. Auxin (red) diffuses down from the shoot tip to cells on its shaded side.

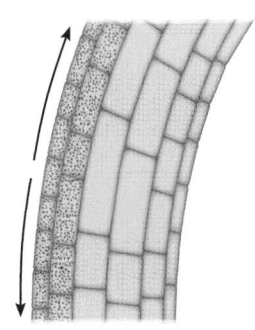

d. The auxin-stimulated cells elongate more quickly, causing the seedling to bend.

Figure 35.19

Phototropism in seedlings. (a) Tomato seedlings grown in darkness; their right side was illuminated for a few hours before they were photographed. **(b–d)** Hormone-mediated differences in the rates of cell elongation bring about the bending toward light. (Auxin is shown in red.)

a. Root oriented vertically

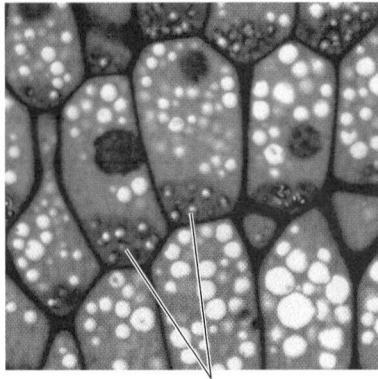

Statoliths

b. Root oriented horizontally

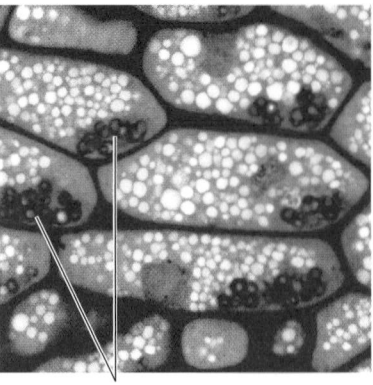

Statoliths

Figure 35.20

Evidence that supports the statolith hypothesis. When a corn root was laid on its side, amyloplasts—statoliths—in cells from the root cap settled to the bottom of the cells within 5 to 10 minutes. Statoliths may be part of a gravity-sensing mechanism that redistributes auxin through a root tip.

(Micrographs courtesy of Randy Moore, from "How Roots Respond to Gravity," M. L. Evans, R. Moore, and K. Hasenstein, *Scientific American*, December 1986.)

called phototropin can absorb blue wavelengths, and it may play a role in stimulating the initial lateral transport of IAA to the dark side of a shoot tip. Studies with *Arabidopsis* suggest there is more than one blue light receptor, however. One is a light-absorbing protein called **cryptochrome**, which is sensitive to blue light and may also be an important early step in the various light-based growth responses. As you will read later, cryptochrome appears to have a role in other plant responses to light as well.

Gravitropism Orients Plant Parts to the Pull of Gravity

Plants show growth responses to Earth's gravitational pull, a phenomenon called **gravitropism**. After a seed germinates, the primary root curves down, toward the "pull" (positive gravitropism), and the shoot curves up (negative gravitropism).

Several hypotheses seek to explain how plants respond to gravity. The most widely accepted hypothesis

proposes that plants detect gravity much as animals do—that is, particles called **statoliths** in certain cells move in the direction gravity pulls them. In the semicircular canals of human ears, tiny calcium carbonate crystals serve as statoliths; in most plants the statoliths are amyloplasts, modified plastids that contain starch grains (see Chapter 5). In eudicot angiosperm stems, amyloplasts often are present in one or two layers of cells just outside the vascular bundles. In monocots such as cereal grasses, amyloplasts are located in a region of tissue near the base of the leaf sheath. In roots, amyloplasts occur in the root cap. If the spatial orientation of a plant cell is shifted experimentally, its amyloplasts sink through the cytoplasm until they come to rest at the bottom of the cell **(Figure 35.20)**.

How do amyloplast movements translate into an altered growth response? The full explanation appears to be fairly complex, and there is evidence that somewhat different mechanisms operate in stems and in roots. In stems, the sinking of amyloplasts may provide a mechanical stimulus that triggers a gene-guided redistribution of IAA. **Figure 35.21** shows what happens when a potted sunflower seedling is turned on its side in a dark room. Within 15–20 minutes, cell elongation decreases markedly on the upper side of the growing horizontal stem, but increases on the lower side. With the adjusted growth pattern, the stem curves upward, even in the absence of light. Using different types of tests, researchers have been able to document the shifting of IAA from the top to the bottom side of the stem. The changing auxin gradient correlates with the altered pattern of cell elongation.

In roots, a high concentration of auxin has the opposite effect—it inhibits cell elongation. If a root is placed on its side, amyloplasts in the root cap accumulate near the side wall that now is the bottom side of the cap. In some way this stimulates cell elongation in the opposite wall, and within a few hours the root once again curves downward. In root tips of many plants, however, especially eudicots, researchers have not been able to detect a shift in IAA concentration that correlates with the changing position of amyloplasts. One hypothesis is that IAA is redistributed over extremely short distances in root cells, and therefore is difficult to measure. Root cells are much more sensitive to IAA than are cells in stem tissue, and even a tiny shift in IAA distribution could significantly affect their growth.

Along with IAA, calcium ions (Ca^{2+}) appear to play a major role in gravitropism. For example, if Ca^{2+} is added to an otherwise untreated agar block that is then placed on one side of a root cap, the root will bend toward the block. In this way, experimenters have been able to manipulate the direction of growth so that the elongating root forms a loop. Similarly, if an actively bending root is deprived of Ca^{2+}, the gravitropic response abruptly stops. By contrast, the negative gravitropic response of a shoot tip is inhibited when the tissue is exposed to excess calcium.

Figure 35.21

Gravitropism in a young shoot. A newly emerged sunflower seedling was grown in the dark for 5 days. Then it was turned on its side and marked at 0.5 cm intervals. Negative gravitropism turned the stem upright in 2 hours.

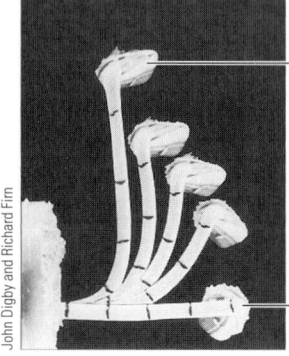

Position 2 hours later

Position 30 minutes after turn

Just how Ca^{2+} interacts with IAA in gravitropic responses is unknown. One hypothesis posits that calcium functions as an activator. Calcium binds to a small protein called *calmodulin,* activating it in the process. Activated calmodulin in turn can activate a variety of key cell enzymes in many organisms, both plants and animals. One possibility is that calcium-activated calmodulin stimulates cell membrane pumps that enhance the flow of both IAA and calcium through a gravity-stimulated plant tissue.

Some of the most active research in plant biology focuses on the intricate mechanisms of gravitropism. For example, there is increasing evidence that in many plants, cells in different regions of stem tissue are more or less sensitive to IAA, and that gravitropism is linked in some fundamental way to these differences in auxin sensitivity. In a few plants, including some cultivated varieties of corn and radish, the direction of the gravitropic response by a seedling's primary root is influenced by light. Clearly there is much more to be learned.

Thigmotropism and Thigmomorphogenesis Are Responses to Physical Contact

Varieties of peas, grapes, and some other plants demonstrate **thigmotropism** (*thigma* = touch), which is growth in response to contact with a solid object. Thigmotropic plants typically have long, slender stems and cannot grow upright without physical support. They often have *tendrils,* modified stems or leaves that can rapidly curl around a fencepost or the sturdier stem of a neighboring plant. If one side of a grape vine stem grows against a trellis, for example, specialized epidermal cells on that side of the stem tendril shorten while cells on the other side of the tendril rapidly elongate. Within minutes the tendril starts to curl around the trellis, forming tight coils that provide strong support for the vine stem. **Figure 35.22** shows thigmotropic twisting in the passionflower *(Passiflora).* Auxin and ethylene may be involved in thigmotropism, but most details of the mechanism remain elusive.

The rubbing and bending of plant stems caused by frequent strong winds, rainstorms, grazing animals, and even farm machinery can inhibit the overall growth of plants and can alter their growth patterns. In this phenomenon, called **thigmomorphogenesis,** a stem stops elongating and instead adds girth when it is regularly subjected to mechanical stress. Merely shaking some plants daily for a brief period will inhibit their upward growth **(Figure 35.23).** But although such plants may be shorter, their thickened stems will be stronger. Thigmomorphogenesis helps explain why plants growing outdoors are often shorter, have somewhat thicker stems, and are not as easily blown over as plants of the same species grown indoors. Trees growing near the snowline of windswept mountains

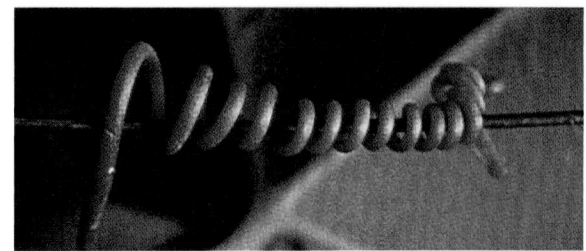

Figure 35.22
Thigmotropism in a passionflower *(Passiflora)* tendril, which is twisted around a support.

show an altered growth pattern that reflects this response to wind stress.

Research on the cellular mechanisms of thigmomorphogenesis has begun to yield tantalizing clues. In one study, investigators repeatedly sprayed *Arabidopsis* plants with water and imposed other mechanical stresses, then sampled tissues from the stressed plants. The samples contained as much as double the usual amount of mRNA for at least four genes, which had been activated by the stress. The mRNAs encoded calmodulin and several other proteins that may have roles in altering *Arabidopsis* growth responses. The test plants were also short, generally reaching only half the height of unstressed controls.

Nastic Movements Are Nondirectional

Tropisms are responses to directional stimuli, such as light striking one side of a shoot tip, but many plants also exhibit **nastic movements** (*nastos* = pressed close together)—reversible responses to nondirectional stimuli, such as mechanical pressure or humidity. We see nastic movements in leaves, leaflets, and even flowers. For instance, certain plants exhibit nastic sleep movements, holding their leaves (or flower petals) in roughly horizontal positions during the day but folding

a.   b.   c.

Cary Mitchell

Figure 35.23
Effect of mechanical stress on tomato plants *(Lycopersicon esculentum).* **(a)** This plant was the control; it was grown in a greenhouse, protected from wind and rain. **(b)** Each day for 28 days this plant was mechanically shaken for 30 seconds at 280 rpm. **(c)** This plant received the same shaking treatment, but twice a day for 28 days.

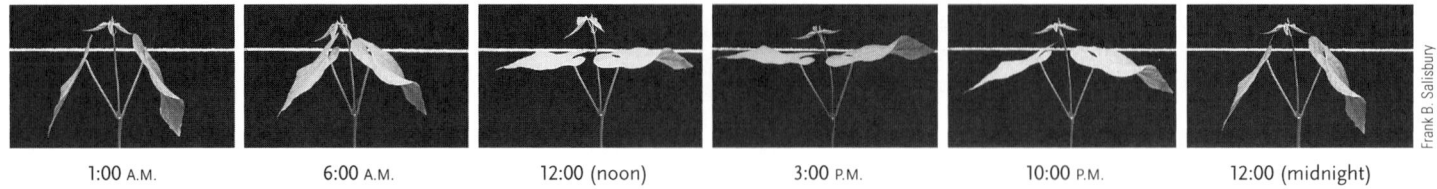

| 1:00 A.M. | 6:00 A.M. | 12:00 (noon) | 3:00 P.M. | 10:00 P.M. | 12:00 (midnight) |

Frank B. Salisbury

Figure 35.24

Nastic sleep movements in leaves of a bean plant. Although this plant was kept in constant darkness for 23 hours, its sleep movements continued independently of sunrise (6 A.M.) and sunset (6 P.M.). Folding the leaves closer to the stem may prevent phytochrome from being activated by bright moonlight, which could interrupt the dark period necessary to trigger flowering. Or perhaps it helps slow heat loss from leaves otherwise exposed to the cold night air.

them closer to the stem at night **(Figure 35.24).** Tulip flowers "go to sleep" in this way.

Many nastic movements are temporary and result from changes in cell turgor. For example, the daily opening and closing of stomata in response to changing light levels are nastic movements, as is the traplike closing of the lobed leaves of the Venus flytrap when an insect brushes against hairlike sensory structures on the leaves. The leaves of *Mimosa pudica,* the sensitive plant, also close in a nastic response to mechanical pressure. Each *Mimosa* leaf is divided into pairs of leaflets **(Figure 35.25a).** Touching even one leaflet at the leaf tip triggers a chain reaction in which each pair of leaflets closes up within seconds **(Figure 35.25b).**

In many turgor-driven nastic movements, water moves into and out of the cells in **pulvini** (*pulvinus =* cushion), thickened pads of tissue at the base of a leaf or petiole. Stomatal movements depend on changing concentrations of ions within guard cells, and pulvinar cells drive nastic leaf movements in *Mimosa* and numerous other plants by the same mechanism **(Figure 35.25c).**

How is the original stimulus transferred from cells in one part of a leaf to cells elsewhere? The answer lies in the polarity of charge across cell plasma membranes (see Chapter 6). Touching a *Mimosa* leaflet triggers an **action potential**—a brief reversal in the polarity of the membrane charge. When an action potential occurs at the plasma membrane of a pulvinar cell, the change in polarity causes potassium ion (K^+) channels to open, and ions flow out of the cell, setting up an osmotic gradient that draws water out as well. As water leaves by osmosis, turgor pressure falls, pulvinar cells become flaccid, and the leaflets move together. Later, when the process is reversed, the pulvinar cells regain turgor and the leaflets spread apart. Action potentials travel between parenchyma cells in the pulvini via plasmodesmata at the rate of about 2 cm/sec. Animal nerves conduct similar changes in membrane polarity along their plasma membranes (see Chapter 37). These changes in polarity, which are also called action potentials, occur much more rapidly—at velocities between 1 and 100 m/sec.

Stimuli other than touch also can trigger action potentials leading to nastic movements. Cotton, soybean, sunflower, and some other plants display *solar tracking,* nastic movements in which leaf blades are oriented toward the east in the morning, then steadily change their position during the day, following the sun across the sky.

a.

b.

David Sieren/Visuals Unlimited

c.

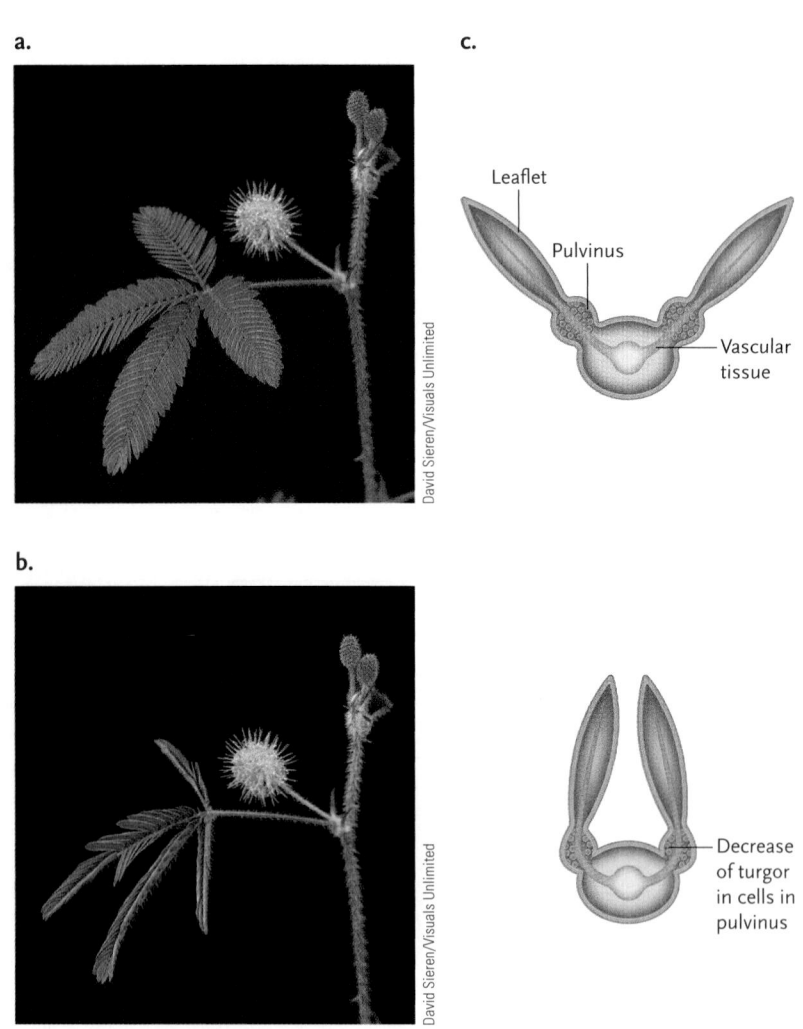

Leaflet

Pulvinus

Vascular tissue

Decrease of turgor in cells in pulvinus

Figure 35.25

Nastic movements in leaflets of *Mimosa pudica,* the sensitive plant. (a) In an undisturbed plant the leaflets are open. If a leaflet near the leaf tip is touched, changes in turgor pressure in pulvini at the base cause the leaf to fold closed. **(b, c).** The diagram sketches this folding movement in cross section. Other leaflets close in sequence as action potentials transmit the stimulus along the leaf.

Such movements maximize the amount of time that leaf blades are perpendicular to the sun, which is the angle at which photosynthesis is most efficient.

35.4 Plant Responses to the Environment: Biological Clocks

Like all eukaryotic organisms, plants have internal time-measuring mechanisms called **biological clocks** that adapt the organism to recurring environmental changes. In plants biological clocks help adjust both daily and seasonal activities.

Circadian Rhythms Are Based on 24-Hour Cycles

Some plant activities occur regularly in cycles of about 24 hours, even when environmental conditions remain constant. These are **circadian rhythms** (*circa* = around, *dies* = day). In Chapter 32, we noted that stomata open and close on a daily cycle, even where plants are kept in total darkness. Nastic sleep movements, described earlier, are another example of a circadian rhythm. Even when such a plant is kept in constant light or darkness for a few days, it folds its leaves into the "sleep" position at roughly 24-hour intervals. In some way, the plant measures time without sunrise (light) and sunset (darkness). Such experiments demonstrate that internal controls, rather than external cues, largely govern circadian rhythms.

Circadian rhythms and other activities regulated by a biological clock help ensure that plants of a single species do the same thing, such as flowering, at the same time. For instance, flowers of the aptly named four-o'clock plant *(Mirabilis jalapa)* open predictably every 24 hours—in nature, in the late afternoon. Such coordination can be crucial for successful pollination. Although some circadian rhythms can proceed without direct stimulus from light, many biological clock mechanisms are influenced by the relative lengths of day and night.

Photoperiodism Involves Seasonal Changes in the Relative Length of Night and Day

Obviously, environmental conditions in a 24-hour period are not the same in summer as they are in winter. In North America, for instance, winter temperatures are cooler and winter day length is shorter. Experimenting with tobacco and soybean plants in the early 1900s, two American botanists, Wightman Garner and Henry Allard, elucidated a phenomenon they called **photoperiodism**, in which plants respond to changes in the relative lengths of light and dark periods in their environment during each 24-hour period. Through photoperiodism, the biological clocks of plants (and animals) make seasonal adjustments in their patterns of growth, development, and reproduction.

In plants, we now know that a blue-green pigment called **phytochrome** often serves as a switching mechanism in the photoperiodic response, signaling the plant to make seasonal changes. Plants synthesize phytochrome in an inactive form, P_r, which absorbs light of red wavelengths. Sunlight contains relatively more red light than far-red light. During daylight hours when red wavelengths dominate, P_r absorbs red light. Absorption of red light triggers the conversion of phytochrome to an active form designated P_{fr}, which absorbs light of far-red wavelengths. At sunset, at night, or even in shade, where far-red wavelengths predominate, P_{fr} reverts to P_r **(Figure 35.26)**.

In nature a high concentration of P_{fr} "tells" a plant that it is exposed to sunlight, an adaptation that is vital given that over time sunlight provides favorable conditions for leaf growth, photosynthesis, and flowering. The exact mechanism of this crucial transfer of environmental information still is not fully understood. Phytochrome activation may stimulate plant cells to take up Ca^{2+} ions, or it may induce certain plant organelles to release them. Either way, when free calcium ions combine with calcium-binding proteins (such as calmodulin), they may initiate at least some responses to light. Botanists suspect that P_{fr} controls the types of

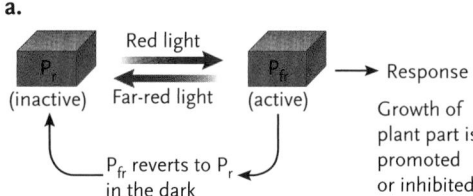

a.

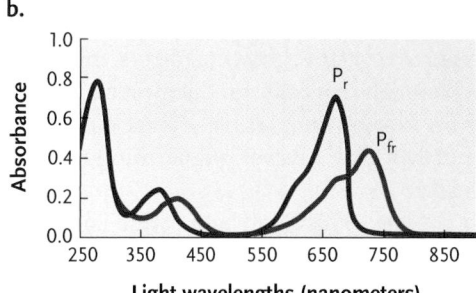

b.

Figure 35.26

The phytochrome switching mechanism, which can promote or inhibit growth of different plant parts. **(a)** Interconversion of phytochrome from the active form (P_{fr}) to the inactive form (P_r). **(b)** The absorption spectra associated with the interconversion of P_r and P_{fr}.

Figure 35.27
Effects of the absence of light on young bean plants *(Phaseolus)*. The two plants at the right, the control group, were grown in a greenhouse. The other two were grown in darkness for 8 days. Note that the dark-grown plants are yellow; they could form carotenoid pigments but not chlorophyll in darkness. They have longer stems, smaller leaves, and smaller root systems than the controls.

enzymes being produced in particular cells—and different enzymes are required for seed germination, stem elongation and branching, leaf expansion, and the formation of flowers, fruits, and seeds. When plants adapted to full sunlight are grown in darkness, they put more resources into stem elongation and less into leaf expansion or stem branching **(Figure 35.27)**.

Cryptochrome—which, recall, is sensitive to blue light and appears to influence light-related growth responses—also interacts with phytochromes in producing circadian responses. Researchers have recently discovered that cryptochrome occurs not only in plants but also in animals such as fruit flies and mice. Does it act as a circadian photoreceptor in both kingdoms? Only further study will provide the answer.

Cycles of Light and Dark Often Influence Flowering

Photoperiodism is especially apparent in the flowering process. Like other plant responses, flowering is often keyed to changes in day length through the year and to the resulting changes in environmental conditions. Corn, soybeans, peas, and other annual plants begin flowering after only a few months of growth. Roses and other perennials typically flower every year or after several years of vegetative growth. Carrots, cabbages, and other biennials typically produce roots, stems, and leaves the first growing season, die back to soil level in autumn, then grow a new flower-forming stem the second season.

In the late 1930s Karl Hamner and James Bonner grew cocklebur plants *(Xanthium strumarium)* in chambers in which the researchers could carefully control environmental conditions, including photoperiod. And they made an unexpected discovery: Flowering occurred only when the test plants were exposed to at least a single night of 8.5 hours of uninterrupted darkness. The length of the "day" in the growth chamber did not matter, but if light interrupted the dark period for even a minute or two, the plant would not flower at all. Subsequent research confirmed that for most angiosperms, it is the length of darkness, not light, that controls flowering.

Kinds of Flowering Responses. The photoperiodic responses of flowering plants are so predictable that botanists have long used them to categorize plants **(Figure 35.28)**. The categories, which refer to day length, reflect the fact that scientists recognized the phenomenon of photoperiodic flowering responses long before they understood that darkness, not light, was the cue. **Long-day plants,** such as irises, daffodils, and corn, usually flower in spring when dark periods become shorter and day length becomes longer than some critical value—usually 9–16 hours. **Short-day plants,** including cockleburs, chrysanthemums, and potatoes, flower in late summer or early autumn when dark periods become longer and day length becomes shorter than some critical value. **Intermediate-day plants,** such as sugarcane, flower only when day length falls in between the values for long-day and short-day plants. **Day-neutral plants,** such as dandelions and roses, flower whenever they become mature enough to do so, without regard to photoperiod.

Experiments demonstrate what happens when plants are grown under the "wrong" photoperiod regimes. For instance, spinach, a long-day plant, flowers and produces seeds only if it is exposed to no more than 10 hours of darkness each day for two weeks (see Figure 35.28). **Figure 35.29** illustrates the results of an experiment to test the responses of short-day and long-day plants to night length. In this experiment, bearded iris plants *(Iris* species), which are long-day plants, and chrysanthemums, which are short-day plants, were exposed to a range of light conditions. In each case, when the researchers interrupted a critical dark period

— Flowers

Figure 35.28
Effect of day length on spinach *(Spinacia oleracea)*, a long-day plant.

with a pulse of red light, the light reset the plants' clocks. The experiment provided clear evidence that short-day plants flower only when nights are longer than a critical value—and long-day plants flower only when nights are shorter than a critical value.

Chemical Signals for Flowering.

When photoperiod conditions are right, what sort of chemical message stimulates a plant to develop flowers? In the 1930s botanists began postulating the existence of "florigen," a hypothetical hormone that served as the flowering signal. In a somewhat frustrating scientific quest, researchers spent the rest of the twentieth century seeking this substance in vain. Recently, however, molecular studies using *Arabidopsis* plants have defined a sequence of steps that may collectively provide the internal stimulus for flowering. Here again, we see one of the recurring themes in plant development—major developmental changes guided by several interacting genes.

Figure 35.30 traces the steps of the proposed flowering signal. To begin with, a gene called *CONSTANS* is expressed in a plant's leaves in tune with the daily light/dark cycle, with expression peaking at dusk (step 1). The gene encodes a regulatory protein called CO (not to be confused with carbon monoxide). As days lengthen in spring, the concentration of CO rises in leaves, and as a result a second gene is activated (step 2). The product of this gene, a regulatory protein called FT, travels in

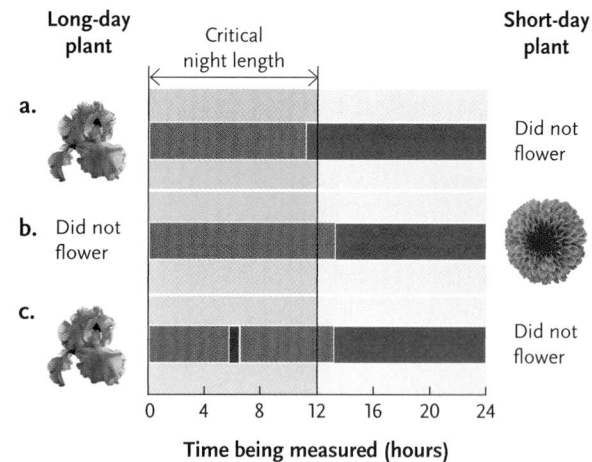

Figure 35.29

Experiments showing that short-day and long-day plants flower by measuring night length. Each horizontal bar signifies 24 hours. Blue bars represent night, and yellow bars day. **(a)** Long-day plants such as bearded irises flower when the night is shorter than a critical length, while **(b)** short-day plants such as chrysanthemums flower when the night is longer than a critical value. **(c)** When an intense red flash interrupts a long night, both kinds of plants respond as if it were a short night; the irises flowered but the chrysanthemums did not.

(Long-day plant photos: Clay Perry/Corbis; short-day plant photo: Eric Chrichton/Corbis.)

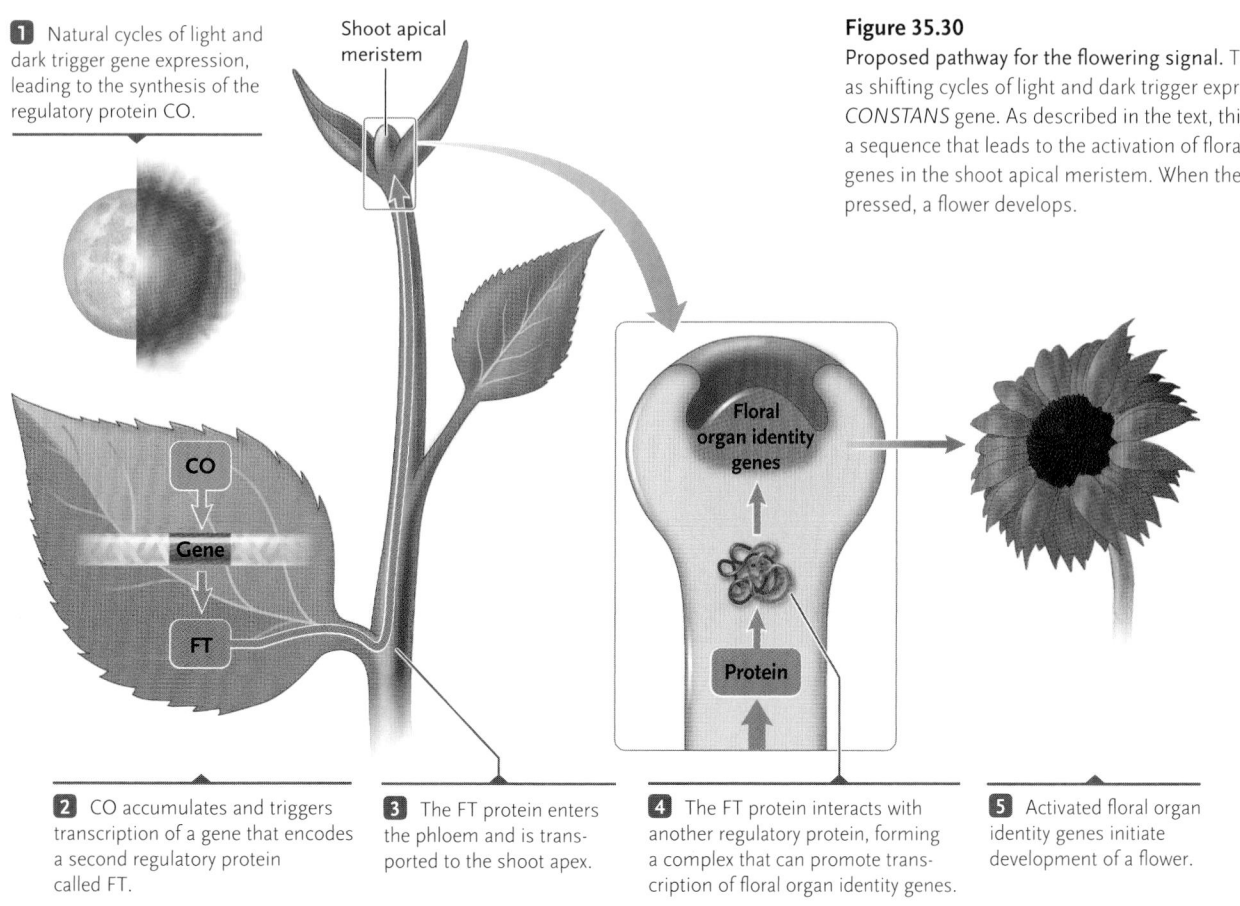

1 Natural cycles of light and dark trigger gene expression, leading to the synthesis of the regulatory protein CO.

Shoot apical meristem

Figure 35.30

Proposed pathway for the flowering signal. The pathway starts as shifting cycles of light and dark trigger expression of the *CONSTANS* gene. As described in the text, this step is the first in a sequence that leads to the activation of floral organ identity genes in the shoot apical meristem. When these genes are expressed, a flower develops.

CO

Gene

FT

Floral organ identity genes

Protein

2 CO accumulates and triggers transcription of a gene that encodes a second regulatory protein called FT.

3 The FT protein enters the phloem and is transported to the shoot apex.

4 The FT protein interacts with another regulatory protein, forming a complex that can promote transcription of floral organ identity genes.

5 Activated floral organ identity genes initiate development of a flower.

the phloem to shoot tips (step 3). Once there, the mRNA is translated into a second regulatory protein (step 4) that in some way interacts with yet a third regulatory protein that is synthesized only in shoot apical meristems (step 5). The encounter apparently sparks the development of a flower (step 6) by promoting the expression of floral organ identity genes in the meristem tissue (see Section 34.5). Key experiments that uncovered this pathway all relied on analysis of DNA microarrays, a technique introduced in Chapter 18 and featured in this chapter's *Focus on Research*.

Vernalization and Flowering. Flowering is more than a response to changing night length. Temperatures also change with the seasons in most parts of the world, and they too influence flowering. For instance, unless buds of some biennials and perennials are exposed to low winter temperatures, flowers do not form on stems in spring. Low-temperature stimulation of flowering is called **vernalization** ("making springlike").

In 1915 the plant physiologist Gustav Gassner demonstrated that it was possible to influence the flowering of cereal plants by controlling the temperature of seeds while they were germinating. In one case, he maintained germinating seeds of winter rye *(Secale cereale)* at just above freezing (1°C) before planting them. In nature, winter rye seeds in soil germinate during the winter, giving rise to a plant that flowers months later, in summer. Plants grown from Gassner's test seeds, however, flowered the same summer even when the seeds were planted in the late spring. Home gardeners can induce flowering of daffodils and tulips by putting the bulbs (technically, *corms*) in a freezer for several weeks before early spring planting. Commercial growers use vernalization to induce millions of plants, such as Easter lilies, to flower just in time for seasonal sales.

Dormancy Is an Adaptation to Seasonal Changes or Stress

As autumn approaches and days grow shorter, growth slows or stops in many plants even if temperatures are still moderate, the sky is bright, and water is plentiful. When a perennial or biennial plant stops growing under conditions that seem (to us) quite suitable for growth, it has entered a state of **dormancy**. Ordinarily, its buds will not resume growth until early spring.

Short days and long nights—conditions typical of winter—are strong cues for dormancy. In one experiment, in which a short period of red light interrupted the long dark period for Douglas firs, the plants responded as if nights were shorter and days were longer; they continued to grow taller **(Figure 35.31)**. Conversion of P_r to P_{fr} by red light during the dark period prevented dormancy. In nature, buds may enter dormancy because less P_{fr} can form when day length shortens in late summer. Other environmental cues are at work also. Cold nights, dry soil, and a deficiency of nitrogen apparently also promote dormancy.

The requirement for multiple dormancy cues has adaptive value. For example, if temperature were the only cue, plants might flower and seeds might germinate in warm autumn weather—only to be killed by winter frost.

A dormancy-breaking process is at work between fall and spring. Depending on the species, breaking

Figure 35.31
Effect of the relative length of day and night on the growth of Douglas firs *(Pseudotsuga menziesii)*. The young tree at the left was exposed to alternating periods of 12 hours of light followed by 12 hours of darkness for a year; its buds became dormant because day length was too short. The tree at the right was exposed to a cycle of 20 hours of light and 4 hours of darkness; its buds remained active and growth continued. The middle plant was exposed each day to 12 hours of light and 11 hours of darkness, with a 1-hour light in the middle of the dark period. This light interruption of an otherwise long dark period also prevented buds from going dormant.

Potted plant grown inside a greenhouse did not flower.   Branch exposed to cold outside air flowered.

Figure 35.32
Effect of cold temperature on dormant buds of a lilac *(Syringa vulgaris)*. In this experiment, a plant was grown in winter inside a warm greenhouse with one branch growing out of a hole. Only the buds on the branch exposed to low outside temperatures resumed growth in spring. This experiment suggests that low-temperature effects are localized.

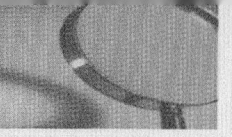

Research Methods: Using DNA Microarray Analysis to Track Down "Florigen"

The more plant scientists learn about plant genomes, the more they are relying on DNA microarray assays to elucidate the activity of plant genes.

Recall from Section 18.3 that a DNA microarray, also called a DNA chip, allows an investigator to explore questions such as how the expression of a particular gene differs in different types of cells. To quantify the expression of specific genes in particular types of cells, mRNA transcripts are isolated from the cells; then a cDNA library is created from each mRNA sample, using nucleotides labeled with fluorescent dyes. Probes (nucleotide sequences) representing every gene in the organism's genome are fixed onto a slide; when the labeled cDNAs are added to the slide, each will hybridize to the gene that expressed the mRNA from which it was made. Next, the DNA microarray is scanned with a laser that can detect fluorescence. When a gene is expressed in a cell, the dye fluoresces and gives a color that accords with the degree of its expression. The procedure can be manipulated to reveal the relative amounts of expression of more than one of a cell's genes.

Philip A. Wigge and his colleagues used this method to learn more about the signaling pathway that causes a plant's apical meristem to give rise to flowers. Previous research had established that in leaves, lengthening spring days coincided with rising concentrations of CO, a regulatory protein

encoded by the *CONSTANS* gene. But what did CO regulate? Working with *Arabidopsis thaliana*, Wigge's group was able to narrow down the field to four genes, and using microarray analysis of DNA from leaf cells they pinpointed one called FT (for flowering locus T). The researchers found that in leaves, CO causes strong expression of FT: When enough CO is present, FT mRNA is rapidly transcribed, then enters the phloem. (The transport of mRNA in phloem is not unusual.) By contrast, when they tested CO's effects in shoot apex cells, they found that it triggers far less gene expression there. Clearly, CO was not directly triggering the development of flowers. However, FT mRNA moves in the phloem to the shoot apex, where it is translated into protein. Was that protein the direct flowering signal? Other studies had implicated a regulatory protein called FD, which microarray analysis had shown was expressed *only*—but very strongly—in the shoot apex.

To sort out this final piece of the puzzle, the Wigge team examined flowering responses in normal *A. thaliana* plants as well as in mutants having a normal FT protein but a defective *fd*, and vice versa. Flowering was abnormal in both types of mutants, possibly because the mutated "partner" suppressed some aspect of the functioning of the normal protein. On the other hand, in wild-type plants, which had a functioning FD protein, expression of FT triggered

a marked increased in the expression of the floral organ gene *APETALA1* **(Figure a).** These results have two major implications. First, they support the hypothesis that FT and FD interact in a normal flowering response. Second, the study suggests that FT, the CO-induced signal from leaves, conveys the environmental signal that it is time for a plant to flower. In that sense, FT may be the long sought "florigen." However, only by interacting with FD does FT "know" where to deliver its flowering signal—in the apical meristems of shoots.

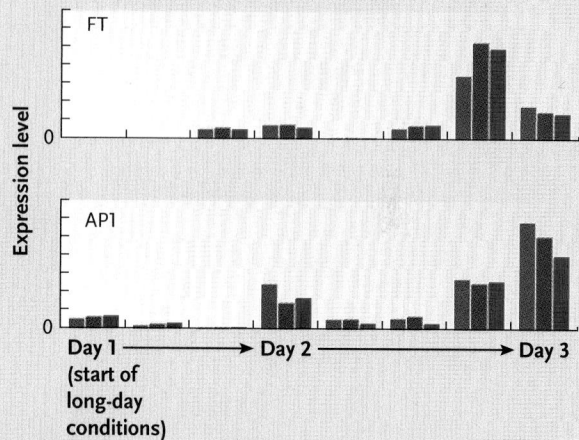

Figure a

Effect of the FT protein on expression of the *APETALA1 (AP1)* floral organ identity gene. In nature, *Arabidopsis thaliana* is a long-day plant, and the experiment was carried out under long-day (that is, short-night) conditions. Three groups of replicates shown here in yellow, orange, and red respectively, were monitored for both AP1 and FT. After a brief delay, the expression of AP1 closely tracked the appearance of the FT regulatory protein, which had been activated by its interaction with the FD protein.

---

dormancy probably involves gibberellins and abscisic acid, and it requires exposure to low winter temperatures for specific periods **(Figure 35.32).** The temperature needed to break dormancy varies greatly among species. For example, the Delicious variety of apples grown in Utah requires 1230 hours near 43°F (6°C); apricots grown there require only 720 hours at that temperature. Generally, trees growing in the southern United States or in Italy require less cold exposure than those growing in Canada or in Sweden.

STUDY BREAK

1. Summarize the switching mechanism that operates in plant responses to changes in photoperiod.
2. Give some examples of how relative lengths of dark and light can influence flowering.
3. Explain why dormancy is an adaptive response to a plant's environment.

35.5 Signal Responses at the Cellular Level

Environmental stimuli such as changing light, temperature, or chemicals on the surface of an attacking pathogen are cues that signal a plant to alter its growth or physiology. For decades plant physiologists have looked avidly for clues about how those signals are converted into a chemical message that produces a change in cell metabolism or growth. As Chapter 7 describes, research on the ways animal cells respond to external signals has revealed some basic mechanisms, and at least some of these mechanisms also apply to plant cells.

Figure 35.33
Signal response pathways in plant cells.

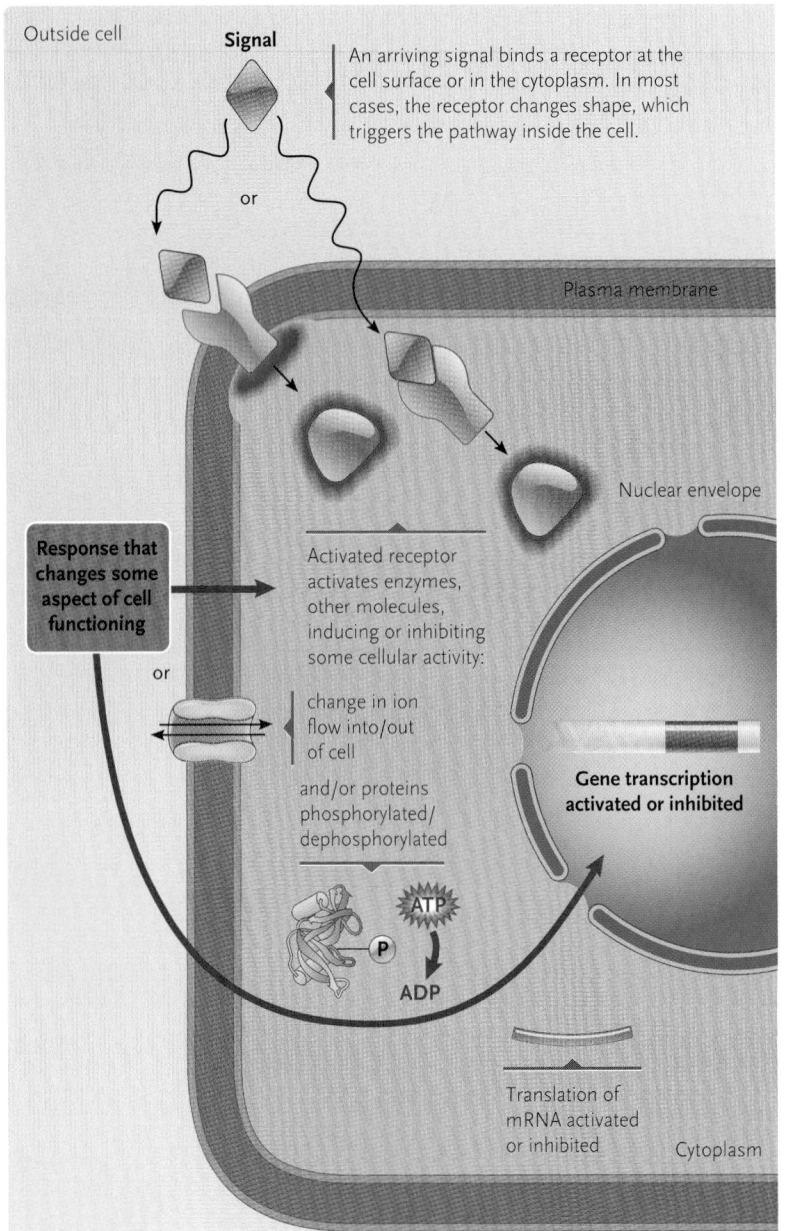

An arriving signal binds a receptor at the cell surface or in the cytoplasm. In most cases, the receptor changes shape, which triggers the pathway inside the cell.

Outside cell

Signal

or

Plasma membrane

Nuclear envelope

Response that changes some aspect of cell functioning

or

Activated receptor activates enzymes, other molecules, inducing or inhibiting some cellular activity:

change in ion flow into/out of cell

and/or proteins phosphorylated/ dephosphorylated

ATP

P

ADP

Gene transcription activated or inhibited

Translation of mRNA activated or inhibited

Cytoplasm

Several Signal Response Pathways Operate in Plants

Hormones and environmental stimuli alter the behavior of target cells, which have receptors to which specific signal molecules can bind and elicit a cellular response. By means of a response pathway, a signal can induce changes in the cell's shape and internal structure or influence the transport of ions and other substances into and out of cells. Some signals cause cells to alter gene activation and the rate of protein synthesis; others set in motion events that modify existing cell proteins. Section 7.1 presents some general features of the signaling process. Here, we'll briefly consider how signal molecules may operate in plants.

Certain hormones and growth factors bind to receptors at the target cell's plasma membrane, on its endoplasmic reticulum (ER), or in the cytoplasm. For example, ethylene receptors are on the ER, and the auxin receptor is a protein in the cytoplasm called TIR1. Research in several laboratories recently confirmed that auxin binds directly to TIR1, setting in motion events that break down protein factors that inhibit transcription. As a result, previously repressed genes are turned on. In many cases (although not with TIR1), binding causes the receptor to change shape. Regardless, binding of a hormone or growth factor triggers a complex pathway that leads to the cell response—the opening of ion channels, activation of transport proteins, or some other event.

Only some cells can respond to a particular signaling molecule, because not all cells have the same types of receptors. For example, particular cells in ripening fruits and developing seeds have ethylene receptors, but few if any cells in stems do. Different signals also may have different effects on a single cell, and may exert those effects by way of different response pathways. One type of signal might stimulate transcription, and another might inhibit it. In addition, as we've seen, some genes controlled by particular receptors encode proteins that regulate still *other* genes.

In plants we know the most about pathways involving auxin, ethylene, salicylic acid, and blue light. **Figure 35.33** diagrams a general model for these response pathways in plant cells. As the figure shows, the response may lead to a change in the cell's structure, its metabolic activity, or both, either directly or by altering the expression of one or more genes.

Second Messenger Systems Enhance the Plant Cell's Response to a Hormone's Signal

We can think of plant hormones and other signaling molecules as external first messengers that deliver the initial physiological signal to a target cell. Often, as with salicylic acid, binding of the signal molecule triggers the synthesis of internal second messengers (introduced in Section 7.4). These intermediary molecules diffuse rapidly through the cytoplasm and provide the main chemical signal that alters cell functioning.

Second messengers usually are synthesized in a sequence of chemical reactions that converts an external signal into internal cell activity. For many years the details of plant second-messenger systems were sketchy and hotly debated. More recently, however, reaction sequences that occur in the cells of animals and some fungi have also been found in plants. The following example describes reactions that close plant stomata in response to a signal from abscisic acid.

As discussed in Section 35.1, abscisic acid helps regulate several responses in plants, including the maturation of seeds and the closing of stomata (see Section 32.3). ABA's role in stomatal closure—triggered by water stress or some other environmental cue—begins when the hormone activates a receptor in the plant cell plasma membrane. Experiments have shown that this binding activates G proteins that in turn activate phospholipase C (see Figure 7.12). This enzyme stimulates the synthesis of second messengers, such as inositol triphosphate (IP_3).

The second messenger diffuses through the cytoplasm and binds with calcium channels in cell structures such as the endoplasmic reticulum, vacuole, and plasma membrane. The bound channels then open, releasing calcium ions that activate protein kinases in the cytoplasm. In turn, the activated enzymes activate their target proteins (by phosphorylating them). Each protein kinase can convert a large number of substrate molecules into activated enzymes, transport proteins, open ion channels, and so forth. Soon the number of molecules representing the final cellular response to the initial signal is enormous.

Recent experimental evidence indicates that, in similar fashion, auxin's hormonal signal is conveyed by cAMP (cyclic adenosine monophosphate), another major second messenger in cells of animals and other organisms.

In addition to the basic pathways described here, other routes may exist that are unique to plant cells.

UNANSWERED QUESTIONS

Do plants have a "backup" copy of their genome?

As described at the beginning of this chapter, land plants manifest adaptations that allow them to survive and reproduce in unfavorable or hostile conditions. Adaptations include changes in growth and development reflecting responses to environmental fluctuations that occur naturally during the normal life cycle of plants. Being physically anchored in one place has driven plant adaptation so that changes in the plant body can facilitate survival. It is also possible, however, that the sessile existence of land plants may have selected for unusual adaptive strategies. Might plants have devised a strategy to utilize previously unknown genetic resources and thereby expand their potential repertoire of adaptive responses? Recent findings demonstrating the existence of a previously unknown mechanism of genetic instability suggest that such a strategy may indeed have been in place during the evolution of land plants. These findings suggest that, at least in *Arabidopsis thaliana*, a "backup" copy of the genome exists that can be accessed under unfavorable conditions.

Why have a backup copy of the genome? Simply put, if the system crashes, it can be restored. By analogy, the genome could be considered the "operating system" stored on the "hard drive" of the organism. If that operating system becomes corrupted, for example by a devastating power surge or a computer virus, a global systems failure might occur. However, if a backup copy were maintained at least in a subset of the population, then, under conditions that might lead to extinction, the backup copy could be used to "restore" the system and increase the chances of survival for that organism or population. In other words, the genome could adapt using the stored information.

An intriguing possibility is that such a backup genome might exist in the form of RNA. The fact that backup copies have not been found using conventional DNA-based detection methods or classical genetic approaches leaves open the exciting possibility that RNA might serve as the storage medium for this information. In a sense, having the information stored in an alternative chemical form (analogous to a different computer language or code) might also make it less susceptible to corruption. Furthermore, it may be that this backup genome is a remnant of an ancestral condition where the genome was RNA-based.

How would you go about testing these different possibilities? First, the findings would have to be independently verified and the existence of a "restoration" mechanism would have to be confirmed by other research groups working on *Arabidopsis* or other plant species. Second, the source and chemical nature of the backup information would need to be identified. Where is it, and is it RNA, DNA, protein, or a combination of these? The question of mechanism would also need to be addressed. How is the system restored, and when does it happen? The question of how widespread this phenomenon is would also need to be considered. Do all plant species maintain a backup copy and do organisms outside of the plant kingdom have a backup genome?

Susan Lolle is associate professor of biology at the University of Waterloo in Waterloo, Canada. Her research interests include plant development, genetics and genome biology. To learn more go to http://www.biology.uwaterloo.ca.

Light is the driving force for photosynthesis, and it may not be farfetched to suppose that plants have evolved other unique light-related biochemical pathways as well. For instance, exciting experiments are extending our knowledge of how plant cells respond to blue light, which, as we have discussed, triggers some photoperiod responses such as the opening and closing of stomata. In all likelihood, much remains to be discovered about this and many other aspects of plant functioning.

STUDY BREAK

1. Summarize the various ways that chemical signals reaching plant cells are converted to changes in cell functioning.
2. What basic task does a second messenger accomplish?
3. Thinking back to Chapter 7, can you describe parallels between signal transduction mechanisms in the cells of plants and animals?

Review

Go to ThomsonNOW™ at www.thomsonedu.com/login to access quizzing, animations, exercises, articles, and personalized homework help.

35.1 Plant Hormones

- At least seven classes of hormones govern flowering plant development, including germination, growth, flowering, fruit set, and senescence (Table 35.1).
- Auxins, mainly IAA, promote elongation of cells in the coleoptile and stem (Figures 35.2–35.7).
- Gibberellins promote stem elongation and help seeds and buds break dormancy (Figures 35.1, 35.8, and 35.9).
- Cytokinins stimulate cell division, promote leaf expansion, and retard leaf aging (Figure 35.10).
- Ethylene promotes fruit ripening and abscission (Figures 35.11 and 35.12).
- Brassinosteroids stimulate cell division and elongation (Figure 35.13).
- Abscisic acid (ABA) promotes stomatal closure and may trigger seed and bud dormancy (Figure 35.14).
- Jasmonates regulate growth and have roles in defense.

Animation: Plant development

Animation: Auxin's effects

35.2 Plant Chemical Defenses

- Plants have diverse chemical defenses that limit damage from bacteria, fungi, worms, or plant-eating insects (Figure 35.15).
- The hypersensitive response isolates an infection site by surrounding it with dead cells (Figure 35.16). During the response salicylic acid (SA) induces the synthesis of PR (pathogenesis-related) proteins.
- Oligosaccharins can trigger the synthesis of phytoalexins, secondary metabolites that function as antibiotics.
- Gene-for-gene recognition enables a plant to chemically recognize a specific pathogen and mount a defense (Figure 35.17).
- Systemic acquired resistance provides long-term protection against some pathogens. Salicylic acid passes from the infected organ to newly forming organs such as leaves, which then synthesize PR proteins (Figure 35.18).
- Heat-shock proteins can reversibly bind enzymes and other proteins in plant cells and prevent them from denaturing when the plant is under heat stress.
- Some plants can synthesize "antifreeze" proteins that stabilize cell proteins when cells are threatened with freezing.

35.3 Plant Responses to the Environment: Movements

- Plants adjust their growth patterns in response to environmental rhythms and unique environmental circumstances. These responses include tropisms.
- Phototropisms are growth responses to a directional light source. Blue light is the main stimulus for phototropism (Figures 35.3 and 35.19).
- Gravitropism is a growth response to Earth's gravitational pull. Stems exhibit negative gravitropism, growing upward, while roots show positive gravitropism (Figures 35.20 and 35.21).
- Some plants or plant parts demonstrate thigmotropism, growth in response to contact with a solid object (Figure 35.22).
- Mechanical stress can cause thigmomorphogenesis, which causes the stem to add girth (Figure 35.31).
- In nastic leaf movements, water enters or exits cells of a pulvinus, a pad of tissue at the base of a leaf or petiole, in response to action potentials (Figures 35.24 and 35.25).

Animation: Gravitropism

Animation: Gravity and statolith distribution

Animation: Phototropism

35.4 Plant Responses to the Environment: Biological Clocks

- Plants have biological clocks, internal time-measuring mechanisms with a biochemical basis. Environmental cues can "reset" the clocks, enabling plants to make seasonal adjustments in growth, development, and reproduction.
- In photoperiodism, plants respond to a change in the relative length of daylight and darkness in a 24-hour period. A switching mechanism involving the pigment phytochrome promotes or inhibits germination, growth, and flowering and fruiting.
- Phytochrome is converted to an active form (P_{fr}) during daylight, when red wavelengths dominate. It reverts to an inactive form (P_r) at sunset, at night, or in shade, when far-red wavelengths predominate. P_{fr} may control the types of metabolic pathways that operate under specific light conditions (Figure 35.26).
- Long-day plants flower in spring or summer, when day length is long relative to night. Short-day plants flower when day length is relatively short, and intermediate-day plants flower when day length falls in between the values for long-day and short-day

plants. Flowering of day-neutral plants is not regulated by light. In vernalization, a period of low temperature stimulates flowering (Figures 35.27–35.29).

- The direct trigger for flowering may begin in leaves, when the regulatory protein CO triggers the expression of the FT gene. The resulting mRNA transcripts move in phloem to apical meristems where translation of the mRNAs yields a second regulatory protein, which in turn interacts with a third. This final interaction activates genes that encode the development of flower parts (Figure 35.30).
- Senescence is the sum of processes leading to the death of a plant or plant structure.
- Dormancy is a state in which a perennial or biennial stops growing even though conditions appear to be suitable for continued growth (Figures 35.31 and 35.32).

Animation: Phytochrome conversions

Animation: Vernalization

Animation: Day length and dormancy

Animation: Flowering response experiments

35.5 Signal Responses at the Cellular Level

- Hormones and environmental stimuli alter the behavior of target cells, which have receptors to which signal molecules can bind. By means of a response pathway that ultimately alters gene expression, a signal can induce changes in the cell's shape or internal structure or influence its metabolism or the transport of substances across the plasma membrane (Figure 35.33).
- Some plant hormones and growth factors may bind to receptors at the target cell's plasma membrane, changing the receptor's shape. This binding often triggers the release of internal second messengers that diffuse through the cytoplasm and provide the main chemical signal that alters gene expression.
- Second messengers usually act by way of a reaction sequence that amplifies the cell's response to a signal. An activated receptor activates a series of proteins, including G proteins and enzymes that stimulate the synthesis of second messengers (such as IP_3) that bind ion channels on endoplasmic reticulum.
- Binding releases calcium ions, which enter the cytoplasm and activate protein kinases, enzymes that activate specific proteins that produce the cell response.

Questions

Self-Test Questions

1. Which of the following plant hormones does *not* stimulate cell division?
   a. auxins
   b. cytokinins
   c. ethylene
   d. gibberellins
   e. abscisic acid

2. Which is the correct pairing of a plant hormone and its function?
   a. salicylic acid: triggers synthesis of general defense proteins
   b. brassinosteroids: promote responses to environmental stress
   c. cytokinins: stimulate stomata to close in water-stressed plants
   d. gibberellins: slow seed germination
   e. ethylene: promotes formation of lateral roots

3. A characteristic of auxin (IAA) transport is:
   a. IAA moves by polar transport from the base of a tissue to its apex.
   b. IAA moves laterally from a shaded to an illuminated side of a plant.
   c. IAA enters a plant cell in the form of IAAH, an uncharged molecule that can diffuse across cell membranes.
   d. IAA exits one cell and enters the next by means of transporter proteins clustered at both the apical and basal ends of the cells.
   e. All of the above are characteristics of auxin transport in different types of cells.

4. Hanging wire fruit baskets have many holes or open spaces. The major advantage of these spaces is that they:
   a. prevent gibberellins from causing bolting or the formation of rosettes on the fruit.
   b. allow the evaporation of ethylene and thus slow ripening of the fruit.
   c. allow oxygen in the air to stimulate the production of ethylene, which hastens the abscission of fruits.
   d. allow oxygen to stimulate brassinosteroids, which hasten the maturation of seeds in/on the fruits.
   e. allow carbon dioxide in the air to stimulate the production of cytokinins, which promotes mitosis in the fruit tissue and hastens ripening.

5. Which of the following is *not* an example of a plant chemical defense?
   a. ABA inhibits leaves from budding if conditions favor attacks by sap-sucking insects.
   b. Jasmonate activates plant genes encoding protease inhibitors that prevent insects from digesting plant proteins.
   c. Acting against fungal infections, the hypersensitive response allows plants to produce highly reactive oxygen compounds that kill selected tissue, thus forming a dead tissue barrier that walls off the infected area from healthy tissues.
   d. Chitinase, a PR hydrolytic protein produced by plants, breaks down chitin in the cell walls of fungi and thus halts the fungal infection.
   e. Attack by fungi or viruses triggers the release of oligosaccharins, which in turn stimulate the production of phytoalexins having antibiotic properties.

6. Which of the following statements about plant responses to the environment is true?
   a. The heat-shock response induces a sudden halt to cellular metabolism when an insect begins feeding on plant tissue.
   b. In gravitropism, amyloplasts sink to the bottom of cells in a plant stem, causing the redistribution of IAA.
   c. The curling of tendrils around a twig is an example of thigmotropism.
   d. Phototropism results when IAA moves first laterally, then downward in a shoot tip when one side of the tip is exposed to light.
   e. Nastic movements, such as the sudden closing of the leaves of a Venus flytrap, are examples of a plant's ability to respond to specific directional stimuli.

7. In nature the poinsettia, a plant native to Mexico, blooms only in or around December. This pattern suggests:
   a. the long daily period of darkness (short day) in December stimulates the flowering.
   b. vernalization stimulates the flowering.
   c. the plant is dormant for the rest of the year.
   d. phytochrome is not affecting the poinsettia flowering cycle.
   e. a circadian rhythm is in effect.

8. Which of the following steps is *not* part of the sequence that is thought to trigger flowering?
   a. Cycles of light and dark stimulate the expression of the *CONSTANS* gene in a plant's leaves.
   b. CO proteins accumulate in the leaves and trigger expression of a second regulatory gene.
   c. mRNA transcribed during expression of a second regulatory gene moves via the phloem to the shoot apical meristem.
   d. Interactions among regulatory proteins promote the expression of floral organ identity genes in meristem tissue.
   e. CO proteins in the floral meristem interact with florigen, a so-called flowering hormone, which provides the final stimulus for expression of floral organ identity genes.

9. Damage from an infectious bacterium, fungus, or worm may trigger a plant defensive response when the pathogen or a substance it produces binds to:
   a. a receptor encoded by the plant's *avirulence (Avr)* gene.
   b. an *R* gene in the plant cell nucleus.
   c. a receptor encoded by a dominant R gene.
   d. PR proteins embedded in the plant cell plasma membrane.
   e. salicylic acid molecules released from the besieged plant cell.

10. In the sequence that unfolds after molecules of a hormone such as ABA bind to receptors at the surface of a target plant cell:
   a. first messenger molecules in the cytoplasm are mobilized, then G proteins carry the signal to second messengers such as protein kinases, which alter the activity of cell proteins such as IP$_3$.
   b. binding activates G proteins, which in turn activate second messengers such as IP$_3$; subsequent steps are thought to involve activation of genes that encode protein kinases.
   c. binding activates phospholipase C, which in turn activates G proteins, which then activate molecules of IP$_3$, a step that leads to the synthesis of protein kinases.
   d. binding stimulates G proteins to activate protein kinases, which then bind calcium channels in ER; the flux of calcium ions activates second messenger molecules that alter the activity of cell proteins or enter the cell nucleus and alter the expression of target genes.
   e. binding activates G proteins, which in turn activate phospholipase C; this substance then stimulates the synthesis of second messenger molecules, the second messengers bind calcium channels in the cell's ER, and finally protein kinases alter the activity of proteins by phosphorylating them.

Questions for Discussion

1. You work for a plant nursery and are asked to design a special horticultural regimen for a particular flowering plant. The plant is native to northern Spain, and in the wild it grows a few long, slender stems that produce flowers each July. Your boss wants the nursery plants to be shorter, with thicker stems and more branches, and she wants them to bloom in early December in time for holiday sales. Outline your detailed plan for altering the plant's growth and reproductive characteristics to meet these specifications.

2. Synthetic auxins such as 2,4-D can be weed killers because they cause an abnormal growth burst that kills the plant within a few days. Suggest reasons why such rapid growth might be lethal to a plant.

3. In some plant species, an endodermis is present in both stems and roots. In experiments, the shoots of mutant plants lacking differentiated endodermis in their root and shoot tissue don't respond normally to gravity, but roots of such plants do respond normally. Explain this finding, based on your reading in this chapter.

4. In *A. thaliana* plants carrying a mutation called *pickle (pkl)*, the primary root meristem retains characteristics of embryonic tissue—it spontaneously regenerates new embryos that can grow into mature plants. However, when the mutant root tissue is exposed to a gibberellin (GA), this abnormal developmental condition is suppressed. Explain why this finding suggests that additional research is needed on the fundamental biological role of GA.

Experimental Analysis

Tiny, thornlike trichomes on leaves are a common plant adaptation to ward off insects. Those trichomes develop very early on, as outgrowths of a seedling's epidermal cells. Biologists have observed, however, that many mature plants develop more leaf trichomes after the fact, as a *response* to insect damage. Researchers at the University of Chicago decided to study this phenomenon, and specifically wanted to determine the effects, if any, of jasmonate, salicylic acid, and gibberellin in stimulating trichome development. Keeping in mind that plant hormones often interact, how many separate experiments, at a minimum, would the research team have had to carry out in order to obtain useful initial data? Do you suppose they used mutant plants for some or all of the tests? Why or why not?

Evolution Link

Cryptochrome occurs in plants and animals. If it was inherited from their shared ancestor, what other major groups of organisms might also have it?

How Would You Vote?

1-Methylcyclopropene, or MCP, is a gas that keeps ethylene from binding to cells in plant tissues. It is used to prolong the shelf life of cut flowers and the storage time for fruits. Should produce that is treated this way be labeled to alert consumers? Go to www.thomsonedu.com/login to investigate both sides of the issue and then vote.

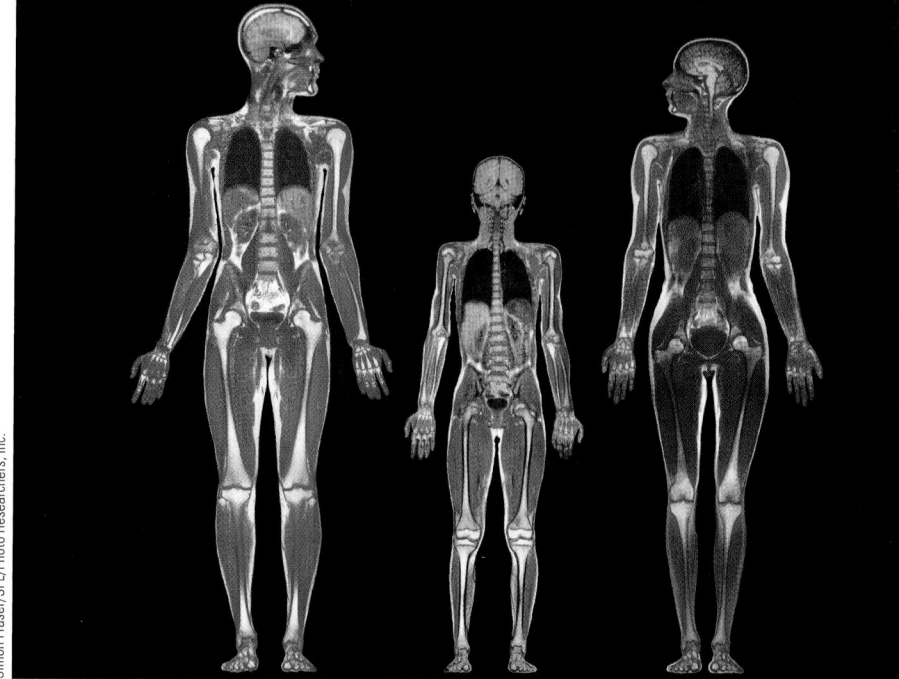

Magnetic resonance imaging (MRI) whole body scans of a man (left), a 9-year-old boy (middle), and a woman. Various organs can be seen in the scans: the whitish skeleton throughout the bodies, the brains within the skulls, lungs (dark) in the chests, lobes of the liver (green and brown ovals) in the abdomens, and bladders (dark ovals) in the lower abdomens.

Simon Fraser/SPL/Photo Researchers, Inc.

36 Introduction to Animal Organization and Physiology

WHY IT MATTERS

After a cold night in Africa's Kalahari Desert, gray meerkats (*Surricata suricatta*), a type of mongoose, awaken in their burrows. Although, like all mammals, meerkats regulate their body temperature, their internal temperature falls during cold nights. If the sun is shining in the morning and warms their burrows, the meerkats emerge and stand on their hind legs facing east, warming their bodies in the rays of the sun **(Figure 36.1).** This sunning behavior helps raise their body temperature.

Once the meerkats warm up, they fan out from their burrows looking for food, mainly insects and an occasional lizard. Their highly integrated body systems allow them to move about, sense the presence of prey, react with speed and precision to capture those prey, and consume them. Within their bodies, the food is broken down into glucose and other nutrient molecules, which are transported throughout the body to provide energy for living. At the same time, balancing mechanisms are constantly at work to maintain the animals' internal environment at a level that keeps body cells functioning. The maintenance of the internal environment in a stable state is called **homeostasis** (*homeo* = the same; *stasis* = standing or stopping). The processes and activi-

Figure 36.1
Meerkats lining up to warm themselves in sunlight.

ties responsible for homeostasis are called **homeostatic mechanisms.** These mechanisms compensate both for the external environmental changes that the meerkats encounter as they explore places with differences in temperature, humidity, and other physical conditions, and for changes in their own body systems.

All animals have body systems for acquiring and digesting nutrients to provide energy for life, growth, reproduction, and movement. Biologists are interested in the structures and functions of these systems. **Anatomy** is the study of the structures of organisms, and **physiology** is the study of their functions—the physico-chemical processes of organisms.

In this chapter we begin with the organization of individual cells into tissues, organs, and organ systems, the major body structures that carry out animal activities. Our discussion continues with a look at how the processes and activities of organ systems coordinate to accomplish homeostasis. The other chapters in this unit discuss the individual organ systems that carry out major body functions such as digestion, movement, and reproduction. Although we emphasize vertebrates throughout the unit, with particular reference to human physiology, we also make comparisons with invertebrates, to keep the structural and functional diversity of the animal kingdom in perspective and to understand the evolution of the structures and processes involved.

36.1 Organization of the Animal Body

In Animals, Specialized Cells Are Organized into Tissues, Tissues into Organs, and Organs into Organ Systems

The individual cells of animals have the same requirements as cells of any kind. They must be surrounded by an aqueous solution that contains ions and molecules required by the cells, including complex organic molecules that can be used as an energy source. The concentrations of these molecules and ions must be balanced to keep cells from shrinking or swelling excessively due to osmotic water movement. Most animal cells also require oxygen to serve as the final acceptor for electrons removed in oxidative reactions. Animal cells must be able to release waste molecules and other by-products of their activities, such as carbon dioxide,

to their environment. The physical conditions of the cellular environment, such as temperature, must also remain within tolerable limits.

The evolution of multicellularity (see Section 24.3) made it possible for organisms to create an *internal fluid environment* that supplies all the needs of individual cells, including nutrient supply, waste removal, and osmotic balance. This internal environment allows multicellular organisms to occupy diverse habitats, including dry terrestrial habitats that would be lethal to single cells. Multicellular organisms can also become relatively large because their individual cells remain small enough to exchange ions and molecules with the internal fluid. The fluid occupying the spaces between cells in multicellular animals is called **interstitial fluid,** or **extracellular fluid.**

The evolution of multicellularity also allowed major life functions to be subdivided among specialized groups of cells, with each group concentrating on a single activity. In animals, some groups of cells became specialized for movement, others for food capture, digestion, internal circulation of nutrients, excretion of wastes, reproduction, and other functions. Specialization greatly increases the efficiency by which animals carry out these functions.

In most animals, these specialized groups of cells are organized into tissues, the tissues into organs, and the organs into organ systems **(Figure 36.2).** A **tissue** is a group of cells with the same structure and function, working together as a unit to carry out one or more activities. The tissue lining the inner surface of the intestine, for example, is specialized to absorb nutrients released by digestion of food in the intestinal cavity.

An **organ** integrates two or more different tissues into a structure that carries out a specific function. The eye, liver, and stomach are examples of organs. Thus, the stomach integrates several different tissues into an organ specialized for processing food.

An **organ system** coordinates the activities of two or more organs to carry out a major body function such as movement, digestion, or reproduction. The organ system carrying out digestion, for example, coordinates the activities of organs including the mouth, stomach, pancreas, liver, and small and large intestines. Some organs contribute functions to more than one organ system. For instance, the pancreas forms part of the endocrine system as well as the digestive system.

STUDY BREAK

1. What are some advantages for an organism being multicellular?
2. What is the difference between a tissue, an organ, and an organ system?

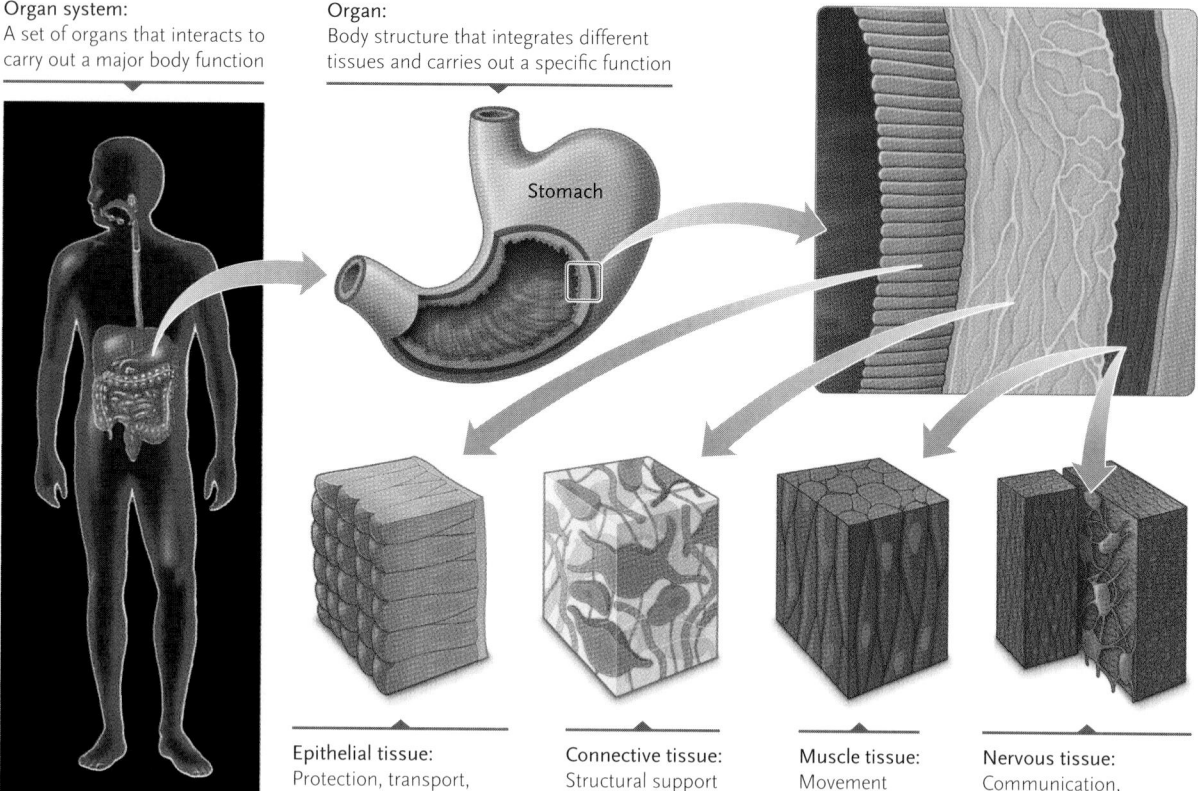

Organ system:
A set of organs that interacts to carry out a major body function

Organ:
Body structure that integrates different tissues and carries out a specific function

Stomach

Epithelial tissue:
Protection, transport, secretion, and absorption

Connective tissue:
Structural support

Muscle tissue:
Movement

Nervous tissue:
Communication, coordination, and control

Figure 36.2
Organization of animal cells into tissues, organs, and organ systems.

36.2 Animal Tissues

Although the most complex animals may contain hundreds of distinct cell types, all can be classified into one of four basic tissue groups: *epithelial, connective, muscle,* and *nervous* (see Figure 36.2). Each tissue type is assembled from individual cells. The properties of those cells determine the structure and, therefore, the function of the tissue. More specifically, the structure and integrity of a tissue depend on the structure and organization of the cytoskeleton within the cell, the type and organization of the extracellular matrix (ECM) surrounding the cell, and the junctions holding cells together (see Section 5.5).

Junctions of various kinds link cells into tissues (see Figure 5.27). *Anchoring junctions* form buttonlike spots or belts that weld cells together. They are most abundant in tissues subject to stretching, such as skin and heart muscle. *Tight junctions* seal the spaces between cells, keeping molecules and even ions from leaking between cells. For example, tight junctions in the tissue lining the urinary bladder prevent waste molecules and ions from leaking out of the bladder into other body tissues. *Gap junctions* open channels between cells in the same tissue, allowing ions and small molecules to flow freely from one to another. For example, gap junctions between muscle cells help muscle tissue to function as a unit.

Let us now consider the structural and functional features that distinguish the four types of tissues, with primary emphasis on the forms they take in vertebrates.

Epithelial Tissue Forms Protective, Secretory, and Absorptive Coverings and Linings of Body Structures

Epithelial tissue (*epi* = over; *thele* = covering) consists of sheetlike layers of cells that are usually joined tightly together, with little ECM material between them **(Figure 36.3)**. Also called *epithelia* (singular, *epithelium*), these tissues cover body surfaces and the surfaces of internal organs, as well as line cavities and ducts within the body. They protect body surfaces from invasion by bacteria and viruses, and secrete or absorb substances. For example, the epithelium covering a fish's gill structures serves as a barrier to bacteria and viruses and exchanges oxygen, carbon dioxide, and ions with the aqueous environment. Some epithelia, such as those lining the capillaries of the circulatory system, act as filters, allowing ions and small molecules to leak from the blood into surrounding tissues while barring the passage of blood cells and large molecules such as proteins.

Because epithelia form coverings and linings, they have one free (or outer) surface, which may be exposed to water, air, or fluids within the body. In internal cavities and ducts, the free surface is often covered with *cilia,* which beat like oars to move fluids through the cavity or duct. The epithelium lining the oviducts in mammals, for example, is covered with

Figure 36.3
Structure of epithelial tissues.

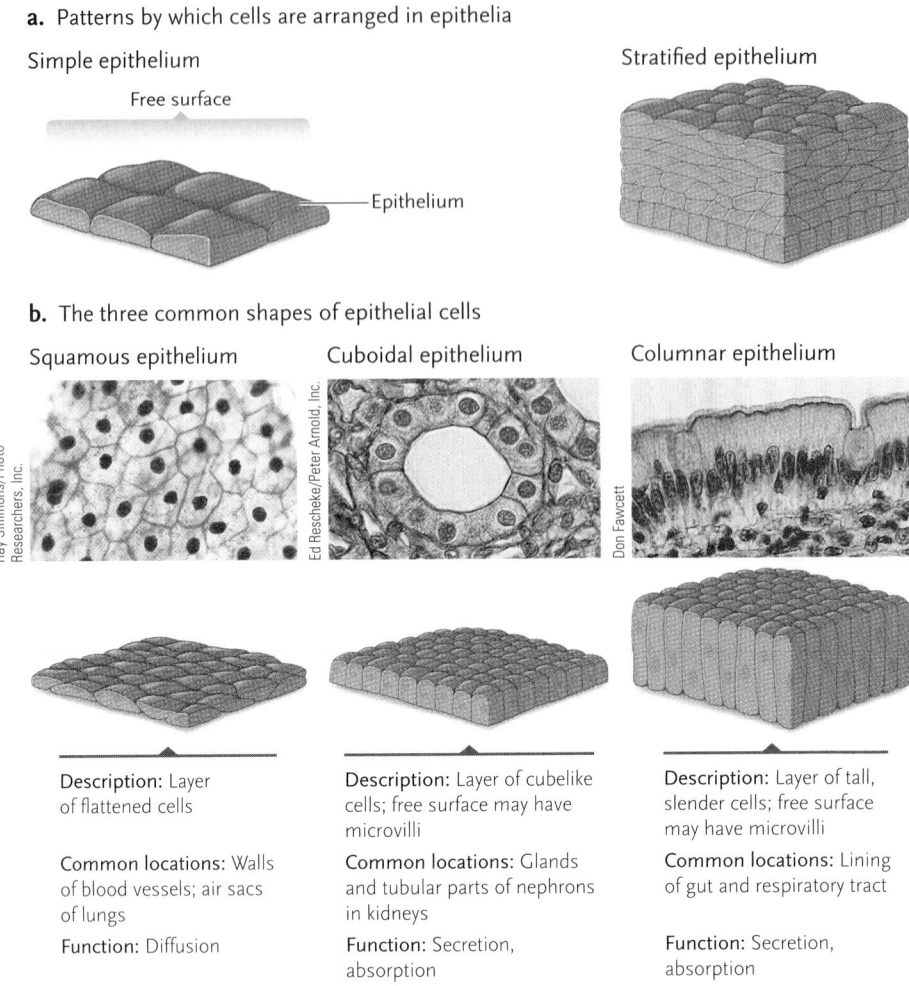

a. Patterns by which cells are arranged in epithelia

Simple epithelium

Free surface

Epithelium

Stratified epithelium

b. The three common shapes of epithelial cells

Squamous epithelium

Cuboidal epithelium

Columnar epithelium

Ray Simmons/Photo Researchers, Inc.

Ed Reschke/Peter Arnold, Inc.

Don Fawcett

Description: Layer of flattened cells

Common locations: Walls of blood vessels; air sacs of lungs

Function: Diffusion

Description: Layer of cubelike cells; free surface may have microvilli

Common locations: Glands and tubular parts of nephrons in kidneys

Function: Secretion, absorption

Description: Layer of tall, slender cells; free surface may have microvilli

Common locations: Lining of gut and respiratory tract

Function: Secretion, absorption

cilia that generate fluid currents to move eggs from the ovaries to the uterus. In some epithelia, including the lining of the small intestine, the free surface is crowded with *microvilli,* fingerlike extensions of the plasma membrane that increase the area available for secretion or absorption.

The inner surface of an epithelium adheres to a layer of glycoproteins secreted by the epithelial cells called the **basal lamina,** which fixes the epithelium to underlying tissues. The basal lamina is secreted by connective tissue cells immediately under the epithelium.

Epithelial Cell Structure. Epithelia are classified as *simple*—formed by a single layer of cells—or *stratified*—formed by multiple cell layers (see Figure 36.3a). The shapes of cells within an epithelium may be *squamous* (mosaic, flattened, and spread out), *cuboidal* (shaped roughly like dice or cubes), or *columnar* (elongated, with the long axis perpendicular to the epithelial layer; see Figure 36.3b). For example, the outer epithelium of mammalian skin is stratified and contains columnar, cuboidal, and squamous cells; the epithelium lining blood vessels is simple and squamous; and the intestinal epithelium is simple and columnar.

The cells of some epithelia, such as those forming the skin and the lining of the intestine, divide constantly to replace worn and dying cells. New cells are produced through division of stem cells in the basal (lowest) layer of the skin. *Stem cells* are undifferentiated (unspecialized) cells in the tissue that divide to produce more stem cells as well as cells that differentiate (that is, become specialized into one of the many cell types of the body). Stem cells are found both in adult organisms and in embryos. Besides the skin, adult stem cells are found in tissues of the brain, bone marrow, blood vessels, skeletal muscle, and liver. (*Insights from the Molecular Revolution* describes an effort to culture embryonic stem cells as a source of replacements for damaged tissues and organs.)

Glands Formed by Epithelia. Epithelia typically contain or give rise to cells that are specialized for secretion. Some of these secretory cells are scattered among nonsecretory cells within the epithelium. Others form structures called **glands,** which are derived from pockets of epithelium during embryonic development.

Some glands, called **exocrine glands** (*exo* = external; *crine* = secretion), remain connected to the epithelium by a duct, which empties their secretion at

Cultured Stem Cells

Stem cells derived from human embryos or fetal tissue have the potential to develop into any tissue, but until recently biomedical researchers had no method for maintaining human stem cells indefinitely in cultures. A reliable supply of cultured stem cells is essential to the growth of tissues for research or for possible use in replacing damaged tissues and organs.

Just a few years ago, James A. Thompson and his coworkers at the University of Wisconsin developed a successful method for culturing stem cells. Their starting point was very early human embryos produced by fertilization of eggs in the test tube. After the embryos had grown for several days, 14 samples of cells were removed and cultured. The challenge was to maintain the cultured cells in the embryonic state and keep them from differentiating into specialized forms.

The researchers' strategy was to place the cells in culture dishes over a bed of mouse fibroblasts. Earlier work had shown that this technique allows mouse and nonhuman primate embryonic cells to survive outside the body, but it had failed to work with some mammalian species. Would it work with human cells?

In 5 of the 14 cultures, the human cells multiplied on the fibroblasts without differentiating as long as the cell masses never contained more than 50 to 100 cells. Larger masses had to be separated and placed in small numbers on a fresh layer of mouse fibroblasts. Using this technique, the five cell cultures were maintained for as long as 8 months in the laboratory with no signs of differentiation or deterioration. They could be frozen, stored, and returned to active cultures at will.

But did they still have full stem-cell function? The investigators ran several tests to answer this question. One experiment tested molecules on the surface of the cultured cells. Stem cells have characteristic surface molecules that change when the cells begin to differentiate into adult tissues. Antibodies against typical stem cell surface molecules all reacted with the cultured cells, indicating that the cells still had the characteristic molecules.

Another experiment tested for telomerase, an enzyme that helps maintain chromosomes at their normal length during rapid cycles of DNA replication and cell division. The enzyme becomes inactive in most cells as they differentiate into adult form. A standard test for the enzyme showed that it was present and remained fully active in the cultured cells.

In the final experiment, the researchers injected samples of the cells into mice to test whether the cultured cells could differentiate into a wide range of tissues. Within these mice, the cells grew into balls of tissue that included skin cells, gut epithelium, cartilage, bone, smooth and striated muscle, and nerve cells. The cultured cells seemed to be able to differentiate into adult tissues when stimulated to do so, and thus to have all the characteristics of stem cells.

The ability to culture stem cells opens many opportunities for future biological and medical discoveries. Observing the differentiation of stem cells into adult types should fill gaps in our knowledge of human development, which until now could be studied only in embryos. These studies may also give clues to the processes that produce birth defects and spontaneous abortion and could indicate means to correct these problems.

Further, if stem cells can be stimulated to differentiate into desired tissues and organs, they may provide an essentially unlimited supply of material for transplants. Many conditions, such as Parkinson disease and juvenile-onset diabetes, result from the death or malfunction of only one or a few cell types. Replacing defective cells from banks of cultured stem cells may be the key to curing the diseases.

These scientific and medical prospects do not provide answers to ethical questions about culturing stem cells derived from embryos. Such questions are the subject of intense scrutiny and debate in the U.S. Congress and among scientists, religious authorities, and the general public. Hopefully, a balance will be found between concerns about the use of human embryonic cells in research and the prospects for significant improvements in human health and scientific knowledge.

---

the epithelial surface. Exocrine secretions include mucus, saliva, digestive enzymes, sweat, earwax, oils, milk, and venom (**Figure 36.4a** shows an exocrine gland in the skin of a poisonous tree frog). Other glands, called **endocrine glands** (*endo* = inside), become suspended in connective tissue underlying the epithelium, with no ducts leading to the epithelial surface. These ductless glands, such as the pituitary gland, adrenal gland, and thyroid gland **(Figure 36.4b),** release their products—called hormones—directly into the interstitial fluid, to be picked up and distributed by the circulatory system.

Some glands act as both exocrine glands and endocrine glands. The pancreas, for instance, has an exocrine function of secreting pancreatic juice through a duct into the small intestine where it plays an important role in food digestion, and an endocrine function of secreting the hormones insulin and glucagon into the bloodstream to help regulate glucose levels in the blood.

Connective Tissue Supports Other Body Tissues

Most animal body structures contain one or more types of **connective tissue.** Connective tissues support other body tissues, transmit mechanical and other forces, and in some cases act as filters. They consist of cells

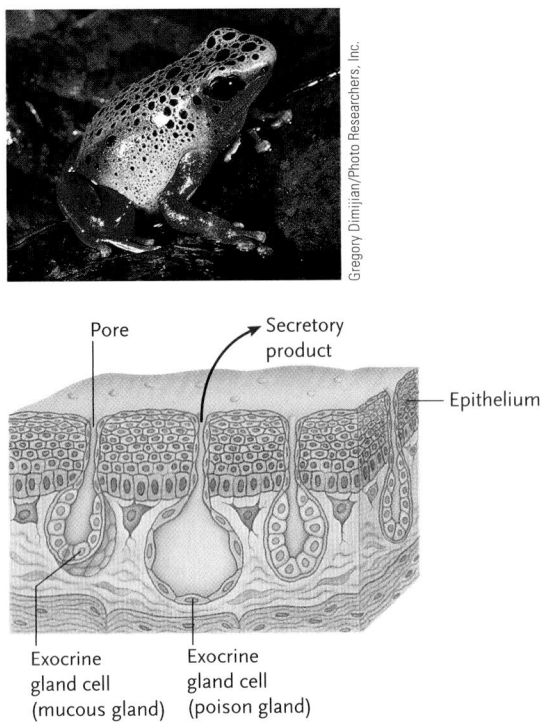

Pore — Secretory product

Epithelium

Exocrine gland cell (mucous gland)   Exocrine gland cell (poison gland)

a. Examples of exocrine glands: The mucus- and poison-secreting glands in the skin of a blue poison frog

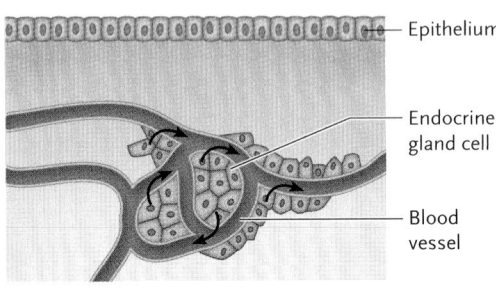

Thyroid

Epithelium

Endocrine gland cell

Blood vessel

b. Example of an endocrine gland: The thyroid gland, which secretes hormones that regulate the rate of metabolism and other body functions

Figure 36.4

Exocrine and endocrine glands.

The poison secreted by the blue poison frog *(Dendrobates azureus)* is one of the most lethal glandular secretions known.

that form networks or layers in and around body structures and that are separated by nonliving material, specifically the ECM secreted by the cells of the tissue (see Section 5.5). Many forms of connective tissue have more ECM material (both by weight and by volume) than cellular material.

The mechanical properties of a connective tissue depend on the type and quantity of its ECM. The consistency of the ECM ranges from fluid (as in blood and lymph), through soft and firm gels (as in tendons), to the hard and crystalline (as in bone). In most connective tissues, the ECM consists primarily of the fibrous glycoprotein **collagen** embedded in a network of proteoglycans—glycoproteins that are very rich in carbohydrates. In bone, the glycoprotein network surrounding the collagen is impregnated with mineral deposits that produce a hard, yet still somewhat elastic, structure. Another class of glycoproteins, **fibronectin**, aids in the attachment of cells to the ECM and helps hold the cells in position.

In some connective tissues another rubbery protein, **elastin**, adds elasticity to the ECM—it is able to return to its original shape after being stretched, bent, or compressed. Elastin fibers, for example, help the skin return to its original shape when pulled or stretched, and give the lungs the elasticity required for their alternating inflation and deflation.

Vertebrates have six major types of connective tissue: *loose connective tissue, fibrous connective tissue, cartilage, bone, adipose tissue,* and *blood.* Each type has a

characteristic function correlated with its structure **(Figure 36.5).**

Loose Connective Tissue. **Loose connective tissue** consists of sparsely distributed cells surrounded by a more or less open network of collagen and other glycoprotein fibers (see Figure 36.5a). The cells, called **fibroblasts,** secrete most of the collagen and other proteins in this connective tissue.

Loose connective tissues support epithelia and form a corsetlike band around blood vessels, nerves, and some internal organs; they also reinforce deeper layers of the skin. Sheets of loose connective tissue, covered on both surfaces with epithelial cells, form the **mesenteries,** which hold the abdominal organs in place and provide lubricated, smooth surfaces that prevent chafing or abrasion between adjacent structures as the body moves.

Fibrous Connective Tissue. In **fibrous connective tissue,** fibroblasts are sparsely distributed among dense masses of collagen and elastin fibers that are lined up in highly ordered, parallel bundles (see Figure 36.5b). The parallel arrangement produces maximum tensile strength and elasticity. Examples include **tendons,** which attach muscles to bones, and **ligaments,** which connect bones to each other at a joint. The cornea of the eye is a transparent fibrous connective tissue formed from highly ordered collagen molecules.

a. Loose connective tissue

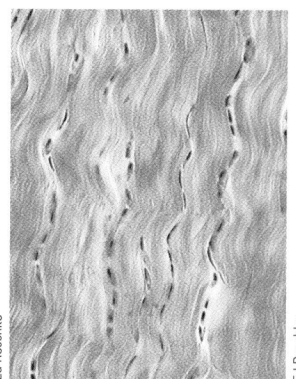

Ed Reschke

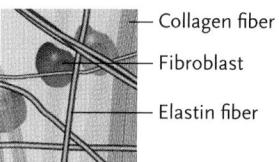

— Collagen fiber

— Fibroblast

— Elastin fiber

Description: Fibroblasts and other cells surrounded by collagen and elastin fibers forming a glycoprotein matrix

Common locations: Under the skin and most epithelia

Function: Support, elasticity, diffusion

b. Fibrous connective tissue

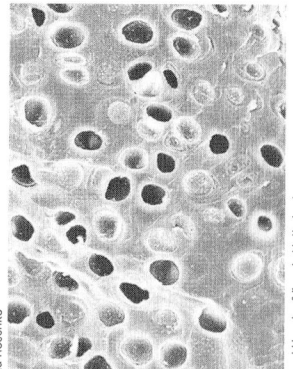

Ed Reschke

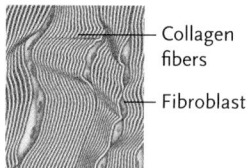

— Collagen fibers

— Fibroblast

Description: Long rows of fibroblasts surrounded by collagen and elastin fibers in parallel bundles with a dense extracellular matrix

Common locations: Tendons, ligaments

Function: Strength, elasticity

c. Cartilage

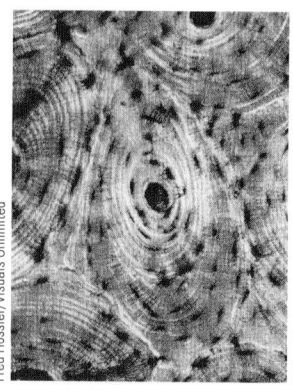

Fred Hossler/Visuals Unlimited

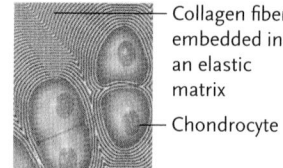

— Collagen fibers embedded in an elastic matrix

— Chondrocyte

Description: Chondrocytes embedded in a pliable, solid matrix of collagen and chondroitin sulfate

Common locations: Ends of long bones, nose, parts of airways, skeleton of vertebrate embryos

Function: Support, flexibility, low-friction surface for joint movement

d. Bone tissue

Ed Reschke

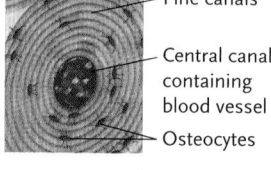

— Fine canals

— Central canal containing blood vessel

— Osteocytes

Description: Osteocytes in a matrix of collagen and glycoproteins hardened with hydroxyapatite

Common locations: Bones of vertebrate skeleton

Function: Movement, support, protection

Figure 36.5
The six major types of connective tissues.

e. Adipose tissue

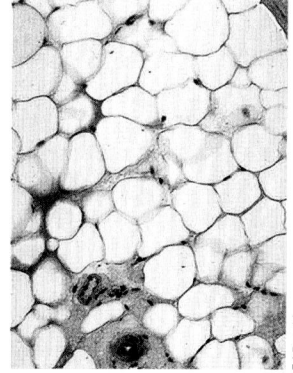

Ed Reschke

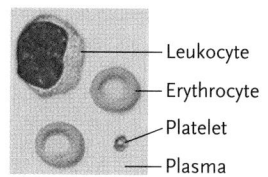

— Nucleus

— Fat deposit

Description: Large, tightly packed adipocytes with little extracellular matrix

Common locations: Under skin; around heart, kidneys

Function: Energy reserves, insulation, padding

f. Blood

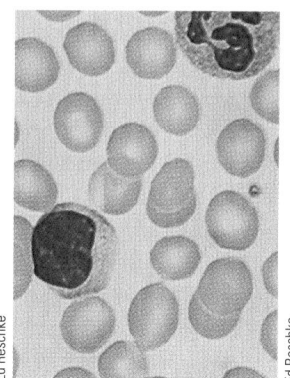

Ed Reschke

— Leukocyte

— Erythrocyte

— Platelet

— Plasma

Description: Leukocytes, erythrocytes, and platelets suspended in a plasma matrix

Common locations: Circulatory system

Function: Transport of substances

Cartilage. **Cartilage** consists of sparsely distributed cells called **chondrocytes**, surrounded by networks of collagen fibers embedded in a tough but elastic matrix of the glycoprotein *chondroitin sulfate* (see Figure 36.5c). Elastin is also present in some forms of cartilage.

The elasticity of cartilage allows it to resist compression and stay resilient, like a piece of rubber. Bending your ear or pushing the tip of your nose, which are supported by a core of cartilage, gives a good idea of the flexible nature of this tissue. In humans, cartilage also supports the larynx, trachea, and smaller air passages in the lungs. It forms the disks cushioning the vertebrae in the spinal column and the smooth, slippery capsules around the ends of bones in joints such as the hip and knee. Cartilage also serves as a precursor to bone during embryonic development; in sharks and rays and their relatives, almost the entire skeleton remains as cartilage in adults.

Bone. The densest form of connective tissue, **bone** forms the skeleton, which supports the body, protects softer body structures such as the brain, and contributes to body movements.

Mature bone consists primarily of cells called **osteocytes** (*osteon* = bone) embedded in an ECM containing collagen fibers and glycoproteins impregnated with *hydroxyapatite,* a calcium-phosphate mineral (see Figure 36.5d). The collagen fibers give bone tensile strength and elasticity; the hydroxyapatite resists compression and allows bones to support body weight. Cells called **osteoblasts** (*blast* = bud or sprout) produce the collagen and mineral of bone—as much as 85% of the weight of bone is mineral deposits. Osteocytes, in fact, are osteoblasts that have become trapped and surrounded by the bone materials they themselves produce. **Osteoclasts** (*clast* = break) remove the minerals and recycle them through the bloodstream. Bone is not a stable tissue; it is reshaped continuously by the bone-building osteoblasts and the bone-degrading osteoclasts.

Although bones appear superficially to be solid, they are actually porous structures, consisting of a system of microscopic spaces and canals. The structural unit of bone is the **osteon.** It consists of a minute central canal surrounded by osteocytes embedded in concentric layers of mineral matter (see Figure 36.5d). A blood vessel and extensions of nerve cells run through the central canal, which is connected to the spaces containing cells by very fine, radiating canals filled with interstitial fluid. The blood vessels supply nutrients to the cells with which the bone is built, and the nerve cells hook up the bone cells to the body's nervous system.

Adipose Tissue. The connective tissue called **adipose tissue** mostly contains large, densely clustered cells called *adipocytes* that are specialized for fat storage (see Figure 36.5e). It has little ECM. Adipose tissue also cushions the body and, in mammals, forms an especially important insulating layer under the skin.

The animal body stores limited amounts of carbohydrates, primarily in muscle and liver cells. Excess carbohydrates are converted into the fats stored in adipocytes. The storage of chemical energy as fats offers animals a weight advantage. For example, the average human would weigh about 45 kg (100 pounds) more if the same amount of chemical energy was stored as carbohydrates instead of fats. Adipose tissue is richly supplied with blood vessels, which move fats or their components to and from adipocytes.

Blood. **Blood** (see Figure 36.5f) is considered a connective tissue because its cells are suspended in a fluid ECM, plasma. The straw-colored **plasma** is a solution of proteins, nutrient molecules, ions, and gases.

Blood contains two primary cell types, **erythrocytes** (red blood cells; *erythros* = red) and **leukocytes** (white blood cells; *leukos* = white). Erythrocytes are packed with hemoglobin, a protein that can bind and transport oxygen. There are several types of leukocytes—all help to protect the body against invading viruses, bacteria, and other disease-causing agents. The blood plasma also contains **platelets,** membrane-bound fragments of specialized blood cells, which take part in the reactions that seal wounds with blood clots.

Blood is the major transport vehicle of the body. It carries oxygen and nutrients to body cells, removes wastes and by-products such as carbon dioxide, and maintains the internal fluid environment, including the osmotic balance between cells and the interstitial fluid. Blood also transports hormones and other signal molecules that coordinate body responses. (The components and roles of blood are described in Chapter 42.)

Muscle Tissue Produces the Force for Body Movements

Muscle tissue consists of cells that have the ability to contract (shorten). The contractions, which depend on the interaction of two proteins—*actin* and *myosin*—move body limbs and other structures, pump the blood, and produce a squeezing pressure in organs such as the intestine and uterus. Three types of muscle tissue, *skeletal, cardiac,* and *smooth,* produce body movements in vertebrates **(Figure 36.6).** In all types of muscle tissue, the cells are densely packed, leaving little room for ECM.

Skeletal Muscle. **Skeletal muscle** is so called because most muscles of this type are attached by tendons to the skeleton. Skeletal muscle cells are also called **muscle fibers** because each is an elongated cylinder (see Figure 36.6a). These cells contain many nuclei and are packed with actin and myosin molecules arranged in highly ordered, parallel units that give the tissue a banded or striated appearance when viewed under a microscope. Muscle fibers packed side by side into parallel bundles surrounded by sheaths of connective tissue form many body muscles, such as the biceps.

Skeletal muscle contracts in response to signals carried by the nervous system. The contractions of skeletal muscles, which are characteristically rapid and powerful, move body parts and maintain posture. The contractions also release heat as a by-product of cellular metabolism. This heat helps mammals, birds, and some other vertebrates maintain their body temperatures when environmental temperatures fall. (Skeletal muscle is discussed further in Chapter 41.)

Cardiac Muscle. **Cardiac muscle** is the contractile tissue of the heart (see Figure 36.6b). Cardiac muscle has a striated appearance because it contains actin and myosin molecules arranged like those in skeletal muscle. However, cardiac muscle cells are short and branched, with each cell connecting to several neighboring cells; the joining point between two such cells is called an *intercalated disk.* Cardiac muscle cells thus form an interlinked network, which is stabilized by anchoring junctions and gap junctions. This network makes heart muscle contract in all directions, producing a squeezing or pumping action rather than the

a. Skeletal muscle   **b.** Cardiac muscle   **c.** Smooth muscle   **Figure 36.6**
Structure of skeletal, cardiac, and smooth muscle.

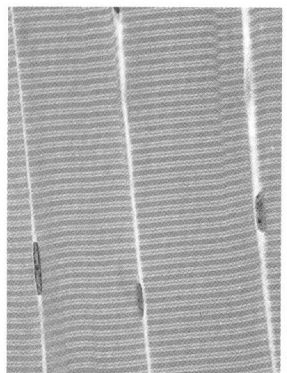 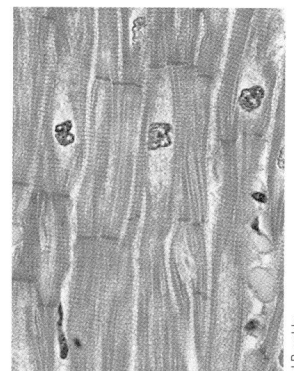 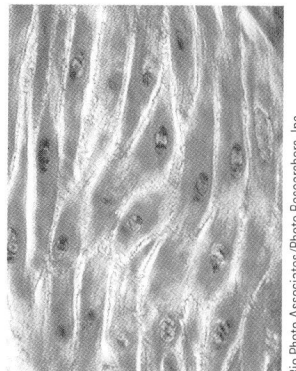

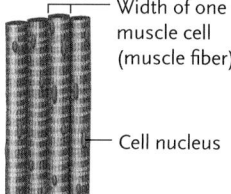

Width of one muscle cell (muscle fiber)

Cell nucleus

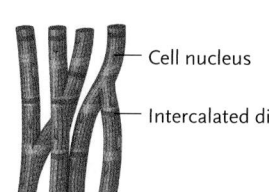

Cell nucleus

Intercalated disk

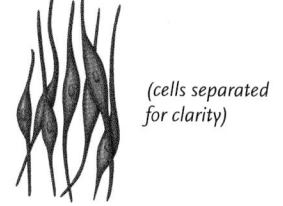

(cells separated for clarity)

Description: Bundles of long, cylindrical, striated, contractile cells called muscle fibers

Typical location: Attached to bones of skeleton

Function: Locomotion, movement of body parts

Description: Cylindrical, striated cells that have specialized end junctions

Location: Wall of heart

Function: Pumping of blood within circulatory system

Description: Contractile cells with tapered ends

Typical location: Wall of internal organs, such as stomach

Function: Movement of internal organs

lengthwise, unidirectional contraction characteristic of skeletal muscle.

Smooth Muscle. **Smooth muscle** is found in the walls of tubes and cavities in the body, including blood vessels, the stomach and intestine, the bladder, and the uterus. Smooth muscle cells are relatively small and spindle-shaped (pointed at both ends), and their actin and myosin molecules are arranged in a loose network rather than in bundles (see Figure 36.6c). This loose network makes the cells appear smooth rather than striated when viewed under a microscope. Smooth muscle cells are connected by gap junctions and enclosed in a mesh of connective tissue. The gap junctions transmit ions that make smooth muscles contract as a unit, typically producing a squeezing motion. Although smooth muscle contracts more slowly than skeletal and cardiac muscle do, its contractions can be maintained at steady levels for a much longer time. These contractions move and mix the stomach and intestinal contents, constrict blood vessels, and push the infant out of the uterus during childbirth.

Nervous Tissue Receives, Integrates, and Transmits Information

Nervous tissue contains cells called **neurons** (also called *nerve cells*) that serve as lines of communication and control between body parts. Billions of neurons are packed into the human brain; others extend throughout the body. Nervous tissue also contains **glial cells** (*glia* = glue), which physically support and provide nutrients to neurons, provide electrical insulation between them, and scavenge cellular debris and foreign matter.

A neuron consists of a *cell body,* which houses the nucleus and organelles, and two types of cell extensions, dendrites and axons **(Figure 36.7).** *Dendrites* receive chemical signals from other neurons or from body cells of other types, and convert them into an electrical signal that is transmitted to the cell body of the receiving neuron. Dendrites are usually highly branched. *Axons* conduct electrical signals away from the cell body to the axon terminals, or endings. At their terminals, axons convert the electrical signal to a chemical signal that stimulates a response in nearby muscle cells, gland cells, or other neurons. Axons are usually unbranched except at their terminals. Depending on the type of neuron and its location in the body, its axon may extend from a few micrometers or millimeters to more than a meter. (Neurons and their organization in body structures are discussed further in Chapters 37, 38, and 39.)

All four major tissue types—epithelial, connective, muscle, and nervous—combine to form the organs and organ systems of animals. The next section depicts the major organs and organ systems of vertebrates, and outlines their main tasks.

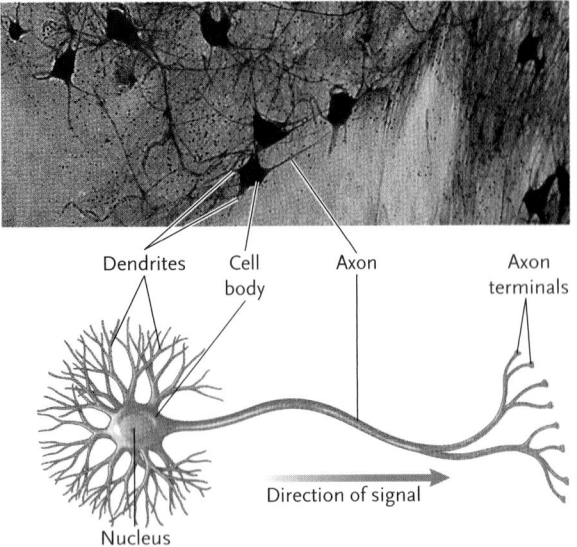

Dendrites · Cell body · Axon · Axon terminals

Direction of signal

Nucleus

Figure 36.7

Neurons and their structure. The micrograph shows a network of motor neurons, which relay signals from the brain or spinal cord to muscles and glands. (Micrograph: Lennart Nilsson from Behold Man, © 1974 Albert Bonniers Forlag and Little, Brown and Company, Boston.)

STUDY BREAK

1. Distinguish between exocrine and endocrine glands. What is the tissue type of each of these glands?
2. What are the six major types of connective tissue in vertebrates?
3. What three types of muscle tissue produce body movements?

36.3 Coordination of Tissues in Organs and Organ Systems

Organs and Organ Systems Function Together to Enable an Animal to Survive

In the tissues, organs, and organ systems of an animal, each cell engages in the basic metabolic activities that ensure its own survival, and performs one or more functions of the system to which it belongs. All verte-

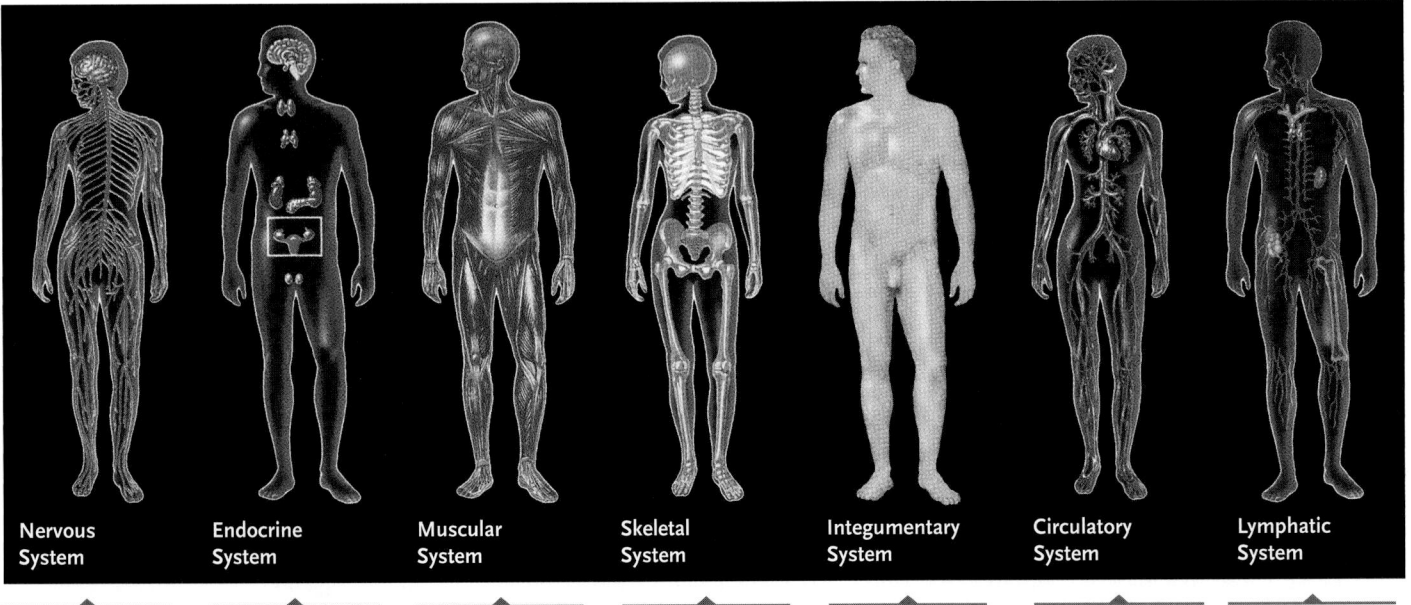

Nervous System	Endocrine System	Muscular System	Skeletal System	Integumentary System	Circulatory System	Lymphatic System
Main organs: Brain, spinal cord, peripheral nerves, sensory organs	**Main organs:** Pituitary, thyroid, adrenal, pancreas, and other hormone-secreting glands	**Main organs:** Skeletal, cardiac, and smooth muscle	**Main organs:** Bones, tendons, ligaments, cartilage	**Main organs:** Skin, sweat glands, hair, nails	**Main organs:** Heart, blood vessels, blood	**Main organs:** Lymph nodes, lymph ducts, spleen, thymus
Main functions: Principal regulatory system; monitors changes in internal and external environments and formulates compensatory responses; coordinates body activities	**Main functions:** Regulates and coordinates body activities through secretion of hormones	**Main functions:** Moves body parts; helps run bodily functions; generates heat	**Main functions:** Supports and protects body parts; provides leverage for body movements	**Main functions:** Covers external body surfaces and protects against injury and infection; helps regulate water content and body temperature	**Main functions:** Distributes water, nutrients, oxygen, hormones, and other substances throughout body and carries away carbon dioxide and other metabolic wastes; helps stabilize internal temperature and pH	**Main functions:** Returns excess fluid to the blood; defends body against invading viruses, bacteria, fungi, and other pathogens as part of immune system

brates (and most invertebrates) have eleven major organ systems, which are summarized in **Figure 36.8,** and discussed in the rest of this unit of the book.

The functions of all these organ systems are coordinated and integrated to accomplish collectively a series of tasks that are vital to all animals, whether a flatworm, a salmon, a meerkat, or a human. These functions include:

1. Acquiring nutrients and other required substances such as oxygen, coordinating their processing, distributing them throughout the body, and disposing of wastes.
2. Synthesizing the protein, carbohydrate, lipid, and nucleic acid molecules required for body structure and function.
3. Sensing and responding to changes in the environment, such as temperature, pH, and ion concentrations.
4. Protecting the body against injury or attack from other animals, and from viruses, bacteria, and other disease-causing agents.

5. Reproducing and, in many instances, nourishing and protecting offspring through their early growth and development.

Together these tasks maintain homeostasis, preserving the internal environment required for survival of the body. Homeostasis is the topic of the next section.

STUDY BREAK

What are the major functions of each of the eleven organ systems of the vertebrate body?

36.4 Homeostasis

Homeostasis is the process by which animals maintain their internal environment in a steady state (constant level) or between narrow limits. Homeostasis depends on a number of the body's organ systems, with the

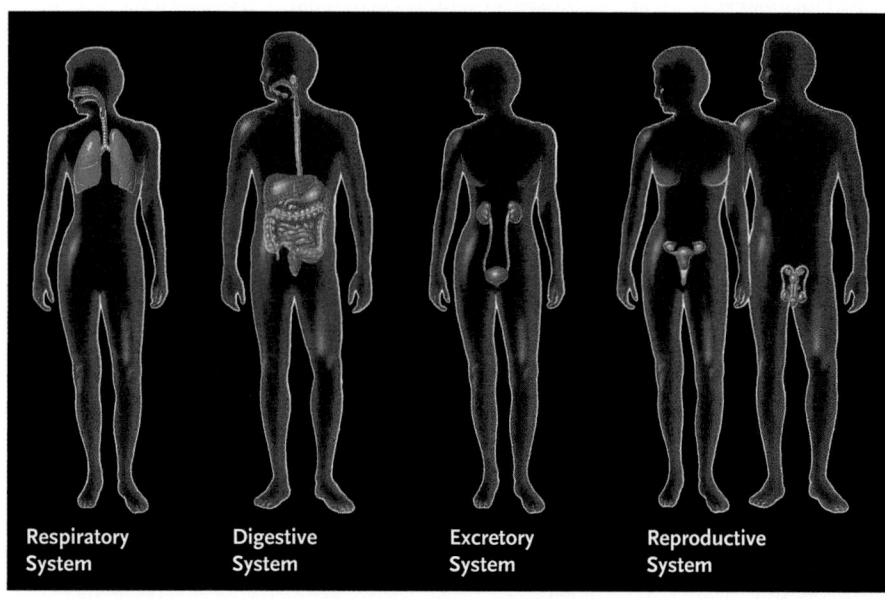

Figure 36.8
Organ systems of the human body. The immune system, which is primarily a cellular system, is not shown.

Respiratory System	**Digestive System**	**Excretory System**	**Reproductive System**
Main organs: Lungs, diaphragm, trachea, and other airways	**Main organs:** Pharynx, esophagus, stomach, intestines, liver, pancreas, rectum, anus	**Main organs:** Kidneys, bladder, ureter, urethra	**Main organs:** *Female*: ovaries, oviducts, uterus, vagina, mammary glands *Male*: testes, sperm ducts, accessory glands, penis
Main functions: Exchanges gases with the environment, including uptake of oxygen and release of carbon dioxide	**Main functions:** Converts ingested matter into molecules and ions that can be absorbed into body; eliminates undigested matter; helps regulate water content	**Main functions:** Removes and eliminates excess water, ions, and metabolic wastes from body; helps regulate internal osmotic balance and pH	**Main functions:** Maintains the sexual characteristics and passes on genes to the next generation

nervous system and endocrine system being the most important. For example, blood pH is controlled by both the nervous and endocrine systems, blood glucose by the endocrine system, internal temperature by the nervous and endocrine systems, and oxygen and carbon dioxide concentrations by the nervous system.

Although the *stasis* part of homeostasis might suggest a static, unchanging process, homeostasis is actually a dynamic process, in which internal adjustments are made continuously to compensate for environmental changes. For example, internal adjustments are needed for homeostasis during exercise or hibernation. The factors controlled by homeostatic mechanisms all require energy that must be constantly acquired from the external environment.

Homeostasis Is Accomplished by Negative Feedback Mechanisms

The primary mechanism of homeostasis is **negative feedback**, in which a stimulus—a change in the external or internal environment—triggers a response that compensates for the environmental change **(Figure 36.9)**. Homeostatic mechanisms typically include three elements: a sensor, an integrator, and an effector. The **sensor** consists of tissues or organs that detect a change in external or internal factors such as pH, temperature, or the concentration of a molecule such as glucose. The **integrator** is a control center that compares the detected environmental change with a **set point,** the level at which the condition controlled by the pathway is to be maintained. The **effector** is a system, activated by the integrator, that returns the condition to the set point if it has strayed away. In most animals, the integrator is part of the central nervous system or endocrine system, while effectors may include parts of essentially any body tissue or organ.

The Thermostat as a Negative Feedback Mechanism. The concept of negative feedback may be most familiar in systems designed by human engineers. The thermostat maintaining temperature at a chosen level in a house provides an example. A sensor within the thermostat measures the temperature. If the room temperature changes more than a degree or so from the set point—the temperature that you set in the thermostat—an integrator circuit in the thermostat activates an effector that returns the room temperature to the set point. If the temperature has fallen below the set point, the effector is the furnace, which adds heat to the house until the temperature rises to the set point. If the temperature has risen above the set point, the effector is the air conditioner, which removes heat from the room until the temperature falls to the set point.

Negative Feedback Mechanisms in Animals. Mammals and birds—warm-blooded vertebrates—also have a homeostatic mechanism that maintains body temperature within a relatively narrow range around a set point. The integrator (thermostat) for this mechanism is located in a brain center called the *hypothalamus*. A group of neurons in the hypothalamus detects changes in the temperature of the brain and the rest of the body, and compares it with a set point. For humans, the set point has a relatively narrow range centered at about 37°C.

One or more effectors are activated in humans if the temperature varies beyond the limits of the set point. If the temperature falls below the lower limit, the hypothalamus activates effectors that constrict the blood vessels in the skin. The reduction in blood flow means that less heat is conducted from the blood through the skin to the environment; in short, heat loss from the skin is reduced. Other effectors may induce shivering, a physical mechanism to generate body heat. Also, integrating neurons in the brain, stimulated by signals from the hypothalamus, make us consciously sense a chill, which we may counteract behaviorally by putting on more clothes or moving to a warmer area.

Conversely, if blood temperature rises above the set point, the hypothalamus triggers effectors that dilate the blood vessels in the skin, increasing blood flow and heat loss from the skin. Other effectors induce sweating, which cools the skin and the blood flowing through it as the sweat evaporates. And again, through integrating neurons in the brain, we may consciously sense being overheated, which we may counteract by shedding clothes, moving to a cooler location, or taking a dip in a pool.

Sometimes the temperature set point changes, and the negative feedback mechanisms then operate to maintain body temperature at the new set point. For example, if you become infected by certain viruses and bacteria, the temperature set point increases to a higher level, producing a fever to help overcome the infection. Once the infection is combated, the set point is readjusted down again to its normal level.

All other mammals have similar homeostatic mechanisms that maintain or adjust body temperature. Dogs and birds pant to release heat from their bodies **(Figure 36.10)** and shiver to increase internal heat production. Many terrestrial animals

Figure 36.9
Components of a negative feedback mechanism maintaining homeostasis. The integrator coordinates a response by comparing the level of an environmental condition with a set point that indicates where the level should be.

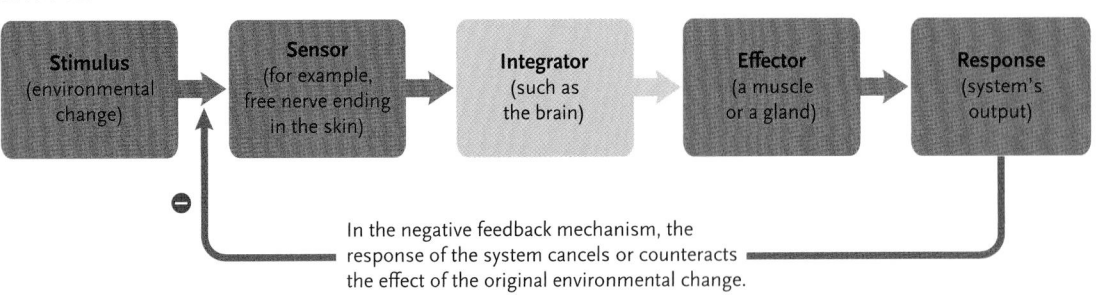

In the negative feedback mechanism, the response of the system cancels or counteracts the effect of the original environmental change.

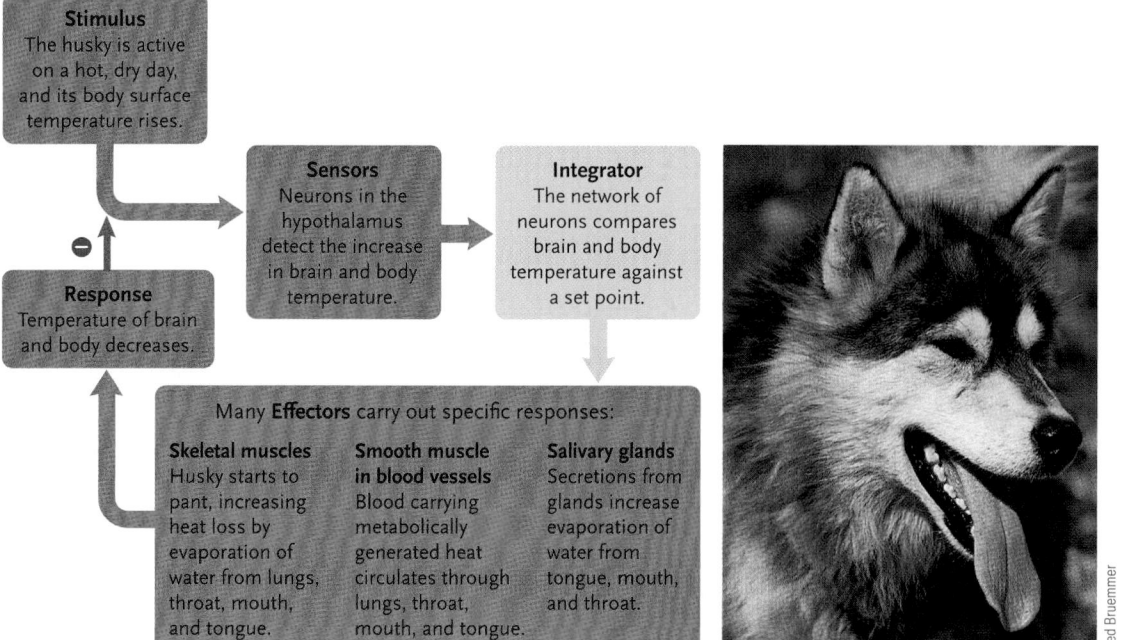

Stimulus
The husky is active on a hot, dry day, and its body surface temperature rises.

Sensors
Neurons in the hypothalamus detect the increase in brain and body temperature.

Integrator
The network of neurons compares brain and body temperature against a set point.

Response
Temperature of brain and body decreases.

Many **Effectors** carry out specific responses:

Skeletal muscles
Husky starts to pant, increasing heat loss by evaporation of water from lungs, throat, mouth, and tongue.

Smooth muscle in blood vessels
Blood carrying metabolically generated heat circulates through lungs, throat, mouth, and tongue.

Salivary glands
Secretions from glands increase evaporation of water from tongue, mouth, and throat.

Fred Bruemmer

Figure 36.10
Homeostatic mechanisms maintaining the body temperature of a husky when environmental temperatures are high.

enter or splash water over their bodies to cool off. Also, recall from the beginning of the chapter how meerkats use behavioral mechanisms to regulate their body temperature.

Whereas mammals regulate their internal body temperature within a narrow range around a set point, certain other vertebrates regulate over a broader range. These vertebrates use other, less precise negative feedback mechanisms for their temperature regulation. Snakes and lizards, for example, respond behaviorally to compensate for variations in environmental temperatures. They may absorb heat by basking on sunny rocks in the cool early morning and move to cooler, shaded spots in the heat of the afternoon. Some fishes, such as the tuna, generate enough heat by contraction of the swimming muscles to maintain body temperature well above the temperature of the surrounding water.

Some invertebrates, such as dragonflies, moths, and butterflies, use muscular contractions equivalent to shivering when their body temperature falls below the level required for flight. The shivering contractions warm the muscles to flying temperature. All of these physiological and behavioral responses depend on negative feedback mechanisms involving sensors, integrators, and effectors.

Animals Also Have Positive Feedback Mechanisms That Do Not Result in Homeostasis

Under certain circumstances, animals respond to a change in internal or external environmental condition by a **positive feedback** mechanism that intensifies or adds to the change. Such mechanisms, with some exceptions, do not result in homeostasis. They operate when the animal is responding to life-threatening conditions (an attack, for instance), or as part of reproductive processes.

The birth process in mammals is a prime example. During human childbirth, initial contractions of the uterus push the head of the fetus against the cervix, the opening of the uterus into the vagina. The pushing causes the cervix to stretch. Sensors that detect the stretching signal the hypothalamus to release a hormone, oxytocin, from the pituitary gland. Oxytocin increases the uterine contractions, intensifying the squeezing pressure on the fetus and further stretching the cervix. The stretching results in more oxytocin release and stronger uterine contraction, repeating the positive feedback circuit and increasing the squeezing pressure until the fetus is pushed entirely out of the uterus.

Because positive feedback mechanisms such as the one triggering childbirth do not result in homeostasis, they occur less commonly than negative feedback in animals. They also operate as part of larger, more inclusive negative feedback mechanisms that ultimately shut off the positive feedback pathway and return conditions to normal limits.

In conclusion, we learned in this chapter about the various tissues and organ systems of the body, and of the involvement of organ systems in homeostasis. Next, we begin a series of chapters describing the organ systems in detail, starting with the nervous system.

STUDY BREAK

What are the components of a negative feedback mechanism that results in homeostasis?

Why do so many strokes and heart attacks occur in the morning?

Stroke and heart attack occur most frequently at a particular time of the day—in the morning—exhibiting a profound circadian variation. Circadian rhythms are generated through a discrete set of molecular interactions including the Bmal1, Clock, NPAS2, Cry, and Per proteins. We have recently shown that the biological clock is expressed and oscillating in blood vessels; however, it remains unknown if this "vascular clock" acts to modulate the function of blood vessels. Moreover, if the vascular clock does influence normal vascular function, might a broken clock contribute to the onset of heart attack and stroke? Current research in my laboratory and others is addressing these questions.

How could a vascular clock influence vascular function?

Circadian rhythms are seen in endothelial function, blood pressure, vascular resistance, and blood flow. Could the circadian clock play a role in the regulation of blood vessel homeostasis? What targets within vascular cells might the clock control? One possibility is that the vascular clock may act to regulate production of signaling molecules in endothelial cells, which comprise the inner lining of blood vessels. In addition, direct actions on vascular smooth muscle cells, which contain the nerve and muscle elements critical for constriction and relaxation of blood vessels, may also be under circadian control. To assess these questions, the use of mice with genetic disruption of the circadian clock (knockout or mutant mice) has been invaluable. Garret FitzGerald and colleagues at the University of Pennsylvania demonstrated that *Bmal1* knockout mice and *Clock* mutant mice lack circadian rhythms in blood pressure, in part due to a blunted sympathetic drive. However, the contribution of parasympathetic outflow remains unknown. Might the clock directly induce or inhibit transcription of genes important to vas-

cular function? One approach is to introduce molecular clock components into cultured cells and to assess promoter regulation of a gene of interest. This has already proved useful in identifying the PAI-1 (plasminogen activator inhibitor-1) protein as a target of the molecular clock. PAI-1 inhibits plasminogen activator, an enzyme that breaks up clots. So PAI-1 promotes clot formation. Future studies implementing tissue-specific knockout mice of molecular clock components will determine more directly how and where these targets are regulated by the circadian clock.

Might a dysfunctional vascular clock contribute to chronic vascular disease?

Chronic impairments in blood pressure and blood flow rhythms are sensed by blood vessels, which causes them to respond by changing architecture through a process called vascular remodeling. Vascular remodeling is an extremely intensive and active area of vascular biology research that is important to understanding the progression to blood vessel disease. Using models of blood vessel ligation in mice with a disrupted molecular clock, we are currently assessing the impact of the biological clock on vascular remodeling.

Research to address the role of the vascular clock may ultimately change the way we understand and treat arteriosclerosis, hypertension, and heart attack.

 R. Daniel Rudic is an assistant professor in the Department of Pharmacology and Toxicology at the Medical College of Georgia. To learn more about his research on circadian rhythms and vascular biology, go to http://www.mcg.edu/som/phmtox/RudicLab/index.asp.

Review

Go to **ThomsonNOW™** at www.thomsonedu.com/login to access quizzing, animations, exercises, articles, and personalized homework help.

36.1 Organization of the Animal Body

- In most animals, cells are specialized and organized into tissues, tissues into organs, and organs into organ systems. A tissue is a group of cells with the same structure and function, working as a unit to carry out one or more activities. An organ is an assembly of tissues integrated into a structure that carries out a specific function. An organ system is a group of organs that carry out related steps in a major physiological process.

36.2 Animal Tissues

- Animal tissues are classified as epithelial, connective, muscle, or nervous (Figure 36.2). The properties of the cells of these tissues determine the structures and functions of the tissues.

- Various kinds of junctions link cells in a tissue. Anchoring junctions "weld" cells together. Tight junctions seal the cells into a leak-proof layer. Gap junctions form direct avenues of commu-

nication between the cytoplasm of adjacent cells in the same tissue.

- Epithelial tissue consists of sheetlike layers of cells that cover body surfaces and the surfaces of internal organs, and line cavities and ducts within the body (Figure 36.3).

- Glands are secretory structures derived from epithelia. They may be exocrine (connected to an epithelium by a duct that empties on the epithelial surface) or endocrine (ductless, with no direct connection to an epithelium) (Figure 36.4).

- Connective tissue consists of cell networks or layers and an extracellular matrix (ECM). It supports other body tissues, transmits mechanical and other forces, and in some cases acts as a filter (Figure 36.5).

- Loose connective tissue consists of sparsely distributed fibroblasts surrounded by an open network of collagen and other glycoproteins. It supports epithelia and organs of the body and forms a covering around blood vessels, nerves, and some internal organs.

- Fibrous connective tissue contains sparsely distributed fibroblasts in a matrix of densely packed, parallel bundles of collagen and elastin fibers. It forms high tensile-strength structures such as tendons and ligaments.

- Cartilage consists of sparsely distributed chondrocytes surrounded by a network of collagen fibers embedded in a tough but highly elastic matrix of branched glycoproteins. Cartilage provides support, flexibility, and a low-friction surface for joint movement.

- In bone, osteocytes are embedded in a collagen matrix hardened by mineral deposits. Osteoblasts secrete collagen and minerals for the ECM; osteoclasts remove the minerals and recycle them into the bloodstream.

- Adipose tissue consists of cells specialized for fat storage. It also cushions and rounds out the body and provides an insulating layer under the skin.

- Blood consists of a fluid matrix, the plasma, in which erythrocytes and leukocytes are suspended. The erythrocytes carry oxygen to body cells; the leukocytes produce antibodies and initiate the immune response against disease-causing agents.

- Muscle tissue contains cells that have the ability to contract forcibly (Figure 36.6). Skeletal muscle, containing long cells called muscle fibers, moves body parts and maintains posture.

- Cardiac muscle, which contains short contractile cells with a branched structure, forms the heart.

- Smooth muscle consists of spindle-shaped contractile cells that form layers surrounding body cavities and ducts.

- Nervous tissue contains neurons and glial cells. Neurons communicate information between body parts in the form of electrical and chemical signals (Figure 36.7). Glial cells support the neurons or provide electrical insulation between them.

Animation: Cell junctions

Animation: Structure of an epithelium

Animation: Types of simple epithelium

Animation: Soft connective tissues

Animation: Specialized connective tissues

Animation: Muscle tissues

Animation: Functional zones of a motor neuron

Animation: Structure of human skin

Practice: Differences between cell and tissue types

36.3 Coordination of Tissues in Organs and Organ Systems

- Organs and organ systems are coordinated to carry out vital tasks, including maintenance of internal body conditions; nutrient acquisition, processing, and distribution; waste disposal; molecular synthesis; environmental sensing and response; protection against injury and disease; and reproduction.

- In vertebrates and most invertebrates, the major organ systems that accomplish these tasks are the nervous, endocrine, muscular, skeletal, integumentary, circulatory, lymphatic, immune, respiratory, digestive, excretory, and reproductive systems (Figure 36.8).

Animation: Human organ systems

36.4 Homeostasis

- Homeostasis is the process by which animals maintain their internal fluid environment under conditions their cells can tolerate. It is a dynamic state, in which internal adjustments are made continuously to compensate for environmental changes.

- Homeostasis is accomplished by negative feedback mechanisms that include a sensor, which detects a change in an external or internal condition; an integrator, which compares the detected change with a set point; and an effector, which returns the condition to the set point if it has varied (Figure 36.9).

- Animals also have positive feedback mechanisms, in which a change in an internal or external condition triggers a response that intensifies the change, and typically does not result in homeostasis.

Questions

Self-Test Questions

1. Which organ or tissue is an early major defense against viruses and bacteria?
   a. kidneys
   b. skin
   c. stomach
   d. skeletal muscle
   e. heart

2. Which tissue is a constant source of adult stem cells in a mammal?
   a. bone marrow
   b. pancreas
   c. basal lamina
   d. heart muscle
   e. kidneys

3. A flexible, rubbery protein in connective tissue is called ___, whereas a more fibrous, less flexible glycoprotein is called ___.
   a. adipose; cartilage
   b. endocrine; exocrine
   c. sweat; hormones
   d. chondroitin sulfate; hydroxapatite
   e. elastin; collagen

4. Adipose tissue:
   a. gives elasticity under epithelium.
   b. gives strength to tendons.
   c. insulates and is an energy reserve.
   d. provides movement, support, and protection.
   e. supports the nose and airways.

5. The bones of an elderly woman break more easily than those of a younger person. You would surmise that with aging, the cell type that diminishes in activity is the:
   a. osteocyte.
   b. osteoblast.
   c. osteoclast.
   d. chondrocyte.
   e. fibroblast.

6. The enormous mass of weight lifters is due to an increase in the size of:
   a. skeletal muscle.
   b. smooth muscle.
   c. cardiac muscle.
   d. involuntary muscle.
   e. interlinked, branched muscle.

7. Which muscle types appear striated under a microscope?
   a. skeletal muscles only
   b. cardiac muscles only
   c. skeletal muscles and cardiac muscles
   d. smooth muscles only
   e. skeletal muscles and smooth muscles

8. Which of the following is *not* a homeostatic response?
   a. In a contest, a student eats an entire chocolate cake in 10 minutes. Due to hormonal secretions, his blood glucose level does not change dramatically.
   b. The basketball players are dripping sweat at half time.
   c. The pupils in the eyes constrict when looking at a light.

d. Slower breathing in sleep changes carbon dioxide and oxygen blood levels, which affect blood pH.

e. The brain is damaged when a fever rises above 105°F.

8. The pituitary gland secretes a hormone that in turn stimulates the thyroid to secrete hormones. When the thyroid hormones are no longer needed, the pituitary stops or reduces its stimulus. This is an example of:

a. osmolarity.

b. environmental sensing.

c. integration.

d. positive feedback.

e. negative feedback.

10. The system that coordinates other organ systems is the:

a. skeletal system.
d. nervous system.

b. reproductive system.
e. integumentary system.

c. muscular system.

Questions for Discussion

1. Blood is often described as an atypical connective tissue. If you had to argue that blood is a connective tissue, what reasons would you include? What reasons would you include if you had to argue that blood is not a connective tissue?

2. What effect do you think a program of lifting weights would have on the bones of the skeleton? How would you design an experiment to test your prediction?

3. Positive feedback mechanisms are rare in animals compared with negative feedback mechanisms. Why do you think this is so?

4. Near the time of childbirth, collagen fibers in the connective tissue of the cervix break down, and gap junctions between the smooth muscle cells of the uterus increase in number. What do you think is the significance of these tissue changes?

5. Explain how, when driving, you control the car's speed by a typical negative feedback mechanism.

Experimental Analysis

The regulation of temperature in mammals and birds is an example of homeostasis. Design an experiment to observe and measure processes involved in temperature homeostasis in sedentary versus athletic humans during exercise.

Evolution Link

Steroid hormones are similar in structure and function across a wide array of animal species. For example, estradiol, which plays a critical role in reproductive and sexual functioning, is chemically identical in turtles and humans. What do these observations suggest about the time when steroid hormones evolved?

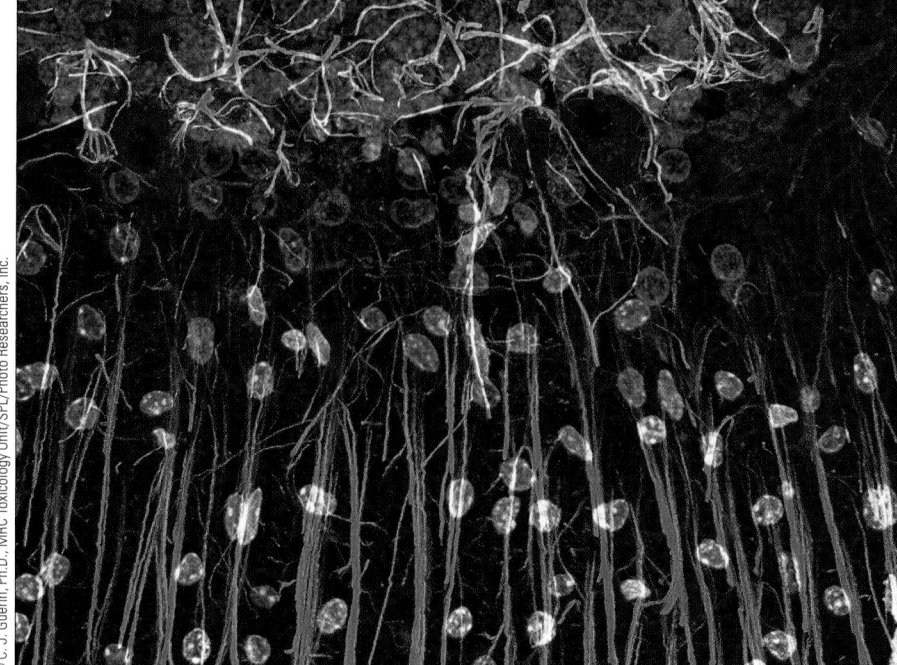

Section through the cerebellum, a part of the brain that integrates signals coming from particular regions of the body (confocal light micrograph). Neurons, the cells that send and receive signals, are red; glial cells, which provide structural and functional support for neurons, are yellow; and nuclei are purple.

Study Plan

37 Information Flow and the Neuron

Why It Matters

The dog stands alert, muscles tense, motionless except for a wagging tail. His eyes are turned toward his master, a boy poised to throw a Frisbee for him to catch. Even before the Frisbee is released, the dog has anticipated the direction of its flight from the eyes and stance of the boy.

With a snap of the wrist, the boy throws the Frisbee, and the dog springs into action. Legs churning, eyes following the Frisbee, the dog runs beneath its track, closing the distance as the Frisbee reaches the peak of its climb and begins to descend. All this time, parts of the dog's brain have been processing information received through various sensory inputs. The eyes report his travel over the ground and the speed and arc of the Frisbee. Sensors in the inner ears, muscles, and joints detect the position of the dog's body, and his brain sends out signals that keep his movements on track and in balance. Other parts of the brain register inputs from sensors monitoring body temperature and carbon dioxide levels in the blood, and send signals that adjust heart and breathing rate accordingly.

At just the right instant, a burst of signals from the dog's brain causes trunk and leg muscles to contract in a coordinated pattern, and

Figure 37.1
With perfect timing, a dog leaps to catch a Frisbee. The coordinated leap involves processing and integration of information by the dog's nervous system.

© Gary Gervat/Masterfile

the dog leaps to intercept the Frisbee in midair with an assured snap of his jaws **(Figure 37.1)**. Now the animal twists, turning his head and eyes toward the ground as his brain calculates the motions required to land on his feet and in balance. The dog makes a perfect landing and trots happily back to his master, ready to repeat the entire performance.

The functions of the dog's nervous system in the chase and capture are astounding in the amount and variety of sensory inputs, the rate and complexity of the brain's analysis and integration of incoming information, and the flurry of signals the brain sends to make compensating adjustments in body activities. Yet they are ordinary in the sense that the same activities take place countless times each day in the nervous system of all but the simplest animals.

All these activities, no matter how complex, depend on the functions of only two major cell types: *neurons* and *glial cells*. In most animals, these cells are organized into complex networks called *nervous systems*. This chapter describes neuron structure and tells how neurons send and receive signals with the aid and support of glial cells. The next chapter considers the neural networks of the brain and its associated structures. Chapter 39 discusses the sensory receptors that detect environmental changes and convert that information into signals for integration by the nervous system.

37.1 Neurons and Their Organization in Nervous Systems

An animal constantly receives stimuli from both internal and external sources. **Neural signaling**, communication by neurons, is the process by which an animal responds appropriately to a stimulus **(Figure 37.2)**. In most animals, the four components of neural signaling are *reception, transmission, integration,* and *response*. **Reception**, the detection of a stimulus, is performed by **neurons,** the cellular components of nervous systems, and by specialized sensory receptors such as those in the eye and skin. **Transmission** is the sending of a message along a neuron, and then to another neuron or to a muscle or gland. **Integration** is the sorting and interpretation of neural messages and the determination of the appropriate response(s). **Response** is the "output" or action resulting from the integration of neural messages. For a dog catching a Frisbee, for instance, sensors in the eye receive light stimuli from the environment, and internal sensors receive stimuli from all the animal's organ systems. The neural messages generated are transmitted through the nervous system and integrated to determine the appropriate response, in this case stimulating the muscles so the dog jumps into the air and catches the Frisbee.

Neurons Are Cells Specialized for the Reception and Transmission of Informational Signals

Neural signaling involves three functional classes of neurons (the blue boxes in Figure 37.2). **Afferent neurons** (also called **sensory neurons**) transmit stimuli collected by their sensory receptors to **interneurons,** which integrate the information to formulate an appropriate response. In humans and some other primates, 99% of neurons are interneurons. **Efferent neurons** carry the signals indicating a response away from the interneuron networks to the **effectors,** the muscles and glands. Efferent neurons that carry signals to skeletal muscle are called **motor neurons.** The information-processing steps in the nervous system can be summarized, therefore, as: (1) sensory receptors on

Figure 37.2
Neural signaling: the information-processing steps in the nervous system.

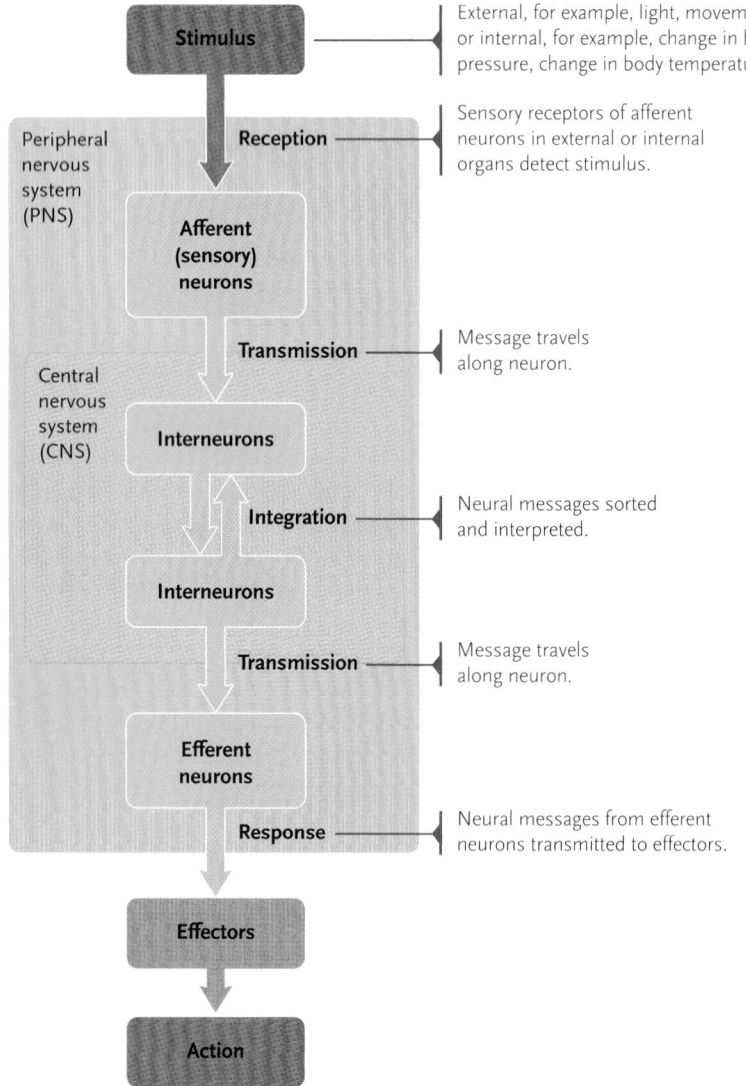

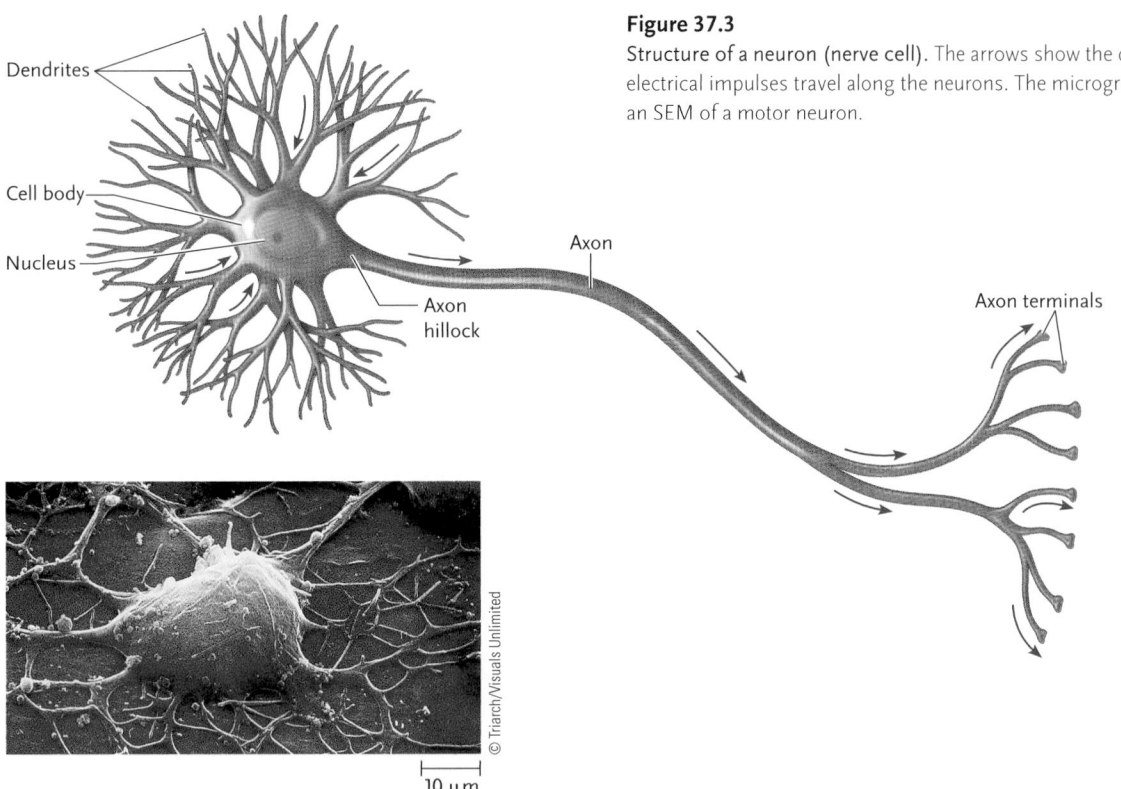

Figure 37.3

Structure of a neuron (nerve cell). The arrows show the direction electrical impulses travel along the neurons. The micrograph is an SEM of a motor neuron.

Dendrites

Cell body

Nucleus

Axon hillock

Axon

Axon terminals

© Triarch/Visuals Unlimited

10 μm

afferent neurons receive a stimulus; (2) afferent neurons transmit the information to interneurons; (3) interneurons integrate the neural messages; and (4) efferent neurons transmit the neural messages to effectors, which act in a way appropriate to the stimulus.

Neurons vary widely in shape and size. All have an enlarged cell body and two types of extensions or processes, called dendrites and axons **(Figure 37.3)**. The **cell body,** which contains the nucleus and the majority of cell organelles, synthesizes most of the proteins, carbohydrates, and lipids of the neuron. Dendrites and axons conduct electrical signals, which are produced by ions flowing down concentration gradients through channels in the plasma membrane of the neuron. **Dendrites** receive the signals and transmit them toward the cell body. Dendrites are generally highly branched, forming a treelike outgrowth at one end of the neuron (*dendros* = tree). **Axons** conduct signals away from the cell body to another neuron or an effector. Neurons typically have a single axon, which arises from a junction with the cell body called an **axon hillock.** The axon has branches at its tip that end as small, buttonlike swellings called **axon terminals.** The more terminals contacting a neuron, the greater its capacity to integrate incoming information.

Connections between axon terminals of one neuron and the dendrites or cell body of a second neuron form **neuronal circuits.** A typical neuronal circuit contains an afferent (sensory) neuron, one or more interneurons, and an efferent neuron. The circuits combine into networks that interconnect the parts of the nervous system. In vertebrates, the afferent neurons and efferent neurons collectively form the *peripheral nervous system (PNS)*. The interneurons form the brain and spinal cord, called the *central nervous system (CNS)*. As depicted in Figure 37.2, afferent (*afferre* = carry toward) information is ultimately transmitted to the CNS where efferent (*efferre* = carry away) information is initiated. The nervous systems of most invertebrates are also divided into central and peripheral divisions.

Neurons Are Supported Structurally and Functionally by Glial Cells

Glial cells are nonneuronal cells that provide nutrition and support to neurons. One type, called **astrocytes** because they are star-shaped **(Figure 37.4),** occurs only in the vertebrate CNS, where they closely cover the surfaces of blood vessels. Astrocytes provide physical support to neurons and help maintain the concentrations of ions in the interstitial fluid surrounding them. Two other types of glial cells— **oligodendrocytes** in the CNS and **Schwann cells** in the PNS—wrap around axons in a jelly roll fashion to form myelin sheaths **(Figure 37.5).** Myelin sheaths have a high lipid content because of the

Figure 37.4

Astrocytes (orange), a type of glial cell, and a neuron (yellow) in brain tissue.

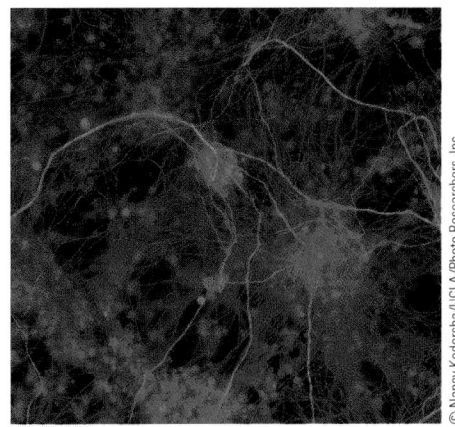

© Nancy Kedersha/UCLA/Photo Researchers, Inc.

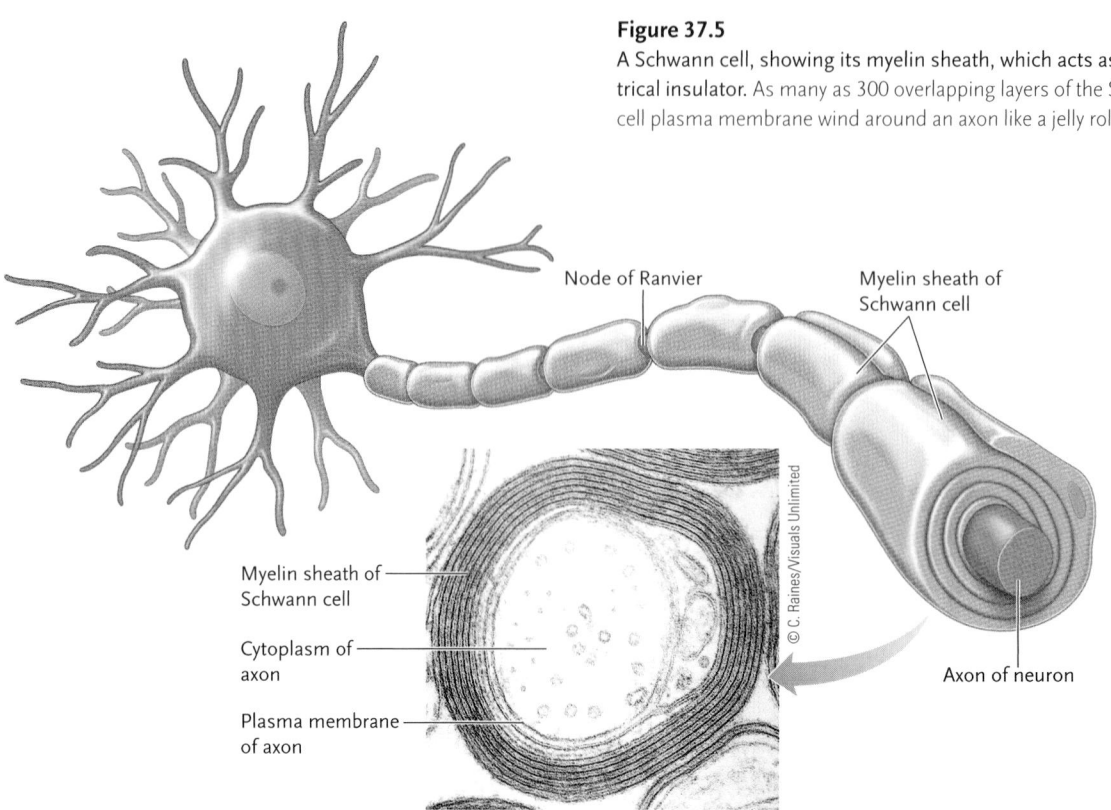

Figure 37.5
A Schwann cell, showing its myelin sheath, which acts as an electrical insulator. As many as 300 overlapping layers of the Schwann cell plasma membrane wind around an axon like a jelly roll.

Node of Ranvier

Myelin sheath of
Schwann cell

Myelin sheath of
Schwann cell

Cytoplasm of
axon

Plasma membrane
of axon

© C. Raines/Visuals Unlimited

Axon of neuron

many layers of plasma membranes of the myelin-forming cells. Because of their high lipid content, the myelin sheaths act as electical insulators. The gaps between Schwann cells, called **nodes of Ranvier**, expose the axon membrane directly to extracellular fluids. This structure speeds the rate at which electrical impulses move along the axons covered by glial cells.

Unlike most neurons, glial cells retain the capacity to divide throughout the life of the animal. This capacity allows glial tissues to replace damaged or dead cells, but also makes them the source of almost all brain tumors, produced when regulation of glial cell division is lost.

Neurons Communicate via Synapses

A **synapse** (*synapsis* = juncture) is a site where a neuron makes a communicating connection with another neuron or with an effector such as a muscle fiber or gland. On one side of the synapse is an axon terminal of the **presynaptic cell**, the neuron that transmits the signal. On the other side is the cell body or a dendrite of the **postsynaptic cell**, the neuron or the surface of an effector that receives the signal. Communication across a synapse may occur by the direct flow of an electrical signal or by means of a **neurotransmitter**, a chemical released by an axon terminal at a synapse. The vast majority of vertebrate neurons communicate by means of neurotransmitters.

In **electrical synapses**, the plasma membranes of the presynaptic and postsynaptic cells are in direct con-

tact **(Figure 37.6a)**. When an electrical impulse arrives at the axon terminal, gap junctions (see Section 36.2) allow ions to flow directly between the two cells, leading to unbroken transmission of the electrical signal. Although electrical synapses allow the most rapid conduction of signals, this type of connection is essentially "on" or "off" and unregulated. In humans, electrical synapses occur in locations such as the pulp of a tooth, where they contribute to the almost instant and intense pain we feel if the pulp is disturbed.

In **chemical synapses**, the plasma membranes of the presynaptic and postsynaptic cells are separated by a narrow gap, about 25 nm wide, called the **synaptic cleft (Figure 37.6b).** When an electrical impulse arrives at an axon terminal, it causes the release of a neurotransmitter into the synaptic cleft. The neurotransmitter diffuses across the synaptic cleft and binds to a receptor in the plasma membrane of the postsynaptic cell. If enough neurotransmitter molecules bind to these receptors, the postsynaptic cell generates a new electrical impulse, which travels along its axon to reach a synapse with the next neuron or effector in the circuit. A chemical synapse is more than a simple on-off switch because many factors can influence the generation of a new electrical impulse in the postsynaptic cell, including neurotransmitters that inhibit that cell rather than stimulating it. The balance of stimulatory and inhibitory effects in chemical synapses contributes to the integration of incoming information in a receiving neuron.

a. Electrical synapse

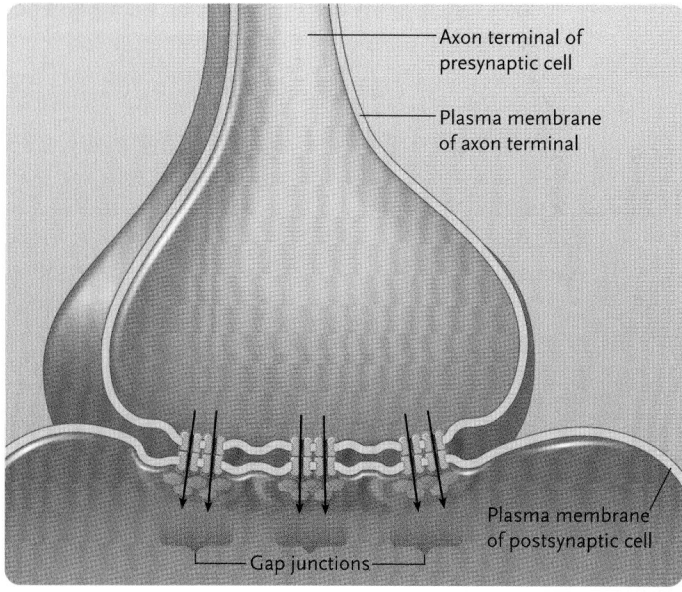

Axon terminal of
presynaptic cell

Plasma membrane
of axon terminal

Plasma membrane
of postsynaptic cell

Gap junctions

In an electrical synapse, the plasma membranes of the presynaptic and post-synaptic cells make direct contact. Ions flow through gap junctions that connect the two membranes, allowing impulses to pass directly to the postsynaptic cell.

b. Chemical synapse

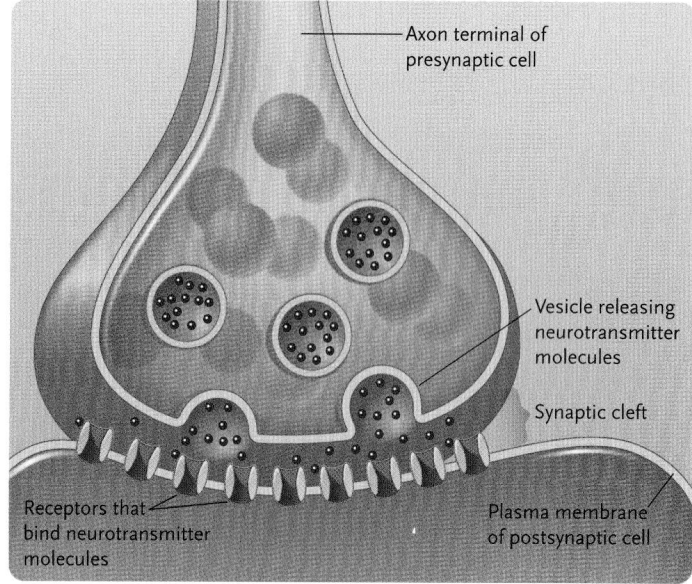

Axon terminal of
presynaptic cell

Vesicle releasing
neurotransmitter
molecules

Synaptic cleft

Receptors that
bind neurotransmitter
molecules

Plasma membrane
of postsynaptic cell

In a chemical synapse, the plasma membranes of the presynaptic and post-synaptic cells are separated by a narrow synaptic cleft. Neurotransmitter molecules diffuse across the cleft and bind to receptors in the plasma membrane of the postsynaptic cell. The binding opens channels to ion flow that may generate an impulse in the postsynaptic cell.

Figure 37.6
The two types of synapses by which neurons communicate with other neurons or effectors.

STUDY BREAK

1. Distinguish between a dendrite and an axon.
2. Distinguish between the functions and locations of afferent neurons, efferent neurons, and interneurons.
3. What is the difference between an electrical synapse and a chemical synapse?

37.2 Signal Conduction by Neurons

All cells of an animal have a **membrane potential**, a separation of positive and negative charges across the plasma membrane. Outside the cell the charge is positive, and inside the cell it is negative. This charge separation produces *voltage*—an electrical potential difference—across the plasma membrane.

The membrane potential is caused by the uneven distribution of Na^+ and K^+ inside and outside the cell. As you learned in Chapter 6, plasma membranes are *selectively* permeable in that they allow some ions but not others to move across the membrane through protein channels embedded in the phospholipid bilayer. Plasma membrane-embedded Na^+/K^+ active transport pumps use energy from ATP hydrolysis to pump simultaneously three Na^+ out of the cell for every two K^+ pumped in. This exchange generates a higher Na^+ concentration outside the cell than inside, and a higher K^+ concentration inside the cell than outside, explaining the positive charge outside the cell. The inside of the

cell is negatively charged because the cell also contains many negatively charged molecules (anions) such as proteins, amino acids, and nucleic acids.

In most cells, the membrane potential does not change. However, neurons and muscle cells use the membrane potential in a specialized way. That is, in response to electrical, chemical, mechanical, and certain other types of stimuli, their membrane potential changes rapidly and transiently. Cells with this property are said to be *excitable cells*. Excitability, produced by a sudden flow across the plasma membrane, is the basis for nerve impulse generation.

Resting Potential Is the Unchanging Membrane Potential of an Unstimulated Neuron

The membrane of a neuron that is not being stimulated is not conducting an impulse—exhibits a steady negative membrane potential, called the **resting potential** because the neuron is at rest. The resting potential has been measured at between -50 and -60 millivolts (mV) for neurons in the body, and at about -70 mV in isolated neurons **(Figure 37.7)**. A neuron exhibiting the resting potential is said to be *polarized*.

The distribution of ions inside and outside an axon that produces the resting potential is shown in **Figure 37.8**. As described earlier in this section, the Na^+/K^+ pump is responsible for creating the imbalance of Na^+ and K^+ inside and outside of the cell, and the concentration of negatively charged molecules within the cell results in the inside being negatively charged and the

Figure 37.7 Research Method

Measuring Membrane Potential

PURPOSE: To determine the membrane potentials of unstimulated and stimulated neurons and muscle cells.

PROTOCOL: Prepare a microelectrode by drawing out a glass capillary tube to a tip with a diameter much smaller than that of a cell and filling it with a salt solution that can conduct an electric current. Under a microscope, use a micromanipulator (mechanical positioning device) to insert the tip of the microelectrode into an axon. Place a reference electrode in the solution outside the cell. Use an oscilloscope or voltmeter to measure the voltage between the microelectrode tip in the axon and the reference electrode outside the cell.

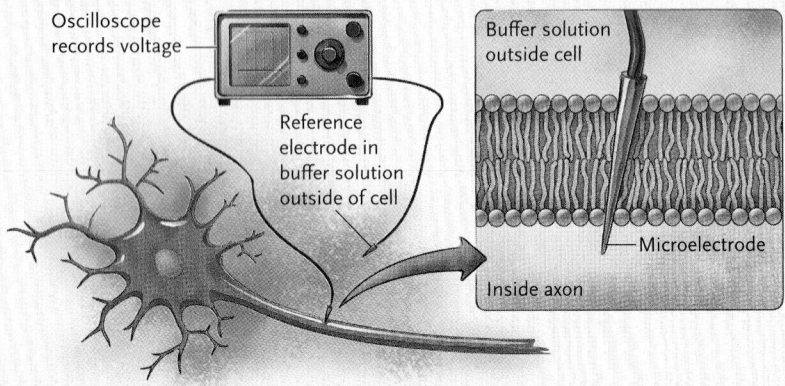

INTERPRETING THE RESULTS: The oscilloscope or voltmeter indicates the membrane potential in volts. Changes in membrane potential caused by stimuli or chemical treatments can be measured and recorded. For an isolated, unstimulated neuron (shown above) the membrane potential is typically about −70 mV.

threshold potential, about −50 to −55 mV in isolated neurons. Once the threshold is reached, the action potential fires—and the membrane potential suddenly increases. In less than 1 msec (millisecond, one-thousandth of a second), it rises so high that the inside of the plasma membrane becomes positive due to an influx of positive ions across the cell membrane, momentarily reaching a value of +30 mV or more. The potential then falls again, in many cases dropping to about −80 mV before rising again to the resting potential. When the potential is below the resting value, the membrane is said to be **hyperpolarized.** The entire change, from initiation of the action potential to the return to the resting potential, takes less than 5 msec in the fastest neurons. Action potentials take the same basic form in neurons of all types, with differences in the values of the resting potential and the peak of the action potential, and in the time required to return to the resting potential.

All stimuli cause depolarization of a neuron, but an action potential is produced only if the stimulus is strong enough to cause depolarization to reach the threshold. This is referred to as the **all-or-nothing principle;** once triggered, the changes in membrane potential take place independently of the strength of the stimulus.

Beginning at the peak of an action potential, the membrane enters a **refractory period** of a few milliseconds during which the threshold required for generation of an action potential is much higher than normal. The refractory period lasts until the membrane has stabilized at the resting potential. As we shall see, the refractory period keeps impulses traveling in a one-way direction in neurons.

outside being positively charged. As we will see in the following description of the changes in a neuron that occur when it is stimulated, the *voltage-gated ion channels* for Na$^+$ and K$^+$ open and close when the membrane potential changes.

The Membrane Potential Changes from Negative to Positive during an Action Potential

When a neuron conducts an electrical impulse, an abrupt and transient change in membrane potential occurs; this is called the **action potential.** An action potential begins as a stimulus that causes positive charges from outside the neuron to flow inward, making the cytoplasmic side of the membrane less negative (**Figure 37.9**). As the membrane potential becomes less negative, the membrane (which was polarized at rest) becomes **depolarized.** Depolarization proceeds relatively slowly until it reaches a level known as the

The Action Potential Is Produced by Ion Movements through the Plasma Membrane

The action potential is produced by movements of Na$^+$ and K$^+$ through the plasma membrane. The movements are controlled by specific **voltage-gated ion channels,** membrane-embedded proteins that open and close as the membrane potential changes (see Figure 37.8). Voltage-gated Na$^+$ channels have two gates, an *activation gate* and an *inactivation gate*, whereas voltage-gated K$^+$ channels have one gate, an *activation gate*.

How the two voltage-gated ion channels operate to generate an action potential is shown in **Figure 37.10.** When the membrane is at the resting potential, the activation gates of both the Na$^+$ and K$^+$ channels are closed. As a depolarizing stimulus raises the membrane potential to the threshold, the activation gate of the Na$^+$ channels opens, allowing a burst of Na$^+$ ions to flow into the axon along their concentration gradient. Once above the threshold, more Na$^+$ channels open, causing a rapid inward flow of positive charges that raises the membrane potential to-

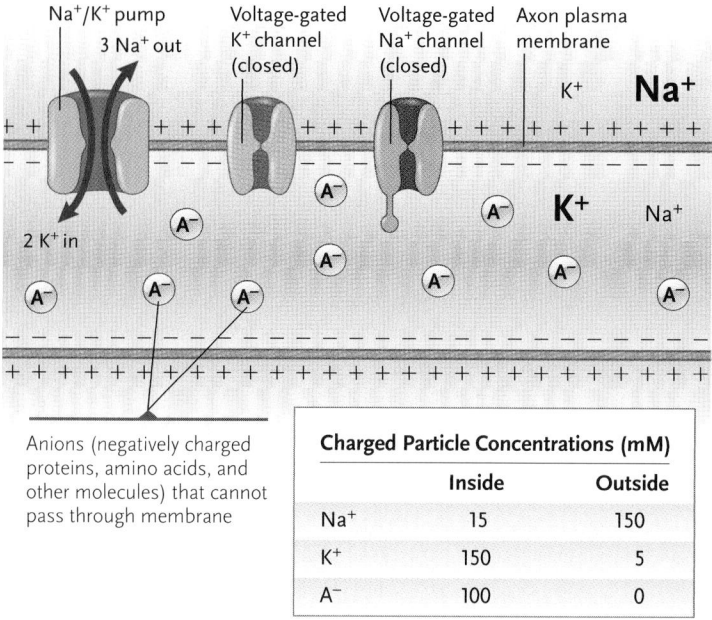

Charged Particle Concentrations (mM)		
	Inside	Outside
Na⁺	15	150
K⁺	150	5
A⁻	100	0

Anions (negatively charged proteins, amino acids, and other molecules) that cannot pass through membrane

Figure 37.8
The distribution of ions inside and outside an axon that produces the resting potential, −70 mV. The distribution of ions that do not directly affect the resting potential, such as Cl⁻, is not shown. The voltage-gated ion channels open and close when the membrane potential changes.

ward the peak of the action potential. As the action potential peaks, the inactivation gate of the Na⁺ channel closes (resembling putting a stopper in the sink), which stops the inward flow of Na⁺. The refractory period now begins.

At the same time, the activation gates of the K⁺ channels begin to open, allowing K⁺ ions to flow rapidly outward in response to their concentration gradient. The K⁺ ions contribute to the refractory period and compensate for the inward movement of Na⁺ ions, returning the membrane to the resting potential. As the resting potential is reestablished, the activation gates of the K⁺ channels close, as do those of the Na⁺ channels, and the inactivation gates of the Na⁺ channels open. These events end the refractory period and ready the membrane for another action potential.

In some neurons, closure of the gated K⁺ channels lags, and K⁺ continues to flow outward for a brief time after the membrane returns to the resting potential. This excess outward flow causes the hyperpolarization shown in Figure 37.9, in which the membrane potential dips briefly below the resting potential.

At the end of an action potential, the membrane potential has returned to its resting state, but the ion distribution has changed slightly. That is, some Na⁺ ions have entered the cell, and some K⁺ ions have left the cell—but not many, relative to the total number of ions, and the distribution is not altered enough to prevent other action potentials from occurring. In the long term, the Na⁺/K⁺ active transport pumps restore the Na⁺ and K⁺ to their original locations.

Some of what is known about how ion flow through channels can change membrane potential has come from experiments using the *patch-clamp* technique. In the patch part of the technique, a micropipette with a

tip 1 to 3 μm in diameter is touched to the plasma membrane of a neuron (or other cell type). The contact seals the membrane to the micropipette and, when the micropipette is pulled away, a patch of membrane with one or a few ion channels comes with it. The clamp part of the technique refers to a voltage clamp, in which an electronic device holds the membrane potential of the patch at a steady value chosen by the investigator. The investigator can add a stimulus that is expected to open or close ion channels. The amount of current the clamping device needs to keep the voltage constant is directly related to the number and charge of the ions moving through the channels and, hence, measures channel activity.

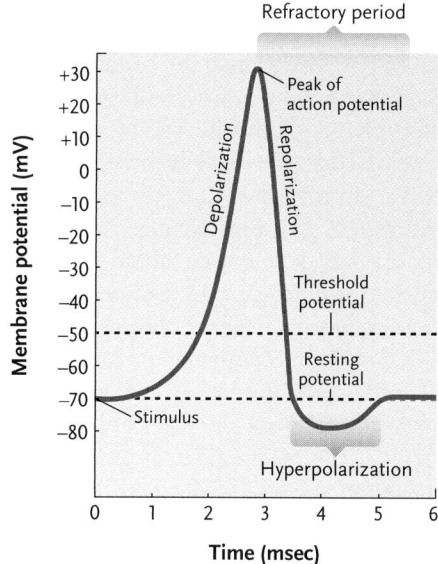

Figure 37.9
Changes in membrane potential during an action potential.

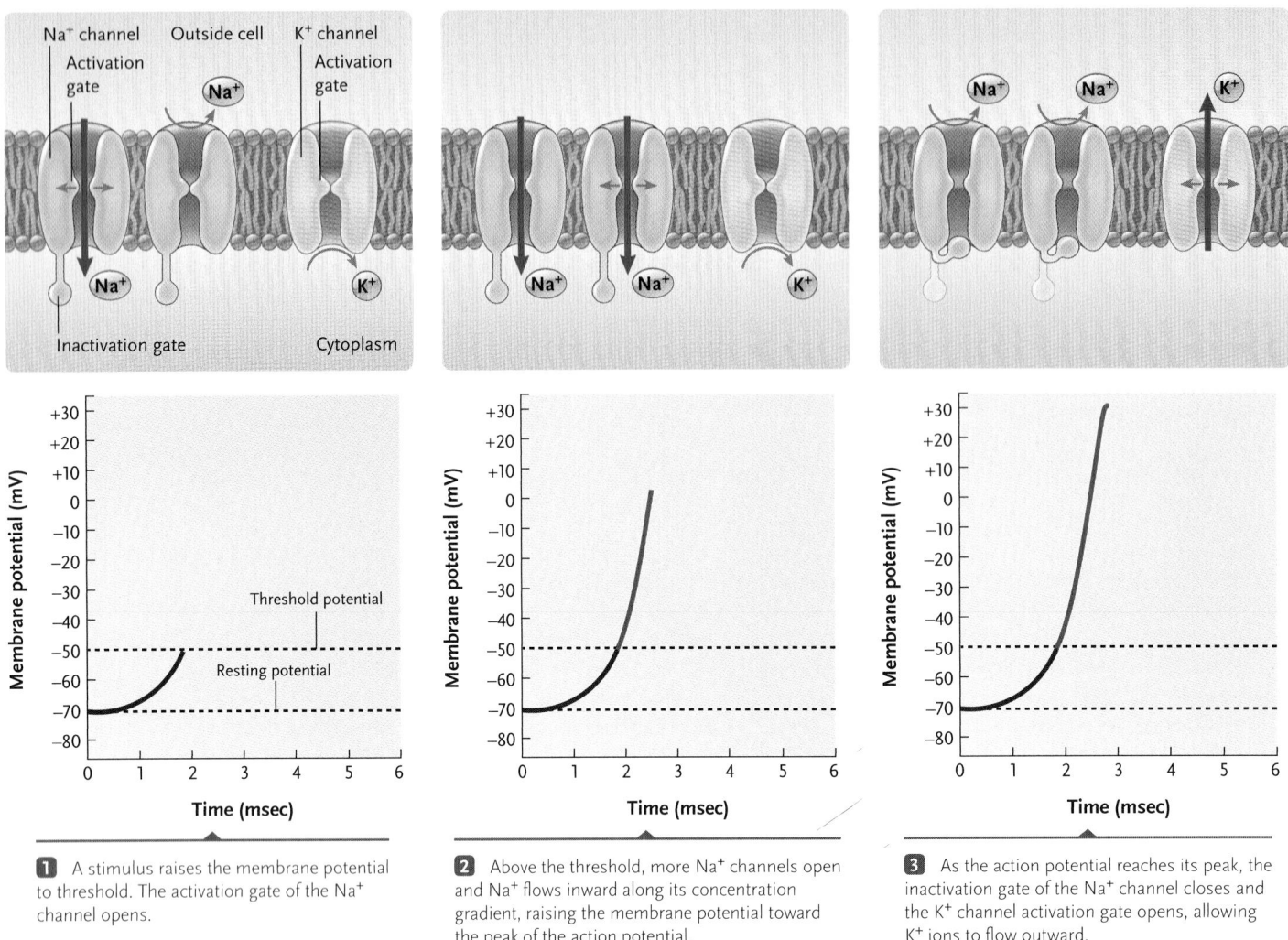

1 A stimulus raises the membrane potential to threshold. The activation gate of the Na⁺ channel opens.

2 Above the threshold, more Na⁺ channels open and Na⁺ flows inward along its concentration gradient, raising the membrane potential toward the peak of the action potential.

3 As the action potential reaches its peak, the inactivation gate of the Na⁺ channel closes and the K⁺ channel activation gate opens, allowing K⁺ ions to flow outward.

Figure 37.10
Changes in voltage-gated Na^+ and K^+ channels that produce the action potential.

Neural Impulses Move by Propagation of Action Potentials

Once an action potential is initiated at the dendrite end of the neuron, it passes along the surface of a nerve or muscle cell as an automatic wave of depolarization traveling away from the stimulation point **(Figure 37.11)**. The action potential does not need further trigger events in order for it to be propagated along the axon to the terminals. In a segment of an axon that is generating an action potential, the outside of the membrane becomes temporarily negative and the inside positive. Because opposites attract, as the region outside becomes negative, local current flow occurs between the area undergoing an action potential and the adjacent downstream inactive area both inside and outside the membrane (arrows, Figure 37.11). This current flow makes nearby regions the axon membrane less positive on the outside and more positive on the inside; in other words, they depolarize the membrane.

The depolarization is large enough to push the membrane potential past the threshold, opening the

voltage-gated Na^+ and K^+ channels and starting an action potential in the downstream adjacent region. In this way, each segment of the axon stimulates the next segment to fire, and the action potential moves rapidly along the axon as a nerve impulse.

The refractory period keeps an action potential from reversing direction at any point along an axon; only the region in front of the action potential can fire. The refractory period results from the properties of the voltage-gated ion channels. Once they have been opened to their activated state, the upstream voltage-gated ion channels need time to reset to their original positions before they can open again. Therefore, only downstream voltage-gated ion channels are able to open, ensuring the one-way movement of the action potential along the axon toward the axon tips. By the time the refractory period ends in a membrane segment that has just fired an action potential, the action potential has moved too far away to cause a second action potential to develop in the same segment.

The magnitude of an action potential stays the same as it travels along an axon, even where the axon

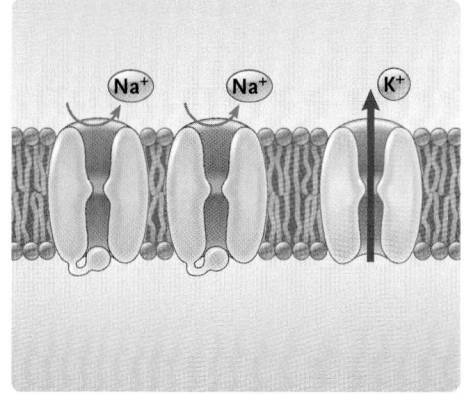

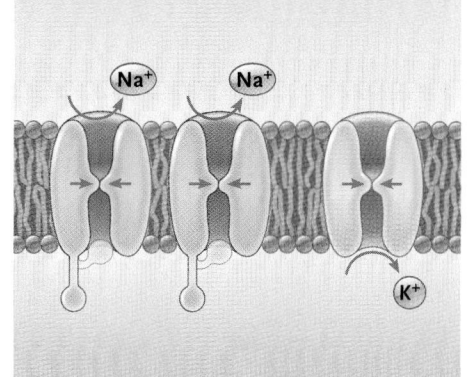

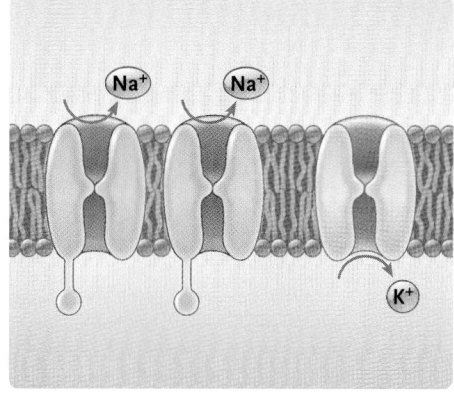

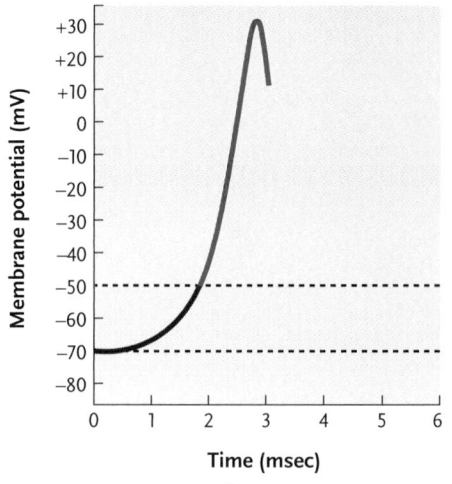

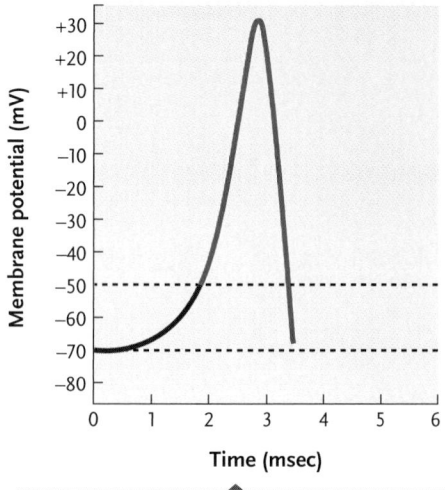

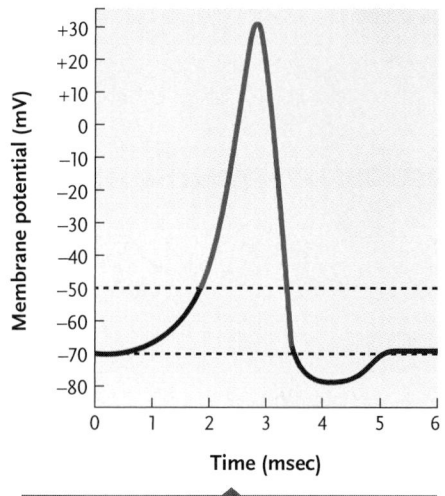

4 The outward flow of K⁺ along its concentration gradient causes the membrane potential to begin to fall.

5 As the membrane potential reaches the resting value, the activation gate of the Na⁺ channel closes and the inactivation gate opens. The K⁺ activation gate also closes.

6 Closure of the K⁺ activation gate stabilizes the membrane potential at the resting value.

branches at its tips. Thus the propagation of an action potential resembles a burning fuse, which burns with the same intensity along its length, and along any branches, once it is lit at one end. Unlike a fuse, however, an axon can fire another action potential of the same intensity within a few milliseconds after an action potential passes through.

Due to the all-or-nothing principle of action potential generation, the intensity of a stimulus is reflected in the *frequency* of action potentials—the greater the stimulus, the more action potentials per second, up to a limit depending on the axon type—rather than by the change in membrane potential. For most neuron types, the limit lies between 10 and 100 action potentials per second.

Both natural and synthetic substances target specific parts of the mechanism generating action potentials. Local anesthetics, such as procaine and lidocaine, bind to voltage-gated Na⁺ channels and block their ability to transport ions; thus, sensory nerves in the anesthetized region cannot transmit pain signals. The potent poison of the pufferfish, tetrodotoxin, also blocks voltage-gated Na⁺ channels in neurons, poten-

tially causing muscle paralysis and death. The pufferfish is highly prized as a delicacy in Japan, eaten after careful preparation to remove organs carrying the tetrodotoxin. A mistake can kill the diners, however, making pufferfish sashimi a kind of culinary Russian roulette.

Saltatory Conduction Increases Propagation Rate in Small-Diameter Axons

In the propagation pattern shown in Figure 37.11, an action potential spreads along every segment of the membrane along the length of the axon. For this type of action potential propagation, the rate of conduction increases with the diameter of the axon. Axons with a very large diameter have evolved in invertebrates such as lobsters, earthworms, and squids as well as a few marine fishes. Giant axons typically carry signals that produce an escape or withdrawal response, such as the sudden flexing of the tail (abdomen) in lobsters that propels the animal backward. The largest known axons, 1.7 mm in diameter, occur in fanworms *(Myxicola)*.

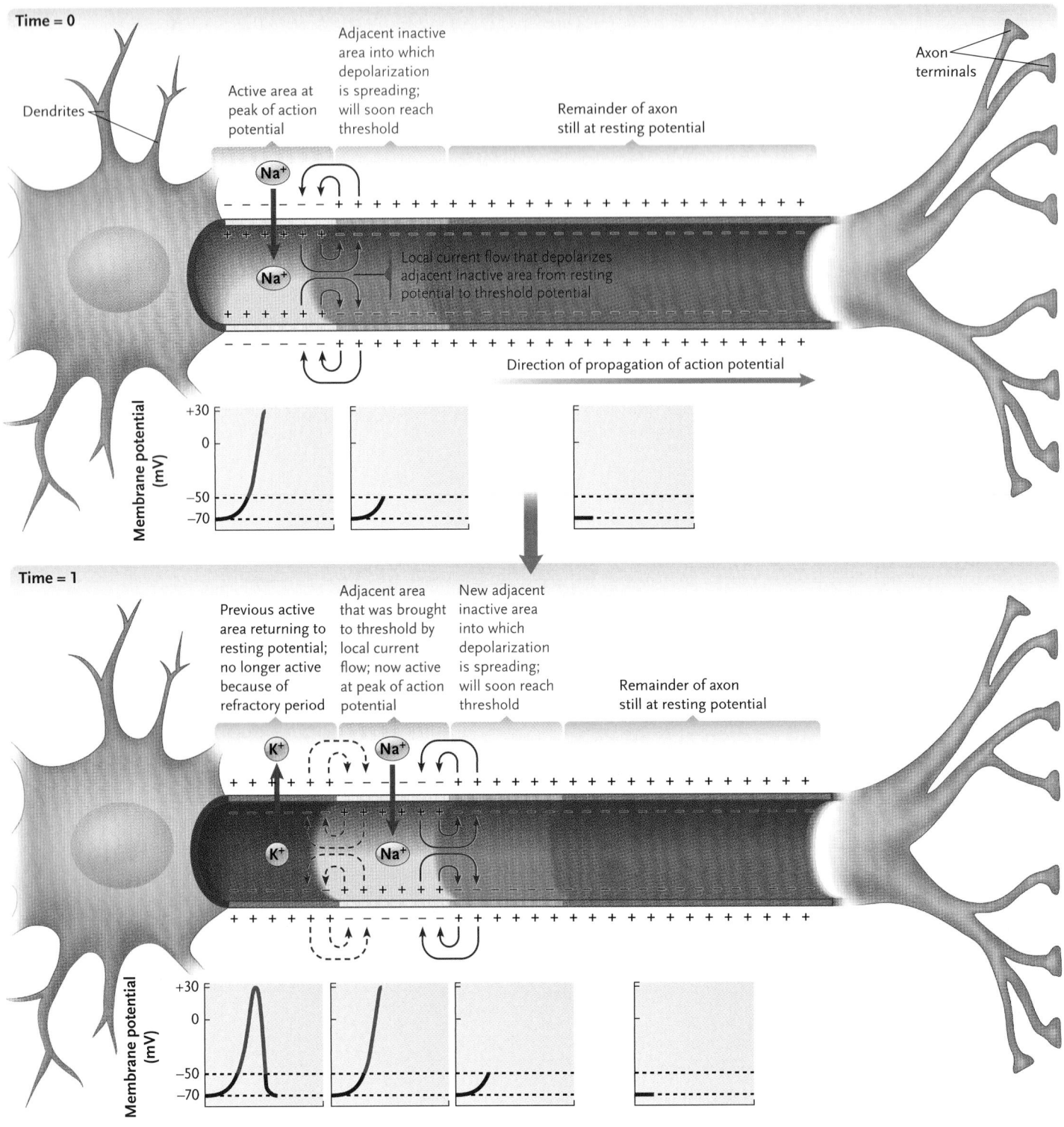

Figure 37.11

Propagation of an action potential along an unmyelinated axon by ion flows between a firing segment and an adjacent unfired region of the axon. Each firing segment induces the next to fire, causing the action potential to move along the axon.

The signals they carry contract a muscle that retracts the fanworm's body into a protective tube when the animal is threatened.

Although large-diameter axons can conduct impulses as rapidly as 25 m/sec (over twice the speed of the world record 100-meter dash), they take up a great deal of space. In complex vertebrates, natural selection has led to a mechanism that allows small-diameter axons to conduct impulses rapidly. The mechanism, called **saltatory conduction** (*saltere* = to leap), allows action potentials to "hop" rapidly along axons instead of burning smoothly like a fuse.

Saltatory conduction depends on the insulating myelin sheath that forms around some axons and in particular on the nodes of Ranvier exposing the axon membrane to extracellular fluids. Voltage-gated Na^+ and K^+ channels crowded into the nodes allow action potentials to develop at these positions **(Figure 37.12)**. The inward movement of Na^+ ions produces depolarization, but the excess positive ions are unable to leave the axon through the membrane regions covered by the myelin sheath. Instead, they diffuse rapidly to the next node where they cause depolarization, inducing an action potential at that node. As this mechanism repeats, the action potential jumps rapidly along the axon from node to node. Saltatory conduction proceeds at rates up to 130 m/sec while an unmyelinated axon of the same diameter conducts action potentials at about 1 m/sec.

Saltatory conduction allows thousands to millions of fast-transmitting axons to be packed into a relatively small diameter. For example, in humans the optic nerve leading from the eye to the brain is only 3 mm in diameter but is packed with more than a million axons. If those axons were unmyelinated, each would have to be about 100 times thicker to conduct impulses at the same velocity, producing an optic nerve about 300 mm (12 inches) in diameter.

The disease *multiple sclerosis* (*sclero* = hard) underscores the importance of myelin sheaths to the operation of the vertebrate nervous system. In this disease, myelin is progressively lost from axons and replaced by hardened scar tissue. The changes block or slow the transmission of action potentials, producing numbness, muscular weakness, faulty coordination of movements, and paralysis that worsens as the disease progresses.

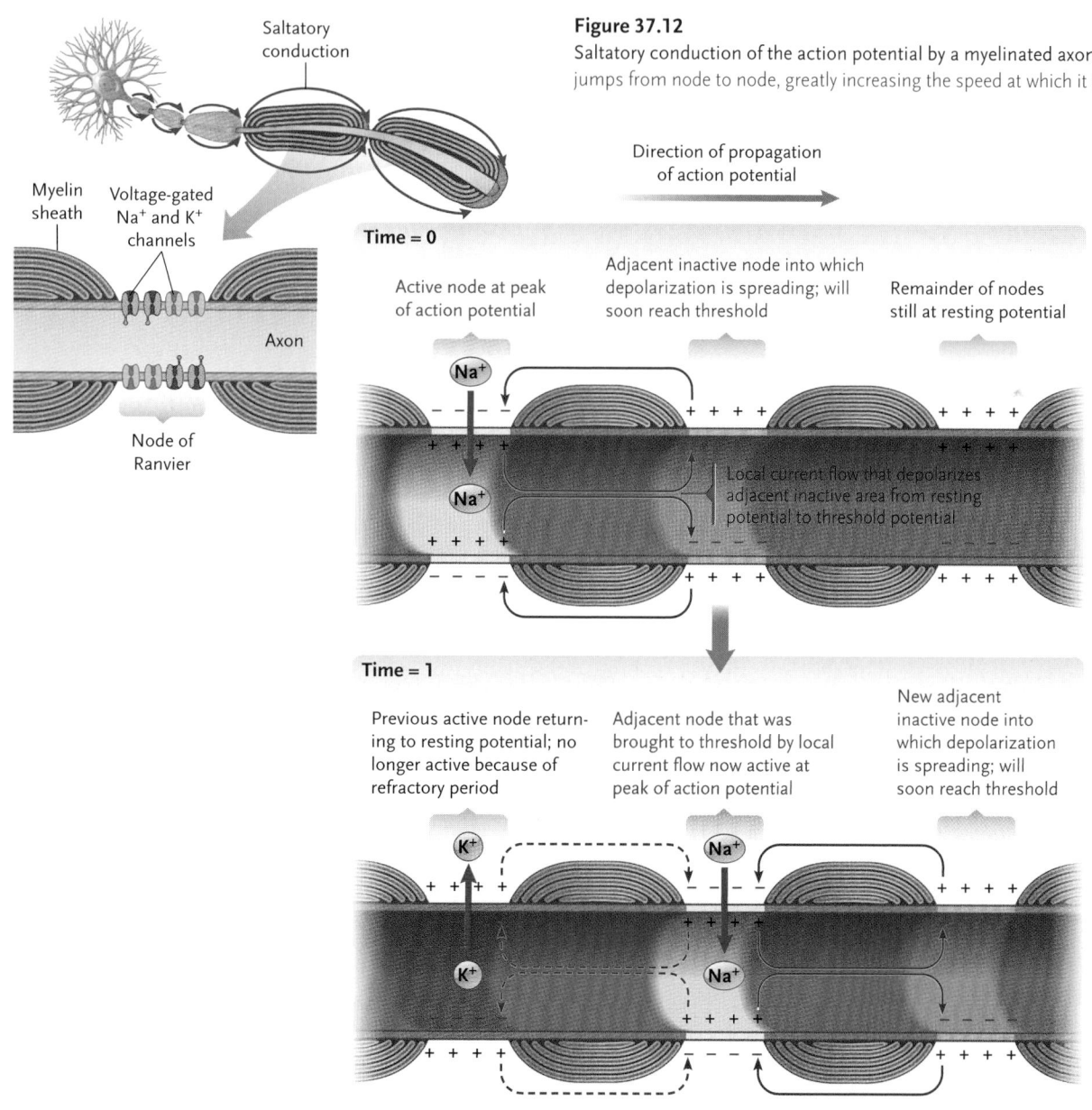

Figure 37.12
Saltatory conduction of the action potential by a myelinated axon. The action potential jumps from node to node, greatly increasing the speed at which it travels along the axon.

1. What mechanism ensures that an electrical impulse in a neuron is conducted in only one direction down the axon?
2. How does having a myelin sheath affect the conduction of impulses in neurons?

37.3 Conduction across Chemical Synapses

Action potentials are transmitted directly across electrical synapses, but they cannot jump across the cleft in a chemical synapse. Instead, the arrival of an action potential causes neurotransmitter molecules—which are synthesized in the cell body of the neuron—to be released by the plasma membrane of the axon terminal, called the **presynaptic membrane (Figure 37.13).** The neurotransmitter diffuses across the cleft and al-

ters ion conduction by activating *ligand-gated ion channels* in the **postsynaptic membrane**, the plasma membrane of the postsynaptic cell. **Ligand-gated ion channels** are channels that open or close when a specific chemical, such as a neurotransmitter, binds to the channel. Neurotransmitter communication from presynaptic to postsynaptic cells is a specialized case of cell-to-cell communication and signal transduction, which you learned about in Chapter 7. Neurobiologists study this phenomenon from the standpoint of understanding the function of neurons, but some of the details being learned are similar in many other types of cells.

Neurotransmitters work in one of two ways. **Direct neurotransmitters** bind directly to a ligand-gated ion channel in the postsynaptic membrane, which opens or closes the channel gate and alters the flow of a specific ion or ions in the postsynaptic cell. The time between arrival of an action potential at an axon terminal and alteration of the membrane potential in the postsynaptic cell may be as little as 0.2 msec.

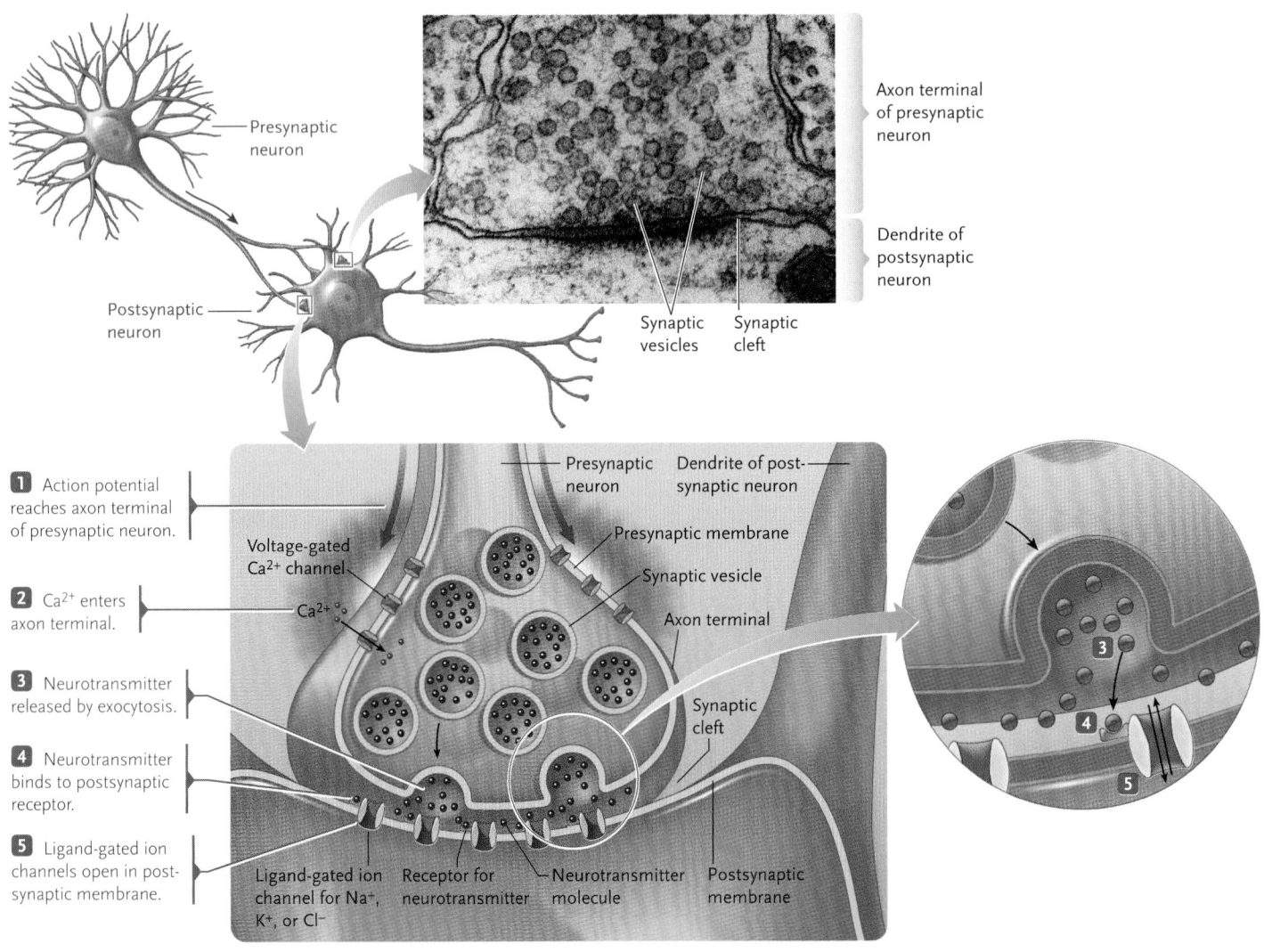

1 Action potential reaches axon terminal of presynaptic neuron.

2 Ca²⁺ enters axon terminal.

3 Neurotransmitter released by exocytosis.

4 Neurotransmitter binds to postsynaptic receptor.

5 Ligand-gated ion channels open in postsynaptic membrane.

Presynaptic neuron

Postsynaptic neuron

Axon terminal of presynaptic neuron

Dendrite of postsynaptic neuron

Synaptic vesicles

Synaptic cleft

Presynaptic neuron

Dendrite of postsynaptic neuron

Presynaptic membrane

Synaptic vesicle

Axon terminal

Synaptic cleft

Postsynaptic membrane

Neurotransmitter molecule

Receptor for neurotransmitter

Ligand-gated ion channel for Na⁺, K⁺, or Cl⁻

Voltage-gated Ca²⁺ channel

Ca²⁺

Figure 37.13
Structure and function of chemical synapses.

(Micrograph: © Dennis Kunkel/Visuals Unlimited.)

Indirect neurotransmitters work more slowly (on the order of hundreds of milliseconds). They act as *first messengers,* binding to G-protein-coupled receptors in the postsynaptic membrane, which activates the receptor and triggers generation of a *second messenger* such as cyclic AMP or other processes (see Section 7.4). The cascade of second messenger reactions opens or closes ion-conducting channels in the postsynaptic membrane. Indirect neurotransmitters typically have effects that may last for minutes or hours. Some substances can act as either direct or indirect neurotransmitters, depending on the types of receptors they bind in the receiving cell.

The time required for the release, diffusion, and binding of neurotransmitters across chemical synapses delays transmission as compared with the almost instantaneous transmission of impulses across electrical synapses. However, communication through chemical synapses allows neurons to receive inputs from hundreds to thousands of axon terminals at the same time. Some neurotransmitters have stimulatory effects, while others have inhibitory effects. All of the information received at a postsynaptic membrane is integrated to produce a response. Thus, by analogy, communication by electrical synapses resembles the effect of simply touching one wire to another; communication by direct and indirect neurotransmitters resembles the integration of multiple inputs by a computer chip.

Neurotransmitters Are Released by Exocytosis

Neurotransmitters are stored in secretory vesicles called **synaptic vesicles** in the cytoplasm of an axon terminal. The arrival of an action potential at the terminal releases the neurotransmitters by *exocytosis:* the vesicles fuse with the presynaptic membrane and release the neurotransmitter molecules into the synaptic cleft.

The release of synaptic vesicles depends on voltage-gated Ca^{2+} channels in the plasma membrane of an axon terminal (see Figure 37.13). Ca^{2+} ions are constantly pumped out of all animal cells by an active transport protein in the plasma membrane, keeping their concentration higher outside than inside. As an action potential arrives, the change in membrane potential opens the Ca^{2+} channel gates in the axon terminal, allowing Ca^{2+} to flow back into the cytoplasm. The rise in Ca^{2+} concentration triggers a protein in the membrane of the synaptic vesicle that allows the vesicle to fuse with the plasma membrane, releasing neurotransmitter molecules into the synaptic cleft.

Each action potential arriving at a synapse typically causes approximately the same number of synaptic vesicles to release their neurotransmitter molecules. For example, arrival of an action potential at one type of synapse causes about 300 synaptic vesicles to release

a neurotransmitter called acetylcholine. Each vesicle contains about 10,000 molecules of the neurotransmitter, giving a total of some 3 million acetylcholine molecules released into the synaptic cleft by each arriving action potential.

When a stimulus is no longer present, action potentials are no longer generated and a response is no longer needed. In this case a series of events prevents continued transmission of the signal. When action potentials stop arriving at the axon terminal, the voltage-gated Ca^{2+} channels in the axon terminal close and the Ca^{2+} in the axon cytoplasm is quickly pumped to the outside. The drop in cytoplasmic Ca^{2+} stops vesicles from fusing with the presynaptic membrane, and no further neurotransmitter molecules are released. Any free neurotransmitter molecules remaining in the cleft quickly diffuse away, are broken down by enzymes in the cleft, or are pumped back into the axon terminals or into glial cells by active transport. Transmission of impulses across the synaptic cleft ceases within milliseconds after action potentials stop arriving at the axon terminal.

Most Neurotransmitters Alter Ion Flow through Na⁺ or K⁺ Channels

Neurotransmitters work by opening or closing membrane-embedded ligand-gated ion channels; most of these channels conduct Na^+ or K^+ across the postsynaptic membrane, although some regulate chloride ions (Cl^-). The altered ion flow in the postsynaptic cell that results from the opening or closing of the gates may stimulate or inhibit the generation of action potentials by that cell. For example, if Na^+ channels are opened, the inward Na^+ flow brings the membrane potential of the postsynaptic cell toward the threshold (the membrane becomes depolarized). If K^+ channels are opened, the outward flow of K^+ has the opposite effect (the membrane becomes hyperpolarized). The combined effects of the various stimulatory and inhibitory neurotransmitters at all the chemical synapses of a postsynaptic neuron or muscle cell determine whether the postsynaptic cell triggers an action potential. (*Insights from the Molecular Revolution* describes experiments that worked out the structure and function of an ion channel gated directly by a neurotransmitter.)

Many Different Molecules Act as Neurotransmitters

In all, nearly 100 different substances are now known or suspected to be neurotransmitters. Most of them are relatively small molecules that diffuse rapidly across the synaptic cleft. Some axon terminals release only one type of neurotransmitter while others release several types. Depending on the type of receptor to which it binds, the same neurotransmitter may stimulate or inhibit the generation of action potentials in the post-

INSIGHTS FROM THE MOLECULAR REVOLUTION

Dissecting Neurotransmitter Receptor Functions

Many receptors for direct neurotransmitters are part of an ion channel that is opened or closed by the binding of a neurotransmitter molecule. Each of these receptors has two regions: a large, hydrophilic portion on the outside surface of the plasma membrane that binds the neurotransmitter and a hydrophobic transmembrane portion that anchors the receptor in the plasma membrane and forms the ion-conducting channel.

Jean-Luc Eiselé and his coworkers at the National Center of Scientific Research and the Central Medical University in Switzerland were interested in determining whether the two primary activities of these receptors—binding neurotransmitters and conducting ions—depend on parts of the protein that work independently or reflect an integration of the entire protein structure.

To find out, Eiselé and his colleagues constructed artificial receptors using regions of the receptors for two different neurotransmitters, acetylcholine and serotonin. These two receptors, although related in amino acid sequence and structure, bind different neurotransmitters and react differently to calcium ions. Ion conduction by the acetylcholine receptor is enhanced by Ca^{2+}, while Ca^{2+} ions block the channel of the serotonin receptor and stop ion conduction.

To create the artificial receptors, the investigators broke the genes encoding the acetylcholine and serotonin receptors into two parts. They then reassembled the parts so that in the protein encoded by the composite gene, the part of the acetylcholine receptor located on the membrane surface was joined to the transmembrane channel of the serotonin receptor. Five versions of the artificial gene, encoding proteins in which the two parts were joined at different positions in the amino acid sequence, were then cloned to increase their quantity and injected into oocytes of the clawed frog, *Xenopus laevis*. Once in the oocytes, the genes were translated into the artificial receptor proteins, which were inserted into the oocyte plasma membranes.

Of the five artificial receptors, all were able to bind acetylcholine, but only two were able to conduct ions in response to binding the neurotransmitter, as measured by an increase in the electrical current flowing across the plasma membrane of the oocytes. Agents that inhibit the normal acetylcholine receptor, such as curare, also inhibited the artificial receptors. Serotonin, in contrast, was not bound and did not open the receptor channels, and agents that inhibit the normal serotonin receptor had no effect on the artificial receptors. However, elevated Ca^{2+} concentrations blocked the channel, as in the normal serotonin receptor.

The remarkable research by Eiselé and his coworkers indicates that the parts of a receptor binding a neurotransmitter and conducting ions function independently. Their work also demonstrates the feasibility of constructing composite receptors as a means for dissecting the functions of subregions of the receptors.

---

synaptic cell. **Figure 37.14** depicts some examples of neurotransmitters.

Acetylcholine acts as a neurotransmitter in both invertebrates and vertebrates. In vertebrates, it acts as a direct neurotransmitter between neurons and muscle cells and as an indirect neurotransmitter between neurons carrying out higher brain functions such as memory, attention, perception, and learning. Acetylcholine-releasing neurons in the brain degenerate in people who develop Alzheimer disease, in which memory, speech, and perceptual abilities decline.

Acetylcholine is the target of many natural and artificial poisons. Curare, a plant extract used as an arrow poison by some indigenous peoples of South America, blocks muscle contraction and produces paralysis by competing directly with acetylcholine for binding sites in synapses that control muscle cells. Atropine, an ingredient of the drops an eye doctor uses to dilate your pupils, is also a plant extract; it relaxes the iris muscles by blocking their acetylcholine receptors. Nicotine also binds to acetylcholine receptors, but acts as a stimulant by turning the receptors on rather than off.

Several amino acids operate as direct neurotransmitters in the CNS of vertebrates and in nerve-muscle synapses of insects and crustaceans. *Glutamate* and *aspartate* stimulate action potentials in postsynaptic cells. They are directly involved in vital brain functions such as memory and learning. *Gamma aminobutyric acid (GABA)*, a derivative of glutamate, acts as an inhibitor by opening Cl^- channels in postsynaptic membranes. *Glycine* is also an inhibitor.

Other substances can block the operation of these neurotransmitters. For example, tetanus toxin, released by the bacterium *Clostridium tetani*, blocks GABA release in synapses that control muscle contraction. The body muscles contract so forcibly that the body arches painfully and the teeth become tightly clenched, giving the condition its common name of lockjaw. Once the effects extend to respiratory muscles, the victim quickly dies.

The biogenic amines, which are derived from amino acids, act primarily as indirect neurotransmitters in the CNS. *Norepinephrine, epinephrine,* and *dopamine,* all derived from tyrosine, function as neurotransmitters between interneurons involved in such diverse brain and body functions as consciousness, memory, mood, sensory perception, muscle movements, maintenance of blood pressure, and sleep. Norepinephrine

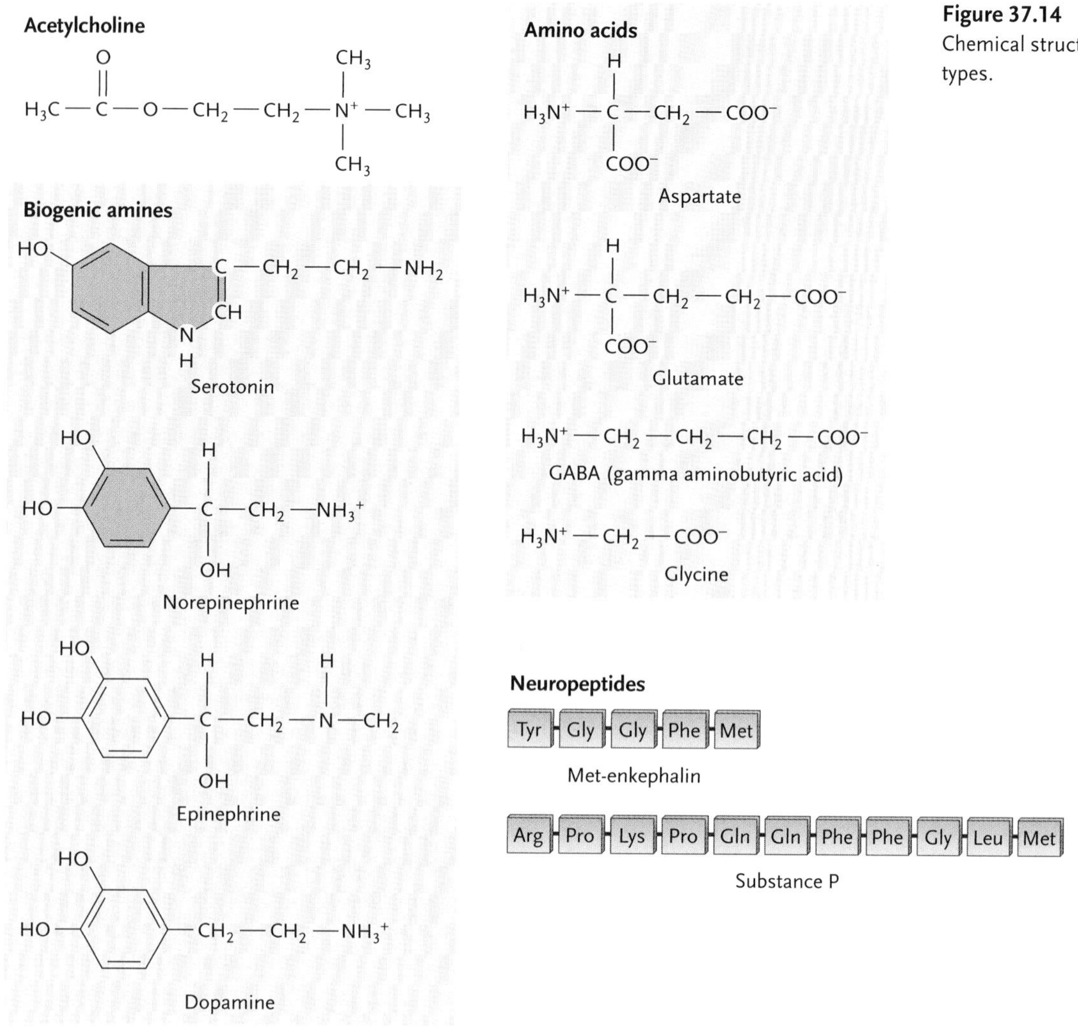

Acetylcholine

$$H_3C - \overset{\overset{\displaystyle O}{\|}}{C} - O - CH_2 - CH_2 - \overset{\overset{\displaystyle CH_3}{|}}{\underset{\underset{\displaystyle CH_3}{|}}{N^+}} - CH_3$$

Biogenic amines

Serotonin

Norepinephrine

Epinephrine

Dopamine

Amino acids

$$H_3N^+ - \overset{\overset{\displaystyle H}{|}}{\underset{\underset{\displaystyle COO^-}{|}}{C}} - CH_2 - COO^-$$

Aspartate

$$H_3N^+ - \overset{\overset{\displaystyle H}{|}}{\underset{\underset{\displaystyle COO^-}{|}}{C}} - CH_2 - CH_2 - COO^-$$

Glutamate

$$H_3N^+ - CH_2 - CH_2 - CH_2 - COO^-$$

GABA (gamma aminobutyric acid)

$$H_3N^+ - CH_2 - COO^-$$

Glycine

Neuropeptides

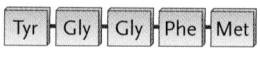

Met-enkephalin

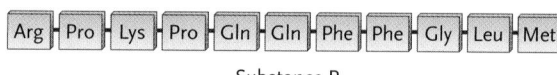

Substance P

Figure 37.14
Chemical structures of the major neurotransmitter types.

and epinephrine are also released into the general body circulation as hormones. Parkinson disease, in which there is a progressive loss of muscle control, results from degeneration of dopamine-releasing neurons in regions of the brain that coordinate muscular movements. *Serotonin,* which is derived from tryptophan, is released by interneurons in the pathways regulating appetite, reproductive behavior, muscular movements, sleep, and emotional states such as anxiety.

Several drugs enhance or inhibit the action of biogenic amines. For example, cocaine binds to the transporters for active reuptake of certain neurotransmitters such as norepinephrine, dopamine, and serotonin from the synaptic cleft, thereby preventing them from being reabsorbed by the neurons that released them. As a result, the concentrations of the neurotransmitters increase in the synapses, leading to amplification of their natural effects. That is, the affected neurons produce symptoms characteristic of cocaine use, namely high energy from the norepinephrine, euphoria from the dopamine, and feelings of confidence from the serotonin.

Neuropeptides, which are short chains of two or more amino acids, act as indirect neurotransmitters in the central and peripheral nervous systems of both ver-tebrates and invertebrates. More than 50 neuropeptides are now known. Neuropeptides are also released into the general body circulation as peptide hormones.

Neuropeptides called *endorphins* ("endogenous morphines") are released during periods of pleasurable experience such as eating or sexual intercourse, or physical stress such as childbirth or extended physical exercise. These neurotransmitters have the opiatelike property of reducing pain and inducing euphoria, well known to exercise buffs as a pleasant by-product of their physical efforts. Most endorphins act on the PNS and effectors such as muscles, but *enkephalins,* a sub-class of the endorphins, bind to particular receptors in the CNS. Morphine, a potent drug extracted from the opium poppy, blocks the sensation of pain and produces a sensation of well-being by binding to the same receptors in the brain.

Another neuropeptide associated with pain response is *substance P,* which is released by special neurons in the spinal cord. Its effect is to increase messages associated with intense, persistent, or severe pain. For example, suppose you put your hand on a hot barbecue grill. You will snatch your hand away immediately by reflex action, and you will feel the "ouch" of the pain a little later. Why do events occur in this order? The reflex action is driven by rapid nerve impulse conduction along myelinated neurons. The neurons that release substance P are not myelinated, however, so their signal is conducted more slowly and the feeling of pain is delayed. The action of endorphins is antagonistic to substance P, reducing the perception of pain.

In mammals and probably other vertebrates, some neurons synthesize and release dissolved carbon monoxide and nitric oxide as neurotransmitters. For example, in the brain, carbon monoxide regulates the release of hormones from the hypothalamus. Nitric oxide contributes to many nervous system functions such as learning, sensory responses, and muscle movements. By relaxing smooth muscles in the walls of blood vessels, nitric oxide causes the vessels to dilate, increasing the flow of blood. For example, when a male is sexually aroused, neurons release nitric oxide into the erectile tissues in the penis. Relaxation of the muscles increases blood flow into the tissues, causing them to fill with blood and produce an erection. The impotency drug Viagra aids erection by inhibiting an enzyme that normally reduces nitric oxide concentration in the penis.

STUDY BREAK

1. What features characterize a substance as a neurotransmitter?
2. Describe how a direct neurotransmitter in a presynaptic neuron controls action potentials in a postsynaptic neuron.

37.4 Integration of Incoming Signals by Neurons

Most neurons receive a multitude of stimulatory and inhibitory signals carried by both direct and indirect neurotransmitters. These signals are integrated by the postsynaptic neuron into a response that reflects their combined effects. The integration depends primarily on the patterns, number, types, and activity of the synapses the postsynaptic neuron makes with presynaptic neurons. Inputs from other sources, such as indirect neurotransmitters and other signal molecules, can modify the integration. The response of the postsynaptic neuron is elucidated by the frequency of action potentials it generates.

Integration at Chemical Synapses Occurs by Summation

As mentioned earlier, depending on the type of receptor to which it binds, a neurotransmitter may stimulate or inhibit the generation of action potentials in the postsynaptic neuron. If a neurotransmitter opens a ligand-gated Na^+ channel, Na^+ enters the cell, causing a depolarization. This change in membrane potential pushes the neuron closer to threshold; that is, it is excitatory and is called an **excitatory postsynaptic potential**, or **EPSP**. On the other hand, if a neurotransmitter opens a ligand-gated ion channel that allows Cl^- to flow into the cell and K^+ to flow out, hyperpolarization occurs. This change in membrane potential pushes the neuron farther from threshold; that is, it is inhibitory and is called an **inhibitory postsynaptic potential**, or **IPSP**. In contrast to the all-or-nothing operation of an action potential, EPSPs and IPSPs are **graded potentials**, in which the membrane potential increases or decreases without necessarily triggering an action potential. And there are no refractory periods for EPSPs and IPSPs.

A neuron typically has hundreds to thousands of chemical synapses formed by axon terminals of presynaptic neurons contacting its dendrites and cell body **(Figure 37.15)**. The events that occur at a single synapse produce either an EPSP or an IPSP in that postsynaptic neuron. But how is an action potential produced if a single EPSP is not sufficient to push the postsynaptic neuron to threshold? The answer involves the summation of the inputs received through those many chemical synapses formed by presynaptic neurons. At any given time, some or many of the presynaptic neurons may be firing, producing EPSPs and/or IPSPs in the postsynaptic neuron. The sum of all the EPSPs and IPSPs at a given time determines the total potential in the postsynaptic neuron and, therefore, how that neu-

Cell body of postsynaptic neuron

Axon terminals

Figure 37.15
The multiple chemical synapses relaying signals to a neuron. The drying process used to prepare the neuron for electron microscopy has toppled the axon terminals and pulled them away from the neuron's surface.

E. R. Lewis, T. E. Everhart, Y. Y. Zevi/Visuals Unlimited

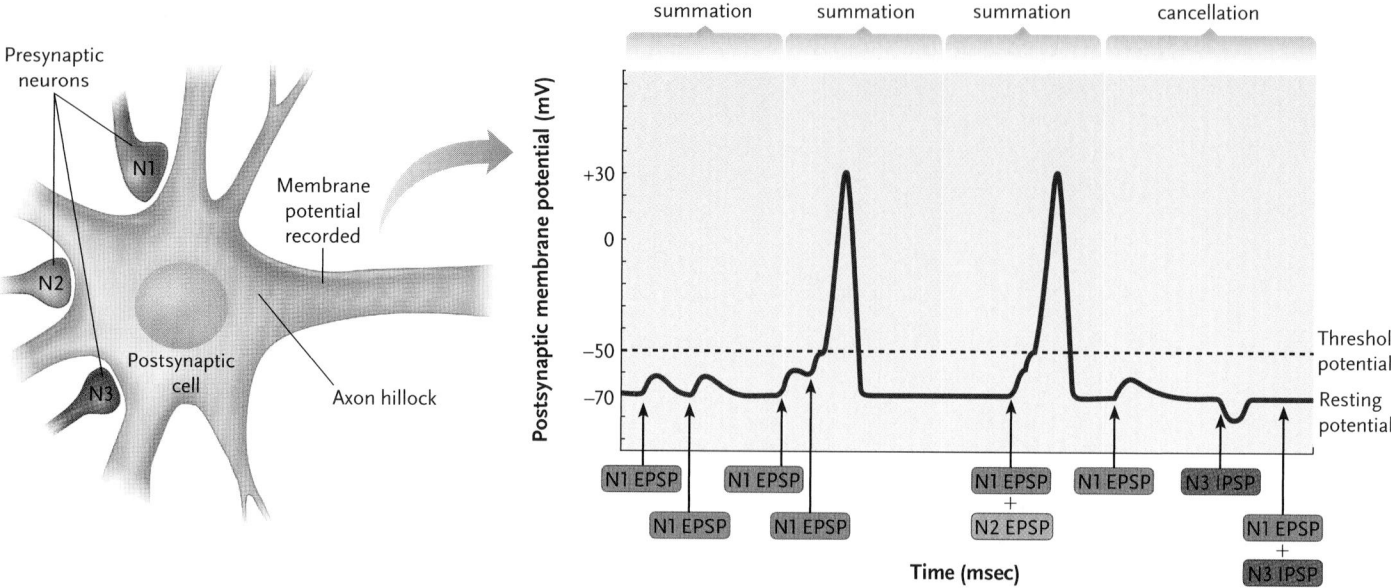

Figure 37.16
Summation of EPSPs and IPSPs by a postsynaptic neuron.

ron responds. **Figure 37.16** shows, in a greatly simplified way, the effects of EPSPs and IPSPs on membrane potential, and how summation of inputs brings a postsynaptic neuron to threshold.

The postsynaptic neuron in the figure has three neurons, N1–N3, forming synapses with it. Suppose that the axon of N1 releases a neurotransmitter, which produces an EPSP in the postsynaptic cell (see Figure 37.16a). The membrane depolarizes, but not enough to reach threshold. If N1 input causes a new EPSP after the first EPSP has died down, it will be of the same magnitude as the first EPSP and no progression toward threshold has taken place—no summation has occurred. If instead, N1 input causes a new EPSP before the first EPSP has died down, the second EPSP will sum with the first and a greater depolarization will have taken place (see Figure 37.16b). This summation of two (or more) EPSPs produced by successive firing of a single presynaptic neuron over a short period of time is called **temporal summation.** If the total depolarization achieved in this way reaches threshold, an action potential will be produced in the postsynaptic neuron. The postsynaptic cell may also be brought to threshold by **spatial summation,** the summation of EPSPs produced by the firing of different presynaptic neurons, such as N1 and N2 (see Figure 37.16c). Lastly, EPSPs and IPSPs can cancel each other out. In the example shown in Figure 37.16d, firing of N1 alone produces an EPSP, firing of N3 alone produces an IPSP, while firing of N1 and N3 simultaneously produces no change in the membrane potential.

The summation point for EPSPs and IPSPs is the axon hillock of the postsynaptic neuron. The greatest density of voltage-gated Na^+ channels occurs in that region, resulting in the lowest threshold potential in the neuron.

The Patterns of Synaptic Connections Contribute to Integration

The total number of connections made by a neuron may be very large—some single interneurons in the human brain, for example, form as many as 100,000 synapses with other neurons. The synapses are not absolutely fixed; they can change through modification, addition, or removal of synaptic connections—or even entire neurons—as animals mature and experience changes in their environments. The combined activities of all the neurons in the nervous system, constantly integrating information from sensory receptors and triggering responses by effectors, control the internal activities of animals and regulate their behavior. This behavior ranges from the simple reflexes of a flatworm to the complex behavior of mammals, including consciousness, emotions, reasoning, and creativity in humans.

Although researchers do not yet understand how processes such as ion flow, synaptic connections, and neural networks produce complex mental activities, they continue to find correspondences between them and the types of neuronal communication described in this chapter. In the next chapter we learn about how nervous systems of animals are organized, and how higher functions such as memory, learning, and consciousness are produced.

STUDY BREAK

How does a postsynaptic neuron integrate signals carried by direct and indirect neurotransmitters?

What is the basic wiring diagram of the brain?

In this chapter, you've learned that neurons communicate with each other through synaptic actions that can directly or indirectly cause a change in the membrane potential of a postsynaptic neuron. In order to understand how the brain processes information, it is important first to understand the wiring that handles this information and what forms that communication can take. There are about 100,000,000,000 neurons in the human brain. This is a daunting number, but it can be managed by grouping the neurons into classes or categories. It has been estimated that there are fewer than 10,000 different neuronal classes in the entire brain. Still, each neuron makes and receives synaptic contacts with about 1,000 other neurons on average, making the wiring diagram quite complex. We do not have an adequate means to represent this complexity, nor do we have a means of understanding how information flows through such a complex network. Just as modern sequencing technology allowed a revolution in the field of genomics (the categorization of all of the genes in an organism); similar breakthroughs will need to occur in the field of neuromics (the categorization of all of the neurons and their interactions). These breakthroughs will have to be in data management, computational simulations, and multisite recording techniques.

In our lab, we are working on a way to represent our knowledge of the brain's wiring with an online knowledge base called NeuronBank. To test NeuronBank, we are using the simple nervous systems of sea slugs, especially *Tritonia diomedea*. A sea slug brain has only 10,000 neurons total, many of which are individually identifiable from animal to animal. Eventually, different branches of NeuronBank will represent our knowledge about the basic wiring of the nervous systems of different animals, allowing the neurons and their connections to be compared across species. Having ready access to this information will allow researchers to better design drugs that target specific neurons. This may aid in treatments for neurological conditions ranging from Parkinson's disease to some forms of blindness.

The way information is conveyed between neurons is not yet understood fully. Much of neuroscience has focused on classical neurotransmission. However, the brain uses many other signaling devices. In our lab, we are also using sea slugs to study neuromodulatory signaling by neurons that release the neurotransmitter serotonin, which regulates appetite, reproductive behavior, muscular movements, sleep, and emotional states such as anxiety. We have found that these neurons can change the strength of connections made by other neurons, and that the effects of a serotonin-releasing neuron depend upon the state of the neuron that it is modulating. So, signaling in the nervous system is not a simple matter of summating excitatory and inhibitory inputs; it involves complex, state-dependent actions. Understanding the complexities of neuronal signaling will allow researchers to understand better how the brain processes information. Another unanswered question is simply, "What are all of the different ways that neurons use for communicating information?"

The ultimate question is "How does all of this processing in the brain lead to self-awareness or consciousness?" We do not have an answer for this question as yet. We know that blocking the activity in parts of the brain, through injury, disease, or drugs, can decrease or alter consciousness. But we do not understand how this activity gives rise to the sensation of "being." That is the ultimate question about how the brain works.

Paul Katz is a professor of biology and the director of the Center for Neuromics at Georgia State University. His research interests include neuromics, neuromodulation, and the evolution of neuronal circuits. Learn more about his work at http://www2.gsu.edu/~biopsk.

Review

Go to **ThomsonNOW** at www.thomsonedu.com/login to access quizzing, animations, exercises, articles, and personalized homework help.

37.1 Neurons and Their Organization in Nervous Systems: An Overview

- The nervous system of an animal (1) receives information about conditions in the internal and external environment, (2) transmits the message along neurons, (3) integrates the information to formulate an appropriate response, and (4) sends out signals to muscles or glands that accomplish the response (Figure 37.2).

- Neurons have dendrites, which receive information and conduct signals toward the cell body, and axons, which conduct signals away from the cell body to another neuron or an effector (Figure 37.3).

- Afferent neurons conduct information from sensory receptors to interneurons, which integrate the information into a response. The response signals are passed to efferent neurons, which activate the effectors carrying out the response (Figure 37.2).

- The combination of an afferent neuron, an interneuron, and an efferent neuron makes up a basic neuronal circuit. The circuits combine into networks that interconnect the peripheral and central nervous systems.

- Glial cells help maintain the balance of ions surrounding neurons and form insulating layers around the axons (Figure 37.5).

- Neurons make connections by two types of synapses. In an electrical synapse, impulses pass directly from the sending to the receiving cell. In a chemical synapse, neurotransmitter molecules released by the presynaptic cell diffuse across a narrow synaptic cleft and bind to receptors in the plasma membrane of the postsynaptic cell (Figure 37.6).

Animation: Neuron structure and function

Animation: Nerve structure

Animation: Impulse travelling through a nerve

37.2 Signal Conduction by Neurons

- The membrane potential of a cell depends on the unequal distribution of positive and negative charges on either side of the membrane, which establishes a potential difference across the membrane.

- Three primary conditions contribute to the resting potential of neurons: (1) an Na^+/K^+ active transport pump that sets up concentration gradients of Na^+ ions (higher outside) and K^+ ions (higher inside); (2) an open channel that allows K^+ to flow out freely; and (3) negatively charged proteins and other molecules inside the cell that cannot pass through the membrane (Figure 37.8).

- An action potential is generated when a stimulus pushes the resting potential to the threshold value at which voltage-gated Na^+ and K^+ channels open in the plasma membrane. The inward flow of Na^+ changes membrane potential abruptly from negative to a positive peak. The potential falls to the resting value again as the gated K^+ channels allow this ion to flow out (Figure 37.10).

- Action potentials move along an axon as the ion flows generated in one segment depolarize the potential in the next segment (Figure 37.11).

- Action potentials are prevented from reversing direction by a brief refractory period, during which a segment of membrane that has just generated an action potential cannot be stimulated to produce another for a few milliseconds.

- In myelinated axons, ions can flow across the plasma membrane only at nodes where the myelin sheath is interrupted. As a result, action potentials skip rapidly from node to node by saltatory conduction (Figure 37.12).

Animation: Ion concentrations

Animation: Ion flow in myelinated axons

Animation: Action potential propagation

Animation: Measuring membrane potential

Animation: Stretch reflex

37.3 Conduction across Chemical Synapses

- Neurotransmitters released into the synaptic cleft bind to receptors in the plasma membrane of the postsynaptic cell, altering the flow of ions across the plasma membrane of the postsynaptic cell and pushing its membrane potential toward or away from the threshold potential (Figure 37.13).

- A direct neurotransmitter binds to a receptor associated with a ligand-gated ion channel in the postsynaptic membrane; the binding opens or closes the channel.

- An indirect neurotransmitter binds to a receptor in the postsynaptic membrane and triggers generation of a second messenger, which leads to the opening or closing of a gated channel.

- Neurotransmitters are released from synaptic vesicles into the synaptic cleft by exocytosis, which is triggered by entry of Ca^{2+} ions into the cytoplasm of the axon terminal through voltage-gated Ca^{2+} channels opened by the arrival of an action potential.

- Neurotransmitter release stops when action potentials cease arriving at the axon terminal. Neurotransmitters remaining in the synaptic cleft are broken down by enzymes or taken up by the axon terminal or glial cells.

- Types of neurotransmitters include acetylcholine, amino acids, biogenic amines, neuropeptides, and gases such as NO and CO (Figure 37.14). Many of the biogenic amines and neuropeptides are also released into the general body circulation as hormones.

Animation: Chemical synapse

37.4 Integration of Incoming Signals by Neurons

- Neurons carry out integration by summing excitatory postsynaptic potentials (EPSPs) and inhibitory postsynaptic potentials (IPSPs); the summation may push the membrane potential of the postsynaptic cell toward or away from the threshold for an action potential (Figure 37.16).

- The combined effects of summation in all the neurons in the nervous system control behavior in animals and underlie complex mental processes in mammals.

Animation: Synaptic integration

Questions

Self-Test Questions

1. Nerve signals travel in the following manner:
   a. A dendrite of a sensory neuron receives the signal; its cell body transmits the signal to a motor neuron's axon, and the signal is sent to the target.
   b. An axon of a motor neuron receives the signal; its cell body transmits the signal to a sensory neuron's dendrite, and the signal is sent to the target.
   c. Efferent neurons conduct nerve impulses toward the cell body of sensory neurons, which send them on to interneurons and ultimately to afferent motor neurons.
   d. A dendrite of a sensory neuron receives a signal; the cell's axon transmits the signal to an interneuron; the signal is then transmitted to dendrites of a motor neuron and sent forth on its axon to the target.
   e. The axons of oligodendrocytes transmit nerve impulses to the dendrites of astrocytes.

2. Glial cells:
   a. are unable to divide after an animal is born.
   b. in the PNS called Schwann cells form the insulating myelin sheath around axons.
   c. called astrocytes form the nodes of Ranvier in the brain.
   d. called oligodendrocytes are star-shaped cells in the PNS.
   e. are neuronal cells that connect to interneurons.

3. An example of a synapse could be the site where:
   a. neurotransmitters released by an axon travel across a gap and are picked up by receptors on a muscle cell.

   b. an electrical impulse arrives at the end of a dendrite causing ions to flow onto axons of presynaptic neurons.
   c. postsynaptic neurons transmit a signal across a cleft to a presynaptic neuron.
   d. oligodendrocytes contact the dendrites of an afferent neuron directly.
   e. an on-off switch stimulates an electrical impulse in a presynaptic cell to stimulate, not inhibit, other presynaptic cells.

4. The resting potential in neurons requires:
   a. membrane transport channels to be constantly open for Na^+ and K^+ flow.
   b. the inside of neurons to be positive relative to the outside.
   c. a slow movement of K^+ outward with a charge difference in the neural membrane set up by this movement of K^+.
   d. an active Na^+/K^+ pump, which pumps Na^+ and K^+ into the neuron.
   e. three Na^+ ions to be pumped through three Na^+ gates and two K^+ ions to be pumped through two K^+ gates.

5. The major role of the sodium potassium pump is to:
   a. cause a rapid firing of the action potential so the inside of the membrane becomes momentarily positive.
   b. decrease the resting potential to zero.
   c. hyperpolarize the membrane above resting value.
   d. increase a high action potential to enter a refractory period.
   e. maintain the resting potential at a constant negative value.

6. In the propagation of a nerve impulse:
    a. the refractory period begins as the K^+ channel opens, allowing K^+ ions to flow outward with their concentration gradient.
    b. Na^+ ions rush with their concentration gradient out of the axon.
    c. positive charges lower the membrane potential to its lowest action potential.
    d. gated K^+ channels open at the same time as the activation gate of Na^+ channels closes.
    e. the depolarizing stimulus lowers the membrane potential to open the Na^+ gates.

7. Which of the following does not contribute to propagation of action potentials?
    a. As the area outside the membrane becomes negative, it attracts ions from adjacent regions; as the inside of the membrane becomes positive, it attracts negative ions from nearby in the cytoplasm. These events depolarize nearby regions of the axon membrane.
    b. The refractory period allows the impulse to travel in only one direction.
    c. Each segment of the axon prevents the adjacent segments from firing.
    d. The magnitude of the action potential stays the same as it travels down the axon.
    e. Increasing the intensity of the stimulus increases the number of action potentials up to a limit.

8. Which of the following statements best describes saltatory conduction?
    a. It inhibits direct neurotransmitter release.
    b. It transmits the action potential at the nodes of Ranvier and thus speeds up impulses on myelinated axons.
    c. It increases neurotransmitter release at the presynaptic membrane.
    d. It decreases neurotransmitter uptake at chemically gated postsynaptic channels.
    e. It removes neurotransmitters from the synaptic cleft.

9. Transmission of a nerve impulse to its target cell requires:
    a. endocytosis of neurotransmitters by the excitatory presynaptic vesicles.
    b. thousands of molecules of neurotransmitter that had been stored in the postsynaptic cell to be released into the synaptic cleft.
    c. Ca^{2+} ions to diffuse through voltage-gated Ca^{2+} channels.
    d. the fall in Ca^{2+} to trigger a protein that causes the presynaptic vesicle to fuse with the plasma membrane.
    e. an action potential to open the Ca^{2+} gates so that Ca^{2+} ions, in higher concentration outside the axon, can flow back into the cytoplasm of the neuron.

10. Autopsy reports reveal that above a certain threshold, brain size is not related to intelligence. A possible explanation is that the brains of:
    a. gifted people have a much vaster network of neural synapses than do the brains of people with normal intelligence.
    b. people with normal intelligence release far more NO and CO neurotransmitters than do those of the gifted.
    c. people with normal intelligence contain more glutamate and aspartate than do those of the gifted.
    d. gifted people have excessive quantities of gamma aminobutyric acid.
    e. people with normal intelligence contain more glycine than do those of the gifted.

Questions for Discussion

1. In some cases of ADHD (attention deficit hyperactivity disorder) the impulsive, erratic behavior typical of affected people can be calmed with drugs that *stimulate* certain brain neurons. Based on what you have learned about neurotransmitter activity in this chapter, can you suggest a neural basis for this effect?

2. Most sensory neurons form synapses either on interneurons in the spinal cord or on motor neurons. However, in many vertebrates, certain sensory neurons in the nasal epithelium synapse directly on brain neurons that activate behavioral responses to odors. Suggest at least one reason why natural selection might favor such an arrangement.

3. How did evolution of chemical synapses make higher brain functions possible?

4. Use an Internet search engine with the term "Pediatric Neurotransmitter Disease" and, for one such disease, explain how the symptoms relate to neurotransmitter function.

Experimental Analysis

Design an experiment to test whether neurons are connected via electrical or chemical synapses.

Evolution Link

A biologist hypothesized that the mechanism for the propagation of action potentials down a neuron evolved only once. What evidence would you collect from animals living today to support or refute that hypothesis?

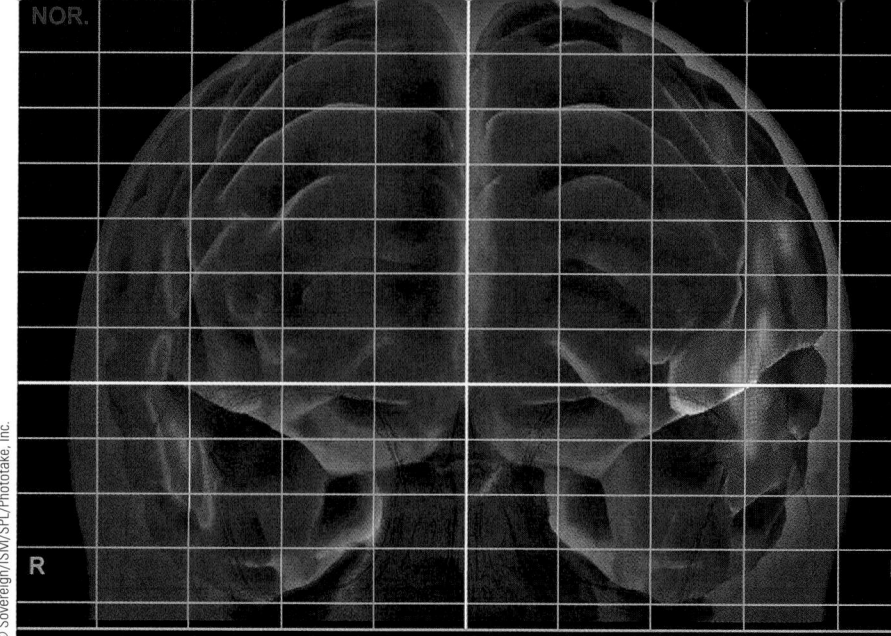

Activity in the human brain while reading aloud. The image combines an MRI of a male brain with a PET scan, which shows that blood circulation increases in the language, hearing, and vision areas of the brain, especially in the left hemisphere.

© Sovereign/ISM/SPL/Phototake, Inc.

38 Nervous Systems

WHY IT MATTERS

The conductor's baton falls and the orchestra plays the first notes of a Mozart symphony. Unaware of the complex interactions of their nervous systems, the musicians translate printed musical notation into melodious sounds played on their instruments. Although their fingers and arms move to produce precise harmonies, the musicians are only vaguely conscious of these movements, learned through years of practice. Their only conscious endeavor is to interpret the music in line with the conductor's directions.

From the back of the hall, a common housefly, *Musca domestica*, moves in random twists and turns that bring it toward the stage. Although far less complex than that of a human, the fly's nervous system contains networks of neurons that work in the same way, in patterns adapted to its lifestyle.

The fly does not register the sounds reverberating through the hall as a significant sensory input. However, some of its receptors are exquisitely sensitive to the presence of potential food molecules, including those in the sweat on the conductor's face. The fly's swoops and turns bring it closer to the conductor; soon it alights on the tip of his nose. When sensory receptors in the fly's footpads detect organic matter on the surface of the nose, they trigger an automated feeding response: the fly's proboscis lowers and its gut begins contractions that suck up the nutrients.

The conductor's eyes notice the insect's approach, and sensory receptors in his skin pinpoint the spot where it lands. Without missing a beat, the conductor's hand flicks toward that exact spot. But his nervous system and effectors, although highly sophisticated, are no match for the escape reflexes of the fly. The fly's sensory receptors detect the motion of the fingers, sending impulses to the fly's leg and wing muscles that launch it into flight long before the fingers reach the nose.

The fly wanders into the orchestra, attracted to potential nutrients on various musicians, who respond with flicking movements that are no more successful than those of the conductor. At last, the fly lands on the left hand of the timpanist, who is listening with pleasure to the music while he awaits his entrance late in the first movement. His right hand holds a mallet. With a skill born of long practice in hitting drums, gongs, and bells with speed and precision, the timpanist deftly swings his mallet and dispatches the fly, ending the latest contest between mammalian and arthropod nervous systems.

The nervous systems underlying these behaviors are one of the features that set animals apart from other organisms. As animals evolved, the need to find food, living space, and mates, and to escape predators and other dangers, provided a powerful selection pressure for increasingly complex and capable nervous systems. Neurons, described in the previous chapter, provide the structural and functional basis for all these systems. We can trace some of the developments along this extended evolutionary pathway by examining the nervous systems of living animals, from invertebrates to mammals, and especially humans.

38.1 Invertebrate and Vertebrate Nervous Systems Compared

The nervous systems of most invertebrates are relatively simple, typically containing fewer neurons, arranged in less complex networks, than vertebrate systems. As animal groups evolved, their nervous systems became more elaborate, providing the ability to integrate more sensory information and to formulate more complex responses. Our comparative survey of nervous systems begins with the simplest invertebrates.

Cnidarians and Echinoderms Have Nerve Nets

Cnidarians and echinoderms are radially symmetrical animals with body parts arranged regularly around a central axis like the spokes of a wheel. Their nervous systems, called **nerve nets,** are loose meshes of neurons organized within that radial symmetry.

The nerve nets of cnidarians such as sea anemones extend into each "spoke" of the body (**Figure 38.1a**). Their neurons lack clearly differentiated dendrites and axons. When part of the animal is stimulated, impulses are conducted through the nerve net in all directions from the point of stimulation. Although there is no cluster of neurons that plays the coordinating role of a brain, nerve cells may be more concentrated in some regions. For example, in scyphozoan jellyfish, which swim by rhythmic contractions of their bells, neurons are denser in a ring around the margin of the bell, in the same area as the contractile cells that produce the swimming movements.

In echinoderms, including sea stars, the nervous system is a modified nerve net, with some neurons organized into **nerves,** bundles of axons enclosed in connective tissue and following the same pathway. A *nerve ring* surrounds the centrally located mouth, and a *radial nerve* that is connected to nerve nets branches throughout each arm (**Figure 38.1b**). If the radial nerve serving an arm is cut, the arm can still move, but not in coordination with the other arms.

More Complex Invertebrates Have Cephalized Nervous Systems

More complex invertebrates have neurons with clearly defined axons and dendrites, and more specialized functions. Some neurons are concentrated into functional clusters called **ganglia** (singular, *ganglion*). A key evolutionary development in invertebrates is a trend toward *cephalization,* the formation of a distinct head region containing both ganglia that constitute a **brain,** the control center of the nervous system, and major sensory structures. One or more solid **nerve cords**—bundles of nerves—extend from the central ganglia to the rest of the body; they are connected to smaller nerves. Another evolutionary trend is toward bilateral symmetry of the body and the nervous system, in which body parts are mirror images on left and right sides. These trends toward cephalization and bilateral symmetry are illustrated here in flatworms, arthropods, and mollusks.

In flatworms, a small brain consisting of a pair of ganglia at the anterior end is connected by two or more longitudinal nerve cords to nerve nets in the rest of the body (**Figure 38.1c**). The brain integrates inputs from sensory receptors, including a pair of anterior eyespots with receptors that respond to light. The brain and longitudinal nerve cords constitute the flatworm's **central nervous system (CNS),** the simplest one known, while the nerves from the CNS to the rest of the body constitute the **peripheral nervous system (PNS).**

a. Cnidarian (sea anemone)

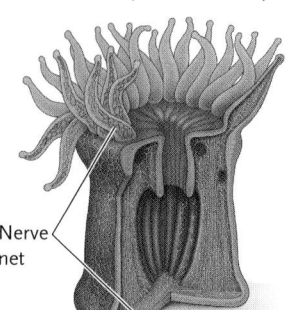

Nerve net

b. Echinoderm (sea star)

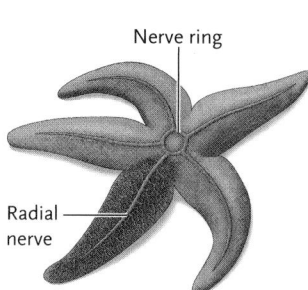

Nerve ring

Radial nerve

c. Planarian (flatworm)

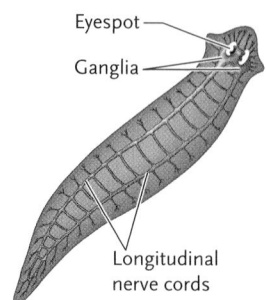

Eyespot

Ganglia

Longitudinal nerve cords

d. Arthropod (grasshopper)

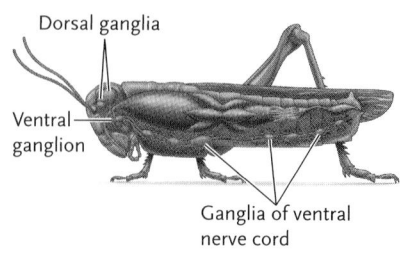

Dorsal ganglia

Ventral ganglion

Ganglia of ventral nerve cord

e. Mollusk (octopus)

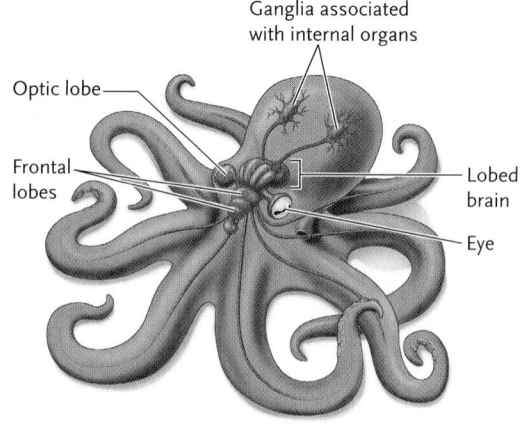

Ganglia associated with internal organs

Optic lobe

Frontal lobes

Lobed brain

Eye

f. Chordate (salamander)

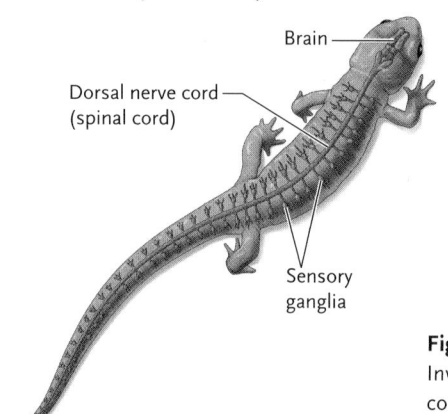

Brain

Dorsal nerve cord (spinal cord)

Sensory ganglia

Figure 38.1
Invertebrate and vertebrate nervous systems compared, showing increasing cephalization. The diagrams are not drawn to the same scale.

Arthropods such as insects have a head region that contains a brain consisting of dorsal and ventral pairs of ganglia, and major sensory structures, usually including eyes and antennae **(Figure 38.1d)**. The brain exerts centralized control over the remainder of the animal. A ventral nerve cord enlarges into a pair of ganglia in each body segment. In arthropods with fused body segments, as in the thorax of insects, the ganglia are also fused into larger masses forming secondary control centers.

Although different in basic plan from the arthropod system, the nervous systems of mollusks (such as clams, snails, and octopuses) also rely on neurons clustered into paired ganglia and connected by major nerves. Different mollusks have varying degrees of cephalization, with cephalopods having the most pronounced cephalization of any invertebrate group. In the head of an octopus, for example, a cluster of ganglia fuses into a complex, lobed brain with clearly defined sensory and motor regions. Paired nerves link different lobes with muscles and sensory receptors, including prominent optic lobes linked by nerves to large, complex eyes **(Figure 38.1e).** Octopuses are capable of rapid movement to hunt prey and to escape from predators, behaviors that rely on rapid, sophisticated processing of sensory information.

Vertebrates Have the Most Specialized Nervous Systems

In vertebrates, the CNS consists of the brain and spinal cord, and the PNS consists of all the nerves and ganglia that connect the brain and spinal cord to the rest of the body **(Figure 38.1f).** All vertebrate nervous systems are highly cephalized, with major concentrations of neurons in a brain located in the head. In contrast to invertebrate nervous systems, which have solid nerve cords located ventrally, the brain and nerve cord of vertebrates are hollow, fluid-filled structures located dorsally. The head contains specialized sensory organs, which are connected directly to the brain by nerves. Compared with invertebrates, the ganglia are greatly reduced in mass and functional activity except in the gut, which contains extensive interneuron networks.

The structure of the vertebrate nervous system reflects its pattern of development. The nervous system of a vertebrate embryo begins as the hollow **neural tube,** the anterior end of which develops into the brain and the rest into the **spinal cord.** The cavity of the neural tube becomes the fluid-filled **ventricles** of the brain and the **central canal** through the spinal cord. Adjacent tissues give rise to nerves that connect the brain and spinal cord with all body regions.

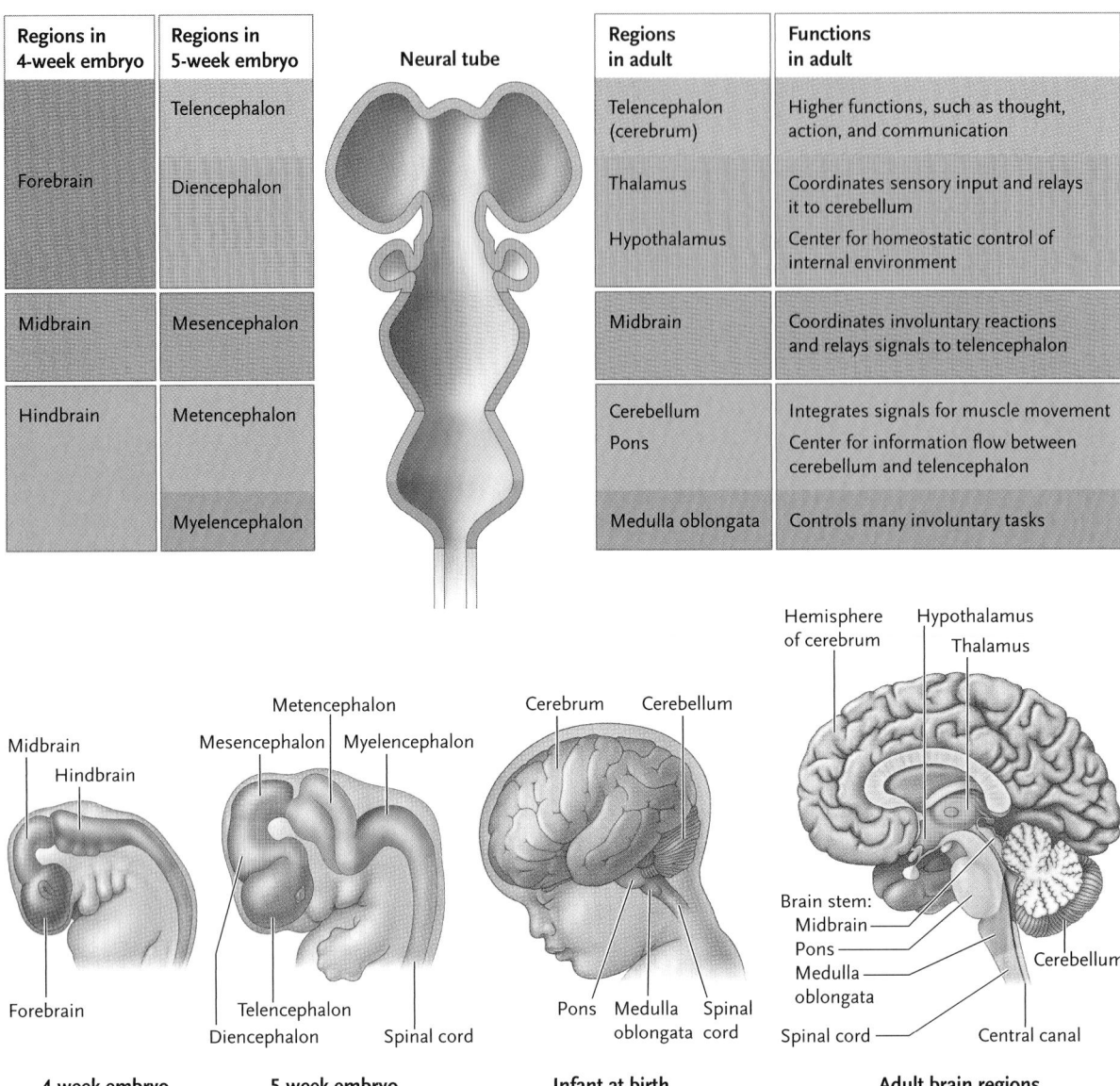

Regions in 4-week embryo	Regions in 5-week embryo
Forebrain	Telencephalon
	Diencephalon
Midbrain	Mesencephalon
Hindbrain	Metencephalon
	Myelencephalon

Neural tube

Regions in adult	Functions in adult
Telencephalon (cerebrum)	Higher functions, such as thought, action, and communication
Thalamus	Coordinates sensory input and relays it to cerebellum
Hypothalamus	Center for homeostatic control of internal environment
Midbrain	Coordinates involuntary reactions and relays signals to telencephalon
Cerebellum	Integrates signals for muscle movement
Pons	Center for information flow between cerebellum and telencephalon
Medulla oblongata	Controls many involuntary tasks

4-week embryo

5-week embryo

Infant at birth

Adult brain regions

Figure 38.2
Development of the human brain from the anterior end of an embryo's neural tube.

Early in embryonic development, the anterior part of the neural tube enlarges into three distinct regions: the **forebrain, midbrain,** and **hindbrain (Figure 38.2).** A little later, the embryonic hindbrain subdivides into the *metencephalon* and *myelencephalon;* the midbrain develops into the *mesencephalon;* and the forebrain subdivides into the *telencephalon* and *diencephalon.*

The metencephalon gives rise to the *cerebellum,* which integrates sensory signals from the eyes, ears, and muscle spindles with motor signals from the telencephalon, and the *pons,* a major traffic center for information passing between the cerebellum and the higher integrating centers of the adult telencephalon. The myelencephalon gives rise to the *medulla oblongata* (commonly shortened to medulla), which controls many vital involuntary tasks such as respiration and blood circulation. The mesencephalon gives rise to the (adult) midbrain, which with the pons and the medulla constitutes the brain stem. The midbrain has centers for coordinating reflex responses (involuntary reac-

tions) to visual and auditory (hearing) input and relays signals to the telencephalon.

The embryonic telencephalon develops into the *cerebrum* (or adult telencephalon), the largest part of the brain. The cerebrum controls higher functions such as thought, memory, language, and emotions, as well as voluntary movements. The diencephalon gives rise to the *thalamus,* a coordinating center for sensory input and a relay station for input to the cerebellum, and to the *hypothalamus,* the primary center for homeostatic control over the internal environment. In fishes, the cerebrum is little more than a relay station for olfactory (sense of smell) information. In amphibians, reptiles, and birds, it becomes progressively larger and contains greater concentrations of integrative functions. In mammals, the cerebrum is the major integrative structure of the brain.

In the following sections, we examine vertebrate nervous systems, and the human nervous system in particular, beginning with the peripheral nervous system.

38.2 The Peripheral Nervous System

Afferent neurons in the peripheral nervous system transmit signals to the CNS, and signals from the CNS are sent via efferent neurons in the peripheral nervous system to the effectors that carry out responses **(Figure 38.3)**. The afferent part of the system includes all the neurons that transmit sensory information from their receptors. The efferent part of the system consists of the axons of neurons that carry signals to the muscles and glands acting as effectors. In mammals, 31 pairs of **spinal nerves** carry signals between the spinal cord and the body trunk and limbs, and 12 pairs of **cranial nerves** connect the brain directly to the head, neck, and body trunk. The efferent part of the PNS is further divided into somatic and autonomic systems (see Figure 38.3).

The Somatic System Controls the Contraction of Skeletal Muscles, Producing Body Movements

The **somatic system** controls body movements that are primarily conscious and voluntary. Its neurons, called motor neurons, carry efferent signals from the CNS to the skeletal muscles. The dendrites and cell bodies of motor neurons are located in the spinal cord; their axons extend from the spinal cord to the skeletal muscle cells they control. As a result, the somatic portions of the cranial and spinal nerves consist only of axons.

Although the somatic system is primarily under conscious, voluntary control, some contractions of skeletal muscles are unconscious and involuntary. These include the reflexes, shivering, and the constant muscle contractions that maintain body posture and balance.

The Autonomic System Is Divided into Sympathetic and Parasympathetic Divisions

The **autonomic nervous system** controls largely involuntary processes including digestion, secretion by sweat glands, circulation of the blood, many functions of the reproductive and excretory systems, and contraction of smooth muscles in all parts of the body. It is organized into *sympathetic* and *parasympathetic* divisions, which

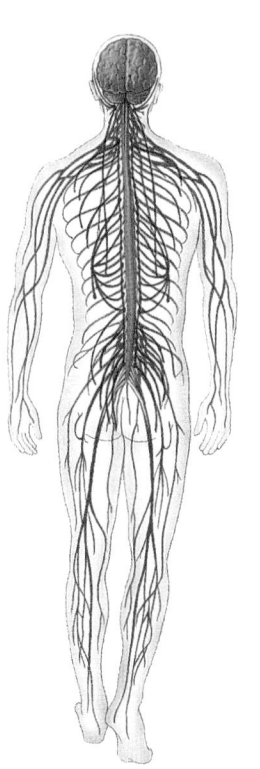

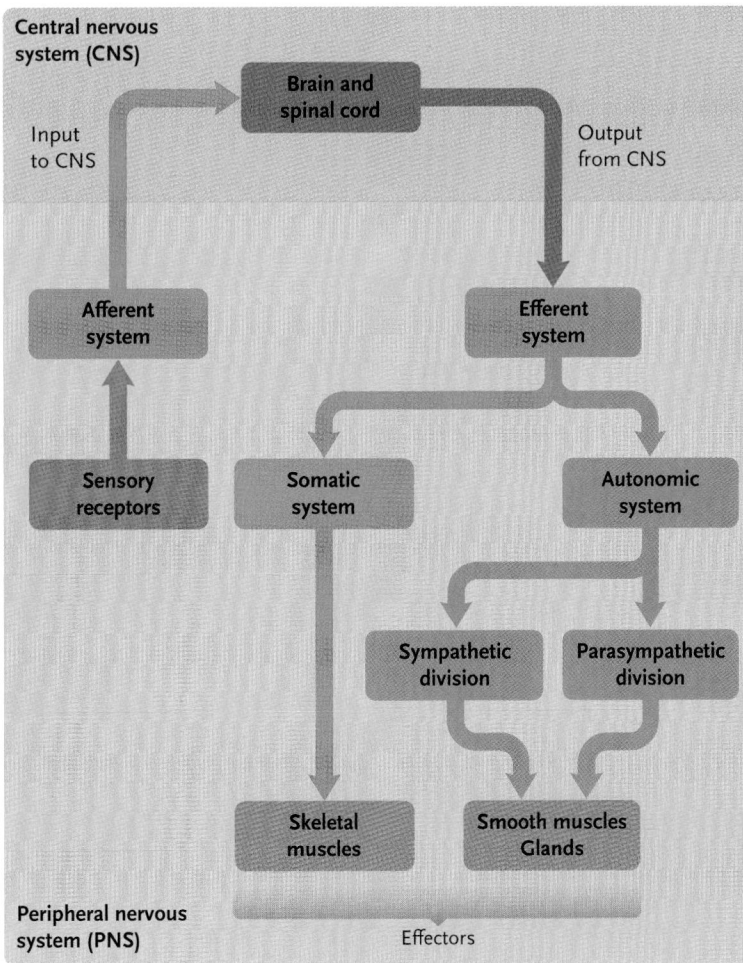

Figure 38.3

The central nervous system (CNS) and peripheral nervous system (PNS), and their subsystems.

Central nervous system (CNS)

Brain and spinal cord

Input to CNS

Output from CNS

Afferent system

Efferent system

Sensory receptors

Somatic system

Autonomic system

Sympathetic division

Parasympathetic division

Skeletal muscles

Smooth muscles Glands

Effectors

Peripheral nervous system (PNS)

KEY

- Central nervous system
- Peripheral nervous system
- Afferent and somatic systems
- Autonomic system

are always active, and have opposing effects on the organs they affect, thereby enabling precise control **(Figure 38.4)**. For example, in the circulatory system, sympathetic neurons stimulate the force and rate of the heartbeat, and parasympathetic neurons inhibit these activities. In the digestive system, sympathetic neurons inhibit the smooth muscle contractions that move materials through the small intestine, and parasympathetic neurons stimulate the same activities. These opposing effects control involuntary body functions precisely.

The pathways of the autonomic nervous system include two neurons. The first neuron has its dendrites and cell body in the CNS, and its axon extends to a ganglion outside the CNS in the PNS. There it synapses with the dendrites and cell body of the second neuron in the pathway. The axon of the second neuron extends from the ganglion to the effector carrying out the response.

The **sympathetic division** predominates in situations involving stress, danger, excitement, or strenuous physical activity. Signals from the sympathetic division increase the force and rate of the heartbeat, raise the blood pressure by constricting selected blood vessels, dilate air passages in the lungs, induce sweating, and open the pupils wide. Activities that are less important in an emergency, such as digestion, are suppressed by the sympathetic system. The **parasympathetic division**, in contrast, predominates during quiet, low-stress situations, such as while relaxing. Under its influence the effects of the sympathetic division, such as rapid heartbeat and elevated blood pressure, are reduced and "housekeeping" (maintenance) activities such as digestion predominate.

STUDY BREAK

Which of the two autonomic nervous system divisions predominates in the following scenarios? (a) You are hiking on a trail and suddenly a bear appears in your path. (b) It is a hot sunny day. You find a shady tree and sit down. Leaning against its trunk, you feel your eyes becoming heavy.

38.3 The Central Nervous System (CNS) and Its Functions

The central nervous system integrates incoming sensory information from the PNS into compensating responses, thus managing body activities. Our examina-

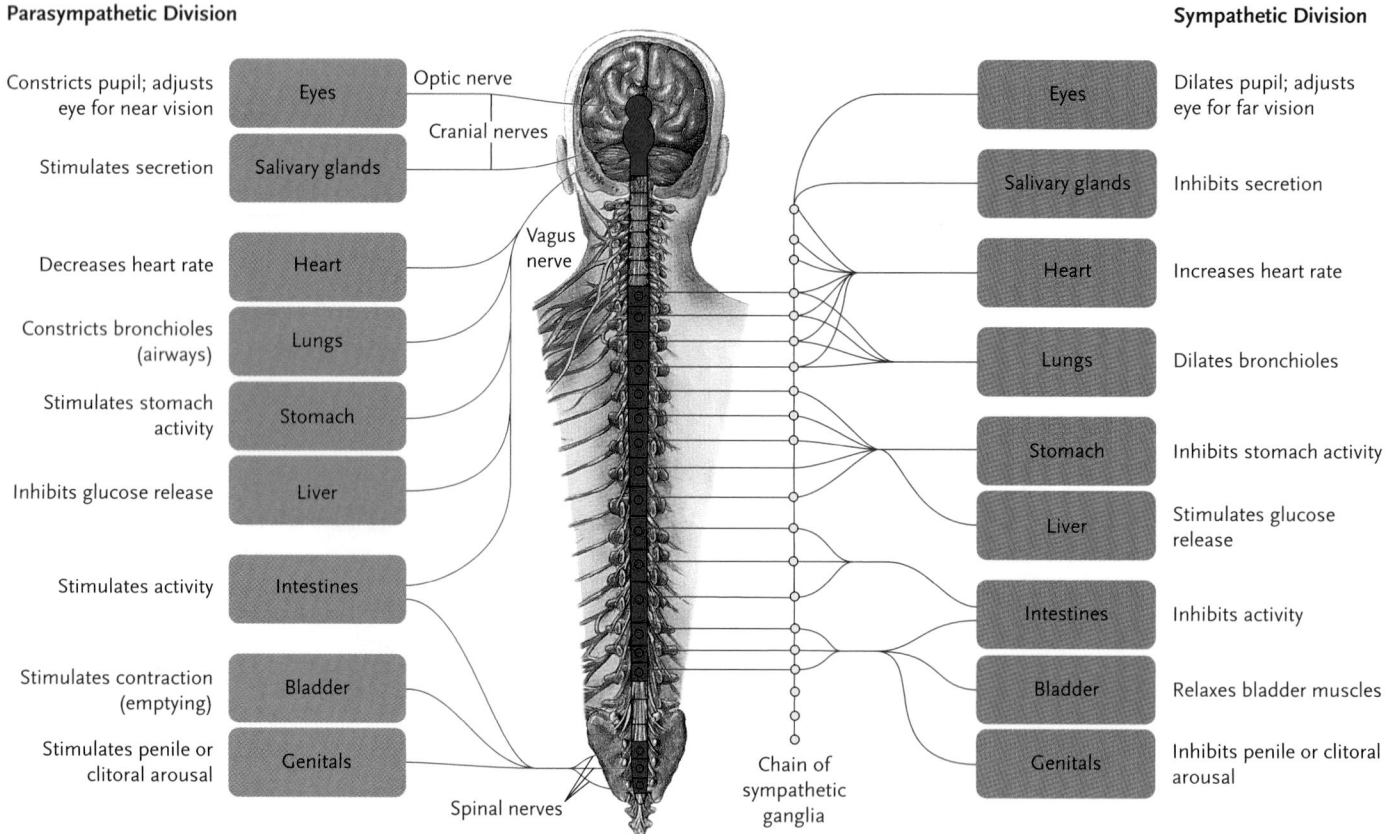

Parasympathetic Division   **Sympathetic Division**

Parasympathetic	Organ	Sympathetic
Constricts pupil; adjusts eye for near vision	Eyes	Dilates pupil; adjusts eye for far vision
Stimulates secretion	Salivary glands	Inhibits secretion
Decreases heart rate	Heart	Increases heart rate
Constricts bronchioles (airways)	Lungs	Dilates bronchioles
Stimulates stomach activity	Stomach	Inhibits stomach activity
Inhibits glucose release	Liver	Stimulates glucose release
Stimulates activity	Intestines	Inhibits activity
Stimulates contraction (emptying)	Bladder	Relaxes bladder muscles
Stimulates penile or clitoral arousal	Genitals	Inhibits penile or clitoral arousal

Optic nerve
Cranial nerves
Vagus nerve
Spinal nerves
Chain of sympathetic ganglia

Figure 38.4
Effects of the sympathetic and parasympathetic divisions of the central nervous system on organ and gland function. Only one side of each division is shown; both are duplicated on the left and right sides of the body.

tion of the vertebrate CNS begins with the spinal cord, and then considers the brain and its functions.

The Spinal Cord Relays Signals between the PNS and the Brain and Controls Reflexes

The spinal cord, which extends dorsally from the base of the brain, carries impulses between the brain and the PNS and contains the interneuron circuits that control motor reflexes.

The spinal cord and brain are surrounded and protected by three layers of connective tissue, the **meninges** (*meninga* = membrane), and by **cerebrospinal fluid,** which circulates through the central canal of the spinal cord, through the ventricles of the brain, and between two of the meninges. The fluid cushions the brain and spinal cord from jarring movements and impacts, and it both nourishes the CNS and protects it from toxic substances.

In cross section, the spinal cord has a butterfly-shaped core of **gray matter,** consisting of nerve cell bodies and dendrites. This is surrounded by **white matter,** consisting of axons, many of them surrounded by myelin sheaths. Pairs of spinal nerves connect with the spinal cord at spaces between the vertebrae (**Figure 38.5**).

The afferent axons entering the spinal cord make synapses with interneurons in the gray matter, which send axons upward through the white matter of the spinal cord to the brain. Conversely, axons from interneurons of the brain pass downward through the white matter of the cord and make synapses with the dendrites and cell bodies of efferent neurons in the gray matter of the cord. The axons of these efferent neurons exit the spinal cord through the spinal nerves.

The gray matter of the spinal cord also contains interneurons of the pathways involved in **reflexes,** programmed movements that take place without conscious effort, such as the sudden withdrawal of a hand from a hot surface (see Figure 38.5). When your hand touches the hot surface, the heat stimulates an afferent neuron, which makes connections with at least two interneurons in the spinal cord. One of these interneurons stimulates an efferent neuron, causing the *flexor* muscle of the arm to contract, which bends the arm and withdraws the hand almost instantly from the hot surface. The other interneuron synapses with an efferent neuron connected to an *extensor* muscle, relaxing it so that the flexor can move more quickly. Interneurons connected to the reflex circuits also send signals to the brain, making you aware of the stimulus causing the reflex. You know from experience that when a reflex movement withdraws your hand from a hot surface or other damaging stimulus, you feel the pain shortly *after* the hand is withdrawn. This is the extra time required for impulses to travel from the neurons of the reflex to the brain (see discussion of the neurotransmitter substance P in Section 37.3).

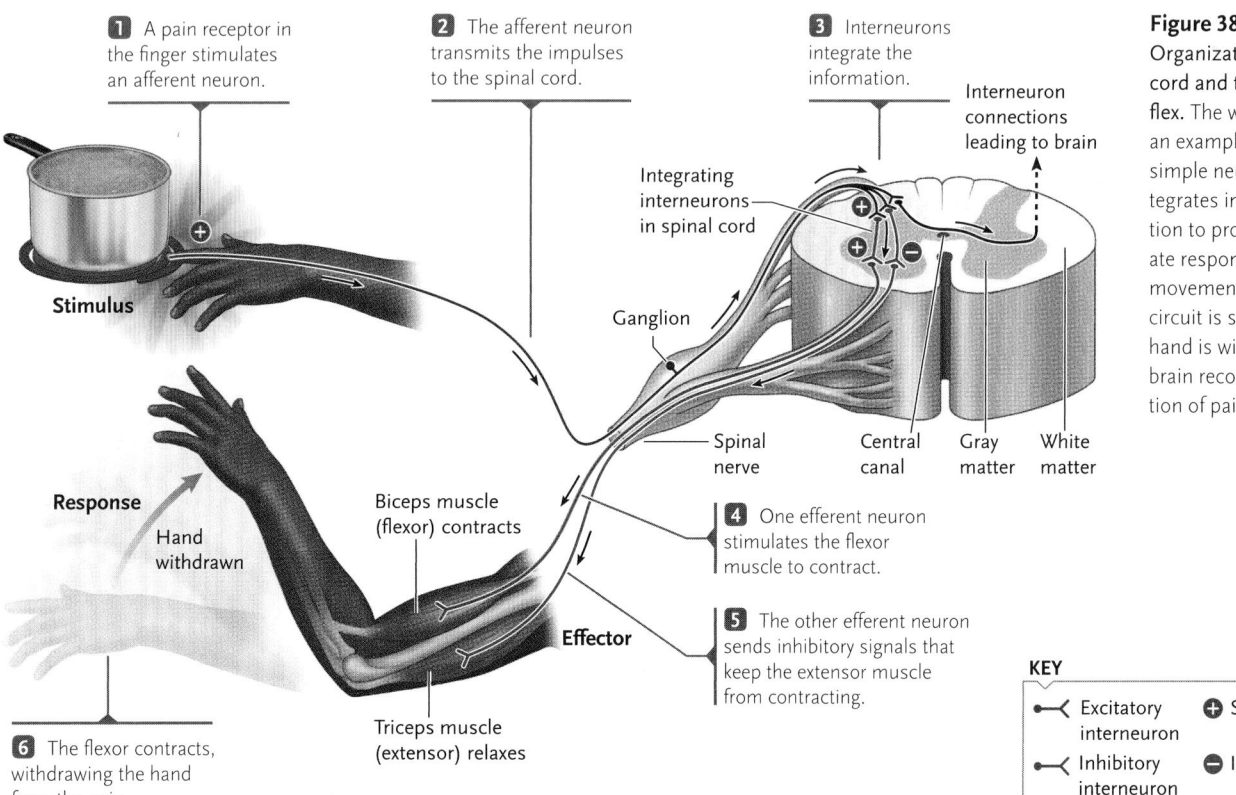

1 A pain receptor in the finger stimulates an afferent neuron.

2 The afferent neuron transmits the impulses to the spinal cord.

3 Interneurons integrate the information.

Interneuron connections leading to brain

Integrating interneurons in spinal cord

Ganglion

Stimulus

Response

Hand withdrawn

Biceps muscle (flexor) contracts

Effector

Triceps muscle (extensor) relaxes

Spinal nerve

Central canal

Gray matter

White matter

4 One efferent neuron stimulates the flexor muscle to contract.

5 The other efferent neuron sends inhibitory signals that keep the extensor muscle from contracting.

6 The flexor contracts, withdrawing the hand from the pain.

KEY
⊱ Excitatory interneuron
⊕ Stimulates
⊱ Inhibitory interneuron
⊖ Inhibits

Figure 38.5
Organization of the spinal cord and the withdrawal reflex. The withdrawal reflex is an example of a relatively simple neuron circuit that integrates incoming information to produce an appropriate response. The reflex movement produced by this circuit is so rapid that the hand is withdrawn before the brain recognizes the sensation of pain.

The Brain Integrates Sensory Information and Formulates Compensating Responses

The brain is the major center that receives, integrates, stores, and retrieves information in vertebrates. Its interneuron networks generate responses that provide the basis for our voluntary movements, consciousness, behavior, emotions, learning, reasoning, language, and memory, among many other complex activities.

Major Brain Structures. We have noted that the three major divisions of the embryonic neural tube—forebrain, midbrain, and hindbrain—give rise to the structures of the adult brain. Like the spinal cord, each brain structure contains both gray matter and white matter and is surrounded by meninges and circulating cerebrospinal fluid **(Figure 38.6).**

The hindbrain develops into three major structures in the adult brain: the *cerebellum,* the *pons,* and the *medulla oblongata* (the *medulla*) (see Figure 38.2).

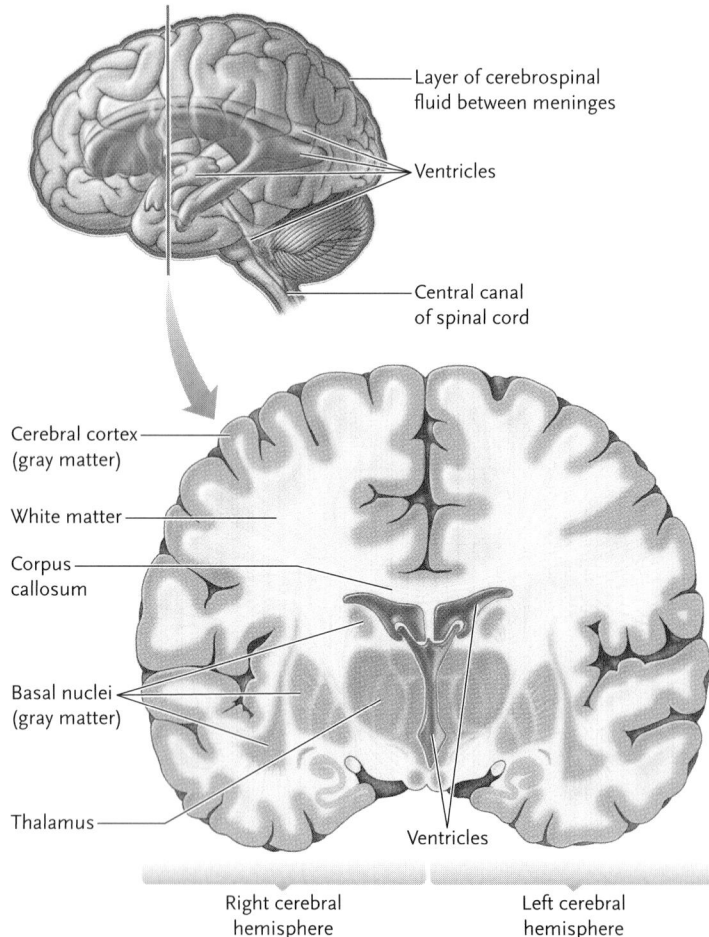

Cerebral cortex (gray matter)

White matter

Corpus callosum

Basal nuclei (gray matter)

Thalamus

Layer of cerebrospinal fluid between meninges

Ventricles

Central canal of spinal cord

Ventricles

Right cerebral hemisphere

Left cerebral hemisphere

Figure 38.6
The human brain, illustrating the distribution of gray matter and white matter, and the locations of the four ventricles (in blue) with their connection to the central canal of the spinal cord.

The pons and medulla, along with the midbrain, form a stalklike structure known as the **brain stem,** which connects the forebrain with the spinal cord. All but two of the twelve pairs of cranial nerves also originate from the brain stem. The cerebellum, with its deeply folded surface, is an outgrowth of the pons.

The forebrain, which makes up most of the mass of the brain in humans, forms the *telencephalon (cerebrum).* The cerebrum, the largest part of the brain in humans, is organized into the left and right *cerebral hemispheres,* which have many fissures and folds (see Figure 38.6). Each hemisphere consists of **cerebral cortex,** a thin outer shell of gray matter covering a thick core of white matter. The basal nuclei, consisting of several regions of gray matter, are located deep within the white matter.

The Blood-Brain Barrier. Unlike the epithelial cells forming capillary walls elsewhere in the body, which allow small molecules and ions to pass freely from the blood to surrounding fluids, those forming capillaries in the brain are sealed together by tight junctions (see Figure 5.27). The tight junctions set up a **blood-brain barrier** that prevents most substances dissolved in the blood from entering the cerebrospinal fluid and thus protects the brain and spinal cord from viruses, bacteria, and toxic substances that may circulate in the blood.

A few types of molecules and ions, such as oxygen, carbon dioxide, alcohol, and anesthetics, can move directly across the lipid bilayer of the epithelial cell membranes by diffusion. A few other substances—most significantly glucose, the only molecule that brain and spinal cord cells can oxidize for energy—are moved across the plasma membrane by highly selective transport proteins.

The Brain Stem Regulates Many Vital Housekeeping Functions of the Body

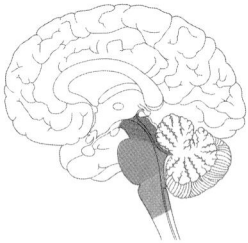

Physicians and scientists have learned much about the functions of various brain regions by studying patients with brain damage from stroke, infection, tumors, or mechanical disturbance. Techniques such as *functional magnetic resonance imaging (fMRI)* and *positron emission tomography (PET)* allow researchers to identify the normal functions of specific brain regions in noninvasive ways. The instruments record a subject's brain activity during various mental and physical tasks by detecting minute increases in blood flow or metabolic activity in specific regions **(Figure 38.7).**

From such medical and experimental analyses, we know that gray-matter centers in the brain stem control

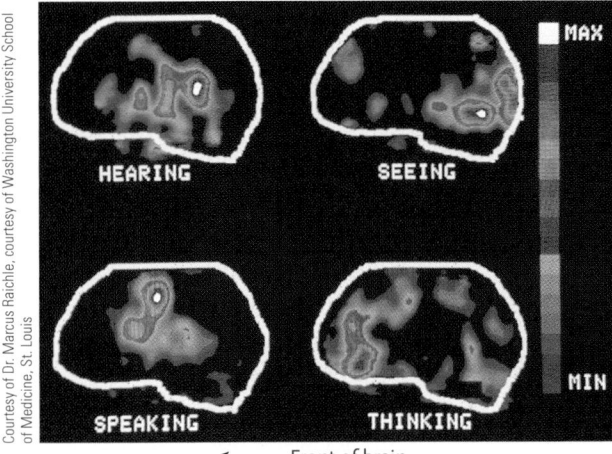

HEARING   SEEING

SPEAKING   THINKING

← Front of brain

Figure 38.7
PET scans showing regions of the brain active when a person performs specific mental tasks. The colors show the relative activity of the sections, with white the most active.

many vital body functions without conscious involvement or control by the cerebrum. Among these functions are the heart and respiration rates, blood pressure, constriction and dilation of blood vessels, coughing, and reflex activities of the digestive system such as vomiting. Damage to the brain stem has serious and sometimes lethal consequences.

A complex network of interconnected neurons known as the **reticular formation** (*reticulum* = netlike structure) runs through the length of the brain stem, connecting to the thalamus at the anterior end and to the spinal cord at the posterior end **(Figure 38.8).** All incoming sensory input goes to the reticular formation, which integrates the information and then sends signals to other parts of the CNS. The reticular formation has two parts. The ascending reticular formation, also called the *reticular activating system,* contains neurons that convey stimulatory signals via the thalamus to arouse and activate the cerebral cortex. It is responsible for the sleep-wake cycle; depending on the level of stimulation of the cortex, various levels of alertness and consciousness are produced. Lesions in this part of the brain stem result in coma. The other part, the descending reticular formation, receives information from the hypothalamus and connects with interneurons in the spinal cord that control skeletal muscle contraction, thereby controlling muscle movement and posture. The reticular formation filters incoming signals, helping to discriminate between important and unimportant ones. Such filtering is necessary because the brain is unable to process every one of the signals from millions of sensory receptors. For example, the action of the reticular formation enables you to sleep through many sounds but waken to specific ones, such as a cat meowing to be let out or a baby crying.

The Cerebellum Integrates Sensory Inputs to Coordinate Body Movements

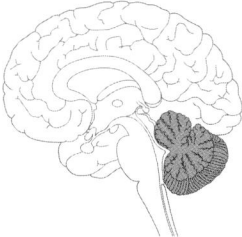

Although the cerebellum is an outgrowth of the pons (see Figure 38.8), it is separate in structure and function from the brain stem. Through its extensive connections with other parts of the brain, the **cerebellum** receives sensory input from receptors in muscles and joints, from balance receptors in the inner ear, and from the receptors of touch, vision, and hearing. These signals convey information about how the body trunk and limbs are positioned, the degree to which different muscles are contracted or relaxed, and the direction in which the body or limbs are moving. The cerebellum integrates these sensory signals and compares them with signals from the cerebrum that control voluntary body movements. Outputs from the cerebellum to the cerebrum, brain stem, and spinal cord modify and fine-tune the movements to keep the body in balance and directed toward targeted positions in space. The cerebellum of all mammals has essentially the same capabilities and works in the same way. The human cerebellum also contributes to the learning and memory of motor skills such as typing.

Gray-Matter Centers Control a Variety of Functions

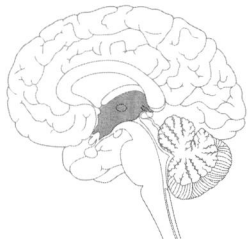

Gray-matter centers derived from the embryonic forebrain include the thalamus, hypothalamus, and basal nuclei **(Figure 38.9).** These centers contribute to the control and integration of voluntary movements, body temperature and glandular secretions, osmotic balance of the blood and extracellular fluids, wakefulness, and the

Figure 38.8
Location of the reticular formation (in blue) in the brain stem.

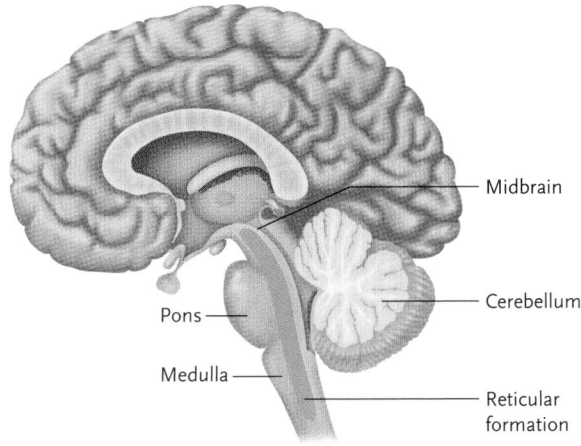

Midbrain

Pons

Cerebellum

Medulla

Reticular formation

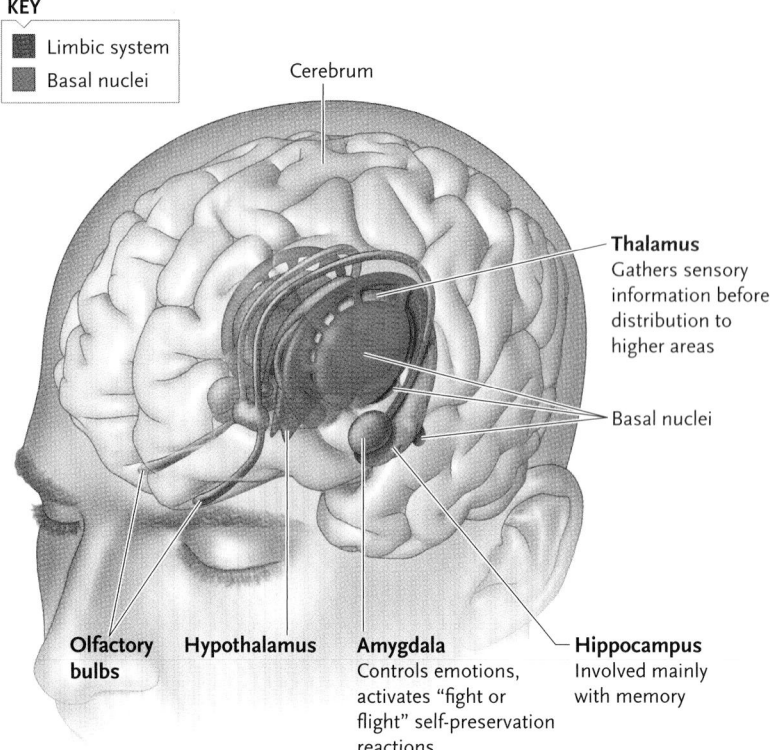

KEY
- Limbic system
- Basal nuclei

Cerebrum

Thalamus
Gathers sensory information before distribution to higher areas

Basal nuclei

Olfactory bulbs

Hypothalamus

Amygdala
Controls emotions, activates "fight or flight" self-preservation reactions

Hippocampus
Involved mainly with memory

Figure 38.9
Basal nuclei, thalamus, and hypothalamus gray-matter centers. The centers shown in this view are those in the left hemisphere.

emotions, among other functions. Some of the gray-matter centers route information to and from the cerebral cortex, and between the forebrain, brain stem, and cerebellum.

The **thalamus** (see Figure 38.9) forms a major switchboard that receives sensory information and relays it to the regions of the cerebral cortex concerned with motor responses to sensory information of that type. Part of the thalamus near the brain stem cooperates with the reticular formation in alerting the cerebral cortex to full wakefulness, or in inducing drowsiness or sleep.

The **hypothalamus** (see Figure 38.9) contains centers that regulate basic homeostatic functions of the body and contribute to the release of hormones. Some centers set and maintain body temperature by triggering reactions such as shivering or sweating. Others constantly monitor the osmotic balance of the blood by testing its composition of ions and other substances. If departures from normal levels are detected, the hypothalamus triggers responses such as thirst or changes in urine output that restore the osmotic and fluid balance.

The centers of the hypothalamus that detect blood composition and temperature are directly exposed to the bloodstream—they are the only parts of the brain *not* protected by the blood-brain barrier. Parts of the hypothalamus also coordinate responses triggered by the autonomic system, making it an important link in such activities as control of the heartbeat, contraction

of smooth muscle cells in the digestive system, and glandular secretion. Some regions of the hypothalamus establish a biological clock that sets up daily metabolic rhythms, such as the regular changes in body temperature that occur on a daily cycle.

The **basal nuclei** are gray-matter centers that surround the thalamus on both sides of the brain (see Figure 38.9). They moderate voluntary movements directed by motor centers in the cerebrum. Damage to the basal nuclei can affect the planning and fine-tuning of movements, leading to stiff, rigid motions of the limbs and unwanted or misdirected motor activity, such as tremors of the hands and inability to start or stop intended movements at the intended place and time. Parkinson disease, in which affected individuals exhibit all of these symptoms, results from degeneration of centers in and near the basal nuclei.

Parts of the thalamus, hypothalamus, and basal nuclei, along with other nearby gray-matter centers— the amygdala, hippocampus, and olfactory bulbs— form a functional network called the **limbic system** (*limbus* = arc), sometimes called our "emotional brain" (see Figure 38.9). The **amygdala** works as a switchboard, routing information about experiences that have an emotional component through the limbic system. The **hippocampus** is involved in sending information to the frontal lobes, and the **olfactory bulbs** relay inputs from odor receptors to both the cerebral cortex and the limbic system. The olfactory connection to the limbic system may explain why certain odors can evoke particular, sometimes startlingly powerful emotional responses.

The limbic system controls emotional behavior and influences the basic body functions regulated by the hypothalamus and brain stem. Stimulation of different parts of the limbic system produces anger, anxiety, fear, satisfaction, pleasure, or sexual arousal. Connections between the limbic system and other brain regions bring about emotional responses such as smiling, blushing, or laughing.

The Cerebral Cortex Carries Out All Higher Brain Functions in Humans

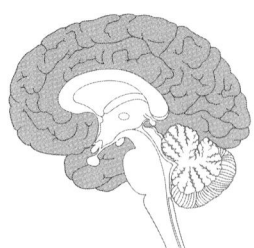

The gray matter of each hemisphere, the cerebral cortex, contains the processing centers for the integration of neural input and the initiation of neural output. The white matter of the cerebral hemispheres, by contrast, contains the neural routes for signal transmission between parts of the cerebral cortex, or from the cerebral cortex to other parts of the CNS. No information processing occurs in the white matter.

Over the course of evolution, the surface area of the cerebral cortex increased by continually folding in

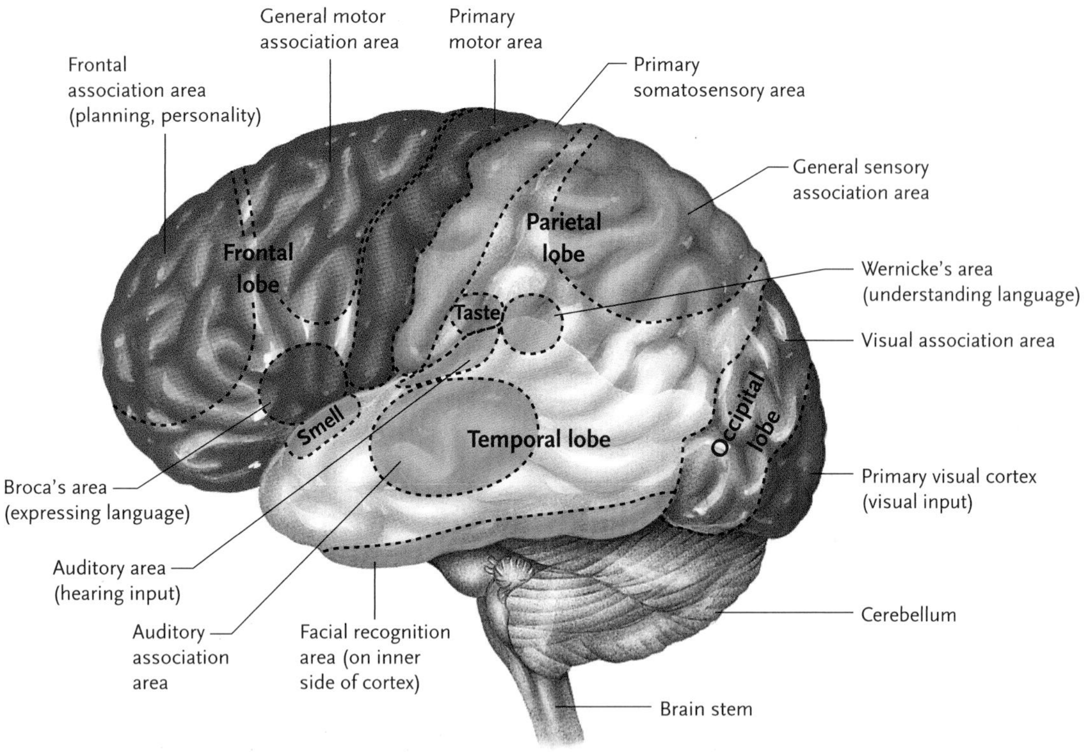

Figure 38.10
The lobes of the cerebrum, showing major regions and association areas of the cerebral cortex.

Frontal association area (planning, personality)

General motor association area

Primary motor area

Primary somatosensory area

General sensory association area

Wernicke's area (understanding language)

Visual association area

Parietal lobe

Frontal lobe

Taste

Smell

Temporal lobe

Occipital lobe

Broca's area (expressing language)

Auditory area (hearing input)

Auditory association area

Facial recognition area (on inner side of cortex)

Brain stem

Primary visual cortex (visual input)

Cerebellum

on itself, thereby expanding the structure into sophisticated information encoding and processing centers. Primates have cerebral cortexes with the largest number of convolutions. In humans, each cerebral hemisphere is divided by surface folds into *frontal, parietal, temporal,* and *occipital* lobes **(Figure 38.10).** Uniquely in mammals, the cerebral cortex of the cerebral hemispheres is organized into six layers of neurons; these layers are the newest part of the cerebral cortex in an evolutionary sense.

The two cerebral hemispheres can function separately, and each has its own communication lines internally and with the rest of the CNS and the body. The left cerebral hemisphere responds primarily to sensory signals from, and controls movements in, the right side of the body. The right hemisphere has the same relationships to the left side of the body. This opposite connection and control reflects the fact that the nerves carrying afferent and efferent signals cross from left to right within the spinal cord or brain stem. Thick axon bundles, forming a structure called the **corpus callosum,** connect the two cerebral hemispheres and coordinate their functions.

Sensory Regions of the Cerebral Cortex. Areas that receive and integrate sensory information are distributed over the cerebral cortex. In each hemisphere, the **primary somatosensory area,** which registers information on touch, pain, temperature, and pressure, runs in a band across the parietal lobes of the brain (see Figure 38.10). Experimental stimulation of this band in one hemisphere causes prickling or tingling sensations in specific parts on the opposite side of the body,

beginning with the toes at the top of each hemisphere and running through the legs, trunk, arms, and hands, to the head **(Figure 38.11).**

Other sensory regions of the cerebral cortex have been identified with hearing, vision, smell, and taste (see Figure 38.10). Regions of the temporal lobes on both sides of the brain receive auditory inputs from the ears, while inputs from the eyes are processed in the primary visual cortex in both occipital lobes. Olfactory input from the nose is processed in the olfactory bulbs, located on the ventral side of the temporal lobes. Regions in the parietal lobes receive inputs from taste receptors on the tongue and other locations in the mouth.

Motor Regions of the Cerebral Cortex. The **primary motor area** of the cerebral cortex runs in a band just in front of the primary somatosensory area (see Figure 38.10). Experimental stimulation of points along this band in one hemisphere causes movement of specific body parts on the opposite side of the body, corresponding generally to the parts registering in the primary somatosensory area at the same level (see Figure 38.11). Other areas that integrate and refine motor control are located nearby.

In both the primary somatosensory and primary motor areas, some body parts, such as the lips and fingers, are represented by large regions, and others, such as the arms and legs, are represented by relatively small regions. As shown in Figure 38.11, the relative sizes produce a distorted image of the human body that is quite different from the actual body proportions. The differences are reflected in the precision of

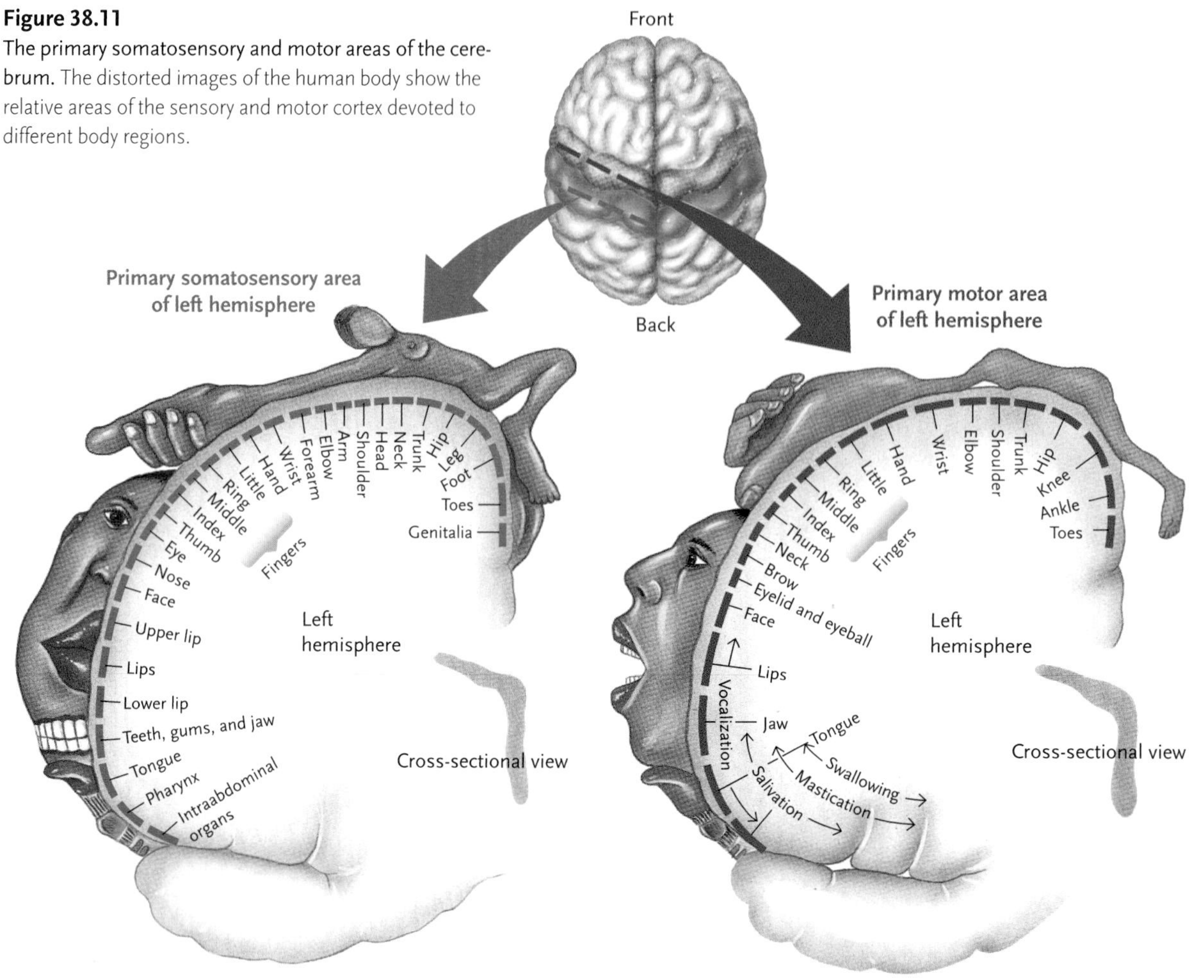

Figure 38.11
The primary somatosensory and motor areas of the cerebrum. The distorted images of the human body show the relative areas of the sensory and motor cortex devoted to different body regions.

touch and movement in structures such as the lips, tongue, and fingers.

Association Areas. The sensory and motor areas of the cerebral cortex are surrounded by **association areas** (see Figure 38.10), which integrate information from the sensory areas, formulate responses, and pass them on to the primary motor area. Two of the most important association areas are *Wernicke's area* and *Broca's area* (see Figure 38.10), which function in spoken and written language. They are usually present on only one side of the brain—in the left hemisphere in 97% of the human population. Comprehension of spoken and written language depends on Wernicke's area, which coordinates inputs from the visual, auditory, and general sensory association areas. Interneuron connections lead from Wernicke's area to Broca's area, which puts together the motor program for coordination of the lips, tongue, jaws, and other structures producing the sounds of speech, and passes the program to the primary motor area. The brain-scan images in Figure 38.7 dramatically illustrate how these brain regions participate as a person performs different linguistic tasks.

People with damage to Wernicke's area have difficulty comprehending spoken and written words,

even though their hearing and vision are unimpaired. Although they can speak, their words usually make no sense. People with damage to Broca's area have normal comprehension of written and spoken language, and know what they want to say, but are unable to speak except for a few slow and poorly pronounced words. Often, such people are also unable to write. Other areas of the brain are also involved in language functions.

Some Higher Functions Are Distributed in Both Cerebral Hemispheres; Others Are Concentrated in One Hemisphere

Most of the other higher functions of the human brain—such as abstract thought and reasoning; spatial recognition; mathematical, musical, and artistic ability; and the associations forming the basis of personality—involve the coordinated participation of many regions of the cerebral cortex. Some of these regions are equally distributed in both cerebral hemispheres, and some are more concentrated in one hemisphere.

Among the functions more or less equally distributed between the two hemispheres is the ability to recognize faces. This function is concentrated along the bottom margins of the occipital and temporal lobes (see Figure 38.10). People with damage to these lobes are often unable to recognize even close relatives by sight but can recognize voices immediately. Functions such as consciousness, the sense of time, and recognizing emotions also seem to be distributed in both hemispheres.

Typically some brain functions are more localized in one of the two hemispheres, a phenomenon called **lateralization.** The unequal distribution of these functions was originally worked out in the 1960s by Roger Sperry and Michael S. Gazzaniga of the California Institute of Technology (Sperry received a Nobel Prize for his research in 1981) in subjects who had had their corpus callosum cut surgically **(Figure 38.12).**

Studies of people with split hemispheres as well as surveys of brain activity by PET and fMRI have confirmed that, for the vast majority of people, the left hemisphere specializes in spoken and written language, abstract reasoning, and precise mathematical calculations. The right hemisphere specializes in nonverbal conceptualizing, intuitive thinking, musical and artistic abilities, and spatial recognition functions such as fitting pieces into a puzzle. The right hemisphere also handles mathematical estimates and approximations that can be made by visual or spatial representations of numbers. Thus the left hemisphere in most people is verbal and mathematical, and the right hemisphere is intuitive, spatial, artistic, and musical.

STUDY BREAK

1. Human newborn babies, as well as premature babies, have an incompletely developed blood-brain barrier. Should this condition influence what food and medications are given to them?
2. Distinguish the structure and functions of the cerebellum from those of the cerebral cortex.

38.4 Memory, Learning, and Consciousness

We set memory, learning, and consciousness apart from the other functions because they appear to involve coordination of structures from the brain stem to the cerebral cortex. **Memory** is the storage and retrieval of a sensory or motor experience, or a thought. **Learning** involves a change in the response to a stimulus based on information or experiences stored in memory. **Consciousness** may be defined as awareness of ourselves, our identity, and our surroundings, and an understanding of the significance and likely consequences of events that we experience.

Memory Takes Two Forms, Short Term and Long Term

Psychology research and our everyday experience indicate that humans have at least two types of memory. **Short-term memory** stores information for seconds, minutes, or at most an hour or so. **Long-term memory** stores information from days to years or even for life. Short-term memory, but not long-term memory, is usually erased if a person experiences a disruption such as a sudden fright, a blow, a surprise, or an electrical shock. For example, a person knocked unconscious by an accident typically cannot recall the accident itself or the events just before it, but long-standing memories are not usually disturbed.

To explain these differences, investigators propose that short-term memories depend on transient changes in neurons that can be erased relatively easily, such as changes in the membrane potential of interneurons caused by EPSPs and IPSPs (excitatory and inhibitory postsynaptic potentials) and the action of indirect neurotransmitters that lead to reversible changes in ion transport (see Section 37.3). By contrast, storage of long-term memory is considered to involve more or less permanent molecular, biochemical, or structural changes in interneurons, which establish signal pathways that cannot be switched off easily.

All memories probably register initially in short-term form. They are then either erased and lost, or committed to long-term form. The intensity or vividness of an experience, the attention focused on an event, emotional involvement, or the degree of repetition may all contribute to the conversion from short-term to long-term memory.

The storage pathway typically starts with an input at the somatosensory cortex that then flows to the amygdala, which relays information to the limbic system, and to the hippocampus, which sends information to the frontal lobes, a major site of long-term memory storage. People with injuries to the hippocampus cannot remember information for more than a few minutes; long-term memory is limited to information stored before the injury occurred.

How are neurons and neuron pathways permanently altered to create long-term memory? One change that has been much studied is **long-term potentiation:** a long-lasting increase in the strength of synaptic connections in activated neural pathways following brief periods of repeated stimulation. The synapses become increasingly sensitive over time, so that a constant level of presynaptic stimulation is converted into a larger postsynaptic output that can last hours, weeks, months, or years. (*Insights from the Molecular Revolution* describes experiments investigating the basis of long-term potentiation in neurons of the hippocampus.) Other changes consistently noted as part of long-term memory include more or less permanent alterations in the number and

Figure 38.12 Experimental Research

Investigating the Functions of the Cerebral Hemispheres

QUESTION: Do the two cerebral hemispheres have different functions?

EXPERIMENT: Roger Sperry and Michael Gazzaniga studied split-brain individuals, in whom the corpus callosum connecting the two cerebral hemispheres had been surgically severed to relieve otherwise uncontrollable epileptic convulsions. In one experiment, they tested how subjects perceived words that were projected onto a screen in front of them.

The retinas of the eyes gather visual information and send signals via the optic nerves to the cerebral hemispheres (Figure 38.12a). Light from the *left* half of the visual field reaches light receptors on the *right* sides of the retinas, and parts of the two optic nerves carry signals to the *right* cerebral hemisphere. Light from the *right* half of the visual field reaches light receptors on the *left* sides of the retinas, and signals are sent to the *left* cerebral hemisphere.

The researchers projected words such as COWBOY in such a way that the subjects could see only the left half of the word (COW) with the left eye and the right half of the word (BOY) with the right eye (Figure 38.12b). Sperry asked the subjects to say what word they saw, and he asked them to write the perceived word with the left hand—a hand that was deliberately blocked from the subject's view.

a. Pathway of visual information from eyes to cerebral hemisphere

b. Experimental set up—COW seen by right sides of retinas and BOY by left sides

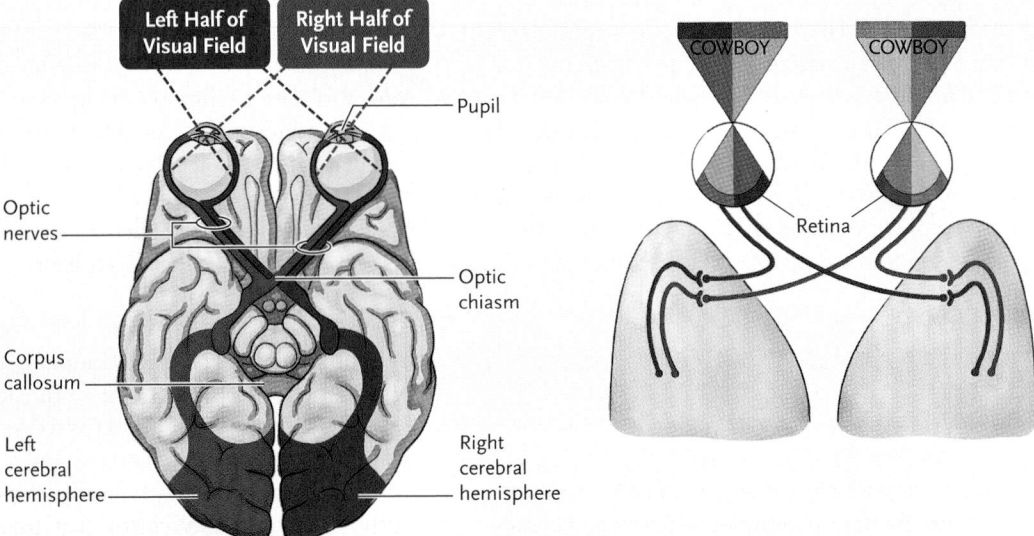

RESULTS: The split-brain subjects said the word in the right half of the visual field (BOY), but wrote the word in the left half of the visual field (COW).

CONCLUSIONS: The studies showed that the left and right hemispheres are specialized in different tasks. The left hemisphere processes language and was able to recognize BOY but received no information about COW. The right hemisphere directs motor activity on the left side of the body and was able to direct the left hand to write COW. However, the subjects could not say what word they wrote. That is, cutting the corpus callosum interrupted communication between the two halves of the cerebrum. In effect, one cerebral hemisphere did not know what the other was doing, and information stored in the memory on one side was not available to the other. In normal individuals, information is shared across the corpus callosum; they would see COWBOY and be able to speak and write the entire word.

Knocked-Out Mice with a Bad Memory

Long-term potentiation (LTP) in neurons of the hippocampus is thought to be central to the conversion of short-term to long-term memory. An indirect neurotransmitter, glutamate, is most often involved in the process. One of the receptors that binds glutamate in postsynaptic membranes of the hippocampus is the *NMDA receptor;* a prolonged series of stimuli causes glutamate to bind to the receptor, which opens an ion channel that forms part of the receptor's structure. Among other effects, Ca^{2+} ions flowing inward through the channel activate a protein kinase in the cytoplasm called *CaMKII*. (Protein kinases are enzymes that, when activated, add phosphate groups to certain proteins; addition of the phosphate groups increases or reduces activity of the proteins.)

One of the target proteins for the activated CaMKII is itself. After it adds phosphate groups to its own structure, CaMKII no longer needs to be activated by calcium—it remains turned on at elevated levels. Its more or less permanent activation makes the neuron more sensitive to incoming signals and increases the number of the signals it sends to other neurons. Researchers hypothesize that this change is a major contributor to LTP in the neuron. Among the supporting evidence is the observation that chemical blockage of the NMDA receptor impairs spatial learning and long-term memory in mice.

A research group at the Massachusetts Institute of Technology led by a Nobel Prize–winning scientist, Susumu Tonegawa, applied molecular techniques to test this hypothesis. Tonegawa's group created a strain of "knockout" mice (see Section 18.2), from which the gene encoding the NMDA receptor has been eliminated.

To test the effects of the knockout on LTP in the neurons, the investigators dissected out brain slices and used microelectrodes to stimulate neurons in the hippocampus. Unlike the response of neurons with an intact NMDA receptor, repeated stimulation was not followed by any potentiation in the knockout neurons.

Would the lack of potentiation in the hippocampal neurons result in the failure of long-term memory storage in the knockout mice? The researchers tested this outcome by placing the knockout mice in a pool in which they had to swim until they could find a submerged platform on which to rest. The knockout mice were much slower than normal mice in finding the platform in initial trials, and, unlike normal mice, were unable to remember its location in later trials.

The results thus support the hypothesis that the NMDA receptor is involved in the conversion from short-term to long-term memory, and that the hippocampus has a central role in this activity. The novel approach used by the Tonegawa group also provides a new, molecular research method by which to analyze higher brain function.

the area of synaptic connections between neurons, in the number and branches of dendrites, and in gene transcription and protein synthesis in interneurons.

Experiments have shown that protein synthesis is critical to long-term memory storage in animals as varied as *Drosophila* and rats. For example, goldfish were trained to avoid an electrical shock by swimming to one end of an aquarium when a light was turned on. The fish could remember the training for about a month under normal conditions; if exposed to a protein synthesis inhibitor while being trained, they forgot the training within a day.

Learning Involves Combining Past and Present Experiences to Modify Responses

As with memory, all animals appear to be capable of learning to some degree. Learning involves three sequential mechanisms, (1) storing memories, (2) scanning memories when a stimulus is encountered, and (3) modifying the response to the stimulus in accordance with the information stored as memory.

One of the simplest forms of memory is **sensitization**—increased responsiveness to mild stimuli after experiencing a strong stimulus. The process was nicely illustrated by Eric Kandel of Columbia University and his associates in experiments with a shell-less marine snail known as the Pacific sea hare, *Aplysia californica,* which is frequently used in research involving reflex behavior, memory, and learning. Many of its neuron circuits have been completely worked out, allowing investigators to follow the reactions of each neuron active in pathways such as learning. The first time the researchers administered a single sharp tap to the siphon (which admits water to the gills), the slug retracted its gills by a reflex movement. However, at the next touch, whether hard or gentle, the siphon retracted much more quickly and vigorously. Sensitization in *Aplysia* has been shown to involve changes in synapses, which become more reactive when more serotonin is released by action potentials. Kandel received the Nobel Prize in 2000 for his research.

Learning skills or procedures, such as tying one's shoes, typing, or playing a musical instrument, involve additional regions of the brain, particularly the cerebellum, where motor activity is coordinated. As we learn such skills, the process gradually becomes automated so that we do not think consciously about each step. (Learning and its relationship to animal behavior are considered further in Chapter 54.)

Consciousness Involves Different States of Awareness

The spectrum of human consciousness ranges from alert wakefulness to daydreaming, dozing, and sleep. Even during sleep there is some degree of awareness, because sleepers can respond to stimuli and waken, unlike someone who is unconscious. Moving between the states of consciousness has been found to involve changes in neural activity over the entire surface of the telencephalon. These changes can be seen using an *electroencephalogram* (EEG), which records voltage changes detected by electrodes placed on the scalp.

When an individual is fully awake, the EEG records a pattern of rapid, irregular *beta waves* (**Figure 38.13**). With mind at rest and eyes closed, the person's EEG pattern changes to slower and more regular *alpha waves*. As drowsiness and light sleep come on, the wave trains gradually become larger, slower, but again less regular; these slower pulsations are called *theta waves*. During the transition from drowsiness to deep sleep, the EEG pattern shifts to even slower *delta waves*. The heart and breathing rates become slower and the skeletal muscles increasingly relaxed, although the sleeper may still change position and move the arms and legs.

Periodically during deep sleep, the delta wave pattern is replaced by the rapid, irregular beta waves characteristic of the waking state. The person's heartbeat and

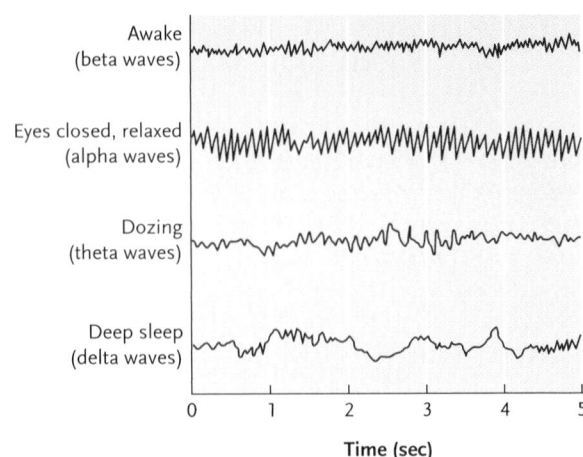

Figure 38.13
Brain waves characteristic of various states of consciousness.

breathing rate increase, the limbs twitch, and the eyes move rapidly behind the closed eyelids, giving this phase its name of **rapid-eye-movement (REM) sleep.** The REM sleep phase occurs about every 1.5 hours while a healthy adult is sleeping, and lasts for 10 to 15 minutes. Sleepers do most of their dreaming during REM sleep, and most research subjects awakened from REM sleep report they were experiencing vivid dreams.

As mentioned earlier, the reticular activating system controls the sleep-wake cycle. It sends signals to the spinal cord, cerebellum, and cerebral cortex, and receives signals from the same locations. The flow of

Unanswered Questions

What gives humans their unique brain capacity?
The human brain is larger relative to our body size than the brains of other mammals and has more functions. How did that come to be the case? At the simplest conceptual level, the answer must lie in the human genome. Researchers led by David Haussler of the University of California, Santa Cruz, have found evidence for unique human DNA that appears to play a central role in giving humans their unique brain capacity. Haussler's group compared the sequences of the human genome with the sequences of the genomes of other primates and other vertebrates. They looked for regions in the human genome that show significantly accelerated rates of base-pair changes since divergence from our common ancestor with the chimpanzee. The investigators found 49 such regions and dubbed them "human accelerated regions," or HARs.

One of the regions—*HAR1*—had dramatic changes that made it stand out from the rest. Haussler's group therefore focused on learning more about the region, looking specifically to see if it contained a gene and, if it did, what that gene encoded. Computer analysis of *HAR1* indicated the presence of what seemed to be a gene, and biochemical analysis confirmed that conclusion by detecting the presence of an RNA encoded by the gene. Interestingly the RNA is not an mRNA, meaning

that the gene does not encode a protein. Rather, it is a structural RNA; that is, when it is expressed, it functions in the cell as an RNA. The researchers next investigated where and when the *HAR1* gene was expressed by looking for the RNA product in human embryonic brain tissue samples taken from different stages of development. The results showed that *HAR1* is expressed in a particular type of neuron in the developing human cerebral cortex at 7 to 19 weeks of development. That period of development is crucial for neuron specification in the cerebral cortex.

The researchers conclude that *HAR1* is a highly promising candidate for a gene involved in uniquely human biology. Work is now continuing to determine how the RNA encoded by *HAR1* functions in the cell. Does it interact with a protein, or with another RNA? How does it play a role in neuron specification in the developing cerebral cortex? As the project continues, the investigators hope to be able to use the mouse model to examine key aspects of *HAR1* function, because human experiments are not possible for ethical reasons. Haussler's group is also investigating other HARs to see what genes they might contain and what functions those genes might have in human brain development and function.

Peter J. Russell

signals along these circuits determines whether we are awake or asleep.

Many other animals also alternate periods of wakefulness and sleep or inactivity. Although sleep obviously has restorative effects on mental and physical functions, the physiological basis of these effects remains unknown.

In the previous chapter we learned about neurons, and in this chapter we have discussed the organization of neurons into nervous systems, as well as the structures of the brain and their functions. In the next chapter we consider the sensory systems that provide input for the brain to process.

STUDY BREAK

An aging person often experiences a progressive decline in cognitive function. This typically begins with short-term memory loss and the inability to learn new information. What brain changes might be occurring?

Review

Go to **ThomsonNOW™** at www.thomsonedu.com/login to access quizzing, animations, exercises, articles, and personalized homework help.

38.1 Invertebrate and Vertebrate Nervous Systems Compared

- The simplest nervous systems are the nerve nets of cnidarians. Echinoderms have modified nerve nets, with some neurons grouped into nerves (Figure 38.1a–b).

- Flatworms, arthropods, and mollusks have a simple central nervous system (CNS), consisting of ganglia in the head region (a brain), and a peripheral nervous system (PNS), consisting of nerves from the CNS to the rest of the body (Figure 38.1c–e).

- In vertebrates, the CNS consists of a large brain located in the head and a hollow spinal cord, and the PNS consists of all the nerves and ganglia connecting the CNS to the rest of the body (Figure 38.1f).

- In the vertebrate embryo, the anterior end of the hollow neural tube develops into the brain, and the rest develops into the spinal cord. The embryonic brain enlarges into the forebrain, midbrain, and hindbrain, which develop into the adult structures (Figure 38.2).

  Animation: Comparisons of animal nervous systems

  Animation: Bilateral nervous systems

  Animation: Vertebrate nervous system divisions

38.2 The Peripheral Nervous System

- Afferent neurons in the PNS conduct signals to the CNS, and signals from the CNS travel via efferent neurons to the effectors—muscles and glands—that carry out responses (Figure 38.3).

- The somatic system of the PNS controls the skeletal muscles, producing voluntary body movements as well as involuntary muscle contractions that maintain balance, posture, and muscle tone.

- The autonomic system of the PNS, which controls involuntary functions, is organized into the sympathetic division and the parasympathetic division (Figure 38.4).

  Animation: Autonomic nerves

38.3 The Central Nervous System (CNS) and Its Functions

- The spinal cord carries signals between the brain and the PNS. Its neuron circuits control reflex muscular movements and some autonomic reflexes (Figure 38.5).

- The medulla, pons, and midbrain form the brain stem, which connects the cerebrum, thalamus, and hypothalamus with the spinal cord.

- The cerebrum is divided into right and left cerebral hemispheres, which are connected by a thick band of nerve fibers, the corpus callosum. Each hemisphere consists of the cerebral cortex, a thin layer of gray matter, covering a thick core of white matter. Other collections of gray matter, the basal nuclei, are deep in the telencephalon (Figure 38.6).

- Cerebrospinal fluid provides nutrients and cushions the CNS (Figure 38.6). A blood-brain barrier allows only selected substances to enter the cerebrospinal fluid.

- Gray-matter centers in the pons and medulla control involuntary functions. Centers in the midbrain coordinate responses to visual and auditory sensory inputs.

- The reticular formation receives sensory inputs from all parts of the body and sends outputs to the cerebral cortex that help maintain balance, posture, and muscle tone. It also regulates states of wakefulness and sleep (Figure 38.8).

- The cerebellum integrates sensory inputs on the positions of muscles and joints, along with visual and auditory information, to coordinate body movements.

- The telencephalon's subcortical gray-matter centers control many functions. The thalamus receives, filters, and relays sensory and motor information to and from regions of the cerebral cortex. The hypothalamus regulates basic homeostatic functions of the body and contributes to the endocrine control of body functions. The basal nuclei affect the planning and fine-tuning of body movements (Figure 38.9).

- The limbic system includes parts of the thalamus, hypothalamus, and basal nuclei, as well as the amygdala and hippocampus. It controls emotions and influences the basic body functions controlled by the hypothalamus and brain stem (Figure 38.9).

- The primary somatosensory areas of the cerebral cortex register incoming information on touch, pain, temperature, and pressure from all parts of the body. In general, the right cerebral hemisphere receives sensory information from the left side of the body and vice versa (Figures 38.10 and 38.11).

- The primary motor areas control voluntary movements of skeletal muscles (Figures 38.10 and 38.11).

- The association areas integrate sensory information and formulate responses that are passed on to the primary motor areas. Wernicke's area integrates visual, auditory, and other sensory information into the comprehension of language; Broca's area coordinates movements of the lips, tongue, jaws, and other structures to produce the sounds of speech (Figure 38.10).

- Long-term memory and consciousness are equally distributed between the two cerebral hemispheres. Spoken and written language, abstract reasoning, and precise mathematical calculations are left hemisphere functions; nonverbal conceptualizing, mathematical estimation, intuitive thinking, spatial recognition, and artistic and musical abilities are right hemisphere functions (Figure 38.12).

Animation: Organization of the spinal cord

Animation: Regions of the vertebrate brain

Animation: Human brain development

Animation: Sagittal view of a human brain

Animation: Primary motor cortex

Animation: Receiving and integrating areas

Animation: Path to visual cortex

38.4 Memory, Learning, and Consciousness

- Memory is the storage and retrieval of a sensory or motor experience or a thought. Short-term memory involves temporary storage of information, whereas long-term memory is essentially permanent.
- Learning involves modification of a response through comparisons made with information or experiences that are stored in memory.
- Consciousness is the awareness of ourselves, our identity, and our surroundings. It varies through states from full alertness to sleep and is controlled by the reticular activating system (Figure 38.13).

Animation: Structures involved in memory

Questions

Self-Test Questions

1. Ganglia first became enlarged and fused into a lobed brain in the evolution of:
   a. vertebrates.
   b. annelids.
   c. flatworms.
   d. cephalopods.
   e. mammals.

2. The metencephalon develops into the:
   a. spinal cord.
   b. cerebellum.
   c. mesencephalon.
   d. medulla oblongata.
   e. cerebrum.

3. The autonomic nervous system is subdivided into:
   a. afferent and efferent systems.
   b. sympathetic and parasympathetic divisions.
   c. skeletal and smooth muscle innervations.
   d. voluntary and involuntary controls.
   e. peripheral and central systems.

4. People with severe insect-sting allergies carry an *epipen* containing medication that they can inject in an emergency. The medication causes smooth muscles in the lung passages to relax so they can breathe but causes their hearts to pound rapidly. This is an example of stimulation of the:
   a. parasympathetic system.
   b. sympathetic system.
   c. somatic nervous system.
   d. limbic system.
   e. voluntary system.

5. Which one of the following structures participates in a reflex?
   a. the gray matter of the brain
   b. the white matter of the brain
   c. the gray matter of the spinal cord
   d. an interneuron that stimulates an afferent neuron
   e. an interneuron that inhibits an afferent neuron

6. Which of the following statements about the blood-brain barrier is incorrect?
   a. It is formed of capillary walls composed of tight junctions.
   b. It transports glucose to brain cells by means of transport proteins.
   c. It allows alcohol to pass through its lipid bilayer.
   d. It moves oxygen through the lipid bilayer.
   e. It reduces blood supply to brain cells compared with other body cells.

7. A segment of the brain stem that coordinates spinal reflexes with higher brain centers and regulates breathing and wakefulness is the:
   a. reticular formation.
   b. white matter of the pons.
   c. white matter of the medulla.
   d. hypothalamus.
   e. cerebellum.

8. Cushioning and nourishing the brain and spinal cord and filling the ventricles of the brain is (are):
   a. meninges.
   b. myelin.
   c. cerebrospinal fluid.
   d. ganglia.
   e. astrocytes.

9. Which structure and function are correctly paired below?
   a. thalamus: relays emotion signals through the limbic system
   b. basal nuclei: relay inputs from odor receptors to the cerebrum
   c. hypothalamus: releases hormones; sets up daily rhythms
   d. amygdala: relays sensory information to the cerebrum
   e. olfactory bulbs: moderate motor centers in the cerebrum

10. A patient had a tumor in Wernicke's area. It was initially diagnosed when he could not:
    a. understand his morning newspaper.
    b. hear his child crying.
    c. see the traffic light turn red.
    d. speak.
    e. feel if the car heater was on.

Questions for Discussion

1. Meningitis is an inflammation of the meninges, the membranes that cover the brain and spinal cord. Diagnosis involves using a needle to obtain a sample of cerebrospinal fluid to analyze for signs of infection. Why analyze this fluid and not blood?

2. An accident victim arrives at the emergency room with severe damage to the reticular formation. Based on information in this chapter, describe some of the symptoms that the examining physician might discover.

3. In the 1930s and 1940s prefrontal lobotomy, in which neural connections in the frontal lobes of both cerebral hemispheres were severed, was used to treat behavioral conditions such as extreme anxiety and rebelliousness. Although the procedure calmed patients, it had side effects such as apathy and a seriously disrupted personality. In view of the information presented in this chapter, why do you think the operation had these effects?

Experimental Analysis

How would you demonstrate that gene activity in the brain altered with aging in mice?

Evolution Link

How do paleontologists contribute to our understanding of the evolution of the brain?

A greater horseshoe bat *(Rhinolophus ferrumequinum)* hunting a moth. The bat uses its sensory system to pursue prey, and the moth uses its sensory system in attempting to evade capture.

© Stephen Dalton/Animals, Animals—Earth Scenes

39 Sensory Systems

WHY IT MATTERS

An insectivorous bat leaves its cave after a good day's sleep to look for food. As it flies, the bat emits a steady stream of ultrasonic clicking noises. Receptors in the bat's ears detect echoes of the clicks bouncing off objects in the environment and send signals to the brain, where they are integrated into a sound map that the animal uses to avoid trees and other obstacles. This ability to detect objects by *echolocation* is so keenly developed that a bat can detect and avoid a thin wire in the dark.

Besides recognizing obstacles, the bat's sensory system is keenly tuned to the distinctive pattern of echoes reflected by the fluttering wings of its favorite food, a moth. Although the slow-flying moth would seem doomed to become a meal for the foraging bat, natural selection has provided some species of moths with an astoundingly sensitive and efficient auditory sense as well as a programmed escape mechanism. On each side of its abdomen is an "ear," a thin membrane that resonates at the frequencies of the clicks emitted by the bat. The moth's ears register the clicks while the bat is still about 30 m away and initiate a response that turns its flight path directly away from the source of the clicks, giving the moth an early advantage in the nocturnal dance of life and death.

In spite of the moth's evasive turn, the bat's random flight pattern carries it in the direction of its prey. At a distance of about 6 m, when echoes from the moth begin to register in the bat's auditory system, the bat increases the frequency of its clicks, enabling it to pinpoint the moth's position.

The moth has not exhausted its evasive tactics, however. As the bat closes in, the increased frequency of the clicks sets off another programmed response that alters the moth's flight into sudden loops and turns, ending with a closed-wing, vertical fall toward the ground. After dropping a few feet, the moth resumes its fluttering flight.

Although the moth escapes for a moment, the bat also alters its path, turning back toward the moth as its echolocation again locks on the prey. The contest goes on, but the bat's sensory system finally leads it to intercept the moth an instant before its vertical drop, and the bat retires to a branch to eat its meal.

Natural selection has produced highly adaptive sensory receptors in moths, bats, and all other animals. These systems, the subject of this chapter, provide animals with a steady stream of information about their internal and external environments. After integrating the information in the central nervous system (CNS), animals respond in ways that enable them to survive and reproduce. We begin this chapter with a survey of animal sensory systems and the ways in which they work.

39.1 Overview of Sensory Receptors and Pathways

Information about an animal's external and internal environments is picked up by **sensory receptors**, formed by the dendrites of afferent neurons, or by specialized receptor cells making synapses with afferent neurons **(Figure 39.1)**. The receptors associated with eyes, ears, skin, and other surface organs detect stimuli from the external environment. Sensory receptors associated with internal organs detect stimuli arising in the body interior.

Sensory receptors respond to stimuli by undergoing a change in membrane potential, caused in most receptors by changes in the rate at which channels conduct positive ions such as Na^+, K^+, or Ca^{2+} across the plasma membrane. Examples of stimuli are light, heat, sound waves, mechanical stress, and chemicals; the conversion of a stimulus into a change in membrane potential is called **sensory transduction**. The change in membrane potential may generate one or more action potentials, which travel along the axon of an afferent neuron to reach the interneuron networks of the CNS. These interneurons integrate the action potentials, and the brain formulates a compensating response, that is, a response appropriate for the stimulus (see Section 38.3). In animals with complex nervous systems, the

Figure 39.1
Sensory receptors, formed **(a)** by the dendrites of an afferent neuron or **(b)** by a separate cell or structure that communicates with an afferent neuron via a neurotransmitter.

a. Sensory receptor formed by dendrites of an afferent neuron

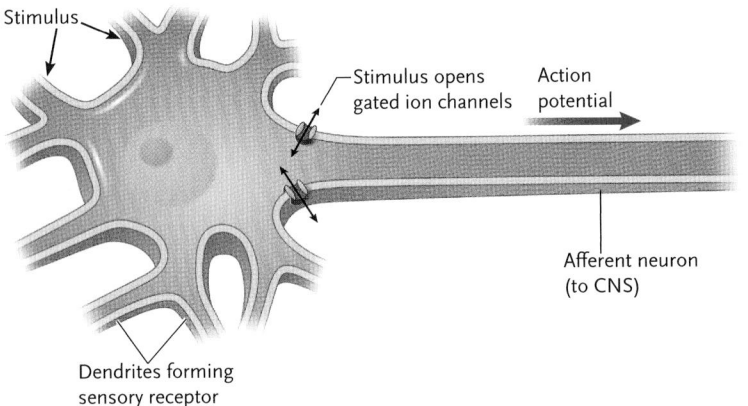

In sensory receptors formed by the dendrites of afferent neurons, a stimulus causes a change in membrane potential that generates action potentials in the axon of the neuron. Temperature and pain receptors are among the receptors of this type.

b. Sensory receptor formed by a cell that synapses with an afferent neuron

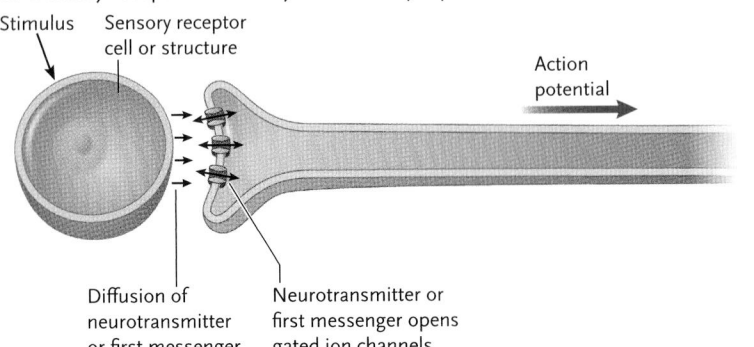

In sensory receptors consisting of separate cells, a stimulus causes a change in membrane potential that releases a neurotransmitter from the cell. The neurotransmitter triggers an action potential in the axon of a nearby afferent neuron. Mechanoreceptors, photoreceptors, and chemoreceptors are examples of receptors of this type.

interneuron networks may also produce an awareness of a stimulus in the form of a conscious sensation or perception.

Five Basic Types of Receptors Are Common to Almost All Animals

Many sensory receptors are positioned individually in body tissues. Others are part of complex sensory organs, such as the eyes or ears, that are specialized for reception of physical or chemical stimuli. Commonly, sensory receptors are classified into five major types, based on the type of stimulus that each detects:

1. **Mechanoreceptors** detect mechanical energy, such as changes in pressure, body position, or acceleration. The auditory receptors in the ears are examples of mechanoreceptors.
2. **Photoreceptors** detect the energy of light. In vertebrates, photoreceptors are mostly located in the retina of the eye.
3. **Chemoreceptors** detect specific molecules, or chemical conditions such as acidity. The taste buds on the tongue are examples of chemoreceptors.
4. **Thermoreceptors** detect the flow of heat energy. Receptors of this type are located in the skin, where they detect changes in the temperature of the body surface.
5. **Nociceptors** detect tissue damage or noxious chemicals; their activity registers as pain. Pain receptors are located in the skin, and also in some internal organs.

In addition to these major types, some animals have receptors that can detect electrical or magnetic fields.

Although humans are traditionally said to have five senses—vision, hearing, taste, smell, and touch—our sensory receptors actually detect more than twice as many kinds of environmental stimuli. Among these are external heat, internal temperature, gravity, acceleration, the positions of muscles and joints, body balance, internal pH, and the internal concentration of substances such as oxygen, carbon dioxide, salts, and glucose.

Afferent Neurons Link Receptors to the CNS

Sensory pathways begin at a sensory receptor and proceed by afferent neurons to the CNS. Because of its wiring, each type of receptor produces a specific kind of response. For example, action potentials arising in the retina of the eye travel along the optic nerve to the visual cortex, where they are interpreted by the brain as differences in the pattern, color, and intensity of light. If you receive a blow to the eye, the stimulus is still interpreted in the visual cortex as differences in the color and intensity of light detected by the eyes—you "see stars"—even though the stimulus is mechanical.

One way in which the intensity and extent of a stimulus is registered is by the frequency (number per unit time) of action potentials traveling along each axon of an afferent pathway. That is, the stronger the stimulus, the more frequently afferent neurons fire action potentials (see Section 37.2). A light touch to the hand, for example, causes action potentials to flow at low frequencies along the axons leading to the primary somatosensory area of the cerebral cortex. As the pressure increases, the number of action potentials per second rises in proportion; in the brain, the increase is interpreted as greater pressure on the hand. Maximum stimulus input is interpreted as pain in the sensory cortex.

The second way in which the intensity and extent of a stimulus is registered is by the number of afferent neurons that the stimulus activates to generate action potentials in the pathway. This way reflects the number of afferent neurons carrying signals from a stimulated region to the brain. The more sensory receptors that are activated, the more axons carry information to the brain. A light touch activates a relatively small number of receptors in a small area near the surface of the finger, for example. As the pressure increases, the resulting indentation of the finger's surface increases in area and depth, activating more receptors. In the appropriate somatosensory area of the brain, the larger number of axons carrying action potentials is interpreted as an increase in pressure spread over a greater area of the finger.

Many Receptor Systems Reduce Their Response When Stimuli Remain Constant

In many systems, the effect of a stimulus is reduced if it continues at a constant level. The reduction, called **sensory adaptation**, reduces the frequency of action potentials generated in afferent neurons when the intensity of a stimulus remains constant. Some receptors adapt quickly and broadly; other receptors adapt only slightly.

For example, when you go to bed, you are initially aware of the touch and pressure of the covers on your skin. Within a few minutes, the sensations lessen or are lost even though your position remains the same. The loss reflects adaptation of mechanoreceptors in your skin. If you move, so that the stimulus changes, the mechanoreceptors again become active. In contrast, nociceptors adapt only slightly, or not at all, to painful stimuli.

In some sensory receptors, biochemical changes in the receptor cell contribute to adaptation. For example, when you move from a dark movie theater into bright sunshine, the photoreceptors of the eye adapt to bright light partly through breakdown of some of the pigments that absorb light.

Sensory adaptation is crucial to animal survival. The adaptation of photoreceptors in our eyes keeps us from being blinded indefinitely as we pass from a darkened room into bright sunlight. Sensory adaptation also increases the sensitivity of receptor systems to *changes* in environmental stimuli, which may be more important to survival than keeping track of environmental factors that remain constant. You may have noticed a cat sitting motionless, focused on its prey, a mouse. As long as the environmental stimuli are constant, the cat's position remains fixed. However, if the mouse moves, the cat will respond rapidly and attempt to capture and kill it.

Many prey animals take advantage of adaptation in predators as a means for concealment or defense. These animals instinctively become motionless when they sense a predator in their environment, which frequently allows them to remain undetected by the adapted senses of their predator.

Nonadapting receptors, such as those detecting pain, are also essential for survival. Pain signals a potential danger to some part of the body, and the signals are maintained until a response by the animal compensates for the stimulus causing the pain.

We now examine the individual receptor types and their characteristics.

39.2 Mechanoreceptors and the Tactile and Spatial Senses

Mechanical stimuli such as touch and pressure are detected by mechanoreceptors. The mechanical forces of a stimulus distort proteins in the plasma membrane of receptors, altering the flow of ions through the membrane. The changed ion flows generate action potentials in afferent neurons leading to the CNS. Sensory information from the receptors informs the brain of the body's contact with objects in the environment, provides information on the movement, position, and balance of body parts, and underlies the sense of hearing.

Receptors for Touch and Pressure Occur throughout the Body

In vertebrates, mechanoreceptors detecting touch and pressure are embedded in the skin and other surface tissues, in skeletal muscles, in the walls of blood vessels, and in internal organs. In humans, touch receptors in the skin are concentrated in greatest numbers in the fingertips, lips, and tip of the tongue, giving these regions the greatest sensitivity to mechanical stimuli. In other areas, such as the skin of the back, arms, and legs, the receptors are more widely spaced.

You can compare the spacing of receptors by pressing two toothpicks lightly against a fingertip and then against the skin of your arm or leg. On your fingertip, the toothpicks can be quite close together—separated by only a millimeter or so—and still be discerned as two separate points. On your arm or leg, they must be nearly 5 cm (almost 2 inches) apart to be distinguished.

Human skin contains several types of touch and pressure receptors **(Figure 39.2)**. Some are free nerve endings, the dendrites of afferent neurons with no specialized structures surrounding them. Others, such as Pacinian corpuscles, have structures surrounding the nerve endings that contribute to reception of stimuli. Free nerve endings wrapped around hair follicles respond when the hair is bent, making you instantly aware, for example, of a spider exploring your arm or leg.

Proprioceptors Provide Information about Movements and Position of the Body

Mechanoreceptors called **proprioceptors** (*proprius* = one's own) detect stimuli that are used in the CNS to maintain body balance and equilibrium and to monitor

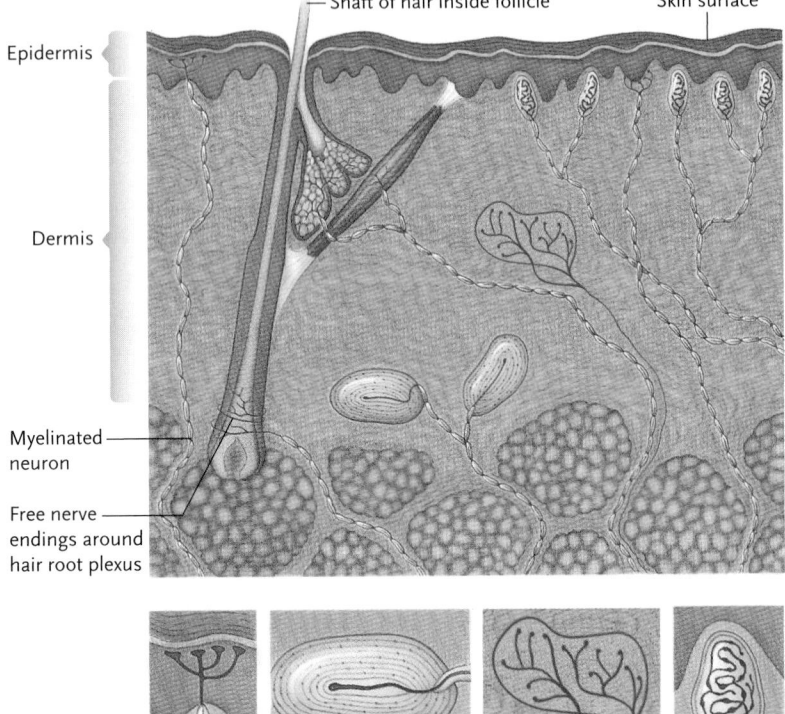

Epidermis

Dermis

Myelinated neuron

Free nerve endings around hair root plexus

Shaft of hair inside follicle — Skin surface

Free nerve endings: light touch

Pacinian corpuscle: deep pressure and vibrations

Ruffini endings: deep pressure

Meissner's corpuscle: light touch, surface vibrations

Figure 39.2
Types of mechanoreceptors detecting tactile stimuli in human skin.

the position of the head and limbs. The activity of these receptors allows you to touch the tip of your nose with your eyes closed, for example, or reach and scratch an itch on your back precisely. Here we consider examples of proprioceptors found in various animals.

Statocysts in Invertebrates. Many aquatic invertebrates, including jellyfishes, some gastropods, and some arthropods, have organs of equilibrium called **statocysts** (*statos* = standing; *kystis* = bag). Most statocysts are fluid-filled chambers with walls that contain **sensory hair cells** enclosing one or more movable stonelike bodies called **statoliths (Figure 39.3).** For example, lobsters have statoliths consisting of sand grains stuck together by mucus. When the animal moves, the statoliths lag behind the movement, bending the sensory hairs and triggering action potentials in afferent neurons. In this way, the statocysts signal the brain about the body's position and orientation with respect to gravity.

The Lateral Line System in Amphibians and Fish. Fishes and some aquatic amphibians detect vibrations and currents in the water through mechanoreceptors along the length of the body called the **lateral line system (Figure 39.4).** In fish, the mechanoreceptors, known as *neuromasts,* also provide information about the fish's orientation with respect to gravity and its swimming velocity. In some fishes, neuromasts are exposed on the body surface; in others, they are recessed in water-filled canals with porelike openings to the outside (as in Figure 39.4). Each dome-shaped neuromast has sensory hair cells clustered in its base. One surface of the hair cell is covered with **stereocilia**, which are actually

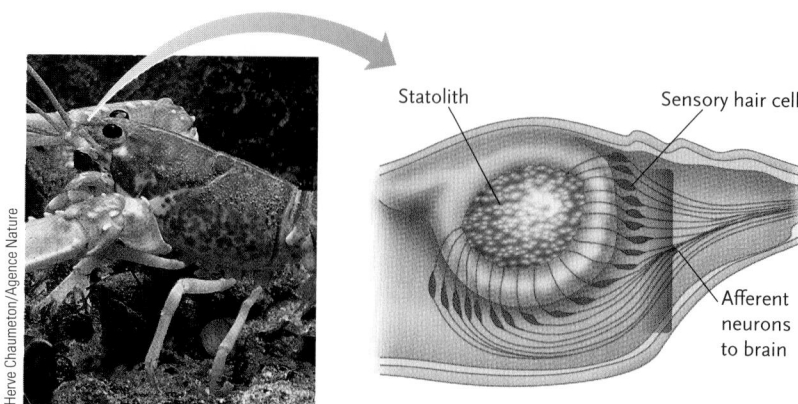

Figure 39.3
A statocyst, an invertebrate organ of equilibrium, and its location at the base of an antenna in a lobster. The statoliths inside are usually formed from fused grains of sand, as they are in the lobster, or from calcium carbonate.

microvilli (cell processes reinforced by bundles of microfilaments). The stereocilia extend into a gelatinous structure, the **cupula** (*cupule* = little cup), which moves with pressure changes in the surrounding water. Movement of the cupula bends the stereocilia, which causes the hair cell's plasma membrane to become depolarized and release neurotransmitter molecules; the neurotransmitters then generate action potentials in associated afferent neurons.

Vibrations detected by the lateral line enable fishes to avoid obstacles, orient in a current, and monitor the presence of other moving objects in the water. The system is also responsible for the ability of schools of fish to move in unison, turning and diving in what appears to be a perfectly synchronized aquatic ballet. In actual-

Figure 39.4
The lateral line system of fishes. The sensory receptor of the lateral line, the neuromast, has a gelatinous cupula that is pushed and pulled by vibrations and currents transmitted through the lateral line canal. As the cupula moves, the stereocilia of the sensory hair cells are bent, generating action potentials in afferent neurons that lead to the brain.

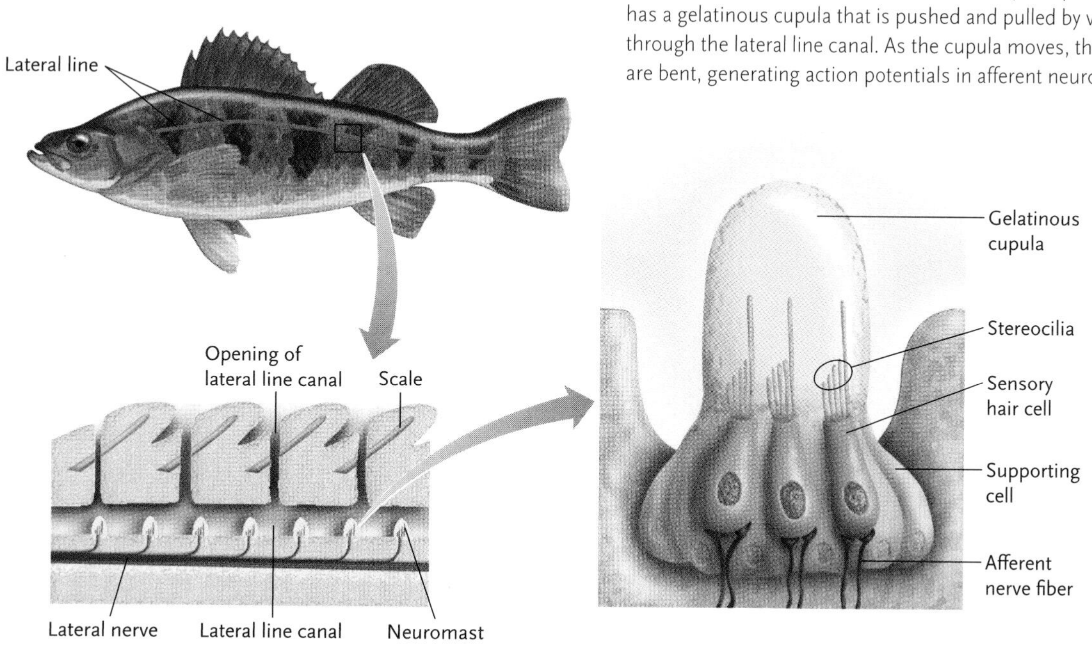

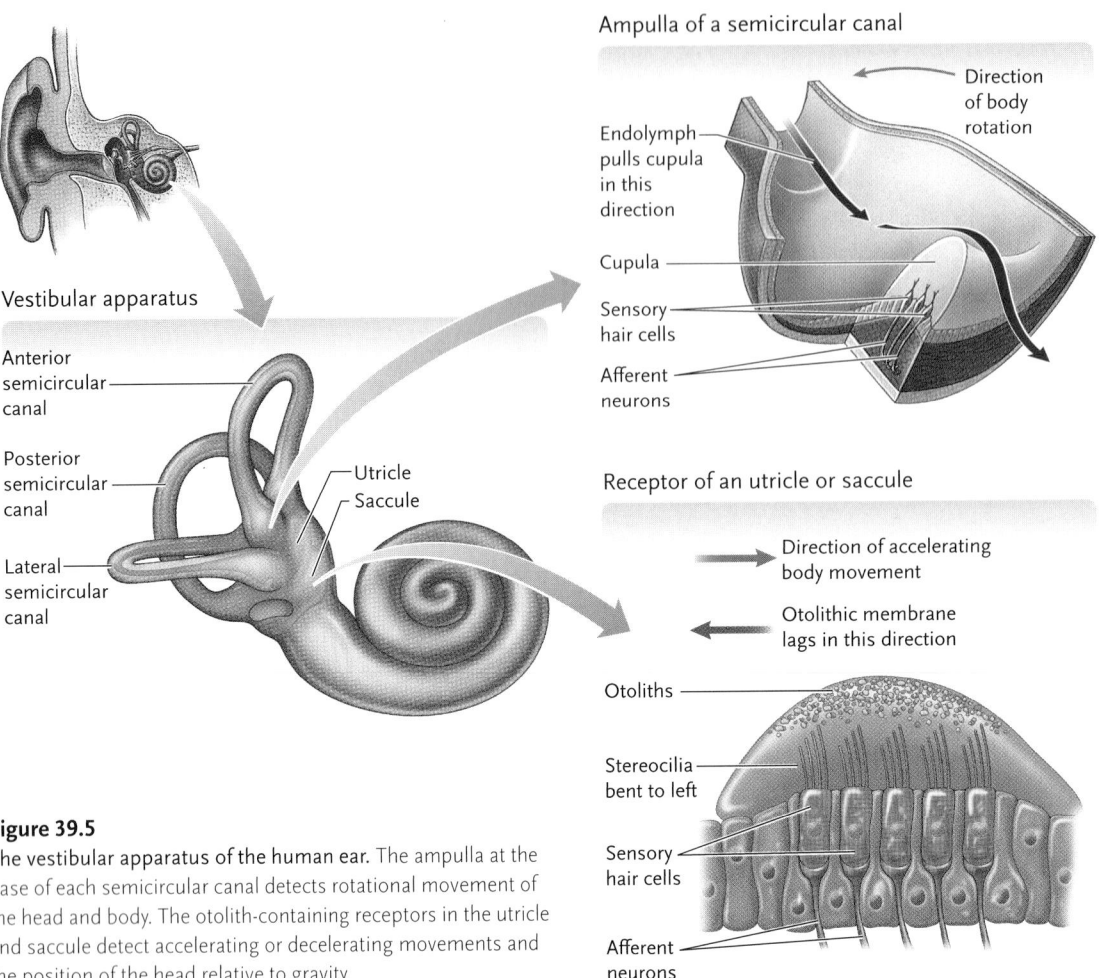

Vestibular apparatus

Anterior semicircular canal

Posterior semicircular canal

Lateral semicircular canal

Utricle
Saccule

Ampulla of a semicircular canal

Direction of body rotation

Endolymph pulls cupula in this direction

Cupula

Sensory hair cells

Afferent neurons

Receptor of an utricle or saccule

Direction of accelerating body movement

Otolithic membrane lags in this direction

Otoliths

Stereocilia bent to left

Sensory hair cells

Afferent neurons

Figure 39.5
The vestibular apparatus of the human ear. The ampulla at the base of each semicircular canal detects rotational movement of the head and body. The otolith-containing receptors in the utricle and saccule detect accelerating or decelerating movements and the position of the head relative to gravity.

ity, the movement of each fish creates a pressure wave in the water that is detected by the lateral line systems of other fishes in the school. Schooling fish can still swim in unison even if blinded, but if the nerves leading from the lateral line system to the brain are severed, the ability to school is lost.

The Vestibular Apparatus in Vertebrates. The inner ear of most terrestrial vertebrates has two specialized sensory structures, the *vestibular apparatus* and the *cochlea*. The **vestibular apparatus** is responsible for perceiving the position and motion of the head and, therefore, is essential for maintaining equilibrium and for coordinating head and body movements. The cochlea is used in hearing, which we discuss later in this chapter.

The vestibular apparatus **(Figure 39.5)** consists of three **semicircular canals** and two chambers, the **utricle** and the **saccule**, filled with a fluid called *endolymph*. The semicircular canals, which are positioned at angles corresponding to the three planes of space, detect rotational (spinning) motions. Each canal has a swelling at its base called an *ampulla,* which is topped with sensory hair cells embedded in a cupula similar to that found in lateral line systems. The cupula protrudes into the endolymph of the canals. When the body or

head rotates horizontally, vertically, or diagonally, the endolymph in the semicircular canal corresponding to that direction lags behind, pulling the cupula with it. The displacement of the cupula bends the sensory hair cells and generates action potentials in afferent neurons making synapses with the hair cells.

The utricle and saccule provide information about the position of the head with respect to gravity (up versus down), as well as changes in the rate of linear movement of the body. The utricle and saccule, which are oriented approximately 30° to each other, each contain sensory hair cells with stereocilia. The hair cells are covered with a gelatinous *otolithic membrane* (which is similar to a cupula) in which **otoliths** (*oto* = ear; *lithos* = stone), small crystals of calcium carbonate, are embedded (see Figure 39.5); otoliths are similar to invertebrate statoliths.

When an animal is upright, the sensory hairs in the utricle are oriented vertically, and those in the saccule are oriented horizontally. When the head is tilted in any direction other than straight up and down, or when there is a change in linear motion of the body, the otolithic membrane of the utricle moves and bends the sensory hairs. Depending on the direction of movement, the hair cells release more or less neurotransmitter, and the brain integrates the signals it receives and generates a perception of the movement. The saccule responds to the tilting of the head away from the horizontal (such as in diving) and to a change in movement up and down (such as jumping up to dunk a basketball). The ultricle and saccule adapt quickly to the body's motion, decreasing their response when there is no change in the rate and direction of movement. In other words, the body adapts to the new position. For instance, when you move your head to the left, that new position becomes the "norm." Then, if you move your head again in any direction, signals from the utricle and saccule tell your brain that your head is moving to a new position.

Stretch Receptors in Vertebrates. In the muscles and tendons of vertebrates, proprioceptors called **stretch receptors** detect the position and movement of the limbs. The stretch receptors in muscles are **muscle spindles,** bundles of small, specialized muscle cells wrapped with the dendrites of afferent neurons and enclosed in connective tissue **(Figure 39.6).** When the muscle stretches, the spindle stretches also, stimulating the dendrites and triggering the production of action potentials. The strength of the response of stretch receptors to stimulation depends on how much and how fast the muscle is stretched. The proprioceptors of tendons, called **Golgi tendon organs,** are dendrites that branch within the fibrous connective tissue of the tendon (see Figure 39.6). These nerve endings measure stretch and compression of the tendon as the muscles move the limbs.

Proprioceptors allow the CNS to monitor the body's position and help keep the body in balance. They also allow muscles to apply constant force under a constant load, and to adjust almost instantly if the load changes. When you hold a cup while someone fills it with coffee, for example, the muscle spindles in your biceps muscle detect the additional stretch as the cup becomes heavier. Signals from the spindles allow you to compensate for the additional weight by increasing the contraction of the muscle, keeping your arm level with no conscious effort on your part. Proprioceptors are typically slow to adapt, so that the body's position and balance are constantly monitored.

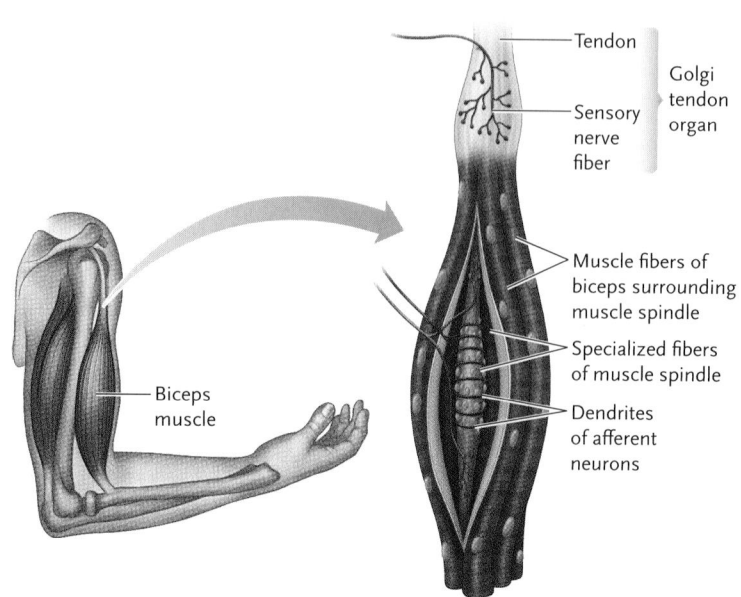

Figure 39.6

Muscle spindles, which detect the stretch and tension of muscles, and Golgi tendon organs, which detect the stretch of tendons.

STUDY BREAK

1. What is the function of proprioceptors?
2. What properties qualify proprioceptors as mechanoreceptors?

39.3 Mechanoreceptors and Hearing

In most animals, the receptors that detect sound are closely related to the receptors that detect body movement.

Sounds are vibrations that travel as waves produced by the alternating compression and decompression of the air. Although sound waves travel through air rapidly—at speeds of about 340 meters per second (700 miles per hour) at sea level—the individual air molecules transmitting the waves move back and forth over only a short distance as the wave passes. The vibrations of sound waves travel through water and solids by a similar mechanism.

The loudness, or *intensity,* of a sound depends on the amplitude (height) of the wave. The *pitch* of a sound—whether a musical tone, for example, is a high note or a low note—depends on the frequency of the waves, measured in cycles per second. The more cycles per second, the higher the pitch. Some animals, such as the bat in the introduction to this chapter, can hear sounds well above 100,000 cycles per second. Humans can hear sounds between about 20 and 20,000 cycles per second, which is why we cannot hear the bat's sonar clicks.

Invertebrates Have Varied Vibration-Detecting Systems

Most invertebrates detect sound and other vibrations through mechanoreceptors in their skin or other surface structures. An earthworm, for example, quickly retracts into its burrow at the smallest vibration of the surrounding earth, even though it has no specialized structures serving as ears. Cephalopods such as squids and octopuses have a system of mechanoreceptors on their head and tentacles, similar to the lateral line of fishes, which detects vibrations in the surrounding water. Many insects have sensory receptors in the form

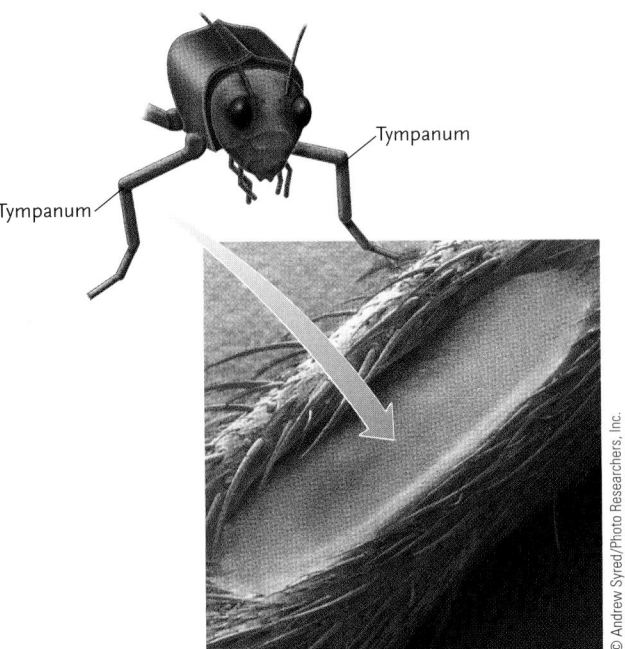

Figure 39.7
The tympanum or eardrum of a cricket, located on the front walking legs.

Tympanum

Tympanum

© Andrew Syred/Photo Researchers, Inc.

of hairs or bristles that vibrate in response to sound waves, often at particular frequencies.

Some insects, such as moths, grasshoppers, and crickets, have complex auditory organs on either side of the abdomen or on the first pair of walking legs **(Figure 39.7)**. These "ears" consist of a thinned region of the insect's exoskeleton that forms a tympanum (*tympanum* = drum) over a hollow chamber. Sounds reaching the tympanum cause it to vibrate; mechanoreceptors connected to the tympanum translate the vibrations into nerve impulses. Some insect ears respond to sounds only at certain frequencies, such as to the pitch of a cricket's song or, in certain moths, to the frequencies of the echolocation sounds emitted by foraging bats.

Human Ears Are Representative of the Auditory Structures of Mammals

The auditory structures of terrestrial vertebrates transmit the vibrations of sound waves to sensory hair cells, which respond by triggering action potentials. Here, we describe the human ear, which is representative of mammalian auditory structures **(Figure 39.8)**.

The **outer ear** has an external structure, the **pinna** (*pinna* = wing or leaf), which concentrates and focuses sound waves. Most mammals can turn the pinnae to help focus sounds, but human pinnae are inefficient compared with the ears of dogs, rabbits, or horses. The sound waves enter the auditory canal, which leads from the exterior, and strike a thin sheet of tissue, the **tympanic membrane**, or eardrum, which vibrates back and forth in response.

Behind the eardrum is the **middle ear**, an air-filled cavity containing three small, interconnected bones: the **malleus** (hammer), the **incus** (anvil), and the **stapes**

(stirrup). The stapes is attached to a thin, elastic membrane, the **oval window.** The vibrations of the eardrum, transmitted by the malleus and incus, push the stapes back and forth against the oval window. The levering action of the bones, combined with the much larger size of the eardrum as compared with the oval window, amplifies the vibrations transmitted to the oval window by more than 20 times.

Inside the oval window is the **inner ear.** It contains several fluid-filled compartments, including the semicircular canals, utricle, saccule, and a spiraled tube, the **cochlea** (*kochlias* = snail). The cochlea twists through about two and a half turns; if stretched out flat, it would be about 3.5 cm long. Thin membranes divide the cochlea into three longitudinal chambers, the *vestibular canal* at the top, the *cochlear duct* in the middle, and the *tympanic canal* at the bottom (see Figure 39.8). The vestibular canal and the tympanic canal join at the outer tip of the cochlea, so that the fluid within them is continuous. Within the cochlear duct is the **organ of Corti**; it contains the sensory hair cells that detect sound vibrations transmitted to the inner ear (see Figure 39.8).

The vibrations of the oval window pass through the fluid in the vestibular canal, make the turn at the end, and travel back through the fluid in the tympanic canal. At the end of the tympanic canal, they are transmitted to the **round window**, a thin membrane that faces the middle ear.

The vibrations traveling through the inner ear cause the *basilar membrane* to vibrate in response. The basilar membrane, which forms part of the floor of the cochlear duct, anchors the sensory hair cells in the organ of Corti. The stereocilia of these cells are embedded in the *tectorial membrane*, which extends the length of the cochlear canal. Vibrations of the basilar membrane cause the hair cells to bend, stimulating them to release a neurotransmitter that triggers action potentials in afferent neurons leading from the inner ear.

The basilar membrane is narrowest near the oval window and gradually widens toward the outer end of the cochlear duct. The high-frequency vibrations produced by high-pitched sounds vibrate the basilar membrane most strongly near its narrow end while vibrations of lower frequency vibrate the membrane nearer the outer end. Thus each frequency of sound waves causes hair cells in a different segment of the basilar membrane to initiate action potentials.

More than 15,000 hair cells are distributed in small groups along the basilar membrane. Each group of hairs is connected by synapses to afferent neurons, which in turn are bundled together in the *auditory nerve,* a cranial nerve that leads to the thalamus. From there, the signals are routed to specific regions in the auditory center of the temporal lobe. As sound waves of a particular frequency and intensity stimulate a specific segment of the basilar membrane, the region of

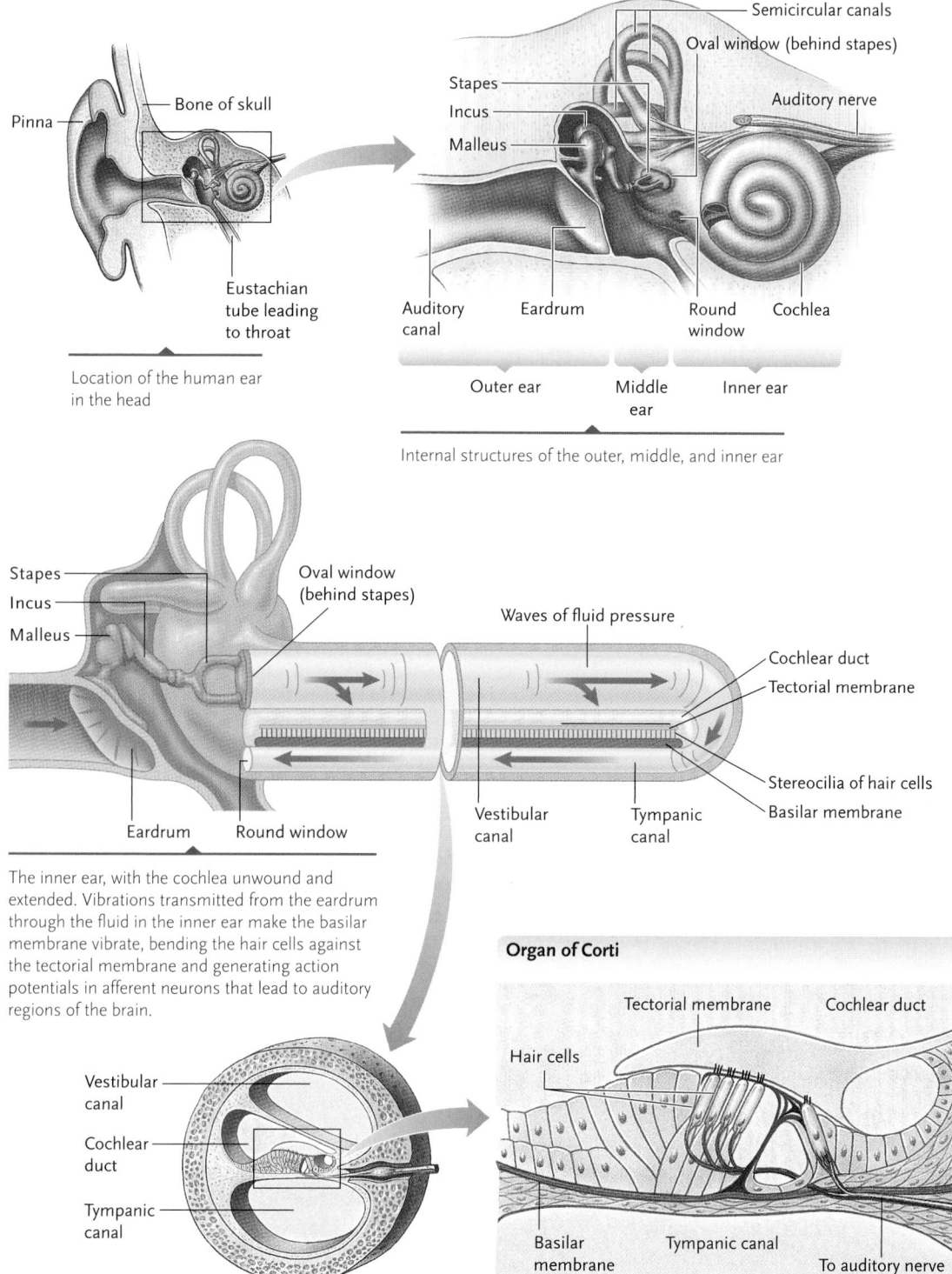

Figure 39.8
Structures of the human ear.

Pinna

Bone of skull

Semicircular canals

Oval window (behind stapes)

Stapes

Incus

Malleus

Auditory nerve

Eustachian tube leading to throat

Auditory canal

Eardrum

Round window

Cochlea

Outer ear

Middle ear

Inner ear

Location of the human ear in the head

Internal structures of the outer, middle, and inner ear

Stapes

Incus

Malleus

Oval window (behind stapes)

Waves of fluid pressure

Cochlear duct

Tectorial membrane

Stereocilia of hair cells

Basilar membrane

Eardrum

Round window

Vestibular canal

Tympanic canal

The inner ear, with the cochlea unwound and extended. Vibrations transmitted from the eardrum through the fluid in the inner ear make the basilar membrane vibrate, bending the hair cells against the tectorial membrane and generating action potentials in afferent neurons that lead to auditory regions of the brain.

Vestibular canal

Cochlear duct

Tympanic canal

Organ of Corti

Tectorial membrane

Cochlear duct

Hair cells

Basilar membrane

Tympanic canal

To auditory nerve

the auditory center to which the signals are sent integrates the information into the perception of sound at a corresponding pitch and loudness.

The sounds we hear are usually a rich combination of vibrations at different frequencies and intensities, which result in hundreds to thousands of different keys being struck simultaneously, with different degrees of force, in the cochlear keyboard. The combination of signals transmitted from the cochlea to the auditory centers is integrated into the perception of a human voice, the song of a sparrow, the roar of a jet plane, or a Mozart symphony.

Another system protects the eardrum from damage by changes in environmental atmospheric pressure. The system depends on the *Eustachian tube,* a duct that leads from the air-filled middle ear to the throat (see Figure 39.8). As we swallow or yawn, the tube opens, allowing air to flow into or out of the middle ear to equalize the pressure on both sides of the eardrum. When swelling or congestion due to infections prevents the tube from admitting air, we complain of having stopped-up ears—we can sense that a pressure difference between the outer and middle ear is bulging the eardrum inward or outward and interfering with the transmission of sounds.

Many Vertebrates Keep Track of Obstacles and Prey by Echolocation

Many vertebrates, like the bat in the introduction to this chapter, locate prey or avoid obstacles by **echolocation**—by making squeaking or clicking noises, and then listening for the echoes that bounce back from objects in their environment. By sensing the direction of the echoes and the time between the squeak or click and the returning echo, the animal can pinpoint the locations of barriers or prey animals.

Porpoises and dolphins locate food fishes in murky water by echolocation, and whales also use echolocation to keep track of the sea bottom and rocky obstacles. Two bird species, the oilbird and the cave swiftlet, use echolocation to avoid obstacles and find their nests in the dark of caves. Even humans can learn to use echolocation—a blind person, for example, listens for the echoes from a tapping cane to avoid posts, walls, and other barriers.

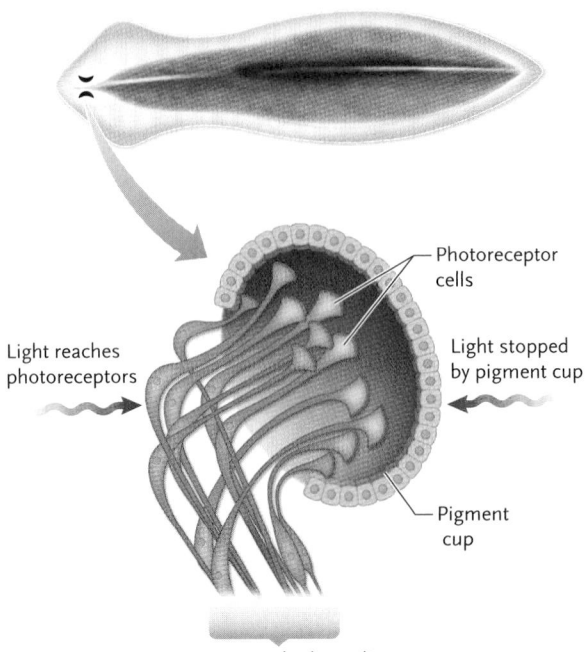

Figure 39.9
The ocellus of *Planaria*, a flatworm, and the arrangement of pigment cells on which its orientation response is based.

STUDY BREAK

1. What vibration-detecting systems are found in cephalopods and insects?
2. How are sounds of particular frequencies distinguished and "heard" by humans?

39.4 Photoreceptors and Vision

Virtually all animals have receptors that can detect and respond to light. As animals evolved and became more complex, the complexity of their visual sensory receptors increased, leading to the highly developed eyes of vertebrates.

We begin our discussion of these remarkable receptors by examining the basic visual structures and their functions in animals. We then compare vision in representative invertebrates and vertebrates.

Vision Involves Detection and Perception of Radiant Energy

Photoreceptors detect light at particular wavelengths, while centers in a brain or central ganglion integrate signals arriving from the receptors into a perception of light. All animals use different forms of a single lipid-like pigment, *retinal* (synthesized from vitamin A), in the photoreceptors to absorb light energy. The absorbed energy generates action potentials in afferent neurons leading to visual centers in the CNS. The organ of vision that detects light is the *eye*. The simplest eyes are capable only of distinguishing light from dark, while the most complex eyes distinguish shapes and colors and focus an accurate image of objects being viewed onto a layer of photoreceptors. Signals originating from the photoreceptors are integrated in the brain into an accurate, point-by-point perception of the object being viewed.

Invertebrate Eyes Take Many Forms

Some invertebrates, such as earthworms, do not have visual organs; instead, photoreceptors in their skin allow them to sense and respond to light. Earthworms respond negatively to light, as you can easily discover by shining a flashlight on an earthworm outside its burrow at night.

The eyes of other invertebrates are diverse, ranging from collections of photoreceptors with no lens and no image-forming capability to eyes remarkably like those of vertebrates. The photoreceptors of invertebrates are depolarized when they absorb light, and generate action potentials or increase their release of neurotransmitter molecules when they are stimulated. Vertebrate photoreceptors function differently, as we will see.

The simplest eye is the **ocellus** (plural, *ocelli;* also called an *eyespot* or *eyecup*). An ocellus, which detects light but does not form an image, consists of fewer than 100 photoreceptor cells lining a cup or pit. In planarians, for example, photoreceptor cells in a cuplike depression below the epidermis are connected to the dendrites of afferent neurons, which are bundled into nerves that travel from the ocelli to the cerebral ganglion **(Figure 39.9).** Each ocellus is covered on one side by a layer of pigment cells that blocks most of the light rays arriving from the opposite side of the animal. As a result, most of the light received by the pigment cells enters the ocellus from the side that it faces. Through integration of information transmitted to the cerebral ganglion from the eyecups, planarians orient themselves so that the amount of light falling on the two ocelli is equal and diminishes as they swim. This reaction carries them directly away from the source of the light and towards darker areas where the chance of a predator catching them is smaller. Similar ocelli are found in a variety of animals, including a number of insects, arthropods, and mollusks.

Two main types of image-forming eyes have evolved in invertebrates: compound eyes and single-lens eyes. The **compound eye** of insects, crustaceans, and a few annelids and mollusks contains hundreds to thousands of faceted visual units called **ommatidia** (*omma* = eye) fitted closely together **(Figure 39.10).** In insects, light entering an ommatidium is focused by a transparent **cornea** and a *crystalline cone* (just below the cornea) onto a bundle of photoreceptor cells. Microvilli of these cells interdigitate like the fingers of clasped hands, forming a central axis that contains rhodopsin, a **photopigment** (light-absorbing pigment) also found in the rods of vertebrate eyes. Absorption of light by rhodopsin causes action potentials to be generated in afferent neurons connected to the base of the ommatidium. Each ommatidium of a compound eye samples a small part of the visual field. From these signals, the brain receives a mosaic image of the world. Because even the slightest motion is detected simultaneously by many ommatidia, compound eyes are extraordinarily adept at detecting movement—a lesson soon learned by fly-swatting humans.

The **single-lens eye** of cephalopods **(Figure 39.11)** resembles a vertebrate eye in that both types operate like a camera. In the cephalopod eye, light enters through the transparent cornea, a **lens** concentrates the light, and a layer of photoreceptors at the back of the eye, the **retina**, records the image. Behind the cornea is the **iris**, which surrounds the **pupil**, the opening through which light enters the eye. Muscles in the iris adjust the size of the pupil to vary the amount of light entering the eye. When the light is bright, circular muscles in the iris contract, shrinking the size of the pupil and reducing the amount of light that enters the eye. In dim light, radial muscles contract and enlarge the pupil, increasing the amount of light that enters the eye. Muscles move the lens forward and back with respect to the retina to focus the image. This is an ex-

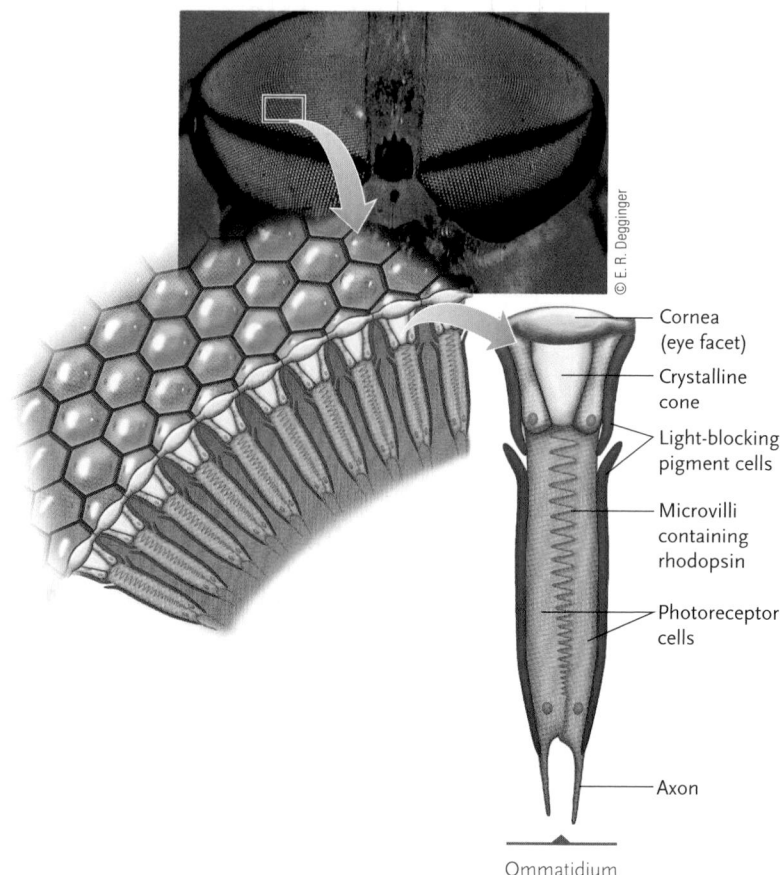

© E. R. Degginger

Cornea (eye facet)
Crystalline cone
Light-blocking pigment cells
Microvilli containing rhodopsin
Photoreceptor cells
Axon

Ommatidium

Figure 39.10

The compound eye of a deer fly. Each ommatidium has a cornea that directs light into the crystalline cone; in turn, the cone focuses light on the photoreceptor cells. A light-blocking pigment layer at the sides of the ommatidium prevents light from scattering laterally in the compound eye.

ample of **accommodation**, a process by which the lens changes to enable the eye to focus on objects at different distances.

A neural network lies under the retina, meaning that light rays do not have to pass through the neurons to reach the photoreceptors. The vertebrate eye has the opposite arrangement. This and other differences in structure and function indicate that cephalopod and vertebrate eyes evolved independently.

Figure 39.11

The eye of an octopus, a cephalopod mollusk.

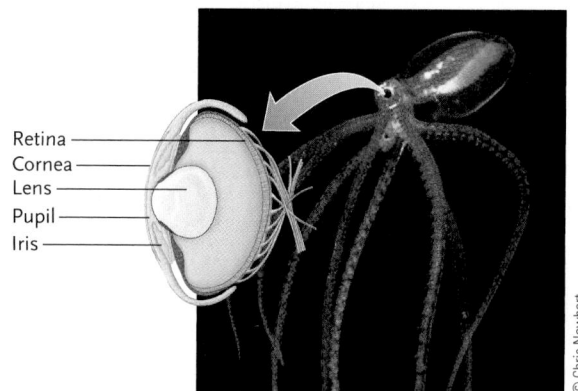

Retina
Cornea
Lens
Pupil
Iris

© Chris Newbert

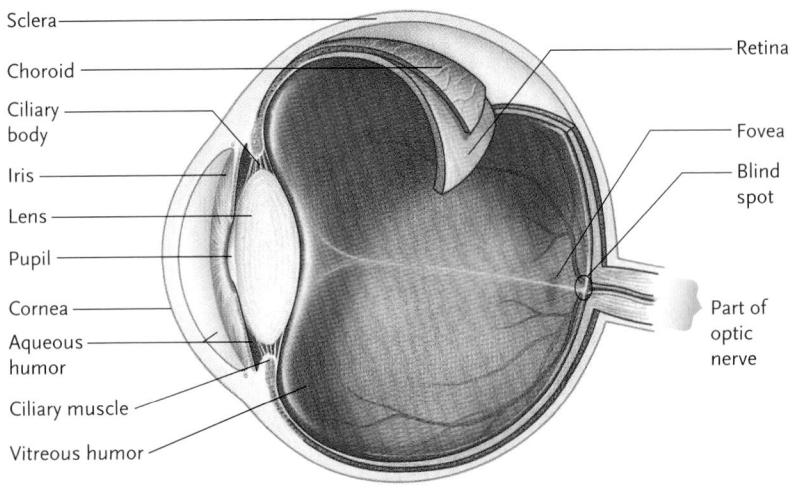

Figure 39.12
Structures of the human eye.

Vertebrate Eyes Have a Complex Structure

The human eye **(Figure 39.12)** has similar structures—cornea, iris, pupil, lens, and retina—to those of the cephalopod eye just described. Light entering the eye through the cornea passes through the iris and then the lens. The lens focuses an image on the retina, and the axons of afferent neurons originating in the retina converge to form the optic nerve leading from the eye to the brain.

A clear fluid called the **aqueous humor** fills the space between the cornea and lens. This fluid carries nutrients to the lens and cornea, which do not contain any blood vessels. The main chamber of the eye, located between the lens and the retina, is filled with the jellylike **vitreous humor** (*vitrum* = glass). The outer wall of the eye contains a tough layer of connective tissue (the *sclera*). Inside it is a darkly pigmented layer (the *choroid*) that prevents light from entering except through the pupil. It also contains the blood vessels nourishing the retina.

Two types of photoreceptors, rods and cones, occur in the retina along with layers of neurons that carry out an initial integration of visual information before it is sent to the brain. The **rods** are specialized for detection of light at low intensities; the **cones** are specialized for detection of different wavelengths (colors).

Accommodation does not occur by forward and back movement of the lens, as described for cephalopods. Rather, the lens of most terrestrial vertebrates is focused by changing its shape. The lens is held in place by fine ligaments that anchor it to a surrounding layer of connective tissue and muscle, the **ciliary body.** These ligaments keep the lens under tension when the ciliary muscle is relaxed. The tension flattens the lens, which is soft and flexible, and focuses light from distant objects on the retina **(Figure 39.13a)**. When the ciliary muscles contract, they relieve the tension of the ligaments, allowing the lens to assume a more spherical shape and focusing light from nearby objects on the retina **(Figure 39.13b)**.

The Retina of Mammals and Birds Contains Rods and Cones and a Complex Network of Neurons

The retina of a human eye contains about 120 million rods and 6 million cones organized into a densely packed, single layer. Neural networks of the retina are layered on top of the photoreceptor cells, so that light rays focused by the lens on the retina must pass through the neurons before reaching the photoreceptors. The light must also pass through a layer of fine blood vessels that covers the surface of the retina.

In mammals and birds with eyes specialized for daytime vision, cones are concentrated in and around a small region of the retina, the **fovea** (see Figure 39.12). The image focused by the lens is centered on

Figure 39.13
Accommodation in terrestrial vertebrates: the lens changes shape rather than moving forward and back to focus on **(a)** distant and **(b)** near objects.

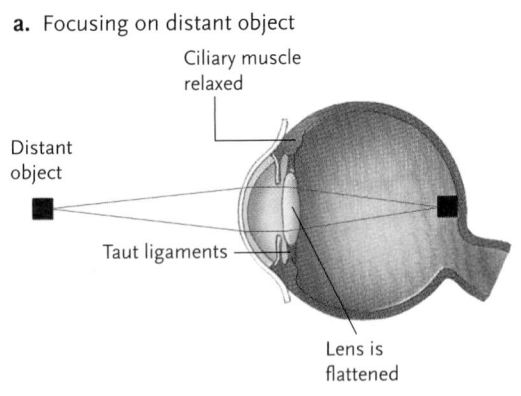

a. Focusing on distant object

When the eye focuses on a distant object, the ciliary muscles relax, allowing the ligaments that support the lens to tighten. The tightened ligaments flatten the lens, bringing the distant object into focus on the retina.

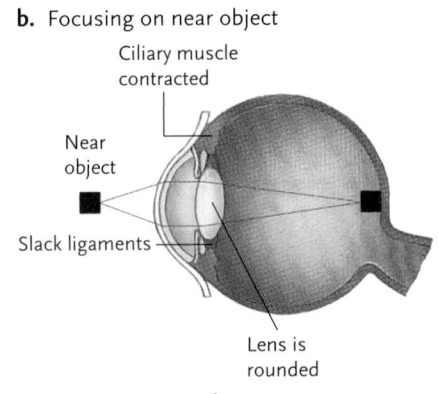

b. Focusing on near object

When the eye focuses on a near object, the ciliary muscles contract, loosening the ligaments and allowing the lens to become rounder. The rounded lens focuses a near object on the retina.

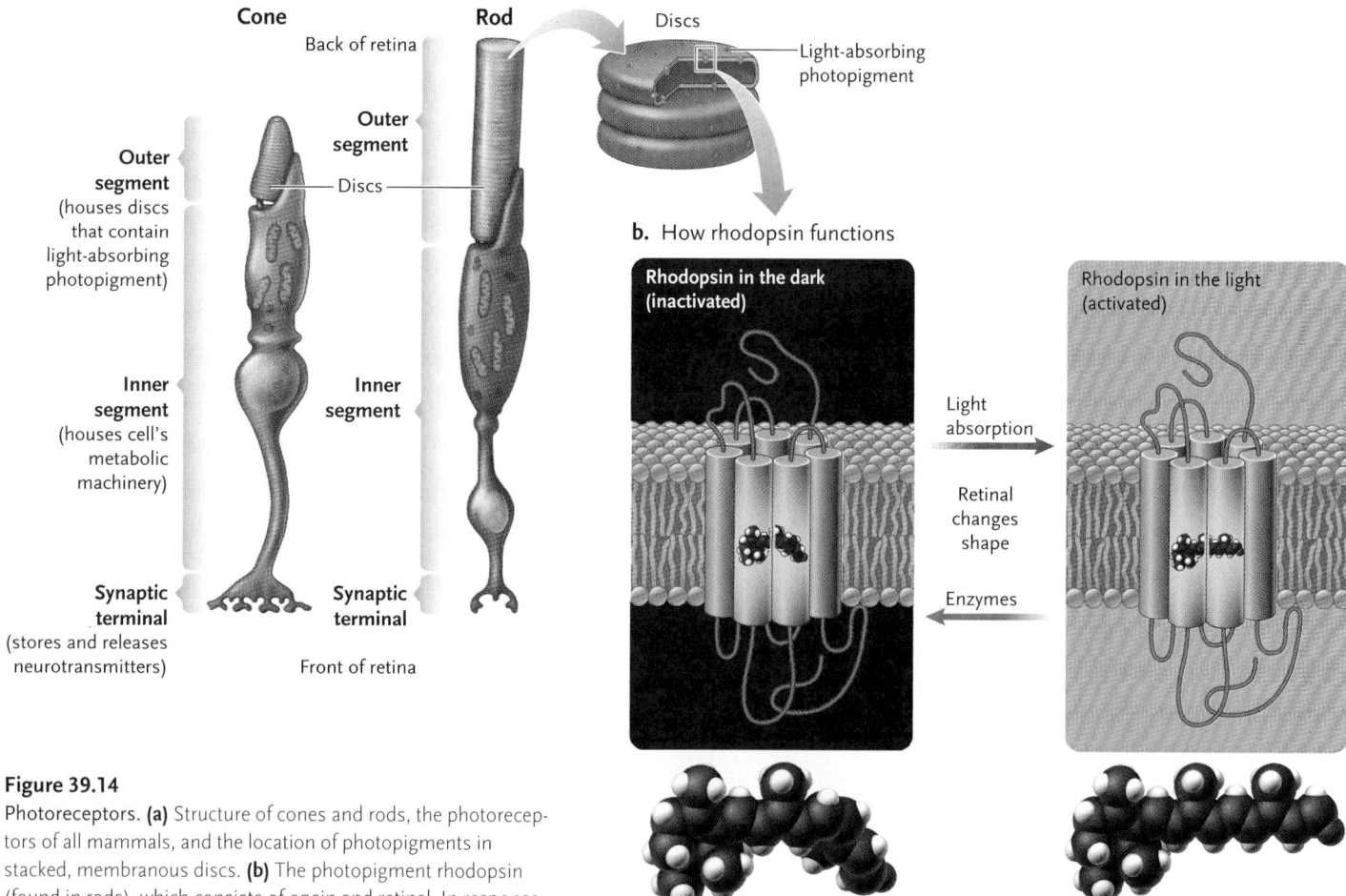

a. Structure of cones and rods

Cone

Rod

Back of retina

Outer segment (houses discs that contain light-absorbing photopigment)

Outer segment

Discs

Inner segment (houses cell's metabolic machinery)

Inner segment

Synaptic terminal (stores and releases neurotransmitters)

Synaptic terminal

Front of retina

Discs

Light-absorbing photopigment

b. How rhodopsin functions

Rhodopsin in the dark (inactivated)

Rhodopsin in the light (activated)

Light absorption

Retinal changes shape

Enzymes

cis-Retinal

trans-Retinal

Figure 39.14

Photoreceptors. **(a)** Structure of cones and rods, the photoreceptors of all mammals, and the location of photopigments in stacked, membranous discs. **(b)** The photopigment rhodopsin (found in rods), which consists of opsin and retinal. In response to light, the retinal changes from a bent to a straight structure.

the fovea, which is circular and less than a millimeter in diameter in humans. The rods are spread over the remainder of the retina. We can see distinctly only the image focused on the fovea; the surrounding image is what we term *peripheral vision*. Mammals and birds with eyes specialized for night vision have retinas containing mostly rods, without a clearly defined fovea. Some fishes and many reptiles have cones generally distributed throughout their retina and very few rods.

The rods of mammals are much more sensitive than the cones to light of low intensity; in fact, they can respond to a single photon of light. This is why, in dim light, we can see objects better by looking slightly to the side of the object. This action directs the image away from the cones in the fovea to the highly light-sensitive rods in surrounding regions of the retina.

Sensory Transduction by Rods and Cones. Photoreceptors have three parts: an outer segment consisting of stacked, flattened, membranous discs; an inner segment where the cell's metabolic activities occur; and the synaptic terminal, where neurotransmitter molecules are stored and released **(Figure 39.14a).** The light-absorbing pigment of rods and cones, retinal, is bonded covalently in the photoreceptors with one of several different pro-

teins called **opsins** to produce **photopigments.** The photopigments are embedded in the membranous discs of the photoreceptors' outer segments **(Figure 39.14b).** The retinal-opsin photopigment in rods is called **rhodopsin.** Let us see how light stimulating a rod photoreceptor is transduced; the mechanism is essentially the same in the cone photoreceptors.

In the dark, the retinal segment of the unstimulated rhodopsin is in an inactive form known as *cis*-retinal (see Figure 39.14b), and the rods steadily release the neurotransmitter glutamate. When rhodopsin absorbs a photon of light, retinal converts to its active form, *trans*-retinal (see Figure 39.14b), and the rods *decrease* the amount of glutamate they release. This will be discussed in the next section.

Rhodopsin is a membrane-embedded G-protein-coupled receptor (see Section 7.4). Recall that an extracellular signal received by a G-protein-coupled receptor activates the receptor, which triggers a signal transduction pathway within the cell, leading to a cellular response. Here, activated rhodopsin triggers a signal transduction pathway that leads to the closure of Na^+ channels in the plasma membrane **(Figure 39.15).** Closure of the channels hyperpolarizes the photoreceptor's membrane, thereby decreasing neuro-

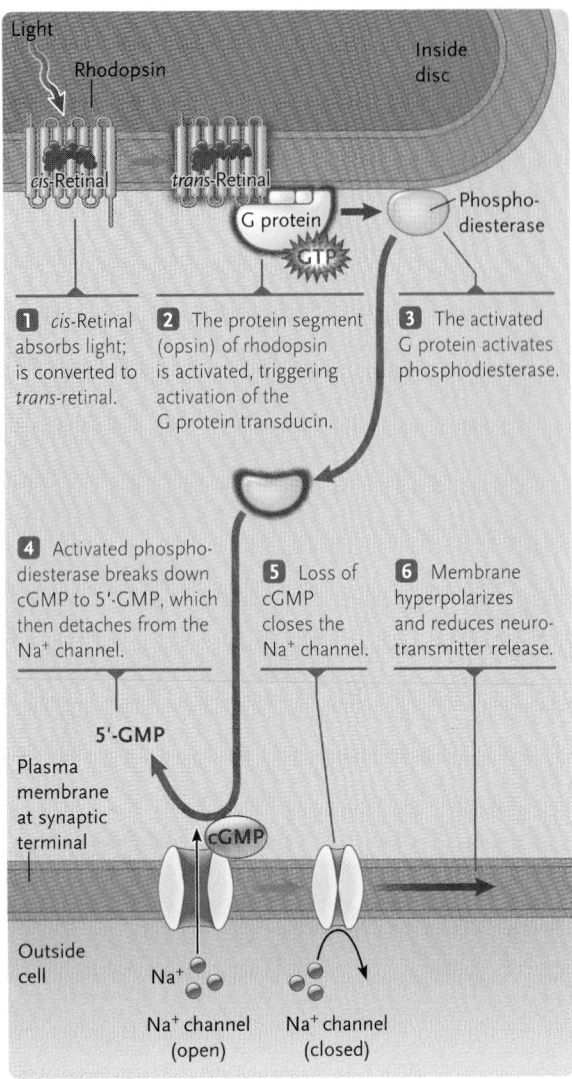

1 *cis*-Retinal absorbs light; is converted to *trans*-retinal.

2 The protein segment (opsin) of rhodopsin is activated, triggering activation of the G protein transducin.

3 The activated G protein activates phosphodiesterase.

4 Activated phosphodiesterase breaks down cGMP to 5'-GMP, which then detaches from the Na⁺ channel.

5 Loss of cGMP closes the Na⁺ channel.

6 Membrane hyperpolarizes and reduces neurotransmitter release.

Figure 39.15
The signal transduction pathway that closes Na⁺ channels in photoreceptor plasma membranes when rhodopsin absorbs light.

transmitter release. The response is graded in the sense that as light absorption by photopigment molecules increases, the amount of neurotransmitter released is reduced proportionately; if light absorption decreases, neurotransmitter release by the photoreceptor increases proportionately. Note that transduction in rods works in the opposite way from most sensory receptors, in which a stimulus increases neurotransmitter release.

Visual Processing in the Retina. In the retina of all vertebrates, the two types of photoreceptors are linked to a network of neurons that carries out initial integration and processing of visual information. The retina of mammals contains four types of neurons **(Figure 39.16).** Just over the rods and cones is a layer of **bipolar cells.** These neurons make synapses with the rods or cones at one end and with a layer of neurons called **ganglion cells** at the other end. The axons of ganglion cells extend over the retina and collect at the back of the eyeball to form the optic nerve, which transmits action potentials to the brain. The point where the optic nerve exits the

eye lacks photoreceptors, resulting in a *blind spot* several millimeters in diameter. Two other types of neurons form lateral connections in the retina: **horizontal cells** connect photoreceptor cells, and **amacrine cells** connect bipolar cells and ganglion cells.

In the dark, the steady release of glutamate from rods and cones depolarizes some of the postsynaptic bipolar cells and hyperpolarizes others, depending on the type of receptor those cells have. In the light, the decrease in neurotransmitter release from rods and cones results in the polarized bipolar cells becoming hyperpolarized, and hyperpolarized bipolar cells becoming polarized. These membrane potential changes in response to light are transmitted to the brain for processing.

Signals from the rods and cones may move vertically or laterally in the retina. Signals move vertically from the photoreceptors to bipolar cells and then to ganglion cells. However, while the human retina has over 120 million photoreceptors, it has only about 1 million ganglion cells. This disparity is explained by the fact that each ganglion cell receives signals from a clearly defined set of photoreceptors that constitute the *receptive field* for that cell. Therefore, stimulating numerous photoreceptors in a ganglion cell's receptive field results in only a single message to the brain from that cell. Receptive fields are typically circular and are of different sizes. Smaller receptive fields result in sharper images because they send more precise information to the brain regarding the location in the retina where the light was received.

Signals that move laterally from a rod or cone proceed to a horizontal cell and continue to bipolar cells with which the horizontal cell makes inhibitory connections. To understand this, consider a spot of light falling on the retina. Photoreceptors detect the light and send a signal to bipolar cells and horizontal cells. The horizontal cells inhibit more distant bipolar cells that are outside the spot of light, causing the light spot to appear lighter and its surrounding dark area to appear darker. This type of visual processing is called **lateral inhibition** and serves both to sharpen the edges of objects and enhance contrast in an image.

Three Kinds of Opsin Pigments Underlie Color Vision

Many invertebrates and some species in each class of vertebrates have color vision. Color vision depends on the cones in the retina. Most mammals have only two types of cones, making their color vision limited, while humans and other primates have three types. Each human or primate cone cell contains one of three different photopigments, collectively called **photopsins,** in which retinal is combined with different opsins. The three photopsins absorb light over different, but overlapping, wavelength ranges, with peak absorptions at 445 nm (blue light), 535 nm (green light), and 570 nm

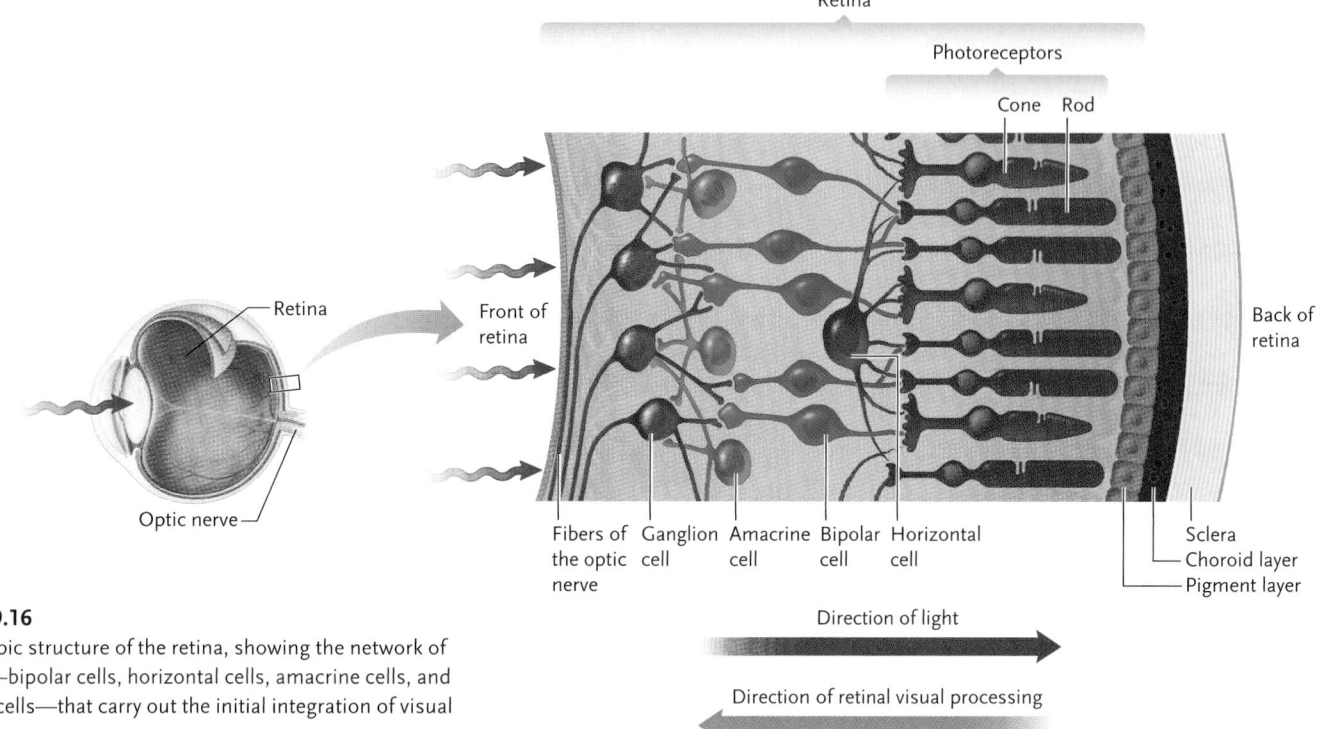

Figure 39.16
Microscopic structure of the retina, showing the network of neurons—bipolar cells, horizontal cells, amacrine cells, and ganglion cells—that carry out the initial integration of visual information.

(red light). The farther a wavelength is from the peak color absorbed, the less strongly the cone responds.

Having overlapping wavelength ranges for the three photoreceptors means that light at any visible wavelength will stimulate at least two of the three types of cones. However, because the maximal absorption of each type of cone is a different wavelength, it is stimulated to a different extent by light at a given wavelength. The differences, relayed to the visual centers of the brain, are integrated into the perception of a color corresponding to the particular wavelength absorbed. Light stimulating all three receptor types equally is seen as white.

The Visual Cortex Processes Visual Information

Just behind the eyes, the optic nerves converge before entering the base of the brain. A portion of each optic nerve crosses over to the opposite side, forming the **optic chiasm** (*chiasma* = crossing place). Most of the axons enter the **lateral geniculate nuclei** in the thalamus, where they make synapses with interneurons leading to the visual cortex **(Figure 39.17)**.

Because of the optic chiasm, the left half of the image seen by both eyes is transmitted to the visual cortex in the right cerebral hemisphere, and the right half of the image is transmitted to the left cerebral hemisphere. The right hemisphere thus sees objects to the left of the center of vision, and the left hemisphere sees objects to the right of the center of vision. Communication between the right and left hemi-

spheres integrates this information into a perception of the entire visual field seen by the two eyes.

If you look at a nearby object with one eye and then the other, you will notice that the point of view is slightly different. Integration of the visual field by the brain creates a single picture with a sense of distance and depth. The greater the difference between the images seen by the two eyes, the closer the object appears to the viewer.

The two optic nerves together contain more than a million axons, more than all other afferent neurons of the body put together. Almost one-third of the cerebral cortex is devoted to visual information. These numbers give some idea of the complexity of the information integrated into the visual image formed by the brain.

STUDY BREAK

For vertebrate photoreception, define: (a) photopigment; (b) cone; (c) receptive field.

39.5 Chemoreceptors

Chemoreceptors form the basis of taste (gustation) and smell (olfaction), and measure the levels of internal body molecules such as oxygen, carbon dioxide, and hydrogen ions. All chemoreceptors probably work through membrane receptor proteins that are stimulated when they bind with specific molecules in the

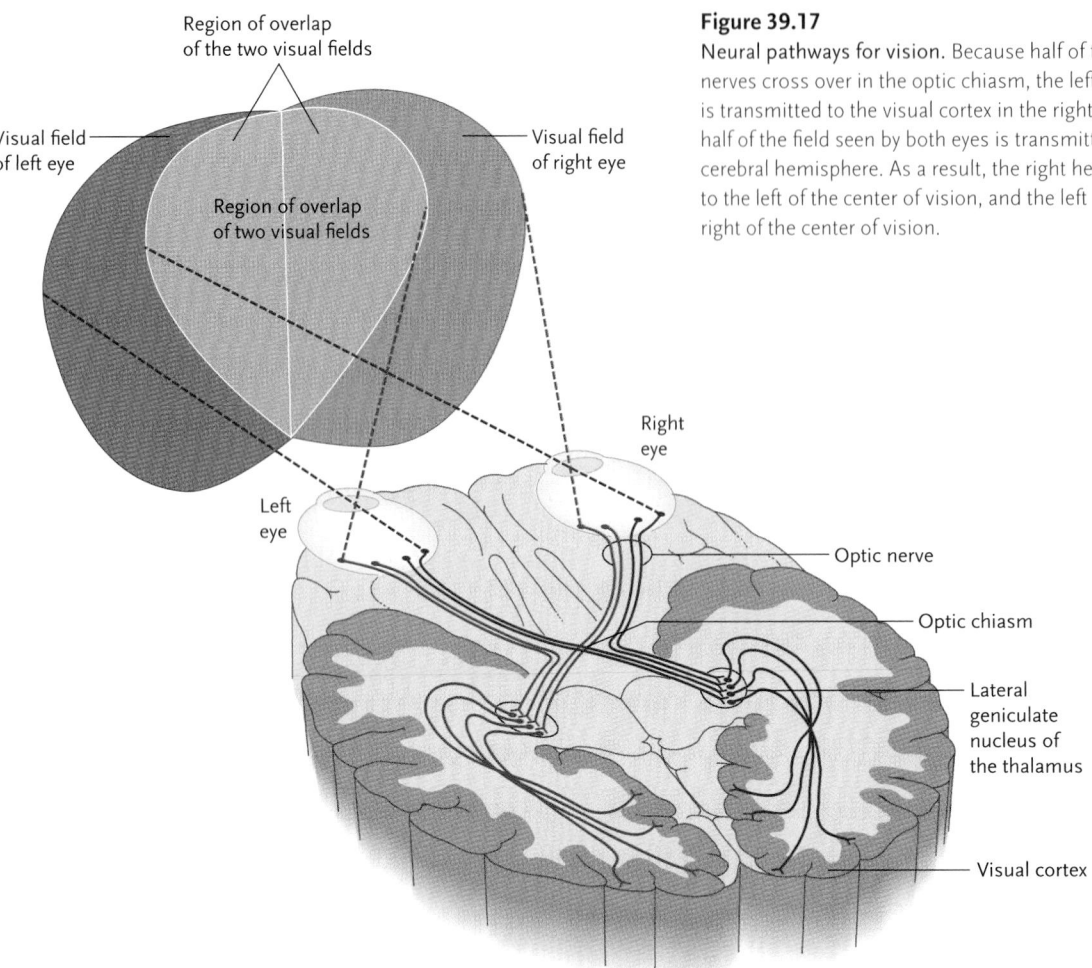

Figure 39.17

Neural pathways for vision. Because half of the axons carried by the optic nerves cross over in the optic chiasm, the left half of the field seen by both eyes is transmitted to the visual cortex in the right cerebral hemisphere. The right half of the field seen by both eyes is transmitted to the visual cortex in the left cerebral hemisphere. As a result, the right hemisphere of the brain sees objects to the left of the center of vision, and the left hemisphere sees objects to the right of the center of vision.

environment, generating action potentials in afferent nerves leading to the CNS.

Invertebrates Have Either the Same or Different Receptors for Taste and Smell

In many invertebrates the same receptors serve for the senses of smell and taste. These receptors may be confined to certain locations or distributed over the body surface. For example, the cnidarian *Hydra* has chemoreceptor cells around its mouth that respond to glutathione, a chemical released from prey organisms ensnared in the cnidarian's tentacles. Stimulation of the chemoreceptors by glutathione causes the tentacles to retract, resulting in ingestion of the prey. By contrast, earthworms have taste/smell receptors distributed over the entire body surface.

Some terrestrial invertebrates, particularly insects, have clearly differentiated taste and smell receptors. In insects, taste receptors occur inside hollow sensory bristles called *sensilla* (singular, *sensillum*), which may be located on the antennae, mouthparts, or feet **(Figure 39.18).** Pores in the sensilla admit molecules from potential food to the chemoreceptors, which are specialized to detect sugars, salts, amino acids, or other chemicals. Many female insects have chemoreceptors on

their ovipositors, which allow them to lay their eggs on food appropriate for the hatching larvae.

Insect olfactory receptors detect airborne molecules. Some insects use odor as a means of communication, as with the pheromones released into the air as sexual attractants by female moths. Olfactory receptors in the bristles of male silkworm moth antennae **(Figure 39.19)** have been shown experimentally to be able to detect pheromones released by a female of the same species in concentrations as low as one attractant molecule per 10^{17} air molecules; when as few as 40 of the 20,000 receptor cells on its antennae have been stimulated by pheromone molecules, the male moth responds by fluttering its wings rapidly to attract the female's attention. Ants, bees, and wasps may identify members of the same hive or nest or communicate by means of odor molecules.

Taste and Smell Receptors Are Differentiated in Terrestrial Animals

In terrestrial animals, taste involves the detection of potential food molecules in objects that are touched by a receptor, while smell involves the detection of airborne molecules. Although both taste and smell receptors have hairlike extensions containing the proteins that bind environmental molecules, the hairs of taste receptors are derived from microvilli and contain microfilaments, while the hairs of smell receptors are derived from cilia and contain microtubules. Another significant difference between taste and smell is that information from taste receptors is typically processed in the parietal lobes, while information from smell receptors is processed in the olfactory bulbs and the temporal lobes.

In Vertebrates, Taste Receptors Are Located in Taste Buds

The taste receptors of most vertebrates form part of a structure called a taste bud, a small, pear-shaped capsule with a pore at the top that opens to the exterior **(Figure 39.20).** The sensory hairs of the taste receptors pass through the pore of a taste bud and project to the exterior. The opposite end of the receptor cells forms synapses with dendrites of an afferent neuron.

The taste receptors of terrestrial vertebrates are concentrated in the mouth. Humans have about 10,000 taste buds, each 30 to 40 μm in diameter, scattered over the tongue, roof of the mouth, and throat. Those on the tongue are embedded in outgrowths called *papillae* (*papula* = pimple), which give the surface of the tongue its rough or furry texture.

Taste receptors on the human tongue are thought to respond to five basic tastes: sweet, sour, salty, bitter, and umami (savory). Some of the receptors for umami respond to the amino acid glutamate (familiar as monosodium glutamate or MSG). Recent research indicates that the classes of receptors may all have many subtypes, each binding a specific molecule within that class.

Signals from the taste receptors are relayed to the thalamus. From there, some signals lead to gustatory centers in the cerebral cortex, which integrate them into the perception of taste, while others lead to the brain stem and limbic system, which links tastes to involuntary visceral and emotional responses. Through these connections, a pleasant taste may lead to salivation, secretion of digestive juices,

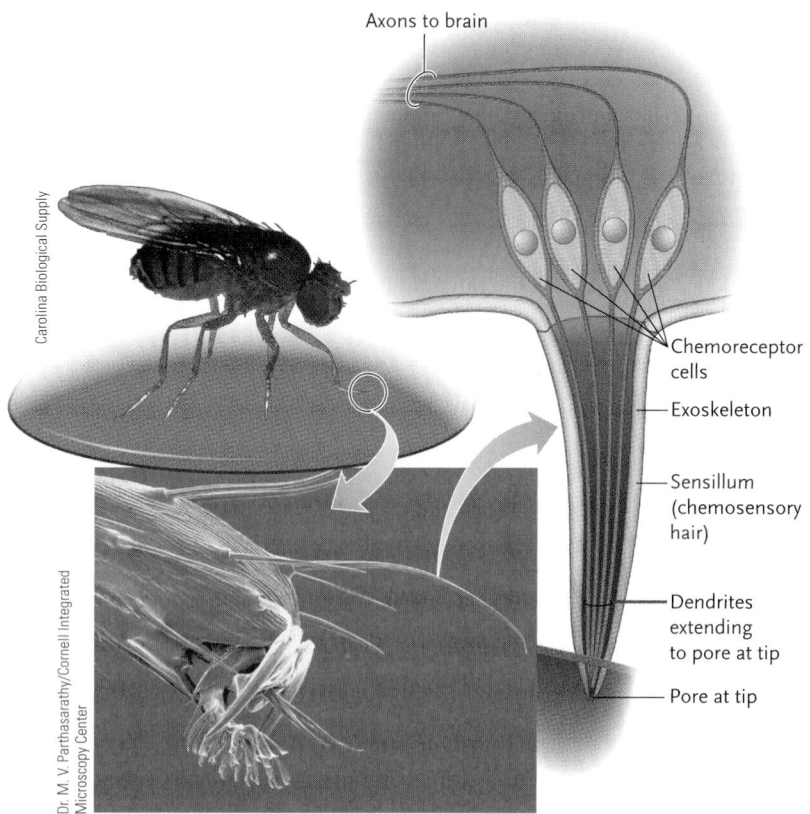

Carolina Biological Supply

Dr. M. V. Parthasarathy/Cornell Integrated Microscopy Center

- Chemoreceptor cells
- Exoskeleton
- Sensillum (chemosensory hair)
- Dendrites extending to pore at tip
- Pore at tip

Axons to brain

Figure 39.18
Taste receptors on the foot of a fruit fly, *Drosophila*.

sensations of pleasure, and even sexual arousal, while an unpleasant taste may produce revulsion, nausea, and even vomiting.

Olfactory Receptors Are Concentrated in the Nasal Cavities in Terrestrial Vertebrates

Receptors that detect odors are located in the nasal cavities in terrestrial vertebrates. Bloodhounds have more than 200 million olfactory receptors in patches of olfactory epithelium in the upper nasal passages; humans have about 5 million olfactory receptors.

On one end, each olfactory receptor cell has 10 to 20 sensory hairs that project into a layer of mucus covering the olfactory area in the nose **(Figure 39.21).** To be

© A. Shay/OSF/Animals Animals—Earth Scenes

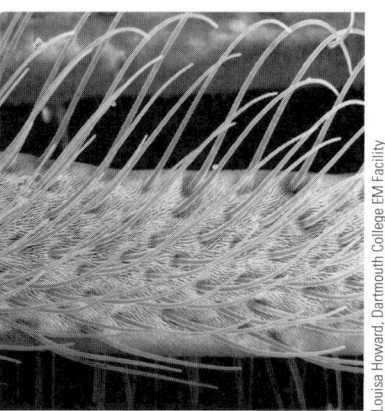

Louisa Howard, Dartmouth College EM Facility

25 μm

Figure 39.19
The brushlike antennae of a male silkworm moth. Fine sensory bristles containing olfactory receptor cells cover the filaments of the antennae.

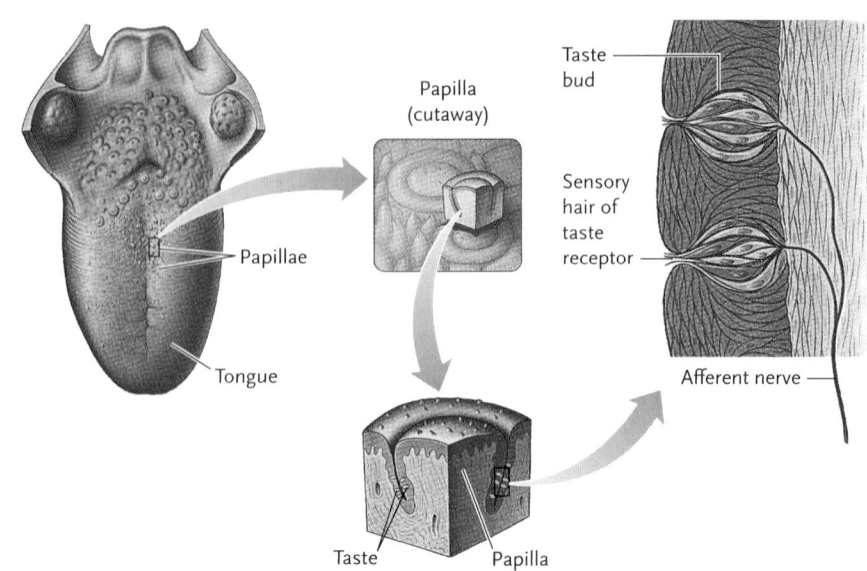

Figure 39.20

Taste receptors in the human tongue. The receptors occur in microscopic taste buds that line the sides of the furry papillae.

detected, airborne molecules must dissolve in the watery mucus solution. On the other end, the olfactory receptors make synapses with interneurons in the olfactory bulbs. Olfactory receptors are the only receptor cells that make direct connections with brain interneurons, rather than via afferent neurons.

From the olfactory bulbs, nerves conduct signals to the olfactory centers of the cerebral cortex, where they are integrated into the perception of tantalizing or unpleasant odors from a rose to a rotten egg. Most odor

perceptions arise from a combination of different olfactory receptors. In the early 1990s, Richard Axel and Linda Buck discovered that about 1000 different human genes give rise to an equivalent number of olfactory receptor types, each of which responds to a different class of chemicals. Axel and Buck received the Nobel Prize in 2004 in recognition of their research.

Other connections from the olfactory bulbs lead to the limbic system and brain stem where the signals elicit emotional and visceral responses similar to those caused by pleasant and unpleasant tastes (see Section 38.3). As a result, different odors, like tastes, may give rise to a host of involuntary responses, from salivation to vomiting, as well as conscious responses.

Olfaction contributes to the sense of taste because vaporized molecules from foods are conducted from the throat to the olfactory receptors in the nasal cavities. Olfactory input is the reason why anything that dulls your sense of smell—such as a head cold, or holding your nose—diminishes the apparent flavor of food.

Many mammals use odors as a means of communication. Individuals of the same family or colony are identified by their odor; odors are also used to attract mates and to mark territories and trails. Dogs, for example, use their urine to mark home territories with identifying odors. Humans use the fragrances of perfumes and colognes as artificial sex attractants.

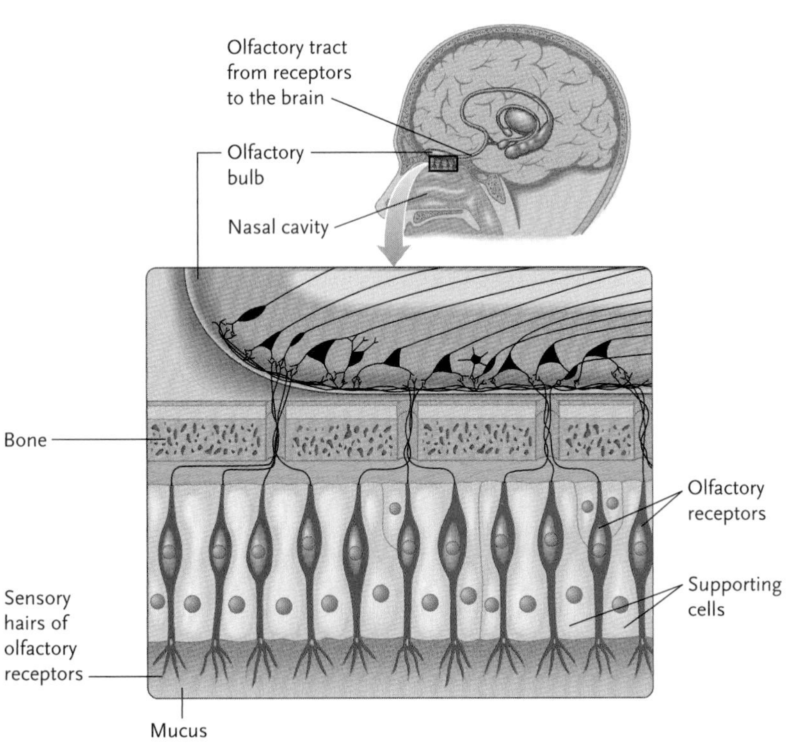

Figure 39.21

Olfactory receptors in the roof of the nasal passages in humans. Axons from these receptors pass through holes in the bone separating the nasal passages from the brain, where they make synapses with interneurons in the olfactory bulbs.

STUDY BREAK

1. How do we distinguish different kinds of smells?
2. For terrestrial vertebrates, describe the pathway by which a signal generated by taste receptors leads to a response.

39.6 Thermoreceptors and Nociceptors

Thermoreceptors detect changes in the surrounding temperature. Nociceptors respond to stimuli that may potentially damage their tissues. Both types of receptors consist of free nerve endings formed by the dendrites of afferent neurons, with no specialized receptor structures surrounding them.

Thermoreceptors Can Detect Warm and Cold Temperatures and Temperature Changes

Most animals have thermoreceptors. Some invertebrates such as mosquitoes and ticks use thermoreceptors to locate their warm-blooded prey. Some snakes, including rattlesnakes and pythons, use thermoreceptors called *pit organs* to detect the body heat of warm-blooded prey animals **(Figure 39.22)**.

In mammals, distinct thermoreceptors respond to heat and cold. Researchers have shown that three members of the *transient receptor potential* (TRP) gated Ca^{2+} channel family act as heat receptors. One responds when the temperature reaches 33°C and another responds above 43°C, where heat starts to be painful; these two receptors are believed to be involved in thermoregulation. The third receptor responds at 52°C and above, in this case producing a pain response rather than being involved in thermoregulation.

Two cold receptors are known in mammals. One responds between 8 and 28°C, and is thought to be involved in thermoregulation. The second responds to temperatures below 8°C and appears to be associated with pain rather than thermoregulation. The molecular mechanisms that control the opening and closing of heat and cold receptor channels are not currently known.

Some neurons in the hypothalamus of mammals also function as thermoreceptors, an ability that has only recently been investigated. Not only do these neurons sense changes in brain temperature, but they also receive afferent thermal information. These neurons are highly sensitive to shifts from the normal body temperature, and trigger involuntary responses such as sweating, panting, or shivering, which restore normal body temperature.

Nociceptors Protect Animals from Potentially Damaging Stimuli

The signals from nociceptors—receptors in mammals, and possibly other vertebrates, that detect damaging stimuli—are interpreted by the brain as pain. Pain is a protective mechanism; in humans, it prompts us to do something immediately to remove or decrease the damaging stimulus. Often pain elicits a reflex response—such as withdrawing the hand from a hot

Pit organs

Figure 39.22
The pit organs of an albino Western diamondback rattlesnake *(Crotalus atrox)*, located in depressions on both sides of the head below the eyes. These thermoreceptors detect infrared radiation emitted by warm-bodied prey animals such as mice.

stove—that proceeds before we are even consciously aware of the sensation.

Various types of stimuli cause pain, including mechanical damage such as a cut, pinprick, or blow to the body, and temperature extremes. Some nociceptors are specific for a particular type of damaging stimulus, while others respond to all kinds.

The axons that transmit pain are part of the somatic system of the PNS (see Section 38.2). They synapse with interneurons in the gray matter of the spinal cord, and activate neural pathways to the CNS by releasing the neurotransmitters glutamate or substance P (see Section 37.3). Glutamate-releasing axons produce sharp, prickling sensations that can be localized to a specific body part—the pain of stepping on a tack, for example. Substance P–releasing axons produce dull, burning, or aching sensations, the location of which may not be easily identified—the pain of tissue damage such as stubbing your toe.

As part of their protective function, pain receptors adapt very little, if at all. Some pain receptors, in fact, gradually intensify the rate at which they send out action potentials if the stimulus continues at a constant level.

The CNS also has a pain-suppressing system. In response to stimuli such as exercise, hypnosis, and stress, the brain releases *endorphins,* natural painkillers that bind to membrane receptors on substance P neurons, reducing the amount of neurotransmitter released.

Nociceptors contribute to the taste of some spicy foods, particularly those that contain hot peppers. In fact, researchers who study pain often use *capsaicin,* the organic compound that gives jalapeños and other peppers their hot taste, to identify nociceptors. To some, the burning sensation from capsaicin is addictive. Here is the reason. Nociceptors in the mouth, nose, and throat immediately transmit pain messages to the brain when they detect capsaicin. The brain re-

Hot News in Taste Research

Biting into a jalapeño pepper (a variety of *Capsicum annuum*) can produce a burning pain in your mouth strong enough to bring tears to your eyes. This painfully hot sensation is due primarily to *capsaicin*, a chemical that probably evolved in pepper plants as a defense against foraging animals. The defense is obviously ineffective against the humans who relish peppers and foods containing capsaicin (such as buffalo wings).

Research by David Julius and his coworkers at the University of California, San Francisco, revealed the molecular basis for detection of capsaicin by nociceptors. They designed their experiments to test the hypothesis that the responding nociceptors have a cell surface receptor that binds capsaicin. Binding the chemical opens a membrane channel in the receptor that admits calcium ions and initiates action potentials interpreted as pain.

The Julius team isolated the total complement of messenger RNAs from nociceptors able to respond to capsaicin and made complementary DNA (cDNA) clones of the mRNAs. The cDNAs, which represented sequences encoding proteins made in the nociceptors, contained thousands of different sequences. The cDNAs were transformed individually into embryonic kidney cells (which do not normally respond to capsaicin), and the transformed cells were screened with capsaicin to identify which took in calcium ions; presumably, these cells had received a cDNA encoding a capsaicin receptor. Messenger RNA transcribed from the identified cDNA clone was injected into both frog oocytes and cultured mammalian cells. Tests showed that both the oocytes and the cultured cells responded to capsaicin by admitting calcium ions, which confirmed that the researchers had found the capsaicin receptor cDNA.

Among the effects noted when the receptor was introduced into oocytes was a response to heat. Increasing the temperature of the solution surrounding the oocytes from 22°C to about 48°C produced a strong calcium inflow. In short, capsaicin and heat produce the same response in cells containing the receptor. Therefore the feeling that your mouth is on fire when you eat a hot pepper probably results from the fact that, as far as your nociceptors and CNS are concerned, it *is* on fire.

---

sponds by releasing endorphins, which act as a painkiller and create temporary euphoria—a natural high, if you will. *Insights from the Molecular Revolution* describes a series of experiments investigating the molecular basis of the pain caused by capsaicin.

STUDY BREAK

> What distinguishes thermoreceptors and nociceptors from the other types of sensory receptors discussed previously?

39.7 Magnetoreceptors and Electroreceptors

Some animals have poorly developed visual systems but can gain information about their environment by sensing magnetic or electrical fields. In so doing, they directly sense stimuli that humans can detect only with scientific instruments.

Magnetoreceptors Are Used for Navigation

Some animals that navigate long distances, including migrating butterflies, beluga whales, sea turtles, homing pigeons, and foraging honeybees, have **magnetoreceptors** that allow them to detect and use Earth's magnetic field as a source of directional information (experiments with sea turtles are described in **Figure 39.23**).

The pattern of Earth's magnetic field differs from region to region yet remains almost constant over time, largely unaffected by changing weather and day and night. As a result, animals with magnetic receptors are able to monitor their location reliably. Although little is known about the receptors that detect magnetic fields, they may depend on the fact that moving a conductor, such as an electroreceptor cell, through a magnetic field generates an electric current.

Some magnetoreceptors may depend on the effect of Earth's magnetic field on the mineral *magnetite*. Magnetite is found in the bones or teeth of many vertebrates, including humans, and also in insects—in the abdomen of honeybees and in the heads and abdomens of certain ants, for example.

Other animals, including homing pigeons, which are famous for their ability to find their way back to their nests even when released far from home, navigate by detecting their position with reference to both Earth's magnetic field and the sun. Magnetite is located in the bills of these birds, which is where research indicates magnetoreception likely occurs.

Electroreceptors Are Used for Location of Prey or for Communication

Many sharks and bony fishes, some amphibians, and even some mammals (such as the star-nosed mole and duckbilled platypus) have specialized **electroreceptors**

Figure 39.23 Experimental Research

Demonstration That Magnetoreceptors Play a Key Role in Loggerhead Sea Turtle Migration

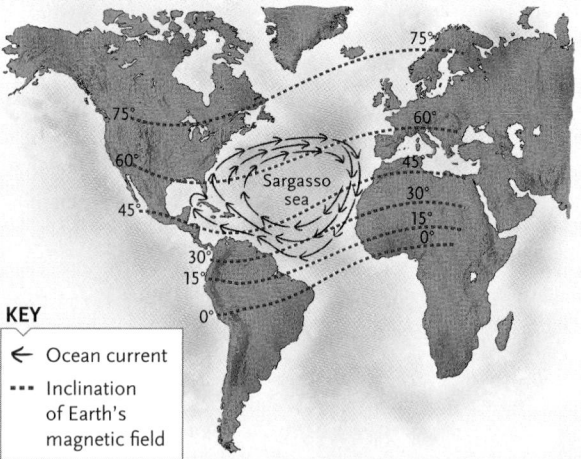

KEY

← Ocean current

··· Inclination of Earth's magnetic field

QUESTION: Do loggerhead sea turtles use a magnetoreceptor system for migration?

EXPERIMENT: Loggerhead sea turtles *(Caretta caretta)* that hatch along the east coast of Florida spend much of their lives traveling the North Atlantic current system around the Sargasso Sea, a pool of warm water with a unique seaweed ecosystem. Eventually and unerringly, the turtles return to their hatching beach for the mating season. Kenneth Lohmann of the University of North Carolina hypothesized that magnetoreception, likely involving magnetite, plays a central role in loggerhead migration. Lohmann tested his hypothesis using an experimental system in which the direction hatchling turtles swam was analyzed in different magnetic fields.

1. Lohmann placed each turtle hatchling he tested in a harness and tethered it to a swiveling, electronic system in the center of a circular pool of water. The pool was surrounded by a large electromagnetic coil system that allowed the researchers to reverse the direction of the magnetic field. The direction the turtle swam was recorded by the tracking system and relayed to a computer.

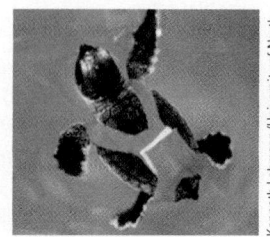

Kenneth Lohmann/University of North Carolina

2. Lohmann allowed the turtles to swim under two experimental conditions: half of the turtles swam in Earth's magnetic field, and the other half swam in a reversed magnetic field.

RESULTS: The turtle hatchlings tested in Earth's magnetic field swam in an east-to-northeast direction on average, mimicking the direction they follow normally when migrating at sea. The turtle hatchlings tested in the reversed magnetic field on average swam in a direction 180° opposite that of the hatchlings swimming in Earth's magnetic field.

CONCLUSION: The results indicate that loggerhead sea turtle hatchlings have the ability to detect Earth's magnetic field and use it as a way to orient their migration. Their direction of migration, east to northeast, matches the inclination of Earth's magnetic field in the Atlantic Ocean where they migrate (see map figure). Lohmann believes that the magnetoreception system in the turtles involves magnetite.

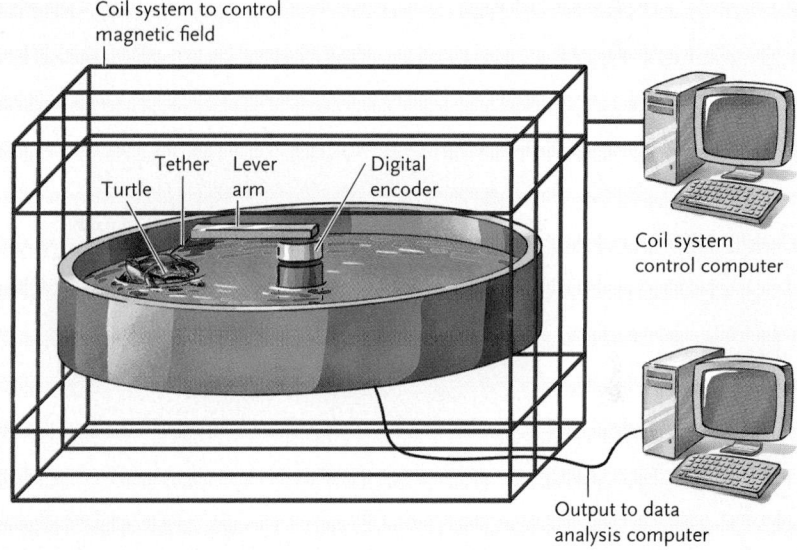

Coil system to control magnetic field

Tether Lever arm Digital encoder

Turtle

Coil system control computer

Output to data analysis computer

---

that detect electrical fields. The plasma membrane of an electroreceptor cell is depolarized by an electrical field, leading to the generation of action potentials. The electrical stimuli detected by the receptors are used to locate prey or navigate around obstacles in muddy water, or, by some fishes, to communicate. Some electroreception systems are passive—they detect electric fields in the environment, not the animal's own electric currents. Passive systems are used mainly to find prey. For example, the electroreceptors of sharks and rays can locate fish buried under the sand from the electrical currents generated by their prey's heartbeat or by the muscle contractions that move water over their gills.

Other electroreception systems are active—the animal emits and receives low voltage electrical signals, either to locate prey or to communicate with members of the same species. The electrical signals

What happens when the senses get scrambled—when listening to music causes you to "see" colors, or when you "taste" certain words?

Synesthesia (joined senses) occurs when two senses, normally separate, are perceived together. For the most part, people with synesthesia are born with it, and it tends to run in families. A recent study by Michael Esterman and his colleagues at the University of California, Berkeley, showed that the posterior parietal cortex, a region of the brain thought to be involved in sensory integration, appears to be crucial to sensory commingling. Some researchers think that this commingling is how the senses function early in development, when the nervous system is still immature. They believe that the senses normally separate from one another around four months after birth. In synesthetes, however, this separation is incomplete and two of their senses remain mingled.

London's Science Museum collaborated with Jamie Ward of University College London in an experiment that paired sounds and music. They wanted to determine if volunteers visiting the museum would prefer combinations of sound and vision as described by synesthetes over combinations randomly generated by a computer. Interestingly, people found the synesthetic combinations more pleasing than the computer-generated ones. Thus, it is possible that everyone may have a built-in understanding of what sounds and colors go together.

In an evolutionary context, which "sense" developed first?

The descriptions of the senses in this chapter focus primarily on vertebrate sensory systems, but the nervous systems of many invertebrates can be quite complex. Indeed, squids, sea hares, leeches, horseshoe crabs, lobsters, and cockroaches have been quite instrumental in helping scientists understand the nervous system. As for senses, squids have sensitive eyes and accentuated smell and taste, and they respond to touch and vibration. Clearly, vertebrates aren't the only multicellular organisms to develop senses.

However, is a nervous system necessary for organisms to have senses? Can single-celled organisms (which, of course, do not have nervous systems) respond to stimuli? Clearly, the answer is yes. Consider *Paramecium tetraurelia*, a single-celled organism covered with cilia that lives in water. It can detect substances in its environment and swim toward certain chemicals while avoiding others. It also responds to solid objects by turning when it runs into one. Thus, it has a "chemical sense" similar to taste or smell, and it responds to a type of "touch." We will probably never know which sense developed first, but some type of touch or chemical sense seems the most likely.

Rona Delay is an associate professor in the Department of Biology at the University of Vermont. Her research centers on understanding how sensory receptors change or transduce information about the external world into a language the brain can understand. The focus of her research is the sense of smell. To learn more go to http://www.uvm.edu/~biology/Faculty/Delay/Delay.html.

---

are generated by special electric organs. A few species, such as the electric eel and the electric catfish, produce discharges on the order of several hundred volts. These discharges are used to stun or kill prey. The voltage is high enough to stun, but not kill, a human.

STUDY BREAK

What are three ways electroreceptors are used in aquatic vertebrates?

Review

Go to **ThomsonNOW**™ at www.thomsonedu.com/login to access quizzing, animations, exercises, articles, and personalized homework help.

39.1 Overview of Sensory Receptors and Pathways

- Sensory receptors are formed by the endings of afferent neurons or specialized cells adjacent to the neurons. They detect stimuli such as mechanical pressure, sound waves, light, or specific chemicals. Action potentials generated by the receptors are carried by the axons of afferent neurons to pathways leading to specific parts of the brain, where signals are processed into sensory sensations (Figure 39.1).

- Receptors are specialized as mechanoreceptors, photoreceptors, chemoreceptors, thermoreceptors, and nociceptors. Some animals have receptors that detect electrical or magnetic fields.

- The routing of information from sensory receptors to particular regions of the brain identifies a specific stimulus as a sensation. The intensity of a stimulus is determined by the frequency of action potentials traveling along the neural pathways and the number of afferent neurons carrying action potentials.

- Many sensory systems show sensory adaptation, in which the frequency of action potentials decreases while a stimulus remains constant. Some sensory receptors, such as those related to pain, show little or no sensory adaptation.

Animation: Action potentials

39.2 Mechanoreceptors and the Tactile and Spatial Senses

- Mechanoreceptors detect touch, pressure, acceleration, or vibration. Touch and pressure receptors are free nerve endings or encapsulated nerve endings of sensory neurons (Figure 39.2).

- Mechanoreceptors called proprioceptors detect stimuli used by the CNS to monitor and maintain body and limb positions.

- Proprioceptors based on sensory hair cells generate action potentials when the hairs are moved (Figures 39.3–39.5).

- Receptors in muscles, tendons, and joints of vertebrates detect changes in stretch and tension of body parts (Figure 39.6).

Animation: Dynamic equilibrium

39.3 Mechanoreceptors and Hearing

- Many invertebrates have mechanoreceptors in their skin or other surface structures that detect sound and other vibrations.
- Hearing relies on sensory hair cells in organs that respond to the vibrations of sound waves.
- In terrestrial vertebrates, the ear consists of three parts. The outer ear directs sound to the eardrum. Vibrations of the eardrum are transmitted through one or more bones in the middle ear to the fluid-filled inner ear. In the inner ear, the vibrations are transmitted through membranes that bend the stereocilia of the hair cells, leading to bursts of action potentials that are reflected in the frequency of the sound waves (Figure 39.8).

Animation: Ear structure and function

Animation: Properties of sound

39.4 Photoreceptors and Vision

- Invertebrates possess many forms of eyes, from the simplest, an ocellus, to single-lens eyes that are similar to vertebrate eyes (Figures 39.9–39.11).
- The photoreceptors of all animal eyes contain the pigment retinal, which absorbs the energy of light and uses it to generate changes in membrane potential.
- The transparent cornea admits light into the vertebrate eye. Behind the cornea, the iris controls the diameter of the pupil, regulating the amount of light that strikes the lens. The lens focuses an image on the retina lining the back of the eye, where photoreceptors and neurons carry out the initial integration of information detected by the photoreceptors (Figure 39.12).
- In terrestrial vertebrates, the lens is focused by adjusting its shape (Figure 39.13). The retina contains two types of photoreceptors, rods and cones. Rods are specialized for detecting light of low intensity; cones are specialized for detecting light of different wavelengths, which are perceived as colors.
- The light-absorbing pigment in photoreceptor cells consists of retinal combined with an opsin protein. When it absorbs light, retinal changes form, initiating reactions that alter the amount of neurotransmitter released by the photoreceptor cells (Figures 39.14 and 39.15).
- Rods and cones are linked to neurons in the retina that perform the initial processing of visual information. The processed signal is sent via the optic nerve through the lateral geniculate nuclei to the visual cortex (Figures 39.16 and 39.17).

Animation: Eye structure

Animation: Visual accommodation

Animation: Organization of cells in the retina

Animation: Receptive fields

Animation: Pathway to visual cortex

Animation: Focusing problems

39.5 Chemoreceptors

- Chemoreceptors respond to the presence of specific molecules in the environment. In vertebrates, they form parts of receptor organs for taste (gustation) and smell (olfaction).
- Taste receptors detect molecules from food or other objects that come into direct contact with the receptor and are used primarily to identify foods (Figures 39.18 and 39.20).
- Olfactory receptors detect molecules from distant sources; besides identifying food, they are used to detect predators and prey, identify family and group members, locate trails and territories, and communicate (Figures 39.19 and 39.21).

Animation: Olfactory pathway

Animation: Taste receptors

39.6 Thermoreceptors and Nociceptors

- Thermoreceptors, which consist of free nerve endings located at the body surface and in limited numbers in the body interior, detect changes in body temperature.
- Nociceptors, located on both the body surface and interior, detect stimuli that can damage body tissues. Information from these receptors is integrated in the brain into the sensation of pain.

Animation: Sensory receptors in the human skin

Animation: Referred pain

39.7 Electroreceptors and Magnetoreceptors

- Some vertebrates have electroreceptors that detect electrical currents and fields, or magnetoreceptors that detect magnetic fields (Figure 39.23).

Questions

Self-Test Questions

1. The frequency of a blast from a nearby ambulance siren can cause a dog to howl in pain. Activated under this circumstance are:
   a. thermoreceptors and chemoreceptors.
   b. photoreceptors and nociceptors.
   c. mechanoreceptors and nociceptors.
   d. chemoreceptors and mechanoreceptors.
   e. photoreceptors and chemoreceptors.

2. Two common side effects of Hansen's disease (leprosy) are a permanent numbness in the hands, feet, and buttocks of affected people and a loss of perception of their spatial position. Affected are:
   a. mechanoreceptors.
   b. adapting receptors.
   c. pH change receptors.
   d. the vestibular apparatus.
   e. vibration detecting systems.

3. Neuromasts are best described as:
   a. nonadapting pain receptors.
   b. components of the fish lateral line system.
   c. statoliths that detect motion.
   d. motor axons that activate motion.
   e. cupulas that detect vibrations.

4. Structures are activated by sound waves in the vertebrate ear in the following order:
   a. oval window, tympanum, semicircular canals, Golgi tendon organ, incus, malleus, stapes.
   b. organ of Corti, malleus, incus, stapes, auditory nerve, eardrum.
   c. eustachian tube, round window, vestibular canal, tympanic canal, cochlear canal, oval window, pinna.
   d. basilar membrane, tectorial membrane, otoliths, utricle, saccule, malleus, cochlea.
   e. pinna, tympanic membrane, malleus, incus, stapes, oval window, cochlear duct.

5. The following situation is associated with movement and position in the human body:
   a. Statoliths in statocysts bend sensory hairs and trigger action potentials.
   b. If sensory hairs in the utricle are oriented horizontally and those in the saccule are oriented vertically, the person is lying down.
   c. When the head rotates, the endolymph in the semicircular canal pulls the cupula with it to activate sensory hair cells.
   d. Displacement of the utricle and saccule generates action potentials.
   e. If the body is spinning at a constant rate and direction, the cupula is displaced and action potentials are initiated.

6. The difference between the vertebrate eye and the cephalopod eye is that the vertebrate eye has:
   a. an iris surrounding the pupil, whereas in cephalopods the pupil surrounds the iris.
   b. a lens that changes shape when focusing, whereas in cephalopods the lens moves back and forth to focus.
   c. a retina that moves in the socket when recording the image, whereas in cephalopods the retina changes shape when stimulated.
   d. a pupil that shrinks in size in bright light, whereas cephalopods have a pupil that enlarges in bright light.
   e. retinal synthesized from vitamin A, whereas cephalopods lack retinal.

7. Which of the following events does not occur during light absorption in the vertebrate eye?
   a. The retinal component of rhodopsin changes from *cis* to *trans* form.
   b. Rhodopsin, a G membrane-embedded protein, triggers a signal transduction pathway to close Na^+ channels in the plasma membrane.
   c. The light stimulus passes from rods and cones to bipolar cells and horizontal cells and then to ganglion cells, whose axons compose the optic nerve.
   d. As light absorption increases, the rhodopsin response causes an increase in the release of neurotransmitters.
   e. When integrating information across the retina, horizontal cells connect the rods and cones, and amacrine cells join with the bipolar cells and ganglion cells.

8. The variety of color seen by humans is directly dependent upon the:
   a. activation of three different photopsins in cones.
   b. transmission of an image to separate brain hemispheres by the optic chiasm.
   c. transmission of impulses from rods across the lateral geniculate nuclei.
   d. lateral inhibition by amacrine cells.
   e. light stimulation of all photoreceptor types equally.

9. In terrestrial animals:
   a. the hairs of taste receptors are derived from cilia and contain microtubules.
   b. the hairs of smell receptors are derived from microvilli and contain microfilaments.
   c. signals from taste receptors are relayed to the temporal lobes.
   d. information from olfactory receptors is processed in the parietal lobes.
   e. connections from the olfactory bulbs lead to the limbic system.

10. In the human response to temperature or pain:
   a. all three transient receptor potential (TRP) gated Ca^{2+} channels act as pain receptors.
   b. cold receptors are activated between 27°C and 37°C.
   c. pain receptors decrease the rate at which they send out action potentials if the pain is constant.
   d. nociceptors, activated by capsaicin in the mouth and nose, can sense pain.
   e. the CNS releases glutamate or substance P to dull the pain sensation.

Questions for Discussion

1. Humans have about 200 million photoreceptors in two eyes, and about 32,000 sensory hair cells in two ears. About 3% of the somatosensory cortex is devoted to hearing, whereas roughly 30% of it is devoted to visual processing. Suggest an explanation for these differences from the perspective of natural selection and adaptation.

2. In owls and many other birds of prey, the fovea is located toward the top of the retina rather than at the center as in humans. This arrangement correlates with the birds' hunting behavior, in which they look down when they fly, scanning the ground for a meal. With this arrangement in mind, why do you think the standing owl in the picture is turning its head upside down?

Chase Smith

3. A patient made an appointment with her doctor because she was experiencing recurrent episodes of dizziness. Her doctor asked questions to distinguish whether she had sensations of lightheadedness, as if she were going to faint, or vertigo, as if she or objects near her were spinning around. Why was this clarification important in the evaluation of her condition?

Experimental Analysis

The fruit fly *Drosophila melanogaster* can distinguish a large repertoire of odors in the environment. Their response may be to move toward food or away from danger. Moreover, particular odors play an important role in their mating behavior. The olfactory organs of a fruit fly are the antennae and an elongated bulge on the head called the maxillary pulp. Because of the ease with which fruit fly genes can be manipulated, identifying and studying their olfactory receptors likely would contribute significantly to our understanding of neural pathways of odor recognition more generally. How could you identify candidate fruit fly genes that encode components of olfactory receptors?

Evolution Link

In 2005, researchers took saliva and blood samples from six cats, including domestic cats, a tiger, and a cheetah, and found that all have a defective gene for one of the two chemoreceptor proteins needed to identify food as sweet. (The scientists conjecture that the lack of a sweet tooth may explain why cats are finicky eaters.) What are the evolutionary implications of the finding?

How Would You Vote?

Noise pollution from commercial shipping and other human activities generates low-frequency sounds that are believed to interfere with the acoustical signals that whales use for navigation, location of food, and communication. To what extent should we limit these activities to protect whales against potential harm? Would you support banning activities that exceeded a certain noise level from U.S. territorial waters? If so, how would you get other nations to do the same? Go to www.thomsonedu.com/login to investigate both sides of the issue and then vote.

Two North American bull elks contesting for cows in Yellowstone National Park. The shorter days of autumn trigger hormone production, battling, and reproductive behavior.

© Mark Wallner

40 The Endocrine System

WHY IT MATTERS

Every September, as the days grow shorter and autumn approaches, bull elks *(Cervus canadensis)* begin to strut their stuff. Although they have grazed peacefully together at high mountain elevations from the Yukon to Arizona, they now become testy with each other. They also rasp at tree branches and plow the ground with their antlers. Soon, they descend to lower elevations, where the cow elks have been feeding in large nursery groups with their calves and yearlings.

The bulls move in among the cows and chase away the male yearlings. As part of the mating ritual, the bulls bugle, square off, strut, and circle; then they clash their antlers together, attempting to drive each other from the cows. The winning males claim harems of about 10 females each, a major prize.

After the mating season ends, tranquility returns. The cows again graze in herds; the males form now-friendly bachelor groups that also feed quietly in the meadows. Eating is their major occupation, storing nutrients in preparation for the snowy winter. The young will be born eight to nine months later, when summer returns.

The next year, the shortening days of late summer and fall again trigger the transition to mating behavior. Detected by the eyes and registered in the brain, reduced daylength initiates changes in the secretion of long-distance signaling molecules called **hormones** (*hormaein* = to excite). Hormones are released from one group of cells and are transported through the circulatory system to other cells, their target cells, whose activities they change. Among the changes will be a rise in the concentration of hormones responsible for mating behavior.

We too are driven by our hormones. They control our day-to-day sexual behavior—often as outlandish as that of a bugling bull elk—as well as a host of other functions, from the concentration of salt in our blood, to body growth, to the secretion of digestive juices. Along with the central nervous system, hormones coordinate the activities of multicellular life.

The best-known hormones are secreted by cells of the **endocrine system** (*endo* = within; *krinein* = to separate), although hormones actually are produced by almost all organ systems in the body. The endocrine system, like the nervous system, regulates and coordinates distant organs. The two systems are structurally, chemically, and functionally related, but they control different types of activities. The nervous system, through its high-speed electrical signals, enables an organism to interact rapidly with the external environment, while the endocrine system mainly controls activities that involve slower, longer-acting responses. Typical responses to hormones may persist for hours, weeks, months, or even years.

The mechanisms and functions of the endocrine system are the subjects of this chapter. As in other chapters of this unit, we pay particular attention to the endocrine system of humans and other mammals.

40.1 Hormones and Their Secretion

Cells signal other cells using neurotransmitters, hormones, and local regulators. Recall from Chapters 37 through 39 that a neurotransmitter is a chemical released by an axon terminal at a synapse which affects the activity of a postsynaptic cell. Our focus in this chapter is hormones and local regulators, molecules that act locally rather than over long distances.

The Endocrine System Includes Four Major Types of Cell Signaling

Four types of cell signaling occur in the endocrine system: classical endocrine signaling, neuroendocrine signaling, paracrine regulation, and autocrine regulation. In *classical endocrine signaling*, hormones are secreted into the blood or extracellular fluid by the cells of ductless secretory organs called **endocrine glands (Figure 40.1a).** (In contrast, *exocrine glands,* such as the sweat and salivary glands, release their secretions into ducts that lead outside the body or into the cavities of the digestive tract, as described in Section 36.2.) The

hormones are circulated throughout the body in the blood and, as a result, most body cells are constantly exposed to a wide variety of hormones. (The cells of the central nervous system are sequestered from the general circulatory system by the blood–brain barrier, described in Section 38.3.) Only *target cells* of a hormone, those with *receptor proteins* recognizing and binding that hormone, respond to it. Through these responses, hormones control such vital functions as digestion, osmotic balance, metabolism, cell division, reproduction, and development. The action of hormones may either speed or inhibit these cellular processes. For example, growth hormone stimulates cell division, whereas glucocorticoids inhibit glucose uptake by most cells in the body.

Hormones are cleared from the body at a steady rate by enzymatic breakdown in their target cells or in either the liver or kidneys. Breakdown products are excreted by the digestive and excretory systems; depending on the hormone, the breakdown takes minutes to days.

In *neuroendocrine signaling,* specialized neurons called **neurosecretory neurons** release a hormone called a *neurohormone* into the circulatory system when appropriately stimulated **(Figure 40.1b).** The neurohormone is distributed by the circulatory system and elicits a response in target cells that have receptors for the hormone. Note that both neurohormones and neurotransmitters are secreted by neurons. Neurohormones are distinguished from neurotransmitters in that neurohormones affect distant target cells, whereas neurotransmitters affect adjacent cells. However, both neurohormones and neurotransmitters function in the same way—they cause cellular responses by interacting with specific receptors on target cells. For instance, gonadotropin-releasing hormone, a neurohormone secreted by the hypothalamus, controls the release of luteinizing hormone from the pituitary.

In *paracrine regulation,* a cell releases a signaling molecule that diffuses through the extracellular fluid and acts on nearby cells—regulation is *local* rather than at a distance, as is the case with hormones and neurohormones **(Figure 40.1c).** In some cases the local regulator acts on the same cells that produced it; this is called *autocrine regulation* **(Figure 40.1d).** For example, many of the growth factors that regulate cell division and differentiation act in both a paracrine and autocrine fashion.

Hormones and Local Regulators Can Be Grouped into Four Classes Based on Their Chemical Structure

More than 60 hormones and local regulators have been identified in humans. Many human hormones are either identical or very similar in structure and function to those in other animals, but other vertebrates as well as invertebrates have hormones not found in humans.

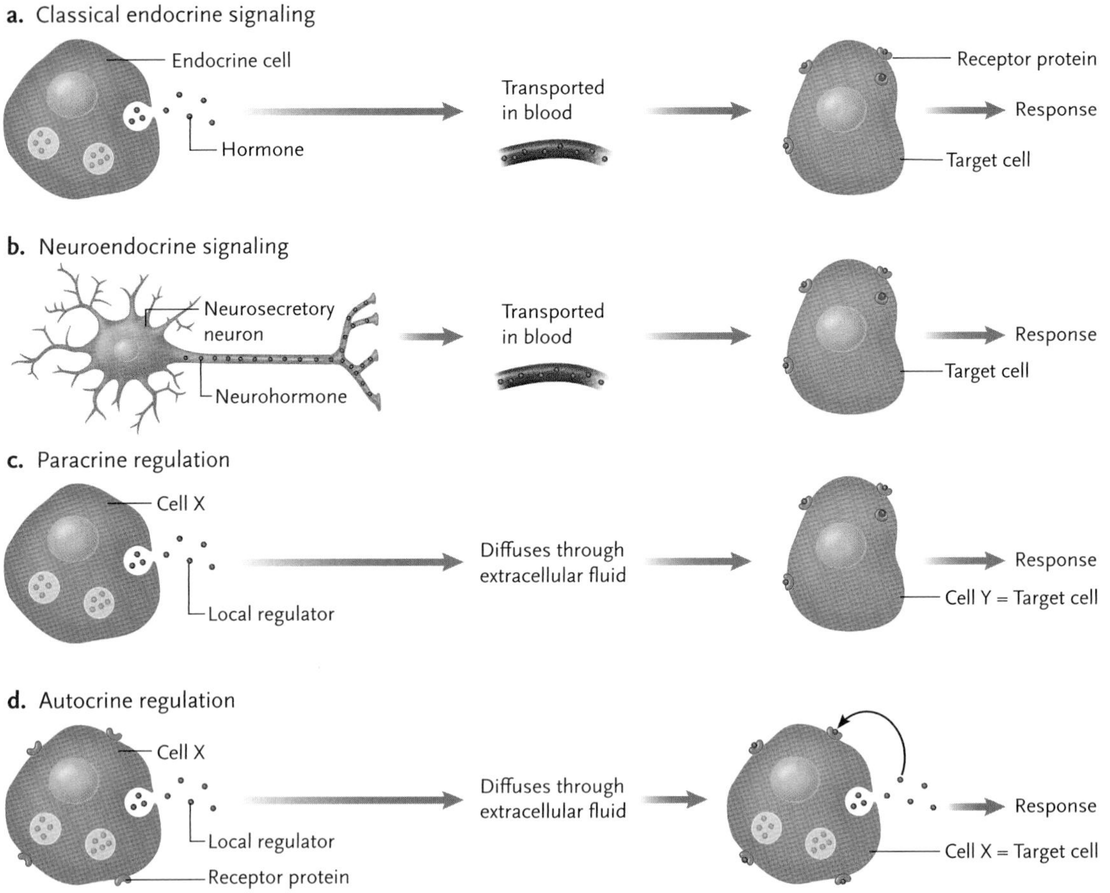

a. Classical endocrine signaling

Endocrine cell

Hormone

Transported in blood

Receptor protein

Response

Target cell

b. Neuroendocrine signaling

Neurosecretory neuron

Neurohormone

Transported in blood

Response

Target cell

c. Paracrine regulation

Cell X

Local regulator

Diffuses through extracellular fluid

Response

Cell Y = Target cell

d. Autocrine regulation

Cell X

Local regulator

Receptor protein

Diffuses through extracellular fluid

Response

Cell X = Target cell

Figure 40.1
The four major types of cell signaling in the endocrine system.

Most of these chemicals can be grouped into four molecular classes: amine, peptide, steroid, and fatty acid–derived molecules.

Amine hormones are involved in classical endocrine signaling and neuroendocrine signaling. Most amine hormones are based on tyrosine. With one major exception, they are hydrophilic molecules, which diffuse readily into the blood and extracellular fluids. On reaching a target cell, they bind to receptors at the cell surface. The amines include epinephrine and norepinephrine, already familiar as neurotransmitters released by some neurons (see Section 37.3). The exception is thyroxine, a hydrophobic amine hormone secreted by the thyroid gland. This hormone, based on a pair of tyrosines, passes freely through the plasma membrane and binds to a receptor inside the target cell, as do steroid hormones (see following discussion and Section 7.5).

The *peptide hormones* consist of amino acid chains, ranging in length from as few as three amino acids to more than 200. Some have carbohydrate groups attached. They are involved in classical endocrine signaling and neuroendocrine signaling. Mostly hydrophilic hormones, peptide hormones are released into the blood or extracellular fluid by exocytosis when cytoplasmic vesicles containing the hormones fuse with the plasma membrane. One large group of peptide hormones, the **growth factors,** regulates the division and differentiation of many cell types in the body. Many growth factors act in both a paracrine and autocrine manner as well as in classical endocrine signaling. Because they can switch cell division on or off, growth factors are an important focus of cancer research.

Steroid hormones are involved in classical endocrine signaling. All are hydrophobic molecules derived from cholesterol and are insoluble in water. They combine with hydrophilic carrier proteins to form water-soluble complexes that can diffuse through extracellular fluids and enter the bloodstream. On contacting a cell, the hormone is released from its carrier protein, passes through the plasma membrane of the target cell, and binds to internal receptors in the nucleus or cytoplasm. Steroid hormones include aldosterone, cortisol, and the sex hormones. Steroid hormones may vary little in structure, but produce very different effects. For example, testosterone and estradiol, two major sex hormones responsible for the development of mammalian male and female characteristics, respectively, differ only in the presence or absence of a methyl group.

Fatty acid–derived molecules are involved in paracrine and autocrine regulation. **Prostaglandins,** for example, are important as local regulators. First discovered in the 1930s in seminal fluid, prostaglandins were so named because they were thought to be secreted by the prostate gland, although actually they are secreted by the seminal vesicles. Scientists later discovered that virtually every cell can secrete prostaglandins, and they are present at essentially all times. In semen, they enhance the transport of sperm through the female reproductive tract by increasing the contractions of smooth muscle cells, particularly in the uterus. During childbirth, prostaglandins secreted by the placenta work with a peptide hormone called oxytocin to stimulate labor contractions. Other prostaglandins induce contraction or relaxation of smooth muscle cells in many parts of the body, including blood vessels and air passages in the lungs. When released as a product of membrane breakdown in injured cells, prostaglandins may also intensify pain and inflammation.

Many Hormones Are Regulated by Feedback Pathways

The secretion of many hormones is regulated by feedback pathways, some of which operate partially or completely independent of neuronal controls. Most of these pathways are controlled by negative feedback—that is, a product of the pathway inhibits an earlier step in the pathway (see Section 36.4). For example, in some mammals, secretion by the thyroid gland is regulated by a negative feedback loop **(Figure 40.2)**. Neurosecretory neurons in the hypothalamus secrete thyroid-releasing hormone (TRH) into a vein connecting the hypothalamus to the pituitary gland. In response, the pituitary releases thyroid-stimulating hormone (TSH) into the blood, which stimulates the thyroid gland to release thyroid hormones. As the thyroid hormone concentration in the blood increases, it begins to inhibit TSH secretion by the

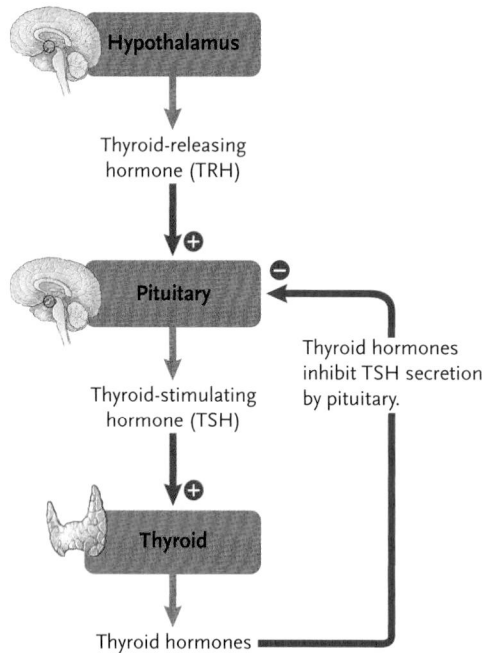

Figure 40.2
A negative feedback loop regulating secretion of the thyroid hormones. As the concentration of thyroid hormones in the blood increases, the hormones inhibit an earlier step in the pathway (indicated by the negative sign).

pituitary; this action is the negative feedback step. As a result, secretion of the thyroid hormones is reduced.

Body Processes Are Regulated by Coordinated Hormone Secretion

Although we will talk mostly about individual hormones in the remainder of the chapter, body processes are affected by more than one hormone. For example, the blood concentrations of glucose, fatty acids, and ions such as Ca^{2+}, K^+, and Na^+ are regulated by the coordinated activities of several hormones secreted by different glands. Similarly, body processes such as oxidative metabolism, digestion, growth, sexual development, and reactions to stress are all controlled by multiple hormones.

In many of these systems, negative feedback loops adjust the levels of secretion of hormones that act in antagonistic (opposing) ways, creating a balance in their effects that maintains body homeostasis. For example, consider the regulation of fuel molecules such as glucose, fatty acids, and amino acids in the blood. We usually eat three meals a day and fast to some extent between meals. During these periods of eating and fasting, four hormone systems act in coordinated fashion to keep the fuel levels in balance: (1) insulin and glucagon, secreted by the pancreas; (2) growth hormone, secreted by the anterior pituitary; (3) epinephrine and norepinephrine, released by the sympathetic nervous system and the adrenal medulla; and (4) glucocorticoid hormones, released by the adrenal cortex.

The entire system of hormones regulating fuel metabolism resembles the failsafe mechanisms designed by human engineers, in which redundancy, overlapping controls, feedback loops, and multiple safety valves ensure that vital functions are maintained at constant levels in the face of changing and even extreme circumstances.

STUDY BREAK

1. Distinguish between a hormone and a neurohormone.
2. Distinguish among the four major types of cell signaling.

40.2 Mechanisms of Hormone Action

Hormones control cell functions by binding to receptor molecules in their target cells. Small quantities of hormones can typically produce profound effects in cells and body functions due to **amplification**. In amplification, an activated receptor activates many proteins, which then activate an even larger number of proteins for the next step in the cellular reaction pathway, and so on, increasing in magnitude for each subsequent step in the pathway.

Hydrophilic Hormones Bind to Surface Receptors, Activating Protein Kinases Inside Cells

Hormones that bind to receptor molecules in the plasma membrane—primarily hydrophilic amine and peptide hormones—produce their responses through signal transduction pathways (see Section 7.2). In brief, when a surface receptor binds a hormone, the receptor is activated and transmits a signal through the plasma membrane. There are two kinds of surface receptors; the cytoplasmic reactions they control when they become activated are described in detail in Sections 7.3 and 7.4. Within the cell, the signal is transduced, changed into a form that causes the cellular response **(Figure 40.3a).** Typically, the reactions of signal transduction pathways involve protein kinases, enzymes that add phosphate groups to proteins. Adding a phosphate group to a pro-

tein may activate it or inhibit it, depending on the protein and the reaction. The particular response produced by a hormone depends on the kinds of protein kinases activated, the type of cell that can respond, and the types of target proteins they phosphorylate. Importantly, a small amount of hormone can elicit a large response because of amplification (see Figure 7.6).

The peptide hormone glucagon illustrates the mechanisms triggered by surface receptors. When glucagon binds to surface receptors on liver cells, it triggers the breakdown of glycogen stored in those cells into glucose. The glucose then is released into the circulatory system.

Hydrophobic Hormones Bind to Receptors Inside Cells, Activating or Inhibiting Genetic Regulatory Proteins

After passing through the plasma membrane, the hydrophobic steroid and thyroid hormones bind to internal receptors in the nucleus or cytoplasm **(Figure 40.3b,**

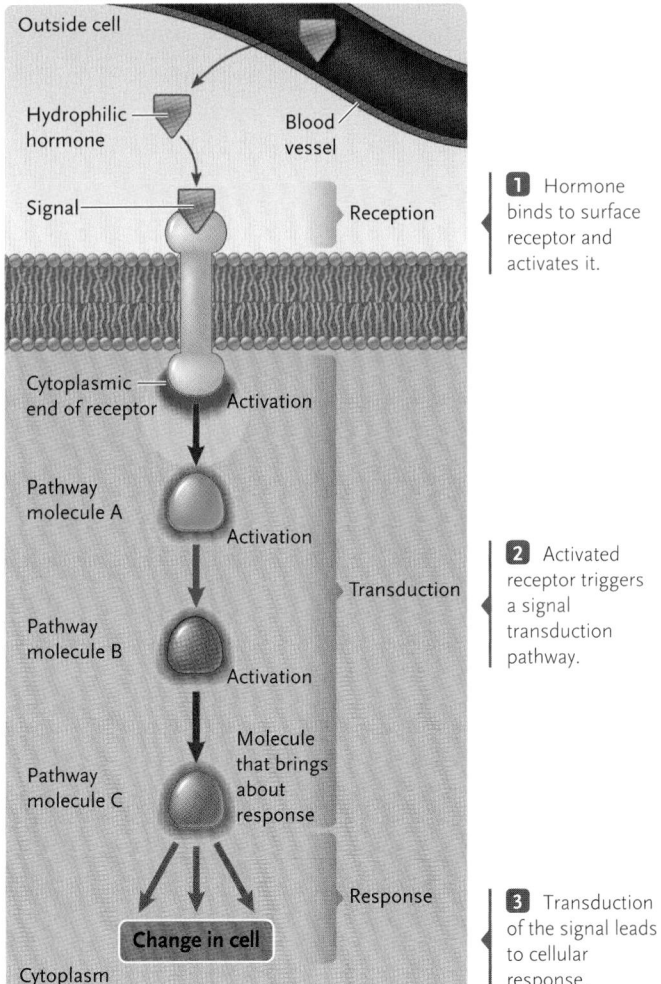

a. Hormone binding to receptor in the plasma membrane

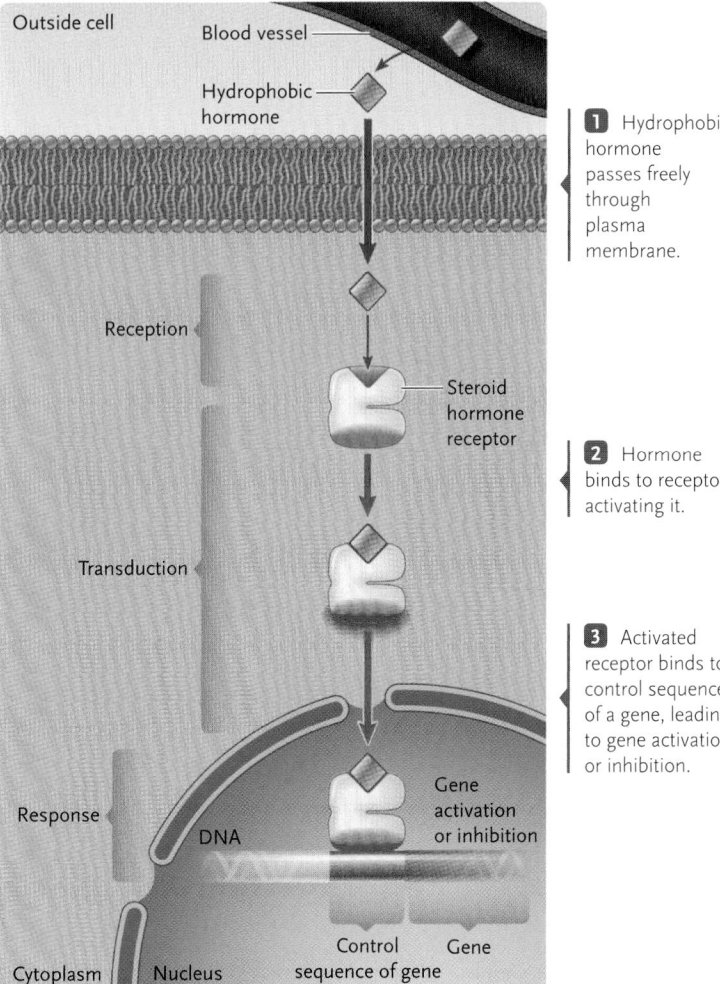

b. Hormone binding to receptor inside the cell

Figure 40.3
The reaction pathways activated by hormones that bind to receptor proteins in the plasma membrane **(a)** or inside cells **(b).** In both mechanisms, the signal—the binding of the hormone to its receptor—is transduced to produce the cellular response.

and described in detail in Section 7.5). Binding of the hormone activates the receptor, which then binds to a control sequence of specific genes. Depending on the gene, binding the control sequence either activates or inhibits its transcription, leading to changes in protein synthesis that accomplish the cellular response. The characteristics of the response depend on the specific genes controlled by the activated receptors, and on the presence of other proteins that modify the activity of the receptor.

One of the actions of the steroid hormone aldosterone illustrates the mechanisms triggered by internal receptors (Figure 40.4). If blood pressure falls below optimal levels, aldosterone is secreted by the adrenal glands. The hormone affects only kidney cells that contain the aldosterone receptor in their cytoplasm. When activated by aldosterone, the receptor binds to the control sequence of a gene, leading to the synthesis of proteins that increase reabsorption of Na^+ by the kidney cells. The resulting increase in Na^+ concentration in body fluids increases water retention and, with it, blood volume and pressure.

Target Cells May Respond to More Than One Hormone, and Different Target Cells May Respond Differently to the Same Hormone

A single target cell may have receptors for several hormones and respond differently to each hormone. For example, vertebrate liver cells have receptors for the pancreatic hormones insulin and glucagon. Insulin increases glucose uptake and conversion to glycogen, which decreases blood glucose levels, while glucagon stimulates the breakdown of glycogen into glucose, which increases blood glucose levels.

Conversely, particular hormones interact with different types of receptors in or on a range of target cells. Different responses are then triggered in each target cell type because the receptors trigger different transduction pathways. For example, the amine hormone epinephrine secreted by the adrenal medulla prepares the body for handling stress (including dangerous situations) and physical activity. (Epinephrine is discussed in more detail in Section 40.4.) In mammals, epinephrine can bind to three different plasma membrane-

Figure 40.4

The action of aldosterone in increasing Na^+ reabsorption in the kidneys when concentration of the ion falls in the blood.

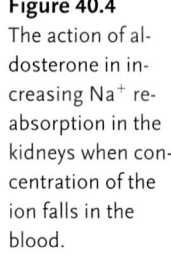

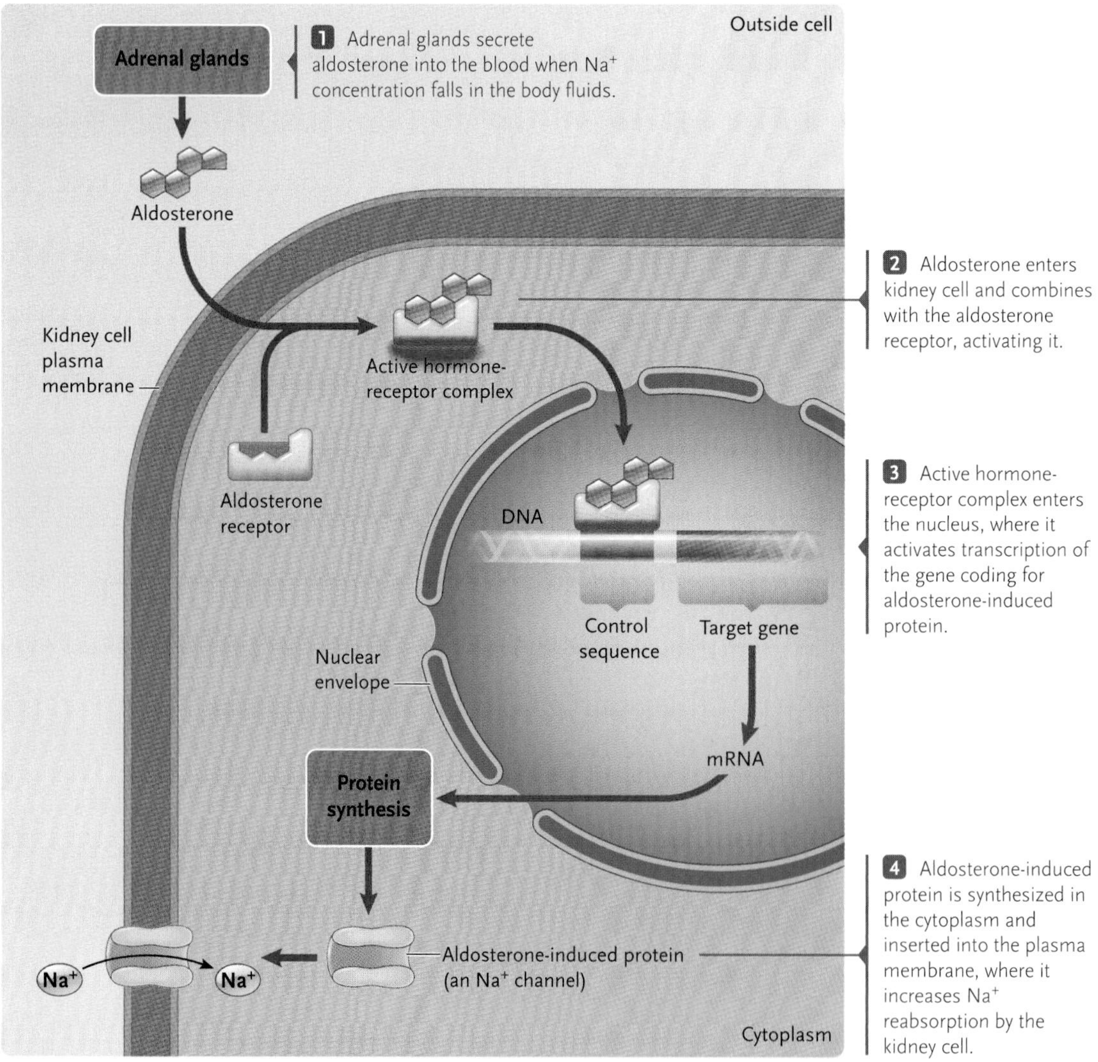

Adrenal glands

1 Adrenal glands secrete aldosterone into the blood when Na^+ concentration falls in the body fluids.

Outside cell

Aldosterone

Kidney cell plasma membrane

Active hormone-receptor complex

Aldosterone receptor

2 Aldosterone enters kidney cell and combines with the aldosterone receptor, activating it.

DNA

3 Active hormone-receptor complex enters the nucleus, where it activates transcription of the gene coding for aldosterone-induced protein.

Nuclear envelope

Control sequence   Target gene

mRNA

Protein synthesis

Aldosterone-induced protein (an Na^+ channel)

Na^+ Na^+

4 Aldosterone-induced protein is synthesized in the cytoplasm and inserted into the plasma membrane, where it increases Na^+ reabsorption by the kidney cell.

Cytoplasm

Figure 40.5 Experimental Research

Demonstration That Binding of Epinephrine to β Receptors Triggers a Signal Transduction Pathway within Cells

QUESTION: Is binding of epinephrine to β receptors necessary for triggering a signal transduction pathway within cells?

EXPERIMENT: It was known that epinephrine triggers a signal transduction pathway within cells. First, activation of adenylyl cyclase causes the level of the second messenger cAMP to increase, and then cAMP activates protein kinases in a signaling cascade that generates a cellular response (see Section 7.4 for specific details of such pathways). Richard Cerione and his colleagues at Duke University Medical Center performed experiments to show whether the signal transduction pathway is stimulated by binding of epinephrine to β receptors.

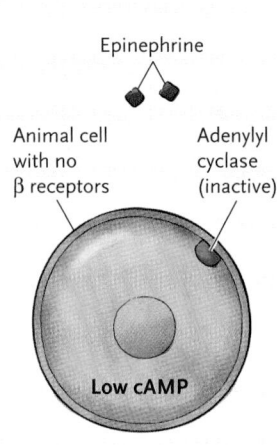

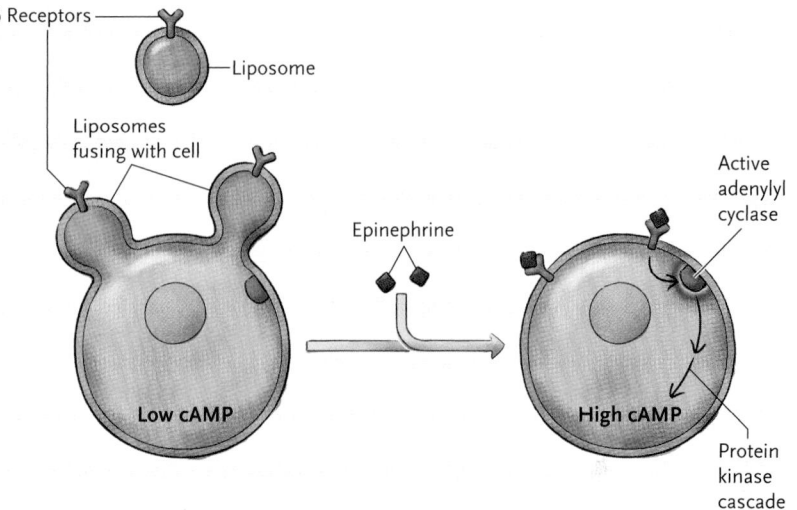

1. Epinephrine was added to animal cells lacking β receptors.

RESULT: No change occurred to the low level of cAMP in those cells. This result demonstrated that epinephrine alone was not able to trigger an increase in cAMP.

2. Liposomes—artificial spherical phospholipid membranes—containing purified β receptors were fused with the animal cells, and then epinephrine was added.

RESULT: When the liposomes fused with the animal cells, β receptors became part of the fused cell's plasma membrane. Then, adding epinephrine triggered synthesis of cAMP, resulting in high levels of cAMP in the cells. This result demonstrated that β receptors must be present in the membrane for epinephrine to trigger an increase in cAMP in the cell. The simplest interpretation was that epinephrine bound to the β receptors, activating adenylyl cyclase within the cell.

CONCLUSION: The cellular response depended upon binding of the hydrophilic hormone to a specific plasma membrane-embedded receptor.

---

embedded receptors: α, β₁, and β₂ receptors. (The experimental demonstration that the binding of epinephrine to a specific receptor triggers a cellular response is described in **Figure 40.5.**) When epinephrine binds to α receptors on smooth muscle cells, such as those of the blood vessels, it triggers a response pathway that causes the cells to constrict, cutting off circulation to peripheral organs. When epinephrine binds to β₁ receptors on heart muscle cells, the contraction rate of the cells increases, which in turn enhances blood supply. When epinephrine binds to β₂ receptors on liver cells, it stimulates the breakdown of glycogen to glucose, which is released from the cell. The overall

effect of these, and a number of other, responses to epinephrine secretion is to supply energy to the major muscles responsible for locomotion—the body is now prepared for handling stress or for physical activity.

Different receptors binding hydrophobic hormones also may generate diverse responses. *Insights from the Molecular Revolution* describes an investigation that tested the cellular responses produced by different receptors binding the same steroid hormone.

In summary, the mechanisms by which hormones work have four major features. First, only the cells that contain surface or internal receptors for a particular hormone respond to that hormone. Second, once

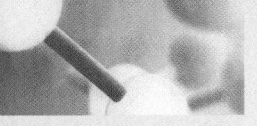

Two Receptors for Estrogens

Estrogens have many effects on female sexual development, behavior, and the menstrual cycle. One negative effect is to stimulate the growth of tumors in breast and uterine cancer. This cancer-enhancing effect can be reduced by administering *antiestrogens*, estrogen-like chemicals that bind competitively to estrogen receptors and block the sites that would normally be bound by the hormone. The antiestrogen *tamoxifen*, for example, inhibits the growth of breast tumors by blocking the activity of estrogen in breast tissue, but patients receiving it are at increased risk of developing uterine cancer.

How can tamoxifen have opposite effects in two different tissues? A group of investigators led by Thomas Scanlan and Peter Kushner at the University of California, San Francisco, joined by others at the Karolinska Institute and the Karo Bio Company in Sweden, had discovered that humans have two highly similar estrogen receptors, ERα and ERβ. Could differences between them account for the opposing effects of tamoxifen in breast and uterine tissues?

To find out, the researchers constructed two pairs of recombinant DNA plasmids. One pair consisted of either the ERα receptor gene or the ERβ receptor gene, adjacent to a promoter for continuous transcription of the receptor gene in human tissue culture cells. The other pair consisted of the firefly luciferase gene, which catalyzes a light-producing reaction, adjacent to one of two gene control sequences, AP1 or ERE, which act in estrogen-regulated systems.

One receptor plasmid and one luciferase plasmid were introduced together into human cell lines that do not normally make estrogen receptors in four possible combinations (see **figure**), with two cell lines making the ERα receptor and two making the ERβ receptor. Estrogen receptors are produced in all four of the resulting cell lines because the gene for the receptor is transcribed from the introduced plasmid.

With this experimental design, the researchers could test whether the ERα or ERβ receptor is activated by binding estrogen, and also whether the activated receptor would bind to the luciferase plasmid. If these two conditions were met, luciferase would be synthesized, and its activity could be measured using a special apparatus. (In this experiment, the luciferase gene acts as a *reporter* for the biological reactions that are occurring—luciferase has nothing to do with estrogen or estrogen activity.)

When the researchers added estrogen to cell lines containing luciferase plasmid with ERE (combinations 1 and 2 in the figure), all the cells produced luciferase. This indicated that both ERα and ERβ could bind the hormone, were activated, and could combine with ERE. When they added tamoxifen alone or along with the estrogen, the cells did not produce luciferase, indicating that tamoxifen acts as an antiestrogen and could combine with either receptor type to block the action of estrogen.

The results were different when the experiment was conducted using cell lines containing luciferase plasmid with AP1 (combinations 3 and 4 in the figure). If the estrogen and tamoxifen were added either separately or together to cells containing gene combination 3, the cells produced luciferase. This result demonstrated that ERα was activated and could combine with the AP1 control sequence, whether it was bound to estrogen or tamoxifen. Thus, in these cells the tamoxifen acted the same as an estrogen.

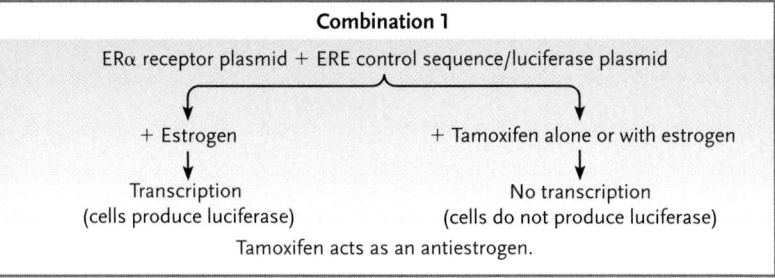

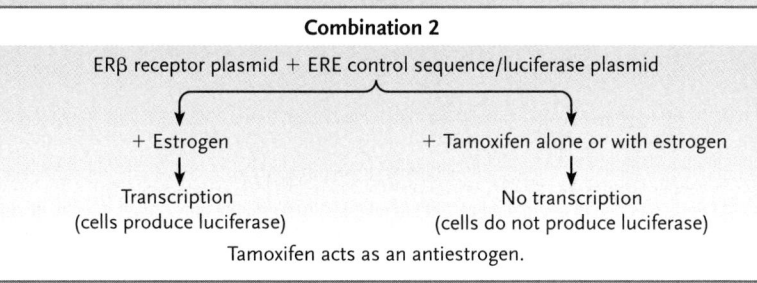

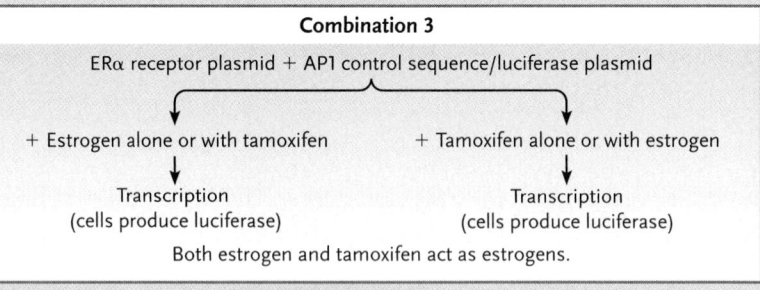

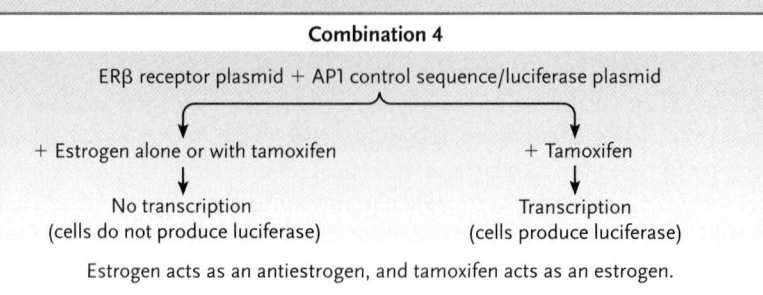

When estrogen was added, alone or with tamoxifen, to cells containing combination 4, no luciferase was produced. However, tamoxifen, if added alone, caused the cells to produce luciferase. These results indicate that ERβ combined with estrogen does not bind and activate AP1. ERβ combined with tamoxifen, however, can bind to AP1 and induce transcription of the lucifer-

ase gene. Evidently, estrogen actually acted as an antiestrogen in cells with combination 4—when added along with tamoxifen, the cells did not produce luciferase, indicating that the hormone blocked the action of tamoxifen.

The experiments indicate that the previously baffling and opposing effects of tamoxifen on breast and uterine tissues occur because different es-

trogen receptors are present, acting on genes controlled by either the ERE or AP1 control sequences. The results emphasize the fact that hormones can have distinct effects in different cell types depending on the types of receptors present. The research also opens the possibility of new cancer treatments that take advantage of the receptor differences.

bound by their receptors, hormones may produce a response that involves stimulation or inhibition of cellular processes through the specific types of internal molecules triggered by the hormone action. Third, because of the amplification that occurs in both the surface and internal receptor mechanisms, hormones are

effective in very small concentrations. Fourth, the response to a hormone differs among target organs.

In the next two sections we discuss the major endocrine cells and glands of vertebrates. The locations of these cells and glands in the human body and their functions are summarized in **Figure 40.6** and **Table 40.1**. Pep-

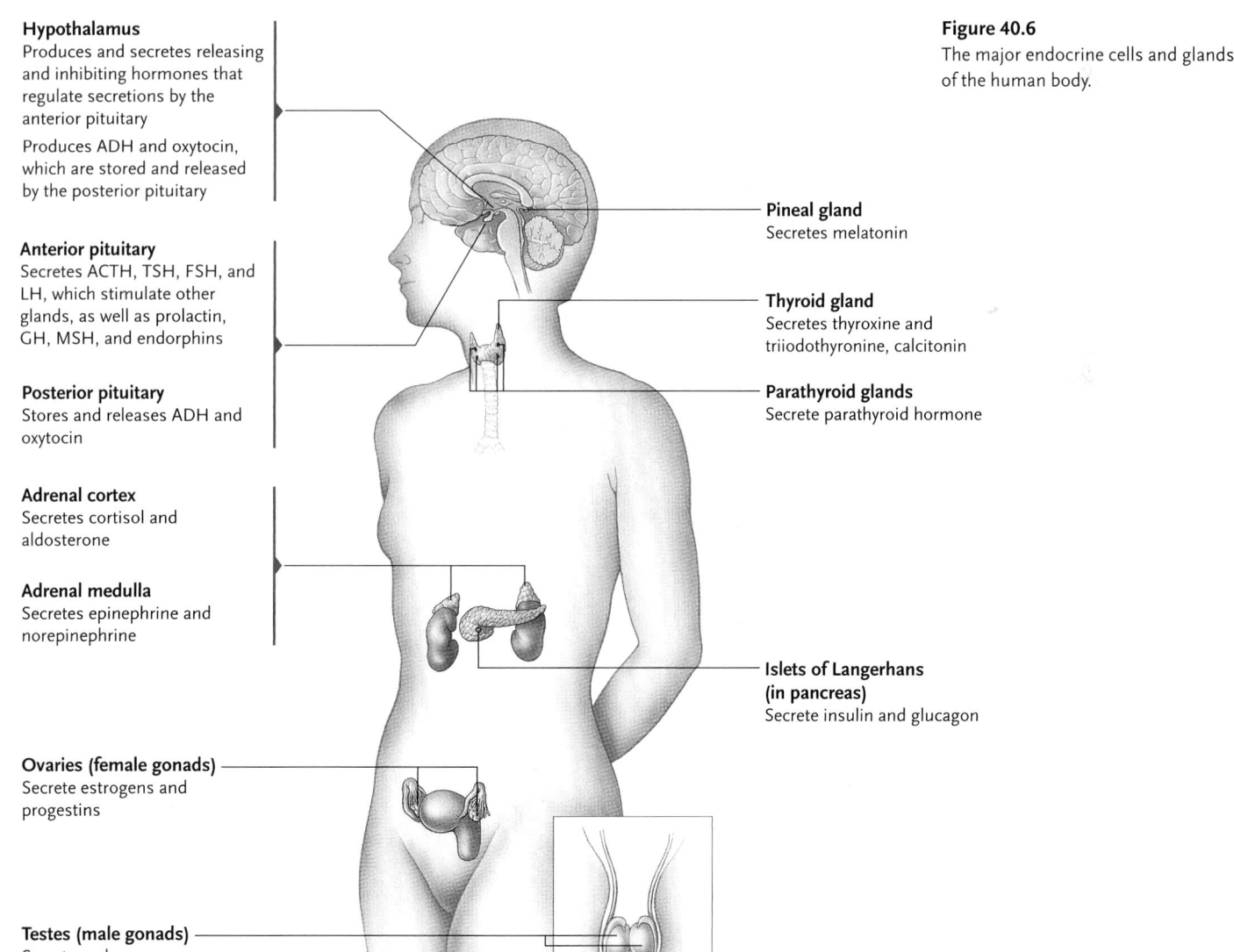

Hypothalamus
Produces and secretes releasing and inhibiting hormones that regulate secretions by the anterior pituitary
Produces ADH and oxytocin, which are stored and released by the posterior pituitary

Anterior pituitary
Secretes ACTH, TSH, FSH, and LH, which stimulate other glands, as well as prolactin, GH, MSH, and endorphins

Posterior pituitary
Stores and releases ADH and oxytocin

Adrenal cortex
Secretes cortisol and aldosterone

Adrenal medulla
Secretes epinephrine and norepinephrine

Ovaries (female gonads)
Secrete estrogens and progestins

Testes (male gonads)
Secrete androgens

Figure 40.6
The major endocrine cells and glands of the human body.

Pineal gland
Secretes melatonin

Thyroid gland
Secretes thyroxine and triiodothyronine, calcitonin

Parathyroid glands
Secrete parathyroid hormone

Islets of Langerhans (in pancreas)
Secrete insulin and glucagon

Table 40.1 | **The Major Human Endocrine Glands and Hormones**

Secretory Tissue or Gland	Hormones	Molecular Class	Target Tissue	Principal Actions
Hypothalamus	Releasing and inhibiting hormones	Peptide	Anterior pituitary	Regulate secretion of anterior pituitary hormones
Anterior pituitary	Thyroid-stimulating hormone (TSH)	Peptide	Thyroid gland	Stimulates secretion of thyroid hormones and growth of thyroid gland
	Adrenocorticotropic hormone (ACTH)	Peptide	Adrenal cortex	Stimulates secretion of glucocorticoids by adrenal cortex
	Follicle-stimulating hormone (FSH)	Peptide	Ovaries in females, testes in males	Stimulates egg growth and development and secretion of sex hormones in females; stimulates sperm production in males
	Luteinizing hormone (LH)	Peptide	Ovaries in females, testes in males	Regulates ovulation in females and secretion of sex hormones in males
	Prolactin (PRL)	Peptide	Mammary glands	Stimulates breast development and milk secretion
	Growth hormone (GH)	Peptide	Bone, soft tissue	Stimulates growth of bones and soft tissues; helps control metabolism of glucose and other fuel molecules
	Melanocyte-stimulating hormone (MSH)	Peptide	Melanocytes in skin of some vertebrates	Promotes darkening of the skin
	Endorphins	Peptide	Pain pathways of PNS	Inhibit perception of pain
Posterior pituitary	Antidiuretic hormone (ADH)	Peptide	Kidneys	Raises blood volume and pressure by increasing water reabsorption in kidneys
	Oxytocin	Peptide	Uterus, mammary glands	Promotes uterine contractions; stimulates milk ejection from breasts
Thyroid gland	Calcitonin	Peptide	Bone	Lowers calcium concentration in blood
	Thyroxine and triiodothyronine	Amine	Most cells	Increase metabolic rate; essential for normal body growth
Parathyroid glands	Parathyroid hormone (PTH)	Peptide	Bone, kidneys, intestine	Raises calcium concentration in blood; stimulates vitamin D activation
Adrenal medulla	Epinephrine and norepinephrine	Amine	Sympathetic receptor sites throughout body	Reinforce sympathetic nervous system; contribute to responses to stress
Adrenal cortex	Aldosterone (mineralocorticoid)	Steroid	Kidney tubules	Helps control body's salt-water balance by increasing Na^+ reabsorption and K^+ excretion in kidneys
	Cortisol (glucocorticoid)	Steroid	Most body cells, particularly muscle, liver, and adipose cells	Increases blood glucose by promoting breakdown of proteins and fats
Testes	Androgens, such as testosterone[*]	Steroid	Various tissues	Control male reproductive system development and maintenance; most androgens are made by the testes
	Oxytocin	Peptide	Uterus	Promotes uterine contractions when seminal fluid ejaculated into vagina during sexual intercourse
Ovaries	Estrogens, such as estradiol[**]	Steroid	Breast, uterus, other tissues	Stimulate maturation of sex organs at puberty, and development of secondary sexual characteristics
	Progestins, such as progesterone[**]	Steroid	Uterus	Prepare and maintain uterus for implantation of fertilized egg and the growth and development of embryo

[*]Small amounts secreted by ovaries and adrenal cortex.
[**]Small amounts secreted by testes.

Table 40.1 | The Major Human Endocrine Glands and Hormones (Continued)

Secretory Tissue or Gland	Hormones	Molecular Class	Target Tissue	Principal Actions
Pancreas (islets of Langerhans)	Glucagon (alpha cells)	Peptide	Liver cells	Raises glucose concentration in blood; promotes release of glucose from glycogen stores and production from noncarbohydrates
	Insulin (beta cells)	Peptide	Most cells	Lowers glucose concentration in blood; promotes storage of glucose, fatty acids, and amino acids
Pineal gland	Melatonin	Amine	Brain, anterior pituitary, reproductive organs, immune system, possibly others	Helps synchronize body's biological clock with day length; may inhibit gonadotropins and initiation of puberty
Many cell types	Growth factors	Peptide	Most cells	Regulate cell division and differentiation
	Prostaglandins	Fatty acid	Various tissues	Have many diverse roles

tide hormones secreted by other body regions, including the stomach and small intestine, the thymus gland, the kidneys, and the heart will be described in the chapters in which these tissues and organs are discussed.

STUDY BREAK

1. Compare and contrast the mechanisms by which glucagon and aldosterone cause their specific responses.
2. Explain how one type of target cell could respond to different hormones, and how the same hormone could produce different effects in different cells.

40.3 The Hypothalamus and Pituitary

The hormones of vertebrates work in coordination with the nervous system. The action of several hormones is closely coordinated by the hypothalamus and its accessory gland, the pituitary.

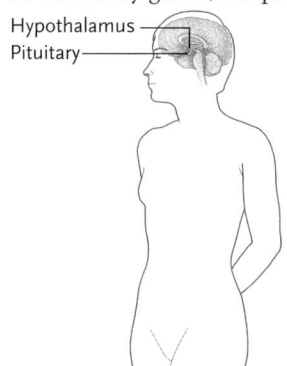

Hypothalamus
Pituitary

The hypothalamus is a region of the brain located in the floor of the cerebrum (see Section 38.3). The **pituitary gland,** consisting mostly of two fused lobes, is suspended just below it by a slender stalk of tissue that contains both neurons and blood vessels **(Figure 40.7).** The **posterior pituitary** contains axons and nerve endings of neurosecretory neurons that originate in the hypothalamus. The **anterior pituitary** contains nonneuronal endocrine cells that form a distinct gland. The two lobes are separate in structure and embryonic origins.

Under Regulatory Control by the Hypothalamus, the Anterior Pituitary Secretes Eight Hormones

The secretion of hormones from the anterior pituitary is controlled by peptide neurohormones called **releasing hormones (RHs)** and **inhibiting hormones (IHs),** which are released by the hypothalamus. These neurohormones are carried in the blood from the hypothalamus to the anterior pituitary in a *portal vein,* a special vein that connects the capillaries of the two glands. The portal vein provides a critical link between the brain and the endocrine system, ensuring that most of the blood reaching the anterior pituitary first passes through the hypothalamus.

RHs and IHs are **tropic hormones** (*tropic* = stimulating, not to be confused with *trophic,* which means "nourishing"), hormones that regulate hormone secretion by another endocrine gland. RHs and IHs regulate the anterior pituitary's secretion of another group of hormones; those hormones in turn control many other endocrine glands of the body, and also control some body processes directly.

Secretion of hypothalamic RHs is controlled by neurons containing receptors that monitor the blood to detect changes in body chemistry and temperature. For example, TRH, discussed earlier, is secreted in response to a drop in body temperature. Input to the hypothalamus also comes through numerous connections from control centers elsewhere in the brain, including the brain stem and limbic system. Negative feedback pathways regulate secretion of the releasing hormones, such as the pathway regulating TRH secretion.

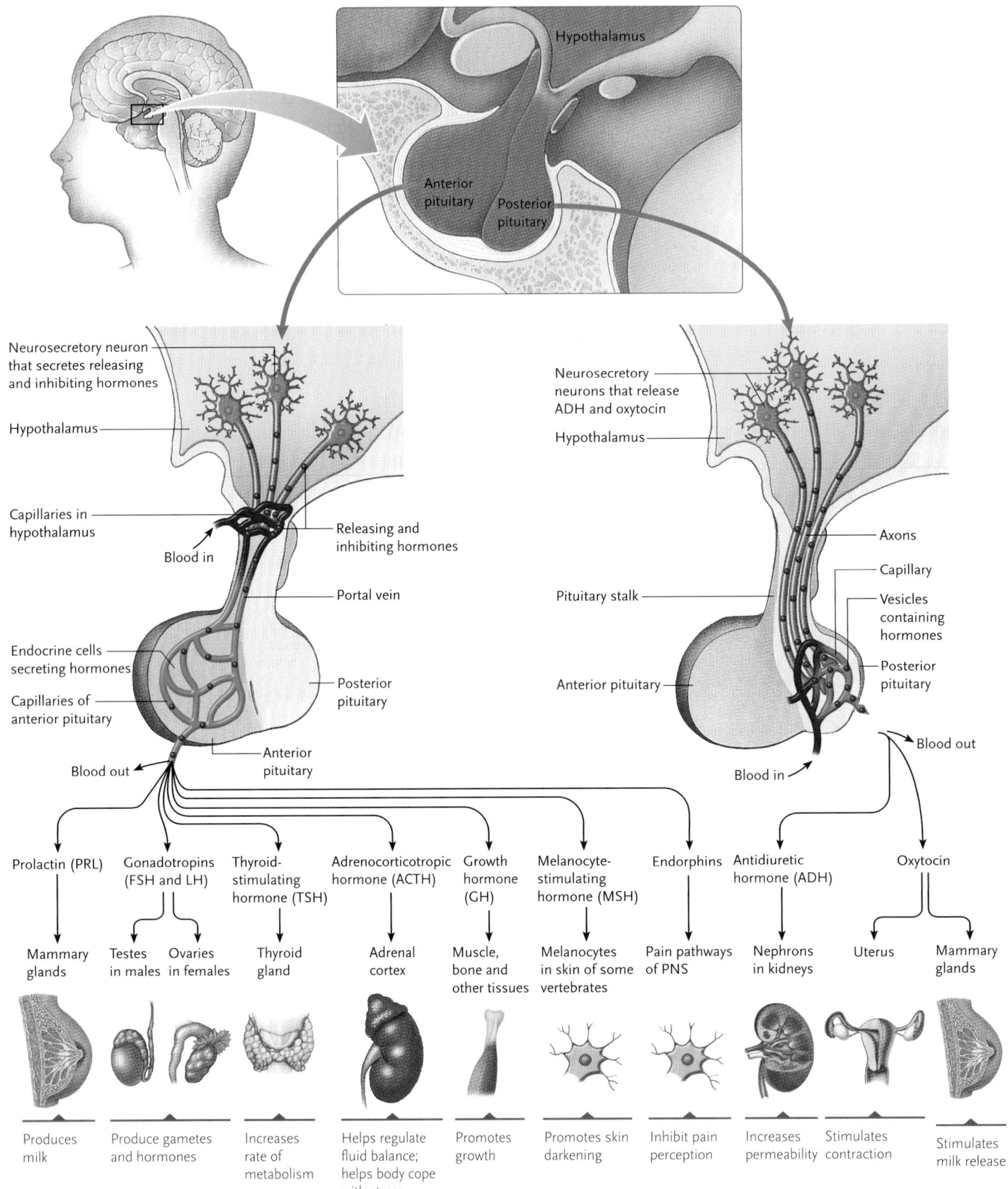

Figure 40.7

The hypothalamus and pituitary. Hormones secreted by the anterior and posterior pituitary are controlled by neurohormones released in the hypothalamus.

Under the control of the hypothalamic RHs, the anterior pituitary secretes six major hormones into the bloodstream: prolactin, growth hormone, thyroid-stimulating hormone, adrenocorticotropic hormone, follicle-stimulating hormone, and luteinizing hormone, and two other hormones, melanocyte-stimulating hormone (MSH) and endorphins. **Prolactin (PRL),** a *nontropic hormone* (a hormone that does not regulate hormone secretion by another endocrine gland), influences reproductive activities and parental care in vertebrates. In mammals, PRL stimulates development of the secretory cells of mammary glands during late pregnancy, and stimulates milk synthesis after a female mammal gives birth. Stimulation of the mammary glands and the nipples, as occurs during suckling, leads to PRL release.

Growth hormone (GH) stimulates cell division, protein synthesis, and bone growth in children and adolescents, thereby causing body growth. GH also stimulates protein synthesis and cell division in adults. For these actions, GH acts as a tropic hormone by binding to target tissues, mostly liver cells, causing them to release insulin-like growth factor **(IGF),** a peptide that directly stimulates growth processes. GH also acts as a nontropic hormone to control a number of major metabolic processes in mammals of all ages, including the conversion of glycogen to glucose and fats to fatty acids as a means of regulating their levels in the blood. In addition, GH stimulates body cells to take up fatty acids and amino acids and limits the rate at which muscle cells take up glucose. These actions help maintain the availability of glucose and fatty acids to tissues and organs between feedings; this is particularly important for the brain. In humans, deficiencies in GH secretion during childhood produce *pituitary dwarfs,* who remain small in stature **(Figure 40.8).** Overproduction of GH during childhood or adolescence, often due to a tumor of the anterior pituitary, produces *pituitary giants,* who may grow above seven feet in height.

The other four major hormones secreted by the anterior pituitary are tropic hormones that control endocrine glands elsewhere in the body. **Thyroid-stimulating hormone (TSH)** stimulates the thyroid gland to grow in size and secrete thyroid hormones. **Adrenocorticotropic hormone (ACTH)** triggers hormone secretion by cells in the adrenal cortex. **Follicle-stimulating hormone (FSH)** controls egg development and the secretion of sex hormones in female mammals, and sperm production in males. **Luteinizing hormone (LH)** regulates part of the menstrual cycle in human females and the secretion of sex hormones in males. FSH and LH are grouped together as **gonadotropins** because they regulate the activity of the gonads (ovaries and testes). The roles of the gonadotropins and sex hormones in the reproductive cycle are described in Chapter 47.

Melanocyte-stimulating hormone (MSH) and **endorphins** are nontropic hormones produced by the anterior pituitary. MSH is named because of its effect in

Figure 40.8
The results of overproduction and underproduction of growth hormone by the anterior pituitary. The man on the left is of normal height. The man in the center is a pituitary giant, whose pituitary produced excess GH during childhood and adolescence. The man on the right is a pituitary dwarf, whose pituitary produced too little GH.

some vertebrates on melanocytes, skin cells that contain the black pigment melanin. For example, an increase in secretion of MSH produces a marked darkening of the skin of fishes, amphibians, and reptiles. The darkening is produced by a redistribution of melanin from the centers of the melanocytes throughout the cells. In humans, an increase in MSH secretion also causes skin darkening, although the effect is by no means as obvious as in the other vertebrates mentioned. For example, MSH secretion increases in pregnant women. That, with the effects of increased estrogens, results in increased skin pigmentation; the effects fade after birth of the child.

Endorphins, nontropic peptide hormones produced by the hypothalamus and pituitary, are also released by the anterior pituitary. In the peripheral nervous system (PNS), endorphins act as neurotransmitters in pathways that control pain, thereby inhibiting the perception of pain. Hence, endorphins are often called "natural painkillers."

The Posterior Pituitary Secretes Two Hormones into the Circulatory System

The neurosecretory neurons in the posterior pituitary secrete two nontropic peptide hormones, antidiuretic hormone and oxytocin, directly into the circulatory system (see Figure 40.7).

Antidiuretic hormone (ADH) stimulates kidney cells to absorb more water from urine, thereby increasing the volume of the blood. The hormone is released when sensory receptor cells of the hypothalamus detect an increase in the blood's Na^+ concentration during periods of body dehydration or after a salty meal. Ethyl alcohol and caffeine inhibit ADH secretion, explaining in part why alcoholic drinks and coffee increase the volume of urine excreted. Nicotine and emotional stress, in contrast, stimulate ADH secretion and water retention. After severe stress is relieved, the return to normal ADH secretion often makes a trip to the bathroom among our most pressing needs. The hypothalamus also releases a flood of ADH when an injury results in heavy blood loss or some other event triggers a severe drop in blood pressure. ADH helps maintain blood pressure by reducing water loss and also by causing small blood vessels in some tissues to constrict.

Hormones with structure and action similar to ADH are also secreted in fishes, amphibians, reptiles, and birds. In amphibians, these ADH-like hormones increase the amount of water entering the body through the skin and from the urinary bladder.

We have noted that **oxytocin** stimulates the ejection of milk from the mammary glands of a nursing mother. Stimulation of the nipples in suckling sends neuronal signals to the hypothalamus, and leads to release of oxytocin from the posterior pituitary. The released oxytocin stimulates more oxytocin secretion by a positive feedback mechanism. Oxytocin causes the smooth muscle cells surrounding the mammary glands to contract, forcibly expelling the milk through the nipples. The entire cycle, from the onset of suckling to milk ejection, takes less than a minute in mammals. Oxytocin also plays a key role in childbirth, as we discussed in Section 36.4.

In males, oxytocin is secreted into the seminal fluid by the testes. Like prostaglandins, when the seminal fluid is ejaculated into the vagina during sexual intercourse, oxytocin stimulates contractions of the uterus that aid movement of sperm through the female reproductive tract.

STUDY BREAK

1. Summarize the functional interactions between the hypothalamus and the anterior pituitary gland.
2. Distinguish between how tropic hormones and nontropic hormones produce responses.

40.4 Other Major Endocrine Glands of Vertebrates

Besides the hypothalamus and pituitary, the body has seven major endocrine glands or tissues, many of them regulated by the hypothalamus-pituitary connection. These glands are the thyroid gland, parathyroid glands,

adrenal medulla, adrenal cortex, gonads, pancreas, and pineal gland (shown in Figure 40.6 and summarized in Table 40.1).

The Thyroid Hormones Stimulate Metabolism, Development, and Maturation

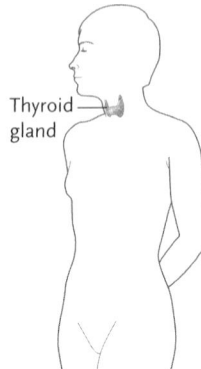

Thyroid gland

The **thyroid gland**, which is located in the front of the throat in humans, has a shape similar to that of a bowtie. It secretes the same hormones in all vertebrates. The primary thyroid hormone, **thyroxine**, is known as T_4 because it contains four iodine atoms. The thyroid also secretes smaller amounts of a closely related hormone, **triiodothyronine** or T_3, which contains three iodine atoms. A supply of iodine in the diet is necessary for production of these hormones. Normally, their concentrations are kept at finely balanced levels in the blood by negative feedback loops such as that described in Figure 40.2.

Both T_4 and T_3 enter cells; however, once inside, most of the T_4 is converted to T_3, the form that combines with internal receptors. Binding of T_3 to receptors alters gene expression, which brings about the hormone's effects.

The thyroid hormones are vital to growth, development, maturation, and metabolism in all vertebrates. They interact with GH for their effects on growth and development. Thyroid hormones also increase the sensitivity of many body cells to the effects of epinephrine and norepinephrine, hormones released by the adrenal medulla as part of the "fight or flight response" (discussed further later).

In amphibians such as frogs, thyroid hormones trigger **metamorphosis,** or change in body form from tadpole to adult **(Figure 40.9).** Thyroid hormones also contribute to seasonal changes in the plumage of birds and coat color in mammals.

In human adults, low thyroid output, *hypothyroidism,* causes affected individuals to be sluggish mentally and physically; they have a slow heart rate and weak pulse, and often feel confused and depressed. Hypothyroidism in infants and children leads to cretinism, that is, stunted growth and diminished intelligence. Overproduction of thyroid hormones in human adults, *hyperthyroidism,* produces nervousness and emotional instability, irritability, insomnia, weight loss, and a rapid, often irregular heartbeat. The most common form of hyperthyroidism is *Graves' disease,* characterized by inflamed, protruding eyes in addition to the other symptoms mentioned.

Insufficient iodine in the diet can cause *goiter,* enlargement of the thyroid. Without iodine, the thyroid cannot make T_3 and T_4 in response to stimulation by

TSH. Because the thyroid hormone concentration remains low in the blood, TSH continues to be secreted, and the thyroid grows in size. Dietary iodine deficiency has been eliminated in developed regions of the world by the addition of iodine to table salt.

In mammals, the thyroid also has specialized cells that secrete **calcitonin**, a nontropic peptide hormone. The hormone lowers the level of Ca^{2+} in the blood by inhibiting the ongoing dissolution of calcium from bone. Calcitonin secretion is stimulated when Ca^{2+} levels in blood rise above the normal range and inhibited when Ca^{2+} levels fall below the normal range.

The Parathyroid Glands Regulate Ca^{2+} Level in the Blood

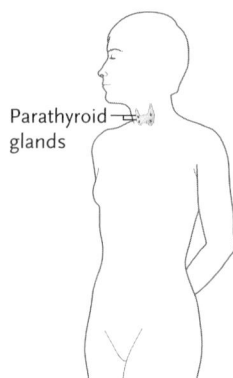

The **parathyroid glands** occur only in tetrapod vertebrates—amphibians, reptiles, birds, and mammals. Each is a spherical structure about the size of a pea. Mammals have four parathyroids located on the posterior surface of the thyroid gland, two on each side. The single hormone they produce, a nontropic hormone called **parathyroid hormone (PTH)**, is secreted in response to a fall in blood Ca^{2+} levels. PTH stimulates bone cells to dissolve the mineral matter of bone tissues, releasing both calcium and phosphate ions into the blood. The released Ca^{2+} is available for enzyme activation, conduction of nerve signals across synapses, muscle contraction, blood clotting, and other uses. How blood Ca^{2+} levels control PTH and calcitonin secretion is shown in **Figure 40.10.**

PTH also stimulates enzymes in the kidneys that convert **vitamin D**, a steroidlike molecule, into its fully active form in the body. The activated vitamin D increases the absorption of Ca^{2+} and phosphates from ingested food by promoting the synthesis of a calcium-binding protein in the intestine; it also increases the release of Ca^{2+} from bone in response to PTH.

PTH underproduction causes Ca^{2+} concentration to fall steadily in the blood, disturbing nerve and muscle function—the muscles twitch and contract uncontrollably, and convulsions and cramps occur. Without treatment, the condition is usually fatal, because the severe muscular contractions interfere with breathing. Overproduction of PTH results in the loss of so much calcium from the bones that they become thin and fragile. At the same time, the elevated Ca^{2+} concentration in the blood causes calcium deposits to form in soft tissues, especially in the lungs, arteries, and kidneys (where the deposits form kidney stones).

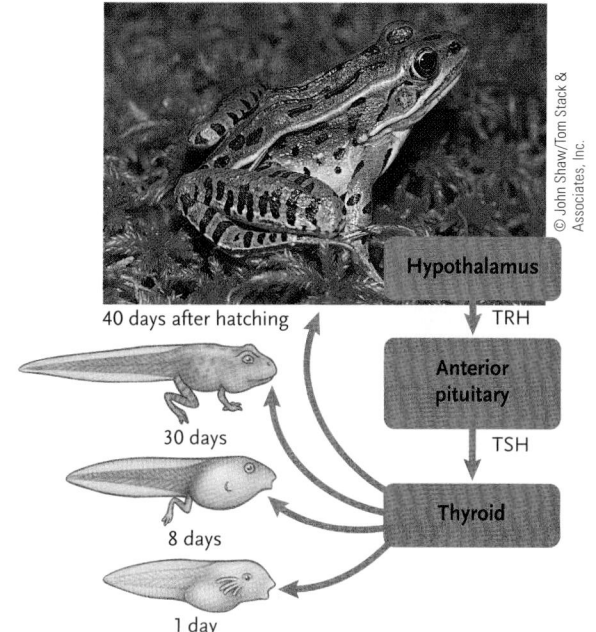

Figure 40.9
Metamorphosis of a tadpole into an adult frog, under the control of thyroid hormones. As a part of the metamorphosis, changes in gene activity lead to a change from an aquatic to a terrestrial habitat. TRH, thyroid-releasing hormone; TSH, thyroid-stimulating hormone.

The Adrenal Medulla Releases Two "Fight or Flight" Hormones

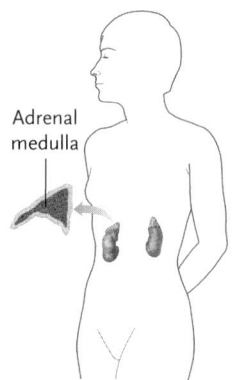

The adrenal glands (*ad* = next to, *renes* = kidneys) of mammals consist of two distinct regions. The central region, the **adrenal medulla**, contains neurosecretory neurons; the tissue surrounding it, the **adrenal cortex**, contains endocrine cells. The two regions secrete hormones with entirely different functions. Nonmammalian vertebrates have glands equivalent to the adrenal medulla and adrenal cortex of mammals, but they are separate. Most of the hormones produced by these glands have essentially the same functions in all vertebrates. The only major exception is aldosterone, which is secreted by the adrenal cortex or its equivalent only in tetrapod vertebrates.

In most species, the adrenal medulla secretes two nontropic amine hormones, **epinephrine** and **norepinephrine**, which are **catecholamines**, chemical compounds derived from the amino acid tyrosine that circulate in the bloodstream. They bind to receptors in the plasma membranes of their target cells. (Epinephrine is also secreted by some cells of the CNS, and norepinephrine is also secreted by some cells of the CNS and neurons of the sympathetic nervous system. In these cases, epinephrine and norepinephrine

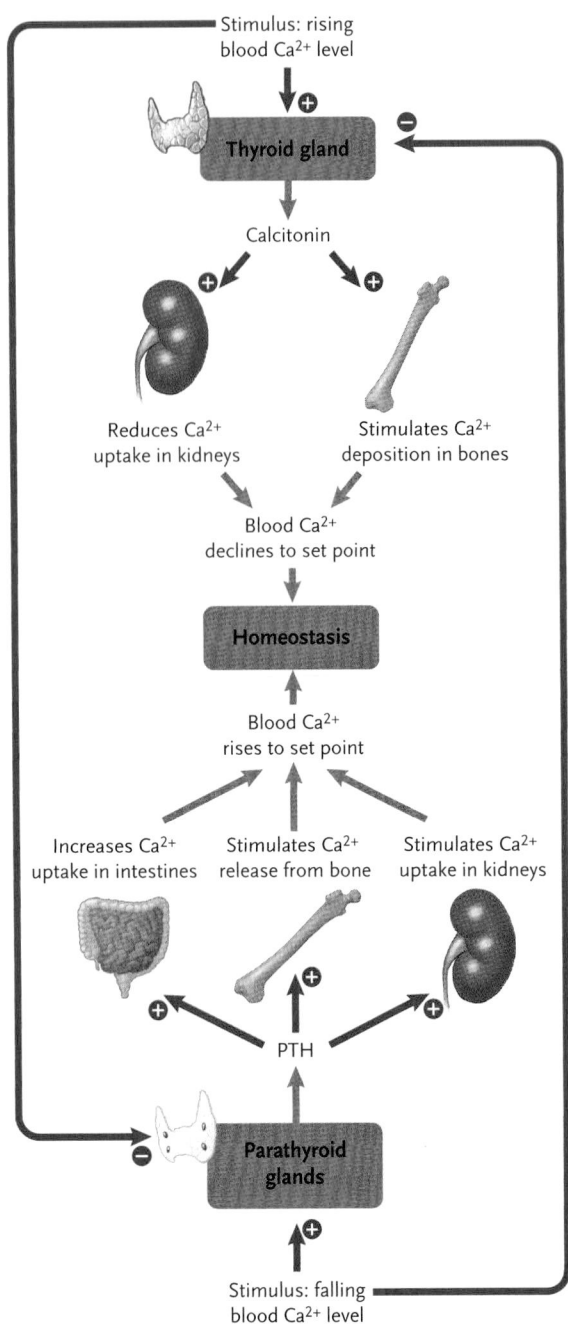

Stimulus: rising blood Ca²⁺ level

Thyroid gland

Calcitonin

Reduces Ca²⁺ uptake in kidneys

Stimulates Ca²⁺ deposition in bones

Blood Ca²⁺ declines to set point

Homeostasis

Blood Ca²⁺ rises to set point

Increases Ca²⁺ uptake in intestines

Stimulates Ca²⁺ release from bone

Stimulates Ca²⁺ uptake in kidneys

PTH

Parathyroid glands

Stimulus: falling blood Ca²⁺ level

Figure 40.10
Negative feedback control of PTH and calcitonin secretion by blood Ca²⁺ levels.

function as neurotransmitters between interneurons involved in a diversity of brain and body functions; see Section 37.3.)

Epinephrine and norepinephrine, which reinforce the action of the sympathetic nervous system, are secreted when the body encounters stresses such as emotional excitement, danger (fight-or-flight situations), anger, fear, infections, injury, even midterm and final exams. Epinephrine in particular prepares the body for handling stress or physical activity. The heart rate increases. Glycogen and fats break down, releasing glucose and fatty acids into the blood as fuel molecules. In the heart, skeletal muscles, and lungs, the blood vessels dilate to increase blood flow. Elsewhere in the body, the blood vessels constrict, raising blood pressure, reducing blood flow to the intestine and kidneys, and inhibiting smooth muscle contractions, which reduces water loss and slows down the digestive system. Airways in the lungs also dilate, helping to increase the flow of air.

The effects of norepinephrine on heart rate, blood pressure, and blood flow to the heart muscle are similar to those of epinephrine. However, in contrast to epinephrine, norepinephrine causes blood vessels in skeletal muscles to constrict. This antagonistic effect is largely canceled out because epinephrine is secreted in much greater quantities.

No known human diseases are caused by underproduction of the hormones of the adrenal medulla, as long as the sympathetic nervous system is intact. Overproduction of epinephrine and norepinephrine, which can occur if there is a tumor in the adrenal medulla, leads to symptoms duplicating a stress response.

The Adrenal Cortex Secretes Two Groups of Steroid Hormones That Are Essential for Survival

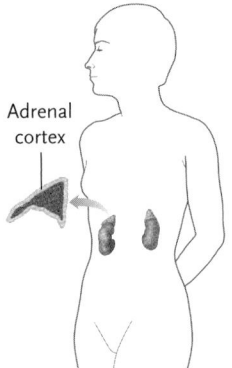

Adrenal cortex

The adrenal cortex of mammals secretes two major types of steroid hormones: **glucocorticoids** help maintain the blood concentration of glucose and other fuel molecules, and **mineralocorticoids** regulate the levels of Na⁺ and K⁺ ions in the blood and extracellular fluid.

The Glucocorticoids. The glucocorticoids help maintain glucose levels in the blood by three major mechanisms: (1) stimulating the synthesis of glucose from noncarbohydrate sources such as fats and proteins, (2) reducing glucose uptake by body cells except those in the central nervous system, and (3) promoting the breakdown of fats and proteins, which releases fatty acids and amino acids into the blood as alternative fuels when glucose supplies are low. The absence of down-regulation of glucose uptake to the CNS keeps the brain well supplied with glucose between meals and during periods of extended fasting. **Cortisol** is the major glucocorticoid secreted by the adrenal cortex.

Secretion of glucocorticoids is ultimately under control of the hypothalamus **(Figure 40.11)**. Low glucose concentrations in the blood, or elevated levels of epinephrine secreted by the adrenal medulla in response to stress, are detected in the hypothalamus, leading to secretion of the tropic hormone ACTH by the anterior pituitary. ACTH promotes the secretion of glucocorticoids by the adrenal cortex.

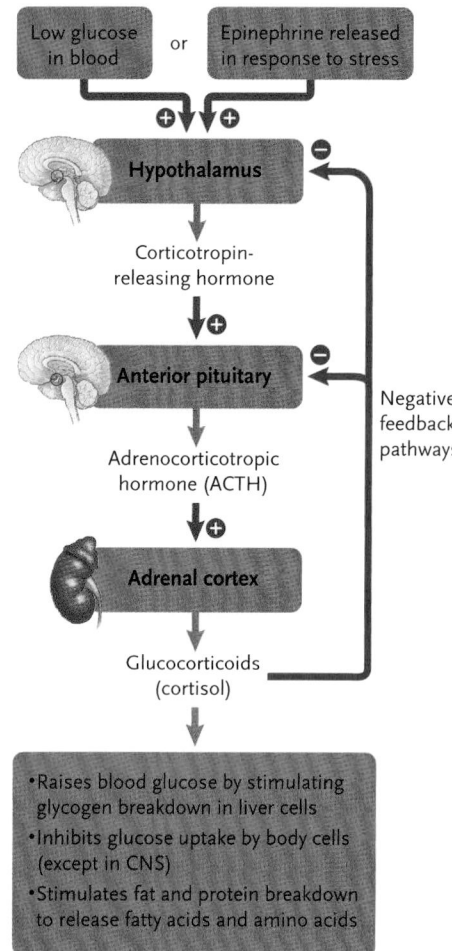

Figure 40.11
Pathways linking secretion of glucocorticoids to low blood sugar and epinephrine secretion in response to stress.

Overproduction of glucocorticoids makes blood glucose rise and increases fat deposition in adipose tissue and protein breakdown in muscles and bones. The loss of proteins from muscles causes weakness and fatigue; loss of proteins from bone, particularly collagens, makes the bones fragile and susceptible to breakage. Underproduction of glucocorticoids causes blood glucose concentration to fall below normal levels in the blood and diminishes tolerance to stress.

Glucocorticoids have anti-inflammatory properties and, consequently, they are used clinically to treat conditions such as arthritis or dermatitis. They also suppress the immune system and are used in the treatment of autoimmune diseases such as rheumatoid arthritis.

The Mineralocorticoids. In tetrapods, the mineralocorticoids, primarily **aldosterone**, increase the amount of Na^+ reabsorbed from the fluids processed by the kidneys and absorbed from foods in the intestine. They also reduce the amount of Na^+ secreted by salivary and sweat glands and increase the rate of K^+ excretion by the kidneys. The net effect is to keep Na^+ and K^+ bal-

anced at the levels required for normal cellular functions, including those of the nervous system. Relatedly, secretion of aldosterone is tightly linked to blood volume and indirectly to blood pressure (see Section 40.2 and Chapter 46).

Moderate overproduction of aldosterone causes excessive water retention in the body, so that tissues swell and blood pressure rises. Conversely, moderate underproduction can lead to excessive water loss and dehydration. Severe underproduction is rapidly fatal unless mineralocorticoids are supplied by injection or other means.

The adrenal cortex also secretes small amounts of androgens, steroid sex hormones responsible for maintenance of male characteristics, which are synthesized primarily by the gonads. These hormones have significant effects only if they are overproduced, as can occur with some tumors in the adrenal cortex. The result is altered development of primary or secondary sex characteristics.

The Gonadal Sex Hormones Regulate the Development of Reproductive Systems, Sexual Characteristics, and Mating Behavior

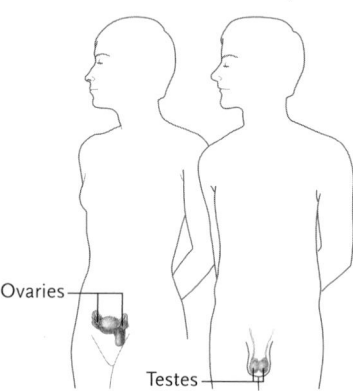

The **gonads**, the testes and ovaries, are the primary source of sex hormones in vertebrates. The steroid hormones they produce, the **androgens**, **estrogens**, and **progestins**, have similar functions in regulating the development of male and female reproductive systems, sexual characteristics, and mating behavior. Both males and females produce all three types of hormones, but in different proportions. Androgen production is predominant in males, while estrogen and progestin production is predominant in females. An outline of the actions of these hormones is presented here, and a more complete picture is given in Chapter 47.

The **testes** of male vertebrates secrete androgens, steroid hormones that stimulate and control the development and maintenance of male reproductive systems. The principal androgen is **testosterone**, the male sex hormone. In young adult males, a jump in testosterone levels stimulates puberty and the development of secondary sexual characteristics, including the growth of facial and body hair, muscle development,

Basic Research: Neuroendocrine and Behavioral Effects of Anabolic–Androgenic Steroids in Humans

Anabolic–androgenic steroids (AAS) are synthetic derivatives of the natural steroid hormone testosterone. They were designed to have potent anabolic (tissue building) activity and low androgenic (masculinizing) activity in therapeutic doses. Overall, there are about 60 AAS that vary in chemical structure and, therefore, in their physiological effects.

AAS are used for treating conditions such as delayed puberty and subnormal growth in children, as well as for therapy in chronic conditions such as cancer, AIDS, severe burns, liver and kidney failure, and anemias. AAS are not used exclusively for medical purposes, however. Because of their anabolic effects, which include an increase in muscle mass, strength, and endurance, as well as acceleration of recovery from injuries, AAS are used by athletes such as bodybuilders, weight lifters, baseball players, and football players. This use is actually abuse, because the doses typically administered are far higher than therapeutic doses. AAS abuse is significant: in the early 1990s, about one million Americans had used or were using AAS to increase strength, muscle mass, or athletic ability. While originally limited to elite athletes, use has

trickled down to average athletes, including adolescents. It is estimated that perhaps 4% of high school students have used AAS. The greatest increase in AAS abuse over the past decade has been by adolescent girls.

Are AAS harmful at high doses? When researchers gave rodents doses of AAS comparable to those associated with human AAS abuse, they observed significant increases in aggression, anxiety, and sexual behaviors. These changes occur as a result of alterations in the neurotransmitters and other signaling molecules associated with those behaviors. All of these changes have been hypothesized to occur in human AAS abusers.

To study the effect of high doses of AAS on the human endocrine system, R. C. Daly and colleagues at the National Institute of Mental Health, in Bethesda, Maryland, administered the AAS methyltestosterone (MT) to normal (medication-free) human volunteers over a period of time in an inpatient clinic. The subjects were examined for the effects of MT on pituitary–gonadal, pituitary–thyroid, and pituitary–adrenal hormones, and the researchers attempted to correlate endocrine changes with psychological symptoms caused by the MT.

The researchers found, for instance, that high doses of MT caused a significant decrease in the levels of gonadotropins and gonadal steroid hormones in the blood. At the same time, thyroxine and TSH levels increased. No significant increases were seen in pituitary–adrenal hormones.

The decrease in testosterone levels correlated significantly with cognitive problems, such as increased distractibility and forgetfulness. The increase in thyroxine correlated significantly with a rise in aggressive behavior, notably anger, irritability, and violent feelings. There were no changes in activities associated with pituitary–adrenal hormones—energy, disturbed sleep, and sexual arousal—as was expected by the lack of change in those hormones.

In sum, behavioral changes associated with high doses of an AAS suggest that AAS-induced hormonal changes may well contribute to the adverse behavioral and mood changes that occur during AAS abuse. Clearly, there is every reason to believe that taking high doses of AAS for athletic gain alters the normal hormonal balance in humans, as it does in rodents.

---

changes in vocal cord morphology, and development of normal sex drive. The synthesis and secretion of testosterone by cells in the testes is controlled by the release of luteinizing hormone (LH) from the anterior pituitary, which in turn is controlled by **gonadotropin releasing hormone (GnRH)**, a tropic hormone secreted by the hypothalamus.

Androgens are natural types of **anabolic steroids**, hormones that stimulate muscle development. Natural and synthetic anabolic steroids have been in the news over the years because of their use by bodybuilders and other athletes from sports in which muscular strength is important. *Focus on Research* discusses the potential adverse effects of anabolic–androgenic steroids, synthetic derivatives of testosterone, in humans.

The **ovaries** of females produce estrogens, steroid hormones that stimulate and control the development and maintenance of female reproductive systems. The principal estrogen is **estradiol**, which stimulates maturation of sex organs at puberty and the development of secondary sexual characteristics. Ovaries also produce progestins, principally **progesterone**, the steroid hormone that prepares and maintains the uterus for implantation of a fertilized egg and the subsequent growth and development of an embryo. The synthesis and secretion of progesterone by cells in the ovaries is controlled by the release of follicle-stimulating hormone (FSH) from the anterior pituitary, which in turn is controlled by the same GnRH as in males.

The Pancreatic Islet of Langerhans Hormones Regulate Glucose Metabolism

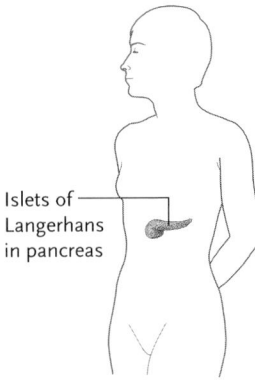

Islets of Langerhans in pancreas

Most of the **pancreas,** a relatively large gland located just behind the stomach, forms an exocrine gland that secretes digestive enzymes into the small intestine (see Chapter 45). However, about 2% of the cells in the pancreas are endocrine cells that form the **islets of Langerhans.** Found in all vertebrates, the islets secrete the peptide hormones insulin and glucagon into the bloodstream.

Insulin and glucagon regulate the metabolism of fuel substances in the body. **Insulin,** secreted by *beta cells* in the islets, acts mainly on cells of nonworking skeletal muscles, liver cells, and adipose tissue (fat). (Brain cells do not require insulin for glucose uptake.) Insulin lowers blood glucose, fatty acid, and amino acid levels and promotes their storage. That is, the actions of insulin include stimulation of glucose transport into cells, glycogen synthesis from glucose, uptake of fatty acids by adipose tissue cells, fat synthesis from fatty acids, and protein synthesis from amino acids. Insulin also inhibits glycogen degradation to glucose, fat degradation to fatty acids, and protein degradation to amino acids.

Glucagon, secreted by *alpha cells* in the islets, has effects opposite to those of insulin: it stimulates glycogen, fat, and protein degradation. Glucagon also uses amino acids and other noncarbohydrates as the input for glucose synthesis; this aspect of glucagon function operates during fasting. Negative feedback mechanisms that are keyed to the concentration of glucose in the blood control secretion of both insulin and glucagon to maintain glucose homeostasis **(Figure 40.12).**

Diabetes mellitus, a disease that afflicts more than 14 million people in the United States, results from problems with insulin production or action. The three classic diabetes symptoms are frequent urination, increased thirst (and consequently increased fluid intake), and increased appetite. Frequent urination occurs because without insulin, body cells are not stimulated to take up glucose, leading to abnormally high glucose concentration in the blood; excretion of the excess glucose in the urine requires water to carry it, which causes increased fluid loss and frequent trips to the bathroom. The need to replace the excreted water causes increased thirst. Increased appetite comes about because cells have low glucose levels and, therefore, proteins and fats are broken down as energy sources. Food intake is necessary to offset the negative energy balance or else weight loss will occur. Two of these classic symptoms gave the disease its name: *diabetes* is derived from a Greek word meaning "siphon," referring to the frequent urination, and *mellitus,* a Latin word meaning "sweetened with honey," refers to the sweet taste of a diabetic's urine. (Before modern blood or urine tests were developed, physicians tasted a patient's urine to detect the disease.)

The disease occurs in two major forms called *type 1* and *type 2.* Type 1 diabetes (insulin-dependent diabetes), which occurs in about 10% of diabetics, results from insufficient insulin secretion by the pancreas. This type of diabetes is usually caused by an autoimmune reaction in which an antibody destroys pancreatic beta cells. To survive, type 1 diabetics must receive regular insulin injections (typically, a genetically engineered human insulin called Humulin); careful dieting and exercise also have beneficial effects, because active skeletal muscles do not require insulin to take up and utilize glucose.

In type 2 diabetes (non-insulin-dependent diabetes), insulin is usually secreted at or above normal levels, but target cells have altered receptors that make

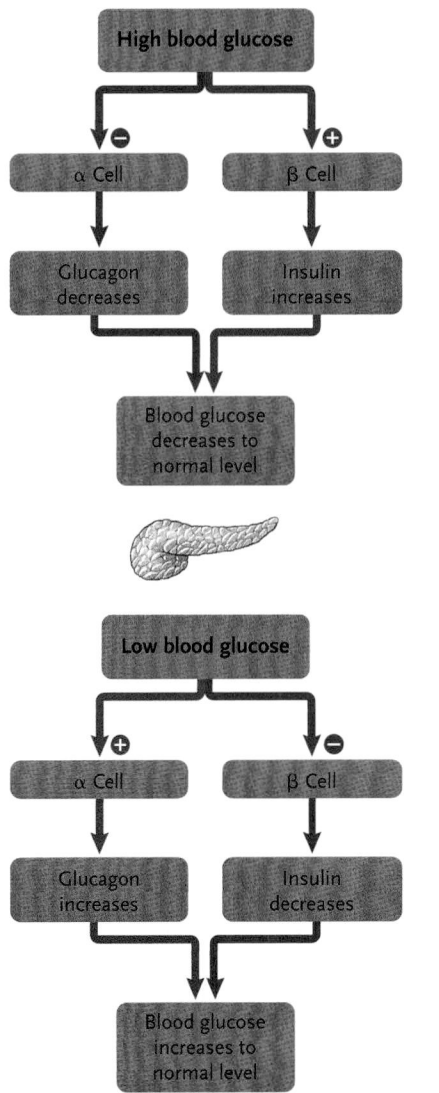

Figure 40.12
The action of insulin and glucagon in maintaining the concentration of blood glucose at an optimal level.

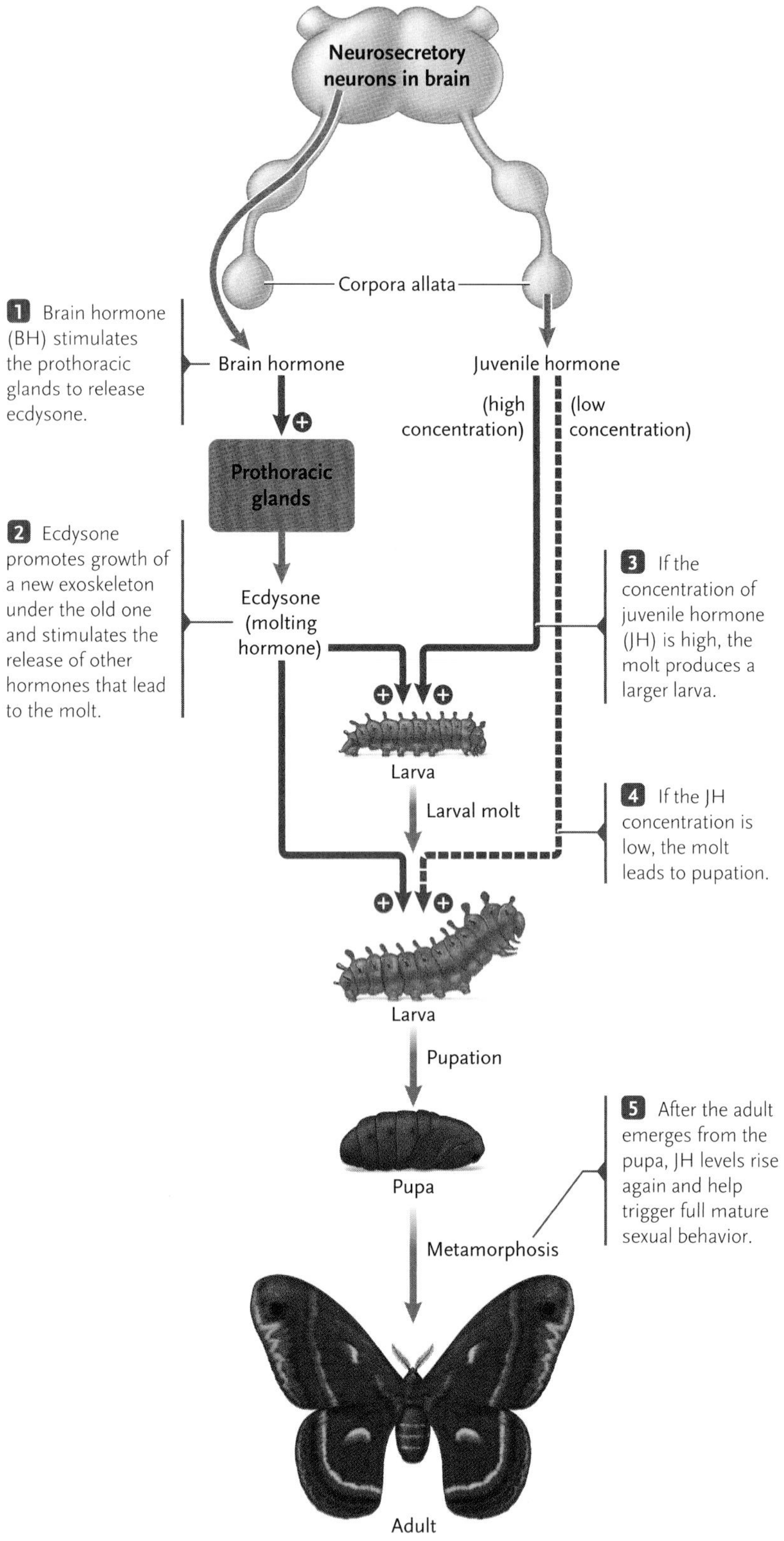

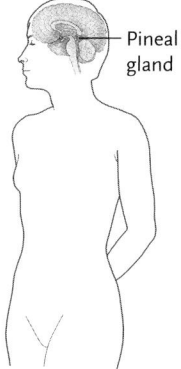

1 Brain hormone (BH) stimulates the prothoracic glands to release ecdysone.

2 Ecdysone promotes growth of a new exoskeleton under the old one and stimulates the release of other hormones that lead to the molt.

Neurosecretory neurons in brain

Corpora allata

Brain hormone

Juvenile hormone

(high concentration)   (low concentration)

Prothoracic glands

Ecdysone (molting hormone)

3 If the concentration of juvenile hormone (JH) is high, the molt produces a larger larva.

Larva

Larval molt

4 If the JH concentration is low, the molt leads to pupation.

Larva

Pupation

Pupa

5 After the adult emerges from the pupa, JH levels rise again and help trigger full mature sexual behavior.

Metamorphosis

Adult

Figure 40.13

The roles of brain hormone, ecdysone, and juvenile hormone in the development of a silkworm moth.

start breaking down proteins and fats to generate energy. The protein breakdown weakens blood vessels throughout the body, particularly in the arms and legs and in critical regions such as the kidneys and retina of the eye. The circulation becomes so poor that tissues degenerate in the arms, legs, and feet. Bleeding in the retina causes blindness at advanced stages of the disease. The breakdown of circulation in the kidneys can lead to kidney failure. In addition, in type 1 diabetes, acidic products of fat breakdown (ketones) are produced in abnormally high quantities and accumulate in the blood. The lowering of blood pH that results can disrupt heart and brain function, leading to coma and death if the disease is untreated.

The Pineal Gland Regulates Some Biological Rhythms

Pineal gland

The **pineal gland** is found at different locations in the brains of vertebrates—for example, in mammals, it is at roughly the center of the brain, while in birds and reptiles, it is on the surface of the brain just under the skull. The pineal gland regulates some biological rhythms.

The earliest vertebrates had a third, light-sensitive eye at the top of the head, and some species, such as lizards and tuataras (New Zealand reptiles), still have an eyelike structure in this location. In most vertebrates, the third eye became modified into a pineal gland, which in many groups retains some degree of photosensitivity. In mammals it is too deeply buried in the brain to be affected directly by light; nonetheless, specialized photoreceptors in the eyes make connections to the pineal gland.

In mammals, the pineal gland secretes a peptide hormone, **melatonin**, which helps to maintain daily biorhythms. Secretion of melatonin is regulated by an inhibitory pathway. Light hitting the eyes generates signals that inhibit melatonin secretion; consequently, the hormone is secreted most actively during periods of darkness. Melatonin targets a part of the hypothalamus called the *suprachiasmatic nucleus,* which is the primary biological clock coordinating body activity to a daily cycle. The nightly release of melatonin may help synchronize the biological clock with daily cycles of light and darkness. The physical and mental discomfort associated with jet lag may reflect the time required for melatonin secretion to reset a traveler's daily biological clock to match the period of daylight in a new time zone.

Melatonin also plays a role in other vertebrates. In some fishes, amphibians, and reptiles, melatonin and other hormones produce changes in skin color through their effects on *melanophores,* the pigment-containing

them less responsive to the hormone than cells in normal individuals. About 90% of patients in the developed world who develop type 2 diabetes are obese. A genetic predisposition can also be a factor. Most affected people can lead a normal life by controlling their diet and weight, exercising, and taking drugs that enhance insulin action or secretion.

Diabetes has long-term effects on the body. Its cells, unable to utilize glucose as an energy source,

cells of the skin. Skin color may vary with the season, the animal's breeding status, or the color of the background.

STUDY BREAK

1. What effect does parathyroid hormone have on the body?
2. What hormones are secreted by the adrenal medulla, and what are their functions?
3. What are the two types of hormones secreted by the adrenal cortex, and what are their functions?
4. To what molecular class of hormones do estradiol and progesterone belong, and what are their functions?

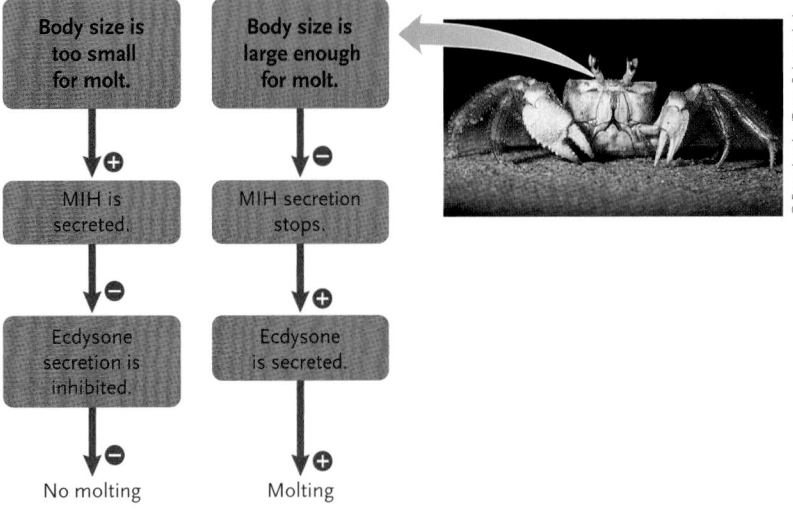

Figure 40.14

Control of molting by molt-inhibiting hormone (MIH), which is secreted by a gland in the eye stalks of crustaceans such as this crab.

40.5 Endocrine Systems in Invertebrates

Invertebrates have fewer hormones, regulating a narrower range of body processes and responses, than vertebrates do. However, in even the simplest animals, such as the cnidarian *Hydra,* hormones produced by neurosecretory neurons control reproduction, growth, and development of some body features. In annelids, arthropods, and mollusks, endocrine cells and glands produce hormones that regulate reproduction, water balance, heart rate, and sugar levels.

Some hormones occur in related forms in invertebrates and vertebrates. For example, both fruit flies and humans have insulin-like hormones and receptors, even though molecular studies suggest that their last common ancestor existed more than 800 million years ago. Both invertebrates and vertebrates secrete peptide and steroid hormones, but most of the hormones have different structures in the two groups, and therefore most have no effect when injected into members of the other group. However, the reaction pathways stimulated by the hormones are the same in both groups, suggesting that these regulatory mechanisms appeared very early in animal evolution.

Hormones Regulate Development in Insects and Crustaceans

Hormones have been studied in detail in only a few invertebrate groups, with the most extensive studies focusing on regulation of metamorphosis in insects. Butterflies, moths, and flies undergo the most dramatic changes as they mature into adults. They hatch from the egg as a caterpillar-like *larva*. During the larval stage, growth is accompanied by one or more *molts,*

in which an old exoskeleton is shed and a new one forms. The insects then enter an inactive stage, the *pupa,* in which the body forms a thick, resistant coating, and finally is transformed into an adult.

Three major hormones regulate molting and metamorphosis in insects: **brain hormone** (BH), a peptide hormone secreted by neurosecretory neurons in the brain; **ecdysone** (*ekdysis* = emerging from), a steroid hormone secreted by the *prothoracic glands;* and **juvenile hormone** (JH), a peptide hormone secreted by the *corpora allata,* a pair of glands just behind the brain **(Figure 40.13).** The outcome of the molt depends on the level of JH. If it is high, the molt produces a larger larva; if it is low, the molt leads to pupation and the emergence of the adult.

Hormones that control molting have also been detected in crustaceans, including lobsters, crabs, and crayfish. Before growth reaches the stage at which the exoskeleton is shed, **molt-inhibiting hormone (MIH),** a peptide neurohormone secreted by a gland in the eye stalks, inhibits ecdysone secretion **(Figure 40.14).** As body size increases to the point requiring a molt, MIH secretion is inhibited, ecdysone secretion increases, and the molt is initiated.

In the next chapter we discuss the structure and functions of muscles, and their interactions with the skeletal system to cause movement. Muscle function depends primarily on the action of the nervous system, but the endocrine system plays a role in the control of smooth muscle contraction.

STUDY BREAK

How do hormones compare structurally and functionally in invertebrates and vertebrates?

What are the cellular mechanisms for insulin resistance in patients with type 2 diabetes?

Insulin resistance is the condition in which the normal physiological levels of insulin are inadequate to produce a normal insulin response in the body. It plays an important role in the development of type 2 diabetes. Gerald Shulman and his research group at Yale Medical School have a long-term goal of elucidating the cellular mechanisms of insulin resistance. Once the mechanisms are known, therapeutic agents can then be developed to reverse insulin resistance in patients with this type of diabetes. In their research, Shulman's group studies patients with type 2 diabetes as well as transgenic mouse models of insulin resistance.

Recall from the chapter that one of the effects of insulin is the conversion of glucose to glycogen. In one set of experiments, Shulman's group studied the rate of glucose incorporation into muscle glycogen. They discovered that muscle glycogen synthesis plays a major role in causing insulin resistance in patients with type 2 diabetes. More detailed studies showed that defects in insulin-stimulated glucose transport and glucose phosphorylation activity in muscles correlate with the early stages in the onset of type 2 diabetes.

Could the defect in glucose transport and phosphorylation activity be reversed? Shulman's group answered this question in a study of lean offspring of type 2 diabetes parents. The offspring examined were insulin-resistant and synthesized insulin-stimulated muscle glycogen at a level only 50% that of normal individuals but, in contrast with their parents, they showed normal blood glucose levels. The potential for these individuals to develop type 2 diabetes later in life is high; that is, they are considered to be prediabetic. After six weeks of following a four-times-a-week aerobic exercise regime on a StairMaster, their insulin-stimulated muscle glycogen synthesis rates returned to normal due to correction of the glucose transport and glucose phosphorylation defects. Thus, the results suggest that regular aerobic exercise potentially could be useful in reversing insulin resistance in prediabetic individuals such as these offspring and, hence, that it might prevent the development of type 2 diabetes. More research is needed to see if that is the case.

Peter J. Russell

Review

Go to **ThomsonNOW**™ at www.thomsonedu.com/login to access quizzing, animations, exercises, articles, and personalized homework help.

40.1 Hormones and Their Secretion

- Hormones are substances secreted by cells that control the activities of cells elsewhere in the body. The cells that respond to a hormone are its target cells. The best-known hormones are secreted by the endocrine system.

- The endocrine system includes four major types of cell signaling: classical endocrine signaling, in which endocrine glands secrete hormones; neuroendocrine signaling, in which neurosecretory neurons release neurohormones into the circulatory system; paracrine regulation, in which cells release local regulators that diffuse through the extracellular fluid to regulate nearby cells; and autocrine regulation, in which cells release local regulators that regulate the same cells that produced it (Figure 40.1).

- Most hormones and local regulators fall into one of four molecular classes: amines, peptides, steroids, and fatty acids.

- Many hormones are controlled by negative feedback mechanisms (Figure 40.2).

Animation: Major human endocrine glands

40.2 Mechanisms of Hormone Action

- Hormones typically are effective in very low concentrations in the body fluids because of amplification.

- Hydrophilic hormones bind to receptor proteins embedded in the plasma membrane, activating them. The activated receptors transmit a signal through the plasma membrane, triggering signal transduction pathways that cause a cellular response. Hydrophobic hormones bind to receptors in the cytoplasm or nucleus, activating them. The activated receptors control the expression of specific genes, the products of which cause the cellular response (Figure 40.3).

- As a result of the types of receptors they have, target cells may respond to more than one hormone, or they may respond differently to the same hormone.

- The major endocrine cells and glands of vertebrates are the hypothalamus, pituitary gland, thyroid gland, parathyroid glands, adrenal medulla, adrenal cortex, testes, ovaries, islets of Langerhans of the pancreas, and pineal gland. Hormones are also secreted by endocrine cells in the stomach and intestine, thymus gland, kidneys, and heart. Most body cells are capable of releasing prostaglandins (Figure 40.6).

Animation: Hormones and target cell receptors

40.3 The Hypothalamus and Pituitary

- The hypothalamus and pituitary together regulate many other endocrine cells and glands in the body (Figure 40.7).

- The hypothalamus produces tropic hormones (releasing hormones and inhibiting hormones) that control the secretion of eight hormones by the anterior pituitary: prolactin (PRL), growth hormone (GH), thyroid-stimulating hormone (TSH), adrenocorticotropic hormone (ACTH), follicle-stimulating hormone (FSH), luteinizing hormone (LH), melanocyte-stimulating hormone (MSH), and endorphins.

- The posterior pituitary secretes antidiuretic hormone (ADH), which regulates body water balance, and oxytocin, which stimulates the contraction of smooth muscle in the uterus as a part of childbirth and triggers milk release from the mammary glands during suckling of the young.

Animation: Posterior pituitary function

Animation: Anterior pituitary function

40.4 Other Major Endocrine Glands of Vertebrates

- The thyroid gland secretes the thyroid hormones and, in mammals, calcitonin. The thyroid hormones stimulate the oxidation of carbohydrates and lipids, and coordinate with growth hormone to stimulate body growth and development. Calcitonin lowers the Ca^{2+} level in the blood by inhibiting the release of Ca^{2+} from bone. In amphibians, such as the frog, thyroid hormones trigger metamorphosis (Figure 40.9).

- The parathyroid glands secrete parathyroid hormone, which stimulates bone cells to release Ca^{2+} into the blood. PTH also stimulates the activation of vitamin D, which promotes Ca^{2+} absorption into the blood from the small intestine (Figure 40.10).

- The adrenal medulla secretes epinephrine and norepinephrine, which reinforce the sympathetic nervous system in responding to stress. The adrenal cortex secretes glucocorticoids, which help maintain glucose at normal levels in the blood, and mineralocorticoids, which regulate Na^+ balance and extracellular fluid volume. The adrenal cortex also secretes small amounts of androgens (Figure 40.11).

- The gonadal sex hormones—androgens, estrogen, and progestins—play a major role in regulating the development of reproductive systems, sexual characteristics, and mating behavior.

- The islet of Langerhans cells of the pancreas secrete insulin and glucagon, which together regulate the concentration of fuel substances in the blood. Insulin lowers the concentration of glucose in the blood and inhibits the conversion of noncarbohydrate molecules into glucose. Glucagon raises blood glucose by stimulating glycogen, fat, and protein degradation (Figure 40.12).

- The pineal gland secretes melatonin, which interacts with the hypothalamus to set the body's daily rhythms.

Animation: Parathyroid hormone action

Animation: Hormones and glucose metabolism

40.5 Endocrine Systems in Invertebrates

- Hormones control development and function of the gonads, manage salt and water balance in the body fluids, and control molting in insects and crustaceans.

- Three major hormones—brain hormone (BH), ecdysone, and juvenile hormone (JH)—control molting and metamorphosis in insects. Hormones that control molting are also present in crustaceans (Figures 40.13 and 40.14).

Questions

Self-Test Questions

1. Amine hormones are usually:
   a. hydrophilic when secreted by the thyroid gland.
   b. based on tyrosine.
   c. paracrine but not autocrine.
   d. not transported by the blood.
   e. repelled by the plasma membrane.

2. Prostaglandins would be best described as inducers of:
   a. male and female characteristics.
   b. cell division.
   c. nerve transmission.
   d. smooth muscle contractions.
   e. cell differentiation.

3. When the concentration of thyroid hormone in the blood increases, it:
   a. inhibits TRH secretion by the hypothalamus.
   b. stimulates a secretion by the hypothalamus.
   c. stimulates the pituitary to secrete TRH.
   d. stimulates the pituitary to secrete TSH.
   e. activates a positive feedback loop.

4. Which of the following statements about endocrine targeting and reception is correct?
   a. The idea that one hormone affects one type of tissue is illustrated when epinephrine binds to smooth muscle cells in blood vessels as well as to beta cells in heart muscle.
   b. The idea that one hormone affects one type of tissue is shown when epinephrine cannot activate both the receptors on liver cells and the beta receptors of heart muscle.
   c. The idea that a target cell can respond to more than one hormone is seen when a vertebrate liver cell can respond to insulin and glucagon.
   d. The idea that a minute concentration of hormone can cause widespread effects demonstrates the specificity of cells for certain hormones.
   e. The idea that the response to a hormone is the same among different target cells is shown when different liver cells are activated by insulin.

5. The posterior pituitary secretes:
   a. tropic hormones, which control the hypothalamus.
   b. IGF, which simulates cell division and protein synthesis.
   c. ADH, which increases water absorption by the kidneys.
   d. oxytocin, which controls egg and sperm development.
   e. prolactin, which stimulates milk synthesis.

6. Blood levels of calcium are regulated directly by:
   a. insulin synthesized by the alpha cells of the pancreas.
   b. PTH made by the pituitary.
   c. vitamin D activated in the liver.
   d. prolactin synthesized by the anterior pituitary.
   e. calcitonin secreted by specialized thyroid cells.

7. If the human body is stressed, glucocorticoids:
   a. promote the breakdown of proteins in the muscles and bones.
   b. increase the amount of sodium reabsorbed from urine in the kidneys.
   c. decrease potassium secretion from the kidneys.
   d. decrease glucose uptake by cells in the nervous system.
   e. inhibit the synthesis of glucose from noncarbohydrate sources.

8. When blood glucose rises:
   a. the alpha cells increase glucagon secretion.
   b. the beta cells increase insulin secretion.
   c. in uncontrolled Type I diabetes, urination decreases.
   d. glucagon uses amino acids as an energy source.
   e. target cells decrease their insulin receptors.

9. In mammals:
   a. the suprachiasmatic nucleus of the pineal gland controls both male and female reproductive systems.
   b. estradiol is produced by the hypothalamus to control ovulation.
   c. melatonin controls anabolic steroid production.
   d. GnRH stimulates LH to control testosterone production.
   e. progesterone increases the secretion of LH from the posterior pituitary.

10. Insect development is regulated by:
    a. ecdysone, a peptide secreted by the brain.
    b. juvenile hormone, a peptide secreted by the corpora allata near the brain.
    c. molt-inhibiting hormone, a steroid secreted by the prothoracic glands.
    d. brain hormone, a steroid secreted by the hypothalamus.
    e. melatonin, a peptide secreted by the brain in the larval stage.

Questions for Discussion

1. A physician sees a patient whose symptoms include sluggishness, depression, and intolerance to cold. What disorder do these symptoms suggest?

2. Cushing's syndrome occurs when an individual overproduces cortisol; this rare disorder is also known as hypercortisolism. In children and teenagers, symptoms include extreme weight gain, retarded growth, excess hair growth, acne, high blood pressure, tiredness and weakness, and either very early or late puberty. Adults with the disease may also exhibit extreme weight gain, excess hair growth, and high blood pressure, and in addition may show muscle and bone weakness, moodiness or depression, sleep disorders, and reproductive disorders. Propose some hypotheses for the overproduction of cortisol in individuals with Cushing's syndrome.

3. A 20-year-old woman with a malignant brain tumor has her pineal gland removed. What kinds of side effects might this loss have?

4. In integrated pest management, a farmer uses a variety of tools to combat unwanted insects. These include applications of either hormones or hormone-inhibiting compounds to prevent insects from reproducing successfully. How might each of these hormone-based approaches disrupt reproduction?

Experimental Analysis

The Environmental Protection Agency (EPA) defines endocrine disruptors as chemical substances that can "interfere with the synthesis, secretion, transport, binding, action, or elimination of natural hormones in the body that are responsible for the maintenance of homeostasis (normal cell metabolism), reproduction, development, and/or behavior." The chemicals, sometimes called environmental estrogens, come from both natural and man-made sources. A simple hypothesis is that endocrine disruptors act by mimicking hormones in the body. Many endocrine disruptors affect sex hormone function and, therefore, reproduction.

Examples of endocrine disruptors are the synthetic chemicals DDT (a pesticide) and dioxins, and natural chemicals such as phytoestrogens (estrogen-like molecules in plants), which are found in high levels in soybeans, carrots, oats, onions, beer, and coffee.

Design an experiment to investigate whether a new synthetic chemical (pick your own interesting scenario) is an endocrine disruptor. (Hint: You probably want to work with a model organism.)

Evolution Link

Which endocrine system evolved earlier, endocrine glands or neurosecretory neurons? Support your conclusion with information obtained from online research.

How Would You Vote?

Crop yields that sustain the human population currently depend on agricultural pesticides, some of which may disrupt hormone function in frogs and other untargeted species. Should chemicals that may cause problems remain in use while researchers investigate them? Go to www.thomsonedu.com/login to investigate both sides of the issue and then vote.

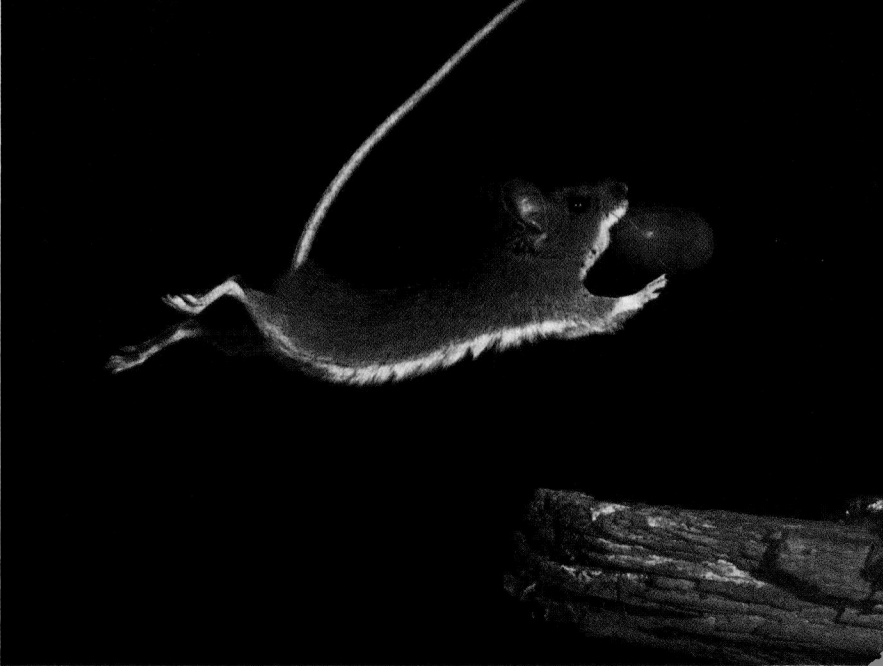

Movement in a long-tailed field mouse *(Apodemus sylvaticus)*. Movement of vertebrates occurs as a result of contractions and relaxations of skeletal muscles. When stimulated by the nervous system, actin filaments in the muscles slide over myosin filaments to cause muscle contractions.

G. Delpho/Peter Arnold, Inc.

41 Muscles, Bones, and Body Movements

WHY IT MATTERS

A Mexican leaf frog *(Pachymedusa dacnicolor)* sits motionless, its prominent eyes staring into space **(Figure 41.1).** But when the frog detects an approaching cricket, it lunges forward at just the right moment, thrusts out its sticky tongue, and captures the prey. This sequence of events, from the beginning of the movement until the frog's mouth closes, sealing the cricket's fate, requires only 260 milliseconds (ms)—about one quarter of a second. How does the frog move so swiftly, and so surely?

As its prey draws near, neuronal signals travel from the frog's brain to the muscles that extend the frog's hind legs, causing the muscles to contract and propel the frog forward on its forelimbs toward the cricket. Within 50 ms after the jump begins, other signals contract the muscles of the lower jaw, opening the mouth. Then, a muscle on the upper surface of the tongue contracts, which raises the tongue and flips it out of the mouth. As the tongue shoots forward, muscle contractions along the ventral side of the trunk arch the body and direct the head downward toward the prey. Within 80 ms after the lunge begins, the tip of the frog's tongue contacts the cricket. Completion of the lunge folds the tongue—and the cricket—into the frog's mouth, aided by contraction of a muscle

933

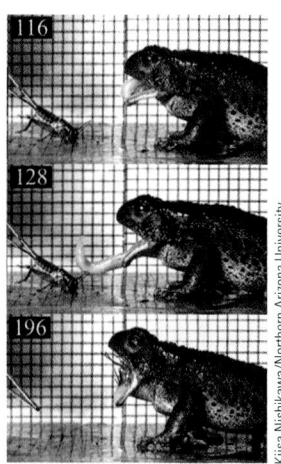

Figure 41.1
A Mexican leaf frog *(Pachymedusa dacnicolor)* capturing a grasshopper.

on the bottom of the tongue. After the mouth closes, further muscle contractions pull the legs forward and fold them under the body.

We know this because Kiisa Nishikawa, Lucie Gray, and James O'Reilly of Northern Arizona University recorded the frog's movements using a high-speed video camera linked to a millisecond timer, with a grid in the background that allowed precise measurement of the distances body parts traveled during the capture. Nishikawa's research group uses the camera's record to study movement in frogs in particular and animals in general.

In Section 36.2 you learned that there are three types of muscle tissue: skeletal, cardiac, and smooth. Skeletal muscle is so named because most muscles of this type are attached by tendons to the skeleton of vertebrates. Cardiac muscle is the contractile muscle of the heart, and smooth muscle is found in the walls of tubes and cavities of the body, including blood vessels and the intestines. In this chapter we describe the structure and function of skeletal muscles, the skeletal systems found in invertebrates and vertebrates, and how muscles bring about movement.

41.1 Vertebrate Skeletal Muscle: Structure and Function

Vertebrate **skeletal muscles** connect to bones of the skeleton. The cells forming skeletal muscles are typically long and cylindrical, and contain many nuclei (shown in Figure 36.6a). Skeletal muscle is controlled by the somatic nervous system.

Most skeletal muscles in humans and other vertebrates are attached at both ends across a joint to bones of the skeleton. (Some, such as those that move the lips, are attached to other muscles or connective tissues under skin.) Depending on its points of attachment, contraction of a single skeletal muscle may extend or bend body parts, or may rotate one body part with respect to another. The human body has more than 600 skeletal muscles, ranging in size from the small muscles that move the eyeballs to the large muscles that move the legs.

Skeletal muscles are attached to bones by cords of connective tissue called *tendons* (see Section 36.2). Tendons vary in length from a few millimeters to some, such as those that connect the muscles of the forearm to the bones of the fingers, that are 20 to 30 cm long.

The Striated Appearance of Skeletal Muscle Fibers Results from a Highly Organized Internal Structure

A skeletal muscle consists of bundles of elongated, cylindrical cells called **muscle fibers**, which are 10 to 100 μm in diameter and run the entire length of the muscle **(Figure 41.2)**. Muscle fibers contain many nuclei, reflecting their development by fusion of smaller cells. Some very small muscles, such as some of the muscles of the face, contain only a few hundred muscle fibers; others, such as the larger leg muscles, contain hundreds of thousands. In both cases, the muscle fibers are held in parallel bundles by sheaths of connective tissue that surround them in the muscle and merge with the tendons that connect muscles to bones or other structures. Muscle fibers are richly supplied with nutrients and oxygen by an extensive network of blood vessels that penetrates the muscle tissue.

Muscle fibers are packed with **myofibrils**, cylindrical contractile elements about 1 μm in diameter that run lengthwise inside the cells. Each myofibril consists of a regular arrangement of **thick filaments** (13–18 nm in diameter) and **thin filaments** (5–8 nm in diameter) (see Figure 41.2). The thick and thin filaments alternate with one another in a stacked set.

The thick filaments are parallel bundles of myosin molecules; each myosin molecule consists of two protein subunits that together form a *head* connected to a long double helix forming a *tail*. The head is bent toward the adjacent thin filament to form a *crossbridge*. In vertebrates, each thick filament contains some 200 to 300 myosin molecules and forms as many crossbridges. The thin filaments consist mostly of two linear chains of actin molecules twisted into a double helix, which creates a groove running the length of the molecule. Bound to the actin are *tropomyosin* and *troponin* proteins. Tropomyosin molecules are elongated fibrous proteins that are organized end to end next to the groove of the actin double helix. Troponin is a three-subunit globular protein that binds to tropomyosin at intervals along the thin filaments.

The arrangement of thick and thin filaments forms a pattern of alternating dark bands and light bands, giving skeletal muscle a striated appearance under the microscope (see Figure 41.2). The dark bands, called *A bands,* consist of stacked thick filaments along with the parts of thin filaments that overlap both ends. The lighter-appearing middle region of an A band, which contains only thick filaments, is the *H zone.* In the center of the H zone is a disc of proteins called the *M line,* which holds the stack of thick filaments together. The light bands, called *I bands,* consist of the parts of the thin filaments not in the A band. In the center of each I band is a thin *Z line,* a disc to which the thin filaments are anchored. The region between two adjacent Z lines is a **sarcomere** (*sarco* = flesh; *meros* = segment); sarcomeres are the basic units of contraction in a myofibril.

At each junction of an A band and an I band, the plasma membrane folds into the muscle fiber to form a **T (transverse) tubule (Figure 41.3)**. Encircling the sarcomeres is the **sarcoplasmic reticulum**, a complex system of vesicles modified from the smooth endoplasmic reticulum. Segments of the sarcoplasmic re-

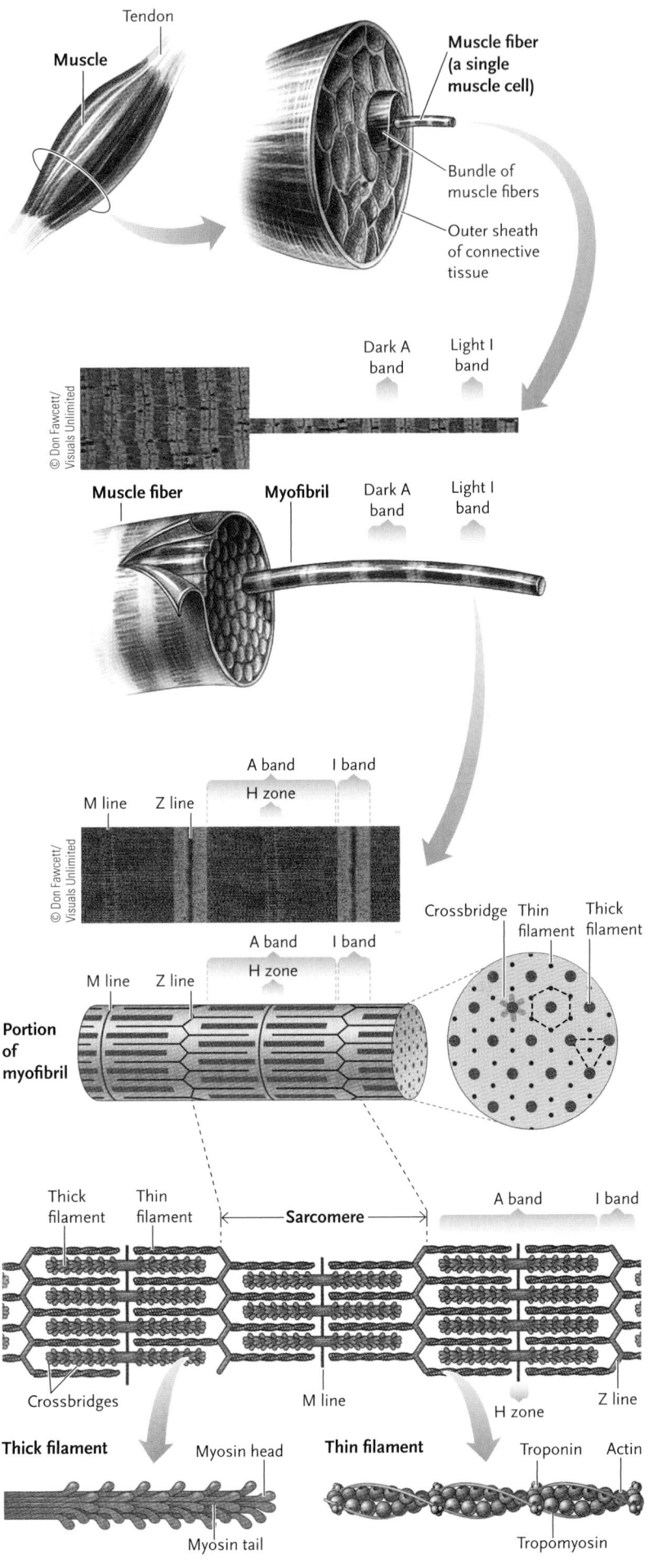

Figure 41.2

Skeletal muscle structure. Muscles are composed of bundles of cells called muscle fibers; within each muscle fiber are longitudinal bundles of myofibrils. The unit of contraction within a myofibril, the sarcomere, consists of overlapping myosin thick filaments and actin thin filaments. The myosin molecules in the thick filaments each consist of two subunits organized into a head and a double-helical tail. The actin subunits in the thin filaments form twisted, double helices, with tropomyosin molecules arranged head-to-tail in the groove of the helix and troponin bound to the tropomyosin at intervals along the thin filaments.

ticulum are wrapped around each A band and I band, and are separated from the T tubules in those regions by small gaps.

An axon of an efferent neuron leads to each muscle fiber. The axon terminal makes a single, broad synapse with a muscle fiber called a **neuromuscular junction** (see Figure 41.3). The neuromuscular junction, T tubules, and sarcoplasmic reticulum are key components in the pathway for stimulating skeletal muscle contraction by neural signals—which starts with action potentials traveling down the efferent neuron—as will be described next.

During Muscle Contraction, Thin Filaments on Each Side of a Sarcomere Slide over Thick Filaments

The precise control of body motions depends on an equally precise control of muscle contraction by a signaling pathway that carries information from nerves to muscle fibers. An action potential arriving at the neuromuscular junction leads to an increase in the concentration of Ca^{2+} in the cytosol of the muscle fiber. The increase in Ca^{2+} triggers a process in which the thin filaments on each side of a sarcomere slide over the thick filaments toward the center of the A band, which brings the Z lines closer together, shortening the sarcomeres and contracting the muscle **(Figure 41.4)**. This *sliding filament mechanism* of muscle contraction depends on dynamic interactions between actin and myosin proteins in the two filament types. That is, the myosin crossbridges make and break contact with actin and pull the thin filaments over the thick filaments—the action is similar to rowing, or a ratcheting process. A model for muscle contraction is shown in **Figure 41.5.**

Conduction of an Action Potential into a Muscle Fiber. Like neurons, skeletal muscle fibers are *excitable,* meaning that the electrical potential of their plasma membrane can change in response to a stimulus. When an action potential arrives at the neuromuscular junction, the axon terminal releases a neurotransmitter, *acetylcholine,* which triggers an action potential in the muscle fiber (see Figure 41.5, step 1). The action potential travels in all directions over the muscle fiber's

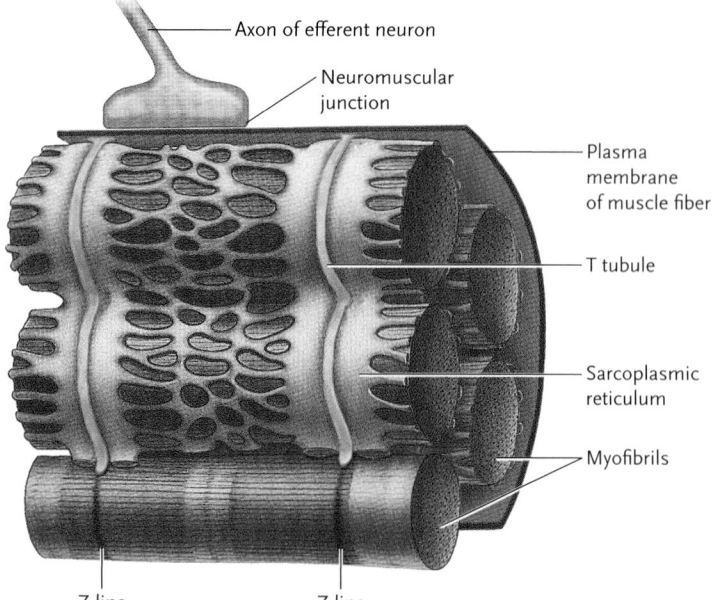

Axon of efferent neuron

Neuromuscular junction

Plasma membrane of muscle fiber

T tubule

Sarcoplasmic reticulum

Myofibrils

Z line   Z line

Figure 41.3
Components in the pathway for the stimulation of skeletal muscle contraction by neural signals. T (transverse) tubules are infoldings of the plasma membrane into the muscle fiber originating at each A band–I band junction in a sarcomere. The sarcoplasmic reticulum encircles the sarcomeres and segments of it end in close proximity to the T tubules.

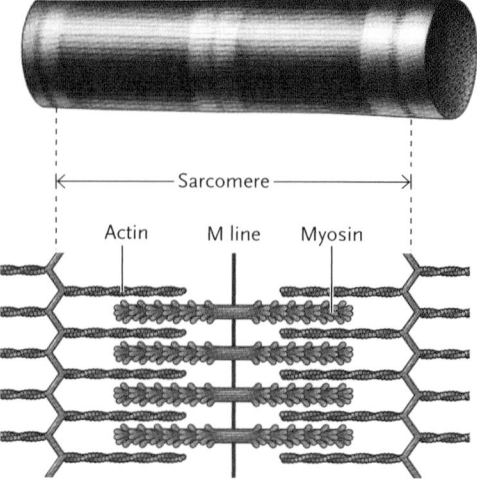

a. Relaxed sarcomere

Sarcomere

Actin   M line   Myosin

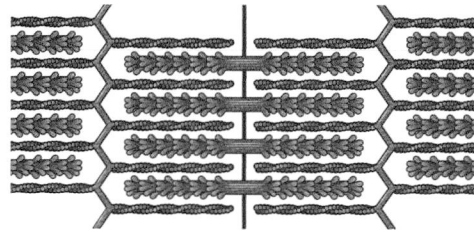

b. Contracted sarcomere

Figure 41.4
Shortening of sarcomeres by the sliding filament mechanism, in which the thin filaments are pulled over the thick filaments.

surface membrane, and also penetrates into the interior of the fiber through the T tubules.

Release of Calcium into the Cytosol of the Muscle Fiber. In the absence of a stimulus, the Ca^{2+} concentration is kept high inside the sarcoplasmic reticulum by active transport proteins that continuously pump Ca^{2+} out of the cytosol and into the sarcoplasmic reticulum. (The active transport proteins are Ca^{2+} pumps, discussed in Section 6.4.) When an action potential reaches the end of a T tubule, it opens ion channels in the sarcoplasmic reticulum that allow Ca^{2+} to flow out into the cytosol (see Figure 41.5, step 2).

When Ca^{2+} flows into the cytosol, the troponin molecules of the thin filament bind the calcium and undergo a conformational change that causes the tropomyosin fibers to slip into the grooves of the actin double helix. The slippage uncovers the actin's binding sites for the myosin crossbridge (see Figure 41.5, step 3). At this point in the process, the myosin crossbridge has a molecule of ATP bound to it, and is not in contact with the thin filament.

The Crossbridge Cycle. Using the energy of ATP hydrolysis, the myosin crossbridge bends away from the tail and binds to a newly exposed myosin crossbridge binding site on an actin molecule (see Figure 41.5, step 4). In effect, this bending compresses a molecular spring in the myosin head. The binding of the crossbridge to actin triggers release of the molecular spring in the crossbridge, which snaps back toward the tail

producing the power stroke (motor) that pulls the thin filament over the thick filament (step 5).

The crossbridge now binds another ATP and myosin detaches from actin (see Figure 41.5, step 6). The cycle repeats again, starting with ATP hydrolysis (step 4). Contraction ceases when action potentials stop: Ca^{2+} is pumped back into the sarcoplasmic reticulum, and its effect on troponin is reversed, leading to tropomyosin again blocking myosin crossbridge binding sites on actin. Contraction ceases and the actin thin filaments slide back over the myosin thick filaments to their original relaxed positions (step 7). Crossbridge cycles based on actin and myosin power movements in all living organisms, from cytoplasmic streaming in plant cells and amoebae to muscle contractions in animals.

Although the force produced by a single myosin crossbridge is comparatively small, it is multiplied by the hundreds of crossbridges acting in a single thick filament, and by the billions of thin filaments sliding in a contracting sarcomere. The force, multiplied further by the many sarcomeres and myofibrils in a muscle fiber, is transmitted to the plasma membrane of a muscle fiber by the attachment of myofibrils to elements of the cytoskeleton. From the plasma membrane, it is transmitted to bones and other body parts

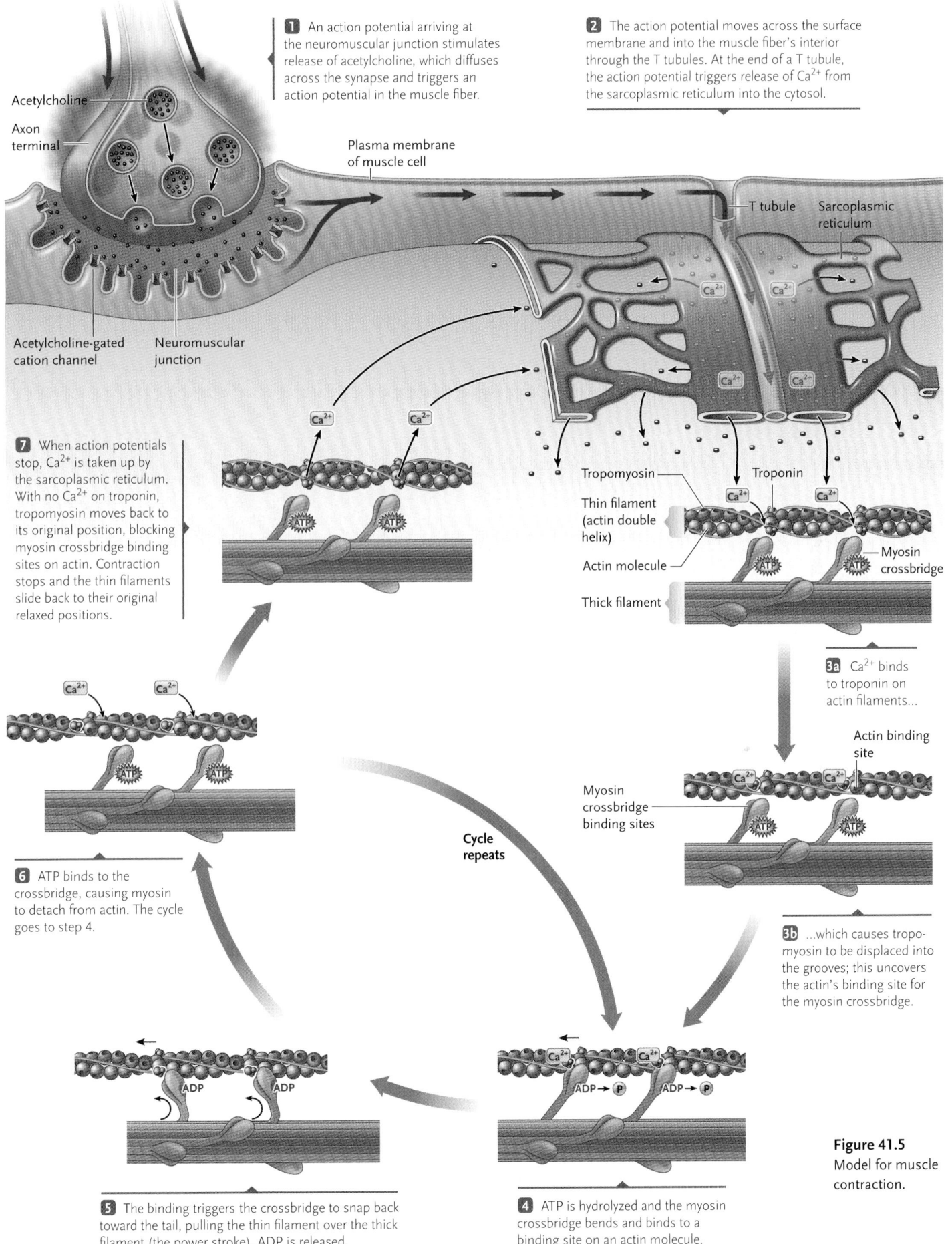

1 An action potential arriving at the neuromuscular junction stimulates release of acetylcholine, which diffuses across the synapse and triggers an action potential in the muscle fiber.

2 The action potential moves across the surface membrane and into the muscle fiber's interior through the T tubules. At the end of a T tubule, the action potential triggers release of Ca^{2+} from the sarcoplasmic reticulum into the cytosol.

Acetylcholine

Axon terminal

Plasma membrane of muscle cell

T tubule

Sarcoplasmic reticulum

Acetylcholine-gated cation channel

Neuromuscular junction

7 When action potentials stop, Ca^{2+} is taken up by the sarcoplasmic reticulum. With no Ca^{2+} on troponin, tropomyosin moves back to its original position, blocking myosin crossbridge binding sites on actin. Contraction stops and the thin filaments slide back to their original relaxed positions.

Tropomyosin

Troponin

Thin filament (actin double helix)

Actin molecule

Myosin crossbridge

Thick filament

3a Ca^{2+} binds to troponin on actin filaments...

Actin binding site

Myosin crossbridge binding sites

6 ATP binds to the crossbridge, causing myosin to detach from actin. The cycle goes to step 4.

Cycle repeats

3b ...which causes tropomyosin to be displaced into the grooves; this uncovers the actin's binding site for the myosin crossbridge.

5 The binding triggers the crossbridge to snap back toward the tail, pulling the thin filament over the thick filament (the power stroke). ADP is released.

4 ATP is hydrolyzed and the myosin crossbridge bends and binds to a binding site on an actin molecule.

Figure 41.5
Model for muscle contraction.

A Substitute Player That May Be a Big Winner in Muscular Dystrophy

Duchenne muscular dystrophy (DMD) is an inherited disease, characterized by progressive muscle weakness, that primarily affects males—about 1 out of every 3500 males is born with the disease. When DMD patients are 3 to 5 years old, their muscle tissue begins to break down, and by the time they are in their teens most can walk only with braces. They usually die of complications from degeneration of the heart and diaphragm muscle by their early 20s. Currently, there is no effective treatment for DMD.

The gene that causes DMD, which is located on the X chromosome, was isolated and identified in 1985. In its normal form, the gene encodes the protein *dystrophin*, which anchors a glycoprotein complex in the plasma membrane of a muscle fiber to the underlying actin cytoskeleton (see Section 5.3). In most people with DMD, segments of DNA are missing from the coding sequence of the gene, so the protein cannot function. Without functional dystrophin, the plasma membrane of the muscle fibers is susceptible to tearing during contraction, which leads to muscle destruction. Creatine kinase (CK), an enzyme found predominantly in muscles and in the brain, leaks out of the damaged muscles and accumulates in the blood, which normally contains little CK. Elevated CK in the blood, then, is diagnostic of muscle damage such as that found in DMD.

Many researchers are working to develop a gene-therapy cure for DMD. For example, Kay E. Davies and her colleagues at Oxford University in England have identified a protein that is structurally similar to dystrophin and appears to have a highly similar function. That protein, called *utrophin*, is made in small quantities in muscle fibers and normally functions only in neuromuscular junctions. The utrophin gene and its protein function normally in DMD patients.

The Davies team reasoned that utrophin might be able to substitute for the missing dystrophin in DMD patients if a means could be found to increase its quantity in muscle cells. For their research, they used *mdx* mice, a strain that has the dystrophin gene deleted and is therefore a mouse model of human DMD. First, they introduced an artificial gene (consisting of the mouse utrophin gene under the control of a strong promoter) into fertilized oocytes; the resulting transgenic mice produced much more than the usual amount of utrophin. The researchers were excited to find that CK levels in the blood of the transgenic mice were reduced to 25% of the level in *mdx* mice without the added gene, indicating that muscle damage was markedly decreased. This was confirmed by microscopic examination. Other techniques showed that utrophin, instead of being concentrated in neuromuscular junctions as it is

normally, was now distributed throughout the muscle plasma membranes. In short, in these experiments the elevated level of utrophin was able to substitute for dystrophin, and decreased significantly the onset of disease symptoms. Moreover, no deleterious side effects from the overproduction of utrophin could be detected in the genetically engineered mice.

Promising as these results are, germline gene therapy of humans is not allowed, so this approach cannot be used with human patients. Davies's group looked for another way to increase utrophin production, and suggested that upregulating the utrophin gene in all cells of the body could be a strategy to treat DMD. In experiments again using transgenic *mdx* mice, they showed that moderate overproduction of utrophin beginning as late as 10 days after birth caused improvements in muscle appearance compared with controls. Overall, the results show that utrophin overproduction therapy, initiated after birth, can be effective, but that both the timing of therapy and the amount of utrophin expressed are important. Davies's group is now searching for a chemical compound that would increase the levels of utrophin already present in DMD patients. However, much work remains before this can be an effective therapy in humans.

by the connective tissue sheaths surrounding the muscle fibers and by the tendons.

Several mutations affecting muscle and nerve tissues interrupt the transmission of force and cause severe disabilities. Duchenne muscular dystrophy (DMD), for example, is caused by a mutation that weakens the cytoskeleton of the muscle fiber, causing the cells to rupture when contractile forces are generated. *Insights from the Molecular Revolution* describes experiments that may lead to a cure for this debilitating disease.

From Contraction to Relaxation. As long as action potentials continue to arrive at the neuromuscular junc-

tion, Ca^{2+} is released in response, and ATP is available, the crossbridge cycle continues to run, shortening the sarcomeres and contracting the muscle fiber.

When action potentials stop, excitation of the T tubules ceases, and the Ca^{2+} release channels in the sarcoplasmic reticulum close. The active transport pumps quickly remove the remaining Ca^{2+} from the cytosol. In response, troponin releases its Ca^{2+} and the tropomyosin fibers are pulled back to cover the myosin binding sites in the thin filaments. The crossbridge cycle stops, and contraction of the muscle fiber ceases. In a muscle fiber that is not contracting, ATP is bound to the myosin head and the crossbridge is not bound to the actin filament (see Figure 41.5, step 7).

Deadly Interruptions of the Crossbridge Cycle. The mechanism controlling vertebrate muscle contraction can be blocked by several toxins and poisons. For example, the bacterium *Clostridium botulinum,* which grows in improperly preserved food, produces a toxin that blocks acetylcholine release in neuromuscular junctions. Many of the body muscles are unable to contract, including the diaphragm, the muscle that is essential for inflating the lungs. As a result, the victim dies from respiratory failure. The toxin is so poisonous that 0.0000001 g is enough to kill a human; 600 g could wipe out the entire human population. This same toxin, under the brand name Botox, is injected in low doses as a cosmetic treatment to remove or reduce wrinkles—if muscles cannot contract, then wrinkles cannot form.

The venom of black widow spiders (genus *Latrodectus*) causes massive release of acetylcholine, leading to convulsive contractions of body muscles; the diaphragm becomes locked in position, causing respiratory failure. Curare, extracted from the bark and sap of some South American trees, blocks acetylcholine from binding to its receptors in muscle fibers. The body muscles, including the diaphragm, become paralyzed and the victim dies of respiratory failure. Some native peoples in South America took advantage of these effects by using curare as an arrow and dart poison.

In a natural process, within a few hours after an animal dies, Ca^{2+} diffuses into the cytoplasm of muscle cells and initiates the crossbridge cycle, producing *rigor mortis,* a strong tension of essentially all the skeletal muscles that stiffens the entire body. As part of rigor mortis, the crossbridges become locked to the thin filaments because ATP production stops (remember that ATP is required to release the crossbridges from actin). The stiffness reverses as actin and myosin are degraded.

The Response of a Muscle Fiber to Action Potentials Ranges from Twitches to Tetanus

A single action potential arriving at a neuromuscular junction usually causes a single, weak contraction of a muscle fiber called a **muscle twitch (Figure 41.6a).** After a muscle twitch begins, the tension of the muscle fiber increases in magnitude for about 30 to 40 ms, and then peaks as the action potential runs its course through the T tubules and the Ca^{2+} channels begin to close. Tension then decreases as the Ca^{2+} ions are pumped back into the sarcoplasmic reticulum, falling to zero in about 50 ms after the peak.

If a muscle fiber is restimulated after it has relaxed completely, a new twitch identical to the first is generated (see Figure 41.6a). However, if a muscle fiber is restimulated before it has relaxed completely, the second twitch is added to the first, producing what is called *twitch summation,* which is basically a summed, stronger contraction **(Figure 41.6b).** And, if action potentials arrive so rapidly (about 25 ms apart) that the fiber cannot relax at all between stimuli, the Ca^{2+} channels remain open continuously and twitch summation produces a peak level of continuous contraction called **tetanus (Figure 41.6c).** (This is not to be confused with the disease of the same name, in which a bacterial toxin causes uncontrolled and con-

Figure 41.6
The relationship of the tension produced in a muscle fiber to the frequency of action potentials.

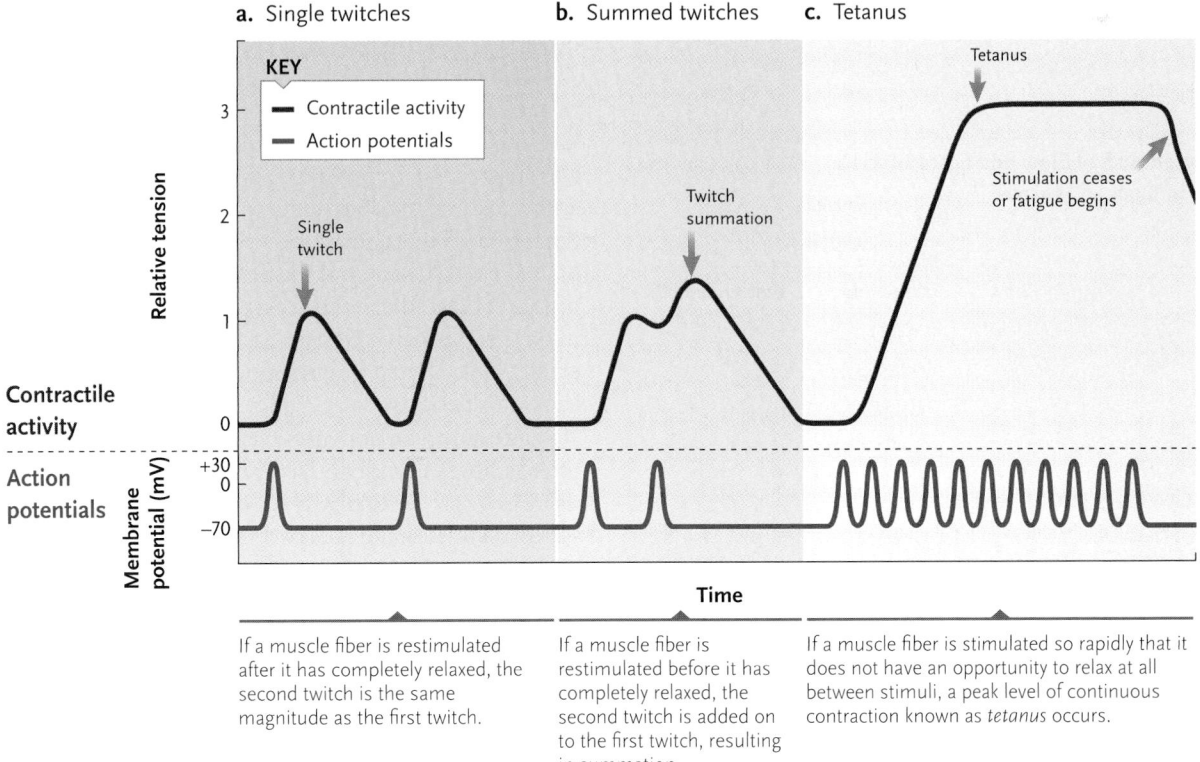

a. Single twitches   **b.** Summed twitches   **c.** Tetanus

KEY
— Contractile activity
— Action potentials

Relative tension

Single twitch

Twitch summation

Tetanus

Stimulation ceases or fatigue begins

Contractile activity

Action potentials

Membrane potential (mV)

+30
0
−70

Time

If a muscle fiber is restimulated after it has completely relaxed, the second twitch is the same magnitude as the first twitch.

If a muscle fiber is restimulated before it has completely relaxed, the second twitch is added on to the first twitch, resulting in summation.

If a muscle fiber is stimulated so rapidly that it does not have an opportunity to relax at all between stimuli, a peak level of continuous contraction known as *tetanus* occurs.

tinuous muscle contraction.) Contractile activity will then decrease if either the stimuli cease or the muscle fatigues.

Tetanus is an essential part of muscle fiber function. If we lift a moderately heavy weight, for example, many of the muscle fibers in our arms enter tetanus and remain in that state until the weight is released. Even body movements that require relatively little effort, such as standing still but in balance, involve tetanic contractions of some muscle fibers.

Muscle Fibers Differ in Their Rate of Contraction and Susceptibility to Fatigue

Muscle fibers differ in their rate of contraction and resistance to fatigue, and thus can be classified as slow, fast aerobic, and fast anaerobic muscle fibers. Their properties are summarized in **Table 41.1.** The proportions of the three types of muscle fibers tailor the contractile characteristics of each muscle to suit its function within the body.

Slow muscle fibers contract relatively slowly and the intensity of contraction is low because their myosin crossbridges hydrolyze ATP relatively slowly. They can remain contracted for relatively long periods without fatiguing. Slow muscle fibers typically contain many mitochondria and make most of their ATP by oxidative phosphorylation (aerobic respiration). They have a low capacity to make ATP by anaerobic glycolysis. They also contain high concentrations of the oxygen-storing protein **myoglobin**, which greatly enhances their oxygen supplies. Myoglobin is closely related to hemoglobin, the oxygen-carrying protein of red blood cells. Myoglobin gives slow muscle fibers, such as those in the legs

of ground birds such as quail, chickens, and ostriches, a deep red color. In sharks and bony fishes, strips of slow muscles concentrated in a band on either side of the body are used for slow, continuous swimming and maintaining body position.

Fast muscle fibers contract relatively quickly and powerfully because their myosin crossbridges hydrolyze ATP faster than those of slow muscle fibers. Fast aerobic fibers have abundant mitochondria, a rich blood supply, and a high concentration of myoglobin, which makes them red in color. They have a high capacity for making ATP by oxidative phosphorylation, and an intermediate capacity for making ATP by anaerobic glycolysis. They fatigue more quickly than slow fibers, but not as quickly as fast anaerobic fibers. Fast aerobic muscle fibers are abundant in the flight muscles of migrating birds such as ducks and geese.

Fast anaerobic fibers typically contain high concentrations of glycogen, relatively few mitochondria, and a more limited blood supply than fast aerobic fibers. They generate ATP mostly by anaerobic respiration (glycolysis) and have a low capacity to produce ATP by oxidative respiration. Fast anaerobic fibers produce especially rapid and powerful contractions but are more susceptible to fatigue. Because their myoglobin supply is limited and they contain few mitochondria, they are pale in color. Some ground birds have flight muscles consisting almost entirely of fast anaerobic muscle fibers. These muscles can produce a short burst of intensive contractions allowing the bird to escape a predator, but they cannot produce sustained flight. Most muscles of lampreys, sharks, fishes, amphibians, and reptiles also contain fast anaerobic muscle fibers, allowing the animals to move quickly to capture prey and avoid danger.

The muscles of humans and other mammals are mixed, and contain different proportions of slow and fast muscle fibers, depending on their functions. Muscles specialized for prolonged, slow contractions, such as the postural muscles of the back, have a high proportion of slow fibers and are a deep red color. The muscles of the forearm that move the fingers have a higher proportion of fast fibers and are a paler red than the back muscles. These muscles can contract rapidly and powerfully, but they fatigue much more rapidly than the back muscles.

The number and proportions of slow and fast muscle fibers in individuals are inherited characteristics. However, particular types of exercise can convert some fast muscle fibers between aerobic and anaerobic types. Endurance training, such as long-distance running, converts fast muscle fibers from the anaerobic to the aerobic type, and regimes such as weight lifting induce the reverse conversion. If the training regimes stop, most of the fast muscle fibers revert to their original types.

Table 41.1	Characteristics of Slow and Fast Muscle Fibers in Skeletal Muscle		
	Fiber Type		
Property	Slow	Fast Aerobic	Fast Anaerobic
Contraction speed	Slow	Fast	Fast
Contraction intensity	Low	Intermediate	High
Fatigue resistance	High	Intermediate	Low
Myosin–ATPase activity	Low	High	High
Oxidative phosphorylation capacity	High	High	Low
Enzymes for anaerobic glycolysis	Low	Intermediate	High
Mitochondria	Many	Many	Few
Myoglobin content	High	High	Low
Fiber color	Red	Red	White
Glycogen content	Low	Intermediate	High

Skeletal Muscle Control Is Divided among Motor Units

The control of muscle contraction extends beyond the simple ability to turn the crossbridge cycle on and off. We can adjust a handshake from a gentle squeeze to a strong grasp, or exactly balance a feather or dumbbell in the hand. How are entire muscles controlled in this way? The answer lies in activation of the muscle fibers in blocks called **motor units.**

The muscle fibers in each motor unit are controlled by branches of the axon of a single efferent neuron **(Figure 41.7).** As a result, all those fibers contract each time the neuron fires an action potential. All the muscle fibers in a motor unit are of the same type—either slow, fast aerobic, or fast anaerobic. When a motor unit contracts, its force is distributed throughout the entire muscle because the fibers are dispersed throughout the muscle rather than being concentrated in one segment.

For a delicate movement, only a few efferent neurons carry action potentials to a muscle, and only a few motor units contract. For more powerful movements, more efferent neurons carry action potentials, and more motor units contract.

Muscles that can be precisely and delicately controlled, such as those moving the fingers in humans, have many motor units in a small area, with only a few muscle fibers—about 10 or so—in each unit. Muscles that produce grosser body movements, such as those moving the legs, have fewer motor units in the same volume of muscle but thousands of muscle fibers in each unit. In the calf muscle that raises the heel, for example, most motor units contain nearly 2000 muscle fibers. Other skeletal muscles fall between these extremes, with an average of about 200 muscle fibers per motor unit.

Invertebrates Move Using a Variety of Striated Muscles

Invertebrates also have muscle cells in which actin-based thin filaments and myosin-based thick filaments produce movements by the same sliding mechanism as in vertebrates. Muscles that are clearly striated, which occur in virtually all invertebrates except sponges, have thick and thin filaments arranged in sarcomeres remarkably similar to those of vertebrates, except for variations in sarcomere length and the ratio of thin to thick filaments.

In invertebrates, an entire muscle is typically controlled by one or a few motor neurons. Nevertheless, invertebrate muscles are capable of finely graded contractions because individual neurons make large numbers of synapses with the muscle cells. As action potentials arrive more frequently at the synapses, more Ca^{2+} is released into the cells, and the muscles contract more strongly.

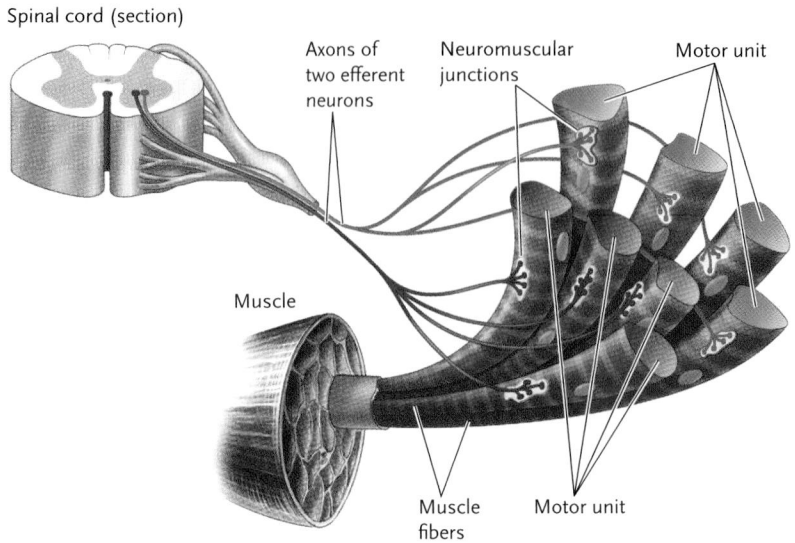

Figure 41.7

Motor units in vertebrate skeletal muscles. Each motor unit consists of groups of muscle fibers activated by branches of a single efferent (motor) neuron.

STUDY BREAK

1. Muscle contraction occurs in response to a stimulus from the nervous system. How does this occur?
2. Outline the molecular events that take place in the sliding filament mechanism of muscle contraction.

41.2 Skeletal Systems

Animal skeletal systems provide physical support for the body and protection for the soft tissues. They also act as a framework against which muscles work to move parts of the body or the entire organism. There are three main types of skeletons found in both invertebrates and vertebrates: hydrostatic skeletons, exoskeletons, and endoskeletons.

A Hydrostatic Skeleton Consists of Muscles and Fluid

A **hydrostatic skeleton** (*hydro* = water; *statikos* = causing to stand) is a structure consisting of muscles and fluid that, by themselves, provide support for the animal or part of the animal; no rigid support, like a bone, is involved. A hydrostatic skeleton consists of a body compartment or compartments filled with water or body fluids, which are incompressible liquids. When the muscular walls of the compartment contract, they pressurize the contained fluid. If muscles in one part of the compartment are contracted while muscles in another part are relaxed, the pressurized fluid will

a. Resting position   **b.** Feeding position

Linda Pitkin/Planet Earth Pictures

Figure 41.8
Sea anemones in **(a)** the resting and **(b)** the feeding position. In **(a)**, longitudinal muscles in the body wall are contracted, and circular muscles are relaxed. In **(b)**, the longitudinal muscles are relaxed, and the circular muscles are contracted. Both sets of muscles work against a hydrostatic skeleton.

move to the relaxed part of the compartment, distending it. In short, the contractions and relaxations of the muscles surrounding the compartments change the shape of the animal.

Hydrostatic skeletons are the primary support systems of cnidarians, flatworms, roundworms, and annelids. In all these animals, compartments containing fluids under pressure make the body semirigid and provide a mechanical support on which muscles act. For example, sea anemones have a hydrostatic skeleton consisting of several fluid-filled body cavities. The body wall contains longitudinal and circular muscles that work against that skeleton. Between meals, longitudinal muscles are contracted (shortened), while

the circular ones are relaxed, and the animal looks short and squat **(Figure 41.8a)**. It lengthens into its upright feeding position by contracting the circular muscles and relaxing the longitudinal ones **(Figure 41.8b)**. In flatworms, roundworms, and annelids, striated muscles in the body wall act on the hydrostatic skeleton to produce creeping, burrowing, or swimming movements. Among these animals, annelids have the most highly developed musculoskeletal systems, with an outer layer of circular muscles surrounding the body, and an inner layer of longitudinal muscles running its length **(Figure 41.9)**. Contractions of the circular muscles reduce the diameter of the body and increase the length of the worm; contractions of the longitudinal muscles shorten the body and increase its diameter. Annelids move along a surface or burrow by means of alternating waves of contraction of the two muscle layers that pass along the body, working against the fluid-filled body compartments of the hydrostatic skeleton.

Many arthropods have hydrostatic skeletal elements. In the larvae of flying insects, internal fluids held under pressure by the muscular body wall provide some body support. In spiders, the legs are extended from the bent position by muscles exerting pressure against body fluids.

Some structures of echinoderms are supported by hydrostatic skeletons. The tube feet of sea stars and sea urchins, for example, have muscular walls enclosing the fluid of the water vascular system (see Figure 29.46).

In vertebrates, the erectile tissue of the penis is a fluid-filled hydrostatic skeletal structure.

An Exoskeleton Is a Rigid External Body Covering

An **exoskeleton** (*exo* = outside) is a rigid external body covering, such as a shell, that provides support. In an exoskeleton, the force of muscle contraction is applied against that covering. An exoskeleton also protects delicate internal tissues such as the brain and respiratory organs.

Many mollusks, such as clams and oysters, have an exoskeleton consisting of a hard calcium carbonate shell secreted by glands in the mantle. Arthropods, such as insects spiders, and crustaceans, have an external skeleton in the form of a chitinous cuticle, secreted by underlying tissue, that covers the outside surfaces of the animals. Like a suit of armor, the arthropod exoskeleton has movable joints, flexed and extended by muscles that extend across the inside surfaces of the joints **(Figure 41.10)**. The exoskeleton protects against dehydration, serves as armor against predators, and provides the levers against which muscles work. In many flying insects, elastic flexing of the exoskeleton contributes to the movements of the wings.

In vertebrates, the shell of a turtle or tortoise is an exoskeletal structure, as are the bony plates, abdominal

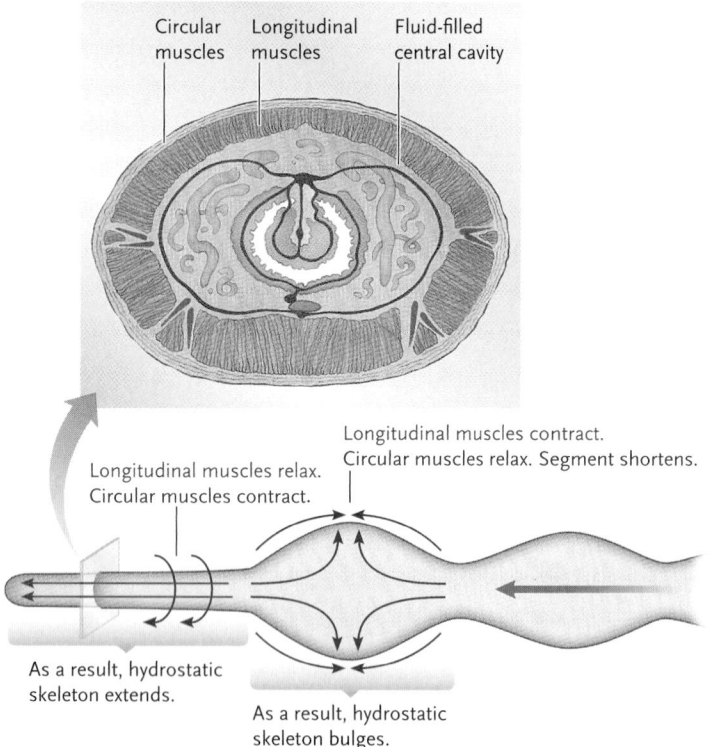

Circular muscles   Longitudinal muscles   Fluid-filled central cavity

Longitudinal muscles contract. Circular muscles relax. Segment shortens.

Longitudinal muscles relax. Circular muscles contract.

As a result, hydrostatic skeleton extends.

As a result, hydrostatic skeleton bulges.

Figure 41.9
Movement of an earthworm, showing how muscles in the body wall act on its hydrostatic skeleton. Contraction of the circular muscles reduce body diameter and increase body length, while contraction of the longitudinal muscles decrease body length and increase body diameter.

ribs, collar bones, and most of the skull of the American alligator.

An Endoskeleton Consists of Supportive Internal Body Structures Such as Bones

An **endoskeleton** (*endon* = within) consists of internal body structures, such as bones, that provide support. In an endoskeleton, the force of contraction is applied against those structures. Like exoskeletons, endoskeletons also protect delicate internal tissues such as the brain and respiratory organs.

In mollusks, the mantle of squids and cuttlefish is reinforced by an endoskeletal element commonly called a "pen" (in squid) or the "cuttlebone" in cuttlefish (see Figure 29.22). Squids also have an internal case of cartilage that surrounds and protects the brain; other segments of cartilage support the gills and siphon in squids and octopuses.

Echinoderms have an endoskeleton consisting of *ossicles* (*ossiculum* = little bone), formed from calcium carbonate crystals. The shells of sand dollars and sea urchins are the endoskeletons of these animals.

The endoskeleton is the primary skeletal system of vertebrates. An adult human, for example, has an endoskeleton consisting of 206 bones arranged in two structural groups **(Figure 41.11)**. The **axial skeleton**, which includes the skull, vertebral column, sternum, and rib cage, forms the central part of the structure (shaded in red in Figure 41.11). The **appendicular skeleton** (shaded in green) includes the shoulder, hip, leg, and arm bones.

Bones of the Vertebrate Endoskeleton Are Organs with Several Functions

The vertebrate endoskeleton supports and maintains the overall shape of the body and protects key internal organs. In addition, the skeleton is a storehouse for calcium and phosphate ions, releasing them as required to maintain optimal levels of these ions in body fluids. Bones are also sites where new blood cells form.

Bones are complex organs built up from multiple tissues, including bone tissue with cells of several kinds, blood vessels, nerves, and in some, stores of adipose tissue. Bone tissue is distributed between dense, compact bone regions, which have essentially no spaces other than the microscopic canals of the osteons (see Figure 36.5d), and spongy bone regions, which are opened by larger spaces (see Figure 41.11). Compact bone tissue generally forms the outer surfaces of bones, and spongy bone tissue the interior. The interior of some flat bones, such as the hip bones and the ribs, are filled with *red marrow,* a tissue that is the primary source of new red blood cells in mammals and birds. The shaft of long bones such as the femur is opened by a large central canal filled with adipose tissue called *yellow marrow,* which is a source of some white blood cells.

Throughout the life of a vertebrate, calcium and phosphate ions are constantly deposited and withdrawn from bones. Hormonal controls maintain the concentration of Ca^{2+} ions at optimal levels in the blood and extracellular fluids (see Figure 40.10), ensuring that calcium is available for proper functioning of the nervous system, muscular system, and other physiological processes.

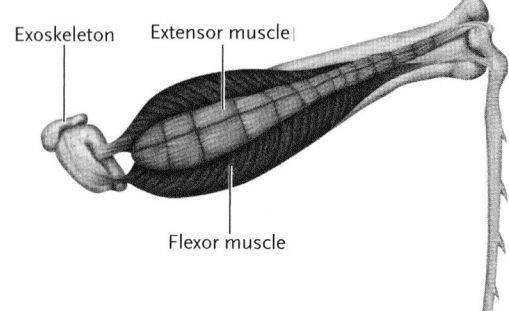

Figure 41.10
Muscles are attached to the inside surfaces of the exoskeleton in a typical insect leg such as this one.

Exoskeleton Extensor muscle Flexor muscle

STUDY BREAK

1. How do hydrostatic skeletons, exoskeletons, and endoskeletons provide support to the body? Give an example of each of these types in echinoderms and vertebrates.
2. What are the functions of the bones of the vertebrate endoskeleton?

41.3 Vertebrate Movement: The Interactions between Muscles and Bones

The skeletal systems of all animals act as a framework against which muscles work to move parts of the body or the entire organism. In this section, the muscle–bone interactions that are responsible for the movement of vertebrates are described.

Joints of the Vertebrate Endoskeleton Allow Bones to Move and Rotate

The bones of the vertebrate skeleton are connected by joints, many of them movable. The most-movable joints, including those of the shoulders, elbows, wrists, fingers, knees, ankles, and toes, are *synovial joints,* consisting of the ends of two bones enclosed by a fluid-filled capsule of connective tissue **(Figure 41.12a)**. Within the joint, the ends of the bones are covered by a smooth layer of cartilage and lubricated by synovial fluid, which makes the bones slide easily as the joint moves. Synovial joints are held together by straps of connective tissue called *ligaments,* which extend across the joints outside the capsule **(Figure 41.12b)**. The ligaments restrict the motion of the joint and help prevent it from buckling or twisting under heavy loads.

In other, less movable joints, called *cartilaginous joints,* the ends of bones are covered with layers of cartilage, but have no fluid-filled capsule surrounding them. Fibrous connective tissue covers and connects

Figure 41.11

Major bones of the human body. The inset shows the structure of a limb bone, with the location of red and yellow marrow. The internal spaces lighten the bone's structure. The cartilage layer forms a smooth, slippery cushion between bones in a joint.

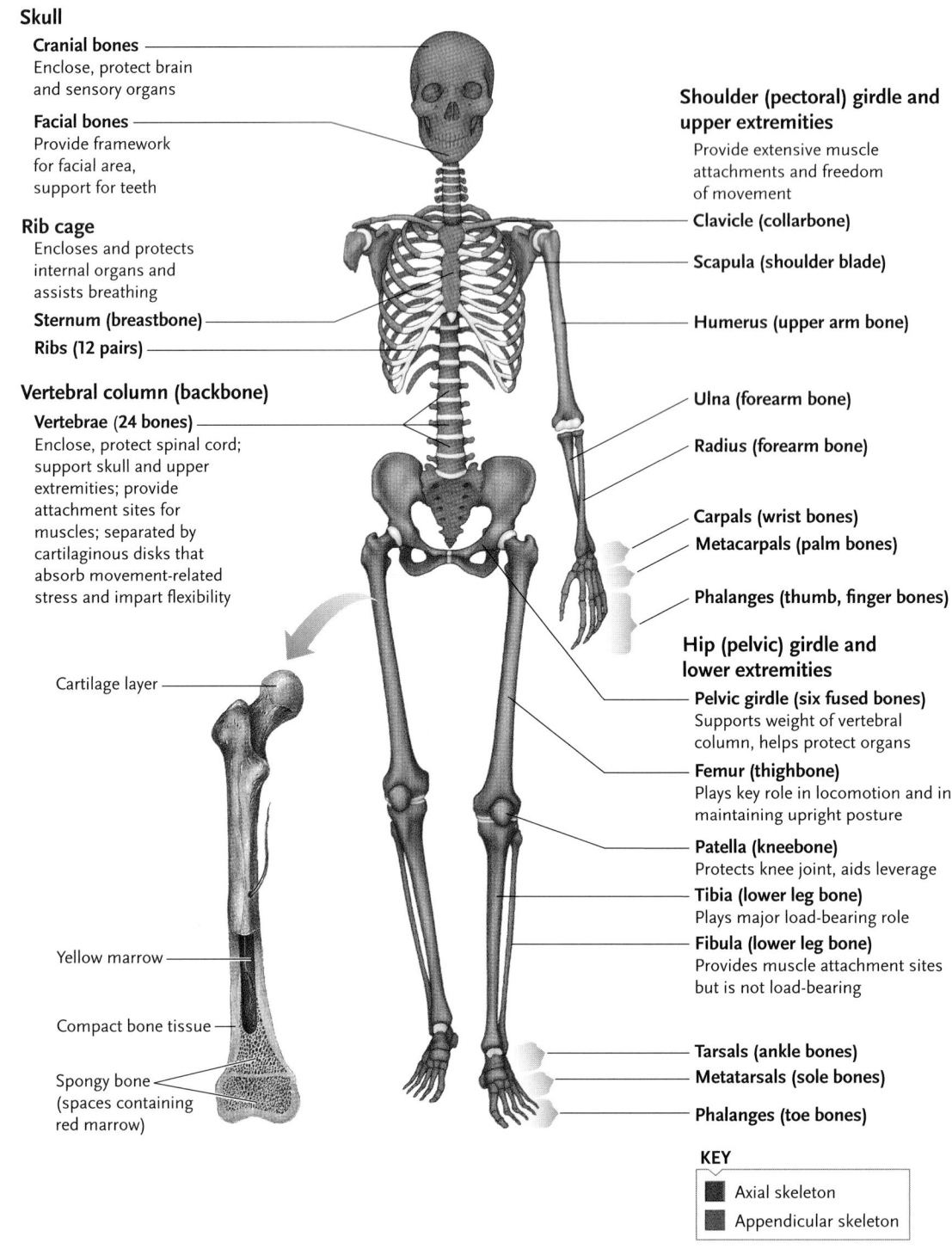

Skull

Cranial bones
Enclose, protect brain and sensory organs

Facial bones
Provide framework for facial area, support for teeth

Rib cage
Encloses and protects internal organs and assists breathing

Sternum (breastbone)

Ribs (12 pairs)

Vertebral column (backbone)

Vertebrae (24 bones)
Enclose, protect spinal cord; support skull and upper extremities; provide attachment sites for muscles; separated by cartilaginous disks that absorb movement-related stress and impart flexibility

Cartilage layer

Yellow marrow

Compact bone tissue

Spongy bone (spaces containing red marrow)

Shoulder (pectoral) girdle and upper extremities
Provide extensive muscle attachments and freedom of movement

Clavicle (collarbone)

Scapula (shoulder blade)

Humerus (upper arm bone)

Ulna (forearm bone)

Radius (forearm bone)

Carpals (wrist bones)
Metacarpals (palm bones)

Phalanges (thumb, finger bones)

Hip (pelvic) girdle and lower extremities

Pelvic girdle (six fused bones)
Supports weight of vertebral column, helps protect organs

Femur (thighbone)
Plays key role in locomotion and in maintaining upright posture

Patella (kneebone)
Protects knee joint, aids leverage

Tibia (lower leg bone)
Plays major load-bearing role

Fibula (lower leg bone)
Provides muscle attachment sites but is not load-bearing

Tarsals (ankle bones)
Metatarsals (sole bones)

Phalanges (toe bones)

KEY
Axial skeleton
Appendicular skeleton

the bones of these joints, which occur between the vertebrae and some rib bones.

In still other joints, called *fibrous joints,* stiff fibers of connective tissue join the bones and allow little or no movement. Fibrous joints occur between the bones of the skull and hold the teeth in their sockets.

The bones connected by movable joints work like levers. A lever is a rigid structure that can move around a pivot point known as a *fulcrum.* Levers differ with re-

spect to where the fulcrum is along the lever and where the force is applied. The most common type of lever system in the body—exemplified by the elbow joint—has the fulcrum at one end, the load at the opposite end, and the force applied at a point between the ends **(Figure 41.13).** For this lever, the force applied must be much greater than the load, but it increases the distance the load moves as compared with the distance over which the force is applied. This allows small muscle movements to produce large body movements, and also allows movements such as running or throwing to be carried out at high speed.

At a joint, a muscle that causes movement in the joint when it contracts is called an **agonist**. In many cases, other muscles that assist the action of an agonist are involved in the movement of a joint. For instance, deltoid and pectoral muscles assist the biceps brachii muscle in lifting a weight.

Most of the bones of vertebrate skeletons are moved by muscles arranged in **antagonistic pairs**: *extensor muscles* extend the joint, meaning increasing the angle between the two bones, while *flexor muscles* do the opposite. (Antagonistic muscles are also used in invertebrates for movement of body parts—for example, the limbs of insects and arthropods.) In humans, one such pair is formed by the biceps brachii muscle at the front of the upper arm and the triceps brachii muscle at the back of the upper arm **(Figure 41.14)**. When the biceps muscle contracts, the bone of the lower arm is bent (flexed) around the elbow joint, and the triceps muscle is passively stretched (see Figure

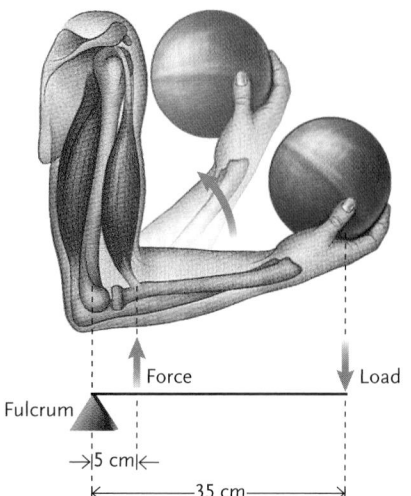

Figure 41.13
A body lever: The lever formed by the bones of the forearm. The fulcrum (the hinge or joint) is at one end of the lever, the load is placed on the opposite end, and the force is exerted at a point on the lever between the fulcrum and the load.

41.14a); when the triceps muscle contracts, the lower arm is straightened (extended) and the biceps muscle is passively stretched (see Figure 41.14b).

Vertebrates Have Muscle–Bone Interactions Optimized for Specific Movements

Vertebrates differ widely in the patterns by which muscles connect to bones, and in the length and mechanical advantage of the levers produced by these connections. These differences produce limbs and other body parts that are adapted for either power or speed, or the most advantageous compromise between these characteristics. Among burrowing mammals such as the mole, for example, the limb bones are short, thick, and heavy, and the point at which muscles attach produce levers that are slow to move

a. Synovial joint cross section

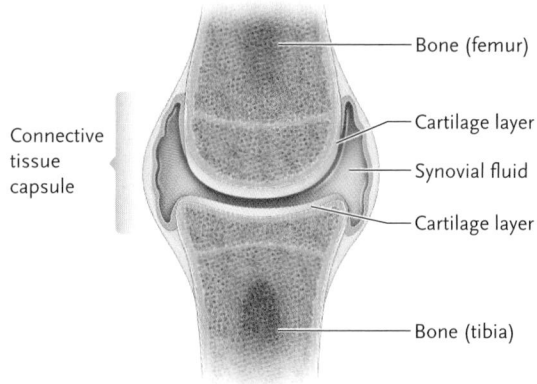

b. Knee joint ligaments

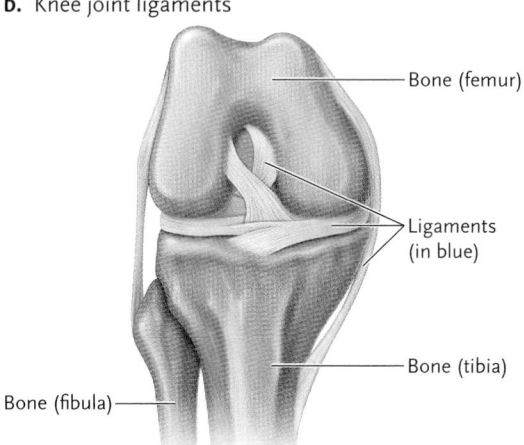

Figure 41.12
A synovial joint. **(a)** Cross section of a typical synovial joint. **(b)** Ligaments reinforcing the knee joint.

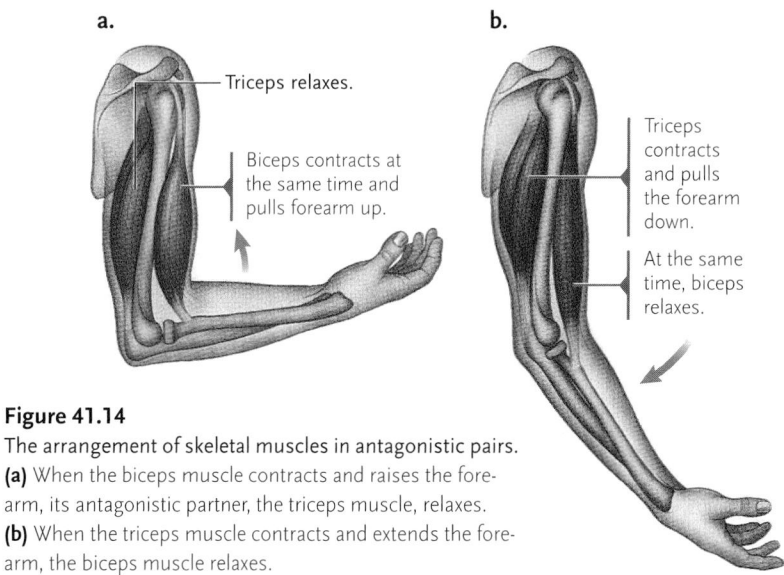

Figure 41.14
The arrangement of skeletal muscles in antagonistic pairs.
(a) When the biceps muscle contracts and raises the forearm, its antagonistic partner, the triceps muscle, relaxes.
(b) When the triceps muscle contracts and extends the forearm, the biceps muscle relaxes.

but that need to apply smaller forces to move a load compared with a human biceps. In contrast, a mammal such as the deer has relatively light and thin bones with muscle attachments producing levers that can produce rapid movement, moving the body easily over the ground.

STUDY BREAK

1. Distinguish synovial joints, cartilaginous joints, and fibrous joints.
2. What are antagonistic muscle pairs?

UNANSWERED QUESTIONS

How can muscle growth processes be controlled to improve the clinical treatment of muscular dystrophy and related disorders?

The mechanisms by which different muscle types develop and their impact on organismal metabolism is not well known. However, we do know that one of the proteins produced in muscle cells, myostatin, is a growth factor that inhibits skeletal muscle growth and development. Thus, myostatin is a potential therapeutic target for treating some of the most debilitating types of muscular dystrophy, which is a degenerative and fatal disease associated with the progressive loss of skeletal muscle mass. Animals with mutations that knock out myostatin gene function completely (for example, the Belgian Blue and Piedmontese cattle breeds) or myostatin knockout mice, in which the gene has been removed experimentally (see Section 18.2 and *Focus on Research* in Chapter 43), have significantly enhanced musculature that is commonly referred to as *double muscling*. Such mutations and enhanced skeletal muscle mass have also been described in a racing dog breed, the whippet, and recently in a young boy.

Our laboratory studies focus on developing novel technologies that introduce protein inhibitors of myostatin activity—in essence, inhibiting the inhibitor—and thereby stimulate skeletal muscle growth in both clinical and agricultural settings. We have recently determined that myostatin can also negatively regulate cardiac muscle growth. Thus, disrupting myostatin production or availability may also help heart attack patients. Replacing damaged skeletal and cardiac muscle using adult or embryonic stem cells engineered to match either tissue type is another highly promising technique for treating these disorders; it could be improved by using "antimyostatin" technologies that enhance growth of the transplanted cells.

How much do the metabolic processes of skeletal muscle specifically contribute to energy storage and whole body form?

Complications associated with obesity, particularly type 2 diabetes mellitus, have reached near-epidemic proportions worldwide. Type 2 diabetes differs from type 1 and is caused not by a lack of the pancreatic hormone insulin but rather by insulin resistance, in which an individual's physiological levels of insulin are inadequate to produce a normal insulin response in the tissues. Both types, however, result in the body's inability to properly process and store metabolites, mostly glucose. Type 2 diabetes can be a debilitating and fatal disease if poorly managed and often aggravates other diseases as well. Scientists now recognize that growth and metabolic processes are integrated and controlled by the same hormones, growth factors, and cytokines. Indeed, skeletal muscle is the largest consumer of metabolites and has the greatest potential to impact their circulating levels. Recent studies suggest that increasing muscle mass can significantly reduce fat mass as growing muscle is supported by energy from fat metabolism. Enhancing skeletal muscle growth and/or the ability of the tissue to consume blood metabolites in obese patients with type 2 diabetes could therefore improve treatments for both. The same antimyostatin technologies used to treat muscle growth disorders could also be used to treat severe cases of obesity and type 2 diabetes with the goals of increasing muscle mass, decreasing fat mass, and improving insulin sensitivity.

How do the extremely complex electrical properties of cardiac muscle develop, and how can they be controlled for biomedical purposes?

An ischemic event that blocks blood flow to a region of the heart and ultimately deprives the muscle of oxygen often results in a heart attack and can damage or destroy significant amounts of cardiac muscle. The surviving muscle, however, compensates by increasing the specific force generated by individual myofibers. Scientists have recently determined that these changes are due to the remodeling of electrical properties—changes in the amount and relative distribution as well as the activity of different classes of ion channels—within the surviving muscle itself. Although the specific channels and the mechanisms of regulation are unknown, a better understanding and ultimately control of these processes could help heart attack patients survive.

Buel (Dan) Rodgers is an assistant professor and assistant animal scientist at Washington State University, studying molecular endocrinology and animal genomics, specifically skeletal muscle growth and development. To learn more about his research, visit http://www.ansci.wsu.edu/People/rodgers/faculty.asp.

Review

Go to **ThomsonNOW**™ at www.thomsonedu.com/login to access quizzing, animations, exercises, articles, and personalized homework help.

41.1 Vertebrate Skeletal Muscle: Structure and Function

- Skeletal muscles move the joints of the body. They are formed from long, cylindrical cells called muscle fibers, which are packed with myofibrils, contractile elements consisting of myosin thick filaments and actin thin filaments. The two types of filaments are arranged in an overlapping pattern of contractile units called sarcomeres (Figure 41.2).

- Infoldings of the plasma membrane of the muscle fiber form T tubules. The sarcomeres are encircled by the sarcoplasmic reticulum, a system of vesicles with segments separated from T tubules by small gaps (Figure 41.3).

- In the sliding filament mechanism of muscle contraction, the simultaneous sliding of thin filaments on each side of sarcomeres over the thick filaments shortens the sarcomeres and the muscle fibers, producing the force that contracts the muscle (Figure 41.4).

- The sliding motion of thin and thick filaments is produced in response to an action potential arriving at the neuromuscular junction. The action potential causes the release of acetylcholine, which triggers an action potential in the muscle fiber that spreads over its plasma membrane and stimulates the sarcoplasmic reticulum to release Ca^{2+} into the cytosol. The Ca^{2+} combines with troponin, inducing a conformational change that moves tropomyosin away from the myosin-binding sites on thin filaments. Exposure of the sites allows myosin crossbridges to bind and initiate the crossbridge cycle in which the myosin heads of thick filaments attach to a thin filament, pull, and release in cyclic reactions powered by ATP hydrolysis (Figure 41.5).

- When action potentials stop, Ca^{2+} is pumped back into the sarcoplasmic reticulum, leading to Ca^{2+} release from troponin, which allows tropomyosin to cover the myosin-binding sites in the thin filaments, thereby stopping the crossbridge cycle (Figure 41.5).

- A single action potential arriving at a neuromuscular junction causes a muscle twitch. Restimulation of a muscle fiber before it has relaxed completely causes a second twitch, which is added to the first, causing a summed, stronger contraction. Rapid arrival of APs causes the twitches to sum to a peak level of contraction called tetanus. Normally, muscles contract in a tetanic mode (Figure 41.6).

- Muscle fibers occur in three types. Slow muscle fibers contract relatively slowly, but do not fatigue rapidly. Fast aerobic fibers contract relatively quickly and powerfully, and fatigue more quickly than slow fibers. Fast anaerobic fibers can contract more rapidly and powerfully than fast aerobic fibers, but fatigue more rapidly. The fibers differ in their number of mitochondria and capacity to produce ATP (Table 41.1).

- Skeletal muscles are divided into motor units, consisting of a group of muscle fibers activated by branches of a single motor neuron. The total force produced by a skeletal muscle is determined by the number of motor units that are activated (Figure 41.7).

- Invertebrate muscles contain thin and thick filaments arranged in sarcomeres, and contract by the same sliding filament mechanism that operates in vertebrates.

Animation: Structure of skeletal muscle

Animation: Sliding filament model

Animation: Nervous system and muscle contraction

Animation: Troponin and tropomyosin

Animation: Energy sources for contraction

Animation: Types of contractions

41.2 Skeletal Systems

- A hydrostatic skeleton is a structure consisting of a muscle-surrounded compartment or compartments filled with fluid under pressure. Contraction and relaxation of the muscles changes the shape of the animal (Figures 41.8 and 41.9).

- In an exoskeleton, a rigid external covering provides support for the body. The force of muscle contraction is applied against the covering. An exoskeleton can also protect delicate internal tissues (Figure 41.10).

- In an endoskeleton, the body is supported by rigid structures within the body, such as bones. The force of muscle contraction is applied against those structures. Endoskeletons also protect delicate internal tissues. In vertebrates, the endoskeleton is the primary skeletal system. The vertebrate axial skeleton consists of the skull, vertebral column, sternum, and rib cage, while the appendicular skeleton includes the shoulder bones, the forelimbs, the hip bones, and the hind limbs (Figure 41.11).

- Bone tissue is distributed between compact bone, with no spaces except the microscopic canals of the osteons, and spongy bone tissue, which has spaces filled by red or yellow marrow (Figure 41.11).

- Calcium and phosphate ions are constantly exchanged between the blood and bone tissues. The turnover keeps the Ca^{2+} concentration balanced at optimal levels in body fluids.

Animation: Vertebrate skeletons

Animation: Human skeletal system

Animation: Structure of a femur

Animation: Long bone formation

41.3 Vertebrate Movement: The Interactions between Muscles and Bones

- The bones of a skeleton are connected by joints. A synovial joint, the most movable type, consists of a fluid-filled capsule surrounding the ends of the bones forming the joint. A cartilaginous joint, which is less movable, has smooth layers of cartilage between the bones with no surrounding capsule. The bones of a fibrous joint are joined by connective tissue fibers that allow little or no movement (Figure 41.12).

- The bones moved by skeletal muscles act as levers, with a joint at one end forming the fulcrum of the lever, the load at the opposite end, and the force applied by attachment of a muscle at a point between the ends (Figure 41.13).

- At a joint, an agonist muscle, perhaps assisted by other muscles, causes movement. Most skeletal muscles are arranged in antagonistic pairs, in which the members of a pair pull a bone in opposite directions. When one member of the pair contracts, the other member relaxes and is stretched (Figure 41.14).

- Vertebrates have a variety of patterns in which muscles connect to bones, giving different properties to the levers produced. Those properties are specialized for the activities of the animal.

Animation: Opposing muscle action

Animation: Human skeletal muscles

Questions

Self-Test Questions

1. Vertebrate skeletal muscle:
   a. is attached to bone by means of ligaments.
   b. may bend but not extend body parts.
   c. may rotate one body part with respect to another.
   d. is found in the walls of blood vessels and intestines.
   e. is usually attached at each end to the same bone.

2. In a resting muscle fiber:
   a. sarcomeres are regions between two H zones.
   b. discs of M line proteins called the A band separate the thick filaments.
   c. I bands are composed of the same thick filaments seen in the A bands.
   d. Z lines are adjacent to H zones, which attach thick filaments.

    e.   dark A bands contain overlapping thick and thin filaments with a central thin H zone composed only of thick filaments.

3.   The sliding filament contractile mechanism:
    a.   causes thick and thin filaments to slide toward the center of the A band, bringing the Z lines closer together.
    b.   is inhibited by the influx of Ca^{2+} into the muscle fiber cytosol.
    c.   lengthens the sarcomere to separate the I regions.
    d.   depends on the isolation of actin and myosin until a contraction is completed.
    e.   uses myosin crossbridges to stimulate delivery of Ca^{2+} to the muscle fiber.

4.   During contraction of skeletal muscle:
    a.   ATP stimulates Ca^{2+} to move out of the cytosol, which allows tropomyosin to bind myosin causing contraction of the thin filament.
    b.   myosin crossbridges use ATP to relax the molecular spring in the myosin head, which pulls the thick filaments away from the thin actin filaments.
    c.   actin binds ATP, allowing troponin in the thick filaments to form the myosin crossbridge.
    d.   action potentials cause the release of Ca^{2+} into the sarcoplasmic reticulum allowing tropomyosin fibers to uncover the actin binding sites needed for the myosin crossbridge.
    e.   botulinum toxin could increase the release of acetylcholine at the contracting muscle site.

5.   When a trained marathoner is running, most likely his:
    a.   muscles have low concentrations of myoglobin.
    b.   slow muscle fibers will do most of the work for the run.
    c.   slow muscle fibers will remain in constant tetanus over the length of the run.
    d.   fast muscle fibers will be employed in the middle of his run.
    e.   slow muscle fibers are using ATP obtained primarily by anaerobic respiration.

6.   Which description is characteristic of a motor unit?
    a.   A single motor unit's muscle fibers vary among the slow/fast aerobic and slow/fast anaerobic forms.
    b.   When receiving an action potential, a motor unit is controlled by a single efferent axon that causes all its fibers to contract.
    c.   When a motor unit contracts, certain sections of the muscle as a whole remain relaxed.
    d.   If a motor unit controls walking, it is found in large numbers in the same volume of muscle.
    e.   If a motor unit controls finger movement, it contains a large number of muscle fibers that are stimulated over a large area.

7.   Which of the following is *not* an example of a hydrostatic skeletal structure?
    a.   the tube feet of sea urchins
    b.   the body wall of annelids
    c.   the trunk of an elephant
    d.   the body wall of cnidarians
    e.   the penis of mammals

8.   Endoskeletons:
    a.   protect internal organs and provide structures against which the force of muscle contraction can work.
    b.   differ from exoskeletons in that endoskeletons do not support the external body.
    c.   cannot be found in mollusks and echinoderms.
    d.   are composed of appendicular structures that form the skull.
    e.   compose the arms and legs, which are part of the axial skeleton.

9.   Connecting the vertebrate skeleton are:
    a.   nonmovable synovial joints.
    b.   ligaments holding together connective tissue of fibrous joints.
    c.   cartilaginous joints found in the shoulders and elbows.
    d.   synovial joints lubricated by synovial fluid.
    e.   fibrous joints that move around a fulcrum allowing small muscle movements to produce large body movements.

10.   The movement of vertebrate muscles is:
    a.   agonistic when it extends the joint.
    b.   antagonistic when it causes movement in the joint.
    c.   caused by extensor muscles that flex the joint.
    d.   caused by flexor muscles that extend the joint.
    e.   most efficient when the biceps and triceps contract simultaneously.

Questions for Discussion

1.   A coach must train young athletes for the 100-meter sprint. They need muscles specialized for speed and strength, rather than for endurance. What kinds of muscle characteristics would the training regimen aim to develop? How would it be altered to train marathoners?

2.   What kind of exercise program might the coach in question 1 recommend to an older person developing osteoporosis? Why?

3.   Based on material in this chapter and in Chapter 40 on endocrine controls, outline some possible causes and physiological effects of calcium deficiency in an active adult.

Experimental Analysis

Design an experiment with rats to determine whether endurance training alters the proportion of slow, fast aerobic, and fast anaerobic muscle fibers.

Evolution Link

What characteristics of vertebrate muscle suggest that the genes for muscle structure were inherited from invertebrate ancestors?

How Would You Vote?

Dietary supplements are largely unregulated. Should they be placed under the jurisdiction of the Food and Drug Administration, which could subject them to more stringent testing for effectiveness and safety? Go to www.thomsonedu.com/login to investigate both sides of the issue and then vote.

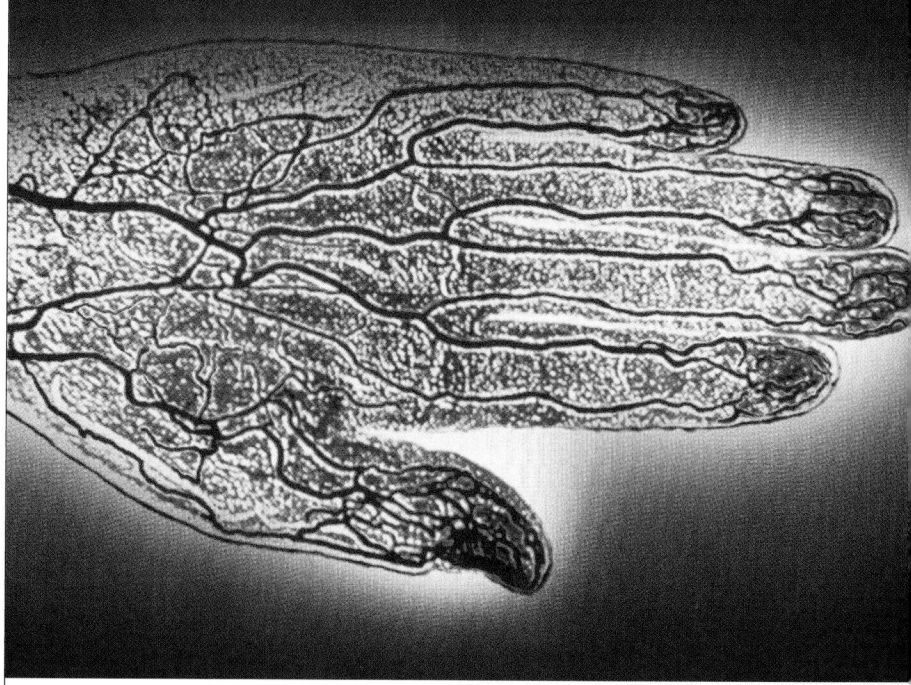

Arteries of the human hand (colorized X-ray). Arteries are vessels of the circulatory system that transport molecules, such as oxygen, nutrients, hormones, and wastes, as well as certain cells, from one tissue to another.

GJLP/CNRI/SPL/Photo Researchers, Inc.

42 The Circulatory System

WHY IT MATTERS

Jimmie the bulldog stood on the stage of a demonstration laboratory at a meeting of the Royal Society in London in 1909, with one front paw and one rear paw in laboratory jars containing salt water (**Figure 42.1**). Wires leading from the jars were connected to a galvanometer, a device that can detect electrical currents. Jimmie's master, Dr. Augustus Waller, a physician at St. Mary's Hospital, was relating his experiments in the emerging field of *electrophysiology*. Among other discoveries, Waller had found that his apparatus detected the electrical currents produced each time the dog's heart beat.

Waller originally made his finding by experimenting on himself. He already knew that the heart creates an electrical current as it beats; other scientists had found this out by attaching electrodes directly to the heart of experimental animals. Looking for a painless alternative to that procedure, Waller reasoned that because the human body can conduct electricity, his arms and legs might conduct the currents generated by the heart if they were connected to a galvanometer. Accordingly, Waller set up two metal pans containing salt water and connected wires from the pans to a galvanometer. He put his bare left foot in one of the pans, and his right hand in the other one. The tech-

From A. D. Waller, Physiology, The Servant of Medicine, Hitchcock Lectures, University of London Press, 1910

a. Jimmie the bulldog

b. Electrocardiogram

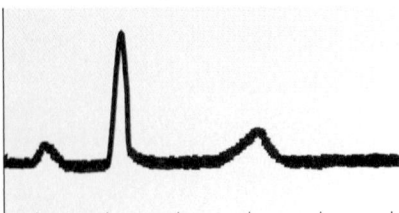

Figure 42.1

The first electrocardiograms. **(a)** Jimmie the bulldog standing in laboratory jars containing salt water, with wires leading to a galvanometer that recorded the electrical currents produced by his heartbeat. **(b)** One of Waller's early electrocardiograms.

nique worked; the indicator of the galvanometer jumped each time his heart beat. And, it worked with Jimmie.

Waller also invented a method for recording the changes in current, which became the first electrocardiogram (ECG). He constructed a galvanometer by placing a column of mercury in a fine glass tube, with a conducting salt solution layered above the mercury. Changes in the current passing through the tube caused corresponding changes in the surface tension of the mercury, which produced movements that could be detected by reflecting a beam of light from the mercury surface. By placing a moving photographic plate behind the mercury tube, Waller could record the movements of the reflected light on the plate (Figure 42.1b shows one of his records).

The beating of Jimmie's heart, recorded as an electrical trace by Augustus Waller, is part of the actions of the **circulatory system**, an organ system consisting of a fluid, a heart, and vessels for moving important molecules, and often cells, from one tissue to another. Examples of transported molecules are oxygen (O_2), nutrients, hormones, and wastes.

We study these systems in this chapter, with emphasis on the circulatory system of humans and other mammals. We also discuss the **lymphatic system**, an accessory system of vessels and organs that helps balance the fluid content of the blood and surrounding tissues and participates in the body's defenses against invading disease organisms.

42.1 Animal Circulatory Systems: An Introduction

The least complex animals, including sponges, cnidarians, and flatworms, function with no circulatory system. All of these animals are aquatic or, like parasitic flatworms, live surrounded by the body fluids of a host animal. Their bodies are structured as thin sheets of cells that lie close to the fluids of the surrounding environment. Substances diffuse between the cells and the environment through the animal's external surface, or through the surfaces of internal channels and cavities that are open to the environment.

In sponges and cnidarians, flagella or cilia help to circulate the external fluid through the internal spaces. As the water passes over the cells, they pick up O_2 and nutrients, and release wastes and CO_2.

In the sponges, water carrying nutrients and O_2 enters through pores in body walls surrounding a central cavity, passes through the cavity, and leaves through a large exit pore **(Figure 42.2a).** Hydras, jellyfish, sea anemones, and other cnidarians have a central *gastrovascular cavity* with a mouth that opens to the outside and extensions that radiate into the tentacles and all other body regions **(Figure 42.2b).** Water enters and leaves through the mouth, serving both digestion and circulation.

a. Sponge **b.** *Hydra;* a cnidarian

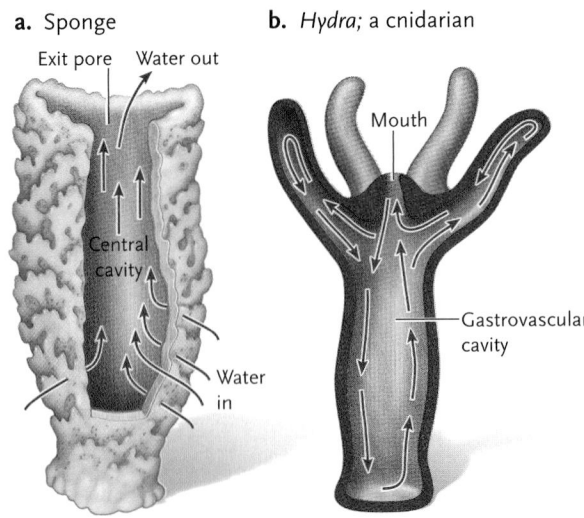

Figure 42.2

Invertebrates with no circulatory system.

Animal Circulatory Systems Share Basic Elements

In larger and more complex animals, most cells lie in cell layers too deep within the body to exchange substances directly with the environment via diffusion. Instead, the animals have a set of tissues and organs—a circulatory system—that conducts O_2, CO_2, nutrients, and wastes between body cells and specialized regions of the animal where substances are exchanged with the external environment. In terrestrial vertebrates, for example, oxygen from the environment is absorbed in the lungs and is carried by the blood to all parts of the body; CO_2 released from body cells is carried by the blood to the lungs, where it is released to the environment. Wastes are conducted from body cells to the kidneys, which remove wastes from the circulation and excrete them into the environment.

The animal circulatory systems carrying out these roles share certain basic features:

1. A specialized fluid medium, such as the blood of mammals and other vertebrates, that carries O_2, CO_2, nutrients, and wastes, and plays a major role in homeostasis (see Section 36.4).
2. A muscular heart that pumps the fluid through the circulatory system.
3. Tubular vessels that distribute the fluid pumped by the heart.

Words associated with the heart often include *cardi(o)*, from *kardia*, the Greek word for heart.

Animal circulatory systems take one of two forms, either *open* or *closed* **(Figure 42.3)**. In an **open circulatory system**, vessels leaving the heart release bloodlike fluid termed **hemolymph** directly into body spaces, called **sinuses**, that surround organs. Thus there is no distinction between hemolymph and *interstitial fluid*, the fluid immediately surrounding body cells. After flowing through the sinuses, the hemolymph reenters the heart through valves in the heart wall. In a **closed circulatory system**, the fluid—blood—is confined to blood vessels and is distinct from the interstitial fluid. Substances are exchanged between the blood and the interstitial fluid and then between the interstitial fluid and cells.

Most Invertebrates Have Open Circulatory Systems

Arthropods and most mollusks have open circulatory systems with one or more muscular hearts **(Figure 42.4)**. In an open system, most of the fluid pressure generated by the heart dissipates when the hemolymph is released into the sinuses. As a result, hemolymph flows relatively slowly. Open systems operate efficiently in these animals because they lead relatively sedentary lives and, as a consequence, their tissues do not require O_2 and nutrients at the rate and quantities required by

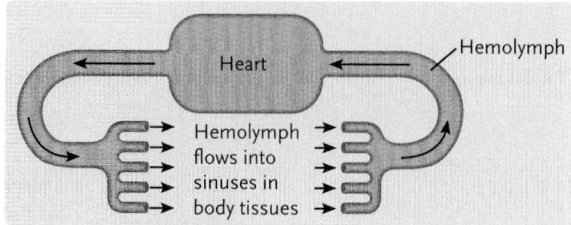

a. Open circulatory system: no distinction between hemolymph and interstitial fluid

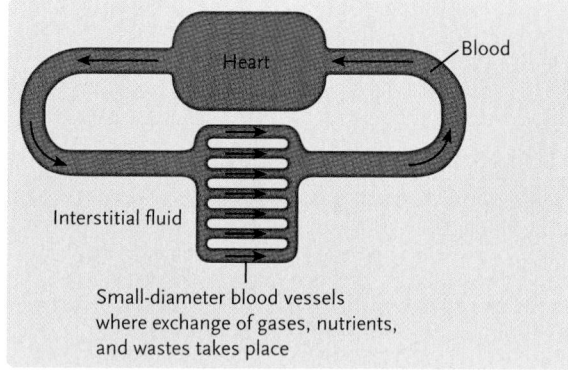

b. Closed circulatory system: blood separated from interstitial fluid

Figure 42.3
Open and closed circulatory systems.

more active species. Among highly mobile and active species with open systems, such as insects and crustaceans, other adaptations compensate for the relatively slow distribution of hemolymph. In insects, for example, O_2 and CO_2 are exchanged efficiently with the environment by specialized air passages that branch throughout the body rather than by the hemolymph (these air passages, called *tracheae*, are discussed in Section 44.2).

Some Invertebrates and All Vertebrates Have Closed Circulatory Systems

Annelids, cephalopod mollusks such as squids and octopuses, and all vertebrates have closed circulatory systems. In these systems, vessels called **arteries** conduct blood away from the heart at relatively high pressure. From the arteries, the blood enters highly branched networks of microscopic, thin-walled vessels called **capillaries** that are well adapted for diffusion of substances. Nutrients and wastes are exchanged between the blood and body tissues as the blood moves through the capillaries. The blood then flows at relatively low

Figure 42.4
Open circulatory system of a grasshopper.

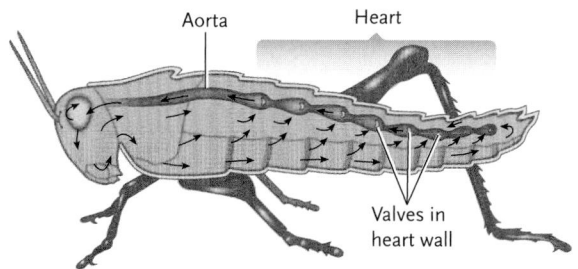

a. Circulatory system of fishes

Capillary networks of gills

- Artery
- Ventricle
- Heart
- Atrium
- Vein

Capillary networks in other body tissues

In fishes, a heart consisting of a series of two chambers pumps blood into one circuit. Blood picks up oxygen in the gills and delivers it to the rest of the body. Deoxygenated blood flows back to the heart.

b. Circulatory system of amphibians

Lungs and skin capillaries

PULMOCUTANEOUS CIRCUIT

Right atrium — Left atrium

Ventricle — Heart — Tissue flap

SYSTEMIC CIRCUIT

Capillary networks in other body tissues

In amphibians, the heart pumps blood through two circuits. Oxygenated blood from the lungs and skin and deoxygenated blood from the rest of the body are kept partially separate by a smooth pattern of flow and a flap of tissue in the large artery leaving the heart.

c. Circulatory system of turtles, lizards, and snakes

Lung capillaries

PULMONARY CIRCUIT

Right atrium — Left atrium

Ventricles — Septum

SYSTEMIC CIRCUIT

Capillary networks in other body tissues

In turtles, lizards, and snakes, a wall of tissue, the septum, improves the separation of oxygenated blood from the lungs and deoxygenated blood from the rest of the body in the single ventricle.

d. Circulatory system of crocodilians, birds, and mammals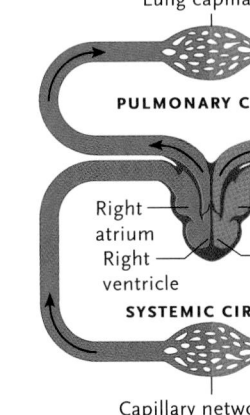

Lung capillaries

PULMONARY CIRCUIT

Right atrium — Left atrium
Right ventricle — Left ventricle

SYSTEMIC CIRCUIT

Capillary networks in other body tissues

In the four-chambered heart of crocodilians, birds, and mammals, a complete septum forms two ventricles and keeps the flow of oxygenated blood from the lungs and deoxygenated blood entirely separate from the rest of the body.

KEY

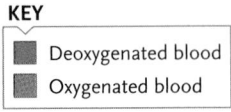

■ Deoxygenated blood
■ Oxygenated blood

Figure 42.5
Evolutionary developments in the heart and circulatory system of major vertebrate groups.

---

pressure from the capillaries to larger vessels, the **veins**, which carry the blood back to the heart.

Typically, the blood is maintained at higher pressure and moves more rapidly through the body in closed systems than in open systems. Closed systems are highly efficient in the distribution of O_2 and nutrients and in the clearance of CO_2 and wastes.

In many animals, but particularly the vertebrates, closed systems allow precise control of the distribution and rate of blood flow to different body regions by means of muscles that contract or relax to adjust the diameter of the blood vessels.

Vertebrate Circulatory Systems Have Evolved from Single to Double Blood Circuits

A comparison of the different vertebrate groups reveals several evolutionary trends that accompanied the invasion of terrestrial habitats. Among the most striking is a trend from an effectively single circuit of blood pumped by the heart in sharks and bony fishes to the two completely separate circuits of birds and mammals. As part of this development, the heart evolved from one series of chambers to a double pump acting in two parallel series. In other words, depending on the vertebrate, the heart may consist of one or two **atria** (singular, *atrium*), the chambers that receive blood returning to the heart, and one or two **ventricles**, the chambers that pump blood from the heart.

The Single Blood Circuit of Fishes. In fishes, the heart consists of two chambers arranged in a single line **(Figure 42.5a).** The ventricle of the heart pumps blood into arteries leading to capillary networks in the gills, where the blood releases CO_2 and picks up O_2. The oxygenated blood—now moving more slowly and at lower pressure after passing through the gill capillaries—flows through another series of arteries and is delivered to capillary networks in other body tissues where it delivers O_2 and picks up CO_2. The comparatively slow movement of the blood through this segment of the circulatory system is accelerated by the contractions of skeletal muscles as the animal swims. In some fishes, a vein in the tail is expanded and surrounded by a mass of skeletal muscle that contracts rhythmically, forming an accessory *caudal heart,* which pumps venous blood toward the anterior end of the body. Eventually, the deoxygenated blood enters veins that carry it back to the atrium of the heart, and the single circuit is repeated.

The circulatory system of fishes is highly suited to the environments in which these animals live. Their bodies, supported by water and adapted for swimming, do not use as much energy in locomotion as do animals of an equivalent size moving on the land or in the air. Hence, fishes require less O_2 for their activities than terrestrial vertebrates do, and their relatively simple circulatory systems fully meet their O_2 requirements.

Double Blood Circuits. Vertebrate hearts changed significantly when the first air-breathing fishes such as the lungfish evolved. In these fishes, the lung evolved as a respiratory organ in addition to gills. The lung necessitated a separate circuit because it is an additional organ for oxygenating the blood and, unlike other organs, does not need to receive oxygenated blood from the gills.

In amphibians, the separation into two circuits was accomplished by division of the atrium into two parallel chambers, the left and right atria, to produce a three-chambered heart, and by adaptations that keep oxygenated and deoxygenated blood partially separate as they are pumped by the single ventricle **(Figure 42.5b).** Amphibians obtain O_2 from gas exchange across their moist skin as well as in their gills or lungs. Oxygenated blood from these organs enters veins that lead to the left atrium, while O_2-depleted blood from the rest of the body enters the right atrium. The atria contract simultaneously, pumping the oxygenated and deoxygenated blood into the single ventricle. The two types of blood remain mostly separated because they differ in density, although some mixing occurs. As the blood is pumped from the ventricle, most of the oxygenated blood enters the branch that leads to the **systemic circuit** of the body, which provides the blood supply for most of the tissues and cells of the body. The deoxygenated blood is directed into the other branch, which leads to the skin and lungs or gills, called the **pulmocutaneous circuit.** Because the blood flows through separate systemic and pulmocutaneous circuits in the amphibians, the blood leaving the heart flows through only one capillary network in each circuit before returning to the heart. This separation greatly increases the blood pressure and flow in the systemic circuit as compared with that of fishes.

Reptiles also have two atria and a single ventricle. The ventricle is divided into right and left halves by a flap of connective tissue called the *septum,* which keeps the flow of oxygenated and deoxygenated blood almost completely separate in a systemic circuit and a **pulmonary circuit,** a branch leading to the lungs **(Figure 42.5c).** The septum is incomplete in turtles, lizards, and snakes, allowing some mixing of oxygenated and deoxygenated blood. In crocodilians (crocodiles and alligators), the septum is complete **(Figure 42.5d).**

Birds (which share ancestry with crocodilians) and mammals have a double heart consisting of two atria and two ventricles (as in Figure 42.5d). In effect, each half of the heart operates as a separate pump, restricting the blood circulation to completely separate pulmonary and systemic circuits. Blood is pumped by a ventricle in each circuit, so that both operate at relatively high pressure.

STUDY BREAK

1. What are the three basic features of animal circulatory systems?
2. Distinguish between an open circulatory system and a closed circulatory system. Why do you think humans could not function with an open circulatory system?
3. Distinguish among atria, ventricles, arteries, and veins.
4. Distinguish among the systemic, pulmocutaneous, and pulmonary circuits.

42.2 Blood and Its Components

In vertebrates, blood is a complex connective tissue containing blood cells suspended in a liquid matrix called the *plasma.* In addition to transporting molecules, blood helps stabilize the internal pH and salt composition of body fluids, and serves as a highway for cells of the immune system and the antibodies produced by some of these cells. It also helps regulate body temperature by transferring heat between warmer and cooler body regions, and between the body and the external environment (the role of the blood in temperature regulation is discussed in Chapter 46).

For an average-sized adult human, the total blood volume is about 4 to 5 liters—more than a gallon—and makes up about 8% of body weight. The *plasma,* a clear, straw-colored fluid, makes up about 55% of the volume of blood in human males and 58% in human females on average. Suspended in the plasma are three main types of blood cells—*erythrocytes, leukocytes,* and *platelets*—which make up the remainder of the blood volume; on average about 45% and 42% for human males and females, respectively. The typical components of human blood are shown in **Figure 42.6.**

In humans, blood cells develop in red bone marrow primarily in the vertebrae, sternum (breastbone), ribs, and pelvis. Blood cells originate in a single type of cells called *pluripotent (plura* = multiple; *potens* = power) *stem cells,* which retain the embryonic capacity to divide **(Figure 42.7).** Pluripotent stem cells differentiate into two other types of stem cells, myeloid stem cells and lymphoid stem cells. The myeloid stem cells give rise to erythrocytes, platelets, and several types of leukocytes, namely neutrophils, basophils, eosinophils, and monocyte/macrophages. The lymphoid stem cells give rise to other types of leukocytes, namely the natural killer cells, T lymphocytes, and B lymphocytes.

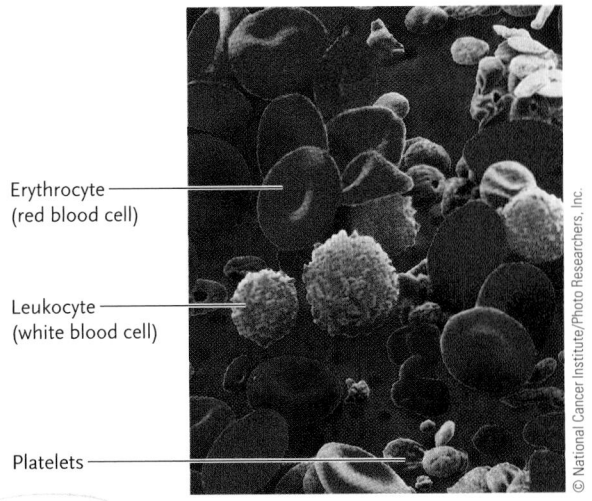

Erythrocyte (red blood cell)

Leukocyte (white blood cell)

Platelets

© National Cancer Institute/Photo Researchers, Inc.

Figure 42.6
Typical components of human blood. The colorized scanning electron micrograph shows the three major cellular components. The sketch of the test tube shows what happens when you centrifuge a blood sample. The blood separates into three layers: a thick layer of straw-colored plasma on top, a thin layer containing leukocytes and platelets, and a thick layer of erythrocytes. The table shows the relative amounts and functions of the various components of blood.

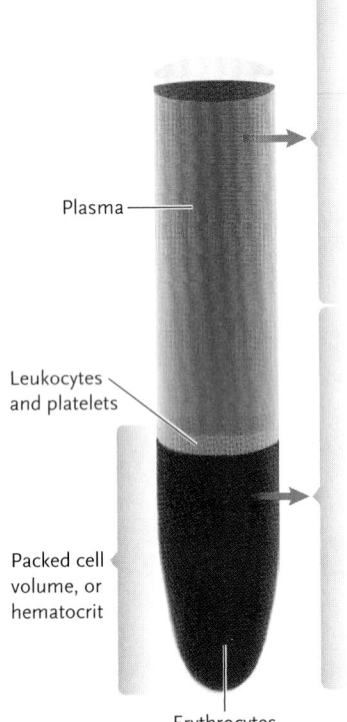

Plasma

Leukocytes and platelets

Packed cell volume, or hematocrit

Erythrocytes

Plasma Portion (55%–58% of total volume):

Components	Relative Amounts	Functions
1. Water	91%–92% of plasma volume	Solvent
2. Plasma proteins (albumin, globulins, fibrinogen, etc.)	7%–8%	Defense, clotting, lipid transport, roles in extracellular fluid volume, etc.
3. Ions, sugars, lipids, amino acids, hormones, vitamins, dissolved gases	1%–2%	Roles in extracellular fluid volume, pH, etc.

Cellular Portion (45%–42% of total volume):

Components	Relative Amounts	Functions
1. Erythrocytes (red blood cells)	4,800,000–5,400,000 per microliter	Oxygen, carbon dioxide transport
2. Leukocytes (white blood cells) 　Neutrophils 　Lymphocytes 　Monocytes/macrophages 　Eosinophils 　Basophils	 3,000–6,750 1,000–2,700 150–720 100–360 25–90	 Phagocytosis during inflammation Immune response Phagocytosis in all defense responses Defense against parasitic worms Secrete substances for inflammatory response and for fat removal from blood
3. Platelets	250,000–300,000	Roles in clotting

Plasma Is an Aqueous Solution of Proteins, Ions, Nutrient Molecules, and Gases

Plasma is so complex that its complete composition is unknown. Among its known components are water (91%–92% of its volume), glucose and other sugars, amino acids, plasma proteins, dissolved gases (mostly O_2, CO_2, and nitrogen), ions, lipids, vitamins, hormones and other signal molecules, and metabolic wastes, including urea and uric acid.

The plasma proteins fall into three classes, the *albumins*, the *globulins*, and *fibrinogen*. The **albumins**, the most abundant proteins of the plasma, are important for osmotic balance and pH buffering. They also transport a wide variety of substances through the circulatory system, including hormones, therapeutic drugs, and metabolic wastes. The **globulins** transport lipids (including cholesterol) and fat-soluble vitamins; a specialized subgroup of globulins, the *immunoglobulins,* constitute antibodies and other molecules contributing to the immune response. Some globulins are also enzymes. **Fibrinogen** plays a central role in the mechanism clotting the blood.

The ions of the plasma include Na^+, K^+, Ca^{2+}, Cl^-, and HCO_3^- (bicarbonate) ions. The Na^+ and Cl^- ions—the components of common table salt—are the most abundant ions. Some of the ions, particularly the bicarbonate ion, help maintain arterial blood at its characteristic pH, which in humans is slightly on the basic side at pH 7.4 (the bicarbonate ion and its role in pH balance are discussed further in Chapter 44).

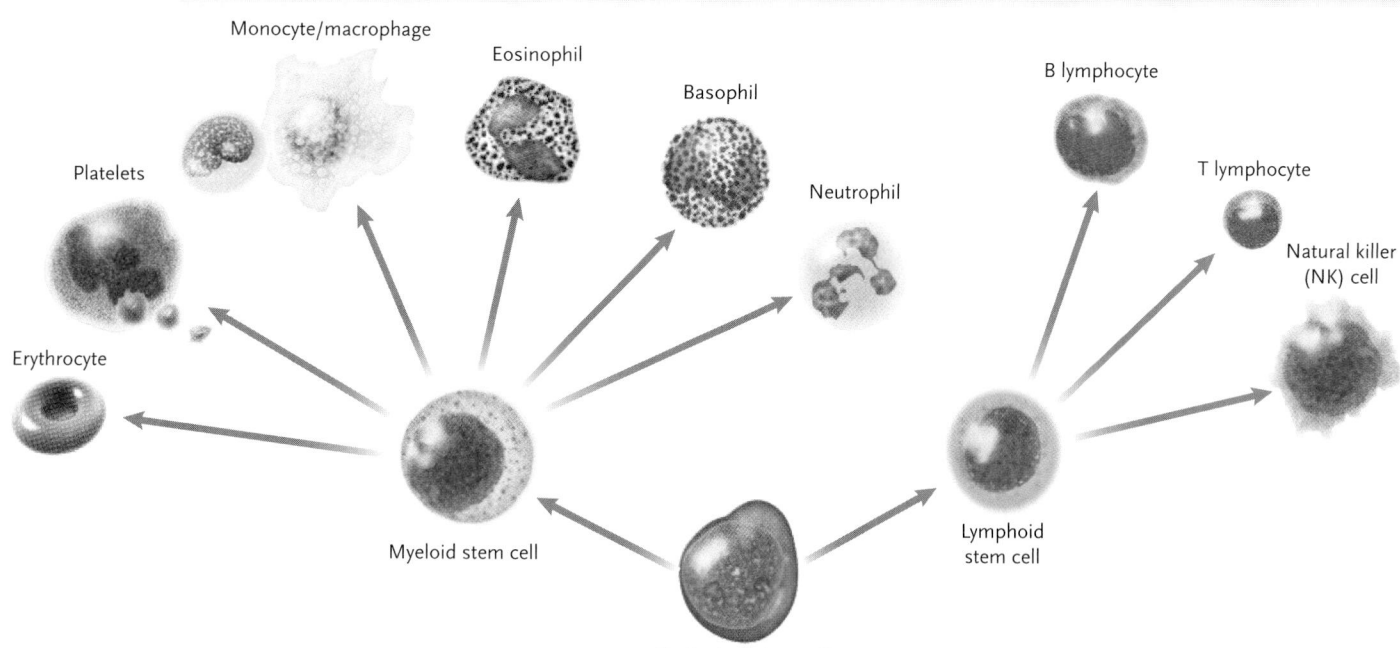

Leukocytes

Monocyte/macrophage

Eosinophil

Basophil

B lymphocyte

Neutrophil

T lymphocyte

Platelets

Natural killer (NK) cell

Erythrocyte

Myeloid stem cell

Lymphoid stem cell

Pluripotent stem cell

Figure 42.7
Major cellular components of mammalian blood and their origins from stem cells.

Erythrocytes Are the Oxygen Carriers of the Blood

Erythrocytes—the red blood cells—carry O_2 from the lungs to body tissues. Each microliter of human blood normally contains about 5 million erythrocytes, which are small, flattened, and disclike. Indentations on each flattened surface make them *biconcave*—thinner in the middle than at the edges (see Figure 42.6). They measure about 7 μm in diameter and 2 μm in thickness. Erythrocytes are highly flexible cells, able to squeeze through narrow capillaries.

As they mature, mammalian erythrocytes lose their nucleus, cytoplasmic organelles, and ribosomes, thereby limiting their metabolic capabilities and life span. The remaining cytoplasm contains enzymes, which carry out glycolysis, and large quantities of *hemoglobin,* the O_2-carrying protein of the blood. Most of the ATP produced by glycolysis is used to power active transport mechanisms that move ions in and out of erythrocytes.

Hemoglobin, the molecule that gives erythrocytes and the blood its red color, consists of four polypeptides, each linked to a nonprotein *heme* group that contains an iron atom in its center (see Figure 3.18). The iron atom binds O_2 molecules as blood circulates through the lungs and releases the O_2 as blood flows through other body tissues.

Hemoglobin can also combine with carbon monoxide (CO), which is common in the exhaust gases of cars and boats, and may be produced by appliances fueled with natural gas or oil. The combination, which is essentially irreversible, blocks hemoglobin's ability

to combine with O_2, which can quickly lead to death if exposure continues. CO is also present in cigarette smoke.

Some 2 million to 3 million erythrocytes are produced in the average human each *second*. The life span of an erythrocyte in the circulatory system is about 120 days. At the end of their useful life, erythrocytes are engulfed and destroyed by *macrophages (makros = big; phagein = to eat)*, a type of large leukocyte, in the spleen, liver, and bone marrow.

A negative feedback mechanism keyed to the blood's O_2 content stabilizes the number of erythrocytes in blood. If the O_2 content drops below the normal level, the kidneys synthesize **erythropoietin** (EPO), a hormone that stimulates stem cells in bone marrow to increase erythrocyte production. Erythropoietin is also secreted after blood loss and when mammals move to higher altitudes. As new red blood cells enter the bloodstream, the O_2-carrying capacity of the blood rises. If the O_2 content of the blood rises above normal levels, erythropoietin production falls in the kidneys and red blood cell production drops. The gene encoding human erythropoietin has been cloned, allowing researchers to produce this protein in large quantities. It can then be injected into the body to stimulate erythrocyte production, for example, in patients with anemia (lower-than-normal hemoglobin levels) caused by kidney failure or chemotherapy. It can also supplement or even replace blood transfusions. Some endurance athletes such as triathletes, bicycle racers, marathon runners, and cross-country skiers have used EPO to increase their erythrocyte levels in order to enhance performance. Such blood doping is deemed illegal by the governing organizations of most endurance sports and, as a result, many athletes have been sanctioned or banned in recent years.

Disorders of erythrocytes are responsible for a number of human disabilities and diseases. The *anemias*, which result from too few or malfunctioning erythrocytes, prevent O_2 from reaching body tissues in sufficient amounts. Shortness of breath, fatigue, and chills are common symptoms of anemia. Anemia can be produced, for example, by blood loss from a wound or bleeding ulcer, or by certain infections.

Leukocytes Provide the Body's Front Line of Defense against Disease

Leukocytes eliminate dead and dying cells from the body, remove cellular debris, and provide the body's first line of defense against invading organisms. They are called white cells because they are colorless, in contrast to the strongly pigmented red blood cells. Also unlike red blood cells, leukocytes retain their nuclei, cytoplasmic organelles, and ribosomes as they mature, and hence are fully functional cells.

Like red blood cells, leukocytes arise from the division of stem cells in red bone marrow (see Figure 42.7). As they mature, they are released into the bloodstream, from which they enter body tissues in large number. Some types of leukocytes are capable of continued division in the blood and body tissues. The specific types of leukocytes and their functions in the immune reaction are discussed in Chapter 43.

Platelets Induce Blood Clots That Seal Breaks in the Circulatory System

Blood **platelets** are oval or rounded cell fragments about 2 to 4 μm in diameter, each enclosed in its own plasma membrane. They are produced in red bone marrow by the division of stem cells (see Figure 42.7).

Platelets contain enzymes and other factors that take part in blood clotting. When blood vessels are damaged, collagen fibers in the extracellular matrix are exposed to the leaking blood. Platelets in the blood then stick to the collagen fibers and release signaling molecules that induce additional platelets to stick to them. The process continues, forming a plug that helps seal off the damaged site. As the plug forms, the platelets release other factors that convert the soluble plasma protein, fibrinogen, into long, insoluble threads of **fibrin**. Cross-links between the fibrin threads form a meshlike network that traps blood cells and platelets and further seals the damaged area **(Figure 42.8)**. The entire mass is a blood clot.

Mutations or diseases that interfere with the enzymes and factors taking part in the clotting mechanism can have serious effects and lead to uncontrolled bleeding. In the most common form of hemophilia, for example, a mutation in a single protein (called *clot-*

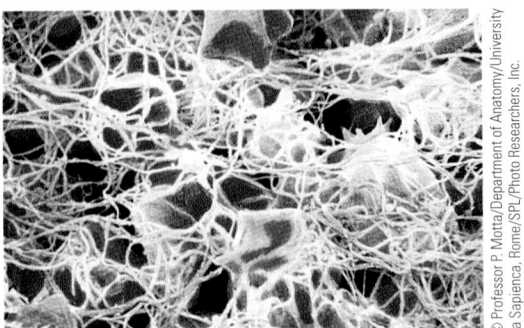

Figure 42.8
Red blood cells caught in a meshlike network of fibrin threads during formation of a blood clot.

ting factor VIII) interferes with the clotting reaction. Bleeding is uncontrolled in afflicted individuals; even small cuts and bruises can cause life-threatening blood loss.

STUDY BREAK

1. Outline the life cycle of an erythrocyte.
2. How does the body compensate for a lower-than-normal level of oxygen in the blood?
3. What are the roles of leukocytes and platelets?

42.3 The Heart

In mammals, the heart is structured from cardiac muscle cells (see Figure 36.6) forming a four-chambered pump, with two atria at the top of the heart pumping blood into two ventricles at the bottom of the heart **(Figure 42.9)**. The powerful contractions of the ventricles push the blood at relatively high pressure into arteries leaving the heart. This arterial pressure is responsible for the blood circulation. Valves between the atria and the ventricles and between the ventricles and the arteries leaving the heart keep the blood from flowing backward.

The heart of mammals pumps the blood through two completely separate circuits of blood vessels: the systemic circuit and the pulmonary circuit **(Figure 42.10)**. The right atrium (toward the right side of the body) receives blood returning to the heart in vessels coming from the entire body, except for the lungs: the *superior vena cava* conveys blood from the head and forelimbs, and the *inferior vena cava* conveys blood from the abdominal organs and the hind limbs. This blood is depleted of O_2 and has a high CO_2 content. The right atrium pumps blood into the right ventricle, which contracts to push the blood into the *pulmonary arteries* leading to the lungs. In the capillaries of the lungs, the blood releases CO_2 and picks up O_2. The oxygenated blood completes this pulmonary circuit by returning in *pulmonary veins* to the heart.

Blood returning from the pulmonary circuit enters the left atrium, which pumps the blood into the left ventricle. This ventricle, the most thick-walled and powerful of the heart's chambers, contracts to send the oxygenated blood coursing into a large artery, the **aorta,** which branches into arteries leading to all body regions except the lungs. In the capillary networks of these body regions, the blood releases O_2 and picks up CO_2. The O_2-depleted blood collects in veins, which complete the systemic circuit. The blood from the veins enters the right atrium. The amount of blood pumped by the two halves of the heart is normally balanced so that neither side pumps more than the other.

The heart also has its own circulation, called the *coronary circulation*. The aorta gives off two *coronary arteries* that course over the heart. The coronary arteries branch extensively, leading to dense capillary beds that serve the cardiac muscle cells. The blood from the capillary networks collects into veins that empty into the right atrium.

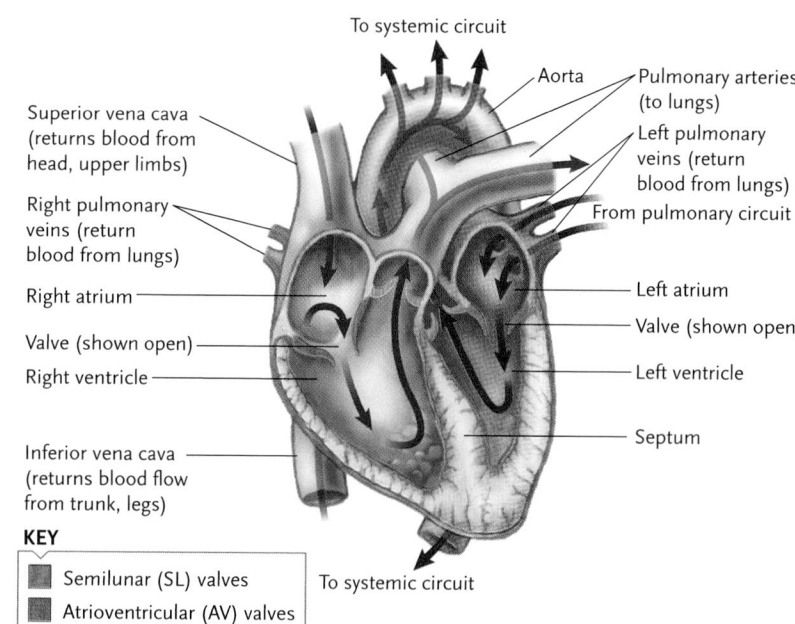

Right pulmonary veins (return blood from lungs)

Left atrium

Valve (shown open)

Left ventricle

Septum

KEY

■ Semilunar (SL) valves

■ Atrioventricular (AV) valves

To systemic circuit

Figure 42.9
Cutaway view of the human heart showing its internal organization.

The Heartbeat Is Produced by a Cycle of Contraction and Relaxation of the Atria and Ventricles

Average heart rates vary among mammals (and among vertebrates generally), depending on body size and the overall level of metabolic activity. A human heart beats 72 times each minute, on average, with each beat lasting about 0.8 second. The heart beat rate of a trained endurance athlete is typically much lower. The heart of a flying bat may beat 1200 times a minute, while that of an elephant beats only 30 times a minute. The period of contraction and emptying of the heart is called the **systole** and the period of relaxation and filling of the heart between contractions is called the **diastole.** The systole-diastole sequence of the heart is called the **cardiac cycle (Figure 42.11).** The following discussion goes through one cardiac cycle.

Starting when both the atria and ventricles are relaxed in diastole, the atria begin filling with blood (see Figure 42.11, step 1). At this point, the **atrioventricular valves (AV valves)** between each atrium and ventricle, and the **semilunar valves (SL valves)** between the ventricles and the aorta and pulmonary arteries, are closed. As the atria fill, the pressure pushes open the AV valves and begins to fill the relaxed ventricles (step 2). When the ventricles are about 80% full, the atria contract and completely fill the ventricles with blood (step 3). Although there are no valves where the veins open into the atria, the atrial contraction compresses the openings, sealing them so that little backflow occurs into the veins.

As the ventricles begin to contract, the rising pressure in the ventricular chambers forces the AV valves shut (step 4). As they continue to contract, the pressure in the ventricular chambers rises above that in the ar-

teries leading away from the heart, forcing open the SL valves. Blood now rushes from the ventricles into the aorta and the pulmonary arteries (step 5).

The completion of the contraction squeezes about two-thirds of the blood in the ventricles into the arteries. Now the ventricles relax, lowering the pressure in the ventricular chambers below that in the arteries. This reversal of the pressure gradient reverses the direction of blood flow in the regions of the SL valves, causing them to close. For about half a second, both the atria and ventricles remain in diastole and blood flows into the atria and ventricles. Then, the blood-filled atria contract, and the cycle repeats.

In an adult human at rest, each ventricle pumps roughly 5 liters of blood per minute—an amount roughly equivalent to the entire volume of blood in the body. At maximum rate and strength the human heart pumps about five times the resting amount, or more than 25 liters per minute.

The heart makes a "lub-dub" sound when it beats, which you can hear by placing an ear against a person's chest or listening to the heart through a stethoscope. The sound is produced mostly by vibrations created when the valves close. The "lub" sound occurs when the AV valves are pushed shut by the contraction of the ventricles; the "dub" sound is made when the SL valves are forced shut as the ventricles relax. *Heart murmurs* are abnormal sounds produced by turbulence created in the blood when one or more of the valves fails to open or close completely and blood flows backward.

The Cardiac Cycle Is Initiated within the Heart

Contraction of cardiac muscle cells is triggered by action potentials that spread across the muscle cell membranes. Some crustaceans, such as crabs and lobsters,

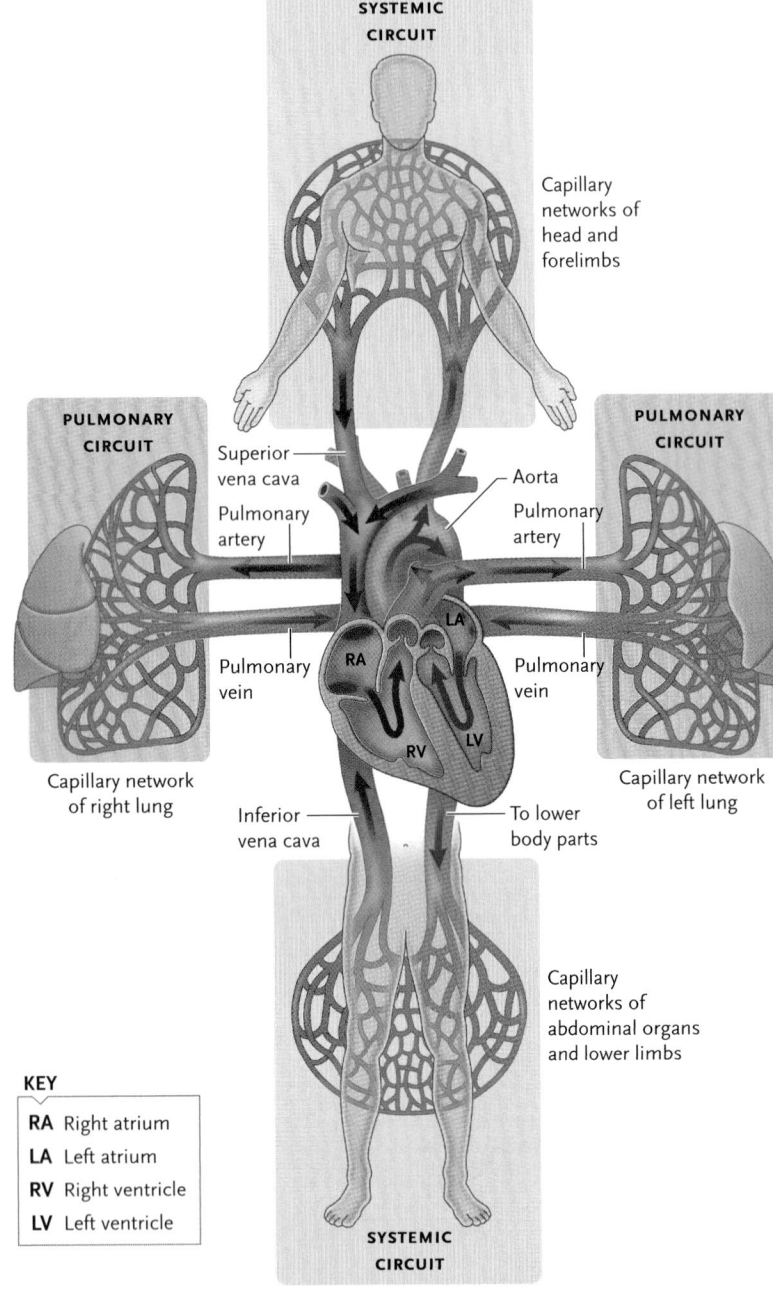

SYSTEMIC CIRCUIT

Capillary networks of head and forelimbs

PULMONARY CIRCUIT

Superior vena cava

Pulmonary artery

Pulmonary vein

Capillary network of right lung

Inferior vena cava

PULMONARY CIRCUIT

Aorta

Pulmonary artery

Pulmonary vein

Capillary network of left lung

To lower body parts

LA

RA

RV

LV

Capillary networks of abdominal organs and lower limbs

SYSTEMIC CIRCUIT

KEY

RA	Right atrium
LA	Left atrium
RV	Right ventricle
LV	Left ventricle

Figure 42.10

The pulmonary and systemic circuits of mammals. The right half of the heart pumps blood into the pulmonary circuit, and the left half of the heart pumps blood into the systemic circuit.

region of the heart called the **sinoatrial node (SA node)**, which controls the rate and timing of cardiac muscle cell contraction. The SA node consists of **pacemaker cells**, which are specialized cardiac muscle cells in the upper wall of the right atrium near where the blood enters the heart from the systemic circuit **(Figure 42.12)**. Ion channels in these cells open in a cyclic, self-sustaining pattern that alternately depolarizes and re-polarizes their plasma membranes. The regularly timed depolarizations initiate waves of contraction that travel over the heart (step 1). The first effect of a wave is to cause the cells of the atria to contract in unison and fill the ventricles with blood.

A layer of connective tissue separates the atria from the ventricles, in effect placing a layer of electrical insulation between the top and bottom of the heart. The insulating layer keeps a contraction signal from the SA node from spreading directly from the atria to the ventricles (step 2). Instead, the atrial wave of contraction excites cells of the **atrioventricular node (AV node)**, located in the heart wall between the right atrium and right ventricle, just above the insulating layer of connective tissue. The signal produced travels from the AV node to the bottom of the heart via *Purkinje fibers* (step 3). These fibers follow a path downward through the insulating layer to the bottom of the heart, where they branch through the walls of the ventricles. The signal carried by the Purkinje fibers induces a wave of contraction that begins at the bottom of the heart and proceeds upward, squeezing the blood from the ventricles into the aorta and pulmonary arteries (step 4). The transmission of a signal from the AV node to the ventricles takes about 0.1 second; this delay gives the atria time to finish their contraction before the ventricles contract. Cardiac muscle cells have a relatively long refractory period, about 0.25 second, which keeps the signals or contractions from reversing at any point.

As Augustus Waller found, the electrical signals passing through the heart can be detected by attaching electrodes to different points on the surface of the body. The signals change in a regular pattern corresponding to the electrical signals that trigger the cardiac cycle, producing what is known as an **electrocardiogram (ECG;** also **EKG,** from German *elektrocardiogramm)*. The highlighted region of the ECG under each stage of the cardiac cycle in Figure 42.12 indicates the electrical activity measured in those stages. Many malfunctions of the heart alter the ECG pattern in characteristic ways, providing clues to the location and type of heart disease.

The spontaneous, rhythmic signal set up by the SA node is the foundation of the normal heartbeat. Because of this internal signaling system, a myogenic heart will continue beating if all the nerves leading to the heart are severed.

The SA node normally dominates the AV node and other conductive regions of the heart, keeping the atria

have **neurogenic hearts,** that is, hearts that beat under the control of signals from the nervous system. This type of heart contraction regulation allows for the heart to be stopped completely in these animals, which may not need continuous hemolymph circulation in some situations. Other animals, including all insects and all vertebrates, have **myogenic hearts,** that is, hearts that maintain their contraction rhythm with no requirement for signals from the nervous system. The advantage of a myogenic heart is that blood flow is ensured in the event of serious trauma to the nervous system.

The contraction of individual cardiac muscle cells in mammalian myogenic hearts is coordinated by a

Figure 42.11
The cardiac cycle.

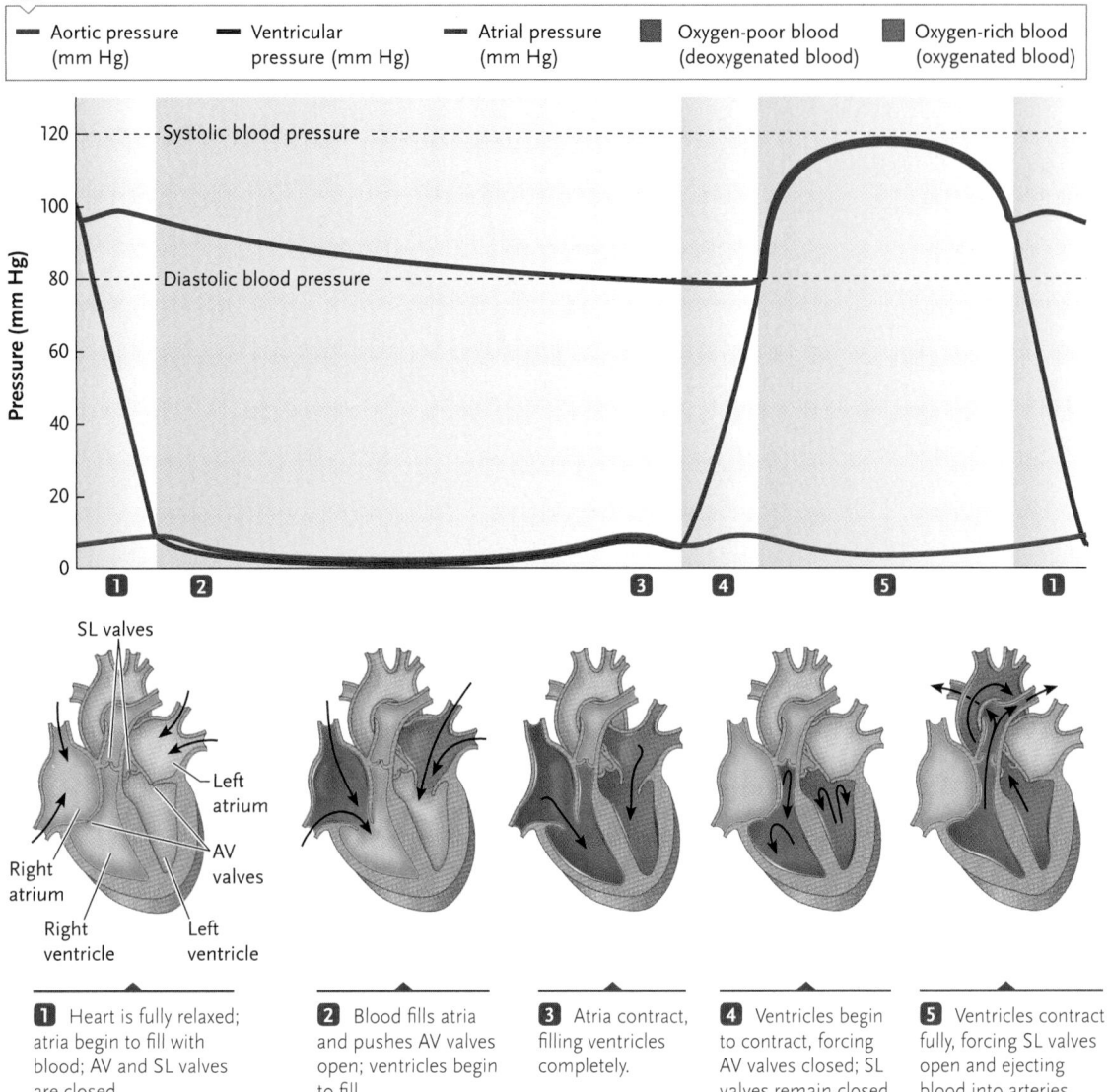

KEY

— Aortic pressure (mm Hg) ▬ Ventricular pressure (mm Hg) ▬ Atrial pressure (mm Hg) ■ Oxygen-poor blood (deoxygenated blood) ■ Oxygen-rich blood (oxygenated blood)

1 Heart is fully relaxed; atria begin to fill with blood; AV and SL valves are closed.

2 Blood fills atria and pushes AV valves open; ventricles begin to fill.

3 Atria contract, filling ventricles completely.

4 Ventricles begin to contract, forcing AV valves closed; SL valves remain closed.

5 Ventricles contract fully, forcing SL valves open and ejecting blood into arteries.

and ventricles beating in a fully coordinated fashion. At times, however, parts of the conductive system outside the SA node may generate signals independently and produce uncoordinated contractions known as *arrhythmias*. Depending on their source and characteristics, arrhythmias range from harmless to life threatening. Most commonly, the ventricles beat prematurely, and then fill more slowly for the next beat, which is proportionately more powerful. This arrhythmia, called a *premature ventricular contraction (PVC)*, feels like a skipped beat but is usually harmless. Consumption of too much caffeine, chocolate, or alcohol can increase the frequency of PVCs. Other arrhythmias, such as those produced when the AV node becomes the dominant pacemaker, can be more dangerous. Some of the more threatening arrhythmias are corrected by surgically implanting an artificial pacemaker that produces an overriding, regular electrical impulse to keep the heart beating at a normal rhythm and rate.

Arterial Blood Pressure Cycles between a High Systolic and a Low Diastolic Pressure

The pressure that a fluid in a confined space exerts is called *hydrostatic pressure*. That is, fluid in a container exerts some pressure on the wall of the container. Blood vessels are essentially tubular containers that are part of a closed system filled with fluid. Hence, the blood in vessels exerts hydrostatic pressure against the walls of the vessels. *Blood pressure* is the measurement of that hydrostatic pressure on the walls of the arteries as the heart pumps blood through the body. Blood pressure is determined by the force and amount of blood pumped by the heart and the size and flexibility of the arteries. In any person, blood pressure changes continually in response to activity, temperature, body position, emotional state, diet, and medications being taken.

Figure 42.12

The electrical control of the cardiac cycle. The top part of the figure shows how a signal originating at the SA node leads to ventricular contraction. The bottom part of the figure shows the electrical activity for each of the stages as seen in an ECG. The colors in the hearts show the location of the signal at each step and correspond to the colors in the ECG.

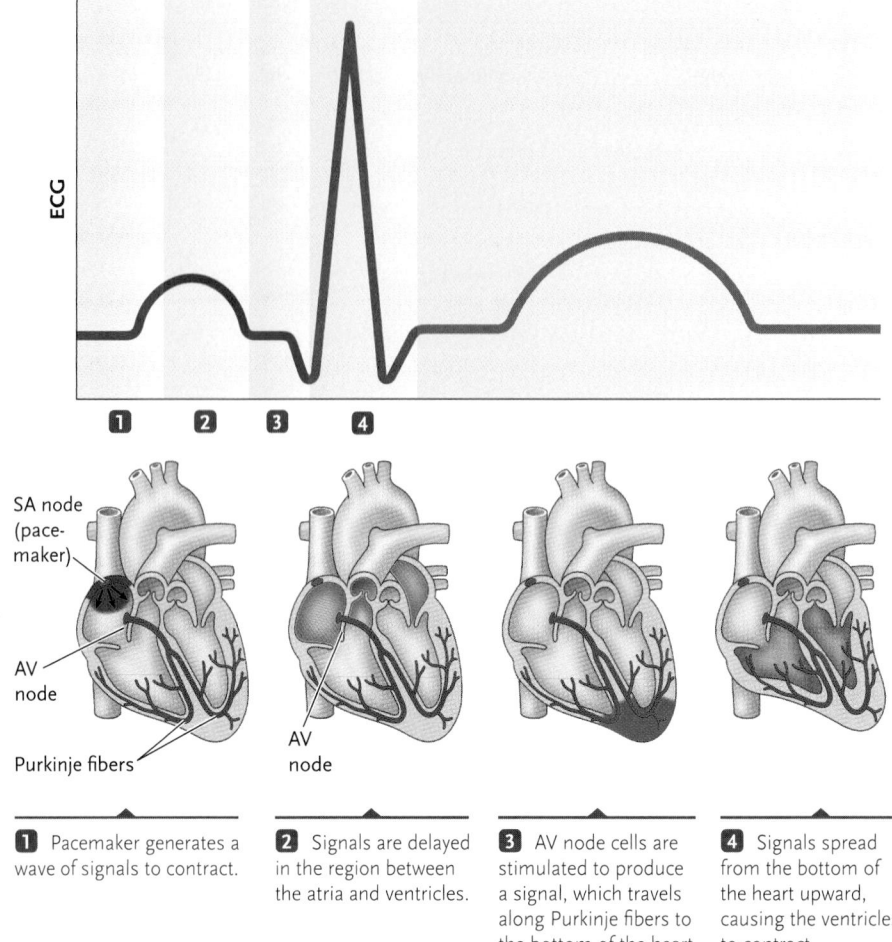

SA node (pace-maker)

AV node

Purkinje fibers

AV node

1 Pacemaker generates a wave of signals to contract.

2 Signals are delayed in the region between the atria and ventricles.

3 AV node cells are stimulated to produce a signal, which travels along Purkinje fibers to the bottom of the heart.

4 Signals spread from the bottom of the heart upward, causing the ventricles to contract.

Two blood pressure measurements are typically made in clinical medicine, representing different parts of the cardiac cycle. As the ventricles contract, a surge of high-pressure blood moves outward through the arteries leading from the heart. This peak of high pressure, called the *systolic blood pressure,* can be felt as a *pulse* by pressing a finger against an artery that lies near the skin, such as the arteries of the neck or the artery that runs along the inside of the wrist. Between ventricular contractions, the arterial blood pressure reaches a low point called the *diastolic blood pressure.*

For most healthy adults at rest, the systolic pressure measured at the upper arm is between 90 and 120 mm Hg, and the diastolic pressure is between 60 and 80 mm Hg (**Figure 42.13** shows how these pressures are measured). The numbers, written in the form 120/80 mm Hg and stated verbally as "120 over 80," refer to the height of a column of mercury in millimeters that would be required to balance the pressure exactly. The systolic and diastolic blood pressures in the pulmonary arteries are typically much lower, about 24/8 mm Hg.

The blood pressure in the systemic and pulmonary circuits is highest in the arteries leaving the heart, and drops as the blood passes from the arteries into the capillaries. By the time the blood returns to the heart, its pressure has dropped to 2 to 5 mm Hg, with no differentiation between systolic and diastolic pressures.

The reduction in pressure occurs because the blood encounters resistance as it moves through the vessels, produced primarily by the friction created when blood cells and plasma proteins move over each other and over vessel walls.

Some people have **hypertension,** commonly called high blood pressure, a medical condition in which blood pressure is chronically elevated above normal values; that is, at least 140/90. In some cases, no specific medical cause can be found to explain the hypertension. In other cases, the hypertension results from another medical condition, such as kidney disease, or diseases or cancers affecting the adrenal cortex. Hypertension can also be caused by certain medications, such as ibuprofen and steroids. Age is also a contributor to hypertension because over time the walls of blood vessels become stiffer as more collagen fibers are added, and this causes decreased elasticity of the arteries. During systole, these arteries cannot expand as much as they once could, and this results in a higher arterial blood pressure.

Hypertension is rarely severe enough to cause symptoms. However, in the long term, the increased pressure in the arteries can cause damage to organs. Hence, hypertension is treated because of the correlation with an increased risk for a number of medical conditions, including myocardial infarction (heart attack), cardiovascular accident (stroke), chronic renal failure, and retinal damage. Treatments to reduce hypertension typically involve life style changes, such as weight loss and regular exercise; in the case of moderate to severe hypertension, drugs are also prescribed.

What happens to blood pressure during exercise? The answer depends on the type of exercise. During static exercise involving a sustained contraction of a muscle group or groups, such as weight lifting or Nautilus machine routines, both systolic and diastolic pressure increase. In elite weight lifters, for instance, blood pressure during lifts can reach 300/150 mm Hg. However, during dynamic exercise involving intermittent and rhythmical muscle contractions, such as running, bicycling, and swimming, only the systolic pressure increases.

STUDY BREAK

1. Explain the role of each of the four chambers of the mammalian heart in blood circulation.
2. Distinguish systole and diastole.
3. Distinguish neurogenic hearts and myogenic hearts.
4. Describe the electrical events that occur during the cardiac cycle in a mammalian myogenic heart.

42.4 Blood Vessels of the Circulatory System

Both the systemic and pulmonary circuits consist of a continuum of different blood vessel types that begin and end at the heart **(Figure 42.14)**. From the heart, large arteries carry blood and branch into progressively smaller arteries, which deliver the blood to the various parts of the body. When a small artery reaches the organ it supplies, it branches into yet smaller vessels, the **arterioles.** Within the organ, arterioles branch into capillaries, the smallest vessels of the circulatory system. The capillaries form a network in the organ that is used to exchange substances between the blood and the surrounding cells. Capillaries rejoin to form small **venules,** which merge into the small veins that leave the organ. The small veins progressively join to form larger veins that eventually become the large veins that enter the heart.

Arteries Transport Blood Rapidly to the Tissues and Serve as a Pressure Reservoir

Arteries have relatively large diameters and, therefore, provide little resistance to blood flow. Structurally, they are adapted to the relatively high pressure of the blood passing through them. The walls of arteries consist of

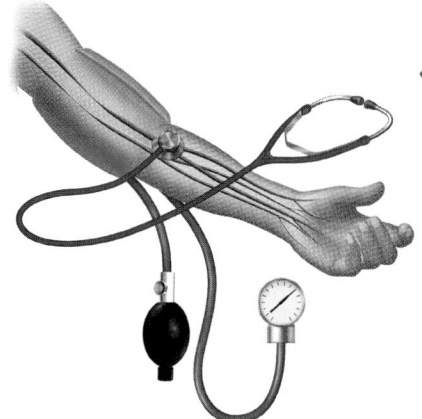

1 The cuff is pumped up until its pressure is higher than the arterial blood pressure, cutting off the flow of blood through the large artery serving the arm.

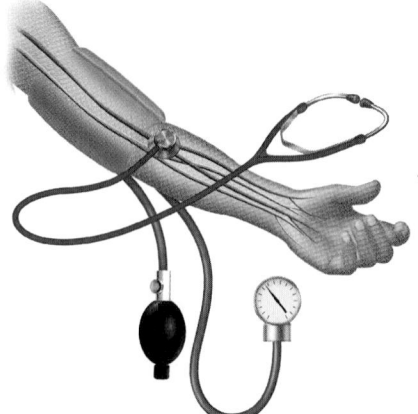

2 A stethoscope is placed over the artery just below the cuff, and the pressure in the cuff is slowly released. When cuff pressure drops to the point that blood just begins to flow through the artery, a faint thumping sound is heard in the stethoscope. The sound is produced by turbulence created as spurts of blood pass through the narrowed artery. The pressure read on the meter when the thumping begins is the systolic pressure.

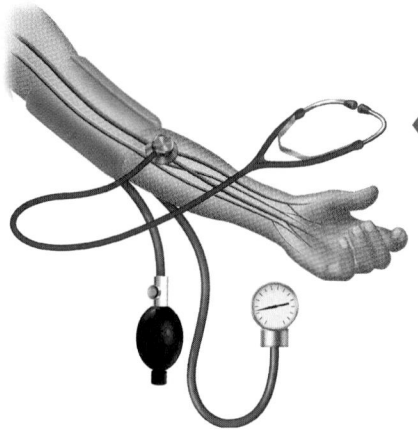

3 The pressure in the cuff is released further until the thumping sound just disappears. At this point, there is no turbulence because the artery is fully open. The pressure read on the meter at this point is the diastolic pressure.

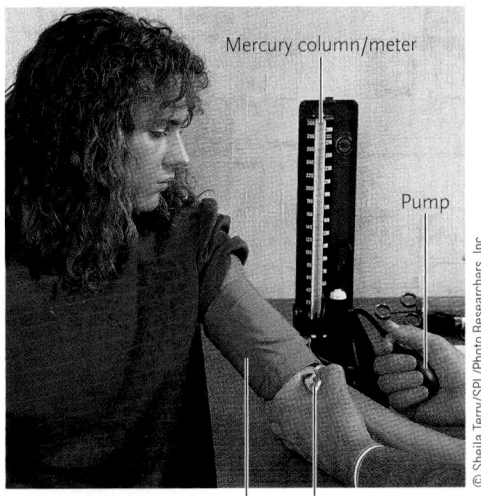

Mercury column/meter

Pump

Cuff   Stethoscope

© Sheila Terry/SPL/Photo Researchers, Inc.

Figure 42.13

Taking blood pressure with a **sphygmomanometer.** The device consists of a rubber bladder (called a cuff) that is wrapped around an arm or leg and connected to an air pump and a mercury column, which records the pressure inside the bladder in millimeters of mercury (mm Hg).

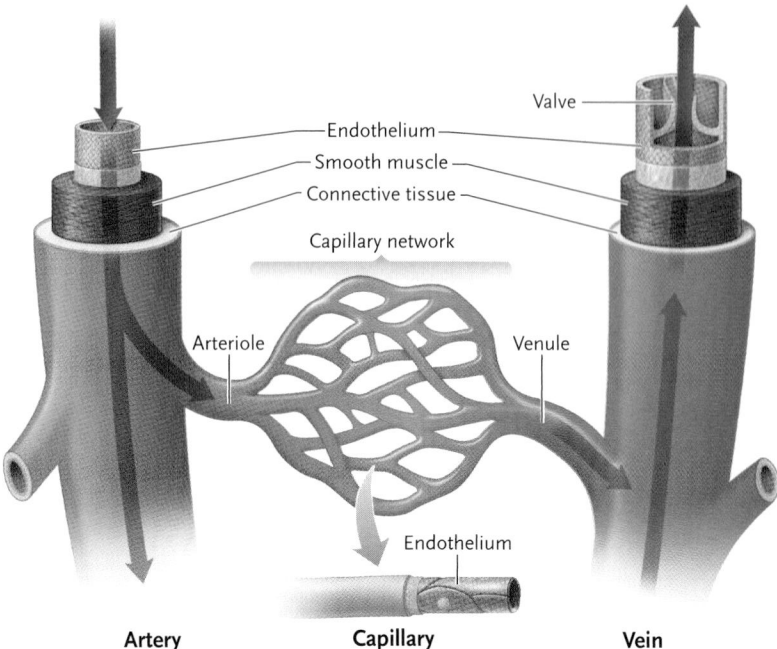

Figure 42.14
The structure of arteries, capillaries, and veins and their relationship in blood circuits.

is relaxing. When contraction of the ventricles pumps blood into the arteries, the amount of blood flowing into the arteries is greater than the amount flowing out into the smaller vessels downstream because of the higher resistance to blood flow in those smaller vessels. The arteries can accommodate the excess volume of blood because their elastic walls allow the arteries to expand in diameter. When the heart then relaxes and blood is no longer being pumped into the arteries, the arterial walls recoil passively back to their original state. The re-coil pushes the excess blood from the arteries into the smaller downstream vessels. As a result, blood flow to tissues is continuous during systole and diastole.

Capillaries Are the Sites of Exchange between the Blood and Interstitial Fluid

Capillaries thread through nearly every tissue in the body, arranged in networks that bring them within 10 μm of most body cells. They form an estimated 2600 km² of total surface area for the exchange of gases, nutrients, and wastes with the interstitial fluid. Capillary walls consist of a single layer of endothelial cells.

Control of Blood Flow through Capillaries. Blood flow through the capillary networks is controlled by contraction of smooth muscle in the arterioles **(Figure 42.15)**. The capillaries themselves do not have smooth muscle but, in many cases, a small ring of smooth muscle called a *precapillary sphincter* is present at the junction between an arteriole and a capillary.

When the sphincter is relaxed, blood flows readily through the arterioles and capillary networks (see Figure 42.15a). In the most contracted state, blood flow

three major tissue layers: (1) an outer layer of connective tissue containing collagen fibers mixed with fibers of the protein elastin, which gives the vessel recoil ability; (2) a relatively thick middle layer of vascular smooth muscle cells also mixed with elastin fibers; and (3) a one-cell-thick inner layer of flattened cells that forms an *endothelium,* a specialized type of epithelial tissue that lines the entire circulatory system (see Figure 42.14).

In addition to being the conduits for blood traveling to the tissues, arteries also act as a pressure reservoir to generate the force for blood movement when the heart

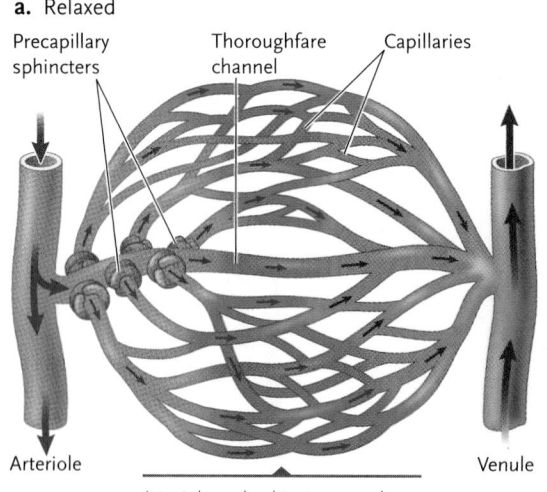

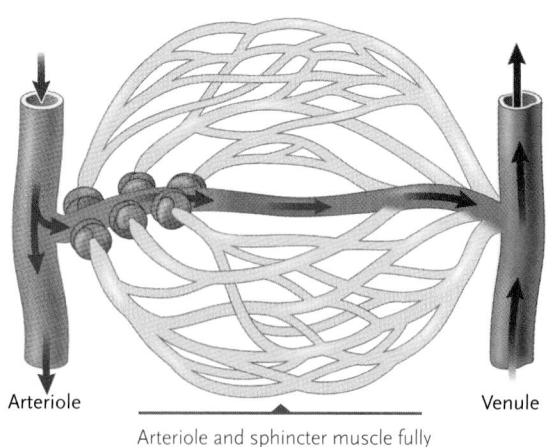

Figure 42.15
Control of blood flow through capillary networks. **(a)** Maximal blood flow when arteriole and sphincter muscles are fully relaxed. **(b)** Minimal blood flow when the arteriole and sphincter muscles are fully contracted.

through the arterioles and capillary networks is limited (see Figure 42.15b). Variation in the contraction of arteriole and sphincter smooth muscles adjusts the rate of flow through the capillary networks between the two flow limits. For example, during exercise, flow of blood through the capillary networks is increased severalfold over the resting state by relaxation of the precapillary sphincters.

Control of Blood Volume to Capillaries by Arterioles.

The volume of blood flowing through an organ is adjusted by regulating the internal diameter of the arterioles of the organ. The blood leaving the arterioles enters the capillaries that branch from them. Although their total surface area is astoundingly large, the diameter of individual capillaries is so small that red blood cells must squeeze through most of them in single file **(Figure 42.16)**. As a result, each capillary presents a high resistance to blood flow. Yet there are so many billions of capillaries in the networks that their combined diameter is about 1300 times greater than the cross-sectional area of the aorta. As a result, blood slows considerably as it moves through the capillaries **(Figure 42.17)**. This is analogous to the slowing that occurs when a narrow, swiftly running streams widens into a broad pool. The slow movement of blood through the capillaries maximizes the time for exchange of substances between blood and tissues. As they leave the tissues, the capillaries rejoin to form veins. Veins have a reduced total cross-sectional area compared with capillaries, so the rate of flow increases as blood returns to the heart (see Figure 42.17).

Exchange of Substances across Capillary Walls.

In most body tissues, narrow spaces between the capillary endothelial cells allow water, ions, and small molecules such as glucose to pass freely between the blood and interstitial fluid. Erythrocytes, platelets, and most plasma proteins are too large to pass between the cells and are retained inside the capillaries, except for molecules that are transported through the epithelial cells by specific carriers. Leukocytes, however, are able to squeeze actively between the cells and pass from the blood to the interstitial fluid.

There are exceptions to these general properties in some tissues; in the brain, for example, the capillary endothelial cells are tightly sealed together, preventing essentially all molecules and even ions from passing between them. The tight seals set up the *blood–brain barrier*, which limits the exchange between capillaries and brain tissues to molecules and ions that are specifically transported through the capillary endothelial cells (see Section 38.3 for additional discussion of the blood–brain barrier).

Forces Driving the Exchange.

Two major mechanisms drive the exchange of molecules and ions between the capillaries and the interstitial fluid: (1) diffusion along concentration gradients and (2) bulk flow.

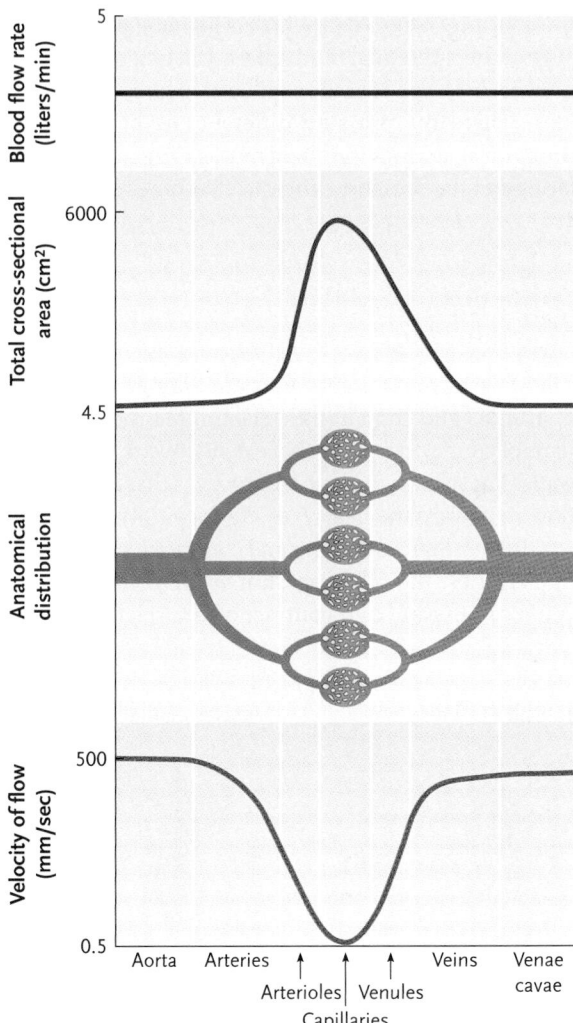

Figure 42.17
Blood flow rate and velocity of flow in relation to total cross-sectional area of the blood vessels. The blood flow rate is identical throughout the circulatory system and is equal to the cardiac output. The velocity of flow in the different types of blood vessels is inversely related to the total cross-sectional area of all the vessels of a particular type: for example, the velocity is highest in the aorta, which has the smallest cross-sectional area, and lowest in the capillaries, which collectively have the largest total cross-sectional area.

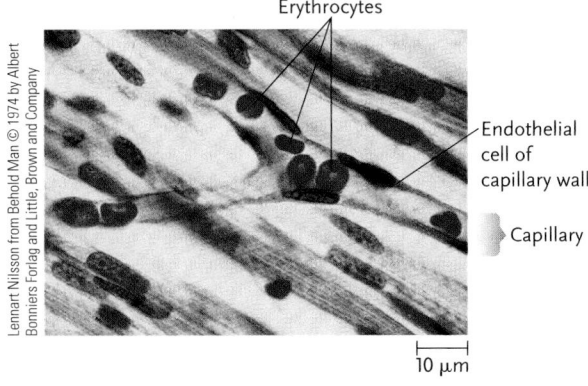

Lennart Nilsson from Behold Man © 1974 by Albert Bonniers Förlag and Little, Brown and Company

Figure 42.16
Erythrocytes moving through a capillary that is just wide enough to admit the cells in single file.

Diffusion along concentration gradients occurs both through the spaces between the capillary endothelial cells and through their plasma membranes. Ions or molecules such as O_2 and glucose, which are more concentrated inside the capillaries, diffuse outward along their gradients; other ions or molecules that are more concentrated outside, such as CO_2, diffuse inward. Some molecules, such as O_2 and CO_2, diffuse directly through the lipid bilayer of the endothelial cell plasma membranes; others, such as glucose, pass by facilitated diffusion through transport proteins. The total diffusion is greatest at the ends of capillaries nearest the arterioles, where the concentration differences between blood plasma and interstitial fluid are highest, and lowest at the ends nearest the venules, where the inside/outside concentrations of diffusible substances are almost equal.

Bulk flow, which carries water, ions, and molecules out of the capillaries, occurs through the spaces between capillary endothelial cells. The flow is driven by the pressure of the blood, which is higher than the pressure of the interstitial fluid. Like diffusion, bulk flow is greatest in the ends of the capillaries nearest the arterioles, where the pressure difference is highest, and drops off steadily as the blood moves through the capillaries and the pressure difference becomes smaller.

Venules and Veins Serve as Blood Reservoirs in Addition to Conduits to the Heart

The walls of venules and veins are thinner than those of arteries and contain little elastin. Many veins have flaps of connective tissue that extend inward from their walls. These flaps form one-way valves that keep blood flowing toward the heart (see Figure 42.14).

Rather than stretching and contracting elastically, like arteries, the relatively thin walls of venules and veins can expand and contract over a relatively wide range, allowing them to act as blood reservoirs as well as conduits. At times, the venules and veins may contain from 60% to 80% of the total blood volume of the body. The stored volume is adjusted by skeletal muscle contraction and by the valves, in response to metabolic conditions and signals carried by hormones and neurotransmitters.

Although blood pressure in the venous system is relatively low, several mechanisms assist the movement of blood back to the heart. As skeletal muscles contract, they compress nearby veins, increasing their internal pressure (Figure 42.18). The one-way valves in the veins, especially numerous in the larger veins of the limbs, keep the blood from flowing backward when the muscles relax.

When you sit without moving for long periods of time, as you might during an airline flight, the lack of skeletal muscular activity greatly reduces the return of venous blood to the heart. As a result, the blood pools in the veins of the body below the heart, making the hands, legs, and feet swell. The motionless blood can also form clots, particularly in the veins of the legs, a condition called *deep vein thrombosis*. Deep vein thrombosis often does not cause symptoms, but can cause serious medical problems if a clot breaks loose and moves elsewhere in the body, such as to the lungs. Raising the arms and getting up at intervals to exercise or contracting and relaxing the leg muscles as you sit can relieve this condition.

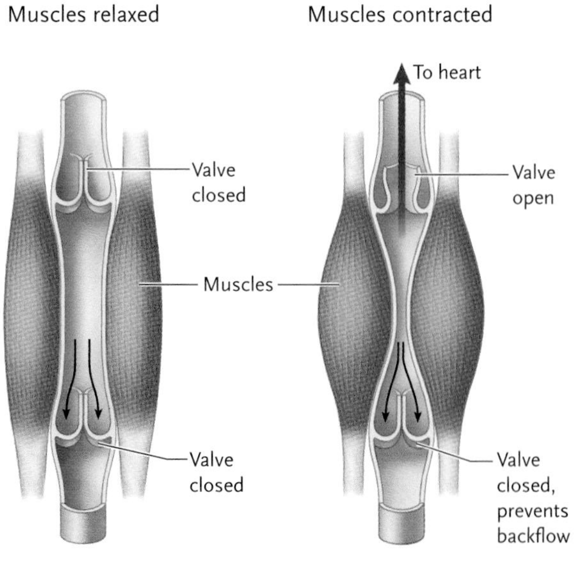

Disorders of the Circulatory System Are Major Sources of Human Disease

The layer of endothelial cells lining the arteries and veins is normally smooth and does not impede blood flow. However, several conditions, including bacterial and viral infections, chronic hypertension, smoking, and a diet high in fats, can damage the endothelial cells, exposing the underlying smooth muscle tissue, which begins a cycle of injury and repair leading to lesions. (*Focus on Research* in Chapter 3 discussed the relationship between fats and cholesterol and coronary artery disease.) Thickened deposits of material called *atherosclerotic plaques* may form at the damaged sites (Figure 42.19). The plaques, which consist of cholesterol-rich fatty substances, smooth muscle cells, and collagen deposits, reduce the diameter of the blood vessel and impede blood flow. Worse, the damaged endothelial lining may stimulate platelets to adhere and trigger the formation of blood clots. The clots further reduce the vessel diameter and flow and may break loose, along with segments of plaque material, to block finer vessels in other regions of the body.

Atherosclerosis has its most serious effects in the smaller arteries of the body, particularly in the fine coronary arteries that serve the heart muscle. Here, the plaques and clots reduce or block the flow of blood to the heart muscle cells. Serious blockage can cause a heart attack—the death of cardiac muscle cells de-

Figure 42.18
How skeletal muscle contraction, and the valves inside veins, help move blood toward the heart.

Muscles relaxed   Muscles contracted

To heart

Valve closed — Valve open

Muscles

Valve closed — Valve closed, prevents backflow

prived of blood flow. The blockage of arteries in the brain by plaque material or blood clots released from atherosclerotic arteries is also a common cause of stroke—a loss of critical brain functions due to the death of nerve cells in the brain. Heart attacks and strokes are the most common causes of death in North America and in Europe.

The risk of heart attacks and stroke can be reduced by avoiding the conditions that damage the blood vessel endothelium. A diet low in fats and cholesterol, avoidance of cigarette smoke, and a program of exercise can reduce epithelial damage and plaque deposition. Medication and exercise, or exercise alone, can also reduce the effects of hypertension. There are good indications that these preventive programs can also reduce the size of existing atherosclerotic plaques.

STUDY BREAK

1. How is blood flow through capillary networks controlled?
2. Explain how, in contrast to most body tissues, the brain does not allow exchange of molecules and ions with blood.
3. Describe the two major mechanisms that drive the exchange of molecules and ions between the capillaries and the interstitial fluid.
4. What are atherosclerotic plaques? Indicate, in general, how they form, and provide some examples of factors contributing to their formation.

42.5 Maintaining Blood Flow and Pressure

Arterial blood pressure is the principal force moving blood to the tissues. Blood pressure must be regulated carefully so that the brain and other tissues receive adequate blood flow, but not so high that the heart is overburdened, risking damage to blood vessels. The three main mechanisms for regulating blood pressure are controlling *cardiac output* (the pressure and amount of blood pumped by the left and right ventricles), the degree of constriction of the blood vessels (primarily the arterioles), and the total blood volume. The sympathetic division of the autonomic nervous system and the endocrine system interact to coordinate these mechanisms. The system is effective in counteracting the effects of constantly changing internal and external conditions, such as movement from rest to physical activity or ending a period of fasting by eating a large meal. For example, the heart responds to the initiation of moderate physical activity by increasing blood flow to the heart itself by 367%, to the muscles of the skin by 370%, and to the skeletal muscle by 1066%, while decreasing flow to the digestive

a. Normal artery **b.** Clogged artery

Wall of artery, cross section
Unobstructed lumen

Atherosclerotic plaque
Blood clot sticking to plaque
Narrowed lumen

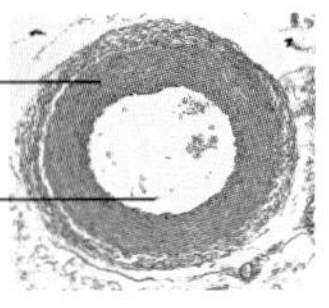

Figure 42.19
Atherosclerosis.
(a) A normal coronary artery. **(b)** A coronary artery that is partially clogged by an atherosclerotic plaque.
(a: Ed Reschke; b: © Biophoto Associates/Photo Researchers, Inc.)

tract and liver by 56%, to the kidneys by 45%, and to the bone and most other tissues by 30%. Only the blood flow to the brain remains unchanged.

Cardiac Output Is Controlled by Regulating the Rate and Strength of the Heartbeat

Regulation of the strength and rate of the heartbeat starts at stretch receptors called *baroreceptors* (a type of mechanoreceptor; see Section 39.2), located in the walls of blood vessels. By detecting the amount of stretch of the vessel walls, baroreceptors constantly provide information about blood pressure. The baroreceptors in the cardiac muscle, the aorta, and the carotid arteries (which supply blood to the brain), are the most crucial. Signals sent by the baroreceptors go to the medulla within the brain stem. In response, the brain stem sends signals via the autonomic nervous system that adjust the rate and force of the heartbeat: the heart beats more slowly and contracts less forcefully when arterial pressure is above normal levels, and it beats faster and contracts more forcefully when arterial pressure is below normal levels.

The O_2 content of the blood, detected by chemoreceptors in the aorta and carotid arteries, also influences cardiac output. If the O_2 concentration falls below normal levels, the brain stem integrates this information with the signals sent from baroreceptors and issues signals that increase the rate and force of the heartbeat. Too much O_2 in the blood has the opposite effect, reducing the cardiac output.

Hormones Regulate both Cardiac Output and Arteriole Diameter

Hormones secreted by several glands contribute to the regulation of blood pressure and flow. As part of the stress response, the adrenal medulla reinforces the action of the sympathetic nervous system by secreting epinephrine and norepinephrine into the bloodstream (see Section 40.4). Epinephrine in particular raises the blood pressure by increasing the strength and rate of the heartbeat and stimulating vasoconstriction of arterioles in some parts of the body, including the skin, gut, and kidneys. At the same time, by inducing the vasodilation of arterioles that deliver blood to the heart, skeletal muscles, and lungs, epinephrine increases the blood flow to these structures. *Insights from the Molecular Revolution* shows how gene manipulation experi-

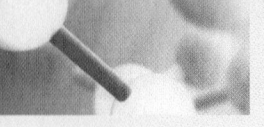

Identifying the Role of a Hormone Receptor in Blood Pressure Regulation Using Knockout Mice

The sympathetic division of the autonomic nervous system brings about a temporary increase in blood pressure when the body experiences stress. That is, the sympathetic nervous system stimulates the adrenal medulla to release the catecholamines epinephrine and norepinephrine. These hormones stimulate the strength and rate of the heartbeat and change arteriole diameter. Epinephrine and norepinephrine exert their effects by binding to specific membrane-embedded receptors on target cells, activating them and triggering a cellular response via a signal transduction pathway. The receptors, which are G-protein–coupled receptors (see Section 7.4), are known as *adrenergic receptors* because of the adrenal origin of the hormones that bind to them.

Different types of adrenergic receptors are responsible for different responses to epinephrine and norepinephrine. With respect to blood pressure regulation, two key receptors are the α_1 receptor and the β_2 receptor. α_1 Receptors are found on cells of all arteriolar smooth muscle, but not in the brain, whereas β_2 receptors are found only on cells of arteriolar smooth muscle in the heart and skeletal muscles. Norepinephrine has strong affinity for the α_1 receptors, while epinephrine has less affinity for this type of receptor. Binding of norepinephrine, and to a lesser extent, epinephrine, to α_1 receptors on arteriolar

smooth muscle causes vasoconstriction of arterioles, thereby contributing to an increase in blood pressure.

Researchers have identified three subtypes of α_1-adrenergic receptors: α_{1A}, α_{1B}, and α_{1D}. Tissues that express α_1 receptors can express all subtypes. Each of the subtypes responds to catecholamines, but the contribution of each subtype to the physiological responses caused by the hormones has not been characterized well. For example, despite attempts, scientists have not been able to develop drugs that are completely specific for inhibiting one subtype, making it impossible to use that approach to look at the effects of loss of activity of one subtype. Now, Gozoh Tsujimoto and colleagues at the Tokyo University of Pharmacy and Life Sciences, and the National Children's Medical Research Center, Tokyo, Japan, have used a different approach—making a gene knockout for the gene that encodes the α_{1D} receptor (the technique for making a gene knockout is described in Section 18.2). Using molecular techniques, the researchers showed that, in the knockout mice, no mRNA transcripts of the α_{1D} receptor gene was produced in any of the tissues examined, and there was no change in the expression of the α_{1A} and α_{1B} receptor subtypes. These results indicated that these mice were a highly useful model system for their physiological studies.

Next, the researchers compared the α_{1D} receptor knockout mice with normal mice to determine what had changed in the knockout mice with respect to cardiovascular function. They found that the knockout mice were modestly hypotensive (had slightly lower-than-normal blood pressure) but had a normal heart rate. They also had normal levels of circulating catecholamines. The investigators tested whether norepinephrine would stimulate an increase in blood pressure. They found that increasing doses of norepinephrine progressively increased blood pressure in both knockout and normal mice but that the response was markedly reduced in the knockout mice. Then, to determine whether the α_{1D} receptor is directly involved in vascular smooth muscle contraction, the researchers measured the effect of norepinephrine on contraction of segments of the aortas isolated from knockout and normal mice. Their results showed that norepinephrine induced concentration-dependent contraction of aortal segments from both types of mice but that the contraction response was considerably reduced in the knockout mice. The researchers concluded from their results that the α_{1D}-adrenergic receptor participates directly in sympathetic nervous system–driven regulation of blood pressure by vasoconstriction.

ments illuminated the role of a receptor for these hormones in regulating blood pressure.

The adrenal cortex and the posterior pituitary also release hormones that regulate blood pressure. Those hormones and their effects are described in Chapter 46.

Local Controls Also Regulate Arteriole Diameter

Several automated mechanisms also operate locally to increase the flow of blood to body regions engaged in increased metabolic activity, such as the muscles of your legs during an extended uphill bike ride. Low O_2 and high CO_2 concentrations, produced by the increased oxidation of glucose and other fuels, induce vasodilation

of the arterioles serving muscles. The vasodilation increases the flow of blood and the O_2 supply. Nitric oxide (NO) released by arterial endothelial cells in body regions engaged in increased metabolic activity also works as a potent vasodilator. NO is broken down quickly after its release, ensuring that its effects are local.

STUDY BREAK

1. Why is it important to regulate arterial blood pressure?
2. What are the three main mechanisms for regulating blood pressure?
3. How does epinephrine affect blood pressure?

42.6 The Lymphatic System

Under normal conditions, a little more fluid from the blood plasma in the capillaries enters the tissues than is reabsorbed from the interstitial fluid into the plasma. The **lymphatic system** is an extensive network of vessels that collect the excess interstitial fluid and return it to the venous blood **(Figure 42.20).** The interstitial fluid picked up by the lymphatic system is called **lymph.** This system also collects fats that have been absorbed from the small intestine and delivers them to the blood circulation (see Chapter 45). The lymphatic system is also a key component of the immune system.

Vessels of the Lymphatic System Extend throughout Most of the Body

Vessels of the lymphatic system collect the lymph and transport it to *lymph ducts* that empty into veins of the circulatory system. The *lymph capillaries,* the smallest vessels of the lymphatic system, are distributed throughout the body, intermixed intimately with the capillaries of the circulatory system. Although they are several times larger in diameter than the blood capillaries, the walls of lymph capillaries also consist of a single layer of endothelial cells surrounded by a thin network of collagen fibers. Interstitial fluid enters the lymph capillaries—becoming lymph—at sites in their walls where the endothelial cells overlap, forming a flap that is forced open by the higher pressure of the interstitial fluid. The openings are wide enough to admit all components of the interstitial fluid, including infecting bacteria, damaged cells, cellular debris, and lymphocytes.

The lymph capillaries merge into *lymph vessels,* which contain one-way valves that prevent the lymph from flowing backward. The lymph vessels lead to the thoracic duct and the right lymphatic duct (see Figure 42.20), which empty the lymph into a vein beneath the clavicles (collarbones), adding it to the plasma in the vein.

Movements of the skeletal muscles adjacent to the lymph vessels and breathing movements help move the lymph through the vessels, just as they help move the blood through veins. Over a day's time, the lymphatic system returns about 3 to 4 liters of fluid to the bloodstream.

Lymphoid Tissues and Organs Act as Filters and Participate in the Immune Response

The tissues and organs of the lymphatic system include the *lymph nodes,* the *spleen,* the *thymus,* and the *tonsils.* They play primary roles in filtering viruses, bacteria, damaged cells, and cellular debris from the lymph and bloodstream, and in defending the body

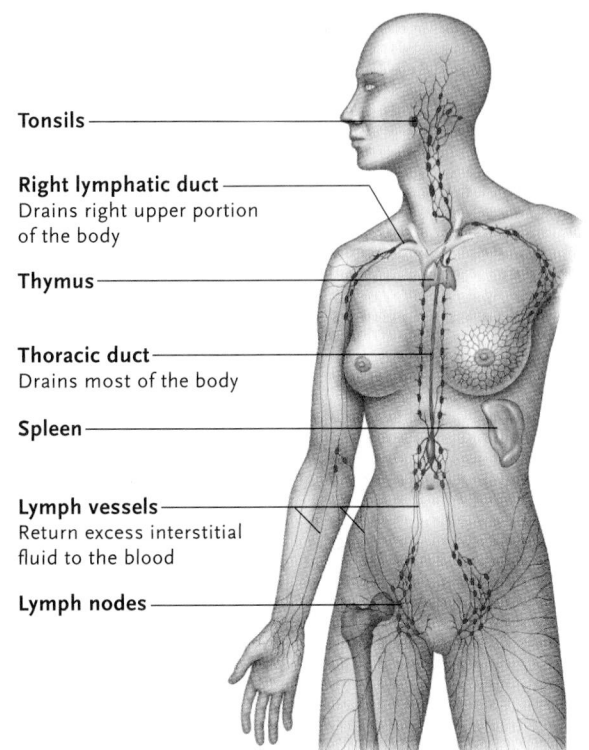

Tonsils

Right lymphatic duct
Drains right upper portion of the body

Thymus

Thoracic duct
Drains most of the body

Spleen

Lymph vessels
Return excess interstitial fluid to the blood

Lymph nodes

Figure 42.20
The human lymphatic system. Patches of lymphoid tissue in the small intestine and in the appendix also are part of the lymphatic system.

against infection and cancer. Patches of lymphoid tissue are also scattered in other regions of the body, such as the small intestine and the appendix.

The **lymph nodes** are small, bean-shaped organs spaced along the lymph vessels and clustered along the sides of the neck, in the armpits and groin, and in the center of the abdomen and chest cavity (see Figure 42.20). Spaces in the nodes contain macrophages, a type of leukocyte that engulfs and destroys cellular debris and infecting bacteria and viruses in the lymph. The lymph nodes also contain other leukocytes, which produce antibodies that aid in the destruction of invading pathogens (discussed more in Chapter 43). Cancer cells that lodge in the nodes may be destroyed or may remain to grow and divide, forming new tumors within the nodes. Therefore, to reduce the risk of cancer spread, lymph nodes near a tumor typically are inspected and may be removed during surgery to excise that tumor.

The lymph nodes may become enlarged and painful if large numbers of bacteria or viruses carried by the lymph become trapped inside them. A doctor usually checks for swollen nodes, particularly in the neck, armpits, and groin, as indicators of an infection in the region of the body served by the nodes.

STUDY BREAK

What is lymph, and how does it enter the lymph capillaries?

How did vertebrate blood clotting evolve?

Vertebrate blood clotting is a complex process involving about two dozen different proteins found in the blood plasma. The system has the properties of a biochemical amplifier in that the exposure of a tiny amount of tissue initiates a series of proteolytic events, one protease activating another successively, with the climax being a large amount of localized thrombin that transforms fibrinogen into a fibrin clot.

Many years ago when I was a graduate student working in a laboratory devoted to blood proteins, I asked myself the question, How could blood clotting ever have evolved? The process seemed much too complicated to have been concocted in one fell swoop, so, I reasoned, it must have begun in a simpler fashion and gradually become more complicated. Certainly, other people had been asking similar questions about complex organs—the evolution of the eye, for example, had been considered by many scientists, including Charles Darwin—but facts about the evolution of individual proteins were just beginning to emerge.

In particular, Vernon Ingram, a scientist working at MIT, had just determined the amino acid sequences of the alpha and beta chains of hemoglobin and found that these two proteins were about 45% identical. They must be the products of a gene duplication, he reported. Given his observation, it seemed to me that the proteases involved in blood clotting ought also to be the products of gene duplication. At the time, none of their amino acid sequences was known, but several had unique properties that distinguished them from other proteases, so it stood to reason that they were related. Yet, some nonproteolytic proteins were also involved, and they must have been added to the process independently.

So where to start? I decided to compare the blood clotting process in a wide range of animals. As it happens, most animals, invertebrate and vertebrate alike, have a kind of blood (see Section 42.2), and in most cases it can be coagulated by various stressful events. But the vertebrate process looked unique to me and must have evolved independently of the system that occurs in lobsters (Crustacea), for example. Among the vertebrates, the most primitive (early diverging) creature I could get my hands on was the lamprey (see Section 30.4, where its place in the vertebrate lineage is discussed).

Because I was hoping to find a simple, predecessor scheme, I was a little disappointed to find that lampreys have a rather sophisticated coagulation process that involved many of the proteins observed in mammals. Certainly, a small amount of tissue factor provoked a thrombin generation that converted fibrinogen to fibrin, just as in humans, at least in a general way. At the time there was no way to determine whether lampreys use the equivalent of *all* the clotting factors on the way to generating thrombin. Some of the most important proteins in the human system occur in minute amounts, and there was no possibility of ever isolating them from the lamprey, short of collecting several barrels of lamprey blood.

Many years later, after the sequences for many of the clotting factors had been reported for humans, my colleague Da-Fei Feng and I were able to align the sequences with a computer and make a phylogenetic tree. It fit the notion of a series of gene duplications very well. Moreover, it implied that most of the duplications had occurred a long time ago, at the very dawn of the vertebrates.

In 2003, the complete genomic sequence of a modern bony fish, the pufferfish, was determined, and one could scan through it and see what genes it has. It has all but a few of the more peripheral clotting proteins found in humans, the central theme being the same. However, the amount of sequence difference between human and pufferfish proteins compared with the degree of difference observed between the duplicated genes suggested some of the gene duplications had occurred not long before the appearance of bony fish. Why not look at the lamprey genome, which diverged 50 million to 100 million years before the appearance of bony fish, to see if the preduplication genes were there? The reason is that the lamprey genome has not been totally sequenced.

Recently, various genome centers around the world have begun maintaining "trace databases." These are uncurated collections of raw DNA sequences determined by random shotgun methods and robotized sequencers. The data are not assembled in any way, and each entry is at best a fragment of a gene. Several hundred organisms, from bacteria to monkeys, are being logged automatically. Among them is the lamprey! So now one can get a glimpse of what genes the lamprey has. We have been scrutinizing this database, even while recognizing its limitations. The reason is that I am getting on in years and can't wait for some mammoth operation with the resources to assemble all the lamprey DNA fragments into a complete genome.

At this point it looks like the lamprey may lack at least two of the mainline clotting factors (Factors VIII and IX), each of which in other vertebrates is the result of a gene duplication. Although the step-by-step evolution of vertebrate blood clotting factors is still technically an "unanswered question," I'm hoping our efforts with the limited trace database will spur others to go for a more convincing, fully assembled lamprey genome.

Russell Doolittle is professor emeritus at the University of California at San Diego. Finding out how blood clotting evolved is one of his major research interests. To learn more about Dr. Doolittle, go to http://www-biology.ucsd.edu/faculty/doolittle.html.

Review

Go to **ThomsonNOW** at www.thomsonedu.com/login to access quizzing, animations, exercises, articles, and personalized homework help.

42.1 Animal Circulatory Systems: An Introduction

- Only the simplest invertebrates—the sponges, cnidarians, and flatworms—have no circulatory systems (Figure 42.2).

- Animals with circulatory systems have a muscular heart that pumps a specialized fluid, such as blood, from one body region to another through tubular vessels. The blood carries O_2 and nutrients to body tissues, and carries away CO_2 and wastes.

- Most invertebrates have an open circulatory system, in which the heart pumps hemolymph into vessels that empty into body spaces called sinuses before returning to the heart. Some invertebrates and all vertebrates have a closed system, in which the blood is confined in blood vessels throughout the body and does not mix directly with the interstitial fluid (Figure 42.3).

- In invertebrates, open circulatory systems occur in arthropods and most mollusks, while closed circulatory systems occur in annelids and in mollusks such as squids and octopuses. In vertebrates, the circulatory system has evolved from a heart with a single series of chambers, pumping blood through a single circuit, to a double heart that pumps blood through separate pulmonary and systemic circuits (Figures 42.4 and 42.5).

Animation: Types of circulatory systems

Animation: Circulatory systems

42.2 Blood and Its Components

- Mammalian blood is a fluid connective tissue consisting of erythrocytes, leukocytes, and platelets, suspended in a fluid matrix, the plasma (Figure 42.6).
- Plasma contains water, ions, dissolved gases, glucose, amino acids, lipids, vitamins, hormones, and plasma proteins. The plasma proteins include albumins, globulins, and fibrinogen.
- Erythrocytes contain hemoglobin, which transports O_2 between the lungs and all body regions (Figure 42.7).
- Leukocytes defend the body against infecting pathogens.
- Platelets are functional cell fragments that trigger clotting reactions at sites of damage to the circulatory system.

Animation: White blood cells

Animation: ABO compatibilities

Animation: Rh factor and pregnancy

Animation: Hemostasis

42.3 The Heart

- The mammalian heart is a four-chambered pump. Two atria at the top of the heart pump the blood into two ventricles at the bottom of the heart, which pump blood into two separate pulmonary and systemic circuits of blood vessels (Figures 42.9 and 42.10).
- In both circuits, the blood leaves the heart in large arteries, which branch into smaller arteries, the arterioles. The arterioles deliver the blood to capillary networks, where substances are exchanged between the blood and the interstitial fluid. Blood is collected from the capillaries in small veins, the venules, which join into larger veins that return the blood to the heart (Figure 42.10).
- Contraction of the ventricles pushes blood into the arteries at a peak (systolic) pressure. Between contractions, the blood pressure in the arteries falls to a minimum (diastolic) pressure. The systole–diastole sequence is the cardiac cycle (Figure 42.11).
- Contraction of the atria and ventricles is initiated by signals from the SA node (pacemaker) of the heart (Figure 42.12).

Animation: Human blood circulation

Animation: Major human blood vessels

Animation: The human heart

Animation: Cardiac cycle

Animation: Cardiac conduction

Animation: Examples of ECGs

42.4 Blood Vessels of the Circulatory System

- Blood is carried from the heart to body tissues in arteries; small branches of arteries, the arterioles, deliver blood to the capillaries, where substances are exchanged with the interstitial fluid. The blood is collected from the capillaries in venules and then returned to the heart in veins (Figure 42.14).
- The walls of arteries consist of an inner endothelial layer, a middle layer of smooth muscle, and an outer layer of elastic fibers. The smallest arteries, the arterioles, constrict and dilate to regulate blood flow and pressure into the capillaries.
- Capillary walls consist of a single layer of endothelial cells. Blood flow through capillaries is controlled by variation in contraction of the smooth muscles of arterioles and precapillary sphincters (Figure 42.15).
- In the capillary networks, the rate of blood flow is considerably slower than that in arteries and veins. This maximizes the time for exchange of substances between blood and tissues. Diffusion along concentration gradients and bulk flow drive the exchange of substances (Figure 42.17).
- Venules and veins have thinner walls than arteries, allowing the vessels to expand and contract over a wide range. As a result, they act as blood reservoirs as well as conduits.
- The return of blood to the heart is aided by pressure exerted on the veins when surrounding skeletal muscles contract and by respiratory movements. One-way valves in the veins prevent the blood from flowing backward (Figure 42.18).

Animation: Capillary forces

Animation: Vessel anatomy

42.5 Maintaining Blood Flow and Pressure

- Blood pressure and flow are regulated by controlling cardiac output, the degree of blood vessel constriction (primarily arterioles), and the total blood volume. The autonomic nervous system and the endocrine system coordinate these mechanisms.
- Regulation of cardiac output starts with baroreceptors, which detect blood pressure changes and send signals to the medulla. In response, the brain stem sends signals via the autonomic nervous system that alter the rate and force of the heartbeat.
- Hormones secreted by several glands contribute to the regulation of blood pressure and flow.
- Local controls respond primarily to O_2 and CO_2 concentrations in tissues. Low O_2 and high CO_2 concentration cause dilation of arteriole walls, increasing the arteriole diameter and blood flow. High O_2 and low CO_2 concentrations have the opposite effects. NO released by arterial endothelial cells acts locally to increase arteriole diameter and blood flow.

Animation: Measuring blood pressure

Animation: Vein function

42.6 The Lymphatic System

- The lymphatic system, a key component of the immune system, is an extensive network of vessels that collect excess interstitial fluid—which becomes lymph—and returns it to the venous blood (Figure 42.20).
- The tissues and organs of the lymphatic system include the lymph nodes, the spleen, the thymus, and the tonsils. They remove viruses, bacteria, damaged cells, and cellular debris from the lymph and bloodstream, and defend the body against infection and cancer.

Animation: Human lymphatic system

Animation: Lymph vascular system

Questions

Self-Test Questions

1. Compared with vertebrates, most invertebrates:
   a. lead more mobile lives.
   b. require a higher level of oxygen.
   c. have more complex layers of cells.
   d. have slower distribution of blood.
   e. require faster delivery and greater quantities of nutrients.

2. Which circulatory system best describes the animal?
   a. Squids and octopuses have open circulatory systems with ventricles that pump blood away from the heart.
   b. Fishes have a single-chambered heart with an atrium that pumps blood through gills for oxygen exchange.
   c. Amphibians have the most oxygenated blood in the pulmocutaneous circuit and the most deoxygenated blood in the systemic circuit.
   d. Amphibians and reptiles use a two-chambered heart to separate oxygenated and deoxygenated blood.
   e. Birds and mammals pump blood to separate pulmonary and systemic systems from two separate ventricles in a four-chambered heart.

3. A healthy student from the coastal city of Boston enrolls at a college in Boulder, Colorado, a mile above sea level. An analysis of her blood in her first months at college would show:
   a. decreased macrophage activity.
   b. increased secretion of erythropoietin by the kidneys.
   c. increased signaling ability of platelets.
   d. anemia caused by malfunctioning erythrocytes.
   e. increased mitosis of leukocytes.

4. A characteristic of blood circulation through or to the heart is that:
   a. the superior vena cava conveys blood to the head.
   b. the inferior vena cava conveys blood to the right atrium.
   c. the pulmonary arteries convey blood from the lungs to the left atrium.
   d. the pulmonary veins convey blood into the left ventricle.
   e. the aorta branches into two coronary arteries that convey blood from heart muscle.

5. The heartbeat includes:
   a. the systole when the heart fills.
   b. the diastole when the heart muscle contracts.
   c. pressure that causes the AV valves to open, filling the ventricles.
   d. rising pressure in the ventricles to open the AV valves and close the SL valves.
   e. the "lub" sound when the SL valves open and the "dub" sound when the AV valves close.

6. Keeping the mammalian cardiac cycle balanced is/are:
   a. an AV node between the right atria and right ventricle, which signals the Purkinje fibers.
   b. pacemaker cells, which compose the AV node and signal the SA node.
   c. an insulating layer that isolates the SA node from the right atrium.
   d. ion channels in pacemaker cells, which close to depolarize their plasma membranes.
   e. neurogenic stimuli from the nervous system.

7. Hydrostatic pressure is best described as:
   a. the uncoordinated contractions that occur during heart attacks.
   b. a premature ventricular contraction that signifies a skipped beat.
   c. a high point of pressure called diastolic blood pressure.
   d. hypertension, which decreases with age.
   e. the pressure of blood on the walls of arteries.

8. Characteristics of veins and venules are:
   a. thick walls.
   b. large muscle mass in walls.
   c. a large quantity of elastin in the walls.
   d. low blood volume compared with arteries.
   e. one-way valves to prevent backflow of blood.

9. When capillaries exchange substances:
   a. red blood cells move through the capillary lumens in double file.
   b. blood flow resistance is lower than it is in arteries and veins.
   c. water, ions, glucose, and erythrocytes pass freely between blood and tissues.
   d. diffusion along a concentration gradient and bulk flow are operating.
   e. diffusion is greatest closest to the arterioles.

10. To increase cardiac output:
    a. the adrenal medulla and sympathetic nervous system secrete epinephrine and norepinephrine.
    b. baroreceptors in the brain signal the sympathetic nerves.
    c. the brain stem signals the baroreceptors, causing the heart to beat faster.
    d. the autonomic nervous system responds to low oxygen on chemoreceptors and decreases the force of the heartbeat.
    e. chemoreceptors, stimulated by excessive blood oxygen, increase the rate of the heartbeat.

Questions for Discussion

1. *Aplastic anemia* develops when certain drugs or radiation destroy red bone marrow, including the stem cells that give rise to erythrocytes, leukocytes, and platelets. Predict some symptoms a person with aplastic anemia would be likely to develop. Include at least one symptom related to each type of blood cell.

2. In addition to the engine exhaust of boats and cars, carbon monoxide is also a component of cigarette smoke. What might be the impact of this phenomenon on a smoker's health?

3. In some people, the pressure of the blood pooling in the legs leads to a condition called *varicose veins,* in which the veins stand out like swollen, purple knots. Explain why this might happen, and why veins closer to the leg surface are more susceptible to the condition than those in deeper leg tissues.

Experimental Analysis

Mice in which the apolipoprotein E gene has been knocked out (deleted) by genetic engineering methods have high levels of plasma cholesterol and readily develop atherosclerosis, particularly on diets high in cholesterol. The immunosuppressant drug rapamycin is being touted also to be a drug that can affect atherosclerosis. Design an experiment to determine whether and at what dose rapamycin is effective in reducing atherosclerosis caused by dietary cholesterol.

Evolution Link

What is the evolutionary advantage of closure of the septum between the two ventricles to create a double circulatory system?

How Would You Vote?

Cardiopulmonary resuscitation (CPR) can make the difference between life and death after a cardiac arrest or a heart attack. Should public high schools in your state require all students to take a course in CPR? Is such a course worth diverting time and resources from the basic curriculum? Go to www.thomsonedu.com/login to investigate both sides of the issue and then vote.

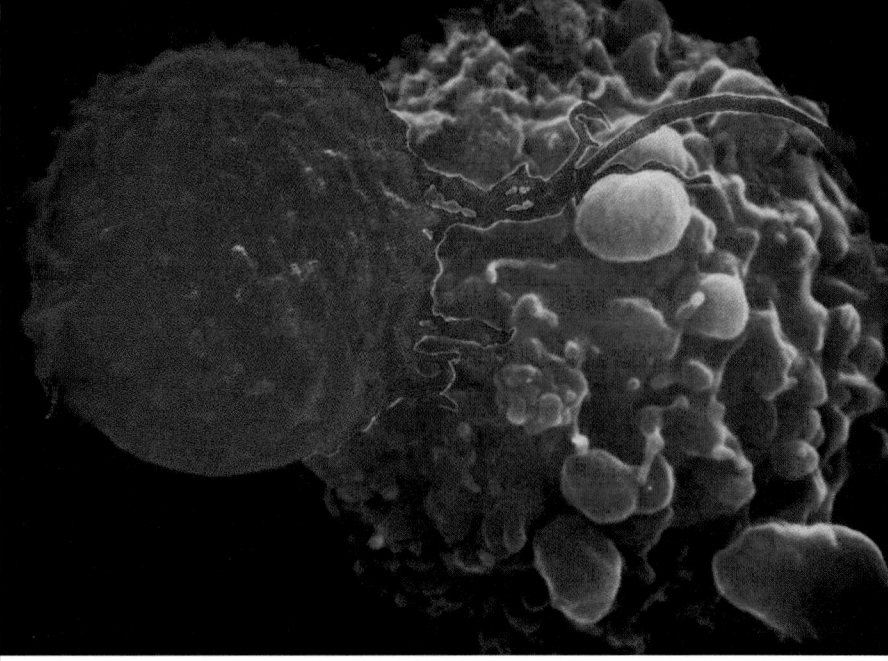

Death of a cancer cell. A cytotoxic T cell (orange) induces a cancer cell (mauve) to undergo apoptosis (programmed cell death). Cytotoxic T cells are part of the body's immune response system programmed to seek out, attach themselves, and kill cancer cells and pathogen-infected host cells.

© Dr. Andrejs Liepins/Science Photo Library/Photo Researchers, Inc.

43 Defenses against Disease

WHY IT MATTERS

Acquired immune deficiency syndrome (AIDS), which was first identified in the early 1980s, now infects about 40 million people worldwide, and continues to spread. Thousands of health-care workers, physicians, and researchers have joined the effort to control AIDS and develop effective treatments. Their primary aim is to develop an anti-AIDS *vaccine*—a substance that, when swallowed or injected, provides protection against infection by HIV (human immunodeficiency virus), which causes the disease.

The development of vaccines began with efforts to control smallpox, a dangerous and disfiguring viral disease that once infected millions of people worldwide. As early as the twelfth century, healthy individuals in China sought out people who were recovering from mild smallpox infections, ground up scabs from their lesions, and inhaled the powder or pushed it into their skin. Variations on this treatment were effective in protecting many people against smallpox infection.

In 1796, an English country doctor, Edward Jenner, used a more scientific approach. He knew that milkmaids never got smallpox if they had contracted cowpox, a similar but mild disease of cows that can be transmitted to humans. Jenner decided to see if a deliberate

971

infection with cowpox would protect humans from smallpox. He scratched material from a cowpox sore into a boy's arm, and 6 weeks later, after the cowpox infection had subsided, he scratched fluid from human smallpox sores into the boy's skin. (Jenner's use of the boy as an experimental subject would now be considered unethical.) Remarkably, the boy remained free from smallpox. Jenner carried out additional, carefully documented case studies with other patients with the same results. His technique became the basis for worldwide **vaccination** (*vacca* = cow) against smallpox. With improved vaccines, smallpox has now been eradicated from the human population.

Vaccination takes advantage of the **immune system** (*immunis* = exempt), the natural protection that is our main defense against infectious disease. This chapter focuses on the immune system and other defenses against infection, such as the skin. Our description emphasizes human and other mammalian systems, in which most of the scientific discoveries revealing the structure and function of the immune system have been made. At the end of the chapter we compare mammalian systems with the protective systems of nonmammalian vertebrates and invertebrates.

43.1 Three Lines of Defense against Invasion

Every organism is constantly exposed to *pathogens,* disease-causing viruses or organisms such as infectious bacteria, protists, fungi, and parasitic worms. Humans and other mammals have three lines of defense against these threats. The first line of defense involves physical barriers that prevent infection; it is not part of the immune system. The second line of defense is the *innate immunity system,* the inherited mechanisms that protect the body from many kinds of pathogens in a nonspecific way. The third line of defense, the *adaptive immunity system,* involves inherited mechanisms leading to the synthesis of molecules that target pathogens in a specific way. Reaction to an infection takes minutes in the case of the innate immunity system versus days for the adaptive immunity system.

Epithelial Surfaces Are Anatomical Barriers That Help Prevent Infection

An organism's first line of defense is the body surface—the skin covering the body exterior and the epithelial surfaces covering internal body cavities and ducts, such as the lungs and intestinal tract. The body surface forms a barrier of tight junctions between epithelial cells that keeps most pathogens (as well as toxic substances) from entering the body.

Many epithelial surfaces are coated with a mucus layer secreted by the epithelial cells that protects against pathogens as well as toxins and other chemicals. In the respiratory tract, ciliated cells constantly sweep the mucus, with its trapped bacteria and other foreign matter, into the throat, where it is coughed out or swallowed.

Many of the body cavities lined by mucous membranes have environments that are hostile to pathogens. For example, the strongly acidic environment of the stomach kills most bacteria and destroys many viruses that are carried there, including those trapped in swallowed mucus from the respiratory tract. Most of the pathogens that survive the stomach acid are destroyed by the digestive enzymes and bile secreted into the small intestine. The vagina, too, is acidic, which prevents many pathogens from surviving there. The mucus coating in some locations contains the enzyme lysozyme, which was secreted by the epithelial cells. Lysozyme breaks down the walls of some bacteria, causing them to lyse.

Two Immunity Systems Protect the Body from Pathogens That Have Crossed External Barriers

The body's second line of defense is a series of generalized internal chemical, physical, and cellular reactions that attack pathogens that have breached the first line. These defenses include inflammation, which creates internal conditions that inhibit or kill many pathogens, and specialized cells that engulf or kill pathogens or infected body cells.

Innate immunity is the term for this initial response by the body to eliminate cellular pathogens, such as bacteria and viruses, and prevent infection. You are born with an innate immune system. Innate immunity provides an immediate, *nonspecific* response; that is, it targets any invading pathogen and has no memory of prior exposure to the pathogen. It provides some protection against invading pathogens while a more powerful, specific response system is mobilized.

The third and most effective line of defense, **adaptive** (also called **acquired**) **immunity,** is *specific:* it recognizes individual pathogens and mounts an attack that directly neutralizes or eliminates them. It is so named because it is stimulated and shaped by the presence of a specific pathogen or foreign molecule. This mechanism takes several days to become protective. Adaptive immunity is triggered by specific molecules on pathogens that are recognized as being foreign to the body. The body retains a memory of the first exposure to a foreign molecule, enabling it to respond more quickly if the pathogen is encountered again in the future.

Innate immunity and adaptive immunity together constitute the immune system, and the defensive reactions of the system are termed the **immune response.** Functionally, the two components of the immune system interconnect and communicate at the chemical and cellular levels. The immune system is the product of a long evolutionary history of compensating adaptations by both pathogens and their targets. Over millions of years of vertebrate history, the mechanisms by which pathogens attack and invade have become more efficient, but the defenses of animals against the invaders have kept pace.

Study Break

1. What features of epithelial surfaces protect against pathogens?
2. What are the key differences between innate immunity and adaptive immunity?

43.2 Nonspecific Defenses: Innate Immunity

In most cases, the body needs 7 to 10 days to develop a fully effective immune response against a new pathogen, one that is invading the body for the first time. Innate immunity holds off invading pathogens in the meantime, killing or containing them until adaptive immunity comes fully into play. We have already discussed the body's anatomical barriers. Now let us look at the internal mechanisms of innate immunity: secreted molecules and cellular components. As you will see, cellular pathogens (such as bacteria) and viral pathogens elicit different responses.

Innate Immunity Provides an Immediate, General Defense against Invading Cellular Pathogens

Cellular pathogens—typically microorganisms—usually enter the body when injuries break the skin or epithelial surfaces. How does the host body recognize the pathogen as foreign? The answer is that the host has mechanisms to distinguish self from nonself. The innate immune system recognizes particular molecules that are common to many pathogens but absent in the host. An example is the lipopolysaccharide of gram-negative bacteria. The host then responds immediately to combat the pathogen.

Several types of specific cell-surface receptors in the host recognize the various types of molecules on microbial pathogens. The response depends on the receptor. For some receptors, the response is secretion of *antimicrobial peptides,* which kill the microbial pathogen. Other receptors include those that trigger the host cell to engulf, and destroy the pathogen, initiating inflammation, and the soluble receptors of the *complement system.*

Antimicrobial Peptides. All of our epithelial surfaces, namely skin, the lining of the gastrointestinal tract, the lining of the nasal passages and lungs, and the lining of the genitourinary tracts, are protected by antimicrobial peptides called *defensins.* Epithelial cells of those surfaces secrete defensins upon attack by a microbial pathogen. The defensins attack the plasma membranes of the pathogens, eventually disrupting them, thereby killing the cells. In particular, defensins play a significant role in innate immunity of the mammalian intestinal tract.

Inflammation. A tissue's rapid response to injury, including infection by most pathogens, involves **inflammation** (*inflammare* = to set on fire), the heat, pain, redness, and swelling that occur at the site of an infection.

Several interconnecting mechanisms initiate inflammation **(Figure 43.1)**. Let us consider bacteria entering a tissue as a result of a wound. **Monocytes** (a type of leukocyte) enter the damaged tissue from the bloodstream through the endothelial wall of the blood vessel. Once in the damaged tissue, the monocytes differentiate into **macrophages** ("big eaters"), which are phagocytes that are usually the first to recognize pathogens at the cellular level. (**Table 43.1** lists the major types of leukocytes such as macrophages; see also Figure 42.7.) Cell-surface receptors on the macrophages recognize and bind to surface molecules on the pathogen, activating the macrophage to phagocytize (engulf) the pathogen (see Figure 43.1, step 1). Activated macrophages also secrete **cytokines,** molecules that bind to receptors on other host cells and, through signal transduction pathways, trigger a response. Usually, not enough macrophages are present in the area of a bacterial infection to handle all of the bacteria.

The death of cells caused by the pathogen at the infection site activates cells dispersed through connective tissue called **mast cells,** which then release histamine (step 2). This histamine, along with the cytokines from activated macrophages, dilates local blood vessels around the infection site and increases their permeability, which increases blood flow and leakage of fluid from the vessels into body tissues (step 3). The response initiated by cytokines directly causes the heat, redness, and swelling of inflammation.

Cytokines also make the endothelial cells of the blood vessel wall stickier, causing circulating **neutrophils** (another type of phagocytic leukocyte) to attach to it in massive numbers. From there, the neutrophils are attracted to the infection site by **chemokines,** proteins also secreted by activated macrophages (step 4). To get to the infection site, the neutrophils pass between endothelial cells of the blood vessel wall. Neutrophils may also be attracted to the pathogen directly by molecules released from the pathogens themselves. Like macrophages, neutrophils have cell-surface recep-

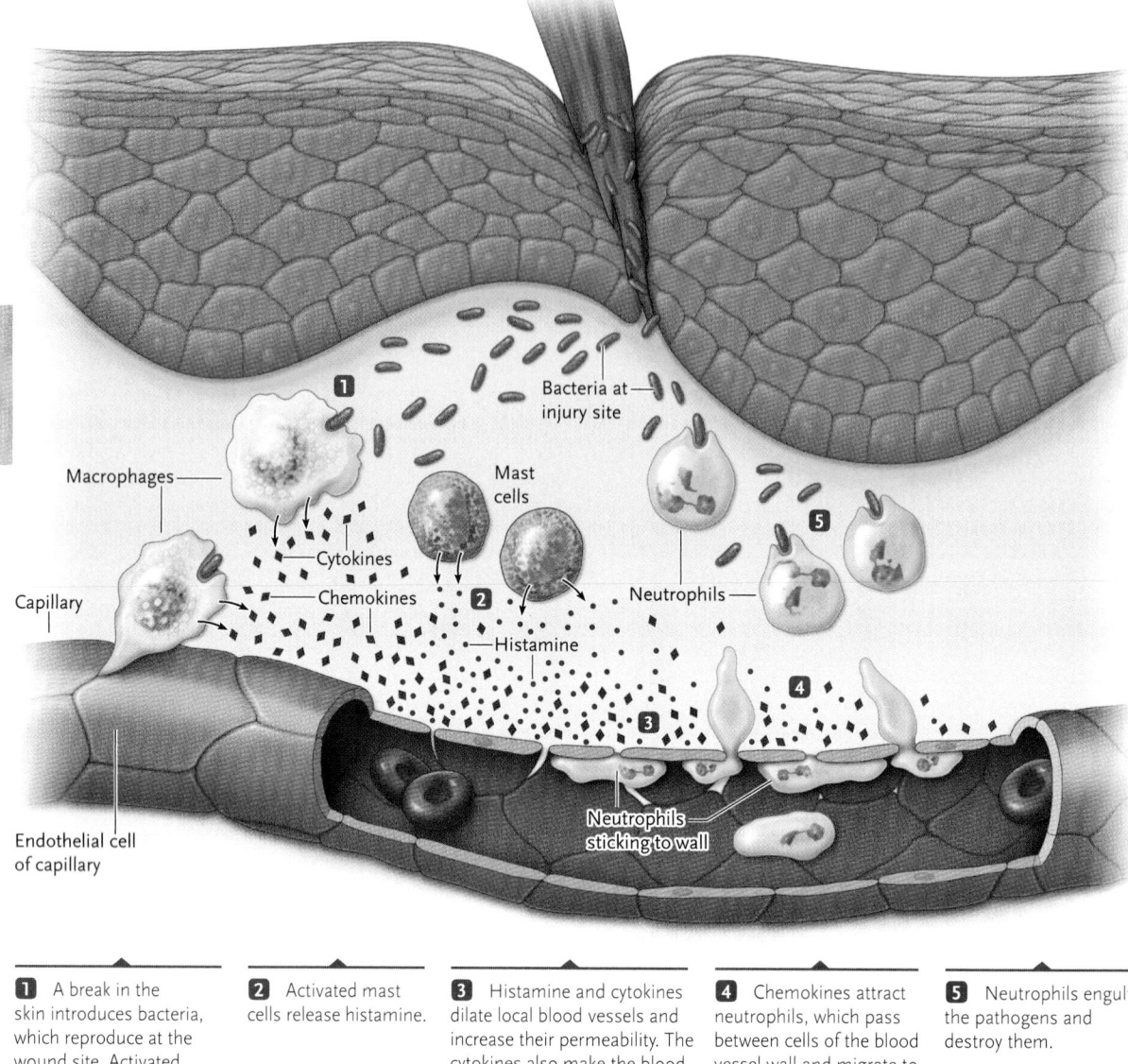

Bacteria at injury site

Macrophages

Mast cells

Cytokines

Chemokines

Neutrophils

Capillary

Histamine

Endothelial cell of capillary

Neutrophils sticking to wall

1 A break in the skin introduces bacteria, which reproduce at the wound site. Activated macrophages engulf the pathogens and secrete cytokines and chemokines.

2 Activated mast cells release histamine.

3 Histamine and cytokines dilate local blood vessels and increase their permeability. The cytokines also make the blood vessel wall sticky, causing neutrophils to attach.

4 Chemokines attract neutrophils, which pass between cells of the blood vessel wall and migrate to the infection site.

5 Neutrophils engulf the pathogens and destroy them.

Figure 43.1
The steps producing inflammation. The colorized micrograph on the left shows a macrophage engulfing a yeast cell.

tors that enable them to recognize and engulf pathogens (step 5).

Once a macrophage or neutrophil has engulfed the pathogen, it uses a variety of mechanisms to destroy it. These mechanisms include attacks by enzymes and defensins located in lysosomes and the production of toxic chemicals. The harshness of these attacks usually kills the neutrophils as well, while macrophages usually survive to continue their pathogen-scavenging activities. Dead and dying neutrophils, in fact, are a major component of the pus formed at infection sites. The pain of inflammation is caused by the migration of macrophages and neutrophils to the infection site and their activities there.

Some pathogens, such as parasitic worms, are too large to be engulfed by macrophages or neutrophils. In that case, macrophages, neutrophils, and

eosinophils (another type of leukocyte) cluster around the pathogen and secrete lysosomal enzymes and defensins in amounts that are often sufficient to kill the pathogen.

The Complement System. Another nonspecific defense mechanism activated by invading pathogens is the **complement system,** a group of more than 30 interacting soluble plasma proteins circulating in the blood and interstitial fluid **(Figure 43.2).** The proteins are normally inactive; they are activated when they recognize molecules on the surfaces of pathogens. Activated complement proteins participate in a cascade of reactions on pathogen surfaces, producing large numbers of different complement proteins, some of which assemble into **membrane attack complexes.** These complexes insert into the plasma membrane of many types of bacterial cells and create pores that allow ions and small molecules to pass readily through the membrane. As a result, the bacteria can no longer maintain osmotic balance, and they swell and lyse. For the other types of bacterial cells, the cascade of reactions coats the pathogen with fragments of the complement proteins. Cell-surface receptors on phagocytes then recognize these fragments, and engulf and destroy the pathogen.

Several activated proteins in the complement cascade also act individually to enhance the inflammatory response. For example, some of the proteins stimulate mast cells to enhance histamine release, while others cause increased blood vessel permeability.

Combating Viral Pathogens Requires a Different Innate Immune Response

You have learned that specific molecules on cellular pathogens such as bacteria are key to initiating innate immune responses. By contrast, the innate immunity system is unable to distinguish effectively the surface molecules of viral pathogens from those of the host. The host must, therefore, use other strategies to provide some immediate protection against viral infec-

Table 43.1	Major Types of Leukocytes and Their Functions
Type of Leukocyte	Function
Monocyte	Differentiates into a macrophage when released from blood into damaged tissue
Macrophage	Phagocyte that engulfs infected cells, pathogens, and cellular debris in damaged tissues; helps activate lymphocytes carrying out immune response
Neutrophil	Phagocyte that engulfs pathogens and tissue debris in damaged tissues
Eosinophil	Secretes substances that kill eukaryotic parasites such as worms
Lymphocyte	Main subtypes involved in innate and adaptive immunity are natural killer (NK) cells, B cells, plasma cells, helper T cells, and cytotoxic T cells. NK cells function as part of innate immunity to kill virus-infected cells and some cancerous cells of the host. The other cell types function as part of adaptive immunity: they produce antibodies, destroy infected and cancerous body cells, and stimulate macrophages and other leukocyte types to engulf infected cells, pathogens, and cellular debris
Basophil	Located in blood, responds to IgE antibodies in an allergic response by secreting histamine, which stimulates inflammation

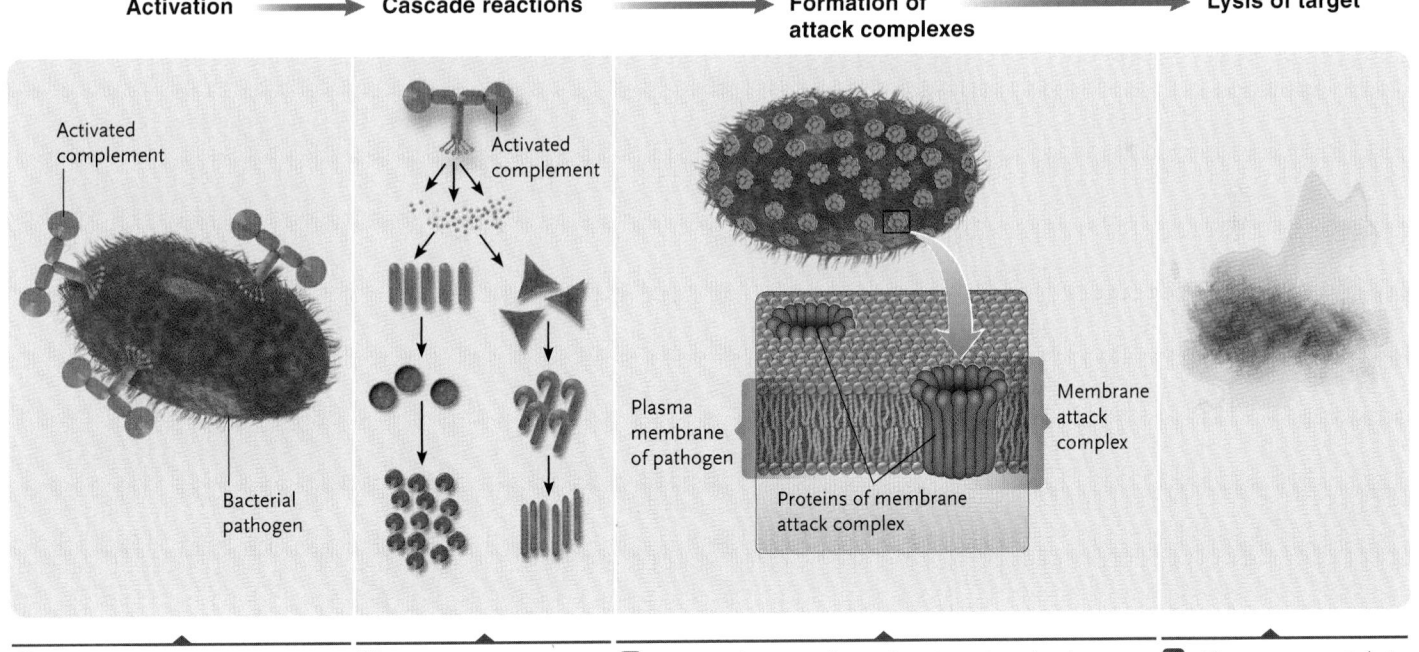

Activation → Cascade reactions → Formation of attack complexes → Lysis of target

Activated complement

Bacterial pathogen

Activated complement

Plasma membrane of pathogen

Proteins of membrane attack complex

Membrane attack complex

1 Complement proteins are activated by binding directly to a bacterial surface.

2 Cascading reactions produce huge numbers of different complement proteins. These assemble into many molecules, which form many membrane attack complexes.

3 The membrane attack complexes insert into the plasma membrane of the pathogen. Each forms a large pore across the membrane.

4 The pores promote lysis of the pathogen, which dies because of the severe disruption of its structure.

Figure 43.2
The role of complement proteins in combating microbial pathogens.

tions until the adaptive immunity system, which can discriminate between viral and host proteins, is effective. Three main strategies involve RNA interference, interferon, and natural killer cells.

RNA Interference. *RNA interference (RNAi)* is a cellular mechanism that is triggered by a virus's double-stranded (ds) RNA molecules (see Section 16.3). Such molecules are a natural part of the life cycles of a number of viruses. The RNAi system destroys dsRNA molecules, thereby inhibiting the virus's life cycle.

Interferon. Viral dsRNA also causes the infected host cell to produce two cytokines, called interferon-α and interferon-β. **Interferons** can be produced by most cells of the body. These proteins act both on the infected cell that produces them, an autocrine effect, and on neighboring uninfected cells, a paracrine effect (see Section 40.1). They work by binding to cell-surface receptors, triggering a signal transduction pathway that changes the gene expression pattern of the cells. The key changes include activation of a ribonuclease enzyme that degrades most cellular RNA and inactivation of a key protein required for protein synthesis, thereby inhibiting most protein synthesis in the cell. These effects on RNA and protein synthesis inhibit replication of the viral genome, while putting the cell in a weakened state from which it often can recover.

Natural Killer Cells. Cells that have been infected with virus must be destroyed. That is the role of **natural killer (NK) cells**. NK cells are a type of **lymphocyte**, a leukocyte that carries out most of its activities in tissues and organs of the lymphatic system (see Figure 42.20). NK cells circulate in the blood and kill target host cells—not only cells that are infected with virus, but also some cells that have become cancerous.

NK cells can be activated by cell-surface receptors or by interferons secreted by virus-infected cells. NK cells are not phagocytes; instead, they secrete granules containing *perforin,* a protein that creates pores in the target cell's membrane. Unregulated diffusion of ions and molecules through the pores causes osmotic imbalance, swelling, and rupture of the infected cell. NK cells also kill target cells indirectly through the secretion of *proteases* (protein-degrading enzymes) that pass through the pores. The proteases trigger **apoptosis**, or programmed cell death (see *Insights from the Molecular Revolution* in Chapter 7). That is, the proteases activate other enzymes that cause the degradation of DNA which, in turn, induces pathways leading to the cell's death.

How does an NK cell distinguish a target cell from a normal cell? The surfaces of most vertebrate cells contain particular *major histocompatibility complex (MHC) proteins.* You will learn about the role of these proteins in adaptive immunity in the next section; for now, just consider them to be tags on the cell surface. NK cells monitor the level of MHC proteins

and respond differently depending on their level. An appropriately high level, as on normal cells, inhibits the killing activity of NK cells. Viruses often inhibit the synthesis of MHC proteins in the cells they infect, lowering the levels of those proteins and identifying them to NK cells. Cancer cells also have low or, in some cases, no MHC proteins on their surfaces, which makes them a target for destruction by NK cells as well.

STUDY BREAK

1. What are the usual characteristics of the inflammatory response?
2. What processes specifically cause each characteristic of the inflammatory response?
3. What is the complement system?
4. Why does combating viral pathogens require a different response by the innate immunity system than combating bacterial pathogens? What are the three main strategies a host uses to protect against viral infections?

43.3 Specific Defenses: Adaptive Immunity

Adaptive immunity is a defense mechanism that recognizes specific molecules as being foreign and clears those molecules from the body. The foreign molecules recognized may be free, as in the case of toxins, or they may be on the surface of a virus or cell, the latter including pathogenic bacteria, cancer cells, pollen, and cells of transplanted tissues and organs. Adaptive immunity develops only when the body is exposed to the foreign molecules and, hence, takes several days to become effective. This would be a significant problem in the case of invading pathogens were it not for the innate immune system, which combats the invading pathogens in a nonspecific way within minutes after they enter the body. There are two key distinctions between innate and adaptive immunity: innate immunity is nonspecific whereas adaptive immunity is specific, and innate immunity retains no memory of exposure to the pathogen whereas adaptive immunity retains a memory of the foreign molecule that triggered the response, thereby enabling a rapid, more powerful response if that pathogen is encountered again at a later time.

In Adaptive Immunity, Antigens Are Cleared from the Body by B Cells or T Cells

A foreign molecule that triggers an adaptive immunity response is called an **antigen** (meaning "*anti*body *gen*erator"). Antigens are macromolecules; most are large proteins (including glycoproteins and lipoproteins) or polysaccharides (including lipopolysaccharides). Some types of nucleic acids can also act as antigens, as can various large, artificially synthesized molecules.

Antigens may be *exogenous,* meaning they enter the body from the environment, or *endogenous,* meaning they are generated within the body. Exogenous antigens include antigens on pathogens introduced beneath the skin, antigens in vaccinations, and inhaled and ingested macromolecules, such as toxins. Endogenous antigens include proteins encoded by viruses that have infected cells, and altered proteins produced by mutated genes, such as those in cancer cells.

Antigens are recognized in the body by two types of lymphocytes, B cells and T cells. **B cells** differentiate from stem cells in the bone marrow (see Section 42.2). It is easy to remember this as "B for bone." However, the "B" actually refers to the *bursa of Fabricus,* a lymphatic organ found only in birds; B cells were first discovered there. After their differentiation, B cells are released into the blood and carried to capillary beds serving the tissues and organs of the lymphatic system. Like B cells, **T cells** are produced by the division of stem cells in the bone marrow. Then, they are released into the blood and carried to the **thymus,** an organ of the lymphatic system (the "T" in "T cell" refers to the thymus).

The role of lymphocytes in adaptive immunity was demonstrated by experiments in which all of the leukocytes in mice were killed by irradiation with X rays. These mice were then unable to develop adaptive immunity. Injecting lymphocytes from normal mice into the irradiated mice restored the response; other body cells extracted from normal mice could not restore the response. (For more on the use of mice as an experimental organism in biology, see *Focus on Research Organisms.*)

There are two types of adaptive immune responses: **antibody-mediated immunity** (also called *humoral immunity*) and **cell-mediated immunity.** In antibody-mediated immunity, B-cell derivatives called **plasma cells** secrete **antibodies,** highly specific soluble protein molecules that circulate in the blood and lymph recognizing and binding to antigens and clearing them from the body. In cell-mediated immunity, a subclass of T cells becomes activated and, with other cells of the immune system, attacks foreign cells directly and kills them.

The steps involved in the adaptive immune response are similar for antibody-mediated immunity and cell-mediated immunity:

1. Antigen encounter and recognition: Lymphocytes encounter and recognize an antigen.
2. Lymphocyte activation: The lymphocytes are activated by binding to the antigen and proliferate by cell division to produce large clones of identical cells.
3. Antigen clearance: The large clones of activated lymphocytes are responsible for clearing the antigen from the body.
4. Development of immunological memory: Some of the activated lymphocytes differentiate into **memory cells** that circulate in the blood and lymph, ready to initiate a rapid immune response upon subsequent exposure to the same antigen.

These steps will be expanded upon in the following discussions of antibody-mediated immunity and cell-mediated immunity.

Antibody-Mediated Immunity Involves Activation of B Cells, Their Differentiation into Plasma Cells, and the Secretion of Antibodies

An adaptive immune response begins as soon as an antigen is encountered and recognized in the body.

Antigen Encounter and Recognition by Lymphocytes. Exogenous antigens are encountered by lymphocytes in the lymphatic system. As already mentioned, the two key lymphocytes that recognize antigens are B cells and T cells. Each B cell and each T cell is specific for a particular antigen, meaning that the cell can bind to only one particular molecular structure. The binding is so specific because the plasma membrane of each B cell and T cell is studded with thousands of identical receptors for the antigen; in B cells they are called **B-cell receptors (BCRs)** and in T cells they are called **T-cell receptors (TCRs) (Figure 43.3).** Considering the entire populations of B cells and T cells in the body, there are multiple cells that can recognize each antigen but, most importantly, the populations (in normal persons) contain cells capable of recognizing any antigen. For example, each of us has about 10 trillion B cells that collectively have about 100 million different kinds of BCRs. And, these cells are present *before* the body has encountered the antigens.

The binding between antigen and receptor is an interaction between two molecules that fit together like an enzyme and its substrate. A given BCR or TCR typically does not bind to the whole antigen molecule, but to small regions of it called **epitopes** or *antigenic determinants.* Therefore, several different B cells and T cells may bind to the population of a particular antigen encountered in the lymphatic system.

BCRs and TCRs are encoded by different genes and thus have different structures. The BCR on a B cell (see Figure 43.3a) corresponds to the antibody secreted by that particular B cell when it is activated and differentiates into a plasma cell. As you will learn in more detail shortly, an antibody molecule is a protein consisting of four polypeptide chains. At one end, it has two identical *antigen-binding sites,* regions that bind to a specific antigen. In the case of the BCR, at the opposite end of the

a. B-cell receptor (BCR)

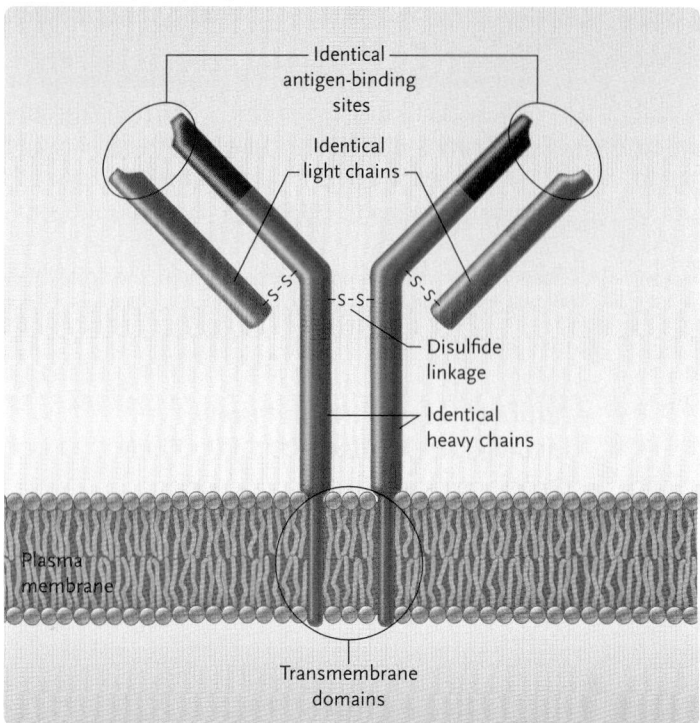

b. T-cell receptor (TCR)

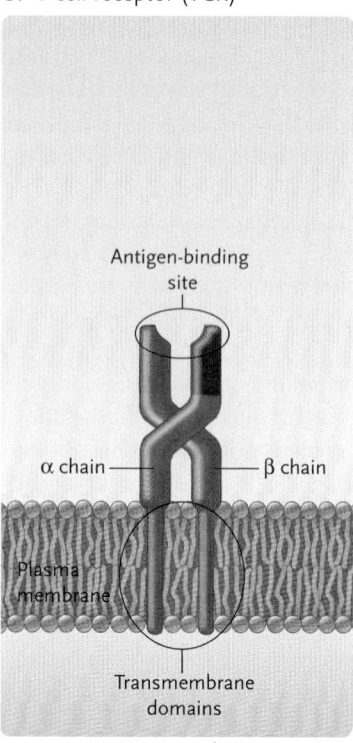

Figure 43.3
Antigen-binding receptors on B cells and T cells.

molecule from the antigen-binding sites are *transmembrane domains*, which embed in the plasma membrane. TCRs are simpler than BCRs, consisting of a protein made up of two different polypeptides (see Figure 43.3b). Like BCRs, TCRs have an antigen-binding site at one end and transmembrane domains at the other end.

Antibodies. Antibodies are the core molecules of antibody-mediated immunity. Antibodies are large, complex proteins that belong to a class of proteins known as **immunoglobulins** (Ig). Each antibody molecule consists of four polypeptide chains: two identical

light chains and two identical **heavy chains** about twice or more the size of the light chain **(Figure 43.4).** The chains are held together in the complete protein by disulfide (—S—S—) linkages and fold into a Y-shaped structure. The bonds between the two arms of the Y form a hinge that allows the arms to flex independently of one another.

Each polypeptide chain of an antibody molecule has a *constant region* and a *variable region*. The constant region of each antibody type has the same amino acid sequence for that part of the heavy chain, and likewise for that part of the light chain. The variable region of

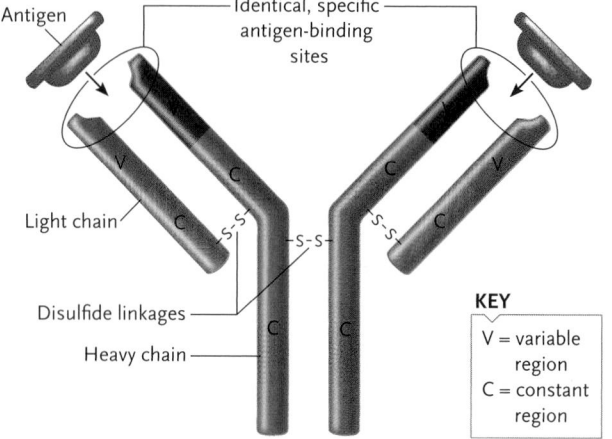

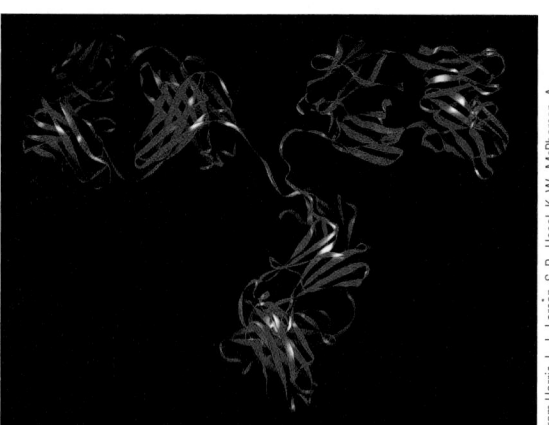

From Harris, L. J., Larson, S. B., Hasel, K. W., McPherson, A., *Biochemistry* 36, p. 1581 (1997). Structure rendered with RIBBONS.

KEY

V = variable region
C = constant region

Figure 43.4
The arrangement of light and heavy polypeptide chains in an antibody molecule. As shown, two sites, one at the tip of each arm of the Y, bind the same antigen.

FOCUS ON RESEARCH

Research Organisms: The Mighty Mouse

The "wee, sleekit, tim'rous, cowrin' beastie," as the poet Robert Burns called the mouse *(Mus musculus)*, has a much larger stature among scientists. The mouse and its cells have been used to great advantage as models for research on mammalian developmental genetics, immunology, and cancer. The availability of the mouse as a research tool enables scientists to carry out experiments with a mammal that would not be practical or ethical with humans.

Mice are grown by the millions in laboratories all over the world. Its small size makes the mouse relatively inexpensive and easy to maintain in the laboratory, and its short generation time, compared with most other mammals, allows genetic crosses to be carried out within a reasonable time span. Mice can be mated when they are 10 weeks old; within 18 to 22 days the female gives birth to a litter of about 5 to 10 offspring. A female may be rebred within little more than a day after giving birth.

© Peter Skinner/Photo Researchers, Inc.

Mice have a long and highly productive history as experimental animals. Gregor Mendel, the founder of genetics, is known to have kept mice as part of his studies. Toward the end of the nineteenth century, August Weissmann helped disprove an early evolutionary hypothesis, the inheritance of acquired characters, by cutting off the tails of mice for 22 successive generations and finding that it had no effect on tail length. The first example of a lethal allele was also found in mice, and pioneering experiments on the transplantation of tissues between individuals were conducted with mice. During the 1920s, Fred Griffith laid the groundwork for the research showing that DNA is the hereditary molecule in his work with pneumonia-causing bacteria in mice (see Section 14.1). More recently, genetic experiments with mice have revealed more than 500 mutants that cause hereditary diseases, immunological defects, and cancer in mammals including humans.

The mouse has also been the mammal of choice for experiments that introduce and modify genes through genetic engineering. One of the most spectacular results of this research was the production of giant mice by introducing a human growth hormone gene into a line of dwarfed

mice that were deficient for this hormone.

Genetic engineering has also produced "knockout" mice, in which a gene of interest is completely nonfunctional (see Section 18.2). The effects of the lack of function of a gene in the knockout mice often allow investigators to determine the role of the normal form of the gene. Some knockout mice have been developed to be defective in genes homologous to human genes that cause serious diseases, such as cystic fibrosis. This allows researchers to study the disease in mice with the goal of developing cures or therapies.

The revelations in developmental genetics from studies with the mouse have been of great interest and importance in their own right. But more and more, as we find that much of what applies to the mouse also applies to humans, the findings in mice have shed new light on human development and opened pathways to the possible cure of human genetic diseases.

In 2002 the sequence of the mouse genome was reported. The sequence is enabling researchers to refine and expand their use of the mouse as a model organism for studies of mammalian biology and mammalian diseases.

---

both the heavy and light chains, by contrast, has a different amino acid sequence for each antibody molecule in a population. Structurally, the variable regions are the top halves of the polypeptides in the arms of the Y-shaped molecule. The three-dimensional folding of the heavy chain and light chain variable regions of each arm creates the antigen-binding site. The antigen-binding site is identical for the two arms of the same antibody molecule because of the identity of the two heavy chains and the two light chains in a molecule, as mentioned earlier. However, the antigen-binding sites are different from antibody molecule to antibody molecule because of the amino acid differences in the variable regions of the two chain types.

The constant regions of the heavy chains in the tail part of the Y-shaped structure determine the *class* of the antibody. Humans have five different classes of anti-

bodies—*IgM, IgG, IgA, IgE,* and *IgD* **(Table 43.2).** Due to differences in their heavy chain constant regions, they have specific structural and functional differences.

IgM remains bound to the cells that make it due to a region at the end opposite from the antigen-binding end that inserts into the plasma membrane of the cell. BCRs on B cells for antigen recognition are IgM molecules; as we shall see, IgM is also the first type of antibody secreted from plasma cells in the early stages of an antibody-mediated response. (When secreted, they exist as a pentamer.) IgM antibodies activate the complement system when they bind an antigen, and stimulate the phagocytic activity of macrophages.

IgG is the most abundant antibody circulating in the blood and lymphatic system, where it stimulates phagocytosis and activates the complement system when it binds an antigen. IgG is produced in large

Table 43.2   **Five Classes of Antibodies**

Class	Structure (Secreted Form)	Location	Functions
IgM		Surfaces of unstimulated B cells (as monomer); free in circulation (as pentamer)	First antibodies to be secreted by B cells in primary response. When bound to antigen, promotes agglutination reaction, activates complement system, and stimulates phagocytic activity of macrophages.
IgG		Blood and lymphatic circulation	Most abundant antibody in primary and secondary responses. Crosses placenta, conferring passive immunity to fetus; stimulates phagocytosis and activates complement system.
IgA		Body secretions such as tears, breast milk, saliva, and mucus	Blocks attachment of pathogens to mucous membranes; confers passive immunity for breastfed infants.
IgE		Skin and tissues lining gastrointestinal and respiratory tracts (secreted by plasma cells)	Stimulates mast cells and basophils to release histamine; triggers allergic responses.
IgD		Surface of unstimulated B cells	Membrane receptor for mature B cells; probably important in B-cell activation (clonal selection).

amounts when the body is exposed a second time to the same antigen.

IgA is found mainly in body secretions such as saliva, tears, breast milk, and the mucus coating of body cavities such as the lungs, digestive tract, and vagina. In these locations, the antibodies bind to surface groups on pathogens and block their attachment to body surfaces. Breast milk transfers IgA antibodies and thus immunity to a nursing infant.

IgE is secreted by plasma cells of the skin and tissues lining the gastrointestinal tract and respiratory tract. IgE binds to basophils and mast cells where it mediates many allergic responses, such as hay fever, asthma, and hives. Binding of a specific antigen to IgE stimulates the cell to which it is bound to release histamine, which triggers an inflammatory response. IgE also contributes to mechanisms that combat infection by parasitic worms.

IgD occurs with IgM as a receptor on the surfaces of B cells; its function is uncertain.

The Generation of Antibody Diversity. The human genome has approximately 20,000–25,000 genes, far fewer than necessary to encode 100 million different antibodies if two genes encoded one antibody, one gene for the heavy chain and one for the light chain. Antibody diversity is generated in a different way from one gene per chain, however, instead involving three rear-

rangements during B-cell differentiation of DNA segments that encode parts of the light and heavy chains. Let us consider how this process produces light-chain genes for the B-cell receptor **(Figure 43.5)**; the production of heavy-chain genes is similar. The genes for the two different subunits of the TCR undergo similar rearrangements to produce the great diversity in antigen-binding capability of those receptors.

An undifferentiated B cell has three types of light-chain DNA segments: V, J, and C. One of each type is needed to make a complete, functional light-chain gene (see Figure 43.5, top). In humans, there are about 40 different V segments encoding most of the variable region of the chain, 5 different J (joining) segments encoding the rest of the variable region, and only one copy of the segment for the constant (C) part of the chain.

During B-cell differentiation, a DNA rearrangement occurs in which one random V and one random J segment join with the C segment to form a functional light-chain gene (see Figure 43.5, step 1). The rearrangement involves deletion of the DNA between the V and J segments. The positions at which the DNA breaks and rejoins in the V and J joining reaction occur randomly over a distance of several nucleotides, which adds greatly to the variability of the final gene assembly. The DNA between the J segment and the C segment is an intron in the final assembled gene.

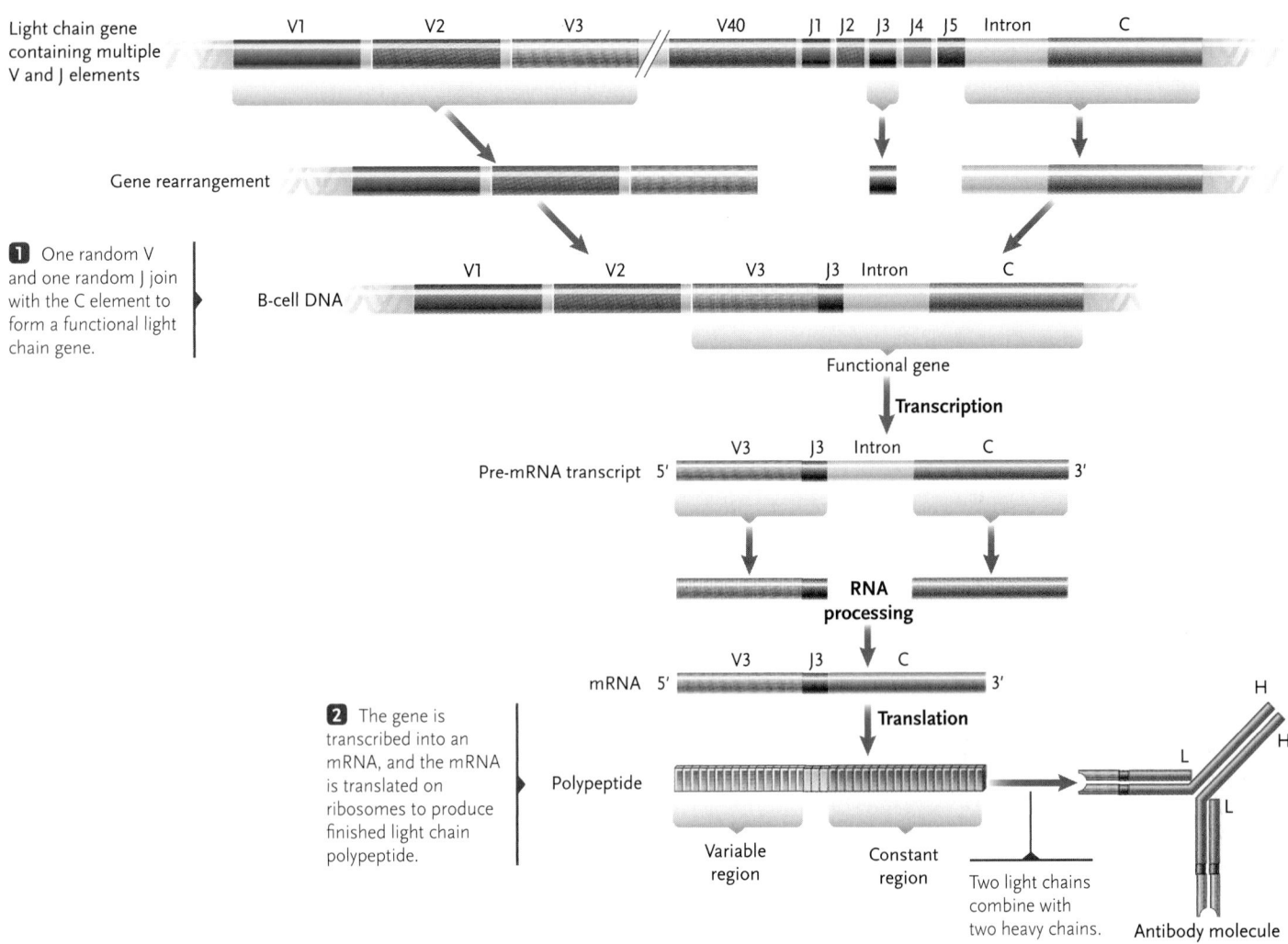

Light chain gene containing multiple V and J elements

V1 V2 V3 // V40 J1 J2 J3 J4 J5 Intron C

Gene rearrangement

1 One random V and one random J join with the C element to form a functional light chain gene.

B-cell DNA

V1 V2 V3 J3 Intron C

Functional gene

Transcription

Pre-mRNA transcript 5′ V3 J3 Intron C 3′

RNA processing

mRNA 5′ V3 J3 C 3′

2 The gene is transcribed into an mRNA, and the mRNA is translated on ribosomes to produce finished light chain polypeptide.

Translation

Polypeptide

Variable region   Constant region   Two light chains combine with two heavy chains.

H H L L

Antibody molecule

Figure 43.5
The DNA rearrangements producing a functional light-chain gene, in simplified form.

Transcription of this newly assembled gene produces a typical pre-mRNA molecule (see Section 15.3). The intron between the J and C segments is removed during the production of the mRNA by RNA processing. Translation of the mRNA produces the light chain with variable and constant regions (step 2).

As noted earlier, the assembly of functional heavy-chain genes occurs similarly. However, whereas light-chain genes have one C segment, heavy-chain genes have five types of C segments, each of which encodes one of the constant regions of IgM, IgD, IgG, IgE, and IgA. The inclusion of one of the five C segment types in the functional heavy-chain gene therefore specifies the class of antibody that will be made by the cell. Thus, the various DNA rearrangements producing the various light chain and heavy chain genes, along with the various combinations of light and heavy chains, generates the 100 million different antibodies.

T-Cell Activation. Let us now follow the development of an antibody-mediated immune response by linking the recognition of an antigen by lymphocytes, the activation of lymphocytes by antigen binding, and the production of antibodies. Typically, the pathway begins when a type of T cell becomes activated, following the steps outlined in **Figure 43.6.** Let us learn about this pathway by considering the fate of pathogenic bacteria that have been introduced under the skin. Circulating viruses in the blood follow the same pathway.

First, a type of phagocyte called a **dendritic cell** engulfs a bacterium in the infected tissue by phagocytosis **(Figure 43.7,** step 1). Dendritic cells are so named because they have many surface projections resembling dendrites of neurons. They have the same origin as leukocytes, and recognize a bacterium as foreign by the same recognition mechanism used by macrophages in the innate immunity system. In essence, the dendritic cell is part of the innate immunity system, but its primary role is to stimulate the development of an adaptive immune response.

Engulfment of a bacterium activates the dendritic cell; the cell now migrates to a nearby lymph node. Then, within the dendritic cell, the endocytic vesicle containing the engulfed bacterium fuses with a lysosome. In the lysosome, the bacterium's proteins are

Antibody-mediated immune response: T-cell activation

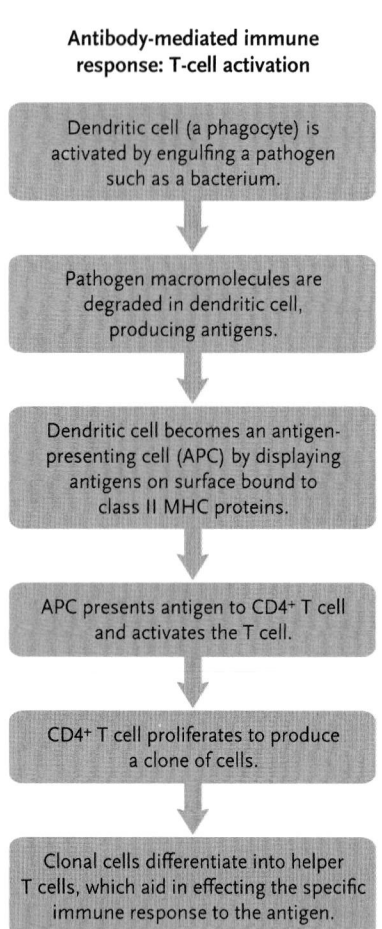

Dendritic cell (a phagocyte) is activated by engulfing a pathogen such as a bacterium.

↓

Pathogen macromolecules are degraded in dendritic cell, producing antigens.

↓

Dendritic cell becomes an antigen-presenting cell (APC) by displaying antigens on surface bound to class II MHC proteins.

↓

APC presents antigen to CD4⁺ T cell and activates the T cell.

↓

CD4⁺ T cell proliferates to produce a clone of cells.

↓

Clonal cells differentiate into helper T cells, which aid in effecting the specific immune response to the antigen.

Figure 43.6
An outline of T-cell activation in antibody-mediated immunity.

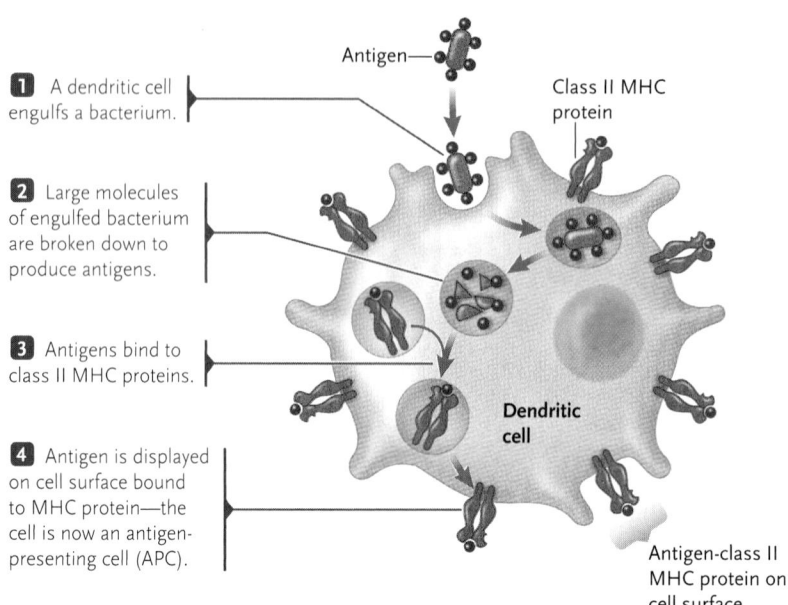

1 A dendritic cell engulfs a bacterium.

2 Large molecules of engulfed bacterium are broken down to produce antigens.

3 Antigens bind to class II MHC proteins.

4 Antigen is displayed on cell surface bound to MHC protein—the cell is now an antigen-presenting cell (APC).

Figure 43.7
Generation of an antigen-presenting cell when a dendritic cell engulfs a bacterium.

degraded into short peptides, which function as antigens (step 2). The antigens bind intracellularly to **class II major histocompatibility complex (MHC)** proteins (step 3); the interacting molecules then migrate to the cell surface where the antigen is displayed (step 4). These steps, which occur in the dendritic cell after it has migrated to the lymph node, convert the cell into an **antigen-presenting cell (APC)**, ready to present the antigen to T cells in the next step of antibody-mediated immunity. The process is recapped in **Figure 43.8.**

Antibody-mediated immune response

T-cell activation

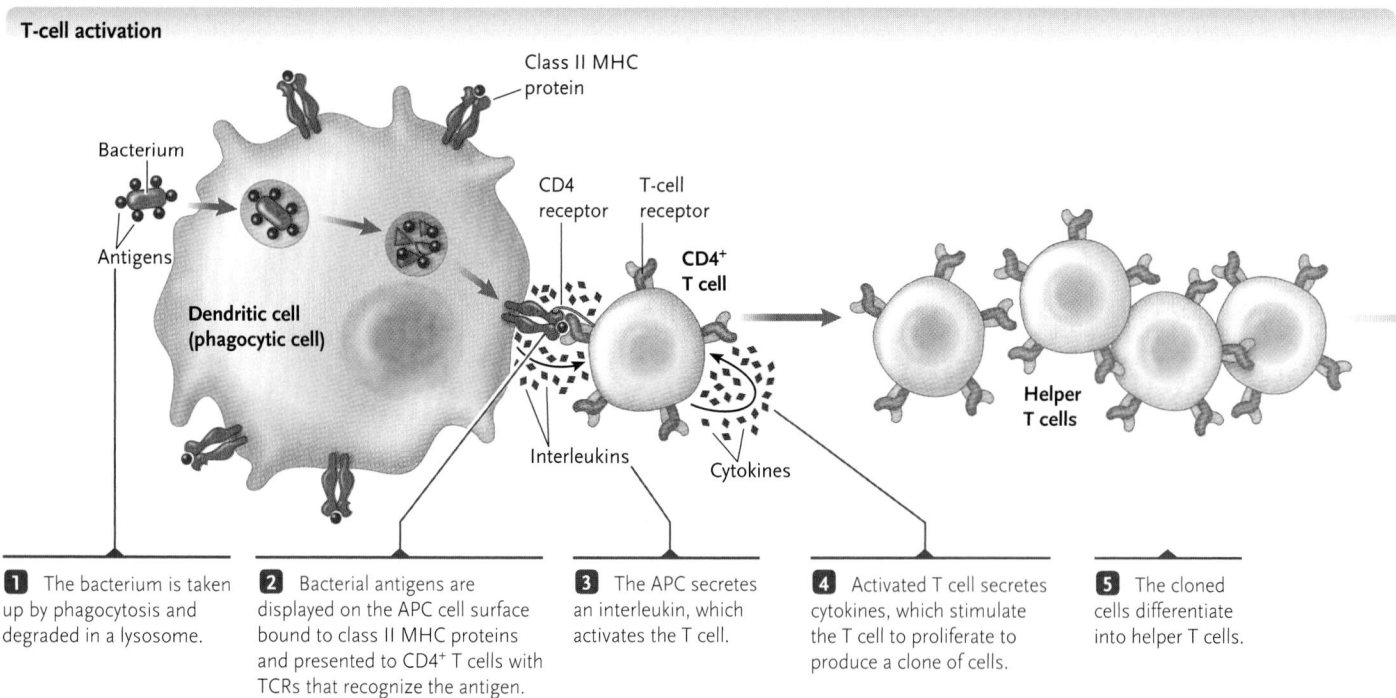

1 The bacterium is taken up by phagocytosis and degraded in a lysosome.

2 Bacterial antigens are displayed on the APC cell surface bound to class II MHC proteins and presented to CD4⁺ T cells with TCRs that recognize the antigen.

3 The APC secretes an interleukin, which activates the T cell.

4 Activated T cell secretes cytokines, which stimulate the T cell to proliferate to produce a clone of cells.

5 The cloned cells differentiate into helper T cells.

MHC proteins are named for a large cluster of genes encoding them, called the **major histocompatibility complex.** The complex spans 4 million base pairs and contains 128 genes. Many of these genes play important roles in the immune system. Each individual of each vertebrate species has a unique combination of MHC proteins on almost all body cells, meaning that no two individuals of a species except identical siblings are likely to have exactly the same MHC proteins on their cells. There are two classes of MHC proteins, class I and class II, which have different functions in adaptive immunity, as we will see.

The key function of an APC is to present the antigen to a lymphocyte. In the antibody-mediated immune response, the APC presents the antigen, bound to a class II MHC protein, to a type of T cell in the lymphatic system called a **CD4$^+$ T cell** because it has receptors named CD4 on its surface. A specific CD4$^+$ T cell having a TCR with an antigen-binding site that recognizes the antigen (epitope, actually) binds to the antigen on the APC (see Figure 43.8, step 2). The CD4 receptor on the T cell helps link the two cells together.

When the APC binds to the CD4$^+$ T cell, the APC secretes an *interleukin* (meaning "between leukocytes"), a type of cytokine, which activates the associated T cell (step 3). The activated T cell then secretes cytokines (step 4), which act in an autocrine manner (see Section 40.1) to stimulate **clonal expansion,** the proliferation of the activated CD4$^+$ T cell by cell division to produce a clone of cells. These clonal cells differentiate into **helper T cells,** so named because they assist with the activation of B cells (step 5). A helper T cell is an example of an **effector T cell,** meaning that it is involved in effecting—bringing about—the specific immune response to the antigen.

B-Cell Activation. Antibodies are produced in and secreted by activated B cells. The activation of a B cell requires the B cell to present the antigen on its surface, and then to link with a helper T cell that has differentiated as a result of encountering and recognizing the same antigen. The process is outlined in **Figure 43.9,** and diagrammed in the second phase of Figure 43.8).

The process of antigen presentation on a B-cell surface begins when BCRs on the B cell interact directly with soluble bacterial (in our example) antigens in the blood or lymph. Once the antigen binds to a BCR, the complex is taken into the cell and the antigen is processed in the same way as in dendritic cells, culminating with a presentation of antigen pieces on the B-cell surface in a complex with class II MHC proteins (see Figure 43.8, step 6).

When a helper T cell encounters a B cell displaying the same antigen, usually in a lymph node or in the spleen, the T and B cells become tightly linked together (step 7). The linkage depends on the TCRs, which recognize and bind the antigen fragment displayed by class II MHC molecules on the surface of the B cell, and on CD4, which stabilizes the binding as it did for T-cell binding to the dendritic cell. The linkage between the cells first stimulates the helper T cell to secrete interleukins that activate the B cell and then stimulates the B cell to proliferate, producing a clone of those B cells with identical B-cell receptors (step 8). Some of the cloned cells differentiate into relatively short-lived **plasma cells,** which now secrete the same antibody that

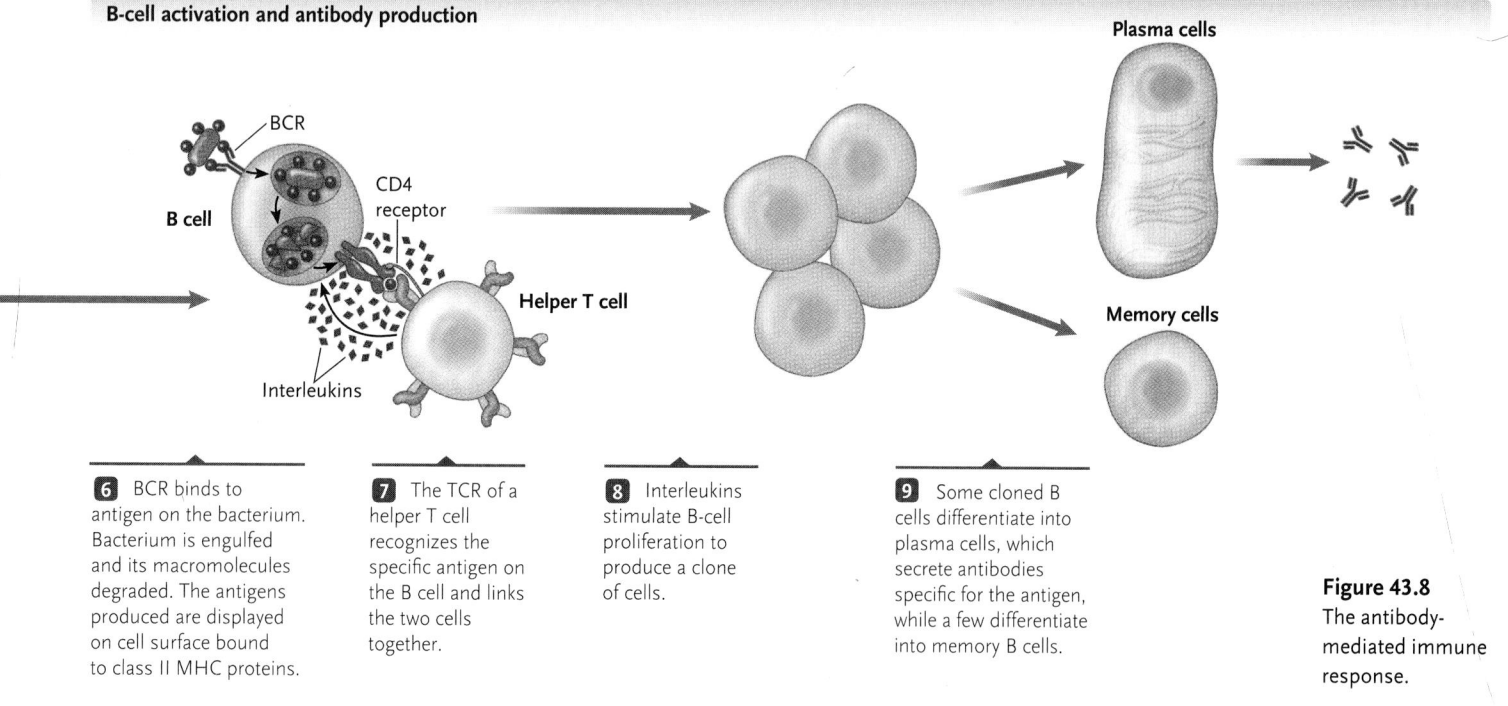

B-cell activation and antibody production

6 BCR binds to antigen on the bacterium. Bacterium is engulfed and its macromolecules degraded. The antigens produced are displayed on cell surface bound to class II MHC proteins.

7 The TCR of a helper T cell recognizes the specific antigen on the B cell and links the two cells together.

8 Interleukins stimulate B-cell proliferation to produce a clone of cells.

9 Some cloned B cells differentiate into plasma cells, which secrete antibodies specific for the antigen, while a few differentiate into memory B cells.

Figure 43.8
The antibody-mediated immune response.

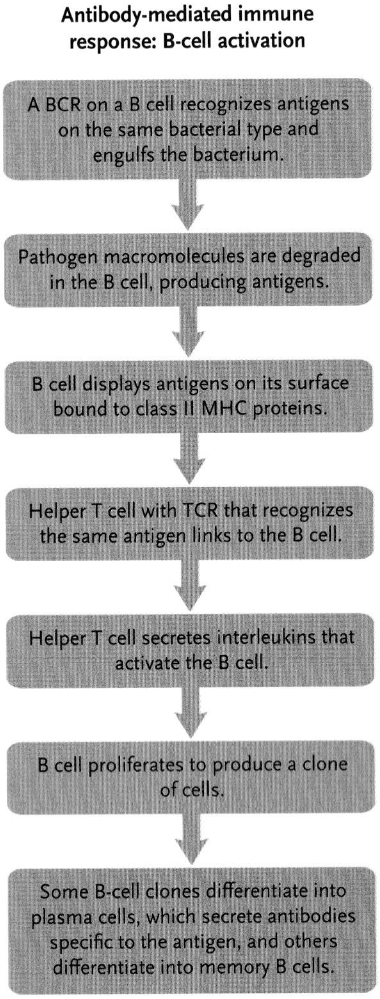

Antibody-mediated immune response: B-cell activation

A BCR on a B cell recognizes antigens on the same bacterial type and engulfs the bacterium.

↓

Pathogen macromolecules are degraded in the B cell, producing antigens.

↓

B cell displays antigens on its surface bound to class II MHC proteins.

↓

Helper T cell with TCR that recognizes the same antigen links to the B cell.

↓

Helper T cell secretes interleukins that activate the B cell.

↓

B cell proliferates to produce a clone of cells.

↓

Some B-cell clones differentiate into plasma cells, which secrete antibodies specific to the antigen, and others differentiate into memory B cells.

Figure 43.9
An outline of B-cell activation in antibody-mediated immunity.

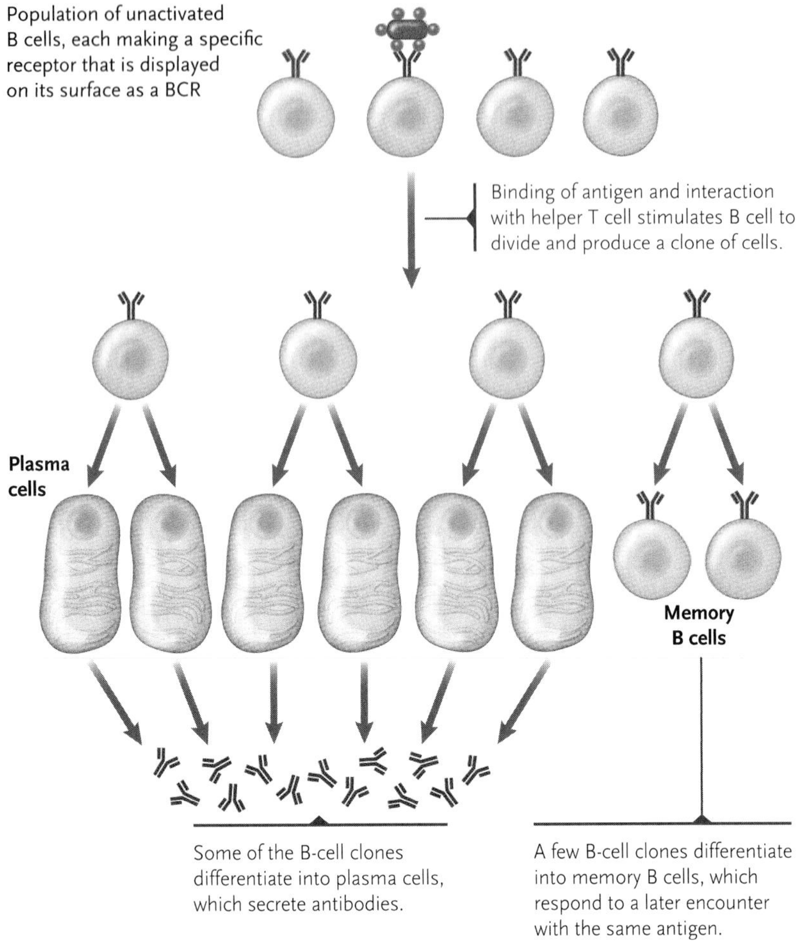

Population of unactivated B cells, each making a specific receptor that is displayed on its surface as a BCR

Binding of antigen and interaction with helper T cell stimulates B cell to divide and produce a clone of cells.

Plasma cells

Memory B cells

Some of the B-cell clones differentiate into plasma cells, which secrete antibodies.

A few B-cell clones differentiate into memory B cells, which respond to a later encounter with the same antigen.

Figure 43.10
Clonal selection. The binding of an antigen to a B cell already displaying a specific antibody to that antigen stimulates the B cell to divide and differentiate into plasma cells, which secrete the antibody, and memory cells, which remain in the circulation ready to mount a response against the antigen at a later time.

was displayed on the parental B cell's surface to circulate in lymph and blood. Others differentiate into **memory B cells**, which are long-lived cells that set the stage for a much more rapid response should the same antigen be encountered later (step 9).

Clonal selection is the process by which a particular lymphocyte is specifically selected for cloning when it recognizes a particular foreign antigen **(Figure 43.10)**. Remember that there is an enormous diversity of randomly generated lymphocytes, each with a particular receptor that may potentially recognize a particular antigen. The process of clonal selection was proposed in the 1950s by several scientists, most notably F. Macfarlane Burnet, Niels Jerne, and David Talmage. Their proposals, made long before the mechanism was understood, described clonal selection as a form of natural selection operating in miniature: antigens select the cells recognizing them, which reproduce and become dominant in the B-cell

population. Burnet received the Nobel Prize in 1960 for his research in immunology.

Clearing the Body of Foreign Antigens. How do the antibodies produced in an antibody-mediated immune response clear foreign antigens from the body? Let us consider some examples concerning bacteria and viruses.

Toxins produced by invading bacteria, such as tetanus toxin, can be *neutralized* by antibodies **(Figure 43.11a)**. The antibodies bind to the toxin molecules, preventing them from carrying out their damaging action. For intact bacteria at an infection site or in the circulatory system, antibodies will bind to antigens on their surfaces. Because the two arms of an antibody molecule bind to different copies of the antigen molecule, an antibody molecule may bind to two bacteria with the same antigen. A population of antibodies against the bacterium, then, link many bacteria to-

a. Neutralization

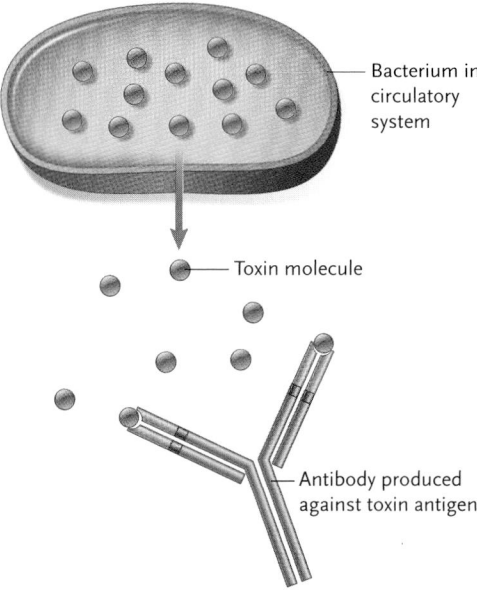

Bacterium in circulatory system

Toxin molecule

Antibody produced against toxin antigen

b. Agglutination

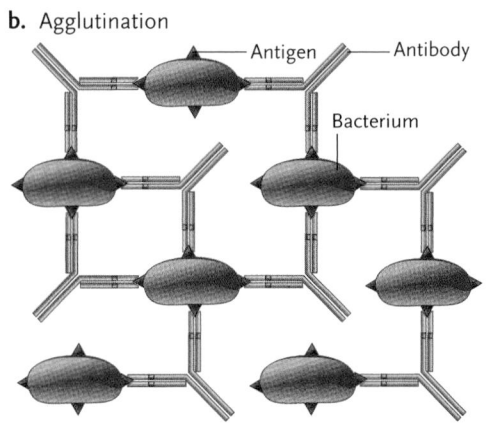

Antigen — Antibody

Bacterium

Figure 43.11
Examples of clearing antigens from the body.

gether into a lattice causing *agglutination,* that is, clumping of the bacteria **(Figure 43.11b).** Agglutination immobilizes the bacteria, preventing them from infecting cells. Antibodies can also agglutinate viruses, thereby preventing them from infecting cells.

More importantly, antibodies aid the innate immune response set off by the pathogens. That is, antibodies bound to antigens stimulate the complement system. Membrane attack complexes are formed and insert themselves into the plasma membranes of the bacteria, leading to their lysis and death. In the case of viral infections, membrane attack complexes can insert themselves into the membranes surrounding enveloped viruses, which disrupts the membrane and prevents the viruses from infecting cells.

Antibodies also enhance phagocytosis of bacteria and viruses. Phagocytic cells have receptors on their surfaces that recognize the heavy-chain end of antibodies (the end of the molecule opposite the antigen-binding sites). Antibodies bound to bacteria or viruses

therefore bind to phagocytic cells, which then engulf the pathogens and destroy them.

For simplicity, the adaptive immune response has been described here in terms of a single antigen. Pathogens have many different types of antigens on their surfaces, which means that many different B cells are stimulated to proliferate and many different antibodies are produced. Pathogens therefore are attacked by many different types of antibodies, each targeted to one antigen type on the pathogen's surface.

Immunological Memory. Once an immune reaction has run its course and the invading pathogen or toxic molecule has been eliminated from the body, division of the plasma cells and T-cell clones stops. Most or all of the clones die and are eliminated from the bloodstream and other body fluids. However, long-lived memory B cells and **memory helper T cells** (which differentiated from helper T cells), derived from encountering the same antigen, remain in an inactive state in the lymphatic system. Their persistence provides an **immunological memory** of the foreign antigen.

Immunological memory is illustrated in **Figure 43.12.** When exposed to a foreign antigen for the first time, a **primary immune response** results, following the steps already described. The first antibodies appear in the blood in 3 to 14 days and, by week 4, the primary response has essentially gone away. IgM is the main antibody type produced in a primary immune response. The primary immune response curve is followed whenever a new foreign antigen enters the body.

When a foreign antigen enters the body for a second or subsequent time, a **secondary immune response** results, while any new antigen introduced at the same time produces a primary response (see Figure 43.12). The secondary response is more rapid than a primary response because it involves the memory B cells and memory T cells that have been stored in the meantime, rather than having to initiate the clonal selection of a new B cell and T cell. Moreover, less antigen is needed

Figure 43.12
Immunological memory: primary and secondary responses to the same antigen.

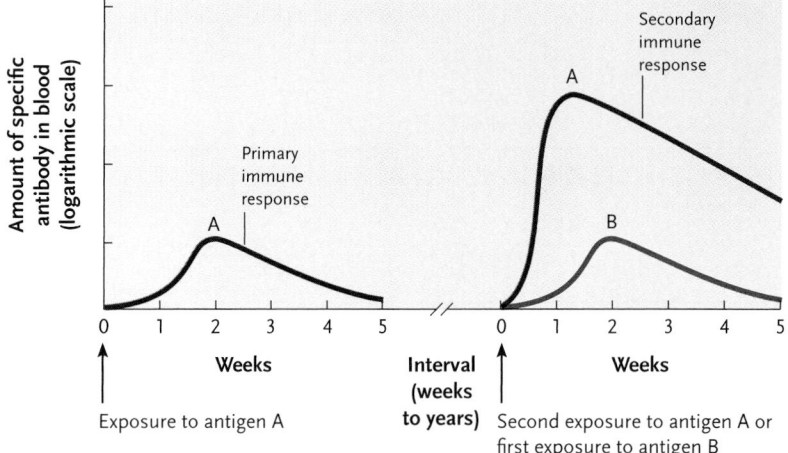

to elicit a secondary response than a primary response, and many more antibodies are produced. The predominant antibody produced in a secondary immune response is IgG; the switch occurs at the gene level in the memory B cells.

Immunological memory forms the basis of vaccinations, in which antigens in the form of living or dead pathogens or antigenic molecules themselves are introduced into the body. After the immune response, memory B cells and memory T cells remaining in the body can mount an immediate and intense immune reaction against similar antigens in the dangerous pathogen. In Edward Jenner's technique, for example, introducing the cowpox virus, a related, less virulent form of the smallpox virus, into healthy individuals initiated a primary immune response. After the response ran its course, a bank of memory B cells and memory T cells remained in the body, able to recognize quickly the similar antigens of the smallpox virus and initiate a secondary immune response. Similarly, a polio vaccine developed by Jonas Salk uses polio viruses that have been inactivated by exposing them to formaldehyde. Although the viruses are inactive, their surface groups can still act as antigens. The antigens trigger an immune response, leaving memory B and T cells able to mount an intense immune response against active polio viruses.

Active and Passive Immunity. **Active immunity** is the production of antibodies in the body in response to exposure to a foreign antigen—the process that has been described up until now. **Passive immunity** is the acquisition of antibodies as a result of direct transfer from another person. This form of immunity provides immediate protection against antigens that the antibodies recognize without the person receiving the antibodies having developed an immune response. Examples of passive immunity include the transfer of IgG antibodies from the mother to the fetus through the placenta and the transfer of IgA antibodies in the first breast milk fed from the mother to the baby. Compared with active immunity, passive immunity is a short-lived phenomenon with no memory, in that the antibodies typically break down within a month. However, in that time, the protection plays an important role. For example, a breast-fed baby is protected until it is able to mount an immune response itself, an ability that is not present until about a month after birth.

Drug Effects on Antibody-Mediated Immunity. Several drugs used to reduce the rejection of transplanted organs target helper T cells. Cyclosporin A, used routinely after organ transplants, blocks the activation of helper T cells and, in turn, the activation of B cells. Unfortunately, cyclosporin and other immunosuppressive drugs also leave the treated individual more susceptible to infection by pathogens.

In Cell-Mediated Immunity, Cytotoxic T Cells Expose "Hidden" Pathogens to Antibodies by Destroying Infected Body Cells

In **cell-mediated immunity**, cytotoxic T cells directly destroy host cells infected by pathogens, particularly those infected by a virus **(Figure 43.13)**. The killing pro-

Figure 43.13
The cell-mediated immune response.

Cell-mediated immune response

T-cell activation

Antigen
Class I MHC protein
CD8 receptor
T-cell receptor (TCR)
Cytotoxic T cells

Virus-infected cell
CD8⁺ T cell

Destruction of infected cells by cytotoxic T cells

CD8⁺ T cell
Perforins
Virus-infected cell

Lennart Nilsson/Bonnier Fakta AB

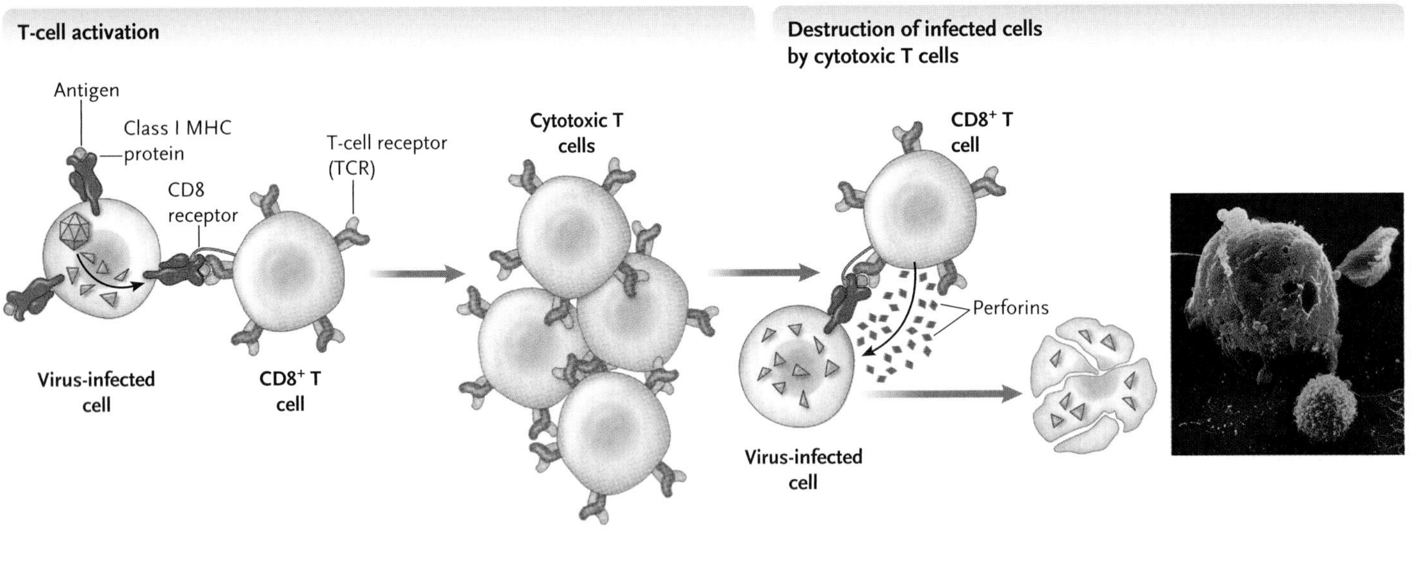

1 Viral proteins are degraded into fragments that act as antigens. The antigens are displayed on the cell surface bound to class I MHC proteins.

2 A TCR on a CD8⁺ T cell recognizes an antigen bound to a class I MHC protein on an infected cell, and the two cells link together. The interaction activates the T cell.

3 The CD8⁺ T cell proliferates and forms a clone. The cloned cells differentiate into cytotoxic T cells and memory cytotoxic T cells.

4 A TCR on a cytotoxic T cell recognizes the antigen bound to a class I MHC protein on the infected cell. The T cell releases perforins.

5 The perforins insert into the membrane of the infected cell, forming pores. Leakage of ions and other molecules, along with other events, causes the cell to lyse.

Some Cancer Cells Kill Cytotoxic T Cells to Defeat the Immune System

Among the arsenal of weapons employed by cytotoxic T cells to eliminate infected and cancerous body cells is the *Fas–FasL* system. Fas is a receptor that occurs on the surfaces of many body cells; FasL is a signal molecule that is displayed on the surfaces of some cell types, including cytotoxic T cells. If a cell carrying the Fas receptor contacts a cytotoxic T cell with the FasL signal displayed on its surface, the effect for the Fas-bearing cell is something like stepping on a mine. When FasL is bound by the Fas receptor, a cascade of internal reactions initiates apoptosis and kills the cell with the Fas receptor.

Surprisingly, cytotoxic T cells also carry the Fas receptor, so they can kill each other by displaying the FasL signal. This mutual killing plays an important role in reducing the level of an immune reaction after a pathogen has been eliminated. In addition, cells in some regions of the body, such as the eye, the nervous system, and the testis, can be severely damaged by inflammation; these cells can make and display FasL, which kills cytotoxic T cells and reduces the severity of inflammation.

Molecular research by a group of Swiss investigators at the Universities of Lausanne and Geneva now shows that some cancer cells survive elimination by the immune system by making and displaying the FasL signal, and thus killing any cytotoxic T cells attacking the tumor. The researchers were led to their discovery by the observation that many patients with malignant melanoma, a dangerous skin cancer, had a breakdown product associated with FasL in their bloodstream.

The investigators extracted proteins from melanoma cells and tested them with antibodies against FasL. The test was positive, showing that FasL was in the tumor cells. The investigators were also able to detect an mRNA encoding FasL in the tumor cells, showing that the gene encoding FasL was active. As a final confirmation, they tagged antibodies against the FasL protein with a dye molecule to make them visible in the light microscope and added them to sections of melanoma tissue from patients. Intense staining of the cells with the dye showed that FasL was indeed present. Tests for the presence of the Fas receptor were negative, showing that Fas synthesis was turned off in the melanoma cells.

Thus the FasL in melanoma cells kills cytotoxic T cells that invade the tumor. At the same time, the absence of Fas receptors ensures that the tumor cells do not kill each other. The presence of FasL and absence of the Fas receptor may explain why melanomas are rarely destroyed by an immune reaction, and also why many other types of cancer also escape immune destruction.

Melanoma cells originate from pigment cells in the skin called *melanocytes*. Normal melanocytes do not contain FasL, indicating that synthesis of the protein is turned on as a part of the changes transforming normal melanocytes into cancer cells.

The findings of the Swiss group could lead to an effective treatment for cancer using the Fas–FasL system. If melanoma cells could be induced to make Fas as well as FasL, for example, they might eliminate a tumor by killing each other!

---

cess begins when some of the pathogens break down inside the infected host cells, releasing antigens that are fragmented by enzymes in the cytoplasm. The antigen fragments bind to class I MHC proteins, which are delivered to the cell surface by essentially the same mechanisms as in B cells (step 1). At the surface, the antigen fragments are displayed by the class I MHC protein and the cell then functions as an APC.

In a cell-mediated immune response, the APC presents the antigen fragment to a type of T cell in the lymphatic system called a **CD8+ T cell** because it has receptors named **CD8** on its surface in addition to TCRs. The presence of a CD8 receptor distinguishes this type of T cell from that involved in antibody-mediated immunity. A specific CD8+ T cell having a TCR with an antigen-binding site that recognizes the antigen fragment binds to that fragment on the APC (step 2). The CD8 receptor on the T cell helps the two cells link together.

The interaction between the APC and the CD8+ T cell activates the T cell, which then proliferates to form a clone. Some of the cells differentiate to become **cytotoxic T cells** (step 3), while a few differentiate into *memory cytotoxic T cells*. Cytotoxic T cells are another type of effector T cell. TCRs on the cytotoxic T cells again recognize the antigen fragment bound to class I MHC proteins on the infected cells (the APCs) (step 4). The cytotoxic T cell then destroys the infected cell in mechanisms similar to those used by NK cells. That is, an activated cytotoxic T cell releases perforin, which creates pores in the membrane of the target cell. The leakage of ions and other molecules through the pores causes the infected cell to rupture. The cytotoxic T cell also secretes proteases that enter infected cells through the newly created pores and cause it to self-destruct by apoptosis (step 5, and photo inset). Rupture of dead, infected cells releases the pathogens to the interstitial fluid, where they are open to attack by antibodies and phagocytes.

Cytotoxic T cells can also kill cancer cells if their class I MHC molecules display fragments of altered cellular proteins that do not normally occur in the body. Another mechanism used by cytotoxic T cells to kill cells, and a process used by some cancer cells to defeat the mechanism, is described in *Insights from the Molecular Revolution*.

Figure 43.14 Research Method

Production of Monoclonal Antibodies

PURPOSE: Injecting an antigen into an animal produces a collection of different antibodies that react against different parts of the antigen. Monoclonal antibodies are produced to provide antibodies that all react against the same epitope of a single antigen.

PROTOCOL:

1. Inject antigen into mouse.

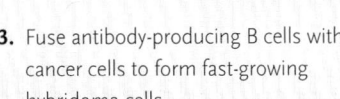

2. Extract activated B cells from spleen.

Activated B cells   Myeloma (cancer) cells

3. Fuse antibody-producing B cells with cancer cells to form fast-growing hybridoma cells.

Hybridoma cell

4. Grow clone from single hybridoma cells; test antibodies produced by clone for reaction against antigen.

5. Grow clone producing antibodies against antigen to large size.

6. Extract and purify antibodies.

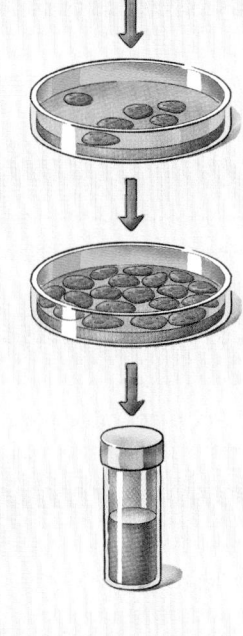

Antibodies Have Many Uses in Research

The ability of the antibody-mediated immune system to generate antibodies against essentially any antigen provides an invaluable research tool to scientists, who can use antibodies to identify biological molecules and to determine their locations and functions in cells. To obtain the antibodies, a molecule of interest is injected into a test animal such as a mouse, rabbit, goat, or sheep. In response, the animal develops antibodies capable of binding to the molecule. The antibodies are then extracted and purified from a blood sample.

To identify the cellular location of a molecule, antibodies made against the molecule are combined with a visible marker such as a dye molecule or heavy metal atom. When added to a tissue sample, the marked antibodies can be seen in the light or electron microscope localized to cellular structures such as membranes, ribosomes, or chromosomes, showing that the molecule forms part of the structure.

Antibodies can also be used to "grab" a molecule of interest from a preparation containing a mixture of all kinds of cellular molecules. For such studies, the antibodies are often attached to plastic beads that are packed into a glass column. When the mixture passes through the column, the molecule is trapped by attachment to the antibody and remains in the column. It is then released from the column in purified form by adding a reagent that breaks the antigen-antibody bonds.

Injecting a molecule of interest into a test animal typically produces a wide spectrum of antibodies that react with different parts of the antigen. Some of the antibodies also cross-react with other, similar antigens, producing false results that can complicate the research. These problems have been solved by producing **monoclonal antibodies**, each of which reacts only against the same segment (epitope) of a single antigen.

Georges Kohler and Cesar Milstein pioneered the production of monoclonal antibodies in 1975. In their technique, a test animal (usually a mouse) is injected with a molecule of interest **(Figure 43.14).** After the animal has developed an immune response, fully activated B cells are extracted from the spleen and placed in a cell culture medium. Because B cells normally stop dividing and die within a week when cultured, they are induced to fuse with cancerous lymphocytes called *myeloma cells,* forming single, composite cells called **hybridomas.** Hybridomas combine the desired characteristics of the two cell types—they produce antibodies like fully activated B cells, and they divide continuously and rapidly like the myeloma cells.

Single hybridoma cells are then separated from the culture and used to start clones. Because all the cells of a clone are descended from a single hybridoma cell, they all make the same, highly specific antibody, able to bind the same part of a single antigen. In addition to their use in scientific research, monoclonal antibodies are also widely used in medical applications such as pregnancy tests, screening for prostate cancer, and testing for AIDS and other sexually transmitted diseases.

1. How, in general, do the antibody-mediated and cell-mediated immune responses help clear the body of antigens?
2. Describe the general structure of an antibody molecule.
3. What are the principles of the mechanism used for generating antibody diversity?
4. What is clonal selection?
5. How does immunological memory work?

43.4 Malfunctions and Failures of the Immune System

The immune system is highly effective, but it is not foolproof. Some malfunctions of the immune system cause the body to react against its own proteins or cells, producing *autoimmune disease*. In addition, some viruses and other pathogens have evolved means to avoid destruction by the immune system. A number of these pathogens, including the AIDS virus, even use parts of the immune response to promote infection. Another malfunction causes the *allergic reactions* that most of us experience from time to time.

An Individual's Own Molecules Are Normally Protected against Attack by the Immune System

B cells and T cells are involved in the development of **immunological tolerance**, which protects the body's own molecules from attack by the immune system. Although the process is not understood, molecules present in an individual from birth are not recognized as foreign by circulating B and T cells, and do not elicit an immune response. Evidently, during their initial differentiation in the bone marrow and thymus, any B and T cells that are able to react with self molecules carried by MHC molecules are induced to kill themselves by apoptosis, or enter a state in which they remain in the body but are unable to react if they encounter a self molecule. The process of excluding self-reactive B and T cells goes on throughout the life of an individual.

Evidence that immunological tolerance is established early in life comes from experiments with mice. For example, if a foreign protein is injected into a mouse at birth, during the period in which tolerance is established, the mouse will not develop antibodies against the protein if it is injected later in life. Similarly, if mutant mice are produced that lack a given complement protein, so that the protein is absent during embryonic development, they will produce antibodies against that protein if it is injected during adult life. Normal mice do not produce antibodies if the protein is injected.

Autoimmune Disease Occurs When Immunological Tolerance Fails

The mechanisms setting up immunological tolerance sometimes fail, leading to an **autoimmune reaction**—the production of antibodies against molecules of the body. In most cases, the effects of such anti-self antibodies are not serious enough to produce recognizable disease. However, in some individuals, about 5% to 10% of the human population, anti-self antibodies cause serious problems.

For example, type 1 diabetes (see Section 40.4) is an autoimmune reaction against the pancreatic beta cells producing insulin. The anti-self antibodies gradually eliminate the beta cells until the individual is incapable of producing insulin. *Systemic lupus erythematosus (lupus)* is caused by production of a wide variety of anti-self antibodies against blood cells, blood platelets, and internal cell structures and molecules such as mitochondria and proteins associated with DNA in the cell nucleus. People with lupus often become anemic and have problems with blood circulation and kidney function because antibodies, combined with body molecules, accumulate and clog capillaries and the microscopic filtering tubules of the kidneys. Lupus patients may also develop anti-self antibodies against the heart and kidneys. *Rheumatoid arthritis* is caused by a self-attack on connective tissues, particularly in the joints, causing pain and inflammation. *Multiple sclerosis* results from an autoimmune attack against a protein of the myelin sheaths insulating the surfaces of neurons. Multiple sclerosis can seriously disrupt nervous function, producing such symptoms as muscle weakness and paralysis, impaired coordination, and pain.

The causes of most autoimmune diseases are unknown. In some cases, an autoimmune reaction can be traced to injuries that expose body cells or proteins that are normally inaccessible to the immune system, such as the lens protein of the eye, to B and T cells. In other cases, as in type 1 diabetes, an invading virus stimulates the production of antibodies that can also react with self proteins. Antibodies against two viruses, the Epstein-Barr and hepatitis B viruses, can react against myelin basic protein, the protein attacked in multiple sclerosis. Sometimes, environmental chemicals, drugs, or mutations alter body proteins so that they appear foreign to the immune system and come under attack.

Some Pathogens Have Evolved Mechanisms That Defeat the Immune Response

Several pathogens regularly change their surface groups to avoid destruction by the immune system. By the time the immune system has developed antibodies

Applied Research: HIV and AIDS

Acquired immune deficiency syndrome (AIDS) is a constellation of disorders that follows infection by the **human immunodeficiency virus, HIV (Figure a).** First reported in various countries in the late 1970s, HIV now infects more than 40 million people worldwide, 64% of them in Africa. AIDS is a potentially lethal disease, although drug therapy has reduced the death rate for HIV-infected individuals in many countries, including the United States.

HIV is transmitted when an infected person's body fluids, especially blood or semen, enter the blood or tissue fluids of another person's body. The entry may occur during vaginal, anal, or oral intercourse, or via contaminated needles shared by intravenous drug users. HIV can also be

transmitted from infected mothers to their infants during pregnancy, birth, and nursing. AIDS is rarely transmitted through casual contact, food, or body products such as saliva, tears, urine, or feces.

The primary cellular hosts for HIV are macrophages and helper T cells, which are ultimately destroyed in large numbers by the virus. The infection makes helper T cells unavailable for the stimulation and proliferation of B cells and cytotoxic T cells. The assault on lymphocytes and macrophages cripples the immune system and makes the body highly vulnerable to infections and to development of otherwise rare forms of cancer.

In 1996 researchers confirmed the process by which HIV initially infects its primary target, the helper T cells.

First, a glycoprotein of the viral coat, called *gp120*, attaches the virus to a helper T cell by binding to its CD4 receptor. Then, another viral protein triggers fusion of the viral surface membrane with the T-cell plasma membrane, releasing the virus into the cell **(Figure b).** Once inside, a viral enzyme, *reverse transcriptase*, uses the viral RNA as a template for making a DNA copy. (The genetic material of HIV is RNA rather than DNA when it is outside a host cell.) Another viral enzyme, *integrase*, then splices the viral DNA into the host cell's DNA. Once it is part of the host cell DNA, the viral DNA is replicated and passed on as the cell divides. As part of the host cell DNA, the virus is effectively hidden in the helper T cell and protected from attack by the immune system.

The viral DNA typically remains dormant until the helper T cell is stimulated by an antigen. At that point, the viral DNA is copied into new viral RNA molecules, and into mRNAs that direct host cell ribosomes to make viral proteins. The viral RNAs are added to the viral proteins to make infective HIV particles, which are released from the host cell by budding **(Figure c).** The viral particles may infect more body cells or another person. The viral infection also leads uninfected helper T cells to destroy themselves in large numbers by apoptosis, through mechanisms that are still unknown.

At the time of initial infection, many people suffer a mild fever and other symptoms that may be mistaken for the flu or the common cold. The symptoms disappear as antibodies against viral proteins appear in the body, and the number of viral particles

Figure a
Structure of a free HIV viral particle.

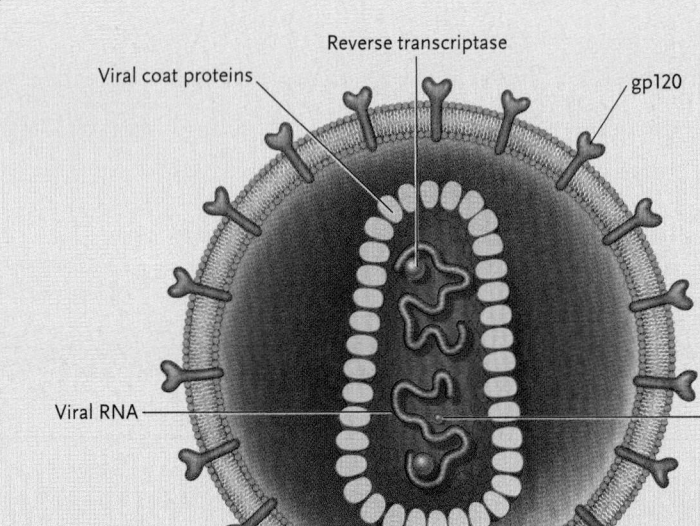

Reverse transcriptase
Viral coat proteins
gp120
Viral RNA
Integrase
Membrane coat derived from host cell

against one version of the surface proteins, the pathogens have switched to different surface proteins that the antibodies do not match. These new proteins take another week or so to stimulate the production of specific antibodies; by this time, the surface groups change again. The changes continue indefinitely, always keeping the pathogens one step ahead of the immune system. Pathogens that use these mechanisms to sidestep the immune system include the protozoan causing African sleeping sickness, the bacterium causing gonorrhea, and the viruses causing influenza, the common cold, and AIDS.

Some viruses use parts of the immune system to get a free ride to the cell interior. For example, the

drops in the bloodstream. However, the virus's genome is still present, integrated into the DNA of T cells, and the virus steadily spreads to infect other T cells. An infected person may remain apparently healthy for years, yet can infect others. Both the transmitter and recipient of the virus may be unaware that the disease is present, making it difficult to control the spread of HIV infections.

With time, more and more helper T cells and macrophages are destroyed, eventually wiping out the body's immune response. The infected person becomes susceptible to opportunistic, secondary infections, such as a pneumonia caused by a fungus (*Pneumocystis carinii*); drug-resistant tuberculosis; persistent yeast (*Candida albicans*) infections of the mouth, throat, rectum, or vagina; and infection by many common bacteria and viruses that rarely infect healthy humans. These secondary infections signal the appearance of full-blown AIDS. Steady debilitation and death typically follow within a period of years in untreated persons.

As yet, there is no cure for HIV infection and no vaccine that can protect against infection. HIV coat proteins mutate constantly, making a vaccine developed against one form of the virus useless when the next form appears. Most of these mutations occur during replication of the virus, when reverse transcriptase makes a DNA copy of the viral RNA.

The development of AIDS can be greatly slowed by drugs that interfere with reverse transcription of the viral genomic RNA into the DNA copy that integrates into the host's genome.

Treatment with a "cocktail" of drugs called *reverse transcriptase inhibitors (RTIs)* inhibits viral reproduction and destruction of helper T cells and extends the lives of people with the AIDS virus. The inhibiting cocktails are not a cure for AIDS, however, because the virus is still present in dormant form in helper T cells. If the therapy is stopped, the virus again replicates and the T-cell population drops.

Presently, the only certain way to avoid HIV infection is to refrain from unprotected sex with people whose HIV status is unknown, and from the use of contaminated needles of the type used to administer drugs intravenously.

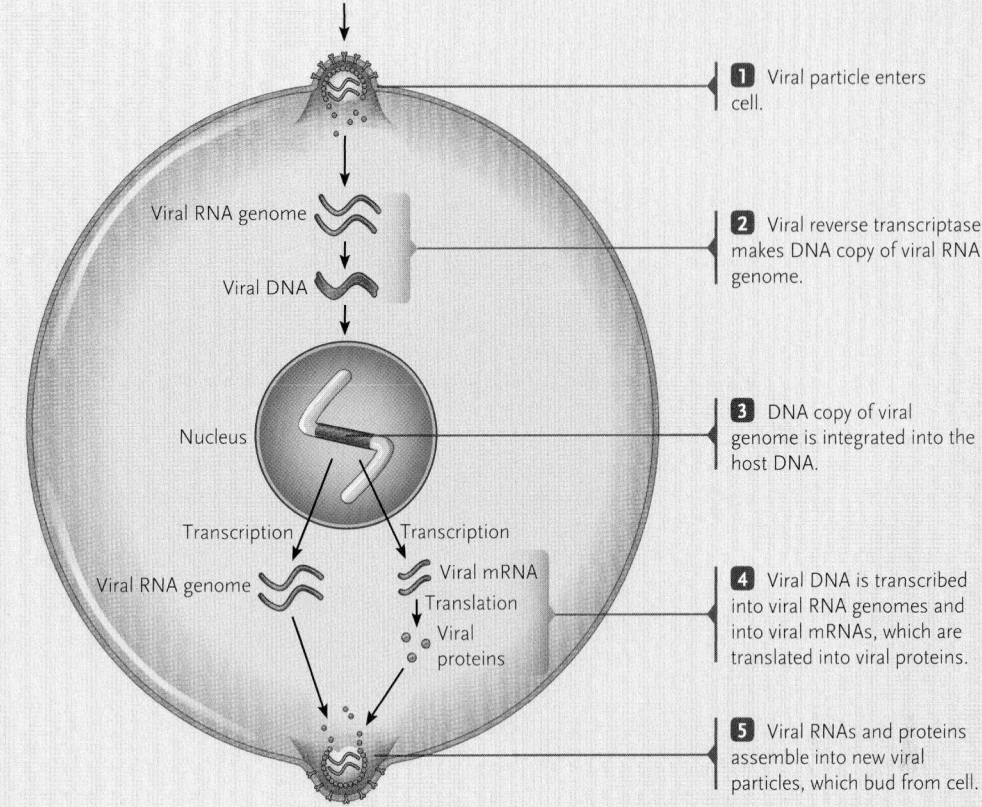

1 Viral particle enters cell.

2 Viral reverse transcriptase makes DNA copy of viral RNA genome.

3 DNA copy of viral genome is integrated into the host DNA.

4 Viral DNA is transcribed into viral RNA genomes and into viral mRNAs, which are translated into viral proteins.

5 Viral RNAs and proteins assemble into new viral particles, which bud from cell.

Figure b
The steps in HIV infection of a host cell.

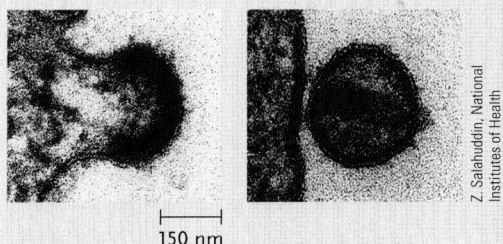

150 nm

Z. Salahuddin, National Institutes of Health

Figure c
An HIV particle budding from a host cell. As it passes from the host cell, it acquires a membrane coat derived from the host cell plasma membrane.

AIDS virus has a surface molecule that is recognized and bound by the CD4 receptor on the surface of helper T cells. Binding to CD4 locks the virus to the cell surface and stimulates the membrane covering the virus to fuse with the plasma membrane of the helper T cell. (The protein coat of the virus is wrapped in a membrane derived from the plasma membrane of the host cell in which it was produced.) The fusion introduces the virus into the cell, initiating the infection and leading to destruction and death of the T cell. (Further details of HIV infection and AIDS are presented in *Focus on Applied Research*.)

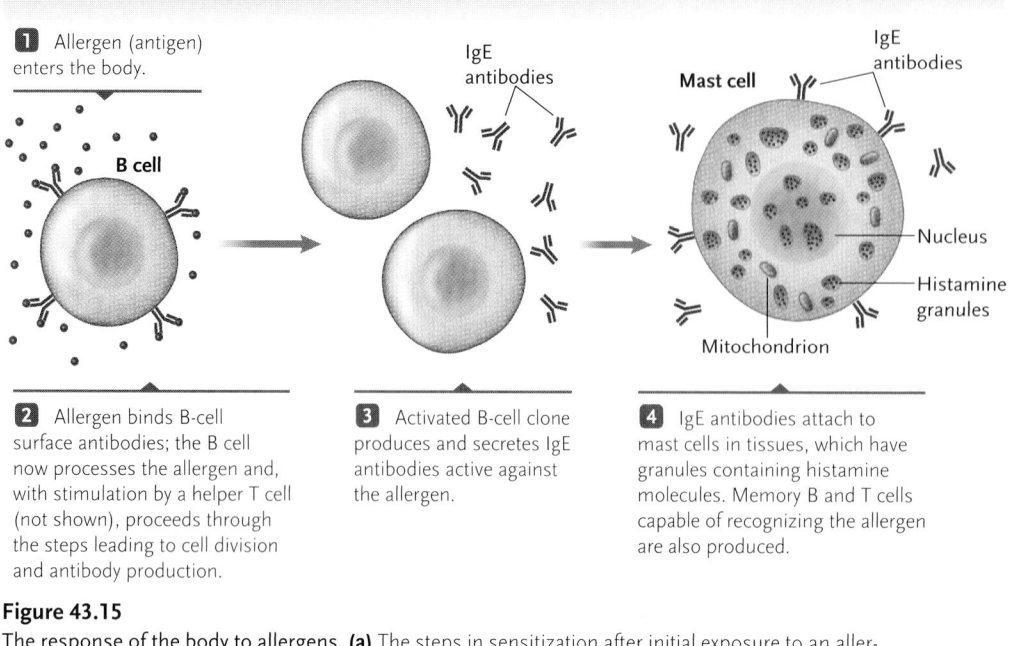

a. Initial exposure to allergen

1 Allergen (antigen) enters the body.

B cell

IgE antibodies

Mast cell

IgE antibodies

Nucleus

Histamine granules

Mitochondrion

2 Allergen binds B-cell surface antibodies; the B cell now processes the allergen and, with stimulation by a helper T cell (not shown), proceeds through the steps leading to cell division and antibody production.

3 Activated B-cell clone produces and secretes IgE antibodies active against the allergen.

4 IgE antibodies attach to mast cells in tissues, which have granules containing histamine molecules. Memory B and T cells capable of recognizing the allergen are also produced.

b. Further exposures to allergen

Allergen

Histamine release

5 After the first exposure, when the allergen enters the body, it binds with IgE antibodies on mast cells; binding stimulates the mast cell to release histamine and other substances.

Figure 43.15

The response of the body to allergens. **(a)** The steps in sensitization after initial exposure to an allergen. **(b)** Production of an allergic response by further exposures to the allergen.

Allergies Are Produced by Overactivity of the Immune System

The substances responsible for allergic reactions form a distinct class of antigens called **allergens,** which induce B cells to secrete an overabundance of IgE antibodies **(Figure 43.15).** The IgE antibodies, in turn, bind to receptors on mast cells in connective tissue and on **basophils,** a type of leukocyte in blood (see Table 43.1), inducing them to secrete histamine, which produces a severe inflammation. Most of the inflammation occurs in tissues directly exposed to the allergen, such as the surfaces of the eyes, the lining of the nasal passages, and the air passages of the lungs. Signal molecules released by the activated mast cells also stimulate mucosal cells to secrete floods of mucus and cause smooth muscle in airways to constrict (histamine also causes airway constriction). The resulting allergic reaction can vary in severity from a mild irritation to serious and even life-threatening debilitation. *Asthma* is a severe response to allergens involving constriction of airways in the lungs. Antihistamines (substances that block histamine receptors) are usually effective in countering the effects of the histamine released by mast cells.

An individual is *sensitized* by a first exposure to an allergen, which may produce only mild allergic symptoms or no reaction at all (see Figure 43.15a). However, the sensitization produces memory B and T cells; at the next and subsequent exposures, the system is poised to produce a greatly intensified allergic response (see Figure 43.15b).

In some persons, inflammation stimulated by an allergen is so severe that the reaction brings on a life-threatening condition called **anaphylactic shock.** Extreme swelling of air passages in the lungs interferes with breathing, and massive leakage of fluid from capillaries causes the blood pressure to drop precipitously. Death may result in minutes if the condition is not treated promptly. In persons who have become sensitized to the venom of wasps and bees, for example, a single sting may bring on anaphylactic shock within minutes. Allergies developed against drugs such as penicillin and certain foods can have the same drastic effects. Anaphylactic shock can be controlled by immediate injection of epinephrine (adrenaline), which reverses the condition by constricting blood vessels and dilating air passages in the lungs.

STUDY BREAK

1. What is immunological tolerance?
2. Explain how a failure in the immune system can result in an allergy.

43.5 Defenses in Other Animals

Other Vertebrate Groups Have Defenses against Infections

This chapter has emphasized the mammalian immune system, the focus of most immunology research. We know relatively little about defenses against infections in most other vertebrate groups. Yet like all other physiological systems, the mammalian defenses against pathogens are the result of evolution, and evidence of their functions can be seen in other vertebrate groups and also in invertebrates.

For example, molecular studies in sharks and rays have revealed DNA sequences that are clearly related to the sequences coding for antibodies in mammals. If injected with an antigen, sharks produce antibodies, formed from light- and heavy-chain polypeptides, capable of recognizing and binding the antigen. Although embryonic gene segments for the two polypeptides are arranged differently in sharks than they are in mammals, antibody diversity is produced by the same kinds of genetic rearrangements in both. Sharks also mount nonspecific defenses, including the production of a steroid that appears to kill bacteria and neutralize viruses nonspecifically and with high efficiency.

Invertebrates lack specific immune defenses equivalent to antibodies and the activities of B and T cells, so their reactions to invading pathogens most closely resemble the nonspecific defenses of humans and other vertebrates. However, all invertebrates have phagocytic cells, which patrol tissues and engulf pathogens and other invaders. Some of the signal molecules that stimulate phagocytic activity, such as interleukins, appear to be similar in invertebrates and vertebrates.

Antibodies do not occur in invertebrates, but proteins of the immunoglobulin family are widely distributed. In at least some invertebrates, these Ig proteins have a protective function. In moths, for example, an

UNANSWERED QUESTIONS

How does gene expression in a bacterial pathogen change during infection?

Characterizing the genetic events involved in the interactions between pathogens and hosts is important for understanding the development of an infectious disease in the host, and for producing effective therapeutic treatments. You learned in this chapter that pathogenic bacteria are first combated by the innate immunity system. Researchers in James Musser's lab at Baylor College of Medicine in collaboration with researchers at Rocky Mountain Laboratories, National Institute of Allergy and Infectious Diseases in Montana, have now obtained information about changes in gene expression for a bacterial pathogen, group A *Streptococcus* (GAS), during infection of a mammal.

GAS is a Gram-positive bacterium that is the cause of pharyngitis ("strep throat"; 2 million cases annually in the United States) and various other infections, including rheumatic heart disease. The researchers studied how gene expression in GAS changed during an 86-day period following infection that caused pharyngitis in macaque monkeys. They are an excellent model for the study because the progression of pharyngitis caused by GAS in these monkeys is highly similar to that seen in human infections. There are three distinct phases, during which GAS can be detected by culturing throat swabs. In the first phase, colonization, GAS establishes infection of host cells, producing only mild pharyngitis. In the second phase, the acute phase, pharyngitis symptoms peak, as does the number of GAS bacteria. In the third phase, the asymptomatic phase, symptoms decrease and disappear along with a decrease in GAS bacteria.

Experimentally, the researchers analyzed gene expression from the entire genome of GAS in the three phases of pharyngitis using DNA microarrays (see Section 18.3). They found that the pattern of GAS gene expression changed over the course of the disease. Significantly, they saw characteristic gene expression patterns for each of the three phases of pathogen–host interaction. These results indicate that GAS regulates expression of its genes extensively as it establishes an infection, and as

the host mounts an innate immune response against it. This work will help direct future research efforts to control infections caused by GAS. More broadly, genomic studies of this kind are likely to provide insights into genetic events contributing to pathogenesis in other pathogen–host interactions.

How does HIV evade the adaptive immunity system?

In cell-mediated immunity, a pathogen-infected antigen-presenting cell (APC) presents an antigen fragment bound to a class I MHC protein to a CD8$^+$ T cell, stimulating the T cell to differentiate into cytotoxic T cells. The cytotoxic T cells then bind to the infected APCs and destroy them. Cytotoxic T cells act particularly against host cells infected by viral pathogens. HIV infects host cells but, rather than being eliminated by the host, this virus establishes a chronic infection that leads to the development of AIDS. That is, HIV evades the adaptive immunity system.

Kathleen Collins and her group at the University of Michigan Medical School have investigated the mechanism of this evasion. They have learned that the virus down-regulates the display of class I MHC proteins on the surfaces of HIV-infected APCs, which thereby limits the presentation of viral antigens by those cells. The down-regulation occurs by the action of the HIV Nef (*negative factor*) protein. Nef binds to class I MHC molecules and inhibits them from moving through the Golgi complex to the cell surface. Without class I MHC molecules on the cell surface to present antigens, the immune response is compromised. The action of Nef, therefore, enhances the ability of HIV to induce AIDS. Research now being pursued is directed toward characterizing more completely the role of Nef in disrupting class I MHC movement from the ER through the Golgi complex to the cell surface, identifying and characterizing other proteins that may interact with Nef in an HIV-infected cell, and developing pharmaceutical reagents aimed at blocking Nef action.

Peter J. Russell

Ig-family protein called *hemolin* binds to the surfaces of pathogens and marks them for removal by phagocytes.

Many invertebrates produce antimicrobial proteins such as lysozyme that are able to kill bacteria and other invading cells. Insects, for example, secrete lysozyme in response to bacterial infections.

STUDY BREAK

Compare invertebrate and mammalian immune defenses.

Review

Go to **ThomsonNOW** at www.thomsonedu.com/login to access quizzing, animations, exercises, articles, and personalized homework help.

43.1 Three Lines of Defense against Invasion

- Humans and other vertebrates have three lines of defense against pathogens. The first, which is nonspecific, is the barrier set up by the skin and mucous membranes.

- The second line of defense, also nonspecific, is innate immunity, an innate system that defends the body against pathogens and toxins penetrating the first line.

- The third line of defense, adaptive immunity, is specific: it recognizes and eliminates particular pathogens and retains a memory of that exposure so as to respond rapidly if the pathogen is encountered again. The response is carried out by lymphocytes, a specialized group of leukocytes.

43.2 Nonspecific Defenses: Innate Immunity

- In the innate immunity system, molecules on the surfaces of pathogens are recognized as foreign by receptors on host cells. The pathogen is then combated by the inflammation and complement systems.

- Epithelial surfaces secrete defensins, a type of antimicrobial peptide, in response to attack by a microbial pathogen. Defensins disrupt the plasma membranes of pathogens, killing them.

- Inflammation is characterized by heat, pain, redness, and swelling at the infection site. Several interconnecting mechanisms initiate inflammation, including pathogen engulfment, histamine secretion, cytokine release, and local blood vessel dilation and permeability increase. (Figure 43.1).

- Large arrays of complement proteins are activated when they recognize molecules on the surfaces of pathogens. Some complement proteins form membrane attack complexes, which insert into the plasma membrane of many types of bacteria and cause their lysis. Fragments of other complement proteins coat pathogens, stimulating phagocytes to engulf them (Figure 43.2).

- Three nonspecific defenses are used to combat viral pathogens: RNA interference, interferons, and natural killer cells.

Animation: Innate defenses

Animation: Complement proteins

Animation: Inflammatory response

Animation: Immune responses

Animation: Human lymphatic system

43.3 Specific Defenses: Adaptive Immunity

- Adaptive immunity, which is carried out by B and T cells, targets particular pathogens or toxin molecules.

- Antibodies consist of two light and two heavy polypeptide chains, each with variable and constant regions. The variable regions of the chains combine to form the specific antigen-binding site (Figure 43.4).

- Antibodies occur in five different classes: IgM, IgD, IgG, IgA, and IgE. Each class is determined by its constant region (Table 43.2).

- Antibody diversity is produced by genetic rearrangements in developing B cells that combine gene segments into intact genes encoding the light and heavy chains. The rearrangements producing heavy-chain genes and T-cell receptor genes are similar. The light and heavy chain genes are transcribed into precursor mRNAs, which are processed into finished mRNAs, which are translated on ribosomes into the antibody polypeptides (Figure 43.4).

- The antibody-mediated immune response has two general phases: T-cell activation, and B-cell activation and antibody production. T-cell activation begins when a dendritic cell engulfs a pathogen and produces antigens, making the cell an antigen-presenting cell (APC). The APC secretes interleukins, which activate the T cell. The T cell then secretes cytokines, which stimulate the T cell to proliferate, producing a clone of cells. The clonal cells differentiate into helper T cells (Figures 43.3, 43.6, and 43.8).

- B-cell receptors (BCRs) on B cells recognize antigens on a pathogen and engulf it. The B cells then display the antigens. The TCR on a helper T cell activated by the same antigen binds to the antigen on the B cell. Interleukins from the T cell stimulate the B cell to produce a clone of cells with identical BCRs. The clonal cells differentiate into plasma cells, which secrete antibodies specific for the antigen, and memory B cells, which provide immunological memory of the antigen encounter (Figures 43.3, 43.8, and 43.9).

- Clonal expansion is the process of selecting a lymphocyte specifically for cloning when it encounters an antigen from among a randomly generated, large population of lymphocytes with receptors that specifically recognize the antigen (Figure 43.10).

- Antibodies clear the body of antigens by neutralizing or agglutinating them, or by aiding the innate immune response (Figure 43.11).

- In immunological memory, the first encounter of an antigen elicits a primary immune response and later exposure to the same antigen elicits a rapid secondary response with a greater production of antibodies (Figure 43.12).

- Active immunity is the production of antibodies in the body in response to an antigen. Passive immunity is the acquisition of antibodies by direct transfer from another person.

- In cell-mediated immunity, cytotoxic T cells recognize and bind to antigens displayed on the surfaces of infected body cells, or to cancer cells. They then kill the infected body cell (Figure 43.13).

- Antibodies are widely used in research to identify, locate, and determine the functions of molecules in biological systems.

- Monoclonal antibodies are made by isolating fully active B cells from a test animal, fusing them with cancer cells to produce hybridomas, and using single hybridomas to start clones of cells, all of which make highly specific antibodies against the same epitope of an antigen (Figure 43.14).

Animation: Antibody structure

Animation: Gene rearrangements

Animation: Clonal selection of a B cell

Animation: Antibody-mediated response

Animation: Cell-mediated response

43.4 Malfunctions and Failures of the Immune System

- In immunological tolerance, molecules present in an individual at birth normally do not elicit an immune response.
- In some people, the immune system malfunctions and reacts against the body's own proteins or cells, producing autoimmune disease.

- The first exposure to an allergen sensitizes an individual by leading to the production of memory B and T cells, which cause a greatly intensified response at the next and subsequent exposures.
- Most allergies result when antigens act as allergens by stimulating B cells to produce IgE antibodies, which leads to the release of histamine. Histamine produces the symptoms characteristic of allergies (Figure 43.15).

Animation: HIV replication cycle

43.5 Defenses in Other Animals

- Antibodies, complement proteins, and other molecules with defensive functions have been identified in all vertebrates.
- Invertebrates rely on nonspecific defenses, including surface barriers, phagocytes, and antimicrobial molecules.

Questions

Self-Test Questions

1. Which of the following most directly affects a cell harboring a virus?
   a. $CD8^+$ T cells that bind class I MHC proteins holding viral antigen
   b. $CD4^+$ T cells that bind free viruses in the blood
   c. B cells secreting perforin
   d. antibodies that bind the viruses with their constant ends
   e. natural killer cells secreting antiviral antibodies

2. Components of the inflammatory response include all *except*:
   a. macrophages.
   b. neutrophils.
   c. B cells.
   d. mast cells.
   e. eosinophils.

3. When a person resists infection by a pathogen after being vaccinated against it, this is the result of:
   a. innate immunity.
   b. immunological memory.
   c. a response with defensins.
   d. an autoimmune reaction.
   e. an allergy.

4. One characteristic of a B cell is that it:
   a. has the same structure in both invertebrates and vertebrates.
   b. recognizes antigens held on class I MHC proteins.
   c. binds viral infected cells and directly kills them.
   d. makes many different BCRs on its surface.
   e. has a BCR on its surface, which is the IgM molecule.

5. Antibodies:
   a. are each composed of four heavy and four light chains.
   b. display a variable end, which determines the antibody's location in the body.
   c. belonging to the IgE group are the major antibody class in the blood.
   d. found in large numbers in the mucous membranes belong to class IgG.
   e. function primarily to identify and bind antigens free in body fluids.

6. The generation of antibody diversity includes the:
   a. joining of V to C to J segments to make a functional light chain gene.
   b. choice from several different types of C segments to make a functional light chain gene.

   c. deletion of the J segment to make a functional light chain gene.
   d. joining of V to J to C segments to make a functional light chain gene.
   e. initial generation of IgG followed later by IgM on a given cell.

7. An APC:
   a. can be a $CD8^+$ T cell.
   b. derives from a phagocytic cell and is lymphocyte-stimulating.
   c. secretes antibodies.
   d. cannot be a B cell.
   e. cannot stimulate helper T cells.

8. Antibodies function to:
   a. deactivate the complement system.
   b. neutralize natural killer cells.
   c. clump bacteria and viruses for easy phagocytosis by macrophages.
   d. eliminate the chance for a secondary response.
   e. kill viruses inside of cells.

9. After Jen punctured her hand with a muddy nail, in the emergency room she received both a vaccine and someone else's antibodies against tetanus toxin. The immunity conferred here is:
   a. both active and passive.
   b. active only.
   c. passive only.
   d. first active; later passive.
   e. innate.

10. Medicine attempts to enhance the immune response when treating:
    a. organ transplant recipients.
    b. anaphylactic shock.
    c. rheumatoid arthritis.
    d. HIV infection.
    e. Type I diabetes.

Questions for Discussion

1. HIV wreaks havoc with the immune system by attacking helper T cells and macrophages. Would the impact be altered if the virus attacked only macrophages? Explain.

2. Given what you know about how foreign invaders trigger immune responses, explain why mutated forms of viruses, which have altered surface proteins, pose a monitoring problem for memory cells.

3. Cats, dogs, and humans may develop myasthenia gravis, an autoimmune disease in which antibodies develop against acetylcholine receptors in the synapses between neurons and skeletal muscle fibers. Based on what you know of the biochemistry of muscle contraction (see Section 41.1), explain why people with this disease typically experience severe fatigue with even small levels of exertion, drooping of facial muscles, and trouble keeping their eyelids open.

Experimental Analysis

Space, the final frontier! Indeed, but being in space has some problems. Astronauts in space show a decline in their ability to mount an immune response and, consequently, develop a decreased resistance to infection. Two potentially important differences in physiology in space versus on Earth are more fluid flowing to the head and a lack of weight-bearing on the lower limbs. Could they be involved somehow in the deleterious effect on the immune system? Design an experiment to be done on Earth to answer this question.

Evolution Link

Defensins are found in a wide range of organisms, including plants as well as animals. What are the evolutionary implications of this observation?

How Would You Vote?

Drugs are available that can extend the life of patients with AIDS, but their high cost is more than people in most developing countries can afford to pay. Should the federal government offer incentives to companies to discount the drugs for developing countries? What about AIDS patients at home? Who should pay for their drugs? Go to www.thomsonedu.com/login to investigate both sides of the issue and then vote.

Lining of the trachea (windpipe) shown in a colorized SEM, with mucus-secreting cells (white) and epithelial cells with cilia (pink). The trachea is positioned between the larynx and the lungs, providing a conduit for air entering and leaving the body.

© Steve Gschmeissner/Science Photo Library/Photo Researchers, Inc.

44 Gas Exchange: The Respiratory System

WHY IT MATTERS

On October 25, 1999, at 9:19 A.M., the captain lined up Learjet N47BA on the runway at Orlando International Airport and opened the throttles. Within seconds, the sleek corporate jet was airborne and climbing; 2 minutes later, at 9:21 EDT, the pilots reported passing through 9500 feet.

As the jet continued its climb, the pressure of the outside air dropped steadily and with it the availability of the oxygen (O_2) that all animal life requires, including the two pilots and three passengers on the jet. Normally, in aircraft, the cabin pressure is maintained at a level equivalent to an altitude of 8000 feet, more than sufficient to keep O_2 available to all on board. But, unknown to the pilots, the pressurization system was not functioning normally.

At 9:27 EDT, the controller at the Jacksonville Control Center instructed the jet to climb to 39,000 feet. The first officer acknowledged the instruction, her voice strong and clear. Her acknowledgment was the last radio transmission anyone was to hear from N47BA.

When humans experience increasingly higher altitudes, each breath brings less O_2 into the body. Of all the cells affected by reduced O_2, the ones most sensitive are those of the eyes and brain. Without an O_2 supply at 25,000 feet, most people progress from fully alert to unconscious in about 3 minutes; at 40,000 feet, the progression takes only 15 seconds.

The jet continued its climb, eventually reaching an altitude of 46,000 feet. When the pilots stopped responding to communications, military jets were sent to investigate. The military pilots could see no movement in the Learjet cabin and there was no response to their transmissions. The forward windshields of the Learjet were frosted over, indicating that warm air from the engines was not ventilating the cabin correctly. Evidently, the aircraft was maintaining its course through the autopilot, without conscious human direction.

Many hours later, at 12:11 P.M. CDT, one of the two engines failed: the aircraft, now unbalanced, rolled over and entered a steep, spiraling descent that ended in a shattering impact in a field near Aberdeen, South Dakota. The subsequent investigation pointed to faulty operation of a single valve controlling cabin pressurization as a likely cause of the accident. This tragic loss of life emphasizes the vital importance of O_2 to the survival of humans and other animals. In this chapter we discuss the respiratory system, the system that allows an animal to exchange CO_2 produced in the body for O_2 from the surroundings. The respiratory systems of animals reflect the environmental conditions under which they live, and this general principle has resulted in a truly remarkable array of adaptations.

44.1 The Function of Gas Exchange

Physiological respiration is the process by which animals exchange gases with their surroundings—how they take in O_2 from the outside environment and deliver it to body cells, and remove CO_2 from body cells and deliver it to the environment **(Figure 44.1)**. The absorbed O_2 is used as the final electron acceptor for the oxidative reactions that produce ATP in mitochondria (see Section 8.4). The CO_2 released to the environment is a product of those oxidative reactions. Because they use O_2 and release CO_2, these ATP-producing reactions are called *cellular respiration*.

How gas exchange occurs in an animal depends on its respiratory medium—air or water—and the nature of its respiratory surface. The **respiratory medium** is the environmental source of O_2 and the "sink" for released CO_2. For aquatic animals, of course, the respiratory medium is water; for terrestrial animals, it is air. Amphibians and some fishes use both water and air as respiratory media. The exchange of gases with the respiratory medium by animals is called **breathing**, whether the medium is air or water.

The **respiratory surface**, formed by a layer of epithelial cells, provides the interface between the body and the respiratory medium. Oxygen is absorbed across the respiratory surface, and CO_2 is released. In all animals, the exchange of gases across the respiratory surface occurs by simple diffusion, movement of molecules from a region of higher concentration to a region of lower concentration (see Section 6.2).

Generally, the concentration of O_2 is higher in the respiratory medium than on the internal side of the respiratory surface, and thus the net diffusion of O_2 is inward. Carbon dioxide moves in the opposite direction because the CO_2 concentration is higher on the internal side of the respiratory surface than in the respiratory medium.

Respiratory surfaces typically have two structural properties that favor a high rate of diffusion: they are thin, and they have large surface areas. The rate of diffusion is inversely proportional to the square of the distance over which the diffusion occurs; diffusion rates are therefore higher through thin surfaces such as the single layer of epithelial cells forming many respiratory surfaces. And, the rate of diffusion is directly proportional to the surface area across which diffusion occurs, meaning that large surface areas allow for higher rates of gas exchange than small surface areas. In addition, the rate of diffusion becomes higher with larger concentration gradients and with increasing temperature.

In some relatively small animals, such as sponges, ctenophores, roundworms, flatworms, and some annelids, the entire body surface serves as the respiratory surface. All these animals are invertebrates that live in aquatic or moist environments.

In larger animals, specialized structures, *gills* and *lungs*, form the primary respiratory surface for exchanging gases with water and air, respectively. In insects, a **tracheal system**, an extensive system of branching tubes, channels air from the outside to the internal organs and most individual cells of the animal.

Because gases must dissolve in water to enter and leave epithelial cells, the respiratory surface must be wetted to function in gas exchange, either directly by

Figure 44.1
The relationship between cellular respiration and physiological respiration.

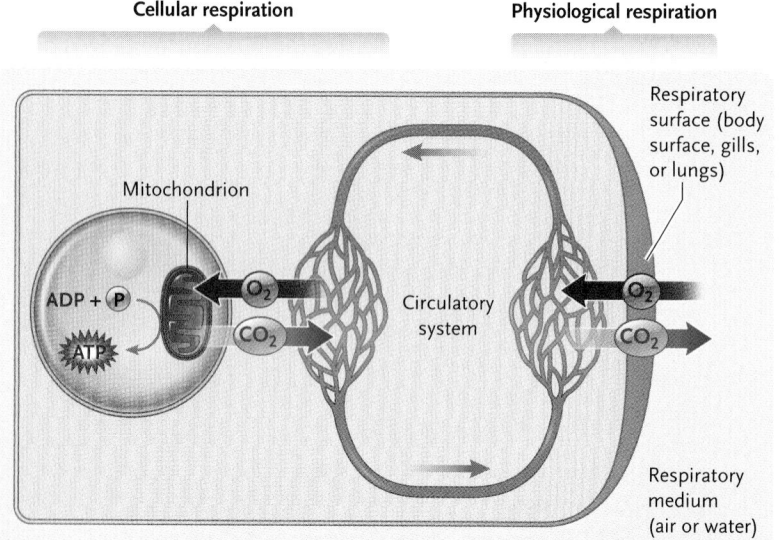

the respiratory medium or by a thin film of water. For this reason, in water-breathing animals, **gills** are *evaginations* of the body: they extend outward into the respiratory medium. In terrestrial animals, **lungs** are typically pockets or *invaginations* of the body surface, buried deeply in the body interior where they are less susceptible to drying out. Also, terrestrial animals have adaptations that moisten dry air before it reaches the respiratory surface. For example, in humans and other mammals, moisture is added to air as it passes through the mouth, nasal passages, throat, and air passages leading to the lungs.

The organ system responsible for gas exchange is termed the **respiratory system.** The respiratory system consists of all the parts of the body involved in exchanging air between the external environment and the blood. In mammals, this includes the airways leading to and into the lungs, the lungs themselves, and the structures of the chest used to move air through the airways into and out of the lungs.

Adaptations That Increase Ventilation and Perfusion of the Respiratory Surface Maximize the Rate of Gas Exchange

Two primary adaptations help animals maintain the difference in concentration between gases outside and inside the respiratory surface, thereby keeping the rate of gas exchange at maximal levels. One is **ventilation,** the flow of the respiratory medium (air or water, depending on the animal) over the external side of the respiratory surface. The second is **perfusion,** the flow of blood or other body fluids on the internal side of the respiratory surface.

Ventilation. As they respire, animals remove O_2 from the respiratory medium and replace it with CO_2. Without ventilation, the concentration of O_2 would fall in the respiratory medium close to the respiratory surface, and the concentration of CO_2 would rise, gradually reducing the concentration gradients and dropping the rate of gas exchange below the minimum level required to sustain life. Examples of ventilation include the one-way flow of water over the gills in fish and many other aquatic animals and the in-and-out flow of air in the lungs of most vertebrates and in the tracheal system of insects.

Perfusion. The constant replacement of blood or another fluid on the internal side of the respiratory surface helps to keep the inside/outside concentration differences of O_2 and CO_2 at a maximum. In animals without a circulatory system, such as roundworms and flatworms, body movements help circulate body fluids beneath the skin. Most animals without a circulatory system are small or have thin, greatly flattened bodies, because all body cells must be located close to the respiratory surface to exchange O_2 and CO_2 adequately. In

animals with a circulatory system, the circulatory system brings blood to the internal side of the respiratory surface, transporting CO_2 from all cells of the body—no matter how far they are from the respiratory surface—to exchange for O_2, which is then taken to all cells of the body.

Adaptations That Increase the Area of the Respiratory Surface Maximize the Quantity of Gases Exchanged

Most animals have adaptations that increase the quantity of gases exchanged by increasing the area of the respiratory surface. In animals whose skin serves as the respiratory surface, an elongated or flattened body form increases the area of the respiratory surface **(Figure 44.2a).**

In animals with gills, the respiratory surface is increased by highly branched structures that include many fingerlike or platelike projections **(Figure 44.2b).** Similarly, in animals with lungs or tracheae, the respiratory surface is increased by a multitude of branched tubes, folds, or pockets **(Figure 44.2c).**

Water and Air Have Advantages and Disadvantages as Respiratory Media

Because their respiratory surfaces are exposed directly to the environment, water breathers have no problem keeping the respiratory surface wetted. However, aquatic animals face two main challenges in obtaining O_2 from water compared with terrestrial animals. First, water contains approximately one-thirtieth as much O_2 as air

a. Extended body surface: flatworm

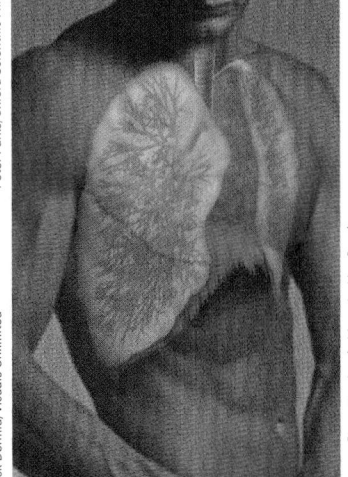
c. Lungs: human

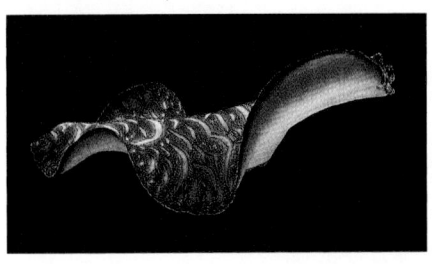
b. External gills: mudpuppy

Peter Parks/Oxford Scientific Films

Jack Dermid/Visuals Unlimited

© 2000 Photodisc, Inc. (with art by Lisa Starr)

Figure 44.2
Adaptations increasing the area of the respiratory surface. **(a)** The flattened and elongated body surface of a flatworm. **(b)** The highly branched, feathery structure of the external gills in an amphibian, the mudpuppy *(Necturus)*. **(c)** The many branches and pockets expanding the respiratory surface in the human lung.

does (at 15°C). Therefore, to obtain the same amount of O_2, an aquatic animal must process 30 times as much of its respiratory medium as a terrestrial animal does. Second, water is about 1000 times as dense as air and about 50 times as viscous. Therefore, it takes significantly more energy to move water than air over a respiratory surface. For this reason, ventilation in most aquatic animals takes place in a one-way direction. In bony fishes, for instance, water enters the mouth, flows over the gills, and exits through the gill covers, all in one direction.

In addition, temperature and solutes affect the O_2 content of water. That is, as either the temperature or the amount of solutes increases, the amount of gas that can dissolve in water decreases. Therefore, with respect to obtaining O_2, aquatic animals that live in warm water are at a disadvantage compared with those that live in cold water. And, because levels of solutes (such as sodium chloride) are higher in seawater than in freshwater, aquatic animals living in seawater are at a disadvantage.

The relatively high O_2 content, low density, and low viscosity of air greatly reduce the energy required to ventilate the respiratory surface. These advantages allow animals with lungs to breathe in and out, re-versing the direction of flow of the respiratory medium, without a large energy penalty. As you will see later, reversing the direction of flow decreases the efficiency of gas exchange.

Another advantage of breathing air is that gas molecules diffuse nearly 10,000 times faster through air than through water. This increases the rate at which molecules of the gases at the respiratory surface exchange with those located farther away in the air, and reduces the requirement for ventilation as compared with water.

A major disadvantage of air is that it constantly evaporates water from the respiratory surface unless the air is saturated with water vapor. Therefore, except in an environment with 100% humidity, animals lose water by evaporation during breathing and must replace the water to keep the respiratory surface from drying and causing the death of the surface cells.

We next turn to the adaptations that allow water-breathing and air-breathing animals to obtain O_2 and release CO_2 in aquatic and terrestrial environments. These adaptations allow animals to exploit the advantages and circumvent the disadvantages of water and air as respiratory media.

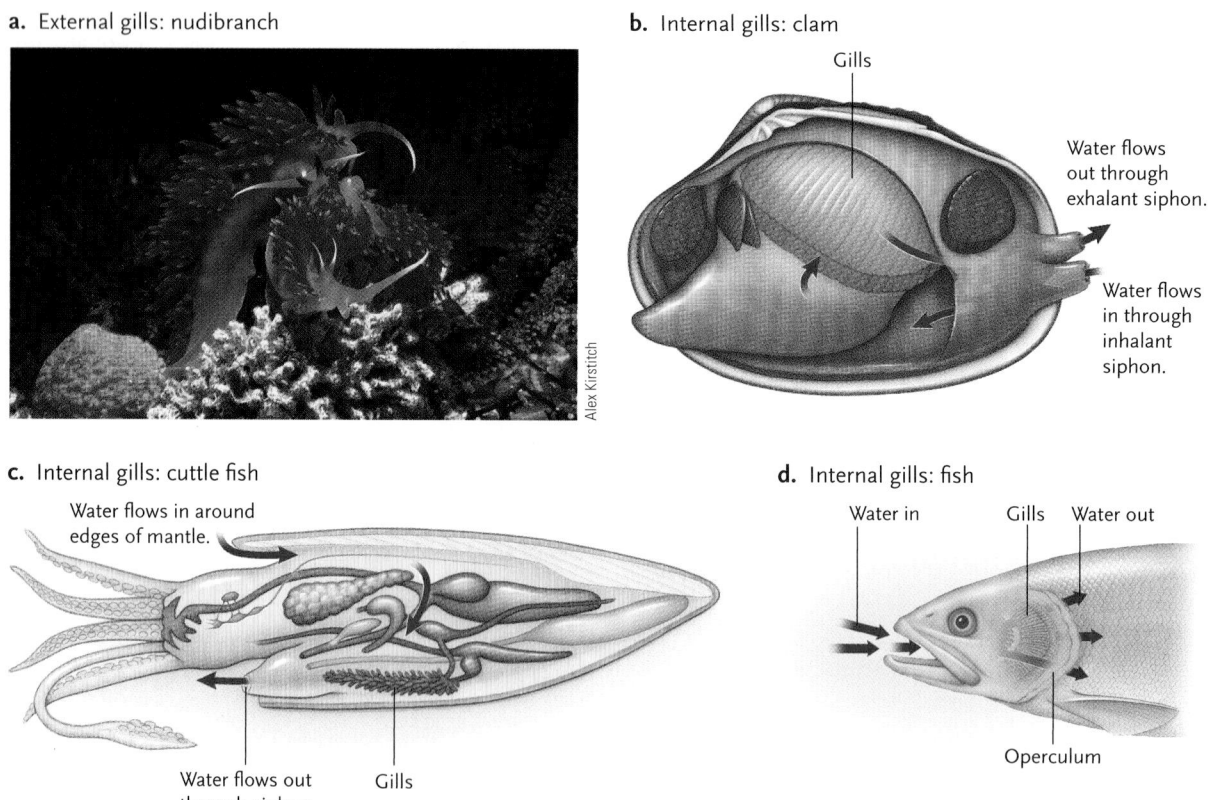

a. External gills: nudibranch

b. Internal gills: clam

Gills

Water flows out through exhalant siphon.

Water flows in through inhalant siphon.

Alex Kirstitch

c. Internal gills: cuttle fish

Water flows in around edges of mantle.

Water flows out through siphon.

Gills

d. Internal gills: fish

Water in   Gills   Water out

Operculum

Figure 44.3

External and internal gills. **(a)** The external gills of a nudibranch *(Flabellina iodinea)*. **(b)** The internal gills in a clam. **(c)** The internal gills in a cuttlefish. **(d)** Internal gills of a bony fish. Water enters through the mouth and passes over the filaments of the gills before exiting through an opening at the edges of the flaplike protective covering, the operculum.

1. Distinguish between the roles of the respiratory medium and the respiratory surface in respiratory systems.
2. What is an advantage of water over air as a respiratory medium? What are two key advantages of air over water as a respiratory medium?

44.2 Adaptations for Respiration

Although most animals that live in water exchange gases through the skin or gills, some, such as whales, seals, and dolphins, exchange gases through lungs (which originally evolved in aquatic creatures). And although most animals that live on land exchange gases through lungs, some, such as sow bugs and land crabs, exchange gases through gills, and others, such as insects, exchange gases using a tracheal system.

Aquatic Gill Breathers Exchange Gases More Efficiently Than Skin Breathers

Gills provide water-breathing animals, and a few air-breathers, with more efficient gas exchange than skin breathers have. In combination with the organized circulatory system common to these animals, gills also allow animals to live in more diverse habitats, and to achieve greater body mass, than animals that breathe primarily or exclusively through the skin.

External and Internal Gills. Gills are respiratory surfaces that are branched and folded evaginations of the body surface. **External gills** are gills that do not have protective coverings; they extend out from the body and are in direct contact with the water. **Internal gills,** by contrast, are located within chambers of the body that have a cover providing physical protection for the gills. Water must be brought to internal gills.

Because external gills have no protective coverings, they are exposed to mechanical damage and must be immersed in water to keep them from collapsing or drying. For these reasons, animals with external gills, including some annelids and mollusks **(Figure 44.3a),** aquatic insects, the larval forms of some bony fishes, and some amphibians, are limited to relatively protected aquatic environments.

The coverings of internal gills protect them from mechanical damage and drying. Covered internal gills allow animals to live in highly diverse habitats, ranging from small streams and ponds to rivers, lakes, and the open seas, and even in moist terrestrial habitats. Most crustaceans, mollusks, sharks, and bony fishes have internal gills. Some invertebrates, such as clams and oysters, use beating cilia to circulate water over their internal gills **(Figure 44.3b).** Others, such as the cuttlefish, use contractions of the muscular mantle to pump water over their gills **(Figure 44.3c).** In adult bony fishes, the gills extend into a chamber covered by gill flaps or *opercula* (singular, *operculum* = little lid) on either side of the head. The operculum also serves as part of a one-way pumping system that ventilates the gills **(Figure 44.3d).**

Many Animals with Internal Gills Use Countercurrent Flow to Maximize Gas Exchange

Sharks, fishes, and some crabs take advantage of one-way flow of water over the gills to maximize the amounts of O_2 and CO_2 exchanged with water. In this mechanism, called **countercurrent exchange,** the water flowing over the gills moves in a direction opposite to the flow of blood under the respiratory surface.

Figure 44.4 illustrates countercurrent exchange in the uptake of O_2. At the point where fully oxygenated water first passes over a gill filament in countercurrent flow, the blood flowing beneath it in the opposite direction is also almost fully oxygenated. However, O_2 concentration is still higher in the water than in the blood, and the gas diffuses from the water into the blood, raising the concentration of O_2 in the blood almost to the level of the fully oxygenated water. At the opposite end of the filament, much of the O_2 has been removed from the water, but the blood flowing under the filament, which has just arrived from body tissues and is fully deoxygenated, contains even less O_2. As a result, O_2 also diffuses from the water to the blood at this end of the filament. All along the gill filament, the same relationship exists, so that at any point, the water is more highly oxygenated than the blood, and O_2 diffuses from the water into the blood across the respiratory surface.

The overall effect of countercurrent exchange is the removal of 80% to 90% of the O_2 content of water as it flows over the gills. In comparison, by breathing in and out and constantly reversing the direction of air flow, mammals manage to remove only about 25% of the O_2 content of air. Efficient removal of O_2 from water is important because of the much lower O_2 content of water compared with air.

Insects Use a Tracheal System for Gas Exchange

Insects breathe air by a respiratory system consisting of air-conducting tubes called **tracheae** (singular *trachea,* or "windpipe"). The tracheae are invaginations of the outer epidermis of the animal, reinforced by rings of chitin, the material of the insect exoskeleton. They lead from the body surface and branch so extensively inside the animal that almost every cell is served

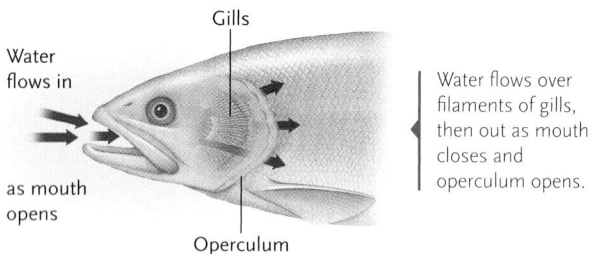

a. The flow of water around the gill filaments

b. Countercurrent flow in fish gills, in which the blood and water move in opposite directions

Water flows in as mouth opens

Gills

Operculum

Water flows over filaments of gills, then out as mouth closes and operculum opens.

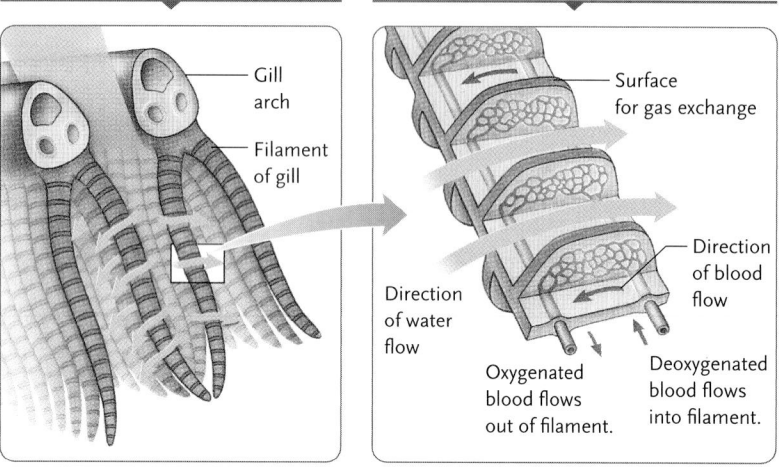

Gill arch

Filament of gill

Surface for gas exchange

Direction of blood flow

Direction of water flow

Oxygenated blood flows out of filament.

Deoxygenated blood flows into filament.

c. In countercurrent exchange, blood leaving the capillaries has the same O₂ content as fully oxygenated water entering the gills

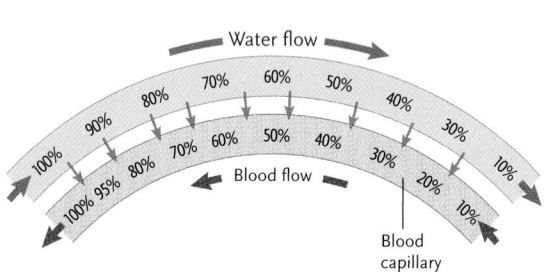

Water flow

Blood flow

Blood capillary

Figure 44.4

Ventilation and countercurrent exchange in bony fishes. **(a)** Water flows around the gill filaments. **(b)** Water and blood flow in opposite directions through the gill filaments. **(c)** Countercurrent exchange: oxygen from the water diffuses into the blood, raising its oxygen content. The percentages indicate the degree of oxygenation of water (blue) and blood (red).

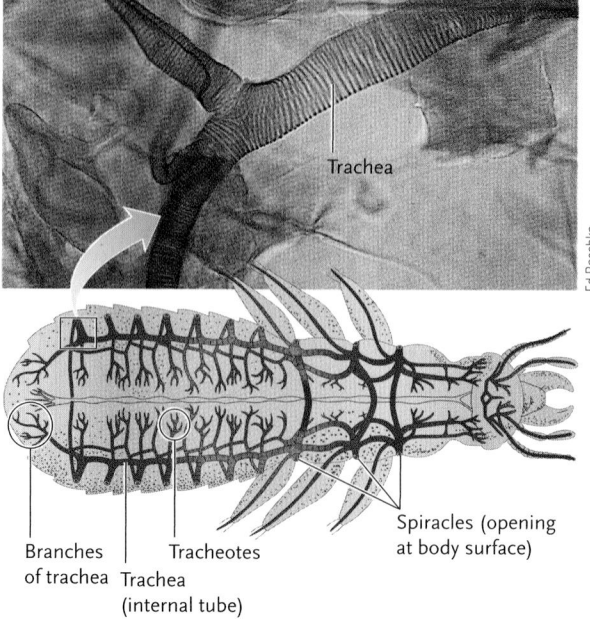

Trachea

Branches of trachea

Trachea (internal tube)

Tracheotes

Spiracles (opening at body surface)

Figure 44.5

The tracheal system of insects. Chitin rings, visible in the photomicrograph, reinforce many of the tracheae.

by a microscopic branch **(Figure 44.5)**. Some of the branches even penetrate inside larger cells, such as those of insect flight muscles. The finest branches of the tracheae, called *tracheoles,* form the respiratory surface of the insect system. Tracheoles are dead-end tubes with very small fluid-filled tips that are in contact with cells of the body. Air is transported by the tracheal system to those tips, and gas exchange occurs directly across the plasma membranes of the body cells in contact with the tips. At places within the body, the tracheae expand into internal air sacs that act as reservoirs to increase the volume of air in the system.

Air enters and leaves the tracheal system at openings in the insect's chitinous exoskeleton called **spiracles** (*spiraculum* = airhole). In adult insects, the spiracles are located in a row on either side of the thorax and abdomen. The spiracles open and close in coordination with body movements to compress and expand the air sacs and pump air in and out of the tracheae. During insect flight, alternating compression and expansion of the thorax by the flight muscles also pump air through the tracheal system.

Lungs Allow Animals to Live in Completely Terrestrial Environments

Lungs are one of the primary adaptations that allowed animals to fully invade terrestrial environments. Some fishes and amphibians have lungs, as do all reptiles, birds, and mammals. All lungs are invaginated structures located internally in the body.

In some fishes, such as lungfishes, lungs and air breathing evolved as adaptations to survive in oxygen-poor water or temporarily in air when the water level dropped and exposed them. The lungs of these fishes consist of thin-walled sacs, which branch off from the mouth, pharynx, or parts of the digestive system; air is obtained by **positive pressure breathing**, a gulping or swallowing motion that forces air into the lungs.

The lungs of mature amphibians such as frogs and salamanders are also thin-walled sacs with relatively

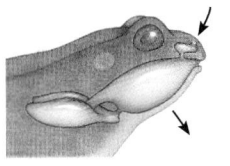

1 The frog lowers the floor of its mouth and inhales through its nostrils.

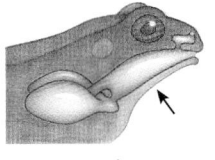

2 It closes its nostrils, opens the glottis, and elevates the floor of the mouth, forcing air into the lungs.

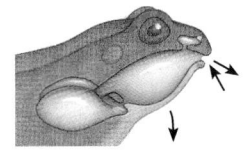

3 Rhythmic ventilation assists in gas exchange.

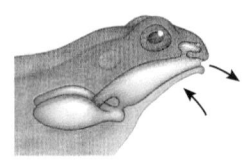

4 Air is forced out when muscles in the body wall above the lungs contract and the lungs recoil elastically.

Figure 44.6
Positive pressure breathing in an amphibian (a frog).

little folding or pocketing. Amphibians also fill their lungs by positive pressure breathing, in this case using a rhythmic motion of the floor of the mouth as the pump, in coordination with opening and closing of the nostrils **(Figure 44.6)**.

The lungs of reptiles, birds, and mammals have many pockets and folds that increase the area of the respiratory surface, which contains dense, highly branched capillary networks. Mammalian lungs consist of millions of tiny air pockets, the **alveoli**, each surrounded by dense capillary networks. Reptiles and mammals fill their lungs by **negative pressure breathing**—by muscular contractions that expand the lungs, lowering the pressure of the air in the lungs and causing air to be pulled inward. (Mammalian negative pressure breathing is described in more detail in the next section.)

In birds, a countercurrent exchange system provides the most complex and efficient vertebrate lungs **(Figure 44.7)**. In addition to paired lungs, birds have nine pairs of air sacs that branch off the respiratory tract. The air sacs, which collectively contain several times as much air as the lungs, set up a pathway that allows air to flow in one direction through the lungs, rather than in and out as in other vertebrates. Within the lungs, air flows through an array of fine, parallel tubes that are surrounded by a capillary network. The blood flows in the direction opposite to the air flow, setting up a countercurrent exchange. The countercurrent exchange allows bird lungs to extract about one-third of the O_2 from the air as compared with about one-fourth in the lungs of mammals.

STUDY BREAK

1. What advantages do gills confer upon a water-breathing animal over skin breathing?
2. What is countercurrent exchange, and how is it beneficial for gas exchange?
3. How does the tracheal system of insects facilitate gas exchange with the cells of the body?
4. Distinguish between positive pressure breathing and negative pressure breathing in animals with lungs.

a. Lungs and air sacs of a bird

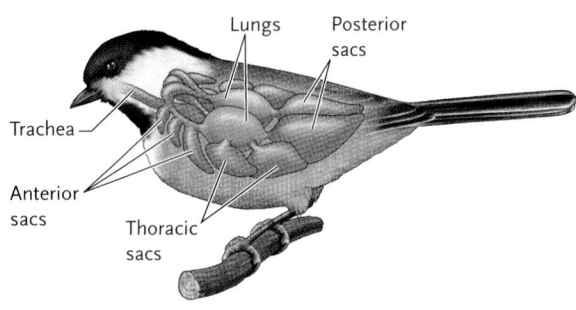

b. Countercurrent exchange

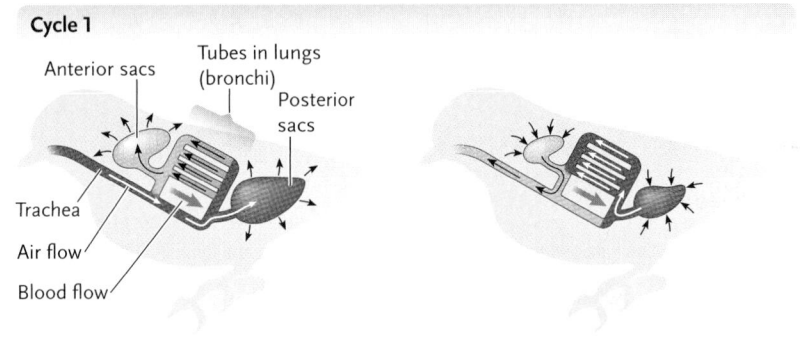

1 During the first inhalation, most of the oxygen flows directly to the posterior air sacs. The anterior air sacs also expand but do not receive any of the newly inhaled oxygen.

2 During the following exhalation, both anterior and posterior air sacs contract. Oxygen from the posterior sacs flows into the gas-exchanging tubes (bronchi) of the lungs.

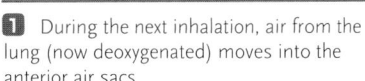

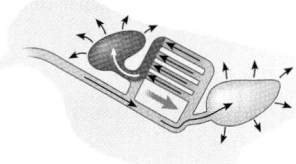

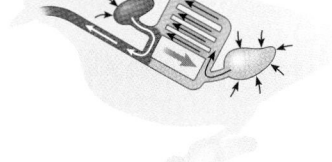

1 During the next inhalation, air from the lung (now deoxygenated) moves into the anterior air sacs.

2 In the second exhalation, air from anterior sacs is expelled to the outside through the trachea.

Figure 44.7
Countercurrent exchange in bird lungs. **(a)** Unlike mammalian lungs, bird lungs do not expand and contract. Changes in pressure in the expandable air sacs move air in and out. **(b)** Air flows in one direction through the tubes of the lungs; blood flows in the opposite direction in the surrounding capillary network. Two cycles of inhalation and exhalation are needed to move a specific volume of air through the bird respiratory system.

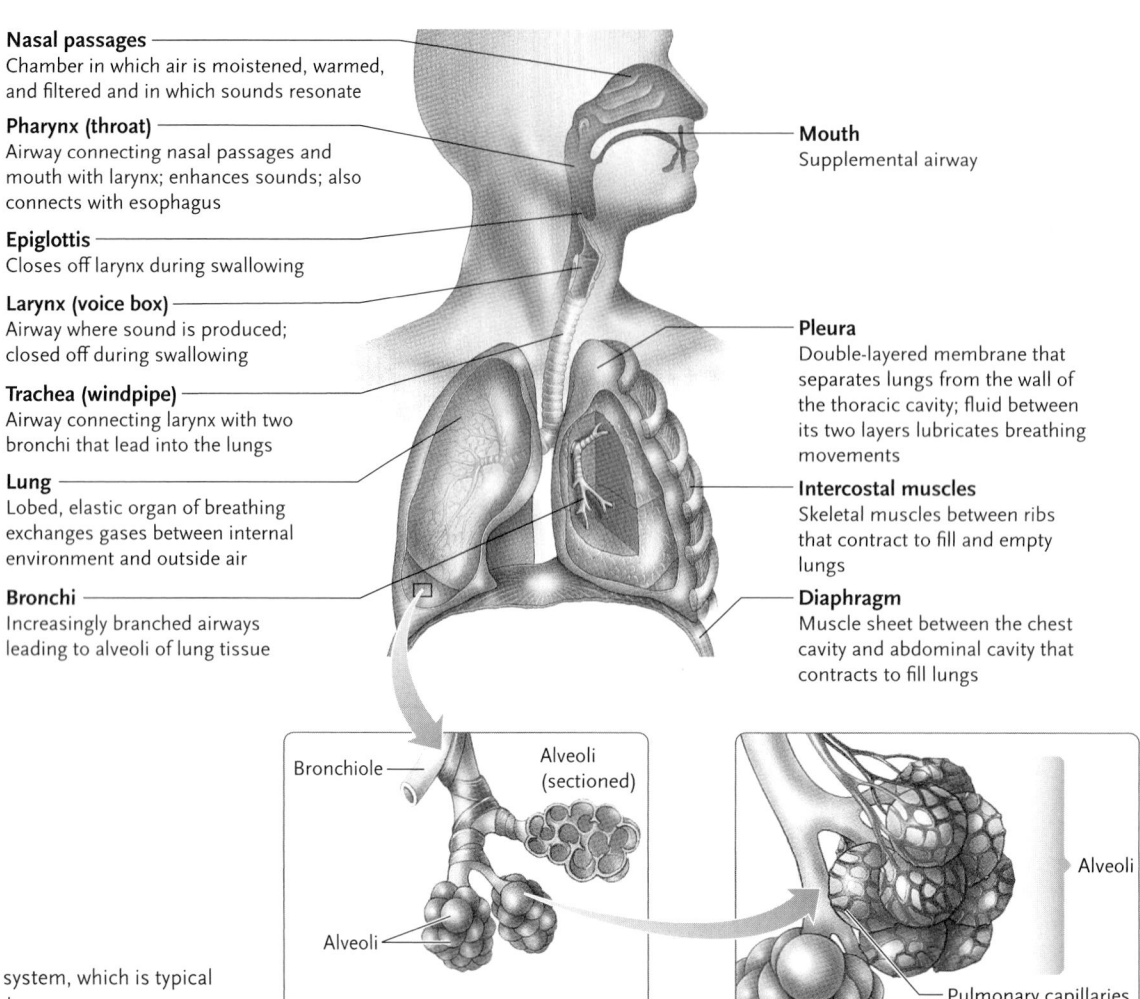

Nasal passages
Chamber in which air is moistened, warmed, and filtered and in which sounds resonate

Pharynx (throat)
Airway connecting nasal passages and mouth with larynx; enhances sounds; also connects with esophagus

Epiglottis
Closes off larynx during swallowing

Larynx (voice box)
Airway where sound is produced; closed off during swallowing

Trachea (windpipe)
Airway connecting larynx with two bronchi that lead into the lungs

Lung
Lobed, elastic organ of breathing exchanges gases between internal environment and outside air

Bronchi
Increasingly branched airways leading to alveoli of lung tissue

Mouth
Supplemental airway

Pleura
Double-layered membrane that separates lungs from the wall of the thoracic cavity; fluid between its two layers lubricates breathing movements

Intercostal muscles
Skeletal muscles between ribs that contract to fill and empty lungs

Diaphragm
Muscle sheet between the chest cavity and abdominal cavity that contracts to fill lungs

Bronchiole

Alveoli (sectioned)

Alveoli

Alveoli

Pulmonary capillaries

Figure 44.8
The human respiratory system, which is typical for a terrestrial mammal.

44.3 The Mammalian Respiratory System

All mammals have a pair of lungs and a diaphragm in the chest cavity that plays an important role in negative pressure breathing. Rapid ventilation of the respiratory surface and perfusion by blood flow through dense capillary networks maximizes gas exchange.

The Airways Leading from the Exterior to the Lungs Filter, Moisten, and Warm the Entering Air

The human respiratory system is typical for a terrestrial mammal **(Figure 44.8).** Air enters and leaves the respiratory system through the nostrils and mouth. Hairs in the nostrils and mucus covering the surface of the airways filter out and trap dust and other large particles. Inhaled air is moistened and warmed as it moves through the mouth and nasal passages.

Next, air moves into the throat, or **pharynx**, which forms a common pathway for air entering the **larynx** (or "voice box") and food entering the esophagus,

which leads to the stomach. The airway through the larynx is open except during swallowing.

From the larynx, air moves into the trachea (or "windpipe"), which branches into two airways, the **bronchi** (singular, *bronchus*). The bronchi lead to the two elastic, cone-shaped lungs, one on each side of the chest cavity. Inside the lungs, the bronchi narrow and branch repeatedly, becoming progressively narrower and more numerous. The terminal airways, the **bronchioles**, lead into cup-shaped pockets, the alveoli (singular, *alveolus;* shown in Figure 44.8 insets).

Each of the 150 million alveoli in each lung is surrounded by a dense network of capillaries. By the time inhaled air reaches the alveoli, it has been moistened to the saturation point and brought to body temperature. The many alveoli provide an enormous area for gas exchange. If the alveoli of an adult human were flattened out in a single layer, they would cover an area approaching 100 square meters, about the size of a tennis court!

The larynx, trachea, and larger bronchi are non-muscular tubes encircled by rings of cartilage that prevent the tubes from compressing. The largest of the rings, which reinforces the larynx, stands out at the front of the throat as the Adam's apple; smaller supporting rings can be felt at the front of the throat just below the larynx. The walls of the smaller bronchi and the bronchioles contain smooth muscle cells that contract or relax to control the diameter of these passages, and with it, the amount of air flowing to and from the alveoli.

The epithelium lining each bronchus contains cilia and mucus-secreting cells. Bacteria and airborne particles such as dust and pollen are trapped in the mucus and then moved upward and into the throat by the beating of the cilia lining the airways. Infection-fighting macrophages also patrol the respiratory epithelium.

Tobacco smoke, by paralyzing the cilia lining the respiratory tract, interferes with the processes that clear bacteria and airborne particles from the lungs. The bacteria and foreign matter persisting in the lungs can cause infections and smoker's cough.

Contractions of the Diaphragm and Muscles between the Ribs Ventilate the Lungs

The lungs are located in the rib cage above the *diaphragm,* a dome-shaped sheet of skeletal muscle separating the chest cavity from the abdominal cavity. The lungs are covered by a double layer of epithelial tissue called the **pleura.** The inner pleural layer is attached to the surface of the lungs, and the outer layer is attached to the surface of the chest cavity. A narrow space between the inner and outer layers is filled with slippery fluid, which allows the lungs to move within the chest cavity without rubbing or abrasion as they expand and contract.

Contraction of muscles between the ribs and the diaphragm brings air into the lungs by a negative pressure mechanism. As an inhalation begins, the diaphragm contracts and flattens, and one set of muscles between the ribs, the *external intercostal* muscles, contracts, pulling the ribs upward and outward **(Figure 44.9).** These movements expand the chest cavity and lungs, lowering the air pressure in the lungs below that of the atmosphere. As a result, air is drawn into the lungs, expanding and filling them.

The expansion of the lungs is much like filling two rubber balloons. Like balloons, the lungs are elastic, and resist stretching as they are filled. Also like balloons, the stretching stores energy, which can be released to expel air from the lungs. When a person at rest exhales, the diaphragm and muscles between the ribs relax, and the elastic recoil of the lungs expels the air.

When physical activity increases the body's demand for O_2, other muscles help expel the air by forcefully reducing the volume of the chest cavity. Contractions of abdominal wall muscles increase abdominal pressure, exerting an upward-directed force on the dia-

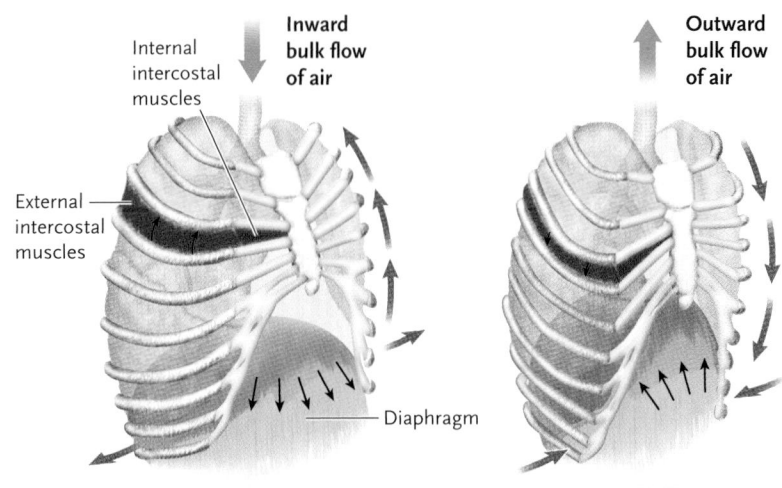

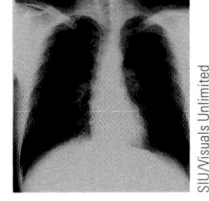 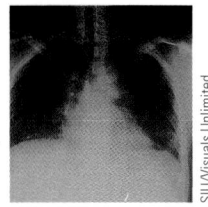

Inhalation. Diaphragm contracts and moves down. The external intercostal muscles contract and lift rib cage upward and outward. The lung volume expands.

SIU/Visuals Unlimited

Exhalation during breathing or rest. Diaphragm and external intercostal muscles return to the resting positions. Rib cage moves down. Lungs recoil passively.

SIU/Visuals Unlimited

Figure 44.9

The respiratory movements of humans during breathing at rest. The movements of the rib cage and diaphragm fill and empty the lungs. Inhalation is powered by contractions of the external intercostal muscles and diaphragm, and exhalation is passive. During exercise or other activities characterized by deeper and more rapid breathing, contractions of the internal intercostal muscles and the abdominal muscles add force to exhalation. The X-ray images show how the volume of the lungs increases and decreases during inhalation and exhalation.

phragm and thus pushing it upward. Contractions of *internal intercostal* muscles pull the chest wall inward and downward, causing it to flatten. As a result, the dimensions of the chest cavity decrease.

The Volume of Inhaled and Exhaled Air Varies over Wide Limits

The volume of air entering and leaving the lungs during inhalation and exhalation is called the **tidal volume.** In a person at rest, the tidal volume amounts to about 500 mL. As physical activity increases, the tidal volume increases to match the body's needs for O_2; at maximal levels, the tidal volume reaches about 3400 mL in females and 4800 mL in males. The maximum tidal volume is called the **vital capacity** of an individual.

Even after the most forceful exhalation, about 1200 mL of air remains in the lungs in males, and about 1000 mL in females; this is the **residual volume** of the lungs. In fact, the lungs cannot be deflated completely because small airways collapse during forced exhalation, blocking further outflow of air. Because air cannot be removed from the lungs completely, some gas exchange can always occur between blood flowing through the lungs and the air in the alveoli.

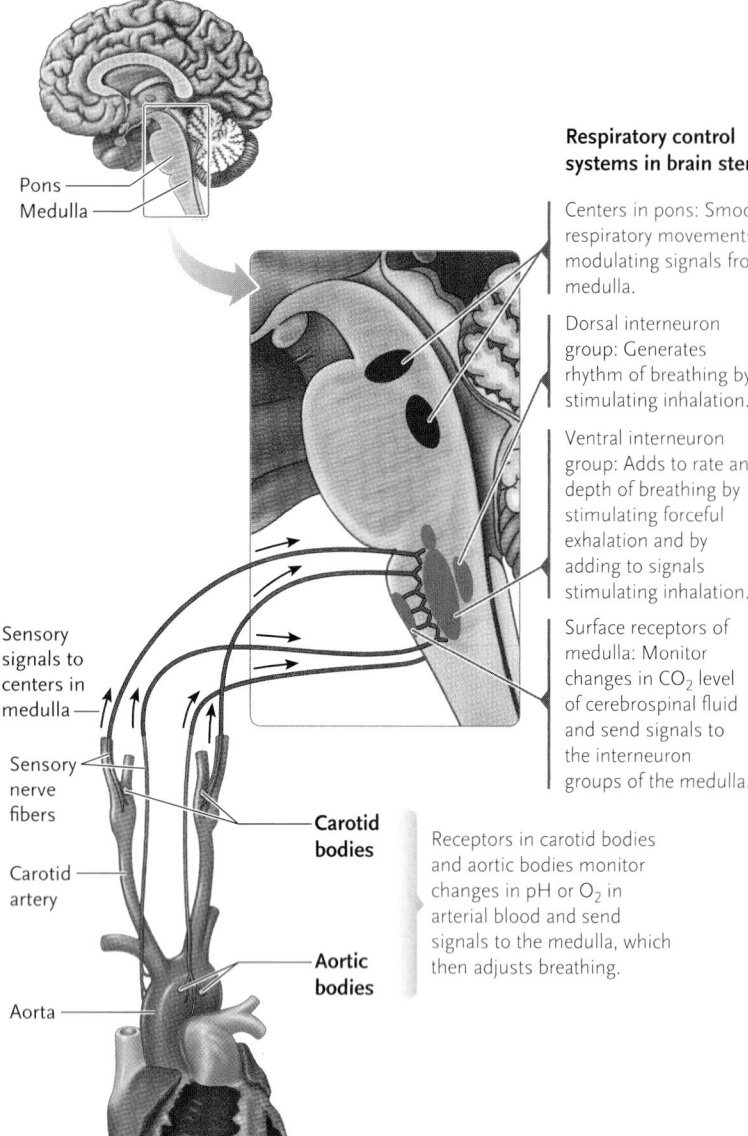

Pons
Medulla

**Respiratory control
systems in brain stem**

Centers in pons: Smooth
respiratory movements by
modulating signals from
medulla.

Dorsal interneuron
group: Generates
rhythm of breathing by
stimulating inhalation.

Ventral interneuron
group: Adds to rate and
depth of breathing by
stimulating forceful
exhalation and by
adding to signals
stimulating inhalation.

Surface receptors of
medulla: Monitor
changes in CO_2 level
of cerebrospinal fluid
and send signals to
the interneuron
groups of the medulla.

Sensory
signals to
centers in
medulla

Sensory
nerve
fibers

Carotid
artery

Aorta

Heart

**Carotid
bodies**

**Aortic
bodies**

Receptors in carotid bodies
and aortic bodies monitor
changes in pH or O_2 in
arterial blood and send
signals to the medulla, which
then adjusts breathing.

Figure 44.10

Control of breathing. Centers in the pons and medulla control the rhythm, rate, and
depth of breathing. Receptors in the carotid arteries and aorta detect changes in the levels
of O_2 and CO_2 in blood and body fluids. Signals from these receptors are integrated in the
respiratory centers of the medulla and pons.

The Centers That Control Breathing
Are Located in the Brain Stem

Breathing is controlled by centers in the medulla and
pons, which form part of the brain stem **(Figure 44.10)**.
Groups of interneurons in the centers regulate the rate
and depth of breathing, ranging from shallow, slow
breathing when the body is at rest to the deep and rapid
breathing of intense physical exercise, excitement, or
fear. Over these extremes, the air entering and leaving
lungs of a human male varies from as little as 5 to 6 L

per minute to (for a brief time only) as much as 150 L
per minute.

Interneurons That Regulate Breathing. Signals from
interneurons in the medulla carried by efferent (motor) neurons of the autonomic system produce the
breathing movements. A set of signals from a dorsal
group of interneurons acts as the primary stimulator
of inhalation by causing the diaphragm and the external intercostal muscles to contract, which expands the
chest cavity and produces an inhalation. In a person at
rest, the signal is switched off as the lungs become
moderately full: the rib muscles and the diaphragm
relax, and a passive exhalation occurs. These signals
act as the primary generator of breathing rhythm.

A ventral group of interneurons in the medulla
can send signals for both inhalation and exhalation.
These neurons become active only during physical exercise, fear, or other situations that require more oxygen when active rather than passive exhalation is
needed. In that case, some of the ventral neurons send
signals that stimulate the abdominal and internal intercostal muscles to contract, thereby causing active
exhalation. Other neurons in the ventral group become
stimulated by signals from the dorsal group, and then
help increase inhalation activity when faster and deeper
breathing is required.

Two interneuron groups in the pons modulate the
signals originating from the medulla, fine-tuning and
smoothing the muscle contractions so that inhalations
and exhalations are gradual and controlled rather than
sudden and abrupt. Signals sent from higher brain
centers in the cerebrum can override the control of
respiratory rate and depth by the brain stem. For example, as we speak or sing, or hold our breath, we can
consciously alter or stop breathing to match the demands of these activities. Breathing rate and depth are
also modified by emotional states, controlled by centers in the limbic system of the brain (see Section 38.3).
Thus breathing is altered as we laugh, gasp, groan, cry,
and sigh.

Receptors That Send Information to the Brain Centers.
The brain centers controlling the rate and depth of
breathing integrate sensory information sent by receptors that monitor O_2 and CO_2 levels in the blood and
body fluids. The integration of sensory information
serves to match breathing rate to the metabolic demands of the body. These *chemoreceptors* are located
centrally on the surface of the medulla, and peripher-
ally in **carotid bodies** in the carotid arteries leading to
the brain and in **aortic bodies** in the large arteries leaving the heart (see Figure 44.10).

The receptors of the medulla detect changes in pH
in the cerebrospinal fluid; the pH is determined mostly
by the CO_2 concentration in the blood. (Remember that
pH decreases as CO_2 levels increase.) The receptors in

the carotid and aortic bodies detect changes in CO_2 and O_2 concentrations in the blood.

The CO_2 receptors in the medulla have the greatest effects on breathing. If increased body activities cause the CO_2 concentration to rise in the blood, the medulla receptors trigger interneuron groups in the medulla that increase the rate and depth of breathing. If CO_2 concentration falls, the receptors send signals to the medulla that lead to a slowing of the rate and depth of breathing.

The peripheral receptors in the carotid and aortic bodies detect changes in pH or O_2 concentration in arterial blood. When these receptors detect a rise in blood pH they send signals to the medulla that cause the medulla to increase the rate and depth of breathing. Although the receptors in the carotid and aortic bodies also detect the O_2 level in arterial blood, the receptors do not respond until blood O_2 level falls below 60% of normal. This reaction makes the O_2 receptors act as a backup system that comes into play only when blood O_2 concentration falls to critically low levels.

Thus, the level of CO_2 in the blood and body fluids is much more closely monitored, and has a much greater effect on breathing, than the O_2 level. This reflects the fact that small fluctuations in blood pH have much greater effects on the ability of hemoglobin to carry oxygen, and on enzyme activity in the blood and interstitial fluid, than fluctuations in the O_2 level.

Local Controls. Other, automated controls within the lungs match the rates of ventilation and perfusion by responding to O_2 concentrations in the blood. If air flow lags behind capillary blood flow, so that the O_2 level falls in the blood, the reduced O_2 concentration causes smooth muscles in the walls of arterioles in the lungs to contract. This reduces the flow of blood, thereby giving it more time to pick up O_2. Conversely, if blood flow lags behind, the rising blood O_2 concentration causes the smooth muscle cells in arteriole walls to relax, dilating the arterioles and increasing the rate of blood flow through lung capillaries. These local controls, in combination with the neural controls that regulate rate and depth of breathing, ensure that the respiratory system meets the body's varying need to obtain O_2 and release CO_2.

STUDY BREAK

1. Explain how inhalation and exhalation occur in a mammal at rest.
2. You can consciously initiate and sustain an exhalation. What is going on muscularly in this case?
3. What is the most important feedback stimulus for breathing?
4. What is the role of the chemoreceptors in the medulla?

44.4 Mechanisms of Gas Exchange and Transport

In both the lungs and body tissues, gas exchange occurs when the gas diffuses from an area of higher concentration to an area of lower concentration. In this section, we consider the mechanics of gas exchange between air and the blood in mammals, and the means by which gases are transported between the lungs and other body tissues. A major part of this story involves hemoglobin, the vertebrate respiratory pigment.

The Proportion of a Gas in a Mixture Determines Its Partial Pressure

For gases, it is often more accurate and convenient to consider concentration differences as differences in pressure. When gases are present in a mixture, the pressure of each individual gas, called its *partial pressure,* is determined by its proportion in the mixture. Air, water, and blood all contain mixtures of gases, including oxygen, carbon dioxide, nitrogen, and other gases, so each gas exerts only a part of the total gas pressure. For example, the proportion of O_2 in dry air is about 21%, or 21/100. In dry air at sea level, the total atmospheric pressure under standard conditions is 760 mm Hg. The partial pressure of O_2, written as P_{O_2}, is equivalent to $760 \times 21/100$, or about 160 mm Hg. The proportion of CO_2 in dry air is about 0.04%, so its partial pressure, P_{CO_2}, is equivalent to $760 \times 0.04/100$, or about 0.3 mm Hg. For O_2 to diffuse inward across a respiratory surface, its partial pressure outside the surface must be greater than inside; for CO_2 to diffuse outward, its partial pressure inside must be greater inside than outside.

In the lungs, even though the P_{O_2} is reduced by mixing with the air in the residual volume, it is still much higher than the P_{O_2} in deoxygenated blood entering the network of capillaries in the lungs **(Figure 44.11)**. As a result, O_2 readily diffuses from the alveolar air into the plasma solution in the capillaries.

Hemoglobin Greatly Increases the O_2-Carrying Capacity of the Blood

After entering the plasma, O_2 diffuses into erythrocytes, where it combines with hemoglobin. The combination with hemoglobin removes O_2 from the plasma, lowering the P_{O_2} of the plasma and allowing additional O_2 molecules to diffuse from alveolar air to the blood.

Recall from Section 42.2 that a mammalian hemoglobin molecule has four heme groups, each containing an iron atom that can combine reversibly with an O_2 molecule. A hemoglobin molecule can therefore bind a total of four molecules of O_2. The combination of O_2 with hemoglobin allows blood to carry about

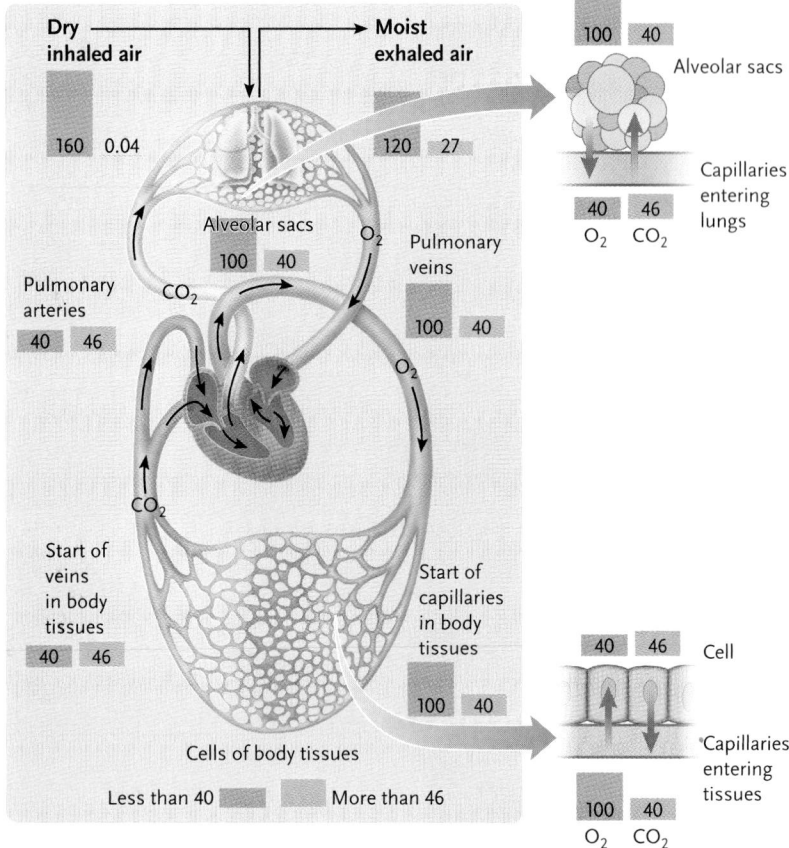

Figure 44.11
The partial pressures of O_2 (pink) and CO_2 (blue) in various locations in the body.

60 times more O_2 (about 200 mL per liter) than it could if the O_2 simply dissolved in the plasma (about 3 mL per liter). About 98.5% of the O_2 in blood is carried by hemoglobin and about 1.5% is carried in solution in the blood plasma.

The reversible combination of hemoglobin with O_2 is related to the partial pressure of O_2 in a pattern shown by the *hemoglobin-O_2 dissociation curve* in **Figure 44.12.** (The curve is generated by measuring the amount of hemoglobin saturated at a given P_{O_2}.) The curve is S-shaped with a plateau region, rather than linear. The top, plateau part of the curve above 60 mm Hg is in the blood P_{O_2} range found in the pulmonary capillaries where O_2 is binding to hemoglobin. For this part of the curve, the blood remains highly saturated with O_2 over a relatively large range of P_{O_2}. Even at P_{O_2} levels much higher than shown on the graph (P_{O_2} theoretically can go up to 760 mm Hg), only a small extra amount of O_2 will bind to hemoglobin. The steep part of the curve between 0 and 60 mm Hg is in the blood P_{O_2} range found in the capillaries in the rest of the body. For this part of the curve, small changes in P_{O_2} result in a large change in the amount of O_2 bound to hemoglobin.

Because the partial pressure of O_2 in alveolar air is about 100 mg Hg, most of the hemoglobin molecules are fully saturated in the blood leaving the alveolar networks, meaning that most of the hemoglobin molecules are bound to four O_2 molecules (see Figure 44.12a). The P_{O_2} of the O_2 in solution in the blood plasma has risen to approximately the same level as in the alveolar air, about 100 mm Hg. The blood has also changed color, reflecting the bright red color of oxygenated hemoglobin as compared with the darker red color of deoxygenated hemoglobin.

The oxygenated blood exiting from the alveoli collects in venules, which merge into the pulmonary veins leaving the lungs. These veins carry the blood to the heart, which pumps the blood through the systemic circulation to all parts of the body.

As the oxygenated blood enters the capillary networks of body tissues, it encounters regions in which the P_{O_2} in the interstitial fluid and body cells is lower than that in the blood, ranging from about 40 mm Hg downward to 20 mm Hg or less (see Figure 44.12b). As a result, O_2 diffuses from the blood plasma into the interstitial fluid, and from the fluid into body cells. As O_2 diffuses from the blood plasma into body tissues, it is replaced by O_2 released from hemoglobin.

Several factors contribute to the release of O_2 from hemoglobin, including increased acidity (lower pH) in active tissues. The acidity increases because oxidative reactions release CO_2, which combines with water to form carbonic acid (H_2CO_3). The lowered pH alters hemoglobin's conformation, reducing its affinity for O_2, which is released and used in cellular respiration.

The net diffusion of O_2 from blood to body cells continues until, by the time the blood leaves the capillary networks in the body tissues, much of the O_2 has been removed from hemoglobin. The blood, now with a P_{O_2} of 40 mm Hg or less, returns in veins to the heart, which pumps it through the pulmonary arteries to the lungs for another cycle of oxygenation.

Carbon Dioxide Diffuses down Concentration Gradients from Body Tissues into the Blood and Alveolar Air

The CO_2 produced by cellular oxidations diffuses from active cells into the interstitial fluid, where it reaches a partial pressure of about 46 mm Hg. Because this P_{CO_2} is higher than the 40 mm Hg P_{CO_2} in the blood entering the capillary networks of body tissues (see Figure 44.10), CO_2 diffuses from the interstitial fluid into the blood plasma **(Figure 44.13a).**

Some of the CO_2 remains in solution as a gas in the plasma. However, most of the CO_2, about 70%, combines with water to produce carbonic acid (H_2CO_3), which dissociates into bicarbonate (HCO_3^-) and H^+ ions. The reaction takes place both in the blood plasma and inside erythrocytes, where an enzyme, *carbonic anhydrase,* greatly speeds the reaction.

Most of the H^+ ions produced by the dissociation of carbonic acid combine with hemoglobin or with proteins in the plasma. The combination, by removing excess H^+ from the blood solution, *buffers* the

a. Hemoglobin saturation level in lungs

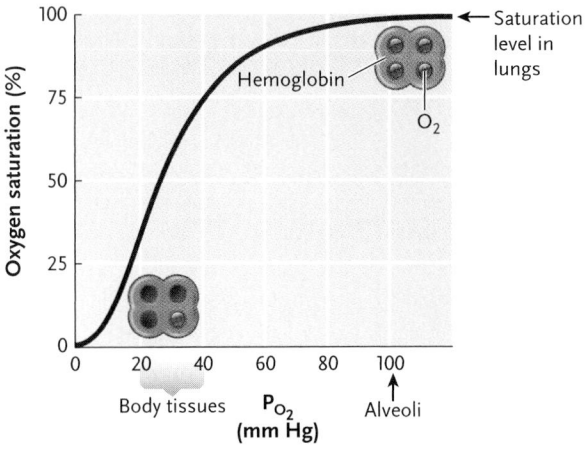

In the alveoli, in which the P_{O_2} is about 100 mm Hg and the pH is 7.4, most hemoglobin molecules are 100% saturated, meaning that almost all have bound four O_2 molecules.

b. Hemoglobin saturation range in body tissues

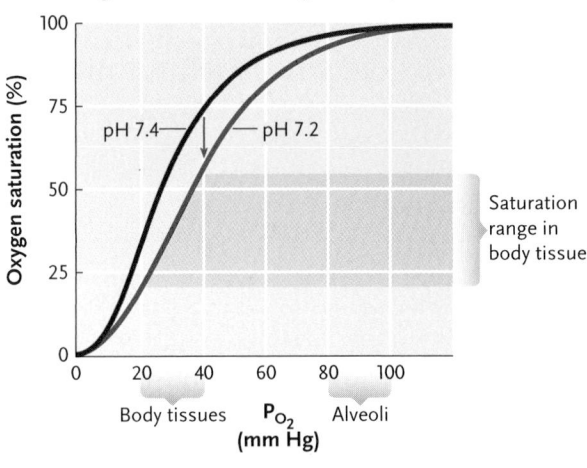

In the capillaries of body tissues, where the P_{O_2} varies between about 20 and 40 mm Hg depending on the level of metabolic activity and the pH is about 7.2, hemoglobin can hold less O_2. As a result, most hemoglobin molecules release two or three of their O_2 molecules to become between 25% and 50% saturated. Note that the drop in pH to 7.2 (red line) in active body tissues reduces the amount of O_2 hemoglobin can hold as compared with pH 7.4. The reduction in binding affinity at lower pH increases the amount of O_2 released in active tissues.

Figure 44.12
Hemoglobin–O_2 dissociation curves, which show the degree to which hemoglobin is saturated with O_2 at increasing P_{O_2}.

blood pH, helping to maintain it at its near-neutral set point of 7.4. (Buffers are discussed in Section 2.5.) The combined pathways absorbing CO_2 in the blood—solution in the plasma, conversion to bicarbonate, and combination with hemoglobin—help maintain the concentration gradient for gaseous CO_2 and keep its diffusion from the interstitial fluid into the blood at optimal levels.

The blood leaving the capillary networks of body tissues is collected in venules and veins and returned to

a. Body tissues

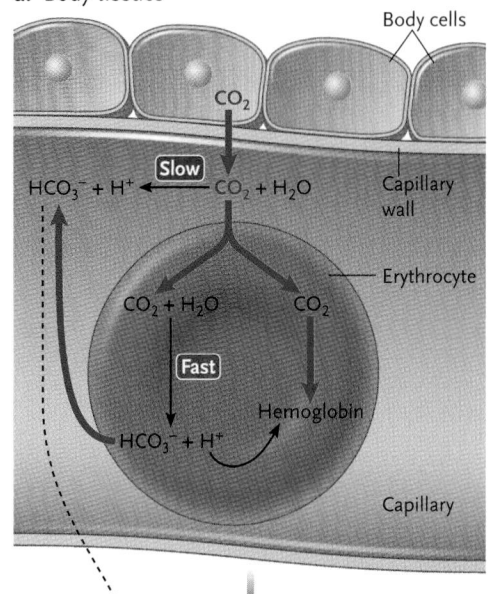

In body tissues, some of the CO_2 released into the blood combines with water in the blood plasma to form HCO_3^- and H^+. However, most of the CO_2 diffuses into erythrocytes, where some combines directly with hemoglobin and some combines with water to form HCO_3^- and H^+. The H^+ formed by this reaction combines with hemoglobin; the HCO_3^- is transported out of erythrocytes to add to the HCO_3^- in the blood plasma.

b. Lungs

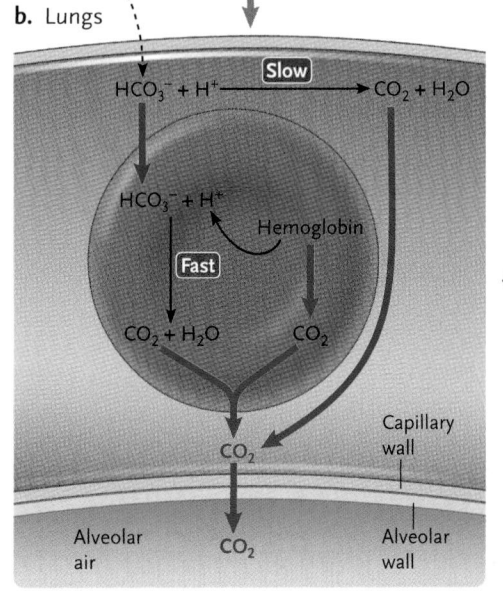

In the lungs, the reactions are reversed. Some of the HCO_3^- in the blood plasma combines with H^+ to form CO_2 and water. However, most of the HCO_3^- is transported into erythrocytes, where it combines with H^+ released from hemoglobin to form CO_2 and water. CO_2 is released from hemoglobin. The CO_2 diffuses from the erythrocytes and, with the CO_2 in the blood plasma, diffuses from the blood into the alveolar air.

Figure 44.13
The reactions occurring during the transfer of CO_2 from body tissues to alveolar air.

the heart, which pumps it through the pulmonary arteries into the lungs. As the blood enters the capillary networks surrounding the alveoli, the entire process of CO_2 uptake is reversed **(Figure 44.13b)**. The P_{CO_2} in the blood, now about 46 mm Hg, is higher than the P_{CO_2} in the alveolar air, about 40 mm Hg (see Figure 44.10). As a result, CO_2 diffuses from the blood into the air. The diminishing CO_2 concentrations in the plasma, along with the lower pH encountered in the lungs, promote the release of CO_2 from hemoglobin. As CO_2 diffuses away, bicarbonate ions in the blood combine with H^+ ions, forming carbonic acid molecules that break down into water and additional CO_2. This CO_2 adds to the quantities diffusing from the blood into the alveolar air. By the time the blood leaves the capillary networks in the lungs, its P_{CO_2} has been reduced to the same level as that of the alveolar air, about 40 mm Hg (see Figure 44.10).

Giving Hemoglobin and Myoglobin Air

Carbon monoxide, like O_2, combines directly with the heme group in the proteins myoglobin and hemoglobin. The heme group by itself, unassociated with the polypeptide chain of hemoglobin or myoglobin, has an affinity for CO some 10,000 times greater than for O_2. Combination of the heme group with the proteins reduces its affinity for CO to only 250 times greater than O_2 for hemoglobin and 30 times greater for myoglobin.

How does combination with the proteins reduce the affinity of the heme group for CO so effectively? One hypothesis is that the reduction depends on a stabilizing hydrogen bond between O_2 and an amino acid, histidine, located in a protein pocket formed by a fold in the amino acid chain near the iron atom of the heme group. According to this proposal, the hydrogen bond allows O_2 to displace a water molecule that occupies the pocket when neither

O_2 nor CO is bound. Because carbon monoxide cannot form this hydrogen bond, it displaces the water molecule less readily than O_2 does.

Research with myoglobin, conducted by John S. Olson and George Phillips and their colleagues at Rice University, supports the hydrogen-bond hypothesis. For their study, the team used a myoglobin gene isolated from a sperm whale. They chemically altered the DNA of the myoglobin gene so that one of seven other amino acids was substituted for the histidine in the pocket. A highly active bacterial promoter was added to the altered genes, which were then introduced one at a time into *Escherichia coli* bacteria. The bacteria expressed the genes, producing the altered myoglobin molecules in quantity, thereby providing the researchers with seven different forms of myoglobin to test for binding affinity for O_2.

The seven amino acids substituted for the histidine—glycine, alanine, leucine, phenylalanine, threonine, valine, and glutamine—are all nonpolar, or uncharged (histidine is positively charged), and thus unlikely to form a hydrogen bond with O_2. Further, none of these amino acids except glycine and glutamine should be able to hold a water molecule stably in the binding pocket. If the hydrogen-bond hypothesis is correct, the affinity of O_2 for the altered myoglobin should be greatly reduced in the mutant forms of the molecule.

Binding tests showed that the substitutions indeed reduced the affinity for O_2 by a factor that varied between 10 and 100 times. The greatest reduction was produced by the most nonpolar of the amino acids, leucine and phenylalanine. The smallest reduction was observed for glycine and glutamine. Thus the results strongly support the hydrogen-bond hypothesis.

Carbon monoxide (CO), a colorless, odorless gas produced when fuels are incompletely burned, as in automobile exhaust and in faulty furnaces, gas appliances, or space heaters, also binds to hemoglobin if it is inhaled into the lungs. It binds so strongly that it displaces O_2 from hemoglobin and drastically reduces the amount of O_2 carried to body tissues. If CO is inhaled in high quantity for even a few minutes, the reduction in oxygen delivered to the brain can lead to unconsciousness and brain damage. Sustained exposure leads to death by hypoxia (lack of oxygen). Because the brain regulates breathing based on CO_2 levels in blood rather than on O_2 levels, victims breathing CO can die from hypoxia without noticing anything amiss up to the point of unconsciousness. Interestingly, the combination of CO with hemoglobin, carboxyhemoglobin, is bright red. This has led to a myth often seen in textbooks that victims of CO poisoning turn a "classic cherry red" in color. However, this actually occurs in less than 2% of cases. *Insights from the Molecular Revolution* describes recent research testing the molecular basis for the binding of CO to hemoglobin and to *myoglobin,* a muscle protein with structure and properties similar to hemoglobin (myoglobin is discussed in Section 41.1).

STUDY BREAK

1. Explain the role of hemoglobin in gas exchange.
2. Why is carbon monoxide potentially lethal?

44.5 Respiration at High Altitudes and in Ocean Depths

This chapter's introduction described some challenges to respiration that arise when humans travel to high altitude. In this concluding section we look more closely at the effects of high altitude on respiration, along with the effects of increased pressures when humans and other mammals dive under water.

High Altitudes Reduce the P_{O_2} of Air Entering the Lungs

As altitude increases, atmospheric pressure decreases, and with it, the P_{O_2} of alveolar air and the concentration gradient of O_2 across the respiratory surface. At 20,000 feet, where most people become unconscious unless they have supplemental O_2, the dry air pressure is about 380 mm Hg and the P_{O_2} is only $380 \times 21/100 =$ about 80 mm Hg, half that at sea level.

Humans who travel from sea level to elevations of 6000 feet or more often experience one or more unpleasant symptoms, including headache, blurred vision, dizziness, nausea, and fatigue. However, after a few weeks at higher elevation, the body adjusts by increasing the number of erythrocytes in the blood. The increase in erythrocyte production is stimulated primarily by a hormone, *erythropoietin* (EPO), which the kidneys secrete in greater quantities in response to a drop in blood O_2. Erythrocyte production slows when people return to lower altitudes. However, the erythrocyte count remains high for several weeks after high-altitude exposure. Athletes often train at high altitudes to increase their erythrocyte count, with the idea that it will improve their stamina and endurance at lower altitudes.

People who live at high altitudes from childhood develop more permanent changes, including an increase in the number of alveoli and more extensive capillary networks in the lungs. These developments are retained if they move to lower altitudes.

Some mammals evolutionarily adapted to high altitudes show genetically determined changes that are present throughout life. For example, llamas, which customarily live at altitudes as high as 4500 m (14,000 feet), have hemoglobin molecules with greater affinity for O_2 than does the hemoglobin of sea level-dwelling mammals. As a result, hemoglobin becomes saturated with O_2 at the lower partial pressures typical of high altitudes. The same adaptation occurs in birds adapted to life at high altitudes, such as the bar-headed goose *(Anser indicus)*. These birds have been observed flying over the peaks of the Great Himalayas, which have altitudes greater than 6000 m.

Diving Mammals Are Adapted to Survive the High Partial Pressures of Gases at Extreme Depths

As a mammal such as a seal or whale dives from the surface, each additional 10 m of depth increases the partial pressure of dissolved gases by about 1 atmo-

sphere. Below about 25 m or so, the pressure becomes so great that the lungs collapse and cease to function. Adaptations of diving mammals such as seals and whales allow these animals to survive the extreme pressure and lack of lung function, in some species for over an hour at ocean depths of more than a mile.

Among these adaptations are more blood per unit of body weight and more red blood cells, which are stored in the spleen and released during a dive. In addition, the muscles of these animals contain much greater quantities of the O_2-binding protein myoglobin than the muscles of land-dwelling mammals do. In all, the adaptations pack about twice as much O_2 per kilogram of body weight into a seal, for example, than into a human.

Other adaptations decrease O_2 consumption during a deep and prolonged dive. The heart rate slows by about 80% to 90% and the circulation of blood to internal organs and muscles is cut by as much as 95%, leaving only the brain with its normal blood supply. Even though most of the blood supply to muscles is cut off, the muscles continue to work by shifting to anaerobic oxidation. The lactic acid produced by anaerobic respiration in the muscles is not released into the blood until the animal returns to the surface.

These combined adaptations give seals and whales an amazing ability to dive to great depths and remain under water for extended periods. Although average dives are on the order of 10 to 20 minutes, some sperm whales, tracked by sonar, have reached depths of 2250 m (more than 7000 feet) and remained under water for as long as 82 minutes.

STUDY BREAK

List the key adaptations that diving mammals use to survive at significant ocean depths.

Review

Go to ThomsonNOW™ at www.thomsonedu.com/login to access quizzing, animations, exercises, articles, and personalized homework help.

44.1 The Function of Gas Exchange

- Physiological respiration is the process by which animals exchange O_2 and CO_2 with the environment (Figure 44.1).

- The two primary operating features of gas exchange are the respiratory medium, either air or water, and the respiratory surface, a wetted epithelium over which gas exchange takes place.

- In some invertebrates, the skin serves as the respiratory surface. In other invertebrates and all vertebrates, gills or lungs provide the primary respiratory surface (Figure 44.2).

- Simple diffusion of molecules from regions of higher concentration to regions of lower concentration drives the exchange of gases across the respiratory surface. The area of the respiratory surface determines the total quantity of gases exchanged by diffusion.

- The concentration gradients of O_2 and CO_2 across the respiratory surface are kept at optimal levels by ventilation and perfusion.

Animation: Examples of respiratory surfaces

44.2 Adaptations for Respiration

- Animals breathing water keep the respiratory surface wetted by direct exposure to the environment. The high density and viscosity of water, and its relatively low O_2 content as compared with air, requires water-breathing animals to expend significant energy to keep their respiratory surface ventilated.

- Air is high in O_2 content, allowing air-breathing animals to maintain higher metabolic levels than water breathers. The low density and viscosity of air as compared with water allows air breathers to ventilate the respiratory surface with relatively little energy. To accommodate water loss by evaporation, lungs typically are invaginations of the body surface, allowing air to become saturated with water before it reaches the respiratory surface.

- Gills are evaginations of the body surface. Water moves over the gills by the beating of cilia or is pumped over the gills by contractions of body muscles (Figures 44.2b and 44.3a–d).

- Water moves in a one-way direction over the gills of sharks, bony fishes, and some crabs, allowing these animals to use countercurrent exchange to maximize the exchange of gases over the respiratory surface (Figure 44.4).

- Insects breathe by means of tracheae, air-conducting tubes that lead from the body surface and send branches to essentially every cell in the body. Gas exchange takes place in the fluid-filled tips at the ends of the branches (Figure 44.5).

- Lungs consist of an invaginated system of branches, folds, and pockets. They may be filled by positive pressure breathing, in which air is forced into the lungs by muscle contractions, or by negative pressure breathing, in which muscle contractions expand the lungs, lowering the air pressure inside them and allowing air to be pulled into the lungs (Figures 44.6 and 44.9).

Animation: Bony fish respiration

Animation: Frog respiration

Animation: Vertebrate lungs

Animation: Bird respiration

44.3 The Mammalian Respiratory System

- Air enters the respiratory system through the nose and mouth and passes through the pharynx, larynx, and trachea. The trachea divides into two bronchi, which lead to the lungs. Within the lungs, the bronchi branch into bronchioles, which lead into the alveoli, which are surrounded by dense networks of blood capillaries (Figure 44.8).

- Mammals inhale by a negative pressure mechanism. Air is exhaled passively by relaxation of the diaphragm and the external intercostal muscles between the ribs, and elastic recoil of the lungs. During deep and rapid breathing, the expulsion of air is forceful, driven by contraction of the internal intercostal muscles (Figure 44.9).

- The tidal volume of the lungs is the air moved in and out of the lungs during an inhalation and exhalation. The vital capacity is the total volume of air a person can inhale and exhale by breathing as deeply as possible. The air remaining in the lungs after as much air as possible is exhaled is the residual volume of the lungs.

- Breathing is controlled by a combination of local chemical controls and regulation by centers in the brain stem. These controls match the rate of air and blood flow in the lungs, and link the rate and depth of breathing to the body's requirements for O_2 uptake and CO_2 release (Figure 44.10).

- The basic rhythm of breathing is produced by interneurons in the medulla. When more rapid breathing is required, another group of interneurons in the medulla sends signals reinforcing inhalation and producing forceful exhalation. Two interneuron groups in the pons smooth and fine-tune breathing by stimulating or inhibiting the inhalation center in the medulla.

- Sensory receptors in the medulla, the carotid bodies, and the aortic bodies detect changes in the levels of O_2 and CO_2 in the blood and body fluids. The control centers in the medulla and pons adjust the rate and depth of breathing to compensate for changes in the blood gases.

    Animation: Human respiratory system

    Animation: Structure of an alveolus

    Animation: Respiratory cycle

    Animation: Changes in lung volume and pressure

    Animation: Partial pressure gradients

    Animation: Pressure-gradient changes during respiration

44.4 Mechanisms of Gas Exchange and Transport

- The partial pressure of O_2 is higher in the alveolar air than in the blood in the capillary networks surrounding the alveoli causing O_2 to diffuse from the alveolar air into the blood. Most of the O_2 entering the blood combines with hemoglobin inside erythrocytes (Figure 44.11).

- A hemoglobin molecule can combine with four O_2 molecules. The large quantities of O_2 that combine with hemoglobin maintain a large gradient in partial pressure between O_2 in the alveolar air and in the blood (Figure 44.12).

- In body tissues outside the lungs, the O_2 concentration in the interstitial fluid and body cells is lower than in the blood plasma. As a result, O_2 diffuses from the blood into the interstitial fluid, and from the fluid into body cells.

- The partial pressure of CO_2 is higher in the tissues than in the blood. About 10% of this CO_2 dissolves in the blood plasma; 70% is converted into H^+ and HCO_3^- (bicarbonate) ions. The remaining 20% combines with hemoglobin (Figures 44.11 and 44.13a).

- In the lungs, the partial pressure of CO_2 is higher in the blood than in the alveolar air. As a result, the reactions packing CO_2 into the blood are reversed, and the CO_2 is released from the blood into the alveolar air (Figure 44.13b).

    Animation: Globin and hemoglobin structure

44.5 Respiration at High Altitudes and in Ocean Depths

- In mammals that move to high altitudes, the number of red blood cells and the amount of hemoglobin per cell increase. These changes are reversed if the animals return to lower altitudes.

- Humans living at higher altitudes from birth develop more alveoli and capillary networks in the lungs.

- Some mammals and birds adapted to high altitudes have forms of hemoglobin with greater affinity for O_2, allowing saturation at the lower P_{O_2} typical of high altitudes.

- Marine mammals adapted to deep diving have a greater blood volume per unit of body weight, and their blood contains more red blood cells, with a higher hemoglobin content, than other mammals. Their muscles also contain more myoglobin than those of land mammals, allowing more O_2 to be stored in muscle tissues. During a dive, the heartbeat slows, and circulation is reduced to all parts of the body except the brain.

Questions

Self-Test Questions

1. Which of the following describes a respiratory medium?
    a. In the liver the rate of diffusion is high.
    b. In the brain CO_2 moves from the neurons to the blood.
    c. In the big toe O_2 moves from blood to tissues.
    d. Epithelial cells form thin surfaces in the lungs.
    e. A running brook provides O_2 to fish.

2. Which of the following describes a respiratory surface?
    a. a surface consisting of multiple layers of epithelial cells
    b. the exoskeleton of an insect
    c. the nasal passages of a mammal
    d. a thin surface consisting of a single layer of epithelial cells
    e. the outer membrane of a mitochondrion.

3. At the end of a basketball game, the opposing teams line up and file past each other and shake hands. This efficient exposure of the teams to each other is analogous to:
    a. countercurrent exchange of gases in fish gills and bird lungs.
    b. diffusion of O_2 from blood to cells in shark tissues.
    c. diffusion of CO_2 from cells to blood in crabs.

    d. utilization of O_2 in cells in insects.
    e. excretion of CO_2 from mammalian cells.

4. Tracheal systems are characterized by:
    a. closed circulatory tubes that move gases.
    b. spiracles that move gases between cells and body fluids.
    c. body movements that compress and expand air sacs to pump air.
    d. positive pressure breathing, which swallows air into the body.
    e. negative pressure breathing, which lowers air pressure at the respiratory surfaces.

5. The structures at which one third of O_2 in the atmosphere moves into the blood of humans are:
    a. alveoli.   d. tracheae.
    b. bronchi.   e. pharynges.
    c. bronchioles.

6. A speed skater is finishing his last lap. At this time:
    a. the diaphragm and rib muscles contract when he exhales.
    b. positive pressure brings air into his lungs.
    c. his lungs undergo an elastic recoil when he inhales.

d. his tidal volume is at vital capacity.

e. his residual volume momentarily reaches zero.

7. A teenager is frightened when she is about to step onto the stage but then remembers to breathe deeply and slowly as she faces the audience. What is occurring here?

a. Interneurons in the medulla cause the rib muscles to relax, followed later by stimulation and contraction of the intercostal muscles.

b. Signals from the pons override the initial brain stem stimuli.

c. The limbic system stabilized her emotional state, so there is no change in the mechanical movement of air.

d. The brain signals the aortic bodies in the carotid arteries to adjust the breathing rate.

e. Initial low CO_2 blood levels causing high pH are followed by increased CO_2 levels that lower pH.

8. Oxygen enters the blood in the lungs because relative to alveolar air:

a. the CO_2 concentration in the blood is high.

b. the CO_2 concentration in the blood is low.

c. the O_2 concentration in the blood is high.

d. the O_2 concentration in the blood is low.

e. the process is independent of gas concentrations in the blood.

9. The hemoglobin O_2 dissociation curve:

a. reflects about 50% saturation of hemoglobin in the alveoli.

b. shifts to the left when pH rises.

c. demonstrates that hemoglobin holds less O_2 when the pH is higher.

d. proves lack of dependence on CO_2 levels.

e. explains how hemoglobin can bind O_2 at high pH in the lungs and release it at lower pH in the tissues.

10. The majority of CO_2 in the blood:

a. is in the form of carbonic acid and bicarbonate ions

b. dissociates to add H^+ to the blood to raise its pH to 7.4.

c. has a lower P_{CO_2} than the P_{CO_2} in the alveolar air.

d. increases in the lung capillaries, which have a higher pH than the tissue capillaries.

e. can be displaced on the hemoglobin molecule by CO if CO is inhaled.

Questions for Discussion

1. Smoking has traditionally been considered to reduce the ability of athletes to run without becoming exhausted. Why might this be true?

2. People are occasionally found unconscious from breathing too much CO_2 (as from a charcoal heater placed indoors) or too much CO (as from auto exhaust in a closed garage). Would it be more advantageous to give pure O_2 to a person breathing too much CO_2 than simply moving the person to fresh air? Why? Which—pure O_2 or fresh air—would be best for a person unconscious from breathing CO? Why?

3. Hyperventilation, or overbreathing, is breathing faster or deeper than necessary to meet the body's needs. Hyperventilation reduces the CO_2 content of blood, but does not significantly increase the amount of O_2 available to tissues. Why might this be so?

Experimental Analysis

Propose a hypothesis for the effect of zero gravity on respiration, and design an experiment to test the hypothesis.

Evolution Link

From what you have learned in this chapter and in Chapter 30, do you think lungs evolved once, or on several occasions? Justify your answer.

How Would You Vote?

Tobacco is a worldwide threat to health and a profitable product for American companies. As tobacco use by its citizens declines, should the United States encourage international efforts to reduce tobacco use around the globe? Go to www.thomsonedu.com/login to investigate both sides of the issue and then vote.

An Alaskan brown bear *(Ursus arctos)* catching a sockeye salmon *(Oncorynchus nerka)*. Animals obtain nutrients by eating other organisms. Their digestive systems break down macromolecules in the food to produce simple organic molecules that are used for fuels and as building blocks for more complex molecules.

45 Animal Nutrition

WHY IT MATTERS

Invisible in the inky darkness, a deep-sea anglerfish (a member of the Order Lophiiformes) lies in wait, its gaping mouth lined with sharp teeth. Just above the mouth dangles a glowing lure suspended from a fishing-rod-like spine that projects from the fish's dorsal fin **(Figure 45.1).** The lure resembles a tiny fish; it even wiggles back and forth in imitation of swimming movements. Its glow is produced by bioluminescent bacteria that live symbiotically in the lure's tissues.

A hapless fish is attracted to the lure. As it comes within range, the anglerfish's mouth expands suddenly, creating a powerful suction that whips the prey in. The backward-angling fangs keep the prey from escaping. The strike takes only 6 ms (milliseconds), among the fastest of any known fishes.

Contractions of throat muscles send the prey to the anglerfish's stomach, which can expand to accommodate a meal as large as the anglerfish itself. In the fish's digestive tract, acids and enzymes dissolve the body of the prey, gradually breaking it into molecules small enough to be absorbed. In this function, the digestive system of the anglerfish is the same that of any other vertebrate, including humans—it provides nutrients that allow the animal to live. And the

Figure 45.1
A deep sea angler-fish, with its rod and lure lit and ready to attract prey.

anglerfish's adaptations for feeding, although bizarre to human sensibilities, are no more remarkable than those of many other animals.

Animal **nutrition**—which includes the processes by which food is ingested, digested, and absorbed into body cells and fluids—is the subject of this chapter. Our discussion begins with the basic categories of animal foods and **ingestion**, the feeding methods used to take food into the digestive cavity. Then we examine the process of **digestion**: the splitting of carbohydrates, proteins, lipids, and nucleic acids in foods into chemical subunits small enough to be absorbed into an animal's body fluids and cells. The chapter also presents the main structural and functional features of digestive systems, with special emphasis on humans and other mammals. The adaptations animals use to obtain and digest food are among their most strongly defining anatomical and functional characteristics.

45.1 Feeding and Nutrition

All organisms require sources of matter and energy for metabolism, homeostasis (maintaining their internal environment in a stable state; see Section 36.4), growth, and reproduction. For animals, meeting these nutritional requirements involves *feeding*, the uptake of food from the surroundings. Animals employ various feeding methods ranging from the ingestion of molecules in liquid solutions to eating entire organisms in one gulp. Once the food is ingested, digestive processes convert its molecules into absorbable subunits. In this section, we survey animal nutritional requirements and feeding methods as an introduction to animal digestive processes.

Animals Require both Organic and Inorganic Molecules for Nutrition

Plants and other photosynthesizers need only sunlight as an energy source and a supply of simple inorganic precursors such as water, carbon dioxide, and minerals to make all the organic molecules they require. In contrast, animals require a constant diet of organic molecules as a source of both energy and nutrients that they cannot make for themselves.

Animals are classified according to their sources of organic molecules. **Herbivores** such as antelopes, horses, bison, giraffes, kangaroos, manatees, and grasshoppers obtain organic molecules primarily by eating plants. **Carnivores**—cats, Tasmanian devils, penguins, sharks, and spiders, for example—primarily eat other animals. We say "primarily" because many herbivores eat animal matter at times, and a number of carnivores occasionally eat plant material. An antelope will eat insects as it grazes, and a grizzly bear, although primarily carnivorous, also eats berries. **Omnivores,** such as crows, cockroaches, and humans, eat both plants and animals and, in fact, any source of organic matter.

Organic molecules are the basis for two of the most fundamental processes of life: they act as fuels for oxidative reactions supplying energy and as building blocks for making complex biological molecules.

Energy supplies and requirements are usually described in terms of calories. A *calorie* (with a lowercase c) is the amount of heat energy required to raise 1 mL of pure water 1°C, from 14.5°C to 15.5°C. In animal nutrition, calories are usually considered in units of 1000 as kilocalories (kcal; the units listed on food packages in the United States) or Calories (with an uppercase C). One Calorie thus equals 1000 calories. Carbohydrates contain about 4.2 kcal per gram, fats about 9.5 kcal per gram, and proteins about 4.1 kcal per gram. At rest, a human female of average size expends about 1300 to 1500 kcal per day, and a human male about 1600 to 1800 kcal per day. Exercise and physical labor can increase these daily totals.

Carbohydrates and fats are the primary organic molecules used as fuels. Animals whose intake of organic fuels is inadequate, or whose assimilation of such fuels is abnormal, suffer from **undernutrition.** Undernutrition is a form of **malnutrition,** which is a condition resulting from an improper diet. **Overnutrition,** the condition caused by excessive intake of specific nutrients, is another main type of malnutrition.

An animal suffering from undernutrition essentially is starving for one or more nutrients, taking in fewer calories than needed for daily activities. Animals with chronic undernutrition lose weight because they have to use molecules of their own bodies as fuels. Mammals use stored fats and glycogen (animal starch) first. Once those stores have been used up, proteins are metabolized as fuels. The use of proteins as fuels leads to muscle wastage and, in the long term, to organ and brain damage and, therefore, eventually to death.

Organic molecules also serve as building blocks for carbohydrates, lipids, proteins, and nucleic acids. Animals can synthesize many of the organic molecules that they do not obtain directly in the diet by converting one type of building block into another. Typically, however, they cannot make certain amino acids and fatty acids from other organic molecules. These required organic building blocks are called **essential amino acids** and **essential fatty acids** because they must be obtained in the diet. If they are not obtained in the diet over a period of time, there may be serious consequences. For instance, protein synthesis cannot continue unless all 20 amino acids are present. In the absence of essential amino acids in the diet, the animal would have to break down its own proteins to provide them for new protein synthesis.

Animals must also take in **vitamins**, organic molecules required in small quantities that the animal cannot synthesize for itself. Many vitamins are *coenzymes,* nonprotein organic subunits associated with enzymes that assist in enzymatic catalysis (see Section 4.4).

Individual species differ in the vitamins and essential amino acids and fatty acids they require. Various species also have differing dietary requirements for inorganic elements such as calcium, iron, and magnesium. These required inorganic elements are known collectively as **essential minerals.**

The essential amino acids, fatty acids, vitamins, and minerals are known collectively as an animal's **essential nutrients.** The list of essential nutrients differs from animal to animal. For domesticated animals, this means that specific feed formulations must be given to each type of animal. For instance, the essential nutrients for cats and dogs are different, which is why there are specific cat foods and dog foods, and why it is does not make good sense, nutritionally speaking, to feed cats and dogs human food.

Animals Obtain Nutrients in Fluid, Particle, or Bulk Form

All animals display adaptations that allow them to obtain the food they need in particular environments. Although these adaptations are amazingly varied, animals can be classified into one of four groups according to overall feeding methods and the physical state of the organic molecules they consume. These four groups are fluid feeders, suspension feeders, deposit feeders, and bulk feeders **(Figure 45.2).**

a. Fluid feeder

b. Suspension feeder

Baleen

c. Deposit feeder

d. Bulk feeder

Figure 45.2
Grouping of animals with respect to overall feeding methods and the physical state of the organic molecules they consume. **(a)** Fluid feeders, exemplified by a hummingbird, which obtains nectar from deep within a flower using its long bill and tongue. **(b)** Suspension feeders, exemplified by the northern right whale *(Balaena glacialis),* which gulps tons of water containing plankton into its mouth, pushes the water out through the sievelike baleen, and swallows the remaining plankton. **(c)** Deposit feeders, exemplified by a fiddler crab *(Uca* species), which sifts edible material from the sediment it takes into its mouth. **(d)** Bulk feeders, exemplified in an extreme way by a python, which ingests its prey (here, a gazelle) whole. Elastic ligaments connecting the jaws allow the snake's mouth to open wide enough to swallow large prey.

Fluid feeders obtain nourishment by ingesting liquids that contain organic molecules in solution. Among the invertebrates, aphids, mosquitoes, leeches, and spiders are examples of fluid feeders. Vertebrate fluid feeders include birds such as hummingbirds (see Figure 45.2a), which feed on flower nectar; parasitic fishes such as lampreys, which feed on body fluids of their hosts; and some bats, which feed on nectar or blood. Many fluid feeders have mouthparts specialized to reach the source of their nourishment. For example, mosquitoes, bedbugs, and aphids have needlelike mouthparts that pierce body surfaces. Nectar-feeding birds and bats have long tongues that can extend deep within flowers. Some fluid feeders use enzymes or other chemicals to liquefy their food or to keep it liquid during feeding. For example, spiders inject digestive enzymes that liquefy tissues inside their victim and then suck up the liquid. The saliva of mosquitoes, leeches, and vampire bats includes an anticoagulant that keeps blood in liquid form during a feeding by inhibiting the clotting reaction.

Suspension feeders ingest small organisms suspended in water, such as bacteria, protozoa, algae, and small crustaceans, or fragments of these organisms. Among the suspension feeders are aquatic invertebrates such as clams, mussels, and barnacles; many fishes; and even some birds and whales (see Figure 45.2b). These animals strain food particles suspended in water through a body structure covered with sticky mucus or through a filtering network of bristles, hairs, or other body parts. The trapped particles are then funneled into the animal's mouth. Bits of organic matter are trapped by the gills of bivalves such as clams and oysters, and plankton is filtered from water by the sievelike fringes of horny fiber hanging in the mouths of baleen whales (see Figure 45.2b).

Deposit feeders pick up or scrape particles of organic matter from solid material they live in or on. Earthworms are deposit feeders that eat their way through soil, taking the soil into their mouth and digesting and absorbing any organic material it contains. Some burrowing mollusks and tube-dwelling polychaete worms use body appendages to gather organic deposits from the sand or mud around them. Mucus on the appendages traps the organic material, and cilia move it to the mouth. The fiddler crab (*Uca* species) is also a deposit feeder (see Figure 45.2c). This animal has claws of markedly different sizes. The small claw picks up sediment and moves it to the mouth where the contents are sifted. The edible parts of the sediment are ingested, and the rest is put back on the sediment as a small ball. The feeding-related movement of the small claw over the larger claw looks like the crab is playing the large claw like a fiddle and hence gives the crab its name.

Bulk feeders are animals that consume sizeable food items whole or in large chunks. Most mammals eat this way, as do reptiles, most birds and fishes, and adult amphibians. Depending on the animal, adaptations for bulk feeding include teeth for tearing or chewing, claws and beaks for holding large food items, and jaws that are hinged or otherwise modified to permit a food mass to enter the mouth (see Figure 45.2d).

We now take up the processes by which animals, having fed, undertake the mechanical and chemical breakdown of food into absorbable molecular subunits.

STUDY BREAK

1. What are carnivores, herbivores, and omnivores?
2. What are essential nutrients, and are they the same for all animals?
3. What is the difference between deposit feeders and suspension feeders?

45.2 Digestive Processes

Digestive processes break food molecules into molecular subunits that can be absorbed into body fluids and cells. The breakdown occurs by **enzymatic hydrolysis,** in which chemical bonds are broken by the addition of H^+ and OH^-, the components of a molecule of water (see Section 3.2). Specific enzymes speed these reactions: *amylases* catalyze the hydrolysis of starches, *lipases* break down fats and other lipids, *proteases* hydrolyze proteins, and *nucleases* digest nucleic acids. Depending on the animal, the enzymatic hydrolysis of food molecules may take place inside or outside the body cells.

Intracellular Digestion Takes Place within Cells; Extracellular Digestion Occurs in an Internal Pouch or Tube

In **intracellular digestion,** cells take in food particles by endocytosis (described in Section 6.5). Inside the cell, the endocytic vesicle containing the food particles fuses with a lysosome, a vesicle containing hydrolytic enzymes. The molecular subunits produced by the hydrolysis pass from the vesicle to the cytosol. Any undigested material remaining in the vesicle is released to the outside of the cell by exocytosis (also discussed in Section 6.5). Only a few animals, primarily sponges and some cnidarians, break down food exclusively by intracellular digestion. In sponges, water containing particles of organic matter and microorganisms enter the animal's saclike body through pores in the body wall (see Figure 29.8). In the body cavity, individual *choanocytes* (collar cells) lining the body wall trap the food particles, take them in by endocytosis, and transport them to amoeboid cells, which digest them intracellularly.

Extracellular digestion takes place outside body cells, in a pouch or tube enclosed within the body. Epithelial cells lining the pouch or tube secrete enzymes that digest the food. Processing food in specialized compartments in this way prevents the animal from digesting its own body tissues.

Most invertebrates and all vertebrates digest food primarily by extracellular digestion. From an adaptive standpoint, extracellular digestion greatly expands the range of available food sources by allowing animals to digest much larger food items than single cells can take in. Extracellular digestion also allows animals to eat large batches of food, which can be stored and digested while the animal continues other activities.

Saclike Digestive Systems Have a Single Opening through Which Food Enters and Undigested Matter Exits

Some animals, including flatworms and cnidarians such as hydras, corals, and sea anemones, have a saclike digestive system with a single opening, a mouth, that serves both as the entrance for food and the exit for undigested material. In some of these animals, such as the flatworm *Dugesia,* the digestive cavity is called a **gastrovascular cavity** because it contributes to circulation as well as digestion. Food is brought to the mouth by a protrusible **pharynx** (a throat that can be stuck out) and then enters the gastrovascular cavity **(Figure 45.3)**; glands in the cavity wall secrete enzymes that begin the digestive process. Cells lining the cavity then take up the partially digested material by endocytosis and complete digestion intracellularly. Undigested matter is released to the outside through the pharynx and mouth.

Digestive Tubes Typically Process Nutrients in Five Successive Steps

Most invertebrates and all vertebrates have a tubelike digestive system with two openings that form a separate mouth and anus; the digestive contents move in one direction through specialized regions of the tube, from the mouth to the anus. This type of digestive system is called a **digestive tube**, *gut, alimentary canal, digestive tract,* or *gastrointestinal (GI) tract.* Structurally, the inside of the digestive tube—called the **lumen**—is external to all body tissues. In other words, the lumen is *outside* of the body.

In most animals with a digestive tube, digestion occurs in five successive steps, with each step taking place in a specialized region of the tube. The tube thus acts as a sort of biological disassembly line, with food entering at one end and passing through as many as five areas in which food processing occurs.

1. **Mechanical processing:** Chewing, grinding, and tearing breaks food chunks into smaller pieces, increasing their mobility and the surface area exposed to digestive enzymes.
2. **Secretion of enzymes and other digestive aids:** Enzymes and other substances that aid the process of digestion, such as acids, emulsifiers, and lubricating mucus, are released into the tube.
3. **Enzymatic hydrolysis:** Food molecules are broken down through enzyme-catalyzed reactions into absorbable molecular subunits.
4. **Absorption:** The molecular subunits are absorbed from the digestive contents into body fluids and cells.
5. **Elimination:** Undigested materials are expelled through the anus.

The material being digested is pushed along by muscular contractions of the wall of the digestive tube. During its progress through the tube, the digestive contents may be stored temporarily at one or more locations. The storage allows animals to take in larger quantities of food than they can process immediately, so that feedings can be spaced in time rather than continuous.

Digestion in an Annelid. The earthworm (genus *Lumbricus,* **Figure 45.4a**) is a deposit feeder. As it burrows, it pushes soil particles into its mouth. The particles pass from the mouth through a connecting passage, the **esophagus**, into the **crop**, an enlargement of the digestive tube where the contents are stored and mixed with lubricating mucus. This mixture enters the **gizzard**, which contains grains of sand, and is ground into fine particles by muscular contractions of the wall. The pulverized mixture then enters a long **intestine**, where the organic matter is hydrolyzed by enzymes secreted into the digestive tube. As muscular contractions of the intestinal wall move the mixture along, cells lining the intestine absorb the molecular subunits produced by digestion. The absorptive surface of the intestine is increased by folds of the wall called *typhlosoles*. At the end of the intestine, the undigested residue is expelled through the anus.

Digestion in an Insect. Herbivorous insects such as the grasshopper **(Figure 45.4b)** tear leaves and other plant parts into small particles with hard external

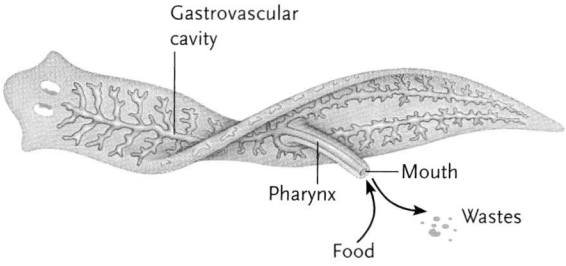

Figure 45.3

The digestive system of the flatworm *Dugesia*. The gastrovascular cavity (in blue) is a blind sac, with one opening to the exterior through which food is ingested and wastes are expelled.

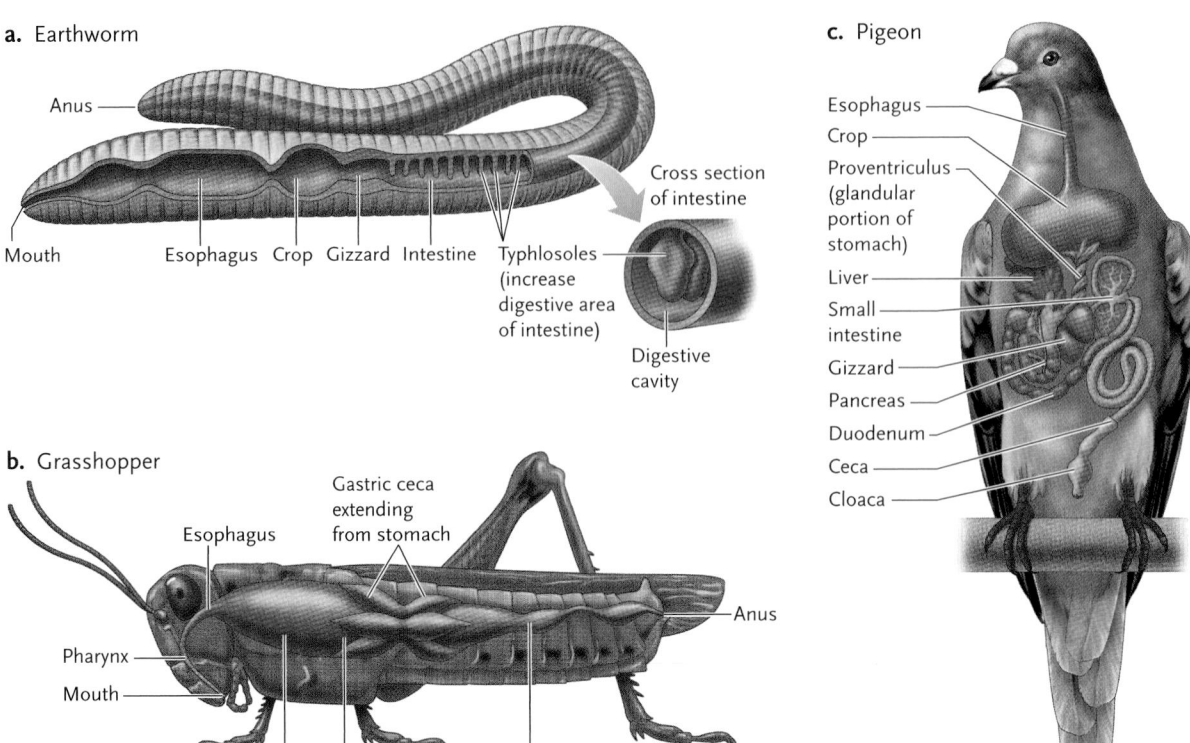

a. Earthworm

Anus

Mouth Esophagus Crop Gizzard Intestine Typhlosoles (increase digestive area of intestine)

Cross section of intestine

Digestive cavity

b. Grasshopper

Gastric ceca extending from stomach

Esophagus

Pharynx

Mouth

Crop Gizzard Intestine

Anus

c. Pigeon

Esophagus
Crop
Proventriculus (glandular portion of stomach)
Liver
Small intestine
Gizzard
Pancreas
Duodenum
Ceca
Cloaca

Figure 45.4
The digestive systems of an annelid, the earthworm **(a)**, an insect, the grasshopper **(b)**, and a bird, the pigeon *(Columba)* **(c)**.

mouth parts. From the mouth, the food particles pass through the pharynx, where salivary secretions moisten the mixture before it enters the esophagus and passes into the crop. These secretions begin the process of chemical digestion. From the crop, the food mass enters the gizzard, which grinds it into smaller pieces. These food particles enter the **stomach**, in which food is stored and digestion begins. Insect stomachs have saclike outgrowths, the *gastric ceca* (*caecus* = blind), where enzymes hydrolyze the digestive contents; the products of digestion are absorbed through the walls of the ceca. The undigested contents then move into the intestine for further digestion and absorption. At the end of the intestine, water is absorbed from the undigested matter and the remnants are expelled through the anus. The digestive systems of other arthropods are similar to the insect system.

Digestion in a Bird. A pigeon (*Columba,* **Figure 45.4c**) picks up seeds with its bill. The bird's tongue moves the seeds into its mouth, where they are moistened by mucus-filled saliva and swallowed whole (birds have no teeth). The seeds then pass through the pharynx into the esophagus. (In some cases, birds crack open seeds with their bills and ingest seed kernels in a similar fashion.) The anterior end of the esophagus is tubelike; at the posterior end is the pouchlike crop, in which the bird can store large quantities of food. From the crop, the food passes into the anterior glandular portion of the stomach, called the *proventriculus,* which secretes digestive enzymes and acids. The posterior end is the gizzard, in which the seeds are ground into

fine particles, aided by ingested bits of sand and rock. The food particles are released into the intestine, where the liver secretes bile and the pancreas adds digestive enzymes. The molecular subunits produced by enzymatic digestion are absorbed as the mixture passes along the intestine, and the undigested residues are expelled through the anus.

Many of the structures of the pigeon's digestive system, including the mouth, pharynx, esophagus, stomach, intestine, liver, and pancreas, occur in almost all vertebrates.

STUDY BREAK

1. Distinguish between extracellular digestion and intracellular digestion.
2. What are the five steps of food processing in a digestive tube?

45.3 Digestion in Humans and Other Mammals

Mammals digest foods using the same five steps as other animals with a digestive tube: mechanical processing, secretion of enzymes and other digestive aids, enzymatic hydrolysis, absorption of molecular subunits, and elimination. The mammalian digestive system is a series of specialized digestive regions that perform these steps, including the mouth, pharynx,

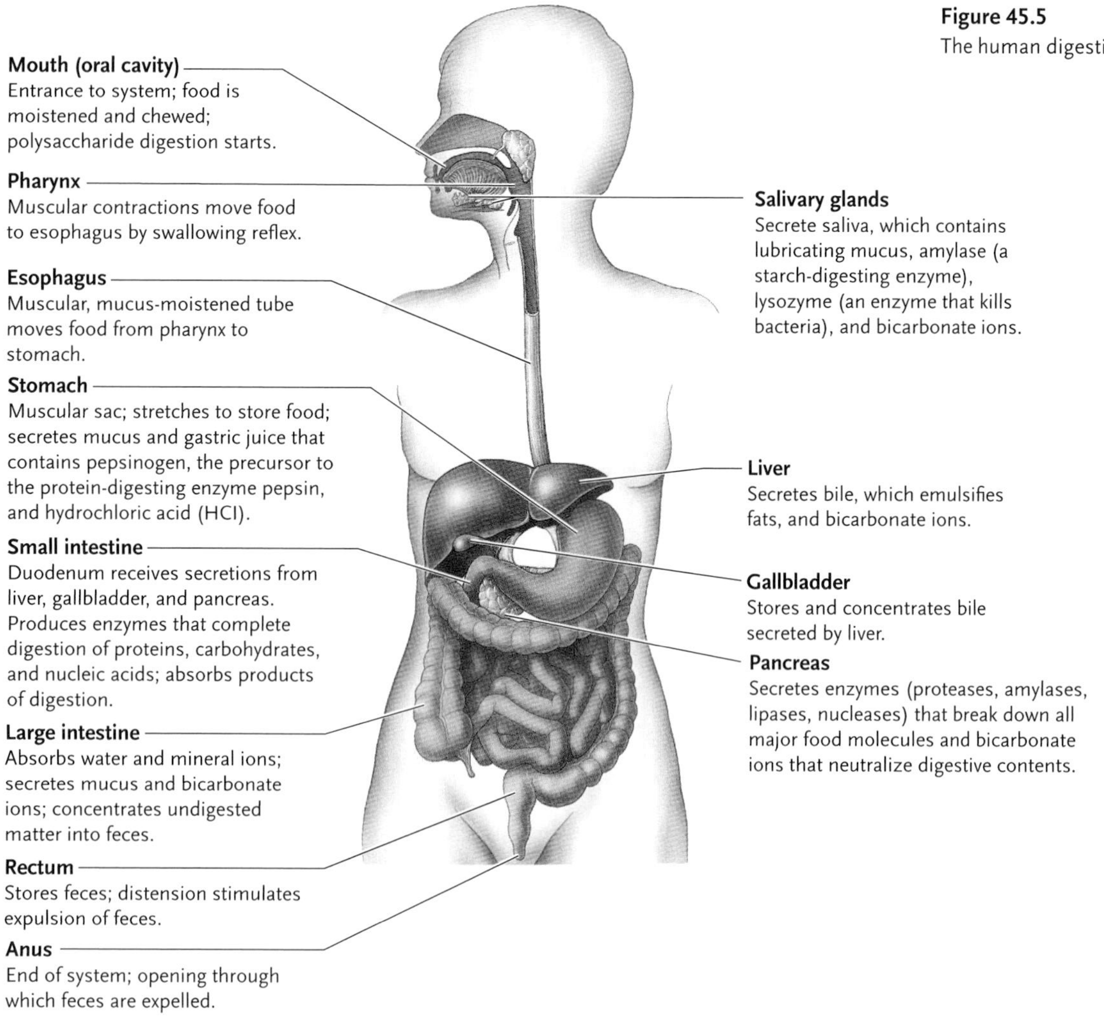

Figure 45.5
The human digestive system.

Mouth (oral cavity)
Entrance to system; food is moistened and chewed; polysaccharide digestion starts.

Pharynx
Muscular contractions move food to esophagus by swallowing reflex.

Esophagus
Muscular, mucus-moistened tube moves food from pharynx to stomach.

Stomach
Muscular sac; stretches to store food; secretes mucus and gastric juice that contains pepsinogen, the precursor to the protein-digesting enzyme pepsin, and hydrochloric acid (HCl).

Small intestine
Duodenum receives secretions from liver, gallbladder, and pancreas. Produces enzymes that complete digestion of proteins, carbohydrates, and nucleic acids; absorbs products of digestion.

Large intestine
Absorbs water and mineral ions; secretes mucus and bicarbonate ions; concentrates undigested matter into feces.

Rectum
Stores feces; distension stimulates expulsion of feces.

Anus
End of system; opening through which feces are expelled.

Salivary glands
Secrete saliva, which contains lubricating mucus, amylase (a starch-digesting enzyme), lysozyme (an enzyme that kills bacteria), and bicarbonate ions.

Liver
Secretes bile, which emulsifies fats, and bicarbonate ions.

Gallbladder
Stores and concentrates bile secreted by liver.

Pancreas
Secretes enzymes (proteases, amylases, lipases, nucleases) that break down all major food molecules and bicarbonate ions that neutralize digestive contents.

esophagus, stomach, small and large intestines, and anus **(Figure 45.5)**. These regions are under the control of the nervous and endocrine systems.

Humans Require Specific Essential Amino Acids, Fatty Acids, Vitamins, and Minerals in Their Diet

The human digestive system meets our basic needs for fuel molecules and for a wide range of nutrients, including the molecular building blocks of carbohydrates, lipids, proteins, and nucleic acids. If the diet is adequate, the digestive system also absorbs the essential nutrients—the amino acids, fatty acids, vitamins, and minerals that cannot be synthesized within our bodies.

Essential Amino Acids and Fatty Acids. There are eight essential amino acids for adult humans: lysine, tryptophan, phenylalanine, threonine, valine, methionine, leucine, and isoleucine. Infants and young children also require histidine. The proteins in fish, meat, egg whites, milk, and cheese supply all the essential amino acids, provided those foods are eaten in adequate quantities. In contrast, the proteins of many plants are deficient in one or more of the essential amino acids.

Corn, for example, contains inadequate amounts of lysine, and beans contain little methionine. Vegetarians, and especially vegans who eat a diet with no animal-derived nutrients, must choose their foods carefully to obtain all of the essential amino acids **(Figure 45.6)**. Such diets typically include combinations of foods, each of which provides some amino acids, and that together contain all of the essential amino acids. An example is including in the diet rice or corn (low in lysine but high in methionine) with legumes such as lentils or with soybeans, perhaps in the form of tofu (low in methionine but high in lysine).

Figure 45.6
Obtaining essential amino acids in a human vegetarian diet.

Eight essential amino acids

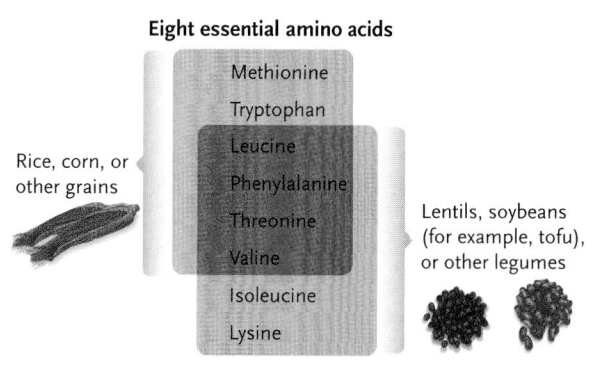

Rice, corn, or other grains

Methionine
Tryptophan
Leucine
Phenylalanine
Threonine
Valine
Isoleucine
Lysine

Lentils, soybeans (for example, tofu), or other legumes

If the diet lacks one or more essential amino acids, many enzymes and other proteins cannot be synthesized in sufficient quantities. The resulting protein deficiency is most damaging to the young, who must rapidly synthesize proteins for development and growth. Even mild protein starvation during pregnancy or for some months after birth can retard a child's mental and physical development.

Only two fatty acids, linoleic acid and linolenic acid, are essential in the human diet. Both are required for synthesis of phospholipids forming parts of biological membranes and certain hormones. Because almost all foods contain these fatty acids, most people have no problem obtaining them. However, people on a low-fat diet deficient in linoleic acid and linolenic acid are at serious risk for developing coronary heart disease. That is, there is an inverse correlation between the concentration of these essential fatty acids in the diet and the incidence of coronary heart disease. This is illustrated in the case of Hindu vegetarians from India. Their diet consists mainly of low-fat grains and legumes—clearly a low-fat diet—yet their rate of coronary heart disease is higher than that in the United States and Europe, where dietary fat content is higher.

Vitamins. Humans require 13 known vitamins in their diet. Many metabolic reactions depend on vitamins, and the absence of one vitamin can affect the functions of the others. These essential nutrients fall into two classes: **water-soluble** (hydrophilic) **vitamins** and **fat-soluble** (hydrophobic) **vitamins** (summarized in **Table 45.1**). The body stores excess fat-soluble vitamins in adipose tissues, but any amount of water-soluble vitamins above daily nutritional requirements is excreted in the urine. Thus, meeting the daily minimum requirements of water-soluble vitamins is critical. The body can tap its stores of fat-soluble vitamins to meet daily requirements; however, these stores are quickly depleted, so that prolonged deficiencies of the fat-soluble vitamins also become critical to health.

Most of us get all the vitamins we need through a normal and varied diet that includes meats, fish, eggs, cheese, and vegetables. Vitamin supplements are usually necessary only for strict vegetarians, newborns, the elderly, and individuals who are taking medication that affects the body's uptake of nutrients.

Vitamin D (calciferol) differs from other essential vitamins because humans can actually synthesize it themselves, through the action of ultraviolet light on lipids in the skin. However, many people are not exposed to enough sunlight to make sufficient quantities of the vitamin, and so must rely on dietary sources. And, although we cannot make vitamin K, much of our requirement for this vitamin is supplied through the metabolic activity of bacteria living in our large intestine. Vitamin K deficiency, therefore, is exceedingly rare in healthy persons. Vitamin K plays a role in blood clotting, so individuals with vitamin K deficiency will bruise easily and show increased blood clotting times. Vitamin K deficiency can be caused in persons on long-term antibiotic therapy because the antibiotics kill intestinal bacteria.

Other mammals have essentially the same vitamin requirements as humans, with some differences. For example, most other mammals, with the exception of primates, guinea pigs, and fruit bats, can synthesize vitamin C. So far as is known, no animal can synthesize B vitamins, but ruminants such as cattle and deer are supplied with these vitamins by microorganisms that live in the digestive tract (see Section 45.5).

Minerals. Many minerals are essential in the human diet (**Table 45.2**). Some of them, called **macronutrients**, are required in amounts ranging from 50 mg to more than a gram per day; others, such as zinc, are **micronutrients**, or **trace elements**, required only in small amounts, some less than 1 mg per day. All of the minerals, although listed as elements, are ingested as compounds or as ions in solution.

A normal and varied diet supplies adequate amounts of the essential minerals. Supplements may be required for those on a strict vegetarian diet, the very young, and the aged. Overdoses of some minerals can cause problems; ingesting excess iron, for example, has been linked to liver, heart, and blood vessel damage; too much sodium can lead to elevated blood pressure and excess water retention in tissues.

We now turn to the structures that extract nutrients from ingested foods. We begin with a survey of digestive structures common to all vertebrates.

Four Major Layers of the Gut Each Have Specialized Functions in Digestion

The wall of the gut in mammals and other vertebrates contains four major layers, each with specialized functions. These layers are shown for the stomach in **Figure 45.7**.

1. The **mucosa**, which contains epithelial and glandular cells, lines the inside of the gut. The epithelial cells absorb digested nutrients and seal off the digestive contents from body fluids. The glandular cells secrete enzymes, aids to digestion such as lubricating mucus, and substances that adjust the pH of the digestive contents.
2. The **submucosa** is a thick layer of elastic connective tissue that contains neuron networks and blood and lymph vessels. The neuron networks provide local control of digestive activity and carry signals between the gut and the central nervous system. The lymph vessels carry absorbed lipids to other parts of the body.

Table 45.1 | **Vitamins: Sources, Functions, and Effects of Deficiencies in Humans**

Vitamin	Common Sources	Main Functions	Effects of Chronic Deficiency
Fat-Soluble Vitamins			
A (retinol)	Yellow fruits, yellow or green leafy vegetables; also in fortified milk, egg yolk, fish liver	Used in synthesis of visual pigments, bone, teeth; maintains epithelial tissues	Dry, scaly skin; lowered resistance to infections; night blindness
D (calciferol)	Fish liver oils, egg yolk, fortified milk; manufactured when body exposed to sunshine	Promotes bone growth and mineralization; enhances calcium absorption from gut	Bone deformities (rickets) in children; bone softening in adults
E (tocopherol)	Whole grains, leafy green vegetables, vegetable oils	Antioxidant; helps maintain cell membrane and red blood cells	Lysis of red blood cells; nerve damage
K (napthoquinone)	Intestinal bacteria; also in green leafy vegetables, cabbage	Promotes synthesis of blood clotting protein by liver	Abnormal blood clotting, severe bleeding (hemorrhaging)
Water-Soluble Vitamins			
B_1 (thiamine)	Whole grains, green leafy vegetables, legumes, lean meats, eggs, nuts	Connective tissue formation; folate utilization; coenzyme forming part of enzyme in oxidative reactions	Beriberi; water retention in tissues; tingling sensations; heart changes; poor coordination
B_2 (riboflavin)	Whole grains, poultry, fish, egg white, milk, lean meat	Coenzyme	Skin lesions
Niacin	Green leafy vegetables, potatoes, peanuts, poultry, fish, pork, beef	Coenzyme of oxidative phosphorylation	Sensitivity to light; contributes to pellagra (damage to skin, gut, nervous system, etc.)
B_6 (pyridoxine)	Spinach, whole grains, tomatoes, potatoes, meats	Coenzyme in amino acid and fatty acid metabolism	Skin, muscle, and nerve damage
Pantothenic acid	In many foods (meats, yeast, egg yolk especially)	Coenzyme in carbohydrate and fat oxidation; fatty acid and steroid synthesis	Fatigue, tingling in hands, headaches, nausea
Folic acid	Dark green vegetables, whole grains, yeast, lean meats; intestinal bacteria produce some folate	Coenzyme in nucleic acid and amino acid metabolism; promotes red blood cell formation	Anemia; inflamed tongue; diarrhea; impaired growth; mental disorders; neural tube defects and low birth weight in newborns
B_{12} (cobalamin)	Poultry, fish, eggs, red meat, dairy foods (not butter)	Coenzyme in nucleic acid metabolism; necessary for red blood cell formation	Pernicious anemia; impaired nerve function
Biotin	Legumes, egg yolk; colon bacteria produce some	Coenzyme in fat and glycogen formation, and amino acid metabolism	Scaly skin (dermatitis), sore tongue, brittle hair, depression, weakness
C (ascorbic acid)	Fruits and vegetables, especially citrus, berries, cantaloupe, cabbage, broccoli, green pepper	Vital for collagen synthesis; antioxidant	Scurvy, delayed wound healing, impaired immunity

3. In most regions of the gut, the **muscularis** is formed by two smooth muscle layers, a *circular layer* that constricts the diameter of the gut when it contracts and a *longitudinal layer* that shortens and widens the gut. The stomach also has an *oblique layer* running diagonally around its wall. The circular and longitudinal muscle layers of the muscularis coordinate their activities to push the digestive contents through the gut **(Figure 45.8).** In this mechanism, called **peristalsis**, the circular muscle layer contracts in a wave that passes along the gut, constricting the gut and pushing the digestive contents onward. Just in front of the advancing constriction, the longitudinal layer contracts, shortening and expanding the tube and making space for the contents to advance.

4. The outermost gut layer, the **serosa**, consists of connective tissue that secretes an aqueous, slippery fluid. The fluid lubricates the areas between the digestive organs and other organs, reducing friction between them as they move together as a result of muscle movement. Along much of the length of the digestive system, the serosa is continuous with the *mesentery,* a tissue that suspends the digestive system from the inner wall of the abdominal cavity.

Table 45.2 | **Major Minerals: Sources, Functions, and Effects of Deficiencies in Humans**

Mineral	Sources	Functions	Effects of Deficiencies
Calcium (Ca)	Dairy products, leafy green vegetables, legumes, whole grains, nuts	Bone, tooth formation; blood clotting; neural and muscle action	Stunted growth; diminished bone mass (osteoporosis)
Chlorine (Cl)	Table salt, meat, eggs, dairy products	HCl formation in stomach, contributes to body's acid-base balance; neural function, water balance	Muscle cramps; impaired growth; poor appetite
Chromium (Cr)*	Meat, liver, cheese, whole grains, brewer's yeast, peanuts	Roles in carbohydrate metabolism	Impaired response to insulin; increases risk of type 2 diabetes mellitus
Cobalt (Co)*	Meat, liver, fish, milk	Constituent of vitamin B_{12} (required for red blood cell maturation)	Same as for vitamin B_{12} (see Table 45.1)
Copper (Cu)*	Nuts, legumes, seafood, drinking water, whole grains, nuts	Used in synthesis of melanin, hemoglobin, and some electron transport chain components in mitochondria	Anemia, changes in bone and blood vessels
Fluorine (F)*	Fluoridated water, tea, seafood	Bone, tooth maintenance	Tooth decay
Iodine (I)*	Marine fish, shellfish, iodized salt	Thyroid hormone formation	Goiter (enlarged thyroid), with metabolic disorders
Iron (Fe)	Liver, whole grains, green leafy vegetables, legumes, nuts, eggs, lean meat, molasses, dried fruit, shellfish	Component of hemoglobin, cytochrome, myoglobin	Iron-deficiency anemia
Magnesium (Mg)	Whole grains, green vegetables, legumes, nuts, dairy products	Required for action of many enzymes; roles in muscle, nerve function	Weak, sore muscles; impaired neural function
Manganese (Mn)*	Whole grains, nuts, legumes, many fruits	Activates many enzymes, including ones with roles in synthesis of urea, fatty acids	Abnormal bone and cartilage
Molybdenum (Mo)*	Dairy products, whole grains, green vegetables, legumes	Component of some enzymes	Impaired nitrogen excretion
Phosphorus (P)	Whole grains, legumes, poultry, red meat, dairy products	Component of bones and teeth, nucleic acids, ATP, phospholipids	Muscular weakness; loss of minerals from bone
Potassium (K)	Meat, milk, many fruits, vegetables	Muscle and neural function; roles in protein synthesis	Muscular weakness
Selenium (Se)*	Meat, seafood, cereal grains, poultry, garlic	Constituent of several enzymes; antioxidant	Muscle pain
Sodium (Na)	Table salt, dairy products, meats, eggs	Acid-base balance, water balance; roles in muscle and neural function	Muscle cramps
Sulfur (S)	Meat, eggs, dairy products	Component of body proteins	Same as protein deficiencies
Zinc (Zn)*	Whole grains, legumes, nuts, meats, seafood	Component of digestive enzymes and transcription factors; roles in normal growth, wound healing, sperm formation, taste and smell	Impaired growth, scaly skin, impaired immune function

*Required in trace amounts in diet

Powerful rings of smooth muscle called **sphincters** form valves between major regions of the digestive tract. By contracting and relaxing, sphincters control the passage of the digestive contents from one region to the next, and ultimately through the anus.

The specialized regions of the gut that perform the sequential processes of digestion in humans allow us to extract nutrients efficiently from the highly varied foods we ingest.

Food Begins Its Travel through the Digestive System in the Mouth, Pharynx, and Esophagus

The human digestive system in its normal contracted state in a living adult is about 4.5 m long. Fully relaxed, it is about twice as long. Food begins its travel through this tract in the mouth, where the teeth cut, tear, and crush food items into small pieces. While chewing is

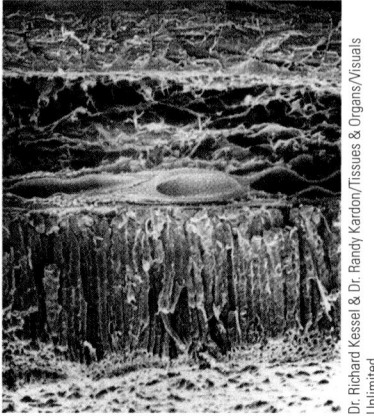

Gastroesophageal sphincter

Esophagus

Stomach

Pyloric sphincter

Duodenum

Serosa

Longitudinal muscle

Circular muscle

Muscularis

Oblique muscle

Submucosa

Mucosa

Figure 45.7
Layers of the gut wall in vertebrates, as seen in the stomach wall.

Dr. Richard Kessel & Dr. Randy Kardon/Tissues & Organs/Visuals Unlimited

in progress, three pairs of **salivary glands** secrete saliva through ducts that open on the inside of the cheeks and under the tongue.

Saliva, which is more than 99% water, moistens the food. Saliva contains **salivary amylase,** which hydrolyzes starches to the disaccharide maltose. It also contains mucus, which lubricates the food mass, and bicarbonate ions (HCO_3^-), which neutralize acids in the food and keep the pH of the mouth between 6.5 and 7.5, which is the optimal range for salivary amylase to function. Another component of saliva is *lysozyme,* an enzyme that kills bacteria by breaking open their cell walls. Some 1 to 2 L of saliva are secreted into the mouth each day.

After a suitable period of chewing, the food mass, called a **bolus,** is pushed by the tongue to the back of the mouth, where touch receptors detect the pressure and trigger the *swallowing reflex* **(Figure 45.9).** The reflex is an involuntary action produced by contractions of muscles in the walls of the pharynx that direct food into

Pyloric sphincter

Chyme

1 The circular layer of the muscularis contracts in a wave, constricting the gut and pushing the digestive contents onward.

2 The longitudinal layer contracts, shortening and expanding the gut and making space for the contents to advance.

3 Partially processed food (chyme) enters the small intestine.

Figure 45.8
The waves of peristaltic contractions moving food through the stomach.

the esophagus. Peristaltic contractions of the esophagus, aided by mucus secreted by the esophagus, propel a bolus towards the stomach. The passage of a bolus down the esophagus stimulates the *gastroesophageal sphincter* at the junction between the esophagus and

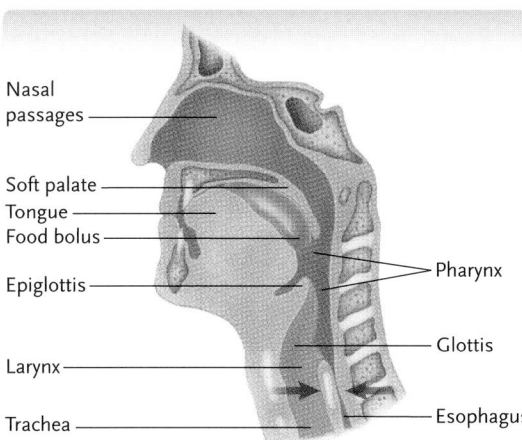

Structures of the mouth, pharynx, and esophagus involved in the swallowing reflex

Nasal passages

Soft palate
Tongue
Food bolus

Epiglottis

Pharynx

Glottis

Larynx

Trachea

Esophagus

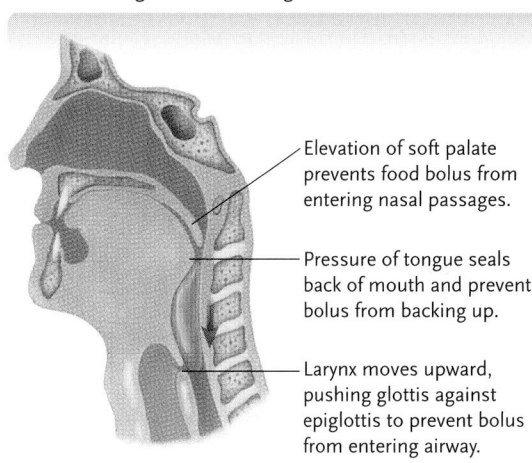

Motions that seal the nasal passages, mouth, and trachea during the swallowing reflex

Elevation of soft palate prevents food bolus from entering nasal passages.

Pressure of tongue seals back of mouth and prevents bolus from backing up.

Larynx moves upward, pushing glottis against epiglottis to prevent bolus from entering airway.

Figure 45.9
The swallowing reflex.

the stomach (see Figure 45.7) to open and admit the bolus to the stomach. After the bolus enters the stomach, the sphincter closes tightly. If the closure is imperfect, the acidic stomach contents can enter the esophagus and produce the irritation and pain we recognize as *acid reflux* or heartburn.

We can consciously initiate the swallowing reflex. However, once the swallowing reflex has begun, we cannot voluntarily stop it, as you might have noticed when you get that feeling of a piece of food or a pill being stuck in the throat or chest. This is because the muscles of the pharynx and upper esophagus are skeletal muscles, which you can control, while the muscles below are smooth muscles, which you cannot control.

Involuntary movements of the tongue and soft palate at the back of the mouth prevent food from backing into the mouth or nasal cavities. Entry into the trachea (the airway to the lungs) is blocked by closure of the *glottis* (the space between the vocal cords) and an upward movement of the *larynx* (the voice box) at the top of the trachea, which closes against a flaplike valve, the **epiglottis.** You can feel the larynx and the front of the epiglottis bob upward if you place your hand on your throat while you swallow. If these blocking mechanisms fail, touch receptors in the nasal passages and larynx trigger coughing and sneezing reflexes that clear these passages.

The Stomach Stores Food and Continues Digestion

The stomach is a muscular, elastic sac that stores food and adds secretions that further the process of digestion. The mucosal layer of the stomach is an epithelium covered with tiny *gastric pits* that are entrances to millions of *gastric glands*. These glands extend deep into the stomach wall and contain cells that secrete some of the products needed to digest food.

The entry of food into the stomach activates stretch receptors in its wall. Signals from the stretch receptors stimulate the secretion of **gastric juice (Figure 45.10),** which contains **pepsinogen**, the precursor for the digestive enzyme pepsin, hydrochloric acid (HCl), and lubricating mucus. The stomach secretes about 2 L of gastric juice each day.

Pepsinogen is secreted by *chief cells* in the gastric pits. It is an inactive precursor molecule that is converted to the digestive enzyme **pepsin** by the highly acid conditions of the stomach. Once produced, pepsin itself can catalyze the reaction that converts more pepsinogen to pepsin. Pepsin begins the digestion of proteins by introducing breaks in polypeptide chains. The activation of pepsinogen illustrates a common theme in the digestive system: powerful hydrolytic enzymes that would be dangerous to the cells secreting them are synthesized in the form of inactive precursors and are not converted into active form until they are exposed to the digestive contents.

Parietal cells secrete H^+ and Cl^-, which combine to form HCl in the lumen of the stomach. The HCl lowers the pH of the digestive contents to pH 2 or lower, the level at which pepsin reaches optimal activity. To put this pH in perspective, lemon juice is pH 2.4, and sulfuric acid or battery acid is approximately pH 1. The acidity of the stomach also helps break up food particles and causes proteins in the digestive contents to unfold, exposing their peptide linkages to hydrolysis by pepsin. The acid also kills most of the bacteria that reach the stomach and stops the action of salivary amylase.

A thick coating of alkaline mucus, secreted by *mucous cells*, protects the stomach's mucosal layer from attack by pepsin and HCl. Behind the mucous barrier, tight junctions between cells prevent gastric juice from seeping into the stomach wall. Even so, some break-

Figure 45.10
Cells that secrete mucus, pepsin, and HCl in the stomach lining.

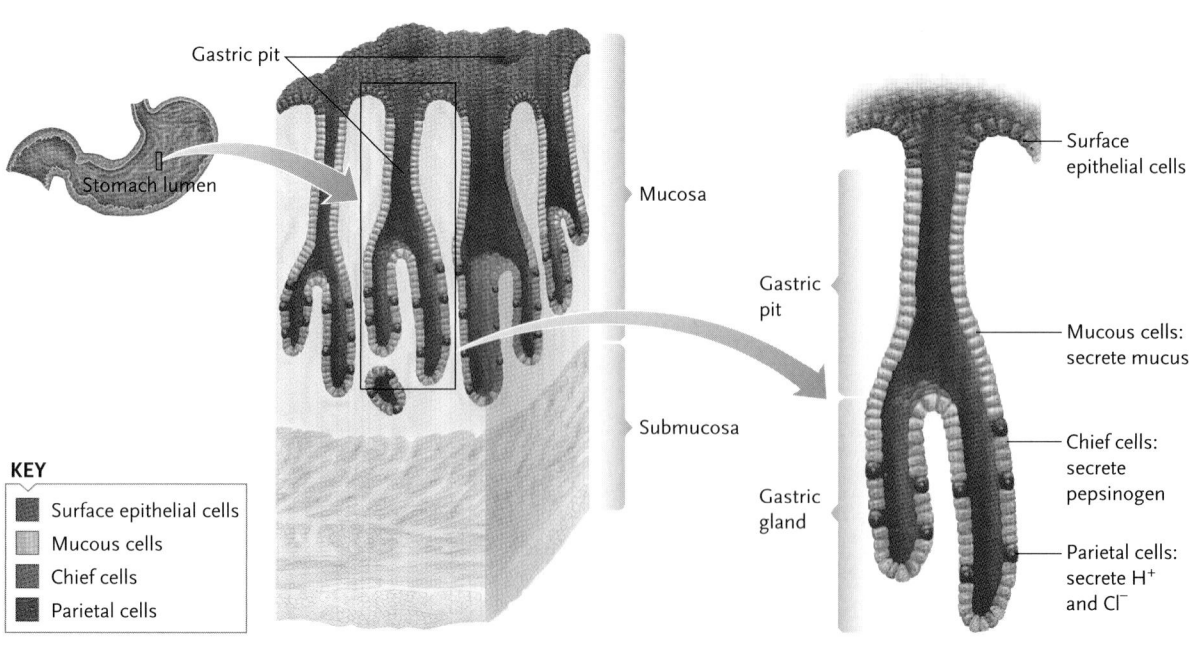

KEY

- Surface epithelial cells
- Mucous cells
- Chief cells
- Parietal cells

Gastric pit

Stomach lumen

Mucosa

Submucosa

Gastric pit

Gastric gland

Surface epithelial cells

Mucous cells: secrete mucus

Chief cells: secrete pepsinogen

Parietal cells: secrete H^+ and Cl^-

down of the stomach's mucosal layer does occur. However, the damage is normally repaired quickly by the rapid division of mucosal cells. Even without damage, the rapid cell division leads to complete replacement of the stomach's mucosal layer.

Most bacteria cannot survive the highly acid environment of the stomach, but one, *Helicobacter pylori,* thrives there. In some people, the bacterium breaks down the mucous barrier, exposing the stomach wall to attack by HCl and pepsin. The resulting lesion, known as a *peptic* or *stomach ulcer,* causes bleeding and pain. If untreated, an ulcer can become so deep that it perforates the entire stomach wall, with potentially fatal consequences. Ulcers are treated by taking an antibiotic that kills *H. pylori.* Barry J. Marshall of the University of Western Australia and J. Robin Warren of Royal Perth Hospital received a 2005 Nobel Prize for their discovery that a bacterium is responsible for most human ulcers.

As part of the digestive process, contractions of the stomach walls continually mix and churn the contents, which can amount to as much as 2 L when the stomach is full. Peristaltic contractions of the stomach wall move the digestive contents toward the *pyloric sphincter* (*pylorus* = gatekeeper) at the junction between the stomach and small intestine. The arrival of a strong stomach contraction relaxes and opens the valve briefly, releasing a pulse of the stomach contents, now called **chyme,** into the small intestine (see Figure 45.8).

Depending on the volume and composition of the stomach contents, it can take from one to six hours for the stomach to empty after a meal. Feedback controls that regulate the rate of gastric emptying tend to match it to the rate of digestion, so that food is not moved along faster than it can be chemically processed. In particular, when chyme with high fat content and high acidity enters the first part of the small intestine, it stimulates the mucosal layer to secrete hormones that slow stomach emptying. Fat is digested more slowly than other nutrients, and it is digested only in the lumen of the small intestine. Therefore, further emptying of the stomach is prevented until the processing of fat has been completed in the small intestine. This is why a fatty meal, such as a greasy pizza, feels so heavy in the stomach. Highly acidic chyme must be neutralized by bicarbonate in the small intestine. Unneutralized stomach acid inactivates digestive enzymes secreted in the small intestine and, therefore, such acid inhibits further emptying of the stomach until it is neutralized.

The Small Intestine Completes Digestion and Begins the Absorption of Nutrients

No absorption of nutrients occurs in the mouth, pharynx, or esophagus; with the exception of a few substances, such as alcohol, aspirin, caffeine, and water, little absorption occurs in the stomach. Most absorption begins, and digestion is completed, as the contents move through the small intestine.

The "small" in the small intestine refers to its diameter, about 4 cm. It is roughly 6 m long and complexly coiled within the abdominal cavity. The lining of the small intestine folds into ridges that are densely covered by microscopic, fingerlike extensions, the **intestinal villi** (singular, *villus*). In addition, the epithelial cells covering the villi have a *brush border* consisting of fingerlike projections of the plasma membrane called **microvilli (Figure 45.11).** The intestinal villi and microvilli are estimated to increase the absorptive surface area of the small intestine to as much as 300 square meters, about the size of a doubles tennis court.

Secretions of the Pancreas, Liver, and Intestinal Mucosa. Digestion in the small intestine depends on enzymes and other substances secreted by the intestine itself and by the pancreas and liver. The secretions from the pancreas and liver enter a common duct that empties into the lumen of the first segment of the small intestine, a short region about 20 cm long called the **duodenum (Figure 45.12).**

In all, about 7 to 9 L of fluid from the stomach, liver, pancreas, and intestinal glandular cells enter the small intestine each day. About 95% of this amount is reabsorbed as water and nutrients as the digestive contents travel along the small intestine. Movement of the contents from the duodenum to the end of the small intestine takes about 3 to 5 hours.

In humans, the **pancreas** is an elongated, flattened gland located between the stomach and duodenum (see Figures 45.5 and 45.12). Exocrine cells in the pancreas secrete bicarbonate ions ($H_2CO_3^-$) and digestive enzymes—*pancreatic enzymes*—into ducts that empty into the lumen of the duodenum. The bicarbonate ions neutralize the acid in the chyme, bringing the digestive contents to a slightly alkaline pH. The alkaline pH allows optimal activity of the enzymes secreted by the pancreas, which include proteases, an amylase, nucleases, and lipases. All of these enzymes act in the lumen of the small intestine. Like pepsin, the proteases released by the pancreas are secreted in an inactive precursor form; contact with the digestive solution activates them. Among the active forms of these enzymes are *trypsin,* which hydrolyzes bonds within polypeptide chains, and *carboxypeptidase,* which cuts amino acids from polypeptide chains one at a time.

The **liver** secretes bicarbonate ions and **bile,** a mixture of substances including *bile salts,* cholesterol, and *bilirubin.* Bile salts are derivatives of cholesterol and amino acids that aid fat digestion through their detergent action. They form a hydrophilic coating around fats and other lipids, which allows the churning motions of the small intestine to *emulsify* the fats—break them into tiny droplets called *micelles,* as in mixing oil and water in making a salad dressing. Lipase, a pancreatic enzyme,

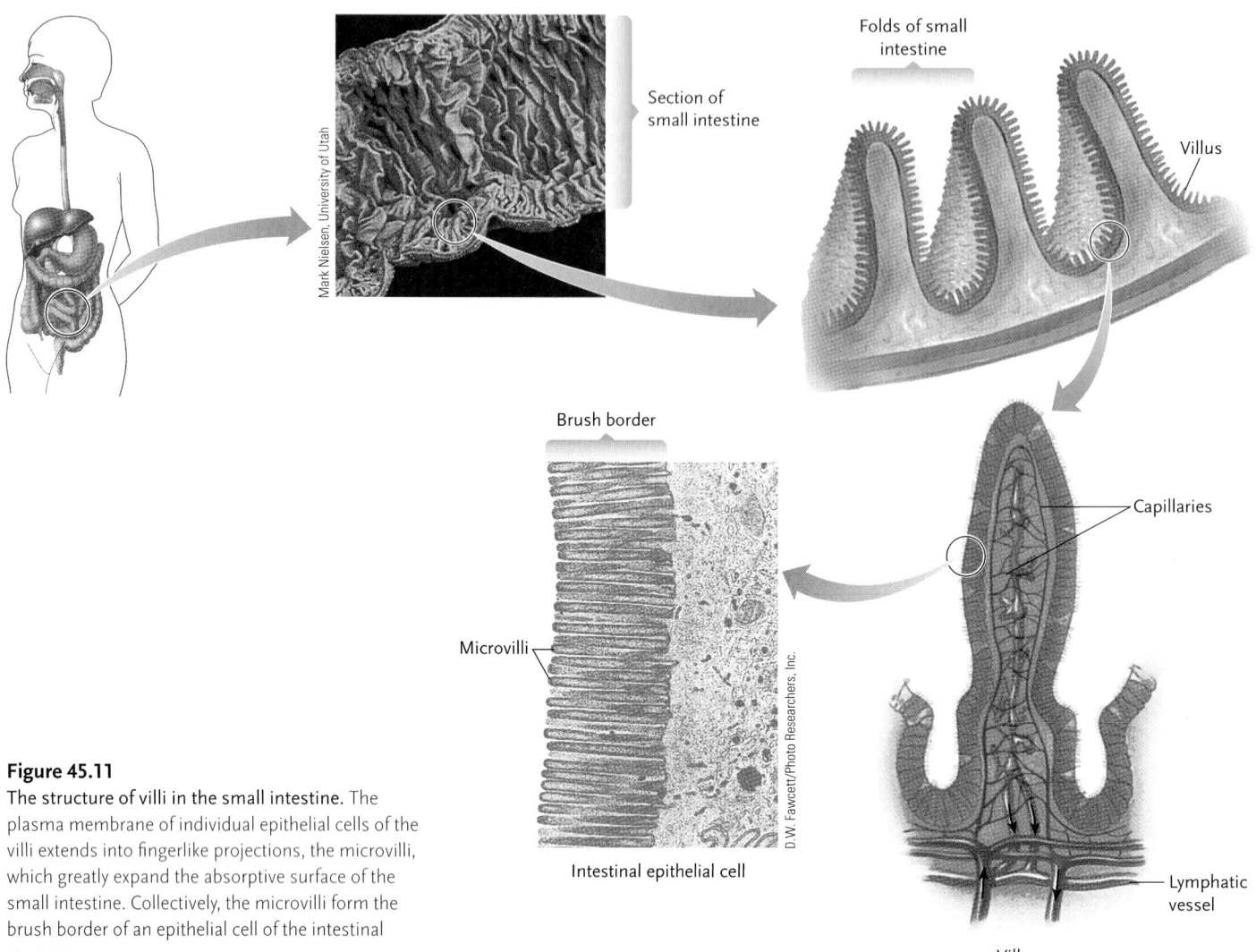

Figure 45.11

The structure of villi in the small intestine. The plasma membrane of individual epithelial cells of the villi extends into fingerlike projections, the microvilli, which greatly expand the absorptive surface of the small intestine. Collectively, the microvilli form the brush border of an epithelial cell of the intestinal mucosa.

Section of small intestine

Mark Nielsen, University of Utah

Folds of small intestine

Villus

Brush border

Microvilli

Intestinal epithelial cell

D. W. Fawcett/Photo Researchers, Inc.

Capillaries

Lymphatic vessel

Villus

can then hydrolyze the fats in the micelles to produce monoglycerides and free fatty acids. Bilirubin, a waste product derived from worn-out red blood cells, is the yellow pigment that gives the bile its color. Bacterial enzymes in the intestines modify the pigment, resulting in the characteristic brown color of feces.

The liver secretes bile continuously. Between meals, when no digestion is occurring, bile is stored in the **gallbladder,** where it is concentrated by the removal of water. After a meal, entry of chyme into the small intestine stimulates the gallbladder to release the stored bile into the small intestine.

Brush-border epithelial cells on the villi of the small intestine secrete water and mucus into the intestinal contents. They also produce enzymes that complete the digestion of carbohydrates, proteins, and nucleic acids. Substrates for those enzymes are breakdown products produced by enzyme activity elsewhere in the digestive system—disaccharides, large peptides, dipeptides, and nucleotides—that are transported across the plasma membranes of the brush-border epithelial cells **(Figure 45.13).** Different *disaccharidases* break maltose, lactose, and sucrose into individual monosac-

charides. Two proteases complete protein digestion: an *aminopeptidase* cuts amino acids from the end of a polypeptide, and a *dipeptidase* splits dipeptides into individual amino acids. Nucleases and other enzymes digest the nucleic acids: *nucleotidases* break them down into nucleosides, and *nucleosidases* convert the nucleosides to nitrogenous bases, five-carbon sugars, and phosphates.

Many adults lose the capacity to synthesize lactase, the enzyme that breaks down the milk sugar lactose. The lactose remaining in the intestine is broken down by bacteria, producing excess methane and CO_2. The accumulating gases distend the large intestine, producing pain, discomfort, and other symptoms of *lactose intolerance.* For many people lactose intolerance can be relieved by taking tablets containing lactase before eating milk products. One estimate is that 70% of the world's population is lactose intolerant. In the United States, between 30 and 50 million people are lactose intolerant, with some ethnic and racial groups affected more than others. For instance, over 80% of Native Americans, up to 80% of African Americans, and over 90% of Asian Americans are lactose intolerant. Generally, lactose intolerance is far less common among people of northern European descent.

Absorption by the Brush-Border Cells of the Intestinal Mucosa. The water-soluble products of digestion enter the intestinal mucosa cells by active transport or facilitated diffusion **(Figure 45.14a),** and water follows by osmosis. The nutrients are then transported from the mucosal cells into the extracellular fluids, from where they enter the bloodstream in the capillary networks of the submucosa.

The micelles formed by bile salts assist in the absorption of fatty acids, monoglycerides, fat-soluble vitamins, and cholesterol and other products of lipid breakdown by lipase **(Figure 45.14b).** When a micelle contacts the plasma membrane of a mucosal cell, the hydrophobic molecules within the droplet penetrate through the membrane and enter the cytoplasm.

In the mucosal cells, the fatty acids and monoglycerides are combined into fats (specifically, triglycerides) and packaged into **chylomicrons,** small droplets covered by a protein coat. Cholesterol absorbed in the small intestine is also packed into the chylomicrons. The protein coat of the chylomicrons provides a hydrophilic surface that keeps the droplets suspended in the cytosol. After traveling across the mucosal cells, the chylomicrons are secreted into the interstitial fluid of

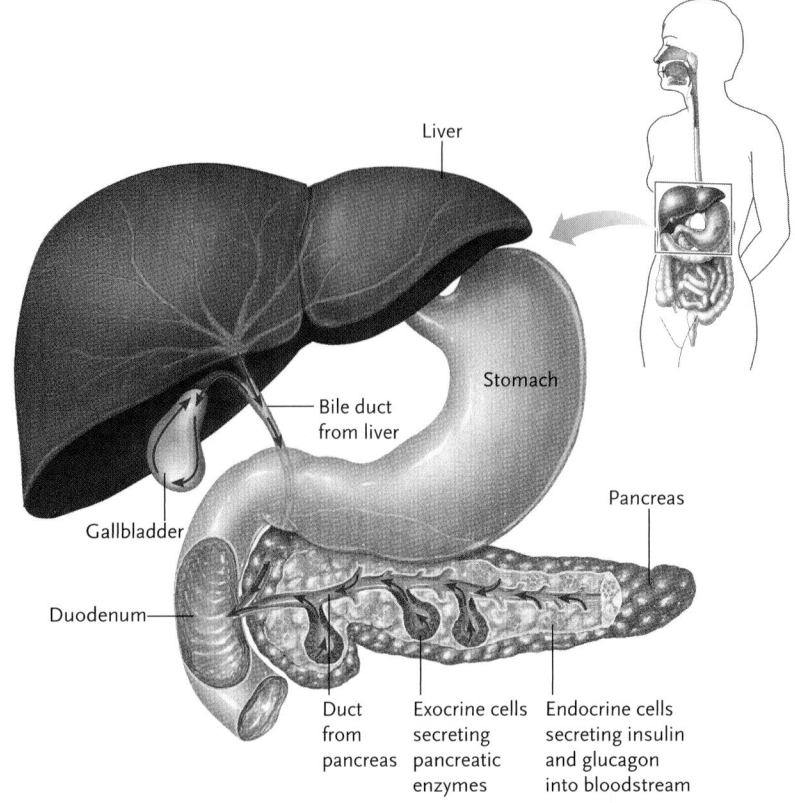

Figure 45.12
The ducts delivering bile and pancreatic juice to the duodenum of the small intestine.

Figure 45.13
Enzymatic digestion of carbohydrates, proteins, fats, and nucleic acids in the human digestive system.

	Carbohydrates	Proteins	Fats	Nucleic acids
Mouth	Polysaccharides ↓ **Salivary amylase** Smaller polysaccharides, disaccharides			
Stomach		Proteins ↓ **Pepsin** Peptides		
Lumen of small intestine	Polysaccharides ↓ **Pancreatic amylase** Disaccharides	Proteins ↓ **Trypsin, chymotrypsin** Peptides Large peptides ↓ **Carboxypeptidase** Amino acids	Triglycerides and other lipids ↓ **Lipase** Fatty acids, monoglycerides	DNA, RNA ↓ **Pancreatic nucleases** Nucleotides
Epithelial cells (brush border) of small intestine	Disaccharides (maltose, sucrose, lactose) ↓ **Disaccharidases** Monosaccharides (for example, glucose)	Large peptides Dipeptides ↓ **Amino-peptidase** ↓ **Dipeptidase** Amino acids Amino acids		Nucleotides ↓ **Nucleotidases, nucleosidases, phosphatases** Nitrogeneous bases, five-carbon sugars, and phosphates

a. Absorption of water-soluble products of digestion by intestinal mucosa cells

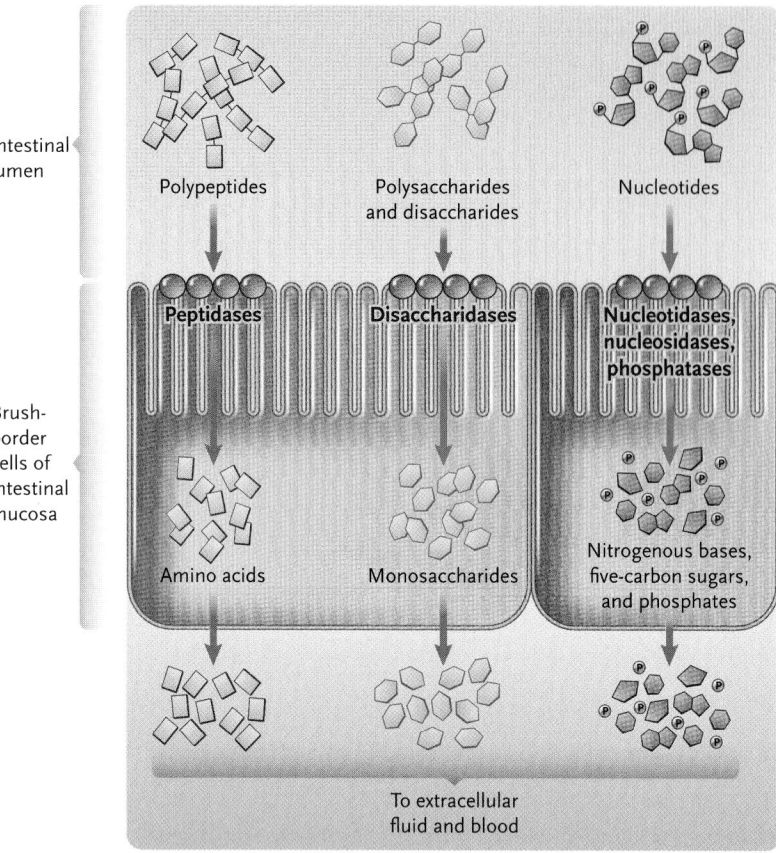

Intestinal lumen

Brush-border cells of intestinal mucosa

To extracellular fluid and blood

Water-soluble molecules are broken into absorbable subunits at brush borders of mucosal cells and transported inside; the subunits are transported on the other side to extracellular fluid and blood.

b. Absorption of fat-soluble products of digestion by intestinal mucosa cells

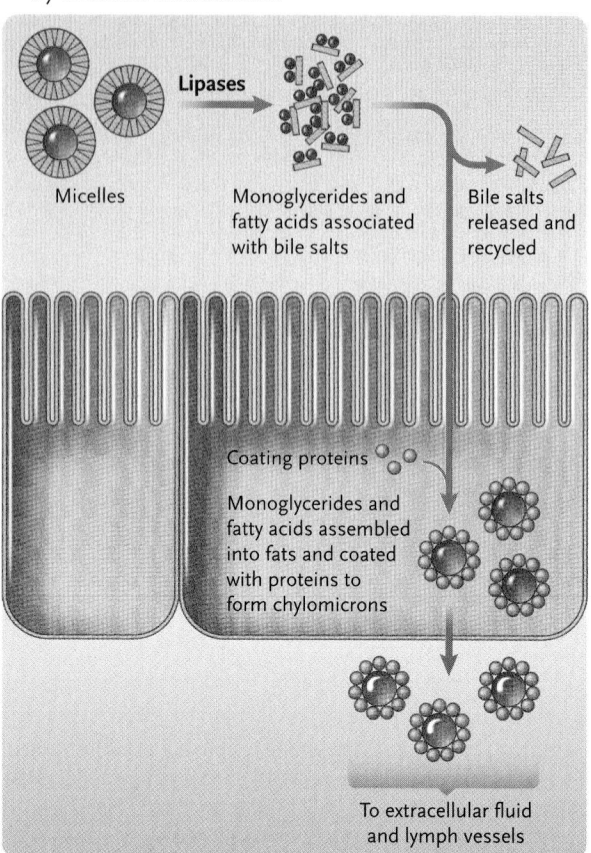

To extracellular fluid and lymph vessels

Micelles (fats coated with bile salts) are digested to monoglycerides and fatty acids, which penetrate into cells and are assembled into fats. The fats are coated with proteins to form chylomicrons, which are released by exocytosis to extracellular fluids, where they are picked up by lymph vessels.

Figure 45.14
Absorption of digestive products by the brush-border cells of the intestinal mucosa.

the submucosa, where they are taken up by lymph vessels. Eventually, they are transferred with the lymph into the blood circulation.

The small intestine reabsorbs all but about 1 L of the 7 to 9 L of fluid released from the stomach. By the time the digestive contents reach the large intestine, almost all nutrients have been hydrolyzed and absorbed.

Many Nutrients Absorbed in the Small Intestine Are Processed in the Liver

The capillaries absorbing nutrient molecules in the small intestine collect into veins that join to form a larger blood vessel, the **hepatic portal vein**, which leads to capillary networks in the liver. In the liver, some of the nutrients leave the bloodstream and enter liver cells for chemical processing. Among the reactions taking place in the liver is the combination of excess glucose units into glycogen, which is stored in the liver cells. This reaction reduces the glucose concentration in the blood exiting the liver to about 0.1%. If the glu-

cose concentration in the blood entering the liver falls below 0.1% during a period of fasting between meals, the reaction reverses. The reversal adds glucose to return the blood concentration to the 0.1% level before it exits the liver.

The liver also synthesizes the lipoproteins that transport cholesterol and fats in the bloodstream, detoxifies ethyl alcohol and other toxic molecules, and inactivates steroid hormones and many types of drugs.

As a result of the liver's activities, the blood leaving the liver has a markedly different concentration of nutrients than the blood carried into the liver by the hepatic portal vein. From the liver, blood is carried to the heart and then pumped by the heart to deliver nutrients to all parts of the body.

The Large Intestine Primarily Absorbs Water and Mineral Ions from Digestive Residues

From the small intestine, the contents move on to the large intestine. A sphincter at the junction between the small and large intestines controls the passage of material and prevents backward movement of the contents. The large intestine is several times larger in diameter than the small intestine, but it is relatively short, about 1.2 m long in humans, as compared with the 6 m length of the small intestine. The inner surface of the large intestine is relatively smooth and contains no villi.

The large intestine has several distinct regions **(Figure 45.15)**. At the junction with the small intestine, a part of the large intestine forms a blind pouch called the **cecum.** A fingerlike sac, the **appendix**, extends from the cecum. The appendix is on average 100 mm long and 7 mm in diameter; it is a vestigial structure with no known function. It certainly has no function in digestion, but it does contain patches of lymphoid tissue, suggesting that at one time it may have functioned as part of the immune system. The cecum merges with the **colon**, the main part of the large intestine, which forms an inverted U. At its distal end, the colon connects with the final segment of the large intestine, the **rectum.**

The large intestine secretes mucus and bicarbonate ions and absorbs water and other ions, primarily sodium and chloride. The absorption of water condenses and compacts the digestive contents into solid masses, the **feces.** Normally, by the time the fecal matter reaches the rectum, it contains less than 200 mL of the fluid entering the digestive tract each day. Diarrhea, by contrast, is an abnormal condition in which the fecal matter is highly fluid. The most common cause of diarrhea is a higher-than-normal rate of movement of materials through the small intestine, which does not leave adequate time for absorption of water to occur. The higher rate of movement can occur as a result of irritation of the small intestine wall in bacterial and viral infections, or because of emotional stress.

About 30% to 50% of the dry matter of feces in humans and other vertebrates consists of more than 500 species of bacteria that live as essentially permanent residents in the large intestine. Most common is the bacterium *Escherichia coli,* which also lives in the intestine of many other mammals. Intestinal bacteria metabolize sugars and other nutrients remaining in the digestive residue, and produce useful fatty acids and vitamins (such as vitamin K, the B vitamins, folic acid, and biotin), some of which are absorbed in the large intestine. Their activity also produces large quantities of gas—*flatus*—primarily CO_2, methane, and hydrogen sulfide. Most of the gas is absorbed through the intestinal mucosa, and the rest is expelled through the anus in the process of *flatulence.* The amount and composition of the gas produced depends on the type of food being ingested and the particular population of bacteria present in the large intestine. Some foods, such as beans, contain carbohydrates that humans cannot digest but that can be metabolized by the gas-producing intestinal bacteria. After eating such foods,

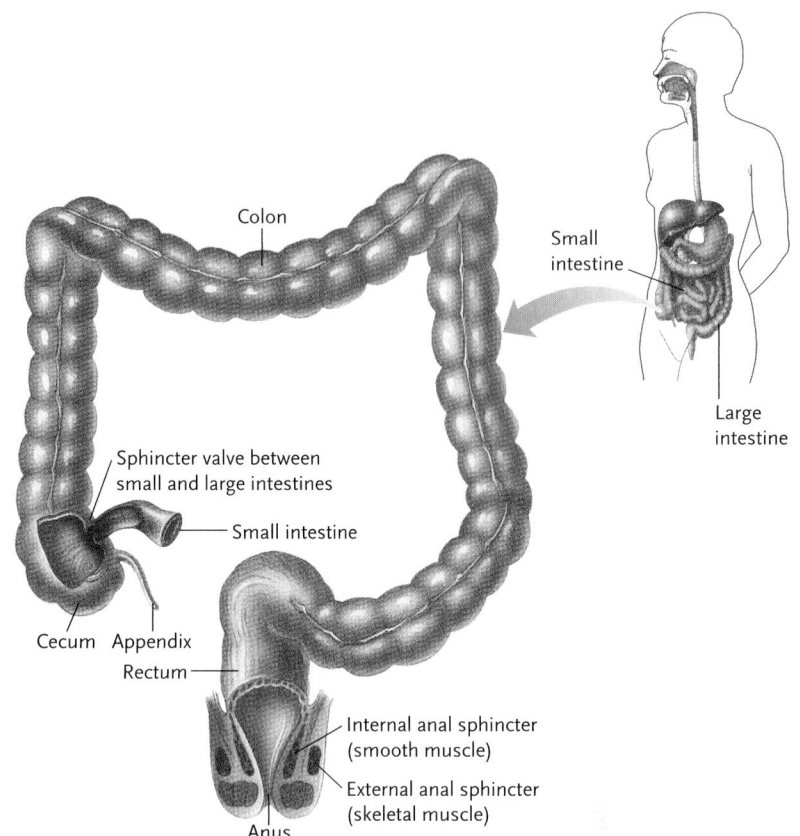

Figure 45.15
The human large intestine.

humans may produce more gas than usual, and flatulence is more likely.

When feces enter the rectum, they stretch the rectal wall. The stretching triggers a *defecation reflex* that opens the *anal sphincter* and expels the feces through the anus. Because the anal sphincter contains rings of voluntary skeletal muscle as well as involuntary smooth muscle (see Figure 45.15), we can resist the defecation reflex by voluntarily tightening the striated muscle ring.

Having completed the journey of ingested food through the digestive tract, we now turn our attention to the mechanisms that regulate the activities of the digestive system. These mechanisms coordinate one region of the digestive tract with another, and help match the production of nutrients with the body's needs.

STUDY BREAK

1. What are the two classes of vitamins? Which of the two types is more critical in the diet and why?
2. What are the four layers of the mammalian gut? Which layer is responsible for peristalsis?
3. How is pepsin produced? What is its function?
4. Distinguish between the functions of the small and large intestines in digestion.

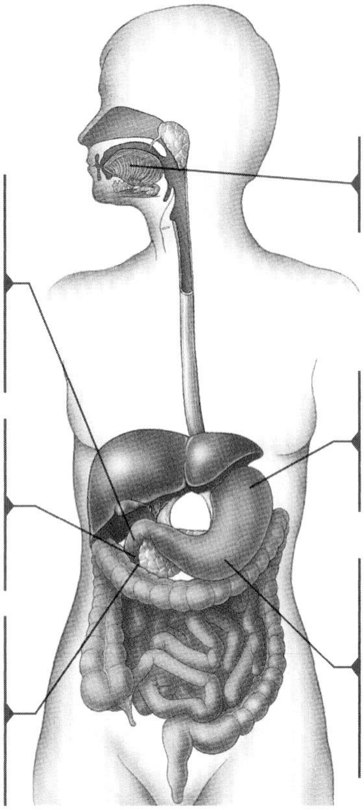

Hormone controls

Acidic chyme stimulates release of the hormone secretin in the small intestine. Secretin inhibits gastric emptying and gastric secretion and stimulates HCO_3^- secretion into the duodenum.

Fat (mostly) in chyme stimulates release of the hormone cholecystokinin (CCK). CCK inhibits gastric activity and stimulates secretion of pancreatic enzymes.

A meal entering the digestive tract stimulates GIP secretion, which triggers insulin release. Insulin stimulates the uptake and storage of glucose from the digested food.

Receptor controls

Receptors in the mouth respond to food by increasing salivary secretion.

Stretch receptors in the stomach respond to food, signaling neuron networks to increase stomach contractions.

Chemoreceptors in the stomach respond to food, signaling neuron networks to stimulate the stomach to secrete the hormone gastrin, which in turn stimulates the stomach to secrete HCl and pepsinogen.

Figure 45.16
Control of digestion by receptors and hormones in the digestive system.

45.4 Regulation of the Digestive Process

The digestive process is regulated and coordinated at many steps by controls that are largely automated. The autonomic nervous system, local neuron networks in the gut wall, and endocrine glands interact in these controls, in response to sensory information gathered by receptors in the digestive tract. The integration of the controls speeds or slows digestion to produce maximum efficiency in the breakdown of food molecules and the absorption of the products.

Much of the control of the digestive system originates in the neuron networks of the submucosa. Other controls, particularly those regulating appetite and oxidative metabolism, originate in the brain, in control centers forming part of the hypothalamus.

The Digestive Tract Itself Has a Number of Control Systems

The movement of food through the digestive system is controlled by receptors in and hormones secreted by various parts of the system **(Figure 45.16)**. Control starts with the mouth. Saliva is secreted constantly into the mouth. The presence of food activates receptors that increase the rate of salivary secretion by as much as 10-fold.

Swallowed food expands the stomach and sets off signals from stretch receptors in the stomach walls. Chemoreceptors in the stomach respond to the presence of food molecules, particularly proteins. Signals from these two types of receptors are integrated in neuron networks in both the stomach and the autonomic nervous system to produce several reflex responses. One is an increase in the rate and strength of stomach contractions. Another is secretion of a hormone, *gastrin*, into the blood leaving the stomach. After traveling through the circulatory system, gastrin returns to the stomach, where it stimulates the secretion of HCl and pepsinogen. These molecules are then used in the digestion of the protein that, as part of the swallowed meal, was responsible for their secretion. Gastrin also stimulates stomach and intestinal contractions, activities that serve to keep the digestive contents moving through the digestive system when a new meal arrives.

Three hormones secreted when food is present in the duodenum also participate in regulating the digestive processes. When chyme is emptied into the duodenum, its acidic nature stimulates the release of the hormone *secretin*. Secretin inhibits further gastric emptying to prevent further acid from entering the duodenum until the newly arrived chyme is neutralized. It also inhibits gastric secretion to reduce acid production in the stomach, and it stimulates HCO_3^- secretion into the lumen of the duodenum to neutralize the acid. If the acid is not neutralized, the duodenal wall will become damaged.

Fat, and to a lesser extent protein, in the chyme entering the duodenum stimulates the release of the hormone *cholecystokinin (CCK)*. CCK inhibits gastric activity, thereby allowing time for nutrients in the duodenum to be digested and absorbed. It also stimulates the secretion of pancreatic enzymes, used to digest the macromolecules in the chyme.

The hormone *glucose-dependent insulinotrophic peptide (GIP)* acts primarily to stimulate insulin release by the pancreas. When a meal is ingested, the body must change its metabolic state to use and store the new nutrients absorbed. Those activities are mostly under the control of insulin. Therefore, when a meal enters the digestive tract, GIP secretion is stimulated to trigger the release of insulin. Insulin is particularly important in stimulating the uptake and storage of glucose and so, not surprisingly, glucose in the duodenum causes an increase in GIP secretion.

The Hypothalamus Exerts Overall Controls

The hypothalamus contains two interneuron centers that work in opposition to control appetite and oxidative metabolism. One center stimulates appetite and reduces oxidative metabolism; the other center stimulates the release of a peptide hormone called *α-melanocyte-stimulating hormone (α-MSH)*, which inhibits appetite.

A major link in the control pathways is the peptide hormone *leptin* (*leptos* = thin), discovered in mice by Jeffrey Friedman and his coworkers at the

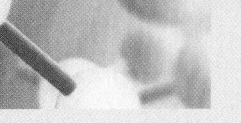

Food for Thought on the Feeding Response

A number of small proteins called *neuropeptides* regulate various physiological responses in humans and other mammals, including appetite and feeding, pain reception, and blood pressure regulation. They exert their effects by binding to receptors on the surfaces of neurons and other cells. For example, the neuropeptide NPY (neuropeptide Y) strongly stimulates appetite and food uptake when it binds to neurons in the hypothalamus and other locations in the brain.

A group of researchers at the Synaptic Pharmaceutical Corporation in New Jersey and at Ciba-Geigy in Switzerland was one of several teams that tried to identify the receptor binding NPY in the hypothalamus. The investigators extracted all the messenger RNAs made in the rat hypothalamus and converted them into DNA copies (cDNAs). The cDNAs were inserted into plasmid cloning vectors for cloning of the sequences in *E. coli*. Next, the plasmids were transferred into mammalian cells that do not normally make neuropeptide receptors, and the researchers screened for cells that were now able to bind the neuropep-

tide. The cells that did must have received an active gene encoding the receptor.

The investigators isolated the rat gene that encoded the receptor and obtained its sequence. The gene sequence showed that it encoded a previously unknown neuropeptide receptor, which they called Y5. An equivalent gene for this receptor was then identified in the human genome.

Next, different rat tissues were tested with a probe that could pair with the mRNA transcribed from the *Y5* gene. The result showed that *Y5* is expressed only in the brain, with the strongest expression in the hypothalamus and the amygdala, a center associated with emotional responses.

After testing the ability of various peptides to bind the Y5 receptor, and the effects of the binding on feeding behavior, the authors proposed that the binding of NPY initiates signals that stimulate hunger and the feeding response. Binding to receptors in the amygdala may add an emotional dimension to the craving for food. The authors conjectured that studies of Y5 could help further our understanding

of eating disorders and perhaps aid in the development of drugs to combat those disorders.

Indeed, the Y5 receptor became a focus of considerable drug discovery efforts. However, a recent study has burst the bubble—drugs designed to inhibit the receptor are not going to be effective in combating obesity. In this study, Andrew Turnbull and his colleagues at AstraZeneca in the United Kingdom, tested whether one such drug affected feeding in rats. Unfortunately, the drug had no significant effect on the increase in food intake induced by NPY itself either in normal or genetically obese rats. Further, the drug had no effect on food intake or body weight in normal rats or in rats that were obese due to their diet. In short, the Y5 receptor is not a significant regulator of feeding behavior. The difference between these results and those of the previous study may reflect the experimental design—the compounds used in the earlier study may well have had other activities that were responsible for their effects on feeding behavior.

Rockefeller University. Fat-storing cells secrete leptin when the deposition of fat increases in the body. Leptin travels in the bloodstream and binds to receptors in both centers in the hypothalamus. Binding stimulates the center that reduces appetite and inhibits the center that stimulates appetite. At the same time, leptin binds to receptors on body cells, triggering reactions that oxidize fatty acids rather than converting them into fats. When fat storage is reduced, leptin secretion drops off, and signals from other pathways activate the appetite-stimulating center in the hypothalamus and turn off the appetite-inhibiting center.

These controls closely match the activity of the digestive system to the amount and types of foods that are ingested, and they coordinate appetite and oxidative metabolism with the body's needs for stored fats. *Insights from the Molecular Revolution* describes recent investigations identifying signal molecules and receptors that regulate appetite and feeding behavior in mammals.

STUDY BREAK

How does the hypothalamus regulate the digestive process?

45.5 Digestive Specializations in Vertebrates

Natural selection has modified the basic vertebrate digestive system into a multitude of structural and functional variations. The most common modifications are in the form of the mouth, teeth, and jaws; the structure and function of the esophagus, stomach, and cecum; and the length of the digestive tract. Vertebrates also vary in the types of enzymes secreted by the digestive system. For example, humans secrete an amylase into the saliva, but cats and pigs secrete a salivary lipase.

Teeth Are Adapted to Feeding Methods

To anthropologists and paleontologists, dentition (the number, kind, and arrangement of teeth) opens a window to an animal's diet and feeding method—and hence reveals a great deal about its habitat and life style. For example, snakes have sharp, pointed teeth that curve backward into the mouth, which helps to ensure that prey (dead or living) does not slip out of the animal's mouth as muscles contract to swallow. The dentition is combined with specializations in jaw structure; many snakes have jaws with elastic connections that allow them to open wide enough to swallow prey whole.

Tooth specialization is especially evident among mammals **(Figure 45.17).** Typically, mammals have four types of upper and lower teeth. **Incisors,** located at the front of the mouth, are flattened, chisel-shaped teeth used to nip or cut food. Horses use their prominent incisors to clip off blades of grasses. Pointed **canines** at the sides of the incisors are specialized for biting and piercing. Carnivores such as wolves and tigers use their long, sharp canines to pierce and kill prey, but the canines are minimally developed or absent in many herbivores. The blocky teeth at the sides of the mouth, the **premolars** and **molars,** have surface bumps, or *cusps,* that are used in crushing, grinding, and shearing food. Large premolars and molars with a ridged surface are characteristic of animals, such as deer, that consume fibrous plant material. The premolars and molars of some carnivores, such as cats, have sharp shearing surfaces that can slice meat efficiently. All four types of teeth are typically well developed in omnivores, such as humans.

The Length of the Intestine and Specializations of the Digestive Tract Reflect Feeding Patterns

There is a strong correlation between diet and the length of the digestive system **(Figure 45.18).** Vertebrates that feed primarily on nutrient-rich foods such as meat, blood, nectar, or insects, including carnivores (such as the dog), generally have a relatively short intestine. In contrast, herbivores (such as the rabbit) have a long intestinal tract and specializations of the esophagus, stomach, and cecum, or other structures that can store large volumes of plant material. Both the longer intestinal tract and greater storage capacity allow an herbivore to extract more nutrients from plant matter, which is relatively difficult to digest. Both types of intestine appear during the life cycle of frogs: frog tadpoles, which primarily eat algae, have a relatively long, coiled intestine; after metamorphosis, the adult frogs, which primarily eat insects, have a short intestine.

Symbiotic Microorganisms Aid Digestion in a Number of Organisms

Many herbivores use the hydrolytic capabilities of microorganisms such as bacteria, protists, and fungi to aid digestion of plant material, housing them in specialized structures of the esophagus, stomach, or cecum. Unlike vertebrates, the microorganisms can synthesize *cellulase,* the enzyme that hydrolyzes the cellulose of plant cell walls into glucose subunits. The arrangement is a classic example of symbiosis; the herbivores benefit from the digestive capabilities of the microorganisms, and the microorganisms benefit from an ideal habitat and an abundant supply of nutrients.

The most remarkable adaptations for symbiotic digestion of plant matter among vertebrates occur in the **ruminants,** which include cattle, deer, goats, sheep, and antelopes. These animals have a complex, four-chambered stomach **(Figure 45.19).** The first three chambers are derived from the esophagus. After the ruminant's teeth tear, cut, and grind plant matter, it is swallowed and arrives in the *reticulum.* In the reticulum, and in the next and largest chamber, the *rumen,* symbiotic microorganisms hydrolyze cellulose in the plant matter into fuels for fermentation reactions (the oxygen

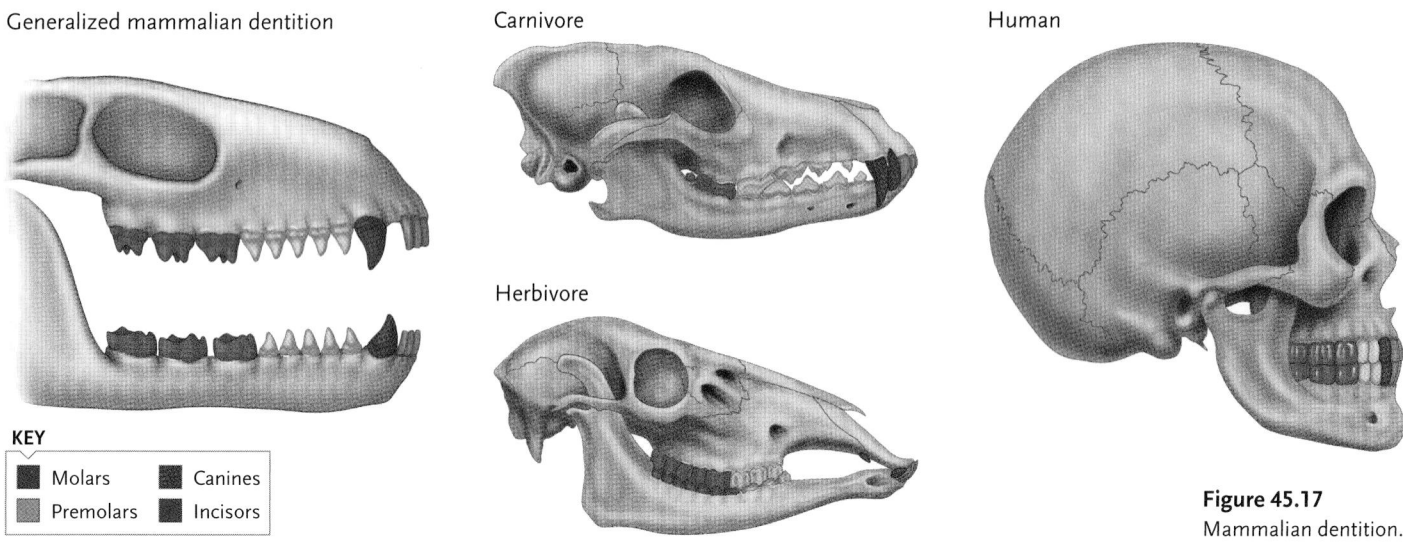

Generalized mammalian dentition

Carnivore

Herbivore

Human

KEY

■ Molars ■ Canines
■ Premolars ■ Incisors

Figure 45.17
Mammalian dentition.

level in the chambers is too low to support mitochondrial reactions). The fermentations generate various products, including alcohols, amino acids, and fatty acids, which are used as nutrients by the ruminants. Methane, another product, collects in the fermentation chambers. Ruminants belch the gas in huge quantities; a cow potentially can release more than 400 L of methane per day. In fact, cattle are estimated to contribute 20% of the methane polluting our atmosphere.

As part of the digestive process, a ruminant "chews its cud"—it regurgitates material from the reticulum and rumen, rechews it, and swallows it again. This process crushes the plants into smaller fragments, exposing more surface area to the microbial enzymes, and gives the enzymes more time to act.

Matter that has been digested and liquefied by the microorganisms moves to the *omasum,* where water is absorbed from the mass, and then to the *abomasum* (the ruminant's true stomach). There, the addition of acids and pepsin to the food mass kills the microorganisms and starts the process of typical vertebrate digestion. As the food mass moves to the small intestine, the dead microorganisms, which are a rich source of proteins, vitamins, and other nutrients, are digested and absorbed along with other hydrolyzable molecules in the digestive contents.

Although the ruminant digestive system is uniquely specialized, many other vertebrate species also have esophageal or gastric chambers containing plant-digesting symbiotic microorganisms. These include the camel, sloth, and langur monkey, and marsupials such as kangaroos and wallabies. One bird, the South American hoatzin *(Opisthocomus hoazin),* is known to have a crop in which microorganisms break down plant matter.

Many vertebrates house symbiotic, plant-digesting microorganisms in the cecum. Horses, elephants, rhinos, rabbits, koalas, many rodents, and some reptiles and birds, including the iguana and chicken, are all examples. Even humans benefit from microbial symbionts living in the cecum. However, because the

Carnivore (a dog)

Herbivore (a rabbit)

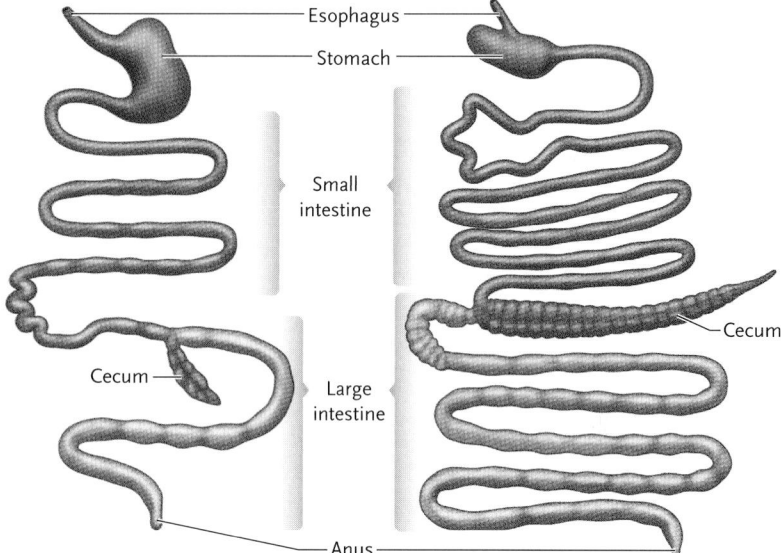

Figure 45.18

Comparison of the length of the digestive tract in a carnivore and an herbivore. The carnivore has a relatively short digestive tract, while the herbivore's digestive tract is much longer.

cecum and the remainder of the large intestine have little capacity to absorb nutrients, microbial digestion in the cecum is not as productive as microbial digestion in the stomach.

We have seen that animals use various strategies to extract the available nutrients from foods. These nu-

Figure 45.19

A ruminant, the pronghorn *(Antilocapra americanus),* and its four-chambered system that digests plant matter with the aid of symbiotic microorganisms.

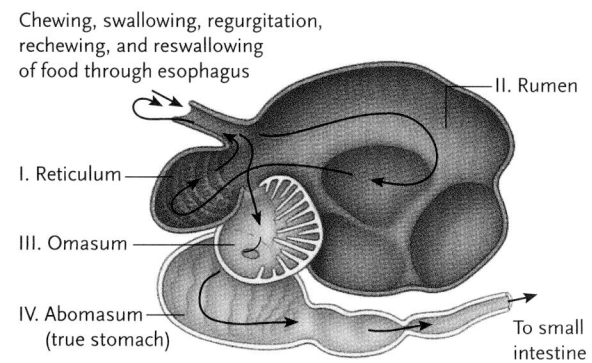

tritional strategies involve multiple steps combining both mechanical and chemical processing, which convert complex foodstuffs into the absorbable subunits that animals need to sustain life. Obtaining food is costly in terms of energy and risk, and animals have evolved many ways to make the most of it.

STUDY BREAK

1. How are the different types of teeth used in feeding?
2. What roles do symbiotic microorganisms play in digestion?

UNANSWERED QUESTIONS

How is energy partitioned in animals?

In this chapter you learned that ingested food molecules are broken down into smaller units that can be readily absorbed into body fluids and cells. The ingested energy is assimilated into the organism and partitioned into four main categories: maintenance metabolism, growth, reproduction, and storage. The top priority for energy allocation is maintenance metabolism—the energy required to seek and digest food and to support essential life processes. Researchers are studying the regulation of energy partitioning among the other categories when available energy exceeds that required for maintenance. The findings indicate that priorities for energy allocation vary between species as well as during an individual's life history. In the case of growth, for example, some species grow for a short period (determinant growth), while other species grow throughout their lives (indeterminant growth). Animal growth is controlled by genetic, environmental, and nutritional factors, but how such cues are integrated with each other and with other processes, such as reproduction and storage, remain to be determined. For example, the cues used to shut down growth and reproduction during fasting are not fully known. (In humans, malnourished juveniles become growth retarded and adult females stop menstruating.) The extent to which severe energy restriction may result in metabolic adaptation (reduced metabolic rate) also needs to be examined.

How are foraging and feeding regulated?

In this chapter you also learned about the control of appetite and oxidative metabolism. While some of the players are known—among them hunger centers in the brain, appetite-stimulating hormones such as neuropeptide Y, and appetite-suppressing or satiety hormones such as leptin—a number of questions remain. For example, how are various sensory inputs, such as smell, integrated to initiate feeding? Similarly, how are other inputs, such as gut contents or increasing nutrient concentration in the blood, integrated to terminate feeding? Also, how is an animal's feeding strategy matched to nutrient quantity and quality in the environment and to the animal's metabolic pattern: for example, sluggish or active (tortoise or hare)? Inevitably, a complex of integrating chemicals—produced by the brain, gut, fats cells, and other tissues—will prove to be involved. Knowledge of this complex system will provide insight not only into obesity but into eating disorders such as anorexia as well.

How are lipids taken up and utilized?

Adipose and other cells take up lipids (mostly triglycerides) that have been hydrolyzed from chylomicrons and very-low-density lipoprotein (VLDL) by the enzyme lipoprotein lipase (LPL). Low-density lipoproteins (LDLs) are VLDL remnants and are the major means of delivering cholesterol to tissues. Normally, LDLs are taken up by receptor-mediated endocytosis. Cholesterol can be scavenged by high-density lipoproteins (HDLs) through the action of the enzyme lecithin cholesterol acyltransferase (LCAT), taken up by the liver, and removed from the body. Researchers have found that defects in LPL, LCAT, and LDL disrupt normal lipid metabolism and lead to severe health risks. For example, genetic defects in the LDL receptor are associated with familial hypercholesterolemia, in which excess cholesterol is left in the blood and causes heart disease and other abnormalities. Continued research on lipid and lipoprotein metabolism will be important for understanding obesity, type 2 diabetes mellitus, and cardiovascular disease.

Mark Sheridan is the James A. Meier Professor of Biological Sciences at North Dakota State University, where he studies the control of growth, development, and metabolism in vertebrates. His current research focuses on somatostatin signaling, growth regulation, and environmental endocrinology. Find out more about his research at http://www.ndsu.nodak.edu/ndsu/msherida/research.interests.

Review

45.1 Feeding and Nutrition

- Animals obtain organic molecules by eating other organisms. Herbivores primarily eat plants, carnivores primarily eat other animals, and omnivores eat animals, plants, and other sources of organic nutrients. The organic molecules are used as fuels for oxidative reactions providing energy and as building blocks for making complex biological molecules.

- Animals require essential substances in their diets—amino acids, fatty acids, vitamins, and minerals—that they cannot make for themselves.

- Animals may be classified with respect to feeding methods and the physical state of the organic molecules they eat. Fluid feeders ingest liquids containing organic molecules in solution. Suspension feeders eat small particles of organic matter or small organisms in suspension in fluids. Deposit feeders ingest small organic particles or organisms that are part of solid matter that the feeders live in or on. Bulk feeders consume large pieces of organisms, or entire large organisms (Figure 45.2).

45.2 Digestive Processes

- Digestion is the process of mechanical and chemical breakdown of food into molecular subunits small enough to be absorbed into body fluids and cells.

- Digestion may be intracellular or extracellular. Extracellular digestion allows food to be eaten in large batches, stored, and broken down while the animal carries out other activities.

- In animals with extracellular digestion, the digestive processes take place in an internal body cavity that is either a pouch or sac with one opening that serves as both mouth and anus, or a tube with two openings forming a mouth on one end and an anus on the other end (Figures 45.3 and 45.4).

- In animals with a digestive tube, digestion occurs in five stages: (1) mechanical processing, including chewing and grinding of food; (2) secretion of enzymes and other digestive aids into the digestive tract; (3) enzymatic hydrolysis of food molecules into molecular subunits; (4) absorption of the molecular subunits across cell membranes; and (5) elimination of undigested matter.

- Food particles and molecules are pushed through the digestive tube by muscular contractions of its wall. Storage of food at various locations in the tube allows animals to digest food while engaged in other activities.

Animation: Examples of digestive systems

45.3 Digestion in Humans and Other Mammals

- Adult humans require eight essential amino acids, 13 vitamins (Table 45.1) and a large number of essential minerals (Table 45.2).

- The mouth, pharynx, esophagus, stomach, intestine, and anus are common to the digestive system of mammals, including humans, and most vertebrates (Figure 45.5).

- The wall of the vertebrate gut is formed from four layers of tissues: the mucosa, the submucosa, the muscularis, and the serosa (Figure 45.7).

- Coordinated contractions of the circular and smooth muscles produce peristaltic waves that move the digestive contents from the mouth to the anus (Figure 45.8).

- Digestion begins in the mouth, where the teeth break the food into smaller bits. Salivary amylase, an enzyme that digests starch, is secreted into the food in the mouth. After chewing, the food is swallowed and travels through the pharynx and esophagus to reach the stomach (Figure 45.9).

- In the stomach, hydrochloric acid, the protein-digesting enzyme pepsin, and mucus are added to the food mass. The stomach churns the acid contents into chyme, which is released in pulses into the small intestine (Figure 45.10).

- Absorption of nutrients begins in the small intestine. Specializations of the small intestine to optimize absorption are the intestinal villi and microvilli (Figure 45.11).

- In the small intestine, digestive juices from the pancreas and liver add enzymes and digestive aids to the food mass (Figure 45.12). The pancreatic juice contains digestive enzymes and bi-carbonate ions that neutralize the acidity of the digestive contents. The liver secretion, bile, contains bile salts, which emulsify fats, cholesterol, bilirubin, and additional bicarbonate ions.

- The small intestine secretes enzymes that complete most digestion. The mucosal cells of the small intestine absorb the molecular subunits created by digestion (Figures 45.13 and 45.14).

- Absorbed nutrients are delivered to the liver, where excess glucose is converted into glycogen and fats, and some of the amino acids are converted into plasma proteins or sugars. The liver also synthesizes cholesterol from lipids, carbohydrates, and other substances.

- The large intestine absorbs water and mineral ions from the digestive contents. At the end of the large intestine the undigested remnants, the feces, are expelled from the anus (Figure 45.15).

Animation: Human digestive system

Animation: Peristalsis

Animation: Structure of the small intestine

Animation: Structure of the large intestine

Animation: Vitamins

45.4 Regulation of the Digestive Process

- Digestion is regulated by signals from the autonomic nervous system, by the activity of neuron networks in the digestive tube wall, and by hormones secreted by the digestive system. The regulatory mechanisms operate in response to signals from sensory receptors that monitor the volume and composition of the digestive contents (Figure 45.16).

Animation: Body mass index

Animation: Caloric requirements

Animation: Chronology of leptin research

45.5 Digestive Specializations in Vertebrates

- Common variations in vertebrate digestive systems include modifications of the teeth, length of the digestive tract, and structure and function of the stomach.

- Mammals have four basic types of teeth—incisors for cutting, canines for piercing, and premolars and molars for cutting, grinding, and smashing food (Figure 45.17).

- Carnivores and vertebrates that eat other nutrient-rich foods have a relatively short digestive tract. Herbivores, which eat nutrient-poor foods, typically have a relatively long digestive tract that includes extensive storage regions (Figure 45.18).

- Many herbivores have digestive chambers in which symbiotic microorganisms digest plant matter into molecules that can be absorbed by the host (Figure 45.19).

Animation: Human teeth

Animation: Ruminant stomach function

Questions

Self-Test Questions

1. Required molecules that animals cannot synthesize are called:
   a. nutrients.
   b. essential nutrients.
   c. enzymes.
   d. proteins.
   e. carbohydrates.

2. Which of the following accurately describes a feeding style?
   a. Deposit feeders obtain nutrients from organic molecules in solution.
   b. Deposit feeders scrape organic matter from solid material on which they live.
   c. Fluid feeders digest organisms suspended in water.

d. Fluid feeders strain food with networks of mucus or bristles and hairs.
e. Suspension feeders consume sizable food whole or in chunks.

3. The order of successive steps in digestion is:
a. absorption follows enzymatic hydrolysis.
b. secretion of enzymes follows absorption of digestive material.
c. mechanical processing follows enzyme secretion.
d. mechanical processing follows enzymatic hydrolysis.
e. enzymatic hydrolysis precedes secretion of digestive aids.

4. The esophagus, crop, gizzard, and intestine are found in:
a. birds and mammals.
b. insects and mammals.
c. flatworms and birds.
d. earthworms and birds.
e. sponges and cnidarians.

5. All of the following are essential nutrients in humans *except:*
a. vitamin B.
b. calcium.
c. glycogen.
d. linoleic acid.
e. vitamin K.

6. A specialized region of the gut is/are the:
a. submucosa formed by circular and longitudinal layers.
b. serosa lining the gut for absorption.
c. mucosa composed of thick elastic connective tissue for movement.
d. muscularis, an outer layer that secretes a slippery material to prevent friction with other organs.
e. sphincters, which form valves between major digestive organs.

7. If the fat in whole milk is ingested:
a. the stomach, with its high pH, will stimulate cells of the duodenum to hasten stomach emptying.
b. parietal cells in the stomach will absorb it.
c. in the small intestine, bile salts emulsify the fats and then lipase hydrolyzes them.
d. lactase deficiency in the small intestine would prevent its digestion.
e. microvilli will absorb the fat in the form of chylomicrons directly into the blood of the hepatic portal vein.

8. The liver's role in digestion is to:
a. synthesize aminopeptidase and dipeptidase to digest polypeptides.
b. synthesize lipase to form free fatty acids.
c. secrete trypsin to break the bonds in polypeptides.
d. secrete bile and bicarbonate ions to help emulsify fats.
e. store bile between meals.

9. Which of the following best describes regulation of digestion?
a. GIP inhibits insulin release from the pancreas.
b. Gastrin stimulates pancreatic secretion of HCl and pepsinogen.
c. Secretin stimulates gastric emptying into the duodenum.
d. CCK stimulates gastric activity to activate the duodenum.
e. Leptin binds different hypothalamic receptors to stimulate or inhibit appetite.

10. An example of a digestive specialization is seen in:
a. the long intestines characteristic of herbivores.
b. the incisors being the dominant teeth in wolves.
c. the canine teeth being the dominant teeth in deer.
d. salivary lipase being made by humans.
e. cellulose being made by humans.

Questions for Discussion

1. As a person ages, the number of cells in the body steadily decreases and their energy needs decline. If you were planning an older person's diet, what kind(s) of nutrients would you emphasize, and why? Which ones would you recommend less of? Include vitamins and minerals in your answer.

2. Formulate a healthy diet for a young, actively growing 7-year-old, and explain why you have included each part of the diet. Refer to Question 1, above, for some issues to consider.

3. A baby develops symptoms of protein deficiency, and the attending physician suggests the cause is a genetic defect leading to a nonfunctional enzyme associated with digestion. Name at least three enzymes that might be likely suspects, and for each one explain how the defect would result in a protein deficiency.

Experimental Analysis

Design experiments to test whether cigarette smoke affects the functioning of the various parts of the digestive system.

Evolution Link

What is the advantage of a tubelike digestive system over a saclike digestive system?

How Would You Vote?

Many nutritionists suspect that increasing consumption of "fast foods" is contributing to rising levels of obesity. Should fast-food labels carry consumer warnings, as alcohol and cigarette labels do? Go to www.thomsonedu.com/login to investigate both sides of the issue and then vote.

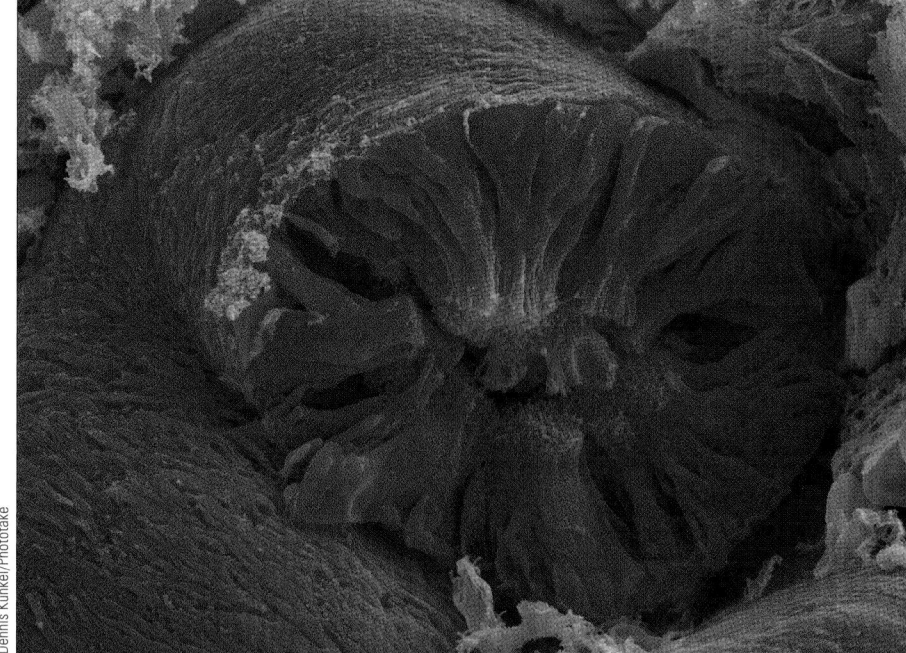

A nephron in a human kidney (colorized SEM). Nephrons are the specialized tubules in kidneys that filter the blood to conserve nutrients and water, balance salts in the body, and concentrate wastes for excretion from the body.

Continued on next page

46 Regulating the Internal Environment

WHY IT MATTERS

The crew of the World War II bomber *Lady Be Good* was assigned to fly a night mission to Naples, Italy, from a base on the North African coast on August 4, 1943. But trouble dogged the mission, forcing the crew to turn back before reaching their target. Navigational errors and a cloud layer led them to miss their home base and continue south over the hostile Sahara Desert. Some 440 miles from the coastline, with the fuel running out, the nine crew members parachuted from the aircraft. The bomber remained airborne for a few more minutes and then crashed, leaving its crew miles behind.

The eight men who survived began a northward trek with only half a canteen of water among them, in desert heat that reached 130°F during the day. In a testimony to the physiological mechanisms that conserve water and cool the body, they continued onward for eight days. But then, one by one, they succumbed as the merciless heat and dehydration exceeded their capacity to survive. Rescue teams searched the desert for weeks after their disappearance, but no trace was found of the crew or their airplane.

The fate of the *Lady Be Good* remained unknown until 1958, when an oil exploration team flying over the desert spotted the aircraft, sitting largely intact in the desert sands. A 2-year search finally led to the remains of the crew, some of them more than a hundred miles north of the downed bomber. Diaries found among the scattered effects told the poignant story of the flight and the futile struggle against the dehydrating desert environment.

This story illustrates only too clearly the trials of animal life under changing environmental conditions. Water and required nutrients may become more or less abundant. Temperatures may rise or fall. Animals have evolved an astounding capacity to compensate for fluctuating external conditions and to maintain the internal environment of their bodies within the relatively narrow limits that cells can tolerate.

These limits, and the compensating mechanisms that maintain them, are the subjects of this chapter. First we examine **osmoregulation**, the regulation of water and ion balance, and the closely related topic of **excretion**, which helps maintain the body's water and ion balance while ridding the body of metabolic wastes. We then consider **thermoregulation**, the control of body temperature.

46.1 Introduction to Osmoregulation and Excretion

Living cells contain water, are surrounded by water, and constantly exchange water with their environment. For the simplest animals, the water of the external environment directly surrounds cells. For more complex animals, an aqueous extracellular fluid surrounds the cells, and is separated from the external environment by a body covering. In animals with a circulatory system, the extracellular fluid includes both the interstitial fluid immediately surrounding cells and the blood or other circulated fluid; these are commonly called body fluids.

In this section, we review the mechanisms cells use to exchange water and solutes with the surrounding fluid through *osmosis*. We also look at how animals harness osmosis to maintain *water balance,* the equilibrium in inward and outward flow of water.

Osmosis Is a Form of Passive Diffusion

In osmosis (see Section 6.3), water molecules move across a selectively permeable membrane from a region where they are more highly concentrated to a region where they are less highly concentrated. The difference in water concentration is produced by differing numbers of solute molecules or ions on the two sides of the membrane. The side of the membrane with a *lower* solute concentration has a *higher* concentration of water molecules, so water will move osmotically to the other side, where water concentration is *lower.*

Selective permeability is a key factor in osmosis because it helps maintain differences in solute concentration on either side of biological membranes. Proteins are among the most important solutes in establishing the conditions that produce osmosis.

The total solute concentration of a solution, called its **osmolarity,** is measured in *osmoles*—the number of solute molecules and ions (in moles)—per liter of solution. Because the total solute concentration in the body fluids of most animals is less than 1 osmole, osmolarity is usually expressed in thousandths of an osmole, or *milliosmoles* (mOsm). As shown in **Figure 46.1,** the osmolarity of body fluids in humans and other mammals is about 300 mOsm/L; osmolarity in a flounder, a marine teleost (bony fish), is about 330 mOsm/L, and in a goldfish, a freshwater teleost, it is about 290 mOsm/L. By contrast, sharks and many marine invertebrates such as lobsters have osmolarities close to that of seawater, about 1000 mOsm/L, and freshwater invertebrates have an osmolarity of about 225 mOsm/L.

Considering solutions on either side of a selectively permeable membrane, a solution of higher osmolarity is said to be *hyperosmotic* to a solution of lower osmolarity, and a solution of lower osmolarity is said to be *hypoosmotic* to a solution of higher osmolarity. If the solutions on either side of a membrane have the same osmolarity, they are said to be *isoosmotic*. Water moves across the membrane between solutions that differ in osmolarity (see Figure 6.9), whereas when two solutions are isoosmotic, no net water movement occurs.

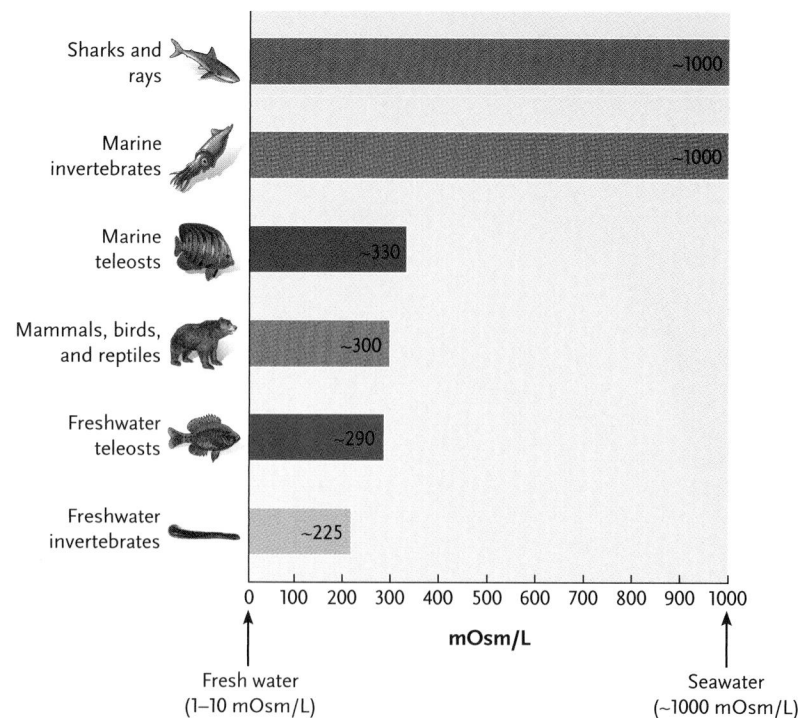

Figure 46.1
Osmolarity of body fluids in some animal groups.

Animals Use Different Approaches to Keep Osmosis from Swelling or Shrinking Their Cells

Because even small differences in osmolarity can cause cells to swell or shrink, animals must keep their cellular and extracellular fluids isoosmotic. In some animals, called **osmoconformers,** the osmolarity of the cellular and extracellular solutions simply matches the osmolarity of the environment. Most marine invertebrates are osmoconformers. Other animals, called **osmoregulators,** use control mechanisms to keep the osmolarity of cellular and extracellular fluids the same, but at levels that may differ from the osmolarity of the surroundings. Most freshwater and terrestrial invertebrates, and almost all vertebrates, are osmoregulators.

For terrestrial animals, one of the greatest challenges to osmoregulation is the limited supply of water in the environment—if the crew of the *Lady Be Good* had had an adequate supply of water, for example, they could probably have reached safety at the North African coast even without food.

Excretion Is Closely Tied to Osmoregulation

Control over osmolarity is partly maintained by removing certain molecules and ions from cells and body fluids and releasing them into the environment; thus, excretion is closely related to osmoregulation. Animals excrete H^+ ions to keep the pH of body fluids near the neutral levels required by cells for survival. They also excrete toxic products of metabolism, such as nitrogenous (nitrogen-containing) compounds resulting from the breakdown of proteins and nucleic acids, and

breakdown products of poisons and toxins. Excretion of ions and metabolic products is accompanied by water excretion since water serves as a solvent for those molecules. Animals that take in large amounts of water may also excrete water to maintain osmolarity.

Microscopic Tubules Form the Basis of Excretion in Most Animals

Except in the simplest animals, minute tubular structures carry out osmoregulation and excretion **(Figure 46.2).** The tubules are immersed in body fluids at one end (called the *proximal end* of the tubules), and open directly or indirectly to the body exterior at the other end (called the *distal end* of the tubules). The tubules are formed from a **transport epithelium**—a layer of cells with specialized transport proteins in their plasma membranes. The transport proteins move specific molecules and ions into and out of the tubule by either active or passive transport, depending on the particular substance and its concentration gradient.

Typically, the tubules function in a four-step process:

- **Filtration.** Filtration is the nonselective movement of some water and a number of solutes—ions and small molecules, but not large molecules such as proteins—into the proximal end of the tubules through spaces between cells. In animals with an open circulatory system, the water and solutes come from body fluids, with movement into the tubules driven by the higher pressure of the body fluids compared with the fluid inside the tubule. In animals with a closed circulatory system, such

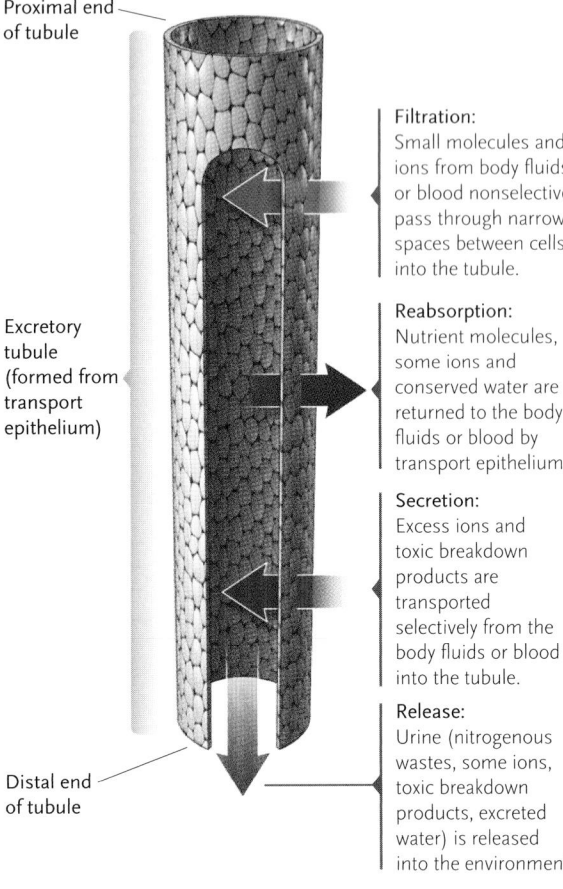

Proximal end of tubule

Filtration:
Small molecules and ions from body fluids or blood nonselectively pass through narrow spaces between cells into the tubule.

Reabsorption:
Nutrient molecules, some ions and conserved water are returned to the body fluids or blood by transport epithelium.

Secretion:
Excess ions and toxic breakdown products are transported selectively from the body fluids or blood into the tubule.

Release:
Urine (nitrogenous wastes, some ions, toxic breakdown products, excreted water) is released into the environment.

Excretory tubule (formed from transport epithelium)

Distal end of tubule

Figure 46.2

Common structures and operations of the tubules carrying out osmoregulation and excretion in animals. The tubules are typically formed from a single layer of cells with transport functions.

as humans, the water and solutes come from the blood in capillaries that surround the tubules, with the movement into the tubules driven by blood pressure. (Open and closed circulatory systems are described in Section 42.1.)

- **Reabsorption.** In reabsorption, some molecules (for example, glucose and amino acids) and ions are transported by the transport epithelium back into the body fluid (animals with open circulatory systems) or into the blood in capillaries surrounding the tubules (animals with closed circulatory systems) as the filtered solution moves through the excretory tubule.

- **Secretion.** Secretion is a selective process in which specific small molecules and ions are transported from the body fluids (animals with open circulatory systems) or blood (animals with closed circulatory systems) into the tubules. Secretion is the second and more important route for eliminating particular substances from the body fluid or blood, filtration being the first. The difference between the two processes is that filtration is nonselective whereas secretion is selective for substances transported. The same substances are transported into the tubule by secretion as in filtration; those sub-

stances are added, therefore, to substances already in the tubule as a result of filtration.

- **Release.** The fluid containing waste materials—urine—is released into the environment from the distal end of the tubule. In some animals the fluid is concentrated into a solid or semisolid form.

The tubules may number from hundreds to millions depending on the species. In combination, they expand the transport epithelium to a total surface area large enough to accomplish the osmoregulatory and excretory functions of the animal. In all vertebrates and many invertebrates, the excretory tubules are concentrated in specialized organs, the *kidneys*, which are discussed in later sections.

Animals Excrete Nitrogen Compounds as Metabolic Wastes

The metabolism of ingested food is a source of both energy and molecules for the biosynthetic activities of an animal. Importantly, metabolism of ingested food produces water—called *metabolic water*—that is used in chemical reactions and is involved in physiological processes such as the excretion of wastes.

The proteins, amino acids, and nucleic acids in food are broken down as part of digestion. The same molecules are broken down in body cells as a result of the normal processes of synthesis and replacement. The nitrogenous products of this breakdown are excreted by most animals as *ammonia, urea,* or *uric acid,* or a combination of these substances **(Figure 46.3).** The particular molecule or combination of molecules produced depends on a balance among toxicity, water conservation, and energy requirements.

Ammonia. Ammonia (NH_3) is the result of a series of biochemical steps beginning with the removal of amino groups ($—NH_3^+$) from amino acids as a part of protein breakdown. Ammonia is readily soluble in water, but it is also highly toxic. Therefore, ammonia must either be excreted or be converted to a nontoxic derivative. However, because of its toxicity, ammonia can be excreted from the body only in dilute solutions, making this path possible only in animals with a plentiful supply of water. Those animals include aquatic invertebrates, teleosts, and larval amphibians; ammonia is the primary nitrogenous waste for them. Terrestrial animals, and some aquatic animals, instead detoxify ammonia, converting it either into urea or uric acid.

Urea. All mammals, most amphibians, some reptiles, some marine fishes, and some terrestrial invertebrates combine ammonia with HCO_3^- and convert the product in a series of steps to *urea,* a soluble and relatively nontoxic substance. Although producing urea requires more energy than forming ammonia, excreting urea instead of ammonia requires only about 10% as much water.

Uric Acid. Water is conserved further in some animals, including terrestrial invertebrates, reptiles, and birds, by the formation of uric acid instead of ammonia or urea. Uric acid is nontoxic, and so insoluble that it precipitates in water as a crystal. (The white substance in bird droppings is uric acid.) The embryos of reptiles and birds, which develop within leathery or hard-shelled eggs that are impermeable to liquids, also conserve water by forming uric acid, which is stored as a waste product.

Although making uric acid requires even more energy than urea, molecule for molecule it contains four times as much nitrogen as ammonia. And, because uric acid precipitates from water as a crystal, it can be excreted as a concentrated paste. These factors conserve about 99% of the water that would be required to excrete an equivalent amount of nitrogen as ammonia.

We have now covered the basics of osmoregulation and excretion. In the sections that follow, we look at the specifics of these processes in different animal groups, beginning with the invertebrates.

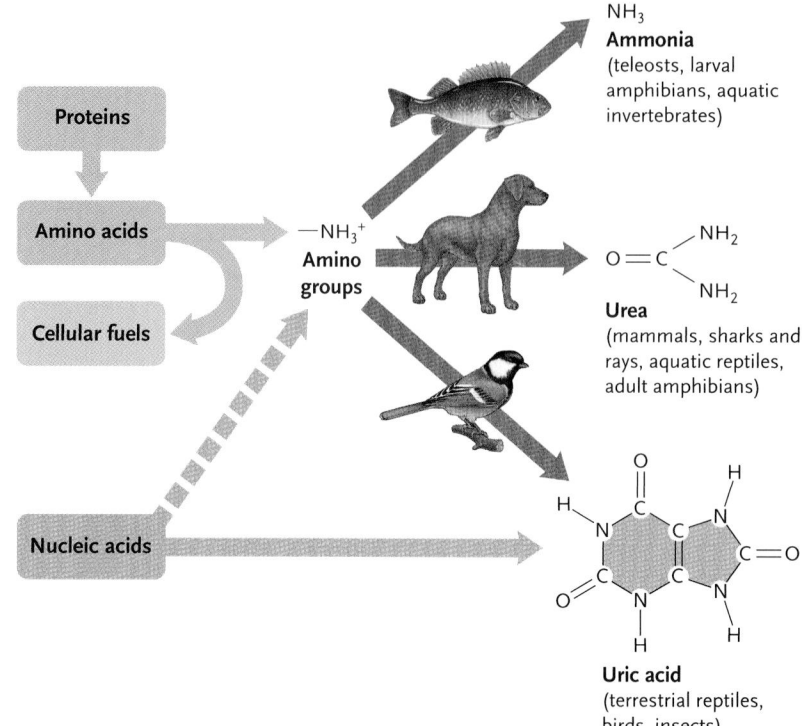

Figure 46.3

Nitrogenous wastes excreted by different animal groups. Although humans and other mammals primarily excrete urea, they also excrete small amounts of ammonia and uric acid.

STUDY BREAK

Define the terms osmosis, osmolarity, hypoosmotic, osmoregulator, and transport epithelium.

46.2 Osmoregulation and Excretion in Invertebrates

Both osmoconformers and osmoregulators occur among the invertebrates. Except for the simplest groups, most invertebrates, whether osmoconformers or osmoregulators, carry out excretion by specialized excretory tubules.

Most Marine Invertebrates Are Osmoconformers; All Freshwater and Terrestrial Invertebrates Are Osmoregulators

Most marine invertebrates are osmoconformers. All these animals release water, certain ions, and nitrogenous wastes—usually in the form of ammonia—directly from body cells to the surrounding seawater. The cells of these animals do not swell or shrink because the osmolarity of their intracellular and extracellular fluids and the surrounding seawater is the same, about 1000 mOsm/L. Therefore, they do not have to expend energy to maintain their osmolarity. However, osmoconformers do expend energy to keep some ions,

such as Na^+, at lower concentrations inside cells than in the surroundings.

In contrast, all freshwater invertebrates are osmoregulators because their cells could not survive if their internal ion concentrations were reduced to freshwater levels. Terrestrial invertebrates are osmoregulators as well. These animals must expend energy to keep their internal fluids hyperosmotic to their surroundings. Although osmoregulation is energetically expensive, these invertebrates can live in more varied habitats than osmoconformers can.

The internal hyperosmoticity of freshwater osmoregulators such as flatworms and mussels causes water to move constantly from the surroundings into their bodies. This excess water must be excreted, at a considerable cost in energy, to maintain internal hyperosmoticity. These animals must also obtain the salts required to keep their body fluids hyperosmotic to fresh water. The salts are obtained from foods, and by actively transporting salt ions from the water into their bodies (even fresh water contains some dissolved salts). This active ion transport occurs through the skin or gills.

Among terrestrial osmoregulators are annelids (earthworms), arthropods (insects, spiders and mites, millipedes, and centipedes), and mollusks (land snails and slugs). Like their freshwater relatives, these invertebrates must obtain salts from their surroundings, usually in their foods. While they do not have to excrete water entering by osmosis, they must constantly replace water lost from their bodies by evaporation.

In Invertebrate Osmoregulators, Specialized Excretory Tubules Participate in Osmoregulation and Carry Out Excretion

Invertebrate osmoregulators typically use specialized tubules for carrying out excretion. Three common types of these specialized tubules that differ in which body fluids are processed and how are *protonephridia*, found in flatworms and larval mollusks; *metanephridia*, found in annelids and most adult mollusks; and *Malpighian tubules*, found in insects and other arthropods.

Protonephridia. The flatworm *Dugesia* provides an example of the simplest form of invertebrate excretory tubule, the **protonephridium** (*protos* = first; *nephros* = kidney). In *Dugesia*, two branching networks of protonephridia run the length of the body **(Figure 46.4)**. The smallest branches of the tubule network end with a large cell containing a bundle of cilia that reach into the tubule and that beat to move fluid through the tubule. This cell is called a *flame cell* because the movement of its cilia resembles a flickering flame. The plasma membrane of the flame cell interdigitates with the plasma membrane of the tubular cell with which it connects. Hemolymph enters the tubule through the membranes in the area where the two membranes interdigitate. As the fluids pass through the protonephridia, some molecules and ions are reabsorbed and others, including nitrogenous wastes, are secreted into the tubules; the urine resulting from this filtration system is released through pores at the ends of the tubules where the tubules reach the body surface.

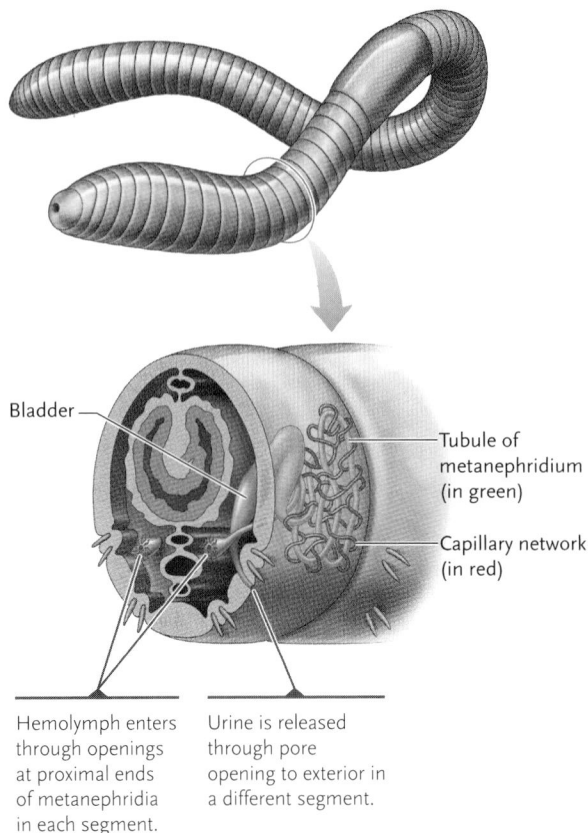

Bladder — Tubule of metanephridium (in green)

Capillary network (in red)

Hemolymph enters through openings at proximal ends of metanephridia in each segment.

Urine is released through pore opening to exterior in a different segment.

Figure 46.5
The metanephridium of an earthworm.

Figure 46.4
The protonephridia of the planarian *Dugesia*, showing a flame cell.

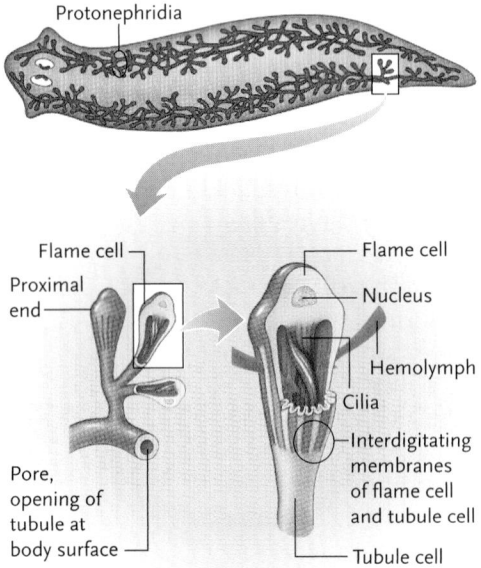

Protonephridia

Flame cell — Flame cell
Proximal end — Nucleus
Hemolymph
Cilia
Interdigitating membranes of flame cell and tubule cell
Pore, opening of tubule at body surface — Tubule cell

Metanephridia. The excretory tubule of most annelids and adult mollusks, the **metanephridium** (*meta* = between), has a funnel-like proximal end surrounded with cilia that admits hemolymph. Like protonephridia, metanephridia are filtration systems. As hemolymph moves through the tubule, some molecules and ions are reabsorbed, and other ions and nitrogenous wastes are secreted into the tubule and excreted from the body surface.

Figure 46.5 shows the arrangement of metanephridia in an earthworm. The proximal ends of a pair of metanephridia are located in each body segment, one on either side of the animal. Each tubule of the pair extends into the following segment, where it bends and folds into a convoluted arrangement surrounded by a network of blood vessels. Reabsorption and secretion take place in the convoluted section. Urine from the distal end of the tubule collects in a saclike storage organ, the *bladder*, from where it is released through a pore in the surface of the segment. Samples taken with a microneedle from various regions of a metanephridium show that the fluid entering the tubule contains all the smaller molecules and ions of the body fluid; as the fluid moves through the tubules, specific molecules and ions are removed by reabsorption and added by secretion.

Malpighian Tubules. The excretory tubule of insects, the **Malpighian tubule**, has a closed proximal end that is immersed in the hemolymph **(Figure 46.6)**. The distal ends of the tubules empty into the gut. In contrast to protonephridia and metanephridia, Malpighian tubules do not filter body fluids; instead, they are excretory systems that use secretion to generate the fluid for release from the body. In particular, uric acid and several ions, including Na^+ and K^+, are actively secreted into the tubules. As the concentration of these substances rises, water moves osmotically from the hemolymph into the tubule. The fluid then passes into the hindgut (intestine and rectum) of the insect as dilute urine. Cells in the hindgut wall actively reabsorb most of the Na^+ and K^+ back into the hemolymph; water follows by osmosis. The uric acid left in the gut precipitates into crystals, which mix with the undigested matter in the rectum and are released with the feces.

STUDY BREAK

Describe protonephridia, metanephridia, and Malpighian tubules. In which animal groups are each of these excretory tubules found?

46.3 Osmoregulation and Excretion in Mammals

In all vertebrates, specialized excretory tubules contribute to osmoregulation and carry out excretion. The excretory tubules, called **nephrons**, are located in a specialized organ, the kidney. We begin our survey of vertebrate osmoregulation and excretion with a description of the structure and function of the mammalian kidney.

The Kidneys and Ureters, the Bladder, and the Urethra Constitute the Urinary System

Mammals have a pair of kidneys, located on either side of the vertebral column at the back of the abdominal cavity **(Figure 46.7)**. Internally, the mammalian kidney is divided into an outer **renal cortex** surrounding a central region, the **renal medulla.**

Body fluids are carried in blood through the **renal artery** to a kidney, where metabolic wastes and excess ions are moved into the nephrons and where urine is formed. The filtered blood is routed away from the kidney by the **renal vein.** The urine leaving individual nephrons is processed further in **collecting ducts** and then drains into a central cavity in the kidney called the **renal pelvis.**

From the renal pelvis, the urine flows through a tube called the **ureter** to the **urinary bladder,** a storage sac located outside the kidneys. Urine leaves the bladder through another tube, the **urethra,** which (in most mammals) opens to the outside. In human females, the opening of the urethra is just in front of the vagina; in males, the urethra opens at the tip of the penis. The two kidneys and ureters, the urinary bladder, and the urethra constitute the mammalian urinary system.

Two sphincter muscles control the flow of urine from the bladder to the urethra. In human infants, urination is an autonomic reflex triggered by stretch receptors in the bladder wall. When the bladder be-

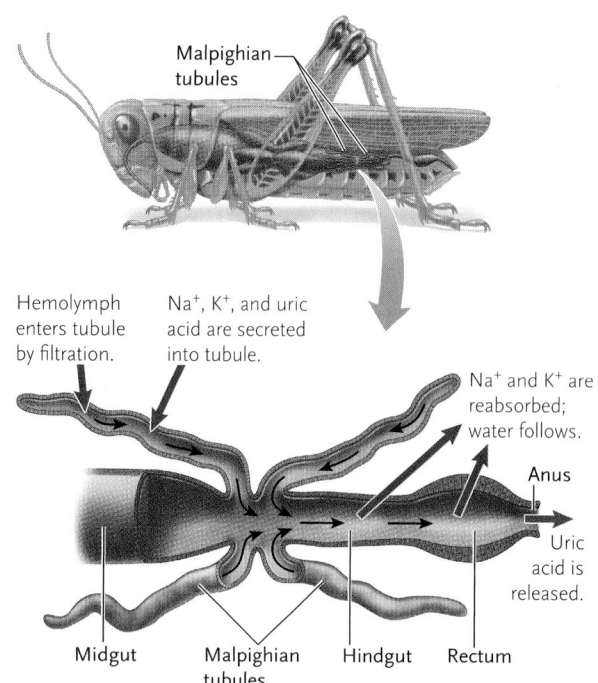

Figure 46.6
Excretion through Malpighian tubules in a grasshopper.

comes full, the sphincters relax, smooth muscles in the bladder wall contract, and the urine is forced to the exterior. At about the age of two years, children learn to override the autonomic reflex by consciously keeping the striated sphincter contracted until urination is convenient.

Mammalian Nephrons Are Differentiated into Regions with Specialized Functions

As in all mammals, human nephrons are differentiated into regions that perform successive steps in excretion. At its proximal end, a human nephron forms the **Bowman's capsule,** an infolded region that cups around a ball of blood capillaries called the **glomerulus (Figure 46.8).** The capsule and glomerulus are located in the renal cortex. Filtration takes place as body fluids are forced into Bowman's capsule from the capillaries of the glomerulus.

Following Bowman's capsule, the nephron forms a **proximal convoluted tubule** in the renal cortex, which descends into the renal medulla in a U-shaped bend called the **loop of Henle** and then ascends again to form a **distal convoluted tubule.** The distal tubule drains the urine into a collecting duct that leads to the renal pelvis. As many as eight nephrons may drain into a single collecting duct. The combined activities of the proximal convoluted tubule, the loop of Henle, the distal convoluted tubule, and the collecting duct convert the filtrate entering the nephron into urine.

Unlike most capillaries in the body, the capillaries in the glomerulus do not lead directly to venules. In-

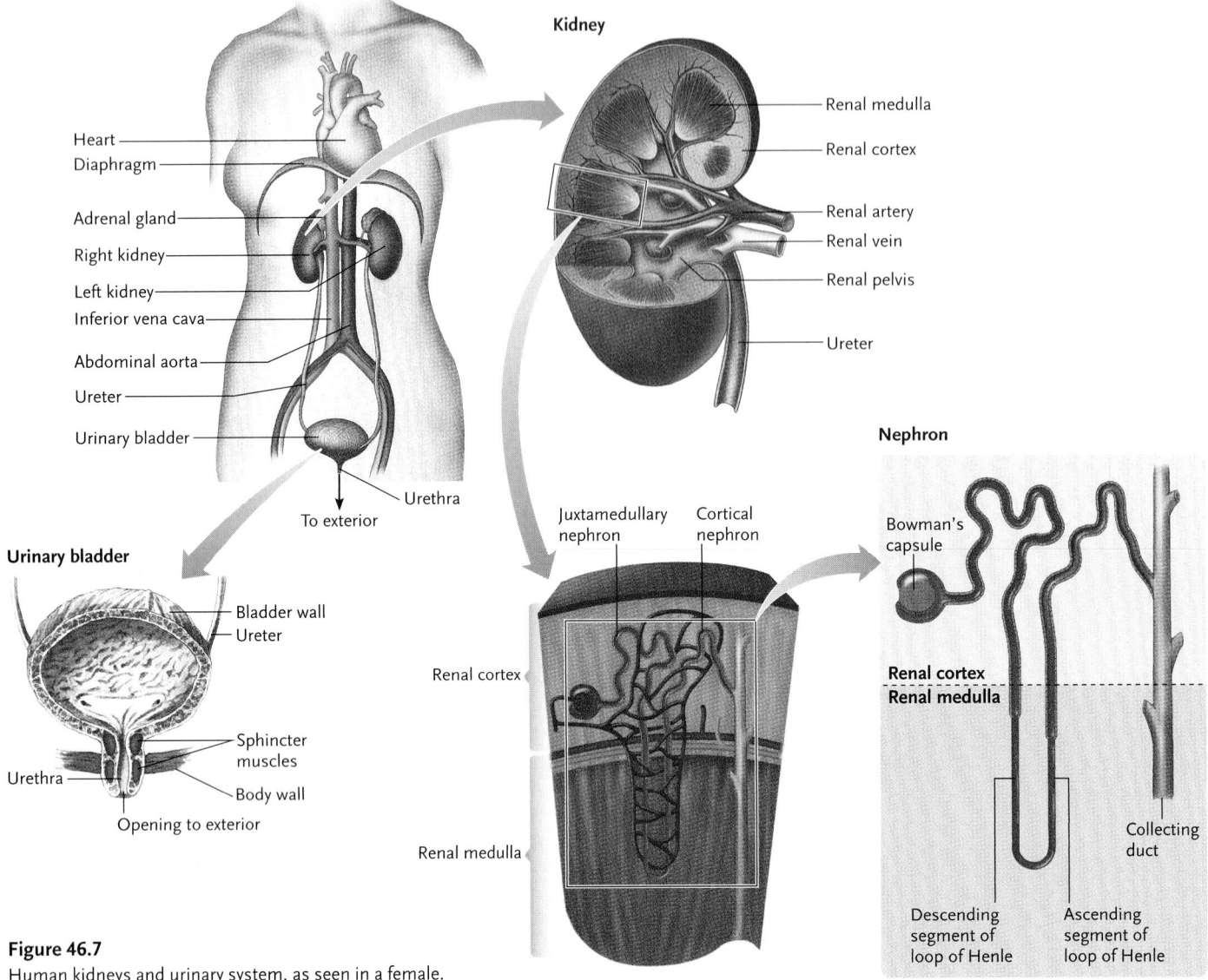

Figure 46.7
Human kidneys and urinary system, as seen in a female.

stead, they form another arteriole that branches into a second capillary network called the **peritubular capillaries.** These capillaries thread around the proximal and distal convoluted tubules and the loop of Henle. Molecules and ions that were reabsorbed during excretion are transferred between the nephron and the peritubular capillaries. However, because the capillaries and the tubules are not in physical contact due to the interstitial fluid between them, this transfer is not direct. Instead, the molecules or ions pass through the wall of the tubule, which is one cell layer thick; diffuse through the interstitial fluid; and then pass into the capillary through its wall, also one cell thick.

Each human kidney has more than a million nephrons. Of these, about 20% (the *juxtamedullary nephrons*) have long loops that descend deeply into the me-

dulla of the kidney. The remaining 80% (the *cortical nephrons*) have shorter loops, most of which are located entirely in the cortex, and the remainder of which extend only partway into the medulla.

Mammalian Nephrons Interact with Surrounding Kidney Structures to Produce Hyperosmotic Urine

In mammals, urine is hyperosmotic to body fluids. All other vertebrates except for a few aquatic bird species produce urine that is hypoosmotic to body fluids, or is at best isoosmotic. Production of hyperosmotic urine is a water-conserving adaptation that is primarily a mammalian characteristic. The production of hyperosmotic urine involves the activities of the mammalian

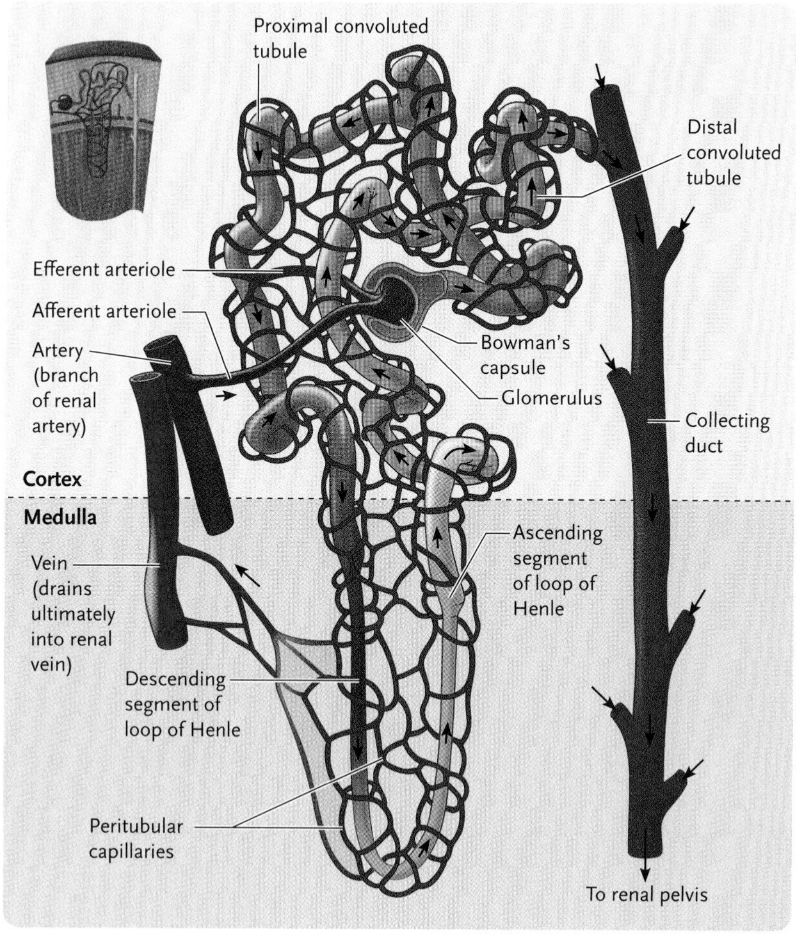

Figure 46.8
A nephron and its blood circulation.

Proximal convoluted tubule

Distal convoluted tubule

Efferent arteriole

Afferent arteriole

Artery (branch of renal artery)

Bowman's capsule

Glomerulus

Collecting duct

Cortex

Medulla

Vein (drains ultimately into renal vein)

Ascending segment of loop of Henle

Descending segment of loop of Henle

Peritubular capillaries

To renal pelvis

nephron itself and an interaction between nephrons and the highly ordered structure of the mammalian kidney. Three features underlie this interaction:

- The arrangement of the loop of Henle, which descends through the medulla and returns to the cortex again.
- Differences in the permeability of successive regions of the nephron, established by a specific group of membrane transport proteins in each region.
- A gradient in the concentration of molecules and ions in the interstitial fluid of the kidney, which increases gradually from the renal cortex to the deepest levels of the renal medulla.

These features interact to conserve nutrients and water, balance salts, and concentrate wastes for excretion from the body.

Researchers determined the transport activities of specific regions of nephrons by dissecting segments of nephrons out of an animal and experimentally manipulating them in vitro. They placed segments in different buffered solutions and passed solutions containing various components of filtrates through the

segment. By labeling specific molecules or ions radioactively, the scientists followed the movements of molecules in the solution surrounding the nephron segment or in the filtrate.

Filtration in Bowman's Capsule Begins the Process of Excretion

The mechanisms of excretion (shown in **Figure 46.9** and summarized in **Table 46.1**) begin in Bowman's capsule. The endothelial cells of the glomerulus capillaries and the cells of the Bowman's capsule are separated by spaces just wide enough to admit water, ions, small nutrient molecules such as glucose and amino acids, and nitrogenous waste molecules, primarily urea. The higher pressure of the blood drives fluid containing these molecules and ions from the capillaries of the glomerulus into the capsule. A thin net of connective tissue between the capillary and Bowman's capsule epithelia contributes to the filtering process. Blood cells and plasma proteins are too large to pass and are retained inside the capillaries.

Two factors help maintain the pressure driving fluid into Bowman's capsule. First, the diameters of

Figure 46.9

The movement of ions, water, and other molecules to and from nephrons and collecting tubules in the human kidney. Nephrons in other mammals and in birds work in similar fashion. The numbers are osmolarity values in mOsm/L.

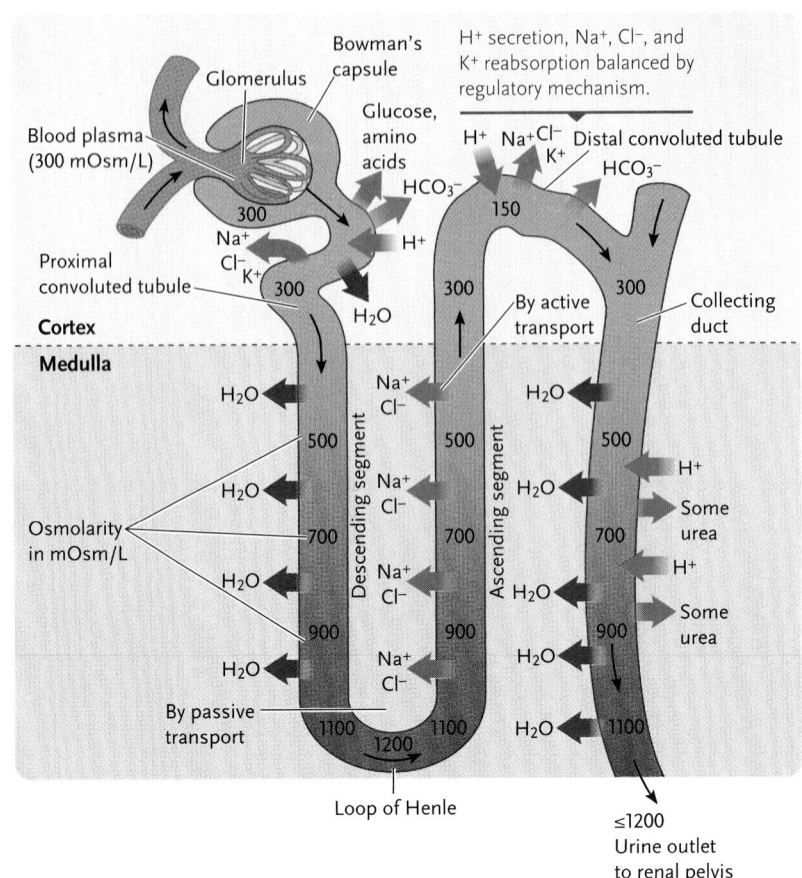

Table 46.1		Filtration, Reabsorption, and Secretion in Nephrons and Collecting Ducts		
Segment	Location	Permeability and Movement	Osmolarity of Filtrate and Urine	Result of Passage
Bowman's capsule	Cortex	Water, ions, small nutrients, and nitrogenous wastes move through spaces between epithelia	300 mOsm/L, same as surrounding interstitial fluid	Water and small substances, but not proteins, pass into nephron
Proximal convoluted tubule	Cortex	Na$^+$ and K$^+$ actively reabsorbed, Cl$^-$ follows; water leaves through aquaporins; H$^+$ actively secreted; HCO$_3^-$ reabsorbed into plasma of peritubular capillaries; glucose, amino acids, and other nutrients actively reabsorbed	300 mOsm/L	67% of ions, 65% of water, 50% of urea, and all nutrients return to interstitial fluid; pH maintained
Descending segment of loop of Henle	Cortex into medulla	Water leaves through aquaporins; no movement of ions or urea	From 300 mOsm/L at top to 1200 mOsm/L at bottom of loop	Additional water returned to interstitial fluid
Ascending segment of loop of Henle	Medulla into cortex	Na$^+$ and Cl$^-$ actively transported out; no entry of water; no movement of urea	From 1200 mOsm/L at bottom to 150 mOsm/L at top of loop	Additional ions returned to interstitial fluid
Distal convoluted tubule	Cortex	K$^+$ and Na$^+$ secreted via active transport into urine; Na$^+$ and Cl$^-$ reabsorbed; water moves into urine through aquaporins; HCO$_3^-$ reabsorbed into plasma of peritubular capillaries	From 150 mOsm/L at beginning to 300 mOsm/L at junction with collecting duct	Ion balance, pH balance
Collecting ducts	Cortex through medulla, empties into renal pelvis	Water moves out via aquaporins; no movement of ions; some urea leaves at bottom of duct	From 300 mOsm/L to 1200 mOsm/L at junction with renal pelvis	More water and some urea returned to interstitial fluid; some H$^+$ added to urine

the arteriole delivering blood to the glomerulus (called the **afferent arteriole**) and the capillaries of the glomerulus itself are larger than that of arterioles and capillaries elsewhere in the body. The larger diameter maintains blood pressure by presenting less resistance to blood flow. Second, the diameter of the arteriole that receives blood from the glomerulus (called the **efferent arteriole**) is smaller than the diameter of the afferent arteriole, producing a damming effect that backs up the blood in the glomerulus and helps keep the pressure high.

In humans, Bowman's capsules collectively filter about 180 L (47.5 gallons) of fluid each day, from a daily total of 1400 L (369.5 gallons) of blood that pass through the kidneys. The human body contains only about 2.75 L of blood plasma, meaning that the kidneys filter a fluid volume equivalent to 65 times the volume of the blood plasma each day. On average, more than 99% of the filtrate, mostly water, is reabsorbed in the nephrons, leaving about 1.5 L to be excreted daily as urine.

Reabsorption and Secretion Take Place in the Remainder of the Nephron

The fluid filtered into Bowman's capsule contains water, other small molecules, and ions at essentially the same concentrations as the blood plasma. By the time the fluid reaches the distal end of the tubules and passes through the collecting ducts, reabsorption out of the tubules and secretion into them have markedly altered the concentrations of all components of the filtrate.

The Proximal Convoluted Tubule. Reabsorption of water, ions, and nutrients back into the interstitial fluid is the main function of the proximal convoluted tubule. Na^+/K^+ pumps in the epithelium of the proximal convoluted tubule move Na^+ and K^+ from the filtrate into the interstitial fluid surrounding the tubule (see Figure 46.9). The movement of positive charges sets up a voltage gradient that causes Cl^- ions to be reabsorbed from within the tubule with the positive ions. Specific active transport proteins reabsorb essentially all the glucose, amino acids, and other nutrient molecules from the filtrate into the interstitial fluid, making the filtrate hypoosmotic to the interstitial fluid surrounding the tubule. As a result, water moves from the tubule into the interstitial fluid by osmosis. The osmotic movement is aided by *aquaporins*, transport proteins that form passages for water molecules in the transport epithelium of the tubule cells. The nutrients and water that entered the interstitial fluid move into the capillaries of the peritubular network.

Some substances are also secreted into the tubule, however: primarily H^+ ions by active transport and the products of detoxified poisons by passive secretion (detoxification takes place in the liver). The secretion of H^+ ions into the filtrate helps balance the acidity constantly generated in the body by metabolic reactions. H^+ secretion is coupled with HCO_3^- reabsorption from the filtrate in the tubule to the plasma in the peritubular capillaries. Small amounts of ammonia are also secreted into the tubule.

In all, the proximal convoluted tubule reabsorbs about 67% of the Na^+, K^+, and Cl^- ions, 65% of the water, 50% of the urea, and essentially all the glucose, amino acids, and other nutrient molecules in the filtrate. The ions, nutrients, and water reabsorbed by the tubule are transported into the interstitial fluid, and then into capillaries of the peritubular network. Although half of the urea is reabsorbed, the constant flow of filtrate through the tubules keeps the concentration of nitrogenous wastes low in body fluids.

The proximal convoluted tubule has structural specializations that fit its function. The epithelial cells that make up its walls are carpeted on their inner surface by a brush border of microvilli. Like the brush border of epithelial cells in the small intestine (see Section 45.3), these microvilli greatly increase the surface area available for reabsorption and secretion.

The Descending Segment of the Loop of Henle. The filtrate leaving the proximal convoluted tubule enters the descending segment of the loop of Henle, where water is reabsorbed. As this tubule segment descends, it passes through regions of increasingly higher solute concentrations in the interstitial fluid of the medulla (see Figure 46.9). (The generation of this concentration gradient is described later.) As a result, more water moves out of the tubule by osmosis as the fluid travels through the descending segment.

The descending segment has aquaporins, which allow the rapid transport of water. The outward movement of water concentrates the molecules and ions inside the tubule, gradually increasing the osmolarity of the fluid to a peak of about 1200 mOsm/L at the bottom of the loop. This is the same as the osmolarity of the interstitial fluid at the bottom of the medulla.

The Ascending Segment of the Loop of Henle. The fluid then moves into the ascending segment of the loop of Henle, where Na^+ and Cl^- are reabsorbed into the interstitial fluid. As this segment ascends, it passes through regions of gradually lessening osmolarity in the interstitial fluid of the medulla. The ascending segment has membrane proteins that transport salt ions, but no aquaporins. Because water is trapped in the ascending segment, the osmolarity of the urine is reduced as salt ions, primarily Na^+ and Cl^-, move out of the tubule.

In the part of the ascending segment immediately following the loop, the ion concentrations in the tubule filtrate are still high enough to move Na^+ and Cl^- out of the tubule by passive transport. Toward the top of the segment, they are moved out by active transport.

An Ore Spells Relief for Osmotic Stress

Almost all cells respond to osmotic stress—osmotic imbalance with the surroundings—by adjusting the cytoplasmic concentration of small organic molecules called *osmolytes*. When cells are surrounded by a hyperosmotic solution, for example, osmolytes accumulate in the cytoplasm, raising its osmolarity to match that of the surroundings. The almost universal occurrence of osmolytes means that they must have appeared very early in the evolution of life.

In humans, cells in the renal medulla are regularly exposed to high solute concentrations in the interstitial fluid. These cells would quickly die from osmotic water loss if they were not protected by osmolytes. In these cells, as well as in many other types of mammalian cells, one of the primary osmolytes is *sorbitol,* made from glucose in a reaction catalyzed by the enzyme *aldose reductase.* In some unknown way, placing cells in a hyperosmotic medium activates the gene encoding and synthesizing aldose reductase in the cytoplasm.

Joan D. Ferraris and her colleagues at the National Institutes of Health in Bethesda, Maryland, were interested in the molecular steps leading to activation of the aldose reductase gene. To begin their research, the investigators extracted DNA from cells in the renal medulla of a rabbit, and cloned the DNA to increase its quantity. They then probed the DNA with a radioactive DNA segment that could pair with the DNA of aldose reductase genes previously isolated from humans and rats. The probe marked the rabbit version of the gene with radioactivity so that it could be separated from the sample.

The researchers were particularly interested in the promoter sequences controlling the gene, which might contain a region activating transcription in cells exposed to a hypertonic medium. To identify the region, they constructed a composite gene from a segment of the separated DNA containing the promoter and surrounding sequences attached to the coding portion of a gene for luciferase (the firefly enzyme, which catalyzes a cytoplasmic reaction emitting light). The composite gene was then increased in quantity by the polymerase chain reaction.

The composite gene was introduced into cultured renal medulla cells from the rabbit, which were then divided into two groups. One group was maintained in an isotonic medium; the other one was exposed to a medium made hypertonic by added NaCl. The cells exposed to the hypertonic solution glowed with light, showing that the luciferase gene had been turned on by having some part of the promoter segment derived from the aldose reductase gene.

The next step was to isolate the particular control sequence. To accomplish this, the researchers broke the promoter segment into fragments, attached them one at a time to the luciferase gene, and tested them by the same experimental procedure. Eventually, they identified the smallest fragment capable of activating the luciferase gene in cells placed in hypertonic medium; it proved to be the sequence CGGAAAATCAC, beginning 1105 base pairs in advance of the site where transcription begins. The investigators termed the activating sequence an *osmotic response element (ORE).*

Presumably, placing the renal medulla cells under osmotic stress triggers a series of reactions that culminates in synthesis or activation of a nuclear regulatory protein that binds the ORE, leading to activation of the aldose reductase gene. The next step in the investigation is to find the nuclear regulatory protein, and then work backwards one step at a time until the entire series of reactions leading from the first cell receptor to activation of the gene is traced out. If successful, the research will reveal an evolutionarily ancient mechanism that is critical to the survival of virtually all living cells.

Besides reducing the osmolarity of the filtrate in the ascending segment, the reabsorption of salt ions from the tubule into the interstitial fluid helps establish the concentration gradient of the medulla, high near the renal pelvis and low near the renal cortex. The energy required to transport NaCl from higher levels of the ascending segment makes the kidneys one of the major ATP-consuming organs of the body.

By the time the fluid reaches the cortex at the top of the ascending loop, its osmolarity has dropped to about 150 mOsm/L. During the travel of fluid around the entire loop of Henle, water, nutrients, and ions have been conserved and returned to body fluids, and the total volume of the filtrate in the nephron has been greatly reduced. Urea and other nitrogenous wastes have been concentrated in the filtrate. Little secretion occurs in either the descending or ascending segments of the loop of Henle.

The Distal Convoluted Tubule. The transport epithelium of the distal convoluted tubule removes additional water from the fluid in the tubule, and works to balance the salt and bicarbonate concentrations of the tubule fluid against body fluids. In response to hormones triggered by changes in the body's salt concentrations (described in Section 46.4), varying amounts of K^+ and H^+ ions are secreted into the fluid, and varying amounts of Na^+ and Cl^- ions are reabsorbed. Bicarbonate ions are reabsorbed from the filtrate as in the proximal tubule.

In total, more ions move outward than inward and, as a consequence, water moves out of the tubule by osmosis, through aquaporins in the distal tubule. The amounts of urea and other nitrogenous wastes remain the same. By the time the fluid—now urine— enters the collecting ducts at the end of the nephron, its osmolarity is about 300 mOsm/L.

The Collecting Ducts. The collecting ducts concentrate the urine. These ducts, which are permeable to water but not salt ions, descend downward from the cortex through the medulla of the kidney. As the ducts descend, they encounter the gradient of increasing solute concentration in the medulla. This increase makes water move osmotically out of the ducts and greatly increases the concentration of the urine, which can become as high as 1200 mOsm/L at the bottom of the medulla. Near the bottom of the medulla, the walls of the collecting ducts contain passive urea transporters that allow a portion of this nitrogenous waste to pass from the duct into the interstitial fluid. This urea adds significantly to the concentration gradient of solutes in the medulla.

In addition to these mechanisms, H^+ ions are actively secreted into the fluid by the same mechanism as in the proximal and distal convoluted tubules. The balance of the H^+ and bicarbonate ions established in the urine, interstitial fluid, and blood, achieved by secretion of H^+ into the urine by the nephrons and collecting ducts, is important for regulating the pH of blood and body fluids. The kidneys thus provide a safety valve if the acidity of body fluids rises beyond levels that can be controlled by the blood's buffer system (see Section 44.4).

At its maximum value of 1200 mOsm/L, reached when water conservation is at its maximum, the urine at the bottom of the collecting ducts is about four times more concentrated than body fluids. It can also be as low as 50 to 70 mOsm/L, when very dilute urine is produced in response to conditions such as excessive water intake.

The high osmolarity of the interstitial fluid toward the bottom of the medulla would damage the medulla cells if they were not protected against osmotic water loss. The protection comes from high concentrations of otherwise inert organic molecules called *osmolytes* in these cells. The osmolytes, primarily a sugar alcohol called *sorbitol*, raise the osmolarity of the cells to match that of the surrounding interstitial fluid. *Insights from the Molecular Revolution* describes research that identified the genetic controls leading to sorbitol production in kidney medulla cells and other cells subjected to osmotic stress.

Urine flows from the end of the collecting ducts into the renal pelvis, and then through the ureters into the urinary bladder where it is stored. From the bladder, urine exits through the urethra to the outside.

Terrestrial Mammals Have Additional Water-Conserving Adaptations

Terrestrial mammals have other adaptations that complement the water-conserving activities of the kidneys. One is the location of the lungs deep inside the body, which reduces water loss by evaporation during breathing (see Section 44.1). Another is a body covering of keratinized skin. Skin is so impermeable that it almost eliminates water loss by evaporation, except for the controlled loss through evaporation of sweat in mammals with sweat glands.

Among mammals, water-conserving adaptations reach their greatest efficiency in desert rodents such as the kangaroo rat **(Figure 46.10).** The proportion of nephrons with long loops extending deep into the kidney medulla of kangaroo rats is very high, allowing them to excrete urine that is 20 times more concentrated than body fluids. Further, most of the water in the feces is absorbed in the large intestine and rectum. Lacking sweat glands, kangaroo rats lose little water by evaporation from the body surface. Much of the moisture in their breath is condensed and recycled by specialized passages in the nasal cavities. They stay in burrows during daytime, and come out to feed only at night.

About 90% of the kangaroo rat's daily water supply is generated from oxidative reactions in its cells. (Humans, in contrast, can make up only about 12% of their daily water needs from this source.) The remaining

	Kangaroo Rat	Human
Water gain (milliliters)		
From ingesting food	6.0	850
From drinking liquids	0.0	1400
By metabolism	54.0	350
	60.0	2600
Water loss (milliliters)		
In urine	13.5	1500
In feces	2.6	200
By evaporation	43.9	900
	60.0	2600

Figure 46.10

A comparison of the sources of water for a human and a kangaroo rat (genus *Dipodymus*). Water conservation in the kangaroo rat is so efficient that the animal never has to drink water.

10% of the kangaroo rat's water comes from its food. These structural and behavioral adaptations are so effective that a kangaroo rat can survive in the desert without ever drinking water.

Marine mammals, including whales, seals, and manatees, eat foods that are high in salt content and never drink fresh water. They are able to survive the high salt intake because they produce urine that is more concentrated than seawater. As a result, they are easily able to excrete all the excess salt they ingest in their diet.

We now turn to the regulatory mechanisms that integrate kidney function with body functions as a whole.

STUDY BREAK

1. Describe the structure of a human nephron from the proximal end to the distal end.
2. The urine entering the collecting ducts at the end of the nephron has an osmolarity essentially the same as that of fluids in other parts of the body. How is the urine subsequently made more concentrated?

46.4 Regulation of Mammalian Kidney Function

Mammalian excretory functions are integrated into overall body functions by three primary control systems, which link kidney functions to blood pressure, to the osmolarity and pH of body fluids, and to the body's water balance. An *autoregulation system* located entirely within the kidney keeps glomerular filtration constant during relatively small variations in blood pressure, as when we move from sitting to standing. Two other systems involve hormonal controls that compensate for excessive loss of salt and body fluids and that adjust the rate of water uptake in the kidneys to compensate for excessive water intake or loss. These two hormonal systems regulate interactions between the kidneys and the rest of the body.

Autoregulation Involves Interactions between the Glomerulus and the Nephron

The autoregulation system responds almost instantly to keep the filtration rate constant during small variations in blood pressure. The system depends on signals from receptors in the **juxtaglomerular apparatus** (*juxta* = near) **(Figure 46.11)**, which is located at a point where the distal convoluted tubule contacts the afferent arteriole carrying blood to the glomerulus. The receptors, located in the tubule wall, monitor the pressure and flow of fluid through the distal tubule. If a rise in blood pressure increases the filtration rate, the receptors release chemical signals that trigger constriction of the afferent arteriole. The constriction reduces blood flow through the glomerulus and lowers the filtration rate. A drop in filtration rate due to a fall in blood pressure has the opposite effect: signals from the receptors cause the arterioles to dilate, blood flow increases in the glomerulus, and the filtration rate rises.

The RAAS Responds to Na^+ by Triggering Na^+ Reabsorption

Major changes in blood volume and pressure occur when the body loses or gains Na^+ in excessive amounts. Excessive Na^+ loss may result from prolonged and heavy sweating, repeated vomiting, severe diarrhea, or insufficient Na^+ uptake in the diet. The Na^+ loss reduces the osmolarity of body fluids, which causes less water to be reabsorbed in the kidneys. The water loss reduces the volume of blood and interstitial fluid and causes the blood pressure to drop. Excessive Na^+ intake in salty foods may have the opposite effects. The body must compensate for significant changes in Na^+.

The **renin-angiotensin-aldosterone system (RAAS)** is the most important hormonal system involved in regulating Na^+ (see Figure 46.11). At normal body salt concentrations, the RAAS allows about 10 g of salt to be excreted in the urine each day. If excessive Na^+ is lost in the excreted salt, blood pressure and body fluid volume drop, and the glomerular filtration rate falls below levels that can be restored by the juxtaglomerular apparatus. In response, cells in the juxtaglomerular apparatus secrete the enzyme **renin** into the bloodstream. (The RAAS also is activated to promote renin secretion when blood pressure or blood volume decreases independently of Na^+ levels, as in the case of a hemorrhage.) Renin converts a blood protein into the peptide hormone **angiotensin.** Angiotensin quickly raises blood pressure by constricting arterioles in most parts of the body; it also stimulates the release of the steroid hormone **aldosterone** from the adrenal cortex. Aldosterone increases Na^+ reabsorption in the kidneys, which raises the osmolarity of body fluids. As a result, water moves from the tubules into the extracellular fluid, which conserves water. Angiotensin also stimulates secretion of **antidiuretic hormone (ADH)** by the posterior pituitary (antidiuretic means "against urine output"). ADH increases water absorption in the kidneys. And, angiotensin stimulates thirst so that more water will be brought into the body. Overall, the combined effects of angiotensin act to raise the blood pressure back to normal levels.

In the opposite situation, when salt intake is too high, both body fluid volume and blood pressure rise above normal levels. Under these conditions, renin secretion is inhibited and, as a result, angiotensin production and aldosterone secretion are not stimulated. The reduction in angiotensin lowers blood pressure by al-

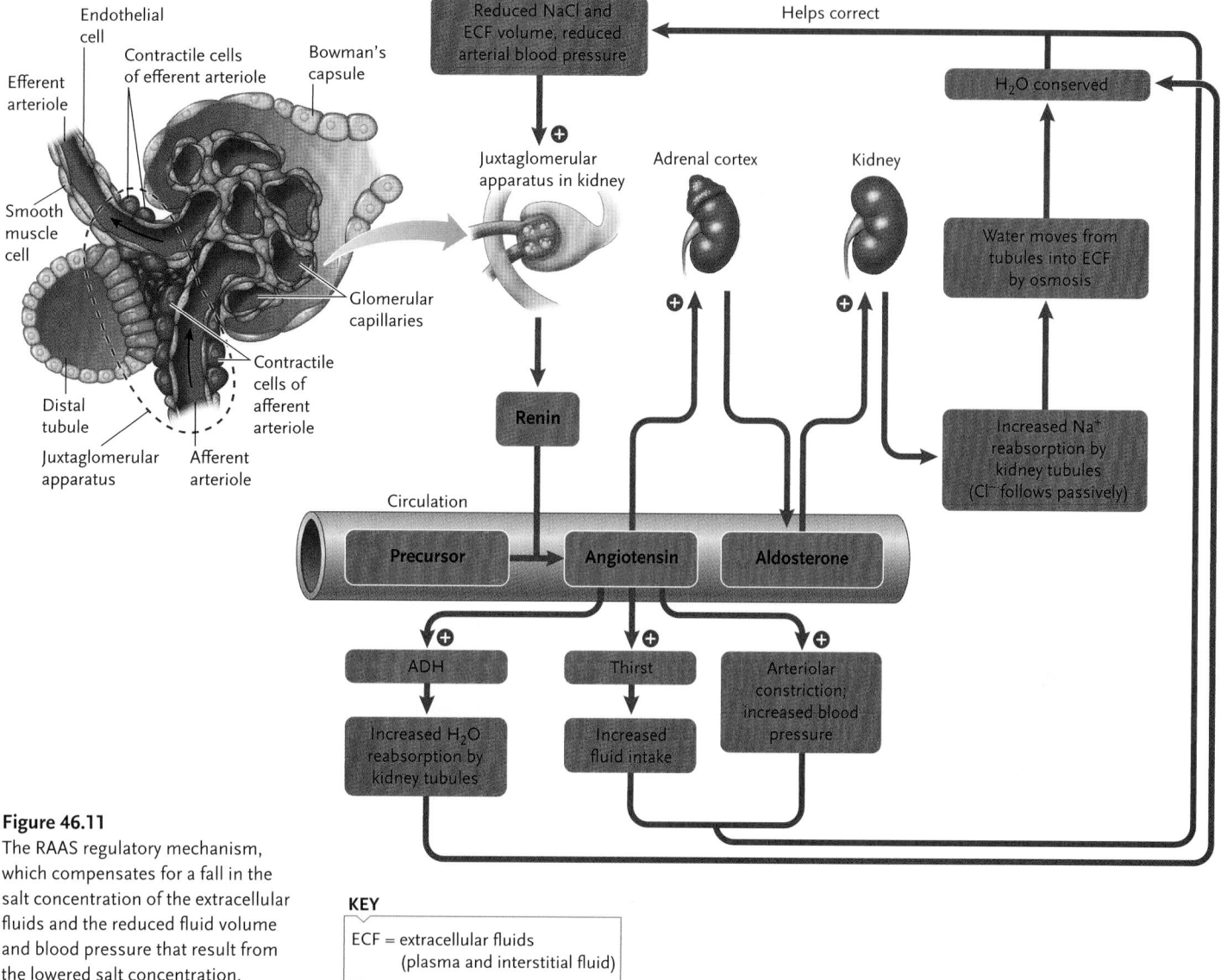

Figure 46.11
The RAAS regulatory mechanism, which compensates for a fall in the salt concentration of the extracellular fluids and the reduced fluid volume and blood pressure that result from the lowered salt concentration.

KEY

ECF = extracellular fluids
(plasma and interstitial fluid)

lowing arterioles to dilate; the reduction in aldosterone increases Na⁺ release in the urine by retarding the reabsorption of Na⁺ and Cl⁻ from the kidney tubules.

As a backup to these controls, elevated blood pressure stimulates specialized cells in the heart to release **atrial natriuretic factor (ANF),** a peptide hormone that also inhibits renin release. ANF also increases the filtration rate by dilating the arterioles that deliver blood to glomeruli and by inhibiting aldosterone release. As less Na⁺ is reabsorbed and urine volume increases, both plasma volume and blood pressure fall to normal levels.

The ADH System Also Regulates Osmolarity and Water Balance

You have just learned that the ADH system is stimulated by angiotensin of the RAAS in response to an increase in Na⁺. The ADH system regulates osmolarity and water balance—and, therefore, urinary output—by

increasing water reabsorption in the kidneys without changing the usual excretion of salt. Independently of being stimulated by angiotensin, the ADH system is triggered by **osmoreceptors,** chemoreceptors in the hypothalamus that respond to changes in the osmolarity of the fluid surrounding them, which reflects the osmolarity generally of the body fluids **(Figure 46.12).**

When an animal becomes dehydrated, the osmolarity of its body fluids increases and its need for water conservation increases. In this situation, osmoreceptors detect the increase in concentration of salts and other dissolved substances in the extracellular fluid (see Figure 46.12, step 1). Signals from the osmoreceptors are routed to the brain stem, where they trigger thirst (step 2). The resulting increase in water ingestion helps compensate for water loss (step 3).

In addition, neurons of the hypothalamus stimulate the posterior pituitary to secrete ADH (step 4). ADH makes the otherwise impermeable distal convoluted tubules and collecting ducts permeable to water.

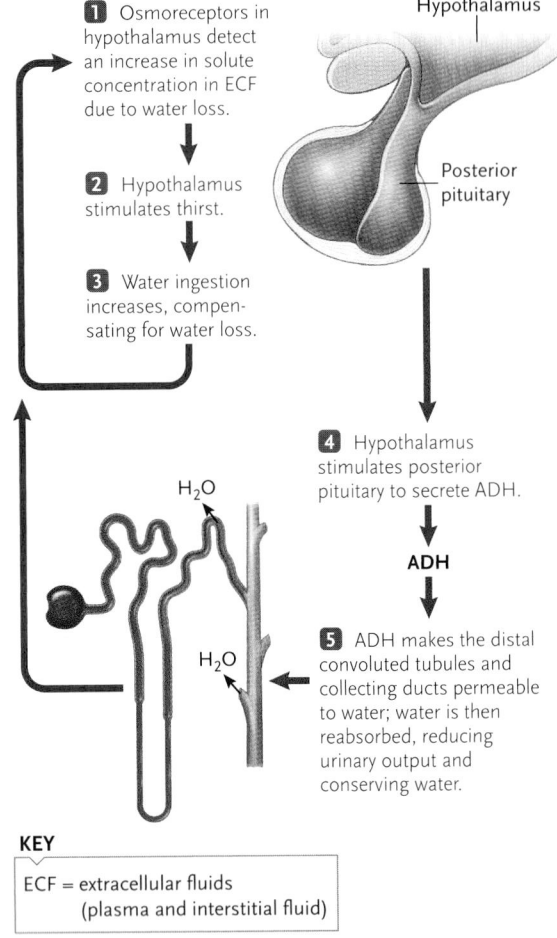

1. Osmoreceptors in hypothalamus detect an increase in solute concentration in ECF due to water loss.

2. Hypothalamus stimulates thirst.

3. Water ingestion increases, compensating for water loss.

Hypothalamus

Posterior pituitary

4. Hypothalamus stimulates posterior pituitary to secrete ADH.

ADH

H_2O

H_2O

5. ADH makes the distal convoluted tubules and collecting ducts permeable to water; water is then reabsorbed, reducing urinary output and conserving water.

KEY

ECF = extracellular fluids (plasma and interstitial fluid)

Figure 46.12
The ADH regulatory system, which stimulates water reabsorption to compensate for a loss in the fluid volume of the extracellular fluids due to excessive water loss from the body.

As a result, water is reabsorbed into those tubules and ducts so that urinary output is reduced and water is conserved (step 5).

By stimulating thirst and water reabsorption in the kidneys, the body's depleted stock of water is restored. The newly added water dilutes the solutes in the body fluids to normal concentrations.

In the opposite condition, when there is a water excess in extracellular fluids, the osmolarity of those fluids drops below normal levels. Here, there is no stimulation of the osmoreceptors in the hypothalamus. Consequently, there is no sensation of thirst, and no ADH release from the posterior pituitary. (In going from water deficiency to water excess, there is a gradual change in both of these parameters as the body adjusts to match its needs, meaning that the sensation of thirst decreases and ADH release is reduced as water in body fluids increases.) Without ADH, the distal convoluted tubules and collecting ducts again become impermeable to water. The animal excretes large volumes of dilute urine until the osmolarity of the extracellular fluids returns to normal. Alcohol also causes frequent urination by inhibiting ADH release.

Although the RAAS and ADH systems interact to regulate the body's water balance over a wide range of conditions, their regulatory mechanisms cannot compensate for water losses for more than a few days if water is unavailable. Dehydration becomes fatal when water loss amounts to about 12% of the normal fluid volume of the body.

Unlike mammals, most other vertebrates cannot conserve water by producing highly concentrated, hyperosmotic urine. In the next section, we consider some of the adaptations that nonmammalian vertebrates use to maintain the osmolarity of body fluids and water balance while excreting hypoosmotic urine.

STUDY BREAK

Outline the roles of the RAAS and ADH system in regulating mammalian kidney function.

46.5 Kidney Function in Nonmammalian Vertebrates

Among nonmammalian vertebrates, only a few species of aquatic birds produce urine that is hyperosmotic to body fluids. The particular adaptations that maintain osmolarity and water balance among these animals vary depending on whether retention of water or salts is the major issue.

Marine Fishes Conserve Water and Excrete Salts

Marine teleosts live in seawater, which is strongly hyperosmotic to their body fluids. As a result, they continually lose water to their environment by osmosis and must replace it by continual drinking. The kidneys of marine teleosts play little role in regulating salt in their body fluids because they cannot produce hyperosmotic urine that would both remove salt and conserve water. Instead, excess Na^+, K^+, and Cl^- ions are eliminated from the body by specialized cells in the gills, called *chloride cells*, which actively transport Cl^- into the surrounding seawater; the Na^+ and K^+ ions are also actively transported to maintain electrical neutrality **(Figure 46.13a)**. Certain other ions in the ingested seawater, such as Ca^{2+} and Mg^{2+}, are removed by the kidneys in an isoosmotic urine. On balance, a marine teleost is able to retain most of the water it drinks and eliminate most of the salt, allowing its body fluids to remain hypoosmotic to the surrounding water with no need to secrete hyperosmotic urine. Nitrogenous wastes are released from the gills, primarily as ammonia, by simple diffusion. The kidneys play little role in nitrogenous-waste removal.

Sharks and rays have a different adaptation to seawater—the osmolarity of their body fluids is maintained close to that of seawater by retaining high levels of urea in body fluids, along with another nitrogenous waste, *trimethylamine oxide (TMAO)*. The match in osmolarity keeps sharks and rays from losing water to the surrounding sea by osmosis, and they do not have to drink seawater continually to maintain their water balance. Excess salts ingested with food are excreted in the kidney and by specialized secretory cells in a *rectal salt gland* located near the anal opening.

Freshwater Fishes and Amphibians Excrete Water and Conserve Salts

The body fluids of freshwater fishes and aquatic amphibians (no amphibians live in seawater) are hyperosmotic to the surrounding water, which usually ranges from about 1 to 10 mOsm/L. Water therefore moves osmotically into their tissues. Such animals rarely drink, and they excrete large volumes of dilute urine to get rid of excess water **(Figure 46.13b)**. In freshwater fishes, salt ions lost with the urine are replaced by salt in foods and by active transport of Na^+ and K^+ into the body by the gills; Cl^- follows to maintain electrical neutrality. Aquatic amphibians obtain salt in the diet and by active transport across the skin from the surrounding water. Nitrogenous wastes are excreted from the gills as ammonia in both freshwater fishes and aquatic amphibians.

Terrestrial amphibians must conserve both water and salt, which is obtained primarily in foods. In these animals, the kidneys secrete salt into the urine, causing water to enter the urine by osmosis. In the bladder, the salt is reclaimed by active transport and returned to body fluids. The water remains in the bladder, making the urine very dilute; during times of drought, it is reabsorbed as a water source. Terrestrial amphibians also have behavioral adaptations that help minimize water loss, such as seeking shaded, moist environments and remaining inactive during the day.

Larval amphibians, which are completely aquatic, excrete nitrogenous wastes from their gills as ammonia. Adult amphibians excrete nitrogenous wastes through their kidneys as urea.

Reptiles and Birds Excrete Uric Acid to Conserve Water

Terrestrial reptiles conserve water by secreting nitrogenous wastes in the form of an almost water-free paste of uric acid crystals. Further water conservation occurs as the epithelial cells of the cloaca, the common exit for the digestive and excretory systems, absorb water from feces and urine before those wastes are excreted. Most birds conserve water by the same processes—they excrete nitrogenous wastes as uric acid and absorb water from the urine and feces in the cloaca. In reptiles, the scales covering the skin allow almost no water to escape through the body surface.

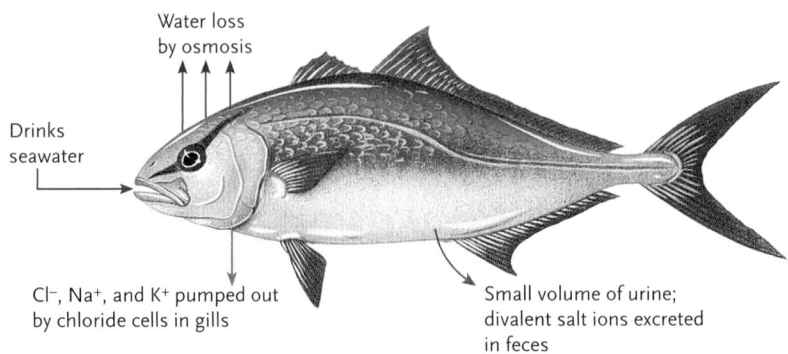

a. Marine teleosts

Water loss by osmosis

Drinks seawater

Cl^-, Na^+, and K^+ pumped out by chloride cells in gills

Small volume of urine; divalent salt ions excreted in feces

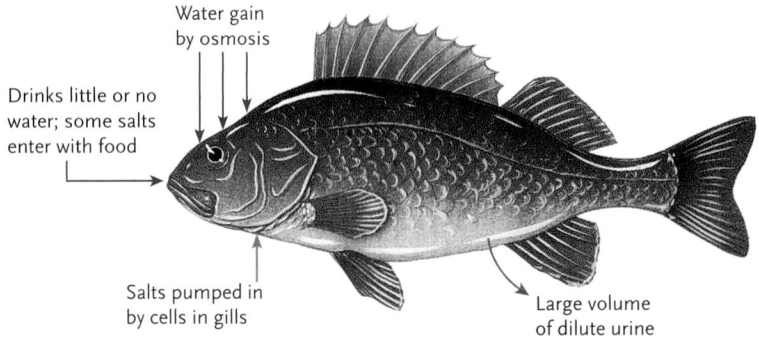

b. Freshwater teleosts

Water gain by osmosis

Drinks little or no water; some salts enter with food

Salts pumped in by cells in gills

Large volume of dilute urine

Figure 46.13
The mechanisms balancing the water and salt content of **(a)** marine teleosts and **(b)** freshwater teleosts.

Reptiles and birds that live in or around seawater, including reptiles such as crocodilians, sea snakes, and sea turtles and birds such as seagulls, penguins, and pelicans, take in large quantities of salt with their food and rarely or never drink fresh water. These animals typically excrete excess salt through specialized *salt glands* located in the head **(Figure 46.14)**, which remove salts from the blood by active transport. The salts are secreted to the environment as a water solution in which salts are two to three times more concentrated

Salt glands in skull

Ducts leading to nostrils and beak

Droplets of concentrated salt solution

Figure 46.14
Salt glands in a bird living on a seacoast.

than in body fluids. The secretion exits through the nostrils of birds and lizards, through the mouth of marine snakes, and as salty tears from the eye sockets of sea turtles and crocodilians. Neural and hormonal controls, essentially the same as those regulating osmolarity in mammals, control the rate of fluid secretion and its salt concentration.

The adaptations described in this section permit excretion of toxic wastes and allow animals to maintain the concentration of body fluids at levels that keep cells from swelling or shrinking. Animals also have mechanisms that address an equally vital challenge—maintaining their internal environment at temperatures that can be tolerated by body cells. We take up these processes in the next section.

STUDY BREAK

1. How do marine and freshwater teleosts differ in water, salt, and nitrogenous-waste regulation?
2. Reptiles and birds excrete nitrogenous wastes in the form of uric acid. Is there an advantage to doing this as opposed to the mammalian process of excreting nitrogenous wastes as urea?

46.6 Introduction to Thermoregulation

Environmental temperatures vary enormously across Earth's surface. However, animal cells can survive only within a temperature range from about 0°C to 45°C (32°F to 113°F). Not far below 0°C, the lipid bilayer of a biological membrane changes from a fluid to a frozen gel, which disrupts vital cell functions, and ice crystals destroy the cell's organelles. At the other extreme, as temperatures approach 45°C, the kinetic motions of molecules become so great that most proteins and nucleic acids unfold from their functional form. Either condition leads quickly to cell death. Animals therefore usually maintain internal body temperatures somewhere within the 0°C to 45°C limits.

Temperature regulation—thermoregulation—is based on negative feedback pathways in which temperature receptors (thermoreceptors) detect changes from a temperature set point. Signals from the receptors trigger physiological and behavioral responses that return the temperature to the set point (thermoreceptors are discussed in Section 39.6; negative feedback mechanisms and set points are discussed in Section 36.4). All of the responses triggered by negative feedback mechanisms involve adjustments in the rate of heat-generating oxidative reactions within the body, coupled with adjustments in the rate of heat gain or loss at the body surface. The particular adaptations accomplishing these responses vary widely among species, however. And, while body temperature is closely regulated around a set point in all endotherms, the set point itself may vary over the course of a day and between seasons.

In this section, we describe the structures, mechanisms, and behavioral adaptations that enable animals to regulate their temperature.

Thermoregulation Allows Animals to Reach Optimal Physiological Performance

Within the 0°C to 45° range of tolerable temperatures, an animal's *organismal performance*—the rate and efficiency of its biochemical, physiological, and whole-body processes—varies greatly. For example, the speed at which the Middle Eastern lizard *Agama stellio* can sprint is low when the animal's body temperature is cold, rises smoothly with body temperature until it levels to a fairly broad plateau, and then drops off dramatically with further increases in body temperature (Figure 46.15a). Similar patterns of temperature dependence are observed for numerous other body functions (Figure 46.15b). The temperature range that provides good organismal performance varies from one species to another, however.

Animals that maintain body temperature within a fairly narrow optimal range can run quickly, digest food efficiently, and carry out necessary activities and processes rapidly and effectively (see Figure 46.15b). Besides keeping body temperatures within tolerable limits, thermoregulation allows animals to achieve this level of performance.

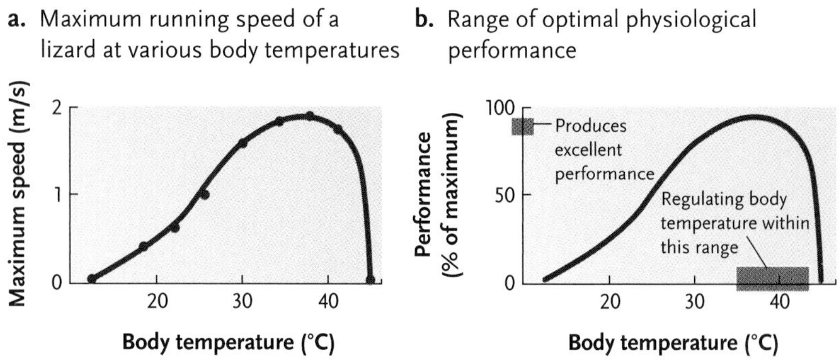

Animals Exchange Heat with Their Environments by Conduction, Convection, Radiation, and Evaporation

As part of thermoregulation, animals exchange heat with their environment. Virtually all heat exchange occurs at surfaces where the body meets the external en-

a. Maximum running speed of a lizard at various body temperatures

b. Range of optimal physiological performance

Figure 46.15

Body temperature and organismal performance. **(a)** The maximum sprint speed of a lizard *(Agama stellio)* changes dramatically with body temperature. **(b)** An animal's other behavioral and physiological processes respond to temperature changes in similar ways. The advantage of regulating body temperature within the range indicated by the bar on the horizontal axis is a high level of organismal performance, indicated by the bar on the vertical axis.

vironment. As with all physical bodies, heat flows into animals if they are cooler than their surroundings and flows outward if they are warmer. This heat exchange occurs by four mechanisms: *conduction, convection, radiation,* and *evaporation* **(Figure 46.16).**

Conduction is the flow of heat between atoms or molecules in direct contact. An animal loses heat by conduction when it contacts a cooler object, and gains heat when it contacts an object that is warmer. **Convection** is the transfer of heat from a body to a fluid, such as air or water, that passes over its surface. The movement maximizes heat transfer by replacing fluid that has absorbed or released heat with fluid at the original temperature. **Radiation** is the transfer of heat energy as electromagnetic radiation. Any object warmer than absolute zero (−273°C) radiates heat; as the object's temperature rises, the amount of heat it loses as radiation increases as well. Animals also gain heat through radiation, particularly by absorbing radiation from the sun. **Evaporation** is heat transfer through the energy required to change a liquid to a gas. Evaporation of water from a surface is an efficient way to transfer heat; when the water in sweat evaporates from the body surface, the body cools down because heat is being transferred to the evaporated water in the surrounding air.

All animals gain or lose heat by a combination of these four mechanisms. A marathon runner or a bicycle racer struggling with the heat on a sunny summer day, for example, loses heat by the evaporation of sweat from the skin, by convection as air flows over the skin, and by outward infrared radiation. The runner gains heat from internal biochemical reactions (especially oxidations), by absorbing infrared and solar radiation, and by conduction as the feet contact the hot ground. To maintain a constant body temperature, the heat gained and lost through these pathways must balance.

Ectothermic and Endothermic Animals Rely on Different Heat Sources to Maintain Body Temperature

Different animals use one of two major strategies to balance heat gain and loss. Animals that obtain heat primarily from the external environment are known as **ectotherms** (*ecto* = outside); those obtaining most of their heat from internal physiological sources are called **endotherms** (*endo* = inside). All ectotherms generate at least some heat from internal reactions, however, and endotherms can obtain heat from the environment under some circumstances.

Virtually all invertebrates, fishes, amphibians, and reptiles are ectotherms. Although these animals are popularly described as cold-blooded, the body temperature of some, such as an active lizard, may be as high as or higher than ours on a sunny day. Ectotherms regulate body temperature by controlling the rate of heat exchange with the environment. Through behav-

Figure 46.16

Heat flow to (in red) and from (in blue) a marathon runner on a hot, sunny day. Unlike conduction, convection, and evaporation, which take place through the kinetic movement of molecules, electromagnetic radiation is transmitted through space as waves of energy. (Photo: Rafael Winer/Corbis.)

ioral and physiological mechanisms, they adjust body temperature toward a level that allows optimal physiological performance. However, most ectotherms are unable to maintain optimal body temperature when the temperature of their surroundings departs too far from that optimum, particularly when environmental temperatures fall. As a result, the body temperatures of ectotherms fluctuate with environmental temperatures, and they typically are less active when it is cold. Nevertheless, ectotherms are highly successful, particularly in warm environments.

The endotherms—birds, mammals, some fishes, sea turtles, and some invertebrates—keep their bodies at an optimal temperature by regulating two processes: (1) the amount of heat generated by internal oxidative reactions and (2) the amount of heat exchanged with the environment. Because endotherms use internal heat sources to maintain body temperature at optimal levels, they can remain active over a broader range of environmental temperatures than ectotherms, and they can inhabit a wider range of habitats. However, endotherms require a nearly constant supply of energy to maintain their body temperatures. And because that energy is provided by food, endotherms typically consume much more food than ectotherms of equivalent size.

The difference between ectotherms and endotherms is reflected in their metabolic responses to environmental temperature (Figure 46.17). For example, the metabolic rate of a resting mouse *increases* steadily as the environmental temperature falls from 25°C to 10°C (77°F to 50°F). This increase reflects the fact that in order to maintain a constant body temperature in a colder environment, endotherms must process progressively more food and generate more heat to compensate for their increased rate of heat loss. In this respect, an endotherm can be likened to a house in winter. To maintain a constant internal temperature, the homeowner must burn more oil or gas on a cold day than on a warm day.

By contrast, the metabolic rate of a resting lizard typically *decreases* steadily over the same temperature range. Because ectotherms don't maintain a constant body temperature, their biochemical and physiological functions, including oxidative reactions, slow down as environmental and body temperatures decrease. Thus, an ectotherm consumes and uses less energy when it is cold than when it is warm. This difference between ectotherms and endotherms is so fundamental that even samples of living tissue extracted from an ectotherm consume energy more slowly than equivalent samples from an endotherm.

Ectothermy and endothermy represent different strategies for coping with the variations in environmental temperature that all animals encounter; neither strategy is inherently superior to the other. Endotherms can remain fully active over a wide temperature range. Cold weather does not prevent them from foraging, mating, or escaping from predators, but it does increase their energy and food needs—and, to satisfy their need for food, they may not have the option of staying curled up safely in a warm burrow. Ectotherms do not have the capacity to be active when environmental temperatures drop too low; they move sluggishly and are unable to capture food or escape from predators. However, because their metabolic rates are lower under such circumstances, so are their food needs, and they do not have to actively look for food and expose themselves to danger to the extent that endotherms do.

Having laid the ground rules of heat transfer and weighed the relative advantages and disadvantages of ectothermy and endothermy, we now begin a more detailed examination of how animals actually regulate their body temperatures within these overall strategies.

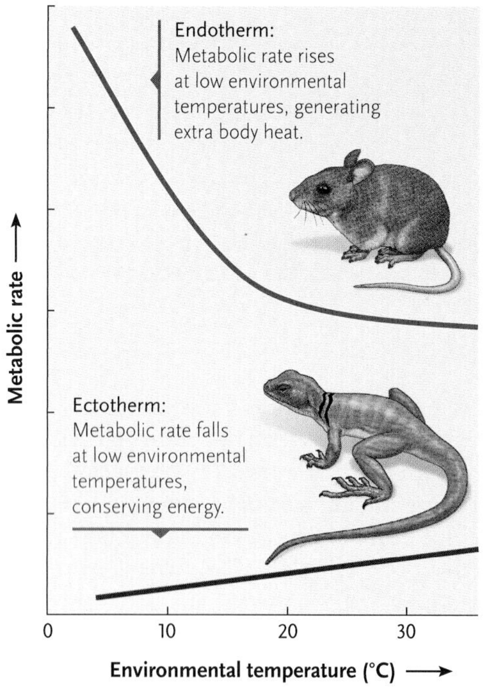

STUDY BREAK

Distinguish between ectothermy and endothermy. Give one advantage and one disadvantage for each form of thermoregulation.

46.7 Ectothermy

Ectotherms vary widely in their ability to regulate internal body temperatures. For example, most aquatic invertebrates have such limited ability to thermoregulate that their body temperatures closely match those of the surrounding environment. These species live in or seek warm or temperate environments, where temperatures fall within a range that produces optimal physiological performance. Ectotherms with a greater ability to thermoregulate may occupy more varied habitats.

Ectotherms Are Found in All Invertebrate Groups

Most aquatic invertebrates are limited thermoregulators whose body temperature closely follows the temperature of their surroundings. However, even among

Figure 46.17
Metabolic responses of ectotherms and endotherms to cooling environmental temperatures. At any temperature, the metabolic rates of endotherms are always higher than those of endotherms of comparable size.

Endotherm: Metabolic rate rises at low environmental temperatures, generating extra body heat.

Ectotherm: Metabolic rate falls at low environmental temperatures, conserving energy.

Metabolic rate →

Environmental temperature (°C) →

0   10   20   30

these animals, some use behavioral responses to regulate body temperature. For example, a South American intertidal mollusk, *Echinolittorina peruviana*, is longer than it is wide. Researchers have shown that this animal orients itself as a means of thermoregulation. On sunny, summer days, it faces the sun, offering a smaller surface area for the sun's rays. On overcast summer days, or during the winter, it orients itself with a lateral side—which has the larger surface area—toward the sun's rays.

Invertebrates living in terrestrial habitats regulate body temperatures more closely. Many also use behavioral responses, such as moving between shaded and sunny regions, to regulate body temperature. Some winged arthropods, including bees, moths, butterflies, and dragonflies, use a combination of behavioral and heat-generating physiological mechanisms for thermoregulation. For example, in cool weather, these animals warm up before taking flight by rapidly vibrating the large flight muscles in the thorax, in a mechanism similar to shivering in humans. The tobacco hawkmoth *(Manduca sexta)* vibrates its flight muscles until its thoracic temperature reaches about 36°C before flying. During flight, metabolic heat generated by the flight muscles sustains the elevated thoracic temperature, so much so that a flying sphinx moth produces more heat per gram of body weight than many mammals.

Most Fishes, Amphibians, and Reptiles Are Ectotherms

Vertebrate ectotherms—fishes, amphibians, and reptiles—also vary widely in their ability to thermoregulate. Most aquatic species have a more limited thermoregulatory capacity than that found among terrestrial species, particularly the reptiles. Some fishes, however, are highly capable thermoregulators.

Fishes. The body temperatures of most fishes remain within one or two degrees of their aquatic environment. However, many fishes use behavioral mechanisms to keep body temperatures at levels allowing good physiological performance. Freshwater species, for example, may use opportunities provided by the thermal stratification of lakes and ponds (see Figure 52.22). During hot summer days they remain in deep, cool water, moving to the shallows to feed only during early morning and late evening when air and water temperatures fall.

Amphibians and Reptiles. The body temperatures of most amphibians also closely match environmental temperatures. Some, such as the tadpoles of foothill yellow-legged frogs *(Rana boylii)*, regulate their body temperature to some degree by changing their location in ponds and lakes to take advantage of temperature differences between deep and shallow water, or be-

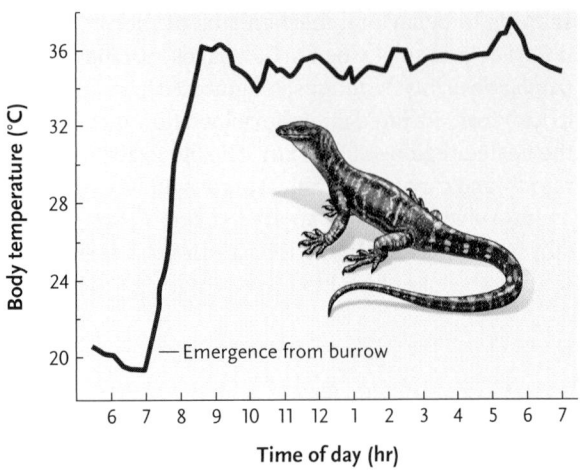

Figure 46.18

An example of excellent thermoregulation in ectotherms. The body temperature of the Australian lizard *Varanus varius* rises quickly after the animal emerges from its burrow and remains relatively stable throughout the day.

tween sunny and shaded regions. Some terrestrial amphibians bask in the sun to raise their body temperature, and seek shade to lower body temperature. However, basking can be dangerous to amphibians because they lose water rapidly through their permeable skin. One South American hylid frog *(Phyllomedusa sauvagei)*, which often basks in sunlight, avoids this problem by coating itself with waterproofing lipids secreted by glands in its skin.

Thermoregulation is more pronounced among terrestrial reptiles. Some lizard species can maintain temperatures that are nearly as constant as those of endotherms **(Figure 46.18)**. For small lizards, the most common behavioral thermoregulatory mechanism is shuttling between sunny (warmer) and shady (cooler) regions; in the deserts, lizards and other reptiles retreat into burrows during the hottest part of summer days. Some, such as the desert iguana *(Dipsosaurus dorsalis)*, lose excess heat by *panting*—rapidly moving air in and out of the airways. The air movement increases heat loss by convection and by evaporation of water from the respiratory tract.

Lizards also frequently adjust their posture to foster heat exchange with the environment, and control the angle of their body relative to the rays of the sun. For example, horned lizards (genus *Phrynosoma*) often warm up by flattening themselves against warm, sunlit rocks to maximize their rate of heat gain by conduction from the rock and radiation from the sun. Snakes and lizards can often be found on large rocks and on roads on chilly nights, taking advantage of the heat retained by the stone or concrete. *Agama savignyi*, a lizard that lives in the Negev Desert in Israel, cools off at midday by climbing into shady bushes, moving away from the hot sand and catching a cooling breeze.

Researchers have demonstrated experimentally that several lizard species couple physiological responses to behavioral mechanisms of thermoregulation. For example, when a Galapagos marine iguana (*Amblyrhynchus cristatus;* see Figure 19.8c) is exposed to heat from infrared radiation, blood flow increases in the heated regions of the skin. The blood absorbs heat rapidly and carries it to critical organs in the core of the body. Conversely, when an area of skin is experimentally cooled, blood flow to it is restricted, thereby preventing the loss of heat to the external environment.

Ectotherms Can Compensate for Seasonal Variations in Environmental Temperature

Many ectotherms undergo physiological changes, called **thermal acclimatization,** in response to seasonal shifts in environmental temperature. These changes allow the animals to attain good physiological performance at both winter and summer temperatures.

For example, in the summer a bullhead catfish (*Ameiurus* species) can survive water temperatures as high as 36°C (97°F), but it cannot tolerate temperatures below 8°C (46°F). In the winter, however, the bullhead cannot survive water temperatures above 28°C (82°F), but can tolerate temperatures near 0°C (32°F). Scientists have hypothesized that the production of different versions of the same enzyme (perhaps encoded by different genes, or produced as a result of alternative splicing; see Section 16.3), each having optimal activity at cooler or warmer temperatures, underlies such acclimatization.

Another acclimatizing change involves the phospholipids of biological membranes (see *Focus on Research* in Chapter 6). For example, membrane phospholipids have higher proportions of double bonds in carp living in colder environments than in carp living in warmer environments. The higher proportion of double bonds makes it harder for the membrane to freeze. A higher proportion of cholesterol also protects membranes from freezing.

When seasonal temperatures fall below 0°C (32°F), some ectotherms add molecules to their body fluids that act as antifreeze molecules to depress their freezing point and retard ice crystal formation. For example, glycerol added to the cellular and extracellular fluids of a parasitic wasp (*Bracon* species) keeps the insect from freezing at temperatures as low as −45°C (−49°F). Similarly, antifreeze proteins allow fishes such as the winter flounder to remain active in seawater as cold as −1.8°C (29°F) (see *Insights from the Molecular Revolution* in Chapter 52).

Ectotherms thus primarily control body temperature by regulating heat exchange with the environment; internal-heat generating mechanisms contribute to the control mechanisms in some species, but are rarely the primary source of body heat. The opposite conditions occur among endotherms: although these animals also regulate heat exchange with the environment, their primary sources of body heat are internal.

46.8 Endothermy

Endotherms—mostly birds and mammals—have the most elaborate and extensive thermoregulatory adaptations of all animals. Highly specialized features of body structure interact with both physiological and behavioral mechanisms to keep the body temperature constant within a narrow range. Typically, the body temperatures of fully active individuals are held constant at levels between about 39° to 42°C (102° to 108°F) in birds, and 36° to 39°C (97° to 102°F) in mammals. These internal temperatures are maintained in the face of environmental temperatures that may range over much greater extremes, from as low as −42°C to as high as +48°C (−45°F to +120°F). Some highly specialized endotherms can even survive temperatures beyond these limits.

We begin by describing the basic feedback mechanisms that maintain body temperature, with primary emphasis on the human system. Later sections discuss variations in the responses of other mammals and of birds, and daily and seasonal variations in the temperature set point.

Information from Thermoreceptors Located in the Skin and Internal Structures Is Integrated in the Hypothalamus

Thermoreceptors are found in various locations in the human body, including the **integument** (skin; introduced in Chapter 36), spinal cord, and hypothalamus. Two types of thermoreceptors occur in human skin. One, called a *warm receptor,* sends signals to the hypothalamus as skin temperature rises above 30°C (86°F), and reaches maximum activity when the temperature rises to 40°C (104°F). The other type, the *cold receptor,* sends signals when skin temperature falls below about 35°C (95°F) and reaches maximum activity at 25°C (77°F). By contrast, the highly sensitive thermoreceptors in the hypothalamus send signals when the blood temperature shifts from the set point by as little as 0.01°C (0.02°F).

Signals from the thermoreceptors are integrated in the hypothalamus and other regions of the brain to bring about compensating physiological and behavioral responses **(Figure 46.19).** The responses keep body temperature close to the set point, which varies normally in humans between 35.5° and 37.7°C (96.0° to 99.9°F) for the head and trunk. The appendages typically vary more widely in temperature; in freezing weather, for example, our arms, hands, legs, and feet are typically lower in temperature than the body core—and the ears and nose especially so.

The hypothalamus was identified as a major thermoreceptor and response integrator in mammals by experiments in which various regions of the brain were heated or cooled with a temperature probe. Within the brain, only the hypothalamus produced thermoregulatory responses such as shivering or panting. Later experiments revealed a similar response when regions of the spinal cord were cooled, indicating that thermoreceptors also occur in this location. The hypothalamus is also a major thermoreceptor and response integrator in fishes and reptiles. In birds, thermoreceptors in the spinal cord appear to be most significant in thermoregulation.

Responses When Core Temperature Falls below the Set Point.

When thermoreceptors signal a fall in core temperature below the set point, the hypothalamus triggers compensating responses by sending signals through the autonomic nervous system. Among the immediate responses is constriction of the arterioles in the skin (vasoconstriction), which reduces the flow of blood to capillary networks in the skin. The reduced flow cuts down the amount of heat delivered to the skin and lost from the body surface. The reduction in flow is most pronounced in the skin covering the extremities, where blood flow may be reduced by as much as 99% when core temperature falls.

Another immediate response is contraction of the smooth muscles erecting the hair shafts in mammals and feather shafts in birds, which traps air in pockets over the skin, reducing convective heat loss. The response is minimally effective in humans because hair is sparse on most parts of the body—it produces the goose bumps we experience when the weather gets chilly. However, in mammals with fur coats or in birds, erection of the hair or feather shafts significantly increases the thickness of the insulating layer that covers the skin.

Immediate behavioral responses triggered by a reduction in skin temperature also help reduce heat loss from the body. Mammals may reduce heat loss by moving to a warmer locale, curling into a ball, or huddling together. We have all seen puppies huddled together to keep warm; birds such as penguins also keep warm by huddling. We humans may also put on more clothes or slip into a tub of hot water.

If these immediate responses do not return body temperature to the set point, the hypothalamus triggers further responses, most notably the rhythmic tremors of skeletal muscle we know as shivering. The heat released by the muscle contractions and the oxidative reactions powering them can raise the total heat production of the body substantially. At the same time, the hypothalamus triggers secretion of *epinephrine* (from the adrenal medulla) and *thyroid hormone* (see

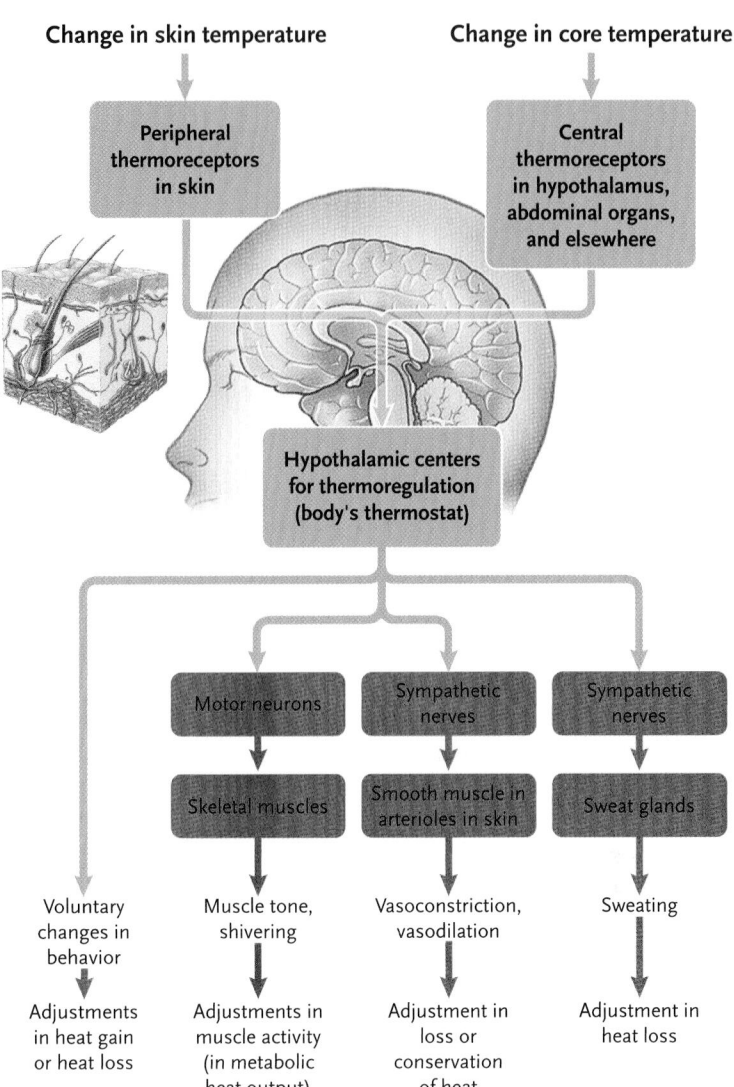

Figure 46.19
The physiological and behavioral responses of humans and other mammals to changes in skin and core temperature.

Section 40.4), both of which increase heat production by stimulating the oxidation of fats and other fuels. The generation of heat by oxidative mechanisms in non-muscle tissue throughout the body is termed **nonshivering thermogenesis.**

In human newborn babies and many other mammals, the most intense heat generation by nonshivering thermogenesis takes place in a specialized **brown adipose tissue** (also called **brown fat**) that can produce heat rapidly. Heat is generated by a mechanism that uncouples electron transport from ATP production in mitochondria (see Section 8.4); the heat is transferred throughout the body by the blood. Animals that hibernate or are active in cold regions, as well as the young of many others, contain brown adipose tissue. In most mammals, brown adipose tissue is concentrated between the shoulders in the back and around the neck. In human newborn babies, this tissue accounts for about 5% of body weight. Typically the tissue shrinks as humans age, until it is absent or essentially so in most adults. However, if exposure to cold is ongoing, the tissue remains. For instance, some Japanese and

Korean divers who harvest shellfish in frigid waters, and male Finlanders who work outside during the year, have significant amounts of brown adipose tissue.

If none of these responses succeeds in raising body temperature to the set point, the result is **hypothermia**, a condition in which the core temperature falls below normal for a prolonged period. In humans, a drop in core temperature of only a few degrees affects brain function and leads to confusion; continued hypothermia can lead to coma and death.

Responses When Core Temperature Rises above the Set Point. When core temperature rises above the set point, the hypothalamus sends signals through the autonomic system that trigger responses lowering body temperature. As an immediate response, the signals relax smooth muscles of arterioles in the skin (vasodilation), increasing blood flow and with it, the heat lost from the body surface. In addition, in humans and other mammals with sweat glands, such as antelopes, cows, and horses, signals from the hypothalamus trigger the secretion of sweat, which absorbs heat as it evaporates from the surface of the skin.

Some endotherms, including dogs (which have sweat glands only on their feet) and many birds (which have no sweat glands), use panting as a major way to release heat. These physiological changes are reinforced by behavioral responses such as seeking shade or a cool burrow, plunging into cold water, wallowing in mud, or taking a cold drink. Elephants typically take up water in their trunk and spray it over their body to cool off in hot weather.

When the heat gain of the body is too great to be counteracted by these responses, **hyperthermia** results. An increase of only a few degrees above normal for a prolonged period is enough to disrupt vital biochemical reactions and damage brain cells. Most adult humans become unconscious if their body temperature reaches 41°C (106°F) and die if it goes above 43°C (110°F) for more than a few minutes.

The Skin Is Highly Adapted to Control Heat Transfer with the Environment

Besides its defensive role against infection described in Section 43.1, the skin of birds and mammals is an organ of heat transfer. The arterioles delivering blood to the capillary networks of the skin constrict or dilate to control blood flow and, with it, the amount of heat transferred from the body core to the surface.

The outermost living tissue of human skin, the **epidermis**, consists of cells that grow and divide rapidly **(Figure 46.20)**, becoming packed with fibers of a highly insoluble protein, *keratin* (see Section 5.3). When fully formed, the epidermal cells die and become compacted into a tough, impermeable layer that limits water loss primarily to evaporation of the fluids secreted by the sweat glands.

The sweat glands and hair follicles are embedded in the layer below the epidermis. Called the **dermis**, it is packed with connective tissue fibers such as collagen, which resist compression, tearing, or puncture of the skin. The dermis also contains thermoreceptors and the dense networks of arterioles, capillaries, and venules that transfer heat between the skin and the environment.

The innermost layer of the skin, the **hypodermis**, contains larger blood vessels and additional reinforcing connective tissue. The hypodermis also contains an insulating layer of fatty tissue below the dermal capillary network, which ensures that heat flows between the body core and the surface primarily through the blood. The insulating layer is thickest in mammals that live in cold environments, such as whales, seals, walruses, and polar bears, in which it is known as *blubber*.

Figure 46.20
The structure of human skin.
(Micrograph: John D. Cunningham/Visuals Unlimited.)

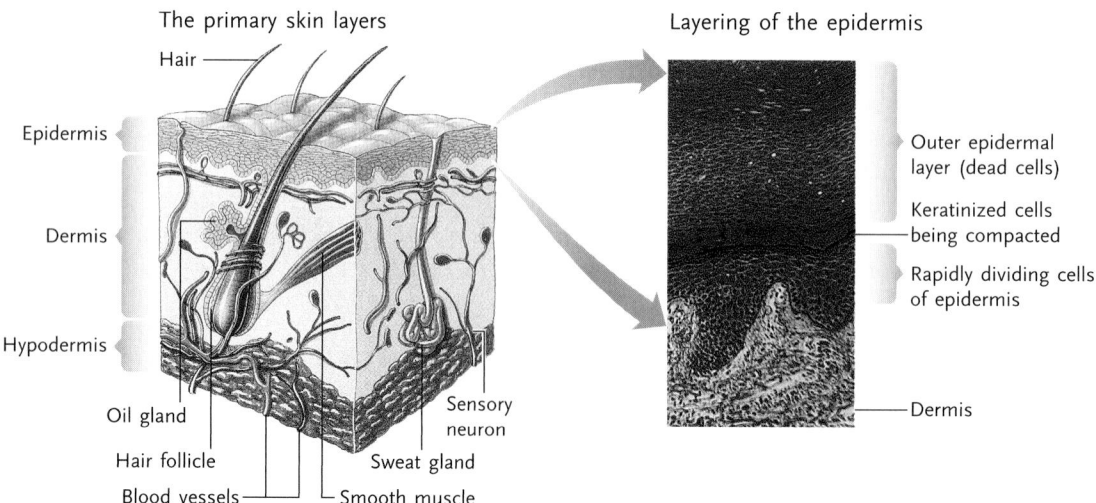

The primary skin layers

Hair
Epidermis
Dermis
Hypodermis
Oil gland
Hair follicle
Blood vessels
Smooth muscle
Sweat gland
Sensory neuron

Layering of the epidermis

Outer epidermal layer (dead cells)
Keratinized cells being compacted
Rapidly dividing cells of epidermis
Dermis

a. Dissipating heat

Joe McDonald/Corbis

b. Conserving heat

Fredrik Broman/Iconica/Getty Images, Inc.

Figure 46.21
Structural and behavioral adaptations controlling heat transfer at the body surface. **(a)** A jackrabbit *(Lepus californicus)* dissipating heat from its ears on a hot summer day. Notice the dilated blood vessels in its large ears. Both the large surface area of the ears and the extensive network of blood vessels promote the dissipation of heat by convection and radiation. **(b)** A husky *(Canis lupus familiaris)* conserving heat by curling up with the limbs under the body and the tail around the nose.

Many Birds and Mammals Have Additional Thermoregulatory Structures and Responses

The thermoregulatory mechanisms we have described to this point are common to many birds and mammals. Many species also have specialized responses that enhance thermoregulation. In hot weather, for example, many birds fly with their legs extended, so that heat flows from their legs into the passing air. Similarly, penguins expose featherless patches of skin under their wings to cool off on days when the weather is too warm. Jackrabbits **(Figure 46.21a)** and elephants dissipate heat from their large ears, which are richly supplied with blood vessels. In times of significant heat stress, kangaroos and rats spread saliva on their fur to increase heat loss by evaporation; some bats coat their fur with both saliva and urine.

Many mammals have an uneven distribution of fur that aids thermoregulation. In a dog, for example, the fur is thickest over the back and sides of the body and the tail, and thinnest under the legs and over the belly. In cold weather, dogs curl up, pull in their limbs, wrap their tail around the body, and bury their nose in the tail, so that only body surfaces insulated by thick fur are exposed to the air **(Figure 46.21b)**. When the weather is hot, dogs spread their limbs, turn on their side or back, and expose the relatively bare skin of the belly, which acts as a heat radiator. These responses are combined with seeking sun or shade or a warm or cool surface to lie on.

In marine mammals such as whales and seals, heat loss is regulated by adjustments in the blood flow through the thick blubber layer to the skin. In cold water, blood flow is minimized by constriction of the vessels, making the skin temperature close to that of the surrounding water while the body temperature remains constant under the insulating blubber. In warmer water, blood flow to the skin increases, bypassing the blubber and allowing excess heat to be lost from the body surface.

In addition, heat loss in whales and seals is controlled by adjustments of the flow of blood to the flippers, which are not insulated by blubber and act as a heat radiator. When a whale generates excessive internal heat through the muscular activity of swimming, the flow of blood from the body core to the flippers increases. In contrast, when heat must be conserved to maintain core temperature at the set point, blood flow to the flippers is reduced.

As with ectotherms, many mammals also undergo thermal acclimatization to adjust to seasonal temperature change. That is, the development of a thick fur coat in winter, which is shed in summer, enables them to adapt to seasonal temperature changes. Note that the trigger for the coat changes in many cases is day length, rather than temperature; there is a general correlation with temperature and day length over the year. Some arctic and subarctic mammals develop a thicker layer of insulating fat in winter.

The Set Point Varies in Daily and Seasonal Rhythms in Many Birds and Mammals

The temperature set point in many birds and mammals varies in a regular cycle during the day. In some, the daily variations are relatively small and not obviously keyed to changes in environmental temperature. In others, larger variations are correlated with daily or seasonal temperature changes.

Humans are among the endotherms for which daily variations in the temperature set point are small. Normally, human core temperature varies from a minimum of about 35.5°C (95.9°F) in the morning to a maximum of about 37.7°C (99.9°F) in the evening. Women also show a monthly variation keyed to the menstrual cycle, with temperatures rising about 0.5°C (0.9°F) from the time of ovulation until menstruation begins. The physiological significance of these variations is unknown.

Camels undergo a daily variation of as much as 7°C (13°F) in set point temperature. During the day, a camel's set point gradually resets upward, an adaptation that allows its body to absorb a large amount of heat. The heat absorption conserves water that would otherwise be lost by evaporation to keep the body at a lower set point. At night, when the desert is cooler, the thermostat resets again, allowing the body temperature to cool several degrees, releasing the excess heat absorbed during the day.

When the environmental temperature is cool, having a lowered temperature set point greatly reduces the energy required to maintain body temperature. In many animals, the lowered set point is accompanied by reductions in metabolic, nervous, and physical activity (including slower respiration and heartbeat), producing a sleeplike state known as **torpor.**

Entry into **daily torpor**—a period of inactivity keyed to variations in daily temperature—is typical of many small mammals and birds. These animals typically expend more energy per unit of body weight to keep warm than larger animals, because the ratio of body surface to volume increases as body size decreases. Hummingbirds, for example, feed actively during the daytime, when their set point is close to 40°C (104°F). During the cool of night, however, the set point drops to as low as 13°C (55°F), which allows the birds to conserve enough energy to survive overnight without feeding. Some nocturnal animals, including bats and small rodents such as the deer mouse, become torpid in cool locations during daylight hours when they do not actively feed. At night, their temperature set point rises and they become fully active **(Figure 46.22).**

Many animals enter a prolonged state of torpor tied to the seasons, triggered in most cases by a change in day length that signals the transition between summer and winter. The importance of day length has been demonstrated by laboratory experiments in which animals have been induced to enter seasonal torpor by changing the period of artificial light to match the winter or summer day length.

Extended torpor during winter, called **hibernation** (*hibernus* = relating to winter), greatly reduces metabolic expenditures when food is unobtainable. Typically, hibernators must store large quantities of fats to serve as energy reserves. The drop in body temperature during hibernation varies with the mammal. In some, such as hedgehogs, woodchucks, and squirrels, body temperature may fall by 20°C (36°F) or more. In certain hedgehogs, for example, body temperature falls from about 38°C (100°F) in the summer to as low as 5° to 6°C (41° to 43°F) during winter hibernation. Body temperature even drops to near 0°C in some small hibernating mammals and, in the Arctic ground squirrel, the body supercools (goes to a below-freezing, unfrozen state) during hibernation, with body temperature dropping to about −3°C. Some ectotherms, including amphibians and reptiles living in northern latitudes and even some insects, also become torpid during winter.

The depth of torpor differs among hibernating mammals. In bears, the core temperature drops only a few degrees. Although sluggish, hibernating bears will waken readily if disturbed. They also waken normally from time to time, as when females wake to give birth during the hibernating season.

Some mammals enter seasonal torpor during summer, called **estivation** (*aestivus* = relating to summer), when environmental temperatures are high and water is scarce. Some ground squirrels, for example, remain inactive in the cooler temperatures of their burrows during extreme summer heat. Many ectotherms, among them land snails, lungfishes, many toads and frogs, and some desert-living lizards, weather such climates by digging into the soil and entering a state of estivation that lasts throughout the hot dry season.

Some Animals Use a Form of Endothermy That Does Not Heat All of Their Cores

In contrast to birds and mammals, some animals exhibit a form of endothermy that does not heat all of their cores. For example, some cold-water marine teleosts (such as tunas and mackerels) and some sharks (such as the great white) use endothermy in their aerobic swimming muscles to maintain a body core temperature as much as 10° to 12°C warmer than their surroundings. These animals have in common the fact that they migrate over long distances, swimming continuously and, therefore, generating constant heat with the swimming muscles. That heat is insufficient to heat the entire body because too much heat is lost at the gill–water interface. These animals have evolved a *countercurrent heat exchanger* system between the swimming muscles and the gills to prevent most of the loss (countercurrent exchange is discussed in Section 44.2).

The system works as follows. Cold blood from the gills is first routed through arteries under the skin,

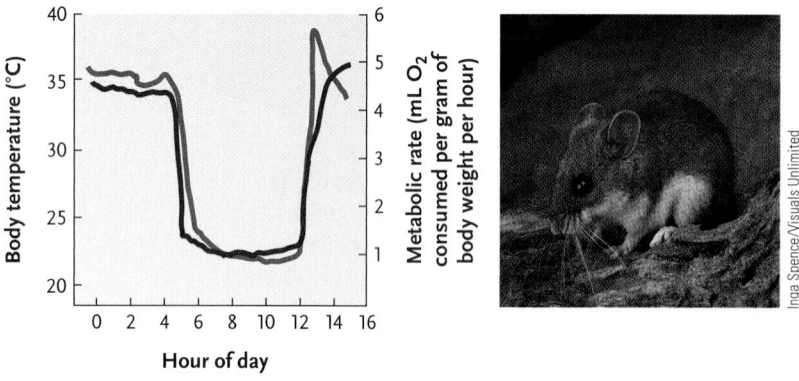

KEY
— Body temperature
— Metabolic rate

Figure 46.22
Cycle of daily torpor in a deer mouse (*Peromyscus maniculatus*).

What is the maximum temperature for life?

As you've read, most nondormant animals cannot live above about 45°C. However, the actual limit remains controversial. Some desert ants, such as *Cataglyphis,* forage on hot Saharan sand with body temperatures exceeding 50°C! In laboratory studies, *Cataglyphis bicolor* remained active at a body temperature of 55°C for short periods. This tolerance for high temperatures may allow it to outcompete others scavengers during the heat of the day.

Most controversial are the alvinellid worms, small tube-building polychaetes that live at hydrothermal vents in the deep sea. Investigators on the submersible *Alvin* (for which the worms were named) found that temperatures in the tubes of *Alvinella* worms were routinely about 60°C, with occasional peaks over 80°C. But other scientists were skeptical, noting that accurate temperature measurements are difficult due to violent currents near the vents, which mix 300°C vent water with the near-freezing water nearby. (At the high pressures of the deep sea, water does not boil at 300°C.) Moreover, laboratory studies of alvinellid enzymes showed that they malfunction above 45°–50°C. To examine tolerance directly, Peter Girguis of Harvard and Raymond Lee of Washington State University placed *Paralvinella* worms (a species related to *Alvinella* and having enzymes with similar thermal limits) in special high-pressure chambers. The chambers have a regulated temperature gradient ranging from 20°C at one end to 61°C at the other. The worms were kept at their natural habitat pressure and observed for seven hours. The animals crawled about and settled around the area at 50°C, where they appeared to behave normally. One worm even survived 55°C water for 15 minutes. These behavioral studies (published in 2006) show that 50°C is *Paralvinella*'s preferred temperature and that it may tolerate higher temperatures briefly. However, the actual limit in nature for these and similar worms remains unknown. What sets the upper limit for animals and other eukaryotes is not known, but is suspected to be fundamental features of gene transcription, RNA processing, and/or translation that cannot be stabilized beyond a certain temperature.

What about limits for archaeans and bacteria? Hyperthermophilic ("high-heat-loving") microbes can be found in abundance at hydrothermal vents. Although none are known to live at the highest temperatures (up to 400°C), one species discovered in 2003 by Derek Lovley and Kazem Kashefi of the University of Massachusetts remained viable in the laboratory after 10 hours at 121°C, a temperature used in autoclaves for sterilizing medical equipment. But the actual upper limit for hyperthermophiles remains speculative. Understanding these aspects of cells is crucial to hypotheses about life's origins and possible life elsewhere in the universe.

How do proteins work in high urea and at high pressure?

You have just read that sharks and their relatives use urea and trimethylamine oxide (TMAO) as osmolytes. Organic osmolytes are often said to be *compatible;* that is, unlike inorganic ions such as Na^+ and Cl^-, they do not perturb proteins even at high concentrations. Thus, organic osmolytes can safely build up in an organism. However, urea is a clear exception. At the urea concentration typical of sharks (300–400 mM),

many proteins—including ones in sharks—are perturbed. Indeed, biochemists often use urea to unfold proteins. How, then, can the sharks function with such high urea concentration? In the 1970s, George Somero and I, then of Scripps Institution of Oceanography, found that TMAO is not simply compatible: it actually stabilizes proteins and can *counteract* urea's effect. In mixtures of TMAO and urea at shark levels, stabilizing and destabilizing effects cancel out. Thus, by using two waste products as osmolytes, sharks maintain water balance while not perturbing proteins.

This shark finding led Robert Balaban, Maurice Burg, and coworkers at the National Institutes of Health to realize that a balancing effect could explain how mammalian kidneys survive the high levels of salt and urea in the medulla (where urea can exceed 1 M). In fact, a study led by Serena Bagnasco, now at Johns Hopkins Hospital, found that kidney cells maintain osmotic balance with sorbitol, inositol, glycerophosphorylcholine (GPC), and betaine (and not with salts, as previously believed). Moreover, GPC and betaine are *methylamines,* which, like TMAO, can counteract urea's effects. How methylamines like TMAO actually stabilize proteins remains uncertain. Wayne Bolen and colleagues at the University of Texas Medical Branch at Galveston recently showed that the peptide backbone of proteins is in a sense repelled by solutions of TMAO, making proteins fold up to avoid contact. The physicochemical properties of TMAO responsible for this effect are under investigation.

Stabilizing osmolytes may also help deep-sea organisms cope with high pressure. Low TMAO levels have long been known in bony fish (it is the source of "fishy odor"), but as osmoregulators, all bony fish were thought to have osmotic pressures of 300–400 mOsm/L, with little need for osmolytes. Recently, I and my students at Whitman College found that the deeper a species lives, the more TMAO it has. Indeed, deep-sea species can have osmotic pressures of 600 mOsm/L or more due to TMAO. Why might this be? Laboratory studies showed that TMAO readily counteracts the destabilizing effects that pressure has on protein structure and function.

Such research may be medically useful. William Welch and colleagues at the University of California at San Francisco hypothesize that stabilizing osmolytes might "repair" disease-causing mutant proteins. For example, cystic fibrosis (CF) arises from a chloride channel protein that does not fold properly, leading to symptoms that include impaired production of sweat and digestive secretions. Marybeth Howard in Welch's laboratory and collaborators recently treated cultured CF cells with organic osmolytes, and the mutant protein indeed folded and worked properly. Whether osmolytes can be used in whole mammals is now being studied.

Paul H. Yancey holds the Carl E. Peterson Endowed Chair of Sciences at Whitman College in Walla Walla, Washington. His main research interests are in areas of animal physiology, especially water stress and osmoregulation. To learn more about his research, go to http://marcus.whitman.edu/~yancey/.

which is at the same temperature as the water. The blood enters the body core through small arteries that form the countercurrent heat exchanger along with small veins that bring warm blood from the swimming muscles in the core. Countercurrent exchange warms the arterial blood, which returns to the heart in veins and is then pumped around the body in arteries, including to the core and the gills. The overall result is that heat is retained within the muscles, so that the core body temperature can remain significantly higher than that of the surrounding water.

STUDY BREAK

Describe how thermoreceptors and negative feedback pathways achieve temperature regulation in endotherms.

Review

Go to **ThomsonNOW**™ at www.thomsonedu.com/login to access quizzing, animations, exercises, articles, and personalized homework help.

46.1 Introduction to Osmoregulation and Excretion

- Solute concentration is measured as osmolarity in milliosmoles per liter of solution (mOsm/L). A solution can be comparatively hyperosmotic, hypoosmotic, or isoosmotic to another solution. Water moving from a region of higher osmolarity to a region of lower osmolarity across a selectively permeable membrane is known as osmosis.

- Osmoregulators keep the osmolarity of body fluids different from that of the environment. Osmoconformers allow the osmolarity of their body fluids to match that of the environment.

- Molecules and ions must be removed from the body to keep cellular and extracellular fluids isoosmotic. In most animals, extracellular fluids are filtered through tubules formed from a transport epithelium and released to the exterior of the animal as urine (Figure 46.2).

- Nitrogenous wastes are excreted as ammonia, urea, or uric acid, or as a combination of these substances (Figure 46.3).

  Animation: Diffusion, osmosis, and countercurrent systems

  Animation: Water and solute balance

46.2 Osmoregulation and Excretion in Invertebrates

- Most marine invertebrates are osmoconformers. Because their body fluids are isoosmotic to seawater, they expend little or no energy on maintaining water balance.

- Freshwater and terrestrial invertebrates are osmoregulators, with body fluids that are hyperosmotic to their surroundings. They must expend energy to excrete water that moves into their cells by osmosis.

- The cells of the simplest marine invertebrates exchange water and solutes directly with the surrounding seawater. More complex invertebrates have specialized excretory tubules (Figures 46.4–46.6).

46.3 Osmoregulation and Excretion in Mammals

- In mammals and other vertebrates, excretory tubules are concentrated in the kidney.

- The mammalian excretory tubule, the nephron, has a proximal end at which filtration takes place, a middle region in which reabsorption and secretion occur, and a distal end that releases urine. A network of capillaries surrounding the nephron takes up ions and water and other molecules absorbed by the nephron. The urine leaving individual nephrons is processed further in collecting ducts and then pools in the renal pelvis. From there it flows through the ureter to the urinary bladder, and through the urethra to the exterior of the animal (Figures 46.7 and 46.8).

- At its proximal end, the nephron forms a cuplike Bowman's capsule around a ball of capillaries, the glomerulus. A filtrate consisting of water, other small molecules, and ions is forced from the glomerulus into Bowman's capsule, from which it travels through the nephron and drains into the collecting ducts and renal pelvis. The proximal convoluted tubule of the nephron secretes H^+ into the filtrate and reabsorbs Na^+, Cl^-, and K^+ along with water, HCO_3^-, and nutrients. In the descending segment of the loop of Henle, water is reabsorbed by osmosis. In the ascending segment of the loop, Na^+ and Cl^- are reabsorbed. In the distal convoluted tubule, the concentrations of H^+ and salts are balanced between the urine and the interstitial fluid surrounding the nephron. In the collecting ducts, additional H^+ is secreted into the urine and water is reabsorbed; some urea is also reabsorbed at the bottom of the ducts (Figure 46.9 and Table 46.1).

  Animation: Human urinary system

  Animation: Human kidney

  Animation: Urine formation

  Animation: Tubular reabsorption

46.4 Regulation of Mammalian Kidney Function

- The kidney's autoregulation system is activated by receptors in the juxtaglomerular apparatus. The receptors trigger constriction or dilation of the afferent arteriole to keep blood flow and filtration constant during small variations in blood pressure.

- When blood volume and blood pressure drop, the hormones of the renin-angiotensin-aldosterone system (RAAS) raise blood pressure by stimulating arteriole constriction and increasing NaCl reabsorption in the kidneys (Figure 46.11).

- ADH, which increases water reabsorption and stimulates thirst, is released from the pituitary when osmoreceptors detect an increase in the osmolarity of body fluids (Figure 46.12).

  Animation: Structure of the glomerulus

46.5 Kidney Function in Nonmammalian Vertebrates

- Marine teleosts continually drink seawater to replace body water lost by osmosis to their hyperosmotic environment. Excess salts and nitrogenous wastes are excreted by the gills (Figure 46.13a).

- The body fluids of sharks and rays are isoosmotic with seawater. They do not lose water by osmosis, and do not drink seawater. Excess salts are excreted in the kidney and by a rectal salt gland.

- Body fluids of freshwater fishes and amphibians are hyperosmotic to their environment, and these animals must excrete the excess water that enters by osmosis. Body salts are obtained from food and, in fishes, through the gills (Figure 46.13b).

Nitrogenous wastes are excreted from the gills of fishes and larval amphibians as ammonia, and through the kidneys of adult amphibians as urea.

- Reptiles and birds conserve water by secreting nitrogenous wastes as uric acid and by absorbing water from urine and feces in the cloaca.

46.6 Introduction to Thermoregulation

- Animals must maintain body temperature at a level that provides optimal physiological performance. Heat flows between animals and their environment by conduction, convection, radiation, and evaporation (Figures 46.15 and 46.16).
- Ectothermic animals obtain heat energy primarily from the environment; endothermic animals obtain heat energy primarily from internal reactions (Figure 46.17).

Animation: Endotherms and ectotherms

46.7 Ectothermy

- Ectotherms obtain heat energy externally and control body temperature primarily by physiological or behavioral methods of regulating heat exchange with the environment (Figure 46.18).
- Many animals undergo thermal acclimatization, a structural or metabolic change in the limits of tolerable temperatures as the environment alternates between warm and cool seasons.

46.8 Endothermy

- Endotherms obtain heat energy primarily from internal reactions and maintain body temperature over a narrow range by balancing internal heat production against heat loss from the body surface.
- Internal heat production is controlled by negative feedback pathways triggered by thermoreceptors. When deviations from the temperature set point occur, signals from the receptors bring about compensating responses such as changes in blood flow to the body surface, sweating or panting, and behavioral modifications (Figure 46.19).
- The skin of endotherms is water-impermeable, reducing heat lost by direct evaporation of body fluids. The blood vessels of the skin regulate heat loss by constricting or dilating. A layer of insulating fatty tissue under the vessels limits losses to the heat carried by the blood. The hair of mammals and feathers of birds also insulate the skin. Erection of the hair or feathers reduces heat loss by thickening the insulating layer (Figures 46.20 and 46.21).
- The temperature set point in many birds and mammals varies in daily and seasonal patterns. During cooler conditions, a lowered set point is accompanied by torpor (Figure 46.22).
- Some animals exhibit a form of endothermy in which part of their core is maintained at a temperature significantly higher than the surrounding environment.

Animation: Human thermoregulation

Questions

Self-Test Questions

1. Which of the following statements about osmoregulation is true?
   a. In freshwater invertebrates, salts move out of the body into the water because the animal is hypoosmotic to the water.
   b. A marine teleost has to fight gaining water because it is isoosmotic to the sea.
   c. Most land animals are osmoconformers.
   d. Vertebrates are usually osmoregulators.
   e. Terrestrial animals can regulate their osmolarity without expending energy.

2. One role of tubules in excretion is to:
   a. absorb H^+ ions to buffer body fluids.
   b. transport proteins across transport epithelium.
   c. reabsorb glucose and amino acids.
   d. move toxic substances from the filtrate into the cells composing the transport tubules.
   e. filter by maintaining a lower pressure in the fluid outside the tubule than inside it.

3. Products of metabolism in humans, as in:
   a. terrestrial amphibians, can include urea, which requires more energy to produce than ammonia.
   b. birds and reptiles, can include uric acid, which is nontoxic and excreted as a paste.
   c. sharks, are primarily excreted as ammonia.
   d. hydra, must be isoosmotic with the water ingested.
   e. other mammals, cannot be water as water comes only from what they drink.

4. Filtration and/or excretion can be carried out by:
   a. ciliated metanephridia in insects.
   b. protonephridia containing flame cells in flatworms.
   c. a nephron and bladder in insects.
   d. Malpighian tubules on the segments of earthworms.
   e. the hindgut, which reabsorbs Na^+ and K^+ into the hemolymph of earthworms.

5. A mammalian nephron contains the:
   a. Bowman's capsule, which delivers the filtrate to the glomerulus.
   b. Bowman's capsule, which filters fluids, 99% of which will be excreted.
   c. proximal convoluted tubule, which moves Na^+ and K^+ into the filtrate of the interstitial fluids.
   d. proximal convoluted tubule, which reabsorbs K^+, Na^+, Cl^-, H_2O, and urea.
   e. proximal convoluted tubule, which lacks microvilli to ease fluid movement through it.

6. Which of the following correctly describes a part of kidney function?
   a. Collecting ducts dilute urine because they are permeable to salt but not water.
   b. In the ascending loop of Henle, Na^+ and Cl^- move into the tubules because the osmolarity of the filtrate is increased.
   c. The descending loop of Henle receives filtrate from the ascending loop.
   d. The distal convoluted tubule pumps water into the tubule by active transport.
   e. The renal pelvis receives urine from the collecting ducts and carries it to the ureters.

7. Which of the following is an example of autoregulation of kidney function?
   a. The RAAS regulates Na^+ by secreting renin when blood pressure or blood volume decreases.
   b. The ADH system regulates water balance by decreasing water reabsorption and increasing excretion of salt.

c. Receptors in the juxtaglomerular apparatus of the distal convoluted tubule detect drops in blood pressure and cause a higher filtration rate.

d. ANF is released by the kidney to increase renin release.

e. Angiotensin lowers blood pressure by constricting arterioles.

8. Deficient water levels in humans are prevented by:

a. osmoreceptors on the hypothalamus that detect decreases in salt concentrations.

b. the hypothalamus stimulating the posterior pituitary to secrete a hormone that allows the collecting ducts and distal convoluted tubules to be permeable to water.

c. inhibiting ADH, which causes a rise in osmolarity of extracellular fluids.

d. producing dilute urine.

e. drinking alcohol, which stimulates aldosterone to raise the osmolarity of body fluids.

9. Which best exemplifies ectothermy?

a. The metabolic rate increases as the temperature decreases.

b. Body temperature remains constant when environmental temperatures change.

c. Food demand increases when temperatures drop.

d. Virtually all invertebrate groups are ectotherms.

e. No vertebrate groups are ectotherms.

10. Unique to endotherms is:

a. torpor.

b. thermal acclimatization.

c. a nonchanging body temperature.

d. response to seasonal temperature changes.

e. thermoregulation by a hypothalamus.

Questions for Discussion

1. A urinalysis reveals glucose, urea, hemoglobin, and sodium. Which of these substances are abnormal in urine, and why?

2. As a person ages, nephron tubules lose some of their ability to concentrate urine. What is the effect of this change?

3. Shivering increases air movement over the body surface. What effect does this air movement have on heat conservation in the shivering animal?

4. What heat transfer processes might account for the change in body temperature when a mammal's body temperature undergoes daily variations?

Experimental Analysis

Design experiments to demonstrate the role of fluid consumption in thermoregulation during endurance exercise.

Evolution Link

Humans produce urea as an excretion product, whereas reptiles and birds produce uric acid. Indeed, human kidneys are not as efficient as those of reptiles and birds. What does this mean in an evolutionary sense?

How Would You Vote?

Many companies use urine testing to screen for drug and alcohol use among prospective employees. Some people say this is an invasion of privacy. Do you think employers should be allowed to require a person to undergo urine testing before being hired? Go to www.thomsonedu.com/login to investigate both sides of the issue and then vote.

A newly fertilized human egg passing down the oviduct on its way to implantation in the wall of the uterus (colorized SEM).

© Clouds Hill Imaging Ltd./Corbis

47 Animal Reproduction

WHY IT MATTERS

It is 7 days after the October full moon and night is falling. All the inhabitants of the Samoan island of Tutuila who have access to a boat are gathered on the island's large lagoon. Some hold lanterns and look into the water; others have nets at the ready. They are awaiting the palolo worm *(Eunice viridis),* which has appeared in the water as the moon rises on this same night of the lunar year for as long as the islanders can remember.

The moon peeks over the horizon and the excitement of the crowd rises. Then, there they are, untold thousands of blue and green worms, squirming in the water like animated spaghetti. The boaters scoop up the worms by the netfull and dump them into buckets. When the buckets are full, the islanders glide toward the shore where steaming pots are waiting, for palolo worms are a delicacy that the islanders savor only once a year. For the islanders, a night of feasting, singing, and dancing will follow as they cook and eat the worms. Their flavor has been described by some as similar to that of caviar; by others like, well, palolo worms.

The worms that squirm to the surface to delight the islanders are actually not complete individuals. They are tail sections about 10 to

1069

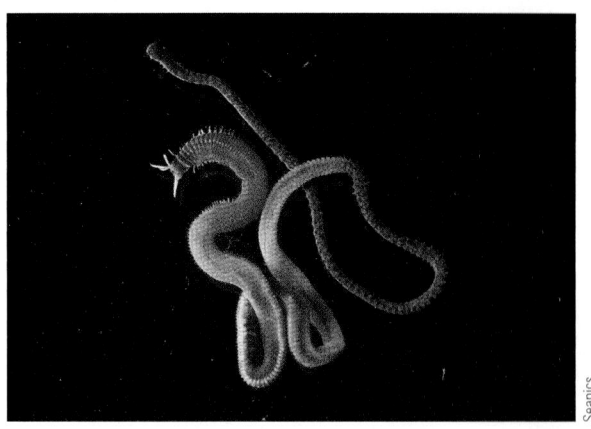

Figure 47.1
The palolo worm. Gametes are packed into segments of the tail section (in blue).

20 cm long that break from adults after they become filled with eggs or sperm. The adults are polychaete annelids that live in burrows in coral reefs of the Samoan and Fiji islands **(Figure 47.1)**. These annelids develop tail segments once a year, just after the October full moon. On the seventh night following the full moon, the tails break off and swim to the surface, where—if Samoan gourmets do not net them first— they disintegrate and release eggs and sperm by the millions, turning the water of the lagoon milky. The anterior ends of the worms, safe in their burrows, will survive to produce tails for next year's mating frenzy. A biological clock in the worms, timed by periods of moonlight, precisely sets both the appearance of the mating swarm and indirectly, the appearance of the islanders with their boats.

The swarm of the palolo worms is only one of many adaptations that accomplish mating in animals. For animals that reproduce by eggs and sperm, the adaptations are as diverse as the number of species on Earth. This diversity allows individuals of the same species to find each other and unite eggs and sperm. Within the diversity, however, are underlying patterns that are shared by all animals.

Both the underlying patterns and the diversity of animal reproduction are the subjects of this chapter. We also discuss the development of eggs and sperm, and the union of egg and sperm that begins the development of a new individual. The next chapter continues with the events of development after eggs and sperm have united.

47.1 Animal Reproductive Modes: Asexual and Sexual Reproduction

Reproduction is part of a life cycle in which individuals grow, develop, and reproduce according to instructions encoded in DNA. Rather than survival of the individual, reproduction is the means of passing on the individual's genes to new generations of the species. As such, it is among the most vital functions of living organisms.

Two basic modes of reproduction operate in the animal kingdom. In **asexual reproduction**, a single individual gives rise to offspring without fusion of **gametes** (egg and sperm); that is, there is no genetic input from another individual. In **sexual reproduction**, male and female parents produce offspring through the union of egg and sperm generated by meiosis (meiosis is discussed in Chapter 11).

Asexual Reproduction Produces Offspring with Genes from Only One Individual

Many aquatic invertebrates and some terrestrial annelids and insects reproduce asexually. Asexual reproduction is rare among vertebrates. In asexual reproduction, one to many cells of a parent's body develop directly into a new individual. In a few animals that undergo asexual reproduction, the cells taking part are genetically varied products of meiosis, but in most they are products of mitosis. The offspring therefore are genetically identical to one another and to the parent: in other words, they are genetic clones of the parent. For this reason, asexual reproduction of this kind is also called *clonal reproduction*.

Genetic uniformity of offspring can be advantageous in environments that remain stable and uniform. Asexual reproduction tends to preserve gene combinations producing individuals that are successful in such environments. Further, individuals do not have to expend energy to produce gametes or find a mate. Asexual reproduction can also bring reproductive advantages to individuals living in sparsely settled populations, or to sessile animals, which cannot move from place to place.

Asexual reproduction involving mitosis occurs in animals by three basic mechanisms: *fission, budding,* and *fragmentation*. In **fission**, the parent separates into two or more offspring of approximately equal size. Planarians (flatworms), for instance, reproduce asexually by fission; depending on the species, they may divide by transverse or longitudinal fission. In **budding**, a new individual grows and develops while attached to the parent. Sponges, tunicates, and some cnidarians reproduce asexually by this mechanism. The offspring may break free from the parent, or remain attached to form a *colony*. In the cnidarian *Hydra*, for example, an offspring buds and grows from one side of the parent's body and then detaches to become a separate individual **(Figure 47.2)**. Among many corals, the buds remain attached when their growth is complete, forming colonies of thousands of interconnected individuals. In **fragmentation**, pieces separate from a parent's body and develop *(regenerate)* into new individuals. Many species of cnidarians, flatworms, annelids, and some echinoderms can reproduce by fragmentation.

Some animals produce offspring by the growth and development of an egg without fertilization. The offspring may be haploid or diploid depending on the species. This form of asexual reproduction is called **parthenogenesis** (*parthenos* = virgin; *genesis* = birth). Because the egg from which a parthenogenetic offspring is produced derives from meiosis in the female parent, the offspring are not genetically identical to the parent or to each other. (How chromosome segregation and genetic recombination during meiosis produces gametes with gene combinations different from the parent is described shortly.)

Parthenogenesis occurs in some invertebrates, including certain aphids, water fleas, bees, and crustaceans. In bees, for instance, haploid male drones are produced parthenogenetically from unfertilized eggs produced by reproductive females (queens) while new queens and sterile workers develop from fertilized eggs. Parthenogenesis also occurs in some vertebrates, for example, in certain fish, salamanders, amphibians, lizards, and turkeys. In these animals, an egg, produced by meiosis, typically doubles its chromosomes to produce a diploid cell that begins development. In single-sex species where females have two identical sex chromosomes, the offspring are female, whereas in single-sex species where males have two identical sex chromosomes, the offspring are male. For instance, all whiptail lizards *(Cnemidophorus)* are females, produced solely by parthenogenesis. Interestingly, these females go through the motions of mating and copulation with each other.

Sexual Reproduction Generates Diversity among Offspring

Animals reproduce sexually by the union of sperm and eggs produced by meiosis. The overriding advantage of sexual reproduction is the generation of genetic diversity among offspring. This diversity increases the chance that, in a changing environment, at least some offspring will grow and reproduce successfully. Diversity also increases the chance that offspring may be able to live and reproduce in environments previously unoccupied by the species.

Two mechanisms that are part of meiosis give rise to the genetic diversity in eggs and sperm: *genetic recombination* (see Section 11.2) and the *independent assortment* of chromosomes of maternal and paternal origin (see Section 12.1). Genetic recombination mixes the alleles of parents into new combinations within chromosomes; independent assortment selects random combinations of maternal and paternal chromosomes to be placed in gamete nuclei. Additional variability is generated at fertilization when eggs and sperm from genetically different individuals fuse together at random to initiate the development of new individuals. To these sources of variability are added random DNA mutations, which are the ultimate source of variability for both sexual and asexual reproduction.

The disadvantages of sexual reproduction include the expenditure of energy and raw materials in produc-

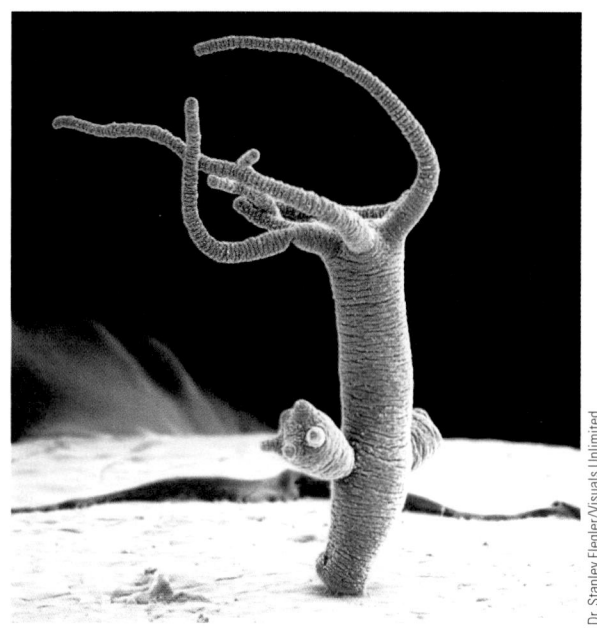

Dr. Stanley Flegler/Visuals Unlimited

Figure 47.2
Asexual reproduction by budding in *Hydra* (colorized SEM).

ing gametes and finding mates. The need to find mates can also expose animals to predation and takes time from finding food and shelter and caring for existing offspring.

With these advantages and disadvantages in mind, we now turn to the mechanisms of sexual reproduction, which include both cellular and whole-organism activities. We begin with the cellular mechanisms in the next section.

STUDY BREAK

What are the advantages and disadvantages of asexual reproduction? Of sexual reproduction?

47.2 Cellular Mechanisms of Sexual Reproduction

The cellular mechanisms of sexual reproduction are **gametogenesis**, the formation of male and female gametes, and **fertilization**, the union of gametes that initiates development of a new individual. The pairing of a male and a female for the purpose of sexual reproduction is **mating**.

Gametogenesis Involves the Coordinated Events of Meiosis and Sperm and Egg Development

Gametes in most animals form from **germ cells**, a cell line that is set aside early in embryonic development and remains distinct from the other, **somatic cells** of the body. During development, the germ cells collect in specialized gamete-producing organs, the **gonads**—the

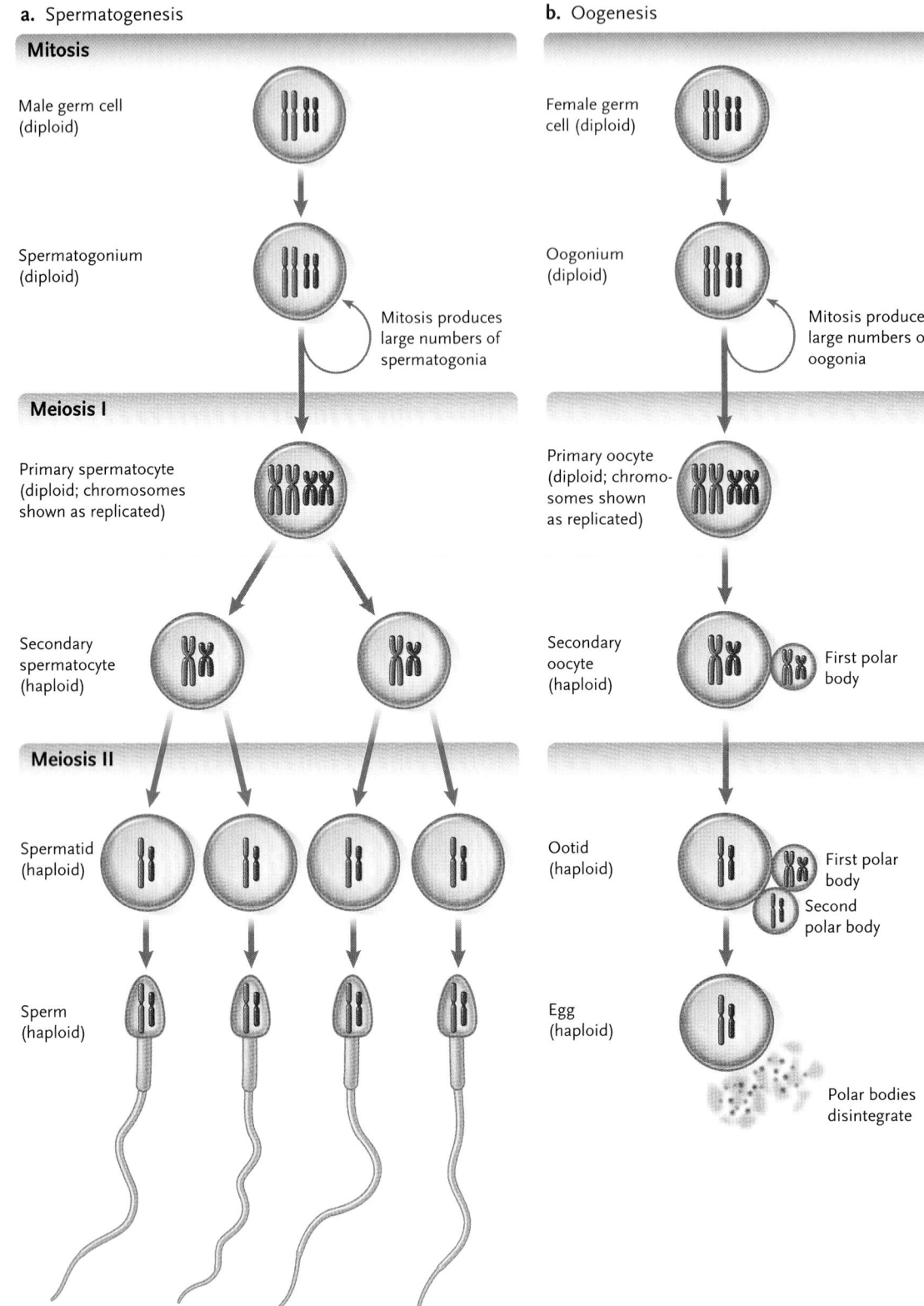

a. Spermatogenesis

Mitosis

Male germ cell (diploid)

Spermatogonium (diploid)

Mitosis produces large numbers of spermatogonia

Meiosis I

Primary spermatocyte (diploid; chromosomes shown as replicated)

Secondary spermatocyte (haploid)

Meiosis II

Spermatid (haploid)

Sperm (haploid)

b. Oogenesis

Mitosis

Female germ cell (diploid)

Oogonium (diploid)

Mitosis produces large numbers of oogonia

Meiosis I

Primary oocyte (diploid; chromosomes shown as replicated)

Secondary oocyte (haploid)

First polar body

Meiosis II

Ootid (haploid)

First polar body

Second polar body

Egg (haploid)

Polar bodies disintegrate

Figure 47.3
The mitotic and meiotic divisions producing eggs and sperm from germ cells. **(a)** Spermatogenesis. **(b)** Oogenesis. The first polar body may or may not divide, depending on the species, so that either two or three polar bodies may be present at the end of meiosis. Two are shown in this diagram.

testes (singular, *testis*) in males and **ovaries** in females. Mitotic divisions of the germ cells produce **spermatogonia** in males and **oogonia** in females; these are the cells that enter meiosis to give rise to gametes **(Figure 47.3)**. In some animals, the germ cells also give rise to families of cells that assist gamete development.

Meiosis reduces the number of chromosomes from the diploid level characteristic of somatic cells of the species, in which there are two copies of each chromosome, to the haploid level of gametes, in which there is one copy of each chromosome. The fusion of a haploid sperm and egg during fertilization restores the diploid number of chromosomes and produces a **zygote**, the first cell of a new individual.

During the meiotic divisions, the developing gametes are known as **spermatocytes** or **oocytes;** at the end of meiosis they become *spermatids* or *ootids.* When meiosis is complete, the haploid cells develop into mature sperm cells, also called **spermatozoa** (singular, *spermatozoon*) or simply *sperm;* and egg cells, also called **ova** (singular, *ovum*) or simply *eggs.* The process of producing sperm is called **spermatogenesis** and the process of producing eggs is called **oogenesis**. The sperm of most animal species are motile cells, driven through a watery medium by the whiplike beating of a flagellum that extends from the posterior end of the cell. The eggs of all animals are nonmotile cells, typically much larger than sperm of the same species.

Spermatogenesis. The events of spermatogenesis produce haploid cells, specialized to deliver their nuclei to eggs of the same species. Two meiotic divisions produce four haploid spermatids (see Figure 47.3a), which develop into mature sperm **(Figure 47.4)**. During maturation, most of the cytoplasm is lost, except for mitochondria, which surround the base of a flagellum. These mitochondria produce the ATP used as the energy source for flagellar beating. At the head of the sperm, a specialized secretory vesicle, the **acrosome**, forms a cap over the nucleus. The acrosome contains enzymes and other proteins that help the sperm attach to and penetrate the surface coatings of an egg of the same species.

Oogenesis. In oogenesis, only one of the cell products of meiosis develops into a functional egg, which retains almost all of the parent cell's cytoplasm. The other products form nonfunctional cells called **polar bodies** (see Figure 47.3b). The unequal cytoplasmic divisions concentrate nutrients and other molecules required for development in the egg. In most species, the polar bodies eventually disintegrate and do not contribute to fertilization or embryonic development.

The oocytes of most animals do not actually complete meiosis until fertilization. For example, mammals follow a complex pattern in which oocytes stop developing at the end of the first meiotic prophase, within a few

a. Human sperm

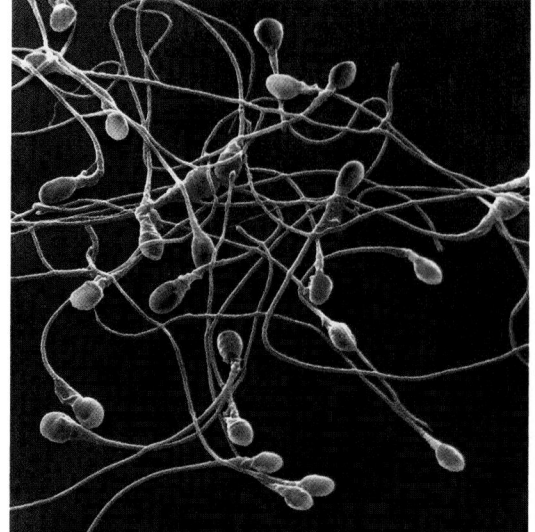

b. Sperm structure

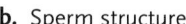
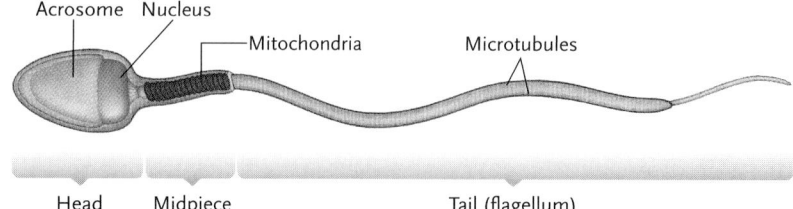

Acrosome   Nucleus

Mitochondria   Microtubules

Head   Midpiece   Tail (flagellum)

Figure 47.4
Spermatozoa. **(a)** Photomicrograph of human sperm. **(b)** Structure of a sperm.

weeks after a female is born. The oocytes remain in the ovary at this stage until the female is sexually mature. In humans, some oocytes may remain in prophase of the first meiotic division for perhaps 50 years. Then, one to several oocytes advance to the metaphase of the second meiotic division and are released from the ovary at intervals ranging from days to months, or at certain seasons, depending on the species. As in other animals, meiosis is completed at fertilization to produce the fully mature egg **(Figure 47.5)**. Mature eggs are the largest cell type of an animal species.

An egg typically has specialized features, which include stored nutrients required for at least the early stages of embryonic development; egg coats of one or more kinds, which protect the egg from mechanical injury and infection and, in some species, protect the embryo after fertilization; and mechanisms that prevent the egg from being fertilized by more than one sperm cell (discussed shortly).

Egg coats are surface layers added during oocyte development or fertilization in many species. The **vitelline coat**, called the **zona pellucida** in mammals (see Figure 47.5) is a gel-like matrix of proteins, glycoproteins, or polysaccharides immediately outside of the plasma membrane of the egg cell. Insect eggs have additional outer protein coats that form a hard, water-

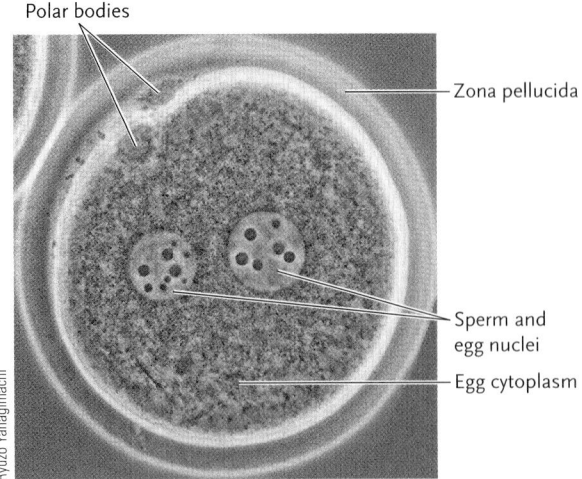

Polar bodies

Zona pellucida

Sperm and egg nuclei

Egg cytoplasm

Ryuzo Yanagimachi

Figure 47.5
A mature hamster egg that has been fertilized.

impermeable layer for preventing desiccation. Amphibians and some echinoderms instead have an additional outer egg jelly layer (see Figure 47.8) that protects the egg from drying.

In birds, reptiles, and one group of egg-laying mammals, the **monotremes**, the egg white, a thick solution of proteins, surrounds the vitelline coat. Outside the white is the *shell* of the egg, flexible and leathery in reptiles and mineralized and brittle in birds. Both the egg white and the shell are added while the egg—fertilized or not—is in transit through the **oviduct**, the tube through which the egg moves from the ovary to the outside of the body. In mammals, the egg is surrounded by **follicle cells** during its development. These cells, which grow from ovarian tissue, nourish the developing egg. They also make up part of the zona pellucida while the egg is in the ovary, and remain as a protective layer after it is released.

The amount of stored nutrients in an egg varies with the animal. Mammalian eggs are microscopic, containing few stored nutrients. In mammals, the embryo develops inside the mother and is supplied with nutrients by the mother's body. In contrast, the relatively huge eggs of birds and reptiles contain all the nutrients required for complete embryonic development: the "yolk" contains the egg cell, and the "white" contains the nutrients. No matter what the size of an animal egg, however, most of the volume is cytoplasm, and the egg nucleus is microscopic or nearly so in all species.

Fertilization Requires an Internal or External Aquatic Medium

Eggs and sperm are delivered from the ovaries and testes to the site of fertilization by oviducts in females and by sperm ducts in males; in many species, external accessory sex organs participate in the delivery. **Figure 47.6** shows examples of invertebrate and vertebrate reproductive systems. The nonmotile eggs move through the oviducts on currents generated by the beating of cilia lining the oviducts, or by contractions of the oviducts or the body wall.

Depending on the species, fertilization may take place externally, in a watery medium outside the body of both parents, or internally, in a watery fluid inside the body of the female. In **external fertilization**, which occurs in most aquatic invertebrates, bony fishes, and amphibians, sperm and eggs are shed into the surrounding water. The sperm swim until they collide with an egg of the same species. The process is helped by synchronization of female and male gamete release, and by the enormous quantities of gametes released. In some animals, such as sea urchins and amphibians, the sperm are attracted to the egg by diffusible attractant molecules released by the egg.

Most amphibians, even terrestrial species such as toads, mate in an aquatic environment. Frogs typically mate by a reflex response called *amplexus*, in which the male clasps the female tightly around the body with his forelimbs **(Figure 47.7)**. The embrace stimulates the female to shed a mass of eggs into the water through the *cloaca*—the cavity in reptiles, birds, amphibians, and many fishes into which both the intestinal and genital tracts empty. As the eggs are released, they are fertilized by sperm released by the male.

Internal fertilization takes place in invertebrates such as annelids, some arthropods, and some mollusks, and in vertebrates such as reptiles, birds, mammals, some fishes, and some salamanders. In these animals, the sperm are released by the male close to or inside the entrance of the reproductive tract of the female. The sperm swim through fluids in the reproductive tract until they reach and fertilize each egg. In some species, molecules released by the egg attract the sperm to its outer coats. The physical act involving the introduction of the male's accessory sex organ (for example, penis) into a female's accessory sex organ (for example, vagina) to accomplish internal fertilization is known as **copulation**. Internal fertilization makes terrestrial life possible by providing the aquatic medium required for fertilization inside the female's body without the danger of gametes drying by exposure to the air.

Sharks and rays have evolved a form of internal fertilization in which the male uses a pair of modified pelvic fins as accessory sex organs to channel sperm directly inside the female's cloaca. Male reptiles, birds, and mammals also have accessory sex organs that place sperm directly inside the reproductive tract of females, where fertilization takes place. In reptiles and birds, sperm fertilize eggs as they are released from the ovary and travel through the oviducts, before the shell is added. In mammals, the male's penis delivers sperm into the female's vagina. Unlike the cloaca, which has both sexual and excretory functions, the vagina is specialized for reproduction. Fertilization takes place when sperm swim into the tubular oviducts containing the eggs.

Fertilization Involves Fusion of a Sperm and an Egg, Which Activates the Egg for Development

Once a sperm touches the outer surface of an egg of the same species **(Figure 47.8a)**, receptor proteins in the sperm plasma membrane bind the sperm to the vitelline coat or zona pellucida. In most animals, only a sperm from the same species as the egg can recognize and bind to the egg surface.

Species recognition is highly important in animals that carry out external fertilization, because the water surrounding the egg may contain sperm of many different species. It is less important in internal fertilization, because structural adaptations and behavioral patterns of mating usually limit sperm transfer from males to females of the same species.

Fertilization. After the initial attachment of sperm to egg, the events of fertilization proceed in rapid succession **(Figure 47.8b)**. The actual attachment event triggers the **acrosome reaction,** in which enzymes contained in the acrosome are released from the sperm and digest a path through the egg coats. The sperm, with its tail still beating, follows the path until its plasma membrane touches and fuses with the plasma membrane of the egg. Fusion introduces the sperm nucleus into the egg cytoplasm and activates the egg to complete meiosis and begin development.

Egg Activation and Blocks to Polyspermy. Two mechanisms can prevent more than one sperm from fertilizing the egg: a *fast block* within seconds of fertilization, and a *slow block* within minutes.

In many invertebrate species, such as the sea urchin, the fusion of egg and sperm opens ion channels in the egg's plasma membrane, spreading a wave of electrical depolarization over the egg surface, much like the nerve impulse traveling along a neuron. The depolarization alters the egg plasma membrane so that it cannot fuse with any additional sperm, thereby eliminating the possibility that more than one set of paternal chromosomes enters the egg. Because it occurs within a few seconds after fertilization, the barrier set up by the wave of depolarization is called the **fast block to polyspermy.**

The fast block depends on a change in the egg's membrane potential from negative to positive. For example, Laurinda Jaffe, of the University of Connecticut Health Center, found that if the membrane potential of a sea urchin egg was artificially kept at a negative value, no fast block was set up, and additional sperm could fuse with the plasma membrane. If the membrane was instead kept positive before sperm contact, fertilization was entirely blocked.

In vertebrates, the wave of membrane depolarization following sperm–egg fusion is not as pronounced, and does not prevent additional sperm from fusing

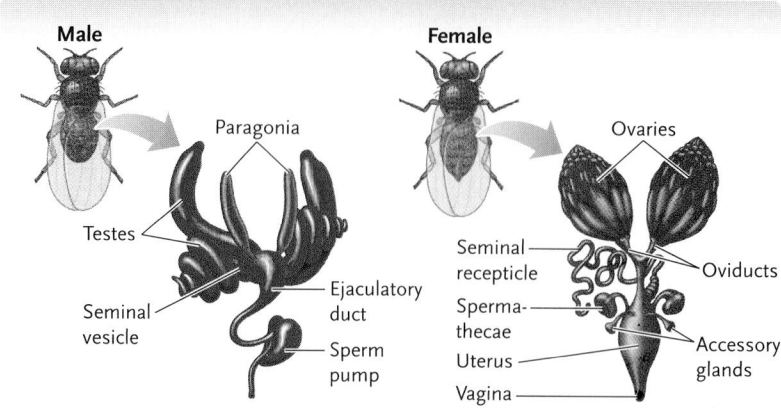

a. *Drosophila* (fruit fly)

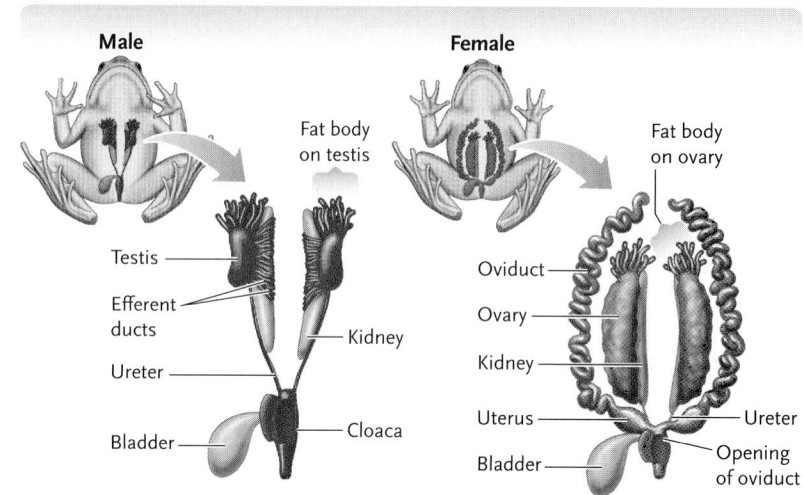

b. Amphibian (frog)

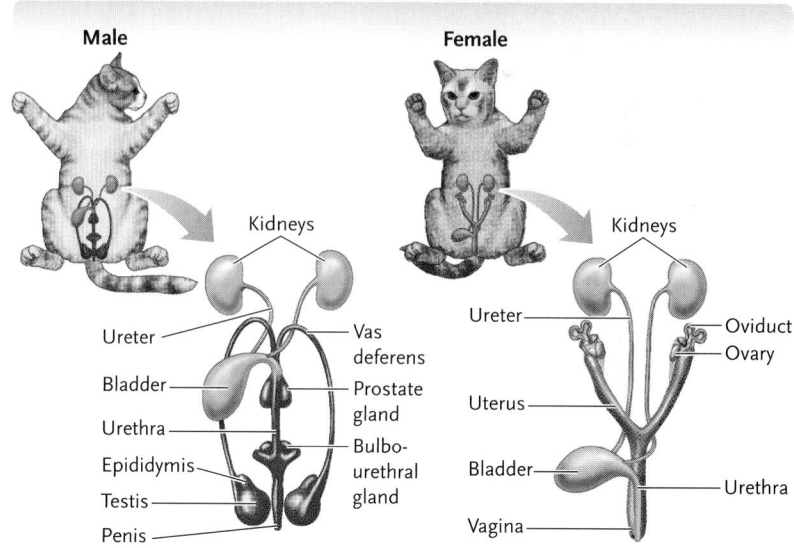

c. Mammal (cat)

Figure 47.6

Some reproductive systems. **(a)** An insect, *Drosophila* (fruit fly). **(b)** An amphibian, a frog. **(c)** A mammal, a cat. Female systems are shown in blue, and male systems in yellow.

— Eggs

Figure 47.7

A male leopard frog *(Rana pipiens)* clasping a female in a mating embrace known as amplexus. The tight squeeze by the male frog stimulates the female to release her eggs, which can be seen streaming from her body, embedded in a mass of egg jelly. Sperm released by the male fertilize the eggs as they pass from the female.

with the egg. However, any additional sperm nuclei entering the egg cytoplasm usually break down and disappear, so that only the first sperm nucleus to enter fuses with the egg nucleus.

In both invertebrates and vertebrates, fusion of egg and sperm triggers the release of stored calcium (Ca^{2+}) ions from the endoplasmic reticulum into the cytosol. The Ca^{2+} ions activate control proteins and enzymes that initiate intense metabolic activity in the fertilized egg, including a rapid increase in cellular oxidations and synthesis of proteins and other molecules.

The Ca^{2+} ions also trigger the **cortical reaction**, in which **cortical granules**, secretory vesicles just under the plasma membrane, fuse with the egg's plasma membrane and release their contents to the outside (see Figure 47.8b). Enzymes released from the cortical granules alter the egg coats within minutes after fertilization, so that no further sperm can attach and penetrate to the egg. Once this barrier, termed the **slow block to polyspermy**, is set up, no further sperm can reach the egg plasma membrane in any animal species.

The importance of Ca^{2+} to cortical granule release has been demonstrated experimentally: if Ca^{2+} is added to the cytoplasm, the granules are released in

a. Sperm adhering to egg

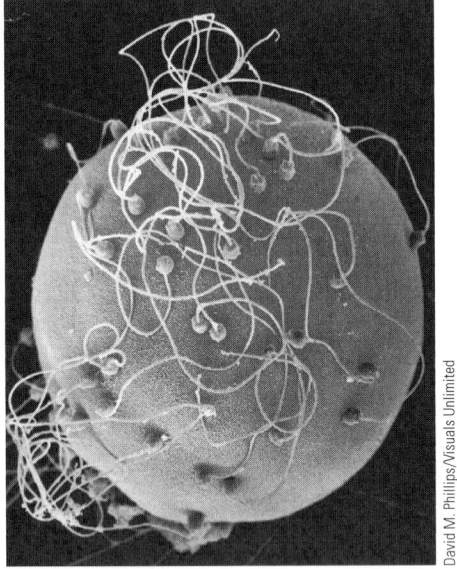

b. Steps in fertilization

1 A sperm contacts the jelly layer of the egg.

2 The acrosomal reaction begins.

3 Acrosomal enzymes dissolve a path through the jelly layer.

4 Proteins in its plasma membrane bind the sperm to the vitelline coat.

5 The sperm lyses a hole in the vitelline coat. The sperm and egg plasma membranes fuse.

6 Membrane depolarization produces the fast block to polyspermy.

7 The sperm nucleus and centriole enter the egg. The sperm nucleus then fuses wtih the egg nucleus.

8 Cortical granules discharge their contents, producing the slow block to polyspermy.

Sperm

Nucleus

Centriole

Actin

Acrosome

Acrosomal process

Jelly layer

Vitelline coat

Plasma membrane

Cortical granule

Egg

Figure 47.8

Fertilization. **(a)** Sperm adhering to the surface coat of a sea urchin egg. Of the many sperm that may initially adhere to the outer surface of an egg, usually only one accomplishes fertilization. **(b)** Steps of fertilization in a sea urchin.

unfertilized eggs; conversely, if Ca^{2+}-binding chemicals are added to the cytoplasm of unfertilized eggs, so that the Ca^{2+} concentration cannot rise, cortical granule release does not occur after fertilization.

After the sperm nucleus enters the egg cytoplasm, microtubules move the sperm and egg nuclei together in the egg cytoplasm and they fuse. The chromosomes of the egg and sperm nuclei then assemble together and enter mitosis. The subsequent, highly programmed events of embryonic development, which convert the fertilized egg into an individual capable of independent existence, are described in the next chapter.

Of the structures in a sperm cell, only the paternal chromosomes, the microtubule organizing center, and one or two centrioles (see Section 10.3 and Figure 10.11) survive in the egg. Therefore, with the exception of the microtubule organizing center and centrioles, all the cytoplasmic structures of the embryo, and of the new individual, are maternal in origin. The centrioles of the new individual are normally paternal in origin.

Reproductive Systems May Be Oviparous or Viviparous in Animals with Internal Fertilization

In animals with internal fertilization, three major types of support for embryonic development have evolved: *oviparity,* meaning egg laying; *viviparity,* meaning live bearing; and *ovoviviparity,* meaning live bearing from eggs that hatch internally. **Oviparous** animals (*ovum* = egg; *parere* = to give birth to) lay eggs that contain the nutrients needed for development of the embryo outside the mother's body. Examples are insects, spiders, most reptiles, and birds. The only oviparous mammals are the *monotremes:* the echidnas and *Ornithorhynchus anatinus* (the duck-billed platypus), both of which inhabit Australia.

Viviparous animals (*vivus* = alive) retain the embryo within the mother's body and nourish it during at least early embryo development. All mammals except the monotremes are viviparous. Viviparity is seen also in all other vertebrate groups except for the crocodiles, turtles, and birds.

In viviparous animals, development of the embryo takes place in a specialized saclike organ, the **uterus** *(womb).* Among mammals, one group, called the *placental mammals* or *eutherians,* has a specialized temporary structure, the **placenta,** that connects the embryo with the uterus. The placenta facilitates the transfer of nutrients from the mother's blood to the embryo and of wastes in the opposite direction. Humans are placental mammals. The other group of mammals, the *marsupials* or *metatherians,* originally were called nonplacental mammals because of a belief that they lacked a placenta. In fact, they do have a placenta, but it derives from a different tissue than that of eutherians and does not connect the embryo and the uterus. Instead

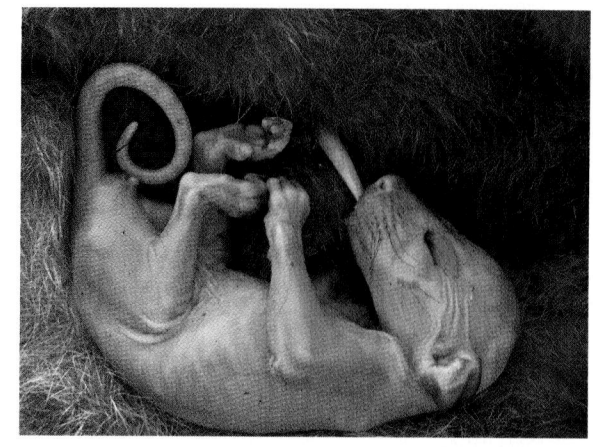

Figure 47.9
Developing offspring of a marsupial mammal, an opossum, attached to nipples in the marsupium (pouch) of the mother.

it provides nutrients to the embryo from an attached membranous sac containing yolk for only the early stages of its development. In many metatherians, the embryo is then born and crawls over the mother's fur to reach the **marsupium,** an abdominal pouch in which it attaches to nipples and continue its development **(Figure 47.9).** Kangaroos, koalas, wombats, and opossums are marsupials.

In some animals, such as some fishes, lizards, and amphibians, many snakes, and many invertebrates, fertilized eggs are retained within the body and the embryo develops using the nutrients provided by the egg. There is no uterus or placenta involved. When development is complete the eggs hatch inside the mother and the young are released to the exterior. Animals showing this form of reproduction are known as **ovoviviparous** animals.

Hermaphroditism Is a Variation on Sexual Reproduction

Some animals have evolved modified mechanisms that they use as their normal sexual reproduction process. One of these mechanisms is **hermaphroditism** (from *Hermes* + *Aphrodite,* a Greek god and goddess), in which both mature egg-producing and mature sperm-producing tissue is present in the same individual. That is, hermaphroditic individuals are able to produce both eggs and sperm. Most flatworms, earthworms, land snails, and numerous other invertebrates are hermaphroditic; in humans and other mammals, hermaphroditism is a rare, abnormal condition.

Most hermaphroditic animals do not fertilize themselves. In those animals, self-fertilization is prevented by anatomical barriers that prevent individuals from introducing sperm into their own body, or by mechanisms in which the egg and sperm mature at different times. The prevention of self-fertilization maintains the genetic variability of sexual reproduction.

Hermaphroditism takes two forms: **simultaneous hermaphroditism,** in which individuals develop functional ovaries and testes at the same time, and

a.

b. Sex organs

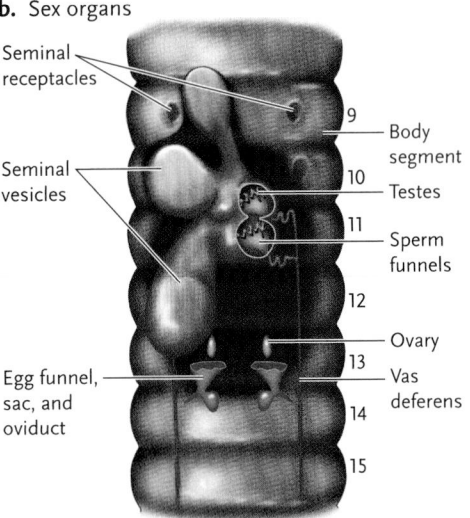

Seminal receptacles

Seminal vesicles

Egg funnel, sac, and oviduct

9 — Body segment

10 — Testes

11 — Sperm funnels

12

Ovary

13 — Vas deferens

14

15

Figure 47.10
Simultaneous hermaphroditism in the earthworm. **(a)** Copulation by a mating pair of earthworms, in which each earthworm releases sperm that fertilizes eggs in its partner. **(b)** Sex organs in the earthworm.

sequential hermaphroditism, in which individuals change from one sex to the other. The two earthworms shown in **Figure 47.10** provide a common example of simultaneous hermaphroditism. The only known vertebrate simultaneous hermaphrodites are hamlets (genus *Hypoplectrus*), a group of predatory sea basses. Sequential hermaphroditism is seen among a number of invertebrates (for example, the gastropod, the slipper shell *Crepidula fornicata*) and some ectothermic vertebrates, notably fishes (for example, the clownfish, genus *Amphiprion*). In some species the initial sex is male (as with the slipper shell and the clownfish), and in others it is female.

STUDY BREAK

1. What are egg coats, and what is their function? What egg coats do mammalian and bird eggs have?
2. How is the slow block to polyspermy brought about?

47.3 Sexual Reproduction in Humans

Except for structural details, human reproduction is typical of that of eutherian (placental) mammals. Internally, these mammals have a pair of gonads, either ovaries or testes. The gonads have a dual function in mammals, as they do in all vertebrates: they both produce gametes and secrete hormones responsible for sexual development and mating behavior (see Section 40.4). Males have ducts that carry sperm from the testes to the exterior. Females have an oviduct that leads from each ovary to the uterus, in which fertilized eggs implant and proceed through embryonic development. Nutrients from the mother and wastes from the embryo are exchanged through the placenta. After birth, the newborn offspring is nourished with milk secreted by the mother's mammary glands.

In this section we survey reproductive structures and functions in humans as representative of eutherian mammals. Our story of human development continues in the next chapter, which traces the process from fertilization to birth.

Human Female Sexual Organs Function in Oocyte Production, Fertilization, and Embryonic Development

Human females have a pair of ovaries suspended in the abdominal cavity **(Figure 47.11)**. An oviduct leads from each ovary to the uterus, which is a hollow, saclike organ with walls containing smooth muscle. The uterus is lined by the endometrium, formed by layers of connective tissue with embedded glands and richly supplied with blood vessels. If an egg is fertilized and begins development, it must implant in the endometrium to continue developing. The lower end of the uterus, the **cervix**, opens into a muscular canal, the **vagina**, which leads to the exterior. Sperm enter the female reproductive tract via the vagina and, at birth, the baby passes from the uterus to the outside through the vagina.

At the birth of a female, each ovary contains about 1 million oocytes, arrested at the end of the first meiotic prophase. Of these oocytes, about 200,000 to 400,000 survive until a female becomes sexually mature; about 400 are **ovulated**—released into the oviducts as immature eggs—during a woman's lifetime. The egg is released into the abdominal cavity and pulled into the nearby oviduct by the current produced by the beating of the cilia lining the oviduct. The cilia also propel the egg through the oviduct and into the uterus. Fertilization of the egg occurs in the oviduct.

The external female sex organs, collectively called the **vulva**, surround the opening of the vagina. Two folds of tissue, the **labia minora**, run from front to rear on either side of the opening to the vagina. These folds are partially covered by a pair of fleshy, fat-padded

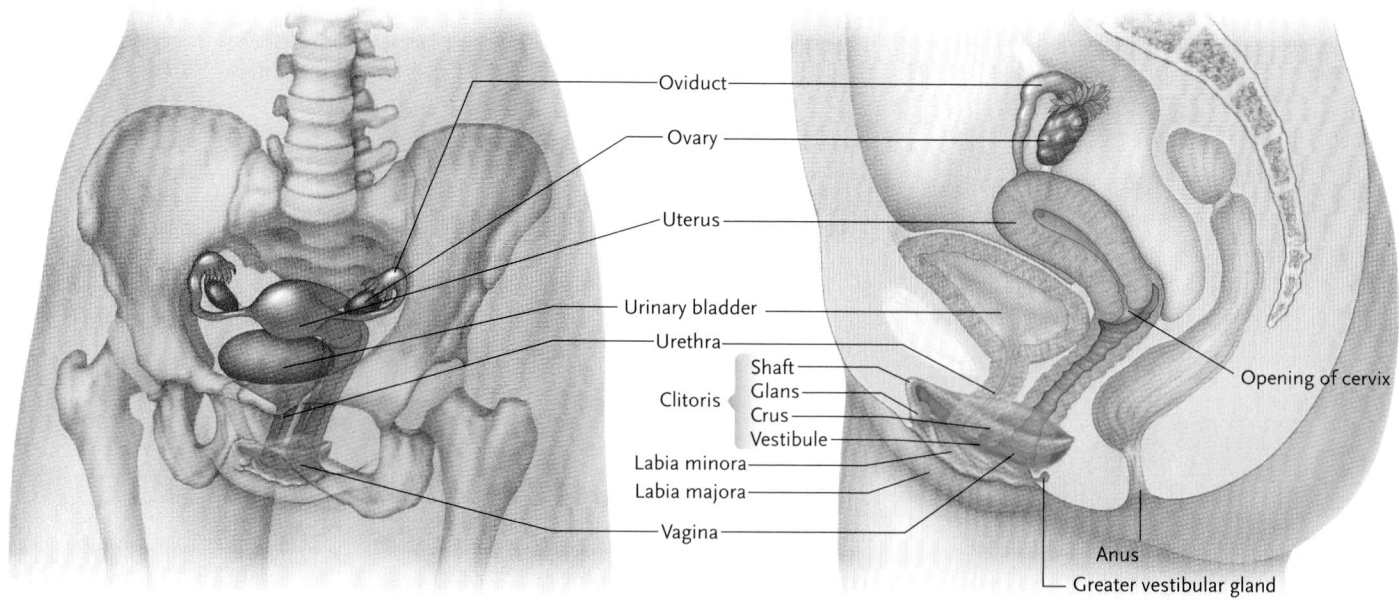

Labels in figure:
Oviduct
Ovary
Uterus
Urinary bladder
Urethra
Clitoris { Shaft, Glans, Crus, Vestibule }
Labia minora
Labia majora
Vagina
Opening of cervix
Anus
Greater vestibular gland

Figure 47.11
The reproductive organs of a human female.

folds, the **labia majora**, which also run from front to rear on either side of the vagina. At the anterior end of the vulva, the labia minora join to partly cover the head of the **clitoris**. The rest of the clitoris is within the body. The clitoris contains erectile tissue and has the same embryonic origins as the penis. A pair of **greater vestibular glands**, with openings near the entrance to the vagina, secretes a mucus-rich fluid that lubricates the vulva. The opening of the urethra, which conducts urine from the bladder, is located between the clitoris and the vaginal opening. Most nerve endings associated with erotic sensations are concentrated in the clitoris, in the labia minora, and around the opening of the vagina. When a human female is born, a thin flap of tissue, the **hymen**, partially covers the opening of the vagina. This membrane, if it has not already been ruptured by physical exercise or other disturbances, is broken by the first sexual intercourse.

Ovulation in Human Females Occurs in a Monthly Cycle

Reproduction in human females is under neuroendocrine control, involving complex interactions between the hypothalamus, pituitary, ovaries, and uterus. Under this control, approximately every 28 days from puberty to menopause, a female releases an egg from one of her ovaries. The cyclic events in the ovary leading to ovulation are known as the **ovarian cycle**. This cycle is coordinated with the **uterine cycle**, or **menstrual cycle** (*menstruus* = monthly), events in the uterus that prepare it to receive the egg if fertilization occurs.

The Ovarian Cycle. The ovarian cycle produces a mature egg **(Figure 47.12)**. The starting point for the cycle is a primary oocyte in prophase of meiosis division I. The beginning of the cycle is triggered by an increase

in the release of **gonadotropin-releasing hormone (GnRH)** by the hypothalamus. This hormone stimulates the pituitary to release **follicle-stimulating hormone (FSH)** and **luteinizing hormone (LH)** into the bloodstream **(Figure 47.13a)**. FSH stimulates 6 to 20 primary oocytes in the ovaries to be released from prophase of meiosis I and continue through the meiotic divisions. As the primary oocytes develop into secondary oocytes—which arrest in metaphase of meiosis II—they become surrounded by cells that form a **follicle** (day 2 of the cycle; **Figure 47.13b**). During this follicular phase, the follicle grows and develops and, at its largest size, becomes filled with fluid and may reach 12 to 15 mm in diameter. Usually only one follicle develops to maturity with release of the egg (secondary oocyte) by ovulation. If two or more follicles develop and their eggs are ovulated, multiple births can result.

As the follicle enlarges, FSH and LH interact to stimulate the follicular cells to secrete **estrogens** (female sex hormones), primarily **estradiol** (see Section 40.4) **(Figure 47.13c)**. Initially, the estrogens are secreted in low amounts; at this level, the estrogens have a negative feedback effect on the pituitary, inhibiting its secretion of FSH. As a result, FSH secretion declines briefly. However, estrogen secretion increases steadily, and its level peaks at about 12 days after follicle development begins (day 14 of cycle). The high estrogen level now has a positive feedback effect on the hypothalamus and pituitary, increasing the release of GnRH and stimulating the pituitary to release a burst of FSH and LH. The increased estrogen levels also convert the mucus secreted by the uterus to a thin and watery consistency, making it easier for sperm to swim through the uterus.

The burst in LH secretion stimulates the follicle cells to release enzymes that digest away the wall of the

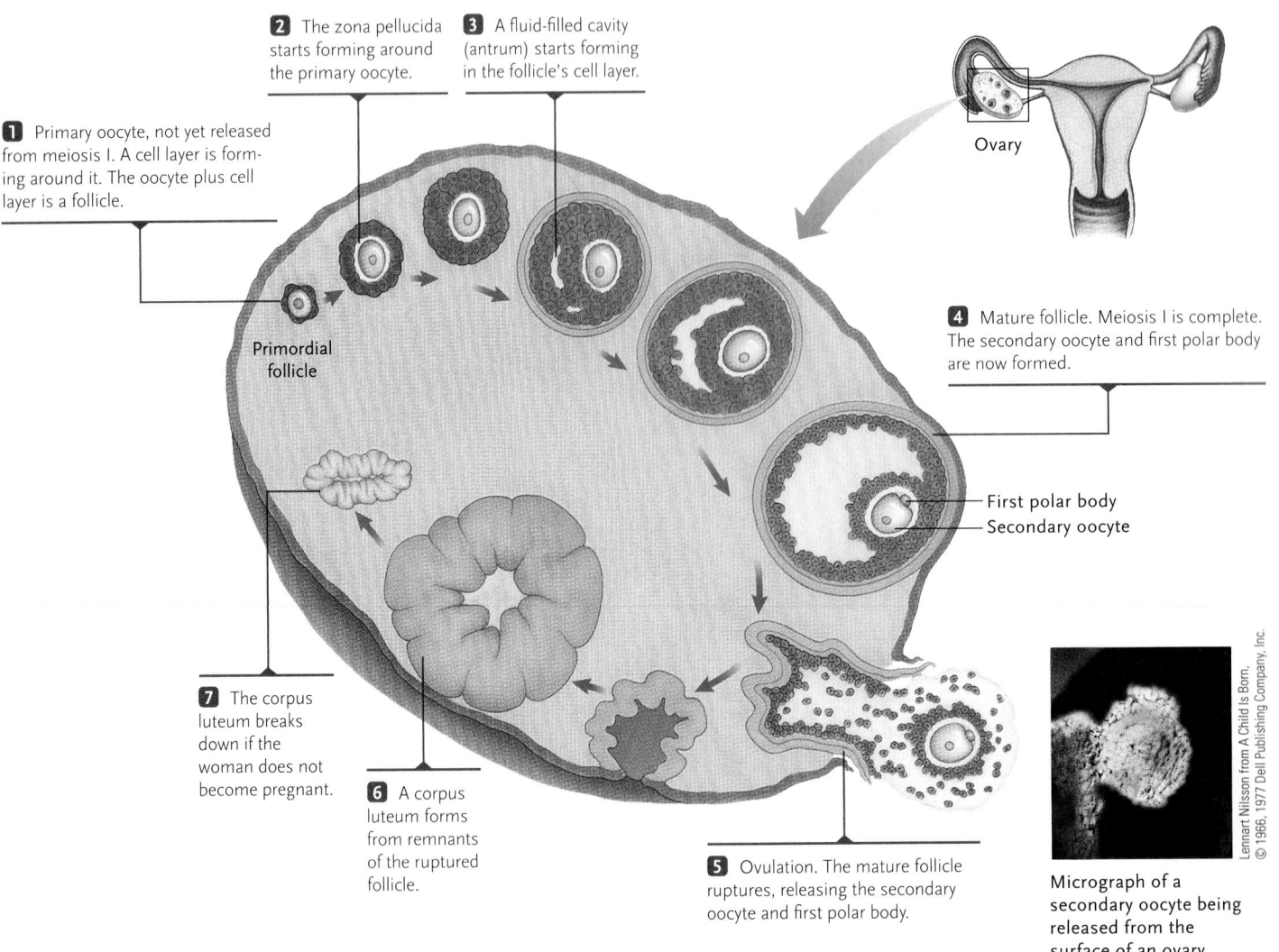

1 Primary oocyte, not yet released from meiosis I. A cell layer is forming around it. The oocyte plus cell layer is a follicle.

2 The zona pellucida starts forming around the primary oocyte.

3 A fluid-filled cavity (antrum) starts forming in the follicle's cell layer.

Primordial follicle

4 Mature follicle. Meiosis I is complete. The secondary oocyte and first polar body are now formed.

First polar body
Secondary oocyte

7 The corpus luteum breaks down if the woman does not become pregnant.

6 A corpus luteum forms from remnants of the ruptured follicle.

5 Ovulation. The mature follicle ruptures, releasing the secondary oocyte and first polar body.

Ovary

Micrograph of a secondary oocyte being released from the surface of an ovary

Lennart Nilsson from A Child Is Born.
© 1966, 1977 Dell Publishing Company, Inc.

Figure 47.12
The growth of a follicle, ovulation, and formation of the corpus luteum in a human ovary.

follicle, causing it to burst and release the egg (see Figure 47.12); this is ovulation. LH also initiates the last phase of the menstrual cycle, the *luteal phase*. That is, LH causes the follicle cells remaining at the surface of the ovary to grow into an enlarged, yellowish structure, the **corpus luteum** (*corpus* = body; *luteum* = yellow; see Figure 47.12). Acting as an endocrine gland, the corpus luteum secretes several hormones: estrogens, large quantities of **progesterone**, and **inhibin**. Progesterone, a female sex hormone, stimulates growth of the uterine lining and inhibits contractions of the uterus. Both progesterone and inhibin have a negative feedback effect on the hypothalamus and pituitary. Progesterone inhibits the secretion of GnRH. Without GnRH, the pituitary does not release FSH and LH. FSH secretion from the pituitary is also inhibited directly by inhibin. The fall in FSH and LH levels diminishes the signal for follicular growth, and no new follicles begin to grow in the ovary.

If fertilization does not occur, the corpus luteum gradually degenerates as cells are phagocytized and blood supply is cut off. By about 10 days after ovulation, little tissue remains, meaning that estrogen, proges-terone, and inhibin are no longer secreted. In the absence of progesterone, *menstruation* begins (described in the next section). As progesterone and inhibin levels decrease, FSH and LH secretion is no longer inhibited, and a new monthly cycle begins.

The Uterine (Menstrual) Cycle. The hormones that control the ovarian cycle also control the uterine (menstrual) cycle **(Figure 47.13d),** keeping the processes connected physiologically. Day 0 of the monthly cycle in the figure is the beginning of follicular development in the ovary (see Figure 47.13b); in the uterus, this correlates with the time at which menstrual flow begins.

Menstrual flow results from the breakdown of the endometrium, which releases blood and tissue breakdown products from the uterus to the outside through the vagina. When the flow ceases, at day 4 to 5 of the cycle, the endometrium begins to grow again; this is the proliferative phase. As the endometrium gradually thickens, the oocytes in both ovaries begin to develop further, eventually leading to ovulation at about 14 days after the beginning of the cycle, as already described.

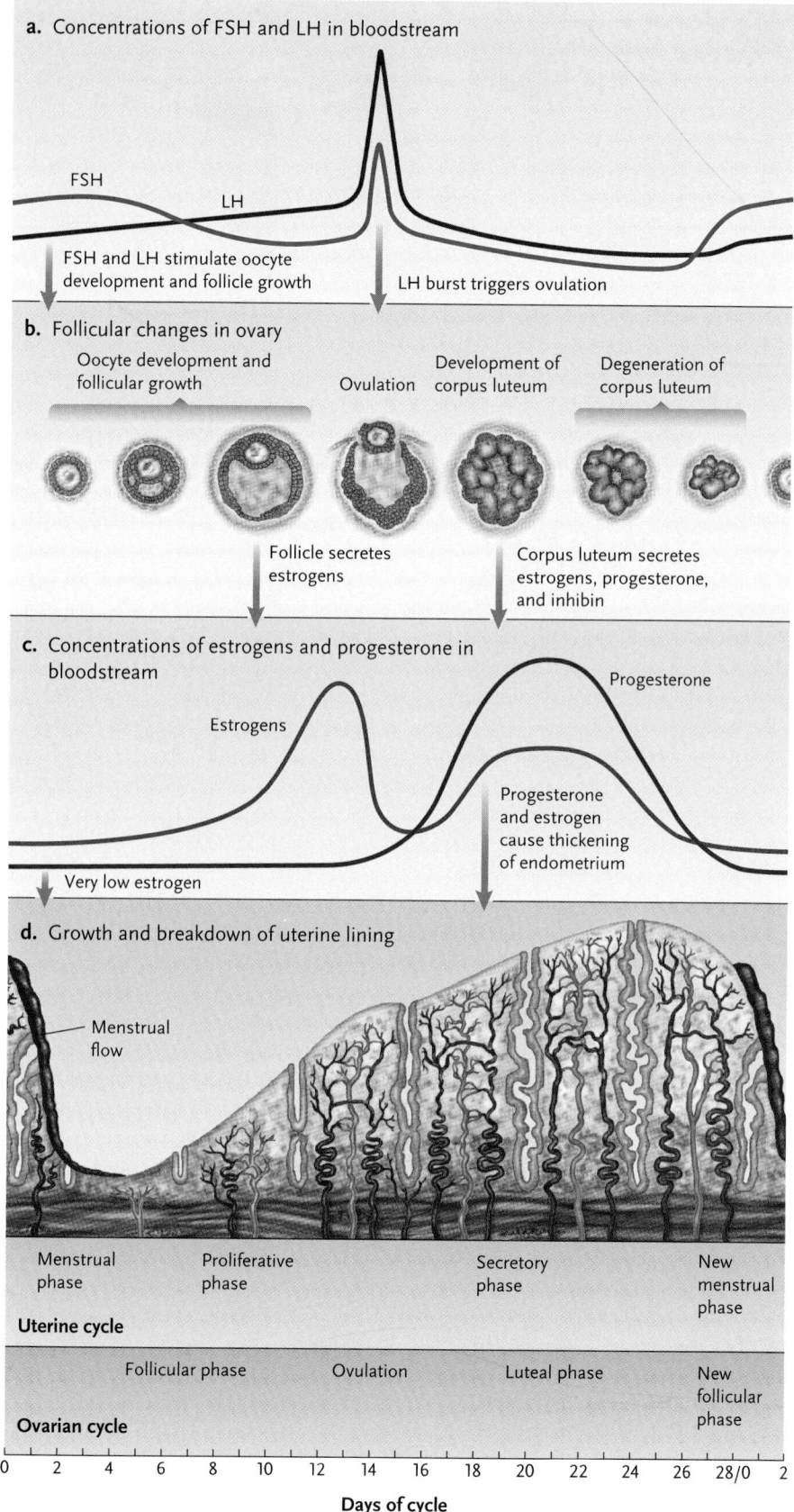

a. Concentrations of FSH and LH in bloodstream

FSH

LH

FSH and LH stimulate oocyte development and follicle growth

LH burst triggers ovulation

b. Follicular changes in ovary

Oocyte development and follicular growth

Ovulation

Development of corpus luteum

Degeneration of corpus luteum

Follicle secretes estrogens

Corpus luteum secretes estrogens, progesterone, and inhibin

c. Concentrations of estrogens and progesterone in bloodstream

Progesterone

Estrogens

Progesterone and estrogen cause thickening of endometrium

Very low estrogen

d. Growth and breakdown of uterine lining

Menstrual flow

| Menstrual phase | Proliferative phase | Secretory phase | New menstrual phase |

Uterine cycle

| Follicular phase | Ovulation | Luteal phase | New follicular phase |

Ovarian cycle

0 2 4 6 8 10 12 14 16 18 20 22 24 26 28/0 2

Days of cycle

Figure 47.13

The ovarian and uterine (menstrual) cycles of a human female. **(a)** The changing concentrations of FSH and LH in the bloodstream, triggered by GnRH secretion by the hypothalamus. **(b)** The cycle of follicle development, ovulation, and formation of the corpus luteum in the ovary. **(c)** The concentrations of estrogens and progesterone in the bloodstream. **(d)** The growth and breakdown of the uterine lining. The days of the monthly cycle are given in the scale at the bottom of the diagram.

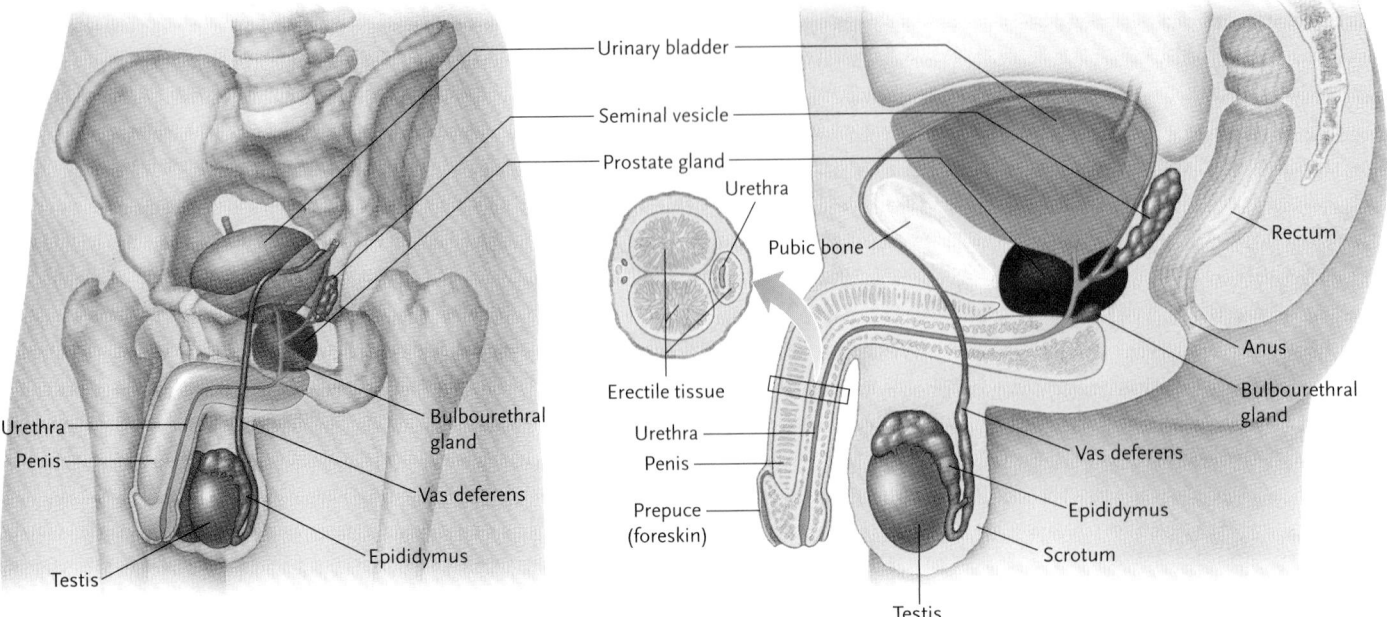

Figure 47.14
The reproductive organs of a human male.

If fertilization does not take place, the uterine lining continues to grow for another 14 days after ovulation; this is the secretory phase. At the end of that time, the absence of progesterone results in the contraction of arteries supplying blood to the uterine lining, shutting down the blood supply and causing the lining to disintegrate. The menstrual flow begins. Contractions of the uterus, no longer inhibited by progesterone, help expel the debris. Prostaglandins released by the degenerating endometrium add to the uterine contractions, making them severe enough to be felt as the pain of "cramps," and also sometimes causing other effects such as nausea, vomiting, and headaches.

Menstruation—the menstrual flow—occurs only in human females and our closest primate relatives, gorillas and chimpanzees. In other mammals, the uterine lining is completely reabsorbed if a fertilized egg does not implant during the period of reproductive activity. The uterine cycle in those mammals is called the *estrous* cycle.

Human Male Sexual Organs Function in Sperm Production and Delivery

Organs that produce and deliver sperm make up the male reproductive system **(Figure 47.14)**. The testes are located outside the abdominal cavity; sperm produced by the testes pass through tubules that enter the abdominal cavity and join with the urethra, the duct that carries urine from the bladder to an opening at the tip of the penis.

Male Reproductive Structures. Human males have a pair of testes, suspended in the baglike **scrotum**. Suspension in the scrotum keeps the testes cooler than the body core, at a temperature that provides an optimal environment for sperm development. Some land mammals such as elephants and monotremes have relatively low body temperatures and have internal testes, that is, testes carried within the body. Marine mammals such as whales and dolphins also have internal testes despite relatively high body temperatures. In these animals particular blood vessel networks serve to lower the temperature in the testes to allow for normal function. A testis is packed with about 125 meters of **seminiferous tubules**, in which sperm proceed through all the stages of spermatogenesis **(Figure 47.15)**. The entire process, from spermatogonium to sperm, takes about 9 to 10 weeks. The testes produce about 130 million sperm each day.

Supportive cells called **Sertoli cells** completely surround the developing spermatocytes in the seminiferous tubules. They supply nutrients to the spermatocytes and seal them off from the body's blood supply. Other cells located in the tissue surrounding the developing spermatocytes, the **Leydig cells**, produce the male sex hormones, known as **androgens**, particularly **testosterone** (see Figure 47.15).

Mature sperm flow from the seminiferous tubules into the **epididymis**, a coiled storage tubule attached to the surface of each testis. Rhythmic muscular contractions of the epididymis move the sperm into a thick-walled, muscular tube, the **vas deferens** (plural, *vasa deferentia*), which leads into the abdominal cavity. Just below the bladder, the vasa deferentia empty into the urethra. During ejaculation, muscular contractions force the sperm into the urethra and out of the penis. The sperm are activated and become motile as they come in contact with alkaline secretions added to the ejaculated fluid by accessory glands.

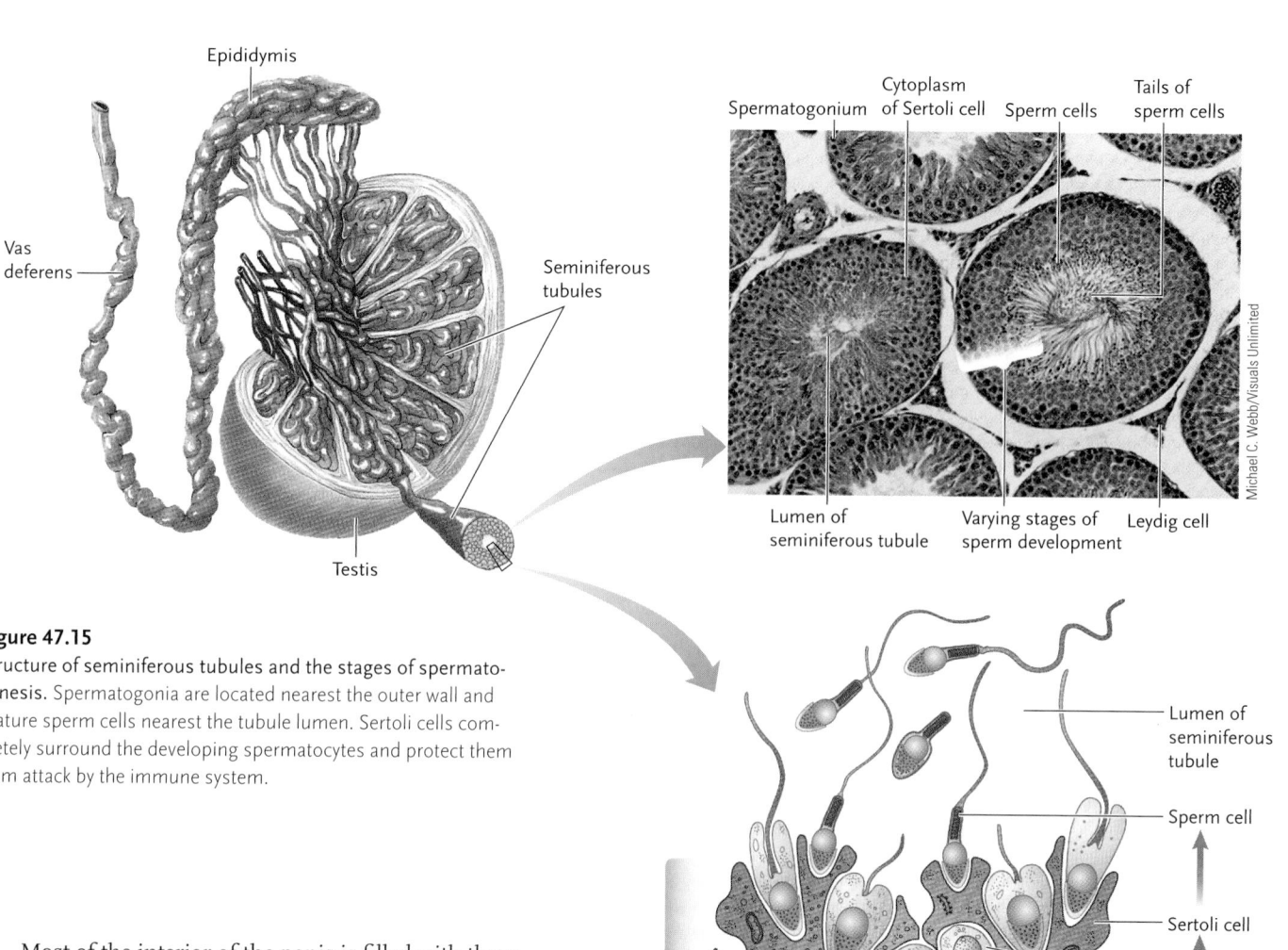

Figure 47.15
Structure of seminiferous tubules and the stages of spermato-genesis. Spermatogonia are located nearest the outer wall and mature sperm cells nearest the tubule lumen. Sertoli cells completely surround the developing spermatocytes and protect them from attack by the immune system.

Most of the interior of the penis is filled with three cylinders of spongelike tissue that become filled with blood and cause erection during sexual arousal. Although the human penis depends solely on engorgement of spongy tissue for erection, the males of many mammalian species, including bats, rodents, walruses, and most other primates, have a bone in the penis, the *baculum,* which helps maintain the penis in an erect state.

The penis ends in a soft, caplike structure, the **glans.** Most of the nerve endings producing erotic sensations are crowded into the glans and the region of the penile shaft just behind the glans. A loose fold of skin, the **prepuce** or **foreskin,** covers the glans (see Figure 47.14). In many cultures the prepuce is removed for hygienic, religious, or other ritualistic reasons by the procedure called **circumcision** ("around cut"). In 2007, the World Health Organization stated that male circumcision is an important strategy to prevent heterosexually acquired HIV infection in males.

Accessory Glands and the Semen. About 150 million to 350 million sperm are released in a single ejaculation. Before they leave the body, these cells are mixed with the secretions of several accessory glands, forming the fluid known as **semen.** In humans, about two-thirds of the volume is produced by a pair of **seminal vesicles,** which secrete a thick, viscous liquid, the **seminal fluid,** into the vasa deferentia near the point where they join with the urethra. The seminal fluid contains prostaglandins that, when ejaculated into the female, trigger contractions of the female reproductive tract that help move the sperm into and through the uterus.

The large **prostate gland,** which surrounds the region where the vasa deferentia empty into the urethra, adds a thin, milky fluid to the semen. The alkaline prostate secretion, which makes up about one-third of the volume of the semen, raises the pH of the semen, and of the vagina, to about pH 6, the level of acidity best tolerated by sperm. The raised pH also activates motility of the sperm. As part of the prostate secretion, a fast-acting enzyme converts the semen to a thick gel when it is first ejaculated. The thickened

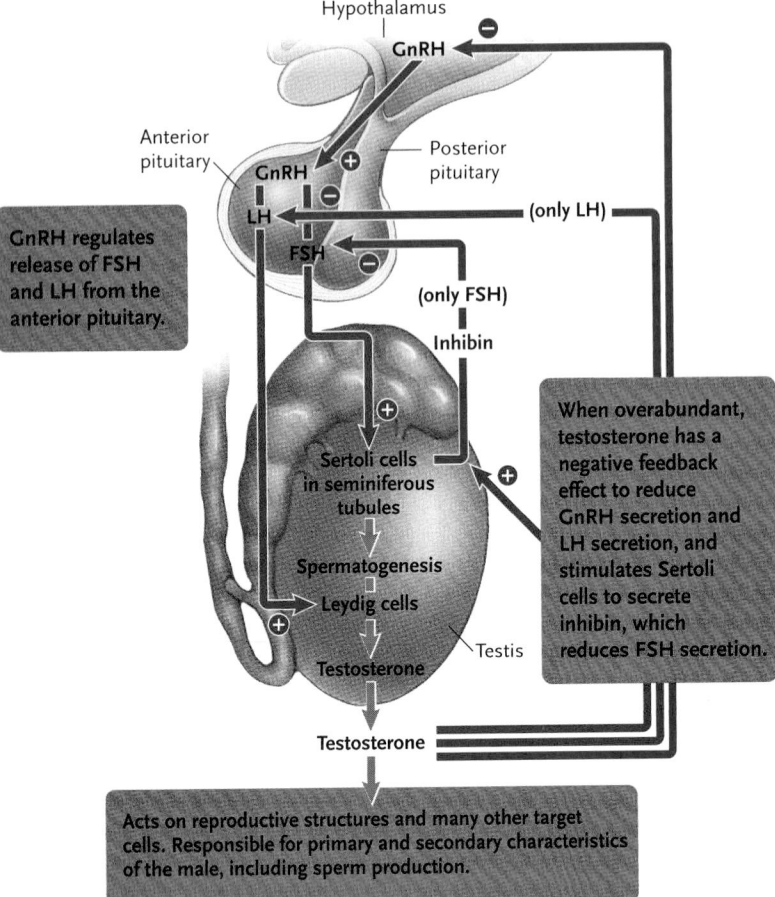

Figure 47.16
Hormonal regulation of reproduction in the male, and the negative feedback systems controlling hormone levels.

Within the figure:

Hypothalamus
GnRH
Anterior pituitary
Posterior pituitary
GnRH
LH
FSH
(only LH)
(only FSH)
Inhibin
Sertoli cells in seminiferous tubules
Spermatogenesis
Leydig cells
Testosterone
Testis
Testosterone

GnRH regulates release of FSH and LH from the anterior pituitary.

When overabundant, testosterone has a negative feedback effect to reduce GnRH secretion and LH secretion, and stimulates Sertoli cells to secrete inhibin, which reduces FSH secretion.

Acts on reproductive structures and many other target cells. Responsible for primary and secondary characteristics of the male, including sperm production.

controls the growth and function of male reproductive structures. FSH stimulates Sertoli cells to secrete a protein and other molecules that are required for spermatogenesis.

The concentrations of these hormones are maintained by negative feedback mechanisms. If the concentration of testosterone falls in the bloodstream, the hypothalamus responds by increasing GnRH secretion. If the concentration of testosterone rises too high, the overabundance inhibits GnRH secretion by the hypothalamus and LH secretion by the anterior pituitary. An overabundance of testosterone also stimulates Sertoli cells to secrete inhibin, which inhibits FSH secretion by the anterior pituitary. As a result, testosterone secretion by the Leydig cells drops off, returning the concentration to optimal levels in the bloodstream.

Human Copulation Follows a Typical Mammalian Pattern

When the male is sexually aroused, sphincter muscles controlling the flow of blood to the spongy erectile tissue of the penis relax, allowing the tissue to become engorged with blood. (The penis is a hydrostatic skeleton structure; see Section 41.2.) As the spongy tissue swells, it maintains the pressure by compressing and almost shutting off the veins draining blood from the penis. The engorgement produces an erection in which the penis lengthens, stiffens, and enlarges. During continued sexual arousal, lubricating fluid secreted by the bulbourethral glands may be released from the tip of the penis.

Female sexual arousal results in enlargement and erection of the clitoris, in a process analogous to erection of the penis in males. The labia minora also become engorged with blood and swell in size, and lubricating fluid is secreted onto the surfaces of the vulva by the greater vestibular glands. In addition to these changes, the nipples become erect by contraction of smooth muscle cells, and the breasts swell due to engorgement with blood.

Insertion of the penis into the vagina and the thrusting movements of copulation lead to the reflex actions of ejaculation, including spasmodic contractions of muscles surrounding the vasa deferentia, accessory glands, and urethra. During ejaculation, the sphincter muscles controlling the exit from the bladder close tightly, preventing urine from being released from the bladder and mixing with the ejaculate. Ejaculation is usually accompanied by *orgasm*, a sensation of intense physical pleasure that is the peak—climax—of excitement for sexual intercourse, followed by feelings of relaxation and gratification.

The motions of copulation stretch the vagina and stimulate the clitoris. The stretching and stimulation can also induce orgasm in females. The vaginal stretching also stimulates the hypothalamus to secrete oxyto-

consistency helps keep the semen from draining from the vagina when the penis is withdrawn. A second, slower-acting enzyme in the prostate secretion then gradually breaks down the semen clot and releases the sperm to swim freely in the female reproductive tract.

Finally, a pair of **bulbourethral glands** secretes a clear, mucus-rich fluid into the urethra before and during ejaculation. This fluid lubricates the tip of the penis and neutralizes the acidity of any residual urine in the urethra. In total, the secretions of the accessory glands make up more than 95% of the volume of semen; less than 5% is sperm.

Hormones Also Regulate Male Reproductive Functions

Many of the hormones regulating the menstrual cycle, including GnRH, FSH, LH, and inhibin, also regulate male reproductive functions. Testosterone, secreted by the Leydig cells in the testes, also plays a key role **(Figure 47.16)**.

In sexually mature males, the hypothalamus secretes GnRH in brief pulses every 1 to 2 hours. The GnRH, in turn, stimulates the pituitary to secrete LH and FSH. LH stimulates the Leydig cells to secrete testosterone, which stimulates sperm production and

Egging on the Sperm

Whether human eggs release attractants to draw sperm near has long been a subject of speculation and research. Now molecular investigations by Marc Spehr and his colleagues at Ruhr University in Germany indicate that sperm can detect and swim toward attractant chemicals.

Other investigators had found that human sperm cells have receptors able to bind to chemical substances classified as odorants, aroma molecules that can be specifically recognized ("smelled"). In vertebrates, more than a thousand genes encode odorant receptors, most of them olfactory receptors associated with the senses of smell and taste. However, odorant receptors are also located on cell types that do not function in taste and smell, including sperm.

But do the odorant receptors function in sperm–egg attraction? The Spehr team began their investigation of this possibility by testing testicular tissue for odorant-receptor-gene activity, using probes for mRNAs with sequences typical of odorant receptor genes. Only two active odorant receptor genes were found in the testes: hOR17-2 (hOR = human olfactory receptor), which had been discovered by others; and hOR17-4, which was not previously known to be active.

The researchers molecularly cloned (see Section 18.1) the hOR17-4 gene of testicular cells and inserted the gene in a line of cultured human embryonic kidney (HEK) cells. Previous work had shown that when an odorant receptor combines with the chemical it recognizes, it triggers cytoplasmic reactions that lead to Ca^{2+} release in the cytoplasm (the IP_3 pathway, described in Section 7.4). Accordingly, the investigators tested the genetically engineered HEK cells for a Ca^{2+} response to any of the chemicals in a mixture containing 100 different chemicals. Only one of the chemicals, *cyclamal,* caused the HEK cells to respond. HEK cells that did not receive the hOR17-4 gene did not respond to cyclamal. The researchers then tested chemicals closely related in chemical structure to cyclamal and found that they also elicit a Ca^{2+} response in the engineered HEK cells. They used one of these chemicals, *bourgonal,* in further testing because it triggered a stronger response than cyclamal.

Next, the investigators tested human sperm cells to see if they would respond to bourgonal. The sperm cells responded to the chemical by an increase in cytoplasmic Ca^{2+} concentration, indicating that the hOR17-4 gene was active during spermatogenesis, leading to synthesis and insertion of the hOR17-4 receptor in the sperm plasma membrane.

In a final experiment, human sperm cells were exposed to gradients of bourgeonal solutions in micropipettes. The sperm swam consistently toward the regions of highest concentration, and swam faster and more directly as the concentration increased.

These experiments indicate that human sperm can detect and respond to chemical attractants by swimming toward the source of the attractant. Whether human eggs actually release such attractants remains to be determined. If so, the egg attractant detected by the hOR17-4 receptor is likely to resemble cyclamal and bourgeonal in chemical structure, because odorant receptors are highly specific in their responses.

As part of their research, the Spehr team found that another chemical, *undecanal,* strongly inhibits the binding of the hOR17-4 receptor to cyclamal and bourgeonal. With chemicals at hand that can both stimulate and eliminate sperm attraction, the system might provide a method for either contraception or procreation.

---

cin, which induces contractions of the uterus. The contractions keep the sperm in suspension and aid their movement through the reproductive tract. Uterine contractions are also induced by the prostaglandins in the semen.

Sperm reach the site of fertilization in the oviducts within 30 minutes after their ejaculation into the vagina. Of the millions of sperm released in a single ejaculation, only a few hundred actually reach the oviducts. After orgasm, the penis, clitoris, and labia minora gradually return to their unstimulated size. Females can experience additional orgasms within minutes or even seconds of a first orgasm, but most males enter a *refractory period* that lasts for 15 minutes or longer before they can regain an erection and have another orgasm.

A Human Egg Can Be Fertilized Only in the Oviduct

A human egg can be fertilized only during its passage through the third of the oviduct nearest the ovary. If the egg is not fertilized during the 12- to 24-hour period that it is in this location, it disintegrates and dies. However, sperm do not swim randomly for a chance encounter with the egg. Rather, they first swim up the cervical canal to reach the oviduct, and then are propelled up the oviduct by contractions of the oviduct's smooth muscles. Further, researchers have found evidence that eggs release chemical attractant molecules that the sperm recognize, causing them to swim directly toward the egg. (*Insights from the Molecular Revolution* describes some of this research.)

To reach the egg, the fertilizing sperm must penetrate the layer of follicle cells surrounding the egg, and then pass through the zona pellucida coating the egg surface **(Figure 47.17)**. Enzymes built into the plasma membrane of the sperm cells aid penetration through the follicle cells. Once through the follicle cells, the sperm binds to receptor molecules on the surface of the zona pellucida. The binding triggers the acrosome reaction in which hydrolytic enzymes are released from the acrosome and digest a pathway to the egg. As soon as the first sperm cell reaches the egg through the pathway digested by the released acrosomal enzymes, the sperm and egg plasma membranes fuse, and the sperm cell enters the cytoplasm of the egg. Although only one sperm fertilizes the egg, the combined release of acrosomal enzymes from many sperm greatly increases the chance that a complete channel will be opened through the zona pellucida. Partially for this reason, a low sperm count is often a source of male infertility. Low sperm count has a number of causes, including infection, heat, frequent intercourse, smoking, and excess alcohol consumption.

a. Sperm attached to zona pellucida

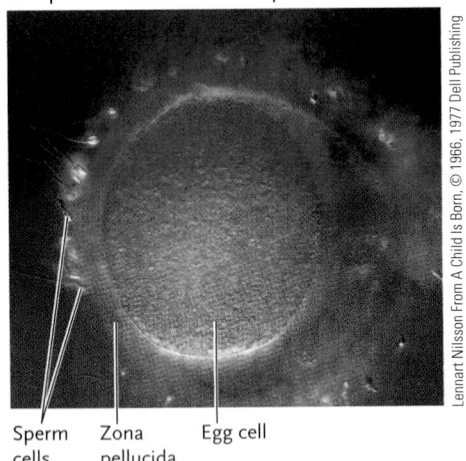

Lennart Nilsson From A Child Is Born, © 1966, 1977 Dell Publishing Coompany, Inc.

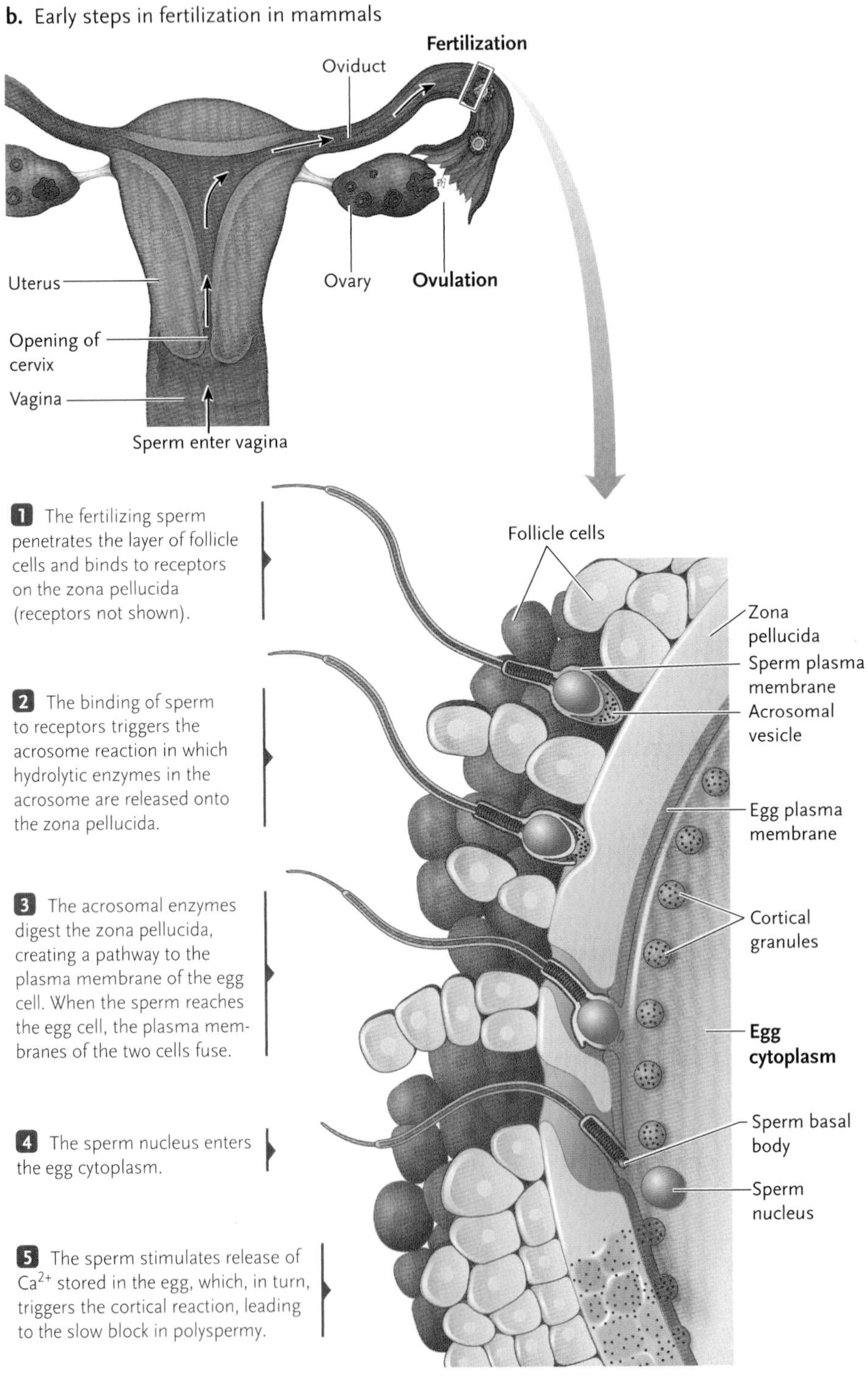

Sperm cells   Zona pellucida   Egg cell

b. Early steps in fertilization in mammals

Fertilization

Oviduct

Uterus

Opening of cervix

Vagina

Ovary   **Ovulation**

Sperm enter vagina

1 The fertilizing sperm penetrates the layer of follicle cells and binds to receptors on the zona pellucida (receptors not shown).

2 The binding of sperm to receptors triggers the acrosome reaction in which hydrolytic enzymes in the acrosome are released onto the zona pellucida.

3 The acrosomal enzymes digest the zona pellucida, creating a pathway to the plasma membrane of the egg cell. When the sperm reaches the egg cell, the plasma membranes of the two cells fuse.

4 The sperm nucleus enters the egg cytoplasm.

5 The sperm stimulates release of Ca^{2+} stored in the egg, which, in turn, triggers the cortical reaction, leading to the slow block in polyspermy.

Follicle cells

Zona pellucida
Sperm plasma membrane
Acrosomal vesicle

Egg plasma membrane

Cortical granules

Egg cytoplasm

Sperm basal body

Sperm nucleus

Figure 47.17
Fertilization in mammals. **(a)** Sperm attached to the zona pellucida of a human egg cell. **(b)** Early steps in fertilization process.

The membrane fusion activates the egg. The sperm that has entered the egg releases nitric oxide, which stimulates the release of stored Ca^{2+} in the egg. The Ca^{2+} triggers cortical granule release to the outside of the egg. Enzymes from the cortical granules crosslink molecules in the zona pellucida, hardening it and sealing the channels opened by acrosomal enzymes. The enzymes also destroy the receptors that bind sperm to the surface of the zona pellucida. As a result, no further sperm can bind to the zona or reach the plasma membrane of the egg. The Ca^{2+} also triggers the completion of meiosis of the egg (recall that, up to that point, it is a secondary oocyte arrested in metaphase of meiosis II). The sperm and egg nuclei then fuse and the cell is now the zygote. Mitotic divisions of the zygote soon initiate embryonic development.

The first cell divisions of embryonic development take place while the fertilized egg is still in the oviduct. By about 7 days after ovulation, the embryo passes from the oviduct and implants in the uterine lining. During and after implantation, cells associated with the embryo secrete **human chorionic gonadotropin (hCG)**, a hormone that keeps the corpus luteum in the ovary from breaking down. Excess hCG is excreted in the urine; its presence in urine or blood provides the basis of pregnancy tests.

The continued activity of the corpus luteum keeps estrogen and progesterone secretion at high levels, maintaining the uterine lining and preventing menstruation. The high progesterone level also thickens the mucus secreted by the uterus, forming a plug that seals the opening of the cervix from the vagina. The plug keeps bacteria, viruses, and sperm cells from further copulation from entering the uterus.

Later in development, about 10 weeks after implantation, the placenta takes over the secretion of progesterone, hCG secretion drops off, and the corpus luteum regresses. However, the corpus luteum continues to secrete the hormone *relaxin,* which inhibits contraction of the uterus until the time of birth is near.

Study Break

Outline the roles of follicle-stimulating hormone (FSH) and luteinizing hormone (LH) in the ovarian cycle of a human female.

47.4 Methods for Preventing Pregnancy: Contraception

In human society, pregnancy can be a blessing or a disaster. An unwanted pregnancy can be inconvenient at the least, or at the worst can have serious physical and social repercussions, particularly for the mother. Many methods exist for achieving contraception—the prevention of pregnancy—some old and others relatively new.

The oldest method of contraception is total abstinence from sex. Unfortunately, millions of years of animal evolution have stacked the cards against total abstinence by making the sex drive among the most powerful of compulsions. Literally millions of unwanted children attest to the failures of this method. Other methods of preventing pregnancy include techniques for (1) preventing the sperm from reaching the site of fertilization, (2) preventing ovulation, or (3) interfering with implantation if fertilization does occur. **Table 47.1** lists the most common contraceptive techniques and their reliability, based on 1 year of use. Two values are given: (1) the lowest expected rate of pregnancy, meaning the rate of pregnancy when the birth control method was used correctly every time; and (2) the typical use rate of pregnancy, meaning the rate of pregnancy when the method was used in a typical manner, meaning that it may not always have been used correctly every time.

Of Methods Preventing Fertilization, Vasectomy and Tubal Ligation Are Most Effective

A natural technique for preventing fertilization is the *rhythm method,* which consists of avoiding intercourse during the time of the month when the egg can be fertilized. Because sperm can survive for as long as 5 days

Table 47.1	Pregnancy Rates for Birth Control Methods	
Method	Lowest Expected Rate of Pregnancy[a]	Typical-Use Rate of Pregnancy[b]
Rhythm method	1%–9%	25%
Withdrawal	4%	19%
Condom (male)	3%	14%
Condom (female)	5%	21%
Diaphragm and spermicidal jelly	6%	20%
Vasectomy (male sterilization)	0.1%	0.15%
Tubal ligation (female sterilization)	0.5%	0.5%
Contraceptive pill (combination estrogen/progestin)	0.1%	5%
Contraceptive pill (progestin only)	0.5%	5%
Implant (progestin)	0.09%	0.09%
Intrauterine device (IUD) (copper T)	0.6%	0.8%

[a]Rate of pregnancy when the birth control method was used correctly every time.
[b]Rate of pregnancy when the method was used typically, meaning that it may not have been always used correctly every time.
Source: U.S. Food and Drug Administration, http://www.fda.gov/fdac/features/1997/conceptbl.html. Data reported in 1997 for effectiveness of methods in a 1-year period.

in the female reproductive tract, intercourse should be avoided from 5 days before ovulation and, for safety's sake, for another 4 or 5 days after ovulation. Although conceptually straightforward, the method is difficult to apply because of the unpredictability of the time of ovulation (and the power of the sex drive). The lowest expected rate of pregnancy for this method is 1% to 9%, while the typical rate is 25%.

Another natural method to prevent fertilization is *withdrawal*—starting sexual intercourse, but withdrawing the penis before ejaculation. Unfortunately, once ejaculation begins, it proceeds as a series of reflexes that is extremely difficult to interrupt; in addition, some sperm may be present in lubrication produced prior to ejaculation. The lowest expected rate of pregnancy for this method is 4%, while the typical rate is 19%.

The *condom,* a thin, close-fitting sheath of latex, lambskin, or polyurethane worn over the penis, is one of the traditional methods of preventing ejaculated sperm from entering the vagina. Condoms made from latex may also provide a barrier to the transmission of disease between sexual partners (condoms made from natural skin do not block viruses such as HIV). Pouch-like "female condoms," inserted into the vagina, prevent ejaculated sperm from entering the uterus. The lowest rate of pregnancy for male condoms is 3%, while the typical rate is 14%. The lowest and typical rates for female condoms are 5% and 21%, respectively.

The *diaphragm* is a cuplike rubber device that blocks the cervix in females. (The similar *cervical cap* is smaller and fits more closely over the cervix.) Typically a spermicidal jelly or cream is also used. To be most effective, a diaphragm and the spermicidal jelly must be inserted no more than an hour before intercourse, and left in place for the recommended time afterward. The lowest rate of pregnancy for a diaphragm used with a spermicide is 6%, while the typical rate is 20%.

Fertilization can also be prevented surgically, by cutting and closing off either the vasa deferentia in males or the oviducts in females. In *vasectomy,* the procedure carried out in males, an incision is made in the scrotum and each vas deferens is severed and tied off. After vasectomy, the seminal fluid is still produced and ejaculated, but it does not contain sperm. In *tubal ligation,* the procedure for females, the oviducts are cut and tied off, or seared with heat (cauterized) to close them. The ligation prevents eggs from being fertilized or reaching the uterus. Neither vasectomy nor tubal ligation interferes with the production of sex hormones by the ovaries or testes, or results in any change in sexual behavior. Both operations are highly effective in preventing pregnancy. Although they can be reversed, the procedures are difficult and not always successful. The lowest rate of pregnancy for vasectomy is 0.1%, while the typical rate is 0.15%. The lowest and typical rates for tubal ligation are both 0.5%.

Of Methods Preventing Ovulation, the Oral Contraceptive Pill Is Most Effective

The primary method used to prevent ovulation is the *oral contraceptive pill,* or simply "the pill," containing a combination of estrogen and *progestin* (a synthetic form of progesterone) or progestin alone. In this highly effective method, the pill is taken daily for 20 to 21 days after the end of the menstrual flow and then stopped (actually, placebo pills are taken for the remaining days of the cycle to maintain the routine of pill taking) to allow menstruation; then the next month's course is begun. If pregnancy is desired, the pill is simply not taken after the menstrual flow.

The pill works by inhibiting the secretion of FSH and LH by the pituitary; without these hormones, ovulation does not occur. When the pill is stopped after 20 to 21 days, the resulting drop in progestin concentration causes the uterine lining to break down and initiates the menstrual flow. Since ovulation does not occur, fertilization and pregnancy are not possible.

The lowest rate of pregnancy for the estrogen/progestin pill is 0.1%, while the typical rate is 5%. The lowest and typical rates for the progestin pill are 0.5% and 5%, respectively. Most pregnancies among women taking the pill result from failure to take it on schedule—often simply by forgetting to take the pill for a day or two at the wrong time of the month. Some women, about one in four, experience unpleasant side effects, such as nausea, tenderness of the breasts, irritability, nervousness, or changes in skin color or texture. Modern versions of the pill have almost eliminated the more serious side effects, such as increased incidence of breast cancer and formation of blood clots. However, cigarette smoking significantly increases the risk of heart attacks and strokes for women taking the pill. This risk increases with age and with the number of cigarettes smoked per day.

As an alternative to the pill, progestin is also injected in a time-release form that prevents ovulation throughout the period of release. In one method, plastic tubes containing progestin are implanted under the skin, usually in the upper arm. The tubes release progestin for up to 5 years, making the method effective for women in countries or situations in which obtaining and taking the pill on a daily basis is impractical. The lowest rate and the typical rate of pregnancy for the implant are both 0.09%.

The IUD and the Morning-After Pill Are Effective in Preventing Implantation

A commonly used method for preventing implantation if fertilization occurs is insertion of an *intrauterine device (IUD),* a small plastic or copper device, into the uterus just inside the cervix. The IUD remains in place as a long-term preventive measure; depending on the type, a single IUD is approved for 5 to 10 years of use. It is not clear how the IUD works; presumably, it causes a mild inflammation of the uterine lining that makes it unreceptive to implantation. The IUD is a refinement of a method used by women since ancient times, in which small pebbles were inserted in the uterus to prevent conception.

The IUD is effective as long as it is not deflected from its correct position in the uterus; unfortunately, this may happen without warning or the user's awareness. A few women also experience unpleasant side effects from the IUD such as cramps, uterine infections, or excessive menstrual bleeding. The lowest rate of pregnancy for the copper T types of IUD is 0.6%, while the typical rate is 0.8%.

Whatever the method of birth control, its effectiveness is improved if sex partners are highly motivated and careful in its use. The effectiveness of condoms, for example, is greatly improved if the penis is withdrawn immediately after ejaculation (before the semen has time to spread under the condom and leak into the vagina) and is not reinserted, with or without a condom, for several hours. Similarly, high motivation in use of the rhythm method, which might require abstaining from intercourse for most of the month except for a few days just after the menstrual flow, considerably improves the percentage of success with this method.

Another method used to prevent pregnancy is the so-called *emergency contraception pill,* commonly referred to as the "morning-after pill." These pills are administered after intercourse has occurred as a means to prevent pregnancy. A high dosage synthetic progestin emergency contraception pill called Plan B is available in the United States without prescription to women who are 18 or older. This pill is highly effective if taken within 72 hours after unprotected sexual intercourse. Pregnancy tests do not work until significantly after this time. Research data show that Plan B works by blocking ovulation; there is no effect of the hormone on implantation of a fertilized egg. But, because sperm can survive in the female reproductive tract for a few days, blocking ovulation can obviously be effective in preventing pregnancy.

Another emergency contraception pill is *mifepristone (RU-486),* which contains a molecule that binds to and blocks progesterone receptors in the uterine lining. The blockage prevents the lining from responding to progesterone and causes it to break down (that is, a menstrual period is initiated), taking with it any embryo that may have implanted. Mifepristone is approved in the United States for terminating pregnancies up to 49 days post-conception; the time period is

UNANSWERED QUESTIONS

Why do male mammals have "female" hormones?

Given the paramount importance of reproduction in biology, there is a surprising lack of knowledge regarding even its basic features. For example, both sexes make the same set of reproductive hormones, but for some of these, we are only beginning to understand their roles in males. For example, oxytocin, released by the pituitary, has long been known to control labor and milk ejection in females. But oxytocin is also found in semen, where it may boost sperm counts and aid sperm transport in the female reproductive tract. And it may also have a role in social behavior. Several studies have shown a role for oxytocin in male–female and mother–child bonding; and recently research groups led by Markus Heinrichs at the University of Zurich, Switzerland, and Ernst Fehr at Collegium Helveticum, Zurich, reported that men who sniffed a nasal spray with oxytocin in the laboratory were much more likely to trust other males in engaging in risky financial transactions. Thus oxytocin has been dubbed the "trust hormone." However, its roles (and how it is regulated) in real social interactions are not known.

Prolactin, also from the pituitary gland, stimulates milk production in females. Tillmann Krüger and colleagues at the Swiss Federal Institute of Technology have found that prolactin levels surge during orgasms in both sexes and that prolactin is associated with a feeling of relaxation and satisfaction. In males, it may also inhibit erection during post-coitus recovery. In 2006, Stuart Brody of the University of Paisley in Scotland and Dr. Krüger found the surge to be much higher after intercourse than masturbation, which may explain why the latter is generally less satisfying. They suggest that excessive prolactin could play a role in male impotence, a possibility now being investigated.

What is the anatomy of the human clitoris?

There is also surprising ignorance on basic reproductive anatomy. Diagrams of the human clitoris are quite misleading, showing it simply as a small external bulb in the female vulva. However, dissections and magnetic resonance images in studies led by Dr. Helen O'Connell, an Australian urology surgeon, revealed the clitoris to be much larger, rivaling the penis. Most of the clitoral tissue consists of large, highly vascular internal bulbs that surround the urethra and vagina. The role of these bulbs is uncertain, but they engorge during intercourse, perhaps to squeeze the urethra closed to prevent infections, to support the vaginal wall during penile penetration, and/or to increase pleasure signaling. Knowledge of this anatomy is crucial for performing safe pelvic surgeries on women.

Paul H. Yancey holds the Carl E. Peterson Endowed Chair of Sciences at Whitman College in Walla Walla, Washington. His main research interests are in the areas of animal physiology, especially water stress and osmoregulation. To learn more about his research, go to http://marcus.whitman .edu/~yancey/.

longer in some foreign countries. It is available only by prescription.

In this chapter we have focused on animal reproduction up to the point of the fertilized egg. In the next chapter, we address the final stage of reproduction in sexually reproducing organisms, the development of a new individual from the fertilized egg.

Review

47.1 Animal Reproductive Modes: Asexual and Sexual Reproduction

- In asexual reproduction, a single parent gives rise to offspring without genetic input from another individual. In sexual reproduction, offspring are produced by the union of gametes—eggs and sperm—from two parents.

- Asexual reproduction involving mitosis occurs in animals by fission, budding, or fragmentation (Figure 47.2). In parthenogenesis, a form of asexual reproduction, females produce eggs that develop without being fertilized.

- In sexual reproduction, genetic variability is produced by the meiotic processes of genetic recombination and independent assortment.

47.2 Cellular Mechanisms of Sexual Reproduction

- Sexual reproduction includes two cellular processes, gametogenesis and fertilization, and a whole-organism process, mating. Gametogenesis is the formation of male and female gametes by meiotic cell division, followed by differentiation of the gametes; fertilization is the union of gametes that initiates development of new individuals (Figure 47.3).

- Gametogenesis takes place in the testes of males and in the ovaries of females. Sperm and eggs are delivered to the site of fertilization by sperm ducts in males and oviducts in females. External reproductive structures aid the delivery in many species.

- In male gametogenesis—spermatogenesis—each cell entering meiosis produces four haploid motile sperm cells. In female gametogenesis—oogenesis—each cell entering meiosis produces one haploid egg cell. The meiotic divisions of oogenesis concentrate almost all the cytoplasm in the single egg cell; the other division products are nonfunctional polar bodies (Figures 47.3–47.5).

- The egg contains stored nutrients and information required for at least the early stages of embryonic development. It is covered by one or more protective coats, and it has a mechanism that blocks additional sperm from entering after fertilization (Figure 47.5).

- Fertilization, which follows mating in most animals, may be external or internal. In external fertilization, sperm and eggs are shed into the surrounding water. In internal fertilization, sperm are released close to or inside the female reproductive ducts via copulation (Figure 47.6).

- When a sperm and egg touch during fertilization, their plasma membranes fuse, introducing the sperm nucleus into the egg cytoplasm. The sperm and egg nuclei then fuse to form a diploid zygote nucleus and initiate embryonic development (Figure 47.8).

- Oviparous animals lay eggs in which development of new individuals takes place outside the female's body. In viviparous animals, development takes place inside the female's body. In ovoviviparous animals, fertilized eggs are retained within the body while the embryo develops, the eggs hatch within the mother, and the young are then released from the body.

- In hermaphroditism, single individuals produce both mature egg-producing tissue and mature sperm-producing tissue (Figure 47.10).

Animation: Spermatogenesis

Animation: Fertilization

47.3 Sexual Reproduction in Humans

- In females, eggs released from the ovaries travel through the oviducts to the uterus. The uterus opens into the vagina, the entrance for sperm and the exit for offspring during birth (Figure 47.11).

- The ovarian cycle produces an egg. The cycle begins with the release of GnRH by the hypothalamus, which stimulates the release of FSH and LH from the anterior pituitary. FSH stimulates oocytes in the ovaries to begin meiosis. One oocyte typically develops to maturity surrounded by cells that form a follicle (Figures 47.12 and 47.13).

- The enlarging follicle secretes estrogens, causing a burst in FSH and LH release; at about 14 days, the LH stimulates ovulation, the bursting of the follicle and the release of the egg. The remainder of the follicle forms the corpus luteum, which secretes estrogens, progesterone, and inhibin (Figures 47.12 and 47.13).

- Day 0 of the monthly uterine (menstrual) cycle correlates with the beginning of follicular development in the ovary and the beginning of the menstrual flow. Secretion of estrogen from the developing follicle stimulates the growth of a new endometrium. If fertilization does not occur, progesterone and inhibin maintain the endometrium until the 28th day of the cycle, when the corpus luteum regresses. Without progesterone, the endometrium breaks down and is released as the menstrual flow (Figure 47.13).

- In males, sperm develop in seminiferous tubules in the testes and are released into the epididymis. When a male ejaculates, sperm travel from the epididymis to the vas deferens, and then through the urethra and the penis. The seminal vesicles, prostate gland, and bulbourethral glands add fluids to the sperm traveling to the outside (Figures 47.14 and 47.15).

- Sperm production in males is also controlled by LH and FSH. LH stimulates Leydig cells in the testes to secrete testosterone, which stimulates sperm production. FSH stimulates Sertoli cells in the testes to secrete molecules needed for spermatogenesis (Figure 47.16).

- During copulation, sperm are ejaculated into the vagina of the female. The sperm then swim through the female reproductive tract, aided by contractions of the oviduct and guided by molecules released by the egg. Upon contact with the egg in the oviduct, the acrosomes of sperm release enzymes that digest a path through the egg coats. As the fertilizing sperm contacts the egg, the sperm and egg plasma membranes fuse, releasing the sperm nucleus into the egg cytoplasm and activating the egg.

The egg completes meiosis, and the sperm and egg nuclei fuse, producing the zygote (Figure 47.17).

- As the embryo implants, the hormone hCG sustains the corpus luteum, which continues to secrete estrogen and progesterone at high levels. These hormones maintain the uterine lining and prevent menstruation.

  Animation: Male reproductive system

  Animation: Route sperm travel

  Animation: Hormonal control of sperm production

  Animation: Female reproductive system

  Animation: Ovarian function

  Animation: Hormones and the menstrual cycle

  Animation: Menstrual cycle summary

47.4 Methods for Preventing Pregnancy: Contraception

- Methods of contraception work by preventing sperm from reaching the site of fertilization, by preventing ovulation, or by interfering with implantation (Table 47.1).
- Methods for preventing the sperm from reaching the site of fertilization include the condom, the diaphragm or cervical cap, and the rhythm method, as well as vasectomy or tubal ligation.
- The oral contraceptive pill prevents ovulation. It contains a combination of estrogen and the progesterone-like progestin, which inhibiting the secretion of FSH and LH and follicle formation.
- Methods for preventing implantation include the IUD and the morning-after pill.

Questions

Self-Test Questions

1. Asexual reproduction is most successful in:
   a. changing environments.
   b. sessile animals.
   c. densely settled populations.
   d. land animals.
   e. genetically varied individuals.

2. Which of the following processes does *not* increase genetic diversity?
   a. parthenogenesis.
   b. random DNA mutations.
   c. genetic recombination.
   d. independent assortment.
   e. random combinations of paternal and maternal chromosomes.

3. Gametogenesis has parallel stages in egg and sperm formation. The stage in eggs that is equivalent to spermatids is the:
   a. primary oocyte.
   b. oogonium.
   c. ovum.
   d. ootid and polar bodies.
   e. secondary oocyte and polar body.

4. The animal group that exhibits external fertilization is the:
   a. amphibians.   d. reptiles.
   b. birds.   e. mammals.
   c. sharks.

5. The slow block to polyspermy:
   a. is caused by a change in membrane potential from negative to positive.
   b. triggers the movement of Ca^{2+} from the cytosol to the endoplasmic reticulum.
   c. triggers a decrease in egg oxidation and protein synthesis.
   d. describes the fusion of egg and sperm nuclei.
   e. includes the fusion of cortical granules with the egg's plasma membrane.

6. Some placental animals provide nutrients to their embryos from an attached membranous yolk-containing sac. They are called:
   a. oviparous animals.   d. eutherians.
   b. ovoviviparous animals. e. mammals.
   c. metatherians.

7. Which activity is a step in the ovarian cycle?
   a. FSH stimulates the pituitary to release GnRH.
   b. When FSH and LH levels fall, the corpus luteum shrinks and the uterine lining breaks down.
   c. Luteinizing hormone stimulates the uterus to make progesterone.
   d. Estrogen levels initially have a positive feedback effect on the pituitary, which is followed by higher estrogen levels causing negative feedback.
   e. A fully developed corpus luteum inhibits uterine lining growth.

8. During spermatogenesis in mammals, sperm travels from the:
   a. Sertoli cells past the epididymis and urethra, through the vas deferens to the prepuce.
   b. seminal vesicles past the prostate gland, through the glans and prepuce to the bulbourethral glands.
   c. vestibular glands past the Leydig cells, through the accessory glands and epididymis to the vas deferens.
   d. labia past the bulbourethral glands, through the vas deferens and urethra to the epididymis.
   e. seminiferous tubules past the Leydig cells, through the epididymis and vas deferens to the urethra.

9. The human egg is fertilized in the:
   a. uterus.   d. cervical canal.
   b. vagina.   e. ovary.
   c. oviduct.

10. The most effective method to prevent fertilization is:
    a. the oral contraceptive.
    b. the IUD.
    c. the morning-after pill.
    d. vasectomy and tubal ligation.
    e. the rhythm method.

Questions for Discussion

1. Currently under development is an "anti-pregnancy vaccine" that stimulates a woman's immune system to develop antibodies against human chorionic gonadotropin (hCG). How would this method prevent pregnancy?

2. Men sometimes have reduced fertility because of *testicular varioceles*, varicose veins in the testes in which blood pools. Based on what you now know of the conditions under which sperm develop properly, how do think this condition might impair sperm development?

3. Spermatogenesis produces four sperm for each spermatocyte, but oogenesis produces only one egg for each oocyte. Why might these different outcomes be adaptive?

4. Sertoli cells protect spermatocytes from attack by antibodies during their development in the human male. What structures

might protect the oocyte and egg from attack by antibodies in the human female?

5. Compare the advantages and disadvantages of sexual and asexual reproduction for an aphid and a parasitic worm.

6. It may be possible to develop a birth control drug that would prevent conception by interfering with fertilization. Outline the design for such a drug and explain exactly how it would work. (There may be more than one design that, in theory, would be effective.)

Experimental Analysis

Design experiments to determine if, and at what dose, vitamin E can decrease menstrual cramping significantly.

Evolution Link

The nematode species *Caenorhabditis elegans* and *Caenorhabditis briggsae* are both hermaphroditic. Phylogenetic evidence indicates that the last common ancestor of these two species had a normal male-female mechanism of reproduction. What does this evidence suggest about their hermaphroditism?

How Would You Vote?

Fertility drugs induce multiple ovulations at the same time and increase the likelihood of high-risk multiple pregnancies. Should the use of such drugs be restricted to conditions that limit the number of embryos formed? Go to www.thomsonedu.com/login to investigate both sides of the issue and then vote.

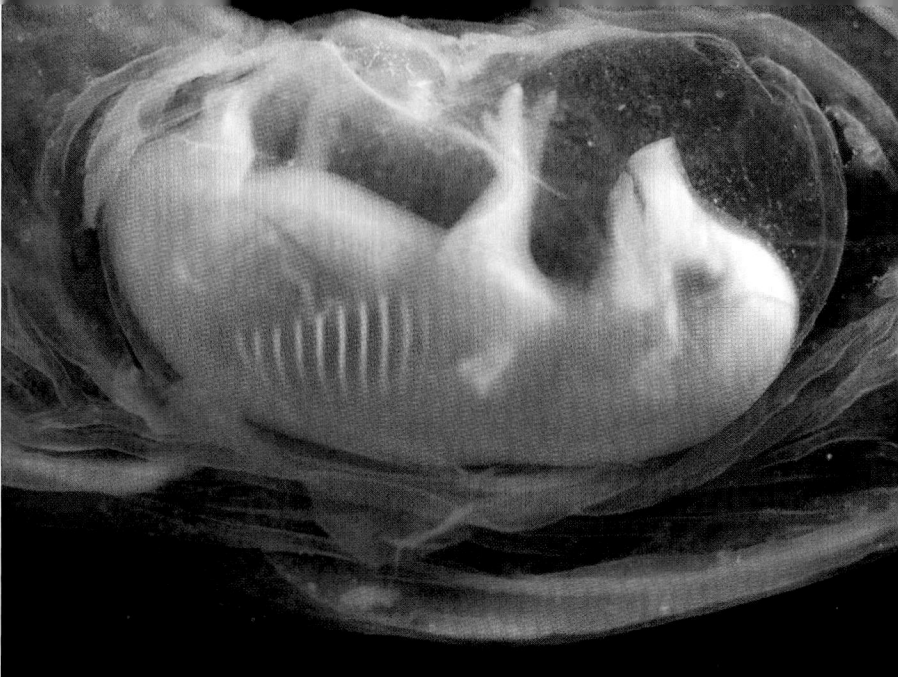

Embryo pig *(Sus scrofa domestica)* after 33 days of development. The embryo is about 16 mm long and is surrounded by several membranous sacs, including the fluid-filled amnion (closest to the embryo), which cushions and protects it.

Daniel Sambraus/SPL/Photo Researchers, Inc.

48 Animal Development

WHY IT MATTERS

The uterine contractions announcing birth are taking place at shorter intervals and with greater intensity. The mother-to-be endures the discomfort and apprehension with the knowledge that the child that has been growing in her body will soon come into the world. It began as a fertilized egg, about the size of a period on this page, and grew through a program of cell divisions, complex cell movements, and molecular interactions. She was unaware of these complexities except for movements of the fetus that became apparent about 14 weeks after she became pregnant.

Her baby's development required no conscious attention on her part: human development, like that of all animals, is programmed to proceed inexorably from fertilized egg to free-living offspring. Even childbirth is the result of programmed events that, once started, normally move to conclusion without requiring deliberate input from the mother.

Over the course of its development, the baby's body formed all the organ systems required for independent existence, and at its birth they are already working to sustain its life. Most astonishing, perhaps, is the baby's brain. It began as a tube of nerve tissue that bulged outward and enlarged, continually adding nerve cells and connecting them into circuits until it attained what may well be some of the most complexly organized matter in the universe—all as part of the automated events of development. Still, the human brain is unique only in the degree of its complexity and integrative capacity; the brains of other mammals are basically similar and develop through the same embryonic pathways.

The baby enters the outside world passing head first through the cervix, and then the vagina. Soon the rest of the body slips through, aided and lubricated in its passage by release of the fluid that surrounded and cushioned it in the uterus. In the first indignity of life, the baby is briefly held upside down to drain fluid from its lungs. This action triggers its first breath, followed by a satisfyingly loud cry.

The baby is proudly displayed to the mother, who greets it with love, relief, joy, and realization of the responsibilities the baby will bring. It is a girl, who with further luck and good care will continue developing through childhood, puberty, adult life, and old age, all through programs built into her hereditary molecules. As part of these passages, she may bring her own child into the world.

People have tried since ancient times to understand how development and birth take place. The scientific quest began with Aristotle, who observed chick development and correctly interpreted the functions of the placenta and umbilical cord in humans. The investigators who followed Aristotle concentrated on describing developmental changes in **morphology**, which is the form or shape of an organism, or of a part of an organism. More recently, investigators began to trace the molecular underpinnings of the morphological events.

In this chapter we survey the results of these investigations. We take up the story of animal development where the previous chapter left off, with the fertilized egg. We continue with the early events leading from the fertilized egg to the primary tissues of the embryo, and then trace the development of organs from these tissues. Next, we describe human development as representative of the process in mammals. Then, we survey the cellular and molecular bases of these mechanisms. At the cellular level, the development of an adult animal from a fertilized egg involves cell division, in which more cells are produced by mitosis; **cell differentiation**, in which changes in gene expression establish cells with specialized structure and function; and **morphogenesis** ("form creation"), the generation of the body form of the animal as differentiated cells end up in their appropriate sites. Finally, we discuss the genetic and molecular mechanisms that are largely responsible for directing the course of development.

48.1 Mechanisms of Embryonic Development

Fertilization of an egg by a sperm cell produces a zygote. Embryonic development begins at this point and ultimately produces a free-living individual. All the instructions required for development are packed into the fertilized egg.

Developmental Information Is Located in both the Nucleus and Cytoplasm of the Fertilized Egg

Mitotic divisions of the zygote formed when egg and sperm nuclei fuse are the beginning of developmental activity (see Section 47.2).

Information Storage in the Egg. The information that directs the initiation of development is stored in two locations in the fertilized egg. Part of the information is stored in the zygote nucleus, in the DNA derived from the egg and sperm nuclei. This information directs development as individual genes are activated or turned off in a highly ordered manner. The rest of the information is stored in the egg cytoplasm, in the form of messenger RNA (mRNA) and protein molecules.

Because the fertilizing sperm contributes essentially no cytoplasm, nearly all the cytoplasmic information of the fertilized egg is maternal in origin. The mRNA and proteins stored in the egg cytoplasm are known as **cytoplasmic determinants**. They direct the first stages of animal development, in the period before genes of the zygote become active. Depending on the animal group, the control of early development by cytoplasmic determinants may be limited to the first few divisions of the zygote, as in mammals, or it may last until the actual tissues of the embryo are formed, as in most invertebrates.

Other Components of the Egg. In addition to cytoplasmic determinants, the oocyte cytoplasm also contains ribosomes and other cytoplasmic components required for protein synthesis and the early cell divisions of embryonic development. For example, the egg cytoplasm contains all the tubulin molecules required to form the spindles for early cell divisions. It also contains mitochondria, nutrients stored in granules in the yolk and in lipid droplets, and, in many animals, pigments that color the egg or regions of it.

The **yolk** contains nutrients. When the egg itself supplies all the nutrients for development of the embryo, as in the eggs laid by insects, reptiles, and birds, it contains large amounts of yolk. When the mother supplies most of the nutrients, as in the placental mammals, the egg has a small quantity of yolk that is used only for the earliest stages of development.

Depending on the species, the yolk may be concentrated at one end or in the center of the egg, or distributed evenly throughout the cytoplasm. Its distribution influences the rate and location of cell division during early embryonic development. Typically, cell division proceeds more slowly in the region of the egg containing the yolk. In the large, yolky eggs of birds and reptiles, cell division takes place only in a small, yolk-free patch at the surface of the egg.

Unequal distribution of yolk and other components in a mature egg is termed **polarity**. For example, in most species the egg nucleus is located toward one end of the egg. This end of the egg, called the **animal pole**, typically gives rise to surface structures and the anterior end of the embryo. The opposite end of the egg, the **vegetal pole**, typically gives rise to internal structures such as the gut and the posterior end of the embryo. Yolk, when unequally distributed in the egg cytoplasm, is most frequently concentrated in the vegetal half of the egg. The egg's polarity contributes to the generation of body axes. For example, egg polarity plays a role in setting the three body axes of bilaterally symmetrical animals (such as humans and dogs): the anterior–posterior axis, the dorsal–ventral (back–front) axis, and the left–right axis **(Figure 48.1)**.

Cleavage, Gastrulation, and Organogenesis Are Early Events in Development

Soon after fertilization, the zygote begins a series of mitotic **cleavage** divisions, so called because cycles of DNA replication and division occur without the production of new cytoplasm. As a result, the cytoplasm of the egg is partitioned into successively smaller cells without increasing the overall size or mass of the embryo **(Figure 48.2)**. These cells are called **blastomeres** (*blastos* = bud or offshoot; *meros* = part or division). In the frog *Xenopus laevis,* for example, a sequence of twelve cleavage divisions produces an embryo of about 4000 cells, which collectively occupy about the same volume and mass as did the original zygote.

Cleavage is the first of three major developmental processes that, with modifications, are common to the early development of most animals (described in detail for particular animals in the next section). Following cleavage, the second major process, **gastrulation**, produces an embryo with three distinct primary tissue layers. Following gastrulation, the development of the major organ systems, called **organogenesis**, gives rise to a free-living individual with the body organization characteristic of its species. Organogenesis involves the same mechanisms used in gastrulation—cell division, cell movements, and cell rearrangements. **Figure 48.3** outlines these stages in the life cycle of a frog.

The cleavage divisions lead to three successive developmental stages that are common to the early development of most animals. The first stage, called a **morula** (*morula* = mulberry), is a solid ball or layer of blastomeres. As cleavage divisions continue, the ball or layer hollows out to form the second stage, the **blastula** (*ula* = small), in which the blastomeres enclose a fluid-filled cavity, the **blastocoel** (*koilos* = hollow).

Once cleavage is complete, the cells of the blastula migrate and divide to produce the **gastrula** (*gaster* = gut or belly). This morphogenetic process, gastrulation, dramatically rearranges the cells of the blastula into the three primary cell layers of the embryo: the outer **ectoderm** (*ecto* = outside; *derma* = skin), the inner **endoderm** (*endo* = inside), and the **mesoderm** (*meso* = middle) between the ectoderm and the endoderm. Gastrulation establishes body pattern; that is,

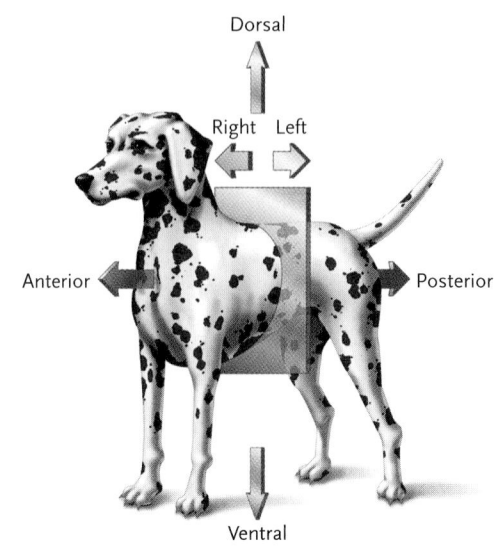

Figure 48.1
Body axes: anterior–posterior, dorsal–ventral, and left–right.

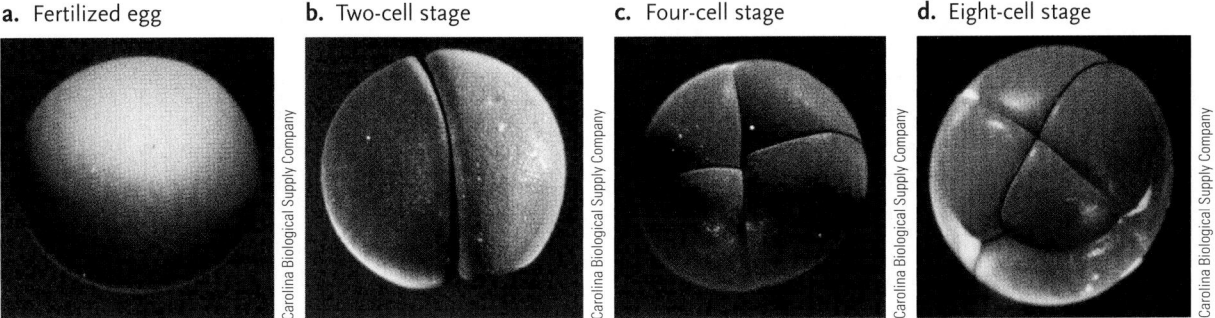

a. Fertilized egg   **b.** Two-cell stage   **c.** Four-cell stage   **d.** Eight-cell stage

Carolina Biological Supply Company

Figure 48.2
The first three cleavage divisions of a frog embryo, which convert the fertilized egg into the eight-cell stage. Note that the cleavage divisions cut the volume of the fertilized egg into successively smaller cells.

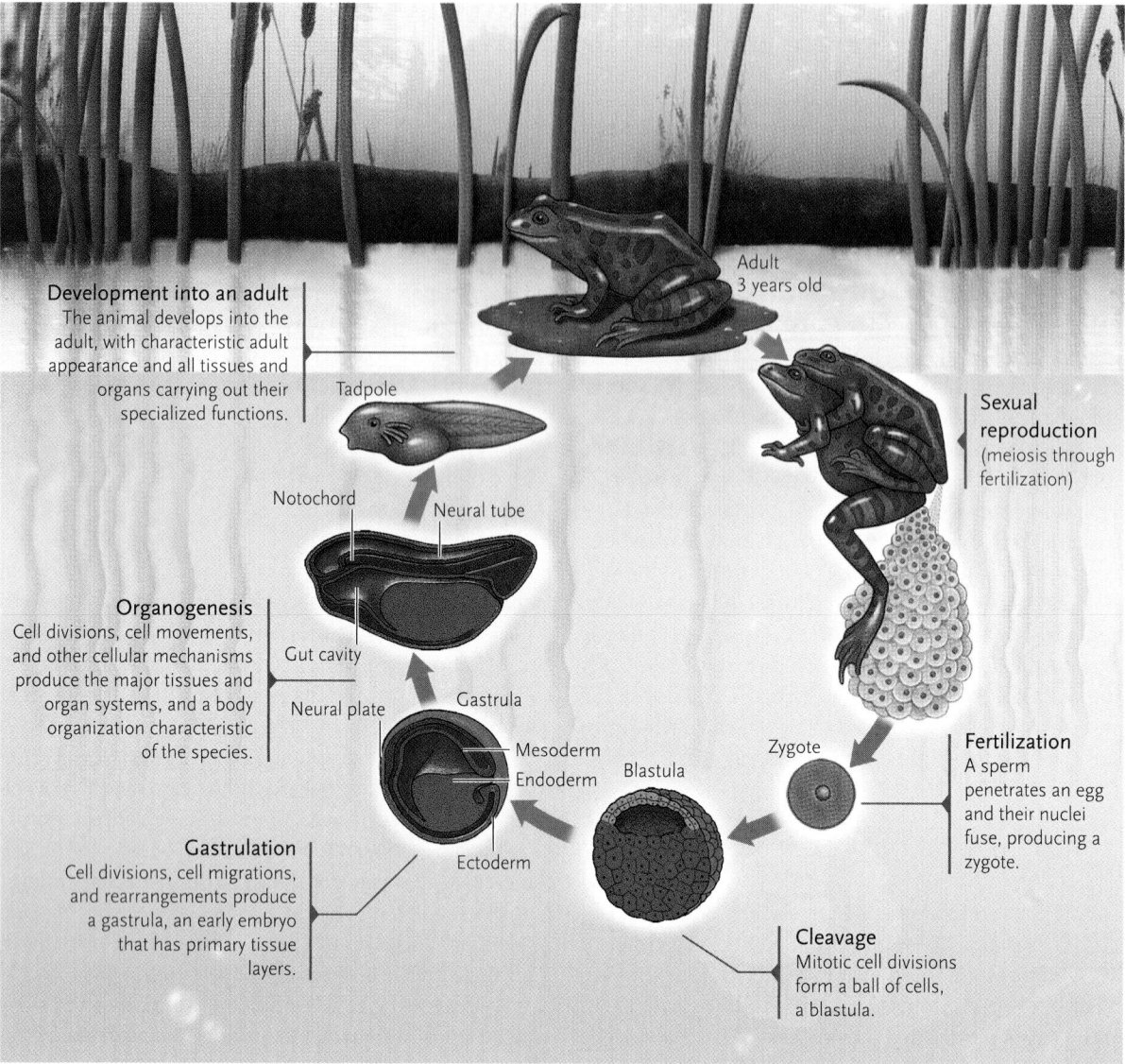

Figure 48.3
Stages of animal development shown in a frog.

Within the figure:

Development into an adult
The animal develops into the adult, with characteristic adult appearance and all tissues and organs carrying out their specialized functions.

Adult 3 years old

Tadpole

Sexual reproduction (meiosis through fertilization)

Notochord

Neural tube

Organogenesis
Cell divisions, cell movements, and other cellular mechanisms produce the major tissues and organ systems, and a body organization characteristic of the species.

Gut cavity

Neural plate

Gastrula

Mesoderm

Endoderm

Blastula

Zygote

Fertilization
A sperm penetrates an egg and their nuclei fuse, producing a zygote.

Gastrulation
Cell divisions, cell migrations, and rearrangements produce a gastrula, an early embryo that has primary tissue layers.

Ectoderm

Cleavage
Mitotic cell divisions form a ball of cells, a blastula.

each tissue and organ of the adult animal originates in one of the three primary cell layers of the gastrula **(Table 48.1).**

The cell movements also form a new cavity within the embryo, the **archenteron** (*arche* = beginning; *enteron* = intestine or gut), which is lined with endoderm. The archenteron forms the primitive gut of the embryo; an opening at one end, the **blastopore**, gives rise to the anus or mouth of the embryo, depending on the animal group (see Section 29.2). In the protostomes, which include annelids, arthropods, and mollusks, the blastopore develops into the mouth, and the anus forms at the opposite end of the embryonic gut. In the deuterostomes, which include echinoderms and chordates, the blastopore develops into the anus and the mouth forms at the opposite end of the embryonic gut. By the time gastrulation is complete, the embryo has clearly defined anterior, posterior, dorsal, and ventral regions.

As the blastula develops into the gastrula, embryonic cells begin to differentiate: they become recognizably different in biochemistry, structure, and function.

The developmental potential of the cells also becomes more limited than that of the fertilized egg from which they originated. For example, although a fertilized egg is capable of developing into a complete embryo, a mesoderm cell may develop into muscle or bone but not normally into outside skin or brain. This restriction of developmental potential does not occur, as was once thought, because the cells have lost all their genes except those for the structure and function of the cell type they will become. Rather, the differentiating cells actually all contain complete genomes of the organism, but each type of cell has a different program of gene expression.

Development in all animals is accomplished by a number of mechanisms that are under genetic control but are influenced to some extent by the environment (for example, temperature affects the rate of cell division). The mechanisms are

1. Mitotic cell divisions.
2. Cell movements.

3. **Selective cell adhesions**, in which cells make and break specific connections to other cells or to the extracellular matrix.

4. **Induction**, in which one group of cells (the inducer cells) causes or influences another nearby group of cells (the responder cells) to follow a particular developmental pathway. The key to induction is that only certain cells can respond to the signal from the inducer cells. Induction typically involves signal transduction events (see Chapter 7). These events are triggered either by direct cell-to-cell contact involving interaction between a membrane-embedded protein on the inducer cell and a receptor protein on the surface of the responder cell, or by a signal molecule released by the inducer cell that interacts with a receptor on the responder cell. (The latter is an example of paracrine signaling; see Section 40.1.)

5. **Determination**, in which the developmental fate of a cell is set. Prior to determination, a cell has the potential to become any cell type of the adult but, after determination, that property is lost as the cell commits to becoming a particular cell type. Typically, determination is the result of induction, but in some cases it results from the asymmetric segregation of cellular determinants.

6. **Differentiation**, which follows determination, involves the establishment of a cell-specific developmental program in cells. Differentiation results in cell types with clearly defined structures and functions; those features derive from specific patterns of gene expression in cells.

You will see examples of these mechanisms in the examples of development discussed in the following three sections.

STUDY BREAK

1. How do cleavage divisions differ from cell division in an adult organism?
2. What are the primary cell layers of the embryo, and what process is responsible for producing them?

48.2 Major Patterns of Cleavage and Gastrulation

With the principles of early embryonic development established, we describe cleavage and gastrulation in three animal groups that have been models in *embryology* (the study of embryos and their development): sea urchins, amphibians, and birds. Later in the chapter, we describe cleavage and gastrulation in humans and other mammals, which resemble the pattern in birds.

Table 48.1	Origins of Adult Tissues and Organs in the Three Primary Tissue Layers
Primary Tissue Layer	Adult Tissues and Organs
Ectoderm	Skin and its elaborations, including hair, feathers, scales, and nails; nervous system, including brain, spinal cord, and peripheral nerves; lens, retina, and cornea of eye; lining of mouth and anus; sweat glands, mammary glands, adrenal medulla, and tooth enamel
Mesoderm	Muscles; most of skeletal system, including bones and cartilage; circulatory system, including heart, blood vessels, and blood cells; internal reproductive organs; kidneys and outer walls of digestive tract
Endoderm	Lining of digestive tract, liver, pancreas, lining of respiratory tract, thyroid gland, lining of urethra, and urinary bladder

Sea Urchin Gastrulation Follows a Symmetrical Pattern That Reflects an Even Distribution of Yolk

Cleavage divisions proceed at approximately the same rate in all regions of a sea urchin embryo (**Figure 48.4,** step 1), reflecting the uniform distribution of yolk in the sea urchin egg. These divisions continue until a blastula containing about a thousand cells is formed (step 2).

Gastrulation begins at the vegetal pole of the blastula. As a result of induction, some cells in the middle of that region become elongated and cylindrical, causing the region to flatten and thicken. Then, some cells break loose and migrate into the blastocoel (see Figure 48.4, step 3). These cells, called *primary mesenchyme cells* (mesenchyme means "middle juice"), move around inside the blastocoel, making and breaking adhesions, until eventually they attach along the ventral sides of the blastocoel. These cells will eventually become the mesoderm (see step 7), which will give rise to skeletal elements of the embryo. Next, the flattened vegetal pole of the blastula invaginates, pushing gradually into the interior (steps 4 and 5). The cells that invaginate are future endoderm cells. The inward movement, in effect much like pushing in the side of a hollow rubber ball, generates a new cavity, the archenteron. The opening of the archenteron is the blastopore.

As the archenteron is forming, the cells of the invaginated cell layer send out extensions that stretch across the blastocoel and contact the inside of the ectoderm (step 6). These extensions make tight adhesions and then contract, pulling the invaginated cell layer inward with them and thereby eliminating most of the blastocoel.

At this point the embryo has two complete cell layers. The outer layer remaining from the original blastula surface makes up the ectoderm of the embryo. The second, inner layer, derived from the cells forming the

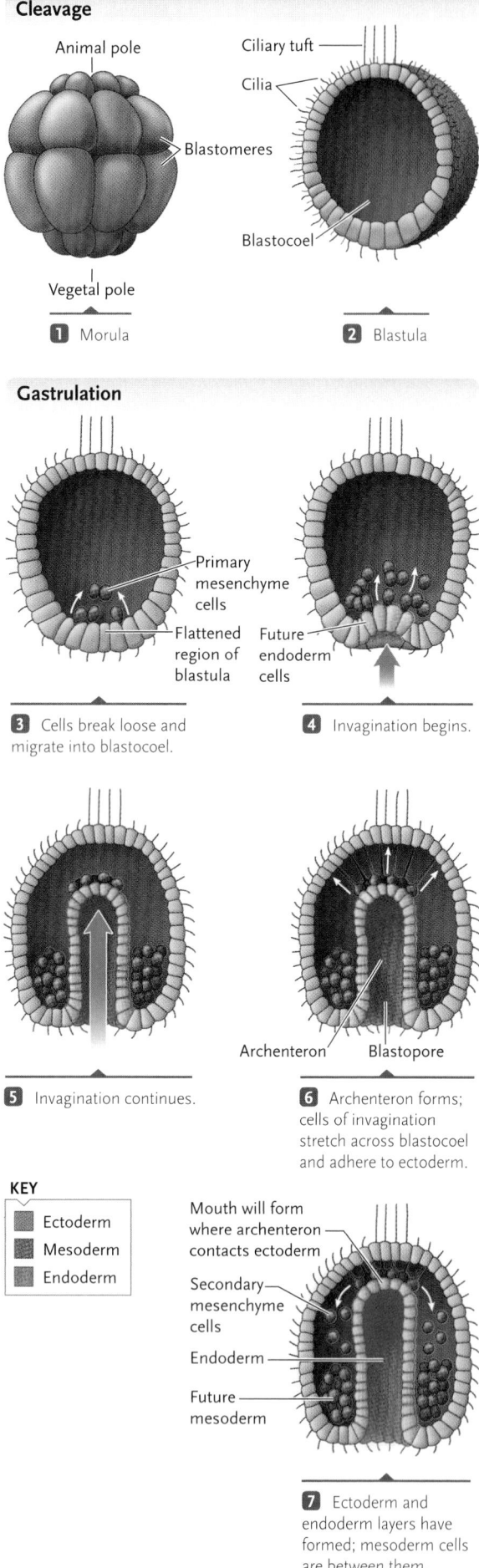

Figure 48.4
Cleavage and gastrulation in the sea urchin.

Cleavage

Animal pole

Blastomeres

Vegetal pole

1 Morula

Ciliary tuft

Cilia

Blastocoel

2 Blastula

Gastrulation

Primary mesenchyme cells

Flattened region of blastula

Future endoderm cells

3 Cells break loose and migrate into blastocoel.

4 Invagination begins.

Archenteron   Blastopore

5 Invagination continues.

6 Archenteron forms; cells of invagination stretch across blastocoel and adhere to ectoderm.

KEY

Ectoderm
Mesoderm
Endoderm

Mouth will form where archenteron contacts ectoderm

Secondary mesenchyme cells

Endoderm

Future mesoderm

7 Ectoderm and endoderm layers have formed; mesoderm cells are between them.

archenteron, makes up the endoderm. Mesodermal cells are also beginning to form a third layer, the mesoderm. Some are derived from the primary mesenchyme cells and others from *secondary mesenchyme cells,* cells that migrated into the space between ectoderm and endoderm (step 7). When the mesoderm layer is complete, the embryo has three complete layers: ectoderm, mesoderm, and endoderm. At this point, cells within each layer begin to differentiate, as evidenced by the synthesis of different proteins in each layer.

As the ectoderm, mesoderm, and endoderm develop, the embryo lengthens into an ellipsoidal shape with the blastopore marking the posterior end of the embryo. From this point on, organ systems differentiate through further cell division, cell movements, selective cell adhesions, induction, and differentiation. The blastopore forms the anus; a mouth will form at the opposite, anterior end of the gut.

Amphibian Cleavage and Gastrulation Are Influenced by an Unequal Distribution of Yolk

In amphibian eggs, such as those of frogs, yolk is concentrated in the vegetal half, which gives it a pale color. The animal half is darkly colored by a layer of pigment granules just below the surface. The sperm normally fertilizes the egg in the animal half (**Figure 48.5,** step 1). After fertilization, the pigmented layer of cytoplasm rotates toward the site of sperm entry, exposing a crescent-shaped region of the underlying cytoplasm at the side opposite the point of sperm entry (step 2). This region, called the **gray crescent**, establishes the dorsal–ventral axis of the embryo, with the gray crescent marking the future dorsal side.

Normally, the first cleavage division runs perpendicular to the long axis of the gray crescent and divides the crescent equally between the resulting cells (step 3). If one of the first two blastomeres does not receive gray crescent material, and the cells are separated experimentally, the cell without gray crescent divides to produce a disordered mass that stops developing. The cell receiving the gray crescent produces a normal embryo. Thus cytoplasmic material localized in the gray crescent is essential to normal development in frog embryos.

As cleavage of a frog embryo continues, cell divisions proceed more rapidly in the animal half, producing smaller and more numerous cells in this region than in the yolky vegetal half. By the time cleavage has produced an embryo with 15,000 cells, the animal half of the embryo has hollowed out, forming the blastula (**Figure 48.6,** step 1, and **Figure 48.7a**).

Gastrulation begins when cells from the animal pole move across the embryo surface and reach the region derived from the gray crescent. This site is marked by a crescent-shaped depression rotated clockwise 90°

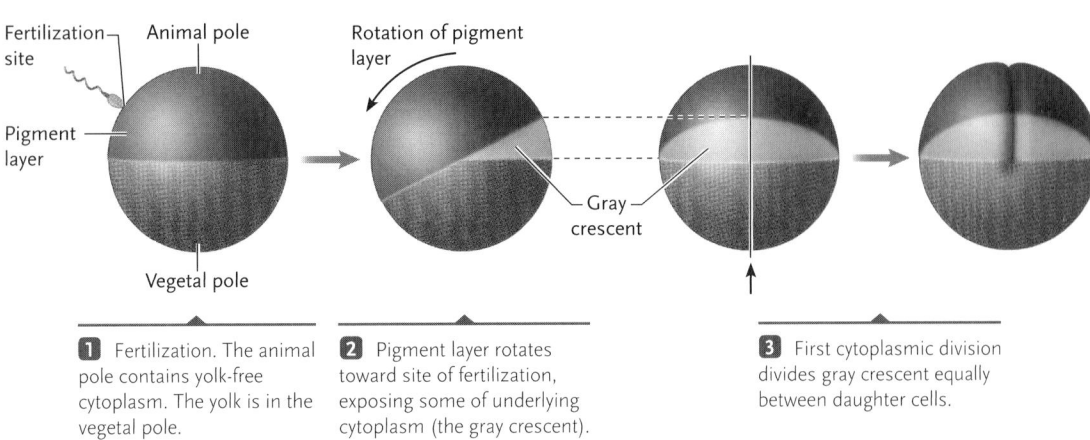

Figure 48.5

Rotation of the pigment layer and development of the gray crescent after fertilization in a frog egg. The gray crescent marks the site where gastrulation of the embryo will begin.

Animal pole

Fertilization site

Pigment layer

Vegetal pole

Rotation of pigment layer

Gray crescent

1 Fertilization. The animal pole contains yolk-free cytoplasm. The yolk is in the vegetal pole.

2 Pigment layer rotates toward site of fertilization, exposing some of underlying cytoplasm (the gray crescent).

3 First cytoplasmic division divides gray crescent equally between daughter cells.

called the **dorsal lip of the blastopore.** Cells changing shape and pushing inward from the surface in a process called **invagination** produce the depression. With continued inward movement of additional cells, the depression eventually forms a complete circle (Figure 48.6, step 2, and **Figure 48.7b**), which is the blastopore.

As cells migrate into the blastopore by a process is called **involution,** the pigmented cell layer of the animal half expands to cover the entire surface of the embryo (Figure 48.6, step 3). The cells of the vegetal half are enclosed by the movement, and show on the outside as a yolk plug in the blastopore (Figure 48.6,

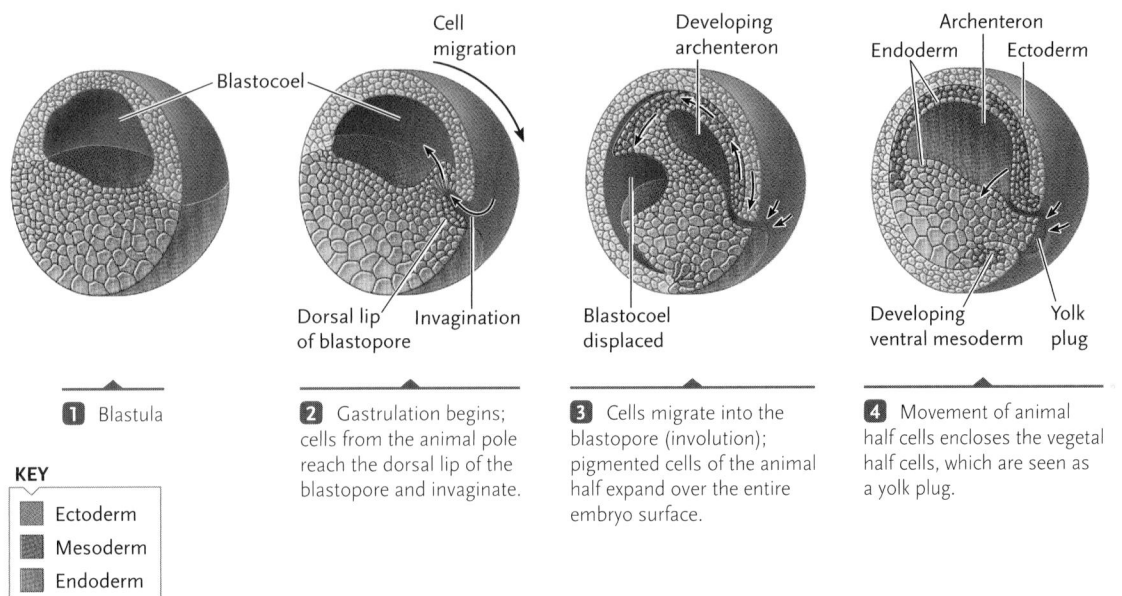

Figure 48.6

Gastrulation in a frog embryo. Yolk cells are shown in darker yellow.

Blastocoel

Cell migration

Dorsal lip of blastopore   Invagination

Developing archenteron

Blastocoel displaced

Archenteron

Endoderm   Ectoderm

Developing ventral mesoderm   Yolk plug

1 Blastula

2 Gastrulation begins; cells from the animal pole reach the dorsal lip of the blastopore and invaginate.

3 Cells migrate into the blastopore (involution); pigmented cells of the animal half expand over the entire embryo surface.

4 Movement of animal half cells encloses the vegetal half cells, which are seen as a yolk plug.

KEY

Ectoderm
Mesoderm
Endoderm

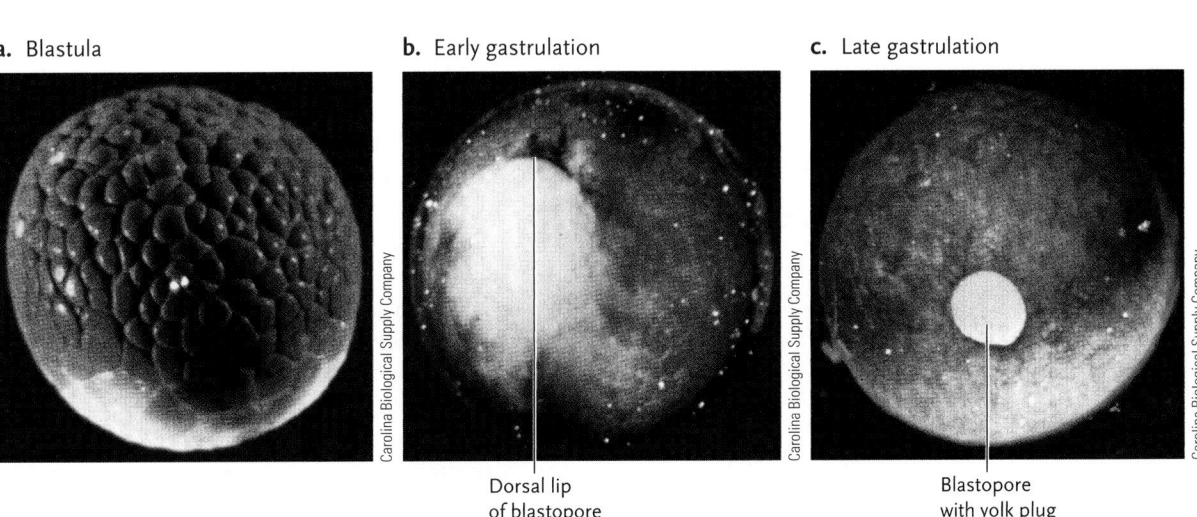

a. Blastula

b. Early gastrulation

c. Late gastrulation

Dorsal lip of blastopore

Blastopore with yolk plug in center

Figure 48.7

Formation of the dorsal lip of the blastopore and the completed blastopore, closed by a yolk plug, in photomicrographs of a frog embryo.

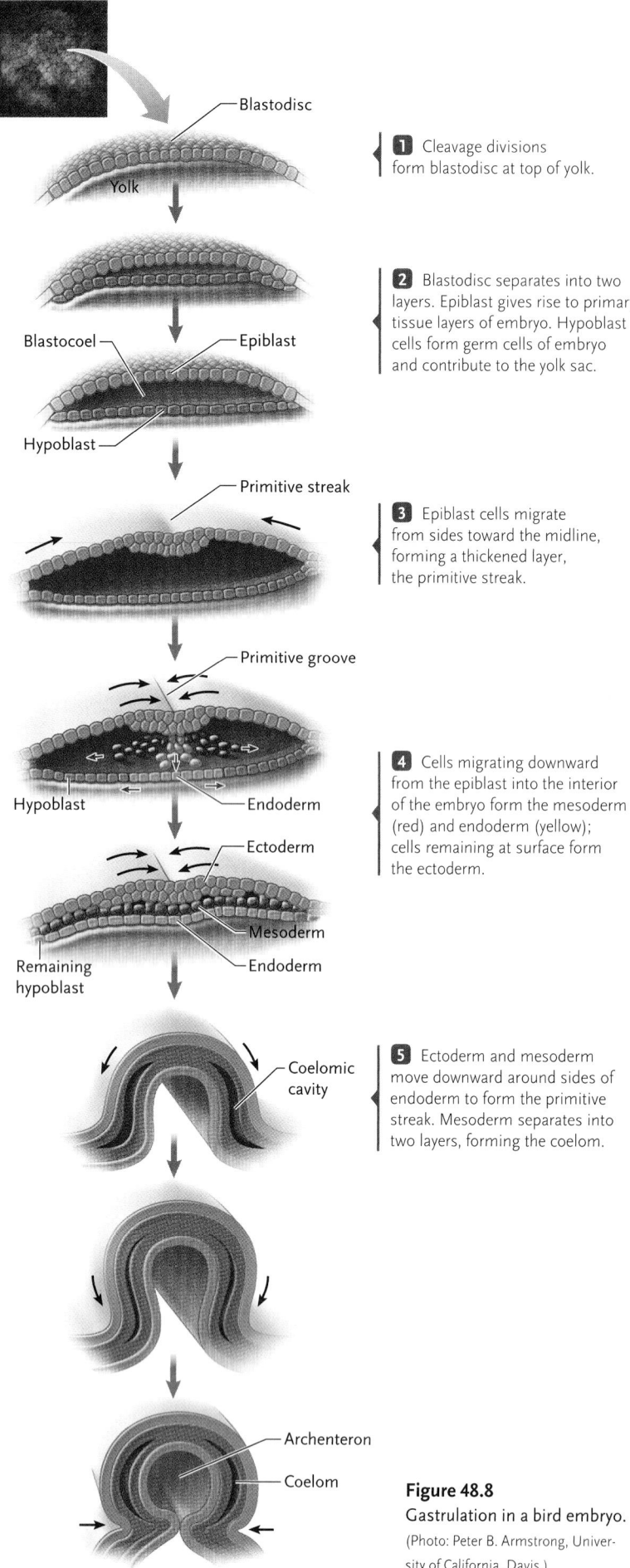

1 Cleavage divisions form blastodisc at top of yolk.

2 Blastodisc separates into two layers. Epiblast gives rise to primary tissue layers of embryo. Hypoblast cells form germ cells of embryo and contribute to the yolk sac.

3 Epiblast cells migrate from sides toward the midline, forming a thickened layer, the primitive streak.

4 Cells migrating downward from the epiblast into the interior of the embryo form the mesoderm (red) and endoderm (yellow); cells remaining at surface form the ectoderm.

5 Ectoderm and mesoderm move downward around sides of endoderm to form the primitive streak. Mesoderm separates into two layers, forming the coelom.

Figure 48.8

Gastrulation in a bird embryo.

(Photo: Peter B. Armstrong, University of California, Davis.)

step 4, and **Figure 48.7c**). As in other vertebrates, the blastopore gives rise to the anus.

Within the embryo, continuing involution moves cells into the interior and upward (see Figures 48.7b and c), forming two layers that line the inside top half of the embryo. The uppermost of these layers is induced to become the dorsal mesoderm (shown in red). The layer beneath it, which contains cells originating from both the outer surface of the embryo and the yolky interior, becomes the endoderm (shown in lighter yellow). The pigmented cells remaining at the surface of the embryo form the ectoderm (shown in blue). The ventral mesoderm begins to be induced near the vegetal pole.

As the mesoderm and endoderm form, the depression created by the inward cell movements gradually deepens and extends inward as the archenteron (see Figures 48.6c and d), which displaces the blastocoel. The cells of the three primary cell layers continue to increase in number by further movements and divisions as development proceeds.

During frog gastrulation, cells of the dorsal lip of the blastopore are inducer cells that control blastopore formation; if the cells in the dorsal lip are removed and transplanted elsewhere in the egg, they cause a second blastopore—and a second embryo—to form in this region (see Section 48.5).

The events of gastrulation in frogs thus include the same developmental mechanisms as in sea urchins—cell divisions, cell movements, selective adhesions, induction, and differentiation.

Gastrulation in Birds Proceeds at One Side of the Yolk

The pattern of gastrulation in birds and reptiles is modified by the distribution of yolk, which occupies almost the entire volume of the egg. (Birds and reptiles, as well as mammals, are all amniotes; see Section 30.7.) The portion of the cytoplasm that divides to give rise to the primary tissues of the embryo is confined to a thin layer at the egg surface. Although mammalian eggs have relatively little yolk, gastrulation follows a similar pattern in mammals, as discussed in Section 48.4.

Cleavage and Gastrulation in Birds. The early cleavage divisions in birds produce a disclike layer of cells at the surface of the yolk called the **blastodisc (Figure 48.8,** step 1). When blastodisc formation is complete, the layer contains about 20,000 cells. The cells of the blastodisc then separate into two layers, called the **epiblast** (top layer) and **hypoblast** (bottom layer). The flattened cavity between them is the blastocoel (step 2).

Gastrulation begins as cells in the epiblast stream toward the midline of the blastodisc, thickening the epiblast in this region. The thickened layer—the **primitive streak**—begins forming in the posterior end of the embryo and extends toward the anterior end as more cells of the epiblast move into it (step 3). The thickening at the anterior end of the primitive streak, called the *primitive knot,* is the functional equivalent of the amphibian dorsal lip of the blastopore. The primitive streak initially defines the axes of the embryo: it extends from posterior to anterior, the region where the streak forms is the dorsal while beneath it is the ventral side, and it defines left and right sides of the embryo.

As the primitive streak forms, its midline sinks, forming the **primitive groove.** The primitive groove is a conduit for migrating cells to move into the blastocoel. The first cells to migrate through the primitive groove are epiblast cells (step 4), which will form the endoderm. Cells migrating laterally between the epiblast and the endoderm form the mesoderm. The epiblast cells left at the surface of the blastodisc form the ectoderm (see step 4). Thus all the primary tissue layers of the chick embryo arise from the epiblast.

Of the cells in the hypoblast, only a few, near the posterior end of the embryo, contribute directly to the embryo. These hypoblast cells form the *germ cells* that, later in development, migrate to the developing gonads and found the cell line leading to eggs and sperm (see Section 47.2).

Initially, the ectoderm, mesoderm, and endoderm are located in three more or less horizontal layers. During gastrulation, the endoderm pushes upward along its midline. At the same time, its left and right sides fold downward, forming a tube oriented parallel to the primitive streak (step 5). The central cavity of the tube is the archenteron, the primitive gut. The mesoderm separates into two layers, forming the coelom, a fluid-filled body cavity (see Section 29.2 and Figure 29.4c). These movements complete formation of the gastrula.

Formation of Extraembryonic Membranes. Each of the primary tissue layers of a bird embryo extends outside the embryo to form four **extraembryonic membranes (Figure 48.9),** which conduct nutrients from the yolk to the embryo, exchange gases with the environment outside the egg, and store metabolic wastes removed from the embryo. The **yolk sac** consists of extensions of mesoderm and endoderm that enclose the yolk. Although the yolk sac remains connected to the gut of the embryo by a stalk, yolk does not directly enter the embryo by this route. Instead, it is absorbed by blood vessels in the membrane, which transport the nutrients to the embryo. The **chorion,** produced from ectoderm and mesoderm, is the outermost membrane, which surrounds the embryo and yolk sac completely, and lines the inside of the shell. This membrane exchanges oxygen and carbon dioxide with the environment through the shell of the egg. The **amnion** is the innermost membrane, which closes over the embryo to form the *amniotic cavity.* The cells of the amnion secrete *amniotic fluid* into the cavity, which bathes the embryo and provides an aquatic environment in which it can develop. Reptilian and mammalian embryos are also surrounded by an amnion and amniotic fluid. By

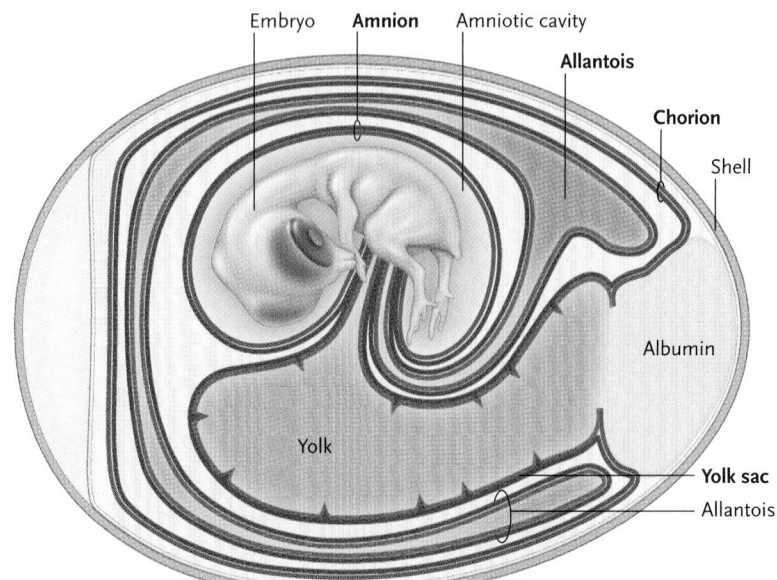

Figure 48.9
The four extraembryonic membranes in a bird embryo (in bold).

providing the embryo with an aquatic environment, this adaptation made possible the development of fully terrestrial vertebrates. The evolutionary importance of the amnion to the fully terrestrial vertebrates is recognized by classifying them together as **amniotes.** A membrane derived from mesoderm and endoderm that has bulged outward from the gut forms a sac called the **allantois.** This sac closely lines the chorion and fills much of the space between the chorion and the yolk sac. The allantois stores nitrogenous wastes (primarily uric acid) removed from the embryo. In addition, the part of the allantoic membrane that lines the chorion forms a rich bed of blood capillaries that is connected to the embryo by arteries and veins. This circulatory system delivers carbon dioxide to the chorion and picks up the oxygen that is absorbed through the shell and chorion.

STUDY BREAK

1. What is the role of the gray crescent in amphibian development?
2. What evidence indicates that cells of the dorsal lip of the blastopore act as inducer cells?
3. What are the extraembryonic membranes in birds, and what are their functions?

48.3 From Gastrulation to Adult Body Structures: Organogenesis

Following gastrulation, organogenesis—the process by which the ectoderm, mesoderm, and endoderm develop into organs—gives rise to an individual with the body organization characteristic of its species. Organo-

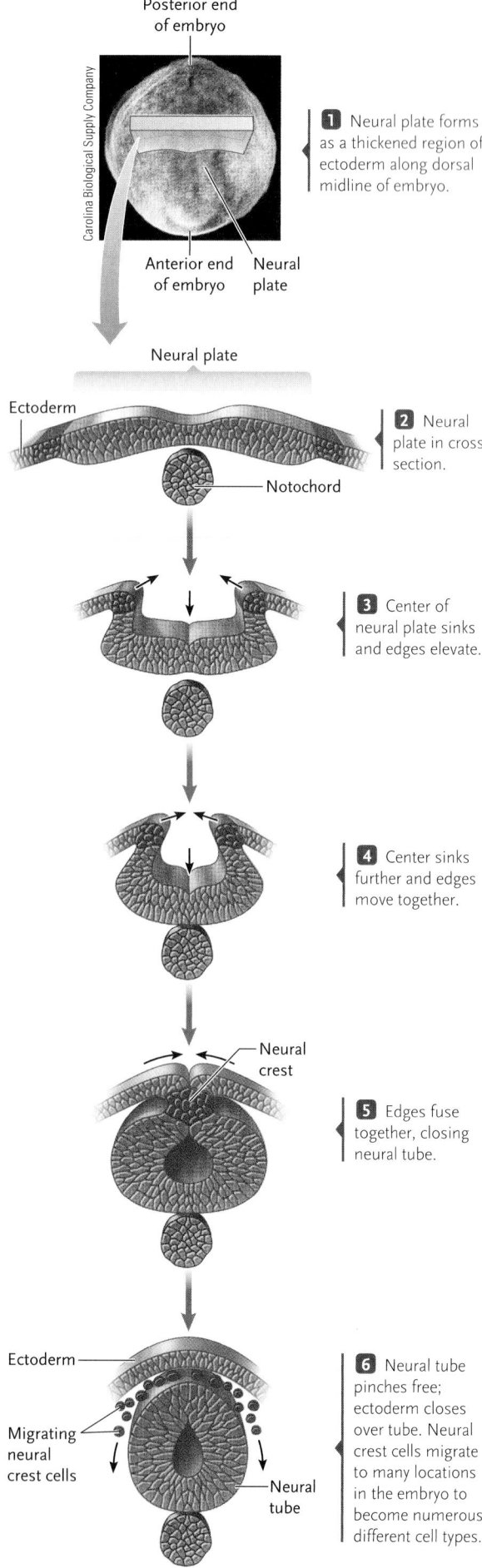

Figure 48.10
Development of the neural tube and neural crest cells in vertebrates. Photo is of an amphibian embryo; drawings show steps in a bird embryo.

Carolina Biological Supply Company

Posterior end of embryo

Anterior end of embryo

Neural plate

1 Neural plate forms as a thickened region of ectoderm along dorsal midline of embryo.

Neural plate

Ectoderm

Notochord

2 Neural plate in cross section.

3 Center of neural plate sinks and edges elevate.

4 Center sinks further and edges move together.

Neural crest

5 Edges fuse together, closing neural tube.

Ectoderm

Migrating neural crest cells

Neural tube

6 Neural tube pinches free; ectoderm closes over tube. Neural crest cells migrate to many locations in the embryo to become numerous different cell types.

genesis involves the same mechanisms used in gastrulation—cell division, cell movements, selective cell adhesion, induction, and differentiation—plus an additional mechanism, *apoptosis,* in which certain cells are programmed to die (apoptosis is also discussed in Section 43.2). To illustrate how the cellular mechanisms of development interact in organogenesis, we follow the formation of major organ systems in the frog embryo. Then we describe the generation of one organ, the eye, which follows a pathway typical of eye development in all vertebrates.

The Nervous System Develops from Ectoderm

In vertebrates, organogenesis begins with development of the nervous system from ectoderm, a process called **neurulation.** As a preliminary to neurulation, cells of the mesoderm form a solid rod of tissue, the **notochord,** which extends the length of the embryo under the dorsal ectoderm. Notochord cells carry out a major induction, in which they cause the overlying ectoderm to thicken and flatten into a longitudinal band called the **neural plate** (**Figure 48.10,** steps 1 and 2). Experiments have shown that if the notochord is removed, the neural plate does not form.

Once induced, the neural plate sinks downward along its midline (steps 2 and 3), creating a deep longitudinal groove. At the same time, ridges elevate along the sides of the neural plate. The ridges move together and close over the center of the groove (steps 4 and 5), converting the neural plate into a **neural tube** that runs the length of the embryo. The neural tube then pinches off from the overlying ectoderm, which closes over the tube (step 6). The central nervous system, including the brain and spinal cord, develops directly from the neural tube.

During formation of the neural tube, cells of the **neural crest**—the region where the neural tube pinches off from the ectoderm (shown in blue in Figure 48.10)—migrate to many locations in the developing embryo and become numerous different types of cells which contribute to a variety of organ systems. (The neural crest is one of the defining features of vertebrates.) Some cells develop into cranial nerves in the head; others contribute to the bones of the inner ear and skull, the cartilage of facial structures, and the teeth. Yet others form ganglia of the autonomic nervous system, peripheral nerves leading from the spinal cord to body structures, and nerves of the developing gut. Still others move to the skin, where they form pigment cells, or to the adrenal glands, where they form the medulla of these glands. The migration of neural crest cells occurs in the development of all vertebrates.

Other structures differentiate in the embryo while the neural tube is forming. On either side of the notochord, the mesoderm separates into blocks of cells called **somites,** spaced one after the other along both sides of the notochord **(Figure 48.11)**. The somites give rise to the vertebral column, the ribs, the repeating sets of muscles associated with the ribs and vertebral column, and muscles of the limbs. The mesoderm outside the somites, which extends around the primitive gut (lateral mesoderm in Figure 48.11), splits into two layers, one covering the surface of the gut, and the other lining the body wall. The space between the layers is the coelom of the adult (see Section 29.2).

Sequential Inductions and Differentiation Are Central to Eye Development

We now take up the development of the eye, to show how cellular mechanisms interact in organogenesis. Eyes develop by the same pathway in all vertebrates.

The brain forms at the anterior end of the neural tube from a cluster of hollow vesicles that swell outward from the neural tube **(Figure 48.12,** step 1). One paired set of vesicles, the *optic vesicles*, develop into the eyes. The figure depicts the optic vesicles in the brain of a frog embryo; note that the morphology of the forebrain, midbrain, and hindbrain in embryos differs among vertebrates.

The optic vesicles grow outward until they contact the overlying ectoderm, inducing a series of developmental responses in both tissues. The outer surface of the optic vesicle thickens and flattens at the region of contact and then pushes inward, transforming the optic vesicle into a double-walled *optic cup,* which ultimately becomes the retina. The optic cup induces the overlying ectoderm to thicken into a disclike swelling, the *lens placode* (step 2). The center of the lens placode sinks inward toward the optic cup, and its edges eventually fuse together, forming a ball of cells, the *lens vesicle* (step 3).

The developing lens cells begin to synthesize *crystallin,* a fibrous protein that collects into clear, glassy deposits. The lens cells finally lose their nuclei and form the elastic, crystal-clear lens.

As the lens develops, it contacts the overlying ectoderm, which has closed over it. In response, the ec-

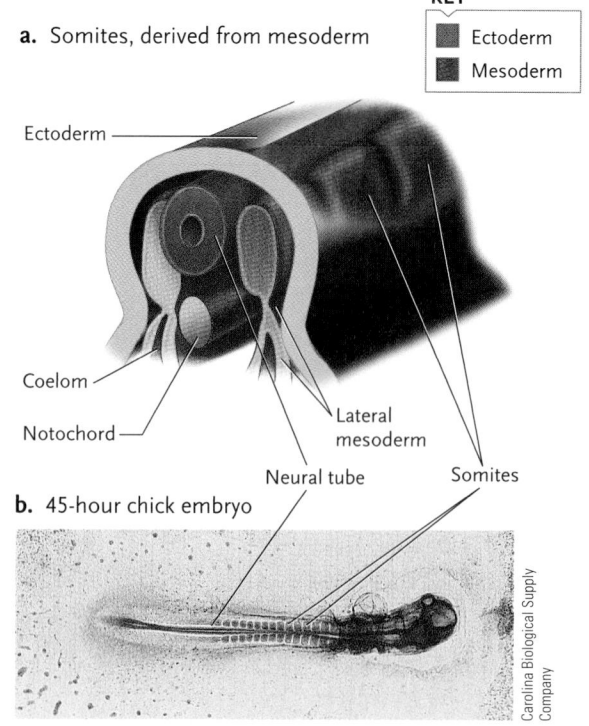

Figure 48.11

Later development of the mesoderm. **(a)** The somites develop into segmented structures such as the vertebrae, the ribs, and the musculature between the ribs. The lateral mesoderm gives rise to other structures, such as the heart and blood vessels and the linings of internal body cavities. **(b)** The somites in a 45-hour chick embryo.

toderm cells lose their pigment granules and become clear, developing into the cornea. Eventually, the developing cornea joins with the edges of the optic cup to complete the primary structures of the eye (step 4). Other cells contribute to accessory structures of the eye; for example, mesoderm and neural crest cells contribute to the reinforcing tissues in the wall of the eye and the muscles that move the eye. Figure 48.12, step 5, shows a fully developed vertebrate eye.

Many experiments have shown that the initial induction by the optic vesicle is necessary for develop-

Figure 48.12
Stages in the formation of the vertebrate eye from the optic vesicle of the brain and the overlying ectoderm.

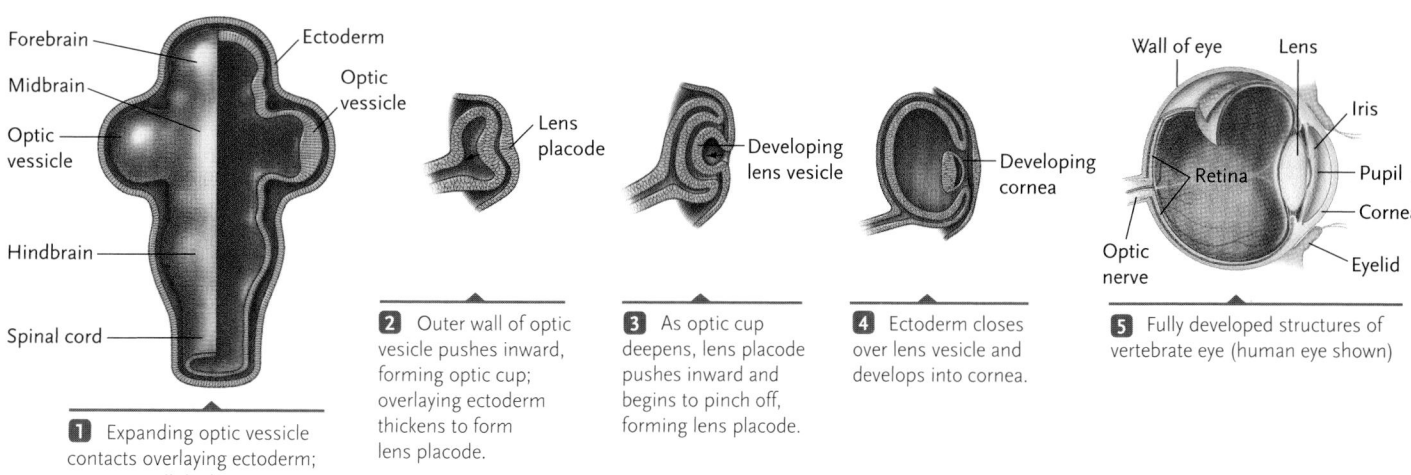

1 Expanding optic vessicle contacts overlaying ectoderm; its outer wall thickens.

2 Outer wall of optic vesicle pushes inward, forming optic cup; overlaying ectoderm thickens to form lens placode.

3 As optic cup deepens, lens placode pushes inward and begins to pinch off, forming lens placode.

4 Ectoderm closes over lens vesicle and develops into cornea.

5 Fully developed structures of vertebrate eye (human eye shown)

ment of the eye. For example, if an optic vesicle is removed before lens formation, the ectoderm fails to develop a lens placode and vesicle. Moreover, placing a removed optic vesicle under the ectoderm in other regions of the head causes a lens to form in the new location. Or, if the ectoderm over an optic vesicle is removed and ectoderm from elsewhere in the embryo is grafted in its place, a normal lens will develop in the grafted ectoderm, even though in its former location it would not differentiate into lens tissue.

Eye development also demonstrates differentiation. Ectoderm cells that are induced to form the lens of the eye synthesize crystallin; in other locations, ectoderm cells typically synthesize a different protein, *keratin,* as their predominant cell product. Keratin is a component of surface structures such as skin, hair, feathers, scales, and horns. In other words, as a response to induction by the optic vesicle, the genes of the ectoderm cells coding for crystallin are activated, while genes coding for keratin are not expressed.

Apoptosis Eliminates Tissues That Are No Longer Required

Induction and differentiation build complex, specialized organs from the three fundamental tissue types. Complementing these processes is *apoptosis,* programmed cell death, which in this case removes tissues present during development but not in the fully formed organ. Apoptosis plays an important role in the

development of animals, both invertebrates and vertebrates. The best example of apoptosis in frog development occurs during metamorphosis, in which the tadpole changes into an adult frog. The tail of the tadpole becomes progressively smaller and finally disappears because its cells disintegrate and their components are absorbed and recycled by other cells. Cells that are eliminated by apoptosis, like those of a tadpole's tail, are typically parts of structures required at one stage of development but not for later stages.

In the next section, we describe cleavage, gastrulation, and organogenesis in human and other mammalian embryos.

STUDY BREAK

1. What is the outcome of organogenesis?
2. What tissues or organs develop from the neural tube and neural crest cells?

48.4 Embryonic Development of Humans and Other Mammals

Human embryonic development is representative of the placental mammals, in which the embryo develops in the uterus of the mother. In the uterus, the embryo is nourished by the placenta, which supplies oxygen and nutrients and carries away carbon dioxide and nitrogenous wastes.

The period of mammalian development that is called **pregnancy** or **gestation** varies in different species. Larger mammals generally have longer gestation periods; for example, gestation takes 600 days in elephants, about 1 year in blue whales, and a mere 21 days in hamsters.

In humans, gestation takes an average of 266 days from the time of fertilization, or about 38 weeks. Because the date of fertilization may be difficult to establish, human gestation is usually calculated from the beginning of the menstrual cycle in which fertilization takes place, giving a period of about 9 months. On this basis, human gestation is divided into three **trimesters,** each 3 months long.

The major developmental events in human gestation—cleavage, gastrulation, and organogenesis—take place during the first trimester. By the fourth week, the embryo's heart is beating, and by the end of the eighth week, the major organs and organ systems have formed. From this point until birth, the developing human is called a **fetus.** Only 5 cm long by the end of the first trimester, the fetus grows during the second and third trimesters to an average length of 50 cm and an average weight of 3.5 kg (or about 19.7 inches and 7.7 pounds). The period of gestation ends with birth.

Figure 48.13
Early stages in the development of the human embryo.

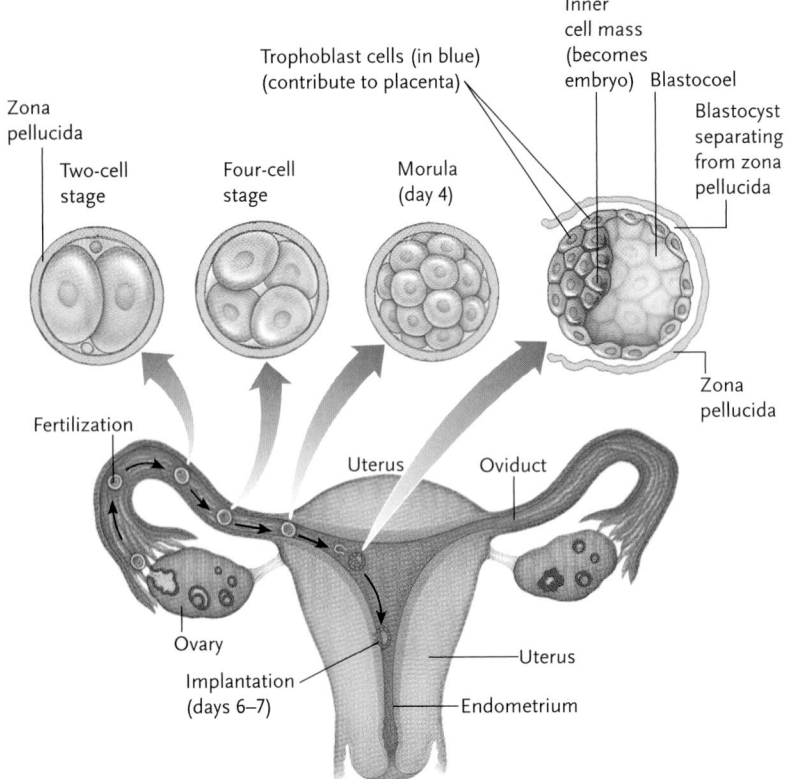

Cleavage and Implantation Occupy the First 2 Weeks of Development

We noted in Section 47.3 that human fertilization occurs when the egg is in the first third of the oviduct leading from the ovary to the uterus. After fertilization, cleavage divisions take place during passage of the developing embryo down the fallopian tube and while it is still enclosed in the zona pellucida—the original coat of the egg **(Figure 48.13).**

By day 4, the morula, a 16- or 32-cell ball, has been produced. By the time the endometrium (uterine lining) is ready for implantation (about 7 days after ovulation; see Section 47.3), the morula has reached the uterus and has undergone further cell divisions and differentiation into a blastocyst. At this time, the **blastocyst** is a single-cell-layered hollow ball of about 120 cells with a fluid-filled cavity, the blastocoel, in which a dense mass of cells is localized to one side. This **inner cell mass** will become the embryo itself, while the rest of the blastocyst will become tissues that support the development of the embryo in the uterus. The outer single layer of cells of the blastocyst is the **trophoblast.**

When it is ready to implant, the blastocyst breaks out of the zona pellucida and sticks to the endometrium on its inner cell mass side **(Figure 48.14a).** Implantation begins when the trophoblast cells overlying the inner cell mass secrete proteases that digest pathways between endometrial cells. Dividing trophoblast cells fill in the digested spaces, appearing like finger-like projections into the endometrium. These cells continue to digest the nutrient-rich endometrial cells, serving both to produce a hole in the endometrium for the blastocyst and to release nutrients that the developing embryo can use after the small amount of yolk contained in the egg cytoplasm is used up. While the blastocyst is burrowing into the endometrium, the inner cell mass separates into the *embryonic disc,* which consists of two distinct cell layers (see Figure 48.14a). The layer farther from the blastocoel is the epiblast, which gives rise to the embryo proper, and the layer nearer the blastocoel is the hypoblast, which gives rise to part of the extraembryonic membranes. When implantation is complete, the blastocyst has completely burrowed into the endometrium and is covered by a layer of endometrial cells **(Figure 48.14b).**

Mammalian Gastrulation and Neurulation Resemble the Reptilian–Bird Pattern

Gastrulation proceeds as in birds (see Figure 48.8), with the formation of a primitive streak in the epiblast. Some epiblast cells remain in place, becoming the ectoderm, while others enter the streak to form the endoderm and mesoderm. The ectoderm, mesoderm, and endoderm are located initially in three layers; from this initial arrangement, the endoderm folds to form the primitive gut, and becomes surrounded with ectoderm and mesoderm. Neurulation in human and other mammalian embryos takes place essentially as in birds (see Figure 48.10).

Extraembryonic Membranes Give Rise to the Amnion and Part of the Placenta

Soon after the inner cell mass separates into the epiblast and hypoblast, a layer of cells separates from the epiblast along its top margin (see Figure 48.14b). The fluid-filled space created by the separation becomes the amniotic cavity, and the layer of ectodermal cells forming its roof becomes the amnion, the extraembryonic membrane surrounding the cavity. The amnion expands until eventually it completely surrounds the embryo and suspends it in amniotic fluid. As in birds, the hypoblast develops into the yolk sac. However, in mammals, the mesoderm of the yolk sac gives rise to the blood vessels in the embryonic portion of the placenta.

While the amnion is expanding around the embryonic disc, blood-filled spaces form in maternal tissue, and trophoblast cells grow rapidly around both the embryo and amnion to form the chorion **(Figure 48.14c).** Next, a connecting stalk forms between the embryonic disc and the chorion, while the chorion begins to grow into the endometrium as fingerlike or treelike extensions called **chorionic villi (Figure 48.14d).** The chorionic villi greatly increase the surface area of the chorion. Where these villi grow into the endometrium is the area of the future placenta. As the chorion develops, mesodermal cells of the yolk sac grow into it and form a rich network of blood vessels, the embryonic circulation of the placenta. At the same time, the expanding chorion stimulates the blood vessels of the endometrium to grow into the maternal circulation of the placenta **(Figure 48.14e).**

Within the placenta of humans, apes, monkeys, and rodents, the maternal circulation opens into spaces in which the maternal blood directly bathes the capillaries coming to the placenta from the embryo **(Figure 48.14f).** (Different types of placentas are found in other mammals.) The embryonic circulation remains closed, however, so that the embryonic and maternal blood do not mix directly. This prevents the mother from developing an immune reaction against cells of the embryo, which may be recognized as foreign by the mother's immune system. Eventually, the placenta and its blood circulation grow to cover about a quarter of the inner surface of the enlarged uterus and reach the size of a dinner plate.

As the embryonic blood circulation develops, this connecting stalk between the embryo and placenta develops into the **umbilical cord,** a long tissue with blood vessels linking the embryo and the placenta. The vessels in the umbilical cord are derived from the extraembryonic membrane, the allantois. They conduct

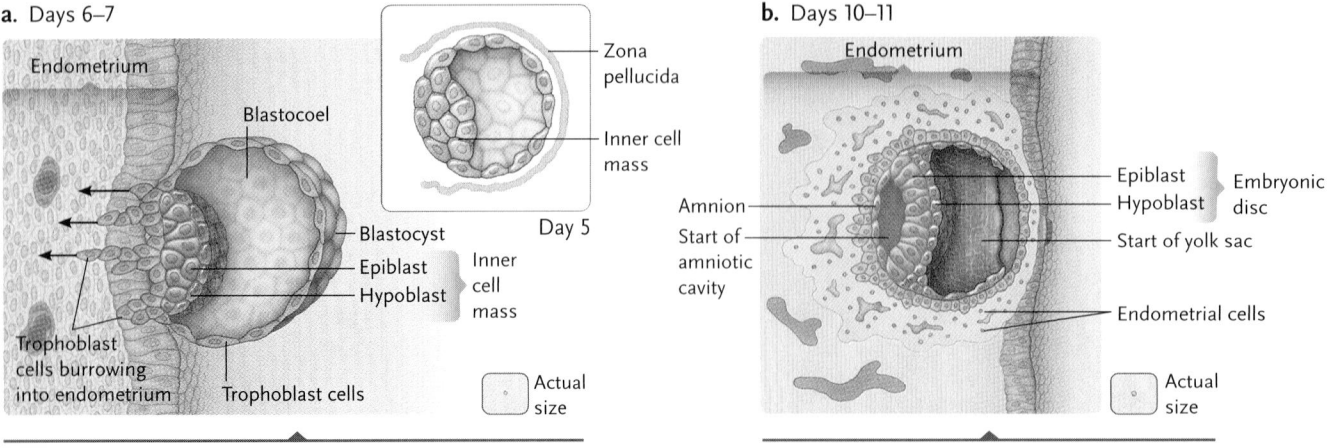

a. Days 6–7

Endometrium

Blastocoel

Blastocyst

Epiblast
Hypoblast
} Inner cell mass

Trophoblast cells burrowing into endometrium

Trophoblast cells

Actual size

Zona pellucida

Inner cell mass

Day 5

Surface cells of the blastocyst attach to the endometrium and start to burrow into it. Implantation is under way.

b. Days 10–11

Endometrium

Amnion

Start of amniotic cavity

Epiblast
Hypoblast
} Embryonic disc

Start of yolk sac

Endometrial cells

Actual size

A layer of epiblast cells separates, producing the amniotic cavity. The cells above the cavity become the amnion, which eventually surrounds the embryo. The hypoblast begins to form around the yolk sac.

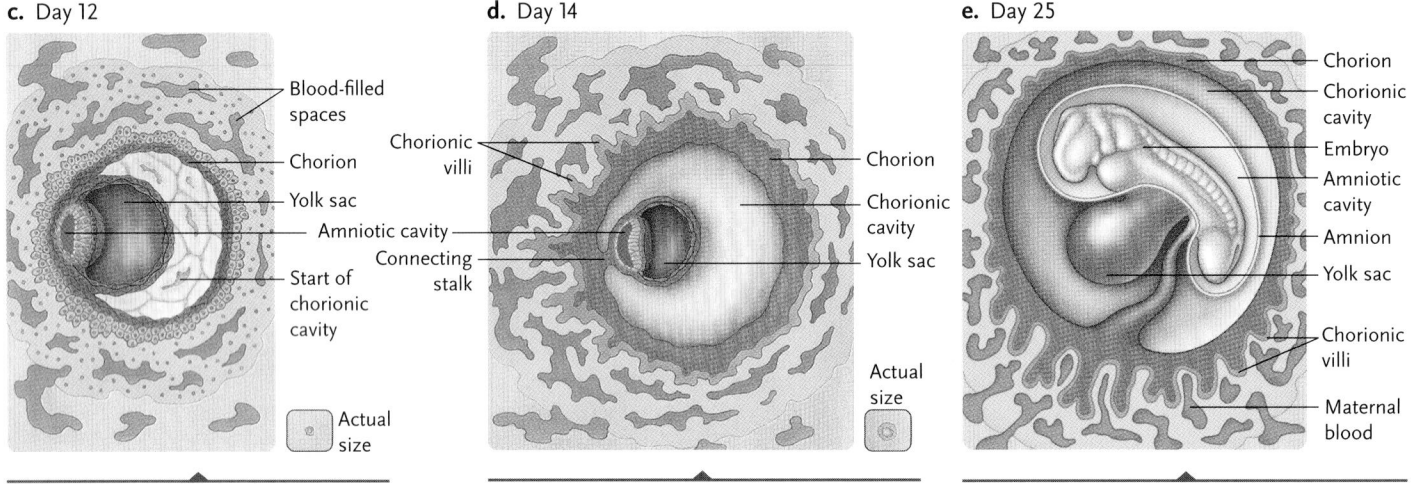

c. Day 12

Blood-filled spaces

Chorion

Yolk sac

Amniotic cavity

Start of chorionic cavity

Actual size

Blood-filled spaces form in maternal tissue. The chorion forms, derived from trophoblast cells, and encloses the chorionic cavity.

d. Day 14

Chorionic villi

Chorion

Chorionic cavity

Yolk sac

Connecting stalk

Actual size

A connecting stalk has formed between the embryonic disk and chorion. Chorionic villi, which will be features of a placenta, start to form.

e. Day 25

Chorion

Chorionic cavity

Embryo

Amniotic cavity

Amnion

Yolk sac

Chorionic villi

Maternal blood

The chorion continues to grow into the endometrium, producing the chorionic villi. The chorion growth stimulates blood vessels of the endometrium to grow into the maternal circulation of the placenta.

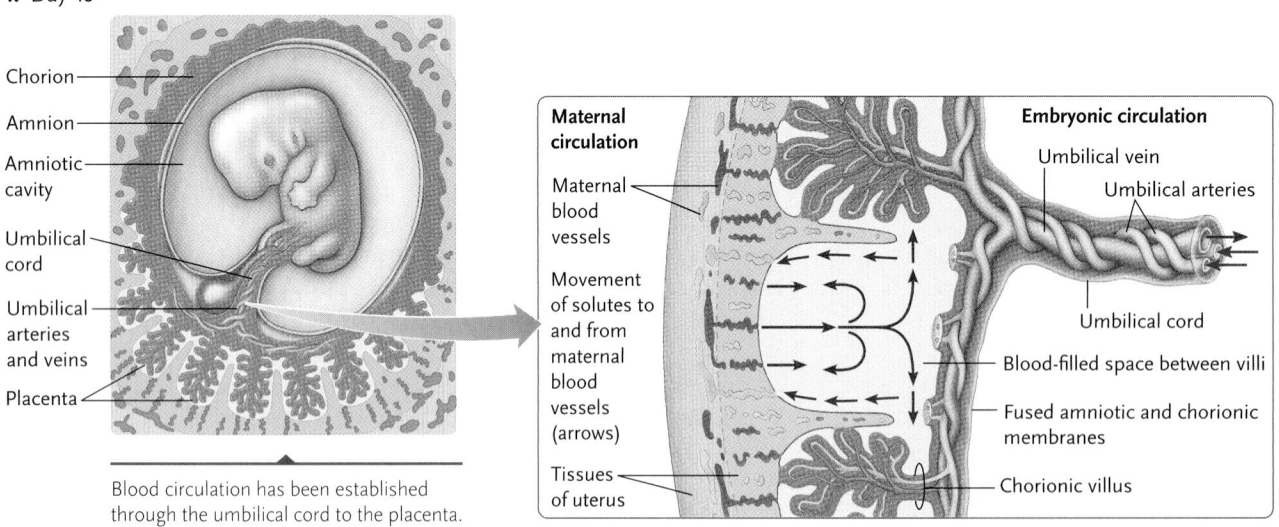

f. Day 45

Chorion

Amnion

Amniotic cavity

Umbilical cord

Umbilical arteries and veins

Placenta

Blood circulation has been established through the umbilical cord to the placenta.

Maternal circulation

Maternal blood vessels

Movement of solutes to and from maternal blood vessels (arrows)

Tissues of uterus

Embryonic circulation

Umbilical vein

Umbilical arteries

Umbilical cord

Blood-filled space between villi

Fused amniotic and chorionic membranes

Chorionic villus

Figure 48.14

Implantation of a human blastocyst in the endometrium of the uterus and the establishment of the placenta.

blood between the embryo and the placenta (shown in the inset for Figure 48.14f).

Within the placenta, nutrients and oxygen pass from the mother's circulation into the circulation of the embryo. Besides nutrients and oxygen, many other substances taken in by the mother—including alcohol, caffeine, drugs, and toxins in cigarette smoke—can pass from mother to embryo. Carbon dioxide and nitrogenous wastes pass from the embryo to the mother, and are disposed of by the mother's lungs and kidneys.

If the presence of a genetic disease such as cystic fibrosis or Down syndrome is suspected, tests can be carried out on cells removed from the embryonic portion of the placenta or from the amniotic fluid, which contains cells derived from the embryo. The test using cells of the placenta is called *chorionic villus sampling;* the test using cells derived from the amniotic fluid is called *amniocentesis* (*centesis* = puncture, referring to the use of a needle, which is pushed through the abdominal wall, to obtain fluid from the amniotic cavity). Chorionic villus sampling can be carried out as early as the eighth week, compared with 14 weeks for amniocentesis. Both tests carry some degree of risk to the embryo.

Further Growth of the Fetus Culminates in Birth

By the end of its fourth week, a human embryo is 3–5 mm long, 250–500 times the size of the zygote **(Figure 48.15a).** It has a tail and pharyngeal arches, which are embryonic features of all vertebrates (see Section 30.2). The pharyngeal arches contribute to the formation of the face, neck, mouth, nasal cavities, larynx, and pharynx. After 5 to 6 weeks, most of the tail has disappeared and the embryo is beginning to take on recognizable human form **(Figure 48.15b).** At 8 weeks, the embryo, now a fetus, is about 2.5 cm long **(Figure 48.15c).** Its organ systems have formed, and its limbs, with fingers or toes at their ends, have developed **(Figure 48.15d).**

Figure 48.16 shows the hormonal events and associated physical events of birth. As the period of fetal growth comes to a close, the fetus typically turns so that its head is downward, pressed against the cervix. A steep rise in the levels of estrogen secreted by the placenta at this time causes cells of the uterus to express the gene for the receptor of the hormone *oxytocin.* The receptors become inserted into the plasma membranes of those cells. Oxytocin—which is secreted by the pituitary gland—binds to its receptor, triggering the smooth muscle cells of the uterine wall to contract and begin the rhythmic contractions of labor. These contractions mark the beginning of **parturition** (*parturire* = to be in labor), the process of giving birth.

The contractions push the fetus further against the cervix and stretch its walls (see Figure 48.16, step 1). In response, stretch receptors in the walls send nerve signals to the hypothalamus, which responds by stimulating the pituitary to secrete more oxytocin. In turn, the oxytocin stimulates more forceful contractions of the uterus, pressing the fetus more strongly against the cervix, and further stretching its walls. The positive feedback cycle continues, steadily increasing the strength of the uterine contractions.

As the contractions force the head of the fetus through the cervix (step 2), the amniotic membrane bursts, releasing the amniotic fluid. Usually, within 12 to 15 hours after the onset of uterine contractions, the head passes entirely through the cervix. Once the head is through, the rest of the body follows quickly and the entire fetus is forced through the vagina to the exterior, still connected to the placenta by the umbilical cord (step 3).

After the baby takes its first breath the umbilical cord is cut and tied off by the birth attendant. Contractions of the uterus continue, expelling the placenta and any remnants of the umbilical cord and embryonic membranes as the afterbirth, usually within 15 minutes to an hour after the infant's birth. The short length of umbilical cord still attached to the infant dries and shrivels within a few days. Eventually, it separates entirely and leaves a scar, the **umbilicus** or navel, to mark its former site of attachment during embryonic development.

The Mother's Mammary Glands Become Active after Birth

Before birth of the fetus, estrogen and progesterone secreted by the placenta stimulate the growth of the mammary glands in the mother's breasts. However, the high levels of these hormones prevent the mammary glands from responding to *prolactin,* the hormone secreted by the pituitary that stimulates the glands to produce milk. After birth of the fetus and release of the placenta, the levels of estrogen and progesterone fall steeply in the mother's bloodstream, and the breasts begin to produce milk (stimulated by prolactin) and secrete it (stimulated by oxytocin).

Continued milk secretion depends on whether the infant is suckled by the mother. If the infant is suckled, stimulation of the nipples sends nerve impulses to the hypothalamus, which responds by signaling the pituitary to release a burst of prolactin and oxytocin. Hormonal stimulation of milk production and secretion continues as long as the infant is breastfed.

So far, we have followed the development of a generic human, but certain aspects of development differ depending on the offspring's sex. Next we look at the specifics of male and female development.

A Gene on the Y Chromosome Determines the Development of Male or Female Sex Organs

The gonads and their ducts begin to develop during the fourth week of gestation. Until the seventh week, male and female embryos have the same set of inter-

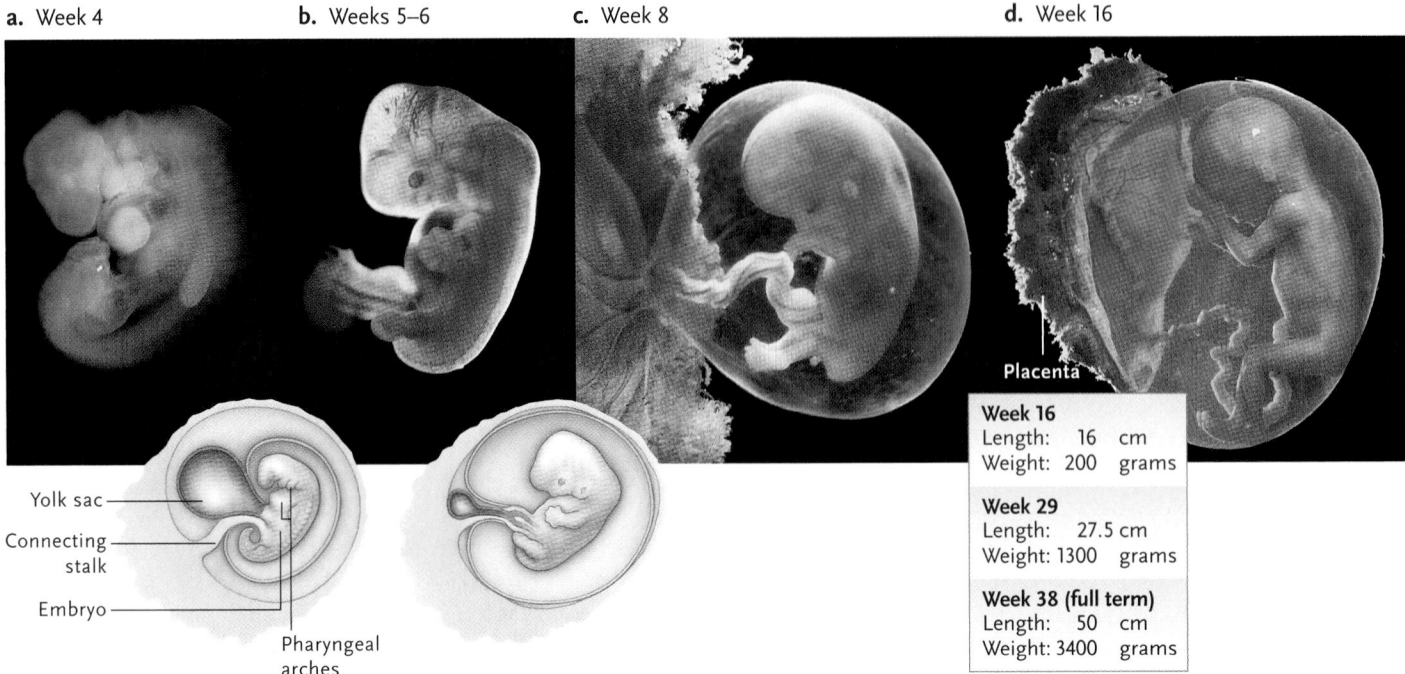

a. Week 4   **b.** Weeks 5–6   **c.** Week 8   **d.** Week 16

Yolk sac

Connecting stalk

Embryo

Pharyngeal arches

Placenta

Week 16		
Length:	16	cm
Weight:	200	grams

Week 29		
Length:	27.5	cm
Weight:	1300	grams

Week 38 (full term)		
Length:	50	cm
Weight:	3400	grams

Figure 48.15

The human embryo at various stages of development, beginning at week 4. The chorion has been pulled aside to reveal the embryo in the amnion at week 8 and week 16. By week 16, movements begin as nerves make functional connections with the forming muscles.

(Photos: Lennart Nilsson, A Child Is Born, © 1966, 1977, Dell Publishing Company, Inc.)

nal structures derived from mesoderm, including a pair of gonads **(Figure 48.17a).** Each gonad is associated with two primitive ducts, the **Wolffian duct** and the **Müllerian duct**, which lead to a cloaca. These internal structures are *bipotential:* they can develop into either male or female sexual organs.

The presence or absence of a Y chromosome determines whether the internal structures develop into male or female sexual organs. If the fetus has the XY combi-

nation of sex chromosomes, a single gene on the Y chromosome, *SRY (Sex-determining Region of the Y),* becomes active in the seventh week. The protein encoded by the gene sets a molecular switch that causes the primitive gonads to develop into testes. The fetal testes then secrete two hormones, testosterone and the *anti-Müllerian hormone (AMH).* The testosterone stimulates development of the Wolffian ducts into the male reproductive tract, including the epididymis, vas deferens,

Figure 48.16

Birth of the fetus. Hormonal events of birth are at the top, and physical events of birth are at the bottom.

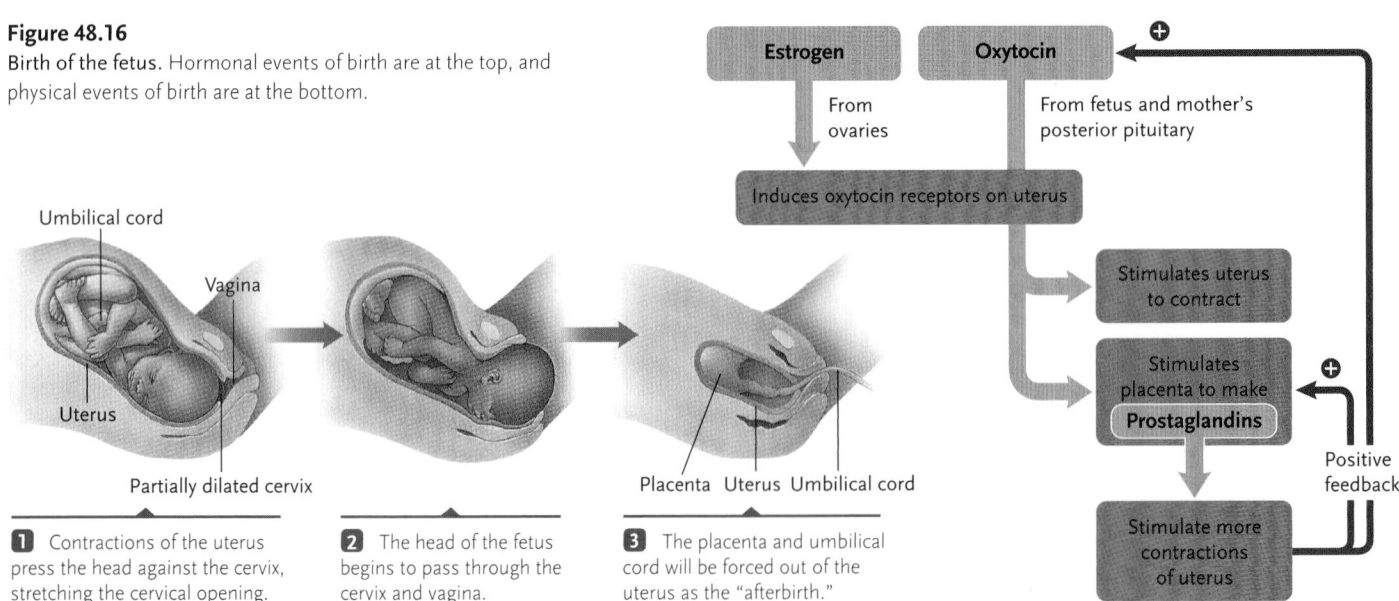

Umbilical cord

Vagina

Uterus

Partially dilated cervix

Placenta   Uterus   Umbilical cord

1 Contractions of the uterus press the head against the cervix, stretching the cervical opening.

2 The head of the fetus begins to pass through the cervix and vagina.

3 The placenta and umbilical cord will be forced out of the uterus as the "afterbirth."

Estrogen

Oxytocin

From ovaries

From fetus and mother's posterior pituitary

Induces oxytocin receptors on uterus

Stimulates uterus to contract

Stimulates placenta to make **Prostaglandins**

Stimulate more contractions of uterus

Positive feedback

and seminal vesicles **(Figure 48.17b)**. AMH causes the Müllerian ducts to degenerate and disappear. (*Insights from the Molecular Revolution* describes experiments that traced the activity of the *SRY* gene and its encoded protein in male development.) Testosterone additionally stimulates the development of the male genitalia.

If the fetus has the XX combination of chromosomes, no SRY protein is produced and the primitive gonads, under the influence of the estrogens and progesterone secreted by the placenta, develop into ovaries. The Müllerian ducts develop into the oviducts, uterus, and part of the vagina, and the Wolffian ducts degenerate and disappear **(Figure 48.17c)**. The female sex hormones additionally stimulate the development of the female external genitalia.

Development Continues after Birth

Once fetal development is over, humans and other mammals, and indeed most other animals, follow a prescribed course of further growth and development that leads to the adult, the sexually mature form of the species. In humans, the internal and external sexual organs mature and secondary sexual characteristics appear at puberty. Similar changes occur in most mammals. There are, in fact, many examples among different animal groups of developmental changes that take place after hatching or birth. In some cases, offspring hatch that are distinctly different in structure from the adult. Examples among invertebrates include insects such as *Drosophila* and butterflies, in which eggs hatch to produce larva that undergo metamorphosis into the adult. Frogs similarly hatch as tadpoles, which undergo metamorphosis to produce the adult.

We have now described embryonic development in animals from a morphological perspective. In the rest of the chapter, we will focus on the cellular and molecular mechanisms that underlie development.

STUDY BREAK

1. Distinguish between the roles of the trophoblast and inner cell mass of the blastocyst in mammalian development.
2. What hormone would you use to induce labor in a pregnant woman?

48.5 The Cellular Basis of Development

In the preceding sections, you learned about the processes of development from a mainly structural point of view. Underlying those developmental processes are specific cellular and molecular events. In this section,

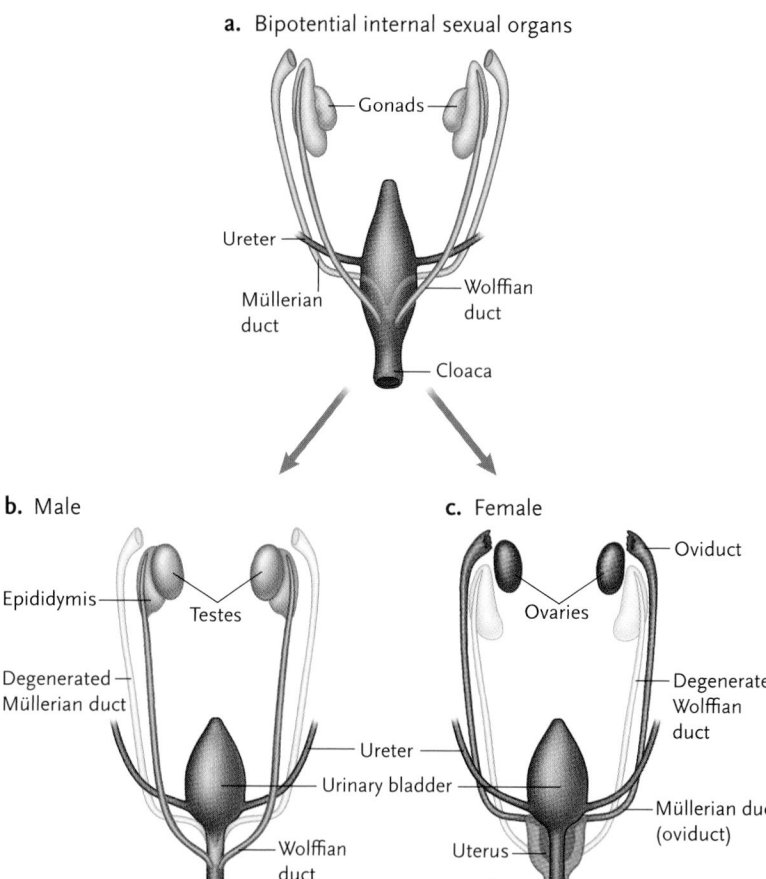

Figure 48.17
Development of the internal sexual organs of males and females from common bipotential origins.

you will learn about all the cellular events that underlie the stages of development.

Cell Division Varies in Orientation and Rate during Embryonic Development

The *orientation* and *rate* of mitotic cell division have special significance in the development of the shape, size, and location of the organ systems of the embryo. Regulation of these two features of mitotic cell division occurs at all stages of development.

The orientation of cell division refers to the angles at which daughter cells are added to older cells as development proceeds. It is determined by the location of a furrow that separates the cytoplasm after mitotic division of the nucleus (furrowing is discussed in Section 10.2). The furrow forms in alignment with the spindle midpoint. Therefore, when the spindle is centrally positioned in the cell, the furrow leads to symmetric division of the cell. However, when the spindle is displaced to one end of the cell, the furrow leads to asymmetric division of the cell into a smaller and a larger cell. Little is known about how spindle positioning is regulated.

The rate of cell division primarily reflects the time spent in the G_1 period of interphase (see Section 10.2); once DNA replication begins, the rest of the cell cycle is usually of uniform length in all cells of the same species. As an embryo develops and cells differentiate, the time spent in interphase increases and varies in length in different cell types. As a result, different cell types proliferate at various rates, giving rise to tissues and organs with different cell numbers. Some cells, when fully differentiated, remain fixed in interphase and stop replicating their DNA or dividing. Nerve cells in the mammalian brain and spinal cord, for example, stop dividing once the nervous system is fully formed. Ultimately, the rate of cell division is under genetic control.

Frog egg cleavage provides examples of how both changes in orientation and rate of mitotic division affect development. The first two cleavages start at the animal pole and extend to the vegetal pole, producing four equal blastomeres (see Figure 48.2). The third cleavage occurs equatorially. However, because there is yolk in the vegetal region of the embryo, this cleavage furrow forms not at the equator but up higher toward the animal pole. The result is an eight-cell embryo with four small blastomeres in the animal region of the embryo, and four large blastomeres in the vegetal region. The blastomeres in the animal region of the embryo proceed to divide rapidly, while the blastomeres in the vegetal part of the embryo divide more slowly because division is inhibited by yolk. As a result, the morula produced consists of an animal region with many small cells, and a vegetal region with relatively few blastomeres.

Cell-Shape Changes and Cell Movements Depend on Microtubules and Microfilaments

We have seen that embryonic cells undergo changes in shape that generate movements, such as the infolding of surface layers to produce endoderm or mesoderm. Entire cells also move during the embryonic growth of animals, both singly and in groups. Both the shape changes and the whole-cell movements are produced by microtubules, powered by dyneins and kinesins, and microfilaments, powered by myosins (see Section 5.3). Movements are also produced by changes in the rate of growth or by the breakdown of microtubules and microfilaments. Generally speaking, changes in both cell shape and cell movement play important roles in cleavage, gastrulation, and organogenesis.

Changes in Cell Shape. Changes in cell shape typically result from reorganization of the cytoskeleton. For example, during the development of the neural plate in frogs, the ectoderm flattens and thickens; that is, microtubules within cells in the ectoderm layer lengthen and slide farther apart, causing the cells to change from a cubelike to a columnar shape (**Figure 48.18a**).

Once formed, the neural plate sinks downward along its midline. This change is produced by a change in cell shape from columnar to wedgelike (**Figure**

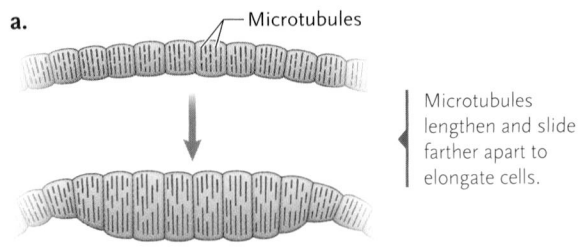

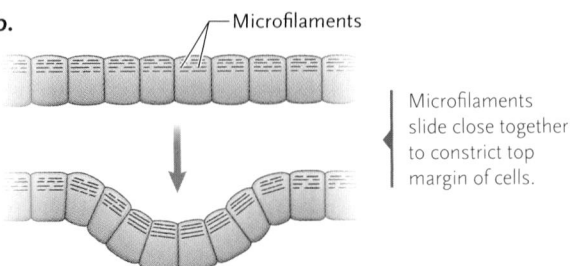

Figure 48.18
The roles of microtubules **(a)** and microfilaments **(b)** in the changes in cell shape that produce developmental movements.

48.18b). As the tops of the cells narrow, the entire cell layer is forced inward—it invaginates. How does this occur? Each wedge-shaped cell contains a group of microfilaments arranged in a circle at its top. Research suggests that the microfilaments slide over each other, tightening the ring like a drawstring and narrowing the top of the cell. This mechanism is supported by experiments in which cytochalasin, a chemical that interferes with microfilament assembly, was added to the cells. As a result, the microfilament circle was dispersed, and no invagination of the ectoderm occurred.

Whole-Cell Movements. Among the most striking examples of whole-cell movements in embryonic development are the cell movements during gastrulation and the often long-distance migrations of neural crest cells. These whole-cell movements involve the coordinated activity of microtubules and microfilaments. The typical pattern of movement is a repeating cycle of steps that resemble how an amoeba moves. First, a cell attaches to the substrate (**Figure 48.19,** step 1) and moves forward by elongating from the point of attachment (step 2). The cell next makes a new attachment at the advancing tip (step 3), and then contracts until the rearmost attachment breaks (step 4). The front attachment now serves as the base for another movement.

How do the cells know where to go? Typically, cells migrate over the surface of stationary cells in one of the embryo's layers. In many developmental systems, migrating cells follow tracks formed by molecules of the extracellular matrix (ECM), secreted by the cells along the route over which they travel. An important

Turning On Male Development

The switch to male development in mammalian embryos is triggered by the protein encoded in the *SRY* gene, carried on the Y chromosome. Individuals with a mutation in which *SRY* encodes a faulty, inactive protein develop into females, even though they have the XY combination of sex chromosomes.

Molecular studies revealed that the mutant SRY proteins have changes in single amino acids or have a missing segment. All the single amino acid changes are concentrated in a region of the SRY protein known as the *HMG box*, which can bind to DNA. This discovery suggested that SRY is a regulatory protein that binds to the control regions of genes such as *AMH*, which encodes the anti-Müllerian hormone, and turns them on.

A group of investigators led by Michael Weiss at the Harvard Medical School and Massachusetts General Hospital in Boston carried out molecular studies testing whether SRY directly turns on the *AMH* gene. For their experiments, the researchers attached the control region of the *AMH* gene to the coding portion of a lucifer-ase gene. Luciferase is the firefly enzyme that catalyzes a reaction with the substrate luciferin to produce light. In this experiment, luciferase was used as a reporter; measuring its activity indirectly informed the researchers about the molecular reactions occurring with the *AMH* gene. Luciferase activity is measured by breaking open cells to produce cell extracts, adding the substrate luciferin to samples of the extract, and quantifying the light emitted from the reaction using a special photodetector system. The composite *AMH*-luciferase gene was introduced into embryonic cells removed from the developing gonads of XY rat embryos, taken at the time when differentiation into a testis or ovary would normally begin.

A normal human *SRY* gene was then introduced into the gonad cells containing the artificial gene. High luciferase activity was seen, confirming that the normal SRY protein activates the *AMH* gene. The experiment was repeated with a mutant *SRY* gene isolated from a human patient who had developed into a female even though she had the XY combination of sex chromosomes. In her case, the mutation resulted from a change of a single amino acid in the HMG box of the SRY protein. When her *SRY* gene was added to the embryonic rat gonad cells, there was no luciferase activity, indicating that her altered SRY protein could not turn on the *AMH* gene.

Adding a normal SRY protein to the *AMH* gene in a test tube showed that the protein binds directly to the gene. Tests with DNA-digesting enzymes showed that combination with SRY protects a segment of the control region of the *AMH* gene from attack by the enzymes. This protection indicates that SRY binds in this region, as expected for a regulatory protein.

Current goals include finding the genes activated by SRY that direct development toward the male. The research promises to reveal the complete sequence of molecular events directing male development. It may also lead to treatments for developmental abnormalities produced when the sex-determining system goes awry.

---

track molecule is *fibronectin,* a fibrous, elongated protein of the ECM. Migrating cells recognize and adhere to the fibronectin; in response, internal changes in the cells trigger movement in a direction based on the alignment of the fibronectin molecules.

Some migrating cells follow concentration gradients instead of molecular tracks. The gradients are created by the diffusion of molecules (often proteins) released by cells in one part of an embryo. Cells with receptors for the diffusing molecule follow the gradient toward its source, or move away from the source.

Selective Cell Adhesions Underlie Cell Movements

Selective cell adhesion, the ability of an embryonic cell to make and break specific connections to other cells, is closely related to cell movement. As development proceeds, many cells break their initial adhesions and move, and then form new adhesions in different locations. Final cell adhesions hold the embryo in its correct shape and form. Junctions of various kinds, including tight, anchoring, and gap junctions, reinforce the final adhesions (see Section 5.5).

Figure 48.19
The cycle of attachments, stretching, and contraction by which cells move over other cells or extracellular materials in embryos.

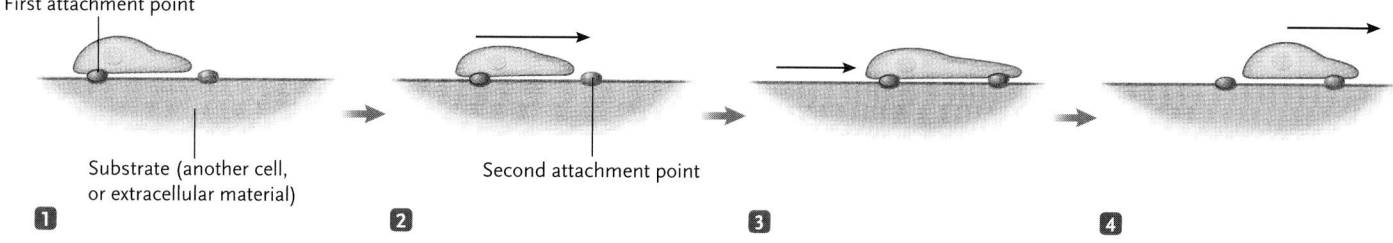

First attachment point
Substrate (another cell, or extracellular material)
Second attachment point

1   2   3   4

The selective nature of cell adhesions was first demonstrated in a classic experiment by Johannes Holtfreter of the University of Rochester and his student P. L. Townes. In this experiment, the researchers removed pieces of ectoderm, mesoderm, and endoderm from living amphibian embryos in the neurulation stage, separated them into individual cells, and added the cells in various combinations to a culture medium. Initially the cells clumped together at random into a ball. After a few hours, they sorted themselves out and moved into arrangements resembling their normal locations in the gastrula **(Figure 48.20)**.

Further research has identified many cell surface proteins responsible for selective cell adhesions, including **cell adhesion molecules** (CAMs; see Section 5.5) and **cadherins** *(calcium-dependent adhesion molecules)*. The cadherins are so named because they require calcium ions to set up adhesions. As cells develop, different types of CAMs or cadherins may appear or disappear from their surfaces as they make and break cell adhesions. The changes reflect alterations in gene activity, often in response to molecular signals arriving from other cells. For example, in the neural plate stage of neurulation in the frog, N-cadherin is on neural plate cells, keeping those cells together, while E-cadherin is on the adjacent ectodermal cells, keeping those cells together. The neural tube is produced when the neural plate cells separate from the ectodermal cells, while both cell types retain their respective cadherin type. The neural crest cells have neither cadherin bound to them, so they do not bind to each other and they disperse (as described earlier). However, if N-cadherin is expressed in the ectodermal cells through experimental manipulation, the forming neural tube does not separate from the flanking ectodermal cells because all of the cells are held together by N-cadherin.

Induction Depends on Molecular Signals Made by Inducing Cells

Recall from Section 48.1 that induction is the process in which a group of cells (the inducer cells) causes or influences a nearby group of cells (the responder cells) to follow a particular developmental pathway. Recall also that induction is the major process responsible for determination, in which the developmental fate of a cell is set. Many experiments have shown that induction occurs through the interaction of signal molecules with surface receptors on the responding cells. The signal molecules may be located on the surface of the inducing cells, or they may be released by the inducing cells. The surface receptors are activated by binding the signal molecules; in the activated form, they trigger internal response pathways that produce the developmental changes (surface receptors and their associated signal transduction pathways are discussed in Sections 7.3 and 7.4). Often, the responses include changes in gene activity.

A German scientist, Hans Spemann of the University of Freiburg, carried out the first experiments identifying induction in embryos in the 1920s. He and his doctoral student, Hilde Mangold, found that if the dorsal lip of a newt embryo was removed and grafted into a different position on another newt embryo, on the ventral side for instance, cells moving inward from the dorsal lip induced a neural plate, a neural tube, and eventually an entire embryo to form in the new location **(Figure 48.21)**. On the basis of his pioneering research, Spemann proposed that the dorsal lip is an *organizer,* acting on other cells to alter the course of development. This action is now known as *induction,* and the cells responsible for induction are known generally as inducer cells. Spemann received the Nobel Prize in 1935 for his research. (In 1924, the year their research paper was published, Mangold died in an accident when her kitchen gasoline heater exploded. She would likely have also received the Nobel Prize, but they are never awarded posthumously.)

Spemann's findings touched off a search for the inducing molecules that must pass from the inducing cells to the responding cells. It took many years to achieve success. Finally, molecular techniques led to the identification of inducing molecules. For example, in 1992, researchers constructed a DNA library from *Xenopus* gastrulas by isolating and cloning the cellular DNA in gene-size pieces. They made mRNA transcripts of the cloned genes and injected them into early *Xenopus* embryos in which the inducing ability of the mesoderm had been destroyed by exposure to ultraviolet light. Some of the injected mRNAs, translated into proteins in the embryos, were able to induce formation of a neural plate and tube and lead to a normal embryo. More than 10 other proteins acting as inducing molecules have been identified in the *Xenopus* system.

Differentiation Produces Specialized Cells without Loss of Genes

Differentiation is the process by which cells that have committed to a particular developmental fate by the determination process (see Section 48.1) now develop into specialized cell types with distinct structures and functions. As part of differentiation, cells concentrate on the production of molecules characteristic of the specific types. For example, 80% to 90% of the total protein that lens cells synthesize is crystallin.

Research into differentiation confirmed that as cells specialize, they retain all the genes of the original egg cell; except in rare instances, differentiation does not occur through selective gene loss. Several definitive experiments supporting this conclusion were carried out several decades ago by Robert Briggs and Thomas King of Lankenau Hospital Research Institute in Philadelphia (now Fox Chase Cancer Center),

Figure 48.20 Experimental Research

Demonstrating the Selective Adhesion Properties of Cells

QUESTION: Do cells make specific connections to other cells?

EXPERIMENT: Johannes Holtfreter and P. L. Townes demonstrated that cells make specific connections to other cells, that is, that cells have selective adhesion properties.

1. Holtfreter and Townes separated ectoderm, mesoderm, and endoderm tissue from amphibian embryos soon after the neural tube had formed. They used embryos from amphibian species that had cells of different colors and sizes, so they could follow under the microscope where each cell type ended up. (The colors shown here are for illustrative purposes only.)

2. The researchers placed the tissues individually in alkaline solutions, which caused the tissues to break down into single cells.

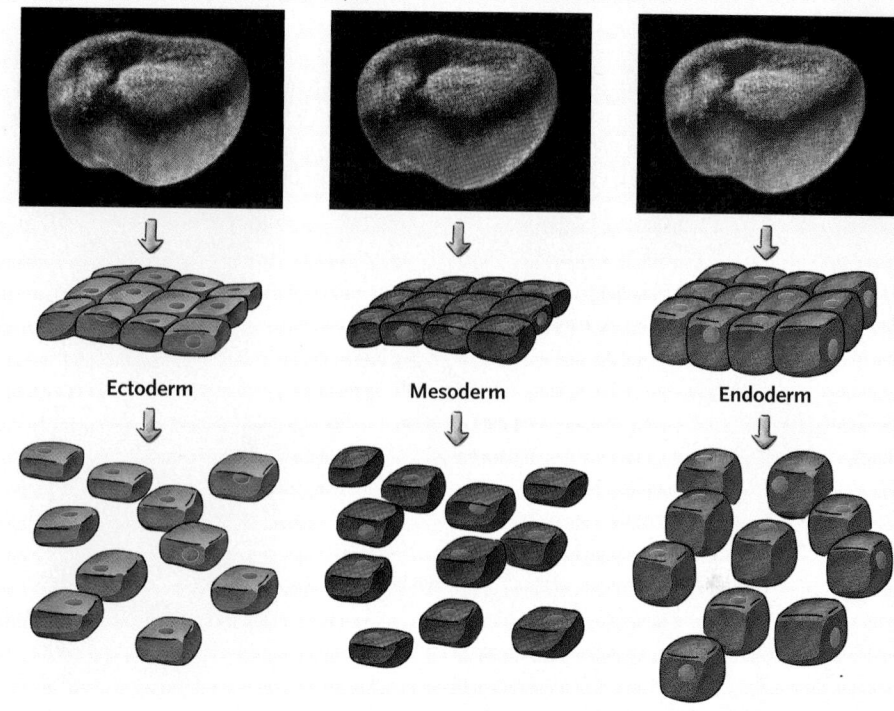

Amphibian embryos of different species

Ectoderm · Mesoderm · Endoderm

3. Holtfreter and Townes then combined suspensions of single cells in various ways. Shown here are ectoderm + mesoderm, and ectoderm + mesoderm + endoderm. When the pH was returned to neutrality, the cells formed aggregates. Through a microscope, the researchers followed what happened to the aggregates on agar-filled petri dishes.

RESULTS: In time the reaggregated cells sorted themselves with respect to cell type; that is, instead of the cell types remaining mixed, each cell type became separated spatially. That is, in the ectoderm + mesoderm mixture, the ectoderm moved to the periphery of the aggregate, surrounding mesoderm cells in the center. In no case did the two cell types remain randomly mixed. The ectoderm + mesoderm + endoderm aggregate showed further that cell sorting in the aggregates generated cell positions reflecting the positions of the cell types in the embryo. That is, the endoderm cells separated from the ectoderm and mesoderm cells and became surrounded by them. In the end, the ectoderm cells were located on the periphery, the endoderm cells were internal, and the mesoderm cells were between the other two cell types.

KEY: Ectoderm · Mesoderm · Endoderm

Ectoderm + Mesoderm

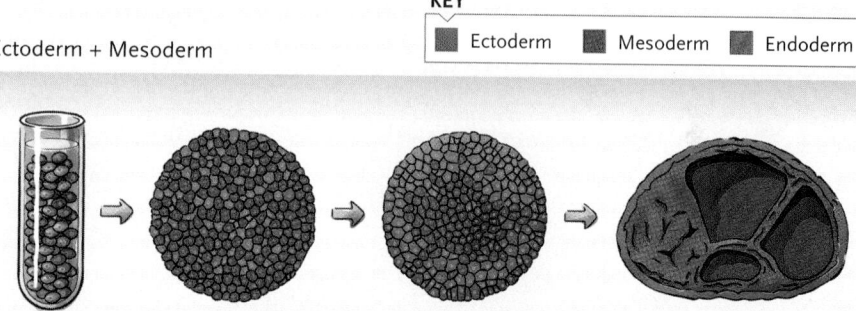

Ectoderm + Mesoderm + Endoderm

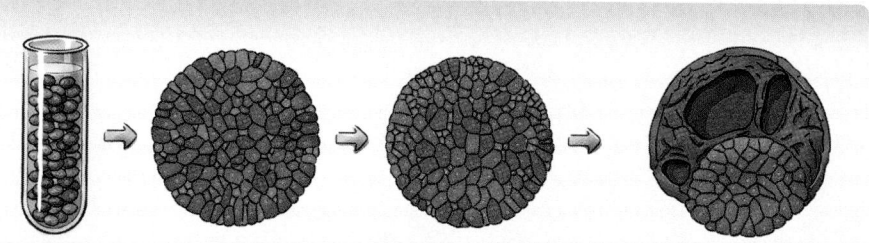

CONCLUSION: Holtfreter interpreted the results to mean that cells have selective affinity for each other; that is, cells have selective adhesion properties. Specifically, he proposed that ectoderm cells have positive affinity for mesoderm cells but negative affinity for endoderm cells, while mesoderm cells have positive affinity for both ectoderm cells and endoderm cells. In modern terms, these properties result from cell surface molecules that give cells specific adhesion properties.

Figure 48.21   Experimental Research

Spemann and Mangold's Experiment Demonstrating Induction in Embryos

QUESTION: Does induction occur in embryonic development?

EXPERIMENT: Hans Spemann and Hilde Mangold performed transplantation experiments with newt embryos, the results of which demonstrated that specific induction of development occurs in the embryos. The researchers removed the dorsal lip of the blastopore from one newt embryo and grafted it onto a different position—the ventral side—of another embryo. The two embryos were from different newt species that differed in pigmentation, allowing them to follow the fate of the tissue easily. The embryo with the transplant was allowed to develop.

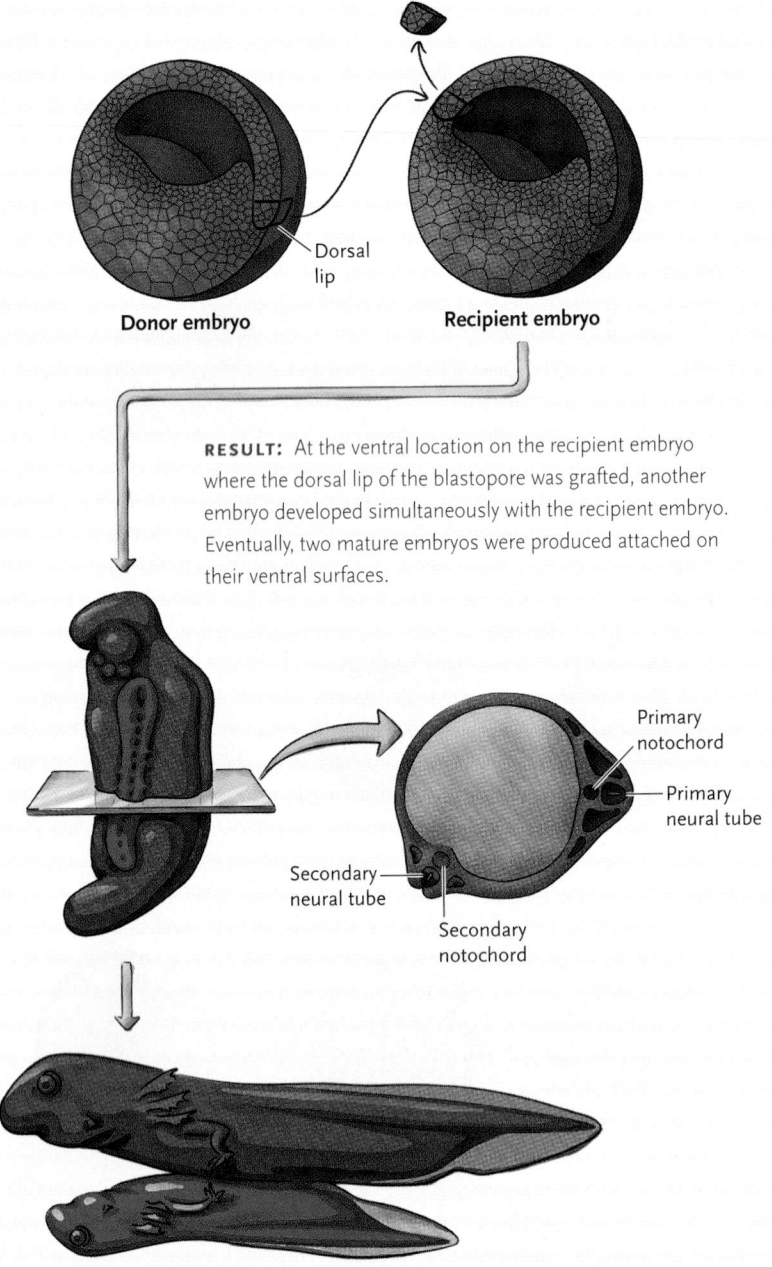

Donor embryo   Recipient embryo

RESULT: At the ventral location on the recipient embryo where the dorsal lip of the blastopore was grafted, another embryo developed simultaneously with the recipient embryo. Eventually, two mature embryos were produced attached on their ventral surfaces.

Primary notochord

Primary neural tube

Secondary neural tube

Secondary notochord

CONCLUSION: The grafted dorsal lip of the blastopore induced a second gastrulation and subsequent development in the ventral region of the recipient embryo. The result demonstrated the ability of particular cells to induce the development of other cells.

and extended by John B. Gurdon of the University of Cambridge, United Kingdom. In a typical experiment, the nucleus of a fertilized frog egg was destroyed by ultraviolet light. A micropipette was then used to transfer a nucleus from a fully differentiated tissue, intestinal epithelium, to the enucleated egg. Some of the eggs receiving the transplanted nuclei subsequently developed into normal tadpoles and adult frogs. This outcome was possible only if the differentiated intestinal cells still retained their full complement of genes. This conclusion was extended to mammals in 1997 when Ian Wilmut and his colleagues successfully cloned a sheep—Dolly—starting with an adult cell nucleus. (This experiment is described in Section 18.2.)

Fate Mapping Maps Adult Structures onto Regions of the Embryos from Which They Developed

From the early days of studying development, embryologists have focused on describing not only how embryos form and develop, but exactly how adult tissues and organs are produced from the cells of the embryo. Thus, an important goal in embryology was to trace cell lineages from embryo to adult. For most organisms it is not possible to trace lineages at the individual cell level, primarily because of the complexity of the developmental process and the typical opacity of embryos. However, it has been possible to map adult or larval structures onto the region of the embryo from which each structure developed. This type of study is called *fate mapping,* and the result is called a **fate map.** Experimentally, fate mapping is done by following development of living embryos under the microscope, either using species in which the embryo is transparent, or by marking cells so they can be followed. Cells may be marked with vital dyes (dyes that do not kill cells), fluorescent dyes, or radioactive labels. Fate maps have been produced for a number of organisms, including the chick, *Xenopus,* and *Drosophila.*

In most cases a fate map is not detailed enough to relate how particular cells in the embryo gave rise to cells of the adult. The exception is the fate map of the nematode *Caenorhabditis elegans,* an organism that has a fixed, reproducible developmental pattern. This animal has a transparent body, and scientists have been able to map the fate—trace the **cell lineage**—of every somatic and germ-line cell as the zygote divides and the resulting embryo differentiates into the 959-cell adult hermaphrodite or the 1031-cell adult male **(Figure 48.22).** They found that all somatic cells of the adult can be traced from five somatic *founder cells* produced during early development. Knowing the cell lineages of *C. elegans* has been a valuable tool for research into the genetic and molecular control of development in this organism, for mutants affecting development have an easily visualized effect.

a. Founder cells

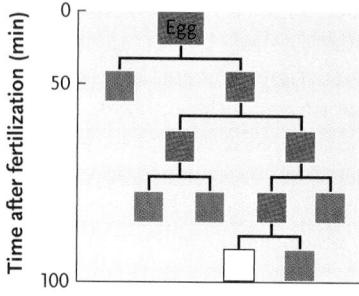

b. Cell lineage for intestinal cells

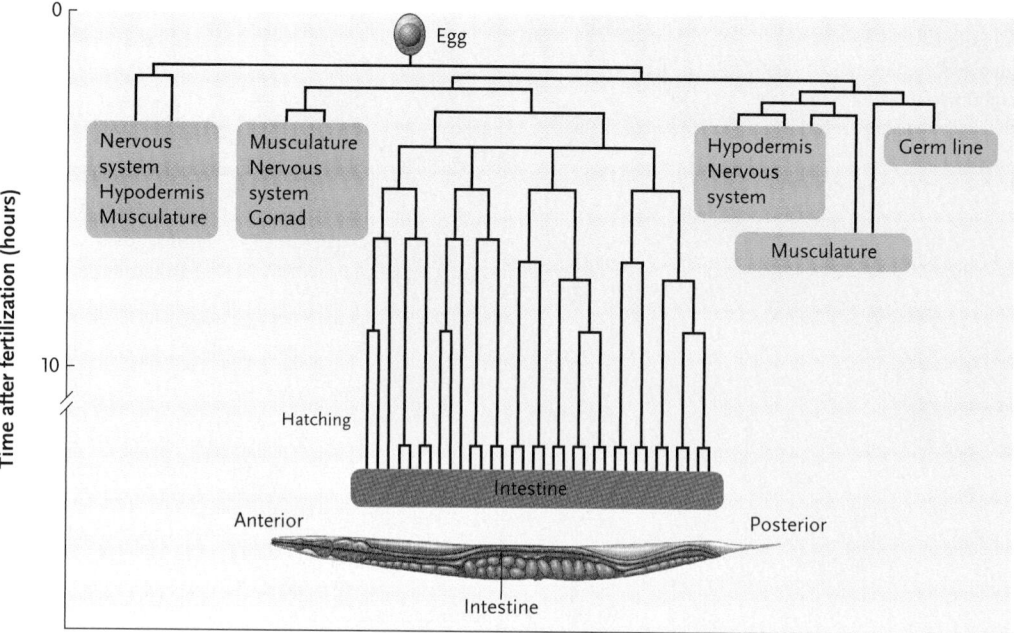

Figure 48.22

Cell lineages of *C. elegans.* **(a)** The founder cells (blue) produced in early cell divisions from which all adult somatic cells are produced. The cell in white gives rise to germ-line cells. **(b)** The cell lineage for cells that form the intestine. The detailed lineages for the other parts of the adult are not shown.

STUDY BREAK

1. What are the key cellular events that contribute to morphogenesis in animals?
2. What is induction? What molecules are involved?

48.6 The Genetic and Molecular Control of Development

We have now looked at development at the level of the whole organism, from the fertilized egg to the fully formed individual, and then at the cellular changes and movements that underlie this progression. We now turn to genetic and molecular mechanisms which, to a large extent, determine the course of development. In particular, these include the molecular mechanisms that control gene expression (see Chapter 16).

Developmental biologists are very interested in identifying and characterizing the genes involved in development, and defining how the products of the genes regulate and bring about the elaborate events we see. One productive research approach has been to isolate mutants that affect developmental processes. Researchers can then identify the genes involved, clone these genes, and analyze them in detail to build models for the molecular functions of the gene products in development. A number of model organisms are used for these studies because of the relative ease with which mutants can be made and studied and the ease of performing molecular analyses. These organisms include the fruit fly *(Drosophila melanogaster)* and *C. elegans*

among invertebrates, and the zebrafish *(Danio rerio)* and the mouse *(Mus musculus)* among vertebrates. *Focus on Research* describes why the zebrafish is a valuable model organism for genetic and molecular studies of development.

Genes Control Cell Determination and Differentiation

As you have learned, determination, the setting of the developmental fate of a cell, in many cases is the result of induction. The end result of determination is differentiation, which produces cell types of particular kinds, such as skin cells or nerve cells. Both determination and differentiation involve specific, regulated changes in gene expression.

One well-studied example of the genetic control of determination and differentiation is the production of skeletal muscle cells from somites in mammals **(Figure 48.23)**. Recall that somites are blocks of mesoderm cells that form along both sides of the notochord (see Figure 48.11). Under genetic control, particular cells of a somite differentiate into skeletal muscle cells. First, paracrine signaling from nearby cells induces those somite cells to express the master regulatory gene, *myoD*. The product of *myoD* is the transcription factor MyoD. By turning on specific muscle-determining genes, the action of MyoD brings about the determination of those cells, converting them to undifferentiated muscle cells known as **myoblasts.** Among the genes that MyoD regulates are the myogenin and MEF genes. These genes are also regulatory genes, expressing transcription factors in the myoblasts that turn on yet another set of genes. The products of those genes—which in-

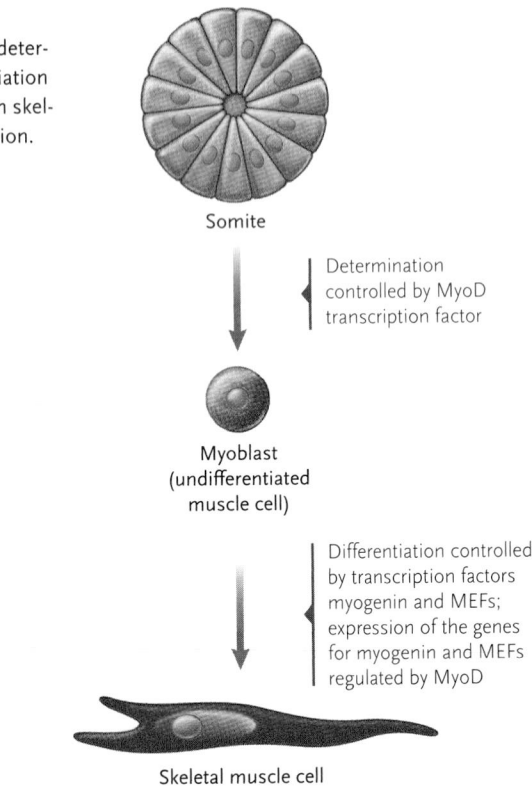

Figure 48.23
The genetic control of determination and differentiation involved in mammalian skeletal muscle cell formation.

Somite

Determination controlled by MyoD transcription factor

Myoblast (undifferentiated muscle cell)

Differentiation controlled by transcription factors myogenin and MEFs; expression of the genes for myogenin and MEFs regulated by MyoD

Skeletal muscle cell

clude myosin, a major protein involved in muscle contraction—are needed for the differentiation of myoblasts into skeletal muscle cells.

Generally speaking, the molecular mechanisms involved in determination and differentiation depend on regulatory genes that encode regulatory proteins controlling the expression of other genes. In essence, the regulatory genes act as master regulators; expression of the regulatory genes is controlled by induction in most cases.

Genes Control Pattern Formation during Development

As a part of the signals guiding differentiation, cells receive positional information that tells them where they are in the embryo. The positional information is vital to **pattern formation**: the arrangement of organs and body structures in their proper three-dimensional relationships. Positional information is laid down primarily in the form of concentration gradients of regulatory molecules produced under genetic control. In most cases, gradients of several different regulatory molecules interact to tell a cell, or a cell nucleus, where it is in the embryo. Below, we describe in brief the results of studies of the genetic control of pattern formation during the development of the fruit fly, *Drosophila melanogaster*. The developmental principles discovered from these studies apply to many other animal species, including humans.

Embryogenesis in *Drosophila*. The production of an adult fruit fly from a fertilized egg occurs in a sequence of genetically controlled development events. Following fertilization, division of the nucleus begins by mitosis, but the cytoplasm does not divide in the early embryo (cytokinesis does not occur) **(Figure 48.24).** The result is a multinucleate blastoderm. At the tenth nuclear division, the nuclei migrate to the periphery of the embryo where, after three more divisions, the 6000 or so nuclei are organized into separate cells. At this stage, the embryo is a *cellular blastoderm,* corresponding to a late blastula stage in the animals we discussed earlier. The cellular blastoderm develops into a segmented embryo (an embryo with distinct segments); at that point, 10 hours have passed since the egg was fertilized. About 24 hours after fertilization, the egg hatches into a larva, which undergoes three molts, then becoming a pupa. The pupa undergoes metamorphosis to produce the adult fly, which emerges about 10–12 days after fertilization. As illustrated by the color usage in Figure 48.24, the segments of the embryo can be mapped to the segments of the adult fly.

Genetic Analysis of *Drosophila* Development. The study of developmental mutants by a large number of researchers has given us important information about

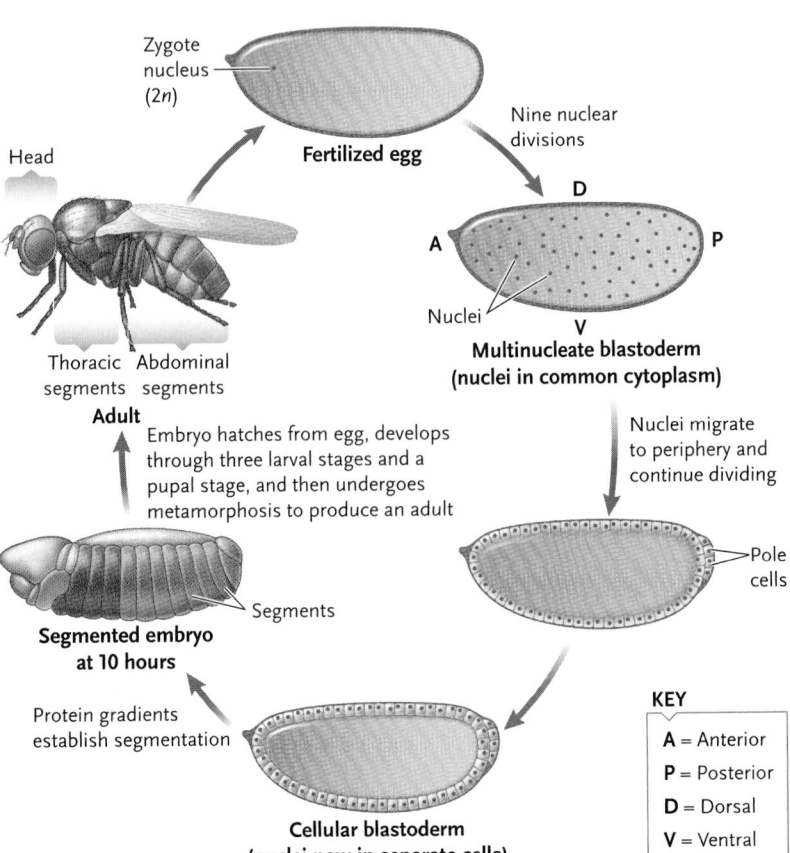

Zygote nucleus (2n)

Fertilized egg

Nine nuclear divisions

Head

D

A — P

Nuclei

V

Multinucleate blastoderm (nuclei in common cytoplasm)

Thoracic Abdominal segments segments

Adult

Embryo hatches from egg, develops through three larval stages and a pupal stage, and then undergoes metamorphosis to produce an adult

Nuclei migrate to periphery and continue dividing

Segments

Pole cells

Segmented embryo at 10 hours

Protein gradients establish segmentation

Cellular blastoderm (nuclei now in separate cells)

KEY

A = Anterior
P = Posterior
D = Dorsal
V = Ventral

Figure 48.24
Embryogenesis in *Drosophila* and the relationship between segments of the embryo and segments of the adult.

Model Research Organisms: The Zebrafish Makes a Big Splash as the Vertebrate Fruit Fly

David M. Parichy

The zebrafish (*Danio rerio*) is a small (3 cm) freshwater fish that gets its name from the black and white stripes running along its body. Native to India, it has spread around the world as a favorite aquarium fish. Beginning about 30 years ago, it began also to be used in scientific laboratories as a model vertebrate organism for studying the roles of genes in development. Its use is now so widespread that it has been dubbed the "vertebrate fruit fly."

The zebrafish brings many advantages as a model research organism. It can be maintained easily in an ordinary aquarium on a simple diet. Although its generation time is relatively long (3 months for the zebrafish as compared with $1\frac{1}{2}$ months for the mouse), a female zebrafish produces about 200 offspring at a time, as compared with an average of 10 for the mouse.

Embryonic development of the zebrafish takes place in eggs released to the outside by the female. The embryos develop rapidly, taking only 3 days from egg laying to hatching. Best of all, the eggs and embryos are transparent, providing an open window that allows researchers to observe developmental stages directly, with little or no disturbance to the embryo. Observational conditions are so favorable that the origin and fate of each cell can be traced from the fertilized egg to the hatchling. Individual nerve cells can be traced, for example, as

they grow and make connections in the brain, spinal cord, and peripheral body regions. Removing or transplanting cells and tissues is also relatively easy. Biochemical and molecular studies can be carried out by techniques ranging from the simple addition of reactants to the water surrounding the embryos to injection of chemicals into individual cells.

The zebrafish has some advantages for developmental studies compared with other vertebrate organisms used as developmental models, including the amphibian *Xenopus*, and the mouse. The early developmental stages of a zebrafish are remarkably like those of mammals, and adult structures such as the eye and skeletal system are typically vertebrate. *Xenopus* takes years to become developmentally mature and produce offspring, and it is not readily amenable to genetic analysis. Although the mouse is a mammal with development and anatomy closely related to those of humans, mouse embryos develop inside the mother and can be observed only by removing them from the mother's body; outside the body, they can be maintained only by demanding and elegant experimental techniques. Chemical studies while the embryo is inside the mother are difficult to perform. Additionally, maintaining colonies of mice is expensive.

The advantages of working with the zebrafish have spurred efforts to investigate its genetics, with particular interest in genes that regulate embryonic development. This work has already identified mutants of more than

2000 genes, including more than 400 genes that influence development. Most of the mechanisms controlled by the developmental genes resemble their counterparts in humans and other mammals.

For some of the zebrafish genes, developmental and physiological studies have revealed functions that were previously unknown for their mammalian equivalents. For example, Nancy Hopkins, a developmental geneticist at MIT, found a gene necessary for normal liver and gut development in the zebrafish. The gene is 80% identical in nucleotide sequence to a human gene; identification of the gene's role in zebrafish gave the first clues to its function in mammals. Other zebrafish mutants have been identified that affect development of the brain and spinal cord, the eyes and ears, the skeletal and digestive systems, and the circulatory system, including the heart, blood vessels, and blood cells. The mutants open these systems to biochemical and molecular study and experimentation.

Genetic studies with the zebrafish have been reinforced by a project to obtain the DNA sequence of its entire genome. The sequencing project, which was completed in 2005, will allow all the zebrafish genes to be located, identified, and correlated with their equivalents in humans and other model research organisms, including the mouse, *C. elegans*, and *Drosophila*. Undoubtedly, the zebrafish will continue to be a valuable model in developmental biology research.

Drosophila development. Three researchers performed key, pioneering research with developmental mutants: Edward B. Lewis of the California Institute of Technology, Christiane Nüsslein-Volhard of the Max Planck Institute for Developmental Biology in Tübingen, Germany, and Eric Wieschaus of Princeton University. The three shared a Nobel Prize in 1995 "for their discoveries concerning the genetic control of early embryonic development."

Nüsslein-Volhard and Wieschaus studied early embryogenesis. They searched for *every* gene required for early pattern formation in the embryo. They did this by looking for recessive *embryonic lethal* mutations. These mutations, when homozygous, result in the death of the embryo during development. By examining at what stage of development an embryo died, and how development was disrupted, they gained insights into the role of the particular genes in embryogenesis.

Lewis studied mutants that changed the fates of cells in particular regions in the embryo, producing structures in the adult that normally were produced by other regions. His work was the foundation of research identifying master regulatory genes that control the development of body regions in a wide range of organisms.

Maternal-Effect Genes and Segmentation Genes for Establishing the Body Plan in the Embryo. A number of genes control the establishment of the embryo's body plan. These genes regulate the expression of other genes. There are two classes: *maternal-effect genes*, and *segmentation genes* that work sequentially **(Figure 48.25)**.

Many **maternal-effect genes** are expressed by the mother during oogenesis. These genes control the polarity of the egg and, therefore, of the embryo. Some control the formation of the anterior structures of the embryo, others control the formation of the posterior structures, and yet others control the formation of the terminal end.

The *bicoid* gene is the key maternal-effect gene responsible for head and thorax development. The *bicoid* gene is transcribed in the mother during oogenesis, and the resulting mRNAs are deposited in the egg, localizing near the anterior pole **(Figure 48.26)**. After the egg is fertilized, translation of the mRNAs produces BICOID protein, which diffuses through the egg to form a gradient with its highest concentration at the anterior end of the egg, and fading to none at the posterior end of the egg. The BICOID protein is a transcription factor that activates some genes and represses others along the anterior–posterior axis of the embryo. Embryos with mutations in the *bicoid* gene have no thoracic structures, but have posterior structures at each end. Researchers concluded, therefore, that the *bicoid* gene in normal embryos is a master regulator gene controlling the expression of genes for the development of anterior structures (head and thorax).

A number of other maternal-effect genes, through the activities of their products in gradients in the embryo, are also involved in axis formation. The *nanos* gene, for instance, is the key maternal-effect gene for the posterior structures. When the *nanos* gene is mutated, embryos lack abdominal segments.

Once the axis of the embryo is set, the expression of at least 24 **segmentation genes** progressively subdivides the embryo into regions, determining the segments of the embryo and the adult (see Figure 48.25). Gradients of BICOID and other proteins encoded by maternal-effect genes regulate expression of the embryo's segmentation genes differentially. That is, each segmentation gene is expressed at a particular time and in a particular location during embryogenesis.

Three sets of segmentation genes are regulated in a cascade of gene activations. The first set to be expressed is the **gap genes**, for example, *hunchback* and *tailless*. These genes are activated based on their positions in the maternally directed anterior–posterior axis of the egg by reading the concentrations of BICOID and other proteins. Gap genes, through their activation of the next genes in the regulatory cascade, control the subdivision of the embryo along the anterior–posterior axis into several broad regions. Mutations in gap genes result in the loss of one or more body segments in the embryo **(Figure 48.27a)**.

The products of gap genes are transcription factors that activate **pair-rule genes.** The actions of the prod-

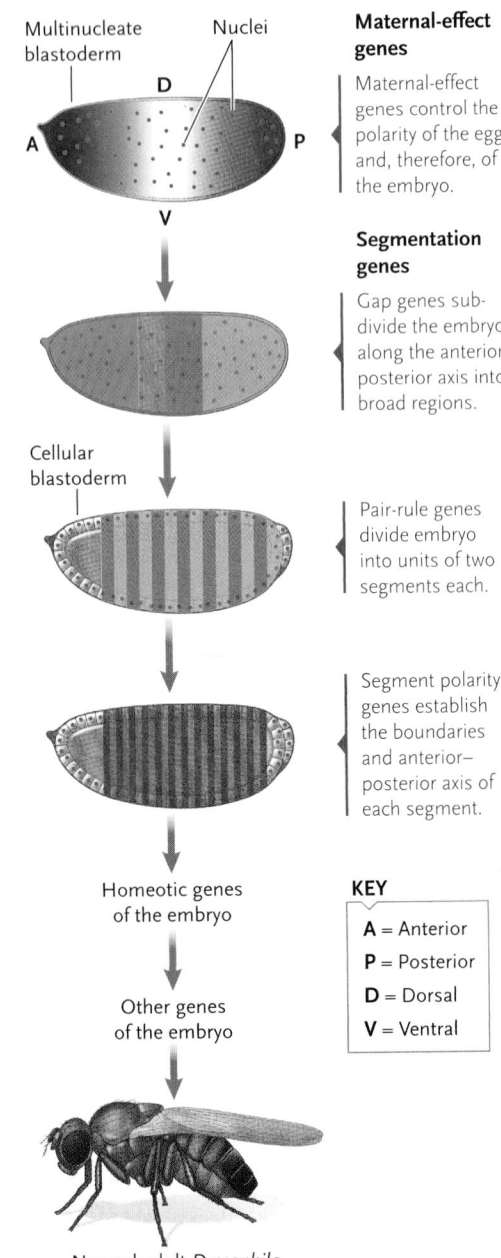

Maternal-effect genes

Maternal-effect genes control the polarity of the egg and, therefore, of the embryo.

Segmentation genes

Gap genes subdivide the embryo along the anterior–posterior axis into broad regions.

Pair-rule genes divide embryo into units of two segments each.

Segment polarity genes establish the boundaries and anterior–posterior axis of each segment.

KEY

A = Anterior
P = Posterior
D = Dorsal
V = Ventral

Multinucleate blastoderm   Nuclei

Cellular blastoderm

Homeotic genes of the embryo

Other genes of the embryo

Normal adult *Drosophila*

Figure 48.25
Maternal-effect genes and segmentation genes and their role in *Drosophila* embryogenesis.

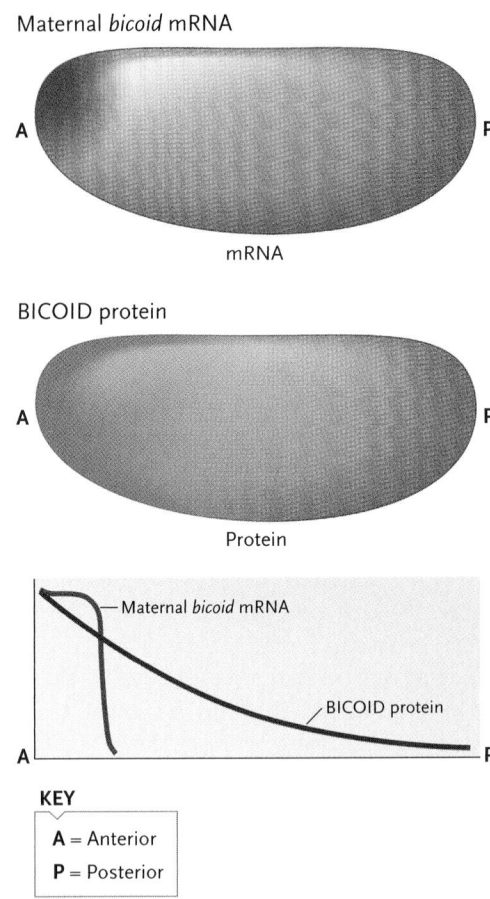

Maternal *bicoid* mRNA

A — P

mRNA

BICOID protein

A — P

Protein

KEY

A = Anterior
P = Posterior

Figure 48.26
Gradients of *bicoid* mRNA and BICOID protein in the *Drosophila* egg.

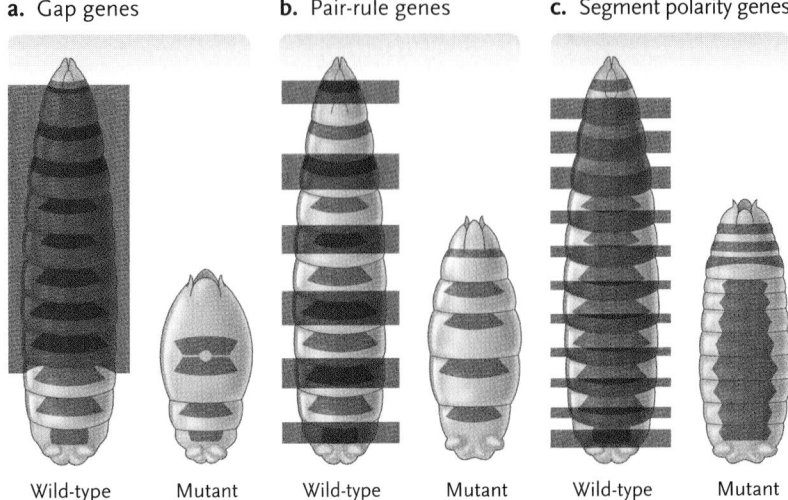

a. Gap genes   **b.** Pair-rule genes   **c.** Segment polarity genes

Wild-type   Mutant   Wild-type   Mutant   Wild-type   Mutant

Figure 48.27
Examples of effects of mutations in the different types of segmentation genes of *Drosophila*. Blue highlights indicate wild-type segments that are mutated. **(a)** Gap gene mutants lack one or more segments. **(b)** Pair-rule gene mutants are missing every other segment. **(c)** Segment polarity gene mutants have segments with one part missing and the other part duplicated as a mirror image.

ucts of pair-rule genes divide the embryo into units of two segments each. Mutations in pair-rule genes delete every other segment of the embryo **(Figure 48.27b).**

The products of pair-rule genes are transcription factors that regulate the expression of the last set of genes in the series, the **segment polarity genes.** The actions of the products of segment polarity genes set the boundaries and anterior–posterior axis of each segment in the embryo. Mutations in segment polarity genes produce segments in which one part is missing and the other part is duplicated as a mirror image **(Figure 48.27c).** The products of segment polarity genes are transcription factors and other molecules that regulate other genes involved in laying down the pattern of the embryo.

Homeotic Genes for Specifying the Developmental Fate of Each Segment. Once the segmentation pattern has been set, **homeotic** (structure-determining) **genes** of the embryo specify what that segment will become after metamorphosis. In normal flies, homeotic genes are master regulatory genes that control the development of structures such as eyes, antennae, legs, and wings on particular segments (see Figure 48.24). Researchers discovered the role of homeotic genes from

the study of mutations in these genes; such mutations alter the developmental fate of a segment in the embryo in a major way. For example, in flies with a mutation in the *Antennapedia* gene, legs develop in place of antennae **(Figure 48.28).**

How do homeotic genes regulate development? Homeotic genes encode transcription factors that regulate expression of genes responsible for the development of adult structures. Each homeotic gene has a common region called the **homeobox** that is key to its function. A homeobox corresponds to an amino acid section of the encoded transcription factor called the **homeodomain.** The homeodomain of each protein binds to a region in the promoters of the genes whose transcription it regulates.

Homeobox-containing genes are called *Hox* genes. There are eight *Hox* genes in *Drosophila* and, interest-

Figure 48.28
Antennapedia, a homeotic mutant of *Drosophila*, in which legs develop in place of antennae.

Normal

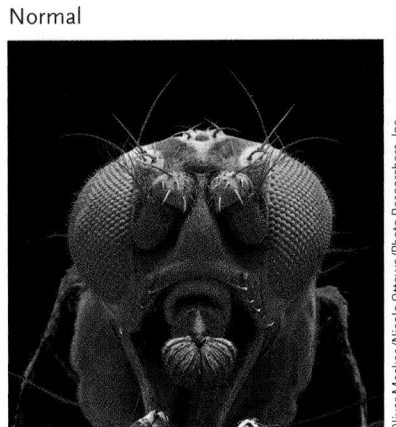

Antennapedia mutant

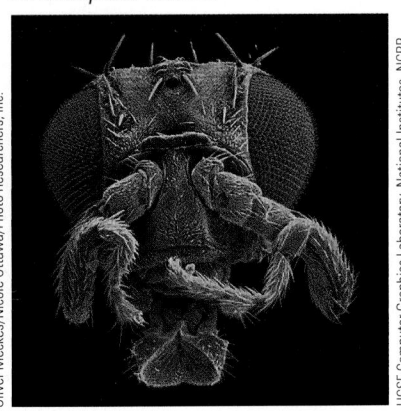

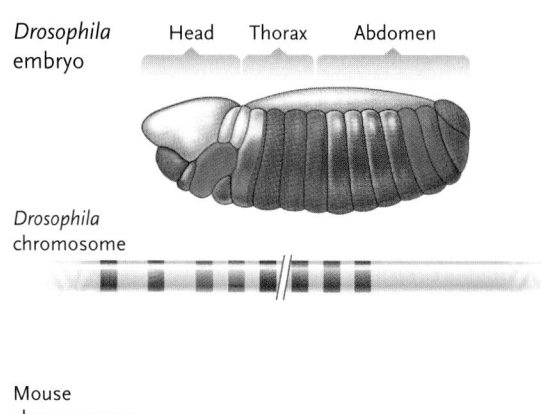

Drosophila embryo — Head   Thorax   Abdomen

Drosophila chromosome

Mouse chromosomes

Mouse embryo

Figure 48.29

The *Hox* genes of the fruit fly and the corresponding regions of the embryo they affect. The mouse has four sets of *Hox* genes on four different chromosomes. Their relationship to the fruit fly genes is shown by the colors.

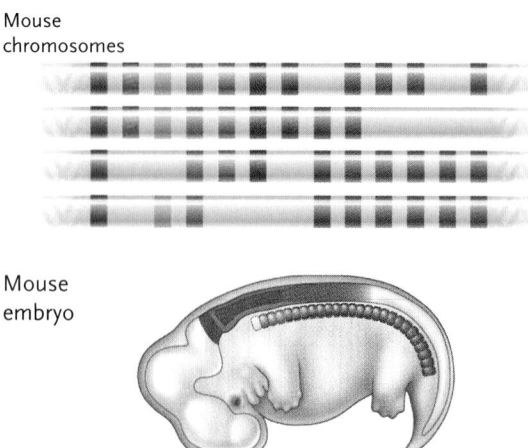

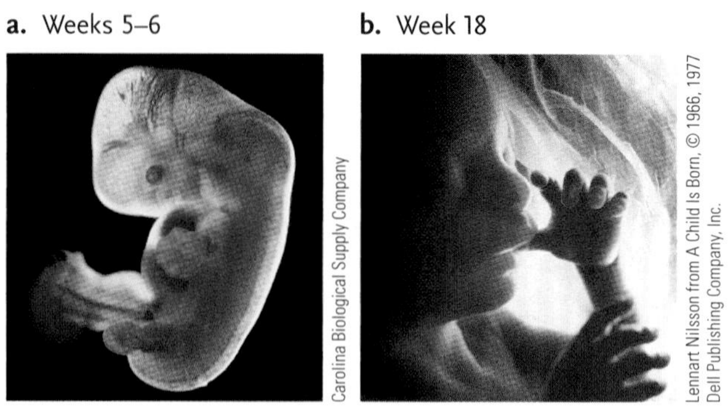

a. Weeks 5–6   **b.** Week 18

Figure 48.30

An illustration of apoptosis in humans: the removal of tissue between developing fingers and toes to produce the free fingers and toes later in development.

ingly, they are organized along a chromosome in the same order as they are expressed along the anterior–posterior body axis **(Figure 48.29).**

The discovery of *Hox* genes in *Drosophila* led to a search for equivalent genes in other organisms. The result of that search has shown that *Hox* genes are present in all major animal phyla. In each case, the genes control the development of the segments/regions of the body and are arranged in order in the genome. (This is a good example of how the results of studies in a model research organism are broadly applicable.) The homeobox sequences in the *Hox* genes are highly conserved,

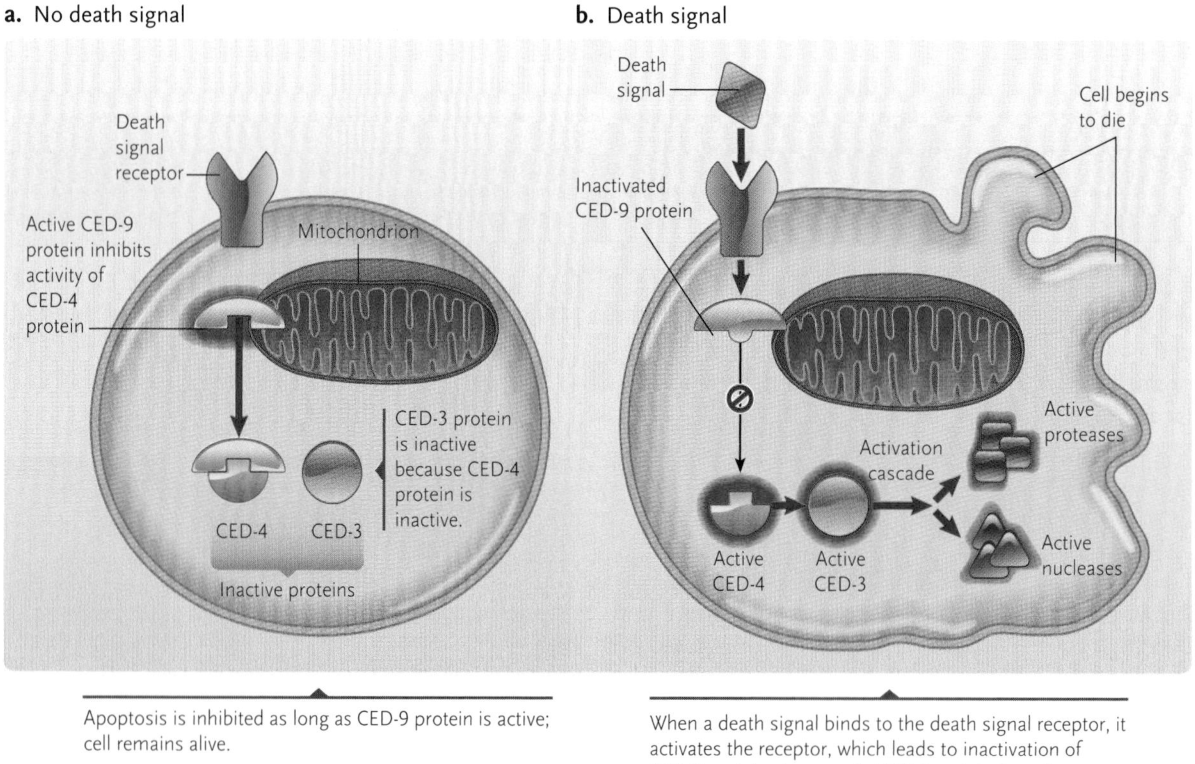

a. No death signal

Death signal receptor

Active CED-9 protein inhibits activity of CED-4 protein

Mitochondrion

CED-3 protein is inactive because CED-4 protein is inactive.

CED-4   CED-3

Inactive proteins

Apoptosis is inhibited as long as CED-9 protein is active; cell remains alive.

b. Death signal

Death signal

Cell begins to die

Inactivated CED-9 protein

Activation cascade

Active proteases

Active CED-4   Active CED-3

Active nucleases

When a death signal binds to the death signal receptor, it activates the receptor, which leads to inactivation of CED-9 protein. As a result, CED-4 protein is no longer inhibited and becomes active, activating CED-3 protein. Active CED-3 triggers a cascade of activations producing active proteases and nucleases, which cause the changes seen in apoptotic cells and eventually to cell death.

Figure 48.31

The molecular basis of apoptosis in *C. elegans.* **(a)** In the absence of a death signal, no apoptosis occurs. **(b)** In the presence of a death signal, activation of CED-4 and CED-3 proteins triggers a pathway that leads to the cell's death.

indicating common function in the wide range of animals in which they are found. For example, the homeobox sequences of mammals are the same or very similar to those of the fruit fly (see Figure 48.29).

Homeotic genes are also found in plants. For example, many homeotic mutations that affect flower development have been identified and analyzed in *Arabidopsis* (see Section 34.5).

Apoptosis Is Triggered by Cell-Death Genes

We have noted the role of apoptosis in the breakdown of a tadpole's tail; there are many other examples of apoptosis in both vertebrate and invertebrate development. In humans, for example, the developing fingers and toes are initially connected by tissue, forming a paddle-shaped structure; later in development, cells of the tissue die by apoptosis resulting in separated fingers and toes **(Figure 48.30)**. Like many other mammals, kittens and puppies are born with their eyes sealed shut by an unbroken layer of skin. Just after birth, cells die in a thin line across the middle of each eyelid, freeing the eyelids to open. During the pupation stage that converts a caterpillar to a butterfly, many tissues of the larva break down by apoptosis and are replaced by newly formed adult tissues.

Apoptosis results from gene activation, in response to molecular signals received by receptors at the surfaces of the marked cells. In effect, the signals amount to a death notice, delivered at a specific time during embryonic development. For example, in the nematode *C. elegans,* division of the fertilized egg leads to a total of 1090 cells. Of these, exactly 131 die at prescribed times to produce a total of 959 cells in the adult hermaphrodite.

The molecular basis of apoptosis in *C. elegans* involves a molecule that can be considered a death signal binding to a receptor in the plasma membrane of a target cell for apoptosis. The receptor is activated and this leads to activation of proteins that kill the cell. In the absence of the death signal, the killing proteins remain inactive.

Let's walk through the two situations. In the absence of a death signal, the membrane receptor is inactive **(Figure 48.31a)**. This allows a protein associated with the outer mitochondrial membrane, CED-9 (encoded by the *ced-9 cell death* gene) to inhibit CED-4 (encoded

by the *ced-4* gene) and CED-3 (encoded by the *ced-3* gene), the two proteins that are needed to turn on the cell death program. Cells in this situation, with the *ced-9* gene being expressed and its product CED-9 being active, are those that normally survive to form the adult nematode. If a death signal binds to the receptor, however, the receptor becomes activated and the events that follow are typical of signal transduction pathways **(Figure 48.31b)** (see Sections 7.1–7.4). In this case, the activated receptor leads to inactivation of CED-9. Because CED-9 no longer is inhibiting them, CED-4 is activated and, in turn, activates CED-3. Activated CED-3 triggers a cascade of reactions, including the activation of proteases and nucleases that degrade cell structures and chromosomes as part of the cell death program.

Studies of mutants helped understand the role of the cell death genes in *C. elegans*. In mutants lacking a normal *ced-3* or *ced-4* gene, the 131 marked cells fail to die, producing a highly disorganized embryo. In the nervous system, for example, 103 cells that die by apoptosis in normal embryos live to form neurons in the mutants. These extra neurons, which are inserted at random in the embryo, lead to a disorganized and nonfunctional nervous system.

Genes related to *ced-3* and *ced-4* have been found in all animals that have been tested for their presence. In humans and other mammals, the equivalent of *ced-3* is the *caspase-9* gene, which encodes a protease that degrades cell structures. The *caspase-9* gene becomes active, for example, in the cells that form the webbing between the fingers and toes, and causes the webbing to break down. The equivalent of *ced-4* is the *Apaf* gene (for *Apoptotic protease-activating factor*). Mammalian cells are saved from death by the *Bcl* family of genes, which are the equivalent of *ced-9* in *C. elegans*. The genes are so closely related that they retain their effects if they are exchanged between *C. elegans* and human cells.

STUDY BREAK

1. In general, how are determination and differentiation controlled?
2. How do the segmentation genes and homeotic genes of *Drosophila* differ in function?

Review

Go to **ThomsonNOW** at www.thomsonedu.com/login to access quizzing, animations, exercises, articles, and personalized homework help.

48.1 Mechanisms of Embryonic Development

- Developmental information is stored in both the nucleus and cytoplasm of the fertilized egg. The mRNA and protein molecules that direct the first stages of development are the cytoplasmic determinants.

- The unequal distribution of yolk and other components makes eggs polar. The animal pole typically gives rise to surface structures and the anterior end of the embryo, while the vegetal pole typically gives rise to internal structures of the embryo such as the gut (Figure 48.2).

- Following fertilization, cleavage divisions produce the morula. The morula hollows out to form the blastula, which develops into the gastrula, the stage in which rearrangements of cells produce the ectoderm, mesoderm, and endoderm. Gastrulation establishes the body pattern, in that the organs and other structures of embryo arise from these three tissue layers (Figure 48.3; Table 48.1).

- Development proceeds as a result of cell division, cell movements, selective adhesions, induction, determination, and differentiation.

Animation: Where embryos develop

Animation: Stages of development

Animation: Cytoplasmic localization

48.2 Major Patterns of Cleavage and Gastrulation

- In sea urchins eggs, yolk is distributed evenly. As a result, cleavage divisions take place at the same rate in all regions of the embryo, and gastrulation follows a symmetrical pattern (Figure 48.4).

- In amphibian eggs, yolk is distributed unequally, with most in the vegetal pole. As a result, the rate of cell division is more rapid in the animal pole, and gastrulation shows an asymmetric pattern (Figures 48.5 and 48.7).

- In bird and reptile embryos, the cleavage divisions give rise to a flat disc of cells at the top of the yolk, which divides into the epiblast and the hypoblast. In gastrulation, cells of the epiblast migrate to the interior to form the endoderm and the mesoderm. The epiblast cells left at the surface form the ectoderm (Figure 48.8).

- In birds and reptiles, the yolk sac, chorion, amnion, and allantoic membrane form from extensions of the primary tissue layers. These extraembryonic membranes conduct nutrients from the yolk to the embryo, exchange gases with the environment, and store metabolic wastes (Figure 48.9).

Animation: Process of gastrulation

48.3 From Gastrulation to Adult Body Structures: Organogenesis

- In organogenesis, the three primary tissues give rise to the tissues and organs of the embryo. Organogenesis begins with neurulation, the development of the nervous system from ectoderm (Figures 48.10 and 48.11).

- The mesoderm splits into somites, which give rise to the vertebral column and to the muscles of the ribs, vertebral column, and limbs (Figure 48.11).

- Development of the eye from optic vesicles is illustrative of the inductions and differentiations common to organogenesis in vertebrates (Figure 48.12).

- Apoptosis—programmed cell death—plays an important role in development by removing tissues present during development but not in the adult organ.

Animation: Neural tube formation

Animation: Embryonic induction

Animation: AER transplant

Animation: Formation of human fingers

48.4 Embryonic Development of Humans and Other Mammals

- In humans, as in other placental mammals, cleavage divisions produce a morula that differentiates into a blastocyst. The blastocyst implants into the endometrium of the uterus, and its inner cell mass separates into the epiblast and hypoblast. The epiblast produces the ectoderm, mesoderm, and endoderm of the embryo (Figures 48.13 and 48.14a, b).

- Gastrulation, neurulation, differentiation of cell layers, and formation of extraembryonic membranes occur by mechanisms similar to those of bird and reptile embryos. Differentiation of ectoderm, mesoderm, and endoderm into their final tissues and organs also occurs in a similar way to birds and reptiles.

- Extraembryonic membranes form in mammals by processes that are also similar to the reptilian–bird pattern. However, some of the membranes have altered functions, reflecting the minimal amount of yolk in mammalian embryos, and maintenance of the embryo by the placenta (Figure 48.14c–e).

- The placenta is connected to the embryo by the umbilical cord, which conducts blood between the embryo and the placenta (Figure 48.14f).

- Fetal growth proceeds until birth, when the fetus is forced from the uterine cavity and through the vagina by contractions of the uterus, stimulated by oxytocin (Figures 48.15 and 48.16).

- The mother's mammary glands secrete milk once the offspring is born. Suckling by the offspring stimulates prolactin and oxytocin release from the pituitary, which stimulates milk production and secretion from the glands, respectively.

- Embryos develop internal male or female sex organs from the same primitive structures. The presence or absence of a Y chromosome, which carries the key *SRY* gene, determines whether the internal structures develop into male or female sexual organs (Figure 48.17).

- Most animals continue development after hatching or birth, leading to the adult, the sexually mature form of the species.

Animation: Fertilization

Animation: Cleavage and implantation

Animation: First 2 weeks of development

Animation: Weeks 3 to 4 of development

Animation: Proportional changes during development

Animation: Structure of the placenta

Animation: Fetal development

Animation: Birth

Animation: Anatomy of the breast

48.5 The Cellular Basis of Development

- Development in animals involves the regulation of specific cellular events, including cell division, cell movement, and cell adhesion.

- Cell division in development varies in orientation and rate.

- Cell movements in development occur through changes in cell shape or the migrations of entire cells. Shape changes are produced by microtubules or microfilaments. In cell migrations, cells follow tracks in the embryo or move in response to gradients of signal molecules (Figures 48.18 and 48.19).

- Selective cell adhesions, which depend on surface glycoproteins including CAMs and cadherins, underlie many cell movements. The final cell adhesions hold the embryo in its correct shape and form (Figure 48.20).

- Induction results from the effects of signaling molecules of the inducing cells on the responding cells (Figure 48.21).

- In differentiation, cells change from embryonic form to specialized types with distinct structures and functions. Differentiation occurs by differential gene activation.

- For some organisms, the origins of adult or larval structures have been mapped to regions of the embryo from which each structure derived (Figure 48.22).

48.6 The Genetic and Molecular Control of Development

- Pattern formation derives from the positions of cells in the embryo. Typically, positional information is detected by the cells in the form of concentration gradients of regulatory molecules encoded by genes (Figures 48.24–48.27).

- *Hox* genes are evolutionarily conserved regulatory genes that control the development of the segments or regions of the body (Figures 48.28 and 48.29).

- Apoptosis, programmed cell death, typically eliminates structures required for earlier but not later stages of development (Figures 48.30 and 48.31).

Questions

Self-Test Questions

1. Major contributors to the axes of the animal body are the:
   a. sperm and egg cytoplasm.
   b. sperm and egg chromosomes.
   c. ribosomes and mitochondria
   d. egg nucleus and yolk.
   e. pigments.

2. The process by which cells undergo mitosis without a corresponding increase in cytoplasm is called:
   a. polarity.
   b. cleavage.
   c. gastrulation.
   d. organogenesis.
   e. induction.

3. Which of the following mechanisms does *not* contribute to zygote development?
   a. meiosis
   b. mitosis
   c. selective cell adhesions
   d. determination
   e. induction

4. A major event during gastrulation is:
   a. the outward movement of cells at the dorsal lip of the blastopore.
   b. the displacement of the archenteron by the blastocoel.
   c. the formation of the coelom from the endoderm.
   d. the extension of ectoderm and endoderm to form the yolk sac.
   e. the development of ectoderm to form epidermal and neural tissues.

5. To contribute to the formation of a nervous system:
   a. the neural crest develops into motor neurons.
   b. the neural tube is converted into a neural plate.
   c. the notochord induces the overlying ectoderm to become a neural plate.
   d. the roof of the archenteron induces the formation of the neural tube.
   e. somites give rise to the autonomic nervous system.

6. In mammalian development:
   a. the morula develops into a trophoblast.
   b. the chorionic villi allow the blastocyst to move down the oviduct.
   c. the allantois takes over the work of the amnion.
   d. the pharyngeal arches transform into the pharynx, larynx, and nasal cavities.
   e. prolactin stimulates parturition.

7. In the development of the female sex organs:
   a. all ducts in the 7-week embryo become Wolffian ducts.
   b. the *SRY* gene is activated.
   c. anti-Mullerian hormone is secreted.
   d. the Mullerian ducts develop into oviducts.
   e. the mother secretes oxytocin.

8. In the embryonic development of the eye:
   a. the optic vesicle cells permanently adhere to each other to prevent movement, whereas the optic cup cells are very motile.
   b. signals from the optic cup trigger surface receptors on the lens placode.
   c. gradients determine that the ectoderm overlying the lens vesicle develops into the optic vesicle.
   d. microtubules powered by myosins and microfilaments powered by dyneins move the eye components around in the head region.
   e. cadherins function in the presence of calcium to allow the lens placode and optic cup to break apart.

9. In mammals, the nose is located on the anterior end and the heart in the center. These positions are the result of activation of:
   a. *Hox* genes arranged along a number of chromosomes in the same order as they are expressed along the anterior–posterior axis.
   b. maternal-effect genes after somites differentiate into muscle.
   c. *Hox* genes scattered randomly among different chromosomes.
   d. a transcription factor called the homeobox.
   e. a homeodomain that binds ribosomes.

10. During embryonic development in many humans, the cells of the lower earlobe die resulting in an unattached earlobe. For this to occur one would deduce that:
    a. *CED-3* genes are inhibited.
    b. *CED-4* genes are inhibited.
    c. caspase-9 is deactivated.
    d. *CED-9* is actively expressed.
    e. the proteins encoded by *Bcl* genes are inactivated.

Questions for Discussion

1. Experimentally, it is possible to divide an amphibian egg so that the gray crescent is wholly within one of the two cells formed. If the two cells are separated, only the cell with the gray crescent will form an embryo with a long axis, notochord, nerve cord, and back musculature. The other forms a shapeless mass of immature gut and blood cells. Propose an explanation of these outcomes.

2. The renowned developmental biologist Lewis Wolpert once observed that birth, marriage, and death are not the most important events in human life; rather, gastrulation is. In what sense was he correct?

3. Arguably, in sexually reproducing animals development begins when eggs and sperm form in the parents. In a paragraph, explain the rationale for this idea.

4. Investigators discovered a *Drosophila* protein that triggers development of the nerve cord on the ventral side of the embryos. When an mRNA encoding the protein was injected into cells on the ventral side of *Xenopus* embryos, dorsal structures were formed on the ventral side, including incomplete heads. What do these findings suggest about the evolution of embryonic development?

Experimental Analysis

As you have learned in this chapter, embryogenesis in *Drosophila* has been well described. In Chapter 18 you also learned that the genome of *Drosophila* has been completely sequenced, allowing each gene in the genome to be cataloged. Design an experiment to identify all the genes that are activated during embryogenesis, and when they are activated.

Evolution Link

Every one of more than a million species of insects has six legs, a pair on each of the three thoracic segments. By contrast, other arthropods, such as crustaceans, have a variable number of limbs; some species have limbs on every segment in both the thorax and abdomen. Propose a molecular mechanism by which limbs might have been lost during the evolution of the insects.

How Would You Vote?

Sanitation and medical advances have greatly extended the average human life span, especially in developed countries. Some researchers are now looking for ways to extend the human life span even further. Do you think research into life extension should be supported by federal research funding? Go to www.thomsonedu.com/login to investigate both sides of the issue and then vote.

A population of Caribbean flamingos *(Phaenicopterus ruber)*. Each pair of flamingos in this breeding colony incubates a single egg in a mud nest.

© Gerry Ellis/Minden Pictures

49 Population Ecology

WHY IT MATTERS

When humans immigrate to new places, they often transport familiar plants and animals from home, introducing them into their new gardens, fields, and forests. Some organisms fail to survive in the new environments. But other species—like the European starlings *(Sturnus vulgaris)* and house sparrows *(Passer domesticus)* that are now so common in North America—flourish and sometimes become pests.

In 1859, an Australian rancher released a few pairs of European rabbits *(Oryctolagus cuniculus)* for sport hunting in the state of Victoria. The rabbits bred rapidly, sometimes producing litters of four or five offspring every month. They had no natural predators in Australia, and by 1900, an estimated 20 million rabbits had overrun much of the continent; their advance was limited only by extreme climates, clay soil, and lack of food or water. The rabbits destroyed natural vegetation and the pastures that supported a large sheep industry. The government tried in vain to poison the rabbits. Ranchers introduced predators, hoping that they would eat rabbits faster than the rabbits could reproduce. But the rabbits continued to multiply. Eventually, the government built a "rabbit-proof fence" that stretched more than

1125

Figure 49.1
Introduced organisms. European rabbits multiplied so rapidly and destroyed so much vegetation in Australia that the government built a fence across the country to prevent their spread.

Peter Bird/Australian Picture Library/Westlight/Corbis

3200 km (2000 miles) to keep the rabbits out of the rich pasture lands in Western Australia **(Figure 49.1)**.

In 1950, scientists tackled the devastating problems caused by the introduced rabbits. Biologists collected myxoma virus (a relative of smallpox) from infected rabbits in South America and released it among the European rabbits in Australia. The virus was lethal to European rabbits, which had never evolved resistance to it. The first epidemic of myxomatosis killed more than 99% of infected rabbits. But in the following season, the virus killed only 90% of infected rabbits, and within a few years, the virus was killing only half the rabbits it infected. Clearly, some rabbits were becoming more resistant to the virus. Resistant rabbits survived and reproduced, comprising a larger percentage of the population over time (see Section 20.3 to review natural selection). Subsequent research showed that the virus had also become less virulent. Today, wildlife-control agents develop and release more deadly viruses to control the rabbit population.

This brief history of an environmental disruption introduces our unit on **ecology**, the study of interactions between organisms and their environments. All environments have both **abiotic** (nonbiological) and **biotic** (biological) components. The abiotic environment includes temperature, moisture, soil chemistry, and other physical factors; the biotic environment includes all the organisms found in a particular place.

This story also identifies several ecological phenomena that we consider in this chapter. For example, some species produce large numbers of young at frequent intervals; their numbers may increase rapidly for a period of time and then drop precipitously. Moreover, a species' abundance is often governed by the presence of other species—its food, predators, parasites, and disease-causing microorganisms. Finally, over time, ecological interactions foster adaptation and evolutionary change.

49.1 The Science of Ecology

The subject matter of ecology is so vast and so diverse that research in ecology is often linked to work in genetics, physiology, anatomy, behavior, paleontology, and evolution as well as in geology, geography, and environmental science. Many ecological phenomena occur over huge areas and long time spans. Ecologists

must devise ways to determine how environments influence organisms and how organisms change the environments in which they live. Today, the science of ecology encompasses two related disciplines. The major research questions of *basic ecology* relate to the distribution and abundance of species and how they interact with each other and the physical environment. Using these data as a baseline, workers in *applied ecology* develop conservation plans and amelioration programs to limit, repair, and mitigate ecological damage caused by human activities.

Ecologists Study Levels of Organization Ranging from Individual Organisms to the Biosphere

Ecology can be divided into five increasingly complex and inclusive levels of organization. In **organismal ecology** researchers study the genetic, biochemical, physiological, morphological, and behavioral adaptations of organisms to the abiotic environment. We have described many such adaptations in Units V and VI; we describe the evolution of animal behavior in Chapter 55.

Population ecology, the subject of this chapter, focuses on **populations**, groups of individuals of the same species that live together. Population ecologists study how the size and other characteristics of populations change in space and time. Research in **community ecology** examines groups of populations that occur together in one area. Community ecologists study interactions between species, analyzing how predation, competition, and environmental disturbances influence a community's development, organization, and structure. We address major issues in community ecology in Chapter 50. Ecologists studying **ecosystem ecology** explore the cycling of nutrients and the flow of energy between the biotic components of an ecological community and the abiotic environment. We consider this topic in Chapter 51. Finally, some ecological studies focus on the **biosphere,** the total of all ecosystems on Earth. In Chapter 52, we examine global patterns in abiotic factors and their effects on populations, communities, and ecosystems. We discuss biodiversity and conservation biology in Chapter 53.

Ecologists Test Hypotheses with Observational and Experimental Data

Ecology has its roots in descriptive natural-history studies that date back to the ancient Greeks. Modern ecology was born in 1870 when the German biologist Ernst Haeckel coined the term (*oikos* = house). Contemporary researchers still gather descriptive information about ecological relationships, but these observations are only a starting point for more rigorous studies.

Most ecologists create hypotheses about ecological relationships and how they change through time or differ from place to place. Like other scientists, some ecologists formalize these ideas in mathematical models that express clearly defined, but hypothetical, relationships among important variables in a system. Manipulation of a model, usually with the help of a computer, can allow researchers to ask what would happen if some of the variables or their relationships change. Thus, researchers can simulate natural events and large-scale experiments before investing time, energy, and money in field and laboratory work. Bear in mind, however, that mathematical models are no better than the ideas and assumptions they embody, and useful models are constructed only after basic observations define the relevant variables.

Ecologists often conduct field or laboratory studies to test the predictions of their hypotheses. In controlled experiments, researchers compare data from an experimental treatment (in which one or more variables are artificially manipulated) with data from a control (in which nothing is changed). Sometimes the distributions of species create "natural experiments," eliminating the need to manipulate variables. For example, two species of fishes, cutthroat trout *(Oncorhynchus clarki)* and Dolly Varden char *(Salvelinus malma)*, live in coastal lakes of British Columbia, Canada. Some lakes have either trout or char, but other lakes contain both species. The natural distributions of these fishes allowed researchers to measure the effect of each species on the other. In lakes where both species live, each restricts its activities to fewer areas and feeds on a smaller variety of prey than it does in lakes where it occurs alone.

Study Break

1. Why are studies of ecosystems more "inclusive" than studies of populations?
2. In what ways are mathematical models useful in ecological research?

49.2 Population Characteristics

Populations have characteristics that transcend those of the individuals they comprise. For example, every population has a **geographical range**, the overall spatial boundaries within which it lives. Geographical ranges vary enormously. A population of snails might inhabit a small tidepool, whereas a population of marine phytoplankton might occupy an area orders of magnitude larger. Every population also occupies a **habitat**, the specific environment in which it lives, as characterized by its biotic and abiotic features. Ecologists also mea-

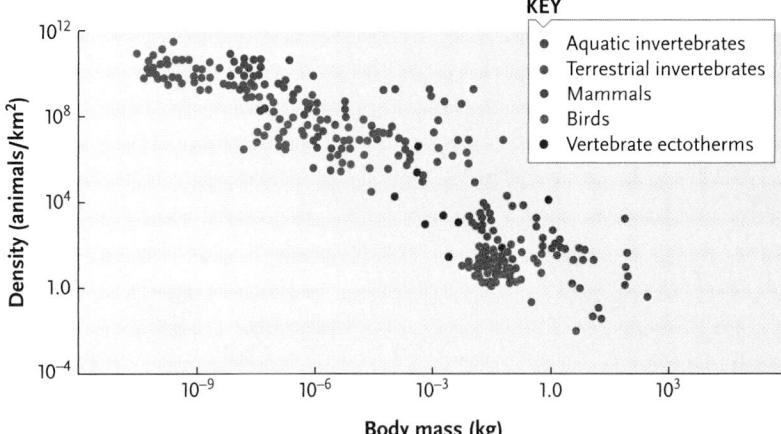

Figure 49.2
Population density and body size. Population density generally declines with increasing body size among animal species. Similar trends exist for other types of organisms.

sure other population characteristics, such as size, distribution in space, and age structure.

A Population's Size and Density Determine the Amount of Resources It Uses

Population size is simply the number of individuals in a population at a specified time. **Population density** is the number of individuals per unit area or per unit volume of habitat. Species with large body size generally have lower population densities than those with small body size **(Figure 49.2)**. Although population size and density are related measures, knowing a population's density provides more information about its relationship to the resources it uses. For example, if a population of 200 oak trees occupies 1 hectare (10,000 m²), its population density is 200/10,000 m² or one tree per 50 m². But if a population of 200 oaks is spread over 5 hectares, its density is one tree per 250 m². Clearly, the second population is less dense than the first, and its members will have greater access to sunlight, water, and other resources.

Ecologists measure population size and density to monitor and manage populations of endangered species, economically important species, and agricultural pests. For large-bodied species, a simple head count provides accurate information. For example, ecologists survey the size and density of African elephant *(Loxodonta africana)* populations by flying over herds and counting individuals. Researchers use a variation on that technique to estimate population size in tiny organisms that live at high population densities. To estimate the density of aquatic phytoplankton, for example, you might collect water samples of known volume from representative areas in a lake and use a microscope to count the organisms; you could then extrapolate their population size and density based on the estimated volume of the entire lake. In other cases, researchers use the mark-release-recapture sampling technique **(Figure 49.3)**.

Figure 49.3 Research Method

Using Mark-Release-Recapture to Estimate Population Size

PURPOSE: Ecologists use the mark-release-recapture technique to estimate the population size of mobile animals that live within a restricted geographic range.

PROTOCOL: A sample of organisms is captured, marked in some permanent but harmless way, and released. Insects and reptiles are marked with ink or paint, birds with rings on their legs, and mammals with ear tags or collars. Some time later, a second sample of organisms is captured, and the researcher notes what proportion of the second sample carries the mark. That proportion tells us what percentage of the total population was captured and marked at the first sampling. The total population size is estimated as (number marked) × (number in the second sample/number of marked recaptures).

Michael C. Singer, University of Texas

INTERPRETING THE RESULTS: Imagine that you capture 120 butterflies, mark each with a black spot on its wing, and release them. A week later, you capture a second sample of 150 butterflies and find that 30 of them have the black mark. Thus, you had marked one out of every five butterflies (30/150) on your first field trip. Because you captured 120 individuals on that first excursion, you would estimate that the total population size is 120 × (150/30) = 600 butterflies.

The technique is based on several assumptions that are critical to its accuracy: (1) that being marked has no effect on survival; (2) that marked and unmarked animals mix randomly in the population; (3) that no migration into or out of the population takes place during the estimating period; and (4) that marked individuals are just as likely to be captured as unmarked individuals. (Sometimes animals become "trap shy" or "trap happy," a violation of the fourth assumption.)

Populations Differ in How They Are Distributed in Space

Populations also vary in their **dispersion**, the spatial distribution of individuals within the geographical range. Ecologists define three theoretical patterns of dispersion: *clumped, uniform,* and *random* **(Figure 49.4).**

Three reasons explain why a **clumped dispersion**—with individuals grouped together—is extremely common in nature. First, suitable conditions often have a patchy distribution. For example, certain pasture plants may be clumped in small, scattered areas where cowpats fell months before, locally enriching the soil. Second, some animals live in social groups (see Section 55.5). Mates are easy to locate within groups, and individuals may cooperate in rearing offspring, feeding, or defending themselves from predators. Third, some organisms are clumped because of their reproductive pattern. Plants and animals that produce asexual clones, such as aspen trees and sea anemones, often occur in large aggregations (see Chapters 34 and 47).

In other species, seeds, eggs, or larvae lack dispersal mechanisms, and offspring grow near their parents.

Organisms are evenly spaced in their habitat, a pattern called **uniform dispersion**, when individuals repel each other because resources are in short supply. For example, creosote bushes *(Larrea tridentata)* are uniformly distributed in the dry scrub deserts of the American Southwest. Mature bushes deplete the surrounding soil of water and secrete toxic chemicals, making it impossible for seedlings to grow. Moreover, seed-eating ants and rodents that live at the base of mature bushes consume any seeds that fall nearby. Territorial behavior, the defense of an area and its resources, produces uniform dispersion in animals (see Section 55.2).

For some populations, environmental conditions don't vary much within a habitat, and individuals are neither attracted to nor repelled by others of their species. These populations exhibit **random dispersion**, which has a formal statistical definition that serves as a theoretical baseline for assessing whether organisms are clumped or uniformly distributed. In cases of random dispersion, individuals are distributed unpredictably. Some spiders, burrowing clams, and rainforest trees exhibit random dispersion.

Whether the spatial distribution of a population appears to be clumped, uniform, or random depends partly on how large an area an ecologist studies. Oak seedlings may be randomly dispersed on a spatial scale of a few square meters, but over an entire mixed hardwood forest, they are clumped under the parent trees.

In addition, the dispersion of animal populations often varies through time in response to natural environmental rhythms. Few habitats provide a constant supply of resources throughout the year, and many animals move from one habitat to another on a seasonal cycle. For example, tropical birds and mammals are often widely dispersed in deciduous forests during the wet season. But during the dry season, they crowd into narrow "gallery forests" along watercourses where evergreen trees provide food and shelter.

A Population's Age Structure, Generation Time, and Sex Ratio Influence How Quickly It Will Grow

All populations have an **age structure**, a statistical description of the relative numbers of individuals in each age class (see Section 49.7). Individuals can be roughly categorized as prereproductive (younger than the age of sexual maturity), reproductive, or postreproductive (older than the maximum age of reproduction). A population's age structure reflects its recent growth history and predicts its future growth potential. Populations that include many prereproductive individuals grew rapidly in the recent past and will continue to grow larger as the young individuals mature and reproduce.

Another characteristic that influences a population's growth is its **generation time**, the average time between the birth of an organism and the birth of its offspring. Generation time is usually short in species that reach sexual maturity at a small body size **(Figure 49.5)**. Their populations often grow rapidly because of the speedy accumulation of reproductive individuals.

Populations also vary in their **sex ratio**, the relative proportions of males and females. In general, the number of females in a population has a bigger impact on population growth than the number of males because only females actually produce offspring. Moreover, in many species, one male can mate with several females, and the number of males may have little effect on the population's reproductive output. In northern elephant seals *(Mirounga angustirostris),* for example, mature bulls fight for dominance on the beaches where the seals mate, and only a few males may ultimately inseminate a hundred or more females. Thus, the presence of other males in the group has little effect on the size of future generations. In animals that form lifelong pair bonds, such as geese and swans, the number of males does influence reproduction in the population.

Population ecologists often try to determine the proportion of individuals in a population that are reproducing. This issue is particularly relevant to the conservation of any species in which individuals are rare or widely dispersed in the habitat (see Section 53.4). As *Insights from the Molecular Revolution* describes, ecologists now use DNA analysis to address this question. In the next section we consider factors that influence the age structure of a population and its potential for future growth.

STUDY BREAK

1. What is the difference between a population's size and its density?
2. What do the three patterns of dispersion imply about the relationships between individuals in a population?

49.3 Demography

Populations grow larger through the birth of individuals and the **immigration** (movement into the population) of organisms from neighboring populations. Conversely, death and **emigration** (movement out of the population) reduce population size. **Demography** is the statistical study of the processes that change a population's size and density through time.

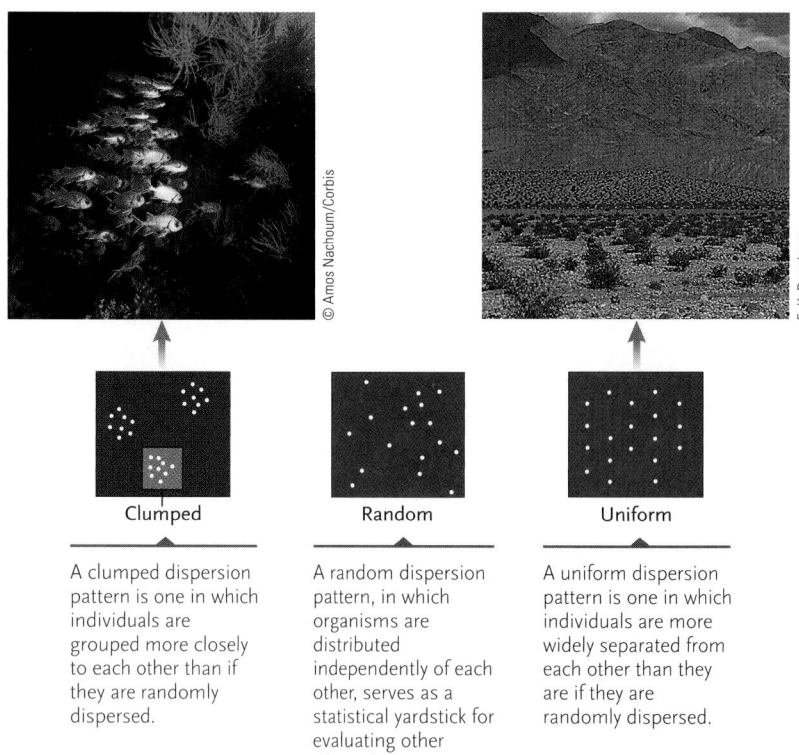

Clumped	Random	Uniform
A clumped dispersion pattern is one in which individuals are grouped more closely to each other than if they are randomly dispersed.	A random dispersion pattern, in which organisms are distributed independently of each other, serves as a statistical yardstick for evaluating other dispersion patterns.	A uniform dispersion pattern is one in which individuals are more widely separated from each other than they are if they are randomly dispersed.

Figure 49.4

Dispersion patterns. Schooling fishes, like these sabre squirrelfish *(Sargocentron spiniferum)* from the Maldives in the Indian Ocean, exhibit a clumped pattern of dispersion. A random pattern of dispersion, which is fairly rare in nature, occurs in organisms that are neither attracted to nor repelled by each other. Creosote bushes *(Larrea tridentata)* near Death Valley, California, exhibit uniform dispersion.

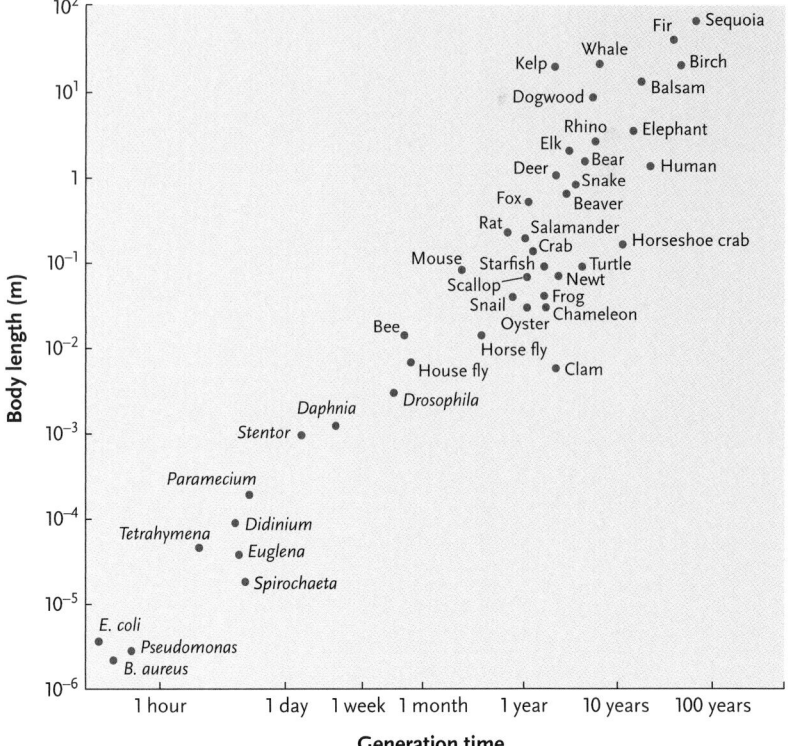

Figure 49.5

Generation time and body size. Generation time increases with body size among bacteria, protists, plants, and animals. The logarithmic scale on both axes compresses the data into a straight line.

Tracing Armadillo Paternity and Migration

Nine-banded armadillos *(Dasypus novemcinctus)* are slow moving and solitary animals. Given their almost completely asocial behavior, what proportion of an armadillo population successfully mates? And, given their slow movements, how have they spread from Mexico and southern Texas through most of the southern United States in only 100 years?

Paulo A. Prodöhl and his colleagues at the University of Georgia, University of Washington, and Valdosta State University used molecular techniques to answer these questions. Their previous work had identified seven short tandem repeat (STR) loci in the armadillo genome. An STR locus consists of a segment of a chromosome with a short sequence repeated in series (see Section 18.2). Here the loci are approximately 200 bp long with 2- and 4-bp tandemly repeated sequences. Alleles of each locus vary in the number of copies of the repeated sequences, resulting in variable lengths for the locus. Because alleles of STR loci are inherited in the same way as alleles of genes, the allelic variations of the

STR loci made it possible for the researchers to trace parentage and migration patterns, in a manner analogous to human DNA fingerprinting.

Prodöhl and his coworkers collected tissue samples by clipping small pieces of tissue from the ears of 290 armadillos living in the Tall Timbers Research Station near Tallahassee, Florida. They extracted genomic DNA from the tissue sample and used the polymerase chain reaction (PCR) to amplify alleles for each of the seven STR loci. They determined the sizes of the fragments amplified by PCR using gel electrophoresis. (These molecular techniques are discussed in Chapter 18.)

The investigators used statistical methods to compare the gel patterns produced by DNA from adults with those from juveniles to determine their relatedness. Adult males and females with gel patterns most similar to those of a given juvenile were considered to be its parents. When the data identified more than two possible parents, the male and female living closest to a juvenile were scored as the most likely candidates. These techniques allowed

the investigators to assign parents to 69 sets of juveniles. Only seven juveniles could not be assigned parents, possibly because the parents had died, avoided capture, or emigrated from the population.

The results from 4 years of study suggest that 36% to 46% of adult armadillos reproduced at least once, despite their asocial habits—a moderately successful reproductive rate. In general, parents and offspring lived between 800 and 1500 m of each other, and individuals were usually captured within a 200-m radius of the same spot from season to season and from one year to the next. Thus, migration appears to be very limited, leaving the basis of their rapid spread unexplained.

© Fred Whitehead/Animals, Animals-Earth Scenes

Ecologists use demographic analysis to predict a population's future population growth. For human populations, these data help governments anticipate the need for social services such as schools and hospitals. Demographic data also allow conservation ecologists to develop plans to protect endangered species. For example, demographic data on the northern spotted owl *(Strix occidentalis caurina)* helped convince the courts to restrict logging in the owl's primary habitat, the old growth forests of the Pacific Northwest. *Life tables* and *survivorship curves* are among the tools ecologists use to analyze demographic data.

Life Tables Summarize a Population's Survival and Reproductive Rates

Although every species has a characteristic life span, few individuals survive to the maximum age possible. Mortality results from starvation, disease, accidents, predation, or the inability to find a suitable habitat. Life insurance companies first developed techniques for measuring mortality rates, but ecologists adapted these approaches to the study of nonhuman populations.

A **life table** summarizes the demographic characteristics of a population **(Table 49.1)**. To collect life-table data for short-lived organisms, demographers typically mark a **cohort**, a group of individuals of similar age, at birth and monitor their survival until all members of the cohort die. For organisms that live more than a few years, a researcher might sample the population for 1 or 2 years, recording the ages at which individuals die, and then extrapolate those results over the species' life span.

In any life table, the life span of the organisms is divided into age intervals of convenient length: days, weeks, or months for short-lived species; years or groups of years for longer-lived organisms. Mortality can be expressed in two complementary ways. **Age-specific mortality** is the proportion of individuals alive at the start of an age interval that died during that age interval. Its more cheerful reflection, **age-specific survivorship**, is the proportion of individuals alive at the start of an age interval that survived until the start of the next age interval. Thus, for the data shown in Table 49.1, the age-specific mortality rate during the 3-to-6-month age interval is 195/722 = 0.270, and the

Table 49.1 — Life Table for a Cohort of 843 Individuals of the Grass *Poa annua* (Annual Bluegrass)

Age Interval (in months)	Number Alive at Start of Age Interval	Number Dying During Age Interval	Age-Specific Mortality Rate	Age-Specific Survivorship Rate	Proportion of Original Cohort Alive at Start of Age Interval	Age-Specific Fecundity (Seed Production)
0–3	843	121	0.144	0.856	1.000	0
3–6	722	195	0.270	0.730	0.856	300
6–9	527	211	0.400	0.600	0.625	620
9–12	316	172	0.544	0.456	0.375	430
12–15	144	90	0.625	0.375	0.171	210
15–18	54	39	0.722	0.278	0.064	60
18–21	15	12	0.800	0.200	0.018	30
21–24	3	3	1.000	0.000	0.004	10
24–	0	—	—	—	—	—

Source: Begon, M., and M. Mortimer. *Population Ecology.* Sunderland, MA: Sinauer Associates, 1981. Adapted from R. Law. 1975.

age-specific survivorship rate is 527/722 = 0.730. For any age interval, the sum of age-specific mortality and age-specific survivorship always equals 1. Life tables also summarize the proportion of the cohort that survived to a particular age, a statistic that identifies the probability that any randomly selected newborn will still be alive at that age. For the 3-to-6-month interval in Table 49.1, this probability is 722/843 = 0.856.

Life tables also include data on **age-specific fecundity**, the average number of offspring produced by surviving females during each age interval. Table 49.1 shows, for example, that plants in the 3-to-6-month age interval each produced an average of 300 seeds. In some species, including humans, fecundity is highest in individuals of intermediate age. Younger individuals have not yet reached sexual matu-

rity, and older individuals are past their reproductive prime. However, in some plants and animals fecundity increases steadily with age.

Survivorship Curves Depict Changes in Survival Rate over the Life Span

Survivorship data are depicted graphically in a **survivorship curve**, which displays the rate of survival for individuals over the species' average life span. Ecologists have identified three generalized survivorship curves (blue lines in **Figure 49.6**), although most organisms exhibit survivorship patterns that fall between these idealized patterns.

Type I curves reflect high survivorship until late in life, when mortality takes a great toll. Type I curves are

Figure 49.6
Survivorship curves. The survivorship curves of many organisms (pink data points) roughly match one of three idealized patterns (blue curves).

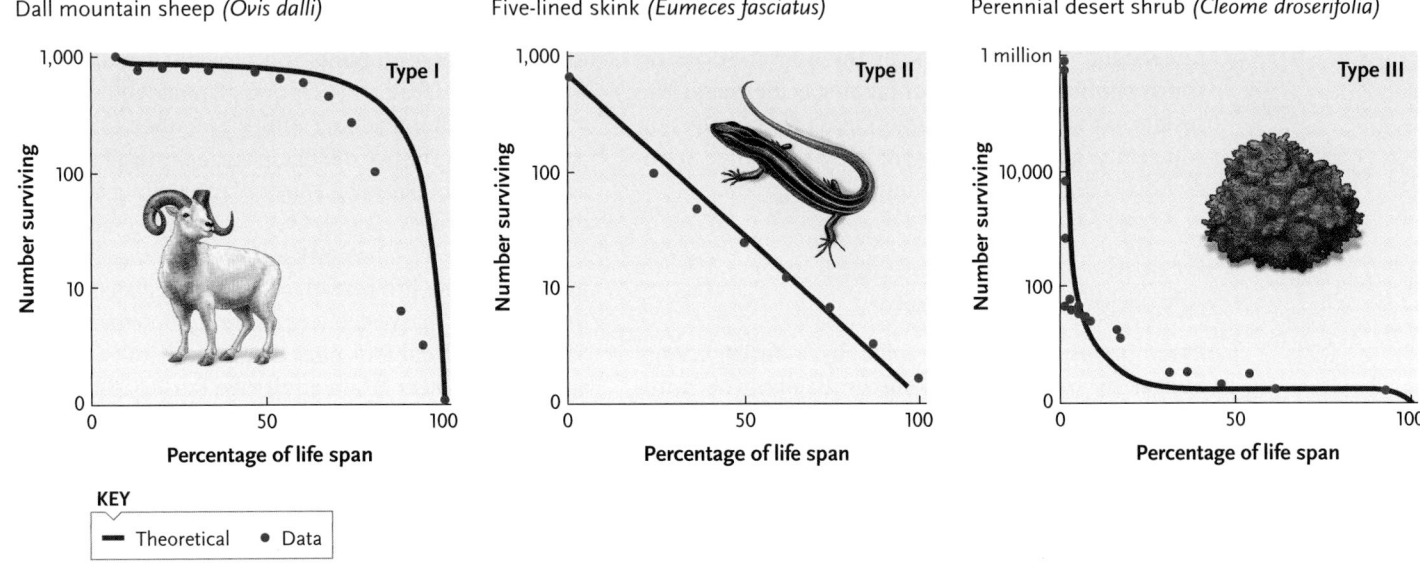

KEY
— Theoretical   • Data

typical of large animals that produce few young and provide them with extended care, which reduces juvenile mortality. For example, large mammals, such as the Dall mountain sheep *(Ovis dalli)*, produce only one or two offspring at a time and nurture them through their vulnerable first year.

Type II curves reflect a relatively constant rate of mortality in all age classes, a pattern that produces steadily declining survivorship. Many lizards, such as the five-lined skink *(Eumeces fasciatus)*, as well as songbirds and small mammals, face a constant probability of mortality from predation, disease, and starvation.

Type III curves reflect high juvenile mortality, followed by a period of low mortality once the offspring reach a critical age and size. For example, *Cleome droserifolia,* a desert shrub from the Middle East, experiences extraordinarily high mortality in its seed and seedling stages. Researchers estimate that for every 1 million seeds produced, fewer than 1000 germinate, and only about 40 individuals survive their first year. Once a plant becomes established, however, its likelihood of future survival is higher, and the survivorship curve flattens out. Many plants, insects, marine invertebrates, and fishes exhibit type III survivorship.

STUDY BREAK

> 1. What statistics are usually included in a life table?
> 2. Which type of survivorship curve is characteristic of humans in industrialized countries? Explain your answer.

49.4 The Evolution of Life Histories

The analysis of life tables reveals how natural selection affects the **life histories**—the lifetime patterns of growth, maturation, and reproduction—that characterize different species. Ecologists study life histories to understand how trade-offs in the allocation of resources influence the evolution of specific traits. The results of their research suggest that natural selection adjusts the allocation of resources to maximize an individual's number of surviving offspring.

Organisms Face Trade-Offs in Their Allocation of Resources

Every organism is constrained by a finite **energy budget,** the total amount of energy that it can accumulate and use to fuel its activities. An organism's energy budget is like a savings account. When the individual accumulates more energy than it needs, it makes deposits to this account—energy is stored as starch, glycogen, or fat. When it expends more energy than it

harvests, it makes withdrawals from its energy stores. But unlike a bank account, an organism's energy budget cannot be overdrawn, and no loans against future "earnings" are possible.

Organisms use the energy they harvest for three broadly defined functions: maintenance (the preservation of good physiological condition), growth, and reproduction. And when an organism devotes energy to any one of these functions, the balance in its energy budget is reduced, leaving less energy for the other functions.

Life History Patterns Vary Dramatically among Species

A fish, a deciduous tree, and a mammal illustrate the dramatic variations that exist in life history patterns. Larval coho salmon *(Oncorhynchus kisutch)* hatch in the headwaters of a stream, where they feed and grow for about a year before assuming their adult body form. After swimming downstream to the ocean, they remain at sea for a year or two, feeding voraciously and growing rapidly. Eventually, salmon use sun-compass, geomagnetic, and chemical cues to return to the rivers and streams where they hatched. The fishes swim upstream, and each female lays hundreds or thousands of relatively small eggs. After spending all of their energy reserves on the upstream journey and reproduction, their condition deteriorates and they die.

Most deciduous trees in the temperate zone, such as oaks (genus *Quercus*) and maples (genus *Acer*), begin their lives as seeds in late summer. The seeds remain metabolically inactive until the following spring or a later year. After germinating, trees collect nutrients and energy and continue to grow throughout their lives. Once they achieve a critical size, they may produce thousands of seeds annually for many years. Thus, growth and reproduction occur simultaneously through much of the trees' life.

European red deer *(Cervus elaphus)* are born in spring, and young remain with their mothers for an extended period, nursing and growing rapidly. After weaning, they feed on their own. Female red deer begin to breed after reaching adult size in their third year, producing one or two offspring annually until they are about 16 years old, when they reach their maximum life span and die.

How can we summarize the similarities and differences in the life histories of these organisms? All three species harvest energy throughout their lives. Salmon and deciduous trees continue to grow until old age, whereas deer reach adult size fairly early in life. Salmon produce many offspring in a single reproductive episode, whereas deciduous trees and deer reproduce repeatedly. However, most trees produce thousands of seeds annually, whereas deer produce only one or two young each spring.

What factors have produced these variations in life history patterns? Life history traits—like all population characteristics—are modified by natural selection. Thus, organisms exhibit evolutionary adaptations that increase the fitness of individuals. Each species' life history is, in fact, a highly integrated "strategy"—not in the human sense of planning ahead, but as a suite of selection-driven adaptations.

Ecologists Analyze the Individual Components of Life Histories

In analyzing life histories, ecologists compare the number of offspring with the amount of care provided to each by the parents; they look at the number of reproductive episodes in the organism's lifetime; and they look at the timing of first reproduction. Because these characteristics evolve together, a change in one trait is likely to influence the success of the others.

Fecundity versus Parental Care. If a female has a fixed amount of energy for reproduction, she can package that energy in various ways. By way of illustration, a female duck with 1000 units of energy for reproduction might lay 10 eggs that each contain 100 units of energy per egg. A salmon, which has higher fecundity, might lay 1000 eggs, each endowed with 1 unit of energy. The amount of energy invested in each offspring *before* it is born represents the **passive parental care** that the female provides. Passive parental care is provided through yolk in an egg, endosperm in a seed, or, in mammals, nutrients that cross the placenta.

Many animals, especially birds and mammals, also provide **active parental care** to offspring *after* their birth. In general, species that produce many offspring in a reproductive episode—such as the coho salmon— provide relatively little active parental care *to each offspring*. In fact, female coho salmon, which produce 2400 to 4500 eggs, die before their eggs even hatch. Conversely, species that produce only a few offspring at a time—such as the European red deer—provide a lot of care to each. A red deer doe nurses its single fawn for up to 8 months before weaning it.

One Reproductive Episode versus Several. A second life history characteristic adjusted by natural selection is the number of reproductive episodes in an organism's lifetime. Some organisms, like the coho salmon, devote all of their stored energy to a single reproductive event. Any adult that survives the upstream migration is likely to leave some surviving offspring. Other species, like deciduous trees and red deer, reproduce multiple times. In contrast to salmon, individuals of these species devote only some of their energy budget to reproduction at any time, with the balance allocated to maintenance and growth. Moreover, in some plants, invertebrates, fishes, and reptiles, larger individuals produce more offspring than small ones do. Thus, one advantage of using only part of the energy budget for reproduction is that continued growth may result in greater fecundity at a later age. However, if an organism does not survive until the next breeding season, the potential advantage of putting energy into maintenance and growth is lost.

Early Reproduction versus Late Reproduction. Individuals that first reproduce at the earliest possible age may stand a good chance of leaving some surviving offspring. But the energy devoted to reproduction is no longer available for maintenance and growth. Thus, early reproducers may be smaller and less healthy than individuals that delay reproduction in favor of these other functions. Conversely, an individual that delays reproduction may increase its chance of survival and its future fecundity by becoming larger or more experienced. But there is always some chance that it will die before the next breeding season, leaving no offspring at all. Thus, a finite energy budget and the risk of mortality establish a trade-off in the timing of first reproduction. Mathematical models suggest that delayed reproduction will be favored by natural selection if a sexually mature individual has a good chance of surviving to an older age, if organisms grow larger as they age, and if larger organisms have higher fecundity. Early reproduction will be favored if adult survival rates are low, if animals don't grow larger as they age, or if larger size does not increase fecundity.

Life history characteristics not only vary from one species to another, but they also vary among populations of a single species. *Focus on Research* describes how predation influences life history characteristics in natural populations of guppies *(Poecilia reticulata)* in Trinidad.

STUDY BREAK

1. To what two broad categories of activities do children devote their energy budget?
2. Why do fecundity and the amount of parental care devoted to each offspring exhibit an inverse relationship?

49.5 Models of Population Growth

We now examine mathematical models of population growth that describe very different responses to changes in a population's density. *Exponential* models apply when populations experience unlimited growth. The *logistic* model applies when population growth is limited, often because available resources are finite. These simple models are tools that help ecologists refine their hypotheses, but neither provides entirely accurate predictions of population growth in nature. In the simplest versions of these models, ecologists define births as the production of offspring by any form

Basic Research: The Evolution of Life History Traits in Guppies

Some years ago, drenched with sweat and with fishnets in hand, two ecologists were engaged in fieldwork on the Caribbean island of Trinidad. They were after guppies *(Poecilia reticulata)*— small fish that bear live young in shallow mountain streams **(Figure a)**. John Endler and David Reznick, then of the University of California at Santa Barbara, were studying the environmental variables that influence the evolution of life history patterns in guppies.

Male guppies are easy to distinguish from females. Males, which stop growing at sexual maturity, are smaller, and their scales have bright colors that serve as visual signals in intricate courtship displays. The drably colored females continue to grow larger throughout their lives.

In the mountains of Trinidad, guppies living in different streams—and even in different parts of the same stream—are eaten by one of two other fish species **(Figure b)**. In some streams, a large pike-cichlid *(Crenicichla alta)* prefers mature guppies and tends not to spend time hunting small, immature ones. In other streams, a small killifish *(Rivulus hartii)* preys on immature guppies but does not have much success with the larger adults.

Reznick and Endler found that the life history patterns of guppies vary among streams with different preda-

tors. In streams with pike-cichlids, both male and female guppies mature faster and begin to reproduce at a smaller size and a younger age than their counterparts in streams where killifish live **(Figure c)**. In addition, female guppies from pike-cichlid streams reproduce more often and produce smaller and more numerous young **(Figure d)**. These differences allow guppies to avoid some predation. Those in pike-cichlid streams begin to reproduce when they are smaller than the size preferred by that predator. And those from killifish streams grow

Figure a
David Reznick surveys a shallow stream in the mountains of Trinidad.

David Reznick/University of California, Riverside

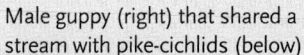
Male guppy (right) that shared a stream with pike-cichlids (below)

Male guppy (right) that shared a stream with killifish (below)

Figure b
Male guppies from streams where pike-cichlids live (top) are smaller, more streamlined, and have duller colors than those from streams where killifish live (bottom). The pike-cichlid prefers to eat large guppies, and the killifish feeds on small guppies. Guppies are shown approximately life-size, adult pike-cichlids grow to 16 cm in length, and adult killifish grow to 10 cm. (Guppy photos: David Reznick/University of California, Riverside: computer enhanced by Lisa Starr; predator photos: Hippocampus Bildarchiv.)

of reproduction, and ignore the effects of immigration and emigration.

Models of Exponential Population Growth Describe Growth without Limitation

Sometimes populations increase in size for a period of time with no apparent limits on their growth. In models of exponential growth, population size increases steadily by a constant ratio. Bacterial populations provide the most obvious examples, but multicellular organisms also sometimes exhibit exponential population growth.

Bacterial Population Growth. Bacteria reproduce by binary fission. A parent cell divides in half, producing two daughter cells, which each divide to produce two

granddaughter cells. Generation time in a bacterial population is simply the time between successive cell divisions. And if no bacteria in the population die, the population doubles in size each generation.

Bacterial populations grow quickly under ideal temperatures and with unlimited space and food. Consider a population of the human intestinal bacterium *Escherichia coli,* for which the generation time can be as short as 20 minutes. If we start with a population of one bacterium, the population doubles to two cells after one generation, to four cells after two generations, and to eight cells after three generations **(Figure 49.7)**. After only 8 hours, or 24 generations, the population will number more than 16 million. And after a single day, or 72 generations, the population will number nearly 5×10^{21} cells. Although other bacteria grow more slowly than *E. coli,* it is no wonder that patho-

quickly to a size that is too large to be consumed by killifish.

Although these life history differences were correlated with the distributions of the two predatory fishes, they might result from some other, unknown differences between the streams. Endler and Reznick investigated this possibility with controlled laboratory experiments. They shipped groups of guppies to California, where they bred guppies from each kind of stream for two generations. Both types of experimental populations were raised under identical conditions in the absence of predation. Even when predators were absent, the two types of experimental populations retained their life history differences. These results provided evidence of a heritable genetic basis for the observed life history differences.

Endler and Reznick also examined the role of predators in the *evolution* of the size differences. They raised guppies for many generations in the laboratory under three experimental conditions—some alone, some with killifish, and some with pike-cichlids. As predicted, the guppy lineage that was subjected to predation by killifish became larger at maturity. Individuals that were small at maturity were frequently eaten, and their reproduction was limited. The lineage that was

raised with pike-cichlids showed a trend toward earlier maturity. Individuals that matured at a larger size faced a greater likelihood of being eaten before they had reproduced.

Finally, when they first visited Trinidad, Endler and Reznick had introduced guppies from a pike-cichlid stream to another stream that contained killifish but no pike-cichlids or guppies. Eleven years later, the guppy populations had changed. As the researchers predicted, the guppies had become larger in size and reproduced more slowly, characteristics that are typical of natural guppy populations that live and die with killifish.

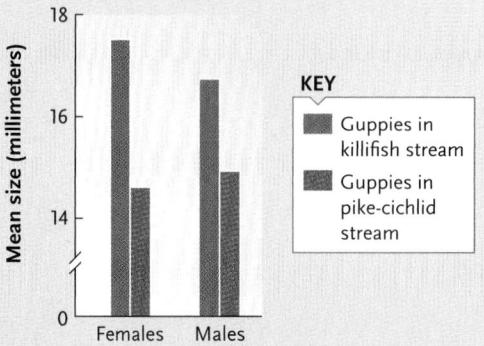

Figure c
Guppies in streams occupied by pike-cichlids are smaller than those in streams occupied by killifish.

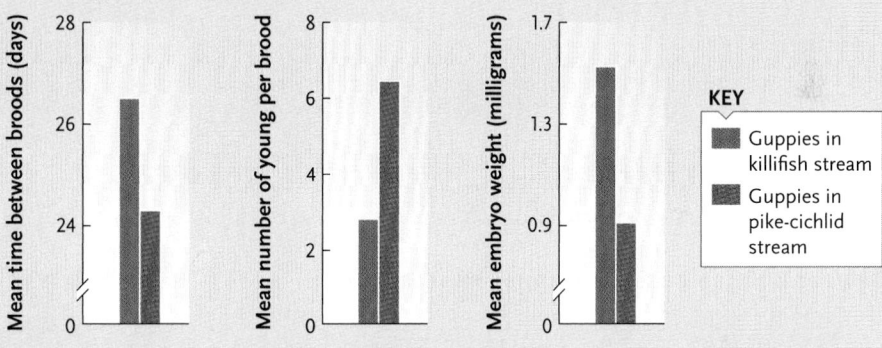

Figure d
Female guppies from streams occupied by pike-cichlids reproduce more often (shorter time between broods) and produce more young per brood and smaller young (lower embryo weight) than females living in streams occupied by killifish.

genic bacteria, such as those causing cholera or plague, can quickly overtake the defenses of an infected animal.

Exponential Population Growth in Other Organisms.
By contrast to bacteria, many plants and animals live side-by-side with their offspring. In these populations, births increase a population's size and deaths decrease it. Over a given time period:

change in population size =
        number of births − number of deaths

We express this relationship mathematically by defining N as the population size; ΔN (pronounced "delta N") as the change in population size; Δt as the time period during which the change occurs; and B and D as the numbers of births and deaths, respectively, *dur-*

ing that time period. Thus, $\Delta N/\Delta t$ symbolizes the change in population size over time, and

$$\Delta N/\Delta t = B - D$$

The preceding equation applies to any population for which we know the exact numbers of births and deaths.

Ecologists usually express births and deaths as *per capita* (per individual) rates, allowing them to apply the model to a population of any size. The per capita birth rate, symbolized b, is simply the number of births in the population during the specified time period divided by the population size: $b = (B/N)$. Similarly, the per capita death rate, d, is the number of deaths divided by the population size: $d = (D/N)$. If, for example, in a population of 2000 field mice, 1000 mice are born and 200 mice die during 1 month's time, $b = 1000/2000 =$

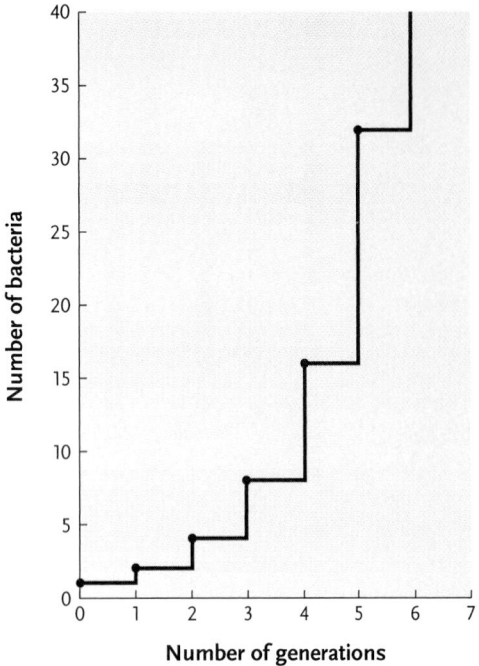

Figure 49.7

Bacterial population growth. If all members of a bacterial population divide simultaneously, a plot of population size over time forms a stair-stepped curve in which the steps get larger as the number of dividing cells increases.

0.5 births per individual per month, and $d = 200/2000 = 0.1$ deaths per individual per month. Of course, no mouse can give birth to half an offspring, and no individual can die one-tenth of a death. But these rates tell us the per capita birth and death rates *averaged over all mice in the population*. Per capita birth and death rates are always expressed over a specified time period. For long-lived organisms, such as humans, time is measured in years; for short-lived organisms, such as fruit

flies, time is measured in days. We can calculate per capita birth and death rates from data in a life table.

We can now revise the population growth equation to use per capita birth and death rates instead of the actual numbers of births and deaths. The change in a population's size during a given time period ($\Delta N/\Delta t$) depends on the per capita birth and death rates, as well as on the number of individuals in the population. Mathematically, we can write

$$\Delta N/\Delta t = B - D = bN - dN = (b - d)N$$

or, in the notation of calculus,

$$dN/dt = (b - d)N$$

This equation describes the **exponential model of population growth.** (Note that in calculus, dN/dt is the notation for the population growth rate; the "d" in dN/dt is *not* the same "d" that we use to symbolize the per capita death rate.)

The difference between the per capita birth rate and the per capita death rate, $b - d$, is the **per capita growth rate** of the population, symbolized by r. Like b and d, r is always expressed per individual per unit time. Using the per capita growth rate, r, in place of $(b - d)$, the exponential growth equation is written

$$dN/dt = rN$$

If the birth rate exceeds the death rate, r has a positive value ($r > 0$), and the population is growing. In our example with field mice, $r = 0.5 - 0.1 = 0.4$ mice per mouse per month. If, on the other hand, the birth rate is lower than the death rate, r has a negative value ($r < 0$), and the population is getting smaller. In populations where the birth rate equals the death rate, r is exactly zero, and the population's size is not changing—a situation known as **zero population**

Figure 49.8

Exponential population growth. Exponential population growth produces a J-shaped curve of population size plotted against time. Although the per capita growth rate (*r*) remains constant, the increase in population size gets larger every month, because more individuals are reproducing.

Month	Old Population Size		Net Monthly Increase		New Population Size
1	2,000	+	800	=	2,800
2	2,800	+	1,120	=	3,920
3	3,920	+	1,568	=	5,488
4	5,488	+	2,195	=	7,683
5	7,683	+	3,073	=	10,756
6	10,756	+	4,302	=	15,058
7	15,058	+	6,023	=	21,081
8	21,081	+	8,432	=	29,513
9	29,513	+	11,805	=	41,318
10	41,318	+	16,527	=	57,845
11	57,845	+	23,138	=	80,983
12	80,983	+	32,393	=	113,376
13	113,376	+	45,350	=	158,726
14	158,726	+	63,490	=	222,216
15	222,216	+	88,887	=	311,103
16	311,103	+	124,441	=	435,544
17	435,544	+	174,218	=	609,762
18	609,762	+	243,905	=	853,667
19	853,677	+	341,467	=	1,195,134

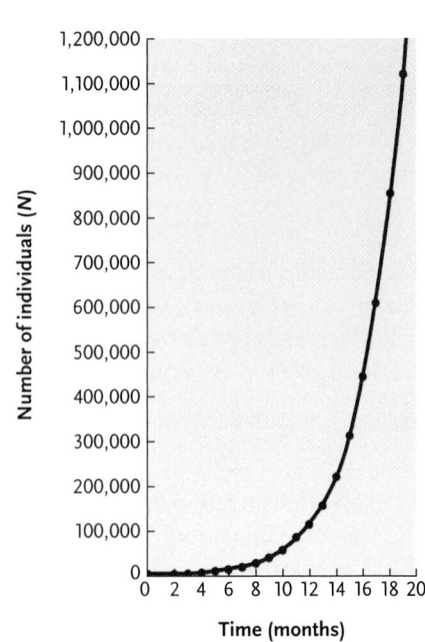

growth, or ZPG. Even under conditions of ZPG, births and deaths still occur, but the numbers of births and deaths cancel out.

As long as a population's per capita growth rate is positive ($r > 0$), the population will increase in size. In our hypothetical population of field mice, we started with $N = 2000$ mice, and calculated a per capita growth rate of 0.4 mice per individual per month. In the first month, the population grows by $0.4 \times 2000 = 800$ mice **(Figure 49.8).** At the start of the second month, $N = 2800$ and r still $= 0.4$. Thus, in the second month, the population grows by $0.4 \times 2800 = 1120$ mice. Notice that even though r remains constant, the *increase* in population size gets larger each month simply because more individuals are reproducing. In less than 2 years, the mouse population will grow to more than 1 million! A graph of exponential population growth has a characteristic J shape, getting steeper through time. The population grows at an ever-increasing pace because the change in a population's size depends on the number of individuals in the population as well as its per capita growth rate.

Population Growth under Ideal Conditions. Imagine a hypothetical population living in an ideal environment—one with unlimited food and shelter; no predators, parasites, or disease; and a comfortable abiotic environment. Under such circumstances, which are admittedly unrealistic, the per capita birth rate is very high, the per capita death rate is very low, and the per capita growth rate, r, is as high as it can possibly be. This maximum per capita growth rate, symbolized r_{max}, is the population's **intrinsic rate of increase.** Under these ideal conditions, our exponential growth equation is

$$dN/dt = r_{max}N$$

When populations are growing at their intrinsic rate of increase, population size increases very rapidly. Across a wide variety of protists and animals r_{max} varies inversely with generation time: species with short generation time have higher intrinsic rates of increase than those with long generation time **(Figure 49.9).**

The Logistic Model Describes Population Growth When Resources Are Limited

The exponential model predicts unlimited population growth. But we know from even casual observations that the population sizes of most species are somehow limited—we are not knee-deep in bacteria, rosebushes, or garter snakes. What factors limit the growth of populations? As a population gets larger, it uses more vital resources, and a shortage of resources may eventually develop. As a result, individuals may have less energy available for maintenance and reproduction, causing per capita birth rates to decrease and per capita death rates to increase. Changes in these rates reduce the

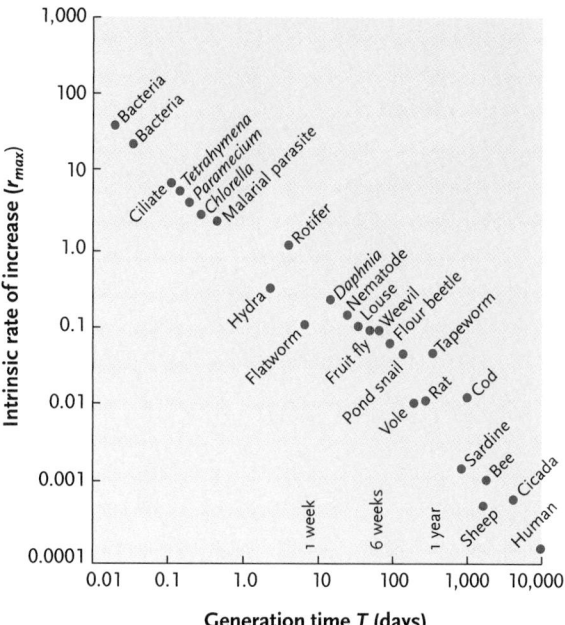

Figure 49.9
Generation time and r_{max}. The intrinsic rate of increase (r_{max}) is high for protists and animals with short generation time and low for those with long generation time. The logarithmic scale on both axes compresses the data into a straight line.

population's per capita growth rate, causing population growth to slow down or to stop altogether.

The Logistic Model. Environments provide enough resources to sustain only a finite population of any species. The maximum number of individuals that an environment can support indefinitely is termed its **carrying capacity,** symbolized K. The carrying capacity, which is defined for each population, is a property of the environment, and it varies from one habitat to another and in a single habitat through time. For example, the spring and summer flush of insects in temperate habitats supports large populations of insectivorous birds. But fewer insects are available in autumn and winter, causing a seasonal decline in the carrying capacity *for birds.* Many birds then migrate to habitats that provide more food and better weather.

The **logistic model of population growth** assumes that a population's per capita growth rate, r, decreases as the population gets larger **(Figure 49.10a).** In other words, population growth slows as the population size approaches the carrying capacity. The mathematical expression ($K - N$) tells us how many individuals can be added to a population before it reaches carrying capacity. And the expression ($K - N$)/K indicates what *percentage* of the carrying capacity is still available.

To create the logistic model, we factor the impact of carrying capacity into the exponential model by letting $r = r_{max}(K - N)/K$. This calculation reduces the per capita growth rate (r) from its maximum value (r_{max}) as N increases:

$$dN/dt = r_{max}N(K - N)/K$$

The calculation of how r varies with population size is straightforward **(Table 49.2).** In a very small population (N much smaller than K), plenty of resources are still

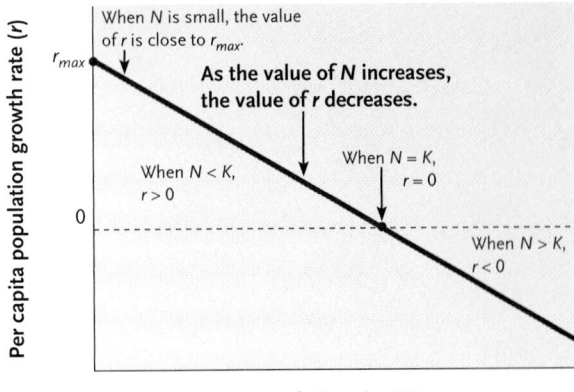

a. The predicted effect of *N* on *r*

When *N* is small, the value of *r* is close to r_{max}.

As the value of *N* increases, the value of *r* decreases.

When *N* < *K*, *r* > 0

When *N* = *K*, *r* = 0

When *N* > *K*, *r* < 0

Per capita population growth rate (*r*)

Population size (*N*)

b. Population size through time

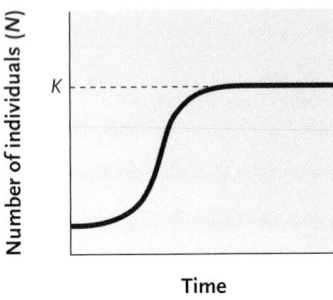

Number of individuals (*N*)

Time

Figure 49.10

The logistic model of population growth. **(a)** The logistic model assumes that the per capita population growth rate *(r)* decreases linearly as population size *(N)* increases. **(b)** The logistic model predicts that population size increases quickly at first, but then slowly approaches the carrying capacity *(K)*.

Table 49.2	The Effect of *N* on *r* and ΔN^* in a Hypothetical Population Exhibiting Logistic Growth in which $K = 2000$ and $r_{max} = 0.04$ per capita per year		
N (population size)	$(K - N)/K$ (% of *K* available)	$r = r_{max}(K - N/K)$ (per capita growth rate)	$\Delta N = rN$ (change in *N*)
50	0.990	0.0396	2
100	0.950	0.0380	4
250	0.875	0.0350	9
500	0.750	0.0300	15
750	0.625	0.0250	19
1000	0.500	0.0200	20
1250	0.375	0.0150	19
1500	0.250	0.0100	15
1750	0.125	0.0050	9
1900	0.050	0.0020	4
1950	0.025	0.0010	2
2000	0.000	0.0000	0

*ΔN rounded to the nearest whole number.

available; the value of $(K - N)/K$ is close to 1, and the per capita growth rate *(r)* is therefore close to the maximum possible (r_{max}). Under these conditions, population growth is close to exponential. If a population is large (*N* close to *K*), few additional resources are available; the value of $(K - N)/K$ is small, and the per capita growth rate *(r)* is very low. When the size of the population exactly equals the carrying capacity, $(K - N)/K$ becomes zero, and so does the population growth rate—the situation defined as ZPG.

The logistic model of population growth predicts an S-shaped graph of population size over time, with the population slowly approaching its carrying capacity and remaining at that level **(Figure 49.10b)**. According to this model, the population grows slowly when the population size is small, because there are few individuals reproducing. It also grows slowly when the population size is large because the per capita population growth rate is low. The population grows quickly (*dN/dt* is highest) at intermediate population sizes, when a sizable number of individuals are breeding and the per capita population growth rate *(r)* is still fairly high (see Table 49.2).

Intraspecific Competition. The logistic model assumes that vital resources become increasingly limited as a population grows larger. Thus, the model is a mathematical portrait of **intraspecific** (within species) **competition,** the dependence of two or more individuals in a population on the same limiting resource. For animals, limiting resources can be food, water, nesting sites, refuges from predators, and, for sessile species (those permanently attached to a surface), space. For plants, sunlight, water, inorganic nutrients, and growing space can be limiting. The pattern of uniform dispersion described earlier often reflects intraspecific competition for limited resources.

In some very dense populations, the accumulation of poisonous waste products may also reduce survivorship and reproduction. Most natural populations live in open systems where wastes are consumed by other organisms or flushed away. But the buildup of toxic wastes is common in laboratory cultures of microorganisms. For example, yeast cells ferment sugar and produce ethanol as a waste product. Thus, the alcohol content of wine rarely exceeds 13% by volume, the ethanol concentration that poisons winemaking yeasts.

Logistic Growth in the Laboratory and in Nature. How well do species conform to the predictions of the logistic model? In simple laboratory cultures, relatively small organisms, such as *Paramecium,* some crustaceans, and flour beetles, often show an S-shaped pattern of population growth **(Figure 49.11)**. Moreover, large animals that have been introduced into new environments sometimes exhibit a pattern of population

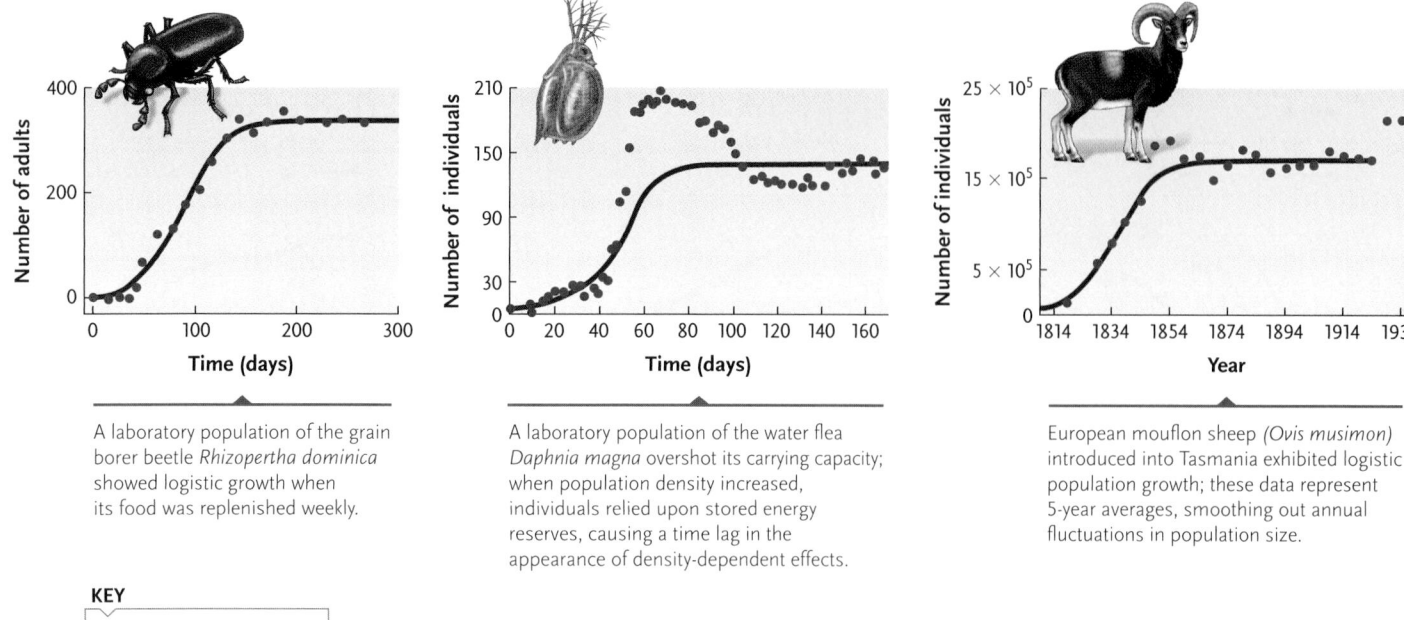

A laboratory population of the grain borer beetle *Rhizopertha dominica* showed logistic growth when its food was replenished weekly.

A laboratory population of the water flea *Daphnia magna* overshot its carrying capacity; when population density increased, individuals relied upon stored energy reserves, causing a time lag in the appearance of density-dependent effects.

European mouflon sheep *(Ovis musimon)* introduced into Tasmania exhibited logistic population growth; these data represent 5-year averages, smoothing out annual fluctuations in population size.

KEY

— Theoretical • Data

Figure 49.11
Examples of logistic population growth.

growth that matches the predictions of the logistic model (see Figure 49.11).

Nevertheless, some assumptions of the logistic model are unrealistic. For example, the model predicts that survivorship and fecundity respond immediately to changes in a population's density. But many organisms exhibit a delayed response, called a **time lag.** Some time lags occur because fecundity is usually determined by the availability of resources at some time in the past, when individuals were adding yolk to eggs or endosperm to seeds. Moreover, when food resources become scarce, individuals may use stored energy reserves to survive and reproduce, and the effects of crowding may not be felt until those reserves are depleted. As a result, the population size may temporarily overshoot its carrying capacity (see Figure 49.11b). Deaths may then outnumber births, causing the population size to drop below the carrying capacity, at least temporarily. Time lags often cause a population to oscillate around its carrying capacity.

Another unrealistic assumption of the logistic model is that the addition of new individuals to a population always decreases survivorship and fecundity, no matter how small the population is. But in small populations, modest population growth probably doesn't have much effect on these processes. In fact, most organisms probably require a minimum population density to survive and reproduce. For example, some plants flourish in small clumps that buffer them from physical stresses, whereas a single individual living in the open would suffer adverse effects. And in some animal populations, a minimum population density is necessary for individuals to find mates—an important issue in conservation biology (see Chapter 53).

STUDY BREAK

1. How does the prediction of the exponential model of population growth differ from that of the logistic model?
2. What is carrying capacity? Is it a property of a habitat or of a population?
3. What is a time lag?

49.6 Population Regulation

As you have seen, the population sizes of some species change from month to month or from year to year, whereas others remain fairly stable. What environmental factors influence population growth rates and control fluctuations in population size?

Density-Dependent Factors Often Regulate Population Size

Some factors that affect population size are **density-dependent:** their influence increases or decreases with the density of the population. Examples of density-dependent environmental factors include intraspecific competition and predation. The logistic model includes the effects of density-dependence in its assumption that per capita birth and death rates change with a population's density.

The Effects of Crowding. Numerous laboratory and field studies show that crowding (high population density) decreases the individual growth rate, adult size,

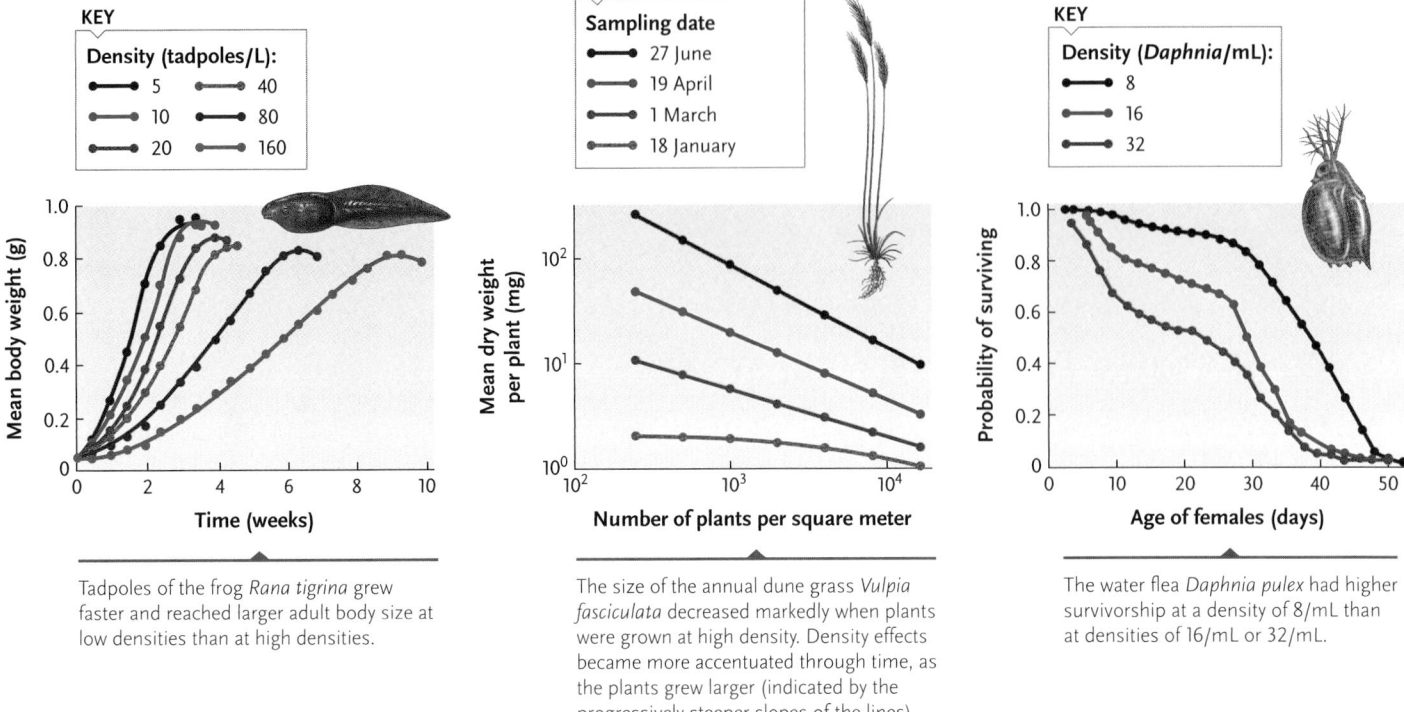

KEY
Density (tadpoles/L):
- 5
- 10
- 20
- 40
- 80
- 160

KEY
Sampling date
- 27 June
- 19 April
- 1 March
- 18 January

KEY
Density (*Daphnia*/mL):
- 8
- 16
- 32

Tadpoles of the frog *Rana tigrina* grew faster and reached larger adult body size at low densities than at high densities.

The size of the annual dune grass *Vulpia fasciculata* decreased markedly when plants were grown at high density. Density effects became more accentuated through time, as the plants grew larger (indicated by the progressively steeper slopes of the lines).

The water flea *Daphnia pulex* had higher survivorship at a density of 8/mL than at densities of 16/mL or 32/mL.

Figure 49.12
Effects of crowding on individual growth, size, and survival.

Figure 49.13
Effects of crowding on fecundity.

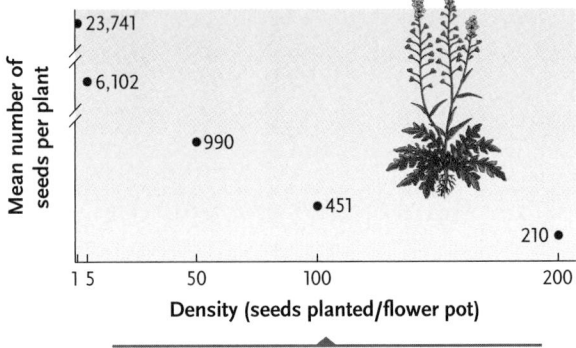

The number of seeds produced by shepherd's purse (*Capsella bursa-pastoris*) decreased dramatically with increasing density in experimental plots.

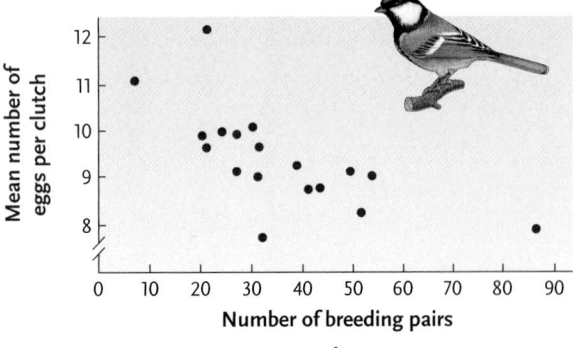

The mean number of eggs produced by the great tit (*Parus major*), a woodland bird, declined as the number of breeding pairs in Marley Wood increased.

and survival of plants and animals **(Figure 49.12)**. Organisms living in extremely dense populations are unable to harvest enough resources; they grow slowly and tend to be small, weak, and less likely to survive. Gardeners understand this relationship, thinning their plants to a density that maximizes the number of vigorous individuals.

Crowding also has a negative effect on reproduction **(Figure 49.13)**. When resources are in short supply, each individual has less energy available for reproduction after meeting its basic maintenance needs. Hence, females in crowded populations produce either fewer offspring or smaller offspring that are less likely to survive.

In some species, crowding stimulates developmental and behavioral changes that may influence the density of a population. For example, migratory locusts (*Locusta migratoria*) can develop into either solitary or migratory forms in the same population. Migratory individuals have longer wings and more body fat, characteristics that allow them to disperse great distances. High population density increases the frequency of the migratory form, and huge numbers of locusts move away from the area of high density **(Figure 49.14)**, reducing the size of the original population.

These studies confirm the assumptions of the logistic equation, but they don't prove that natural populations are regulated by density-dependent factors. A convincing demonstration requires experimental evi-

Figure 49.14

A swarm of locusts. Migratory locusts *(Locusta migratoria),* moving across an African landscape, can devour their own weight in plant material every day.

dence that an increase in population density causes population size to decrease, and that a decrease in density causes it to increase. In one study conducted in the 1960s, Robert Eisenberg of the University of Michigan experimentally increased the numbers of aquatic snails in some ponds, decreased them in others, and maintained natural densities in control ponds. Although adult survivorship did not differ between experimental and control treatments, snails in the high-density ponds produced fewer eggs, and those in the low-density ponds produced more eggs, than those living at the control density. In addition, the survival rates of young snails declined as density increased. After 4 months, the densities in the two experimental groups converged on those in the control, providing strong evidence of density-dependent population regulation.

Other Density-Dependent Factors. Our discussion of the logistic equation described intraspecific competition as the primary density-dependent factor regulating population size. Competition between populations of different species also exerts density-dependent effects on population growth, a topic we consider in Section 50.1.

Predation can also cause density-dependent population regulation. As a particular prey species becomes more numerous, predators may consume more of it because it is easier to find and catch. Once a prey species has exceeded a threshold density, predators may consume a *larger percentage* of the prey population, which is a density-dependent effect (see Figure 20.16). For example, on rocky shores in California, sea stars concentrate their feeding on the most abundant of several invertebrate species. When one prey species becomes common, predators feed on it disproportionately, drastically reducing its numbers.

Like predation, parasitism and disease cause density-dependent regulation of plant and animal populations. Infectious microorganisms spread quickly

in a crowded population. In addition, if crowded individuals are weak or malnourished, they are more susceptible to infection and may die from diseases that healthy organisms would survive.

Density-Independent Factors Can Limit Population Size

Some populations are affected by **density-independent** factors that reduce population size regardless of its density. If an insect population is not physiologically adapted to high temperature, a sudden hot spell may kill 80% of the insects whether they number 100 or 100,000. Fires, earthquakes, storms, and other natural disturbances may contribute directly or indirectly to density-independent mortality. But because such factors do not cause a population to fluctuate around its carrying capacity, density-independent factors do not *regulate* population size, although they may reduce it.

Density-independent factors have a particularly strong effect on populations of small-bodied species that cannot buffer themselves against environmental change. Their populations grow exponentially for a time, but shifts in climate or random events cause high mortality before populations reach a size at which density-dependent factors regulate their numbers. When conditions improve, populations grow exponentially—at least until another density-independent factor causes them to crash again. For example, a small Australian insect, *Thrips imaginis,* feeds on pollen and flowers of plants in the rose family; they are frequently abundant enough to damage the blooms. *Thrips* populations grow exponentially in spring, when many flowers are available and the weather is warm and moist **(Figure 49.15).** But populations crash predictably during summer because *Thrips* do not tolerate extremely hot

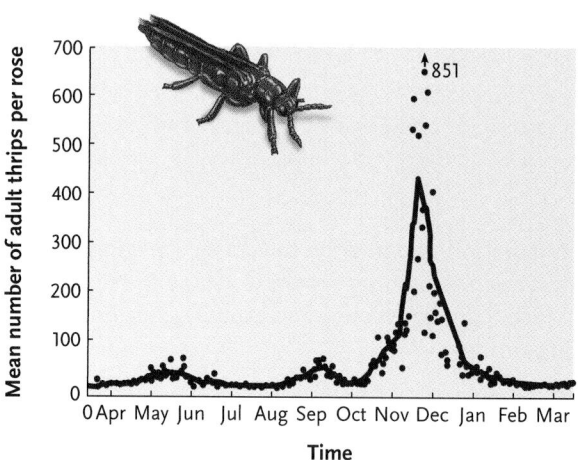

Figure 49.15

Booms and busts in a *Thrips* population. Populations of the Australian insect *Thrips imaginis* grow exponentially when conditions are favorable during spring (which begins in September in the southern hemisphere). The populations crash in summer, however, when hot and dry conditions cause high mortality.

Figure 49.16 Experimental Research

Evaluating Density-Dependent Interactions between Species

QUESTION: Does the population density of lizards on Caribbean islands have any effect on the population density of spiders?

EXPERIMENT: Spiller and Schoener built fences around a series of study plots on a small island in the Bahamas. They excluded all individuals of three lizard species from the experimental plots, but left resident lizards undisturbed in the control plots. They then made monthly measurements of population densities of the web-building spider *Metepeira datona* in both experimental plots and control plots.

RESULTS: Over the 20-month course of the experiment, spider densities were as much as five times higher in the experimental plots than in the control plots.

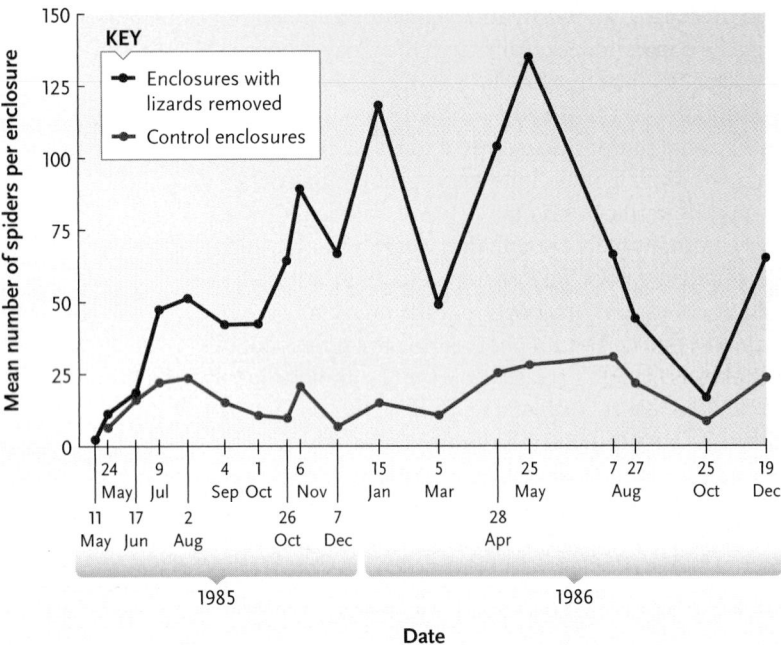

CONCLUSION: Spiller and Schoener concluded that the presence of lizards has a large impact on spider populations. The lizards not only compete with the spiders for insect food, but they also appear to prey on the spiders.

Interacting Environmental Factors Often Limit Population Sizes

Sometimes several density-dependent factors influence a population at the same time. For example, on small islands in the West Indies, the spider *Metepeira datona* is rare wherever lizards (*Ameiva festiva, Anolis*

carolinensis, and *Anolis sagrei*) are abundant, but common where the lizards are rare or absent. To test whether the presence of lizards limits the abundance of spiders, David Spiller and Tom Schoener of the University of California, Davis, built fences around plots on islands where these species occur. They eliminated lizards from experimental plots, but left them in control plots. After 2 years, spider populations in some experimental plots were five times denser than those in control plots **(Figure 49.16).** In this case, lizards had two density-dependent effects on spider populations: they preyed upon spiders, and they competed with them for food.

Density-dependent factors can also interact with density-independent factors, limiting population growth. For example, food shortage caused by high population density (a density-dependent factor) may lead to malnourishment; in turn, malnourished individuals may be more likely to succumb to the stress of extreme weather (a density-independent factor).

Populations can also be affected by density-independent factors in a density-dependent manner. For example, animals often retreat into shelters to escape environmental stresses, such as floods or severe heat. If a population is small, most individuals can fit into a limited number of available refuges. But if a population is large, only a small proportion will find suitable shelter; and the larger the population is, the greater the percentage of individuals that will experience the stress. Thus, although the density-independent effects of weather limit *Thrips* populations, it is the availability of flowers in summer—clearly a density-dependent factor—that regulates the size of the *Thrips* starting stock the following spring. Hence, both types of factors influence the size of *Thrips* populations.

The Life History Characteristics of a Species Govern Fluctuations in Its Population Size through Time

Even casual observation reveals tremendous variation in how rapidly population size changes in different species. For example, new weeds often appear in a vegetable garden overnight, whereas the number of oak trees in a forest may remain relatively stable for years. Why do some species have the potential for explosive population growth, but others do not? The answer lies in how natural selection has molded life history strategies that are adapted to different ecological conditions. Ecologists describe two divergent life history patterns—**r-selected** species and **K-selected** species—with very different characteristics **(Table 49.3, Figure 49.17).** These strategies represent extremes on a continuum of possible patterns, and the life histories of most species actually fall somewhere between them.

Species with an *r*-selected life history are adapted to function well in rapidly changing environments. They are generally small, have short generation times,

and dry conditions. After the crash, a few individuals survive in remaining flowers, forming the stock from which the population grows exponentially the following spring.

and produce numerous, tiny offspring, often in a single reproductive event. The offspring receive little or no parental care of any kind. Because species with short generation times tend to have high r_{max} (see Figure 49.9), their populations grow exponentially when environmental conditions are favorable—hence the name *r*-selected. Although their numerous offspring disperse and rapidly colonize available habitats, most die before reaching sexual maturity (Type III survivorship). Thus, the success of an *r*-selected life history depends on flooding the environment with a *large quantity* of young, only a few of which may be successful.

Because they have small body size, *r*-selected species lack physiological mechanisms to buffer them from environmental variation. Thus, as described earlier for the Australian thrips living in roses, survivorship and fecundity are often greatly influenced by density-independent factors, and population size fluctuates markedly. In good years, survivorship and fecundity may be high, and the population explodes. In bad years, survivorship and reproduction may be low, and the population crashes. Populations of *r*-selected species are often so greatly reduced by changes in abiotic environmental factors, such as temperature or moisture, that they never grow large enough to face a shortage of limiting resources; thus, their carrying capacity cannot be estimated, and changes in their population size cannot be described by the logistic model of population growth.

By contrast, *K*-selected species thrive in more stable environments. They are generally large, have long generation times, and produce offspring repeatedly during their lifetimes. Their offspring receive substantial parental care, either as energy reserves in an egg or seed or as active care, ensuring that most survive the early stages of life (Type I or Type II survivorship). Because *K*-selected species typically have a low r_{max}, their populations often grow slowly. The success of a *K*-selected life history therefore depends on the production of a relatively small number of *high quality* offspring that join an already well-established population.

The large body size of *K*-selected species allows them to use behavioral and physiological mechanisms to buffer themselves against environmental change, so that survivorship and fecundity do not fluctuate wildly in response to environmental variations. Instead, their populations are often affected by density-dependent factors, which regulate population size near their carrying capacity—hence the name *K*-selected. For these species, natural selection has favored life history characteristics that result in stable population sizes: the production of relatively few offspring, extensive parental care, good competitive ability, a long life span, and repeated reproductions. Many large terrestrial vertebrates are examples of *K*-selected species.

Table 49.3	Characteristics of *r*-Selected and *K*-Selected Species	
Characteristic	***r*-Selected Species**	***K*-Selected Species**
Maturation time	Short	Long
Life span	Short	Long
Mortality rate	Usually high	Usually low
Reproductive episodes	Usually one	Usually several
Time of first reproduction	Early	Late
Clutch or brood size	Usually large	Usually small
Size of offspring	Small	Large
Active parental care	Little or none	Often extensive
Population size	Fluctuating	Relatively stable
Tolerance of environmental change	Generally poor	Generally good

Some Species Exhibit Regular Cycles in Population Size

The population densities of many insects, birds, and mammals in the northern hemisphere fluctuate between species-specific lows and highs in a multiyear cycle. Arctic populations of small rodents vary in size over a 4-year cycle, whereas snowshoe hares, ruffed grouse, and lynxes have 10-year cycles. Ecologists documented such cyclic fluctuations more than a century ago, but none of the general hypotheses so far proposed explains the cycles in all species. The availability and quality of food, the abundance of predators, the prevalence of disease-causing microorganisms, and variations in weather may influence population growth. Furthermore, a cycling population's food supply and

a. An *r*-selected species

b. A *K*-selected species

Figure 49.17

Life history differences. **(a)** An *r*-selected species, like quinoa (*Chenopodium quinoa*), matures in one growing season and produces many tiny seeds, which were a traditional food staple for the indigenous people of North and South America. **(b)** A *K*-selected species, like the coconut palm (*Cocos nucifera*), grows slowly and produces a few large seeds repeatedly during its long life.

predators are themselves influenced by the population's size.

Theories of *intrinsic control* suggest that as an animal population grows, individuals undergo hormonal changes that increase aggressiveness, reduce reproduction, and foster dispersal to other areas. The dispersal phase of the cycle may be dramatic. For example, when populations of the Norway lemming *(Lemmus lemmus),* a rodent that lives in the Scandinavian arctic, reach their peak density, aggressive interactions drive younger and weaker individuals away from their place of birth. The exodus of many thousands of lemmings, scrambling over rocks and even cliffs, is sometimes incorrectly portrayed in nature films as a suicidal mass migration. Researchers do not yet know how widespread these hormonal and behavioral changes are among different species or exactly what regulates them.

Other explanations focus on *extrinsic control,* such as the relationship between a cycling species and its food or predators. A dense population may exhaust its food supply, increasing mortality and decreasing reproduction. But experimental food supplementation does not always prevent the decline in mammal populations, indicating that other factors are also at work.

Some researchers have suggested that the cycles of predators and their prey are induced by time lags in each population's response to changes in density of the other **(Figure 49.18).** The 10-year cycles of snowshoe hares *(Lepus americanus)* and their feline predators, Canada lynxes *(Lynx canadensis),* were often cited as a classic example of such an interaction. But recent research has cast doubt on this straightforward explanation. Hare populations exhibit a 10-year fluctuation even on islands where lynxes are absent. Thus, the lynx cannot be solely responsible for the hare's cycle, although cycles in the hare populations may trigger cycles in populations of their predators.

Charles Krebs and his colleagues at the University of British Columbia studied hare and lynx interactions with a large-scale, multiyear experiment in the southern Yukon. They fenced experimental areas where they added food for the hares, excluded mammalian predators, or applied both experimental treatments; unmanipulated plots served as controls. Where mammalian

© Ed Cesar/Photo Researchers, Inc.

Figure 49.18

The predator-prey model. Predator-prey interactions may contribute to density-dependent regulation of both populations. **(a)** A mathematical model predicts cycles in the numbers of predators and prey because of time lags in each species' responses to changes in the density of the other. (Predator population size is exaggerated in this graph: predators are usually less common than prey.) **(b)** The interaction between the Canada lynx *(Lynx canadensis)* and the snowshoe hare *(Lepus americanus)* was often described as a cyclic predator-prey interaction. The abundances of lynx and hare are based on counts of pelts that trappers sold to Hudson's Bay Company over a 90-year period. Recent research has shown that population cycles in snowshoe hares are caused by complex interactions between the hare, its food plants, and its predators.

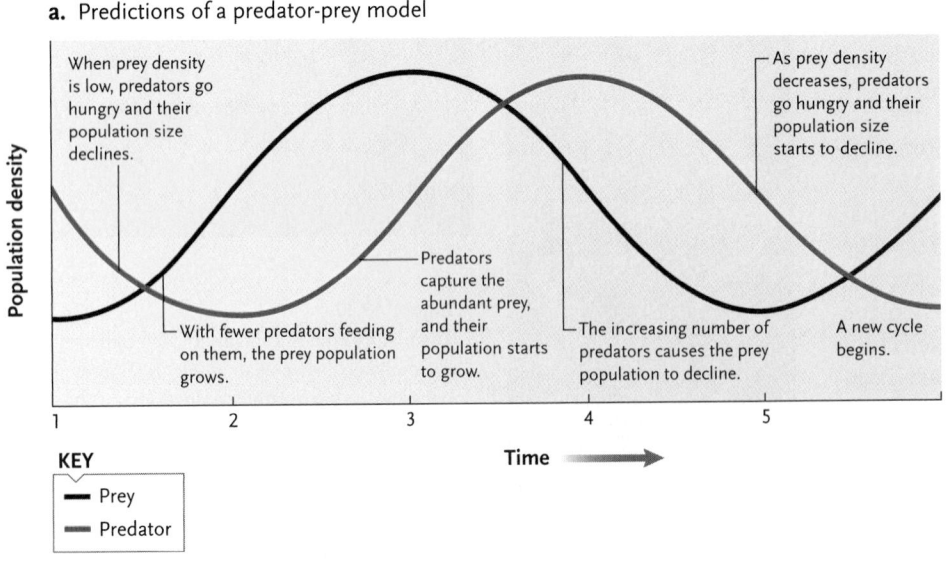

a. Predictions of a predator-prey model

When prey density is low, predators go hungry and their population size declines.

With fewer predators feeding on them, the prey population grows.

Predators capture the abundant prey, and their population starts to grow.

The increasing number of predators causes the prey population to decline.

As prey density decreases, predators go hungry and their population size starts to decline.

A new cycle begins.

Population density

Time

KEY
— Prey
— Predator

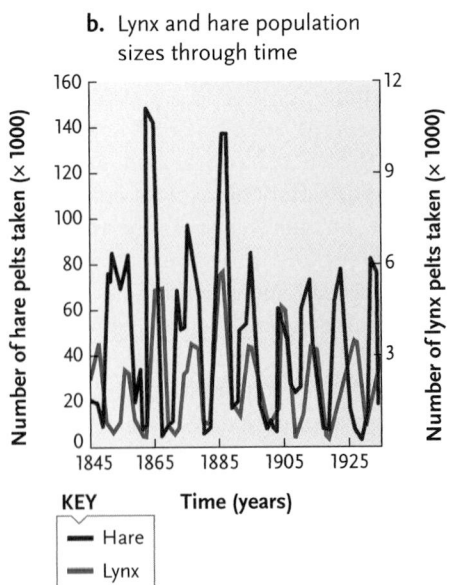

b. Lynx and hare population sizes through time

Number of hare pelts taken (× 1000)

Number of lynx pelts taken (× 1000)

Time (years)

KEY
— Hare
— Lynx

predators were excluded, hare densities approximately doubled relative to the controls. Where food was added, hare densities tripled. But in plots where food was added *and* predators were excluded, the hare densities increased 11-fold. Krebs and his colleagues concluded that neither food availability nor predation alone is solely responsible for arctic hare population cycles; instead, complex interactions between the hares, their food plants, and their predators create the cyclic fluctuations in hare population size.

STUDY BREAK

1. How can you tell whether an environmental factor causes density-dependent or density-independent effects on a population?
2. Are the effects of infectious diseases on populations more likely to be density-dependent or density-independent?

49.7 Human Population Growth

How do human populations compare with those of other species we have studied? The worldwide human population surpassed 6 billion on October 12, 1999. Like many other species, humans live in somewhat isolated populations, which vary in their demographic traits and access to resources. Although many of us live comfortably, at least a billion people are malnourished or starving, lack clean drinking water, and live without adequate shelter or health care. Even if it were possible to double the food supply, increased agricultural production would inevitably increase pollution and contribute to spoiled croplands, deforestation, and desertification, which are described in Chapter 53.

Human Populations Have Sidestepped the Usual Density-Dependent Controls

For most of human history, our population grew slowly; but over the past two centuries, the worldwide human population has grown exponentially **(Figure 49.19)**. Demographers have identified three ways in which humans have avoided the effects of density-dependent regulating factors.

First, humans have expanded their geographical range into virtually every terrestrial habitat. Our early ancestors lived in tropical and subtropical grasslands, but by 40,000 years ago, they had dispersed through much of the world (see Section 30.13). Their success resulted from their ability to solve ecological problems by building fires, assembling shelters, making clothing and tools, and planning community hunts. Vital survival skills spread from generation to generation and from one population to another because language allowed the communication of complex ideas and knowledge.

Second, humans have increased the carrying capacities of habitats they occupy. About 11,000 years ago, many populations shifted from hunting and gathering to agriculture. They cultivated wild grasses, diverted water to irrigate crops, and used domesticated animals for food and labor. Such innovations increased the availability of food, raising both the carrying capacity and the population growth rates. In the mid-eighteenth century, people harnessed the energy in fossil fuels, and industrialization began in Western Europe and North America. Food supplies and the carrying capacity increased again, at least in the industrialized countries, through the use of synthetic fertilizers, pesticides, and efficient methods of transportation and food distribution.

Third, advances in public health have reduced the effects of critical population-limiting factors such as malnutrition, contagious diseases, and poor hygiene. Over the past 300 years, modern plumbing and sewage

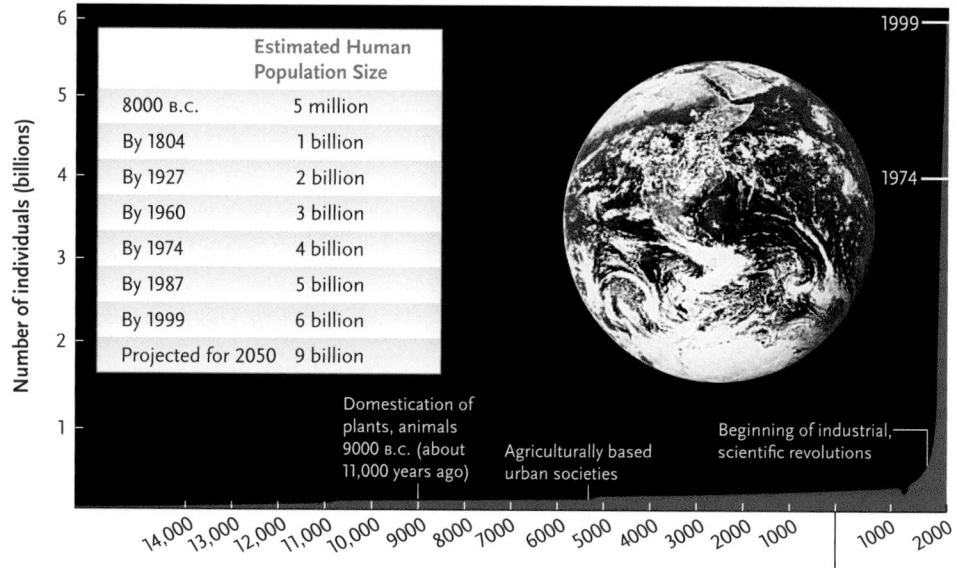

Figure 49.19

Human population growth. The worldwide human population grew slowly until 200 years ago, when it began to increase explosively. The dip in the mid-fourteenth century represents the death of 60 million Asians and Europeans from the bubonic plague. The table shows the years when the human population reached each additional billion people. (Photo: NASA.)

Estimated Human Population Size	
8000 B.C.	5 million
By 1804	1 billion
By 1927	2 billion
By 1960	3 billion
By 1974	4 billion
By 1987	5 billion
By 1999	6 billion
Projected for 2050	9 billion

treatment, improvements in food handling and processing, and medical discoveries have reduced death rates sharply. Births now greatly exceed deaths, especially in less industrialized countries, resulting in rapid population growth.

Age Structure and Economic Development May Now Control Our Population Growth

Where have our migrations and technological developments taken us? It took about 2.5 *million* years for the human population to reach 1 billion, 123 years to reach the second billion, and only 13 years to jump from 5 billion to 6 billion (see the inset table in Figure 49.19). Rapid population growth may now be an inevitable consequence of our age structure and economic development.

Population Growth and Age Structure. On a worldwide scale, the annual growth rate for the human population averaged nearly 1.2% ($r = 0.012$ new individuals per individual per year) between 2000 and 2005. Population experts expect that rate to decline, but even so, the human population will probably exceed 9 *billion* by 2050.

The population growth rates of individual nations vary widely, however, ranging from much less than 1% to more than 3% in 2001 **(Figure 49.20a).** The industrialized countries of Western Europe have achieved nearly zero population growth, but other countries—notably those in Africa, Latin America, and Asia—will experience huge increases over the next 20 or 25 years **(Figure 49.20b).**

For all long-lived species, differences in age structure are a major determinant of differences in population growth rates **(Figure 49.21).** The uniform age structure of countries with zero growth—with approximately equal numbers of people of reproductive and prereproductive ages—suggests that individuals have just been replacing themselves and that these populations will not experience a growth spurt when today's children mature. By contrast, the narrow-based age structure of countries with negative growth illustrates a continuing decrease in population size. Reproductives have been producing very few offspring, and the small group of prereproductives may not even replace themselves. Countries with rapid growth have a broad-based age structure, with many youngsters born during the previous 15 years. Worldwide, more than one-third of the human population falls within this prereproductive base. This age class will soon reach sexual maturity. Even if each woman produces only two offspring, populations will continue to grow rapidly because so many individuals are reproducing.

The age structure of the United States falls between those for countries with zero growth and countries with rapid growth. The average number of children per family has declined to the two that are necessary to replace their parents in the population. Nevertheless, the U.S. population will continue to grow slowly for the next couple of generations largely because of continued immigration.

Population Growth and Economic Development. The relationship between a country's population growth and its economic development can be depicted by the **demographic transition model (Figure 49.22).** This model describes historical changes in demographic patterns in the industrialized countries of Western Europe; we do not know if it accurately predicts the future for developing nations today.

a. Mean annual population growth rates

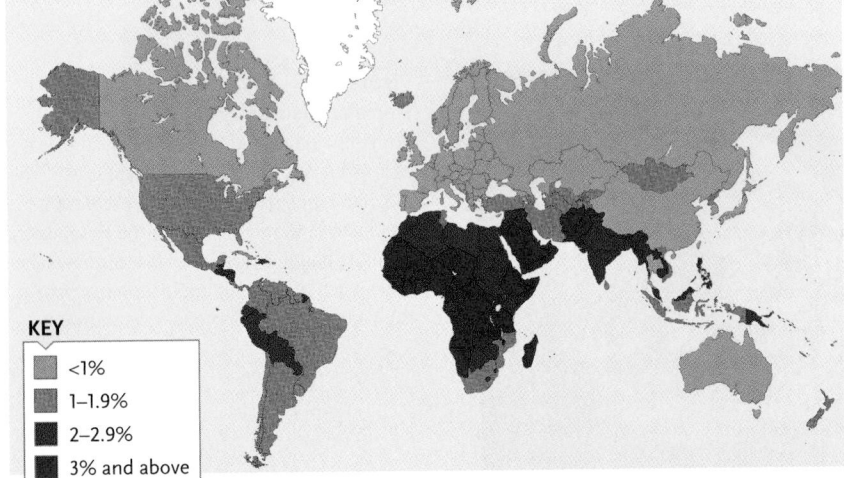

KEY
- <1%
- 1–1.9%
- 2–2.9%
- 3% and above

b. Projected population sizes for 2025

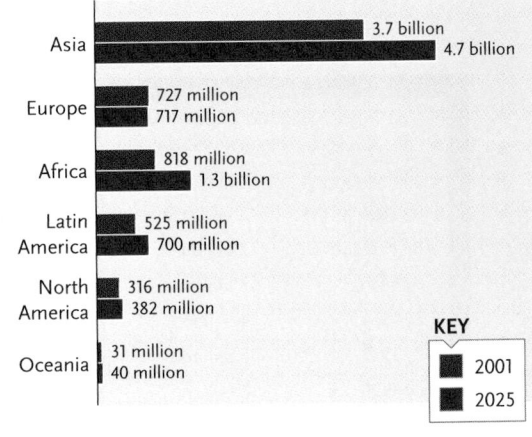

Asia — 3.7 billion / 4.7 billion
Europe — 727 million / 717 million
Africa — 818 million / 1.3 billion
Latin America — 525 million / 700 million
North America — 316 million / 382 million
Oceania — 31 million / 40 million

KEY
- 2001
- 2025

Figure 49.20
Local variation in human population growth rates. **(a)** Average annual population growth rates varied among countries and continents in 2001. **(b)** In some regions, the population is projected to increase greatly by 2025 (red) as compared with the population size in 2001 (orange); the population of Europe will likely decline.

a. Hypothetical age distributions for populations with different growth rates

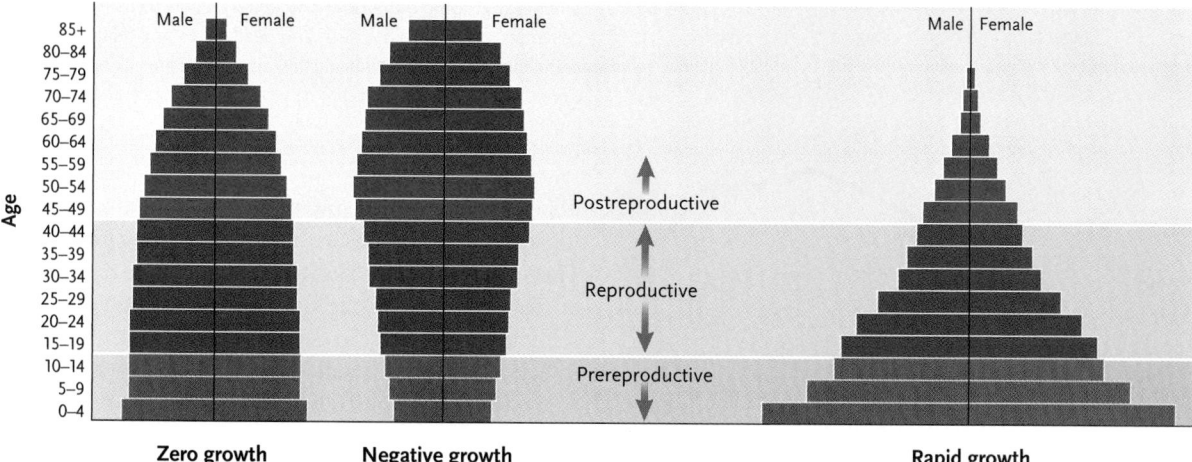

Zero growth   Negative growth   Rapid growth

b. Age pyramids for the United States and Mexico in 2000

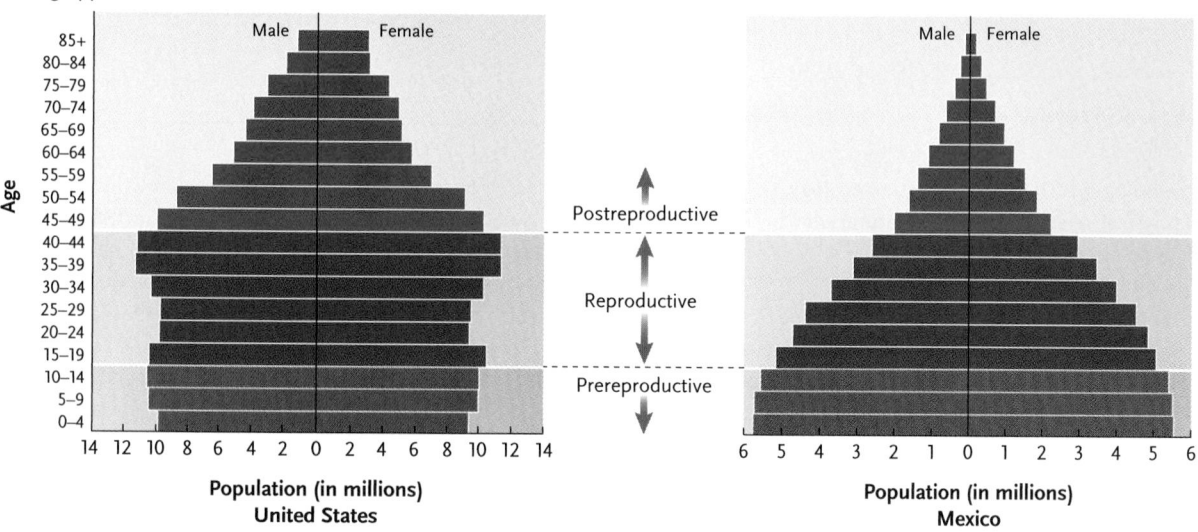

Population (in millions)
United States

Population (in millions)
Mexico

Figure 49.21

Age-structure diagrams. **(a)** Age-structure diagrams differ for countries with zero, negative, and rapid population growth rates. The width of each bar represents the proportion of the population in each age class. **(b)** Age-structure diagrams for the United States and Mexico in 2000 (measured in millions of people) suggest that these countries would experience different population growth rates.

According to this model, during a country's *preindustrial* stage, birth and death rates are high, and the population grows slowly. Industrialization begins a *transitional* stage, when food production rises, and health care and sanitation improve. The death rate declines, resulting in an increased rate of population growth. Later, as living conditions improve, the birth rate also declines, causing the population growth rate to drop. When the *industrial* stage is in full swing, population growth slows dramatically. People move from the countryside to cities, and urban couples often choose to accumulate material goods instead of having large families. Zero population growth is reached in the *postindustrial* stage. Eventually, the birth rate falls below the death rate, *r* falls below zero, and population size begins to decrease.

Today, the United States, Canada, Australia, Japan, Russia, and most of Western Europe are in the industrial stage. Their growth rates are slowly decreasing. In

Bulgaria, Germany, Hungary, and Sweden, birth rates are lower than death rates, and populations are getting smaller, indicating their entry into the postindustrial stage. Kenya and other less industrialized countries are in the transitional stage, but they may not have enough skilled workers or enough capital to make the transition to an industrialized economy. Thus, many poorer nations may be stuck in the transitional stage.

Limiting Population Growth. Most governments realize that increased population size is now the major factor causing resource depletion, excessive pollution, and an overall decline in the quality of life. The principles of population ecology demonstrate that a slowing of population growth—or an actual decline in population size—can be achieved only by decreasing the birth rate or increasing the death rate. And because increasing mortality is neither a rational nor humane means

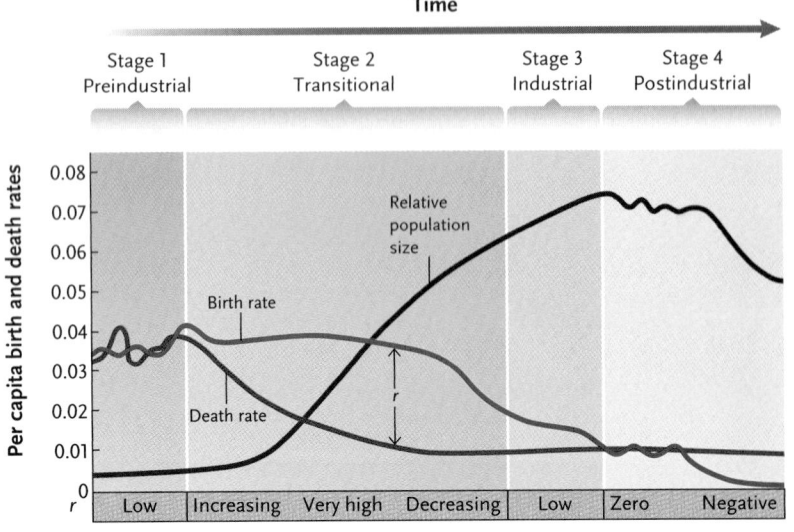

Figure 49.22

The demographic transition. The demographic transition model describes changes in the birth and death rates and relative population size as a country passes through four stages of economic development. The bottom bar describes the net population growth rate, *r*.

of population control, most governments are attempting to lower birth rates with **family planning programs.** These programs educate people about ways to produce an optimal family size on an economically feasible schedule. Programs vary in their details, but all provide information on methods of birth control (see Section 47.6). When thoughtfully developed and carefully administered, family planning programs cause birth rates to decline significantly.

All species face limits to their population growth. We have postponed the action of most factors that limit

UNANSWERED QUESTIONS

Are there universal governing principles in population ecology, similar to the laws of physical sciences? Or is the natural world so complex that each population must be considered individually, leaving us with just a series of case studies?

These types of broad questions motivated the founders of modern ecological studies, such as G. Evelyn Hutchinson and Robert MacArthur. Many ecologists have attempted to codify aspects of population ecology in terms of specific principles, sometimes imposing artificial dichotomies in the process. For example, this chapter considered whether or not natural populations are subject to either density-dependent or density-independent regulation. Ecologists have also attempted to uncover basic patterns in community ecology, which are described in the next chapter. We do know that some general principles are often important in governing the structure of populations or natural communities but, as yet, we cannot apply any of them to a specific system without also including a detailed study of that system.

What is the importance of scale in ecology?

Although an individual population can be a meaningful object of study, ecologists often collect data from multiple populations of the same species to compare results among the "replicates." However, although the separate populations may appear to be replicates, they are often quite different in appearance, age structure, life history, or other characteristics. Ecologists confronted with such variation might seek explanations in the differences between the populations' environments, including both abiotic and biotic factors. However, such local variation may also be attributable to the larger context, such as the landscape or surrounding communities.

One important manifestation of this question applies to how the populations of a species are distributed in space. In many cases, discrete populations are widely separated from one another. Such separation is easy to imagine in terms of fish that live in lakes or organisms that live on islands, but it also applies in a diversity of other organisms. For example, many plants and animals are found only on chemically distinct patches of soil that are distributed like islands across a terrestrial environment. Many

lizards prefer to live in rock outcrops that dot the landscape. Other organisms live on cool, wet mountaintops surrounded by desert. Such isolation is often exaggerated by human modification of the landscape, which progressively fragments and isolates habitable environments from one another. How does such subdivision change the dynamics of the individual populations? How does it change the way they evolve? These are questions that have challenged population and evolutionary biologists for decades. The effects of humans on the environment are making the answers to these questions more than a theoretical concern.

What is the importance of evolution in ecological interactions?

Most research in ecology, ranging from formal models of population growth and regulation to empirical studies, treats populations as if they were unchanging—as if they were not evolving. This implicit perspective does not deny that evolution is happening, but treats it as if it happens on such a long time scale that it need not be considered in contemporary studies. However, many recent studies have shown that populations may evolve quickly, often on a year-to-year basis. If this observation is generally true, then ecological studies that do not include evolutionary change may be compromised. For example, the monitoring and management of commercially exploited fish populations are based entirely on models of population growth, demography, and life histories similar to those considered in this chapter. Commercial fisheries often capture a large proportion of a population every year, focusing on the largest adults. Although research has clearly shown that these practices are likely to select for earlier maturity at a smaller size, these findings have not yet been incorporated into fisheries-management policy. More generally, ecologists have not yet included sufficient emphasis on the interaction between evolution as it occurs on the scale of our day-to-day existence and the modeling and empirical study of ecological processes.

David Reznick is Professor of Biology at the University of California, Riverside. He studies natural selection both from an experimental perspective and by testing evolutionary theory in natural populations. He works primarily with guppies on the island of Trinidad. Learn more about his work at http://www.biology.ucr.edu/people/faculty/Reznick.html.

population growth, but no amount of invention can expand the ultimate limits set by resource depletion and a damaged environment. We now face two options for limiting human population growth: we can make a global effort to limit our population growth, or we can wait until the environment does it for us.

STUDY BREAK

1. How have humans sidestepped the controls that regulate populations of other organisms?
2. How does the age structure of a population influence its future population growth?

Review

Go to **ThomsonNOW** at www.thomsonedu.com/login to access quizzing, animations, exercises, articles, and personalized homework help.

49.1 The Science of Ecology

- Ecology is the study of the interactions between organisms and their environments. Basic ecology focuses on undisturbed natural systems, whereas applied ecology considers the effects of human disturbance (Figure 49.1).
- Ecologists do research at five levels of organization: organisms, populations, communities, ecosystems, and the biosphere.
- Ecologists test hypotheses about ecological relationships with experimental or observational data. They sometimes frame hypotheses in mathematical models.

49.2 Population Characteristics

- A population's size and density can be measured directly or with sampling techniques (Figures 49.2 and 49.3).
- Organisms within a population may be clumped, uniformly distributed, or randomly distributed within their habitat (Figure 49.4). Clumped dispersion is the most common, but animals may change their dispersion pattern seasonally.
- The relative numbers of individuals of different ages determines a population's age structure. Generation time is the average time between an individual's birth and the birth of its offspring (Figure 49.5). A population's sex ratio is the relative proportion of males and females.

Animation: Distribution patterns

Animation: Mark-recapture method

Animation: Age structure diagrams

49.3 Demography

- Demography is the study of the survivorship, reproduction, immigration, and emigration patterns that influence population characteristics.
- Life tables summarize age-specific mortality, survivorship, and age-specific fecundity of surviving individuals (Table 49.1).
- Survivorship curves depict a population's survival pattern over its life span. Ecologists define three general patterns of survivorship: high survivorship until late in life, a constant mortality level at all ages, and high juvenile mortality (Figure 49.6).

49.4 The Evolution of Life Histories

- An organism's energy budget mandates trade-offs in the allocation of energy to maintenance, growth, and reproduction.
- Natural selection has molded several interacting components of life history variation based upon the allocation of resources to growth, maintenance, and reproduction: the trade-off between fecundity and parental care; whether to reproduce once versus multiple times; and the age of first reproduction.

Animation: Life history patterns

Animation: Guppy characteristics

49.5 Models of Population Growth

- Bacteria reproduce by binary fission, and their populations double in size each generation (Figure 49.7).
- The exponential growth model, $dN/dt = rN$, describes unlimited population growth. A graph of exponential growth is J-shaped (Figure 49.8).
- The logistic model, $dN/dt = r_{max}N(K - N/K)$, includes the effects of resource limitation. The carrying capacity, K, is the maximum population size that an environment can sustain. The per capita population growth rate, r, decreases as N approaches K. A graph of logistic growth is S-shaped (Figures 49.9 and 49.10, Table 49.2).
- Some populations exhibit logistic growth in the laboratory and in nature, but time lags in responses to increased density may cause N to oscillate around K (Figure 49.11).

Animation: Exponential growth

Animation: Effect of death on growth

Practice: Comparison of exponential and logistic population growth

49.6 Population Regulation

- Density-dependent factors regulate population size by reducing individual growth rates, adult size, survivorship, and fecundity (Figures 49.12 and 49.13). Competition within populations or between species, predator–prey interactions, parasites, and infectious diseases can cause density-dependent population regulation (Figure 49.14).
- Abiotic environmental factors, which affect a population regardless of its size, cause density-independent limitation of population size (Figure 49.15).
- Interactions between density-dependent and density-independent factors often influence population size (Figure 49.16).
- The life history patterns of most organisms fall between two extremes: r-selected species and K-selected species (Figure 49.17). They differ in many life history characteristics (Table 49.3).
- Some animal populations exhibit cyclic fluctuations in size (Figure 49.18). No general model has successfully explained all population cycles.

49.7 Human Population Growth

- Human populations have sidestepped density-dependent population regulation by expanding into most terrestrial habitats, increasing carrying capacity, and reducing death rates with improved medical care and sanitation (Figures 49.19 and 49.20).
- Age structure may now control human population growth rates (Figure 49.21). In countries with large numbers of young people, populations will continue to grow rapidly as they reach sex-

ual maturity. The populations of countries with a uniform age structure will not experience much growth in the foreseeable future.

- The demographic transition model describes the influence of economic development on population growth (Figure 49.22).

- Many governments encourage population control through family planning programs.

  Animation: Current and projected population sizes by region

  Animation: U.S. age structure

  Animation: Demographic transition model

Questions

Self-Test Questions

1. Ecologists sometimes use mathematical models to:
   a. avoid conducting laboratory studies or field work.
   b. simulate natural events before conducting detailed field studies.
   c. make basic observations about ecological relationships in nature.
   d. collect survivorship and fecundity data to construct life tables.
   e. determine the geographical ranges of populations.

2. The number of individuals per unit area or volume of habitat is called the population's:
   a. geographical range.   d. size.
   b. dispersion pattern.   e. age structure.
   c. density.

3. One day you caught and marked 90 butterflies in a population. A week later, you returned to the population and caught 80 butterflies, including 16 that had been marked previously. What is the size of the butterfly population?
   a. 170   b. 450   c. 154   d. 186   e. 106

4. A uniform dispersion pattern implies that members of a population:
   a. cooperate in rearing their offspring.
   b. work together to escape from predators.
   c. use resources that are patchily distributed.
   d. may experience intraspecific competition for vital resources.
   e. have no ecological interactions with each other.

5. The model of exponential population growth predicts that the per capita population growth rate (r):
   a. does not change as a population gets larger.
   b. gets larger as a population gets larger.
   c. gets smaller as a population gets larger.
   d. is always at its maximum level (r_{max}).
   e. fluctuates on a regular cycle.

6. A population of 1000 individuals experiences 462 births and 380 deaths in 1 year. What is the value of r for this population?
   a. 0.842/individual/year   d. 0.820/individual/year
   b. 0.462/individual/year   e. 0.082/individual/year
   c. 0.380/individual/year

7. According to the logistic model of population growth, the absolute number of individuals by which a population grows during a given time period:
   a. gets steadily larger as the population size increases.
   b. gets steadily smaller as the population size increases.
   c. remains constant as the population size increases.
   d. is highest when the population is at an intermediate size.
   e. fluctuates on a regular cycle.

8. Which example might reflect density-dependent regulation of population size?
   a. An exterminator uses a pesticide to eliminate carpenter ants from a home.
   b. Mosquitoes disappear from an area after the first frost.
   c. The lawn dies after a month-long drought.
   d. Storms blow over and kill all willow trees along a lake.
   e. The size of a clam population declines as the number of predatory herring gulls explodes.

9. A K-selected species is likely to exhibit:
   a. a Type I survivorship curve and a short generation time.
   b. a Type II survivorship curve and a short generation time.
   c. a Type III survivorship curve and a short generation time.
   d. a Type I survivorship curve and a long generation time.
   e. a Type II survivorship curve and a long generation time.

10. One reason that human populations have sidestepped factors that usually control population growth is:
    a. The carrying capacity for humans has remained constant since humans first evolved.
    b. Agriculture and industrialization have increased the carrying capacity for our species.
    c. The population growth rate (r) for the human population has always been small.
    d. The age structure of human populations has no impact on its population growth.
    e. Plagues have killed off large numbers of humans at certain times in the past.

Questions for Discussion

1. Choose an animal or plant species that lives in your environment and identify the density-dependent and density-independent factors that might influence its population size. How could you demonstrate conclusively that the factors work in either a density-dependent or density-independent fashion?

2. Many city-dwellers have noted that the density of cockroaches in apartment kitchens appears to vary with the habits of the occupants: people who wrap food carefully and clean their kitchen frequently tend to have fewer arthropod roommates than those who leave food on kitchen counters and clean less often. Interpret these observations from the viewpoint of a population ecologist.

3. How could you define the worldwide carrying capacity for humans? What factors would you have to take into account?

Experimental Analysis

Design an experiment using fruit flies or some other small laboratory animal to test the hypothesis that delaying the age of first reproduction will decrease a population's per capita birth rate. Your experimental design should include experimental and control groups as well as details about your experimental methods and the data you would collect.

Evolution Link

Many animals, including humans and other primates, live long beyond their reproductive years. Develop an evolutionary hypothesis to explain this observation, and design a study that might test it.

How Would You Vote?

Some people oppose any deer hunting, whereas others see hunters as a logical substitute for an absence of natural predators. Do you support encouraging hunting in areas where the presence of too many deer is harming the habitat? Go to www.thomsonedu.com/login to investigate both sides of the issue and then vote.

Three interacting populations. Ladybird beetles *(Coccinella septempunctata)* feed on aphids (order Hemiptera), which consume the sap of plants.

© Claude Nuridsany & Marie Perennou/SPL/Photo Researchers, Inc.

50 Population Interactions and Community Ecology

WHY IT MATTERS

In some open woodlands in Central America, flocks of chestnut-headed oropendolas *(Zarhynchus wagleri)*, members of the blackbird family, build hanging nests in isolated trees **(Figure 50.1)**. Female giant cowbirds *(Scaphidura oryzivora)* often bully their way into a colony, laying an egg or two in each oropendola nest. Cowbirds are *brood parasites* on oropendolas, tricking them into caring for cowbird young. The cowbird chicks grow faster than oropendola chicks, and they consume much of the food that the oropendolas bring to their own offspring. Because cowbird chicks take food away from their oropendola nest mates, we might expect adult oropendolas to eject cowbird eggs and chicks from their nests—but often they don't.

Why do some oropendolas care for offspring that are not their own? In an ingenious study conducted in the 1960s, Neal Smith of the Smithsonian Tropical Research Institute determined that cowbird chicks could actually increase the number of offspring that some oropendolas raise. Oropendola chicks are frequently parasitized by botfly larvae, which feed on their flesh. The aggressive cowbird chicks snap at adult botflies and pick fly larvae off their nest mates. Although cowbird chicks eat food meant for oropendola chicks, they also protect

Cortez C. Austin

Figure 50.1
Potential victims of brood parasitism. Chestnut-headed oropendolas (*Zarhynchus wagleri*) rear their young in elaborate hanging nests. Some populations of oropendolas are subject to brood parasitism by giant cowbirds (*Scaphidura oryzivora*).

them from potentially lethal parasites; twice as many young oropendolas survive in nests with cowbird chicks as in nests without them.

In other areas of Central America, oropendolas build nests near the hives of bees or wasps. These oropendolas chase cowbirds from their colonies, and when a cowbird does manage to sneak an egg into one of their nests, the oropendolas frequently eject it. Why do oropendolas in these colonies reject cowbird eggs when others do not? Smith determined that the swarms of bees and wasps keep botflies away from the oropendola colonies. At these sites, twice as many oropendola chicks survive in nests without cowbirds as in those that include them. Thus the oropendolas derive no benefit from having cowbird chicks in their nests, and natural selection has favored discriminating behavior in oropendolas that nest near bees and wasps.

The story of the oropendolas, cowbirds, botflies, bees, and wasps provides an example of the population interactions that characterize life in an **ecological community**, an assemblage of species living in the same place. And as this story reveals, the presence or absence of certain species may alter the effects of such interactions in almost unimaginably complex ways. We begin this chapter with a description of some of the many ways that populations in a community interact. We then examine how population interactions and other factors, such as the kinds of species present and

the relative numbers of each species, influence a community's characteristics.

50.1 Population Interactions

Population interactions usually provide benefits or cause harm to the organisms engaged in the interaction **(Table 50.1)**. And because interactions with other species often affect the survival and reproduction of individuals, many of the relationships that we witness today are the products of long-term evolutionary modification. Before examining several general types of population interactions, we briefly consider how natural selection has shaped the relationships between interacting species.

Coevolution Produces Reciprocal Adaptations in Species That Interact Ecologically

Population interactions change constantly. New adaptations that evolve in one species exert selection pressure on another, which then evolves adaptations that exert selection pressure on the first. The evolution of genetically based, reciprocal adaptations in two or more interacting species is described as **coevolution.**

Some coevolutionary relationships are straightforward. For example, ecologists describe the coevolutionary interactions between some predators and their prey as a race in which each species evolves adaptations that temporarily allow it to outpace the other. When antelope populations suffer predation by cheetahs, natural selection fosters the evolution of faster speed in the antelopes. Cheetahs then experience selection for increased speed so that they can overtake and capture antelopes. Other coevolved interactions provide bene-

Table 50.1	Population Interactions and Their Effects		
Interaction	**Effects on Interacting Populations**		
Predation	+/−	Predators gain nutrients and energy; prey are killed or injured.	
Herbivory	+/−	Herbivores gain nutrients and energy; plants are killed or injured.	
Competition	−/−	Both competing populations lose access to some resources.	
Commensalism	+/0	One population benefits; the other population is unaffected.	
Mutualism	+/+	Both populations benefit.	
Parasitism	+/−	Parasites gain nutrients and energy; hosts are injured or killed.	

fits to both partners. For example, the flower structures of different monkey-flower species have evolved characteristics that allow them to be visited by either bees or hummingbirds (see Figure 21.7).

Although one can hypothesize a coevolutionary relationship between any two interacting species, documenting the evolution of reciprocal adaptations is difficult. As our introductory story about oropendolas and their parasites illustrated, coevolutionary interactions often involve more than two species. Indeed, most organisms experience complex interactions with numerous other species in their communities, and the simple portrayal of coevolution as taking place between two species rarely does justice to the complexity of these relationships.

Predation and Herbivory Define Many Relationships in Ecological Communities

Because animals acquire nutrients and energy by consuming other organisms, **predation** (the interaction between predatory animals and the animal prey they consume) and **herbivory** (the interaction between herbivorous animals and the plants they eat) are often the most conspicuous relationships in ecological communities.

Adaptations for Feeding. Both predators and herbivores have evolved remarkable characteristics that allow them to feed effectively. Carnivores use sensory systems to locate animal prey and specialized behaviors and anatomical structures to capture and consume it. For example, a rattlesnake (genus *Crotalus*) uses heat sensors on its head (see Figure 39.22) and chemical sensors in the roof of its mouth to find rats or other endothermic prey. Its hollow fangs inject toxins that kill the prey and begin to digest its tissues even before the snake consumes it. And elastic ligaments connecting the bones of its jaws and skull allow a snake to swallow prey that is larger than its head. Herbivores have comparable adaptations for locating and processing their food plants. Insects use chemical sensors on their legs and heads to identify edible plants and sharp mandibles or sucking mouthparts to consume plant tissues or sap. Herbivorous mammals have specialized teeth to harvest and grind tough vegetation (see Section 45.5).

All animals must select their diets from a variety of potential food items. Some species, described as *specialists,* feed on one or just a few types of food. Among birds, for example, the Everglades kite (*Rostrhamus sociabilis*) consumes just one prey species, the apple snail (*Pomacea paludosa*). Other species, described as *generalists,* have broader tastes. Crows (genus *Corvus*) consume food ranging from grain to insects to carrion.

How does an animal select what type of food to eat? Some mathematical models, collectively described as **optimal foraging theory,** predict that an animal's diet is a compromise between the costs and benefits associated with different types of food. Assuming that animals try to maximize their energy intake in a given feeding time, their diets should be determined by the time and energy it takes to pursue, capture, and consume a particular kind of food compared with the energy that food provides. For example, a cougar (*Puma concolor*) will invest more time and energy hunting a mountain goat (*Oreamnos americanus*) than a jackrabbit (*Lepus townsendii*), but the payoff for the cougar is a bigger meal.

Food abundance also affects food choice. When prey are scarce, animals often take what they can get, settling for food that has a low benefit-to-cost ratio. But when food is abundant, they may specialize, selecting types that provide the largest energetic return. Bluegill sunfish (*Lepomis macrochirus*), for example, feed on *Daphnia* and other small crustaceans. When crustacean density is high, the fish hunt mostly large *Daphnia,* which provide more energy for their effort; but when prey density is low, bluegills feed on *Daphnia* of all sizes (**Figure 50.2**).

Defenses against Herbivory and Predation. Because herbivory and predation have a negative impact on the organisms being consumed, plants and animals have evolved mechanisms to avoid being eaten. Some plants

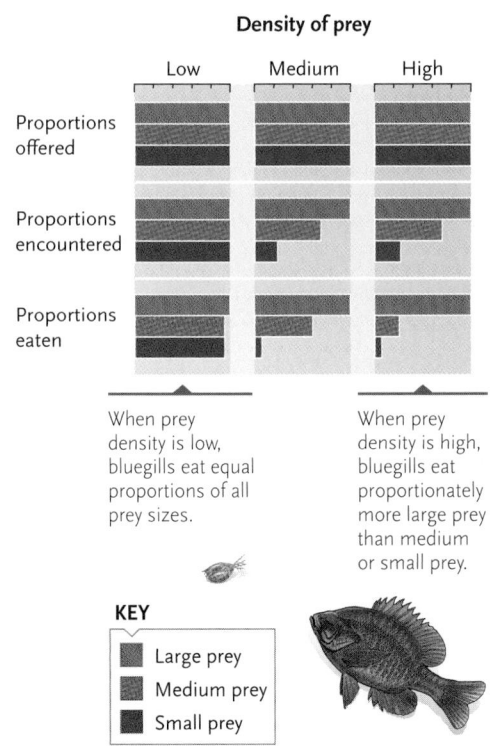

Figure 50.2

An experiment demonstrating that prey density affects predator food choice. Researchers tested the food size preferences of captive bluegill sunfish (*Lepomis macrochirus*) by offering them equal numbers of small, medium, and large-sized prey (*Daphnia magna*) at three different prey densities. Because large prey are the easiest to find, bluegills encountered them more frequently than small or medium-sized prey, especially at the highest prey density. The bluegills' selection of prey varied with prey density; they strongly preferred large prey when prey of all sizes were abundant.

Figure 50.3

Hiding in plain sight. Some animals, such as **(a)** giant swallowtail butterfly *(Papilio cresphontes)* larvae that resemble bird droppings and **(b)** some katydids *(Mimetica* species) that resemble insect-damaged leaves, do not attract the attention of predators.

a. Bird dropping mimic

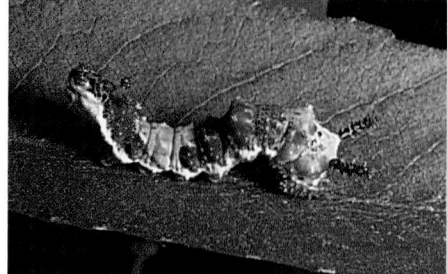

Edward S. Ross

b. Damaged leaf mimic

Frans Lanting/Minden Pictures

use spines, thorns, and irritating hairs to protect themselves from herbivores. Many plant tissues also contain poisonous chemicals that deter herbivores from feeding. For example, plants in the milkweed family (Asclepiadeceae) exude a milky, irritating sap that contains cardiac glycosides, even small amounts of which are toxic to vertebrate heart muscle. Other compounds mimic the structure of insect hormones, disrupting the development of insects that consume them. Most of these poisonous compounds are volatile, giving plants their typical aromas; some herbivores have co-evolved the ability to recognize these odors and avoid the toxic plants. Recent research indicates that some plants increase their production of toxic compounds in response to herbivore feeding. For example, potato and tomato plants that have been damaged by herbivores produce higher levels of protease-inhibiting chemicals; these compounds prevent herbivores from digesting proteins they have just consumed, reducing the food value of these plant tissues.

Many animals have evolved an appearance that provides a passive defense against predation **(Figure 50.3)**. Caterpillars that look like bird droppings, for example, may not attract much attention from a hungry predator. And as you learned in Chapter 1 (see Figure 1.9), **cryptic coloration** helps some prey (as well as some predators) to blend in with their surroundings.

Once discovered by a predator, many animals first try to run away. When cornered, they may try to startle or intimidate the predator with a display that increases their apparent size or ferocity **(Figure 50.4)**. Such a display might confuse the predator just long enough to allow the potential victim to escape. Other species seek shelter in protected sites. For example, flexible-shelled African pancake tortoises *(Malacochersus tornieri)* retreat into rocky crevices and puff themselves up with air, becoming so tightly wedged between rocks that predators cannot extract them.

Other animals defend themselves actively. North American porcupines (genus *Erethizon*) release hairs modified into sharp, barbed quills that stick in a predator's mouth, causing severe pain and swelling. Other species fight back by biting, charging, or kicking an attacking predator. Chemical defenses also provide effective protection. Skunks release a noxious spray when threatened, and some frogs and toads produce neurotoxic skin secretions that paralyze and kill mammals. Some insects even protect themselves with poisons acquired from plants. The caterpillars of monarch butterflies *(Danaus plexippus)* are immune to the cardiac glycosides in the milkweed leaves they eat. They store these chemicals at high concentration, even through metamorphosis, making adult monarchs poisonous to vertebrate predators.

Poisonous or repellant species often advertise their unpalatability with bright, contrasting patterns, called **aposematic coloration (Figure 50.5)**. Although a predator might attack a black-and-white skunk, a

Photo Researchers, Inc.

Figure 50.4

Startle defenses. A short-eared owl *(Asio flammeus)* increases its apparent size when threatened by a predator.

Courtesy of Ken Nemuras

Figure 50.5

Aposematic coloration. Poisonous animals, like the harlequin toad *(Atelopus varius)* from Central America often have bright warning coloration.

yellow-banded wasp, or an orange monarch butterfly once, it quickly learns to associate the gaudy color pattern with pain, illness, or severe indigestion—and rarely attacks these easily recognized animals again.

Mimicry, in which one species evolves an appearance resembling that of another **(Figure 50.6),** is also a form of defense. In **Batesian mimicry,** named for English naturalist Henry W. Bates, a palatable or harmless species, the **mimic,** resembles an unpalatable or poisonous one, the **model.** Any predator that eats the poisonous model will subsequently avoid other organisms that resemble it. In **Müllerian mimicry,** named for German zoologist Fritz Müller, two or more unpalatable species share a similar appearance, which reinforces the lesson learned by a predator that attacks any species in the mimicry complex.

Despite the effectiveness of many antipredator defenses, coevolution has often molded the responses of predators to overcome them. For example, when threatened by a predator, the beetle *Eleodes longicollis* raises its rear end and sprays a noxious chemical from a gland at the tip of its abdomen. Although this behavior deters many would-be predators, grasshopper mice (genus *Onychiomys*) of the American southwest circumvent this defense: they grab the beetles and shove their abdomens into the ground, rendering the beetle's spray ineffective **(Figure 50.7).**

Interspecific Competition Occurs When Different Species Depend on the Same Limiting Resources

Populations of different species often use the same limiting resources, causing **interspecific competition** (competition between species). The competing populations may experience increased mortality and decreased reproduction, responses that are similar to the effects of intraspecific competition (see Section 49.5). Interspecific competition reduces the size and population growth rate of one or more of the competing populations.

Community ecologists identify two main forms of interspecific competition. In **interference competition,** individuals of one species harm individuals of another species directly. Animals may fight for access to resources, as when lions chase smaller scavengers like hyenas and jackals from their kills. Similarly, many plant species, including creosote bushes (see Figure 49.4), release toxic chemicals, which prevent other plants from growing nearby. In **exploitative competition,** two or more populations use ("exploit") the same limiting resource; the presence of one species reduces resource availability for the others, even in the absence of snout-to-snout or root-to-root confrontations. For example, in the deserts of the American Southwest, many bird and ant species feed largely on seeds. Thus, each seed-eating species may deplete the food supply available to others.

a. Batesian mimicry

Drone fly *(Eristalis tenax)*, the mimic   Honeybee *(Apis mellifera)*, the model

b. Müllerian mimicry

Heliconius erato   *Heliconius melpone*

Figure 50.6

Mimicry. **(a)** Batesian mimics are harmless animals that mimic a dangerous one. The harmless drone fly *(Eristalis tenax)* is a Batesian mimic of the stinging honeybee *(Apis mellifera)*. **(b)** Müllerian mimics are poisonous species that share a similar appearance. Two distantly related species of butterfly, *Heliconius erato* and *Heliconius melpone*, have nearly identical patterns on their wings.

Competitive Exclusion and the Niche Concept. In the 1920s, the Russian mathematician Alfred J. Lotka and the Italian biologist Vito Volterra independently proposed a model of interspecific competition, modifying the logistic equation (see Section 49.5) to describe the effects of competition between two species. In their model, an increase in the size of one population reduces the population growth rate of the other.

a. *Eleodes* bettle   **b.** Grasshopper mouse

Figure 50.7

Coevolution of predators and prey. **(a)** When disturbed by a predator, the beetle *Eleodes longicollis* sprays a noxious chemical from its posterior end. **(b)** Grasshopper mice (genus *Onychiomys*) overcome this defense by shoving a beetle's rear end into the soil and dining on it headfirst.

Figure 50.8 Experimental Research

Gause's Experiments on Interspecific Competition in Paramecium

QUESTION: Can two species of *Paramecium* coexist in a simple laboratory environment?

EXPERIMENT: Gause grew populations of two *Paramecium* species, *Paramecium aurelia* and *Paramecium caudatum*, alone (single species cultures) or together (mixed culture) in small bottles in his laboratory. To determine whether the growth of these populations followed the predictions of the logistic equation, Gause had to maintain a reasonably constant carrying capacity in each culture. Thus, he fed the cultures a broth of bacteria, and he eliminated their waste products (by centrifuging the cultures and removing some of the culture medium) on a regular schedule. He then monitored their population sizes through time.

RESULTS: When grown separately, *P. caudatum* **(a)** and *P. aurelia* **(b)** each exhibited logistic population growth. But when the two species were grown together in a mixed culture **(c)**, *P. aurelia* persisted and *P. caudatum* was nearly eliminated from the culture.

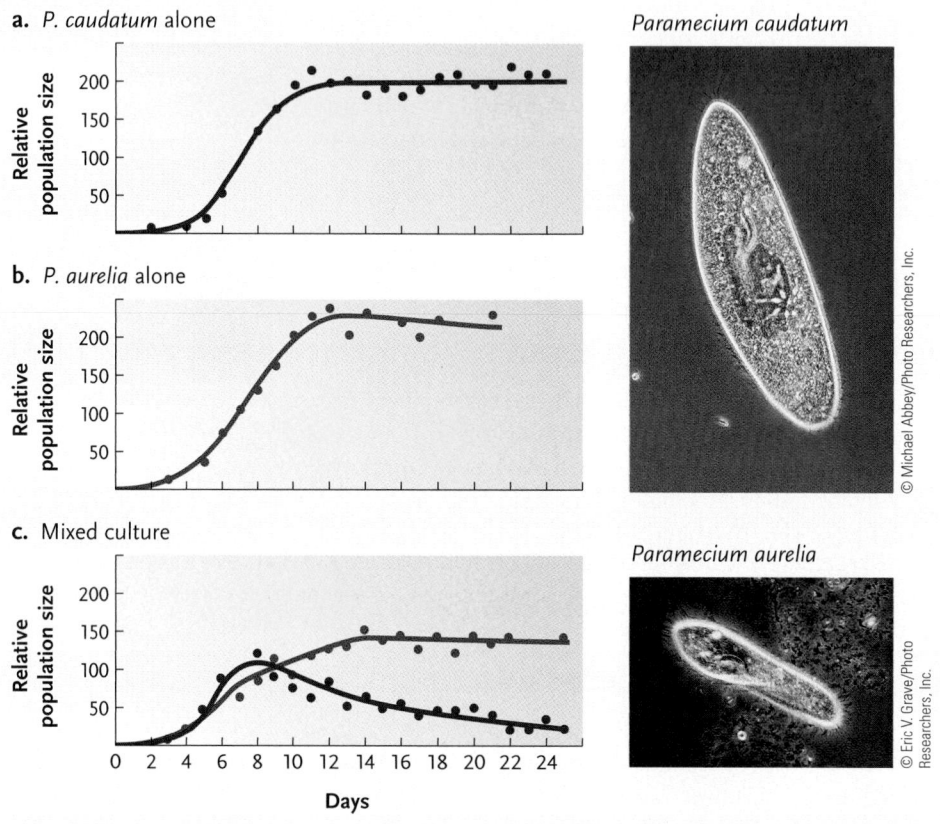

a. *P. caudatum* alone

b. *P. aurelia* alone

c. Mixed culture

Paramecium caudatum

© Michael Abbey/Photo Researchers, Inc.

Paramecium aurelia

© Eric V. Grave/Photo Researchers, Inc.

CONCLUSION: Because one species was almost always eliminated from mixed species cultures, Gause formulated the competitive exclusion principle: populations of two or more species cannot coexist indefinitely if they rely on the same limiting resources and exploit them in the same way.

A Russian biologist, G. F. Gause, tested the model experimentally in the 1930s. He grew cultures of two *Paramecium* species (ciliate protozoans) under constant laboratory conditions, regularly renewing food and removing wastes. Both species feed on bacteria suspended in the culture medium. When grown alone, each species exhibited logistic growth; but when grown together in the same dish, *Paramecium aurelia* persisted at high density, but *Paramecium caudatum* was nearly eliminated **(Figure 50.8)**. These results inspired Gause to define the **competitive exclusion principle:** populations of two or more species cannot coexist indefinitely if they rely on the same limiting resources and exploit them in the same way. One species inevi-

tably harvests resources more efficiently and produces more offspring than the other.

Ecologists developed the concept of the **ecological niche** as a tool for visualizing resource use and the potential for interspecific competition in nature. We define a population's niche by the resources it uses and the environmental conditions it requires over its lifetime. In this context, the niche includes food, shelter, and nutrients as well as abiotic conditions, such as light intensity and temperature, which cannot be depleted. In theory, one could identify an almost infinite variety of conditions and resources that contribute to a population's niche. In practice, ecologists usually analyze a few critical resources for which populations might compete. Sunlight, soil moisture, and inorganic nutrients are important resources for plants. Food type, food size, and nesting sites are important for animals.

Ecologists distinguish the **fundamental niche** of a population, the range of conditions and resources that it can possibly tolerate and use, from its **realized niche**, the range of conditions and resources that it actually uses in nature. Realized niches are smaller than fundamental niches, partly because all tolerable conditions are not always present in a habitat, and partly because some resources are used by other species. We can visualize competition between two populations by plotting their fundamental and realized niches with respect to one or more resources **(Figure 50.9)**. If the fundamental niches of two populations overlap, they *might* compete in nature.

Evaluating Competition in Nature. The observation that several populations use the same resource does not demonstrate that competition occurs. For example, all terrestrial animals consume oxygen, but they don't compete for oxygen because it is usually plentiful. Nevertheless, two general observations provide *indirect* evidence that interspecific competition may have important effects. The first is the extremely common observation of **resource partitioning**, the use of different resources or the use of resources in different ways, by species living in the same place. For example, weedy plants might compete for water and dissolved nutrients in abandoned fields. But they avoid competition by partitioning these resources, collecting them from different depths in the soil **(Figure 50.10)**.

A second phenomenon that suggests the importance of competition is observed in comparisons of species that are sometimes sympatric (that is, living in the same place) and sometimes allopatric (that is, living in different places). In several studies of animals, researchers have documented **character displacement**: allopatric populations are morphologically similar and use similar resources, but sympatric populations are morphologically different and use different resources. The differences between the sympatric populations allow them to coexist without competing. Differences in bill size among sympatric finch species on the Galápagos Islands (see Sections 19.2 and 20.3) may be the product of character displacement **(Figure 50.11)**.

Data on resource partitioning and character displacement merely suggest the possible importance of interspecific competition in nature. To demonstrate *conclusively* that interspecific competition limits natural populations, one must show that the presence of one population reduces the population size or distribution of its presumed competitor. In a classic field experiment, Joseph Connell of the University of California, Santa Barbara, determined that competition between two barnacle species caused the realized niche of one species to be smaller than its fundamental niche

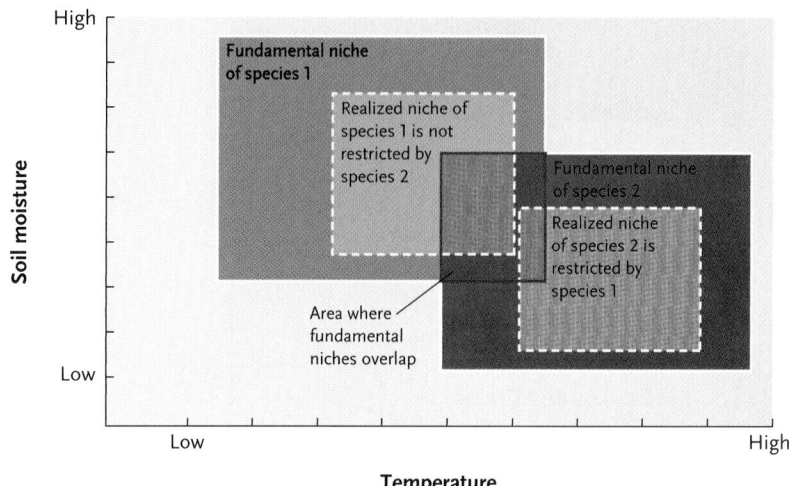

Figure 50.9

Fundamental versus realized niches. In this hypothetical example, both species 1 and species 2 can survive intermediate temperature and soil moisture conditions, as indicated by the shading where their fundamental niches overlap. Because species 1 actually occupies most of this overlap zone, its realized niche is not much affected by the presence of species 2. By contrast, the realized niche of species 2 is restricted by the presence of species 1, and species 2 occupies warmer and dryer parts of the habitat.

(Figure 50.12). Connell first observed the distributions of barnacles in undisturbed habitats. *Chthamalus stellatus* is generally found in shallow water on rocky coasts, where it is periodically exposed to air. *Balanus balanoides* typically lives in deeper water, where it is usually submerged.

Connell determined the fundamental niche of each species by removing either *Chthamalus* or *Balanus* from rocks and monitoring the distribution of each

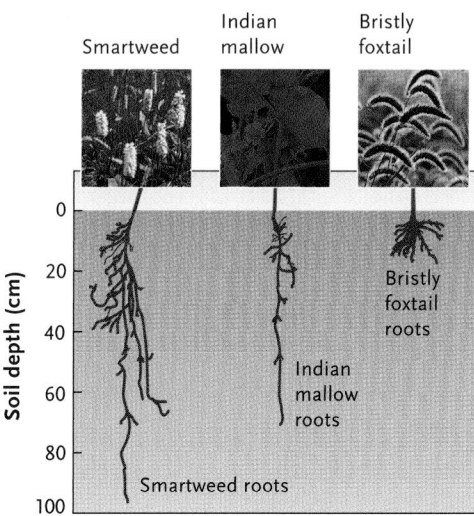

Figure 50.10

Resource partitioning. The root systems of three plant species that grow in abandoned fields partition water and nutrient resources in soil. Bristly foxtail grass *(Setaria faberii)* has a shallow root system; Indian mallow *(Abutilon theophraste)* has a moderately deep taproot; and smartweed *(Polygonum pensylvanicum)* has a deep taproot that branches at many depths.

(Photos: left, © Tony Wharton, Frank Lane Picture Agency/Corbis; middle, © Hal Horwitz/Corbis; right, © Joe McDonald/Corbis.)

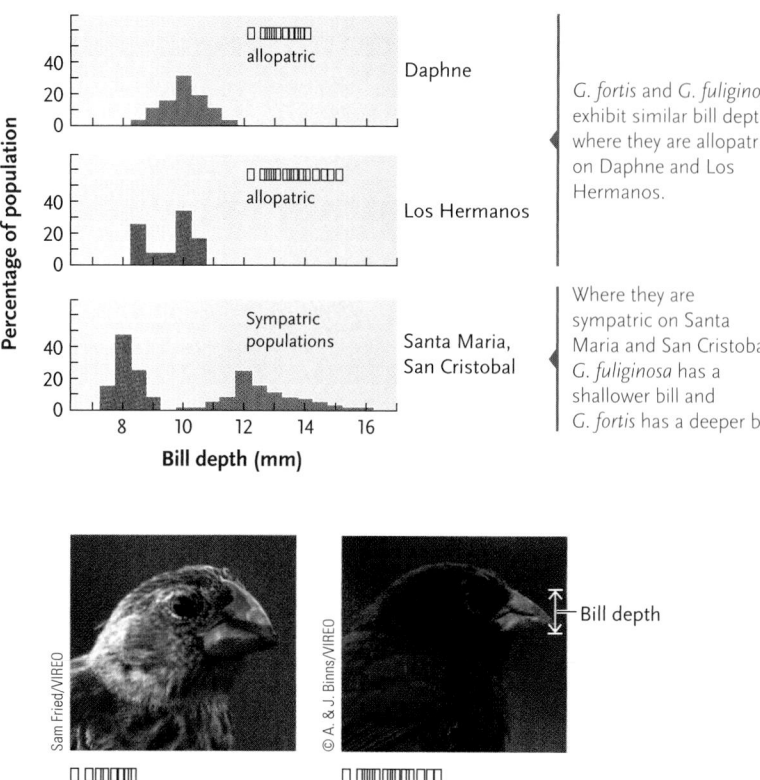

Figure 50.11

Character displacement. *Geospiza fortis* and *Geospiza fuliginosa* exhibit character displacement in the depth of their bills, a trait that is correlated with the sizes of seeds they eat.

G. fortis and G. fuliginosa exhibit similar bill depths where they are allopatric on Daphne and Los Hermanos.

Where they are sympatric on Santa Maria and San Cristobal, G. fuliginosa has a shallower bill and G. fortis has a deeper bill.

species in the absence of the other. When Connell removed *Balanus* from rocks in deep water, larval *Chthamalus* colonized the area and produced a flourishing population of adults. Connell observed that *Balanus* physically displaced *Chthamalus* from these rocks. Thus, interference competition from *Balanus* prevents *Chthamalus* from occupying areas where it would otherwise live. By contrast, the removal of *Chthamalus* from rocks in shallow water did not result in colonization by *Balanus*. *Balanus* is apparently unable to live in habitats that are frequently exposed to air. Connell therefore concluded that competition from *Chthamalus* does not affect the distribution of *Balanus*. Thus, the competitive interaction between these two species is asymmetrical: *Balanus* has a substantial effect on *Chthamalus*, but *Chthamalus* has virtually no effect on *Balanus*.

In Symbiotic Associations, the Lives of Two or More Species Are Closely Intertwined

Some species have a physically close ecological association called **symbiosis** (*sym* = together; *bio* = life; *sis* = process). Biologists define three types of symbiotic interactions—*commensalism, mutualism,* and *parasitism*—that differ in their effects.

Commensalism, in which one species benefits and the other is unaffected, is rare in nature, because few species are unaffected by their interactions with another. One possible example is the relationship between cattle egrets *(Bubulcus ibis)*, birds in the heron family, and the large grazing mammals with which they associate **(Figure 50.13)**. Cattle egrets feed on insects and other small animals that their commensal partners flush from grass. Feeding rates of egrets are higher when they associate with large grazers than when they do not. The birds clearly benefit from this interaction, but the presence of birds has no apparent positive or negative impact on the mammals.

Mutualism, in which both partners benefit, is extremely common. The coevolved relationships between flowering plants and animal pollinators are largely mutualistic. Animals that feed on a plant's nectar or pollen carry its gametes from one flower to another **(Figure 50.14)**. Similarly, animals that eat the fruits of flowering plants disperse the seeds, "planting" them in a pile of nutrient-rich feces. These mutualistic relationships between plants and animals do not require active cooperation. Each species simply exploits the other for its own benefit.

Some associations between bacteria and plants are also mutualistic. One of the most important of these associations is between *Rhizobium* and leguminous plants, such as peas, beans, and clover (see Section 33.3). *Insights from the Molecular Revolution* describes how the genes responsible for the association were identified and their possible evolutionary origin.

Mutualistic relationships between animal species are also common. For example, some small marine fishes feed on parasites that attach to the mouths and gills of large predatory fishes **(Figure 50.15)**. Parasitized fishes hover motionless while the "cleaners" scour their tissues. The relationship is mutualistic because the cleaner fishes get a meal, and the larger fishes are relieved of parasites.

The relationship between the bull's horn acacia tree (*Acacia cornigera*) of Central America and a small ant species (*Pseudomyrmex ferruginea*) is one of the most highly coevolved mutualisms known **(Figure 50.16)**. Each acacia is inhabited by an ant colony that lives in the tree's swollen thorns. The ants swarm out of the thorns to sting—and sometimes kill—herbivores that touch the tree. The ants also clip any vegetation that grows nearby. Thus, acacia trees that are colonized by ants grow in a space free of herbivores and competitors, and occupied trees grow faster and produce more seeds than unoccupied trees. In return, the plants produce sugar-rich nectar consumed by adult ants and protein-rich structures that the ants feed to their larvae. Ecologists describe the coevolved mutualism between these species as *obligatory,* at least for the ants; they cannot subsist on any other food sources.

Figure 50.12 Experimental Research

Demonstration of Competition between Two Species of Barnacles

QUESTION: Do two barnacle species limit one another's realized niche in habitats where they coexist?

EXPERIMENT: Connell observed a difference in the distributions of two barnacle species on a rocky coast: *Chthamalus stellatus* occupies shallow water, and *Balanus balanoides* lives in deeper water. He then determined the fundamental niche of each species by removing either *Chthamalus* or *Balanus* from rocks and monitoring the distribution of each species in the absence of the other.

RESULTS: When Connell removed *Balanus* from rocks in deep water, larval *Chthamalus* colonized the area and produced a flourishing population of adults. By contrast, the removal of *Chthamalus* from rocks in shallow water did not result in colonization by *Balanus*.

Realized niches before experimental treatments

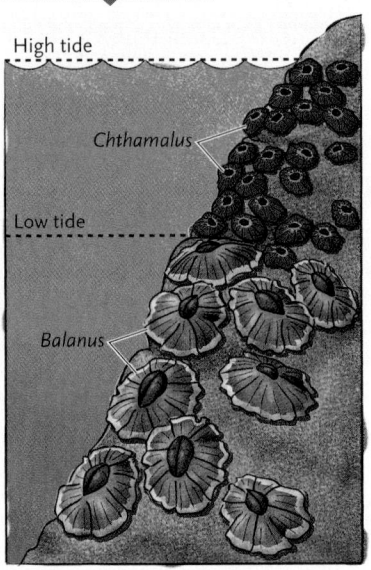

Treatment 1: Remove *Balanus*
In the absence of *Balanus*, *Chthamalus* occupies both shallow water and deep water.

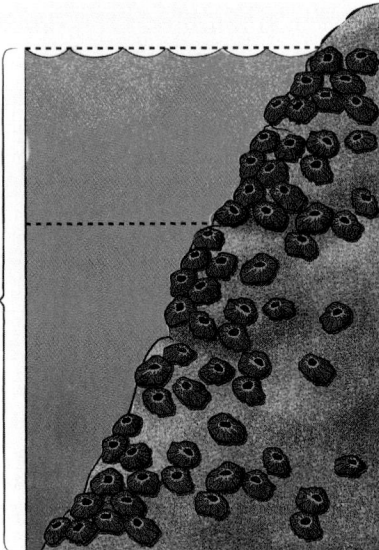

Treatment 2: Remove *Chthamalus*
In the absence of *Chthamalus*, *Balanus* still occupies only deep water.

CONCLUSION: In habitats where *Balanus* and *Chthamalus* coexist, the realized niche of *Chthamalus* is smaller than its fundamental niche because of competition from *Balanus*. The realized niche of *Balanus* is similar to its fundamental niche because it is not affected by the competitive interaction.

Parasitism is a type of interaction in which one species, the **parasite**, uses another, the **host**, in a way that is harmful to the host. Parasite–host relationships are like predator–prey relationships: one population of organisms feeds on another. But parasites rarely kill their hosts quickly because a dead host is useless as a continuing source of nourishment.

Tapeworms and other parasites that live *within* a host are **endoparasites.** Many endoparasites acquire their hosts passively, when a host accidentally ingests the parasite's eggs or larvae (see *Focus on Research,* Chapter 29). Endoparasites generally complete their

Figure 50.13
Commensalism. Cattle egrets *(Bubulcus ibis)* feed on insects and other small animals flushed by the movements of large grazing mammals, like this African buffalo *(Sycerus coffer).*

Fritz Polking/Frank Lane Picture Agency

Figure 50.14

Mutualism between plants and animals. Several species of yucca plants (*Yucca* species) are each pollinated exclusively by one species of yucca moth (*Tegeticula* species). The adult stage of each moth appears at the time of year when its yucca plant flowers. These species are so mutually interdependent that the larvae of each moth species can feed on only one type of yucca, and the flowers of each yucca can be fertilized by only one species of moth. Most plant-pollinator mutualisms are much less specific.

a. Flowering yucca plant

b. Female yucca moth

c. Yucca moth larva

A female yucca moth uses highly modified mouthparts to gather the sticky pollen and roll it into a ball. She carries the pollen to another flower, and after piercing its ovary wall, she lays her eggs. She then places the pollen ball into the opening of the stigma.

When moth larvae hatch from the eggs, they eat some of the yucca seeds and gnaw their way out of the ovary to complete their life cycle. Enough seeds remain undamaged to produce a new generation of yuccas.

life cycle in one or two host individuals. By contrast, leeches, aphids, mosquitoes, and other parasites that feed on the *exterior* of a host are **ectoparasites**. Most animal ectoparasites have elaborate sensory and behavioral mechanisms that allow them to locate specific hosts, and they feed on numerous host individuals during their lifetimes. Some plants, such as mistletoes (genus *Phoradendron*), live as ectoparasites on the trunks and branches of trees; their roots penetrate the host's xylem and extract water and nutrients.

Not all parasites feed directly on a host's tissues. The giant cowbirds described earlier are brood parasites, as are other species of cowbirds and cuckoos. Although oropendolas sometimes benefit from the presence of cowbirds, most brood parasites have negative effects on their hosts. For example, brood parasitism by the brown-headed cowbird *(Molothrus ater)* has played a large role in the near-extinction of Kirtland's warbler *(Dendroica kirtlandii)*.

The feeding habits of some insects, called **parasitoids**, fall somewhere between true parasitism and predation. A female parasitoid lays eggs in the larva or pupa of another insect species, and her young consume the tissues of the living host. Because the hosts chosen by most parasitoids are highly specific, agricultural ecologists often release parasitoids to control populations of insect pests.

STUDY BREAK

1. Why are some carnivores willing to spend more time and energy capturing large prey than small prey?
2. What are the differences between cryptic coloration, aposematic coloration, and mimicry? Can a mimic ever have aposematic coloration?
3. How can field experiments demonstrate conclusively that two species compete for limiting resources?

Cleaner wrasse

Figure 50.15
Mutualism between animal species. A large potato cod (*Epinephelus tukula*) from the Great Barrier Reef in Australia remains nearly motionless in the water while a striped cleaner wrasse (*Labroides dimidiatus*) carefully removes and eats ectoparasites attached to its lip. The potato cod is a predator, and the striped cleaner wrasse is a potential prey—but their mutualistic interaction supersedes a possible predator–prey interaction.

50.2 The Nature of Ecological Communities

Ecologists have often debated the nature of ecological communities, asking if they have emergent properties that transcend the interactions among the populations they contain.

Most Ecological Communities Blend into Neighboring Communities

How do complex population interactions affect the organization and functioning of ecological communities? In the 1920s, ecologists in the United States de-

Finding a Molecular Passport to Mutualism

The mutualistic association between *Rhizobium* bacteria and leguminous plants is established through a complex signaling process. When roots of one of these plants are invaded by *Rhizobium*, the plants respond by developing *root nodules* that house the bacteria and supply them with carbohydrates. In return, the bacteria fix atmospheric nitrogen into ammonia, which the plants use as a nitrogen source. This mutualistic association fixes about 120 million metric tons of nitrogen annually into ammonia, and greatly reduces farmers' need to use nitrogen-containing chemical fertilizers.

Proteins encoded in several sets of *Rhizobium* genes (called *nod*, *nif*, and *fix*) promote the mutualistic association with legumes. Enzymes encoded in the *nod* genes catalyze the synthesis of polysaccharides stimulating growth of a tubelike *infection thread*, which admits the bacteria to the root tissue. Once inside, the same polysaccharides promote development of the root nodule. The *nif* and *fix* genes encode enzymes involved in nitrogen fixation.

Most of these genes are carried on a single plasmid in *Rhizobium*. (Plasmids are small circles of DNA located outside the main bacterial chromosome.) The DNA sequence of the plasmid carrying the *nod*, *nif*, and *fix* genes was revealed by Cristoph Freiberg and his colleagues at the Institute for Molecular Biotechnology in Jena, Germany, and the University of Geneva in Switzerland.

The investigators studied the large plasmid of the *Rhizobium* species designated NGR234, which can invade an unusually large selection of legumes—more than 110 genera—and even one nonleguminous plant. They isolated the bacterium from root nodules and extracted the plasmid DNA. Sequencing showed that the plasmid contains an astounding 416 coding sequences. A computer search identified 277 sequences as relatives of known genes with established functions, including relatives of *nod*, *nif*, and *fix*. The remaining 139 genes have no known counterparts in any other living organism.

Among the known genes, close similarities were found to genes of a plasmid in *Rhizobium radiobacter*, another bacterium able to invade plant hosts. *Rhizobium radiobacter* invades various deciduous plants and promotes growth of large masses of tissue called *crown gall tumors* (see Figure 18.14). The similarities between the *Rhizobium* NGR234 and *Rhizobium radiobacter* plasmids suggest that the mechanisms by which they invade their host plants may have originated in a common evolutionary ancestor.

This research sequencing the *Rhizobium* NGR234 plasmid may help to reveal the molecular and biochemical basis of the mutualistic relationship between nodule-inducing *Rhizobium* and legumes. As a practical matter, the plasmid and its genes may provide a "genetic passport" that could be adapted to allow nodule-inducing *Rhizobium* to invade nonleguminous plants. If successful, this adaptation might allow the equivalent of nitrogen-fixing root nodules to be developed in many nonleguminous crops, eliminating their need for nitrogenous fertilizers and reducing both the cost of growing food crops and pollution by fertilizer runoff.

---

veloped two extreme hypotheses about the nature of ecological communities. Frederic Clements of the University of Minnesota championed an *interactive* view of communities. He described communities as "superorganisms," assemblages of species bound together by complex population interactions. According to this view, each species in a community requires interactions with a set of ecologically different species, just as every cell in an organism requires services that other types of cells provide. Clements believed that once a mature community was established, its **species composition**—the particular combination of species that occupy the site—was at *equilibrium*. If a fire or some other environmental factor disturbed the community, it would return to its predisturbance state.

Henry A. Gleason of the University of Michigan proposed an alternative, *individualistic* view of ecological communities. He believed that population interactions do not always determine species composition. Instead, a community is just an assemblage of species that are individually adapted to similar environmental conditions. According to Gleason's hypothesis, communities do not achieve equilibrium; rather, they constantly change in response to disturbance and environmental variation.

In the 1960s, Robert Whittaker of Cornell University suggested that ecologists could determine which

a. Ants patrolling an acacia

b. Cleared area around an acacia

Figure 50.16

A highly coevolved mutualism. **(a)** Bull's horn acacia trees *(Acacia cornigera)* provide colonies of small ants *(Pseudomyrmex ferruginea)* with homes in hollow enlarged thorns as well as other resources. Although individual ants are small, they are numerous and aggressive. **(b)** Because the ants attack herbivores and remove vegetation near their tree, acacias occupied by ants grow in a space that is free of herbivores and competitors.

a. Interactive hypothesis

The interactive hypothesis predicts that species within communities exhibit similar distributions along environmental gradients (indicated by the close alignment of several curves over each section of the gradient) and that boundaries between communities (indicated by arrows) are sharp.

b. Individualistic hypothesis

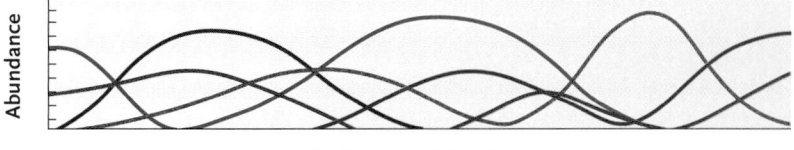

The individualistic hypothesis predicts that species distributions along the gradient are independent (indicated by the lack of alignment of the curves) and that sharp boundaries do not separate communities.

c. Siskiyou Mountains

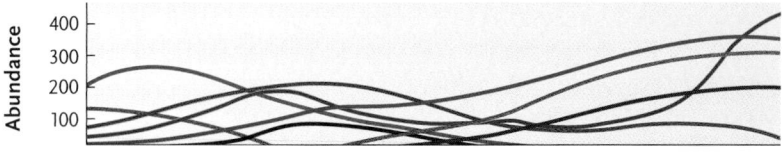

d. Santa Catalina Mountains

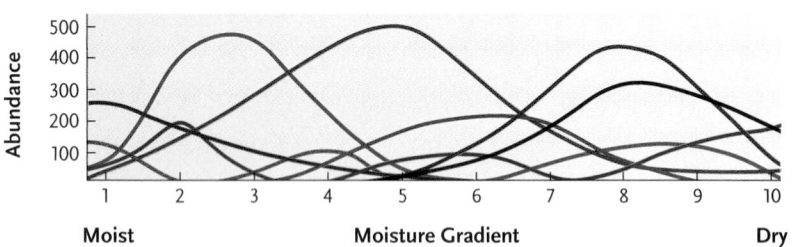

Most gradient analyses support the individualistic hypothesis, as illustrated by distributions of tree species along moisture gradients in Oregon's Siskiyou Mountains and Arizona's Santa Catalina Mountains.

Figure 50.17
Two views of ecological communities.

Figure 50.18
Sharp community boundaries. Soils derived from serpentine rock have high magnesium and heavy metal content, which many plants cannot tolerate. Although native California wildflowers (bright yellow in this photograph) thrive on serpentine soil at the Jasper Ridge Preserve of Stanford University, introduced European grasses (green in this photograph) competitively exclude them from adjacent soils derived from sandstone.

hypothesis was correct by analyzing communities along environmental gradients, such as temperature or moisture **(Figure 50.17)**. According to Clements' interactive hypothesis, species that typically occupy the same communities should always occur together. Thus, their distributions along the gradient would be clustered in discrete groups with sharp boundaries between groups (see Figure 50.17a). According to Gleason's individualistic hypothesis, each species is distributed over the section of an environmental gradient to which it is adapted. Different species would have unique distributions, and species composition would change continuously along the gradient. In other words, communities would not be separated by sharp boundaries (see Figure 50.17b).

Most gradient analyses support Gleason's individualistic view of ecological communities. Environmental conditions vary continuously in space, and most plant distributions match these patterns (see Figure 50.17c, d). Species occur together in assemblages because they are adapted to similar conditions, and the species compositions of the assemblages change gradually across environmental gradients.

Nevertheless, the individualistic view does not fully explain all patterns observed in nature. Ecologists recognize certain assemblages of species as distinctive communities and name them accordingly—redwood forests and coral reefs are good examples. But the borders between adjacent communities are often wide transition zones, called **ecotones**. Ecotones are generally rich with species because they include plants and animals from both neighboring communities as well as some species that thrive only under transitional conditions. In some places, however, a discontinuity in a critical resource or some important abiotic factor produces a sharp community boundary. For example, chemical differences between soils derived from serpentine rock and sandstone establish sharp boundaries between communities of native California wildflowers and introduced European grasses **(Figure 50.18)**.

STUDY BREAK

1. Which view of communities suggests that they are just chance assemblages of species that happen to be adapted to similar abiotic environmental conditions?
2. Why would you often find more species living in an ecotone than you would in the communities on either side of it?

50.3 Community Characteristics

Although the species composition of an ecological community may vary somewhat over geographical gradients, every community has certain characteristics that define its overall appearance and structure.

The Growth Forms of Plants Establish a Community's Overall Appearance

The growth forms—sizes and shapes—of plants vary markedly in different environments. Warm, moist environments support complex vegetation with multiple vertical layers. For example, tropical forests include a canopy, formed by the tallest trees; an understory of shorter trees and shrubs; an herb layer under openings in the canopy; vinelike lianas; and epiphytes, which grow on the trunks and branches of trees **(Figure 50.19)**. By contrast, physically harsh environments are occupied by low vegetation with simple structure. For example, trees on mountaintops buffeted by cold winds are short, and the plants below them cling to rocks and soil. Other environments support growth forms between these extremes (see Chapter 52).

Communities Differ in Species Richness and the Relative Abundance of Species They Contain

Communities differ greatly in their **species richness**, the number of species that live within them. For example, the harsh environment on a low desert island may support just a few species of microorganisms, fungi, algae, plants, and arthropods. By contrast, tropical forests, which grow under milder physical conditions, include many thousands of species. Ecologists have studied global patterns of species richness (described below in Section 50.7) for decades. Today, as human disturbance of natural communities has reached a crisis point, conservation biologists focus on such studies to determine which regions of Earth are most in need of preservation (see Chapter 53).

Within every community, populations differ in their commonness or the **relative abundance** of individuals. Some communities have just one or two abundant species and a number of rare species; in other communities, species are represented by more equal numbers of individuals. For example, in a temperate deciduous forest in West Virginia, tulip poplar *(Liriodendron tulipifera)* and sassafras *(Sassafras albidum)*

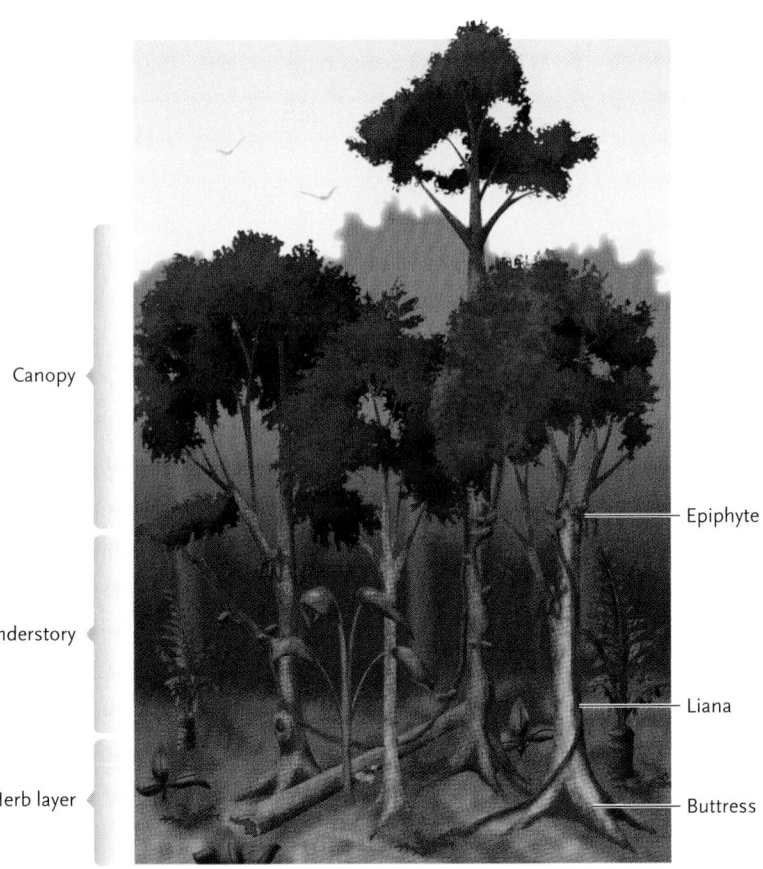

Figure 50.19
Layered forests. Tropical forests include a canopy of tall trees and an understory of short trees and shrubs. Huge vines (lianas) climb through the trees, eventually reaching sunlight in the canopy; and epiphytic plants grow on trunks and branches, increasing the structural complexity of the habitat.

Forest A   Forest B   Forest C

Figure 50.20

Species diversity. In this hypothetical example, each of three forests contains 50 trees. Forest A and forest B each include 10 tree species, but forest C includes only two tree species. Because forest A is dominated by one tree species, but forest B is not, ecologists would say that forest B is more diverse. Forest C, with only two tree species, is less diverse than the others.

might together account for nearly 85% of the trees. By contrast, a tropical forest in Costa Rica may include more than 200 tree species, each making up only a small percentage of the total.

Species richness and relative abundance together contribute to a community characteristic that ecologists call **species diversity.** To demonstrate species diversity, we will compare two hypothetical forest communities, each with 50 trees distributed among 10 species **(Figure 50.20).** In Forest A, the dominant species is represented by 39 individuals, two species by two individuals each, and seven species by one individual each. In Forest B, each of the 10 species is represented by five individuals. Although both communities have the same species richness (10 species), Forest A is less diverse than Forest B, because most of its trees are of the same species. A forest with only two tree species (Forest C in Figure 50.20) would be less diverse than either of the others.

Feeding Relationships within a Community Determine Its Trophic Structure

All ecological communities, regardless of their species richness, also have a trophic structure (*troph* = nourishment) that comprises all of the plant–herbivore, predator–prey, host–parasite, and potential competitive interactions **(Figure 50.21).**

Trophic Levels. We can visualize the trophic structure of a community as a hierarchy of **trophic levels,** defined by the feeding relationships among its species (see Figure 50.21a). Photosynthetic organisms are the **primary producers,** the first trophic level. Primary producers are often described as **autotrophs** (*auto* = self) because they capture sunlight and convert it into chemical energy, using simple inorganic molecules acquired from the environment to build larger organic molecules that

other organisms can use. Plants are the dominant primary producers in terrestrial communities. Multicellular algae and plants are the major primary producers in shallow freshwater and marine environments, but photosynthetic protists and cyanobacteria play that role in deep, open water.

Animals, by contrast, are **consumers.** Herbivores, which feed directly on plants, form the second trophic level, the **primary consumers.** Carnivores that feed on herbivores are the third trophic level, or **secondary consumers;** and carnivores that feed on other carnivores form the fourth trophic level, the **tertiary consumers.** For example, songbirds feeding on herbivorous insects are secondary consumers, and falcons feeding on songbirds are tertiary consumers. Some organisms, like humans and some bears, are **omnivores,** feeding at several trophic levels simultaneously.

A separate and distinct trophic level includes organisms that extract energy from the organic detritus (refuse) produced at other trophic levels. Scavengers, or **detritivores,** are animals such as earthworms and vultures that ingest dead organisms, digestive wastes, and cast-off body parts such as leaves and exoskeletons. **Decomposers** are small organisms, such as bacteria and fungi, that feed on dead or dying organic material. As described in Chapter 51, detritivores and decomposers serve a critical ecological function because their activity reduces organic material to small inorganic molecules that producers can assimilate.

All of the consumers in a community—the animals, fungi, and diverse microorganisms—are described as **heterotrophs** (*hetero* = other) because they acquire energy and nutrients by eating other organisms or their remains.

Food Chains and Webs. Ecologists depict the trophic structure of a community in a **food chain,** a portrait

a. Trophic levels   **b.** Marine food web

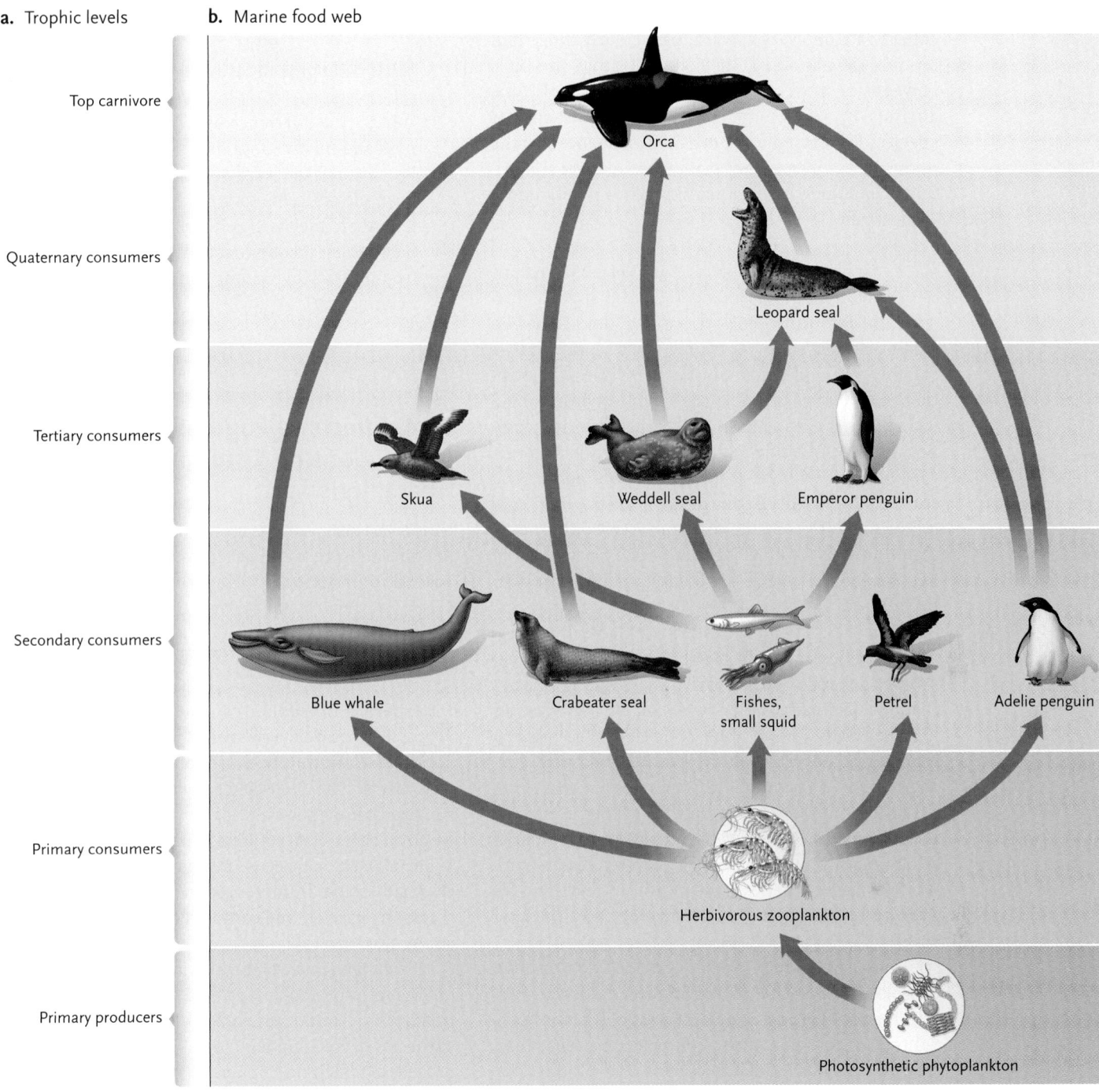

Figure 50.21
The marine food web off the coast of Antarctica.

of who eats whom. Each link in a food chain is represented by an arrow pointing from the food to the consumer. Simple, straight-line food chains are rare in nature because most consumers feed on more than one type of food, and because most organisms are eaten by more than one type of consumer. These complex relationships are portrayed as a **food web**, a set of interconnected food chains with multiple links.

In the food web for the waters off the coast of Antarctica (see Figure 50.21b), the primary producers and primary consumers are small organisms that occur in vast numbers. Microscopic diatoms (phytoplankton) are responsible for most photosynthesis, and small shrimplike krill (zooplankton) are the major primary consumers. These tiny organisms are eaten by larger species, such as fishes, seabirds, and suspension-feeding baleen whales. Some of the secondary consumers are themselves eaten by birds and mammals at higher trophic levels. The top carnivore in this ecosystem, the orca *(Orcinus orca)*, feeds on carnivorous birds and mammals.

Ideally, depictions of food webs would include all species in a community, from microorganisms to the top consumer. But most ecologists simply cannot collect data on every species, particularly those that are rare or very small. Instead, they study the links between the most important species and simplify the analysis by grouping together trophically similar species. For example, Figure 50.21b categorizes the many different species of primary producers and primary consumers as phytoplankton and zooplankton, respectively.

Food-Web Analysis. In the late 1950s, Robert MacArthur of Princeton University pioneered the analysis of food webs to determine how the many links between trophic levels may contribute to a community's **stability**—its ability to maintain its species composition and relative abundances when environmental disturbances eliminate some species from the community. MacArthur hypothesized that in species-rich communities, where animals feed on many food sources, the absence of one or two species would have only minor effects on the structure and stability of the community as a whole. He therefore proposed a connection between species diversity, food-web complexity, and community stability.

Recent research has confirmed MacArthur's reasoning. For example, the average number of links per species generally increases with increasing species richness. Comparative food-web analysis also reveals that the relative proportions of species at the highest, middle, and lowest trophic levels are reasonably constant across communities. When researchers compared the number of prey species to the number of predator species in food webs from 92 communities of freshwater invertebrates, they discovered that, regardless of species richness, a community includes between two and three prey species for every predator species.

Interactions among species in a food web are often complex, indirect, and hard to unravel. In desert communities of the American Southwest, for example, rodents and ants potentially compete for seeds, their main food source. And the plants that produce the seeds compete for water, nutrients, and space. Rodents generally prefer to eat large seeds, but ants prefer small seeds. Thus, feeding by rodents reduces the potential population sizes of plants that produce large seeds. As a result, the population sizes of plants that produce small seeds may increase, ultimately providing more food for ants.

Some analyses of food webs focus on interactions in which predators or prey have significant influence on the growth rates and sizes of other populations in the community; these *strong interactions* can affect overall community structure. In the next section we provide examples of strong interactions when we describe how consumers influence the competitive interactions among populations of their prey.

STUDY BREAK

1. What plant growth forms are common in tropical forests?
2. What is the difference between species richness and relative abundance?
3. Peregrine falcons are predatory birds that have been introduced into many North American cities, where they feed primarily on pigeons. The pigeons eat mostly vegetable matter. To what trophic level do pigeons and peregrine falcons belong?

50.4 Effects of Population Interactions on Community Characteristics

Numerous studies have shown that interspecific competition and predation can influence a community's species composition.

Interspecific Competition Can Reduce Species Richness within Communities

Interspecific competition can cause the local extinction of species or prevent new species from becoming established in a community, thus reducing its species richness. During the 1960s and early 1970s, ecologists emphasized competition as the primary factor structuring communities. Observations of resource partitioning and character displacement suggested that some process had fostered differences in resource use among coexisting species, and competition provided the most straightforward explanation of these patterns.

Seeking to uncover direct evidence of competition, ecologists undertook many field experiments on competition in natural populations. The experiment on barnacles depicted in Figure 50.12 is typical of this approach, in which researchers determine whether adding or removing a species changes the distribution or population size of its presumed competitors. In the early 1980s, two independent reviews of the literature on these field experiments, one by Joseph Connell and the other by Thomas W. Schoener of the University of California at Davis, suggested that competition is sometimes a potent force. Connell's survey, which included 527 published experiments on 215 species, identified competition in roughly 40% of the experiments and more than 50% of the species. Schoener's review, which used different criteria to evaluate 164 experiments on approximately 400 species, found that competition affected more than 75% of the species.

Although these reviews confirm the importance of competition, the ecological literature upon which they were based probably contains several significant biases. First, ecologists who set out to study competition are more likely to study interactions in which they think competition occurs, and they are more likely to publish research that documents its importance. Thus, the literature includes more studies of competition in *K*-selected species than in *r*-selected species. Recall that populations of *r*-selected species, such as herbivorous insects, rarely reach carrying capacity, and competition may not limit their population sizes (review Section 49.6). Thus, the Connell and Schoener surveys may *overestimate* the importance of competition. (Nevertheless, a more recent survey suggests that interspecific competition may be common even among populations of herbivorous insects.) Another bias, which Connell called "the ghost of competition past," *underestimates* the importance of competition. If, as many ecologists believe, resource partitioning and character displacement are the results of past competition, we are unlikely to witness much competition today, even though it was once important in structuring those population interactions.

Ecologists have not yet reached consensus about whether interspecific competition strongly influences the species composition and structure of most communities. Plant ecologists and vertebrate ecologists, who often study *K*-selected species, generally believe that competition has a profound effect on species distributions and resource use. Insect ecologists and marine ecologists, who often study *r*-selected species, argue that competition is not the major force governing community structure, pointing instead to predation or parasitism and physical disturbance.

Predators Can Boost Species Richness by Stabilizing Competitive Interactions among Their Prey

Predators can influence the species richness and structure of communities by reducing the population sizes of their prey. On the rocky coast of the American Northwest, for example, algae and sessile invertebrates compete for attachment sites on rocks, a requirement for life on a wave-swept shore. Mussels *(Mytilus californianus)* are the strongest competitors for space, eliminating other species from the community. But at some sites, predatory sea stars *(Pisaster ochraceus)* preferentially feed on mussels, reducing their numbers and creating space for other species to grow. Because the interaction between *Pisaster* and *Mytilus* affects other species as well, it qualifies as a strong interaction.

In the 1960s, Robert Paine of the University of Washington conducted removal experiments to evaluate the effects of *Pisaster* predation **(Figure 50.22)**. In predator-free experimental plots, mussels outcompeted barnacles, chitons, limpets, and other invertebrate herbivores, reducing species richness from 18 species to 2 or 3. In control plots that contained preda-

tors, however, all 18 species persisted. Ecologists describe predators like *Pisaster* as **keystone species**, species that have a greater effect on community structure than their numbers might suggest.

Herbivores May Counteract or Reinforce Competition among Their Food Plants

Herbivores also exert complex effects on communities. In the 1970s, Jane Lubchenco, then of Harvard University, studied herbivory in a periwinkle snail *(Littorina littorea)*, a keystone species on rocky shores in Massachusetts **(Figure 50.23)**. Periwinkles preferentially graze on the tender green alga *Enteromorpha*. In tidepools, which are usually submerged, *Enteromorpha* outcompetes other algae. Moderate feeding by periwinkles, however, eliminates some *Enteromorpha*, allowing less competitive algal species to grow. Moderate herbivory by periwinkles therefore increases algal species richness in tidepools. But on high rocks, which are exposed to air during low tide, the dehydration-resistant red alga *Chondrus* is competitively dominant. Periwinkles don't eat the tough *Chondrus*, however, feeding instead on the less abundant and competitively inferior *Enteromorpha*. Thus, on exposed rocks, feeding by the snails reduces algal species richness.

STUDY BREAK

1. How is the scientific literature on interspecific competition biased?
2. What are keystone species, and how do they influence species richness in communities?

50.5 Effects of Disturbance on Community Characteristics

Recent research tends to support the individualistic view that many communities are not in equilibrium and that their species composition changes frequently. Environmental disturbances—storms, landslides, fires, floods, and cold spells—often eliminate some species, providing opportunities for others to become established.

Frequent Disturbances Keep Some Communities in a Constant State of Flux

Physical disturbances are common in some environments. For example, lightning-induced fires commonly sweep through grasslands, powerful hurricanes routinely demolish patches of forest, and waves wash over communities that live at the edge of the sea.

Joseph Connell and his colleagues conducted an ambitious long-term study of the effects of disturbance

Figure 50.22 Experimental Research

Effect of a Predator on the Species Richness of Its Prey

QUESTION: Does feeding by a predator influence the species richness and relative abundances of the species on which it feeds?

EXPERIMENT: The predatory sea star *Pisaster ochraceus* preferentially feeds on mussels *(Mytilus californianus)*, which is the strongest competitor for space in rocky intertidal habitats in Washington State. Paine removed *Pisaster* from caged experimental study plots, but left control study plots undisturbed. He then monitored the species richness of *Pisaster*'s invertebrate prey over many years.

RESULTS: Paine documented an increase in mussel populations in the experimental plots as well as complex changes in the feeding relationships among species in the intertidal food web. The overall effect of removing *Pisaster*, the top predator in this food web, was a rapid decrease in the species richness of invertebrates and algae. By contrast, control plots maintained their species richness over the course of the experiment.

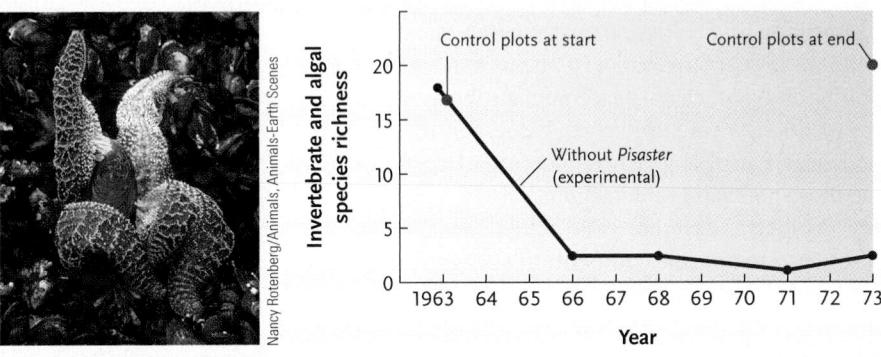

CONCLUSION: Predation by the sea star *Pisaster ochraceus* maintains the species richness of its prey by preventing mussels from outcompeting other invertebrates and algae on rocky shores.

---

on coral reefs, shallow tropical marine habitats that are among the most species-rich communities on Earth. In some parts of the world, reefs are routinely battered by violent storms, which wash corals off the substrate, creating bare patches in the reef. The scouring action of storms creates opportunities for coral larvae to settle on bare substrates and start a new colony; ecologists use the word *recruitment* to describe the process in which young individuals join a population.

From 1963 to 1992, Connell and his colleagues tracked the fate of the Heron Island Reef at the south end of Australia's Great Barrier Reef **(Figure 50.24).** The inner flat and protected crests of the reef are sheltered from severe wave action during storms, whereas some pools and crests are routinely exposed to physical disturbance. Because corals live in colonies of variable size, the researchers monitored coral abundance by measuring the percentage of the substrate (that is, the seafloor) that colonies covered. They revisited marked study plots at intervals, photographing and identifying individual coral colonies.

Five major cyclones crossed the reef during the 30-year study period. Coral communities in the exposed areas of the reef were in a nearly continual state of flux. In exposed pools, four of the five cyclones re-

duced the percentage of cover, often drastically. On exposed crests, the cyclone of 1972 eliminated virtually all of the corals, and subsequent storms slowed the recovery of these areas for more than 20 years. By contrast, corals in sheltered areas suffered much less storm damage. Nevertheless, their coverage also declined steadily during the study as a natural consequence of the corals' growth. As colonies grew taller and closer to the ocean's surface, their increased exposure to air resulted in substantial mortality.

Connell and his colleagues also documented recruitment, the growth of new colonies from settling larvae, in their study plots. They discovered that the rate at which new colonies developed was almost always higher in sheltered areas than in exposed areas. However, recruitment rates were extremely variable, depending in part on the amount of space that storms or coral growth had made available.

This long-term study of coral reefs illustrates that frequent disturbances prevent some communities from reaching an equilibrium determined by interspecific interactions. Changes in the coral reef community at Heron Island result from the combined effects of external disturbances that remove coral colonies from the reef and internal processes (growth and recruit-

Figure 50.23 Experimental Research

The Complex Effects of an Herbivorous Snail on Algal Species Richness

QUESTION: How does feeding by periwinkle snails *(Littorina littorea)* influence the species richness of algae in intertidal communities?

EXPERIMENT: Lubchenco manipulated the densities of periwinkle snails in tidepools and on exposed rocks in a rocky intertidal habitat by creating enclosures that prevented snails from either entering or leaving her study plots. She then monitored the species composition of algae in the study plots and examined those data by plotting them against periwinkle density.

RESULTS: The effects of periwinkle density on algal species richness varied dramatically between study plots in tidepools and on exposed rocks.

Periwinkle snails *(Littorina littorea)*

© Jane Burton/BruceColman, Ltd.

Enteromorpha growing in tidepools

© Jane Burton/BruceColman, Ltd.

Chondrus growing on exposed rocks

Heather Angel/Natural Visions

In tidepools

Periwinkles per square meter

In tidepools, snails at low densities eat little algae and *Enteromorpha* competitively excludes other algal species, reducing species richness. At high snail densities, heavy feeding on all species reduces algal species richness. At intermediate snail densities, grazing eliminates some *Enteromorpha*, allowing other species to grow.

On exposed rocks

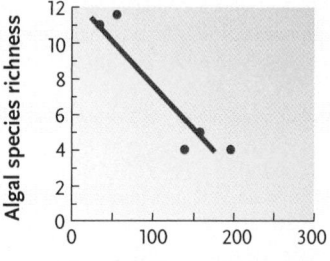

Periwinkles per square meter

On exposed rocks, periwinkles never eat much *Chondrus*, but they consume the tender, less successful competitors. Thus, feeding by periwinkles reinforces the competitive superiority of *Chondrus:* as periwinkle density increases, algal species richness declines.

CONCLUSION: Grazing by periwinkle snails has complex effects on the species richness of competing algae. In tidepools, where periwinkle snails preferentially feed on *Enteromorpha*, the competitively dominant alga, snails at an intermediate density remove some *Enteromorpha*, which allows weakly competitive algae to grow, increasing species richness. Feeding by snails at either low or high densities reduces algal species richness. On exposed rocks, where periwinkle snails rarely eat the competitively dominant alga *Chondrus*, feeding by snails reduces algal species richness.

---

ment) that either eliminate colonies or establish new ones. In this community, growth and recruitment are slow processes, and disturbances are frequent. Thus, the community never attains equilibrium.

Moderate Levels of Disturbance May Foster High Species Richness

According to the **intermediate disturbance hypothesis,** proposed by Connell in 1978, species richness is greatest in communities that experience fairly frequent disturbances of moderate intensity. Moderate distur-

bances create some openings for *r*-selected species to arrive and join the community, but they allow *K*-selected species to survive. Thus, communities that experience intermediate levels of disturbance contain a rich mixture of species. Where disturbances are severe and frequent, communities include only *r*-selected species that complete their life cycles between catastrophes. Where disturbances are mild and rare, communities are dominated by long-lived *K*-selected species that competitively exclude other species from the community.

Several studies in diverse habitats have confirmed the predictions of the intermediate disturbance hy-

Figure 50.24

The effects of storms on corals. Five tropical cyclones (marked by gray arrows) damaged corals on the Heron Island Reef during a 30-year period. Storms reduced the percentage cover of corals in exposed parts of the reef **(a)** much more than in sheltered parts of the reef **(b)**.

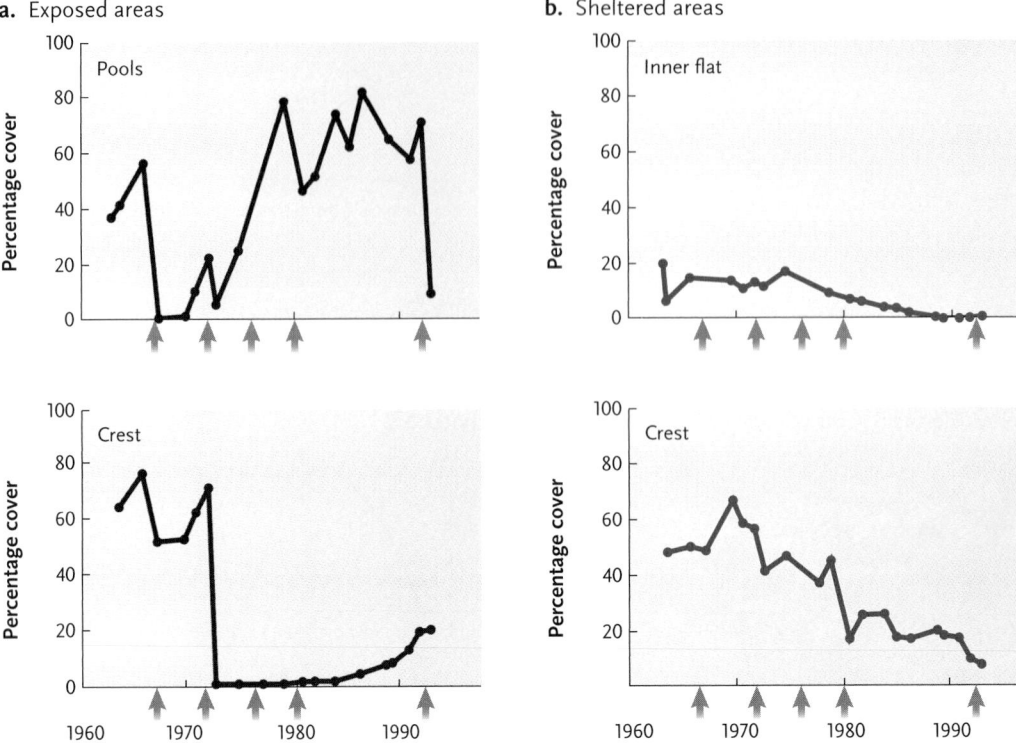

Figure 50.24

The effects of storms on corals. Five tropical cyclones (marked by gray arrows) damaged corals on the Heron Island Reef during a 30-year period. Storms reduced the percentage cover of corals in exposed parts of the reef **(a)** much more than in sheltered parts of the reef **(b)**.

pothesis. For example, Colin R. Townsend and his colleagues at the University of Otago studied the effects of disturbance at 54 stream sites in the Taieri River system in New Zealand. Disturbance occurs in these communities when water flow from heavy rains moves the rocks, soil, and sand in the streambed, disrupting the habitats where animals live. Townsend and his colleagues measured how much of the substrate moved in different streambeds to index the intensity of the disturbance. Their results indicate that species richness is highest in areas that experience intermediate levels of disturbance **(Figure 50.25)**.

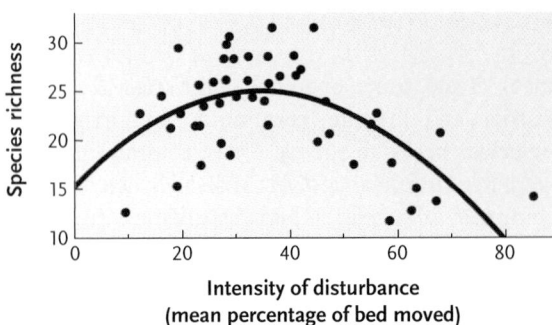

Figure 50.25

An observational study that supports the intermediate disturbance hypothesis. In the Taieri River system in New Zealand, species richness was highest in stream communities that experienced an intermediate level of disturbance.

Some ecologists have also suggested that species-rich communities recover from disturbances more readily than do less diverse communities. For example, David Tilman and his colleagues at the University of Minnesota conducted large-scale experiments in midwestern grasslands on the relationship between species number and the ability of communities to recover from disturbance. Their results demonstrate that grassland plots with high species richness recover from drought faster than plots with fewer species.

STUDY BREAK

1. How might disturbances from storms allow coral reefs to be rejuvenated by the recruitment of young individuals?
2. How do moderately severe and moderately frequent disturbances influence a community's species richness?

50.6 Ecological Succession: Responses to Disturbance

In response to disturbance, communities undergo **ecological succession**, a somewhat predictable series of changes in species composition over time.

Succession Begins after Disturbance Alters a Landscape or Changes the Species Composition of an Existing Community

Primary succession begins when organisms first colonize habitats without soil, such as those created by erupting volcanoes and retreating glaciers **(Figure 50.26)**. Lichens (see Section 28.3), which derive nutrients from rain and bare rock, are usually the first visible colonizers of such inhospitable habitats. They secrete mild acids that erode rock surfaces, initiating the slow development of soil, which is enriched by the organic material lichens produce. After lichens modify a site, mosses (see Section 27.2) colonize patches of soil and grow quickly.

As soil accumulates, hardy opportunistic plants—grasses, ferns, and broad-leaved herbs—colonize the site from surrounding areas. Their roots break up rock, and as they die, their decaying remains enrich the soil. Detritivores and decomposers facilitate these processes. As the soil gets deeper and richer, increased moisture and nutrients support bushes and, eventually, trees. Late successional stages are often dominated by *K*-selected species with woody trunks and branches that position leaves in sunlight and large root systems that acquire water and nutrients from soil.

In the classical view of ecological succession, long-lived species, which replace themselves over time, eventually dominate a community, and new species join it only rarely. This relatively stable, late successional stage is called a **climax community** because the dominant vegetation replaces itself and persists until an environmental disturbance eliminates it, allowing other species to invade. Local climate and soil conditions, the surrounding communities where colonizing species originate, and chance events determine the species composition of climax communities. However, recent research suggests that even "climax communities" change slowly in response to environmental fluctuations, as described below.

Secondary succession occurs after existing vegetation is destroyed or disrupted by an environmental disturbance, such as a fire, a storm, or human activity. The presence of soil makes the disturbed sites ripe for colonization. Moreover, the soil may contain numerous seeds that germinate after the disturbance. The early stages of secondary succession proceed rapidly, but later stages parallel those of primary succession.

Secondary succession in the North Temperate Zone is well studied in abandoned farms, called "old fields," where forests were cleared centuries earlier. Because the transformation from old field back to forest takes at least a hundred years, ecologists use historical records to find the age of different stands of vegetation and reconstruct the successional sequence by comparing stands of different ages. In the Piedmont region of southeastern North America, an abandoned field is covered by crabgrass (genus *Digitaria*), an an-

nual plant, during the first growing season. The following year, crabgrass is replaced by horseweed *(Conyza canadensis)*, which cannot persist because it secretes substances that inhibit the germination of its own seeds. Ragweed *(Ambrosia artemisiifolia)*, another annual, dominates during the third year, but it is gradually replaced by perennial asters (genus *Erigeron*) and broomsedges (genus *Andropogon*), which are, in turn, replaced by shrubs. Ten to fifteen years after the field was abandoned, pine (genus *Pinus*) seedlings germinate. Growing pines cast substantial shade and their fallen needles acidify the soil, making the site unsuitable for the plants from earlier successional stages. Because pines are intolerant of shade, pine seedlings don't flourish under mature pine trees. Thus, after 50 to 100 years, pines are replaced by a taller mixed hardwood forest of oaks (genus *Quercus*) and hickories (genus *Carya*), which develops in the thick, moist soil. The hardwood forest forms the climax community after more than a century of successional change.

Similar climax communities sometimes arise from alternative successional sequences. For example, hardwood forests also develop in sites that were once ponds. During **aquatic succession**, debris from rivers and runoff accumulates in a body of water, causing it to fill in at its margins. The pond is transformed into a swamp, inhabited by plants adapted to a semisolid substrate. As larger plants get established, their high transpiration rates dry the soil, allowing other plant species to colonize. Given enough time, the site may become a meadow or forest, where an area of moist, low-lying ground is the only remnant of the original pond.

Community Characteristics Change during Succession

Several characteristics undergo directional change as succession proceeds. First, because *r*-selected species are short-lived and *K*-selected species long-lived, species composition changes rapidly in the early stages, but slowly in the late stages of succession. Second, species richness increases rapidly during the early stages because new species join the community faster than resident species become extinct; as succession proceeds, however, species richness stabilizes or may even decline. Third, in terrestrial communities that receive sufficient rainfall, the maximum height and total mass of the vegetation increase steadily as large species replace small ones, creating the complex structure of the climax.

Because plants influence the physical environment below them, the community itself increasingly moderates the microclimate. The shade cast by a forest canopy retains soil moisture and reduces temperature fluctuations. The trunks and canopy also reduce wind speed. By contrast, the short vegetation in an early successional stage does not effectively shelter the space below it.

1 The glacier has retreated about 8 m per year since 1794.

2 This site was covered with ice less than 10 years before this photo was taken. When a glacier retreats, a constant flow of melt water leaches minerals, especially nitrogen, from the newly exposed substrate.

3 Once lichens and mosses have established themselves, mountain avens (genus *Dryas*) grows on the nutrient-poor soil. This pioneer species benefits from the activity of mutualistic nitrogen-fixing bacteria, spreading rapidly over glacial till.

4 Within 20 years, shrubby willows (genus *Salix*), cottonwoods (genus *Populus*), and alders (genus *Alnus*) take hold in drainage channels. These species are also symbiotic with nitrogen-fixing microorganisms.

5 In time, young conifers, mostly hemlocks (genus *Tsuga*) and spruce (genus *Picea*), join the community.

6 After 80 to 100 years, dense forests of Sitka spruce (*Picea sichensis*) and western hemlock (*Tsuga heterophylla*) have crowded out the other species.

Figure 50.26
Primary succession following glacial retreat. The retreat of glaciers at Glacier Bay, Alaska, has allowed ecologists to document primary succession on newly exposed rocks and soil.

Although ecologists usually describe succession in terms of vegetation, animals undergo succession, too. As the vegetation shifts, new resources become available, and animal species replace each other over time. Herbivorous insects, which often have strict food preferences, undergo succession along with their food plants. And as the herbivores change, so do their predators, parasites, and parasitoids. In old-field succession in eastern North America, different vegetation stages harbor a changing assortment of bird species (Figure 50.27).

Several Hypotheses Help to Explain the Processes Underlying Succession

Differences in dispersal abilities, maturation rates, and life spans among species are at least partly responsible for ecological succession. Early successional stages harbor many *r*-selected species because they produce numerous small seeds that colonize open habitats and grow quickly. Mature successional stages are dominated by *K*-selected species because they are long-lived. Nevertheless, coexisting populations inevitably affect one another. Although the role of population interactions in succession is generally acknowledged, ecologists debate the relative importance of processes that either facilitate or inhibit the turnover of species in a community.

The **facilitation hypothesis** suggests that species modify the local environment in ways that make it less suitable for themselves but more suitable for colonization by species typical of the next successional stage. For example, when lichens first colonize bare rock, they produce a small quantity of soil, which is required by mosses and grasses that grow there later. According to this hypothesis, changes in species composition are both orderly and predictable because the presence of each stage facilitates the success of the next. Facilitation is very important in primary succession, but it may not be the best model of interactions that influence secondary succession.

The **inhibition hypothesis** suggests that new species are prevented from occupying a community by whatever species are already present. According to this hypothesis, succession is neither orderly nor predictable because each stage is dominated by whichever species happen to colonize the site first. Species replacements occur only when individuals of the dominant species die of old age or when an environmental disturbance reduces their numbers. Eventually, long-lived species replace short-lived species, but the precise species composition of a mature community is up for grabs. Inhibition appears to play a role in some secondary successions. For example, the interactions among early successional species in an old field are highly competitive. Horseweed inhibits the growth of asters, which follow them in succession, by shading the aster seedlings and by releasing toxic substances from their roots. The experimental removal of horseweed enhances the growth of asters, confirming the inhibitory effect.

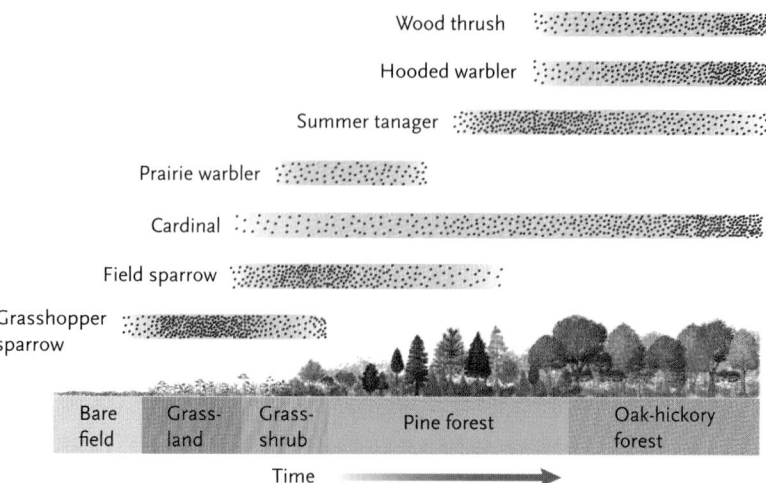

Figure 50.27

Succession in animals. Successional changes in bird species composition in an abandoned agricultural field in eastern North America parallel the changes in plant species composition. Residence times of several representative species are illustrated. The density of stippling inside each bar illustrates the density of each species through time.

The **tolerance hypothesis** asserts that succession proceeds because competitively superior species replace competitively inferior ones. According to this model, early-stage species neither facilitate nor inhibit the growth of later-stage species. Instead, as more species arrive at a site and resources become limiting, competition eliminates species that cannot harvest scarce resources successfully. In the Piedmont region of North America, for example, hardwood trees are more tolerant of shade than pine trees are, and hardwoods gradually replace pines during succession. Thus, the climax community includes only strong competitors. Tolerance may explain the species composition of many transitional and mature communities.

At most sites, succession probably results from a combination of facilitation, inhibition, and tolerance, coupled with interspecific differences in dispersal, growth, and maturation rates. Moreover, within a community, the patchiness of abiotic factors also strongly influences plant distributions and species composition. In the deciduous forests of eastern North America, maples (genus *Acer*) predominate on wet, low-lying ground, but oaks (genus *Quercus*) are more abundant at higher and drier sites. Thus, a mature deciduous forest is often a mosaic of species and not a uniform stand of trees.

Disturbance and density-independent factors also play important roles, in some cases speeding successional change. In northern forests, for example, moose prefer to feed on deciduous shrubs, accelerating the rate at which conifers replace them. In other cases, disturbance inhibits successional change, establishing a *disturbance climax* or **disclimax community**. In many grassland communities, grazing by large mammals and periodic fires kill the seedlings of trees that would otherwise become established. Thus, disturbance prevents the succession from grassland to forest, and grassland persists as a disclimax community.

On a local scale, disturbances often destroy small patches of vegetation, returning them to an earlier successional stage. A hurricane may knock over trees in a forest, creating small, sunny patches of open ground. Locally occurring *r*-selected species take advantage of the resources that are suddenly available and quickly colonize the openings. These local patches then undergo succession that is out of step with the immediately surrounding forest. Thus, moderate disturbance, accompanied by succession in local patches, can increase species richness in many communities.

STUDY BREAK

1. What is the difference between primary succession and secondary succession?
2. How does a climax community differ from early successional stages?
3. How do the three hypotheses about the causes of ecological succession view the role of population interactions in the successional process?

50.7 Variations in Species Richness among Communities

Species richness often varies among communities according to a recognizable pattern. Two large-scale patterns of species richness—latitudinal trends and island patterns—have captured the attention of ecologists for more than a century.

Many Types of Organisms Exhibit Latitudinal Gradients in Species Richness

Ever since Darwin and Wallace traveled the globe (see Section 19.2), ecologists have recognized broad latitudinal trends in species richness. For many, but not all, plant and animal groups, species richness follows a latitudinal gradient, with the most species in the tropics and a steady decline in numbers toward the poles **(Figure 50.28)**. Several general hypotheses may explain these striking patterns.

Some hypotheses propose historical explanations for the *origin* of high species richness in the tropics. The benign climate in tropical regions allows some tropical organisms to have more generations per year than their temperate counterparts. And, given the small seasonal changes in temperature, tropical species may be less likely than temperate species to migrate from one habitat to another, thus reducing gene flow between geographically isolated populations (see Section 21.3). These factors may have fostered higher speciation rates in the tropics, accelerating the accumulation of species. Tropical communities may also have experienced severe disturbance less often than communities at higher latitudes, where periodic glaciations have caused repeated extinctions. Thus, new species may have accumulated in the tropics over longer periods of time.

Other hypotheses focus on ecological explanations for the *maintenance* of high species richness in the tropics. Some resources are more abundant, predictable, and diverse in tropical communities. Tropical regions experience more intense sunlight, warmer temperatures in most months, and higher annual rainfall than temperate and polar regions (see Chapter 52). These factors provide a long and predictable growing season for the lush tropical vegetation, which supports a rich assemblage of herbivores, and through them many carnivores and parasites. Furthermore, the abundance, predictability, and year-round availability of resources allow some tropical animals to have specialized diets. For example, tropical forests support many species of fruit-eating bats and birds, which could not survive in temperate forests where fruits are not available year-round.

Species richness may therefore be a self-reinforcing phenomenon in tropical communities. Complex webs of population interactions and interdependency

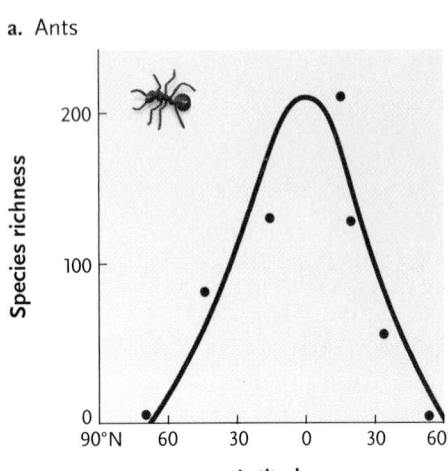

a. Ants

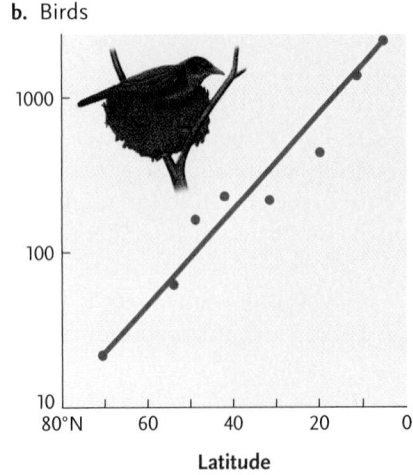

b. Birds

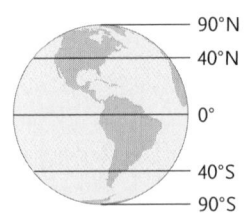

Figure 50.28
Latitudinal trends in species richness. The species richness of many animals and plants varies with latitude, as illustrated here **(a)** for ants and **(b)** for birds of North and Central America. The species-richness data for birds is based on records of where the species breed.

have coevolved in relatively stable and predictable tropical climates. Predator–prey, competitive, and symbiotic interactions may prevent individual species from dominating communities and reducing species richness.

The Theory of Island Biogeography Explains Variations in Species Richness

Although the species richness of communities may be stable over time, species composition is often in flux as new species join a community and others drop out. In the 1960s, Robert MacArthur of Princeton University and Edward O. Wilson of Harvard University addressed the question of why communities vary in species richness, using islands as model systems. Islands provide natural laboratories for studying ecological phenomena, just as they do for evolution (see *Focus on Research* in Chapter 21). Island communities are often small, have well-defined boundaries, and are isolated from surrounding communities.

In developing the **equilibrium theory of island biogeography**, MacArthur and Wilson sought to explain variations in species richness on islands of different size and different levels of isolation from other landmasses **(Figure 50.29)**. They hypothesized that the number of species on any island was governed by a give and take between two processes: the immigration of new species to an island and the extinction of species already there (see Figure 50.29a).

According to the MacArthur–Wilson model, the mainland harbors a *species pool* from which species immigrate to offshore islands. Seeds and small arthropods are carried by wind or floating debris; some animals, such as birds, arrive under their own power. When few species are already on an island, the rate at which new species immigrate to the island is high. But

as more species inhabit the island over time, the immigration rate declines because there are fewer species left in the mainland pool that can still arrive on the island as *new* colonizers.

Once a species immigrates to an island, its population grows and persists for some time. But as the number of species on the island increases, the rate at which those species go extinct also rises. The extinction rate increases through time partly because there are more species that can go extinct there. In addition, as the number of species on the island increases, competition and predator–prey interactions can reduce the population sizes of some species and drive them to extinction.

According to MacArthur and Wilson's theory, an equilibrium between immigration and extinction determines the number of species that ultimately occupy an island. In other words, once equilibrium is reached, the number of species remains relatively constant because one species already on the island goes extinct in about the same time it takes a new species to immigrate to the island. The model does not specify which species immigrate to the island or which ones already on the island go extinct. It simply predicts that the number of species on the island is in equilibrium, although species composition is not. The ongoing processes of immigration and extinction establish a constant turnover in the roster of species that live on any island.

The MacArthur–Wilson model explains why some islands harbor more species than others. Large islands have higher immigration rates than small islands do because they present a larger target for dispersing organisms. Moreover, large islands have lower extinction rates because they can support larger populations and provide a greater range of habitats and resources. Thus, at equilibrium, large islands have more species than small islands (see Figure 50.29b). Similarly, islands

Figure 50.29
Predictions of the theory of island biogeography.

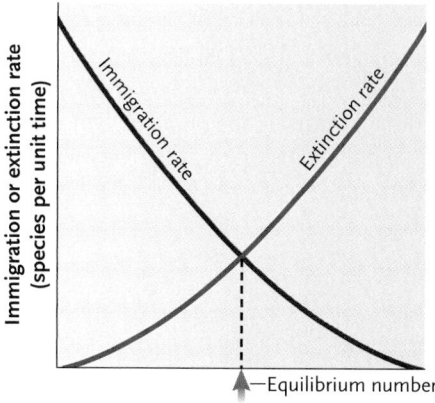

a. Immigration and extinction rates

The number of species on an island at equilibrium (indicated by the arrow) is determined by the rate at which new species immigrate and the rate at which species already on the island go extinct.

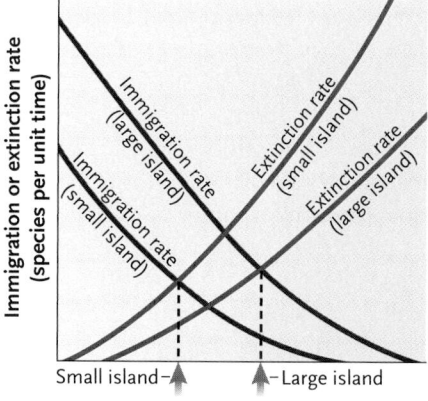

b. Effect of island size

Immigration rates are higher and extinction rates lower on large islands than on small islands. Thus, at equilibrium, large islands have more species.

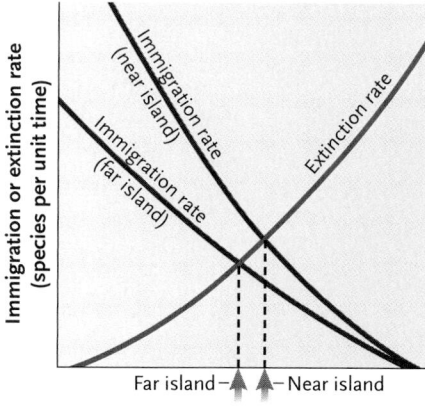

c. Effect of distance from mainland

Organisms leaving the mainland locate nearby islands more easily than distant islands, causing higher immigration rates on near islands. Thus, near islands support more species than far ones.

a. Distance effect

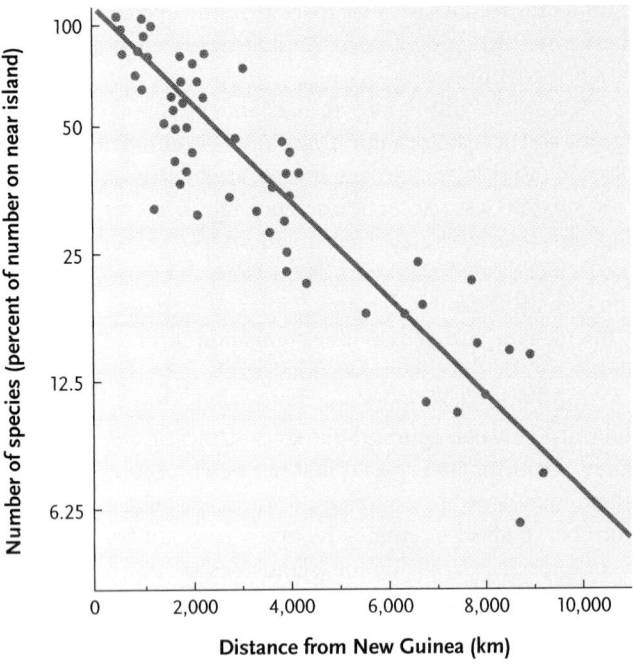

The number of lowland bird species on islands of the South Pacific declines with the islands' distance from the species source, the large island of New Guinea. Data in this graph were corrected for differences in the sizes of the islands. The number of bird species on each island is expressed as a percentage of the number of bird species on an island of equivalent size close to New Guinea.

b. Area effect

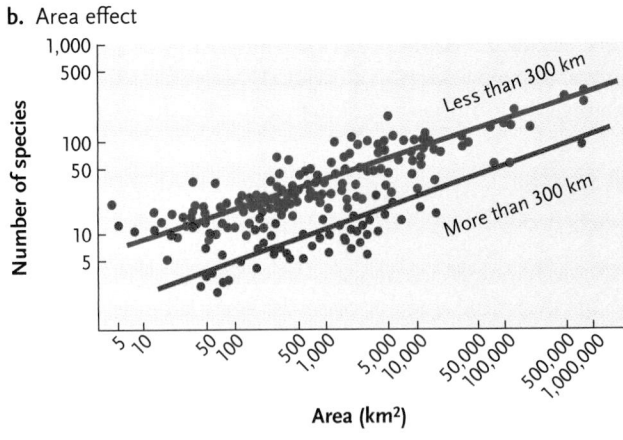

The number of bird species on tropical and subtropical islands throughout the world increases dramatically with island area. The data for islands near to a source and islands far from a mainland source are presented separately to minimize the effect of distance. Notice that the "distance effect" reduces the number of bird species on islands that are more than 300 km from a mainland source.

Figure 50.30
Factors that influence bird species richness on islands.
(a) Fewer bird species colonize islands that are distant from the mainland source. **(b)** More bird species colonize large islands than small ones.

near the mainland have higher immigration rates than distant islands do, because dispersing organisms are more likely to locate islands that are close to their point of departure. Distance does not affect extinction rates. Thus, at equilibrium, near islands have more species than far islands (see Figure 50.29c).

The equilibrium theory's predictions about the effects of area and distance are generally supported by data on plants and animals **(Figure 50.30)**. Experimental work has also verified some of its basic assumptions. For example, Amy Schoener of the University of Washington found that, within 30 days, more than 200 species of marine organisms colonized tiny artificial "islands" (plastic kitchen scrubbers) that she placed in a Bahaman lagoon. Her research confirmed that immigration rate increases with island size. In another ambitious study, Daniel Simberloff and Edward O. Wilson exterminated insects on small islands in the Florida Keys and monitored subsequent immigration and extinction (see the *Focus on Research*). Their research also confirmed the equilibrium theory's predictions that an island's size and distance from the mainland influence how many species will occupy it.

The equilibrial view of species richness also applies to mainland communities, which exist as islands in a metaphorical sea of dissimilar habitat. Lakes are "islands" in a "sea" of dry land, and mountaintops are habitat "islands" in a "sea" of low terrain. Species richness in these communities is partly governed by the immigration of new species from distant sources and the extinction of species already present. As human activities disrupt environments across the globe, undisturbed sites function as islandlike refuges for threatened and endangered species. Conservation biologists now apply the general lessons of MacArthur and Wilson's theory to the design of nature preserves (see Chapter 53).

In the next chapter we examine ecosystems, which include ecological communities interacting with their abiotic environments, focusing on the movements of energy and nutrients.

STUDY BREAK

1. What factors may foster the maintenance of high species richness in tropical communities?
2. According to the equilibrium theory of island biogeography, what are the effects of an island's size and its distance from the mainland on the number of species that can occupy it?

Basic Research: Testing the Theory of Island Biogeography

Shortly after Robert MacArthur and Edward O. Wilson published the equilibrium theory of island biogeography in the 1960s, Daniel Simberloff, one of Wilson's graduate students at Harvard University, and Wilson himself undertook one of the most ambitious experiments ever attempted in community ecology. Simberloff reasoned that the best way to test the theory's predictions was to monitor immigration and extinction on barren islands.

Simberloff and Wilson devised a system for removing all the animals from individual red mangrove trees in the Florida Keys. The trees, with canopies that spread from 11 to 18 m in diameter, grow in shallow water and are isolated from their neighbors; thus, each tree is an island that harbors an arthropod community. The species pool on the Florida mainland includes about 1000 arthropod species, but each mangrove island contains no more than 40 species at one time.

After cataloging the species on each island, Simberloff and Wilson hired an extermination company to erect large tents and fumigate the islands to eliminate all arthropods on them **(Figure a).** The exterminators used methyl bromide, a pesticide that doesn't harm trees or leave any residue. Simberloff then monitored both the immigration of arthropods to the islands and the extinction of species that became established on them. He surveyed six islands regularly for 2 years and at intervals thereafter.

The results of this experiment confirm several predictions of MacArthur and Wilson's theory **(Figure b).** Arthropods recolonized the islands rapidly, and within 8 or 9 months the number of species living on each island had reached an equilibrium that was near the original species number. In addition, the island nearest to the mainland had more species than the most distant island. However, immigration

Figure a

After cataloguing the arthropods, Simberloff and Wilson hired an exterminating company to erect a tent over each mangrove island. Once the islands were fully covered, exterminators used methyl bromide to eliminate all living arthropods.

and extinction were incredibly rapid, and Simberloff and Wilson suspected that some species went extinct even before they had noted their presence. The researchers also discovered that 3 years after the experimental treatments, the species composition of the islands was still changing constantly and did not remotely resemble the species composition in the islands before they were defaunated.

Simberloff and Wilson's research was a landmark study in ecology because it tested the predictions of an important theory using a field experiment. Although such efforts are now almost routine in ecological studies, this project was one of the first to demonstrate that large-scale experimental manipulations of natural systems are feasible and that they often produce clear results.

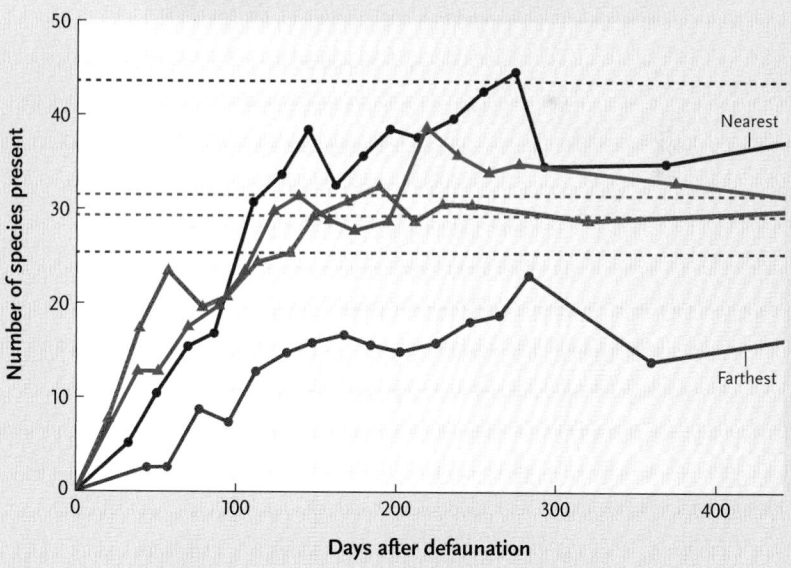

Figure b

On three of four islands, species richness gradually returned to the predefaunation level (indicated by color-coded dashed lines on the graph). The most distant island had not reached its predefaunation species richness after 2 years.

Do species interactions change predictably across environments?

As we learned in this chapter, the population interactions that occur between species range from mutualistic to parasitic. Some biologists have suggested that we should expect more competitive interactions between species in some kinds of environments, but more positive interactions in others. Community ecology will become a more quantitative and predictive discipline if researchers focus on how abiotic and biotic environmental factors—such as the presence of particular community members, environmental gradients, or global climate change—influence the strength of the interactions between species. For example, as physical environments become more stressful, the abundance and distribution of species should be determined less by resource limitation and more by the stress itself. Accordingly, plants tend to compete far less with each other in stressful environments than they do under ideal growing conditions. Scientists are now engaged in the intellectual feedback of theory development and experimental testing aimed at generating a predictive framework for particular types of interactions and their consequences for community structure.

What is the relative importance of positive versus negative interactions for community structure?

It was once suggested that ecologists in capitalist societies, like the United States, tend to more often study competition and predation, but ecologists in socialist societies tend to study mutualism. Although the truth of this anecdote is unclear, it is remarkable that ecologists still do not agree on the relative importance of positive interactions (for example, mutualism or commensalism) versus negative interactions (such as predation or competition) in generating community structure. Advances in this area of study may result from "factorial" experiments, in which two or more types of interactions are manipulated. For example, one might examine the relative effects of excluding pollinators versus excluding herbivores on the success of a plant population. In factorial experiments, the researcher can conclude that one factor has a bigger effect than the other, because all other factors were controlled. These sorts of experiments may eventually lead to an emerging picture of the relative importance of positive versus negative population interactions.

How does the evolutionary history of a species influence its ecology today?

The great evolutionary biologist Theodosius Dobzhansky once noted that "Nothing in biology makes sense except in the light of evolution." Although we know a great deal about both ecology and evolutionary biology, researchers are only beginning to explore the impact of an organism's evolutionary history on its ecology. This very active area of research includes the use of phylogenetic information (see Chapter 23), selection experiments (see Chapter 21), and a knowledge of the genetic basis of particular traits (see Chapter 12). For example, are closely related species more likely to compete with each other than more distantly related species are? Do organisms that are well adapted to particular environments fare poorly in other environments? Why do some organisms specialize in their resource use? Are the population dynamics that species experience shaped by past evolutionary events? These questions are currently being addressed, and the answers uncovered by researchers may unravel many current mysteries about the ecology of populations and communities.

Anurag Agrawal is an associate professor in the Departments of Entomology and Ecology and Environmental Biology at Cornell University. He studies the evolutionary and community ecology of plant–insect interactions. To learn more about Dr. Agrawal's research go to http://www.herbivory.com.

Review

Go to **ThomsonNOW**‾ at www.thomsonedu.com/login to access quizzing, animations, exercises, articles, and personalized homework help.

50.1 Population Interactions

- Coevolution is the evolution of reciprocal adaptations in species that interact ecologically (Figure 50.1).
- Predators and herbivores use diverse adaptations to select, locate, capture, and ingest an appropriate diet (Figure 50.2). Plants have both structural and chemical defenses against herbivores. Animal prey may try to hide or escape from predators, defend themselves actively, or advertise their unpalatability (Figures 50.3–50.5); some species mimic the appearance of poisonous species (Figure 50.6). Predators may evolve adaptations to counter prey defenses (Figure 50.7).
- Interspecific competition results if two or more populations use the same limiting resources; competition may lead to the extinction of one competitor (Figure 50.8). Ecologists use the ecological niche concept to visualize a population's resource use (Figure 50.9). Observations of resource partitioning (Figure 50.10) and character displacement (Figure 50.11) suggest that competition may be important, but only field experiments can demonstrate that competition occurs (Figure 50.12).
- Symbiosis is a close ecological association between species. In commensal interactions, one species benefits and the other is unaffected (Figure 50.13). In mutualistic interactions, both partners benefit (Figures 50.14–50.16). In parasitic interactions, one species benefits and the other is harmed.

Animation: Predator–prey interactions

Animation: Competitive exclusion

Animation: Hairston's experiment

Animation: Resource partitioning

Animation: Wasp and mimics

Practice: Understanding the major types of species interactions: competition, predation, parasitism, and mutualism

50.2 The Nature of Ecological Communities

- An interactive view suggests that species in a community are bound together in a complex web of necessary biotic interactions; an individualistic view recognizes communities as loose assemblages of organisms that have similar physical requirements (Figure 50.17).

- Ecotones occur where adjacent communities grade into one another; sharp boundaries occur between communities where a critical resource or an important abiotic factor is discontinuous (Figure 50.18).

50.3 Community Characteristics

- In benign environments, vegetation is tall and has a complex physical structure (Figure 50.19). In stressful environments, vegetation is short and has a simple physical structure.

- Communities differ in species richness and the relative abundances of species. Both characteristics contribute to a community's species diversity (Figure 50.20).

- Organisms are classified as producers, consumers, detritivores, or decomposers. Ecologists depict the trophic structure (feeding relationships) of communities in food webs (Figure 50.21). Food-web analyses seek to identify generalities about trophic structure and its relationship to community stability.

Animation: Trophic levels in a simple food chain

Animation: Rain forest food web

50.4 Effects of Population Interactions on Community Characteristics

- Interspecific competition often affects the species composition and structure of communities.

- Predators may increase species richness by reducing the population size of the competitively most successful prey, thus allowing other prey species to occupy the community (Figure 50.22).

- Herbivores sometimes increase species richness and sometimes decrease it (Figure 50.23).

Animation: Effect of keystone species on diversity

50.5 Effects of Disturbance on Community Characteristics

- Environmental disturbances may eliminate populations from a community. Some communities, such as coral reefs, experience such frequent disturbance that their species composition is never at equilibrium (Figure 50.24).

- Disturbances of intermediate intensity and frequency allow both r-selected and K-selected species to occupy a site, increasing species richness (Figure 50.25).

50.6 Ecological Succession: Responses to Disturbance

- Ecological succession is a somewhat predictable change in species composition over time.

- Primary succession occurs on bare ground or rock (Figure 50.26). Secondary succession occurs where a community existed in the past (Figure 50.27).

- Species composition changes quickly and species richness rises rapidly during early successional stages. Early stages include short-lived r-selected species; later stages include long-lived K-selected species. Some communities eventually achieve a relatively stable climax state.

- Most communities include a mosaic of species that reflect patchiness in environmental conditions and the mixture of relatively undisturbed and recently disturbed sites.

Animation: Succession

50.7 Variations in Species Richness among Communities

- Communities near the equator have higher species richness than those near the poles (Figure 50.28). Explanations for this latitudinal gradient focus on either the origin or the maintenance of high species richness in the tropics.

- The equilibrium theory of island biogeography predicts that the number of species on an island represents a balance between the immigration of new species and the extinction of species already present (Figure 50.29). Studies show that large islands harbor more species than small islands and that islands near a mainland source have more species than distant islands (Figure 50.30).

Animation: Species diversity by latitude

Animation: Area and distance effects

Questions

Self-Test Questions

1. According to optimal foraging theory, predators:
   a. always feed on the largest prey possible.
   b. always feed on the prey that are easiest to catch.
   c. choose prey based on the costs of capturing and consuming it compared with the energy it provides.
   d. feed on plants when animal prey are scarce.
   e. have coevolved mechanisms to overcome prey defenses.

2. The use of the same limiting resource by two species is called:
   a. brood parasitism.
   b. interference competition.
   c. exploitative competition.
   d. mutualism.
   e. optimal foraging.

3. The range of resources that a population can possibly use is called:
   a. its fundamental niche.
   b. its realized niche.
   c. character displacement.
   d. resource partitioning.
   e. its relative abundance.

4. Differences in bill size of finch species living on the same island in the Galápagos may be caused by:
   a. predation.
   b. character displacement.
   c. mimicry.
   d. interference competition.
   e. cryptic coloration.

5. Bacteria that live in the human intestine assist digestion and feed on nutrients the human consumed. This relationship might best be described as:
   a. commensalism.
   b. mutualism.
   c. endoparasitism.
   d. ectoparasitism.
   e. predation.

6. The table below shows how many individuals were recorded for each of five species in five separate communities (a–e). Which community has the highest species diversity?

Community	Species 1	Species 2	Species 3	Species 4	Species 5
a.	90	10	0	0	0
b.	80	10	10	0	0
c.	25	25	25	25	0
d.	2	4	6	8	80
e.	20	20	20	20	20

7. A keystone species:
   a. is usually a primary producer.
   b. has a critically important role in determining the species composition of its community.
   c. is always a predator.
   d. usually reduces the species diversity in a community.
   e. usually exhibits aposematic coloration.

8. Species richness is often highest in communities where disturbances are:
   a. very frequent and severe.
   b. very frequent and of moderate intensity.
   c. very rare and severe.
   d. of intermediate frequency and moderate intensity.
   e. very rare and mild.

9. The change in the species composition of a community from bare and lifeless rock to climax vegetation is called:
   a. disturbance.
   b. competition.
   c. secondary succession.
   d. primary succession.
   e. facilitation.

10. The equilibrium theory of island biogeography predicts that the number of species found on an island:
   a. increases steadily until it equals the number in the mainland species pool.
   b. is greater on large islands than on small ones.
   c. is smaller on islands near the mainland than on distant islands.
   d. can never reach an equilibrium number.
   e. is greater for islands near the equator than for islands near the poles.

Questions for Discussion

1. Using the terms and concepts introduced in this chapter, describe the interactions that humans have with ten other species. Try to pick at least eight species that we do not eat.

2. After reading about the two potential biases in the scientific literature on competition, describe how future studies of competition might avoid such biases.

3. Humans are destroying natural communities at an ever-increasing pace. Using the predictions of the theory of island biogeography, develop hypotheses about what might happen as patches of natural habitats get smaller and smaller. How would you test these hypotheses?

Experimental Analysis

Chaparral, a community of woody shrubs that is fairly common in California, often grows adjacent to grassland. The two communities are consistently separated by a "bare zone," usually less than 1 m wide, where no vegetation of either type grows. Ecologists have proposed two possible explanations for this strip of bare soil: (1) that the leaves of chaparral shrubs release harmful, water-soluble chemicals that keep the grass seeds from germinating in the adjacent soil; and (2) that small mammals living in the dense cover provided by chaparral consume the grass seeds before they germinate; the animals don't venture very far from the shrubs because they would be easy targets for predatory hawks. Design a set of field experiments to test the two hypotheses.

Evolution Link

Five processes can foster microevolutionary change: gene flow, genetic drift, mutation, natural selection, and nonrandom mating (see Section 20.3). Which of those processes might contribute to the evolution of Batesian mimicry in two butterfly species? Would the same processes affect both the mimic and the model similarly? Which processes might have contributed to the evolution of the mutualistic relationship between ants and acacia trees, and how would their action on the two mutualists differ?

How Would You Vote?

Currently, only a fraction of the crates being imported into the United States are inspected for the inadvertent or deliberate presence of exotic species. Would the cost of added inspections be worth it? Go to www.thomsonedu.com/login to investigate both sides of the issue and then vote.

© Mark J. Barrett 2005 www.markjbarrett.com

Silver Springs, Florida. This small river was the site of one of the earliest comprehensive studies of ecosystem structure and function.

51 Ecosystems

WHY IT MATTERS

Poor Lake Erie, the shallowest of the Great Lakes. Several major industrial cities, including Toledo, Cleveland, Erie, and Buffalo, sprawl along its shoreline. Most of its water comes from the Detroit River, which flows past Detroit; the other rivers that flow into Lake Erie carry runoff from agricultural fields in Canada and the United States.

When Europeans first settled along its shores roughly 300 years ago, Lake Erie was a wetland paradise. Fishes and waterfowl reproduced in marshes and bays. Even after steel mills and oil refineries were built nearby in the 1860s and 1870s, the lake supported a busy fishing industry and was famous as a recreation area.

By 1970, wetlands had been filled for building; bays had been dredged for shipping lanes; and the shoreline had been converted to beaches. Worst of all, household sewage, industrial effluent, and agricultural runoff had so polluted the lake that it no longer supported the activities that had made it famous **(Figure 51.1).** The water was murky with algae and cyanobacteria; dead fishes washed up on the shore; local health departments closed beaches; and the fishing industry collapsed.

Figure 51.1
Pollution of Lake Erie. A steel mill in Lackawanna, New York, discharged industrial wastes into Lake Erie until 1983, when the mill was closed.

How can a vibrant natural resource become a foul smelling dump? The answer lies in the human activities that disrupt an **ecosystem,** a biological community and the physical environment with which it interacts. Between the 1930s and the 1970s, Lake Erie's concentration of phosphorus, which had been a limiting nutrient, tripled, largely from household detergents and agricultural fertilizers. High phosphorus concentrations encouraged the growth of photosynthetic algae, changing the phytoplankton community. The density of coliform bacteria, which originate in the human gut and serve as indicators of organic pollution, also skyrocketed as a result of the surge in sewage and nutrients entering the lake.

Increased phytoplankton and bacterial populations depleted oxygen in the lake's waters, contributing to changes elsewhere in the lake. Mayflies (*Hexagenia* species), whose larvae live in well-oxygenated bottom sediments, had once been so abundant that their aerial breeding swarms were a public nuisance. But they became nearly extinct in the polluted lake, replaced by oligochaete worms, snails, and other invertebrates. Along with overfishing, changes in the bottom fauna shifted the composition of the fish community; the catch of desirable food fishes declined to almost zero by the mid-1960s.

In 1972, Canada and the United States began efforts to restore the lake. They spent billions of dollars to reduce the influx of phosphates and limited fishing of the most vulnerable native species. Nonnative salmon (*Onchorhynchus* species) and other predatory fishes were introduced in the hope that they could bring the lake back to its original condition. Even the accidental introduction of zebra mussels *(Dreissina polymorpha)*, an aquatic pest, inadvertently helped the effort because they feed on phytoplankton.

But, although somewhat improved, Lake Erie will never return to its former glory. Some native species are now extinct there, and the introduced species that replaced them function differently within the ecosystem. The lake still suffers periods of uncontrolled algal growth, fish kills, and high levels of harmful bacteria.

This story of an ecological disaster and partial recovery introduces ecosystem ecology, the branch of ecology that analyzes the flow of energy and the cycling of materials between an ecosystem's living and nonliving components. These processes make the resident organisms highly dependent on each other and on their physical surroundings. Ultimately, the Lake Erie ecosystem unraveled because human activities disrupted the flow of energy and the cycling of materials upon which the organisms depended.

51.1 Energy Flow and Ecosystem Energetics

Ecosystems receive a steady input of energy from an external source, which in virtually all cases is the sun. Energy flows through an ecosystem, but, as dictated by the laws of thermodynamics (see Section 4.1), much of it is lost without being used by organisms.

Food webs define the pathways by which energy moves through an ecosystem's biotic components (see Section 50.3). In most ecosystems, energy moves simultaneously through a *grazing food web* and a *detrital food web* **(Figure 51.2).** The grazing food web includes the producer, herbivore, and carnivore trophic levels. The detrital food web includes detritivores and decomposers. Because detritivores and decomposers subsist on the remains and waste products of organisms at every trophic level, the two food webs are closely interconnected. Detritivores also contribute to the grazing food web when carnivores eat them.

All of the organisms in a trophic level are the same number of energy transfers from the ecosystem's ultimate energy source. Plants are one energy transfer removed from sunlight; herbivores are two transfers away; carnivores feeding on herbivores are three transfers away; and carnivores feeding on other carnivores are four transfers away. In this section, we consider the details of energy flow and the efficiency of energy transfer from one trophic level to another.

Sunlight Provides the Energy Input for Practically All Ecosystems

Virtually all life on Earth depends on the input of solar energy. Every minute of every day, the atmosphere intercepts roughly 19 kcal of energy per square meter. (Recall from Chapter 2 that 1 kcal = 1000 calories.) About half that energy is absorbed, scattered, or reflected by gases, dust, water vapor, and clouds without ever reaching the planet's surface (see Chapter 52). Most energy that reaches the surface falls on bodies of water or bare ground, where it is absorbed as heat or reflected back into the atmosphere; reflected energy warms the atmosphere, as we discuss later in this chapter. Only a small percentage contacts primary producers, and most of that energy evaporates water, driving transpiration in plants (see Section 32.3).

Ultimately, photosynthesis converts less than 1% of the solar energy that arrives at Earth's surface into chemical energy. But primary producers capture enough energy to create an average of several kilograms of dry plant material per square meter per year. On a global scale, they produce more than 150 billion metric tons of new biological material annually. Some of the solar energy that producers convert into chemical energy is transferred to consumers at higher trophic levels.

The rate at which producers convert solar energy into chemical energy is an ecosystem's **gross primary productivity.** But like all other organisms, producers also use energy for their own maintenance. After deducting the energy used for these functions, which are collectively called *cellular respiration* (see Section 8.1), whatever chemical energy remains is the ecosystem's **net primary productivity.** In most ecosystems, net primary productivity is between 50% and 90% of gross primary productivity. In other words, producers use between 10% and 50% of the energy they capture for their own respiration.

Ecologists generally measure primary productivity in units of energy captured ($kcal/m^2/yr$) or in units of biomass created ($g/m^2/yr$). *Biomass* is the dry weight of biological material per unit area or volume of habitat. (We measure biomass as the *dry* weight of organisms because their water content, which fluctuates with water uptake or loss, has no energetic or nutritional value.) You should not confuse an ecosystem's productivity with its **standing crop biomass,** the total dry weight of plants present at a given time. Net primary productivity is the *rate* at which the standing crop produces *new* biomass.

The energy captured by plants is stored in biological molecules—mostly carbohydrates, lipids, and proteins. Ecologists can convert units of biomass into

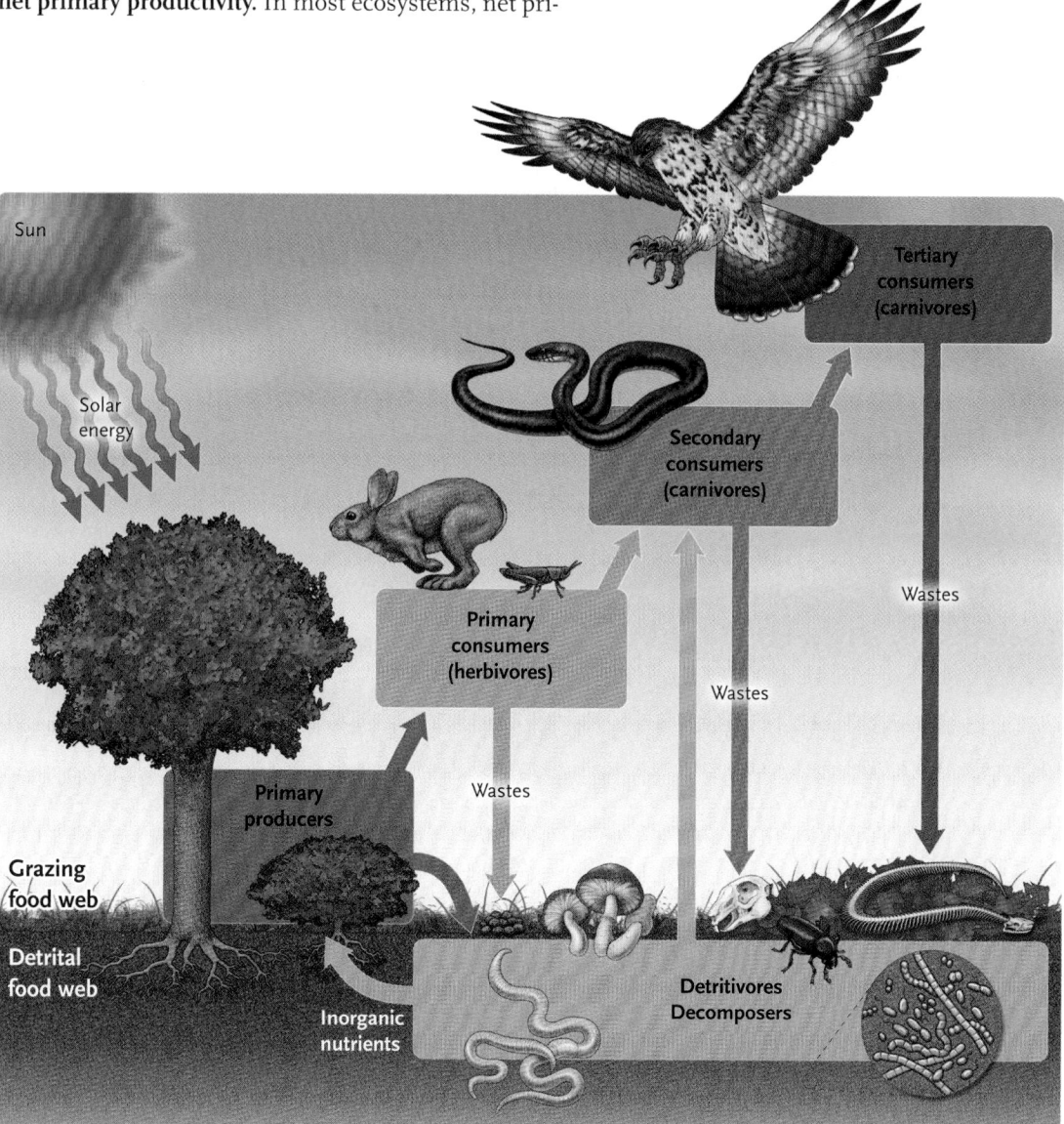

Figure 51.2
Grazing and detrital food webs. Energy and nutrients move through two parallel food webs in most ecosystems. The grazing food web includes producers, herbivores, and carnivores. The detrital food web includes detritivores and decomposers. Each box in this diagram represents many species, and each arrow represents many arrows.

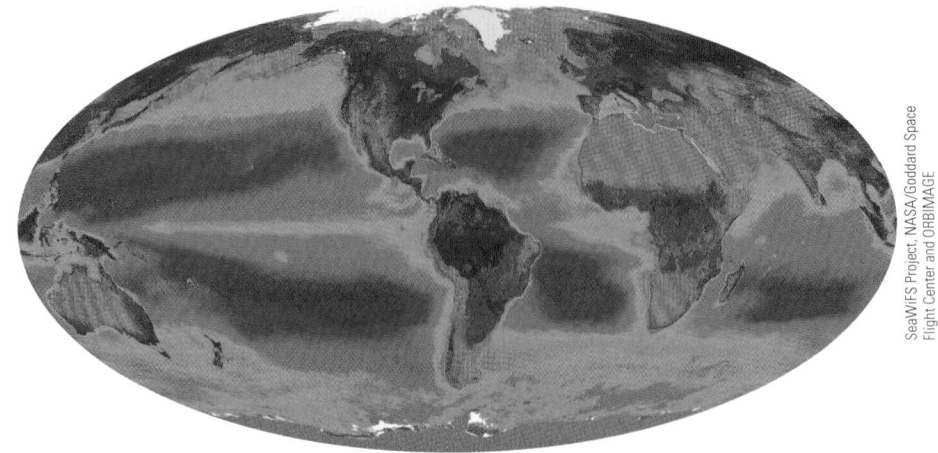

units of energy or vice versa as long as they know how much carbohydrate, protein, and lipid a sample of biological material contains (4.2 kcal/g of carbohydrate; nearly 4.1 kcal/g of protein; and 9.5 kcal/g of lipid). Thus, net primary productivity is a measure of the rate at which producers accumulate energy as well as the rate at which new biomass is added to an ecosystem. Because it is far easier to measure biomass than energy content, ecologists usually measure changes in biomass to estimate productivity. New biomass takes several forms: the growth of existing producers; the creation of new producers by reproduction; and the storage of energy as carbohydrates. Because herbivores eat all three forms of new biomass, net primary productivity also measures how much new energy is available for primary consumers.

Primary Productivity Varies Greatly on Global and Local Scales

The potential rate of photosynthesis in any ecosystem is proportional to the intensity and the duration of sunlight, which vary geographically and seasonally (see Chapter 52). Sunlight is most intense and day length least variable near the equator. By contrast, light intensity is weakest and day length most variable near the poles. Thus, producers at the equator can photosynthesize nearly 12 hours a day, every day of the year. Near the poles, photosynthesis is virtually impossible during the long, dark winter; in summer, however, plants can photosynthesize around the clock.

Sunlight is not the only factor that influences the rate of primary productivity, however; temperature as well as the availability of water and nutrients also have big effects. For example, many of the world's deserts receive plenty of sunshine but have low rates of productivity because water is in short supply and the soil is nutrient-poor. Thus, mean annual net primary productivity varies greatly on a global scale **(Figure 51.3)**, reflecting variations in these environmental factors (see Chapter 52).

On a finer geographical scale, within a particular terrestrial ecosystem, mean annual net productivity often increases with the availability of water **(Figure 51.4)**. In systems with sufficient water, a shortage of mineral nutrients may be limiting. All plants need specific ratios of macronutrients and micronutrients for maintenance and photosynthesis (see Section 33.1). But plants withdraw nutrients from soil, and if nutrient concentration drops below a critical level, photosynthesis may decrease or stop altogether. In every ecosystem, one nutrient inevitably runs out before the supplies of other nutrients are exhausted. The element in short supply is called a **limiting nutrient** because its absence limits productivity. Productivity in agricultural fields is subject to the same constraints as productivity in natural ecosystems. Farmers increase productivity by irrigating (adding water to) and fertilizing (adding nutrients to) their crops.

In freshwater and marine ecosystems, where water is always readily available, the depth of the water and the combined availability of sunlight *and* nutrients govern the rate of primary productivity. Productivity is high in near-shore ecosystems where sunlight pene-

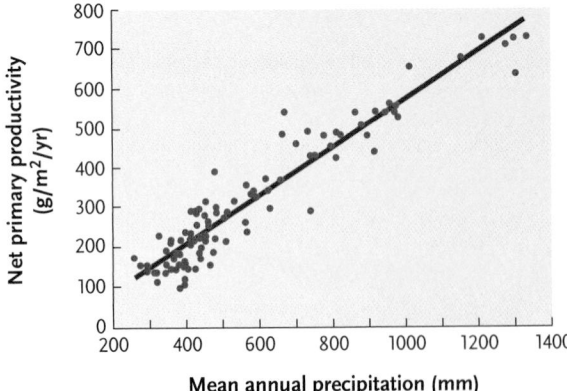

Figure 51.4

Water and net primary productivity. Mean annual net primary productivity increases with mean annual precipitation among 100 sites in the Great Plains of North America. These data include only aboveground productivity.

trates shallow, nutrient-rich waters. Kelp beds and coral reefs, for example, which occur along temperate and tropical coastlines respectively, are among the most productive ecosystems on Earth **(Table 51.1).** By contrast, productivity is low in the open waters of a large lake or ocean: sunlight penetrates only the upper layers, and nutrients sink to the bottom. Thus, the two requirements for photosynthesis, sunlight and nutrients, are available in different places.

Although ecosystems vary in their net primary productivity, the differences are not always proportional to variations in their standing crop biomass (see Table 51.1). For example, biomass in temperate deciduous forests and temperate grasslands differs by a factor of 20, but the difference in their rates of net primary productivity is only twofold. Most biomass in trees is present in non-photosynthetic tissues such as wood. As a result, their ratio of productivity to biomass is low (1200 g/m^2 ÷ 30,000 g/m^2 = 0.040). By contrast, grasslands don't accumulate much biomass because annual mortality, herbivores, and fires remove plant material as it is produced; and their productivity to biomass ratio is much higher (600 g/m^2 ÷ 1600 g/m^2 = 0.375).

Some ecosystems contribute more than others to overall net primary productivity **(Figure 51.5).** Ecosystems that cover large areas make substantial contributions, even if their productivity is low. Conversely, geographically restricted ecosystems make large contributions if their productivity is high. For example, the open ocean and tropical rain forests contribute about equally to total global productivity, but for different reasons. Open oceans have low productivity, but they cover nearly two-thirds of Earth's surface. Tropical rain forests cover only a small area, but they are highly productive.

Some Stored Energy Is Always Lost before It Is Transferred from One Trophic Level to the Next

Net primary productivity ultimately supports all the consumers in grazing and detrital food webs. Consumers in the grazing food web eat some of the biomass at every trophic level except the highest; uneaten biomass eventually dies and passes into detrital food webs. However, consumers assimilate only a portion of the material they ingest, and unassimilated material is passed as feces, which also supports detritivores and decomposers.

As energy is transferred from producers to consumers, some is stored in new consumer biomass, called **secondary productivity.** Nevertheless, two factors cause energy to be lost from the ecosystem every time it flows from one trophic level to another. First, animals use much of the energy they assimilate for maintenance or locomotion rather than the production of new biomass. Second, as dictated by the second law of thermodynamics, no biochemical reaction is 100% effi-

Table 51.1	Standing Crop Biomass and Net Primary Productivity of Different Ecosystems	
Ecosystem	**Mean Standing Crop Biomass (g/m^2)**	**Mean Net Primary Productivity (g/m^2/yr)**
Terrestrial Ecosystems		
Tropical rain forest	45,000	2,200
Tropical deciduous forest	35,000	1,600
Temperate rain forest	35,000	1,300
Temperate deciduous forest	30,000	1,200
Savanna	4,000	900
Boreal forest (taiga)	20,000	800
Woodland and shrubland	6,000	700
Agricultural land	1,000	650
Temperate grassland	1,600	600
Tundra and alpine tundra	600	140
Desert and thornwoods	700	90
Extreme desert, rock, sand, ice	20	3
Freshwater Ecosystems		
Swamp and marsh	15,000	2,000
Lake and stream	20	250
Marine Ecosystems		
Open ocean	3	125
Upwelling zones	20	500
Continental shelf	10	360
Kelp beds and reefs	2,000	2,500
Estuaries	1,000	1,500
World Average	**3,600**	**333**

From Whittaker, R.H. 1975. *Communities and Ecosystems.* 2nd ed. Macmillan.

cient; thus, some of the chemical energy liberated by cellular respiration is always converted to heat, which most organisms do not use.

Ecological efficiency is the ratio of net productivity at one trophic level to net productivity at the trophic level below it. For example, if the plants in an ecosystem have a net primary productivity of 100 g/m^2/year of new tissue and the herbivores that eat those plants produce 10 g/m^2/year, the ecological efficiency of the herbivores is 10%. The efficiencies of three processes—harvesting food, assimilating ingested energy, and producing new biomass—determine the ecological efficiencies of consumers.

Harvesting efficiency is the ratio of the energy content of food consumed to the energy content of food available. Predators harvest food efficiently when prey are abundant and easy to capture (see Section 50.1).

Assimilation efficiency is the ratio of the energy absorbed from consumed food to the food's total energy content. Because animal prey is relatively easy to digest,

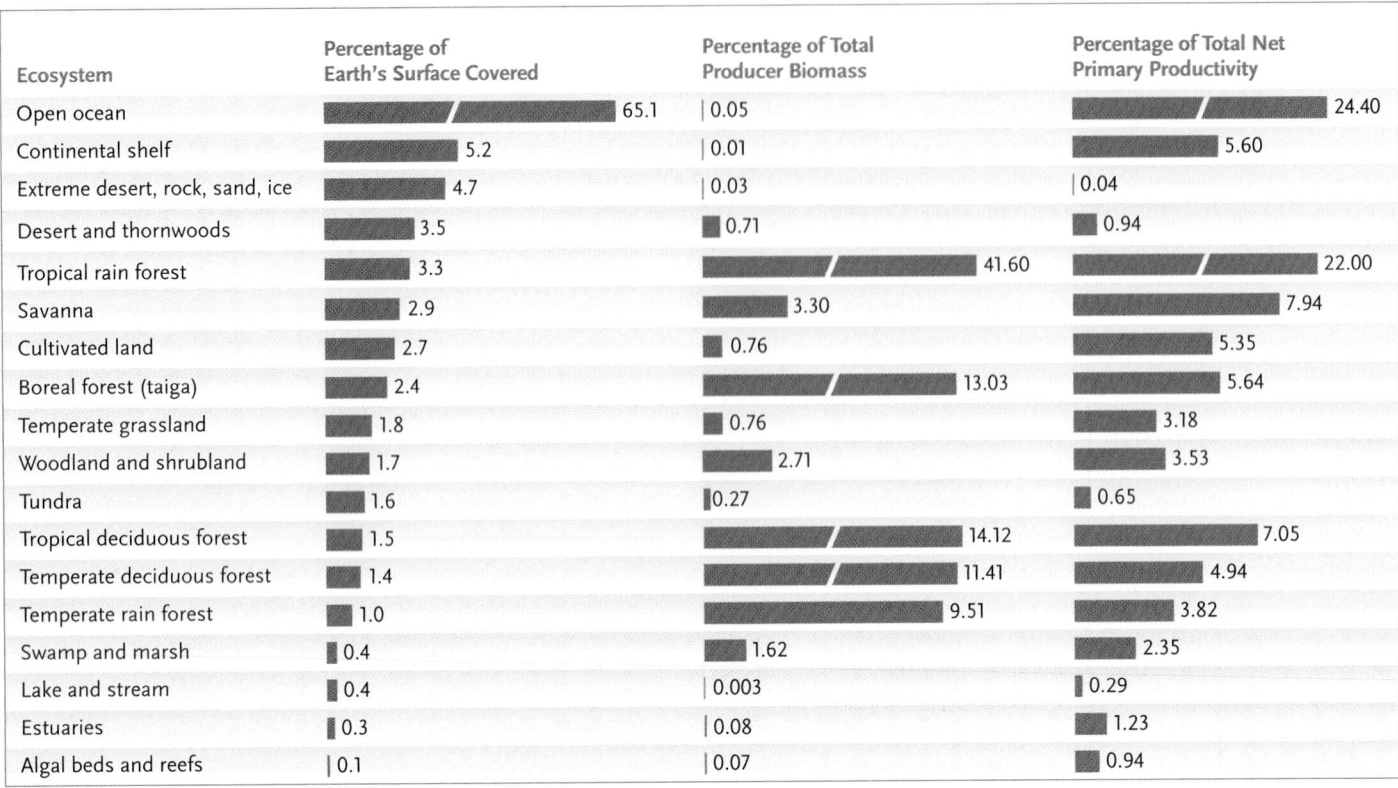

Ecosystem	Percentage of Earth's Surface Covered	Percentage of Total Producer Biomass	Percentage of Total Net Primary Productivity
Open ocean	65.1	0.05	24.40
Continental shelf	5.2	0.01	5.60
Extreme desert, rock, sand, ice	4.7	0.03	0.04
Desert and thornwoods	3.5	0.71	0.94
Tropical rain forest	3.3	41.60	22.00
Savanna	2.9	3.30	7.94
Cultivated land	2.7	0.76	5.35
Boreal forest (taiga)	2.4	13.03	5.64
Temperate grassland	1.8	0.76	3.18
Woodland and shrubland	1.7	2.71	3.53
Tundra	1.6	0.27	0.65
Tropical deciduous forest	1.5	14.12	7.05
Temperate deciduous forest	1.4	11.41	4.94
Temperate rain forest	1.0	9.51	3.82
Swamp and marsh	0.4	1.62	2.35
Lake and stream	0.4	0.003	0.29
Estuaries	0.3	0.08	1.23
Algal beds and reefs	0.1	0.07	0.94

Figure 51.5

Biomass and net primary productivity. The percentage of Earth's surface that an ecosystem covers is not proportional to its contribution to the total biomass of producers or its contribution to the total net primary productivity.

carnivores absorb between 60% and 90% of the energy in their food; assimilation efficiency is lower for prey with indigestible parts like bones or exoskeletons. Herbivores assimilate only 15% to 80% of the energy they consume because cellulose is not very digestible.

Production efficiency is the ratio of the energy content of new tissue produced to the energy assimilated from food. Production efficiency varies with maintenance costs. For example, endothermic animals often use less than 10% of their assimilated energy for growth and reproduction, because they use energy to generate body heat (see Section 46.8). Ectothermic animals, by contrast, channel more than 50% of their assimilated energy into new biomass.

The overall ecological efficiency of most organisms is between 5% and 20%. As a rule of thumb, only about 10% of the energy accumulated at one trophic level is converted into biomass at the next higher trophic level, as illustrated by energy transfers at Silver Springs, Florida **(Figure 51.6)**. Producers in the Silver Springs ecosystem convert 1.2% of the solar energy they intercept into chemical energy (represented by 20,810 kcal of gross primary productivity). However, they use about two-thirds of this energy for respiration, leaving only one-third to be included in new plant biomass, the net primary productivity. All consumers in the grazing food web (on the right in Figure 51.6) ultimately depend on this energy source, which dwindles with each transfer between trophic levels. Energy is lost to respiration and export (that is, the transport of energy-containing materials out of the ecosystem by flowing water) at each trophic level. In addition, substantial energy, represented in organic wastes and uneaten biomass, flows into the detrital food web (on the left in Figure 51.6). To determine the ecological efficiency of any trophic level, we divide its productivity by the productivity of the level below it. For example, the ecological efficiency of midlevel carnivores at Silver Springs is 111 kcal/yr ÷ 1103 kcal/yr = 10.06%.

As energy works its way up a food web, energy losses are multiplied in successive energy transfers, greatly reducing the energy available to support the highest trophic levels. Consider a hypothetical example in which ecological efficiency is 10% for all consumers. Assume that the plants in a small field annually produce new tissues containing 100 kcal of energy. Because only 10% of that energy is transferred to new herbivore biomass, the 100 kcal in plants produces only 10 kcal of new herbivorous insects; only 1 kcal of new songbirds, which feed on insects; and only 0.1 kcal of new falcons, which feed on songbirds. Thus, after three energy transfers, only 0.1% of the energy from primary productivity remains at the highest trophic levels. If the energy available to each trophic level is depicted graphically, the result is a **pyramid of energy** with primary producers on the bottom and higher-level consumers on the top. We discuss ecological pyramids in detail in the next section.

The low ecological efficiencies that characterize most energy transfers illustrate one advantage of eating "lower on the food chain." Even though humans digest and assimilate meat more efficiently than vegetables, we might be able to feed more people if we all ate more vegetables directly instead of first passing

Figure 51.6 Observational Research

Energy Flow in the Silver Springs Ecosystem

HYPOTHESIS: Only a small percentage of the energy present in a trophic level is transferred to the next higher trophic level in the ecosystem.

PREDICTION: The energy content of the organisms present in each trophic level will decline steadily from the lowest to highest trophic levels.

METHOD: Howard T. Odum and his research team analyzed energy flow in an aquatic ecosystem at Silver Springs, Florida. The producers in this small spring are mostly aquatic plants. The herbivores include snails, shrimp, insects, fishes, and turtles. The carnivores include a variety of invertebrates and fishes. The top carnivores are large fish. Sunlight is available as an energy source all year round. After defining the food web in this ecosystem, researchers estimated the biomass and energy content (kcal/g) of each trophic level. They then constructed a diagram that illustrates how much energy is present at each trophic level and how much energy is lost as it works its way through the food web.

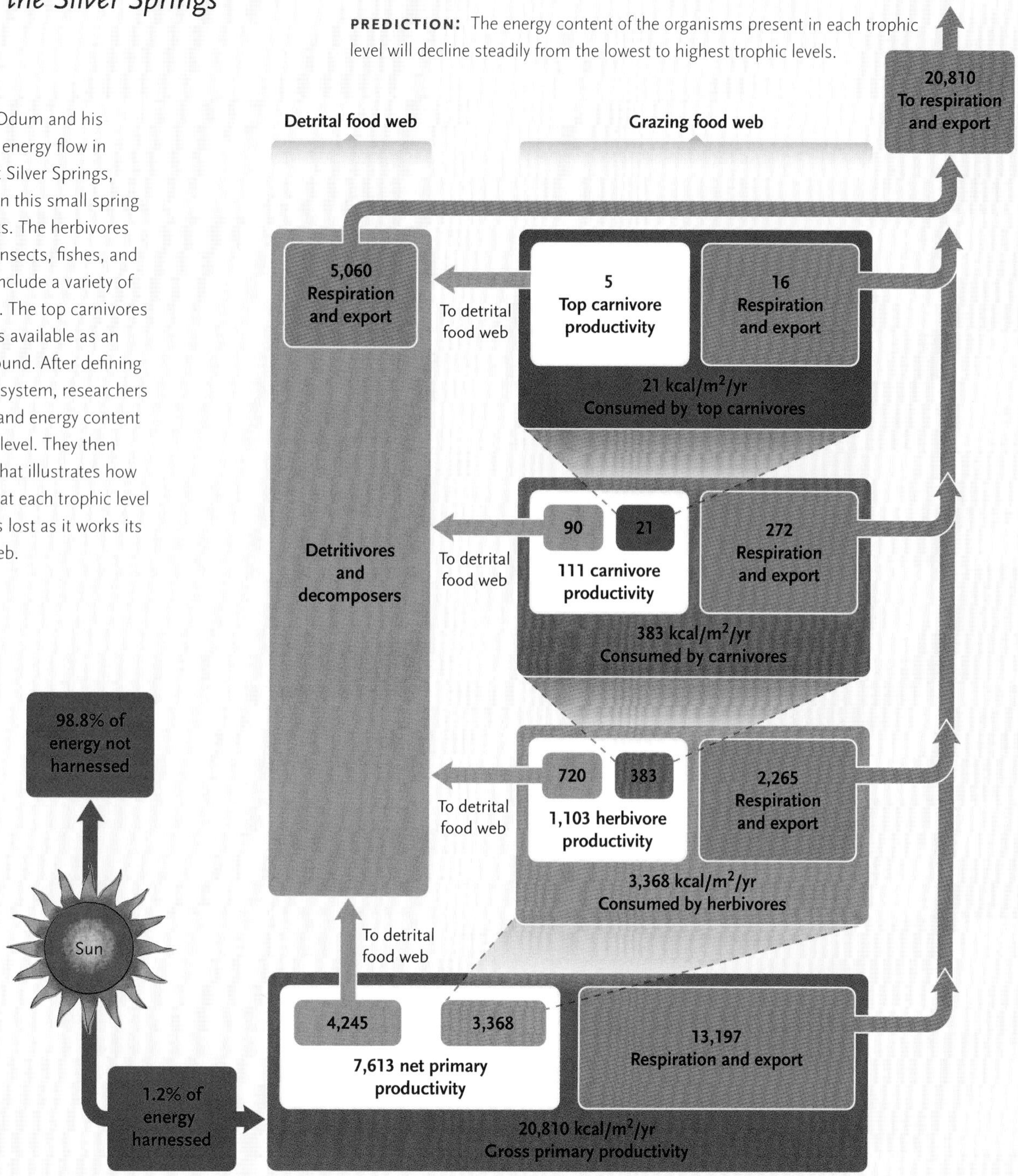

RESULTS: The diagram illustrates annual energy flow for the spring ecosystem at Silver Springs, Florida. Numbers on the diagram indicate the quantity of energy (kcal/m²/yr). Because the ecosystem is based on flowing water, small quantities of energy arrive from other ecosystems and small quantities are exported in material carried away by stream flow.

CONCLUSION: The study confirmed the hypothesis that only a small proportion of the energy present at a trophic level is transferred to the next higher trophic level. Ultimately, all of the energy that passes through the grazing and detrital food webs is released as metabolically generated heat.

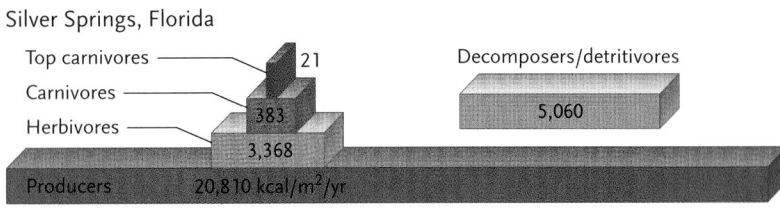

Silver Springs, Florida

Figure 51.7
Pyramids of energy. The pyramid of energy for Silver Springs, Florida, shows that the amount of energy passing through each trophic level decreases as it moves up the food web.

these crops through another trophic level, such as cattle or chickens, to produce meat. The production of animal protein is costly because much of the energy fed to livestock is used for their own maintenance rather than the production of new biomass. But despite the economic—not to mention health-related—logic of a more vegetarian diet, a change in our eating habits alone won't eliminate food shortages or the frequency of malnutrition. Many regions of Africa, Australia, North America, and South America support vegetation that is suitable only for grazing by large herbivores. These areas could not produce significant quantities of edible grains and vegetables.

Ecological Pyramids Illustrate the Effects of Energy Losses

The inefficiency of energy transfer from one trophic level to the next has profound effects on ecosystem structure. Ecologists illustrate these effects in diagrams called **ecological pyramids.** Trophic levels are drawn as stacked blocks, with the size of each block proportional to the energy, biomass, or numbers of organisms present. We mentioned the pyramid of energy in the previous section. Pyramids of energy typically have wide bases and narrow tops **(Figure 51.7)** because each trophic level contains only about 10% as much energy as the trophic level below it.

The progressive reduction in productivity at higher trophic levels, as illustrated in Figure 51.6, usually establishes a **pyramid of biomass (Figure 51.8).** The biomass at each trophic level is proportional to the chemical energy temporarily stored there. Thus, in terrestrial ecosystems, the total mass of producers is generally greater than the total mass of herbivores, which is, in turn, greater than the total mass of predators (see Fig-

ure 51.8a). Populations of top predators—animals like mountain lions or alligators—contain too little biomass and energy to support another trophic level; thus, they have no nonhuman predators.

Freshwater and marine ecosystems sometimes exhibit inverted pyramids of biomass (see Figure 51.8b). In the open waters of a lake or ocean, primary consumers (zooplankton) eat the primary producers (phytoplankton) almost as soon as they are produced. As a result, the standing crop of primary consumers at any moment in time is actually larger than the standing crop of primary producers. Food webs in these ecosystems are stable, however, because the producers have exceptionally high **turnover rates.** In other words, the producers divide and their populations grow so quickly that feeding by zooplankton doesn't endanger their populations or reduce their productivity. And on an annual basis, the *cumulative total* biomass of primary producers far outweighs that of primary consumers.

The reduction of energy and biomass also affects the population sizes of organisms at the top of a food web. Top predators are often relatively large animals. Thus, the limited biomass present in the highest trophic levels is concentrated in relatively few animals **(Figure 51.9).** The extremely narrow top of this **pyramid of numbers** has grave implications for conservation biology. Top predators tend to be large animals with small population sizes. And because each individual must patrol a large area to find sufficient food, the members of a population are often widely dispersed within their habitats. As a result, they are subject to genetic drift (see Section 20.3) and are highly sensitive to hunting, habitat destruction, and random events, which can lead to extinction (see Chapter 53). Top predators may also suffer from the accumulation of poisonous materials that move through food webs (see *Focus on Research* on biological magnification in Chapter 53). Even predators that feed below the top trophic level often suffer the ill effects of human activities. *Insights from the Molecular Revolution* describes how researchers determined that fishing diminishes fragile populations of loggerhead sea turtles *(Caretta caretta),* a predator that routinely travels from one ecosystem to another.

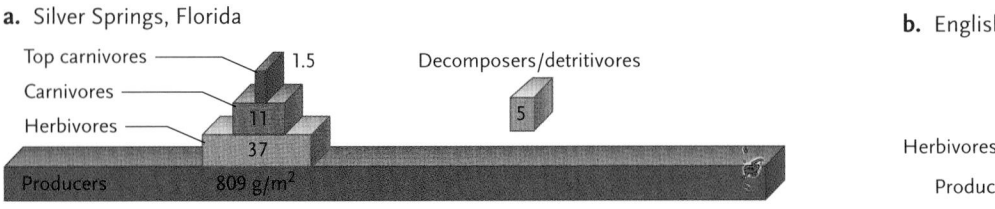

a. Silver Springs, Florida

b. English Channel

Figure 51.8
Pyramids of biomass. **(a)** The pyramid of standing crop biomass for Silver Springs is bottom heavy, as it is for most ecosystems. **(b)** Some marine ecosystems, such as that in the English Channel, have an inverted pyramid of biomass because producers are quickly eaten by primary consumers. Only the producer and herbivore trophic levels are illustrated here. The data for both pyramids are given in grams of dry biomass per square meter.

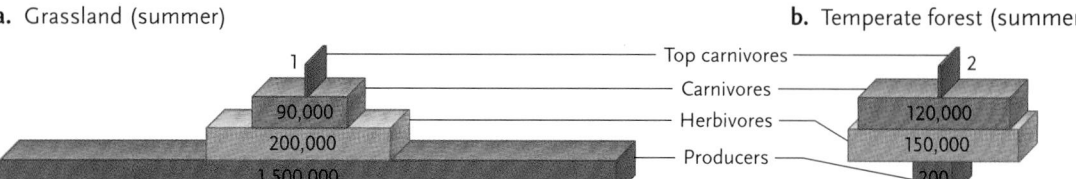

a. Grassland (summer)

Top carnivores — 1
Carnivores
Herbivores — 90,000
Producers — 200,000
1,500,000

b. Temperate forest (summer)

Top carnivores — 2
Carnivores — 120,000
Herbivores — 150,000
Producers — 200

Figure 51.9

Pyramids of numbers. **(a)** The pyramid of numbers (number of individuals per 1000 m^2) for temperate grasslands is bottom-heavy because individual producers are small and very numerous. **(b)** The pyramid of numbers for forests may have a narrow base because herbivorous insects often outnumber the producers, which are large trees. Data for both pyramids were collected in summer. Detritivores and decomposers (soil animals and microorganisms) are not included because they are difficult to count.

Consumers Sometimes Regulate Ecosystem Processes

As you know from the preceding discussion, numerous abiotic factors—the intensity and duration of sunlight, rainfall, temperature, and the availability of nutrients—have significant effects on primary productivity. Primary productivity, in turn, has profound effects on populations of herbivores and the predators that feed on them. But what effect does feeding by these consumers have on primary productivity?

Research conducted in the 1990s suggests that consumers may sometimes influence rates of primary productivity, especially in ecosystems with low species diversity and relatively few trophic levels. For example, food webs in lake ecosystems depend primarily on the productivity of phytoplankton **(Figure 51.10).** These producers are consumed by herbivorous zooplankton, which are in turn eaten by predatory invertebrates and fishes. The top nonhuman carnivore in these food webs is usually a predatory fish.

Herbivorous zooplankton play a central role in the regulation of lake ecosystems. Small zooplankton species consume only small phytoplankton. Thus, when small zooplankton are especially abundant, the large phytoplankton escape predation and survive, and the lake's primary productivity is high. By contrast, large zooplankton are voracious, eating both small and large phytoplankton. When *large* zooplankton are especially abundant, they reduce the overall biomass of phytoplankton, lowering the ecosystem's primary productivity.

In what has been termed a **trophic cascade**—predator–prey effects that reverberate through the population interactions at two or more trophic levels in an ecosystem—feeding by plankton-eating invertebrates and fishes has a *direct* impact on herbivorous zooplankton populations and an *indirect* impact on phytoplankton populations and the ecosystem's primary productivity. Invertebrate predators prefer small zooplankton. And when the invertebrates that eat small zooplankton are the dominant carnivores in the ecosystem, large zooplankton become more abundant;

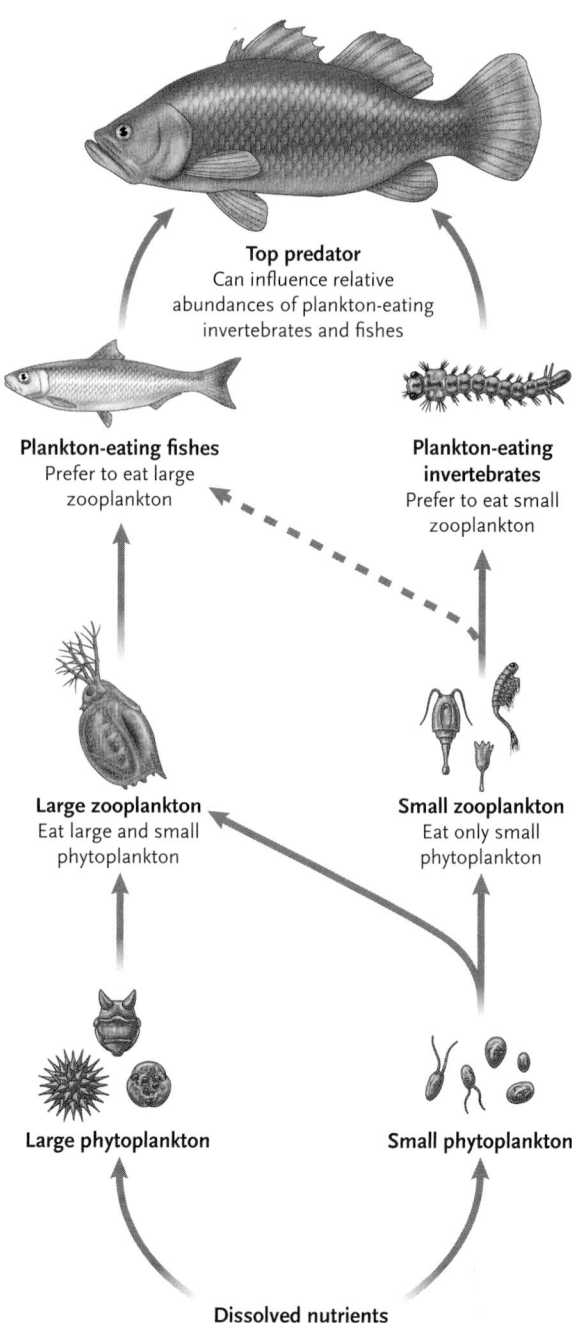

Top predator
Can influence relative abundances of plankton-eating invertebrates and fishes

Plankton-eating fishes
Prefer to eat large zooplankton

Plankton-eating invertebrates
Prefer to eat small zooplankton

Large zooplankton
Eat large and small phytoplankton

Small zooplankton
Eat only small phytoplankton

Large phytoplankton

Small phytoplankton

Dissolved nutrients

Figure 51.10

Consumer regulation of primary productivity. A simplified food web illustrates that lake ecosystems have relatively few trophic levels. The effects of feeding by top carnivores can cascade downward, exerting an indirect effect on phytoplankton and, thus, on primary productivity.

Fishing Fleets at Loggerheads with Sea Turtles

Populations of loggerhead sea turtles (*Caretta caretta*) that nest on Western Pacific beaches in Australia and Japan have been in decline. A surprising recent discovery indicates that the explanation may lie many thousands of miles away. Loggerhead sea turtles hatch from eggs that females bury on sandy beaches. The hatchlings then scurry to the surf and migrate to distant feeding grounds. The turtles mature at the feeding grounds, and eventually return to their hatching beaches to lay eggs.

Recently, a population of loggerhead sea turtles was discovered feeding along the coast of Baja California. Nesting grounds for these turtles are known only in the western Pacific, in Australia and Japan; none had been identified in the eastern Pacific. Did these turtles really migrate across 10,000 km of open ocean from Japan and Australia to Baja California? If so, the trip would be the longest open ocean migration known for any marine animal.

In addition, this long journey might explain the decline of the turtles in Japan and Australia. Scientists know that as many as 4000 loggerhead turtles drown in fishing nets in the north Pacific each year. Are these turtles intercepted on their way to Baja California from the Australian or Japanese feeding grounds? If so, the large numbers caught in fishing nets may contribute to the decline of loggerhead populations in the western Pacific.

Brian W. Bowen and his colleagues at the University of Florida, Gainesville, used mitochondrial DNA (mtDNA) sequences to answer these questions. One 350-base-pair segment of mtDNA was particularly useful because it includes sequence variations that are characteristic of different loggerhead populations.

The investigators took DNA samples from nesting populations in Australia and Japan, from feeding populations in Baja California, and from turtles drowned in fishing nets in the north Pacific. They used the polymerase chain reaction (PCR) to amplify the mtDNA segment from the DNA samples. Sequencing of the amplified segments revealed three major variants of mtDNA, which the researchers designated sequences A, B, and C. The sequences were distributed among loggerhead turtles as shown in the accompanying table.

The mtDNA of most turtles found in Baja California and in fishing nets in the north Pacific match that of turtles from the Japanese nesting areas, supporting the idea that loggerhead turtles hatched in Japan make the long migration across the north Pacific to Baja California. The data also indicate that a few turtles hatched in Australia may follow the same migratory route.

The investigators propose that the North Pacific Current, which moves from west to east, aids the migration. The return trip from Baja to Japan could be made via the North Equatorial Current, which runs from east to west just north of the equator. Loggerhead turtles have been found in this current; further tests will reveal whether they have the mtDNA sequence characteristic of the individuals nesting in Japan and feeding in Baja California.

Because only 2000 to 3000 female loggerhead turtles nest in Japan, it is uncertain whether the Japanese nesting population can survive the loss of thousands of offspring to fishing in the north Pacific. The number of female loggerhead turtles nesting in Australia has declined by 50% to 80% in the last decade; the loss of only a few individuals in fishing nets could have a drastic impact on this population as well. To save the loggerhead turtles, wildlife managers and international agencies must establish and enforce limits on the number of migrating individuals trapped and killed in the ocean fisheries.

Location	Sequence A	Sequence B	Sequence C
Australian nesting areas	26 turtles	0 turtles	0 turtles
Japanese nesting areas	0 turtles	23 turtles	3 turtles
Baja California feeding grounds	2 turtles	19 turtles	5 turtles
North Pacific	1 turtle	28 turtles	5 turtles

they consume many phytoplankton, causing productivity to decrease. By contrast, zooplankton-eating fishes prefer to eat large zooplankton (see Figure 50.2). Thus, when plankton-eating fishes are abundant, small zooplankton become the dominant herbivores. As a result, large phytoplankton become more numerous and the lake's productivity rises.

Large predatory fishes may add an additional level of control to the system because they feed on and regulate the population sizes of plankton-eating invertebrates and fishes. Thus, the effects of feeding by the top predator can cascade downward through the food web, affecting the densities of plankton-eating invertebrates and fishes, herbivorous zooplankton, and phytoplankton.

STUDY BREAK

1. What is the difference between gross primary productivity and net primary productivity?
2. What environmental factors influence rates of primary productivity in terrestrial and aquatic ecosystems?
3. Why is energy lost from an ecosystem at every transfer from one trophic level to the trophic level above it?
4. How can the presence of a top predator influence the interactions of organisms at lower trophic levels and an ecosystem's productivity?

51.2 Nutrient Cycling in Ecosystems

The availability of nutrients is as important to ecosystem function as the input of energy. Photosynthesis—the conversion of solar energy into chemical energy—requires carbon, hydrogen, and oxygen, which producers acquire from water and air. Producers also need nitrogen, phosphorus, and other minerals (see Table 33.1). A deficiency in any of these minerals can reduce primary productivity.

Earth is essentially a closed system with respect to matter. Thus, unlike energy, for which there is a constant cosmic input, virtually all the nutrients that will ever be available for biological systems are already present. Nutrient ions or molecules constantly circulate between the abiotic environment and living organisms in what ecologists describe as **biogeochemical cycles.** And unlike energy, which flows through ecosystems and is gradually lost as heat, matter is conserved in biogeochemical cycles. Although there may be local shortages of specific nutrients, Earth's overall supplies of these chemical elements are never depleted.

Ecologists Describe Nutrient Cycling with a Generalized Compartment Model

Nutrients take various forms as they pass through biogeochemical cycles. Some materials, such as carbon, nitrogen, and oxygen, form gases, which move through global *atmospheric cycles.* Geological processes move other materials, such as phosphorus, through local *sedimentary cycles,* carrying them between dry land and the seafloor. Rocks, soil, water, and air are the reservoirs where mineral nutrients accumulate, sometimes for many years.

Ecologists use a **generalized compartment model** to describe nutrient cycling (**Figure 51.11**). Two criteria divide ecosystems into four compartments where nutrients accumulate. First, nutrient molecules and ions are described as either *available* or *unavailable,* depending upon whether or not they can be assimilated by organisms. Second, nutrients are present either in *organic* material, the living or dead tissues of organisms, or in *inorganic* material, such as rocks and soil. For example, minerals in dead leaves on the forest floor are in the available-organic compartment because they are in the remains of organisms that can be eaten by detritivores. But calcium ions in limestone rocks are in the unavailable-inorganic compartment because they exist in a nonbiological form that producers cannot assimilate.

Nutrients move rapidly within and between the available compartments. Living organisms are in the available-organic compartment, and whenever heterotrophs consume food, they recycle nutrients within that reservoir (indicated by the oval arrow in the upper left of Figure 51.11). Producers acquire nutrients

from the air, soil, and water of the available-inorganic compartment. Consumers also acquire nutrients from the available-inorganic compartment when they drink water or absorb mineral ions through the body surface. Several processes routinely transfer nutrients from organisms to the available-inorganic compartment. As one example, respiration releases carbon dioxide, moving both carbon and oxygen from the available-organic compartment to the available-inorganic compartment.

By contrast, the movement of materials into and out of the unavailable compartments is generally slow. Sedimentation, a long-term geological process, converts ions and particles of the available-inorganic compartment into rocks of the unavailable-inorganic compartment. Materials are gradually returned to the available-inorganic compartment when rocks are uplifted and eroded or weathered. Similarly, over millions of years, the remains of organisms in the available-organic compartment were converted into coal, oil, and peat of the unavailable-organic compartment.

Except for the input of solar energy, we have described energy flow and nutrient cycling as though ecosystems were closed systems. In fact, most ecosystems exchange energy and nutrients with neighboring ecosystems. For example, rainfall carries nutrients into a forest ecosystem, and runoff carries nutrients from a forest into a lake or river. Ecologists have mapped the biogeochemical cycles of important elements, often by

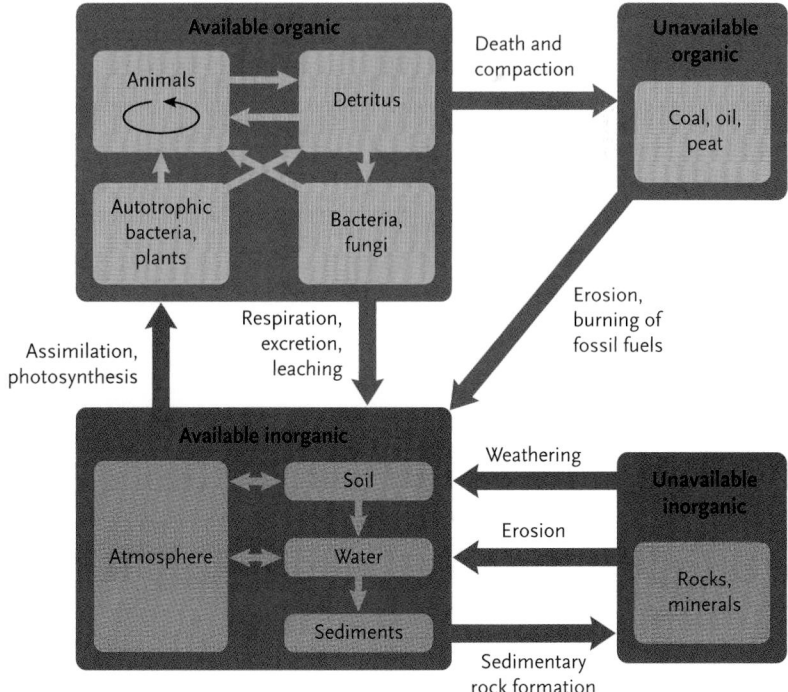

Figure 51.11

A generalized compartment model of nutrient cycling. Nutrients cycle through four major compartments within ecosystems. Processes that move nutrients from one compartment to another are indicated on the arrows. The oval arrow in the upper left corner of the figure represents animal predation on other animals.

using radioactively labeled molecules that they can follow in the environment. As you study the details of the four biogeochemical cycles described below, try to understand them in terms of the generalized compartment model of nutrient cycling.

The Hydrologic Cycle Recirculates All the Water on Earth

Although it is not a mineral nutrient, water is the universal intracellular solvent for biochemical reactions. Nevertheless, only a fraction of 1% of Earth's total water is present in biological systems at any time.

The cycling of water, called the **hydrologic cycle**, is global, with water molecules moving from the ocean into the atmosphere, to the land, through freshwater ecosystems, and back to the ocean **(Figure 51.12)**. Solar energy causes water to evaporate from oceans, lakes, rivers, soil, and living organisms, entering the atmosphere as a vapor and remaining aloft as a gas, as droplets in clouds, or as ice crystals. It falls as precipitation, mostly in the form of rain and snow. When precipitation falls on land, water flows across the surface or percolates to great depth in the soil, eventually reentering the ocean reservoir through the flow of streams and rivers.

The hydrologic cycle maintains its global balance because the total amount of water that enters the atmosphere is equal to the amount that falls as precipitation. Most water that enters the atmosphere evaporates from the ocean, which represents the largest reservoir on the planet. A much smaller fraction evaporates from terrestrial ecosystems, and most of that results from transpiration in green plants.

The constant recirculation provides fresh water to terrestrial organisms and maintains freshwater ecosystems such as lakes and rivers. Water also serves as a transport medium that moves nutrients within and between ecosystems, as demonstrated in a series of classic experiments in the Hubbard Brook Experimental Forest, described in *Focus on Basic Research*.

The Carbon Cycle Includes a Large Atmospheric Reservoir

Carbon atoms provide the backbone of most biological molecules, and carbon compounds store the energy captured by photosynthesis (see Section 9.1). Carbon enters food webs when producers convert atmospheric carbon dioxide (CO_2) into carbohydrates. Heterotrophs acquire carbon by eating other organisms or detritus.

a. The water cycle

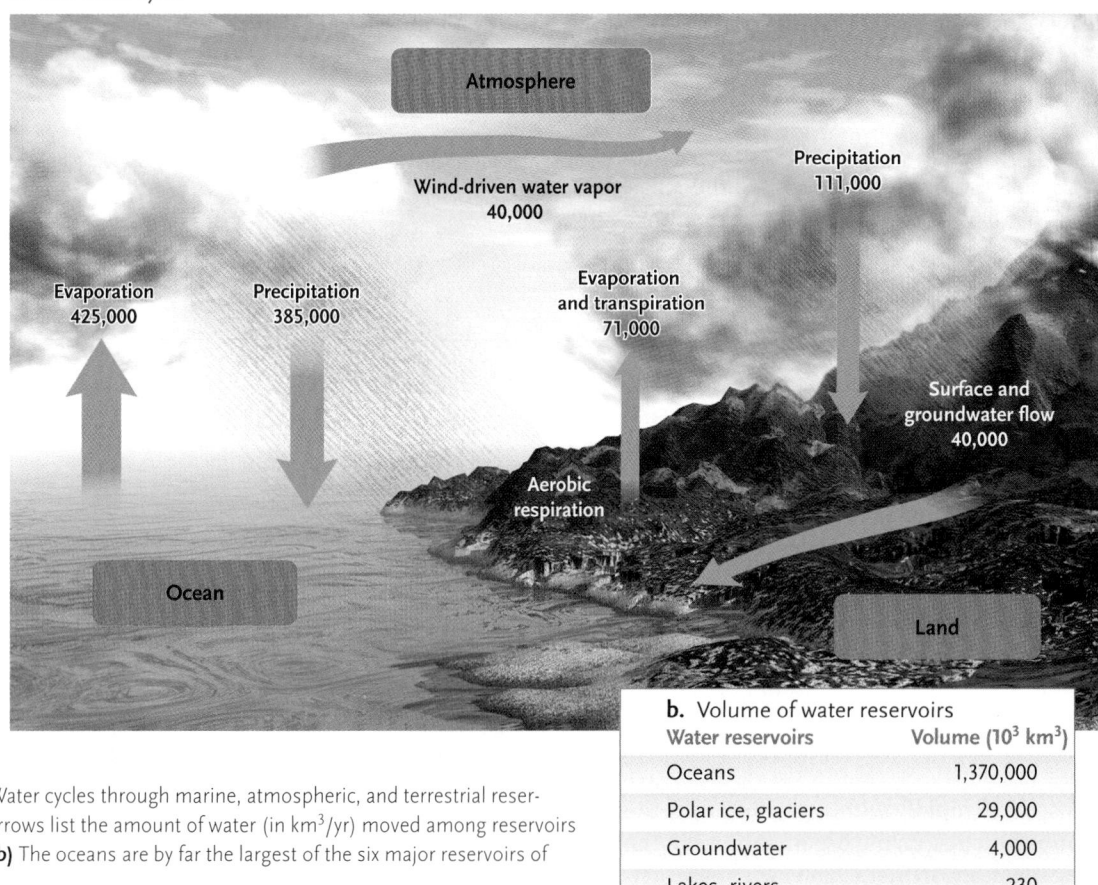

Figure 51.12
The hydrologic cycle. Water cycles through marine, atmospheric, and terrestrial reservoirs. **(a)** Data on the arrows list the amount of water (in km³/yr) moved among reservoirs by various processes. **(b)** The oceans are by far the largest of the six major reservoirs of water on Earth.

b. Volume of water reservoirs

Water reservoirs	Volume (10³ km³)
Oceans	1,370,000
Polar ice, glaciers	29,000
Groundwater	4,000
Lakes, rivers	230
Soil moisture	67
Atmosphere (water vapor)	14

FOCUS ON RESEARCH

Basic Research: Studies of the Hubbard Brook Watershed

Because water always flows downhill, local topography affects the movement of dissolved nutrients in terrestrial ecosystems. A **watershed** is an area of land from which precipitation drains into a stream or river system. Thus, each watershed represents a part of an ecosystem from which nutrients exit through a single outlet, much the way a bathtub empties through a single drain. When several streams join to form a river, the watershed drained by the river encompasses all of the smaller watersheds drained by the streams. For example, the Mississippi River watershed covers roughly one-third of the United States, and it includes watersheds drained by the Illinois, Missouri, and Tennessee Rivers as well as many other watersheds drained by smaller streams and rivers.

Because watersheds are relatively self-contained units, they are ideal for large-scale field experiments about nutrient flow in ecosystems. Herbert Bormann of Yale University and Gene Likens of Cornell University have conducted a classic experiment on this topic since the 1960s. Bormann and Likens manipulated small watersheds of temperate deciduous forest in the Hubbard Brook Experimental Forest in the White Mountain National Forest of New Hampshire. They measured precipitation and nutrient input into the watersheds, the uptake of nutrients by vegetation, and the amount of nutrients leaving the watershed via streamflow. Nutrients exported in streamflow were monitored in water samples collected from V-shaped concrete weirs built into bedrock below the streams that drained the watersheds **(Figure a).** Impermeable bedrock underlies the soil, preventing water from leaving the system by deep seepage.

After collecting several years of baseline data on six undisturbed watersheds, the researchers cut all the trees in one small watershed in 1965 and 1966. They also applied herbicides to prevent regrowth. After establishing

this experimental treatment, they monitored the output of nutrients in streams that drained experimental and control watersheds. They attributed differences in nutrient export between undisturbed watersheds (controls) and the clear-cut watershed (experimental treatment) to the effects of deforestation.

Bormann and Likens determined that vegetation absorbed substantial water and conserved nutrients in undisturbed watersheds. Plants used about 40% of the precipitation for transpiration. The rest contributed to runoff and groundwater. Control watersheds lost only about 8–10 kg of calcium per hectare each year, an amount that was replaced by the erosion of bedrock and input from rain. Moreover, control watersheds actually accumulated about 2 kg of nitrogen per hectare per year and slightly smaller amounts of potassium.

By contrast, the experimentally deforested watershed experienced a 40% annual increase in runoff. During a 4-month period in the summer, runoff increased 300%. Some mineral losses were similarly large. The net loss of calcium was 10 times higher than in the control watersheds **(Figure b)** and the loss of potassium 21 times higher. Phosphorus losses did not increase;

Figure a
Weir used to measure the volume and nutrient content of water leaving a watershed by streamflow.

this mineral was apparently retained by the soil. However, the loss of nitrogen was an astronomical 120 kg per hectare per year. So much nitrogen entered the stream draining the experimental watershed that the stream became choked with algae and cyanobacteria. Thus, the results of the Hubbard Brook experiment suggest that deforestation increases flooding and decreases the fertility of ecosystems.

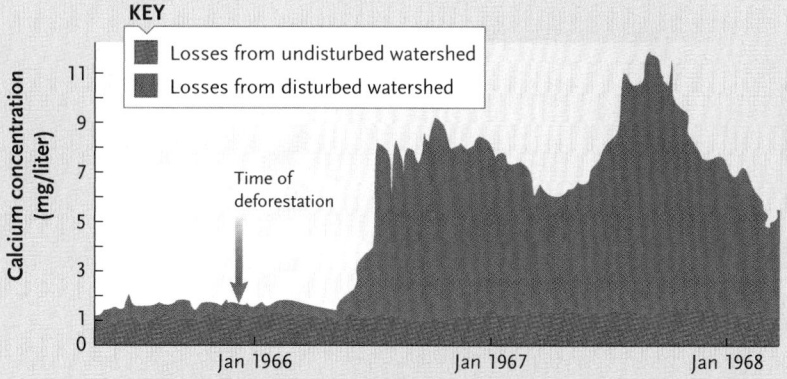

Figure b
Calcium losses from a deforested watershed were much greater than those from controls. The arrow indicates the time of deforestation in early winter. Mineral losses did not increase until after the ground thawed the following spring; increased runoff also caused large water losses from the watershed.

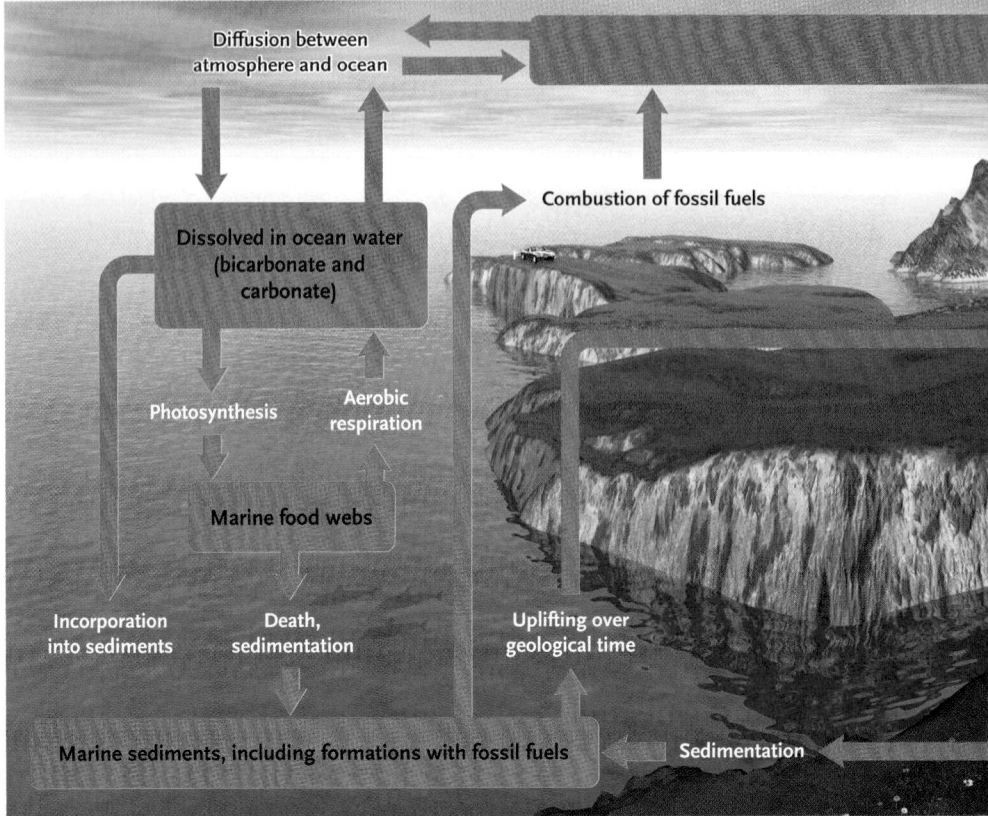

a. Amount of carbon in major reservoirs

Carbon reservoirs	Mass (10^{15} g)
Sediments and rocks	77,000,000
Ocean (dissolved forms)	39,700
Soil	1,500
Atmosphere	750
Biomass on land	715

b. Annual global carbon movement between reservoirs

Direction of movement	Mass (10^{15} g)
From atmosphere to plants (carbon fixation)	120
From atmosphere to ocean	107
To atmosphere from ocean	105
To atmosphere from plants	60
To atmosphere from soil	60
To atmosphere from burning fossil fuel	5
To atmosphere from burning plants	2
To ocean from runoff	0.4
Burial in ocean sediments	0.1

c. The global carbon cycle

Figure 51.13

The carbon cycle. Marine and terrestrial components of the global carbon cycle are linked through an atmospheric reservoir of carbon dioxide. **(a)** By far, the largest amount of Earth's carbon is found in sediments and rocks. **(b)** Earth's atmosphere mediates most movements of carbon. **(c)** In this illustration of the carbon cycle, boxes identify major reservoirs, and labels on the arrows identify the processes that cause carbon to move between reservoirs.

Although carbon moves somewhat independently in the sea and on land, a common atmospheric pool of CO_2 creates a global **carbon cycle (Figure 51.13).**

The largest reservoir of carbon is sedimentary rock, such as limestone or marble. Rocks are in the unavailable-inorganic compartment, and they exchange carbon with living organisms at an exceedingly slow pace. Most *available* carbon is present as dissolved bicarbonate ions (HCO_3^-) in the ocean. Soil, the atmosphere, and plant biomass form other significant, but much smaller, reservoirs of available carbon. Atmospheric carbon is mostly in the form of molecular CO_2, a product of aerobic respiration. Volcanic eruptions also release CO_2 into the atmosphere.

Sometimes carbon atoms leave the organic compartments for long periods of time. Some organisms in marine food webs build shells and other hard parts by incorporating dissolved carbon into calcium carbonate ($CaCO_3$) and other insoluble salts. When shelled organisms die, they sink to the bottom and are buried in sediments. The insoluble carbon that accumulates as rock in deep sediments may remain buried for millions of years before tectonic uplifting brings it to the surface, where erosion and weathering dissolve sedimentary rocks and return carbon to an available form.

Carbon atoms were also transferred to the unavailable-organic compartment when soft-bodied organisms were buried in habitats where low oxygen concentration prevented decomposition. Under suitable geological conditions, these carbon-rich tissues were slowly converted to gas, petroleum, or coal, which humans now use as fossil fuels. Human activities, especially the burning of fossil fuels, are transferring carbon into the atmosphere at a high rate. The resulting change in the worldwide distribution of carbon is having profound consequences for Earth's atmosphere and climate, including a general warming of the climate and a rise in sea level, as described in *Focus on Applied Research.*

The Nitrogen Cycle Depends upon the Activity of Diverse Microorganisms

All organisms require nitrogen to construct nucleic acids, proteins, and other biological molecules. Earth's atmosphere had a high nitrogen concentration long

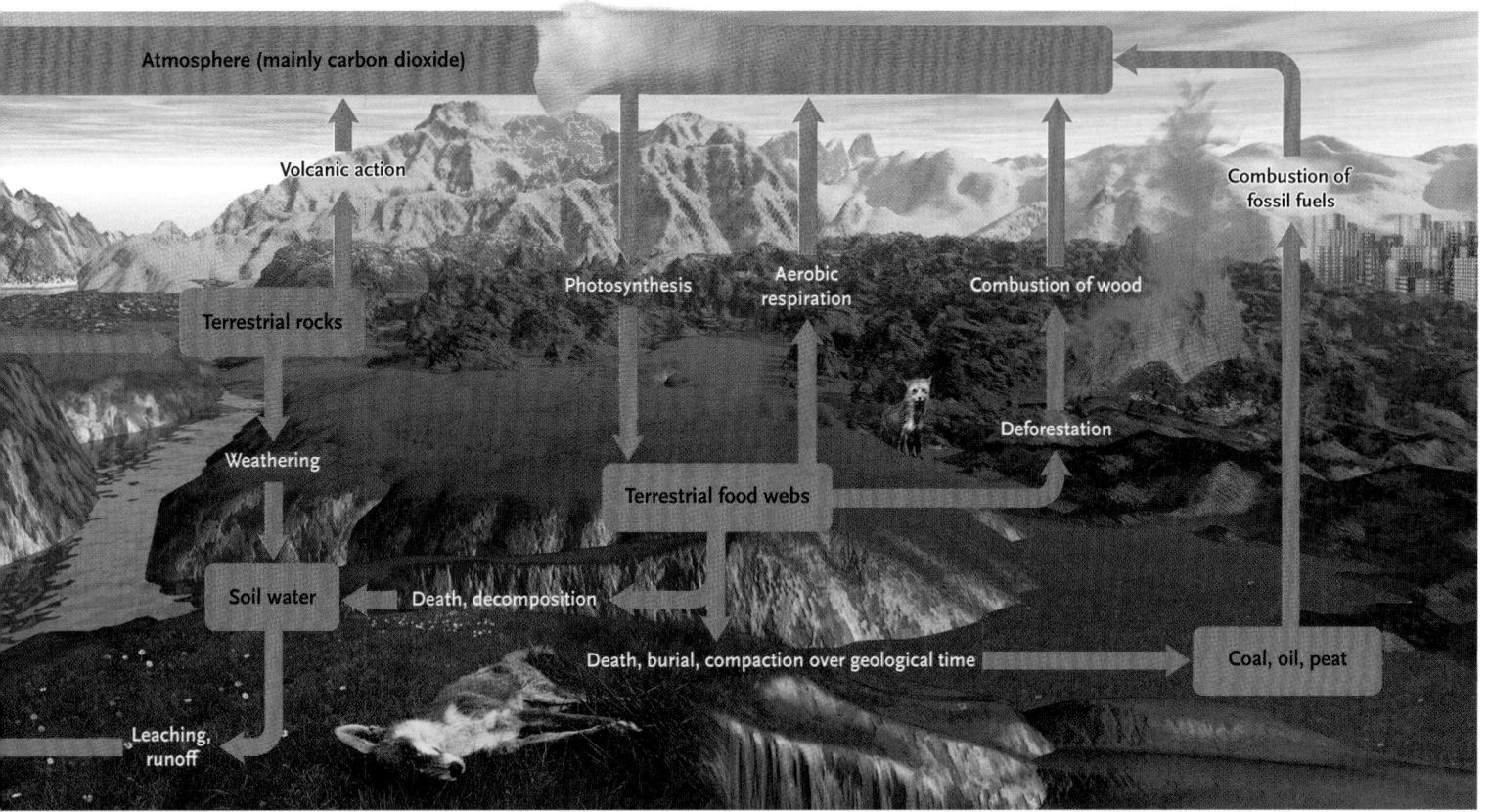

Atmosphere (mainly carbon dioxide)

Volcanic action

Combustion of fossil fuels

Photosynthesis

Aerobic respiration

Combustion of wood

Terrestrial rocks

Weathering

Deforestation

Terrestrial food webs

Soil water

Death, decomposition

Death, burial, compaction over geological time

Coal, oil, peat

Leaching, runoff

before life originated. Today, a global **nitrogen cycle** moves this element between the huge atmospheric pool of gaseous molecular nitrogen (N_2) and several much smaller pools of nitrogen-containing compounds in soils, marine and freshwater ecosystems, and living organisms **(Figure 51.14).**

Nitrogen Cycling within Ecosystems. Molecular nitrogen is abundant in the atmosphere, but triple covalent bonds bind its two atoms so tightly that most organisms cannot use it. However, three biochemical processes—nitrogen fixation, ammonification, and nitrification **(Table 51.2)**—convert nitrogen into nitrogen compounds that primary producers can incorporate into biological molecules such as proteins and nucleic acids. Secondary consumers obtain their nitrogen by consuming primary producers, thereby initiating the movement of nitrogen through the food webs of an ecosystem.

In **nitrogen fixation** (see Section 33.3), molecular nitrogen (N_2) is converted into ammonia (NH_3) and ammonium ions (NH_4^+). Certain bacteria, including *Azotobacter* and *Rhizobium,* which collect molecular nitrogen from the air between soil particles, are the major nitrogen fixers in terrestrial ecosystems. The cyanobacteria partners in some lichens (see Section 28.3) also fix molecular nitrogen. Other cyanobacteria, such as *Anabaena* and *Nostoc,* are important nitrogen fixers in

aquatic ecosystems; the water fern (genus *Azolla*) plays that role in rice paddies. Collectively, these organisms fix an astounding 200 million metric tons of nitrogen each year; nitrogen fixation can also result from lightning and volcanic action. Plants and other primary producers assimilate and use this nitrogen in the biosynthesis of amino acids, proteins, and nucleic acids, which then circulate through food webs.

Some plants, including legumes (such as beans and clover), alders (*Alnus* species), and some members of the rose family (Rosaceae), are mutualists with nitrogen-fixing bacteria. These plants acquire nitrogen from soils much more readily than plants that lack such mutualists. Although these plants have the competitive edge in nitrogen-poor soil, nonmutualistic species often displace them in nitrogen-rich soil.

In addition to nitrogen fixation, several other biochemical processes make large quantities of nitrogen available to producers. **Ammonification** of detritus by bacteria and fungi converts organic nitrogen into ammonia (NH_3), which dissolves into ammonium ions (NH_4^+) that plants can assimilate; some ammonia escapes into the atmosphere as a gas. **Nitrification** by certain bacteria produces nitrites (NO_2^-) that are then converted by other bacteria to usable nitrates (NO_3^-). All of these compounds are water-soluble, and water rapidly leaches them from soil into streams, lakes, and oceans.

Applied Research: Disruption of the Carbon Cycle

Concentrations of gases in the lower atmosphere have a profound effect on global temperature, which in turn has enormous impact on global climate. Molecules of carbon dioxide (CO_2), water, ozone, methane, nitrous oxide, and other compounds collectively act like a pane of glass in a greenhouse (hence the term *greenhouse gases*). They allow the short wavelengths of visible light to reach Earth's surface; but they impede the escape of longer, infrared wavelengths back into space, trapping much of that energy as heat. In short, greenhouse gases foster the accumulation of heat in the lower atmosphere, a warming action known as the **greenhouse effect,** which prevents Earth from being a cold and lifeless planet.

Since the late 1950s, scientists have measured atmospheric concentrations of CO_2 and other greenhouse gases at remote sampling sites, which are free of local contamination and reflect the average concentrations of these gases in the atmosphere. Results indicate that concentrations of greenhouse gases have increased steadily for as long as they have been monitored.

The graph for atmospheric CO_2 concentration **(Figure a)** has a regular zigzag pattern that follows the annual cycle of plant growth in the northern hemisphere. Photosynthesis withdraws so much CO_2 from the atmospheric available-inorganic pool during the northern hemisphere summer that

its concentration falls. The concentration is higher during the northern hemisphere winter, when aerobic respiration continues, returning carbon to the atmospheric available-inorganic pool, and photosynthesis slows. The zigs and the zags in the data for CO_2 represent seasonal highs and lows, but the midpoint of the annual peaks and troughs has increased steadily for 40 years. Many scientists interpret these data as evidence of a rapid buildup of atmospheric CO_2, which represents a shift in the distribution of carbon in the major reservoirs on Earth. The best estimates suggest that CO_2 concentration has increased by 35% in the last 150 years and by more than 10% in the last 30 years.

What has caused the increase in the atmospheric concentration of CO_2? Burning of fossil fuels and wood is the largest contributor, because CO_2 is a combustion product of this process. Today, humans burn more wood and fossil fuels than ever before. Vast tracts of tropical forests are being cleared and burned (see Section 53.2). To make matters worse, deforestation reduces the world's biomass of plants, which assimilate CO_2 and help maintain the carbon cycle as it existed before human activities disrupted it.

Why is an increase in the atmospheric CO_2 concentration so alarming? Recent research suggests that plants with C_3 metabolism will respond to increased CO_2 concentrations with increased growth rates, but that C_4 plants will not (review Section 9.4 on C_3 and C_4 plants). Thus, rising atmospheric levels of CO_2 will probably alter the relative abundances of many plant species, changing the composition and dynamics of their communities.

Simulation models by scientists who study the global climate suggest that increasing concentrations of any greenhouse gas may also intensify the greenhouse effect, contributing to a trend of global warming. Should we be alarmed about the prospect of a warmer planet? Some models predict that the mean temperature of the lower atmosphere will rise by 4° C,

enough to increase ocean surface temperatures. Water expands when heated, and global sea level could rise as much as 0.6 m just from this expansion. In addition, atmospheric temperature is rising fastest near the poles. Thus, global warming may also foster melting of glaciers and the Antarctic ice sheet, which might raise sea level much more, inundating low coastal regions. Waterfronts in Vancouver, Los Angeles, San Diego, Galveston, New Orleans, Miami, New York, and Boston could be submerged. So might agricultural lands in India, China, and Bangladesh, where much of the world's rice is grown. Moreover, global warming could disturb regional patterns of precipitation and temperature. Areas that now produce much of the world's grains, including parts of Canada and the United States, would become arid scrub or deserts, and the now-forested areas to their north would become dry grasslands.

Many scientists believe that atmospheric levels of greenhouse gases will continue to increase at least until the middle of the twenty-first century and that global temperature may rise by several degrees. At the Earth Summit in 1992, leaders of the industrialized countries agreed to try to stabilize CO_2 emissions by the end of the twentieth century. We have already missed that target, and some countries, including the United States, which is the largest producer of greenhouse gases, have now abandoned that goal as too costly. Stabilizing emissions at current levels will not reverse the damage already done, nor will it stop the trend toward global warming. Many scientists agree that we should begin preparing for the consequences of global warming now. For example, we might increase reforestation efforts because a large tract of forest can withdraw significant amounts of CO_2 from the atmosphere. We might also step up genetic engineering studies to develop heat-resistant and drought-resistant crop plants, which may provide crucial food reserves in regions of climate change.

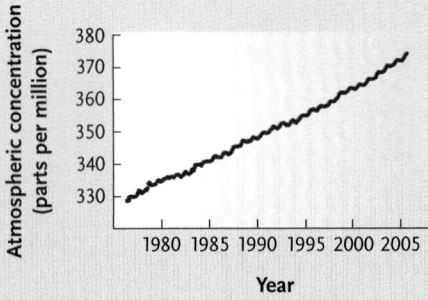

Figure a

Increases in atmospheric concentration of carbon dioxide, mid-1970s through 2004. The data were collected at a remote monitoring station in Australia (Cape Grim, Tasmania) and compiled by scientists at the Commonwealth Scientific and Industrial Research Organization, an agency of the Australian government.

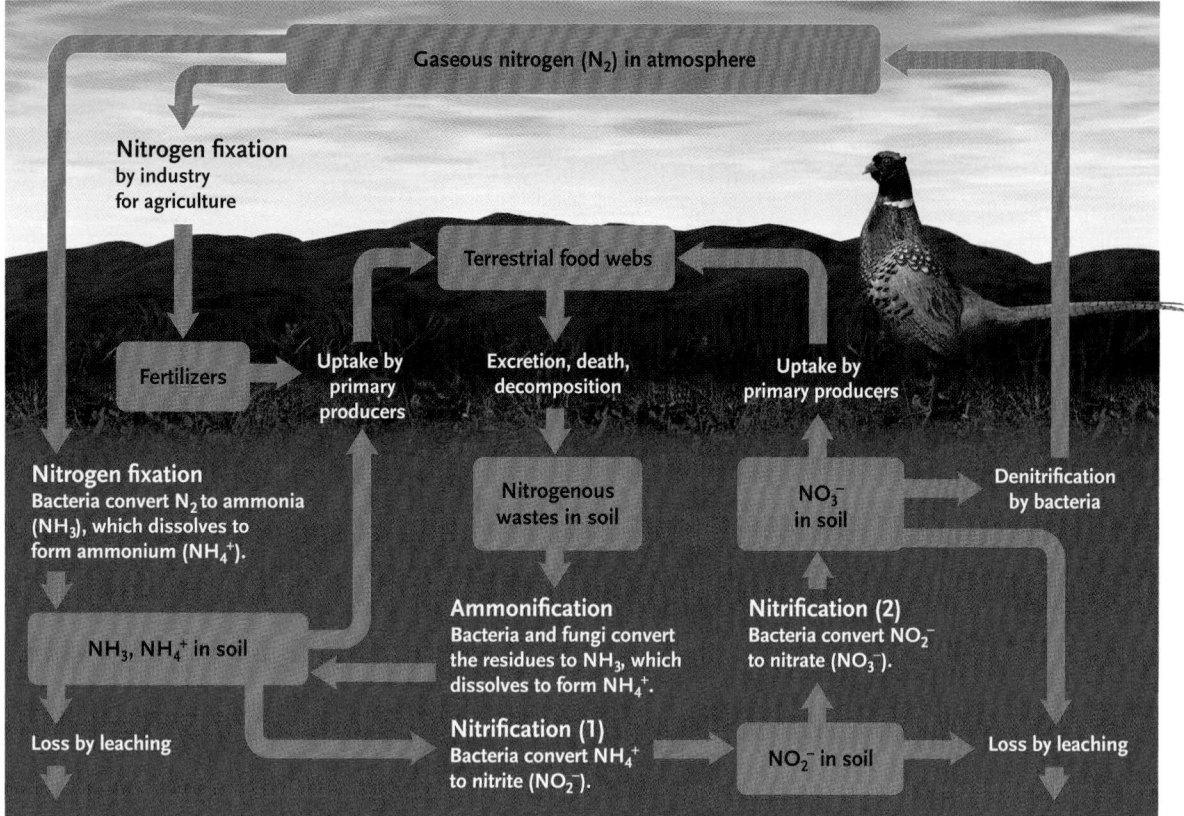

Figure 51.14

The nitrogen cycle in terrestrial ecosystems. Nitrogen cycles through terrestrial ecosystems when unavailable molecular nitrogen is made available through the action of nitrogen-fixing bacteria. Other bacteria recycle nitrogen within the available organic compartment through ammonification and two types of nitrification, converting organic wastes into ammonium ions and nitrates. Denitrification converts nitrate to molecular nitrogen, which returns to the atmosphere. Runoff carries various nitrogen compounds from terrestrial ecosystems into oceans, where it is recycled in marine food webs.

Under conditions of low oxygen availability, **denitrification** by still other bacteria converts nitrites or nitrates into nitrous oxide (N_2O) and then into molecular nitrogen (N_2), which enters the atmosphere, completing the cycle. This action can deplete supplies of soil nitrogen in waterlogged or otherwise poorly aer-ated environments, such as bogs and swamps. In an interesting twist on the usual predator–prey relation-ships, several species of flowering plants that live in nitrogen-poor soils, such as Venus' fly trap *(Dionaea muscipula)*, capture and digest small insects as their primary nitrogen source.

Table 51.2 **Biochemical Processes That Influence Nitrogen Cycling in Ecosystems**

Process	Organisms Responsible	Products	Outcome
Nitrogen fixation	Bacteria: *Rhizobium, Azotobacter, Frankia* Cyanobacteria: *Anabaena, Nostoc*	Ammonia (NH_3), ammonium ions (NH_4^+)	Assimilated by primary producers
Ammonification of organic detritus	Soil bacteria and fungi	Ammonia (NH_3), ammonium ions (NH_4^+)	Assimilated by primary producers
Nitrification			
(1) Oxidation of NH_3	Bacteria: *Nitrosomonas, Nitrococcus*	Nitrite (NO_2^-)	Used by nitrifying bacteria
(2) Oxidation of NO_2^-	Bacteria: *Nitrobacter*	Nitrate (NO_3^-)	Assimilated by primary producers
Denitrification of NO_3^-	Soil bacteria	Nitrous oxide (N_2O), molecular nitrogen (N_2)	Released to atmosphere

Human Disruption of the Nitrogen Cycle. Human activities are altering the nitrogen cycle, primarily through the application of nitrogen-containing fertilizers. Of all nutrients required for primary production, nitrogen is often the least abundant. Agriculture routinely depletes soil nitrogen: with each harvest, nitrogen is removed from fields through the harvesting of plants that have accumulated nitrogen. Soil erosion and leaching remove more. Traditionally, farmers rotated their crops, alternately planting legumes and other crops in the same fields. In combination with other soil-conservation practices, crop rotation stabilized soils and kept them productive, sometimes for thousands of years.

Until 50 years ago, nearly all the nitrogen in living systems was made available by nitrogen-fixing microorganisms. Today, however, agriculture relies on the application of synthetic fertilizers. Some yields have quadrupled over the past 50 years. But 50 years is just an instant in the history of agriculture, and such high yields may not be sustainable for very long. Moreover, the production of synthetic fertilizers is expensive. It uses fossil fuels both as a raw material and as an energy source, so that fertilizer becomes increasingly costly as supplies of fossil fuels dwindle. Furthermore, rain and runoff leach excess fertilizer from agricultural fields and carry it into aquatic ecosystems. Like the phosphorus in Lake Erie, nitrogen has become a major pollutant of freshwater ecosystems, artificially enriching the waters and allowing producers to expand their populations.

The Phosphorus Cycle Includes a Large Sedimentary Reservoir

Phosphorus compounds lack a gaseous phase, and this element moves between terrestrial and marine ecosystems in a sedimentary cycle **(Figure 51.15)**. Earth's crust is the main reservoir of phosphorus, as it is for other minerals such as calcium and potassium that undergo sedimentary cycles.

Phosphorus is present in terrestrial rocks in the form of phosphates (PO_4^{3-}). In the **phosphorus cycle**, weathering and erosion carry phosphate ions from rocks to soil and into streams and rivers, which eventually transport them to the ocean. Once there, some phosphorus enters marine food webs, but most of it precipitates out of solution and accumulates for millions of years as insoluble deposits, mainly on continental shelves. When parts of the seafloor are uplifted and exposed, weathering releases the phosphates.

Plants absorb and assimilate dissolved phosphates directly, and phosphorus moves easily to higher trophic levels. All heterotrophs excrete some phosphorus as a waste product in urine and feces, which are decom-

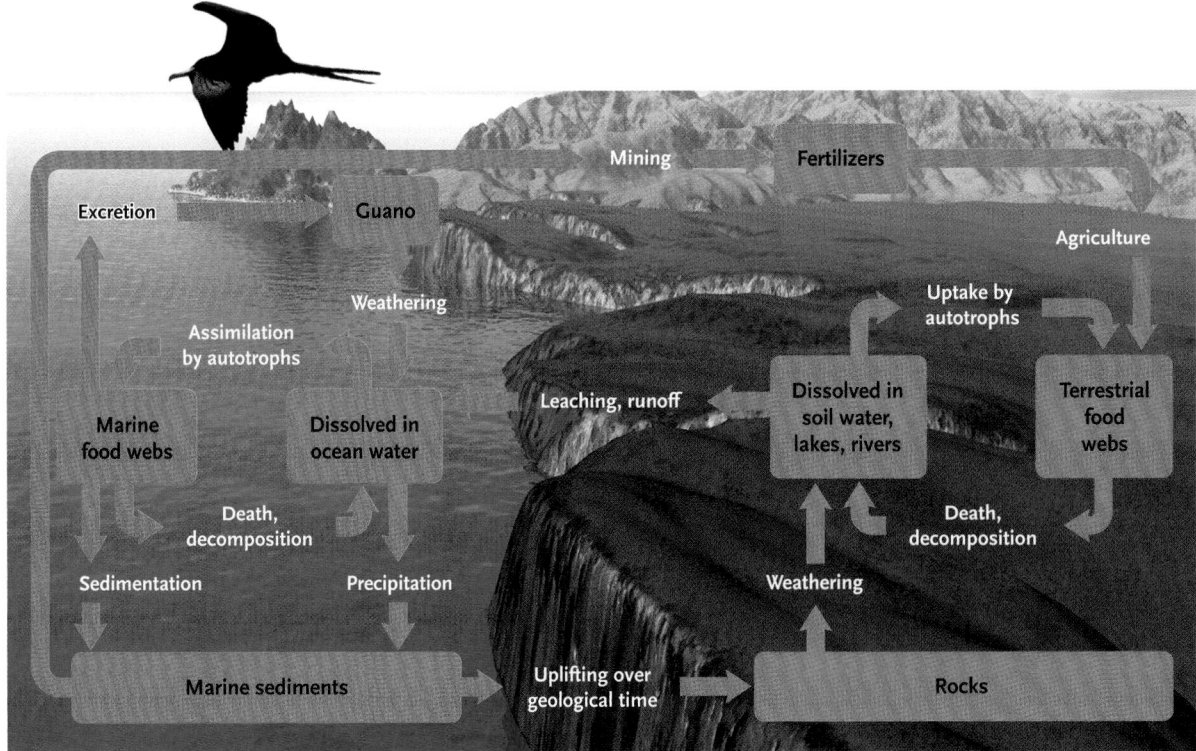

Figure 51.15

The phosphorus cycle. Phosphorus becomes available to biological systems when wind and rainfall dissolve phosphates in rocks and carry them into adjacent soil and freshwater ecosystems. Runoff carries dissolved phosphorus into marine ecosystems, where it precipitates out of solution and is incorporated into marine sediments.

posed, and producers readily absorb the phosphate ions that are released. Thus, phosphorus cycles rapidly *within* terrestrial communities.

Supplies of available phosphate are generally limited, however, and plants acquire it so efficiently that they reduce soil phosphate concentration to extremely low levels. Thus, like nitrogen, phosphorus is a common ingredient in agricultural fertilizers, and excess phosphates are pollutants of freshwater ecosystems. For many years, phosphate for fertilizers was obtained from *guano* (the droppings of seabirds that consume phosphorus-rich food), which was mined on small islands off the Pacific coast of South America. Most phosphate for fertilizer now comes from phosphate rock mined in Florida and other places with abundant marine deposits.

STUDY BREAK

1. In the generalized compartment model of biogeochemical cycling, how are the compartments where nutrients accumulate classified?
2. How does the global hydrologic cycle maintain its balance?
3. What process moves large quantities of carbon from an organic compartment to an inorganic compartment?
4. What microorganisms drive the global nitrogen cycle, and how do they do it?
5. What is Earth's main reservoir for phosphorus, and why is it recycled at such a slow rate from that reservoir?

51.3 Ecosystem Modeling

Ecologists Use Conceptual Models and Simulation Models to Understand Ecosystem Dynamics

To make predictions about how an ecosystem will respond to specific changes in physical factors, energy flow, or nutrient availability, ecologists turn to ecosystem modeling. Analyses of energy flow and nutrient cycling allow us to create a *conceptual model* of how ecosystems function **(Figure 51.16)**. Energy that enters ecosystems is gradually dissipated as it flows through a food web. By contrast, nutrients are conserved and recycled among the system's living and nonliving components. This very general model does not include processes that carry nutrients and energy out of one ecosystem and into another.

Note that the conceptual model ignores the nuts-and-bolts details of exactly how specific ecosystems function. Although it is a useful tool, a conceptual

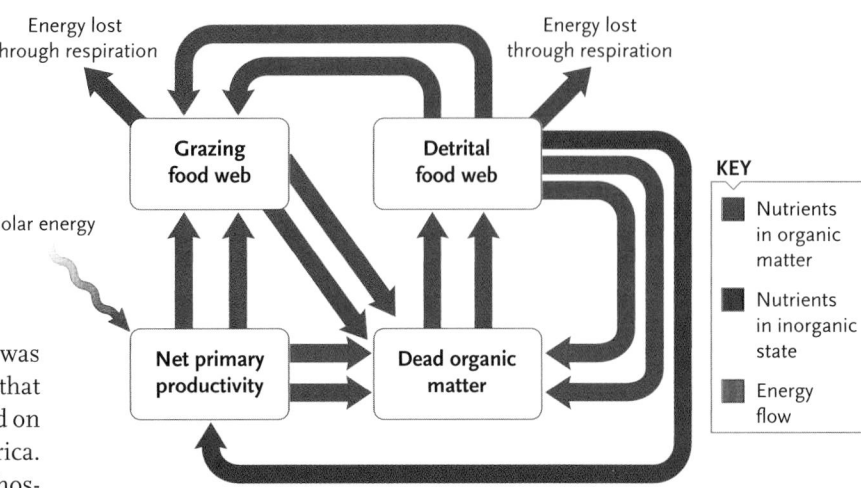

KEY

■ Nutrients in organic matter

■ Nutrients in inorganic state

■ Energy flow

Figure 51.16
A conceptual ecosystem model. A simple conceptual model of an ecosystem illustrates how energy flows through the system and is lost from both detrital and grazing food webs. Nutrients are recycled and conserved.

model doesn't really help us predict what would happen, say, if we harvested 10 million tons of introduced salmon from Lake Erie every year. We could simply harvest the fishes and see what happens. But ecologists prefer less intrusive approaches to study the potential effects of disturbances.

One approach to predicting "what would happen if . . ." is **simulation modeling**. Using this method, researchers gather detailed information about a specific ecosystem. They then create a series of mathematical equations that define its most important relationships. For example, one set of equations might describe how nutrient availability limits productivity at various trophic levels. Another might relate population growth of zooplankton to the productivity of phytoplankton. Other equations would relate the population dynamics of primary carnivores to the availability of their food, and still others would describe how the densities of primary carnivores influence reproduction in populations at both lower and higher trophic levels. Thus, a complete simulation model is a set of interlocking equations that collectively predict how changes in one feature of an ecosystem might influence others.

Creating a simulation model is no easy task, because the relationships within every ecosystem are complex. First, you must identify the important species, estimate their population sizes, and measure the average energy and nutrient content of each. Next, you would describe the food webs in which they participate, measure the quantity of food each species consumes, and estimate the productivity of each population. And, for the sake of completeness, you would determine the ecosystem's energy and nutrient gains and losses caused by erosion, weathering, precipitation, and runoff. You would repeat these measurements seasonally to identify annual variation in these factors. Finally, you might repeat the measurements over several years to determine the effects of year-to-year variation in climate and chance events.

After collecting these data, you would write equations that quantify the relationships in the ecosystem, including information about how temperature and other abiotic factors influence the ecology of each species. Having completed that job, you could begin to predict—possibly in great detail—the effects of harvesting 10 million or even 50 million tons of salmon annually from Lake Erie. Of course, you would have to refine the model whenever new data became available.

Some ecologists devote their professional lives to the study of ecosystem processes. The long-term initiative at the Hubbard Brook Forest provides a good example. As we attempt to understand larger and more complex ecosystems—and as we create larger and more complex environmental problems—modeling becomes an increasingly important tool. If a model is based on well-defined ecological relationships and good empirical data, it can allow us to make accurate predictions about ecosystem changes without the need for costly and environmentally damaging experiments. But like all ideas in science, a model is only as good as its assumptions, and models must constantly be adjusted to incorporate new ideas and recently discovered facts.

In the next chapter we examine how interactions among ecosystems establish the global phenomena that characterize the biosphere.

STUDY BREAK

1. What are the advantages and disadvantages of relying on conceptual models that describe ecosystem function?
2. What data must ecologists collect before constructing a simulation model of an ecosystem?

UNANSWERED QUESTIONS

How does the carbon cycle of a forest respond to climate change and urbanization?

As you've read in this chapter, human influences on the environment can have dramatic unforeseen consequences for ecosystems, altering energy flow and nutrient cycling. Given the complexity of ecosystems—the myriad scales of influence and multiple interactions among the organisms, the physical environment, and climate change variables—the precise response of an ecosystem is difficult to predict, even with advanced ecosystem models. We do known with certainty, however, that the carbon, nitrogen, and water cycles of forested ecosystems in the northeastern United States are changing, and they are likely to continue to do so.

Carbon cycle research in forested ecosystems often entails building an ecosystem model from quantitative data on the various pools and fluxes of carbon in the ecosystem and how these change with time. Scientists then correlate these changes with the environmental conditions and derive a mechanistic understanding that they can use to make predictions about how the ecosystem will respond to future changes. In theory, *gross primary productivity* (GPP) should be predictable from a basic understanding of photosynthesis and a general description of the ambient environmental conditions. In practice, however, the complexity of canopy architecture and leaf positioning, the timing of recurring natural phenomena, and the effects of herbivory and leaf losses from abiotic factors all make accurate predictions more difficult. Furthermore, the problem is dynamic because age-related changes in stand structure, disturbance, invasion, drought, seasonality, and pests or pathogens all add spatial and temporal complexities. Scientists should be able to predict *net primary productivity* (NPP), a key parameter used by ecologists to classify the world's ecosystems, from measurements of the cellular respiration and the relative abundances of representative organisms from the ecosystem.

Quantifying GPP and NPP on a large spatial scale can be challenging, and discovering the underlying mechanisms that control ecosystem responses to changes in environmental conditions is difficult. For example, studies at Black Rock Forest, a deciduous-oak-dominated forest in New York State, revealed that temporal heterogeneity (seasonal variation in leaf and stem respiration) and spatial heterogeneity (variations in canopy and hill slope position) are important factors that must be included in models of canopy respiration. Nevertheless, some simplifications may be possible. For example, while the basal rate of respiration is quite variable and subject to acclimation, it may be predictable from basic plant properties such as their nitrogen concentrations. Furthermore, the temperature coefficient of respiration is relatively constant, greatly simplifying the construction of an ecosystem model. To consider the impact of tree respiration on ecosystem form and function fully, my research team experimented with models that explicitly consider physiological linkages between photosynthesis and respiration, as mediated by leaf carbohydrate pools. We found that when we included direct linkages to carbon gain in the analysis, the model correctly predicted a large (23%) decrease in the estimated nighttime canopy respiration during the growing season. This result emphasizes the need for a process-based modeling approach when estimating forest productivity.

Our research at Black Rock Forest has also demonstrated that human activities in New York City (60 miles to the south) may be influencing tree growth in both urban and rural areas, with significant changes in seedling size, biomass allocation, herbivory, stomatal densities, nutrient concentrations, efficiency of water use, and rates of key physiological processes such as photosynthesis and respiration. Urbanization has a clear effect on the land area developed, but current research is showing that human activities in urban areas also influence forested ecosystems in the surrounding rural areas. Understanding how human activity, climate change, and forest ecosystems interact is crucial if we are to make prudent and sustainable development decisions, preserving the health of the ecosystems and the services they provide.

 Kevin Griffin is an associate professor at Columbia University's Lamont-Doherty Earth Observatory. His research centers on processes in plant and ecosystem ecology, the goal of which is to increase our understanding of both the role of vegetation in the global carbon cycle and the interactions between the carbon cycle and Earth's climate system. To learn more about Dr. Griffin's research, go to http://www.ldeo.columbia.edu/.

Review

Go to **ThomsonNOW**™ at www.thomsonedu.com/login to access quizzing, animations, exercises, articles, and personalized homework help.

51.1 Energy Flow and Ecosystem Energetics

- Ecosystems include biological communities and the abiotic environmental factors with which they interact (Figure 51.1).

- Food webs define the pathways along which energy and nutrients move through the biological components of an ecosystem. Ecosystems include both grazing and detrital food webs, which are closely interconnected (Figure 51.2).

- Only a small portion of the solar energy that reaches Earth is converted into chemical energy through the process of photosynthesis.

- An ecosystem's gross primary productivity is the rate at which producers convert solar energy into chemical energy. Producers use some energy for respiration; some is converted to heat; and some remains in the ecosystem as net primary productivity.

- Primary productivity is measured in units of energy captured or biomass produced per unit area per unit time. Net primary productivity indexes the energy available to support heterotrophs. Ecosystems vary in productivity and in their contributions to Earth's total productivity (Figure 51.3, Table 51.1).

- On land, primary productivity is limited by the availability of sunlight, water, and nutrients; temperature; and how much photosynthetic tissue is present. In marine and aquatic ecosystems, primary productivity is limited when sunlight and nutrients are not available in the same place (Figures 51.4 and 51.5).

- Only a fraction of the energy at any trophic level is converted into biomass at higher trophic levels. Ecological efficiencies generally range from 5% to 20%. As energy passes through a food web, an average of 90% is lost at each transfer between trophic levels, limiting the number of trophic levels that a food web can support (Figure 51.6).

- Ecological pyramids portray the effects of energy losses. For terrestrial ecosystems, pyramids of energy, biomass, and numbers generally have broad bases and narrow tops (Figures 51.7–51.9).

- The food preferences of consumers can influence primary productivity through a trophic cascade (Figure 51.10).

  Animation: The role of organisms in an ecosystem

  Animation: Food webs

  Animation: Energy flow at Silver Springs

51.2 Nutrient Cycling in Ecosystems

- Earth is a closed system with respect to matter.

- Nutrients circulate in biogeochemical cycles between living organisms and nonliving reservoirs. Nutrients accumulate in four compartments, defined by whether the nutrients are available or unavailable and whether they are in organic or inorganic material (Figure 51.11). Nutrients move rapidly between available compartments. Exchange rates for the unavailable compartments are slow. Some biogeochemical cycles are atmospheric; others are sedimentary.

- Water circulates through the atmosphere, oceans, and terrestrial and freshwater ecosystems in a global hydrologic cycle. Water evaporates from the oceans and continents and falls as precipitation. Runoff and streamflow return excess precipitation from the land to the oceans (Figure 51.12).

- The carbon cycles in terrestrial and aquatic ecosystems are linked through an atmospheric pool of CO_2, which primary producers assimilate. Respiration returns carbon to the atmosphere as CO_2. Earth's largest reservoir of carbon is unavailable in sedimentary rock. Other large reservoirs include coal, oil, and peat as well as dissolved bicarbonate and carbonate ions in seawater (Figure 51.13).

- Nitrogen is cycled between living organisms and an atmospheric pool of nitrogen gas. Bacteria and cyanobacteria make nitrogen available to the food web through the processes of nitrogen fixation, ammonification, and nitrification. Denitrification converts nitrogen compounds to molecular nitrogen, which enters the atmosphere (Figure 51.14, Table 51.2). The use of synthetic fertilizers disrupts the nitrogen cycle.

- Phosphorus undergoes a sedimentary cycle. Weathering and erosion of rock make phosphorus available; it is leached from soil and carried to the ocean. Dissolved phosphates precipitate out of seawater, forming insoluble deposits, which are eventually uplifted by tectonic processes (Figure 51.15).

  Animation: Hydrologic cycle

  Animation: Hubbard Brook experiment

  Animation: Carbon cycle

  Animation: Greenhouse effect

  Animation: Greenhouse gases

  Animation: Carbon dioxide and temperature

  Animation: Nitrogen cycle

  Animation: Phosphorus cycle

51.3 Ecosystem Modeling

- Conceptual models describe energy flow and nutrient cycling in ecosystems (Figure 51.16).

- Simulation models are interlocking mathematical equations that define the relationships between populations and between populations and the physical environment. They allow users to predict the effects of changes in ecosystem structure and function.

Questions

Self-Test Questions

1. Which of the following events would move energy and material from a detrital food web into a grazing food web?
   a. A beetle eats the leaves of a living plant.
   b. An earthworm eats dead leaves on the forest floor.
   c. A robin catches and eats an earthworm.
   d. A crow eats a dead robin.
   e. A bacterium decomposes the feces of an earthworm.

2. The total dry weight of plant material in a forest is a measure of the forest's:
   a. gross primary productivity.
   b. net primary productivity.
   c. cellular respiration.

d. standing crop biomass.

e. ecological efficiency.

3. Which of the following ecosystems has the highest rate of net primary productivity?

a. open ocean

b. temperate deciduous forest

c. tropical rain forest

d. desert and thornwoods

e. agricultural land

4. Endothermic animals exhibit a lower ecological efficiency than ectothermic animals because:

a. endotherms are less successful hunters than ectotherms.

b. endotherms eat more plant material than ectotherms.

c. endotherms are larger than ectotherms.

d. endotherms produce fewer offspring than ectotherms.

e. endotherms use more of their energy to maintain body temperature than ectotherms.

5. The amount of energy available at the highest trophic level in an ecosystem is determined by:

a. only the gross primary productivity of the ecosystem.

b. only the net primary productivity of the ecosystem.

c. the gross primary productivity and the standing crop biomass.

d. the net primary productivity and the ecological efficiencies of herbivores.

e. the net primary productivity and the ecological efficiencies at all lower trophic levels.

6. Some freshwater and marine ecosystems exhibit an inverted pyramid of:

a. biomass.

b. energy.

c. numbers.

d. turnover.

e. ecological efficiency.

7. Which process moves nutrients from the available-organic compartment to the available-inorganic compartment?

a. respiration

b. erosion

c. assimilation

d. sedimentation

e. photosynthesis

8. Which of the following materials has a sedimentary cycle?

a. water

b. oxygen

c. nitrogen

d. phosphorus

e. carbon

9. Which of the following statements is supported by the results of studies at the Hubbard Brook Experimental Forest?

a. Most of the energy captured by primary producers is lost before it reaches the highest trophic level in an ecosystem.

b. Deforested watersheds experience significantly less run-off than undisturbed watersheds.

c. Deforested watersheds lose more calcium and nitrogen in runoff than undisturbed watersheds.

d. Nutrients generally move through biogeochemical cycles very quickly.

e. Deforested watersheds generally receive more rainfall than undisturbed watersheds.

10. Nitrogen fixation converts:

a. atmospheric molecular nitrogen to ammonia.

b. nitrates to nitrites.

c. ammonia to molecular nitrogen.

d. ammonia to nitrates.

e. nitrites to nitrates.

Questions for Discussion

1. A lake near your home became overgrown with algae and pondweeds a few months after a new housing development was built nearby. What data would you collect to determine whether the housing development might be responsible for the changes in the lake?

2. Some politicians question whether recent increases in atmospheric temperature result from our release of greenhouse gases into the atmosphere. They argue that atmospheric temperature has fluctuated widely over Earth's history, and the changing temperature is just part of an historical trend. What information would allow you to refute or confirm their hypothesis? In addition, describe the pros and cons of reducing greenhouse gases as soon as possible versus taking a "wait and see" approach to this question.

3. If you could design the ideal farm animal—one that was grown as food for humans—from scratch, what characteristics would it have?

4. If you were growing a vegetable garden, identify the factors that might affect its primary productivity. How would you increase productivity? Identify some of the possible consequences of your gardening activities to nearby ecosystems.

Experimental Analysis

Design an experiment to test the hypothesis that the top predator in an aquatic ecosystem regulates the ecosystem's productivity. Establish as many experimental ponds as you wish, and imagine stocking them with organisms at different trophic levels. If the hypothesis is correct, describe the results you would expect to record from each of your experimental treatments.

Evolution Link

In the discussion of trophic cascades, we described how herbivorous zooplankton of different sizes eat phytoplankton of different sizes and how different types of predators preferentially feed on different sizes of zooplankton. Develop hypotheses about how these feeding preferences might establish different patterns of natural selection on the phytoplankton and zooplankton. How could you test your hypotheses?

How Would You Vote?

Emissions from motor vehicles are a major source of greenhouse gases. Many people buy large vehicles that use more fuel but are viewed as safer and more useful. Should such vehicles be taxed extra to discourage sales and offset their environmental costs? Can we expect the emergence of better fuels as well as more of the fuel-efficient, larger vehicles that are becoming available? Go to www.thomsonedu.com/login to investigate both sides of the issue and then vote.

Stormy weather in the biosphere. Atmospheric disturbances like Hurricane Fran, seen here in a satellite photograph taken on September 4, 1996, often have a dramatic impact on living systems.

52 The Biosphere

WHY IT MATTERS

The winter of 1997–1998 was one for the books. Record rainfall caused mudslides in California and flooding along the normally arid coast of Ecuador and Peru. But the annual rains never arrived in Asia and Australia, and fires consumed tropical rain forests in Indonesia and Malaysia. What caused these major climatic dislocations? Every 3 to 7 years, interactions between the upper layers of the Pacific Ocean and the atmosphere produce El Niño, a climatic event with global consequences **(Figure 52.1).** The 1997–1998 El Niño altered weather patterns worldwide, killing more than 2000 people and causing at least $30 billion in property damage.

In most years, air flows from a high pressure system over the eastern Pacific toward a low pressure system over the western Pacific. These winds move surface water from east to west and bring heavy rains to parts of Asia and Australia. Winds also usually blow from the poles toward the equator along the western sides of continents, and Earth's rotation causes these winds to push ocean surface water westward, away from the coast. The displaced surface water is replaced by cold, deep, nutrient-rich water carried by vertical currents called *upwellings* (see Figure 52.1a). The nutrients support complex marine

a. Usual pattern of Pacific Ocean currents

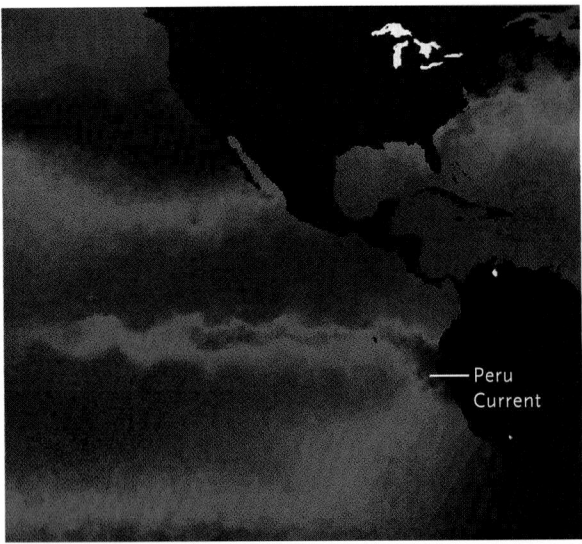

In most years, the powerful Peru Current carries cold water from the ocean bottom to the surface off the west coast of South America. The cold surface water then flows westward along the equator toward a large pool of warm water in the western Pacific Ocean. In this satellite photo taken on May 31, 1988, dark red indicates the warmest water and dark green the coldest upwelled water.

b. Pacific Ocean currents in an El Niño year

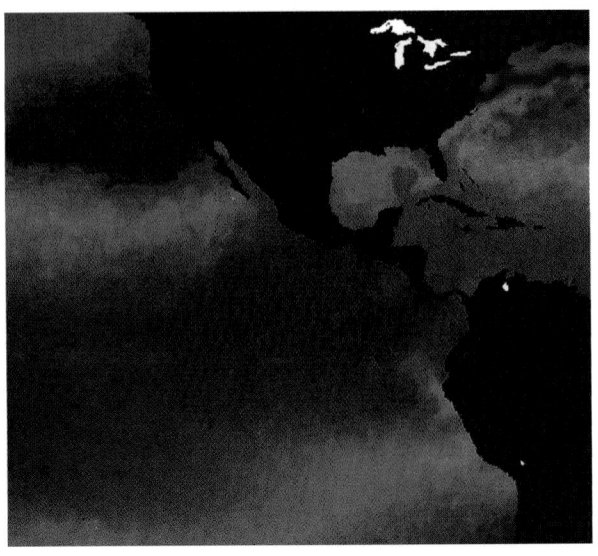

During an El Niño event, equatorial winds reverse directions and warm surface water flows eastward along the equator from the western Pacific Ocean toward South America. In this satellite photo taken on May 13, 1992, the warm water (red) spreads up and down the west coast of North and South America, suppressing the upwelling of cold water by the Peru Current.

Figure 52.1
El Niño and Pacific Ocean currents.

food webs in the shallow water above the continental shelf. For example, the Peru Current along the west coast of South America once supported a rich anchovy fishery.

Ocean currents vary seasonally, however, and in late December or early January, a warm, nutrient-poor current flows eastward along the equator and then north and south along the coastlines of Central and South America. Peruvian fishermen call this warm current El Niño (Spanish for "the child"), because it reaches their coast around Christmas. It usually persists for only a few weeks.

In strong El Niño years, atmospheric pressure systems change over the Pacific, altering the prevailing winds and ocean currents. Equatorial winds weaken; surface currents reverse direction, flowing from west to east; and a huge pool of warm ocean water accumulates in the eastern Pacific. During these shifts, the heavy rain that usually falls on Asia and Australia is instead delivered to the central and eastern Pacific. Thus, in El Niño years, Asia and Australia receive less rain than usual, and the west coasts of the Americas receive more. In the United States, winter temperatures are unusually high in the north central states and unusually low in the southern states.

El Niño episodes also alter sea surface temperature. When the warm current flowing from west to east reaches the continental shelf, it displaces the cold water of the Peru Current and prevents the usual upwelling (see Figure 52.1b). These changes in ocean currents have catastrophic effects on marine food webs. Lacking sufficient nutrients, phytoplankton die, followed by fishes that eat phytoplankton, and seabirds

that eat fishes. In combination with overfishing, the El Niño of 1972 drove the Peruvian anchovy population to the brink of extinction.

Some El Niño years are followed by a weather pattern called La Niña: the low pressure system over the western Pacific is accentuated, pulling air and ocean surface water from east to west. Low ocean surface temperatures extend from the coast of South America to Samoa. La Niña's effect on winter weather is opposite that of El Niño: parts of Asia and Australia are unusually wet; and the northern United States experiences periods of cold, wet weather, whereas the southern region is unusually warm and dry.

El Niño and La Niña are two extremes of a global climate cycle called the El Niño Southern Oscillation, or ENSO (the name refers to fluctuations in air pressure over the tropical Pacific). ENSO is a product of large-scale interactions between the ocean and atmosphere and has a major impact on the **biosphere**, all the places on Earth where organisms live. The biosphere has three abiotic components, which surround Earth's geological bulk like a skin. The **hydrosphere** encompasses all the water, including oceans and polar ice caps. The **lithosphere** includes the rocks, sediments, and soils of the crust. Finally, the **atmosphere** includes gases and airborne particles that envelop the planet.

In this chapter, we survey the biosphere with a wide-angle lens. First, we examine its environmental diversity and how organisms cope with it. We then consider how variations in the physical environment influence the large-scale distributions of ecosystems on land, in fresh water, and in the sea.

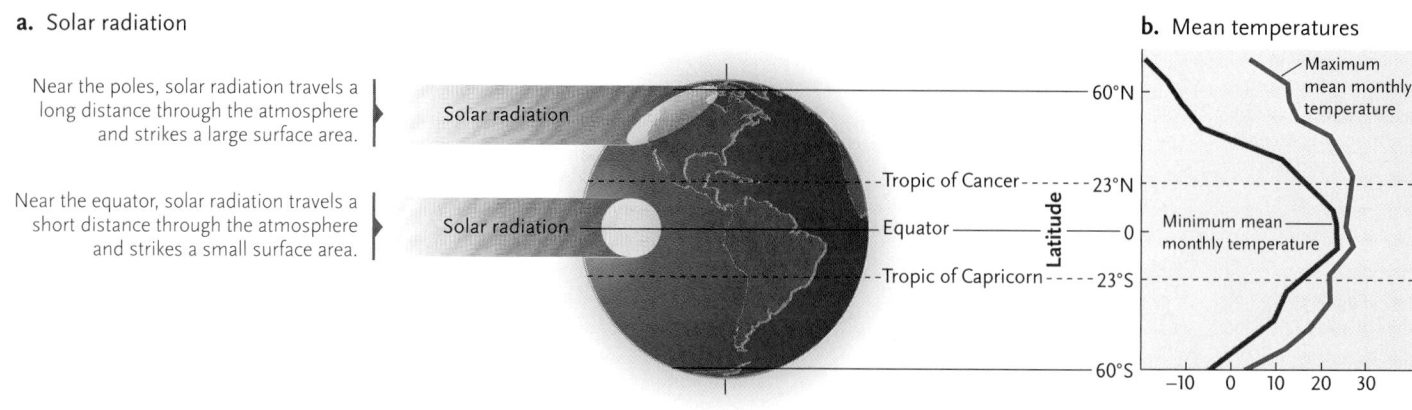

a. Solar radiation

Near the poles, solar radiation travels a long distance through the atmosphere and strikes a large surface area.

Solar radiation

Near the equator, solar radiation travels a short distance through the atmosphere and strikes a small surface area.

Solar radiation

60°N
Tropic of Cancer -- 23°N
Equator -- 0
Tropic of Capricorn -- 23°S
60°S

b. Mean temperatures

Maximum mean monthly temperature

Minimum mean monthly temperature

Temperature (°C)

Figure 52.2
Latitudinal variation in solar radiation and temperature. **(a)** Solar radiation is more intense near the equator than near the poles. **(b)** Minimum and maximum mean monthly temperatures as well as the range of mean monthly temperatures vary with latitude.

52.1 Environmental Diversity of the Biosphere

Numerous abiotic factors—sunlight, temperature, humidity, wind speed, cloud cover, and rainfall—contribute to a region's **climate**, the weather conditions prevailing over an extended period of time. Climates vary on global, regional, and local scales, and they undergo seasonal changes almost everywhere.

Variations in Incoming Solar Radiation Create Global Climate Patterns

A global pattern of environmental diversity results from latitudinal variation in incoming solar radiation, Earth's rotation on its axis, and its orbit around the sun.

Solar Radiation. Earth's spherical shape causes the intensity of incoming solar radiation to vary from the equator to the poles **(Figure 52.2).** When sunlight strikes Earth directly at a 90° angle, as it does near the equator, it travels the shortest possible distance through the radiation-absorbing atmosphere and falls on the smallest possible surface area (see Figure 52.2a). When sunlight arrives at an oblique angle, as it does near the poles, it travels a longer distance through the atmosphere and shines on a larger area. Thus, solar radiation is more concentrated near the equator than it is at higher latitudes, causing latitudinal variation in temperature (see Figure 52.2b).

Seasonality. Earth is tilted on its axis at a fixed position of 23.5° from the perpendicular to the plane on which it orbits the sun **(Figure 52.3).** This tilt produces seasonal variation in the duration and intensity of incoming solar radiation. The Northern Hemisphere receives its maximum illumination—and the Southern Hemisphere its minimum—on the June solstice (around

June 22), when the sun shines directly over the Tropic of Cancer (23.5° N latitude). The reverse is true on the December solstice (around December 22), when the sun shines directly over the Tropic of Capricorn (23.5° S latitude). Twice each year, on the vernal and autumnal equinoxes (around March 21 and September 23, respectively), the sun shines directly over the equator.

Earth's tilt is permanent, and only the **tropics**—the latitudes between the Tropics of Cancer and Capricorn—ever receive intense solar radiation from directly overhead. Moreover, the tropics experience only small seasonal changes in temperature and day length: environmental temperature is high and days last approximately 12 hours throughout the year. (Tropical seasonality is reflected in the alternation of wet and dry periods, rather than warm and cold seasons.) Seasonal variation in temperature and day length increases steadily toward the poles. Polar winters are long and cold with periods of continuous dark-

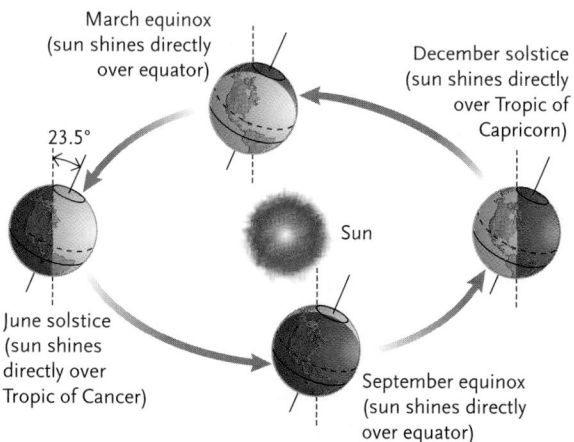

March equinox (sun shines directly over equator)

December solstice (sun shines directly over Tropic of Capricorn)

23.5°

Sun

June solstice (sun shines directly over Tropic of Cancer)

September equinox (sun shines directly over equator)

Figure 52.3
Seasonal variation in solar radiation. Earth's fixed tilt on its axis causes the Northern Hemisphere to receive more sunlight in June and the Southern Hemisphere to receive more in December. These differences are reflected in seasonal variations in day length and temperature, which are more pronounced at the poles than at the equator.

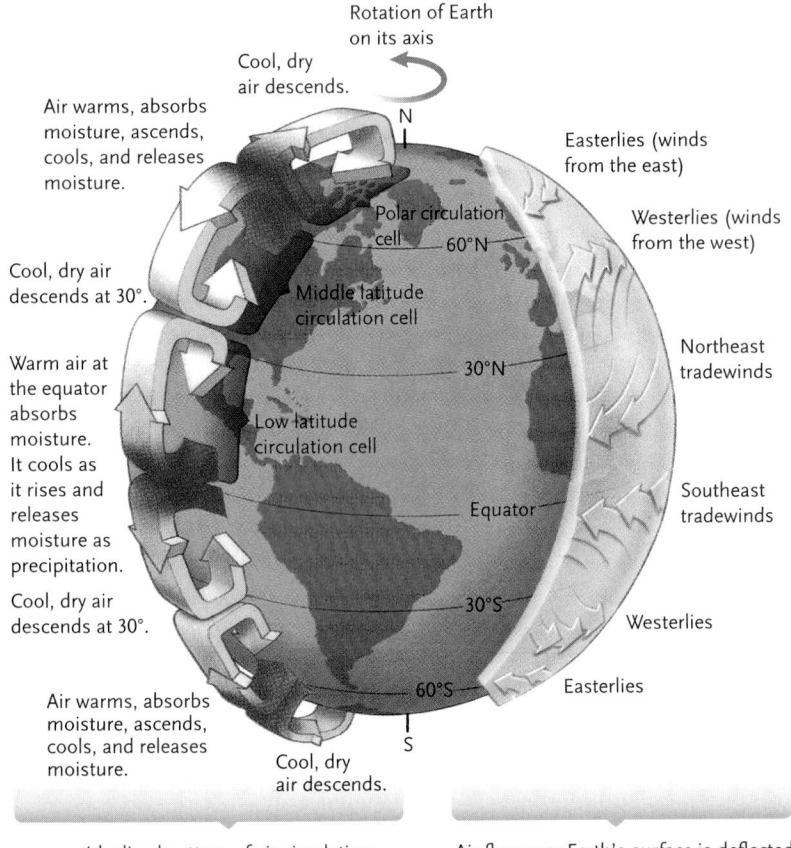

Rotation of Earth
on its axis

Cool, dry
air descends.

Air warms, absorbs
moisture, ascends,
cools, and releases
moisture.

Easterlies (winds
from the east)

Westerlies (winds
from the west)

Cool, dry air
descends at 30°.

Polar circulation
cell

60°N

Middle latitude
circulation cell

Warm air at
the equator
absorbs
moisture.
It cools as
it rises and
releases
moisture as
precipitation.

30°N

Low latitude
circulation cell

Northeast
tradewinds

Equator

Southeast
tradewinds

Cool, dry air
descends at 30°.

30°S

Westerlies

Air warms, absorbs
moisture, ascends,
cools, and releases
moisture.

60°S

Easterlies

S

Cool, dry
air descends.

Idealized pattern of air circulation.

Air flow near Earth's surface is deflected
from a strictly north–south direction.

Figure 52.4

Global air circulation. Latitudinal variations in the intensity of solar radiation cause equatorial air masses to warm and rise, initiating a global pattern of air movement in three circulation cells in each hemisphere. Air masses moving near Earth's surface create easterly and westerly winds, which are deflected from a strictly north–south flow by the planet's rotation.

ness, and polar summers are short with periods of continuous light.

Air Circulation. Sunlight warms air masses, causing them to expand, lose pressure, and rise in the atmosphere. The unequal heating of air at different latitudes initiates global air movements, producing three circulation cells in each hemisphere **(Figure 52.4)**. Warm equatorial air masses rise to high altitude before spreading north and south. They eventually sink back to Earth at about 30° N and S latitude. At low altitude, some air masses flow back toward the equator, completing low-latitude circulation cells. Others flow toward the poles, rise at 60° latitude, and divide at high altitude. Some air flows toward the equator, completing the pair of middle-latitude circulation cells. The rest moves toward the poles, where it descends and flows toward the equator, forming the polar circulation cells.

The flow of air masses at low altitude creates winds near the planet's surface. But the surface ro-

tates beneath the atmosphere, moving rapidly near the equator, where Earth's diameter is greatest, and slowly near the poles. Latitudinal variation in the speed of rotation deflects the movement of the rising and sinking air masses from a strictly north–south path into belts of easterly and westerly winds (see Figure 52.4); this deflection is called the Coriolis effect. Winds near the equator are called the trade winds; those further from the equator are the temperate westerlies and easterlies, named for the direction from which they blow.

Precipitation. Differences in solar radiation and global air circulation create latitudinal variations in rainfall **(Figure 52.5)**. Warm air holds more water vapor than cool air does. As air near the equator heats up, it absorbs water, primarily from the oceans. However, the warm air masses expand as they rise, and their heat energy is distributed over a larger volume, causing their temperature to drop. A decrease in temperature without the actual *loss* of heat energy is called **adiabatic cooling**. After cooling adiabatically, the rising air masses release moisture as rain. Torrential rainfall is characteristic of warm equatorial regions, where rising, moisture-laden air masses cool as they reach high altitude.

As cool, dry air masses descend at 30° latitude, increased air pressure at low altitude compresses them, concentrating their heat energy, raising their temperature, and increasing their capacity to hold moisture. The descending air masses absorb water from the land, so these latitudes are typically dry. Some air masses continue moving poleward in the lower atmosphere. When they rise at 60° latitude, they cool adiabatically and release precipitation (see Figure 52.4), creating moist habitats in the northern and southern temperate zones.

Ocean Currents. Latitudinal variations in solar radiation also warm the oceans' surface water unevenly. Because the volume of water increases as it warms, sea level is about 8 cm higher at the equator than at the poles. The volume of water associated with this "slope" is enough to cause surface water to move in response to gravity. The trade winds and temperate westerlies also contribute to the mass flow of water at the ocean surface. Thus, surface water flows in the direction of prevailing winds, forming major currents. Earth's rotation, the positions of landmasses, and the shapes of ocean basins also influence their movement.

Oceanic circulation is generally clockwise in the Northern Hemisphere and counterclockwise in the Southern **(Figure 52.6)**. The trade winds push surface water toward the equator and westward until it contacts the eastern edge of a continent. Swift, narrow, and deep currents of warm, nutrient-poor water run toward the poles, parallel to the east coasts of continents. For ex-

Figure 52.5

Variations in precipitation. The tropics receive high annual rainfall, whereas regions near 30° latitude are usually dry. Local topographic features and ocean currents also influence precipitation patterns.

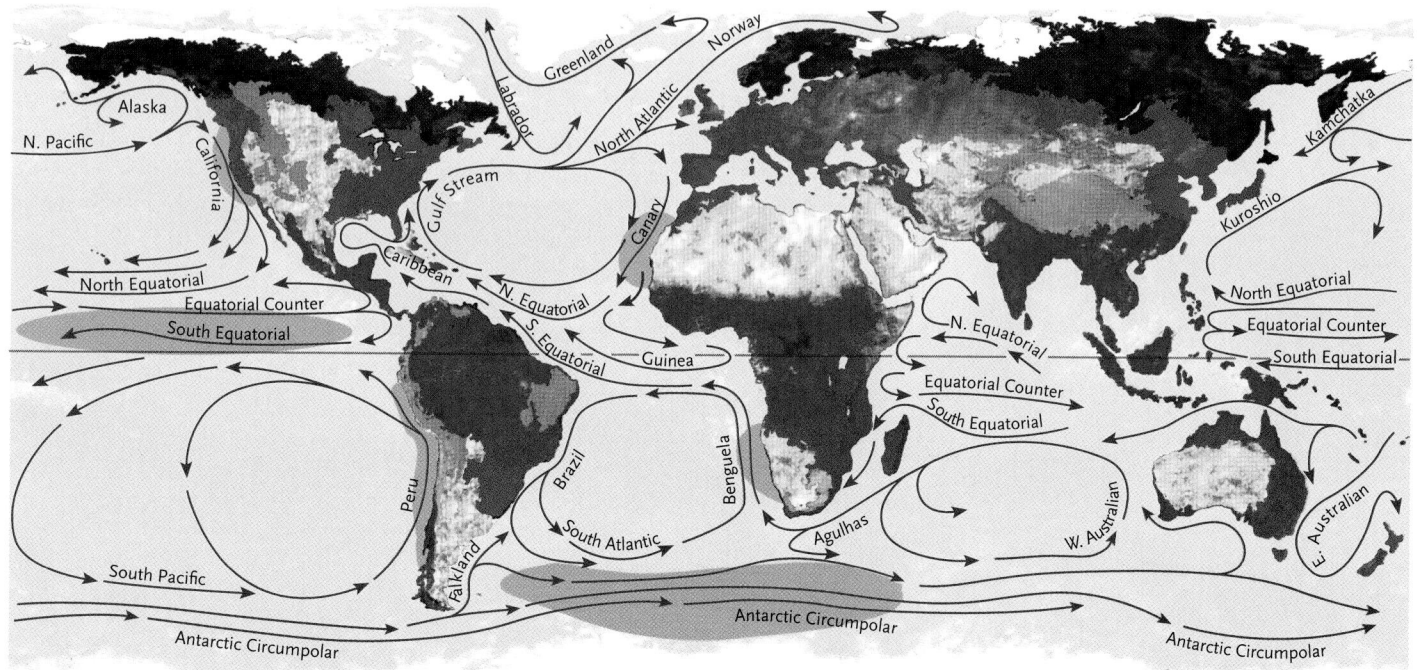

Figure 52.6

Ocean currents. Prevailing winds, Earth's rotation, gravity, the shape of ocean basins, and the positions of landmasses establish the direction and intensity of surface currents in the oceans. In general, warm currents flow away from the equator, and cold currents flow toward it.

ample, the Gulf Stream flows northward along the east coast of North America, carrying warm water toward northwestern Europe. Cold water returns from the poles toward the equator in slow, broad, and shallow currents, such as the California Current, that parallel the west coasts of continents.

Regional and Local Effects Overlay Global Climate Patterns

Although global and seasonal patterns determine a site's climate, regional and local effects also influence abiotic conditions.

Proximity to the Ocean. Currents running along seacoasts exchange heat with air masses flowing above them, moderating the temperature over nearby land. Breezes often blow from the sea toward the land during the day and in the opposite direction at night **(Figure 52.7).** These local effects sometimes override latitudinal variations in temperature. For example, the climate in London is much milder than that in Minneapolis, even though Minneapolis is slightly further south. Minneapolis has a **continental climate** that is not moderated by the distant ocean, but London has a **maritime climate,** tempered by winds that cross the nearby North Atlantic Current.

a. Daytime: land warmer than sea

b. Nighttime: sea warmer than land

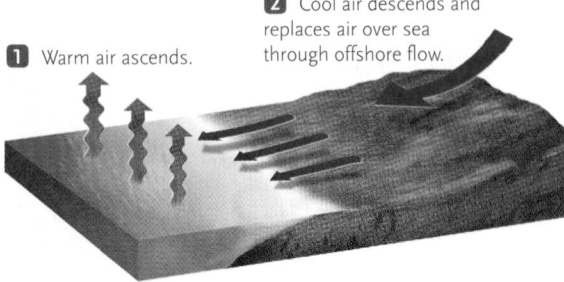

Figure 52.7
Sea breezes and land breezes. On a summer afternoon **(a),** warm air rises over the land, and a cool sea breeze blows inland from the ocean. At night **(b),** when the ocean is warmer than the land, the pattern is reversed.

Ocean currents also affect moisture conditions in coastal habitats. For example, air masses absorb water as they move from west to east across the Pacific Ocean. They cool as they cross the cold California Current, and when they reach land in northern California and Oregon during winter, their water vapor condenses into heavy fog and rain. During summer, however, land is warmer than the adjacent ocean. The air masses heat up as they cross the land, and they accumulate water, creating dry conditions.

Some regions experience **monsoon cycles** caused by seasonal reversals of wind direction. In the North American southwest, for example, summer heat causes air masses over land to rise, creating a zone of low pressure. Moist air from the nearby Gulf of California flows inland, where it rises and cools adiabatically, releasing substantial precipitation. Summer monsoon rains deliver one-third to one-half of the annual rainfall in Arizona and New Mexico. During the winter, when land is cooler than the nearby ocean, low-pressure systems form over the ocean, and winds blow from the land to the sea; thus, winters in the southwest are generally dry. Seasonal monsoon cycles also deliver torrential rainfall to parts of Africa, Asia, and South America.

The Effects of Topography. Mountains, valleys, and other topographic features also influence regional climates. In the Northern Hemisphere, south-facing slopes are warmer and drier than north-facing slopes, because they receive more solar radiation. In addition, adiabatic cooling causes air temperature to decline 3° to 6°C for every 1000 m increase in altitude.

Mountains also establish regional and local rainfall patterns. For example, after a warm air mass picks up moisture from the Pacific Ocean, it moves inland and reaches the Sierra Nevada, which parallels the California coast. As it rises to cross the mountains, the air cools adiabatically and loses moisture, releasing heavy rainfall on the windward side **(Figure 52.8).** After the now-dry air crosses the peaks, it descends and warms, absorbing moisture and forming a **rain shadow.** Habitats on the leeward side of mountains, such as the Great Basin Desert in western North America, are typically drier than those on the windward side.

Microclimate. Although climate influences the overall distributions of organisms, the abiotic conditions that immediately surround them—the **microclimate**—have the greatest effect on survival and reproduction. For example, a fallen log on the forest floor creates a microclimate in the underlying soil that is shadier, cooler, and moister than surrounding soil exposed to sun and wind. Many animals, including some insects, worms, salamanders, and snakes, occupy these sheltered sites and avoid the effects of prolonged exposure to the elements.

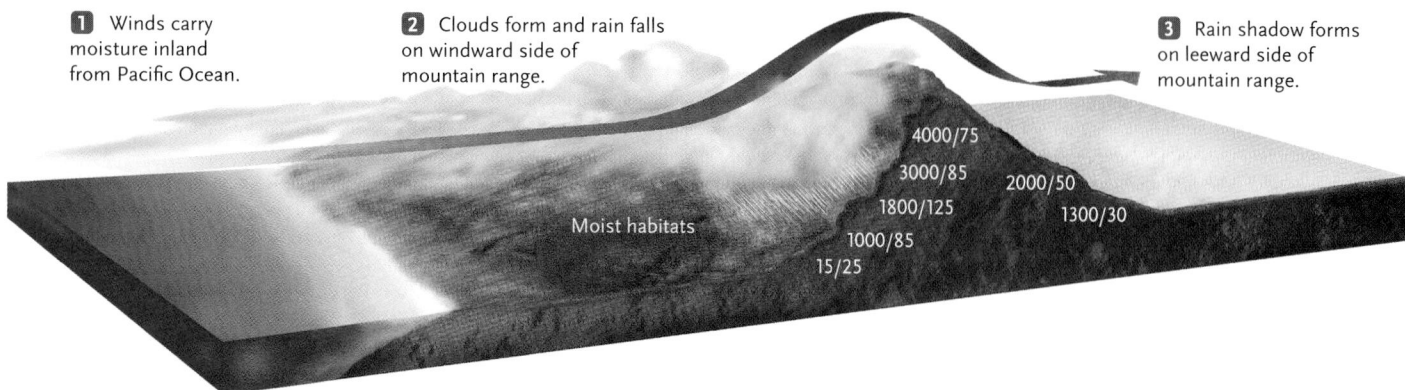

1. Winds carry moisture inland from Pacific Ocean.

2. Clouds form and rain falls on windward side of mountain range.

3. Rain shadow forms on leeward side of mountain range.

4000/75
3000/85
1800/125
1000/85
Moist habitats
15/25
2000/50
1300/30

Figure 52.8
Formation of a rain shadow. White numbers indicate altitude in meters followed by mean annual precipitation in centimeters for the Sierra Nevada of California.

STUDY BREAK

1. How does Earth's spherical shape influence temperature and air movements at different latitudes?
2. What causes seasonality of the climate in the temperate zone?
3. Why do dry conditions occur at 30° N and S latitude?
4. Briefly describe how mountains influence local precipitation.

52.2 Organismal Responses to Environmental Variation

Daily and seasonal variations in physical factors have profound effects on the biology of individual organisms. Moreover, large-scale variations in environmental conditions often influence the distributions of populations.

Organisms Use Homeostatic Responses to Cope with Environmental Variation

Animals in particular exhibit diverse homeostatic responses—biochemical, behavioral, physiological, and morphological—that enable them to maintain relatively constant conditions within their cells and tissues. Although the ability to use these responses almost certainly has a genetic basis, only some responses to environmental variation are *obligate* (that is, they must always be used). *Insights from the Molecular Revolution* describes one such evolutionary response at the biochemical level. Many behavioral and physiological responses are *facultative*. In other words, animals may use them or not, as their immediate conditions demand. Here we provide two brief examples of facultative behavioral and physiological responses to variations in environmental temperature.

Like many ectothermic animals, lizards often use behaviors to regulate body temperature (see Figure 1.15 and Section 46.6). They commonly *bask* in sunny spots to raise body temperature and seek shaded places to cool off. Many *Anolis* lizard species (see *Focus on Research Organisms* in Chapter 30) are distributed over broad altitudinal ranges, and populations living at high altitude encounter cooler environments than do those at low altitude. While they were graduate students at Harvard University, Paul E. Hertz, now of Barnard College, and Raymond B. Huey, now of the University of Washington, hypothesized that *Anolis* populations living at cool, high altitudes would bask more frequently than those living at warm, low altitudes. Hertz and Huey tested their hypothesis by observing *Anolis cybotes* and its close relative *Anolis shrevei* along an altitudinal gradient in the Dominican Republic **(Figure 52.9).** Their results indicate that basking frequency increases steadily with altitude. Moreover, the body temperatures of the lizards vary much less with altitude than do air temperatures at the same localities. The researchers therefore concluded that increased basking frequency by lizards at high altitude partially compensates for the lower environmental temperatures they encounter.

The state of extreme physiological sluggishness called *torpor* is a facultative response to daily variations in environmental temperature. Endothermic animals use the heat generated by the metabolic breakdown of food to maintain high body temperature (see Section 46.6). However, small endotherms, such as hummingbirds, have a large relative surface area through which they lose body heat. When environmental temperature is low, they may lose heat faster than they can generate it, risking the total depletion of their energy reserves and death by starvation. The problem is particularly acute at night, when hummingbirds cannot feed to replenish their energy stores. F. Reed Hainsworth and Larry Wolf of Syracuse University discovered that the purple-throated carib *(Eulampis jugularis),* a West Indian hummingbird, often becomes torpid at night, lowering its body

INSIGHTS FROM THE MOLECULAR REVOLUTION

Fish Antifreeze Proteins

Polar-dwelling fishes, such as winter flounder, Alaskan plaice, and Arctic sculpin, have "antifreeze proteins" that prevent their bodies from freezing into solid ice at the extremely low environmental temperatures they encounter. As ice crystals begin to form within a fish's cells and tissues, the antifreeze proteins bind to the crystals and cover them with a protein coat that prevents further crystal growth and fusion. As a result, the fishes freeze only to an ultrafine slush that allows continued activity, including movement and feeding. The antifreeze proteins are small molecules containing between 30 and 50 amino acids.

Researchers do not fully understand how the antifreeze proteins bind to ice crystals. Frank Sicheri and D. S. C. Yang at McMaster University in Hamilton, Ontario, and the Bio-Crystallography Laboratory of the VA Medical Center in Pittsburgh, Pennsylvania, sought a molecular solution to this problem. They used X-ray diffraction (see Section 14.2) to work out the molecular structure of the antifreeze protein from the winter flounder *(Pseudopleuronectes americanus)*. They grew protein crystals in a solution at 4°C and then examined them by X-ray diffraction at 4°C and −180°C.

The X-ray diffraction data indicated that the 37 amino acids of the antifreeze protein wind into a single, linear alpha helix. Along one side of the helix, side groups of two polar amino acids, threonine and asparagine, extend from the surface at four evenly spaced locations in a flat plane, one at either end of the molecule and two within the helix. Sicheri and Yang propose that these locations, which are spaced 1.65 nm apart, are *ice-binding motifs*.

The four motifs would fit nicely to the tips of ridges formed by water molecules on the surface of an ice crystal, which are spaced at intervals of 1.67 nm **(Figure a)**. Hydrogen bonds between the polar amino acid side groups and water molecules along the ridges would hold the proteins tightly to the surface of the ice crystal. The tight fit would prevent more water molecules from adding to the ice surface and thereby prevent further crystal growth.

Understanding how the fish antifreeze proteins work is not just a fascinating scientific issue. The description and characterization of the antifreeze proteins could lead to medical and industrial applications in situations where procedures and equipment must tolerate freezing conditions. In fact, the U.S. Food and Drug Administration recently approved the use of an antifreeze protein—originally discovered in an arctic fish, but now produced by genetically engineered yeast—as an ingredient in ice cream to enhance its creamy texture.

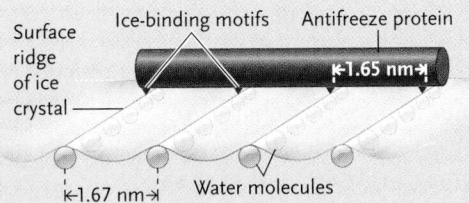

Figure a

How winter flounder antifreeze protein may bind to the surface ridges of an ice crystal. Only the water molecules forming the tips of the ridges are shown.

---

temperature from 40° to 20°C. Because torpor reduces the temperature difference between their bodies and the environment, torpid birds lose heat less rapidly. At the nighttime environmental temperatures they usually encounter, the torpid hummingbirds may use 80% less energy than they would if they had not entered a temporarily dormant state.

Global Warming Is Changing the Ecology of Many Organisms

As described in Chapter 51's *Focus on Applied Research,* most scientists agree that the atmosphere is getting warmer. What effect will global warming have on biological systems? Biologists hypothesize that, on the spatial scale of the biosphere, rising temperatures will affect the geographical distributions of populations, species, and communities. Models of climate change predict that the distributions of polar species will contract to even higher latitudes, and the ranges of temperate and tropical species will expand or shift toward the poles. The models also predict that global warming will change the timing of important biological events. For example, plants whose flowering is triggered by warm

springtime temperatures will flower earlier in the season; similarly, migratory animals will return from their wintering grounds and begin reproducing earlier in the year.

Camille Parmesan of the University of Texas at Austin and Gary Yohe of Wesleyan University tested these predictions with a massive literature review. They surveyed studies of changes in the geographical distributions and timing of springtime activities in a wide variety of herbaceous plants, trees, invertebrates, and vertebrates over roughly the past 100 years. Their analysis, published in 2003, suggests that the geographical ranges of 99 species of butterflies, birds, and alpine herbs in the Northern Hemisphere have shifted dramatically into habitats that had previously been too cold for them. Some species have expanded their distributions northward an average of 6.1 km per decade. Other species have shifted their distributions to higher altitude, an average of 6.1 m per decade. Their analysis also indicated that for 172 species of plants, butterflies, amphibians, and birds, springtime growth and reproduction has occurred on average 2.3 days earlier per decade. If these trends continue at the same rate, spring flowering and animal reproduction will occur

Figure 52.9 Observational Research

How Lizards Compensate for Altitudinal Variations in Environmental Temperature

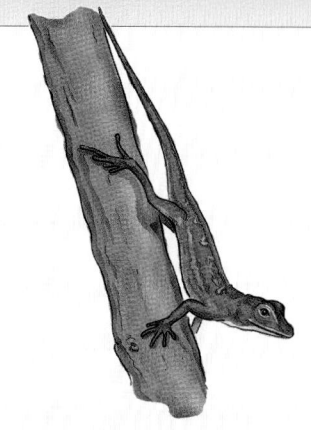

HYPOTHESIS: Lizards living at high altitude can use behaviors to compensate for the low environmental temperatures they encounter.

PREDICTION: The percentage of lizards observed basking in the sun will increase with altitude, and mean air temperatures will vary more than lizard body temperatures among study sites distributed along an altitudinal gradient.

METHOD: Hertz and Huey measured the basking behavior as well as air temperatures and body temperatures of two closely related species of *Anolis* lizards distributed along an altitudinal gradient in the Dominican Republic. They surveyed populations of lizards at sea level, 550 m, 1100 m, and 2200 m altitude. They then compared the percentages of lizards basking and the mean air and lizard body temperatures at the four study sites.

RESULT: The percentage of lizards basking increased steadily with altitude. Mean air temperature differed by as much as 8°C among study sites, but mean body temperature differed by only 2°C.

CONCLUSION: Lizards living at high altitude bask in patches of sun more frequently, partially compensating for the low environmental temperatures in their habitats.

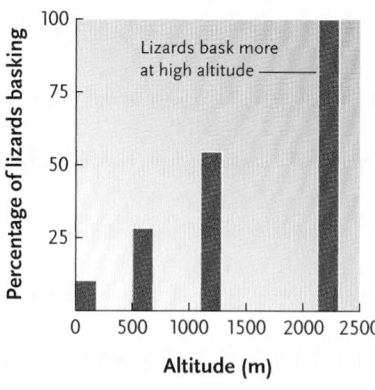

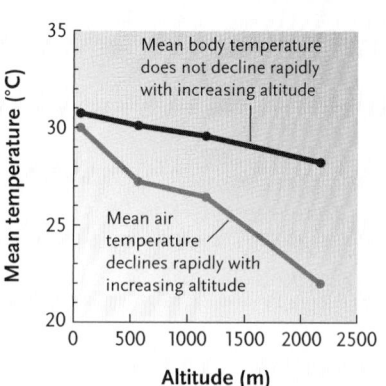

one full month earlier in the year 2130 than it did in 2000.

A parallel analysis of less detailed data on 677 species of plants and animals suggests that 62% of the species surveyed showed trends toward earlier flowering, breeding, or growth. And for 434 species in which researchers documented a change in geographical distribution, 80% of the shifts were in the direction predicted by climate change models. Parmesan and Yohe noted that geographical distributions change rapidly, and that species respond to both cooling and warming trends. Marine species in Europe expanded their ranges northward during two warming periods in the twentieth century (1930–1945 and 1975–1999), but shifted their ranges southward during a cooling period (1950–1970).

Global warming is also changing species composition and relative abundance within ecological communities. For example, among invertebrates and fishes on the California coast, cold-adapted species have become less abundant and warm-adapted species more abundant. Comparable changes have been noted in communities from Antarctica to the Arctic.

The geographical distributions of species and communities have often changed with climate shifts over evolutionary time, but the rate of global warming

has accelerated in your lifetime. As you know from preceding chapters, the factors that govern the structures of communities and ecosystems are complex, and scientists are far from being able to predict all of the consequences of these changes in detail. In the next section, we describe how today's climate affects species and community distributions on a biosphere-wide scale. You can be certain that biology texts in the twenty-second century will paint a very different portrait of these large-scale associations.

STUDY BREAK

1. How does the behavior of *Anolis* lizards in the Dominican Republic change over altitude?
2. What effect is global warming likely to have on the geographical distributions of organisms?

52.3 Terrestrial Biomes

In Section 22.2 we described how convergent evolution produces morphological and physiological similarities in species that occupy similar environments. Early

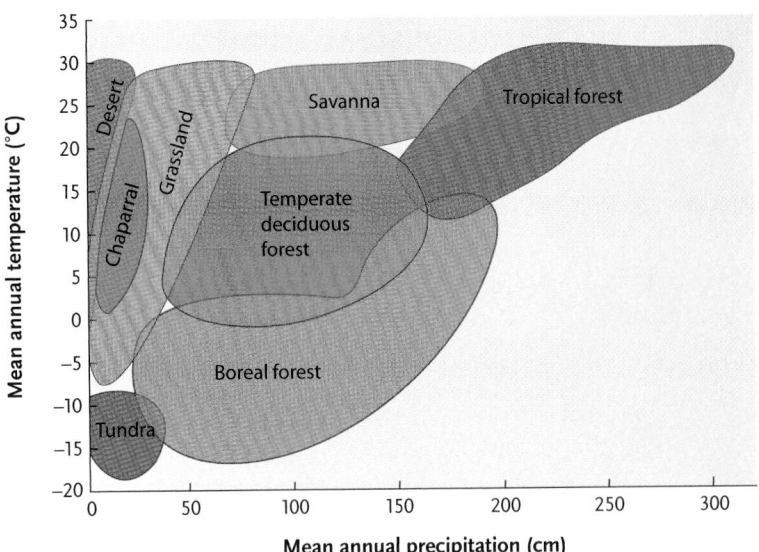

Figure 52.10

Climograph. Each of the major terrestrial biomes occupies a characteristic combination of temperature and moisture conditions.

example, in eastern North America, the temperate deciduous forest biome includes beech-maple forests in the north and oak-hickory forests in the south. Before surveying eight major terrestrial biomes—*tropical forests, savannas, deserts, chaparral, temperate grasslands, temperate deciduous forests, evergreen coniferous forests,* and *tundra*—we consider how environmental factors influence their overall distribution.

Environmental Variation Governs the Distribution of Terrestrial Biomes

Because organisms—and the communities they form—are sensitive to abiotic factors, climate is the main determinant of biome distribution. A **climograph** portrays the particular combination of temperature and rainfall conditions where each terrestrial biome occurs **(Figure 52.10)**. For example, some deserts, grasslands, savannas, and tropical forests occur in areas that have comparable mean annual temperatures but vastly different rainfall. Conversely, some biomes, such as boreal forests, temperate deciduous forests, and savannas, are found under similar moisture conditions but different temperature regimes.

Although the climograph provides a general portrait of the temperature and moisture conditions where the different biomes occur, it does not address the de-

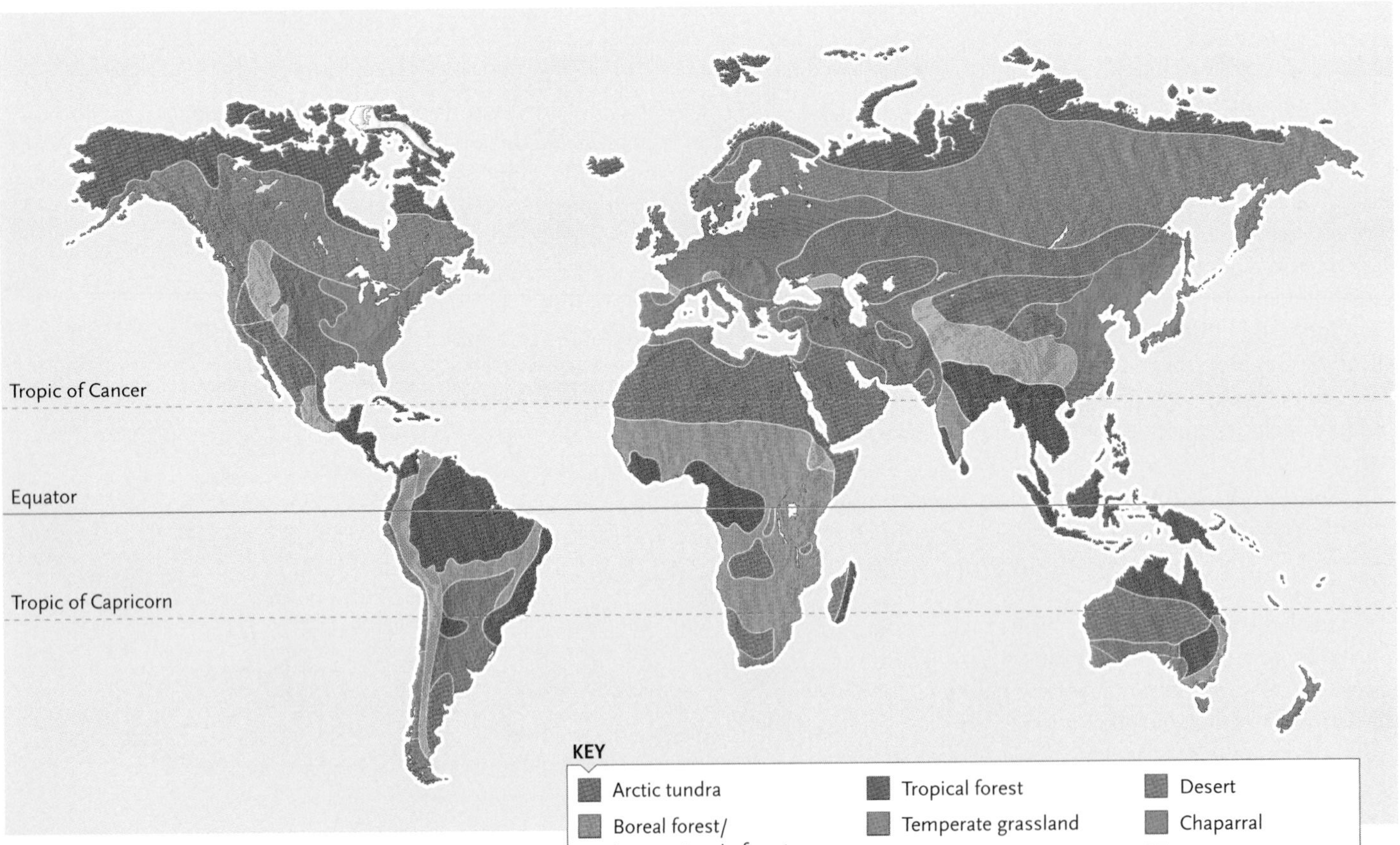

in the twentieth century, two American ecologists, Frederic Clements of the Carnegie Institution in Washington and Victor Shelford of the University of Illinois, generalized this observation within a larger perspective by defining the **biome** as a vegetation type plus its associated microorganisms, fungi, and animals. Although vegetation is superficially similar throughout a biome, its species composition varies from place to place. For

KEY

- Arctic tundra
- Boreal forest/ temperate rain forest
- Temperate deciduous forest
- Tropical forest
- Temperate grassland
- Savanna and thorn forest
- Desert
- Chaparral
- Mountains (complex zonation)

Tropic of Cancer

Equator

Tropic of Capricorn

Figure 52.11

Terrestrial biomes. Climate governs the distributions of the world's major terrestrial biomes.

tails of environmental variation. For example, the climograph includes only mean annual temperature and rainfall, not seasonal variation in these factors. Two regions may have the same mean temperature even though one experiences blazingly hot summers and bitterly cold winters and the other has moderate temperature throughout the year; we would expect them to harbor different organisms. Moreover, the distributions of communities are also influenced by nonclimatic factors, such as regional variations in soil structure and mineral composition (see Section 33.2).

Because temperature and rainfall exhibit latitudinal patterns (displayed in Figures 52.2 and 52.5), the distributions of some terrestrial biomes appear as bands on a world map **(Figure 52.11)**. But regional and local climatic variations influence these broad patterns. For example, chaparral is common in certain coastal habitats, whereas grasslands occur further inland at similar latitudes. Comparable bands of distinct vegetation form on mountainsides because temperature and moisture conditions also change with altitude.

Tropical Forests Include the Most Species-Rich Communities on Earth

Three types of **tropical forests**—rain forest, deciduous forest, and montane forest—sweep across the parts of Africa, Asia, Australia, and Central and South America that receive intense solar radiation and heavy rainfall.

Tropical rain forests grow where some rain falls every month, mean annual rainfall exceeds 250 cm, mean annual temperature is at least 25°C, and humidity is above 80%. Limited by neither temperature nor water, the productivity of a tropical rain forest is exceptionally high (see Section 51.1). Trees replace their leaves throughout the year, producing a continuous rain of detritus that ants, land crabs, and other detritivores quickly consume. Decomposers are also active in the hot, moist environment, and almost no litter accumulates on the ground. Because nutrients released by decomposition are promptly absorbed by vegetation or leached by rain, soil in tropical rain forests is nutrient-poor, with low humus content (see Section 33.2).

Tropical rain forests are usually layered (see Figure 50.19). The crowns of tall trees form a dense, tangled canopy that intercepts most incoming sunlight 40 to 45 m above the ground **(Figure 52.12)**. Even the largest trees grow only shallow roots in the thin soil, but many have wide *buttresses,* woody lateral extensions of their trunks, that stabilize them in the ground. Shade-tolerant shrubs and small trees form understory layers below the canopy. The woody stems of lianas climb through both layers, and epiphytes, such as bromeliads and orchids, cover the trunks and branches of trees, especially in sunlit openings. In mature rain forests, the ground is surprisingly bare of leafy vegetation, because very little sunlight reaches the forest floor.

© 1991 Gary Braasch

Figure 52.12
Tropical forest. Many tropical rain forest trees are covered with lianas and epiphytes.

Tropical rain forests probably harbor more plant and animal species than all other terrestrial biomes combined. Ecologists have proposed numerous hypotheses to explain both the evolution and the maintenance of high species richness in these communities (see Section 50.7), but no single hypothesis explains the pattern adequately. In fact, we do not even have a complete species list for any rain forest community, largely because most animals live in the highly productive canopy, which ecologists have only recently begun to study in detail (see *Focus on Research*). The most extensive tracts of tropical rain forest occur in South America, central and western Africa, and Southeast Asia. Unfortunately, they are being cleared at an alarming rate (see Chapter 53); some experts predict that this biome will all but disappear before the middle of the twenty-first century.

Habitats centered at 20° north and south of the equator experience a pronounced summer rainy season and winter dry season. **Tropical deciduous forests** occur where winter drought reduces photosynthesis, and most trees drop their leaves. For example, the monsoon forests of Southeast Asia, which harbor teak and other tropical hardwoods, are as lush as tropical rain forests in the rainy season; but many trees are bare in the dry season.

High altitudes in the tropics support distinctive **tropical montane forests,** or "cloud forests," which are frequently enveloped in mist. The trees, often no more

Basic Research: Exploring the Rain Forest Canopy

Biological diversity in tropical rain forests has fascinated naturalists for centuries. Sadly, most of its organisms live beyond our reach. The forest canopy extends from 9 or 10 m above the ground to heights as great as 45 m, making the canopy inaccessible and largely unexplored. Early ecologists were able to study canopy-dwelling species only when they found a fallen tree or followed loggers into the forest. In the 1930s, a clever botanist trained monkeys to retrieve plants from the canopy, but these efforts provided little data about the ecological interactions that govern life in the treetops.

Many ecologists still study canopy-dwelling organisms from the safety of the ground. Binoculars provide a good view of fairly large vertebrates. And a hike along a ridge top can provide a canopy-level view of trees growing in an adjacent valley or ravine. Some researchers use ropes to hoist nets or traps into the canopy, lowering them periodically to see what they have caught. Others spray a fog of insecticide into the canopy to kill small invertebrates, which then rain down onto plastic sheets spread below the trees. These ground-based techniques have led to the discovery of hundreds—perhaps thousands—of new arthropod species. Ecologists now collect huge samples of arthropods to study the species composition and structure of communities and to monitor changes in these communities over time. But distant observations and mass sampling techniques don't provide detailed data about which insects are feeding on a tree, how often hummingbirds pollinate a flower, or when a tiny lizard hunts its prey.

Today many ecologists routinely risk life and limb to collect detailed ecological data in the rain forest canopy. They climb trees and crawl along stout branches. Many build stable observation decks with walkways, allowing study on either side of the "trail."

What does this newfound access to the rain forest canopy add to our knowledge of organisms that live there? Researchers can measure the physical environment of the canopy and observe the physiological and behavioral adaptations of its plants and animals. For example, researchers are gathering data on the feeding habits and behavior of small animals that never venture to the ground, such as fruit-eating bats and birds. When coupled with information about the movement patterns of these animals, the data provide insight into the dispersal of seeds in the fruits. And an understanding of seed dispersal provides information for studies of the population ecology of rain forest trees.

Canopy ecologists have also discovered fascinating relationships between plants and their animal pollinators. For example, Donald Perry, a freelance biologist, discovered that birds are attracted to the sweet nectar of the vine *Norantea sessilis*. Feeding birds step on the vine's sturdy flowers; their feet become covered with the plant's pollen, which is embedded in a gummy substance. When the birds visit another vine of the same species, they transfer the pollen to that plant's flowers, providing cross-pollination, which appears to be necessary for the vine's reproduction.

Research in the tropical rain forest canopy promises exciting discoveries about ecological relationships in this unique biome, which is the most threatened on Earth (see Chapter 53). Such research is essential for developing a public appreciation of tropical forests and for creating conservation plans to preserve them.

A platform in the canopy of a tropical rain forest in Costa Rica provides a comfortable perch for Donald Perry to survey the pollinating activity of birds and bees.

than 3 m tall, are densely covered with epiphytes, which thrive in the moisture-laden air. Cloud-forest plants grow slowly because productivity is limited by low temperatures, high humidity (making transpiration difficult), and sunlight-blocking clouds.

Savannas Grow Where Moderate Rainfall Is Highly Seasonal

Grasslands with few trees, the biome called **savanna**, grow in areas adjacent to tropical deciduous forests **(Figure 52.13).** Seasonality in tropical and subtropical savannas is determined by the availability of water; although annual rainfall averages 90 to 150 cm, droughts typically last for months. Grasses are successful in semiarid conditions because their shallow roots harvest water efficiently. With the onset of seasonal rains, they grow quickly, reaching a height of 2 to 3 m. During the dry season, grasses die back and frequently burn, but their underground parts remain alive and resprout when water again becomes available. Shrubby trees outcompete grasses in moist, low-lying areas or on rocky ground, but periodic fires and grazing mammals eliminate most trees as seedlings.

The largest savannas stretch across eastern and southern Africa; smaller patches occur in India, Aus-

Figure 52.13

Savanna. The African savanna, a warm grassland with scattered stands of shrubby trees, has an enormous concentration of large ungulates (hoofed, herbivorous mammals), such as these wildebeests *(Connochaetes taurinus)*.

tralia, and South America. African savannas are home to large herbivorous mammals, including antelopes, zebras, giraffes, and elephants, some of which fall prey to savanna predators, such as lions, leopards, cheetahs, and wild dogs. Grazing mammals follow the seasonal cycle of grasses, migrating away from dry areas to greener pastures.

Thorn forests grow at the arid borders of true savanna, where large mammals are less abundant. Grasses and other plants that store energy in large underground root systems grow among scrubby trees. Thorn forests are also highly seasonal, growing dramatically in the rainy season and dying back during the annual dry season, which may last for 8 to 9 months.

Deserts Develop in Places Where Little Precipitation Falls

Deserts form where rainfall averages less than 25 cm per year. The hot deserts of the American Southwest, northern Chile, Australia, northern and southern Africa, and Arabia occur near 30° latitude, where descending air masses create very dry conditions. Cool deserts, such as the Gobi and Kyzyl-Kum of Asia and the Great Basin of North America, form in massive rain shadows at higher latitudes.

Desert conditions are often extreme. Rainfall arrives infrequently in heavy, brief pulses; and sudden runoff erodes topsoil, which often has high mineral content but little organic matter. Dry air and scant cloud cover allow most sunlight to reach the ground, raising daytime air and ground temperatures as high as 45°C and 70°C, respectively. At night, the surface loses heat quickly; in some deserts, temperatures drop below freezing in winter.

Desert vegetation is always sparse because arid environments do not favor large, leafy plants. Some deserts, such as the Namib of Africa and the Atacama-Sechura of South America, receive so little rainfall that

large areas are practically devoid of vegetation. By contrast, the hot Sonoran Desert in northern Mexico, southeastern California, and southern Arizona harbors a diverse flora, including deep-rooted shrubs and shallow-rooted cacti **(Figure 52.14)**. Mesquite and cottonwood trees grow deep taproots into the permanent water supply below streambeds. Perennial plants often protect their tissues from herbivores with spines or toxic chemicals, and many use CAM photosynthesis to conserve water (see Section 9.4). After seasonal rains, annual plants germinate, mature, flower, and produce seeds before brutally dry conditions resume.

Deserts also support abundant animals, most of them fairly small. Ants, birds, and rodents often subsist on seeds. Some seed-eating mammals survive on the water they extract from food. Insects, some lizards, and mammals consume the sparse vegetation. Scorpions, lizards, and birds feed primarily on insects; snakes, owls, and foxes prey on other animals. Most

Figure 52.14

Desert. The warm Sonoran Desert near Tucson, Arizona, is home to columnar saguaro cacti *(Carnegiea gigantea)* and other drought-adapted plants.

Figure 52.15
Chaparral. Chaparral covers broad expanses of hills in coastal areas of central and southern California.

central and southern California, central Chile, southwestern Australia, southern Africa, and the Mediterranean region.

Chaparral shrubs are dense, with hard, tough, evergreen leaves **(Figure 52.15)**. They build woody stems above ground and large root systems in the soil. Many species, such as sages (genus *Salvia*), produce toxic, aromatic compounds that inhibit the germination and growth of potential competitors. Just after the winter rains, the shrubs are covered with new leaves and flowers, and the vegetation teems with insects and breeding birds. During the hot, dry summers, however, most plants are dormant, and lightning sparks frequent fires. The aromatic oils and resins of many species, such as eucalyptus, make them highly flammable. Their aboveground parts burn swiftly, but they quickly resprout from large root crowns. Other species release seeds from fire-resistant cones or pods, and their seedlings grow in ash-enriched soil.

Temperate Grasslands Are Held in a Disclimax State by Periodic Disturbance

Temperate grasslands include the prairies of North America, the steppes of central Asia, the pampas of South America, and the veldt of southern Africa **(Figure 52.16)**. They stretch across the interiors of continents, where winters are cold and snowy and summers are warm and fairly dry. Only 25 to 100 cm of rain falls unevenly through the year. Temperate grasslands are disclimax communities: seasonal drought, periodic fires, and grazing by mammals inhibit succession, preventing shrubs and trees from displacing perennial grasses and herbaceous plants (see Section 50.6). Grassland soil is rich in organic matter because the aboveground parts of most plants die and decompose annually.

desert animals avoid the midday heat and dehydrating conditions; many retreat into underground burrows, where water vapor from their respiration cools and moistens the air. Many species are nocturnal or active only in the early morning and late afternoon.

Chaparral Grows Where Winters Are Cool and Wet and Summers Are Hot and Dry

A scrubby mix of short trees and low shrubs called **chaparral** dominates narrow sections of coastal land between 30° and 40° latitude, where winters are cool and wet and summers hot and dry. Seasonal rainfall averages only 25 to 60 cm per year. Chaparral occurs in

a. Shortgrass prairie

b. Tallgrass prairie

Figure 52.16
Temperate grassland. **(a)** The western plains of North America were once covered with shortgrass prairie, as shown here east of the Rocky Mountains. Bison were the dominant large herbivores. **(b)** Tallgrass prairie, like this lush patch in eastern Kansas, once covered the eastern plains.

In North America (see Figure 52.16a) shortgrass prairie covers much of the west, where winds are strong, rainfall light and infrequent, and evaporation rapid. Drought-tolerant perennials have deep roots, and their underground rhizomes, which store energy, resprout quickly after a fire. Tallgrass prairie (see Figure 52.16b) once occupied moister regions to the east of the shortgrass prairie. It boasted an abundance of legumes and sunflowers, often 3 m tall, but most of it was converted to farmland long ago; small patches still exist in nature preserves and in glades within eastern deciduous forests.

North American grasslands are still occupied by large grazing mammals, including pronghorns and bison, which once numbered in the millions. The most familiar burrowing mammal is the prairie dog, a rodent, but pocket gophers, ground squirrels, and jackrabbits are also common. Wolves were the primary large predators until they were hunted nearly to extinction. Coyotes, foxes, ferrets, hawks, and owls still take small prey today.

Temperate Deciduous Forests Experience Seasonal Dormancy

At temperate latitudes, with warm summers, cold winters, and annual precipitation between 75 and 250 cm, **temperate deciduous forests** grow at low to middle altitudes. In winter, low temperatures reduce photosynthetic rates, and snow and ice can damage leaves. Thus, most plants shed their leaves and grow new ones in spring **(Figure 52.17)**. The thick layer of leaf litter, which releases mineral nutrients as it decomposes, enriches the soil. Decomposition is slow, however, because the growing season is only about 7 months long.

Temperate deciduous forests have much lower species richness than tropical forests. Trees form a canopy 10 to 35 m high, and woody shrubs form an understory below it. Herbaceous plants and a ground layer of mosses or liverworts grow below the shrubs. Many herbaceous plants, including some terrestrial orchids, flower early in spring, before trees produce sunlight-blocking leaves; others flower near the end of the growing season.

Forests of ash, beech, birch, chestnut, elm, and oak stretched unbroken across eastern North America, Europe, and eastern Asia before farmers cleared the land. In North America, introduced diseases and insects have nearly eliminated the once dominant species, such as American chestnut and American elm. Today, beech, birch, and maple predominate in the Northeast; oak–hickory forests dominate farther south and west; and oak woodlands merge into tallgrass prairie to the west. Before the arrival of Europeans, deer, bison, bears, and pumas roamed the forests with many smaller species of animals. Today, small mammals such as voles, mice, chipmunks, squirrels, rabbits,

Summer   Winter

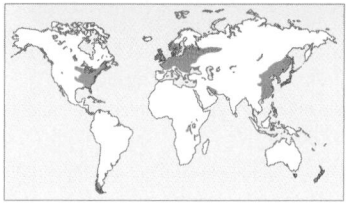

Figure 52.17
Temperate deciduous forest. Seasonal variations in temperature and water change the character of this forest south of Nashville, Tennessee.

opossums, and raccoons predominate, although deer and bears have recently surged in abundance.

Evergreen Coniferous Forests Predominate at High Northern Latitudes

The **boreal forest**, or **taiga** (Russian for "swamp forest"), is a circumpolar expanse of evergreen coniferous trees in Europe, Asia, and North America **(Figure 52.18)**. Snow blankets the ground during long and extremely cold winters, and most precipitation falls during the short summer. In the northernmost taiga, plants grow quickly during long (18-hour) summer days.

Stands of white spruce and balsam fir dominate North America's boreal forest. Their needle-shaped leaves have a thick cuticle and recessed stomata that conserve water during winter, when ground water is frozen. Fallen needles acidify the thin soil, which speeds the leaching of most nutrients, and few shrubs and herbaceous plants grow beneath the conifers. Lightning-sparked fires are common; some deciduous trees grow in areas opened by fire, but conifers eventually replace them. Cold streams, marshes, ponds, and lakes often dot the landscape; at flat, poorly drained sites, peat mosses, shrubs, and stunted trees dominate acidic bogs, called muskegs.

Most taiga is relatively undisturbed by humans, and it still harbors its native animals. Moose, elk, and deer are the dominant large herbivores. Hare as well as squirrels, porcupines, and other rodents also feed on plants. Some small animals are active all winter in runways they dig beneath the snow. Wolves, lynx, and

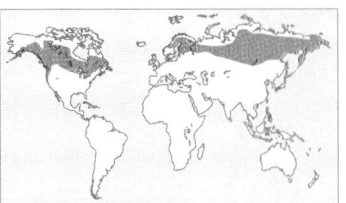

Figure 52.18
Boreal forest. Single-species stands of spruce dominate this boreal forest, the predominant forest at high latitudes in the Northern Hemisphere.

wolverines prey on herbivores. Grizzly bears and black bears roam the forest, devouring seeds, berries, fishes, and small animals. Mosquitoes, black flies, and gnats are superabundant near bogs and lakes in summer.

Other types of coniferous forest grow in more southerly coastal lowlands where winters are mild and wet and the summers are cool. For example, a **temperate rain forest**, supported by heavy rain and fog, parallels the coast from Alaska into northern California. In western Washington State, the rain forest

on the Olympic Peninsula receives 500 cm of rainfall per year, as much as some tropical forests. This temperate rain forest harbors some of the world's tallest trees, including Douglas fir and Sitka spruce to the north and coast redwoods to the south.

Tundra Comprises a Vast, Treeless Plain in the Northernmost Habitats

The treeless **arctic tundra** stretches from the boreal forests to the polar ice cap in Europe, Asia, and North America. Covering almost 5% of the land, this biome is windswept and wet. Winter temperatures are consistently below freezing. The 2-month summer is so cool that only the topmost layer of soil ever thaws, leaving the ground below perpetually frozen; in some areas, this **permafrost** is more than 500 m thick. Although less than 25 cm of precipitation falls each year, evaporation is slow, and permafrost is impermeable; thus, low-lying soil remains permanently waterlogged, forming bogs **(Figure 52.19a)**. Anaerobic conditions and low temperatures retard decomposition, and soggy masses of detritus accumulate.

Plants in the tundra are short because the weak sunlight and minimal growing season provide barely enough energy and warmth for net primary productivity; moreover, strong winter winds shred any plants with a high profile. The vegetation consists of low-growing lichens, mosses, grasses, perennial herbs, dwarf shrubs, and a few stunted trees, usually less than 1 m tall. During summer's nearly continuous sunlight, plants flower profusely, and their fruits ripen fast.

Some animals, including herbivorous arctic hares, lemmings, and willow ptarmigans as well as predatory

a. Arctic tundra

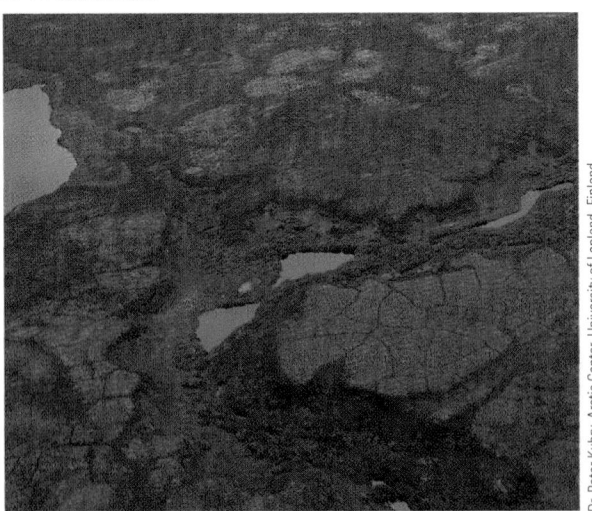

b. Alpine tundra

Figure 52.19
Tundra. **(a)** Rain and snowmelt cannot percolate through the arctic tundra's permafrost. In summer water accumulates in ponds and bogs as shown in this aerial photograph of the tundra in northern Russia. **(b)** Compact, short plants form the alpine tundra, which grows on mountaintops at temperate latitudes, such as the Cascade Range of Washington state.

snowy owls, wolves, foxes, and lynx, are permanent tundra residents. In summer, herds of herbivorous musk oxen, caribou, and reindeer migrate there from boreal forests, and migratory shorebirds and waterfowl arrive to breed. Flying insects abound in summer, especially mosquitoes and black flies, which reproduce in boggy habitats.

A similar biome, called **alpine tundra**, occurs on high mountaintops throughout the world **(Figure 52.19b)**. Dominant plants form cushions and mats that withstand the buffeting of strong winds. Winter temperatures are well below freezing, and shaded patches of snow persist even in summer. The thin, fast-draining soil is nutrient-poor, and primary productivity is low.

STUDY BREAK

1. Which terrestrial biomes occur in habitats that receive the greatest amount of rainfall?
2. Which terrestrial biomes are renewed by periodic fires?
3. Which terrestrial biomes have the tallest vegetation? Which ones have the shortest?
4. In which terrestrial biomes are the trees usually evergreen?

52.4 Freshwater Biomes

Aquatic biomes comprise several distinctive habitats in either freshwater or marine environments. Freshwater biomes occur where water with a salt concentration below 0.5% accumulates or moves through a landscape. Ecologists distinguish between *lotic* biomes, where water flows through channels, and *lentic* biomes, where water stands in an open basin. All freshwater biomes interact with surrounding land, because runoff carries a nearly constant input of nutrients. Highly productive ecotones, called **wetlands**, often define the borders of freshwater biomes. These marshes and swamps may harbor an astounding array of microorganisms, algae, plants, invertebrates, and vertebrates.

Streams and Rivers Carry Water Downhill to a Lake or the Sea

The flowing-water biomes start as seeps on high ground. As they flow downhill, they grow into narrow streams, which merge to form wide rivers **(Figure 52.20)**. Streams and rivers include three habitats. *Riffles* are shallow, fast-moving, turbulent stretches over a rough bottom of pebbles or rocks. *Pools* are deep, slow-moving areas with a smooth sand or mud bottom. *Runs* are deep, fast-moving stretches over smooth bedrock or sand. Streams generally have high flow rate, low volume, and lots of riffles and pools. As they merge into rivers, flow rate declines, but flow volume increases, and runs and pools predominate. Flow rate and volume also vary seasonally with the rate of water input from rainfall and snowmelt and geographically with altitude and topography.

Physical factors change over the length of a flowing-water system. The concentration of suspended particulate material is low in streams, but high in rivers, which are often turbid with silt. Temperature also increases as water flows downstream to warmer lowland habitats. Because oxygen is more soluble in cold water than in warm water, dissolved oxygen is usually higher in streams than in rivers. Erosion of the streambed and surrounding land provides the solute content of flowing water. Today, agricultural runoff and industrial and municipal wastes provide major input. In unpolluted streams, organic detritus provides more than 95% of the nutrients and energy entering aquatic food webs. This input is particularly important in streams flowing through dense forests, where vegetation blocks the sunlight necessary for primary productivity.

The flow of water affects every aspect of life in streams and rivers. In swift-moving riffles, primary producers cling permanently to fixed substrates, because phytoplankton are swept away by the current. Insect larvae and other invertebrates attach to the un-

a. A stream

b. A river

Figure 52.20
Stream and river habitats. **(a)** In streams, such as this one in Virginia, water flows quickly through narrow channels, often with a rocky bottom. **(b)** In rivers, like the Rio Napo in Ecuador, water flows more slowly through broad channels, and suspended sediments often make the water murky.

Figure 52.21

Lakes. A lake in Torres del Paine National Park, a biosphere reserve in Chile.

dersides of rocks, and many species are flattened, maintaining a low profile in the current. By contrast, large rivers have dense populations of algae and cyanobacteria, which attach to rocks and other substrates, and rooted aquatic plants at the river's edge.

Lakes Are Bodies of Standing Water That Accumulates in Basins

Lakes and other standing-water biomes are generally fed by rainfall and by streams and rivers that drain surrounding watersheds **(Figure 52.21).** Because the availability of light affects a lake's primary productivity, ecologists often distinguish between the **photic zone** of a lake, the surface water that sunlight penetrates, and the deeper **aphotic zone**, which is always dark.

Lake Zonation. Every lake includes zones, defined by depth and distance from the shore, that provide distinctive environments **(Figure 52.22).** In the **littoral zone,** the shallow water near the shore, sunlight penetrates to the bottom. Enriched by nutrients made available by decomposers and runoff, the littoral zone has high productivity and species richness. Rooted aquatic plants, such as cattails and water lilies, grow above the surface, and "floating aquatics," such as duckweed, are common. Submerged vegetation harbors a rich community of microorganisms, epiphytes, and invertebrates. Numerous animals—insects, worms, snails, crayfish, fishes, frogs, turtles, and water birds—use the littoral zone to feed and reproduce.

The **limnetic zone,** the sunlit water beyond the littoral, supports plankton communities: the primary producers are phytoplankton—cyanobacteria, diatoms, and green algae; the primary consumers are zooplankton—rotifers, copepods, and other tiny heterotrophs. Small fishes, which feed on plankton, are themselves consumed by larger fishes, such as bass.

Photosynthesis is impossible, however, in the **profundal zone**, the perpetually dark water below the limnetic zone. Nevertheless, a constant rain of detritus from the limnetic zone supports a community of bacterial decomposers and animal detritivores, including worms, clams, insect larvae, and catfish.

Seasonal Changes in Temperate Lakes. In temperate areas, seasonal temperature variations induce changes in the vertical zonation of lakes **(Figure 52.23).** Like other liquids, water gets denser as it cools. But water has a unique property: it reaches maximum density at 4°C, with the density declining as it gets colder. Thus, water at 4°C sinks below water that is either warmer or colder; ice floats because it is less dense than very cold water.

During winter, ice forms on the surface of temperate zone lakes. Water temperature varies from near freezing just below the ice to 4°C at the bottom. Differences in the density of water at 0° and 4°C maintain this thermal stratification. In spring, as the ice melts, the warmer, denser water sinks; and the surface temperature gradually rises to 4°C. For a brief time, the temperature is uniform at all depths. Winds blowing across the lake create vertical currents that cause a **spring overturn**, mixing surface water with deep water. Oxygen at the surface moves to the bottom, and nutrients from the bottom move to the surface.

By midsummer, sunlight heats the top layer of the limnetic zone, called the **epilimnion**, to temperatures above 4°C. In large lakes, the epilimnion may be more than 10 m deep. In the deep water of the lake's profundal zone, called the **hypolimnion**, the temperature remains near 4°C. However, at the boundary between the epilimnion and the hypolimnion, water temperature changes abruptly over a narrow depth range, called the **thermocline.** The thermocline prevents vertical mixing because warm surface water floats above the thermocline, and cool deep water stays below it. During summer, nutrient-rich detritus sinks to the bottom of the lake, where decomposition depletes the oxygen dissolved in the hypolimnion. In autumn, declining sunlight and winds cause the epilimnion to cool, and as the water becomes denser, it sinks, eliminating the thermocline. Winds then mix the water vertically once again during an **autumn overturn,** and dissolved gases and nutrients are equalized at all depths.

Primary productivity in the limnetic zone varies with the seasonal overturns. In spring, increased sunlight, warm temperatures, and the sudden

Figure 52.22

Lake zonation. The zonation in a lake is based upon the water's depth and its distance from shore.

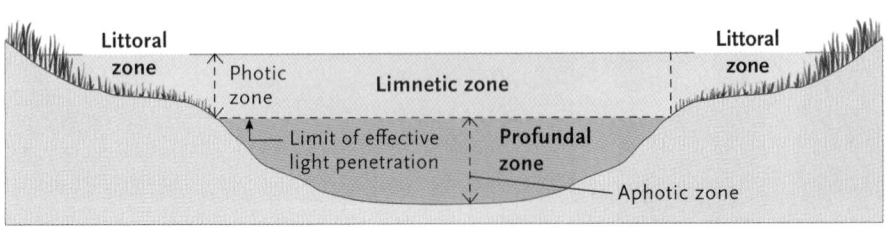

availability of nutrients induce a bloom of productivity. As the season progresses and the thermocline prevents vertical mixing, nutrient levels dwindle in the epilimnion, and primary productivity declines. By late summer, nutrient shortages limit photosynthesis. After the autumn overturn, nutrient cycling drives a short burst of primary productivity. But as days get shorter and temperature declines, primary productivity remains low until spring.

Trophic Nature of Lakes. Ecologists classify lakes by their nutrient content and rates of productivity. **Oligotrophic lakes** are poor in nutrients and organic matter, but rich in oxygen. Their low primary productivity keeps the water crystal clear, making them popular recreational sites. By contrast, **eutrophic lakes** are rich in nutrients and organic matter. The decomposition of organic matter depletes oxygen in the hypolimnion when the lake is stratified, and high primary productivity in the epilimnion often chokes the water with seasonal blooms of cyanobacteria and filamentous algae. Eutrophic lakes are often thick and "soupy," making them unattractive for recreation. Over long periods of time, as sediments accumulate, lakes naturally change from oligotrophic to eutrophic; their basins eventually fill with sediments, and terrestrial plants invade.

As you learned in the description of changes in Lake Erie at the beginning of Chapter 51, the addition of nutrients to a lake often disrupts its trophic condition. In a classic experiment conducted in the late 1960s, David Schindler and his colleagues at The Experimental Lakes Project in Ontario, Canada, experimentally separated the two basins of a lake with a plastic curtain. The researchers added phosphates to one basin and used the other basin as a control. Within 2 months, the artificially enriched basin sported a bloom of cyanobacteria, a sign of eutrophication; the control basin remained oligotrophic and crystal clear **(Figure 52.24).**

STUDY BREAK

1. How does the availability of dissolved oxygen vary from the headwaters of a stream to the mouth of a river?
2. What factors cause the seasonal overturns in lakes?
3. Why are oligotrophic lakes better for recreational purposes than eutrophic lakes?

52.5 Marine Biomes

Marine biomes, in which salinity (salt concentration) averages about 3%, cover nearly three-fourths of Earth's surface and account for a large fraction of its primary

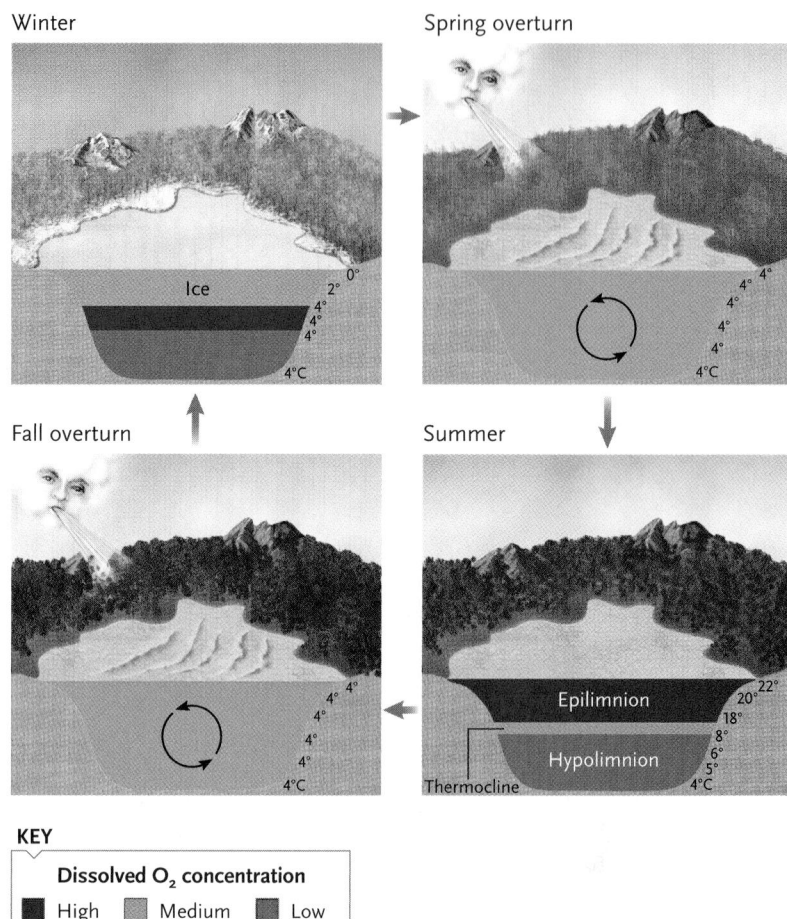

Figure 52.23

Seasonal overturns in lakes. The waters of shallow temperate-zone lakes mix twice each year. During the spring and autumn overturns, temperature is equalized at all depths; nutrients are carried upward from the bottom; and oxygen is carried downward from the surface.

productivity. They also mediate important global processes: marine phytoplankton process large amounts of carbon dioxide, generating oxygen and moderating the greenhouse effect.

As with standing freshwater biomes, depth and distance from shore govern the physical characteristics of marine habitats. Ecologists describe ocean zonation in several ways **(Figure 52.25),** including the distinction between the photic and aphotic zones. Another major distinction is between the **pelagic province,** the water, and the **benthic province,** the bottom sediments. The pelagic province includes the **neritic zone,** the shallow water above the continental shelves, and the **oceanic zone,** the deep water beyond them. The benthic province is divided into the **intertidal zone,** the shoreline that is alternately submerged and exposed by tides, and the **abyssal zone,** the bottom sediments that lie permanently below deeper water. Here we describe five marine biomes—*estuaries, rocky and sandy coasts, continental shelves and oceanic banks, open ocean,* and *benthic regions*—that represent particular associations of organisms occupying different marine zones and provinces.

Figure 52.24 Experimental Research

Artificial Eutrophication of a Lake

QUESTION: Does the addition of excess phosphorus to a lake encourage the growth of primary producers, such as cyanobacteria?

EXPERIMENT: Schindler and his colleagues experimentally separated the two basins of a lake in Ontario, Canada, with a plastic curtain. The researchers added phosphates to one basin and used the other basin as a control.

RESULTS: Within 2 months, the artificially enriched basin (in the upper left of the photo) sported a pale green bloom of cyanobacteria, a sure sign of eutrophication; the control basin remained oligotrophic and crystal clear.

D. W. Schindler, *Science*, 897–899

CONCLUSION: The addition of excess phosphorus to a lake encourages blooms of cyanobacteria, causing the lake to change from oligotrophic to eutrophic.

Estuaries Form Where Rivers Meet the Sea

Estuaries are coastal regions where seawater mixes with fresh water from rivers, streams, and runoff **(Figure 52.26).** Salinity is low where fresh water enters the estuary and high on the tidal side. After heavy rainfall, fresh water floods into the habitat, reducing salinity and raising water temperature. At high tide, cold, salty water flows in from the sea. All estuarine organisms must tolerate these variable conditions.

Variations in local topography influence an estuary's physical features. Chesapeake Bay in Maryland, Mobile Bay in Alabama, and San Francisco Bay in California are broad, shallow estuaries. The estuaries in Alaska and British Columbia are narrow and deep, as are Norway's fjords. Many estuaries are bordered by **salt marshes,** tidal wetlands dominated by emergent grasses and reeds (see Figure 52.26a). In tropical estuaries, the roots of densely packed mangrove trees penetrate the muddy bottom, accumulating sediments and slowly adding land to the shoreline (see Figure 52.26b).

The constant input of nutrients and removal of wastes by the tides contribute to exceptionally high productivity in estuaries. Primary producers include phytoplankton, salt-tolerant grasses and reeds that can withstand submergence at high tide, and algae that grow in mud and on plant surfaces. Roots and stems trap organic matter, which enters detrital food webs. The detritus (and bacteria clinging to it) supports nematodes, snails, crabs, and fishes; suspension-feeding mollusks and arthropods capture edible particles in the slowly moving water. Many marine arthropods and fishes breed in calm, shallow estuaries, where their young find abundant food and refuge from predators in the complex vegetation. Migratory birds use estuaries as rest stops, and shore birds and waterfowl use their muddy bottoms as rich feeding grounds, particularly at low tide.

Rocky and Sandy Coasts Experience Cyclic Periods of Exposure and Submergence

The intertidal zone, the area between low and high tide marks, is one of the most stressful habitats on Earth. On rocky shores, residents are battered by waves and floating debris. Sessile species, such as mussels and barnacles, attach to substrates with special structures or cement. Motile species, such as limpets and sea stars, simply hang onto rocks. Organisms that live high on the shore dry out at low tide, freeze in winter, and bake in summer. Exposed animals often seal themselves inside shells, and intertidal algae have thick polysaccharide coats that adsorb water and prevent dehydration.

Biotic interactions also take their toll. Organisms throughout the intertidal zone compete for attachment sites to avoid being washed away (see Figure 50.12). At low tide, predatory birds and mammals attack from above; at high tide, predatory fishes move in from the sea. Because the tides often scour detritus from the rocky intertidal, grazing food webs predominate.

Rocky shores often have three zones **(Figure 52.27).** The *upper intertidal* is submerged only during the highest tide of the lunar cycle. It is sparsely populated by barnacles, sturdy algae, and grazing and predatory snails. The *middle intertidal* is submerged daily during the highest regular tide and exposed during the lowest. Its tide pools are occupied by red, brown, and green algae, grazing and predatory mollusks, sponges, sea anemones, worms from several phyla, hermit crabs, echinoderms, and small fishes. Biodiversity is greatest in the *lower intertidal,* which is exposed only during the lowest tide of the lunar cycle. It is occupied by dense beds of algae, tunicates, echinoderms, other invertebrates, and fishes.

Sandy shores are composed of loose sediments that waves and currents constantly rearrange. Large plants cannot grow on such unstable substrates, so grazing food webs are rare. Organic debris imported from offshore or from nearby land supports detrital food webs. Animals live in burrows, which they must frequently repair as the substrate shifts. Crabs and shorebirds live as scavengers or predators above the high tide mark. At night, beach hoppers and ghost crabs leave their burrows, seeking food. Marine worms, clams, crabs, and other invertebrates live in the sand between the high and low tide marks.

Light Penetrates the Shallow Water over Continental Shelves and Oceanic Banks

The neritic zone includes the shallow water over continental shelves and oceanic banks, underwater landmasses that rise to within 300 m of the surface. Although small in area, the neritic zone is highly productive and species-rich **(Figure 52.28).** Runoff from the land brings a steady inflow of nutrients; and upwelling and waves circulate nutrients from the bottom to the photic zone.

In temperate regions, giant kelp forests, which are among Earth's most productive ecosystems, occupy some continental shelves and banks (see Figure 52.28a). Kelp are enormous algae that attach to the bottom with giant holdfasts; their stipes ("stems") reach upward with fronds fanning out into the water. Sea anemones, snails, echinoderms, lobsters, and other invertebrates live in the kelp, where fishes and other predators consume them. Even where kelp does not grow, continental shelves and banks teem with life. Most of the important fisheries in the temperate zone occur there.

In the tropics, the warm but nutrient-poor water above continental shelves is often occupied by **coral reefs** (see Figure 52.28b). Sunlight penetrates the clear water all the way to the bottom. Photosynthetic dinoflagellates, living as endosymbionts of the coral animals (see Section 26.2), and coralline algae are largely responsible for primary productivity. Coral animals also feed on microscopic organisms and suspended particles. The reefs are the remains of corals, algae, and

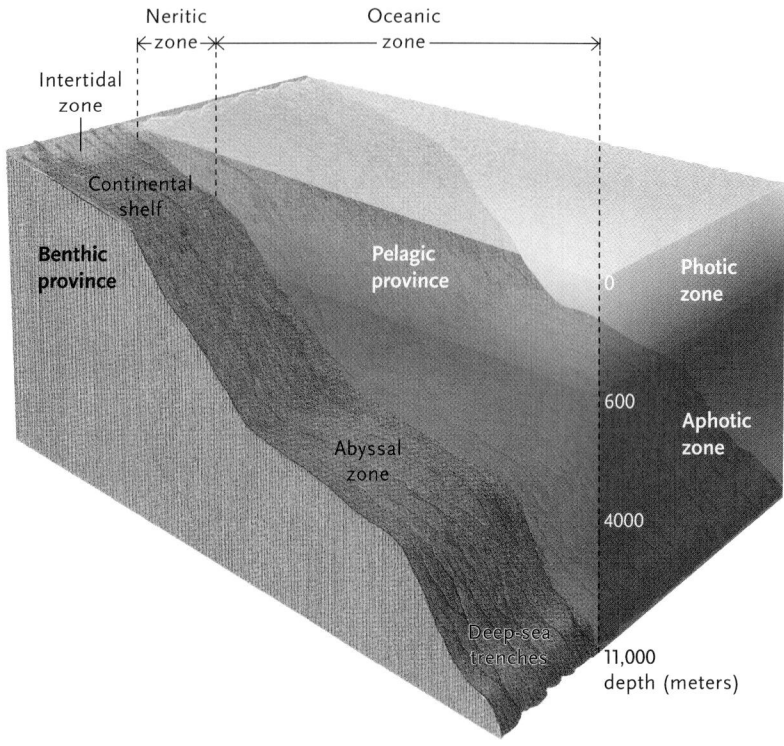

Figure 52.25

Oceanic zonation. Ecologists divide the ocean into the pelagic province (the water) and the benthic province (the ocean bottom). Zones are defined according to the depth of water (photic versus aphotic zones) and distance from shore (neritic versus oceanic zones in the pelagic province, intertidal versus abyssal zones in the benthic province). The different zones are not drawn to scale.

other organisms, and their structural complexity rivals that of tropical rain forests. Tides and currents carve ledges and caverns; and storms frequently disturb the reefs, creating openings in which new coral colonies can grow (see Section 50.5). A reef may be festooned with as many as 750 species of corals and a dizzying variety of algae. The diversity of coral skeletons provides a complex structure that is used by invertebrates from nearly every phylum and by a host of herbivorous and carnivorous fishes.

a. Salt marsh grasses

b. Mangroves

Figure 52.26

Estuaries. **(a)** The salt marsh grass (*Spartina* species) is the major producer in a South Carolina estuary. **(b)** Red mangroves (*Rhizophora mangle*) are abundant in Florida's Everglades.

Upper intertidal

Middle intertidal

Lower intertidal

Figure 52.27
Vertical zonation in the intertidal. A rocky shore in the Pacific Northwest clearly exhibits the characteristic vertical zonation. The distance between low and high tide marks on this rocky shore is about 3 m.

In the Open Ocean, Photosynthesis Occurs Only in the Upper Layers

The oceanic zone lies beyond the continental shelves. Though generally low in nutrients, it is locally enriched by runoff from land and by upwelling bottom waters. The open ocean is typically cold, except in the tropics. The surface water is illuminated by sunlight, which warms it somewhat and allows photosynthesis. Most primary productivity is restricted to a depth of about 50 m, however, because seawater filters light. Photosynthetic activity varies seasonally, as it does on land.

"Pastures" of phytoplankton are eaten by zooplankton, including copepods, shrimplike krill, small worms, cnidarians, and the larvae of invertebrates and fishes. Consumers that can actively swim against the currents, such as squids, fishes, marine turtles, and whales, are called **nekton (Figure 52.29).** Some consumers feed on plankton, and some prey on other nekton. Low light levels in water between about 50 and 600 m allow little photosynthesis, but many fishes and some mobile invertebrates are active at these depths, traveling into the sunlit zone to feed on organisms near the surface.

No sunlight ever penetrates the deepest part of the oceanic zone below 600 m. Some of these abyssal regions, such as the Marianas Trench, are more than 9 km below the surface. Scientists have explored the deepest water in the ocean only during the past few decades, but we know that it is a cold (2° to 3°C), dark environment, where organisms live under tremendous pressure from the ocean above. Abyssal communities are surprisingly diverse, although population densities tend to be low. The denizens of the abyssal zone include invertebrates, bony fishes, and sharks. Some fishes and invertebrates are bioluminescent, producing spots of light that may serve for communication or as lures to entice prey within reach of their large jaws **(Figure 52.30).**

The Benthic Province Includes the Rocks and Sediments of the Ocean Bottom

The benthic province extends from the intertidal zone to the deep-sea trenches. In the oceanic zone, bottom sediments are composed of soft mud, fine particles of silt, detritus, and the shells of dead microscopic organ-

a. Kelp forest **b.** Coral reef

Figure 52.28
Neritic zone. (a) Kelp forests, such as this one off the coast of California, often grow in the neritic zone along the coast at temperate latitudes. **(b)** This coral reef in the Raja Ampat islands of Indonesia illustrates the structural complexity and biological diversity found in reef communities.

isms. Species living in and on the bottom are collectively called **benthos.** Sunlight never strikes the benthic province of the open ocean, which is inhabited by bacteria, fungi, and a variety of animals. Sessile invertebrates, such as sponges, sea anemones, and clams, live amidst the sediments, and many motile animals, including worms, mollusks, crustaceans, echinoderms, and fishes, form detrital food webs supported by organic remains that sink from pelagic communities.

In 1976, researchers found communities thriving near hydrothermal vents at a depth of 3000 m near the Galápagos Rift, a volcanically active boundary between two crustal plates. Near-freezing water seeps into fissures where it is heated to temperatures of 350°C or higher. Pressure forces the heated water upward, and minerals are leached from porous rocks as the water spews out through vents in the seafloor. This hydrothermal outpouring releases hydrogen sulfide, which serves as an energy source for chemoautotrophic bacteria, the primary producers in hydrothermal vent communities. Some of these bacteria live as endosymbionts of giant clams and tube-dwelling worms **(Figure 52.31).** Deep-sea food webs also include sea anemones, crustaceans, and fishes. Researchers have located hydrothermal vent ecosystems in the South Pacific, near Easter Island; in the North Pacific, off the coast of British Columbia; the Gulf of California, about 150 miles south of the tip of Baja California, Mexico; and the Atlantic.

Recent research in the deepest reaches of the ocean reveals that communities also exist in areas far from hydrothermal vents. These "cold seep" communities thrive on broad expanses of the seafloor, where extremely salty water percolates upward from the underlying rocks and sediment, carrying abundant minerals, hydrogen sulfide, and methane to areas that are accessible to organisms. Chemosynthetic bacteria, which grow in large mats, can metabolize these molecules, forming the base of food webs that also include sponges, worms, and bivalve mollusks.

STUDY BREAK

1. What is the difference between the benthic and pelagic provinces of the ocean?
2. Which marine biomes experience the largest fluctuations in salinity (salt concentration) over time?
3. Which marine biomes or regions within marine biomes receive abundant energy input from sunlight?
4. What is the source of nutrients and energy for the benthos of the oceanic zone?
5. What organisms are the primary producers in hydrothermal vent and cold seep communities, and how do they differ from the primary producers in the photic zone?

Figure 52.29

Open ocean. A humpback whale (*Megaptera novaeangliae*) breaches (leaps out of the water) near a British Columbia coastline.

Figure 52.30

Deep sea. A deep-sea anglerfish (*Himantolophus* species) uses a bioluminescent lure to attract prey to its formidable jaws.

Figure 52.31

Deep benthos. Giant tube-dwelling worms are common in the hydrothermal vent communities on the deep ocean floor.

How will biomes change in response to anthropogenic (human-induced) global warming?

Will biomes remain largely intact and simply shift their geographical distributions northward or upward (to higher altitudes)? Or will the biomes we recognize today become disrupted, and new biomes arise as species associate in different combinations? Parmesan and Yohe estimated that 59% of wild species around the world have already shown some change in their geographical distributions in response to the relatively small level of global warming—a 0.7°C rise in average temperature—over the past 100 years. Documented responses to global warming vary from species to species, however. For example, only 20% of butterfly species in Spain, France, and North Africa have shifted their southern range boundaries northward, but 70% of butterfly species in the United Kingdom and Scandinavia have expanded their northern range borders further northward, sometimes by as much as 300 km over the past 30 years.

Thus, although the distributions of some species appear stable, the ranges of others are showing strong responses to global warming. At least two hypotheses may explain these patterns. First, some species may be stressed by rising temperatures but have not yet shown a measurable response. Second, the geographical distributions of some species may not be governed primarily by climate. Whatever the reason, the fact that we observe large variation in the response of different species suggests that not all species in a community are moving together. Thus, the existing communities of birds, butterflies, and trees are being disrupted—with some species moving and others not. Biologists have also noted differences in the response of different taxonomic groups. Butterflies in Europe and North America seem to be shifting their distributions northward and upward at about the same rate that temperatures are changing, but plants appear to lag behind. Alpine herbs in Switzerland, for example, have shifted their distributions upward at about half the rate that one might expect from the rate of regional warming, and it was 30 years after warming began before tree seedlings in Sweden started to colonize alpine habitats at higher elevations, shifting the treeline upward.

Even bigger questions remain. Will the vegetation that currently lives in the tropics expand into what is now the temperate zone and cover more of the planet? Some studies suggest that tropical lowland trees are already at their physiological limit—already showing signs of stress by shutting down photosynthesis on the hottest and driest days. Furthermore, climate model projections from a 2007 report of the Intergovernmental Panel on Climate Change consistently show substantial drying as well as warming in midlatitudes. If this projection is correct, many plants and animals now living in the wet tropics will be unable to shift northward into what may become an extreme desert climate. And what will happen to the arctic tundra? Researchers have already collected strong evidence that shrubs and trees are encroaching northward into the tundra of Alaska and Canada. The permafrost is melting, and the soil is drying. How do these observed changes in plant and animal distributions relate to the future? Human activity has already caused Earth's mean annual temperature to rise by 0.7°C in the past 100 years. Climate model projections suggest that further increases between 1.8°C and 4.0°C are likely; some models suggest the rise will be over 6.0°C. Can the tundra biome survive even the lowest projections—more than twice the warming it has already experienced?

How will evolution shape the ways that wild species respond to climate change? Which groups of organisms are likely to adapt, and which are likely to become extinct?

Populations are evolving all the time in response to changing selection regimes. Global warming is one of many human-driven environmental changes that could foster genetic change. Biologists have known for decades that organisms are locally adapted to the climatic conditions under which they routinely live. Scientists have documented local genetic changes toward more warm-adapted genotypes in fruit flies, mosquitoes, and the algal symbionts of corals. Do the observed genetic changes suggest that these species are adapting to anthropogenic global warming? Will other species follow suit? The fossil record suggests that during the Pleistocene glaciations, when Earth's temperature shifted between glacial periods (4°–8°C colder than now) and interglacial periods (today's temperatures), very few species became extinct and few experienced substantial morphological evolution. But, before the Pleistocene, Earth was much hotter than it is today, and the atmosphere had higher levels of CO_2. During the transition from these very warm, high CO_2 conditions to the colder, low CO_2 conditions of the Pleistocene, a large proportion of species became extinct. To how much climate change can organisms adapt? At what point is climate change extreme enough that species come to the limit of their genetic variation, can no longer adapt, and become extinct?

Camille Parmesan is an associate professor in the Section of Integrative Biology at the University of Texas at Austin. Her recent work has focused on current impacts of climate change on wildlife, and especially on butterfly range shifts. To learn more about her research, go to http://cluster3 .biosci.utexas.edu/research/parmesanLab.

Review

Go to ThomsonNOW™ at www.thomsonedu.com/login to access quizzing, animations, exercises, articles, and personalized homework help.

52.1 Environmental Diversity of the Biosphere

- The biosphere encompasses all the regions on Earth where organisms live, including the atmosphere, hydrosphere, and lithosphere.

- Latitudinal variations in solar radiation establish global climate patterns (Figure 52.2). Earth's tilt on its axis causes seasonal variation in solar radiation and climate (Figure 52.3). Seasonal variations in day length and temperature increase steadily from tropical latitudes toward the poles.

- Unequal heating of the atmosphere causes air masses to flow in circulation cells that create worldwide wind and precipitation patterns (Figures 52.4 and 52.5). Ocean currents generally flow clockwise in the Northern Hemisphere and counterclockwise in the Southern Hemisphere (Figure 52.6).

- The oceans and local topographical features influence regional and local climates. Proximity to the ocean has a moderating effect on terrestrial climates (Figure 52.7). Habitats are generally wetter on the windward sides of mountains than on the leeward sides (Figure 52.8).

Animation: Global air circulation patterns

Animation: Air circulation and climate

Animation: Major climate zones and ocean currents

Animation: Rain shadow effect

Animation: El Niño Southern Oscillation

52.2 Organismal Responses to Environmental Variation

- Organisms use homeostatic responses to cope with environmental variation. Animals often use facultative behavioral and physiological mechanisms to respond to environmental temperature (Figure 52.9).

- Global warming is affecting the ecology of many organisms. Many species are experiencing changes in their geographical ranges or in the timing of their reproduction.

52.3 Terrestrial Biomes

- Biomes are general types of vegetation and other associated organisms. Climate is the major determinant of terrestrial biome distributions (Figures 52.10 and 52.11).

- Tropical forest occurs at low latitudes where seasonality is determined by variations in rainfall rather than by day length and temperature (Figure 52.12). Tropical rain forests are the most species-rich terrestrial biome, but they grow on nutrient-poor soils.

- Savanna is tropical and subtropical grassland with scattered trees (Figure 52.13). Long dry seasons, fires, and grazing by large mammals prevent trees from replacing perennial grasses.

- Deserts form in arid regions where precipitation is low and temperature varies widely on a daily and seasonal basis (Figure 52.14).

- Chaparral is a coastal biome dominated by dense, woody shrubs and trees that resprout after periodic fires (Figure 52.15). Chaparral occurs where winters are mild and wet and summers hot and dry.

- Temperate grassland grows where winters are cold, summers are warm, and rainfall is moderate (Figure 52.16). Tree seedlings are eliminated from grasslands by droughts, periodic fires, and grazing by mammals. Grassland soils are rich and deep.

- Temperate deciduous forest flourishes at middle latitudes with abundant rainfall. The seasonality of the climate is reflected in the annual loss and regrowth of leaves (Figure 52.17).

- The boreal forest, or taiga, includes dense stands of coniferous trees at high latitudes, where winters are long and cold (Figure 52.18).

- Tundra is the northernmost biome, where plants grow in shallow topsoil over a layer of permafrost (Figure 52.19). The brief growing season and winter winds cause tundra plants to be very short.

Animation: Terrestrial biomes

Animation: Soil profiles

52.4 Freshwater Biomes

- Freshwater biomes include both flowing-water and standing-water systems.

- The physical characteristics of flowing-water biomes change from the headwaters of a stream to the mouth of a river (Figure 52.20), and their food webs are largely detrital.

- The physical characteristics of standing-water biomes change with the depth of water and distance from shore (Figures 52.21 and 52.22). Lakes exhibit marked vertical zonation and, in the temperate zone, undergo a seasonal mixing of their waters (Figure 52.23). Lakes are generally classified by their nutrient status and productivity (Figure 52.24).

Animation: Lake zonation

Animation: Lake turnover

Animation: Trophic nature of lakes

52.5 Marine Biomes

- The world's oceans exhibit marked zonation based on water depth and distance from shore (Figure 52.25).

- Estuaries are highly productive tidal biomes where rivers provide a constant input of nutrients and freshwater, and the tides carry away wastes (Figure 52.26).

- The intertidal zone is a stressful environment that is alternately submerged and exposed (Figure 52.27).

- Highly productive and diverse shallow-water biomes grow on continental shelves and oceanic banks. Kelp forests predominate at high latitudes, whereas coral reefs occur in the tropics (Figure 52.28).

- The open ocean is highly stratified because photosynthesis is possible only in the uppermost 50 m of water. Plankton are the primary producers in the uppermost layers; they support grazing food webs (Figure 52.29). The deep sea includes many predatory species (Figure 52.30).

- Organisms of the sea floor occupy the benthic province. Falling detritus supports most benthic communities, but chemoautotrophic bacteria support communities near deep-sea hydrothermal vents (Figure 52.31).

Animation: Rocky intertidal zones

Animation: Three types of reefs

Animation: Oceanic zones

Animation: Coastal upwelling

Animation: Hydrothermal vent community

Questions

Self-Test Questions

1. The lithosphere includes all:
   a. oceans.
   b. ice caps.
   c. rocks, soils, and sediments.
   d. gases and airborne particles.
   e. places where organisms live.

2. Earth's 23.5° tilt on its axis directly causes:
   a. latitudinal variation in average annual rainfall.
   b. ocean currents to rotate clockwise in the Northern Hemisphere.
   c. microclimates to vary dramatically over short distances.
   d. low rainfall on the leeward side of mountain ranges.
   e. seasonal variation in the amount of solar radiation.

3. Adiabatic cooling causes rising air masses to:
   a. absorb moisture from Earth's surface.
   b. release precipitation.
   c. change the direction of the El Niño current.
   d. flow toward the equator from the poles.
   e. be deflected from a strictly northward or southward flow.

4. The term "rain shadow" describes the:
   a. low rainfall that is typical on the leeward side of mountains.
   b. low rainfall that is typical at 30° latitude.
   c. high rainfall that is typical on the windward side of mountains.
   d. blocking of rain by vegetation in dense tropical forests.
   e. low rainfall that is typical in the interior of continents.

5. The major climatic factors that govern the distributions of terrestrial biomes are:
   a. temperature only.
   b. rainfall only.
   c. wind speed only.
   d. temperature and rainfall.
   e. temperature, rainfall, and wind speed.

6. Which biome experiences the highest annual rainfall?
   a. tropical rain forest
   b. tropical savanna
   c. chaparral
   d. temperature grassland
   e. arctic tundra

7. From which biome are trees excluded by periodic fires and grazing herbivores?
   a. tropical rain forest
   b. thorn forest
   c. chaparral
   d. temperate grassland
   e. arctic tundra

8. The major source of nutrients in the headwaters of a small stream is from:
   a. dead leaves and other organic matter from adjacent land.
   b. photosynthesis by phytoplankton.
   c. photosynthesis by floating aquatic plants.
   d. the activity of chemoautotrophic bacteria.
   e. minerals from the underlying bedrock.

9. During the spring overturn in a temperate zone lake:
   a. oxygen is carried from the surface to the bottom, and nutrients are carried from the bottom to the surface.
   b. nutrients are carried from the surface to the bottom, and oxygen is carried from the bottom to the surface.
   c. nutrients and oxygen are carried from the bottom waters to the surface waters.
   d. nutrients and oxygen are carried from the surface waters to the bottom waters.
   e. oxygen concentration remains constant at all depths, and nutrients sink to the bottom.

10. In which habitat must organisms adjust regularly to changing salinity?
    a. salt marsh
    b. coral reef
    c. benthic province
    d. estuary
    e. riffle

Questions for Discussion

1. Temperate grassland and chaparral often burn in lightning-induced fires, which stimulate the germination of seeds and regrowth of existing vegetation. Do you think that companies or the government should sell fire insurance to people who build expensive homes in places where periodic fires are virtually inevitable?

2. Boreal forests generally harbor many fewer species of trees than tropical forests do. Develop three hypotheses to explain this pattern. What data would you collect to test your hypotheses?

3. Describe the biome in which you live, noting the prevailing climate and microclimates and any other factors that govern the characteristics of the organisms that naturally occur there.

4. Many regions on Earth have been developed for agriculture, industry, and human habitation. Have our activities created new biomes? What physical environments are created by development, and what plants and animals occupy developed areas?

Experimental Analysis

Design an experiment to test the hypothesis that streams receive much of their nutrients and energy from material that falls into them from overhanging vegetation.

Evolution Link

If the geographical ranges of species change in response to global warming, what new selection pressures will organisms face as they move into ecological communities where they have not previously occurred? Your answer should address the effects of novel species interactions as well as the effects of encountering different physical environments.

How Would You Vote?

We cannot stop an El Niño from happening, but we might be able to minimize its environmental, social, and economic impacts. Would you support the use of taxpayer dollars to fund research into the causes and effects of El Niño? Go to www.thomsonedu.com/login to investigate both sides of the issue and then vote.

Florida panther *(Puma concolor coryi)*. Fewer than 100 individuals of this endangered subspecies survive.

JH Pete Carmichael/Getty Images Inc.

53 Biodiversity and Conservation Biology

WHY IT MATTERS

Someone seems to be missing. Investigators thoroughly checked the subject's known haunts, but found no trace. They questioned others in the neighborhood, but came up with few leads. The case is especially difficult because the subject was last seen alive in 1978. With so cold a trail to follow, investigators reluctantly marked the case file "Missing and Presumed Extinct."

The subject in this case was Miss Waldron's red colobus monkey, *Procolobus badius waldroni* **(Figure 53.1).** Named for a traveling companion of the taxonomist who first described it in 1933, this distinctively colored subspecies lived in large and noisy social groups in a remote forest on the border between Ivory Coast and Ghana in West Africa.

John Oates of the City University of New York recently led a research team that tried to locate Miss Waldron's red colobus. They used every imaginable method, including visual and auditory censuses, searching for scat (dung) in natural habitats, interviewing local people, and looking in marketplaces where monkey meat is commonly traded. In 2000, more than 20 years after the last confirmed sighting, the researchers concluded that this monkey is probably extinct. A later

Figure 53.1
Miss Waldron's red colobus. *Procolobus badius waldroni*, which weighed about 10 kg, may be the first primate subspecies to become extinct in more than 100 years.

search by a member of the team, William S. McGraw of Ohio State University, did find the skin of one monkey that a hunter had shot 6 months before. But McGraw searched in vain for a living monkey, and he concluded that even if a few are still alive, the population is so small that continued hunting will surely eliminate it.

Procolobus badius waldroni may be the first primate subspecies to become extinct in more than 100 years—and only the second in the last 500 years. Monkeys and other primates are among the most closely monitored and protected species on Earth. Nonetheless, Oates

Figure 53.2
The Pacific yew tree. The slow-growing Pacific yew *(Taxus brevifolia)* is the original source of Taxol, a compound that effectively fights several cancers.

and his colleagues concluded, these monkeys probably became extinct because they were hunted locally for food by a growing human population and because humans have destroyed their natural habitats.

Miss Waldron's red colobus is just one of many species driven to extinction every year. Current threats to biodiversity, all of which ultimately result from human activities, are massive. The likely loss of this monkey should warn us that many taxa are at risk, even those that are most rigorously protected.

When ecologists speak of **biodiversity**, they are referring to the richness of living systems. At the most fundamental level of biological organization, biodiversity encompasses the *genetic variation* that is raw material for adaptation, speciation, and evolutionary diversification (see Chapters 20 and 21). At a higher level of organization, biodiversity includes *species richness* within communities (see Section 50.3). The number and variety of species within a community influences its overall characteristics, population interactions, and trophic structure. Finally, biodiversity exists at the *ecosystem level*. Complex networks of interactions bind species in an ecosystem together, and because different ecosystems interact within the biosphere, damage to one ecosystem can reverberate through others.

In this chapter we reflect on the importance of biodiversity and describe how human activities threaten it. We also consider theoretical and practical approaches to conservation biology, the scientific discipline that focuses on preserving Earth's biological resources.

53.1 The Benefits of Biodiversity

What is the value of biodiversity, and why should humans preserve it? Arguments for conserving biodiversity fall into three general groups: its direct benefit to humans, its indirect benefit to all living systems, and its intrinsic worth.

Biodiversity Benefits Humans Directly

Scientists constantly search for natural products that might provide humans with better food, clothing, or medicine. The development of a new medicine often begins when a scientist analyzes a traditional folk remedy or screens naturally occurring compounds for curative properties. Chemists then isolate and purify the active ingredient and devise a way to synthesize it in the laboratory. More than half of the 150 most commonly prescribed drugs were developed from natural products in this manner.

For example, *Taxol,* a drug treatment for breast and ovarian cancer, was isolated from the narrow strip of vascular cambium beneath the bark of the Pacific yew tree, *Taxus brevifolia* **(Figure 53.2).** Unfortunately, a fully grown, 100-year-old tree produces only a tiny amount of Taxol, and six trees must be destroyed to extract enough to treat one patient. Pacific yew trees are not abundant, and they grow slowly. Harvesting them for Taxol extraction could quickly lead to their extinction—and an end to the natural source of this life-saving compound. However, after much research, scientists can now synthesize this widely used drug in the laboratory.

Wild plants and animals also serve as sources of genetic traits that may improve agricultural crops and domesticated livestock. For example, corn *(Zea mays)* is an annual plant. Its cultivation requires yearly tilling of the soil, a labor-intensive activity that leads to erosion and loss of topsoil. Farmers have yearned for a perennial strain of corn, one that would produce grain for years after a single planting. In 1978, botanists discovered teosinte *(Zea diploperennis)* a perennial plant closely related to corn, in the mountains of western Mexico. Researchers crossed the two species, producing a *perennial* corn. If they can increase the yield of this hybrid, it may prove to be an economically valuable crop **(Figure 53.3)**.

Today, many agricultural researchers use genetic engineering, the transfer of selected genes from one species into another (see Section 18.2), to alter crop plants more precisely than they can using hybridization. The transferred genes may be chosen to increase resistance to pests or environmental stress, promote faster growth, or increase shelf life after harvesting. However, many scientists and environmentalists fear that genetically modified crops may create environmental hazards that will inadvertently endanger biodiversity. For example, a genetically modified plant or animal that escaped into a natural habitat might compete with naturally occurring species. Or a genetically modified plant might poison harmless animals as well as insect pests.

Figure 53.3
Teosinte and domesticated corn. Ears of domesticated corn (*Zea mays*, right) are much larger than those of its wild relative teosinte (*Zea diploperennis*, left). Scientists crossed the two species in the hope of producing a perennial corn; the hybrids produce ears of an intermediate size (middle).

Ecosystem Services Benefit All Forms of Life

Humans and other species derive indirect benefits when ecosystems perform the ecological processes on which all life depends. These **ecosystem services**, as they are called, include the decomposition of wastes, nutrient recycling, oxygen production, maintenance of fertile topsoil, and air and water purification.

Some ecosystem services can even mitigate environmental damage caused by humans. As you may recall from *Focus on Applied Research* in Chapter 51, the combustion of fossil fuels produces CO_2 and other waste products that accumulate in the atmosphere, increasing the greenhouse effect and fostering global warming. Photosynthetic organisms use CO_2 for essential metabolic processes; thus, forests and, even more importantly, communities of marine phytoplankton withdraw CO_2 from the atmosphere and incorporate it into living organisms (see Figure 51.13), a phenomenon called *carbon sequestration*. Recent research indicates that these organisms are essential for limiting the damage caused by the burning of fossil fuels. In the long run, biodiversity's indirect benefits, provided in the form of ecosystem services, may be even more valuable to humans than the direct benefits.

Biodiversity Has Intrinsic Worth beyond Its Utility to Humans

Some ethicists argue that we should preserve biodiversity because it has intrinsic worth, independent of its direct or indirect value to humans. They note that humans are just one species among millions in the remarkable network of life. Countering this position is the view that our immediate needs should always rank above those of other species and that we should use them to maximize our own welfare. The latter view inevitably leads to the disruption of natural environments and the loss of biodiversity. Framed in this way, the debate lies more within the realms of philosophy and public policy than biology. Nevertheless, many people feel an emotional or spiritual connection to natural landscapes and the plants and animals they harbor. Thus, biodiversity enhances human existence in intangible ways.

Figure 53.4 Experimental Research

Predation on Songbird Nests in Forests and Forest Fragments

QUESTION: Are songbird nests in small forest fragments more likely to be found by predators than nests in large forest patches?

EXPERIMENT: Wilcove placed between 13 and 50 artificial bird nests, each containing three quail eggs, in three habitat types: large areas of intact forest, rural forest fragments, and suburban forest fragments. He placed about half the nests at each study site on the ground at the base of a tree or shrub and half the nests 1 to 2 m above the ground in a sapling or shrub. He checked the nests after 7 days to determine what proportion of the nests had been subjected to predation.

RESULT: Predators generally found a larger proportion of the artificial bird nests in small forest fragments than they did in large forest patches.

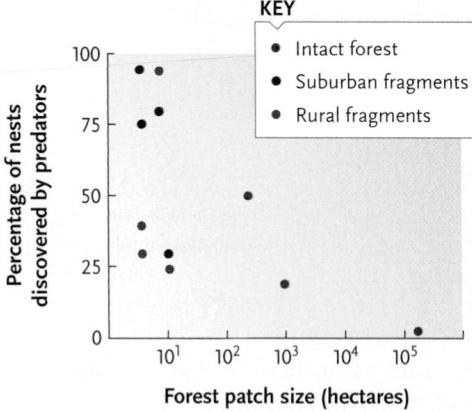

CONCLUSION: Songbirds' nests are much more likely to suffer from predation in small forest fragments than they are in large patches of intact forest.

STUDY BREAK

1. How does biodiversity serve as a storehouse of genetic information that is potentially useful to humans?
2. How do naturally occurring organisms provide humans with ecosystem services?

53.2 The Biodiversity Crisis

Earth's biodiversity is currently declining dramatically. Although the proximate causes of the decline may vary from one group of organisms to another, the ultimate cause is always the same: human disruption of natural communities and ecosystems.

Human Activities Disturb and Fragment Habitats

When humans first enter undisturbed habitats, they typically build roads to gain access to resources, such as oil, wood, or game animals, or to begin agricultural development. The roads bring in settlers, who clear isolated areas for specific uses. Nonnative organisms are often introduced by humans or migrate into the now-disturbed area under their own power. These invaders then consume, parasitize, or compete with the native plants and animals. As the land is further changed and degraded, the habitat is altered dramatically, possibly forever. Although this pattern of development initially affects only locally distributed species, the negative effects spread rapidly to a regional scale. The remaining areas of *intact* habitat are inevitably reduced to small, isolated patches, a phenomenon that ecologists describe as **habitat fragmentation.**

Habitat fragmentation is a threat to biodiversity because small habitat patches can sustain only small populations. As you learned in Section 49.5, a habitat's *carrying capacity,* the maximum population size that it can support, varies with available resources. Populations that occupy small habitat patches inevitably experience low carrying capacities, a problem that is especially acute for species at the higher trophic levels (see Section 51.1). Furthermore, fragmented habitat patches are often separated by unsuitable habitat that organisms may be unable or unwilling to cross. As a result, individuals from one isolated population are unlikely to migrate into another, reducing gene flow between them. The combination of small population size and genetic isolation fosters genetic drift, which reduces genetic variability and fosters extinction (see Section 20.3).

Habitat fragmentation not only reduces the amount of undisturbed habitat; it also jeopardizes the quality of the habitat that remains. Human activities create noise and pollution that spread into nearby areas. The removal of natural vegetation disrupts the local physical environment, exposing the borders of the remaining habitat to additional sunlight, wind, and rainfall. Increased runoff compacts the soil and makes it waterlogged. These phenomena are collectively described as **edge effects.**

The effects of habitat fragmentation are often profound. For example, populations of forest-dwelling, migratory songbirds have declined markedly in eastern North America since the late 1940s, largely because of habitat fragmentation in their North American breeding grounds and in their Caribbean and South American wintering grounds.

In 1994, Scott K. Robinson of the Illinois Natural History Survey and David S. Wilcove of the Environmental Defense Fund identified three factors that decrease populations of migratory songbirds in fragmented breeding habitats. First, small forest patches often lack specific habitat types—such as streams, cool ravines, or dense ground cover—that many songbird species require.

Second, songbirds breeding in forest patches are more likely to suffer from brood parasitism (described in the opening of Chapter 50) by brown-headed cowbirds (*Molothrus ater*) than are those breeding in intact forests. Brown-headed cowbirds, which prefer open habitats, were rare in eastern North America before European settlers converted forests to farmland. Today, cowbirds are abundant in open agricultural fields and suburban gardens, and they locate the nests of unwitting "foster parents" in nearby forest fragments where the host species breed. Parasitized songbirds rear fewer than half as many young as they might otherwise raise, and their populations decline accordingly.

The third factor that reduces songbird numbers in forest fragments is increased nest predation by blue jays (*Cyanocitta cristata*), American crows (*Corvus brachyrhynchus*), common grackles (*Quiscalus quiscula*), squirrels (genus *Sciurus*), raccoons (*Procyon lotor*), and domestic dogs and cats. These predators, which feed on songbird eggs and young, are now superabundant in rural and suburban areas, and they enter adjacent forest fragments in search of an easy meal. Wilcove tested the predation hypothesis experimentally by placing artificial nests with quail eggs in intact forests and in forest fragments. Although he did not observe predation directly, he found that predators discovered only 2% of the nests in the largest intact forest, but they often found 50% or more of the nests placed in small, suburban forest fragments **(Figure 53.4)**.

Deforestation May Lead to Desertification

Forests are among the habitats that humans most frequently clear and convert. According to the United Nations Forest Resources Assessment released in 2005, global deforestation is occurring at a rate of about 13 million hectares per year, or 25 hectares per minute. In other words, an area of forest equivalent to 42 football fields is cleared of all trees every minute of every day.

Deforestation does not occur uniformly across the globe. Today, more than 90% of the deforestation occurs in tropical regions, mostly to clear land for grazing. Brazil has experienced the most extensive recent damage, accounting for 25% of all deforestation during the late twentieth century **(Figure 53.5)**. This assessment is particularly troubling because Brazil contains approximately 27% of the planet's total aboveground woody biomass. Compounding the environmental damage, most tropical forests are burned as they are cleared, a process that adds CO_2 to the atmosphere, enhancing the greenhouse effect and increasing the rate of global warming (see *Focus on Applied Research* in Chapter 51).

Once a forest has been cut, heavy grazing or farming drains nutrients from the soil. To remain productive, even the best agricultural or grazing lands require either the application of fertilizers or long periods during which the land is fallow, allowing plants to replenish the soil naturally. Unfortunately, the soil where tropical forests grow is often of marginal value right from the start (for reasons described in Chapter 52), and it is rapidly degraded; it becomes hard, even more nutrient-poor, unable to retain water, and likely to wash away.

When large tracts of subtropical forest are cleared and overused, the land often undergoes **desertification**: the groundwater table recedes to deeper levels; less surface water is available for plants; soil accumulates high concentrations of salts (a process called *salinization*); and topsoil is eroded by wind and water. In other words, the habitat is converted to desert.

Desertification speeds the loss of biodiversity locally and can eliminate entire ecosystems. For example, desertification has decimated habitats in the Sahel re-

1975   2001

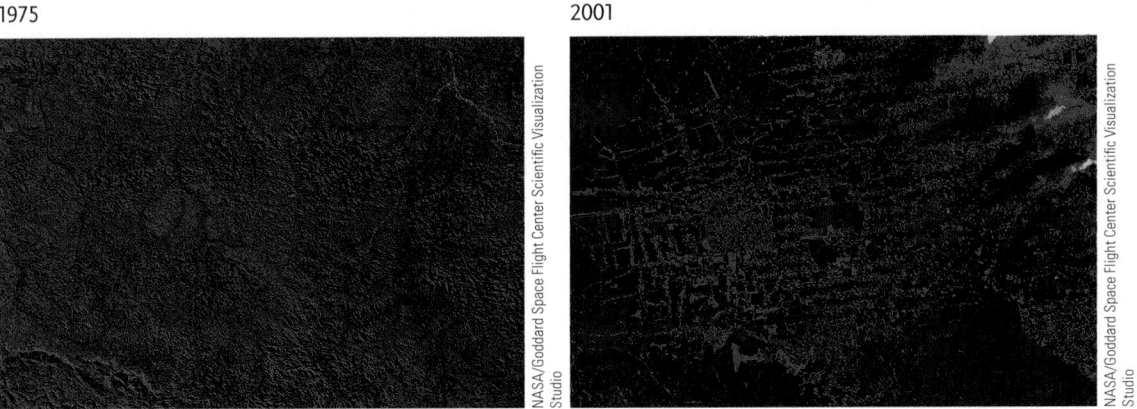

NASA/Goddard Space Flight Center Scientific Visualization Studio

Figure 53.5
Deforestation in the Amazon Basin. Satellite photos of Rondonia, in the Brazilian Amazon, show how much of the Amazon forest was cut (light green) between 1975 and 2001. Each photo illustrates an area approximately 60 by 85 km.

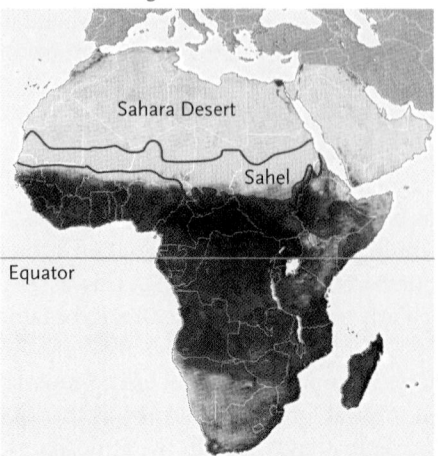 

Figure 53.6

Desertification in the Sahel. **(a)** A satellite photo taken near the end of the dry season in June 2005 illustrates the severe desertification in parts of the Sahel region of Africa. Dark green areas are densely vegetated; light green areas are sparsely vegetated, and sand-colored areas are barren. **(b)** People who live in this region can barely eke out a living on the land.

gion of Africa, just south of the Sahara Desert **(Figure 53.6).** Excessive grazing of cattle and goats by an ever-expanding human population is the main reason for the Sahara's southward expansion at a rate of 5.5 to 8 km per year. Because the sand dunes of the expanding desert shift constantly, agriculture and grazing are nearly impossible, resulting in frequent famines among the people of the Sahel.

Desertification and salinization have also begun in the Everglades, a unique, shallow "river of grass" that covers much of southern Florida. The amount of fresh water flowing through South Florida to the Everglades has decreased approximately 70% since 1948, when an extensive network of canals and levees was built to reduce flooding. The rapidly growing human population in South Florida contributes directly to desertification, as groundwater is tapped for domestic use and to irrigate lawns, golf courses, and agricultural fields. Salt water from the Gulf of Mexico now intrudes into the water table, causing salinization of the soil. The Comprehensive Everglades Restoration Plan (CERP), approved by the U.S. Congress in 2000, seeks to restore the natural flow of the Everglades over the next 30 years. This project may halt or reverse the desertification process.

Sadly, deforestation, desertification, and global warming reinforce each other in a positive feedback cycle (see *Focus on Applied Research* in Chapter 51). If scientists' projections are correct, desertification will lead to an increase in the average global temperature, speeding evaporation and the retreat of forests, which, in turn, will increase rates of desertification. If deforestation and desertification continue, we will soon lose a large proportion of Earth's forests and face a decrease in the area of habitable land.

Many Forms of Pollution Overwhelm Species and Ecosystems

The release of **pollutants**—materials or energy in forms or quantities that organisms do not usually encounter—poses another major threat to biodiversity.

Although chemical pollutants, the by-products or waste products of agriculture and industry, are released locally, many spread in water or air, sometimes on a continental or global scale. Within North America, for example, winds carry airborne pollutants from coal-burning power plants to the Northeast **(Figure 53.7).** Sulfur dioxide (SO_2), which dissolves in water vapor in the air and forms sulfuric acid, falls as **acid precipitation,** acidifying soil and bodies of water. Many lakes in northeastern North America have experienced a precipitous drop in pH from historical readings near 6 to values that are now well below 5—a 10-fold increase in acidity. Although the lakes once harbored lush aquatic vegetation and teemed with fishes, they are now crystal clear and nearly devoid of life.

As residents of major cities and industrial areas know all too well, carbon wastes from factories and automobile engines cause terrible local pollution, increasing rates of asthma and other respiratory ailments. Some airborne pollutants, notably CO_2, also join the general atmospheric circulation, where they contribute to the greenhouse effect and global warming.

Like air pollution, water pollution originates locally but has a much broader impact. Oil spills, for example, disrupt local ecosystems, killing most organisms near the spill. Because oil floats on water, it spreads rapidly to nearby areas. The wreck of an oil tanker off the coast of Spain in 2002 destroyed many fertile fishing grounds within a few weeks. Scientists

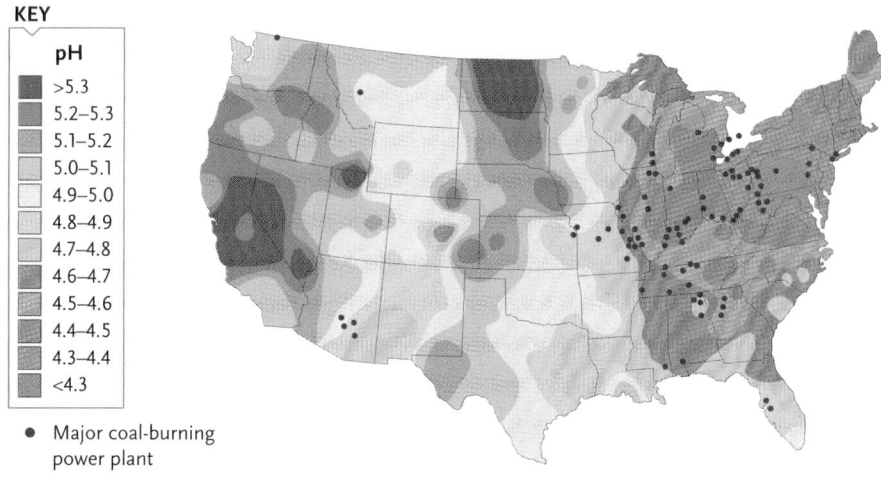

KEY

pH
- >5.3
- 5.2–5.3
- 5.1–5.2
- 5.0–5.1
- 4.9–5.0
- 4.8–4.9
- 4.7–4.8
- 4.6–4.7
- 4.5–4.6
- 4.4–4.5
- 4.3–4.4
- <4.3

● Major coal-burning power plant

Figure 53.7

Acid precipitation. Coal-burning power plants (indicated by red dots) release air pollution that is carried northeast, where it falls as acid precipitation. The map shows the average pH of rainfall.

expect oil to continue leaking from the sunken ship for another 50 years; the effects of the long-term leakage may linger for centuries.

Pollution can also have serious effects on terrestrial ecosystems. As a recent disaster in India, Nepal, and Pakistan illustrates, the application of synthetic compounds to agricultural fields or livestock can have dire and far-reaching consequences. For thousands of years, gigantic populations of vultures (several *Gyps* species)—estimated at more than 40 million birds—performed an important ecosystem service by consuming the abandoned carcasses of farm animals across South Asia. In the early 1990s, however, farmers began to administer diclofenac, a new and inexpensive anti-inflammatory drug, to injured livestock. Within a few years, vultures began to disappear; in 2006, scientists estimated that their populations had declined by more than 97%. Recent research revealed that diclofenac, which causes fatal kidney failure in birds, was responsible for the deaths: vultures were ingesting substantial doses of the drug from the livestock carcasses they ate. All vulture species in South Asia are now on the verge of extinction, and although governments in the region have banned the sale of diclofenac, wildlife experts say that the vulture populations are unlikely to recover soon, if ever.

The decline in vulture populations has had a disastrous impact on urban and rural communities in South Asia. Livestock carcasses are now consumed by growing populations of wild dogs, many of which carry rabies. India has the world's highest human death toll from rabies—30,000 per year—and two-thirds of the cases are caused by dog bites. Populations of rats and flies also appear to be increasing. *Focus on*

Research (p. 1236) describes another example of how pesticides and other chemicals accumulate at lethal concentrations in organisms living at higher trophic levels.

Some forms of pollution have more subtle effects. Light and noise pollution disrupt the activities of nocturnal animals or those that rely on vision or hearing for orientation. For example, light pollution in beachfront communities disrupts the reproduction of marine turtles, all species of which are declining in numbers **(Figure 53.8)**. Female turtles crawl up on beaches at night to lay their eggs in the sand; after the eggs hatch, the young dig their way out of the nest and head for the ocean. But female turtles are reluctant to come ashore on beaches with artificial light. And lights may later confuse and misdirect their hatchlings, making them even easier prey for predators, or cause them to stay too long on shore, where they dehydrate and die.

Exotic Species Often Eliminate Native Species

As humans travel from one habitat to another, we inevitably carry other species with us. Seeds cling to our legs, insects accompany us in our food and possessions, and some organisms hitch a ride on boats or cars. The introduction of nonnative organisms, called

Figure 53.8

Light pollution disrupts green turtle reproduction. **(a)** Female green turtles (*Chelonia mydas*) are reluctant to nest on beaches affected by light and noise pollution. **(b)** Artificial light confuses hatchling turtles, hindering their escape from eager predators like this great blue heron (*Ardea herodias*).

a. Female green turtle digging a nest

b. Heron eating hatchling green turtle

Applied Research: Biological Magnification

The synthetic organic pesticide DDT (dichloro-diphenyl-trichloroethane) was first used widely during World War II. In the tropical Pacific, it killed the mosquitoes that transmitted malarial parasites (*Plasmodium* species) to soldiers. In war-ravaged European cities, it controlled body lice that carried the bacteria causing typhus (*Rickettsia rickettsii*). After the war, people started using DDT to kill agricultural pests, disease vectors, and insects in homes and gardens.

Although DDT is a stable hydrocarbon compound that is nearly insoluble in water, it is more mobile than its users expected. Winds carry it as a vapor, and water transports it as fine particles. DDT is also highly soluble in fats, accumulating in animal tissues—and it travels with animals wherever they go.

Unfortunately, consumers accumulate the DDT from all of the organisms they eat in their lifetimes. Primary consumers, like herbivorous insects, may ingest relatively small quantities. But a songbird that eats many insects will accumulate a moderate amount, and a predator that feeds on songbirds will accumulate even more. Thus, DDT and other nondegradable poisons be-

come concentrated in organisms at higher trophic levels, a phenomenon called **biological magnification** (see **figure**). Although many organisms can partially metabolize DDT to other compounds, these products are also toxic or physiologically disruptive.

After the war, DDT moved rapidly through ecosystems, affecting organisms in ways that no one had predicted. In cities where DDT controlled Dutch elm disease, songbirds died after eating contaminated insects and seeds. In streams flowing through forests where DDT killed spruce budworms, salmon died because runoff carried the pesticide into their habitat. And in croplands around the world, new pests flourished because DDT indiscriminately killed the natural predators that had kept their populations in check.

Eventually, the effects of biological magnification began to show up in places far removed from the sites of DDT application. Top carnivores in some food webs were pushed to the brink of extinction. The reproduction of bald eagles, peregrine falcons, ospreys, and brown pelicans was disrupted because one DDT breakdown product interferes with the deposition

of calcium in their eggshells. When birds tried to incubate their eggs, the shells cracked beneath the parents' weight. Even today, traces of DDT are found in the bodies of nearly all species, including in human fat and breast milk.

Since the 1970s, DDT has been banned in the United States, except for restricted applications to protect public health. Many hard-hit species have partially recovered, but some birds still lay thin-shelled eggs because they pick up DDT at their winter ranges in Latin America. As recently as 1990, the California State Department of Health recommended that a fishery off the coast of California be closed; DDT from industrial waste discharged 20 years earlier was still moving through that ecosystem. Moreover, DDT is still used in other countries, and some enters the United States on imported fruit and vegetables.

Biological magnification is a problem that applies to many compounds that humans release into the environment. For example, polychlorinated biphenyls (PCBs), commonly used in the manufacture of plastics and electrical insulation, enters aquatic ecosystems in factory wastes. Their use has been banned in the United States since the 1970s. But these compounds break down very slowly, and vast deposits have accumulated in the bottom sediments of rivers and lakes. Once bottom-feeding organisms ingest them, the toxins work their way up food webs, accumulating at higher and higher concentrations in consumers. The effects on humans can be severe; pregnant women who regularly eat fish from the Great Lakes often give birth to children with below-average weight and neonatal behavioral problems. The pollution in some areas of New York State was so severe that the Department of Health advised people to avoid eating freshwater fish more than once a month. PCBs can be removed from aquatic ecosystems by dredging, but the dredging activity itself stirs up the polluted sediments, releasing the toxins into the water that flows above.

In this food web near Long Island Sound, New York, DDT concentration (measured in parts per million, ppm) was magnified nearly 10 million times between zooplankton and the osprey.

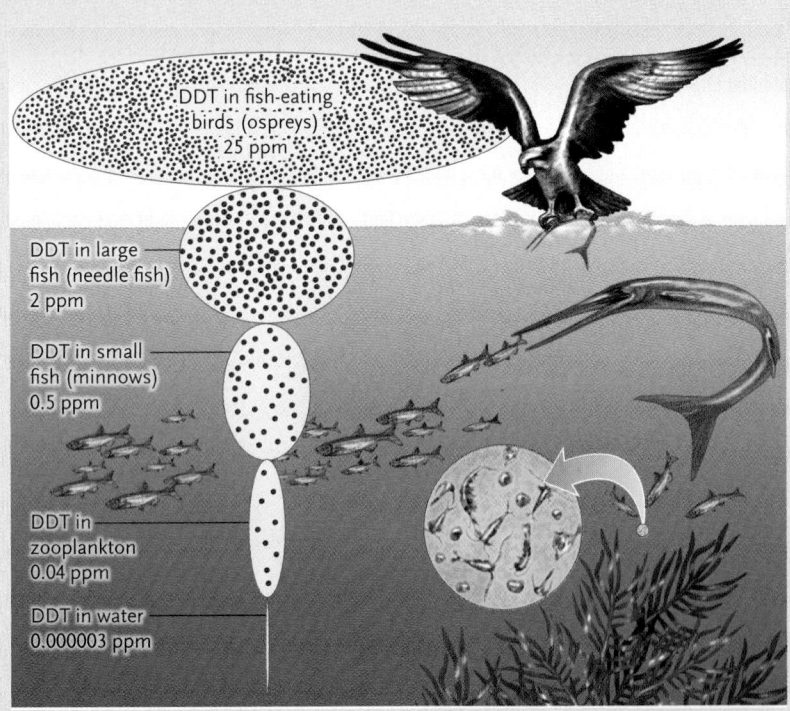

DDT in fish-eating birds (ospreys) 25 ppm

DDT in large fish (needle fish) 2 ppm

DDT in small fish (minnows) 0.5 ppm

DDT in zooplankton 0.04 ppm

DDT in water 0.000003 ppm

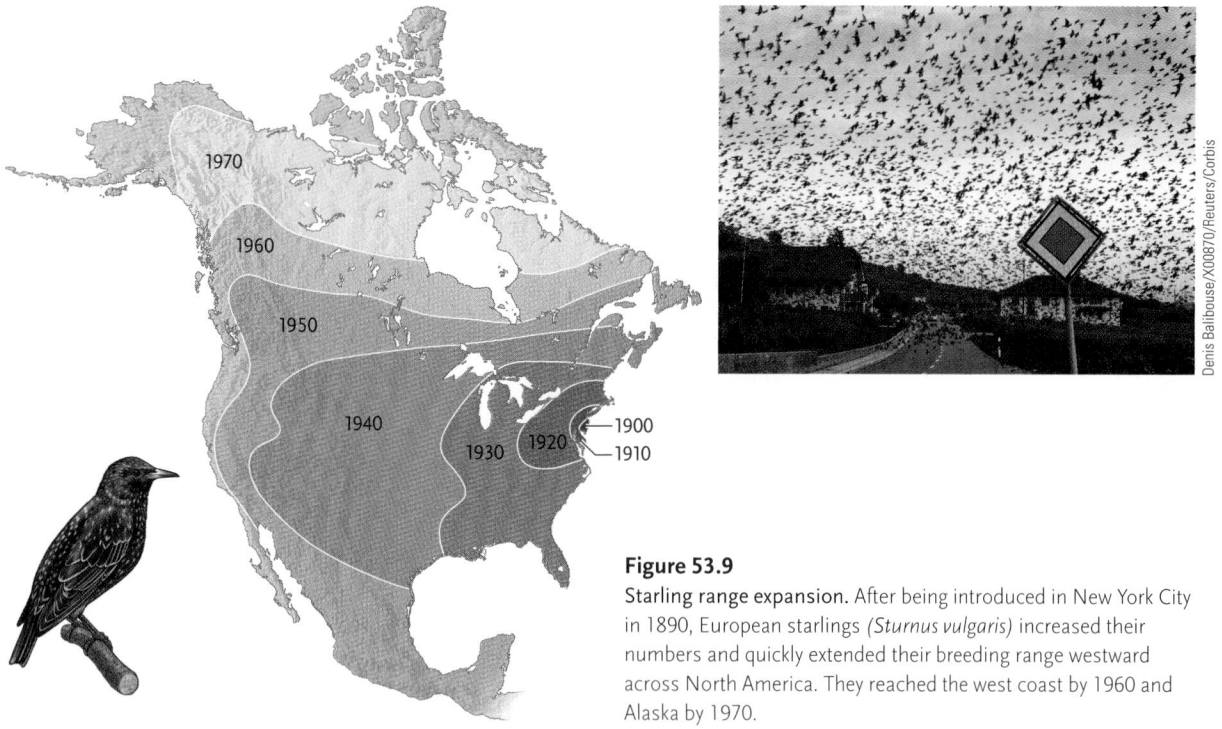

Figure 53.9

Starling range expansion. After being introduced in New York City in 1890, European starlings *(Sturnus vulgaris)* increased their numbers and quickly extended their breeding range westward across North America. They reached the west coast by 1960 and Alaska by 1970.

exotic species, into new habitats poses one of the most serious threats to biodiversity.

Exotic species often prey upon, parasitize, or outcompete native species, leading to their extinction. Many have *r*-selected life histories (see Section 49.6); they mature quickly and reproduce prodigiously, and they thrive in the degraded habitats that humans so frequently create. In the absence of natural checks on population growth—such as competitors, predators, and parasites—exotics often experience exponential population growth (see Section 49.5).

The European starling *(Sturnus vulgaris)* provides an example of the explosive population growth and range expansion of an exotic species. These birds were released in North America in 1890 when a misguided individual, who wanted to introduce all of the bird species mentioned by Shakespeare into North America, imported them into Brooklyn, New York. Within 70 years, they had spread across the continent **(Figure 53.9);** their population size is now estimated at 200 million. Starlings pose a serious threat to native birds, including several woodpecker species, because they successfully compete with them for nesting sites in natural cavities in trees.

Introduced plants often transform entire ecosystems. One of the best-known examples is kudzu *(Pueraria lobata),* a fast-growing species from Asia. In the early 1900s, it was widely planted in the southeastern United States as a source of animal feed. Later, a government agency promoted it as a plant that could stabilize soils and decrease erosion on deforested hillsides; we now know that it does not perform those functions effectively. But when kudzu has access to abundant nutrients and water, its branches can grow up to 30 cm per day. It spread quickly across the South, literally overgrowing almost all native plants **(Figure 53.10).**

Exotic insects often become pests of agricultural crops and native plants. The hemlock woolly adelgid *(Adelges tsugae)* was accidentally introduced into North America from Asia. The adelgid kills eastern hemlocks *(Tsuga canadensis)* by feeding on their sap. It now threatens the trees from North Carolina to Massachusetts **(Figure 53.11).** But adelgids endanger far more than these evergreen trees. Hemlocks buffer the physical

Figure 53.10
Kudzu, the vine that ate the South. *Kudzu (Pueraria lobata),* an introduced vine, grows so quickly that it often covers living trees or even abandoned buildings.

a. Woolly adelgids

b. Hemlocks killed by woolly adelgids

c. Eastern hemlock and woolly adelgid ranges

Adelgids

KEY

Native range of eastern hemlock

Hemlock woolly adelgid reported

Figure 53.11

Hemlock woolly adelgid. **(a)** The aphidlike woolly adelgid *(Adelges tsugae)* feeds on the sap of **(b)** eastern hemlock *(Tsuga canadensis)*, often killing the tree. This insect pest is spreading northward **(c)** and may someday endanger hemlocks throughout their geographical range.

conditions below them: hemlock stands are cool in summer and warm in winter, sustaining a unique community of organisms that includes ruffed grouse *(Bonasa umbellus)*, turkey *(Meleagris gallopavo)*, white-tailed deer *(Odocoileus virginianus)*, and snowshoe hare *(Lepus americanus)*. Infested stands rarely survive more than a few years, and the communities established under pure stands of eastern hemlock will likely become extinct because of feeding by the adelgid.

Overexploitation Greatly Reduces Population Sizes

Many local extinctions result from **overexploitation**, the excessive harvesting of an animal or plant species. At a minimum, overexploitation leads to declining population sizes in the harvested species. In the most extreme cases, a species may be wiped out completely. Overexploitation also can foster evolutionary changes in the exploited population, much the way guppies respond to natural predators in the streams of Trinidad (described in *Focus on Research* in Chapter 49).

The fishery on the Grand Banks off the coast of Newfoundland, Canada, provides a sad example of overexploitation **(Figure 53.12)**. For hundreds of years, fishermen used traditional line and small-net fishing to harvest a large but sustainable catch. During the twentieth century, however, new technology allowed them to locate and exploit schools of fishes more efficiently. As a result, 45% of the fish species harvested there are now overfished. Haddock *(Melanogrammus aegelfinus)* and yellowtail flounder *(Limanda ferruginea)*

a. The Grand Banks

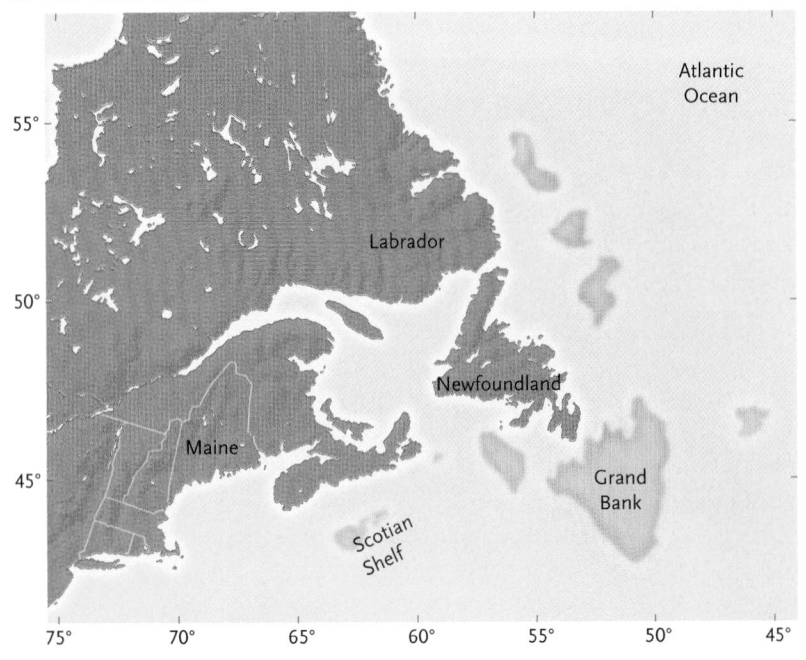

b. Atlantic cod

Figure 53.12

Overexploitation of North Atlantic fisheries. **(a)** The Grand Banks (sand-colored shading) were severely overfished in the late twentieth century, leading to the near extinction of many species, including the **(b)** Atlantic cod *(Gadus morhua)*.

have been essentially eliminated from the Grand Banks, and their populations will probably never recover. And because fishermen preferentially harvest the oldest and largest individuals, which fetch a higher market price, Atlantic cod *(Gadus morhua)* now mature at a younger age (3 years compared with 5 or 6 years) and smaller size.

As a consequence of overfishing, the average yield of the Grand Banks has declined to less than 10% of the highest historic levels. In the mid-1960s, Atlantic cod yielded a minimum of 350,000 tons per year. By the mid-1970s, the catch dropped to 50,000 tons per year. The Canadian government finally closed the fishery in 1993, after the cod catch fell below 20,000 tons for several consecutive years. But the damage had already been done: the most heavily exploited species are less marketable because of their smaller size, fish populations have decreased to dangerously low levels, and the fishing industry is itself imperiled. This sequence of events has been replicated in fisheries around the world. Indeed, in a report published in 2003, Ransom A. Myers and Boris Worm of Dalhousie University in Nova Scotia estimated that modern fishing techniques have reduced the biomass of large predatory fishes by about 90% in marine ecosystems.

Overexploitation is not inevitable; careful management of fisheries can achieve sustainable harvests. Many approaches are possible, such as providing supplemental food or shelter; maintaining captive breeding populations, from which individuals are introduced into the wild; limiting the times of harvest to avoid disrupting reproductive cycles; and limiting the size and character of the catch. Similar strategies can be devised for other resource populations.

Human Activities Are Causing a Dramatic Increase in Extinction Rates

As you may remember from Section 22.5, extinction has been common in the history of life: roughly 10% of the species alive at any time in the past became extinct within 1 million years. These *background extinction rates* eliminated perhaps seven or eight species per year. Paleobiologists have also documented at least five *mass extinctions,* during which extinction rates increased greatly above the background rate for short periods of geological time (see Figure 22.18).

At present, Earth appears to be experiencing the greatest mass extinction of all time. According to Edward O. Wilson of Harvard University, extinction rates today may be 1000 times the historical background rate, meaning that thousands of species are being driven to extinction each year. The vast majority of extinctions are a direct result of habitat fragmentation, desertification, rising levels of pollution, the introduction of exotic species, and the overexploitation of natural populations.

If humans are the cause of the current mass extinction, why has it taken so long to occur? Why didn't the mass extinction begin long ago? The answer lies in our increased rate of population growth (see Section 49.7). During the nineteenth and twentieth centuries, improvements in food production, sanitation, and health care increased human life expectancy. Our ever-increasing population consumes resources and produces wastes at an escalating rate. As global population continues to increase, so will the habitat destruction that inevitably accompanies population growth.

STUDY BREAK

1. How has habitat fragmentation affected breeding songbird populations in eastern North America?
2. What factors have increased the likelihood of desertification in southern Florida?
3. What are the consequences of the overexploitation of fish populations?
4. How do extinction rates today compare with the background extinction rate evident in the fossil record?

53.3 Biodiversity Hotspots

Given the detrimental effects of human activities on biodiversity and natural environments, conservation biologists are constantly seeking ways to minimize or reverse the damage.

Conservation Biologists Focus Their Efforts in Areas Where Biodiversity Is both Concentrated and Endangered

If we are to limit the effects of human activities and preserve biodiversity, we must know how and where biodiversity is distributed. Although species richness generally increases from the poles to the tropics within many communities (see Section 50.7), these large global patterns do not help biologists pinpoint those areas where conservation efforts will have the greatest impact.

In a survey published in 2000, Norman Myers of Oxford University and his colleagues in England and the United States pinpointed 25 **biodiversity hotspots**, areas where biodiversity is both concentrated and endangered **(Figure 53.13)**. As defined by the Endangered Species Act, adopted by the U.S. Congress in 1973, an **endangered species** is one that is "in danger of extinction throughout all or a significant portion of its range." (Species that are likely to become endangered in the near future are designated as *threatened*.) Thus, to qualify as a biodiversity hotspot, an

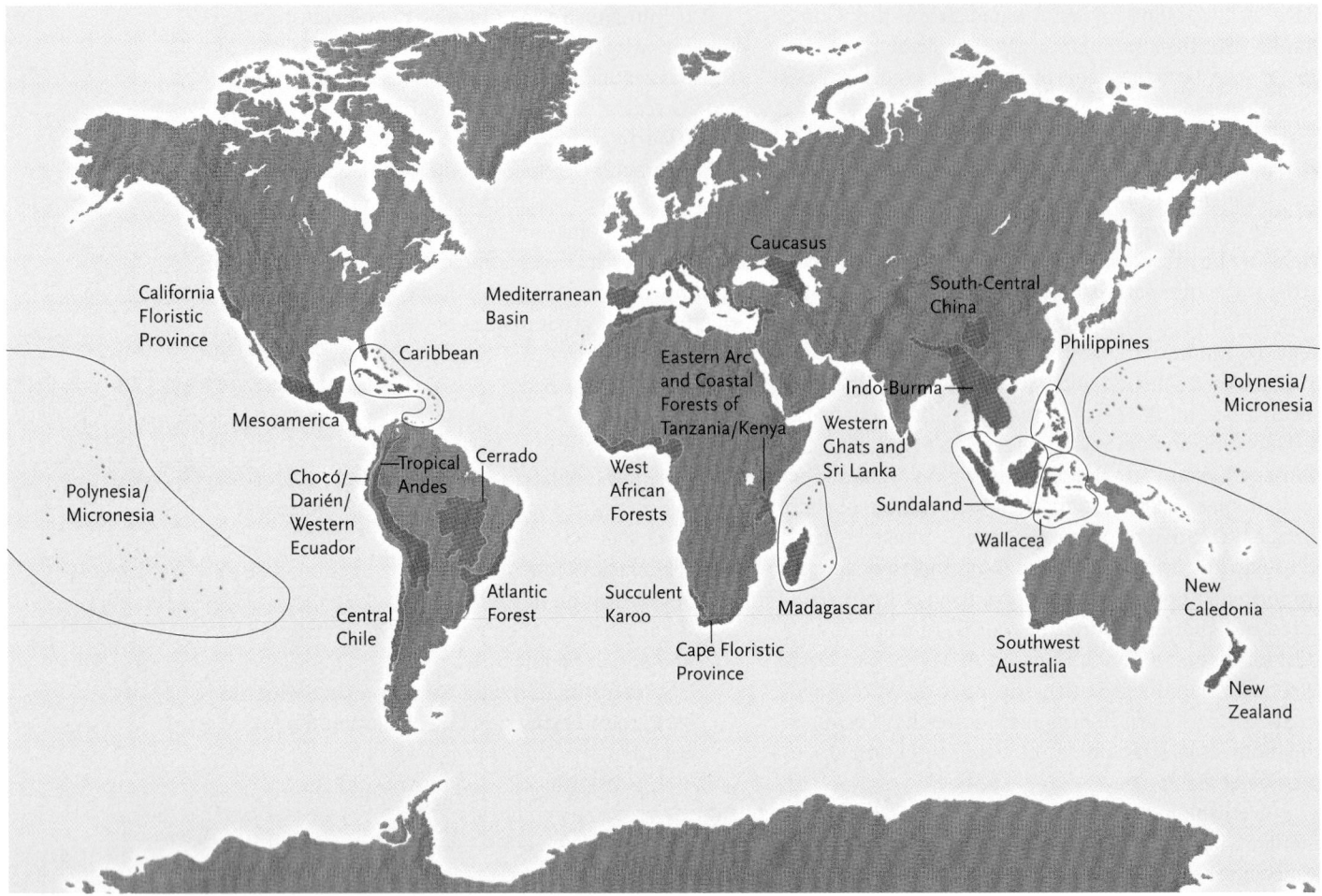

Figure 53.13

Biodiversity hotspots. Norman Myers and his colleagues identified 25 places that harbor many endemic species that are threatened by human encroachment.

area must harbor a large number of **endemic species** (those that are found nowhere else), and it must have already lost much of its natural vegetation to human encroachment.

Endemic species tend to have highly specific habitat or dietary requirements, low dispersal ability, and restricted geographical distributions. Myers used the number of endemic species as a criterion for identifying hotspots because locally distributed species account for much of Earth's biodiversity; and if the local habitats where these species occur are at risk of development, the species are also at risk. Although the 25 hotspots occupy only 1.4% of Earth's land surface, they include the only remaining habitat for approximately 45% of all terrestrial plant species and 35% of all terrestrial vertebrate species.

Sixteen of the 25 hotspots are in the tropics, where humans have already cut much of the natural vegetation. For example, until the mid-1900s, the Brazilian Atlantic Forest stretched undisturbed along the southern coast of Brazil and parts of Paraguay and Argentina. Since then, 93% of the forest has been cleared for agriculture and grazing, making it one of the most endangered ecosystems on Earth. Today more than 70% of Brazil's population lives within the historical distribution of the Atlantic Forest, and most of its endemic species are threatened. Yet the Atlantic Forest still harbors more than 5% of Earth's butterfly species, 7% of the primate species, and more than 430 tree species per hectare.

Nearly all tropical islands fall within one of the designated hotspots, and 9 of the 25 hotspots are mostly or completely made up of tropical islands. As you may recall from Chapter 21, island clusters often harbor many species because their geography fosters adaptive radiations. By definition, because island-dwelling species have limited geographical ranges, their population sizes tend to be small; and small populations always face a high likelihood of extinction. Because most tropical islands also house dense human populations, it is not surprising that they are well represented on the hotspot list.

1. What criteria do ecologists use to identify biodiversity hotspots?
2. Why are conservation biologists especially concerned about the rapid rate of deforestation in the tropics?

53.4 Conservation Biology: Principles and Theory

Conservation biology is an interdisciplinary science that focuses on the maintenance and preservation of biodiversity. Conservation biologists use theoretical concepts from systematics, population genetics, behavior, and ecology to develop ways to protect threatened wildlife. We introduce theoretical aspects of conservation biology in this section and practical applications in the next.

Systematics Organizes Our Knowledge of the Biological World

To develop a conservation plan for any habitat, scientists must start with an inventory of its species. Their primary tool is systematics, the branch of biology that discovers, describes, and organizes our knowledge of biodiversity (see Chapter 23).

Cataloguing the diversity of life may be the most daunting task that biologists face. After more than 200 years of work, systematists have described and named approximately 1.6 million species. However, they realize that this number represents only a fraction of existing species.

In 1982, Terry Erwin of the Smithsonian Institution studied beetle biodiversity at the Tambopata National Reserve in Southern Peru. He sprayed biodegradable insecticide into the canopy of one large tree and collected 15,869 individual beetles, which he sorted into 3429 species. More than 90% of the individual beetles he collected belonged to species that had not yet been described. Erwin used this astounding result and a complex mathematical model to predict that approximately 30 million species currently exist.

Nigel Stork of the Natural History Museum in London later questioned Erwin's conclusions. Using additional data and a modified set of assumptions, he estimated that the actual number of living species was closer to 100 million. If his figure is correct, more than 98% of species—most of them arthropods, nematodes, bacteria, and archaeans—are still unknown to science. Regardless of whether biodiversity encompasses 30 million species or 100 million, systematists clearly have much work to do.

Recently, conservation biologists and systematists have begun to develop a new technology that will simplify the identification of species in the field, thereby facilitating the creation of a catalog of biodiversity. *Insights from the Molecular Revolution* describes the effort to develop a "DNA barcode scanner."

Population Genetics Informs Strategies for Species Preservation

When populations are reduced to small size, genetic drift inevitably reduces their genetic variability (see Section 20.3) and the evolutionary potential to adapt to changing environments. Thus, the loss of even a small fraction of a species' genetic diversity reduces its survival potential. To avoid this problem, conservationists strive not only to increase the population sizes of threatened or endangered species but to maintain or increase their genetic variation, both within and between populations.

For example, the whooping crane *(Grus americana)* was once an abundant bird in wet grassland environments through much of central North America **(Figure 53.14).** By the early 1940s, excessive hunting and habitat destruction had caused their numbers to decline to just 21 individuals in two isolated populations. This population bottleneck and the resultant loss of genetic variability apparently contributed to developmental deformities of the spine and trachea that had not been seen previously.

During the 1970s, biologists began an aggressive conservation program. In addition to preserving habitats in the crane's summer and winter ranges, they initiated a carefully controlled captive breeding program designed to minimize the effects of inbreeding. Although more than 300 whooping cranes now survive in several wild and captive populations, recent research reveals that they still have a remarkably low level of genetic variability. As expected, the genetic effects of a severe population bottleneck may persist long after a population begins to increase in size.

Figure 53.14
Whooping cranes. Endangered whooping cranes *(Grus americana)* winter in the Aransas National Wildlife Refuge in Corpus Christi, Texas.

Brian K. Miller/Animals, Animals—Earth Scenes

Developing a DNA Barcode System

Everyone is familiar with the checkout scanners at supermarkets and other stores. The cashier quickly passes an item's barcode over the scanner, and the register identifies it and records its price. The system works because the barcode on every item contains unique identifying information. Some biologists have proposed an analogous method, called DNA barcoding, for identifying animal and plant species quickly and accurately. The researchers envision using a handheld device to rapidly analyze DNA in the field; the resulting data would be sent to a database by cell phone, and minutes later an identification and a description of the species would appear on the instrument's screen.

While the analytical device is not yet ready for use in the field, DNA barcoding is now being tested. This technique is the brainchild of Paul Hebert, a population geneticist at the University of Guelph, Ontario, Canada. His idea has caught on, and in 2004 a consortium of major natural history museums and herbariums started the Barcode of Life Initiative, with the goal of creating a database of DNA barcodes linked to specimens already identified in their collections. The approach potentially could replace the traditional methods of systematic analysis using organismal and genetic characters to identify species.

Hebert proposed using the first part of the *COI* (cytochrome oxidase 1) gene—a sequence of about 500 nucleotides—as the DNA barcode to distinguish animal species. This mitochondrial gene tends to vary greatly between species. Moreover, it appears to have no inserted or deleted DNA

segments in most animal species, making the alignment and comparison of sequences straightforward. Hebert's hope is that any *COI* gene sequence obtained in the field will provide a unique identifier for the species from which the DNA sample was obtained.

Early tests of Hebert's barcode approach have been promising. He and his collaborators first analyzed the *COI* gene sequence in the skipper butterflies of Costa Rica. Although adult skippers look pretty much alike, their caterpillars vary in appearance and in their food plant preferences, leading researchers to wonder if butterflies that had been assigned to one species (*Astraptes fulgerator*) might actually represent several. Analyses of the *COI* gene sequence sampled from 484 adults allowed Hebert and his colleagues to identify 10 distinctive DNA barcodes, suggesting that there are at least 10 species of skipper in Costa Rica rather than just one.

In early 2007, Hebert and his colleagues reported that they had used the DNA barcode to analyze 2500 specimens of 643 North American bird species. The results were impressive: barcode differences between species were an order of magnitude greater than the differences within species, allowing the unambiguous identification of species from a short DNA sequence. Interestingly, the barcode analysis identified 15 probable new species that had not been previously identified and revealed that 8 supposed species of gull may be variants of just one species.

Taken together, the results of the two research studies provide support for the use of DNA barcodes, and for

using the *COI* gene sequence specifically for the barcode analysis, as a means of identifying animal species.

Can DNA barcodes be used to identify plant species? The mitochondrial *COI* gene sequence used for barcoding animals is not suitable for barcoding plants because the gene has evolved much more slowly in plants and therefore exhibits less variability among species. However, in 2005, researchers reported on a study that used two different DNA sequences, one from the nuclear genome and the other from the chloroplast genome, to barcode flowering plants. Trials involving 53 plant families, with a total of 99 species from 80 genera, suggested that the two sequences could distinguish a large number of flowering plant species, making barcoding of flowering plants a feasible proposition.

Despite these early successes, many skeptics believe that DNA barcoding will prove to be inaccurate and may, in fact, produce false conclusions about species designations and incorrect counts of biodiversity. The skeptics argue that the approach has not been tested sufficiently in closely related species, which may exhibit only small differences in the barcode sequence. They also point out that the assumption that organisms have a fixed genetic characteristic—like the barcodes on items at the supermarket—contradicts fundamental ideas about genetic variability that are at the core of contemporary evolutionary theory. The DNA barcoding efforts continue nonetheless, and new data will continue to fuel the debate between the supporters of this approach and the naysayers.

Studies of Population Ecology and Behavior Are Essential Elements of Conservation Plans

Conservation programs also require data about target species' ecology and behavior, including their feeding habits, movement patterns, and rates of reproduction.

Sea otters (*Enhydra lutris*) are predatory marine mammals that live along the coastline of the North Pacific Ocean. In the early 1700s, they numbered ap-

proximately 300,000 individuals (**Figure 53.15**), but commercial hunting reduced their numbers to about 3000 individuals by the start of the twentieth century. Sea otters are keystone predators (see Section 50.4), and the destruction of sea otter populations had profound effects on the communities in which they lived. As the numbers of sea otters plummeted, populations of sea urchins, one of their favored prey, exploded; burgeoning sea urchin populations decimated local kelp

a. Sea otter

Malcolm Schuyl/Peter Arnold, Inc.

b. Geographical range of sea otters

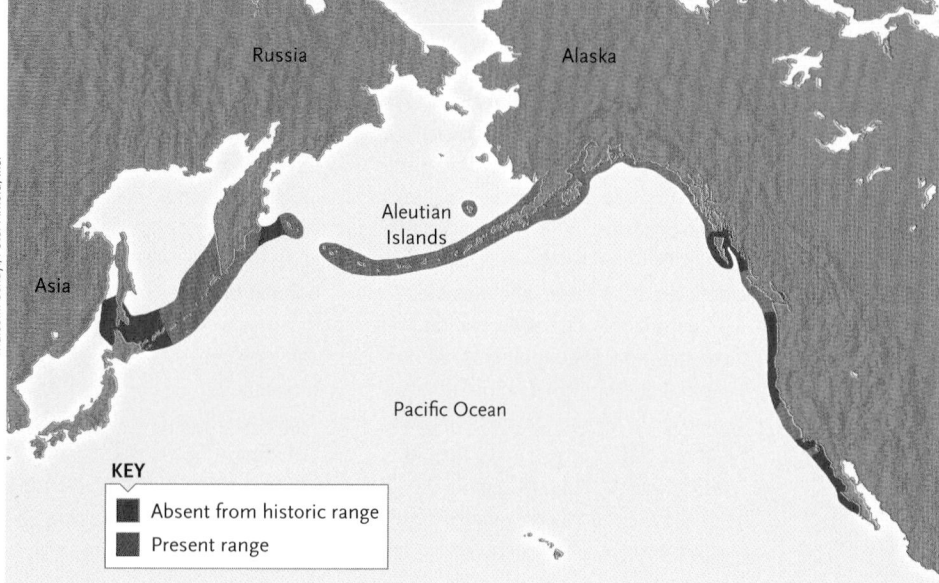

Russia

Alaska

Aleutian
Islands

Asia

Pacific Ocean

KEY

Absent from historic range

Present range

Figure 53.15
Sea otters. After being hunted nearly to extinction, **(a)** sea otters *(Enhydra lutris)* have been reintroduced in many parts of their historical range **(b)**.

beds, disrupting the communities of animals that live among these giant algae.

International treaties ended nearly all hunting of sea otters in 1911, and the populations subsequently recovered to about one-third of their original levels. Conservation biologists facilitated the recovery by reintroducing otters into southeastern Alaska, British Columbia, Washington, and California. Before deciding where otters should be reintroduced, scientists had to assess the resources available at different sites and determine how far individual otters would move, how rapidly they would reproduce, and how quickly their populations would spread. The reintroduction effort was successful at first. However, populations in California have experienced high mortality since the mid-1990s, and nearly half of those dying have been adults in their reproductive prime. Researchers have identified parasitic infections and heart disease as leading causes of death, suggesting that some coastal environments are so badly degraded that they may no longer support populations of this species.

Given the complexities of the ecological relationships in natural communities and ecosystems, conservation biologists have developed two sophisticated types of population analysis, *population viability analysis* and *metapopulation dynamics,* to design effective conservation plans.

Population Viability Analysis. Using complex mathematical models, conservation biologists can conduct a **population viability analysis** (PVA) to determine how large a population must be to ensure its long-term survival. PVAs evaluate phenomena that may influence the longevity of the population or species: habitat suitability, the likelihood of catastrophic events, and other factors that may cause fluctuations in demo-

graphics, population size, or genetic variability. When conducting a PVA, researchers must decide what level of risk is acceptable for a given survival time. For example, should a conservation plan attempt to ensure a 95% probability that the species will survive for 100 years, or should it specify a 99% survival probability? An increase in either the survival probability or the survival time requires an increase in the size of the population that must be conserved. The **minimum viable population size** identifies the smallest population that fits the desired specifications of the conservation plan. *Focus on Research* describes how biologists used PVA in the conservation of an Australian marsupial, the yellow-bellied glider.

Metapopulation Dynamics. In many species, individuals move frequently from one local population to another. To describe the dynamics of such movements, ecologists define a **metapopulation** as a group of neighboring populations that exchange individuals. Local populations within a metapopulation are not all equal: they often differ in size, population growth rates, the suitability of their habitats, their exposure to predators, and other factors. Moreover, some may decline steadily in size, while others may increase.

Under favorable circumstances, a population may produce numerous offspring, some of which emigrate and join nearby populations, where they breed, providing a genetic connection between local populations (see the discussion of gene flow in Section 20.3). Thus, dispersal and gene flow between local populations maintain the metapopulation.

Populations that are either stable or increasing in size are described as **source populations** because they are a possible source of immigrants to other populations. Those that decline in size are called **sink**

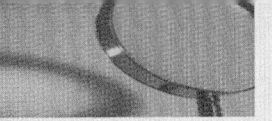

Applied Research: Preserving the Yellow-Bellied Glider

Predicting the future is never easy, especially the future of a threatened species. But population viability analysis (PVA) allows conservation biologists to predict how a species will fare under a range of possible scenarios. An effective PVA for an animal species requires detailed information about its diet, predators, mating habits, habitat preferences, space requirements, demography, geographical distribution, responses to climatic fluctuations and human disturbances, and a host of other aspects of its biology.

The Australian yellow-bellied gliding marsupial, *Petaurus australis*, better known as the yellow-bellied glider, provides an example of how PVA is essential for a conservation effort. This mammal, about the size of a squirrel, lives in small family groups in undisturbed *Eucalyptus* forests along Australia's eastern coast. Each glider family maintains a home range (the area it uses for feeding and other activities) of 25 to 85 hectares; the home ranges of neighboring families do not overlap. As a result, the population density of gliders has never been high. But glider populations have declined precipitously as forests have been cleared, and the species is now considered threatened.

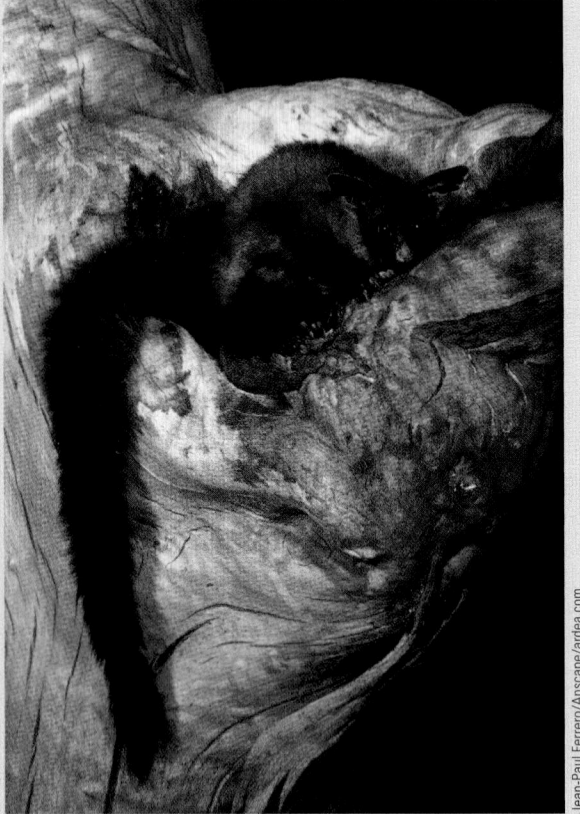

Jean-Paul Ferrero/Anscape/ardea.com

Using data from nearly 20 published papers, two Australian conservation biologists, Russ Goldingay of the University of Wollongong and Hugh Possingham of the University of Adelaide, conducted a PVA for this species. They estimated age distributions in glider populations as well as survival probabilities, litter sizes, sex ratios, lifespan, and home range sizes. They analyzed these data using a mathematical model that predicts the viability for populations of various sizes. In most PVAs, a population is considered viable if it has a 95% probability of surviving for 100 years. Goldingay and Possingham introduced additional complexity to their analysis by assessing the effects of unpredictable environmental events, such as drought, on breeding success. They also conducted sensitivity analyses to examine how changing the values of specific parameters—such as litter size, mortality rates of the different age classes, or the frequency and severity of droughts—might influence the general predictions of the viability model.

Once Goldingay and Possingham had completed many thousands of these calculations, they concluded that a viable population of gliders would require at least 150 family groups. They also suggested that a population of that size would need approximately 18,000 hectares (roughly 70 square miles). Currently, only 1 of the 15 existing conservation reserves is that large.

Goldingay and Possingham did not factor some common environmental disturbances—fire, disease, or predation by introduced species—into their analyses. Such disturbances could decimate a small glider population in short order. Thus, the outlook for gliders may be bleaker than the researchers suggest, because their estimates of minimum viable population size and minimum necessary habitat size are almost certainly too low. Given only this information, we might predict that the glider will inevitably become extinct.

However, there is some hope for the yellow-bellied glider. Goldingay and Possingham assumed that gliders don't move between populations, a behavior that promotes gene flow. They ignored this aspect of metapopulation dynamics because they had no data on gene flow in this species. The movement of individuals between populations could reduce the required minimum viable population size by decreasing the likelihood of genetic drift and the extinction of local populations. Biologists may even be able to transplant gliders from one population to another, effectively creating source and sink populations. This procedure might increase population size and genetic diversity in the most endangered populations. If successful, such an approach could stave off extinction.

As a result of this PVA, conservation biologists can determine which of the remaining forest tracts are large enough to sustain a yellow-bellied glider population. Thus, they now know where to concentrate their limited resources to secure the future survival of this species. Although predicting the future is difficult, PVAs allow conservation biologists to make accurate and reliable recommendations for selective transplants that will contribute to the conservation of threatened species.

Figure 53.16 Observational Research

Metapopulation Structure of the Bay Checkerspot Butterfly

Paul Ehrlich

Map of serpentine habitat patches near Morgan Hill

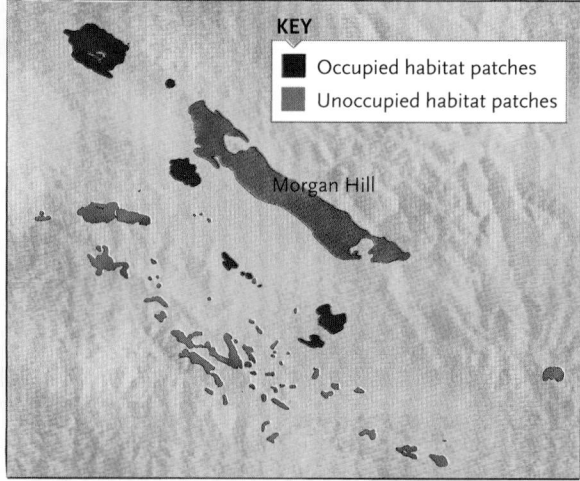

KEY
- ■ Occupied habitat patches
- ▨ Unoccupied habitat patches

Morgan Hill

10 km

HYPOTHESIS: Populations of the bay checkerspot butterfly *(Euphydryas editha bayensis)* living on small patches of suitable habitat are "sink" populations that frequently become extinct. Populations in large habitat patches can serve as a "source" of individuals to recolonize small habitat patches nearby.

PREDICTION: Because the bay checkerspot butterfly is a weak flyer, small patches of suitable habitat that are close to a large source population will be recolonized frequently. Patches of suitable habitat that are far from a large source population will be recolonized only rarely.

METHOD: Susan Harrison, Dennis D. Murphy, and Paul R. Ehrlich of Stanford University surveyed 59 small patches of serpentine grassland near San Jose, California, in 1986 and 1987. They estimated each patch's "quality" based on the presence or absence of food plants on which bay checkerspots depend and on aspects of the physical environment that are important to these butterflies. They also measured each patch's distance from Morgan Hill, a very large patch of suitable habitat that had sustained a bay checkerspot population for years. In patches where they found butterflies, they estimated bay checkerspot population sizes.

RESULTS: A complex statistical analysis revealed that both distance from the Morgan Hill population and habitat patch quality were important factors in determining whether bay checkerspots would be present or absent in a small habitat patch. The authors noted that only the nine high-quality habitat patches near Morgan Hill (red on the map) were occupied by bay checkerspots. Of 50 unoccupied habitat patches, 6 were near Morgan Hill but of low quality; 18 were of high quality but far from Morgan Hill; and 26 were too far from Morgan Hill and of too low quality to support a population of bay checkerspots.

CONCLUSION: Populations of bay checkerspot butterflies that occupy large patches of suitable habitat serve as source populations for individuals that recolonize small patches of suitable habitat where butterfly populations frequently become extinct. However, because the bay checkerspot is a weak flyer, it recolonizes small patches of suitable habitat only if they are close to a source population.

populations because they represent a drain on the supply of available immigrants. Individuals usually move from source populations to sink populations, and sink populations persist because they receive immigrants from source populations in the metapopulation.

The bay checkerspot butterfly, *Euphydryas editha bayensis* (**Figure 53.16**), provides an example of metapopulation dynamics. This species is restricted to serpentine grassland in the San Francisco Bay area (see Figure 50.18) because its larvae eat plants that grow only in that community. Human disturbance has fragmented much of the butterfly's natural habitat into patches of varying size, each of which may support a local butterfly population. The life cycle of these butterflies is always a race against time, because the larvae must feed and mature before dry summer weather kills their food plants. Populations in small patches often become extinct, but those occupying larger patches, where food plants stay alive longer, generally survive the seasonal drought. Butterfly populations in larger

habitat patches therefore serve as source populations for emigrants that repopulate small habitat patches the following year. But the bay checkerspot is a poor flyer, and it cannot disperse long distances. Thus, small patches of suitable habitat harbor bay checkerspots only if they are close to a larger patch that serves as a source. A conservation plan for this butterfly would therefore aim to preserve habitat patches of sufficient size to serve as sources for nearby smaller patches.

Community and Landscape Ecology Help Large-Scale Preservation Projects

Many conservation efforts focus on the preservation of entire communities or ecosystems. These projects often depend on the work of community and landscape ecologists.

Species/Area Relationships. As you know from Chapter 50, community composition is dynamic: some species become extinct and others join the community

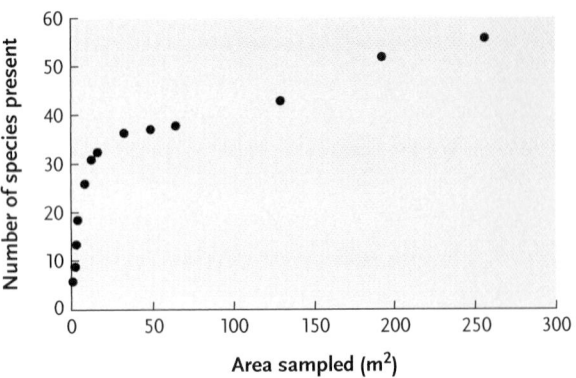

through immigration. If we view fragmented patches of intact habitat as islands in a sea of unsuitable terrain, we can apply the predictions of the theory of island biogeography (see Section 50.7) to the design of protected areas. For example, we might expect that the number of species a patch will support depends on its size and proximity to larger patches.

Indeed, ecologists recognized long ago that large habitat patches sustain more species than small patches do **(Figure 53.17).** When plotted on an arithmetic scale, the relationship between species richness and habitat area increases sharply at first and then flattens. In other words, for relatively small habitat patches, even minor increases in area allow a large increase in the number of resident species; but as habitat patches get larger, the number of species present eventually levels off. You encountered an example of this relationship in our discussion of bird species richness on islands of different sizes (see Figure 50.30b).

As habitats become increasingly fragmented, edge effects exaggerate the species/area relationship in mainland habitat patches **(Figure 53.18).** Consider two hypothetical patches of habitat: one is 100 m on a side, with a total area of 10,000 m²; the other is 200 m on a side, with a total area of 40,000 m². Now, imagine that edge-effect disturbances penetrate 20 m into each patch from all directions. The small patch contains only 3600 m² of intact habitat, but the large patch con-

tains 25,600 m² of intact habitat. Although the large patch is only four times larger than the small patch, the large patch contains more than seven times as much *intact* habitat.

Landscape Ecology. Researchers in the field of **landscape ecology** determine how large-scale ecological factors—such as the distribution of plants, topography, and human activity—influence local populations and communities. Knowing that larger protected areas will preserve more species, conservation biologists have debated whether nature preserves should comprise one large habitat patch or several smaller patches. Ecologists identify this debate with the acronym **SLOSS** (*S*ingle *L*arge *O*r *S*everal *S*mall). Jared Diamond of the University of California, Los Angeles, initiated the SLOSS debate in 1975. Applying the lessons of island biogeography, Diamond concluded that a single large preserve was preferable to several smaller ones, even if they encompassed an equivalent area.

Conservation biologists have since concluded that no single design is best for all organisms. For large animals, such as predatory cats, one large preserve may be best, because individuals must patrol large areas to search for food. For smaller animals, such as insects, several small preserves, each providing a slightly different environment that supports one population, is preferable; if a population in one preserve becomes extinct, individuals from elsewhere in the metapopulation can recolonize the area.

Diamond also suggested that small preserves would function better if corridors of intact habitat connected them. Individuals could move between preserves, reviving any local populations that experienced a decline. These landscape corridors might effectively join the smaller constituent populations into one larger population, which would avoid some of the genetic difficulties encountered by small populations.

However, some conservation biologists argued that landscape corridors connecting small preserves may actually threaten biodiversity. Corridors are usually narrow and thus subject to strong edge effects. In some environments, they are drier and more susceptible to fires that could spread into the preserves they connect. Corridors might also provide entry points for exotic species and disease-causing organisms. Finally, species that don't enter habitat edges would be unlikely to use the corridors at all.

Ellen I. Damschen of North Carolina State University and several colleagues conducted an ambitious long-term field experiment on the effect of landscape corridors on plant species richness **(Figure 53.19).** Their results, published in late 2006, suggest that habitat patches connected by corridors retain more native plant species than isolated patches and that corridors did not promote the entry of exotic species. Thus, based on limited experimental evidence, corridors appear to be a useful feature in the design of nature preserves.

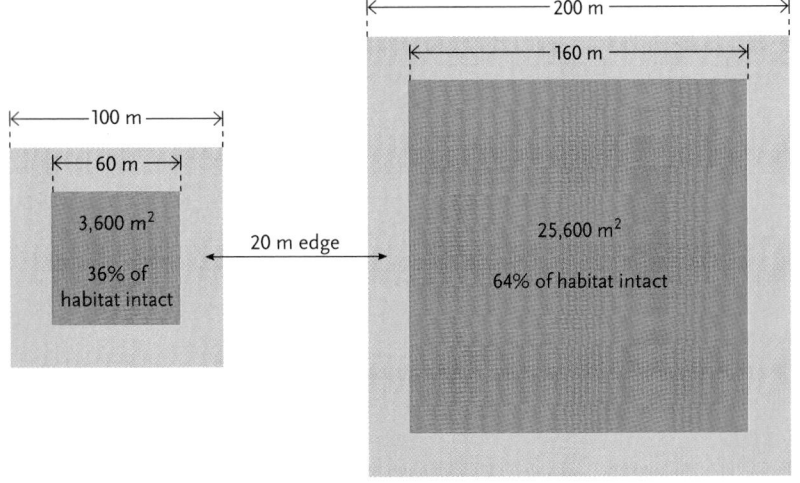

Landscape corridors may also allow large animals to move freely between patches of suitable habitat. For example, the Florida panther *(Puma concolor coryi)*, shown on page 1229, is critically endangered: only 70 to 100 individuals of this subspecies remain from a population that once ranged throughout the southeastern United States; other panther subspecies still inhabit the western states. Panthers are large predators, and each female requires nearly 20,000 hectares (more than 75 square miles) for hunting and breeding; males each require more than twice as much space.

Although the state and federal governments have set aside several panther conservation areas in Florida, 52% of the habitat panthers occupy is privately owned, and most of it is highly fragmented. Panthers frequently cross roads, and most panther deaths in Florida are caused by accidents with motor vehicles. Protected landscape corridors might enable panthers to move more safely between conservation areas. A preliminary study found that panthers already use such corridors, typically along wooded riverbanks, when they are available. The Florida Fish and Wildlife Service has proposed the creation of an ambitious 6100-hectare network of such corridors alongside the Caloosahatchee River to link several significant habitat fragments in neighboring counties.

STUDY BREAK

1. How does a population bottleneck change the likelihood that a species will become extinct?
2. How does a population viability analysis assist in the development of a conservation plan for a species?
3. Would a single large nature preserve or several small preserves experience greater edge effects?

53.5 Conservation Biology: Practical Strategies and Economic Tools

Conservation biology seeks to protect native species, communities, and ecosystems from the effects of human activity. Meeting that goal and reversing some of the existing damage requires the integration of biological research with economic and social realities.

Conservation Efforts Aim to Preserve, Conserve, and Restore Habitats

Conservation groups often highlight efforts to preserve individual animal species, such as the giant panda *(Ailuropoda melanoleuca)* or California condor *(Gymnogyps californianus)*. The preservation of "charismatic megavertebrates," as these large animals are sometimes described, attracts substantial public sup-

Figure 53.19 Experimental Research

Effect of Landscape Corridors on Plant Species Richness in Habitat Fragments

QUESTION: Do landscape corridors connecting habitat patches influence the species richness of native and exotic plants within the habitat patches?

EXPERIMENT: Damschen and her colleagues studied changes in the community composition and species richness of the plants in open habitat patches within a longleaf pine *(Pinus palustris)* forest in South Carolina. Their experimental design included both isolated patches and patches that were connected to one another by a landscape corridor. All patches included the same land area, and their large size (1.375 ha each, including the landscape corridors) allowed the researchers to make a realistic assessment of the effects of landscape corridors. After creating the patches of open habitat within the forest in 2000, the researchers catalogued all plant species occurring in the patches through 2005, although they were unable to collect data in 2004.

RESULTS: Over the course of the study, habitat patches that were connected by landscape corridors harbored increasingly more plant species than did isolated habitat patches. The researchers also noted that the difference in species richness between the two experimental treatments was caused by a difference in the number of native plant species present. The number of exotic species in connected and isolated habitat patches was similar.

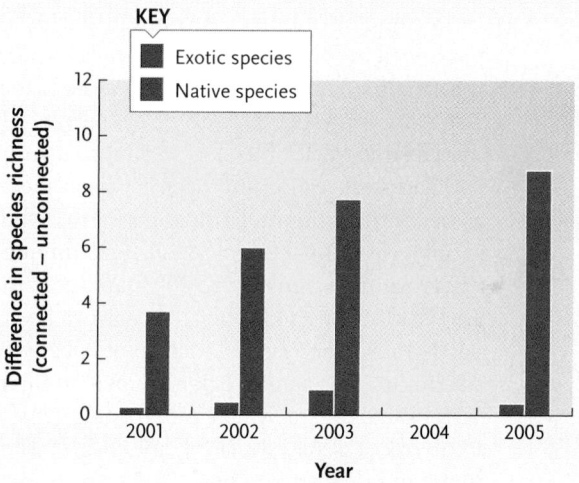

CONCLUSION: Landscape corridors between patches of open habitat in longleaf pine forests increase the species richness of native species in open habitat patches, but they do not foster the entry of exotic species.

port. Nonetheless, there is little point in trying to preserve natural populations of individual species if their habitats are in jeopardy. An alternative to species-based conservation focuses on the preservation of intact habitats; individual species are conserved as a consequence of preserving the habitats on which they depend. Conservation biologists approach this goal with a continuum of approaches, which fall into three general categories: *preservation, mixed-use conservation,* and *restoration.*

a. Albany Pine Bush

b. Karner Blue butterfly

Figure 53.20

The Albany Pine Bush habitat. **(a)** The Pine Bush lies entirely within the city limits of Albany, New York. It is home to about 50 threatened or endangered plant and animal species, including **(b)** the Karner Blue butterfly *(Lycaeides melissa samuelis)*.

Conservation through Preservation. In many countries, habitats are preserved when an individual or organization purchases them and enforces strict standards of land use. In sensitive habitats, people may be excluded altogether; in other cases, access is restricted and the exploitation of resources is controlled. This approach works well in countries with efficient law enforcement and a tradition of private land ownership. In the United States, for example, the Nature Conservancy has purchased large tracts of land to preserve native species.

The preservation approach has been successful in preserving portions of the Pine Bush habitat near Albany, New York **(Figure 53.20)**. This unique ecosystem arose approximately 11,000 years ago at the end of the last glacial period, when a massive deposit of sand was left near the western margin of Albany's current city limits. This sandy region formed an inland pine-barrens habitat in which pitch pine *(Pinus rigida)*, scrub oak *(Quercus ilicifolia)*, and dwarf chestnut oak *(Quercus prinoides)* are now the dominant vegetation. The Pine Bush is home to more than 50 plant and animal species that the state and federal government list as threatened or endangered. The habitat itself was once vulnerable because it lies within Albany's city limits; however, since 1988, the Pine Bush has been jointly owned and protected by New York state, local municipalities, and the Nature Conservancy.

Mixed-Use Conservation. The preservation approach does not work under all circumstances. Where outright preservation is impractical, conservation biologists advocate mixed-use con-

servation, which combines the protection of some land parcels with the controlled development of others.

The Ngorongoro Conservation Area (NCA) in Tanzania provides an example of mixed-use conservation. The NCA covers 829,000 hectares of grassland and borders the Serengeti National Park. Because it houses a high concentration of wildlife, the NCA is one of the most heavily visited tourist destinations in eastern Africa. For the past several hundred years, the Maasai people have herded cattle, goats, and sheep in the Serengeti and Ngorongoro **(Figure 53.21)**. The Maasai are nomadic pastoralists who frequently move their relatively small herds to new grazing areas in the region. As a result, their activities do not degrade the land or exclude native wildlife. In 1959 the Maasai agreed to vacate the Serengeti, which was converted into a national park, in return for retaining the rights to live and herd livestock within the NCA. The government of Tanzania helped create the necessary infrastructure within the NCA, including a constant water supply as well as social services. Under this agreement, 40,000 indigenous residents, most of them Maasai, live in this large and valuable conservation area.

Conservation through Restoration. Conservation biologists sometimes create restoration plans to reestablish the vitality of a previously disrupted community or ecosystem. This effort requires the removal of contaminants, impediments to the natural flow of water, and barriers to animal movement as well as the restoration of natural processes, such as periodic fires or floods. Most restoration projects also require replanting key plant communities and long-term management once restoration is complete.

Not all degraded habitats can be restored, and not all potential restoration projects are equally feasible. When making project decisions, restoration ecologists

Figure 53.21

Mixed-use conservation. The Maasai use the Ngorongoro Conservation Area to graze cattle and goats.

consider a number of factors: Will the restored habitat be suitable for rare or endangered species, and will its creation increase endemic biodiversity? Would the restoration reunite previously fragmented land parcels? Will the restored habitat experience the periodic disturbances, such as fires or floods, that are essential for its continued existence? What are the costs of implementing the plan and maintaining the area? Finally, would the restored land be valued by local residents, and will they support and maintain it?

A successful restoration project is currently underway in the Brazilian Atlantic Forest, sponsored by the Instituto de Pesquisas Ecológicas (IPÊ), a Brazilian nongovernmental organization. In western São Paulo state, near the Morro do Diabo state park, IPÊ is trying to recreate the natural Brazilian Atlantic Forest ecosystem by planting native trees in habitat corridors between remaining forest fragments. These corridors of native tree species should facilitate the preservation of species in those forest patches and supply valuable botanical resources for endemic wildlife and local residents.

Successful Conservation Plans Must Incorporate Economic Factors

Biologists can almost always develop a plan to conserve a species, community, or ecosystem. But to be successful, a plan must be economically feasible, and it must provide direct benefits to local residents whose lives it will affect.

Local Involvement. Early conservation efforts simply set aside protected areas in which most human activities were banned. Local people were denied access to resources within the preserve—resources that were sometimes essential for their survival. Not surprisingly, these plans generated antipathy towards conservationists and the organisms they were trying to preserve.

For example, the northern spotted owl (*Strix occidentalis caurina*) lives only in old growth coniferous forests of the Pacific Northwest, where many local residents worked in forestry or supporting industries. The suggestion that the owl be listed as an endangered species triggered a bitter political battle between conservationists and local residents because the conservation plan for the owls required closing large tracts of forest to logging. Washington State listed the owl as an endangered species in 1988, but local residents, who lost jobs when logging was reduced, remain hostile to these conservation efforts.

Conservation plans are more successful if they provide local residents with benefits that depend on the existence of a preserve. Royal Chitwan National Park provides an excellent example. For more than 100 years, this area, located in south central Nepal near the northern border of India, was a privately owned "preserve" used for big game hunting by royalty. These

activities decimated local populations of large mammals, especially the Bengal tiger (*Panthera tigris*) and one-horned or Indian rhinoceros (*Rhinoceros unicornis*). Populations of both species dwindled to approximately 100 individuals by the mid-1960s. The area was subsequently opened for settlement, and, as immigrants swarmed into the fragile grassland, its human population exploded.

The area was converted into Royal Chitwan National Park in 1973. Today, humans are excluded from the park for most of the year. But each January, after the monsoon rains end and the grasses have dried, local residents are welcomed into the park for their annual harvest festival. They cut the grass and carry it away to thatch roofs, make mats, and feed domestic animals **(Figure 53.22)**. The local people value Chitwan and argue for its preservation. Today, more than 600 one-horned rhinos survive in Nepal, most of them in Royal Chitwan National Park. And the Bengal tiger population of Chitwan has increased to approximately 250 individuals.

Ecotourism. In some preserves, governments enlist local residents in park development and operations, providing them with a viable livelihood. The most successful approach has been the development of **ecotourism**, in which visitors, often from wealthier countries, pay a fee to visit a nature preserve. Local people work as guides, cooks, and logistical and support staff.

Not everyone agrees that ecotourism is helpful. Critics note that increased human traffic may degrade habitats, and unregulated ecotourism can eventually lead to overdevelopment. For example, several million people visit national parks in the western United States annually. Traffic jams, automobile accidents, and long lines routinely plague visitors at the most popular sites. Cranky ecotourists call for the construction of more roads and parking lots, which are inconsistent with the purpose of a national park because cars increase local air pollution and occasionally kill wildlife. In 2006, the government began charging a $20 fee for each automobile entering Yosemite Na-

Figure 53.22
Conservation and the local economy. Local residents support conservation efforts at the Royal Chitwan National Park, Nepal, because officials open the park for a grass harvest each year.

Ed Degginger/Color-Pic

tional Park in California, hoping to limit the number of visitors arriving in private vehicles and to increase reliance on public transportation.

Countrywide Economic Approaches. In the mid-1990s, conservation biologists and economists developed the concept of **ecosystem valuation**, in which ecosystem services—such as carbon dioxide processing or water retention and purification, which are best provided by intact ecosystems—are assigned an economic value. These estimated values are used to negotiate contracts in which a private company or conservation organization pays a community, state, or country to maintain intact ecosystems. By one 1997 estimate, the gross global ecosystem valuation is roughly 18 trillion U.S. dollars per year. If less obvious benefits provided by nature are tallied—soil formation, crop pollination, and nutrient cycling—the total value of ecosystem ser-

vices rises to 33 trillion U.S. dollars—almost twice the value of all goods produced by all humans on the planet!

The implementation of ecosystem valuation exchanges is determined on a case-by-case basis, depending on what ecosystem services the paying organization wants to preserve. Costa Rica is leading the way in this effort by creating valuation contracts with several corporations. For example, in 1998, the Monteverde Conservation League signed a contract with a local electrical company to ensure the continued flow of water from the Bosque Eternal de los Niños, a forest preserve. The company had plans to build a hydroelectric dam on the Rio Esperanza, and feared that deforestation upstream would disrupt water flow through the dam. The contract specifies that the electrical company will pay the people who live upstream to preserve their forests rather than cutting them.

UNANSWERED QUESTIONS

Are there general patterns in networks of interacting species? What causes those patterns, and what are their consequences for biodiversity conservation?

Biodiversity is a buzzword referring to the diversity of life at its different levels of organization—from molecules and genes to organisms and ecosystems. The term describes not only the component parts but also the way the parts are assembled and how they function. In any ecosystem, organisms interact with each other through a variety of beneficial and detrimental interactions. Ecologists usually describe and summarize these complex interactions as *networks,* in which species are represented as nodes and interspecific interactions as links. Ecologists have long sought to uncover the emergent properties of interaction networks, their causes, and their consequences.

Some early theoretical models of food webs (networks depicting who eats whom in an ecosystem) suggested that species-rich, highly connected networks would be less stable than smaller, simpler networks. This finding astounded many ecologists because it contradicted a widely accepted hypothesis that larger, more complex food webs were more stable and thus more resistant to disturbance, such as invasions by exotic species. Supporters of the "complex food webs are more stable" hypothesis noted that many large, complex food webs—such as those in tropical forests—exist in nature, and their persistence over time had to be explained.

Later theoretical research suggested that the solution to the contradiction might lie in the strength of species interactions (that is, the impact that one species has on others). Food webs with a few very strong interactions and many weaker ones tend to be more stable than food webs with many interactions of medium strength. In other words, food webs that include one or two keystone species tend to be more stable than those without keystone species. Further research has revealed that many real food webs include a few strong interactions and many weaker ones, but ecologists still do not know what factors determine interaction strength and why we often observe only a few strong interactions and many weak ones.

For many years, research on ecological interaction networks has focused almost exclusively on food webs—that is, on predator–prey interactions. More recently, however, other types of interactions (including those between plants and their mutualistic pollinators and seed dispersers, or between hosts and their parasites) have started to receive more attention. This new research has uncovered some intriguing general patterns. For example, mutualistic networks include asymmetric interactions of two general types: (1) species that have few links to other species ("specialists") tend to interact with species that have many links to other species ("generalists"); and (2) species that exhibit weak interactions tend to be associated with species that affect them strongly. These two types of asymmetry in mutualistic networks seem to result from the fact that only a few species have many links and only a few interactions are strong. As was the case for the analysis of food webs, the unresolved issue is *why* only a few species have many links or have strong interactions.

Answering these questions is important not only because they improve our understanding of the complexity of ecological systems, but also because they may have important implications for biodiversity conservation. For example, the widespread existence of asymmetric interactions makes interaction networks highly resistant to perturbations (such as habitat modification and species invasions) that could result in the local extinction of some species. The removal of highly linked species with strong interactions will affect many other species in the community, but because most species have few links and weak interactions, most extinctions will have only minor effects on the overall structure and functioning of the network.

Diego Vázquez is a researcher with the National Research Council of Argentina at the Argentine Institute of Dryland Research in Mendoza, Argentina. He received his *licenciatura* (undergraduate degree) in biology from the University of Buenos Aires and his Ph.D. in ecology and evolutionary biology from the University of Tennessee. His interests include community ecology, plant-animal interactions, mutualism, biological invasions, and conservation biology. You can learn more about his research at http://www.cricyt.edu.ar/interactio.

Thus, both the forests and water flow are preserved, maintaining the forest ecosystem and generating badly needed electricity.

Biodiversity is a precious resource that is disappearing rapidly throughout the world. It can still be conserved through a monumental effort to catalog the diversity of living organisms and develop an understanding of their ecological relationships. Perhaps the major challenge for conservation biologists is the education of the human population about the value of biodiversity and the development of conservation plans that will enlist the support of people who live among the threatened species.

STUDY BREAK

1. Is the Pine Bush habitat in New York State an example of preservation, mixed-use conservation, or restoration?
2. How has the establishment of the Royal Chitwan National Park in Nepal been a successful conservation effort? How do conservation biologists measure its success?
3. How can the concept of ecosystem services be used to foster conservation of threatened habitats and species?

Review

Go to **ThomsonNOW**™ at www.thomsonedu.com/login to access quizzing, animations, exercises, articles, and personalized homework help.

53.1 The Benefits of Biodiversity

- Biodiversity provides direct benefits to humans because natural populations of organisms can be sources of useful natural products as well as genetic resources that can improve domesticated crops and animals (Figures 53.2 and 53.3).
- Biodiversity provides indirect benefits to humans by maintaining normal ecosystem processes, some of which help to counteract the harmful effects of human activities.
- Ethicists and environmentalists argue that biodiversity should be preserved simply because of its intrinsic worth.

53.2 The Biodiversity Crisis

- Human disruption of a habitat usually begins with the construction of a road that provides access to resources; the disruption spreads rapidly. Habitat fragmentation reduces the size of intact habitat patches, and edge effects diminish the quality of remaining habitat (Figure 53.4). Only small populations, which are subject to genetic drift and an increased likelihood of extinction, can inhabit small habitat patches.
- Deforestation is occurring at an alarming rate, especially in tropical regions (Figure 53.5). Excessive deforestation may lead to desertification and the loss of entire ecosystems (Figure 53.6). Deforestation, desertification, and global warming reinforce each other in a positive feedback cycle.
- Although pollution is released locally, it often spreads to regional and global scales, especially in bodies of water and the atmosphere (Figure 53.7). Pollution can take many forms (Figure 53.8).
- Exotic species often contribute to the extinction of native species through competition, predation, or parasitism (Figures 53.9–53.11). Humans frequently introduce exotics into communities either intentionally or inadvertently.
- Overexploitation of natural populations reduces their sizes and may induce evolutionary responses in the exploited populations (Figure 53.12).
- Although extinction has been common in the history of life, human activities have recently initiated what may be the greatest mass extinction of all time. Some biologists estimate that extinction rates today may be 1000 times the background extinction rate.

Animation: Five major extinctions

Animation: Effects of deforestation

Animation: Effect of air pollution in forests

53.3 Biodiversity Hotspots

- Biodiversity hotspots harbor large numbers of endemic species and are threatened by human activities (Figure 53.13). Although hotspots encompass only 1.4% of the land, a much larger proportion of biodiversity inhabits these areas.
- More than half the identified hotspots are in the tropics, and nearly all tropical islands are included within the hotspot designation.
- Preserving the hotspots will conserve a substantial part of Earth's biodiversity.

Animation: Global crises by region and habitat

Animation: Three types of reefs

53.4 Conservation Biology: Principles and Theory

- Conservation biology draws its theoretical foundation from systematics, population genetics, population ecology, behavior, community ecology, and landscape ecology.
- Systematists provide taxonomic inventories of biodiversity that are helpful for establishing conservation priorities.
- Conservation biologists design breeding programs to maintain or increase the genetic variability of species being preserved (Figure 53.14).
- Besides studying the population ecology and behavior of targeted species (Figure 53.15), conservation biologists use population viability analyses to determine the minimum viable population size necessary to conserve threatened species. Analyses of metapopulation dynamics can help conservation biologists understand the interactions among small populations of threatened species (Figure 53.16).
- Studies in community ecology have established the generality of the species/area effect: large habitat patches harbor more species than small habitat patches do (Figure 53.17).
- From the perspective of landscape ecology, biologists have debated the advantages and disadvantages of establishing one large reserve versus several smaller ones that are connected by habitat corridors (Figures 53.18 and 53.19).

53.5 Conservation Biology: Practical Strategies and Economic Tools

- Efforts to conserve communities or ecosystems follow one of three general strategies. *Preservation* requires the restriction or prohibition of human access to the area (Figure 53.20). *Mixed-use conservation,* an approach that balances the conflicting demands of habitat preservation and development, allows local residents to use the protected area in limited ways (Figure

53.21). *Restoration* attempts to recreate natural communities and ecosystems in places that have already been degraded by human activities.

- Conservation plans must also incorporate economic and social factors to win local support. Most conservation plans now include the involvement of local residents to generate revenue for their communities (Figure 53.22). Ecosystem valuation also encourages the preservation of ecosystems by assigning them a significant economic value.

Animation: Sustainable resource management

Questions

Self-Test Questions

1. The greatest extinction in the history of life on Earth:
   a. occurred at the end of the Permian period.
   b. occurred at the end of the Cretaceous period.
   c. occurred at the end of the Ordovician period.
   d. occurred at the end of the Cambrian era.
   e. may be occurring now.

2. Which of the following is usually the first step in the disruption of a natural habitat by humans?
   a. establishment of small villages
   b. planting of crops
   c. building of a road
   d. invasion by exotic species
   e. overexploitation of resources

3. Habitat fragmentation has damaged populations of breeding birds in North America because:
   a. the remaining habitat patches rarely contain enough food for birds to rear their offspring.
   b. the nests of birds in small habitat patches are frequently attacked by predators.
   c. pairs of breeding birds cannot easily move from one habitat patch to another.
   d. female birds cannot locate potential mates in small habitat patches.
   e. small habitat patches do not have enough edges to provide adequate hiding places.

4. Deforestation:
   a. is a problem only in the tropics.
   b. may speed desertification.
   c. is slowed by grazing and farming.
   d. permanently enriches the soil.
   e. leads to the formation of lush grasslands.

5. Chemical pollutants:
   a. can spread rapidly from the places they are released.
   b. do not appear to influence global climate change.
   c. have contributed to global mass extinctions.
   d. rarely affect natural bodies of water.
   e. rarely influence animals feeding at higher trophic levels.

6. Which of the following is most likely to be a biodiversity hotspot?
   a. a patch of forest in the middle of North America that is 500 km from the nearest big city
   b. a series of uninhabitable sand dunes in the Sahara Desert
   c. a botanical garden that houses representatives of 25,000 plant species
   d. a tropical island with many endemic species and a growing human population
   e. a suburban neighborhood where fields have been converted to backyards and playgrounds

7. Population viability analyses allow conservation biologists to:
   a. identify the source population from which an individual dispersed to a sink population.
   b. determine how large an area must be preserved for the protection of a threatened species.
   c. identify whether individuals of a threatened species are reproductively mature.
   d. predict the minimum population size of a threatened species that is likely to survive.
   e. predict whether a threatened species will use habitat corridors.

8. Metapopulations are defined as:
   a. neighboring populations that exchange individuals.
   b. populations that steadily decrease in size.
   c. populations that steadily increase in size.
   d. populations that produce numerous fertile offspring.
   e. populations that never receive immigrants.

9. For which of the following species has the use of habitat corridors been proposed as an important conservation tool?
   a. sea otters
   b. bay checkerspot butterflies
   c. Florida panthers
   d. whooping cranes
   e. Eastern hemlocks

10. The main goal of restoration ecology is the reestablishment of:
    a. natural patterns of water flow.
    b. the vitality of a degraded ecosystem.
    c. the historical corridors linking forest fragments.
    d. the natural barriers to animal movement.
    e. ecotourism.

Questions for Discussion

1. National parks are often established in ecologically sensitive areas. In many places they have become so popular that visitors endanger the ecosystems the parks were originally designed to preserve. How can the goals of conservationists, who work to maintain intact ecosystems, be balanced with those of citizens who wish to visit intact ecosystems? In other words, how would you regulate domestic ecotourism?

2. How do the principles of population genetics and the principles of metapopulation dynamics apply to the SLOSS debate? Do they suggest different ideal designs for nature preserves?

3. Imagine that you are a conservation biologist who has been asked to develop a conservation plan for a species of lizard that lives in the deserts of the American Southwest. What sorts of data would you collect before developing a final plan?

Experimental Analysis

Devise a field study to determine whether the species/area relationship applies to aquatic ecosystems, such as ponds and lakes, as it does to terrestrial habitats.

Evolution Link

Overexploitation of marine fish stocks has depleted natural populations and caused a reduction in the age and size at which many fish species become reproductively mature. What sort of government regulations of fishing might reverse the current trend toward smaller adult size? Explain your answer in terms of the selection pressures that fishing places on targeted species.

How Would You Vote?

Material goods can be manufactured in ways that protect biodiversity but often are more expensive than comparable goods produced without regard for the environment. As a consumer, are you willing to pay extra for the first kind? Go to www.thomsonedu.com/login to investigate both sides of the issue and then vote.

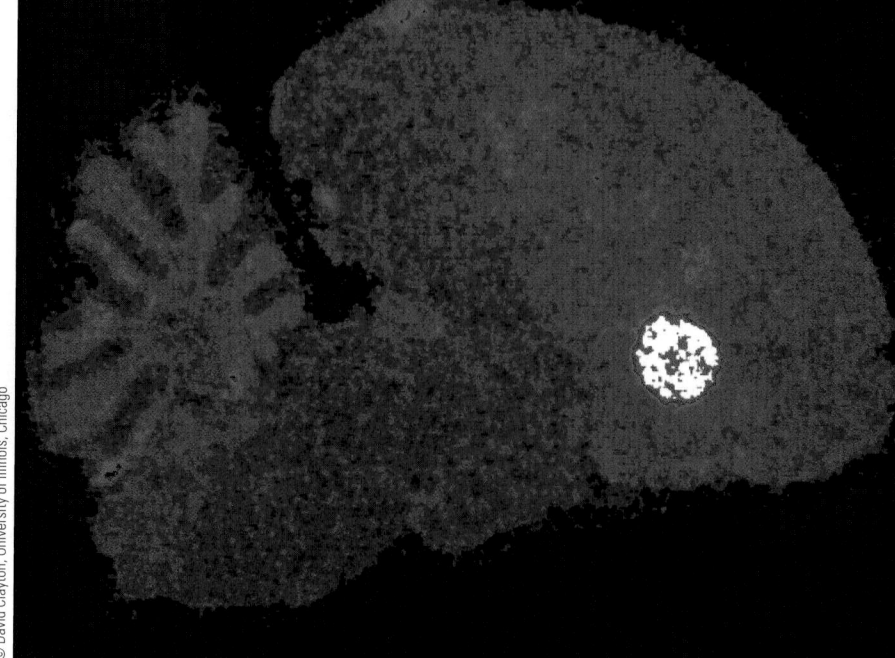

A section of zebra finch *(Taeniopygia guttata)* brain, stained to illuminate expression of the *zenk* gene, which helps a male bird reproduce his species' song.

© David Clayton, University of Illinois, Chicago

54 The Physiology and Genetics of Animal Behavior

WHY IT MATTERS

Male white-crowned sparrows *(Zonotrichia leucophrys)* are handsome birds with a song that birdwatchers describe as a "plaintive whistle" followed by a "husky trilled whistle." This distinctive song is a critical part of a male white-crown's **behavioral repertoire**, the set of actions that it can perform in response to stimuli in its environment. An adult male sparrow's song is one of the ways he struts his stuff. The song not only announces his presence to rival males, but it also signals to females that he is available as a potential mate. Experienced birders easily recognize this song, which differs from that of song sparrows *(Melospiza melodia)* and swamp sparrows *(Melospiza georgiana)*, as sound spectrograms illustrate **(Figure 54.1)**. In fact, every songbird species produces vocal signals that are characteristic of its species and its species alone.

The study of **animal behavior** involves discovering how animals respond to specific stimuli and why they respond in predictable and characteristic ways. A comprehensive approach to animal behavior studies first crystallized in the 1930s, when European researchers—notably Konrad Lorenz, Niko Tinbergen, and Karl von Frisch, who shared a Nobel Prize for their work in 1973—developed the discipline

1253

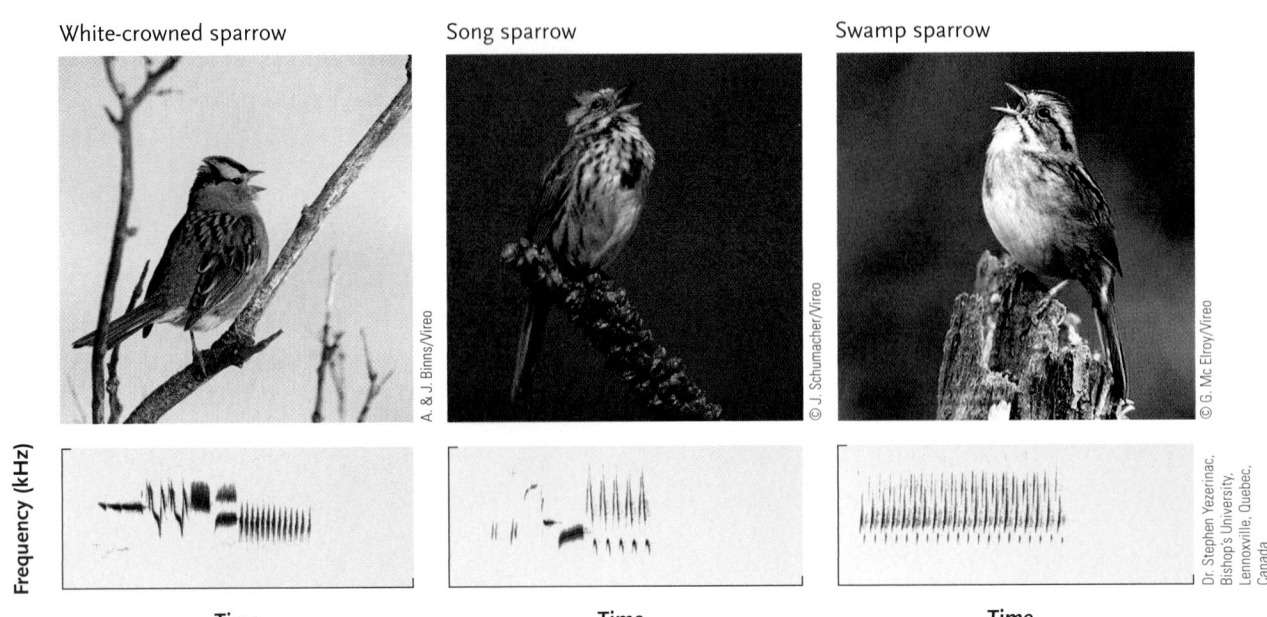

White-crowned sparrow   Song sparrow   Swamp sparrow

Frequency (kHz)

Time   Time   Time

Figure 54.1

Songbirds and their songs. Sound spectrograms (visual representations of sound graphed as frequency versus time) illustrate differences in the songs of the white-crowned sparrow *(Zonotrichia leucophrys)*, the song sparrow *(Melospiza melodia)*, and the swamp sparrow *(Melospiza georgiana)*.

of **ethology**, which focuses on how animals behave in their natural environments. They analyzed how evolutionary processes shape inherited behaviors and the ways that animals respond to specific stimuli. Tinbergen identified four basic questions that any broad study of animal behavior should address: (1) What mechanisms trigger a specific behavioral response? (2) How does the expression of a behavior develop as an animal matures? (3) What is the behavior's function and how does it increase an animal's chances of surviving and reproducing? (4) How did the behavior evolve?

Advances in **neuroscience**—the integrated study of the structure, function, and development of the nervous system—now allow researchers to explore the first and second questions in detail. Comparable advances in genetic analysis and evolutionary theory enable scientists to address the third and fourth questions. In this chapter, we examine the *proximate causes* of behavior—the genetic, cellular, physiological, and anatomical mechanisms that underlie an animal's ability to detect internal stimuli and environmental cues and react to them in species-specific ways. In Chapter 55, we consider the *ultimate causes* of animal behavior—its adaptive value and evolution.

54.1 Genetic and Environmental Contributions to Behavior

For many years, animal behaviorists debated whether animals are born with the ability to perform most behaviors completely or whether experience is necessary

to shape their actions. However, extensive research in neuroscience has demonstrated that no behavior is determined entirely by genetics or entirely by environmental factors. Instead, behaviors develop through complex gene–environment interactions. We illustrate such an interaction below with a detailed description of the process through which male white-crowned sparrows learn their adult song.

Most Behaviors Have both Instinctive and Learned Components

Why do adult male white-crowned sparrows sing a song that no other species sings? One possible explanation is that they possess an innate (inborn) ability to produce their particular song, an ability so reliable that young males sing the "right" song the first time they try. According to this hypothesis, their distinctive song would be an example of an **instinctive behavior**, a genetically "programmed" response that appears in complete and functional form the first time it is used. An alternative hypothesis is that they acquire the song as a result of certain experiences, such as hearing the songs of adult male white-crowns that live nearby. In other words, this species' distinctive song might be an example of a **learned behavior**, one that is dependent upon having a particular kind of experience during development.

How can we determine which of these two hypotheses is correct? If the white-crowned sparrow's song is instinctive, isolated male nestlings that have never heard other members of their species should be able to sing their species' song when they mature. But if the learning hypothesis is correct, young birds deprived of certain essential experiences should not sing "properly" when they become adults.

In a set of pioneering experiments conducted at Rockefeller University, Peter Marler tested these alternative hypotheses. He took newly hatched white-crowns from nests in the wild and reared them individually in soundproof cages in his laboratory. Some of the chicks listened to recordings of a male white-crowned sparrow's song when they were 10 to 50 days old; others did not. The juvenile males in both groups first started to vocalize when they were about 150 days old. For many days, they produced whistles and twitters that only vaguely resembled the songs of adults. But gradually the young males that had listened to tapes of their species' song began to sing better and better approximations of that song. At about 200 days of age, they were right on target, producing a song that was nearly indistinguishable from the one they had heard months before. By contrast, males in the group that had not heard tape-recorded white-crown songs never came close to singing the way wild males do.

These results revealed that learning is essential for a young male white-crowned sparrow to acquire the full song of its species. Although birds isolated as nestlings did sing instinctively, they needed the acoustical experience of listening to their species' song early in life if they were to reproduce it months later. We can therefore reject the hypothesis that white-crowned sparrows hatch from their eggs with the ability to produce the "right" song. Their species-specific song—and presumably those of other songbirds—has both instinctive and learned components.

Although early researchers generally classified behaviors as *either* instinctive *or* learned, we now know that most behaviors include both instinctive and learned components. Nevertheless, some behaviors have a strong instinctive component, whereas others are mostly learned.

STUDY BREAK

1. What is the difference between an instinctive behavior and a learned behavior?
2. How did the isolation of young male sparrows in soundproof cages allow Marler to conclude that learning was important to song acquisition?

54.2 Instinctive Behaviors

Instinctive behaviors—which are often grouped into functional categories, such as feeding behaviors, defensive responses, mating behaviors, and parental care activities—can be performed without the benefit of prior experience. We therefore assume that they have a strong genetic basis and that natural selection has preserved them as adaptive behaviors.

Many Instinctive Behaviors Are Highly Stereotyped

Many instinctive behaviors are highly stereotyped; in other words, when triggered by a specific cue, they are performed over and over in almost exactly the same way. Such behaviors are called **fixed action patterns**, and the simple cues that trigger them are called **sign stimuli**. For example, sign stimuli and fixed action patterns govern the transfer of food from herring gull *(Larus argentatus)* parents to their offspring. Researchers found that very young chicks secure food from their parents through a begging response (the fixed action pattern), which is triggered when they see a red spot on the lower bill of an adult (the sign stimulus). This cue "releases" the begging behavior of hungry baby gulls, which peck at the spot on the parent's bill. In turn, the tactile stimulus delivered by the pecking chick serves as a sign stimulus that induces the adult bird to regurgitate food stored in its crop. The baby gulls then feed on the chunks of fish, clams, or other food that lie before them. We know that the spot on the parent's bill releases the begging response of the young gull because the same response is triggered by an artificial bill that looks only vaguely like a herring gull's bill, provided it has a dark contrasting spot near the tip **(Figure 54.2)**. Thus, even very simple cues can activate fixed action patterns.

Human infants often respond innately to the facial expressions of adults **(Figure 54.3)**. For example, researchers can trigger smiling in even very young babies simply by moving a mask toward the infant, as long as the mask possesses two simple, diagrammatic eyes. Clearly the infant, like a nestling herring gull, is not reacting to every feature of a face; instead it focuses on simple cues, which function as sign stimuli that release a fixed behavioral response.

Natural selection has molded the behavior of some parasitic species to exploit the relationship between sign stimuli and fixed action patterns for their own benefit. For example, birds that are brood parasites lay their eggs in the nests of other species (see *Why It Matters* at the beginning of Chapter 50). When the brood parasite's egg hatches, the alien nestling mimics and even exaggerates sign stimuli that are ordinarily exhibited by its hosts' own chicks: opening its mouth, bobbing its head, and calling vigorously. These exaggerated behaviors elicit feeding by the foster parents, and the young brood parasite often receives more food than the hosts' own young **(Figure 54.4)**.

Although instinctive behaviors are often performed completely the first time an animal responds to a stimulus, they can by modified by an individual's experiences. For example, the fixed action patterns of a young herring gull change through time. Although

Figure 54.2 Experimental Research

The Role of Sign Stimuli in Parent-Offspring Interactions

QUESTION: What feature of the parent's head triggers pecking behavior in young herring gulls?

EXPERIMENT: Niko Tinbergen and A. C. Perdeck tested the responses of young herring gull *(Larus argentatus)* chicks to cardboard cutouts of an adult herring gull's head and bill. They waved these models in front of the chicks and recorded how often a particular model elicited a pecking response from the chicks. One cutout included an entire gull's head with a red spot near the tip of the bill; another cutout included just the bill with the red spot; the third cutout included the entire head but lacked the red spot.

RESULT: Young herring gulls pecked at the model of the bill with a red spot almost as often as they pecked at the model of an entire head with a red spot, but they pecked much less frequently at the model of an entire head that lacked a red spot.

Herring gulls *(Larus argentatus)*

© Marie Read Natural History Photography

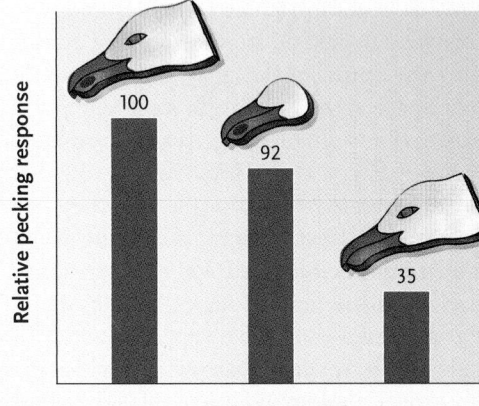

Relative pecking response

100   92   35

Model presented

CONCLUSION: Begging behavior by young herring gulls is triggered by a simple sign stimulus, the red spot on the parent's bill. Experimental tests revealed that herring gull chicks respond more to the presence of the contrasting spot than they do to the outline of an adult's head.

Evan Cerasoli

Figure 54.3
Instinctive responses in humans. The smiling face of an adult is a sign stimulus that triggers smiling behavior in very young infants.

the youngster initially begs by pecking at almost anything remotely similar to an adult gull's bill, it eventually learns to recognize the distinctive visual and vocal features associated with its parents. The chick uses this information to become increasingly selective about the stimuli that will elicit its begging behavior. Thus, instinctive behaviors can be modified in response to particular experiences during their early performances.

Behavioral Differences between Individuals May Reflect Underlying Genetic Differences

Because the performance of instinctive behaviors does not depend on prior experience, behavioral differences between individuals may reflect genetic differences between them. Stevan Arnold, then at the University of Chicago, tested that hypothesis by studying the innate responses of captive newborn garter snakes *(Thamnophis elegans)* to the olfactory stimuli provided by potential food items that they had never before encountered. Arnold measured the snakes' responses to cotton swabs that had been dipped in a smelly extract of banana slug *(Ariolimax columbianus),* a shell-less mollusk. A snake "smells" by tongue-flicking, which draws volatile chemicals into a special sensory organ in the roof of its mouth. If the young snake had been born to a mother captured in coastal California, where adult garter snakes regularly eat banana slugs, it almost always began tongue-flicking at the slug-scented cotton swab **(Figure 54.5).** By contrast, newborn snakes whose parents came from inland California, where banana slugs do not occur, rarely tongue-flicked at the swabs. Thus, although the coastal and inland snakes belong to the same species, their instinctive responses to the volatile chemicals associated with banana slugs were markedly different.

In another experiment, Arnold tested whether newborn snakes would feed on bite-sized chunks of slug. After a brief flick of the tongue, 85% of the newborn snakes from a coastal population routinely struck at the slug and swallowed it, despite having had no prior experience with this prey. By contrast, only 17% of newborn snakes from the inland population ate slugs consistently, even when no other food was available. Arnold hypothesized that coastal and inland garter snakes possess different alleles at one or more gene loci controlling their odor-detection mechanisms, leading to differences in their behavior. To test this hypothesis, Arnold crossbred coastal and inland snakes. If genetic differences contribute to the different food preferences of the two snake populations, then hybrid offspring, which receive genetic information from each parent, should behave in an intermediate fashion. Results of the experiment confirmed his prediction: when presented with bite-sized chunks of slug, 29% of the newborn snakes of mixed parentage consumed them every time.

Many additional experiments have confirmed that genetic differences between individuals can translate into behavioral differences between them. *Insights from the Molecular Revolution* describes a striking example of a single gene that influences the grooming behavior of mice. Bear in mind that single genes do not control complex behavior patterns directly. Instead, the alleles present affect the kinds of enzymes that cells can produce, influencing the biochemical pathways involved in the development of an animal's nervous system. The resulting neurological differences can translate into a behavioral difference between individuals that have certain alleles and those that do not.

STUDY BREAK

1. How do the chicks of brood parasites stimulate unwitting foster parents to feed them?
2. How did Arnold demonstrate that the receptiveness of garter snakes to a meal of banana slugs had a genetic basis?

54.3 Learned Behaviors

Unlike instinctive behaviors, learned behaviors are not performed completely the first time an animal responds to a specific stimulus. Instead, they change in response to environmental stimuli that an individual experiences as it develops.

Learned Behaviors Are Modified by an Animal's Prior Experiences

Behavioral scientists generally define **learning** as a process in which experiences change an animal's behavioral responses. Different types of learning occur un-

Figure 54.4

Exploitation of a releaser. This young European cuckoo *(Cuculus canorus)*, a brood parasite, stimulates feeding behavior by its foster parent, a hedge sparrow *(Prunella modularis)*. It secures food by displaying exaggerated versions of the sign stimuli used by the host offspring to release feeding behavior by the parents.

a. Banana slug

b. Adult coastal garter snake eating a banana slug

c. Newborn coastal garter snake "smelling" slug extract

Figure 54.5

Genetic control of food preference. (a) Banana slugs *(Ariolimax columbianus)* are a preferred food of **(b)** an adult garter snake *(Thamnophis elegans)* from coastal California. **(c)** A newborn garter snake from a coastal population flicks its tongue at a cotton swab drenched with tissue fluids from a banana slug.

A Knockout by a Whisker

Almost all eukaryotic organisms share a series of developmental interactions called the *wingless/Wnt* pathway. The name comes from the original discovery of the pathway in the fruit fly *Drosophila melanogaster*, in which mutant genes of the pathway cause alterations in the wings and other segmental structures. Recently, three genes closely related to *disheveled (dsh)*, one of the genes of the *Drosophila wingless/Wnt* pathway, were isolated and identified in mice. No functions have yet been identified for the proteins encoded in the three mouse disheveled genes, but tests show that they are highly active in both embryos and adults. Their function must be important, but what could it be?

Nardos Lijam and his coworkers in several laboratories, including Case Western University, the Universities of Colorado and Maryland, and the National Institutes of Health in Bethesda, Maryland, decided to seek an answer to this question by developing a line of mice that totally lacked one of the disheveled genes, called *Dvl-1* in genetic shorthand. First they constructed an artificial copy of the *Dvl-1* gene with the central section scrambled so that no functional proteins could be made from its encoded directions. Next they introduced the artificial gene into em-bryonic mouse cells. Cells that successfully incorporated the gene were then injected into very early mouse embryos. Some of the mice grown from these embryos were heterozygotes, with one normal copy of the *Dvl-1* gene and one nonfunctional copy. Interbreeding of the heterozygotes produced some individuals that carried two copies of the altered *Dvl-1* gene and no normal copies. Such individuals, in which the normal gene is eliminated, are called knockout mice for the missing gene. (Making knockout mice is described in Section 18.2.)

Surprisingly, the knockout mice grew to maturity with no apparent morphological defects in any tissue examined, including the brain. Their motor skills, sensitivity to pain, cognition, and memory all appeared to be normal. Their social behavior was a different story, however. When housed with normal mice, the knockouts failed to take part in the common activities of mouse social groups: social grooming, tail pulling, mounting, and sniffing. Rather than building nests and sleeping in huddled groups, as normal mice do in the cages, the knockouts tended to sleep alone, without constructing full nests from cage materials. Mice heterozygous for the *Dvl-1* gene—that is, with one normal and one altered copy of the gene—behaved normally in all these social activities.

The knockout mice also jumped around wildly in response to an abrupt, startling sound while the response of normal mice was less extreme. Since it is known that a neural circuit of the brain inhibits the startle response of normal mice, the reaction of the knockout mice suggested that this inhibitory circuit was probably altered. Humans with schizophrenia, obsessive-compulsive disorders, Huntington disease, and some other brain dysfunctions also show an intensified startle reflex similar to that of the *Dvl-1* knockout mice.

The researchers' analysis revealed that the *Dvl-1* gene modifies developmental pathways affecting complex social behavior in mice, and probably in other mammals. It is one of the first genes affecting mammalian behavior to be identified. The similarity in startle-reflex intensity between the knockout mice and humans with neurological or psychiatric disorders also suggests that mutations in the *Dvl* genes and the *wingless* developmental pathway may underlie some human mental diseases. If so, further studies of the *Dvl* genes may give us clues to the molecular basis of these diseases, and a possible means to their cure.

© Nina Leen/Time and Life Pictures/ Getty Images

Figure 54.6
Imprinting. Having imprinted on him shortly after hatching, young greylag geese *(Anser anser)* frequently joined Konrad Lorenz for a swim.

der different environmental circumstances. In this section we consider *imprinting, classical conditioning, operant conditioning, insight learning,* and *habituation.*

Some animals learn the identity of a caretaker or the key features of a suitable mate during a **critical period**, a restricted stage of development early in life. This type of learning is called **imprinting**. For example, newly hatched geese imprint on their mother's appearance and identity, staying near her for months. And when they reach sexual maturity, they try to mate with other geese, which exhibit the visual and behavioral stimuli on which they had imprinted as youngsters. When Konrad Lorenz, one of the founders of ethology, tended a group of newly hatched greylag geese *(Anser anser),* they imprinted on him instead of an adult of their own species **(Figure 54.6)**. The male geese not only followed Lorenz about, but they also courted humans when they achieved sexual maturity.

Other forms of learning can occur throughout an animal's lifetime. Russian physiologist Ivan Pavlov's classic experiments with dogs explored **classical conditioning**, a type of learning in which animals develop a mental association between two phenomena that are usually unrelated. Dogs generally salivate when they eat. The food is called an *unconditioned stimulus* because the dogs respond to it instinctively; no learning is required for the stimulus (food) to elicit the response (salivation). In his experiment, Pavlov rang a bell just before offering food to dogs. After about 30 trials in which dogs received food immediately after the bell rang, the dogs associated the bell with feeding time, and they drooled profusely whenever it rang—even when no food was forthcoming. Thus, the bell became a *conditioned stimulus,* one that elicited a particular learned response. In classical conditioning, an animal learns to respond to a conditioned stimulus when it precedes an unconditioned stimulus that normally triggers the response. For example, your cat may become exceptionally friendly whenever she hears the sound of a can opener, another example of classical conditioning.

In another form of associative learning, called trial-and-error learning or **operant conditioning,** animals learn to link a voluntary activity, called an *operant,* with its favorable consequences, called a *reinforcement.* For example, a laboratory rat will explore a new cage randomly. If the cage is equipped with a bar that releases food when it is pressed, the rat will eventually lean on the bar by accident (the operant) and immediately receive a morsel of food (the reinforcement). After just a few such experiences, a hungry rat will learn to press the bar in its cage more frequently—as long as bar-pressing behavior is followed by access to food. Laboratory rats have also learned to press bars to turn off disturbing stimuli, such as bright lights.

A few animal species can abruptly solve problems without apparent trial-and-error attempts at the solution; researchers call this **insight learning.** For example, captive chimpanzees *(Pan troglodytes)* were able to solve a novel problem that their keepers devised: how to get bananas hung far out of reach. The chimps studied the situation, then stacked and stood on several boxes, and used a stick to knock the fruit to the floor.

Animals typically lose their responsiveness to frequent stimuli that are not quickly followed by the usual reinforcement. This learned loss of responsiveness, called **habituation,** saves the animal the time and energy of responding to stimuli that are no longer important. For example, the sea hare *Aplysia,* a shell-less mollusk, typically responds to a touch on the side of its body by retracting its delicate gills, a response that helps protect it from approaching predators. But if an *Aplysia* is touched repeatedly over a short period of time with no harmful consequences, it stops retracting its gills.

STUDY BREAK

1. Dogs typically wag their tails when they see their owners pick up a leash. What kind of learning does this demonstrate?
2. What type of learning allows you to sleep through your alarm clock when it rings to awaken you for biology class?

54.4 The Neurophysiological Control of Behavior

Research in neuroscience has shown that all behavioral responses, even those that are either mostly instinctive or mostly learned, depend on an elaborate physiological foundation provided by the biochemistry and structure of neurons (nerve cells). The neurons that regulate an innate response as well as those that make it possible for an animal to learn something are products of a complex developmental process in which genetic information and environmental contributions are intertwined. Although the anatomical and physiological basis for some behaviors is present at birth, an individual's experiences alter cells of its nervous system in ways that produce particular patterns of behavior. In this section we use examples from research on the singing behavior of songbirds to explore general principles about the physiological basis of behavior that apply to many other kinds of animals.

Discrete Neural Circuits in Specific Brain Regions Control Singing Behavior in Songbirds

Marler's experiments (see Section 54.1) help explain the physiological underpinnings of singing behavior in male white-crowned sparrows. If acoustical experience shapes this behavior, a sparrow chick's brain must be able to acquire and store information present in the songs of other males. Then, months later, when the young male starts to sing, its nervous system must have special features that enable the bird to match its vocal output to the stored memory of the song that it had heard earlier. Eventually, when it achieves a good match, the sparrow's brain must "lock" on the now complete song and continue to produce it when the bird is singing.

Additional experiments have provided detailed information about the nature of the sparrow's nervous system. Young birds that did not hear taped song during their critical period, between 10 and 50 days old, never produced the full song of their species, even if they heard it later in life. In addition, young birds that heard recordings of *other* bird species' songs during the critical period never generated replicas of those songs as they matured. These and other findings suggested that certain neurons in the young male's brain are influenced only by appropriate stimuli, namely the acoustical signals from individuals of its own species, and only during the critical period. Neuroscientists have identified the neurons clusters, called *nuclei* (singular, *nucleus*), that make song learning and song production possible.

Moreover, every behavioral trait appears to have its own neural basis. For example, a male zebra finch, *Tae-niopygia guttata* **(Figure 54.7),** another songbird, can

Figure 54.7
Zebra finches.
Native to Indonesia, zebra finches (*Taeniopygia guttata*), have played an important role in studies of the physiological basis of song learning. The male has a striped throat.

discriminate between the songs of strangers and the songs of established neighbors on adjacent **territories.** (In many bird species, territories are plots of land, defended by individual males or breeding pairs, within which the territory holders have exclusive access to food and other necessary resources. Territories are discussed further in Chapter 55.) The ability to discriminate between the songs of neighbors and those of strangers also involves a nucleus in the forebrain. Cells in this nucleus fire frequently the first time that the song of a new zebra finch is played to a test subject. But as the song is played again and again, these cells cease to respond, indicating that the bird becomes habituated to a now familiar song, although it still reacts to the songs of strangers. The neurophysiological networks that make this selective learning possible enable male zebra finches to behave differently toward familiar neighbors, which they largely ignore, and unfamiliar singers, which they attack and drive away.

The Activation of Specific Genes Fosters the Development of Nuclei That Regulate a Bird's Song

The role of genes in learning has been identified by research using new molecular and cellular techniques that reveal when a specific gene is active in neurons. When a bird is exposed to relevant acoustical stimuli, such as the songs of potential rivals of its own species, certain genes are "turned on" within neurons in the song-controlling nuclei of the bird's brain. For example, when a zebra finch hears the elements of its species' song, a gene called *zenk* becomes active in the brain, producing an enzyme that changes the structure and function of the neurons (see photo on p. 1253). In effect, the ZENK enzyme programs the neurons of the bird's brain to "anticipate" key acoustical events of potential biological importance. When these events occur, they trigger additional changes in the bird's brain that affect its actions. As a result, a territory owner habituates to (that is, learns to ignore) a singing neighbor with which it has already adjusted territorial boundaries; but it retains the ability to detect and repel new in-

truders of its own species, which represent a real threat to its continued control of its territory.

STUDY BREAK

1. What research results suggest that certain neurons in the young male bird's brain are influenced only by acoustical signals from members of its own species and only during a critical period?
2. What happens to cells in the nucleus in the forebrain of a zebra finch after it hears a neighboring bird's song many times?
3. What is the role of the ZENK enzyme in song learning?

54.5 Hormones and Behavior

Research on many animal species has revealed that hormones are the chemical signals triggering the performance of specific behaviors. They often accomplish this function by regulating the development of neurons and neural networks or by stimulating the cells within endocrine glands to release chemical signals.

Hormones Regulate the Development of Cells and Networks That Form the Neural Basis of Behavior

How did the neurons in an adult zebra finch acquire the remarkable capacity to change in response to specific stimuli? In zebra finches, only males produce courtship songs. Very early in a male songbird's life, certain cells in its brain produce the hormone estrogen, which affects target neurons in an area of the developing brain called the *higher vocal center*. The presence of this hormone leads to a complex series of biochemical changes that result in the production of more neurons in parts of the brain that regulate singing. By contrast, the brains of developing females do not produce estrogen, and in the absence of this hormone, the number of neurons in the higher vocal center of females *declines* over time **(Figure 54.8).** Experiments have shown that when young female zebra finches are given estrogen, they produce more neurons in the higher vocal center; but the treated females do not sing later in life unless they are also treated with androgens (male hormones).

Thus, genetically induced hormone production contributes to song learning and singing behavior in male zebra finches by regulating the numbers and types of neurons in the brain centers that produce those behaviors. The development of these neurons primes them for additional changes in response to specific acoustical experiences during the bird's development. Moreover, specific stimuli, such as the songs of either familiar or unfamiliar males, can alter the genetic activity of the neurons that control the behavior of adult birds.

Changing Hormone Concentrations Alter the Behavior of Animals as They Mature

Just as estrogen influences the development of singing ability in zebra finches, other hormones mediate the development of the nervous system in other species. Indeed, a change in the concentration of a certain hormone is often the physiological trigger that induces important changes in an animal's behavior as it matures.

In honeybees *(Apis mellifera)*, worker bees perform different tasks for the colony's welfare as they grow older: bees that are less than 15 days old tend to care for larvae and maintain the hive, whereas those that are more than 15 days old often make foraging excursions from the hive to collect the nectar and pollen that bees eat **(Figure 54.9)**. These behavioral changes are induced by rising concentrations of juvenile hormone (see Section 40.5), which is released by a gland near the bee's brain. Despite its name, circulating levels of juvenile hormone actually increase as a honeybee gets older.

Juvenile hormone may exert its effect on the bee's behavior by stimulating genes in certain brain cells to produce proteins that affect nervous system function. One such chemical, *octopamine,* stimulates neural transmissions and reinforces memories. It is concentrated in the antennal lobes, a part of the bee's brain that contributes to the analysis of chemical scents in the bee's external environment. Octopamine is found at higher concentrations in the older, foraging bees that have higher levels of juvenile hormone. And when extra juvenile hormone is administered to bees experimentally, their production of octopamine increases. Thus, increased octopamine levels in the antennal lobes may help a foraging bee home in on the odors of flowers where it can collect nectar and pollen.

The honeybee example illustrates how genes and hormones interact in the development of behavior. Genes code for the production of hormones, which change the intracellular environment of assorted target cells. The hormones then directly or indirectly change the genetic activity and enzymatic biochemistry in their targets. If the cells in question are neurons, the changes in their biochemistry translate into changes in the animal's behavior.

Hormone Levels Affect Reproductive Activity in Many Animals

The African cichlid fish *(Haplochromis burtoni)* provides an example of how hormones regulate reproductive behavior. Some adult males maintain nesting ter-

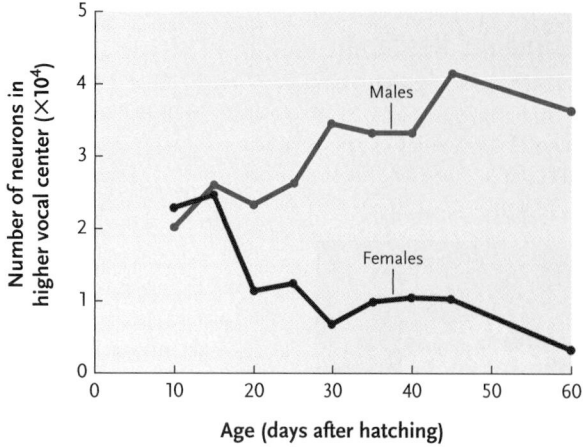

Figure 54.8

Hormonally induced changes in brain structure. The brains of young male zebra finches secrete estrogen, which stimulates the production of additional neurons in the higher vocal center. Lacking this hormone, the brains of young female zebra finches lose neurons in this brain region.

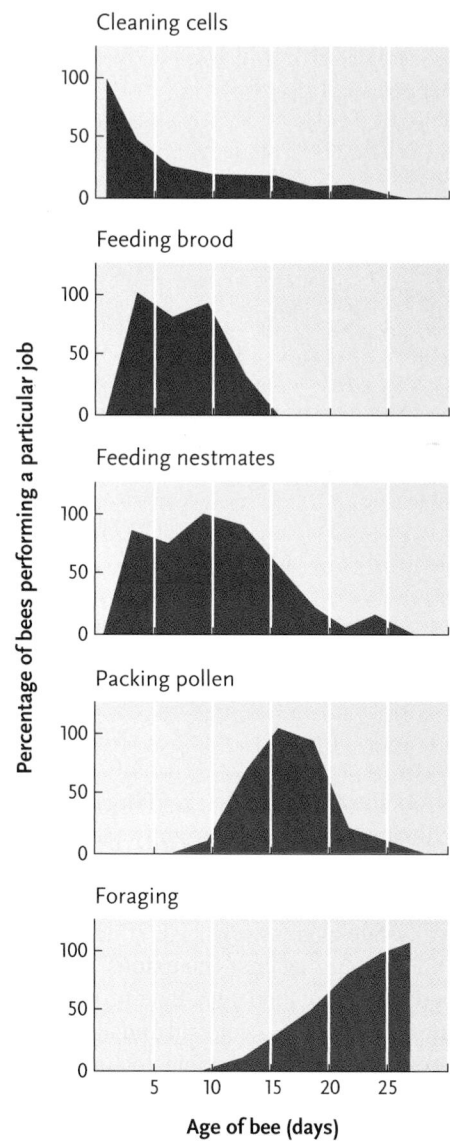

Figure 54.9

Age and task specialization in honeybee *(Apis mellifera)* workers. Young bees typically clean cells and feed the brood; older workers leave the hive to forage for food.

Figure 54.10 Experimental Research

Effects of the Social Environment on Brain Anatomy and Chemistry

a. African cichlid fish *(Haplochromis burtoni)*

Nonterritorial male

Territorial male

Russell Fernald, Stanford University

QUESTION: How does the acquisition or the loss of a territory affect the brain anatomy and chemistry of an African cichlid fish *(Haplochromis burtoni)?*

EXPERIMENT: Fernald and his students housed groups of male cichlids in aquariums in their laboratory. Males that established and maintained territories in the aquariums were brightly colored, whereas those that could not hold territories were pale and drab. The researchers then moved some small territorial males into tanks where larger males had already established territories. The newly introduced males could not establish and maintain territories under these experimental conditions, and therefore changed status from territorial to nonterritorial. The researchers also moved some large nonterritorial males into tanks with smaller territorial males. Under these experimental conditions, the newly introduced males quickly established and maintained territories, changing their status from nonterritorial to territorial. Other males, left in their original tanks so that their territorial status did not change, served as controls. Four weeks later, the researchers examined the brains of the experimental and control fish and measured the size of the neurons that produce GnRH, a hormone that stimulates bright coloration as well as aggressive behavior and mating behavior in males.

RESULT: The GnRH-producing cells in the brains of experimental males that had lost their territories were much smaller than those in the brains of control males that had maintained their territories. By contrast, the GnRH-producing cells in the brains of experimental males that had gained territories were much larger than those of control males that had never held territories.

b. GnRH-secreting cells

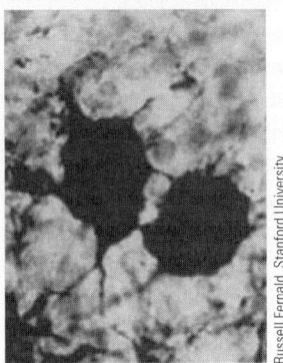 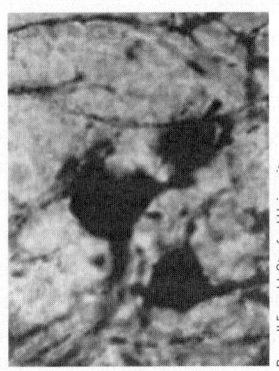 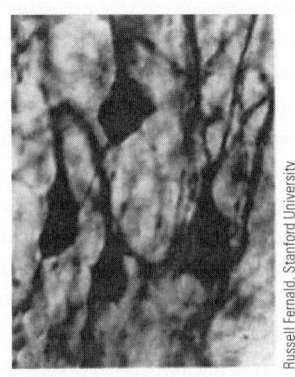 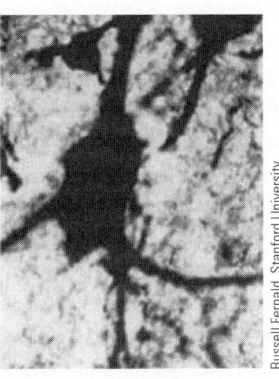

Territorial control

Territorial to nonterritorial experimentals

Nonterritorial control

Nonterritorial to territorial experimentals

CONCLUSION: Changes in social status influence the size of brain cells producing hormones that influence the color and behavior of males.

---

ritories on the bottom of Lake Tanganyika in East Africa. Territory holders are brightly colored, and they exhibit elaborate behavioral displays that attract egg-laden females to their territories. These males defend their real estate aggressively against neighboring territory holders and against incursions by males that have no territories of their own. By contrast, nonterritorial males are much less colorful and aggressive; they do not control a patch of suitable nesting habitat, and they make no effort to court females.

The behavioral differences between the two types of males are caused by differences in their levels of circulating sex hormones. Recall from Section 47.3 that gonadotropin-releasing hormone (GnRH) stimulates the testes to produce testosterone and sperm. When the circulating testosterone is carried to the brain, it modulates the activity of neurons that regulate sexual and aggressive behavior. In territorial fish the GnRH-producing neurons in the hypothalamus are large and biochemically active, but in nonterritorial fish they are small and inactive. In the absence of GnRH, the testes do not produce testosterone; the testosterone-deficient fish do not court females with sexual displays, nor do they usually attack other males.

What causes the differences in the neuronal and hormonal physiology of the two types of male fish? Russell Fernald and his students at Stanford University conducted laboratory experiments in which they manipulated the territorial status of males: some territorial males were changed into nonterritorial males; some nonterritorial males were changed into territorial males; and the territorial status of other males was left unchanged as a control **(Figure 54.10)**. Four weeks later, they compared the coloration and behavior as well as the size of the GnRH-producing cells in the brains of the experimental fishes with those of the control males that had retained their original status. Males that had held territories in the past, but had then been defeated by another male, quickly lost their bright colors and stopped being combative. Moreover, their GnRH-producing cells were smaller than those of the successful territory-holding controls. Conversely, males that gained a territory in the experiment quickly developed bright colors and displayed aggressive behaviors towards other males. And the GnRH-producing cells in their brains were larger than those of fishes that had maintained their status as non-territory-holding controls.

The neuronal, hormonal, and behavioral differences between the two experimental groups of males are therefore correlated with a key environmental variable: success or failure in the acquisition and maintenance of a territory. The fish can detect and store information about their aggressive interactions. The neurons that process this information transmit their input to the hypothalamus where it affects the size of the GnRH cells, which in turn dictates the hormonal state of the male. A decrease in GnRH production can turn a feisty territorial male into a subdued drifter, biding his time and building his energy reserves for a future attempt at defeating a weaker male and taking over his territory. If successful in regaining territorial status, the male's GnRH levels will increase again, and the once-peaceful male will revert to vigorous sexual and aggressive behavior.

Note the general similarity of these processes to those described for the white-crowned sparrow's song learning: the fish's brain possesses cells that can change their biochemistry, structure, and function in response to well-defined social stimuli. These physiological changes make it possible for the fish to modify its behavior, depending on its social circumstances. In the next section, we examine how the structure of the nervous system allows animals to respond to important environmental stimuli.

STUDY BREAK

1. What is the effect of estrogen on the development of neurons in the higher vocal center of young zebra finches?
2. How might juvenile hormone production influence a bee's ability to recognize and locate appropriate food sources?
3. How does the loss of its territory change the brain chemistry of an African cichlid fish?

54.6 Nervous System Anatomy and Behavior

Although many behaviors result from gene–environment interactions and changes in hormone concentrations, some specific behaviors are produced by the anatomical structure of an animal's nervous system. Studies on a wide range of animal species demonstrate that the nervous systems of many animals provide rapid responses to key stimuli. In other species, sensory systems are structured to acquire a disproportionately large amount of information about those stimuli that are most important to survival and reproductive success.

Hard-Wired Connections between Sensory and Motor Systems Provide Rapid Behavioral Responses to Life-Threatening Stimuli

In some animals, important information acquired by the senses is relayed directly to motor neurons. Such a system provides crickets with a potentially lifesaving predator avoidance behavior. Crickets and some other insects fly mainly at night, a behavior that allows them to avoid day-flying predatory birds. But flying crickets aren't safe even at night, because insect-eating bats can detect them in pitch darkness.

Bats detect potential prey by echolocation (see *Why It Matters* at the beginning of Chapter 39). They call almost continuously while flying at night, and the sound waves they produce bounce off items in their path, creating echoes that the bats hear and use to track their prey. Their vocalizations are of such high frequency (up to 100,000 hertz) that they lie outside the upper limit (20,000 hertz) of unaided human hearing. However, a bat's auditory apparatus and brain not only can hear ultrasound, as these high frequency sounds are called, but also can analyze ultrasonic echoes in a way that permits the bat to identify, approach, and capture flying insect prey. With enemies of this sort in its environment, a cricket flying at night is in real danger of being intercepted and eaten.

Crickets are not defenseless, however. Black field crickets *(Teleogryllus oceanicus)*, for example, hear ultrasound through ears in their front legs (see Figure 39.7). The approach of a calling bat causes sensory neurons connected to the ears to fire. However, to be of any use to the cricket, this information must be translated immediately into evasive action—and crickets have the anatomical and physiological equipment to do exactly that.

Imagine that a bat is zeroing in for the kill, rushing toward the left side of a flying cricket. The cricket's left ear will be bombarded with more intense ultrasound than the right ear, and the neurons that receive input from the ears will also be stimulated unequally. The cricket's nervous system is structured to relay

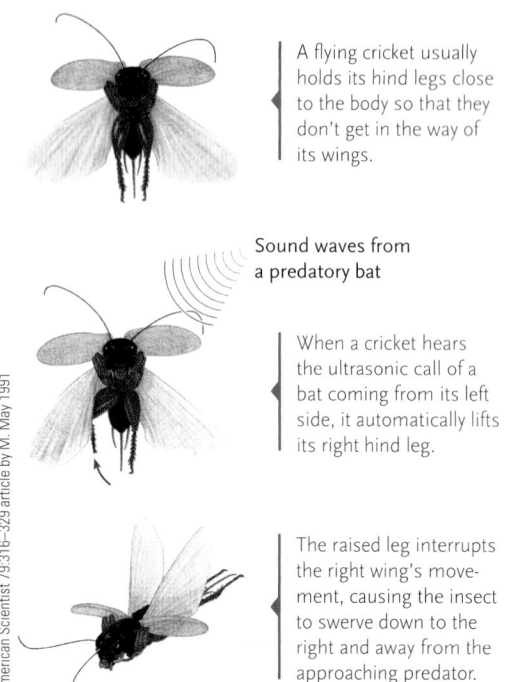

A flying cricket usually holds its hind legs close to the body so that they don't get in the way of its wings.

Sound waves from a predatory bat

When a cricket hears the ultrasonic call of a bat coming from its left side, it automatically lifts its right hind leg.

The raised leg interrupts the right wing's movement, causing the insect to swerve down to the right and away from the approaching predator.

American Scientist 79:316–329 article by M. May 1991

Figure 54.11
A neural mechanism for escape behavior in the black field cricket (*Teleogryllus oceanicus*).

incoming messages from the *left* ear to the motor neurons (muscle-regulating nerve cells) that control the *right* hind leg. Sufficient ultrasonic stimulation on the left side of the body will induce the motor neurons for the right hind leg to fire, causing muscle contractions in that leg. As the right hind leg jerks up, it blocks movement of the right hind wing, reducing the flight power generated on the right side of the cricket's body. The flying cricket then swerves sharply to the right and loses altitude, diving down and away from the approaching bat **(Figure 54.11)**. Thus, the anatomical structure of the cricket's nervous system produces a behavioral response that takes the cricket out of harm's way.

If all goes well, the cricket will gain the safety of foliage or leaf litter on the ground before the bat can reach it. Once there, echoes bouncing off the materials all around it will mask any ultrasonic echoes coming from its body. The thwarted bat will be forced to look elsewhere for prey that responds less rapidly.

The Structure of Sensory Systems Allows Animals to Respond Appropriately to Different Stimuli

In some animals, the structure and neural connections of sensory systems allow them to distinguish potentially life-threatening stimuli from those that are more mundane. For example, fiddler crabs *(Uca pugilator)* live and feed on mud flats where they build burrows that provide safe refuge from predators, including crab-hunting shorebirds. But to use its burrow wisely,

a crab must be able to distinguish between predatory gulls and its fellow fiddler crabs. Otherwise, it would dash for cover whenever anything moved in its field of vision.

Fiddler crabs possess long-stalked eyes that they hold above their carapace perpendicularly to the ground. John Layne, a neurophysiologist at Duke University, wondered whether a crab might distinguish between dangerous predators and fellow crabs by having a divided field of vision. A large predatory gull sailing in for the kill would stimulate receptors on the upper part of the eye, whereas a fellow crab, whose movements would be slightly below the midpoint of the eyes, would stimulate a lower set of visual receptors. If the receptors above and below the retinal equator relayed their signals to different groups of neurons, the crab's nervous system could be "wired" to provide different responses to the different stimuli.

Layne hypothesized that receptors above the midline of the eye activate neurons that control an escape response, so that stimulation from above would reliably trigger a dash for the burrow. By contrast, a moving stimulus at or below eye level, as when one crab approached another, would provide input to the neurons that allow a crab to behave appropriately to a male or female of its species.

Layne tested this hypothesis by placing crabs on an elevated platform in a glass jar. He then presented the same moving stimulus, a black square, to each crab at two heights; sometimes the stimulus circled the jar above the crab's eyes and sometimes below. Stimuli that activated the upper part of the retina did indeed induce escape behavior, but if the stimuli were below the retinal equator, the animal generally ignored the moving objects altogether **(Figure 54.12)**. Thus, specific nervous system connections between a fiddler crab's eyes and brain provide appropriate responses to different specific stimuli.

The Amount of Brain Tissue Devoted to Analyzing Sensory Information Varies from One Sensory System to Another

The match between the structure of an animal's nervous system and the real-world challenges it faces extends beyond the ability to avoid predators. For example, the star-nosed mole (*Condylura cristata*), which lives in wet tunnels in North American marshlands, spends almost all of its life in complete darkness. Like nocturnal insect-eating bats, the mole must find food without benefit of visual cues; and, like the bats, it has a receptor-perceptual system that enables it to feed effectively. The star-nosed mole subsists largely on earthworms, and it uses its nose to locate them—but not by smell. Instead, as the mole proceeds down a tunnel, 22 fingerlike tentacles from its nose sweep the area di-

Figure 54.12 Experimental Research

Nervous System Structure and Appropriate Behavioral Responses

QUESTION: Do fiddler crabs respond differently to stimuli that are presented above the midline of their visual field than they do to stimuli presented below it?

EXPERIMENT: Layne investigated this question by placing crabs on an elevated platform in a glass jar. He then presented the same moving stimulus, a black square, to each crab at varying heights; sometimes the stimulus circled the jar above the crab's eyes and sometimes below.

RESULTS: Stimuli that activated the upper part of the retina did indeed induce escape behavior, but when the stimuli were below the retinal equator, the animal generally ignored the moving objects altogether.

Fiddler crab *(Uca pugilator)*

© Jeff Foott/Dcom/DRK Photo

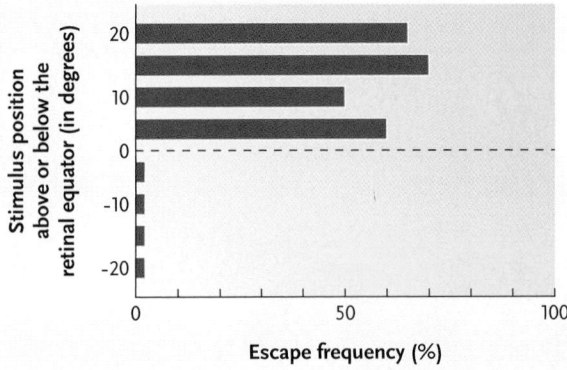

CONCLUSION: Specific nervous system connections between a fiddler crab's eyes and brain provide appropriate responses to different specific stimuli.

How does an animal choose which behavior to perform?

Animals must perform many types of behaviors in their lifetimes. But what determines which behavior is most appropriate in any particular situation? In some cases, it appears that an animal performs whichever behavior it is most provoked to do at the time. For example, when the Eimer's organs of a star-nosed mole tell it that its tentacles have contacted an earthworm, the mole responds by biting into whatever is in front of its mouth. In other cases, an animal may perform whichever behavior will bring it the most reward. Researchers are studying how animals choose behaviors, using a combination of methods including theoretical modeling, neuroscience, and molecular biology. Male fruit flies often must choose between aggression and courtship, and molecular manipulations that result in the inability of a fly to discriminate between males and females often result in the fly making the wrong choice. Understanding the behavioral choices that animals make will help us understand both the mechanisms and evolution of behavior.

How does experience change the brain to affect behavior?

Some behaviors develop only after an animal has had certain experiences. How does experience translate into a new or improved behavior? Researchers are tracing the paths from stimulus perception, to activation of specific neural circuits and molecular pathways in the brain, to changes in brain structure and chemistry that then lead to behavioral change. This research is helped enormously by the burgeoning field of molecular neuroscience. Knowing how experience translates into a new or improved behavior has broad implications for our understanding of learning and memory, brain plasticity, brain disease, and recovery from stroke.

How do genes influence behavior?

Genetic differences between individuals can translate into behavioral differences between them. But genes encode proteins, and the road between the transcription of a gene and the performance of a specific behavior is a "long and winding" one. Researchers are studying how genetic differences between individuals influence the biochemical pathways that shape the development of the nervous system or later modify its function. Insights into how genes influence behavior have profound implications for our understanding of brain function and perhaps even policy decisions about screening people for genes that may influence their behavior.

John Dixon The Champaign–Urbana News–Gazette

Gene E. Robinson is the G. William Arends Professor of Integrative Biology at the University of Illinois at Urbana-Champaign, where he studies social behavior and genomics using the honeybee. To learn more about his research, go to http://www.life.uiuc.edu/robinson.

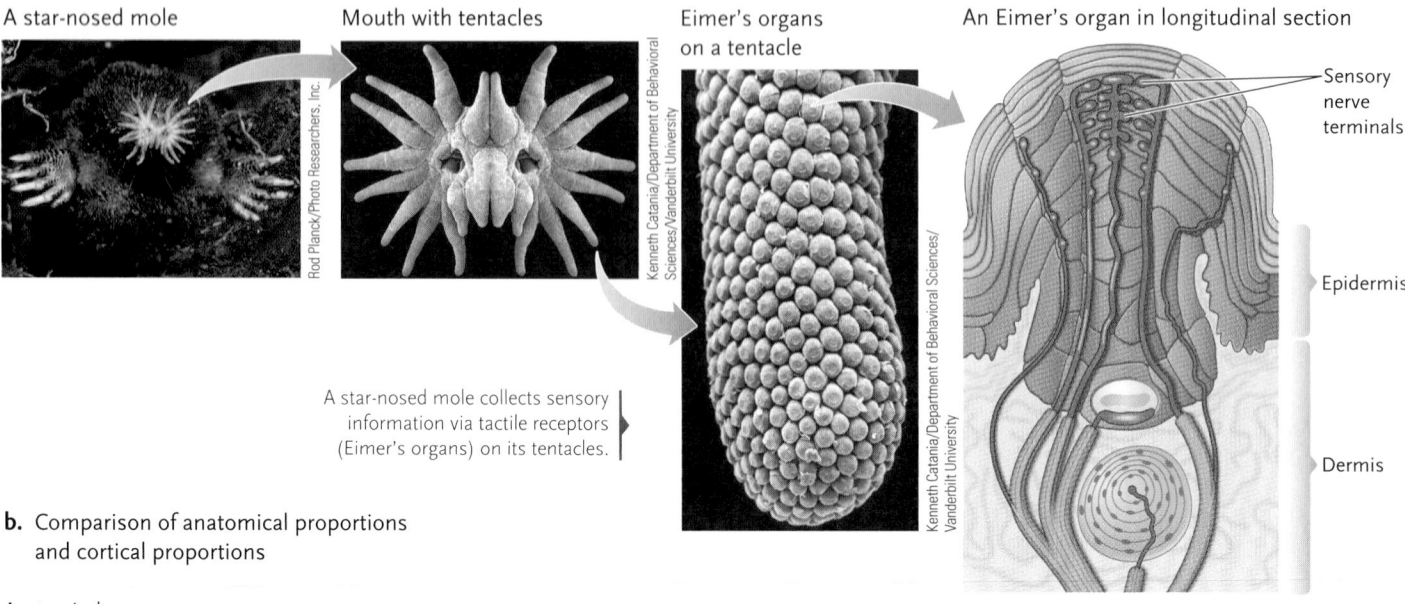

a. Sensory organs on the tentacle of a star-nosed mole

A star-nosed mole

Mouth with tentacles

Eimer's organs on a tentacle

An Eimer's organ in longitudinal section

Sensory nerve terminals

Epidermis

Dermis

A star-nosed mole collects sensory information via tactile receptors (Eimer's organs) on its tentacles.

Rod Planck/Photo Researchers, Inc.

Kenneth Catania/Department of Behavioral Sciences/Vanderbilt University

Kenneth Catania/Department of Behavioral Sciences/ Vanderbilt University

b. Comparison of anatomical proportions and cortical proportions

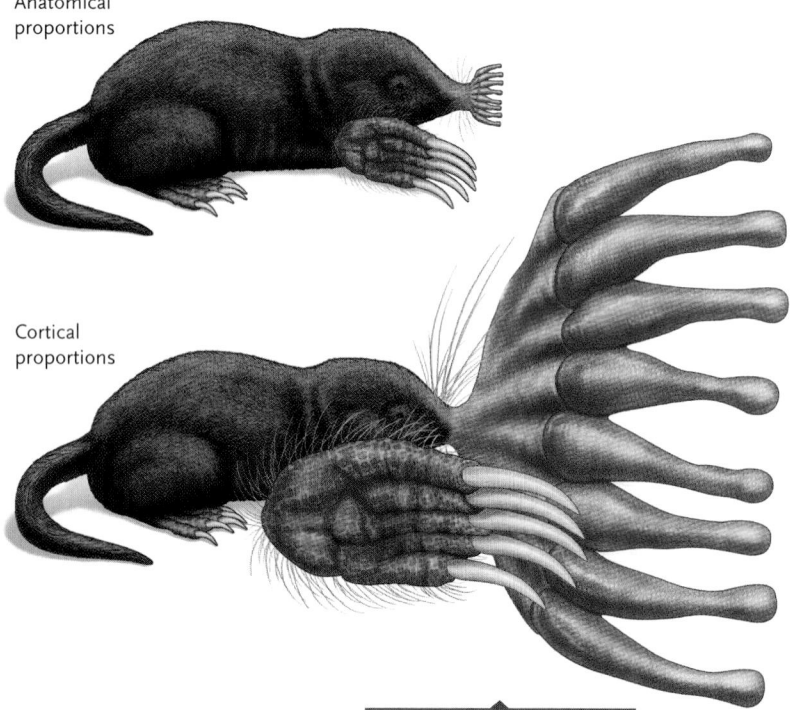

Anatomical proportions

Cortical proportions

Most of the mole's cereberal cortex is devoted to the tentacles and front, digging feet.

Figure 54.13

The collection and analysis of sensory information by the star-nosed mole *(Condylura cristata)*. **(a)** The mole's nose has 22 fleshy tentacles covered with cylindrical tactile receptors called Eimer's organs, each containing sensory nerve terminals. **(b)** The star-nosed mole's cerebral cortex devotes far more space and neurons to the analysis of tactile inputs from the tentacles than from elsewhere on the body. These drawings compare the mole's actual body proportions with the relative amount of cortical tissue that processes sensory information from the various parts of the body.

rectly ahead. These tentacles are covered with thousands of tactile (touch) receptors called Eimer's organs **(Figure 54.13a)**. Sensory nerve terminals in the Eimer's organs generate complex and detailed patterns of signals about the objects they contact. These messages are relayed by neurons to the cortex of the mole's brain, much of which is devoted to the analysis of information from the nose's tactile receptors.

The structural basis of the mole's sensory analysis is apparent when we consider that the amount of brain tissue responding to signals from the mole's nose contains many more cells than do the tissues

that decode tactile signals from all other parts of the animal's body combined **(Figure 54.13b)**. Moreover, the brain does not treat inputs from all 22 of the mole's "fingers" equally. Instead, the brain devotes more cells to the tentacles closest to the mole's mouth, and fewer cells to analyzing messages from tentacles that are farther away.

The processing of tactile information in this species is clearly related to the importance of finding food in totally dark underground tunnels. Moreover, the extra attention given to signals from certain tentacles almost certainly helps the mole locate prey that are close to its mouth, allowing it to bite worms before they can move away after being touched.

As these examples illustrate, animal nervous systems do not offer neutral and complete pictures of the environment. Instead, distorted and unbalanced perceptions of the world are advantageous because certain types of information are far more important for survival and reproductive success than others. In the next chapter, we examine the ecological circumstances and selective forces that have promoted the evolution of specific behaviors.

STUDY BREAK

1. How does the anatomy of the cricket's nervous system helps it avoid an approaching bat?
2. What behavioral response is elicited in a fiddler crab when its eyes detect movement above the midline of its visual field?
3. Explain the sensory mechanism that allows a star-nosed mole to locate earthworms in its tunnels.

Review

Go to **ThomsonNOW** at www.thomsonedu.com/login to access quizzing, animations, exercises, articles, and personalized homework help.

54.1 Genetic and Environmental Contributions to Behavior

- Most behaviors have both instinctive and learned components. Some behaviors can be produced only if the animal's nervous system acquires inputs from specific experiences during a critical stage of its development (Figure 54.1).

54.2 Instinctive Behaviors

- Instinctive behaviors are those that an animal performs completely the first time it is presented with a stimulus.
- Fixed action patterns are highly stereotyped behaviors that animals exhibit in response to simple cues called sign stimuli (Figures 54.2–54.4). Fixed action patterns often change through time in response to an animal's experiences.
- Behavioral differences between individuals often reflect underlying genetic differences. Research on garter snakes suggests that certain food preferences are genetically based (Figure 54.5).

  Animation: Instinctive behavior in infants

  Animation: Adaptive behavior in starlings

  Animation: Snake taste preference

  Animation: Cuckoo and foster parent

54.3 Learned Behaviors

- Learned behaviors develop only after an animal has had certain experiences in its environment. The different forms of learning include imprinting (Figure 54.6), classical conditioning, operant conditioning, and insight learning. Habituation is a learned loss of responsiveness to specific stimuli.

54.4 The Neurophysiological Control of Behavior

- Animal behavior requires an anatomical, physiological, and biochemical foundation based in the nervous system. An individual's experience alters cells of the nervous system in ways that produce particular patterns of behavior.

- The physiological basis of bird singing behavior resides in specific neuron clusters, called nuclei, that communicate with each other in the bird's brain.
- Bird song and some other behaviors develop only after specific genes are activated within the neurons that produce the behavior.

  Video: Development and elicitation of bird song

54.5 Hormones and Behavior

- Hormones can mediate the expression of specific behaviors by activating genes that change the biochemistry, morphology, and number of neurons in specific nuclei. Estrogen stimulates the production of neurons in the higher vocal center of male zebra finches (Figures 54.7 and 54.8).
- Age-related changes in hormone levels can alter the behavior of animals over the course of their lives. Changes in juvenile hormone concentration are correlated with changes in task specialization in honeybees (Figure 54.9).
- Behavioral interactions with other individuals can alter an animal's hormone levels, inducing changes in its behavior. Research on male cichlid fishes suggests that variations in coloration and aggressive behavior associated with territorial status and social interactions are mediated by the production of certain hormones (Figure 54.10).

  Animation: Hormonal control of behavior

54.6 Nervous System Anatomy and Behavior

- Sensory information enables animals to make adaptive behavioral responses to their environments. In crickets and some other animals, sensory systems relay information to motor systems, inducing an almost instantaneous response (Figure 54.11).
- Nervous systems can generate prompt and effective responses to those environmental stimuli that may have a large impact on an animal's survival and reproductive success. Fiddler crabs respond differently to movements that occur either above or below the midline of their visual field (Figure 54.12).
- Sensory systems are often structured to provide more information about important environmental factors. A large fraction of a star-nosed mole's brain is devoted to analyzing input from tactile receptors on its nose (Figure 54.13).

Questions

Self-Test Questions

1. Marler concluded that white-crowned sparrows can learn their species' song only:
   a. after receiving hormone treatments.
   b. during a critical period of their development.
   c. under natural conditions.
   d. from their genetic father.
   e. if they are reared in isolation cages.

2. Instinctive behaviors are:
   a. performed completely the first time they are used.
   b. always modified by an animal's experiences.
   c. performed only by very young animals.
   d. often the product of habituation.
   e. sometimes called trial-and-error responses.

3. A stimulus that always causes an animal to behave in a highly stereotyped way is called:
   a. a fixed action pattern.
   b. an instinct.
   c. habituation.
   d. a sign stimulus.
   e. a reinforcement.

4. Arnold's experiments on the feeding preferences of garter snakes demonstrated that food choice is largely governed by a snake's:
   a. early experiences.
   b. genetics.
   c. size and color.
   d. diet while it was developing inside its mother.
   e. trial-and-error learning.

5. Learning in which an animal associates two phenomena that it experiences at approximately the same time is called:
   a. imprinting.
   b. operant conditioning.
   c. classical conditioning.
   d. insight learning.
   e. habituation.

6. The development of the song system in male songbirds depends on:
   a. direct connections between sensory neurons and motor neurons.
   b. a decrease in the number of neurons in the song system.
   c. the behaviors of females, which stimulate hormone production.
   d. the successful defense of a territory.
   e. the production of estrogen early in life.

7. One of the functions of octopamine in foraging honeybees is to:
   a. increase the production of juvenile hormone.
   b. decrease the production of juvenile hormone.
   c. make the bees defend their territory more aggressively.
   d. stimulate neural transmissions and reinforce memories.
   e. increase the time they spend caring for larvae.

8. In cichlid fishes, high levels of the hormone GnRH:
   a. make females more receptive to male attention.
   b. cause males to be sexually aggressive but not territorial.
   c. stimulate a male to defend its territory.
   d. cause males to abandon their territories.
   e. cause males to lose their bright colors.

9. Sensory bias in the nervous system of a cricket ensures that ultrasound perceived on one side of the body will cause:
   a. a movement in a leg on the same side of the body.
   b. a movement in a leg on the opposite side of the body.
   c. the cricket to respond with a vocalization.
   d. the cricket to stop vocalizing.
   e. the cricket to fly toward the sound.

10. In the brain of a star-nosed mole, more cells decode:
    a. tactile information from its feet than from all other parts of its body.
    b. tactile information from the tentacles on its nose than from all other parts of its body.
    c. tactile information from its mouth than from all other parts of its body.
    d. visual information from the top part of its visual field than the bottom part.
    e. visual information from the bottom part of its visual field than the top part.

Questions for Discussion

1. One day, while walking in the country, you see a rooster wade into a pond and begin to court a female mallard duck. What probably happened to the rooster early in life?

2. Using an example from your own experience, explain why habituation to a frequent stimulus might be beneficial. Also describe an example in which habituation might be harmful or even dangerous.

3. Is learning always superior to instinctive behavior? If you think so, why do so many animals react instinctively to certain stimuli? Are there some environmental circumstances in which being able to respond "correctly" the first time would have a big payoff?

4. Cockroaches have two small projections called *cerci* at the tip of the abdomen. You suspect that the cerci might be responsible for the insects' ability to detect predators, such as lizards, rushing toward them from behind. Under the microscope you see that each cercus is covered with fine hairs. What properties should these hairs have if they are part of a system that detects moving air pushed ahead by an approaching predator? How might the roach determine whether the danger was coming from the right or left side? How quickly should cercal information be processed compared with information about the chemicals in a food item?

Experimental Analysis

You find that some fruit flies in your lab are quick to come to a dish containing citrus oils, but others are not as responsive. How could you test whether these behavioral differences are caused by genetic differences among the flies? What should happen if you performed an artificial selection experiment (see Section 19.2) in which you tried to select for quick versus slow responses to citrus oils?

Evolution Link

Some birds that are frequently kept as pets, such as parrots and myna birds, have the uncanny ability to imitate human speech. Develop a hypothesis that explains why the ability to be a good mimic might have evolved in these species. What features of the birds' brains might be involved in this behavior?

Musk oxen *(Ovibos moschatus).* The social behavior of a herd of musk oxen includes encircling their young to protect them from predators.

© Paul Nicklen/National Geographic/Getty Images

55 The Ecology and Evolution of Animal Behavior

WHY IT MATTERS

In early spring, male white-crowned sparrows leave their wintering grounds in Mexico and fly thousands of kilometers to their northern breeding range. There, they select patches of habitat that contain the resources necessary for breeding—suitable cover, potential nesting sites, and abundant food. Then, they start to sing and sing, repeating their song thousands of times every day. The songs are a form of communication through which males announce their presence to rival males and to females. Males also perform elaborate courtship behaviors. And once the young hatch, they communicate with their parents, eliciting the care they need before leaving the nest.

All of these behaviors carry significant costs and risks. For example, migration requires enormous energy expenditure, and many migrating birds die before completing their trip. Moreover, singing males are conspicuous, and they may attract the attention of a hawk or some other predator. Given the costs and dangers associated with these behaviors, what benefits do the birds gain from performing them? The ultimate evolutionary benefit is obvious: with luck, individuals performing these complex and diverse behaviors may leave surviving offspring **(Figure 55.1).**

Figure 55.1

Reproductive success. Parental care is just one of many behaviors required for successful reproduction in white-crowned sparrows *(Zonotrichia leucophrys)* and many other animal species. The number of surviving nestlings will determine the reproductive success of their parents.

Questions about ultimate benefits are fundamentally different from the questions we considered in the previous chapter, where we focused on *how* underlying physiological and genetic mechanisms enable animals to behave. In this chapter, we try to explain *why* animals behave as they do. Why do sparrows migrate to their breeding grounds, breed in certain habitats but not others, and expose themselves to predation by singing? The behavior of animals is closely tied to ecological circumstances, and evolutionary biologists view most behaviors as an individual's responses to its environment. Moreover, like morphological traits, behaviors are subject to microevolutionary change (see Chapter 20). If particular alleles contribute even slightly to the development of a behavior that enhances an animal's fitness, natural selection will cause the frequency of those alleles to increase in the next generation.

Behavioral biologists apply ecological and evolutionary analyses to all forms of animal behavior, including those described above. In this chapter, we examine the ecology and evolution of several categories of animal behavior: orientation, navigation, and migration; habitat selection and territoriality; communication; reproductive behavior and mating systems; and social behavior, including behaviors described as altruistic. We close the chapter with a brief look at human behavior.

55.1 Migration and Wayfinding

Most animals move through their environments at some stage of their life cycles. Although some species move only short distances to find suitable environmen-

tal conditions, many others undertake large-scale movements on a seasonal schedule.

Migrating Animals Make Long Round-Trips on a Seasonal Cycle

Many animal species undertake a seasonal **migration**, traveling from the area where they were born to a distant and initially unfamiliar destination, and returning to their birth site later. The Arctic tern *(Sterna paradisaea)*, a seabird, makes an annual round-trip migration of 40,000 km **(Figure 55.2).** Many other vertebrate species, including gray whales and salmon, also undertake long and predictable journeys. Even some arthropods migrate long distances. For example, spiny lobsters *(Panulirus* species) form long conga lines as they move between coral reefs and the open ocean floor on a seasonal cycle **(Figure 55.3).**

Animals Use Wayfinding Mechanisms to Guide Their Movements

Moving animals use various wayfinding mechanisms to arrive at their destination. Biologists group these mechanisms into three general categories: *piloting, compass orientation,* and *navigation.* Many species probably use a combination of these mechanisms to guide their movements.

The simplest wayfinding mechanism is **piloting,** in which animals use familiar landmarks to guide their journey. For example, gray whales *(Eschrichtius robustus)* migrate from Alaska to Baja California and back using visual cues provided by the Pacific coastline of North America. When it is time to breed and lay eggs, Pacific salmon (genus *Oncorhynchus*) use olfactory cues to pilot their way from the ocean back to the stream where they themselves hatched.

Animals that do not undertake long migrations also use specific landmarks to identify their nest site or places where they have stashed food. In a famous experiment published in 1938, Niko Tinbergen showed that female digger wasps *(Philanthus triangulum),* which nest in soil, use visual landmarks to find their nests after flying off in search of food **(Figure 55.4).** Tinbergen arranged pinecones in a circle around one nest while the female was still inside. As she left, she flew around the area, apparently noting nearby landmarks. Tinbergen then moved the circle of pinecones a short distance away. Each time a female returned, she searched for her nest within the pinecone circle—and never once found it unless the pinecones were returned to their original position. In a follow-up study, Tinbergen rearranged the circle of pinecones into a triangle after females left their nests and added a ring of stones nearby. The returning females looked for their nest in the stone circle. Tinbergen concluded that digger wasps respond to the general outline or geometry of landmarks around their nests and not to the specific objects that create those landmarks.

A more sophisticated wayfinding mechanism, **compass orientation**, allows animals to move in a particular direction, often over a specific distance or for a prescribed length of time. Some day-flying migratory birds, for example, orient themselves using the sun's position in the sky in conjunction with an internal biological clock (see Section 40.4). The internal clock allows the bird to use the sun as a compass, compensating for changes in its position through the day; the clock may also allow some birds to estimate how far they have traveled since beginning their journey. Other migratory animals use polarized light or Earth's magnetic field as a compass.

Some birds that migrate at night use the positions of stars to determine their direction. The indigo bunting *(Passerina cyanea)*, for example, flies about 3500 km from the northeastern United States to the Caribbean or Central America each fall and makes the return journey each spring. Stephen Emlen of Cornell University demonstrated that these birds use celestial cues to direct their migration **(Figure 55.5)**. Emlen confined individual buntings in cone-shaped test cages. He lined the sides of the cages with blotting paper, placed inkpads on the bottom, and kept the cages in an outdoor enclosure so that the birds had a full view of the sky. Whenever a bird made a directed movement, its inky footprints indicated the direction in which it was trying to fly. Emlen found that on clear nights in fall, the footprints pointed to the south; on clear nights in spring, they pointed north. On cloudy nights, when the buntings could not see the stars, their footprints were evenly distributed in all directions, indicating that their compass required a view of the stars.

The most complex wayfinding mechanism is **navigation**, in which an animal moves toward a specific destination, using both a compass and a "mental map" of where it is in relation to the destination. Human hikers in unfamiliar surroundings routinely use navigation to find their way home: they use a map to determine their current position and the necessary direction of movement and a compass to orient themselves in that direction. Scientists have documented true navigation in only a few animal species. Perhaps the most notable is the homing pigeon *(Columba livia)*, which can navigate to its home coop from any direction. Recent research suggests that homing pigeons probably use the sun's position as their compass and olfactory cues as their map.

Environmental Cues Trigger Hormonal Changes That Induce Seasonal Migration

For white-crowned sparrows and many other species, researchers have shown that decreasing (or increasing) day length, a correlate of the approaching autumn (or

Figure 55.2

Long-distance migration. Arctic terns *(Sterna paradisaea)* migrate from the high Arctic to Antarctica each year, a round-trip journey of 40,000 km. This species' summer breeding range is shaded on the map.

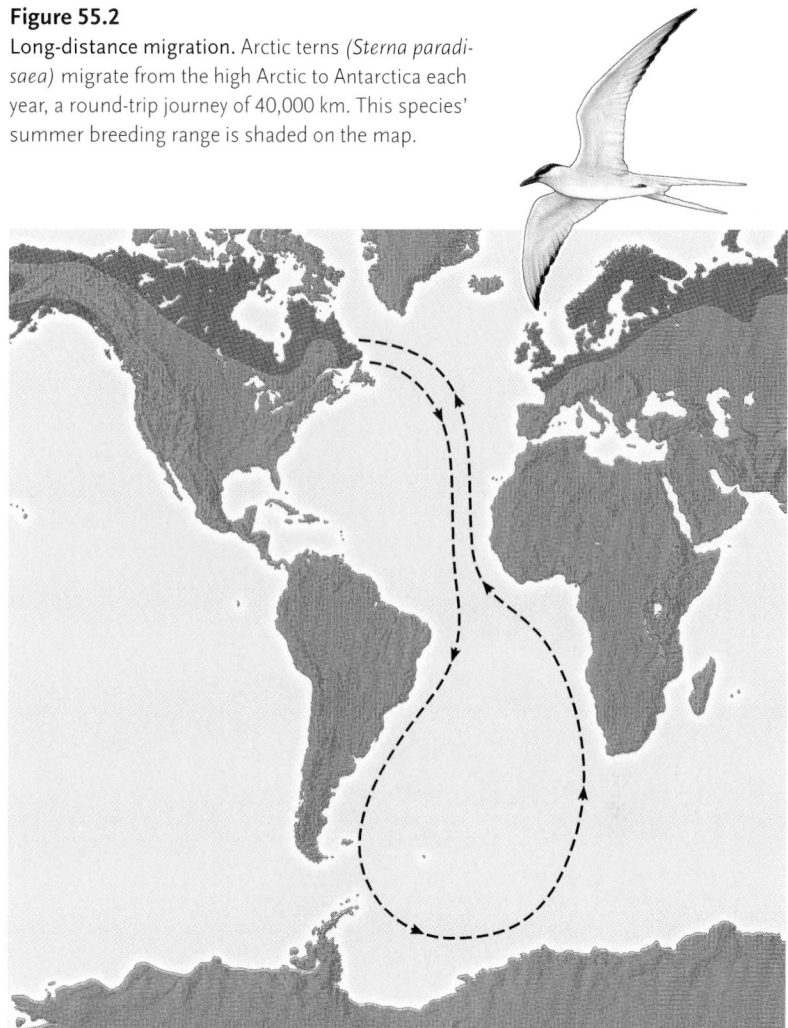

Figure 55.3

Migrating arthropods. Spiny lobsters *(Panulirus argus)* make seasonal migrations between coral reefs and the open ocean floor. As many as 50 individuals march in single file for several days.

Figure 55.4 Experimental Research

Using Landmarks to Find the Way Home

QUESTION: How do female digger wasps *(Philanthus triangulum)* relocate their nests after flying off to search for food?

EXPERIMENT: Tinbergen arranged pinecones in a circle around the nest of a female digger wasp while she was still inside. After leaving the nest, she circled the area a few times, apparently noting nearby landmarks. Tinbergen then moved the circle of pinecones a short distance away.

RESULT: Each time the female returned, she searched for her nest within the pinecone circle. She was unable to find the nest unless Tinbergen replaced the pinecones in their original position.

Wasp's flight pattern on leaving nest

Wasp's return, looking for nest

Nest

CONCLUSION: Female digger wasps use the location of local landmarks to find the entrances to their underground nests.

---

spring), stimulates the anterior pituitary of the bird's brain to generate a series of hormonal changes. The birds then feed heavily and accumulate the fat reserves necessary to fuel the long journey. Sparrows also become increasingly restless at night, until one evening they launch themselves into their nocturnal migration. Their ability to adopt and maintain a southerly orientation in autumn (and a northerly one in spring) rests in part on their capacity to use the positions of stars to provide them with directional information.

Seasonal Variation in Food Supply May Explain the Evolution of Migratory Behavior

Migratory behavior entails obvious costs, such as the time and energy devoted to the journey and the risk of death from exhaustion or predator attack. Why then do some species migrate? What benefits accrue to an individual that undertakes a costly migration?

For migratory birds, the most widely accepted hypothesis focuses on seasonal changes in food supplies. The amount of insect food available in northern forests increases explosively during the warm spring and summer, providing abundant resources to produce eggs and rear offspring. Then, during the late fall and winter, insects all but disappear. A few bird species that forage on seeds and dormant insects do not head south. However, energy supplies are more predictably available in tropical overwintering grounds, and migratory birds may have a better chance of surviving there. The following spring they return north to exploit the food bonanza on their summer breeding grounds.

The two-way migratory journeys may provide other benefits as well. Avoiding the northern winter is probably adaptive because endotherms must increase their metabolic rates just to stay warm in cold climates (see Section 46.8). But in summer the days are longer at high latitudes than they are in the tropics (see Section 52.1), giving adult birds more time to collect enough food to rear a brood.

Seasonal changes in food supply also underlie the migration of monarch butterflies *(Danaus plexippus)*, which eat milkweed *(Asclepias* species) leaves as larvae and the nectar of milkweeds and other plants as adults.

Figure 55.5 Experimental Research

Experimental Analysis of the Indigo Bunting's Star Compass

QUESTION: Do indigo buntings *(Passerina cyanea)* use the positions of stars in the night sky to orient their migrations?

EXPERIMENT: Emlen placed individual buntings in cone-shaped test cages. He lined the sides of the cages with blotting paper, placed inkpads on the bottom, and kept the cages in an outdoor enclosure so that the birds had a full view of the sky. Whenever a bird made a directed movement, its inky footprints indicated the direction in which it was trying to fly. Emlen predicted that the footprints would show the buntings' inclination to migrate south in autumn and north in spring.

Indigo bunting

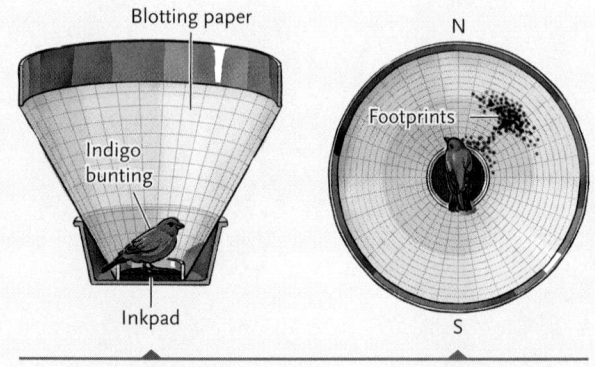

R. & N. Bowers/VIREO

Side (left) and overhead (right) views of the test cage with blotting paper on the sides and an inkpad on the bottom

RESULTS: On clear nights in autumn, the footprints pointed to the south; on clear nights in spring, they pointed north. On cloudy nights, when buntings could not see the stars, their footprints were evenly distributed in all directions.

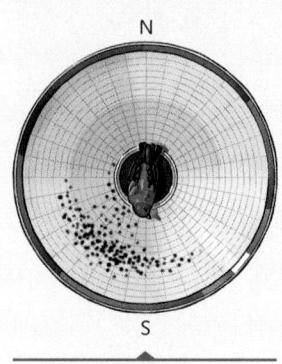

In autumn, the bunting footprints indicated that they were trying to fly south.

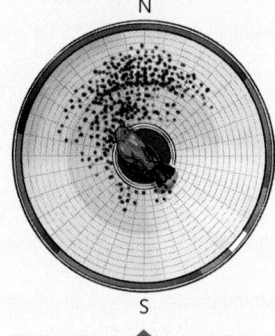

In spring, the bunting footprints indicated that they were trying to fly north.

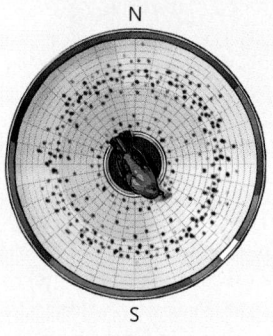

On cloudy nights, when buntings could not see the stars, their footprints indicated a random pattern of movement.

CONCLUSION: Indigo buntings use the positions of the stars to direct their seasonal migrations. When they could see the stars above their test cages, they moved in the predicted direction; but when clouds obscured their view of the stars, they moved in random directions.

In eastern North America, milkweed plants grow only during spring and summer. Many adult monarchs head south in late summer, when milkweeds are beginning to die, migrating as much as 4000 km from eastern and central North America to central Mexico, where they cluster in spectacular numbers **(Figure 55.6)**. Unlike migrant birds, these insects do not feed on their overwintering grounds. Instead, their metabolic rate decreases in the cool mountain air, and the butterflies become inactive for months, thereby conserving precious energy reserves. When spring arrives, the butterflies become active again and begin the return migration to northern breeding habitats. The northward migration is slow, however, and many individuals stop along the way to feed and lay eggs. But their offspring, and their offspring's offspring, continue the northward

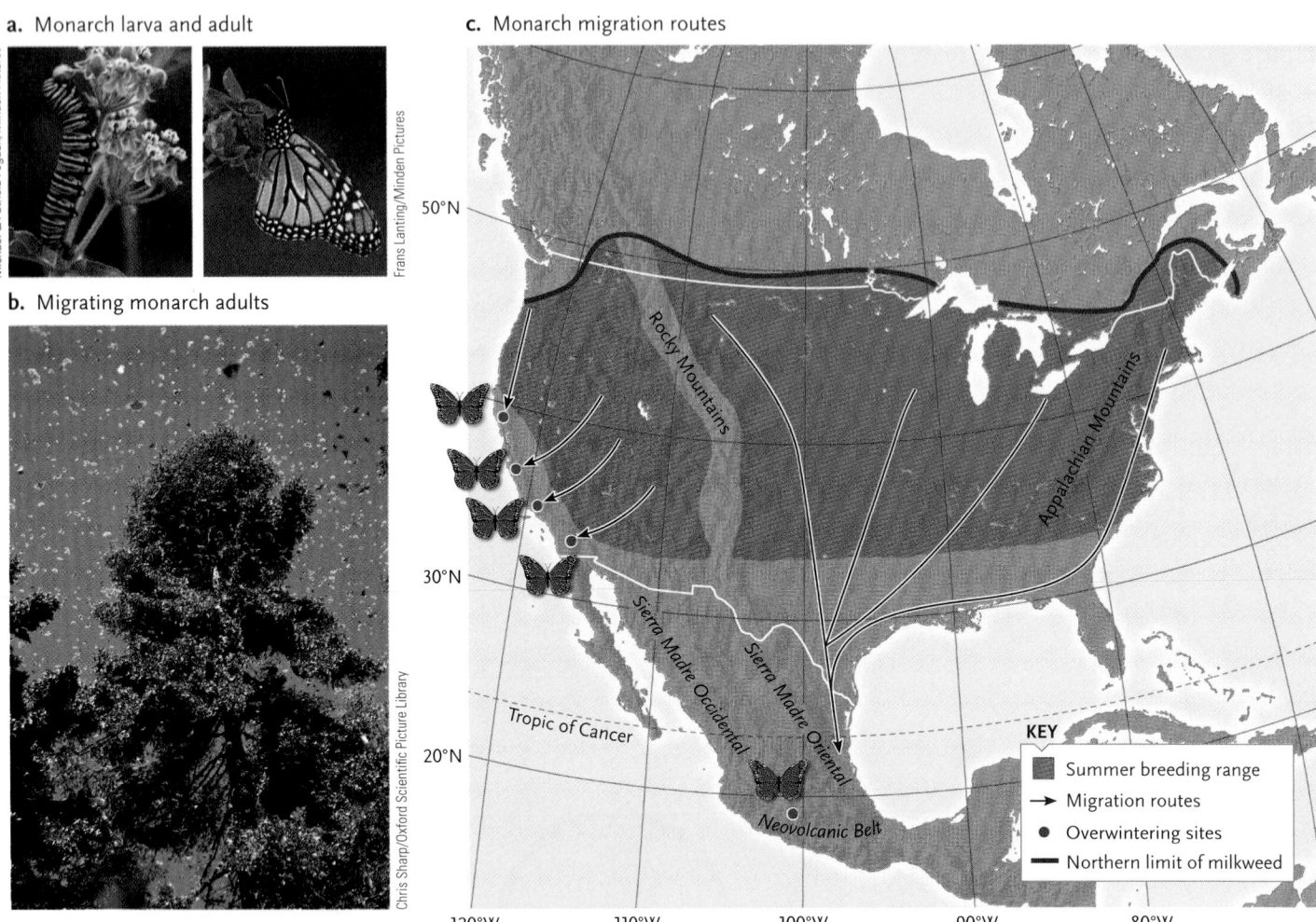

a. Monarch larva and adult

b. Migrating monarch adults

c. Monarch migration routes

KEY

- Summer breeding range
- → Migration routes
- ● Overwintering sites
- ▬ Northern limit of milkweed

50°N
30°N
20°N

Rocky Mountains
Appalachian Mountains
Sierra Madre Occidental
Sierra Madre Oriental
Tropic of Cancer
Neovolcanic Belt

120°W 110°W 100°W 90°W 80°W

Figure 55.6

Monarch butterfly migrations. **(a)** Monarch butterflies *(Danaus plexippus)* feed primarily on milkweed plants. **(b)** When milkweed plants in their breeding range die back at the end of summer, millions of monarchs begin a southward migration. **(c)** Butterflies that live and breed east of the Rocky Mountains migrate to Mexico. After passing the winter in a semidormant state, they migrate northward the following spring. Monarchs living west of the Rocky Mountains winter in coastal California.

migration through the summer; some descendants eventually reach Canada for a final round of breeding. The summer's last generation then returns south to the spot where their ancestors, two to five generations removed, spent the previous winter.

For other animals, the migration to breeding grounds may provide special conditions necessary for reproduction. For example, gray whales migrate south to breeding grounds in quiet, shallow lagoons where predators are rare and warm water temperatures will not stress their calves.

STUDY BREAK

1. What is the difference between piloting, compass orientation, and navigation?
2. What is the most probable selection pressure that has fostered seasonal migrations in birds?

55.2 Habitat Selection and Territoriality

The geographical range of nearly every animal species includes a mosaic of habitat types. The breeding range of white-crowned sparrows, for example, encompasses forests, meadows, housing developments, and city dumps. An animal's choice of habitat is critically important because the habitat provides food, shelter, nesting sites, and the other organisms with which it interacts. If an animal chooses a habitat that does not provide appropriate resources, it will not survive and reproduce.

Animals Use Multiple Criteria for Selecting Habitats

On a large spatial scale, animals almost certainly use multiple criteria to select the habitats they occupy, but no research has yet established any general principles about how animals make these choices. When a migrating bird arrives at its breeding range, for example, it probably cues on large-scale geographical features, such as a pond or a patch of large trees. If it does not find the food or nesting resources it needs—or if other individuals have already depleted those resources—it may move to another habitat patch.

On a very fine spatial scale, basic responses to physical factors enable some animals to find suitable habitats. The simplest such mechanism is called a **kinesis** (*kinesis* = movement), a change in the rate of movement or the frequency of turning movements in response to environmental stimuli. For example, the terrestrial crustaceans known as wood lice (Isopoda) typically live under rocks and logs or in other damp places. Although these arthropods are not attracted to moisture per se, laboratory experiments have shown that when a wood louse encounters dry soil, it exhibits a kinesis, scrambling around and turning frequently; when it reaches a patch of moist soil, it moves much less. As a result, these animals accumulate in moist habitats. Biologists infer that this behavior is adaptive because wood lice exposed to dry soil quickly dehydrate and die. Other animals may exhibit a **taxis** (*taxis* = ordered movement), a response that is directed either toward or away from a specific stimulus. For example, cockroaches (order Blattodea) exhibit negative phototaxis: they actively avoid light and seek darkness, a behavior that makes them harder for visually oriented predators to detect.

Genetics and Learning Influence Habitat Selection

Biologists generally assume that habitat selection is adaptive and has been shaped by natural selection in most animal species. For example, some animals instinctively select habitats where they are well camouflaged, a means of avoiding detection by predators (see Figure 50.3); predators would discover and eliminate any individual that fails to select a matching background—along with any alleles responsible for the mismatch. Many insects have a genetically determined preference for the plants that they eat during their larval stage. Adults often restrict their mating and egg-laying activities to these food plants, effectively selecting the habitats where their offspring will live and feed, as described in the discussion of sympatric speciation (see Section 21.3).

Even vertebrates sometimes exhibit such innate preferences, as demonstrated by two closely related European bird species, blue tits *(Parus caeruleus)* and coal tits *(Parus ater)*. Adult blue tits feed mostly in oak trees, whereas coal tits prefer to feed in pines. When researchers reared the young of both species in cages without any vegetation at all and then offered them a choice between oak branches and pine branches, coal tits immediately gravitated toward pines and blue tits toward oaks, strongly suggesting that the preference is

Figure 55.7

Habitat selection by birds. Wild blue tits *(Parus caeruleus)* show a strong preference for oak trees, and coal tits *(Parus ater)* show a strong preference for pine trees. Hand-reared birds that were raised in a vegetation-free environment showed identical, though slightly weaker, preferences.

innate **(Figure 55.7)**. Further research demonstrated that each species feeds most successfully in the tree species it prefers. Thus, natural selection probably fostered these preferences.

Habitat preferences can be molded by experiences early in life, however. For example, the tadpoles of red-legged frogs *(Rana aurora)* usually live in aquatic habitats cluttered with sticks, strands of algae, and plant stems; when given a choice in the laboratory, they prefer striped backgrounds to plain ones. By contrast, tadpoles of the closely related cascade frog *(Rana cascadae)* live over gravel bottoms, and they prefer plain substrates over striped ones. However, when red-legged frogs are reared over plain substrates and cascade frogs over striped substrates, they no longer exhibit preferences for their usual substrates.

Animals Sometimes Defend Patches of Habitat for Their Exclusive Use

Under some circumstances, animals may defend a **territory** from other members of their species, retaining more or less exclusive use of the resources it contains. Territorial behavior occurs in all major groups of vertebrates, many insects, and some other invertebrates, but it is by no means universal. In many organisms, territorial behavior occurs only during the breeding season.

Animals establish and defend territories only when some critical resource is in short supply. Moreover, the resource must be fixed in space so that the area around it can be defended. For example, during the breeding season, most songbirds defend a territory within which they build a nest and collect food for their young. By contrast, many sea bird species, such as terns and penguins, do not defend a feeding territory. They catch fish in the ocean and build nests on the shore. Although they defend a tiny area around the nest, they never attempt to defend a section of ocean; fishes come and go at will and thus do not constitute a defendable resource.

Territorial defense is always a costly activity. Patrolling territory borders, performing displays hundreds of times per day, and chasing intruders take time and energy. Moreover, territorial displays increase an animal's likelihood of being injured or detected and captured by a predator.

Experiments conducted by Catherine Marler and Michael Moore of Arizona State University illustrate the cost of territorial behavior in Jarrow's spiny lizard (*Sceloporus jarrovi*). Male lizards ordinarily exhibit strong territoriality only during the autumn mating season, when elevated blood levels of testosterone stimulate their aggressive behavior. The researchers implanted small doses of testosterone under the skin of experimental animals in June and July, during the nonmating season; controls received a placebo treatment. Testosterone-enhanced males were more active and displayed more often than control males. But experimental males spent less time feeding, even though they used about 30% more energy per day than control males. Over the course of about 7 weeks, a significantly higher percentage of experimental males died—a clear sign that engaging in territorial behavior is costly.

On the other hand, the benefits of maintaining a territory include having access to nesting sites, food supplies, and refuges from predators. For example, the surgeonfish *(Acanthurus lineatus)*, which lives in the coral reefs around American Samoa, may engage in as many as 1900 chases per day to defend a small territory from other algae-eating fish species. But territory holders may consume up to five times as much food as nonterritory holders.

Figure 55.8
Visual displays. The courtship display of a male wandering albatross *(Diomedea exulans)* includes ritualized postures and movements of the wings and body.

© E. Mickleburgh/Ardea, London

55.3 The Evolution of Communication

When resident animals advertise their presence in their territories, they are communicating information to nearby animals. In the formal language of animal behavior studies, all communication systems involve an interaction between a *signaler,* the animal that transmits information, and a *signal receiver,* the animal that intercepts the information and makes a behavioral response. Natural selection has adjusted the ability of signalers to transmit information and the ability of receivers to get the message.

Animal Signals Can Activate Different Sensory Receptors in Receivers

Biologists categorize animal signals according to the sensory receptors, or "channels," through which the signal acts: *acoustical, visual, chemical, tactile,* or *electrical.* Each channel has specific advantages.

Bird songs are examples of **acoustical signals;** a signaler produces a sound that is heard by a signal receiver. Many animals use the acoustical channel, including a host of nocturnal and burrow-dwelling insects and amphibians. These signals reach distant receivers, even at night and in cluttered environments where visual signals are less effective.

Because humans frequently use facial expressions and body language to send messages, **visual signals** are a familiar form of communication. In many animals, visual signals are *ritualized;* in other words, they have become exaggerated and stereotyped over evolutionary time, forming an easily recognized visual display **(Figure 55.8).** Visual displays can even be useful at night or in the darkness of the deep sea; some animals, such as fireflies and certain fishes, send bioluminescent signals to distant receivers.

Many species release **chemical signals,** which carry messages to signal receivers through the olfactory channel. Scent marking (spraying) by male cats is an example. In particular, mammals and insects often communicate through **pheromones,** distinctive volatile chemicals released in minute amounts to influence the behavior of members of the same species. For example, a worker ant's body contains a battery of glands, each releasing a different pheromone **(Figure 55.9).** One set of pheromones recruits fellow workers to battle colony invaders; another set stimulates workers to collect food that has been discovered outside the colony. Other animals release pheromones to attract mates. Female silkworm moths *(Bombyx mori)* produce bombykol, a single molecule of which can generate a message in specialized receptors on the antennae of any male silkworm moth that is downwind (see Figure 39.19).

In many species, touch conveys important messages from a signaler to a receiver. **Tactile signals** can operate only over very short distances, but for social animals living in close company, they play a significant role in the development of friendly bonds between individuals **(Figure 55.10)**.

Some freshwater fish species, especially those that occupy murky tropical rivers where visual signals could not be seen, use weak **electrical signals** to communicate. These fishes have electric organs that can release charges of variable intensity, duration, and frequency, allowing substantial modulation of the message that a signaler sends. Among the New World knifefishes (order Gymnotiformes), including the electric eel *(Electrophorus electricus)*, electrical discharges can signal threats, submission, or a readiness to breed.

Honeybees Use Several Communication Channels to Transmit Complex Messages

When animals need to convey a complex message, they may use several channels of communication simultaneously. For example, as Karl von Frisch demonstrated, the famous dance of the honeybee *(Apis mellifera)* involves tactile, acoustical, and chemical communication **(Figure 55.11)**. When a foraging honeybee discovers pollen or nectar, it returns to its colony and performs a complex dance on the vertical surface of the honeycomb in the complete darkness of the hive. The dancer moves in a circle, attracting a crowd of workers, some of which follow and maintain physical contact with the dancer. From the dance, they acquire information about the distance and direction they will need to fly to locate the food source.

When the food source is less than about 75 m from the hive, the bee performs a "round dance" (see Figure 55.11a). It moves in tight circles, swinging its abdomen back and forth. Bees surrounding the dancer produce a brief acoustical signal, which stimulates the dancer to regurgitate a sample of the food it discovered. The regurgitated sample serves as a chemical cue to other workers, which then leave the hive to search for that type of food.

If the food source is more distant, the forager performs what von Frisch described as the "waggle dance." The bee dances a half circle in one direction and then dances in a straight line while waggling its abdomen before dancing a half circle in the other direction (see Figure 55.11b). With each waggle, the dancer produces a brief buzzing sound. Von Frisch determined that the angle of the straight run relative

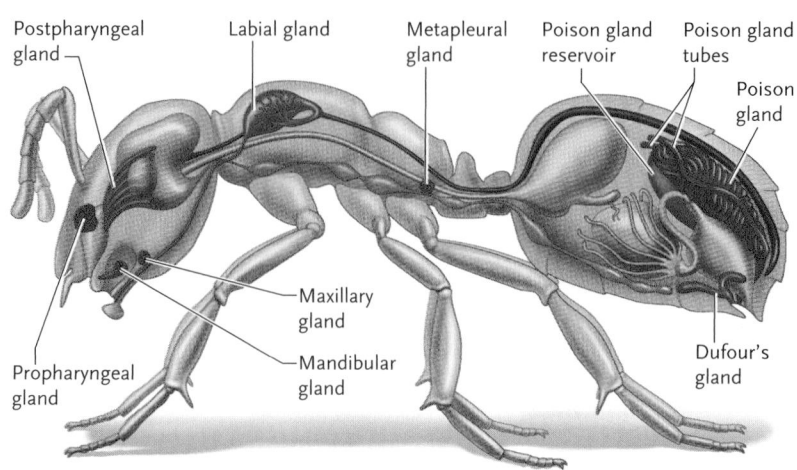

Figure 55.9
Chemical signals. An ant's body contains a host of pheromone-producing glands, each of which manufactures and releases its own volatile chemical or chemicals.

to the vertical honeycomb indicates the direction of the food source relative to the position of the sun (see Figure 55.11c). The duration of the waggles and buzzes that the bee makes on the straight run carries information about the distance to the food: the longer the time spent waggling and buzzing, the further the food is from the hive.

Figure 55.10
Tactile signals. Grooming by hyacinth macaws *(Anodorhynchus hyacinthinus)* removes parasites and dirt from feathers. The close physical contact also promotes friendly relations between groomer and groomee.

a. Round dance

b. Waggle dance

c. Coding direction in the waggle dance

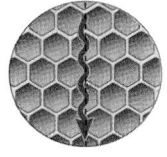 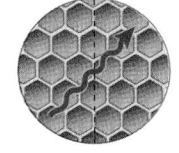 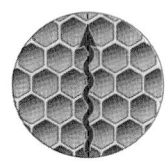

When the bee moves straight down the comb, other bees fly to the source directly away from the sun.

When the bee moves 45° to the right of vertical, other bees fly at a 45° angle to the right of the sun.

When the bee moves straight up the comb, other bees fly straight toward the sun.

Figure 55.11

Dance communication by honeybees. Foraging honey bees *(Apis mellifera)* transmit information about the location of a food source by dancing on the vertical honeycomb. **(a)** If the food source is close to the hive, the forager performs a "round dance." **(b)** If the food source is more than about 75 m from the hive, the forager performs a "waggle dance," which indicates the distance to the food source. **(c)** The dancing bee indicates the direction to a distant food source by the angle of the waggle run.

Figure 55.12

Threat displays. The threat display of a dominant male mandrill *(Mandrillus sphinx)*, used to drive away rival males, features exposed canines.

Tom and Pat Leeson

Biologists Use Evolutionary Hypotheses to Analyze Communication Systems

Signal receivers often respond to communication from signalers in predictable ways. For example, a male white-crowned sparrow generally avoids entering a neighboring territory simply because it hears the song of the resident male. Similarly, young male baboons often retreat without a fight when they see an older male's visual threat display **(Figure 55.12)**, even though they may lose the chance to mate with a female. Why do these receivers behave in ways that appear to be beneficial to their rivals, but not to themselves?

When biologists try to explain behavioral interactions, their hypotheses focus on how an animal's actions may allow it to contribute more offspring to the next generation. In our first example, the retreating male sparrow avoids wasting time and energy on a battle he is likely to lose—as well as the possibility of being injured or killed by another male. Moreover, ousting the current resident might be more tiring and risky than finding a suitable unoccupied breeding site. This hypothesis predicts that resident males should almost always win physical contests. In cases when an intruder wins a territory from a resident, it may do so only after a prolonged series of exhausting clashes. Observations of territorial species—whether birds, lizards, frogs, fish, or insects—generally support these predictions.

Applying a similar argument to competition among male baboons, we can predict that smaller or younger males will concede females to threatening older rivals without fighting. The signal receiver retreats after receiving the threat because he judges that he would be demolished in real combat—after all, a male baboon's canine teeth are not just for show. Evolutionary analyses therefore suggest that both the signaler and the signal receiver benefit from the transfer of information in their communication system.

An evolutionary analysis also helps to explain the strange yell of ravens *(Corvus corax)*, which scavenge carcasses of deer, elk, or moose in northern forests during winter. When one of these large birds comes across a food bonanza, it may call loudly, attracting a crowd of hungry ravens. The calling behavior puzzled Bernd Heinrich of the University of Vermont. Wouldn't a quiet raven eat more, survive longer, and produce more offspring than a noisy bird? If natural selection favored the raven's calling behavior, we might expect that the cost of calling (in terms of lost food) would be offset by a reproductive benefit for the individual caller. Heinrich noticed that paired, territory-owning adults did not yell loudly when they found goat carcasses that he had hauled into the Maine woods; instead, they fed quietly. Only young, wandering ravens that happened upon a carcass in another bird's territory advertised their discovery. The signals of these birds attracted other nonterritorial ravens, which collectively overwhelmed the resident pair's attempts to defend their territory. Only then was a wanderer likely to have a chance to feed in an area that would otherwise be off-limits.

STUDY BREAK

1. Which channels do humans consciously use to communicate with each other?
2. How does a honeybee tell its hive-mates that it has discovered a distant food source?
3. Why do ravens sometimes announce their discovery of food?

55.4 The Evolution of Reproductive Behavior and Mating Systems

In many animal species, communication coordinates the reproductive activities of males and females and governs the interactions between parents and offspring. In this section, we examine how several elements of behavior contribute to the reproductive success of individuals.

Males and Females Use Different Reproductive Strategies

In sexually reproducing species, males and females often differ in their overall **reproductive strategies,** the set of behaviors that lead to reproductive success. This difference arises in part from a fundamental difference in the amount of **parental investment,** the time and energy devoted to the production and rearing of offspring, provided by the two sexes. Because eggs are much larger than sperm, females almost always contribute more energy than males to the production of a gamete.

A male might increase the number of offspring that carry his alleles simply by mating with multiple females, especially if he does not spend time and energy providing parental care to his offspring. Thus, in many animal species, males compete intensely for access to females, and any trait that increases a male's access or attractiveness to females has a big reproductive payoff.

Entirely different selection pressures operate on females, whose reproductive output is generally limited by the number of eggs they can produce. Mating with multiple males will not increase that number. But the success of her offspring may depend on the attributes of their father or the territory he holds. Thus, the females of many species choose their mates carefully. In some cases, females mate with males whose territories include abundant resources, ensuring an ample food supply for their young. In other cases, females choose robust males that will contribute "good genes" (that is, alleles that confer a high likelihood of surviving and reproducing) to her offspring, increasing their chances of long-term success.

Male Competition for Females and Female Mate Choice Foster Sexual Selection

Male competition for access to females coupled with the females' choice of mates establishes a form of natural selection called **sexual selection,** that is, selection for mating success (see Section 20.3). As a result of sexual selection, males are larger than females in many species, and males have ornaments and weapons, such as horns and antlers, that are useful for attracting fe-

males as well as for butting, stabbing, or intimidating rival males. Males typically show off these elaborate structures in complex **courtship displays** to attract the attention of females. For example, male peafowl *(Pavo cristatus)* strut in front of females while spreading a gigantic fan of tail feathers, which they shake, rattle, and roll.

Why should females choose males with exaggerated structures that they display conspicuously? Biologists have developed several hypotheses to explain the attraction. First, a male's large size, bright feathers, or large horns might indicate that he is particularly healthy, that he can harvest resources efficiently, or simply that he has managed to survive to an advanced age. These traits are, in effect, signals of male quality; and if they reflect a male's genetic makeup, he is likely to fertilize a female's eggs with sperm containing successful alleles. In some cases, big, showy males hold large, rich territories, and females that choose them gain access to the resources their territories contain.

The degree to which females *actively* choose genetically superior mates varies among species. In the northern elephant seal *(Mirounga angustirostris),* for example, female choice is more or less passive. Large numbers of females gather on beaches to give birth to their pups before becoming sexually receptive again. Males locate these clusters of females and fight to keep other males away (see Figure 20.8). Males that win have exceptional reproductive success, but only after engaging in violent and relentless combat with rival males. In this kind of mating system, females are practically guaranteed to receive sperm from large and powerful males in superb physiological condition, attributes that may well be associated with alleles that will increase their offspring's chances of living long enough to reproduce.

In other species, females exercise more active mate choice, copulating only after inspecting a group of potential partners. Among birds, active female mate choice is most apparent at **leks,** display grounds where each male possesses a small territory from which it courts attentive females. Male sage grouse *(Centrocercus urophasianus),* a lekking bird of western North America, gather in open areas among stands of sagebrush. Each male defends just a few square meters, where it struts in circles while emitting booming calls and showing off its elegant tail feathers and big neck pouches **(Figure 55.13).** Females wander among the displaying males, presumably analyzing the males' visual and acoustical displays. Eventually, each female selects a mate from among the dozens of males present. Females repeatedly favor males that come to the lek daily, defend their small area vigorously, and display more frequently than the average lek participant. In other words, favored males can sustain their territorial defense and high display rate over long periods, an ability that may correlate with useful genetic traits.

Figure 55.13

Lekking behavior. Male sage grouse *(Centrocercus urophasianus)* use their ornamental feathers in visual courtship displays performed at a lek, where each male has his own small territory. The smaller brown females observe the prancing males before choosing a mate.

Experimental studies of peafowl suggest that the top peacocks at a lek may indeed supply advantageous alleles to their offspring. In nature, peahens prefer males whose tails have many ornamental eyespots **(Figure 55.14)**. In an experiment on captive birds, some females were mated to males with highly attractive tails, but others were paired with males whose tails were less impressive. The offspring of both groups were reared under uniform conditions for several months and then released into an English woodland. After 3 months on their own, the offspring of fathers with impressive tails survived better and weighed significantly more than did those whose fathers had less

Figure 55.14

Sexual selection for ornamentation. The attractiveness of a peacock *(Pavo cristatus)* to females depends in part on the number of eyespots on his extraordinary tail. The offspring of males with elaborate tails are more successful than the offspring of males with plainer tails.

attractive tails. Apparently, a peahen's mate choice does provide her offspring with a survival advantage.

Another hypothesis argues that females select showy males even though their ornate structures may impede their locomotion or their elaborate displays may attract the attention of a predator. According to this hypothesis, any male that survives *despite* carrying such a handicap must have a very strong constitution indeed, and he will pass those successful alleles—as well as the alleles responsible for the ornamental handicap—to the female's offspring.

Patterns of Parental Care and Territoriality Influence Mating Systems

In the examples of mate choice just described, successful males inseminate many females, increasing their reproductive success dramatically. But one male mating with many females is only one of several **mating systems,** the ways in which males and females pair up. Some species are **promiscuous:** individuals do not form close pair bonds, and both males and females mate with multiple partners. Other species are **monogamous:** one male and one female form a long-term association. Finally, some species are **polygamous:** *either* males *or* females may have many mating partners. If one male mates with many females, the relationship is called **polygyny;** if one female mates with multiple males, it is described as **polyandry.**

Mating systems appear to have evolved to maximize reproductive success, partly in response to the amount of parental care that offspring require and partly in response to other aspects of a species' ecology. For example, the young of most songbird species, like the white-crowned sparrow, are helpless upon hatching; all they can do is open their mouths and peep, signaling to their parents that they are ready to be fed. These young require lots of parental care, and they are more likely to flourish if both parents bring food to the nest. As you might expect, nearly all songbirds are monogamous, and males and females team up to provide parental care to their offspring.

In some other bird species, such as red-winged blackbirds *(Agelaius phoenecius),* males establish large, resource-filled territories, and females select mates largely by the quality of the real estate a male holds. Any male with an exceptionally fine territory will be desirable, even if another female has already established herself there. A second female may judge that more resources are available in his territory than in a neighboring one, despite competition with the other female. However, if many females have already settled in a male's territory, intense competition from them may make it less attractive. Given this pattern of habitat and mate choice by females, red-winged blackbirds have a polygynous mating system; males may fertilize the eggs of multiple females and provide little if any direct care to their offspring.

1. For monogamous species, what characteristics of males should increase their attractiveness to females?
2. What activities do male and female sage grouse perform at a lek?
3. Why might a female red-winged blackbird settle on a territory that was already occupied by another female?

55.5 The Evolution of Social Behavior

Social behavior, the interactions that animals have with other members of their species, has profound effects on an individual's reproductive success. Some animals are solitary, getting together only briefly to mate (rhinoceroses and leopards); others spend most of their lives in small family groups (gorillas); still others live in groups with thousands of relatives (termites and honey bees). Some species, such as some African antelopes and humans, live in large social units composed primarily of nonrelatives.

Group Living Carries both Benefits and Costs

Ecological factors have a large impact on the reproductive benefits and costs of social living. Groups of cooperating predators frequently capture prey more effectively than they would on their own. For example, white pelicans *(Pelecanus erythrorhynchos)* often encircle a school of fish before moving in for the kill. Conversely, prey that are subject to intense predation often gain safety in numbers. Those living in groups have more watchful eyes to detect an approaching predator. In ad-

dition, a predator may be confused when multiple prey scatter in many directions. Finally, few predators have the capacity to capture every individual in a prey cluster, so that some prey escape while the predator pursues others.

Some prey species, such as musk oxen *(Ovibos moschatus),* join forces to defend themselves actively (see the photo that opens this chapter). Even some insects, such as Australian sawfly caterpillars *(Perga dorsalis),* exhibit cooperative defensive behavior **(Figure 55.15).** When predators disturb the caterpillars, all members of the group rear up and writhe about, regurgitating sticky, pungent oils that they have collected from the eucalyptus leaves they eat. Although the caterpillars can store these oils safely, they are toxic and repellent to bird predators.

A group of sawflies regurgitates more repellent eucalyptus oils than a single individual, which may explain why these insects form their simple societies. If this hypothesis is correct, solitary individuals should be at greater risk of being eaten than those that live communally. Birgitta Sillén-Tullberg of the University of Stockholm, Sweden, tested this prediction by offering sawfly caterpillars to young great tits *(Parus major),* a songbird species. Birds that received caterpillars one at a time consumed an average of 5.6, but those that received them in groups of 20 ate an average of only 4.1 caterpillars. As Sillén-Tullberg had predicted, the caterpillars were somewhat safer in a group than on their own.

In some environments, the costs of social clumping can be significant. These costs may include increased competition for food. For example, when thousands of royal penguins *(Eudyptes schlegeli)* crowd together in huge colonies **(Figure 55.16),** the pressure

Figure 55.15
Social defensive behavior. Australian sawfly *(Perga dorsalis)* caterpillars clump together on tree branches. These larvae each regurgitate yellow blobs of sticky, aromatic fluid. The accumulation of fluid from a large group of caterpillars successfully deters some predators.

John Alcock/Arizona State University

Figure 55.16
Colonial living. Royal penguins *(Eudyptes schlegeli)* on Macquarie Island, between New Zealand and Antarctica, experience both benefits and costs from living together in huge colonies.

A. E. Zuckerman/Tom Stack & Associates

on the local food supplies is great, increasing the risk of starvation. Communal living also facilitates the spread of contagious diseases and parasites. Nestlings in large colonies of cliff swallows *(Petrochelidon pyrrhonota)* are often stunted in growth because their nests are swarming with blood-feeding parasites, which move easily from nest to nest under crowded conditions. Such costs are probably why the vast majority of animals do not live in large, complex societies.

Fitness Varies among the Members of a Dominance Hierarchy

Recognizing the costs as well as the benefits of social living, biologists have examined features of social living that appear to reduce the fitness of some individuals. For example, some animal species form **dominance hierarchies**, social systems in which each individual's behavior is governed by its place in a highly structured social ranking. In a typical dominance hierarchy, the dominant or *alpha* individual rules the roost; subordinate individuals typically concede valuable resources to more dominant animals without so much as a peep of protest.

Although dominant individuals gain first access to resources, they also incur costs. Frequent challenges from lower ranking individuals may induce a stress response in dominant animals, which must constantly defend their status. For example, in some primates, wild dogs, and other mammals, dominant males have higher blood levels of cortisol and other stress-related hormones (see Section 40.4) than do subordinates. Elevated cortisol levels may induce high blood pressure, the disruption of sugar metabolism, and other pathological conditions.

Why does a subordinate remain in the group when dominant companions reduce its chances for reproductive success? A possible explanation is that survival rates and reproductive success may be even lower for animals that live by themselves: a solitary baboon surely quickens the pulse of a passing leopard **(Figure 55.17)**. A subordinate member of a group gains the benefits, such as protection against predators, that come from being part of the group. Low-ranking males may even have the chance to copulate with one of the group's females when dominant males are not watching, thus ensuring some representation of their alleles in the next generation. And if a low-ranking individual can live long enough, its social superiors may be toppled by predation, accidents, or old age, and a one-time subordinate may find itself high on the social register with food and mates galore.

In Some Animal Societies, Individuals Exhibit Altruistic Behavior

In some species, group members appear to sacrifice their own reproductive success to help individuals that are not their direct descendants; such behaviors are collectively called **altruism.** For example, subordinate members of a wolf pack do not reproduce, but they share captured prey with the dominant pair and that pair's offspring. Altruistic behavior, by its very definition, appears to contradict a basic premise of Darwinian evolutionary theory, namely that natural selection favors traits that increase an *individual's* relative fitness. Why don't subordinate wolves simply save the energy spent on helping, bide their time until they can become dominant, and then produce their own offspring?

Behavioral ecologist William D. Hamilton of University College, London, provided a solution to this puzzle. He recognized that alleles favoring altruism could be propagated indirectly if altruistic individuals sacrificed personal reproduction to help their relatives reproduce. Helping relatives in this way can propagate the helper's own genes because the family shares alleles inherited from their ancestors.

We can quantify the average percentage of alleles that relatives are likely share by calculating their degree of relatedness **(Figure 55.18)**. We start by considering half siblings who, by definition, share only one genetic parent. Half siblings share on average 25% of their alleles by inheritance from their shared parent, making their degree of relatedness 0.25. By contrast, full siblings, who share the same genetic mother *and* father, share 25% of their alleles through the mother and 25% of their alleles through the father, for a total, on average, of 25% + 25% = 50% of their alleles. In other words, the degree of relatedness for full siblings is 0.50. The degree of relatedness between a nephew or niece and an aunt or uncle is 0.25, and the degree of relatedness between first cousins is 0.125. Thus, individuals should be more likely to help close relatives because, by increasing a close relative's fitness, the individual is helping to propagate some of its own alleles.

If altruistic behavior reduces the reproductive success of an individual exhibiting that behavior, how could an allele that promotes altruistic behavior persist

Figure 55.17

The cost of living alone. A solitary olive baboon *(Papio anubis)* confronts a leopard *(Felis pardalis)* bravely but without much chance of survival.

Figure 55.18 Research Method

Calculating Degrees of Relatedness

PURPOSE: The kin-selection hypothesis suggests that the extent of altruistic behavior exhibited by one individual to another is directly proportional to the percentage of alleles that they share. The hypothesis therefore predicts that individuals are more likely to help close relatives because, by increasing a close relative's fitness, the individual is helping to propagate some of its own alleles. Researchers calculate the degree of relatedness between individuals to test this prediction.

PROTOCOL: To calculate the degree of relatedness between any two individuals, we first draw a family tree that shows all of the genetic links between them. The alleles of a parent are shuffled by recombination and independent assortment in the gametes they produce, so we can calculate only the average percentage of a parent's alleles that offspring are likely to share.

We start by considering *half* siblings, those who share only one genetic parent. Each sibling receives half of its alleles from its mother. Because a parent has only two alleles at each gene locus, the probability of sibling A getting a particular allele from its mother is 0.5 (decimal notation for 50%). Similarly, the probability of sibling B getting the same allele from its mother is also 0.5. Statistically, the probability that two independent events—in this case, the transfer of an allele to sibling A and the transfer of *the same* allele to sibling B—will both occur is the product of their separate probabilities. Thus, the likelihood that both siblings receive the same allele from their mother is $0.5 \times 0.5 = 0.25$.

Now consider two *full* siblings, who share the same genetic mother *and* father. They share 25% of their alleles through the mother *plus* 25% of their alleles through the father, for a total of 50% (half their alleles). In other words, the degree of relatedness for full siblings is 0.50.

INTERPRETING THE RESULTS: Each link drawn between a parent and an offspring or between full siblings indicates that those two individuals share, on average, 50% of their alleles. We can calculate the total relatedness between any two individuals by multiplying out the probabilities across all of the links between them. Thus, the degree of relatedness between a niece and an uncle is 0.25, and the degree of relatedness between first cousins is 0.125.

Half Siblings

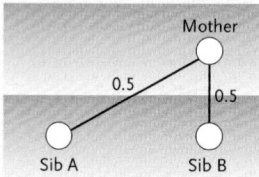

Relatedness = (0.5)(0.5) = 0.25

Full Siblings

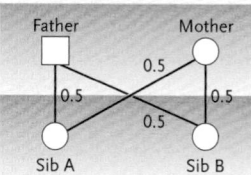

Relatedness
Through mother = (0.5)(0.5) = 0.25
Through father = (0.5)(0.5) = 0.25
Total relatedness = 0.25 + 0.25 = 0.5

First Cousins

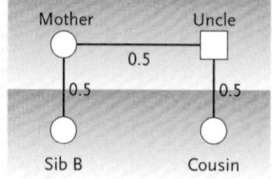

Relatedness = (0.5)(0.5)(0.5) = 0.125

---

or even increase in frequency in a population? The answer depends on the overall number of offspring that carry the allele in the next generation. Altruistic behavior may increase the survival of the altruist's relatives—and they may share the allele in question. If the altruistic behavior allows the assisted relatives to produce proportionately more offspring than the altruist might have produced without helping them, the allele for altruism can increase in frequency in the population. This form of natural selection is aptly called **kin selection.**

For example, suppose a male wolf helps his parents rear four pups to adulthood, pups that would have died without the extra assistance provided by the altruist. Because the pups are his siblings, they share 50% of his genes; thus, on average the helper wolf has created "by proxy" two ($0.50 \times 4 = 2$) copies of any allele that contributed to his altruistic behavior. The costs of his altruism must be measured against this indirect reproductive success. If he had abstained from altruism, the helper wolf might have raised, say, two surviving offspring of his own. Each of his offspring would

carry half of his alleles, preserving just one ($0.50 \times 2 = 1$) copy of a given allele. Under these hypothetical circumstances, reproducing on his own would have produced fewer copies of his alleles in the next generation than helping to raise his siblings.

Although our example of the altruistic wolf is hypothetical, biologists have observed sibling helpers in many bird and mammal species. The phenomenon is especially common among animals in which inexperienced young adults are unable to control sufficient resources to reproduce successfully on their own. Their altruistic behavior not only assists reproduction by their close relatives, but it may also provide useful practice for rearing their own future offspring.

Hamilton's kin-selection hypothesis explains altruistic behavior between closely related individuals, but behavioral biologists have also observed examples of altruism between nonrelatives. For example, the common vampire bat (*Desmodus rotundus*), which feeds on the blood of sleeping mammals, must consume a meal every 2 days to avoid starving to death. Bats that have consumed a large meal often share

a. Queen with sterile workers

b. Workers sharing food and passing pheromones

Kenneth Lorenzen

Kenneth Lorenzen

Figure 55.19
Life in a honeybee *(Apis mellifera)* colony. **(a)** A court of sterile worker daughters surrounds a queen bee, the only female of the colony that reproduces. **(b)** Worker bees routinely share food and transfer pheromones to one another.

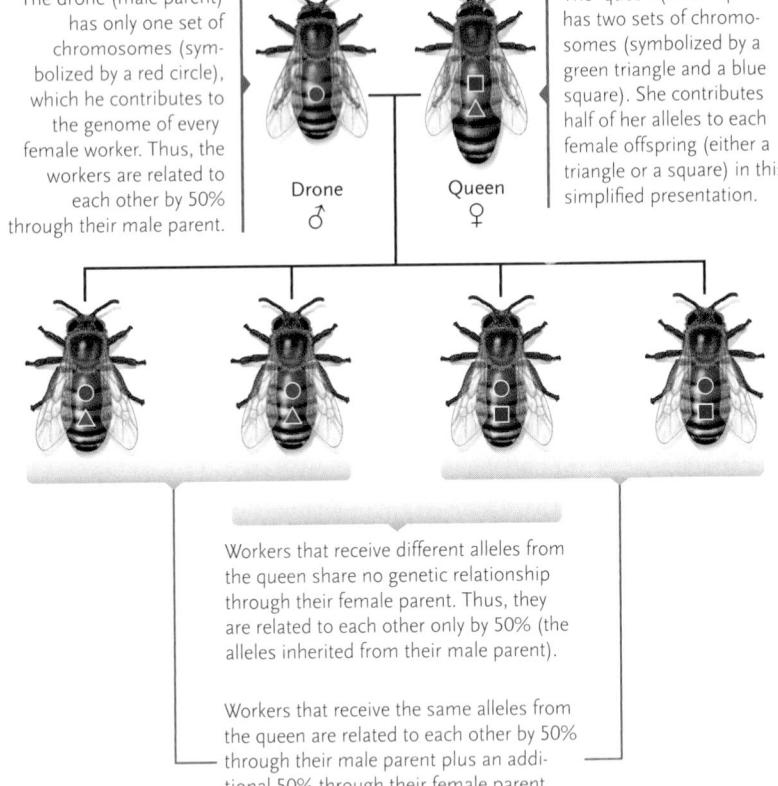

The drone (male parent) has only one set of chromosomes (symbolized by a red circle), which he contributes to the genome of every female worker. Thus, the workers are related to each other by 50% through their male parent.

The queen (female parent) has two sets of chromosomes (symbolized by a green triangle and a blue square). She contributes half of her alleles to each female offspring (either a triangle or a square) in this simplified presentation.

Drone ♂

Queen ♀

Workers that receive different alleles from the queen share no genetic relationship through their female parent. Thus, they are related to each other only by 50% (the alleles inherited from their male parent).

Workers that receive the same alleles from the queen are related to each other by 50% through their male parent plus an additional 50% through their female parent.

Figure 55.20
Haplodiploidy. The genetic system of eusocial insects produces full siblings that have an exceptionally high degree of relatedness. Although this simplified model ignores recombination between the queen's two sets of chromosomes, it demonstrates how half the workers are related to each other by 50% and half are related to each other by 100%. Thus, the average degree of relatedness among workers is 75%. Including recombination would complicate the illustration, but the conclusion would be the same.

their bounty with unrelated members of their group. Why would one bat share its resources with a nonrelative? Robert Trivers, then of Harvard University, proposed that individuals will help nonrelatives if they are likely to return the favor in the future. Trivers called this form of altruistic behavior **reciprocal altruism**, because each member of the partnership can potentially benefit from the relationship. Trivers hypothesized that reciprocal altruism would be favored by natural selection as long as individuals that do *not* reciprocate—called "cheaters" by behavioral biologists—are denied future aid. Observations of vampire bats and some other animals have confirmed Trivers' hypothesis: when a vampire bat accepts a "blood donation" from another bat, but then refuses to share food that it has collected, the other bats refuse to share their food with it in the future.

An Unusual Genetic System May Explain Altruism in Eusocial Insects

Hamilton's insights lead to a critical prediction about the occurrence of self-sacrificing behavior: altruism should usually be directed to close relatives. The evidence from many animal species overwhelmingly supports this prediction, but some species of ants, bees, wasps, and termites, those known as eusocial insects, provide a truly remarkable example. In **eusocial** insects, thousands of related individuals—a large percentage of them sterile female workers—live and work together in a colony for the reproductive benefit of a single queen and her mate(s). The workers may even die in defense of their colonies. How did this self-sacrificing social behavior evolve, and why does it persist over time? The failure of altruistic workers to reproduce should doom any alleles that promote altruism to early extinction.

For example, in a honeybee *(Apis mellifera)* colony, which may contain 30,000 to 50,000 related individuals, the only fertile female is the queen bee; all of the workers are her daughters **(Figure 55.19)**. The queen's role in the colony is to reproduce. The workers perform all other tasks in maintaining the hive, from feeding the queen and her larvae to constructing new honeycomb and foraging for nectar and pollen. They also transfer food to one another and sometimes guard the entrance to the hive. Some pay the ultimate sacrifice when they sting intruders: this act of defense tears open the bee's abdomen, leaving the stinger and the poison sac behind in the intruder's skin, but killing the bee.

Why do bees and other eusocial insects devote their entire lives to helping their mother produce hundreds of thousands of eggs? One answer may lie in a genetic phenomenon called **haplodiploidy**, an unusual pattern of sex determination in these insects **(Figure 55.20)**. Like many other organisms, female bees are diploid, receiving one set of chromosomes from each parent. But male bees are haploid: they hatch from unfertilized eggs. When a queen bee mates with one drone (a male), all of the sperm he delivers are genetically identical because males have only one set of chromosomes. Thus, all workers inherit exactly the same set of alleles from their male parent, producing a 50% degree of relatedness among them. Like other diploid organisms, the workers are also related to each other by an average of 25% through their female parent. Adding these two components of relatedness, we see that workers are related to each other by an average of 75%, a higher degree of relatedness than they would have to any offspring they produced if they were fertile.

This extremely high degree of relatedness among the workers in some eusocial insect colonies may explain their exceptional level of cooperation. When Hamilton first worked out this explanation of social behavior in these insects, he suggested that the workers devote their lives to caring for their siblings—the queen's other offspring—because a few of those siblings, which carry 75% of the workers' alleles, may become future queens producing enormous numbers of offspring themselves.

Nonbreeding workers also exist in a mammalian species, the naked mole-rat *(Heterocephalus glaber)*, a small, almost hairless animal that lives in underground colonies of 70 to 80 individuals in eastern Africa. As described in *Insights from the Molecular Revolution*, recent studies have shown that the highly cooperative individuals occupying a colony share an exceptionally high proportion of their alleles.

STUDY BREAK

1. What do the social behaviors of musk oxen and sawfly larvae have in common?
2. Which animals in a dominance hierarchy are most likely to reproduce?
3. Why might the genetic system of many eusocial insects promote altruistic behavior?

55.6 An Evolutionary View of Human Social Behavior

Evolutionary Analyses May Help to Explain Human Social Behavior

If we can analyze the evolutionary basis of the behavior of honeybees, naked mole-rats, and other animals, perhaps we can do the same for human behavior. According to Hamilton's kin selection hypothesis, we would expect human altruism toward nonrelatives to be rare. And it is true that *most* acts of human altruism are directed toward family members; huge sacrifices to help nonrelatives are relatively uncommon. But why, from an evolutionary perspective, do such charitable acts toward strangers occur at all?

Many behavioral biologists believe that reciprocal altruism can explain why humans have an evolved willingness to engage in low-cost acts of charity. Such behavior demonstrates their capacity for cooperation, and generosity is a socially approved trait that may confer benefits on those who exhibit it. This hypothesis yields the prediction that people who engage in charity will usually let others know about it. That prediction is supported by data showing that when organizers of blood drives offer small participation pins to donors, more people sign up to give blood.

Sometimes researchers employ an evolutionary perspective to study difficult or painful societal issues, such as the occurrence of child abuse within families. Evolutionary theory leads us to predict that family members should generally help, not harm, one another. Margo Wilson and Martin Daly of McMaster University wondered if child abuse might be more common in reconstituted families, those with stepparents who are *not* genetically related to all the children in their care. To test this hypothesis, they examined data on criminal child abuse within families, made available by the police department of a Canadian city. In this city, the chance that a young child would be subject to criminal abuse was 40 times higher for children living with one stepparent and one genetic parent than for children who lived with both genetic parents **(Figure 55.21).**

This example illustrates the sort of insights that an evolutionary analysis of human behavior can provide. Wilson and Daly are not justifying or excusing child abusers. Neither are they claiming that abusive stepparenting is evolutionarily adaptive. Instead, their point is that humans may have some genetic characteristic that makes it more difficult to invest in children that they know are not their own, particularly if they also care for their own genetic children. These results are not just academic. Although a large majority of stepparents cope well with the difficulties of their role, a few do not. Knowledge of familial circumstances under which child abuse is more likely to occur may allow us to provide social assistance that would prevent some children from being abused in the future.

In recent years, the application of evolutionary thinking to human behavior has produced research on all sorts of questions. Sometimes the questions are interesting or even profound. Why do some tightly knit ethnic groups discourage intermarriage with members of other groups? At other times the issues may seem frivolous. Why do men often find women with certain physical characteristics attractive? Although evolution-

Unadorned Truths about Naked Mole-Rat Workers

Naked mole-rats (*Heterocephalus glaber*) are sightless and essentially hairless burrowing mammals (see the accompanying photo) that live in mazes of subterranean tunnels in parts of Ethiopia, Somalia, and Kenya. Mole-rat colonies, which may include from 25 to several hundred individuals, contain a single "queen" and one to three males as the only breeding individuals. The remaining males and females are nonbreeding workers that, like the worker bees, ants, and termites of insect colonies, do all the labor: digging and defending the tunnels and caring for the queen and her mates.

One of the many unanswered questions about these colonial mammals is the genetic structure of a colony. Is close kinship one of the relationships underlying the altruistic behavior of the workers? In other words, do they cooperate because they are all brothers and sisters?

H. Kern Reeve and his colleagues at Cornell University investigated this question using molecular techniques resembling the DNA fingerprinting analysis often used to determine human kinship. The technique (see Section 18.2) depends on a group of repeated DNA sequences that vary to a greater or lesser extent among individuals; that is, they are *polymorphic*. No two individuals (except identical twins) are likely to have exactly the same combination of sequences. Brothers and sisters with the same parents have the most closely related sequences; as relationships become more distant, the differences in the sequences increase.

The researchers began their work by capturing mole-rats living in four colonies in Kenya. Individuals from the same colony were placed together in a system of artificial tunnels. Samples of DNA, taken from individuals that died naturally in the artificial colonies, were subjected to DNA fingerprinting analysis (see Section 18.2). First, the extracted DNA was fragmented by treating with a restriction endonuclease. The DNA fragments were separated by agarose gel electrophoresis, then transferred to a membrane filter by the Southern blot technique (see Figure 18.9). Next, the DNA fragments on the filter were hybridized independently with three radioactively labeled probes that identify three distinct groups of polymorphic sequences in the mole-rat DNA. The hybridization patterns were visualized by autoradiography. The pattern of bands, different for each individual (other than twins), is the DNA fingerprint.

The fingerprint of each mole-rat was compared with the fingerprints of other members of the same and other colonies. In the comparisons, bands that were the same in two individuals were scored as "hits." The number of hits was then analyzed to assign relatedness by noting which individuals shared the greatest number of bands.

The comparisons revealed that individuals in the same mole-rat colony were indeed closely related—they shared an unusually high number of bands, higher than human siblings and approaching the band similarity of identical twins. The number of bands shared between individuals of different colonies was significantly lower, but still higher than that noted between unrelated individuals of other vertebrate species. The close relatedness of even separate colonies, as the investigators point out, may be due to similar selection pressures or to recent common ancestry among colonies in the same geographical region.

On the basis of their results, the researchers propose that the close genetic relatedness among individuals in a colony, which is assumed to increase the degree of altruistic behavior, is one of two major factors underlying the evolution and maintenance of the nonbreeding worker caste in the colonies. The second major factor, they propose, is the chance of survival, which is greater for mole-rats remaining in colonies than for those that attempt to live and breed on their own.

Naked mole-rats (*Heterocephalus glaber*) live in colonies containing many workers that are effectively sterile.

© Gregory D. Dimijian/Photo Researchers, Inc.

ary hypotheses about the adaptive value of behavior can be tested, helping us to understand why we behave as we do, they should never be used to *justify* behavior that is harmful to other individuals. Understanding why we get along or fail to get along with each other and the ability to make moral judgments about our behavior are uniquely human characteristics that set us apart from other animals.

STUDY BREAK

1. How might evolutionary biologists explain altruistic behavior that people exhibit to nonrelatives?
2. Why might stepparents provide fewer resources to their children than birth parents do?

Figure 55.21 Observational Research

An Evolutionary Analysis of Human Cruelty

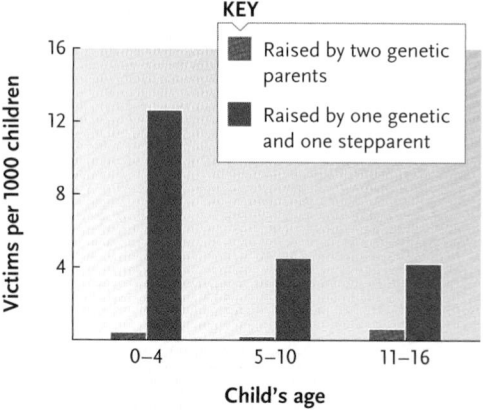

KEY

■ Raised by two genetic parents

■ Raised by one genetic and one stepparent

Victims per 1000 children

0–4   5–10   11–16

Child's age

HYPOTHESIS: Wilson and Daly hypothesized that child abuse would be more common in families in which parents and children were not genetically related than in families in which parents raise their biological offspring.

PREDICTION: Stepparents abuse their stepchildren more frequently than birth parents abuse their biological children.

METHOD: The researchers analyzed data on criminal child abuse within families that had been collected by the police department of a large Canadian city.

RESULT: The data indicated that children living with one stepparent and one genetic parent were 40 times more likely to suffer criminal abuse than children living with two genetic parents.

CONCLUSION: Children raised by one genetic parent and one stepparent are significantly more likely to suffer abuse than those raised by two genetic parents.

UNANSWERED QUESTIONS

Who else is watching and listening?

Studies of communication have typically concentrated on the signaler and the intended receiver. But others are lurking in the background—eavesdroppers who are also attending to these signals. These third parties often use the signal to the signaler's detriment. Some flies, for example, listen for calling male crickets and deposit their larvae on the caller; the larvae eventually kill the cricket as they use him for a food source. Sometimes the signal that makes a male more attractive to females has the same effect on eavesdroppers. Male túngara frogs, for example, can add a "chuck" to the "whine" component of their call. These more complex calls make them more attractive to female frogs but also increase their risk of being captured by frog-eating bats or found by blood-sucking flies. Other animals circumvent this cruel bind by evolving signals in "private channels" to which intended receivers, but not predators, are privy. Ultraviolet signals in swordtail fishes and electrical signals in weakly electric fish are two examples. Increasingly, communication systems are being analyzed from the perspective of a complex communication network rather than the more simple two-way interactions.

Why do females prefer attractive mates?

One reason that females prefer attractive mates is that females can use the ornaments of males to judge their quality in terms of performance and survivorship. A superior male might provide better resources for the female and her offspring, or even pass on better genes to their young. Alternatively, male courtship traits must stand out against a sometimes chaotic background of environmental noise and signals from competing males. To stand out more than others, males might evolve signals that are better at stimulating the female's sensory, neural, and cognitive systems. If a species has evolved visual pigments that allow individuals to locate orange fruit, for example, males should

evolve orange colors, because they will better stimulate the female's visual system. Such a scenario has been suggested for guppies. Researchers who study *sensory drive* try to understand how selection on sensory systems in one context, such as feeding, can influence functions in other contexts, such as mate choice.

To what degree is human behavior influenced by natural selection?

Cooperative and altruistic behavior sometimes evolve in animal societies. Since all societies face similar basic challenges, it is not surprising that cooperation can also be important in human societies. Strong evidence suggests that cooperation has evolved under selection in animal societies, but does logic demand that it has evolved under selection in humans as well? If not, what data would provide strong evidence for the evolution of human social behavior?

The field of evolutionary psychology, which poses these questions, is constrained because the invasive experimental approaches that have been so successful in animal biology cannot be applied to humans. Here the cross-cultural, comparative approach has made some important advances. These questions are being pursued by researchers in biology and psychology as well as anthropology and sociology. This mix of researchers with different approaches guarantees excitement and controversy about these very basic questions that ask why we are who we are.

Michael J. Ryan is the Clark Hubbs Regents Professor in Zoology at the University of Texas at Austin, where he studies the evolution and function of animal behavior. Most of his work has addressed sexual selection and communication in frogs and fish. To learn more about his research, go to http://www.sbs.utexas.edu/ryan/.

Review

Go to ThomsonNOW™ at www.thomsonedu.com/login to access quizzing, animations, exercises, articles, and personalized homework help.

55.1 Migration and Wayfinding

- Some animals—including some arthropods, fishes, birds, and mammals—migrate seasonally, traveling from their birthplace to a distant locality and back again (Figures 55.2 and 55.3).

- Migrating animals may use piloting, compass orientation, or navigation to find their way. In piloting, animals use familiar landmarks to guide their journey (Figure 55.4). In compass orientation, animals use the position of the sun or stars, polarized light, or the Earth's magnetic field as a guide (Figure 55.5). In navigation, animals use mental maps of their position to find their destination.

- Biologists frequently interpret migratory behavior as an adaptive response to changing food supplies (Figure 55.6). Some animals occupy northern habitats when food is plentiful during the spring and summer breeding season. They generally head south to seasonally more productive habitats before the onset of winter.

55.2 Habitat Selection and Territoriality

- Animals use multiple criteria when selecting their habitats.

- Kineses and taxes help animals orient to appropriate portions of the habitats they occupy.

- Habitat selection often has a largely genetic basis, but learning and prior experience influence habitat selection in some species (Figure 55.7).

- Animals may establish and defend territories to gain exclusive use of defendable resources that are in short supply. The costs of territoriality include the time and energy devoted to territory defense and the risk of injury from fights or exposure to predators.

55.3 The Evolution of Communication

- Animal communication occurs between a signaler, which sends a message, and a signal receiver, which receives and interprets the message.

- Animals communicate using acoustical, visual, chemical, tactile, or electrical signals (Figures 55.8–55.10). Each sensory channel provides specific advantages. Animals may use more than one channel simultaneously.

- Honeybees use a combination of tactile, acoustical, and chemical channels to share information about the location of food sources (Figure 55.11).

Animation: Honeybee dances

55.4 The Evolution of Reproductive Behavior and Mating Systems

- Males and females exhibit different reproductive strategies. Males can increase their reproductive success by inseminating the eggs of many females. Females generally seek mates that provide successful alleles to offspring, have access to abundant resources, or help care for young.

- Males often compete for access to females (Figure 55.12). Sexual selection has produced elaborate structures that males use for displays and for aggressive interactions with other males (Figures 55.13 and 55.14). Females may choose to mate with males that have showy structures and great stamina, which may function as signs that they possess successful alleles.

- The type of mating system a species uses is tied to its pattern of territoriality and the amount of parental care the male parent provides.

55.5 The Evolution of Social Behavior

- Social interactions between individuals of the same species provide both benefits and costs. Group living may provide better protection from predators, more efficient feeding, and communal care of young (Figures 55.15 and 55.17). The costs of living in a group include increased competition for scarce resources and an increase in the spread of contagious diseases (Figure 55.16).

- Dominance hierarchies are highly structured societies in which some individuals have high status and first access to resources.

- Altruistic behavior appears to contradict a basic premise of Darwinian evolutionary theory, because altruistic individuals sacrifice their own fitness for the benefit of others. However, individuals generally display altruistic behavior to close relatives that share some of their alleles (Figures 55.18).

- An unusual mechanism of sex determination, haplodiploidy, makes the workers in some eusocial insect colonies more closely related to each other than siblings are in most species (Figures 55.19 and 55.20). Haplodiploidy may have fostered the evolution of highly altruistic behavior.

Animation: Sawfly defense

55.6 An Evolutionary View of Human Social Behavior

- Although humans are more likely to provide assistance to close relatives than to nonrelatives, acts of charity to strangers are common, especially if the altruist can advertise the generosity.

- An analysis of child abuse suggests that humans are more likely to abuse children to whom they are not genetically related than they are to abuse their biological children (Figure 55.21).

Questions

Self-Test Questions

1. Which of the following statements about animal migration is true?
   a. Piloting animals use the position of the sun to acquire information about their direction of travel.
   b. Animals migrating by compass orientation use mental maps of their position in space.
   c. Navigating animals use familiar landmarks to guide their journey.
   d. Navigating animals use a compass and a mental map of their position to reach a destination.
   e. Most migrating birds use olfactory cues to return to the place where they hatched from eggs.

2. In Marler and Moore's experiment with Jarrow's spiny lizard, what evidence from males that had received testosterone implants suggested that engaging in territorial behavior carries a heavy cost?
   a. They had to consume more water than control males.
   b. They mated with fewer females than control males.

c. They ate more frequently than control males.
d. They had higher death rates than control males.
e. They weighed more than control males.

3. Which signal type would provide the fastest communication between bats flying in a dark forest?
   a. chemical signals
   b. acoustical signals
   c. visual signals
   d. tactile signals
   e. electrical signals

4. Squashing an ant on a picnic blanket often attracts many other ants to its "funeral." What kind of signal did squashing the ant likely produce?
   a. an electrical signal
   b. a visual signal
   c. an acoustical signal
   d. a chemical signal
   e. a tactile signal

5. Which of the following behaviors might have been produced by sexual selection?
   a. A male frog calls loudly and clearly from a pond during the breeding season.
   b. A young male goat bleats plaintively when left by its mother.
   c. A hen clucks to call its chicks closer when a predator approaches.
   d. A female lion ignores the sexual advances of a young male.
   e. A male dog is attracted to the odor of a female dog.

6. In comparison to males, the females of many animal species:
   a. compete for mates.
   b. choose mates that are well camouflaged in their habitats.
   c. choose to mate with many partners.
   d. are always monogamous.
   e. choose their mates carefully.

7. Social behavior:
   a. is exhibited *only* by animals that live in groups with close relatives.
   b. cannot evolve in animals that maintain territories.
   c. evolved because group living provides benefits to individuals in the group.
   d. is never observed in insects and other invertebrate animals.
   e. can be explained only by the hypothesis of kin selection.

8. Altruism is a behavior that:
   a. cannot evolve.
   b. has been observed only in insects.
   c. increases the number of offspring an individual produces.
   d. can indirectly spread the altruist's alleles.
   e. can evolve only in animals with a haplodiploid genetic system.

9. The degree of relatedness between a parent and its biological offspring:
   a. is the same as that between brother and sister.
   b. is less than that between brother and sister.

c. depends on how many siblings the parent has.
d. promotes an individual's reproductive success.
e. is the same as between first cousins.

10. The tendency for humans to be charitable to perfect strangers can be explained by the hypothesis of:
    a. sexual selection.
    b. kin selection.
    c. reciprocal altruism.
    d. polyandry.
    e. navigation.

Questions for Discussion

1. In Chapter 53, you learned about some of the environmental changes associated with global warming. What effects might global warming have on animal species that undertake seasonal migrations?

2. The yellow-rumped whippersnapper, an imaginary species of songbird, always established breeding territories in forests where trees are interspersed among many small ponds. Design an experiment to determine what features of the environment this species uses to select its habitat.

3. Although females provide parental care far more often than males in the animal kingdom as a whole, exceptions exist, especially among birds and fishes. Develop three evolutionary hypotheses to explain why male birds are so likely to involve themselves in caring for their broods.

Experimental Analysis

You discover that a particular butterfly species almost always lives in open meadows and almost never lives in nearby shaded forests. Design an experiment to test whether or not habitat selection by this species is adaptive.

Evolution Link

African honeyguides (family Indicatoridae) are birds that call to humans and other mammals, leading them to honeybee colonies in woodlands. The mammals then open the hives to extract the honey, and the honeyguide feeds on the beeswax. How could a communication system between two species evolve?

How Would You Vote?

Africanized bees are slowly expanding their range in North America. Some researchers think the more we know about them, the better we will be able to protect ourselves. Should we fund more research into the genetic basis of their behavior? Go to www.thomsonedu.com/login to investigate both sides of the issue and then vote.

Appendix A
Answers

Chapter 1

Study Break 1.1
1. The major levels in the hierarchy of life and some of their emergent properties are: cells—life; organisms—learning; populations—birth and death rates; communities, ecosystems, biosphere—diversity and stability.
2. Organisms use energy collected from the external environment for growth (including the production of new molecules and cells), maintenance and repair of body parts, and reproduction.
3. A life cycle is the series of structurally and functionally distinct developmental stages through which multicellular organisms pass.

Study Break 1.2
1. In artificial selection, humans selectively breed individuals who possess desirable heritable characteristics to enhance those traits in the next generation. In natural selection, genetically based characteristics that increase survival and reproduction become more common in the next generation.
2. Random changes in DNA—mutations—may change the structure of proteins that contribute to the physical appearance and internal functions of an organism.
3. Being camouflaged may make an animal less likely to be noticed by a predator.

Study Break 1.3
1. In the cells of prokaryotic organisms, DNA is not separated from other parts of the cell. In the cells of eukaryotic organisms, DNA is enclosed within a nucleus.
2. Humans are classified in Domain Eukarya and Kingdom Animalia.

Study Break 1.4
1. A scientific hypothesis must be falsifiable. In other words, we must be able to imagine what sort of data would demonstrate that the hypothesis is incorrect.
2. The copper lizard models told the researchers how frequently lizards would perch in the sun just by chance and what the temperatures of nonthermoregulating lizards would be.
3. Model organisms are usually easy to maintain and study, and they have been so well studied that researchers already know a lot about their biology.
4. When scientists describe a set of ideas as a "theory," they recognize that the ideas have already withstood many scientific tests.

Self-Test Questions
1. c 2. b 3. c 4. d 5. b 6. d 7. c 8. a 9. e 10. d

Chapter 2

Study Break 2.1
An element is a pure substance that consists of one type of atom. An atom is the smallest unit of an element that retains its chemical and physical properties.

A molecule is a collection of atoms chemically combined in fixed numbers and ratios. Molecules can consist of the same atoms, as is seen for the two oxygen atoms in the molecule oxygen, or they can consist of a mixture of different atoms, as in the combination of two hydrogen atoms and one oxygen atom in a molecule of water. Molecules with component atoms that are different, such as water, are compounds.

Study Break 2.2
1. Protons and neutrons are found in the nucleus of an atom. Electrons are found in orbitals located in energy levels (shells) that surround the nucleus.
2. Carbon 11 has six protons and five neutrons. Oxygen 15 has eight protons and seven neutrons.
3. The number of valence electrons—the electrons in the outermost shell of an atom—determines its chemical reactivity. If the outermost shell is not completely filled with electrons, the atom tends to be chemically reactive, whereas if that shell is completely filled, the atom is nonreactive.

Study Break 2.3
1. An ionic bond forms between atoms when those atoms gain or lose electrons completely. An example is the ionic bond in NaCl.
2. A covalent bond forms when atoms share a pair of valence electrons rather than gaining or losing them completely.
3. Electronegativity is a measure of an atom's attraction for the electrons it shares in a chemical bond with another atom. When electrons are shared equally, the atoms remain uncharged and the result is a nonpolar covalent bond. When electrons are shared unequally, one atom carries a partial negative charge and the other atom carries a partial positive charge. The molecule then has polarity, and the bond is a polar covalent bond.
4. In a chemical reaction, atoms or molecules interact to form new chemical bonds or break old ones. Atoms are added to or removed from molecules, or linkages of atoms in molecules are rearranged as a result of bond formation.

Study Break 2.4
1. Hydrogen bonds between neighboring water molecules produce a water lattice. The constant breakage and re-formation of hydrogen bonds in the lattice allows water to flow easily. The polarity of water molecules also contributes to the properties of water. That is, in liquid water, the lattice resists invasion by other molecules unless the invading molecules also contain polar regions that can form competing attractions with water molecules. In that case, the water lattice opens, forming a cavity in which the polar or charged molecule can move. However, nonpolar molecules are unable to affect the water lattice. Hydrogen bonds also give water its unusual ability to resist changes in temperature by absorbing or releasing heat, its unusually high boiling point, and its unusually high internal cohesion and surface tension.
2. A solute is a dissolved substance. A solvent is a substance capable of dissolving another substance. A solution is a solute dissolved in a solvent. For example, salt (NaCl) is a solute that can dissolve in the solvent water.

Study Break 2.5
1. Acids are hydrogen ion (proton, H^+) donors, whereas bases are proton acceptors. An acid dissociates in water to produce a hydrogen ion and an anion. Most bases dissociate in water to give hydroxide ions, which then accept protons to produce water.
2. Buffers act to control the pH of a solution. In living organisms, buffers keep the pH of body and cell fluids within a narrow range, enabling normal cell and body functions to occur. Outside of the normal pH range, the functions of proteins can be affected, thereby adversely affecting the functions of the organism.

Self-Test Questions
1. d 2. e 3. a 4. b 5. a 6. d 7. d 8. c 9. e 10. b

Chapter 3

Study Break 3.1
1. Organic molecules are molecules based on carbon. Hydrocarbons are a type of organic molecule that consists of carbon linked only to hydrogen atoms.
2. The maximum number of bonds that a carbon atom can form is four.

Study Break 3.2

1. In a dehydration synthesis reaction, components of water (—H and —OH) are removed. In hydrolysis, the components of a water molecule are added to functional groups as molecules are broken down into smaller subunits.
2. Carboxyl groups donate a hydrogen ion in water and therefore act as acids. Amino groups accept a hydrogen ion in water and therefore act as bases. Phosphate groups donate hydrogen ions in water and therefore act as acids.

Study Break 3.3

A monosaccharide is the structural unit of carbohydrate molecules. Monosaccharides are simple sugars such as trioses, pentoses, and hexoses. Glucose, galactose, and fructose are hexoses.

A disaccharide is a molecule assembled from two monosaccharides linked by a dehydration synthesis reaction. Lactose, sucrose, and maltose are disaccharides.

A polysaccharide is a polymer of monosaccharide subunits. The subunits are identical or different, depending on the particular polysaccharide. Polysaccharides, as large polymers, are examples of macromolecules. Glycogen, starch, cellulose, and chitin are polysaccharides.

Study Break 3.4

The three most common lipids found in living organisms are neutral lipids, phospholipids, and steroids. Most neutral lipids consist of a three-carbon backbone chain formed from glycerol, with each carbon linked to a fatty acid side chain. In the most common phospholipids, glycerol is the backbone, with two of its binding sites linked to fatty acids. The third binding site is linked to a polar phosphate group. Steroids have structures based on a framework of four carbon rings. Differences in side groups attached to the rings distinguish the different types of steroids.

Study Break 3.5

1. Differences in the side groups (R in the figures) give the amino acids their individual properties.
2. A peptide bond is the bond between the C of the carboxyl group of one amino acid and the N of the amino group of the adjacent amino acid (see Figure 3.17a). The bond is formed in a dehydration synthesis reaction between an amino group of one amino acid and a carboxyl group of another amino acid.
3. In their final folded form, proteins have distinct, structural subdivisions called domains. The domains are the result of the amino acid sequence of the protein (the primary structure of the protein) and the secondary, tertiary, and quaternary (if more than one polypeptide is involved) structures of the protein.

Study Break 3.6

1. Nucleic acids are polymers of nucleotides. A nucleotide consists of a nitrogenous base, a five-carbon sugar, and a phosphate.
2. DNA has the five-carbon sugar deoxyribose, whereas RNA has ribose. DNA has the pyrimidine nitrogenous base T (thymine), and RNA has U (uracil).

Self-Test Questions

1. a 2. d 3. c 4. e 5. c 6. d 7. b 8. d
9. b 10. a

Chapter 4

Study Break 4.1

1. Kinetic energy is the energy of motion, whereas potential energy is stored energy.
2. Catabolic reactions involve the breakdown of complex molecules into simpler molecules, whereas anabolic reactions involve the synthesis of complex molecules from simpler molecules.
3. An open system can exchange both energy and matter with its environment. A closed system can exchange energy, but not matter, with its environment.

Study Break 4.2

1. Many individual reactions found in living cells are not spontaneous because they have a positive ΔG. By joining that reaction to another reaction with a large negative ΔG, the reaction can be completed. The combined reaction is called a coupled reaction.
2. ATP contains the five-carbon sugar ribose, with the nitrogenous base adenine linked to one of the carbons and a chain of three phosphate bonds to another carbon. In a phosphorylation reaction, one phosphate of ATP is transferred to another molecule, leaving ADP.

Study Break 4.3

The greater the negative value of ΔG, the further a reaction will proceed toward completion and therefore the greater the concentration of product molecules versus reactant molecules.

Study Break 4.4

Enzymes accelerate reactions by reducing the activation energy of a reaction, the initial input of energy required to start a reaction. Enzymes lower activation energy by altering reacting molecules to an intermediate arrangement of atoms and bonds that both the reactants and the products of a reaction can assume. This arrangement is called the transition state, an activated state that is highly unstable. With relatively little change in energy, the transition state can move forward toward products or backward toward reactants.

Study Break 4.5

1. As the temperature increases, the increasing kinetic motions of the amino acid chains of an enzyme eventually result in a disruption of the enzyme's three-dimensional structure, causing it to unfold and denature. At that point, there is no enzyme activity.
2. There is a maximum rate at which an enzyme can combine with substrates and release products. Beyond that point, increasing substrate concentration does not increase the reaction rate any further.
3. In competitive inhibition, the inhibitor competes with the normal substrate molecule for binding to the active site of the enzyme, whereas in noncompetitive inhibition, the inhibitor does not compete directly with the substrate for binding to the active site.

Study Break 4.6

A ribozyme is an RNA molecule that accelerates the rate of a biological reaction. A ribozyme qualifies as an enzyme because it remains unchanged after the reaction is complete; that is, it is a true catalyst.

Self-Test Questions

1. c 2. b 3. c 4. e 5. d 6. a 7. a 8. b
9. d 10. e

Chapter 5

Study Break 5.1

The plasma membrane is a bilayer of lipid and suspended protein molecules that bounds the cytoplasm of a cell. The lipid bilayer is hydrophobic; therefore, it is a barrier to the passage of water-soluble substances. The membrane has protein channels through which selected water-soluble substances are able to pass.

Study Break 5.2

The DNA of a prokaryotic cell is located in the nucleoid region, a central area of the cell that has no membrane around it to separate it from the cytoplasm. In most prokaryotes, the DNA is a folded mass. In its unfolded state, it is a circular DNA molecule. In bacteria, this DNA molecule is called the bacterial chromosome.

Study Break 5.3

1. The DNA of a eukaryotic cell is mostly found within a nucleus, which is located roughly centrally in the cell. The nucleus is bounded by a membrane—the nuclear envelope—which separates its contents from those of the cytoplasm. The DNA is complexed with proteins and organized into several linear chromosomes.
2. The nucleolus is an area within the nucleus; it looks like a mass of fibers and granules in the electron microscope. A nucleolus forms around the genes for ribosomal RNA in the chromosomes and is, in fact, the location where the information in those genes is copied into the ribosomal RNA. The ribosomal RNA combines with proteins in the nucleolus to form the large and small ribosomal subunits. A large and a small ribosomal

subunit function together in the cytoplasm to synthesize proteins.

3. The endomembrane system is a collection of organelles, membranous channels, and vesicles within the eukaryotic cell. The synthesis, distribution, and storage of major biological molecules is conducted by this system, which includes the endoplasmic reticulum (ER) and the Golgi complex. The rough ER has ribosomes on its outer surface. Proteins synthesized by those ribosomes enter the ER lumen, where they fold into their final shape. Often those proteins are modified chemically. Then they are delivered to other regions of the cell within vesicles that pinch off from the rough ER. For most of the proteins made on rough ER, the next destination is the Golgi complex.

The smooth ER consists of membranes that lack ribosomes. The smooth ER has various functions, including synthesis of lipids that become parts of cell membranes.

The Golgi complex is a stack of flattened, membranous sacs. This organelle receives vesicles that contain proteins released from the rough ER and continues their chemical modifications. After their modification, the proteins are sorted into vesicles that pinch off from the margins of the Golgi sacs. Some of the released vesicles remain in the cytoplasm as storage vesicles of various types, whereas other vesicles, called secretory vesicles, release their contents to the outside of the cell.

In summary, the endomembrane system is a major traffic network for proteins and other molecules within the cell.

4. A mitochondrion is an organelle enclosed by two membranes. The outer mitochondrial membrane is smooth and covers the outside of the organelle. The inner mitochondrial membrane is highly folded into cristae. The interior of the mitochondrion, bounded by the inner membrane, is the mitochondrial matrix that contains DNA and ribosomes. Most of the energy required for eukaryotic cellular activities is generated by reactions in the mitochondria. Those reactions break down sugars, fats, and other fuel molecules into water and carbon dioxide, releasing energy mostly in the form of ATP. The ATP-generating reactions are located in the cristae and in the matrix.

5. The cytoskeleton is an internal cytoplasmic network of protein filaments and tubules. The main proteins involved are actins and tubulins. The function of the cytoskeleton is to maintain the shape of the cell, reinforce the plasma membrane, and organize internal structures. Changes in the cytoskeleton are responsible for movements of cell organelles, movements of parts of the cell, or movements of the whole cell itself.

Study Break 5.4

1. A chloroplast has two membranes: an outer boundary membrane and an inner boundary membrane. The latter, similar to the cristae of the mitochondrion, is highly folded. The two membranes enclose an inner compartment known as the stroma. Within the stroma is a membrane system composed of flattened, closed sacs called thylakoids. Chloroplasts are the sites of photosynthesis in plant cells. The thylakoid membranes contain chlorophyll and other molecules that absorb light energy and convert it into chemical energy. Enzymes in the stroma use the chemical energy to make carbohydrates and other complex organic molecules from water, carbon dioxide, and other simple inorganic precursors.

2. The tonoplast, the membrane surrounding the central vacuole, contains transport proteins for the movement of substances into and out of the vacuole. Central vacuoles also store organic and inorganic salts, organic acids, sugars, storage proteins, pigments, and, in some cells, waste products. Chemical defense molecules are found in the central vacuoles of some plants.

Study Break 5.5

1. Anchoring junctions are spots, or belts that run entirely around cells, effectively sticking adjacent cells together. Microfilaments (in adherens junctions) or intermediate filaments (in desmosomes) anchor the junction in the underlying cytoplasm.

Tight junctions involve fusion of a network of junction proteins in the outer halves of the plasma membranes of adjacent cells, forming a tight seal between the cells that can keep even ions from moving between the cells.

Gap junctions open direct channels between adjacent cells through which ions and small molecules can pass directly. The gap junctions are formed by aligned hollow protein cylinders in the plasma membranes of two cells.

2. The extracellular matrix (ECM) is a complex of proteins and polysaccharides secreted by the cells that it surrounds. Depending on the nature of the network of proteoglycans (carbohydrate-rich glycoproteins) in the ECM, the consistency ranges from soft and jellylike to hard and elastic.

Self-Test Questions

1. d 2. e 3. a 4. b 5. c 6. b 7. d 8. c
9. b 10. e

Chapter 6

Study Break 6.1

1. The fluid mosaic model proposes that the membrane consists of a fluid phospholipid bilayer in which proteins are embedded and float freely.

2. Integral proteins include transport proteins, receptor proteins, recognition proteins, and cell adhesion proteins. Peripheral proteins include microtubules, microfilaments, intermediate filaments, and proteins that link the cytoskeleton together.

Study Break 6.2

1. In passive transport, molecules and ions move across the membrane from the side with the higher concentration to the side with the lower concentration; that is, the difference in concentration provides the energy for passive transport. In active transport, molecules and ions move across the membrane from the side with the lower concentration to the side with the higher concentration, that is, against the concentration gradient. The energy for active transport comes from the hydrolysis of ATP.

2. The transport of substances through membranes based solely on molecular size and lipid solubility is simple diffusion. The diffusion of polar and charged molecules across membranes with the help of transport proteins is facilitated diffusion.

Study Break 6.3

1. Osmosis is the passive transport of water across a membrane. This movement follows concentration gradients. For osmosis to occur, there must be a selectively permeable membrane, that is, a membrane that will allow water molecules, but not molecules of the solute, to pass. As long as the solute is at different concentrations on the two sides of the membrane, water movement will occur; that is, it is not necessary for pure water to be present on one side of the membrane.

2. If animal cells are in a hypertonic solution, water molecules will move by osmosis from within the cells to the surrounding solution. If the outward movement of water exceeds the capacity of the cells to replace the lost water, the cells will shrink.

Study Break 6.4

1. Active transport is the movement of substances across membranes against their concentration gradients by pumps; the energy for active transport comes from ATP hydrolysis. In primary active transport, ATP hydrolysis directly drives the process; that is, the same protein that transports a substance also hydrolyzes ATP. In secondary active transport, ATP hydrolysis indirectly drives the process; that is, the transport proteins themselves do not hydrolyze ATP. Rather, the transporters use a favorable concentration gradient of ions, generated by primary active transport (where ATP hydrolysis is used), as their energy source for active transport of a different ion or molecule.

2. A membrane potential is a voltage difference across a membrane. Ion transport by membrane pumps contributes to this voltage difference. The sodium–potassium pump in the plasma membrane pushes three sodium ions out of the cell and two potassium ions into the cell with each turn of the pump. This leads to an accumulation of positive charges outside the membrane, causing the inside of the cell to become negatively charged with respect to the outside of the cell. In addition, an unequal distribution of ions across the membrane is created by passive transport. The electrical potential difference (voltage) across the plasma membrane is the membrane potential.

Study Break 6.5
1. In exocytosis, secretory vesicles in the cytoplasm contact and fuse with the plasma membrane, releasing their contents to the outside of the cell.
2. Endocytosis is a mechanism by which substances are brought into the cell from the exterior. The substances become trapped in pitlike depressions that bulge inward from the plasma membrane. The depression pinches off as an endocytic vesicle. In bulk-phase endocytosis, no binding by surface receptors is involved. Extracellular water is taken in together with any other molecules that are in solution in the water. This is the simplest form of endocytosis. In receptor-mediated endocytosis, molecules to be taken in become bound to the outer cell surface by receptor proteins. The receptor proteins are specific in that they recognize and bind only certain molecules from the solution that surrounds the cell. The molecules recognized are mostly proteins or other molecules carried by proteins. Once the receptors have bound their target molecules, the receptors collect into a coated pit, a depression in the plasma membrane. The pits, with the contained target molecules, pinch off from the plasma membrane to form endocytic vesicles.

Self-Test Questions
1. d 2. b 3. a 4. c 5. e 6. d 7. e 8. b
9. a 10. c

Chapter 7

Study Break 7.1
Specificity of a cellular response depends on the signal–receptor interaction. Specificity starts with the signal molecule; that is, the specific signal molecule is the messenger that elicits a specific cellular response. For example, the hormone epinephrine causes glucose release into the bloodstream. Specificity also depends on the target cells; that is, only target cells respond to the signal molecule because they exclusively have receptors for the signal molecule.

Study Break 7.2
1. Protein kinases are enzymes that add phosphate groups to other proteins. The result of phosphorylation is that the protein will be either stimulated or inhibited in its activity. The cellular responses of signal transduction pathways are produced through the actions of protein kinases.
2. Amplification is the phenomenon of an increase in the magnitude of each step of a signal transduction pathway. Amplification typically occurs because the proteins conducting each step of the pathway are enzymes. That is, each enzyme, when it is becomes activated, activates large numbers of molecules entering the next step of the pathway.

Study Break 7.3
1. For a receptor tyrosine kinase to become activated, first, the signal molecule binds to the receptor, which then assembles into a dimer. The receptor adds phosphate groups to tyrosines on the cytoplasmic side of itself, which activates the receptor.
2. The insulin receptor is different from general receptor tyrosine kinases in that it is permanently assembled into a dimer. Generally, receptor tyrosine kinases assemble into dimers only after the signal molecule binds to them.

Study Break 7.4
1. The first messenger in a G-protein–coupled receptor–controlled pathway is the extracellular signal molecule. When it binds to the G-protein–coupled receptor, it activates a site on the cytoplasmic side of the receptor; the activated receptor, in turn, activates the G protein next to it.
2. The effector is activated by the G protein. The effector is a plasma-associated enzyme that generates a nonprotein signal molecule called the second messenger. The second messenger leads to the activation of protein kinases leading to the cellular responses triggered by the signal molecule.
3. A main way the pathway is turned off is by the conversion of cAMP to 5′-AMP by phosphodiesterase. As long as the receptor is bound by the signal molecule, cAMP is being generated by the activated effector. The continued synthesis of cAMP balances the degradation of cAMP by phosphodiesterase, ensuring that the pathway continues to run. However, if the signal molecule no longer is bound to the receptor, the effector again becomes inactive, cAMP therefore is not generated, and existing cAMP is rapidly degraded by phosphodiesterase. As a result, the protein kinase cascade is shut down and no cellular responses occur.

Study Break 7.5
1. The steroid receptor is within the cell, whereas the receptor tyrosine kinase and G-protein–coupled receptors are in the membrane or associated with the membrane. Also, the activated steroid receptor directly activates genes, whereas the other two receptors, when active, are just the first steps in pathways that may or may not activate genes.
2. A steroid hormone brings about a specific cellular response because whether a cell responds to a steroid hormone depends on whether it has the internal receptor. Then within the cells with the receptor, the specific genes controlled depend on the genes with the regulatory sequences recognized by the activated receptor.

Study Break 7.6
Signal transduction pathways, cellular response systems triggered by cell adhesion molecules, and communication pathways that involve gap junctions between adjacent cells might be integrated in a cross-talk network.

Self-Test Questions
1. b 2. a 3. d 4. c 5. d 6. c 7. e 8. b
9. d 10. e

Chapter 8

Study Break 8.1
1. Oxidation is the removal of electrons from a substance; reduction is the addition of electrons to a substance.
2. Cellular respiration refers to the reactions in which oxygen is used as final electron acceptor; it includes the reactions that transfer electrons from organic molecules to oxygen and the reactions that make ATP. Oxidative phosphorylation is the process by which ATP is synthesized using the energy released by electrons as they are transferred to oxygen.

Study Break 8.2
1. The initial steps of glycolysis, which convert glucose to a phosphorylated derivative, require 2 ATP. The later steps, leading to two molecules of pyruvate, involve electron removal from the glucose derivatives and release 4 ATP.
2. If ATP is in excess in the cytosol, it binds to and inhibits the activity of phosphofructokinase. As a result, the concentration of fructose-1,6-bisphosphate, the product of the phosphofructokinase reaction, decreases, and the following reactions of glycolysis are slowed or stopped. This is reversed when ATP levels in the cytosol decrease. In the end, this control mechanism helps prevent the needless oxidation of fuel molecules when ATP is at adequate levels in the cell.

Study Break 8.3
The three-carbon pyruvate molecules are transported from the cytosol into the mitochondria, where they are converted into two-carbon acetyl units through pyruvate oxidation. The citric acid cycle oxidizes the acetyl

units completely to carbon dioxide with the transfer of electrons to NAD$^+$ or FAD.

Study Break 8.4
1. Each complex contains a unique combination of nonprotein carriers that pick up and release electrons.
2. The proton pumps push protons (H$^+$) from the mitochondrial matrix to the intermembrane compartment, increasing the proton concentration there. The resulting proton gradient produces an electrical gradient across the inner mitochondrial membrane with the matrix negatively charged with respect to the intermembrane compartment. The charge and proton concentration differences together provide energy for ATP synthesis in what is called proton-motive force.

Study Break 8.5
Fermentation occurs when oxygen is absent or limited. The electrons carried by the NADH produced in glycolysis are transferred to an organic molecule instead of the electron transfer system. In lactate fermentation, the end product of glycolysis, pyruvate, is converted to lactate. The lactate stores electrons temporarily, transferring them to the mitochondrial electron transfer system when the oxygen content of cells returns to normal. In alcoholic fermentation, pyruvate is converted into ethyl alcohol.

Self-Test Questions
1. b 2. a 3. e 4. c 5. b 6. e 7. a 8. d
9. a 10. c

Chapter 9

Study Break 9.1
1. The two stages of photosynthesis are light-dependent reactions, in which the energy of sunlight is absorbed and converted into chemical energy in the form of ATP and NADPH, and light-independent (dark) reactions, in which electrons carried by NADPH are used as a source of energy to convert carbon dioxide from inorganic to organic form.
2. Chloroplast. The light-dependent reactions are carried out on the thylakoid membranes and stromal lamellae. The light-independent reactions are carried out in the stroma.

Study Break 9.2
1. The chlorophyll a molecules in the antenna complexes are normal molecules of the pigment, consisting of a carbon ring structure with an attached hydrophobic side chain. A magnesium atom is bound at the center of the ring structure. These chlorophyll a pigments absorb light. The chlorophyll a molecules in the reaction centers have modified, specific light absorption properties that result from interactions with particular proteins of the photosystems. The special

chlorophyll a of photosystem II is P680; that of photosystem I is P700. These pigment molecules receive light energy from normal pigments via inductive resonance. The energy is used to push electrons to an excited state.
2. Electrons derived from water splitting and pushed to higher energies by light absorption in photosystem II begin an electron transfer process. The high-energy electrons pass to a primary acceptor in photosystem II and are passed down an electron transfer system to P700 in photosystem I, losing energy along the way. Light energy absorbed by photosystem I again excites the electrons, which pass to different electron carriers, ending with ferredoxin. The ferredoxin transfers high-energy electrons to NADP$^+$, which is reduced to NADPH by NADP$^+$ reductase.
3. In the noncyclic electron flow pathway, electrons run through the entire set of photosystems and electron carriers, producing both NADPH and ATP. NADPH is produced as described in the answer to the previous question. ATP is made from ADP and P$_i$ by ATP synthase. An H$^+$ gradient is used to power ATP synthesis, a process called chemiosmosis.

In the cyclic electron flow pathway, electrons flow cyclically around photosystem I; photosystem II is not involved. The cycle of electrons is through the cytochrome complex and plastocyanin to photosystem I, to ferredoxin, but then back to the cytochrome complex rather than on to NADP$^+$ reductase. The cyclic flow of electrons pumps protons across the thylakoid membranes into the compartment and fuels ATP synthesis as for the noncyclic electron flow pathway. Only ATP is produced by this pathway.

Study Break 9.3
1. Rubisco catalyzes a reaction combining carbon dioxide with RuBP to form two molecules of 3-phosphoglycerate (3PGA). Rubisco, an enzyme unique to photosynthetic organisms, is the key enzyme for producing the world's food because it is responsible for carbon dioxide fixation, a process that ultimately provides organic molecules for most of the world's organisms. Rubisco is the key regulatory site of the Calvin cycle for the following reason: During the daytime, sunlight powers the light-dependent reactions, and the NADPH and ATP produced by those reactions stimulate rubisco, which, in turn, keeps the Calvin cycle running. In darkness, however, NADPH and ATP levels are low, and as a result, rubisco's activity is inhibited and the Calvin cycle slows down or stops.
2. For each carbon atom that is released from the Calvin cycle in a carbohydrate molecule, one carbon dioxide molecule must enter the cycle. Therefore, to produce a molecule containing 12 carbon

atoms, 12 molecules of carbon dioxide must enter the cycle.

Study Break 9.4
1. Photorespiration occurs when oxygen concentrations are high relative to CO$_2$ concentrations. In that condition, rubisco acts as an oxygenase rather than a carboxylase. As an oxygenase, rubisco catalyzes the combination of RuBP with O$_2$ rather than CO$_2$, forming toxic products that cannot be used in photosynthesis. The toxic products are eliminated by reactions that release carbon in inorganic form as CO$_2$, greatly reducing the efficiency of photosynthesis. The entire process is called photorespiration because it uses oxygen and releases CO$_2$.

Photorespiration uses energy to salvage the carbons from phosphoglycolate. Overall, therefore, photorespiration reduces the efficiency of energy use in photosynthesis. This can be seen in the reduced growth of plants grown under photorespiration conditions.
2. In C$_4$ plants, a C$_4$ pathway bypasses the oxygenase activity of rubisco. In the pathway, initial fixation of CO$_2$ is catalyzed by a carboxylase that has no oxygenase activity in specific locations or at times within the plant when oxygen is overabundant. In later steps, the CO$_2$ is released in relatively oxygen-free regions or at times for final fixation in the reactions using RuBP in the Calvin cycle.

Self-Test Questions
1. d 2. b 3. a 4. c 5. e 6. a 7. b 8. d
9. d 10. c

Chapter 10

Study Break 10.1
In mitosis, DNA replication is followed by the equal separation of the replicated DNA molecules and their delivery to daughter cells. The process ensures that the two cell products of a division have the same DNA content and the same genetic information as the parent cell entering division.

Study Break 10.2
1. The order of the stages of mitosis are prophase, prometaphase, metaphase, anaphase, and telophase.
2. Each eukaryotic chromosome has a specialized region known as a centromere. The centromere is where a complex of several proteins, called a kinetochore, forms. During mitosis, some spindle microtubules attach to each kinetochore. These connections determine the outcome of mitosis, because they attach the sister chromatids of each chromosome to microtubules leading to the opposite spindle poles; during anaphase, the spindle separates sister chromatids and pulls them to opposite spindle poles. In brief, the centromeres are key to chromosome

segregation during mitosis. Although not mentioned in the chapter, this is also apparent when problems occur in which a chromosome fragment without a centromere breaks off from a chromosome. The fragment without a centromere cannot connect to the spindle and, hence, is not segregated properly.

3. Joined sister chromatid pairs attach to kinetochore microtubules during prometaphase and begin their migration to the metaphase plate. The spindle microtubules are also necessary for segregating the sister chromatids to opposite poles of the cell during anaphase. If colchicine is present, no spindle will form and no kinetochore microtubules are present to attach to the sister chromatids. Therefore, the cell will be stuck in mitosis with the condensed pairs of sister chromatids in an unorganized array.

Study Break 10.3

1. In animal cells with centrosomes, the spindle forms through division of the cell center. As the dividing centrosome separates into two parts, the microtubules of the spindle form between them. In plant cells that have no centrosome, the spindle microtubules simply assemble around the nucleus. In either case, the microtubules assemble in a parallel array that creates two poles in the dividing cell.

2. Chromosomes (sister chromatids) move apart during anaphase. During the anaphase movements, the kinetochores move along the kinetochore microtubules, which become shorter as anaphase progresses. The nonkinetochore microtubules slide over each other, decreasing the degree of overlap and pushing the poles farther apart. The total distance traveled by the chromosomes is the sum of the two movements.

Study Break 10.4

1. A CDK can become active only once it has complexed with a cyclin protein. Cyclin proteins fluctuate in concentration throughout the cell cycle, and only when the concentration is high enough will a cyclin/CDK complex form. Thus, the point in the cycle at which at a particular CDK can become active depends on when the concentration of its associated cyclin is at a sufficient level.

2. When the kinase of the CDK becomes active, it phosphorylates target proteins in the cell, activating them. Those activated proteins are responsible for moving the cell on in the cell cycle, such as from G_1 to S or from G_2 to M.

3. An oncogene is an altered gene in an organism that contributes to the development of cancer—that is, uncontrolled cell division. Some of the genes that become oncogenes encode components of the cyclin/CDK system that regulates cell division, whereas others encode proteins that regulate gene activity, form

cell-surface receptors, or make up elements of the systems controlled by the receptors.

Study Break 10.5

1. Prokaryotic cell division begins with replication of the bacterial chromosome, starting with duplication of the origin of replication. Once the origin of replication is duplicated, the two origins actively migrate to the two ends of the cells, a process that separates the two replicating chromosomes in the cell. Division of the cytoplasm then occurs by means of a partition of cell wall material that grows inward until the cell is separated into two parts. The cytoplasmic division divides the replicated DNA molecules and cytoplasmic structures between the daughter cells.

2. Present in eukaryotic cell division, but absent from prokaryotic cell division, is the process of mitosis, any form of microtubules for chromosome segregation, a spindle apparatus, and cyclin/CDK control proteins.

Self-Test Questions

1. c 2. a 3. e 4. b 5. d 6. d 7. a 8. b
9. b 10. c

Chapter 11

Study Break 11.1

1. Mitosis produces daughter cells that are genetically identical to the parent cell. Either a haploid or a diploid cell can undergo mitosis. Meiosis starts with a diploid cell. There is one round of DNA replication but two rounds of cell division, with the result that four haploid cells are produced from the parent diploid cell.

2. Recombination is the physical exchange of segments between the chromatids of homologous chromosomes. Recombination occurs in prophase I, when homologous chromosomes have each duplicated to produce sister chromatids and are aligned fully in an organization called a tetrad.

3. Meiosis II is the meiotic division that is similar to a mitotic division.

Study Break 11.2

1. There are three ways in which sexual reproduction generates genetic variability. First, recombination, which involves the physical exchange of segments between homologous chromatids in prophase I of meiosis, generates new combinations of alleles. Second, the random separation of homologous chromosomes during meiosis generates genetic variability. That is, in metaphase I, for each homologous pair of chromosomes, one chromosome makes spindle connections leading to one pole and the other chromosome connects to the opposite pole. This pro-

cess operates independently for each homologous pair of chromosomes; thus, for each meiosis, random combinations of maternal and paternal chromosomes move to the poles during anaphase I. Third, the random joining of male and female gametes produces additional genetic variability.

2. The proportion of gametes that will have chromosomes that originate from the animal's female parent is: $(1/2)^6 = 1/64$.

Study Break 11.3

In animals, the diploid phase dominates the life cycle; mitotic divisions occur only in this phase. Meiosis in the diploid phase gives rise to products that develop directly into egg and sperm cells without undergoing mitosis.

In most plants, the life cycle alternates between haploid and diploid generations, both of which grow by mitotic divisions. Fertilization produces the diploid sporophyte generation; after growth by mitotic divisions, cells of the sporophyte undergo meiosis and produce haploid spores. The spores germinate and grow by mitotic divisions into the gametophyte generation. After growth of the gametophyte, cells develop directly into egg or sperm nuclei, which fuse in fertilization to produce the diploid sporophyte generation again.

Self-Test Questions

1. b 2. a 3. d 4. c 5. b 6. b 7. b 8. d
9. a 10. b

Chapter 12

Study Break 12.1

1. The genotypes of the two parents are Aa Bb × Aa Bb.

2. The genotypes of the parents are Aa Bb × aa bb. This is a testcross.

Study Break 12.2

1. The color pattern involved is an incompletely dominant trait.

2. The fur colors here involve multiple alleles of a single gene. The allele symbols are C for wild type, c^{ch} for chinchilla, c^h for Himalayan, and c for albino, with dominance in the order $C \rightarrow c^{ch} \rightarrow c^h \rightarrow c$. That is, the C allele is completely dominant to the c^{ch} allele, the c^{ch} allele is completely dominant to the c^h allele, and so on. Therefore, we have these genotypes and phenotypes:

C with c^{ch}, c^h, or c = agouti (CC, Cc^{ch}, Cc^h, Cc)

c^{ch} with c^{ch}, c^h, or c = chinchilla ($c^{ch}c^{ch}$, $c^{ch}c^h$, $c^{ch}c$)

c^h with c^h or c = Himalayan (c^hc^h, c^hc)

c with c = albino (cc)

Genetics Problems

1. (a) The CC parent produces all C gametes, and the Cc parent produces 1/2 C and 1/2 c gametes. All offspring would

have colored seeds—half homozygous *CC* and half heterozygous *Cc*. (b) Both parents produce 1/2 *C* and 1/2 *c* gametes. Of the offspring, three-fourths would have colored seeds (1/4 *CC* + 1/2 *Cc*) and one-fourth would have colorless seeds (1/4 *cc*). (c) The *Cc* parent produces 1/2 *C* gametes and 1/2 *c* gametes, and the *cc* parent produces all *c* gametes. Half of the offspring are colored (1/2 *Cc*) and half are colorless (1/2 *cc*).

2. The genotypes of the parents are *Tt* and *tt*.

3. The taster parents could have a nontaster child, but nontaster parents are not expected to have a child who can taste PTC. The chance that they might have a taster child is 3/4. The chance of a nontaster child being born to the taster couple would be 1/4. Because each combination of gametes is an independent event, the chance of the couple having a second child, or any child, who cannot taste PTC is expected to be 1/4.

4. (a) All *A B*. (b) 1/2 *A B* + 1/2 *a B*. (c) 1/2 *A b* + 1/2 *a b*. (d) 1/4 *A B* + 1/4 *A b* + 1/4 *a B* + 1/4 *a b*.

5. (a) All *Aa BB*. (b) 1/4 *AA BB* + 1/4 *AA Bb* + 1/4 *Aa BB* + 1/4 *Aa Bb*. (c) 1/4 *Aa Bb* + 1/4 *Aa bb* + 1/4 *aa Bb* + 1/4 *aa bb*. (d) 1/4 *Aa Bb* + 1/8 *AA Bb* + 1/8 *AA BB* + 1/8 *Aa bb* + 1/8 *aa Bb* + 1/16 *AA BB* + 1/16 *AA bb* + 1/16 *aa BB* + 1/16 *aa bb*.

6. (a) All *A B C*. (b) 1/2 *A B c* + 1/2 *a B c*. (c) 1/4 *A B C* + 1/4 *A B c* + 1/4 *a B C* + 1/4 *a B c*. (d) 1/8 *A B C* + 1/8 *A B c* + 1/8 *A b C* + 1/8 *A b c* + 1/8 *a B C* + 1/8 *a B c* + 1/8 *a b C* + 1/8 *a b c*.

7. Because the man can produce only 1 type of allele for each of the 10 genes, he can produce only 1 type of sperm cell with respect to these genes. The woman can produce 2 types of alleles for each of her 2 heterozygous genes, so she can produce $2 \times 2 = 4$ different types of eggs with respect to the 10 genes. In general, as the number of heterozygous genes increases, the number of possible types of gametes increases as 2^n, where $n =$ the number of heterozygous genes.

8. Use a standard testcross; that is, cross the guinea pig with rough, black fur with a double recessive individual, *rr bb* (smooth, white fur). If your animal is homozygous *RR BB*, you would expect all the offspring to have rough, black fur.

9. One gene probably controls pod color. One allele, for green pods, is dominant; the other allele, for yellow pods, is recessive.

10. The cross *RR* × *Rr* will produce 1/2 *RR* and 1/2 *Rr* offspring. The cross *Rr* × *Rr* will produce 1/4 *RR*, 1/2 *Rr*, and 1/4 *rr* as combinations of alleles. However, the 1/4 *rr* combination is lethal, so it

does not appear among the offspring. Therefore, the offspring will be born with only two types, *RR* and *Rr*, with twice as many *Rr* as *rr* in a 1 : 2 ratio (or 1/3 *RR* + 2/3 *Rr*).

11. The parental cross is *GG TT RR* × *gg tt rr*. All offspring of this cross are expected to be tall plants with green pods and round seeds, or *Gg Tt Rr*. When crossed, this heterozygous F₁ generation is expected to produce eight different phenotypes among the offspring: green-tall-round, green-dwarf-round, yellow-tall-round, green-tall-wrinkled, yellow-dwarf-round, green-dwarf-wrinkled, yellow-tall-wrinkled, yellow-dwarf-wrinkled, in a 27 : 9 : 9 : 9 : 3 : 3 : 3 : 1 ratio.

12. The genotypes are: bird 1, *Ff Pp*; bird 2, *FF PP*; bird 3, *Ff PP*; bird 4, *Ff Pp*.

13. Yes, it can be determined that the child is not hers, because the father must be AB to have both an A and B child with a type O wife; none of the woman's children could have type O blood with an AB father.

14. The cross is expected to produce white, tabby, and black kittens in a 12 : 3 : 1 ratio.

15. The mother is homozygous recessive for both genes, and the father must be heterozygous for both genes. The child is homozygous recessive for both genes. The chance of having a child with normal hands is 1/2, and that of having a child with woolly hair is 1/2. Using the product rule of probability, the probability of having a child with normal hands and woolly hair is 1/2 × 1/2 = 1/4.

Chapter 13

Study Break 13.1
The cross to use is the testcross. Here, the testcross would be *Aa Bb* × *aa bb*. A testcross is used so that you can follow the meiotic events in the dihybrid parent (including the consequences of crossing-over between linked genes), because all of the gametes from the testcross parent carry recessive alleles for the genes in the cross. A testcross shows linkage when the ratio of 1 : 1 : 1 : 1 for the four possible phenotypes is not seen. That is, the 1 : 1 : 1 : 1 ratio result occurs when two genes assort independently. However, if two genes are linked, there will be excess of the two parental classes of progeny compared with the two recombinant classes.

Study Break 13.2
The differences between sex-linked inheritance and autosomal inheritance are seen clearly when reciprocal crosses are made and followed through to the F₂ generation. If sex-linked inheritance is involved, a cross of miniature-winged female × normal-winged male flies will give an F₁ generation of all normal-winged female and all miniature-winged

male flies. (This result would not be found with autosomal inheritance. Instead, all F₁ flies—both males and females—would have normal wings.) Selfing the F₁ flies will give an F₂ generation with 1 : 1 normal-winged : miniature-winged flies in both sexes. (For autosomal inheritance, you would see a 3 : 1 ratio of normal-winged : miniature-winged flies in both sexes.)

In the reciprocal cross of true-breeding normal-winged female × miniature-winged male, the F₁ flies will all have normal wings if sex-linked inheritance is involved. (This result is the same as for autosomal inheritance.) Selfing the F₁ flies will give an F₂ generation in which all females will have normal wings and the males will be 1/2 normal-winged and 1/2 miniature-winged. (For autosomal inheritance, you would see a 3 : 1 ratio of normal-winged : miniature-winged flies in both sexes.)

In summary, reciprocal crosses show different segregation patterns of phenotypes for sex-linked inheritance and autosomal inheritance. The two modes of inheritance are easiest to distinguish in a cross of a mutant female × wild-type male because then, in the F₁ generation, all males show the mutant phenotype when sex-linked inheritance is involved.

Study Break 13.3
(a) Duplication of a chromosome segment occurs when a segment breaks from one chromosome and is inserted into its homolog.

(b) A Down syndrome individual results when nondisjunction of chromosome 21 during meiosis occurs (usually in females), producing gametes with two copies of chromosome 21 and one copy of every other chromosome. When such a gamete fuses with a normal gamete, the result is a zygote with three copies of chromosome 21 and two copies of the other chromosomes. This individual will have Down syndrome. Aneuploidy is the term for the condition of extra or missing chromosomes.

(c) A translocation occurs when a broken segment of a chromosome becomes attached to a different, nonhomologous chromosome.

(d) Polyploidy means that there are more sets of chromosomes than the typical diploid set. Polyploidy may result if the spindle fails to function properly in mitosis of cell lines leading to gametes. Cells affected in this way will have twice the normal number of sets of chromosomes. When meiosis subsequently occurs, the gametes produced will have two sets of chromosomes. Fusion of these gametes with, for instance, a gamete with one set of chromosomes will produce a zygote with three sets of chromosomes—a triploid cell.

Study Break 13.4
1. Autosomal recessive inheritance: For a child to exhibit an autosomal recessive

trait, he or she must inherit one recessive allele from each parent. For autosomal recessive inheritance to explain Simpson syndrome in the family, the father must be homozygous for the Simpson syndrome allele, ss, and the mother must be heterozygous, Ss. The expectation would be that 1/2 of the children would be Ss and 1/2 would be ss, regardless of sex, and that is what is found. Therefore, on the assumption that the mother is heterozygous, the syndrome could be an autosomal recessive trait. Sex-linked recessive inheritance: One characteristic of sex-linked recessive inheritance is that affected females pass on the trait to all their sons. Here, we start with an affected male, and he would have to be X^sY if it is a sex-linked recessive trait. To explain the children, we would have to assume that the mother is heterozygous, X^SX^s. The cross of $X^SX^s \times X^sY$ is expected to give 1/2 females with the syndrome and 1/2 males with the syndrome, which is what is described. Therefore, the syndrome could be a sex-linked trait.

2. Autosomal recessive inheritance: In pedigrees of autosomal recessive traits, the appearance of progeny with a trait when both parents do not have the trait is one common feature. If we assume that both parents are heterozygous, Ww, then the children can be explained. That is, you can get both wiggly-eared children (ww, expected frequency 1/4) and non–wiggly-eared children (WW or Ww, combined expected frequency 3/4), regardless of sex. Therefore, wiggly ears could be an autosomal recessive trait based on this family. Sex-linked recessive inheritance: Because a male individual has only one X chromosome, it is not possible for two nonwiggler parents to produce a wiggler daughter. To get a wiggler daughter, the male would have to be X^wY (wiggler) and the female would have to be X^WX^w (nonwiggler), which is not the case here. Therefore, we cannot conclude that ear wiggling is a sex-linked recessive trait based on this family.

Study Break 13.5

A mutant trait that shows cytoplasmic inheritance is caused by an alteration in the DNA of an organelle, either the mitochondrion or the chloroplast. A key property of cytoplasmic inheritance is that a trait is transmitted by a parent to all offspring, regardless of sex. The most common form of this is maternal inheritance, in which the progeny inherit the trait from their mother, paralleling the inheritance of mitochondria and mitochondrial DNA from the female parent and not from the male parent. This pattern of inheritance would not be seen for genes on chromosomes in the nucleus (as explained in Chapter 11).

Genetics Problems

1. All sons will be color-blind, but none of the daughters will be. However, all

daughters will be heterozygous carriers of the trait.
2. The chance that her son will be color-blind is 1/2, regardless of whether she marries a normal or color-blind male.
3. All these questions can be answered from the pedigree. Polydactyly is caused by a dominant allele, and the trait is not sex linked. The genotypes of each person are:

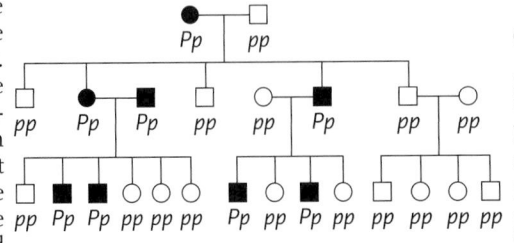

4. The sequence of the genes is ADBC.
5. Let the allele for wild-type gray body color $= b^+$, and the allele for black body $= b$. Let the allele for wild-type red eye color $= p^+$, and the allele for purple eyes $= p$. Then the parents are:

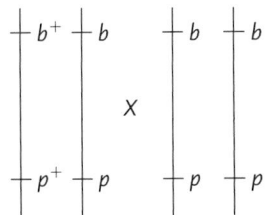

The F$_1$ flies with black bodies and red eyes are:

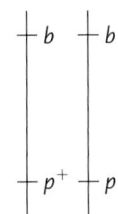

and the flies with gray bodies and purple eyes are:

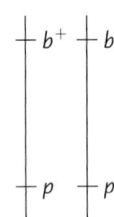

6. The genes are linked by their presence on the same chromosome (an autosome), but they are not sex linked. Because the F$_1$ females must have produced 600 gametes to give these 600 progeny, and because 42 + 30 of these were recombinant, the percentage of recombinant gametes is 72/600, or 12%, which implies that 12 map units separate the two genes.
7. Because this trait is probably carried on the Y chromosome, which a man transmits to all his sons, all will have hairy ears. None of the daughters will have

hairy ears because they do not have a Y chromosome.
8. You might suspect that a recessive allele is sex linked and is carried on one of the two X chromosomes of the female parent in the cross. When present on the single X of the male (or if present on both Xs of a female) the gene is lethal.

Chapter 14

Study Break 14.1

^{35}S-labeled phages in this scenario will have labeled protein coats and labeled DNA. When these phages infect bacteria, radioactivity enters the cell and is found in the progeny phages. In addition, radioactivity is found in the phage material removed by the blender. ^{32}P-labeled phages in this scenario are like the phages in Hershey and Chase's experiment; they have labeled DNA but unlabeled protein. When these phages infect bacteria, radioactivity enters the cell and is found in the progeny phages. No radioactivity is found in the phage material removed by the blender.

Study Break 14.2

1. Adenine and guanine are purines. Thymine and cytosine are pyrimidines.
2. Complementary base pairs are held together by hydrogen bonds. Each base is attached to the deoxyribose sugar by a covalent bond.
3. Watson and Crick described the right-handed double helix that consists of two sugar–phosphate backbones on the outside and complementary base pairs between the two backbones. A complementary base pair is a purine paired with a pyrimidine, more specifically, an A with a T, and a G with a C. The two strands of DNA are antiparallel. The key dimensions of the molecule are: diameter = 2 nm; 1 base pair = 0.34 nm; 1 turn of the helix = 10 base pairs = 3.4 nm.
4. The question focuses on the complementary base-pairing rules: A = T and G = C. If A = 20%, then T = 20%, giving 40% of the DNA as A-T base pairs. Therefore, 60% of the base pairs in this DNA molecule are G-C, and the percentage of C is 30%.

Study Break 14.3

1. Complementary base pairing ensures that the new DNA double helix is a faithful copy of the parental DNA double helix. For whatever base is exposed on the template strand, the DNA polymerase inserts the nucleotide with the complementary base.
2. DNA polymerases cannot initiate a DNA strand; they can add DNA nucleotides only to the 3' end of an existing strand. The primer serves to provide a short stretch of nucleic acid that can be extended by DNA polymerase. The primer consists of RNA, rather than DNA, and is made by primase.

3. One DNA polymerase begins DNA synthesis at the ends of primers. This enzyme does most of the synthesis of new DNA during replication. (In *E. coli*, this enzyme is DNA polymerase III.) The other DNA polymerase removes the RNA primer, replacing the RNA nucleotides with DNA nucleotides. This enzyme takes over after the first enzyme finishes and removes nucleotides from the RNA primer ahead of it one by one, replacing each with a DNA nucleotide.

4. Telomeres are buffers against the progressive loss of the ends of chromosomes by repeated rounds of replication. Only when the hundreds to thousands of copies of the telomere repeats have been lost are genes exposed. When those genes are lost by continued chromosome shortening and/or when chromosomes break down in the absence of telomeres, the cell is severely damaged.

Study Break 14.4

Proofreading prevents errors from being introduced into the DNA sequence. The DNA sequence in an organism's genome specifies everything about that organism—most notably, its function and reproduction. If significant errors occur during replication, gene sequences could be changed and the function of the organism could be adversely affected. Particularly if there is a high rate of errors, as there would be in the absence of proofreading, these errors would have potentially drastic consequences.

As part of the proofreading process, DNA polymerase reverses and removes the mispaired nucleotide. The enzyme then continues to move forward, inserting the correct nucleotide.

DNA repair mechanisms then look for and correct any errors that were not detected by proofreading. For example, repair enzymes remove a section of the newly synthesized DNA with the mismatch, and DNA polymerase synthesizes a replacement section with the correct base pairing.

Study Break 14.5

1. The nucleosome consists of two molecules each of histones H2A, H2B, H3, and H4 assembled into a nucleosome core particle wrapped with almost two turns of DNA. The diameter of the nucleosome is 10 nm.

2. Histone H1 is responsible for the next level of chromosome packing above the nucleosome. H1 binds to the exit/entry point of DNA on the nucleosome and to the linker DNA and brings about a coiling of the chromatin into the 30-nm chromatin fiber. The coiled structure is called the solenoid.

Self-Test Questions

1. b 2. d 3. a 4. a 5. d 6. c 7. b 8. a
9. c 10. b

Chapter 15

Study Break 15.1

1. While most enzymes are proteins, not all proteins are enzymes. And, some proteins consist of more than one polypeptide subunit. Each different polypeptide is encoded by a different gene, hence the one gene–one polypeptide hypothesis.

2. There are four different letters in the code (A, U, G, C), so a five-letter code would have 4^5 possible combinations = 1024 codons.

Study Break 15.2

1. 5'-GUUUAACCGAAUAAUGGCCUAC-3'

2. The promoter determines where transcription of a gene will begin. In prokaryotes, RNA polymerase binds to the nucleotide sequence of the promoter and orients in the correct way to transcribe the associated gene. In eukaryotes, transcription factors bind to the promoter and then recruit RNA polymerase, which then orients properly for transcription from the transcription start point.

Study Break 15.3

1. Both pre-mRNAs and mRNAs have a 5' cap and a 3' poly(A) tail. Only pre-mRNAs have introns, which are removed from pre-mRNAs to produce mRNAs.

2. Particular snRNPs bind to the ends of an intron using their contained RNAs to recognize the boundary sequences of the intron. Other snRNPs then bind, causing the intron to loop out, and completing the active spliceosome. Cleavage at each intron-exon junction, looping back of the intron on itself, and joining the two exons together completes the splicing event; the intron and snRNPs are then released.

Study Break 15.4

1. In eukaryotes, a complex of the small ribosomal subunit, initiator tRNA, initiation factors, and GTP binds to the 5' cap of the mRNA and scans along the mRNA until it reaches the first AUG codon, which is the start codon. The anticodon of the initiator tRNA binds to the start codon, the large ribosomal subunit binds, and the initiation factors are released when GTP is hydrolyzed.

In prokaryotes, a complex of the small ribosomal subunit, initiator tRNA, initiator factors, and GTP binds to the region of the mRNA where the AUG start codon is located, directed by a specific RNA sequence upstream of the start codon. The other steps are the same as those in eukaryotes.

2. The P site is where the tRNA with the growing polypeptide is located. Downstream of the P site is the A site. An incoming aminoacyl-tRNA enters the A site and its anticodon base pairs with the codon of the mRNA in that site. When the polypeptide is transferred to the amino acid on the tRNA in the A site, the ribosome translocates one codon along the mRNA. As translocation takes place, the empty tRNA that was in the P site is moved to the E site. It remains there, blocking a new aminoacyl-tRNA entering the A site until translocation is finished. Then the empty tRNA is released from the ribosome.

3. Proteins found in the cytoplasm are made on free ribosomes. Proteins sorted via the ER have signal sequences at the beginning end that direct it and the ribosome to dock with a receptor on the ER. Continued translation inserts the growing polypeptide into the lumen of the ER, the signal sequence is then removed, and processing and sorting of the polypeptide proceeds. Special sorting signals are also found on proteins that are intended for the nucleus, mitochondria, and chloroplasts.

Self-Test Questions

1. b 2. a 3. e 4. d 5. b 6. c 7. d 8. b
9. a 10. e

Chapter 16

Study Break 16.1

1. The Lac repressor is active when it is made. In normal cells, in the absence of lactose, the Lac repressor binds to the operator, blocking transcription. In the mutant, the Lac repressor is not made, so transcription can never be blocked because no repressor is available to bind to the operator. As a consequence, the lactose metabolizing enzymes will be made both in the presence and absence of lactose in the medium. The mutation involved is an example of a regulatory mutant. Mutations such as this were valuable to Jacob and Monod in developing the operon model for gene regulation.

2. The Trp repressor is inactive when it is made. In normal cells, in the presence of tryptophan, the Trp repressor is activated, binding to the operator and blocking transcription. In the mutant, the Trp repressor is not made, so the operon cannot be turned off when tryptophan is present. This means that the tryptophan biosynthesis enzymes will be produced both in the presence and absence of tryptophan.

Study Break 16.2

1. Histones are general negative regulators of gene expression. When DNA is complexed with histones in normal chromatin, gene promoters typically are not very accessible to the transcription machinery. By acetylating histones, the chromatin is remodeled, making the promoter now accessible to the transcription machinery. Acetylation of histones occurs in response to the binding of an activator to

a regulatory sequence associated with the gene.

2. General transcription factors bind to the promoter and recruit RNA polymerase II, orienting the enzyme so that it will begin transcription at the beginning of the gene. Activators bind to regulatory sequences associated with genes and increase the rate of transcription. Activators that bind to regulatory sequences in the proximal promoter region interact directly with the general transcription factors at the promoter to exert their action. Activators that bind to regulatory sequences in the enhancer stimulate transcription indirectly. These latter activators bind to a coactivator that also binds to the complex of proteins at the promoter, and transcription then occurs at the maximal possible rate.

Study Break 16.3

1. A microRNA, in a complex with particular proteins, binds to an mRNA by complementary base pairing. Either the proteins cut the mRNA in the region of pairing, thereby destroying that molecule, or the double-stranded RNA region blocks translation.

2. Removal of the poly(A) tail would result in the mRNA not being translated.

Study Break 16.4

1. A tumor-suppressor gene encodes a product that has an inhibitory role in the cell division cycle. If both alleles of a tumor-suppressor gene are mutated so that the product is nonfunctional, then those particular "brakes" on cell division are lost.

2. A proto-oncogene encodes a product that has a stimulatory role in cell division. Mutations that lead to increased levels of that product have converted a proto-oncogene to an oncogene. The increased amount of product is stimulatory to cell division.

Self-Test Questions
1. d 2. c 3. a 4. b 5. e 6. b 7. d 8. a
9. d 10. b

Chapter 17

Study Break 17.1
An F$^+$ cell contains the F factor plasmid in addition to the chromosomal DNA. The F factor enables an F$^+$ cell to conjugate with an F$^-$ cell, which lacks the F factor. By a special replication mechanism, a copy of the F factor is transferred to the recipient F$^-$ cell in an F$^+$ × F$^-$ mating, so the recipient is converted to an F$^+$ cell.

An Hfr cell has the F factor integrated into the chromosomal DNA. When an Hfr cell mates with an F$^-$ cell, the F factor begins its replicative transfer into the recipient as in an F$^+$ × F$^-$ mating and, by that transfer mechanism brings in chromosomal genes from the Hfr donor. Those chromosomal genes can recombine with the genes in the recipient. Because replication of the F factor begins in the middle of the plasmid, the entire F factor cannot be transferred to the recipient unless the entire chromosome is transferred to the recipient. This occurs only rarely because conjugating pairs typically fall apart before the 90 to 100 minutes required for complete transfer. Hence, the recipient remains F$^-$ in this mating.

Study Break 17.2
A virulent phage always enters the lytic cycle when it infects a bacterial cell. The end result is the assembly of many progeny phages and their release into the surroundings when the cell breaks open.

A temperate phage follows one of two paths when it infects a cell. It can enter either the lytic cycle or the lysogenic cycle. The lytic cycle is the same as that for a virulent phage. The lysogenic cycle involves integration of the phage's chromosome into the bacterial chromosome. In the integrated state the phage—now called the prophage—is inactive and replicates only when the bacterial chromosome replicates. In response to an adverse environmental signal to the cell, the phage chromosome can excise itself from the bacterial chromosome and enter the lytic cycle.

Study Break 17.3
A transposon is a mobile genetic element that moves from one location to another in the genome as a DNA molecule.

A retrotransposon is a mobile genetic element that moves from one location to another in the genome using an RNA intermediate. That is, the integrated DNA element is transcribed to produce an RNA copy. The RNA copy is reverse-transcribed into DNA, which then integrates at a new location in that genome.

A retrovirus is a virus with an RNA genome. When a retrovirus infects a cell, the RNA genome is reverse-transcribed into DNA and the DNA copy integrates into the genome to produce the provirus. New virus RNA genomes are made by transcription of the integrated DNA copy. Clearly there are similarities between retroviruses and retrotransposons with respect to how they integrate into chromosomal DNA.

Self-Test Questions
1. e 2. a 3. b 4. c 5. d 6. a 7. c 8. e
9. d 10. b

Chapter 18

Study Break 18.1
1. Each restriction enzyme recognizes a specific sequence in DNA, typically in the range of 4–6 base pairs, and cuts both strands of the DNA within the sequence. Restriction enzymes differ with respect to the DNA sequences—the restriction sites—they recognize and cut. The enzymes most useful for cloning produce sticky ends.

2. Replication origin so that the plasmid will replicate in *E. coli*, an antibiotic resistance gene to allow selection of bacteria containing the plasmid, a cluster of restriction sites (a multiple cloning site) to provide choices for inserting fragments of DNA, and the *lacZ$^+$* gene to use in blue-white screening, in which colonies containing recombinant plasmids (white) are distinguished from colonies with vectors lacking inserted DNA (blue).

3. A cDNA library is a collection of clones in which DNAs that are complementary to the mRNAs in the cell have been inserted into a cloning vector. The array of particular clones in a cDNA library is directly related to the genes being expressed in the cells from which the mRNAs are isolated. Therefore, not all genes are represented in any given cDNA library. By contrast, a genomic library is a collection of clones containing all the sequences of an organism inserted into a cloning vector. A genomic library contains, therefore, all of the genes of an organism, as well as noncoding sequences found between and within genes.

4. PCR is a method to amplify a specific segment of DNA. The amplification process depends on DNA replication and, therefore, requires primers. A limitation of PCR, then, is that DNA sequence information must be available for the segment of DNA the amplification of which is desired. Otherwise, the primers cannot be synthesized. The ingredients of a polymerase chain reaction are the template DNA containing the sequence to be amplified, the two primers, a buffer, and a DNA polymerase such as *Taq* polymerase, which is tolerant to the high temperature used to denature the DNA repeatedly during the cycles of the reaction.

Study Break 18.2
1. Each human (or other organism) has unique combinations and variations of DNA sequences. DNA fingerprinting exploits these combinations and variations to distinguish between different individuals (with the exception of identical twins). That is, through the use of DNA technologies to analyze the particular regions of the genome showing sequence variation, it is possible to compare two DNA samples to see if they are from the same or different person. DNA fingerprinting is used in a number of ways, including molecular testing for changes associated with certain genetic diseases, forensics, paternity testing, and basic research.

2. A transgenic organism is one into which a gene or genes from an external source have been introduced as a means to modify the organism genetically.

3. Germ-line cells develop into reproductive cells, so modifying this cell type genetically will lead to the genetic modification being passed on to offspring. In somatic cell gene therapy, germ-line cells and their products are not involved, so the inserted genes remain with the individual and are not passed on to offspring.

Study Break 18.3

1. To find protein-coding genes in bacteria, computer algorithms search a genome sequence for open reading frames (ORFs), which are defined as the DNA equivalent of a start codon (ATG) in frame (a multiple of three away) from a stop codon (TAG, TAA, or TGA). Such a segment of DNA could potentially produce an mRNA that could be translated into a protein. Other analyses would be required to show if any given ORF identified in this way is an actual protein-coding gene.

   To find protein-coding genes in a mammal, in essence one also needs to find ORFs. However, the presence of introns in eukaryotic protein-coding genes complicates matters. Therefore, more sophisticated computer algorithms are needed, in this case ones that attempt to locate exon-intron boundaries while they search for ORFs.

2. A genome-wide analysis of gene expression here could involve DNA microarray assays. mRNAs could be isolated from untreated tissue culture cells and from cells treated with a steroid hormone. Convert each mRNA preparation to cDNAs using reverse transcriptase, using nucleotide precursors with different fluorescent labels for the two batches, for example, green for the untreated (reference) sample and red for the treated (experimental) sample. Mix the two cDNAs and pump them through a DNA chip prepared to have spots with DNA representing every gene in the genome. Allow hybridization to occur, and analyze the hybridization by laser detection. The colors of the spots indicate which genes are affected by the steroid hormone. Purely red spots represent genes that are active only in hormone-treated cells. Purely green spots represent genes that are active only in untreated cells. Spots that are a mixture of green and red represent genes that are active in both types of cells. Based on controls, it would be possible to see if any of the genes have higher or lower expression levels in the treated cells.

Self-Test Questions
1. e 2. c 3. d 4. a 5. b 6. b 7. e 8. a
9. e 10. d

Chapter 19

Study Break 19.1
1. Buffon did not understand how "anatomically perfect" animals could have useless structures.
2. Lamarck proposed that all species change through time, the changes are inherited by the next generation, the changes arise in response to environmental conditions, and specific mechanisms caused the changes.
3. If the geological features that are seen on Earth today were produced by the very slow processes seen today, acting over long periods of time, they must have taken more than 6000 years.

Study Break 19.2
1. Darwin observed that living organisms often resemble fossils found in the same area; that organisms found on South America resembled one another, even if they occupied different environments; and that many species found on the Galápagos Islands resembled species from the South American mainland.
2. Darwin realized that the effects of competition for resources in nature were similar to the action of a plant or animal breeder who used only certain individuals as parents for the next generation.
3. Darwin's theory relied on physical explanations for the origin of biodiversity; he recognized that evolutionary change takes place within a population rather than in individuals, he recognized that natural selection is a multistage process, and he emphasized the importance of environmental conditions to the process of natural selection.

Study Break 19.3
1. Two problems that slowed the acceptance of Darwin's theory are that Mendel's genetic studies focused on simple traits and that these traits often changed in just a few generations.
2. Microevolution refers to small, genetically based changes within populations; macroevolution refers to large-scale changes in patterns of biodiversity.
3. Evidence for evolution comes from studies of adaptation, the fossil record, historical biogeography, comparative morphology, comparative embryology, and comparative molecular biology.

Self-Test Questions
1. c 2. e 3. d 4. b 5. c 6. b 7. a 8. b
9. d 10. d

Chapter 20

Study Break 20.1
1. The variation in skunks is qualitative.
2. The researchers used artificial selection to change the activity levels of the mice.
3. Genetic variation, differing environmental effects on individuals, and interactions between genes and the environment affect phenotypic variation in a population.

Study Break 20.2
1. Genotype frequencies specify how alleles are combined in individuals, and allele frequencies specify how common the alleles are.
2. The Hardy-Weinberg model identifies the conditions under which evolution will *not* occur.
3. If genotype frequencies are not already in equilibrium, they will stop changing after one generation of random mating.

Study Break 20.3
1. Mutation and gene flow tend to increase genetic variation within populations, and natural selection and genetic drift tend to decrease it.
2. Stabilizing selection increases the representation of the average phenotype in a population.
3. Sexual selection, like directional selection, favors extreme phenotypes.

Study Break 20.4
1. Diploidy protects harmful recessive alleles because dominant alleles mask their effects in heterozygotes.
2. A balanced polymorphism is one in which two or more phenotypes are maintained in fairly stable proportions over many generations.
3. The sickle-cell allele is rare in Northern Europe because, in the absence of the malarial parasite, it confers no advantage on individuals that carry it.

Study Break 20.5
1. The adaptive value of a trait can be evaluated by comparing closely related species that live in different environments.
2. Natural selection preserves traits that were useful when the organisms subject to selection were alive and reproducing.

Self-Test Questions
1. c 2. b 3. c 4. d 5. b 6. e 7. a 8. b
9. c 10. d

Chapter 21

Study Break 21.1
1. The morphological species concept defines species based on morphological differences between them. The biological species concept defines species as populations that can successfully interbreed under natural conditions.
2. Clinal variation is a pattern of smooth variation along a geographical gradient.

Study Break 21.2
1. Prezygotic isolating mechanisms either prevent individuals of different species

from mating or prevent sperm from one species from fertilizing the eggs of another. Postzygotic isolating mechanisms limit the survivorship or reproductive capability of hybrid individuals.
2. The scenario illustrates a behavioral isolating mechanism.

Study Break 21.3
1. In the first stage of allopatric speciation, populations become geographically separated. In the second stage, they become reproductively isolated.
2. Some populations of bent grass survive better on unpolluted soil, whereas others survive better on polluted soil.
3. Insects from different host races spend most of their time on different host plant species. Thus, they rarely encounter each other and would be unlikely to mate under natural conditions.

Study Break 21.4
1. Natural selection cannot promote reproduction isolation in allopatric populations directly, but it can lead to genetic divergence that results in reproductive isolation.
2. Polyploidy has frequently led to speciation in flowering plants.

Self-Test Questions
1. a 2. e 3. c 4. d 5. b 6. e 7. b 8. c
9. a 10. b

Chapter 22

Study Break 22.1
1. Hard parts, such as the shells or bones of animals, are the materials most likely to fossilize.
2. The fossil record provides an incomplete portrait of life in the past because not all organisms are equally likely to form fossils; fossils do not form in all types of habitats; and fossils are often destroyed by geological processes and erosion.
3. The fossil record provides information about the morphology of ancient organisms; how structures changed over time; and the proliferation and extinction of evolutionary lineages. It also offers indirect evidence about the behavior, ecology, and physiology of organisms that lived in the past.

Study Break 22.2
1. Continental drift caused large-scale geographical separation of populations and lineages that subsequently evolved in isolation.
2. Distantly related species that live in widely separated parts of the world may resemble each other because convergent evolution fosters similar adaptations to the environments they occupy.

Study Break 22.3
1. The horse lineage was highly branched, and it included some species that were larger and some that were smaller than their ancestors.
2. The gradualist hypothesis predicts slow, continuous morphological changes in a lineage. The punctuated equilibrium hypothesis predicts that morphological changes occur rapidly as new species form and that most species experience little morphological change for long periods of time.

Study Break 22.4
1. In general, the average size of organisms has increased since life first appeared.
2. The processes that can produce evolutionary novelties include preadaptation, the differential growth of body parts (allometry), and changes in the timing of developmental events (heterochrony).

Study Break 22.5
1. A population of organisms may occupy a new adaptive zone after the evolution of a key morphological innovation that allows it to use the environment in a unique way or after a once-successful group of organisms declines.
2. The mass extinction at the end of the Cretaceous period took place over tens of thousands of years.
3. The first major adaptive radiation of animals took place in the Cambrian era.

Study Break 22.6
1. Similar developmental control genes are present in a wide variety of animals, plants, fungi, and prokaryotes. Their widespread distribution suggests that they were present in the ancestor of all these organisms and have been conserved through countless generations.
2. The *Pitx1* gene is expressed in fin buds that later produce spines, and it is not expressed in those that fail to produce spines.

Self-Test Questions
1. a 2. e 3. b 4. d 5. c 6. c 7. a 8. d
9. e 10. b

Chapter 23

Study Break 23.1
1. A phylogenetic tree is a formal hypothesis about the evolutionary relationships among species. A classification is an arrangement of organisms into hierarchical groups that reflect their relatedness.
2. A phylogenetic tree and classification allow biologists to conduct research on just one species or a group of related species that share a genetic history.

Study Break 23.2
1. The system of binomial nomenclature avoids ambiguity in the naming of species because it assigns a unique two-part name to each species.
2. The taxonomic category immediately above the family is the order. The category immediately below family is the genus.

Study Break 23.3
1. Morphological traits are often useful for tracing the evolutionary relationships within a group of organisms because they can be observed and measured in fossils as well as in living organisms.
2. Prezygotic isolating mechanisms are often useful characters in systematic studies of animals because they are the characteristics that the animals use to identify the species of a potential mate.

Study Break 23.4
1. Systematists use homologous characters in their analyses because similarities in homologous characters indicate genetic relatedness and shared ancestry.
2. Outgroup comparison is a technique that compares the group under study to more distantly related organisms to identify the ancestral and derived versions of characters.

Study Break 23.5
1. A monophyletic taxon contains an ancestor and all of its descendants. A polyphyletic taxon includes species from different evolutionary lineages.
2. The traditional definition of Reptilia (including turtles, lizards and snakes, and crocodilians) is paraphyletic because it includes only some descendants of ancestral Sauropsida. It does not include birds, which are part of the sauropsid lineage.
3. In a cladistic analysis, organisms that share derived homologous characters are grouped together.

Study Break 23.6
1. Molecular characters provide several advantages in systematic analyses: (1) they provide abundant data; (2) molecular sequences can be compared between distantly related organisms that share no organismal characteristics or between closely related species with only minor morphological differences; and (3) proteins and nucleic acids are not directly affected by developmental or environmental factors that cause nongenetic morphological variation.
2. Mutations in some types of DNA appear to arise at a relatively constant rate. Thus, differences in the DNA sequences of two species can index their time of divergence. Large differences imply divergence in the distant past, whereas small differences suggest a more recent common ancestor.
3. Phylogenetic analyses of prokaryotes based on morphological data were not very successful because prokaryotes do not have many morphological features. Analyses based on molecular sequence data have been more successful because

researchers can identify many molecular differences among the lineages of prokaryote.

Self-Test Questions
1. c 2. b 3. a 4. d 5. b 6. e 7. a 8. a 9. e 10. d

Chapter 24

Study Break 24.1
1. Only in a reducing atmosphere can amino acids be produced from simpler chemicals and energy. Without amino acids there can be no life. Thus, any theory for the origin of life must consider the necessity of reducing conditions.
2. Current thinking is that early Earth's atmosphere per se was not reducing; hence, scientists have looked for localized regions where conditions were of a reducing nature, such as ocean floor hydrothermal vents. Another hypothesis is that some amino acids had an extraterrestrial origin.

Study Break 24.2
Lipid bilayers in the form of membranes are present in all present-day cellular organisms. Therefore, it is more intuitive to conceive of a path from a simple lipid bilayer system to a membrane-bound cell than it is to conceive of a transition of a clay-based system to a cellular system.

Study Break 24.3
The basic tenet of the theory of endosymbiont origins for mitochondria and chloroplasts is that organelles such as mitochondria and chloroplasts originated from symbiotic relationships between two prokaryotic organisms. An anaerobic prokaryotic is proposed to have ingested an aerobic prokaryote, which persisted in the cytoplasm, continuing to respire aerobically. A gradual process of mutual adaptation transformed the cytoplasmic aerobes into mitochondria.

The same basic mechanism is believed to have led to the appearance of membrane-bound plastids (including chloroplasts) at a later time. In this case, nonphotosynthetic aerobic cells with mitochondria are proposed to have ingested photosynthetic prokaryotes that were perhaps similar to present-day cyanobacteria. Again, through mutual adaptation, the photosynthetic prokaryotes changed into plasmids.

Self-Test Questions
1. d 2. e 3. d 4. a 5. b 6. b 7. c 8. e 9. c 10. e

Chapter 25

Study Break 25.1
1. Prokaryotes have no major cytoplasmic organelles equivalent to the endoplasmic reticulum, Golgi complex, mitochondria, or chloroplasts of eukaryotes, nor does it have lysosomes.

The genetic material of a prokaryote generally is a single, circular DNA molecule localized in a non-membrane-bound central region of the cell called a nucleoid. By contrast, the genetic material of a eukaryote is distributed among a number of linear chromosomes, which consist of DNA complexed with basic proteins known as histones.

Most prokaryotes are surrounded by a cell wall located outside the plasma membrane. Animal cells do not have a cell wall, but plants, fungi, and some other eukaryotes do. The compositions of the eukaryotic cell walls are chemically different from bacterial cell walls.

2. A chemoheterotroph oxidizes organic molecules as its energy source and obtains carbon in organic form. A photoautotroph uses light as its energy source and carbon dioxide as its carbon source.
3. Obligate anaerobes are poisoned by oxygen. They survive either by fermentation or by a form of respiration in which inorganic molecules are used as final electron acceptors. Facultative anaerobes use oxygen when it is present but live by fermentation when conditions are anaerobe.
4. Nitrogen fixation is the reduction of atmospheric nitrogen (N_2) to ammonia (NH_3). Nitrification is the conversion of ammonium (NH_4^+) to nitrate (NO_3^-).

Nitrogen fixation, an exclusively prokaryotic process, is the only means of replenishing the nitrogen sources used by most microorganisms, and by all animals and plants.
5. A biofilm is a complex aggregation of microorganisms, many or all of them prokaryotes, attached to a surface. Biofilms are used in a variety of beneficial applications, including bioremediation of toxic organic chemicals contaminating the groundwater. On the other hand, biofilms can have adverse effects on human health. For instance, biofilms can result in antibiotic-resistant infections if they adhere to surgical materials, such as catheters and implants. Other beneficial and detrimental examples are in the chapter text.

Study Break 25.2
1. Comparing DNA sequences, RNA sequences, and protein sequences are the main methods used to classify prokaryotes. As they become available, genomic sequences are also being used for evolutionary comparisons.

Ribosomal RNA sequences have been used extensively for the classification of prokaryotes. Of particular value to evolutionary studies are rRNA sequences. The amount of sequence divergence gives researchers an estimate of how much time has passed since any two species shared the same ancestor. Turning this around, an investigator uses the differences between rRNA sequences as an indicator of evolutionary relatedness.
2. The hypothesized evolutionary ancestor of present-day Proteobacteria is a purple, photosynthesizing bacterium.
3. Photosynthetic Proteobacteria carry out photosynthesis as either photoautotrophs (the purple sulfur bacteria) or photoheterotrophs (the purple nonsulfur bacteria). The photosynthesis process used does not use water as an electron donor and does not release oxygen as a by-product of photosynthesis. Their photosynthetic pigment is a type of chlorophyll distinct from that of plants.

Cyanobacteria are photoautotrophs that carry out photosynthesis by the same pathways as eukaryotic algae and plants. They use the same chlorophyll as plants are their main photosynthetic pigment, and release oxygen as a by-product of photosynthesis.
4. An exotoxin is a toxic protein that leaks from or is secreted from the bacterium that makes it. An endotoxin is a normal lipopolysaccharide component of the outer membrane of Gram-negative bacteria; it is released when bacteria die and lyse. An exoenzyme is an enzymatic protein that is released from cells.

An exotoxin interferes with biochemical processes of body cells. An endotoxin overstimulates the immune system, often causing inflammation. Depending on the bacterium, the endotoxin release has different effects, that may include organ failure and death. An exoenzyme digests plasma membranes, causing cells of the infected host to rupture and die. Exoenzymes may also digest extracellular materials and red and white blood cells.

Study Break 25.3
1. See Table 25.1 for a comparison of the properties of organisms in each of the three domains. The classification of Archaea as a distinct domain was based on comparisons of DNA and rRNA sequences. Unique to Archaea are chemical features of the plasma membrane and cell wall.
2. A methanogen lives in reducing environments, generating energy by converting substrates such as carbon dioxide, hydrogen gas, methanol or acetate into methane gas. All known methogens belong to the Euryarchaeota.
3. Extreme halophilic Archaea live in high-salt environments, requiring at least 1.5 M NaCl in order to live. Most of these organisms are aerobic chemoheterotrophs, obtaining energy from sugars, alcohols, and amino acids using pathways similar to those of bacteria. All known extreme halophilic Archaea belong to the Euryarchaeota.
4. Extreme thermophiles live in extremely hot environments such as thermal hot springs and hydrothermal ocean floor

vents. Psychrophiles grow optimally in temperatures in the range −10 to −20°C, such as in the Antarctic and Arctic oceans.

Study Break 25.4

A virus is a nonliving entity that consists of genetic material (DNA or RNA) surrounded by a layer of protein (the coat or capsid). When a virus enters a cell, the virus's genetic material directs the replication of its genetic material and the production of progeny viruses. In other words, a virus depends upon a host cell for its life cycle, while directing that life cycle through the action of its own genes.

Viroids are infective plant pathogens consisting only of RNA.

A prion is an infectious agent capable of replication, but containing no nucleic acid molecule. Consisting only of a protein molecule, prions invade nerve cells of a variety of animals, causing fatal degeneration of the nervous system. The prion protein is an misfolded form of a normal cellular protein that is encoded by a nuclear gene. The misfolded prion protein typically forms aggregates, which in part are responsible for their adverse effects. The prion protein "replicates" by converting normal proteins to misfolded prion proteins.

Self-Test Questions

1. c 2. d 3. a 4. b 5. e 6. a 7. d 8. b 9. e 10. a

Chapter 26

Study Break 26.1

A protist is distinguished from a prokaryote by having typical eukaryotic cell features like a nuclear envelope surrounding its genetic material, and cell organelles such as mitochondria (in most protists), chloroplasts (in some protists), endoplasmic reticulum, Golgi complex, and so on.

Distinguishing protists from fungi, animals, and plants is more blurry. Fungi are nonmotile at all stages of their life cycles, while most protists are motile or have motile stages in their life cycles. Cell wall structure is also different from fungi, and from plants. Protists differ from both animals and plants by lacking highly differentiated structures and by not having complex developmental stages. Collagen, the extracellular support protein of animals, is absent in protists.

Study Break 26.2

1. Excavates' nuclear genomes contain genes that are of mitochondrial origin, arguing that they once had mitochondria.
2. The chloroplast will have two membranes: one derived from the plasma membrane of the engulfing eukaryote and the other from the plasma membrane of the cyanobacterium.

Self-Test Questions

1. e 2. d 3. d 4. b 5. b 6. c 7. b 8. a 9. b 10. c

Chapter 27

Study Break 27.1

1. Evolution of a root system gave land plants access to minerals and water in soil and provided physical support for aerial parts. The evolving shoot system of land plants, including lignified tissues in stems, allowed vascular plants to grow taller and stay erect, thereby gaining better access to sunlight for photosynthesis. Reproductive structures borne on aerial stems (such as flowers) might serve as platforms for more efficient dispersal of spores from the parent plant. Vascular tissues were innovations for distributing water (xylem) and sugars (phloem) through the plant body.

2. Homosporous plants produce a single type of sexual spore and are in effect bisexual, with each gametophyte capable of producing both sperm and eggs. Heterosporous species, including angiosperms and gymnosperms, produce two types of spores, which develop into sexually different gametophytes that produce either sperm or eggs. Plant scientists associate the evolution of heterospory with several key reproductive innovations in land plant evolution, including the protection of male gametes inside pollen grains and the protection of plant embryos inside seeds.

Study Break 27.2

1. Like aquatic plants, bryophytes produce flagellated sperm that must swim through water to reach eggs, and they lack a complex vascular system (although some have a primitive type of conducting tissue). Bryophytes do have parts that are rootlike, stemlike, and leaflike, although the "roots" are rhizoids, and bryophyte "stems" and "leaves" did not evolve from the same structures that vascular plant stems and leaves did. Sporophytes of some species have a water-conserving cuticle and stomata. Like most plants, bryophytes also have both sexual and asexual reproductive modes.

2. In general, mosses are the bryophytes that most closely resemble vascular plants. Some species produce structurally complex gametophytes that have a central strand of primitive water-conducting tissue that resembles the xylem of vascular plants, and in a few species the water-conducting cells are surrounded by sugar-conducting tissue resembling the phloem of vascular plants.

Study Break 27.3

1. In bryophytes, the gametophyte is much larger than the sporophyte and obtains its nutrition independently. The comparatively tiny sporophyte remains attached to the gametophyte and depends on the gametophyte for much of its nutrition. In modern lycophytes (club mosses and their close relatives), the gametophyte is free-living—though it is nourished by mycorrhizae instead of carrying out photosynthesis—and it is smaller than the sporophyte, which is a photosynthetic autotroph.

2. Ferns leaves often take the form of feathery fronds, and roots extend from underground stems called rhizomes. Whisk ferns lack true leaves and roots; instead, small leaflike scales dot an upright, green, branching stem, which arises from a horizontal rhizome system anchored by rhizoids. Horsetail sporophytes typically have underground rhizomes and roots that anchor the rhizome to the soil. The scalelike leaves are arranged in whorls about a photosynthetic stem.

3. In horsetails, the sporangia that produce spores are borne in strobili, and spores are carried away from the plant by air currents. In ferns, sporangia are produced on the lower surface or margin of leaves and spores are forcefully dispersed from the parent plant when contraction of a beltlike annulus rips open the sporangium and ejects the spores.

Study Break 27.4

1. The four major reproductive adaptations that evolved in gymnosperms include pollen and the ovule, both of which shelter spores; pollination rather than dispersal of swimming sperm; and the seed as a "package" that protects and often nourishes the embryo.

2. The three basic parts of a seed are the embryo sporophyte, endosperm, and outer seed coat. The endosperm nourishes the embryo sporophyte, and the seed coat protects it.

3. Features that make conifer sporophytes structurally more complex than other gymnosperms include anatomically complex needlelike or scalelike leaves that are adapted to aridity and the production of resins as metabolic by-products.

Study Break 27.5

1. The lack of fossil early angiosperms, including obvious transitional forms, has made it difficult to trace a clear evolutionary path for flowering plants. As result plant scientists have proposed several, often conflicting, classification schemes for angiosperms.

2. Seeds leaves (cotyledons) and pollen morphology are two major features used to distinguish monocots and eudicots. Monocots have a single seed leaf and pollen grain with a single groove. Eudicots have two seed leaves and pollen grains with three grooves. Monocots (such as grasses, lilies, and palms) also generally have fibrous root systems, leaves with

parallel veins, flower parts in multiples of three, and scattered vascular bundles in stems. Eudicots (most flowering trees and shrubs, roses, sunflowers, beans) usually have netlike leaf venation, a primary taproot, flower parts in fours or fives, and vascular tissues arranged in a ring.

3. Adaptations that have contributed to the evolutionary success of angiosperms include vascular tissue modifications that make transport of water and nutrients more efficient; double fertilization, which results in enhanced nutrition (endosperm) for embryos; physical protection of embryos within ovaries and seeds; and coevolution with animal pollinators, which increases the likelihood that pollination will occur.

Self-Test Questions
1. c 2. b 3. a 4. d 5. d 6. e 7. c 8. c 9. e 10. d

Chapter 28

Study Break 28.1

1. Some fungi are multicellular while others, the yeasts, are single cells. (Some species alternate between these two forms at different life cycle stages.) The cells of all fungi are surrounded by a hardened wall; in most cases the hardener is the polysaccharide chitin. The body of a multicellular fungus consists of a dense mesh of filaments called hyphae, which in some groups are separated into cell-like compartments by cross walls (septa). Aggregations of hyphae are the structural foundation for all other parts that develop as part of a multicellular fungus. For example, in some species modified hyphae form rhizoids that anchor the fungus to its substrate.

2. Fungal spores are microscopic, usually nonmotile reproductive cells in which haploid nuclei are surrounded by a tough outer wall. They are produced sexually or asexually. Sexual spores are produced by genetically different parent fungi and may unite in a sexual process that gives rise to a diploid life stage; asexual spores are genetically identical to the parent fungus and may give rise to a new, haploid individual.

3. Many fungal species have a life cycle stage called a dikaryon, which contains two haploid nuclei (a condition expressed as $n + n$). A dikaryon forms as the result of plasmogamy, a sexual stage in which the cytoplasms of two genetically different partners fuse. This fusion ensures genetic diversity in new individuals. At some point after a dikaryon forms, the nuclei fuse (karyogamy) to form a short-lived zygote. Meiosis in the zygote produces haploid nuclei that become packaged into sexual spores.

Study Break 28.2

1. The main phyla of fungi are the Chytridiomycota, Zygomycota, Glomeromycota, Ascomycota, and Basidiomycota. Chytrids are the only fungi that produce motile, flagellated spores. Zygomycetes often reproduce asexually, but sometimes reproduce sexually by way of hyphae that occur in + and − mating strains; haploid nuclei in the hyphae function as gametes. Following plasmogamy, further development produces zygospores in which karyogamy gives rise to diploid zygotes ($2n$ nuclei), which then undergo meiosis as sexual spores form. Glomeromycetes reproduce asexually, by way of spores that form at the tips of hyphae. In ascomycetes, chains of asexual spores called conidia, each containing a haploid nucleus, develop during asexual reproduction. Ascomycetes produce haploid sexual spores in pouchlike cells called asci. Most basidiomycetes reproduce only sexually: Club-shaped basidia develop on a basidiocarp (for example, a "mushroom") and bear sexual spores on their outer surface. When dispersed, the spores may germinate and give rise to a haploid mycelium. Cytoplasmic fusion may occur between hyphae of two compatible mating strains, producing a dikaryotic mycelium from which basidiocarps may grow. Microsporidia, single-celled parasites that may be related to zygomycetes, resemble spores but lack mitochondria. Fungi for which no sexual life stage has been identified are placed in a convenience grouping called "conidial fungi."

2. Anatomically, the simplest fungi are chytrids, which are microscopic, and zygomycetes, which have aseptate hyphae. Ascomycetes, basidiomycetes, and glomeromycetes all form septate (walled) hyphae.

3. Most chytrids are aquatic, but some are parasites on insects, plants, and some animals. Other are symbiotic partners in the gut of cattle and some other herbivores.

Many zygomycetes are saprobes in soil, feeding on plant detritus. Their metabolic activities release mineral nutrients that plant roots can take up. Some zygomycetes are parasites of insects or spoil stored grains, bread, fruits, and vegetables such as sweet potatoes. Others are used in manufacturing products such as industrial pigments and pharmaceuticals. Many more are highly destructive plant pathogens and several can be human pathogens, causing athlete's foot, ringworm infections, and more serious illnesses. The pink bread mold *Neurospora crassa* has been crucial in genetic research.

Some basidiomycetes participate in vital mutualistic associations (mycorrhizae) with the roots of forest trees. Others produce prized edible mushrooms.

Rusts and smuts are parasites that cause serious diseases in wheat, rice, and other plants.

Glomeromycetes are all specialized to form mycorrhizae with plant roots.

Study Break 28.3

1. A lichen is a communal life form representing a symbiosis between a photosynthetic green alga or species of cyanobacteria (the photobiont) and a nonphotosynthetic fungus (the mycobiont). The algal cells supply the lichen's carbohydrates, most of which are absorbed by the fungus. In some cases, the alga is protected from desiccation or some other environmental threat.

2. A mycorrhiza is a symbiotic association between a fungus and plant roots. The fungal hyphae make mineral ions and sometimes water available to the plant's roots, and in exchange the fungus absorbs carbohydrates, amino acids, and possibly other growth-enhancing substances provided by the plant. Mycorrhizae greatly enhance the plant's ability to extract various nutrients, especially phosphorus and nitrogen, from soil, and they are crucial to the survival of many plant species.

3. In endomycorrhizae the hyphae of a fungus (typically a glomeromycete) enter plant roots, where their tips branch into clusters called arbuscules where exchanges of nutrients take place. In ectomycorrhizae the fungal partners are basidiomycetes. Their hyphae surround plant roots but do not penetrate them.

Self-Test Questions
1. b 2. d 3. e 4. b 5. a 6. d 7. a 8. e 9. a 10. d

Chapter 29

Study Break 29.1

1. Several characteristics distinguish animals from plants: animal cells lack cell walls; animals are heterotrophic and, at some stage of the life cycle, motile.

2. The ability of animals to move through the environment allows them to search for and pursue the food items that supply them with nutrients and energy.

Study Break 29.2

1. A tissue is a group of cells that share a common structure and function. The three primary tissue layers that contribute to the bodies of most animals are endoderm, mesoderm, and ectoderm.

2. Humans are bilaterally symmetrical.

3. The coelom is a space within which internal organs can move independently of the body wall muscles. The fluid within it provides protection for internal organs. In some animals the coelom functions as a hydrostatic skeleton.

4. The advantages of having a segmented body include the redundancy of vital organ systems in different segments, allowing an animal to survive damage to some segments, and improved control over body movements.

Study Break 29.3
1. Molecular sequence studies have confirmed the distinctions between the Parazoa and the Eumetazoa, between the Radiata and the Bilateria, and between the Protostomia and the Deuterostomia.
2. A schizocoelom appears to be the ancestral body cavity among the protostomes.

Study Break 29.4
1. Sponges do not exhibit any kind of body symmetry.
2. A sponge gathers food from its environment by drawing water into its body through numerous small pores and harvesting particulate matter from the water with its choanocytes, or collar cells.

Study Break 29.5
1. Cnidarians capture animal prey by stinging it with their nematocysts and using their tentacles to pull it into their mouths.
2. The anthozoans, including sea anemones and corals, have only a polyp stage in their life cycle.
3. Ctenophores capture microscopic plankton in sticky filaments on their two tentacles, which are then drawn across the mouth.

Study Break 29.6
1. Free-living flatworms have digestive, excretory, nervous, and reproductive systems. Tapeworms lack a digestive system.
2. Ectoprocts, brachiopods, and phoronid worms all have a circular or U-shaped feeding structure called a lophophore, a characteristic that reveals their close evolutionary relationship.
3. The anatomical and physiological systems that allow squids and other cephalopods to be more active than other types of mollusks include a closed circulatory system with accessory hearts and a complex nervous system with giant nerve fibers. Many cephalopods use their excurrent siphon to expel jet of water, allowing them to move rapidly through the environment.
4. The organ systems that exhibit segmentation in most annelid worms include respiratory surfaces; parts of the nervous, circulatory, and excretory systems; and the body wall and coelom.

Study Break 29.7
1. The cuticle protects a nematode from the digestive enzymes of its host.
2. Although the rigid exoskeletons of arthropods do not expand, these animals grow a new, soft exoskeleton inside the existing one. After shedding the old exoskeleton, they grow to a larger size by expanding the new exoskeleton with either water or air before it hardens.
3. The body regions of the four living subphyla of arthropods differ in how they have become fused. Chelicerates have a fused cephalothorax and an abdomen. Crustaceans show variable patterns, but many have a fused cephalothorax and an abdomen. Myriapods have a head and a trunk. Hexapods have a separate head, thorax, and abdomen.
4. Insects with incomplete metamorphosis hatch from their eggs as wingless nymphs, which vary in how closely they resemble adults; nymphs then undergo metamorphosis into the adult form. Insects with complete metamorphosis hatch from eggs as larvae, which are always very different from adults. After becoming a pupa, their cells and tissues are reorganized into the adult form.

Self-Test Questions
1. b 2. d 3. c 4. a 5. c 6. e 7. e 8. a
9. d 10. b

Chapter 30

Study Break 30.1
1. Echinoderms have a water vascular system that operates their tube feet.
2. Water enters the pharynx of a hemichordate through its mouth, and it exits the pharynx through the gill slits. The animal extracts oxygen and particulate food from the water as it passes through the pharynx.

Study Break 30.2
1. The animal is not a chordate, because chordates have a dorsal nerve cord.
2. Vertebrates have an internal bony skeleton, including a cranium and vertebral column in most groups, as well as structures derived from neural crest cells.

Study Break 30.3
1. Vertebrates have multiple *Hox* gene complexes, which provide them with several copies of each *Hox* gene. Cephalochordates have just one *Hox* gene complex.
2. Of the three groups listed, Gnathostomata has the most species, and Amniota has the fewest.

Study Break 30.4
1. Hagfishes lack bone, paired fins, and scales in their skin. Hagfishes have neither a cranium nor a vertebral column, and lampreys have only rudimentary traces of vertebrae. These observations suggest that their lineages arose before these structures appeared in the vertebrates.
2. The derived traits possessed by conodonts and ostracoderms include structures made of bone or a bonelike material and, in some ostracoderms, a brain divided into three regions.

Study Break 30.5
1. Sharks are more efficient predators than acanthodians or placoderms were because they have well-developed sensory systems to detect prey; their lightweight skeletons and absence of heavy body armor allow them to pursue prey rapidly; and they have numerous teeth that are replaced when damaged or worn and a loosely attached upper jaw that permits them to suck in large chunks of food.
2. The air bladders of ray-finned bony fishes increase their locomotor abilities by allowing them to rise or sink easily in the water. Their fin rays allow them to engage in precise movements during locomotion.
3. The lungs of lungfish allow them to survive in environments with low oxygen content because they can acquire oxygen from the air.

Study Break 30.6
1. For the first tetrapods, the advantages of moving onto land included abundant food resources, an absence of predators, and readily available oxygen. The disadvantages included the need for more skeletal support against gravity, mechanisms to prevent dehydration in air, and modifications of sensory systems so that they would function in air.
2. The parts of the amphibian life cycle that are most dependent on water are the egg and larval stages.

Study Break 30.7
1. The amniote egg freed amniotes from a dependence on standing water because it can survive on land. The shells of amniote eggs mediate gas exchange and water exchange with the environment.
2. Among the three amniote lineages, the Synapsida includes the mammals, the Anapsida includes turtles, and the Diapsida includes lizards, snakes, crocodilians, and birds.
3. Because lizards are lepidosaurs and both birds and crocodilians are archosaurs, crocodilians are more closely related to birds than they are to lizards.

Study Break 30.8
The overall structure of turtles differs from other amniotes in that their bodies are enclosed within a bony, keratin-covered shell.

Study Break 30.9
1. Besides their loss of legs, snakes differ from their lizard ancestors in having smaller skull bones, and the connections between them are more elastic.
2. Several characteristics reveal the close evolutionary relationship of crocodilians and birds, including a four-chambered heart and maternal care of offspring.

Study Break 30.10

1. The specific adaptations that allow birds to fly either reduce their weight or increase their muscle power. Weight-reducing adaptations include a lightweight skeleton, the absence of teeth and a urinary bladder, and the habit of laying an egg as soon as it has a shell. Power-promoting adaptations include large wing muscles; efficient digestive, respiratory, and circulatory systems; and a high metabolic rate.

2. The structure of a bird's bill reflects its diet. For example, hummingbirds, which drink nectar, have long thin bills, and parrots, which eat hard nuts, have stout sharp bills. Wings and feet are adapted to birds' flying habits and habitats. For example, ducks have webbed feet that allow them to paddle in water, and albatrosses have long thin wings that work efficiently for long distance flight.

Study Break 30.11

1. Most mammals were probably active at night during the Mesozoic era to avoid competition with and predation by dinosaurs, which were active during the day.

2. The key adaptations that allow mammals to be active under many types of environmental conditions include insulating fur and fat and a high metabolic rate that generates lots of body heat.

3. The major groups of living mammals are distinguished on the basis of their reproductive habits. Monotremes lay eggs. Marsupials give birth to relatively undeveloped young after a short period of gestation. Placentals give birth to more developed young after a long period of gestation.

Study Break 30.12

1. The characteristics that allow many species of primates to spend a lot of time in trees include flexible shoulder and hip joints, grasping hands, and excellent depth perception.

2. The lowest taxonomic group that includes monkeys, apes, and humans is the Anthropoidea. The lowest taxonomic group that includes only apes and humans is the Hominoidea.

3. Gorillas, chimpanzees, and bonobos are the apes that spend the most time on the ground.

Study Break 30.13

1. Researchers usually use the criterion of bipedality to distinguish between humans and apes. Humans (that is, hominids) are bipedal, and apes are not.

2. The strongest evidence suggesting that Neanderthals and modern humans belong to different species comes from mtDNA sequence data: differences between gene sequences of Neanderthals and humans are three times greater than the differences between pairs of modern humans.

3. The African Emergence Hypothesis proposes that modern humans arose in Africa and then migrated to various other regions; all modern humans are descended from that wave of immigrants. The Multiregional Hypothesis argues that modern humans evolved simultaneously in many different regions from archaic human ancestors that had migrated from Africa.

Self-Test Questions

1. a 2. c 3. d 4. b 5. e 6. c 7. e 8. d 9. b 10. a

Chapter 31

Study Break 31.1

1. A land plant's shoot system consists of its photosynthetic tissues and organs—stems, leaves, and buds. Stems are frameworks for upright growth and favorably position leaves for light exposure and flowers for pollination. Leaves increase a plant's surface area and thus its exposure to sunlight. Buds eventually extend the shoot or give rise to a new, branching shoot. The shoot system of a flowering plant also includes flowers and fruits. Parts of the shoot system store carbohydrates manufactured during photosynthesis.

   The root system usually grows belowground. It anchors the plant, and sometimes structurally supports its upright parts. It also absorbs water and dissolved minerals from soil and stores carbohydrates.

2. Meristem tissue is self-perpetuating embryonic tissue. Apical meristems, at the tips of shoots and roots, gives rise to a young plant's stems, buds, roots, and other primary tissues. In plants that show secondary growth, cylinders of lateral meristem tissue give rise to (often woody) secondary tissues that increase the diameter of older stems and roots.

Study Break 31.2

1. The ground tissue system makes up most of the plant body. It includes three types of structurally simple tissues—parenchyma, collenchyma, and sclerenchyma—each of which is composed mainly of one type of cell. Parenchyma makes up most of a plant's primary tissue and typically has air spaces between its cells, which are alive at maturity and can continue to divide. Subgroups of parenchyma cells are specialized for photosynthesis, secretion, and storage (of starch). Collenchyma is flexible ground tissue that contains cellulose. Its cells remain alive and metabolically active at maturity. They provide mechanical support for parenchyma and often collectively form strands or a sheathlike cylinder under the dermal tissue of growing shoot regions and leaf stalks. Cells of sclerenchyma are dead at maturity, but while alive they develop thick secondary walls that typically are lignified and provide additional support and protection in mature plant parts.

2. Xylem and phloem are the tissues of the vascular tissue system. The two types of xylem cells, called tracheids and vessel members, both develop thick, lignified secondary cell walls and die at maturity. The empty cell walls of abutting cells serve as pipelines for water and minerals. The conducting cells of phloem, called sieve tube members, form sieve tubes that conduct solutes, mainly sugars made during photosynthesis, throughout a plant.

3. The dermal tissue system serves as a skinlike protective covering for the plant body. Cells of the epidermis are tightly packed and cover the primary plant body. They secrete a cuticle that coats all plant parts except the very tips of the shoot and most absorptive parts of roots. Some epidermal cells become modified for specialized functions; examples are guard cells, which form stomata; root hairs, which absorb water and minerals; and hairlike trichomes, which function in defense against herbivory or secrete sugars that attract pollinators.

Study Break 31.3

1. Stems have four main functions: (1) they provide mechanical support for body parts involved in growth, photosynthesis, and reproduction; (2) they house the vascular tissues (xylem and phloem), which transport products of photosynthesis, water and dissolved minerals, hormones, and other substances throughout the plant; (3) they often are modified to store water and food; and (4) they have specific stem regions that contain meristematic tissue, which gives rise to new cells of the shoot.

   A plant stem is divided into modules, each consisting of a node, where leaves are attached, and an internode, the space between nodes. New primary growth occurs in buds—a terminal bud at the apex of the main shoot, and lateral buds, which produce branches (lateral shoots), in the leaf axils. Meristem tissue in buds gives rise to leaves, flowers, or both.

   In eudicots, most primary growth in a stem's length occurs directly below the shoot apical meristem. When a meristematic cell divides, one of its daughter cells becomes an initial, a cell that remains as part of the meristem. The other daughter cell becomes a derivative, which typically divides once or twice and then enters on the path to differentiation. As derivatives differentiate, they give rise to three primary meristems: protoderm, procambium, and ground meristem. These primary meristems produce cells that differentiate into specialized cells and tissues. In eudicots, the primary meristems are also responsible for elongation of the plant body. Each primary meristem occupies a different position in the shoot tip. Outermost is protoderm, which gives rise to the stem's epidermis. Inward from the protoderm the ground meristem gives rise to ground tissue (mostly parenchyma). Procambium, which produces the primary vascular tissues, is sandwiched between ground meristem layers. In most plants, inner procambial cells give rise to xylem and outer procambial cells to phloem. The developing vascular tissues become organized into vascular bundles that are wrapped in sclerenchyma and thread lengthwise through the parenchyma. In the stems and roots of most eudicots and some conifers, the vascular bundles form a stele (vascular cylinder) that vertically divides the column of ground tissue into an outer cortex and an inner pith.

2. Leaves are organs specialized for photosynthesis. In both eudicots and monocots, the leaf blade provides a large surface area for absorbing sunlight and carbon dioxide. Many eudicot leaves have a broad, flat blade attached to the stem by a petiole. Unless a petiole is very short, it holds a leaf away from the stem and helps prevent individual leaves from shading one another. In most monocot leaves, such as those of rye grass or corn, the blade is longer and narrower and its base simply forms a sheath around the stem.

3. Leaves develop on the sides of the shoot apical meristem. Initially, meristem cells near the apex divide and their derivatives elongate. The resulting bulge enlarges into a thin, rudimentary leaf, or leaf primordium. As the plant grows and internodes elongate, the leaves become spaced at intervals along the length of the stem or its branches. Leaf tissues typically form several layers. Uppermost is epidermis, with cuticle covering its outer surface. Just beneath the epidermis is mesophyll composed of loosely packed parenchyma cells that contain chloroplasts. Leaves of many plants, especially eudicots, contain two layers of mesophyll. Palisade mesophyll cells contain more chloroplasts and are arranged in compact columns with smaller air spaces between them, typically toward the upper leaf surface. Spongy mesophyll, which tends to be located toward the underside of a leaf, consists of irregularly arranged cells with a network of air spaces that enhance the uptake of carbon dioxide and release of oxygen during photosynthesis and account for 15% to 50% of a leaf's volume. Below the mesophyll is another cuticle-covered epidermal layer. Except in grasses and a few other plants, this layer contains most of the stomata through which water vapor exits the leaf and gas exchange occurs. Vascular bundles form a network of veins throughout the leaf.

4. Plants that live many years may spend part of their lives in a juvenile phase, then shift to a mature or adult phase. The differences between juveniles and adults often are reflected in leaf size and shape, in the arrangement of leaves on the stem, or in a change from vegetative growth to a reproductive stage. Most woody plants must attain a certain size before their meristem tissue can respond to the hormonal signals that govern flower development.

Study Break 31.4
1. Most eudicots have a taproot system—a single main root, or taproot, that is adapted for storage and smaller branching lateral roots. As the main root grows downward, its diameter increases, and the lateral roots emerge along the length of its older, differentiated regions. Grasses and many other monocots develop a fibrous root system in which several main roots branch to form a dense mass of smaller roots. Fibrous root systems are adapted to absorb water and nutrients from the upper layers of soil, and tend to spread out laterally from the base of the stem.

2. The root apical meristem and the actively dividing cells behind it form the zone of cell division. Cells in the center of the root tip become the procambium; those just outside the procambium become ground meristem; and those on the periphery of the apical meristem become protoderm. The zone of cell division merges into the zone of elongation, most of the increase in a root's length occurs. Above the zone of elongation, cells may differentiate further and take on specialized roles in the zone of maturation.

3. Primary root growth produces a system of vascular pipelines extending from root tip to shoot tip. The root procambium produces cells that mature into the root's xylem and phloem. Ground meristem gives rise to the root's cortex, its ground tissue of starch-storing parenchyma cells that surround the stele. In many flowering plants, the outer root cortex cells give rise to an exodermis, a thin band of cells that may limit water losses from roots and help regulate the absorption of ions. The innermost layer of the root cortex is the thin endodermis, which helps control the movement of water and dissolved minerals into the stele. Between the stele and the endodermis is the pericycle, which gives rise to lateral roots. In some cells in the developing root epidermis the outer surface becomes extended into root hairs.

Study Break 31.5
1. Secondary growth processes add girth to roots and stems over two or more growing seasons. In plant species that have secondary growth, older stems and roots become more massive and woody through the activity of two types of lateral meristems. One of these meristems, the vascular cambium, produces secondary xylem and phloem. The other, the cork cambium, produces cork, a secondary epidermis that is one element of bark.

2. Vascular cambium consists of two types of cells—fusiform initials and ray initials. Fusiform initials are derived from cambium inside the vascular bundles and give rise to secondary xylem and phloem cells. Secondary xylem forms on the inner face of the vascular cambium, and secondary phloem forms on the outer face. Ray initials are derived from the parenchyma cells between vascular bundles. Their descendants form spoke-like rays of parenchyma cells—horizontal channels that carry water sideways through the stem. As the mass of secondary xylem inside the ring of vascular cambium increases, it forms hard tissue known as wood. Bark encompasses all the living and nonliving tissues between the vascular cambium and the stem surface. It includes the secondary phloem and the periderm, the outermost portion of bark that consists of cork, cork cambium, and secondary cortex.

3. Roots of some plant species also undergo secondary growth, but the ring of vascular cambium develops differently than it does in stems. When their primary growth is complete, these roots have a layer of residual procambium between the xylem and phloem of the stele. The vascular cambium arises in part from this residual cambium, and in part from the pericycle. Eventually, the cambial tissues arising from the procambium and those arising from the pericycle merge into a complete cylinder of vascular cambium. As in stems, the vascular cambium gives rise to secondary xylem to the inside and secondary phloem to the outside. As secondary xylem accumulates, older roots can become thick and woody.

Self-Test Questions
1. d 2. c 3. b 4. a 5. c 6. a 7. d 8. e 9. c 10. b

Chapter 32

Study Break 32.1
1. After an H^+ gradient is established (by proton pumping out of the cell), the resulting inward flow of H^+ down its concentration gradient provides the energy to actively transport other substances into the cell.

2. In symport two substances move in the same direction through a cell membrane; in antiport two substances cross the cell membrane in opposite directions. Plants take up organic substances and important anions such as potassium and nitrate by symport. Antiport commonly moves ions of calcium and sodium out of plant cells.

3. Water potential is potential energy stored in water. It is the driving force for osmo-

sis, which in turn is responsible for the moment of water into and out of plant cells, including root cells.

Study Break 32.2

1. In the apoplastic pathway, water and dissolved substances don't pass through living root cells but instead move through the continuous network of adjoining cell walls and air spaces. When apoplastic water (and solutes) reach the endodermis, however, they must detour around the impermeable Casparian strip and pass through cells in order to move into the stele. The symplastic pathway passes through living cells. Water that diffuses into root cells moves in this pathway from cell to cell through plasmodesmata.
2. Epidermal cells of root hairs actively transport most mineral ions into root epidermal cells. These ions travel inward via the transmembrane pathway. Other ions may be dissolved in apoplastic water. They ultimately travel to the xylem in the symplast after crossing into and through endodermal cells of the Casparian strip. Once an ion reaches the stele it enters the xylem.

Study Break 32.3

1. In the cohesion–tension mechanism, water transport begins as water evaporates from the walls of mesophyll cells inside leaves and into the intercellular spaces. This water vapor escapes by transpiration through open stomata. As water molecules exit the leaf, they are replaced by others from the mesophyll cell cytoplasm. The water loss gradually reduces the water potential in a transpiring cell below the water potential in the leaf xylem. Water from the xylem in the leaf veins then follows the gradient into cells, replacing the water lost in transpiration.
2. Stomata open and close in response to changing environmental cues, such as light levels (detected via blue-light receptors), CO_2 concentration in the air spaces inside leaves, and the amount of water available to the plant. Stomata open when hydrogen ions are pumped out of guard cells, setting up the symport of H^+ and K^+ into the guard cells through ion channels. Water then follows by osmosis. Stomata close when H^+ pumping in guard cells ceases and K^+ moves out of guard cells, with water again following by osmosis. Through their ability to open and close, stomata help regulate water loss by plants and the uptake of carbon dioxide for photosynthesis.

Study Break 32.4

1. Translocation is the long-distance transport of substances in plants. The term generally applies to the transport of organic compounds, mainly sucrose, in phloem. Transpiration is the evaporation of water from a plant's aerial parts, mainly leaves. This water moves from roots upward to aerial parts in the xylem.

2. The mechanism of pressure flow moves sucrose from a source (such as a leaf or stem) into sieve tubes. Pressure builds up at the source end of a sieve tube system as sucrose enters sieve tubes at sources and water follows by osmosis. Under high pressure, sucrose moves by bulk flow toward a sink (plant parts that take up sucrose as metabolic fuel), where the sugar is unloaded.

Self-Test Questions

1. b 2. d 3. a 4. e 5. a 6. b 7. e 8. d 9. c 10. b

Chapter 33

Study Break 33.1

1. Plants require relatively large amounts of macronutrients such as nitrogen, sulfur, potassium, and calcium, and trace amounts of micronutrients such as iron, chlorine, zinc, nickel, and copper.
2. Plants vary in their nutritional requirements. For example, as described in the text, leafy plants require more nitrogen and magnesium than other plant types do, and alfalfa, a grass, requires significantly more potassium than lawn grasses do. An adequate amount of an essential element for one plant also may be toxic for another. For these reasons, the nutrient content of soils is an important factor determining which plants grow well in a given location.

Study Break 33.2

1. Humus is important in soil because it generally contains nutrient-rich organic material and because it absorbs water, which contributes to the water-holding capacity of soil.
2. The amount of water that is available in soil to be taken up by plant roots depends primarily on the relative proportions of different soil components. Water moves quickly through sandy soils, while soils rich in clay and humus tend to hold the most water.
3. A plant's ability to absorb soil minerals depends partly on cation exchange, in which one cation, usually H^+, replaces a soil cation. As H^+ enters the soil solution, it displaces adsorbed mineral cations attached to clay and humus, freeing them to move into roots. Anions in the soil solution, such as nitrate (NO_3^-), sulfate (SO_4^{2-}), and phosphate (PO_4^-), generally move more readily into root hairs. Soil pH also affects the availability of some mineral ions because chemical reactions in very acid (pH < 5.5) or very alkaline (pH > 9.5) soils can trigger chemical reactions that bind various mineral cations in compounds that are insoluble in soil water.

Study Break 33.3

1. As described in this chapter and in Section 28.3, a mycorrhiza is a symbiotic

association between a fungus and plant roots. Most plants form mycorrhizal associations, which facilitate the plant's ability to extract soil nutrients such as nitrogen and phosphorus. As with plant roots, mineral ions enter fungal hyphae by way of transport proteins. Some of the plant's sugars and nitrogenous compounds nourish the fungus, and as the root grows, it takes up a portion of the minerals that the fungus has secured. In some types of mycorrhizae the fungus actually lives inside cells of the root cortex.
2. Nitrogen fixation refers to the incorporation of atmospheric nitrogen into compounds, especially nitrate (NO_3^-), that plants can readily take up. Ammonification is a process in which soil bacteria known as ammonifying bacteria break down decaying organic matter and convert it to ammonium (NH_4^+). In nitrification, nitrifying bacteria oxidize NH_4^+ to NO_3^-. Inside root cells, absorbed NO_3^- is converted by a multistep process back to NH_4^+. In this form, it is rapidly used to synthesize organic molecules, mainly amino acids.
3. Associations with bacteria supply nitrogen to certain types of plants, such as legumes. The host plant provides organic molecules that the bacteria use for cellular respiration, and the bacteria supply NH_4^+ that the plant uses to produce nitrogenous molecules. In legumes the nitrogen-fixing bacteria reside in root nodules. Usually, a single species of nitrogen-fixing bacteria colonizes a single legume species, drawn to the plant's roots by chemical attractants (mainly flavonoids) that the roots secrete. By way of exchanged molecular signals, bacteria then are able to penetrate a root hair and form a colony inside the root cortex. Each cell in a root nodule may contain several thousand bacteria (now called bacteroids). The plant takes up some of the nitrogen fixed by the bacteroids, and the bacteroids utilize some compounds produced by the plant.

Self-Test Questions

1. e 2. c 3. d 4. c 5. b 6. c 7. a 8. d 9. a 10. e

Chapter 34

Study Break 34.1

1. The two alternating generations of plants are the sporophyte (spore-producing) and gametophyte (gamete-producing) generations.
2. Sporophytes produce spores that give rise to gametophytes. Gametophytes then may produce gametes; male gametophytes produce sperms, and female gametophytes produce eggs. In all seed plants the sporophyte is much larger and longer-lived than the gametophyte, and

the gametophyte is protected within sporophyte tissues for all or part of its life. Gametophytes also are dependent upon the sporophyte for their nutrition.

Study Break 34.2
1. Flowers are specialized for reproduction. Before an angiosperm can produce a flower, biochemical signals (triggered in part by environmental cues such as day length and temperature) travel to the apical meristem of a shoot. In response, cells there change their activity: Instead of continuing vegetative growth, the shoot is modified into a floral shoot that will give rise to floral organs.
2. Pollen grains are the mature male gametophytes. They arise by the following steps:
   a. Spores that give rise to male gametophytes are produced in a flower bud's anthers.
   b. Diploid microsporocytes inside an anther's pollen sacs undergo meiosis; eventually each one produces four small haploid microspores.
   c. Microspores then divide by mitosis.
   d. One of the two resulting nuclei divides again by mitosis, yielding a three-celled immature gametophyte: two haploid sperm cells and a third cell that will control the development of a pollen tube after pollen lands on a receptive stigma.
   e. A mature male gametophyte consists of the pollen tube and sperm cells.
3. Female gametophytes develop inside ovules in a flower's carpels. They arise by the following steps:
   a. In an ovule, a diploid megasporocyte divides by meiosis, forming four haploid megaspores.
   b. Three (usually) of these megaspores disintegrate.
   c. The remaining megaspore undergoes three rounds of mitosis without cytokinesis. The result is a single large cell with eight nuclei arranged in two groups of four.
   d. One nucleus in each group migrates to the center of the cell
   e. After the cell undergoes cytokinesis a cell wall forms around these two polar nuclei, forming a large "central cell."
   f. A wall also forms around each of the remaining nuclei, and three of them, including an egg cell, cluster near the micropyle.
   g. The result is an embryo sac containing seven cells and eight nuclei. This sac is the mature female gametophyte.

Study Break 34.3
1. A pollen grain that lands on a compatible stigma absorbs moisture and germinates a pollen tube, which burrows through the stigma and style toward an ovule. Chemical cues from the two synergids cells help guide the pollen tube toward the egg. Before or during these events, the pollen grain's haploid sperm-producing cell divides by mitosis, forming two haploid sperm. When the pollen tube reaches the ovule, it enters through the micropyle and an opening forms in its tip. The two sperm are released into the cytoplasm of a disintegrating synergid. Next double fertilization occurs: typically, one sperm nucleus fuses with the egg to form a diploid (2n) zygote. The other sperm nucleus fuses with the central cell, forming a cell with a triploid (3n) nucleus. Tissues derived from the 3n cell are called endosperm.
2. As a seed matures, the embryo inside it develops a root–shoot axis with root apical meristem at one end and shoot apical meristem at the other end. Depending on the plant group, one or two cotyledons also develop. In monocots a single large cotyledon develops and stores endosperm; protective tissues arise around the root and shoot apical meristems. In eudicots, two endosperm-storing cotyledons form. Near the micropyle the radicle (embryonic root) attaches to the cotyledon at a region called the hypocotyl. Beyond the hypocotyl is the epicotyl, which has the shoot apical meristem at its tip and which often bears a cluster of tiny foliage leaves, the plumule. At germination, when the root and shoot first elongate and emerge from the seed, the cotyledons are positioned at the first stem node with the epicotyl above them and the hypocotyl below them.
3. Imbibition causes the seed coat to split, and water and oxygen move more easily into the seed. Metabolism switches into high gear as cells divide and elongate to produce the seedling. Enzymes that were synthesized before dormancy become active; other enzymes are produced as the genes encoding them begin to be expressed. The increased gene activity and enzyme production mobilizes the seed's food reserves in cotyledons or endosperm. Nutrients released by the enzymes sustain the developing seedling sporophyte until its root and shoot systems are established.

Study Break 34.4
1. Flowering plants may reproduce asexually (vegetatively) by fragmentation, in which cells in a piece of the parent plant dedifferentiate and then regenerate a whole plant; by apomixis, in which a diploid embryo develops from an unfertilized egg or from diploid cells in ovule tissue; or by the production of structures such as rhizomes or suckers from a nonreproductive plant part, typically meristem tissues in a bud on a root or stem.
2. Totipotency is the capacity of fully differentiated cells to dedifferentiate, return to an unspecialized embryonic state, and then develop into a fully functional mature plant. Plant tissue culture procedures trigger the development of a mass of dedifferentiated cells (a callus), some of which regain totipotency and develop into plantlets.

Study Break 34.5
1. A homeotic gene is a regulatory gene in the genome of an organism that encodes a transcription factor. In *Arabidopsis*, a eudicot, homeotic genes govern the development of the root and shoot tissue systems, as well as of floral organs.
2. The two basic mechanisms of plant morphogenesis are oriented cell division and cell expansion. Oriented cell division establishes the general shape of a plant organ, and cell expansion enlarges the cells in a developing organ in particular directions. They increase in circumference (girth) when new cell walls form parallel to the nearest plant surface (such as the surface of a stem or tree trunk) or when cell walls form at right angles both to the nearest surface and to the transverse plane.
3. Leaves arise through a developmental program that begins with gene-regulated activity in meristematic tissue. Hormones or other signals may arrive at target cells via the stem's vascular tissue, activating genes that regulate development. Small phloem vessels penetrate a young leaf primordium almost immediately after it begins to bulge out from the underlying meristematic tissue, followed by xylem. A growing primordium becomes cone-shaped (wider at its base than at its tip). Rapid mitosis in a particular plane in cells along the flanks of the cone (perpendicular to the surface in eudicots and parallel to the surface in monocots) produce the leaf blade characteristic of the particular species. Leaf tip cells typically are the first to stop dividing. By the time a leaf has expanded to its mature size, mitosis has ended and the leaf is a fully functional photosynthetic organ.

Self-Test Questions
1. c 2. a 3. b 4. e 5. d 6. c 7. b 8. e 9. c 10. c

Chapter 35

Study Break 35.1
1. Auxins, gibberellins, cytokinins, and brassinosteroids all promote the growth of plant parts, and ethylene stimulates cell division in seedlings. Abscisic acid is the major growth-inhibiting plant hormone.
2. Ethylene is a good example of a hormone that can stimulate or inhibit growth at various stages of the plant life cycle. In seedlings, it simultaneously slows elongation of the stem and stimulates cell divisions that increase stem girth. In mature plants of deciduous species it governs senescence (including fruit ripening) and the abscission of flowers,

fruits, and leaves. Studies of brassino-steroids have revealed that this family of steroid hormones have different effects in different tissues—for example, promoting the elongation of vascular tissue and pollen tubes, but inhibiting elongation in roots.

Study Break 35.2

1. General plant responses to attack include mobilization of jasmonates and salicylic acid, systemin (in tomato), the hypersensitive response, PR proteins, and secondary metabolites (phytoalexins). Gene-for-gene recognition is a pathogen-specific response.
2. Salicylic acid (SA) is considered to be a general systemic response to damage because experiments show that when a plant is wounded, soon thereafter SA can be detected in a variety of its tissues.
3. While the hypersensitive response is underway, SA also is synthesized and operates in other defensive chemical pathways in a plant. This effect includes the synthesis of PR (pathogenesis-related) proteins that attack pathogenic cells.

Study Break 35.3

1. Directional light of blue wavelengths is the direct stimulus for phototropism. The most widely accepted scientific explanation for gravitropism is the sinking of amyloplasts in cells surrounding vascular bundles in response to gravity. Sinking amyloplasts may provide a mechanical stimulus that triggers a gene-guided redistribution of IAA. The changing auxin gradient in turn adjusts a plant's growth pattern.
2. Unlike tropisms, nastic movements occur in response to nondirectional stimuli, such as mechanical pressure resulting from an insect brushing against hairlike sensory structures in the leaves of a Venus flytrap plant.

Study Break 35.4

1. Plant responses to changes in photoperiod rely on different chemical forms of the blue-green pigment phytochrome. Daylight converts the inactive phytochrome (P_r) to an active form, (P_{fr}). When light levels fall, P_{fr} reverts to P_r. This switching mechanism helps regulate light-related processes such as photosynthesis.
2. Dormancy is an adaptive response because it attunes a plant's growth to the most favorable environmental conditions for survival.

Study Break 35.5

1. Some response pathways may directly stimulate gene transcription while others might inhibit gene expression. Some pathways have an intermediary step, in which the original signal triggers the synthesis of regulatory proteins that in turn promote or inhibit the expression of still other genes.

2. Second messengers boost cellular responses to a hormonal signal by setting in motion of cascade of activated protein kinases, which in turn activate cell proteins that carry out the cell's response.
3. Although Chapter 7 focuses on signal transduction pathways in animals, in fact the basics of some transduction pathways are similar in animals and plants. In both groups we see examples of one-step signal transduction (in which a signal exerts its effect by binding a cell receptor) and of signaling cascades such as second messenger systems.

Self-Test Questions

1. e 2. a 3. c 4. b 5. b 6. b 7. a 8. d
9. c 10. d

Chapter 36

Study Break 36.1

1. Multicellularity made it possible for animals to create an internal fluid environment for fulfilling the nutrient supply, waste removal, and osmotic balance needs of individual cells. As a result, multicellular organisms could evolve to occupy a variety of habitats, including dry terrestrial environments, in which single cells cannot survive. Multicellularity also allowed major life functions to be distributed among specialized groups of cells, with each group having a single activity. The specialized groups of cells are typically organized into tissues, the tissues into organs, and the organs into organ systems.
2. A tissue is a group of cells with the same structure and function. Cells in the tissue work together to perform one or more activities. An organ integrates two or more different tissues into a structure that carries out a specific function. An organ system coordinates the activities of two or more organs to carry out a major body function.

Study Break 36.2

1. Exocrine and endocrine glands are formed by epithelia. An exocrine gland remains connected to the epithelium by a duct, whereas an endocrine gland is suspended in connective tissue underlying the epithelium, with no ducts leading to the epithelial surface.
2. Loose connective tissue, fibrous connective tissue, cartilage, bone, adipose tissue, and blood.
3. Skeletal, cardiac, and smooth.

Study Break 36.3

See Figure 36.11.

Study Break 36.4

A stimulus—a change in the external or internal environment—starts the homeostatic mechanism. The stimulus is detected by a *sensor*, an *integrator* compares the environ-

mental change with a set point, and the *effector* becomes activated by the integrator and functions to return the environmental parameter to the set point.

Self-Test Questions

1. b 2. a 3. e 4. c 5. b 6. a 7. c 8. e
9. e 10. d

Chapter 37

Study Break 37.1

1. A dendrite receives signals and conducts them toward the cell body. An axon conducts signals away from the cell body toward another neuron or effector.
2. An afferent neuron conducts information from its sensory receptors to interneurons. Efferent neurons conduct signals from interneuron networks to effectors, the muscles and glands that carry out the response. The afferent and efferent neurons constitute the peripheral nervous system (PNS). Interneurons process information received from afferent neurons and send a response to the efferent neurons. Interneurons form the brain and spinal cord, the central nervous system (CNS).
3. In an electrical synapse, the plasma membrane of the axon terminal of the presynaptic cell is in direct contact with the postsynaptic cell, allowing ions to pass directly between the cells when an electrical impulse arrives. In a chemical synapse, the presynaptic and postsynaptic cells are separated by a small gap. Neurotransmitters released from the presynaptic cell diffuse across the gap and bind to receptors in the plasma membrane of the postsynaptic cell. If enough neurotransmitter molecules bind to those receptors, the postsynaptic cell generates a new electrical impulse, which travels along its axon to the next neuron or effector in the circuit.

Study Break 37.2

1. At the peak of an action potential, the plasma membrane of the neuron enters a short refractory period, in which the threshold for generation of an action potential is much higher than normal. Only the region in front of the action potential can fire, meaning that the impulse can only move in one direction, that is, toward the axon tip. The refractory period remains in effect until the membrane again reaches the resting potential. By that time, the action potential has moved too far away to cause a second action potential in the same region.
2. Neurons insulated with myelin sheaths have gaps in the sheaths called nodes of Ranvier, where the axon membrane is exposed to extracellular fluids. The inward movement of sodium ions at a node produces depolarization and an action potential, but the adjacent myelin

sheath prevents the excess positive ions from exiting through the membrane. Instead, they diffuse rapidly to the next node where they cause depolarization, inducing an action potential there. Continuation of this process allows the action potential to jump rapidly along the axon from node to node, at a faster rate than a nonmyelinated neuron of the same diameter.

Study Break 37.3

1. A neurotransmitter is synthesized in a neuron, is released into the synaptic cleft from a presynaptic axon terminal, and binds to receptors in the plasma membrane of the postsynaptic cell. Depending on the type of receptor, a neurotransmitter either stimulates or inhibits the generation of action potentials.

2. A direct neurotransmitter, like all neurotransmitters, is stored in synaptic vesicles in the cytoplasm at an axon terminal. When an action potential arrives at the terminal, the change in membrane potential opens voltage-gated Ca^{2+} channels in the axon terminal, allowing Ca^{2+} to flow back into the cytoplasm. The rise in Ca^{2+} concentration triggers the release of the neurotransmitters into the synaptic cleft by exocytosis. Direct neurotransmitters diffuse across the synaptic cleft and open or close ligand-gated ion channels in the postsynaptic neuron's membrane. Most of the channels regulate Na^+ or K^+ movement through the membrane, although some regulate Cl^-. Depending on the ion flow, action potentials are either stimulated or inhibited in the postsynaptic neuron.

Study Break 37.4

One way a postsynaptic neuron integrates signals is through summation of EPSPs and IPSPs that alter the neuron's membrane potential. EPSPs move the membrane potential toward the threshold for an action potential, while IPSPs move the membrane potential away from the threshold for an action potential. The final change in membrane potential depends on the particular array of EPSPs and IPSPs received. The patterns of synaptic connections made by a neuron also contribute to integration.

Self-Test Questions
1. d 2. b 3. a 4. c 5. e 6. a 7. c 8. b
9. e 10. a

Chapter 38

Study Break 38.1

1. A nerve net is a loose meshwork of neurons organized in a radial pattern to reflect the radial symmetry of the animal in which they are found. Nerves are bundles of axons surrounded by connective tissue. Nerve cords are bundles of nerves.

2. Cephalization is the formation of a distinct head region containing ganglia, which form a major central control center or brain, and major sensory structures. Cephalization is found in more complex invertebrates and in all vertebrates.

3. The embryonic hindbrain subdivides into the metencephalon, which gives rise to the cerebellum and the pons, and the myelencephalon, which gives rise to the medulla. The cerebellum integrates sensory signals from eyes, ears, and muscle spindles with motor signals from the forebrain. The pons is a center for information transfer between the cerebellum and the higher integrating centers of the forebrain. The medulla controls vital tasks such as respiration and blood circulation.

Study Break 38.2

(a) Sympathetic nervous system; this is the classic "fight or flight" scenario for this autonomic nervous system division.

(b) Parasympathetic nervous system.

Study Break 38.3

1. The blood-brain barrier prevents most substances dissolved in the blood from entering the cerebrospinal fluid and protects the brain and spinal cord from infectious agents, such as bacteria and viruses, and from toxic substances that may be in the blood. With an incompletely developed blood-brain barrier, it would be important to take precautions that infectious agents, toxic substances, or any chemicals that could affect brain development do not reach the brain.

2. The cerebellum is an outgrowth of the pons. However, it is structurally and functionally separate from the brain stem. The cerebellum has extensive connections with other parts of the brain, through which it receives sensory inputs from receptors in muscles and joints, from balance receptors in the inner ear, and from the receptors of touch, vision, and hearing. The cerebellum integrates the various sensory signals and compares them with signals from the cerebrum that control voluntary body movements. Information flow from the cerebellum to the cerebrum, brain stem, and spinal cord modifies and fine-tunes movements of the body. In humans, the cerebellum also is involved in the learning and memory of motor skills.

The cerebral cortex is a thin layer of gray matter that forms the surface of the cerebrum. The more evolutionarily advanced an animal is, the more folded its cerebral cortex. The cerebral cortex contains sensory areas that receive and integrate sensory information of many kinds, including touch, pain, temperature, pressure, hearing, vision, smell, and taste. In addition there are motor areas that are involved in controlling body movements and position, and association areas, which integrate informa-

tion from the sensory areas and send responses to the motor area. Most higher functions of the human brain, including critical thinking, abstract thought, musical ability, and aspects of personality, involve activities of many regions of the cerebral cortex.

Study Break 38.4

Short-term memory involves transient changes in neurons. Short-term memory loss resulting from aging may be caused by a loss of control of the mechanisms involved, or, more likely, by a loss or degeneration of the neurons that constitute the short-term memory system.

Learning involves first storing memories. If the short-term memory system is faulty, perhaps due to loss or degeneration of neurons, then the transfer of short-term memories to the long-term memory system will be impaired. Another possibility is a loss or degeneration of the neurons responsible for long-term memory.

Self-Test Questions
1. d 2. b 3. b 4. b 5. c 6. e 7. a 8. c
9. c 10. a

Chapter 39

Study Break 39.1

1. Sensory transduction is the conversion of a stimulus into a change in membrane potential.

2. Signals from sensory receptors proceed by afferent neurons to the CNS. The afferent neurons from particular receptors are routed to specific regions of the CNS. Processing of the incoming signals by those regions gives the "sense" of the stimulus—a smell, pain, and so forth.

Study Break 39.2

1. Proprioceptors detect stimuli that are processed to provide the animal with information about movements and position of the body.

2. All proprioceptors are stimulated by a mechanical force, hence they are mechanoreceptors. For example, when a vertebrate moves, fluid moving in the vestibular apparatus bends sensory hairs, which generate action potentials in afferent neurons that synapse with the hair cells. The signal is sent to the brain, which integrates the information into a perception of movement.

Study Break 39.3

1. Cephalopods have mechanoreceptors on their head and tentacles, similar to the lateral line of fishes. These mechanoreceptors detect vibrations in the water.

Many insects have sensory receptors in the form of hairs or bristles that vibrate in response to sound waves, while some insects (for example, moths, grasshoppers, and crickets) have auditory organs on either side of the abdomen or on the first pair of walking legs.

2. Vibrations representing sound frequencies are transmitted into the fluid-filled inner ear. They travel through the inner ear and cause the basilar membrane to vibrate in response, bending the sensory hair cells and stimulating them to release a neurotransmitter that triggers action potentials in afferent neurons leading from the inner ear. The key to "hearing" different frequencies of sound is that the vibrations from a particular sound frequency cause the basilar membrane to vibrate maximally at one particular location. Each frequency of sound waves causes hair cells in a specific region of the basilar membrane to initiate action potentials, and that information is sent to the brain, which integrates it into a perception of the sound stimulus.

Study Break 39.4

(a) A photopigment is an association of retinal with one of several different opsin proteins.

(b) A photoreceptor is a receptor specialized for detection of colors (different wavelengths of light).

(c) A ganglion cell's receptive field is the specific set of photoreceptors that send signals to that cell. Receptive fields are usually circular and vary in size; the smaller the receptive field, the more precise the information sent to the brain, and the sharper the image.

Study Break 39.5

1. Most odor perceptions arise from a combination of different olfactory receptors, which are located in the nasal cavities in humans. We have about 1000 different olfactory receptor types, each of which responds to a different class of chemicals. The stimulated olfactory receptors send signals via the olfactory bulbs to the olfactory centers of the cerebral cortex, where the signals are interpreted as particular smells.

2. Chemicals binds to taste receptors and generate signals. Signals from taste receptors are relayed to the thalamus. From there, some signals go to gustatory centers in the cerebral cortex where they are integrated to produce taste perception. Other signals go to the brain stem and limbic system, which links tastes to involuntary visceral and emotional responses, such as sensations of pleasure or revulsion.

Study Break 39.6

The other sensory receptors involve specialized receptor structures. Afferent neurons synapse with the receptors. When a receptor is stimulated, the change in membrane potential (sensory transduction) is transmitted to the afferent neuron, which transmits the signal to the interneuron networks of the CNS. By contrast, both thermoreceptors and nociceptors consist of free nerve endings formed by the dendrites of afferent neurons, with no specialized receptor structures involved.

Study Break 39.7

Electroreceptors are used in aquatic vertebrates for electrolocation (locating other animals such as prey), electrocommunication (communicating with other members of the same species), and killing prey (involving high voltage discharge).

Self-Test Questions

1. c 2. a 3. b 4. e 5. c 6. b 7. d 8. a 9. e 10. d

Chapter 40

Study Break 40.1

1. A *hormone* is a signaling molecule secreted by one group of cells and transported through the circulatory system to other, target cells, whose activities they change. Specific target cells react to the hormone because they carry receptors for the hormone. The best-known hormones are secreted by cells of the endocrine system and elicit a response in target cells that have receptors for the hormone. A *neurohormone* is a type of hormone. Specifically, it is a signaling molecule released into the circulatory system from stimulated neurosecretory neurons. Like other hormones, it moves around the body in the circulatory system and causes a response in target cells with receptors for the neurohormone.

2. In classical endocrine signaling, hormones are secreted by endocrine glands into the blood or extracellular fluid and exert their effects on distant target cells.

   In neuroendocrine signaling, neurosecretory neurons respond to and conduct signals, but instead of synapsing with target cells, release neurohormones into the circulatory system. The neurohormones exert their effects on distant target cells.

   In paracrine regulation, a cell releases a signaling molecule—a local regulator—that diffuses through the extracellular fluid and acts on nearby cells, rather than on target cells at a distance.

   Autocrine regulation is similar to paracrine regulation, but the released local regulator acts on the same cells that produced it.

Study Break 40.2

1. Glucagon is a peptide hormone, which triggers a response by binding to a surface receptor. Glucagon binding activates the receptor, which triggers a signal transduction pathway inside the cell, leading to phosphorylation of target proteins. The altered activities of the phosphorylated proteins produce responses in the target cells, in this case the breakdown of glycogen in liver cells to glucose.

Aldosterone is a steroid hormone, a hydrophobic molecule that passes though the plasma membrane and binds to an internal receptor in the cytoplasm or nucleus, activating it. The hormone-activated receptor complex binds to control sequences in the DNA that the receptor recognizes and either activates or inhibits transcription of the associated target genes. In short, through binding to a receptor, the hormone affects transcription of specific genes in target cells.

2. A target cell could respond to different hormones if it carries receptors for those different hormones. Turning this around, a target cell will respond only to one hormone if it just has the receptor for that hormone. The same hormone could produce different effects in different cells if there are different receptors for that hormone, each triggering a distinct response pathway.

Study Break 40.3

1. The hypothalamus and anterior pituitary gland are connected by nerve and vascular tissues. The hypothalamus releases peptide neurosecretory hormones into the linking blood vessels. These hormones are tropic hormones that regulate the secretion of peptide hormones by the anterior pituitary gland. The pituitary hormones regulate several key body systems.

2. A tropic hormone regulates hormone secretion by another endocrine gland. The hormone or hormones produced by that gland produces the response. A nontropic hormone acts directly on target cells and is directly responsible for the response.

Study Break 40.4

1. Parathyroid hormone stimulates the dissolution of calcium and phosphate ions from bone and their release into the bloodstream.

2. The adrenal medulla secretes epinephrine and norepinephrine. Epinephrine prepares the body for handling stress or physical activity by, among other actions, (1) increasing heart rate; (2) breaking down glycogen and fats, thereby releasing glucose and fatty acids into the blood for fuel; (3) dilating blood vessels in the heart, skeletal muscles, and lungs to increase blood flow; (4) constricting blood vessels elsewhere, thereby raising blood pressure, reducing blood flow to the intestine and kidneys, and inhibiting smooth muscle contraction, which reduces water loss and slows down the digestive system; and (5) dilating airways in the lungs, thereby increasing air flow.

   Norepinephrine has similar effects to epinephrine on heart rate, blood pressure, and blood flow to the heart muscle. In contrast to epinephrine, norepinephrine causes blood vessels in skeletal muscles to contract.

3. Glucocorticoids and mineralocorticoids are the two major classes of steroid hormones secreted by the adrenal cortex. Glucocorticoids help maintain the concentration of glucose and other fuel molecules in the blood, and mineralocorticoids regulate the levels of Na^+ and K^+ in blood and the extracellular fluid.

4. Estradiol is an estrogen and progesterone is a progestin; both are steroid hormones Estradiol produced by the ovaries stimulates maturation of sex organs at puberty and development of secondary sexual characteristics. Progesterone, also produced by the ovaries, prepares and maintains the uterus for implantation of a fertilized egg and the growth and development of an embryo.

Study Break 40.5

In general, invertebrates have fewer hormones, regulating fewer body processes and responses, than vertebrates do. Peptide and steroid hormones are produced in both invertebrates and vertebrates. However, most of those hormones are different in structure and molecular function in the two groups, even though the reaction pathways stimulated by the hormones are the same.

Self-Test Questions

1. b 2. d 3. a 4. c 5. c 6. e 7. a 8. b
9. d 10. b

Chapter 41

Study Break 41.1

1. The tip of an axon of an efferent neuron makes a synapse with a muscle fiber called a neuromuscular junction. When an action potential arrives at that junction, the axon tip releases acetylcholine, which triggers an action potential in the muscle fiber that moves in all directions over its surface, and penetrates to the interior of the fiber through T tubules. When the action potential reaches the end of the T tubules, ion channels are opened in the sarcoplasmic reticulum that allow calcium ions to flow from the sarcoplasmic reticulum into the cytosol. Troponin then binds the ion and undergoes a conformation change that allows tropomyosin to enter the grooves in the actin helix of the thin filaments. As a result, the myosin-binding sites are uncovered, and the crossbridge cycle is turned on, leading to muscle contraction.

2. In the sliding filament mechanism of muscle contraction, the thin filaments on each side of a sarcomere slide over the thick filaments toward the center of the A band, thereby bringing the Z lines closer together. A crossbridge cycle is responsible for the contraction. At the beginning of the cycle, the myosin crossbridge has an ATP bound to it and is not in contact with the actin of the thin filament. The ATP is hydrolyzed, causing the myosin crossbridge to bend away from the tail and bind to a myosin-binding site on the thin filament that was uncovered in response to the release of calcium ions into the cytosol from the sarcoplasmic reticulum. When the myosin crossbridge binds to the actin, the crossbridge snaps back toward the myosin tail to produce the power stroke that pulls the thin filament over the thick filament. ADP and phosphate are released from the crossbridge in this step. A new molecule of ATP now binds to the crossbridge, causing the myosin to detach from the actin. The cycle then repeats.

Study Break 41.2

1. A hydrostatic skeleton provides support to the body or body part through muscles acting on compartments filled with fluid under pressure. In mollusks, the tube feet of sea stars and sea urchins are hydrostatic skeletal structures. In vertebrates, the penis is a hydrostatic skeletal structure.

An exoskeleton is a rigid external body covering. The force of muscle contraction against the covering provides support to the body or part of the body. Many mollusks, such as clams and oysters, have exoskeletons consisting of a shell secreted by glands in the mantle. In vertebrates, the shell of a turtle or tortoise, and the bony plates in the skin, the abdominal ribs, the collar bones, and most of the bony skull of the American alligator are exoskeletal structures.

An endoskeleton consists of internal body structures such as bones. The force of muscle contraction against the internal body structures provides support. In mollusks, the mantle of squids and cuttlefish, the case of cartilage surrounding and protecting the squid brain, and segments of cartilage supporting the gills and siphon in squids and octopuses are endoskeletal structures. In vertebrates, the endoskeleton is the primary skeletal system.

2. The bones of the vertebrate endoskeleton provide support for the body and body parts, protect key internal organs, store calcium and phosphate ions, and are the sites where new blood cells form.

Study Break 41.3

1. Synovial joints consist of the ends of two bones enclosed by a fluid-filled capsule of connective tissue. Fluid within the capsule, and a smooth layer of cartilage over the ends of the bones, enable the bones to slide easily as the joint moves. In these joints, ligaments extend across the joints across the capsule. The ligaments serve to confine the motion of the joint and protect it to an extent from deleterious effects of heavy loads.

Cartilaginous joints have no fluid-filled capsule surrounding them, and the bones involved are covered with cartilage. The bones of the joint are covered by and connected with fibrous connective tissue. Cartilaginous joints are less movable than synovial joints.

Fibrous joints have stiff fibers of connective tissue joining the bones. As a result, the bones show little or no movement.

2. Antagonistic muscle pairs are muscles arranged so that bones can be extended, flexed, or rotated in opposite directions around a joint. For example, in humans, the biceps and triceps muscles are antagonistic muscle pairs.

Self-Test Questions

1. c 2. e 3. a 4. d 5. b 6. b 7. c 8. a
9. d 10. b

Chapter 42

Study Break 42.1

1. The three basic features of animal circulatory systems are (1) a specialized fluid medium, exemplified by the blood of vertebrates, for transporting molecules; (2) a muscular heart for pumping the fluid; and (3) tubular vessels for distributing the fluid pumped by the heart.

2. In an open circulatory system, there is no distinction between blood and interstitial fluid. Vessels from the heart release hemolymph directly into body spaces and the fluid is subsequently collected and reenters the heart. In a closed circulatory system, blood is channeled in blood vessels leading to and from the heart and is distinct from the interstitial fluid.

A limitation of an open circulatory system is that most of the fluid pressure generated by the heart dissipates when the blood is released into the body spaces. Consequently, blood flows relatively slowly. Humans could not function as we do with such a system, because we would not be able to distribute oxygen efficiently throughout the body; nor would we be able to eliminate the wastes we produce.

3. Atria are chambers of the heart that receive blood returning to the heart. Ventricles are chambers of the heart that pump blood from the heart. Arteries are vessels of circulatory systems that conduct blood away from the heart at relatively high pressure. Veins are vessels of circulatory systems that carry blood back to the heart.

4. The systemic circuit of the body's circulatory system is the circuit from the heart to most of the tissues and cells of the body and back to the heart. The pulmocutaneous circuit goes from the heart to the skin and lungs or gills in amphibians and back to the heart. The pulmonary circuit goes from the heart to the lungs and back to the heart.

Study Break 42.2

1. Pluripotent stem cells in the red bone marrow are the origin of erythrocytes. In

humans and other mammals, erythrocytes lose their nucleus, cytoplasmic organelles, and ribosomes as they mature, becoming essentially a membrane-bound hemoglobin reservoir that is not capable of protein synthesis. At the end of their life span—about 4 months—erythrocytes are engulfed by macrophages, a type of leukocyte, in the spleen, liver, and bone marrow.

2. Low oxygen content triggers a negative feedback mechanism to increase erythrocytes in the blood. The kidneys are stimulated to synthesize the hormone erythropoietin (EPO), which stimulates stem cells in the bone marrow to increase erythrocyte production. EPO synthesis is stopped when the oxygen content of the blood rises above normal levels.

3. Leukocytes are the first line of defense against invading organisms, eliminate dead and dying cells from the body, and remove cellular debris.

    Platelets assist in blood clotting. When blood vessels are damaged, platelets stick to the collagen fibers exposed to leaking blood and recruit other platelets to the site. Eventually a plug forms at the site, sealing off the damaged area.

Study Break 42.3

1. The right atrium receives blood in the systemic circuit returning to the heart in vessels coming from the entire body, with the exception of the lungs. The superior vena cava is the vessel draining blood from the head and forelimbs, and the inferior vena cava is the vessel draining blood from the abdominal organs and the hind limbs. The right atrium pumps blood into the right ventricle.

    The right ventricle receives blood from the right atrium, and pumps it into the pulmonary arteries going to the lungs, beginning the pulmonary circuit.

    The left atrium receives blood returning from the lungs in the pulmonary veins, completing the pulmonary circuit. The left atrium pumps this blood into the left ventricle.

    The left ventricle pumps the blood received from the left atrium into the aorta where the blood begins its path in the systemic circuit.

2. Systole is the period of contraction and emptying of the heart. Diastole is the period of relaxation and filling of the heart between contractions.

3. Neurogenic hearts beat under the control of signals from the nervous system. If the signals cease, this type of heart stops beating. Myogenic hearts maintain a contraction rhythm without signals from the nervous system. In the event of a serious trauma to the nervous system, this type of heart keeps beating.

4. Pacemaker cells of the SA node undergo regularly timed depolarizations which initiate waves of contraction that travel over the heart. The waves stop before they can go to the bottom part of the heart as they encounter an insulating layer of connective tissue. The contraction signals at this point excite cells of the atrioventricular node, which is located in the heart wall between the right atrium and right ventricle. The signal from the AV node passes to the bottom of the heart along Purkinje fibers where it induces a wave of contraction that begins at the bottom of the heart and moves upwards, expelling blood from the ventricles into the aorta and pulmonary arteries.

Study Break 42.4

1. Blood flow through capillary beds is controlled by contraction of smooth muscles in arterioles, and contraction of precapillary sphincters that are at the junctions of capillaries and arterioles.

2. In most body tissues, there are small spaces between capillary endothelial cells, but in the brain the capillary endothelial cells are tightly sealed together, forming a blood–brain barrier.

3. The two major mechanisms are diffusion along concentration gradients and bulk flow. Diffusion along concentration gradients occurs both through the spaces between the capillary endothelial cells and through the plasma membranes. The direction of movement of the molecule or ion depends, then, on the concentration gradient. For instance, oxygen and glucose, which are at higher concentrations in the capillaries, diffuse into the interstitial fluid and then into the cells of the tissues. Carbon dioxide, by contrast, is at a higher concentration in the interstitial and therefore diffuses into the capillaries.

    Bulk flow carries water, ions, and molecules out of the capillaries. Driven by the pressure of the blood, which is higher than the pressure of the interstitial fluid, bulk flow occurs through the spaces between capillary endothelial cells.

4. Atherosclerotic plaques are thickened deposits of cholesterol-rich fatty substances, smooth muscle cells, and collagen that form at damaged sites (lesions) within arteries and veins.

    In general, they form when the endothelial cells lining the blood vessels become damaged, leading to exposure of the underlying smooth muscle tissue. A cycle of injury and repair leads to lesions in the blood vessels which are the potential sites for atherosclerotic plaque formation. Contributing factors to atherosclerotic plaque formation are conditions that cause lesions, such as bacterial and viral infections, hypertension, smoking, and a diet that is high in fats.

Study Break 42.5

1. Arterial blood pressure must be regulated within limits to provide sufficient blood flow for the brain and other tissues, and to prevent damage to blood vessels, tissues, and organs that would occur at high blood pressures.

2. The three main mechanisms for regulating blood pressure are (1) controlling cardiac output (the pressure and amount of blood pumped by the ventricles), (2) controlling the degree of the blood vessels (mostly the arterioles), and (3) controlling the total blood volume.

3. Epinephrine raises blood pressure. The hormone does so by increasing the strength and rate of the heart rate, and stimulating vasoconstriction of arterioles in certain parts of the body.

Study Break 42.6

Lymph is interstitial fluid, an aqueous solution containing molecules, ions, infecting bacteria, damaged cells, cellular debris, and lymphocytes.

Lymph capillaries consist of a single layer of endothelial cells. Interstitial fluid enters these capillaries at sites where the endothelial cells overlap when the pressure of the interstitial forces flaps open. The openings produced are large enough for cells to enter.

Self-Test Questions

1. d   2. e   3. b   4. b   5. c   6. a   7. e   8. e
9. d   10. a

Chapter 43

Study Break 43.1

1. Mucus layers are physical barriers to pathogens. Mucus layers may contain toxins and other chemicals that kill pathogens. The aqueous environment in contact with an epithelial surface may inhibit or kill pathogens by being acidic, or by containing enzymes or bile juices.

2. Innate immunity provides a nonspecific, immediate response to a pathogen. It is the first response system that comes into play when a pathogen is encountered, but it retains no memory of the encounter. The innate immune responses include inflammation and specialized cells that attack pathogens.

    Adaptive immunity, by contrast, provides a specific response to a pathogen, and it retains a memory of exposure to the pathogen so that it can respond more quickly to future attacks. The adaptive immune response takes several days to become protective, in contrast to minutes for an innate immune response.

Study Break 43.2

1. The inflammatory response is characterized by heat, pain, redness, and swelling at the infection site.

2. Heat, redness, and swelling: Inflammation-mediating molecules released at the infection site cause the dilation of local blood vessels and increase their permeability. As a result, blood flow increases and fluid leaks from the blood vessels.

Heat, redness, and swelling are direct consequences of these effects.

Pain: Pain is caused by the migration of macrophages and neutrophils to the infection site, and their activities at the site.

3. The complement system is a nonspecific defense mechanism involving a group of more than 30 interacting soluble plasma proteins. The proteins are activated by molecules on the surface of pathogens. Activated complement system proteins participate in a cascade of reactions, producing membrane attack complexes that insert into the plasma membrane of many types of bacterial cells and coating other types of cells with fragments of the complement proteins. The membrane attack complexes create pores that lead to loss of osmotic balance, which causes the pathogens to lyse. Pathogens coated with complement protein fragments are recognized by phagocytes that then engulf and destroy the pathogens.

4. Viral pathogens do not have surface molecules that are distinct from those of the host. The three main strategies used by the host in innate immunity are RNA interference, interferon, and natural killer cells.

Study Break 43.3

1. The antibody-mediated immune response uses antibodies secreted by plasma cells (differentiated from activated B cells) to target antigens in various body fluids. The cell-mediated immune response uses cytotoxic T cells to target and kill cells infected with a pathogen.

2. An antibody is a multi-subunit protein. It is a member of the immunoglobulin (Ig) family of proteins. It consists of four polypeptides, two of which are identical heavy chains and two of which are identical light chains. Structurally an antibody looks like a Y. Each of the two arms of the Y involves pairing of a light chain with part of the heavy chain. The tail of the Y consists of the remaining parts of the heavy chains paired together. Each heavy chain and each light chain has a variable region and a constant region. The variable regions represent the top half of the arms of the Y and constitute two identical binding sites the antibody has for the antigen with which it can react.

3. DNA rearrangements of genes during differentiation are responsible for generating diverse light-chain genes and heavy-chain genes for the B-cell receptor and for the T-cell receptor. In brief, there are many different V DNA segments, one C DNA segment, and a few different J (joining) DNA segments. During differentiation, a light-chain gene is assembled from one V segment, one J segment and the C segment, creating a gene that can be transcribed. Transcription of the gene produces a pre-mRNA that is spliced to remove introns and generate the translatable mRNA. The various combinations of segments plus imprecision in the joining at the DNA level of the V and J segments produces many types of light-chain genes. A similar DNA rearrangement process occurs for heavy-chain genes. Together, this results in tremendous variability in the antibody molecules that can be made.

4. Clonal selection is the process by which an antigen stimulates the production of a clone of B-cell-derived plasma cells that secrete antibodies against that antigen. Clonal selection accounts for the rapid response, specific action, and diversity of antibodies seen for the adaptive immune system. It also accounts for immunological memory through the production of memory B cells and memory T cells.

5. Immunological memory is the aspect of adaptive immunity that allows for a rapid, intense immune reaction to develop if the body encounters an antigen it has seen before.

Study Break 43.4

1. Immunological tolerance, a feature of the adaptive immune system, protects the body's own molecules from attack by its immune system. B cells and T cells are involved in the development of immunological tolerance, but the exact process is not understood.

2. Basically, an allergy results from an overreaction of the immune system to a particular antigen. The substances responsible for allergic reactions are a class of antigens known as allergens. Allergens stimulate B cells to secrete IgE antibodies in high amounts. IgE binds to mast cells, which are then induced to oversecrete histamine. Histamine contributes to inflammation, and at high amounts, the histamine-induced inflammation can be severe and even life threatening.

Study Break 43.5

Invertebrates lack the adaptive immunity system seen in mammals; they have no B or T cells, for instance. The invertebrate immune defenses, like the innate immunity system of mammals, are nonspecific. For example, all invertebrates have phagocytic cells that engulf pathogens, and many invertebrates produce antimicrobial proteins such as lysozyme. Many invertebrates produce proteins of the immunoglobulin family. While these are not antibody molecules, in some invertebrates they provide a protective function against some pathogens.

Self-Test Questions

1. a 2. c 3. b 4. e 5. e 6. d 7. b 8. c
9. a 10. d

Chapter 44

Study Break 44.1

1. The respiratory medium is the environmental source of oxygen, and the repository for released carbon dioxide. For aquatic animals, water is the respiratory medium and for terrestrial animals it is air. The respiratory surface is the layer of epithelial cells between the body and the respiratory medium. Gas exchange occurs across the respiratory surface—oxygen in and carbon dioxide out of the body. In certain small animals, the body surface itself is the respiratory surface. Among larger animals, aquatic animals use gills, insects use tracheal systems, and terrestrial animals use lungs as the respiratory surface.

2. The advantage of water over air as a respiratory medium is that it enables the respiratory surface readily to remain wet at all times. Two key advantages of air over water as a respiratory medium are (1) there is much more oxygen in air than in water; and (2) air is much less dense and less viscous than water, so significantly less energy is needed to move air over the respiratory surface than to move water.

Study Break 44.2

1. The advantages of gills to water-breathing animals over skin breathing are greater efficiency of gas exchange, the ability to live in more diverse habitats, and the potential to achieve a greater body mass.

2. In countercurrent exchange, the respiratory medium flows in the opposite direction of the blood flow under the respiratory surface. Examples are the flow of water over the gills in sharks, fishes, and some crabs and the one-way flow of air through the lungs of birds; all these are opposite to the flow of blood. The advantage of this mechanism is that it maximizes the amounts of oxygen and carbon dioxide exchanged with the respiratory medium.

3. In the tracheal system, fine branches called tracheoles end in tips that are in contact with body cells. The tracheole tips are filled with fluid, and gas exchange occurs through the fluid and the plasma membrane of the body cells in contact with the tips. Air enters the tracheal system through spiracles on the body surface. Movement of air through the tracheal system occurs by muscle-driven contractions and expansions of air sacs within the system.

4. In positive pressure breathing, gulping, swallowing, or pumping action forces air into the lungs. In negative pressure breathing, muscular activity expands the lungs, lowering the pressure of air in the lungs and thereby causing air to be pulled inward.

Study Break 44.3

1. Contraction of the diaphragm and the external intercostal muscles pulls the ribs upward and outward, expanding the chest and lungs. By this negative pressure mechanism, the air pressure within the lungs is lower than outside of the

body. The higher outside pressure drives air into the lungs, expanding and filling them. Relaxation of the diaphragm and the intercostal muscles reverses the pressure condition, and the elastic recoil of lungs expels the air.

2. For conscious exhalation, you contract the internal intercostal muscles between the ribs to pull the diaphragm down while simultaneously contracting abdominal muscles to compress the abdominal organs and force them upward against the diaphragm to expel air from the lungs.

3. The most important feedback stimulus for breathing is the level of carbon dioxide in the blood.

4. The chemoreceptors in the medulla respond to carbon dioxide levels in the blood. If carbon dioxide levels rise, the medulla responds to trigger an increase in the rate and depth of breathing. The opposite response is triggered if carbon dioxide levels fall.

Study Break 44.4

1. Hemoglobin is present in red blood cells. Oxygen diffuses from alveoli into the plasma solution in the capillaries and then into the red blood cells. In the red blood cells, oxygen binds to hemoglobin, thereby lowering the partial pressure of oxygen in the plasma. This leads to more oxygen molecules diffusing down the oxygen concentration gradient from alveolar air to blood.

   The binding of oxygen to hemoglobin is reversible. Further, the affinity of hemoglobin for oxygen increases as the partial pressure of oxygen increases and vice versa. This property is important for determining the release of oxygen from hemoglobin keyed to tissue requirements.

2. Carbon monoxide has much greater affinity for hemoglobin than oxygen does, and so it displaces oxygen from hemoglobin. This leads to a reduction in the amount of oxygen carried in the blood. Since the brain does not monitor oxygen levels, and other receptors do not respond until blood oxygen levels are critically low, victims can easily lapse into unconsciousness and death.

Study Break 44.5

- More blood per unit body weight than land animals
- Additional red blood cells, many stored in the spleen and released during a dive
- More myoglobin in muscles than land animals
- Slowing of the heart by as much as 90%
- Reduction of blood circulation to internal organs and muscles by up to 95%
- Retention of lactic acid in the muscles with no release into the blood until the animal returns to the surface

Self-Test Questions

1. e 2. d 3. a 4. c 5. a 6. d 7. b 8. d
9. e 10. a

Chapter 45

Study Break 45.1

1. Carnivores primarily eat other animals as their source of organic materials. Herbivores primarily eat plants as their source of organic materials. Omnivores obtain their organic materials from any source.

2. Essential nutrients are the amino acids, fatty acids, vitamins, and minerals that the animal cannot make itself and must obtain from its diet. The list of essential nutrients varies among animal types.

3. Deposit feeders pick up or scrape particles of organic matter from solid material they live in or on. Suspension feeders ingest small organisms that are suspended in water. Depending on the animal, the ingested organisms may be bacteria, protozoa, algae, or small crustaceans, or fragments of those organisms.

Study Break 45.2

1. Extracellular digestion takes place outside body cells, either in a pouch or a tube enclosed by the body, whereas intracellular digestion takes place within cells.

2. (1) Mechanical processing, to break up the food; (2) secretion of enzymes and other substances that aid digestion; (3) enzymatic hydrolysis of food molecules into simpler molecular subunits; (4) absorption of the molecular subunits into body fluids and cells; and (5) elimination of undigested materials.

Study Break 45.3

1. The two classes of vitamins are water-soluble vitamins and fat-soluble vitamins. Fat-soluble vitamins ingested in the diet that are in excess of bodily needs can be stored in adipose tissues. In contrast, excess water-soluble vitamins in the diet are excreted in the urine. Hence, it is critical that humans meet their daily requirements for water-soluble vitamins.

2. The four layers of the gut are the mucosa, the submucosa, the muscularis, and the serosa. The muscularis layer, which is formed from two smooth muscle layers, is responsible for peristalsis.

3. Pepsin is secreted from chief cells into the stomach lumen as an inactive precursor molecule, pepsinogen. The pepsinogen is converted to pepsin by the highly acid conditions of the stomach. Pepsin is a digestive enzyme that begins the digestion of proteins by making breaks in polypeptide chains.

4. In the small intestine, the digestion of macromolecules into their molecular subunits occurs, and those subunits are absorbed. In the large intestine, water and mineral ions are absorbed from the remaining digestive contents; this leaves undigested remnants, the feces, which are expelled from the body.

Study Break 45.4

The hypothalamus has two interneuron centers that play a central role in regulating the digestive processes. One center stimulates appetite and reduces oxidative metabolism. The other center works in opposition to the first center by stimulating the release of a peptide hormone that inhibits appetite.

Study Break 45.5

1. Incisors nip or cut food. Canines bite and pierce food. Premolars and molars crush, grind, and shear food.

2. Symbiotic microorganisms aid digestion in many herbivores by assisting in the breakdown of plant material. The microorganisms synthesize cellulase, an enzyme vertebrates cannot make, which hydrolyzes the cellulose of plant cell walls into glucose subunits.

Self-Test Questions

1. b 2. b 3. a 4. d 5. c 6. e 7. c 8. d
9. e 10. a

Chapter 46

Study Break 46.1

Osmosis is a process in which water molecules move across a selectively permeable membrane from a region where they are more highly concentrated to one where they are less highly concentrated.

Osmolarity is the total solute concentration of a solution. Osmolarity is measured in osmoles per liter of solution, where an osmole is the number of solute molecules and ions (in moles).

A solution that is hypoosmotic has lower osmolarity than the solution on the other side of a selectively permeable membrane. The solution with the higher osmolarity is said to be hyperosmotic.

An osmoregulator is an organism (animal) that uses active control mechanism to keep the osmolarity of cellular and extracellular fluids the same. This osmolarity value may differ from the osmolarity of the surroundings.

Tubules that carry out osmoregulation and excretion are formed from transport epithelium, a layer of cells with specialized transport proteins in their plasma membranes that move specific molecules and ions into and out of the tubule.

Study Break 46.2

Protonephridia are found in flatworms and larval mollusks. They are the simplest invertebrate excretory tubule. Body fluids enter the blind end of a protonephridium, and are propelled through the tube by cilia movement on the flame cell, a large cell at the blind end. As the fluids move through the protonephridium, some molecules and ions are reabsorbed, and others, including nitrogenous wastes, are secreted into the tubules. Excess fluid is released through pores connecting the network of protonephridia to the body surface.

Metanephridia are found in annelids and most adult mollusks. Body fluids enter the funnel-like proximal end, driven by cilia surrounding that end. Some molecules and ions are reabsorbed as the fluids move through the tubule, and other ions and nitrogenous wastes are secreted into the tubule and excreted from the body surface.

Malpighian tubules are found in insects and other arthropods. Body fluids enter the tubules through spaces between the tubule cells. The distal ends of the tubules empty into the gut. Uric acid and several ions, including sodium ions and potassium ions, are actively secreted into the tubules. The concentration of these substances causes water to move from the body fluids into the tubule by osmosis. The fluid then passes into the hindgut as dilute urine. In the hindgut, cells in the gut wall transport most of the ions back into the body fluids; water follows by osmosis. The uric acid remaining is ultimately excreted with the feces.

Study Break 46.3

1. At the proximal end, a human nephron forms a cuplike region called Bowman's capsule. Bowman's capsule surrounds the glomerulus, a complex of blood capillaries. The capsule and the glomerulus are located in the renal cortex.

   Next comes the proximal convoluted tubule in the renal cortex. The tubule descends in a U-shaped bend called the loop of Henle, and ascends again to form the distal convoluted tubule. Up to eight distal convoluted tubules drain into one collecting duct.

2. The collecting ducts are permeable to water but not to salt ions. The ducts begin in the cortex and extend into the medulla of the kidney. As the ducts descend, they become surrounded by an ever-increasing solute concentration in the medulla. As the urine passes down the collecting ducts, water moves osmotically out of the ducts, causing an increase in the concentration of the urine. At the bottom of the collecting ducts, the urine is about four times more concentrated than body fluids.

Study Break 46.4

The RAAS is a regulatory system that compensates for excessive loss of salt and body fluids. The ADH system is a regulatory system that compensates for excessive water intake or loss. Combined, the two systems play an important role in regulating the interactions between the kidneys and the rest of the body.

In brief, the RAAS works as follows: When the sodium ion concentration of body fluids falls, the volume of extracellular fluids and blood pressure also drop. This activates the RAAS. Receptors in the juxtaglomerular apparatus and the heart wall detect the changes and trigger the secretion of renin which, in turn, triggers the production of angiotensin. Angiotensin stimulates constriction of arterioles in many parts of the body,

thereby increasing blood pressure. It also stimulates the adrenal cortex to secrete aldosterone, which increases Na^+ reabsorption in the kidneys. The RAAS is suppressed if NaCl concentration in body fluids is higher than normal.

In the ADH system, ADH is released from the pituitary when osmoreceptors in the hypothalamus detect an increase in the osmolarity of body fluids. ADH increases water reabsorption in the kidney; as a result, urinary output is reduced and water is conserved. In the case of a decrease in osmolarity, ADH release from the pituitary is inhibited, thereby decreasing water reabsorption in the kidney.

Study Break 46.5

1. Marine teleosts live in seawater, which is hyperosmotic to their body fluids. These fishes drink seawater continuously to replace water lost to the environment by osmosis. Sodium ions, potassium ions, and chloride ions from the seawater they drink are excreted by the gills. Nitrogenous wastes are released from the gills, primarily as ammonia, by simple diffusion.

   Freshwater fishes live in fresh water, which is hypoosmotic to their body fluids. They take in water by osmosis and excrete excess water. Salts needed for bodily functions are obtained from food and by active transport through the gills from the water. Nitrogenous wastes are excreted from the gills as ammonia.

2. The excretion of urea in mammals involves expelling the urea in solution—urine. Thus, there is loss of water with excretion of nitrogenous waste as urea. Excretion of nitrogenous waste in the form of uric acid conserves water because uric acid crystals are almost water free.

Study Break 46.6

Ectothermy applies to animals that obtain heat energy primarily from the environment, whereas endothermy applies to animals that obtain heat energy primarily from internal reactions.

Generally speaking, ectotherms are highly successful in warm environments, but their bodily functions slow down as the temperature decreases. Endotherms can remain active over a broader range of environmental temperatures than ectotherms do, but they require an almost constant supply of energy to maintain their body temperature.

Study Break 46.7

1. The thermoregulatory responses shown by ectotherms can be physiological, such as regulating blood flow to internal or external body organisms, or behavioral, such as physically moving to a location in the environment suitable for their heat energy needs at the time.

2. Thermal acclimatization refers to the physiological changes that many ectotherms make to compensate for seasonal shifts in environmental temperature.

Study Break 46.8

Temperature regulation in endothermic animals involves mechanisms that balance internal heat production against heat loss from the body. Internal heat production is controlled by negative feedback pathways that are triggered by thermoreceptors in the skin, the hypothalamus, and the spinal cord. When a deviation from the set point occurs, this system operates to return the core temperature to that set point through changes in metabolic processes, behavioral activities, and control of heat loss at the skin surface.

Self-Test Questions

1. d 2. c 3. a 4. b 5. d 6. e 7. a 8. b
9. d 10. e

Chapter 47

Study Break 47.1

Recall that asexual reproduction produces offspring with genes from only one parent. In most animals that undergo asexual reproduction, the offspring are produced by mitosis and, therefore, are genetically identical to one another and to the parent. Such genetic homogeneity can be advantageous in stable and uniform environments. Another advantage is that individuals do not need to use energy to produce gametes or to find and select a mate. Further, for individuals in sparse populations, or for sessile animals, asexual reproduction can be an advantage. A disadvantage is that a genetically homogenous population may not adapt readily to new environments.

By contrast, sexual reproduction always generates genetic diversity among offspring. This provides a population the opportunity to adapt to changing environments, and perhaps to move to and colonize new environments. Disadvantages of sexual reproduction include the expenditure of energy and raw materials to produce gametes and to find and select mates.

Study Break 47.2

1. Egg coats are surface coats around the egg. They are added during oocyte development or fertilization. Egg coats provide protection against mechanical injury, infection by microorganisms, and, in some species, loss of water.

   Mammalian eggs have an egg coat called the zona pellucida immediately surrounding the egg. This gel-like coating is called the vitelline coat in other organisms. In birds, a thick solution of proteins—the egg white—surrounds a vitelline coat. Bounding the egg white is a hard shell.

2. The slow block to polyspermy occurs in many organisms, including mammals. The fusion of egg and sperm triggers an increase in calcium ions in the cytosol. The calcium ions activate proteins that initiate a high level of metabolic activity in the egg. The released calcium ions also cause the cortical granules to fuse with

the egg's plasma membrane and release their contents to the outside. Enzymes released from those granules alter the egg coats in a matter of minutes after fertilization, and this blocks further sperm from attaching and penetrating the egg.

Study Break 47.3
- FSH stimulates a number of oocytes to begin meiosis. A follicle forms around each oocyte.
- FSH and LH interact to stimulate follicle cells to secrete estrogens.
- Later in the cycle, an increased level of estrogen leads to a new burst of release of FSH and LH. This new LH stimulates follicle cells to release enzymes that digest away the follicle wall, leading to egg release.
- The new LH also causes the remaining follicle cells to grow into the corpus luteum.
- FSH and LH levels decrease. This removes the stimulatory signal for follicular growth, and no new follicles grow in the ovary.

Study Break 47.4
The oral contraceptive pill inhibits the secretion of FSH and LH by the pituitary. FSH and LH stimulate oocytes in the ovaries to begin meiosis. They also stimulate follicle cells that surround the oocytes to secrete estrogens. Later, LH causes ovulation—the release of an egg. Therefore, without FSH and LH, ovulation cannot occur.

Self-Test Questions
1. b 2. a 3. d 4. a 5. e 6. c 7. b 8. e
9. c 10. d

Chapter 48

Study Break 48.1
1. In cleavage divisions, cycles of DNA replication and division occur without the production of new cytoplasm. Therefore, the cytoplasm of the egg partitions into many cells without increasing in overall size or mass. In cell division in an adult organism, the mitotic cell cycle involves cycles of DNA replication and division interspersed with cell growth. That is the cell grows, then divides, then the two progeny cells grow, then divide and so on. New cytoplasm is produced during these cell cycles so that overall there is an increase in mass of the cells.
2. The three primary cell layers of the embryo are produced by gastrulation. They are ectoderm, mesoderm, and endoderm.

Study Break 48.2
1. The gray crescent establishes the dorsal–ventral axis, with the gray crescent marking the future dorsal side.
2. If cells of the dorsal lip of the blastopore are transplanted to another location in the egg, they cause a second blastopore, and subsequently a second embryo, to form in that region.
3. The extraembryonic membranes in birds and their functions are as follows:
   - Yolk sac: Surrounds the yolk, which provides nutrients to the embryo.
   - Chorion: Exchanges oxygen and carbon dioxide with the environment through the egg shell.
   - Amnion: Encloses the embryo and secretes amniotic fluid into the space that provides an aquatic environment in which the embryo can develop.
   - Allantois: Stores nitrogenous wastes, primarily in the form of uric acid, that are derived from the embryo. Part of the allantoic membrane forms a bed of capillaries that is connected to the embryo and that delivers carbon dioxide from the embryo to the chorion and picks up oxygen absorbed through the shell.

Study Break 48.3
1. The three primary tissues, ectoderm, mesoderm, and endoderm, give rise to the tissues and organs of the embryo.
2. The central nervous system, notably the brain and spinal cord, develops from the neural tube. Neural crest cells give rise to parts of the nervous system (including cranial nerves, ganglia of the autonomic nervous system, peripheral nerves from the spinal cord to body structures, and nerves of the developing gut) and contribute to a variety of other body structures (for example, bones of the inner ear and skull, cartilage of facial structures, teeth, pigment cells in the skin, and the adrenal medulla).

Study Break 48.4
1. Cells of the trophoblast are responsible for implantation of the blastocyst in the endometrium (uterine lining). When the blastocyst is ready for implantation, trophoblast cells secrete proteases that digest pathways between endometrial cells. Dividing trophoblast cells fill the spaces and, through continued digestion and division, the blastocyst burrows into the endometrium and eventually becomes covered by a layer of endometrial cells. From then on, cells originating in the trophoblast support the development of the embryo/fetus in the uterus through the production of the chorion.

   The inner cell mass becomes the embryo/fetus itself. During implantation, the inner cell mass separates into the epiblast and the hypoblast. The epiblast produces the ectoderm, mesoderm, and endoderm of the developing embryo. The hypoblast gives rise to part of the extraembryonic membranes.
2. Oxytocin. This hormone is responsible for stimulating contractions of the uterus.

Study Break 48.5
1. Cell division, cell movement, and cell adhesion.
2. Induction is the process whereby a region of the embryo acts on other cells to alter the course of development. Induction is brought about by proteins; hence, induction is under genetic control.

Study Break 48.6
1. Determination and differentiation are under molecular control. Regulatory genes encode regulatory proteins that bind to promoters of the genes they control, switching the genes on or off depending on the interaction.
2. The segmentation genes subdivide the embryo progressively into regions, thereby determining the segments of the embryo and the adult. In essence, they organize the embryo into segments. The homeotic genes specify the identity of each segment with respect to the body part it will become.

Self-Test Questions
1. d 2. b 3. a 4. e 5. c 6. d 7. d 8. b
9. a 10. e

Chapter 49

Study Break 49.1
1. Studies of ecosystems are more "inclusive" than studies of populations because ecosystems include the populations of many different species.
2. Mathematical models are useful in ecological research because they help scientists formalize hypotheses about the relationships between variables and because they allow researchers to simulate the effects of changing variables before investing time and resources in experiments or observational studies.

Study Break 49.2
1. A population's size is simply the number of individual it contains. Its density is the number of individuals per area or volume of habitat occupied,
2. A clumped pattern of dispersion implies that individuals in the population help each other or that some vital resource also has a clumped distribution in the environment. A uniform pattern of dispersion implies that individuals in the population repel each other. A random pattern of dispersion does not imply either positive or negative interactions among individuals in the population.

Study Break 49.3
1. A life table usually summarizes statistics about the age-specific survival rates, age-specific mortality rates, and age-specific fecundity of a population.
2. Humans in the industrialized countries exhibit Type I survivorship curves because they provide lots of care to their

offspring, thus reducing infant and childhood mortality to low levels.

Study Break 49.4
1. Children spend most of their energy on growth and maintenance, and devote energy to reproduction later in life.
2. Fecundity and the amount of parental care devoted to each offspring exhibit an inverse relationship because organisms that produce few offspring can devote substantial time and energy to each, whereas those that produce many offspring can devote only minimal time and energy to each.

Study Break 49.5
1. The model of exponential population growth predicts unlimited population growth through time, generating a J-shaped curve of population size versus time. The logistic model predicts that population growth slows down as the population approaches its carrying capacity, generating an S-shaped curve of population size versus time.
2. Carrying capacity is the maximum number of individuals in a population that an environment can support. The carrying capacity is thus a property of the environment with reference to a particular population.
3. A time lag is a delay in a population's response to a changing environment. It may cause a population's size to oscillate around its carrying capacity.

Study Break 49.6
1. The effects of density-dependent factors get stronger (that is, they affect a larger percentage of the individuals in the population) as the population's density increases. The effects of density-independent factors do not change (that is, they affect the same percentage of the individuals in a population) as the population's density changes.
2. The affects of infectious diseases are usually density-dependent because disease-causing pathogens spread more quickly through dense populations of the organisms they infect.

Study Break 49.7
1. Humans have sidestepped the controls that regulate the populations of other organisms by expanding their geographical range to include a wide variety of habitats, by increasing the carrying capacity through agricultural production, and by decreasing death rates through the introduction of medical care and sanitation.
2. The age structure of a population influences its future population growth by determining how many individuals will reach reproductive age in the future. Populations with a bottom-heavy age structure (that is, with many young children) will experience a growth spurt when children alive today reach sexual maturity.

Populations with a more even age structure will not experience a dramatic future increase in population size.

Self-Test Questions
1. b 2. c 3. b 4. d 5. a 6. e 7. d 8. e
9. d 10. b

Chapter 50

Study Break 50.1
1. Some carnivores will spend more time and energy capturing large prey than small prey because large prey provide a larger return on their investment of time and energy in the hunt.
2. Cryptic coloration makes an organism inconspicuous, allowing it to blend in with its surroundings. Aposematic coloration makes an organism highly conspicuous, advertising its unpalatability. Mimicry allows one organism, the mimic, to resemble another species, the model; models are usually unpalatable or poisonous. A mimic will have aposematic coloration if it resembles an aposematic model.
3. Field experiments can demonstrate that two species are competing for limiting resources if the removal of one species increases population size or density in the other or if the addition of a potential competitor decreases the population size or density of the other.

Study Break 50.2
1. Gleason's individualistic view of communities suggests that they are just chance assemblages of species that happen to be adapted to similar abiotic environmental conditions.
2. Ecologists find more species living in an ecotone than in the communities on either side of it because ecotones contain species from both neighboring communities as well as species that are adapted to transitional environmental conditions.

Study Break 50.3
1. The plant growth forms found in tropical forests include a canopy of tall trees, an understory of shorter trees and shrubs, an herb layer, vinelike lianas, and epiphytes.
2. The species richness of a community is the number of species it contains. Relative abundance refers to the commonness or rarity of species in the community.
3. Pigeons, which eat grain and other vegetable matter, are included in the second trophic level, primary consumers. Peregrine falcons, which feed on pigeons and other birds, are in the third trophic level, secondary consumers.

Study Break 50.4
1. On the one hand, the ecological literature on competition may overestimate

the importance of competition because ecologists are more likely to study and publish papers on interactions in which competition is important than on interactions in which it is not. On the other hand, the literature may underestimate the importance of competition because, if strong competition between species can not persist for long periods of time, we are unlikely to find populations competing strongly in nature.
2. Keystone species are those that have a substantial effect on community structure even if their populations are not very dense. Keystone species may either increase or decrease species richness in the communities they occupy.

Study Break 50.5
1. Strong storms allow coral communities to be rejuvenated through the recruitment of new individuals because they scour the seafloor, removing existing coral colonies from the community. These openings provide spaces where coral larvae may settle and initiate the growth of new colonies.
2. Moderately severe and moderately frequent disturbances increase a community's species richness by creating opportunities for r-selected species to colonize the habitat while allowing populations of K-selected species to persist.

Study Break 50.6
1. Primary succession occurs in places without soil; secondary succession occurs after a disturbance has destroyed vegetation.
2. A climax community differs from earlier successional stages in having taller, longer-lived vegetation, generally higher species richness, and a buffered physical environment under the vegetation.
3. Three hypotheses about the underlying causes of succession differ in how they view the role of population interactions. The facilitation hypothesis specifies no particular role for population interactions. The inhibition hypothesis suggests that species that are already present prevent other species from joining a community. The tolerance hypothesis suggests that as environmental conditions within the community change during succession, only species that can compete strongly under the changing conditions will persist.

Study Break 50.7
1. Some explanations of the high species richness in the tropics suggest that the benign climate and historically low levels of severe disturbance have fostered more rapid rates of speciation in tropical regions. Other explanations suggest that the year-round availability of food resources and complex food webs allow more species to coexist in tropical regions.
2. According to the equilibrium theory of island biogeography, large islands will

harbor more species than small islands, and islands that are close to the mainland will harbor more species than those that are further away.

Self-Test Questions
1. c 2. c 3. a 4. b 5. b 6. e 7. b 8. d
9. d 10. b

Chapter 51

Study Break 51.1
1. Gross primary productivity is a measure of the total amount of solar energy converted into chemical energy by the producers in an ecosystem. Net primary productivity is the amount of chemical energy that remains after deducting the producers' maintenance costs from the gross primary productivity.
2. In terrestrial ecosystems, primary productivity may be influenced by the availability of light, water, and nutrients and by the environmental temperature. In aquatic ecosystems, primary productivity is often limited by the joint availability of light and nutrients in the same place.
3. Energy is lost from an ecosystem at every transfer between trophic levels because some of the energy is not assimilated; because organisms use some of the energy they assimilate for maintenance costs; and because biological processes are never 100 efficient.
4. The presence of a top predator can influence the interactions of organisms at lower trophic levels and an ecosystem's productivity by changing the relative abundances of organisms at lower trophic levels. These effects can reverberate through an ecosystem in a trophic cascade.

Study Break 51.2
1. In the generalized compartment model of biogeochemical cycling, nutrient pools are classified as either available or unavailable and as either organic or inorganic.
2. The global hydrologic cycle maintains its balance because the amount of water returned to the atmosphere by evaporation and transpiration is equal to the amount that falls as precipitation. Runoff from the land maintains the balance between terrestrial and marine components of the cycle.
3. Respiration, excretion, leaching, and the burning of fossil fuels move large quantities of carbon from an organic compartment to an inorganic compartment of an ecosystem.
4. Bacteria, cyanobacteria, and fungi drive the global nitrogen cycle through their activities in nitrogen fixation, ammonification, nitrification, and denitrification.
5. Marine sediments are Earth's main reservoir for phosphorus, which is recycled slowly after geological uplifting and erosion make it available to producers.

Study Break 51.3
1. The advantage of using conceptual models of ecosystem function is that they are a simplification of the processes that determine ecosystem function in nature. The disadvantage of these models is that they do not include processes that carry nutrients and energy out of one ecosystem and into another; neither do they include the nuts-and-bolts details of exactly how specific ecosystems function. Thus, conceptual models do not provide precise predictions about potential changes in ecosystem function.
2. Before constructing a simulation model of an ecosystem, ecologists must collect data about the population sizes of important species, the average energy and nutrient content of each, the food webs in which they participate, the quantity of food each species consumes, and the productivity of each population; the ecosystem's energy and nutrient gains and losses caused by erosion, weathering, precipitation, and runoff; and seasonal and annual variations in these factors.

Self-Test Questions
1. b 2. d 3. c 4. e 5. e 6. c 7. a 8. d
9. c 10. a

Chapter 52

Study Break 52.1
1. Because of Earth's spherical shape, sunlight striking the planet's surface is more concentrated near the equator than at the poles. As a result, temperatures are higher at low latitudes. The concentrated sunlight near the equator heats the atmosphere, causing air masses near the equator to rise, establishing three circulation cells in the Northern Hemisphere and three in the Southern Hemisphere.
2. Earth's fixed tilt on its axis causes seasonal variation in the amount of sunlight striking the temperate zones as the planet orbits the sun.
3. Dry conditions prevail at 30° north and south latitudes because sinking air masses warm as they descend, causing them to absorb water from the land.
4. Mountains affect local precipitation because rising air masses on the windward side of a mountain cool adiabatically and release moisture. When the air masses descend on the leeward side of a mountain, they warm and absorb moisture, causing a rain shadow.

Study Break 52.2
1. *Anolis* lizards in the Dominican Republic bask more frequently at high elevation than they do at low elevation.
2. Global warming will likely cause the geographical distributions of species to shift or expand to higher latitudes and to higher elevations.

Study Break 52.3
1. Tropical rain forest and temperate rainforest are the terrestrial biomes that occur in habitats that receive the most rainfall.
2. Savannas, chaparral, and temperate grasslands are renewed by periodic fires.
3. Tropical rain forests and temperate rainforests have the tallest vegetation. Arctic and alpine tundra have the shortest vegetation.
4. Trees are usually evergreen in tropical rain forest, temperate rain forest, and taiga.

Study Break 52.4
1. Dissolved oxygen concentration is usually high in the headwaters of a stream, gradually diminishing as water flows into a river.
2. The factors that cause seasonal overturns in lakes include seasonal changes in environmental temperatures, variations in wind velocity, and the fact that water is densest at 4°C.
3. Oligotrophic lakes are better than eutrophic lakes for recreational purposes because the water in oligotrophic lakes is clear, whereas the water in eutrophic lakes is often clogged with strands of algae and cyanobacteria.

Study Break 52.5
1. The benthic province of the ocean includes all of the bottom sediments. The pelagic province includes all of the water.
2. Of all the marine biomes, estuaries experience the largest fluctuations in salinity over time.
3. Estuaries, the intertidal zone, and the upper layer of the oceanic pelagic zone receive substantial energy inputs from sunlight.
4. The benthos of the oceanic pelagic zone receives nutrients and energy from the detritus sinking from the upper layers of water.
5. Chemoautotrophic bacteria are the primary producers of hydrothermal vent communities and cold seep communities. Unlike photosynthetic organisms of the photic zone, they use hydrogen sulfide and other molecules, instead of sunlight, as an energy source for their chemosynthetic activity.

Self-Test Questions
1. c 2. e 3. b 4. a 5. d 6. a 7. d 8. a
9. a 10. d

Chapter 53

Study Break 53.1
1. Living systems are a storehouse of potentially useful genetic information because naturally occurring compounds may prove to be useful in the treatment of disease, in the manufacture of new products, or in agriculture

2. Naturally occurring organisms provide many ecosystem services, such as the sequestration of carbon dioxide, fixation of nitrogen into forms that plants can absorb, recycling of nutrients with ecosystems, and the retention of water in ecosystems.

Study Break 53.2
1. Habitat fragmentation in eastern North America has affected breeding songbirds by reducing the variety of habitats available to them, increasing the frequency of brood parasitism of their nests, and increasing the rate of predation on their eggs and young.
2. The growing human population in south Florida has increased the likelihood of desertification there by withdrawing groundwater for agricultural, recreational, and residential uses faster than it is replenished.
3. Overexploitation of fish populations typically causes fishes to reach reproductive maturity at a smaller size and younger age; decreases population sizes; and sometimes leads to the extinction of populations.
4. By one estimate, extinction rates today may be 1000 times greater than the background extinction rate evident in the fossil record.

Study Break 53.3
1. Ecologists generally identify biodiversity hotspots as areas that include many endemic species that face the threat of extinction.
2. Conservation biologists are especially concerned about the rapid rate of deforestation in the tropics because tropical forests, although not very extensive in area, harbor many terrestrial species.

Study Break 53.4
1. Population bottlenecks—large, temporary reductions in a population's size—inevitably foster genetic drift, thereby reducing a population's genetic variability, which increases its likelihood of becoming extinct.
2. A population viability analysis allows a conservation biologist to identify the minimum population size that is likely to survive both predictable and unpredictable environmental change. It therefore specifies how many individuals must be conserved for the continued survival of the population and species.
3. Several small preserves would collectively experience more edge effects than one large preserve of the same total size.

Study Break 53.5
1. The Pine Bush habitat in New York state is an example of conservation through preservation.
2. The Royal Chitwan National Park has been judged a success because local residents benefit from the park's existence, and therefore support it, and because

populations of many animals, including tigers and rhinoceroses, have increased within its borders.
3. Economists can determine the economic values of specific ecosystem services and convince local governments that it is economically beneficial to preserve ecosystems and the services they provide.

Self-Test Questions
1. e 2. c 3. b 4. b 5. a 6. d 7. d 8. a
9. c 10. b

Chapter 54
Study Break 54.1
1. An instinctive behavior is a genetically or developmentally programmed response that appears in complete and functional form the first time it is used. A learned behavior is one that is dependent on having a particular kind of experience during development.
2. Marler demonstrated that singing the correct species song is a learned behavior by isolating some young male sparrows in soundproof cages, thereby preventing them from hearing their species song. Because the isolated males never learned to sing the correct song, Marler concluded that the experience of hearing the species song was necessary for song learning.

Study Break 54.2
1. The chicks of brood parasites stimulate their foster parents to feed them by engaging in exaggerated behaviors—opening their mouths and begging and peeping vigorously—that serve as sign stimuli, triggering feeding behavior in the parents.
2. Arnold demonstrated that the receptiveness of garter snakes to a meal of banana slugs had a genetic basis by breeding snakes that almost always eat banana slugs with snakes that rarely eat them. The behavior of the hybrid offspring was intermediate between the behaviors of the two parent populations.

Study Break 54.3
1. Tail wagging by a dog when it sees its owner pick up a leash is an example of classical conditioning.
2. Sleeping through an alarm clock is an example of habituation.

Study Break 54.4
1. The conclusion that certain neurons in a young male bird's brain are influenced only by acoustical signals from its own species and only during a critical period is supported by two observations: (1) young birds that did not hear taped songs during the critical period never produced the full song of their species; and (2) young birds that heard recordings of *other* bird species' songs during

the critical period never generated replicas of those songs as they matured.
2. After a zebra finch hears the song of a neighbor many times, cells in a nucleus in its forebrain habituate to that stimulus and stop responding to it.
3. The role of the ZENK enzyme in song learning is to program the nerve cells of the bird's brain to anticipate key acoustical events of potential biological importance.

Study Break 54.5
1. A high estrogen concentration in the brains of young male zebra finches stimulates the production of more neurons in the higher vocal center.
2. Juvenile hormone stimulates the production of octopamine, a protein that may help a foraging bee home in on the odors of flowers where it can collect nectar and pollen.
3. The loss of a territory by a male African cichlid fish causes its brain to produce less GnRH.

Study Break 54.6
1. The anatomy of the cricket's nervous system helps it avoid an approaching bat because sensory neurons on one side of the body send messages to motor neurons on the other side. When the cricket hears a bat approaching from the right, it automatically lifts a leg on the left side, blocking the left wing and causing the cricket to veer away from the oncoming bat's path.
2. When a fiddler crab detects movement above the midline of its visual field, it dashes into its burrow.
3. A star-nosed mole locates earthworms in its tunnels with touch receptors on the tentacles that sprout from its nose.

Self-Test Questions
1. b 2. a 3. d 4. b 5. c 6. e 7. d 8. c
9. b 10. b

Chapter 55
Study Break 55.1
1. When piloting, animals use familiar landmarks to guide their journey. When using compass orientation, animals use external environmental cues such as the position of the sun or stars as a compass to move in a particular direction, often over a specific distance or for a prescribed length of time. When navigating, animals use a compass as well as a mental map of their position in relation to their destination.
2. Seasonal changes in temperature and food availability are the most likely selection pressures responsible for the evolution of migratory behavior in birds.

Study Break 55.2
1. Wood lice accumulate in moist habitats because they move much less in moist habitats than they do in dry habitats.

2. The costs of maintaining a territory include the time and energy needed to defend territory borders, the possibility of being injured or killed during a territorial encounter, and the increased likelihood of being noticed and captured by a predator. The benefits of holding a territory include access to all of the resources found within the territory.

Study Break 55.3

1. Humans consciously use acoustical, visual, and tactile channels to communicate with each other.
2. A honeybee uses the waggle dance to communicate the location of a distant food source to its hive-mates.
3. A young wandering raven that discovers a food source in the territory of other ravens will call vigorously to attract other ravens from outside the territory. Collectively, the non-territory holders can overwhelm the defenses of the resident birds and then consume the food.

Study Break 55.4

1. For monogamous species, the males that are most attractive to females are those that can demonstrate their good genes with large showy morphological characteristics and elaborate behavioral displays and those that hold territories rich in resources.
2. Male sage grouse perform displays at a lek, trying to attract the attention of females. Female sage grouse go to a lek to evaluate the qualities of the males that are displaying there.
3. A female red-winged blackbird might settle in the territory of a male even in the presence of other resident females if the male's territory is very rich in resources.

Study Break 55.5

1. The social behavior of musk oxen and sawfly larvae includes cooperative defense of the group against predators.

2. Among the animals in a dominance hierarchy, the most dominant individuals are the most likely to reproduce.
3. The genetic system in eusocial insects promotes altruistic behavior because most individuals in a colony are more closely related to each other than are siblings in most other animal species.

Study Break 55.6

1. People may exhibit altruistic behavior to nonrelatives because their acts of charity may induce others to be charitable toward them.
2. Stepparents might provide fewer resources to their children than birthparents do because stepparents do not share as many genes with their stepchildren as birthparents share with their biological children.

Self-Test Questions

1. d 2. d 3. b 4. d 5. a 6. e 7. c 8. d
9. a 10. c

Appendix B
Classification System

The classification system presented here is based on a combination of organismal and molecular characters and is a composite of several systems developed by microbiologists, botanists, and zoologists. This classification reflects current trends toward a phylogenetic approach to taxonomy, one that incorporates the ever more detailed information about the relationships of monophyletic lineages provided by new molecular sequence data. In keeping with these trends, we have omitted reference to the traditional taxonomic categories, such as "class" and "order." Instead, we present the major monophyletic lineages in each of the three domains, and we indicate their relationships within a nested hierarchy that parallels that of traditional Linnaean classification.

Although researchers generally agree on the identity of the major monophyletic lineages, the biologists who study different groups have not established universal criteria for identifying the somewhat arbitrary taxonomic categories included in the traditional Linnaean hierarchy. As a result, a "class" or "order" of flowering plants may not be the equivalent of a "class" or "order" of animals. In fact, as described in *Unanswered Questions* at the end of Chapter 23, systematic biologists are shifting toward a more phylogenetic approach to taxonomy and classification, such as the one represented here.

Bear in mind that we include this appendix to introduce the diversity of life and illustrate many of the evolutionary relationships that link monophyletic groups. Like all phylogenetic hypotheses, this classification is open to revision as new information becomes available. Moreover, the classification is incomplete because it includes only those lineages that are described in Unit Four.

Prokaryotes and Eukaryotes

Organisms fall into two groups, prokaryotes and eukaryotes, based on the organization of their cells. Prokaryotes consist of the Domains Bacteria and Archaea and are characterized by a central region, the nucleoid, which has no boundary membrane separating it from the cytoplasm, and by membranes typically limited to the plasma membrane. Most prokaryotes are single-celled, although some are found in simple associations. All other organisms are eukaryotes, which make up the Domain Eukarya. Eukaryotes are characterized by cells with a central, membrane-bound nucleus, and an extensive membrane system. Some eukaryotes are single-celled, while others are multicellular.

Domain Bacteria

The largest and most diverse group of prokaryotes. Includes photoautotrophs, chemoautotrophs, and heterotrophs.

PROTEOBACTERIA Purple sulfur bacteria, purple nonsulfur bacteria, and some chemoheterotrophs

GREEN BACTERIA Green sulfur bacteria and green nonsulfur bacteria

CYANOBACTERIA Photoautotrophic Gram-negative bacteria that use the same chlorophyll as in plants

GRAM-POSITIVE BACTERIA Chemoheterotrophic bacteria with thick cell walls

SPIROCHETES Helically spiraled bacteria that move by twisting in a corkscrew pattern

CHLAMYDIAS Gram-negative intracellular parasites of animals, with cell walls that lack peptidoglycans

Domain Archaea

Prokaryotes that are evolutionarily between eukaryotic cells and the bacteria. Most are chemoautotrophs. None is photosynthetic. Originally discovered in extreme habitats, they are now known to be widely dispersed. Compared with bacteria, the Archaea have a distinctive cell wall structure and unique membrane lipids, ribosomes, and RNA sequences. Some are symbiotic with animals, but none is known to be pathogenic.

EURYARCHAEOTA Includes methanogens, extreme halophiles, and some extreme thermophiles

CRENARCHAEOTA Includes most of the extreme thermophiles, as well as psychrophiles; mesophilic species comprise a large part of plankton in cool, marine waters

KORARCHAEOTA Known only from DNA isolated from hydrothermal pools. As of this writing, none has been cultured and no species have been named.

Domain Eukarya

PROTOCTISTA A collection of single-celled and multicelled lineages, which are almost certainly not a monophyletic group. Some biologists consider the groups listed below to be kingdoms in their own right.

> **Excavates** Single-celled animal parasites that lack mitochondria and move using flagella; most have a hollow, ventral feeding groove
>
> > Diplomonadida (diplomonads)—two nuclei; move by multiple free flagella

> > Parabasala (parabasalids)—move by an undulating membrane and free flagella
>
> **Discicristates** Mostly single-celled, highly motile cells that swim using flagella, have disc-shaped mitochondrial inner membranes
>
> > Euglenoids—free-living photosynthetic autotrophs
> >
> > Kinetoplastids—nonphotosynthetic, heterotrophs that live as animal parasites

Alveolates Characterized by small membrane-bound vesicles called alveoli in a layer under the plasma membrane

> Ciliophora (ciliates)—single-celled heterotrophs; swim by means of cilia

> Dinoflagellata (dinoflagellates)—single-celled marine heterotrophs or autotrophs; shell formed from cellulose plates

> Apicomplexa (apicomplexans)—nonmotile parasites of animals with apical complex for attachment and invasion of host cells

Heterokonts Characterized by two different flagella

> Oomycota (oomycetes)—water molds, white rusts, and mildews

> Bacillariophyta (diatoms)—single-celled; covered by a glassy silica shell

> Chrysophyta (golden algae)—colonial; each cell of the colony has a pair of flagella and is covered by a glassy shell consisting of plates or scales

> Phaeophyta (brown algae)—photoautotrophic protists

Cercozoa Amoebas with stiff, filamentous pseudopodia; some with outer shells

> Radiolaria (radiolarians)—heterotrophic; glassy internal skeleton with projecting raylike strands of cytoplasm

> Foraminifera (forams)—heterotrophic protists with shells consisting of organic matter reinforced by calcium carbonate

> Chlorarachniophyta (chloroarachniophytes)—green, photosynthetic amoebas that also engulf food

Amoebozoa Includes most of the amoebas and the slime molds

> Amoebas—single-celled; use non-stiffened pseudopods for locomotion and feeding

> Cellular slime molds—heterotrophs; primarily individual cells; move by amoeboid motion, or as a multicellular mass

> Plasmodial slime molds—heterotrophs; live as plasmodium, a large composite mass with nuclei in a common cytoplasm, that moves and feeds like a giant amoeba

Opisthokonts A single posterior flagellum at some stage in the life cycle

> Choanoflagellata (choanoflagellates)—motile protists with a single flagellum surrounded by collar of closely packed microvilli; likely ancestor of animals and fungi

Archaeplastida Red algae, green algae, and land plants, photosynthesizers with a common evolutionary origin

> Rhodophyta (red algae)—marine seaweeds, typically multicellular, reddish in color; with plantlike bodies

> Chlorophyta (green algae)—green photosynthetic single-celled, colonial, and multicellular protists that have the same photosynthetic pigments as plants; likely ancestor of land plants

FUNGI Heterotrophic, mostly multicellular organisms with cell wall containing chitin and cell nuclei occurring in threadlike hyphae; life cycle typically includes both asexual and sexual phases, with sexual structures used as the basis for phylum-level classification. Single-celled species are known as yeasts.

> Zygomycota (zygomycetes)—terrestrial; asexual reproduction via nonmotile haploid spores formed in sporangia; sexual spores (zygospores) form in zygosporangia; aseptate hyphae

> Glomeromycota (glomeromycetes)—terrestrial; asexual reproduction via spores at the tips of hyphae; form mycorrhizal associations with plant roots

> Ascomycota (ascomycetes/sac fungi)—terrestrial and aquatic; sexual spores form in asci; asexual reproduc-

tion occurs via conidia (nonmotile spores); septate hyphae

> Basidiomycota (basidiomycetes)—terrestrial; reproduction usually via asexual basidiospores produced by basidia; septate hyphae

>> Basidiomycetes: mushroom-forming fungi and relatives

>> Teliomycetes: rusts

>> Ustomycetes: smuts

> Chytridiomycota (chytrids)—mostly aquatic; asexual reproduction by way of motile zoospores; sexual reproduction via gametes produced in gametangia; hyphae mostly aseptate

> Conidial fungi—not a true phylum but a convenience grouping of species for which no sexual phase is known

> Microsporidia—single-celled sporelike parasites of animals, other groups; phylogeny uncertain

PLANTAE Multicellular autotrophs, mostly terrestrial, and most of which gain energy via photosynthesis; life cycle characterized by alternation of a gametophyte (gamete-producing) generation and sporophyte (spore-producing) generation

Nonvascular plants (bryophytes)—no vessels for transporting water and nutrients; swimming sperm require liquid water for sexual reproduction

> Hepatophyta (liverworts)—leafy or simple flattened thallus with rhizoids; no true leaves, stems, roots, or stomata (porelike openings for gas exchange); spores in capsules

> Anthocerophyta (hornworts)—simple flattened thallus, hornlike sporangia

> Bryophyta (mosses)—feathery or cushiony thallus; some with hydroids; spores in capsules

Seedless vascular plants—plants in which embryos are not housed inside seeds

> Lycophyta (club mosses)—simple leaves, cuticle, stomata, true roots; most species have sporangia on sporophylls; fertilization by swimming sperm

> Pterophyta (ferns, whisk ferns, horsetails)—*Ferns*: Finely divided leaves; sporangia in sori. *Whisk ferns*: Branching stem from rhizomes; sporangia on stem scales. *Horsetails*: hollow stem, scalelike leaves, sporangia in strobili.

Seed plants—vascular plants in which embryos develop within seeds

> Gymnosperms—seeds born on stems, on leaves, or under scales

> Cycadophyta (cycads)—shrubby or treelike with palmlike leaves; male and female strobili on separate plants

> Ginkgophyta (ginkgoes)—lineage with a single living species (*Ginkgo biloba*); tree with deciduous, fan-shaped leaves; male, female reproductive structures on separate plants

> Gnetophyta (gnetophytes)—shrubs or woody vinelike plants; male and female strobili on separate plants

> Coniferophyta (conifers)—predominant extant gymnosperm group; mostly evergreen trees and shrubs with needlelike or scalelike leaves; male and female cones usually on the same plant

> Anthophyta (angiosperms/flowering plants)—reproductive structures in flowers

>> Monocotyledones (monocots)—grasses, palms, lilies, orchids and their relatives; a single cotyledon (seed leaf); pollen grains have one groove

Eudicotyledones (eudicots)—roses, melons, beans, potatoes, most fruit trees, others; two cotyledons; pollen grains have three grooves

Other major angiosperm lineages: magnoliids (magnolias and relatives); water lilies (Family Nymphaeaceae); *Amborella* (Family Amborellaceae)

ANIMALIA Multicellular heterotrophs; nearly all with tissues, organs, and organ systems; motile during at least part of the life cycle; sexual reproduction in most; embryos develop through a series of stages; many with larval and adult stages in life cycle

Parazoa Animals lacking tissues and body symmetry

Porifera (sponges)—multicellular; extract oxygen and particulate food from water drawn into a central cavity

Eumetazoa Animals possessing tissues and either radial or bilateral symmetry

Radiata—acoelomate animals possessing radial symmetry and two tissue layers

Cnidaria (cnidarians)—two tissue layers; single opening into gastrovascular cavity; nerve net; nematocysts for defense and predation; some sessile, some motile; most are predatory, some with photosynthetic endosymbionts; freshwater and marine

Hydrozoa: hydrozoans

Scyphozoa: jellyfishes

Cubozoa: box jellyfishes

Anthozoa: sea anemones, corals

Ctenophora (comb jellies)—two (possibly three) tissue layers; feeding tentacles capture particulate food; beating cilia provide weak locomotion; marine

Bilateria—animals possessing bilateral symmetry and three tissue layers

PROTOSTOMIA—acoelomate, pseudocoelomate, or schizocoelomate; many with spiral, indeterminate cleavage; blastopore forms mouth; nervous system on ventral side

Lophotrochozoa—many with either a lophophore for feeding and gas exchange or a trochophore larva

Ectoprocta (bryozoans)—coelomate; colonial; secrete hard covering over soft tissues; lophophore; sessile; particulate feeders; marine

Brachiopoda (lamp shells)—coelomate; dorsal and ventral shells; lophophore; sessile; particulate feeders; marine

Phoronida (phoronid worms)—coelomate; secrete tubes around soft tissues; sessile; particulate feeders; lophophore

Platyhelminthes (flatworms)—acoelomate; dorsoventrally flattened; complex reproductive, excretory, and nervous systems; gastrovascular cavity in many; free-living or parasitic, often with multiple hosts; terrestrial, freshwater, and marine

Turbellaria: free-living flatworms

Trematoda: flukes

Cestoda: tapeworms

Rotifera (wheel animals)—pseudocoelomate; microscopic; complete digestive system; well-developed reproductive, excretory, and nervous systems; particulate feeders; major components of marine and freshwater plankton

Nemertea (ribbon worms)—schizocoelomate; proboscis housed within rhynchocoel; complete digestive tract; circulatory system; predatory; mostly marine

Mollusca (mollusks)—schizocoelomate; many with trochophore larva; many with shell secreted by mantle; body divided into head–foot, visceral mass, and mantle; well-developed organ systems; variable locomotion; herbivorous or predatory; terrestrial, freshwater, and marine

Polyplacophora: chitons

Gastropoda: snails, sea slugs, land slugs

Bivalvia: clams, mussels, scallops, oysters

Cephalopoda: squids, octopuses, cuttlefish, nautiluses

Annelida (segmented worms)—schizocoelomate; many with trochophore larva; segmented body and organ systems; well-developed organ systems; many use hydrostatic skeleton for locomotion; some predatory, some particulate feeders, some detritivores; terrestrial, freshwater, and marine

Polychaeta: marine worms

Oligochaeta: freshwater and terrestrial worms

Hirudinea: leeches

Ecdysozoa—cuticle or exoskeleton is shed periodically

Nematoda (roundworms)—pseudocoelomate; body covered with tough cuticle that is shed periodically; well-developed organ systems; thrashing locomotion; many are parasitic on plants or animals; mostly terrestrial

Onychophora (velvet worms)—schizocoelomate; segmented body covered with cuticle; locomotion by many unjointed legs; complex organ systems; predatory; terrestrial

Arthropoda (arthropods)—schizocoelomate; jointed exoskeleton made of chitin; segmented body, some with fusion of segments in head, thorax, or abdomen; complex organ systems; variable modes of locomotion, including flight; specialization of numerous appendages; herbivorous, predatory, or parasitic; terrestrial, freshwater, and marine

Trilobita: trilobites (extinct)

Chelicerata: horseshoe crabs, spiders, scorpions, ticks, mites

Crustacea: shrimps, crayfishes, lobsters, crabs, barnacles, copepods, isopods

Myriapoda: centipedes, millipedes

Hexapoda: springtails and insects

DEUTEROSTOMIA—enterocoelomate; many with radial, determinate cleavage; blastopore forms anus; nervous system on dorsal side in many

Echinodermata (echinoderms)—secondary radial symmetry, often organized around five radii; hard internal skeleton; unique water vascular system with tube feet; complete digestive system; simple nervous system; no circulatory or respiratory system; generally slow locomotion using tube feet; predatory, herbivorous, particulate feeders, detritivores; exclusively marine

Asteroidea: sea stars

Ophiuroidea: brittle stars

Echinoidea: sea urchins, sand dollars

Holothuroidea: sea cucumbers

Crinoidea: feather stars, sea lilies

Concentricycloidea: sea daisies

Hemichordata (acorn worms)—pharynx perforated with gill slits; proboscis; complex organ systems; tube-dwelling in soft sediments; particulate or deposit feeders; exclusively marine

Chordata (chordates)—notochord; segmental body wall and tail muscles; dorsal hollow nerve chord; perforated pharynx; complex organ systems; variable modes of locomotion; extremely varied diets; terrestrial, freshwater, and marine

    Urochordata: tunicates, sea squirts

    Cephalochordata: lancelets

    Vertebrata: vertebrates

        Myxinoidea: hagfishes

        Petromyzontoidea: lampreys

        Placodermi: placoderms (extinct)

        Chondrichthyes: sharks, skates, and rays

Acanthodii: acanthodians

Actinopterygii: ray-finned fishes

Sarcopterygii: fleshy-finned fishes

Amphibia: salamanders, frogs, caecelians

Synapsida: mammals

Anapsida: turtles

Diapsida: sphenodontids, lizards, snakes, crocodilians, birds

Appendix C
Annotations to a Journal Article

This journal article reports on the movements of a female wolf during the summer of 2002 in northwestern Canada. It also reports on a scientific process of inquiry, observation and interpretation to learn where, how and why the wolf traveled as she did. In some ways, this article reflects the story of "how to do science" told in section 1.4 of this textbook. These notes are intended to help you read and understand how scientists work and how they report on their work.

1 Title of the journal, which reports on science taking place in Arctic regions.

2 Volume number, issue number and date of the journal, and page numbers of the article.

3 Title of the article: a concise but specific description of the subject of study—one episode of long-range travel by a wolf hunting for food on the Arctic tundra.

4 Authors of the article: scientists working at the institutions listed in the footnotes below. Note #2 indicates that P. F. Frame is the corresponding author—the person to contact with questions or comments. His email address is provided.

5 Date on which a draft of the article was received by the journal editor, followed by date on which a revised draft was accepted for publication. Between these dates, the article was reviewed and critiqued by other scientists, a process called peer review. The authors revised the article to make it clearer, according to those reviews.

6 ABSTRACT: A brief description of the study containing all basic elements of this report. First sentence summarizes the background material. Second sentence encapsulates the methods used. The rest of the paragraph sums up the results. Authors introduce the main subject of the study—a female wolf (#388) with pups in a den—and refer to later discussion of possible explanations for her behavior.

7 Key words are listed to help researchers using computer databases. Searching the databases using these key words will yield a list of studies related to this one.

8 RÉSUMÉ: The French translation of the abstract and key words. Many researchers in this field are French Canadian. Some journals provide such translations in French or in other languages.

9 INTRODUCTION: Gives the background for this wolf study. This paragraph tells of known or suspected wolf behavior that is important for this study. Note that (a) major species mentioned are always accompanied by scientific names, and (b) statements of fact or postulations (claims or assumptions about what is likely to be true) are followed by references to studies that established those facts or supported the postulations.

10 This paragraph focuses directly on the wolf behaviors that were studied here.

11 This paragraph starts with a statement of the hypothesis being tested, one that originated in other studies and is supported by this one. The hypothesis is restated more succinctly in the last sentence of this paragraph. This is the inquiry part of the scientific process—asking questions and suggesting possible answers.

---

1 ARCTIC

2 VOL. 57, NO. 2 (JUNE 2004) P. 196–203

3 Long Foraging Movement of a Denning Tundra Wolf

4 Paul F. Frame,[1,2] David S. Hik,[1] H. Dean Cluff,[3] and Paul C. Paquet[4]

5 (Received 3 September 2003; accepted in revised form 16 January 2004)

6 ABSTRACT Wolves (*Canis lupus*) on the Canadian barrens are intimately linked to migrating herds of barren-ground caribou (*Rangifer tarandus*). We deployed a Global Positioning System (GPS) radio collar on an adult female wolf to record her movements in response to changing caribou densities near her den during summer. This wolf and two other females were observed nursing a group of 11 pups. She traveled a minimum of 341 km during a 14-day excursion. The straight-line distance from the den to the farthest location was 103 km, and the overall minimum rate of travel was 3.1 km/h. The distance between the wolf and the radio-collared caribou decreased from 242 km one week before the excursion to 8 km four days into the excursion. We discuss several possible explanations for the long foraging bout.

7 *Key words:* wolf, GPS tracking, movements, *Canis lupus*, foraging, caribou, Northwest Territories

8 RÉSUMÉ Les loups (*Canis lupus*) dans la toundra canadienne sont étroitement liés aux hardes de caribous des toundras (*Rangifer tarandus*). On a équipé une louve adulte d'un collier émetteur muni d'un système de positionnement mondial (GPS) afin d'enregistrer ses déplacements en réponse au changement de densité du caribou près de sa tanière durant l'été. On a observé cette louve ainsi que deux autres en train d'allaiter un groupe de 11 louveteaux. Elle a parcouru un minimum de 341 km durant une sortie de 14 jours. La distance en ligne droite de la tanière à l'endroit le plus éloigné était de 103 km, et la vitesse minimum durant tout le voyage était de 3,1 km/h. La distance entre la louve et le caribou muni du collier émetteur a diminué de 242 km une semaine avant la sortie à 8 km quatre jours après la sortie. On commente diverses explications possibles pour ce long épisode de recherche de nourriture.

Mots clés: loup, repérage GPS, déplacements, *Canis lupus*, recherche de nourriture, caribou, Territoires du Nord-Ouest

Traduit pour la revue *Arctic* par Nésida Loyer.

9 Introduction

Wolves (*Canis lupus*) that den on the central barrens of mainland Canada follow the seasonal movements of their main prey, migratory barren-ground caribou (*Rangifer tarandus*) (Kuyt, 1962; Kelsall, 1968; Walton et al., 2001). However, most wolves do not den near caribou calving grounds, but select sites farther south, closer to the tree line (Heard and Williams, 1992). Most caribou migrate beyond primary wolf denning areas by mid-June and do not return until mid-to-late July (Heard et al., 1996; Gunn et al., 2001). Conse-quently, caribou density near dens is low for part of the summer.

During this period of spatial separation from the main caribou herds, wolves must either search near the homesite for scarce caribou or alternative prey (or both), travel to where prey are abundant, or use a combination of these strategies.

Walton et al. (2001) postulated that the travel of tundra wolves outside their normal summer ranges is a response to low caribou availability rather than a pre-dispersal exploration like that observed in territorial wolves (Fritts and Mech, 1981; Messier, 1985). The authors postulated this because most such travel was directed toward caribou calving grounds. We report details of such a long-distance excursion by a breeding female tundra wolf wearing a GPS radio collar. We discuss the relationship of the excursion to movements of satellite-collared caribou (Gunn et al., 2001), supporting the hypothesis that tundra wolves make directional, rapid, long-distance movements in response to seasonal prey availability.

[1] Department of Biological Sciences, University of Alberta, Edmonton, Alberta T6G 2E9, Canada
[2] Corresponding author: pframe@ualberta.ca
[3] Department of Resources, Wildlife, and Economic Development, North Slave Region, Government of the Northwest Territories, P.O. Box 2668, 3803 Bretzlaff Dr., Yellowknife, Northwest Territories X1A 2P9, Canada; Dean_Cluff@gov.nt.ca
[4] Faculty of Environmental Design, University of Calgary, Calgary, Alberta T2N 1N4, Canada; current address: P.O. Box 150, Meacham, Saskatchewan S0K 2V0, Canada

196

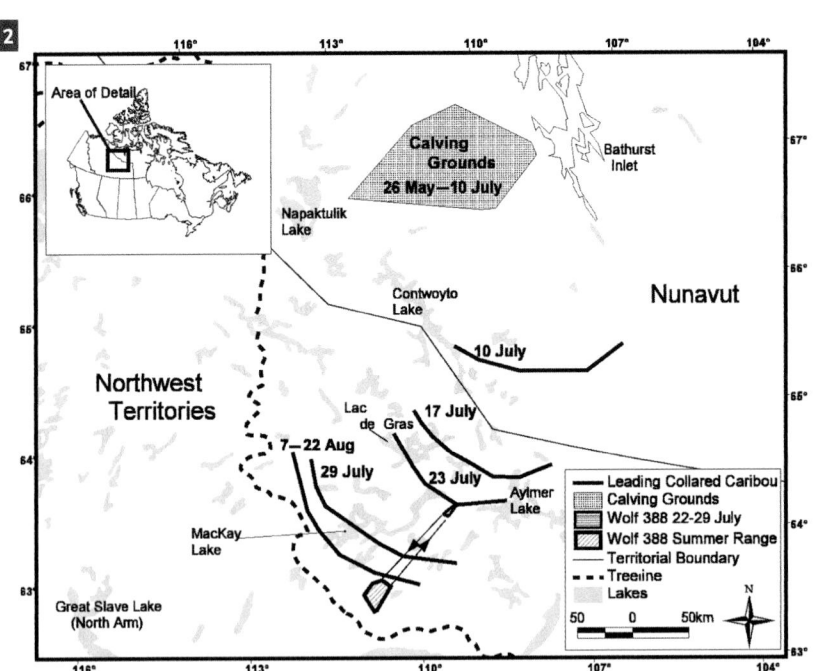

Figure 1. Map showing the movements of satellite radio-collared caribou with respect to female wolf 388's summer range and long foraging movement, in summer 2002.

12 This map shows the study area and depicts wolf and caribou locations and movements during one summer. Some of this information is explained below.

13 STUDY AREA: This section sets the stage for the study, locating it precisely with latitude and longitude coordinates and describing the area (illustrated by the map in Figure 1).

14 Here begins the story of how prey (caribou) and predators (wolves) interact on the tundra. Authors describe movements of these nomadic animals throughout the year.

15 We focus on the denning season (summer) and learn how wolves locate their dens and travel according to the movements of caribou herds.

13 Study Area

Our study took place in the northern boreal forest–low Arctic tundra transition zone (63° 30′ N, 110° 00′ W; Figure 1; Timoney et al., 1992). Permafrost in the area changes from discontinuous to continuous (Harris, 1986). Patches of spruce (*Picea mariana, P. glauca*) occur in the southern portion and give way to open tundra to the northeast. Eskers, kames, and other glacial deposits are scattered throughout the study area. Standing water and exposed bedrock are characteristic of the area.

14 *Details of the Caribou-Wolf System*

The Bathurst caribou herd uses this study area. Most caribou cows have begun migrating by late April, reaching calving grounds by June (Gunn et al., 2001;

Figure 1). Calving peaks by 15 June (Gunn et al., 2001), and calves begin to travel with the herd by one week of age (Kelsall, 1968). The movement patterns of bulls are less known, but bulls frequent areas near calving grounds by mid-June (Heard et al., 1996; Gunn et al., 2001). In summer, Bathurst caribou cows generally travel south from their calving grounds and then, parallel to the tree line, to the northwest. The rut usually takes place at the tree line in October (Gunn et al., 2001). The winter range of the Bathurst herd varies among years, ranging through the taiga and along the tree line from south of Great Bear Lake to southeast of Great Slave Lake. Some caribou spend the winter on the tundra (Gunn et al., 2001; Thorpe et al., 2001).

15 In winter, wolves that prey on Bathurst caribou do not behave territorially. Instead, they follow the herd throughout its winter range (Walton et al., 2001; Musiani, 2003). However, during denning (May–

16 Other variables are considered—prey other than caribou and their relative abundance in 2002.

17 METHODS: There is no one scientific method. Procedures for each and every study must be explained carefully.

18 Authors explain when and how they tracked caribou and wolves, including tools used and the exact procedures followed.

19 This important subsection explains what data were calculated (average distance…) and how, including the software used and where it came from. (The calculations are listed in Table 1.) Note that the behavior measured (traveling) is carefully defined.

20 RESULTS: The heart of the report and the observation part of the scientific process. This section is organized parallel to the Methods section.

21 This subsection is broken down by periods of observation. Pre-excursion period covers the time between 388's capture and the start of her long-distance travel. The investigators used visual observations as well as telemetry (measurements taken using the global positioning system (GPS)) to gather data. They looked at how 388 cared for her pups, interacted with other adults, and moved about the den area.

Table 1. Daily distances from wolf 388 and the den to the nearest radio-collared caribou during a long excursion in summer 2002.

Date (2002)	Mean distance from caribou to wolf (km)	Daily distance from closest caribou to den
12 July	242	241
13 July	210	209
14 July	200	199
15 July	186	180
16 July	163	162
17 July	151	148
18 July	144	137
19 July[1]	126	124
20 July	103	130
21 July	73	130
22 July	40	110
23 July[2]	9	104
29 July[3]	16	43
30 July	32	43
31 July	28	44
1 August	29	46
2 August[4]	54	52
3 August	53	53
4 August	74	74
5 August	75	75
6 August	74	75
7 August	72	75
8 August	76	75
9 August	79	79

[1] Excursion starts.
[2] Wolf closest to collared caribou.
[3] Previous five days' caribou locations not available.
[4] Excursion ends.

August, parturition late May to mid-June), wolf movements are limited by the need to return food to the den. To maximize access to migrating caribou, many wolves select den sites closer to the tree line than to caribou calving grounds (Heard and Williams, 1992). Because of caribou movement patterns, tundra denning wolves are separated from the main caribou herds by several hundred kilometers at some time during summer (Williams, 1990:19; Figure 1; Table 1).

16 Muskoxen do not occur in the study area (Fournier and Gunn, 1998), and there are few moose there (H.D. Cluff, pers. obs.). Therefore, alternative prey for wolves includes waterfowl, other ground-nesting birds, their eggs, rodents, and hares (Kuyt, 1972; Williams, 1990:16; H.D. Cluff and P.F. Frame, unpubl. data). During 56 hours of den observations, we saw no ground squirrels or hares, only birds. It appears that the abundance of alternative prey was relatively low in 2002.

17 Methods

Wolf Monitoring

18 We captured female wolf 388 near her den on 22 June 2002, using a helicopter net-gun (Walton et al., 2001). She was fitted with a releasable GPS radio collar (Merrill et al., 1998) programmed to acquire locations at 30-

minute intervals. The collar was electronically released (e.g., Mech and Gese, 1992) on 20 August 2002. From 27 June to 3 July 2002, we observed 388's den with a 78 mm spotting scope at a distance of 390 m.

Caribou Monitoring

In spring of 2002, ten female caribou were captured by helicopter net-gun and fitted with satellite radio collars, bringing the total number of collared Bathurst cows to 19. Eight of these spent the summer of 2002 south of Queen Maud Gulf, well east of normal Bathurst caribou range. Therefore, we used 11 caribou for this analysis. The collars provided one location per day during our study, except for five days from 24 to 28 July. Locations of satellite collars were obtained from Service Argos, Inc. (Landover, Maryland).

Data Analysis

Location data were analyzed by ArcView GIS soft- **19** ware (Environmental Systems Research Institute Inc., Redlands, California). We calculated the average distance from the nearest collared caribou to the wolf and the den for each day of the study.

Wolf foraging bouts were calculated from the time 388 exited a buffer zone (500 m radius around the den) until she re-entered it. We considered her to be traveling when two consecutive locations were spatially separated by more than 100 m. Minimum distance traveled was the sum of distances between each location and the next during the excursion.

We compared pre- and post-excursion data using Analysis of Variance (ANOVA; Zar, 1999). We first tested for homogeneity of variances with Levene's test (Brown and Forsythe, 1974). No transformations of these data were required.

Results **20**

Wolf Monitoring

Pre-Excursion Period: Wolf 388 was lactating when **21** captured on 22 June. We observed her and two other females nursing a group of 11 pups between 27 June and 3 July. During our observations, the pack consisted of at least four adults (3 females and 1 male) and 11 pups. On 30 June, three pups were moved to a location 310 m from the other eight and cared for by an uncollared female. The male was not seen at the den after the evening of 30 June.

Before the excursion, telemetry indicated 18 foraging bouts. The mean distance traveled during these bouts was 25.29 km (± 4.5 SE, range 3.1–82.5 km). Mean greatest distance from the den on foraging

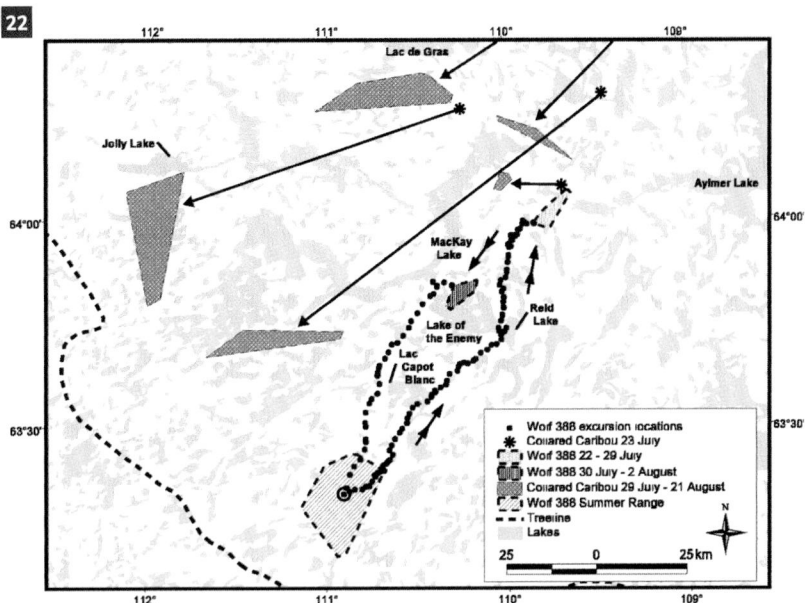

22

Figure 2. Details of a long foraging movement by female wolf 388 between 19 July and 2 August 2002. Also shown are locations and movements of three satellite radio-collared caribou from 23 July to 21 August 2002. On 23 July, the wolf was 8 km from a collared caribou. The farthest point from the den (103 km distant) was recorded on 27 July. Arrows indicate direction of travel.

bouts was 7.1 km (± 0.9 SE, range 1.7–17.0 km). The average duration of foraging bouts for the period was 20.9 h (± 4.5 SE, range 1–71 h).

The average daily distance between the wolf and the nearest collared caribou decreased from 242 km on 12 July, one week before the excursion period, to 126 km on 19 July, the day the excursion began (Table 1).

23   **Excursion Period:** On 19 July at 2203, after spending 14 h at the den, 388 began moving to the northeast and did not return for 336 h (14 d; Figure 2). Whether she traveled alone or with other wolves is unknown. During the excursion, 476 (71%) of 672 possible locations were recorded. The wolf crossed the southeast end of Lac Capot Blanc on a small land bridge, where she paused for 4.5 h after traveling for 19.5 h (37.5

km). Following this rest, she traveled for 9 h (26.3 km) onto a peninsula in Reid Lake, where she spent 2 h before backtracking and stopping for 8 h just off the peninsula. Her next period of travel lasted 16.5 h (32.7 km), terminating in a pause of 9.5 h just 3.8 km from a concentration of locations at the far end of her excursion, where we presume she encountered caribou. The mean duration of these three movement periods was 15.7 h (± 2.5 SE), and that of the pauses, 7.3 h (± 1.5). The wolf required 72.5 h (3.0 d) to travel a minimum of 95 km from her den to this area near caribou (Figure 2). She remained there (35.5 km2) for 151.5 h (6.3 d) and then moved south to Lake of the Enemy, where she stayed (31.9 km^2) for 74 h (3.1 d) before returning to her den. Her greatest distance from the den, 103 km, was recorded 174.5 h (7.3 d) after the excursion

22   The key in the lower right-hand corner of the map shows areas (shaded) within which the wolves and caribou moved, and the dotted trail of 388 during her excursion. From the results depicted on this map, the investigators tried to determine when and where 388 might have encountered caribou and how their locations affected her traveling behavior.

23   The wolf's excursion (her long trip away from the den area) is the focus of this study. These paragraphs present detailed measurements of daily movements during her two-week trip—how far she traveled, how far she was from collared caribou, her time spent traveling and resting, and her rate of speed. Authors use the phrase "minimum distance traveled" to acknowledge they couldn't track every step but were measuring samples of her movements. They knew that she went at least as far as they measured. This shows how scientists try to be exact when reporting results. Results of this study are depicted graphically in the map in Figure 2.

24 Post-excursion measurements of 388's movements were made to compare with those of the pre-excursion period. In order to compare, scientists often use means, or averages, of a series of measurements—mean distances, mean duration, etc.

25 In the comparison, authors used statistical calculations (F and df) to determine that the differences between pre- and post-excursion measurements were statistically insignificant, or close enough to be considered essentially the same or similar.

26 As with wolf 388, the investigators measured the movements of caribou during the study period. The areas within which the caribou moved are shown in Figure 2 by shaded polygons mentioned in the second paragraph of this subsection.

27 This subsection summarizes how distances separating predators and prey varied during the study period.

28 DISCUSSION: This section is the interpretation part of the scientific process.

29 This subsection reviews observations from other studies and suggests that this study fits with patterns of those observations.

30 Authors discuss a prevailing theory (CBFT) which might explain why a wolf would travel far to meet her own energy needs while taking food caught closer to the den back to her pups. The results of this study seem to fit that pattern.

---

began, at 0433 on 27 July. She was 8 km from a collared caribou on 23 July, four days after the excursion began (Table 1).

The return trip began at 0403 on 2 August, 318 h (13.2 d) after leaving the den. She followed a relatively direct path for 18 h back to the den, a distance of 75 km.

The minimum distance traveled during the excursion was 339 km. The estimated overall minimum travel rate was 3.1 km/h, 2.6 km/h away from the den and 4.2 km/h on the return trip.

24 **Post-Excursion Period:** We saw three pups when recovering the collar on 20 August, but others may have been hiding in vegetation.

Telemetry recorded 13 foraging bouts in the post-excursion period. The mean distance traveled during these bouts was 18.3 km (+ 2.7 SE, range 1.2–47.7 km), and mean greatest distance from the den was 7.1 km (+ 0.7 SE, range 1.1–11.0 km). The mean duration of these post-excursion foraging bouts was 10.9 h (+ 2.4 SE, range 1–33 h).

When 388 reached her den on 2 August, the distance to the nearest collared caribou was 54 km. On 9 August, one week after she returned, the distance was 79 km (Table 1).

Pre- and Post-Excursion Comparison

25 We found no differences in the mean distance of foraging bouts before and after the excursion period (F = 1.5, df = 1, 29, p = 0.24). Likewise, the mean greatest distance from the den was similar pre- and post-excursion (F = 0.004, df = 1, 29, p = 0.95). However, the mean duration of 388's foraging bouts decreased by 10.0 h after her long excursion (F = 3.1, df = 1, 29, p = 0.09).

26 *Caribou Monitoring*

Summer Movements: On 10 July, 5 of 11 collared caribou were dispersed over a distance of 10 km, 140 km south of their calving grounds (Figure 1). On the same day, three caribou were still on the calving grounds, two were between the calving grounds and the leaders, and one was missing. One week later (17 July), the leading radio-collared cows were 100 km farther south (Figure 1). Two were within 5 km of each other in front of the rest, who were more dispersed. All radio-collared cows had left the calving grounds by this time. On 23 July, the leading radio-collared caribou had moved 35 km farther south, and all of them were more widely dispersed. The two cows closest to the leader were 26 km and 33 km away, with 37 km between them. On the next location (29 July), the most southerly caribou were 60 km

farther south. All of the caribou were now in the areas where they remained for the duration of the study (Figure 2).

A Minimum Convex Polygon (Mohr and Stumpf, 1966) around all caribou locations acquired during the study encompassed 85 119 km².

27 **Relative to the Wolf Den:** The distance from the nearest collared caribou to the den decreased from 241 km one week before the excursion to 124 km the day it began. The nearest a collared caribou came to the den was 43 km away, on 29 and 30 July. During the study, four collared caribou were located within 100 km of the den. Each of these four was closest to the wolf on at least one day during the period reported.

28 Discussion

Prey Abundance

29 Caribou are the single most important prey of tundra wolves (Clark, 1971; Kuyt, 1972; Stephenson and James, 1982; Williams, 1990). Caribou range over vast areas, and for part of the summer, they are scarce or absent in wolf home ranges (Heard et al., 1996). Both the long distance between radio-collared caribou and the den the week before the excursion and the increased time spent foraging by wolf 388 indicate that caribou availability near the den was low. Observations of the pups' being left alone for up to 18 h, presumably while adults were searching for food, provide additional support for low caribou availability locally. Mean foraging bout duration decreased by 10.0 h after the excursion, when collared caribou were closer to the den, suggesting an increase in caribou availability nearby.

Foraging Excursion

30 One aspect of central place foraging theory (CPFT) deals with the optimality of returning different-sized food loads from varying distances to dependents at a central place (i.e., the den) (Orians and Pearson, 1979). Carlson (1985) tested CPFT and found that the predator usually consumed prey captured far from the central place, while feeding prey captured nearby to dependants. Wolf 388 spent 7.2 days in one area near caribou before moving to a location 23 km back towards the den, where she spent an additional 3.1 days, likely hunting caribou. She began her return trip from this closer location, traveling directly to the den. While away, she may have made one or more successful kills and spent time meeting her own energetic needs before returning to the den. Alternatively, it may have taken several attempts to make a kill,

200 *P.F. Frame, et al*

which she then fed on before beginning her return trip. We do not know if she returned food to the pups, but such behavior would be supported by CPFT.

31 Other workers have reported wolves' making long round trips and referred to them as "extraterritorial" or "pre-dispersal" forays (Fritts and Mech, 1981; Messier, 1985; Ballard et al., 1997; Merrill and Mech, 2000). These movements are most often made by young wolves (1–3 years old), in areas where annual territories are maintained and prey are relatively sedentary (Fritts and Mech, 1981; Messier, 1985). The long excursion of 388 differs in that tundra wolves do not maintain annual territories (Walton et al., 2001), and the main prey migrate over vast areas (Gunn et al., 2001).

Another difference between 388's excursion and those reported earlier is that she is a mature, breeding female. No study of territorial wolves has reported reproductive adults making extraterritorial movements in summer (Fritts and Mech, 1981; Messier, 1985; Ballard et al., 1997; Merrill and Mech, 2001). However, Walton et al. (2001) also report that breeding female tundra wolves made excursions.

Direction of Movement

32 Possible explanations for the relatively direct route 388 took to the caribou include landscape influence and experience. Considering the timing of 388's trip and the locations of caribou, had the wolf moved northwest, she might have missed the caribou entirely, or the encounter might have been delayed.

A reasonable possibility is that the land directed 388's route. The barrens are crisscrossed with trails worn into the tundra over centuries by hundreds of thousands of caribou and other animals (Kelsall, 1968; Thorpe et al., 2001). At river crossings, lakes, or narrow peninsulas, trails converge and funnel towards and away from caribou calving grounds and summer range. Wolves use trails for travel (Paquet et al., 1996; Mech and Boitani, 2003; P. Frame, pers. observation). Thus, the landscape may direct an animal's movements and lead it to where cues, such as the odor of caribou on the wind or scent marks of other wolves, may lead it to caribou.

33 Another possibility is that 388 knew where to find caribou in summer. Sexually immature tundra wolves sometimes follow caribou to calving grounds (D. Heard, unpubl. data). Possibly, 388 had made such journeys in previous years and killed caribou. If this were the case, then in times of local prey scarcity she might travel to areas where she had hunted successfully before. Continued monitoring of tundra wolves may answer questions about how their food needs are met in times of low caribou abundance near dens.

Caribou often form large groups while moving **34** south to the tree line (Kelsall, 1968). After a large aggregation of caribou moves through an area, its scent can linger for weeks (Thorpe et al., 2001:104). It is conceivable that 388 detected caribou scent on the wind, which was blowing from the northeast on 19–21 July (Environment Canada, 2003), at the same time her excursion began. Many factors, such as odor strength and wind direction and strength, make systematic study of scent detection in wolves difficult under field conditions (Harrington and Asa, 2003). However, humans are able to smell odors such as forest fires or oil refineries more than 100 km away. The olfactory capabilities of dogs, which are similar to wolves, are thought to be 100 to 1 million times that of humans (Harrington and Asa, 2003). Therefore, it is reasonable to think that under the right wind conditions, the scent of many caribou traveling together could be detected by wolves from great distances, thus triggering a long foraging bout.

Rate of Travel

Mech (1994) reported the rate of travel of Arctic **35** wolves on barren ground was 8.7 km/h during regular travel and 10.0 km/h when returning to the den, a difference of 1.3 km/h. These rates are based on direct observation and exclude periods when wolves moved slowly or not at all. Our calculated travel rates are assumed to include periods of slow movement or no movement. However, the pattern we report is similar to that reported by Mech (1994), in that homeward travel was faster than regular travel by 1.6 km/h. The faster rate on return may be explained by the need to return food to the den. Pup survival can increase with the number of adults in a pack available to deliver food to pups (Harrington et al., 1983). Therefore, an increased rate of travel on homeward trips could improve a wolf's reproductive fitness by getting food to pups more quickly.

Fate of 388's Pups

Wolf 388 was caring for pups during den observa- **36** tions. The pups were estimated to be six weeks old, and were seen ranging as far as 800 m from the den. They received some regurgitated food from two of the females, but were unattended for long periods. The excursion started 16 days after our observations, and it is improbable that the pups could have traveled the distance that 388 moved. If the pups died, this would have removed parental responsibility, allowing the long movement.

Our observations and the locations of radio-collared caribou indicate that prey became scarce in

31 Here our authors note other possible explanations for wolves' excursions presented by other investigators, but this study does not seem to support those ideas.

32 Authors discuss possible reasons for why 388 traveled directly to where caribou were located. They take what they learned from earlier studies and apply it to this case, suggesting that the lay of the land played a role. Note that their description paints a clear picture of the landscape.

33 Authors suggest that 388 may have learned in traveling during previous summers where the caribou were. The last two sentences suggest ideas for future studies.

34 Or maybe 388 followed the scent of the caribou. Authors acknowledge difficulties of proving this, but they suggest another area where future studies might be done.

35 Authors suggest that results of this study support previous studies about how fast wolves travel to and from the den. In the last sentence, they speculate on how these observed patterns would fit into the theory of evolution.

36 Authors also speculate on the fate of 388's pups while she was traveling. This leads to . . .

37 Discussion of cooperative rearing of pups and, in turn, to speculation on how this study and what is known about cooperative rearing might fit into the animal's strategies for survival of the species. Again, the authors approach the broader theory of evolution and how it might explain some of their results.

38 And again, they suggest that this study points to several areas where further study will shed some light.

39 In conclusion, the authors suggest that their study supports the hypothesis being tested here. And they touch on the implications of increased human activity on the tundra predicted by their results.

40 ACKNOWLEDGEMENTS: Authors note the support of institutions, companies and individuals. They thank their reviewers and list permits under which their research was carried out.

41 REFERENCES: List of all studies cited in the report. This may seem tedious, but is a vitally important part of scientific reporting. It is a record of the sources of information on which this study is based. It provides readers with a wealth of resources for further reading on this topic. Much of it will form the foundation of future scientific studies like this one.

---

the area of the den as summer progressed. Wolf 388 may have abandoned her pups to seek food for herself. However, she returned to the den after the excursion, where she was seen near pups. In fact, she foraged in a similar pattern before and after the excursion, suggesting that she again was providing for pups after her return to the den.

37 A more likely possibility is that one or both of the other lactating females cared for the pups during 388's absence. The three females at this den were not seen with the pups at the same time. However, two weeks earlier, at a different den, we observed three females cooperatively caring for a group of six pups. At that den, the three lactating females were observed providing food for each other and trading places while nursing pups. Such a situation at the den of 388 could have created conditions that allowed one or more of the lactating females to range far from the den for a period, returning to her parental duties afterwards. However, the pups would have been weaned by eight weeks of age (Packard et al., 1992), so nonlactating adults could also have cared for them, as often happens in wolf packs (Packard et al., 1992; Mech et al., 1999).

Cooperative rearing of multiple litters by a pack could create opportunities for long-distance foraging movements by some reproductive wolves during summer periods of local food scarcity. We have recorded multiple lactating females at one or more tundra wolf dens per year since 1997. This reproduc-

38 tive strategy may be an adaptation to temporally and spatially unpredictable food resources. All of these possibilities require further study, but emphasize both the adaptability of wolves living on the barrens and their dependence on caribou.

Long-range wolf movement in response to caribou

39 availability has been suggested by other researchers (Kuyt, 1972; Walton et al., 2001) and traditional ecological knowledge (Thorpe et al., 2001). Our report demonstrates the rapid and extreme response of wolves to caribou distribution and movements in summer. Increased human activity on the tundra (mining, road building, pipelines, ecotourism) may influence caribou movement patterns and change the interactions between wolves and caribou in the region. Continued monitoring of both species will help us to assess whether the association is being affected adversely by anthropogenic change.

40 **Acknowledgements**

This research was supported by the Department of Resources, Wildlife, and Economic Development, Government of the Northwest Territories; the Department of Biological Sciences at the University of Alberta; the Natural Sciences and Engineering Research Council of Canada; the Department of Indian and Northern Affairs Canada; the Canadian Circumpolar Institute; and DeBeers Canada, Ltd. Lorna Ruechel assisted with den observations. A. Gunn provided caribou location data. We thank Dave Mech for the use of GPS collars. M. Nelson, A. Gunn, and three anonymous reviewers made helpful comments on earlier drafts of the manuscript. This work was done under Wildlife Research Permit – WL002948 issued by the Government of the Northwest Territories, Department of Resources, Wildlife, and Economic Development.

41 **References**

BALLARD, W.B., AYRES, L.A., KRAUSMAN, P.R., REED, D.J., and FANCY, S.G. 1997. Ecology of wolves in relation to a migratory caribou herd in northwest Alaska. Wildlife Monographs 135. 47 p.

BROWN, M.B., and FORSYTHE, A.B. 1974. Robust tests for the equality of variances. Journal of the American Statistical Association 69:364–367.

CARLSON, A. 1985. Central place foraging in the red-backed shrike (*Lanius colturio* L.): Allocation of prey between forager and sedentary consumer. Animal Behaviour 33:664–666.

CLARK, K.R.F. 1971. Food habits and behavior of the tundra wolf on central Baffin Island. Ph.D. Thesis, University of Toronto, Ontario, Canada.

ENVIRONMENT CANADA. 2003. National climate data information archive. Available online: http://www.climate.weatheroffice.ec.gc.ca/Welcome_e.html

FOURNIER, B., and GUNN, A. 1998. Musk ox numbers and distribution in the NWT, 1997. File Report No. 121. Yellowknife: Department of Resources, Wildlife, and Economic Development, Government of the Northwest Territories. 55 p.

FRITTS, S.H., and MECH, L.D. 1981. Dynamics, movements, and feeding ecology of a newly protected wolf population in northwestern Minnesota. Wildlife Monographs 80. 79 p.

GUNN, A., DRAGON, J., and BOULANGER, J. 2001. Seasonal movements of satellite-collared caribou from the Bathurst herd. Final Report to the West Kitikmeot Slave Study Society, Yellowknife, NWT. 80 p. Available online: http://www.wkss.nt.ca/HTML/08_ProjectsReports/PDF/Seasonal MovementsFinal.pdf

HARRINGTON, F.H., and ASA, C.S. 2003. Wolf communication. In: Mech, L.D., and Boitani, L., eds. Wolves: Behavior, ecology, and conservation. Chicago: University of Chicago Press. 66–103.

HARRINGTON, F.H., MECH, L.D., and FRITTS, S.H. 1983. Pack size and wolf pup survival: Their relationship under varying ecological conditions. Behavioral Ecology and Sociobiology 13:19–26.

HARRIS, S.A. 1986. Permafrost distribution, zonation and stability along the eastern ranges of the cordillera of North America. Arctic 39(1):29–38.

HEARD, D.C., and WILLIAMS, T.M. 1992. Distribution of wolf dens on migratory caribou ranges in the Northwest

202 *P.F. Frame, et al*

Territories, Canada. Canadian Journal of Zoology 70:1504–1510.

HEARD, D.C., WILLIAMS, T.M., and MELTON, D.A. 1996. The relationship between food intake and predation risk in migratory caribou and implication to caribou and wolf population dynamics. Rangifer Special Issue No. 2:37–44.

KELSALL, J.P. 1968. The migratory barren-ground caribou of Canada. Canadian Wildlife Service Monograph Series 3. Ottawa: Queen's Printer. 340 p.

KUYT, E. 1962. Movements of young wolves in the Northwest Territories of Canada. Journal of Mammalogy 43:270–271.

———. 1972. Food habits and ecology of wolves on barren-ground caribou range in the Northwest Territories. Canadian Wildlife Service Report Series 21. Ottawa: Information Canada. 36 p.

MECH, L.D. 1994. Regular and homeward travel speeds of Arctic wolves. Journal of Mammalogy 75:741–742.

MECH, L.D., and BOITANI, L. 2003. Wolf social ecology. In: Mech, L.D., and Boitani, L., eds. Wolves: Behavior, ecology, and conservation. Chicago: University of Chicago Press. 1–34.

MECH, L.D., and GESE, E.M. 1992. Field testing the Wildlink capture collar on wolves. Wildlife Society Bulletin 20:249–256.

MECH, L.D., WOLFE, P., and PACKARD, J.M. 1999. Regurgitative food transfer among wild wolves. Canadian Journal of Zoology 77:1192–1195.

MERRILL, S.B., and MECH, L.D. 2000. Details of extensive movements by Minnesota wolves (Canis lupus). American Midland Naturalist 144:428–433.

MERRILL, S.B., ADAMS, L.G., NELSON, M.E., and MECH, L.D. 1998. Testing releasable GPS radiocollars on wolves and white-tailed deer. Wildlife Society Bulletin 26:830–835.

MESSIER, F. 1985. Solitary living and extraterritorial movements of wolves in relation to social status and prey abundance. Canadian Journal of Zoology 63:239–245.

MOHR, C.O., and STUMPF, W.A. 1966. Comparison of methods for calculating areas of animal activity. Journal of Wildlife Management 30:293–304.

MUSIANI, M. 2003. Conservation biology and management of wolves and wolf-human conflicts in western North America. Ph.D. Thesis, University of Calgary, Calgary, Alberta, Canada.

ORIANS, G.H., and PEARSON, N.E. 1979. On the theory of central place foraging. In: Mitchell, R.D., and Stairs, G.F., eds. Analysis of ecological systems. Columbus: Ohio State University Press. 154–177.

PACKARD, J.M., MECH, L.D., and REAM, R.R. 1992. Weaning in an arctic wolf pack: Behavioral mechanisms. Canadian Journal of Zoology 70:1269–1275.

PAQUET, P.C., WIERZCHOWSKI, J., and CALLAGHAN, C. 1996. Summary report on the effects of human activity on gray wolves in the Bow River Valley, Banff National Park, Alberta. In: Green, J., Pacas, C., Bayley, S., and Cornwell, L., eds. A cumulative effects assessment and futures outlook for the Banff Bow Valley. Prepared for the Banff Bow Valley Study. Ottawa: Department of Canadian Heritage.

STEPHENSON, R.O., and JAMES, D. 1982. Wolf movements and food habits in northwest Alaska. In: Harrington, F.H., and Paquet, P.C., eds. Wolves of the world. New Jersey: Noyes Publications. 223–237.

THORPE, N., EYEGETOK, S., HAKONGAK, N., and QITIRMIUT ELDERS. 2001. The Tuktu and Nogak Project: A caribou chronicle. Final Report to the West Kitikmeot/Slave Study Society, Ikaluktuuttiak, NWT. 160 p.

TIMONEY, K.P., LA ROI, G.H., ZOLTAI, S.C., and ROBINSON, A.L. 1992. The high subarctic forest-tundra of northwestern Canada: Position, width, and vegetation gradients in relation to climate. Arctic 45(1):1–9.

WALTON, L.R., CLUFF, H.D., PAQUET, P.C., and RAMSAY, M.A. 2001. Movement patterns of barren-ground wolves in the central Canadian Arctic. Journal of Mammalogy 82:867–876.

WILLIAMS, T.M. 1990. Summer diet and behavior of wolves denning on barren-ground caribou range in the Northwest Territories, Canada. M.Sc. Thesis, University of Alberta, Edmonton, Alberta, Canada.

ZAR, J.H. 1999. Biostatistical analysis. 4th ed. New Jersey: Prentice Hall. 663 p.

Glossary

3′ end The end of a polynucleotide chain at which a hydroxyl group is bonded to the 3′ carbon of a deoxyribose sugar.

5′ cap In eukaryotes, a guanine-containing nucleotide attached in a reverse orientation to the 5′ end of pre-mRNA and retained in the mRNA produced from it. The 5′ cap on an mRNA is the site where ribosomes attach to initiate translation.

5′ end The end of a polynucleotide chain at which a phosphate group is bound to the 5′ carbon of a deoxyribose sugar.

10-nm chromatin fiber The most fundamental level of chromatin packing of a eukaryotic chromosome in which DNA winds for almost two turns around an eight-protein nucleosome core particle to form a nucleosome and linker DNA extends between adjacent nucleosomes. The result is a beads-on-a-string type of structure with a 10-nm diameter.

30-nm chromatin fiber Level of chromatin packing of a eukaryotic chromosome in which histone H1 binds to the 10-nm chromatin fiber causing it to package into a coiled structure about 30 nm in diameter and with about six nucleosomes per turn. Also referred to as a *solenoid*.

A site The site where the incoming aminoacyl-tRNA carrying the next amino acid to be added to the polypeptide chain binds to the mRNA.

abdomen The region of the body that contains much of the digestive tract and sometimes part of the reproductive system; in insects, the region behind the thorax.

abiotic Nonbiological, often in reference to physical factors in the environment.

abscisic acid (ABA) A plant hormone involved in the abscission of leaves, flowers, and fruits, dormancy of buds and seeds, and closing of stomata.

abscission In plants, the dropping of flowers, fruits, and leaves in response to environmental signals.

absorption spectrum Curve representing the amount of light absorbed at each wavelength.

abyssal zone The bottom sediments that lie permanently below deep ocean water.

accommodation A process by which the lens changes to enable the eye to focus on objects at different distances.

acid Proton donor that releases H+ (and anions) when dissolved in water.

acid precipitation Rainfall with low pH, primarily created when gaseous sulfur dioxide (SO_2) dissolves in water vapor in the atmosphere, forming sulfuric acid.

acid-growth hypothesis A hypothesis to explain how the hormone auxin promotes growth of plant cells; it suggests that auxin stimulates H+ pumps in the plasma membrane to move H+ from the cell interior into the cell wall, which increases wall acidity, making the wall expandable.

acidity The concentration of H+ in a water solution, as compared with the concentration of OH−.

acoelomate A body plan of bilaterally symmetrical animals that lack a body cavity between the gut and the body wall.

acoustical signaling A means of animal communication in which a signaler produces a sound that is heard by a signal receiver.

acquired immune deficiency syndrome (AIDS) A constellation of disorders that follows infection by the HIV virus.

acrosome A specialized secretory vesicle on the head of an animal sperm, which helps the sperm penetrate the egg.

acrosome reaction The process in which enzymes contained in the acrosome are released from an animal sperm and digest a path through the egg coats.

action potential The abrupt and transient change in membrane potential that occurs when a neuron conducts an electrical impulse.

action spectrum Graph produced by plotting the effectiveness of light at each wavelength in driving photosynthesis.

activation energy The initial input of energy required to start a reaction.

activator A regulatory protein that controls the expression of one or more genes.

active immunity The production of antibodies in the body in response to exposure to a foreign antigen.

active parental care Parents' investment of time and energy in caring for offspring after they are born or hatched.

active site The region of an enzyme that recognizes and combines with a substrate molecule.

active transport The mechanism by which ions and molecules move against the concentration gradient across a membrane, from the side with the lower concentration to the side with the higher concentration.

adaptation Characteristic that helps an organism survive longer or reproduce more under a particular set of environmental conditions.

adaptation, evolutionary The accumulation of adaptive traits over time.

adaptation, sensory *See* sensory adaptation.

adaptive (acquired) immunity A specific line of defense against invasion of the body in which individual pathogens are recognized and attacked to neutralize and eliminate them.

adaptive radiation A cluster of closely related species that are each adaptively specialized to a specific habitat or food source.

adaptive trait A genetically based characteristic, preserved by natural selection, that increases an organism's likelihood of survival or its reproductive output.

adaptive zone A part of a habitat that may be occupied by a group of species exploiting the same resources in a similar manner.

adductor muscle A muscle that pulls inward toward the median line of the body; in bivalve mollusks, it pulls the shell closed.

adenine A purine that base-pairs with either thymine in DNA or uracil in RNA.

adherens junction Animal cell junction in which intermediate filaments are the anchoring cytoskeletal component.

adhesion The adherence of molecules to the walls of conducting tubes, as in plants.

adiabatic cooling A decrease in temperature without the actual loss of heat energy, occurring in air masses that expand as they rise in the atmosphere.

adipose tissue Connective tissue containing large, densely clustered cells called adipocytes that are specialized for fat storage.

adrenal cortex The outer region of the adrenal glands, which contains endocrine cells that secrete two major types of steroid hormones, the glucocorticoids and the mineralocorticoids.

adrenal medulla The central region of the adrenal glands, which contains neurosecretory neurons that secrete the catecholamine hormones epinephrine and norepinephrine.

adrenocorticotropic hormone (ACTH) A hormone that triggers hormone secretion by cells in the adrenal cortex.

adventitious root A root that develops from the stem or leaves of a plant.

aerobe An organism that requires oxygen for cellular respiration.

afferent arteriole The vessel that delivers blood to the glomerulus of the kidney.

afferent neuron A neuron that transmits stimuli collected by a sensory receptor to an interneuron.

African emergence hypothesis An hypothesis proposing that modern humans first evolved in Africa and then dispersed to other continents.

agar A gelatinous product extracted from certain red algae or seaweed used as a culture medium in the laboratory and as a gelling or stabilizing agent in foods.

agarose gel electrophoresis Technique by which DNA, RNA, or protein molecules are separated in a gel subjected to an electric field.

age structure A statistical description or graph of the relative numbers of individuals in each age class in a population.

age-specific fecundity The average number of offspring produced by surviving females of a particular age.

age-specific mortality The proportion of individuals alive at the start of an age interval that died during that age interval.

age-specific survivorship The proportion of individuals alive at the start of an age interval that survived until the start of the next age interval.

aggregate fruit A fruit that develops from multiple separate carpels of a single flower, such as a raspberry or strawberry.

agonist A muscle that causes movement in a joint when it contracts.

albumin The most abundant protein in blood plasma, important for osmotic balance and pH buffering; also, the portion of an egg that serves as the main source of nutrients and water for the embryo.

alcohol A molecule of the form R—OH in which R is a chain of one or more carbon atoms, each of which is linked to hydrogen atoms.

alcoholic fermentation Reaction in which pyruvate is converted into ethyl alcohol and CO_2 in a two-step series that also converts NADH into NAD^+.

aldehyde Molecule in which the carbonyl group is linked to a carbon atom at the end of a carbon chain, along with a hydrogen atom.

aldosterone A mineralocorticoid hormone released from the adrenal cortex that increases the amount of Na^+ reabsorbed from the urine in the kidneys and absorbed from foods in the intestine, reduces the amount of Na^+ secreted by salivary and sweat glands, and increases the rate of K^+ excretion by the kidneys, keeping Na^+ and K^+ balanced at the levels required for normal cellular function.

aleurone The thin layer of cells that separates the endosperm of a seed from the pericarp.

algin Alginic acid, found in the cell walls of brown algae.

allantois In an amniote egg, an extraembryonic membrane sac that fills much of the space between the chorion and the yolk sac and store's the embryo's nitrogenous wastes.

allele One of two or more versions of a gene.

allele frequency The abundance of one allele relative to others at the same gene locus in individuals of a population.

allergen A type of antigen responsible for allergic reactions, which induces B cells to secrete an overabundance of IgE antibodies.

allometric growth A pattern of postembryonic development in which parts of the same organism grow at different rates.

allopatric speciation The evolution of reproductive isolating mechanisms between two populations that are geographically separated.

allopolyploidy The genetic condition of having two or more complete sets of chromosomes from different parent species.

all-or-nothing principle The principle that an action potential is produced only if the stimulus is strong enough to cause depolarization to reach the threshold.

allosteric activator Molecule that converts an enzyme with an allosteric site, a regulatory site outside the active site, from the inactive form to the active form.

allosteric inhibitor Molecule that converts an enzyme with an allosteric site, a regulatory site outside the active site, from the active form to the inactive form.

allosteric regulation Specialized control mechanism for enzymes with an allosteric site, a regulatory site outside the active site, that may either slow or accelerate activity depending on the enzyme.

allosteric site A regulatory site outside the active site.

alpine tundra A biome that occurs on high mountaintops throughout the world, in which dominant plants form cushions and mats.

alternation of generations The regular alternation of mode of reproduction in the life cycle of an organism, such as the alternation between diploid (sporophyte) and haploid (gametophyte) phases in plants.

alternative hypothesis An explanation of an observed phenomenon that is different from the explanation being tested.

alternative splicing Mechanism that joins exons in different combinations to produce different mRNAs from a single gene.

altruism A behavioral phenomenon in which individuals appear to sacrifice their own reproductive success to help other individuals.

alveolus (plural, alveoli) One of the millions of tiny air pockets in mammalian lungs, each surrounded by dense capillary networks.

amacrine cell A type of neuron that forms lateral connections in the retina of the eye, connecting bipolar cells and ganglion cells.

amino acid A molecule that contains both an amino and a carboxyl group.

amino group Group that acts as an organic base, consisting of a nitrogen atom bonded on one side to two hydrogen atoms and on the other side to a carbon chain.

aminoacylation The process of adding an amino acid to a tRNA. Also referred to as *charging*.

aminoacyl-tRNA A tRNA linked to its "correct" amino acid, which is the finished product of charging.

aminoacyl-tRNA synthetase An enzyme that catalyzes aminoacylation.

ammonification A metabolic process in which bacteria and fungi convert organic nitrogen compounds into ammonia and ammonium ions; part of the nitrogen cycle.

amniocentesis Technique of prenatal diagnosis in which cells are obtained from the amniotic fluid.

amnion In an amniote egg, an extraembryonic membrane that encloses the embryo, forming the amniotic cavity and secreting amniotic fluid, which provides an aquatic environment in which the embryo develops.

Amniota The monophyletic group of vertebrates that have an amnion during embryonic development.

amniote egg A shelled egg that can survive and develop on land.

amplification An increase in the magnitude of each step as a signal transduction pathway proceeds.

amygdala A gray-matter center of the brain that works as a switchboard, routing information about experiences that have an emotional component through the limbic system.

amyloplast Colorless plastid that stores starch in plants.

anabolic reaction Metabolic reaction that requires energy to assemble simple substances into more complex molecules.

anabolic steroid A steroid hormone that stimulates muscle development.

anaerobe An organism that does not require oxygen to live.

anagenesis The slow accumulation of evolutionary changes in a lineage over time.

anaphase The phase of mitosis during which the spindle separates sister chromatids and pulls them to opposite spindle poles.

anaphylactic shock A severe inflammation stimulated by an allergen, involving extreme swelling of air passages in the lungs that interferes with breathing, and massive leakage of fluid from capillaries that causes blood pressure to drop precipitously.

anapsid A member of the group of amniote vertebrates having no temporal arches and no spaces on the sides of the skull.

anatomy The study of the structures of organisms.

ancestral character A trait that was present in a distant common ancestor.

anchoring junction Cell junction that forms belts that run entirely around cells, "welding" adjacent cells together.

androgen One of a family of hormones that promote the development and maintenance of sex characteristics.

aneuploid An individual with extra or missing chromosomes.

angiosperm A flowering plant. Its egg-containing ovules mature into seeds within protected chambers called ovaries.

angiotensin A peptide hormone that raises blood pressure quickly by constricting arterioles in most parts of the body; it also stimulates release of the steroid hormone aldosterone.

animal behavior The responses of animals to specific internal and external stimuli.

animal pole The end of the egg where the egg nucleus is located, which typically gives rise to surface structures and the anterior end of the embryo.

Animalia The taxonomic kingdom that includes all living and extinct animals.

anion A negatively charged ion.

annual An herbaceous plant that completes its life cycle in one growing season and then dies.

annulus In ferns, a ring of thick-walled cells that nearly encircles the sporangium and functions in spore release.

antagonistic pair Two skeletal muscles, one of which flexes as the other extends to move joints.

antenna A chemosensory appendage attached to the head of some adult arthropods.

antenna complex (light-harvesting complex) In photosystems, the sites at which light is absorbed and converted into chemical energy during photosynthesis, an aggregate of many chlorophyll pigments and a number of carotenoid pigments that serves as the primary site of absorbing light energy in the form of photons.

anterior Indicating the head end of an animal.

anterior pituitary The glandular part of the pituitary, composed of endocrine cells that synthesize and secrete several tropic and nontropic hormones.

anther The pollen-bearing part of a stamen.

antheridium (plural, antheridia) In plants, a structure in which sperm are produced.

Anthocerophyta The phylum comprising hornworts.

Anthophyta The phylum comprising flowering plants.

Anthropoidea The monophyletic lineage of primates comprising monkeys, apes, and humans.

antibody A highly specific soluble protein molecule that circulates in the blood and lymph, recognizing and binding to antigens and clearing them from the body.

antibody-mediated immunity Adaptive immune response in which plasma cells secrete antibodies.

anticodon The three-nucleotide segment in tRNAs that pairs with a codon in mRNAs.

antidiuretic hormone (ADH) A hormone secreted by the posterior pituitary that increases water absorption in the kidneys, thereby increasing the volume of the blood.

antigen A foreign molecule that triggers an adaptive immunity response.

antigen-presenting cell (APC) A cell that presents an antigen to T cells in antibody-mediated immunity and cell-mediated immunity.

antiparallel Strands of DNA that run in opposite directions.

antiport A secondary active transport mechanism in which a molecule moves through a membrane channel into a cell and powers the active transport of a second molecule out of the cell. Also referred to as *exchange diffusion*.

aorta A large artery from the heart that branches into arteries leading to all body regions except the lungs.

aortic body One of several small clusters of chemoreceptors, baroreceptors, and supporting cells located along the aortic arch, that measures changes in blood pressure and the composition of arterial blood flowing past it.

aphotic zone Deeper water of a lake or ocean where sunlight does not penetrate.

apical dominance Inhibition of the growth of lateral buds in plants due to auxin diffusing down a shoot tip from the terminal bud.

apical meristem A region of unspecialized dividing cells at shoot tips and root tips of a plant.

apomixis In plants, the production of offspring without meiosis or formation of gametes.

apoplastic pathway The route followed by water moving through plant cell walls and intercellular spaces (the apoplast). *Compare* symplastic pathway.

apoptosis Programmed cell death.

aposematic coloration Bright, contrasting patterns that advertise the unpalatability of poisonous or repellant species.

appendicular skeleton The bones comprising the pectoral (shoulder) and pelvic (hip) girdles and limbs of a vertebrate.

appendix A fingerlike sac that extends from the cecum of the large intestine.

applied research Research conducted with the goal of solving specific practical problems.

aquaporin A specialized protein channel that facilitates diffusion of water through cell membranes.

aquatic succession A process in which debris from rivers and runoff accumulates in a body of fresh water, causing it to fill in at the margins.

aqueous humor A clear fluid that fills the space between the cornea and lens of the eye.

Archaea One of two domains of prokaryotes; archaeans have some unique molecular and biochemical traits, but they also share some traits with Bacteria and other traits with Eukarya.

archegonium The flask-shaped structure in which bryophyte eggs form.

archenteron The central endoderm-lined cavity of an embryo at the gastrula stage, which forms the primitive gut.

Archosauromorpha A diverse group of diapsids that comprises crocodilians, pterosaurs, and dinosaurs (including birds).

arctic tundra A treeless biome that stretches from the boreal forests to the polar ice cap in Europe, Asia, and North America.

arteriole A branch from a small artery at the point where it reaches the organ it supplies.

artery A vessel that conducts blood away from the heart at relatively high pressure.

artificial selection Selective breeding of animals or plants to ensure that certain desirable traits appear at higher frequency in successive generations.

ascocarp A reproductive body that bears or contains asci.

ascus (plural, asci) A saclike cell in ascomycetes (sac fungi) in which meiosis gives rise to haploid sexual spores (meiospores).

asexual reproduction Any mode of reproduction in which a single individual gives rise to offspring without fusion of gametes; that is, without genetic input from another individual. See also *vegetative reproduction*.

association area One of several areas surrounding the sensory and motor areas of the cerebral cortex that integrate information from the sensory areas, formulate responses, and pass them on to the primary motor area.

aster Radiating array produced as microtubules extending from the centrosomes of cells grow in length and extent.

astrocyte A star-shaped glial cell that provides support to neurons in the vertebrate central nervous system.

asymmetrical Characterized by lack of proportion in the spatial arrangement or placement of parts.

atmosphere The component of the biosphere that includes the gases and airborne particles enveloping the planet.

atom The smallest unit that retains the chemical and physical properties of an element.

atomic nucleus The nucleus of an atom, containing protons and neutrons.

atomic number The number of protons in the nucleus of an atom.

atomic weight The weight of an element in grams, equal to the mass number.

ATP (adenosine triphosphate) The primary agent that couples exergonic and endergonic reactions.

ATP synthase A membrane-spanning protein complex that couples the energetically favorable transport of protons across a membrane to the synthesis of ATP.

atrial natriuretic factor (ANF) A peptide hormone that inhibits renin release and increases the filtration rate by dilating the arterioles that deliver blood to glomeruli and by inhibiting aldosterone release.

atrial siphon A tube through which invertebrate chordates expel digestive and metabolic wastes.

atriopore The hole in the body wall of a cephalochordate through which water is expelled from the body.

atrioventricular node (AV node) A region of the heart wall that receives signals from the sinoatrial node and conducts them to the ventricle.

atrioventricular valve (AV valve) A valve composed of endocardium and connective tissue between each atrium and ventricle, which prevents backflow of blood from the ventricle to the atrium during emptying of the heart.

atrium (plural, atria) A body cavity or chamber surrounding the perforated pharynx of invertebrate chordates; also, one of the chambers that receive blood returning to the heart.

autoimmune reaction The production of antibodies against molecules of the body.

autonomic nervous system A subdivision of the peripheral nervous system that controls largely involuntary processes including digestion, secretion by sweat glands, circulation of the blood, many functions of the reproductive and excretory systems, and contraction of smooth muscles in all parts of the body.

autopolyploidy The genetic condition of having more than two sets of chromosomes from the same parent species.

autosomal dominant inheritance Pattern in which the allele that causes a trait is dominant, and only homozygous recessives are unaffected.

autosomal recessive inheritance Pattern in which individuals with a trait are homozygous for a recessive allele.

autosome Chromosome other than a sex chromosome.

autotroph An organism that produces its own food using CO_2 and other simple inorganic compounds from its environment and energy from the sun or from oxidation of inorganic substances.

autumn overturn A process in which winds mix the water in a lake vertically, equalizing the concentrations of dissolved gases and nutrients at all depths.

auxin Any of a family of plant hormones that stimulate growth by promoting cell elongation in stems and coleoptiles; inhibit abscission; govern responses to light and gravity, and have other developmental effects.

Avogadro's number The number 6.022×10^{23}, derived by dividing the atomic weight of any element by the weight of an atom of that element.

Avr gene A gene in certain plant pathogens that encodes a product triggering a defensive response in the plant.

axial skeleton The bones comprising the head and trunk of a vertebrate: the cranium, vertebral column, ribs, and sternum (breastbone).

axil The upper angle between the stem and an attached leaf.

axon The single elongated extension of a neuron that conducts signals away from the cell body to another neuron or an effector.

axon hillock A junction with the cell body of a neuron from which the axon arises.

axon terminal A branch at the tip of an axon that ends as a small, buttonlike swelling.

B cell A lymphocyte that recognizes antigens in the body.

bacillus (plural, bacilli) A cylindrical or rod-shaped prokaryote.

background extinction rate The average rate of extinction of taxa through time.

bacteria One of the two domains of prokaryotes; collectively, bacteria are the most metabolically diverse organisms.

bacterial chromosome DNA molecule in bacteria in which hereditary information is encoded.

bacteriophage A virus that infects bacteria. Also referred to as a *phage*.

bacteroid A rod-shaped or branched bacterium in the root nodules of nitrogen-fixing plants.

balanced polymorphism The maintenance of two or more phenotypes in fairly stable proportions over many generations.

bark The tough outer covering of woody stems and roots, composed of all the living and nonliving tissues between the vascular cambium and the stem surface.

Barr body The inactive, condensed X chromosome seen in the nucleus of female mammals.

basal angiosperm Any of the earliest branches of the flowering plant lineage; includes the star anise group and water lilies.

basal body Structure that anchors cilia and flagella to the surface of a cell.

basal lamina The membrane that fixes the epithelium to underlying tissues (also called the basement membrane).

basal nucleus One of several gray-matter centers that surround the thalamus on both sides of the brain and moderate voluntary movements directed by motor centers in the cerebrum.

base Proton acceptor that reduces the H^+ concentration of a solution.

base-pair mismatch An error in the assembly of a new nucleotide chain in which bases other than the correct ones pair together.

base-pair substitution mutation A particular mutation involving a change from one base pair to another in DNA.

basic research Research conducted to search for explanations about natural phenomena in order to satisfy curiosity and to advance collective knowledge of living systems.

basidiocarp A fruiting body of a basidiomycete; mushrooms are examples.

basidiospore A haploid sexual spore produced by basidiomycete fungi.

basidium (plural, basidia) A small, club-shaped structure in which sexual spores of basidiomycetes arise.

basophil A type of leukocyte that is induced to secrete histamine by allergens.

Batesian mimicry The form of defense in which a palatable or harmless species resembles an unpalatable or poisonous one.

B-cell receptor (BCR) The receptor on B cells that is specific for a particular antigen.

behavioral isolation A prezygotic reproductive isolating mechanism in which two species do not mate because of differences in courtship behavior; also known as ethological isolation.

behavioral repertoire The set of actions that an animal can perform in response to stimuli in its environment.

benthic province The bottom sediments in the ocean.

benthos Species living in and on the bottom sediments of the ocean.

beta (β) sheet A type of primary structure in a polypeptide in which the amino acid chain zigzags in a flat plane to form a beta strand, and beta strands then align side by side in the same or opposite direction.

biennial A plant that completes its life cycle in two growing seasons and then dies; limited secondary growth occurs in some biennials.

bilateral symmetry The body plan of animals in which the body can be divided into mirror image right and left halves by a plane passing through the midline of the body.

bilayer A membrane with two molecular layers.

bile A mixture of substances including bile salts, cholesterol, and bilirubin that is made in the liver, stored in the gallbladder, and used in the digestion of fats.

binomial Relating to or consisting of two names or terms.

binomial nomenclature The naming of species with a two part scientific name, the first indicating the genus and the second indicating the species.

biodiversity The richness of living systems as reflected in genetic variability within and among species, the number of species living on Earth, and the variety of communities and ecosystems.

biodiversity hotspot An area where biodiversity is both highly concentrated and endangered.

biofilm A microbial community consisting of a complex aggregation of microorganisms attached to a surface.

biogeochemical cycle Any of several global processes in which a nutrient circulates between the abiotic environment and living organisms.

biogeographical realm A major region of Earth that is occupied by distinct evolutionary lineages of plants and animals.

biogeography The study of the geographical distributions of plants and animals.

bioinformatics Field that fuses biology with mathematics and computer science that is used for the analysis of genome sequences.

biological clock An internal time-measuring mechanism that adapts an organism to recurring environmental changes.

biological evolution The process by which some individuals in a population experience changes in their DNA and pass those modified instructions to their offspring.

biological lineage An evolutionary sequence of ancestral organisms and their descendants.

biological magnification The increasing concentration of nondegradable poisons in the tissues of animals at higher trophic levels.

biological research The collective effort of individuals who have worked to understand how living systems function.

biological species concept The definition of species based upon the ability of populations to interbreed and produce fertile offspring.

bioluminescent An organism that glows or releases a flash of light, particularly when disturbed.

biomass The dry weight of biological material per unit area or volume of habitat.

biome A large scale vegetation type and its associated microorganisms, fungi, and animals.

bioremediation Applications of chemical and biological knowledge to decontaminate polluted environments.

biosphere All regions of Earth's crust, waters, and atmosphere that sustain life.

biota The total collection of organisms in a geographic region.

biotechnology The manipulation of living organisms to produce useful products.

biotic Biological, often in reference to living components of the environment.

bipedalism The habit in animals of walking upright on two legs.

bipolar cell A type of neuron in the retina of the eye that connects the rods and cones with the ganglion cells.

blade The expanded part of a leaf that provides a large surface area for absorbing sunlight and carbon dioxide.

blastocoel A fluid filled cavity in the blastula embryo.

blastocyst An embryonic stage in mammals; a single-cell-layered hollow ball of about 120 cells with a fluid-filled blastocoel in which a dense mass of cells is localized to one side.

blastodisc A disclike layer of cells at the surface of the yolk produced by early cleavage divisions.

blastomere A small cell formed during cleavage of the embryo.

blastopore The opening at one end of the archenteron in the gastrula that gives rise to the mouth in protostomes and the anus in deuterostomes.

blastula The hollow ball of cells that is the result of cleavage divisions in an early embryo.

blending theory of inheritance Theory suggesting that hereditary traits blend evenly in offspring through mixing of the blood of the two parents.

blood A fluid connective tissue composed of blood cells suspended in a fluid extracellular matrix, plasma.

blood-brain barrier A specialize arrangement of capillaries in the brain that prevents most substances dissolved in the blood from entering the cerebrospinal fluid and thus protects the brain and spinal cord from viruses, bacteria, and toxic substances that may circulate in the blood.

bolting Rapid formation of a floral shoot in plant species that form rosettes, such as lettuce.

bolus The food mass after chewing.

bone The densest form of connective tissue, in which living cells secrete the mineralized matrix of collagen and calcium salts that surrounds them; forms the skeleton.

book lungs Pocketlike respiratory organs found in some arachnids consisting of several parallel membrane folds arranged like the pages of a book.

boreal forest A biome that is a circumpolar expanse of evergreen coniferous trees in Europe, Asia, and North America.

Bowman's capsule An infolded region at the proximal end of a nephron that cups around the glomerulus and collects the water and solutes filtered out of the blood.

brachiation A pattern of locomotion among primates in which an individual swings below branches from one handhold to another.

brain A single, organized collection of nervous tissue in an organism's head that forms the control center of the nervous system and major sensory structures.

brain hormone (BH) A peptide hormone secreted by neurosecretory neurons in the brain of insects.

brain stem A stalklike structure formed by the pons and medulla, along with the midbrain, which connects the forebrain with the spinal cord.

brassinosteroid Any of a family of plant hormones that stimulate cell division and elongation and differentiation of vascular tissue.

breathing The exchange of gases with the respiratory medium by animals.

bronchiole One of the small, branching airways in the lungs that lead into the alveoli.

bronchus (plural, bronchi) An airway that leads from the trachea to the lungs.

brown adipose tissue A specialized tissue in which the most intense heat generation by nonshivering thermogenesis takes place.

Bryophyta The phylum of nonvascular plants, including mosses and their relatives.

bryophyte A general term for plants (such as mosses) that lack internal transport vessels.

budding A mode of asexual reproduction in which a new individual grows and develops while attached to the parent.

buffer Substance that compensates for pH changes by absorbing or releasing H^+.

bulbourethral gland One of two pea sized glands on either side of the prostate gland, which secrete a mucous fluid that is added to semen.

bulk feeder An animal that consumes size-able food items whole or in large chunks.

bulk flow The group movement of molecules in response to a difference in pressure between two locations.

bulk-phase endocytosis Mechanism by which extracellular water is taken into a cell together with any molecules that happen to be in solution in the water. Also referred to as *pinocytosis*.

C₄ cycle A reaction series that allows CO_2 to be fixed by a carboxylase that is unaffected by high oxygen concentrations.

Ca²⁺ pump (calcium pump) Pump that pushes Ca^{2+} from the cytoplasm to the cell exterior, and also from the cytosol into the vesicles of the endoplasmic reticulum.

cadherin A cell surface protein responsible for selective cell adhesions that require calcium ions to set up adhesions.

calcitonin A nontropic peptide hormone that lowers the level of Ca^{2+} in the blood by inhibiting the ongoing dissolution of calcium from bone.

callus An undifferentiated tissue that develops on or around a cut plant surface or in tissue culture.

calorie (cal) The amount of heat required to raise 1 g of water by 1°C, known as a "small" calorie; when capitalized, a unit equal to 1000 small calories.

Calvin cycle *See* light-independent reaction.

calyx The outermost whorl of a flower, made up of sepals; early in the development of a flower, it encloses all the other parts, as in an unopened bud.

CAM plant C₄ plant that runs the Calvin and C₄ cycles at different times to circumvent photorespiration. CAM stands for "crassu-lacean acid metabolism."

canines Pointed, conical teeth of a mammal, located between the incisors and the first premolars, that are specialized for biting and piercing.

capillary The smallest diameter blood vessel, with a wall that is one cell thick, which forms highly branched networks well adapted for diffusion of substances.

capsid *See* coat.

capsule An external layer of sticky or slimy polysaccharides coating the cell wall in many prokaryotes.

carapace A protective outer covering that extends backward behind the head on the dorsal side of an animal, such as the shell of a turtle or lobster.

carbon cycle The global circulation of carbon atoms, especially via the processes of photo-synthesis and respiration.

carbonyl group The reactive part of aldehydes and ketones, consisting of an oxygen atom linked to a carbon atom by a double bond.

carboxyl group The characteristic functional group of organic acids, formed by the combination of carbonyl and hydroxyl groups.

cardiac cycle The systole-diastole sequence of the heart.

cardiac muscle The contractile tissue of the heart.

carnivore An animal that primarily eats other animals.

carotenoid Molecule of yellow-orange pigment by which light is absorbed in photosynthesis.

carotid body A small cluster of chemoreceptors and supporting cells located near the bifurcation of the carotid artery that measures changes in the composition of arterial blood flowing through it.

carpel The reproductive organ of a flower that houses an ovule and its associated structures.

carrageenan A chemical extracted from the red alga *Eucheuma* that is used to thicken and stabilize paints, dairy products such as pudding and ice cream, and many other creams and emulsions.

carrier An individual who carries a mutant allele and could pass it on to offspring, but does not display its symptoms.

carrier protein Transport protein that binds a specific single solute and transports it across the lipid bilayer.

carrying capacity The maximum size of a population that an environment can support indefinitely.

Cartagena Protocol on Biosafety An international agreement that promotes biosafety as it relates to genetically modified organisms.

cartilage A tissue composed of sparsely distributed chondrocytes surrounded by networks of collagen fibers embedded in a tough but elastic matrix of the glycoprotein.

Casparian strip A thin, waxy impermeable band that seals abutting cell walls in roots; the strip helps control the type and amount of solutes that enter the stele by blocking the apoplastic pathway at the endodermis and forcing substances to pass through cells (the symplast).

catabolic reaction Cellular reaction that breaks down complex molecules such as sugar to make their energy available for cellular work.

catalyst Substance with the ability to accelerate a spontaneous reaction without being changed by the reaction.

catastrophism The theory that Earth has been affected by sudden, violent events that were sometimes worldwide in scope.

catecholamine Any of a class of compounds derived from the amino acid tyrosine that circulates in the bloodstream, including epinephrine and norepinephrine.

cation A positively charged ion.

cation exchange Replacement of one cation with another, as on a soil particle.

CD4⁺ T cell A type of T cell in the lymphatic system that has CD4 receptors on its surface. This type of T cell binds to an antigen-presenting cell in antibody-mediated immunity.

CD8⁺ T cell A type of T cell in the lymphatic system that has CD8 receptors on its surface. This type of T cell binds to an antigen-presenting cell in cell-mediated immunity.

cDNA library The entire collection of cloned cDNAs made from the mRNAs isolated from a cell.

cecum A a blind pouch formed at the junction of the large and small intestine.

cell Smallest unit with the capacity to live and reproduce.

cell adhesion molecule A cell surface protein responsible for selectively binding cells together.

cell adhesion protein Protein that binds cells together by recognizing and binding receptors or chemical groups on other cells or on the extracellular matrix.

cell body The portion of the neuron containing genetic material and cellular organelles.

cell culture A living cell grown in a laboratory vessel.

cell cycle The sequence of events during which a cell experiences a period of growth followed by nuclear division and cytokinesis.

cell differentiation A process in which changes in gene expression establish cells with specialized structure and function.

cell expansion A mechanism that enlarges the cells in specific directions in a developing organ.

cell fractionation Technique that divides cells into fractions containing a single type of organelle.

cell junction Junction that seals the spaces between cells and provides direct communication between cells.

cell lineage Cell derivation from the undifferentiated tissues of the embryo.

cell plate In cytokinesis in plants, a new cell wall that forms between the daughter nuclei and grows laterally until it divides the cytoplasm.

cell theory Three generalizations yielded by microscopic observations: all organisms are composed of one or more cells; the cell is the smallest unit that has the properties of life; and cells arise only from the growth and division of preexisting cells.

cell wall A rigid external layer of material surrounding the plasma membrane of cells in plants, fungi, bacteria, and some protists, providing cell protection and support.

cell-mediated immunity An adaptive immune response in which a subclass of T cells—cytotoxic T cells—becomes activated and, with other cells of the immune system, attacks host cells infected by pathogens, particularly those infected by a virus.

cellular respiration The process by which energy-rich molecules are broken down to produce energy in the form of ATP.

cellular slime mold Any of a variety of primitive organisms of the phylum Acrasiomycota, especially of the genus *Dictyostelium;* the life cycle is characterized by a slimelike amoeboid stage and a multicellular reproductive stage.

cellulose One of the primary constituents of plant cell walls, formed by chains of carbohydrate subunits.

centimorgan *See* map unit.

central canal The central portion of the vertebral column in which the spinal cord is found.

central nervous system (CNS) One of the two major divisions of the nervous system containing the brain and spinal cord.

central vacuole A large, water-filled organelle in plant cells that maintains the turgor of the cell and controls movement of molecules between the cytosol and sap.

centriole A cylindrical structure consisting of nine triplets of microtubules in the centrosomes of most animal cells.

centromere A specialized chromosomal region that connects sister chromatids and attaches them to the mitotic spindle.

centrosome (cell center) The main microtubule organizing center of a cell, which organizes the microtubule cytoskeleton during interphase and positions many of the cytoplasmic organelles.

cephalization The development of an anterior head where sensory organs and nervous system tissue are concentrated.

cephalothorax The anterior section of an arachnid, consisting of a fused head and thorax.

cerebellum The portion of the brain that receives sensory input from receptors in muscles and joints, from balance receptors in the inner ear, and from the receptors of touch, vision, and hearing.

cerebral cortex A thin outer shell of gray matter covering a thick core of white matter within each hemisphere of the brain; the part of the forebrain responsible for information processing and learning.

cerebrospinal fluid Fluid that circulates through the central canal of the spinal cord and the ventricles of the brain, cushioning the brain and spinal cord from jarring movements and impacts, as well as nourishing the CNS and protecting it from toxic substances.

cervix The lower end of the uterus.

channel protein Transport protein that forms a hydrophilic channel in a cell membrane through which water, ions, or other molecules can pass, depending on the protein.

chaparral A biome comprising a scrubby mix of short trees and shrubs that dominates coastal land between 30° and 40° latitude, where winters are cool and wet and summers hot and dry.

chaperone protein (chaperonin) "Guide" protein that binds temporarily with newly synthesized proteins, directing their conformation toward the correct tertiary structure and inhibiting incorrect arrangements as the new proteins fold.

character A heritable characteristic.

character displacement The phenomenon in which allopatric populations are morphologically similar and use similar resources, but sympatric populations are morphologically different and use different resources; may also apply to characters influencing mate choice.

charging *See* aminoacylation.

charophyte A member of the group of green algae most similar to the algal ancestors of land plants.

checkpoint Internal control of the cell cycle that prevents a critical phase from beginning until the previous phase is complete.

chelicerae The first pair of fanglike appendages near the mouth of an arachnid, used for biting prey and often modified for grasping and piercing.

chemical bond Link formed when atoms of reactive elements combine into molecules.

chemical equation A chemical reaction written in balanced form.

chemical reaction A reaction that occurs when atoms or molecules interact to form new chemical bonds or break old ones.

chemical signal Any secretion from one cell type that can alter the behavior of a different cell that bears a receptor for it; a means of cell communication."

chemical synapse A type of communicating connection between two neurons or a neuron and an effector cell in which an electrical impulse arriving at an axon terminal of the presynaptic cell triggers release of a neurotransmitter that crosses the gap and binds to a receptor on the postsynaptic cell, triggering an electrical impulse in that cell.

chemiosmotic hypothesis Model proposing that mitochondrial electron transfer produces an H^+ gradient and that the gradient powers ATP synthesis by ATP synthase.

chemoautotroph An organism that obtains energy by oxidizing inorganic substances such as hydrogen, iron, sulfur, ammonia, nitrites, and nitrates and uses carbon dioxide as a carbon source.

chemoheterotroph An organism that oxidizes organic molecules as an energy source and obtains carbon in organic form.

chemokine A protein secreted by activated macrophages that attracts other cells, such as neutrophils.

chemoreceptor A sensory receptor that detects specific molecules, or chemical conditions such as acidity.

chemotroph An organism that obtains energy by oxidizing inorganic or organic substances.

chiasmata *See* crossover.

chitin A polysaccharide that contains nitrogen and is present in the cell walls of fungi and the exoskeletons of arthropods.

chlorophyll Molecule of green pigment that absorbs photons of light in photosynthesis.

chloroplast The site of photosynthesis in plant cells.

chlorosis An abnormal yellowing of plant tissues due to lack of chlorophyll; a sign of nutrient deficiency or infection by a pathogen.

choanocyte One of the inner layer of flagellated cells lining the body cavity of a sponge.

Choanoflagellata A group of minute, single-celled protists found in water; the flask-shaped body has a collar of closely packed microvilli that surrounds the single flagellum by which it moves and take in food.

cholesterol The predominant sterol of animal cell membranes.

chondrocyte A cartilage-producing cell.

chorion In an amniote egg, an extraembryonic membrane that surrounds the embryo and yolk sac completely and exchanges oxygen and carbon dioxide with the environment; becomes part of the placenta in mammals.

chorionic villus (plural, villi) One of many treelike extensions from the chorion, which greatly increase the surface area of the chorion.

chorionic villus sampling Technique of prenatal diagnosis in which cells are obtained from portions of the placenta that develop from tissues of the embryo.

chromatin The structural building block of a chromosome, which includes the complex of DNA and its associated proteins.

chromatin remodeling Process in which the state of the chromatin is changed so that the proteins that initiate transcription can bind to their promoters.

chromoplast Plastid containing red and yellow pigments.

chromosomal protein The histone and nonhistone protein associated with DNA structure and regulation in the nucleus.

chromosome The nuclear unit of genetic information, consisting of a DNA molecule and associated proteins.

chromosome segregation The equal distribution of daughter chromosomes to each of the two cells that result from cell division.

chromosome theory of inheritance The principle that genes and their alleles are carried on the chromosomes.

chylomicron A small triglyceride droplet covered by a protein coat.

chyme Digested content of the stomach released for further digestion in the small intestine.

ciliary body A fine ligament in the eye that anchors the lens to a surrounding layer of connective tissue and muscle.

cilium Motile structure, extending from a cell surface, that moves a cell through fluid or fluid over a cell.

circadian rhythm Any biological activity that is repeated in cycles, each about 24 hours long, independently of any shifts in environmental conditions.

circulatory system An organ system consisting of a fluid, a heart, and vessels for moving important molecules, and often cells, from one tissue to another.

circulatory vessel An element of the circulatory system through which fluid flows and carries nutrients and oxygen to tissues and remove wastes.

circumcision Removal of the prepuce for religious, cultural, or hygienic reasons.

cisternae Membranous channels and vesicles that make up the endoplasmic reticulum.

citric acid cycle Series of reactions in which acetyl groups are oxidized completely to carbon dioxide and some ATP molecules are synthesized. Also referred to as *Krebs cycle* and *tricarboxylic acid cycle.*

clade A monophyletic group of organisms that share homologous features derived from a common ancestor.

cladistics An approach to systematics that uses shared derived characters to infer the phylogenetic relationships and evolutionary history of groups of organisms.

cladogenesis The evolution of two or more descendant species from a common ancestor.

cladogram A branching diagram in which the endpoints of the branches represent different species of organisms, used to illustrate phylogenetic relationships.

claspers A pair of organs on the pelvic fins of male crustaceans and sharks, which help transfer sperm into the reproductive tract of the female.

class A Linnaean taxonomic category that ranks below a phylum and above an order.

class II major histocompatibility complex (MHC) A collection of proteins that present antigens on the cell surface of an antigen-presenting cell in an antibody-mediated immune response.

classical conditioning A type of learning in which an animal develops a mental associa-tion between two phenomena that are usually unrelated.

classification An arrangement of organisms into hierarchical groups that reflect their relatedness.

clathrin The network of proteins that coat and reinforce the cytoplasmic surface of cell membranes.

cleavage Mitotic cell divisions of the zygote that produce a blastula from a fertilized ovum.

climate The weather conditions prevailing over an extended period of time.

climax community A relatively stable, late successional stage in which the dominant vegetation replaces itself and persists until an environmental disturbance eliminates it, allowing other species to invade.

climograph A graph that portrays the particular combination of temperature and rainfall conditions where each terrestrial biome occurs.

cline A pattern of smooth variation in a characteristic along a geographical gradient.

clitoris The structure at the junction of the labia minora in front of the vulva, homologous to the penis in the male.

clonal analysis A method of culturing meristematic tissue that contains a mutated embryonic cell having a readily observable trait, such as the absence of normal pigment.

clonal expansion The proliferation of the activated CD4+ T cell by cell division to produce a clone of cells.

clonal selection The process by which a lymphocyte is specifically selected for cloning when it encounters a foreign antigen from among a randomly generated, enormous diversity of lymphocytes with receptors that specifically recognize the antigen.

clone An individual genetically identical to an original cell from which it descended.

closed circulatory system A circulatory system in which the fluid, blood, is confined in blood vessels and is distinct from the interstitial fluid.

clumped dispersion A pattern of distribution in which individuals in a population are grouped together.

cnidocyte A prey-capturing and defensive cell in the epidermis of cnidarians.

CO₂ fixation Process in which electrons are used as a source of energy to convert inorganic CO_2 to an organic form.

coactivator (mediator) In eukaryotes, a large multiprotein complex that bridges between activators at an enhancer and proteins at the promoter and promoter proximal region to stimulate transcription.

coat The protective layer of protein that surrounds the nucleic acid core of a virus in free form; also known as a capsid.

coated pit A depression in the plasma membrane that contains receptors for macromolecules to be taken up by endocytosis.

coccus (plural, cocci) A spherical prokaryote.

cochlea A snail-shaped structure in the inner ear containing the organ of hearing.

codominance Condition in which alleles have approximately equal effects in individuals, making the alleles equally detectable in heterozygotes.

codon Each three-letter word (triplet) of the genetic code.

coelom A fluid-filled body cavity in bilaterally symmetrical animals that is completely lined with derivatives of mesoderm.

coelomate A body plan of bilaterally symmetrical animals that have a coelom.

coenocytic Condition in which a single cell has many nuclei.

coenzymes Organic cofactors that include complex chemical groups of various kinds.

coevolution The evolution of genetically based, reciprocal adaptations in two or more species that interact closely in the same ecological setting.

cofactor An inorganic or organic nonprotein group that is necessary for catalysis to take place.

cohesion The high resistance of water molecules to separation.

cohesion-tension mechanism of water transport A model of how water is transported from roots to leaves in vascular plants; the evaporation of water from leaves pulls water up in xylem by creating a continuous negative pressure (tension) that extends to roots.

cohort A group of individuals of similar age.

coleoptile A protective sheath that covers the shoot apical meristem and plumule of the embryo in monocots, such as grasses, as it pushes up through soil.

coleorhiza A sheath that encloses the radicle of an embryo until it breaks out of the seed coat and enters the soil as the primary root.

collagen Fibrous glycoprotein—very rich in carbohydrates—embedded in a network of proteoglycans.

collecting duct A location where urine leaving individual nephrons is processed further.

collenchyma One of three simple plant tissues. Flexibly supports rapidly growing plant parts. Its elongated cells are alive at maturity and collectively often form strands or a sheathlike cylinder under the dermal tissue of growing shoot regions and leaf stalks.

colon The main part of the large intestine.

colony Multiple individual organisms of the same species living in a group.

combinatorial gene regulation The combining of a few regulatory proteins in particular ways so that the transcription of a wide array of genes can be controlled and a large number of cell types can be specified.

commaless The sequential nature of the words of the nucleic acid code, with no indicators such as commas or spaces to mark the end of one codon and the beginning of the next.

commensalism A symbiotic interaction in which one species benefits and the other is unaffected.

community Populations of all species that occupy the same area.

community ecology The ecological discipline that examines groups of populations occurring together in one area.

companion cell A specialized parenchyma cell that is connected to a mature sieve tube member by plasmodesmata and assists sieve tube members both with the uptake of sugars and with the unloading of sugars in tissues.

comparative morphology Analysis of the structure of living and extinct organisms.

compass orientation A wayfinding mechanism that allows animals to move in a particular direction, often over a specific distance or for a prescribed length of time.

competitive exclusion principle The ecological principle stating that populations of two or more species cannot coexist indefinitely if they rely on the same limiting resources and exploit them in the same way.

competitive inhibition Inhibition of an enzyme reaction by an inhibitor molecule that resembles the normal substrate closely enough so that it fits into the active site of the enzyme.

complement system A nonspecific defense mechanism activated by invading pathogens, made up of more than 30 interacting soluble plasma proteins circulating in the blood and interstitial fluid.

complementary base pairing Feature of DNA in which the specific purine–pyrimidine base pairs A–T (adenine–thymine) and G–C (guanine–cytosine) occur to bridge the two sugar–phosphate backbones.

complementary DNA (cDNA) A DNA molecule that is complementary to an mRNA molecule, synthesized by reverse transcriptase.

complete digestive system A digestive system having a mouth at one end, through which food enters, and an anus at the other end, through which undigested waste is voided.

complete flower A flower in which all four whorls (sepals, petals, stamens, carpels) are present.

complete medium A growth medium containing a full complement of nutrient substances.

complete metamorphosis The form of metamorphosis in which an insect passes through four separate stages of growth: egg, larva, pupa, and adult.

complex virus A bacteriophage with a DNA genome that has a tail attached at one side of a polyhedral head.

compound A molecule whose component atoms are different.

compound eye The eye of most insects and some crustaceans, composed of many faceted, light-sensitive units called ommatidia fitted closely together, each having its own refractive system and each forming a portion of an image.

concentration The number of molecules or ions of a substance in a unit volume of space.

concentration gradient The concentration difference that drives diffusion.

condensation reaction Reaction during which the components of a water molecule are removed, usually as part of the assembly of a larger molecule from smaller subunits. Also referred to as *dehydration synthesis reaction*.

conduction The flow of heat between atoms or molecules in direct contact.

cone In the vertebrate eye, a photoreceptor in the retina that is specialized for detection of different wavelengths (colors). In cone-bearing plants, a cluster of sporophylls.

conformation The overall three-dimensional shape of a protein.

conformational change Alteration in the three-dimensional shape of a protein.

conidiophore A fungal hypha that gives rise to conidia.

conidium (plural, conidia) An asexually produced fungal spore.

Coniferophyta The major phylum of cone-bearing gymnosperms, most of which are substantial trees; includes pines, firs, and other conifers.

conjugation In bacteria, the process by which a copy of part of the DNA of a donor cell moves through the cytoplasmic bridge into the recipient cell where genetic recombination can occur. In ciliate protozoans, a process of sexual reproduction in which individuals of the same species temporarily couple and exchange genetic material.

connective tissue Tissue having cells scattered through an extracellular matrix; forms layers in and around body structures that support other body tissues, transmit mechanical and other forces, and in some cases act as filters.

conodont An abundant, bonelike fossil dating from the early Paleozoic era through the early Mesozoic era, now described as a feeding structure of some of the earliest vertebrates.

consciousness Awareness of oneself, identity, and surroundings, with understanding of the significance and likely consequences of events.

conservation biology An interdisciplinary science that focuses on the maintenance and preservation of biodiversity.

consumer An organism that consumes other organisms in a community or ecosystem.

contact inhibition The inhibition of movement or proliferation of normal cells that results from cell–cell contact.

continental climate Climate not moderated by the distant ocean.

continental drift The long-term movement of continents as a result of plate tectonics.

continuous distribution A geographical distribution in which a species lives in suitable habitats throughout a geographical area.

contractile vacuole A specialized cytoplasmic organelle that pumps fluid in a cyclical manner from within the cell to the outside by alternately filling and then contracting to release its contents at various points on the surface of the cell.

control Treatment that tells what would be seen in the absence of the experimental manipulation.

convection The transfer of heat from a body to a fluid, such as air or water, that passes over its surface.

convergent evolution The evolution of similar adaptations in distantly related organisms that occupy similar environments.

copulation The physical act involving the introduction of the accessory sex organ of a male into the accessory sex organ of a female to accomplish internal fertilization.

coral reef A structure made from the hard skeletons of coral animals or polyps; found largely in tropical and subtropical marine environments.

core The nucleic acid center of a virus in the free form.

corepressor In the regulation of gene expression in bacteria, a regulatory molecule that combines with a repressor to activate it and shut off an operon.

cork A nonliving, impermeable secondary tissue that is one element of bark.

cork cambium A lateral meristem in plants that forms periderm, which in turn produces cork.

cornea The transparent layer that forms the front wall of the eye, covering the iris.

corolla The structure formed collectively by the petals of a flower.

corpus callosum A structure formed of thick axon bundles that connect the two cerebral hemispheres and coordinate their functions.

corpus luteum Cells remaining at the surface of the ovary during the luteal phase; the structure acts as an endocrine gland, secreting several hormones: estrogens, large quantities of progesterone, and inhibin.

cortex Generally, an outer, rindlike layer. In mammals, the outer layer of the brain, the kidneys, or the adrenal glands. In plants, the outer region of tissue in a root or stem lying between the epidermis and the vascular tissue, composed mainly of parenchyma.

cortical granule A secretory vesicle just under the plasma membrane of an egg cell.

cortical reaction The reaction in which cortical granules fuse with the plasma membrane of the egg and release their contents to the outside.

cortisol The major glucocorticoid steroid hormone secreted by the adrenal cortex, which increases blood glucose by promoting breakdown of proteins and fats.

cotransport *See* symport.

countercurrent exchange A mechanism in which the water flowing over the gills moves in a direction opposite to the flow of blood under the respiratory surface.

coupled reaction Reaction that occurs when an exergonic reaction is joined to an endergonic reaction, producing an overall reaction that is exergonic.

courtship display A behavior performed by males to attract potential mates or to reinforce the bond between a male and a female.

covalent bond Bond formed by electron sharing between atoms.

cranial nerve A nerve that connects the brain directly to the head, neck, and body trunk.

cranium The part of the skull that encloses the brain.

crassulacean acid metabolism (CAM) A biochemical variation of photosynthesis that was discovered in a member of the plant family Crassulaceae. Carbon dioxide is taken up and stored during the night to allow the stomata to remain closed during the daytime, decreasing water loss.

Crenarchaeota A major group of the domain Archaea, separated from the other archaeans based mainly on rRNA sequences.

crista Fold that expands the surface area of the inner mitochondrial membrane.

critical period A restricted stage of development early in life during which an animal has the capacity to respond to specific environmental stimuli.

crop Of birds, an enlargement of the digestive tube where the digestive contents are stored and mixed with lubricating mucus.

crossing-over The recombination process in meiosis, in which chromatids exchange segments.

crossover Site of recombination during meiosis. Also referred to as a *chiasmata*.

cross-pollination Fertilization of one plant by a different plant.

cross-talk Interaction by which cell signaling pathways communicate with one another to integrate their responses to cellular signals.

cryptic coloration Coloration that allows an organism to match its background and hence become less vulnerable to predation or recognition by prey.

cryptochrome A light-absorbing protein that is sensitive to blue light and that may also be an important early step in various light-based growth responses.

C-terminal end The end of an amino acid chain with a —COO⁻ group.

cupula In certain mechanoceptors, a gelatinous structure with stereocilia extending into it that moves with pressure changes in the surrounding water; movement of the cupula bends the stereocilia, which triggers release of neurotransmitters.

cuticle The outer layer of plants and some animals, which helps prevent desiccation by slowing water loss.

Cycadophyta A phylum of palmlike gymnosperms known as cycads; the pollen-bearing and seed-bearing cones (strobili) occur on separate plants.

cyclic AMP (cAMP) In particular signal transduction pathways, a second messenger that activates protein kinases, which elicit the cellular response by adding phosphate groups to specific target proteins. cAMP functions in one of two major G-protein–coupled receptor–response pathways.

cyclic electron flow An electron transport pathway associated with photosystem I in photosynthesis that produces ATP without the synthesis of NADPH.

cyclin In eukaryotes, protein that regulates the activity of CDK (cyclin-dependent kinase) and controls progression through the cell cycle.

cyclin-dependent kinase (CDK) A protein kinase that controls the cell cycle in eukaryotes.

cytochrome Protein with a heme prosthetic group that contains an iron atom.

cytokine A molecule secreted by one cell type that binds to receptors on other cells and, through signal transduction pathways, triggers a response. In innate immunity, cytokines are secreted by activated macrophages.

cytokinesis Division of the cytoplasm into two daughter cells following the nuclear division stage of mitosis.

cytokinin A hormone that promotes and controls growth responses of plants.

cytoplasm All the parts of the cell that surround the central nuclear or nucleoid region.

cytoplasmic determinants The mRNA and proteins stored in the egg cytoplasm that direct the first stages of animal development in the period before genes of the zygote become active.

cytoplasmic inheritance Pattern in which inheritance follows that of genes in the cytoplasmic organelles, mitochondria or chloroplasts.

cytoplasmic streaming Intracellular movement of cytoplasm.

cytosine A pyrimidine that base-pairs with guanine in nucleic acids.

cytoskeleton The interconnected system of protein fibers and tubes that extends throughout the cytoplasm of a eukaryotic cell.

cytosol Aqueous solution in the cytoplasm containing ions and various organic molecules.

cytotoxic T cell A T lymphocyte that functions in cell-mediated immunity to kill body cells infected by viruses or transformed by cancer.

daily torpor A period of inactivity and lowered metabolic rate that allows an endotherm to conserve energy when environmental temperatures are low.

dalton A standard unit of mass, about 1.66 $\times 10^{-24}$ grams.

day-neutral plant A plant that flowers without regard to photoperiod.

decomposer A small organism, such as a bacterium or fungus, that feeds on the remains of dead organisms, breaking down complex biological molecules or structures into simpler raw materials.

degeneracy (redundancy) The feature of the genetic code in which, with two exceptions, more than one codon represents each amino acid.

dehydration synthesis reaction *See* condensation reaction.

deletion Chromosomal alteration that occurs if a broken segment is lost from a chromosome.

demographic transition model A graphical depiction of the historical relationship between a country's economic development and its birth and death rates.

demography The statistical study of the processes that change a population's size and density through time.

denaturation A loss of both the structure and function of a protein due to extreme conditions that unfold it from its conformation.

dendrite The branched extension of the nerve cell body that receives signals from other nerve cells.

dendritic cell A type of phagocyte, so called because it has many surface projections that resemble dendrites of neurons, which engulfs a bacterium in infected tissue by phagocytosis.

denitrification A metabolic process in which certain bacteria convert nitrites or nitrates into nitrous oxide and then into molecular nitrogen, which enters the atmosphere.

density-dependent Description of environmental factors for which the strength of their effect on a population varies with the population's density.

density-independent Description of environmental factors for which the strength of their effect on a population does not vary with the population's density.

deoxyribonucleic acid (DNA) The large, double-stranded, helical molecule that contains the genetic material of all living organisms.

deoxyribose A 5-carbon sugar to which the nitrogenous bases in nucleotides of DNA link covalently.

depolarized State of the membrane (which was polarized at rest) as the membrane potential becomes less negative.

deposit feeder An animal that consumes particles of organic matter from the solid substrate on which it lives.

derivative One of the daughter cells produced when a plant cell divides; it typically divides once or twice and then enters on the path to differentiation.

derived character A new version of a trait found in the most recent common ancestor of a group.

dermal tissue system The plant tissue system that comprises the outer tissues of the plant body, including the epidermis and periderm; it serves as a protective covering for the plant body.

dermis The skin layer below the epidermis; it is packed with connective tissue fibers such as collagen, which resist compression, tearing, or puncture of the skin.

descent with modification Biological evolution.

desert A sparsely vegetated biome that forms where rainfall averages less than 25 cm per year.

desertification A process in which large tracts of subtropical forest are cleared and overused, the groundwater table recedes to deeper levels, less surface water is available for plants, soil accumulates high concentrations of salts, and topsoil is eroded by wind and water.

desmosome Anchoring junction for which microfilaments anchor the junction in the underlying cytoplasm.

determinate cleavage A type of cleavage in protosomes in which each cell's developmental path is determined as the cell is produced.

determinate growth The pattern of growth in most animals in which individuals grow to a certain size and then their growth slows dramatically or stops.

determination Mechanism in which the developmental fate of a cell is set.

detritivore An organism that extracts energy from the organic detritus (refuse) produced at other trophic levels.

development A series of programmed changes encoded in DNA, through which a fertilized egg divides into many cells that ultimately are transformed into an adult, which is itself capable of reproduction.

diabetes mellitus A disease that results from problems with insulin production or action.

diacylglycerol (DAG) In particular signal transduction pathways, a second messenger that activates protein kinases, which elicit the cellular response by adding phosphate groups to specific target proteins. DAG is involved in one of two major G-protein–coupled receptor–response pathways.

diapsid A member of a group within the amniote vertebrates having a skull with two temporal arches. Their living descendants include lizards and snakes, crocodilians, and birds.

diastole The period of relaxation and filling of the heart between contractions.

diffusion The net movement of ions or molecules from a region of higher concentration to a region of lower concentration.

digestion The splitting of carbohydrates, proteins, lipids, and nucleic acids in foods into chemical subunits small enough to be absorbed into the body fluids and cells of an animal.

digestive tube A tubelike digestive system with two openings that form a separate mouth and anus; the digestive contents move in one direction through specialized regions of the tube, from the mouth to the anus.

dihybrid A zygote produced from a cross that involves two characters.

dihybrid cross A cross between two individuals that are heterozygous for two pairs of alleles.

dikaryon The life stage in certain fungi in which a cell contains two genetically distinct haploid nuclei.

dioecious Having male flowers and female flowers on different plants of the same species.

diploblastic An animal body plan in which adult structures arise from only two cell layers, the ectoderm and the endoderm.

diploid An organism or cell with two copies of each type of chromosome in its nucleus.

direct neurotransmitter A neurotransmitter that binds directly to a ligand-gated ion channel in the postsynaptic membrane, opening or closing the channel gate and altering the flow of a specific ion or ions in the postsynaptic cell.

directional selection A type of selection in which individuals near one end of the phenotypic spectrum have the highest relative fitness.

Discicristates A protist group of single-celled, highly motile organisms that swim by means of flagella, and that have characteristic disc-shaped mitochondrial cristae (inner mitochondrial membranes).

disclimax community An ecological community in which regular disturbance inhibits successional change.

discontinuous replication Replication in which a DNA strand is formed in short lengths that are synthesized in the direction opposite of DNA unwinding.

disjunct distribution A geographical distribution in which populations of the same species or closely related species live in widely separated locations.

dispersal The movement of organisms away from their place of origin.

dispersion The spatial distribution of individuals within a population's geographical range.

disruptive selection A type of natural selection in which extreme phenotypes have higher relative fitness than intermediate phenotypes.

dissociation The separation of water to produce hydrogen ions and hydroxide ions.

distal convoluted tubule The tubule in the human nephron that drains urine into a collecting duct that leads to the renal pelvis.

disulfide linkage Linkage that occurs when two sulfhydryl groups interact during a linking reaction.

DNA *See* deoxyribonucleic acid.

DNA chip *See* DNA microarray.

DNA fingerprinting Technique in which DNA samples are used to distinguish between individuals of the same species.

DNA helicase An enzyme that catalyzes the unwinding of DNA template strands.

DNA hybridization Technique in which a gene or sequence of interest is identified in a set of clones when it base pairs with a single-stranded DNA or RNA molecule called a nucleic acid probe.

DNA ligase In DNA replication, an enzyme that seals the nicks left after RNA primers are replaced with DNA.

DNA methylation Process in which a methyl group is added enzymatically to cytosine bases in the DNA.

DNA microarray A solid surface divided into a microscopic grid of thousands of spaces each containing thousands of copies of a DNA probe. DNA chips are used commonly for analysis of gene activity and for detecting differences between cell types. Also referred to as a *DNA chip*.

DNA polymerase An enzyme that assembles complementary nucleotide chains during DNA replication.

DNA repair mechanism Mechanism to correct base-pair mismatches that escape proofreading.

DNA technologies Techniques to isolate, purify, analyze, and manipulate DNA sequences.

domain In protein structure, a distinct, large structural subdivision produced in many proteins by the folding of the amino acid chain. In systematics, the highest taxonomic category; a group of cellular organisms with characteristics that set it apart as a major branch of the evolutionary tree.

dominance The masking effect of one allele over another.

dominance hierarchy A social system in which the behavior of each individual is constrained by that individual's status in a highly structured social ranking.

dominant The allele expressed when more than one allele is present.

dormancy A period in the life cycle in which biological activity is suspended.

dorsal Indicating the back side of an animal.

dorsal lip of the blastopore A crescent-shaped depression rotated clockwise 90° on the embryo surface that marks the region derived from the gray crescent, to which cells from the animal pole move as gastrulation begins.

double fertilization The characteristic feature of sexual reproduction in flowering plants. In the embryo sac, one sperm nucleus unites with the egg to form a diploid zygote from which the embryo develops, and another unites with two polar nuclei to form the primary endosperm nucleus.

double helix Two nucleotide chains wrapped around each other in a spiral.

double-helix model Model of DNA consisting of two complementary sugar–phosphate backbones.

duodenum A short region of the small intestine where secretions from the pancreas and liver enter a common duct.

duplication Chromosomal alteration that occurs if a segment is broken from one chromosome and inserted into its homolog.

E site The site where an exiting tRNA binds prior to its release from the ribosome.

ecdysis Shedding of the cuticle, exoskeleton, or skin; molting.

ecdysone A steroid hormone secreted by the prothoracic glands of insects.

echolocation A technique for locating prey by making squeaking or clicking noises, and then listening for the echoes that bounce back from objects in their environment.

ecological community An assemblage of species living in the same place.

ecological efficiency The ratio of net productivity at one trophic level to net productivity at the trophic level below it.

ecological isolation A prezygotic reproductive isolating mechanism in which species that live in the same geographical region occupy different habitats.

ecological niche The resources a population uses and the environmental conditions it requires over its lifetime.

ecological pyramid A diagram illustrating the effects of energy transfer from one trophic level to the next.

ecological succession A somewhat predictable series of changes in the species composition of a community over time.

ecology The study of the interactions between organisms and their environments.

ecosystem Group of biological communities interacting with their shared physical environment.

ecosystem ecology A ecological discipline that explores the cycling of nutrients and the flow of energy between the biotic components of an ecological community and the abiotic environment.

ecosystem services The ecological processes on which all life depends, which include decomposition of wastes, nutrient recycling, oxygen production, maintenance of fertile topsoil, and air and water purification.

ecosystem valuation A process in which ecosystem services are assigned an economic value.

ecotone A wide transition zone between adjacent communities.

ecotourism An activity in which visitors, often from wealthy countries, pay a fee to visit a nature preserve.

ectoderm The outermost of the three primary germ layers of an embryo, which develops into epidermis and nervous tissue.

ectomycorrhiza A mycorrhiza that grows between and around the young roots of trees and shrubs but does not enter root cells.

ectoparasite A parasite that lives on the exterior of its host organism.

ectotherm An animal that obtains its body heat primarily from the external environment.

edge effect A phenomenon in which the removal of natural vegetation disrupts the local physical environment, exposing the borders of the remaining habitat to additional sunlight, wind, and rainfall.

effector In signal transduction, a plasma membrane-associated enzyme, activated by a G protein, that generates one or more second messengers. In homeostatic feedback, the system that returns the condition to the set point if it has strayed away.

effector T cell A cell involved in effecting—bringing about—the specific immune response to an antigen.

efferent arteriole The arteriole that receives blood from the glomerulus.

efferent neuron A neuron that carries the signals indicating a response away from the interneuron networks to the effectors.

egg cell The female reproductive cell.

elastin A rubbery protein in some connective tissues that adds elasticity to the extracellular matrix—it is able to return to its original shape after being stretched, bent, or compressed.

electrical signaling A means of animal communication in which a signaler emits an electric discharge that can be received by another individual.

electrical synapse A mechanical and electrically conductive link between two abutting neurons that is formed at the gap junction.

electrocardiogram (ECG) Graphic representation of the electrical activity within the heart, detected by electrodes placed on the body.

electrochemical gradient A difference in chemical concentration and electric potential across a membrane.

electromagnetic spectrum The range of wavelengths or frequencies of electromagnetic radiation extending from gamma rays to the longest radio waves and including visible light.

electron Negatively charged particle outside the nucleus of an atom.

electron microscope Microscope that uses electrons to illuminate the specimen.

electron transfer system Stage of cellular respiration in which high-energy electrons produced from glycolysis, pyruvate oxidation, and the citric acid cycle are delivered to oxygen by a sequence of electron carriers.

electronegativity The measure of an atom's attraction for the electrons it shares in a chemical bond with another atom.

electroreceptor A specialized sensory receptor that detects electrical fields.

element A pure substance that cannot be broken down into simpler substances by ordinary chemical or physical techniques.

embryo An organism in its early stage of reproductive development, beginning in the first moments after fertilization.

embryo sac The female gametophyte of angiosperms, within which the embryo develops; it usually consists of seven cells: an egg cell, an endosperm mother cell, and five other cells with fleeting reproductive roles.

embryophyte Any plant in which the embryo is retained within maternal tissue.

emergent property Characteristic that depends on the level of organization of matter, but does not exist at lower levels of organization.

emigration The movement of individuals out of a population.

enantiomers Isomers that are mirror images of each other. Also referred to as *optical isomers*.

endangered species A species in danger of extinction throughout all or a significant portion of its range.

endemic species A species that occurs in only one place on Earth.

endergonic reaction Reaction that can proceed only if free energy is supplied.

endocrine gland Any of several ductless secretory organs that secrete hormones into the blood or extracellular fluid.

endocrine system The system of glands that release their secretions (hormones) directly into the circulatory system.

endocytic vesicle Vesicle that carries proteins and other molecules from the plasma membrane to destinations within the cell.

endocytosis In eukaryotes, the process by which molecules are brought into the cell from the exterior involving a bulging in of the plasma membrane that pinches off to form an endocytic vesicle.

endoderm The innermost of the three primary germ layers of an embryo, which develops into the gastrointestinal tract and, in some animals, the respiratory organs.

endodermis The innermost layer of the root cortex; a selectively permeable barrier that helps control the movement of water and dissolved minerals into the stele.

endomembrane system In eukaryotes, a collection of interrelated internal membranous sacs that divide a cell into functional and structural compartments.

endomycorrhiza A mycorrhiza in which the fungal hyphae penetrate into cells of the root.

endoparasite A parasite that lives in the internal organs of its host organism.

endoplasmic reticulum (ER) In eukaryotes, an extensive interconnected network of cisternae that is responsible for the synthesis, transport, and initial modification of proteins and lipids.

endorphin One of a group of small proteins occurring naturally in the brain and around nerve endings that bind to opiate receptors and thus can raise the pain threshold.

endoskeleton A supportive internal body structure, such as bones, that provides support.

endosperm Nutritive tissue inside the seeds of flowering plants.

endospore A small, metabolically inactive, asexual spore that develops within some bacterial cells when environmental conditions become unfavorable.

endosymbiont hypothesis The proposal that the membranous organelles of eukaryotic cells (mitochondria and chloroplasts) may have originated from symbiotic relationships between two prokaryotic cells.

endotherm An animal that obtains most of its body heat from internal physiological sources.

endotoxin A lipopolysaccharide released from the outer membrane of the cell wall when a bacterium dies and lyses.

end-product inhibition *See* feedback inhibition.

energy The capacity to do work.

energy budget The total amount of energy that an organism can accumulate and use to fuel its activities.

energy levels Regions of space within an atom where electrons are found. Also referred to as *shells*.

enhancer In eukaryotes, a region at a significant distance from the beginning of a gene containing regulatory sequences that determine whether the gene is transcribed at its maximum possible rate.

enterocoelom In deuterostomes, the body cavity pinched off by outpocketings of the archenteron.

entropy Disorder, in thermodynamics.

enveloped virus A virus that has a surface membrane derived from its host cell.

enzymatic hydrolysis A process in which chemical bonds are broken by the addition of H^+ and OH^-, the components of a molecule of water.

enzyme Protein that accelerates the rate of a cellular reaction.

enzyme specificity The ability of an enzyme to catalyze the reaction of only a single type of molecule or group of closely related molecules.

eosinophil A type of leukocyte that targets extracellular parasites too large for phagocytosis in the inflammatory response.

epiblast The top layer of the blastodisc.

epicotyl The upper part of the axis of an early plant embryo, located between the cotyledons and the first true leaves.

epidermis A complex tissue that covers an organism's body in a single continuous layer or sometimes in multiple layers of tightly packed cells.

epididymis A coiled storage tubule attached to the surface of each testis.

epiglottis A flaplike valve at the top of the trachea.

epilimnion The top layer of the limnetic zone in a lake.

epinephrine A nontropic amine hormone secreted by the adrenal medulla.

epiphyte A plant that grows independently on other plants and obtains nutrients and water from the air.

epistasis Interaction of genes, with one or more alleles of a gene at one locus inhibiting or masking the effects of one or more alleles of a gene at a different locus.

epithelial tissue Tissue formed of sheetlike layers of cells that are usually joined tightly together, with little extracellular matrix material between them. They protect body surfaces from invasion by bacteria and viruses, and secrete or absorb substances.

epitope The small region of an antigen molecule to which BCRs or TCRs bind.

equilibrium point A state of balance between opposing factors that push a reaction in either direction.

equilibrium theory of island biogeography An hypothesis suggesting that the number of species on an island is governed by a give and take between the immigration of new species to the island and the extinction of species already there.

ER (endoplasmic reticulum) lumen The enclosed space surrounded by a cisterna.

erythrocyte A red blood cell, which contains hemoglobin, a protein that transports O_2 in blood.

erythropoietin (EPO) A hormone that stimulates stem cells in bone marrow to increase erythrocyte production.

esophagus A connecting passage of the digestive tube.

essential amino acid Any amino acid that is not made by the human body but must be taken in as part of the diet.

essential element Any of a number of elements required by living organisms to ensure normal reproduction, growth, development, and maintenance.

essential fatty acid Any fatty acid that the body cannot synthesize but needs for normal metabolism.

essential mineral Any inorganic element such as calcium, iron, or magnesium that is required in the diet of an animal.

essential nutrient Any of the essential amino acids, fatty acids, vitamins, and minerals required in the diet of an animal.

estivation Seasonal torpor in an animal that occurs in summer.

estradiol A form of estrogen.

estrogen Any of the group of female sex hormones.

estuary A coastal habitat where tidal seawater mixes with fresh water from rivers, streams, and runoff.

ethology A discipline that focuses on how animals behave in their natural environments.

ethylene A plant hormone that helps regulate seedling growth, stem elongation, the ripening of fruit, and the abscission of fruits, leaves, and flowers.

euchromatin In eukaryotes, regions of loosely packed chromatin fibers in interphase nuclei.

eudicot A plant belonging to the Eudicotyledones, one of the two major classes of angiosperms; their embryos generally have two seed leaves (cotyledons), and their pollen grains have three grooves.

Eukarya The domain that includes all eukaryotes, organisms that contain a membrane-bound nucleus within each of their cells; all protists, plants, fungi, and animals.

eukaryote Organism in which the DNA is enclosed in a nucleus.

eukaryotic chromosome A DNA molecule, with its associated proteins, in the nucleus of a eukaryotic cell.

euploid An individual with a normal set of chromosomes.

Euryarchaeota A major group of the domain Archaea, members of which are found in different extreme environments. They include methanogens, extreme halophiles, and some extreme thermophiles.

eusocial A form of social organization, observed in some insect species, in which numerous related individuals—a large percentage of them sterile female workers—live and work together in a colony for the reproductive benefit of a single queen and her mate(s).

eutrophic lake A lake that is rich in nutrients and organic matter.

evaporation Heat transfer through the energy required to change a liquid to a gas.

evolutionary developmental biology A field of biology that compares the genes controlling the developmental processes of different animals to determine the evolutionary origin of morphological novelties and developmental processes.

evolutionary divergence A process whereby natural selection or genetic drift causes populations to become more different over time.

exchange diffusion *See* antiport.

excitatory postsynaptic potential (EPSP) The change in membrane potential caused when a neurotransmitter opens a ligand-gated Na^+ channel and Na^+ enters the cell, making it more likely that the postsynaptic neuron will generate an action potential.

excretion The process that helps maintain the body's water and ion balance while ridding the body of metabolic wastes.

exergonic reaction Reaction that has a negative ΔG because it releases free energy.

exocrine gland A gland that is connected to the epithelium by a duct and that empties its secretion at the epithelial surface.

exocytosis In eukaryotes, the process by which a secretory vesicle fuses with the plasma membrane and releases the vesicle contents to the exterior.

exodermis In the roots of some plants, an outer layer of root cortex that may limit water losses from roots and help regulate the absorption of ions.

exoenzyme An enzymatic protein released by a bacterium that digests plasma membranes and causes cells of the infected host to rupture and die.

exon An amino acid–coding sequence present in pre-mRNA that is retained in a spliced mRNA that is translated to produce a polypeptide.

exon shuffling Process by which existing amino acid–coding regions or domains are mixed into novel combinations to create new proteins.

exoskeleton A hard external covering of an animal's body that blocks the passage of water and provides support and protection.

exotic species A nonnative organism.

exotoxin A toxic protein that leaks from or is secreted from a bacterium and interferes with the biochemical processes of body cells in various ways.

experimental data Information that describes the result of a careful manipulation of the system under study.

experimental variable The variable to which any difference in observations of experimental treatment subjects and control treatment subjects is attributed.

exploitative competition Form of competition in which two or more individuals or populations use the same limiting resources.

exponential model of population growth Model that describes unlimited population growth.

external fertilization The process in which sperm and eggs are shed into the surrounding water, occurring in most aquatic invertebrates, bony fishes, and amphibians.

external gill A gill that extends out from the body and lacks a protective covering.

extinction The death of the last individual in a species or the last species in a lineage.

extracellular digestion Digestion that takes place outside body cells, in a pouch or tube enclosed within the body.

extracellular fluid The fluid occupying the spaces between cells in multicellular animals.

extracellular matrix (ECM) A molecular system that supports and protects cells and provides mechanical linkages.

extraembryonic membrane A primary tissue layer extended outside the embryo that conducts nutrients from the yolk to the embryo, exchanges gases with the environment outside the egg, or stores metabolic wastes removed from the embryo.

F pilus Structure on the cell surface that allows an F^+ donor bacterial cell to attach to an F^- recipient bacterial cell. Also referred to as a *sex pilus*.

F^- cell Recipient cell in conjugation between bacteria.

F^+ cell Donor cell in conjugation between bacteria.

F_1 generation The first generation of offspring from a genetic cross.

F_2 generation The second generation of offspring from a genetic cross.

facilitated diffusion Mechanism by which polar and charged molecules diffuse across membranes with the help of transport proteins.

facilitation hypothesis An hypothesis that explains ecological succession, suggesting that species modify the local environment in ways that make it less suitable for themselves but more suitable for colonization by species typical of the next successional stage.

facultative anaerobe An organism that can live in the presence or absence of oxygen, using oxygen when it is present and living by fermentation under anaerobic conditions.

family A Linnaean taxonomic category that ranks below an order and above a genus.

family planning program A program that educates people about ways to produce an optimal family size on an economically feasible schedule.

fast block to polyspermy The barrier set up by the wave of depolarization triggered when sperm and egg fuse, making it impossible for other sperm to enter the egg.

fast muscle fiber A muscle fiber that contracts relatively quickly and powerfully.

fat Neutral lipid that is semisolid at biological temperatures.

fate map Mapping of adult or larval structures onto the region of the embryo from which each structure developed.

fat-soluble vitamin A vitamin that dissolves in liquid fat or fatty oils, in addition to water.

fatty acid One of two components of a neutral lipid, containing a single hydrocarbon chain with a carboxyl group linked at one end.

feather A sturdy, lightweight structure of birds, derived from scales in the skin of their ancestors.

feces Condensed and compacted digestive contents in the large intestine.

feedback inhibition In enzyme reactions, regulation in which the product of a reaction acts as a regulator of the reaction. Also referred to as *end-product inhibition*.

fermentation Process in which electrons carried by NADH are transferred to an organic acceptor molecule rather than to the electron transfer system.

fertilization The fusion of the nuclei of an egg and sperm cell, which initiates development of a new individual.

fetus A developing human from the eighth week of gestation onward, at which point the major organs and organ systems have formed.

fiber In sclerenchyma, an elongated, tapered, thick-walled cell that gives plant tissue its flexible strength.

fibrin A protein necessary for blood clotting; fibrin forms a web-like mesh that traps platelets and red blood cells and holds a clot together.

fibrinogen A plasma protein that plays a central role in the blood-clotting mechanism.

fibroblast The type of cell that secretes most of the collagen and other proteins in the loose connective tissue.

fibronectin A class of glycoproteins that aids in the attachment of cells to the extracellular matrix and helps hold the cells in position.

fibrous connective tissue Tissue in which fibroblasts are sparsely distributed among dense masses of collagen and elastin fibers that are lined up in highly ordered, parallel bundles, producing maximum tensile strength and elasticity.

fibrous root system A root system that consists of branching roots rather than a main taproot; roots tend to spread laterally from the base of the stem.

filament In flowers, the stalk of a stamen, which supports the anther.

filtration The nonselective movement of some water and a number of solutes—ions and small molecules, but not large molecules such as proteins—into the proximal end of the renal tubules through spaces between cells.

first law of thermodynamics The principle that energy can be transferred and transformed but it cannot be created or destroyed.

first messenger The extracellular signal molecule in signal transduction pathways controlled by G-protein–coupled receptors.

fission The mode of asexual reproduction in which the parent separates into two or more offspring of approximately equal size.

fixed action pattern A highly stereotyped instinctive behavior; when triggered by a specific cue, it is performed over and over in almost exactly the same way.

flagellum (plural, flagella) A long, thread-like, cellular appendage responsible for movement; found in both prokaryotes and eukaryotes, but with different structures and modes of locomotion.

flower The reproductive structure of angiosperms, consisting of floral parts grouped on a stem; the structure in which seeds develop.

fluid feeder An animal that obtains nourishment by ingesting liquids that contain organic molecules in solution.

fluid mosaic model Model proposing that the membrane consists of a fluid phospholipid bilayer in which proteins are embedded and float freely.

follicle cell A cell that grows from ovarian tissue and nourishes the developing egg.

follicle-stimulating hormone (FSH) The pituitary hormone that stimulates oocytes in the ovaries to continue meiosis and become follicles. During follicle enlargement, FSH interacts with luteinizing hormone to stimulate follicular cells to secrete estrogens.

food chain A depiction of the trophic structure of a community, a portrait of who eats whom.

food vacuole A membrane-bound sac used for digestion.

food web A set of interconnected food chains with multiple links.

forebrain The largest division of the brain, which includes the cerebral cortex and basal ganglia. It is credited with the highest intellectual functions.

foreskin A loose fold of skin that covers the glans of the penis.

formula The name of a molecule written in chemical shorthand.

fossil The remains or traces of an organism of a past geologic age embedded and preserved in Earth's crust.

founder effect An evolutionary phenomenon in which a population that was established by just a few colonizing individuals has only a fraction of the genetic diversity seen in the population from which it was derived.

fovea The small region of the retina around which cones are concentrated in mammals and birds with eyes specialized for daytime vision.

fragmentation A type of vegetative reproduction in plants in which cells or a piece of the parent break off, then develop into new individuals.

frameshift mutation Mutation in a protein-coding gene that causes the reading frame of an mRNA transcribed from the gene to be altered, resulting in the production of a different, and nonfunctional, amino acid sequence in the polypeptide.

free energy The energy in a system that is available to do work.

freeze-fracture technique Technique in which experimenters freeze a block of cells rapidly, then fracture the block to split the lipid bilayer and expose the hydrophobic membrane interior.

frequency-dependent selection A form of natural selection in which rare phenotypes have a selective advantage simply because they are rare.

fruit A mature ovary, often with accessory parts, from a flower.

fruiting body In some fungi, a stalked, spore-producing structure such as a mushroom.

functional genomics The study of the functions of genes and of other parts of the genome.

functional groups The atoms in reactive groups.

fundamental niche The range of conditions and resources that a population can possibly tolerate and use.

furrow In cytokinesis, a groove that girdles the cell and gradually deepens until it cuts the cytoplasm into two parts.

fusiform initial A cell derived from cambium inside a vascular bundle; gives rise to secondary xylem and phloem cells.

G_0 phase The phase of the cell cycle in eukaryotes in which many cell types stop dividing.

G_1 phase The initial growth stage of the cell cycle in eukaryotes, during which the cell makes proteins and other types of cellular molecules but not nuclear DNA.

G_2 phase The phase of the cell cycle in eukaryotes during which the cell continues to synthesize proteins and to grow, completing interphase.

gallbladder The organ that stores bile between meals, when no digestion is occurring.

gametangium A cell or organ in which gametes are produced.

gamete A haploid cell, and egg or sperm. Haploid cells fuse during sexual reproduction to form a diploid zygote.

gametic isolation A prezygotic reproductive isolating mechanism caused by incompatibility between the sperm of one species and the eggs of another; may prevent fertilization.

gametogenesis The formation of male and female gametes.

gametophyte An individual of the haploid generation produced when a spore germinates and grows directly by mitotic divisions in organisms that undergo alternation of generations.

ganglion A functional concentration of nervous system tissue composed principally of nerve-cell bodies, usually lying outside the central nervous system.

ganglion cell A type of neuron in the retina of the eye that receives visual information

from photoreceptors via various intermediate cells such as bipolar cells, amacrine cells, and horizontal cells.

gap gene In *Drosophila* embryonic development, the first activated set of segmentation genes that progressively subdivide the embryo into regions, determining the segments of the embryo and the adult.

gap junction Junction that opens direct channels allowing ions and small molecules to pass directly from one cell to another.

gastric juice A substance secreted by the stomach that contains the digestive enzyme pepsin.

gastrovascular cavity A saclike body cavity with a single opening, a mouth, which serves both digestive and circulatory functions.

gastrula The developmental stage resulting when the cells of the blastula migrate and divide once cleavage is complete.

gastrulation The second major process of early development in most animals, which produces an embryo with three distinct primary tissue layers.

gated channel Ion transporter in a membrane that switches between open, closed, or intermediate states.

gemma (plural, gemmae) Small cell mass that forms in cuplike growths on a thallus.

gene A unit containing the code for a protein molecule or one of its parts, or for functioning RNA molecules such as tRNA and rRNA.

gene flow The transfer of genes from one population to another through the movement of individuals or their gametes.

gene pool The sum of all alleles at all gene loci in all individuals in a population.

gene therapy Correction of genetic disorders using genetic engineering techniques.

gene-for-gene recognition A mechanism in which plants can detect an attack by a specific pathogen; the product of a specific plant gene interacts with the product of a specific pathogen gene, triggering the plant's defensive response.

general transcription factor (basal transcription factor) In eukaryotes, a protein that binds to the promoter of a gene in the area of the TATA box and recruits and orients RNA polymerase II to initiate transcription at the correct place.

generalized compartment model A model used to describe nutrient cycling in which two criteria—organic *versus* inorganic nutrients and available *versus* unavailable nutrients—define four compartments where nutrients accumulate.

generalized transduction Transfer of bacterial genes between bacteria using virulent phages that have incorporated random DNA fragments of the bacterial genome.

generation time The average time between the birth of an organism and the birth of its offspring.

genetic code The nucleotide information that specifies the amino acid sequence of a polypeptide.

genetic counseling Counseling that allows prospective parents to assess the possibility that they might have a child affected by a genetic disorder.

genetic drift Random fluctuations in allele frequencies as a result of chance events; usually reduces genetic variation in a population.

genetic engineering The use of DNA technologies to alter genes for practical purposes.

genetic equilibrium The point at which neither the allele frequencies nor the genotype frequencies in a population change in succeeding generations.

genetic recombination The process by which the combinations of alleles for different genes in two parental individuals become shuffled into new combinations in offspring individuals.

genetic screening Biochemical or molecular tests for identifying inherited disorders after a child is born.

genetically modified organism (GMO) A transgenic organism.

genomic imprinting Pattern of inheritance in which the expression of a nuclear gene is based on whether an individual organism inherits the gene from the male or female parent.

genomic library A collection of clones that contains a copy of every DNA sequence in a genome.

genotype The genetic constitution of an organism.

genotype frequency The percentage of individuals in a population possessing a particular genotype.

genus A Linnaean taxonomic category ranking below a family and above a species.

geographical range The overall spatial boundaries within which a population lives.

germ cell An animal cell that is set aside early in embryonic development and gives rise to the gametes.

germ-line gene therapy Experiment in which a gene is introduced into germ-line cells of an animal to correct a genetic disorder.

gestation The period of mammalian development in which the embryo develops in the uterus of the mother.

gibberellin Any of a large family of plant hormones that regulate aspects of growth, including cell elongation.

gill A respiratory organ formed as evagination of the body that extends outward into the respiratory medium.

gill arch One of the series of curved supporting structures between the slits in the pharynx of a chordate.

gill slit One of the openings in the pharynx of a chordate through which water passes out of the pharynx.

Ginkgophyta A plant phylum with a single living species, the ginkgo (or maiden-hair) tree.

gizzard The part of the digestive tube that grinds ingested material into fine particles by muscular contractions of the wall.

gland A cell or group of cells that produces and releases substances nearby, in another part of the body, or to the outside.

glans A soft, caplike structure at the end of the penis, containing most of the nerve endings producing erotic sensations.

glial cell A nonneuronal cell contained in the nervous tissue that physically supports and provides nutrients to neurons, provides electrical insulation between them, and scavenges cellular debris and foreign matter.

globulin A plasma protein that transports lipids (including cholesterol) and fat-soluble vitamins; a specialized subgroup of globulins, the immunoglobulins, constitute antibodies and other molecules contributing to the immune response.

glomerulus A ball of blood capillaries surrounded by Bowman's capsule in the human nephron.

glucagon A pancreatic hormone with effects opposite to those of insulin: it stimulates glycogen, fat, and protein degradation.

glucocorticoid A steroid hormone secreted by the adrenal cortex that helps maintain the blood concentration of glucose and other fuel molecules.

glycocalyx A carbohydrate coat covering the cell surface.

glycogen Energy-providing carbohydrates stored in animal cells.

glycolysis Stage of cellular respiration in which sugars such as glucose are partially oxidized and broken down into smaller molecules.

glycosidic bond Bond formed by the linkage of two α-glucose molecules with oxygen as a bridge between a carbon of the first glucose unit and a carbon of the second glucose unit.

Gnathostomata The group of vertebrates with moveable jaws.

Golgi complex In eukaryotes, the organelle responsible for the final modification, sorting, and distribution of proteins and lipids.

Golgi tendon organ A proprioceptor of tendons.

gonad A specialized gamete-producing organ in which the germ cells collect. Gonads are the primary source of sex hormones in vertebrates: ovaries in the female and testes in the male.

gonadotropin A hormone that regulates the activity of the gonads (ovaries and testes).

gonadotropin releasing hormone (GnRH) A tropic hormone secreted by the hypothalamus that causes the pituitary to make luteinizing hormone (LH) and follicle stimulating hormone (FSH).

G-protein–coupled receptor In signal transduction a surface receptor that responds to a signal by activating a G protein.

graded potential A change in membrane potential that does not necessarily trigger an action potential.

gradualism The view that Earth and its living systems changed slowly over its history.

gradualist hypothesis The hypothesis that large changes in either geological features or biological lineages result from the slow, continuous accumulation of small changes over time.

Gram stain technique A technique of staining bacteria to distinguish between types of bacteria with different cell wall compositions.

Gram-negative Describing bacteria that do not retain the stain used in the Gram stain technique.

Gram-positive Describing bacteria that appear purple when stained using the Gram stain technique.

granum Structure in the chloroplasts of higher plants formed by thylakoids stacked one on top of another.

gravitropism A directional growth response to Earth's gravitational pull that is induced by mechanical and hormonal influences.

gray crescent A crescent-shaped region of the underlying cytoplasm at the side opposite the point of sperm entry exposed after fertilization when the pigmented layer of cytoplasm rotates toward the site of sperm entry.

gray matter Areas of densely packed nerve cell bodies and dendrites in the brain and spinal cord.

greater vestibular gland One of two glands located slightly below and to the left and right of the opening of the vagina in women. They secrete mucus to provide lubrication, especially when the woman is sexually aroused.

greenhouse effect A phenomenon in which certain gases foster the accumulation of heat in the lower atmosphere, maintaining warm temperatures on Earth.

gross primary productivity The rate at which producers convert solar energy into chemical energy.

ground meristem The primary meristematic tissue in plants that gives rise to ground tissues, mostly parenchyma.

ground tissue system One of the three basic tissue systems in plants; includes all tissues other than dermal and vascular tissues.

growth factor Any of a large group of peptide hormones that regulates the division and differentiation of many cell types in the body.

growth hormone (GH) A hormone that stimulates cell division, protein synthesis, and bone growth in children and adolescents, thereby causing body growth.

guanine A purine that base-pairs with cytosine in nucleic acids.

guard cell Either of a pair of specialized crescent-shaped cells that control the opening and closing of stomata in plant tissue.

guttation The exudation of water from leaves as a result of strong root pressure.

gymnosperm A seed plant that produces "naked" seeds not enclosed in an ovary.

H$^+$ pump *See* proton pump.

habitat The specific environment in which a population lives, as characterized by its biotic and abiotic features.

habitat fragmentation A process in which remaining areas of intact habitat are reduced to small, isolated patches.

habituation The learned loss of responsiveness to stimuli.

half-life The time it takes for half of a given amount of a radioisotope to decay.

haplodiploidy A pattern of sex determination in insects in which females are diploid and males are haploid.

haploid An organism or cell with only one copy of each type of chromosome in its nuclei.

Hardy-Weinberg principle An evolutionary rule of thumb that specifies the conditions under which a population of diploid organisms achieves genetic equilibrium.

haustorium (plural, haustoria) The hyphal tip of a parasitic fungus that penetrates a host plant and absorbs nutrients from it; likewise in parasitic flowering plants, a root that can penetrate a host's tissues and absorb nutrients.

head The anteriormost part of the body, containing the brain, sensory structures, and feeding apparatus.

head–foot In mollusks, the region of the body that provides the major means of locomotion and contains concentrations of nervous system tissues and sense organs.

heartwood The inner core of a woody stem; composed of dry tissue and nonliving cells that no longer transport water and solutes and may store resins, tannins, and other defensive compounds.

heat of vaporization The heat required to give water molecules enough energy of motion to break loose from liquid water and form a gas.

heat-shock protein (HSP) Any of a group of chaperone proteins that are present in all cells in all life forms. They are induced when a cell undergoes various types of environmental stresses like heat, cold, and oxygen deprivation.

heavy chain The heavier of the two types of polypeptide chains that are found in immunoglobulin and antibody molecules.

helical virus A virus in which the protein subunits of the coat assemble in a rodlike spiral around the genome.

helper T cell A clonal cell that assists with the activation of B cells.

hemolymph The circulatory fluid of invertebrates with open circulatory systems, including mollusks and arthropods.

hepatic portal vein The blood vessel that leads to capillary networks in the liver.

Hepatophyta The phylum that includes liverworts and their bryophyte relatives.

herbicide A compound that, at proper concentration, kills plants.

herbivore An animal that obtains energy and nutrients primarily by eating plants.

herbivory The process in which herbivores consume plants.

hermaphroditism The mechanism in which both mature egg-producing and mature sperm-producing tissue are present in the same individual.

heterochromatin In eukaryotes, regions of densely packed chromatin fibers in interphase nuclei.

heterochrony Changes in the relative rate of development of morphological characters.

heterosporous Producing two types of spores, "male" microspores and "female" megaspores.

heterotroph An organism that acquires energy and nutrients by eating other organisms or their remains.

heterozygote An individual with two different alleles of a gene.

heterozygote advantage An evolutionary circumstance in which individuals that are heterozygous at a particular locus have higher relative fitness than either homozygote.

heterozygous State of possessing two different alleles of a gene.

Hfr cell A special donor cell that can transfer genes on a bacterial chromosome to a recipient bacterium.

hibernation Extended torpor during winter.

hindbrain The lower area of the brain that includes the brain stem, medulla oblongata, and pons.

hippocampus A gray-matter center that is involved in sending information.

histone A small, positively charged (basic) protein that is complexed with DNA in the chromosomes of eukaryotes.

historical biogeography The study of the geographical distributions of plants and animals in relation to their evolutionary history.

homeobox A region of a homeotic gene that corresponds to an amino acid section of the homeodomain.

homeodomain An encoded transcription factor of each protein that binds to a region in the promoters of the genes whose transcription it regulates.

homeostasis A steady internal condition maintained by responses that compensate for changes in the external environment.

homeostatic mechanism Any process or activity responsible for homeostasis.

homeotic gene Any of the family of genes that determines the structure of body parts during embryonic development.

hominid A member of a monophyletic group of primates, characterized by an erect bipedal stance, that includes modern humans and their recent ancestors.

Hominoidea The monophyletic group of primates that includes apes and humans.

homologies Characteristics shared by a set of species because they inherited them from their common ancestor.

homologous traits Characteristics that are similar in two species because they inherited the genetic basis of the trait from their common ancestor.

homoplasies Characteristics shared by a set of species, often because they live in similar environments, but not present in their common ancestor; often the product of convergent evolution.

homosporous Producing only one type of spore.

homozygote An individual with two copies of the same allele.

homozygous State of possessing two copies of the same allele.

horizon A noticeable layer of soil, such as topsoil, having a distinct texture and composition that varies with soil type.

horizontal cell A type of neuron that forms lateral connections among photoreceptor cells in the retina of the eye.

hormone A signaling molecule secreted by a cell that can alter the activities of any cell with receptors for it; in animals, typically a molecule produced by one tissue and transported via the bloodstream to another specific tissue to alter its physiological activity.

host A species that is fed upon by a parasite.

host race A population of insects that may be reproductively isolated from other populations of the same species as a consequence of their adaptation to feed on a specific host plant species.

human chorionic gonadotropin (hCG) A hormone that keeps the corpus luteum in the ovary from breaking down.

human immunodeficiency virus (HIV) A retrovirus that causes acquired immunodeficiency syndrome (AIDS).

humus The organic component of soil remaining after decomposition of plants and animals, animal droppings, and other organic matter.

hybrid breakdown A postzygotic reproductive isolating mechanism in which hybrids are capable of reproducing, but their offspring have either reduced fertility or reduced viability.

hybrid inviability A postzygotic reproductive isolating mechanism in which a hybrid individual has a low probability of survival to reproductive age.

hybrid sterility A postzygotic reproductive isolating mechanism in which hybrid offspring cannot form functional gametes.

hybrid zone A geographical area where the hybrid offspring of two divergent populations or species are common.

hybridoma A B cell that has been induced to fuse with a cancerous lymphocyte called a myeloma cell, forming single, composite cell.

hydration layer A surface coat of water molecules that covers other polar and charged molecules and ions.

hydrocarbon Molecule consisting of carbon linked only to hydrogen atoms.

hydrogen bond Noncovalent bond formed by unequal electron sharing between hydrogen atoms and oxygen, nitrogen, or sulfur atoms.

hydrologic cycle The global cycling of water between the ocean, the atmosphere, land, freshwater ecosystems, and living organisms.

hydrolysis Reaction in which the components of a water molecule are added to functional groups as molecules are broken into smaller subunits.

hydrophilic Polar molecules that associate readily with water.

hydrophobic Nonpolar substances that are excluded by water and other polar molecules.

hydroponic culture A method of growing plants not in soil but with the roots bathed in a solution that contains water and mineral nutrients.

hydrosphere The component of the biosphere that encompasses all the waters on Earth, including oceans, rivers, and polar ice caps.

hydrostatic skeleton A structure consisting of muscles and fluid that, by themselves, provide support for the animal or part of the animal; no rigid support, like a bone, is involved.

hydroxyl group Group consisting of an oxygen atom linked to a hydrogen atom on one side and to a carbon chain on the other side.

hymen A thin flap of tissue that partially covers the opening of the vagina.

hyperpolarized The condition of a neuron when its membrane potential is more negative than the resting value.

hypersensitive response A plant defense that physically cordons off an infection site by surrounding it with dead cells.

hypertension Commonly called high blood pressure, a medical condition in which blood pressure is chronically elevated above normal values.

hyperthermia The condition resulting when the heat gain of the body is too great to be counteracted.

hypertonic Solution containing dissolved substances at higher concentrations than the cells it surrounds.

hypha (plural, hyphae) Any of the threadlike filaments that form the mycelium of a fungus.

hypoblast The bottom layer of a blastodisc.

hypocotyl The region of a plant embryo's vertical axis between the cotyledons and the radicle.

hypodermis The innermost layer of the skin that contains larger blood vessels and additional reinforcing connective tissue.

hypolimnion The deep water of the profundal zone of a lake.

hypothalamus The portion of the brain that contains centers regulating basic homeostatic functions of the body and contributing to the release of hormones.

hypothermia A condition in which the core temperature falls below normal for a prolonged period.

hypothesis A "working explanation" of observed facts.

hypotonic Solution containing dissolved substances at lower concentrations than the cells it surrounds.

ice lattice A rigid, crystalline structure formed when a water molecule in ice forms four hydrogen bonds with neighboring molecules.

imbibition The movement of water into a seed as the water molecules are attracted to hydrophilic groups of stored proteins; the first step in germination.

immigration Movement of organisms into a population.

immune response The defensive reactions of the immune system.

immune system The body's system of defenses against disease, composed of certain white blood cells and antibodies.

immunoglobulin A specific protein substance produced by plasma cells to aid in fighting infection.

immunological memory The capacity of the immune system to respond more rapidly and vigorously to the second contact with a specific antigen than to the primary contact.

immunological tolerance The process that protects the body's own molecules from attack by the immune system.

imperfect flower A type of incomplete flower that has stamens or carpels, but not both.

imprinting The process of learning the identity of a caretaker and potential future mate during a critical period.

inbreeding A special form of nonrandom mating in which genetically related individuals mate with each other.

incisors Flattened, chisel-shaped teeth of mammals, located at the front of the mouth, that are used to nip or cut food.

incomplete dominance Condition in which the effects of recessive alleles can be detected to some extent in heterozygotes.

incomplete flower A flower lacking one or more of the four floral whorls.

incomplete metamorphosis In certain insects, a life cycle characterized by the absence of a pupal stage between the immature and adult stages.

incurrent siphon A muscular tube that brings water containing oxygen and food into the body of an invertebrate.

incus The second of the three sound-conducting middle ear bones in vertebrates, located between the malleus and the stapes.

independent assortment Mendel's principle that the alleles of the genes that govern two characters segregate independently during formation of gametes.

indeterminate cleavage A type of cleavage, observed in many deuterostomes, in which the developmental fates of the first few cells produced by mitosis are not determined as soon as cells are produced.

indeterminate growth Growth that is not limited by an organism's genetic program, so that the organism grows for as long as it lives; typical of many plants. *Compare* determinate growth.

indirect neurotransmitter A neurotransmitter that acts as a first messenger, binding to a G-protein–coupled receptor in the postsynaptic membrane, which activates the receptor and triggers generation of a second messenger such as cyclic AMP or other processes.

inducer Concerning regulation of gene expression in bacteria, a molecule that turns on the transcription of the genes in an operon.

inducible operon Operon whose expression is increased by an inducer molecule.

induction A mechanism in which one group of cells (the inducer cells) causes or influences another nearby group of cells (the responder cells) to follow a particular developmental pathway.

infection thread In the formation of root nodules on nitrogen-fixing plants, the tube formed by the plasma membrane of root hair cells as bacteria enter the cell.

inflammation The heat, pain, redness, and swelling that occur at the site of an infection.

ingestion The feeding methods used to take food into the digestive cavity.

inheritance The transmission of DNA (that is, genetic information) from one generation to the next.

inhibin A peptide that, in females, is an inhibitor of FSH secretion from the pituitary thereby diminishing the signal for follicular growth. In males, inhibin inhibits FSH secretion from the pituitary, thereby decreasing spermatogenesis.

inhibiting hormone (IH) A hormone released by the hypothalamus that inhibits the secretion of a particular anterior pituitary hormone.

inhibition hypothesis An hypothesis suggesting that new species are prevented from occupying a community by whatever species are already present.

inhibitory postsynaptic potential (IPSP) A change in membrane potential caused when hyperpolarization occurs, pushing the neuron farther from threshold.

initial A plant cell that remains permanently as part of a meristem and gives rise to daughter cells that differentiate into specialized cell types.

initiator codon *See* start codon.

initiator RNA The aminoacyl-tRNA used for initiation, with an anticodon to the methionine-specifying AUG start codon.

innate immunity A nonspecific line of defense against pathogens that includes inflammation, which creates internal conditions that inhibit or kill many pathogens, and specialized cells that engulf or kill pathogens or infected body cells.

inner boundary membrane Membrane lying just inside the outer boundary membrane of a chloroplast, enclosing the stroma.

inner cell mass The dense mass of cells within the blastocyst that will become the embryo.

inner ear That part of the ear, particularly the cochlea, that converts mechanical vibrations (sound) into neural messages that are sent to the brain.

inner mitochondrial membrane Membrane surrounding the mitochondrial matrix.

inorganic molecule Molecule without carbon atoms in its structure.

inositol triphosphate (IP₃) In particular signal transduction pathways, a second messenger that activates transport proteins in the endoplasmic reticulum to release Ca^{2+} into the cytoplasm. IP_3 is involved in one of two major G-protein–coupled receptor–response pathways.

insertion sequence A transposable element that contains only genes for its transposition.

insight learning A phenomenon in which animals can solve problems without apparent trial-and-error attempts at the solution.

instinctive behavior A genetically "programmed" response that appears in complete and functional form the first time it is used.

insulin A hormone secreted by beta cells in the islets, acting mainly on cells of nonworking skeletal muscles, liver cells, and adipose tissue (fat) to lower blood glucose, fatty acid and amino acids levels, and promote the storage of those molecules.

insulin-like growth factor (IGF) A peptide that directly stimulates growth processes.

integral protein Protein embedded in a phospholipid bilayer.

integration The sorting and interpretation of neural messages and the determination of the appropriate response(s).

integrator In homeostatic feedback, the control center that compares a detected environmental change with a set point.

integument Skin.

interference competition Form of competition in which individuals fight over resources or otherwise harm each other directly.

interferon A cytokine produced by infected host cells affected by viral dsRNA, which acts both on the infected cell that produces it, an autocrine effect, and on neighboring uninfected cells, a paracrine effect.

interkinesis A brief interphase separating the two meiotic divisions.

intermediate disturbance hypothesis Hypothesis proposing that species richness is greatest in communities that experience fairly frequent disturbances of moderate intensity.

intermediate filament A cytoskeletal filament about 10 nm in diameter that provides mechanical strength to cells in tissues.

intermediate-day plant A plant that flowers only when daylength falls between the values for long-day and short-day plants.

internal fertilization The process in which sperm are released by the male close to or inside the entrance of the reproductive tract of the female.

internal gill A gill located within the body that has a cover providing physical protection for the gills. Water must be brought to internal gills.

interneuron A neuron that integrates information to formulate an appropriate response.

internode The region between two nodes on a plant stem.

interphase The first stage of the mitotic cell cycle, during which the cell grows and replicates its DNA before undergoing mitosis and cytokinesis.

interspecific competition The competition for resources between species.

interstitial fluid The fluid occupying the spaces between cells in multicellular animals.

intertidal zone The shoreline that is alternately submerged and exposed by tides.

intestinal villus A microscopic, fingerlike extension in the lining of the small intestine.

intestine The portion of digestive system where organic matter is hydrolyzed by enzymes secreted into the digestive tube. As muscular contractions of the intestinal wall move the mixture along, cells lining the intestine absorb the molecular subunits produced by digestion.

intracellular digestion The process in which cells take in food particles by endocytosis.

intraspecific competition The dependence of two or more individuals in a population on the same limiting resource.

intrinsic rate of increase The maximum possible per capita population growth rate in a population living under ideal conditions.

intron A non–protein-coding sequence that interrupts the protein-coding sequence in a eukaryotic gene. Introns are removed by splicing in the processing of pre-mRNA to mRNA.

invagination The process in which cells changing shape and pushing inward from the surface produce an indentation, such as the dorsal lip of the blastopore.

inversion Chromosomal alteration that occurs if a broken segment reattaches to the same chromosome from which it was lost, but in reversed orientation, so that the order of genes in the segment is reversed with respect to the other genes of the chromosome.

invertebrate An animal without a vertebral column.

involution The process by which cells migrate into the blastopore.

ion A positively or negatively charged atom.

ionic bond Bond that results from electrical attractions between atoms that have lost or gained electrons.

iris Of the eye, the colored muscular membrane that lies behind the cornea and in front of the lens, which by opening or closing determines the size of the pupil and hence the amount of light entering the eye.

islets of Langerhans Endocrine cells that secrete the peptide hormones insulin and glucagon into the bloodstream.

isomers Two or more molecules with the same chemical formula but different molecular structures.

isotonic Equal concentration of water inside and outside cells.

isotope A distinct form of the atoms of an element, with the same number of protons but different number of neutrons.

jasmonate Any of a group of plant hormones that help regulate aspects of growth and responses to stress, including attacks by predators and pathogens.

juvenile hormone (JH) A peptide hormone secreted by the corpora allata, a pair of glands just behind the brain in insects.

juxtaglomerular apparatus A group of receptors that monitor the pressure and flow of fluid through the distal tubule of the kidney.

karyogamy In plants, the fusion of two sexually compatible haploid nuclei after cell fusion (plasmogamy).

karyotype A characteristic of a species consisting of the shapes and sizes of all the chromosomes at metaphase.

keeled sternum The ventrally extended breastbone of a bird to which the flight muscles attach.

ketone Molecule in which the carbonyl group is linked to a carbon atom in the interior of a carbon chain.

keystone species A species that has a greater effect on community structure than its numbers might suggest.

kilocalorie (kcal) The scientific unit equivalent to a Calorie and equal to 1000 small calories.

kin selection Altruistic behavior to close relatives, allowing them to produce proportionately more surviving copies of the altruist's genes than the altruist might otherwise have produced on its own.

kinesis A change in the rate of movement or the frequency of turning movements in response to environmental stimuli.

kinetic energy The energy of motion.

kinetochore A specialized structure consisting of proteins attached to a centromere that mediates the attachment and movement of chromosomes along the mitotic spindle.

kingdom A Linnaean taxonomic category that ranks below a domain and above a phylum.

kingdom Animalia The taxonomic kingdom that includes all living and extinct animals.

kingdom Fungi The taxonomic kingdom that includes all living or extinct fungi.

kingdom Plantae The taxonomic kingdom encompassing all living or extinct plants.

kingdom Protoctista A diverse and polyphyletic group of single-celled and multicellular eukaryotic species.

Korarchaeota A group of Archaea recognized solely on the basis of rRNA coding sequences in DNA taken from environmental samples.

Krebs cycle *See* citric acid cycle.

***K*-selected species** Long-lived species that thrive in more stable environments.

labia majora A pair of fleshy, fat-padded folds that partially cover the labia minora.

labia minora Two folds of tissue that run from front to rear on either side of the opening to the vagina.

lactate fermentation Reaction in which pyruvate is converted into lactate.

lagging strand A DNA strand assembled discontinuously in the direction opposite to DNA unwinding.

landscape ecology The field that examines how large-scale ecological factors—such as the distribution of plants, topography, and

human activity—influence local populations and communities.

larva A sexually immature stage in the life cycle of many animals that is morphologically distinct from the adult.

larynx The voice box.

latent phase The time during which a virus remains in the cell in an inactive form.

lateral bud A bud on the side of a plant stem from which a branch may grow.

lateral geniculate nuclei Clusters of neurons located in the thalamus that receive visual information from the optic nerves and send it on to the visual cortex.

lateral inhibition Visual processing in which lateral movement of signals from a rod or cone proceeds to a horizontal cell and continues to bipolar cells with which the horizontal cell makes inhibitory connections, serving both to sharpen the edges of objects and enhance contrast in an image.

lateral line system The complex of mechano-receptors along the sides of some fishes and aquatic amphibians that detect vibrations in the water.

lateral meristem A plant meristem that gives rise to secondary tissue growth. *Compare* primary meristem.

lateral root A root that extends away from the main root (or taproot).

lateralization A phenomenon in which some brain functions are more localized in one of the two hemispheres.

lateral-line system The complex of organs and sensory receptors along the sides of many fishes and amphibians that detects vibrations in water.

leaching The process by which soluble materials in soil are washed into a lower layer of soil or are dissolved and carried away by water.

leading strand A DNA strand assembled in the direction of DNA unwinding.

leaf primordium A lateral outgrowth from the apical meristem that develops into a young leaf.

learned behavior A response of an animal that is dependent upon having a particular kind of experience during development.

learning A process in which experiences stored in memory change the behavioral responses of an animal.

leghemoglobin An iron-containing, red-pigmented protein produced in root nodules during the symbiotic association between *Bradyrhizobium* or *Rhizobium* and legumes.

lek A display ground where males each possess a small territory from which they court attentive females.

lens The transparent, biconvex intraocular tissue that helps bring rays of light to a focus on the retina.

Lepidosauromorpha A monophyletic lineage of diapsids that includes both marine and terrestrial animals, represented today by sphenodontids, lizards, and snakes.

leukocyte A white blood cell, which eliminates dead and dying cells from the body, removes cellular debris, and participates in defending the body against invading organisms.

Leydig cell A cell that produces the male sex hormones.

lichen A single vegetative body that is the result of an association between a fungus and a photosynthetic partner, often an alga.

life cycle The sequential stages through which individuals develop, grow, maintain themselves, and reproduce.

life history The lifetime pattern of growth, maturation, and reproduction that is characteristic of a population or species.

life table A chart that summarizes the demographic characteristics of a population.

ligament A fibrous connective tissue that connects bones to each other at a joint.

ligand-gated ion channel A channel that opens or closes when a specific chemical, such as a neurotransmitter, binds to the channel.

light chain The lighter of the two types of polypeptide chains found in immunoglobulin and antibody molecules.

light microscope Microscope that uses light to illuminate the specimen.

light-dependent reaction The first stage of photosynthesis, in which the energy of sunlight is absorbed and converted into chemical energy in the form of ATP and NADPH.

light-independent reaction The second stage of photosynthesis, in which electrons are used as a source of energy to convert inorganic CO_2 to an organic form. Also referred to as the *Calvin cycle*.

lignification The deposition of lignin in plant cell walls; it anchors the cellulose fibers in the walls, making them stronger and more rigid, and protects the other wall components from physical or chemical damage.

lignin A tough, rather inert polymer that strengthens the secondary walls of various plant cells and thus helps vascular plants to grow taller and stay erect on land.

limbic system A functional network formed by parts of the thalamus, hypothalamus, and basal nuclei, along with other nearby gray-matter centers—the amygdala, hippocampus, and olfactory bulbs—sometimes called the "emotional brain".

limiting nutrient An element in short supply within an ecosystem, the shortage of which limits productivity.

limnetic zone The sunlit, open water in a lake, beyond the zone where plants rooted in the bottom can grow.

linkage The phenomenon of genes being located on the same chromosome.

linkage map Map of a chromosome showing the relative locations of genes based on recombination frequencies.

linked genes Genes on the same chromosome.

linker A short segment of DNA extending between one nucleosome and the next in a eukaryotic chromosome.

lipopolysaccharide A large molecule that consists of a lipid and a carbohydrate joined by a covalent bond.

lithosphere The component of the biosphere that includes the rocks, sediments, and soils of the crust.

liver A large organ whose many functions include aiding in digestion, removing toxins from the body, and regulating the chemicals in the blood.

loam Any well-aerated soil composed of a mixture of sand, clay, silt, and organic matter.

locus The particular site on a chromosome at which a gene is located.

logistic model of population growth Model of population growth that assumes that a population's per capita growth rate decreases as the population gets larger.

long-day plant A plant that flowers in spring when dark periods become shorter and day length becomes longer.

long-term memory Memory that stores information from days to years or even for life.

long-term potentiation A long-lasting increase in the strength of synaptic connections in activated neural pathways following brief periods of repeated stimulation.

loop of Henle A U-shaped bend of the proximal convoluted tubule.

loose connective tissue A tissue formed of sparsely distributed cells surrounded by a more or less open network of collagen and other glycoprotein fibers.

lophophore The circular or U-shaped fold with one or two rows of hollow, ciliated tentacles that surrounds the mouth of brachiopods, bryozoans, and phoronids and is used to gather food.

loss of imprinting A phenomenon in which the imprinting mechanism for a gene does not work, resulting in both alleles of the gene being active.

lumen The inside of the digestive tube.

lung One of a pair of invaginated respiratory surfaces, buried in the body interior where they are less susceptible to drying out; the organs of respiration in mammals, birds, reptiles, and most amphibians.

luteinizing hormone (LH) A hormone secreted by the pituitary that stimulates the growth and maturation of eggs in females and the secretion of testosterone in males.

Lycophyta The plant phylum that includes club mosses and their close relatives.

lymph The interstitial fluid picked up by the lymphatic system.

lymph node One of many small, bean-shaped organs spaced along the lymph vessels that contain macrophages and other leukocytes that attack invading disease organisms.

lymphatic system An accessory system of vessels and organs that helps balance the fluid content of the blood and surrounding tissues and participates in the body's defenses against invading disease organisms.

lymphocyte A leukocyte that carries out most of its activities in tissues and organs of the lymphatic system. Lymphocytes play major roles in immune responses.

lysogenic cycle Cycle in which the DNA of the bacteriophage is integrated into the DNA of the host bacterial cell and may remain for many generations.

lysosome Membrane-bound vesicle containing hydrolytic enzymes for the digestion of many complex molecules.

lytic cycle The series of events from infection of one bacterial cell by a phage through the release of progeny phages from lysed cells.

macroevolution Large-scale evolutionary patterns in the history of life, producing major changes in species and higher taxonomic groups.

macromolecule A very large molecule assembled by the covalent linkage of smaller subunit molecules.

macronucleus In ciliophorans, a single large nucleus that develops from a micronucleus but loses all genes except those required for basic "housekeeping" functions of the cell and for ribosomal RNAs.

macronutrient In humans, a mineral required in amounts ranging from 50 mg to more than 1 gram per day. In plants, a nutrient needed in large amounts for the normal growth and development.

macrophage A phagocyte that takes part in nonspecific defenses and adaptive immunity.

magnetoreceptor A receptor found in some animals that navigate long distances which allows them to detect and use Earth's magnetic field as a source of directional information.

magnification The ratio of an object as viewed to its real size.

magnoliids An angiosperm group that includes magnolias, laurels, and avocados; they are more closely related to monocots than to eudicots.

major histocompatibility complex A large cluster of genes encoding the MHC proteins.

malleus The outermost of the sound-conducting bones of the middle ear in vertebrates.

malnutrition A condition resulting from a diet that lacks one or more essential nutrients.

Malpighian tubule The main organ of excretion and osmoregulation in insects, helping them to maintain water and electrolyte balance.

mammary glands Specialized organs of female mammals that produce energy-rich milk, a watery mixture of fats, sugars, proteins, vitamins, and minerals.

mantle One or two folds of the body wall that lines the shell and secretes the substance that forms the shell in mollusks.

map unit The unit of a linkage map, equivalent to a recombination frequency of 1%. Also referred to as a *centimorgan*.

maritime climate Climate tempered by ocean winds.

marsupium An external pouch on the abdomen of many female marsupials, containing the mammary glands, and within which the young continue to develop after birth.

mass extinctions The disappearance of a large number of species in a relatively short period of geological time.

mass number The total number of protons and neutrons in the atomic nucleus.

mast cell A type of cell dispersed through connective tissue that releases histamine when activated by the death of cells, caused by a pathogen at an infection site.

maternal chromosome The chromosome derived from the female parent of an organism.

maternal-effect gene One of a class of genes that regulate the expression of other genes expressed by the mother during oogenesis and that control the polarity of the egg and, therefore, of the embryo.

mating The pairing of a male and a female for the purpose of sexual reproduction.

mating systems The social systems describing how males and females pair up.

mating type A genetically defined strain of an organism (such as a fungus) that can only mate with an organism of the opposite mating type; mating types are often designated + and −.

matter Anything that occupies space and has mass.

mechanical isolation A prezygotic reproductive isolating mechanism caused by differences in the structure of reproductive organs or other body parts.

mechanoreceptor A sensory receptor that detects mechanical energy, such as changes in pressure, body position, or acceleration. The auditory receptors in the ears are examples of mechanoreceptors.

medusa The tentacled, usually bell-shaped, free-swimming sexual stage in the life cycle of a coelenterate.

megapascal A unit of pressure used to measure water potential.

megaspore A plant spore that develops into a female gametophyte; usually larger than a microspore.

meiosis The division of diploid cells to haploid progeny, consisting of two sequential rounds of nuclear and cellular division.

meiosis I The first division of the meiotic cell cycle in which homologous chromosomes pair and undergo an exchange of chromosome segments, and then the homologous chromosomes separate, resulting in two cells, each with the haploid number of chromosomes and with each chromosome still consisting of two chromatids.

meiosis II The second division of the meiotic cell cycle in which the sister chromatids in each of the two cells produced by meiosis I separate and segregate into different cells, resulting in four cells each with the haploid number of chromosomes.

melanocyte-stimulating hormone (MSH) A hormone secreted by the anterior pituitary that controls the degree of pigmentation in melanocytes.

melatonin A peptide hormone secreted by the pineal gland that helps maintain daily biorhythms.

membrane attack complexes (MAC) An abnormal activation of the complement (protein) portion of the blood, forming a cascade reaction that brings blood proteins together, binds them to the cell wall, and then inserts them through the cell membrane.

membrane potential An electrical voltage that measures the potential inside a cell membrane relative to the fluid just outside; it is negative under resting conditions and becomes positive during an action potential.

memory The storage and retrieval of a sensory or motor experience, or a thought.

memory B cell In antibody-mediated immunity, a long-lived cell expressing an antibody on its surface that can bind to a specific antigen. A memory B cell is activated the next time the antigen is encountered, producing a rapid secondary immune response.

memory cell An activated lymphocyte that circulates in the blood and lymph, ready to initiate a rapid immune response upon subsequent exposure to the same antigen.

memory helper T cell In cell-mediated immunity, a long-lived cell differentiated from a helper T cell, which remains in an inactive state in the lymphatic system after an immune reaction has run its course and ready to be activated upon subsequent exposure to the same antigen.

meninges Three layers of connective tissue that surround and protect the spinal cord and brain.

menstrual cycle A cycle of approximately 1 month in the human female during which an egg is released from an ovary and the uterus is prepared to receive the fertilized egg; if fertilization does not occur, the endometrium breaks down, which releases blood and tissue breakdown products from the uterus to the outside through the vagina.

meristem An undifferentiated, permanently embryonic plant tissue that gives rise to new cells forming tissues and organs.

mesenteries Sheets of loose connective tissue, covered on both surfaces with epithelial cells, which suspend the abdominal organs in the coelom and provide lubricated, smooth surfaces that prevent chafing or abrasion between adjacent structures as the body moves.

mesoderm The middle layer of the three primary germ layers of an animal embryo, from which the muscular, skeletal, vascular, and connective tissues develop.

mesohyl The gelatinous middle layer of cells lining the body cavity of a sponge.

mesophyll The ground tissue located between the two outer leaf tissues, composed of loosely packed parenchyma cells that contain chloroplasts.

messenger RNA (mRNA) An RNA molecule that serves as a template for protein synthesis.

metabolism The biochemical reactions that allow a cell or organism to extract energy from its surroundings and use that energy to maintain itself, grow, and reproduce.

metamorphosis A reorganization of the form of certain animals during postembryonic development.

metanephridium The excretory tubule of most annelids and mollusks.

metaphase The phase of mitosis during which the spindle reaches its final form and the spindle microtubules move the chromosomes into alignment at the spindle midpoint.

metapopulation A group of neighboring populations that exchange individuals.

microbody Small, membrane-bound organelle that carries out vital reactions linking metabolic pathways.

microclimate The abiotic conditions immediately surrounding an organism.

microevolution Small-scale genetic changes within populations, often in response to shifting environmental circumstances or chance events.

microfilament A cytoskeletal filament composed of actin.

micronucleus In ciliophorans, one or more diploid nuclei that contains a complete complement of genes, functioning primarily in cellular reproduction.

micronutrient Any mineral required by an organism only in trace amounts.

micropyle A small opening at one end of an ovule through which the pollen tube passes prior to fertilization.

microscope Instrument of microscopy with different magnifications and resolutions of specimens.

microscopy Technique for producing visible images of objects that are too small to be seen by the human eye.

microspore A plant spore from which a male gametophyte develops; usually smaller than a megaspore.

microsporidium (plural, microsporidia) A fungal parasite of animals; many mycologists believe they make up a possible sixth phylum within the Kingdom Fungi.

microtubule A cytoskeletal component formed by the polymerization of tubulin into rigid, hollow rods about 25 nm in diameter.

microtubule organizing center (MTOC) An anchoring point near the center of a eukaryotic cell from which most microtubules extend outward.

microvilli Fingerlike projections forming a brush border in epithelial cells that cover the villi.

midbrain The uppermost of the three segments of the brainstem, serving primarily as an intermediary between the rest of the brain and the spinal cord.

middle ear The air-filled cavity containing three small, interconnected bones: the malleus, incus, and stapes.

middle lamella Layer of gel-like polysaccharides that holds together walls of adjacent plant cells.

migration The predictable seasonal movement of animals from the area where they are born to a distant and initially unfamiliar destination, returning to their birth site later.

mimic The species in Batesian mimicry that resembles the model.

mimicry A form of defense in which one species evolves an appearance resembling that of another.

mineralocorticoid A steroid hormone secreted by the adrenal cortex that regulates the levels of Na^+ and K^+ in the blood and extracellular fluid.

minimal medium A growth medium containing the minimal ingredients that enable a nonmutant organism, such as *E. coli*, to grow.

minimum viable population size The smallest population size that is likely to survive both predictable and unpredictable environmental variation.

mismatch repair Repair system that removes mismatched bases from newly synthesized DNA strands.

missense mutation A base-pair substitution mutation in a protein-coding gene that results in a different amino acid in the encoded polypeptide than the normal one.

mitochondrial electron transfer system Series of electron carriers that alternately pick up and release electrons, ultimately transferring them to their final acceptor, oxygen.

mitochondrial matrix The innermost compartment of the mitochondrion.

mitochondrion Membrane-bound organelle responsible for synthesis of most of the ATP in eukaryotic cells.

mitosis Nuclear division that produces daughter nuclei that are exact genetic copies of the parental nucleus.

model The species in Batesian mimicry that is resembled by the mimic.

model organism An organism with characteristics that make it a particularly useful subject of research because it is likely to produce results widely applicable to other organisms.

modern synthesis A unified theory of evolution developed in the middle of the twentieth century.

molarity (M) The number of moles of a substance dissolved in 1 L of solution.

molars Posteriormost teeth of mammals, with a broad chewing surface for grinding food.

mold Asexual, spore-producing stage of many multicellular fungi.

mole (mol) The atomic weight of an element or the molecular weight of a compound.

molecular clock A technique for dating the time of divergence of two species or lineages, based upon the number of molecular sequence differences between them.

molecular weight The weight of a molecule in grams, equal to the total mass number of its atoms.

molecule A unit composed of atoms combined chemically in fixed numbers and ratios.

molt-inhibiting hormone (MIH) A peptide neurohormone secreted by a gland in the eye stalks of crustaceans that inhibits ecdysone secretion.

monoclonal antibody An antibody that reacts only against the same segment (epitope) of a single antigen.

monocot A plant belonging to the Monocotyledones, one of the two major classes of angiosperms; monocot embryos have a single seed leaf (cotyledon) and pollen grains with a single groove.

monocyte A type of leukocyte that enters damaged tissue from the bloodstream through the endothelial wall of the blood vessel.

monoecious Having both "male" flowers (which possess only stamens) and "female" flowers (which possess only carpels).

monogamy A mating system in which one male and one female form a long-term association.

monohybrid An F_1 heterozygote produced from a genetic cross that involves a single character.

monohybrid cross A genetic cross between two individuals that are each heterozygous for the same pair of alleles.

monomers Identical or nearly identical subunits that link together to form polymers during polymerization.

monophyletic taxon A group of organisms that includes a single ancestral species and all of its descendants.

monosaccharides The smallest carbohydrates, containing three to seven carbon atoms.

monotreme A lineage of mammals that lay eggs instead of bearing live young.

monounsaturated Fatty acids with one double bond.

monsoon cycle A wind pattern that brings seasonally heavy rain to a region by blowing moisture-laden air from the sea to the land.

morphogenesis Orderly, genetically programmed changes in the size, shape, and proportion of body parts of an organism; the process by which specialized tissues and organs form.

morphological species concept The concept that all individuals of a species share measurable traits that distinguish them from individuals of other species.

morphology The form or shape of an organism, or of a part of an organism.

morula The first stage of animal development, a solid ball or layer of blastomeres.

mosaic evolution The tendency of characteristics to undergo different rates of evolutionary change within the same lineage.

motif A highly specialized region in a protein produced by the three-dimensional arrangement of amino acid chains within and between domains.

motile Capable of self-propelled movement.

motor neuron An efferent neuron that carries signals to skeletal muscle.

motor unit A block of muscle fibers that is controlled by branches of the axon of a single efferent neuron.

mRNA splicing Process that removes introns from pre-mRNAs and joins exons together.

mucosa The lining of the gut that contains epithelial and glandular cells.

Müllerian duct The bipotential primitive duct associated with the gonads that leads to a cloaca.

Müllerian mimicry A form of defense in which two or more unpalatable species share a similar appearance.

multicellular organism Individual consisting of interdependent cells.

multiple alleles More than two different alleles of a gene.

multiple fruit A fruit that develops from several ovaries in multiple flowers; examples are pineapples and mulberries.

multiregional hypothesis An hypothesis proposing that after archaic humans migrated from Africa to many regions on Earth, their different populations evolved into modern humans simultaneously.

muscle fiber A bundle of elongated, cylindrical cells that make up skeletal muscle.

muscle spindle A stretch receptor in muscle; a bundlesof small, specialized muscle cells wrapped with the dendrites of afferent neurons and enclosed in connective tissue.

muscle tissue Cells that have the ability to contract (shorten) forcibly.

muscle twitch A single, weak contraction of a muscle fiber.

muscularis The muscular coat of a hollow organ or tubular structure.

mutation A spontaneous and heritable change in DNA.

mutualism A symbiotic interaction between species in which both partners benefit.

mycelium A network of branching hyphae that constitutes the body of a multicellular fungus.

mycobiont The fungal component of a lichen.

mycorrhiza A mutualistic symbiosis in which fungal hyphae associate intimately with plant roots.

myoblast An undifferentiated muscle cell.

myofibril A cylindrical contractile element about 1 μm in diameter that runs lengthwise inside the muscle fiber cell.

myogenic heart A heart that maintains its contraction rhythm with no requirement for signals from the nervous system.

myoglobin An oxygen-storing protein closely related to hemoglobin.

Na^+/K^+ pump Pump that pushes 3 Na^+ out of the cell and 2 K^+ into the cell in the same pumping cycle. Also referred to as the *sodium–potassium pump*.

Nanoarchaeota A group of Archaea that was proposed based on rRNA sequence analysis of a thermophilic archaean found in a symbiotic relationship with an other thermophilic archaean; most probably a subgroup of the Euryarchaeota.

nastic movement In plants, a reversible response to nondirectional stimuli, such as mechanical pressure or humidity.

natural history The branch of biology that examines the form and variety of organisms in their natural environments.

natural killer (NK) cell A type of lymphocyte that destroys virus-infected cells.

natural selection The evolutionary process by which alleles that increase the likelihood of survival and the reproductive output of the individuals that carry them become more common in subsequent generations.

natural theology A belief that knowledge of God may be acquired through the study of natural phenomena.

navigation A wayfinding mechanism in which an animal moves toward a specific destination, using both a compass and a "mental map" of where it is in relation to the destination.

negative feedback The primary mechanism of homeostasis, in which a stimulus—a change in the external or internal environment—triggers a response that compensates for the environmental change.

negative pressure breathing Muscular contractions that expand the lungs, lowering the pressure of the air in the lungs and causing air to be pulled inward.

nekton Animals that can actively swim against water currents.

nematocyst A coiled thread, encapsulated in a cnidocyte, that cnidarians fire at prey or predators, sometimes releasing a toxin through its tip.

nephron A specialized excretory tubule that contributes to osmoregulation and carries out excretion, found in all vertebrates.

neritic zone The shallow water of the oceans above the continental shelves.

nerve A bundle of axons enclosed in connective tissue and all following the same pathway.

nerve cord A bundle of nerves that extends from the central ganglia to the rest of the body, connected to smaller nerves.

nerve net A simple nervous system that coordinates responses to stimuli but has no central control organ, or brain.

nervous tissue Tissue that contains neurons, which serve as lines of communication and control between body parts.

net primary productivity The chemical energy remaining in an ecosystem after a producer's cellular respiration is deducted.

neural crest A band of cells that arises early in the embryonic development of vertebrates near the region where the neural tube pinches off from the ectoderm; later, the cells migrate and develop into unique structures.

neural plate Ectoderm thickened and flattened into a longitudinal band, induced by notochord cells.

neural signaling The process by which an animal responds appropriately to a stimulus.

neural tube A hollow tube in vertebrate embryos that develops into the brain, spinal cord, spinal nerves, and spinal column.

neurogenic heart A heart that beats under the control of signals from the nervous system.

neuromuscular junction The junction between a nerve fiber and the muscle it supplies.

neuron An electrically active cell of the nervous system responsible for controlling behavior and body functions.

neuronal circuit The connection between axon terminals of one neuron and the dendrites or cell body of a second neuron.

neuroscience The integrated study of the structure, function, and development of the nervous system.

neurosecretory neuron A neuron that releases a neurohormone into the circulatory system when appropriately stimulated.

neurotransmitter A chemical released by an axon terminal at a chemical synapse.

neurulation The process in vertebrates by which organogenesis begins with development of the nervous system from ectoderm.

neutral lipid Energy-storing molecule consisting of a glycerol backbone and three fatty acid chains.

neutral variation hypothesis An evolutionary hypothesis that some variation at gene loci coding for enzymes and other soluble proteins is neither favored nor eliminated by natural selection.

neutron Uncharged particle in the nucleus of an atom.

neutrophil A type of phagocytic leukocyte that attaches to blood vessel walls in massive numbers when attracted to the infection site by chemokines.

nitrification A metabolic process in which certain soil bacteria convert ammonia or ammonium ions into nitrites that are then converted by other bacteria to nitrates, a form usable by plants.

nitrogen cycle A biogeochemical cycle that moves nitrogen between the huge atmospheric pool of gaseous molecular nitrogen and several much smaller pools of nitrogen-containing compounds in soils, marine and freshwater ecosystems, and living organisms.

nitrogen fixation A metabolic process in which certain bacteria and cyanobacteria convert molecular nitrogen into ammonia and ammonium ions, forms usable by plants.

nitrogenous base A nitrogen-containing molecule with the properties of a base.

nociceptor A sensory receptor that detects tissue damage or noxious chemicals; their activity registers as pain.

node The point on a stem where one or more leaves are attached.

node of Ranvier The gap between two Schwann cells, which exposes the axon membrane directly to extracellular fluids.

noncompetitive inhibition Inhibition of an enzyme reaction by an inhibitor molecule that binds to the enzyme at a site other than the active site and, therefore, does not compete directly with the substrate for binding to the active site.

noncyclic electron flow Pathway in photosynthesis in which electrons travel in a one-way direction from H_2O to $NADP^+$.

nondisjunction The failure of homologous pairs to separate during the first meiotic division or of chromatids to separate during the second meiotic division.

nonhistone protein All the proteins associated with DNA in a eukaryotic chromosome that are not histones.

nonpolar association Association that occurs when nonpolar molecules clump together.

nonpolar covalent bond Bond in which electrons are shared equally.

nonsense codon *See* stop codon.

nonsense mutation A base-pair substitution mutation in a gene in which the base-pair change results in a change from a sense codon to a nonsense codon in the mRNA. The polypeptide translated from the mRNA is shorter than the normal polypeptide because of the mutation.

nonshivering thermogenesis The generation of heat by oxidative mechanisms in non-muscle tissue throughout the body.

nonvascular plant *See* bryophyte.

norepinephrine A nontropic amine hormone secreted by the adrenal medulla.

notochord A flexible rodlike structure, constructed of fluid-filled cells surrounded by tough connective tissue, which supports a chordate embryo from head to tail.

N-terminal end The end of a polypeptide chain with an —NH_3^+ group.

nuclear envelope In eukaryotes, membranes separating the nucleus from the cytoplasm.

nuclear pore Opening in the membrane of the nuclear envelope through which large molecules, such as RNA and proteins, move between the nucleus and cytoplasm.

nucleoid The central region of a prokaryotic cells with no boundary membrane separating it from the cytoplasm, where DNA replication and RNA transcription occur.

nucleolus The nuclear site of rRNA transcription, processing, and ribosome assembly in eukaryotes.

nucleoplasm The liquid or semiliquid substance within the nucleus.

nucleosome The basic structural unit of chromatin in eukaryotes, consisting of DNA wrapped around a histone core.

nucleosome core particle An eight-protein particle formed by the combination of two molecules each of H2A, H2B, H3, and H4, around which DNA winds for almost two turns.

nucleotide The monomer of nucleic acids consisting of a five-carbon sugar, a nitrogenous base, and a phosphate

nucleus The central region of eukaryotic cells, separated by membranes from the surrounding cytoplasm, where DNA replication and messenger RNA transcription occur.

null hypothesis A statement of what would be seen if the hypothesis being tested were wrong.

null model A conceptual model that predicts what one would see if a particular factor had no effect.

nutrition The processes by which an organism takes in, digests, absorbs, and converts food into organic compounds.

obligate aerobe A microorganism that uses oxygen for cellular respiration and requires oxygen in its surroundings to support growth.

obligate anaerobe A microorganism that cannot use oxygen and can grow only in the absence of oxygen.

observational data Basic information on biological structures or the details of biological processes.

oceanic zone The deep ocean water beyond the continental shelves.

ocellus (plural, ocelli) The simplest eye, which detects light but does not form an image.

oil Neutral lipid that is liquid at biological temperatures.

olfactory bulb A gray-matter center that relay inputs from odor receptors to both the cerebral cortex and the limbic system.

oligodendrocyte A type of glial cell that populates the CNS and is responsible for producing myelin.

oligosaccharin A complex carbohydrate that in plants serves as a signaling molecule and as a defense against pathogens.

oligotrophic lake A lake that is poor in nutrients and organic matter, but rich in oxygen.

ommatidium (plural, ommatidia) A faceted visual unit of a compound eye.

omnivore An animal that feeds at several trophic levels, consuming plants, animals, and other sources of organic matter.

oncogene A gene capable of inducing one or more characteristics of cancer cells.

one gene–one enzyme hypothesis Hypothesis showing the direct relationship between genes and enzymes.

one gene–one polypeptide hypothesis Restatement of the one gene–one enzyme hypothesis, taking into account that some proteins consist of more than one polypeptide and not all proteins are enzymes.

oocyte A developing gamete that becomes an ootid at the end of meiosis.

oogenesis The process of producing eggs.

oogonium A cell that enters meiosis and gives rise to gametes, produced by mitotic divisions of the germ cells in females.

open circulatory system An arrangement of internal transport in some invertebrates in which the vascular fluid, hemolymph, is released into sinuses, bathing organs directly, and is not always retained within vessels.

operant conditioning A form of associative learning in which animals learn to link a voluntary activity, an operant, with its favorable consequences, the reinforcement.

operator A DNA regulatory sequence that controls transcription of an operon.

operculum A lid or flap of the bone serving as the gill cover in some fishes.

operon A cluster of prokaryotic genes and the DNA sequences involved in their regulation.

opposable big toe A big toe that can flex to contact the sole of the foot, allowing the foot to grasp objects in the environment.

opsin One of several different proteins that bond covalently with the light-absorbing pigment of rods and cones (retinal).

optic chiasm Location just behind the eyes where the optic nerves converge before entering the base of the brain, a portion of each optic nerve crossing over to the opposite side.

optical isomers *See* enantiomers.

optimal foraging theory A set of mathematical models that predict the diet choices of animals as they encounter a range of potential food items.

oral hood Soft fleshy structure at the anterior end of a cephalochordate that frames the opening of the mouth.

orbital The region of space where the electron "lives" most of the time.

order A Linnaean taxonomic category of organisms that ranks above a family and below a class.

organ Two or more different tissues integrated into a structure that carries out a specific function.

organ of Corti An organ within the cochlear duct that contains the sensory hair cells detecting sound vibrations transmitted to the inner ear.

organ system The coordinated activities of two or more organs to carry out a major body function such as movement, digestion, or reproduction.

organelles The nucleus and other specialized internal structures and compartments of eukaryotic cells.

organic acid (carboxylic acid) Acid for which the characteristic functional group is a carboxyl group (—COOH).

organic molecule Molecule based on carbon.

organismal ecology An ecological discipline in which researchers study the genetic, biochemical, physiological, morphological, and behavioral adaptations of organisms to their abiotic environments.

organogenesis The development of the major organ systems, giving rise to a free-living individual with the body organization characteristic of its species.

oriented cell division Cell division in different planes; establishes the overall shape of a plant organ.

origin of replication (*ori*) A specific region at which replication of a bacterial chromosome commences.

orthogenesis An obsolete theory that evolution is goal oriented, striving to perfect organisms.

oscula One or more openings in a sponge through which water is expelled.

osmoconformer An animal in which the osmolarity of the cellular and extracellular solutions matches the osmolarity of the environment.

osmolarity The total solute concentration of a solution, measured in osmoles—the number of solute molecules and ions (in moles)—per liter of solution.

osmoreceptor A chemoreceptor in the hypothalamus that responds to changes in the osmolarity of the fluid surrounding it, which reflects the osmolarity generally of the body fluids.

osmoregulation The regulation of water and ion balance.

osmoregulator An animal that uses control mechanisms to keep the osmolarity of cellular and extracellular fluids the same, but at levels that may differ from the osmolarity of the surroundings.

osmosis The passive transport of water across a selectively permeable membrane in response to solute concentration gradients, a pressure gradient, or both.

osmotic pressure A state of dynamic equilibrium in which the pressure of the solution on one side of a selectively permeable membrane exactly balances the tendency of water molecules to diffuse passively from the other side of the membrane due to a concentration gradient.

osteoblast A cell that produces the collagen and mineral of bone.

osteoclast A cell that removes bone minerals and recycles them through the bloodstream.

osteocyte A mature bone cell.

osteon The structural unit of bone, consisting of a minute central canal surrounded by osteocytes embedded in concentric layers of mineral matter.

ostracoderm One of an assortment of extinct, jawless fishes that were covered with bony armor.

otolith One of many small crystals of calcium carbonate embedded in the otolithic membrane of the hair cells.

outer boundary membrane A smooth membrane that surrounds a chloroplast, enclosing the stroma.

outer ear The external structure of the ear, consisting of the pinna and meatus.

outer membrane In Gram-negative bacteria, an additional boundary membrane that covers the peptidoglycan layer of the cell wall.

outer mitochondrial membrane The smooth membrane covering the outside of a mitochondrion.

outgroup comparison A technique used to identify ancestral and derived characters by comparing the group under study to more distantly related species that are not otherwise included in the analysis.

oval window An opening in the bony wall that separates the middle ear from the inner ear.

ovarian cycle The cyclic events in the ovary leading to ovulation.

ovary In animals, the female gonad, which produces female gametes and reproductive hormones. In flowering plants, the enlarged base of a carpel in which one or more ovules develop into seeds.

overexploitation The excessive harvesting of an animal or plant species, potentially leading to its extinction.

overnutrition The condition caused by excessive intake of specific nutrients.

oviduct The tube through which the egg moves from the ovary to the outside of the body.

oviparous Referring to animals that lay eggs containing the nutrients needed for development of the embryo outside the mother's body.

ovoviviparous Referring to animals in which fertilized eggs are retained within the body and the embryo develops using nutrients provided by the egg; eggs hatch inside the mother.

ovulation The process in which oocytes are released into the oviducts as immature eggs.

ovule In plants, the structure in a carpel in which a female gametophyte develops and fertilization takes place.

ovum A female sex cell, or egg.

oxidation The removal of electrons from a substance.

oxidative phosphorylation Synthesis of ATP in which ATP synthase uses an H^+ gradient built by the electron transfer system as the energy source to make the ATP.

oxidized Substance from which the electrons are removed during oxidation.

oxytocin A hormone that stimulates the ejection of milk from the mammary glands of a nursing mother.

P generation The parental individuals used in an initial cross.

P site The site in the ribosome where the tRNA carrying the growing polypeptide chain is bound.

pacemaker cell A specialized cardiac muscle cell in the upper wall of the right atrium that sets the rate of contraction in the heart.

paedomorphosis A common form of heterochrony in which juvenile characteristics are retained in a reproductive adult.

pairing Process in meiosis in which homologous chromosomes come together and pair. Also referred to as *synapsis*.

pair-rule genes In *Drosophila* embryonic development, the set of segmentation regulatory genes activated by gap genes that divide the embryo into units of two segments each.

paleobiology The study of ancient organisms.

pancreas A mixed gland composed of an exocrine portion that secretes digestive enzymes into the small intestine and an endocrine portion, the islets of Langerhans, that secretes insulin and glucagon.

parapatric speciation Speciation between populations with adjacent geographical distributions.

paraphyletic taxon A group of organisms that includes an ancestral species and some, but not all, of its descendents.

parapodia Fleshy lateral extensions of the body wall of aquatic annelids, used for locomotion and gas exchange.

parasite An organism that feeds on the tissues of or otherwise exploits its host.

parasitism A symbiotic interaction in which one species, the parasite, uses another, the host, in a way that is harmful to the host.

parasitoid An insect species in which a female lays eggs in the larva or pupa of another insect species, and her young consume the tissues of the living host.

parasympathetic division The division of the autonomic nervous system that predominates during quiet, low-stress situations, such as while relaxing.

parathyroid gland One of a pair of glands that produce parathyroid hormone (PTH) (found only in tetrapod vertebrates).

parathyroid hormone (PTH) The hormone secreted by the parathyroid glands in response to a fall in blood Ca^{2+} levels.

parental Phenotypes identical to the original parental individuals.

parental investment The time and energy devoted to the production and rearing of offspring.

parthenogenesis A mode of asexual reproduction in which animals produce offspring by the growth and development of an egg without fertilization.

partial diploid A condition in which part of the genome of a haploid organism is diploid. Recipients in bacterial conjugation between an Hfr and an F^- cell become partial diploids for part of the Hfr bacterial chromosome.

parturition The process of giving birth.

passive immunity The acquisition of antibodies as a result of direct transfer from another person.

passive parental care The amount of energy invested in offspring—in the form of the energy stored in eggs or seeds or energy transferred to developing young through a placenta—before they are born.

passive transport The transport of substances across cell membranes without expenditure of energy, as in diffusion.

paternal chromosome The chromosome derived from the male parent of an organism.

pathogenesis-related (PR) protein A hydrolytic enzyme that breaks down components of a pathogen's cell wall.

pattern formation The arrangement of organs and body structures in their proper three-dimensional relationships.

pectoral girdle A bony or cartilaginous structure in vertebrates that supports and is attached to the forelimbs.

pedicellariae Small pincers at the base of short spines in starfishes and sea urchins.

pedigree Chart that shows all parents and offspring for as many generations as possible, the sex of individuals in the different generations, and the presence or absence of a trait of interest.

pelagic province The water in a marine biome.

pellicle A layer of supportive protein fibers located inside the cell, just under the plasma membrane, providing strength and flexibility instead of a cell wall.

pelvic girdle A bony or cartilaginous structure in vertebrates that supports and is attached to the hind limbs.

pepsin An enzyme made in the stomach that breaks down proteins.

pepsinogen The inactive precursor molecule for pepsin.

peptide bond A link formed by a dehydration synthesis reaction between the $-NH_2$ group of one amino acid and the $-COOH$ group of a second.

peptidoglycan A polymeric substance formed from a polysaccharide backbone tied together by short polypeptides, which is the primary structural molecule of bacterial cell walls.

peptidyl transferase An enzyme that catalyzes the reaction in which an amino acid is cleaved from the tRNA in the P site of the ribosome and forms a peptide bond with the amino acid on the tRNA in the A site of the ribosome.

peptidyl-tRNA A tRNA linked to a growing polypeptide chain containing two or more amino acids.

per capita growth rate The difference between the per capita birth rate and the per capita death rate of a population.

perennial A plant in which vegetative growth and reproduction continue year after year.

perfect flower A flower that has both male (stamen) and female (carpel) sexual organs.

perfusion The flow of blood or other body fluids on the internal side of the respiratory surface.

pericarp The fruit wall.

pericycle A tissue of plant roots, located between the endodermis and the phloem, which gives rise to lateral roots.

periderm The outermost portion of bark; consists of cork, cork cambium, and secondary cortex.

peripheral nervous system (PNS) All nerve roots and nerves (motor and sensory) that supply the muscles of the body and transmit information about sensation (including pain) to the central nervous system.

peripheral protein Protein held to membrane surfaces by noncovalent bonds formed with the polar parts of integral membrane proteins or membrane lipids.

peristalsis The rippling motion of muscles in the intestine or other tubular organs characterized by the alternate contraction and relaxation of the muscles that propel the contents onward.

peritoneum The thin tissue derived from mesoderm that lines the abdominal wall and covers most of the organs in the abdomen.

peritubular capillary A capillary of the network surrounding the glomerulus.

permafrost Perpetually frozen ground below the topsoil.

peroxisome Microbody that produces hydrogen peroxide as a by-product.

petal Part of the corolla of a flower, often brightly colored. **petiole** The stalk by which a leaf is attached to a stem.

pH scale The numerical scale used by scientists to measure acidity.

phage *See* bacteriophage.

phagocytosis Process in which some types of cells engulf bacteria or other cellular debris to break them down.

pharynx The throat. In some invertebrates, a protrusible tube used to bring food into the mouth for passage to the gastrovascular cavity; in mammals, the common pathway for air entering the larynx and food entering the esophagus.

phenotype The outward appearance of an organism.

phenotypic variation Differences in appearance or function between individual organisms.

pheromone A distinctive volatile chemical released in minute amounts to influence the behavior of members of the same species.

phloem The food-conducting tissue of a vascular plant.

phloem sap The solution of water and organic compounds that flows rapidly through the sieve tubes of flowering plants.

phosphate group Group consisting of a central phosphorus atom held in four linkages: two that bind —OH groups to the central phosphorus atom, a third that binds an oxygen atom to the central phosphorus atom, and a fourth that links the phosphate group to an oxygen atom.

phosphodiester bond The linkage of nucleotides in polynucleotide chains by a bridging phosphate group between the 5′ carbon of one sugar and the 3′ carbon of the next sugar in line.

phospholipid A phosphate-containing lipid.

phosphorus cycle A biogeochemcial cycle in which weathering and erosion carry phosphate ions from rocks to soil and into streams and rivers, which eventually transport them to the ocean, where they are slowly incorporated into rocks.

phosphorylation The addition of a phosphate group to a molecule.

photic zone Surface water of a lake or ocean that sunlight penetrates.

photoautotroph A photosynthetic organism that uses light as its energy source and carbon dioxide as its carbon source.

photobiont The photosynthetic component of a lichen.

photoheterotroph An organism that uses light as the ultimate energy source but obtains carbon in organic form rather than as carbon dioxide.

photoperiodism The response of plants to changes in the relative lengths of light and dark periods in their environment during each 24-hour period.

photophosphorylation The synthesis of ATP coupled to the transfer of electrons energized by photons of light.

photopigment Light-absorbing pigment.

photopsin One of three photopigments in which retinal is combined with different opsins.

photoreceptor A sensory receptor that detects the energy of light.

photorespiration A process that metabolizes a by-product of photosynthesis.

photosynthesis The conversion of light energy to chemical energy in the form of sugar and other organic molecules.

photosystem A large complex into which the light-absorbing pigments for photosynthesis are organized with proteins and other molecules.

photosystem I In photosynthesis, a protein complex in the thylakoid membrane that uses energy absorbed from sunlight to synthesize NADPH.

photosystem II In photosynthesis, a protein complex in the thylakoid membrane that uses energy absorbed from sunlight to synthesize ATP.

phototroph An organism that obtains energy from light.

phototropism The tendency of a plant shoot to bend toward a source of light.

PhyloCode A formal set of rules governing phylogenetic nomenclature.

phylogenetic species concept A concept that seeks to delineate species as the smallest aggregate population that can be united by shared derived characters.

phylogenetic tree A branching diagram depicting the evolutionary relationships of groups of organisms.

phylogeny The evolutionary history of a group of organisms.

phylum (plural, phyla) A major Linnaean division of a kingdom, ranking above a class.

physiological respiration The process by which animals exchange gases with their surroundings—how they take in oxygen from the outside environment and deliver it to body cells, and remove carbon dioxide from body cells and deliver it to the environment.

physiology The study of the functions of organisms—the physico-chemical processes of organisms.

phytoalexin A biochemical that functions as an antibiotic in plants

phytochrome A blue-green pigmented plant chromoprotein involved in the regulation of light-dependent growth processes.

phytoplankton Microscopic, free-flowing aquatic plants and protists.

phytosterol A sterol that occurs in plant cell membranes.

piloting A wayfinding mechanism in which animals use familiar landmarks to guide their journey.

pilus (plural, pili) A hair or hairlike appendage on the surface of a prokaryote.

pinacoderm In sponges, an unstratified outer layer of cells.

pineal gland A light-sensitive, melatonin-secreting gland that regulates some biological rhythms.

pinna The external structure of the outer ear, which concentrates and focuses sound waves.

pinocytosis *See* bulk-phase endocytosis.

pith The soft, spongelike, central cylinder of the stems of most flowering plants, composed mainly of parenchyma.

pituitary A gland consisting mostly of two fused lobes suspended just below the hypothalamus by a slender stalk of tissue that contains both neurons and blood vessels; it interacts with the hypothalamus to control many physiological functions, including the activity of some other glands.

placenta A specialized temporary organ that connects the embryo and fetus with the uterus in mammals, mediating the delivery of oxygen and nutrients.

plasma The clear, yellowish fluid portion of the blood in which cells are suspended. Plasma consists of water, glucose and other sugars, amino acids, plasma proteins, dissolved gases, ions, lipids, vitamins, hormones and other signal molecules, and metabolic wastes.

plasma cell A large antibody-producing cell that develops from B cells.

plasma membrane The outer limit of the cytoplasm responsible for the regulation of substances moving into and out of cells.

plasmid A DNA molecule in the cytoplasm of certain prokaryotes, which often contains genes with functions that supplement those in the nucleoid and which can replicate independently of the nucleoid DNA and be passed along during cell division.

plasmodesma A minute channel that perforates a cell wall and contains extensions of the cytoplasm that directly connect adjacent plant cells.

plasmodial slime mold A slime mold of the class Myxomycetes.

plasmodium The composite mass of plasmodial slime molds consisting of individual nuclei suspended in a common cytoplasm surrounded by a single plasma membrane.

plasmogamy The sexual stage of fungi during which the cytoplasms of two genetically different partners fuse.

plasmolysis Condition due to outward osmotic movement of water, in which plant cells shrink so much that they retract from their walls.

plastids A family of plant organelles.

plastron The ventral part of the shell of a turtle.

plate tectonics The geological theory describing how Earth's crust is broken into irregularly shaped plates of rock that float on its semisolid mantle.

platelet An oval or rounded cell fragment enclosed in its own plasma membrane, which is found in the blood; they are produced in red bone marrow by the division of stem cells and contain enzymes and other factors that take part in blood clotting.

pleiotropy Condition in which single genes affect more than one character of an organism.

pleura The double layer of epithelial tissue covering the lungs.

ploidy The number of chromosome sets of a cell or species.

plumule The rudimentary terminal bud of a plant embryo located at the end of the hypocotyl, consisting of the epicotyl and a cluster of tiny foliage leaves.

polar association Association that occurs when polar molecules attract and align themselves with other polar molecules and with charged ions and molecules.

polar body A nonfunctional cell produced in oogenesis.

polar covalent bond Bond in which electrons are shared unequally.

polar nucleus In the embryo sac of a flowering plant, one of two nuclei that migrate into the center of the sac, become housed in a central cell, and eventually give rise to endosperm.

polar transport Unidirectional movement of a substance from one end of a cell (or other structure) to the other.

polarity The unequal distribution of yolk and other components in a mature egg.

pollen grain The male gametophyte of a seed plant.

pollen sac The microsporangium of a seed plant, in which pollen develops.

pollen tube A tube that grows from a germinating pollen grain through the tissues of a carpel and carries the sperm cells to the ovary.

pollination The transfer of pollen to a flower's reproductive parts by air currents or on the bodies of animal pollinators.

pollutant Materials or energy in a form or quantity that organisms do not usually encounter.

poly(A) tail The string of A nucleotides added posttranscriptionally to the 3′ end of a pre-mRNA molecule and retained in the mRNA produced from it that enables the mRNA to be translated efficiently and protects it from attack by RNA-digesting enzymes in the cytoplasm.

polyandry A polygamous mating system in which one female mates with multiple males.

polygamy A mating system in which either males or females may have many mating partners.

polygenic inheritance Inheritance in which several to many different genes contribute to the same character.

polygyny A polygamous mating system in which one male mates with many females.

polyhedral virus A virus in which the coat proteins form triangular units that fit together like the parts of a geodesic sphere.

polymerase chain reaction (PCR) Process that amplifies a specific DNA sequence from a DNA mixture to an extremely large number of copies.

polymerization Process in which monomers link together to form a polymer.

polymorphism The existence of discrete variants of a character among individuals in a population.

polyp The tentacled, usually sessile stage in the life cycle of a coelenterate.

polypeptide The chain of amino acids formed by sequential peptide bonds.

polyphyletic taxa A group of organisms that belong to different evolutionary lineages and do not share a recent common ancestor.

polyploid An individual with one or more extra copies of the entire haploid complement of chromosomes.

polyploidy The condition of having one or more extra copies of the entire haploid complement of chromosomes.

polysaccharide Chain with more than 10 linked monosaccharide subunits.

polysome The entire structure of an mRNA molecule and the multiple associated ribosomes that are translating it simultaneously.

polyunsaturated Fatty acid with more than one double bond.

population All the individuals of a single species that live together in the same place and time.

population bottleneck An evolutionary event that occurs when a stressful factor reduces population size greatly and eliminates some alleles from a population.

population density The number of individuals per unit area or per unit volume of habitat.

population ecology The ecological discipline that focuses on how a population's size and other characteristics change in space and time.

population genetics The branch of science that studies the prevalence and variation in genes among populations of individuals.

population size The number of individuals in a population at a specified time.

population viability analysis A mathematical analysis used by conservation biologists to determine the minimum viable population size for threatened or endangered species.

positive feedback A mechanism that intensifies or adds to a change in internal or external environmental condition.

positive pressure breathing A gulping or swallowing motion that forces air into the lungs.

posterior Indicating the tail end of an animal.

posterior pituitary The neural portion of the pituitary, which stores and releases two hormones made by the hypothalamus, antidiuretic hormone and oxytocin.

postsynaptic cell The neuron or the surface of an effector after a synapse that receives the signal from the presynaptic cell.

postsynaptic membrane The plasma membrane of the postsynaptic cell.

postzygotic isolating mechanism A reproductive isolating mechanism that acts after zygote formation.

potential energy Stored energy.

preadaptation A characteristic evolved by an ancestral species that serves an adaptive but different function in a descendant species or population.

predation The interaction between animals and the animal prey they consume.

prediction A statement about what the researcher expects to happen to one variable if another variable changes.

pregnancy The period of mammalian development in which the embryo develops in the uterus of the mother.

premolars Teeth located in pairs on each side of the upper and lower jaws of mammals, positioned behind the canines and in front of the molars.

pre-mRNA (precursor-mRNA) The primary transcript of a eukaryotic protein-coding gene, which is processed to form messenger RNA.

prenatal diagnosis Techniques in which cells derived from a developing embryo or its surrounding tissues or fluids are tested for the presence of mutant alleles or chromosomal alterations.

prepuce Foreskin; a loose fold of skin that covers the glans of the penis.

pressure flow mechanism In vascular plants, pressure that builds up at the source end of a sieve tube system and pushes solutes by bulk flow toward a sink, where they are removed.

presynaptic cell The neuron with an axon terminal on one side of the synapse that transmits the signal across the synapse to the dendrite or cell body of the postsynaptic cell.

presynaptic membrane The plasma membrane of the axon terminal of a presynaptic cell, which releases neurotransmitter molecules into the synapse in response to arrival of an action potential.

prezygotic isolating mechanism A reproductive isolating mechanism that acts prior to the production of a zygote, or fertilized egg.

primary active transport Transport in which the same protein that transports a substance also hydrolyzes ATP to power the transport directly.

primary cell layers The ectoderm, mesoderm, and endoderm layers that form the embryonic tissues.

primary cell wall The initial cell wall laid down by a plant cell.

primary consumer An herbivore, a member of the second trophic level.

primary endosymbiosis In the model for the origin of plastids in eukaryotes, the first event in which a eukaryotic cell engulfed a photosynthetic cyanobacterium.

primary growth The growth of plant tissues derived from apical meristems. *Compare* secondary growth.

primary immune response The response of the immune system to the first challenge by an antigen.

primary meristem Root and shoot apical meristems, from which a plant's primary tissues develop. *Compare* lateral meristem.

primary motor area The area of the cerebral cortex that runs in a band just in front of the primary somatosensory area and is responsible for voluntary movement.

primary plant body The portion of a plant that is made up of primary tissues.

primary producer An autotroph, usually a photosynthetic organism, a member of the first trophic level.

primary somatosensory area The area of the cerebral cortex that runs in a band across the parietal lobes of the brain and registers information on touch, pain, temperature, and pressure.

primary structure The sequence of amino acids in a protein.

primary succession Predictable change in species composition of an ecological community that develops on bare ground.

primary tissue A plant tissue that develops from an apical meristem.

primase An enzyme that assembles the primer for a new DNA strand during DNA replication.

primer A short nucleotide chain made of RNA that is laid down as the first series of nucleotides in a new DNA strand, or made of DNA for use in the polymerase chain reaction (PCR).

primitive groove In development of birds, the sunken midline of the primitive streak that acts as a conduit for migrating cells to move into the blastocoel.

primitive streak In development of birds, the thickened region of the embryo produced by cells of the epiblast streaming toward the midline of the blastodisc.

Principle of Independent Assortment Mendel's principle that the alleles of the genes that govern two characters segregate independently during formation of gametes.

principle of monophyly A guiding principle of systematic biology that defines monophyletic taxa, each of which contains a single ancestral species and all of its descendants.

principle of parsimony A principle of systematic biology that states that a particular trait is unlikely to evolve independently in separate evolutionary lineages.

Principle of Segregation Mendel's principle that the pairs of alleles that control a character segregate as gametes are formed, and that half the gametes carry one allele, and the other half carry the other allele.

prion An infectious agent that contains only protein and does not include a nucleic acid molecule.

probability The possibility that an outcome will occur if it is a matter of chance.

procambium The primary meristem of a plant that develops into primary vascular tissue.

product An atom or molecule leaving a chemical reaction.

product rule Mathematical rule in which the final probability is found by multiplying individual probabilities.

profundal zone The perpetually dark layer below the limnetic zone in a lake.

progesterone A female sex hormone that stimulates growth of the uterine lining and inhibits contractions of the uterus.

progestin A class of sex hormones synthesized by the gonads of vertebrates and active predominantly in females.

prokaryote Organism in which the DNA is suspended in the cell interior without separation from other cellular components by a discrete membrane.

prokaryotic chromosome A single, typically circular DNA molecule.

prokaryotic flagellum A long, threadlike protein fiber that rotates in a socket in the plasma membrane and cell wall to push a prokaryotic cell through a liquid medium.

prolactin (PRL) A peptide hormone secreted by the anterior pituitary that stimulates breast development and milk secretion in mammals.

prometaphase A transition period between prophase and metaphase during which the microtubules of the mitotic spindle attach to the kinetochores and the chromosomes shuffle until they align in the center of the cell.

promiscuity A mating system in which individuals do not form close pair bonds, and both males and females mate with multiple partners.

promoter The site to which RNA polymerase binds for initiating transcription of a gene.

promoter proximal region Upstream of a eukaryotic gene, a region containing regulatory sequences for transcription called promoter proximal elements.

proofreading mechanism Mechanism of DNA polymerase to back up and remove mispaired nucleotides from a newly synthesized DNA strand.

propagation In animal nervous systems, the concept that the action potential does not need further trigger events to keep going.

prophage A viral genome inserted in the host cell DNA.

prophase The beginning phase of mitosis during which the duplicated chromosomes within the nucleus condense from a greatly extended state into compact, rodlike structures.

proprioceptor A mechanoreceptor that detects stimuli used in the CNS to maintain body balance and equilibrium and to monitor the position of the head and limbs.

prostaglandin One of a group of local regulators derived from fatty acids that are involved in paracrine and autocrine regulation.

prostate gland An accessory sex gland in males that adds a thin, milky fluid to the semen and adjusts the pH of the semen to the level of acidity best tolerated by sperm.

protein Molecules that carry out most of the activities of life, including the synthesis of all other biological molecules. A protein consists of one or more polypeptides depending on the protein.

protein chip *See* protein microarray.

protein kinase Enzyme that transfers a phosphate group from ATP to one or more sites on particular proteins.

protein microarray Similar in concept to a DNA microarray, a solid surface with a microscopic grid with thousands of spaces containing probes for analyzing the proteome, the complete set of proteins encoded by the genome of an organism. Also referred to as a *protein chip*.

protein phosphatase Enzyme that removes phosphate groups from target proteins.

proteome The complete set of proteins that can be expressed by the genome of an organism.

proteomics The study of the proteome.

protocell A primitive cell-like structure that has some of the properties of life and that might have been the precursor of cells.

Protoctista The kingdom that includes all the eukaryotes that are not fungi, plants, or animals.

protoderm The primary meristem that will produce stem epidermis.

proton Positively charged particle in the nucleus of an atom.

proton pump Pump that moves hydrogen ions across membranes and pushes hydrogen ions across the plasma membrane from the cytoplasm to the cell exterior. Also referred to as H^+ *pump*.

protonema The structure that arises when a liverwort or moss spore germinates and eventually gives rise to a mature gametophyte.

protonephridium The simplest form of invertebrate excretory tubule.

proton-motive force Stored energy that contributes to ATP synthesis, as well as to the cotransport of substances to and from mitochondria.

proto-oncogene A gene that encodes various kinds of proteins that stimulate cell division.

Mutated proto-oncogenes contribute to the development of cancer.

protoplast The cytoplasm, organelles, and plasma membrane of a plant cell.

protoplast fusion A plant breeding process in which protoplasts are fused into a single cell.

proximal convoluted tubule The tubule between the Bowman's capsule and the loop of Henle in the nephron of the kidney, which carries and processes the filtrate.

pseudocoelom A fluid- or organ-filled body cavity between the gut (a derivative of endoderm) and the muscles of the body wall (a derivative of mesoderm).

pseudocoelomate A body plan of bilaterally symmetrical animals with a body cavity that lacks a complete lining derived from mesoderm.

pseudopod (plural, pseudopodia) A temporary cytoplasmic extension of a cell.

psychrophile An archaean or bacterium that grows optimally at temperatures in the range −10 to −20°C.

Pterophyta The plant phylum of ferns and their close relatives.

pulmocutaneous circuit In amphibians, the branch of a double blood circuit that receives deoxygenated blood and moves it to the skin and lungs or gills.

pulmonary circuit The circuit of the cardiovascular system that supplies the lungs.

pulvinus (plural, pulvini) A jointlike, thickened pad of tissue at the base of a leaf or petiole; flexes when the leaf makes nastic movements.

punctuated equilibrium hypothesis The evolutionary hypothesis that most morphological variation arises during speciation events in isolated populations at the edge of a species' geographical distribution.

Punnett square Method for determining the genotypes and phenotypes of offspring and their expected proportions.

pupa The nonfeeding stage between the larva and adult in the complete metamorphosis of some insects, during which the larval tissues are completely reorganized within a protective cocoon or hardened case.

pupil The dark center in the middle of the iris through which light passes to the back of the eye.

purine A type of nitrogenous base with two carbon-nitrogen rings.

pyramid of biomass A diagram that illustrates differences in standing crop biomass in a series of trophic levels.

pyramid of energy A diagram that illustrates the amount of energy that flows through a series of trophic levels.

pyramid of numbers A diagram that illustrates the number of individual organisms present in a series of trophic levels.

pyrimidine A type of nitrogenous base with one carbon-nitrogen ring.

pyruvate oxidation (pyruvic acid oxidation) Stage of cellular respiration in which the three-carbon molecule pyruvate is converted into a two-carbon acetyl group that is completely oxidized to carbon dioxide.

qualitative variation Variation that exists in two or more discrete states, with intermediate forms often being absent.

quantitative variation Variation that is measured on a continuum (such as height in human beings) rather than in discrete units or categories.

quaternary structure The arrangement of polypeptide chains in a protein that contains more than one chain.

quiescent center A region in a root apical meristem where there is no cell division.

R gene A resistance gene in a plant; dominant R alleles confer enhanced resistance to plant pathogens.

R plasmid A bacterial plasmid containing genes that provide resistance to unfavorable conditions.

radial cleavage A cleavage pattern in deuterostomes in which newly formed cells lie directly above and below other cells of the embryo.

radial symmetry A body plan of organisms in which structures are arranged regularly around a central axis, like spokes radiating out from the center of a wheel.

radiation The transfer of heat energy as electromagnetic radiation.

radicle The rudimentary root of a plant embryo.

radioactivity The giving off of particles of matter and energy by decaying nuclei.

radioisotope An unstable, radioactive isotope.

radiometric dating A dating method that uses measurements of certain radioactive isotopes to calculate the absolute ages in years of rocks and minerals.

radula The tooth-lined "tongue" of mollusks that scrapes food into small particles or drills through the shells of prey.

rain shadow An area of reduced precipitation on the leeward side of a mountain.

random coil An arrangement of the amino acid chain providing flexible regions that allow sections of the chain to bend.

random dispersion A pattern of distribution is which the individuals in a population are distributed unpredictably in their habitat.

rapid eye movement (REM) sleep The period during deep sleep when the delta wave pattern is replaced by rapid, irregular beta waves characteristic of the waking state. The person's heartbeat and breathing rate increase, the limbs twitch, and the eyes move rapidly behind the closed eyelids.

ray initial A cell in vascular cambium that gives rise to spokelike rays of parenchyma cells.

reabsorption The process in which some molecules (for example, glucose and amino acids) and ions are transported by the transport epithelium back into the body fluid (animals with open circulatory systems) or into the blood in capillaries surrounding the tubules (animals with closed circulatory systems) as the filtered solution moves through the excretory tubule.

reactants The atoms or molecules entering a chemical reaction.

reaction center Part of photosystems I and II in chloroplasts of plants. In the light-dependent reactions of photosynthesis, the reaction center receives light energy absorbed by the antenna complex in the same photosystem.

realized niche The range of conditions and resources that a population actually uses in nature.

receptacle The expanded tip of a flower stalk that bears floral organs.

reception In signal transduction, the binding of a signal molecule with a specific receptor in a target cell.

receptor protein Protein that recognizes and binds molecules from other cells that act as chemical signals.

receptor tyrosine kinase In signal transduction, a surface receptor with built-in protein kinase activity.

receptor-mediated endocytosis The selective uptake of macromolecules that bind to cell surface receptors concentrated in clathrin-coated pits.

recessive An allele that is masked by a dominant allele.

reciprocal altruism Form of altruistic behavior in which individuals help nonrelatives if they are likely to return the favor in the future.

recognition protein Protein in the plasma membrane that identifies a cell as part of the same individual or as foreign.

recombinant Phenotype with a different combination of traits from those of the original parents.

recombinant DNA DNA from two or more different sources joined together.

recombination The physical exchange of segments between the chromatids of homologous chromosomes, or between chromosomes of prokaryotic cells or viruses.

recombination frequency In the construction of linkage maps of diploid eukaryotic organisms, the percentage of testcross progeny that are recombinants.

rectum The final segment of the large intestine.

red tide A growth in dinoflagellate populations that causes red, orange, or brown discoloration of coastal ocean waters.

redox reaction Coupled oxidation–reduction reaction in which electrons are removed from a donor molecule and simultaneously added to an acceptor molecule.

reduced Substance that receives electrons during reduction.

reduction The addition of electrons to a substance.

reflex A programmed movement that takes place without conscious effort, such as the sudden withdrawal of a hand from a hot surface.

refractory period A period that begins at the peak of an action potential and lasts a few milliseconds, during which the threshold required for generation of an action potential is much higher than normal.

reinforcement The enhancement of reproductive isolation that had begun to develop while populations were geographically separated.

relative abundance The relative commonness of populations within a community.

relative fitness The number of surviving offspring that an individual produces compared with the number left by others in the population.

release The process in which urine is released into the environment from the distal end of the excretory tubule.

release factor A protein that recognizes stop codons in the A site of a ribosome translating an mRNA and terminates translation. Also referred to as the *termination factor*.

releasing hormone (RH) A peptide neurohormone that control the secretion of hormones from the anterior pituitary.

renal artery An artery that carries bodily fluids into the kidney.

renal cortex The outer region of the mammalian kidney that surrounds the renal medulla.

renal medulla The inner region of the mammalian kidney.

renal pelvis The central cavity in the kidney where urine drains from collecting ducts.

renal vein The vein that routes filtered blood away from the kidney.

renin An enzyme secreted by cells in the juxtaglomerular apparatus into the bloodstream that converts a blood protein into the peptide hormone angiotensin.

renin-angiotensin-aldosterone system (RAAS) The most important hormonal system involved in regulation of Na^+ in mammals.

replica plating Technique for identifying and counting genetic recombinants in conjugation, transformation, or transduction experiments in which the colony pattern on a plate containing solid growth medium is pressed onto sterile velveteen and transferred to other plates containing different combinations of nutrients.

replicates Multiple subjects that receive either the same experimental treatment or the same control treatment.

replication fork The region of DNA synthesis where the parental strands separate and two new daughter strands elongate.

replication origin The site at which DNA replication begins.

repressible operon Operon whose expression is prevented by a repressor molecule.

repressor A regulatory protein that prevents the operon genes from being expressed.

reproduction The process in which parents produce offspring.

reproductive isolating mechanism A biological characteristic that prevents the gene pools of two species from mixing.

reproductive strategy One of a set of behaviors that lead to reproductive success.

residual volume The air that remains in lungs after exhalation.

resolution The minimum distance two points in a specimen can be separated and still be seen as two points.

resource partitioning The use of different resources or the use of resources in different ways by species living in the same place.

respiratory medium The environmental source of O_2 and the "sink" for released CO_2. For aquatic animals, the respiratory medium is water; for terrestrial animals, it is air.

respiratory surface A layer of epithelial cells that provides the interface between the body and the respiratory medium.

respiratory system All the parts of the body involved in exchanging air between the external environment and the blood.

response In signal transduction, the last stage in which the transduced signal causes the cell to change according to the signal and to the receptors on the cell. In the nervous system, the output resulting from the integration of neural messages.

resting potential A steady negative membrane potential exhibited by the membrane of a neuron that is not stimulated—that is, not conducting an impulse.

restriction endonuclease (restriction enzyme) An enzyme that cuts DNA at a specific sequence.

restriction fragment A DNA fragment produced by cutting a long DNA molecule with a restriction enzyme.

restriction fragment length polymorphisms When comparing different individuals, restriction enzyme–generated DNA fragments of different lengths from the same region of the genome.

reticular formation A complex network of interconnected neurons that runs through the length of the brain stem, connecting to the thalamus at the anterior end and to the spinal cord at the posterior end.

retina A light-sensitive membrane lining the posterior part of the inside of the eye.

retrotransposon A transposable element that transposes via an intermediate RNA copy of the transposable element.

retrovirus A virus with an RNA genome that replicates via a DNA intermediate.

reverse transcriptase An enzyme that uses RNA as a template to make a DNA copy of the retrotransposon. Reverse transcriptase is used to make DNA copies of RNA in test tube reactions.

reversible The term indicating that a reaction may go from left to right or from right to left, depending on conditions.

rhizoid A modified hypha that anchors a fungus to its substrate and absorb moisture.

rhizome A horizontal, modified stem that can penetrate a substrate and anchor the plant.

rhodopsin The retinal-opsin photopigment.

ribonucleic acid (RNA) A polymer assembled from repeating nucleotide monomers in which the five-carbon sugar is ribose. Cellular RNAs are mRNA (which is translated to produce a polypeptide), tRNA (which brings an amino acid to the ribosome for assembly into a polypeptide during translation), and rRNA (which is a structural component of ribosomes). The genetic material of some viruses is RNA.

ribose A five-carbon sugar to which the nitrogenous bases in nucleotides link covalently.

ribosomal RNA (rRNA) The RNA component of ribosomes.

ribosome A ribonucleoprotein particle that carries out protein synthesis by translating mRNA into chains of amino acids.

ribosome binding site In translation initiation in prokaryotes, a sequence just upstream of the start codon which directs the small ribosomal subunit to bind and orient correctly for the complete ribosome to assemble and start translating in the correct spot.

ribozyme An RNA-based catalyst that is part of the biochemical machinery of all cells.

ring species A species with a geographic distribution that forms a ring around uninhabitable terrain.

RNA *See* ribonucleic acid.

RNA interference (RNAi) The phenomenon of silencing a gene posttranscriptionally by a small, single-stranded RNA that is complementary to part of an mRNA.

RNA polymerase An enzyme that catalyzes the assembly of nucleotides into an RNA strand.

rod In the vertebrate eye, a type of photoreceptor in the retina that is specialized for detection of light at low intensities.

root An anchoring structure in land plants that also absorbs water and nutrients and (in some plant species) stores food.

root cap A dome-shaped cell mass that forms a protective covering over the apical meristem in the tip of a plant root.

root hair A tubular outgrowth of the outer wall of a root epidermal cell; root hairs absorb much of a plant's water and minerals from the soil.

root nodule A localized swelling on a root in which symbiotic nitrogen-fixing bacteria reside.

root pressure The pressure that develops in plant roots as the result of osmosis, forcing xylem sap upward and out through leaves. *See also* Guttation.

root primordium A rudimentary root.

root system An underground (or submerged) network of roots with a large surface area that favors the rapid uptake of soil water and dissolved mineral ions.

rough ER Endoplasmic reticulum with many ribosomes studding its outer surface.

round window A thin membrane that faces the middle ear.

***r*-selected species** A short-lived species adapted to function well in a rapidly changing environment.

RuBP carboxylase/oxygenase (rubisco) An enzyme that catalyzes the key reaction of the Calvin cycle, carbon fixation, in which CO_2 combines with RuBP (ribulose 1,5-bisphosphate) to form 3-phosphoglycerate.

ruminant An animals that has a complex, four-chambered stomach.

S phase The phase of the cell cycle during which DNA replication occurs.

saccule A fluid-filled chamber in the vestibular apparatus that provides information about the position of the head with respect to gravity (up versus down), as well as changes in the rate of linear movement of the body.

salicylic acid (SA) In plants, a chemical synthesized following a wound that has multiple roles in plant defenses, including interaction with jasmonates in signaling cascades.

salivary amylase A substance that hydrolyzes starches to the disaccharide maltose.

salivary gland A gland that secretes saliva through a duct on the inside of the cheek or under the tongue; the saliva lubricates food and begins digestion.

salt marsh A tidal wetland dominated by emergent grasses and reeds.

saltatory conduction A mechanism that allows small-diameter axons to conduct impulses rapidly.

saprobe An organism nourished by dead or decaying organic matter.

sapwood The newly formed outer wood located between heartwood and the vascular cambium. Compared with heartwood, it is wet, lighter in color, and not as strong.

sarcomere The basic unit of contraction in a myofibril.

sarcoplasmic reticulum In vertebrate muscle fibers, a complex system of vesicles modified from the smooth endoplasmic reticulum that encircles the sarcomeres. The sarcoplasmic reticulum is part of the pathway for the stimulation of muscle contraction by neural signals.

saturated enzymes Enzymes for which increases in substrate concentration have no effect on the reaction rate.

saturated fatty acid Fatty acid with only single bonds linking the carbon atoms.

savanna A biome comprising grasslands with few trees, which grows in areas adjacent to tropical deciduous forests.

schizocoelom In protostomes, the body cavity that develops as inner and outer layers of mesoderm separate.

Schwann cell A type of glial cell in the PNS that wraps nerve fibers with myelin and also secretes regulatory factors.

scientific method An investigative approach in which scientists make observations about the natural world, develop working explanations about what they observe, and then test those explanations by collecting more information.

scientific name A two-part name identifying the genus to which a species belongs and designating a particular species within that genus.

scientific theory A broadly applicable idea or hypothesis that has been confirmed by every conceivable test.

sclereid A type of sclerenchyma cell; sclereids typically are short and have thick, lignified walls.

sclerenchyma A ground tissue in which cells develop thick secondary walls, which commonly are lignified and perforated by pits through which water can pass.

scrotum The baglike sac in which the testes are suspended in many mammals.

scutellum The shield-shaped cotyledon of a grass.

second law of thermodynamics Principle that for any process in which a system changes from an initial to a final state, the total disorder of the system and its surroundings always increases.

second messenger In particular signal transduction pathways, an internal, nonprotein signal molecule that directly or indirectly activates protein kinases, which elicit the cellular response.

secondary active transport Transport indirectly driven by ATP hydrolysis.

secondary cell wall A layer added to the cell wall of plants that is more rigid and may become many times thicker than the primary cell wall.

secondary consumer A carnivore that feeds on herbivores, a member of the third trophic level.

secondary endosymbiosis In the model for the origin of plastids in eukaryotes, the second event, in which a nonphotosynthetic eukaryote engulfed a photosynthetic eukaryote.

Secondary growth Plant growth that originates at lateral meristems and increases the diameter of older roots and stems. *Compare* primary growth.

secondary immune response The rapid immune response that occurs during the second (and subsequent) encounters of the immune system of a mammal with a specific antigen.

secondary plant body The part of a plant made up of tissues that develop from lateral meristems.

secondary productivity Energy stored in new consumer biomass as energy is transferred from producers to consumers.

secondary structure Regions of alpha helix, beta strand, or random coil in a polypeptide chain.

secondary succession Predictable changes in species composition in an ecological community that develops after existing vegetation is destroyed or disrupted by an environmental disturbance.

secondary tissue In plants, the tissue that develops from lateral meristems.

secretion A selective process in which specific small molecules and ions are transported from the body fluids (in animals with open circulatory systems) or blood (in animals with closed circulatory systems) into the excretory tubules.

secretory vesicle Vesicle that transports proteins to the plasma membrane.

seed The structure that forms when an ovule matures after a pollen grain reaches it and a sperm fertilizes the egg.

seed coat The outer protective covering of a seed.

segment polarity genes In *Drosophila* embryonic development, the set of segmentation regulatory genes activated by pair-rule genes that set the boundaries and anterior-posterior axis of each segment in the embryo.

segmentation The production of body parts and some organ systems in repeating units.

segmentation genes Genes that work sequentially, progressively subdividing the embryo into regions, determining the segments of the embryo and the adult.

segregation The separation of the pairs of alleles that control a character as gametes are formed.

selective cell adhesion A mechanism in which cells make and break specific connections to other cells or to the extracellular matrix.

selectively neutral *See* neutral variation hypothesis.

selectively permeable Membranes that selectively allow, impede, or block the passage of atoms and molecules.

self-fertilization (self-pollination) Fertilization in which sperm nuclei in pollen produced by anthers fertilize egg cells housed in the carpel of the same flower.

self-incompatibility In plants, the inability of a plant's pollen to fertilize ovules of the same plant.

semen The secretions of several accessory glands in which sperm are mixed prior to ejaculation.

semicircular canal A part of the vestibular apparatus that detects rotational (spinning) motions.

semiconservative replication The process of DNA replication in which the two parental strands separate and each serves as a template for the synthesis of new progeny double-stranded DNA molecules.

semilunar valve (SL valve) A flap of endocardium and connective tissue reinforced by fibers that prevent the valve from turning inside out.

seminal fluid Fluid secreted by the seminal vesicles that contains prostaglandins, which when ejaculated into the female trigger contractions of the female reproductive tract that help move the sperm into and through the uterus.

seminal vesicle A vesicle that secretes seminal fluid.

seminiferous tubule One of the tiny tubes in the testes where sperm cells are produced, grow, and mature.

senescence The biologically complex process of aging in mature organisms that leads to the death of cells and eventually the whole organism.

sense codon A codon that specifies an amino acid.

sensitization Increased responsiveness to mild stimuli after experiencing a strong stimulus; one of the simplest forms of memory.

sensor A tissue or organ that detects a change in an external or internal factor such as pH, temperature, or the concentration of a molecule such as glucose.

sensory adaptation A condition in which the effect of a stimulus is reduced if it continues at a constant level.

sensory hair cell A hair cell that send impulses along the auditory nerve to the brain when alternating changes of pressure agitate the basilar membrane on which the organ of Corti rests, moving the hair cells.

sensory neuron A neuron that transmits stimuli collected by their sensory receptors to interneurons.

sensory receptor A receptor formed by the dendrites of afferent neurons, or by specialized receptor cells making synapses with afferent neurons that pick up information about the external and internal environments of the animal.

sensory transduction The conversion of a stimulus into a change in membrane potential.

sepal One of the separate, usually green parts forming the calyx of a flower.

septum (plural, septa) A thin partition or cross wall that separates body segments.

sequential hermaphroditism The form of hermaphroditism in which individuals change from one sex to the other.

serosa The serous membrane: a thin membrane lining the closed cavities of the body; has two layers with a space between that is filled with serous fluid.

Sertoli cell One of the supportive cells that completely surrounds developing spermatocytes in the seminiferous tubules. Follicle-stimulating hormone stimulates Sertoli cells to secrete a protein and other molecules that are required for spermatogenesis.

sessile Unable to move from one place to another.

set point The level at which the condition controlled by a homeostatic pathway is to be maintained.

seta (plural, setae) A chitin-reinforced bristle that protrudes outward from the body wall in some annelid worms.

sex chromosomes Chromosomes that are different in male and female individuals of the same species.

sex pilus *See* F pilus.

sex ratio The relative proportions of males and females in a population.

sex-linked gene Gene located on a sex chromosome.

sexual dimorphism Differences in the size or appearance of males and females.

sexual reproduction The mode of reproduction in which male and female parents produce offspring through the union of egg and sperm generated by meiosis.

sexual selection A form of natural selection established by male competition for access to females and by the females' choice of mates.

shells *See* energy levels.

shoot system The stems and leaves of a plant.

short-day plant A plant that flowers in late summer or early autumn when dark periods become longer and light periods become shorter.

short-term memory Memory that stores information for seconds.

sieve tube A series of phloem cells joined end to end, forming a long tube through which nutrients are transported; seen mainly in flowering plants.

sieve tube member Any of the main conducting cells of phloem that connect end to end, forming a sieve tube.

sign stimulus A simple cue that triggers a fixed action pattern.

signal peptide A short segment of amino acids to which the signal recognition particle binds, temporarily blocking further translation. A signal peptide is found on polypeptides that are sorted to the endoplasmic reticulum. Also referred to as *signal sequence*.

signal recognition particle (SRP) Protein-RNA complex that binds to signal sequences and targets polypeptide chains to the endoplasmic reticulum.

signal sequence *See* signal peptide.

signal transduction The series of events by which a signal molecule released from a controlling cell causes a response (affects the function) of target cells with receptors for the signal. Target cells process the signal in the three sequential steps of reception, transduction, and response.

silencing Phenomenon in which methylation of cytosines in eukaryotic promoters inhibits transcription and turns the genes off.

silent mutation A base-pair substitution mutation in a protein-coding gene that does not alter the amino acid specified by the gene.

simple diffusion Mechanism by which certain small substances diffuse through the lipid part of a biological membrane.

simple fruit A fruit that develops from a single ovary; in many of them at least one layer of the pericarp is fleshy and juicy.

simulation modeling An analytical method in which researchers gather detailed information about a system and then create a series of mathematical equations that predict how the components of the system interact and respond to change.

simultaneous hermaphroditism A form of hermaphroditism in which individuals develop functional ovaries and testes at the same time.

single-lens eye An eye type that works by changing the amount of light allowed to enter into the eye and by focusing this incoming light with a lens.

single-stranded binding protein Protein that coats single-stranded segments of DNA, stabilizing the DNA for the replication process.

sink Any region of a plant where organic substances are being unloaded from the sieve tube system and used or stored.

sink population In metapopulation analysis, a population that routinely declines in size after being replenished by immigrants from a source population.

sinoatrial node (SA node) The region of the heart that controls the rate and timing of cardiac muscle cell contraction.

sinus A body space that surrounds an organ.

sister chromatid One of two exact copies of a chromosome duplicated during replication.

skeletal muscle A muscle that connect to bones of the skeleton, typically made up of long and cylindrical cells that contain many nuclei.

slime layer A coat typically composed of polysaccharides that is loosely associated with bacterial cells.

SLOSS (*Single Large Or Several Small*) The debate among conservation biologists about the relative merits of establishing fewer large preserves or more numerous small ones.

slow block to polyspermy The process in which enzymes released from cortical granules alter the egg coats within minutes after fertilization, so that no other sperm can attach and penetrate to the egg.

slow muscle fiber A muscle fiber that contracts relatively slowly and with low intensity.

small interfering RNA (siRNA) A class of single-stranded RNAs that cause RNA interference.

small ribonucleoprotein particle A complex of RNA and proteins.

smooth ER Endoplasmic reticulum with no ribosomes attached to its membrane surfaces. Smooth ER has various functions, including synthesis of lipids that become part of cell membranes.

smooth muscle A relatively small and spindle-shaped muscle cell in which actin and myosin molecules are arranged in a loose network rather than in bundles.

social behavior The interactions that animals have with other members of their species.

sodium–potassium pump *See* Na^+/K^+ pump.

soil solution A combination of water and dissolved substances that coats soil particles and partially fills pore spaces.

solenoid *See* 30-nm chromatin fiber.

solute The molecules of a substance dissolved in water.

solution Substance formed when molecules and ions separate and are suspended individually, surrounded by water molecules.

solvent The water in a solution in which the hydration layer prevents polar molecules or ions from reassociating.

somaclonal selection A procedure in which somatic embryos derived from tissue culture are screened to identify those having desired characteristics, such as disease resistance.

somatic cell Any of the cells of an organism's body other than reproductive cells.

somatic embryo A plant embryo that is genetically identical to the parent because it arose through asexual means.

somatic gene therapy Gene therapy in which genes are introduced into somatic cells.

somatic nervous system A subdivision of the peripheral nervous system controlling body movements that are primarily conscious and voluntary.

somites Paired blocks of mesoderm cells along the vertebrate body axis that form during early vertebrate development and differentiate into dermal skin, bone, and muscle.

soredium (plural, soredia) A specialized cell cluster produced by lichens, consisting of a mass of algal cells surrounded by fungal hyphae; soredia function like reproductive spores and can give rise to a new lichen.

sorus (plural, sori) A cluster of sporangia on the underside of a fern frond; reproductive spores arise by meiosis inside each sporangium.

source In plants, any region (such as a leaf) where organic substances are being loaded into the sieve tube system of phloem.

source population In metapopulation analyses, a population that is either stable or increasing in size.

southern blot analysis Technique in which labeled probes are used to detect specific DNA fragments that have been separated by gel electrophoresis.

spatial summation The summation of EPSPs produced by firing of different presynaptic neurons.

specialized transduction Transfer of bacterial genes between bacteria using temperate phages that have incorporated fragments of the bacterial genome as they make the transition from the lysogenic cycle to the lytic cycle.

speciation The process of species formation.

species A group of populations in which the individuals are so closely related in structure, biochemistry, and behavior that they can successfully interbreed.

species cluster A group of closely related species recently descended from a common ancestor.

species composition The particular combination of species that occupy a site.

species diversity A community characteristic defined by species richness and the relative abundance of species.

species richness The number of species that live within an ecological community.

species selection A type of natural selection that acts upon species rather than upon populations.

specific epithet The species name in a binomial.

specific heat The amount of heat required to increase the temperature of a given quantity of water.

spermatocyte A developing gamete that becomes a spermatid at the end of meiosis.

spermatogenesis The process of producing sperm.

spermatogonium (plural, spermatogonia) A cell that enters meiosis and gives rise to gametes, produced by mitotic divisions of the germ cells in males.

spermatozoan Also called sperm; a haploid cell that develops into a mature sperm cell when meiosis is complete.

sphincter A powerful ring of smooth muscle that forms a valve between major regions of the digestive tract.

spinal cord A column of nervous tissue located within the vertebral column and directly connected to the brain.

spinal nerve A nerve that carries signals between the spinal cord and the body trunk and limbs.

spindle The structure that separates sister chromatids and moves them to opposite spindle poles.

spindle pole One of the pair of centrosomes in a cell undergoing mitosis from which bundles of microtubules radiate to form the part of the spindle from that pole.

spinneret A modified abdominal appendage from which spiders secrete silk threads.

spiracle An opening in the chitinous exoskeleton of an insect through which air enters and leaves the tracheal system.

spiral cleavage The cleavage pattern in many protostomes in which newly produced cells lie in the space between the two cells immediately below them.

spiral valve A corkscrew-shaped fold of mucous membrane in the digestive system of elasmobranchs, which slows the passage of material and increases the surface area available for digestion and absorption.

spirillum (plural, spirilla) Any flagellated aerobic bacterium twisted helically like a corkscrew.

spliceosome A complex formed between the pre-mRNA and small ribonucleoprotein particles, in which mRNA splicing takes place.

spongocoel The central cavity in a sponge.

spontaneous reaction Chemical or physical reaction that occurs without outside help.

sporangium (plural, sporangia) A single-celled or multicellular structure in fungi and plants in which spores are produced.

spore A haploid reproductive structure, usually a single cell, that can develop into a new individual without fusing with another cell; found in plants, fungi, and certain protists.

sporophyll A specialized leaf that bears sporangia (spore-producing structures).

sporophyte An individual of the diploid generation produced through fertilization in organisms that undergo alternation of generations; it produces haploid spores.

sporopollenin A tough polymer in the walls of spores and pollen grains, the presence of which helps such structures resist decay.

spring overturn The mixing of surface water with deep water in a lake or pond, causing oxygen at the surface to move to the bottom, and nutrients from the bottom to move to the surface.

squalene A liver oil found in sharks that is lighter than water, which increases their buoyancy.

SRP (signal recognition particle) receptor A protein on the membrane of the endoplasmic reticulum that binds the signal recognition particle.

stability The ability of a community to maintain its species composition and relative abundances when environmental disturbances eliminate some species from the community.

stabilizing selection A type of natural selection in which individuals expressing intermediate phenotypes have the highest relative fitness.

stamen A "male" reproductive organ in flowers, consisting of an anther (pollen producer) and a slender filament.

standing crop biomass The total dry weight of plants present in an ecosystem at a given time.

stapes The smallest of three sound-conducting bones in the middle ear of tetrapod vertebrates.

starch Energy-providing carbohydrates stored in plant cells.

start codon The first codon read in an mRNA in translation—AUG. Also referred to as the *initiator codon*.

statocyst A mechanoreceptor in invertebrates that senses gravity and motion using statoliths.

statolith A movable starch- or carbonate-containing stonelike body involved in sensing gravitational pull.

stele The central core of vascular tissue in roots and shoots of vascular plants; it consists of the xylem and phloem together with supporting tissues.

stereocilia Microvilli covering the surface of hair cells clustered in the base of neuromasts.

steroid A type of lipid derived from cholesterol.

steroid hormone receptor Internal receptor that turns on specific genes when it is activated by binding a signal molecule.

steroid hormone response element The DNA sequence to which the hormone-receptor complex binds.

sterol Steroid with a single polar —OH group linked to one end of the ring framework and a complex, nonpolar hydrocarbon chain at the other end.

sticky end End of a DNA fragment, with a single-stranded structure that can form hydrogen bonds with a complementary sticky end on any other DNA molecule cut with the same enzyme.

stigma The receptive end of a carpel where deposited pollen germinates.

stoma (plural, stomata) The opening between a pair of guard cells in the epidermis of a plant leaf or stem, through which gases and water vapor pass.

stomach The portion of the digestive system in which food is stored and digestion begins.

stop codon A codon that does not specify amino acids. The three nonsense codons are UAG, UAA, and UGA. Also referred to as the *nonsense codon* and *termination codon*.

stratification Horizontal layering of sedimentary rocks beneath the soil surface.

stretch receptor A proprioceptor in the muscles and tendons of vertebrates that detects the position and movement of the limbs.

strict aerobe Cell with an absolute requirement for oxygen to survive, unable to live solely by fermentations.

strict anaerobe Organism in which fermentation is the only source of ATP.

strobilus *See* cone.

stroma An inner compartment of a chloroplast, enclosed by two boundary membranes and containing a third membrane system.

stromatolite Fossilized remains of ancient cyanobacterial mats that carried out photosynthesis by the water-splitting reaction.

structural genomics The sequencing of genomes and the analysis of the nucleotide sequences to locate genes and other functionally important sequences within the genome.

structural isomers Two molecules with the same chemical formula but atoms that are arranged in different ways.

style The slender stalk of a carpel situated between the ovary and the stigma in plants.

submucosa A thick layer of elastic connective tissue that contains neuron networks and blood and lymph vessels.

subsoil The region of soil beneath topsoil, which contains relatively little organic matter.

subspecies A taxonomic subdivision of a species.

substrate The particular reacting molecule or molecular group that an enzyme catalyzes.

substrate-level phosphorylation An enzyme-catalyzed reaction that transfers a phosphate group from a substrate to ADP.

sugar–phosphate backbone Structure in a polynucleotide chain that is formed when deoxyribose sugars are linked by phosphate groups in an alternating sugar–phosphate–sugar–phosphate pattern.

sulfhydryl group Group that works as a molecular fastener, consisting of a sulfur atom linked on one side to a hydrogen atom and on the other side to a carbon chain.

sum rule Mathematical rule in which final probability is found by summing individual probabilities.

surface tension The force that places surface water molecules under tension, making them more resistant to separation than the underlying water molecules.

survivorship curve Graphic display of the rate of survival of individuals over a species' life span.

suspension feeder An animal that ingests small food items suspended in water.

suspensor In seed plants, a stalklike row of cells that develops from a zygote and helps position the embryo close to the nourishing endosperm.

swim bladder A gas-filled internal organ that helps fish maintain buoyancy.

symbiosis An interspecific interaction in which the ecological relations of two or more species are intimately tied together.

symmetry Exact correspondence of form and constituent configuration on opposite sides of a dividing line or plane.

sympathetic division Division of the autonomic nervous system that predominates in situations involving stress, danger, excitement, or strenuous physical activity.

sympatric speciation Speciation that occurs without the geographic isolation of populations.

symplastic pathway The route taken by water that moves through the cytoplasm of plant cells (the symplast). *Compare* apoplastic pathway.

symport The transport of two molecules in the same direction across a membrane. Also referred to as *cotransport*.

synapse A site where a neuron makes a communicating connection with another neuron or an effector such as a muscle fiber or gland.

synapsid One of a group of amniotes having one temporal arch on each side of the head, which includes living mammals.

synapsis *See* pairing.

synaptic cleft A narrow gap that separates the plasma membranes of the presynaptic and postsynaptic cells.

synaptic vesicle A secretory vesicle in the cytoplasm of an axon terminal of a neuron, in which neurotransmitters are stored.

synaptonemal complex A protein framework that tightly holds together homologous chromosomes as they pair.

systematics The branch of biology that studies the diversity of life and its evolutionary relationships.

systemic acquired resistance A plant defense response to microbial invasion; defensive chemicals including salicylic acid may spread throughout a plant, rendering healthy tissues less vulnerable to infection.

systemic circuit In amphibians, the branch of a double blood circuit that receives oxygenated blood and provides the blood supply for most of the tissues and cells of a body.

systemin A plant peptide hormone that functions in defense responses to wounds.

systems biology An area of biology that studies the organism as a whole to unravel the integrated and interacting network of genes, proteins, and biochemical reactions responsible for life.

systole The period of contraction and emptying of the heart.

T (transverse) tubule The tubule that passes in a transverse manner from the sarcolemma across a myofibril of striated muscle.

T cell A lymphocyte produced by the division of stem cells in the bone marrow and then released into the blood and carried to the thymus. T cells participate in adaptive immunity.

tactile signal A means of animal communication in which the signaler uses touch to convey a message to the signal receiver.

taiga *See* boreal forest.

taproot system A root system consisting of a single main root from which lateral roots can extend; often stores starch.

TATA box A regulatory DNA sequence found in the promoters of many eukaryotic genes transcribed by RNA polymerase II.

taxis A behavioral response that is directed either toward or away from a specific stimulus.

taxon (plural, taxa) A name designating a group of organisms included within a category in the Linnaean taxonomic hierarchy.

taxonomic hierarchy A system of classification based on arranging organisms into ever more inclusive categories.

taxonomy The science of the classification of organisms into an ordered system that indicates natural relationships.

T-cell receptor (TCR) A receptor that covers the plasma membrane of a T-cell, specific for a particular antigen.

telomerase An enzyme that adds telomere repeats to chromosome ends.

telomeres Repeats of simple-sequence DNA that maintain the ends of linear chromosomes.

telophase The final phase of mitosis, during which the spindle disassembles, the chromosomes decondense, and the nuclei re-form.

temperate bacteriophage Bacteriophage that may enter an inactive phase (lysogenic cycle) in which the host cell replicates and passes on the bacteriophage DNA for generations before the phage becomes active and kills the host (lytic cycle).

temperate deciduous forest A forested biome found at low to middle altitudes at temperate latitudes, with warm summers, cold winters, and annual precipitation between 75 and 250 cm.

temperate grassland A non-forested biome that stretches across the interiors of most continents, where winters are cold and snowy and summers are warm and fairly dry.

temperate rain forest A coniferous forest biome supported by heavy rain and fog, which grows where winters are mild and wet and the summers are cool.

template A nucleotide chain used in DNA replication for the assembly of a complementary chain.

template strand The DNA strand that is copied into an RNA molecule during gene transcription.

temporal isolation A prezygotic reproductive isolating mechanism in which species live in the same habitat but breed at different times of day or different times of year.

temporal summation The summation of several EPSPs produced by successive firing of a single presynaptic neuron over a short period of time.

tendon A type of fibrous connective tissue that attaches muscles to bones.

terminal bud A bud that develops at the apex of a shoot.

termination codon *See* stop codon.

termination factor *See* release factor.

terminator Specific DNA sequence for a gene that signals the end of transcription of a gene. Terminators are common for prokaryotic genes.

territory A plot of habitat, defended by an individual male or a breeding pair of animals, within which the territory holders have exclusive access to food and other necessary resources.

tertiary consumer A carnivore that feeds on other carnivores, a member of the fourth trophic level.

tertiary structure The overall three-dimensional folding of a polypeptide chain.

testcross A genetic cross between an individual with the dominant phenotype and a homozygous recessive individual.

testis (plural, testes) The male gonad. In male vertebrates, they secrete androgens and steroid hormones that stimulate and control the development and maintenance of male reproductive systems.

testosterone A hormone produced by the testes, responsible for the development of male secondary sex characters and the functioning of the male reproductive organs.

tetanus A situation in which a muscle fiber cannot relax at all between stimuli, and twitch summation produces a peak level of continuous contraction.

tetrad Homologous pair consisting of four chromatids.

Tetrapoda A monophyletic lineage of vertebrates that includes animals having four feet, legs, or leglike appendages.

T-even bacteriophage Virulent bacteriophages, T2, T4, and T6, that have been valuable for genetic studies of bacteriophage structure and function.

thalamus A major switchboard of the brain that receives sensory information and relays it to the regions of the cerebral cortex concerned with motor responses to sensory information of that type.

thallus A plant body not differentiated into stems, roots, or leaves.

thermal acclimatization A set of physiological changes in ectotherms in response to seasonal shifts in environmental temperature, allowing the animals to attain good physiological performance at both winter and summer temperatures.

thermocline The narrow depth range in a lake where water temperature changes abruptly at the boundary between the epilimnion and the hypolimnion.

thermodynamics The study of the energy flow during chemical and physical reactions.

thermoreceptor A sensory receptor that detects the flow of heat energy.

thermoregulation The control of body temperature.

thick filament A type of filament in striated muscle composed of myosin molecules; they interact with thin filaments to shorten muscle fibers during contraction.

thigmomorphogenesis A plant response to a mechanical disturbance, such as frequent strong winds; includes inhibition of cellular elongation and production of thick-walled supportive tissue.

thigmotropism Growth in response to contact with a solid object.

thin filament A type of filament in striated muscle composed of actin, tropomyosin, and troponin molecules; they interact with thick filaments to shorten muscle fibers during contraction.

thorax The central part of an animal's body, between the head and the abdomen.

thorn forest A forested biome that grows at the arid borders of true savanna, where large mammals are less abundant.

threshold potential In signal conduction by neurons, the membrane potential at which the action potential fires.

thylakoids Flattened, closed sacs that make up a membrane system within the stroma of a chloroplast.

thymine A pyrimidine that base-pairs with adenine.

thymus An organ of the lymphatic system that plays a role in filtering viruses, bacteria, damaged cells, and cellular debris from the lymph and bloodstream, and in defending the body against infection and cancer.

thyroid gland A gland located beneath the voice box (larynx) that secretes hormones regulating growth and metabolism.

thyroid-stimulating hormone (TSH) A hormone that stimulates the thyroid gland to grow in size and secrete thyroid hormones.

thyroxine (T_4) The main hormone of the thyroid gland, responsible for controlling the rate of metabolism in the body.

Ti (tumor inducing) plasmid A plasmid used to make transgenic plants.

tidal volume The volume of air entering and leaving the lungs during inhalation and exhalation.

tight junction Region of tight connection between membranes of adjacent cells.

time lag The delayed response of organisms to changes in environmental conditions.

tissue A group of cells and intercellular substances with the same structure that function as a unit to carry out one or more specialized tasks.

tolerance hypothesis Hypothesis asserting that ecological succession proceeds because competitively superior species replace competitively inferior ones.

tonoplast The membrane that surrounds the central vacuole in a plant cell.

topoisomerase An enzyme that relieves the overtwisting and strain of DNA ahead of the replication fork.

topsoil The rich upper layer of soil where most plant roots are located; it generally consists of sand, clay particles, and humus.

torpor A sleeplike state produced when a lowered set point greatly reduces the energy required to maintain body temperature, accompanied by reductions in metabolic, nervous, and physical activity.

torsion The realignment of body parts in gastropod mollusks that is independent of shell coiling.

totipotent Having the capacity to produce cells that can develop into or generate a new organism or body part.

trace element An element that occurs in organisms in very small quantities (<0.01%); in nutrition, a mineral required by organisms only in small amounts.

tracer Isotope used to label molecules so that they can be tracked as they pass through biochemical reactions.

trachea In insects, an extensively branched, air-conducting tube formed by invagination of the outer epidermis of the animal, and reinforced by rings of chitin. In vertebrates, the windpipe, which branches into the bronchi.

tracheal system A branching network of tubes that carries air from small openings in the exoskeleton of an insect to tissues throughout its body.

tracheid A conducting cell of xylem, usually elongated and tapered.

tracheophyte A plant with xylem, phloem, and usually well-developed roots, stems, and leaves.

traditional evolutionary systematics An approach to systematics that uses phenotypic similarities and differences to infer evolutionary relationships, grouping together species that share both ancestral and derived characters.

trait A particular variation in a genetic or phenotypic character.

transcription The mechanism by which the information encoded in DNA is made into a complementary RNA copy.

transcription initiation complex Combination of general transcription factors with RNA polymerase II.

transcription unit A region of DNA that transcribes a single primary transcript.

transcriptional regulation The processes that directly control gene activity.

transduction In cell signaling, the process of changing a signal into the form necessary to cause the cellular response. In prokaryotes, the process in which DNA is transferred from donor to recipient bacterial cells by an infecting bacteriophage.

transfer cell Any of the specialized cells that form when large amounts of solutes must be loaded or unloaded into the phloem; they facilitate the short-distance transport of organic solutes from the apoplast into the symplast.

transfer RNA (tRNA) The RNA that brings amino acids to the ribosome for addition to the polypeptide chain.

transformation The conversion of the hereditary type of a cell by the uptake of DNA released by the breakdown of another cell.

transgenic An organism that has been modified to contain genetic information from an external source.

transition state An intermediate arrangement of atoms and bonds that both the reactants and the products of a reaction can assume.

translation The use of the information encoded in the RNA to assemble amino acids into a polypeptide.

translocation In genetics, a chromosomal alteration that occurs if a broken segment is attached to a different, nonhomologous chromosome. In vascular plants, the long-distance transport of substances by xylem and phloem.

transmembrane pathway The path followed by water when it enters root cells by crossing across the plasma membrane.

transmission In neural signaling, the sending of a message along a neuron, and then to another neuron or to a muscle or gland.

transpiration The evaporation of water from a plant, principally from the leaves.

transport The controlled movement of ions and molecules from one side of a membrane to the other.

transport epithelium A layer of cells with specialized transport proteins in their plasma membranes.

transport protein A protein embedded in the cell membrane that forms a channel allowing selected polar molecules and ions to pass across the membrane.

transposable element (TE) A sequence of DNA that can move from one place to another within the genome of a cell.

transposase An enzyme that catalyzes some of the reactions inserting or removing the transposable element from the DNA.

transposon A bacterial transposable element with an inverted repeat sequence at each end enclosing a central region with one or more genes.

tricarboxylic acid cycle *See* citric acid cycle.

trichocyst A dartlike protein thread that can be discharged from a surface organelle for defense or to capture prey.

trichome A single-celled or multicellular outgrowth from the epidermis of a plant that provides protection and shade and often gives the stems or leaves a hairy appearance.

triglyceride A nonpolar compound produced when a fatty acid binds by a dehydration synthesis reaction at each of glycerol's three —OH-bearing sites.

triiodothyronine (T_3) A hormone secreted by the thyroid gland that regulates metabolism.

trimester A division of human gestation, three months in length.

triploblastic An animal body plan in which adult structures arise from three primary germ layers, endoderm, mesoderm, and ectoderm.

trochophore The small, free-swimming, ciliated aquatic larva of various invertebrates, including certain mollusks and annelids.

trophic cascade The effects of predator–prey interactions that reverberate through other population interactions at two or more trophic levels in an ecosystem.

trophic level A position in a food chain or web that defines the feeding habits of organisms.

trophoblast The outer single layer of cells of the blastocyst.

tropic hormone A hormone that regulates hormone secretion by another endocrine gland.

tropical deciduous forest A tropical forest biome that occurs where winter drought reduces photosynthesis, and most trees drop their leaves seasonally.

tropical forest Any forest that grows between the Tropics of Capricorn and Cancer, a region characterized by high temperature and rainfall and thin, nutrient-poor topsoil.

tropical montane forest A tropical forest biome of short trees, which are frequently enveloped in mist; also known as a "cloud forest."

tropical rain forest A dense tropical forest biome that grows where some rain falls every month, mean annual rainfall exceeds 250 cm, mean annual temperature is at least 25°C, and humidity is above 80%.

tropics The latitudes between 23.5° N and 23.5° S, the Tropics of Cancer and Capricorn.

tropism The turning or bending of an organism or one of its parts toward or away from an external stimulus, such as light, heat, or gravity.

true-breeding Individual that passes traits without change from one generation to the next.

tumor-suppressor gene A gene that encodes proteins that inhibit cell division.

turgor pressure The internal hydrostatic pressure within plant cells.

turgor pressure The normal fullness or tension produced by the fluid content of plant and animal cells.

turnover rate The rate at which one generation of producers in an ecosystem is replaced by the next.

tympanum A thin membrane in the auditory canal that vibrates back and forth when struck by sound waves.

umbilical cord A long tissue with blood vessels linking the embryo and the placenta.

umbilicus Navel; the scar left when the short length of umbilical cord still attached to the infant after birth dries and shrivels within a few days.

undernutrition A condition in animals in which intake of organic fuels is inadequate, or whose assimilation of such fuels is abnormal.

undulating membrane In parabasalid protists, a finlike structure formed by a flagellum buried in a fold of the cytoplasm that facilitates movement through thick and viscous fluids.an expansion of the plasma membrane in some flagellates that is usually associated with a flagellum.

unicellular organism Individual consisting of a single cell.

uniform dispersion A pattern of distribution in which the individuals in a population are evenly spaced in their habitat.

uniformitarianism The concept that the geological processes that sculpted Earth's surface over long periods of time—such as volcanic eruptions, earthquakes, erosion, and the formation and movement of glaciers—are exactly the same as the processes observed today.

universal A feature of the nucleic acid code, with the same codons specifying the same amino acids in all living organisms.

unreduced gamete A gamete that contains the same number of chromosomes as a somatic cell.

unsaturated fatty acid Fatty acid with one or more double bonds linking the carbons.

ureter The tube through which urine flows from the renal pelvis to the urinary bladder.

urethra The tube through which urine leaves the bladder. In most animals, the urethra opens to the outside.

urinary bladder A storage sac located outside the kidneys.

uterine cycle The menstrual cycle.

uterus A specialized saclike organ, in which the embryo develops in viviparous animals.

utricle A fluid-filled chamber of the vestibular apparatus that provides information about the position of the head with respect to gravity (up versus down), as well as changes in the rate of linear movement of the body.

vaccination The process of administering a weakened form of a disease to patients as a means of giving them immunity to a more serious form of the disease.

vagina The muscular canal that leads from the cervix to the exterior.

valence electron An electron in the outermost energy level of an atom.

Van der Waals forces Weak molecular attractions over short distances.

variable An environmental factor that may differ among places or an organismal characteristic that may differ among individuals.

vas deferens The tube through which sperm travel from the epididymis to the urethra in the male reproductive system.

vascular bundle A cord of plant vascular tissue; often multistranded with both xylem and phloem.

vascular cambium A lateral meristem that produces secondary vascular tissues in plants.

vascular plant *See* tracheophyte.

vascular tissue system One of the three tissue systems in plants that provide the foundation for plant organs; it consists of transport tubes for water and nutrients.

vegetal pole The end of the egg opposite the animal pole, which typically gives rise to internal structures such as the gut and the posterior end of the embryo.

vegetative reproduction Asexual reproduction in plants by which new individuals arise (or are created) without seeds or spores; examples include fragmentation from the parent plant or the use of cuttings by gardeners.

vein In a plant, a vascular bundle that forms part of the branching network of conducting and supporting tissues in a leaf or other expanded plant organ. In an animal, a vessel that carries the blood back to the heart.

ventilation The flow of the respiratory medium (air or water, depending on the animal) over the respiratory surface.

ventral Indicating the lower or "belly" side of an animal.

ventricle In the brain, an irregularly shaped cavity containing cerebrospinal fluid. In the heart, a chamber that pumps blood out of the heart.

venule A capillary that merges into the small veins leaving an organ.

vernalization The stimulation of flowering by a period of low temperature.

vertebrae The series of bones that form the vertebral column of vertebrate animals.

vertebral column The series of vertebrae that surrounds and protects the dorsal nerve cord and forms the supporting axis of the body.

vertebrate A member of the monophyletic group of tetrapod animals that possess a vertebral column.

vesicle A small, membrane-bound compartment that transfers substances between parts of the endomembrane system.

vessel In plants, one of the tubular conducting structures of xylem, typically several centimeters long; most angiosperms and some other vascular plants have xylem vessels.

vessel member Any of the short cells joined end to end in tubelike columns in xylem.

vestibular apparatus The specialized sensory structure of the inner ear of most terrestrial vertebrates that is responsible for perceiving the position and motion of the head and, therefore, for maintaining equilibrium and for coordinating head and body movements.

vestigial structure An anatomical feature of living organisms that no longer retains its function.

vibrio Any of various short, motile, S-shaped or comma-shaped bacteria of the genus *Vibrio*.

vicariance The fragmentation of a continuous geographical distribution by nonbiological factors.

virion A complete virus particle.

viroid A plant pathogen that consists of strands or circles of RNA, smaller than any viral DNA or RNA molecule, that have no protein coat.

virulent bacteriophage Bacteriophage that kills its host bacterial cells during each cycle of infection.

virus An infectious agent that contains either DNA or RNA surrounded by a protein coat.

visceral mass In mollusks, the region of the body containing the internal organs.

visual signal A means of communication in which animals use facial expressions or body language to send messages to other individuals.

vital capacity The maximum tidal volume of air that an individual can inhale and exhale.

vitamin An organic molecule required in small quantities that the animal cannot synthesize for itself.

vitamin D A steroidlike molecule that increases the absorption of Ca^{2+} and phosphates from ingested food by promoting the synthesis of a calcium-binding protein in the intestine; it also increases the release of Ca^{2+} from bone in response to PTH.

vitelline coat A gel-like matrix of proteins, glycoproteins, or polysaccharides immediately outside the plasma membrane of an egg cell.

vitreous humor The jellylike substance that fills the main chamber of the eye, between the lens and the retina.

viviparous Referring to animals that retain the embryo within the mother's body and nourish it during at least early embryo development.

voltage-gated ion channel A membrane-embedded protein that opens and closes as the membrane potential changes.

vulva The external female sex organs.

water lattice An arrangement formed when a water molecule in liquid water establishes an average of 3.4 hydrogen bonds with its neighbors.

water potential The potential energy of water, representing the difference in free energy between pure water and water in cells and solutions; it is the driving force for osmosis.

watershed An area of land from which precipitation drains into a single stream or river.

water-soluble vitamin A vitamin with a high proportion of oxygen and nitrogen able to form hydrogen bonds with water.

wax A substance insoluble in water that is formed when fatty acids combine with long-chain alcohols or hydrocarbon structures.

wetland A highly productive ecotone often at the border between a freshwater biome and a terrestrial biome.

white matter The myelinated axons that surround the gray matter of the central nervous system.

wilting The drooping of leaves and stems caused by a loss of turgor.

wobble hypothesis Hypothesis stating that the complete set of 61 sense codons can be read by fewer than 61 distinct tRNAs because of particular pairing properties of the bases in the anticodons.

Wolffian duct A bipotential primitive duct associated with the gonads that leads to a cloaca.

wood The secondary xylem of trees and shrubs, lying under the bark and consisting largely of cellulose and lignin.

X chromosome Sex chromosome that occurs paired in female cells and single in male cells.

X-linked recessive inheritance Pattern in which displayed traits are due to inheritance of recessive alleles carried on the X chromosome.

X-ray diffraction Method for deducing the position of atoms in a molecule.

xylem The plant vascular tissue that distributes water and nutrients.

xylem sap The dilute solution of water and solutes that flows in the xylem.

Y chromosome Sex chromosome that is paired with an X chromosome in male cells.

yeast A single-celled fungus that reproduces by budding or fission.

yolk The portion of an egg that serves as the main energy source for the embryo.

yolk sac In an amniote egg, an extraembryonic membrane that encloses the yolk.

zero population growth A circumstance in which the birth rate of a population equals the death rate.

zona pellucida A gel-like matrix of proteins, glycoproteins, or polysaccharides immediately outside the plasma membrane of the egg cell.

zone of cell division The region in a growing root that consists of the root apical meristem and the actively dividing cells behind it.

zone of elongation The region in a root where newly formed cells grow and elongate.

zone of maturation The region in a root above the zone of elongation where cells do not increase in length but may differentiate further and take on specialized roles.

zooplankton Small, usually microscopic, animals that float in aquatic habitats.

zygospore A multinucleate, thick-walled sexual spore in some fungi that is formed from the union of two gametes.

zygote A fertilized egg.

Credits

CHAPTER 1 **Page 1** Bryan Allen/Corbis. **1.1** Kevin Schafer. **1.2** (top to bottom) Bryan Allen/Corbis. Jamie and Judy Wild/Danita Delimont.com. Ron Sefton/Bruce Coleman USA. Jamie and Judy Wild/Danita Delimont.com. Edward Snow/Bruce Coleman USA. **1.4** PDB ID: 1BBB; Silva, M. M., Rogers, P. H., Amone, A. A third quaternary structure of human hemoglobin A at 1.7-Å resolution, *J Biol Chem, 267,* p. 17248, 1992. **1.6** (top) Norman Meyers/Bruce Coleman. (center left) Edward S. Ross. (center right) Paul De Greve/FPG/Getty Images. (bottom) Paul de Greve/FPG/Getty Images. **1.7** (all) Photographs by Jack de Coningh. **1.8** Photographs courtesy Derrell Fowler, Tecumseh, Oklahoma. **1.9** (all) Photographs courtesy of Hopi Hoekstra, University of California, San Francisco. **1.12** (a) Courtesy © Dr. G. Cohen-Bazire. (b) Courtesy James Evarts. **1.13** (a) © P. Hawtin, University of Southampton/SPL/Photo Researchers, Inc. (b) R. Robinson/Visuals Unlimited. (c) top left, M. Abbey/Visuals Unlimited. (c) top right, Robert C. Simpson/Nature Stock. (c) bottom left, John Lotter Gurling/Tom Stack & Associates. (c) bottom right, James Carmichael Jr./NHPA. **1.15** (left) Kevin de Queiroz, National Museum of Natural History, Smithsonian Institution. (center) Alejandro Sanchez. (right) Kevin de Queiroz, National Museum of Natural History, Smithsonian Institution

CHAPTER 2 **Page 21** Nigel Cattlin/Holt Studios International/Science Photo Library/Photo Researchers, Inc. **2.1** Photo by Gary Head for Norman Terry, University of California, Berkeley. **2.2** (left) Steve Lissau/Rainbow. (right) Jack Carey. **Page 26** Photo by Gary Head. **2.7** (b) Bruce Iverson. **2.11** (all) Dr. Kellar Autumn, Autumn Lab, Lewis & Clark College. **2.12** Wolfgang Kaehler. **2.15** (b) H. Eisenbeiss/Frank Lane Picture Agency. **2.17** Frederica Georgia/Photo Researchers, Inc.

CHAPTER 3 **Page 41** Hybrid Medical Animation/Science Photo Library/Photo Researchers, Inc. **3.1** Dave Schiefelbein. **3.7** (a) Ed Reschke/Peter Arnold, Inc. (b) Dennis Kunkel/Phototake. (c) © Biophoto Associates/Photo Researchers, Inc. (d) David Scharf/Peter Arnold, Inc. **3.10** Clem

Haagner/Ardea, London. **Page 52** Micrograph, Louis L. Lainey. **3.11** (a) Scott Camazine/Photo Researchers, Inc. (b) Larry Lefever/Grant Heilman Photography. **3.14** Tim Davis/Photo Researchers, Inc.

CHAPTER 4 **Page 71** Carolina Biological Supply/Phototake, Inc. **4.1** David Work. **4.3** (top) NASA. **4.3** (bottom) Manfred Kage/Peter Arnold, Inc. **4.13** (b) Douglas Faulkner/Sally Faulkner Collection.

CHAPTER 5 **Page 91** David Becker/Science Photo Library/Photo Researchers, Inc. **5.1** (a) National Library of Medicine. (b) Armed Forces Institute of Pathology. **5.2** (a) Tony Brain/SPL/Photo Researchers, Inc. (b) M. Abbey/Visuals Unlimited. (c) Wim van Egmond. (d) Manfred Kage/Peter Arnold, Inc. (e) C. E. Jeffree, et al, *Planta,* 172(1):20–37, 1987. Reprinted. **5.4** (top, first, second, and third) © Dennis Kunkel Microscopy, Inc. (top, fourth) Jeremy Pickett-Heaps, University of Colorado. (bottom, first, second, and third) © Dennis Kunkel Microscopy, Inc. (bottom, fourth) Jeremy Pickett-Heaps, University of Colorado. **5.6** Prof. H. Wartenberg from Dr. H. Jastrow's electron microscopic *Atlas,* http://www.unimainz.de/FB/Medizin/Anatomie/workshop/EM/EMAtlas.html. **5.7** Dr. G. Cohen-Bazire. **5.8** (b) G. L. Decker. **5.9** (b) M. C. Ledbetter, Brookhaven National Laboratory. **5.10** A. C. Faberge, *Cell and Tissue Research,* 1515:403–415, 1974. **5.11** (a, b) Don W. Fawcett/Visuals Unlimited. **5.12** Dr. Don Fawcett & R. Bollender/Visuals Unlimited. **5.14** Don Fawcett/Photo Researchers, Inc. **5.16** Keith R. Porter. **5.17** Eldon Newcomb, University of Wisconsin **5.18** (a) J. U. Shuler/Photo Researchers. (b) Courtesy of Mary Osborn, Max Planck Institute for Biophysical Chemistry, Goettingen FRG. (c) Courtesy of Dr. Vincenzo Cirulli, Lab of Developmental Biology, The Whittier Inst. for Diabetes, Univ. of Cal.–San Diego, La Jolla, CA. **5.21** (c) Don Fawcett/Photo Researchers, Inc. **5.22** (a) Lennart Nilsson. (b) CNRI/SPL/Photo Researchers, Inc. **5.23** Dr. Donald Fawcett and H. Bernstet/Visuals Unlimited. **5.24** Dr. Jeremy Burgess/SPL/Photo Researchers, Inc. **5.25** (top) Ray F. Evert. (bottom) Biophoto Associates/Photo Researchers, Inc. **5.26** (left) SPL/Photo Researchers, Inc. (center) G. E. Palade. (right) D. W. Fawcett.

CHAPTER 6 **Page 119** Dr. Alexander Gray/Wellcome Trust Medical Photographic Library. **6.1** Andrew Martinez/Photo Researchers, Inc. **6.7** Don W. Fawcett/Photo Researchers, Inc. **6.10** (all) Micrographs, M. Sheetz, R. Painter, and S. Singer. *Journal of Cell Biology,* 70:493, 1976. By permission of Rockefeller University Press. **6.14** M. M. Perry and A. M. Gilbert. **6.15** (b) Mike Abbey/Visuals Unlimited.

CHAPTER 7 **Page 139** © Russell Kightley Media. **7.1** Visuals Unlimited.

CHAPTER 8 **Page 157** Professors P. Motta and T. Naguro/SPL/Photo Researchers, Inc. **8.13** Ralph Pleasant/FPG/Getty Images. **8.17** David M. Phillips/Visuals Unlimited.

CHAPTER 9 **Page 177** Dr. Kari Lounatmaa/Science Photo Library/Photo Researchers, Inc. **9.3** Craig Tuttle/Corbis. **9.4** (a) Barker Blankenship/FPG/Getty Images. **9.16** (b) © 2001 PhotoDisc, Inc. (c) Chris Heller/Corbis.

CHAPTER 10 **Page 201** Dr. Paul Andrews, University of Dundee/Science Photo Library/Photo Researchers, Inc. **10.1** Chris Huss. **10.2** Conly Rieder. **10.4** (all) Ed Reschke. **10.5** (all) Ed Reschke. **10.7** (left) © Leonard Lessin/Peter Arnold, Inc. (right) Peter Arnold, Inc. **10.8** D. M. Phillips/Visuals Unlimited. **10.9** R. Calentine/Visuals Unlimited. **10.11** Photograph by Dr. Conly L. Rieder, Wadsworth Center, Albany, New York 12201-0509. **Page 213** Adrian Warren/Ardea/London. **10.16** Courtesy of Professor Pierre Chambon, Institut Clinique de la Souris, University of Strasbourg. Reprinted by permission from *Nature* 348:699. Copyright 1990 Macmillan Magazines, Ltd.

CHAPTER 11 **Page 221** Adrian T. Sumner/Science Photo Library/Photo Researchers, Inc. **11.3** (all) Micrographs with thanks to the John Innes Foundation Trustees. **11.6** Courtesy Diter von Wettstein.

CHAPTER 12 **Page 235** Carolyn A. McKeone/Science Photo Library/Photo Researchers, Inc. **12.1** (both) Stanley Flegler/Visuals Unlimited. **12.2** Moravian Museum, Brno. **12.12** (a) Dr. P. Marazzi/Photo Researchers, Inc. (b) St. Bartholomew's Hospital/Photo Researchers, Inc. (c) David Frazier/Photo Researchers, Inc. **12.13** (top, left) William Ferguson. (top, right) William Ferguson. (bottom) Francese Muntada/Corbis. **12.16** (a and c) Michael Stuckey/Comstock, Inc. (b) Bosco Broyer, Photo by Gary Head **12.17** (a) Dan Fairbanks/Brigham Young University.

CHAPTER 13 **Page 255** Regents of University of California 2005/Dr. Uli Weier/Photo Researchers, Inc. **13.1** Eddie Adams/AP Wide World Photos. **Page 258** Herman Eisenbeiss/Photo Researchers, Inc. **13.6** (left) © 2001 PhotoDisc, Inc. (right) © 2001 EyeWire. **13.7** (a) © Carolina Biological/Visuals Unlimited. (b) © Terry Gleason/Carolina Biological/Visuals Unlimited. **13.9** Bettmann/Corbis. **13.10** Ulrike Schanz/Animals, Animals. **13.13** (a) © 1997, Hironao Numabe, M.D., Tokyo Medical University. **13.15** © Abraham Menashe. **Page 276** (left) Carolina Biological/

Visuals Unlimited. (right) Bonnie Kamin/Stuart Kenter Associates.

CHAPTER 14 **Page 277** Kenneth Eward/Photo Researchers, Inc. **14.1** A. C. Barrington Brown © 1968 J. D. Watson. **14.5** SPL/Photo Researchers, Inc. **Page 294** C. J. Harrison. **14.18** (left) O. L. Miller, Jr., Steve McKnight. (right) B. Hamkalo.

CHAPTER 15 **Page 301** © LookatSciences/Phototake. **15.1** (left) Dennis Hallinan/FPG/Getty Images. (right) Bob Evans/Peter Arnold, Inc. **15.18** CNRI/SPL/Photo Researchers, Inc. **15.19** Courtesy Barbara A. Hamkalo.

CHAPTER 16 **Page 329** Abby Dernburg and Terumi Kohwi-Shigematsu/Lawrence Berkeley National Laboratory. **16.1** Lennart Nilsson/Bonnier Fakta. **16.15** Lennart Nilsson, © Boehringer Ingelheim International GmbH.

CHAPTER 17 **Page 351** Hybrid Medical Animation/Photo Researchers, Inc. **Page 352** Dennis Kunkel Microscopy, Inc. **17.2** (a) © Dennis Kunkel. (b) Courtesy of L. G. Caro and Academic Press, Inc. (London) Ltd. from *Journal of Molecular Biology* 16:269, 1966. **17.3** (a) Dr. Huntington Porter and Dr. David Dressler. (b) Prof. Stanley Cohen/SPL/Photo Researchers, Inc. **17.6** (a) Eye of Science/Photo Researchers, Inc. **17.11** P. J. Maugham.

CHAPTER 18 **Page 371** Pasteka/SPL/Photo Researchers, Inc. **18.1** © Marilyn Menotti-Raymond, The National Cancer Institute–Frederick. **18.4** Photo courtesy of Lucigen Corporation, Middleton, WI. **18.7** Damon Biotech, Inc. **18.12** R. Brinster, R. E. Hammer, School of Veterinary Medicine, University of Pennsylvania. **18.13** PA News Photo Library/AP Wide World Photos. **18.14** Stephen Wolfe, *Molecular and Cellular Biology.* **18.16** Kevin V. Wood. **18.17** Dr. Jorge Mayer, Golden Rice Project. **18.20** Courtesy Ludwig Institute for Cancer Research.

CHAPTER 19 **Page 401** Christopher Ralling. **19.1** (left) Courtesy George P. Darwin, Darwin Museum, Down House. (right) Down House and The Royal College of Surgeons of England. **19.2** (left) Wolfgang Kaehler. (center) Kenneth W. Fink/Photo Researchers, Inc. (right) Dave Watts/A. N. T. Photo Library. **19.4** Rich Kirchner/Foto Natura/Photo Researchers, Inc. **19.6** (top) Charles R. Knight painting (negative CK21T), Field Museum of Natural History, Chicago. (bottom) Calvin Larsen/Photo Researchers, Inc. **19.7** (left) Hugo Willcox/Foto Natura/Minden Pictures. (right) Fred Hazelhoff/Foto Natura/Minden Pictures. **19.8** (b) D. Kaleth/Image Bank/Getty Images. (c) William Paton/Foto Natura/Photo Researchers, Inc. (d) Heather Angel/Natural Visions.

19.9 (a) Dr. P. Evans/Bruce Coleman. (b) Kevin Schafer/Corbis. (c) Mark Moffatt/Minden Pictures. (d) Alan Root/Bruce Coleman Ltd. **Page 408** William Perlman/Star Ledger/Corbis. **19.12** P. Morris/Ardea, London.

CHAPTER 20 **Page 419** © Mark Moffett/Foto Natura/Minden Pictures. **20.1** The Advertising Archives, London. **20.2** (a) George Bernard/Foto Natura/Photo Researchers, Inc. (b) Timothy A. Pearce, Ph.D./Section of Mollusks/Carnegie Museum of Natural History. Photograph by Mindy McNaugher. **20.4** Arthur Morris/VIREO. **20.5** (left) Eric Crichton/Bruce Coleman, Inc. (right) William E. Ferguson. **20.7** (top) David Neal Parks. (bottom) W. Carter Johnson. **20.8** Frans Lanting/Minden Pictures. **20.11** (a) Forrest W. Buchanan/Visuals Unlimited. (b) Gregory K. Scott/Photo Researchers, Inc. **20.12** (left) Heather Angel/Natural Visions. **20.13** (left) © 2008 Josef Hlasak.

CHAPTER 21 **Page 443** © Mickey Gibson/Animals, Animals—Earth Scenes. **21.1** Bruce Beehler. **21.6** Courtesy of James E. Lloyd. Miscellaneous Publications of the Museum of Zoology of the University of Michigan, 130: 1–195, 1966. **21.7** (both) Reny Parker. **21.8** Jen and Des Bartlett/Bruce Coleman USA. **21.10** (left) Patrice Geisel/Visuals Unlimited. (center) Tom Van Sant/The Geosphere Project, Santa Monica, CA. (right) Fred Mc Connaughey/Photo Researchers, Inc. **Page 452** (both) Kenneth Y. Kaneshiro, University of Hawaii. **21.12** (left) © H. Clarke, VIREO/Academy of Natural Sciences. (right) Robert C. Simpson/Nature Stock. **21.14** Dr. Jim Smith, Michigan State University. **21.15** (left) Andrew Parkinson/Frank Lane Picture Agency. (right) Eric Soder/Foto Natura/Photo Researchers, Inc.

CHAPTER 22 **Page 463** Courtesy Lowcountry Geologic. **22.1** Rudwick, 1985. The Meaning of Fossils. University of Chicago Pres, Figure 3.1 p. 106 (partial). **22.2** (a) George H. H. Huey/Corbis. (b) Neville Pledge/South Australian Museum. (c) Jack Koivula/Photo Researchers, Inc. (d) Novosti/Photo Researchers, Inc. **22.3** David Noble/FPG/Getty Images. **22.4** (all) © 2001 PhotoDisc. **22.8** (both) Edward S. Ross. **Page 473** (figure e) Jack Dermid. **22.9** (a) Douglas P. Wilson/Eric and David Hosking. (b) Superstock, Inc. (c) E. R. Degginger. **22.12** (all) Alan Cheetham et al, Department of Paleobiology, National Museum of Natural History, Smithsonian Institution, Washington, D.C. **22.15** David Scott/SREL. **22.16** (both) Gary Head. **22.22** © Michael D. Shapiro and David Kingsley.

CHAPTER 23 **Page 491** Courtesy of U.S. Forest Service, Boise National Forest (Kathryn M. Beall photo). **23.1** (bottom) From L. W. Hackett, *Malaria in Europe*, Oxford University Press, 1937. © London School of Tropical Medicine & Hygiene. **23.3** (top) S. L. Collins and J. T. Collins. (bottom) The Amphibians and Reptiles of Missouri, by T. R. Johnson © 1987 by the Conservation Commission of the State of Missouri. Reprinted by permission. **23.6** (a) Nature's Images/Photo Researchers, Inc. (b) Neil Bowman/Frank Lane Picture Agency. (c) Millard Sharp/Photo Researchers, Inc. **23.12** (top) Thomas J. Lemieux, University of Colorado. (bottom) Sandra Floyd, University of Colorado.

CHAPTER 24 **Page 511** Dr. Ken Macdonald/SPL/Photo Researchers, Inc. **24.1** (bottom) Jeff Hester and Paul Scowen, Arizona State University, and NASA. (inset left) Stanley M. Awramik. **24.3** Photo by Chesley Bonestell. **24.5** (left) Dr. W. Hargreaves and D. Deamer. **24.6** Bill Bachmann, Photo Researchers, Inc. **24.9** Robert Trench, Professor Emeritus, University of British Columbia.

CHAPTER 25 **Page 525** © Phototake, Inc**. 25.1** (all) Tony Brian, David Parker/SPL/Photo Researchers, Inc. **25.2** (all) David M. Phillips/Visuals Unlimited. **25.4** (both) T. J. Beveridge/Visuals Unlimited. **25.5** © Frank Dazzo, Michigan State University. **25.7** CNRI/SPL/Photo Researchers, Inc. **25.9** Dr. Terry J. Beveridge, Department of Microbiology, University of Guelph, Ontario, Canada/Biological Photo Service. **25.12** Hans Reichenbach, Gesellschaft für Biotechnologische Forschung, Braunsweig, Germany. **25.13** (a) Dr. Jeremy Burgess/SPL/Photo Researchers, Inc. (b) Tony Brian/SPL/Photo Researchers, Inc. (c) P. W. Johnson and J. McN. Sieburth, University of Rhode Island/Biological Photo Service. **25.14** David M. Phillips/Visuals Unlimited. **25.15** David M. Phillips/Visuals Unlimited. **25.16** (a) Barry Rokeach. (b) © Alan L. Detrick/Science Source/Photo Researchers, Inc. **25.17** R. Robinson/Visuals Unlimited. **25.19** (right) K. G. Murti/Visuals Unlimited. **25.20** © APHIS photo by Dr. Al Jenny.

CHAPTER 26 **Page 549** Steve Gschmeissner/SPL/Photo Researchers, Inc. **26.1** (a) Edward S. Ross. (b) Gary W. Grimes and Steven L'Hernault. (c) Steven C. Wilson/Entheos. (d) Wim van Egmond. **26.3** (top) Frieder Sauer/Bruce Coleman Ltd. (bottom) Redrawn from V. & J. Pearse and M. & R. Buchsbaum, *Living Invertebrates*, The Boxwood Press, 1987. **26.4** (a) © Dennis Kunkel Microscopy, Inc. (b) Dr. Dennis Kunkel/Visuals Unlimited. **26.5** (top) P. L. Walne nd J. H. Arnott, Planta, 77:325–354, 1967. **26.6** (top) Oliver Meckes/Photo Researchers, Inc. **26.7** (bottom right) John Walsh/SPL/Photo Researchers, Inc. **26.8** Dr. David Phillips/Visuals Unlimited. **26.9** (a) Claude Taylor and the University of Wisconsin Dept. of Botany. (b) Heather Angel. (c) W. Merrill. **Page 559** (micrograph) Steven L'Hernault. (left) Sinclair Stammers/Photo Researchers, Inc. **26.10** Dr. John Cunningham/Visuals Unlimited. **26.11** Jan Hinsch/SPL/Photo Researchers, Inc. **26.12** (a) Ron Hoham, Dept. of Biology, Colgate University. (b) Lewis Trusty/Animals, Animals. (c) Jeffrey Levinton, State University of New York, Stony Brook. **26.14** (a) Wim van Egmond. (b) Courtesy of Allen W. H. Be and David A. Caron. (c) John Clegg/Ardea, London. (d) Redrawn from V. & J. Pearse and M. & R. Buchsbaum, *Living Invertebrates*, The Boxwood Press, 1987. **26.15** M. Abbey/Visuals Unlimited. **26.16** (a, b) Carolina Biological Supply. (c) Courtesy Robert R. Kay from R R Kay, et al., Development, 1989 Supplement, pp. 81–90. © The Company of Biologists Ltd., 1989. **26.17** (a) Wim van Egmond. (b) Douglas Faulkner/Sally Faulkner Collection. **26.18** (a) Linda Sims/Visuals Unlimited. (b) Brian Parker/Tom Stack and Associates. (c) Manfrage Kage/Peter Arnold, Inc. **26.20** Dr. John Clayton, National Institute of Water and Atmospheric Research, New Zealand.

CHAPTER 27 **Page 575** Animals, Animals—Earth Scenes. **27.1** (a) Craig Wood/Visuals Unlimited. (b) Robert Potts, California Academy of Sciences. (c) © Craig Allikas/www.orchidworks.com. **27.2** © Courtesy Microbial Culture Collection, National Institute for Environmental Studies, Japan. **27.3** (a) George S. Ellmore. (b) Jeremy Burgess/SPL/Photo Researchers, Inc. **27.4** Reprinted with permission from Elsevier. **27.10** (a) Martin Hutten/National Park Service. (b) Paul Stehrgreen/National Park Service. (c) Wayne P. Armstrong, Professor of Biology and Botany, Palomar College, San Francisco, CA. **27.11** © clive@hiddenforest.co.nz. **27.12** (top left) Jane Burton/Bruce Coleman USA. **27.13** Dr. Judith Jernstedt, University California, Davis. **27.14** (b) Field Museum of Natural History, Chicago. **27.15** (a) © Ed Reschke/Peter Arnold, Inc. (b) Kathleen B. Pigg, Arizona State University. **27.16** (top) A. & E. Bomford/Ardea, London. (bottom) © Hubert Klein/Peter Arnold, Inc. **27.17** Kingsley R. Stern. **27.18** (a) William Ferguson. (b) W. H. Hodges. (c) Kratz/Zefa. **27.21** Carlton Ray/Science Photo Library/Photo Researchers, Inc. **27.22** (a) © Joyce Photographics/Photo Researchers, Inc. (b) Runk/Schoenberger/Grant Heilman. (c) William Ferguson. (d) Kingsley R. Stern. **27.23** (a) Edward S. Ross. (b) William Ferguson. (c) Robert & Linda Mitchell Photography. (d) ©Fletcher and Baylis/Photo Researchers, Inc. **27.24** (left) Robert Potts, California Academy of Science. (top right) Robert and Linda Mitchell Photography. (bottom right) © R. J. Erwin/Photo Researchers, Inc. **27.25** (a) George H. Huey/Corbis. (b) Bill Coster/Peter Arnold, Inc. (c) © John Mason/Ardea, London. (d) © Peter F. Zika/Visuals Unlimited. **27.26** © David Dilcher, Florida Museum of Natural History/Paleobotany Laboratory. **27.28** Electron micrograph courtesy of J. Ward. **27.29** (a) D. Harms/Peter Arnold, Inc. (b) Rob and Ann Simpson/Visuals Unlimited. (c) © Gregory C. Dimijian Photo Researchers, Inc. (d) © Sangtae Kim, University of Florida. **27.30** (a, left)

© Earl Roberge/Photo Researchers, Inc. (a, center) © Darrell Gulin/The Image Bank/Getty Images. (a, right) Dr. John Hilty. (b, left) Robert E. Bayse/Rose Breeding and Genetics Research Program/Department of Horticulture, Texas A & M University. (b, center left) Keenan Ward/Corbis. (b, center right) John McAnulty/Corbis. **27.30** (b, right) John M. Roberts/Corbis. **27.32** Merlin D. Tuttle, Bat Conservation International. (b) Photo by Marcel Lecoufle. (c) Robert A. Tyrrell. (d) Thomas Eisner/Cornell University.

CHAPTER 28 **Page 605** Fritz Polking/Peter Arnold, Inc. **28.1** (left) Robert C. Simpson/Nature Stock. (top right) Robert C. Simpson/Nature Stock. (bottom right) SciMAT/Photo Researchers, Inc. **28.2** (a) Garry T. Cole, University of Texas, Austin/BPS. **28.5** (a) John Taylor/Visuals Unlimited. (b) Centers for Disease Control and Prevention. (c) Courtesy Ken Nemuras. **28.6** (both) Micrographs Ed Reschke. **28.7** (a) J. D. Cunningham/Visuals Unlimited. (b) John Hodgin. **28.8** (b) © North Carolina State University, Department of Plant Pathology. (c) © Michael Wood/mykob.com. (d) © Fred Stevens/mykob.com. **28.9** (a) N. Allin and G. L. Barron. (b) © Dennis Kunkel Microscopy, Inc. **28.10** Gary T. Cole, University of Texas, Austin/BPS. **28.12** (a, b) Robert C. Simpson/Nature Stock. (c) Jeffrey Lepore/Photo Researchers, Inc. (d) Jane Burton/Bruce Coleman Ltd. (e) Robert C. Simpson/Nature Stock. **28.13** Biophoto Associates/Photo Reaseachers, Inc. **Page 621** Mark Mattock/Planet Earth Pictures. **28.15** (b) V. Ahmadjian and J. B. Jacobs. (c) Jane Burton/Bruce Coleman Ltd. (d) Eye of Science/SPL/Photo Researchers Inc. **28.16** (a) Bryce Kendrick. **28.17** (a) Prof. D. J. Read, University of Sheffield. (b) © 1999 Gary Braasch. **28.18** F. B. Reeves.

CHAPTER 29 **Page 627** Mark Moffett/Minden Pictures. **29.1** (all) Dr. Chip Clark, National Museum of Natural History, Smithsonian Institution. **29.7** Marty Snyderman/Planet Earth Pictures. **29.8** (d, left) Don W. Fawcett/Visuals Unlimited. **29.10** (a) Kim Taylor/Bruce Coleman Ltd. **29.12** (a) © Michael Durham/Minden Pictures. (b) Courtesy of Dr. William H. Hamner. **29.13** (a) Christian DellaCorte. (b) F. S. Westmorland. **29.14** © Norbert Wu/Minden Pictures. **29.15** (a) © blickwinkel/Hecker/Alamy. (b) © Andrew J. Martinez/Photo Researchers, Inc. (c) © Lawrence Naylor/Photo Researchers, Inc. **29.17** © Cory Gray. **29.18** © E. R. Degginger/Photo Researchers, Inc. **29.19** (a) Robert and Linda Mitchell Photography. (b) Cath Ellis, University of Hull/SPL/Photo Researchers, Inc. **Page 644** © James Marshall/Corbis. **Page 645** (top right) © Andrew Syred/Photo Researchers, Inc. (bottom left) © L. Jensen/Visuals Unlimited. (bottom right) Dianora Niccolin. **29.20** (b) Herve Chaumeton/Agence Nature. **29.21** (a) Kjell B. Sandved. **29.24** Jeff

Foott/Tom Stack & Associates. **29.25** (b) J. Kottmann/Peter Arnold, Inc. (c) © Joe McDonald/Corbis. **29.26** (b) Herve Chaumeton/Agence Nature. (c) © Tom McHugh/Photo Researchers, Inc. **29.27** (a) Bob Cranston. (b) Grossauer/ZEFA. (c, top) A. Kertitch, © 1992 Sea World of California. All rights reserved. (c, bottom) © E. Webber/Visuals Unlimited. **29.29** (a) Kjell B. Sandved@sandved.com. **29.30** Duncan Mcewan/npl/Minden Pictures. **29.31** (both) J. A. L. Cooke/ Oxford Scientific Films. **29.32** © 2006 Alistair Dove/Image Quest Marine. **Page 654** J. Sulston, MRC Laboratory of Molecular Biology. **29.33** © Cliff B. Frith/Bruce Coleman USA. **29.34** Mitsuhiko Imamori/Minden Pictures. **29.35** Dr. Chip Clark. **29.36** (a, c) P. J. Bryand, University of California–Irvine/ BPS. (d) © Andrew Syred/Photo Researchers, Inc. **29.37** (left) Jane Burton/Bruce Coleman. (right) Angelo Giampiccolo/FPG/Getty Images. **29.38** (a) Jane Burton/Bruce Coleman. (b) Herve Chaumeton/Agence Nature. **29.39** (left) Herve Chaumeton/Agence Nature. **29.40** (left) Runk/Schoenberger/Grant Heilman. **29.41** (a) Steve Martin/Tom Stack & Associates. (b) Z. Leszczynski/Animals, Animals. **29.42** (a) © Arthur Evans/Animals, Animals—Earth Scenes. (b) Edward S. Ross. (c) © BIOS Borrell Bartomeu/ Peter Arnold, Inc. (d, e) Edward S. Ross. (f) © Michael Durham/Minden Pictures. (g) C. P. Hickman, Jr. (h) S. J. Krasemann/Peter Arnold, Inc.

CHAPTER 30 **Page 667** © Ingo Arndt/ Foto Natura/npl/Minden Pictures. **30.1** Tom McHugh/Photo Researchers, Inc. **30.2** (a) Herve Chaumeton/Agence Nature. (b) George Perina, www .seapix.com. (c) Edward Snow/Bruce Coleman USA. (d) Jan Haaga, Kodiak Lab, AFSC/NMFS. (e) Chris Huss/The Wildlife Collection. **30.3** (c) Herve Chaumeton/Agence Nature. **30.6** (a, top) Peter Parks/Oxford Scientific Films/Animal Animals. (b, left) 2002 Gary Bell/Taxi/Getty Images. **30.7** (a) Runk and Schoenberger/Grant Heilman Photography, Inc. **30.11** (b) Heather Angel/Natural Visions. **30.14** Bill Wood/Bruce Coleman. **30.16** (a) © Gido Braase/Deep Blue Productions. (b) Jonathan Bird/Oceanic Research Group, Inc. (c) Alex Kerstitch/ Visuals Unlimited. **30.17** (a) Ken Lucas/Visuals Unlimited. (b) Patrice Ceisel/© 1986 John G. Shedd Aquarium. **30.18** (b) Digital Vision/Getty Images, Inc. (c) Kit Kittle/Corbis. (d) F. Graner/Peter Arnold, Inc. (e) Brandon Cole/Visuals Unlimited. (f) Arthur W. Ambler/Photo Researchers, Inc. **30.19** (a) Norbert Wu/Peter Arnold, Inc. (b) Wernher Krutein/ photovault.com. **30.21** (a) Stephen Dalton/Photo Researchers, Inc. (b) Bill M. Campbell, MD. (c) Juan M. Renjifo/ Animals Animals. **30.24** (b) Paul J. Fusco/Photo Researchers, Inc. **30.25** (a) Pete & Judy Morrin/Ardea London. (b) © Stephen Dalton/Photo Researchers, Inc. (c) Andrew Dennis/ A.N.T. Photo Library. (d) Mary Ann McDonald/Corbis. **Page 691** (all)

Manuel Leal, Duke University. **30.26** (a) Gerard Lacz/ANT Photolibrary. **30.27** (a) Devin Schafer/Corbis. (b) Arthur Morris/Visuals Unlimited. (c) Ron Sanford/Corbis. (d) Wim Klomp/Foto Natura/Minden Pictures. (e) Robert A. Tyrrell. (f) Tim Zurowski/ Corbis. **30.28** (a) D. & V. Blagden/ANT Photo Library. (b) Jean Phillipe Varin/ Jacana/Photo Researchers, Inc. **30.29** Milse, T./Arco Images/Peter Arnold, Inc. **30.30** (a) Theo Allofs/Photonica/ Getty Images Inc. (b) J. Scott Altenbach, University of New Mexico. (c) Leonard Lee Rue III/FPG/Getty Images Inc. (d) Martin Harvey/Peter Arnold, Inc. (e) David Parker/SPL/Photo Researchers, Inc. (f) © Douglas Faulkner/ Photo Researchers, Inc. **30.32** Cagan Sekercioglu/Visuals Unlimited. **30.33** Larry Burrows/Aspect Photolibrary. **30.34** (b) Joseph Van Os/Getty Images, Inc. **30.35** (a) Art Wolfe/Photo Researchers, Inc. (b) Kenneth Garrett/ National Geographic Image Collection. **30.39** (a) Dr. Donald Johanson, Institute of Human Origins. (b) Louise M. Robbins. **30.40** (a) Science VU/NM/ Visuals Unlimited. (b) AAAC/Topham/ The Image Works.

CHAPTER 31 **Page 711** © Mark Bolton/Corbis. **31.1** (a) © Earl Roberge/Photo Researchers, Inc. (b) © W. Percy Conway/Corbis. (c) © Gregory K. Scott/Photo Researchers, Inc. **Table 31.1** (row 1, left) © Bruce Iverson. (row 1, right) © Mike Clayton/University of Wisconsin Department of Botany. (row 2, left) © Ernest Manewal/Index Stock Imagery. (row 2, right) © Darrell Gulin/Corbis. (row 3, left) © Simon Fraser/Photo Researchers, Inc. (row 3, right) Gary Head. (row 4, left and right) © Andrew Syred/Photo Researchers, Inc. **31.5** (right) James D. Mauseth. **31.6** (all) © Biophoto Associates. **31.7** (a) © Kingsley R. Stern. (b) © D. E. Akin and I. L. Rigsby, Richard B. Russel, Agricultural Research Service, U.S. Department of Agriculture, Athens, Georgia. **31.8** (a) Alison W. Roberts, University of Rhode Island. (b) H. A. Cote, W. A. Cote, and A. C. Day, *Wood Structure and Identification,* second edition, Syracuse University Press. **31.9** (a) James D. Mauseth, University of Texas. (b) Courtesy of Professor John Main, Pacific Lutheran University. **31.10** (a) George S. Ellmore. (b) © Dr. Jeremy Burgess/SPL/Photo Researchers, Inc. (c) Courtesy Mark Holland, Salisbury University. **Page 720** (figure a) William E. Ferguson. (figure b) Cathie Martin. **31.11** (b) Jakub Jasinski/ Visuals Unlimited. **31.12** (b) Robert and Linda Mitchell Photography. (c) Richard R. Dute. **31.13** (a, center) Ray F. Evert. (a, right) James W. Perry. (b, center) Carolina Biological Supply. (b, right) James W. Perry. **31.14** (a) Mike Hill/Getty Images Inc. (b) Wally Eberhart/Visuals Unlimited. (c) Joerg Boethling/Peter Arnold, Inc. (d) Alan & Linda Detrick/Photo Researchers, Inc. (e) Michael P. Gadomski/Photo Researchers, Inc. **31.16** (a) Joseph Devenney/Getty

Images Inc. (b) Maxine Adcock/SPL/ Photo Researchers Inc. **31.17** (b) C. E. Jeffree, et al, *Planta* 172(1):20–37, 1987. Reprinted by permission of C. E. Jefree and Springer-Verlag. **31.18** (a, b) Thomas L. Rost. **31.19** (c) © Beth Davidow/Visuals Unlimited. **31.20** (b) John Limbaugh/Ripon Microslides, Inc. **31.21** (a) Chuck Brown. (b) Carolina Biological Supply. **31.22** © Omnikron/Photo Researchers, Inc. **31.23** (b) Alison W. Roberts, University of Rhode Island. **31.26** (b) © George Bernard/SPL/Photo Researchers, Inc. **Page 733** Tony Gibson at the National Science Foundation.

CHAPTER 32 **Page 737** © Steve Gschmeissner/SPL/Photo Researchers, Inc. **32.1** Owaki-Kulla/Corbis. **32.2** Micrograph Chuck Brown. **32.5** (both) © Claude Nuridsany and Marie Perennou/Science Photo Library/ Photo Researchers, Inc. **32.7** (b) Micrograph Chuck Brown. **32.9** Dr. John D. Cunningham/Visuals Unlimited. **32.11** (both) T. A. Masefield. **32.13** (a) BIOS Matt Alexander/Peter Arnold, Inc. (b) Thomas L. Rost. (c) Fritz Polking/ Visuals Unlimited. (d) Fritz Polking/ Visuals Unlimited. **32.14** (a) Martin Zimmerman, *Science,* 1961, 133: 73–79, © AAAS. (b) Martin H. Zimmermann.

CHAPTER 33 **757** © Ellen McKnight/ Alamy. **33.1** Gerry Ellis/The Wildlife Collection. **33.3** (all) E. Epstein, University of California, Davis. **33.4** William Ferguson. **33.8** (a) Adrian P. Davies/Bruce Coleman. (b) NifTAL Project, University of Hawaii, Maui. (c) © Dr. Jeremy Burgess/SPL/Photo Researchers, Inc. (d) Mark E. Dudley and Sharon R. Long. **33.10** (a) David Cavagnaro/Peter Arnold, Inc. (b) © Grant Heilman Photography. (c) Beverly McMillan. (d) © Prem Subrahmanyam/www.premdesign.com.

CHAPTER 34 **Page 775** © Ted Kinsman/SPL/Photo Researchers, Inc. **34.1** (left) Courtesy of Caroline Ford, School of Plant Sciences, University of Reading, UK. (right) ZEFA—Rein. **34.2** (both) Gary Head. **34.4** (a) Janet Jones. (b) Karlene V. Schwartz. **34.6** (a) David M. Phillips/Visuals Unlimited. (b) Dr. Jeremy Burgess/SPL/Photo Researchers, Inc. (c) David Scharf/Peter Arnold, Inc. **34.8** (a) Michael Clayton, University of Wisconsin. (b) Patricia Schulz. (c) Michael Clayton, University of Wisconsin. (d) Dr. Charles Good, Ohio State University–Lima. (e, f) Michael Clayton, University of Wisconsin. **34.9** (c) Dr. John D. Cunningham/Visuals Unlimited. **34.10** (b) Siegel, R./Arco Images/Peter Arnold, Inc. (c, top) Richard H. Gross. (c, bottom) Andrew Syred/SPL/Photo Researchers, Inc. (d) Mark Rieger. (e) R. Carr. **34.12** (c) Herve Chaumeton/Agence Nature. **34.13** (c) Barry L. Runk/Grant Heilman, Inc. (c) James Mauseth. **34.14** Ed Reschke/Peter Arnold, Inc. **34.15** (left) R-R/S/Grant Heilman Photography, Inc. (center top and bottom; right) Professor Dr. Hans Hanks-Ulrich Koop. **Page 791** Courtesy of the Arabidopsis Information

Resource, 2005. **34.16** (a) Kelly Yee and John J. Harada. (b) Damien Lovegrove/SPL/Photo Researchers, Inc. **Page 793** (left) Jonathan Plett and Sharon Regan. (center) Daniel Szymanski, *Plant Cell* 10:2047. (right) © Dr. Daniel Szymanski, Agronomy Department, Purdue University. **34.19** (all) S. M. Wick, *J Cell Biol,* 89:685, 1981, Rockefeller University Press. **34.21** (all) Jose Luis Riechmann. **34.22** (a) Roland R. Dute.

CHAPTER 35 **Page 801** © Garry Black/ Masterfile. **35.1** Nigel Cattlin/Visuals Unlimited. **35.4** Kingsley R. Stern. **35.8** Sylvan H. Wittwer/Visuals Unlimited. **35.9** Sylvan Wittwer/Visuals Unlimited. **35.11** N.R. Lersten. **35.12** Larry D. Nooden. **35.13** Joanne Chory. **35.14** Amanda Darcy/Getty Images Inc. **35.15** David Cavagnaro/Peter Arnold, Inc. **35.16** Nigel Cattlin/Photo Researchers, Inc. **35.19** Cathlyn Melloan/Stone/Getty Images. **35.20** (both) Micrographs courtesy of Randy Moore, from "How Roots Respond to Gravity," M. L. Evans, R. Moore, and K. Hasenstein, *Scientific American,* December 1986. **35.21** (left) Michael Clayton, University of Wisconsin. (right) John Digby and Richard Firn. **35.23** Cary Mitchell. **35.24** (all) Frank B. Salisbury. **35.25** (a, b) David Sieren/ Visuals Unlimited. **35.27** Dwight Kuhn. **35.28** Jan Zeevart. **35.29** (left) Clay Perry/Corbis. (right) Eric Chrichton/ Corbis. **35.31** R. J. Downs. **35.32** (right) Eric Welzel/Fox Hill Nursery, Freeport, Maine. **Page 827** Tony Gibson at the National Science Foundation.

CHAPTER 36 **Page 831** Simon Fraser/ SPL/Photo Researchers, Inc. **36.1** David Macdonald. **36.3** (b, left) Ray Simmons/Photo Researchers, Inc. (b, center) Ed Reschke/Peter Arnold, Inc. (b, right) Don Fawcett. **36.4** (a, top) Gregory Dimijian/Photo Researchers, Inc. **36.5** (a, top) Ed Reschke. (b, top) Ed Reschke. (c, top) Fred Hossler/ Visuals Unlimited. (d, top) Ed Reschke. (e, top) Ed Reschke. (f, top) Ed Reschke. **36.6** (a, top) Ed Reschke. (b, top) Ed Reschke. (c, top) BioPhoto Associates/Photo Researchers, Inc. **36.7** Lennart Nilsson from *Behold Man,* © 1974 Albert Bonniers Forlag and Little, Brown and Company, Boston. **36.10** Fred Bruemmer.

CHAPTER 37 **Page 847** © C. J. Guerin, Ph.D., MRC Toxicology Unit/SPL/ Photo Researchers, Inc. **37.1** © Gary Gerovac/Masterfile. **37.3** © Triarch/ Visuals Unlimited. **37.4** © Nancy Kedersha/UCLA/Photo Researchers, Inc. **37.5** © C. Raines/Visuals Unlimited. **37.13** © Dennis Kunkel/Visuals Unlimited. **37.15** E. R. Lewis, T. E. Everhart, Y. Y. Zevi/Visuals Unlimited.

CHAPTER 38 **Page 867** © Sovereign/ ISM/SPL/Phototake, Inc. **38.7** Courtesy of Dr. Marcus Raichle, courtesy of Washington University School of Medicine, St. Louis.

CHAPTER 39 **Page 885** © Stephen Dalton/Animals, Animals—Earth Scenes. **39.3** (left) Herve Chaumeton/

Index